AA002351

Proceedings of the

ASP-DAC 2006

Asia and South Pacific Design Automation Conference
2006

January 24-January 27, 2006
Pacifico Yokohama
Yokohama, Japan

Sponsored by:
IEEE Circuits and Systems Society
ACM SIGDA
IEICE (Institute of Electronics, Information and
Communication Engineers)
IPSJ (Information Processing Society of Japan)
Supported by:
JEITA (Japan Electronics and Information Technologies
Industry Association)
STARC (Semiconductor Technology Academic
Research Center)
SCAT (Support Center for Advanced Telecommunications
Technology Research, Foundation)
In Cooperation with:
JNTO (Japan National Tourist Organization)
City of Yokohama

Additional Copies may be ordered from:

IEEE Order Dept.
Hoes Lane
P.O. Box 1331
Piscataway, NJ 08854, U.S.A.

Copyright and Reprint Permission: Abstracting is permitted with credit to the source. Libraries are permitted to photocopy beyond the limit of U.S. copyright law for private use of patrons those articles in this volume that carry a code at the bottom of the first page, provided the per-copy fee indicated in the code is paid through Copyright Clearance Center, 222 Rosewood Drive, Danvers, MA 01923. For other copying, reprint or republication permission, write to IEEE Copyrights Manager, IEEE Operations Center, 445 Hoes Lane, P.O. Box 1331, Piscataway, NJ 08855-1331. All rights reserved. Copyright ©2006 by the Institute of Electrical and Electronics Engineers, Inc.

IEEE Catalog Number: 06EX1199 (CD-ROM Version: 06EX1199C)
ISBN: 0-7803-9451-8 (CD-ROM Version: 0-7803-9452-6)
Library of Congress: 2005932176

ASP-DAC 2006 General Chair's Message

On behalf of the Organizing Committee, I would like to welcome you to the Asia and South Pacific Design Automation Conference 2006 (ASP-DAC 2006) being held here at Pacifico Yokohama jointly with the Electronic Design and Solution Fair 2006. ASP-DAC is a sister conference of DAC, DATE and ICCAD, and it is the 11th event of this conference series.

ASP-DAC is the meeting place where researchers and engineers come together to learn and discuss state of the art technologies of system/SoC design, EDA and design methodologies. This year, we put special effort into attracting designers and design industries to produce Designers' Forum.

We have three keynote speakers from academia, the semiconductor industry and the systems industry. Professor Alberto Sangiovanni-Vincentelli, University of California at Berkeley, will explore future system design perspectives in automotive electronics. Satoru Ito, President & CEO of RENESAS Technology Corp. will discuss challenges of device innovation. Yukichi Niwa, Senior Advisory Director of CANON INC. will present the company's key concept of architecting platform based design.

The technical program was selected from 424 papers from 27 countries. The 64 members of the Technical Program Committee chaired by Professor Onodera and helped by over 250 reviewers had to make difficult choices to carefully select 135 papers. It is an outstanding program that covers a variety of key topics from system level design to physical design. Designers' Forum is a new program that shares design experience and solutions of real product designs of the industries whose topics include the CELL and mobile designs, panels of top 10 design issues and system verification. The University Design Contest is also an important event of ASP-DAC, which focuses on a real chip design in academia. On Tuesday, one full-day and six half-day tutorials are scheduled to provide introductions to hot topics like DFM, lowpower, packaging/interconnect, system level design and verification.

An event like ASP-DAC can happen only with the efforts of many people. We wish to express our appreciation to all authors, speakers, reviewers, session organizers, moderators, panelists, session chairs, keynote speakers and committee members. Also my sincere thanks to the dedicated members of the Organizing Committee, the Technical Program Committee, the University LSI Design Contest Committee, the Industry Liaison and the Steering Committee.

Finally, special thanks to all ASP-DAC attendees. I hope you will have a productive and exciting experience at ASP-DAC 2006.

Fumiyasu Hirose

Fumiyasu Hirose
General Chair
ASP-DAC 2006

Message from Technical Program Committee

On behalf of the Technical Program Committee for the Asia and South Pacific Design Automation Conference 2006, we would like to welcome all of you to the conference held from January 24 through 27, 2006 at Pacifico Yokohama Conference Center in Yokohama, Japan.

Hidetoshi Onodera   Yusuke Matsunaga

This year, ASPDAC received 424 paper submissions, which is the second highest number in our history. This is a 46% increase in submissions over the last ASPDAC held in Japan two years ago. The submissions span 27 countries/regions in Asia, North America, South America, Europe, Oceania, and Africa.

The Technical Program Committee was composed of 64 professionals who are experts on EDA, LSI design, and embedded system design, and was organized into 11 sub-committees. The number of committee members is relatively small compared to the previous conferences. However, all the members were committed to make in-depth reviews for all the papers assigned to each sub-committee and physically attend the TPC meeting for paper selection. Based on the result of a rigorous and thorough review followed by a full day face-to-face discussion, we have selected 135 papers and compiled them into a 3-day program of 27 technical sessions in three parallel tracks. The technical program is further enriched by multiple special sessions and panels in one more track, resulting in four parallel tracks of exciting presentations and discussions.

Each day, the technical program starts with a keynote address that is organized through the leadership of Fumiyasu Hirose, General Chair, followed by regular and special sessions. We have 7 special sessions in total. On Wednesday, selected submissions to the University Design Contest will be presented at Session 1D. Session 2D focuses on electro-thermal design and Session 3D discusses flash memory in embedded systems. On Thursday, Session 4D addresses an overview of Open Access and Session 5D focuses on low power design challenges for mobile applications. On Friday, Session 7D discusses H.264 design issues and Session 8D addresses design methodology for "Cell" processor development. It should be mentioned that Sessions 5D and 8D are organized as Designers' Forum that is a brand new event for encouraging mutual exchange between/within designers in industry and EDA researchers. Besides two special sessions, Designers' Forum will organize two panel discussions on functional verification (Session 6D, Thursday) and top-10 design issues (Session 9D, Friday). These special sessions will provide you with a wide variety of hot and exciting topics from system level down to physical level, and from industrial design issues to theoretical fundamentals.

On behalf of the Technical Program Committee, we would like to thank all people involved in the 2006 event. In particular, we would like to thank the members of the Organizing and Technical Program Committees, the members of Industry Liaison, special session organizers, and everyone at JESA for conference management. Also, we would like to express our sincere thanks to all the authors who submitted papers with valuable results, since their contributions form the basis of our technical excellence.

We would be more than happy if you could attend the conference and find something new in the directions of EDA and technologies during ASP-DAC 2006.

Technical Program Chair
Hidetoshi Onodera
Kyoto University

Technical Program Vice Chair
Yusuke Matsunaga
Kyushu University

University LSI Design Contest

The University LSI Design Contest was conceived as a unique program of ASP-DAC Conference. The purpose of the Contest is to encourage education and research in LSI design, and its realization on chips at universities, and other educational organizations by providing opportunities to present and discuss innovative and state-of-the-art designs at the conference. Application areas and types of circuits include (1) Analog and Mixed-Signal Circuits, (2) Digital Signal processing, (3) Microprocessors, and (4) Custom Application Specific Circuits. Methods or technology used for implementation include (a) Full Custom and Cell-Based LSIs, (b) Gate Arrays, and (c) Field Programmable Devices, including FPGA/PLDs.

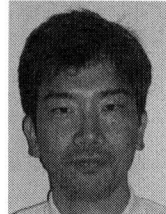 

Kazutoshi Kobayashi   Takahiko Arakawa

This year, nineteen selected designs from seven countries/areas will be disclosed in Session 1D with a short presentations followed by live discussions in front of posters with light meals. Submitted designs were reviewed by the members of the University Design Contest Committee based on the following criteria: Reliability of design and implementation, Quality of implementation, Performance of the design, Novelty, and Additional special features. In the selection process, emphasis was placed more on reliability, quality, and performance. As a result, the nineteen designs were selected. Also, we have instituted one outstanding design award and two special feature awards.

It is with great pleasure that we acknowledge the contributions to the Design Contest, and it is our earnest belief that it will promote and enhance research and education in LSI design in academic organizations. It is also our hope that many people not only in academia but in industry will attend the contest and enjoy the stimulating discussions.

Co-Chairs, University LSI Design Contest Committee

Co-Chair
Kazutoshi Kobayashi
Kyoto University

Co-Chair
Takahiko Arakawa
RENESAS Technology Corp.

Designers' Forum

This brandnew event, Designers' Formum, is conceived as a unique program for ASP-DAC to encourage mutual exchange both between and within designers in industry, researchers in the area of EDAs, and EDA developers. Here, designs will be presented focusing on design styles, design issues, and ways to tackle design issues. Panel discussions will also be held for the latest design issues.

Haruyuki Tago   Makoto Ikeda

This year, we will have 2 special sessions in ASP-DAC 2006 as follows. 4 presentations related to low power designs, power models, and power estimation frameworks for the real SoC designs will be given by Toshiba, Fujitsu, Hitachi and Renesas, and Samsung in session 5D, "Low Power Design." In session 8D, "Cell processor", 4 presentations will be given fucusing on simulations, tests, verifications, power estimation, and design methodology of Cell; and PLL design employed in the Cell processor.

In addition to the special sessions, we will have 2 panel discussions. Session 6D, "Functional Verification -now and future-" will be modareted by Dr. Y. Masubuchi of Toshiba, who has been deeply engaged in the architecture design and verifications of the Cell processor. Three panelists are LSI designers, working for functional verifications in LSI design, and one panelist is from an EDA vendor, developing functional verification tools. Session 9D, "Top 10 design issues seen by LSI designers versus EDA developers", on the other hand, will be moderated by Dr. Y. Hagihara of Sony. Three panelists are managers of SoC and system designs, will focus on the top 10 design issues seen by LSI designers; and 3 panelists are techology officers from 3 major EDA vendors, and they will focus on top 10 design issues seen by EDA developers. Discussions will be led toward the perspectives of the future SoC design issues, comparing with the two top 10 issues seen by both LSI designers and EDA developers.

This Designers' Forum is planned by the Industry Laison Members of ASP-DAC2006. It is with great pleasure that we acknowledge the contributions to the Designers' Forum, and it is our earnest belief that this forum will promote mutual exchange of designers and EDA researchers and developers, toward nano-meter SoC design issues.

Industry Liaison Chair
Haruyuki Tago
TOSHIBA CORPORATION

Designers' Forum Chair
Makoto Ikeda
University of Tokyo

Keynote Address I

Automotive Electronics: Steady Growth for Years to Come!

Alberto Sangiovanni-Vincentelli

The Edgar L. and Harold H. Buttner Chair of Electrical
Engineering and Computer Science, University of California,
Berkeley, and Chief Technology Advisor, Member of the Board
and Co-founder, Cadence Design Systems, United States

The world of electronics is witnessing a revolution in the way products are conceived, designed and implemented. The ever growing importance of the web, the advent of microprocessors of great computational power, the explosion of wireless communication, the development of new generations of integrated sensors and actuators are changing the world in which we live and work. The new key words are:

- Disappearing electronics, i.e., electronics has to be invisible to the user, it has to help unobtrusively.

- Pervasive computing, i.e., electronics is everywhere, all common use objects will have an electronic dimension.

- Ambient intelligence, i.e., the environment will react to us with the use of electronic components. They will recognize who we are and what we like.

- Wearable computing, i.e., the new devices will be worn as a watch or a hat. They will become part of our clothes. Some of these devices will be tags that will contain all important information about us.

- Know more, carry less, i.e., the environment will know more about us so that we will not need to carry all the paraphernalia of keys, credit cards, personal I.D.s, access cards, access codes.

The car as a self-contained microcosm is experiencing a similar revolution: all the key words listed above are going to have a great impact on the automotive world. We need to rethink what a car really is and the role of electronics in it. Electronics is now essential to control the movements of a car, of the chemical and electrical processes taking place in it, to entertain the passengers, to establish connectivity with the rest of the world, to ensure safety. What will an automobile manufacturer's core competence become in the next few years? Will electronics be the essential element in car manufacturing and design? The challenges and opportunities are related to

- how to integrate the mechanical and the electronics worlds, i.e., how to make mechatronics a reality in the automotive world,

- how to integrate the different motion control and power-train control functions so that important synergies can be exploited,

viii

- how to combine entertainment, communication and navigation subsystems,
- how to couple the world of electronics where the life time of a product is around 2 years and shrinking, with the automotive world, where the product life time is 10 years and possibly growing,
- how to develop new services based on electronics technology,
- how to exploit communication among cars and between cars and infrastructure such as Global Positioning Systems and cellular networks,
- how are the markets evolving (for example, what will be the size of the after-market sales for automotive electronics, if any?).

We will pose these questions while reviewing some of the most important technology and product developments of the past few years. We will also present new trends on how the design of electronics of the car should be carried out. We will finally analyze the dynamics of the automotive electronics industry that is bound to produce a major shake-up in the structure of the design chain with particular emphasis on the AUTOSAR consortium.

Keynote Address II

Challenging Device Innovation

Satoru Ito

President & CEO

RENESAS Technology Corp., Japan

The semiconductor industry has continuously transformed our way of life, through a number of underlying technology breakthroughs and innovations over the past years. There are currently two challenges that this industry faces: a limitation of miniaturization technology and a difficulty in maintaining an economy of scale. To cope with these challenges, there is a growing need to work closely with partners and customers who have business related to semiconductors, in addition to semiconductor manufacturers. Especially in the area of semiconductor design, we see a need to create a new EDA methodology that broadens the definition of traditional EDA and re-defines the connection among system designers, SoC designers and development tool designers. As we move closer to the realm driven by the convergence of applications and advancements in miniaturization technology, I'd like to discuss the associated technological challenges as well as economical challenges, and present to you our strategy to overcome these issues.

Keynote Address III

Effective Platform-based Development for Large-scale Systems Design

Yukichi Niwa

Senior Advisory Director,
Group Executive of Platform Technology Development
Headquarters
CANON INC., Japan

Platform-based development (PBD) aims to continuously add new value in both cases of incremental development and product planning based development. By adding new technology to previously existing technology and by storing the technologies as reusable assets, PBD enables high quality, low cost, and short turnaround time development. Furthermore, PBD allows target-oriented development where we can select and concentrate technology to eliminate unnecessary development.

In order to execute effective PBD, it is important to introduce the firm layer structuring of digital/analog technology so that individual professionals in independent layer can maximize their efficiency without any restraint. The act of layer structuring is nothing but the architectural design of the development methodology. Thus, it's no exaggeration to say that success in business profitability management directly depends on the presence of the good architect.

The important thing in the next stage is to optimize the design process by investing in computer resources. For example, it is necessary to thoroughly adapt simulation technology to the development of high quality imaging technology, embedded system (hardware/software) technology, or communication technology. The quantitative evaluation from the early design phase and the workflow based on the accumulated design know-how (IP, methodology) will accelerate technology innovation and strengthen the platform even further. Eventually, management can directly obtain absolute advantage of large-scale system design effectiveness.

ASP-DAC 2005 Best Papers

Best Paper Award

Speed and Voltage Selection for GALS Systems based on Voltage/Frequency Islands
Koushik Niyogi, Diana Marculescu (Carnegie Mellon University)

The Polygonal Contraction Heuristic for Rectilinear Steiner Tree Construction
Yin Wang, Xianlong Hong, Tong Jing, Yang Yang (Tsinghua University, Beijing), Xiaodong Hu, Guiying Yan (Institute of Applied Mathematics, Chinese Academy of Sciences)

Fast PLL Simulation Using Nonlinear VCO Macromodels for Accurate Prediction of Jitter and Cycle-Slipping due to Loop Non-idealities and Supply Noise
Xiaolue Lai, Yayun Wan, Jaijeet Roychowdhury (University of Minnesota)

Design Contest Award

A Bandwidth Efficient Subsampling-based Block Matching Architecture for Motion Estimation
Hao-Yun Chin, Chao-Chung Cheng, Yu-Kun Lin, Tian-Sheuan Chang (Chiao Tung University)

A Low-Power Video Segmentation LSI with Boundary-Active-Only Architecture
Takashi Morimoto, Osamu Kiriyama, Hidekazu Adachi, Zhaomin Zhu, Tetsushi Koide, Hans Jürgen Mattausch (Hiroshima University)

ASP-DAC 2006 Best Papers

Best Paper Award

1B-1 **Constraint-Driven Bus Matrix Synthesis for MPSoC**
Sudeep Pasricha, Nikil Dutt (Univ. of California, Irvine, United States), Mohamed Ben-Romdhane (Conexant, United States)

3C-1 **Post-Routing Redundant Via Insertion for Yield/Reliability Improvement**
Kuang-Yao Lee, Ting-Chi Wang (National Tsing Hua Univ., Taiwan)

Best Paper Candidates

1B-1 **Constraint-Driven Bus Matrix Synthesis for MPSoC**
Sudeep Pasricha, Nikil Dutt (Univ. of California, Irvine, United States), Mohamed Ben-Romdhane (Conexant, United States)

3C-1 **Post-Routing Redundant Via Insertion for Yield/Reliability Improvement**
Kuang-Yao Lee, Ting-Chi Wang (National Tsing Hua Univ., Taiwan)

7B-1 **Equivalent Circuit Modeling of Guard Ring Structures for Evaluation of Substrate Crosstalk Isolation**
Daisuke Kosaka, Makoto Nagata (Kobe Univ., Japan)

7C-2 **Speed Binning Aware Design Methodology to Improve Profit under Parameter Variations**
Animesh Datta (Purdue Univ., United States), Swarup Bhunia (Case Western Reserve Univ., United States), Jung Hwan Choi, Saibal Mukhopadhyay, Kaushik Roy (Purdue Univ., United States)

8A-1 **Fast Substrate Noise-Aware Floorplanning with Preference Directed Graph for Mixed-Signal SOCs**
Minsik Cho, Hongjoong Shin, David Z. Pan (Univ. of Texas, Austin, United States)

8B-1 **Finding Optimal L1 Cache Configuration for Embedded Systems**
Andhi Janapsatya, Aleksandar Ignjatovic, Sri Parameswaran (Univ. of New South Wales, Australia)

9A-1 **TAPHS: Thermal-Aware Unified Physical-Level and High-Level Synthesis**
Zhenyu (Peter) Gu (Northwestern Univ., United States), Yonghong Yang (Queen's Univ., Canada), Jia Wang, Robert P. Dick (Northwestern Univ., United States), Li Shang (Queen's Univ., Canada)

9C-3 **Statistical Leakage Minimization through Joint Selection of Gate Sizes, Gate Lengths and Threshold Voltage**
Sarvesh Bhardwaj, Yu Cao, Sarma Vrudhula (Arizona State Univ., United States)

Design Contest Award

Outstanding Design Award

1D-15 **A 52mW 1200MIPS Compact DSP for Multi-Core Media SoC**
Shih-Hao Ou, Tay-Jyi Lin, Chao-Wei Huang, Yu-Ting Kuo, Chie-Min Chao, Chih-Wei Liu (National Chiao Tung Univ., Taiwan), Chein-Wei Jen (STC, ITRI, Taiwan)

Special Feature Award

1D-1 **A Low Dynamic Power and Low Leakage Power 90-nm CMOS Square-Root Circuit**
Tadayoshi Enomoto, Nobuaki Kobayashi (Chuo Univ., Japan)

1D-8 **Adaptively-Biased Capacitor-Less CMOS Low Dropout Regulator with Direct Current Feedback**
Yat-Hei Lam, Wing-Hung Ki, Chi-Ying Tsui (Hong Kong Univ. of Science and Tech., Hong Kong)

ASP-DAC 2006 Organizing Committee

General Chair

Fumiyasu Hirose
Cadence Design Systems, Japan

SC Chair

Hiroto Yasuura
Kyushu University

Secretary

Atsushi Takahashi
Tokyo Institute of Technology

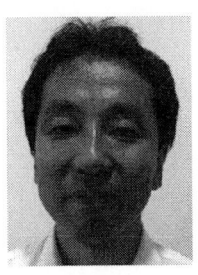

SC Vice Chair

Takeshi Yoshimura
Waseda University

Secretary

Yoshinori Takeuchi
Osaka University

Technical Program Chair

Hidetoshi Onodera
Kyoto University

Assitant Secretary

Asako Kaneko
Cadence Design Systems, Japan

Technical Program Vice Chair

Yusuke Matsunaga
Kyushu University

Past Chair

Masaharu Imai
Osaka University

TPC Secretary

Masanori Hashimoto
Osaka University

Design Contest Co-Chair

Kazutoshi Kobayashi
Kyoto University

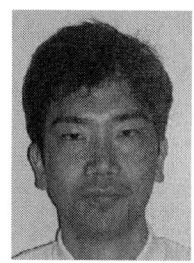

Tutorial Vice Chair

Hiroyuki Higuchi
FUJITSU LABORATORIES LTD.

Design Contest Co-Chair

Takahiko Arakawa
RENESAS Technology Corp.

Finance Co-Chair

Naoya Tohyama
System Fabrication Technologies, Inc

Designers' Forum Chair

Makoto Ikeda
University of Tokyo

Finance Co-Chair

Shinji Kimura
Waseda University

Tutorial Co-Chair

Kazutoshi Wakabayashi
NEC Corporation

Publicity Chair

Masahiro Fukui
Ritsumeikan University

Tutorial Co-Chair

Chung-Kuan Cheng
University of California, San Diego

Publication Chair

Nozomu Togawa
Waseda University

Web Publicity

Bakhtiar Affendi
Tokyo Institute of Technology

ASP-DAC Rep. at DAC

Yusuke Matsunaga
Kyushu University

Industry Liaison Chair

Haruyuki Tago
TOSHIBA CORPORATION

ASP-DAC Rep. at DATE

Masaharu Imai
Osaka University

EDSF Chair

Mitsuru Nadaoka
Oki Electric Industry Co., Ltd.

ASP-DAC Rep. at ICCAD

Kazutoshi Wakabayashi
NEC Corporation

ASP-DAC Japan Council Rep.

Tokinori Kozawa

IEICE/CAS Rep.

Takao Nishitani
Kochi University of Technology

ASP-DAC Japan Council Rep.

Kenji Yoshida
Cadence Design Systems, Japan

IEICE/ICD Rep.

Masao Nakaya
RENESAS Technology Corp.

IEICE/VLD Rep.

Shinji Kimura
Waseda University

Secretariat

Yoshinori Ishizaki
Japan Electronics Show Association

IPSJ/SLDM Rep.

Takashi Kambe
Kinki University

Secretariat

Mieko Mori
Japan Electronics Show Association

JEITA/EDA TC Rep.

Yoshio Okamura
RENESAS Technology Corp.

Secretariat

Fumiaki Yoshinaga
Japan Electronics Show Association

IWCM Rep.

Shigetaka Kumashiro
STARC

Secretariat

Kayoko Oda
Japan Electronics Show Association

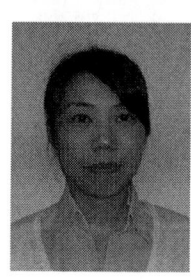

Secretariat

Jiro Irie
Japan Electronics Show Association

ASP-DAC Steering Committee

Chair

Hiroto Yasuura
Kyusyu University
yasuura@c.csce.kyushu-u.ac.jp

ASP-DAC 2006 General Chair

Fumiyasu Hirose
Cadence Design Systems, Japan

Vice Chair

Takeshi Yoshimura
Waseda University
t-yoshimura@waseda.jp

ASP-DAC 2005 General Chair

TingAo Tang
Fudan University

Secretary

Toshihiro Hattori
RENESAS Technology Corp.
hattori.toshihiro@renesas.com

ACM SIGDA Rep.

Nikil Dutt
University of California at Irvine

Secretary

Kazutoshi Wakabayashi
NEC Corporation
wakaba@bl.jp.nec.com

IEEE CAS Rep.

Georges Gielen
The Katholieke Universiteit Leuven

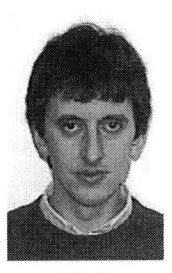

PAST Chair

Tatsuo Ohtsuki
Waseda University

DAC Representative

Steven P. Levitan
University of Pittsburgh

DATE Representative

Georges Gielen
The Katholieke Universiteit Leuven

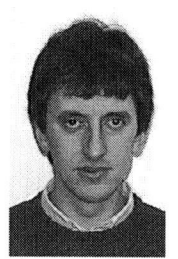

IEICE TGVLD Chair

Shinji Kimura
Waseda University

ICCAD Representative

Kazutoshi Wakabayashi
NEC Corporation

IEICE TGICD Chair

Masao Nakaya
RENESAS Technology Corp.

JEITA Representative

Yoshio Okamura
RENESAS Technology Corp.

IPSJ SIG SLDM Chair

Takashi Kambe
Kinki University

EDSF Chair

Mitsuru Nadaoka
Oki Electric Industry Co., Ltd.

STARC Representative

Takeshi Imamura
Semiconductor Technology Academic Research Center

IEICE TGCAS Chair

Takao Nishitani
Kochi University of Technology

JIEP Representative

Yoichi Shiraishi
Gunma University

International Members

Sunil D. Sherlekar
Tata Consultancy Services

Youn-Long Steve Lin
Tsing Hua University, Hsin-Chu

Richard M M Chen
City University of Hong Kong

Alexander Stempkovsky
Russin Academy of Sciences

Sri Parameswaran
The University of New South Wales

Qianling Zhang
Fudan University

Xian-Long Hong
Tsinghua University, Beijing

Advisory Members

Masaharu Imai
Osaka University

Chong-Min Kyung
Korea Advanced Institute of Science and
Technology

Basant R. Chawla
Genentech

Hideo Fujiwara
Nara Institute of Science and Technology

Hidetoshi Onodera
Kyoto University

Satoshi Goto
Waseda University

Isao Shirakawa
Professor Emeritus of Osaka University

Fumiyasu Hirose
Cadence Design Systems, Japan

Kenji Yoshida
Cadence Design Systems, Japan

Takashi Kambe
Kinki University

Tokinori Kozawa

Hiroaki Kunieda
Tokyo Institute of Technology

ASP-DAC 2006 Technical Program Committee

Chair
Hidetoshi Onodera
Kyoto University
onodera@i.kyoto-u.ac.jp

Vice Chair
Yusuke Matsunaga
Kyushu University
matsunaga@c.csce.kyushu-u.ac.jp

Secretary
Masanori Hashimoto
Osaka University
hasimoto@ist.osaka-u.ac.jp

Subcommittees (∗ indicates the subcommitte chair.)
[1] System Level Design Methodology

∗**Youn-Long Lin** National Tsing Hua University	**Ahmed Jerraya** TIMA	**Yoshinori Takeuchi** Osaka University
Soonhoi Ha Seoul National University	**Tsuneo Nakata** Fujitsu Lab.	

[2] Embedded and Real-Time Systems

∗**Hiroyuki Tomiyama** Nagoya University	**Tei-Wei Kuo** National Taiwan University	**Sungjoo Yoo** Samsung
Pai Chou University of California, Irvine	**Sri Parameswaran** University of New South Wales	

[3] Behavioral/Logic Synthesis and Optimization

∗**Kiyoung Choi** Seoul National University	**Shinji Kimura** Waseda University	**Shigeru Yamashita** AIST Nara
Shih-Chieh Chang National Tsing Hua University	**Diana Marculescu** Carnegie Mellon University	

[4] Validation and Verification for Behavioral/Logic Design

∗**Kiyoharu Hamaguchi** Osaka University	**Shin'ichi Minato** Hokkaido University	**Farn Wang** National Taiwan University
Jin-Young Choi Korea University	**Karem Sakallah** University of Michigan	

xxiii

[5] Physical Design (Routing)

∗Martin D. F. Wong
University of Illinois, Urbana
Champaign
Tong Jing
Tsinghua University

Youichi Shiraishi
Gunma University
Atsushi Takahashi
Tokyo Institute of Technology

Ting-Chi Wang
National Tsing Hua University

[6] Physical Design (Placement)

∗Shin'ichi Wakabayashi
Hiroshima City University
Yao-Wen Chang
National Taiwan University

Jason Cong
University of California, Los
Angeles
Shigetoshi Nakatake
University of Kitakyushu

Evangeline F. Y. Young
Chinese University of Hong
Kong

[7] Timing, Power, Signal/Power Integrity Analysis and Optimization

∗Sachin Sapatnekar
University of Minnesota
Shabbir Batterywala
Synopsys (India)
Jin-Jia Liou
National Taiwan University

Takashi Sato
Renesas
Weiping Shi
Texas A&M University
Youngsoo Shin
KAIST

Sheldon Tan
University of California,
Riverside
Ryuichi Yamaguchi
Matsushita

[8] Interconnect, Device and Circuit Modeling and Simulation

∗Hideki Asai
Shizuoka University
Arun Chandrasekhar
Intel (India)
Charlie Chung-Ping Chen
National Taiwan University

Eli Chiprout
Intel
Yungseon Eo
Hanyang University

Hiroo Masuda
STARC
Jae-Kyung Wee
Soongsil University

[9] Test and Design for Testability

∗Seiji Kajihara
Kyushu Institute of Technology
Masaki Hashizume
Tokushima University

Sungho Kang
Yonsei University
XiaoWei Li
China Academy of Sciences

Prab Varma
Veritable

[10] Analog, RF and Mixed Signal Design and CAD

∗Makoto Nagata
Kobe University
Seijiro Moriyama
PDF Solutions

Hong-June Park
POSTECH
Jaijeet Roychowdhury
University of Minnesota

Chau-Chin Su
National Chao-Tung University

[11] Leading Edge Design Methodology for SOCs and SIPs

∗Hideharu Amano
Keio University

Ing-Jer Huang
National Sun-Yat-Sen University

Satoshi Matsushita
NEC

Borivoje Nikolic
University of California, Berkeley

In-Cheol Park
KAIST

Yulu Yang
Nankai University

University LSI Design Contest Committee

Co-Chairs

Kazutoshi Kobayashi
Kyoto University
kobayasi@kuee.kyoto-u.ac.jp

Takahiko Arakawa
RENESAS Technology Corp.
arakawa.takahiko@renesas.com

Chih-Wei Liu
National Chiao Tung University

Hideo Ohwada
FUJITSU LABORATORIES LTD.

Takao Onoye
Osaka University

Mengtian Rong
Shanghai Jiao Tong University

Hao San
Gunma University

Makoto Takamiya
University of Tokyo

Chi-ying Tsui
The Hong Kong University of Science
and Technology

Tomohisa Wada
University of the Ryukyus

Hideki Yamauchi
SANYO Electric Co., Ltd.

Xiaooyang Zeng
Fudan University

Industry Liaison

Chair
Haruyuki Tago
TOSHIBA CORPORATION
haruyuki.tago@toshiba.co.jp

Design

Yoshiaki Hagihara
Sony Corporation

Sunao Torii
NEC Corporation

Kunio Uchiyama
Hitachi, Ltd.

Shigeru Watari
Matsushita Electric Industrial Co., Ltd.

Takeshi Yamamura
FUJITSU LABORATORIES LTD.

JEITA

Yoshitada Fujinami
NEC Electronics Corporation

Shigemi Saito
Sony Corporation

EDA

Hiromitsu Fujii
Nihon Synopsys Co.,Ltd.

Fumiyasu Hirose
Cadence Design Systems, Japan

Satoshi Kojima
Mentor Graphics Japan Co.,Ltd.

List of Reviewers

Rohit A
Bakhtiar Affendi
Yong-Jin Ahn
Hideharu Amano
Chirayu Amin
Zaher Andraus
Tetsuya Aoyama
Hideki Asai
Shabbir Batterywala
Valeria Bertacco
Sandeep Bhatia
Sambuddha Bhattacharya
Srinivas Bodapati
Zhen Cao
Sourav Chakravarty
Jeremy Chan
Arun Chandrasekhar
Kai-hui Chang
Shih-Chieh Chang
Soon-Jyh Chang
Tsin-Yuan Chang
Yao-Wen Chang
Yuan-Hao Chang
Yun-Nan Chang
Kai-Yuan Chao
Charlie Chung-Ping Chen
De-Sheng Chen
Hsin-Chen Chen
Huang-Yu Chen
Hung-Ming Chen
Jian-Jia Chen
Liang-Bi Chen
Sao-Jie Chen
Tai-Chen Chen
Tung-Chieh Chen
Ya-Shu Chen
Yan-Bin Chen
Yi-An Chen
Yu-Cheng Chen
Yu-Zhi Chen
C.K. Cheng
Li Chia
Charles Chiang
Mei-Fang Chiang
Eli Chiprout
Jih-Ching Chiu

Youngchul Cho
Jin-Young Choi
Kiyoung Choi
Michael Yee-Jern Chong
Pai Chou
Szu-Jui Chou
Chris C.N. Chu
Jason Cong
Florentin Dartu
Liang Deng
Lan-Rong Dung
Yungseon Eo
Jeffrey Fan
Jia-Wei Fang
Yu-Luen Fang
Zhe Feng
Ryuichi Fujimoto
Kunihiro Fujiyoshi
Youxin Gao
Eric Grimme
Ruei-Ting Gu
Hi Annie Guo
Zheng Guo
Soonhoi Ha
Kiyoharu Hamaguchi
Kee-Sung Han
Masaki Hashizume
Atsushi Hatabu
Hao-Chiao Hong
Jin-Hua Hong
Jen-Wei Hsieh
Wang-Jui Hsieh
Pi-Cheng Hsiu
Chin-Hsiung Hsu
Heng-Ruey Hsu
Ren-Chien Hsu
Tien-Chang Hsu
Ching-Chi Hu
Jiang Hu
Wui Hu
Xiaodong Hu
Yu Hu
Chih-Yuan Huang
Chung-Yang (Ric) Huang
Geng-Dian Huang
Ing-Jer Huang

Jiun-Lang Huang
Jun-Dar Huang
Po-Chiun Huang
Shih-Hsu Huang
Szu-Wei Huang
Wen-Kai Huang
Wen-Pin Huang
Chia-Mei Hung
Ting-Ting Hwang
Hiroyuki Igura
Masato Inagi
Shigeto Inui
Takeshi Inuo
Ryosuke Isotani
Tsuyoshi Isshiki
Hiroaki Iwashita
Andhi Janapsatya
Ahmed Jerraya
Hui-Ru Jiang
Jie-Hong Roland Jiang
Rong Jiang
Zhe-Wei Jiang
Tong Jing
Jinyong Jung
Nobuki Kajihara
Seiji Kajihara
Sungho Kang
Hiroshi Kawaguchi
Shoji Kawahara
Hideyuki Kawakita
Masahiro Kawakita
Mahesh Ketkar
Sunil Khatri
Mary Kiemb
Tomohiro Kikuma
Chris H. Kim
Jin-Hyun Kim
Sun-Gyeum Kim
Shinji Kimura
Tomohisa Kimura
Lih Wen Koh
Yukihide Kohira
Tetsushi Koide
Satoshi Komatsu
Michio Komoda
Shiann-Rong Kuang

Yukiko Kubo
Shuichi Kunie
Ko-Chi Kuo
Tei-Wei Kuo
Shin-ya Kuwamura
Chi-Ping Lai
Chun-Hung Lai
Gang-Hee Lee
Imyong Lee
Katherine S.M. Lee
Kuang-Yao Lee
Sung-Hyun Lee
Wan-Ping Lee
Yu-Min Lee
Melinda Ler
Guo-Liang Li
Hang Li
XiaoWei Li
Yih-Lang Li
Guang-Wan Liao
Cheng-Min Lien
John Lillis
Chi-Hung Lin
Chung-Wei Lin
I-Jye Lin
I-Ting Lin
Jiun-Ren Lin
Rung-Bin Lin
Youn-Long Lin
Tsui-Yee Ling
Jin-Jia Liou
Bin Liu
Bo Liu
Chien-Nan Liu
Hung-Yi Liu
Meng-Chung Liu
Pu Liu
Hong-Wen Lu
Yung-Feng Lu
Danny W.S. Luk
Jianfeng Luo
Yuchun Ma
Patrick H. Madden
Amit Majumdar
Arthur W.K. Mak
Wai-Kei Mak
Usama Malik
Junichi Mano
Diana Marculescu

Dejan Markovic
Hiroo Masuda
Osamu Matsumoto
Hidetoshi Matsuoka
Satoshi Matsushita
Hiroki Matsutani
Yasuyuki Matsuya
Noel Menezes
Shin'ichi Minato
Natasa Miskov-Zivanov
Hiroshi Miyashita
Maher Mneimneh
Seijiro Moriyama
Tatsuji Moriyoshi
Rafael Kazumichi Morizawa
Hiroshi Murata
Kouhei Nadehara
Makoto Nagata
Masaki Nakanishi
Koichi Nakashiro
Tsuneo Nakata
Shigetoshi Nakatake
Shogo Nakaya
Takashi Nakayama
Myung-Jin Nam
Borivoje Nikolic
Kengo Nishino
Dongkeun Oh
Michiroh Ohmura
Kenichi Okada
Makiko Okumura
Mustafa Ozdal
Liang-Teck Pang
Sanjay Pant
Sri Parameswaran
Hong-June Park
In-Cheol Park
Priyadarsan Patra
Jorgen Peddersen
Nei-Chiung Perng
Frederic Petrot
Stephen Plaza
Jan Poland
Swarna Radhakrishnan
Roshan Ragel
Subramanian Rajagopalan
Frederic Rousseau
Jaijeet Roychowdhury
Saowanee Saewong

Karem Sakallah
Roy Sanghamitra
Sachin Sapatnekar
Takashi Sato
Prashant Saxena
Muzhou Shao
Seng Lin Shee
Farhana Sheikh
Hossein Sheini
Weiping Shi
Atsufumi Shibayama
Youngsoo Shin
Joseph Shinnerl
Youichi Shiraishi
Alan Su
Chau-Chin Su
Qing Su
Dong-kwan Suh
Pei-Lun Sui
Masaya Sumita
C. N. Sze
Kenton Sze
Akira Tada
Masamichi Takagi
Atsushi Takahashi
Yasuhiro Takashima
Koichiro Takayama
Yoshinori Takeuchi
Emil Talpes
Yutaka Tamiya
Sheldon Tan
Genichi Tanaka
Katsunori Tanaka
Song-Jian Tang
Hiroshi Tanimoto
Yuichi Tanji
Yoshinori Tomita
Hiroyuki Tomiyama
Masahiko Toyonaga
Ming-Chao Tsai
Ren-Song Tsay
Hsueh-Wen Tseng
Shuji Tsukiyama
Yojiro Uchimura
Prab Varma
Ilya Wagner
Shin'ichi Wakabayashi
Chun-Yao Wang
D. H. Wang

Farn Wang
Kuo-Hua Wang
Ting-Chi Wang
Xiaoyi Wang
Takayuki Watanabe
Jae-Kyung Wee
Xinjie Wei
Martin D. F. Wong
Y.C. Wong
Chin-Hsien Wu
David Wu
Min Xie
Jingyu Xu
Ryuichi Yamaguchi

Tatsuya Yamamoto
Shigeru Yamashita
Chuan-Yue Yang
Fu-Ching Yang
Yang Yang
Yulu Yang
Zijiang Yang
Akira Yasuda
Tse-Chen Yeh
Shoko Yonezawa
Jun-hee Yoo
Seung-Mok Yoo
Sungjoo Yoo
Atsushi Yoshikawa

Evangeline F.Y. Young
Ping-Hung Yuh
Nacer eddine Zergainoh
Lizheng Zhang
Yan Zhang
Zhengya Zhang
Xin Zhao
Hai Zhou
Yi Zhou
Haikun Zhu
Qiang Zhu
Radu Zlatanovici

Contents

ASP-DAC 2006 General Chair's Message	iii
Message from Technical Program Committee	iv
University LSI Design Contest	vi
Designers' Forum	vii
Keynote Addresses	viii
ASP-DAC 2005 Best Papers	xii
ASP-DAC 2006 Best Papers	xiii
ASP-DAC 2006 Organizing Committee	xv
ASP-DAC Steering Committee	xix
ASP-DAC 2006 Technical Program Committee	xxiii
University LSI Design Contest Committee	xxvi
Industry Liaison	xxvii
List of Reviewers	xxviii

CONTENTS

Session 1A
Formal Methods for Coverage and Scalable Verification

Chair(s): Kiyoharu Hamaguchi, Valeria Bertacco

1A-1 **Transition-Based Coverage Estimation for Symbolic Model Checking**
Xingwen Xu, Shinji Kimura, Kazunari Horikawa, Takehiko Tsuchiya 1

1A-2 **Word Level Functional Coverage Computation**
Bijan Alizadeh 7

1A-3 **Discovering the Input Assumptions in Specification Refinement Coverage**
Prasenjit Basu, Sayantan Das, Pallab Dasgupta, Partha P Chakrabarti 13

1A-4 **Refinement Strategies for Verification Methods Based on Datapath Abstraction**
Zaher Semon Andraus, Mark Hammond Liffiton, Karem Ahmad Sakallah 19

1A-5 **Generation of Shorter Sequences for High Resolution Error Diagnosis Using Sequential SAT**
Sung-Jui Pan, Kwang-Ting Cheng, John Moondanos, Ziyad Hanna 25

Session 1B
Interconnect for High-End SoC

Chair(s): Yoshinori Takeuchi, Juinn-Dar Huang

1B-1 **Constraint-Driven Bus Matrix Synthesis for MPSoC**
Sudeep Pasricha, Nikil Dutt, Mohamed Ben-Romdhane 30

1B-2 **Improving Routing Efficiency for Network-on-Chip through Contention-Aware Input Selection**
Dong Wu, Bashir M. Al-Hashimi, Marcus T. Schmitz 36

1B-3 **Physical Design Implementation of Segmented Buses to Reduce Communication Energy**
Jin Guo, Antonis Papanikolaou, Pol Marchal, Francky Catthoor 42

1B-4 **Co-Synthesis of a Configurable SoC Platform based on a Network on Chip Architecture**
Mário Pereira Véstias, Horácio Neto 48

1B-5 **Customized SIMD Unit Synthesis for System on Programmable Chip - A Foundation for HW/SW Partitioning with Vectorization**
Muhammad Omer Cheema, Omar Hammami 54

xxxii

Session 1C
Timing Analysis and Optimization

Chair(s): Ryuichi Yamaguchi, Atsushi Kurokawa

1C-1 **Robust Analytical Gate Delay Modeling for Low Voltage Circuits**
Anand Ramalingam, Sreekumar V. Kodakara, Anirudh Devgan, David Z. Pan 61

1C-2 **CGTA: Current Gain-based Timing Analysis for Logic Cells**
Shahin Nazarian, Massoud Pedram, Tao Lin, Emre Tuncer 67

1C-3 **Efficient Static Timing Analysis Using a Unified Framework for False Paths and Multi-Cycle Paths**
Shuo Zhou, Bo Yao, Hongyu Chen, Yi Zhu, Chung-Kuan Cheng, Mike Hutton . 73

1C-4 **Crosstalk Analysis using Reconvergence Correlation**
Sachin Shrivastava, Rajendra Pratap, Harindranath Parameswaran, Manuj Verma 79

1C-5 **Process-Induced Skew Reduction in Nominal Zero-Skew Clock Trees**
Matthew R. Guthaus, Dennis Sylvester, Richard B. Brown 84

Session 1D
University Design Contest

Chair(s): Kazutoshi Kobayashi, Takahiko Arakawa

1D-1 **A Low Dynamic Power and Low Leakage Power 90-nm CMOS Square-Root Circuit**
Tadayoshi Enomoto, Nobuaki Kobayashi 90

1D-2 **A High-Throughput Low-Power Fully Parallel 1024-bit 1/2-Rate Low Density Parity Check Code Decoder in 3-Dimensional Integrated Circuits**
Lili Zhou, Cherry Wakayama, Nuttorn Jangkrajarng, Bo Hu, Richard Shi 92

1D-3 **A 16-Bit, Low-Power Microsystem with Monolithic MEMS-LC Clocking**
Robert M. Senger, Eric D. Marsman, Michael S. McCorquodale, Richard B. Brown 94

1D-4 **Ultra-Low Voltage Power Management Circuit and Computation Methodology for Energy Harvesting Applications**
Chi-Ying Tsui, Hui Shao, Wing-Hung Ki, Feng Su 96

1D-5 **A 0.5-V Sigma-Delta Modulator Using Analog T-Switch Scheme for the Subthreshold Leakage Suppression**
Koichi Ishida, Atit Tamtrakarn, Takayasu Sakurai 98

1D-6 **An Implementation of a CMOS Down-Conversion Mixer for GSM1900 Receiver**
Fangqing Chu, Wei Li, Junyan Ren 100

CONTENTS

1D-7 **Integrated Direct Output Current Control Switching Converter using Symmetrically-Matched Self-Biased Current Sensors**
Yat-Hei Lam, Suet-Chui Koon, Wing-Hung Ki, Chi-Ying Tsui 102

1D-8 **Adaptively-Biased Capacitor-Less CMOS Low Dropout Regulator with Direct Current Feedback**
Yat-Hei Lam, Wing-Hung Ki, Chi-Ying Tsui 104

1D-9 **A Built-in Power Supply Noise Probe for Digital LSIs**
Mitsuya Fukazawa, Koichiro Noguchi, Makoto Nagata, Kazuo Taki 106

1D-10 **A 476-gate-count Dynamic Optically Reconfigurable Gate Array VLSI chip in a standard 0.35um CMOS Technology**
Minoru Watanabe, Fuminori Kobayashi 108

1D-11 **Measurement Results of Within-Die Variations on a 90nm LUT Array for Speed and Yield Enhancement of Reconfigurable Devices**
Kazuya Katsuki, Manabu Kotani, Kazutoshi Kobayashi, Hidetoshi Onodera .. 110

1D-12 **High-Throughput Decoder for Low-Density Parity-Check Code**
Tatsuyuki Ishikawa, Kazunori Shimizu, Takeshi Ikenaga, Satoshi Goto 112

1D-13 **Hardware Implementation of Super Minimum All Digital FM Demodulator**
Nursani Rahmatullah, Arif Nugroho 114

1D-14 **Designing a Custom Architecture for DCT Using NISC Technology**
Bita Gorjiara, Mehrdad Reshadi, Daniel Gajski 116

1D-15 **A 52mW 1200MIPS Compact DSP for Multi-Core Media SoC**
Shih-Hao Ou, Tay-Jyi Lin, Chao-Wei Huang, Yu-Ting Kuo, Chie-Min Chao, Chih-Wei Liu, Chein-Wei Jen 118

1D-16 **Implementation of H.264/AVC Decoder for Mobile Video Applications**
Suh Ho Lee, Ji Hwan Park, Seon Wook Kim, Sung Jea Ko, Suki Kim 120

1D-17 **A High-Performance Platform-Based SoC for Information Security**
Min Wu, Xiaoyang Zeng, Jun Han, Yongyi Wu, Yibo Fan 122

1D-18 **Configurable Multi-Processor Architecture and its Processor Element Design**
Tsutomu Nishimura, Takuji Miki, Hiroaki Sugiura, Yuki Matsumoto, Masatsugu Kobayashi, Toshiyuki Kato, Tsutomu Eda, Hironori Yamauchi 124

1D-19 **Design and Implementation of Transducer for ARM-TMS Communication**
Hansu Cho, Samar Abdi, Daniel Gajski 126

Session 2A
Software Techniques for Efficient SoC Design

Chair(s): Qiang Zhu, Ahmed Jerraya

2A-1 **Energy Savings through Embedded Processing on Disk System**
Seung Woo Son, Guangyu Chen, Mahmut Kandemir, Fehui Li 128

CONTENTS

2A-2 **Energy-Aware Computation Duplication for Improving Reliability in Embedded Chip Multiprocessors**
Guilin Chen, Mahmut Kandemir, Feihui Li 134

2A-3 **Object Duplication for Improving Reliability**
Guilin Chen, Guangyu Chen, Mahmut Kandemir, Narayanan Vijaykrishnan, Mary Jane Irwin 140

2A-4 **Mapping and Configuration Methods for Multi-Use-Case Networks on Chips**
Srinivasan Murali, Martijn Coenen, Andrei Radulescu, Kees Goossens, Giovanni De Micheli 146

2A-5 **Conversion of Reference C Code to Dataflow Model: H.264 Encoder Case Study**
Hyeyoung Hwang, Taewook Oh, Hyunuk Jung, Soonhoi Ha 152

Session 2B
Application Examples with Leading Edge Design Methodology

Chair(s): In-Cheol Park, Hideharu Amano

2B-1 **SAVS: A Self-Adaptive Variable Supply-Voltage Technique for Process - Tolerant and Power-Efficient Multi-issue Superscalar Processor Design**
Hai Li, Yiran Chen, Kaushik Roy, Cheng-Kok Koh 158

2B-2 **The Design and Implementation of a Low-Latency On-Chip Network**
Robert Mullins, Andrew West, Simon Moore 164

2B-3 **A Near Optimal Deblocking Filter for H.264 Advanced Video Coding**
Shen-Yu Shih, Cheng-Ru Chang, Youn-Long Lin 170

2B-4 **Image Segmentation and Pattern Matching Based FPGA/ASIC Implementation Architecture of Real-Time Object Tracking**
Kousuke Yamaoka, Takashi Morimoto, Hidekazu Adachi, Tetsushi Koide, Hans Juergen Mattausch 176

2B-5 **Prefetching-Aware Cache Line Turnoff for Saving Leakage Energy**
Ismail Kadayif, Mahmut Kandemir, Feihui Li 182

Session 2C
Placement

Chair(s): Evangeline F.Y. Young, Shin'ichi Wakabayashi

2C-1 **A Robust Detailed Placement for Mixed-Size IC Designs**
Jason Cong, Min Xie 188

2C-2 **FastPlace 2.0: An Efficient Analytical Placer for Mixed-Mode Designs**
Natarajan Viswanathan, Min Pan, Chris Chu 195

CONTENTS

2C-3 **Timing-Driven Placement Based on Monotone Cell Ordering Constraints**
Chanseok Hwang, Massoud Pedram 201

2C-4 **Constraint Driven I/O Planning and Placement for Chip-package Co-design**
Jinjun Xiong, Yiu-Chung Wong, Egino Sarto, Lei He 207

2C-5 **Simultaneous Block and I/O Buffer Floorplanning for Flip-Chip Design**
Chih-Yang Peng, Wen-Chang Chao, Yao-Wen Chang, J.-H. Wang 213

Session 2D
Special Session: Electrothermal Design of Nanoscale Integrated Circuits

Chair(s): Dennis Sylvester, Mongkol Ekpanyapong

2D-1 **Electrothermal Analysis and Optimization Techniques for Nanoscale Integrated Circuits**
Yong Zhan, Brent Goplen, Sachin S. Sapatnekar 219

2D-2 **Electrothermal Engineering in the Nanometer Era: From Devices and Interconnects to Circuits and Systems**
Kaustav Banerjee, Sheng-Chih Lin, Navin Srivastava 223

2D-3 **Area Optimization for Leakage Reduction and Thermal Stability in Nanometer Scale Technologies**
Ja Chun Ku, Yehea Ismail 231

2D-4 **Compact Thermal Models for Estimation of Temperature-dependent Power/Performance in FinFET Technology**
Aditya Bansal, Mesut Meterelliyoz, Siddharth Singh, Jung Hwan Choi, Jayathi Murthy, Kaushik Roy 237

Session 3A
Logic Synthesis

Chair(s): Shinji Kimura, Shih-Chieh Chang

3A-1 **An Anytime Symmetry Detection Algorithm for ROBDDs**
Neil Kettle, Andy King 243

3A-2 **High Level Equivalence Symmetric Input Identification**
Ming-Hong Su, Chun-Yao Wang 249

3A-3 **Fast Multi-Domain Clock Skew Scheduling for Peak Current Reduction**
Shih-Hsu Huang, Chia-Ming Chang, Yow-Tyng Nieh 254

3A-4 **Low Area Pipelined Circuits by Multi-clock Cycle Paths and Clock Scheduling**
Bakhtiar Affendi Rosdi, Atsushi Takahashi 260

CONTENTS

3A-5 **A Transduction-based Framework to Synthesize RSFQ Circuits**
Shigeru Yamashita, Katsunori Tanaka, Hideyuki Takada, Koji Obata, Kazuyoshi Takagi 266

Session 3B
Future Technical Directions for Design Automation

Chair(s): Makoto Nagata, Ryuichi Fujimoto

3B-1 **Fast Simulation of Large Networks of Nanotechnological and Biochemical Oscillators for Investigating Self-Organization Phenomena**
Xiaolue Lai, Jaijeet Roychowdhury 273

3B-2 **Newton: A Library-Based Analytical Synthesis Tool for RF-MEMS Resonators**
Michael S. McCorquodale, James L. McCann, Richard B. Brown 279

3B-3 **Jitter Decomposition in Ring Oscillators**
Qingqi Dou, Jacob Abraham 285

3B-4 **A Fast Methodology for First-Time-Correct Design of PLLs Using Nonlinear Phase-Domain VCO Macromodels**
Prashant Goyal, Xiaolue Lai, Jaijeet Roychowdhury 291

3B-5 **Double Edge Triggered Feedback Flip-Flop in Sub 100nm Technology**
Seid Hadi Rasouli, Amir Amirabadi, Azam Seyedi, Ali Afzali-Kusha 297

Session 3C
Routing and Interconnect Optimization

Chair(s): Youichi Shiraishi, Tong Jing

3C-1 **Post-Routing Redundant Via Insertion for Yield/Reliability Improvement**
Kuang-Yao Lee, Ting-Chi Wang 303

3C-2 **Temperature-Aware Routing in 3D ICs**
Tianpei Zhang, Yong Zhan, Sachin S. Sapatnekar 309

3C-3 **Closed Form Solution for Optimal Buffer Sizing Using The Weierstrass Elliptic Function**
Sebastian Vogel, Martin D.F. Wong 315

3C-4 **An O(mn) Time Algorithm for Optimal Buffer Insertion of Nets with m Sinks**
Zhuo Robert Li, Weiping Shi 320

3C-5 **Spec-based Flip-Flop and Latch Repeater Planning**
Man Chung Hon 326

xxxvii

CONTENTS

Session 3D
Special Session: Flash Memory in Embedded Systems

Chair(s): Tohru Ishihara, Hiroyuki Tomiyama

3D-1 **Current Trends in Flash Memory Technology**
Sang Lyul Min, Eyee Hyun Nam 332

3D-2 **Configurability of Performance and Overheads in Flash Management**
Tei-Wei Kuo, Jen-Wei Hsieh, Li-Pin Chang, Yuan-Hao Chang 334

Session 4A
Resolving Timing Issues: Design and Test

Chair(s): Masaki Hashizume, Kazumi Hatayama

4A-1 **Delay Defect Screening for a 2.16GHz SPARC64 Microprocessor**
Noriyuki Ito, Akira Kanuma, Daisuke Maruyama, Hitoshi Yamanaka, Tsuyoshi Mochizuki, Osamu Sugawara, Chihiro Endoh, Masahiro Yanagida, Takeshi Kono, Yutaka Isoda, Kazunobu Adachi, Takahisa Hiraide, Shigeru Nagasawa, Yaroku Sugiyama, Eizo Ninoi 342

4A-2 **A Dynamic Test Compaction Procedure for High-quality Path Delay Testing**
Masayasu Fukunaga, Seiji Kajihara, Xiaoqing Wen, Toshiyuki Maeda, Shuji Hamada, Yasuo Sato 348

4A-3 **Delay Variation Tolerance for Domino Circuits**
Kai-Chiang Wu, Cheng-Tao Hsieh, Shih-Chieh Chang 354

4A-4 **Efficient Identification of Multi-Cycle False Path**
Kai Yang, Tim Cheng 360

4A-5 **IEEE Standard 1500 Compatible Interconnect Diagnosis for Delay and Crosstalk Faults**
Katherine Shu-Min Li, Yao-Wen Chang, Chauchin Su, Chung-Len Lee, Jwu E Chen 366

Session 4B
Leading Edge Design Methodology for SoCs and SiPs

Chair(s): Satoshi Matsushita

4B-1 **High-Level Architecture Exploration for MPEG4 Encoder with Custom Parameters**
Marius Bonaciu, Aimen Bouchhima, Wassim Youssef, Xi Chen, Wander Cesario, Ahmed Jerraya 372

4B-2 **Programmable Numerical Function Generators Based on Quadratic Approximation: Architecture and Synthesis Method**
Shinobu Nagayama, Tsutomu Sasao, Jon Butler 378

CONTENTS

4B-3 **An Automated Design Flow for 3D Microarchitecture Evaluation**
Jason Cong, Ashok Jagannathan, Yuchun Ma, Glenn Reinman, Jie Wei, Yan Zhang 384

4B-4 **Optimal Topology Exploration for Application-Specific 3D Architectures**
Ozcan Ozturk, Feng Wang, Mahmut Kandemir, Yuan Xie 390

4B-5 **Task Placement Heuristic Based on 3D-Adjacency and Look-Ahead in Reconfigurable Systems**
Jesus Tabero, Julio Septien, Hortensia Mecha, Daniel Mozos 396

Session 4C
Advanced Circuit Simulation

Chair(s): Hideki Asai, C.J. Richard Shi

4C-1 **A Quasi-Newton Preconditioned Newton-Krylov Method for Robust and Efficient Time-Domain Simulation of Integrated Circuits with Strong Parasitic Couplings**
Zhao Li, Richard Shi 402

4C-2 **An Efficient and Globally Convergent Homotopy Method for Finding DC Operating Points of Nonlinear Circuits**
Kiyotaka Yamamura, Wataru Kuroki 408

4C-3 **Optimization of Circuit Trajectories: An Auxiliary Network Approach**
Baohua Wang, Pinaki Mazumder 416

4C-4 **SASIMI: Sparsity-Aware Simulation of Interconnect-Dominated Circuits with Non-Linear Devices**
Jitesh Jain, Stephen F Cauley, Cheng-Kok Koh, Venkataramanan Balakrishnan 422

4C-5 **An Unconditional Stable General Operator Splitting Method for Transistor Level Transient Analysis**
Zhengyong Zhu, Rui Shi, Chung-Kuan Cheng, Ernest S. Kuh 428

Session 4D
Special Session: Open Access Overview

Chair(s): John Darringer

4D-1 **An Introduction to OpenAccess -An Open Source Data Model and API for IC Design-**
Michaela Guiney, Eric Leavitt 434

4D-2 **Open Access Overview "Industrial Experience"**
Yoshio Inoue 437

4D-3 **EDA Vendor Adoption**
Hillel Ofek 439

CONTENTS

4D-4   **Utility of the OpenAccess Database in Academic Research**
        David Papa, Igor Markov, Philip Chong 440

Session 5A
Advances in Simulation Technologies

Chair(s): Shin'ichi Minato, Karem Sakallah

5A-1   **Depth-Driven Verification of Simultaneous Interfaces**
        Ilya Wagner, Valeria Bertacco, Todd Austin 442

5A-2   **FSM-Based Transaction-Level Functional Coverage for Interface Compliance Verification**
        Man-Yun Su, Che-Hua Shih, Juinn-Dar Huang, Jing-Yang Jou 448

5A-3   **Hardware Debugging Method Based on Signal Transitions and Transactions**
        Nobuyuki Ohba, Kohji Takano 454

5A-4   **Cycle Error Correction in Asynchronous Clock Modeling for Cycle-Based Simulation**
        Junghee Lee, Joonhwan Yi 460

5A-5   **A Fast Logic Simulator Using a Look Up Table Cascade Emulator**
        Hiroki Nakahara, Tsutomu Sasao, Munehiro Matsuura 466

Session 5B
Scheduling for Embedded Systems

Chair(s): Sri Parameswaran, Sang Lyul Min

5B-1   **Power-Aware Scheduling and Dynamic Voltage Setting for Tasks Running on a Hard Real-Time System**
        Peng Rong, Massoud Pedram 473

5B-2   **Optimal TDMA Time Slot and Cycle Length Allocation for Hard Real-Time Systems**
        Ernesto Wandeler, Lothar Thiele 479

5B-3   **POSIX modeling in SystemC**
        Hector Posadas, Jesus Adamez, Pablo Sanchez, Eugenio Villar, Francisco Blasco 485

5B-4   **PARLGRAN: Parallelism Granularity Selection for Scheduling Task Chains on Dynamically Reconfigurable Architectures**
        Sudarshan Banerjee, Elaheh Bozorgzadeh, Nikil Dutt 491

5B-5   **Memory Optimal Single Appearance Schedule with Dynamic Loop Count for Synchronous Dataflow Graphs**
        Hyunok Oh, Nikil Dutt, Soonhoi Ha 497

CONTENTS

Session 5C
High Frequency Interconnect Effects in Nanometer Technology

Chair(s): Charlie Chung-Ping Chen, Noel Menezes

5C-1 **Wire Sizing with Scattering Effect for Nanoscale Interconnection**
Sean X. Shi, David Z. Pan 503

5C-2 **Adaptive Admittance-based Conductor Meshing for Interconnect Analysis**
Ya-Chi Yang, Cheng-Kok Koh, Venkataramanan Balakrishnan 509

5C-3 **Interconnect RL Extraction at a Single Representative Frequency**
Akira Tsuchiya, Masanori Hashimoto, Hidetoshi Onodera 515

5C-4 **An Efficient Algorithm for 3-D Reluctance Extraction Considering High Frequency Effect**
Mengsheng Zhang, Wenjian Yu, Yu Du, Zeyi Wang 521

5C-5 **Macromodelling Oscillators Using Krylov-Subspace Methods**
Xiaolue Lai, Jaijeet Roychowdhury 527

Session 5D
Designers' Forum: Low Power Design

Chair(s): Haruyuki Tago, Makoto Ikeda

5D-1 **Low-Power Design Methodology for Module-wise Dynamic Voltage and Frequency Scaling with Dynamic De-skewing Systems**
Takeshi Kitahara, Hiroyuki Hara, Shinichiro Shiratake, Yoshiki Tsukiboshi, Tomoyuki Yoda, Tetsuaki Utsumi, Fumihiro Minami 533

5D-2 **Single-Chip Multi-Processor Integrating Quadruple 8-Way VLIW Processors with Interface Timing Analysis Considering Power Supply Noise**
Satoshi Imai, Atsuki Inoue, Motoaki Matsumura, Kenichi Kawasaki, Atsuhiro Suga 541

5D-3 **A System-level Power-estimation Methodology based on IP-level Modeling, Power-level Adjustment, and Power Accumulation**
Masafumi Onouchi, Tetsuya Yamada, Kimihiro Morikawa, Isamu Mochizuki, Hidetoshi Sekine 547

5D-4 **PowerViP: SoC Power Estimation Framework at Transaction Level**
Ikhwan Lee, Hyunsuk Kim, Peng Yang, Sungjoo Yoo, Eui-Young Chung, Kyu-Myung Choi, Jeong-Taek Kong, Soo-Kwan Eo 551

xli

CONTENTS

Session 6A
Power Optimization of Large-Scale Circuits

Chair(s): Sheldon Tan, David Z. Pan

6A-1 **Mathematically Assisted Adaptive Body Bias (ABB) for Temperature Compensation in Gigascale LSI Systems**
Sanjay V Kumar, Chris H Kim, Sachin S Sapatnekar 559

6A-2 **Analysis and Optimization of Gate Leakage Current of Power Gating Circuits**
Hyung-Ock Kim, Youngsoo Shin 565

6A-3 **Delay Modeling and Static Timing Analysis for MTCMOS Circuits**
Naoaki Ohkubo, Kimiyoshi Usami 570

6A-4 **Switching-Activity Driven Gate Sizing and Vth Assignment for Low Power Design**
Yu-Hui Huang, Po-Yuan Chen, TingTing Hwang 576

6A-5 **Power Driven Placement with Layout Aware Supply Voltage Assignment for Voltage Island Generation in Dual-Vdd Designs**
Bin Liu, Yici Cai, Qiang Zhou, Xianlong Hong 582

Session 6B
Advanced Memory and Processor Architectures for MPSoC

Chair(s): Soonhoi Ha, Youn-Long Lin

6B-1 **Reusable Component IP Design using Refinement-based Design Environment**
Sanggyu Park, Sang-Yong Yoon, Soo-Ik Chae 588

6B-2 **An Interface-Circuit Synthesis Method with Configurable Processor Core in IP-Based SoC Designs**
Shunitsu Kohara, Naoki Tomono, Jumpei Uchida, Yuichiro Miyaoka, Nozomu Togawa, Masao Yanagisawa, Tatsuo Ohtsuki 594

6B-3 **A Real-Time and Bandwidth Guaranteed Arbitration Algorithm for SoC Bus Communication**
Chien-Hua Chen, Geeng-Wei Lee, Juinn-Dar Huang, Jing-Yang Jou 600

6B-4 **Hierarchical Memory Size Estimation for Loop Fusion and Loop Shifting in Data-Dominated Applications**
Qubo Hu, Arnout Vandecappelle, Martin Palkovic, Per Gunnar Kjeldsberg, Erik Brockmeyer, Francky Catthoor 606

6B-5 **A Novel Instruction Scratchpad Memory Optimization Method based on Concomitance Metric**
Andhi Janapsatya, Aleksandar Ignjatovic, Sri Parameswaran 612

xlii

CONTENTS

Session 6C
New Routing Techniques

Chair(s): Ting-Chi Wang, Vijay Pitchumani

6C-1 **DraXRouter: Global Routing in X-Architecture with Dynamic Resource Assignment**
Zhen Cao, Tong Jing, Yu Hu, Yiyu Shi, Xianlong Hong, Xiaodong Hu, Guiying Yan 618

6C-2 **Diagonal Routing in High Performance Microprocessor Design**
Noriyuki Ito, Hideaki Katagiri, Ryoichi Yamashita, Hiroshi Ikeda, Hiroyuki Sugiyama, Hiroaki Komatsu, Yoshiyasu Tanamura, Akihiko Yoshitake, Kazuhiro Nonomura, Kinya Ishizaka, Hiroaki Adachi, Yutaka Mori, Yutaka Isoda, Yaroku Sugiyama 624

6C-3 **CDCTree: Novel Obstacle-Avoiding Routing Tree Construction based on Current Driven Circuit Model**
Yiyu Shi, Tong Jing, Lei He, Zhe Feng, Xianlong Hong 630

6C-4 **A Novel Framework for Multilevel Full-Chip Gridless Routing**
Tai-Chen Chen, Yao-Wen Chang, Shyh-Chang Lin 636

6C-5 **Monotonic Parallel and Orthogonal Routing for Single-Layer Ball Grid Array Packages**
Yoichi Tomioka, Atsushi Takahashi 642

Session 7A
Minimization of Test Cost and Power

Chair(s): Seiji Kajihara, Satoshi Ohtake

7A-1 **A Routability Constrained Scan Chain Ordering Technique for Test Power Reduction**
Xuan-Lun Huang, Jiun-Lang Huang 648

7A-2 **FCSCAN: An Efficient Multiscan-based Test Compression Technique for Test Cost Reduction**
Youhua Shi, Nozomu Togawa, Shinji Kimura, Masao Yanagisawa, Tatsuo Ohtsuki 653

7A-3 **Compaction of Pass/Fail-based Diagnostic Test Vectors for Combinational and Sequential Circuits**
Yoshinobu Higami, Kewal K. Saluja, Hiroshi Takahashi, Shin-ya Kobayashi, Yuzo Takamatsu 659

7A-4 **Low-Overhead Design of Soft-Error-Tolerant Scan Flip-Flops with Enhanced-Scan Capability**
Ashish Goel, Swarup Bhunia, Hamid Mahmoodi, Kaushik Roy 665

xliii

CONTENTS

7A-5　**A Memory Grouping Method for Sharing Memory BIST Logic**
Masahide Miyazaki, Tomokazu Yoneda, Hideo Fujiwara 671

Session 7B
Substrate Coupling and Analog Synthesis

Chair(s): Jaijeet Roychowdhury, Tomohisa Kimura

7B-1　**Equivalent Circuit Modeling of Guard Ring Structures for Evaluation of Substrate Crosstalk Isolation**
Daisuke Kosaka, Makoto Nagata 677

7B-2　**A New Boundary Element Method for Accurate Modeling of Lossy Substrates with Arbitrary Doping Profiles**
Xiren Wang, Wenjian Yu, Zeyi Wang 683

7B-3　**Parasitics Extraction Involving 3-D Conductors based on Multi-layered Green's Function**
Zuochang Ye, Zhiping Yu 689

7B-4　**Signal-Path Driven Partition and Placement for Analog Circuit**
Di Long, Xianlong Hong, Sheqin Dong 694

7B-5　**An Approach to Topology Synthesis of Analog Circuits Using Hierarchical Blocks and Symbolic Analysis**
Xiaoying Wang, Lars Hedrich 700

Session 7C
Statistical and Yield Analysis

Chair(s): Hiroo Masuda, Seijiro Moriyama

7C-1　**Statistical Corner Conditions of Interconnect Delay (Corner LPE Specifications)**
Kenta Yamada, Noriaki Oda 706

7C-2　**Speed Binning Aware Design Methodology to Improve Profit under Parameter Variations**
Animesh Datta, Swarup Bhunia, Jung Hwan Choi, Saibal Mukhopadhyay, Kaushik Roy 712

7C-3　**Yield-Area Optimizations of Digital Circuits Using Non-dominated Sorting Genetic Algorithm (YOGA)**
Vineet Agarwal, Janet Wang 718

7C-4　**A Probabilistic Analysis of Pipelined Global Interconnect Under Process Variations**
Navneeth Kankani, Vineet Agarwal, Janet M Wang 724

7C-5　**Yield-Preferred Via Insertion Based on Novel Geotopological Technology**
Fangyi Luo, Yongbo Jia, Wayne Wei-Ming Dai 730

Session 7D
Special Session: H.264/AVC Design Challenges and Solutions

Chair(s): Wayne Wolf

7D-1   **Introduction to H.264 Advanced Video Coding**
Jian-Wen Chen, Chao-Yang Kao, Youn-Long Lin 736

7D-2   **Algorithms and DSP Implementation of H.264/AVC**
Hung-Chih Lin, Yu-Jen Wang, Kai-Ting Cheng, Shang-Yu Yeh, Wei-Nien Chen, Chia-Yang Tsai, Tian-Sheuan Chang, Hsueh-Ming Hang 742

7D-3   **Hardware Architecture Design of an H.264/AVC Video Codec**
Tung-Chien Chen, Chung-Jr Lian, Liang-Gee Chen 750

7D-4   **ASIP Approach for Implementation of H.264/AVC**
Sung Dae Kim, Jeong Hoo Lee, Chung Jin Hyun, Myung Hoon Sunwoo 758

Session 8A
Floorplanning

Chair(s): Yao-Wen Chang, Shigetoshi Nakatake

8A-1   **Fast Substrate Noise-Aware Floorplanning with Preference Directed Graph for Mixed-Signal SOCs**
Minsik Cho, Hongjoong Shin, David Z. Pan 765

8A-2   **A Fixed-die Floorplanning Algorithm Using an Analytical Approach**
Yong Zhan, Yan Feng, Sachin S. Sapatnekar 771

8A-3   **A Multi-Technology-Process Reticle Floorplanner and Wafer Dicing Planner for Multi-Project Wafers**
Chien-Chang Chen, Wai-Kei Mak 777

8A-4   **Design Space Exploration for Minimizing Multi-Project Wafer Production Cost**
Rung-Bin Lin, Meng-Chiou Wu, Wei-Chiu Tseng, Ming-Hsine Kuo, Tsai-Ying Lin, Shr-Cheng Tsai 783

8A-5   **SAT-Based Optimal Hypergraph Partitioning with Replication**
Michael G. Wrighton, Andre M. DeHon 789

Session 8B
Memory Optimization for Embedded Systems

Chair(s): Hiroyuki Tomiyama, Preeti Ranjan Panda

8B-1   **Finding Optimal L1 Cache Configuration for Embedded Systems**
Andhi Janapsatya, Aleksandar Ignjatovic, Sri Parameswaran 796

CONTENTS

8B-2 **Memory Size Computation for Multimedia Processing Applications**
Hongwei Zhu, Ilie I. Luican, Florin Balasa 802

8B-3 **Maximizing Data Reuse for Minimizing Memory Space Requirements and Execution Cycles**
Mahmut Kandemir, Guangyu Chen, Feihui Li 808

8B-4 **Compiler-Guided Data Compression for Reducing Memory Consumption of Embedded Applications**
Ozcan Ozturk, Guangyu Chen, Mahmut Kandemir, Ibrahim Kolcu 814

8B-5 **Analysis of Scratch-Pad and Data-Cache Performance Using Statistical Methods**
Javed Absar, Francky Catthoor 820

Session 8C
Inductive Issues in Power Grids and Packages

Chair(s): Takashi Sato, Yehea Ismail

8C-1 **Efficient Early Stage Resonance Estimation Techniques for C4 Package**
Jin Shi, Yici Cai, Shelton X-D Tan, Xianlong Hong 826

8C-2 **Parallel-Distributed Time-Domain Circuit Simulation of Power Distribution Networks with Frequency-Dependent Parameters**
Takayuki Watanabe, Yuichi Tanji, Hidemasa Kubota, Hideki Asai 832

8C-3 **Power Distribution Techniques for Dual VDD Circuits**
Sarvesh Hemchandra Kulkarni, Dennis Sylvester 838

8C-4 **Calculating Frequency-Dependent Inductance of VLSI Interconnect by Complete Multiple Reciprocity Boundary Element Method**
Changhao Yan, Wenjian Yu, Zeyi Wang 844

8C-5 **Controlling Inductive Cross-talk and Power in Off-chip Buses using CODECs**
Brock LaMeres, Kanupriya Gulati, Sunil Khatri 850

Session 8D
Designers' Forum: "Cell" Processor

Chair(s): Haruyuki Tago, Makoto Ikeda

8D-1 **A New Test and Characterization Scheme for 10+ GHz Low Jitter Wide Band PLL**
Kazuhiko Miki, David Boerstler, Eskinder Hailu, Jieming Qi, Sarah Pettengill, Yuichi Goto 856

8D-2 **An SPU Reference Model for Simulation, Random Test Generation and Verification**
Yukio Watanabe, Balazs Sallay, Brad Michael, Daniel Brokenshire, Gavin Meil, Hazim Shafi, Daisuke Hiraoka 860

CONTENTS

8D-3 **A Cycle Accurate Power Estimation Tool**
Rajat Chaudhry, Daniel Stasiak, Stephen Posluszny, Sang Dhong 867

8D-4 **Key Features of the Design Methodology Enabling a Multi-Core SoC Implementation of a First-Generation CELL Processor**
Dac Pham, Hans-Werner Anderson, Erwin Behnen, Mark Bolliger, Sanjay Gupta, Peter Hofstee, Paul Harvey, Charles Johns, Jim Kahle, Atsushi Kameyama, John Keaty, Bob Le, Sang Lee, Tuyen Nguyen, John Petrovick, Mydung Pham, Juergen Pille, Stephen Posluszny, Mack Riley, Joseph Verock, James Warnock, Steve Weitzel, Dieter Wendel 871

Session 9A
High-Level Synthesis

Chair(s): Shigeru Yamashita, Youngsoo Shin

9A-1 **TAPHS: Thermal-Aware Unified Physical-Level and High-Level Synthesis**
Zhenyu (Peter) Gu, Yonghong Yang, Jia Wang, Robert P. Dick, Li Shang 879

9A-2 **An Automated, Efficient and Static Bit-width Optimization Methodology Towards Maximum Bit-width-to-Error Tradeoff With Affine Arithmetic Model**
Yu Pu, Yajun Ha 886

9A-3 **Abridged Addressing: A Low Power Memory Addressing Strategy**
Preeti Ranjan Panda 892

9A-4 **Using Speculative Computation and Parallelizing Techniques to Improve Scheduling of Control based Designs**
Roberto Cordone, Fabrizio Ferrandi, Gianluca Palermo, Marco Domenico Santambrogio, Donatella Sciuto 898

9A-5 **Worst Case Execution Time Analysis for Synthesized Hardware**
Jun-hee Yoo, Xingguang Feng, Kiyoung Choi, Eui-Young Chung, Kyu-Myung Choi 905

Session 9B
Modeling, Compilation and Optimization of Embedded Architectures

Chair(s): Hiroyuki Tomiyama, Lovic Gauthier

9B-1 **Workload Prediction and Dynamic Voltage Scaling for MPEG Decoding**
Ying Tan, Parth Malani, Qinru Qiu, Qing Wu 911

9B-2 **Lazy BTB: Reduce BTB Energy Consumption Using Dynamic Profiling**
Yen-Jen Chang 917

9B-3 **Cache Size Selection for Performance, Energy and Reliability of Time-Constrained Systems**
Yuan Cai, Marcus T. Schmitz, Alireza Ejlali, Bashir M. Al-Hashimi, Sudhakar M. Reddy 923

xlvii

CONTENTS

9B-4 **Reducing Dynamic Compilation Overhead by Overlapping Compilation and Execution**
Priya Unnikrishnan, Mahmut Kandemir, Feihui Li 929

9B-5 **Functional Modeling Techniques for Efficient Sw Code Generation of Video Codec Application**
Sang-Il Han, Soo-Ik Chae, Ahmed Amine Jerraya 935

Session 9C
Statistical Design

Chair(s): Sachin Sapatnekar, Sunil Khatri

9C-1 **Convergence-Provable Statistical Timing Analysis with Level-Sensitive Latches and Feedback Loops**
Lizheng Zhang, Jengliang Tsai, Weijen Chen, Yuhen Hu, Charlie Chungping Chen 941

9C-2 **Parameterized Block-Based Non-Gaussian Statistical Gate Timing Analysis**
Soroush Abbaspour, Hanif Fatemi, Massoud Pedram 947

9C-3 **Statistical Leakage Minimization through Joint Selection of Gate Sizes, Gate Lengths and Threshold Voltage**
Sarvesh Bhardwaj, Yu Cao, Sarma Vrudhula 953

9C-4 **Statistical Bellman-Ford Algorithm With An Application to Retiming**
Mongkol Ekpanyapong, Thaisiri Watewai, Sung Kyu Lim 959

9C-5 **An Exact Algorithm for the Statistical Shortest Path Problem**
Liang Deng, Martin D. F. Wong 965

Author Index A-1

Transition-Based Coverage Estimation for Symbolic Model Checking

Xingwen Xu   Shinji Kimura

Graduate School of IPS
Waseda University
2-7 Hibikino, 808-0135 Japan

Kazunari Horikawa   Takehiko Tsuchiya

System LSI Design Division
Toshiba Corporation
Sawasaki, 210-8520 Japan

Abstract—Lack of complete formal specification is one of the major obstacles for the deployment of model checking. Coverage estimation addresses this issue by revealing the unverified part of the design according to the specified properties. In this paper we propose a new transition-based coverage metric to evaluate the completeness of properties for symbolic model checking. It is more comprehensive and accurate than the existing coverage metrics for model checking. An efficient symbolic algorithm is presented for computing the transition coverage for a subset of ACTL. Our coverage estimator has been applied to the model checking of a cache coherence protocol. We uncovered several coverage holes including one that eventually led to the discovery of a design bug.

I. INTRODUCTION

Model checking [1] proves whether a system satisfies a set of properties under all possible input sequences. However, to model-check a complex design, it is very hard to determine whether sufficient properties have been specified or not [5, 13]. The completeness of formal specification needs to be evaluated for ensuring that there is no unknown behavior in the implementation, assuming the unknown behaviors often contain design bugs. Coverage estimation for model checking compares the given specification with a given implementation and deduces the parts of the implementation that are not covered by the specification. Additional properties then should be specified to close such coverage gap. Assisting the model checking process with coverage estimation can achieve higher degree of confidence in the verification results [10].

One of the most popular coverage methods for model checking is a state-based coverage metric based on state perturbation [2, 3]. Informally, a state is covered by a verified property with respect to an observed signal if changing the value of the signal in that state will cause the property to fail. The accuracy of the coverage computation algorithm is improved in [3]. The state coverage metric has successfully uncovered several meaningful coverage holes in real-word model checking projects [2]. However, one major limitation of the state coverage metric is that it is based on states, not transitions or pathes. A state may be reached via several transitions. Property verification over any of those transitions will cover that state. Consequently the state metric leaves quite a large portion of the design's behaviors unchecked. since design errors usually creep in on the transitions [15], the limitation of the state metric might leave such bugs escaped.

To provide more comprehensive coverage analysis, a transition traversal coverage method is proposed in [4]. Transitions of FSM are covered if they are traversed by verified properties according to the semantics of CTL. The authors provide a novel symbolic approach to compute the transition coverage. However, since the transition traversal dose not consider the signals' values, transition traversal coverage can not precisely reflect the completeness of properties.

Other mutation coverage metrics for model checking are introduced in [6, 7], including metrics based on omitting or replacing transitions (or paths) of the finite state machine. However, the practicality of these metrics has not been full established. Besides perturbations on FSM, the high-level fault model is introduced for generating the perturbed implementation in coverage method of [8]. Farn et. al. also have presented a numerical coverage estimation for the symbolic simulation of real-time systems focusing on safety analysis [9].

In this paper, we propose a new transition-based coverage metric for symbolic model checking based on a novel transition perturbation model. The transition perturbation model is able to pinpoint the transition through which the value of selected observed signal is checked by properties. The transition-based coverage is much more comprehensive than the existing state coverage since the coverage space expands from all states to all transitions of FSM. It avoids the false sense of completeness by the high coverage in the transition traversal method by considering perturbation. We present an efficient symbolic algorithm for computing the transition coverage for a subset of ACTL. The algorithm has the same order of complexity as a model checking algorithm. We have integrated it with the model checking engine of VIS2.0 [14]. The coverage estimator has been applied to the model checking of a cache coherence protocol. We discovered several coverage holes including one that eventually led to the discovery of a design bug. And the computation overhead is less than 20% of the plain model checking in our experiments. We also compare the coverage results and computation time with the state coverage method to demonstrate the advantages of the proposed method.

The rest of the paper is organized as follows. Section II presents the preliminaries including the state coverage metric and the transition representation. The transition coverage metric and symbolic computation algorithm is presented in Section III. In section IV, we discuss our experimental results and Section V concludes this paper.

II. Preliminaries

Let AP be a set of atomic propositions. A Kripke structure over AP is a four tuple $K = <S, S_0, R, L>$, where S is a finite set of states, $S_0 \subseteq S$ is the set of initial states, $R \subseteq S \times S$ is a complete transition relation, $L : S \to 2^{AP}$ is a function that labels each state with the set of atomic propositions true in that state. Any atomic proposition in A is a CTL formula, and if φ and ψ are CTL formula, then so are: $\neg\varphi$, $\varphi \vee \psi$, $\varphi \wedge \psi$, $AX\varphi$ (φ holds at the next time instant), $A\varphi U\psi$ (φ holds until ψ holds), and $AG\varphi$ (φ holds henceforth). The semantics of other CTL operators like $A\varphi R\psi$ can be found in [1].

The state coverage metric is defined as follows [2]:

Definition 1 : *Given a Kripke structure $K = < S, S_0, R, L >$ over AP, a state $s \in S$, and an atomic proposition $q \in AP$. The perturbed Kripke structure for s with respect to q (observed signal) is $K_s^q = < S, S_0, R, L_s^q >$, where for each state $t \in S$:*

$$ L_s^q(t) = \begin{cases} L(t) & \text{if } t \neq s \\ L(s) \setminus \{q\} & \text{if } t = s \text{ and } \{q\} \in L(t) \\ L(s) \cup \{q\} & \text{if } t = s \text{ and } \{q\} \notin L(t) \end{cases} $$

Definition 2 : Given a property f and a Kripke structure K such that $K \models f$, the set of covered states $C \subseteq S$ by f with respect to q satisfies the following condition for any state $s \in S$: $(K_s^q \not\models f) \Leftrightarrow (s \in C)$.

A sequential circuit is usually modeled as a finite state machine (FSM) where input propositions are labeled on the transition edges. We consider the transitions in the circuit FSM when talking about transition coverage. On the other hand, model checking is usually performed on the Kripke structure. Traditionally, a circuit FSM is first translated to a Kripke structure for model checking. For a FSM (Mealy Type) $M = <I, O, S, \delta, \lambda, S_0>$, where S and I are the state space and the input space; δ and λ are the state transition function and the output function, the corresponding Kripke structure can be derived as $K = <S \times I, S_0 \times I, R, L>$, where for any $s, s' \in S$ and any $i, i' \in I$:

- $(<s, i>, <s', i'>) \in R$ iff $\delta(s, i) = s'$,

- $L(<s, i>) = i \cup s \cup \lambda(s, i)$.

For example, the FSM of a simple modulo-3 counter in Fig. 1 is translated to the Kripke structure shown in Fig. 2. From the translation, it is observed that each transition $\delta(s, i) \to s'$ in the circuit FSM corresponds to a state $< s, i >$ in the Kripke structure one-by-one. In the example, the state $S10$ in the Kripke structure represents the transition from $S1$ to $S2$ in the FSM. As a result, we can evaluate the transition coverage of the circuit FSM based on the states of the Kripke structure.

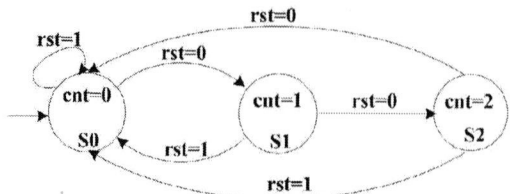

Fig. 1. The FSM of a modulo-3 counter.

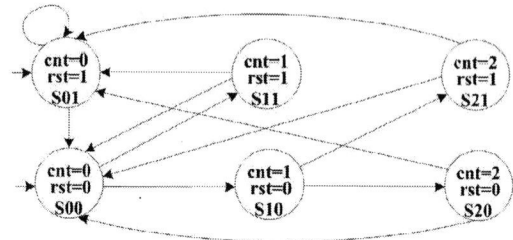

Fig. 2. The Kripke structure model of the module-3 counter.

III. Transition Coverage Estimation

In this section, we first talk about the basic idea of the transition-based coverage method, then we introduce our new transition perturbation model and define the transition coverage metric. We also present a symbolic algorithm to compute the transition coverage for a subset of ACTL.

A. Transition-Based Coverage Method

For a circuit FSM, we can consider two types of coverage metrics: the state coverage and the transition coverage. If the set of states of the FSM is S, then the state coverage space is just S. Deep-hidden design bugs usually creep in on transitions and state-based coverage method is not sufficient to clear such bugs. On the other hand, the transition coverage space is $S \times I$, where I is the set of inputs. Note that by enhancing the coverage space, we can do more precise coverage analysis and higher confidence in the correctness can be achieved [11, 12, 15]. A basic transition-based coverage method for model checking is proposed in [4], where covered transitions are defined as traversed by CTL operators. In this paper, we combine the transition traversal coverage and the state-based one by introducing transition perturbation.

Since the transition coverage method extends the coverage space to $S \times I$, more computation resources might be required. Such drawback should be treated as the tradeoff between completeness and computation efficiency.

B. Coverage Metric Based on Transition Perturbation

A CTL formula usually specifies how particular states should be reached through transitions and the correctness conditions for certain circuit signals on the reached states. The two factors answer the coverage question of what has been verified by a property. In order to catch the verification intent of formal properties on both states and transitions, we propose a novel transition perturbation model. The model is to pinpoint the transition through

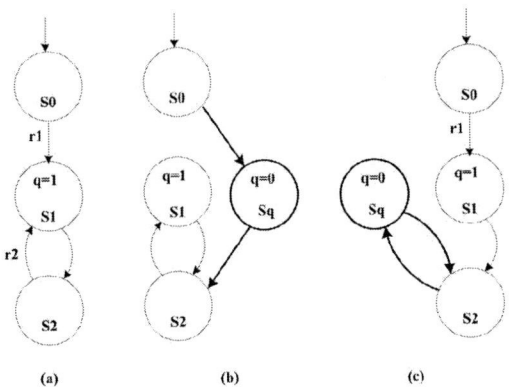

Fig. 3. The transition perturbation model for FSM.

which the correctness condition for the selected observed signal is checked.

We show a simple example to illustrate how a perturbed FSM is constructed. Fig. 3(a) is a simple FSM. To generate a perturbed FSM for transition $r1 = (S0, S1)$ for the selected observed signal q, first a new state Sq is added on which the value of q is different with the state $S1$; then the original transition $r1$ is redirected to the state Sq; finally all transitions starting from the state $S1$ are copied to the state Sq. The perturbed FSM is shown in Fig. 3(b). In the same way, the perturbed FSM for $r2$ is shown in Fig. 3(c).

In the transition perturbation model, there is only one transition reaching at the state where the value of the observed signal q is deliberately changed. On the other hand, all other transition sequences of the original model are maintained. As a result, a property will get failed if and only if the property traverses through the transition to check the value of the observed signal, which provide coverage information for both the state transition and the correctness condition of signals.

Informally, a transition of FSM is covered for the selected observed signal if a proven property gets failed on the perturbed FSM. As in the example of Fig. 3, two properties $AX(q = 1)$ and $AX(AX(AX(q = 1)))$ are satisfied by the original model (Fig. 3(a)). Because $AX(q = 1)$ is no longer satisfied by the model with perturbation on $r1$ (Fig. 3(b)), transition $r1$ is covered by this property with respect to signal q. Similarly, transition $r2$ is covered by $AX(AX(AX(q = 1)))$. According to the state coverage metric, both properties cover state $S1$. With our new transition coverage metric, two transitions reaching at state $S1$ are covered by different properties.

Since model checking is actually performed on the Kripke structure, we give a formal definition of perturbation model and transition coverage metric on the Kripke structure according to the relationship between the FSM and its corresponding Kripke structure (Section II).

Definition 3: *Given a Kripke structure* $K = <S, S_0, R, L>$, *a state* $r_i \in S$ *(it actually represents a transition of the circuit FSM), and an atomic proposition (observed signal)* $q \in AP$. *The perturbed Kripke structure on* r_i *for* q *is* $K_r^q = <S_r^q, S_0, R_r^q, L_r^q>$, *where* $S_r^q = S \cup S^q$, S^q *is a new state set and for each state* r_j *with* $(r_i, r_j) \in R$,

we add a new state r_j^q *in* S^q.
For each state $t \in S_r^q$:

$$L_r^q(t) = \begin{cases} L(t) & \text{if } t \notin S^q; \\ L(r_j) \setminus \{q\} & \text{if } t = r_j^q \text{ and } \{q\} \in L(r_j); \\ L(r_j) \cup \{q\} & \text{if } t = r_j^q \text{ and } \{q\} \notin L(r_j). \end{cases}$$

For each state pair (t_i, t_j), $t_i \in S_r^q$, $t_j \in S_r^q$:

$$(t_i, t_j) \in R_r^q \Leftrightarrow \begin{cases} (r_j, t_j) \in R & \text{if } t_i = r_j^q; \\ true & \text{if } t_i = r_i \text{ and } t_j = r_j^q; \\ false & \text{if } t_i = r_i \text{ and } t_j = r_j; \\ (t_i, t_j) \in R & \text{otherwise.} \end{cases}$$

Definition 4: *Given a property* f *and a Kripke structure* K *such that* $K \models f$. *The set of covered transitions* T *by* f *for the selected observed signal* q *is defined as: for any state* $r \in S$, $(K_r^q \not\models f) \Leftrightarrow (r \in T)$.

$$coverage = \frac{number \ of \ covered \ transitions}{number \ of \ reachable \ transitions}$$

The definition is based on the states of the Kripke structure, but it is different with the state-based coverage method [2]. In the counter example (Fig. 2), consider the property $AG(cnt = 2 \rightarrow AX(cnt = 0))$ and select cnt as observed signal. According to the state coverage method, states $\{S01, S00\}$ are covered, which correspond to the state $S0$ in the FSM (Fig. 1). However, there are 4 transitions arriving at $S0$, but the state coverage metric dose not distinguish which transitions are really concerned by the property. Differently, according to the transition coverage metric, states $\{S21, S20\}$ are covered by the property, which correspond to transitions from $S2$ under input $rst = 1$ and $rst = 0$ in the FSM. The other two transitions arriving at $S0$ in the FSM are not covered. Conceptually, states of the Kripke structure are interpreted as transitions of the circuit FSM in our method, but they are just considered as static states of FSM in the state-based metric.

C. Coverage Computation

The transition coverage metric is general for any property specification language. However, the set of covered transitions may not be easily computed. In this paper, we present a symbolic algorithm which is based on fix point computation and binary decision diagrams (BDDs) [1] to compute the transition coverage for a subset of ACTL.

Definition 5: *The set of formulae acceptable to our algorithm is defined as: if* b *is a propositional formula, then* b *is acceptable; if* f *and* g *are acceptable, then so are* AXf, AGf, $A(fUg)$, $A(fRg)$, $f \wedge g$, $b \rightarrow g$.

Note that AFf can be represented as $A(trueUf)$, so we do not treat it separately. According to industrial experience, the set of formulae is sufficiently expressive to specify most desirable properties for sequential logic circuits.

Fig. 4 is the main function of our coverage computation algorithm $Cov(\varphi, S)$. The algorithm is performed on the Kripke structure of a circuit design. The input

```
Cov(φ, S){
    if (S == empty) return empty;
    if (φ is propositional) return empty;
    switch (φ)
    case f ∧ g : result = Cov(f, S) ∪ Cov(g, S);
    case b → g : result = Cov(g, S ∩ Sat(b));
    case AGf : result = Cov(f, Rch(S));
    case AXf :
        cf = Chk(f, Fwd(S));
        r1 = Bwd(cf) ∩ S; r2 = Cov(f, Fwd(S));
        result = r1 ∪ r2;
    case fUg :
        fTrv = empty; gTrv = empty;
        gS = Sat(g); doS = S;
        do{
            gTrv = gTrv ∪ (doS ∩ gS);
            fS = doS \ gS;
            if(fS ≠ empty){
                fTrv = fTrv ∪ fS;
                doS = Fwd(fS) \ (fTrv ∪ gTrv);
            }
        }while(fS ≠ empty & doS ≠ empty));
        c1 = Chk(f, fTrv); c2 = Chk(g, gTrv);
        r1 = fTrv ∩ Bwd(c1); r2 = fTrv ∩ Bwd(c2);
        r3 = Cov(f, fTrv); r4 = Cov(g, gTrv);
        result = r1 ∪ r2 ∪ r3 ∪ r4;
    case fRg : //similar to fUg
    default : Unacceptable formula;
    return result;
}
```

Fig. 4. Coverage computation algorithm: the main function $Cov(φ, S)$.

is a property and a set of states; the output is a set of covered states which correspond to the transitions of the circuit FSM. For the top-level run, S is the initial state set, and $φ$ is a verified property. As the algorithm proceeds, sub-formulae and states on the traversing pathes are recursively processed by the algorithm. The algorithm finally returns the covered transitions with respect to the given observed signal q. $Chk(φ, S)$ is a sub-function which extracts correctness conditions of each sub-formula and checks their dependency on the observed signal in different states, as shown in Fig. 5. Other predicates used in the algorithm is explained as follows:

$Fwd(S)$: the forward image of a state set S;
$Bwd(S)$: the backward image of a state set S;
$Rch(S)$: all reachable states from a state set S;
$Sat(f)$: the set of states satisfying f;
$\backslash$: the set minus.

The idea of our algorithm is to perform transition traversal for a property, and while traversing a transition, we extract the correctness conditions from the property on its destination state, the transition is identified as covered if the correctness condition depends on the value of the observed signal.

```
Chk(φ, S){
    if (S == empty) return empty;
    if (φ is propositional)
        result = S ∩ Sat(¬φ|_{q→¬q});
        return result;
    switch (φ)
    case f ∧ g : result = Chk(f, S) ∪ Chk(g, S);
    case b → g : result = Chk(g, S ∩ Sat(b));
    case AGf : result = Chk(f, S);
    case AXf : result = empty;
    case fUg :
        r1 = Chk(g, S ∩ Sat(g));
        r2 = Chk(f, S \ Sat(g));
        result = r1 ∪ r2;
    case fRg : //similar to fUg
    default : Unacceptable formula;
    return result;
}
```

Fig. 5. Coverage computation algorithm: the sub function $Chk(φ, S)$.

In our algorithm, we care more about the verification intent of a property formula. As for formula AGf, the set of reachable states are assumed and there are no verification intent for transitions from the initial state to all other states. Thus, in our algorithm, we do not perform traversing for AG but directly compute the coverage of its sub-formula f with all reachable states.

Compared with the state coverage estimation algorithm [2], the main extra computation in our algorithm comes from the backward image computation. This will not increase much computation cost since it is only performed on certain state sets. The algorithm has the same order of complexity as a model checking algorithm, which is exponential in the worst case. The actually computation overhead is quite small in our experiments when we integrate the coverage estimation with model checking itself.

IV. EXPERIMENTAL RESULTS

We have implemented the transition coverage estimator by extending the state-of-the-art model checker VIS2.0 [14]. We integrate the coverage estimation with model checking to save computation cost. The Kripke structure constructed for model checking is also used for coverage estimation. The model checking results for sub-formulas can be used in the coverage estimation. For comparison purpose, we have also implemented the state coverage method [2]. We have applied the coverage method to the model checking of a cache coherence protocol.

One of the most successful application domains for model checking has been multiprocessor cache coherence protocols. This application is commercially very important since almost all high-end servers are now cache-coherence multiprocessors. In our work, we apply coverage metrics to assist the model checking process for a full-

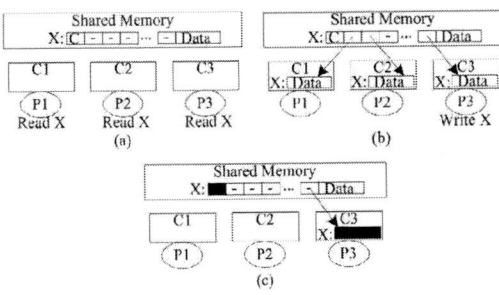

Fig. 6. The three different states in a full-map directory.

map directory-based cache coherence protocol [16]. The protocol uses directory entries with one bit per processor and dirty bit. Each bit represents the status of the block in the corresponding processors's cache (present or absent). If the dirty bit is set, then one and only one processor's bit is set, and that processor has permission to write into the block. Each cache block may be in one of the three states: INVALID, SHARED, or EXCLUSIVE. Fig. 6 illustrates three different states of a full-map directory. In (a), location X is missing in all of the caches; (b) results from three caches requesting copies of location X; (c) results from cache C3 requesting writing to location X. Note the dirty-bit is set to clean in (a) and (b), and to dirty in (c). To simply the complexity, we first configure the protocol model with two processors and two memory entries. The cache for each processor only contains one data block.

The cache protocol should keep the block states in the memory directory and those in the caches consistent. We use invariant properties to check the state consistency. For example, the following two properties check the SHARED and EXCLUSIVE states for cache C1 with address 0:

$$AG(c1\_sta = SHA * c1\_add = 0 \leftrightarrow$$
$$dirty[0] = 0 * share\_c1[0] = 1); \qquad (1)$$
$$AG(c1\_sta = EXC * c1\_add = 0 \leftrightarrow$$
$$dirty[0] = 1 * share\_c1[0] = 1); \qquad (2)$$

Since invariant properties do not contribute transition coverage, we estimate state coverage by selecting the dirty bit $dirty[0]$ as the observed signal. The coverage results show certain states with clean dirty-bit are not covered. After analysis, we found a design error in our initial model. When a cache is full, a write-back operation should be performed when the cache gets a read or write miss. After fixing the bug, additional properties are added to cover all states. For example, dirty bit for address 0 should be clean when cache C1 is in INVALID state and cache C2 is not for address 0:

$$AG(c1\_sta = INV * c2\_add \neq 0 \rightarrow dirty[0] = 0) \qquad (3)$$

At this point, the state coverage metric is of great value since we are mainly concerned for the state consistency. However, we do not consider the state transition behaviors of the protocol. The cache protocol should ensure sequential consistency. Intuitively, each read should return

TABLE I
STATE AND TRANSITION COVERAGE RESULTS.

Observed Signal	Number of Properties	State Coverage	Transition Coverage
ack1	9	100%	100%
c1_o	7	100%	99.72%
dirty[0]	3	100%	95.49%

the value of the most recent write to the same memory location.

We select the data output of cache C1 $c1\_o$ as the observed signal and seven properties are specified. For instance, if any processor reads value 0 from memory address 0, then all read at address 0 should return 0 unless there is a write to address 0 with value 1. This property can be expressed by CTL formula like:

$$AG(r\_a0\_v0\_ack \rightarrow A((w\_a0\_v1)R$$
$$(c1\_r\_a0\_ack \rightarrow c1\_o = 0))); \qquad (4)$$

Three properties are specified for the observed signal $dirty[0]$. For example, the dirty-bit should remain clean until there is a write for its address. And if cache C1 is in EXCLUSIVE state, the dirty-bit for the same address as cache C1 should remain dirty until cache C1 is requested to write back or invalidate its data.

$$AG(dirty[0] = 0 \rightarrow A(wri\_add0)R(dirty[0] = 0)); \qquad (5)$$
$$AG(c1\_sta = EXC * c1\_add = 0 \rightarrow$$
$$A((wri\_bak1 = 1 + inval1 = 1)R(dirty[0] = 1))); \qquad (6)$$

Another 9 properties are specified when taking the acknowledge for processor P1 $ack1$ as the observed signal.

Tab. I compares the state and transition coverage results. All states are covered by these properties but the transition coverage results discover certain conditions which are not considered by these properties.

When a write operation by a processor begins, the read operation by another processor may have not finished. And the outcome of such a read is not verified. The state coverage metric is unable to discover this coverage hole because this state in which the read operation finishes is covered through other transitions. Another corner case is that, when processor P1 begins to write, cache C2 is asked to invalidate its content, then cache C1 writes back its content to the memory because it is full. After that the new value is written to cache C1. The uncovered transition is illustrated in Fig. 7 (the bold edge). The state metric is unable to detect this coverage hole because a read operation by processor P2 will also trigger P1 to write back. The invalidation–write-back sequence is actually a design error since it will degrade performance. A superior sequence is write-back–invalidation for such a condition.

We also compare the computation time of transition and state coverage estimation. We add two other differ-

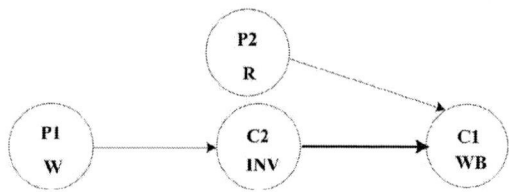

Fig. 7. One uncovered transition in the cache protocol.

TABLE II
COMPARISON OF COMPUTATION TIME.

Protocol Configure	T1 (Seconds)	T2 (Seconds)	T3 (Seconds)
2p2m	51	65	66
3p2m	3956	4065	4071
2p4m	9938	10444	12592

ent types of configuration as two processor with 4 memory entries and 3 processors with 2 memory entries. The computation time for the 7 properties for observed signal c1_o is shown in Tab. II, where T1 is the time of plain model checking; T2 and T3 are respectively the times of model checking along with state and transition coverage estimation. The CPU times are measured on a IBM IntelliStation Z-Pro with 3.0GHZ CPU and 2.3GB RAM. The results confirm our discussion in section III that the computation costs are similar for the transition and state coverage estimation. The computation overhead of coverage estimation compared to plain model checking is also shown in Tab. II, which is less than 20% of the plain model checking.

V. CONCLUSIONS

We have proposed a novel transition perturbation model, and based on this model, we have defined a new transition-based coverage metric for model checking to evaluate the completeness of properties. An efficient symbolic algorithm has been presented to compute the transition coverage for a subset of ACTL. The implemented coverage estimator has been applied to the model checking of a cache coherence protocol. The experiment results have confirmed that the proposed method can identify critical coverage holes which may escape the state-based coverage estimation, while the computation overhead is minor compared with plain model checking. Our future work is to develop automatic techniques for analyzing the coverage results so as to fill coverage holes efficiently.

REFERENCES

[1] E.M. Clarke, O. Grumberg, and D. Peled. *Model Checking*. MIT Press, 1999.

[2] Y. Hoskote, T. Kam, Pei-Hsin Ho, and Xudong Zhao. "Coverage estimation for symbolic model checking". In *Proceedings of the 36th Design Automation Conference (DAC)*, page 300-305, 1999.

[3] N. Jayakumar, M. Purandare, and F. Somenzi. "Do's and don'ts of CTL state coverage estimation". In *Proceedings of*

the *40th Design Automation Conference (DAC)*, page 292-295, 2003.

[4] X. Xu, S. Kimura, K. Horikawa, and T. Tsuchiya. "Transition Traversal Coverage Estimation for Symbolic Model Checking". In *Proceedings of the 3rd ACM/IEEE internatinoal Confernece on Formal Methods and Models for Co-Design (MEMOCODE)*, page 259-260, 2005.

[5] S. Katz, O. Grumberg, and D. Geist. "Have I written enough properties? - A method of comparison between specification and implementation". In *Proceedings of the 10th ACM Advanced Research Working Conference on Correct Hardware Design and Verification Methods (CHARM)*, page 280-297, 1999.

[6] H. Chockler, O. Kupferman, and M.Y. Vardi. "Coverage Metrics for Formal Verification". In *Proceedings of 12th ACM Advanced Research Working Conference on Correct Hardware Design and Verification Methods (CHARM)*, page 111-125, 2003.

[7] H. Chockler, O. Kupferman, R. Kurshan and M.Y. Vardi. "A Practical Approach to Coverage in Model Checking". In *Proceedings of the 2001 International Conference on Computer Aided Verification (CAV)*, page 66-78, 2001.

[8] F. Fummi, G. Pravadelli, A. Fedeli, U. Rossi, and F. Toto. "On the use of a high-level fault model to check properties incompleteness". In *Proceedings of the 1st ACM/IEEE International Conference on Formal Methods and Models for Co-Design (MEMOCODE)*, pages 145-152, 2003.

[9] F. Wang, G.-D. Huang, F. Yu. "Numerical Coverage Estimation for the symbolic simulation of real-time systems". FORTE 2003, LNCS 2767, pages 160-176, 2003.

[10] S. Tasiran, and K. Keutzer. "Coverage metrics for functional validation of hardware designs". In *IEEE Journals of Design & Test of Computers*, Volume 18(4), pages 36-45, 2001.

[11] Y. Hoskote, D. Moundanos, and J.A. Abraham. "Automatic extraction of the control flow machine and application to evaluating coverage of verification vectors". In *Proceedings of the 1995 IEEE Internatinal Conference on Computer Design VLSI in Computers and Processors (ICCD)*, page 532-537, 1995.

[12] R.C. Ho, C. Han Yang, M.A. Horowitz, and D.L. Dill. "Architecture validation for processors". In *Proceedings of the 22nd Annual International Symposium on Computer Architecture*, page 404-413, 1995.

[13] S. Das, et al.. "Formal Verification Coverage: Computing the Coverage Gap between Temoral Specifications". In *Proceedings of the IEEE/ACM International Conference on Compute Aided Design (ICCAD)*, page 198-203, 2004.

[14] R.K. Brayton, et al.. "VIS : A System for Verification and Synthesis". In *Proceedings of the 1996 International Conference on Computer Aided Verification (CAV)*, page 428-432, 1996.

[15] A. Gupta, S. Malik, and P. Ashar. "Toward Formalizing A Validation Methodology Using Simulation Coverage". In *Proceedings of the 34th Design Automation Conference (DAC)*, page 740-745, 1997.

[16] D. Chaiken, C. Fields, K. Kurihara, and A. Agarwal. "Directory-based cache coherence in large-scale multiprocessors Computer". In *Computer*, IEEE Computer Society, Volume 23, Issue 6, page 49-58,1999.

Word Level Functional Coverage Computation

Bijan Alizadeh

Microelectronic Research and Development Center of Iran
(MERDCI)
Tehran, IRAN, Postal Code 1438753645
E-mail: bijan.alizadeh@merdci.com

Abstract— This paper proposes a word-level coverage metric to determine the completeness of a set of properties verified by a word-level method. An algorithm is presented to compute a functionality based coverage metric for a sequence property as specification. Control, intermediate and output signals are represented by a multiplexer based structure of linear integer equations, and RT level properties are directly applied to this representation. A set of integer equations are symbolically simulated based on the specified property in a predictable time. We used a canonical form of linear Taylor Expansion Diagram.

I Introduction

Validating the functionality of digital circuits and systems is an increasingly difficult task because of the growing complexity of designs. Logic simulation was the mainstream approach for the validation of large synchronous systems because of its scalability and flexibility. However, the fraction of the design space that can be explored by simulation is insufficient, especially for large designs.

On the other hand, formal methods provide exhaustive coverage of hardware behavior, which depends on the set of defined properties, and explore the behavior under all the possible input stimuli. In addition, the designer requires automated verification tools at higher levels of abstraction to verify the design at the early stages of the design flow [1]. Therefore, formal verification methods such as symbolic model checking have become important for RT or behavioral level verification.

Most of the formal methods use Binary Decision Diagrams (BDDs) to represent the set of states and the state transition functions [1, 2]. The BDD techniques may still suffer from memory explosion problem when the application is a large datapath. In order to overcome this problem, various solutions have been proposed that try to contain the size of the BDDs involved [3]. There are also some methods that use integer programming to verify datapath circuits [4, 5, 6]. However, these approaches need to use an Integer Linear Programming (ILP) solver and, therefore, are limited to datapath designs. Other high level data structures like Binary Moment Diagram (BMD) [7] and Taylor Expansion Diagram (TED) [8] have also been proposed to check the equivalence between two circuits, but they have not been used to do property checking.

In this paper, we present a novel technique for symbolic simulation that uses a new, parametric, high-level representation for the functions at the inputs of sequential

elements and outputs of the circuit. This representation produces a word-level representation called Linear TED (LTED) which is suitable to compute functional coverage instead of BDDs. For this work, we used VHDL to describe a design and a sequence format to describe its properties. We extract Data Flow Graph (DFG) for the design [9], convert it to LTED, and prove the design property symbolically [10].

A coverage metric can be very useful in achieving a high degree of confidence in the completeness of the verification. We present a functionality based coverage metric which is applicable to our high level model and indicates practical point of view of signal coverage. Two approaches for defining and developing algorithms for coverage metrics in temporal logic model checking have been studied in the literature [11, 12]. The first approach, by Hoskote et al., is to check the influence of small changes in the system on the satisfaction of the specification [11]. Intuitively, if some part of the system can be changed without violating the specifications, this part is *uncovered* by the specification. The second approach, introduced in [12], is suggesting two alternatives to the naive algorithm for specifications in the branching time temporal logic CTL. The first algorithm is symbolic and it computes the set of pairs <w,w'> such that flipping the value of q in w' falsifies φ in w. The second algorithm improves the naive algorithm by exploiting overlaps in the many dual structures that we need to check. Neither one of these algorithms is attractive: the symbolic algorithm doubles the number of BDD's variables, and the second algorithm requires the development of new procedures. Also, these algorithms cannot be extended to specifications in LTL as they heavily use the fixed-point characterization of CTL, which is not applicable to LTL.

The main advantages of our method are as follows: First, our technique has added some parts to TED [8] to represent relational expressions and proposed a simplification process which is based on computing intersection or union areas of two linear equations. The basic idea of our method is to use this simplification instead of solving equations. We computed the union and intersection of two linear equations (equality and nonequality) based on their respective area in a two dimensional space. Second, we propose a practical coverage metric based on the high level model.

Section 2 of this paper presents the way to construct Linear TED as a canonical representation of expressions. In section 3, some algorithms to check the basic properties in our model are given. Section 4 shows a simple method to estimate path coverage and experimental results for some

0-7803-9451-8/06/$20.00 ©2006 IEEE.

examples given in section 5. Last section presents a short conclusion of this work.

II Word Level Representation

The word level representation, used in this paper is called Linear TED (LTED). This structure includes Variable, Constant, Branch, Union and Intersect nodes. The algebraic expression F(x,y,...) will be represented by constant and linear terms of Taylor series expansion [8], Equ(1), where *const* is some part of F(x,y,...) and independent of the variable **x,** while *linear* depends on the variable **x**. The variable **x** is top variable of F(x,y,...) [8].

$$F(x, y,...) = const + x(linear) \qquad (1)$$

For representing relational expression, we have just added relational operators, including E (equal to zero), NE (not equal to zero) and GE (greater or equal to zero), to the LTED node. Each Variable node has a constraint field indicating its respective range. For example consider variable X, as a bit type, so its constraint field indicates $0 \le X \le 1$.

A Branch node has three fields, including *Select*, *InZero* and *InOne*, where *Select* is a relational expression, i.e. CLTED node, and other fields are LTED nodes. The functionality of a Branch node is indicated by Equ(2).

$$F = Select \& InOne + \overline{Select} \& InZero \qquad (2)$$

A Union or Intersect node has two fields; Left child and Right child, which are LTED nodes. In part 2.2, The union and intersection operations will be defined on CLTED nodes.

To support word level arithmetic operations, the **syntax** and **semantics** of word level operations are formally defined as follows:

Syntax:

A word level formula is a list of *term*s. Formally, let *term*, *Var*, and *Const* denote a word level term, a word level variable, and a constant value respectively. Then the syntax is formally defined in Fig. 1.

```
term := Var | Const | Var := term | term1 + term2 | term1 * term2 |
    propBranch term1, term2 | prop1 Intersect prop2 | prop1 Union prop2 |
prop := TRUE | FALSE | term1 = term2 | term1 != term2 | term1 >= term2
Formula := list of terms
```

Fig. 1. Syntax of a word level logic

Semantics:
The interpretation of word level terms, propositions and formulas are defined as in Fig. 2. The *Branch* node is comparable to the *If-Then-Else* operator in BDD package (see Fig. 2). *Intersect/Union* nodes are like the *And/Or* operators, but they have some differences which will be described later.

```
Variable :  [x] = x ∈ {Input ∪ PresentState ∪ NextState ∪ Output}
Constant :  [c] = c ∈ Z
Addition :  [term1 + term2] = addition of term1 and term2
Multiply :  [term1 * term2] = multiplication of term1 and term2
Branch : [prop BR term1, term2] = If (prop = TRUE) term1, Else term2
Intersection :  [prop1 Intersect prop2] = prop1 AND prop2
Union :     [prop1 Union prop2] = prop1 OR prop2
Equal :     [term1 = term2] = If (term1 = term2) TRUE
Inequal :   [term1 ≠ term2] = If (term1 ≠ term2) TRUE
Greater or equal : [term1 ≥ term2] = If (term1 ≥ term2) TRUE
```

Fig. 2. Semantics of word level operations

A. Construction of the LTED

Our method needs two LTEDs called Original LTED (OLTED) and Canonical LTED (CLTED) which are defined as follows:

Definition 1. *An OLTED node is a directed acyclic graph G=(V, E) with vertex set V and edge set E. The vertex set V contains six types of vertices: Branch (B), Union (U), Intersect (I), Variable (V), Relational Variable (RV), and Constant (C) nodes.*

- *A Branch node v has as attributes a select field select(v) ∈ {U, I, RV}, and two children InOne(v), InZero(v) ∈ V.*
- *A Union node v has as attributes two children left(v) ∈ {U, I, RV}, right(v) ∈ {I, RV}. A Union node includes another Union node on its Left-Child sub-term, and there will be an Intersect, or Relational Variable node on its Right-Child sub-term, because of its canonical form.*
- *An Intersect node v has as attributes two children left(v), right(v) ∈ {I, RV}. Relational Variable nodes are ordered from LeftChild to RightChild in each Intersect node.*
- *A Variable node v has as attributes an integer varaible var(v), and two children const(v), linear(v) ∈ {V, C}.*
- *A Relational Variable node v has as attributes an integer variable var(v), a relational operator op(v) ∈ {=, !=, >, >=}, and two children const(v), linear(v) ∈ {V, C}.*
- *A Constant node v has as its attribute a value val(v) ∈ Z.*

The relation between an OLTED and the integer function it represents is straightforward. This leads to the following correspondence between OLTEDs and integer functions:

Definition 2. *A vertex v in an OLTED denotes an integer function f^v defined recursively as:.*
- *If v is a Constant node, then $f^v = val(v)$.*
- *If v is a Relational Variable node, then $f^v = const(v) + var(v).linear(v)\ op(v)\ 0$.*
- *If v is a Variable node, then $f^v = const(v) + var(v).linear(v)$.*
- *If v is an Intersect node, then $f^v = f^{left(v)}$ Intersect $f^{right(v)}$.*
- *If v is a Union node, then $f^v = f^{left(v)}$ Union $f^{right(v)}$.*
- *If v is a Branch node, then $f^v = (f^{InOne(v)}$ Intersect select(v)) Union $(f^{Inzero(v)}$ Intersect Not(select(v))).*

Definition 3. *A CLTED node is formally defined as in Definition 1, when all nodes excluding Branch node are used.*

Example: Fig. 3 shows OLTED node developed for the statement: *If (a) then   X <= b + c   Else   X <= b − c.* As illustrated in this figure, boolean condition *a* is converted to *a − 1 = 0* (Notice: The *E* symbol near the *a* node, in the *Select* field, shows the *Equality* operator, i.e. =).

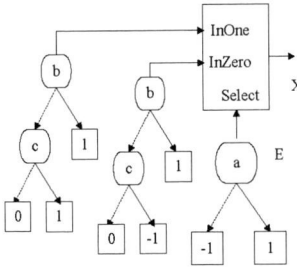

Fig. 3. Example of an OLTED node

B. LTED operations

Now we describe how addition, subtraction, multiplication, union and intersection of two LTEDs are performed.

The addition and multiplication operators are applied similar to TED's ADD and MULT when two OLTEDs are not Branch nodes [8]. Otherwise *InOne* and *InZero* fields of Branch node will be added to (multiplied by) another node as *InOne* and *InZero* fields of result respectively. At this point two Branch nodes with same *Select* fields will be distinguished to make a simpler LTED node.

The union and intersection operators are defined on CLTED. Notice that checking the existence of integer solutions for a conjunction of linear inequalities is an NP-complete problem. But here the intersection operator is used for checking the existence of integer solutions for a conjunction of linear inequalities, when the intersection area in a two dimensional space is considered instead of solving that linear inequalities. The execution time of this method is polynomial. Assuming the two CLTEDs are algebriac expressions with two variables including relational operators, we must consider the following cases:

1. Both nodes are *Relational Variable* nodes (u,v ∈ RV). One of our contribution is related to the way in which linear equations (equality and nonequality) are solved without ILP/SAT solvers. This way, we consider conjunction of two or more integer equations in a two dimentional space, and compute the intersection area which is covered by all equations [10]. For simplifying the problem, we consider the solution of two linear equations of two variables, i.e., I: [$a0*X+b0*Y+c0$ *Op1* 0] and II: [$a1*X+b1*Y+c1$ *Op2* 0], where *Op1* and *Op2* are {=, >, >=, ≠}. To solve these equations, various conditions of coefficients of these equations are considered. These conditions describe positions of the equations in a two dimensional space. For instance, the condition *a0*b1−a1*b0=0* shows that the two equations are parallel. The conditions *b0>0* and *a0*a1+b0*b1>0* indicate that both equations have upward direction when

Op1 and *Op2* are considered *greater* (>). It means that the area above the specified lines are covered by those equations. The condition *c0*b1−c1*b0<0* shows that the first equation is above the second one in a two dimensional space (see Fig. 4). If two linear equations are not parallel, we have to return a *Union (Intersect)* node [u Union (Intersect) v].

2. Otherwise, this procedure is called recursively to compute Conjunction (*Intersect*) or Disjunction (*Union*) of two LTED nodes.

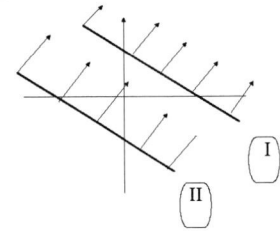

Fig. 4. Comparison of two linear equations in which Op1, Op2 are considered greater (>)

C. DFG to LTED conversion

The first step is extraction of DFG [see reference 9]. After DFG extraction, we will be capable to translate it to LTED. Next state and output functions in DFG have multiplexer based structures, which will be in one-to-one correspondence with Branch node in OLTED. Therefore we can make next state and output functions according to LTEDs and call them *"list of TedState"*. The *list of TedState* includes identifier (Id) of next state variable, Id of related present state variable and value of next state variable as an LTED node. The Id of present state variable will be −1 if there is an output or intermediate variable as the next state variable. Consider Greatest Common Divisor example. TABLE I and Fig. 5 show the *list of TedState* and a LTED node of *nxtX* signal respectively. This LTED node will be regarded as value field of one of the rows in TABLE I that indicates Ids of *nxtX* and *X* signals. In this example relational expressions like "Start = 1" and "X > Y" will be converted to LTED nodes "Start − 1 = 0" and "X − Y −1 ≥ 0", as *Select* field of first and last Branch nodes, respectively.

TABLE I
List of TedState in GCD example

Present State	Next State	Value of Next State
X	nxtX	shown in Fig. 5
Y	nxtY	OLTED structure
Reset	nxtReset	OLTED structure
-1	Out	OLTED structure

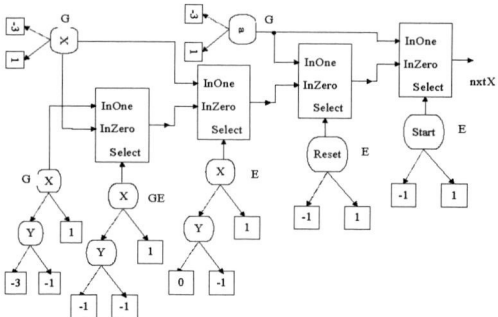

Fig. 5. LTED of $(nxtX - 3 > 0)$

III Property Checking in Design

Properties are described in a linear time logic and subdivided into an assumption part (P1) and a commitment part (P2) where both P1, P2 are defined by the rules below. The assumption part can be specified at different times. This form of property allows us to check output or control signals, safety and liveness properties based on the linear time logic.

```
P::=(P)|P∧P|¬P|P=P|P>P|P>=P|P≠P|
     time=i, P |Variable|IntegerValue
Q ::= {P1=>P2}
```

An overall view of the property checking is shown in Fig. 6. First, we extract LTEDs of next state and output functions from a synthesized design. Afterwards, we extract tree structure of the *P2* part to specify what verification procedures need to be called at each level of the tree. Two procedures, *CheckComb* and *CheckX* perform the task of verification of this flowchart.

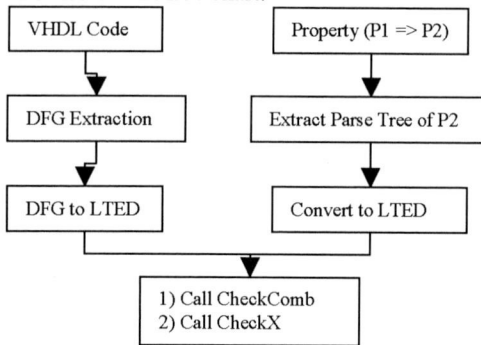

Fig. 6. Flowchart of property Checking mechanism

Fig. 7 shows the *CheckComb* procedure in the flowchart of Fig. 6. When the *P2* part of a property is combinational, i.e. without state operators, we must replace intermediate variables by their values specified in *list of TedState (ALLSTS)*, and convert them to CLTED. Later the assumption part of property (*P1*) will be eliminated from computed CLTED. At the end of the procedure, CLTED equations, that indicate conditions needed to satisfy the property, will be returned.

To *eliminate* one CLTED, e.g. *u*, from another CLTED, e.g. *v*, we recursively perform this procedure till both of

them become Variable nodes. At this point, if intersection of u and v is one of u or v, 1 will be returned. Otherwise intersection result will be returned. If *u* is Intersect or Union node, this procedure is called recursively for u.Left and u.Right.

```
CheckComb (LTED P2;  LTED P1)
  ALLSTS: set of next state, output and
intermediate nodes and their values.
  TopV = Get top variable of P2;
  Linear = ALLSTS.Get()->value;
  Cnst = CheckComb(P2->Cnst;  P1);
  if (P2.kind == E) //Equality Operator
    Result = Linear + Cnst = 0;
  else if (P2.kind == G)
    Result = Linear + Cnst > 0;
  else if (P2.kind == GE)
    Result = Linear + Cnst • 0;
  else if (P2.kind == NE)
    Result = Linear + Cnst • 0;
  else      //no relational expression
    Result = Linear + Cnst;
    Eliminate (P1, Result);
```

Fig. 7. Combinational part

Fig. 8 shows the *CheckX* procedure in the flowchart of Fig. 6. When the *P2* part of a property uses the next-state operator (*X*), correctness of the property is checked in three major steps. These steps are current state variables to next state variables converting, next state variables replacing, and simplifying. Simplification is performed based on Intersect and Union operators which were described in previous sections. If *P1* part is subset of the result, this part of verification is acceptable and the return result is considered as the CLTED node. Else, verification fails.

```
CheckX (LTED P2; LTED P1)
  TopV = Get top variable of P2;
  Linear = ALLSTS.Get()->value;
  Cnst = CheckX(P2->Cnst;  P1);
  if (P2.kind == E)
    Result = Linear + Cnst = 0;
  else if (P2.kind == G)
    Result = Linear + Cnst > 0;
  else if (P2.kind == GE)
    Result = Linear + Cnst • 0;
  else if (P2.kind == NE)
    Result = Linear + Cnst • 0;
  else
    Result = Linear + Cnst;
    Eliminate (P1, Result);
```

Fig. 8. Next State(X) operator

To determine whether a CLTED, e.g. *u*, is *subset* of another one, e.g. *v*, we perform this procedure recursively until *u* and *v* become Variable nodes. In this condition, if union of them is the second one, i.e. *v*, it means that *u* is subset of *v*. If *u* is Intersect node and u.Left and u.Right are subsets of *v*, then *u* will be a subset of *v*. If *u* is Union node and u.Left or u.Right is a subset of *v*, then *u* will be a subset of *v*.

10

IV Coverage in LTED Model

A property specifies the condition on certain circuit signals. It also specifies under which *assumptions*, this condition should be held. One of the signals should be checked is identified as the *observed signal* and coverage is defined on this *observed signal*. The *observed signals* consists of *next state*, *intermediate* and *output* signals which are presented by LTED nodes (see section II). When a property is proved to be true in a circuit, coverage should be defined for the specified signal according to a subset of circuit branches (The circuit branches were satisfied based on the assumptions in the property).

A *covered set of paths* for an *observed signal* is a set of paths in LTED structure of the *observed signal* which is covered based on property assumptions. Here is an example to explain the concept. Suppose that we are to compute the coverage of a simple sequence formula in GCD:

```
at T:      start=0 & reset=0 & X=12 & Y=3
at T+1:    X=9
```

where X at T+1 or *nxtX* at T is considered as *observed signal*. The formula specifies that whenever X=12, Y=3, start=0 and reset=0, X will be 9 at the next clock cycle. Five paths, in *nxtX* signal, are reachable based on different assumptions (see Fig. 9(a)). Also as the bold lines in the Fig. 9(b) show, a path is just specified based on property assumptions. So the coverage of this property is 1/5 = 20% according to *nxtX* signal.

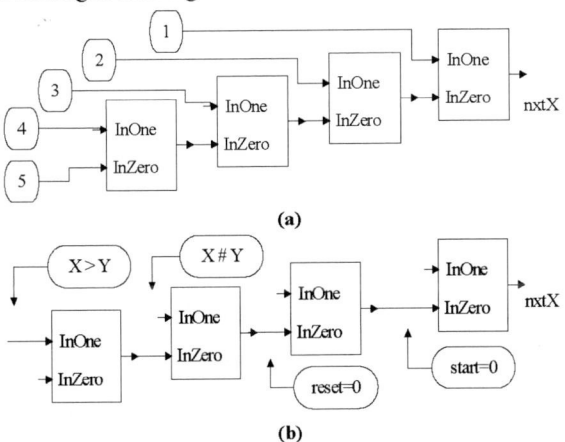

Fig. 9. All paths (a) and covered paths (b) in nxtX signal

Definition 4: *Coverage of a formula for an observed signal on a given high level model is computed as the fraction of paths in the LTED model which are covered based on property assumptions (see Equ(3)). Coverage for a set of properties is simply obtained by adding the coverages of all properties.*

$$cov\ erage = \frac{\#\ cov\ ered\ paths}{\#\ all\ paths} \times 100\% \qquad (3)$$

When a particular *observed signal* is fully (100%) covered that its LTED model is checked based on property assumptions for all paths. It is the best way to determine the full coverage of the properties. On the other hand, the formulation of the coverage metric identifies the partially-covered paths in terms of uncovered paths so that the user can write additional properties to complete the coverage.

We present a recursive algorithm to compute the set of covered paths and all paths in the LTED model of an *observed signal* for a sequence formula (see Fig. 10). *NumberofAllPaths* procedure computes all paths in the LTED structure of an *observed signal*. This function is called recursively to compute all paths in pIn0 (else) and pIn1 (then) parts of a Branch node.

NumCoverPaths procedure specifies number of paths in the LTED structure of an *observed signal*, which are covered when property assumptions are applied to the LTED structure. While a Branch node is processing, the following cases must be checked:
1. Variables in *pSel* field are in assumption part:
   I. If *pSel* is TRUE based on assumptions, the number of the covered paths in *pIn1* field should be returned.
   II. If *pSel* is FALSE based on assumptions, the number of the covered paths in *pIn0* field should be returned.
2. Variables in *pSel* field are not in assumption part, the addition of the number of the covered paths in *pIn0* and *pIn1* fields should be returned.

```
int NumberofAllPaths(LTED In)
if(In->type==Branch)
  if(pIn1->type!=Branch)
    if(pIn0->type!=Branch)  return 2;
    else return NumberofAllPaths(pIn0)+1;
  else
    if(pIn0->type!=Branch)
      return NumberofAllPaths(pIn1)+1;
    else
      return NumberofAllPaths(pIn0)+
             NumberofAllPaths(pIn1)+1;
int NumCoverPaths(LTED source,LTED assum)
if(source->type==Var)
  if(IsIntermediateVar(source))
    tmp = IntermediateVar(source);
    return NumCoverPaths(tmp,assum);
  else      return 1;
if(source->type==Branch)
  if(IsInAssumptions(pSel, assum))
    if(IsConditionTrue(pSel, assum))
      return NumCoverPaths(pIn1,assum);
    else
      return NumCoverPaths(pIn0,assum);
  else
    return NumCoverPaths(pIn1,assum)+
           NumCoverPaths(pIn0,assum)+1;
if(source->type==AndTed)
  return NumCoverPaths(pLeft,assum)*
         NumCoverPaths(pRight,assum);
if(source->type==OrTed)
  return NumCoverPaths(pLeft,assum)+
         NumCoverPaths(pRight,assum);
if(source->type==Const)      return 1;
```

Fig. 10. NumberofAllPaths and NumberofCoveredPaths algorithms

If an Intersect node is processing, the multiplication of the number of the covered paths in Left and Right children will be returned. If a Union node is processing, the addition of the number of the covered paths in Left and Right children will be returned. If a Variable node is processing, we should check whether it is an intermediate signal. If so, its value in the *"list of TedState"* must be checked. Otherwise, 1 will be returned.

V EXPERIMENTAL RESULTS

We verified different properties on five examples including the Traffic Light Control (TLC), Greatest Common Divisor (GCD), Elevator (EL), 2-Client Arbiter (2CA) and a processor named Simple Architecture, Yet Enough Hardware (SAYEH) [see reference 10]. For instance, GCD properties are as follows:

1. *start = 1 & a = 23 => X(x = 23).*
2. *Reset = 1 & a = 13 => X(x = 13).*
3. *start = 0 & Reset = 0 & x =< y => X(x) = x.*
4. *start = 0 & Reset = 0 & x > y => X(x) = x - y.*

TABLE II compares our results with those of the VIS verification tools [13]. Notice that we have used Windows-based VIS in which CPU time pertains to EX, EG or EF functions, not all parts of VIS. To compute the CPU times, we have added appropriate VIS functions to VIS source codes in order to report execution time of EX, EG or EF function calls. The coverage method shows 80% coverage on *hwyl* signal in TLC example. As mentioned before, this method presents how many paths have been considered by the described properties, and in TLC example, our properties could cover 4/5 of paths at signal *hwyl*. In SAYEH example, some properties are not supported by VIS, because they involve both controller and datapath. VIS is not able to construct BDD of the datapath part of the SAYEH because of its large size.

TABLE II
Comparison with VIS

Circuit		TLC	GCD	SAYEH	EL	2CA
SN		hwyl	X	DataBus	door	cntl1
P1	WLM	0.01	0.04	10.9	0.3	0.3
	PC	20%	25%	10%	33.3%	25%
	VIS	0.1	0.2	31.2	2.1	1.5
P2	WLM	0.65	0.1	11.4	0.9	0.4
	PC	20%	25%	15%	33.3%	25%
	VIS	1.2	0.6	NS	3.4	1.5
P3	WLM	12.1	0.03	11.6	0.21	0.1
	PC	20%	25%	10%	33.3%	25%
	VIS	19.2	0.13	NS	4.7	0.9
P4	WLM	0.4	0.03	12.1	---	0.01
	PC	20%	25%	20%	---	25%
	VIS	0.9	0.14	39.8	---	0.1
TC		80%	100%	55%	100%	100%
N	WLM	60	32	1612	87	62
	VIS	974	968442	419062	20418	39381
M	WLM	5.3	4.5	10.3	4.1	5.1
	VIS	10.1	36	26.48	5.2	5.5

SN: Signal Name;   P1-P4: Cpu Time of Property1-4 (seconds)
WLM: our Word Level Method   NS: Not Supported
PC: %Property Coverage   TC: %Total Coverage
N: Number of Nodes(LTED,BDD);   M: Memory Usage (MegaByte)

VI CONCLUSION

In order to overcome problems related to the use of BDDs and other representations [7], we used a high level of representation. As the result, we are able to manipulate complex designs in much less time and memory than BBD-based approaches. Our representation treats data and control units together and is not limited to controller circuits or datapath circuits individually [7]. Also path coverage on LTL properties can be obtained based on this word-level model efficiently unlike of other models [11, 12]. As mentioned before, our approach does not need to solve integer equations or do satisfiability checking despite of other approaches [3, 4, 5, 6, 9].

References

[1] H. Touati, H. Savoj, B. Lin, R.K. Brayton and A. Sangiovanni-Vincentelli, "Implicit State Enumeration of Finite State Machines Using BDDs", in Proceedings ICCAD, pp 130-133, 1990.

[2] K. McMillan, Symbolic Model Checking, Kluwer Academic Publishers, Boston, 1993.

[3] A. Biere, A. Cimatti, E.M. Clarke, M. Fujita and Y. Zhu, "Symbolic Model Checking Using SAT Procedures Instead of BDDs", In Proceedings DAC, pp 317-320, June 1999.

[4] R. Brinkmann and R. Drechsler, "RTL-Datapath Verification using Integer Linear Programming", in Proceedings of IEEE VLSI Design'01 & Asia and South Pacific Design Automation Conference, pp 741-746, 2002.

[5] J. C. Corbett and G. S. Avrunin, "Using Integer Programming to Verify General Safety and Liveness Properties", in Journal of Formal Methods in System Design, Vol. 6, pp 97-123, Jan. 1995.

[6] T. Bultan, R. Gerber, and W. Pugh, "Symbolic Model Checking of Infinite State Systems Using Presburger Arithmetic", in 9th International Conference CAV, pp 400-411, 1997.

[7] R. Drechsler, Formal Verification of Circuits, Kluwer Academic Publishers, 2000.

[8] M. Ciesielski, P. Kalla and Z. Zeng, "Taylor Expansion Diagrams: A Compact Canonical Representation for Arithmetic Expressions", DATE02, pp 285-289, 2002.

[9] B. Alizadeh and M.R. Kakoee, "Using Integer Equations for High Level Formal Verification Property Checking", in ISQED03, pp 69-74, 2003.

[10] B. Alizadeh and Z. Navabi, "Word Level Symbolic Simulation in Processor Verification", in Journal of IEE-Proceedings Computers and Digital Techniques, Vol. 151, No. 5, pp 356-366, Sep. 2004.

[11] Y. Hoskote, T. Kam, P.-H Ho, and X. Zhao. Coverage Estimation for Symbolic Model Checking. In Proc. 36th Design Automation Conference, pp 300-305, 1999.

[12] H. Chockler, O. Kupferman, and M. Y. Vardi, "Coverage Metrics for Temporal Logic Model Checking", in *TACAS*, LNCS 2031, pp 528 – 542, 2001.

[13] Robert K. Brayton, A. Sangiovanni, A. Aziz and et al, "VIS: A system for Verification and Synthesis", in Proceedings of the 8th International Conference on Computer Aided Verification, pp 428-432, 1996.

Discovering the Input Assumptions in Specification Refinement Coverage

Prasenjit Basu   Sayantan Das   Pallab Dasgupta*   P.P. Chakrabarti*

Department of Computer Science & Engineering,
Indian Institute of Technology, Kharagpur, INDIA 721302
{pbasu,sayantan,pallab,ppchak}@cse.iitkgp.ernet.in

Abstract— The design of a large chip is typically hierarchical – large modules are recursively expanded into a collection of sub-modules. Each expansion refines the design due to the addition of level specific details. We believe that a similar approach is necessary to scale the capacity of formal property verification technology – as the design gets refined from one level to another, the formal specification must also be refined to reflect the level specific design decisions. At the heart of this approach we propose a checker that identifies the input assumptions under which the refined specification "covers" the original specification. This enables the validation engineer to focus the verification effort on the remaining input scenarios thereby reducing the number of target coverage points for simulation.

I. INTRODUCTION

In recent times, most leading chip design companies are seriously investigating the possibility of integrating *formal property verification* (FPV) [1] into their pre-silicon validation flows. It is widely accepted that the major challenge in existing FPV technology is in capacity. Unfortunately the theoretical lower bounds on the complexity of model checking problems imply that the technology is unlikely to scale beyond a point, in spite of the advances in the engineering of formal tools. Therefore a new methodology is required to extend the frontiers of FPV technology and enable us to formally verify properties over large designs.

How do we succeed in developing the RTL of large and complex designs? We do so by recursively decomposing the design functionality into the functionality of smaller modules, and by adding more details as we go down the module hierarchy. We continue this process of refinement until the modules are simple enough to be treated as unit level modules. We believe that the future of formal property verification lies in adopting a similar approach for formal property specifications. Whenever, we refine the design functionality into that of its component modules, we need to refine the formal specification of the design into that of its component modules. We refer to this paradigm as *formal specification refinement*.

The key formal method required for enabling specification refinement is a prover that checks whether the refined specification consisting of the specifications of the component modules *covers* the original specification of the design. In other

words we need to verify that *all behaviors that refute the original specification also refute the refined specification*. For example, let $\mathcal{A}$ denote the set of architectural properties of a design. Suppose we implement the design in terms of a set of RTL modules, $M_1 \ldots M_k$. Capacity limitations of existing FPV tools do not allow us to verify $\mathcal{A}$ over the whole design. Today validation engineers write properties, $\mathcal{R}_i$ over each RTL module, M_i and verify them, but this does not guarantee that they together satisfy $\mathcal{A}$. This guarantee can be obtained by a specification refinement proof which verifies that $\mathcal{R}_1, \ldots, \mathcal{R}_k$ together cover $\mathcal{A}$. Since the proof compares the specifications only (and not specification versus implementation – as in existing FPV techniques), we do not have major capacity limitations. In an earlier work, we developed the foundations of specification refinement proofs [2]. In that work we presented the methodology for comparing two temporal logic specifications at different levels of abstraction, and finding whether one covers the other.

What should we do if the refinement checker returns a negative answer? One option is to add more RTL properties to plug the coverage gap. In practice however, it is not always possible to cover system level properties in this way. Therefore the remaining properties in $\mathcal{A}$ must be checked dynamically during simulation.

This brings us to an interesting problem. Typically the coverage gap is a function of the input scenarios. In other words, we often find that the RTL properties succeed in covering the original specification $\mathcal{A}$ under some input constraints and fail for the rest. If we are able to compute the input constraints under which the refinement checker succeeds, then we can target the simulation on the remaining scenarios. This is the main goal of this paper.

In this paper we define a methodology for constraining the coverage gap between temporal specifications to determine the input constraints under which the gap is closed. The negation of these constraints cover the scenarios that must be checked through simulation. We believe that this enhancement to the basic refinement checker is of significant practical value in design validation through specification refinement.

The paper is organized as follows. In Section II, we present the theoretical background of the specification refinement problem. The formalism for computing the input constraint that covers the gap is presented in Section III. Section IV presents the algorithms for finding a legible representation of the input constraint. Section V presents the results of

---

*Pallab Dasgupta and P.P.Chakrabarti acknowledge the partial support of the Dept. of Sci & Tech, Govt. of India.

0-7803-9451-8/06/$20.00 ©2006 IEEE.

our prototype tool on several test cases. We have presented a case study on a realistic example in the Appendix.

II. Specification Refinement Coverage

In this paper we will focus on covering the gap between the formal architectural specification and the combined formal properties of the RTL components through specification refinement.

1. The key architectural properties are specified using a formal property specification language.[1] This forms the *architectural intent*, $\mathcal{A}$.

2. In the first phase of implementation, the design is planned in terms of a set of RTL blocks, and the functionality of each of these RTL blocks are defined. At this stage, formal properties are written to express some of the correctness requirements of each RTL block. The properties of the RTL blocks taken together is called the *RTL specs*, $\mathcal{R}$.

3. Our specification refinement checker checks whether the RTL specs, $\mathcal{R}$, cover the architectural intent, $\mathcal{A}$. We refer to this check as the *primary coverage question*.

The following definition formally describes the notion of coverage of the architectural intent by the RTL specs.

Definition 1 [Coverage Definition:]
The RTL specs cover the architectural intent iff there exists no run that refutes one or more properties of the architectural intent but does not refute any property of the RTL specs. □

In [2], it was shown that *the RTL specs $\mathcal{R}$, cover the architectural intent $\mathcal{A}$, iff the temporal property $\mathcal{R} \Rightarrow \mathcal{A}$ is valid.*

Most model checking tools for LTL [3] and its derivatives already have the capability of performing validity (or satisfiability) checks on temporal specifications, and can therefore be used to check whether $\mathcal{R} \Rightarrow \mathcal{A}$ is valid.

An arsenal of heuristic algorithms had been proposed in [2], for representing the coverage gap as a set of temporal properties when the answer to the primary coverage question is negative. While this feedback is useful in deciding whether and where more properties need to be added, it does not help the validation engineer to find the input scenarios that cover the behaviors in the gap.

III. Covering Input Assumptions

Specification refinement is not always feasible. It relies on the design architect's ability to add properties over individual RTL modules in a way to cover the architectural properties of the whole design. The refinement checker verifies the correctness of the decomposition and shows the gaps, but it does not automate the decomposition of the specification.

It is therefore natural to expect that in spite of adding more RTL properties on component modules, some architectural behaviors will remain uncovered. These behaviors must be targeted by simulation and dynamic property verification techniques.

[1] We shall use Linear Temporal Logic (LTL) in this paper

Our objective in this paper is to partition the architectural behaviors in terms of the input scenarios that trigger those behaviors. We use the specification refinement approach to show that the RTL specs cover the architectural specs under some of the input assumptions (expressed as properties over input variables). Since these scenarios are covered by specification refinement FPV, the remaining scenarios (also expressed as properties over inputs) represent the cases that must be verified through simulation. We believe that this is a very useful feedback to the validation engineer.

The following example demonstrates our intent. This is a toy example, used to demonstrate the main idea intuitively. In the appendix, we present a real example.

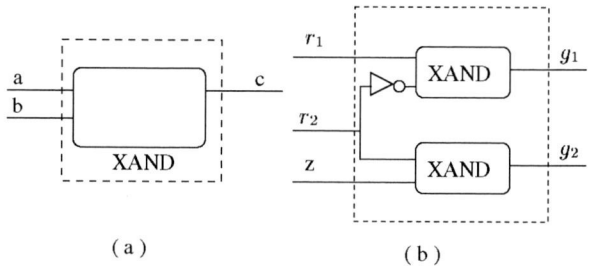

Fig. 1. A sample arbiter

Example 1 Let us consider the design of an arbiter that arbitrates between two request lines r_1 and r_2 from two master devices. Let the corresponding grant lines to the master devices be g_1 and g_2. The arbiter also receives an input z from a slave device, that remains high as long as the slave device is *ready*.

The arbiter specification requires us to treat r_2 as a high-priority request. Whenever r_2 is asserted and the slave is ready (that is, z is high), the arbiter must give the grant, g_2 in the next cycle, and continue to assert g_2 as long as r_2 remains asserted. When r_2 is not high, the arbiter parks the grant on g_1 regardless of whether r_1 is asserted. We are further given, that the request r_2 is fair in the sense that it is de-asserted infinitely often (enabling g_1 to be asserted infinitely often).

The above architectural intent may be expressed in LTL as follows:

$$A_1: \quad G\,F(\,\neg r_2\,)$$
$$A_2: \quad G(\,(\,r_2 \wedge z\,) \Rightarrow X(\,g_2\,U\,\neg r_2\,)\,)$$
$$A_3: \quad G(\,(\,\neg r_2\,) \Rightarrow X\,g_1\,)$$

Let us now consider an implementation of the arbiter using a component called XAND, as shown in Fig 1. The specification of the module XAND is as follows:

$$R_1': \quad G(\,(\,a \wedge b\,) \Rightarrow X\,c\,)\,)$$

It may be noted that we do not require the internal implementation of the RTL module XAND. Property R_1' is part of the RTL specification for XAND. Substituting the signal names of the instances of XAND in Fig 1(b) with r_1, r_2, g_1, g_2 and z, and adding the fairness property on r_2, we have the RTL specification as:

$$R_1: \quad G\,F(\,\neg r_2\,)$$
$$R_2: \quad G(\,(\,r_1 \wedge \neg r_2\,) \;\Rightarrow\; X\,g_1\,)$$
$$R_3: \quad G(\,(\,r_2 \wedge z\,) \;\Rightarrow\; X\,g_2\,)$$

The first property is the same fairness constraint as in the architectural intent. The second property says if r_1 is asserted and r_2 is de-asserted then g_1 is asserted in the next cycle. The third property states that whenever r_2 and z are asserted together, then g_2 is asserted in the next cycle.

The primary coverage problem is to determine whether $(R_1 \wedge R_2 \wedge R_3) \Rightarrow (A_1 \wedge A_2 \wedge A_3)$ is valid. In this case the answer is negative. For example whenever we have a scenario where both r_1 and r_2 are low, the architectural intent requires g_1 to be asserted, but the RTL specification does not have this requirement, which shows that A_3 is not covered. The coverage gap can be accurately represented by the following property:

$$U_1: \quad G(\,(\,\neg r_1 \wedge \neg r_2\,) \;\Rightarrow\; X\,g_1\,)$$

Our previous specification refinement checker will produce the property U_1 as the coverage gap. In the new approach we will produce the input constraint:

$$I_C: \; G(\,r_1 \vee r_2\,)$$

Under assumption I_C, we have $(R_1 \wedge R_2 \wedge R_3) \Rightarrow A_3$. Therefore RTL FPV succeeds in covering A_3 under these cases. Simulation should target input sequences satisfying $\neg I_C$, which is:

$$I_U: \; F(\,\neg r_1 \wedge \neg r_2\,)$$

This is indeed what we want, since the coverage gap U_1 represents the cases where r_1 and r_2 are low together. $\square$

The inputs to our problem are:

1. The *architectural intent* $\mathcal{A}$, as a set of LTL properties over a set $\mathcal{AP}_\mathcal{A}$, of Boolean signals, and

2. The *RTL specification* $\mathcal{R}$, as another set of LTL properties over a set, $\mathcal{AP}_\mathcal{R}$, of Boolean signals,

3. Additionally, we have the interface information of the architectural block; namely the set of inputs $\mathbb{I}_\mathcal{A}$, and the set of outputs $\mathbb{O}_\mathcal{A}$, of the architectural block, where $\mathcal{AP}_\mathcal{A} = \mathbb{I}_\mathcal{A} \cup \mathbb{O}_\mathcal{A}$.

We shall also use $\mathcal{A}$ to denote the conjunction of the properties in the architectural intent, and $\mathcal{R}$ to denote the conjunction of the properties in the RTL specification.

Throughout this paper we assume that $\mathcal{AP}_\mathcal{A} \subseteq \mathcal{AP}_\mathcal{R}$. This assumption essentially means that the RTL specification has the same names for their signals as the corresponding ones in the architectural intent. The RTL specification can have other signals in addition to these. Typically this is not a restrictive assumption within the design hierarchy, since it is generally considered a good practice for designers at a lower level of the design hierarchy to inherit the interface signal names from previous level of hierarchy.

Definition 2 [Strong and weak properties:]
A property $\mathcal{F}_1$ is stronger than a property $\mathcal{F}_2$ iff $\mathcal{F}_1 \Rightarrow \mathcal{F}_2$ and $\mathcal{F}_2 \not\Rightarrow \mathcal{F}_1$. We also say that $\mathcal{F}_2$ is weaker than $\mathcal{F}_1$. $\square$

Definition 3 [Coverage Hole in RTL Spec:]
A coverage hole in the RTL specification is a property $\mathcal{R}_H$ over $\mathcal{AP}_\mathcal{R}$, such that $(\mathcal{R} \wedge \mathcal{R}_H) \Rightarrow \mathcal{A}$, and there exists no property, $\mathcal{R}'_H$, over $\mathcal{AP}_\mathcal{R}$ such that $\mathcal{R}'_H$ is weaker than $\mathcal{R}_H$ and $(\mathcal{R} \wedge \mathcal{R}'_H) \Rightarrow \mathcal{A}$. In other words, this is the weakest property that suffices to close the coverage hole. Adding the weakest property strengthens the RTL specification in a minimal way. $\square$

Since $\mathcal{AP}_\mathcal{A} \subseteq \mathcal{AP}_\mathcal{R}$, each property of the architectural intent is a valid property over $\mathcal{AP}_\mathcal{R}$. In [2], we have shown that the coverage hole in the RTL specification is unique and is given by $\mathcal{A} \vee \neg \mathcal{R}$.

Definition 4 [Covering Input Constraint:]
A covering input constraint is a property $\mathcal{I}_\mathcal{C}$ over $\mathbb{I}_\mathcal{A}$, such that under the assumption $\mathcal{I}_\mathcal{C}$, the RTL specification covers the architectural intent i.e. $\mathcal{I}_\mathcal{C} \Rightarrow (\mathcal{R} \Rightarrow \mathcal{A})$, and there exists no property $\mathcal{I}'_\mathcal{C}$ over $\mathbb{I}_\mathcal{A}$ such that $\mathcal{I}'_\mathcal{C}$ is weaker than $\mathcal{I}_\mathcal{C}$ and $\mathcal{I}'_\mathcal{C} \Rightarrow (\mathcal{R} \Rightarrow \mathcal{A})$. In other words, we find the weakest assumption over $\mathbb{I}_\mathcal{A}$ that suffices to close the coverage gap. $\square$

According to the definition, the covering input constraint $\mathcal{I}_\mathcal{C}$ is a property over $\mathbb{I}_\mathcal{A}$, such that $\mathcal{I}_\mathcal{C} \Rightarrow (\mathcal{A} \vee \neg \mathcal{R})$. That is, it is stronger than the coverage hole in the RTL specification and no other property over $\mathbb{I}_\mathcal{A}$ which is weaker than $\mathcal{I}_\mathcal{C}$ has the same characteristic.

Definition 5 [Uncovered Input Space:]
An uncovered input space is a property $\mathcal{I}_\mathcal{U}$ over $\mathbb{I}_\mathcal{A}$, such that every run that refutes one or more properties of the architectural intent but does not refute any property of the RTL specification, implies $\mathcal{I}_\mathcal{U}$. Intuitively, these are the input scenarios for which the required guarantees have not been specified in the RTL specs. $\square$

Clearly the input scenarios which fall outside the covering input constraint, are the uncovered input scenarios. Thus the uncovered input space is $\neg \, \mathcal{I}_\mathcal{C}$.

IV. Computing the *Uncovered Input Space*

In this section, we present the algorithms for determining the uncovered input space as defined in Definition 5.

A. Coverage Algorithm

Our algorithm takes each formula $\mathcal{F}_\mathcal{A}$ from the architectural intent $\mathcal{A}$ and the interface definition of the architectural block and finds the uncovered input space, $\mathcal{I}_\mathcal{U}$, for $\mathcal{F}_\mathcal{A}$, with respect to the RTL specification $\mathcal{R}$. Since $\mathcal{R}$ is required to cover every property in $\mathcal{A}$ (by definition), we use this natural decomposition of the problem.

Algorithm 1 Coverage Algorithm

Find_Uncovered_Input_Space$(\mathcal{F}_\mathcal{A}, \mathcal{R}, \mathbb{I}_\mathcal{A}, \mathbb{O}_\mathcal{A})$

1. Compute $\mathcal{U} = \mathcal{F}_\mathcal{A} \vee \neg \mathcal{R}$

2. If $\mathcal{U}$ is not valid then

    (a) Unfold $\mathcal{U}$ up to its fixpoint to create two sets of uncovered terms; one is the disjunction of the terms before the fixpoint, $\mathcal{U}_M{}^{BF}$ and the other is the disjunction of the terms at the fixpoint, $\mathcal{U}_M{}^{AF}$;

    (b) Use universal quantification to eliminate signals belonging to $\mathcal{AP}_\mathcal{R} - \mathcal{AP}_\mathcal{A}$ from both the sets;

    (c) $\mathcal{I}_{BF} =$ Call Find_ISpace_BeforeF($\mathcal{U}_M{}^{BF}, \mathbb{I}_\mathcal{A}, \mathbb{O}_\mathcal{A}$);

    (d) $\mathcal{I}_{AF} =$ Call Find_ISpace_AtF($\mathcal{F}_\mathcal{A}, \mathcal{U}_M{}^{AF}, \mathbb{I}_\mathcal{A}, \mathbb{O}_\mathcal{A}$);

    (e) $\mathcal{I}_\mathcal{C} =$ Call Combine_Both_Parts($\mathcal{I}_{BF}, \mathcal{I}_{AF}, k$); [where k is the no. of unfolding required for $\mathcal{U}$ to reach the fixpoint]

    (f) $\mathcal{I}_\mathcal{U} = \neg \mathcal{I}_\mathcal{C}$;

3. Return $\mathcal{I}_\mathcal{U}$;

The first step computes the coverage gap $\mathcal{U}$. If $\mathcal{U}$ is valid then $\mathcal{F}_\mathcal{A}$ is covered. Otherwise we determine the uncovered input space. The second step of the algorithm performs this task. This step is further divided into six steps. The following subsections describe each of these steps in details.

B. Step 2(a): Unfolding of $\mathcal{U}$

In this step we recursively unfold $\mathcal{U}$ and generate two sets of disjunction of terms, namely $\mathcal{U}_M{}^{BF}$ and $\mathcal{U}_M{}^{AF}$, that contain only Boolean subformulas and Boolean subformulas guarded by a finite number of X (next) operators.

Definition 6 [X-depth, X-pushed, X-guarded:]
A formula is said to be X-pushed if all the X operators in the formula are pushed as far as possible to the left. A formula is said to be X-guarded if the corresponding X-pushed formula starts with an X operator whose scope covers the whole formula. The X-depth of an operator within a formula is the number of X operators whose scope covers the operator in the X-pushed form. $\square$

Any LTL property can be recursively unfolded over time steps to create an equivalent properties over Boolean formulas and X-guarded LTL formulas. It is known that for every temporal property such a decomposition begins to produce similar X-guarded subformulas after a well defined number of unfolding. During the unfolding process we check whether such a fixpoint has been reached. Once we reach the fixpoint (say after k steps of unfolding), then we take the formula produced after $(k-1)$ levels of unfolding and abstract out the temporal operators other than X as in [4]. We rewrite it in the form of disjunction of terms to generate $\mathcal{U}_M{}^{BF}$. In case $k = 1$, we consider $\mathcal{U}_M{}^{BF}$ as $True$. $\mathcal{U}_M{}^{AF}$ is obtained by dropping the terms that contain any temporal operator other than X from the one step decomposed form of the fixpoint formula.

Example 2 Let us consider the architectural property A and the RTL properties $R1$ and $R2$ as given below where r_1, r_2, r_3 are inputs to the architectural block and g_1 is the output:

$$A : G(r_1 \Rightarrow X g_1)$$

$$R1 : (r_1 \wedge \neg r_3) \Rightarrow X g_1$$
$$R2 : G((r_1 \wedge r_2) \Rightarrow X g_1)$$

We compute the disjunction of the uncovered terms before the fixpoint, $\mathcal{U}_M{}^{BF}$ and obtain the following formula:

$$\mathcal{U}_M{}^{BF} : \neg r_1 \vee X(g_1) \vee (r_1 \wedge \neg r_3 \wedge \neg X(g_1)) \vee (r_1 \wedge r_2 \wedge \neg X(g_1))$$

Again unfolding the fixpoint part and abstracting out the unbounded temporal operators, we yield the following disjunction of terms as $\mathcal{U}_M{}^{AF}$.

$$\mathcal{U}_M{}^{AF} : \neg r_1 \vee X(g_1) \vee (r_1 \wedge r_2 \wedge \neg X(g_1)) \quad \square$$

C. Step 2(b): Abstraction

In this step we universally eliminate the variables in $\mathcal{AP}_\mathcal{R} - \mathcal{AP}_\mathcal{A}$ from the properties $\mathcal{U}_M{}^{BF}$ and $\mathcal{U}_M{}^{AF}$.

D. Step 2(c): Finding covering input constraint before fixpoint

We treat each different X-guarded Booleans (different in X-guarded Boolean variables and/or in the number of X operators guarding the variables) in $\mathcal{U}_M{}^{BF}$ as separate Boolean variables and characterize them as inputs or outputs depending on whether the X-guarded Boolean variables are inputs or outputs respectively. Now we universally abstract out the outputs from $\mathcal{U}_M{}^{BF}$ and obtain the covering input constraint before fixpoint, namely $\mathcal{I}_{BF}$.

Example 3 Let us again consider the architectural property A and the RTL properties $R1$ and $R2$ as given in Example 2. We consider $X g_1$ as an output variable since g_1 is an output of the architectural block. We universally abstract out the output $X g_1$ from $\mathcal{U}_M{}^{BF}$ to get $\mathcal{I}_{BF}$ as below:

$$\mathcal{I}_{BF} : \neg r_1 \vee (r_1 \wedge \neg r_3) \vee (r_1 \wedge r_2) \quad \square$$

E. Step 2(d): Finding covering input constraint at fixpoint

In this step we first use the algorithm described in [2] to find out the uncovered architectural intent (say $\mathcal{G}$) using $\mathcal{U}_M{}^{AF}$ as the uncovered minterms. We have developed an approximate strategy for automatically identifying the input scenarios that *trigger* non-vacuous interpretations of a temporal property. Given a property, φ, defined over input variables $\mathcal{I}$, and output variables $\mathcal{O}$, we generate a formula S_φ that is stronger than φ and is defined only over $\mathcal{I}$. Since S_φ is stronger than φ and is free from output variables, it follows that S_φ describes input scenarios that make φ vacuously true. Therefore, we restrict the input space by $\neg S_\varphi$, which covers all non-vacuous runs.

Example 4 Consider the property, $\varphi : G[r_1 \Rightarrow X g_1]$. Then $S_\varphi = G(\neg r_1)$ and the constraint that restricts the input space to non-vacuous runs is $\neg S_\varphi = F r_1$, where r_1 as an input and g_1 is an output. $\square$

To find the required property over the inputs we define two operators, namely $\mathcal{S}$ and $\mathcal{W}$ which correspond to strengthening and weakening operations respectively. We present the rules for pushing the operators $\mathcal{S}$ and $\mathcal{W}$ downto the variables.

- **The rules for $\mathcal{S}$ operator:**

  - $\mathcal{S}(\theta\ f) \equiv \theta(\mathcal{S}(f))$ where θ is from $\{X, G, F\}$
  - $\mathcal{S}(f\ \eta\ g) \equiv \mathcal{S}(f)\ \eta\ \mathcal{S}(g)$ where η is from $\{\vee, \wedge, U\}$
  - $\mathcal{S}(f \Rightarrow g) \equiv \mathcal{W}(f) \Rightarrow \mathcal{S}(g)$
  - $\mathcal{S}(\neg f) \equiv \neg\,\mathcal{W}(f)$

- **The rules for $\mathcal{W}$ operator:**

  - $\mathcal{W}(\theta\ f) \equiv \theta(\mathcal{W}(f))$ where θ is from $\{X, G, F\}$
  - $\mathcal{W}(f\ \eta\ g) \equiv \mathcal{W}(f)\ \eta\ \mathcal{W}(g)$ where η is from $\{\vee, \wedge, U\}$
  - $\mathcal{W}(f \Rightarrow g) \equiv \mathcal{S}(f) \Rightarrow \mathcal{W}(g)$
  - $\mathcal{W}(\neg f) \equiv \neg\,\mathcal{S}(f)$

After applying the above rules on $\mathcal{S}(\mathcal{G})$, we yield a form where only the variable instances are preceded by an $\mathcal{S}$ or $\mathcal{W}$ operator. The substitution rule for a variable instance v is to substitute v by False for $\mathcal{S}(v)$, and by True for $\mathcal{W}(v)$.

We substitute the output variable instances of $\mathcal{G}$ according to the actions derived from $\mathcal{S}(\mathcal{G})$ and keep the input variables intact, thus obtaining a formula which is stronger than $\mathcal{G}$ and does not have any output variables. We call this formula, the covering input constraint at fixpoint, namely $\mathcal{I}_{AF}$.

Example 5 Let us consider the same architectural property A and RTL properties $R1$ and $R2$ as given in Example 2. After applying the coverage algorithm of [2] we yield the formula $\mathcal{G}$ as follows:

$$\mathcal{G} : G((r_1 \wedge \neg r_2) \Rightarrow X\ g_1)$$

Now we replace the output variable instances by True/False so as to strengthen $\mathcal{G}$. In this case we substitute g_1 by False. After reducing the substituted formula we get $\mathcal{I}_{AF}$ as shown below:

$$\mathcal{I}_{AF} : G(\neg(r_1 \wedge \neg r_2)) \qquad \square$$

F. Step 2(e): Combining two parts

Here we combine the input constraints $\mathcal{I}_{BF}$ and $\mathcal{I}_{AF}$ to produce the covering input constraint $\mathcal{I}_{\mathcal{C}}$. It may be noted that intuitively $\mathcal{I}_{BF}$ tells about the input constraint before the fixpoint and $\mathcal{I}_{AF}$ tells about the input constraint from the fixpoint under which the RTL properties cover the architectural intent. Hence we augment $\mathcal{I}_{AF}$ with $(k-1)$ X operators in the prefix where k is the fixpoint depth of $\mathcal{U}$ and take conjunction between $\mathcal{I}_{BF}$ and the X-augmented $\mathcal{I}_{AF}$ to yield $\mathcal{I}_{\mathcal{C}}$.

Example 6 Let us again consider the properties of Example 2. The fixpoint depth in this example is 2. After augmenting one X operator in the prefix of $\mathcal{I}_{AF}$ we get the following formula:

$$\mathcal{I}'_{AF} : X\ G(\neg(r_1 \wedge \neg r_2))$$

We produce $\mathcal{I}_{\mathcal{C}}$ by taking the conjunction of $\mathcal{I}_{BF}$ and $\mathcal{I}'_{AF}$.

$$\mathcal{I}_{\mathcal{C}} : (\neg r_1 \vee (r_1 \wedge \neg r_3) \vee (r_1 \wedge r_2)) \wedge$$
$$X\ G(\neg(r_1 \wedge \neg r_2)) \qquad \square$$

G. Step 2(f): Final step

Finally, we present $\mathcal{I}_{\mathcal{U}}$ as the negation of $\mathcal{I}_{\mathcal{C}}$.

Example 7 Continuing with the same example we can see that the uncovered input space in this case is as follows:

$$\mathcal{I}_{\mathcal{U}} : (r_1 \wedge \neg r_2 \wedge r_3) \vee X\ F(r_1 \wedge \neg r_2) \qquad \square$$

V. RESULTS

We have implemented a tool for computing the uncovered input scenarios. In Table I, we give the run times of our tool for some example circuits, including ARM AMBA AHB [5] and two Intel test cases, on a 2.8 GHz Intel Xeon processor with 4GB RAM. The second and third columns of the table present the no. of architectural signals and RTL signals respectively, for the circuit under test (mentioned in the first column). In the fourth and fifth columns, we produce the time (in seconds) taken by our tool to solve the primary coverage question and to compute the uncovered input space respectively. The results are quite promising. In all of the cases, the tool produced the uncovered input space, wherever necessary, in reasonable time.

TABLE I
RUNTIMES OF OUR TOOL

Circuit	$|\mathcal{AP}_{\mathcal{A}}|$	$|\mathcal{AP}_{\mathcal{R}}|$	Time(s) Primary Coverage Question	Time(s) Uncov. Input Space
Memory Arb. Logic	17	26	1.66	26.01
Intel Test Case1	8	16	0.005	0.02
Intel Test Case2	8	16	0.01	0.11
AMBA AHB Arb.	12	12	0.005	0.07
Paper Example	5	5	0.005	0.02

REFERENCES

[1] E. M. Clarke, O. Grumberg, and D. A. Peled, *Model checking*, MIT Press, 2000.

[2] S. Das, P. Basu, A. Banerjee, P. Dasgupta, P. P. Chakrabarti, C.R. Mohan, L. Fix, R. Armoni, "Formal verification coverage: computing the coverage gap between temporal specifications," In *Proc. of ICCAD*, 198-203, 2004.

[3] A. Pnueli, "The temporal logics of programs," In *Proc. of FOCS*, 46-57, 1977.

[4] A. Biere, A. Cimatti, E. M. Clarke, M. Fujita, Y. Zhu, "Symbolic model checking using SAT procedures instead of BDDs," In *Proc. of DAC*, 1999.

[5] *ARM AMBA specification rev 2.0*, http://www.arm.com/

A. CASE STUDY: MEMORY ARBITRATION LOGIC

Fig. 2 shows the architecture of a memory arbitration logic in the presence of a L1 cache. There are four request inputs, $r1, \ldots, r4$, for four independent on-chip requesting modules. Each of these four modules also assert write-enable signals $w1, \ldots, w4$ respectively. When wi is high, it indicates that the

1A-3

2006 Asia and South Pacific Design Automation Conference

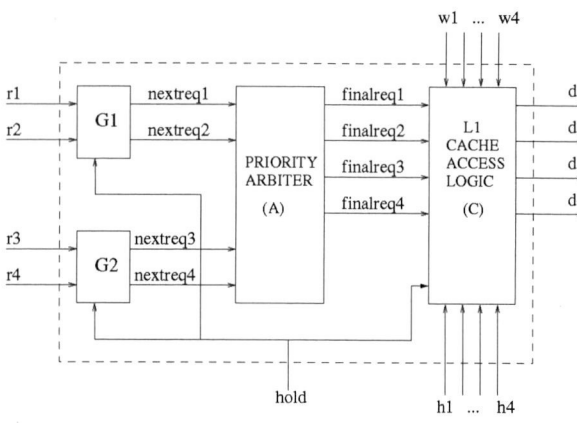

Fig. 2. Memory arbitration with L1 cache

corresponding requesting device intends to perform a write operation, where as the read operation is the default (when wi is low). The signals $h1, \ldots, h4$ are inputs from the cache that indicates whether the page requested by $r1, \ldots, r4$ respectively are present in the cache. The input signal $hold$ signifies that when it is asserted the memory arbitration logic stops accepting any request. The signals $d1, \ldots, d4$ indicates whether the page requested by $r1, \ldots, r4$ respectively is ready.

The architectural intent requires that $r1$ and $r2$ have higher priority than $r3$ and $r4$. Specifically this is translated into the architectural property that-

> If two devices with different priorities make requests for the cache control unit with the higher priority device making the request before the lower priority one, the device with higher priority will always have its page ready at the output, before the device with lower priority.

This architectural level requirement can be expressed by four properties for the four different ways in which a high priority request can come with a low priority one. For example, when we consider $r1$ (high priority) with $r3$ (low priority) we have the property-

$A_1: G(((\ r1 \wedge \neg r2 \wedge \neg r3\ ) \wedge X(\ r1\ U\ r3\ )) \Rightarrow (\ \neg d3\ U\ d1\ ))$

We have the following fairness conditions on the inputs and outputs:

$Q_1: \quad G(\ r1 \Rightarrow (\ r1\ U\ d1\ )\ )$
$Q_2: \quad G(\ r3 \Rightarrow (\ r3\ U\ d3\ )\ )$

which require the requesting device to hold the request line high until the page becomes ready. In addition we have the following two fairness restrictions:

$Q_3: \quad G(\ \neg r1 \Rightarrow (\ \neg d1 \wedge X(\ \neg d1\ )\ )\ )$
$Q_4: \quad G(\ \neg r3 \Rightarrow (\ \neg d3 \wedge X(\ \neg d3\ )\ )\ )$

which states whenever a device is idle (i.e. not requesting), its page ready signal will be low in that cycle and its next cycle. Fig. 2 shows the architecture of this logic in terms of four modules. G1 and G2 are round-robin arbiters. Whenever $hold$ is asserted they ignore the request lines. The variable $lastreq$ represents the state of G1 indicating which device was granted in the last round. Below we present the RTL properties for G1. Those for G2 are similar.

$R_{G1_1}: G((r1 \wedge \neg r2 \wedge \neg hold) \Rightarrow (nextreq1 \wedge X(\neg lastreq)))$
$R_{G1_2}: G((\neg r1 \wedge r2 \wedge \neg hold) \Rightarrow (nextreq2 \wedge X(lastreq)))$
$R_{G1_3}: G((\ r1 \wedge r2 \wedge \neg hold \wedge lastreq\ )$
$\qquad\qquad \Rightarrow (\ nextreq1 \wedge X(\ \neg lastreq\ )\ ))$
$R_{G1_4}: G((\ r1 \wedge r2 \wedge \neg hold \wedge \neg lastreq\ )$
$\qquad\qquad \Rightarrow (\ nextreq2 \wedge X(\ lastreq\ )\ ))$
$R_{G1_5}: G(\ hold \Rightarrow (\ \neg nextreq1 \wedge \neg nextreq2\ )\ )$
$R_{G1_6}: mutex(\ nextreq1,\ nextreq2\ )$

The priority arbiter in Fig 2 selects requests in the priority order, namely G1-highest and G2-lowest. The specification for this module is given below:

$R_{A_1}: G(\ nextreq1 \Rightarrow X\ finalreq1\ )$
$R_{A_2}: G(\ nextreq2 \Rightarrow X\ finalreq2\ )$
$R_{A_3}: G((\neg nextreq1 \wedge \neg nextreq2 \wedge nextreq3) \Rightarrow X\ finalreq3)$
$R_{A_4}: G((\neg nextreq1 \wedge \neg nextreq2 \wedge nextreq4) \Rightarrow X\ finalreq4)$
$R_{A_5}: mutex(\ finalreq1,\ \ldots,\ finalreq4\ )$

The cache access logic in Fig 2 directly interacts with the cache and performs according to the transfer type and whether the transfer is a cache hit or a cache miss. Whenever a cache miss is occurred in a read transfer the cache logic keeps the corresponding request pending in a wait buffer. $bfull$ is an internal signal which notifies the wait buffer full condition. The following three properties are given for device 1. For other devices we have the similar properties.

$R_{C_1}: \quad G(\ (\ finalreq1 \wedge h1\ ) \Rightarrow X\ d1\ )$
$R_{C_2}: \quad G(\ (\ finalreq1 \wedge w1 \wedge \neg h1\ ) \Rightarrow X\ d1\ )$
$R_{C_3}: \quad G(\ (\ finalreq1 \wedge \neg w1 \wedge \neg h1 \wedge \neg bfull\ ) \Rightarrow X\ F\ d1\ )$

Also for the buffer full condition, the following two properties are required.

$R_{C_4}: \quad G(\ \neg hold \Rightarrow \neg bfull\ )$
$R_{C_5}: \quad G(\ bfull \Rightarrow X\ F(\ \neg bfull\ )\ )$

Finally we have the mutual exclusion property for the ready signals.

$R_{C_6}: \quad mutex(\ d1,\ d2,\ d3,\ d4\ )$

The answer to our primary coverage problem for the architectural property A_1 is negative i.e. A_1 is not implied by the RTL specification. On a closer look it can be observed that in the scenario when the higher priority request is a read operation that results in a cache miss, A_1 is not guaranteed by the RTL properties. A counter example scenario that demonstrates this gap is as follows:

> Device 1 has requested for a page read in the present cycle but device 2 and device 3 haven't. In the next cycle device 3 has made a request and device 1's request results in a cache miss. So in the following cycle device 1 does not get the corresponding page ready signal but device 3's request results in a cache hit and hence in the succeeding cycle device 3 has its page ready signal available at the output.

Our tool returns the following property as the uncovered input space:

$A_H: F(\ (\ (\ r1 \wedge \neg r2 \wedge \neg r3\ ) \wedge (\ (\ X(\ \neg w1\ ) \wedge$
$\qquad\qquad X(\ \neg h1\ )\ ) \vee hold\ ) \wedge X(\ r1\ U\ r3\ )\ )\ )$

The property represents the set of scenarios that are not covered by the RTL specification.

18

Refinement Strategies for Verification Methods Based on Datapath Abstraction

Zaher S. Andraus, Mark H. Liffiton, and Karem A. Sakallah

Department of Electrical Engineering and Computer Science
University of Michigan
Ann Arbor, MI 48109-2122

{zandrawi,liffiton,karem}@eecs.umich.edu

Abstract—In this paper we explore the application of Counter-example-Guided Abstraction Refinement (CEGAR) in the context of microprocessor correspondence checking. The approach utilizes automatic datapath abstraction augmented with automatic refinement based on 1) localization, 2) generalization, and 3) minimal unsatisfiable subset (MUS) extraction. We introduce several refinement strategies and empirically evaluate their effectiveness on a set of microprocessor benchmarks. The data suggest that localization, generalization, and MUS extraction from both the abstract *and* concrete models are essential for effective verification. Additionally, refinement tends to converge faster when multiple MUSes are extracted in each iteration.

I. Introduction

Counterexample-Guided Abstraction Refinement (CEGAR for short) has been shown to be an effective paradigm in a variety of hardware and software verification scenarios. Originally pioneered by Kurshan [16], it has since been adopted by several researchers as a powerful means for coping with verification complexity. The widespread use of such a paradigm hinges, however, on the automation of its abstraction *and* refinement phases. Without automation, CEGAR requires laborious user intervention to choose the right abstractions and refinements based on a detailed understanding of the intricate interactions among the components of the design being verified. Clarke et al. [9], Jain et al. [14], and Dill et al. [5] have successfully demonstrated the automation of abstraction and refinement in the context of model checking for safety properties of hardware and software systems. In particular, these approaches create a smaller abstract transition system from the underlying concrete transition system and iteratively refine it with the spurious counterexamples produced by the model checker. The approaches in [9] and [14] are additionally based on the extraction of unsatisfiability explanations derived from the infeasible counterexamples to provide stronger refinement of the abstract model and to significantly reduce the number of refinement iterations. All of these approaches are examples of *predicate abstraction* which, essentially, projects the concrete model onto a given set of relevant predicates to produce an abstraction suitable for model checking a given property. In contrast, Andraus et al. [2] describe a methodology for *datapath*

abstraction that is particularly suited for equivalence checking. In their approach, datapath components in behavioral Verilog models are automatically abstracted to uninterpreted functions and predicates while refinement is performed manually using the ACL2 theorem prover [15].

The use of (near)-minimal explanations of unsatisfiability forms the basis of another class of abstraction methods. These include the work of Gupta et al. [12] and McMillan et al. [20] who employ "proof analysis" techniques to create an abstraction from an unsatisfiable concrete bounded model checking (BMC) instance of a given depth.

In this paper we explore the application of CEGAR in the context of microprocessor correspondence checking. The approach is based on automatic datapath abstraction as in [2] augmented with automatic refinement using minimal unsatisfiable subset (MUS) extraction. One of our main conclusions is the necessity of basing refinement on the extraction of MUSes from both the abstract *and* concrete models. Additionally, refinement tends to converge faster when multiple MUSes are extracted in each iteration. Finally, localization and generalization of the spurious counterexamples are shown to be crucial for fast convergence of the refinement iteration.

The rest of the paper is organized in 4 sections. Section II reviews the basic CEGAR algorithm and describes the various refinement strategies that can be deployed to enhance its performance. Section III briefly describes datapath abstraction and illustrates the abstraction and various refinement steps on a simple example. In Section IV we describe our implementation of these ideas in the Reveal system and discuss the effectiveness of the various refinement options in the verification of a sample benchmark. We conclude in Section V, with a recap of the paper's main contributions and suggestions for further work.

II. Refinement Strategies

A sketch of the basic counterexample-guided abstraction refinement methodology is shown in Algorithm 1. The detailed design is assumed to be characterized by a system of *concrete* constraints $\text{conc}(X) = \bigwedge_{1 \le i \le n} C_i(X)$ where X denotes a suitable vector of variables and

$C_i(X)$ is a Boolean *consistency* constraint that models a particular component in the design. For example, a 32-bit adder with inputs A, B, and output S would be represented by the constraint $C(A, B, S) = (S = A + B)$. The verification objective can also be expressed as a Boolean function of the design's variables. In general we will be concerned with equivalence between signal pairs, but any safety property can be handled similarly. Let $\text{prop}(X)$ denote the condition that we would like to check for on the design. The verification task can be expressed as showing that $\text{conc}(X) \rightarrow \text{prop}(X)$ is valid, i.e., that it is a tautology.

CEGAR starts, in line 1, with the construction of an abstraction $\text{abst}(X) = \bigwedge_{1 \leq i \leq n} A_i(X)$ such that $C_i(X) \rightarrow A_i(X)$. In other words, each constraint in $\text{abst}(X)$ is a *relaxation* of a corresponding constraint in $\text{conc}(X)$[1]. This type of abstraction, is sound, i.e., if $\text{prop}(X)$ holds on $\text{abst}(X)$, then it must also hold on $\text{conc}(X)$ (line 4), but incomplete, i.e., $\text{prop}(X)$ may be violated on $\text{abst}(X)$ but still hold on $\text{conc}(X)$). Completeness is achieved by *refining* $\text{abst}(X)$ to eliminate such cases. Specifically, if X^* denotes an assignment of values to variables such that

$$\text{abst}(X^*) \rightarrow \text{prop}(X^*) = 0 \text{ and } \text{conc}(X^*) = 0$$

then X^* represents a spurious counterexample that must be eliminated from the space of consistent assignments in the abstract model (line 9). Note that this process is guaranteed to converge assuming that the abstraction is finite and noting that it shrinks monotonically with each refinement iteration.

This basic version of CEGAR can be quite inefficient, requiring a large number of refinement iterations, since false counterexamples are eliminated one at a time. The improved version depicted in Algorithm 2 employs several techniques aimed at reducing the number of refinement iterations. The common goal of these techniques is to use the specific counterexample that falsifies the property on the abstract model as a seed to generate a large collection of "related" counterexamples that can then be simultaneously checked on the concrete model and, subsequently, eliminated at the same time from the abstract model. Algorithm 2 employs the following *violation enlargement* methods:

- **Localization** (line 5a) is basically a cone-of-influence (COI) reduction that removes irrelevant (don't-care) assignments from the counterexample by a syntactic traversal of the abstract formula.
- **Generalization** (line 5b) replaces the specific values of the variables in the counterexample with appropriate

---

Algorithm 1 (Basic CEGAR)

```
1.    abst(X) = relax(conc(X))
2.    while (true) {
3.      if abst(X) -> prop(X)
4.        then {"property holds"; exit}
5.        else {// abst(X*) -> prop(X*) == 0
6.          if conc(X*)
7.            then {"property fails"; exit}
8.            else // spurious counter example
9.              abst(X) = abst(X) && !X*;
10.   }// else(line 5)
11. }// while(line 2)
```

Algorithm 2 (CEGAR with enhanced refinement)

```
1.    abst(X) = relax(conc(X))
2.    while (true) {
3.      if abst(X) -> prop(X)
4.        then {"property holds"; exit}
5.        else {// abst(X*) -> prop(X*) = 0
5a.       viol(X) = localize(X*);
5b.       viol(X) = generalize(viol(X));
5c.       expl(X) = MUS(!abst(X) && viol(X));
5d.       viol(X) = viol(X) && expl(X);
6.        if conc(X) && viol(X)
7.          then {"property fails"; exit}
8.          else // spurious violation
8a.           while (expl(X) = newMUS(conc(X) && viol(X)))
9.              abst(X) = abst(X) && !(viol(X) && expl(X));
10.   }// else(line 5)
11. }// while(line 2)
```

equality and dis-equality between pairs of variables. These relations can now be viewed as a conjunction of violation constraints $\text{viol}(X)$ that satisfy

$$[\text{abst}(X) \rightarrow \text{prop}(X)] \wedge \text{viol}(X) = 0$$

- **Minimal Unsatisfiable Subset (MUS) extraction** (line 5c) selects a small subset of the constraints from the unsatisfiable formula $\neg\text{abst}(X) \wedge \text{viol}(X)$[2] that is still sufficient to explain its infeasibility. Denoting this subset as $\text{expl}(X)$, the set of violation constraints is now reduced by eliminating from it those constraints that are not in $\text{expl}(X)$ (line 5d).

At this point, $\text{viol}(X)$ represents not just one but many specific counterexamples that violate $\text{prop}(X)$ but are consistent with $\text{abst}(X)$. The enlargement of X^* to $\text{viol}(X)$ was based solely on information from the abstract model and the property being checked. The check in line 6 is now used to determine if any of these violations is consistent with the concrete model. Failing this check implies that *all* of these violations are spurious and we can refine the abstract model by removing them. However, an additional enhancement (lines 8a and 9) makes it possible to utilize the concrete constraints to enlarge $\text{viol}(X)$ further. Specifically, a small number of MUSes from the unsatisfiable formula $\text{conc}(X) \wedge \text{viol}(X)$ are extracted and used to select subsets of $\text{viol}(X)$ each of which is sufficient to keep the formula unsatisfiable.

---

[1] In general, both design variables and the concrete constraints that relate them are relaxed.

[2] Note that $[\text{abst}(X) \rightarrow \text{prop}(X)] \wedge \text{viol}(X) = 0$ implies that $\neg\text{abst}(X) \wedge \text{viol}(X) = 0$.

```verilog
module example();
    wire [3:0] a, b;
    wire m = a[3]; // msb
    wire l = a[0]; // lsb
    wire c = m? a >> 1 : a;
    wire d = l? b >> 2 : c;
    wire e = m? a : a >> 1;
    wire f = l? {2'b00, b[3:2]} : e;
    wire p = !(a == 0) || (d == f);
endmodule;
```

Fig. 1. Verilog design example used to illustrate abstraction and refinement

III. DATAPATH ABSTRACTION REFINEMENT

The focus of our work is the application of the CEGAR framework, particularly the afore-mentioned enhanced refinement strategies, in the context of hardware correspondence checking. Specifically, we address the task of verifying that an optimized microprocessor implementation is compliant with its functional specification. The implementation and specification are assumed to be given using a hardware description language, such as Verilog, and together are regarded as the concrete model. The correctness criterion, i.e., the property that must hold to insure that the implementation is functionally equivalent to the specification, depends on the nature of the optimizations performed to obtain the implementation. In particular, pipelined implementations require alignment of the programmer-visible implementation and specification states which can be accomplished, for example, by flushing [8]. The specifics of the correctness criterion, while important, are orthogonal to the abstraction refinement flow and can be assumed, for our purposes, to be provided by the user along with the specification and implementation.

The concrete model is relaxed by treating datapath elements as uninterpreted functions (UFs) and uninterpreted predicates (UPs) that operate on unbounded terms [6]. This is justified by the fact that, generally speaking, design optimizations add significant complexity to an implementation's control logic while, mostly, preserving its datapath components. Datapath abstraction, thus, yields a compact representation that preserves the control interactions in the concrete model while maintaining functional consistency of the abstracted datapath elements. The resulting abstract model can now be viewed as a set of constraints—a formula—in CLU, a quantifier-free first-order logic with counter arithmetic and lambda expressions [6], and used in lieu of the detailed bit-level concrete model to check satisfaction of the desired correctness condition.

To illustrate the salient features of datapath abstraction refinement consider the example Verilog "design" in Fig. 1. The verification objective is to prove that signal p is always true, indicating that the design satisfies the condition $(a = 0) \rightarrow (d = f)$. The formula representing the concrete constraints of this design is written by inspection as

$$
\begin{aligned}
\text{conc}\,(a,b,c,d,e,f,l,m,p) = & \\
(m = a[3]) \wedge & \\
(l = a[0]) \wedge & \\
(m \wedge (c = a >> 1) \vee \neg m \wedge (c = a)) \wedge & \\
(l \wedge (d = b >> 2) \vee \neg l \wedge (d = c)) \wedge & \\
(m \wedge (e = a) \vee \neg m \wedge (e = a >> 1)) \wedge & \\
(l \wedge (f = \{\text{2'b00}, b[3:2]\}) \vee \neg l \wedge (f = e)) \wedge & \\
(p = \neg(a = 0) \vee (d = f)) &
\end{aligned}
\tag{1}
$$

Using the semantics of bit vector operations, such as extraction, concatenation, and shifting, along with the standard Boolean connectives, this formula can be translated in a straightforward fashion to propositional conjunctive normal form (CNF) so that it can be checked for satisfiability by standard SAT solvers. In fact, for this simple example it is quite easy for a modern SAT solver to prove that $\text{conc} \wedge \neg p$ is unsatisfiable which is the same as saying that $\text{conc} \rightarrow p$ is valid.

Our objective, however, is to establish this result using CEGAR. A possible abstraction of this design is:

$$
\begin{aligned}
\text{abst}\,(a,b,c,d,e,f,l,m,s,t,u,p,zero) = & \\
(m = \text{EX1}(a)) \wedge & \\
(l = \text{EX2}(a)) \wedge & \\
(s = \text{SR1}(a, \text{succ}(zero))) \wedge & \\
(t = \text{SR2}(b, \text{succ}(\text{succ}(zero)))) \wedge & \\
(u = \text{CT1}(zero, \text{EX3}(b))) \wedge & \\
(c = \text{ite}(m, s, a)) \wedge & \\
(d = \text{ite}(l, t, c)) \wedge & \\
(e = \text{ite}(m, a, s)) \wedge & \\
(f = \text{ite}(l, u, e)) \wedge & \\
(p = \neg(a = zero) \vee (d = f)) &
\end{aligned}
\tag{2}
$$

where detailed bit vector operations have been replaced by UP and UF symbols. For example, EX1 is a UP that corresponds to extracting the most significant bit of a, and SR2 is a UF that corresponds to a right shift of b by two bits. Variables in this abstract formula that correspond to bit vectors in the concrete formula are now considered to be unbounded terms. They can be compared for equality to enforce functional consistency (given two terms t_1 and t_2 and a single-argument UF F, $(t_1 = t_2) \rightarrow (F(t_1) = F(t_2))$) but are otherwise uninterpreted having lost their concrete semantics. On the other hand, variables in the abstract formula that correspond to single bits in the concrete formula (such as m

TABLE I

EXECUTION TRACE OF ALGORITHM 2 ON EXAMPLE OF FIG. 1

		Iteration 1			Iteration 2			
		X^*	Localize	Generalize	Find MUS	X^*	Localize	Generalize	Find MUS
abst(X)		$a = 0$	$a = 0$	$a = 0$	$a = 0$	$a = 0$	$a = 0$	$a = 0$	$a = 0$
		$b = 8$	$b = 8$	$l = 1$	$l = 1$	$b = 16$	$b = 16$	$l = 0$	$l = 0$
		$c = 16$		$t \neq u$	$t \neq u$	$c = 8$	$c = 8$	$m = 1$	$t = u$
		$d = 20$	$d = 20$			$d = 8$	$d = 8$	$t = u$	$a \neq s$
		$e = 0$				$e = 0$	$e = 0$	$a \neq s$	
		$f = 12$	$f = 12$			$f = 0$	$f = 0$		
		$l = 1$	$l = 1$			$l = 0$	$l = 0$		
		$m = 1$				$m = 1$	$m = 1$		
		$s = 16$				$s = 8$	$s = 8$		
		$t = 20$	$t = 20$			$t = 3$	$t = 3$		
		$u = 12$	$u = 12$			$u = 3$	$u = 3$		
conc(X)		$viol_1 = (a = 0) \wedge (a[0] = 1)$ $viol_2 = b \gg 2 \neq \{2\text{'b}00, b[3{:}2]\}$				$viol_3 = (a = 0) \wedge (a \neq a \gg 1)$			

and l) retain their Boolean semantics and can be combined with the standard Boolean connectives. The remaining symbols in the formula represent the CLU built-in functions for counting (succ), decision (if-then-else or ite) and the smallest term (zero).

When Algorithm 2 is invoked on this example, it terminates with a proof of validity after two refinement iterations (see Table I). The counterexample produced in the first iteration is localized by eliminating irrelevant assignments, namely those that correspond to the "else" branches of the the ite operators involving variable l as well as other assignments that depend on them. Next, the remaining relevant assignments are generalized into the violation constraint $(a = 0) \wedge (l = 1) \wedge (t \neq u)$ which, in this case, cannot be enlarged further using MUS extraction from the abstract formula. Upon checking this violation constraint on the concrete formula it is found to be spurious, and leads to the creation of two simpler explanations: 1) the least significant bit of a cannot be 1 when a is 0, and 2) shifting b right by two bits is equivalent to concatenating zeros to the left of b's two most significant bits. The abstract formula is now refined to eliminate these two violations, i.e., at the start of the second iteration the correctness condition is checked against abst $\wedge \neg((a = 0) \wedge (l = 1)) \wedge \neg(t \neq u)$.

In the second iteration, localization is unable to eliminate any assignments from the counterexample. However, generalization retains only three of the assignments as well as the equality between t and u from the first iteration, and deduces that a and s are not equal. MUS extraction identifies that the assignment to m is immaterial to the current violation and can be safely removed. Checking this violation constraint on the concrete formula shows that it is still spurious, and identifies the

minimal explanation "shifting zeros yields zeros!" The concrete formula was thus able to remove the constraints $l = 0$ and $t = u$ as irrelevant to the current violation. When this violation is eliminated, by refining with $\neg((a = 0) \wedge (a \neq s))$, the algorithm terminates proving that p is always true.

IV. IMPLEMENTATION AND EXPERIMENTAL RESULTS

We implemented the above refinement strategies in the Reveal system which performs verification of hardware designs using datapath abstraction. Reveal consists of the following components:

- **Vapor** [2] for abstracting designs written in behavioral Verilog to the UCLID language.
- **UCLID** [6] for converting the abstracted design to a formula in the CLU logic.
- **Wave** for encoding the CLU formula in propositional logic, specifically in the conjunctive normal form suitable for SAT solvers [7].
- **zCore** and **zMinimal** [23] for extracting a single MUS from a CNF instance. To obtain an MUS from the abstract model, we convert the CLU expression in Algorithm 2 (line 5c) to a CNF instance using Ackerman encoding [1].
- **CAMUS** [19] for extracting one or more MUSes from a CNF instance.

We ran the experiments on an 64-bit 2.4-GHz AMD processor with 4GB of RAM running Linux.

To establish a baseline, we ran Algorithm 2 by disabling MUS extraction from both the abstract (line 5c) as well as the concrete (line 8a) formulas. In all cases, the procedure had to be aborted, even for simple designs, suggesting that MUS extraction is essential for refinement. We then ran another set of experiments in which localization and generalization were disabled forcing MUS extraction to be based on the specific counterexamples produced. Again, verification failed to finish in the allotted time in all tested cases, suggesting that localization and generalization are also essential for effective refinement.

We then performed a series of experiments using different combinations of MUS-based refinements with localization and generalization enabled. For ease of exposition, we will use refine(x, y) to denote refinement with:

- $x \in \{0, 1\}$ MUSes extracted using the abstract formula
- $y \in \{0, 1, \text{some}, \text{all}\}$ MUSes extracted using the concrete formula

We used zCore and zMinimal to extract single MUSes, and CAMUS to extract multiple and all MUSes. The number of MUSes produced by CAMUS in the "multiple" mode can be controlled by a user-specified parameter and is typically between 3 and 7.

Table II shows the results of verifying two different properties on the PDLX benchmark [25] with various refinement combinations. This benchmark consists of 686 Verilog lines and 396 latches. In the first set of experi-

TABLE II
PDLX REFINEMENT RESULTS

	refine(0, y)			refine(1, y)		
	T	I	M	T	I	M
refine(x, 0)	T.O.	>9	N/A	T.O.	>7	>7
refine(x, 1)	T.O.	>37	>37	T.O.	>22	>22
refine(x, some)	T.O.	>7	>25	138	2	10
refine(x, all)	T.O.	>1	–	T.O.	>1	–

Buggy PDLX implementation; prop = (RF$_{Spec}$ = RF$_{Impl}$)

	refine(0, y)			refine(1, y)		
	T	I	M	T	I	M
refine(x, 0)	T.O.	>13	N/A	T.O.	>10	N/A
refine(x, 1)	316	25	25	180	8	8
refine(x, some)	404	5	21	64	1	4
refine(x, all)	T.O.	>1	–	T.O.	>1	–

Bug-free PDLX implementation; prop = (PC$_{Spec}$ = PC$_{Impl}$)

ments, the property being checked is the equivalence between the implementation and specification register files using a buggy implementation in which the ALU output is stuck at 0. In the second set of experiments, the correctness condition is the equivalence of the implementation and specification program counters; the implementation in these experiments was bug-free. Both of these criteria are based on the Burch and Dill [8] correspondence checking scheme of pipelined microprocessors. The columns in the table give the total verification time in seconds (T), the number of refinement iterations (I) and, where applicable, the number of MUSes extracted using the concrete model (M). A time out of 600 seconds was used in all experiments.

For the buggy design, the only refinement strategy that did not time out was refine(1, some). For the bug-free design, on the other hand, verification with refinements that employed one or several MUSes extracted from the concrete model finished within the allowed time. In both cases, however, the best performance was obtained with the refine(1, some) combination.

Further analysis of these results reveals that the use of MUSes involves a trade-off between the effort to extract them and their effectiveness in refining the abstraction. This can be seen by comparing refine(x, all), refine(x, 1) and refine(x, some). In the first case, the excessive time needed to extract all MUSes seems to negate their utility for refinement. Comparing the other two scenarios, we note that extracting several MUSes per refinement iteration seems to always yield fewer iterations, and shorter overall verification times, than does the extraction of just one MUS in each iteration.

V. CONCLUSIONS AND FUTURE WORK

In this work we explored the use of MUS extraction for refinement of datapath abstractions in the CEGAR verification flow. We found that extraction of MUSes from both the abstract and concrete models is necessary for faster convergence. We also found that performance tends to improve when refinement is based on the extraction of a small number of MUSes, rather than a single MUS, from the concrete model in each iteration. Additionally, we introduced counterexample localization and generalization, and demonstrated their necessity for speeding up convergence of the refinement iteration.

The robustness and scalability of the verification framework described here can be enhanced further by incorporating several additional improvements. For example, faster MUS extraction from the abstract model may be possible if the extraction algorithm were to operate directly on the abstract CLU formula rather than on its lower-level CNF Boolean encoding. Additionally, extracting multiple MUSes from the abstract model, akin to the extraction of multiple MUSes from the concrete model, may yield a further reduction in the number of refienemnt iterations and, possibly, overall verification time. Finally, by analyzing the *structure* of the MUSes extracted from the concrete model, it may be possible to generalize them into *universal* rules (templates) that can be stored in a rule base that grows with the usage of the system and that can be consulted when verifying other designs.

ACKNOWLEDGMENTS

This work was funded in part by the DARPA/MARCO Gigascale Systems Research Center, and in part by the National Science Foundation under ITR grant No. 0205288.

REFERENCES

[1] W. Ackerman, "Solvable Cases of the Decision Problem." North-Holland, Amsterdam, 1954.

[2] Z. S. Andraus and K. A. Sakallah, "Automatic Abstraction of Verilog Models", In Proceedings of 41st Design Automation Conference 2004, pp. 218-223.

[3] C. Barrett, D. Dill, and J. Levitt, "Validity checking for combinations of theories with equality". In FMCAD96, LNCS 1166, pp. 187-201.

[4] P. Chauhan, E. Clarke, J. Kukula, S. Sapra, H. Veith, and D. Wang, "Automated Abstraction Refinement for Model Checking Large State Spaces using SAT based Conflict Analysis", FMCAD02.

[5] S. Das and D. Dill, "Successive Approximation of Abstract Transition Relations" in 16th Annual IEEE Symposium on Logic in Computer Science (LICS) 2001.

[6] R. E. Bryant, S. K. Lahiri, S. A. Seshia, "Modeling and Verifying Systems using a Logic of Counter Arithmetic with Lambda Expressions and Uninterpreted Functions". In Proc. CAV, July 2002.

[7] R. E. Bryant, S. German, and M. N. Velev, "Exploiting

positive equality in a logic of equality with uninterpreted functions". ACM Transactions on Computational Logic, 2(1):93-134, January 2001.

[8] J. R. Burch and D. L. Dill, "Automatic Verification of Pipelined Microprocessor Control". CAV '94, D. L. Dill, ed., LNCS 818, Springer-Verlag, June 1994, pp. 68-80.

[9] E. Clarke, O. Grumberg. S. Jha, Y. Lu and H. Veith, "Counterexample-Guided Abstraction Refinement," In CAV 2000, pp. 154-169.

[10] N. Een and A. Biere, "Improved Subsumption Techniques for Variables Elimination in SAT," SAT 2005.

[11] S. Graf and H. Saidi, "Construction of abstract state graphs with PVS," In CAV, volume 1254, pp. 72-83, 1997.

[12] A. Gupta, M. Ganai, Z. Yang, and P. Ashar, "Iterative Abstraction Using SAT-based BMC with Proof Analysis." In Proc. of the International Conference on CAD, pp. 416-423, Nov. 2003.

[13] J. L. Hennessy and D. A. Patterson, "Computer Architecture: A Quantitative Approach", Morgan Kaufmann, 1990.

[14] H. Jain, D. Kroening and E. Clarke, "Predicate Abstraction and Verification of Verilog," Technical Report CMU-CS-04-139.

[15] M. Kaufmann and J. Moore, "An Industrial Strength Theorem Prover for a Logic Based on Common Lisp." IEEE Transactions on Software Engineering 23(4), April 1997, pp. 203-213.

[16] R. Kurshan, "Computer-Aided Verification of Coordinating Processes: The Automata-Theoritic Approach," Princ-

eton University Press, 1999.

[17] D. Kroening, E. Clarke and K. Yorav, "Behavioral Consistency of C and Verilog Programs with Bounded Model Checking," in 40th DAC, 2003, pp. 368-342.

[18] S. K. Lahiri, S. A. Seshia, R. E. Bryant, "Modeling and Verification of Out-of-Order Microprocessors in UCLID", FMCAD 2002.

[19] M. H. Liffiton, M. D. Moffitt, M. E. Pollack, and K. A. Sakallah, "Identifying Conflicts in Overconstrained Temporal Problems," in *Proc. 19th International Joint Conference on Artificial Intelligence (IJCAI-05)*, pp. 205-211, Edinburgh, Scotland, 2005.

[20] K. L. McMillan and N. Amla, "Automatic Abstraction without Counterexamples." In International Conference on Tools and Algorithms for Construction and Analysis of Systems (TACAS'03), pp. 2-17, Warsaw, Poland, April, 2003, LNCS 2619.

[21] S. A. Seshia, S. K. Lahiri, R. E. Bryant, "A User's Guide to UCLID version 0.1".

[22] D. E. Thomas and P. R. Moorby, "The Verilog Hardware Description Language", Kluwer Academic Publishers, Nowell, Massachusetts, 1991.

[23] L. Zhang and S. Malik, "Extracting Small Unsatisfiable Cores from Unsatisfiable Boolean Formulas." In SAT, Springer-Verlag, 2003.

[24] www-2.cs.cmu.edu/~uclid

[25] vlsi.colorado.edu/~vis

Generation of Shorter Sequences for High Resolution Error Diagnosis Using Sequential SAT

Sung-Jui (Song-Ra) Pan and Kwang-Ting Cheng

Dept. of Electrical & Computer Engr., U. of California, Santa Barbara, Santa Barbara, CA 93106

{*srpan, timcheng*}*@ece.ucsb.edu*

John Moondanos and Ziyad Hanna

Intel Corporation

{*john.moondanos, ziyad.hanna*}*@intel.com*

Abstract

Commonly used pattern sources in simulation-based verification include random, guided random, or design verification patterns. Although these patterns may help bring the design to those hard-to-reach states for activating the errors and for propagating them to observation points, they tend to be very long, which complicates the subsequent diagnosis process. As a key step in reducing the overall diagnosis complexity, we propose a method of generating a shorter error-sequence based on a given long error-sequence. We formulate the problem as a satisfiability problem and employ a SAT solver as the underlying engine for this task. By heuristically selecting an intermediate state S_i which is reachable by the given long sequence, the task of finding the transfer sequence from the initial state to the target state can be divided into two easier tasks - finding a transfer sequence from the initial state to S_i and one from S_i to the target state. Our preliminary experimental results on public benchmark circuits show that the proposed method can achieve significant reduction in the length of the error sequences.

1 Introduction

Constrained or guided random patterns are still heavily used to detect errors in a modern design, although recent improvements in formal methods such as bounded model checking (BMC) have made it possible to verify certain properties and generate counter-examples for large industrial designs. The counter-examples are often too complex to be manually inspected by the designers. In [11], Ravi and Somenzi propose to guide the bounded model checker to minimize the sizes of counter-examples. However, in current practice, most of the counter-examples are primarily found by simulation-based verification instead of formal methods. Thus, both manual debugging and automatic design-error diagnosis for a sequential circuit often relies on a set of sequences, called *error sequences*, which reveal erroneous responses at the primary outputs. In general, longer error sequences imply higher complexity for diagnosis. An unnecessarily long error sequence (e.g. visiting a state multiple times, not taking the shortest

path from the initial state to the required state for activating and propagating the error, etc.) generally degrades the diagnosis resolution. In contrast, short error sequences minimize the complexity and maximize the resolution of the diagnosis process.

D'Souza and Hsiao in [6] extended the region-based technique proposed in [3] for diagnosing sequential circuits. It attempts to identify the first cycle in which an error appears at either a flip-flop or a primary output to avoid re-simulating the entire sequence. After such a cycle is identified, a shorter sequence can be found for region-based diagnosis. Region-based diagnosis proposed in [3, 6] relaxes the single-error assumption and does not use any specific error models originally proposed in [1]. However, the knowledge of the expected value at each register in each cycle, which is required for these methods, is not typically available in the debugging phase. Usually the expected values are available only for the primary outputs. Therefore, we cannot accurately determine the cycle at which the state of the erroneous circuit first deviates from that of the good circuit.

In [5], the authors propose to eliminate redundant states in an error trace. A new trace can then be found by identifying the shortest path among the remaining unique states in the trace. One main problem with the method is that the new trace may not activate and propagate any errors to the primary outputs. This is because the process of identifying the redundant states in the original error trace and finding the shortest path for the remaining unique states is not based on a "golden model" of the design. Instead, it is based on the design model which contains bugs. Therefore, the new trace may not be a valid error trace. These simulation based techniques have been greatly enhanced in [4].

In this paper, we address the problem of generating a shorter error sequence from a given longer one. Error sequences are often identified by examining the simulation results of the erroneous designs on either functional or random patterns. We focus on formulating this test-generation problem as a satisfiability problem and employ a modern SAT solver [8], for this optimization. By selecting an intermediate state S_i, which is reachable by the given long sequence, the task of finding the transfer sequence from the initial state and the state which reveals the error at the outputs can be divided into two sub-tasks - finding a transfer sequence from the initial state to S_i and one from S_i to the target state. This strategy can be recursively applied until every subproblem becomes solvable by the SAT solver. However, if the transfer sequences are derived this way and then cascaded together to form the final transfer sequence, we may get a "false" sequence that cannot activate the errors or propagate them to the outputs. This could happen if the circuit model used by the SAT solver is erroneous (because it is derived from the circuit under diagnosis). Therefore, each error sequence generated must be simulated to verify the matching. If the generated sequence does not result in the erroneous responses at the outputs, the procedure needs to be called again to generate a new transfer sequence. We propose an algorithm to select states with higher probability of exciting and/or propagating errors as the intermediate states for finding the transfer sequence. The experimental results on ISCAS89 and ITC99 benchmark circuits show that the proposed method can achieve a reduction rate as high as 99.94% in test length.

The rest of the paper is organized as follows. Section 2 gives a detailed statement of the problem. Sections 3 and 4 present the proposed method for generating a shorter test sequence. Section 5 shows the experimental results, and concluding remarks are given in Section 6.

2   Problem Formulation

There often exist multiple transfer sequences between any pair of states; some are significantly shorter than others. In addition, some states visited by an error sequence are not relevant for activating and for propagating errors to the outputs. Figure 1 shows the state diagram of a sequential circuit under diagnosis. Note that our method needs neither the construction nor the use of the state diagram. Our method takes the boolean netlist of the sequential circuit as the input and works primarily at the structural level. The example is mainly used for illustration of the concept. Suppose the transition function from state F to state B is erroneous and the states traversed by the error sequence is $A-C-C-D-E-F-B$. Our objective is to find a shorter sequence that can transfer the circuit from A to B and also propagate errors to the primary outputs. Because transitions $A-C$ and $C-D$ are not relevant for activating and propagating the error, states C and D need not be visited. A shorter sequence such as the one traversing states $A-E-F-B$ would be a better sequence for diagnosis.

We can classify the states traversed by the error sequence into *irrelevant* states (such as states C and D) and *relevant* states (such as states E and F). The objective is to find a new error sequence that will visit all relevant states and avoid irrelevant states. However, because the knowledge of correct values at registers in each clock cycle is not available, we cannot determine precisely whether a state visited by the error sequence is relevant or not. Therefore, we need to simulate

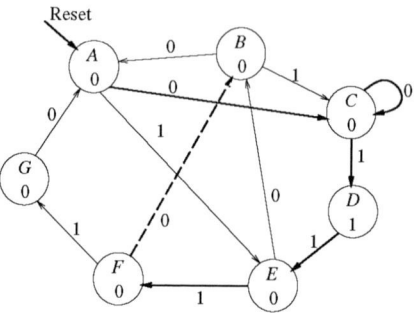

Figure 1: An example of the error sequence, $A-C-C-D-E-F-B$, where the state transition, $F-B$, is erroneous.

the generated sequence to verify whether it indeed activates and propagates the error to the primary outputs.

3   Overview of the Method

For a given error sequence, the initial state S_I and the final state S_F of a circuit C could be derived by simulation. Since our objective is to find a shorter error sequence from S_I to S_F rather than to search for all possible shorter sequences, a modern SAT solver becomes a more attractive method than the BDD-based techniques. We can apply a sequential SAT solver to find the shortest transfer sequence $T'(S_I, S_F)$ between S_I and S_F. However, a sequential SAT solver [8, 9] usually cannot find the shortest solution. In order to find the shortest sequence, one strategy is to follow the principle of bounded model checking. We attempt to find a solution within i time frames where i is initially set to 1. If no solution exists, i is then increased by 1 for another run of the SAT solving. This iterative process continues until a solution is found. Without the exact information of relevant states, the sequence derived this way may be a "false" sequence that cannot activate and propagate the error to the outputs. By selecting intermediate states derived from the simulation results of an erroneous circuit, the SAT solving will be guided to generate a sequence that will visit relevant states. Also, the task of finding a shorter sequence from S_I to S_F can be divided into two sub-tasks that are more amenable to the SAT solving.

The proposed algorithm is shown in Figure 2. In Line 4, we select intermediate states, $S_{targets}$ derived from the simulation result of an erroneous circuit C on a given long error sequence V, as candidates to divide V into two sub-sequences, $T(S_I, S_i)$ and $T(S_i, S_F)$ where $S_i \in S_{targets}$. The detail of the state selection will be explained in Section 4. Then, we use a SAT solver to find the shortest transfer sequence, $T'(S_I, S_i)$, to replace $T(S_I, S_i)$ (Line 5). Note that in finding the shortest sequence from S_I to every $S_i \in S_{targets}$, the learned conflict clauses generated in solving one target can be accumulated and reused in finding the sequences for other targets. However, $T'(S_I, S_i)$ alone may not be able to activate and propagate errors to primary outputs, since errors in the given sequence may only be activated and propagated to primary outputs by $T(S_i, S_F)$. Thus, in Line 9–10, for each $S_i \in S_{targets}$, a new test sequence V' is generated by cascading $T'(S_I, S_i)$ and $T(S_i, S_F)$ followed by re-simulation. If no error appears at the observation points, the new sequence V' is discarded and the shortest transfer sequence from S_I to the next $S_i \in S_{targets}$ is used to form the next test sequence for simulation and verification of its validity. On the other hand, if V' successfully produces errors at primary outputs, a valid new sequence has been found. We then continue to shorten the second part of the sequence, $T(S_i, S_F)$. This can be done by setting S_i as the initial state S_I (Line 13) before calling the process again (Line 3–12). This process continues (by iteratively setting new S_I's) until no reduction can be made.

```
0 # C: Circuit, V: Error Sequence
1 # S_I: Initial State, S_F: Final State
2 push S_F into S_targets.
3 while S_targets is not empty
4     Select intermediates states into S_targets.
5     Apply the sequential SAT solver [8]
6     Sort S_targets in a reverse chronological order, S'_targets
7     for each target S_i ∈ S'_targets
8         if solution of S_i exists
9             V' = Cascade_Sequences(T'(S_I, S_i), T(S_i, S_F)).
10            ErrorObserved = Simulate(V').
11            if ErrorObserved == TRUE
12                break
13    set S_I = S_i.
```

Figure 2: Algorithm for error sequence generation/minimization.

In order to achieve a higher reduction rate in test length, we process the state $S_i \in S_{targets}$ in a reverse chronological order. That is, the first S_i selected for transfer sequence generation is the farthest state S_i in $S_{targets}$ from S_I. If the resulting sequence cannot reveal the errors at observation points, then the next farthest target is selected as the target for transfer sequence generation.

4 Intermediate State Selection

Selection of intermediate states in Figure 2 affects the reduction rate of the test length. Consider the example in Figure 1. If states B and D are selected as the intermediate states to optimize the transfer sequence from state A to state B, then A–E–B and A–C–D–E–B, will be found as the shortest state sequences respectively. However, neither sequence can activate error (because the assumption was that the error only corrupted the transition from F to B). Therefore, both sequences will be rejected and no test length reduction will be achieved. On the other hand, if we select state F as the intermediate state, the valid sequence A–E–F–B will be found which could activate and propagate the error to the outputs. In the following, we propose a heuristic for identifying states with higher probabilities of successful activation and/or propagation of the errors to either primary outputs or registers in the next clock cycle. These states are preferred intermediate states for transfer sequence generation.

Starting from the erroneous outputs E, we trace backwards to find the fan-in registers of E. By analyzing the states visited by the error sequence V, we develop a metric to identify which fan-in registers, denoted as F_1, are irrelevant for activating or propagating the error to E, and which registers, denoted as F_2, are essential for error activation and propagation. For f_i to be in F_2, one of the following two cases must apply : (1) an erroneous value already exists in f_i, or (2) the value in f_i is error-free but is essential for error activation or propagation. Starting from the registers in F_2, we continue to trace backwards for one more cycle to identify their fan-in registers, which are essential for error activation or propagation. The iterative process continues until every state reached by the error sequence has been analyzed. Throughout this process, we can calculate the cycles in which a register is classified as an essential one (i.e. belonging to F_2). Suppose that, in M cycles, a register f_i is classified as F_2. Among these M cycles, assume f_i is at 0(1) for $m_1(m_2)$ cycles. If $m_1 > m_2$, we would guess that f_i being 0 is more likely to excite and/or propagate the errors than f_i being 1. Therefore, we would prefer to select an intermediate state with f_i being 0 rather than with f_i being 1. Likewise, if $m_2 > m_1$, we would prefer to select an intermediate state with f_i being 1.

As shown in Figure 3, for a given error sequence, the outputs with erroneous responses are set as the initial *error candidates*, E (Line 1). In this algorithm, E contains either outputs or registers that might have erroneous values. In Line 3, we trace backwards to find the fan-in registers F of signals in E. In order to determine whether $f_i \in F$ belongs to either category F_1 or category F_2 in each cycle, we complement the value in f_i (and keep value in all other inputs and registers intact - Line 5) and check whether the change can be propagated to E (Line 6). If any value of $e_i \in E$ is changed, we classify f_i as a category F_2 register. Then, we push f_i into E_{next} as a source for back-tracing in the next iteration (Line 12). In Line 8–11, we increment α_i^1 by 1 if the original value of f_i is 1; otherwise, α_i^0 is incremented by 1. α_i^k is the number of cycles that $f_i \in F$ being at value k has been classified as a category F_2 register. For a register whose α_i^0 is larger than α_i^1, heuristically, f_i being 0 has a higher probability to excite and/or propagate the error than that being 1. The whole process (Line 3–11) repeats until all states visited by the error sequence have been examined. After the calculation of α_i^k for each register is completed, the *relevance* β_j of a state S_j for error activation/propagation can be approximated by summing up the α_i^k of register $f_i \in S_j$ based on the value of f_i in S_j (Lines 15 – 20). Then, we sort the states based on their β_j from the highest to the lowest (Line 21). In Line 22, the states S' are selected from the sorted states in a chronological order. The selected states are returned for finding shorter transfer sequences (Line 23).

5 Experimental Results

We implemented our algorithm in the C++ language and tested it on some public benchmark circuits including IS-

```
# every $\alpha_i^k$ and $\beta_i$ is initialized to 0.
# $S$: the simulation result of a given error sequence
1 Identify erroneous outputs as error candidates $E$.
2 repeat
3    Trace back from $E$ to get their fan-in registers $F$.
4    for each register $f_i \in F$.
5        Invert the value of $f_i$ and keep the others.
6        Simulate the circuit.
7        if any value of $e_i \in E$ is changed.
8            if value($f_i$) == 1
9                Increase $\alpha_i^1$ by 1.
10           else if value($f_i$) == 0
11               Increase $\alpha_i^0$ by 1.
12           Push $f_i$ into $E_{next}$.
13   set $E = E_{next}$.
14 until Empty($E$) == TRUE.
15 for each state $S_j \in S$ .
16   for each register $f_i \in S_j$
17       if $value(f_i)$ == 1
18           $\beta_j$ += $\alpha_i^1$.
19       else
20           $\beta_j$ += $\alpha_i^0$.
21 Sort $S$ based on $\beta_i$ (from the highest to the lowest).
22 Select states $S'$ from the sorted states in chronological order.
23 return $S'$.
```

Figure 3: Selection of intermediate states with higher probabilities of activation and/or propagation of the errors.

Ckt. Name	Num. of Registers
s9234	211
s38417	1636
b14	245
b20	490
b21	490
b22	703
or1k	1860

Table 1: Number of registers for each circuit

CAS89, ITC99, and the OpenRisc 1000 (a microprocessor) from the OpenCores organization [10]. The number of registers of each tested circuit is shown in Table 1 where or1k denotes the OpenRisc 1000. In this paper, we do not speculate about what kind of errors will be easier to detect by simulation-based or formal verification methods. Our objective is to show the effectiveness of our method to reduce the error sequence found by simulation. We generated several erroneous circuits from each benchmark circuit listed in Table 1 by randomly replacing several gates with different types of gates. Then, we used the Candence TestBuilder [13] as our random pattern generator to generate input sequences and used Mentor Graphics ModelSim to simulate the erroneous circuits. After a long error sequence had been identified, we applied the proposed method to generate a new shorter error sequence.

As discussed in Section 4, selection of intermediate states is critical to the test generation quality. As shown in Table 2,

we compare our method with the method $Select(k)$ for which the state of every k cycles is selected as an intermediate state. For example, method $Select(20)$ means the reached states in every 20 cycles, starting from the initial state, will be included as the intermediate states. Len denotes the length of the error sequence found by simulation. The numbers before and after "/" in Columns 3, 4, and 5 denote the length of derived test sequence and the reduction rate in percentages with respect to the original test length. For the method $Select(k)$, if each time we fail to generate a transfer sequence for a target intermediate state S_i, the next target intermediate state would be the state reached k cycle earlier, S_{i-k}, under the original error sequence. Even if the transfer sequence for the new target is successfully found, the final error sequence will include $T(S_{i-k}, S_F)$ which is k patterns longer than $T(S_i, S_F)$. Therefore, the length of the final sequence would be increased by k. As shown in Column 4, since the method could not find a short error sequence from the initial state to the final state at which the erroneous responses are present at primary outputs for every single circuit in our experiment, the lengths of sequences for the method $Select(100)$ are always at least 100. In general, the reduction rate of $Select(k)$ can be improved by reducing k. Experimentally, changing k from 100 to 20 increases the reduction rate from 71.32% to 76.68%. In comparison to $Select(k)$, our method achieves average higher average reduction rate (78.49%) than both $Select(100)$ (71.32%) and $Select(20)$ (76.68%). For most cases, the improvement rate is much better than the average. Note that, for some circuits, such as or1k_[1,2], the results of $Select(20)$ are worse than those of $Select(100)$. The reason is that there are more targets for SAT solving for Method $Selct(20)$ than Method $Select(100)$. The sequential SAT solver simply cannot find enough solutions for selected targets to reduce test length within the time limit when time limit for each iteration is set to 1800 seconds.

The reduction rates for all tested circuits are shown in Table 3, where Len, Len', Rate, Ite., and Time denote the length of the original error sequence, the length of generated shorter sequence, the reduction rate in test length, the number of iterations for transfer sequence generation, and the average runtime for SAT solving in each iteration, respectively. As shown in Column 4 of Table 3, the proposed method can achieve more than 80% reduction rate in 20 out of the 36 cases, and the highest reduction rate is 99.94% on b14_9. The average reduction rate is 59.4%. There are 6 out of the 36 cases whose reduction rates are below 10%. One explanation for the low reduction rate is that there may be counters in the circuits, and specific values in the counters are required for activating and propagating the errors to the outputs. For counters, it is unlikely that any shorter sequences exist to reach a specific state.

6   Concluding Remarks

Although long test sequences may help bring the design to those hard-to-reach states for activating the errors and for propagating them to observation points, these sequences com-

Ckt.	Len	Ours	Select(100)	Select(20)
b14_1	369	21/94.31%	115/68.83%	13/96.48%
b14_2	693	77/88.89%	115/83.41%	75/89.18%
b14_3	874	24/97.25%	114/86.96%	32/96.34%
b14_4	2347	29/98.76%	111/95.27%	23/99.02%
b14_5	22796	110/99.52%	112/99.51%	46/99.80%
b14_6	30960	18/99.94%	114/99.63%	46/99.85%
b14_7	4634	24/99.48%	114/97.54%	32/99.31%
b20_1	645	505/21.71%	609/5.58%	127/80.31%
b20_2	1013	163/83.91%	215/78.78%	105/89.63%
s38417_1	1041	103/90.11%	193/81.46%	109/89.53%
s38417_2	2759	457/83.44%	244/91.56%	188/93.19%
s38417_3	1629	53/96.75%	132/91.90%	55/96.62%
s38417_4	1541	83/94.61%	167/89.16%	99/93.58%
s38417_5	2223	99/95.55%	206/90.73%	132/94.06%
s38417_6	1082	86/92.05%	165/84.75%	89/91.77%
or1k_1	2982	246/91.75%	252/91.55%	2865/3.92%
or1k_2	1230	21/98.29%	125/89.84%	360/70.73%
or1k_3	1183	1130/4.48%	1183/0.00%	1183/0.00%
or1k_4	4262	3070/27.97%	4262/0.00%	2661/37.56%
or1k_5	4931	4385/11.07%	4931/0.00%	4400/10.77%
Average	-	-/78.49%	-/71.32%	-/76.68%

Table 2: Comparison of three intermediate state selection methods.

plicate the subsequent diagnosis process. To reduce the complexity of diagnosis, we have proposed a method to generate a shorter error sequence, based on a known error sequence identified by simulation. We have further proposed to simplify the task of generating such a sequence, which brings the circuit from the initial state to the state that reveals the error at the primary outputs. This is accomplished by identifying an intermediate state S_i so that the final sequence is formed by two sub-sequences - one from the initial state to S_i and the other from S_i to the final state. The task of generating each sub-sequence is substantially simpler than the longer original task. We have developed a heuristic to select intermediate states which are more likely to lead to successful generation of error sequences. A sequential SAT solver was used as the underlying engine for finding these sub-sequences. The experimental results on ISCAS89 and ITC99 benchmark circuits show that the proposed method can achieve a very significant reduction rate in test length.

References

[1] Magdy S. Abadir, Jack Ferguson, and Thomas E. Kirkland, "Logic Design Verification via Test Generation," in *IEEE Trans. on Computer-Aided Deisgn*, Jan. 1988, Vol. 7, No. 1, pp. 138–148.

[2] Moayad Fahim Ali, Andreas Veneris, Sean Safarpour, Magdy Abadir, Rolf Drechsler, and Alexander Smith, "Debugging Sequential Circuits Using Boolean Satisfiability," in *Proc. International Conference on Computer Aided Design*, 2004, pp. 204–209.

[3] Vamsi Boppana, Rajarshi Mukherjee, Jawahar Jain, Masahiro Fujita, and Pradeep Bollineni, "Multiple Error Diagnosis Based On Xlists," in *Proc. Design Automation Conference*, 1999, pp. 660–665.

[4] K. Chang, V. Bertacco, and I. Markov, "Simulation based Bug Trace Minimization with BMC-based Refinement," in *Proc. International Conference on Computer Aided Design*, 2005.

[5] Yirng-An Chen and Fang-Sung Chen, "Algorithms for Compacting Error Traces," in *Proc. Asia and South Pacific Design Automation Conference*, 2003, pp. 21–24.

Ckt.	Len	Len	Rate(%)	Ite.	Time
s9234_1	3548	24	99.32	2	287
s9234_2	7114	44	99.38	2	108
s9234_3	3955	18	99.54	2	687
s9234_4	9386	15	99.84	2	5
s38417_1	1041	103	90.10	4	943
s38417_2	2759	457	83.44	3	2530
s38417_3	1629	53	96.75	3	302
s38417_4	1541	83	94.61	3	712
s38417_5	2223	99	95.55	4	942
s38417_6	1082	86	92.05	3	1433
b14_1	369	21	94.31	2	28
b14_2	693	77	88.89	2	60
b14_3	874	24	97.25	3	59
b14_4	2347	29	98.76	3	77
b14_5	22796	110	99.52	2	63
b14_6	30960	18	99.94	4	56
b14_7	4634	24	99.48	4	80
b20_2	1013	163	83.91	4	7520
or1k_1	2982	246	91.75	7	5095
or1k_2	1230	21	98.29	5	1045
b20_1	645	505	22.71	5	2808
b20_3	2130	1902	10.70	5	3518
b20_4	2429	2273	6.42	3	2710
b20_5	939	719	23.43	6	3601
b21_1	433	267	38.34	3	2664
b21_2	2160	2096	3.96	2	3173
b21_3	870	704	19.08	3	2663
b21_4	1231	941	23.56	3	2893
b21_5	955	767	19.69	2	2929
b21_6	4252	3998	5.97	2	2936
b22_1	2138	2080	2.71	3	2792
b22_2	6552	6488	0.98	4	3079
b22_3	1527	1295	15.19	4	3243
or1k_3	1183	1130	4.48	2	3171
or1k_4	4262	3070	27.97	4	5263
or1k_5	4931	4385	11.08	2	4058
Average	-	-	59.4	-	-

Table 3: Experimental results on ISCAS89, ITC99, and or1k benchmark circuits.

[6] Anand L. D'Souza and Michael S. Hsiao, "Error Diagnosis of Sequential Circuits Using Region-Based Model," in *Proc. Intenal Conference on VLSI Design*, 2001, pp. 103–108.

[7] E. Goldberg and Y. Novikov, "BerkMin: A fast and robust SAT-solver," in *Proc. Design, Automation and Test Conference in Europe*, 2002, pp. 142–149.

[8] Feng Lu, M. K. Iyer, G. Parthasarathy, Li.-C. Wang, K.-T. Cheng, and K.C. Chen, "An Efficient Sequential SAT Solver With Improved Search Strategies," in *Proc. Design, Automation, and Test Conference in Europe*, 2005, pp. 1102–1107.

[9] M.W. Moskewicz, C.F. Madigan, Y. Zhao, L. Zhang, and S. Malik, "Chaff: engineering an efficient SAT solver," in *Proc. Design Automation Conference*, 2001, pp. 530–535.

[10] OpenRisc 1000 series, *http://www.opencores.org*

[11] K. Ravi and F. Somenzi, "Minimal assignments for bounded model check," in *Tools and Algorithms for the Construction of Analysis of Systems*, 2004, pp. 31–45.

[12] Alexander Smith, Andreas Veneris, and Anastasios Viglas, "Design Diagnosis Using Boolean Satisfiability," in *Asia and South Pacific Design Automation Conference*, 2004, pp. 218–223.

[13] Cadence TestBuilder, *http://www.testbuilder.net*

1B-1

Constraint-Driven Bus Matrix Synthesis for MPSoC

Sudeep Pasricha[†], Nikil Dutt[†], Mohamed Ben-Romdhane[‡]

†Center for Embedded Computer Systems
University of California, Irvine, CA
{sudeep, dutt}@cecs.uci.edu

‡Conexant Systems Inc.
Newport Beach, CA
m.benromdhane@conexant.com

Abstract – Modern multi-processor system-on-chip (MPSoC) designs have high bandwidth constraints which must be satisfied by the underlying communication architecture. Bus matrix based communication architectures consist of several parallel busses which provide a suitable backbone to support high bandwidth systems, but suffer from high cost overhead due to extensive bus wiring inside the matrix. Manual traversal of the vast exploration space to synthesize a minimal cost bus matrix that also satisfies performance constraints is practically infeasible. In this paper, we address this problem by proposing an automated approach for synthesizing a bus matrix communication architecture which satisfies all performance constraints in the design and minimizes wire congestion in the matrix. To validate our approach, we consider several industrial strength applications from the networking domain and show that our approach results in up to 9× component savings when compared to a full bus matrix and up to 3.2× savings when compared to a maximally connected reduced bus matrix.

I. Introduction

Multi-processor system-on-chip (MPSoC) designs are increasingly being used in today's high performance embedded systems. These systems are characterized by a high level of parallelism, due to the presence of multiple processors, and large bandwidth requirements, due to the massive scale of component integration. The choice of communication architecture in such systems is of vital importance because it supports the entire inter-component data traffic and has a significant impact on the overall system performance.

Traditionally used hierarchical shared bus communication architectures such as those proposed by AMBA [1], CoreConnect [2] and STbus [3] can cost effectively connect few tens of cores but are not scalable to cope with the demands of very high performance systems. Point-to-point communication connection between cores is practical for even fewer components. Network-on-Chip (NoC) based communication architectures [5] have recently emerged as a promising alternative to handle communication needs for the next generation of high performance designs. However, although basic concepts have been proposed, research on NoCs is still in its infancy, and few concrete implementations of complex NoCs exist to date [6].

In this paper we look at *bus matrix* (sometimes also called *crossbar switch*) based communication architectures [7] which are currently being considered by designers to meet the high bandwidth requirements of modern MPSoC systems. Fig. 1 shows an example of a three-master seven-slave AMBA bus matrix architecture for a dual ARM processor based networking subsystem application. A bus matrix consists of several busses in parallel which can support concurrent high bandwidth data streams. The *Input stage* is used to handle

interrupted data bursts, and to register and hold incoming transfers if receiving slaves cannot accept them immediately. The *Decode* stage generates select signal for appropriate slaves. Unlike in traditional shared bus architectures, *arbitration* in a bus matrix is not centralized, but rather distributed so that every slave has its own arbitration.

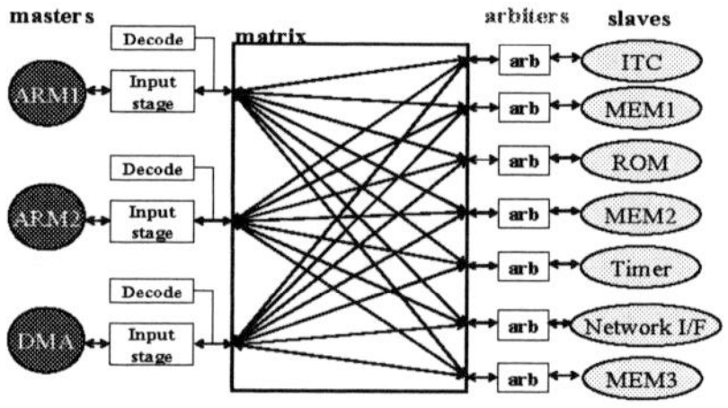

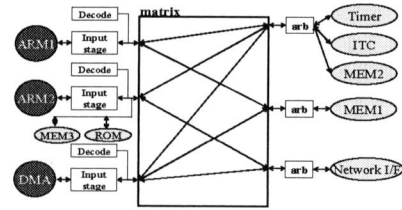

Fig. 1. Full bus matrix architecture

One drawback of the *full bus matrix* structure shown in Fig. 1 is that it connects every master to every slave in the system, resulting in a prohibitively large number of busses. The excessive wire congestion can make it practically impossible to route and achieve timing closure for the design [14]. To overcome this shortcoming, designers tailor a full matrix structure to the particular application at hand, creating a *partial bus matrix*, as shown in Fig. 2. This structure has fewer busses and consequently uses fewer components (arbiters, decoders, buffers), has a smaller area and also utilizes less power.

Fig. 2. Partial bus matrix architecture

The problem of synthesizing a minimal cost (i.e. having the least number of busses) bus matrix for a particular application is complicated by the large number of combinations of possible matrix topologies and bus architecture parameters such as bus widths, clock speeds, out-of-order (OO) buffer sizes and shared slave arbitration schemes. Previous research in the area of bus matrix/crossbar synthesis (discussed in the next section) has been inadequate in addressing the entire problem, and instead has been limited to exploring a small subset of the synthesis problem (such as topology synthesis [8]). Very often, designers end up evaluating the bus matrix design space by creating simulation models annotated with detail based on experience, and manually iterating through

0-7803-9451-8/06/$20.00 ©2006 IEEE.

30

different combinations of topology and communication architecture parameters. Such an effort remains time consuming and produces bus matrix architectures which are generally overdesigned for the application at hand.

Our goal in this paper is to address this problem by presenting an automated approach for synthesizing a bus matrix communication architecture, which generates not only the matrix topology, but also communication parameter values for bus clock speeds, OO buffer sizes and arbitration strategies. Most importantly, our synthesis effort minimizes the number of busses in the matrix and satisfies all performance constraints in the design. To demonstrate the effectiveness of our approach we synthesize a bus matrix architecture for four industrial strength MPSoC case studies from the networking domain and show that our approach significantly reduces wire congestion in a matrix, resulting in up to 9× component savings when compared to a full bus matrix and up to 3.2× savings when compared to a maximally connected reduced bus matrix.

II. Related Work

The need for bus matrix (or crossbar switch) architectures has been emphasized in previous work in the area of communication architecture design. Lahtinen et al. [9] compared the shared bus and crossbar topologies to conclude that the crossbar is superior to a bus for high throughput systems. Ryu et al. [10] compared a full crossbar switch with other bus based topologies and found that the crossbar switch outperformed the other choices due to its superior parallel response. Loghi et al. [11] presented exploration studies with the AMBA and STBus shared bus, full crossbar and partial crossbar topologies, concluding that crossbar topologies are much better suited for high throughput systems requiring frequent parallel accesses. An interesting conclusion from their work is that partial crossbar schemes can perform just as well as the full crossbar scheme, if designed carefully. However, the emphasis of their work was not on the generation of such partial crossbar topologies.

Although a lot of work has been done in the area of hierarchical shared bus architecture synthesis [12-14] and NoC architecture synthesis [15-16], few efforts have focused on bus matrix synthesis. Ogawa et al. [17] proposed a transaction based simulation environment which allows designers to explore and design a bus matrix. But the designer needs to manually specify the communication topology, arbitration scheme and memory mapping, which is too time consuming for the complex systems of today. The automated synthesis approach for STBus crossbars proposed by Murali et al. in [8] is the only work that comes closest to our goal of automated bus matrix synthesis. However, their work primarily deals with automated crossbar topology synthesis – the communication parameters (arbitration schemes, OO buffer sizes, bus widths and speeds) which have considerable influence on system performance [22-23] are not explored or synthesized. Our synthesis effort overcomes this shortcoming and synthesizes both the topology and communication architecture parameters for the bus matrix. Additionally, [8] assumes that critical data streams cannot overlap on the same bus, places a static limit on the maximum number of components that can be attached to a bus and also requires the designer to specify hard-to-determine

threshold values of traffic overlap as an input, based on which components are allocated to separate busses. These are conservative approaches which lead to an overdesigned, sub-optimal system. Our approach carefully selects appropriate arbitration schemes (e.g. TDMA based) that can allow multiple constraint streams to exist on the same bus, and also does not require the designer to specify data traffic threshold values or statically limit the number of components on a bus. Experimental comparison studies (described in Section IV) show that our scheme is more aggressive and obtains greater reduction in bus matrix connections, when compared to [8].

III. Bus Matrix Synthesis

This section describes our approach for automated bus matrix synthesis. First we formulate the problem and present our assumptions. Next, we describe our simulation engine and elaborate on communication parameter constraints, which guide the matrix synthesis process. Finally, we present our automated bus matrix synthesis approach in detail.

A. Problem Formulation

We are given an MPSoC design having several components (IPs) that need to communicate with each other. We assume that hardware/software partitioning has taken place and that the appropriate functionality has been mapped onto hardware and software IPs. These IPs are standard "black box" library components which cannot be modified during the synthesis process, except for the memory components. The target standard bus matrix communication architecture (e.g. AMBA bus matrix [1]) that determines the pins at the IP interface and for which the matrix must be synthesized, is also specified. Typically, all busses within a bus matrix have the same data bus width, which usually depends on the number of data interface pins of the IPs in the design. We assume that this matrix data bus width is specified by the designer, based on the knowledge of the IPs selected for the design.

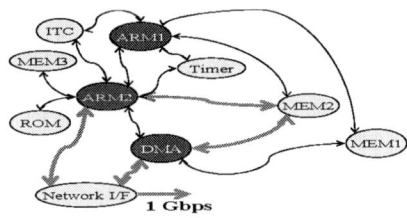

Fig. 3. Communication Throughput Graph (CTG)

Generally, MPSoC designs have performance constraints which are dependent on the nature of the application. The *throughput* of communication between components is a good measure of the performance of a system [12]. To represent performance constraints in our approach, we define a **Communication Throughput Graph** $CTG = G(V,A)$ which is a directed graph, where each vertex v represents a component in the system, and an edge a connects components that need to communicate with each other. A **Throughput Constraint Path** (TCP) is a sub-graph of a CTG, consisting of a single master for which data throughput must be maintained and other masters, slaves and memories which are in the critical path that impacts the maintenance of the throughput. Fig. 3 shows a *CTG* for a network subsystem, with a *TCP* involving the ARM2, MEM2, DMA and 'Network I/F' components, where the rate of data packets streaming out of the 'Network I/F' component must not fall below 1 Gbps.

Problem Definition *A bus B can be considered to be a partition of the set of components V in a CTG, where $B \subset V$. Then the problem is to determine an optimal component to bus assignment for a bus matrix architecture, such that the V is partitioned onto a minimal number of busses N, and satisfies all performance constraints in the design, represented by the TCPs in a CTG.*

B. Simulation Environment

Since communication behavior in a system is characterized by unpredictability due to dynamic bus requests from cores, contention for shared resources, buffer overflows etc., a simulation based approach is necessary for accurate performance estimation. For the simulation part of our flow, we capture behavioral models of components and bus architectures in SystemC [18][24], and keep them in an IP library database. Since we were concerned about the speed of simulation, we chose a fast transaction-based, bus cycle accurate modeling abstraction, which averaged simulation speeds of 150–200 Kcycles/sec [19], while running embedded software applications on processor ISS models. The communication model in this abstraction is extremely detailed, capturing delays arising due to frequency and data width adapters, bridge overheads, interface buffering and all the static and dynamic delays associated with the standard bus architecture protocol being used.

C. Communication Parameter Constraint Set

In the interest of generating a practically realizable system, we allow a designer to specify a discrete set of valid values (constraint set) for communication parameters such as bus clock speeds, OO buffer sizes and arbitration schemes. We allow the specification of two types of constraint sets for components – a global constraint set (Ψ_G) and a local constraint set (Ψ_L). For instance, a designer might set the allowable bus clock speeds for a set of busses locally in a subsystem to multiples of 33 MHz, with a maximum speed of 166 MHz, based on the operation frequency of the cores in the subsystem, while globally, the allowed bus clock speeds are multiples of 50 MHz, up to maximum of 400 MHz. The presence of a local constraint overrides the global constraint, while the absence of it results in the resource inheriting global constraints. This provides a convenient mechanism for the designer to bias the synthesis process based on knowledge of the design and the technology being targeted. Such knowledge about the design is not a prerequisite for using our synthesis framework, but informed decisions can help avoid the synthesis of unrealistic system configurations.

D. Synthesis Approach

We now describe our automated bus matrix synthesis approach. Fig. 4 gives a high level overview of the flow. The

inputs to the flow include a Communication Throughput Graph (CTG), a library of behavioral IP models, a target bus matrix template (e.g. AMBA bus matrix [1]) and a communication parameter constraint set (Ψ) – which includes Ψ_G and Ψ_L. The general idea is to first perform a fast TLM level simulation of the system to get application-specific data traffic statistics. This information is used in a global optimization phase to reduce the full bus matrix architecture, by removing unused busses and local slave components from the matrix. We call the resulting matrix a **maximally connected reduced matrix**. The next step is to perform a static *branch and bound based hierarchical clustering* of slave components in the matrix which further reduces the number of busses in the matrix. We rank the results of the static clustering analysis, from the best case solution (least number of busses) to the worst (most number of busses) and save them in the database. We then use a fast bus cycle accurate simulation engine [19] to validate and select the best solution which meets all the performance constraints, determine slave arbitration schemes, optimize the design to minimize bus speeds and OO buffer sizes and then finally output the optimal synthesized bus matrix architecture.

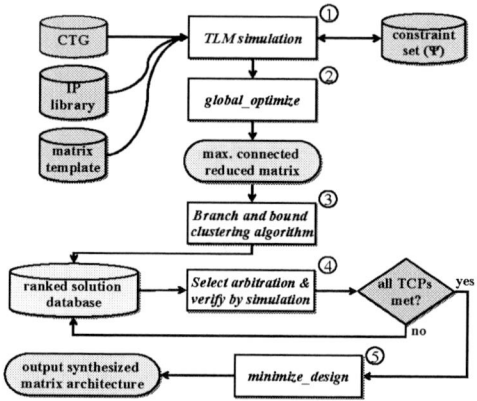

Fig. 4. Automated bus matrix synthesis flow

We now describe the synthesis flow in detail. In the first phase, the IP library is mapped onto a full bus matrix and simulated at the TLM level, with no arbitration contention overhead since there are no shared channels and also because we assume infinite ports at IP interfaces. We also set the OO buffer sizes to the maximum allowed in Ψ. The TLM simulation allows us to obtain application-specific data traffic statistics such as number of transactions on a bus, average transaction burst size on a bus and memory usage profiles. Knowing the bandwidth to be maintained on a channel from the Throughput Constraint Paths (TCPs) in the CTG, we can also estimate the minimum clock speed at which any bus in the matrix must operate, in order to meet its throughput constraint, as follows. The data throughput ($\Gamma_{TLM/B}$) from the TLM simulation, for any bus B in the matrix is given by

$$\Gamma_{TLM/B} = (numT_B \times sizeT_B \times width_B \times \Omega_B) / \sigma$$

where *numT* is the number of data transactions on the bus, *sizeT* is the average size of these data transactions, *width* is the bus width, Ω is the clock speed, and σ is the total number of cycles of TLM simulation for the application. The values of

$numT$, $sizeT$ and σ are obtained from the TLM simulation in phase 1. To meet the throughput constraint $\Gamma_{TCP/B}$ for bus B,

$$\Gamma_{TLM/B} \geq \Gamma_{TCP/B}$$

$$\therefore \quad \Omega_B \geq (\sigma \times \Gamma_{TCP/B}) / (numT_B \times sizeT_B \times width_B)$$

The minimum bus speed thus found is used to create (or update) the local bus speed constraint set $\Psi_{L(speed)}$ for the bus B.

In the next phase (phase 2 in Fig. 4), we perform global optimization (*global_optimize*) on the matrix by using information gathered from the TLM simulation in phase 1. In this phase we first remove all the busses that have no data traffic on them, from the full bus matrix. Next, we analyze the memory usage profile from the simulation run and attempt to split those memory nodes for which different masters access non-overlapping regions. Finally we cluster dedicated slave and memory components with their corresponding masters by migrating them from the matrix to the local busses of the masters, to reduce congestion in the bus matrix. Note that we perform memory splitting before local node clustering because it allows us to generate local memories which can then be clustered with their corresponding masters. After the *global_optimize* phase, the matrix structure obtained is termed as a **maximally connected reduced bus matrix**.

The next phase (phase 3 in Fig. 4) involves static analysis to determine the optimal reduced bus matrix for the given application. We make use of a *branch and bound based hierarchical clustering algorithm* to cluster slave components to reduce the number of busses in the matrix even further. Note that we do not consider merging masters because it adds two levels of contention (one at the master end and another at the slave end) in a data path, which can drastically degrade system performance. Before describing the algorithm, we present a few definitions. A slave cluster $SC = \{s_1...s_n\}$ refers to an aggregation of slaves that share a common arbiter. Let M_{SC} refer to the set of masters connected to a slave cluster SC. Next, let $\Pi_{SC1/SC2}$ be a superset of sets of busses which are merged when slave clusters $SC1$ and $SC2$ are merged. Finally, for a *merged bus set* $\beta = \{b_1...b_n\}$, where $\beta \subset \Pi_{SC1/SC2}$, let K_β refer to the set of allowed bus speeds for the newly created bus when the busses in set β are merged, and is given by

$$K_\beta = \Psi_{L(speed)}(b_1) \cap \Psi_{L(speed)}(b_2) ... \cap \Psi_{L(speed)}(b_n)$$

The branching algorithm starts out by clustering two slave clusters at a time, and evaluating the gain from this operation. Initially, each slave cluster has just one slave. The total number of clustering configurations possible for a bus matrix with n slaves is given by $(n! \times (n-1)!)/2^{(n-1)}$. This creates an extremely large exploration space, which cannot be traversed in a reasonable amount of time. In order to consider only valid clustering configurations and arrive at an optimal solution quickly, we make us of a bounding function. Fig. 5 shows the pseudo code for our bounding function which is called after every clustering operation of any two slave clusters $SC1$ and $SC2$. In Step 1, we use a look up table to see if the clustering operation has already been considered previously, and if so, we discard the duplicate clustering. Otherwise we update the lookup table with the entry for the new clustering. In Step 2, we check to see if the clustering of SC1 and SC2 results in the merging of busses in the matrix, otherwise the clustering is not

beneficial and the solution can be bounded. If the clustering results in bus mergers, we calculate the number of merged busses for the clustering and store the cumulative weight of the clustering operation in the branch solution node. In Step 3, we check to see if the allowed set of bus speeds for every merged bus is compatible or not. If the allowed speeds for any of the busses being merged are incompatible (i.e $K_\beta == \phi$ for any β), the clustering is not possible and we bound the solution. Additionally, we also calculate if the throughput requirement of each of the merged busses can be theoretically supported by the new merged channel. If this is not the case, we bound the solution. The bounding function thus enables a conservative pruning process which quickly eliminates invalid solutions and allows us to rapidly converge on the optimal solution.

> **Step 1:** *if (exists lookupTable(SC1,SC2))* **then** *discard duplicate clustering*
>   *else updatelookupTable(SC1, SC2)*
>
> **Step 2:** *if ($M_{SC1} \cap M_{SC2} == \phi$) then bound clustering*
>   *else cum_weight = cum_weight + | $M_{SC1} \cap M_{SC2}$ |*
>
> **Step 3:** *for each set $\beta \in \Pi_{SC1/SC2}$ do*
>   *if (($K_\beta == \phi$) || ($\sum_{i=1}^{|\beta|} \Gamma_{TCP/i} > (width_B \times max\_speed_B)$)) then*
>   *bound clustering*

Fig. 5. *bound* function

The solutions obtained from the static branch and bound clustering algorithm are ranked from best to worst and stored in a solution database. The next phase (phase 4 in Fig. 4) validates the solutions by simulation. We use a fast transaction-based bus cycle accurate simulation engine [19] to verify that the reduced matrix still satisfies all the constraints in the design. We perform arbitration strategy selection at this stage (from the allowed schemes in the constraint set Ψ). If a static priority based scheme for a shared slave (with priorities distributed among slave ports according to throughput requirements) results in TCP constraint violations, we make use of other arbitration schemes, in increasing order of implementation costs. So we would use a simpler arbitration scheme like round robin (RR) first, before resorting to the more elaborate TDMA/RR scheme like that used in [4]. It is possible that even after using these different arbitration conflict schemes, there are TCP constraint violations. In such a case we remove the solution from the solution database, and proceed to select the next best solution, continuing in this manner till we reach a solution which successfully passes the simulation based verification. This is the minimal cost solution, having the least number of busses in the matrix, while still satisfying all TCP constraints in the design. Once we arrive at such a solution, we call the *minimize_design* procedure (phase 5 in Fig. 4) where we attempt to minimize the bus clock speeds and prune OO buffer sizes. In this procedure, we iteratively select busses in the matrix and attempt to arrive at the lowest value of bus clock speeds (as allowed by Ψ) which does not violate any TCP constraint. We verify any changes made in bus speeds via simulation. After minimizing bus speeds, we prune the OO buffer sizes from the maximum values allowed to their peak traffic buffer count utilization values, obtained from simulation. Finally, we output the synthesized minimal cost bus matrix, with a well defined topology and parameter values.

IV. Case Studies

We applied our automated bus matrix synthesis approach on four MPSoC applications – VIPER, SIRIUS, ORION4 and HNET8 – from the networking domain. While VIPER and SIRIUS are variants of existing industrial strength applications, ORION4 and HNET8 are larger systems which have been derived from the next generation of MPSoC applications currently in development. Table 1 shows the number of components in each of these applications. Note that the *Masters* column includes the processors in the design.

Table 1. Number of cores in MPSoC applications

Applications	Processors	Masters	Slaves
VIPER	2	4	15
SIRIUS	3	5	19
ORION4	4	8	24
HNET8	8	13	29

Table 2. Throughput Constraint Paths (TCPs)

IP cores in Throughput Constraint Path (TCP)	TCP constraint
ARM1, MEM1, DMA, SDRAM1	640 Mbps
ARM1, MEM2, MEM6, DMA, Network I/F2	480 Mbps
ARM2, Network I/F1, MEM3	5.2 Gbps
ARM2, MEM4, DMA, Network I/F3	1.4 Gbps
ASIC1, ARM3, SDRAM1, Acc1, MEM5, Network I/F2	240 Mbps
ARM3, DMA , Network I/F3, MEM5	2.8 Gbps

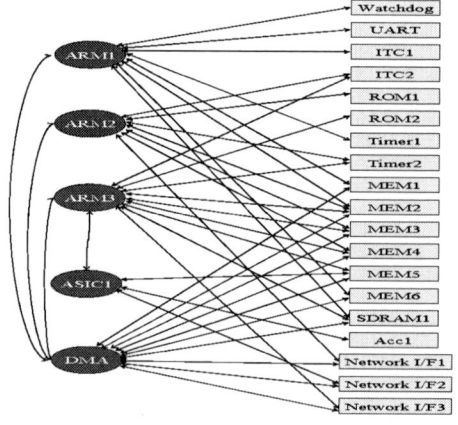

Fig. 6. CTG for SIRIUS application

Fig. 6 shows the CTG for the SIRIUS application. For clarity, the TCPs are presented separately in Table 2. ARM1 is a protocol processor (PP) while ARM2 and ARM3 are network processors (NP). The ARM1 PP is responsible for setting up and closing network connections, converting data from one protocol type to another, generating data frames for signaling, operating and maintenance and exchanging data with NP using shared memory. The ARM2 and ARM3 NPs directly interact with the network ports and are used for assembling incoming packets into frames for the network connections, network port packet/cell flow control, assembling incoming packets/cells into frames, segmenting outgoing frames into packets/cells, keeping track of errors and gathering statistics. The ASIC1 block performs hardware cryptography acceleration for DES, 3DES and AES. The DMA is used to handle fast memory to memory and network interface data transfers, freeing up the processors for more useful work. SIRIUS also has a number of memory blocks, network interfaces and peripherals such as interrupt controllers (ITC1, ITC2), timers (Watchdog, Timer1,

Timer2), UART and a packet accelerator (Acc1).

Table 3. Customizable Parameter Constraint Set

Set	Values
bus speed	25, 50, 100, 200, 300, 400
arbitration strategy	static, RR, TDMA/RR
OO buffer size	1 – 8

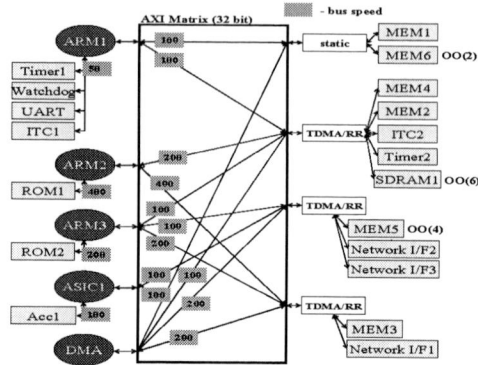

Fig. 7. Synthesized bus matrix for SIRIUS

Table 3 shows the global customizable parameter set Ψ_G. For the synthesis we target an AMBA3 AXI [21] based bus matrix structure. Fig. 7 shows the matrix structure output by our synthesis flow, which satisfies all six throughput constraints in the design (Table 2). The data bus width used in the matrix is 32 bits, and the slave-side arbitration strategies, operating speeds for the busses and OO buffer sizes (for components supporting OO transaction completion) are shown in the figure. While the full bus matrix used 95 busses, after the global optimization phase (Fig. 4) we were able to reduce this number to 34 for the maximally connected reduced matrix. The final synthesized matrix further reduces the number of busses to as few as 16 (this includes the local busses for the masters) which is almost a $6\times$ saving in the number of busses used when compared to the original full matrix. The entire synthesis process took just a few hours to complete instead of the several days or even weeks it would have taken for a manual effort.

We now present two sets of experiments to prove the effectiveness of our approach - the first compares our synthesis results with previous work in the area of bus matrix synthesis, while the second compares the results of our synthesis approach for four MPSoC applications of varying complexity.

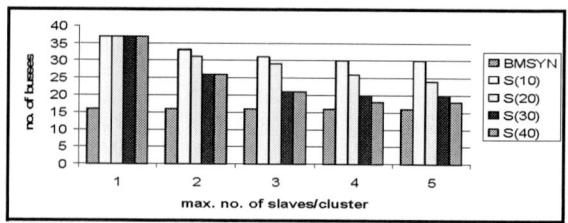

Fig. 8. Comparison with threshold based approach for SIRIUS

To compare the quality of our synthesis results, we chose the closest existing piece of work that deals with automated matrix synthesis with the aim of minimizing number of busses [8]. Since their approach only generates matrix topology (while we generate both topology and parameter values), we restricted our comparison to the number of busses in the final synthesized design. The threshold based approach proposed in [8] requires the designer to statically specify (i) the maximum number of slaves per cluster and (ii) the traffic overlap threshold, which if exceeded prevents two slaves from being assigned to the same bus cluster. The results of our comparison study, performed on the SIRIUS application, are shown in Fig. 8. BMSYN is our bus matrix synthesis approach while the other comparison points are obtained from [8]. $S(x)$, for $x = 10$, 20, 30, 40, represents the threshold based approach where no two slaves having a traffic overlap of greater than x% can be assigned to the same bus, and the X-axis in Fig. 8 varies the maximum number of slaves allowed in a bus cluster for these comparison points. The values of 10 – 40% for traffic overlap are chosen as per recommendations from [8]. It is clear from Fig. 8 that our synthesis approach produces a lower cost system (having lesser number of busses) than approaches which force the designer to statically approximate application characteristics.

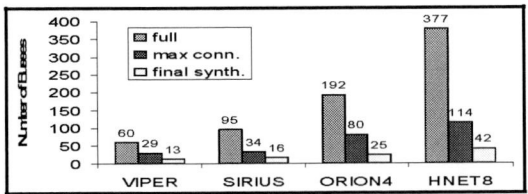

Fig. 9. Comparison of number of busses for MPSoC applications

The number of busses in a full bus matrix, a maximally connected reduced matrix and the final synthesized bus matrix using our approach, for the four applications we considered, are compared in Fig. 9. More detailed experimental results can be found in our technical report [20]. It can be seen that our bus matrix synthesis approach results in significant matrix component savings, ranging from $2.1\times$ to $3.2\times$ when compared to a maximally connected bus matrix, and from $4.6\times$ to $9\times$ when compared with a full bus matrix.

In the present and near future, we believe that the bus matrix communication architecture can efficiently support MPSoC systems with tens to hundreds of cores with several data throughput constraints in the multiple gigabits per second range. However, for very large MPSoC systems in the future, bus-based communication systems will suffer from unpredictable wire cross-coupling effects, significant clock skews on longer wires and serious routability issues for multiple wires crossing the chip in a non-regular manner. Network-on-chip (NoC) based communication architectures, with a regular wire layout and having all links of the same length, offer a predictable model for wire cross-talk and delay. This predictability will permit aggressive clock rates and support much larger data throughputs. Therefore we believe that for very large MPSoC systems in the future having several hundreds of cores, a packet-switched NoC communication backbone would be a more suitable choice.

V. Conclusion

In this paper, we presented an approach for the automated synthesis of a bus matrix communication architecture for MPSoC designs with high bandwidth requirements. Our synthesis approach satisfies all throughput performance constraints in the design, while generating an optimal bus matrix topology having a minimal number of busses, as well as values for parameters such as bus speeds, OO buffer sizes and arbitration strategies. Results from the synthesis of an AMBA3 AXI [21] based bus matrix for four MPSoC applications from the networking domain show a significant reduction in bus count in the synthesized matrix when compared with a full bus matrix (up to $9\times$) and a maximally connected reduced matrix (up to $3.2\times$). Our approach is not restricted to an AMBA3 [21] matrix based architecture and can be easily extended to synthesize CoreConnect [2] and STBus [3] crossbars as well.

Acknowledgements

This research was partially supported by grants from SRC Contract 1330, Conexant Systems, CPCC fellowship and UC Micro (03-029).

References

[1] ARM AMBA Specification (rev2.0), *www.arm.com, 2001*

[2] "IBM On-chip CoreConnect Bus Architecture", *www.chips.ibm.com*

[3] "STBus Communication System: Concepts and Definitions", *Reference Guide, STMicroelectronics, May 2003*

[4] "Sonics Integration Architecture, Sonics Inc", *www.sonicsinc.com*

[5] L.Benini, G.D.Micheli, "Networks on Chips: A New SoC Paradigm", *IEEE Computers, pp. 70-78, Jan. 2002*

[6] J. Henkel, et al, "On-chip networks: A scalable, communication-centric embedded system design paradigm", *VLSI Design, 2004*

[7] M. Nakajima et al. "A 400MHz 32b embedded microprocessor core AM34-1 with 4.0GB/s cross-bar bus switch for SoC", *ISSCC 2002*

[8] S. Murali, G. De Micheli, "An Application-Specific Design Methodology for STbus Crossbar Generation", *DATE 2005*

[9] V. Lahtinen, et al, "Comparison of synthesized bus and crossbar interconnection architectures", *ISCAS 2003*

[10] K.K Ryu, E. Shin, V.J. Mooney, "A Comparison of Five Different Multiprocessor SoC Bus Architectures", *DSS 2001*

[11] M. Loghi, et al "Analyzing On-Chip Communication in a MPSoC Environment", *DATE 2004*

[12] M. Gasteier, M. Glesner "Bus-based communication synthesis on system level", *ACM TODAES, January 1999*

[13] S. Pasricha, N. Dutt, M. Ben-Romdhane, "Automated Throughput-driven Synthesis of Bus-based Communication Architectures", *In Proc of ASPDAC 2005*

[14] S. Pasricha, N. Dutt, E. Bozorgzadeh, M. Ben-Romdhane, "Floorplan-aware Automated Synthesis of Bus-based Communication Architectures", *In Proc. of DAC 2005*

[15] K. Srinivasan, et al, "Linear Programming based Techniques for Synthesis of Network-on-Chip Architectures", *ICCD 2004*

[16] D. Bertozzi et al. "NoC synthesis flow for customized domain specific multiprocessor systems-on-chip", *IEEE TPDS, Feb 2005*

[17] O. Ogawa et al, "A Practical Approach for Bus Architecture Optimization at Transaction Level", *DATE 2003*

[18] SystemC initiative. *www.systemc.org*

[19] S. Pasricha, N. Dutt, M. Ben-Romdhane, "Fast Exploration of Bus-based On-chip Communication Architectures", *In Proc. of CODES+ISSS 2004*

[20] S. Pasricha, N. Dutt, M. Ben-Romdhane, "Bus Matrix Communication Architecture Synthesis", *CECS Technical Report 05-17, October 2005*

[21] ARM AMBA AXI Specification *www.arm.com/armtech/AXI*

[22] S. Pasricha, N. Dutt, M. Ben-Romdhane, "Extending the Transaction Level Modeling Approach for Fast Communication Architecture Exploration", *In Proc. of DAC 2004*

[23] K. Lahiri et al, "Efficient exploration of the SoC communication architecture design space", *ICCAD 2000*

[24] S. Pasricha, "Transaction Level Modeling of SoC with SystemC 2.0" Synopsys User Group Conference (SNUG 2002), Bangalore, May 2002

Improving Routing Efficiency for Network-on-Chip through Contention-Aware Input Selection

Dong Wu, Bashir M. Al-Hashimi, Marcus T. Schmitz
School of Electronics and Computer Science
University of Southampton
Southampton, SO17 1BJ, UK
e-mail: {dw, bmah, ms4}@ecs.soton.ac.uk

Abstract - The performance of Network-on-Chip (NoC) largely depends on the underlying routing techniques, which have two constituencies: *output selection* and *input selection*. Previous research on routing techniques for NoC has focused on the improvement of output selection. This paper investigates the impact of input selection, and presents a novel *contention-aware input selection (CAIS)* technique for NoC that improves the routing efficiency. When there are contentions of multiple input channels competing for the same output channel, CAIS decides which input channel obtains the access depending on the *contention level* of the upstream switches, which in turn removes possible network congestion. Simulation results with different synthetic and real-life traffic patterns show that, when combined with either deterministic or adaptive output selection, CAIS achieves significant better performance than the traditional first-come-first-served (FCFS) input selection, with low hardware overhead (<3%).

I Introduction

As technology scales and chip integrity grows, on-chip communication is playing an increasingly dominant role in System-on-Chip (SoC) design. To meet the performance and design productivity requirements, Network-on-Chip (NoC) [1-4] has been proposed as a solution to provide better modularity, scalability, reliability and higher bandwidth compared to bus-based communication infrastructures. Fig. 1(a) shows a mesh-based NoC, which consists of a grid of 16 cores. Each core is connected to a switch by a network interface. Cores communicate with each other by sending packets via a path consisting of a series of switches and inter-switch links. For each packet, there are several possible paths, which directly influence the time needed for delivery. Therefore, the performance of NoC largely depends on the underlying routing technique, which chooses a path for a packet and decides the routing behaviour of the switches.

Fig. 1(b) shows a block diagram of a switch with n+1 input channels and output channels interconnected by a crossbar. In order to route packets through the network, the switch needs to implement a routing technique. A routing technique has two constituencies: output selection and input selection. A packet coming from an input channel may have a choice of multiple output channels, e.g., a packet p0 of input_0 can be forwarded via output_0, output_1 and so on. The *output selection* chooses one of the multiple output channels to deliver the packet. Similarly, multiple input

channels may request simultaneously the access of the same output channel, e.g., packets p0 of input_0 and p1 of input_1 can request output_0 at the same time. The *input selection* chooses one of the multiple input channels to get the access.

Whilst the impact of the output selection on routing efficiency has been investigated [5-7], no explicit work has been reported on the impact of the input selection, which is the aim of this paper. The main contribution of this paper is a novel *contention-aware input selection (CAIS)*, as part of the routing techniques implemented in switches. With CAIS, each output channel within a switch observes the *contention level* (i.e., the number of request from the input channels, Section III), and transmits this contention level to the input channel of the downstream switch. During the input selection within the downstream switch, CAIS chooses an input channel depending on the contention levels. Input channels with higher contention levels get higher priority. CAIS tries to remove possible network congestion by keeping the traffic flowing even in the paths with heavy traffic load, which in turn improves routing efficiency.

Experimental results with synthetic and real-life examples show that CAIS can be combined with either deterministic or adaptive output selection, and it is capable of decreasing

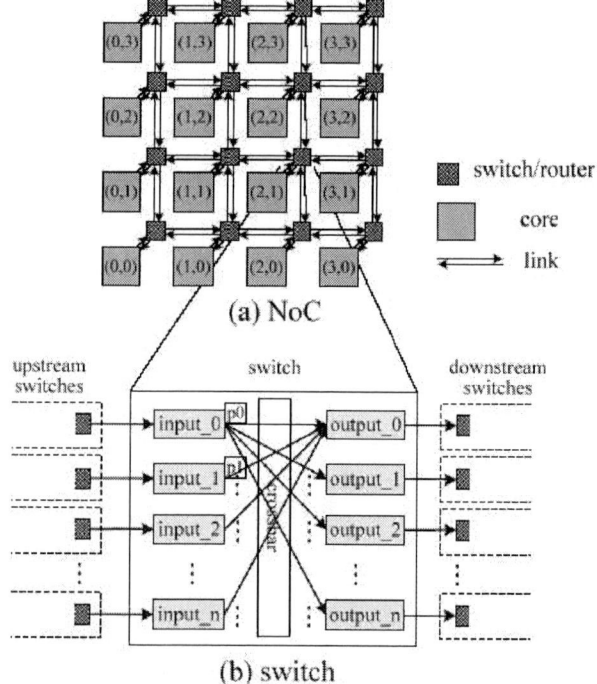

Fig. 1. Block diagram of NoC and switch

---

* This work is supported in part by the EPSRC, UK, under grant EP/C512804, GR/S95770.

0-7803-9451-8/06/$20.00 ©2006 IEEE.

packet latency significantly compared to the traditional first-come-first-served (FCFS) input selection. Furthermore, the synthesis of prototype switches with CAIS shows low hardware overhead compared to FCFS.

The rest of paper is organised as follows. Related work is reviewed in Section II. Section III describes the proposed contention-aware input selection (CAIS) in detail. The experiment results and switch implementation are presented in Section IV. Finally Section V gives the conclusion.

II. Related Work

The idea of NoC is derived from large-scale computer networks and distributed computing [8, 9]. However, the routing techniques for NoC have some unique design considerations besides low latency and high throughput. Due to tight constraints on memory and computing resources, the routing techniques for NoC should be reasonably simple [7]. Several switch architectures have been developed for NoC [10-12], employing XY output selection and wormhole routing (Section III). In [6], a deflective routing technique is proposed to avoid network congestion by spreading the traffic over a larger area. It performs output selection based on the number of packets being handled in the neighbouring switches. Packets are forwarded to switches with less traffic load. The routing technique proposed in [7] is similar to [6] in terms of acquiring information from the neighbouring switches to avoid network congestion, but uses the buffer levels of the downstream switches to perform the output selection. A routing scheme which combines deterministic and adaptive routing is proposed in [5], where the switch works in deterministic mode when the network is not congested, and switches to adaptive mode when the network becomes congested. All the routing techniques [5-7] focused on the output selection. The motivation of this paper is to investigate the impact of input selection and develop a simple yet effective input selection, aiming to improve the routing efficiency with low hardware cost.

III. Proposed Technique

Two input selections have been used in NoC, first-come-first-served (FCFS) input selection [5] and round-robin input selection [10, 11]. In FCFS, the priority of accessing the output channel is granted to the input channel which requested the earliest. Round-robin assigns priority to each input channel in equal portions on a rotating basis. FCFS and round-robin are fair to all channels but do not consider the actual traffic condition. This section presents a contention-aware input selection (CAIS) as part of the routing techniques implemented in switches. CAIS performs more intelligent input selection by considering the actual traffic condition, leading to higher routing efficiency.

A. Preliminaries

In this paper we consider NoCs with 2D mesh topology (Fig. 1(a)). Wormhole switching [9, 13] is employed because of its low latency and low buffer requirement. In wormhole switching, a packet is divided into flits for transmission. The header flit contains the routing information, which is used by the switches to establish the routing path. The remaining flits simply follow the path in a pipeline fashion. A flit is passed to the next switch as soon as enough buffer space is available to store it, even though there is not enough space to store the whole packet. If the header flit encounters a channel already in use, the subsequent flits have to wait at their current locations and are spread over multiple switches, thus blocking the intermediate links.

The proposed contention-aware input selection (CAIS, Section III.B) has been combined with an output selection, either deterministic or adaptive [5], to complete the routing function. In this paper, the XY routing [9] is used as a representative of deterministic output selection for its simplicity and popularity in NoC. In the XY output selection, packets are sent first along the X dimension then along the Y dimension. For example, considering the NoC of Fig. 1(a), a packet from (0, 3) to (2, 2) will take a path as follows: (0, 3) → (1, 3) → (2, 3) → (2, 2). The wormhole switching is sensitive to deadlock [9, 13]. To avoid deadlock, the minimal odd-even (OE) routing [8] is used as a representative of adaptive output selection. In the OE output selection, a packet chooses a path from multiple alternatives, but paths with certain turns are prohibited to avoid deadlock. Considering the previous example again, a packet from (0, 3) to (2, 2) has two alternative paths: (0, 3) → (0, 2) → (1, 2) → (2, 2) and (0, 3) → (1, 3) → (1, 2) → (2, 2). Note the path (0, 3) → (1, 3) → (2, 3) → (2, 2) is invalid because an east-south turn is not allowed in the switch positioned at (2, 3) to avoid deadlock.

B. Contention-Aware Input Selection

To show the influence of input selection and output selection on the routing efficiency, consider the example of Fig. 2, which shows a network of switches (cores are ignored for simplicity). Note the grey scale of the switches indicates the number of packets waiting at the switches. The white colour switches have low number of waiting packets, whilst the grey colour switches have higher number of waiting packets, and the black colour switch at (2, 2) has the highest number of waiting packets. To demonstrate the influence of output selection, consider a packet p0 traveling from (3, 0) to (0, 2), which has a choice of multiple paths. A good path would be to avoid the congested area (i.e., the grey and black switches), as indicated by the dashed line. This shows a suitable output selection can avoid network

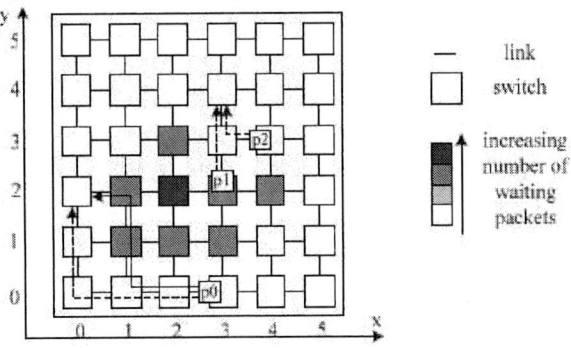

Fig. 2. Motivation of CAIS

congestion. Now consider the input selection. Packets p1 at (3, 2) and p2 at (4, 3) both want to travel through (3, 3). In this case, a good choice would be let p1 take the priority to access (3, 3), because the switch at (3, 2) has more waiting packets than the switch at (4, 3). Such an input selection helps reduce the number of waiting packets in congested areas. This removes possible network congestions and leads to better NoC performance. Based on this observation, a contention-aware input selection (CAIS) is developed.

The basic idea of CAIS is to give the input channels different priorities of accessing the output channels. The priorities are decided dynamically at run-time, based on the actual traffic condition of the upstream switches. More precisely, each output channel within a switch observes the *contention level* (the number of requests from the input channels) and sends this contention level to the input channel of the downstream switch, where the contention level is then used in the input selection. When multiple input channels request the same output channel, the access is granted to the input channel which has the highest contention level acquired from the upstream switch. This input selection removes possible network congestion by keeping the traffic flowing even in the paths with heavy traffic load, which in turn improves routing performance.

Fig. 3 illustrates the detailed architecture of a switch with CAIS. As can be seen, besides wires for data transmission, CAIS requires additional wires to transmit contention levels (CLs) between neighbouring switches. The switch has n+1 input channels and output channels. Input channels contain a buffer to store the incoming flits temporarily before forwarding them to one of the output channels. The output selection (OS) module examines the header flit and decides to which output channel the packet should be passed. The OS sends an access request to the corresponding output channel. In an output channel, once it becomes available, the CAIS module examines the access requests, and sends a selection signal to the MUX module which accordingly

connects one of the input data signals to the output data signal. If there is only one access request, the request is granted. If there are multiple access requests, a selection of which input channel gets the access has to be made. This selection mechanism is explained next.

CAIS performs input selection based on the contention level (CL). The contention level of an output channel is the number of access request received at a certain time. The CL of an input channel is acquired from the output channel of the upstream switch through signal wires. Fig. 4 shows the algorithm of CAIS, which consists of two processes working in parallel. Process *observe_cl* is activated when the status of *req_0..n* changes. It observes the number of request to this output channel (i.e., CL) and puts the CL value at *out_cl_i*. Then the CL is transmitted to the input channel of the downstream switch, where the CL will be used to perform the input selection. For example, considering the switch of Fig. 3, if input channels 0 and 1 are requesting output channel 0, then the CL of output channel 0 (*out_cl_0*) is 2, and the CL of the input channel of the downstream switch is also 2. Note that, to avoid high complexity and hardware cost, CL is only sent to the immediate downstream switches and is not spread any further. For example, considering the previous example again, if the CLs of the input channels 0 and 1 (*in_cl_0* and *in_cl_1*) are 3 and 4 respectively, the CL of output channel 0 is NOT 3+4, but 2. Process *select_input* is activated when the output channel is available and there are requests. It examines all requests and the CLs of the corresponding input channels, and grants the request with the highest CL.

For the input channels connected to the cores, there are no upstream switches transmitting CL to them. The CL value is set to 0 for these input channels. Therefore, the packets already in the network have higher priority than the packets waiting to be injected into the network.

Contention-Aware Input Selection (CAIS)	
req_0..n	request signals from the input channels
out_cl_i	CL of the ith output channel
in_cl_j	CL of the jth input channel acquired from the upstream switch
max_cl	maximum contention level
sel_i	selection signal of the ith output channel
i = 0, 1, 2, ..., n; j = 0, 1, 2, ..., n;	
01	process *observe_cl(req_0..n)*
02	begin
03	*out_cl_i* <= number of request to the ith output channel;
04	end process *observe_cl*
05	
06	process *select_input*
07	begin
08	*max_cl* := 0;
09	for all requests loop
10	if *in_cl_j* >= *max_cl* then
11	*max_cl* := *in_cl_j*;
12	*sel_i* <= j;
13	end if
14	end loop
15	end process *select_input*

Fig. 4. Pseudo VHDL code of the CAIS algorithm

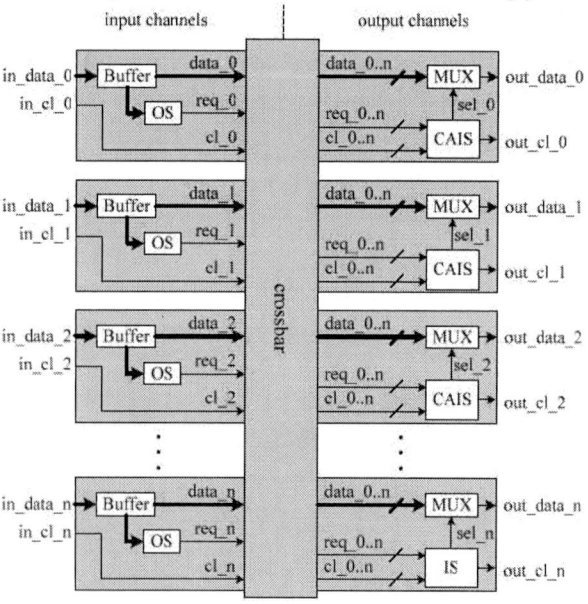

Fig. 3. Switch architecture with CAIS

IV. Experimental Results

Experiments are conducted to evaluate the performance of the contention-aware input selection (CAIS) and give a comparison between CAIS and traditional input selections. Two traditional input selections have been used in NoC, first-come-first-served (FCFS) input selection [5] and round-robin input selection [10, 11]. Due to the advancement of FCFS over round-robin, FCFS is selected to compare with CAIS. Both CAIS and FCFS are combined with a deterministic output selection (XY routing [9]) and an adaptive output selection (OE routing [8]). Four switch models are developed using VHDL to implement the four routing schemes: XY+FCFS, XY+CAIS, OE+FCFS and OE+CAIS. Simulations are carried out on a 6×6 mesh NoC using these four switch models. As in previous work [5, 8, 14], the performance of the routing scheme is evaluated through latency-throughput curves. For a given packet injection rate (i.e., the number of packets injected to the network per cycle), a simulation is conducted to evaluate the average packet latency. It is assumed that the packet latency is the duration from the time when the first flit is created at the source core, to the time when the last flit is delivered to the destination core. For each simulation, the packet latencies are averaged over 50,000 packets. Latencies are not collected for the first 5,000 cycles to allow the network to stabilise. It is assumed that the packets have a fixed length of 5 flits and the buffer size of input channels is 5 flits. Since the network performance is greatly influenced by the traffic pattern, we applied four different traffic patterns, including three synthetic traffic patterns (uniform, transpose and hot spot) and a real-life traffic pattern (GSM voice CODEC).

A. Synthetic Traffic

In the first set of experiments we consider three synthetic traffic patterns: uniform, transpose, and hot spot [8]. In the uniform traffic pattern, a core sends a packet to any other cores with equal probability. In the transpose traffic pattern, a core at (i, j) only send packets to the core at (5-j, 5-i). In the hot spot traffic pattern, the core at (3, 3) is designated as the hot spot, which receives 10% more traffic in addition to the regular uniform traffic.

Fig. 5 shows the performance of the four routing schemes under uniform traffic. The X-axis represents the packet injection rate per node (the packet injection rate for the whole NoC is 36 times higher), and the Y-axis represents the average packet latency. As can be seen from the figure, the four schemes have almost the same performance at low traffic load (<0.038 packets/cycle). As the traffic load increases, the packet latency rises dramatically due to the network congestion. Comparing the curves of OE+FCFS and OE+CAIS, it can be seen that, using the OE output selection, CAIS performs significantly better than FCFS. Similarly, the curves of XY+FCFS and XY+CAIS show that CAIS also outperforms FCFS when using XY output selection. As reported in [5, 8], the XY output selection has better performance than the OE output selection. This is because the XY output selection incorporates global, long-term information about the uniform traffic, leading to

even distribution of traffic. On the other hand, the OE routing is based on local, short-term information, which only benefits the immediate future packets while loses the evenness of uniform traffic in the long run.

Fig. 6 shows the performance of the four routing schemes under transpose traffic. It can be seen that FCFS and CAIS have the same performance when using the XY output

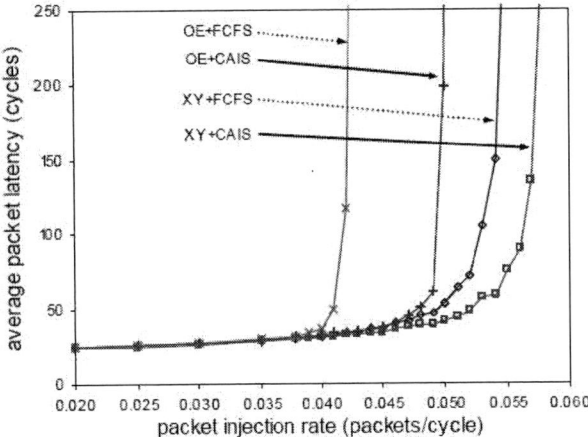

Fig. 5. Routing performance under uniform traffic

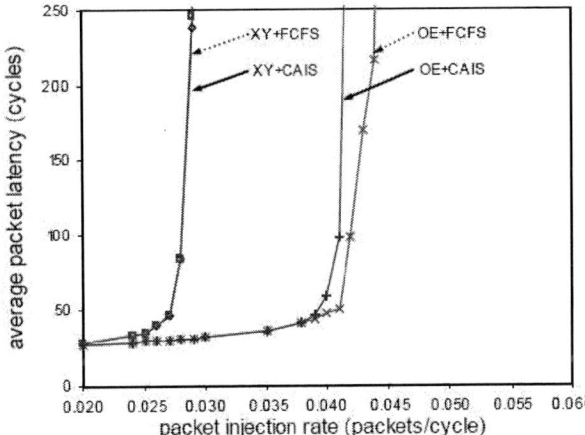

Fig. 6. Routing performance under transpose traffic

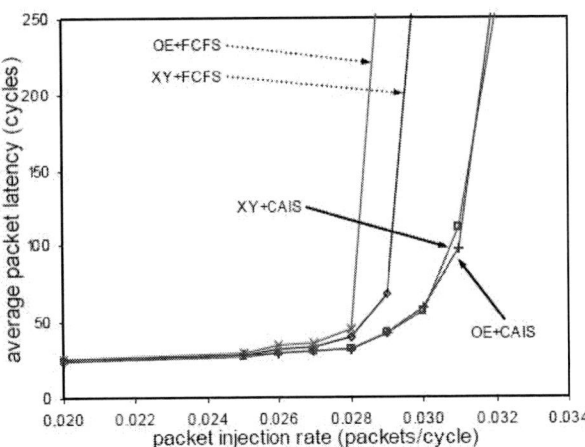

Fig. 7. Routing performance under hot spot traffic

selection; FCFS works slightly better than CAIS when using the OE output selection. This is because with transpose traffic, it is rarely the case that more than one input channels compete for the same output channel. Therefore, the input selection policy has little impact on the routing performance.

Fig. 7 shows the routing performance under hot-spot traffic. Once again, it can be seen that CAIS significantly outperforms FCFS, either using XY or OE output selection. Furthermore, although the OE output selection performs worse than the XY output selection when using the FCFS input selection, it achieves similar performance as XY when using the CAIS input selection.

B. Voice CODEC Traffic

To evaluate the performance of CAIS under more realistic traffic loads, we have conducted simulations using a GSM voice CODEC [15]. The GSM voice CODEC is partitioned into 9 cores. The communication trace between the cores is recorded for an input voice stream of 500 frames (10 seconds of voice). The cores are mapped manually to a 6×6 mesh NoC. Fig. 8 shows the partition and communication trace of the GSM voice CODEC. As can be seen, the encoder and decoder are partitioned into core_0 – core_4 and core_5 – core_8 respectively. The directed edges between the cores represent communications. For example, the edge between core_0 and core_1 means there is a communication from core_0 to core_1, the communication has a size of 320 bytes and occurs at cycle 0. Due to the space limitation, only communications in the first 1000 cycles are shown. Fig. 8 also shows the mapping of the cores to the NoC using the dashed lines. The CODEC cores generate packets according to the recorded communication trace. The other cores in the NoC generate uniform traffic, with the same average packet injection rate as the CODEC cores. The packet injection rate is increased incrementally to get the latency-throughput curve, which is shown in Fig. 9. As can be seen, in both cases of using XY and OE output selection, CAIS achieves better performance than FCFS. Furthermore, although XY has worse performance than OE when using FAFS, it achieves similar performance as OE when the CAIS input selection is used. This observation and the one obtained from Fig. 7 show that the employment of

CAIS can help the otherwise less effective output selection when using FCFS catch up with the more effective output selection. This demonstrates further the importance of input selection in routing efficiency and the effectiveness of CAIS.

One consideration of CAIS is that some packets may experience indefinite waiting. Although theoretically possible, it does not happen in the experiments. To give an insight, Fig. 10 shows the maximum packet latency of the routing schemes under the GSM voice CODEC traffic. Note the curves in Fig. 10 are NOT average latency, but the maximum latency experienced by the packets, thus the curves show some dips and jumps. As can be seen, when the network load is low, CAIS has similar maximum packet latency as FCFS; when the network load is high, CAIS has shorter maximum packet latency than FCFS. CAIS does not cause indefinite waiting. The reason is that, due to the latency introduced by the processing of the header flit, there is a gap between the access requests of two consecutive packets. During this gap, one of the packets waiting in input channels with low contention levels can get the access to the output channel, thus indefinite waiting is avoided.

Overall, the experiments of Sections IV.A and IV.B have demonstrated the importance of input selection, which is in line with that obtained in [14] when applied in distributed computing. The experiments have also shown that CAIS

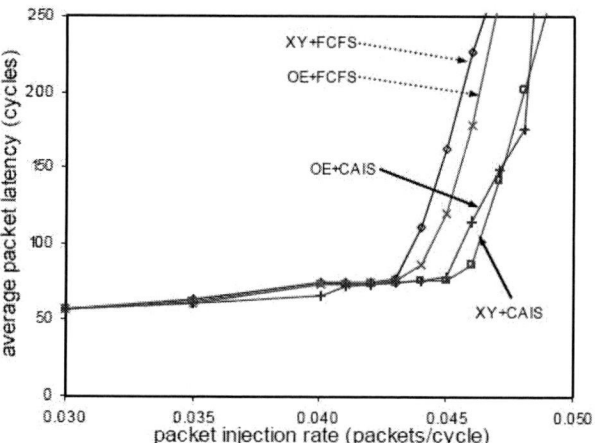

Fig. 9. Routing performance under voice CODEC traffic

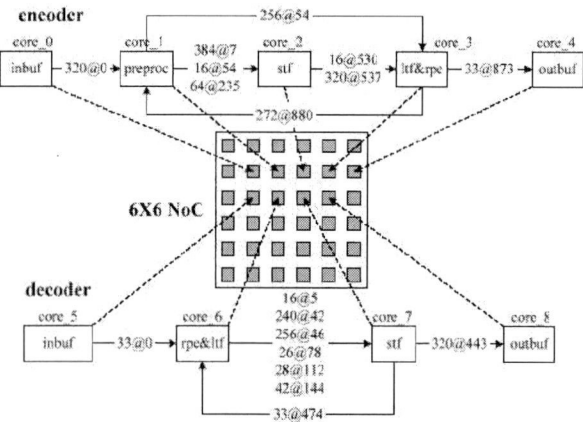

Fig. 8. Partition, communication trace and core mapping of a GSM voice CODEC

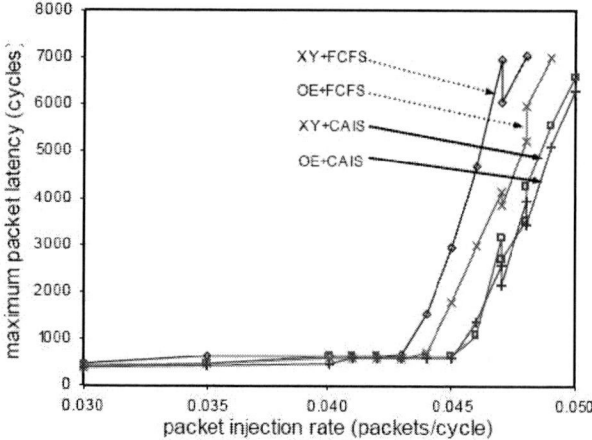

Fig. 10. Maximum packet latency under CODEC traffic

effectively improves the routing efficiency for NoCs.

C. Implementation of Prototype Switch

To evaluate the area overhead of CAIS and show the performance/area trade-off, switches with four different routing schemes have been implemented. The first scheme is XY+FCFS, i.e., XY output selection and FCFS input selection. The other three schemes are XY+CAIS, OE+FCFS and OE+CAIS. The switches were coded in VHDL and synthesized with Synplify ASIC using an ST Microelectronics 0.12 μm standard cell library. For all switches, the data width is set to 32 bits, and each input channel has a buffer size of 5 flits. Fig. 11 shows the area cost of the four switches. As expected, using the deterministic XY output selection and the simple FCFS input selection, XY+FCFS has the lowest area cost of 0.109725mm^2. Due to the relative complexity of the adaptive OE output selection and the CAIS input selection, XY+CAIS and OE+FCFS have slightly higher area costs of 0.111480mm^2 and 0.110355mm^2 respectively, and OE+CAIS has the highest area cost of 0.113580mm^2. Comparing the area costs of XY+FCFS and XY+CAIS, CAIS introduces 1.6% additional overhead than FCFS. Similarly, when comparing OE+FCFS and OE+CAIS, CAIS introduces 2.9% additional overhead than FCFS.

CAIS requires additional wires to transmit the contention levels (CLs). In the case of 2D mesh NoC, each switch have at most 5 input channels, receiving packets from the core and the 4 neighbouring switches. Thus at most 3 ($\lceil \log_2 5 \rceil$) wires are needed to transmit a CL. This is acceptable because NoC have abundant wiring resources [3, 4].

Note although this paper has considered mesh-based NoC, CAIS is flexible enough to support other NoC topologies including irregular topologies. This can be easily done by configuring the value of n (Fig. 3 and Fig. 4), and the number of wires for CL transmission accordingly.

V. Conclusion

This paper has shown the importance of input selection in routing efficiency, and presented a simple yet effective contention-aware input selection (CAIS) as part of the routing techniques implemented in switches. CAIS performs the input selection considering the contention level of the upstream switches. By granting busier input channel higher priority to access the output channel, CAIS keeps the traffic in busy paths flowing, therefore removes possible

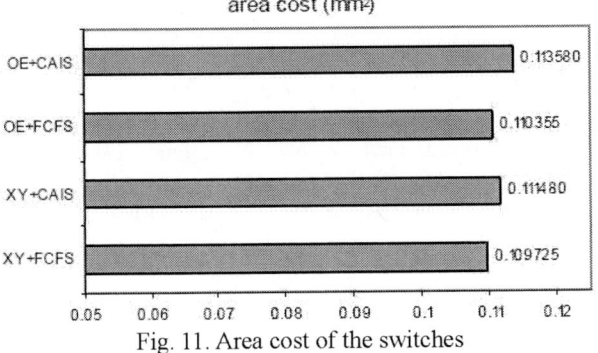

Fig. 11. Area cost of the switches

network congestion. Simulation has been carried out with a number of different traffic patters, including synthetic traffic and realistic voice CODEC traffic. The results shows that, no matter which output selection is used (deterministic XY routing or adaptive odd-even routing), the proposed CAIS achieved better performance than the traditional first-come-first-served (FCFS) input selection for most traffic patters except the transpose traffic, where CAIS has similar performance as FCFS. Furthermore, the employment of CAIS can make the XY routing (low complexity and hardware cost) to achieve similar or even better performance than the higher complexity and hardware cost odd-even routing for some traffic patters, i.e., area saving. The prototype switch with CAIS has been implemented and shows that CAIS has slight hardware overhead compared to FCFS (< 3%). As explained in Section IV.B, although not shown in the experiments, there is a starvation possibility in CAIS, which remains to be addressed in future work.

References

[1] S. Kumar, A. Jantsch, J.-P. Soininen, M. Forsell, M. Millberg, J. Oberg, et al, "A network on chip architecture and design methodology," ISVLSI, pp. 117-24, USA, 2002.

[2] L. Benini and G. De Micheli, "Networks on chips: A new SoC paradigm," Computer, vol. 35, pp. 70-78, 2002.

[3] J. Henkel, W. Wolf, and S. Chakradhar, "On-chip networks: A scalable, communication-centric embedded system design paradigm," VLSI Design, pp. 845-851, India, 2004.

[4] W. J. Dally and B. Towles, "Route packets, not wires: On-chip interconnection networks," DAC, pp. 684-689, USA, 2001.

[5] J. Hu and R. Marculescu, "DyAD - Smart routing for networks-on-chip," DAC, pp. 260-263, USA, 2004.

[6] E. Nilsson, M. Millberg, J. Oberg, and A. Jantsch, "Load distribution with the proximity congestion awareness in a network on chip," DATE, pp. 1126-7, Germany, 2003.

[7] T. T. Ye, L. Benini, and G. De Micheli, "Packetization and routing analysis of on-chip multiprocessor networks," Journal of Systems Architecture, vol. 50, pp. 81-104, 2004.

[8] G.-M. Chiu, "The odd-even turn model for adaptive routing," IEEE Transactions on Parallel and Distributed Systems, vol. 11, pp. 729-38, 2000.

[9] L. M. Ni and P. K. McKinley, "A survey of wormhole routing techniques in direct networks," Computer, vol. 26, pp. 62-76, 1993.

[10] C. A. Zeferino, M. E. Kreutz, and A. A. Susin, "RASoC: A router soft-core for Networks-on-Chip," Designers Forum - DATE, pp. 198-203, France, 2004.

[11] N. Kavaldjiev, G. J. M. Smit, and P. G. Jansen, "A virtual channel router for on-chip networks," IEEE International SOC Conference, pp. 289-93, USA, 2004.

[12] E. Rijpkema, K. Goossens, A. Radulescu, J. Dielissen, J. Van Meerbergen, P. Wielage, and E. Waterlander, "Trade-offs in the design of a router with both guaranteed and best-effort services for networks on chip," IEE Proceedings: Computers and Digital Techniques, vol. 150, pp. 294-302, 2003.

[13] D. Bertozzi and L. Benini, "Xpipes: A network-on-chip architecture for gigascale systems-on-chip," IEEE Circuits and Systems Magazine, vol. 4, pp. 18-31, 2004.

[14] C. J. Glass and L. M. Ni, "Adaptive routing in meshconnected networks," ICDCS, pp. 12-9, Japan, 1992.

[15] M. T. Schmitz, B. M. Al Hashimi, and P. Eles, "Systemlevel design techniques for energy-efficient embedded systems," Kluwer Academic Publishers, 2004.

Physical design implementation of segmented buses to reduce communication energy

Jin Guo[1,2], Antonis Papanikolaou[1], Pol Marchal[1], Francky Catthoor[1,2]

[1]IMEC v.z.w., Kapeldreef 75, 3001 Leuven, Belgium

[2]Katholieke Universiteit Leuven, Kasteelpark Arenberg 10, 3001 Heverlee, Belgium

Abstract— The amount of energy consumed for interconnecting the IP-blocks is increasing significantly due to the suboptimal scaling of long wires. To limit this energy penalty, segmented buses have gained interest in the architectural community. However, the netlist topology and the physical design stage significantly influence the final communication energy cost. We present in this paper an automated way to implement a netlist consisting of hard macro blocks, which are interconnected with heavily segmented buses in an energy optimal fashion for communication. We optimize the network wires energy dissipation in two separate, but related steps: minimizing the number of segments for active communication paths at the first step (block ordering), followed by the activity aware floorplanning step to minimize the physical length of these segments. Energy gains of up to a factor of 4 are achieved compared to a standard system implementation using a shared bus. Especially, the block ordering step contributes significantly to the network energy optimization process.

I. INTRODUCTION

Energy consumption is becoming one of the major optimization targets when designing low-power embedded systems. An important way to reduce the energy cost is to introduce a distributed memory hierarchy [15]. However, due to the increasing number of IP blocks (many dozens), more and longer wires are required to interconnect all of them. The trend toward IP reuse will also push toward the efficient reuse of more hard macro-blocks. Hence, the netlist connecting many hard IP blocks should be efficiently implemented.

The performance and the energy consumption of global wires cannot follow that of the transistors as technology feature sizes scale down. The energy dissipation of those wires is almost not improving anymore, with each new technology node, while transistor capacitance keeps shrinking [2]. As a result, the relative energy consumption of the communication is increasing compared to the computation and the storage system components. The results in [6] show that the energy consumed by the communication network is comparable to the energy consumed by the heavily distributed memories.

At the architectural level a lot of studies exist already for communication energy minimization. Chen et al. [9] have demonstrated how to reduce the energy consumption of the communication by using a segmented bus architecture, which can shut off the unused path via switches. The energy costs of driving the switches for a segmented bus are appreciable lower than the gain obtained from using a segmented bus [16]. The actual wire energy consumption of this architecture, however, depends on the floorplan of the system. The reason is that energy is proportional to the product of wirelength multiplied

by the activity of each wire. The activity can be minimized at higher abstraction levels, but the actual length of each interconnect wire is only decided during physical design. Thus, support is needed from the physical design phase in order to achieve minimal communication energy consumption.

Current physical design flows, however, mostly focus on the optimization of area occupation and total wirelength of the design. Total wirelength is probably a good metric to represent wire congestion, but it is not adequate to reflect the energy consumption since different wires can have very different activities.

The use of heavily segmented buses introduces an additional issue to be solved. The number of segments should indeed increase compared to the industrial bridged or segmented bus architectures [19] [18], in order to reduce the energy cost. Then the order in which the blocks are connected to the bus heavily influences the communication energy. For example, a netlist with optimal connectivity can assure the very active paths are always shorter in length than the less active paths during the physical design, which reduces energy consumption intuitively. Hence, such netlist topology problems together with the activity aware floorplanning techniques need to be efficiently coupled to optimize the communication network energy.

In this paper we show how communication energy optimization can be achieved on real application drivers. We couple an energy-optimal communication architecture, connection ordering to an automated energy-aware physical design flow to show that communication energy can be minimized by up to a factor of 4 compared to shared buses architecture with a standard physical design flow. We also stress that the network energy is very sensitive to the netlist topology in terms of block ordering.

II. RELATED WORK

Reducing the communication power dissipation has increasingly become a key concern in SOC design. Many reported approaches have focused on the savings at architecture level. Chen et al. [9] first demonstrated how the segmented buses can improve communication power and critical path delay. However, they evaluated the wire energy based on the estimation of the wire length, without doing the floorplanning. Hence the results lack of real physical level effects. Furthermore, they calculated the wire energy consumption without buffers to neglect the circuit level issues. But buffers contribute significantly to the overall energy consumption and latency.

A lot of work has been done in the physical design community to optimize chip area and improve circuit performance [10] [11]. To utilize the large impact of physical design on total system energy consumption, researchers have looked into power optimization at the floorplanning stage. Chao et al. [3] have introduced a floorplanner which optimizes the module power consumption and chip area by choosing the specific shape for each module. The wire energy dissipation is neglected. Several other approaches have been introduced to include a low power objective in physical design. Prabhakaran et al. [13] presented a simultaneous scheduling, binding and floorplanning algorithm to minimize interconnect energy dissipation. They contributed to the combination of high-level synthesis and physical design for interconnect power reduction, but not targeting macro blocks.

Jingcao et al. [7] have proposed a new methodology to generate low energy and high performance communication networks at the floorplanning stage. Their methodology is based on a point-to-point connection architecture. This architecture is not suitable for connecting many macro blocks because it can lead to routability and wire congestion problems. Jimenez et al. [12] presented an activity aware placement methodology, which aimed at reducing the power dissipation of macro block-based VLSI design. They mainly introduced how to implement the simulated annealing algorithm efficiently. They did not specify the communication architecture.

To the best of our knowledge, the impact of floorplanning on the energy consumption of heavily segmented buses has not been studied in the past. In sectionV, we will therefore outline an exploration environment for investigating this problem in two steps of optimization. First, we explain the target architecture more in detail.

III. PLATFORM DESCRIPTION

Our target domain is that of application domain specific embedded portable systems. We focus on multimedia and wireless applications. This target implies a number of architectural assumptions that we can exploit. Systems of this domain will follow the System-on-Chip template and they will consist of different tiles, among which one will be the mass storage memory which dominates the system area occupation. In this paper we focus on the implementation of a given tile. The implementation of the complete system can be done in the same manner.

A. Memory Organization

These systems consist of application domain specific processors with their local memories (Fig. 1). The processors can be partitioned into distributed processing elements (PEs). The memories are organized in a hierarchical manner for energy optimal purpose [6]. The local memory layer typically consists of many small memories. This heavily distributed memory organization increases the number of macro blocks, to at least dozens but up to hundreds. And the additional available bandwidth from the memories to the processors increases the number of parallel communication paths among the blocks. Therefore, a hierarchal memory organization results in a more complex communication network architecture and requires novel floorplanning techniques.

The PEs and memories are assumed to be pre-designed IP blocks in order to enable IP reuse and thus limit the system development design efforts. The shapes of the blocks are predefined and assumed to be rectangular. The width and height

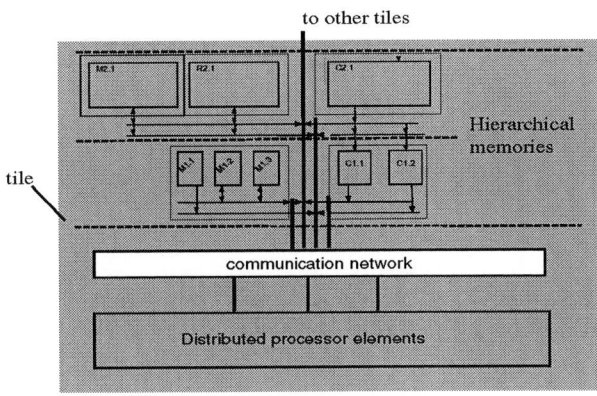

Fig. 1. Memory hierarchy and distributed organization

of the memories are fixed to achieve optimal intra-block energy consumption at the system design phase. During physical design, the physical characteristics of the memories are not changed any more to avoid overruling the already made decisions at higher level.

B. Segmented Buses Network

Due to the large number of IP blocks, interconnecting them becomes a critical issue. Shared buses are conventionally used (Fig. 2-a) instead of the point-to-point connections in order to use less communication wires. The main disadvantage of shared buses is that they are power hungry, since every master has to charge the entire bus for communicating data. In Fig. 2-a, the most frequently accessed path is between memory1 and the processor. During each access the entire bus swings, including connections 1-2-3. Traditionally, segmented buses [9] introduce some switches on the buses (see Fig. 2-b). The switches are 3-port uni-cast or multi-cast components implemented using tri-state buffer chains [14]. They can be programmed in order to configure the needed communication paths and shut off the others. For instance, we can avoid those connections 1-2-3 swing in Fig. 2-b by giving the control signal to the switches to shut off the paths. This way, it reduces the capacitance which toggles when communicating and thus reduces the communication energy consumption. In our case, there are dozens of switching connections to many segments, which implies a large potential energy gain there.

Fixing the netlist of the shared buses is straightforward, where all the blocks connect to the same buses without alternative solutions. When it comes to the segmented buses, additional decisions need to be taken to finalize the netlist of the system. Assuming a linear communication topology, the blocks are connected to the bus one by one in a specific order and no star connections exist. The order of the blocks connected to the bus dictates the activation frequency of the various segments. Hence, the network energy of the segmented buses has a strong dependency on the netlist. This step is actually necessary for any topology of segmented buses. In the following sections we propose a simple way to order the blocks on linear topologies.

1B-3

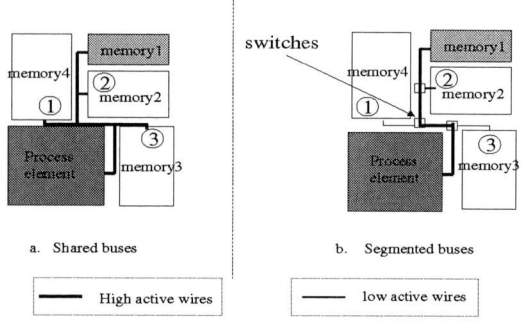

a. Shared buses   b. Segmented buses

Fig. 2. Segmented buses based on shared connections

IV. MOTIVATIONAL EXAMPLE

The energy cost of the segmented bus depends directly on the wirelength and the switching activity of each segment. The energy cost can be computed as follows:

$$E = V_{dd}^2 \times C_l \times \sum_i \left(\alpha_i \times l_i \times bitwidth_i \right) \qquad (1)$$

We realistically assume that only the global metal layers are used for the implementation of the communication network, so V_{dd} and capacitance per unit of length C_l are constants for a specific technology node. α_i is the number of activations of segment i. l_i is the average wirelength among the wires of a given segment and $bitwidth_i$ is the number of parallel wires of that segment. Conventional physical design flows minimize the total wire length $\sum_i (l_i \times bitwidth_i)$. This reduces wire congestion, but it can lead to energy optimization only when α_i is constant or varies little for the different segments.

In typical embedded applications, the required bandwidth on the different connections from processing elements to memories varies significantly. This translates into large differences in activation frequency among the segments. We have measured a ratio of up to ten between the most active and the least active segments [17]. This large range in activities, means that we cannot neglect it in the optimization cost function. To reduce the energy consumption of the segmented bus, we should minimize the sum of the products of activity times total segment wire length, denoted as $\sum_i (l_i \times bitwidth_i \times \alpha_i)$.

This optimization can be decoupled into two steps: (1) heavily active communication paths should utilize the smallest number of segments for the data transfer (block ordering) and (2) the physical lengths of these segments should be minimized (activity aware floorplanning).

A. Block Ordering

As discussed earlier, we have to deal with up to hundreds of connections, leading to many alternative switches organizations and block ordering. The way the switches are connected to each other introduces the order with which the memories are connected on the buses. The number of segments that need to be traversed to communicate between a given PE and a given memory is determined by the connection ordering of the memories. A good ordering decision results in that the very active transfer paths use a minimal number of segments, which implies low cost. For instance, Fig. 3 illustrates two block ordering decisions. The same architecture is shown comprising two processing elements and 4 memories connected with a different order on the linear communication topology. The different order is the only difference between the two netlists. To illustrate the potential for energy optimization we assume for simplicity in this example that all segments have the same wirelength L and we neglect the wirelengths from memory ports to switches. Some memories communicate to 2 PEs like memory3. The values annotated on top of each memory block in the figure are the number of accesses of each memory to the two PEs individually for executing a given functionality. Furthermore, we assume that the bitwidth is the same everywhere, thus we can neglect it in the qualitative calculations that follow.

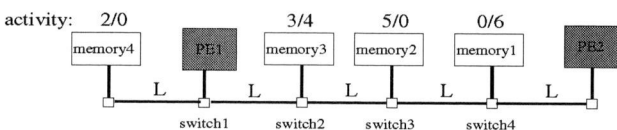

a. Non optimal block ordering decision for energy

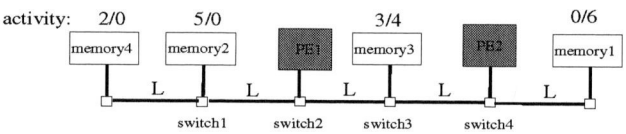

b. Optimal block ordering decision for energy

Fig. 3. Block ordering decisions influence communication cost

Considering the energy related communication cost as the product of total wire length from source to sink of communication and activity, the first block ordering decision has a cost of $E = 2 \times L + 3 \times L + 5 \times 2L + 4 \times 3L + 6 \times L = 33L$. The second netlist has a much lower cost, due to a better block ordering. Its cost is $E = 2 \times 2L + 5 \times L + 3 \times L + 4 \times L + 6 \times L = 22L$, a nearly 40% energy reduction. The intuitive principle for energy optimal ordering is that the connection order is made according to the activity order, that is, the most active memories are closer to PE in terms of the number of switches existing in the communication path between these two blocks and vice versa.

B. Activity Aware Floorplanning

Given a netlist with a good block ordering decision, the physical length of each segment should be minimized (step 2) according to their activity in the physical design steps of floorplanning and placement. An activity aware floorplanning technique has been outlined in [8]. The principle is that very active segments should have minimal length, while not so active segments can tolerate a bit longer length. For instance, in Fig. 4, the two floorplans are using the same netlist. The first floorplan does not consider the activity of each segment and aims at minimizing the total wirelength and area occupation. In the second floorplan, we change the positions of the modules such that the wirelength of the most active link is reduced, which should be energy optimal solution for the network. An area penalty is incurred and the total wirelength might be increased a bit. We quantify these overheads in a later section. However, large main on-chip storage memories (not layer1 memories) will dominate the chip area in the embedded systems, thus small area overheads in the area of the individual tiles for

44

the distributed layer1 memories will be negligible at the level of the entire chip.

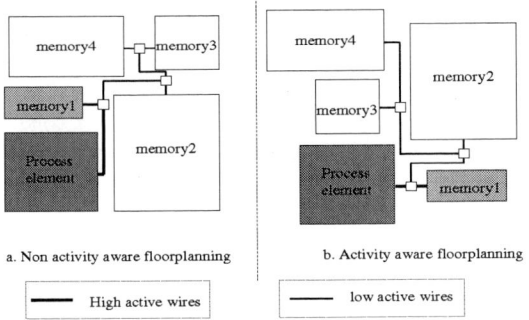

a. Non activity aware floorplanning   b. Activity aware floorplanning

— High active wires   — low active wires

Fig. 4. Floorplanning techniques for segmented buses

In the remainder of this paper we outline an automated methodology, which can produce an optimized implementation of a system consisting of hard macro-blocks interconnected via segmented buses starting from the RTL system description. The main optimization cost is communication energy consumption.

V. PHYSICAL DESIGN FLOW FOR ENERGY OPTIMAL COMMUNICATION

In order to make an automatic physical design implementation for low energy segmented buses architecture interconnecting hard macro blocks, we introduce our approach in Fig. 5, compared to a conventional approach. After the system design step (or high-level synthesis), the system consists of blocks in terms of many memories and a few processing elements. The conventional approach inputs the RTL description into a placer to get the placement and then performs the routing.

For segmented buses, the RTL description should be extended with the information about how the switch blocks are connected. Normally, the switch blocks are much smaller than the memory blocks. It is extremely difficult for most of the current macro cell floorplanning tools to deal with macro blocks of very different sizes. In that case, manual intervention is need. For instance, some extremely large or small blocks are pre-placed manually before automatically floorplanning the other blocks.

Avoiding the manual placement of the switches while keeping the smart strategy made by the floorplanner, we automatically insert the switch blocks after floorplanning, closely to the ports of the memories/PEs. Since the switches are small, the insertion step does not change the relative locations of the large macros, which have been optimized by the floorplanner already. And since the switches are close to the communication ports, the insertion influence on the communication path is slight. Hence the insertion impact on the cost of the activity aware floorplanning is small.

In our approach, we first make the ordering decision according to the memory activities. The ordering decision is necessary to identify the activity of the segments between two switches. Because switches do not appear at floorplanning stage, an intermediate netlist, activity weighted netlist, is produced. This netlist has no switch description, but is annotated with the activities of point-to-point communications between the macros. These activities are equal to the segment activities

after the later switches insertion step. By this way, we use the netlist without switches but still provide enough activity information to physical design for energy optimization. The activity weighted netlist is imported to the public domain floorplanner Parquet [4] to get a network energy optimized placement solution. Then switches are added and the placement is slightly adjusted to accommodate them. In parallel, a new RTL netlist which describes all the macro blocks including the switches in the system after switches insertion is fixed. Finally, the placement and the new RTL netlist are imported to the MAGMA [1] environment for routing. We will explain the major steps in detail.

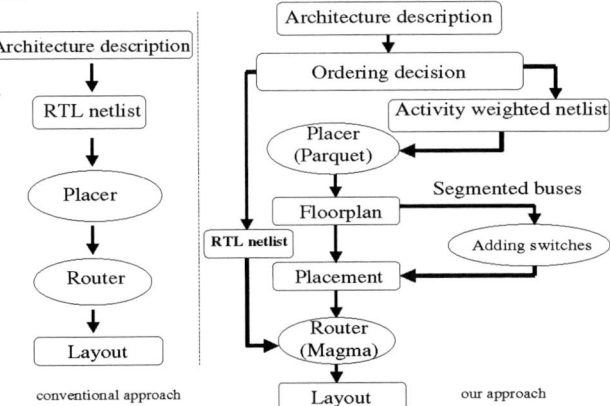

conventional approach   our approach

Fig. 5. Exploration methodology flow-graph

A. Block Ordering Decision

This step aims at minimizing the number of segments those need to be activated for very active communication paths. We currently only deal with linear bus topologies. This is not a severe limitation, since many of the bus-based designs in literature and in industry use linear topologies. This one dimensional topology is already able to illustrate well the gains achieved by the topology decisions, though a two dimensional topology might achieve more gains. Extending the principle to other topologies is future work.

The system components that are involved in most of the data transfers are obviously the processing elements. They fetch or store all the data from and to the individual memories. Thus, we have developed an approach where we start from the connections of the processing elements on the bus and connect the memories close to them based on their number of accesses. The most frequently accessed memories are connected very close to the processing element they communicate with and vice versa. This ordering localizes most of the communication activity in small areas and few segments need to be activated for most of the transfers. Extending this principle to cover multiple processing elements on a single linear bus is trivial. If more than one bus exists in the netlist, we perform the ordering per bus.

B. Floorplanning and Routing

In this stage, we need to implement the activity-aware floorplanning technique in order to minimize the lengths of the active segments in our macro block based netlist. We use Parquet [4] to obtain the floorplan for each design, which is a macro block placement tool. It can minimize the weighted net length and chip area via using a simulated annealing algorithm.

After the ordering decision, the activity of each net can be determined exactly. The activity is annotated to the net as the weight. Normally, the floorplanner minimizes $w1 \times area + w2 \times weighted\_wire\_length$ to achieve a good balance between area cost and network cost. In order to illustrate how much energy gain we will have compared to the shared buses, we set a low weight for area and a large weight for the weighted wire length in the cost function. Due to the fact that the simulated annealing process takes a long time to converge, we run the process for a set of times, with a reasonable CPU time for each run. The best floorplan is selected.

The floorplan generated by Parquet is not routable since the block edges overlap with their neighbors' edges. So it needs to be changed to a legal placement for routing. In parallel, switches need to be added to implement the segmented buses architecture. Again, the switches blocks and memory/PE blocks can not overlap with each other. We do the switch insertion and block separating simultaneously: starting from the left-bottom block, the neighbor block is moved to the right or to the top until a pre-defined space exists between these two blocks. If switches are needed to be inserted there, further movements are going on until enough space is assured for placing the switches.

The MAGMA BlastFusion chip implementation system was used to check the placement and do the detail routing. Only the global metal layers are used for connecting the memories and PEs. Because the bus is segmented, the length of the segments is rather small and buffering is enough. We target low power embedded system whose clock frequencies is several 100 MHZ but less than 1 GHZ. For these two reasons, no repeaters are needed. Hence, the wires can be routed over the blocks. The wirelength of each segment is measured and chip area is reported by the MAGMA tool suite.

VI. EXPERIMENTAL RESULTS

This section explains the way we performed the calculations for communication energy and critical path length. The results of our experiments are also presented here.

A. Network Energy and Critical Path Length

In order to have a better estimation of the real energy consumption of the buses we take into account the required buffers to drive the wires. The wires are buffered in a delay optimal way [5] and all the calculations are made for the 130nm technology node with data coming from the ITRS roadmap [2].

The total network dynamic energy consumption is the sum of all the individual segments energy consumption. We use the critical path wire length as a delay metric. For shared buses, the critical path wire length is equal to the whole bus length in the case of a single bus or the longest bus length for multiple buses. For segmented buses, the communication path length is the sum of the activated segments wire length. So the critical path length is the maximum sum of segment-lengths for any of the communication paths.

B. Results

First, we performed the design on the motion estimation and the motion compensation kernels of the main profile of an MPEG4 encoder. This is an application specific design consisting of two processing elements and 14 local memories.

We have followed two different combinations of communication architecture and physical design approach: shared buses with minimal total HPWL (Half Perimeter Wire Length), and segmented buses with minimal energy. The shared buses implementation follows the conventional design flow shown in Fig. 5. The RTL netlist is imported into Parquet, which generates one floorplan with minimal HPWL. For the second combination, we adopted activity aware floorplanning combined with a netlist which specifies the energy optimal block order for the segmented buses up front.

TABLE I
THREE DESIGN APPROACHES, MPEG4 ENCODER

Buses architecture	area (mm^2)	energy *1e-3(J)	critical path(m)	total WL(m)
Shared buses Min. HPWL	1.3333	0.481	0.0036	0.032
Segmented buses Min. energy	1.423	0.126	0.0019	0.114

Table I presents the chip area, the network energy consumption, critical path wirelength and total wirelength for these two approaches. Compared to shared buses using the conventional approach, significant energy gain is achieved when we use an optimal netlist and the activity aware floorplanning technique based on segmented buses. This combination reduces the wire energy to around one quarter compared to the minimal HPWL shared buses.

We evaluated the segmented buses architecture via our implementation methodology for MPEG4 with two different memory organizations and DAB (Digital Audio Broadcast) receiver with one memory organization in Table II. Similar comparisons of the two designs as in the Table I can be made and the conclusions are the same. In average, the energy optimal implementation of segmented buses can improve communication energy with a factor of 3.1 compared to the shared buses approach.

To assess the impact of the two proposed optimizations in the overall communication energy consumption we have conducted a number of experiments for MPEG4 application. Two different block ordering approaches, a random choice and the optimal choice, were combined with two different floorplanning approaches, a conventional one and the activity aware one. Fig. 6 illustrates the results from the four different experiments. The leftmost bars show the area and energy consumption of a netlist with a random block ordering decision using a conventional floorplanning approach that optimizes area and half-perimeter wirelength with equal weights. The second bar from the left is obtained by combining the same block ordering decision with an activity-aware floorplanning approach. The third set of bars is the result of an optimal block ordering decision (as explained in Section V) and the conventional floorplanning, while the last set of bars are for optimal block ordering and activity-aware floorplanning.

By comparing of the first and third experiments, it is evident that block ordering can have a very significant impact on the resulting communication energy (44.6% reduction). An optimal block ordering decision by itself reduce the energy, by reducing the number of segments that are activated for very

TABLE II
SHARED BUSES VS. SEGMENTED BUSES FOR DIFFERENT APPLICATIONS

Applications	Memory organization	Communication network	area (mm^2)	energy *1e-5(J)	critical path *1e-3(m)	total WL(m)
DAB	13 memories	Shared buses minimal HPWL	6.567	19.6	2.69	0.105
	3 PEs 3 buses	Segmented buses minimal energy	7.145	5.16	1.51	0.158
MPEG4 Memory Mapping1	9 memories	Shared buses minimal HPWL	5.333	2.08	1.92	0.076
	2 PEs 3 buses	Segmented buses minimal energy	5.562	1.16	1.29	0.144
MPEG4 Memory Mapping2	15 memories	Shared buses minimal HPWL	5.193	23.9	2.39	0.101
	2 PEs 2 buses	Segmented buses minimal energy	6.107	6.186	1.76	0.149

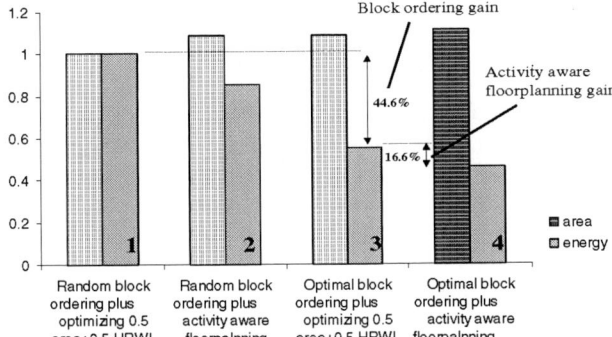

Fig. 6. Impact of block ordering and activity aware floorplanning, MPEG4 application driver

active transfer paths. The comparison of experiments 3 and 4, reveals that impact of activity-aware floorplanning on the final results is relative small (16.6% reduction). Thus, such a floorplanning without an optimized block ordering decision does not provide large energy gains. If the two optimization steps are coupled efficiently, significant energy gains of a factor of 2.16 can be achieved for this design. The area degradation is about 11%. We can conclude that the final network energy cost is much more sensitive to block ordering (netlist topology) than to the activity aware floorplanning techniques.

VII. CONCLUSIONS

We propose a novel automated approach that can implement a hard macro block netlist interconnected by segmented buses while minimizing the energy consumed in the communication network. The results show that this approach can reduce the communication network energy consumption by up to a factor of 4 compared to a conventional physical design stage implemented the netlist interconnected with a shared bus. And we present that the energy consumption of the communication network is high sensitive to the netlist topology decisions.

REFERENCES

[1] "Blast Chip 4.0 User Guide", *Magma Design Automation*, Cupertino, CA 95014, pp.271-351, http://www.magma-da.com.

[2] "International technology roadmap for semiconductors 2001 Edition."

[3] Chao Kai-yuan, D. F. Wong, "Floorplanning for low power designs," *IEEE international Symposium on Circuits and Systems*, Vol. 1, pp. 45-48, May 1995.

[4] S. N. Adya and I. L. Markov, "Fixed-outline floorplanning through better local search", *International Conference On Computer Design (ICCD)*, pp. 328-333, 2001.

[5] J. M. Rabaey, *Digital integrated circuits: a design perspective*, Upper Saddle River (N.J.): Prentive Hall, 2003.

[6] A. Papanikolaou, K. Koppenberger, M. Miranda, F. Catthoor, *Memory communication network exploration for low-power distributed memory organisations*, Proc. IEEE Wsh. on Signal Processing Systems (SIPS), Austin TX, IEEE Press, pp.176-181, Oct. 2004.

[7] Hu Jingcao, Deng Yangdong, R.Marculesu "System-level point-to-point communication synthesis using floorplanning information", *ASP-DAC*, pp. 573-579, 2002.

[8] Hua Wang, A. Papanikolaou, M. Miranda, F.Catthoor, "A global bus power optimization methodology for physical design of memory dominated systems by coupling bus segmentation and activity driven block placement", *ASP-DAC*, pp. 759-761, 2004.

[9] J. Y. Chen et al., "Segmented bus design for low-power system", *IEEE Trans. VLSI Syst* pp. 25-29 Mar 1999.

[10] I. Hui-Ru Jiang, Yao-Wen Chang, Jing-Yang Jou, Kai-Yuan Chao, "Simultaneous floorplanning and buffer block planning", *ASP-DAC*, pp. 431-434, 2003.

[11] Hua Xiang, Xiaoping Tang, M. D. F. Wong, "Bus-Driven floorplanning", *ICCAD*, pp. 66-73, 2003.

[12] M. A. Jimenez and M. Shanblatt, "Integrating a low-power objective into the placement of macro block-based layouts", *Proceedings of the 44th IEEE 2001 Midwest Symposium on Circuits and Systems*, Vol 1 , pp. 14-17 Aug 2001.

[13] P. Prabhakaran P. Banerjee, J. Crenshaw, M. Sarrafzadeh, "Simultaneous scheduling, binding and floorplanning for interconnect power optimization", *Proceedings of VLSI Design*, pp. 423 - 427, 1999 .

[14] A.Papanikolaou, F.Starzer, M.Miranda, F.Catthoor, K.De Bosschere, "Architectural and physical design optimizations for efficient intra-tile communication", *Proc. Intnl. System-on Chip Symp. (SoC)*, Tampere, Finland, pp.-, Nov. 2005.

[15] P. Panda, F. Catthoor, N. Dutt, K. Danckaert, E. Brockmeyer, C. Kulkarni, A.Vandecappelle, P.G.Kjeldsberg, "Data and memory optimizations for embedded systems", *on Design Automation for Embedded Systems (TODAES)*, Vol.6, No.2, pp.142-206, April 2001.

[16] K. Heyrman, A. Papanikolaou, F. Catthoor, P. Veelaert, W. Philips "Energy costs of transporting switch control bits for a segmented bus', *Proceeding of proRISC 2005*.

[17] E. Brockmeyer, J. D'Eer, F. Catthoor, N. Busa, P. Lippens, J. Huisken, *Code transformations for reduced data transfer and storage in low power realization of DAB synchro core* Proc.IEEE Wsh. on Power and Timing Modeling, Optimization and Simulation (PATMOS), Kos, Greece, pp.51-60, Oct. 1999.

[18] STBus specifications http://www.stmcu.com/inchtml-pages-STBus_intro.html

[19] ARM AMBA bus specification http://www.arm.com/armwww.ns4/html/AMBA?OpenDocument

Co-Synthesis of a Configurable SoC Platform based on a Network on Chip Architecture

Mário P. Véstias
mpv@fidelio.inesc-id.pt
INESC-ID, Lisboa
Portugal

Horácio C. Neto
hcn@inesc-id.pt
INESC-ID, Lisboa
Portugal

Abstract - The constant increase of gate capacity and performance of configurable hardware chips made it possible to implement systems-on-chip (SoC) able to tackle the demanding requirements of many embedded systems. In this paper, we propose an approach to the design space exploration of a configurable SoC (CSoC) platform based on a network on chip (NoC) architecture for the execution of dataflow dominated embedded systems. The approach has been validated with the design of a color image compression algorithm in an FPGA.

I Introduction

Configurable hardware is constantly being upgraded with higher working frequencies and gate capacity that allow the implementation of faster complex systems in a single chip, making it a competitive solution for embedded systems. This gate capacity leads to a complexity challenge needing new architectures and design methodologies to increase design productivity. An approach to the design of such complex systems is to reuse hardware and software blocks resulting in a number of interconnected IP cores. Gajski et.al. [1] have proposed an IP-centric embedded system design methodology and Vahid et al. [2] have proposed a platform based methodology which not only allows reuse of components but also of system architectures and topologies.

Many architectural templates have been proposed for hardware platforms for future SoCs with a general emphasis on providing efficient and standardized communication infrastructures for connecting multiple resources on the chip, like the Network-on-Chip (NoC) [3], [4]. The NoC has been introduced as a new interconnection paradigm able to integrate a many cores while keeping a high communication bandwidth. The increased computational power and internal communication bandwidth of NoC can provide better timing performances to the embedded applications than the shared medium in current SoC architectures.

Some works have contributed with concepts in the area of networks on chip, like [5]. Among the few implementations of NoC, we are mainly interested on the configurable hardware implementations of [6] and [7]. Marescaux [6] have implemented a bidirectional torus in a Virtex/VirtexII FPGA. The network uses 16 bits data packets, the XY routing algorithm, virtual output buffers and supports up to 320Mbits/s at 40MHz with two virtual channels time multiplexed for QoS support. HERMES [7] is a NoC mesh topology implemented in a VirtexII FPGA. The network supports up to 500 Mbits/s at 25MHz without QoS.

Many co-synthesis tools will be required to develop NoC based architectures. Tools to choose the best platform configuration and to map applications to the target NoC architecture will be essential. There has been a lot of research on co-synthesis for bus-based architectures [8], [9], [10]. NoC researchers can adapt many of the techniques and ideas from these approaches for NoC tool development.

Since NoC is a novel research area only a few mapping and scheduling approaches have been developed. Lei et al. [11] use a genetic algorithm (GA) for task mapping and list-scheduling (LS) for task scheduling. The communication is neither mapped nor scheduled and delay is estimated as the average distance between processors. Shin et al. [12] proposed a methodology with network assignment and link speed allocation for reducing communication energy. They use GA for mapping and network assignment and LS for task scheduling and link assignment. To our knowledge, there is not a methodology for the development of NoC based SoC considering all aspects of the co-synthesis process.

In this paper, we propose a co-synthesis methodology with the integration of allocation, mapping and scheduling steps for the development of SoC based on a parameterizable NoC for the execution of dataflow-dominated applications, like multimedia. The platform supports hardware/software multitasking and includes hardware support for the operating system. Increased productivity is achieved through orthogonalization of communication and computation and design reuse. A real multimedia example has been simulated and implemented on a Virtex II XC2V6000 FPGA.

II. CSoC Architecture

Our CSoC platform consists of an array of tiles interconnected with a NoC. The NoC consists of an array of routers (R), where a router is connected to at most four neighbor routers and to a local IP core. Among the many interconnection topologies, we use a 2D mesh topology because it fits naturally in a 2-dimensional chip (see example in figure 1).

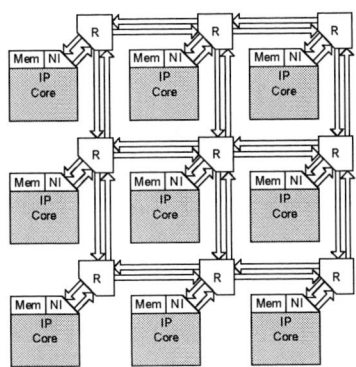

Fig. 1.   SoC architecture

A tile consists of an IP core, local memory and a network interface (NI). An IP core is a piece of configurable hardware or a processor. Each core has direct access to local memory and uses the NoC to exchange data with other cores. The link between a router and a core is established with a NI. The platform connects to the environment using IP cores that implement a particular type of interface.

A NoC can be described by its topology and by the strategies used for routing, flow control, switching, arbitration and buffering. Routing determines how a message chooses a path in this graph, while flow control deals with the allocation of channels and buffers to a message as it traverses this path. Switching is the mechanism that removes data from an input channel of a router and places it on an output channel, while arbitration is responsible for scheduling the access to channels and buffers. Buffering defines the approach used to temporarily store messages.

The communication behavior follows a layered approach similar to the OSI communication architecture (see figure 2).

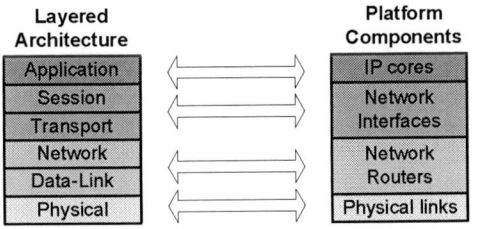

Fig. 2.   Communication architecture of the SoC platform

The application layer includes all tasks implemented by a core that consume/produce data. The session layer includes OS services, namely, memory management and task scheduling. The transport layer manages the identification of task ports for the correct end-to-end delivery of data between tasks on different IP cores and the segmentation of data output into packets and reassembly of packets into input data. The network layer includes services for packet routing. The data-link layer includes protocols for reliable data communication between two routers and between a router and an IP core. Finally, the physical layer models the physical links for transmission of bits.

A. Parameterization

The NoC infrastructure has a set of configurable parameters, including:
1) The size of the 2D mesh topology.
2) The type of IP core of each tile.
3) The data width of point-to-point channels between routers. Supported values are 8, 16 and 32 bits.

III. Router Design

A router forwards packets between IP cores. For each packet received, the router reads the destination address and forwards it to the correct output port. Our router consists of a set of input and output ports with buffering and a set of control blocks for routing, flow control, switching and arbitration (see figure 3).

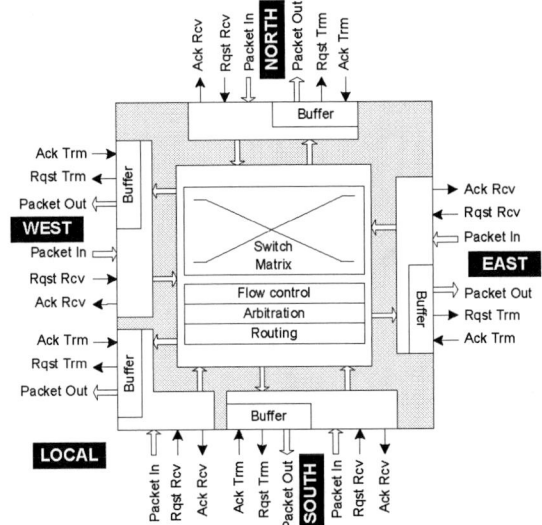

Fig. 3. Architecture of a NoC router

Router ports are used to exchange packets with neighbor routers and with the local IP. A port guarantees the communication reliability through a two-way handshake point-to-point flow control.

The arbitration mechanism uses a round-robin scheme to arbitrate requests from different input ports and grants the output buffer to an input request port. Among the deadlock free routing algorithms for mesh topologies, we implemented the XY algorithm. The XY algorithm routes packets first along the X direction, then along the Y direction until reaching the target. The switching mechanism is based on the store and forward process.

IV. Network Interface Design

The NI consists of an input and an output controller, shared memory ports to connect to the core, a port to connect to the router and an OS memory (see figure 4).

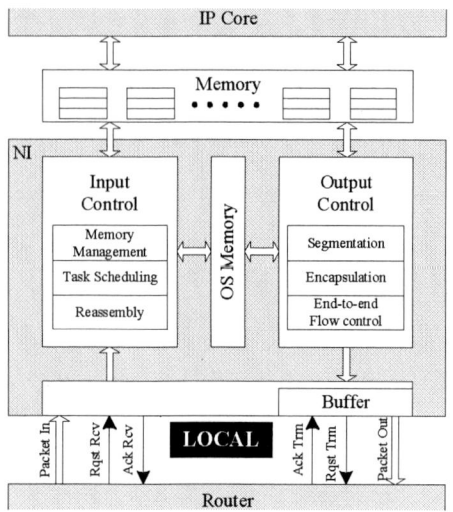

Fig. 4. – Architecture of the network interface

The interface between the NI and the router is identical to the interface between two routers. The interface between the IP core and the NI is implemented with shared memory. The memory may be dual-port RAM or FIFO, depending on the type of the core. For a processor core, the shared memory is always implemented with dual-port RAM. For a hardware core, it may be any of the two kinds of memory. The NI also performs data split and data merge to transmit packets with a data size different than the link size.

The input controller receives data from the router and sends it to the shared memory to an address defined by the port number of the packet. For data transmitted in multiple packets, the controller stores each packet in sequential memory positions until the complete data port is available. When all data of the input ports of a task are available and the corresponding output port is free, the input controller sends a token to the core indicating the task can be executed.

The output controller reads data from the shared memory and sends it in one or more packets. It also implements an end-to-end dataflow control, that is, when the input buffers of a task are available, the output controller sends a token to the producers indicating that they can send more data.

V. Packet Structure

Data to be transmitted is encapsulated in packets at the transmitter and deencapsulated at the receiver. Besides data packets, our NoC implementation uses configuration and token packets. Configuration packets are used to configure the NI and token packets are used for end-to-end flow control.

The highest level of the packet structure is at the application layer where data is produced. If data is too large to fit into one packet it will be divided into several packets. Each packet is encapsulated with its type at the session layer. Next, the transport layer adds the destination port and the network layer adds the destination address.

VI. CSoC Performance Evaluation

We have conducted a simulation of a prototype of the NoC to characterize the following set of parameters (see table 1):

Link latency (LL) - the delay for a packet to move from the output of a router to the output of a neighbor router;

Resource generation latency (RGL) – the delay for a NI to generate a packet;

Resource reception latency (RCL) – the delay for a NI to consume a packet;

Resource to resource latency (R2RL) – the delay for moving a packet from one IP core to a neighbor IP core;

Resource to resource bandwidth (R2RB) - the transmission throughput between IP cores.

TABLE I
NoC Characterization

LL	RGL	RCL	R2RB
1 cycle	4 cycles	5 cycles	$\dfrac{f(frequency)}{5}$ Packets/s

From these parameters, we calculate the communication delay of a packet between two tiles. In a NoC, the communication delay depends on the distance between tiles, the size of transmission data and the network traffic. In the execution of an application, it is possible to have many concurrent transmissions, which conflicts in the use of the communication resources. If tasks have variable execution times, the network traffic is non-deterministic, which makes the analysis more difficult. To simplify, we assume that at any time a link is dedicated to a single data transfer. Hence, the transmission delay of a packet between two tiles separated by NR routers, *EdgeDelay*, is given by:

$$EdgeDelay = \frac{RGL + RCL + LL \times NR}{f(\text{working frequency})} \quad (1)$$

Since the packets are buffered at the routers, the transmission of data requiring more than one packet can be pipelined. In this case, the transmission delay, $EdgeDelay_{pipe}$, of NP packets is given by:

$$EdgeDelay_{pipe} = \frac{1}{R2RB} \times (NR + NP) \quad (2)$$

Besides performance, we have also determined the maximum area occupied by a router and a NI after the synthesis and placement of the components on the target FPGA with the Xilinx ISE 6.2i software (see table 2).

TABLE II
Slice Areas of the Interconnection Components

Block	Size (slices)	BRAM	% XC2V6000
Router (8bits)	189	0	0,56
NI	121	1	0,36

A router can forward five packets per clock cycle at 150MHz. Therefore, a router can forward at most 6Gbps.

VII. CSoC Co-Synthesis

To configure our CSoC architecture for a specific application, we have developed a platform-based co-synthesis methodology. It finds a hardware/software architecture that runs the application with optimized performance and meets the design constraints (see figure 5).

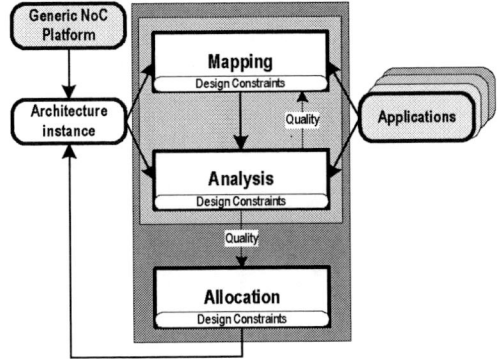

Fig. 5. Co-synthesis flow

It starts with an architecture instance, maps the application onto the architecture and uses the analysis step to determine the quality of the architecture based on cost and performance metrics. The analysis yields quality values that together with design constraints over cost and performance guide the allocation to improve the architecture.

A. Application Model

The applications are modeled with an iterative dataflow graph (IDFG) that can represent iterative behaviors. This model is a directed cyclic graph $G = (V, E)$, where each vertex $v \in V$ represents a task (atomic computation) and each edge $e \in E$ represents intra or inter data dependencies between tasks.

A task vertex has associated its worst and/or average case execution time, data, and program memory size or hardware area in each of the available cores. An edge has an associated vector (v,d), where v is the data size to be transferred and d is the inter delay between two connected tasks.

B. Allocation

The allocation step determines the most appropriate SoC architecture from the generic CSoC platform for the execution of the application. It determines the size of the NoC topology and the type of core associated with each tile. This is a hard problem. Therefore, we have used an heuristic (see figure 6).

From the generic CSoC platform, the algorithm instantiates an initial set of IP cores. From this set, it generates an initial architecture. To generate an architecture from a set of cores, the algorithm reads sequentially the set of tiles with its IP core association (including the NI) and fills the FPGA (including the routers) until it is full or all tiles are associated.

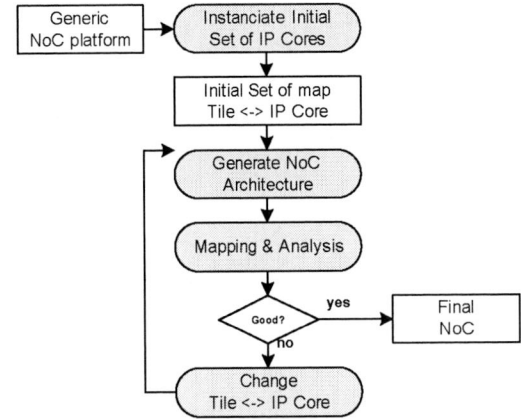

Fig. 6. Allocation algorithm

Next, it maps the application and finds the quality of the architecture with the analysis tool. If the quality is considered acceptable, the algorithm stops. Otherwise, it changes the IP core of a tile and restarts the evaluation flow. The iterative process of the algorithm is controlled with a simulated annealing (SA) algorithm [13] as follows:

Move generation function: generates a new architecture instance from the previous by changing the IP core of a tile.

Cooling schedule: the cooling schedule includes the initial temperature, t0, the decrement rule for the temperature, the stop criterion and the length of the Markov chain. t0 is obtained by incrementing the temperature until the percentage of accepted transitions is higher then 70%. The decrement rule is given by $tk = t0 \times 0.95k$. The algorithm stops when three consecutive Markov chains end with the same value. The length of the Markov chain is equal to the size of the neighborhood.

Cost function: obtained with the mapping and analysis steps.

C. Mapping

Mapping consists on assigning each application object (task, data transfer and variable) onto an architectural element (IP core, link and memory, respectively) in order to maximize the quality of the architecture while satisfying design constraints. Even for small instances, the mapping problem has exponential complexity so that heuristics must be used.

Our mapping approach uses SA to improve an initial solution found with LS while exploring the advantages of both pipelining and unrolling to increase the throughput. Pipelining allows tasks belonging to different iterations to be executed at the same time and unrolling increases the number of tasks within an iteration to explore more parallelism. The NoC structure can easily implement pipelining since the output of a task is buffered and the flow of data is easily controlled by the end-to-end flow control of the NI.

The algorithm starts with the IDFG of the application and a NoC architecture and iteratively runs a mapping design space exploration step with different unrolling values (U) (see figure 7).

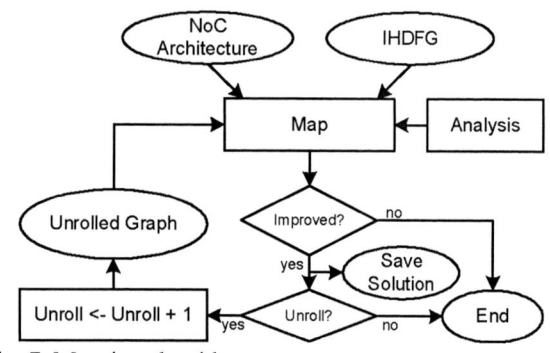

Fig. 7. Mapping algorithm

The iterative process is controlled with SA using the same cooling schedule of the allocation process. Each mapping solution is evaluated with the analysis tool.

D. Analysis

The analysis finds the quality of an architecture based on its performance, cost and memory requirements.

For performance evaluation, it uses LS to order the execution of tasks assigned to a single core to optimize the throughput of the graph, $C_{throughput}$. To schedule tasks on software processors, the algorithm assumes the processor executes one task at a time and the program code of the task is in local memory before starting its execution.

The memory requirements of a CSoC architecture depends on the local memory requirements of each IP core. The local memory of a core is used to store the program instructions of tasks (for a software core), the OS data of the NI and the tasks data. We assume the local memory is implemented with BRAM and the instruction, the OS and data memories use independent BRAM.

To calculate the memory utilization of a core, the analysis uses a table with the instruction memory size of each task on each software core and the size of input and output ports of each task. The analysis process determines the number of local BRAMs, BR, necessary to implement core k as follows:

$$BR(k) = \frac{\sum_{i \in \{coreTasks\}} instrMem(i)}{BRAMSize} + \frac{2 \times \sum_{i \in \{coreTasks\}} \sum_{j \in \{taskPorts\}} portSize(i,j)}{BRAMSize} + 1 \quad (3)$$

where $instrMem(i)$ is the code memory size of task i, $portSize(i, j)$ is the data size of port j of task i. The total number of BRAM used, C_{memory}, is given by:

$$C_{memory} = \sum_{i \in \{setofcores\}} BR(i) \quad (4)$$

The final quality of the architecture is given by $C_t + C_m$ where

$$Cx = \begin{cases} K \times \dfrac{Cmetric(P)}{\overline{Cmetric}}, & \text{w/o constraint} \\ Ka \times \left| \dfrac{(Cmetric(P) - Cmetric\,Constra\,int)}{CmetricCon\,stra\,int} \right|, & \text{w/ constraint} \end{cases} \quad (5)$$

where K and Ka are weighting factors specified by the user

with typical values of 0.5 and 100, respectively. $Cmetric(P)$ is the metric value (throughput or memory) of a solution P and $C_{metricConstraint}$ is a metric (throughput or memory) constraint. For a non-constrained metric, we use the average value ($\overline{Cmetric}$) calculated from the values of the metric in some (<20) previous solutions.

VIII. Design Evaluation

This section describes the design of a JPEG encoder with the proposed CSoC environment. A simulation has been executed and a prototype is under development based on the Xilinx VirtexII XC2V6000.

A. JPEG Encoder

The standard JPEG compression with a block size of 8×8 pixels for color images was implemented. The compression method used is based on the DCT (see IDFG of the JPEG encoder in figure 8).

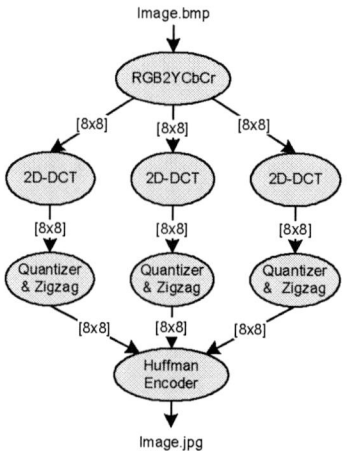

Fig. 8. IDFG of the JPEG encoder

The tasks of the JPEG where characterized considering a hardware implementation (see table 3).

TABLE III
JPEG Hardware Task Characterization

Task	Size (slices)	BRAM	Latency
RGB2YCbCr	204	0	64
2D-DCT	1612	1	168
Quantizer	312	1	64
Huffman	176	1	192

For this application, the co-synthesis tool took less than 5 minutes to found a hardware only solution that can process two blocks of [8×8]×24 bits in 3.8 µs (cores frequency = 100MHz, NoC frequency = 150MHz with 16 bit packets), which is equivalent to a processing capacity of 800 Mbps. With this throughput, we can process color images at the processing times of table 4.

TABLE IV
Execution Time of Hardware Solutions

Image size	HW solution (NoC) seconds (fps)	Pentium 4 at 1.7 GHz	HW solution (bus) seconds
640×480	0.009 (108)	0.046	0,055
800×600	0.015 (67)	0.071	0,086
1024×768	0.024 (42)	0.110	0,14

The JPEG solution was placed & routed and simulated on the FPGA with the Xilinx ISE 6.2i software (see figure 9).

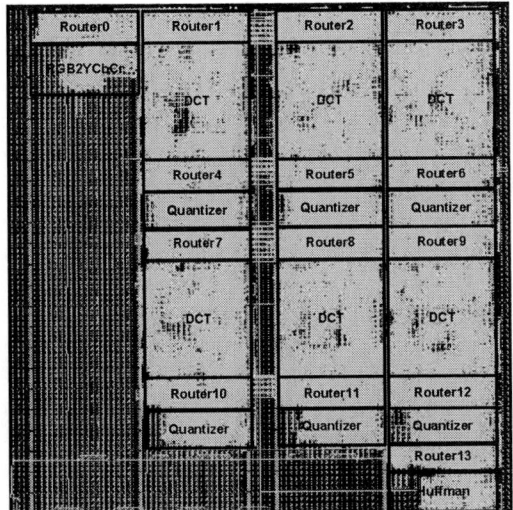

Fig. 9. FPGA implementation of the JPEG case study

In the figure, we have outlined all resources used in the implementation (13877 out of 33792 slices, 45%). We can see the co-synthesis tool have unrolled the graph one time, almost doubling the processing rate. The design was easily routed because of its regularity and the mapping constraints over the IP cores, the routers and block RAM.

From this and other results, we conclude the following:

- Designs can be quickly and easily designed without bus design complications. The same design with a single bus could not achieve the same throughput.
- For certain cores (Huffman), a router uses more slices. This area overhead may become a serious bottleneck for NoC architectures. Other NoC topologies with shared routers and local memory are being considered.
- Many NoC parameters can and should be explored with the cosynthesis tool, including buffer size, switching capacity, routing algorithm and arbitration policy. Adjusting these parameters to specific applications means smaller size routers and consequently less overhead.

IX. Summary and Conclusions

The well-structured design of the CSoC platform and the acceptable computation times of the co-synthesis tool allow the rapid development of SoC architectures.

Our approach has been used to design a JPEG application with a throughput of 800 Mbps. The results are very promising since we where capable of easily integrate several IP cores in a single chip and obtain high quality solutions.

Future research includes developing a more flexible CSoC with different network topologies and including a generic parameterizable router as part of the design space exploration in order to improve the area, performance and energy dissipation of the final SoC architecture.

Acknowledgments

The authors thank the support granted by INESC-ID.

References

[1] D. Gajski, R. Dömer and J. Zhu, "IP-Centric Methodology and Design with the SpecC Language", in *System Level Design*, Nato Science Series 357, 1999.

[2] F. Vahid and T. Givargis, "Platform Tuning for Embedded Systems Design", in *IEEE Computer,* 34, 3.

[3] W. Dally and B. Towles, "Route Packets, Not Wires: On-Chip Interconnection Networks", in *Proc. of DAC*, 2001.

[4] A. Hemani, A. Jantsch, S. Kumar, A. Postula, J. Oberg, M. Millberg and D. Lindqvist, "Network on Chip: An Architecture for Billion Transistor Era", in *Proceedings of the IEEE NorChip Conference*, Nov. 2000.

[5] L. Benini and G. de Micheli, "Networks on Chips: a New SoC Paradigm", in *Computer*, v.35(1), Jan. 2002, pp.70-78.

[6] T. Marescaux, A. Bartic, D. Verkest, S. Vernalde, R. Lauwereins, "Interconnection Networks Enable Fine-Grain Dyanmic Multi-Tasking on FPGAs", in *Field-Programmable Logic and Applications*, 2002, pp. 795-805.

[7] F. Moraes, N. Calazans, A. Mello, L. Möller, L. Ost "HERMES: an Infrastructure for Low Area Overhead Packet-Switching Networks on Chip", in *Integration*, the VLSI Journal 38, 2004, pp. 69-93.

[8] R. Dick and N. Jha, "CORDS: Hardware-Software Co-Synthesis of Reconfigurable Real-Time Distributed Embedded Systems", in *Proc. of ICCAD*, pp. 62-68, 1998.

[9] R. Szymanek and K. Kuchcinski, "Design Space Exploration in System Level Synthesis under Memory Constraints", in *Proceedings EuroMicro*, pp. 8-10, 1999.

[10] U. Shenoy, et al., "A System-Level Algorithm with Guaranteed Solution Quality", in *Proceedings of DATE*, pp. 417-422, 2000.

[11] T. Lei and S. Kumar, "Algorithms and Tools for NoC Based System Design", in *Proceedings of SBCCI*, 2003.

[12] D. Shin and J. Kim, "Power-Aware Communication Optimization for Networks-on-Chips with a Voltage Scalable links", in *Proceedings of CODES+ISSS*, pp. 170-175, 2004.

[13] S. Kirkpatrick, C. Gelatt, and M. Vecchi, "Optimization by Simulated Annealing", in Science, 220(4598): pp. 671-680, May 1983.

Customized SIMD Unit Synthesis for System on Programmable Chip – A Foundation for HW/SW Partitioning with Vectorization

Muhammad Omer Cheema
UEI, ENSTA
Paris, 75739
Tel : 33(0)1 45 52 54 60
Fax : 33(0)1 45 52 83 27
e-mail : cheema@ensta.fr

Omar Hammami
UEI, ENSTA
Paris, 75739
Tel : 33(0)1 45 52 54 60
Fax : 33(0)1 45 52 83 27
e-mail :hammami@ensta.fr

Abstract— **Use of Single Instruction Multiple Data (SIMD) functional units enables multimedia systems to exploit parallelism to a higher degree resulting in significant system performance improvements. While implementation of whole SIMD system functionality for an application results in wastage of area resources, we have observed that for a specific multimedia application, we only need to implement a customized SIMD unit that is a subset of whole SIMD standard implementation. Based on this study, we have proposed an extension to the traditional system design and synthesis flow by integrating a methodology of SIMD unit Synthesis. Our system synthesizes a customized SIMD unit along with an extended instruction set and generates an equivalent version of assembly code for the application using the extended instruction set. The results of area and performance obtained by experimenting over our implementation of AltiVec compatible customized SIMD units show the effectiveness of our approach.**

Index Terms— **SIMD Synthesis, HW/SW Codesign, AltiVec Architecture, Vectorization**

I. INTRODUCTION

Multimedia standards such as MPEG-1, MPEG-2, MPEG-4, MPEG-7, JPEG2000, and H.263 put challenges on both hardware architectures and software algorithms for executing different multimedia processing jobs in real-time. To meet the high computational requirements of emerging media applications, current systems use a combination of general-purpose processors accelerated with DSP (or media) processors and ASICs performing specialized computations. However, benefits offered by general-purpose processors in terms of ease of programming, higher performance growth, easier upgrade paths between generations, and cost considerations argue for increasing use of general purpose processors for media processing applications [1]. The most visible evidence of this trend has been the SIMD-style media instruction-set architecture (ISA) extensions announced for most high-performance general purpose processors (e.g., AMD's 3DNow! [2], Motorola's AltiVec [3], Intel's SSE1/SSE2 [4], Sun's VIS, HP's MAX, Compaq's MVI and MIP's MDMX).

Research work done over the study of area constraints of SIMD shows that implementation of whole SIMD units is very expensive in terms of area and energy requirements [5],[6]. On the other hand, research also indicates that multimedia applications don't use all the components of an SIMD unit and hence implementation of many parts of SIMD units can be avoided to save area and energy without affecting the speed. As a result, synthesis of customized SIMD units to optimize the system resources is suggested. Synthesis of customized SIMD units is somehow equivalent to an Application Specific Instruction Set Processor (ASIP) synthesis problem and recently, an increasing interest in this direction has been observed [7],[8],[9][10],[11],[12]. While ASIP synthesis can be considered a generalized SIMD synthesis problem, a very few methodologies to synthesize SIMD units in particular have been proposed [13],[14]. [14] uses the optimization of Control Data Flow Graphs (CDFG) of the application code for extraction of SIMD instructions. A drawback of this approach is that SIMD pattern recognition through CDFG doesn't completely exploit the possible parallelism of a program hence speedup because of SIMD usage remains very limited. [13] uses step by step SIMD instruction decomposition for a manually vectorized program to get an area efficient processor core, but it doesn't take into account the standard SIMD based systems with complex architectures hence it doesn't represent a real world scenario of a DSP application using SIMD instructions. That's why we have implemented standard AltiVec unit being used as a coprocessor with a PowerPC 405 processor to keep in mind the practical aspects of SIMD while proving the concept behind our work.

There are also some commercial tools available for synthesis of extensible processors. Commercial examples of extensible processors include HP Laboratory and STMicroelectronics' Lx [15], Altera's NIOS [16] and Tensilica's Xtensa [17]. In Altera's NIOS architecture, extensible instruction set is obtained by introducing the instructions in the already existing pipeline of the processor which increases the critical path length of the processor. Just like Xtensa, our methodology emphasizes on the use of coprocessor that extends the existing instruction set with one more advantage that we are using well known Instruction Set Architecture (ISA) based on PowerPC architecture.

0-7803-9451-8/06/$20.00 ©2006 IEEE.

Based on the above discussion, this paper presents a methodology for the application specific synthesis of SIMD units for digital signal processing applications. Given an application program written in C or Assembly language and a set of application data, our methodology synthesizes an RTL description of an SIMD based coprocessor and the extended instruction set along with the modified assembly language program capable of running over synthesized system. As a case study, we experimented over PowerPC architecture based AltiVec [3] units and the results obtained indicate the effectiveness of our methodology. These results testify to the high potential of the SIMD computation paradigm in the synthesis of high performance and low-power application specific hardware architectures.

Rest of the paper is organized as follows: Section II gives an introduction to PowerPC/AltiVec and presents benchmarking results of multimedia applications outlining the motivation behind our work. Section III explains the System Synthesis methodology. Section IV and V describe the experiment environment and results. Section VI and VII present conclusions and future work.

II. STUDY OF UTILIZATION OF ALTIVEC UNITS

AltiVec is a floating point and integer SIMD instruction set designed and owned by Apple Computer, IBM and Motorola (the AIM alliance), and implemented on versions of the PowerPC including Motorola's G4 and IBM's G5 processors. AltiVec is a trademark owned solely by Motorola, so the system is also referred to as Velocity Engine by Apple [18] and VMX by IBM. Fig. 1 explains the various components of an MPC 7400 system that consists of a G4 processor having an AltiVec extension.

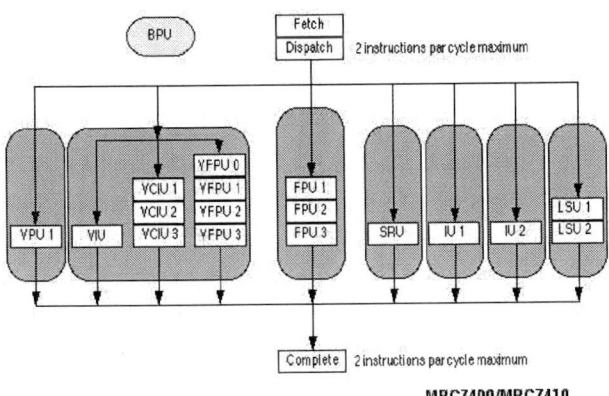

MPC7400/MPC7410

Fig.1 PowerPC G4 Architecture

Vector Permute Unit (VPU) Vector Integer Unit (VIU), Vector Complex Integer Unit (VCIU) and (Vector Floating Point Unit) VFPU are part of the AltiVec unit. Vector permute unit arranges the data to make it usable by vector instructions. VIU, VCIU and VFPU are used to execute vector integer, vector complex integer and vector floating point instructions. Other components shown in Fig.1 are part of general-purpose PowerPC and are used to execute non-SIMD instructions.

Our first experiment was to measure the utilization of these vector units for certain applications. For that we developed a few multimedia test benches for AltiVec enabled G4 system. We used profiling and simulation tools like Shark, Amber, MONster [19] and SimG4 [20] to calculate the usage of various components in the system. We used different versions of filters used in image processing keeping in mind that each of the filters had a different level of vectorization so that it can reflect realistic results on AltiVec unit usage during the application execution.

For an image of size 320x240, when applied to various versions of filter program having different vectorization level, results in Table 1 were obtained. Looking at the Table 1, we can see that even if the branch prediction and cache performances remained in reasonable limits, usage of most of the AltiVec components was very poor. As a matter of fact, for our application that dealt with integers parts only, floating point unit was never used. Looking at the statistics, we can claim that most of the SIMD resources are underutilized (or un-utilized in some cases).

TABLE 1
Statistics of G4 Components Usage for Multimedia Applications

	Filter v1	Filter v2	Filter v3	Filter v4
Instruction/Cycle	0.8615	0.8483	0.6703	0.8465
FXU1 Idle Time	53.28%	58.44%	45.46%	54.09%
FXU2 Idle Time	76.36%	70.41%	64.18%	75.89%
FPU Idle Time	100 %	100 %	100 %	100 %
VAUS IdleTime	99.27%	99.32%	100	99.25%
VAUC Idle Time	93.90%	93.23%	92.83%	93.77%
VAUF Idle Time	100 %	100 %	100 %	100 %
VPU Idle Time	100 %	91.87%	100 %	90.76%
SYS Idle Time	91.90%	92.52%	97.98%	91.74%
LSU Idle Time	56.01%	61.16%	49.73%	67.42%
DL1 Hit Rate	98.52%	98.72%	97.18%	98.36%
IL1 Hit rate	99.82%	99.84%	99.54%	99.82%
Branch Predict.	93.45%	93.45%	94.91%	93.45%

One important thing to note is that researchers in [21] also got the similar results and concluded that in the dynamic instruction stream of media workloads, 85% of the instructions are not performing computation but are load/stores, loop/branches and address generation instructions. They observed an SIMD efficiency ranging only from 1 to 12%.

Using the GCC 4.0.0 [22] and VAST [23] vectorizing tools, we vectorized various benchmarks and obtained even worse results in terms of AltiVec unit utilization due to the facts that vectorizing capabilities of existing tools are limited and also that even if an application is well vectorized, most of the components in SIMD remain underutilized as was observed in Table 1.

Based on above observation, we conclude that it is preferable to include only those components of SIMD in application specific embedded systems that are not underutilized or un-utilized to have a better area and energy consumption of a system: hence the basic motivation behind our work.

III. ADAPTIVE GENERATION OF SIMD UNITS

Our SIMD synthesis flow consists of following sub tasks:
a) Vectorization
b) Static and Dynamic Profiling
c) SIMD AltiVec module Generation
d) Real Time Execution of the Application
e) Repetition of above steps until a set of possible solutions is obtained. Best solution matching the system requirements is chosen.

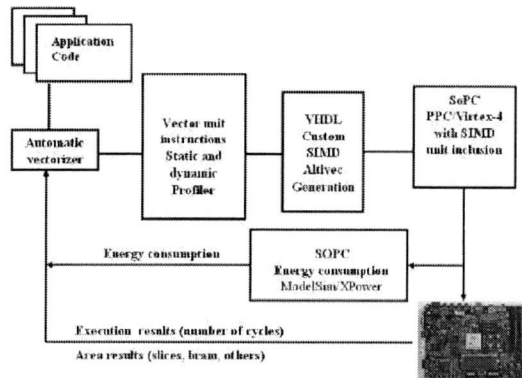

Fig. 2 System Design Flow

a. Vectorization

As mentioned in first section, input to the system is an application written in C or Assembly language. First step in the synthesis process is to vectorize the application. During this phase, vectorizing compiler detects the possible vectorizing options. We have defined equivalence classes of AltiVec and PowerPC instructions. Equivalence classes represent the possible replacement of an instruction with a set of instructions performing the same function. As a result, use of some instructions can be avoided which makes it possible for us to choose alternative components of AltiVec or PowerPC to test the behavior of the system for a program modified using an equivalent class. As a very simple example of the concept of equivalence classes,

let's say that we have a *vadduwm* instruction that adds a vector of four elements having 32-bit size each. RTL description of the SIMD unit is implemented in a module that handles unsigned addition for byte, half word and word elements. Let's suppose that we want to generate a program version that doesn't contain *vadduwm* instruction. An obvious reason for such decision can be that the *vadduwm* is executed only a few times during the whole execution of the program while module inserted inside the system due to its inclusion adds significant amount of energy and area requirements. So we can use the concept of equivalence classes and replace this *vadduwm* instruction with four PowerPC add instructions used for addition of 32 bit unsigned elements to generate such a version. This example mentions the replacement of a vector instruction with its equivalent PowerPC instructions. There are some cases where it seems more beneficial to use another vector instruction to replace a vector instruction (i.e. multiply accumulate operation with two different operations of multiply and then accumulate for vectors). This concept of equivalence classes, when introduced in a vectorizing compiler gives a very large system design space depending on the set of vector instructions being used and their replacement methodology. Ideally, to get an optimal solution, an automatic system design exploration algorithm can be applied. Or alternatively, system can be manually tested for various vectorized versions of the program and the system configuration matching the area and speed constraints can be chosen as is done in this article. Energy based optimization has not been performed in this article although it remains an optional part of the design flow and we are in the process of developing a methodology to have good energy consumption estimates.

b. Static Analysis and Profiling Results

In this phase, we analyze the application and study the various aspects related to instruction set used and its usage. During this process, frequency, timing and repetition patterns of instructions are studied. This helps the system designer to capture the properties of the system and to exploit the inherent parallelism in various ways. Information obtained during this step is also helpful in automatic generation of customized AltiVec module.

c. AltiVec Module Generation

During this phase, the system automatically generates the VHDL description of a suitable AltiVec module that consists of only those components which are needed to execute the specific version of the program generated in step A. The modules not going to be used by program are ignored and kept out of the hardware synthesis process. System is ready to be executed in real time at the end of this step.

d. *Real time execution over Virtex- 4 FPGA*

In this phase, application is run over an FPGA on which customized SIMD unit is synthesized. We preferred actual execution of the application over FPGA instead of simulation to avoid the accuracy limitations of the simulation and to prove our idea in a concrete manner. Execution also speeds up the process as simulation of applications has proven to be very slow in many cases. Results of area and energy consumption and number of cycles taken by application are obtained at the end of this phase.

All of the above steps are repeated several times and results for area, energy and speed are obtained for corresponding configurations. Based on the results obtained and the system requirement, suitable SIMD system and corresponding extended instruction set is chosen for the application.

IV. EXPERIMENTAL SETUP

In this section, we briefly explain the experimental setup for the hardware environment. We have used Xilinx Virtex-4 FX platform devices to execute the application in real time and get the execution results. Virtex-4 consists of a PowerPC 405 processor: a 32-bit IBM RISC processor at its core along with various peripheral component interfaces. Virtex-4 FPGA is the newest of Virtex FPGA [24] series and is the first FPGA that provides an option to connect a coprocessor with PowerPC processors with the help of an APU (Auxiliary Processor Unit). And this feature was the major reason for us to choose Virtex-4 to perform the experiments.

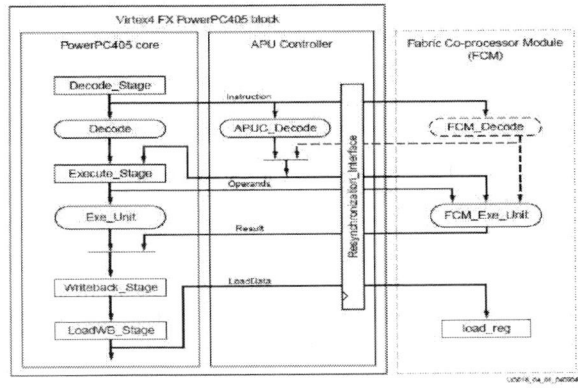

Pipeline Flow Diagram

Fig.3 PowerPC with APU Interface

As shown in the Fig. 3, Virtex-4 APU allows a designer to extend the native PowerPC 405 instruction set with custom instructions for execution by an FPGA Fabric Coprocessor Module (FCM). An APU-enhanced system enables tighter integration between an application-specific function and the processor pipeline, making the APU implementation superior to, for example, a bus peripheral. When an instruction arrives, the processor and the APU decode it simultaneously. If the instruction is meant for the APU and the FCM, the APU relays it to the FCM.

The Embedded Development Kit (EDK) is a widely used tool to program Xilinx FPGAs. EDK 7.1 is the latest version and the only way to develop the Virtex-4 FPGA based APU enabled systems. EDK includes the IPs of Processor Local Bus (PLB), On-Chip Peripheral Bus (OPB), Block RAM (BRAM) controllers that were reused in our system design. (Integrated Software Environment) ISE 7.1 is used to synthesize the system and get the area requirements of the system.

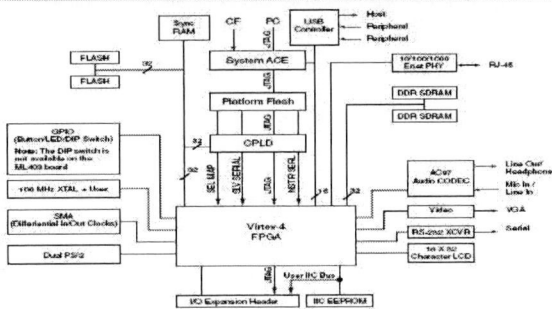

Fig. 4   Xilinx ML403 FPGA Resources

All the experiments have been performed over Xilinx ML403 [25] board that allows designers to investigate and experiment with features of the Virtex™-4 family of FPGA.

V. EVALUATION RESULTS

We tested our methodology over two sets of applications. Smaller application using lesser number of vector instructions was a matrix transpose application. It consisted of only five vector instructions being used including *lvx* and *stvx*. For a larger application, we developed a set of image processing filters which used several vector instructions.

Fig. 5 represents the area taken by various components of AltiVec on a Virtex-4 FPGA. Some components like Vector Permute Units and the modules related to shift instructions take as much as one thousand slices, which is more than 20 % of total ML403 area, while most of the modules are less expensive in terms of area requirements. An obvious reason for this fact is that the shift capabilities in AltiVec instructions are more than that of a "barrel shifter" since every block of the vector can be shifted by a different value. For standard implementation of the VPU, whole "cross bar" functionality has to be implemented to keep it compatible to standards resulting in adding a lot of RTL logic. Area might have been smaller for shift instructions if same shift value

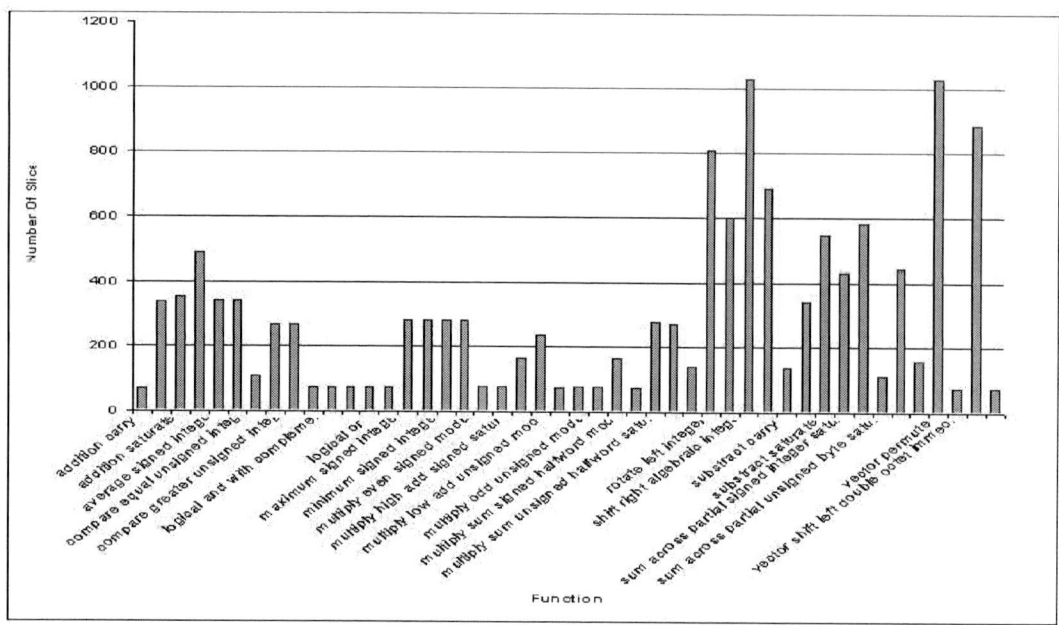

Fig. 5 Area of AltiVec Modules in Virtex-4

was used for every data component in the instruction. Similarly, the instruction with saturation takes up more area because of additional logic for implementation of saturation functionality.

Repeating the methodology mentioned in previous sections, results of area and energy consumption obtained by system synthesis and real time execution of matrix transpose application are summarized in Table 2. Results show that a speed up of up to 5.2 can be obtained with an area cost of 89% of FPGA total area. Configuration 5 is using scalar only code while other configurations use one or more vector instructions. Configuration 1 is using all possible vector instructions in the program resulting in maximum area and maximum speed up.

TABLE 2
Area vs. Speedup for Matrix Transpose Program

Config. No.	FPGA Area	Time (cycle)	Speedup over Non-SIMD Code
Config. 1	89	3 171 944	5.2
Config. 2	83	5 275 383	3.1
Config. 3	75	6 357 824	2.6
Config. 4	70	7 188 232	2.3
Config. 5	28	16 534 108	0

Fig. 6 graphically represents the above table. As expected, in all the configurations, area and execution time tradeoff is clearly visible.

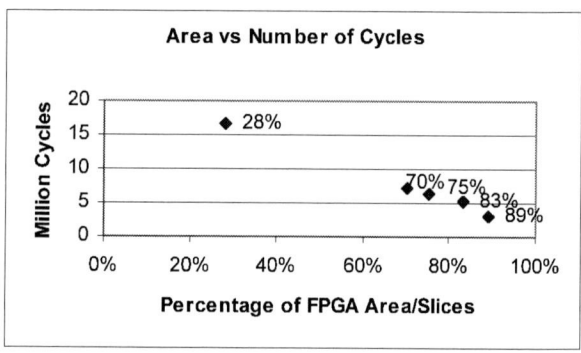

Fig. 6 Area vs. Speedup Tradeoff for Matrix Transpose Program

Similarly, various AltiVec configurations of a filter automatically generated by our customized AltiVec generation tool depending on extended instruction set being used and corresponding area and speedup results are shown in Table 3. An image of size 500x500 was used as data input for the results in Table 3. It is important to note at this point that implementation of vector register bank and vector permute unit took more than 40 percent of the area available over ML 403 board because of the reasons mentioned in the beginning of this section. Rest of the area utilization was dependent on the vectorizing compiler's decision to select/reject certain instructions in a specific SIMD configuration.

TABLE 3
Area vs. Speedup for Average Filters

Config. No.	FPGA Area	Time (cycle)	Speedup over Scalar Code
Config. 1	84%	29 812 080	2
Config. 2	86%	33 087 428	1.8
Config. 3	89%	23 853 953	2.8
Config. 4	89%	34 237 811	1.8
Config. 5	92%	12 388 586	4.9
Config. 6	78%	35 770 207	1.7
Config. 7	28%	60 810 431	0

Fig. 7 shows a tradeoff between area and speedup for the given filter application. We observe that various solutions based on the system requirements are possible. For example, if focus is on the execution speed, configuration 5 is the required solution. If area is to be minimized, among the SIMD based solutions, configuration 6 is the best option. Other configurations represent the tradeoff between these two extremes.

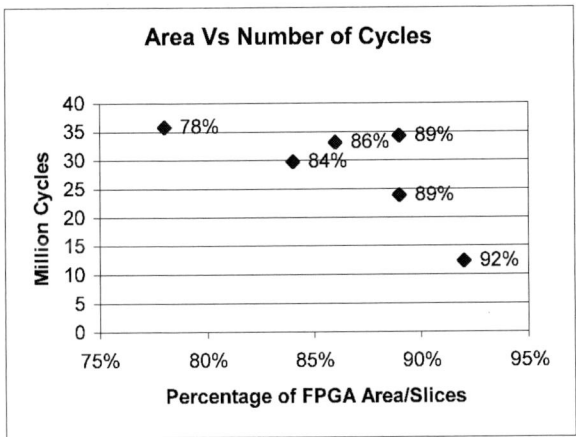

Fig. 7 Area vs. Speedup Tradeoff for Average Filters

To test the impact of system performance for different image sizes, one configuration was chosen and images of various sizes were applied to the application. Results obtained are summarized in the Fig. 8. The results show that optimal solution obtained for one image size might not be optimal for other image sizes and speed up can be lesser if images of smaller sizes are used. In ideal case, for each data size/type, system should be synthesized again to get an optimized solution.

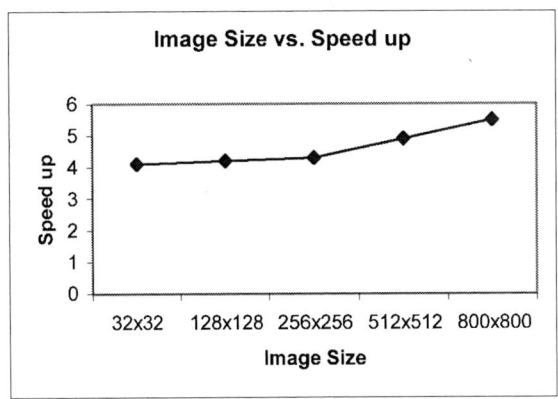

Fig. 8 Image Size vs. Speedup Tradeoff for Average Filters

Needless to say that, generally more speedup has been observed for large data sizes showing the suitability for SIMD for large data applications.

VI. DISCUSSION AND EXTENSIONS

In a broader view, this paper lays out the foundation for a HW/SW partitioning scheme, which includes vectorization. The target platform of such a scheme is described in Fig. 9. This single processor platform is composed of: (1) an IBM PPC 405 processor connected to an IBM CoreConnect infrastructure (2) peripherals (3) a custom Altivec compatible SIMD unit (4) hardware accelerators: this whole platform being the result of a HW/SW partitioning scheme. In such a platform HW/SW partitioning scheme needs to add a new dimension in the design space exploration with the inclusion of vectorization resulting in SIMD units in the system.

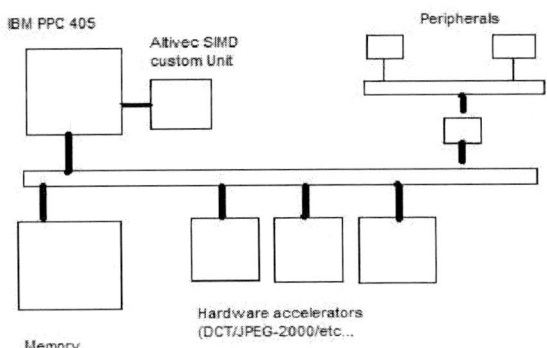

Fig. 9 Target platform with vectorization

This flow accepts as input a C application from which a call graph is extracted and profiled. Compute intensive functions having SIMD like patterns are vectorized. A design space exploration algorithm similar to [26] is applied which partitions each function in either: (1) software without vectorization (2) vectorized software with the custom associated SIMD unit (3) or pure hardware as a hardware accelerator. The resulting configuration is generated, synthesized on an FPGA platform and executed. The actual number of cycles of the execution as well as area values resulting from the synthesis/place and route step are fed back to the DSE engine for a new exploration until the constraints are met.

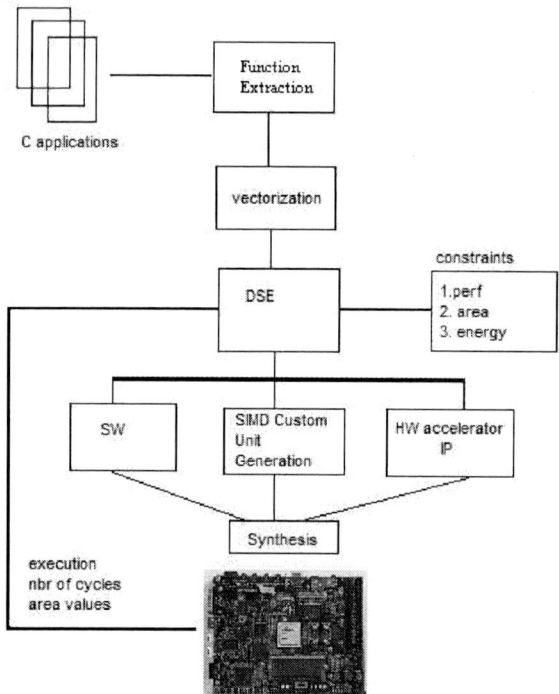

Fig. 10 HW/SW/SIMD Partitioning Flow

VII. CONCLUSIONS

In this paper, we have proposed a methodology for the synthesis of customized SIMD units. Concept of equivalence class between/among SIMD and general-purpose processor instructions has been introduced to create a system design space for synthesis of customized SIMD units. As a case study, we have used AltiVec based SIMD unit along with PowerPC405 and generated a suitable architecture for a specific image processing application along with the extended instruction set and a modified application that can execute itself over the synthesized hardware. Results of area and application execution time show that significant efficiency improvement is achieved through the use of our SIMD based synthesis methodology.

REFERENCES

[1] K. Diefendorff and P. K. Dubey. How multimedia workloads will change processor Ddesign. In IEEE Micro, pages 43–45, Sep 1997.

[2] S. Oberman et al, "AMD 3DNow! Technology and the K6-2 microprocessor", In HOTCHIPS10, 1998.

[3] M. Phillip et al,"AltiVec technology: Accelerating media processing across the spectrum", In HOTCHIPS10, Aug 1998.

[4] S. K. Raman, V. Pentkovski, J. Keshava," Implementing streaming SIMD extensions on the Pentium III processor". In IEEE Micro, volume 20(4), pages 47–57, July-August 2000

[5] Schmookler M, Putrino M, Roth C, Sharma M, Mather A, Tyler J, Nguyen H.V, Pham M.N, Lent J "A low-power, high-speed implementation of a PowerPC microprocessor vector extension",14th IEEE Symposium on Computer Arithmetic, p.12, 1999

[6] Linlay Gwennap, "AltiVec vectorizes PowerPC forthcoming multimedia extensions improve on MMX " Microprocessor report, Volume 12, No. 6, May 11, 1998

[7] T. M. Conte et al., "Challenges to combining general-purpose and multimedia processors.", In IEEE Computer, pages 33–37, Dec 1997.

[8] D. Goodwin and D. Petkov, "Automatic generation of application specific processors", In Proc. of the 2003 International Conference on Compilers, Architecture, and Synthesis for Embedded Systems, pages 137-147, 2003.

[9] Slingerland N, Smith A.J, "Measuring the performance of multimedia instruction sets", IEEE Transactions on Computers, Vol. 51, No. 11, November 2002 page 1317-1332

[10] P. Brisk, A. Kaplan, M. Sarrafzadeh, "Area-efficient instruction set synthesis for reconfigurable system-on-chip designs" Proceedings of Design Automation Conference, 41st Conference on (DAC'04) pp. 395-400,2004

[11] Atasu, K., Pozzi, L., and Ienne, "Automatic application-specific instruction set extensions under microarchitectural constraints.", Design Automation Conf. (DAC), 2003.

[12] P. Ranganatha, S. Adve,N. P. Jouppi , "Performance of image and video processing with general-purpose processors and media ISA extensions", Proceeding of 26th International Symposium on Computer Architecture, pp-124-135, 1999

[13] N. Togawa, M. Yanagisawa, T. Ohtsuki, "A Hardware/Software cosynthesis system for digital signal processor cores" IEICE Trans. Fundamentals, Vol E83-A, NO.11, November 1999

[14] V. Raghunathan, A. Raghunathan, M. B. Srivastave, M. D. Ercegovac, "High level synthesis with SIMD units", Proceedings of the 15th International Conference on VLSI Design (VLSID.02), 2002

[15] P. Faraboschi et al, "Lx: a technology platform for customizable VLIW embedded processing", In ISCA, 2000.

[16] Altera. Nios embedded processor system development. http://www.altera.com/products/ip/processors/nios/nio-index.html.

[17] R. E. Gonzalez, "Xtensa: A configurable and extensible processor." IEEE Micro, 20(2), 2000.

[18] Velocity Engine: http://developer.apple.com/hardware/ve/index.html

[19] CHUD Tools(Amber, Shark, MONster): http://developer.apple.com/tools/performance/overview.html

[20] SIMG4:http://developer.apple.com/Developer/Documentation/CHUD/SimG4_Users_Guide.pdf

[21] Talla D, John L, Burger D "Bottlenecks in multimedia processing with SIMD style extensions and architectural enhancement" IEEE Transactions on Computers, VOL. 52, NO. 8, AUGUST 2003 page 1015 – 1031

[22] Dorit Nicholas, "Autovectorization in GCC", GCC Developers' Summit, pp 105-117, 2004

[23] VAST Code Optimizer, Available: http://www.crescentbaysoftware.com/vast_altivec.html

[24] Virtex-4 FPGA Handbook August 2004

[25] ML 40x Evaluation Platform User Guide, UG080 (v2.0) P/N 0402337 February 28, 2005, 2004-2005 Xilinx, Inc.

[26] K. Ghali, O Hammami, "Multiobjective design of embedded processors on FPGA platform", ICDCS 2004

Robust Analytical Gate Delay Modeling for Low Voltage Circuits

Anand Ramalingam*, Sreekumar V. Kodakara[†], Anirudh Devgan[‡], and David Z. Pan*

* Department of Electrical and Computer Engineering, The University of Texas, Austin, TX 78712
[†] Department of Electrical and Computer Engineering, The University of Minnesota, Minneapolis, MN 55455
[‡] Magma Design Automation, Austin, TX 78759

{anandram,dpan}@cerc.utexas.edu, sreek@ece.umn.edu, and devgan@magma-DA.com

Abstract—Sakurai-Newton (SN) delay metric [1] is a widely used closed form delay metric for CMOS gates because of simplicity and reasonable accuracy. Nevertheless it can be shown that the SN metric fails to provide high accuracy and fidelity when CMOS gates operate at low supply voltages. Thus it may not be applicable in many low power applications with voltage scaling. In this paper, we propose a new closed form delay metric based on the centroid of power dissipation. This new metric is inspired by our key observation and theoretic proof that the SN delay is indeed Elmore delay, which can be viewed as the centroid of current. Our proposed metric has a very high correlation coefficient (≥ 0.98) when correlated with the actual delays got from the HSPICE simulations. Such high correlation is consistent across all major process technologies. In comparison, the SN metric has a correlation coefficient between $(0.70, 0.90)$ depending upon the technology and the CMOS gate, and it is less accurate for lower supply voltages. Since our proposed metric has high fidelity across a wide range of supply voltages yet a simple closed form, it will be very useful to guide low voltage and low power designs.

I. INTRODUCTION

Accurate yet efficient delay modeling is important to guide design optimization, such as transistor and gate sizing, interconnect optimization, placement, and routing. Closed form delay equations with high accuracy is desirable since they are efficient and easy to implement. The alternative to the closed form delay metrics are the lookup tables. The lookup tables though accurate are less attractive since they are computationally expensive to use within an optimization loop and provide little insight [2]. The delay modeling consists of two distinct components, the gate and the interconnect delay modeling.

In the literature, significant amount of work has been devoted to *interconnect delay* characterization. The interconnects are often modeled as RC trees. The widely used Elmore delay is the first moment of the impulse response of the RC tree [3]. To improve the accuracy of the Elmore delay, models based on the higher order moment matching AWE [4] have been proposed. But AWE is expensive to use in optimization since it lacks closed-form expression. To improve the accuracy of Elmore delay and retain its simplicity, several works have proposed delay models that are functions of the higher moments of the impulse response of the RC tree [2], [5], [6]. Another fast approach is the matching the moments of the impulse response to a Probability Density Function (PDF) [7]–[10].

In the literature, the *gate delay* characterization has received lesser attention compared to the interconnect delay characterization. The Sakurai-Newton (SN) delay approximation [1] is a widely used closed-form delay metric for the CMOS gates because of simplicity and reasonable accuracy. Nevertheless the SN metric lacks accuracy when the CMOS gates operate at low supply voltages [11]. But for the nanometer SoC designs, delay modeling needs to address the heterogeneous nature, such as voltage scaling/voltage islands. Thus the delay model needs to be robust across a wide range of operating scenarios.

In this paper, we propose a new, robust closed form gate delay metric based on the centroid of power dissipation. This new model is inspired by our key observation and theoretic proof that the SN metric can be viewed as the centroid of current dissipated by the gate. The proposed metric has a very high correlation coefficient (≥ 0.98) when correlated with the actual delays got from the HSPICE simulations. Such high correlation is consistent across all major process technologies. In comparison, the SN metric has a correlation coefficient between $(0.70, 0.90)$ depending upon the technology and the CMOS gate, and it is less accurate for lower supply voltages. Since our proposed metric has high fidelity across a wide range of supply voltages yet a simple closed form, it will be very useful to guide low voltage and low power designs.

To summarize, we make the following contributions:

- We show that the Elmore delay can be expressed as the centroid of current dissipated.
- We prove that the SN delay approximation is the exact Elmore delay of a CMOS gate.
- We propose a high fidelity closed form metric for the delay of a CMOS gate based on the centroid of the power dissipated by the gate.

The rest of the paper is organized as follows. Section II presents the Sakurai and Newton approximation to the delay. Section III provides the background for the Elmore delay which leads to the proof that the SN delay approximation is the exact Elmore delay of a CMOS gate. In Section IV, we propose a new closed form formula inspired by our observation that the SN delay can be viewed as the centroid of current. The experimental results are presented in Section V, followed by conclusion in Section VI.

II. SAKURAI-NEWTON DELAY APPROXIMATION

The Shockley model for MOSFET [12] fails in the short-channel region because it neglects the velocity saturation effects. Sakurai and Newton proposed a model that takes into account the short-channel behavior while retaining the simplicity of the Shockley model [1], [13]. They modified the quadratic dependence of the drain current on the driving voltage to a α-power dependence, where $1 \leq \alpha \leq 2$ is the called the velocity saturation index.

The drain current i_D according to [1] is,

$$
i_D = \begin{cases} \frac{k}{2}(v_{GS} - V_T)^\alpha & \text{saturation,} \\ k(v_{GS} - V_T)^\alpha \frac{v_{DS}}{V_{DS_{SAT}}} & \text{linear,} \\ 0 & \text{cutoff} \end{cases} \tag{1}
$$

where

- $k = \left(\frac{W}{L}\right)\mu_n C_{ox}$, where μ_n is the mobility of electrons and C_{ox} is the oxide capacitance.
- $V_{DS_{SAT}}$ determines the boundary between linear and saturation regions when $v_{GS} = V_{DD}$.

For the delay *approximation* of the CMOS inverter, we assume a step input to the inverter. Thus we are finding out the inherent delay of the gate ignoring the finite rise time of the input. The delay due to finite rise time can be incorporated using techniques such as PERI [14].

Since we assume a step input, the drain current equation in (1) simplifies to,

$$
i_D = \begin{cases} \frac{k}{2}(V_{DD} - V_T)^\alpha & V_{DD} - V_T < v_{DS} \leq V_{DD}, \\ k(V_{DD} - V_T)^\alpha \frac{v_{DS}}{V_{DD}-V_T} & v_{DS} \leq V_{DD} - V_T \end{cases} \tag{2}
$$

where $(V_{DD} - V_T)$ is the boundary between linear and saturation regions under step input.

The main assumption in the delay approximation is that a constant saturation current I_{D0} discharges the output voltage from $v_{DS} = V_{DD}$ to $\frac{V_{DD}}{2}$.

$$
t_{sn} = \frac{\Delta Q|_{(v_{DS}=V_{DD} \to \frac{V_{DD}}{2})}}{I_{D0}} = \frac{C_L \frac{V_{DD}}{2}}{\frac{k}{2}(V_{DD} - V_T)^\alpha}
$$

Thus the Sakurai-Newton (SN) delay metric is [1],

$$
\boxed{t_{sn} \approx \frac{C_L V_{DD}}{k(V_{DD} - V_T)^\alpha}} \tag{3}
$$

Note that this metric is an *approximation* to the delay since the transistor is assumed to be in *saturation* from $v_{DS} = V_{DD}$ to $\frac{V_{DD}}{2}$. The assumption is *weak*, since under the step input the transistor is in *saturation* region only from $v_{DS} = V_{DD}$ to $(V_{DD} - V_T)$. From $v_{DS} = (V_{DD} - V_T)$ to 0, the transistor is in *linear* region. In this paper, we model the transistor operating in saturation and linear regions as a nonlinear resistor R [11]. Thus the inverter can be modeled as an RC circuit [15] as shown in Figure 1. For an RC tree, the Elmore delay is an upper bound on the actual delay for any input waveform [16]. The theory behind the Elmore delay is discussed in the next section.

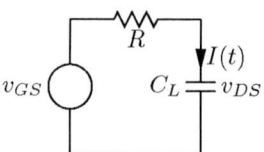

Fig. 1. The RC model of an inverter. Note that R is a nonlinear resistor modeling transistor and C_L is the load capacitance seen by the inverter.

III. CENTROID OF CURRENT BASED DELAY

In this section, we first show that the Elmore delay of a CMOS gate is the centroid of current dissipated by it. Then we prove that the SN metric is the exact Elmore delay of the CMOS gate. This key observation will inspire us to propose a new delay metric in Section IV.

Lemma 1. *The Elmore delay of a CMOS gate is the centroid of the current dissipated by it when it is switching.*

Proof. The Elmore delay is defined as the centroid of the impulse response $h(t)$ of the system [17]. The centroid x_c of the function $f(x)$ is defined as,

$$
x_c = \frac{\int_x x \, f(x) \, dx}{\int_x f(x) \, dx}
$$

Thus the Elmore delay is given by,

$$
t_{elmore} = \frac{\int_0^\infty t \, h(t) \, dt}{\int_0^\infty h(t) \, dt} \tag{4}
$$

since $\int_0^\infty h(t)dt = 1$ for RC circuits with monotonic response [17] we can write (4) as,

$$
t_{elmore} = \int_0^\infty t \, h(t) \, dt \tag{5}
$$

Let $H(s)$ denote the Laplace transform of $h(t)$. The transfer function $H(s)$ is defined as the ratio of output to input voltages [18]. Since we assume a step input, the transfer function reduces to,

$$
H(s) = \frac{V_{DS}(s)}{V_{GS}(s)} = \frac{V_{DS}(s)}{\frac{1}{s}} = s V_{DS}(s)
$$

We apply the Inverse Laplace transform to get the impulse response, $h(t) = \frac{dv_{DS}}{dt}$. We know that the current discharged through the capacitor,

$$
I(t) = C_L \frac{dv_{DS}}{dt}
$$
$$
= C_L h(t)
$$

Hence under the RC model with the assumption of step input,

$$
I(t) \propto h(t) \tag{6}
$$

$$
t_{elmore} = \frac{\int_0^\infty t \, I(t) \, dt}{\int_0^\infty I(t) \, dt} \tag{7}
$$

Thus the Elmore delay is shown as the centroid of the area under the current discharged through the load capacitor. $\square$

We can now show the following result.

Theorem 1. *The Sakurai-Newton delay approximation is the exact Elmore delay of the CMOS gate under the following conditions:*

(i) *A step input is applied;*

(ii) *The CMOS gate is modeled as an RC circuit.*

Proof. We provide the proof when the gate is discharging. The proof is similar when the gate is charging.

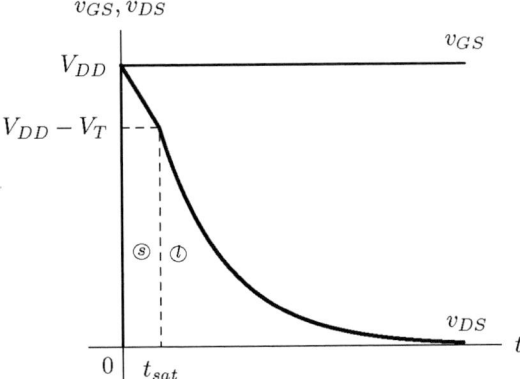

Fig. 2. Inverter waveforms when the output is discharging. The input v_{GS} is a step input. The output v_{DS} decreases linearly in the saturation region (till t_{sat}) and decays exponentially in the linear region (after t_{sat}).

The input and output voltage waveforms associated with the discharging inverter are shown in Figure 2. When a rising step input ($v_{GS} = V_{DD}\, u(t)$) is applied to the inverter, the NMOS is on while the PMOS is off. The NMOS operates in the saturation region when the output discharges from $v_{DS} = V_{DD}$ to $(V_{DD} - V_T)$ and it operates in the linear region when the output discharges from $v_{DS} = (V_{DD} - V_T)$ to 0. The time taken by the output v_{DS} to reach $(V_{DD} - V_T)$ is denoted as t_{sat}, the time at which the NMOS transistor switches from saturation to linear region of operation.

The Elmore delay integral in (7) can be written as,

$$t_{elmore} = \frac{\int_0^{t_{sat}} t\, i_{D_{SAT}}\, dt + \int_{t_{sat}}^{\infty} t\, i_{D_{LIN}}\, dt}{\int_0^{t_{sat}} i_{D_{SAT}}\, dt + \int_{t_{sat}}^{\infty} i_{D_{LIN}}\, dt} \quad (8)$$

To evaluate (8), we need closed form expressions for $i_{D_{SAT}}$, $i_{D_{LIN}}$, and t_{sat}.

When the NMOS is saturated, the output voltage v_{DS} decreases linearly from V_{DD} to $(V_{DD} - V_T)$, shown as ⓢ in Figure 2. The decrease is linear because the current is a constant during that period which is given by,

$$i_{D_{SAT}} = \frac{k}{2}(V_{DD} - V_T)^\alpha \quad (9)$$

When the output voltage v_{DS} goes below $(V_{DD} - V_T)$, the NMOS enters the linear region of operation, shown as ⓛ in Figure 2. The current in the linear region can be written as,

$$\begin{aligned} i_{D_{LIN}} &= k(V_{DD} - V_T)^\alpha \frac{v_{DS}}{V_{DD} - V_T} \\ &= \frac{v_{DS}}{R} \end{aligned}$$

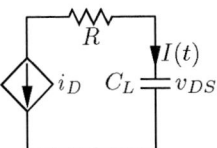

Fig. 3. RC model with discharging current as a controlled current source.

where $\frac{1}{R} = k(V_{DD} - V_T)^{\alpha-1}$ is the resistance through which we discharge the load capacitor C_L as shown in Figure 3. We need an closed form expression for v_{DS} to evaluate $i_{D_{LIN}}$. The output voltage v_{DS} in the linear region is simply the voltage seen at the capacitor of a first order RC circuit under the step input. Thus the output voltage v_{DS} in the linear region can be written as,

$$v_{DS} = (V_{DD} - V_T)e^{\frac{-(t-t_{sat})}{RC_L}} u(t - t_{sat})$$

Thus the current during the linear region of operation can be written as,

$$i_{D_{LIN}} = k(V_{DD} - V_T)^\alpha e^{\frac{-(t-t_{sat})}{RC_L}} u(t - t_{sat}) \quad (10)$$

Finally we need t_{sat}, the time at which the NMOS switches from saturation to the linear region. Applying Kirchhoff current law to the output in Figure 3,

$$-C_L \frac{dv_{DS}}{dt} = \frac{k}{2}(V_{DD} - V_T)^\alpha$$

$$-\int_{V_{DD}}^{V_{DD}-V_T} dv_{DS} = \frac{\frac{k}{2}(V_{DD} - V_T)^\alpha}{C_L} \int_0^{t_{sat}} dt$$

On integrating and simplifying we get,

$$t_{sat} = \frac{2C_L V_T}{k(V_{DD} - V_T)^\alpha} \quad (11)$$

Substituting the unknowns in (8), and evaluating the integrals we get,

$$t_{elmore} = \frac{\frac{C_L^2 V_T^2}{k(V_{DD}-V_T)^\alpha} + \frac{C_L^2(V_{DD}^2 - V_T^2)}{k(V_{DD}-V_T)^\alpha}}{C_L V_T + C_L(V_{DD} - V_T)}$$

$$\boxed{t_{elmore} = \frac{C_L V_{DD}}{k(V_{DD} - V_T)^\alpha}} \quad (12)$$

which is the same as (3). Thus the SN delay *approximation* is the *exact* Elmore delay of the CMOS gate. $\square$

In the nanometer regimes, the velocity saturation constant $\alpha \approx 1$. Thus (12) can be rewritten as,

$$t_{elmore} = \frac{C_L}{k\left(1 - \frac{V_T}{V_{DD}}\right)} \quad (13)$$

The SN metric (13) fails to track the delay when the supply voltages are low [11]. Taur and Ning [11] presented a simple curve fitting metric that works across a wide range of voltages. The Taur-Ning (TN) delay metric is given by,

$$\boxed{t_{tn} \propto \frac{C_L}{\left(0.7 - \frac{V_T}{V_{DD}}\right)}} \quad (14)$$

where 0.7 is a numerical fitting parameter. The TN metric suffers from the drawback of having high absolute errors compared to the actual HSPICE delays. This is further discussed in Section V. Another drawback is that it is applicable only when $\frac{V_T}{V_{DD}} \leq 0.5$ [11]. This means it may not be applied to very low V_{DD} designs.

IV. CENTROID OF POWER BASED DELAY

In this section, we derive a new metric based on the centroid of power (CP) which overcomes the drawbacks of the SN and TN delay metrics.

The SN metric can roughly be thought of as a charge based delay since we integrate over current. The centroid of power can be thought of as an energy based delay since we integrate over power. The delay obtained by taking the centroid of the power at the output can be written as,

$$t_{cp} = \frac{\int_0^\infty t \, v_{DS} \, i_D \, dt}{\int_0^\infty v_{DS} \, i_D \, dt} \qquad (15)$$

Since the NMOS transistor is operating in two different regions namely saturation and linear regions, (15) can be written as,

$$
t_{cp} = \frac{\int_0^{t_{sat}} t \, v_{DS_{SAT}} \, i_{D_{SAT}} \, dt + \int_{t_{sat}}^\infty t \, v_{DS_{LIN}} \, i_{D_{LIN}} \, dt}{\int_0^{t_{sat}} v_{DS} \, i_D \, dt + \int_{t_{sat}}^\infty v_{DS} \, i_D \, dt}
$$
$$
= \frac{\frac{C_L^2 (3V_{DD} - 2V_T) V_T^2}{3k(V_{DD}-V_T)^\alpha} + \frac{C_L^2 (V_{DD}+3V_T)(V_{DD}-V_T)^2}{4k(V_{DD}-V_T)^\alpha}}{\frac{1}{2} C_L (2V_{DD} - V_T) V_T + \frac{1}{2} C_L (V_{DD}-V_T)^2}
$$

which can be simplified to,

$$\boxed{t_{cp} = \frac{C_L (3V_{DD}^3 + 3V_{DD}^2 V_T - 3V_{DD} V_T^2 + V_T^3)}{6k V_{DD}^2 (V_{DD} - V_T)^\alpha}} \qquad (16)$$

The correlation between the centroid of power (CP) delay metric and the HSPICE delay values is better than the correlation between the SN delay metric and the HSPICE delay values. The correlation attains near perfection with a modification in the Taur-Ning spirit.

We found out empirically that $\frac{1}{(V_{DD}-V_T)^2}$ tracks the delay better than $\frac{1}{V_{DD}^2}$. Substituting $(V_{DD}-V_T)^2$ for V_{DD}^2 in the denominator of (16), we get the modified centroid of power (CPM) metric,

$$\boxed{t_{cpm} \propto \frac{C_L (3V_{DD}^3 + 3V_{DD}^2 V_T - 3V_{DD} V_T^2 + V_T^3)}{(V_{DD}-V_T)^2 (V_{DD}-V_T)^\alpha}} \qquad (17)$$

The correlation between the CPM delay metric and the HSPICE delay values is *almost perfect*. Also, the absolute error between the CPM metric and the HSPICE delay values reduces significantly compared to the other metrics discussed in this paper. A possible reason for this near perfect tracking of delay is that the gate overdrive is proportional to $(V_{DD} - V_T)$ and not to V_{DD}. An alternative way to reason about this is the fact that $\frac{1}{(V_{DD}-V_T)^2}$ has a faster rate of change compared with $\frac{1}{V_{DD}^2}$ when V_{DD} varies.

V. EXPERIMENTAL RESULTS

We used the Berkeley Predictive Technology Model [19] for our simulations. The simulations were run on the INV, NAND2, NOR2, XOR2 gates for their worst case input. The load capacitance C_L was varied from $20fF$ to $50fF$. The supply voltage V_{DD} was varied from $2 \times V_{T0}$ to $6 \times V_{T0}$. The threshold voltage V_{T0} was varied within $\pm 10\%$ of its original value. The simulations were run on $45nm$, $65nm$, and $100nm$ technologies. Thus nearly 200 simulations were run on each gate for a given technology under its worst case input.

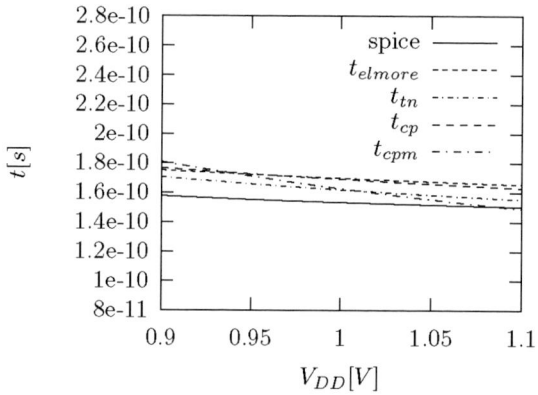

Fig. 4. HSPICE delay and the values predicted by the delay metrics for INV in $65nm$ technology under *nominal* supply voltages. The solid line is the HSPICE delay values and the dotted lines are the delays predicted by the various metrics. The V_{DD} was varied with load capacitance $C_L = 20fF$ and threshold voltage $V_{T0} = 0.22V$. Note that *all* the delay metrics track under *nominal* supply voltages.

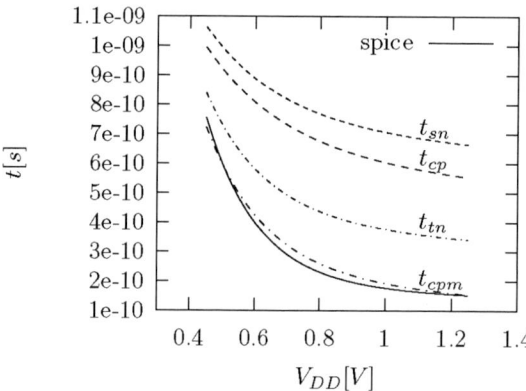

Fig. 5. HSPICE delay and the values predicted by the delay metrics for INV in $65nm$ technology. The solid line is the HSPICE delay values and the dotted lines are the delays predicted by the various metrics. The V_{DD} was varied with load capacitance $C_L = 20fF$ and threshold voltage $V_{T0} = 0.22V$. Note that only CPM can track the delay in the lower voltages while TN can track to quite an extent, the other two metrics SN and CP cannot track it.

The delay values predicted by the metrics were scaled by a constant value c. The constant c is obtained using linear regression. Suppose d_i is the delay obtained from HSPICE during the i th simulation and x_i is the delay predicted by the metric, c is obtained on minimizing $\sum_i (d_i - ax_i)^2$. Note that

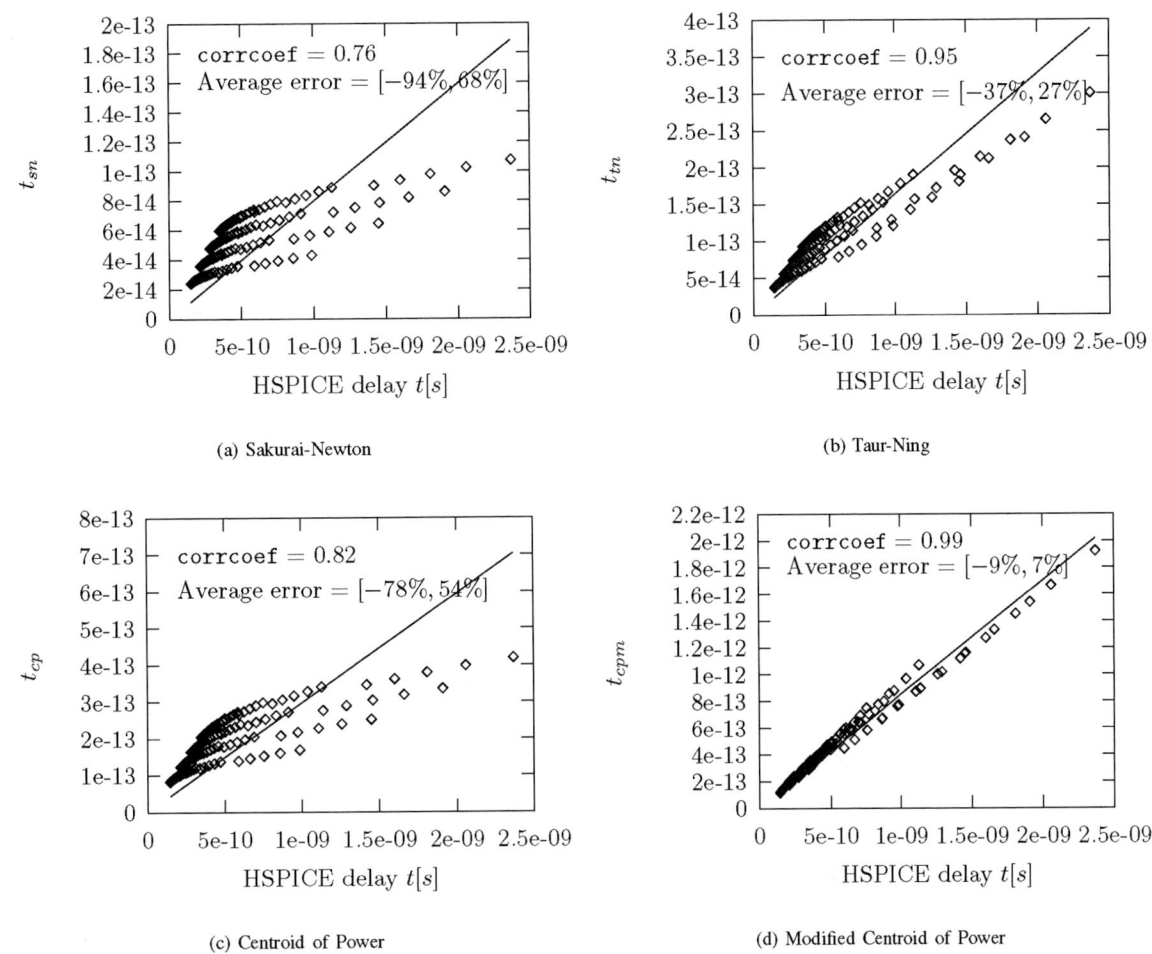

Fig. 6. Scatter plot of different delay metrics with the HSPICE delay for INV in $65nm$ technology. Since we have multiplied by the constant of proportionality, no units are provided for the y-axis.

TABLE I

THE CORRELATION OF HSPICE DELAY VALUES WITH THE DELAY METRICS ACROSS DIFFERENT TECHNOLOGIES AND GATES. THE HSPICE DELAY OF A GATE IS MEASURED FOR ITS WORST CASE INPUT COMBINATION.

Gate	45nm				65nm				100nm			
	SN	TN	CP	CPM	SN	TN	CP	CPM	SN	TN	CP	CPM
INV	0.76	0.97	0.81	0.99	0.76	0.95	0.82	0.99	0.90	0.99	0.94	0.98
NAND2	0.72	0.95	0.76	0.99	0.73	0.91	0.77	0.99	0.83	0.96	0.87	1.00
NOR2	0.73	0.96	0.78	0.99	0.75	0.92	0.80	0.99	0.90	0.99	0.93	0.99
XOR2	0.71	0.95	0.76	0.99	0.71	0.90	0.76	0.98	0.90	0.97	0.93	1.00

TABLE II

THE PERCENTAGE ERROR BETWEEN HSPICE DELAY VALUES AND THE DELAY METRICS ACROSS VARIOUS TECHNOLOGIES AND GATES. A *line* WAS FITTED TO THE DATA POINTS PREDICTED BY THE DELAY METRIC. IN THIS TABLE THE AVERAGE min, max ESTIMATION ERROR PERCENTAGE IS SHOWN.

Gate	45nm (%)				65nm (%)				100nm (%)			
	SN	TN	CP	CPM	SN	TN	CP	CPM	SN	TN	CP	CPM
INV	−161, 97	−41, 26	−139, 76	−14, 10	−94, 68	−37, 27	−78, 54	−9, 7	−24, 22	−7, 5	−19, 16	−7, 9
NAND2	−275, 137	−82, 45	−240, 112	−32, 14	−153, 91	−69, 43	−130, 76	−22, 11	−59, 43	−26, 19	−51, 35	−3, 4
NOR2	−209, 111	−59, 35	−181, 92	−19, 11	−112, 73	−50, 34	−96, 61	−13, 8	−31, 20	−9, 6	−23, 16	−7, 10
XOR2	−271, 141	−80, 47	−236, 115	−31, 15	−151, 94	−68, 45	−129, 79	−22, 13	−57, 43	−24, 19	−49, 35	−3, 4

a changes as we take more samples of the parameters across a wider range. Thus a metric might be able to track the delay across small variations of supply voltage while it may not be able to track delay under large variations of supply voltage. This is illustrated in the Figures 4 and 5.

In Figure 4, the CMOS gates operate under *nominal* supply voltages, $V_{DD} = 4 \times V_{T0}$ to $6 \times V_{T0}$ all the delay metrics correlate to HSPICE reasonably well. However, when the supply voltage drops below $V_{DD} = 4 \times V_{T0}$, only the CPM metric is able to track the delay well shown in Figure 5. The data is taken for an inverter in $65nm$ technology by varying the supply voltage V_{DD} from $2 \times V_{T0}$ to $6 \times V_{T0}$ and fixing the other circuit parameters.

The data obtained from other gates across various technologies and circuit parameters such V_{DD} and V_T have similar results to Figure 5. There are two things to note in this figure:

1) The *correlation* measures the relative error. Intuitively, the relative error gives an estimate of how close the shape of the predicted delay curve is with the actual delay obtained from HSPICE simulations.

2) The *estimation error* gives the absolute difference between the predicted delay and the actual delay obtained from HSPICE simulations.

To visualize the performance of delay metrics with respect to the above two characteristics we use the scatter plot. The scatter plot of different delay metrics versus the actual delay values for INV in $65nm$ technology is shown in Figure 6.

The data points are obtained by varying different circuit parameters. We fitted a line through the data points to find out the constant of proportionality in the delay metrics. Then we find the estimation error between the fitted line and the HSPICE delay values. The correlation is shown as `corrcoef` and the estimation error is shown as 'Average error' in the scatter plot. From the scatter plot it is clear that the CPM delay metric has the highest correlation and the lowest estimation error among all the delay metrics.

Table I summarizes the *correlation* coefficient of different delay metrics for various gates across the technologies. The correlation was taken between the actual HSPICE delays and the delay metric. From the table, we observe that the correlation coefficient of the CPM metric is consistently greater than 0.98, which is not exhibited by the other delay metrics. The estimation errors are tabulated in Table II. The values listed in the table are the *average* of the estimation errors.

VI. Conclusion

In this paper, we proposed a new closed form delay metric based on the modified centroid of power dissipated. This new metric is inspired by our key observation that the SN delay can be viewed as the centroid of current. We also provide a theoretic proof that the SN delay is the Elmore delay of a CMOS gate when a gate is modeled as an RC circuit. The delay due to finite rise time can be incorporated using techniques such as PERI [14].

Our proposed metric has a very high correlation coefficient (≥ 0.98) when correlated with the actual delays got from

the HSPICE simulations. Such high correlation is consistent across all major process technologies. The new metric is both simple and inexpensive to use as compared to the other metrics proposed in the literature. We anticipate its use in low voltage circuits and in the inner optimization of physical design tools where it is necessary to obtain quick and relatively accurate delay estimates.

Acknowledgment

This work is partially sponsored by IBM Faculty Award. We used computers donated by Intel Corporation.

References

[1] T. Sakurai and A. R. Newton, "Alpha-power law MOSFET model and its applications to CMOS inverter delay and other formulas," *IEEE Journal of Solid State Circuits*, vol. 25, no. 2, pp. 584–594, April 1990.

[2] C. J. Alpert, A. Devgan, and C. V. Kashyap, "RC delay metrics for performance optimization," *IEEE Trans. on Computer-Aided Design of Integrated Circuits and Systems*, vol. 20, no. 5, pp. 571–582, May 2001.

[3] J. Rubinstein, P. Penfield, and M. A. Horowitz, "Signal delay in RC tree networks," *IEEE Trans. on Computer-Aided Design of Integrated Circuits and Systems*, vol. 2, no. 3, pp. 202–211, July 1983.

[4] L. T. Pillage and R. A. Rohrer, "Asymptotic waveform evaluation for timing analysis," *IEEE Trans. on Computer-Aided Design of Integrated Circuits and Systems*, vol. 9, no. 4, pp. 352–366, April 1990.

[5] B. Tutuianu, F. Dartu, and L. Pileggi, "An explicit RC-circuit delay approximation based on the first three moments of the impulse response," in *Proc. of Design Automation Conf.*, 1996, pp. 611–616.

[6] A. B. Kahng and S. Muddu, "An analytical delay model for RLC interconnects," *IEEE Trans. on Computer-Aided Design of Integrated Circuits and Systems*, vol. 16, no. 12, pp. 1507–1514, December 1997.

[7] R. Kay and L. Pileggi, "PRIMO: Probability interpretation of moments for delay calculation," in *Proc. of Design Automation Conf.*, 1998, pp. 463–468.

[8] T. Lin, E. Acar, and L. Pileggi, "h-gamma: an RC delay metric based on a gamma distribution approximation of the homogeneous response," in *Proc. of the International Conf. on Computer-Aided Design*, 1998, pp. 19–25.

[9] F. Liu, C. Kashyap, and C. J. Alpert, "A delay metric for RC circuits based on the weibull distribution," in *Proc. of the International Conf. on Computer-Aided Design*, 2002, pp. 620–624.

[10] C. J. Alpert, F. Liu, C. Kashyap, and A. Devgan, "Delay and slew metrics using the lognormal distribution," in *Proc. of Design Automation Conf.*, 2003, pp. 382–385.

[11] Y. Taur and T. H. Ning, *Fundamentals of Modern VLSI Devices*. Cambridge University Press, 1998.

[12] W. Shockley, "A unipolar 'field-effect' transistor," in *Proc. of Institute of Radio Engineers*, 1952, pp. 1365–1376.

[13] T. Sakurai and A. R. Newton, "A simple MOSFET model for circuit analysis," *IEEE Trans. on Electron Devices*, vol. 38, no. 4, pp. 887–894, April 1991.

[14] C. V. Kashyap, C. J. Alpert, F. Y. Liu, and A. Devgan, "Closed-form expressions for extending step delay and slew metrics to ramp inputs for RC trees," *IEEE Trans. on Computer-Aided Design of Integrated Circuits and Systems*, vol. 23, no. 4, pp. 509–516, April 2004.

[15] D. Hodges, H. Jackson, and R. Saleh, *Analysis and Design of Digital Integrated Circuits: In Deep Submicron Technology*. McGraw-Hill, 2003.

[16] R. Gupta, B. Tutuianu, and L. Pileggi, "The elmore delay as a bound for RC trees with generalized input signals," *IEEE Trans. on Computer-Aided Design of Integrated Circuits and Systems*, vol. 16, no. 1, pp. 95–104, January 1997.

[17] W. Elmore, "The transient response of damped linear networks with particular regard to wideband amplifiers," *Journal of Applied Physics*, vol. 19, no. 1, pp. 55–63, January 1948.

[18] A. V. Oppenheim, A. S. Willsky, and S. H. Nawab, *Signals and Systems*. Prentice Hall, 1996.

[19] Y. Cao, T. Sato, M. Orshansky, D. Sylvester, and C. Hu, "New paradigm of predictive MOSFET and interconnect modeling for early circuit simulation," in *Proc. of Custom Integrated Circuits Conf.*, 2000, pp. 201–204.

CGTA: Current Gain-based Timing Analysis for Logic Cells

Shahin Nazarian   Massoud Pedram   Tao Lin   Emre Tuncer

EE-Systems Dept., University of Southern California
Los Angeles, CA 90089

{shahin , pedram@usc.edu}

Magma Design Automation
Santa Clara, CA 95054

{tao , emre @magma-da.com}

ABSTRACT

This paper introduces a new current-based cell timing analyzer, called CGTA, which has a higher performance than existing logic cell timing analysis tools. CGTA relies on a compact lookup table storing the output current gain (sensitivity) of every logic cell as a function of its input voltage and output load. The current gain values are subsequently used by the timing calculator to produce the output current value as a function of the applied input voltage. This current and the output load then uniquely determine the output voltage value. Therefore, CGTA is capable of efficiently and accurately computing the output voltage waveform of a logic cell, which has been subjected to an arbitrary noisy input voltage waveform. Experimental results are presented to assess the quality of CGTA compared to other existing approaches.

1. INTRODUCTION

As the layout geometries in recent technologies scales down, the increase in the package density and operational frequency aggravates the noise sources. To check whether a noise source can create erroneous outputs, the circuit should be analyzed using a timing analysis tool. Input pattern dependent circuit-level timing analysis with tools such as Spice, is very accurate, but requires significant computational resources, which makes this approach impractical for large VLSI circuits. Logic-level timing analysis tools such as static or statistical static timing analysis tools are used as efficient alternatives with an acceptable level of accuracy.

Delay models for both interconnect lines and cells are required to perform timing analysis. The function of an *interconnect delay model* is to take as input the transient waveform at the near-end of an interconnect line and produce as output, the corresponding waveform at the far-end of the line while accounting for the effect of various noise sources that couple to the line. This process is known as *interconnect timing analysis*.

Similarly, the function of a *cell delay model* is to take a noisy input waveform and produce the waveform for the cell output. This process is known as *cell timing analysis*. Conventional timing analysis tools start with arrival time and slope (transition time or slew) at the near-end of an interconnect line and produce the arrival time and slew at the output of a cell that is driven by the far-end of that line.

The fact that the interconnect delay dominates the cell delay in modern VLSI circuits, has made the researchers produce excellent interconnect delay models. However the conventional cell delay models have not improved as much and their deficiencies, especially in handling noisy waveforms have been intensified due to recent technology trend. Consequently cell models are one of the main sources of inaccuracy in existing timing analysis tools. The focus of this paper is on the logic cell timing analysis when noise is present.

Cell delay is conventionally pre-characterized based on input slew and capacitive output load by using a circuit level timing analyzer such as Spice. Therefore the resulting pre-characterized look-up tables are inherently incompatible with the RC/RLC interconnect loads. This incompatibility is dispelled by finding an effective capacitive load, which is in some way equivalent to the more complex RC [1] or RLC load [2]. An iterative or non-iterative approach may be used to calculate the effective capacitance. The goal of cell timing analysis is conventionally stated as: Given a noisy waveform at the input of a cell, find an equivalent input voltage waveform that when is applied to the cell generates an output waveform which is as close as possible to the output waveform in terms of its arrival time and slew.

The interesting fact about the shape of the waveform is that different voltage waveforms with identical arrival time and slew at the input of a cell can result in very different propagation delays through that cell. This is because the exact shape of the input voltage waveform can greatly influence the cell output waveform behavior. Generally speaking, as the crosstalk noise becomes more significant in current technologies, using only a reference point (arrival time) and a constant slope (slew) to convey the timing information for a signal transition adversely impacts the robustness of timing analysis tools. Hence the shape of the waveform should be considered more effectively. We re-state the problem in a more general statement as follows: Given a noisy voltage waveform at the input of a cell, determine the output voltage waveform, which has the minimum error with respect to the actual output waveform.

As the silicon technology is driven to nanometer, conventional voltage-based lookup tables are nearing the end of their useful life. In [3]-[4] the common voltage-based cell timing analyzers are reviewed and their shortcomings are highlighted. In addition to being inefficient in accurately considering the impact of the shape of the noisy waveform, the voltage-based timing analysis tools are inefficient in low power design styles that incorporate two or more logic "islands", each running at a different operating voltage. Traditional library cell characterization that accurately covers a wide range of operating voltages can be prohibitively time consuming.

Current-based has been shown to be more accurate than voltage-based logic cell timing analysis [5]-[7]. In fact some industrial current-based timing analyzers, such as CCSM and ECSM are already in use [8]. Existing current-based approaches may still exhibit large variations from Spice simulation when presented with complex interconnect models or non-monotonic input voltage waveforms. Their complexity is a barrier to apply them in novel design tools.

In this paper, we present CGTA, a current gain-based timing analysis tool for logic cells. The gain (sensitivity) of output current to input voltage is defined as the derivative of output current waveform to the input voltage waveform. The gain is then used to accurately model the impact of the shape of the input voltage waveform on the output current waveform and eventually the voltage waveform. To respond to the more general problem, CGTA is able to directly build the output waveforms without the need for creating an equivalent input waveform as is done by conventional techniques. CGTA is simple and efficient to implement. More precisely, the application of the current sensitivity factor in delay calculation brings together the accuracy of a current-based cell modeling and the efficiency of a voltage-based cell modeling. It will be shown that CGTA can result in a very efficient, yet accurate timing analysis compared to cell timing analysis using existing current-based methods.

The remainder of this paper is organized as follows. In section 2 we review the previous logic cell timing analysis techniques including current-based ones. Section 3 describes CGTA. Section 4 and 5 review the experimental results and conclusions respectively.

2. BACKGROUND

Most of today's logic cell timing analysis techniques used in integrated circuit design flows consist of lookup tables or characteristic equations that rely on linear or ramp voltage waveforms and simplified loads as inputs and create linear or ramp voltage waveform approximations as output. Interested reader may refer to references [3]-[4] that extensively review the various voltage-based cell timing analyzers and discusses their shortcomings and strengths.

Two recently developed approaches, i.e., equation-based and current-based techniques, contend to replace voltage-based lookup tables. Both have the ability to better predict nanometer timing across a range of supply voltage [8].

2.1 Equation-based Techniques

The equation-based timing analyzers generally use a polynomial with multiple coefficients relating timing to a variety of input parameters. The goal is to model delay variation due to environmental factors such as supply voltage and substrate temperature. However, it is difficult to fit the actual non-linear behavior of the timing quantity of interest with a polynomial that has a limited (and relatively small) number of terms.

In practice, the extreme effort to characterize real silicon to the equation-based modeling has made it unpopular. Sophisticated optimization algorithms are required to perform curve fitting of a polynomial to simulation data, and the accuracy and turnaround time of the library creation is limited by the quality of the optimization algorithms.

2.2 Current-based Techniques

Current-based cell timing analyzers generally base their delay calculations on the amount of current flow into or out of a cell. Current-based cell modeling is much easier to characterize than the equation-based one. Rather than a mathematical abstraction, current-based modeling is a physical model patterned after the actual construction of transistors. It improves delay calculation accuracy by modeling a cell's output drive as a current source rather than a voltage source. Current sources are more effective at tracking non-linear transistor switching behavior and permit highly accurate modeling of long complex interconnects, which are common in many of today's largest nanometer low power designs.

One example of a current-based cell timing analysis technique is the Blade in [5]. Blade consists of a voltage-controlled current source, an internal capacitance, and a time shift of the output waveform. First $I_{out}(V_{in}, V_{out})$, the amount of current sourced by a cell in response to DC voltage levels on the input and output pins of interest, is determined and a lookup table (denoted by the cell I-V table) is created for each cell by sweeping the DC values of input and output voltages and measuring the current sourced by the cell output pin. However, a response exclusively derived from the DC-based I-V table results in an overly optimistic timing analysis as the DC sweep of the input and output ignores the effects of parasitic elements. Therefore a calibration procedure is thus performed to consider the cell parasitic effects. This procedure determines an internal capacitive load which, when applied to the Blade model, results in a transient waveform that matches the shape of a Spice-generated waveform for the cell under identical conditions. Once the waveform shapes have been matched, a time shift is calculated by examining the time difference between the 50% points of the Spice output and the calibrated Blade output. A runtime engine consisting 31×31 I-V lookup tables and a secant iteration-based nonlinear solver is used in [5] to compute the output waveforms.

A more complete current-based cell delay technique is presented in [6], where the current drawn by a cell during output switching is computed while considering the miller effect between input and output nodes as well as internal parasitic effects. These effects are modeled by capacitors which are calculated through a series of transient Spice-based simulations. Additionally, I-V tables are generated in the same fashion as the ones in Blade model.

Alternatively, cells under the crosstalk-induced pulse (glitch) attack are studied in [7] by using an analytical current model consisting of four parameters, namely a dc current source, a linear resistance, an output capacitance, and the internal delay of the gate. The DC and transient cell characterization steps and large number of required iterations for the aforementioned techniques are too complex to be utilized in existing CAD tools and flows.

3. CGTA

This section describes CGTA, a new current-based cell delay modeling for the purpose of timing analysis. The key innovation in CGTA originates from its modeling of the output current signal as a function of the input voltage signal. Therefore, we substitute the I-V lookup tables of existing current-based cell timing analyzers with a simpler computational model, while maintaining the accuracy. Unlike the voltage-based methods that first need to find an equivalent linear input waveform, our model directly builds the output voltage waveform.

We define the *current gain*, ρ_c, as the derivative of the output current to the input voltage. Each cell is pre-characterized with a 2-D lookup table with input voltage and effective output capacitance as the input keys and ρ_c as its returned value. Output current waveform is computed by using the lookup table gain information and performing Taylor series expansion. Having the

output current waveform the output voltage waveform can be computed considering the load.

3.1 Intuition Behind Current Gain Utilization

As described in 2.2, the characterization steps in the existing current-based cell timing analyzers are quite involved. The major source of complexity is due to the fact that both input and output voltages should be considered as input parameters to the cell model. These voltages must then be swept during the DC characterization step in order to fill in the I-V lookup tables and compute the parasitic capacitances. Also a series of transient simulations should be performed during which voltage transition are applied to input and output pins. To resolve this issue, we notice that the output voltage of a cell is a function of the input voltage, the parasitic effects, output load, and supply voltage, V_{dd}. Considering the parasitic capacitive values as constant [6], the output voltage may be replaced by the input voltage and output load values. Therefore, having the output load, the output current can be written as a function of input voltage. This is why ρ_c is defined as the sensitivity of the output current to input voltage.

3.2 CGTA Model

Each logic cell in the standard library is pre-characterized with a lookup table, which is used for output voltage calculations of the cell. This table will be referred as $I_{gain}(K\times L)$ where K and L denote the number of input voltage levels and effective capacitance values, respectively. I_{gain} contains $\rho_c(V_{in}^i, C_{eff}^j)$ which is simply the derivative of the cell's output current, i_{out} with respect to its input voltage at voltage value V_{in}^i when the cell output is connected to an effective load with a value of C_{eff}^j:

$$\rho_c(V_{in}^i, C_{eff}^j) = \rho_c(i, j) = \frac{\Delta i_{out}}{\Delta v_{in}}\bigg|_{v_{in}(t)=V_{in}^i} \qquad (1)$$

ρ_c quantitatively shows how sensitive the output current is to the input voltage, at a certain input voltage value and for a certain effective output capacitance value. The $\rho_c(i,j)$ value is stored in row i and column j of the I_{gain} lookup table. Figure 1 depicts an example of such a lookup table. The I_{gain} tables are created per pair of input and output pins by a series of transient Spice-based simulations, in which noiseless (saturated ramp) input waveforms are applied while the output current change is monitored. This process is repeated for different effective load capacitances.

It is shown later in this section that CGTA is able to consider arbitrary loads including simple capacitive, RC-π, or more complex interconnect RC models.

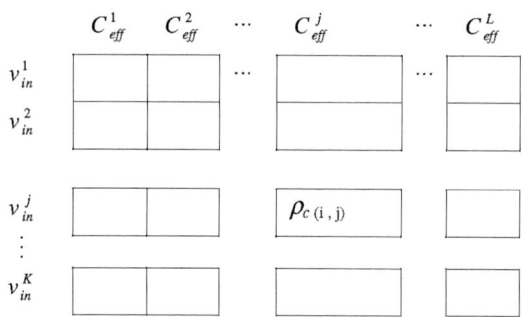

Figure 1. $I_{gain}(K \times L)$, the cell current gain lookup table used in CGTA.

However since ρ_c is a function of the output load, an effective output capacitance, is used to model the output of the load. The iterative effective capacitance calculation technique of [1] is used to determine the effective capacitance.

Effective capacitance is dependent on the input transition time; therefore, given a noisy waveform, the effective capacitance changes for different regions of the waveform due to different slews. We thus divide the noisy waveform into different parts by doing a piecewise linear approximation of the waveform. Each part of the noisy waveform is approximated by a fixed transition time, and therefore, has its own effective capacitance. It is empirically found that the effective capacitance calculation converges in fewer than 3 iterations. Note that the effective capacitance calculation is done only for the purpose of obtaining ρ_c values from the I_{gain} lookup table. Note that when calculating the output voltage, we use the actual load e.g., an RC-π model shown in Figure 2.

The input voltage waveform, v_{in}, is represented by a time-indexed voltage array, i.e., by using P equidistant sample points (t_0, ..., t_{P-1}.) CGTA takes this data and uses the I_{gain} table to find ρ_c values for each point. Figure 3 illustrates ρ_c for a typical crosstalk-induced noisy waveform that was generated from a part of an industrial design.

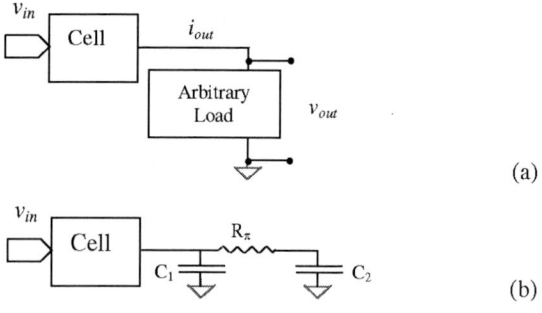

Figure 2. i_{out} is calculated as a function of v_{in} and ρ_c. (a) v_{out} in terms of i_{out} and output load, (b) RC-π modeling of the load.

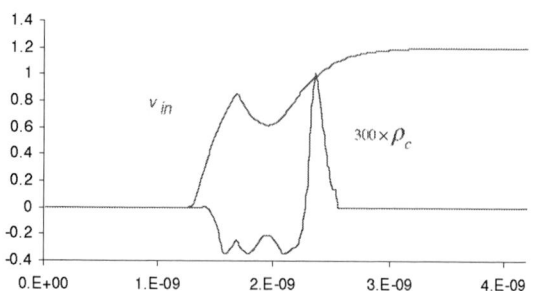

Figure 3. ρ_c for a typical crosstalk-induced noisy waveform. $300 \times \rho_c$ is illustrated for visibility purposes.

We assume that the noisy input voltage waveform, v_{in}, has been characterized by the user (or a timing analysis tool) by having specified the input waveform voltage levels at P equidistant sample points (t_0, ..., t_{P-1}.) CGTA constructs the output waveforms by reporting the output current and voltage levels at P equidistant points. Therefore, it is easy to see that CGTA can be used as the main delay calculation engine in a timing analysis tool which starts from the primary inputs of the circuits and calculates the voltage waveforms for all intermediate signals and the primary outputs during a linear time traversal of the circuit net list. (The goal of this work is to develop CGTA as a cell timing analysis technique; therefore, calculation of the interconnect delay is outside of the scope of the present paper.) To detect noise, P should be selected such that the time between two consecutive sampling points is no larger than one half of the smallest crosstalk noise width. In practice, we have considered a sampling time intervals of at least 50ps, e.g., for an input waveform with a rise time of 1ns, at least 20 sampling points are used.

CGTA builds an equivalent output current waveform in response to the noisy input voltage waveform, v_{in}, using the truncated Taylor series expansion of i_{out}:

$$i_{out}(t_{k+1}) = i_{out}(t_k) + \rho_c(t_k) \cdot \{v_{in}(t_{k+1}) - v_{in}(t_k)\} +$$
$$\frac{1}{2} \cdot \frac{\Delta \rho_c}{\Delta v_{in}}(t_k) \cdot \{v_{in}(t_{k+1}) - v_{in}(t_k)\}^2 \qquad (2)$$

where $i_{out}(t_0)$ is initialized to zero. $\rho_c(t_k)$ is a shorthand notation for $\rho_c(v_{in}(t_k))$. In general, the P computed output values may not be equidistant. This is undesirable when doing the timing analysis of a logic circuit. To avoid this, a set of P equidistant points are computed based on weighted average of the two nearest values found from Equation (2).

As pointed out $\rho_c(v_{in}(t_k))$ is found from the I_{gain} table (if necessary using interpolation.) $\frac{\Delta \rho_c}{\Delta v_{in}}(t_k)$ is found from the I_{gain} table and using Equation (3):

$$\frac{\Delta\rho_c}{\Delta v_{in}}(t_k) = \frac{\Delta\rho_c(t_k)}{\Delta v_{in}(t_k)} = \frac{\Delta\rho_c(v_{in}(t_k))}{\Delta v_{in}(t_k)} \qquad (3)$$

$\frac{\Delta\rho_c}{\Delta v_{in}}(t_k)$ is defined to be zero if the input voltage does not change from the previous sampling point to the current one, i.e., $\Delta v_{in}(t_k)=0$.

A Padé approximation can be used to calculate the output current, instead of the Taylor series expansion of Equation (2). Padé approximations are usually superior to Taylor expansions when functions contain poles, because the use of rational functions allows them to be well-represented [9]. However, our experimental results demonstrate that using the first two terms of the Taylor series to find the output current provides sufficient accuracy, yet it is much more efficient than using the Padé approximation. This makes Equation (2) more suitable than an equivalent Padé formula to be used in a logic cell timing analysis tool.

Having calculated the output current, the output voltage can be found based on the arbitrary load connected to the output. Figure 4 illustrates the equivalent output current waveform and also the resulting output voltage waveform for the noisy input waveform of Figure 3. Both output current and voltage waveforms by CGTA closely match their respective actual waveforms generated by Hspice [10].

The underlying principle of our approach to handle the compound cells (i.e., multi-stage cells, for example an OR gate) is similar to that described in [5]. We repeat the characterization process for each logic function (NOR function and the NOT function.) Therefore two runs of CGTA calculation steps are required for output waveform computation of an OR gate.

Each cell exhibits a kind of low pass filtering effect, which prunes certain amount of input noise. This is not considered in current-based approaches in general.

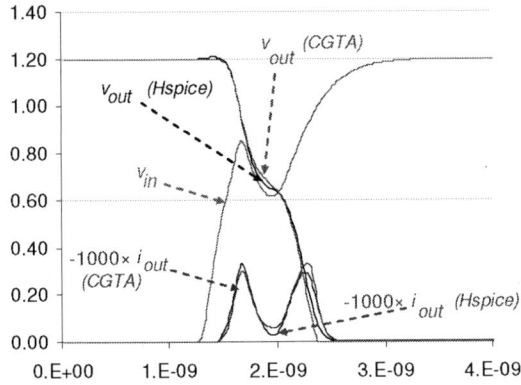

Figure 4. The actual and equivalent waveforms for CGTA. -1000×current waveforms are illustrated for visibility purposes.

To increase the accuracy, similar to [5], a low pass filter may be used on the noisy input waveforms prior to presenting the waveform to the CGTA calculator.

4. EXPERIMENTAL RESULTS

CGTA was written in C and compiled under Sun Blade 1000 machine. The cells used in the experiments are from a 130nm, 1.2V production cell library using parasitically extracted netlists. An automated test system was devised to assess CGTA and compare its delay accuracy and run-time with Hspice. A variety of cells in the production library were tested considering waveforms with a large variety of shapes, from pure ramp to noisy waveforms. The set of experiments included RC-π structure as well as capacitive only loads. The size of I_{gain} for CGTA was set to (20,5) meaning that 20 input voltage values between 0 and 1.2V and 5 output capacitance values are considered. No low pass filters were used in CGTA to generate the results in this paper. Compared with Hspice, the generated output voltage waveforms by CGTA matched the Hspice with 1-3% error. Figure 4 shows a comparison for an example of such output waveforms for CGTA with Hspice. The key advantage is that efficiency has been obtained without compromising the accuracy compared to other cell timing analysis techniques.

The accuracy of CGTA is next demonstrated on realistic circuit configurations that are part of a large high-performance ASIC design obtained from industry. The circuit configurations appraise our model under different scenarios, i.e., for different number of aggressor lines, interconnect lengths, coupling capacitance values, and input slews to create various noisy waveform shapes. Configuration I is a pair of 1000μm coupled interconnect lines running parallel to one another with a total distributed coupling value of 100fF. Both aggressor and victim line inputs have a slew of 150ps. For all configurations we set the arrival time and slew (transition time) of the victim line input to 1000ps to 150ps, respectively. For configuration I we swept the arrival time of the aggressor line input from 500 to 1500ps in steps of 5ps. Configuration II includes two aggressor lines each with 100fF total coupling and a victim, all of which are 500μm long. We maintained a fixed offset of -100ps between signal arrival time of the 1st and 2nd aggressor line inputs, while sweeping that of the 2nd aggressor line input arrival time. The two aggressor inputs have slews 200ps, and 400ps, respectively. Configuration III contains three aggressor lines, each with 50fF total distributed coupling and 300μm long. The victim line is 500μm long. We maintained a fixed offset of -50 between the arrival times of 1st and 3rd aggressor line inputs and -100 between those of 2nd and 3rd. The arrival time of the 3rd aggressor line input was then swept from 500 to 1500ps in steps of 5ps. The slews of the three aggressor lines are 200ps, 350ps, and 400ps respectively.

Table 1 shows the maximum and average delay errors of the existing voltage-based techniques and CGTA compared to Hspice. The cell delays were calculated as the difference between the $0.5V_{dd}$ crossing point of the output waveform and that of the input waveform. In terms of percentage errors, the average and maximum errors for CGTA compared to Hspice are about 1% and 3%, respectively. The average run-time of CGTA output waveform computation for a typical logic cell is less than 100μsec.

Table 1. Absolute errors in calculated delays vs. Spice simulation results for different timing analysis tools

Method	Delay Error (ps) = |Delay(Hspice) – Delay(Method)|					
	Configuration I		Configuration II		Configuration III	
	Max	Avg	Max	Avg	Max	Avg
Noiseless Point-based	81.3	29.3	134.2	48.5	153.4	55.3
Noisy Point-based	82.7	24.5	144.5	51.3	151.6	56.4
Least Square Fitting (LSF)	75.1	30.9	110.8	45.4	124.6	49.4
Elmore-based [3]	82.3	14.5	145.3	33.4	166.3	35.3
Weighted LSF [4]	42.4	10.3	49.3	17.4	48.5	15.6
SDP+Elmore-based [3]	39.5	8.6	46.8	12.8	45.6	11.7
CGTA	11.4	3.7	11.8	3.9	11.9	3.9

5. CONCLUSION

Conventional logic cell timing analysis tools approximate a noisy input waveform by an equivalent saturated ramp input, which then enables them to utilize standard delay look-up tables to report the cell timing information as a function of the input ramp slew and the output load. These techniques can result in significant error because the output waveform corresponding to the actual noisy input tends to be quite different from the one that is produced by assuming a saturated ramp input. Current-based cell timing analyzers have been proposed as an accurate alternative to these voltage-based timing analyzers. Unfortunately, the existing current-based techniques tend to be CPU-intensive in terms of the DC and transient simulations needed to pre-characterize the logic cell output current as a function of its input and output voltages and to calculate the values of internal logic cell capacitances. This paper presented CGTA, an accurate and efficient current-based cell timing analyzer which overcomes the aforementioned shortcoming. In particular, CGTA uses a compact table lookup whereby the output current gain (sensitivity) of a logic cell is pre-characterized as a function of its input voltage and output load. The current gain values are then used as part of a highly efficient timing calculator to provide the output voltage waveform as the logic cell is presented with an arbitrary noisy input voltage waveform. Experimental results demonstrate the high accuracy of CGTA and its efficiency.

REFERENCES

[1] J. Qian, S. Pullela, L. Pillage, "Modeling the effective capacitance for the RC interconnect of CMOS gates," *Transactions On Computer-Aided Design of Integrated Circuits and Systems, Vol. 13*, pp. 1526–1535, 1994.

[2] K. Agarwal, D. Sylvester, D. Blaauw, "An effective capacitance based driver output model for on-chip RLC interconnects," *Proceedings of Design Automation Conference (DAC)*, pp. 376-381, 2003.

[3] S. Nazarian, M. Pedram, E. Tuncer, and T. Lin, "Sensitivity-based gate delay propagation in static timing analysis," *Proceedings of International Symposium on Quality of Electronic Designs (ISQED)*, Mar. 2005, pp. 536-541.

[4] M. Hashimoto, Y. Yamada, H. Onodera, "Equivalent waveform propagation for static timing analysis," *IEEE Trans. Computer-Aided Design of Integ. Circuits & Systems, Vol. 23, No.4*, pp. 498-508, 2004.

[5] J.F. Croix, D.F. Wong, "Blade and razor: cell and interconnect delay analysis using current-based models," *Proceedings of Design Automation Conference (DAC)*, pp. 386-389, 2003.

[6] I. Keller, K. Tseng, N. Verghese, "A robust cell-level crosstalk delay change analysis," *Proceedings of International Conference on Computer-Aided Design (ICCAD)*, pp.147-154, 2004.

[7] A. Korshak, J.C. Lee, "An effective current source cell model for VDSM delay calculation," *Proceedings of International Symposium on Quality of Electronic Designs (ISQED)*, pp. 296–300, 2001.

[8] P. Buch, W.J. Dai "Understanding ECSM and CCSM," http://www.magma-da.com/c/@SQJEgjiSi1J_6/Pages/MWAUnderstandingECSMandCCSM.html.

[9] http://www.mathworld.com.

[10] "Hspice: The golden standard for Accurate Circuit Simulation," http://www.synopsys.com/products/mixedsignal/hspice/hspice.html.

Efficient Static Timing Analysis Using a Unified Framework for False Paths and Multi-Cycle Paths

Shuo Zhou, Bo Yao, Hongyu Chen, Yi Zhu
and Chung-Kuan Cheng
University of California at San Diego
La Jolla, CA, USA
szhou@cs.ucsd.edu

Mike Hutton
Altera Corp., San Jose, CA, USA
mhutton@altera.com

Abstract - We propose a framework to unify the process of false paths and multi-cycle paths in static timing analysis (STA). We use subgraphs attached with timing constraints to represent false paths and multi-cycle paths. The complexity of the subgraph representation is reduced to improve efficiency. Finally, we present theorems to show that the unified framework produces correct timings. The experimental results demonstrate that the minimization is effective for both artificial and industry test cases.

Keywords: static timing analysis, false subgraphs, multi-cycle subgraphs, time shifting, biclique covering.

1. Introduction

Static timing analysis (STA) is widely used in performance driven optimization programs. The overall procedure of STA performs a forward propagation to derive the *arrival times*, and a backward propagation to derive the *required arrival times* and *slacks* at all the points in the circuits [1]. These slacks are useful information for optimization programs.

To derive accurate slacks, STA has to deal with false paths and multi-cycle paths. A *false path* is a path not logically realizable [2]. False path timings must be eliminated from timing analysis. A *multi-cycle path* is a path that signals propagate longer than one clock cycle [8]. The accurate slacks of multi-cycle paths should be computed using multi-cycle arrival times. False paths and multi-cycle paths have to be dealt with efficiently because timing analysis is invoked heavily in the inner loop of optimization programs.

Previous published works in STA focused on false paths [2] [3] [4]. K. P Belkhale et al. [2] used *tags* to distinguish and remove false path timings. Each tag contains a set of false subgraph labels. E. Goldberg et al. [3] proposed to reduce the number of timings with *tags* according to the timing values. Because the timing values may change during the circuit optimization, the reduction is performed together with timing analysis, which induces runtime penalty in the optimization process. D. Blaauw et al. [4] removed the specified false subgraphs and produced a new timing graph using node splitting and edge removal. The optimal set of nodes for splitting is identified. We are unaware of reports that can unify the process of false paths and multi-cycle paths efficiently.

In this paper, we propose a framework to unify the process of false paths and multi-cycle paths with *exceptional rules*, and minimize the number of *rule sets* to improve the efficiency. The contributions of the paper are as follows.
● We represent each *exceptional rule* with a *subgraph* attached with the *setup time* and *hold time*. The rules cover false paths and multi-cycle paths. For the false subgraph, the setup time and hold time are unbounded, i.e., $+\infty$ and $-\infty$, respectively.
● We use *rule sets* to group a set of rules when the timing

information can be shared at a particular vertex. We allow *priority* for the rules. When there is a conflict among the rules in a rule set, the rule with the highest priority dominates.
● We devise time shifting to align the hold and setup times of the rules. By doing so, we can merge the distinguished timings, and collect different rule sets into one *rule collection*. For example, when a 2-cycle path and a 3-cycle path converge at a vertex, we can shift the arrival time of the 2-cycle path by one cycle, and merge the timing information.
● We adopt the *biclique covering* approach in [9] to minimize the number of rule collections at every vertex. We present theorems to guarantee that based on the minimized tags timing analysis produces correct slacks.
● The cost of the rule collection minimization is only incurred at the initializing stage, thus not contributing to the CPU time of the optimization program.
● We test our approach on both artificial and industry test cases. For four industry test cases, the number of rule collections is reduced by 84.6% and the runtime of STA decreases by 38.13% on the average. The runtime of minimization is 383 seconds for the test case containing 533,224 nets.

The remainder of this paper is organized as follows. Section 2 defines terminologies. In Sections 3 and 4, we propose the unified framework and the minimization algorithm. The experimental results are presented in Section 5. In the last section, we give the conclusions.

2. Terminology

Timing analysis is performed on a timing graph, which is a directed acyclic *graph* $G = \{V, E\}$, where V is a set of vertices and E is a set of edges. Each *edge* (u, v) is an ordered pair from vertex u to vertex v. The input degree of vertex v, $d^-(v)$, is the number of edges ending at v. The output degree of v, $d^+(v)$, is the number of edges starting at v. The begin set $B = \{v \mid v \in V, d^-(v) = 0\}$ is the set of primary input vertices. The destination set $D = \{v \mid v \in V, d^+(v) = 0\}$ is the set of primary output vertices.

A *path* p in graph G is a sequence of vertices and edges. We can represent path p by only the edges in the path [5]. Each vertex v separates path p into a *head* and a *tail*. Since graph G is directed acyclic, all the paths in G are simple. If a path starts from a vertex in the begin set B, it is a *prefix path* p^-. If a path ends at a vertex in the destination set D, it is a *suffix path* p^+. A *complete path* is both a prefix and a suffix path, which starts from a vertex in the begin set B, and ends at a vertex in the destination set D. The *prefix cone* $P^-(v)$ of a vertex v contains all the prefix paths ending at v. The *suffix cone* $P^+(v)$ of v contains all the suffix paths starting at v.

In Figure 1, the begin set of graph G contains vertices 1 and 2, and the end set contains vertices 8 and 9. Prefix cone of vertex 5 has two prefix paths $\{(1, 3), (3, 5)\}$ and $\{(2, 4), (4,$

5)}. Vertex 5 separates complete path {(1, 3), (3, 5), (5, 7), (7, 9)} into head {(1, 3), (3, 5)} and tail {(5, 7), (7, 9)}.

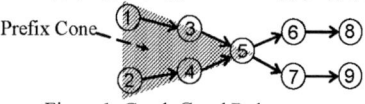

Figure 1. Graph G and Paths.

3. Unified Framework Processing False Paths and Multi-cycle Paths

We represent false paths and multi-cycle paths as *exceptional rules*, and create *rule sets* to unify the process of false paths and multi-cycle paths. We follow the procedure in [2] to compute the rule sets, except that both false and multi-cycle path rules are included in the rule sets. By doing so, we can remove false path arrival times and compute correct slacks for multi-cycle paths according to multi-cycle arrival times.

3.1 General Rule and Exceptional Rules

The *general rule* on graph G is that the complete path p in G satisfies the hold and setup time $[h, s]$, i.e., $h \leq$ delay $(p) \leq s$.

An *exceptional rule* r describes a false or multi-cycle *subgraph*, $G_r = \{V_r, E_r\}$, a pair of hold and setup time $[h_r, s_r]$, and a priority p_r.

● For the multi-cycle subgraph, $[h_r, s_r]$ is the multi-cycle arrival time, and for the false subgraph, the hold time and setup time are unbounded, i.e., $-\infty$ and $+\infty$, respectively.

● Subgraph G_r describes a set of false paths or multi-cycle paths governed by rule r. The begin set of G_r, denoted as B_r, is a set of vertices which have no input edges in E_r.[1] The destination set of G_r, denoted as D_r, is a set of vertices which have no output edges in E_r. A prefix path p_r^- in G_r starts from a vertex in B_r, and a suffix path p_r^+ in G_r ends at a vertex in D_r. The complete path p_r in G_r is both a prefix and a suffix path in G_r. A path p is a false path or multi-cycle path governed by rule r if the intersection of path p and E_r is a complete path in subgraph G_r. All the paths governed by rule r are constrained by an inequality $h_r \leq$ delay $(p) \leq s_r$.

● If a path is governed by various rules, the rule with the highest priority p_r supersedes others.

Figure 2 contains two rules, false subgraph rule 0 and multi-cycle subgraph rule 1. Complete path {(1, 3), (3, 5), (5, 6), (6, 8)} is a false path because it contains complete path {(3, 5), (5, 6)} in subgraph G_0. Another complete path {(2, 4), (4, 5), (5, 7), (7, 9)} is a 2-cycle path because it belongs to the subgraph G_1. Complete path {(1, 3), (3, 5), (5, 7), (7, 9)} belongs to both G_0 and G_1. Because the priority of rule 0, i.e., $p_0 = 2$, is higher than the priority of rule 1, i.e., $p_1 = 1$, the complete path is a false path.

When multiple rules are specified, we map the rules on graph G to formulate rule sets at every vertex v and edge (u, v):

● Rule set $F(v) = \{r| v \in B_r\}$ contains rules starting from v;

● Rule set $T(v) = \{r| v \in D_r\}$ contains rules ending at v;

● Rule set $I(u, v) = \{r| (u, v) \in E_r\}$ contains rules covering edge (u, v).

For example rule set $F(3)$ of vertex 3 contains rule 0. Rule set $T(9)$ of vertex 9 contains both rule 0 and rule 1. Rule set $I(5, 7)$ of edge (5, 7) contains rule 0 and rule 1.

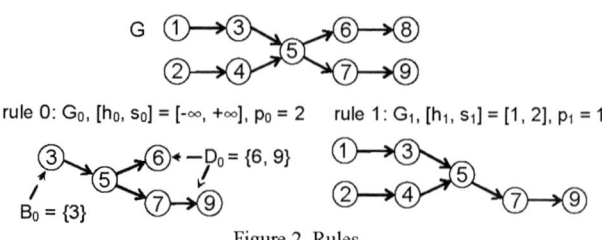

Figure 2. Rules.

3.2 Rule Set Computation

We compute *rule sets* for prefix paths and use prefix *rule sets* to distinguished arrival times.

Definition 3.1: Given a prefix path p^-, the rule set of the prefix path p^- is $R(p^-) = \{r| p^- \cap E_r$ is a prefix path in G_r and the tail of $p^-\}$. Given a suffix path p^+, the rule set of the suffix path p^+ is $R(p^+) = \{r| p^+ \cap E_r$ is a suffix path in G_r) and the head of $p^+\}$.

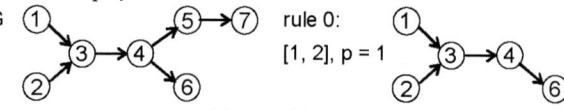

Figure 3. Rule sets.

Conceptually, the prefix or suffix rule set indicates whether the prefix or suffix path belongs to the paths governed by a rule. In Figure 3, the rule set of prefix path p^-_1 = {(1, 3), (3, 4)} contains rule 0 because p^-_1 is the prefix path of {(1, 3), (3, 4), (4, 6)}, which is governed by rule 0. Another prefix path p^-_2 = {(1, 3), (3, 4), (4, 5)} exits from G_0 at vertex 4, which is not an ending vertex of rule 0. Therefore, rule set $R(p^-_2)$ does not contain rule 0, which means prefix path p^-_2 does not belong to the path of rule 0.

The prefix *rule sets* and the arrival times are computed as follows. [2]

Rule Set Computation (v)
I. If vertex v is a primary input
    If $F(v) \cap T(v)$ does not contain false subgraph rule, produce a
    rule set $R = F(v)$;
II. else
    For each edge (u, v)
      For each rule set R *at vertex* u
      1. $R' = (R \cap I(u,v)) \cup F(v)$;
      2. If $R' \cap T(v)$ does not contain false subgraph rule,
        arrival_t(v,R')=max(arrival_t(v,R'),arrival_t(u,R)
        +delay(u,v));

In step *I*, if $F(v) \cap T(v)$ contains false subgraph rules, all the paths through vertex v are false paths. Therefore, we eliminate the arrival times of these false paths by producing no rule set. In step *II.1*, the intersection $R \cap I(u,v)$ means that only rules containing edge (u, v) remain in the rule set. The union with rule set $F(v)$ means that we include rules starting from v in the rule set. In step *II.2*, if the intersection $R' \cap T(v)$ contains false path rules, we eliminate false path arrival times by producing no rule set. Figure 4 illustrates the rule set computation, we compute rule set {0, 1} at vertex 2 by equation ({1} $\cap I(1, 2)) \cup F(2)$, where {1} is the rule set at vertex 1, $I(1, 2) = \{1\}$ is the rule set of edge (1, 2), and $F(2) = \{0\}$ is the starting rule set of vertex 2. At vertex 8, rule set {0, 1} is deleted because the intersection of rule set {0, 1} and ending rule set $T(8) = \{0\}$ contains false subgraph rule 0.

We calculate the required arrival time and slack for each rule set R by a backward sweeping. If the arrival time of R is

---

[1] Since a multi-cycle path is between a pair of flip flops, the vertices in the begin set and end set of a multi-cycle path rule are primary input and output vertices, respectively.

[2] We only compute the maximum arrival times. The minimum arrival times can be computed similarly using min in stead of max operation.

forward propagated to rule set R', the required arrival time of R' is backward propagated to R. For example in Figure 4, the required arrival time for rule set $\{0, 1\}$ at vertex 4, $req(4, \{0, 1\})$, is backward propagated from required arrival times $req(7, \{0, 1\})$ at vertex 7 and $req(6, \varnothing)$ at vertex 6.

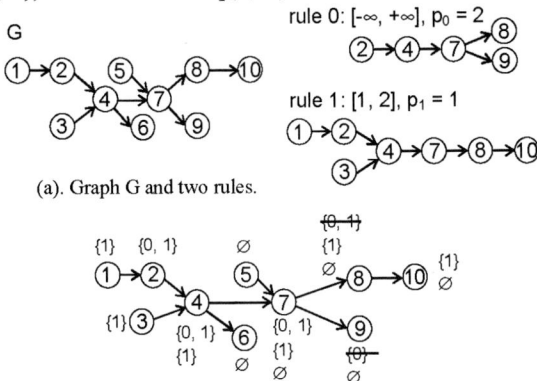

(a). Graph G and two rules.

(b). Prefix rule sets at every vertex.

Figure 4. The rule set computation in a forward propagation.

We show that the unified framework based on rule sets produces correct slacks.

Definition 3.2: A prefix rule set R covers a prefix path p^-, if the arrival time labeled by R is the maximum arrival times of a set of prefix paths including p^-.

For example at vertex 7 in Figure 4, prefix paths $\{(1, 2), (2, 4), (4, 7)\}$, $\{(3, 4), (4, 7)\}$ and $\{(5, 7)\}$ are covered by rule sets $\{0, 1\}$, $\{1\}$, and $\varnothing$, respectively.

Lemma 3.1: At each vertex v, the prefix rule sets produced by forward *Rule Set Computation* cover all the prefix paths in prefix cone $P^-(v)$, and the suffix rule sets produced by backward *Rule Set Computation* cover all the suffix paths in suffix cone $P^+(v)$.

Based on Lemma 3.1, we have Theorem 3.1 as follows.

Theorem 3.1: The slack computation based on prefix rule sets produces correct slacks at every vertex.

4. Rule Collection Minimization

The minimization follows the biclique covering approach in [9], which minimized the number of *rule sets* for false paths. In order to include multi-cycle paths, we devise time shifting to align different hold and setup times, and collects rule sets into *rule collections*. In this section, we first define the rule collection. Then, we use examples to show basic ideas of the biclique covering approach and time shifting. Finally, we propose our minimization algorithm and show that the produced slacks are correct.

Definition 4.1: A rule collection at vertex v is a set of rule sets of prefix paths which ends at v, i.e. $\Re(v) \subseteq \{R(p^-) | p^- \in P^-(v),\}$, where $P^-(v)$ is the prefix cone at v.

Although the rule sets in a rule collection may contain rules with different hold and setup times, we use *time shifting* to align, which is introduced 4.2.3.

4.1 Motivation

We use two examples to show that the arrival times distinguished by different rule sets can be merged; thus the rule sets can be collected.

Example 1: The paths of three false subgraphs converge at vertex 5, and then diverge at vertex 6. The rule set computation produces four rule sets at vertex 5 and 6. According to the rule set propagation relations in Figure 5 (b),

we find that the arrival times $a(5, \{0\})$ and $a(5, \{1\})$ are both propagated through suffix paths $\{(5, 6), (6, 9)\}$ and $\{(5, 6), (6, 10)\}$. Therefore, we can merge the arrival times and collect rule sets $\{0\}$ and $\{1\}$ into one rule collection $\{\{0\}, \{1\}\}$. Similarly, we can merge other arrival times and produce rule collections. As a result, three arrival times distinguished by rule collections are enough to cover all the paths through vertex 5 as shown in Figure 5 (c).

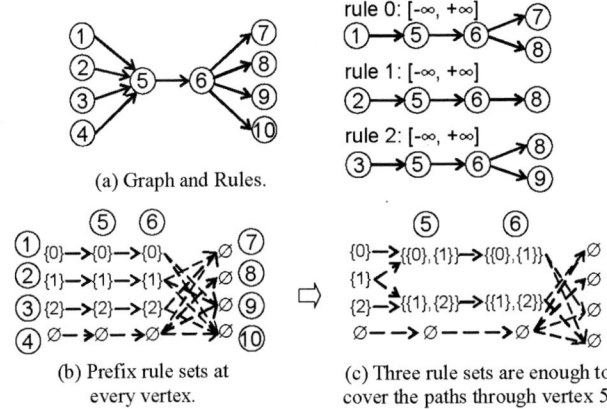

(a) Graph and Rules.

(b) Prefix rule sets at every vertex.

(c) Three rule sets are enough to cover the paths through vertex 5.

Figure 5. Cover complete paths with fewer rule sets.

Example 2: A 2-cycle path $\{(1, 3), (3, 4)\}$ with hold and setup time $[1, 2]$, and a 3-cycle path $\{(2, 3), (3, 4)\}$ with hold and setup time $[2, 3]$ converge at vertex 3. Because the hold and setup times are different, the slack of these paths should be computed separately at vertex 4. Therefore, at vertex 3, we have to distinguish the arrival times by rule sets $\{0\}$ and $\{1\}$. However, if we shift the arrival time of prefix path $\{(1, 3)\}$ forward by 1 cycle, we can merge two arrival times and use hold and setup time $[2, 3]$ to compute the slack for both paths. As a result, we can collect two rule sets into one rule collection.

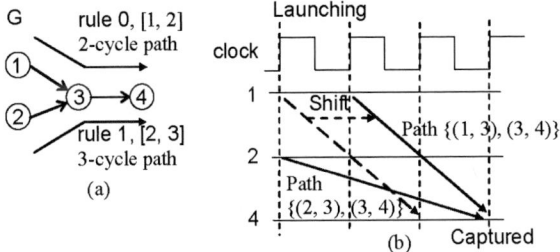

Figure 6. Time shifting: (a) 2-cycle and 3-cycle paths have hold and setup times $[1, 2]$ and $[2, 3]$, respectively; (b) Shift arrival time of path $\{(1, 3), (3, 4)\}$ to align the hold and setup times.

The examples indicate that: 1) the rule set minimization requires information of the complete paths from both forward and backward sweepings, and 2) we can align the hold and setup times of various multi-cycle paths by time shifting, and merge the timings.

4.2 Rule Collection Minimization Algorithm

The basic idea of the minimization is as follows. 1) Gather prefix and suffix rule sets at each vertex. 2) Obtain and align hold and setup times of complete paths. 3) Collect rule sets into rule collections and cover the complete paths.

4.2.1 Main Flow

We first compute suffix rule sets of all the vertices by backward *Rule Set Computation*. Then, the rule collections are minimized and propagated at every vertex in topological

1C-3

order. The main flow is as follows.

Main flow

I. *Produce suffix rule sets by a backward sweeping;*
II. *For each vertex v in topological order*
  1. *If vertex v is a primary input*
    Produce an initial rule collection $\Re(v) = \{F(v)\}$;
  2. *Rule collection minimization(v);*
  3. *For each edge (v, u) Rule Collection Propagation ($\Re(v)$, v, u);*

The rule collection minimization of *II.2* is as follows.

Rule collection minimization (v)

I. *For each rule collection $\Re(v)$*
  For each suffix rule sets $R(p^+)$, where p^+ is the suffix path in the suffix path cone $P^+(v)$
    Intersect rule collection $\Re(v)$ with suffix rule sets $R(p^+)$;
II. *Construct bipartite graph at v based on the intersections;*
III. *Shift the hold and setup times of the intersections, and perform biclique covering on bipartite_graph(v);*

The rule collection propagation in *II.3* propagates rule collection $\Re(v)$ to vertex *u*, which is introduced in Section 4.2.4.

4.2.2 Intersections of the Prefix and Suffix Rule Sets

At each vertex, the intersections of the prefix and suffix rule sets provide the hold and setup times of the complete paths. If rule *r* belongs to intersection $R(p^-) \cap R(p^+)$, the complete path of prefix path p^- plus suffix path p^+ is governed by rule *r*. If there are various rules in intersection $R(p^-_i) \cap R(p^+)$, the rule with the highest priority is dominant.

Definition 4.2: The intersection between a prefix rule collection $\Re(v)$ and a suffix rule set $R(p^+)$ intersects each prefix rule set in $\Re(v)$ with $R(p^+)$, i.e., $Intersect(\Re(v), R(p^+)) = \{R(p^-_i) \cap R(p^+) \mid R(p^-_i) \in \Re(v)\}$.

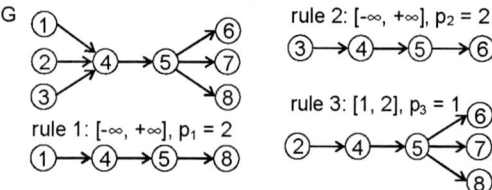

Figure 7. Example for rule collection minimization.

The intersections correspond to the complete paths. For example at vertex 4 in Figure 7, there are three prefix paths and three suffix paths. The first column in Table 4.1 contains the rule collections for the prefix paths, and the first row shows the suffix rule sets. Rule collection {{3}} corresponds to prefix path {(2, 4)}, and suffix rule set {1, 3} corresponds to suffix path {(4, 5), (5, 8)}. *Intersect*({{3}}, {1, 3}) = {{3}} corresponds to complete path {(2, 4), (4, 5), (5, 8)}, which is in rule 3 with hold and setup time [1, 2].

We construct a bipartite graph based on the intersections to cover the complete paths through the vertex. For every intersection, we produce an edge from the rule collection to the suffix rule set and attach the hold and setup time of the intersection on the edge. We eliminate the edge with hold and setup time [-∞, +∞], thus removing false paths.

Table 4.1 Intersections of the prefix rule collections and the suffix rule sets at vertex 4

$R(p^+)$ / $\Re(v)$	{3}	{1, 3}	{2, 3}
{{1[-∞, +∞]}}	{∅[0, 1]}	{{1[-∞, +∞]}}	{∅[0, 1]}
{{2[-∞, +∞]}}	{∅[0, 1]}	{∅[0, 1]}	{{2[-∞, +∞]}}
{{3[1, 2]}}	{{3[1, 2]}}	{{3[1, 2]}}	{{3[1, 2]}}

Figure 8 illustrates the bipartite graph at vertex 4. The bipartite graph does not contain an edge from rule collection {{1}} to suffix rule set {1, 3} because the hold and setup time

of *Intersect* ({{1}}, {1, 3}) = {{1}} is [-∞, +∞].

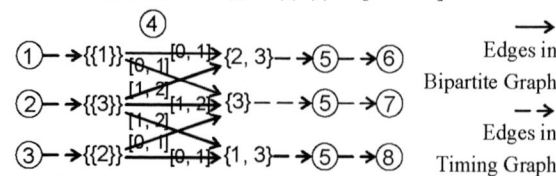

Figure 8. Bipartite graph at vertex 4.

4.2.3 Time Shifting and Biclique Covering

We cover the edges in the bipartite graph by a set of bicliques, i.e. complete bipartite graphs, and collect the prefix rule sets in each biclique into one rule collection. By doing so, we cover the complete paths represented by the edges by rule collections.

All the complete paths covered by one biclique should have aligned hold and setup times. Therefore, we devise time shifting to align different hold and setup times.

Definition 4.3: *Time shifting* on rule collection $\Re(v)$ plus ΔT cycles on the hold and setup time of every rule in $\Re(v)$. The rule collection after time shifting is denoted as $\Re(v)^{+\Delta T}$.

Definition 4.4: *Biclique covering* on a bipartite graph covers the bipartite graph by a set of complete bipartite subgraphs.

We produce aligned bicliques by time shifting and cover the edges of the bipartite graph at vertex *v*. A biclique *b* is *aligned* if for any suffix rule set $R(p^+) \in b$, all the edges to $R(p^+)$ are attached the same hold and setup time. A biclique *b* can be *aligned by time shifting* if there are a set of ΔTs such that after time shifting on each rule collection $\Re(v)_i \in b$ by ΔT_i, and updating the hold and setup times of the edges, biclique *b* becomes aligned.

Biclique Covering (v)

I. *Initialize the biclique set as $B = \varnothing$;*
II. *For every rule collection $\Re(v)$ in minimum degree order*
  a) For every biclique b in the biclique set B
    If b remains a biclique after including a set of edges from $\Re(v)$, and b can be aligned by timing shifting
      Add edges from $\Re(v)$ to b, and do time shifting on b;
  b) If $\exists$ edges from $\Re(v)$ not covered by bicliques in B
    1. *For every biclique b in B*
      If b includes edges from $\Re(v)$, remove the edges from $\Re(v)$ and Shift the hold and setup times back;
    2. *Produce a new biclique containing all the edges from $\Re(v)$;*
    3. *Add the new biclique to the biclique set B;*
III. *For each biclique b in the biclique set B*
  New $\Re(v)' = \cup \Re(v)_i^{\Delta Ti}$, where $\Re(v)_i \in$ biclique b, the hold and setup time of $\Re(v)_i$ is shifted by ΔT_i;

Since computing the minimum biclique covering on the general bipartite graph is NP complete [6, 7], we use a minimum degree order approach. Each time, we try to cover the edges from a rule collection $\Re(v)$ by enlarging smaller bicliques. If all the edges from $\Re(v)$ are covered and the enlarged bicliques are aligned under time shifting, we update the enlarged bicliques by adding the edges and shifting the hold and setup times. Otherwise, we produce a new biclique containing the edges from the rule collection $\Re(v)$. Based on each biclique *b*, the new rule collection is the union of all the rule collections in the biclique, i.e., $\Re(v)' = \cup \Re(v)_i$, $\Re(v)_i \in b$.

Figure 9 shows the biclique covering at vertex 4. Rule collection $\Re'_1$ collects rule collections {{1}} and {{3}}. The hold and setup time of {{1}}, [0, 1], is shifted by 1 cycle to align with the hold and setup time of {{3}}, [1, 2]. After minimization, two rule collections cover the complete paths

76

through vertex 4.

Time shifting 1 cycle: $[0, 1]+1=[1, 2] \Rightarrow \Re'_1 = \{\{1\}^{+1}, \{3\}\}$

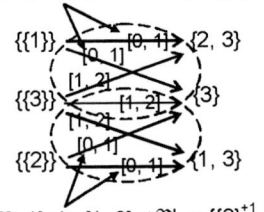

Time shifting 1 cycle: $[0, 1]+1 = [1, 2] \Rightarrow \Re'_2 = \{\{2\}^{+1}, \{3\}\}$

Figure 9. Biclique covering at vertex 4.

4.2.4 Rule Collection Propagation

For edge (v, u), we propagate rule sets in rule collection $\Re(v)$ to vertex u similarly as the *Rule Set Computation* computes the rule sets.

Rule Collection Propagation ($\Re(v)$, v, u)

For each $R \in \Re(v)$

1. $R' = (R \cap I(v, u)) \cup F(u)$;
2. $\Re(u) = \Re(u) \cup \{R'\}$, where $\Re(u)$ is initialized as $\varnothing$;

4.2.5 Hold and setup Time Conflict

The intersections $R(p^-_i) \cap R(p^+)$ in each *Intersect($\Re(v)$, $R(p^+)$)* should produce the same hold and setup time. Otherwise, there is hold and setup time conflict. For example, Table 4.2 shows the intersections at vertex 5. The rule collections are propagated from vertex 4. The intersection of rule collection $\{\{1\}^{+1}, \{3\}\}$, and suffix rule set $\{1, 3\}$ contains two sets, $\{1\}^{+1}$ and $\{3\}$. The hold and setup time of rule 1, i.e., $[-\infty, +\infty]$, conflicts with the hold and setup time of rule 3, i.e., $[1, 2]$.

Table 4.2 Intersections at vertex 5

$\Re(v)$ \ $R(p^+)$	$\{3\}$	$\{1, 3\}$	$\{2, 3\}$
$\{\{1\}^{+1}, \{3\}\}$	$\{\varnothing^{+1}, \{3\}\}$	$\{\{1\}^{+1}, \{3\}\}$	$\{\varnothing^{+1}, \{3\}\}$
$\{\{2\}, \{3\}\}$	$\{\varnothing^{+1}, \{3\}\}$	$\{\varnothing^{+1}, \{3\}\}$	$\{\{2\}^{+1}, \{3\}\}$

The hold and setup time conflict indicates that the complete paths corresponding to the intersection have conflict hold and setup times. For example, *intersect($\{\{1\}^{+1}, \{3\}\}$, $\{1, 3\}$)* at vertex 5 corresponds to two complete paths, $p_1 = \{(1, 4), (4, 5), (5, 8)\}$ which is a false path, and $p_2 = \{(2, 4), (4, 5), (5, 8)\}$ which is multi-cycle path.

Because each rule collection only labeling one arrival time, we cannot use time shifting to align these different hold and setup times. Therefore, in the bipartite graph we do not produce the edge between the rule collection and the suffix rule set if there is a conflict. Theorem and lemmas in Section 4.4 guarantee that all the paths are still covered.

Bipartite graph Construction (v)

I. Collect rule collections propagated from the previous vertices;
II. For each rule collection $\Re(v)$ and suffix rule set $R(p+)$
1. *Intersect($\Re(v)$, $R(p^+)$) = $\{R(p^-)_i{}^{\Delta Ti} \cap R(p^+)| R(p^-)_i \in \Re(v)\}$*;
2. For each $R(p^-)_i{}^{\Delta Ti} \cap R(p^+) \in$ *Intersect($\Re(v)$, $R(p^+)$)*
The hold and setup time is $[h_r + \Delta T_i, s_r + \Delta T_i]$, where r is the rule in $R(p^-)_i{}^{\Delta Ti} \cap R(p^+)$ with the highest priority;
3. If there is no conflict among all $[h_r + \Delta T_i, s_r + \Delta T_i]$s, and $\forall [h_r + \Delta T_i, s_r + \Delta T_i] \neq [-\infty, +\infty]$
Add an edge attached with $[h_r + \Delta T_i, s_r + \Delta T_i]$ from $\Re(v)$ to $R(p^+)$;

Figure 10 illustrates the bipartite graph at vertex 5, which does not contain edges from $\{\{1\}^{+1}, \{3\}\}$ to $\{1, 3\}$ and from $\{\{2\}^{+1}, \{3\}\}$ to $\{2, 3\}$ due to hold and setup time conflicts.

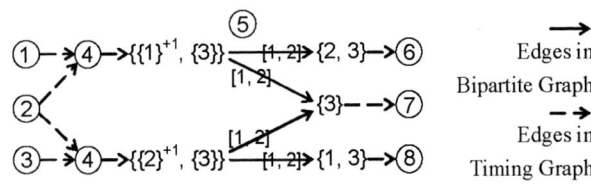

Figure 10. Bipartite graph at vertex 5.

Figure 11 summarizes the rule collections produced by our approach. At vertices 4 and 5, the number of tags is reduced from 3 to 2.

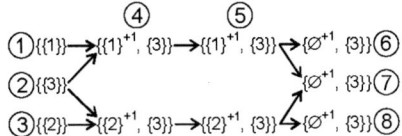

Figure 11. Rule collections at every vertex: $\varnothing^{+1}$'s at vertex 6 and 8 are propagated from $\{1\}^{+1}$ and $\{2\}^{+1}$ at vertex 5, respectively; $\varnothing^{+1}$ at vertex 7 is propagated from $\{1\}^{+1}$ and $\{2\}^{+1}$ at vertex 5.

4.3 Timing Analysis with Rule Collections

We compute the arrival time, required arrival and slack for each rule collection by forward and backward sweepings similar as the timing analysis process based on rule sets. The only difference is that for rule collections with time shifting, $\Re(v)^{+\Delta T}$, we forward and backward shift the arrival times and required arrival times by ΔT. In Figure 11, for the rule collection $\Re'_1 = \{\{1\}^{+1}, \{3\}\}$ at vertex 4, the arrival time of prefix path $\{(1, 4)\}$ is shifted by 1 cycle and merged with the arrival time of $\{(2, 4)\}$. When the required arrival time labeled by $\Re'_1$ is backward propagated to vertex 1, we shift the required arrival time back by 1 cycle.

4.4 Correctness

This section presents lemmas and theorems to guarantee the correctness. We show that all the complete paths are covered by rule collections. Thus, timing analysis based on rule collections produces correct slacks of all the paths. We omit formal proofs due to space constraints.

Definition 4.5: A rule collection $\Re(v)$ covers a complete path p through vertex v, where p is the concatenation of prefix path $p^-\in P^-(v)$ and suffix path $p^+\in P^+(v)$, if 1) p^- is covered by prefix rule set $R(p^-) \in \Re(v)$, 2) p^+ is covered by suffix rule set $R(p^+)$, and 3) the bipartite graph at vertex v contains an edge from $\Re(v)$ to $R(p^+)$.

Lemma 4.1: If rule collection $\Re(u)$ covers a complete path p at vertex u and edge $(u, v) \in p$, after *Rule Collection Propagation*, the produced rule collection $\Re(v)$ covers path p at vertex v.

Theorem 4.1: All the complete paths through a vertex v are covered by the rule collections at v.

Lemma 4.1 and Theorem 4.1 show that rule collections cover all the complete paths. The next lemma and theorem show that timing analysis produces correct slacks.

Lemma 4.2: If rule collection $\Re(v)$ covers path p and path p is governed by rule r, *slack* $(v, \Re(v))$ covers the correct slack of path p, which is computed based on rule r.

Theorem 4.2: Timing analysis with rule collections produces correct slack at each vertex v, which covers the slacks of all the paths through vertex v.

5. Experimental Results

We test our algorithm on both artificial and industry test cases. The algorithm is implemented in C and tested on a

Pentium 4 Linux machine.

We first follow the experiments in [2] to randomly create false and multi-cycle subgraphs on a 100×100 mesh. The average number of edges in subgraphs is 6000. Each test case contains from 9 to 104 rules including 30 percent false subgraph rules. The hold and setup times of multi-cycle paths are in the range from 2-cycle to 4-cycle. We compare the number of rule collections with the number of prefix rule sets in Table 5.1. The number of prefix rule sets equals to the number of tags produced by the approach in [2] if there are only false paths. The average reduction ratio for five test cases is 31.22%. The CPU time for tag minimization increases when the number of rules in the case increases. The largest case including 104 rules consumes 87 seconds.

Table 5.1. *Tag* minimization on a 100 ×100 mesh

# rules	# prefix rule sets	# rule collections	Reduction ratio	Runtime (sec)
9	9129	8281	9.29%	2
34	77102	49321	36.03%	19
69	137581	89987	34.59%	44
88	176384	97124	44.94%	61
104	209718	145484	30.63%	87
average			31.10%	

--Reduction ratio= (#prefix rule sets - #rule collections)/ (#prefix rule sets)

We also test our algorithm on four industry test cases and show the experimental results in Table 5.2. The largest circuit, i.e., atmlcore, contains 533,224 nets, 2 false path rules and 2262 multi-cycle path rules. We use *Rule Set Computation* to produce prefix rule sets, and minimize rule sets into rule collections. The average reduction ratio of four test cases is 84.60%. For the largest circuit, i.e. atmlcore, the runtime of minimization is only 383 seconds, including the CPU time for loading the test cases, mapping the rules on the graph, and minimizing the rule collections.

Table 5.2 also shows runtimes of STA (STA). If STA uses prefix rule sets to deal with false paths and multi-cycle paths, for the largest circuit, i.e., atmlcore, the runtime is 40.33 seconds. If rule collections are used in STA, the runtime is

reduced to 19.5 seconds. Though the reduction is only 20.83 seconds for performing STA once, the reduction ratio is 51.65%. If timing analysis is repeatedly called during performance driven optimization, for example 100 times, the reduction on STA runtime would be 2083 seconds, which is larger than the minimization runtime cost 40.33 seconds.

6. Conclusions

We propose a framework to unify the process of false paths and multi-cycle paths in STA. Furthermore, we improve the efficiency by minimizing the number of distinguished timings created for false paths and multi-cycle paths. Finally, we present theorems to guarantee that our approach produces correct timing information with false paths and multi-cycle paths considered. The experimental results demonstrate that our minimization is effective.

Acknowledgements

This work was supported in part by the California MICRO program and a grant from Altera Corp.

References

[1] R. B. Hitchcock, "Timing Verification and Timing Analysis Program", DAC 1982, pp. 594-604.
[2] K. P. Belkhale, A. J. Suess, "Timing Analysis with known False Sub-Graphs", ICCAD 1995, pp. 736-740.
[3] E. Goldberg, A. Saldanha, "Timing Analysis with Implicitly Specified False Path", Int. Workshop on Timing Issues in the Specification and Synthesis of Digital Designs, T99, 1999.
[4] D. Blaauw, R. Panda, A. Das, "Removing user-specified false paths from timing graphs", DAC 2000, pp. 270-273.
[5] TC Hu, Combinatorial Algorithms, Dover Publication, Second Edition, 2002.
[6] J. Orlin, "Containment in graph theory: Covering graphs with cliques", Nederl. Akad. Wetensch. Indag. Math., 39:211-218, 1977.
[7] H. Muller, "Alternate Cycle-Free Matchings", Order, (7):11-21, 1990.
[8] M. Hutton, D. Karchmer, B. Archell, J. Govig, "Efficient Static Timing Analysis and Applications Using Edge Masks", FPGA 2005, pp. 174-183.
[9] S. Zhou, B. Yao, H. Chen, Y. Zhu, CK Cheng, M. Hutton, et al., "Improving the efficiency of Static Timing Analysis with False Paths", ICCAD 2005, pp. 527-531.

Table 5.2. *Tag* minimization on industry test cases

Cases	# nets	#rules		# prefix rule sets	# rule collections	reduction ratio	Minimization Runtime (sec)	STA runtime		
		False path	Multi-cycle path					Use prefix rule sets (sec)	Use rule collections (sec)	Runtime reduction
tdl	27,555	1	27	158	67	57.59%	1	1.2	1.2	0
cq_mod	38,535	2517	3181	217,456	14,972	93.11%	22	4.2	2	52.38%
pm25c	325,582	7	2574	1,781,400	101,238	94.32%	106	55.33	28.5	48.49%
atmlcore	533,224	2	2262	2,411,892	159,451	93.39%	383	40.33	19.5	51.65%
average						84.60%				38.13%

--Reduction ratio = (#prefix rule sets - #rule collections)/ (#prefix rule sets)

--Reduction of STA runtime = (STA Runtime using prefix rule sets – STA Runtime using rule collections)/ STA Runtime using prefix rule sets

Crosstalk Analysis using Reconvergence Correlation

Sachin Shrivastava, Rajendra Pratap, Harindranath Parameswaran, and Manuj Verma

Cadence Design Systems, India Pvt. Ltd.
Noida Special Economy Zone, Noida, India-201305
Tel : +91-120-2562842 , Fax : +91-120-2562231
e-mail: sachins@cadence.com,pratap@cadence.com,harindra@cadence.com
and mverma@cadence.com

Abstract— In the UDSM era, crosstalk is an area of considerable concern for designers, as it can have a considerable impact on the yield, both in terms of functionality and operating frequency. Methods of crosstalk analysis are pessimistic in nature and the effort is ongoing to come up with techniques that make the analysis as realistic as possible. Using information from timing analysis is one such technique where we use data about overlap in switching among nets to identify those that can potentially switch together. Existing techniques tend to look at the set of a victim and associated aggressor nets in isolation, and select a subset of aggressors based on the absolute timing windows of these nets, thus ignoring the information associated with the fanin of these nets. In reality, however, some of these nets may never switch together because the reconvergence of those nets has not being factored in. Ignoring this correlation can cause false failures being flagged, leading to increased design cycles and conservatism in the design. We propose a technique where the correlation due to reconvergence can be captured in terms of relative switching windows [1]. We apply this technique to real designs and show that this leads to more realistic analysis for crosstalk, and that we can see a reduction in the number of violations reported. We also analyze the effective of the method statistically.

I. INTRODUCTION

With shrinking process dimensions, and incresingly dense designs, crosstalk has been an area of increasing concern to ASIC designers over the past few years. Crosstalk can cause quiescent nets to glitch leading to functional failures, and can cause switching nets to speed up and slow down (depending on the relative directions of switching) thus potentially leading to timing violations in the design. Crosstalk is extremely complex to analyze accurately, so existing techniques make several approximations that can speed up the analysis. These include usage of methods like constraint propagation, generation of noise models for cells, and usage of timing information. All of these techniques tend to make the result of analysis pessimistic, and can lead to the flagging of false violations, where the analysis tool shows a particular net or path to be failing, when in reality the design may perform correctly under the given operating conditions. This can lead to longer design cycles when the designer attempts to fix or verify the failure, and can also lead to overly conservative designs. One common technique used in analysis is to use the timing windows of the set of aggressor net(s) and of the designated victim net so as to find the worst potential noise [1][2][3][4][5]. In crosstalk glitch analysis (where the victim net is static), we look at the set of aggressors to see the strongest (in terms of glitch generated at the victim net receiver input) subset of aggressors that can switch in a common window. Similarly for finding the effect of crosstalk on delay, we look at a scenario where the victim net is switching together or against a set of aggressors, and try to find the combination where the victim signal switches the slowest (victim switching against the aggressor nets) or the fastest (victim switching with the aggressor nets). In general, this problem can be looked at in terms of taking a set of nets and trying to identify a set that maximizes some effect of interest - in this case glitch on victim net or timing change in the victim signal. At a high level, this problem can be decomposed into two parts, one where we identify all possible sets of aggressors that can switch together, and then trying to see which of these sets can have the maximum impact.

Existing approaches take the absolute timing windows of each of these aggressors, and either consider overlap, or take the overlap at the tap point. However, these approaches take the timing windows to be independent of each other. This however, can lead to pessimistic results because it does not consider the case where these nets may not be able to switch together due to reconvergence correlation between the nets. By reconvergence we mean the scenario where nets diverge from a common point and later in the fan-out cone there is need to find if these nets can switch together

Consider the case shown in Fig. 1. Here we have assumed

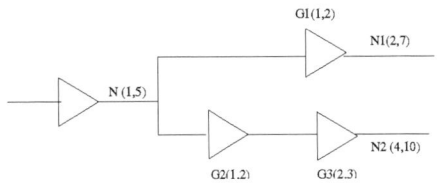

Fig. 1. Absolute timing windows

---

[1] The term "Relative Window" is also used in [8], to avoid any confusion we would like to mention that relative timing windows discussed in this paper is totally different from [8], and should not be confused with the concept defined there.

zero net delays without loss of generality since nets can always be replaced by buffers of equivalent delay. In this figure timing windows of net N is $(1,5)$ and timing windows of net $N1$ and $N2$, can be found by adding gate delays $G1$ and $G2+G3$ respectively. It therefore appears that timing windows of net $N1$ and $N2$ are overlapping, hence in conventional crosstalk analysis they are considered to be switching together. If these two nets are considered as attackers to a common victim net, analysis for worst case crosstalk effect on the victim net will take in the scenario where both of these nets attack the victim together. However, note that if net N switches at time t, net $N1$ switches in time $[t+1, t+2]$, while net $N2$ switches during $[t+3, t+5]$. Hence, in reality $N1$ and $N2$ can never switch together (except for the case when difference between two delays is more than time period, and transition of one net in n^{th} cycle can overlap with transition of other net in $n+1^{th}$ cycle.), and therefore cannot attack the victim net simultaneously. This leads to pessimistic analysis, and can result in a false failure being flagged at the victim net.

This pessimistic analysis happens because the technique does not consider the fact that in this case net $N1$ and $N2$ are correlated due to reconvergence. By reconvergence we mean that nets which diverge from a common point and which later are grouped together in the same set for analysis. In this case it applies to a set of nets which are being examined for a common switching windows. Since these nets diverged from some common point they will exhibit some correlation in their timing windows. Any analysis that ignores this correlation will tend to be inaccurate.

In this paper we define timing windows with reference to the diverging node in the fan-in cone of the net. We propose a method to perform crosstalk analysis using the relative timing windows which utilizes the re-convergence correlation, and thus addresses this source of pessimism present in the current approach.

We also examine this problem analytically, and show using probabilistic models that the probability of a violation being flagged reduces using this method. We have also shown some data on real designs which shows the effectiveness of the method proposed in the paper.

II. RELATIVE TIMING WINDOWS

Let (t_{min}^N, t_{max}^N) be the timing window of a net N, such that net N can switch during time t_{min}^N to t_{max}^N. Similarly (d_{min}^P, d_{max}^P) represents delay of a path P, where d_{min}^P is the minimum delay and d_{max}^P is maximum delay.

In Fig. 2, the timing windows of net N can be found by adding delays to the timing windows.

$$TW(N) = \bigcup_{i=1}^{n} TW(Ni) \oplus D(i) \tag{1}$$

where $TW(Ni) = (t_{min}^{Ni}, t_{max}^{Ni})$ is the timing window of net Ni, and $D(Ni \to N) = (d_{min}^{Ni \to N}, d_{max}^{Ni \to N})$ is the delay from net Ni to N. Here the *and* operation $\oplus$ is defined as $(a,b) \oplus (c,d) = (a+c, b+d)$.

In the absolute timing windows approach, we lose the relative information of the delays between net N and Ni due to this addition of delays to timing windows.

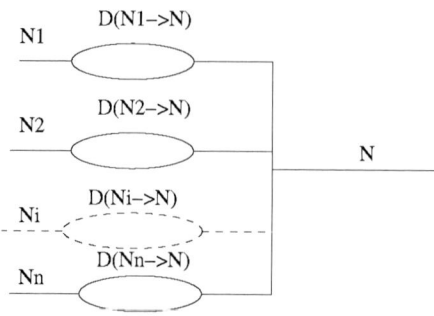

Fig. 2. Relative Timing windows

In the relative timing windows based approach proposed here, instead of adding delays to timing windows, the delays and timing windows of fan-in nets are preserved. Timing window of net N with reference to net Ni can be represented as $TW(N/Ni) = \{TW(N), D(Ni \to N)\}$. Hence,

$$TW(N) = \bigcup_{i=1}^{n} TW(N/Ni) \tag{2}$$

If net N has relative timing windows with reference to $N1$, and net $N1$ has timing windows with reference to net $N2$, timing windows of net N can be found with reference to $N2$. If,

$$TW(N/N1) = \{TW(N1), D(N1 \to N)\} \tag{3}$$
$$TW(N1/N2) = \{TW(N2), D(N2 \to N1)\} \tag{4}$$

Thus, the timing windows of net N with reference to $N2$ will be,

$$TW(N/N2) = \{TW(N2), D(N2 \to N1) \oplus D(N1 \to N)\} \tag{5}$$

III. RELATIVE TIMING WINDOWS BASED CROSSTALK ANALYSIS

In crosstalk analysis (noise or delay), timing windows are used to find if two nets can switch together or not. As we saw in ection I that nets $N1$ and $N2$ in Fig. 1 are considered to be switching together using common timing windows based approach.

If we use relative timing windows in this case, net $N1$ and $N2$ will have relative timing windows $\{TW(N), (1,2)\}$ and $\{TW(N), (3,5)\}$ respectively. Based on relative timing windows, timing windows of net $N1$ and $N2$ with reference to net N are $(1,2)$ and $(3,5)$, hence from use of relative timing windows, it is clear that net $N1$ and $N2$ can never switch together. The approach suggested here applies only to the nets constrained by synchronous clocks, since nets constrained by asynchronous clocks will be uncorrelated and can switch at any time.

Consider the case where we need to find out if two aggressor nets $A1$ and $A2$ can switch together as part of our analysis. Let

$\Phi(N)$ be the set of nets at all latest divergence points in the fan-in cone of net N. Then the relative timing windows of net $A1$ and $A2$ can be represented as,

$$TW(A1) = \bigcup_{i \in \Phi(A1)} \{TW(Ni), D(Ni \rightarrow A1)\} \qquad (6)$$

$$TW(A2) = \bigcup_{i \in \Phi(A2)} \{TW(Ni), D(Ni \rightarrow A2)\} \qquad (7)$$

Note that since aggressors $A1$ and $A2$ are constrained by synchronous clocks, they will always have common divergence points in the fan-in cone. If we trace back in the fan-in cone, we can find latest common points in the fan-in cones using standard graph algorithms. Let the set of latest common point in these fan-in cones be $\Phi(A1, A2)$. Using equations (3,4) we can represent timing windows of net $A1$ and $A2$ with reference to nets in set $\Phi(A1, A2)$.

$$TW(A1) = \bigcup_{i \in \Phi(A1, A2)} \{TW(Ni), D(Ni \rightarrow A1)\} \qquad (8)$$

$$TW(A2) = \bigcup_{i \in \Phi(A1, A2)} \{TW(Ni), D(Ni \rightarrow A2)\} \qquad (9)$$

While finding the overlap relationship of net $A1$ and $A2$, the first term in the relative timing windows (i.e. $TW(Ni)$) can be ignored since it is common for both nets. Now, based on delays $D(Ni \rightarrow A1)$ and $D(Ni \rightarrow A2)$, it can be found whether nets can switch together or not. It is immediately apparent that since delays $D(Ni \rightarrow A1)$ and $D(Ni \rightarrow A2)$ will be significantly less wider than the timing windows, this approach reduces the pessimism caused in the standard approach.

So far we did not consider the direction of switching (rise or fall). Due to different rise and fall delays, rise and fall timing windows can be different, and hence we need to find the rise and fall relative timing window of a net. In the conventional analysis we have absolute timing windows available for all the nets, hence we need to find a way to calculate relative timing windows using the absolute timing windows.

Let the timing window at a divergence point be (R_1, R_2) for rise transition and (F_1, F_2) for fall transition. Also let the delays of the path from this divergence point to the aggressor net under consideration be (r_1, r_2) and (f_1, f_2) for rise and fall transition respectively. Now the timing window at aggressor will depend on the unateness of the path. Hence, the rise and fall timing windows at the aggressor net will be,

- For a non-inverting path

$$(R_1 + r_1, R_2 + r_2), (F_1 + f_1, F_2 + f_2) \qquad (10)$$

- For an inverting path

$$(F_1 + r_1, F_2 + r_2), (R_1 + f_1, R_2 + f_2) \qquad (11)$$

- For a path which can be both inverting and non-inverting.

$$(min(R_1, F_1) + r_1, max(R_2, F_2) + r_2),$$
$$(min(R_1, F_1) + f_1, max(R_2, F_2) + f_2) \qquad (12)$$

If the timing windows at the divergence point are (R'_1, R'_2) for rise transition and (F'_1, F'_2) for fall transition, the relative timing windows will be,

- For a non-inverting path

$$(R'_1 - R_1, R'_2 - R_2)(F'_1 - F_1, F'_2 - F_2) \qquad (13)$$

- For an inverting path

$$(R'_1 - F_1, R'_2 - F_2), (F'_1 - R_1, F'_2 + R_2) \qquad (14)$$

- For a path which can be both inverting and non-inverting

$$(R'_1 - min(R_1, F_1), R'_2 - max(R_2, F_2)),$$
$$(F'_1 - (min(R_1, F_1), F'_2 - max(R_2, F_2)) \qquad (15)$$

We now propose a crosstalk noise analysis method using timing windows below

1. Perform Crosstalk analysis using conventional approach, and find list of violating nets.

2. For each violating net in the design.

   (a) Find all aggressors of this net.

   (b) For each synchronous group of aggressors

      i. Find common divergence points of all the nets in the group.
      ii. Find relative timing windows using equations (13-15)
      iii. Use relative timing windows to find which nets in the group can actually switch together, and modify the aggressor groups.

   (c) Based on the new groups, find glitch caused by the worst group.

Note that in the case of crosstalk delay analysis, we need to compute sets of aggressors that switch with victim net. This means that while finding common divergence point, victim net is also to be considered along with the aggressors.

IV. ANALYTICAL STUDY USING PROBABILISTIC MODELS

In this section we find the effectiveness of the proposed approach by finding the probability that this approach will reduce the glitch on a given net. Consider a victim net which is affected by N aggressors. For simplicity and without loss of generality we assume that all the aggressors are constrained by clocks of same time period T (This assumption can be relaxed by considering LCM of all time periods and then we can represent discontinuous timing windows). In the analysis we are assuming that all switching instances are uniformly distributed in the time period T (Though due to this assumption, the analysis may not be very realistic, but it will be indicative since we are using this analysis only to find the change in violations using our approach). If the width of the timing window of ith aggressor is τ_i, the probability of all the nets switching together is (Refer Appendix A, [7])

81

$$P_{old} = \left(\prod_{i=1}^{N} \frac{\tau_i}{T} \right) \left(\sum_{j=1}^{N} \frac{T}{\tau_j} \right) \qquad (16)$$

If we use our relative timing windows based approach, and if the width of timing window at the common divergence point is d, the probability will be

$$P_{new} = \left(\prod_{i=1}^{N} \frac{(\tau_i - d)}{T} \right) \left(\sum_{j=1}^{N} \frac{T}{(\tau_j - d)} \right) \qquad (17)$$

Hence, given that N nets are switching together using conventional analysis, probability that they will not be considered switching together using our approach,

$$P = \frac{P_{new}}{P_{old}} \qquad (18)$$

If we assume that widths of all the timing windows are same τ, the probability would be

$$P = \left(1 - \frac{d}{\tau} \right)^{N-1} \qquad (19)$$

This shows that by using proposed approach glitch on the net will decrease with a probability P.

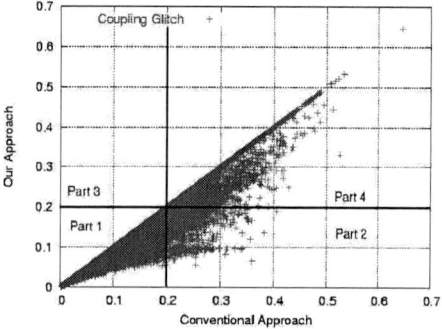

Fig. 3. Coupling glitch comparison

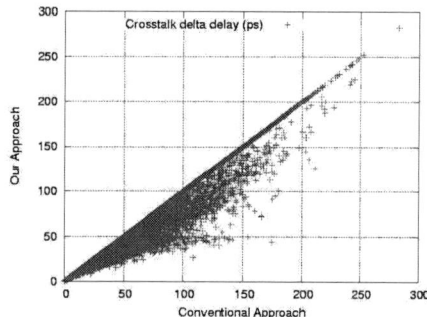

Fig. 4. Crosstalk delay comparison

V. EXPERIMENTAL RESULTS AND CONCLUSIONS

We used this approach on a 65nm design using the method mentioned in section III. Fig. 3 shows the scattered plot comparison of coupling glitch obtained using our method with the existing approach. For simplicity we assume that all the cells have a threshold of 0.2V and glitch is not propagated through cells [3][6]. Though results with these assumptions may not be very accurate but will be indicative since coupling glitch has direct impact on number of violations.

We divide the plot into four partitions as following,

Part 1 : Number of nets below threshold (using existing and proposed approach) = 139634

Part 2 : Number of nets above threshold (using existing approach) and below thresholds (using proposed approach) = 2053

Part 3 : Number of nets below threshold (using existing approach) and above thresholds (using proposed approach) = 0

Part 4 : Number of nets above threshold (using existing and proposed approach) = 14906

This means that Part 2 will have nets which were reported as violation using existing approach, but are filtered using proposed approach. Part 3 will have nets which were not reported as violation using existing approach, but are reported as violation using the proposed approach, but since this approach does not add any pessimism, number of nets in this part will always be 0.

Hence earlier, there were 16959 (Part 2 + Part 4) violations, and our approach filtered 2053 (Part 2) false violations, and now there are only 14906 (Part 3 + Part 4) potential violations left.

It is clear from the results that this approach reduces pessimism in the coupling glitch computation, and reduces the number of violations, making the analysis more accurate.

Fig. 4 shows similar comparison for crosstalk delay analysis. Here we have compared the delta delay at the victim net due to crosstalk. In crosstalk delay analysis victim timing window is also considered [4][5], hence this approach also removes the pessimism in victim-aggressor overlap, as compared to noise analysis where only aggressor-aggressor overlap is considered for pessimism removal.

VI. CONCLUSIONS

In this paper we have proposed the concept of relative timing windows and a method to perform crosstalk analysis using relative timing windows. As compared to conventional analysis where nets are considered independent of each other, our method utilizes the reconvergence correlation between two nets, and reduces the pessimism present in the conventional approach.

We analyze the effectiveness of this approach statistically using simplified probabilistic models. Experimental results have been presented that helps in showing the effectiveness of this approach considering both crosstalk glitch and delay analysis.

Appendix

A. PROBABILITY OF AGGRESSORS SWITCHING TOGETHER

In this section we find the probability of N nets switching together, if there timing windows of width $(\tau_1, \tau_2, \ldots, \tau_N)$ are uniformly distributed in the period T. Since switching will repeat after time period T, we consider time period T as the periphery of circle, and timing windows uniformly distributed around the circle.

We consider the discrete problem first by dividing the time period T in X equal increments, then we can find solution for continuous case by making $X \to \infty$.

Now timing windows will have widths $m_1, m_2, \ldots, m_N$ increments, where $m_i = (\tau/T)X$. All these timing windows can be placed in X^N number of ways. In order to find the probability of all of them switching together, we need to find the number of combinations which contains at least one increments of overlap between all the timing windows.

Each combination of overlap can be translated into X different positions, hence we need to first find number of overlapping combination for just one interval. If we consider one fix interval in the duration, number of combinations in which all timing windows will contain this interval will be $m_1 m_2 \ldots m_N$. But due to the translation, overlap of two increments is counted twice, similarly overlap of three increments is counted thrice and so on. Hence if we fix two particular consecutive increments in the duration, number of combinations which overlap will be $(m_1 - 1)(m_2 - 1) \ldots (m_N - 1)$. This counts overlap of two increments once, overlap of three increments twice and so on. Hence if we subtract this product from the previous product, we will get the number of combinations which counts each overlap exactly once.

$$m_1 m_2 \ldots m_N - (m_1 - 1)(m_2 - 1) \ldots (m_N - 1)$$

Hence probability of N nets switching together,

$$P = \lim_{X \to \infty} \frac{X[m_1 \ldots m_N - (m_1 - 1) \ldots (m_N - 1)]}{X^N} \quad (20)$$

Since $m_i = (\tau_i/T)X$, all the term in numerator of degree less than N drop out, and the probability of N nets switching together for continuous case is,

$$P = \left(\prod_{i=1}^{N} \frac{\tau_i}{T} \right) \left(\sum_{j=1}^{N} \frac{T}{\tau_j} \right) \quad (21)$$

REFERENCES

[1] L.H. Chen and M. Marek-Sadowska, *"Aggressor Alignment for Worst-Case Crosstalk Noise"*, IEEE Trans. on Computer-Aided Design of Integrated Circuits and Systems, vol20, no. 5, May 2001, pp. 612-621.

[2] P. Chen and K. Keutzer, *"Toward True Crosstalk Noise Analysis"*, Proc. Intl. Conf. Computer-Aided Design, 1999.

[3] K.L. Shepherd, V. Narayanan, P.C. Elmendorf, and G.Zheng, *"Global Harmony: Coupled Noise Analysis for Full-Chip RC Interconnect Analysis"*, Proc. Intl. Conf. Computer-Aided Design, Nov. 1997, pp. 139-146.

[4] G. Yee, R. Chandra, V. Ganesan, and C. Sechen, *"Wire Delay in the Presence of Crosstalk"*, Proc. IEEE/ACM Intl. Workshop on Timing Issues in the Specification and Synthesis of Digital Systems, pp. 170-175,1997.

[5] P.D. Gross, R. Arunachalam, K. Rajagopal, and L.T. Pileggi, *"Determination of worst-case aggressor alignment for delay calculation"*, Proc. Intl. Conference Computer-Aided Design, Nov. 1998, pp. 212-219.

[6] V. Zolotov, D. Blaauw, S. Sirichotiyakul, M. Becer, C. Oh, R. Panda, A. Grinshpon, and R. Levy, *"Noise propagation and failure criteria for VLSI designs"*, Proc. IEEE/ACM Intl. Conf. Computer-Aided Design, 2002.

[7] A. Papoulis, *"Probability, Random Variables, and Stochastic Processes"*, 2nd ed. New York: McGraw-Hill, 1984.

[8] Y. Sasaki and G. De Micheli, *"Crosstalk Delay Analysis using Relative Window Method"*, ASIC/SoC Conference, 1999.

Process-Induced Skew Reduction in Nominal Zero-Skew Clock Trees

Matthew R. Guthaus, Dennis Sylvester

Dept. of EECS
University of Michigan
Ann Arbor, MI USA 48109
{mguthaus,dennis}@eecs.umich.edu

Richard B. Brown

Dept. of ECE
University of Utah
Salt Lake City, UT USA 84112
brown@coe.utah.edu

Abstract— This work develops an analytic framework for clock tree analysis considering process variations that is shown to correspond well with Monte Carlo results. The analysis framework is used in a new algorithm that constructs deterministic nominal zero-skew clock trees that have reduced sensitivity to process variation. The new algorithm uses a sampling approach to perform route embedding during a bottom-up merging phase, but does not select the best embedding until the top-down phase. This results in clock trees that exhibit a mean skew reduction of 32.4% on average and a standard deviation reduction of 40.7% as verified by Monte Carlo. The average increase in total clock tree capacitance is less than 0.02%.

I. INTRODUCTION

Integrated circuit performance is largely determined by the efficiency and quality of synchronous clock distribution. A poor clock distribution can consume an inordinate amount of power, limit circuit performance or prevent correct functionality at all speeds. Many high performance designs use clock distribution grids to improve performance, but these consume significantly more power than traditional clocks. Symmetric clock trees such as the H-tree are robust, but they cannot be perfectly balanced since sink loads and locations are rarely uniform and sinks may have differing clock edge requirements.

Many other works have presented algorithms to minimize the deterministic skew of a clock tree during physical design [4, 6, 7, 8, 9]. These algorithms typically use wire jogs that balance the sink delays with minimal routing overhead. However, these works only consider that the delay to balance is the nominal value. Process variability of clock delay elements can have a significant impact on the skew in what is called *process induced skew*. Other works have started to address this issue by improving correlation among sinks [10] or tuning the clock tree after manufacturing [13]. However, reference [10] complicates circuit simulation with cyclic RC graphs and has the potential for DC power consumption. Reference [13] requires improved at-speed testing methodologies and increased testing time.

Nominal, deterministic clock skew can be defined as the largest difference between a fastest and slowest clock sink arrival time:

$$Skew = \max_{\forall i \in S}(AT_i) - \min_{\forall j \in S}(AT_j) \qquad (1)$$

where AT_i is the clock arrival time at sink i and S is the set of clock sinks. Statistical clock tree analysis adds the problem that different manufactured parts will have different slow and fast clock sinks depending on how the sources of variation manifest themselves. Therefore, analysis must be done in a probabilistic manner. Previous works have done this using discretized, numerical integration for only intra-die variation [2] or have focused on non-clock interconnect analysis using a first-order parameterized model [1, 3]. The most similar work [11] to that described in this paper uses a first-order parameterized model but uses heuristic worst-case bounding to minimize skew. This can be overly pessimistic.

This work uses a first-order parameterized model to perform statistical skew analysis and construct robust clock trees. In Section II, we describe the sources of variation and how they are modeled in our experiments. In Section III, we propose an improved statistical analysis framework based on a first-order parameterized Elmore delay computation. In Section IV, deterministic zero-skew clock tree construction is briefly reviewed. Then a new technique is presented for minimizing process-induced skew such that the clock trees remain zero-skew according to traditional deterministic analysis. Experimental results are presented in Section V and conclusions are made in Section VI.

II. PROCESS VARIABILITY

Process variation can be of the inter-die or intra-die type. Both types of variation present challenges to physical design. *Inter-die variation* is fully correlated for a given die, but separate dies can have different amounts of variation. This requires examining an exponential number corners since it is unclear which corner is the worst for interconnect delay. On the other hand, *intra-die variation* can lead to process spread within a single die. This work models both inter- and intra-die variation using correlated and independent sources. All variation is assumed to be Gaussian, but this is not a requirement since numerical integration can be used as shown in [2]. The nominal process values are from a generic 90nm process and the parameters vary according to [5].

This work considers two physical parameters for metal variation: height (H) and width (W). The correlation of CMP variation is on the order of 3.5mm [12] and the tolerance for the range of interconnect density during the fill process is user-defined. Due to these considerations, the interconnect variation is assumed to be fully correlated in this paper. This does not exclude other more advanced models from being used.

A simple extraction model is used for wire resistance and

capacitance. Unit resistance is calculated as

$$R = \frac{\rho}{A} = \frac{\rho}{HW} \qquad (2)$$

where ρ is the resistivity of aluminum ($2.8 \times 10^{-8} \Omega \cdot m$) and A is the cross sectional area. Capacitance estimation uses a parallel plate model for inter-layer and sidewall capacitance and a term for fringe effects. Unit capacitance can then be described as

$$C = \frac{\epsilon W}{T} + \frac{2\pi\epsilon}{log(T/W)} + \frac{\epsilon H}{S} \qquad (3)$$

where $\epsilon = 3.9\epsilon_0$ is the permitivity of SiO_2, S is the space between wires, and T is the space between layers. The method can be easily extended with more accurate 3-D extraction models.

III. Statistical Analysis

Statistical timing analysis performs timing analysis of a circuit considering process variations. An Elmore delay model is used for interconnect modeling. In order to retain sensitivities to the sources of variation, a first-order parameterized form is used to represent all probabilistic quantities including delays, arrival times, skews, resistances and capacitances.

A. First-order Parameterized Form

The fundamental model used in this work is based on a weighted sum of Gaussian distributions which can be expressed as

$$A = a_0 + \sum_{i=1}^{n} a_i X_i + a_r X_r, \qquad (4)$$

where X_r and X_i's are random variables normally distributed with mean zero and variance one, $N(0, 1)$, that represent individual process parameters. The n random variables, X_1 to X_n, represent process parameters that are globally shared among all gates or wires while the X_r random variable represents local, independent variation. The magnitude of the variation is given by the absolute value of coefficients, a_i or a_r. a_0 is the nominal value. For clarity, a capital letter refers to either a parameterized form (e.g., A) or a random variable (e.g., X_i) throughout this paper. We now cover the basic manipulations of parameterized forms that are needed for analytic statistical timing analysis of clock trees.

Addition/Subtraction – The addition operation is

$$A + B = (a_0 + b_0) + \sum_{i=1}^{n}(a_i + b_i)X_i + \sqrt{a_r^2 + b_r^2}X_r$$

and subtraction is similar. It should be noted that the root sum-of-squares of the independent terms is used in both addition and subtraction.

Multiplication – The multiplication operation is

$$
\begin{aligned}
AB \;=\; & (a_0 b_0) + \sum_{i=1}^{n}(a_0 b_i + a_i b_0)X_i \\
& + \sum_{i=1}^{n}\sum_{j=1}^{n}(a_i b_j X_i X_j) + \sum_{i=1}^{n}(a_i b_r X_i X_r) \\
& + \sum_{i=1}^{n}(a_r b_i X_r X_i) + \sqrt{(a_0 b_r)^2 + (b_r a_0)^2}X_r.
\end{aligned}
$$

As can be seen, there are many second order terms. If the signals are not correlated, all cross terms are zero and can be dropped without loss of accuracy as done in [3]. However, in most situations, there will be high levels of correlation. For example, when calculating Elmore delay, the R and C corresponding to a given segment will have highly negative correlation and, therefore, the non-linear terms must be considered. The work in [1] proposed to linearize the squared terms (X_i^2) into a mean and a normally distributed portion (X_i) such that the original first and second moments are preserved. However, they ignore the non-squared terms ($X_i X_j$ when $i \neq j$) and the terms with the random component (X_r). In addition, they adjust the individual sensitivities which can be misleading as they are larger than those in Monte Carlo simulations. *This work considers all cross terms so the multiplication operation is more accurate than previous methods.* The mean term for only the squared components is added to the mean and the other nonlinear components are linearized and added to the random sensitivity to prevent corrupting individual sensitivities. The product (C) is summarized with the following coefficients

$$
\begin{aligned}
c_0 \;=\;& (a_0 b_0) + \alpha \sum_{i=1}^{n} a_i b_i + \alpha a_r b_r \\
c_i \;=\;& a_0 b_i + a_i b_0 \\
c_r^2 \;=\;& (a_0 b_r)^2 + (a_r b_0)^2 \\
& + \beta^2 \sum_{i=1}^{n}\left((a_i b_r)^2 + (a_r b_i)^2 + (a_i b_i)^2\right) + (\beta a_r b_r)^2
\end{aligned}
$$

where α and β are the linear approximation coefficients in $X_i X_j \approx \alpha + \beta X_r$. As in VITA [1], these coefficients are $\alpha = 1$ and $\beta = \sqrt{2}$ for normal distributions. The results using the new procedure are more accurate than VITA, or dropping non-linear terms, in every test case examined.

Maximum/Minimum – The maximum (minimum) operation is calculated as a weighted combination of inputs while matching the analytically calculated first and second moments as in [14]. Fast Monte Carlo with linear regression can be substituted for wide maximum (minimum) operations or in situations where the accuracy is questionable.

B. Process Variation-Induced Skew

Statistical analysis of the clock tree starts by analytically computing the sensitivities of each resistive and capacitive element to obtain the parameterized forms, R and C, respectively. The maximum and minimum delay of each clock subtree and total capacitance are then computed in a bottom-up manner. Given that the addition, multiplication, maximum, and minimum operations were defined in the previous section, the maximum ($\overline{D_i}$) and minimum ($\underline{D_i}$) delay for a node, i, can be calculated in first-order parametric form using the Elmore delay metric and its fanout delays, j,

$$\overline{D_i} = \max_{\forall j \in Sinks(i)} \left(R_{(i,j)}(\frac{1}{2}C_{(i,j)} + C_j) + \overline{D_j} \right) \quad (5)$$

$$\underline{D_i} = \min_{\forall j \in Sinks(i)} \left(R_{(i,j)}(\frac{1}{2}C_{(i,j)} + C_j) + \underline{D_j} \right) \quad (6)$$

where $R_{(i,j)}$ is the resistance and $C_{(i,j)}$ is the capacitance from node i to node j. C_j is the total subtree capacitance in first-order parameterized form that is also calculated bottom-up as a sum of parameterized forms as in

$$C_i = C_{(i,j)} + C_j. \quad (7)$$

At the root of the tree, the bottom-up traversal results in the maximum and minimum delays for all clock branches in the parameterized form. Taking the difference of the maximum and minimum delays when i is the root gives the statistical skew of the clock tree in parameterized form

$$S = \overline{D_i} - \underline{D_i}. \quad (8)$$

To demonstrate the analytic method, zero-skew trees were generated using the approach in [4, 6, 8]. *These trees all have zero skew according to nominal, deterministic Elmore delay.* The proposed statistical analysis shows that the trees actually have poor skew tolerance to process variations. To verify the statistical analysis, the results are compared with Monte Carlo simulations in Table I. The analytic methodology predicts the mean skew value within 1.6% on average, but the standard deviation is systematically underestimated by an average of 45.3%. This is due to the non-Gaussian nature of the skew distribution. Since skew is always positive, the distribution is truncated at zero and is highly asymmetric in situations with near zero skew. The underestimation of the standard deviation may result in less accuracy, but it does not undermine the fidelity of the analytic analysis capability – it is still able to determine which tree is best. In addition, the mean and sensitivities of the maximum ($\overline{D_{root}}$) and minimum ($\underline{D_{root}}$) sink delay match Monte Carlo very well. The fidelity of the analytic analysis is demonstrated by Figure 1 which compares several deterministic zero-skew trees using both the analytic method and Monte Carlo. The relative ordering of the expected skew and standard deviation is identical. Similar results are seen in other test cases.

IV. CLOCK TREE CONSTRUCTION

A. Review of DME

One method of constructing nominal zero-skew clock trees is the Deferred-Merge Embedding (DME) method that was independently proposed in [4, 6, 8]. The DME method requires as input the abstract topology of the clock tree to define the merge order of the sinks. The objective is to produce a clock

TABLE I
PROCESS-INDUCED SKEW IN NOMINAL ZERO-SKEW TREES.

Bench.	Monte Carlo		Analytic		Error (%)	
	μ (ps)	σ (ps)	μ (ps)	σ (ps)	μ	σ
s1423	18.3	9.2	17.9	6.6	-2.2	-28.3
s5378	33.9	18.8	34.6	12.2	3.3	-35.1
s15850	201.6	143.7	207.7	81.3	-3.0	-43.4
r1	74.1	53.1	76.2	32.9	2.8	-38.9
r2	113.7	80.6	114.7	40.0	0.9	-50.4
r3	175.8	126.7	194.7	55.5	10.7	-56.1
r4	371.1	271.5	397.8	115.0	7.2	-57.6
r5	1911.1	1436.7	1930.7	812.8	1.0	-43.4
p1	22.9	13.8	22.7	7.1	-0.9	-48.6
p2	61.2	35.7	63.4	17.6	-3.6	-50.7
				Average	1.6	-45.3

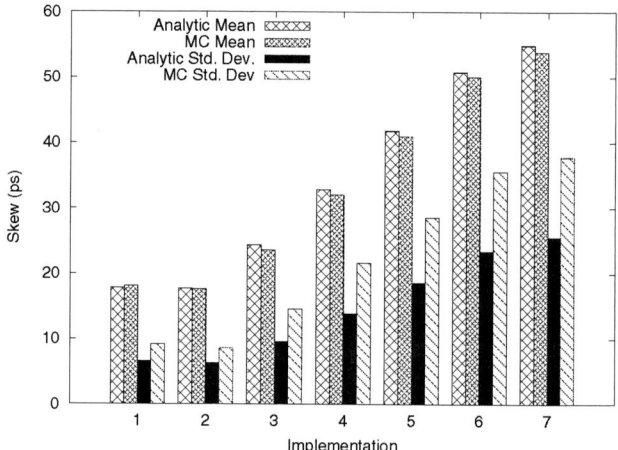

Fig. 1. Fidelity of analytic method on s1423.

tree routing such that all sinks have equal delay while consuming the minimum amount of wire routing to reduce clock power. To do this, it defers choosing where to merge two subtrees until the parent merging point is decided. This allows each zero-skew merging location to be selected such that the overall minimum wire length is used. Figure 2 shows an example with four sinks (A-D) during the bottom-up merging and top-down embedding phase of DME. During the bottom-up phase, the sinks are merged according to the input topology. The merging determines the valid locations where the two subtrees can be joined such that the new subtree has zero skew and minimum wire length. The locus of points that define the area of the zero-skew merge is called the *merging segment* (MS). With rectilinear routing and equal resistance and capacitance on the layers, the MS is an intersection of two Manhattan circles (diamonds). This intersection is always a line with slope of +1 or -1. In Figure 2, sinks A and B are merged into MS1, sinks C and D are merged into MS2, and then MS1 and MS2 are merged into MS3. If the subtrees to be merged have vastly dissimilar delays, extra wire jogs may be required to achieve zero skew. After the bottom-up phase is completed, a single MS remains such as MS3 in Figure 2. The root of the clock

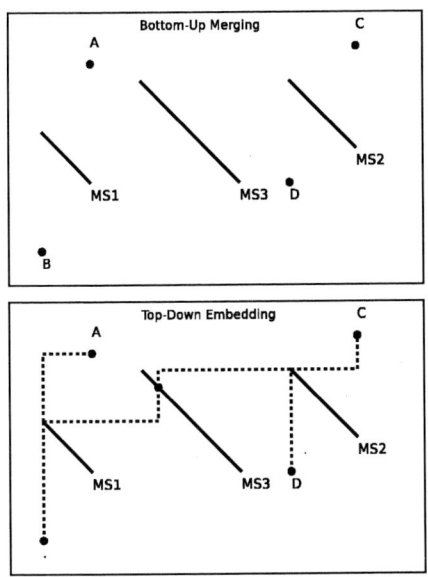

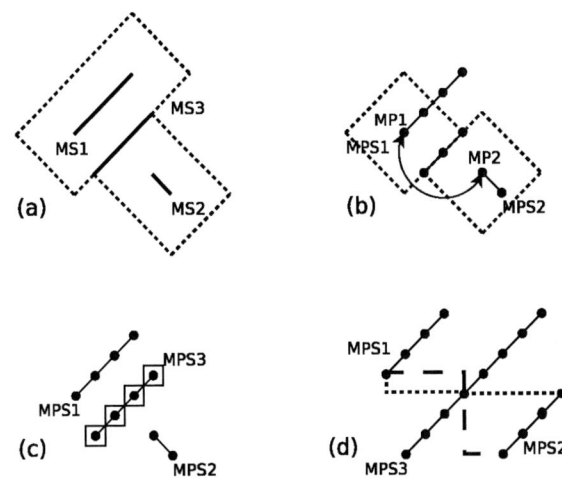

Fig. 3. Example of (a) DME merging, (b) sampled DME closest pair merging, (c) sampled DME pruning, and (d) sampled DME merging with multiple embeddings.

Fig. 2. Example of bottom-up merging and top-down embedding.

tree can be placed anywhere on this MS or the point on the MS closest to the desired root location. In a top-down manner, the location of each child MS that is closest to the parent location is selected. The actual wire route is determined or *embedded* at this stage. The procedure is repeated until the sinks are reached. The resulting tree has minimal wire length and zero skew in nominal process conditions.

B. Sampled Zero-Skew Trees

The method that is presented in the next section relies on a sample-based implementation of traditional DME. Instead of computing the intersection of arbitrary Manhattan arcs as in Figure 3a, a collection of merging points (MPs) is maintained that represents the MS. This set of merging points is referred to as a *merging point set* (MPS). To calculate a MPS, the traditional MS is computed for all closest pairs, like MP1 and MP2 in Figure 3b, and sampled. When there are more than two closest MPs, the union of all the resulting zero-skew MPs is taken and pruned. The pruning divides the area into a proximity grid of a user-specified resolution and each MP is allocated to the appropriate grid location as in Figure 3c. Any single MP in a grid can replace all other MP for nominal optimization since they each represent a zero-skew merging location with minimal wire length. The pruning maintains a linear number of samples for each MS to provide a variety of candidate merge locations for further merging.

Property 1 *All sample-based trees have zero skew in nominal conditions.*

Property 1 is induced from the fact that all MP are created from two nominal zero-skew MP. The two zero-skew MP may not be in the "correct" location according to traditional DME, but the new MP location will be adjusted appropriately to maintain zero skew. Therefore, the resulting tree may not have minimum power, but it is always zero skew. The grid size determines the sub-optimality in power.

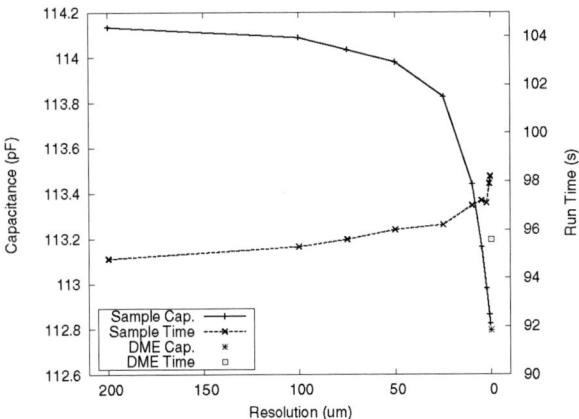

Fig. 4. Resolution impact on optimality and run-time of sample-mode DME compared to traditional DME.

Property 2 *As the grid resolution approaches zero, the sample-based method will obtain the same power zero-skew solution as traditional DME.*

An empirical demonstration of Property 2 can be seen in Figure 4. At the expense of a few seconds of CPU time, the power of the solution quickly approaches the optimal capacitance produced by traditional DME.

C. Reduced Statistical Skew (RSS) Trees

The proposed method for Reduced Statistical Skew (RSS) clock trees is similar to the well known DME approach, but incorporates the previous sample-based technique. The sampled-based technique allows the route embeddings to be stored during the bottom-up merging process. Each MP remembers from which MPs it was constructed so that it can be appropriately embedded in the top-down phase. The actual wires (embedded mergings) allow the process sensitivity to be considered during clock tree construction. In traditional DME, the embedding does not occur until the top-down phase. In the new method, MPs are embedded during the bottom-up phase, but the selection of which MP and corresponding embedding to use is delayed until the root location is known. The actual merging location is chosen according to the deterministic, nominal skew, so the trees generated with the new approach have zero skew in the nominal case, but also exhibit significantly reduced statistical skew.

Consider Figure 3d that shows the merging of MPS1 and MPS2 into MPS3. According to the sample method proposed in Subsection B, all closest pairs of MPs are merged to form a new MPS. Before the results are pruned, there may be several equally viable zero-skew embeddings for a given location. As an example, the dotted and dashed embeddings in Figure 3d are equally viable candidates. In deterministic nominal analysis, each of these solutions is identical and all but one can be dropped from consideration without degrading solution quality. In statistical analysis, however, this is not true because the solutions have different expected skews and process sensitivities. Keeping all embedded MPs in a grid is not viable since the number of MPs can be as much as $m \times n$ for parallel Manhattan arcs with m MPs in MPS1 and n MPs in MPS2. A few properties allow us to determine which criteria are most desirable in choosing which MP embedding to keep; a MP embedding can affect our clock tree skew through its capacitive load, its subtree delay, or its subtree skew.

Subtree Capacitance – With the Elmore delay model and correlated or independent parameters, all sampled DME subtree embeddings rooted at the same location have the same total capacitance with the same parametric sensitivities. This is true because each subtree uses the same (minimum) total amount of horizontal and vertical wire. In the Elmore delay model, a subtree capacitance is a simple sum of these capacitances as in (7)[1]. Since each MP that we wish to compare is in the same grid, they are physically close and have very similar parametric capacitances. The consequences of this are that the choice of an embedding does not significantly effect subsequent mergings through the load capacitance.

Subtree Delay – Typical subtree embeddings show that the delays of the minimum wire length, zero-skew subtrees tend to be highly correlated with similar means. Therefore, the delay does not provide a good differentiating metric.

Subtree Skew – The worst skew of a clock tree is determined by the worst skew of all subtrees. This can be proven by contradiction, but is omitted due to space limitations. The intuition provided by the statement is equally true for clock trees while considering process variation. In addition, since subtree capacitance and subtree delay are not viable for pruning, subtree skew is the most reasonable candidate for differentiating solutions.

Our optimization algorithm, RSS, retains the MP candidate (with embedding) in each grid location that exhibits the least expected skew during bottom-up merging. The merging locations during the bottom-up phase are calculated using deterministic nominal timing analysis. L-shape routes connect the MPs. The order of segments on an L-shape route has an

almost negligible effect on variability so segment order is randomly selected. During the bottom-up phase, the maximum and minimum delays are calculated at each MP as presented in Section III. The bottom-up phase results in a MPS where each MP corresponds to a fully embedded clock tree. The top-down phase of RSS selects MP with the minimum statistical skew tree and uses the corresponding minimal wire-length embedding to construct the tree.

V. Experimental Results

The above algorithm was implemented in C++ and executed on a 3.0Ghz Xeon running Linux. The results of the RSS algorithm are compared with a nominal DME solution by running Monte Carlo simulations on the final trees. The Monte Carlo mean and standard deviation of the final clock tree skew is shown in Table II. The mean skew is reduced by 32.4% on average and the standard deviation is reduced by 40.7% on average. The pruning heuristics preserve the $O(n)$ run-time behavior of the traditional DME algorithm, but the coefficient of the linear run-time is determined by the sample resolution. This tradeoff between run-time and and power optimality was selected to favor power optimality as can be seen by less than 0.02% sub-optimality in total tree capacitance on average. The sample resolution was chosen independently for each benchmark so that there are at least 500 grids in both the horizontal and vertical direction. The largest benchmark, r5, with 3101 sinks has only a 16.3% increase in run-time compared to nominal DME.

Figure 5 compares two trees on a tiny benchmark to illustrate the differences between RSS and traditional DME. The top tree, created by DME, arbitrarily selects the MS endpoint as the clock tree root. This results in a tree that has high sensitivity to metal skew. One branch of the root is primarily on horizontal metal while the other branch is on vertical metal. Picking the mid-point would be a better choice, but not the best. In the bottom of Figure 5, RSS picks a point that minimizes the process induced skew. For clarity, note that the vertical dimension is $2\times$ the horizontal dimension in the figures.

VI. Conclusion

This work developed an analytic framework for clock tree analysis that has high fidelity with Monte Carlo simulations. In addition, the expected skew is very accurate with a typical error of only 1.6%. The analytic analysis framework was used in the implementation of a Reduced Statistical Skew (RSS) clock tree method that minimizes the statistical skew in nominal zero-skew clock trees. The new approach performs route embedding during the bottom-up phase, but does not select the best embedding until the top-down phase. This results in nominal zero-skew trees that exhibit significantly improved statistical properties. On a set of benchmarks, the mean skew was reduced by 32.4% on average and the standard deviation was reduced by 40.7% on average as verified by Monte Carlo. Run-time is preserved through an efficient proximity pruning technique that ensures O(n) run-time and remains efficient for even the largest benchmark. Less than 0.02% power was added on average due to the sub-optimality of the sampling technique.

---

[1]If a higher-order model of interconnect were used, resistive shielding would invalidate this property. However, clock trees tend to use upper layers of metal and wide wires which reduces the effect of resistive shielding.

TABLE II
RSS DME COMPARED WITH TRADITIONAL DME. μ AND σ IN PS, CAPACITANCE IN FF, AND CPU TIMES IN S.

		DME				RSS				Change (%)		
Bench.	Sinks	Cap.	μ	σ	CPU	Cap.	μ	σ	CPU	Cap.	μ	σ
s1423	74	4953.8	18.3	9.2	0.04	4955.3	14.2	5.8	29.6	+0.03	-22.4	-37.0
s5378	179	11002.6	33.9	18.8	0.1	11004.2	23.5	9.8	31.6	+0.01	-30.7	-47.9
s15850	597	35010.2	201.6	143.7	3.0	35018.3	80.4	41.6	43.2	+0.02	-60.1	-71.1
r1	267	15899.5	74.1	53.1	0.2	15901.0	39.1	25.8	13.8	+0.01	-47.2	-52.5
r2	598	35677.0	113.7	80.7	2.5	35682.3	91.7	61.6	35.3	+0.01	-19.3	-23.7
r3	862	51464.8	175.8	126.7	9.1	51476.3	152.3	107.1	66.8	+0.02	-13.4	-15.5
r4	1903	112796.4	371.1	271.5	130.5	112816.7	319.6	228.9	181.3	+0.02	-13.9	-15.7
r5	3101	182224.9	1911.1	1436.7	572.9	182279.4	432.1	285.9	666.3	+0.03	-77.4	-80.1
p1	269	14975.8	22.9	13.8	0.3	14977.4	17.5	8.4	24.1	+0.01	-23.6	-39.1
p2	603	33766.7	61.2	35.7	2.3	33769.3	51.4	27.3	9.0	+0.01	-16.0	-23.5
									Average	+0.017	-32.4	-40.7

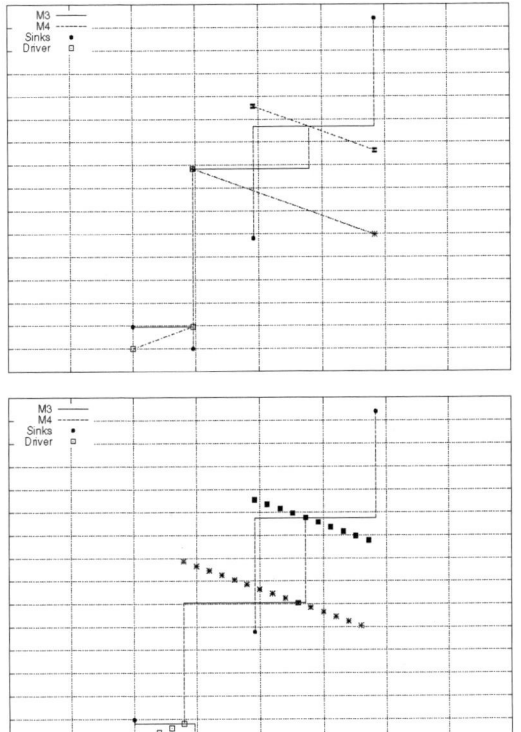

Fig. 5. Traditional DME (top) and RSS DME (bottom) trees.

VII. ACKNOWLEDGMENTS

The authors would like to that Vladimir Zolotov at IBM T.J. Watson Research Center for his many insightful discussions on the topic.

REFERENCES

[1] S. Abbaspour, H. Fantemi, and M. Pedram. VITA: Variation-aware interconnect timing analysis for symmetric and skewed sources of variation considering variational ramp input. In *GLSVLSI*, April 2005.

[2] A. Agarwal, D. Blaauw, and V. Zolotov. Statistical clock skew analysis considering intra-die process variations. In *ICCAD*, pages 914–920, 2003.

[3] K. Agarwal, D. Sylvester, and D. Blaauw. Variational delay metrics for interconnect timing analysis. In *DAC*, 2004.

[4] K. Boese and A. Kahng. Zero-skew clock routing trees with minimum wirelength. In *ASIC Conf.*, pages 1.1.1–1.1.5, 1992.

[5] Y. Cao, P. Gupta, A. B. Kahng, D. Sylvester, and J. Yang. Design sensitivities to variability: Extrapolations and assessments in nanometer VLSI. In *ASIC/SOC*, pages 411–415, September 2002.

[6] T.-H. Chao, Y.-C. Hsu, and J. Ho. Zero skew clock net routing. In *DAC*, pages 518–523, 1992.

[7] J. Cong and C.-K. Koh. Minimum-cost bounded-skew clock routing. In *ISCAS*, pages 215–218, 1995.

[8] M. Edahiro. Minimum path-length equi-distant routing. In *Asia-Pacific Conf. on Circuits and Systems*, pages 41–46, December 1992.

[9] D. J.-H. Huang, A. B. Kahng, and C.-W. A. Tsao. On the bounded-skew clock and steiner routing problems. In *DAC*. 508–513, June 1995.

[10] A. Kapoor, N. Jayakumar, and S. P. Khatri. A novel clock distribution and dynamic de-skewing methodology. In *ICCAD*, pages 626–631, 2004.

[11] B. Lu, J. Hu, G. Ellis, and H. Su. Process variation aware clock tree routing. In *ISPD*, pages 174–181, 2003.

[12] B. E. Stine and D. S. B. et al. The physical and electrical effects of metal-fill patterning practices for oxide chemical-mechanical polishing processes. *IEEE Transactions on Electron Devices*, 45(3):665–679, March 1998.

[13] J.-L. Tsai, D.-H. Baik, C. C. Chen, and K. K. Saluja. A yield improvement methodology using pre- and post-silicon statistical clock scheduling. In *ICCAD*, pages 611–618, 2004.

[14] C. Visweswariah, K. Ravindran, K. Kalafala, S. G. Walker, and S. Narayan. First-order incremental block-based statistical timing analysis. In *DAC*, pages 331–336, 2004.

1D-1

2006 Asia and South Pacific Design Automation Conference

A Low Dynamic Power and Low Leakage Power
90-nm CMOS Square-Root Circuit

Tadayoshi Enomoto and Nobuaki Kobayashi

Chuo University, Graduate School of Science and Engineering

Information and System Engineering Course

1-13-27 Kasuga, Bunkyo-ku, Tokyo 112-0881, Japan

Abstract - To drastically reduce the dynamic power (P_{AT}) and the leakage power (P_{ST}), while to keep speed of a CMOS square-root (SR) circuit, a new algorithm, new architectures and a new leakage reduction circuit were developed. Using these techniques, a 90-nm CMOS LSI was fabricated. The P_{AT} and P_{ST} of the new SR circuit were reduced to about 1/4 and 1/33 those of a conventional SR circuit. Measured results agreed well with simulated results.

1. INTRODUCTION

Low-power circuit techniques are needed for use in battery-driven portable systems. To reduce both the dynamic power ($P_{AT}=GCf_cV_{DD}{}^2$) and the leakage power ($P_{ST}=GI_LV_{DD}$) of the CMOS circuits, we have to reduce the total number of logic gates (G), the supply voltage (V_{DD}), and/or the leakage current (I_L) of an individual logic gate while maintaining a required clock frequency (f_c). Improving both algorithms and architectures can reduce G. Shortening a critical path, that is, decreasing a number of logic gates (G_c) of the critical path can lower V_{DD}. Lowering V_{DD} is also effective for lowering I_L. Furthermore, to drastically decrease I_L, we developed a special leakage current reduction circuit. To examine the effects of the developed low power techniques on both P_{AT} and P_{ST}, we have applied those techniques to a square-root circuit for such uses as in computer graphic application.

2. TECHNIQUE FOR LOWERING SUPPLY VOLTAGE

Let Q ($=.q_1q_2 - - q_m$) be the square root of A ($=.a_1a_2 - - a_{2m-1}a_{2m}$). The mth-bit SR (q_m) is obtained as a carry signal when a mth reminder ($.R_m$) is calculated [1]. R_m is obtained by

$$.R_m = .R_{m-1}a_{2m-1}a_{2m} - .00 - - 0q_1q_2 - - q_{m-2}q_{m-1}01 \quad (1)$$

when q_{m-1} is 1. It is calculated as

$$.R_m = .R_{m-1}a_{2m-1}a_{2m} + .00 - - 0q_1q_2 - - q_{m-2}q_{m-1}11 \quad (2)$$

when q_{m-1} is 0. The above two equations for m of 4 are carried out by the square-root (SR) circuit shown in Fig. 1 [1]. This $2m$-bit conventional SR circuit (C-SR) for m of 4 can be constructed with a 4-stage ripple carry adder that consists of 20 full adders (CASs) with a subtraction function. Bold solid lines indicate the critical path. The C-SR including buffer inverters has G of 189 gates and G_c of 60 gates in the critical path.

Replacing CASs by CAS1s and CAS2s (Fig.2) can drastically reduce G_c. G and G_c of the SR circuit for m of 4 would be reduced to 179 and 40, respectively. To further reduce G and G_c, we have modified Eq. 1 as

$$.R_m = .R_{m-1}a_{2m-1}a_{2m}+.11--1q_{1,B}q_{2,B}--q_{m-2}q_{m-1,B}11, \quad (3)$$

where $q_{1,B}$, $q_{2,B}$, and so on are the inverses of q_1, q_2, and so on. Furthermore, 1 and $q_{m-1,B}$ in Eq. 3 are replaced by q_{m-1}

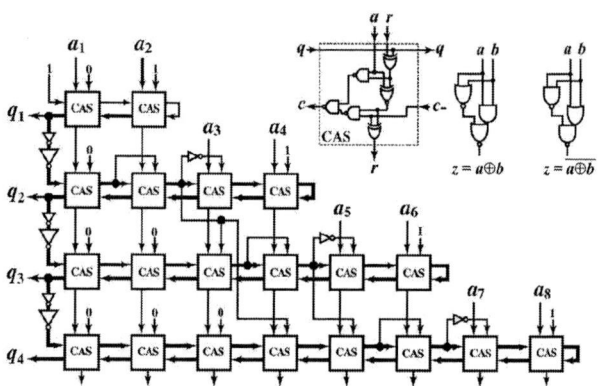

Fig. 1. A conventional square-root circuit (C-SR) for m of 4.

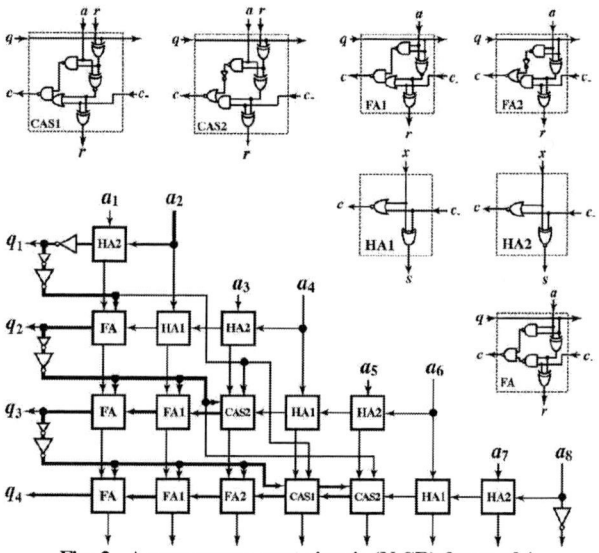

Fig. 2. A new square-root circuit (N-SR) for m of 4.

and 0, respectively. Similarly, 0 and q_{m-1} in Eq. 2 are replaced by q_{m-1} and 0, respectively. Thus, both Eqs. 2 and 3 can be expressed by the same equation as

$$.R_m = .R_{m-1}a_{2m-1}a_{2m}+.q_{m-1}q_{m-1}--q_{m-1}$$
$$(q_1 \oplus q_{m-1})(q_2 \oplus q_{m-1})--(q_{m-2} \oplus q_{m-1})011. \quad (4)$$

G and G_c of the SR circuit (not shown) using Eq. 4 for m of 4 were greatly reduced to 128 and 32, respectively.

The calculation processes of Eq. 4 are mostly carried out by additions, so complicated full adders (CAS-1 and CAS-2) with subtraction functions can mostly be replaced by either smaller full adders (FA, FA-1, FA-2) or simple half adders (HA-1, HA-2). Thus, we were able to significantly simplify the SR circuit (Fig. 2). G and G_c of the new SR circuit (N-SR) were greatly reduced to 95 and 30, respectively. Thus,

0-7803-9451-8/06/$20.00 ©2006 IEEE.

90

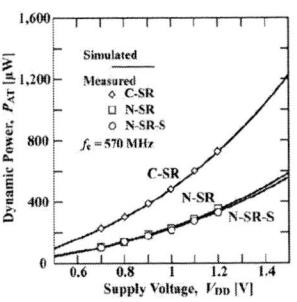
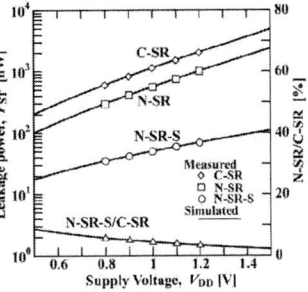

Fig. 3. Dynamic power (P_{AT}) at f_c = 570 MHz. **Fig. 4. Leakage power (P_{ST}).**

G and G_c of N-SR are about 50.0% of those of C-SR, respectively.

At V_{DD} of 1.0 V, the simulated maximum operating clock frequency (f_c) of N-SR was 946 MHz, which was 1.66 times faster than that (= 570 MHz) of C-SR. This great f_c improvement was due to the considerable reduction of G_c. The simulated P_{AT}s of C-SR and N-SR for m of 4 at f_c of 570 MHz are plotted as solid lines in Fig. 3. Between 0.5 V and 1.5 V, P_{AT} of N-SR is less than 50% of that of C-SR. P_{AT} of N-SR at 0.77 V and 570 MHz is 131 μW, which is 27.1% of that (484 μW at 1 V and 570 MHz) of C-SR. The simulated P_{ST}s of C-SR and N-SR for m of 4 are plotted as solid lines in Fig. 4. P_{ST} of N-SR at 0.77 V is 276 nW, which is less than 1/4 of that (1,147 nW) of C-SR. Table 1 summarizes the characteristics of C-SR and N-SR.

3. LEAKAGE CURRENT REDUCTION CIRCUIT

To further reduce P_{ST}, we developed a leakage current reduction circuit called a "self-controllable-voltage-level (SVL)" circuit (Fig. 5). N-SR incorporating the SVL circuits is called N-SR-S. The upper SVL circuit (U-SVL) and the lower SVL circuit (L-SVL) can supply a maximum V_D (= V_{DD}) and a minimum V_S (= V_{SS} = 0 V), respectively to the active N-SR on request (i.e., CLB = 0, CL = 1). The U-SVL and L-SVL can also supply decreased V_D (< V_{DD}) and increased V_S (> 0 V), respectively to the stand-by N-SR when CL is 0 and CLB is 1.

The SVL circuits can simultaneously reduce the drain-to-source voltage (V_{ds}) and increase the substrate bias (V_{sub}) of cut-off MOSFETs. Thus, it decreases the sub-threshold currents of the cut-off MOSFETs [2]. The SVL circuit can also reduce the gate-to-drain electric fields of the cut-off MOSFETs and gate-to-source electric fields of the turn-on MOSFETs; it can reduce not only gate induced drain leakage (GIDL) currents in the cut-off MOSFETs [3], but also gate tunnel currents in the turn-on MOSFETs. Consequently, P_{ST} of the SR circuit is considerably reduced.

At 1.0 V the maximum f_c of N-SR-S was 918 MHz, which was 3% slower than that (= 946 MHz) of N-SR. The simulated P_{AT} of N-SR-S for m of 4 at 570 MHz is plotted in Fig. 3. At 0.78 V P_{AT} was reduced to 132 μW that is 27.3% of that of C-SR. The simulated P_{ST} of N-SR-S is plotted in Fig. 4. P_{ST} of N-SR-S at 0.78 V is 34 nW, a reduction to 3% of C-SR and 12% of N-SR. The SVL

Table 1. Characteristics of C-SR, N-SR and N-SR-S.

SR circuits	C-SR	N-SR (N-SR/C-SR)	N-SR-S (N-SR-S/C-SR)
No. of logic gates G	189	95 (50.3%)	97 (51.3%)
No. of logic gates of critical path G_c	60 (100%)	30 (50.0%)	30 (50.0%)
Supply voltage V_{DD} [V] *	1	0.77	0.78
Dynamic power P_{AT} [μW] f_c of 570 MHz **	484 (100%)	131 (27.1%)	132 (27.3%)
Leakage power P_{ST} [nW]	1,147 (100%)	276 (24%)	34 (3%V)

* Minimum V_{DD} that confirms the 570-MHz operation.
** P_{AT} measured at f_c of 570 MHz.

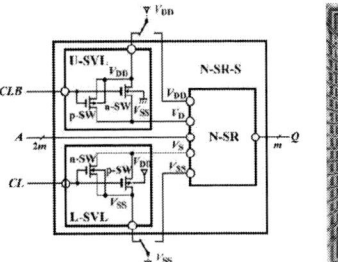

Fig. 5. N-SR and N-SR-S. **Fig. 6. 90-nm CMOS LSI.**

circuit is very effective in reducing P_{ST}, while the speed overhead is negligible.

4. LSI FABRICATION AND EXPERIMENTAL RESULTS

C-SR, N-SR and N-SR-S were fabricated for m of 4 as shown in Fig. 6. The 90-nm, 6-layer Cu CMOS fabrication process was used. The threshold voltage (V_{th}) of n-MOSFETs was 0.22 V and that of p-MOSFETs was -0.24 V. The measured P_{AT}s and P_{ST}s for the three circuits are plotted in Figs. 3 and 4, respectively. Measured results agree well with SPICE simulated results.

5. SUMMARY

We have developed an SR algorithm, small circuit architectures, and a leakage current reduction circuit to reduce P_{AT} and P_{ST}, while maintaining operating speed. Our developed techniques hardly affected the operating speed, while reducing P_{AT} to about 1/4 and P_{ST} to 3% those of the conventional circuit. These power reduction techniques will therefore play a major role in future development of sub-100-nm CMOS circuits.

Acknowledgment - The authors wish to thank our colleagues at the Institute of Science and Engineering, Chuo University for their supported of this work. The VLSI chips used in this study were fabricated in the chip fabrication program of the VLSI Design and Education Centre (VDEC) of the University of Tokyo in collaboration with STARK and ASPLA.

References
[1] Kai Hwang, "Computer Architecture: -Principles, Architecture and Design-," John Wiley & Sons, Inc., Section 11.2, 1979.
[2] H. J. M. Veendrick, "Deep-submicron CMOS ICs", Kuluwer academic publishers, Dordrecht, Netherlands, pp. 73-75, 1998.
[3] M. Rosar, B. Leroy and G. Schweeger, "A New Model for Description of Gate Voltage and Temperature Dependence of Gate Induced Drain Leakage (GIDL) in the Low Electric Field Region", IEEE Tran. on Electron Devices, vol. 47, no. 1, pp. 154 - 159, January 2000.

A High-Throughput Low-Power Fully Parallel 1024-bit ½-Rate Low Density Parity Check Code Decoder in 3-Dimensional Integrated Circuits*

Lili Zhou, Cherry Wakayama, Nuttorn Jangkrajarng, Bo Hu, and C.-J. Richard Shi

Department of Electrical Engineering, University of Washington, Seattle, WA 98195, USA

Abstract - A 1024-bit, ½-rate fully parallel low-density parity-check (LDPC) code decoder has been designed and implemented using a three-dimensional (3D) 0.18μm fully depleted silicon-on-insulator (FDSOI) CMOS technology based on wafer bonding. The taped-out 3D decoder with about 8M transistors was simulated to have a high throughput of 2Gb/s and a low power consumption of only 430mW using 6.4μm by 6.3μm of die area. The 3D implementation is estimated to offer more than 10x power-delay-area product improvement over its corresponding 2D implementation. This first large-scale 3D ASIC with fine-grain (5μm) vertical interconnects is made possible by jointly developing a complete automated 3D design flow from a commercial 2-D design flow combined with the needed 3D-design point tools.

I. Introduction

Low density parity check (LDPC) codes are emerging as standard methods of channel encoding and error correcting for many wireless standards, due to their near Shannon-limit error correction performance [1] and the progress in semiconductor fabrication technologies that allow very large scale integration of circuit functionality. The LDPC block-parallel message passing decoding algorithm and its fully-parallel implementation architecture yield the high-throughput error-correction capacity necessary for large-volume communication and data storage applications. However, the implementation leads to the following interconnect design challenges [2]:

(1) The average wire length can be 3mm for a 7.5mm x 7.0 mm die implementation (half of the die size). Therefore, specific CAD tools have been developed.

(2) The wiring takes more silicon area and power dissipation than the logic itself.

(3) Only 50% area utilization for logic was achieved due to routing congestion.

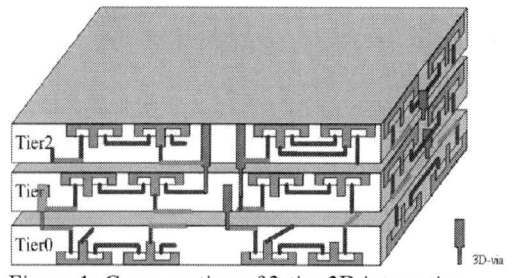

Figure 1: Cross-section of 3-tier 3D integration.

To address this interconnect design challenge, we explore the use of a 3-dimensional integrated circuit process.

---

* This research was sponsored by the U.S. DARPA 3D-IC Program under Grant No. N66001-05-1-8918 monitored by Navy SPAWAR, San Diego.

More specifically, we use MIT Lincoln Lab's wafer bonding 3D process, which stacks 3 wafers each composing of a single layer of transistors with 3 layers of metal wires (called one-tier) formed on fully-depleted silicon-on-insulator (FDSOI) substrates [3]. Figure 1 shows the cross-section view of the 3-tier 3-D IC integration.

We note that all the previous 3D IC designs are limited either to simple logic devices, or to circuits of regular structures such as photo sensors, memories and field-programmable gate arrays (FPGAs). This is primarily due to the lack of 3-D CAD tools and a complete 3D design flow to handle the ASIC design complexity. To accomplish this challenging LDPC ASIC design in 3D, we have developed the needed 3D-CAD tools to automate the 3D IC implementation process. A complete 3D design methodology and design flow is developed based on a commercial 2D design flow augmented with the needed 3D design tools.

II. Fully Parallel LDPC Decoder Architecture

Our 3D LDPC decoder is based on the classical message-passing (belief-propagation) algorithm, which maps extremely well to a parallel decoder architecture that can be represented by a bipartite graph, directly instantiated to hardware [2]. As illustrated in Fig. 2, there are two types of computing nodes (*variable* nodes and *check* nodes) that perform all the logical calculations, edges representing the required interconnect as defined by the very sparse parity check matrix. A 1024-bit ½-rate code decoder requires 1024 variable nodes and 512 check nodes.

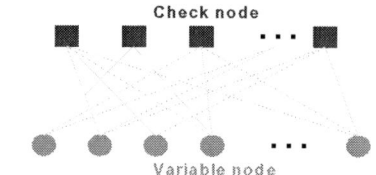

Figure 2. The Tanner graph of a LDPC code.

The main data path is designed as 16 parallel three-stage pipelines (Fig. 3); this allows the decoder to achieve the high throughput of 2Gb/s with the clock frequency at 128MHz (128MHz*16=2Gb/s).

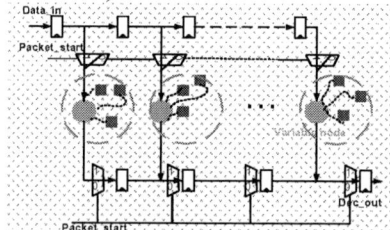

Figure 3. Fully parallel 3-stage pipelined LDPC architecture.

III. 3D LDPC Design

Our 3D design methodology is based on partitioning a 3D-design into 2-D designs at the fine-grain level. An in-house 3D placement tool has been developed that places computing nodes on all the 3 tiers with the objectives to minimize the area, routing density, total wire length and 3D-vias. We have also developed in-house programs for 3D routing, buffer insertion, and circuit-verse-schematic (CVS) checking. Figure 4 shows the 3D LDPC design flow.

IV. Results and 2-D Comparison

The 3D LDPC decoder has been taped out based on the 3D MIT Lincoln Lab FD-SOI process. It contains about 8M transistors using $6.4\mu m$ by $6.3\mu m$ of 3D die area. The top view of 3-tier LDPC layout is shown in Figure 5, where over 10,000 dense 3D vias are used to connect 3 tiers.

The simulated code performance is shown in Fig. 6. The black curve shows the BER vs. SNR performance up to BER of 10^{-5}. The grey curve shows the fast iteration convergence, which yields low-power dissipation of estimated 430mw.

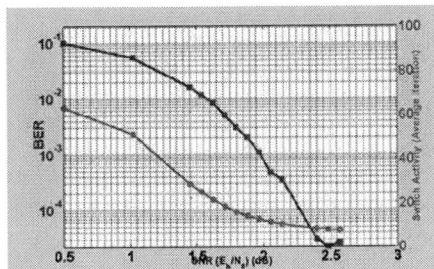

Figure 6. The simulated LDPC decoder performance.

Table 1 summarizes the taped out design's characteristics and a comparison 2D design. The 2D design was accomplished by putting all devices on one tier with the same technology and standard cells. We can see that the 3D implementation achieved a significant advantage over the same technology 2D implementation in terms of wire length, area, clock skew, and buffer size. The improvement in terms

of power-delay-area product is more than an order of magnitude: $2.5 \cdot 3 \cdot 1.75 = 13.125$.

Table 1. The comparison between 3D and 2D designs.

	2D design	3D design
area (mm*mm)	18.238*15.92= **290.35**	(6.4*6.227)*3= **119.56**
total wire length (m)	182.42	22.39+22.57+22.46= **67.42**
max. WL before buffer insertion (mm)	13.82	8.68
max. WL after buffer insertion (mm)	4	4.17
buffer used	32900	24636
clock skew (ns)	2.33	1
power dissipation(mw)	750	430

V. Conclusion

A fully parallel LDPC decoder has been implemented on a 3-tier 3D IC process with 2Gb/s throughput and 430mw power consumption. The significance of this work is three-fold: (1) It is the *first* large-scale 3D ASIC implementation. (2) It is for the first time, by real silicon tape out and simulation, 3D IC process with 3-tier integration was shown to yield an order of magnitude improvement over the corresponding 2D process, in terms of power-delay-area product. (3) It is the first time that an automated 3D design flow has been developed and used to tape out a large-scale silicon ASIC design.

Acknowledgements: Dr. Guoyong Shi, Dr. Sambuddha Bhattacharya, and Dr. Lei Yang contributed to the 3D LDPC design and 3D-CAD tool development. Dr. Craig Keast, Dr. Peter Waytt, and Dr. James Burns of MIT Lincoln Lab contributed to the 3D fabrication.

References

[1] D. Mackay and R.M. Neal, "Near Shannon limit performance LDPC Codes", Electron Letters, 1996.
[2] A. Blanksby and C.J. Howland, "A 690-mw 1-Gb/s 1024b, 1/2 rate LDPC Code Decoder", JSSC, 2002.
[3] J. Burns *et. al.*, "Three-dimensional integrated circuits for low-power, high bandwidth systems on a chip", ISSCC, 2001.

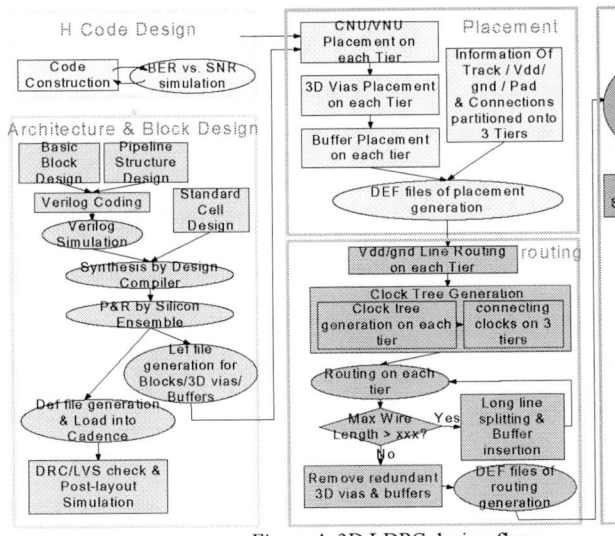

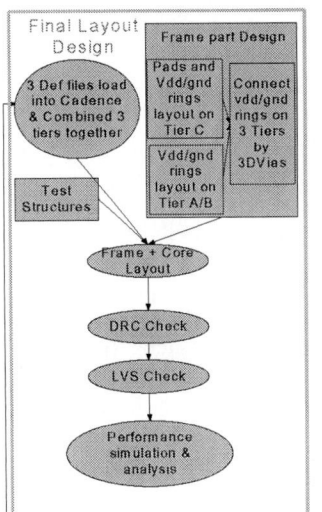

Figure 4. 3D LDPC design flow.

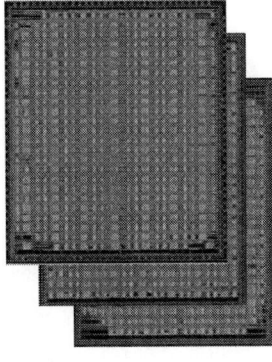

Figure 5. Top view of 3-tier 3D LDPC layout.

93

A 16-Bit, Low-Power Microsystem with Monolithic MEMS-*LC* Clocking

Robert M. Senger, Eric D. Marsman,
Michael S. McCorquodale

Richard B. Brown

Dept. of Electrical Engineering and Computer Science
University of Michigan
Ann Arbor, MI 48109 USA
{rsenger, emarsman, mmccorq}@umich.edu

Dept. of Electrical and Computer Engineering
University of Utah
Salt Lake City, UT 84412 USA
brown@coe.utah.edu

Abstract—**Single-chip systems save the power dissipation that would be required for chip-to-chip communication, resulting in compact, low-power solutions for battery-powered applications. This paper describes the design and measured performance of a fully-functional digital core with a low-jitter, on-chip, MEMS-*LC* clock reference. This chip has been fabricated in TSMC's 0.18μm MM/RF bulk CMOS process. Maximum power consumption of the complete microsystem is 48.78mW operating at 90MHz on a 1.8V power supply.**

I. INTRODUCTION

To satisfy the broad range of workload requirements for microsystems and Systems-on-a-Chip (SoCs), an adaptable microcontroller unit (MCU) must be designed with a wide spectrum of communication capabilities and operating specifications. The size, processing, and power requirements for the embedded MCU in PDAs, cell phones, remote environmental sensors, bio-medical devices, etc. vary significantly with the application. By building an MCU that can satisfy these design requirements and by leveraging an intellectual property (IP) based design methodology [1], manpower and design time can be greatly reduced without sacrificing significant power or performance.

II. MICROSYSTEM ARCHITECTURE

Fig. 1 shows the microsystem architecture consisting of the digital core and the CMOS-MEMS *LC* tank oscillator used as an on-chip clock reference. The digital core includes a 3-stage pipeline, 16-bit data path, a 24-bit unified instruction and data address space, 64KB of on-chip SRAM, and an external memory port supporting up to 64KB. The load-store instruction set architecture (ISA) was custom designed with 77 instructions supporting eight addressing modes and single- and multi-word arithmetic, shift, logical, and control-flow operations [2]. A 3-stage pipeline was chosen to provide adequate performance for remote sensing and bio-medical applications, yet still remain low-power with minimal pipeline hardware overhead. The pipeline utilizes sixteen 16-bit general purpose registers and four 24-bit address registers, divided evenly over two windows. The windowing scheme reduces the size of the register encoding field to enable 16-bit instructions while providing additional registers for temporary storage. [3] gives a detailed analysis of the compiler's efficient utilization of the register windows to achieve up to 19% reduction in power consumption and 30% improvement in performance when compared to a non-windowed architecture. Address register manipulation is enabled through direct memory mapped access or by using address update instructions.

The memory architecture is a banked style with the 64KB of SRAM split into four single-ported 16KB banks. This allows for instruction and data accesses to occur si-

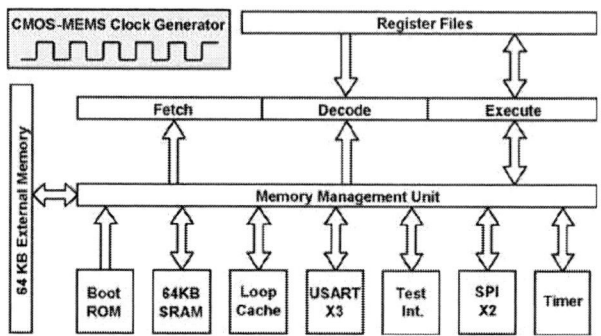

Fig. 1. Microsystem architecture.

multaneously without stalling the machine pipeline as long as they address different banks. To save power, unused banks are deactivated on a cycle-by-cycle basis. Compared to a single 64KB bank, this configuration dissipates 69.2% less energy per access with only a 16.2% area penalty [4]. Additional power savings are enabled by a low-power, 512-byte loop cache. Unlike traditional hardware controlled caches, the loop cache is a tagless bank of low-power memory intelligently managed by the compiler. The cache is filled with commonly executed instructions or accessed data, typically found in program loops.

To generate a stable on-chip clock reference, a complementary, cross-coupled, negative-transconductance MEMS-*LC* oscillator was implemented. A detailed description of this low-jitter, 1.1GHz CMOS compatible reference oscillator is given in [5]. With the proposed *LC* oscillator, neither a PLL/DLL nor off-chip crystal is required, thus reducing system size, cost, and power. Moreover, the clock is significantly more stable and obtains better temperature stability than alternative on-chip clock generation techniques such as ring, relaxation, or phase-shift oscillators [5]. To improve the quality factor, the inductor uses thick top metal that is released from the surrounding oxide when the bond pad openings are etched. The capacitors are metal-insulator-metal. Compensation for frequency deviation due to process or temperature variation can be achieved by modulating the current in the *LC* tank [5]. A buffer amplifier is required to isolate the free-running oscillator from the chained flip-flop frequency divider.

III. TEST RESULTS

Fig. 2 is a die-micrograph of the microsystem fabricated in TSMC's 0.18μm MM/RF bulk CMOS process. The 128 pin die measures 3.54mm per side and contains 3.5 million transistors. The design methodology presented in [6] merges digital, analog, and MEMS domains into a top-down ASIC design flow that was employed to build this chip. The microsystem was verified on an HP82000 digital

0-7803-9451-8/06/$20.00 ©2006 IEEE.

Fig. 2. Die micrograph of the complete microsystem.

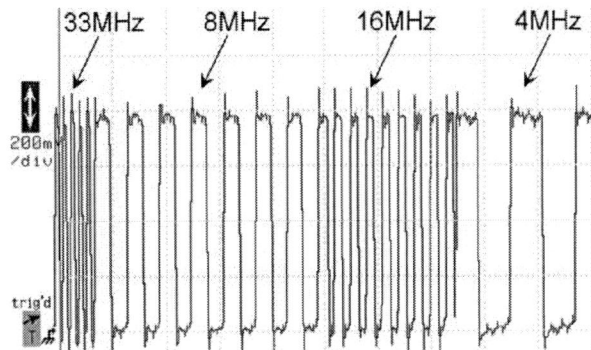

Fig. 4. Low-latency, glitch-free, dynamic clock frequency scaling.

tester using test vectors generated from assembly programs that were used to check the original Verilog model. These programs consisted of hand-written focused test cases, randomly generated test cases, and compiled application code.

The MCU is fully-functional up to a maximum operating frequency of 92.5MHz at 1.8V and consumes a maximum of 33.9mW. At 10MHz and 1.15V, power consumption drops to 1.41mW. When put into a 2kHz low-power idle mode, the core consumes only 740μW from a 1.15V supply [7]. Digital output pins are available to control an off-chip voltage regulator that can modulate the power supply voltage. Fig. 3 shows the measured MCU digital core power as a function of voltage for 10, 50, and 90MHz operation.

A single access to the loop cache consumes 45% of the energy that an access to the SRAM consumes [7]. The custom-built compiler implemented a novel dynamic loop cache filling algorithm that improved power efficiency over more traditional static filling. The dynamic algorithm was simulated and compared against static filling for different size loop caches and was also compared against traditional instruction caches. Across a subset of the embedded benchmarks from MiBench and MediaBench, the dynamic filling algorithm obtained an average energy savings of 43% and

outperformed all other cache configurations [8].

The fabricated *LC* reference oscillates at 1.056GHz with a ±2% precision before trimming. The oscillator achieves a worst case 48/52 duty-cycle. A 1.1% frequency variation was observed over a temperature range of -40 to 100°C. The reference oscillator occupies only $0.3mm^2$ of Si area and consumes 17.28mW from a 1.8V supply. The measured RMS period jitter is 610ppm [5]. Fig. 4 shows oscilloscope traces of the microsystem dynamically selecting different frequency divider outputs without halting the pipeline.

IV. CONCLUSION

This work reports the single-chip integration of a flexible, low-power microsystem that has a custom ISA and C-compiler with a CMOS compatible MEMS-*LC* reference oscillator. Maximum active power consumption of the MCU is 33.9mW at 92.5MHz and 1.8V with an idle-mode drawing only 740μW at 1.15V. The on-chip, $0.3mm^2$ MEMS-*LC* reference supplies a highly accurate, low-jitter clock source while consuming 17.28mW at 1.8V.

ACKNOWLEDGEMENTS

Fabrication of this work at TSMC was supported by the MOSIS Educational Program. The authors wish to thank Artisan for digital cell libraries and memory generators. Work was supported by the Engineering Research Centers program of the NSF under award number EEC-9986866.

REFERENCES

[1] M. McCorquodale, *et al.*, "Microsystem and SoC design with UMIPS," *IFIP International Conf. on VLSI SOC*, pp. 324-329, 2003.

[2] R. Senger, *et al.*, "A 16-bit mixed-signal microsystem with integrated CMOS-MEMS clock reference," in *Proc. Design Automation Conf.*, pp. 520-525, June 2003.

[3] R. Ravindran, *et al.*, "Partitioning variables across register windows to reduce spill code in a low-power processor," *IEEE Trans. Computers*, to be published.

[4] S. Martin, *et al.*, "A low-power microinstrument for chemical analysis of remote environments," *11th NASA Symp. on VLSI Design*, Coeur d' Alene, ID, pp. 1-4, May 2003.

[5] M. McCorquodale, "Monolithic and Top-Down Clock Synthesis with Micromachined RF Reference", Ph.D. Dissertation, Dept. of Elec. Eng. and Comp. Sci., Univ. of Michigan, Ann Arbor, MI, 2004.

[6] M. McCorquodale, F. Gebara, K. Kraver, E. Marsman, R. Senger, and R. Brown, "A top-down microsystems design methodology and associated challenges," in *Design, Automation, and Test in Europe Designers' Forum Proc.*, pp. 292-296, Mar. 2003.

[7] E. Marsman, *et al.*, "A 16-bit low-power microcontroller with monolithic MEMS-*LC* clocking," in *Proc. Intl. Symp. on Circuits and Systems*, pp. 624-627, May 2005.

[8] R. Ravindran, *et al.*, "Compiler managed dynamic instruction placement in a low-power code cache," *Code Generation and Optimization*, pp. 179-190, Mar. 2005.

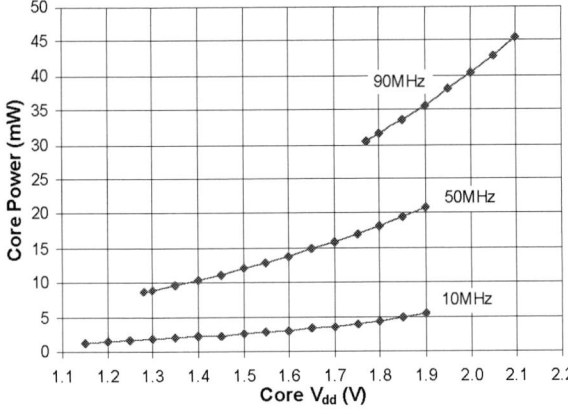

Fig. 3. Measured power as V_{dd} is scaled across frequency ranges.

Ultra-Low Voltage Power Management Circuit and Computation Methodology for Energy Harvesting Applications

Chi-Ying Tsui, Hui Shao, Wing-Hung Ki and Feng Su
Depatment of Electrical and Electronic Engineering
The Hong Kong University of Science and Technology
Fax: (852)2358-1485, Email: eetsui@ee.ust.hk

Abstract- A power management and computation methodology is proposed for ultra-low power energy harvesting applications. An integrated exponential charge pump that accepts an input voltage of around 150mV and provides an unregulated output voltage of more than 1.5V serves as the power supply. To cater with the fluctuated energy source and unregulated power supply, a supply side charge-based computation methodology is proposed, of which the computation activity tracks with the fluctuation of the available energy. The idea is demonstrated in a test chip fabricated using a 0.35µm technology.

I Introduction

Recently, new emerging applications such as RFID and wireless sensors network, which consume very low power and utilize the energy harvested from the environment, are gaining more attention[1][2]. In these applications, the energy are collected from solar, vibration, heat or radioactive decay of matters and at the same time the amount of energy available is limited and the source is unstable.

Previous researches on energy harvesting systems assume that a voltage source of at least 1V is available [3][4]. However for some environment, such as inside a human body, the harvested voltage source may be much lower. At the same time in order to reduce the cost of the device, expensive voltage regulator should be avoided. Unstable low voltage source poses new challenges in the design of the power management circuit and computation paradigm for the applications.

In this work, we propose a novel power management system for energy-limited source applications which can pump the input voltage of ~150mV, a value much lower than the threshold voltage of CMOS transistor, to several volts. We also present the design of the computation module that can operate under the unregulated voltage supply. We propose a supply-side charge-based computation paradigm where computation is carried out only when the energy from the environment is enough to execute a specified atomic operation of the computation. An experimental chip was designed, fabricated and measured to demonstrate the idea.

II. System Design Description

Fig. 1 shows the block diagram of our target system. It contains of 4 blocks, the energy harvesting mechanism, the power management system, the computation module and the charge-based control unit.

A generic energy harvesting mechanism scavenges the energy from the environment and gives out an unregulated source voltage. The voltage is inputted to the power management system where a high conversion ratio charge pump steps up the voltage. The unregulated voltage (V_{out}) is then fed directly to the computation module. The charge-based control unit will make sure the energy available is enough for the atomic operation before triggering the computation.

A. Power Management System

The power management system contains a 4-stage 16x exponential charge pump and a clock generator for control. Figure 2 shows the circuit diagram. Since the voltage source V_s is around 100mv~200mv, the circuit needs a start up circuit, which only functions at the beginning of the circuit running. Once the circuit is started, the generated high voltage source V_{out} will provide the energy to the clock generator and cut the switch between the start-up voltage and the circuit.

This charge pump has a cross-coupled structure that employs a 2-phase non-overlapping clock. The cross-coupled structure consists of two symmetrical branches A and B. When $\phi_1=1$, the kth stage capacitor of Branch A is charged to 2^{k-1} Vs by the $k-1^{th}$ stage capacitor of Branch B. When $\phi_2=1$, the positive plate of the kth stage capacitor of Branch A is pushed by the $k-1^{th}$ stage capacitor of Branch A to 2^kVs and charges the $k+1^{th}$ stage capacitor of Branch B to 2^kVs at the same time. A similar mechanism occurs to the capacitors in Branch B. If the charge pump has 4 stages, an ideal voltage conversion of 16 is obtained.

B. Computation Module

The power supply becomes unstable due to the absence of the voltage regulator, and this will affect the delay of the circuit and may cause timing problems. In order to track the change of supply and automatically adjust the circuit's performance, we propose to use self-time asynchronous pipeline design to implement the computation module [5]. In this case the operation of a pipeline stage gets more robust over various operating conditions for its locally-generated timing signal and it is more suitable for the design to track with the unstable supply voltage

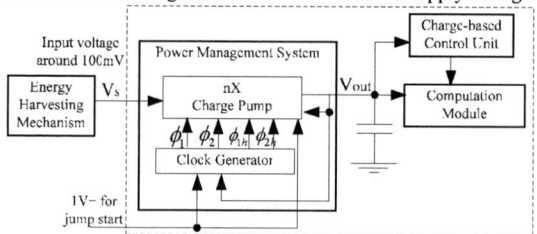

Fig. 1. Targeted system using energy harvesting

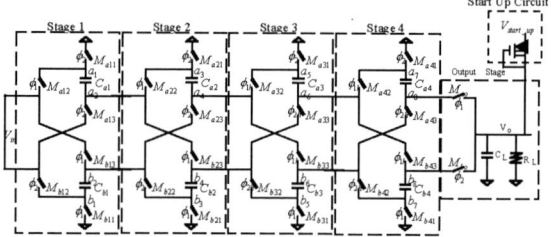

Fig. 2. 16x charge pump with startup circuit

0-7803-9451-8/06/$20.00 ©2006 IEEE.

C. Charge-based Control Unit

Due to the unpredictable nature of the energy source, the energy available at a particular time interval may not be enough for the computation and if we carry it out, the computation may not be able to finish and even worse the voltage may drop to a level that some of the stored data will be lost. Here, we propose a supply side charge-base computation methodology. The charge required at different voltage for a certain atomic computation is estimated and stored in the device. The computation will only be triggered when the harvested energy can provide the charge for it. Moreover, the computation can be further prioritized, both in task level and bit level. Depending on the energy available and priority, different computations will be executed.

III. Experimental Results

To demonstrate the proposed idea, a test chip was designed and fabricated using AMS 0.35μm CMOS technology. A 16x exponential charge pump is used for the power management system with an output capacitance of 0.47μF. For the computation module, we use a self-timed pipelined 4-tap FIR filter as an example where bundle delay is used to generate the local timing signals. An on-chip linear feedback shift register is used to generate the input data to the filter. A simple start-stop control is used to demonstrate the charge-based control where a hysteretic comparator is implemented. The charge required for an atomic computation is converted as a voltage level at the output of the charge pump given the output capacitance. A voltage comparator is used to compare the output voltage with this voltage level for the stopping of the computation. Table 1 shows the chip summary and the die photo is shown in Fig.3.

Experiments were carried out on the chip. First we varied the input voltage to the charge pump between 169mV and 190mV to mimic the unstable harvested energy source. Fig.4 shows the waveform of V_s (input voltage to the charge pump), V_{out} (output voltage at the charge pump), stop signal from the charge based control unit and output data from computation module. Here V_{out} is not regulated when stop signal is disabled and it drops as the computation consumed energy is larger than the converted energy from the power management system. The charge based control unit will stop the computation when the available energy is not enough and activate it again when energy recovers. What's more, the duty cycle under small input voltage is longer due to the lower input power value. Another experiment was conducted using the energy converted from a solar cell as the energy source. The input voltage varies from 80mV to 143mV while the solar cell is under weak light and strong light condition, respectively. Fig.5 shows the corresponding waveforms where Vs is the voltage output from the solar cell. It is shown that under weak light condition, the charge based control unit stops the computation because of small amount of energy available. When the light intensity increases, the computation will be triggered if the available energy can support the atomic operation. To show the reliability of the computation under the unstable energy source and the unregulated power supply, we ran the test chip for days and collect the sample error. About 10^{12} output samples were collected and no error was found when comparing with the correct samples.

IV. Conclusion

A power management circuit and a supply side charge-based computation methodology for energy harvesting applications were proposed. A test chip was designed and fabricated to demonstrate the idea. Future work includes the development of more complicated charge-based control for real applications.

References

[1] S.Jung et.al.,"Enabling technologies for disappearing electronics in smart textiles," *Proceedings of IEEE ISSCC* , Jan.2003

[2] R.Amirtharajah et.al.,"Self-powered signal processing using vibration-based power generation" *IEEE Journal of Solid-state Circuits,* Vol.33,No.5,pp.687-695, May.1998

[3] R.Amirtharajah et.al.,"A micropower programmable DSP powered using a MEMS-based virbration-to-elctric energy converter," *IEEE Int'l Solid-state Ckts. Cof.,* pp.362-363, 2002

[4] B.Warneke et.al.," An autonomous 16mm³ solar-powered node for distributed wireless sensor network," *IEEE Sensors,* pp.1510–1515, Orlando Florida, June 2002

[5] Jung-Lin Yang, E. Brunvand, "Self-timed design with dynamic domino circuits", Proceedings of IEEE Symposium on VLSI, pp. 217-219, Feb 2003

TABLE I   Chip Performance Summary

Technology	0.35μm CMOS 4-Metal 2-Poly
Size of power management system	0.29mmx0.33mm
Switching frequency	100kHz~2MHz
Input voltage range	80mV~200mV
Output voltage	1.12V~2.31V
Maximum Output Current	200μA
Output capacitance	0.47μF
Size of computation module	0.93mmx0.40mm
Size of charge based control unit	0.15mmx0.29mm

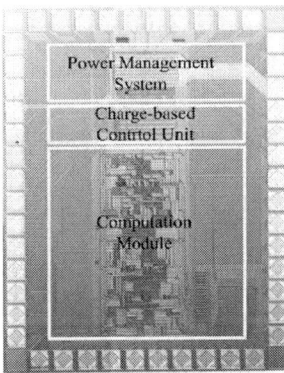

Fig. 3. Die photo of the test chip

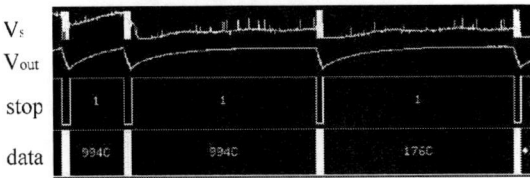

Fig. 4. Measurement result for fluctuated input voltage

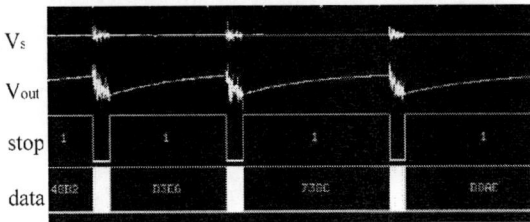

Fig. 5. Measurement result with energy from the solar cell

1D-5

A 0.5-V Sigma-Delta Modulator Using Analog T-Switch Scheme for the Subthreshold Leakage Suppression

Koichi Ishida, Atit Tamtrakarn, and Takayasu Sakurai

Center of Collaborative Research
University of Tokyo
Tokyo, 153-8505, Japan
Tel : +81-3-5452-6253
Fax : -81-3-5452-6632
e-mail : {ishida, atit, tsakukrai} @iis.u-tokyo.ac.jp

Abstract - A 0.5-V sigma-delta modulator implemented in a 0.15-μm FD-SOI process with low V_{TH} of 0.1V using analog T-switch (AT-switch) scheme to suppress subthreshold-leakage problems is presented. The scheme is compared with the conventional circuit, which are also fabricated in the same chip. The measurement result demonstrates that the sigma-delta modulator based on AT-switch realizes 6-bit resolution through reducing non-linear leakage effects while the conventional circuit can achieve 4-bit resolution.

I Introduction

Recently, low-voltage, low-power yet inexpensive VLSI's are getting focus, and analog building blocks tend to be embedded in scaled digital circuits as a part of SoC implemented with advanced VLSI technology. The International Technology Roadmap for Semiconductors (ITRS) predicts that the threshold voltage (V_{TH}) of high-performance logic technology will ever decrease. Hence, techniques for analog circuit implemented with very low-V_{TH} process become more important.

Several sub-1V sigma-delta modulators and ADC's are reported. All are, however, implemented in a high-threshold voltage process [1,2,3]. In this paper, a 0.5-V sigma-delta modulator using analog T-switch (AT-switch) scheme, which can suppress subthreshold leakage in the process with very low V_{TH} of 0.1V, is presented and verified by measurement.

II. Circuit Design

Fig.1 shows a schematic of a sigma-delta modulator using AT-switch scheme. $M_{1a\sim c}$ and $M_{2a\sim c}$ are the AT-switch that consists of two series-connected MOS's and intermediate voltage controlling MOS. The analog ground is set to $V_{DD}/2$. All MOS switches are driven by non-overlapping clocks whose swing is between V_{SS} to V_{DD}. Fig.2 explains why the subthreshold leakage can be suppressed. During the sampling phase, the gate voltage of M_{3b} is set to V_{SS} and that of M_{3a} is set to V_{DD} to cut them off deeply. When the input signal is around V_{DD} the node "A" is set to V_{IN} through M_{2a} and M_{2b}. Although M_{3a} is still leaky, the gate-source of M_{3b}

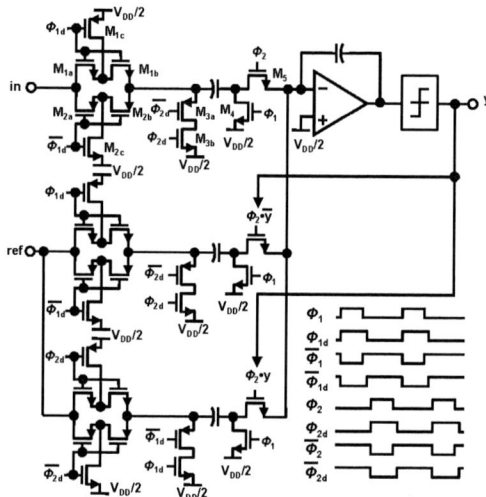

Fig.1. Schematic of a 0.5-V sigma-delta modulator using the AT-switch scheme.

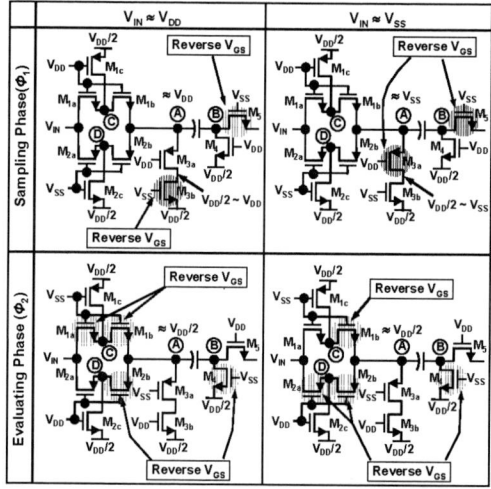

Fig.2. Principle of the AT-switch scheme. Subthreshold leakage is suppressed by reverse-VGS without handling voltage outside the power rails.

0-7803-9451-8/06/$20.00 ©2006 IEEE.

98

is reversely biased by $V_{DD}/2$ and M_{3b} is completely cut off. When the input signal is around V_{SS}, the gate-source of M_{3a} is reversely biased although the M_{3b} is leaky. In this case, M_5 is also reversely biased since the node "B" is always set to $V_{DD}/2$ through M_4. During the evaluation phase, the gates of M_{1a} and M_{1b} are both set to V_{SS} and the gates of M_{2a} and M_{2b} are set to V_{DD} to cut them off. The node "C" and node "D" are connected to $V_{DD}/2$ in this phase through M_{1c} and M_{2c} respectively. Then, both of the V_{GS} of M_{1b} and M_{2b} are reversely biased and they are deeply cut off even though M_{1a} and M_{2a} are leaky. M_4 is also reversely biased since the node "B" is always set to $V_{DD}/2$ through the M_5. Since this scheme is insensitive to parasitic capacitances of MOS switches, added parasitic capacitance introduced by the proposed scheme do not affect the operation [4]

III. Experimental Results

The AT-switch scheme and the conventional scheme are implemented in the same chip. The circuits are operated under 0.5-V supply voltage and at 2-MHz sampling frequency.

Fig.3 demonstrates measured output power spectra. The output spectrum of the conventional scheme is taken at the input level of −7.6dB and the large harmonic tones that degrade SNDR are observed. This is due to the leakage current that introduces non-linear errors. The proposed scheme shows the peak SNDR at the input level of −7.6dB. It is seen that the third order and higher tones are greatly suppressed compared with the conventional circuit.

Fig.4 shows measured SNDR's. The SNDR of the conventional scheme is degraded to 31.5dB, which is below 5-bit resolution, with the power consumption of 71μW. The proposed scheme achieves the peak SNDR of 39.6dB, which realizes more than the 6bit resolution, with the power consumption of 75μW. The peak SNDR and the dynamic range are improved over the conventional approach at the same time. The chip microphotograph of the sigma-delta modulator using the AT-switch scheme is shown in Fig.5 and the area is 130μm×190μm.

IV. Conclusion

A 0.5-V sigma-delta modulator implemented in a 0.1V-V_{TH} 0.15-μm FD-SOI process using AT-switch scheme is experimentally verified. The modulator using AT-switch realizes 6-bit resolution by reducing non-linear leakage effects caused by the leakage and loss of charge through low-V_{TH} transistors.

Acknowledgements

The authors would like to express deep appreciation to Mr. F.Ichikawa, S.Baba, T.Chiba, K.Tani from Oki Electric Industry Co., Ltd. for valuable support and chip fabrication, and Dr. H.Ishikuro from Toshiba Corp. and Prof. M.Takamiya from VLSI Design and Education Center (VDEC), the University of Tokyo for fruitful discussions.

References

[1] V. Peluso, P. Vancorenland, A. Marques, M. Steyaert, and W. Sansen, "A 900-mV Low-Power ΔΣ A/D Converter with 77-dB Dynamic Range," IEEE J. Solid-State Circuits, vol. 33, No. 12, pp. 1887-1897, Dec. 1998.

[2] J.Sauerbrey, M. Wittig, D. Schmitt-Landsiedel, and R. Thewes, "0.65V Sigma-Delta Modulators," Proc. of ISCAS, pp. I1021-I1024, May 2003.

[3] J. Sauerbrey, D. Landsiedel, and R.Thewes, et al., "A 0.5-V 1-μW Successive Approximation ADC," J. Solid State Circuits, vol. 38, pp. 1261-1265, Jul. 2003.

[4] K. Martin, "Improved circuits for the realization of switched-capacitor filters," Trans. on CAS, pp. 237-244, Apr. 1980.

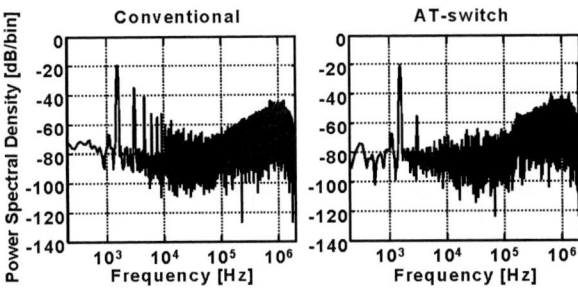

Fig.3. Measured output power spectra of sigma-delta modulators.

Fig.4. Measured SNDR's.

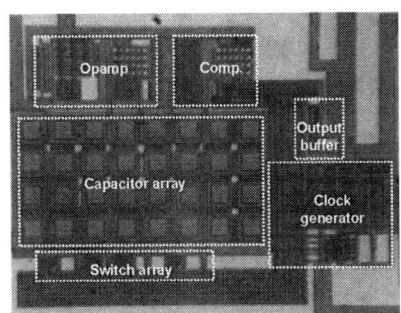

Fig.5. The chip microphotograph of the 0.5-V sigma-delta modulator using AT-switch.

1D-6

An Implementation of a CMOS Down-Conversion Mixer for GSM1900 Receiver

Fangqing Chu, Wei Li, Junyan Ren

State Key Laboratory of ASIC & System, Department of Microelectronics
Fudan University, Shanghai, China
E-mail: fqchu@fudan.edu.cn

Abstract—A 1.9-GHz down-conversion CMOS mixer, intended for the GSM1900 (PCS1900) Low-IF receivers is present with the utilization of novel folded Gilbert Cell fabricated in a RF 0.18-μm CMOS process. The prototype demonstrates a good performance. It achieves a conversion gain of 6dB, SSB Noise Figure of 18.5dB and IIP3 11.5dBm while consuming 7mA current from 3.3V power supply.

I. INTRODUCTION

This paper presents the design and the implementation of a high CMOS frequency down-conversion mixer suitable for a single-chip receiver as shown in Fig. 1. The receiver topology proposed here is designed for the Personal Communications Standard (PCS1900) that operates at central frequency of 1.9GHz designating the receiver band from 1930MHz to 1990MHz.

The proposed mixer here is based on a novel folded Gilbert cell topology shown in Fig.3. This mixer topology is similar to the classical Gilbert cell based mixer (Fig.2). Virtually one of the main advantages of the proposed topology is to use tank circuits to fold the RF signal to the switching pairs, thus retains the same advantages as the Gilbert cell based mixer. Meanwhile with this folded topology we can set the bias current of the trans-conductance stage and the commutate stage independently, thus the noise and linearity performances can be optimized independently, also the P-MOSFET commutate stage suffers less 1/f noise which sometimes dominates the noise performance of the mixer in the Low-IF systems [2] [3] [4]. Section 2 will describe the design of this down-conversion

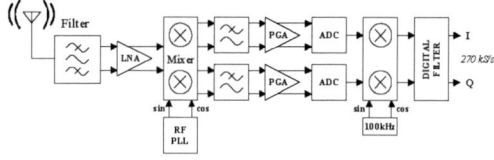

Fig.1 GSM1900 Low-IF Receiver architecture

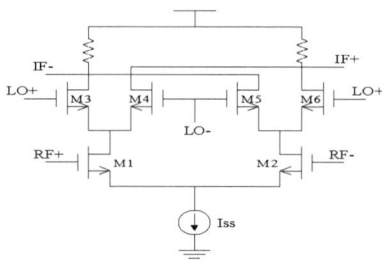

Fig.2 Topology of the traditional Gilbert cell of a mixer

mixer in more details. It is followed by Conclusions in Section 4.

II. DESIGN OF THE MIXER

2.1 Principles and Design considerations

The schematic of the proposed mixer is shown in Fig. 3 and Fig. 4 respectively, including a mixer core, a LO buffer and the bias circuits. In the mixer core design, M3-M6 PMOS cross-coupling cells are designed as mixer current commutation stage and M1, M2 NMOS pairs as trans-conductance stage. Two inductors Ld1, Ld2 co-operate with C1, C2 (including the parasitic capacitances) of point X and Y, which ensure to resonate at the central frequency of the input signal. Two off-chip resistors are employed as output IF loads to facilitate the testing. Ls1 and Ls2 are two inductors used as inductively degenerates that are similarly used in the LNA input stage. These inductors are both good for the conjugate matching of mixer input and the linearity performance [1].

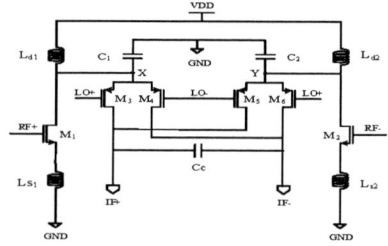

Fig.3. The proposed mixer core circuit with folded topology

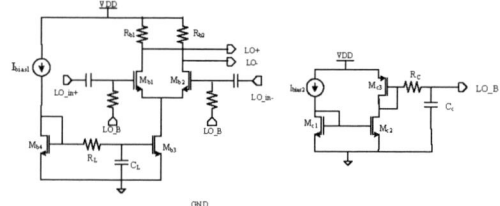

Fig.4 The LO buffer and the bias circuits

III. IMPLEMENTATION AND MEASUREMENTS

The GSM1900 Receiver was fabricated in SMIC 0.18-μm CMOS RF technology. The die micrograph of mixer is shown in Fig 5. The active area of the mixer with LO Buffer is 0.8mm×0.7mm. The testing instruments include Spectrum Analyzer E4440A, Vector Network Analyzer E5071B, RF signal generator E4438C and Noise Source E346C from Agilent Technologies™. The measured performance of the mixer is summarized in Table 1. Fig 6 shows the measured

0-7803-9451-8/06/$20.00 ©2006 IEEE.

output spectrum of the mixer. Considering the losses of the microwave connector and the single to differential Balun, the conversion gain of the mixer at 100kHz IF is around 6dB. In order to evaluate the linearity of the mixer, two-tone inter-modulation measurement was carried out with tone frequencies at 1900MHz and 1900.2 MHz respectively. Fig.7 and Fig.8 show the 1-dB compression point and IIP3 results respectively. Due to the equipment limitation, the SSB noise figure of the mixer can only be measured at 1MHz(instead of 100kHz) IF output and shows 18.5dB(include the noise effects of an off-chip output buffer and the LO buffer).

TABLE I.   SUMMARY OF MIXER MEASUREMENTS

Mixer	Measured Parameters
Supply voltage	3.3V
Current dissipation	7mA
RF frequency	1900MHz
LO frequency(4dBm)	1900.1MHz
SSB(Noise Figure)	18.5dB
Power Conversion Gain	6dB
Input IP3	11.5dBm
Input P- 1dB	1.5dBm
LO-RF feed-through	-53dB
LO-IF feed-through	-48dB

IV.   CONCLUSIONS

A 1.9-GHz down-conversion CMOS Mixer with the utilization of novel folded Gilbert Cell fabricated in a RF 0.18-μm CMOS process has been described. With this folded mixer, the bias current of the trans-conductance stage and the switching stage can be set independently to get an optimization of both linearity and noise performances. The mixer modular, includes a mixer core and a LO buffer, achieves a conversion gain of 6dB, Noise Figure of 18.5dB and IIP3 of 11.5dBm while consuming only 7mA current from 3.3V power supply. The measurement results show that the performance of this mixer has met the requirements of the low-IF GSM1900 receiver system.

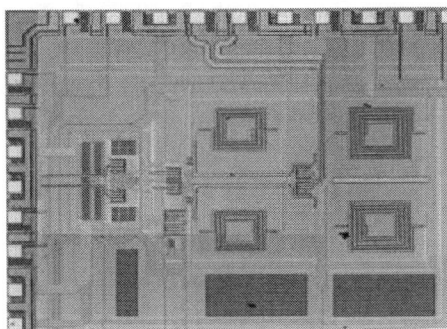

Fig.5 Micrograph of the mixer die

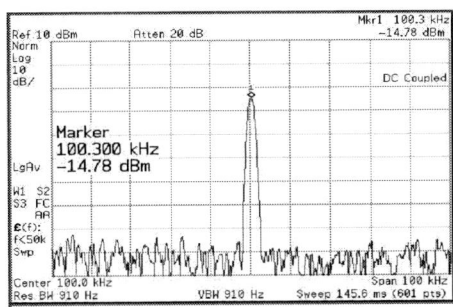

Fig.6 Measured mixer output spectrum: RF input power -20dBm/1900MHz and LO power 4dBm/1900.1MHz

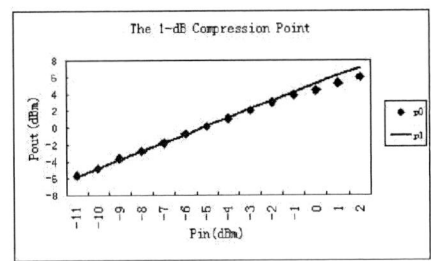

Fig.7 Measured P1-dB Compression Point of the mixer

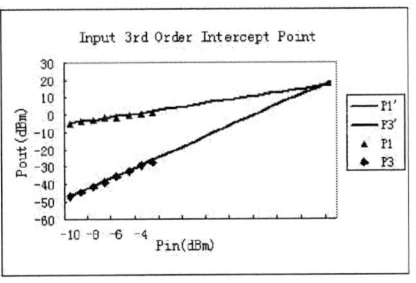

Fig.8 Extrapolation of the mixer IP3

ACKNOWLEDGMENT

The major works were finished at Fudan-Sicomm Union-Laboratory. The authors thank Shanghai Sicomm RF Technology, Inc. for the support on system specification, circuit design, the devices modeling and circuit testing. Besides, the authors thank the team members Junbiao Ding, Fang Yang , Zhisheng Li and Jiasheng Hu for the cooperation.

REFERENCES

[1] Thomas H.Lee. "The Design of CMOS Radio-Frequency Integrated Circuits," Cambridge University Press, 1998.

[2] M.T.Terrovitis, "Analysis and Design of Current-Commutating CMOS Mixers," PhD thesis, UC Berkeley, Berkeley, CA, 2001.

[3] B. Razavi, "Design of analog CMOS integrated circuits," Singapore :McGraw-Hill, 2001.

[4] H. Darabi, A.Abidi, "Noise in RF-CMOS Mixers: A Simple Physical Model," IEEE J. Solid-State Circuit, vol.35, NO.1, January 2000.

Integrated Direct Output Current Control Switching Converter using Symmetrically-Matched Self-Biased Current Sensors

Yat-Hei Lam

Department of EEE
Hong Kong University of
Science & Technology
Hong Kong SAR, China
e-mail: hylas@ee.ust.hk

Suet-Chui Koon

National Semiconductor
Corporation
Hong Kong SAR, China

e-mail: gladys.koon@nsc.com

Wing-Hung Ki

Department of EEE
Hong Kong University of
Science & Technology
Hong Kong SAR, China
e-mail: eeki@ee.ust.hk

Chi-Ying Tsui

Department of EEE
Hong Kong University of
Science & Technology
Hong Kong SAR, China
e-mail: eetsui@ee.ust.hk

Abstract -- A non-inverting flyback converter using an integrated symmetrically-matched self-biased current sensor was fabricated in a 0.35μm CMOS process. It operates in pseudo-continuous conduction mode and employs a direct output current control scheme to achieve excellent line transient response. The converter switches at 1MHz with an input of 1.2V to 2V to give an output of 1.5V and delivers 250mA.

I. INTRODUCTION

The controller of a DC-DC switching converter with voltage feedback is simple, but neither the inductor nor the output current is monitored, and the loop response is slow. Dynamic performance could be enhanced by employing current feedback, but slope compensation for the current loop is required to avoid sub-harmonic oscillation. Switching converters operating in the discontinuous conduction mode (DCM) do not exhibit sub-harmonic oscillation, because the inductor current starts from zero in every switching cycle. By raising the current floor to a non-zero value, the technique of pseudo-continuous conduction mode (PCCM) can be used to enhance the current handling capability [1].

The input voltage of a power converter with multiple sources changes from time to time, and spikes at the output are unavoidable. A control scheme that monitors and controls the output current directly may regulate the output voltage independent of the input voltage. In this paper, a direct output current control non-inverting flyback converter is proposed, and exhibits excellent rejection to abrupt line changes.

II. DIRECT OUTPUT CURRENT CONTROL CONVERTER

The proposed direct output current (DOC) control non-inverting flyback converter operates in PCCM is shown in Fig.1. The supply voltage V_{IN} may vary from 2V down to 1.2V, while the converter maintains an output voltage at V_{OUT}=1.5V. Every switching period is divided into three phases. In Phase 1, switches S_1 and S_3 are closed, charging the inductor at a rate of $di/dt=V_{in}/L$, until it reaches the peak current I_{PK} that is controlled by the output of an error amplifier. In Phase 2, S_2 and S_4 are closed, discharging the inductor at $di/dt=-V_{out}/L$. When the inductor current drops to the predefined freewheeling level I_{FW}, the converter enters Phase 3. In Phase 3, S_2 and S_4 are closed, and the inductor current freewheels at I_{FW} until the period expires. The current flow is shown in Fig.1(a) and the inductor current waveform is shown in Fig.1(b).

Charge is delivered to the output in Phase 2 only, and the average output current is proportional to the shaded area shown in Fig.1(b) that can be controlled by changing I_{PK}

This research is in part supported by Research Grant Council CERG HKUST 6311/04E.

---

through the control loop. Therefore, the output current is directly controlled. If the input voltage changes suddenly, the ramp up slope in Phase 1 is changed. Yet, as both I_{PK} and I_{FW} are not changed in a time frame of one cycle, the shaded area in Phase 2 is not affected (Fig.1(b)). Hence, the output voltage is not disturbed by a sudden change in V_{IN}, and excellent line transient response is achieved.

III. CONTROL SCHEME IMPLEMENTATION

The system block diagram of the proposed converter is shown in Fig.2. All switches are MOS transistors. The output voltage is scaled by a resistor string and compared to the reference voltage through a compensation circuit realized by an op-amp. The output of the error amplifier controls the peak inductor current. The functional block Logics & Drivers controls the switching sequence of the power stage. Inductor current information is extracted by sensing the current that passes through transistors M_{N1} and M_{N2} utilizing a MOS transistor scaling technique [2]. By sensing either M_{N1} or M_{N2}, the inductor current can be sensed all the time. It should be noted that in Phase 3 when the inductor current is freewheeling, both M_{N1} and M_{N2} conduct, but M_{N2} is chosen to be sensed such that there is no hand-over problem at the start of Phase 1.

IV. SM SELF-BIASED CURRENT SENSOR

A symmetrically-matched (SM) self-biased voltage mirror current sensor (Fig.3) [3, 4] is employed in the design. Using a time-multiplexing technique, the voltage-mirror core can be reconfigured in sensing both the switch and diode currents, and hence, the complete inductor current can be sensed without additional loss. The voltage mirror core consists of M_{1A} to M_{4A} and M_{1B} to M_{4B}. Let switches S_{X2} and S_{Y2} be closed, such that the current of M_{N2} is sensed by M_{S2}, with $M_{N2} : M_{S2} = N : 1$. Consider a large inductor current I_L injected into M_{N2} at node V_{SW2}. With $(W/L)_{1A} = (W/L)_{1B}$, $I_{d1A} = I_{d1B}$ are injected into M_{2A} and M_{2B}. Now, M_{2A}, $M_{2B} : M_{4A}$, $M_{4B} = 1 : M$, such that $M_{2A}+M_{4A}$ are matched with $M_{2B}+M_{4B}$, and currents of $(M+1)I_{d2A}$ are injected into M_{N2} and M_{S2}. The matching forces V_Y to be equal to V_X, thus achieving the voltage mirror function, and the current ratio $I_{N2} : I_{S2}$ is $N : 1$, or $I_{S2} = I_L/(N-1)$. This sensed current is mirrored by M_5 for peak current control (using R_{sen}). The X branch and the Y branch are biased with $I_L/(N-1)$, and hence, the larger the I_L, the larger the bias current, and the faster the voltage mirror. The matching is so accurate that even the large-signal analysis using MOS equation with channel length modulation cannot reveal any difference between the corresponding terminal voltages of any paired transistors. Hence, the sensing accuracy surpasses all prior current sensors. For a sensing ratio of 1000 to 1, an inductor current of 1mA can be sensed accurately. By closing switches S_{X1} and S_{Y1}, the switch current of M_{N1} is sensed by M_{S1}, and

---

0-7803-9451-8/06/$20.00 ©2006 IEEE.

the same voltage mirror core is reused. Yet, the connections for M_{N1} and M_{N2} are different, as shown in Fig.3, because M_{N1} is sourcing while M_{N2} is sinking the inductor current.

V. MEASUREMENT RESULTS

The converter was fabricated and tested. Fig.4 shows the inductor current and the corresponding current sensor output voltage. Complete inductor current information was sensed and scaled accurately at a switching frequency of 1MHz. Fig.5 shows the line transient response. The supply voltage changed by 400mV but the output voltage showed no observable changes. Fig.6 shows the load transient response of the DOC control converter. The output voltage settled in 120μs when the load current changed from 50mA to 250mA, and 60μs when the load current changed from 250mA back to 50mA.

VI. CONCLUSIONS

A direct output current control scheme for switching converters in incorporating the current control loop with the pseudo-continuous conduction mode of operation was proposed. The DOC control converter was designed and demonstrated excellent line transient response. An accurate and fully-integrated current sensor is employed. The complete design was realized and verified by measurement results.

REFERENCES

[1] D. Ma, W. H. Ki and C. Y. Tsui, "A Pseudo-CCM / DCM SIMO switching converter with freewheel switching", *IEEE J. of Solid-State Ckts.*, vol.38, No.6, pp.1007-1014, June 2003.

[2] Y. H. Lam, W. H. Ki and D. Ma, "Loop gain analysis and development of high-speed high–accuracy current sensors for switching converters," *IEEE Int'l Symp. on Ckts. & Sys.*, pp.828-831, May 2004.

[3] Y. H. Lam, W. H. Ki, C. Y. Tsui and D. Ma, "Integrated 0.9V charge-control switching converter with self-biased current sensor," *IEEE Int'l Midwest Symp. on Ckts. & Sys.*, Hiroshima, Japan, pp.305–308, July 2004.

[4] Y. H. Lam, W. H. Ki and C. Y. Tsui, "Symmetrically matched voltage mirrors and applications therefor," *US Patent Application No. 11/185,294*, July 20, 2005.

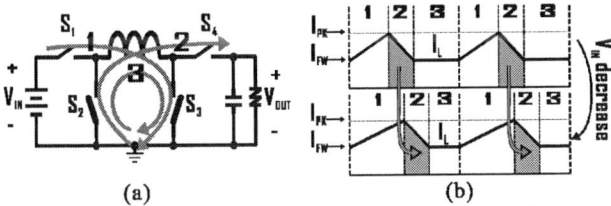

(a)   (b)

Fig.1 (a) PCCM flyback converter current flow and (b) corresponding inductor current waveforms

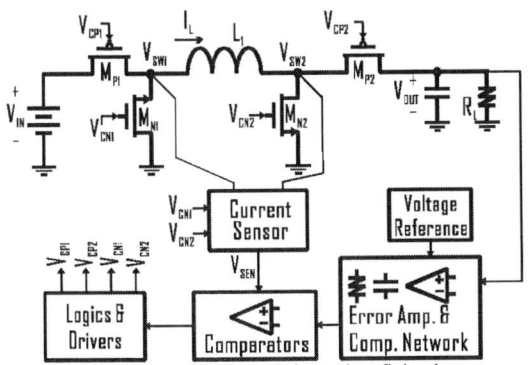

Fig.2 System diagram of the non-inverting flyback converter

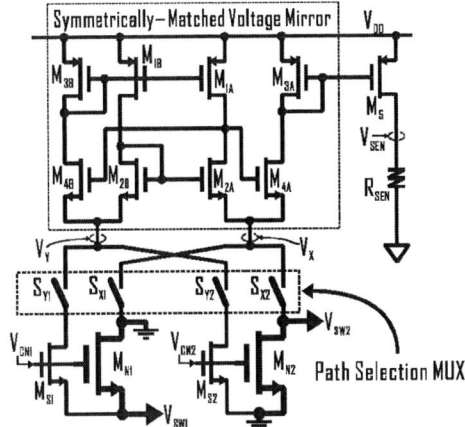

Fig.3 Symmetrically-matched self-biased current sensor

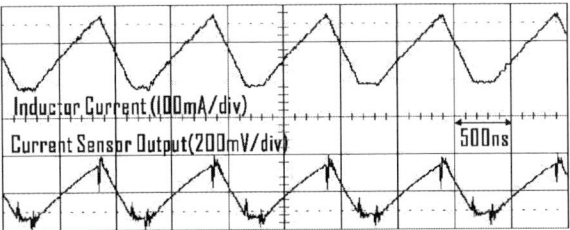

Fig.4 Measured inductor current and sensed current waveforms

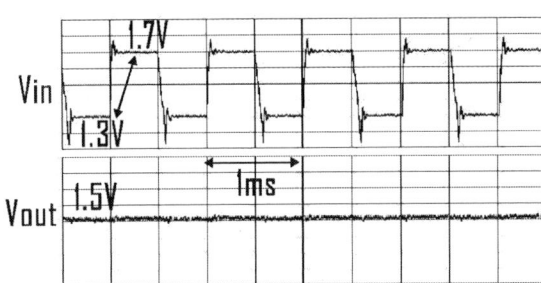

Fig.5 Line transient response of DOC control converter

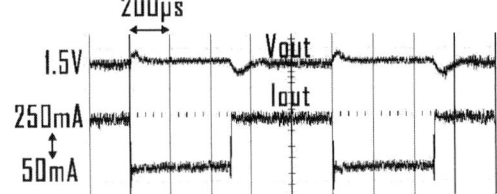

Fig.6 Load transient response of DOC control converter

Die Size	2.46mm × 2.82mm
Technology	0.35μm CMOS 4Metal-2Poly
Inductor	4.7μH (off-chip)
C_{out}	47μF (off-chip)
f_{switch}	1MHz
V_{in}	1.2V to 2V
V_{out}	1.5V
Max. I_{out}	250mA

Fig.7 Converter specifications and chip micrograph

Adaptively-Biased Capacitor-Less CMOS Low Dropout Regulator with Direct Current Feedback

Yat-Hei Lam

Department of EEE
Hong Kong University of
Science & Technology
Hong Kong SAR, China
e-mail: hylas@ee.ust.hk

Wing-Hung Ki

Department of EEE
Hong Kong University of
Science & Technology
Hong Kong SAR, China
e-mail: eeki@ee.ust.hk

Chi-Ying Tsui

Department of EEE
Hong Kong University of
Science & Technology
Hong Kong SAR, China
e-mail: eetsui@ee.ust.hk

Abstract -- A capacitor-less low dropout regulator (LDR) with direct current feedback is proposed. A symmetrically-matched voltage mirror in sensing the load current is employed, and gives excellent line and load regulations. The dynamic biasing results in an LDR with pole-tracking that extends the bandwidth of the loop gain at high load currents. The LDR was fabricated in a 0.35μm CMOS process with an active area of 0.11mm², and measurement results corroborated well with both analysis and simulation.

I. INTRODUCTION

Embedding a low dropout regulator (LDR, also known as LDO) into an IC chip for on-chip power regulation can reduce the circuit board area, but bulky off-chip capacitors are usually unavoidable for adequate filtering. In addition, on-chip compensation capacitors are usually needed and this takes up valuable silicon area.

The performance of the LDR depends much on the design of the error amplifier in the voltage feedback loop. To increase the gain and bandwidth of the loop, and to eliminate the output capacitor, multi-stage amplifier with a rather sophisticated compensation scheme was proposed [1]. However, an on-chip compensation capacitor of even a few pF occupies a large silicon area, and it is better if they could further be reduced, or even eliminated. An LDR needs a large biasing current for the high-gain high-speed error amplifier for control, especially when the load current is high. It would be advantageous if the bias current of the LDR is adaptive to the load, such that at light load, the bias current is low for high efficiency, and at heavy load, the bias current is high for high speed control.

II. LDR WITH DIRECT CURRENT FEEDBACK

The first LDR with a dynamically biased voltage buffer was proposed in 1998 [2] but a BiCMOS process is needed and accurate current sensing was achieved due to the difference in $|V_{ds}|$ between the sensing transistor and the power transistor.

Fig. 1 shows our proposed LDR that employs an accurate current sensor using a symmetrically-matched (SM) voltage mirror [3, 4]. Two feedback loops can be identified. The scaled output voltage is compared to the reference voltage V_{ref} through an error amplifier that is simply a differential pair, and the output of which drives the pass transistor M_o to supply the load cuurent. This is the voltage feedback loop. The current feedback loop consists of the SM voltage mirror that provides a bias current I_{b2} for the error amplifier that drives M_o to complete the loop. The error amplifier is biased by M_{S5} and it sinks a current of $[2P/N] \times I_{do}$ which is the feedback current generated by the current sensor. The fed back bias current is accurately proportional to I_o, and we coined this action as direct current feedback (DCF). The drain current of the power transistor M_o bears a square relation with the gate overdrive voltage:

$$I_{do} = \tfrac{1}{2}\mu_p C_{ox}(W/L)_{M_o}(|V_{gso}| - |V_{tp}|)^2(1+|\lambda_p V_{dso}|) \qquad (1)$$

Suppose the drain current of M_o increases to $I_{do}+i_{do}$ due to an increase in load current. To accommodate this change, the gate overdrive voltage $|V_{gso}| - |V_{tp}|$ has to increase according to equation (1) to keep the output voltage constant. Now, with the ratio assignment of Fig.1, the drain current of M_{A3} increases from $(P/N)I_{do}$ to $(P/N)(I_{do}+i_{do})$, providing just the correct overdrive voltage for M_{A4} and thus, M_o. Consequently, an excellent load regulation is obtained.

For stability and bandwidth, we note that the proposed LDR has two high impedance nodes related to the feedback loops. The first one is at the output of the error amplifier V_a, and is the dominant pole $p_a = 1/(C_a R_a)$, where C_a is mainly the very large gate capacitance of M_o, and R_a is the output resistance at V_a. The second pole is at the output of the regulator V_o, with $p_o = 1/(C_o R_o) \propto I_o$, where C_o is the drain capacitance of M_o plus the parasitic capacitance of the packaging, and $R_o = r_{do}||R_L||(R_1+R_2) \propto 1/I_o$. Since the proposed LDR has no output filtering capacitor, p_o is much larger than p_a. Now, the criteria of design is to make sure that p_o is higher than the unity gain frequency ω_t due to the single pole roll-off of p_a, and $\omega_t = bg_{mA1}g_{mo}R_o/C_a$. M_{A1} and M_{A2} are designed to work in weak inversion region at light to medium load such that $g_{mA1} \propto I_o$ [5]. At light load, M_o works in weak inversion region also and $g_{mo} \propto I_o$, giving $\omega_t \propto I_o$. The unity gain bandwidth tracks with p_o. At medium load, M_o works in strong inversion region such that $g_{mo} \propto \sqrt{I_o}$, while M_{A1} and M_{A2} are designed to stay in weak inversion region, thus $\omega_t \propto \sqrt{I_o}$. The bandwidth ω_t is still extending with the load current but at a slower rate. At heavy load, M_{A1}, M_{A2} and M_o work in strong inversion region, and g_{mA1}, $g_{mo} \propto \sqrt{I_o}$, giving ω_t independent of I_o. Separation between ω_t and p_o becomes larger as I_o increases. As a result, if the LDR is stable at a light load current, it is guaranteed to be stable at a higher load current.

III. SIMULATION AND MEASUREMENT RESULTS

The direct current feedback low dropout regulator was fabricated in a 0.35μm CMOS process. Fig.2 shows the simulated loop gain response. Pole tracking is evident as both p_a and p_o move to higher frequencies at a high load current, with p_o moves faster than p_a, and the unity gain bandwidth is extended. Fig.3(a) shows the measured load transient response of the DCF LDR with a current step of 1mA to 150mA. For a load change from high to low, the initial large bias current gives a fast response and the ripple is small, only 4mV, and is even smaller than 0.5% of the output voltage. For a load change from low to high, the initial bias current of the error amplifier is small and the response is slow and leads to a larger glitch of 50mV at the output. As the bias current increases with the load current, the bandwidth increases and the output voltage is stabilized very quickly. The 1% settling time (18mV of 1.8V) is less than 200ns. Fig.3(b) shows the rejection of switching noise. In this measurement, the LDR is cascaded to a switching converter. Low frequency ripples are filtered out effectively and high frequency glitches are attenuated by approximately 14dB. Fig.4 shows the measured output voltage vs the load current and the supply voltage. For a load current that changes from 0 to 240mA at $V_{dd} = 2.2V$, the load regulation is only

---

This research is in part supported by Research Grant Council CERG HKUST 6311/04E.

0-7803-9451-8/06/$20.00 ©2006 IEEE.

2.77µV/mA (664µV in total). Note that for a conventional LDR, the output voltage decreases monotonically as the load current increases, but for the proposed DCF LDR, the output voltage increases initially at light load current, which is due to the current feedback mechanism using the symmetrically-matched voltage mirror, and the load regulation is not solely controlled by the gain of the error amplifier. Fig.5 shows the measured quiescent current I_q vs the load current I_o. The quiescent current includes the currents consumed by the error amplifier, the voltage mirror and the feedback resistors. At light loads, I_q is closed to I_{b1} (a very small bias current in case the load current goes to zero) plus the current of the potential divider, but as I_o is larger than 1mA, I_q increases linearly, which implies that the current sensor works properly. Table 1 summarizes the specifications of the proposed DCF LDR. Fig.6 shows the chip micrograph.

IV. CONCLUSIONS

We demonstrated a working low dropout regulator that employed adaptive biasing and symmetrical matching techniques. Remarkable performance in both load transient response and load regulation are shown in the measurement results. The resultant LDR needs no filtering and compensation capacitors, and fabrication cost could be much reduced. Therefore, it is suitable for system-on-chip (SOC) applications and as an on-chip power regulator.

REFERENCES

[1] K. N. Leung and P. Mok, "A capacitor-free CMOS low-dropout regulator with damping-factor-control frequency compensation," *IEEE J. of Solid-State Ckts.*, Vol.38, No.10, pp.1691-1702, Oct. 2003.
[2] G. A. Rincon-Mora and P. E. Allen, "A low-voltage, low quiescent current, low drop-out regulator," *IEEE J. of Solid-State Ckts.*, Vol.33, No.1, pp.36–44, Jan. 1998.
[3] Y. H. Lam, W. H. Ki, C. Y. Tsui and D. Ma, "Integrated 0.9V charge-control switching converter with self-biased current sensor," *IEEE Int'l Midwest Symp. on Ckts. & Sys.*, Hiroshima, Japan, pp.305–308, July 2004.
[4] Y. H. Lam, W. H. Ki and C. Y. Tsui, "Symmetrically matched voltage mirrors and applications therefor," *US Patent Application No. 11/185,294*, July 20, 2005.
[5] P. R. Gray, P. J. Hurst, S. H. Lewis and R. G. Meyer, *Analysis and design of Analog Integrated Circuits*, 4th ed. New York: Wiley, 2001.

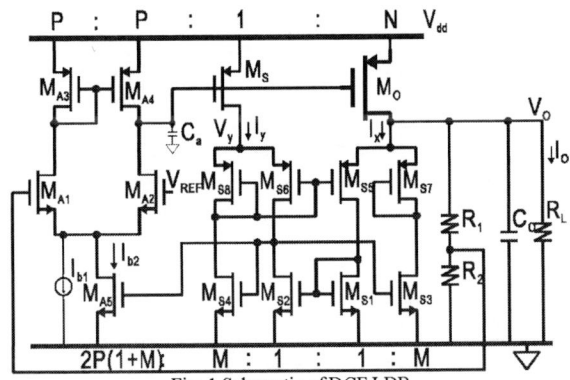

Fig. 1 Schematic of DCF LDR

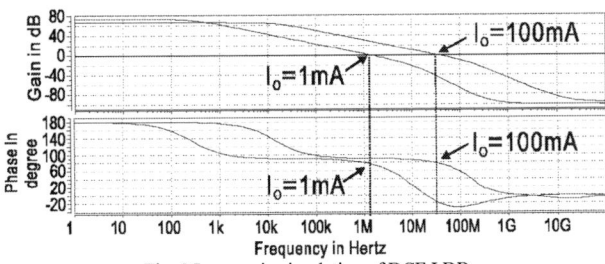

Fig. 2 Loop gain simulation of DCF LDR

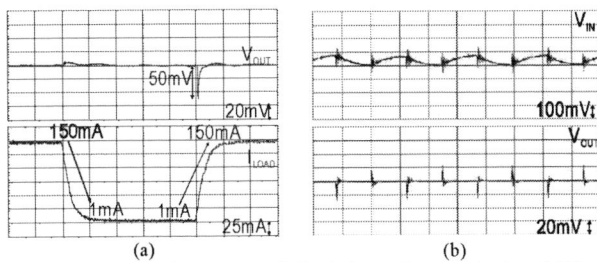

(a)   (b)

Fig. 3 (a) Load transient response in load change between 1mA and 150mA (2µs/div) and (b) Line ripple rejection with high frequency switching noise

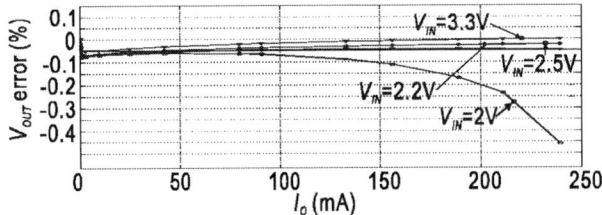

Fig. 4 Load regulation of the DCF LDR

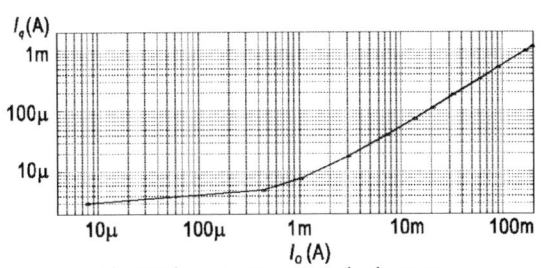

Fig. 5 Quiescent current versus load current

TABLE 1
SPECIFICATIONS OF THE DCF LDO REGULATOR

Technology	0.35µm CMOS process
Chip area (including pads)	0.32 mm²
Chip area (active circuit area)	0.11mm²
V_{dd}	2-3.6V
V_O	1.8V
Maximum I_O	240 mA
Quiescent Current @ I_O=0mA	3 µA
Quiescent Current @ I_O=200mA	1.03 mA
Load regulation @ V_{dd}=2.2V, I_O=0mA to 240mA	2.77µV/mA (664µV total)
Line regulation @ V_{dd}=2V to 3.6V, I_O=100mA	<0.8mV/V (1.28mV total)
Line ripple rejection @ 10kHz	>40 dB
Output impedance @ 1MHz (I_O=100mA)	200mΩ

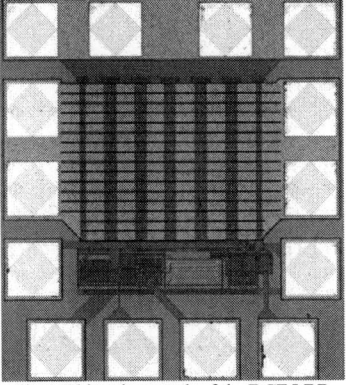

Fig. 6 Chip micrograph of the DCF LDR

A Built-in Power Supply Noise Probe for Digital LSIs

Mitsuya Fukazawa, Koichiro Noguchi, Makoto Nagata, Kazuo Taki

Department of Computer and Systems Engineering, Kobe University
1-1 Rokkodai-cho, Nada-ku, Kobe 657-8501, Japan

1. Introduction

Dynamic power-supply noise has emerged as a critical piece of low voltage designs in current 130-nm and coming sub-100-nm CMOS technologies. On-chip measurements of dynamic power-supply and ground distributions within a large scale digital circuit can provide precious knowledge for establishing reliable design guides of power-supply systems[1]. This paper introduces a design example of a standard logic cell based digital circuit incorporating a built-in noise detection technique[2].

2. Design of built-in noise detector

Figure 1(a) shows a built-in power supply noise probing architecture. Power supply and ground noises in a digital circuit often show strong position dependences arising from the interaction of time-varying digital activity distributions and frequency-domain transfer properties of on-chip and off-chip parasitic impedance notworks. The architecture enables to profile the time-domain variations of noise intensity as well as the frequency-domain emphasis of noise distributions in power-supply/ground grids. A noise detector structure given in Figure 1(b) fits this purpose, consisting of a source follower (SF) that senses noise voltage and a transconductance transistor (Gm) that converts SF's output voltage to a current signal read externally as Iout. Because of the small device count in this simple front-end structure, the size of a detector is comparable to a standard flip-flop cell and thus placed within a cell row of a standard cell-based digital circuit.

Detailed design of noise detector circuits is shown in Figure 2. In addition to a source follower ($M1, M2$) and a single MOSFET common-source amplifier ($M3$), switch MOSFETs for shuttering by signal SHU, $M4$, and for selective activation by signal SEL, $M5, M6, M7$, are provided.

The source followers (SF) of n-channel and p-channel senses voltage fluctuation on the nearest digital power-supply $DVDD$ and ground $DGND$ wirings, respectively, and the common-source amplifier continuously translates it to a current-mode noise signal that is transmitted on a shared current bus $IBUS$ and then read out through a current mirror by an external oscilloscope with a termination resistor. One of the detectors sharing $IBUS$ can be activated at the same time while all the others are cut off.

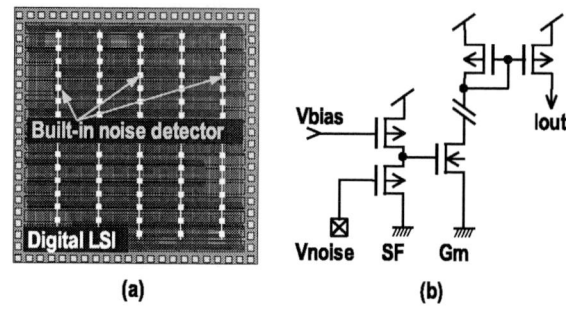

Figure 1: (a) Chip architecture incorporating built-in noise probing technique and (b) SF+Gm noise detector.

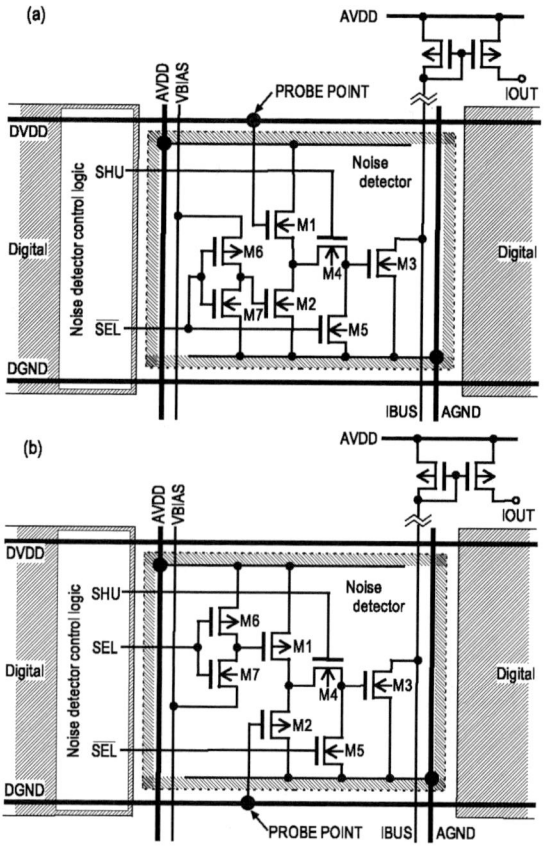

Figure 2: Built-in noise detector circuits for probing (a) power-supply and (b)ground wirings.

0-7803-9451-8/06/$20.00 ©2006 IEEE.

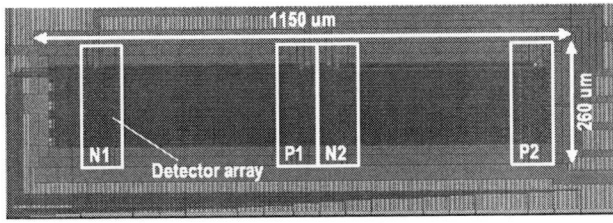

Figure 3: Magnified chip photograph.

The physical layout of the detector can be adjusted to that of a standard logic cell library and its area can be comparable with a few DFF cells.

3. Measurements

A test chip was prepared for the built-in detectors. Two pairs of n-type and p-type detectors for 1.8-V digital circuits were designed with 2.5-V and 1.8-V MOSFETs in a 0.18-μm CMOS, 5LM, multi-oxide, triple-well on P-type bulk substrate technology, where the sizes of MOSFETs were chosen to obtain the noise measurement bandwidth and gain of roughly 1 GHz and 1.0, respectively. A magnified photo of a test chip is given in Figure 3.

Noise waveforms on power-supply and ground rails by the built-in noise detector located in the middle of the 12-th cell row are shown in Figure 4, where a shift register in the test chip was selectively activated. Negative drops on power-supply wirings right after both rise and fall edges of the clock signal are dynamic noises resulting from the switching operations of DFF cells, along with positive counterparts on ground wirings.

Figure 5 extracts the heights of voltage drop at a rise clock edge corresponding to roughly 3.125 ns in Figure 4, where Vdd_drop shows negative peaks measured from 1.8 V while Vgnd_drop gives positive peaks from 0.0V. The location of an active shift register in the 32 logic cell rows is identified by the cell-row number. Both gain-calibrated 2.5-V and 1.8-V detectors show consistent results, which proves the certainty of the built-in noise detection technique.

Observations show that Vdd_drop is larger than 50 mV when a shift register in any cell row is active, and takes the maximum when that in the same 12-th cell row as the active built-in detector is active. On the other hand, Vgnd_drop is normally negligible other than the shift register at 11-th and/or 12-th cell row are active, which both share the ground wiring that the detector is probing. The significantly localized distribution found in ground noise is due to the presence of a substrate.

4. Conclusions

The design of compact noise detector circuitry that could be embedded and arrayed within a high-density large-scale digital circuit was demonstrated. In-depth characterization of dynamic power-supply and ground

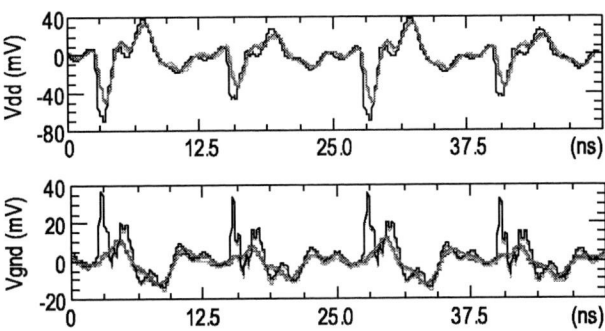

Figure 4: Noise waveforms on power-supply (upper) and ground (lower) wirings measured by detectors on 12-th cell row, where vertical axes show difference from nominal voltage of 1.8 V and 0.0 V, respectively.

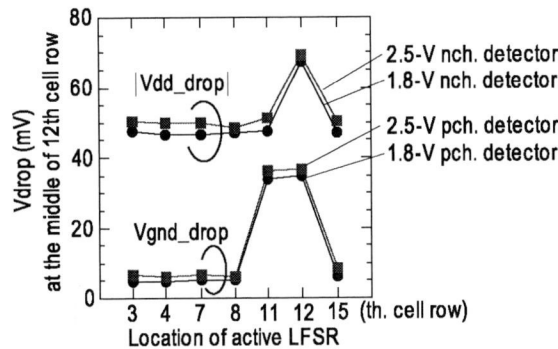

Figure 5: Dynamic voltage drop on power-supply wirings, Vdd_drop, and ground wirings, Vgnd_drop, measured from 1.8 V and 0.0 V, respectively.

noises by the built-in noise detection technique can validate and/or calibrate dynamic power-supply analysis (IR drop) methodologies that are becoming requisite to nanometer scale digital integrated circuits.

Acknowledgments

Test chips were fabricated by Hitachi Ltd. in the chip fabrication program provided by the VLSI Design and Education Center (VDEC) of the University of Tokyo in collaboration with Dai Nippon Printing Corporation.

References

[1] K. Shimazaki, M. Nagata, T. Okumoto, S. Hirano, and H. Tsujikawa, "Dynamic Power-Supply and Well Noise Measurements and Analysis for Low Power Body Biased Circuits," *IEICE Transaction on Electronics*, Vol.E88-C, No.4, pp.589-596, Apr. 2005.

[2] M. Nagata, T. Okumoto, K. Taki, "A Built-in Technique for Probing Power Supply and Ground Noise Distribution within Large-Scale Digital Integrated Circuits," *IEEE J. Solid-State Circuits*, vol.40, No.4, pp.813-819, Apr.2005.

A 476-gate-count Dynamic Optically Reconfigurable Gate Array VLSI chip in a standard $0.35\,\mu m$ CMOS Technology

Minoru Watanabe and Fuminori Kobayashi
Department of Systems Innovation and Informatics
Kyushu Institute of Technology
680-4 Kawazu, Iizuka, Fukuoka, 820-8502, Japan
Email: { watanabe, fkoba }@ces.kyutech.ac.jp

I. INTRODUCTION

High-speed reconfigurable processors have been developed in recent years: they are DAP/DNA chips and DRP chips [1][2]. These devices can be changed from one context to another context at every clock cycle in a few nanoseconds. However, their die size limits the number of reconfiguration contexts of currently available DAP/DNA and DRP chips to 4–16.

In contrast, optically reconfigurable gate arrays (ORGAs) [3][4] enable both fast reconfiguration and numerous reconfiguration contexts using an optical holographic memory and optical wide-band reconfiguration connections. Such devices present the possibility of large virtual gate-count VLSIs.

However, even though the virtual gate-count is extremely large, a high real gate-count is required to increase the amount of working circuits at any moment. For that reason, we previously proposed a dynamic optical reconfiguration circuit that is the smallest optical reconfiguration circuit among all ORGAs. We continue our development of high-gate-count dynamic ORGAs (DORGAs) [5].

This paper presents a new design of a fabricated 476-gate-count DORGA modified from a previously designed 68-gate-count DORGA [6] using standard 0.35 μm three-metal CMOS process technology.

II. DYNAMIC OPTICAL RECONFIGURATION CIRCUIT

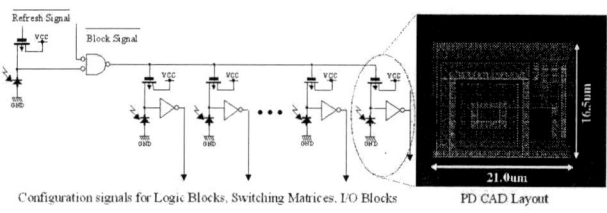

Configuration signals for Logic Blocks, Switching Matrices, I/O Blocks   PD CAD Layout

Fig. 1. Schematic diagram of an array of optical reconfiguration circuits and a CAD layout of a photodiode cell that includes a photodiode and an optical reconfiguration circuit.

A dynamic optical reconfiguration circuit that eliminates a static memory function from the VLSI part consists of only a photodiode and an inverter, as shown in Fig. 1. In that circuit,

TABLE I

SPECIFICATION OF A HIGH-DENSITY DORGA.

Technology	0.35 μm double-poly triple-metal CMOS process
Chip size	4.9 × 4.9 [mm]
Supply Voltage	Core 3.3V, I/O 3.3V
Photodiode size	9.1 × 8.8 [μm]
Distance between Photodiodes	h.=42, v.= 33 [μm]
Number of Photodiodes	3,696
Av. Aperture Ratio	3.1%
Number of Logic Blocks	28
Number of Switching Matrices	36
Number of Wires in a Routing Channel	8
Number of I/O bits	64
Gate Count	476

the gate array information is stored in junction capacitance of photodiodes instead of a static memory function component such as a latch, a flip-flop, or a single memory bit. Photodiodes not only detect light, but also serve as dynamic memory. The photodiode states are connected directly through inverters to the gate array component.

As a result, the dynamic optical reconfiguration circuits can be implemented in less than 10% of the area of all cell areas that comprise an optically reconfigurable logic block. The reconfiguration circuit, including the static memory function, occupied 43% of the chip area in previously proposed ORGAs [7].

III. GATE ARRAY DESIGN

A new 476-gate-count DORGA-VLSI chip was designed using a 0.35 μm standard CMOS process. Table 1 shows those specifications. The respective acceptance surface sizes of the photodiode and the photodiode-cell size, including an optical reconfiguration circuit, are 8.8 μm × 9.1 μm and 21.0 μm × 16.5 μm. The photodiodes were constructed between N+ diffusion and the P-substrate. The photodiode cells were arranged at 42.0 μm horizontal intervals and at 33.0 μm vertical intervals: 3,696 photodiodes were used. The

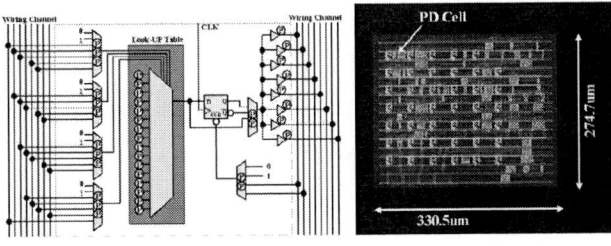 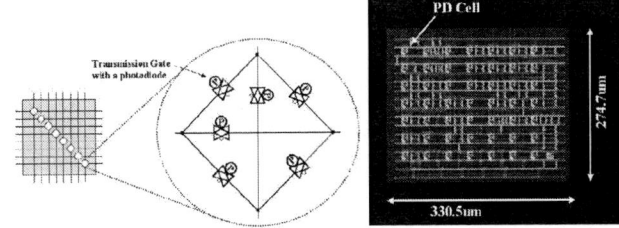

Fig. 2. Block diagram and CAD layout of an optically reconfigurable logic block (ORLB).

Fig. 3. Block diagram and CAD layout of an optically reconfigurable switching matrix (ORSM).

third metal layer was used for shielding transistors from light irradiation; the other two layers were used for wiring. The gate array components were designed using Design Compiler and Apollo (Synopsys Inc.), respectively, for the logic synthesis tool and the place and route tool.

Fig. 2 shows a block diagram and CAD Layout of an ORLB. The ORLB consists of a four-input look-up table (LUT), multiplexers, a D-flip flop, and tri-state buffers, along with the FPGA structure. A point of difference from FPGAs is that all states of the LUTs, multiplexers, and tri-state buffers are optically programmable through 40 photodiodes. One optical reconfiguration circuit is added to an ORLB as a block reconfiguration assignment. Therefore, 41 photodiodes were implemented in this ORLB. Wiring was executed using the first and the second metal layers while avoiding the photodiode-cell aperture area. The cell size is 330.5 μm × 274.7 μm. Fig. 3 shows a block diagram and CAD Layout of an optically reconfigurable switching matrix (ORSM). The ORSM structure is fundamentally identical to that of units sold by Xilinx Inc., but each optical reconfiguration circuit controls a transmission gate. The cell size is 330.5 μm × 274.7 μm.

Fig. 4 shows the CAD layout of a part of the entire gate array. The ORLBs and ORSMs are placed alternately in the horizontal direction. The ORSM cells were placed five in the horizontal direction and eight in the vertical direction. However, the corner cell was removed. The ORLB cells were placed four between ORSMs for the horizontal direction and seven in the vertical direction. In the remaining area, 16 I/O cells with 4 I/O bits were placed.

IV. EXPERIMENTAL RESULT

The photodiode characteristics has been measured by using 633 nm He-Ne Laser. The reconfiguration period is 12 ms when each photodiode receive 505 pW laser power. In the case of reconfiguring DORGA at 100MHz, the required optical power is calculated 2.24 W since the required optical power is inverse proportion to the reconfiguration period

V. CONCLUSION

This paper presented the design of the largest 476-gate count DORGA fabricated by using 0.35 μm three-metal CMOS technology. In the case of reconfiguring DORGA at 100MHz,

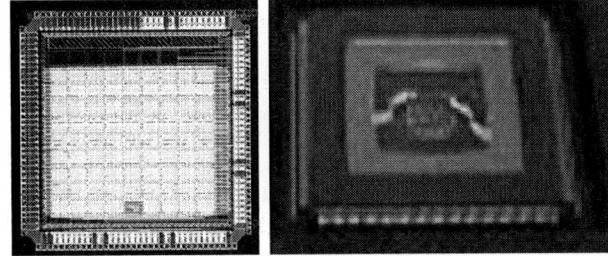

Fig. 4. The CAD layout and chip photograph.

the required optical power was estimated 2.24 W. At that time, the reconfiguration data transfer rate of the DORGA VLSI chip reaches 369.6 Gbit/s.

VI. ACKNOWLEDGMENTS

This research was supported by the project of development of high-density optically and partially reconfigurable gate arrays under Japan Science and Technology Agency. The VLSI chip in this study was fabricated in the chip fabrication program of VLSI Design and Education Center (VDEC), the University of Tokyo in collaboration with Rohm Co. Ltd. and Toppan Printing Co. Ltd.

REFERENCES

[1] H. Nakano, T. Shindo, T. Kazami, M. Motomura, "Development of dynamically reconfigurable processor LSI," NEC Tech. J. (Japan), vol. 56, no. 4, pp. 99–102, 2003.

[2] http://www.ipflex.co.jp

[3] J. Mumbru, G. Panotopoulos, D. Psaltis, X. An, F. Mok, S. Ay, S. Barna, and E. R. Fossum, "Optically Programmable Gate Array," Proc. SPIE - Int. Soc. Opt. Eng., vol. 4089, pp. 763–771, 2000.

[4] J. Mumbru, G. Zhou, X. An, W. Liu, G. Panotopoulos, F. Mok, and D. Psaltis, "Optical memory for computing and information processing," Proc. SPIE - Int. Soc. Opt. Eng., vol. 3804, pp. 14–24, 1999.

[5] M. Watanabe, F. Kobayashi, "A high-density optically reconfigurable gate array using dynamic method," International conference on Field-Programmable Logic and its Applications, pp. 261–269, 2004.

[6] M. Watanabe, F. Kobayashi, "A dynamic optically reconfigurable gate array using dynamic method," International Workshop on Applied Reconfigurable Computing, pp. 50–58, 2005.

1D-11

2006 Asia and South Pacific Design Automation Conference

Measurement Results of Within-Die Variations on a 90nm LUT Array for Speed and Yield Enhancement of Reconfigurable Devices

Kazuya Katsuki, Manabu Kotani, Kazutoshi Kobayashi and Hidetoshi Onodera
Graduate School of Informatics, Kyoto University, Kyoto, Japan.
{katsuki, kotani, kobayasi, onodera}@vlsi.kuee.kyoto-u.ac.jp

Abstract— It is possible to enhance speed and yield of reconfigurable devices utilizing WID variations. An LUT array LSI is fabricated on a 90nm process to measure WID and D2D variations. Performance fluctuations are measured by counting the number of LUTs through which a signal is passing within a certain time. D2D and WID variations are clearly observed by the measurement.

I. INTRODUCTION

Process scaling makes it possible to integrate billions of transistors on a die. It is quite difficult to manufacture such small transistors with similar characteristics. Down-scaling increases variations of transistor performance. Transistor performances are different die-to-die (D2D) and also within-die (WID). [1] reveals that WID variations are apparently observed in a 90nm process, which become dominant according to the process scaling[2].

Degradations of transistor performance by variations impacts gate delay, and finally it degrades the speed and yield of the LSIs. On the other hand, such WID variations can be compensated by reconfiguration. If a circuit is reconfigured according to the process variation, speed and yield can be enhanced. In order to place functional blocks according to the process variations, we must obtain variation data in some way. Once they are obtained, functional blocks can be placed according to the measured variations of each chip and their lengths of critical paths.

We have fabricated an LUT array LSI to confirm whether the WID variations clearly occur in reconfigurable devices. Its structure is shown in section II, measurement results is in section III, and we conclude this paper in section IV.

II. FABRICATED LUT ARRAY

Fig. 1 shows the structure of a logic block (LB) which contains a 4-bit LUT and a scan flip-flop (SDFF). An LUT consists of 16 flip-flops to store an LUT configuration and five MUX4s (4-input multiplexers). The output signal **Mout** from the MUX4 is sent to the adjacent LUT. Fig. 2 shows the array structure of logic blocks in the fabricated chip. They are laid out in a fractal structure to observe

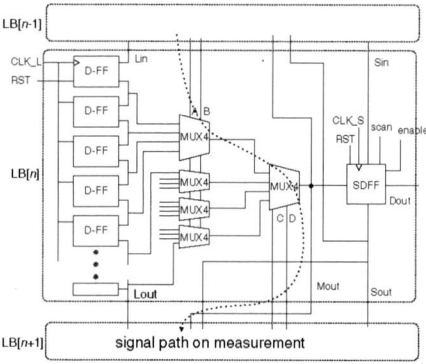

Fig. 1. Structure of a logic block. A signal is transmitted along the dashed arrow through two MUX4s per LB at the measurement.

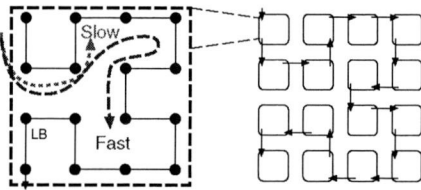

Fig. 2. Structure of the LUT array. LBs are connected in a fractal structure to observe scalable process variations.

scalable process variations. If they are laid out in a line, WID variations may be canceled. The fractal structure makes it possible to measure WID variations in scalable square regions.

On measuring the process variations, a signal is rushing through LUTs from the first LB in a square region, which is captured by the SDFF in each LB. LUTs are configured as follows during the measurement.

- The LUT in the first LB is configured to become true at any input value.

- The LUT in the second LB is configured to become true if the input **B** from the previous Sout becomes true.

- The LUTs in the other LBs are configured to become true only if the input **A** from **Mout** becomes true.

Applying a clock pulse to SDFFs under the above LUT configuration, **Sout** of the first LB becomes true, which is transmitted through LBs. During the transmission, let

0-7803-9451-8/06/$20.00 ©2006 IEEE.

110

Fig. 3. Chip micrograph of a 90nm LUT array LSI including 2,048 logic blocks located at the bottom.

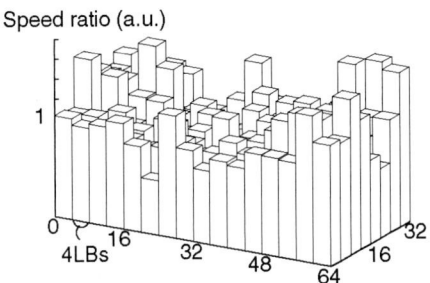

Fig. 4. Statistics of a fabricated dice by regarding the gradient from the least square method as the performance indicator. Peripheral LBs are fast and central ones are slow.

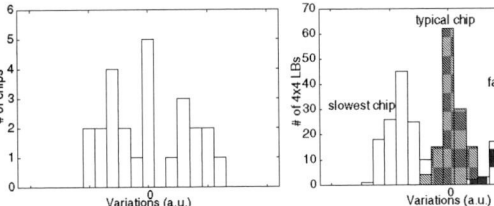

Fig. 5. Observed D2D variations, which is obtained from the average residual error between the measured values and averaged representative ones.

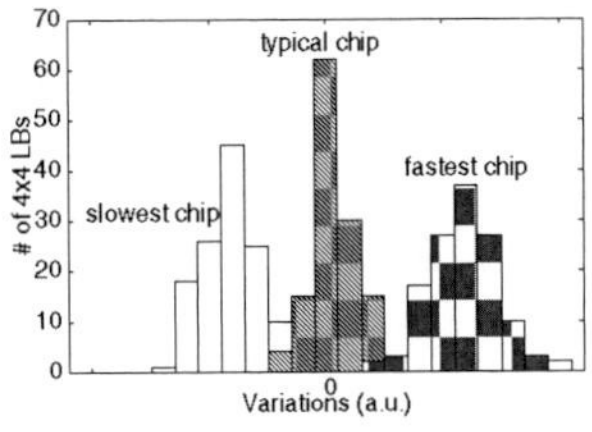

Fig. 6. Extracted WID variations from three distinguished chips. Left: slowest, Center: typical, Right: fastest. They are similar to the Gaussian distribution. The scale of the x axis is same as Fig. 5

us apply another clock pulse to SDFFs. Then the SDFFs in the LBs where the true signal have been transmitted become true. If WID variations are observed, number of transmitted LBs will be different in each square region as shown in Fig. 2. Fig. 3 shows a micrograph of a fabricated LSI.

III. MEASUREMENT RESULTS

In a single measurement, very little variations appear since the transistor speed is quantized as the number of LBs. To avoid the quantization and measure the difference clearly, clock cycle (time to transmission) is varied from 4.0ns to 8.0ns at 0.1ns interval. We repeat it 100 times per cycle at the resolution of 16 (4×4) LBs. The average value of 100 results is regarded as the number of transmissions at the cycle. By setting the clock cycle on the horizontal axis and the average number of transmissions on the vertical axis, the gradient is calculated using the least square method. The gradient depends on the performance of each block of LBs. The ratio of the gradients is equivalent to the ratio of the speeds. We can regard these gradients as the performance indicator.

Fig. 4 shows the statistic of WID variations from a fabricated chip. The peripheral LBs tend to be fast and the central LBs are slow. The other 24 chips have the same tendency. The possible reasons of the concave delay curve is that the central portions degradation caused by IR drop. To distinguish the WID and D2D variations, statistics from the 25 dice are averaged for every 16 LBs. These averaged gradients are called the reference delays. The average value of the residual errors between the measured and reference delays on a die is regarded as the D2D process variation, which is shown in Fig. 5. Fig. 6 shows WID variations of the slowest, typical and fastest chips. Each distribution is obtained to subtract measured delays from the reference delays. The three distributions are very similar to the Gaussian distribution. Therefore the above residual-based method is practical to extract WID variations. Fig. 5 and 6 reveals that the WID and D2D variations have the same order in the 90nm process.

IV. CONCLUSION

We propose compensating WID variations by reconfiguration. An LUT array LSI is fabricated on a 90nm process to measure process variations of reconfigurable devices. D2D and WID variations are cleary observed on the fabricated chip, and they have the same order on 90nm process. So it is likely that WID variations will be dominant in near future. This means that the proposed method is efficient and will be effective.

REFERENCES

[1] S. Ohkawa, M. Aoki, and H. Masuda. Analysis and Characterization of Device Variations in an LSI Chip Using an Integrated Device Matrix Array. *IEEE Transactions on Semiconductor Manufacturing, Vol.17, No.2*, pages 155–165, 2004.

[2] Samie B. Samaan. The Impact of Device Parameter Variations on the Frequency and Performance of VLSI Chips. In *ICCAD2004*, pages 343–346, 2004.

[3] Kazuya Katsuki, Manabu Kotani, Kazutoshi Kobayashi, and Hidetoshi Onodera. A Yield and Speed Enhancement Scheme under Within-die Variations on 90nm LUT Array. In *Proceedings of IEEE 2005 Custom Integrated Circuits Conference*, pages 601–604, 2005.

1D-12

2006 Asia and South Pacific Design Automation Conference

High-Throughput Decoder for Low-Density Parity-Check Code

Tatsuyuki Ishikawa, Kazunori Shimizu, Takeshi Ikenaga and Satoshi Goto

Graduate School of Information, Production and Systems, Waseda University
2-7 Hibikino, Wakamatsuku, Kitakyushushi, Fukuoka 808-0135, Japan
e-mail: i_tatsuyuki@akane.waseda.jp

Abstract— We have designed and implemented the LDPC decoder chip with memory-reduction method to achieve high-throughput and practical chip size. The decoder decodes (3,6)-2304bit regular LDPC codes using modified min-sum algorithm. The decoder achieves a throughput of 530Mb/s at an operating frequency of 147MHz. The chip has been fabricated in a 0.18μm, 6 metal-layer CMOS technology. The chip size is 36mm^2.

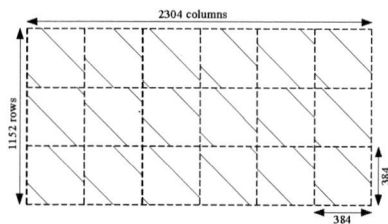

Fig. 1. Parity Check Matrix.

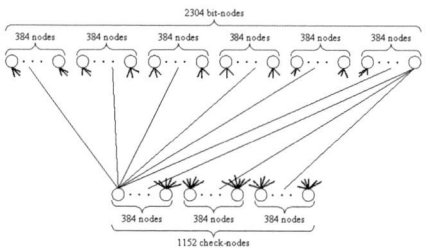

Fig. 2. Tanner Graph of Parity Check Matrix.

I. INTRODUCTION

Low-Density Parity-Check (LDPC) codes achieve good performance for error collecting. In the last few years some work has been done on designing LDPC decoder[1][2]. The decoding algorithm exchanges the messages between the check-nodes and bit-nodes on tanner graph by performing the row operations and column operations iteratively. For future application requires high-throughput over a hundred of Mb/s, it is necessary that row operations and column operations are operated concurrently. However, double amount of memory is required to operate these concurrently compared to serial approach. We have proposed the method to improve this problem. In this method, a row operation outputs minimum absolute value, second minimum absolute value, these flags and signs instead of actual output value. We designed and implemented high-throughput LDPC decoder chip using this method.

II. PARITY CHECK MATRIX

Fig.1 shows the block-structured parity check matrix of LDPC decoder that we have designed. This matrix enables the decoder to perform row and column operations partially in parallel[1]. We specified that block length is 2304 and code rate is 1/2. The diagonal lines represent entries of 1 in the matrix, other entries are 0. The matrix is composed of 3 × 6 sub-blocks of size 384. Each 384 × 384 sub-block is an identity matrix that has been shifted to the right. The shift values have been determined with a heuristic search for the best performance.

Fig.2 shows the tanner graph corresponding to the parity check matrix. Each bit-node corresponds to each column in the parity check matrix. Each check-node corresponds to each row in the parity check matrix. 2304 bit-nodes are divided into 6 groups that contain 384 nodes. 1152 check-nodes are divided into 3 groups that contain 384 nodes.

III. HARDWARE ARCHITECTURE

Fig.3 shows the block diagram of LDPC decoder. We adopted the modified min-sum algorithm to implement the decoder[2]. 6 bit Symbol Log-Likelihood-Ratio (LLR) is fed to the decoder as a input value. Symbol LLRs are stored to the input SRAMs. The decoder contains 6 SRAMs (lram1-6) with 72bit, 64 words for these input values. An input value of same group corresponding to group of bit-nodes is stored to a SRAM. 12 input values are stored to a address (word). The input SRAMs store 2 blocks of code. A block is stored for next decoding while other block is being decoded.

The decoder contains 12 × 3 = 36 Check-node Function Units (CFUs) so that 12 row operations of each group of check-nodes can be operated concurrently. Fig.4 shows the block diagram of CFU. A CFU outputs 5bit minimum absolute value, 5bit second minimum absolute value, 6 flag-bits to select minimum values and 6 sign-bits instead of actual alpha values. Flag-bits and sign-bits are stored to alpha SRAMs. The LDPC decoder contains

0-7803-9451-8/06/$20.00 ©2006 IEEE.

112

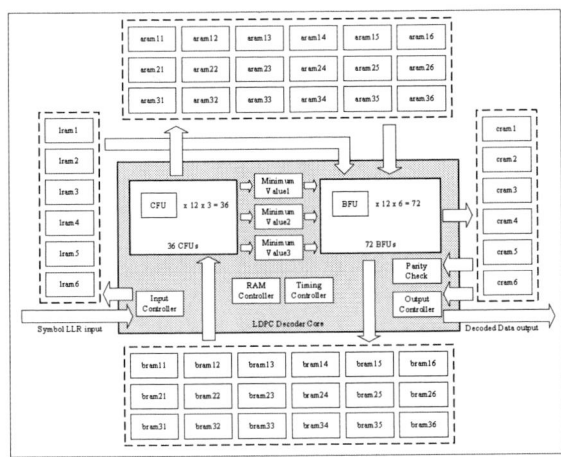

Fig. 3. LDPC Decoder Architecture.

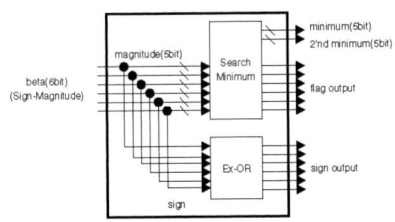

Fig. 4. Check-node Function Unit.

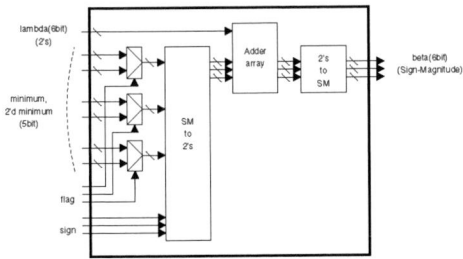

Fig. 5. Bit-node Function Unit.

18 SRAMs (aram11-16, aram21-26 and aram31-36) with 24bit, 32 words for flag-bits and sign-bits. Minimum absolute value and second minimum absolute value are store to shift-registers with $12 \times (5 + 5) = 120$ bit-width.

The decoder also contains $12 \times 6 = 72$ Bit-node Function Units (BFUs) so that 12 column operations of each group of bit-nodes can be operated concurrently. Fig.5 shows the block diagram of BFU. A BFU outputs 6 bit beta values and these values are stored to beta SRAMs. The decoder contains 18 SRAMs (bram11-16, bram21-26 and bram31-36) with 72bit, 32 words for these values.

The decoder contains 6 SRAMs (cram1-6) with 12bit, 32 words for decoded values. Parity check module checks decoded values whether the result of decoding is correct or not. Parity check and decoding operate independently. All of SRAMs are dual-port memory to decode the code consecutively. The LDPC decoder takes $384 / 12 = 32$ cycles to complete an iterative decoding.

TABLE I

SPECIFICATIONS OF LDPC DECODER CHIP.

Design process	0.18μm, 6M, CMOS
Chip size	6.0mm × 6.0mm
Gate count (Decoder core)	206,343 gates
Embedded SRAMs	85,248bits (48 instances)
Clock frequency.	147MHz (Max.)
Throughput	530Mb/s (@147MHz)
Power consumption	3.6W (@147MHz, 10 iterations)

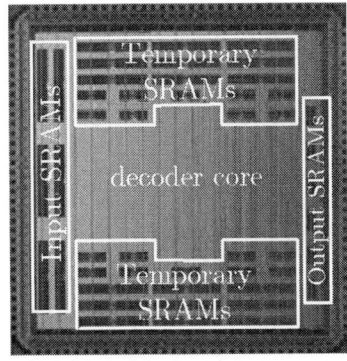

Fig. 6. LDPC decoder microphotograph.

IV. CHIP FABRICATIONS

Fig.6 shows microphotograph of LDPC decoder. A LDPC decoder chip has been fabricated in a 0.18μm, 6 metal-layer CMOS technology. Chip die size is 6.0mm × 6.0mm = 36mm^2. 48 dual-port SRAMs occupy 78.4% of the total synthesized area. Gate count of decoder core is 206,343 gates. The decoder achieves throughput of 530Mb/s and power consumed is 3.6W at an operating frequency of 147MHz with 10 iterative decoding. 48 SRAMs consume 87.8% of the total power. The throughput is comparable to appeared in [2], where the throughput is 127Mb/s at an operationg frequency of 121MHz. Table I shows the specifications of LDPC decoder chip.

V. SUMMARY AND CONCLUSIONS

We have designed and implemented LDPC decoder chip with memory-reduction method. The decoder achieves throughput of 530Mb/s. This architecture will be applied to future application requires high-throughput.

REFERENCES

[1] M. M. Mansour, N. R. Shanbhag, "Low-power VLSI decoder architectures for LDPC codes," *Proceedings of the 2002 International Symposium on Low Power Electronics and Design,* pp.284-289, 2002.

[2] M. Karkooti, J. R. Cavallaro, "Semi-Parallel Reconfigurable Architecutures for Real-Time LDPC Decoding," *Proceeding 2004 IEEE International Conference on Information Technology, ITCC'04,* Volume1, pp.579-585, 2004.

Hardware Implementation of Super Minimum All Digital FM Demodulator

Nursani Rahmatullah

Department of Electrical Engineering
Institut Teknologi Bandung, Indonesia
E-mail: rahmatullah@students.ee.itb.ac.id

Arif E Nugroho

Department of Electrical Engineering
Institut Teknologi Bandung, Indonesia
E-mail: arif_endro@opencores.org

Abstract – We propose improvement of the new architecture of digital FM demodulator. This work enhances signal quality, system clock frequency, and superior than well known PLL technique today. No more multiplier, no more ROM or table, compact size, and very fast in transient or state response. Real implementation in Altera® APEX20K200 EBC652-1X PLD gives 348 logic elements and run up to 224.42 MHz.

I. Introduction

Many efforts have been made to integrate an FM receiver on a single chip using various architectures, but the performance has been limited by analog signal processing accuracy. The main issue of integrating an FM demodulator in a chip is how to accurately discriminate a small frequency deviation of the FM signal from its center frequency. Without doubt, Phase Locked Loop (PLL) type being the most commonly use FM demodulator today. But now a new enhanced architecture which is more superior to PLL technique has been made in this work.

The new algorithm was first explained in [1] which is fall into compact size architecture, a new simple demodulation algorithm without multiplier, very fast, running without ROM or look-up table, and takes an absolute stability structure which has no feedback loop for input phase tracking.

Some improvements have been performed in this work to get better compact architecture, faster system clock, and achieve a good signal quality without FIR filter. The proposed improvement and its PLD implementation can run up to 224.42 MHz system clock using Altera® APEX20K 200EBC652-1X.

II. FM Demodulation

The new algorithm assumes that process in FM demodulation is equivalent with tracking for frequency deviation. Tracking process is performed for each period of cycle, so frequency deviation can be detected as shown in Fig.1. The gap area (D) increases along with frequency deviation magnitude. We assume that each period has 16 sampling points.

Here we try to derive this new concept with our own perception. Assume both carrier (C), and modulating signal (S_N) are harmonic signals, then

$$C(n) = A_C \cos(2\pi f_C n + \theta_C) \quad (1.1)$$

$$S_N(n) = A_S \cos(2\pi f_S n + \theta_S) \quad (1.2)$$

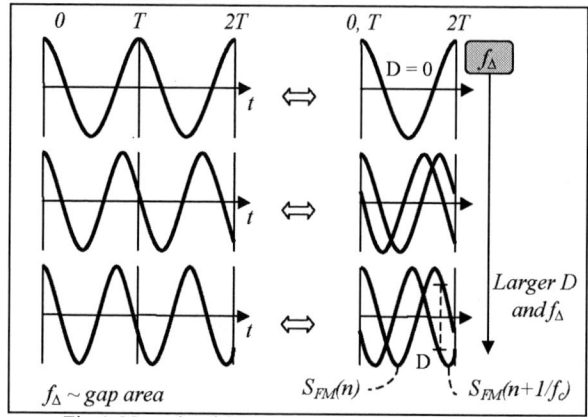

Fig. 1. New algorithm concept for detecting deviation

The modulated signal (S_{FM}) will be

$$S_{FM}(n) = A_C \cos\left[(2\pi f_C + A_S \cos(2\pi f_S n + \theta_S))n + \theta_C\right] \quad (1.3)$$

From Eq.(1.3), S_N is equivalent to frequency deviation in Fig.1, thus tracking for f_Δ means demodulating S_{FM}. For tracking f_Δ, note that gap D between two signals in Fig.1 gives

$$D(n) = \left|S_{FM}(n+T) - S_{FM}(n)\right| \quad (1.4)$$

$$\Delta D(n) = D(n+T+1) - D(n) \quad (1.5)$$

$$S_N(n) = \sum_{k=-\infty}^{n} \Delta D(k) = \sum_{k=-\infty}^{n-1} \Delta D(k) + \Delta D(n)$$
$$= S_N(n-1) + \Delta D(n) \quad (1.6)$$

III. Hardware Implementation

A. Improvement

The basic implementation in [1] did not take into account non linear effect of demodulation when calculating gap, D. It will exhibit high frequency components in the subtraction process in Eq.(1.4) and then accumulated in the end of demodulating process as noise that reduce demodulated signal quality. This problem actually can be solved by subjecting the output signal to low pass filter such as FIR, but it will consume large area and reduce system clock performance.

0-7803-9451-8/06/$20.00 ©2006 IEEE.

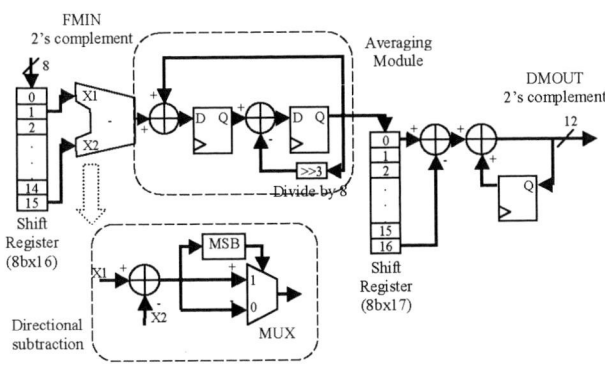

Fig.2. Improvement architecture in proposed method

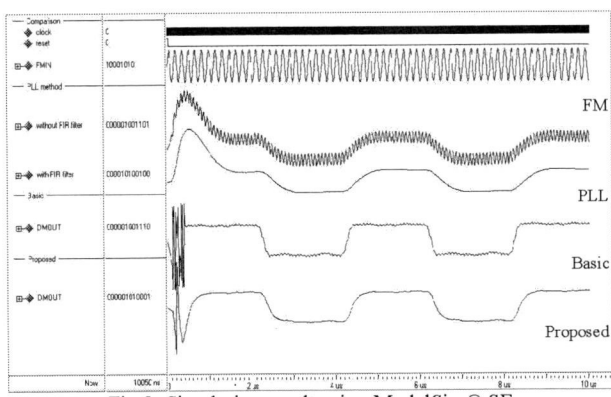

Fig.3. Simulation result using ModelSim® SE

The previous basic implementation in [1] used comparator to detect directional slope in computing gap D, and also need more adders to compensate negative value of subtraction. Our proposed method can eliminate these problems, as shown in Fig.2. Directional subtraction does not need a comparator but only need one adder. Instead of FIR filter, averaging module (loop filter) with a pole of 7/8 is added after directional subtraction to compensate non linear effect of modulation [2].

In Table.1 comparison between PLL method, basic implementation, and proposed method is given. All architectures are synthesized using Xilinx®Spartan3 3S200FT256-4 FPGA, this device is similar to device used in [1] so we can get a clear comparison. PLL has been optimized; it employs Booth's multiplier as phase detector, minimum loop filter, and optimized Numerical Controlled Oscillator (NCO) and FIR filter. Real implementation and measurement performed using Altera® APEX20K200 EBC652-1X. It gives 348 logic elements, minimum period of 4.456 ns, so it can run up to 224.42 MHz system clock frequency.

TABLE I
Synthesized result of each architecture

Architecture using Xilinx®Spartan3 3S200FT256-4	Time		Area		
	Delay (ns)	Frequency (MHz)	Slices	Slices FF	LUT
PLL (optimized)	9.725	102.828	491	548	721
Basic circuit Phase-3 [1]	6.451	155.015	184	311	48
Proposed method	4.658	214.684	195	348	61

B. Simulation and Real Waveform

Simulation in Fig.3 shows superiority of our proposed improvement method that successfully removed the ripple without loosing its quick transient and state response. In this simulation, the modulating signal is square wave with 10 KHz of frequency. We assume carrier frequency of 1 MHz with modulation index 10. System clock frequency and sampling frequency are 16 MHz.

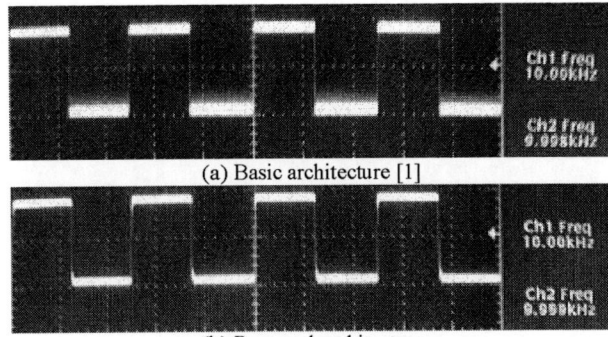

(a) Basic architecture [1]

(b) Proposed architecture

Fig.4. Real waveform captured by oscilloscope

In Fig.4 we can see the real waveform. This waveform is taken by demodulating real FM signal generated by Leader SSG equipment signal generator.

In this real implementation, we use AD9203ARU 10-bit 40-MSPS A/D converter and THS5651AIPW 10-bit 100-MSPS D/A converter. The modulating signal is square wave signal of 10 KHz. FM signal has carrier frequency of 2.5 MHz, and system clock frequency of 40 MHz. Proposed architecture in Fig.4(b) successfully eliminates ripples shown in Fig.4(a). SNR for basic circuit is 27.8 dB while SNR for proposed method is 30.6 dB.

IV. Summary and Conclusions

This work enhanced signal quality and system clock frequency of the new architecture of all digital FM demodulator without significantly increase the area. Signal quality and high system clock frequency has become important issue of today communication systems such as software defined radio.

References

[1] Kenji Yamamoto, Masaya Yokota, *"Development of Super Minimum FM-Demodulator"*, University of the Ryukyus LSI Design Contest, Okinawa, 2005.
[2] John G Proakis, Dimitris G Manolakis, *"Digital Signal Processing"*, Prentice Hall, 1996.

1D-14　　2006 Asia and South Pacific Design Automation Conference

Designing a Custom Architecture for DCT Using NISC Technology

Bita Gorjiara, Mehrdad Reshadi, Daniel Gajski

Center for Embedded System Computers, University of California Irvine

{bgorjiar, reshadi, gajski}@cecs.uci.edu

Abstract– This paper presents design of a custom architecture for Discrete Cosine Transform (DCT) using No-Instruction-Set Computer (NISC) technology that is developed for fast processor customization. Using several software transformations and hardware customization, we achieved up to 10 times performance improvement, 2 times power reduction, 12.8 times energy reduction, and 3 times area reduction compared to an already-optimized soft-core MIPS implementation.

I. Introduction

This paper presents design of a custom architecture for Discrete Cosine Transform (DCT) using No-Instruction-Set Computer (NISC) design flow that is developed for fast processor customization. Processor customization techniques such as designing Application-Specific Instruction-Set Processors (ASIPs) [2] have recently emerged to meet the performance and power constraints of designs starting from high-level languages such as C. A new alternative to ASIP is No-Instruction-Set-Computer (NISC) [3][4][7] in which a *cycle-accurate compiler* generates code to control a given custom datapath at every clock cycle. However, instead of using any abstraction such as instruction-set or microcode, the NISC compiler directly generates the control signal values of every component in the datapath for every clock cycle. A NISC designer needs to only focus on designing the datapath, i.e. selecting the components and connecting them together. Unlike ASIPS, in NISC, there is no need for designing instruction-set and instruction decoder, or updating the compiler. The NISC compiler inputs the datapath as a netlist of RTL components, and automatically analyzes and extracts branch delay and possible operations. The datapath netlist contains components such as bus, multiplexer, register, register-file, memory, and functional unit. After compiling the program onto the given datapath, the compiler generates a string of control values, called Control Word (CW), for each cycle. These control words are stored in a control memory and are applied to the datapath by the controller at every cycle.

In this case-study, first we compile the C code of DCT algorithm on a general-purpose datapath similar to a MIPS processor. Next, we apply several software transformations and hardware customization to improve the performance, power, energy and area.

II. DCT algorithm

The Discrete Cosine Transform (DCT) [1] and Inverse Discrete Cosine Transform (IDCT) are important parts of JPEG and MPEG standards. MPEG encoders use both DCT and IDCT, whereas MPEG decoders only use IDCT. The definition of DCT for a 2-D $N \times N$ matrix of pixels is as follows:

$$F[u,v] = \frac{1}{N^2} \sum_{m=0}^{N-1} \sum_{n=0}^{N-1} f[m,n] \cos \frac{(2m+1)u\pi}{2N} \cos \frac{(2n+1)v\pi}{2N}$$

Where u, v are discrete frequency variables ($0 \le u$, $v \le N-1$), $f[i,j]$ gray level of pixel at position (i, j), and $F[u,v]$ coefficients of point (u, v) in spatial frequency. Assuming $N=8$, matrix C is defined as follows:

$$C[u][n] = \frac{1}{8} \cos \frac{(2n+1)u\pi}{16}$$

Based on matrix C, an integer matrix $C1$ is defined as follows: $C1 = $ round(*factor* $\times C$). The $C1$ matrix is used in calculation of DCT and IDCT: $F = C1 \times f \times C2$, where, $C2 = C1^T$. As a result, DCT can be calculated using two consecutive matrix multiplications. Figure 1(a) shows the C code of multiplying two given matrix A and B using three

nested loops. Using a MIPS M4K ™ Core processor [5], the matrix-multiplication-based DCT takes 13058 cycles to compute [3]. However, given the MIPS datapath, the NISC-style processor takes 10772 cycles to compute DCT. The 20% reduction in number of cycles is because of the finer-grained control that NISC compiler has over the datapath compared to traditional compilers that use instruction-set abstraction. We developed the synthesizable hardware description for our NISC-style MIPS (called NMIPS), and synthesized it using Xilinx ISE 6.3. In our implementation, the bus-width of the datapath is 16-bit for a 16-bit DCT precision, and the datapath does not have any integer divider or floating point unit. The clock frequency of 78.3MHz was achieved after synthesis and Placement-and-Routing (PAR). All of the experiments in this paper are synthesized and mapped on Xilinx FPGA package Virtex2V250-6 using Xilinx ISE 6.3 tool. Two synthesis optimizations of retiming and buffer-to-multiplexer conversions are applied to improve the performance. In these experiments, we set the PAR effort to the highest level possible for maximum clock speed.

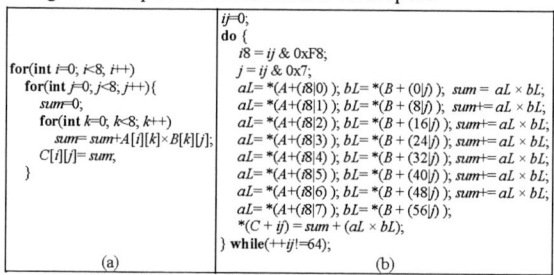

Figure 1. (a) Original and (b) Transformed matrix multiplication

III. Custom DCT implementations

In general, customization of a design involves both software and hardware transformations. To increase the parallelism in code, we unroll the inner-most loop of the matrix multiplication code, merge the two outer loops, and convert some of the costly operations such as addition and multiplication to OR and AND. In DCT, the operation conversions are possible because of the special values of the constants and variables. The transformed code is shown in Figure 1(b). By looking at the body of loop, four steps of computation can be identified: (1) calculation of the memory addresses of the matrix elements; (2) loading the values from data memory; (3) multiplying the two values; (4) accumulating the multiplication results. We design our custom datapath in a way that each of these steps is a pipeline stage. Figure 2(a) shows the proposed custom pipelined datapath called CDCT1. The datapath includes four major pipeline stages that are marked in the figure. In NISC, Comparator (Comp) and Address Generator (AG) are used for handling jumps, while Link Register (LR) and *direct address* are used for supporting function calls. We have used operation chaining to reduce RF file accesses and decrease register pressure. The OR and ALU, as well as the Mul and Adder are chained. Note that the chaining of multiply and add forms a MAC unit in the datapath. After compilation, synthesis and PAR, the total number of cycles of the DCT is reduced to 3080, and the maximum clock frequency is 85.7MHz. Next, we iteratively apply the following datapath refinements to improve the performance, power, and area of DCT implementation:

1) To reduce critical path delay that includes ALU delay and RF setup time, we add an extra register between the output of ALU and the input of RF. Also, LR and *direct address* are removed because, there is no need for a function call (the matrix multiplication code is inlined).

0-7803-9451-8/06/$20.00 ©2006 IEEE.　　116

Additionally, buses are simplified to point-to-point connections that are actually used by DCT. The result architecture is called CDCT2;

2) The unused parts of ALU, Comp and RF are removed. A general-purpose ALU and Comp supports many operations. However, as shown in Figure 1, only Add, Or, And, Multiply, and Not-equal (!=) operations are used in DCT. Therefore, the ALU and Comp can be simplified. Additionally, instead of 32-register RF, we can use a 16-register RF because the rest is not used by the application. We call this new architecture CDCT3.

3) After synthesizing CDCT3, we observe that the critical path includes the Control Memory (CMem) read delay, CW wire delay, and AG's delay. Therefore, to reduce critical path delay, we apply controller pipelining by adding CW register and status register, and call the new architectures CDCT4 and CDCT5, respectively.

4) To decrease the area, we reduce the bit-width of the components in address calculation pipeline stage without affecting the precision of DCT calculations. The result architecture is called CDCT6.

5) After synthesizing CDCT6, we observe that the critical path goes through the Mul. Since Mul is a ASIC multiplier, we cannot reduce the critical path any further. However, if we consider Mul as a two-cycle unit, we can further improve the clock frequency by adding pipeline registers at the outputs of the RF. The final architecture (CDCT7) is shown in Figure 2(b).

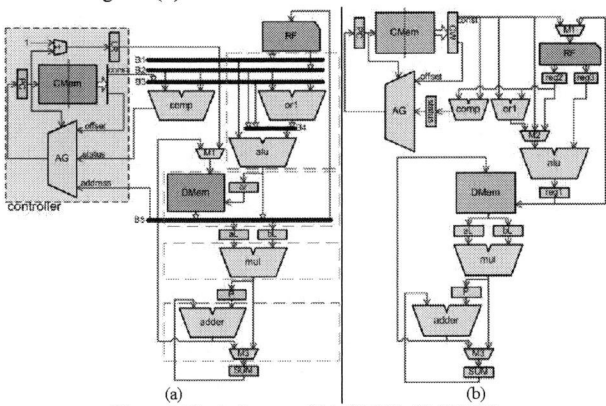

Figure 2. Block diagram of (a) CDCT1, (b) CDCT7

IV. Comparing the DCT implementations

Table 1 compares the performance, power, energy, and area of the all NISC implementations after synthesizing them on FPGA. The fourth column of Table 1 shows the total execution time of the DCT algorithm. Note that although in some cases (CDCT4, CDCT5, and CDCT7) the number of cycles increases, the clock frequency improvement compensates for that. Therefore, the total execution delay maintains a decreasing trend.

Table 1. Performance, dynamic power, energy, and area of the NISCs

	No. of cycles	Clock Freq	DCT exec. time(us)	Power (mW)	Enegy (u.J)	Normalized area
NMIPS	10772	78.3	137.57	177.33	24.40	1.00
CDCT1	3080	85.7	35.94	120.52	4.33	0.81
CDCT2	2952	90.0	32.80	111.27	3.65	0.71
CDCT3	2952	114.4	25.80	82.82	2.14	0.40
CDCT4	3080	147.0	20.95	125.00	2.62	0.46
CDCT5	3208	169.5	18.93	106.00	2.01	0.43
CDCT6	3208	171.5	18.71	104.00	1.95	0.34
CDCT7	3460	250.0	13.84	137.00	1.90	0.35

Dynamic power consumption (column fifth), also decreases as we introduce customization and datapath pipelining. However, in CDCT4, power consumption increases because of extra logic added by retiming algorithm. In general, as frequency increases the clock power of the datapaths increases. The power-breakdown of the designs (Figure 3) confirms this fact.

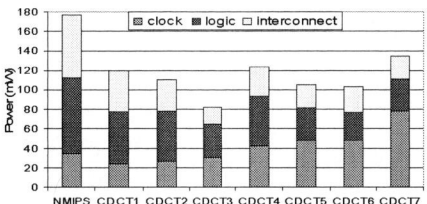

Figure 3. Dynamic power of the DCT implementations

Figure 4 shows the performance, power, energy and area of the designs normalized against NMIPS. As shown in Figure 4, CDCT7 is the best design in terms of delay and energy consumption, while CDCT3 is the best in terms of power, and CDCT6 is the best in terms of area. As a result, CDCT3, CDCT6, and CDCT7 are considered the pareto-optimial solutions. Note that minimum energy and minimum power are achieved by two different designs: CDCT7 and CDCT3, respectively. Compared to NMIPS, CDCT7 runs 10 times faster, consumes 1.3 times less power and 12.8 times less energy. Also, it occupies 2.9 times less area than NMIPS. The minimum power consumption is achieved by the CDCT3, which consumes 2 times less power compared to NMIPS. Note that performance of NMIPS is 20% better than performance of a MIPS core. Also, since NMIPS does not have instruction decoder, its area and power are less than MIPS. In our experiments, we compared the results to NMIPS which is conservative relative to the MIPS core.

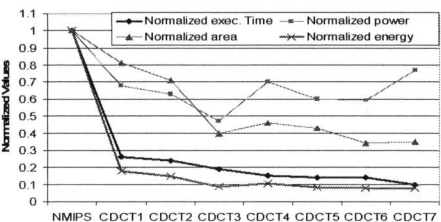

Figure 4. Comparing different DCT implementations

We configured the Xilinx Virtex-II Multimedia development board to run CDCT7. The board has a Virtex-II XC2V2000-FF896 FPGA package and only supports 27MHz, 53MHz, and 108MHz clock frequencies. Although CDCT7 could achieve the maximum clock frequency of 250MHz, we ran it with clock frequency of 108MHz on the board due to unavailability of higher clock frequencies.

Figure 5. Xilinx Virtex-II multimedia board

Figure 6. CDCT7 after placement and routing

For all the DCT implementations, the synthesizable Verilog files, timing constraints, the synthesis scripts, and the Placement-and-Routing results are available for download at [6].

References

[1] N. Ahmed, T. Natarajan, and K.R. Rao, Discrete Cosine Transform, *IEEE Trans. On Computers*, vol. C- 23, pp. 90-93, Jan 1974

[2] M.K. Jain, M. Balakrishnan, and A. Kumar, ASIP Design Methodologies: Survey and Issues, *In Proc. of International Conference on VLSI Design*, 2001.

[3] M. Reshadi, D. Gajski, A Cycle-Accurate Compilation Algorithm for Custom Pipelined Datapaths, *In Proc. ISSS05*, 2005.

[4] M. Reshadi, B. Gorjiara, D. Gajski, Utilizing Horizontal and Vertical Parallelism Using a No-Instruction-Set Compiler and Custom Datapaths, *In Proc. ICCD05*.

[5] MIPS32® M4K™ Core, http://www.mips.com

[6] Design files: http://www.ece.uci.edu/~bgorjiar/projects/NISC/customDCT/

[7] Nisc webpage: http://www.ics.uci.edu/~reshadi/projects/nisc/

A 52mW 1200MIPS Compact DSP for Multi-Core Media SoC

Shih-Hao Ou, Tay-Jyi Lin, Chao-Wei Huang, Yu-Ting Kuo,
Chie-Min Chao, Chih-Wei Liu, and *Chein-Wei Jen*

Department of Electronics Engineering
National Chiao Tung University, Taiwan

Abstract - This paper presents a DSP core for multi-core media SoC, which is optimized to execute a set of signal processing tasks very efficiently. The fully-programmable core has a *data-centric* instruction set and a corresponding *latency-insensitive* micro-architecture, where the hardware design is optimized concurrently with its automatic software generator. The proposed DSP core has 3X performance (in cycles) of those found in commercial dual-core application processors with similar computing resources. The silicon implementation in UMC 0.18μm 1P6M CMOS technology operates at 314MHz and consumes only 52mW average power.

I. Introduction

The computations of next-generation communication devices can no longer be satisfied with a single processor at reasonable cost. Dual-core processor with a RISC and a DSP is a popular solution to meet the computation demands. TI OMAP [1], which consists of an ARM core and a DSP core, is a well-known example. The ARM core is responsible for control-oriented tasks such as the user interfaces, system coordination, and protocol stack, while the DSP core takes care of the computation-intensive tasks such as baseband processing, data transformation and so on. However, most dual-core processors have redundant components because they are constructed with existing processor cores, which are designed for standalone uses. In this project, we design a DSP core from scratch, while the RISC core is kept unchanged for software compatibility. The design goals include multi-core configuration, compactness, and low power. Moreover, automatic software generation from high-level specification is of the same importance. This paper is organized as follows. The processor design is first described in Section II. Section III summarizes the chip specification and silicon implementation of the compact DSP core. Finally, Section IV concludes this work.

II. Processor Design

A. Data-centric ISA

Considering the generic 4-way VLIW processor, conventional ISA (instruction set architecture) [2] specifies operations performed by each functional unit and additionally where to get the corresponding source/destination operands and some control information. We call aforementioned ISA operation-centric. In contrast to the operation-centric ISA which controls the functional units, our data-centric ISA is designed to be responsible for data generation directly. In other words, the instructions specify the required data by each functional unit in each current iteration and take care of the returned results from all functional units in previous iteration. In addition, the operation and other control information are carried with instructions. Take our addition instructions for example. The assembly syntax is described as: `Rd=DS, {Rs>>+Rt>>}>>;` that is interpreted as *"store the DS datum into the register Rd,* and *add the datum in the register Rs and the register Rt together"*. Those ">>" marks enable one-bit shift for input alignment or output normalization.

B. Latency-insensitive microarchitecture

Fig. 1 depicts the DSP core's micro-architecture consisting of the data generator and 4 sites which are the adder, the multiplier, the shifter and the control unit respectively. The data generator is made up of the 4-by-4 crossbar switch and those DRF (distributed register file). Additionally, the DSP is equipped with the banked data memory served as ping-pong buffer to reduce the communication overhead between the DSP core and host processor. Because the micro-architecture is tightly coupled with the data-centric ISA by which the operands is scheduled, the processor absolutely does not require some complex forwarding-path found in conventional VLIW processors between functional units. Therefore, the micro-architecture features latency-insensitive characteristic. Besides, the micro-architecture is so modular and thus simplifies replacement and collocation of functional modules with different latency. Thus, only simple modification in the software generator is required to adapt to different hardware configurations without altering the other hardware blocks. By the way, as the technology advances rapidly, the modular micro-architecture which localizes the interconnection is an effective solution to the increasing wire delay & on-chip interconnection overhead.

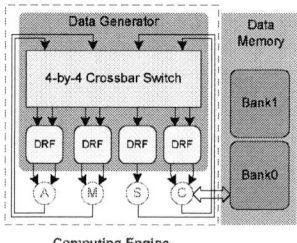

Figure 1.   Computing engine

C. Automated Code Generation

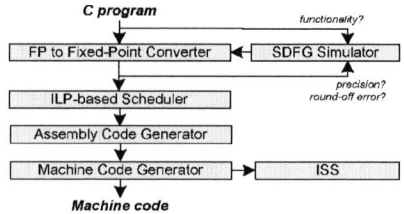

Figure 2.   Code generation flow

Fig. 2 depicts the automatic code generation flow of our DSP core. First, the FP SDFG [3] can be derived from the C descriptions via the SUIF compiler. The SDFG simulator is bit-true and supports both the FP and the fixed-point arithmetic, and the designers can develop and verify their DSP algorithms easily. Once the functionality of the FP SDFG is verified to be correct, the FP-to-fixed-point converter translates the FP SDFG into a fixed-point one by applying static analysis which is based on worst-case range analysis [4]. The operations of the fixed-point SDFG are scheduled

0-7803-9451-8/06/$20.00 ©2006 IEEE.

with the integer linear programming (ILP). Finally, the fixed-point executables can be generated and simulated by the machine code generator and the ISS (instruction set simulator) respectively.

D. Performance Evaluation

Several popular DSP kernels are used to evaluate our proposed DSP core. Table 1 outlines the results. In addition, the 2nd & 3rd rows give the reference performances of two commercial DSP cores, both of which have already been integrated in some dual-core processor designs. ADI ADSP-218x has similar computing resources to our DSP core, while TI C'55 has one more MAC unit. The cycle counts are all excerpted from their application notes. The reasons that our DSP has such improvements over conventional DSP can be summarized as follows. First, its data-driven engine and code generator are developed in parallel based on high-level synthesis to extensively exploit the inherent parallelism of DSP algorithms, and the performance can therefore be very close to that of customized ASIC designs. Then, the data-centric ISA & latency-insensitive micro-architecture enable smooth dataflow with its internal crossbar network and relatively plenty registers (note that the complexity is much less than that of a plain register file in the general-purpose processors). Moreover, the four embedded 1-bit shifters in the fixed-point arithmetic units also help the reduction of the execution cycles. We further have several implementations of the 2-D DCT from the independent JPEG group (IJG) to analyze the round-off error of our proposed fixed-point arithmetic. Table 2 summarizes the comparisons. The proposed 16-bit fixed-point even outperforms the hand-optimized 32-bit integer 2D-DCT from IJG. Moreover, our 24-bit fixed-point has about 64dB PSNR, which has the same maximum precision as the single-precision FP (i.e. with the 23-bit mantissa).

TABLE I.   PERFORMANCE EVALUATION

	Lattice	Biquad	FFT	2D-DCT
ADSP-218x	32	13	874	2,452
TI C'55	12	5	367	1,082
Proposed DSP	**12**	**16**	**268**	**688**

TABLE II.   ROUND-OFF ERROR COMPARISON OF 2D-DCT

	PSNR (dB)	Cycel count
Single-precision FP	--	624
16-bit integer	29.5183	848
32-bit interger	33.8020	656
Proposed 16-bit	**40.0468**	**656**
Proposed 32-bit	**64.1201**	**624**

III. Silicon Implementation

Fig. 3 illustrates the global design flow exercised by the proposed DSP core. First, in the **ISA/Micro-architecture Design** phase, we analyze several popular DSP algorithms to design the ISA and the micro-architecture. The cycle-accurate SystemC model is utilized to be verified with the compiler-generated code. If the resulted performance is not satisfied, the micro-architecture refinement, i.e. latency modification is performed to improve performance. Then in the **RTL Design** phase, we write the fully synthesizable RTL in Verilog and use the SystemC model to cross-check the correctness by HDL simulator. We further manually handcraft assembly codes to account for the corner cases to increase the code coverage. The metrics such as statement, branch, state, and arc all achieve 100%. Finally the **Implementation** phase involves the RTL synthesis and the physical design by means of the *Synopsys Design Compiler* and the *SoC Encounter* respectively. The functional verification of the synthesized gate-level net-list is through the formal equivalence checking with the RTL model. Besides, the timing/area/power consumption estimation is made by the popular estimation tools like *PrimeTime, PrimePower, nanosim* and so on. Others like the

scan-insertion and memory BIST are also implemented to increase testability. If the required performance/cost is not met, one solution is to go back to RTL design, i.e. RTL code quality improvement by *nLint* and the other is again back to micro-architecture refinement.

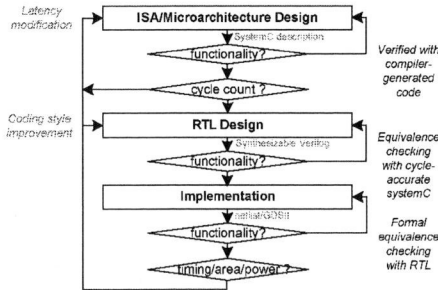

Figure 3.   Design flow

The DSP has been implemented and fabricated in UMC 0.18μm 1P6M CMOS technology. Table III summaries the specification and Fig. 4 shows both the chip layout and the die photo. In the post-layout simulation, the DSP core can operate at 314 MHz and consumes only 52mW average power. The chip has already been fabricated and tested to work correctly at 100MHz. The maximum operating frequency is unknown at the time of paper submission due to the limitation of our available *IMS* test machine.

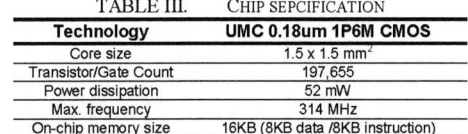

TABLE III.   CHIP SEPCIFICATION

Technology	UMC 0.18um 1P6M CMOS
Core size	1.5 x 1.5 mm^2
Transistor/Gate Count	197,655
Power dissipation	52 mW
Max. frequency	314 MHz
On-chip memory size	16KB (8KB data /8KB instruction)

Figure 4.   Chip layout (left) & die photo (right)

IV. Conclusion

The paper summaries the design of a compact DSP core for multi-core media SoC, from the definition of its instruction set in C++, the micro-architecture exploration in cycle-accurate SystemC, to the synthesizable RTL implementation. The DSP core with compiler-generated software codes has about 3X performance improvements of those found in commercial dual-core application processors. Its silicon implementation in the UMC 0.18μm CMOS technology achieves 314MHz frequency and consumes only 52mW.

References

[1] *OMAP5910 Dual Core Processor – Technical Reference Manual*, Texas Instruments, Jan. 2003

[2] J. L Hennessy, and D. A. Patterson, *Computer Architecture – A Quantitative Approach*, 3rd Edition, Morgan Kaufmann, 2002

[3] K. K. Parhi, *VLSI Digital Signal Processing Systems – Design and Implementation*, John Wiley & Sons, 1999

[4] S. H. Ou, T. J. Lin, H. Y. Lin, C. M. Chao, C. W. Liu and C. W. Jen, "Lightweight arithmetic units for VLSI digital signal processors," in *Proc. VLSI-TSA-DAT*, Apr. 2005

Implementation of H.264/AVC Decoder for Mobile Video Applications

Suh Ho Lee, Ji Hwan Park, Seon Wook Kim, Sung Jea Ko and Suki Kim

Department of Electronics and Computer Engineering
Korea University
Seoul, Korea 136-701
Tel : 82-2-927-1582
Fax : 82-927-2-1582
e-mail : lsh@ulsi.korea.ac.kr

Abstract - This paper presents an H.264/AVC baseline profile decoder based on a SoC platform design methodology. The overall decoding throughput is increased by optimized software and a dedicated hardware accelerator. We minimize the number of bus accesses and use macroblock (MB) level pipeline processing techniques to achieve a real time operation. We implemented and verified a prototype on a SoC platform with a 32-bit RISC CPU core and FPGA module. Our design can process up to 20 frames/sec with QCIF_(176x144). The proposed architecture can be easily applied to many mobile video application areas such as a digital camera and a DMB (Digital Multimedia Broadcasting) phone.

I. Introduction

Recently, H.264/AVC, a new video compression standard, has been developed in partnership with ITU-T and ISO MPEG. The primary achievement is significantly improved coding performance in low and high bit rates as compared with previous coding standards such as H.263, MPEG-2, and MPEG-4 [1][2]. However, H.264/AVC needs large amount of computation, which makes it difficult to achieve real time processing in software manner. As one of the solutions, in this paper, we propose a hardware accelerator for H.264/AVC decoder. H.264/AVC decoder profiling shows that motion compensation, de-blocking filter, inverse integer transform and quantization consumes about 70% of total decoding time [3][4]. We implemented theses components in hardware. Also, we used a parallel processing architecture with more efficient data transmission scheme to achieve real time processing. To speed up inverse quantization, inverse integer transform, de-blocking filter and context adaptive variable length decoding (CAVLD) are implemented in hardware and inter/intra prediction blocks are realized in software. Our experiment shows that our proposed hardware accelerator delivers about 30% speed up over only software execution.

II. Proposed Architecture

For reusability and performance, the decoder includes one RISC CPU, some dedicated image processing hardware blocks, AHB bus interface blocks and internal bus blocks. The RISC CPU is responsible for consulting all other hardware units to make co-operation between software and hardware. The image processing blocks include Inter/Intra prediction unit, CAVLD unit, inverse quantization, inverse 4x4 integer transform unit, and de-blocking filter unit. The

bus interface blocks include AHB master/slave units and direct memory access (DMA) unit. Our proposed H.264/AVC decoder is accessed by control and status registers via an AHB slave interface. For data transfer, the H.264/AVC decoder data from/to external memory buffers with minimum CPU intervention according to control register settings. Fig.1 shows our proposed H.264/AVC decoder architecture. With the new dedicated hardware accelerator, the decoding throughput is increased by about 30% on the average.

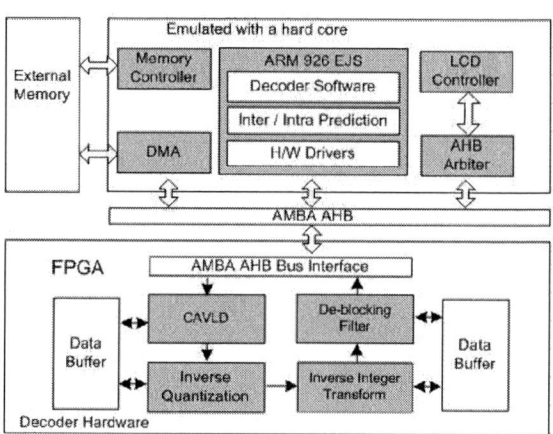

Fig.1. The proposed architecture of H.264/AVC decoder.

III. System Operation and Design

Decoding H.264/AVC video requires fast signal processing. Our designed hardware accelerators for H.264/AVC decoder consist of the following four components: CAVLD, IQ, IDCT and de-blocking filter. We have adopted MB level 4-stage pipeline architecture.

A. CAVLD

The whole CAVLD consist of the following parts: the input data buffer, coeff_token, level, total zeros, run of zeros decoder and VLC table. Together with a barrel shifter and controller, the pipeline architecture can CAVLD every syntax element in on clock cycle. The input data buffer of this CAVLD is to align the input bit stream for decoding the next code word.

0-7803-9451-8/06/$20.00 ©2006 IEEE.

B. IQ, IDCT

The block of IQ, IDCT that is composed of four main stages: input/output buffer stage, the inverse transform stage, the quantization processing stage and the quantization stage. MB data is initially stored in the register file. Then the 4x4 input is passed to the inverse transform block. This block consists of two cascaded sub-blocks. Each of them is responsible of multiplying two 4x4 matrices and is composed of four identical butterfly-adder blocks. Its operation is to perform a group of additions and shifts.

C. De-blocking filter

The architecture includes four components: low bs filter, high bs filter performs the horizontal/vertical filtering in a row-by row manner. 4x4 pixel data register buffers the intermediate results produced by low bs filter, high bs filter and acts as a transposed output buffer. Output buffer is used as a local buffer. Data flow control unit consist of a finite state machine which controls synchronization among low bs filter, high bs filter, 4x4 pixel data register and local output buffer.

IV. Experimental Results

Our design is implemented in the Verilog and synthesized using the Synopsys Design Compiler with Samsung 0.18um standard cell library. The results are shown in Table 1. Table 2 shows the encoder parameters for our designed decoder experiments and Table 3 shows the system performance comparison using the H.264/AVC decoder. In this system, the ARM926EJS CPU is running at 150MHz and the FPGA module is running at 50MHz. As compared to the software implementation, our performance gain mainly comes from the MB-level pipelining architecture and the hardware acceleration. Moreover, we verified our H.264/AVC baseline decoder using FPGA module. Fig.2 shows the photograph of our evaluation system.

TABLE I
Synthesis Results

Target Processing Capability	20 frames/sec @ 50 MHz QCIF(176x144)
Frequency	50 MHz
Power Consumption	113 mW
Functional Block	Gate Count
CAVLC decoder	12,900
IDCT/ Inverse Hadamard	7,380
Inverse Quantization	6,530
De-blocking Filter	24,580
Control Logic	3,400
Total Gate Counts	54,790

TABLE II
Encoder Parameters for Experiments

Frame Size	QCIF
Frame Rate	30fps
QP	I(28) P(31)
Reference Frame Number	5

TABLE III
System Performance Comparison Using H.264 Decoder at ARM926EJS

Sequnece	S/W	Our (S/W+H/W)	Throughput improvment
Foreman	7.56 fps	10.22 fps	35.18 %
News	11.40 fps	15.21 fps	33.42 %
Mother_daughter	13.12 fps	17.22 fps	31.25 %
Tennis	15.21 fps	20.15 fps	32.47 %

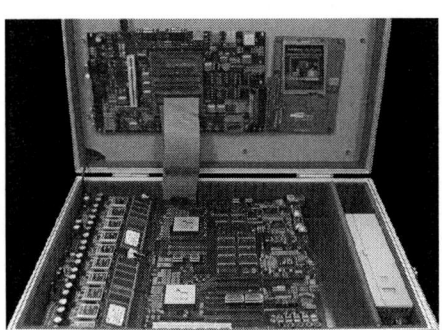

Fig.2. Photography of evaluation system.

V. Summary and Conclusions

In this paper, we have proposed a MB-level pipelining and our bus architecture to minimize the number of bus accesses in order to achieve a real time operation for a H.264/AVC baseline profile decoder. Our design methodology is based on task partitioning and scheduling in MB-level to enhance the overall decoding throughput. We control the hardware components in software and accelerate computationally intensive modules in hardware. The proposed MB-level pipelining improves the overall performance significantly. Our design could process up to 20 frames/sec with QCIF_(176x144), which is 30% more speed up than software only execution.

References

[1] Draft ITU-T Recommendation and Final Draft international Standards of Joint Video Specification (ITU-T Rec. H.264\ISO/IEC 14 496-10 AVC) Joint Video Team (JVT), Mar.2003, Doc. JVT-G050.

[2] Malavar, H.S., Hallapuro, A., Karczewicz, M., and Kerofsky, L.,"Low-complexity transform and quantization in H.264/AVC," IEEE Trans. Circuits Syst. Video Technol vol.13, no.7, pp.598-603, July 2003.

[3] T.Wiegand, G. Sullivan, G. Bjontegaard, and A. Luthra, "Overview of the H.264/AVC video coding standard," IEEE Trans. Circuits Syst. Video Technol., vol. 13, no. 7, pp.560-576, July 2003.

[4] M. Horowitz, A. Joch, F. Kossentini, A. Hallapuro, "H.264/AVC baseline profile decoder complexity analysis," IEEE Trans. Circuits Syst. Video Technol, vol. 13, no. 7, pp.704-716, July 2003.

A High-Performance Platform-Based SoC for Information Security

Min Wu, Xiaoyang Zeng[+], Jun Han, Yongyi Wu, Yibo Fan

State Key Lab of ASIC and System, Fudan University, Shanghai 200433, China

[+] Email: xyzeng@fudan.edu.cn, Tel: 86-21-65104145.

Abstract-A platform-based SoC named as Firebird is presented in this paper, which is used for the applications of information security. Several design aspects, which includes the embedded 32-bit RISC CPU and AMBA controller, the reconfigurable and scalable public-key crypto-coprocessor, high-performance TRNG and several low-power schemes, make Firebird very efficient for the client-end applications of information security. Also the test results of this prototype chip indicate that Firebird can work with all these features efficiently, and has some obvious advantages over other designs in the literatures.

1. Introduction

Recently, Platform-based SoC (System-on-Chip) design has become an efficient solution to meet the continuously increasing requirements of Time-to-Market. In general, a SoC platform consists of embedded CPU and bone bus, DMA controller, memories, and some other assistant modules. With a SoC platform, diverse modules for special applications such as security IPs can be integrated easily.

In this paper, a platform-based SoC named as Firebird for information security is proposed. And it integrates several special sub-modules such as a public-key crypto-coprocessor, a full-customized TRNG (True Random Number Generator) and a USB engine in addition to the necessary modules for the SoC platform.

2. Architecture and merits description of Firebird

Figure 01 shows the hardware-level architecture of the SoC platform used for the design of Firebird, which consists of following several important modules: the embedded 32-bit RISC CPU and it AMBA bus that includes AHB (system bus) and APB (peripheral bus), the reconfigurable and scalable public-key crypto-coprocessor, TRNG (Truly Random Number Generator), USB engine and some other modules. Among these modules, CPU is the main controller for all information security protocols and algorithms; public-key crypto-coprocessor acts as the undertaker of all computations of complex arithmetic, which affects the overall performance of Firebird. And AMBA bus presents a flexible environment for modules' interface and communication.

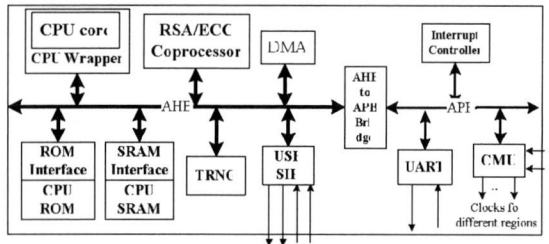

Fig. 01 Platform Architecture for Firebird

Comparing with other design in the literatures, Firebird has several merits such as the reconfigurable and scalable public-key crypto-coprocessor, high performance TRNG,

and efficient low-power schemes, etc. And these merits make Firebird very suitable for the high-performance client-end applications of information security, especially for the fields of portable devices and wireless communications.

Fig.02 shows the architecture of crypto-coprocessor, the core part of Firebird. It can perform the complex arithmetic both of ECC and RSA, i.e., can carry out both point multiplication over elliptic curves and modular multiplication of big integers. The word-based scalable data-path is the most important part of the coprocessor, and it can perform several types of computation abovementioned with a word-based style using the same arithmetic unit without any other hardware modifications. The data-path can also perform two multiplications in finite filed operations in parallel, then speed up point multiplication of ECC, and save about half operating time. It can also be configured to implement up to 2048-bit modular multiplication.

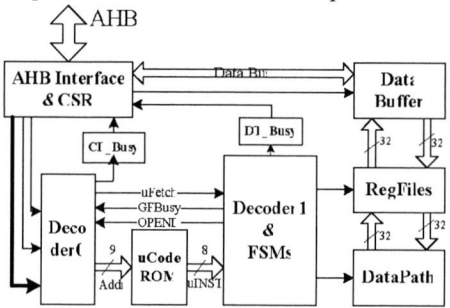

Fig.02 VLSI Architecture of RSA/ECC Coprocessor

TRNG module in Fig.01 shows another merit of Firebird and is mainly used for key generation and one of the countermeasure solutions for anti-attacks. In this design, a thermal noise based scheme is adopted. And the TRNG has passed the strict test with the international standards such as FIPS140-1 and NIST SP800-22. And the TRNG includes the advantages such as high-speed, low power and low cost.

Another important merit of Firebird lies in the low power schemes. A CMU (Clock Management Unit) is created for power saving for the applications such as portable devices and wireless communications. From the operation procedure of Firebird, it is determined that the three modules, e.g. public-key crypto-coprocessor, TRNG and DMA controller, will seldom work simultaneously. Hence a much higher power-efficiency will be achieved if their clocks are closed or waked-up by firmware-controlled CMU dynamically. One scheme of low power is shown in Fig.03. The embedded RISC CPU itself can also be clock-gated by firmware if there are no more tasks, and will be waked-up if any interrupt comes. In this way, the four main power consumers of Firebird are fully controlled by firmware to work or not, and a consequent overall power efficiency will be achieved.

The DMA controller in Firebird is used to improve the

throughput of bulks data transferring. In this paper, DMA Controller can be fully controlled by CPU. And it can be safely and easily clock-gated by firmware-controlled CMU.

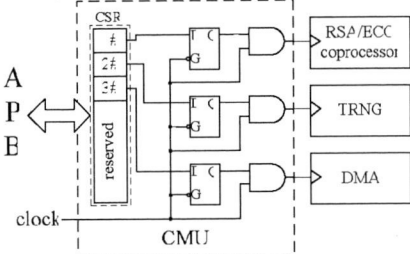

Fig.03 Scheme of Power Efficiency

In Firebird, several modules can share the internal buffers with CPU so as to avoid extra unnecessary operations of data copying. And this kind of architecture shown in Fig.04 is called sharing-memory and very efficient for USB engine and crypto-coprocessor, which must handle bulks of data.

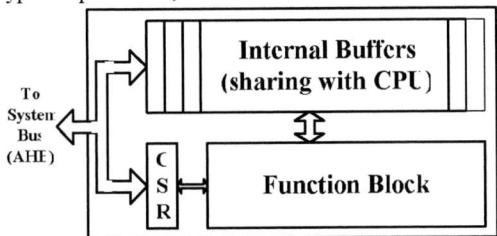

Fig. 04 Scheme for Sharing-Memory

3. Implementation and Test Results

The platform-based SoC, i.e. Firebird, is implemented with TSMC 0.25um CMOS technology and Fig.05 shows its die photo, and this prototype chip is pad-limited. The total die size is about 4.7885*4.3438mm2 (including pads). The test results indicate that: Firebird can work at 60MHz, and the main sub-modules such as crypto-coprocessor can work at 150MHz; the signature rate is about 20 times/sec (@50MHz) with RSA algorithm and 50 times/sec (@50MHz) with ECC algorithm. Fig.06 shows power dissipation of Firebird (@30MHz, 2.5V) during each working-stage, and there exists three kinds of power management level, which are the level of no power management (DPM0), no CPU clock-gated (DPM1) and CPU clock-gated (DPM2). It can be seen that the DPM (Dynamic Power Management) schemes can improve energy efficiency of Firebird and reduce the mean power from about 110mW to 70mW.

All above features are sufficient for the high-performance but low-cost-and-power client-end applications of information security such as the fields of portable devices and wireless communications. And they also have obvious advantages over other related implementations as Table 01 shows.

4. Conclusions

In this paper, a platform-based SoC for information security named as Firebird is presented. And it includes several merits such as reconfigurable and scalable public-key crypto-coprocessor, high-performance TRNG and several low-power schemes. Also the test results of this prototype chip of Firebird indicate that all these features can help Firebird to perform both RSA and ECC computations with rea-

sonable power dissipation, acceptable low-lost but high-performance, which are the key elements for client-end applications of information security.

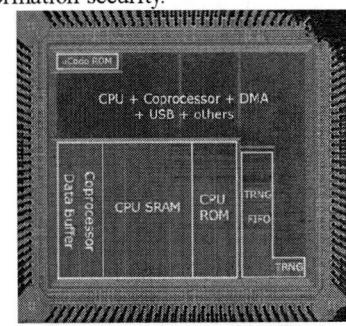

Fig. 05 Die Photo of Firebird

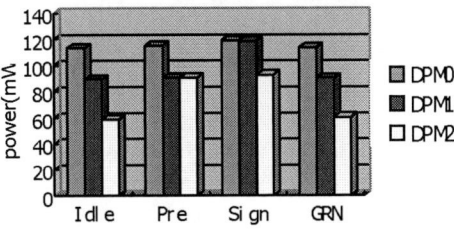

Fig.06 Power Dissipation Test and Comparison

Table 01 Some Comparison for Crypto-coprocessor

items designs	Year	Power (mW)	Kilo gates	Tech (um)	Freq (MHz)	Baud rate (kbps)	Scalable
Ref. [6]	1989	500	-	2	25	8	N
Ref. [5]	1997	330	-	-	45	-	-
Ref. [4]	2003	-	40	0.35	220	69	Y
Ref. [3]	2002	-	77	0.6	300	72	N
Ref. [2]	2004	-	148	0.18	450	214	N
Proposed	2005	41.6	45	0.25	30	14	Y

(*) The work is sponsored by NNSF (National Natural Science Foundation)

Reference

[1]. W. Kim, S. Kim, et al., A Platform-Based SoC Design of a 32-Bit Smart Card, ETRI Journal, Vol.25, No.6, Dec. 2003, PP.510-516.

[2]. Q. Liu, F. Ma, D. Tong, et al., A regular Parallel RSA Processor, IEEE MWSCAS 2004, July 25-July 28,2004,Japan, PP. III467-III470.

[3]. J.H Hong, C.W Wu, Cellular array modular multiplier for the RSA Public-key cryptosystem based on modified Booth's algorithm. IEEE Trans. VLSI Systems, Nol.11, No.3, 2002, PP.474-484.

[4]. M-C. Sun, C-P. Su, et al., Design of a scalable RSA and ECC Crypto-processor, in Proc. ASP-DAC, Kitakyushu, Jan. 2003, PP. 495-498.

[5]. A. Satoh, Y. Kobayashi, et al., A high-speed small RSA encryption LSI with low power dissipation, in Proc. 1st Int. Information Security Workshop (ISW'97), Sept. 1997, Japan, PP.174-187.

[6]. A.Vandemeulebroecke, et al., A Single-chip 1024-bits RSA processor, in Proc. Advances in Cryptology (EUROCRYPT'89), Belgium, PP.219-236.

1D-18

2006 Asia and South Pacific Design Automation Conference

Configurable Multi-Processor Architecture and its Processor Element Design

Tsutomu Nishimura, Takuji Miki, Hiroaki Sugiura, Yuki Matsumoto, Masatsugu Kobayashi,
Toshiyuki Kato, Tsutomu Eda and Hironori Yamauchi

VLSI center

Ritsumeikan University

1-1-1 Nojihigashi, Kusatsu, Shiga 525-8577, Japan

Email: ro010014@se.ritsumei.ac.jp

Abstract— We developed an application specific multi-processor generation system intended for real-time applications. In this system, we adopted a distributed memory type multi-processor architecture with hierarchical tree network as a configurable multi-processor which can be adapted to various scale systems flexibly. We have also developed a configurable multi-processor prototype as LSI chips with the 0.18 μm CMOS standard cell technology.

I. INTRODUCTION

Parallel processing by a multi-processor has been suggested as a solution to the processor performance limit of a single processor [1]. In the traditional multi-processor system, the application program has been scheduled on the fixed hardware. Therefore, the multi-processor's performance has been limited because of structural mismatch between hardware and software.

To solve this problem, we propose the automatic generation system for a multi-processor based on Hardware-Software Co-design [2] [3]. The automatic generation system generates three objectives based on C program and HDL source of PE. These three objectives are PEs, network switches and the objective codes of each PE (see Fig.1).

We have studied the multi-processor hardware ardhitecture for the configurable multi-processor that can be adapted for various scale systems. Based on this study we designed and fab"licated "it" as an LSI chip.

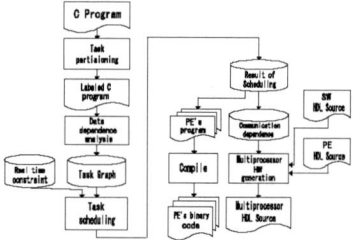

Fig. 1. Automatic generation flow

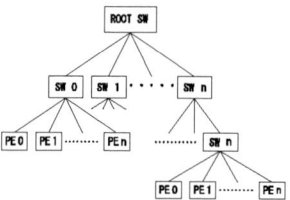

Fig. 2. Multi-processor System

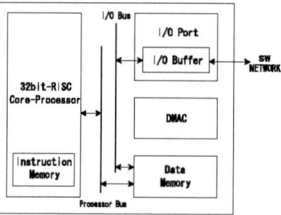

Fig. 3. PE architecture

II. MULTI-PROCESSOR ARCHITECTURE

We designed the multi-processor architecture to be used for the automatic generation. This system has to be flexible enough to adapt itself to various scale multi-processor systems flexibly. Consequently, we adopted the distributed memory type multi-processor that can be a larger scale multi-processor than the shared memory type.

Figure 2 shows our proposed multi-processor system. In this system, the PE network is a hierarchical tree structure composed as a Matrix Switch. Therefore, this system has high scalability and it is able to form the best structure according to C program.

A. PE Architecture

One PE is composed of a 32-bit-RISC Core-processor, Direct Memory Access Controller (DMAC), I/O port,

0-7803-9451-8/06/$20.00 ©2006 IEEE.

124

TABLE I

COMPARISON OF PROCESSING PERFORMANCE BETWEEN
SINGLE-PROCESSOR AND 16-PE MULTI-PROCESSOR

	Single-processor (the number of clocks)	16 PEs Multi-processor (the number of clocks)	Accelerating Ratio (Multi/Single)
1024-FFT	2,283,931	490,950	4.65 times
DFT	1,931,629	138,648	13.93 times

Fig. 4. Layout of the chip

Data-Memory and Instruction-Memory, as shown in Figure 3. The core-processor is based on DLX architecture. In addition to the fundamental instruction set of the RISC architecture, we designed additional instructions that achieve communication processing between PEs. The DMAC and the I/O port are designed to perform communication processing and reduce communication overhead.

III. LSI IMPLEMENTATION

We designed our proposed multi-processor using *Verilog-HDL*, and simulated it with *NC-Verilog* to evaluate its processing performance. In this experiment, we compared the processing performance of the 16-PE multi-processor to that of the single-processor by processing 1024-point FFT program (see table I). We confirmed that the proposed multi-processor processed the FFT and DFT program faster than single processor, up to 14 times as fast as single-processor.

We developed the LSI which is equipped with one PE and one SW as a prytotype for verification and evaluation of the chip. The LSI was fabricated with the 0.18 μm process. The memory is loaded as SRAM (IMEM: 32-bit × 512 words, DMEM: 32-bit × 4K words). About 150K transisters are integrated in each PE.

Figure 4 shows the layout of the chip. The die size is 6166 μm × 6166 μm, and the core size is 3166 μm × 3166 μm. The chip has 208 pins, which include the power supply pins (VDD: 1.8V and 3.3V, VSS). We back-annotated a functional verification with the timing information of the place and route. This chip can operate up to 50 MHz.

The test circuit is equipped in this LSI. We adopted the direct access method as a typical ad-hoc technique. Using multiplexers, drawing out signals, that can not be accessed directly from outside of the chip, we were able to test the each function block. Therefore, this chip can be used as either PE or SW, and we constructed a multi-processor system using many of them on the evaluation boards (see Figure 5). We achieved a multi-processor system with the capability to process the FFT program, and we verified that these chips can operate at 33 MHz.

IV. CONCLUSIONS

In this paper, we have proposed a multi-processor generation system, and designed a configurable multi-processor for this system. The proposed multi-processor

Fig. 5. On board multi-processor system for evaluation

system is so flexible that it is able to form the approate structure for processing various application programs. We confirmed that the proposed multi-processor achieved high parallel processing performance. Following that, we fabricated "it" by using the 0.18 μm CMOS technology and realized a multi-processor system composed of a sufficient number of LSI chips on evaluation boards.

In future work, we will improve our automatic generation system so that it will be applicable to any large scale application programs.

ACKNOWLEDGEMENTS

The authors would like to thank Mr. Daisuke Okawa from Rohm Corporation for his support.

The LSI chip in this research has been fabricated in Rohm Corporation.

REFERENCES

[1] W. H. Wolf, "An architectural co-synthesis algorithm for distributed embedded computer systems," *IEEE Trans.*, Veri Large Scale Integration Systems, Vol. 5, No. 2, pp. 218-229, June 1997.

[2] H. El-Rewini, T. G. Lesis and H. H. Ali, "Task scheduling in parallel and distributed systems," pp. 56-81, Prentice Hall, 1994.

[3] H. Nishikado, A. Nishi and H. Yamauchi, "Fast scheduling method for multi-processor using genetic algorithms," Proc. *IEEE ISPACS'99*, pp. 171-174, Dec. 1999.

Design and Implementation of Transducer for ARM-TMS Communication

Hansu Cho, Samar Abdi and Daniel Gajski
Center for Embedded Computer Systems
University of California, Irvine, CA - 92697
{hscho,abdi,gajski}@cecs.uci.edu

Communication between components, with different interface protocols, requires an extra component that must translate one protocol to another. This component is referred to as a transducer. In this paper we describe the design and implementation of a transducer between AMBA bus and TMS DSP bus. The transducer allows system designers to send data from AMBA compliant components to TMS compliant ones, and vice versa. The transducer was modeled in Verilog and implemented on Xilinx VirtexII FPGA board.

I. INTRODUCTION

With rising design complexity, designers are forced to implement more functionality in a shorter period of time by employing IP reuse. However, different IPs, typically, have different interface protocols and bus widths. Therefore, communication from one IP to another might not be possible using a common bus. This communication must be routed through an additional component known as a transducer.

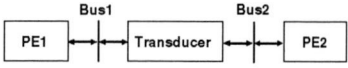

Fig. 1. A transducer for incompatible PE1 and PE2

The transducer is a component with two interfaces as shown in Figure 1. We have two different Processing Elements(PEs), namely PE1 and PE2, using different protocols on Bus1 and Bus2, respectively. We also have a transducer component with two interfaces, one connected to Bus1 and the other to Bus2. Assume that PE1 wants to sent data to PE2. Since PE2 does not support the protocol of Bus1, the communication must go through the transducer component. The data is first sent from PE1 to transducer using Bus1. The transducer locally buffers the data and then sends it to PE2 using Bus2. Thus the transaction is completed.

In this paper, we present the design and implementation of a transducer for a system consisting of and ARM1020T core [1] and a TMS320C50 DSP [2]core. The two processors have different bus interface protocols and are running at different clock speeds. The rest of the paper is organized as follows. In Section II, we give the specification and design of the transducer. In Section III, we discuss the hardware implementation. Experimental setup and results are presented in Section IV.

II. SPECIFICATION AND DESIGN

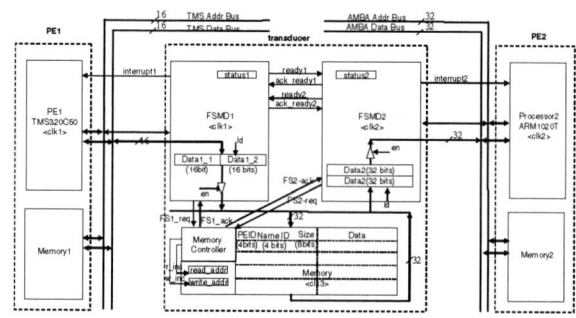

Fig. 2. Block diagram of the AMBA-TMS transducer

Figure 2 shows the architecture of the AMBA-TMS transducer. It is composed of three different subcomponents, namely FSMD1, FSMD2, and FIFO. Each subcomponent is modeled as a finite state machine with datapath (FSMD). FSMD1 implements the protocol of Bus1(TMS) and enables PE1(TMS) to read(write) to(from) the FIFO. FSMD1 runs at the same clock speed as PE1. FSMD2 implements the protocol of Bus2(AMBA) and enables PE2 read(write) from(to) the FIFO. FSMD2 also runs at the same clock speed as PE2. The FIFO consists of a local memory and a controller that communicates with FSMD1 and FSMD2 using double handshake protocol. Mutually exclusive access to the FIFO is guaranteed by the two handshake signals between FSMD1 and FSMD2. The asynchronous handshake between the FIFO and the two FSMDs allows different clock speeds for FSMD1, FSMD2 and the FIFO. As a result, the transducer is capable of interfacing PEs running at different clock speeds.

In order to send a message (size limited by FIFO capacity) from PE1 to PE2, FSMD1 and FSMD2 require the size and the source/destination ID, so that the transducer can identify the number of bus cycles needed for the transaction and the right PE to notify about the transactions. Thus the message must be packetized as follows:

$|PE1id(1byte)|PE2id(1byte)|msgid(2bytes)|$
$size(2bytes)|rawdata...|$

It is assumed that a new message is not initiated unless all the data from the previous message has been sent.

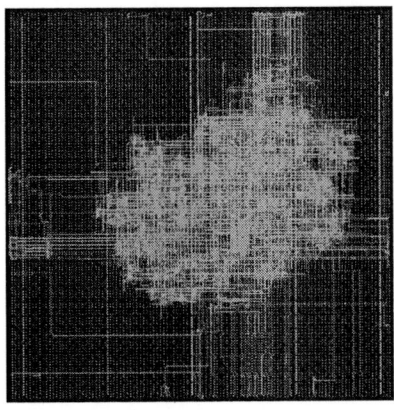

Fig. 3. FPGA implementation of the transducer

III. IMPLEMENTATION

The HDL implementation consisted of 3 top level modules, namely FSMD1, FSMD2 and the FIFO. At the RT level, FSMD1 required 10 states and FSMD2 require 15 states. The FIFO consisted of a controller with 10 states and a 1 kB RAM. The design was synthesized on the Xilinx VirtexII FPGA board. A built-in RAM on the FPGA was used to implement the memory in the FIFO. The FSMD2 block implements the interface to the AMBA bus. It can be seen that a register file (RF) is used to hold the data in transit to(from) the FIFO. This is becase AMBA allows burst transfers of various sizes. The HSIZE/HTRANS signals of AMBA are used to select the number of registers in the RF that are used to hold data for a burst transfer. While sending data to ARM processor, this RF is always filled from the FIFO before initiating a transfer. Figure 3 shows the floorplan of the transducer on a Xilinx VirtexII DS031 [3] board.

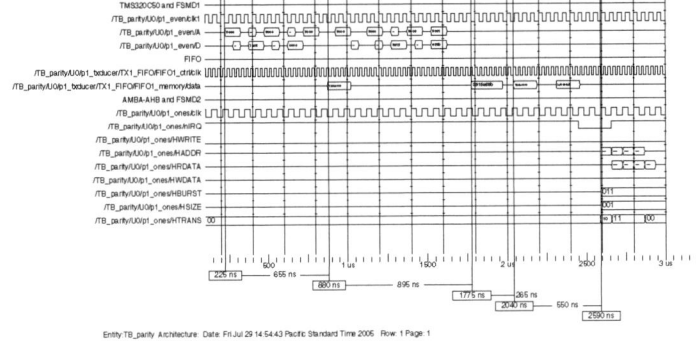

Fig. 4. Simulated waveform showing transaction progress

IV. EXPERIMENTAL RESULTS

Table I shows the synthesis results for the transducer. It can be seen that the ARM processor is allowed to run at a maximum frequency of 183.5 MHz (1/FSMD1 clk), while the TMS processor is allowed to run at a maximum of 182.9 MHz

TABLE I
KEY IMPLEMENTATION METRICS

Feature	Value
V_{dd} core/IO	1.5 V / 3.3 V
No. of F/F (Area)	1561 slices
No. of 4-inp LUTs(Area)	856 slices
Min. clk period (FSMD1)	5.450 ns
Min. clk period (FSMD2)	5.468 ns
Min. clk period (FIFO)	7.322 ns

(1/FSMD2 clk). The waveform showing the progress of a single word going through the transducer is shown in Figure 4. Table II shows the extra number of cycles used to write data through the transducer, as compared to direct write to memory. The delays are shown in sender's cycles for three sets of clock speeds.

TABLE II
TRANSDUCER DELAY

Clk1	Clk2	Clk3	ARM→TMS	TMS→ARM
50ns	70ns	30ns	44	23
70ns	50ns	30ns	35	29
50ns	50ns	50ns	48	35

V. CONCLUSIONS

In this paper we presented the design and implementation of a useful communication element called the transducer that allows communication between AMBA compatible and TMS compatible cores. We observed that such "glue logic" can be automatically generated, given parameters such as message formatting, bus characterstics and protocol specifications. A transducer allows designers to put together different IPs with incompatible interfaces in the same system.

REFERENCES

[1] ARM Inc. ARM1020T Manual. Available http://www.arm.com/pdfs/DDI0135A_1020T.

[2] Texas Instruments Inc. TMS32C5x User's Guide. Avaiable http://focus.ti.com/.

[3] Xilinx Inc. DS031-Virtex-II Platform FPGAs: Complete Data Sheet. Available http://direct.xilinx.com/bvdocs/publications/ds031.pdf.

Energy Savings through Embedded Processing on Disk System*

Seung Woo Son   Guangyu Chen   Mahmut Kandemir   Fehui Li

Department of Computer Science and Engineering
The Pennsylvania State University
University Park, PA, 16802, USA
Tel: 1-814-863-1047
Fax: 1-814-865-3176
e-mail: {sson,gchen,kandemir,feli}@cse.psu.edu

Abstract— Many of today's data-intensive applications manipulate disk-resident data sets. As a result, their overall behavior is tightly coupled with their disk performance. Unfortunately, most of these applications quickly become disk bound since disk I/O times, the communication latencies, and energy consumption required to transfer disk data to the host machine can be very large. A promising solution to this problem is to embed computational power into the disk storage system. This paper concentrates on such a smart disk based architecture and proposes an automated approach that partitions a given application code between the host machine and the smart disk. The main goal is to perform data filterings, identified at compile time, on the smart disk, thereby reducing the energy spent in communicating disk data to the host unit for processing. To achieve this, the proposed approach uses integer linear programming to identify the code fragments that perform significant data filtering and assigns such fragments to the smart disk for execution. In addition to the communication energy benefits of the proposed approach, we show in this paper that this approach can also help us better exploit the low-power management capabilities provided by the system. Our experiments with four data-intensive applications indicate significant energy savings.

I. INTRODUCTION

Applications from different domains that make use of disk-resident data are increasing in both size of the data they manipulate and code complexity. Therefore, disk system performance is critical in shaping both performance and power consumption of such applications. Our analysis of several array-intensive applications shows that a significant fraction of computations that depend on disk data are of *filtering type,* that is, the amount of the input data is larger than that of the output data. Consequently, it is possible to reduce the amount of data to be communicated between the disk system and the host system by executing this filtering type of computations on the disk system. This means performing some form of embedded processing on the disk system, and requires the employment of a processing element on the disk along with its memory. In this paper, we use the term "smart disk" to refer to such a disk storage system equipped with processing capabilities (an embedded CPU) and a memory unit.

Consider, for example, an image processing application such as edge detection. The input to this application is an image (or a series of images) and the output is typically a list of the edges detected. For example, [20] studies real images from IBM Almaden's CattleCam and attempts to detect cows in the landscape above San Jose. The application processes a set of 256 KB images and returns only the edges found in the data

using a fixed 37 pixel mask. The intent is to model a class of image processing applications where only a particular set of features such as edges in an image are important, rather than the entire image. This potentially huge data reduction in transforming input to output presents an important opportunity for reducing the amount of data communication between the disk system and the host system.

The prior efforts from the domain of high-performance computing and databases studied this problem of embedded processing on the disk system from the performance angle. We refer the reader to [1, 3, 4, 11, 15, 19, 20, 24] and the references therein. Code partitioning has also been considered in the context of memories that employ embedded processing capabilities [6, 13]. However, to the best of our knowledge, no past study investigated how this embedded processing on disk system can affect power consumption. Also, none of the prior studies discusses a fully automated mechanism for identifying the computations to be performed on the embedded processor in the smart disk. This is the problem attacked in this paper. Specifically, concentrating on a set of array-intensive applications that make frequent use of the disk system, this paper makes the following two contributions:

• We present an approach which, given a data-intensive application code, determines automatically the parts of the application code that can be executed on the smart disk system. In other words, this approach, which is based on integer linear programming (ILP), partitions the application code between the host system and the smart disk system.

• We study the power consumption behavior of this approach under two different scenarios using a simulation-based platform. In the first scenario, the components of the system do not employ any power-saving mechanisms. In contrast, in the second scenario, we assume that both the host processor and the embedded processor on the disk system have low-power operating modes, which can be activated depending on the current loads of the processors. In addition, the disk itself and the interconnect between the disk and the host processor can be shut-down when they are not in use.

To test our approach, we made experiments with four benchmark codes extracted from the SPEC2000 floating-point and Perfect Club suites, which were modified to operate on disk-resident data sets. These experiments reveal that the proposed approach is very successful, under the scenarios mentioned above, in reducing power consumption without noticeably impacting performance. In the first scenario above, our approach reduces the communication traffic between the smart disk and the host, and this leads to savings in power. In the second scenario, our approach helps to increase the idleness of the host processor, thereby enabling a more effective power management via low-power operating modes.

The remainder of this paper is organized as follows. In the following section we briefly describe the disk architecture

---

*This work is supported in part by NSF Grants 0444158, 0406340 and 0093082 and a grant from GSRC.

2006 Asia and South Pacific Design Automation Conference

2A-1

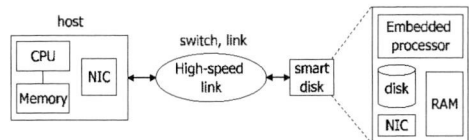

Fig. 1. The smart disk (SD) based system configuration.

equipped with an embedded processor and compare it with a traditional host-based system. In Section III, we present our ILP-based approach and give an example that illustrates how it works. In Section IV, we describe our experimental platform, define the methodology used in the experiments, and present the simulation results. In Section V, we conclude the paper.

II. SMART DISK BASED ARCHITECTURE AND EXECUTION MODEL

Figure 1 shows a smart disk based computing platform. This platform has two major components: a *host* and a *smart disk*. The host is the platform where application execution normally takes place in a system without smart disks (i.e., the one with a conventional disk system). In a smart disk based architecture, on the other hand, the host executes only some parts of the application code (not the entire code), depending on the code partitioning strategy adopted. In general, some parts of the application code are mapped to the smart disk and executed there. This smart disk based storage architecture, which contains a disk system (which can actually be a RAID based disk array), an embedded processor and a memory component, communicates with the host through a communication link, whose exact details are dependent on implementation.

In our execution model, the operation of the smart disk is controlled by the host. Suppose for now that we ran a *code partitioner* and determined the code fragments that are mapped to the host and those mapped to the smart disk (the rest of the paper will discuss our ILP-based code partitioning strategy in detail). When the execution begins, the host first sends the smart disk the code fragments that are mapped to the smart disk. These fragments are typically generated via a cross-compiler at the host system. Following this, the host also sends an *activation signal* to the smart disk, allowing it to start its execution. From this point on, the smart disk and the host can potentially execute in parallel. When the smart disk prepares an output (intermediate or final), it sends the output over the communication link to the host. This requires a synchronization between the host and the smart disk. Similarly, the host can send data to the smart disk during the course of execution.

The main advantage of this type of storage system is to allow the smart disk to execute some parts of the applications that perform data filtering. In this context, a code fragment is said to perform *data filtering* if its output data is much smaller than its input data. For instance, a code fragment that takes a three-dimensional array of size $N \times N \times N$ as input and generates a two-dimensional array of size $N \times N$ as output can be viewed as performing data filtering. If a computation that performs data filtering is executed on the smart disk, one can expect significant reductions in communication energy, as compared to the conventional scenario where the entire computation is executed on the host. Continuing with the example mentioned above, for instance, if we do not employ a smart disk, the three-dimensional input array (a total of N^3 array elements) needs to be transferred from the disk to the host for processing. On the other hand, if we perform filtering on the smart disk, the volume of communicated data is only N^2 elements (i.e., the output array for the code fragment), which represents significant savings in communication energy consumption. There is also a side benefit of employing a smart disk as

far as energy saving is concerned. Many hardware components today support several low-power operating modes. Executing some code fragments on the smart disk allows the host CPU to switch itself to a low-power mode if it needs to wait for the results from the smart disk. This can also help increase the overall energy savings. In our experimental evaluation, we consider both these scenarios: one without any low-power management and one with low-power management.

III. OUR APPROACH

We focus on data-intensive applications that manipulate array data. In this section, we explain how we can reduce power consumption by dividing a given code fragment that accesses array data into two parts, host-resident codes and *disklets*[1], i.e., the code portions assigned to smart disks. We want to emphasize that our approach tries to minimize the overall energy consumption of the given application, i.e., it does not only try to reduce the amount of communication. After all, if our objective was just minimizing communication, we could execute everything on the smart disk. However, this would increase program execution time and leakage energy consumption dramatically.

We assume that each array contains a set of *subarrays*. Each subarray $\mathcal{A}_i$ used by a program $\mathcal{P}$ can be represented using tuple $(Q, \alpha_i, \vec{L}_i, \vec{U}_i)$, where Q is the parent array of subarray $\mathcal{A}_i$ (i.e., the array from which $\mathcal{A}_i$ is extracted), and function α_i maps the each subscript vector of subarray $\mathcal{A}_i$ to a subscript vector of parent array Q. Vectors $\vec{L}_i$ and $\vec{U}_i$ are the lower and the upper bounds for the subscript vectors for subarray $\mathcal{A}_i$, respectively. The set of elements captured by subarray $\mathcal{A}_i$ can be expressed as:

$$A_i = \{Q[\alpha_i(\vec{I})] \mid \vec{L} \preceq \vec{I} \preceq \vec{U}\}. \qquad (1)$$

We use an ILP formulation to find the optimal execution strategy for the given program $\mathcal{P}$. We assume that the program $\mathcal{P}$ accesses m subarrays and consists of n loop nests. Before discussing our ILP formulation, let us first define some variables. The values of the following variables can be determined using a compiler by statically analyzing the source code of a given program and/or through profiling:

- $J_{i,j}$: $J_{i,j} \in \{0, 1\}$. If $J_{i,j} = 1$, this indicates that subarrays $\mathcal{A}_i$ and $\mathcal{A}_j$ share some elements, i.e., $A_i \cap A_j \neq \phi$. On the other hand, we have $J_{i,j} = 0$ if subarrays $\mathcal{A}_i$ and $\mathcal{A}_j$ do not share any data elements.

- N_i: the number of iterations for loop nest $\mathcal{L}_i$.

- X_i, E_i: per iteration execution time and dynamic energy consumption for executing loop nest $\mathcal{L}_i$ on the host processor.

- X_i', E_i': per iteration execution time and dynamic energy consumption for executing loop nest $\mathcal{L}_i$ on the embedded processor (in the smart disk).

- $W_{i,j}$: this is set to 1 if loop nest $\mathcal{L}_j$ updates the values of some elements of subarray $\mathcal{A}_i$.

- $R_{i,j}$: this is set to 1 if loop nest $\mathcal{L}_j$ reads the values of some elements of subarray $\mathcal{A}_i$.

The values of the following variables, on the other hand, are determined by the ILP solver:

- H_i: $H_i \in \{0, 1\}$. If $H_i = 1$, this indicates that loop nest $\mathcal{L}_i$ is assigned to be executed on the host processor. $H_i = 0$ indicates that loop nest $\mathcal{L}_i$ is to be executed on the embedded processor.

---

[1]The term is due to Acharya et al [1].

129

- $M_{i,j}$: $M_{i,j} \in \{0,1\}$. $M_{i,j} = 1$ indicates that subarray $\mathcal{A}_i$ is in the main memory of the host system at the entry of loop nest $\mathcal{L}_j$. $M_{i,j}$ takes the value of 0 if this is not the case.
- $D_{i,j}$: $D_{i,j} \in \{0,1\}$. $D_{i,j} = 1$ indicates that subarray $\mathcal{A}_i$ is dirty at the entry of loop nest $\mathcal{L}_j$. That is, the host processor has updated the values of some of the elements of subarray $\mathcal{A}_i$, and $\mathcal{A}_i$ has not been written back to the disk. Otherwise, we have $D_{i,j} = 0$.

In our formulation, we do not capture the disk energy consumption explicitly, as our approach does not change the original disk I/O activity. However, in our experiments, we also considered the disk power consumption. Let us assume that the total main memory of the host available to the program $\mathcal{P}$ is B. The following expression captures this memory constraint:

$$\sum_{i=1}^{m} M_{i,j}|A_i| \le B, \quad \forall j. \tag{2}$$

Since all the subarrays are initially on the disk (i.e., at the beginning of execution), we have:

$$M_{i,1} = D_{i,1} = 0, \quad \forall i. \tag{3}$$

Since a dirty subarray $\mathcal{A}_i$ must be in the main memory, we have the following constraint:

$$D_{i,j} \le M_{i,j}, \quad \forall i,j. \tag{4}$$

If we execute loop nest $\mathcal{L}_j$ on the embedded processor, all the dirty data that may be accessed by $\mathcal{L}_j$ must have been written back to the disk. Therefore, we have the following constraint:

$$(1 - H_j) + \sum_{k=1}^{m} D_{i,j} J_{i,k}(R_{k,j} + W_{k,j}) \le 1, \quad \forall i,j. \tag{5}$$

If we execute loop nest $\mathcal{L}_j$ on the embedded processor, any data that may be updated by $\mathcal{L}_j$ cannot be in the main memory of the host. Therefore, we have the following constraint:

$$(1 - H_j) + \sum_{k=1}^{m} M_{i,j} J_{i,k} W_{k,j} \le 1, \quad \forall i,j. \tag{6}$$

If we execute loop nest $\mathcal{L}_j$ on the host processor, the size of the data that needs to be loaded into the main memory is:

$$\sum_{i=1}^{m} (1 - M_{i,j}) R_{i,j}|A_i|. \tag{7}$$

After executing loop nest $\mathcal{L}_j$ on the host processor, the size of the data that needs to be written back to the disk is:

$$\sum_{i=1}^{m} W_{i,j}(1 - D_{i,j+1})|A_i|. \tag{8}$$

Assuming the data transfer rate for the communication link is r, the total time required for executing loop nest $\mathcal{L}_j$ on the host processor (including the time to transfer the data over the communication link) can be calculated as:

$$T_j = r\left(\sum_{i=1}^{m}(1 - M_{i,j})R_{i,j}|A_i| + \sum_{i=1}^{m} W_{i,j}(1 - D_{i,j+1})|A_i|\right) + N_j X_j. \tag{9}$$

Assuming that the per byte energy consumption of the link is p, the link energy spent for transferring data for executing loop nest $\mathcal{L}_j$ on the host processor would be:

$$E_j{}^* = p\left(\sum_{i=1}^{m}(1 - M_{i,j})R_{i,j}|A_i| + \sum_{i=1}^{m} W_{i,j}(1 - D_{i,j+1})|A_i|\right). \tag{10}$$

On the other hand, the total time required for executing loop nest $\mathcal{L}_j$ on the embedded processor would be:

$$T_j' = N_j X_j'. \tag{11}$$

Therefore, the execution time for the entire program can be

```
Input:
    H_j, M_{i,j}, and D_{i,j} – the values of binary variables determined by the ILP solver;
    P = {L_1, L_2, ..., L_n} – input program;
Output:
    Transformed program P'

for j = 1 to n {
  if(H_j = 1) {
    // insert code before loop nest L_j
    for i = 1 to m {
      if (M_{i,j} = 0 ∧ M_{i,j-1} = 1) {
        if (dirty_i = 1) {
          dirty_i = 0;  output "write A_i back to disk";
        }
        output "release memory occupied by A_i";
      } else if (M_{i,j} = 1 ∧ M_{i,j-1} = 0)
        output "load A_i into memory";
    }
    < the code for executing loop nest L_j on the host processor >
    // insert code after loop nest L_j
    for i = 1 to m {
      if (W_{i,j} = 1) {
        dirty_i = 1;
        if (D_{i,j+1} = 0) {
          dirty_i = 0;  output "write A_i back to disk";
        }
      }
    }
  } else {
    < the code for executing loop nest L_j on the embedded processor >
  }
}
```

Fig. 2. Rewrite the code of program $\mathcal{P}$ according to the values of H_j, $M_{i,j}$, and $D_{i,j}$.

expressed as follows:

$$T = \sum_{j=1}^{n} (H_j T_j + (1 - H_j) T_j'). \tag{12}$$

Also, the leakage energy for the entire system can be written as:

$$E_{leakage} = P \sum_{j=1}^{n} (H_j T_j + (1 - H_j) T_j'), \tag{13}$$

where P is the leakage power for the entire system.

The dynamic energy for executing the entire program (excluding the energy spent on the communication link) is:

$$E_{dynamic} = \sum_{j=1}^{n} (H_j N_j E_j + (1 - H_j) N_j E_j'). \tag{14}$$

The link energy for the entire program, i.e., the energy spent on the link in communicating data between the host and the smart disk, is:

$$E_{link} = \sum_{j=1}^{n} H_j E_j{}^*. \tag{15}$$

Therefore, the total energy consumed by the system for executing program $\mathcal{P}$ can be expressed as:

$$E = E_{link} + E_{leakage} + E_{dynamic}. \tag{16}$$

We use the Xpress-MP solver [17] to determine the values for binary variables H_j, $M_{i,j}$, and $D_{i,j}$ ($i = 1..m$ and $j = 1..n$) such that all the constraints above are satisfied and the overall energy (E) is minimized[2]. It should be emphasized that our solution considers three energy components: the energy for transmitting data through the I/O link between the host and the smart disk, the dynamic energy consumed by both the host system and the smart disk for executing the code fragments mapped to them, and the leakage energy consumed by the entire system during the execution of the application program.

---

[2]Please note that the right hand side of Equation (16) is not a linear function of variables H_j, $M_{i,j}$, and $D_{i,j}$. However, Xpress-MP [17] allows the target function to contain products of up to two variables.

130

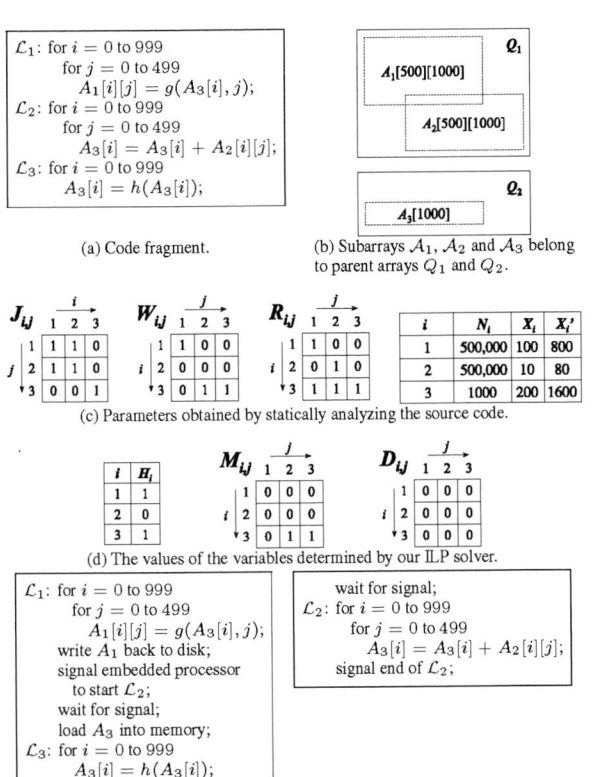

(a) Code fragment.

(b) Subarrays $\mathcal{A}_1$, $\mathcal{A}_2$ and $\mathcal{A}_3$ belong to parent arrays Q_1 and Q_2.

(c) Parameters obtained by statically analyzing the source code.

(d) The values of the variables determined by our ILP solver.

(e) The partitioned code generated by our approach. Left: loop nests $\mathcal{L}_1$ and $\mathcal{L}_3$ executed on the host processor. Right: loop nest $\mathcal{L}_2$ executed on the embedded processor in the smart disk.

Fig. 3. An example application of our approach.

Once the values for binary variables H_j, $M_{i,j}$, and $D_{i,j}$ have been determined, our approach rewrites the original code of program $\mathcal{P}$ using the algorithm shown in Figure 2. Specifically, before each loop nest $\mathcal{L}_j$ that is determined to be executed on the host processor, our approach inserts code to release the memory occupied by each subarray $\mathcal{A}_i$ with $M_{i,j} = 0$. If the subarray to be released is dirty, our approach also generates code to write the dirty value back into the disk. If subarray $\mathcal{A}_i$ is written by loop nest $\mathcal{L}_j$, "$D_{i,j+1} = 0$" means that this subarray might be used by a loop nest that is executed on the embedded processor. In this case, our approach inserts code to write subarray $\mathcal{A}_i$ back to the disk immediately after the execution of loop nest $\mathcal{L}_j$. In addition, we also invoke cross compilation for each loop nest that is determined to be executed on the embedded processor.

Figure 3 gives an example application of our approach described above. The original code fragment shown in Figure 3(a) contains three separate loop nests ($\mathcal{L}_1$, $\mathcal{L}_2$, and $\mathcal{L}_3$), and manipulates three different subarrays ($\mathcal{A}_1$, $\mathcal{A}_2$, and $\mathcal{A}_3$) belonging to two parent arrays (Q_1 and Q_2), as shown in Figure 3(b). By analyzing this code fragment, the compiler extracts the required parameters, given in Figure 3(c), which are subsequently fed to the ILP solver. The ILP solver then determines the H_j values (see Figure 3(d)), which indicate that, in this particular example, loop nests $\mathcal{L}_1$ and $\mathcal{L}_3$ are to be executed on the host machine and loop nest $\mathcal{L}_2$ is to be executed on the smart disk; i.e., loop nest $\mathcal{L}_2$ is a disklet. Since only $\mathcal{L}_2$ performs data filtering in this example, our approach assigns this loop nest to the smart disk. And, according to the $M_{i,j}$ values obtained through the ILP solver, we know that the result of loop nest $\mathcal{L}_2$, $\mathcal{A}_3$, should be transferred to the host machine. Figure 3(e) gives the partitioned code generated by our approach.

IV. EXPERIMENTAL PLATFORM AND EMPIRICAL RESULTS

A. Setup and Benchmarks

To evaluate the effectiveness of our approach, we wrote a trace-driven simulator using CSIM [21][3]. In this setup, all the disk I/O and data communication are assumed to be done at a page granularity (default page size of Solaris is 8KB). The cycle time for each instruction type (e.g., load/store, arithmetic, etc.) is obtained from the processors' manual [9, 10]. The simulator generates energy consumption and performance statistics. The energy statistics are calculated based on the figures extracted from the datasheet of each system component or from the previously-published studies [5, 7, 8, 9, 10, 12, 14, 16, 18, 25], and are given in Table I. More specifically, our power model is as follows. While processing the given input traces, the simulator keeps track of time values (stamps) for all activities involved in processing each trace. These recorded time values indicate the states of each component (i.e., when a particular component is used and how long it is used) throughout the entire simulation time, and then we calculate power values from these determined states of each system component. The required hardware components for this calculation are different for each scheme we considered, which will be explained later in this section. In this study, we assume that each component in our target system has its own power transition state diagram. That is, each component has at least one low-power operating mode (state) as well as active and idle modes, and it takes a certain amount of time and energy to transition from one state to another. Based on these energy and time values, one can determine the *break-even threshold*, i.e., the minimum amount of idle time required to compensate the cost of transitioning a given component into a low-power mode, and this threshold is used in this work in deciding whether it makes sense to put the component into a low-power mode when it becomes idle.

The different components have different low-power modes and usually different names for these modes. For example, a modern server disk has typically only one low-power mode, called shut-down, whereas a DRAM has three low-power modes, named standby, napping, and power-down. So, when there is no confusion, in the rest of the paper, we use the term "low-power mode" to denote only one of the low-power modes that each hardware component provides, specifically, the one which consumes the lowest energy.

Table II gives the set of data-intensive applications, used in this study. We chose these benchmarks from the CFP2000 [22] and Perfect Club benchmarks [2], and made the array data manipulated by the benchmarks disk resident. As a result, each array reference causes a disk access unless the requested block is captured in the buffer cache. Also, to complete the simulation within a reasonable amount of time, we concentrated on the most dominant loop nests in terms of the cumulative I/O times and the amount of data manipulated. The second column of Table II gives the total dataset size manipulated by each benchmark, and the next two columns give the total energy consumption and execution time, respectively, for each application, when all computations are executed in the host (this is the HOST version, as will be described shortly). *The energy and performance numbers presented in the rest of the paper are normalized with respect to the values listed in these two columns of Table II.* The fifth column gives the contribution of the energy consumed in the communication link between the host and the smart disk, which takes a significant fraction of the total energy consumption, as can be seen from the table

---

[3]This should not be confused with the gate-level simulator of the same name [23].

TABLE I

DEFAULT SIMULATION PARAMETERS.

Parameter	Value
Host Processor	
Model	Intel P4 2.0 GHz
Power (active/idle/standby)	100.4/75.3/0.0525 W
Power (standby → active)	0.1 J
Time (standby → active)	1 ms
Power (active → standby)	5.3 uJ
Time (active → standby)	70.38 ns
Embedded Processor	
Model	StrongArm 200 MHz
Power (active/idle/standby)	400/50/0.16 mW
Power (standby → active)	0.064 J
Time (standby → active)	160 ms
Power (active → standby)	0.036 mJ
Time (active → standby)	90 us
Memory	
Model	Rambus DRAM
Capacity	32MB for smart disk and 1GB for host
Power (active/standby)	300/3 mW
Power (standby → active)	15 mW
Time (standby → active)	6000 ns
Power (active → standby)	15 mW
Time (active → standby)	8 memory cycle
Disk	
Model	IBM Ultrastar 36Z15
Storage Capacity	18 GB
RPM	15,000
Average seek time	3.4 msec
Average rotation time	2 msec
Internal transfer rate	55 MB/sec
Power (active/idle/standby)	13.5/10.2/2.5 W
Energy (idle → standby)	13 J
Time (idle → standy)	1.5 sec
Energy (standby → active)	135 J
Time (standby → active)	10.9 sec
Interconnects (Link)	
Model	Infiniband 1X
Bandwidth	2.5G (1X)
Energy	10.21 (pJ/bit)
Power (standby)	ε
Time (standby ↔ active)	800 ns
Energy (standby ↔ active)	0.002 mJ
Switch	
Model	IBM Infiniband 1X
Power (active/standby)	11 / 2 W
Energy (active → standby)	0.18 mJ
Time (active → standby)	0.09 ms
Energy (standby → active)	0.32 J
Time (standby → active)	0.16 sec

TABLE II

BENCHMARKS AND THEIR CHARACTERISTICS.

Name	Total Data (MB)	Base Energy (J)	Execution Time (sec)	Link Energy	% of Code on on Smart Disk
swim	22.1	736.6	4.4	23.9 %	59.0 %
apsi	2.9	101.6	0.6	23.8 %	74.0 %
mgrid	80.7	2707.1	16.2	23.6 %	54.0 %
bmcm	10.3	457.5	2.6	22.3 %	28.3 %

machine occurs only at this point.

- OPT: This is an optimized version in which code fragments to be executed on the host and the smart disk are determined by our ILP-based partitioner described in Section III. The computation occurs in both the host and the smart disk, which requires all system components to execute this scheme. When one side is in use during computation, the other side will remain in idle state. In the general case, some code fragments are executed on the host machine and the others on the smart disk. The communication energy is spent whenever there is a communication between the two code fragments mapped onto the different places.

These three schemes (HOST, SD, and OPT) do not make use of any low-power modes of any component in the system. Apart from these three schemes, we also implemented an energy-optimized version, called EOPT, which can be used in conjunction with the above three schemes. In the EOPT scheme, each system component, e.g., CPU, memory, interconnect, etc., can be in a low-power mode when it is not in use. The decision to place a component in the low-power mode is based on the break-even time of each component, as explained earlier. That is, if the idle period of a given component is longer than its break-even time, the component is placed into the low-power mode. Otherwise, it remains in the idle/active state, i.e., without any power management. Therefore, the schemes in conjunction with EOPT do not change their original execution times. The purpose of our experiments with EOPT is to see how our approach interacts with low-power modes. Combining EOPT with the three schemes described above, we conducted experiments with a total of six different schemes: HOST, SD, OPT, HOST+EOPT, SD+EOPT, and OPT+EOPT. Among the different schemes used in this work, the largest ILP solution time was taken by the OPT+EOPT scheme and was around 172 seconds for the *mgrid* benchmark.

B. Results

The graphs in Figure 4 give the total energy consumption of our benchmarks under the different schemes described earlier. As mentioned earlier, all the results are normalized with respect to the HOST version. One can make several observations from these results. First, for *mgrid* and *bmcm*, executing all the loop nests in the smart disk significantly increases overall energy consumption. This is because the computation power in the embedded processor in the smart disk is much less than that in the host processor. Second, for *swim*, there is not a significant difference whether all code fragments are executed in the host or in the embedded processor. This is because the communication reduction and the increase in computation time when all computations are assigned to the embedded processor balance each other. Lastly, even if we do not employ any shutdown policies in any component, the OPT version results in significant amount of energy savings (23% on average). This shows that our ILP-based approach successfully partitions the computations across the host and the smart disk in an energy-efficient fashion.

It is to be noted, however, that the reduction in the amount of data to be communicated, provided by our approach, might affect the potential energy savings when the system components employ low-power operating modes. The potential en-

(over 20% for all four benchmarks). This suggests that reducing the link power consumption can have a significant impact in practice on overall system power consumption. The last column gives the percentage of the application code mapped to the smart disk (i.e., the disklets) after applying our ILP-based code partitioner.

To quantify the benefits obtained from our approach, we implemented and conducted experiments with different schemes:

- HOST: This is the version where all the computations are performed on the host machine. This scheme requires all system components shown in Table I except the embedded processor and memory in the smart disk, which remain in the idle states. Since the data is stored in the smart disk, this scheme incurs a significant data communication from the smart disk to the host. The energy and performance results with this version are given in Table II.

- SD: This version represents the other extreme, and performs all the computations on the smart disk, that is, the remaining components remain in idle state. Unlike the HOST scheme, it does not incur any communication energy cost due to the disk accesses since the disk is a local resource from the viewpoint of the smart disk. However, this version can increase leakage consumption dramatically over the HOST version, due to the increase in execution time. After finishing execution, the results required by the host are transferred to the host machine, and the communication between the smart disk and the host

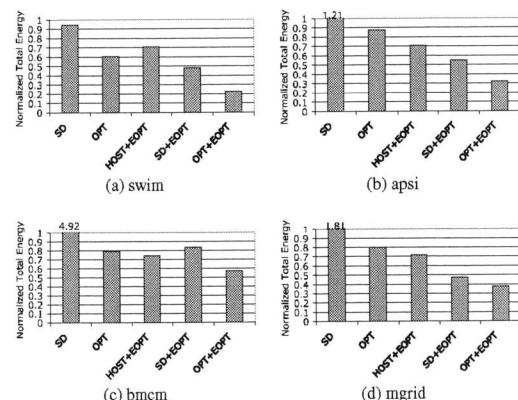

(a) swim   (b) apsi

(c) bmcm   (d) mgrid

Fig. 4. Normalized total energy consumption of each benchmark.

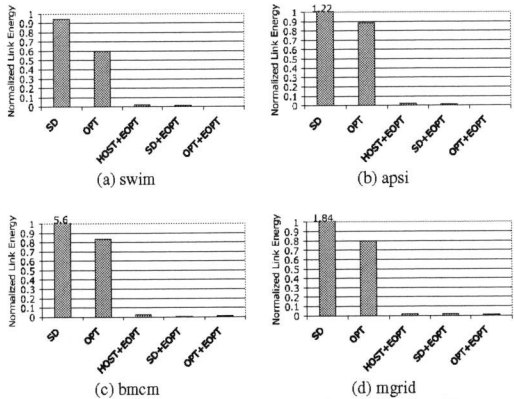

(a) swim   (b) apsi

(c) bmcm   (d) mgrid

Fig. 5. Normalized link energy consumption of each benchmark.

ergy savings in this case are captured by the right three bars, HOST+EOPT, SD+EOPT, and OPT+EOPT, in each bar-chart shown in Figure 4. One can see from these results that, if we can exploit the low-power mode that each component provides whenever possible, the resulting energy savings are really significant. Overall, the results presented in Figure 4 indicate that our approach reduces energy in both the scenarios studied. These benefits are achieved not only due to the reduced volume of communication between the smart disk and the host, but also due to the increased chances for the host to shut down. To see the details of the energy savings in the communication link, we also collected statistics on link energy consumption with these six different schemes we experimented. The results are given in Figure 5. One can see from these results that most of the energy consumption in the communication links can be eliminated if we shut down the links when they are not in use. Overall, our results indicate that the compiler algorithm presented in this paper can be very useful in practice, and the best energy savings are achieved when our approach is used in conjunction with the power saving mechanisms provided by each component. We also measured the impact of our approach on original execution cycles, however, we do not present the detailed results due to lack of space. We found that the OPT and OPT+EOPT schemes improve the original execution cycles by 23% on average.

V. CONCLUSIONS

Many large-scale applications are data-intensive in nature and require manipulation of huge data sets such as multidimensional scientific data, image files, satellite data, database tables, and digital libraries. Apart from the high computational requirements, these applications typically involve the transfer of large amounts of disk-resident data back and forth between the secondary storage devices and the processing units. Observing that a large fraction of these computations are of filtering type, this paper proposes and evaluates an ILP-based approach that partitions an application code between the host system and the disk system (equipped with an embedded processor). We test the behavior of our approach using a set of array-intensive benchmarks that frequently exercise the disk system. Our results show that the proposed partitioning approach reduces power consumption significantly.

REFERENCES

[1] A. Acharya, M. Uysal, and J. H. Saltz. Active Disks: Programming Model, Algorithms and Evaluation. In *Proceedings of the 8th Conference on Architectural Support for Programming Languages and Operating Systems*, pages 81–91, 1998.

[2] M. Berry and et al. The PERFECT Club Benchmarks: Effective Performance Evaluation of Supercomputers. *The International Journal of Supercomputer Applications*, 3(3):5–40, 1989.

[3] G. Chen, G. Chen, M. T. Kandemir, and A. Nadgir. Compiler-Based Code Partitioning for Intelligent Embedded Disk Processing. In *Languages and Compilers for Parallel Computing*, pages 451–465, 2003.

[4] S. Chiu, W. keng Liao, and A. N. Choudhary. Design and Evaluation of Distributed Smart Disk Architecture for I/O-Intensive Workloads. In *International Conference on Computational Science*, pages 230–241, 2003.

[5] C. Eddington. InfiniBridge: An InfiniBand Channel Adapter with Integrated Switch. *IEEE Micro*, 22(2):48–56, 2002.

[6] M. Hall, P. Kogge, J. Koller, P. Diniz, J. Chame, J. Draper, J. LaCoss, J. Granacki, J. Brockman, A. Srivastava, W. Athas, V. Freeh, J. Shin, and J. Park. Mapping irregular applications to DIVA, a PIM-based data-intensive architecture. In *Proceedings of the 1999 ACM/IEEE conference on Supercomputing*, 1999.

[7] IBM hard disk drive. Ultrastar 36Z15, April 2001.

[8] InfiniBand Trade Alliance. The InfiniBand Architecture. http://www.infinibandta.org.

[9] Intel. *Intel StrongARM SA-1100 Microprocessor - Developer's Manual*, June 2000.

[10] Intel. *Intel Pentium-4 Microprocessor at 2GHz - Datasheet*, 2001.

[11] K. Keeton, D. A. Patterson, and J. M. Hellerstein. A Case for Intelligent Disks (IDISKs). *SIGMOD Record*, 27(3):42–52, 1998.

[12] E. J. Kim, K. H. Yum, G. M. Link, N. Vijaykrishnan, M. T. Kandemir, M. J. Irwin, M. S. Yousif, and C. R. Das. Energy Optimization Techniques in Cluster Interconnects. In *Proceedings of the International Symposium on Low Power Electronics and Design*, pages 459–464, 2003.

[13] P. M. Kogge, S. C. Bass, J. B. Brockman, D. Z. Chen, and H. S. E. Pursuing a Petaflop: Point designs for 100TF Computers Using PIM Technologies. In *6th Symp. on Frontiers of Massively Parallel Computation, Annapolis*, pages 25–31, October 1996.

[14] X. Li, Z. Li, F. M. David, P. Zhou, Y. Zhou, S. V. Adve, and S. Kumar. Performance Directed Energy Management for Main Memory and Disks. In *Proceedings of the 11th International Conference on Architectural Support for Programming Languages and Operating Systems*, pages 271–283, 2004.

[15] G. Memik, M. T. Kandemir, and A. N. Choudhary. Design and Evaluation of Smart Disk Architecture for DSS Commercial Workloads. In *Proceedings of the International Conference on Parallel Processing*, pages 335–, 2000.

[16] S. S. Mukherjee, P. J. Bannon, S. Lang, A. Spink, and D. Webb. The Alpha 21364 Network Architecture. *IEEE Micro*, 22(1):26–35, 2002.

[17] D. Optimization. Modeling with Xpress-MP, December 2001.

[18] Rambus. RDRAM. http://www.rambus.com, 1999.

[19] E. Riedel, C. Faloutsos, G. A. Gibson, and D. Nagle. Active Disks for Large-Scale Data Processing. *IEEE Computer*, 34(6):68–74, 2001.

[20] E. Riedel, G. A. Gibson, and C. Faloutsos. Active Storage for Large-Scale Data Mining and Multimedia. In *Proceedings of the International Conference on Very Large Data Bases*, pages 62–73, 1998.

[21] H. Schwetman. CSIM19: A Powerful Tool for Building System Models. In *Proceedings of the 33nd conference on Winter simulation*, pages 250–255, 2001.

[22] SPEC. Specfp 2000. http://www.specbench.org/cpu2000/CFP2000/, 2000.

[23] P. Stenius. Csim - a simple hp48 circuit simulator for educational purposes, 1992.

[24] M. Uysal, A. Acharya, and J. H. Saltz. Evaluation of Active Disks for Decision Support Databases. In *Proceedings of the 6th International Symposium on High-Performance Computer Architecture*, pages 337–348, 2000.

[25] H.-S. Wang, L.-S. Peh, and S. Malik. A Power Model for Routers: Modeling Alpha 21364 and InfiniBand Routers. *IEEE Micro*, 23(1):26–35, 2003.

Energy-Aware Computation Duplication for Improving Reliability in Embedded Chip Multiprocessors *

G. Chen, M. Kandemir, and F. Li
Computer Science and Engineering Department
Pennsylvania State University
e-mail: {guilchen,kandemir,feli}@cse.psu.edu

Abstract— Compilers designed for current embedded systems must be capable of addressing multiple constraints such as low power, high performance, small memory footprint and form factor, and high reliability at the same time. In particular, optimizing for one constraint should be performed carefully, considering its impact on other constraints. Recent trends indicate that transient errors are becoming increasingly important in embedded systems. Focusing on an embedded chip multiprocessor and array-intensive applications, this paper demonstrates how reliability against transient errors can be improved without impacting execution time by utilizing idle processors for duplicating some of the computations of the active processors. It also shows how a balance between power savings and reliability improvement can be struck using a metric called the energy-delay-fallibility product. Our experimental results indicate that the "percentage of duplicated computations" is a useful high-level metric for studying the tradeoffs among performance, power, and reliability.

I. INTRODUCTION

Today's embedded systems are expected to satisfy multiple, and often conflicting, criteria (constraints) such as low power, high performance, small memory footprint and form factor, and high reliability. Therefore, the entire system design cycle employed in the past, which includes software and hardware, must be re-thought based on this multi-criteria requirement. For example, most of the software compilation techniques developed over the years target mainly at optimizing performance. Recently, compiler designers have also focused on memory footprint estimation/reduction [5,20–22] and power/energy optimizations [9, 14]. However, recent increase in transient error rates due to scaling technology and employment of low power techniques made it imperative to consider reliability as a first class optimization metric. The important point though is the reliability measures must be well-balanced against performance and power concerns. In other words, while ensuring reliable execution as much as possible, one also needs to be careful in not increasing execution cycles or power consumption excessively.

One way of addressing growing complexity problem of embedded system designs is chip multiprocessors [2, 3, 13]. The idea is to accommodate multiple simple cores within the same die instead of a complex single core based architecture. Prior research [13, 15] has already discussed several advantages of

chip multiprocessors over complex single processor based solutions, which include appropriateness to high-level code parallelization and design and verification advantages. Performance/verification related issues have been addressed in studies like [16], whereas studies such as [18] and [10] focused on the problem of reducing power consumption. An important observation made by one of the prior studies [10] is that, in executing array-intensive codes in a chip multiprocessor, not all the cores are used all the time. That is, at any given time, certain number of processors are idle and can potentially be switched off or put in a low-power operating mode to save energy.

In this work, we explore an alternate use of such idle processors in an embedded chip multiprocessor. Specifically, focusing on embedded array-intensive applications, we demonstrate that idle processors can be used for *duplicating* some of the computations of the active processors, thereby improving overall reliability against transient errors. In this approach, a computation can be duplicated once to detect transient errors (our focus in this paper) or twice to detect and correct them. It should also be noticed that not all of the idle processors need to be used for duplicating computation. In fact, some of them can still be placed into a low-power mode to save power, which is a particularly promising approach in an environment where both reliability and power consumption are important metrics to consider. Whether an idle processor should be used for saving power or increasing reliability depends on the relative importances of reliability and power as well as allowable power consumption and acceptable error levels.

It is to be emphasized that reliability concerns due to transient errors are becoming an increasingly pressing problem for embedded systems. This is mainly because of two reasons. First, as devices are being pushed into very deep sub-micron technologies (< 250 nano-meter), reliability is becoming an important issue. Some of the growing effects are the so-called "transient errors", which are due to temporary conditions of usage characteristics and the environment. Cross-coupling, ground bounce, external terrestrial radiations create more and more unpredictable transient and soft errors which affect system reliability. Transient errors caused due to radiations especially have been studied very closely in industry. The second reason is that many embedded systems employ several mechanisms that scale voltage or place unused components into low-power modes (sleep states), with the objective of reducing energy consumption. This in turn increases the vulnerability of these systems to transient errors. A recent study [4] investigates the relationship between power-saving strategies and cir-

---

*This work is supported in part by NSF Career Award #0093082 and by a grant from GSRC.

0-7803-9451-8/06/$20.00 ©2006 IEEE.

cuit reliability.

Since we want to measure the impact of our approach on performance, power, and reliability, previous evaluation metrics such as execution cycles, performability, or energy-delay product are not very appropriate for our purposes. Instead, we employ a different metric called the *energy-delay-fallibility* product (EDF), where fallibility in this context is the opposite of reliability. Using this metric, we study how the different divisions (partitionings) of the idle processors between those that execute duplicated computations and those that are placed into a low-power mode to save power can affect the value of the energy-delay-fallibility product. We also study the impact of the duplication granularity for adaptation (i.e., application based versus loop nest based) on the value of the energy-delay-fallibility metric. Note that, using new metrics for characterizing system behavior in terms of energy efficiency, reliability, computation performance and battery lifetime has been a popular research topic recently [19].

We automated our reliability-oriented approach within a parallelizing compiler and tested it using seven array-intensive benchmark codes. Our results, collected using the Simics simulation toolset [17], demonstrate that not just the different applications require different percentages of idle processors to be used for computation duplication, but also even within the same application, the different loop nests work best (from the perspective of the energy-delay-fallibility product) with the different percentages of idle processors being used for duplication. In other words, the amount of duplication should be tuned at a loop nest granularity. While more computation duplication intuitively means better reliability, the latter concept depends also on the frequency of transient errors and the patterns these errors exhibit. The results indicate that the "percentage of computations duplicated" is a reasonable indicator for reliability. In this paper, as long as there is no confusion, we use the terms "percentage of computations duplicated" and "percentage of processors used for duplicating computations" interchangeably.

The remainder of this paper is structured as follows. The next section summarizes the architectural abstraction presented to our approach. Section III discusses the details of our computation duplication strategy. Section IV introduces our experimental setup and reports the experimental results from our implementation. This experimental evaluation considers both the error-free case and the cases with errors. Section V concludes with a summary of our major observations.

II. CHIP MULTIPROCESSOR ARCHITECTURE AND EXECUTION MODEL

We focus on an embedded chip multiprocessor of the shared memory type. In this architecture, multiple processor cores (typically between 4 and 32) reside on the same chip. Each core has private L1 instruction and data caches and there is also a large unified L2 cache shared by all the cores. We also assume existence of an off-chip memory, whose access latency and power consumption are typically much larger than the corresponding values for the on-chip L1 and L2 caches. In this architecture, data sharing and communication among processors is achieved using shared memory components (i.e., through L2 cache and off-chip memory). Note that, several chip multipro-

cessor proposals [13, 15] from academia and industry fit in this abstraction.

We focus on execution of array-intensive applications on this chip multiprocessor. It is to be emphasized that, array-intensive applications frequently appear in many embedded domains where reliability is also a concern. In this work, an array-intensive embedded application is parallelized by considering each loop nest in turn, and distributing its iterations across processors. How exactly the loops are parallelized is orthogonal to the focus of this work. For this purpose, one can employ either user-assisted methods or automatic compiler support. For a given loop nest, the set of loop iterations assigned to a particular processor is called its *local iteration set* or *local iteration space* for that processor with respect to that nest. Our unit of duplication (for reliability purposes) in this paper is a local iteration space. That is, we duplicate the local iteration space of an active processor on an otherwise idle processor. When a processor core is switched to a low-power mode, its L1 cache is also assumed to be put in the low-power mode. While in the low-power mode, a processor/L1 pair consumes a fraction of the energy that they would consume in the full active mode, and the L1 cache maintains its contents. Neither L2 cache nor off-chip memory is turned off during program execution.

III. COMPUTATION DUPLICATION

Figure 1 illustrates our approach to computation duplication. In Figure 1(a), a loop nest is parallelized (either automatically by a parallelizing compiler or through user help), and m out of total n processors (P_1 through P_m) are used to execute the loop nest. In a previous work [10], all the idle processors (P_{m+1} through P_n) along with their L1 caches are placed into a low-power mode to save energy. In our work, on the other hand, we utilize some of these idle processors to improve reliability by executing on them duplicates of the computation performed by the active processors (P_1 through P_m). What we mean by computation in this context is the local iteration space assigned to an active processor as a result of code parallelization. In Figure 1(b), the local iteration spaces of r processors (P_1 through P_r) are duplicated and executed on processors P_{m+1} through P_{m+r} to improve the reliability of the computation. When there is no confusion, we say that P_i ($1 \leq i \leq r$) is a *primary* processor, and P_{m+i} is the *duplicate* of P_i. It is to be noted that, these concepts of primary and duplicate processors are relative to a given loop nest; i.e., a given processor can be primary in one nest and duplicate in another.

To improve reliability of execution for a given primary processor, we have to check its execution with that of its duplicate to see whether their results agree. A straightforward way of achieving this would be letting the primary and its duplicate run in a lock-step fashion, during which they compare their results after executing each and every statement. Although such a lock-step execution can be implemented rather easily, it can also generate a lot of communication and synchronization activities between the primaries and their duplicates, and the overheads it incurs in terms of both performance and energy consumption can be intolerable for an embedded system. This is particularly true considering the fact that communication and synchronization costs can be critical in a chip multi-

2A-2 2006 Asia and South Pacific Design Automation Conference

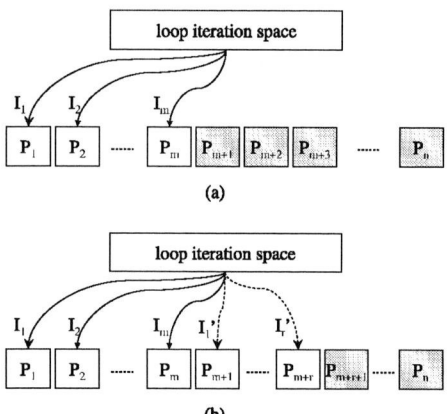

(a)

(b)

Fig. 1. Computation (local iteration space) duplication. P_1 through P_n represent the processors in the chip multiprocessor. The idle processors are shaded. (a) The loop nest is parallelized and set to execute on m processors; I_i ($1 \leq i \leq m$) is the set of iterations assigned to P_i as a result of parallelization, i.e., its local iteration set. (b) I_1 to I_r (where $r \leq m$) are assumed to be duplicated and the duplicated computations are assigned to P_{m+1} through P_r. Dashed curves represent the duplicated iteration sets. I_i' is the duplicate of I_i.

$$A[i] = C[i+1] * D[i] + E[i];$$
$$B[i] = C[i-1] - D[i-1];$$
(a)

$$A[i] = C[i+1] * D[i] + E[i];$$
$$CHECK[prid] += A[i];$$
$$B[i] = C[i-1] - D[i-1];$$
$$CHECK[prid] += B[i];$$
(b)

$$CHECK[prid] += C[i+1] * D[i] + E[i];$$
$$CHECK[prid] += C[i-1] - D[i-1];$$
(c)

Fig. 2. Adding checksum to a code segment. *prid* is the id of a given processor, and $CHECK[prid]$ is the corresponding checksum. (a) Original code (loop body). (b) Transformed code to be executed by the primary processor. (c) Transformed code to be executed by the duplicate processor.

$$A[i] = B[i] + C[i];$$
$$B[i] = C[i] - 10;$$
(a)

$S_{b1}:$ $A[i] = B[i] + C[i];$
$S_{b2}:$ $CHECK[prid] += A[i];$
$S_{b3}:$ $B[i] = C[i] - 10;$
$S_{b4}:$ $CHECK[prid] += B[i];$
(b)

$S_{c1}:$ $CHECK[prid] += B[i] + C[i];$
$S_{c2}:$ $CHECK[prid] += C[i] - 10;$
(c)

$$CHECK[prid] += B'[i] + C[i];$$
$$B'[i] = C[i] - 10;$$
$$CHECK[prid] += B'[i];$$
(d)

Fig. 3. An example code segment that illustrates the need for array duplication. (a) Original code. (b) Transformed code to be executed by the primary processor. (c) Incorrect version of the transformed code to be executed by the duplicate processor. (d) Correct version of the transformed code with array duplication to be executed by the duplicate processor.

processor. Instead, in our work, we use a more efficient way of comparing these two executions, which is as follows. We associate a *checksum* with each processor (i.e., with both primary and its duplicate), and all these checksums are initialized to zero at the beginning. Each time a statement finishes its execution and produces a result, we add it to the corresponding checksum. After all the loop iterations complete, we compare the primary's checksum with its duplicate's checksum to see whether they are equal. Note that, in this approach, we do not need any output (except for the checksum) from the duplicate processor. Therefore, in most cases, in the duplicate, all the results of computations can be directly fed to the checksum without being written into any array or variable, as would be the case in the original code. This also helps reduce the energy overhead associated with the duplicate significantly. Figure 2 gives an example of how a code segment can be transformed to generate checksum. Figure 2(a) is the original code in the loop body (assuming that i is the loop index variable); Figure 2(b) is the transformed code to be executed by the primary; and Figure 2(c) is the transformed code to be executed by the duplicate. It can be seen that the code to be executed by the duplicate does not update arrays A and B due to the reason explained above.

The transformation used in Figure 2 may not work correctly if an array element is both read and written in the iteration space (i.e., when there exists a data dependence involving the array in question). Figure 3 presents such an example. In statement S_{b1}, we use the old value of $B[i]$. To be consistent, S_{c1} should also use the old value of $B[i]$. Otherwise, the checksums computed in Figure 3(b) and Figure 3(c) will be different from each other, even in the absence of any transient error, because the input values used to calculate them are different. Therefore, S_{c1} should be executed before S_{b3} to ensure that the old value of $B[i]$ is not overwritten by S_{b3}. However, doing so means that we need to synchronize the primary processor and its duplicate at the statement level, which is what we want to avoid by introducing checksums at the first place. Our approach to address

this problem is presented in Figure 3(d). In this solution, array B is duplicated, and B' represents the duplicate. We replace B with B' in the code depicted in Figure 3(c), and we obtain the code in Figure 3(d). Notice that, although not shown here explicitly, we need to initialize B' with B's old values before the loop is entered since the old values of B will be used before they are written (updated) within the loop. Using array duplication, the synchronization problem discussed above due to data dependences is solved. It should be observed though that such data duplications bring extra overheads in terms of memory space occupation, performance (execution cycles), and energy consumption. However, our experience with numerous application codes shows that such cases (that require data duplication) do not happen very frequently. In fact, we observed during our experimental evaluation that, in only six out of the twenty four loops in our study need array duplication, and *we take all the associated overheads into account in our experimental evaluation.* The code transformations illustrated in Figures 2 and 3 have been automated within our compiler.

It should be noticed that this computation duplication catches all types of transient errors during loop execution (as long as either the primary and the duplicate executes correctly), not just the memory related errors. That is, it also captures the CPU errors that might occur during the execution of the duplicated loops. Therefore, it is very general. While it is possible that it can fail in cases where a wrong computation still generates a correct checksum, such cases are expected to be very rare in practice. The presented scheme is different from the prior code duplication related studies such as [1,6,8], as we focus on an embedded chip multiprocessor and try to minimize a different metric, which captures performance, energy, and reliability. However, there is still a chance that a transient error can strike during execution of an unduplicated computation, which cannot be detected by our approach. In other words,

136

TABLE I
MAJOR SIMULATION PARAMETERS AND THEIR VALUES.

Parameter	Value
Number of Processors	8
L1 Instruction Cache	8KB 2-way associative 32 byte blocks
L1 Data Cache	8KB 2-way associative 32 byte blocks
L2 Cache	1MB 4-way associative 64 byte blocks
L1 Dynamic Energy Consumption	0.16 nJ/access
L2 Dynamic Energy Consumption	0.65 nJ/access
r	0.1
Reactivation Latency	4 cycles
Off-Chip Memory Energy	6.32 nJ/access
Off-Chip Memory Access Latency	80 cycles
On-Chip Bus Arbitration Delay	5 cycles
Replacement Policy	Strict LRU

while we increase the resilience of execution against soft errors, the number of errors we actually detect is another matter, and depends on error injection rate, error injection pattern, and other factors.

IV. EXPERIMENTS

A. Setup

We used a chip multiprocessor simulator for our experiments, built upon Simics [17]. Simics is a platform for full system simulation that can run actual firmware, complete kernel, and driver codes. It is sufficiently abstract to achieve good performance levels, and it provides both functional accuracy for running commercial workloads and sufficient timing accuracy to interface to detailed hardware models. In particular, it allows us to model the overheads incurred by our approach accurately. Our simulator keeps track of the number of instructions executed by each processor and data/instruction accesses to different memory components (L1, L2, and off-chip memory). We also embedded energy models into this simulator. These energy models are access based, and compute energy of a component by multiplying the number of accesses to that component with a fixed (component specific) per access energy consumption. The per access energy consumption value for each component is obtained through profiling. The necessary code modifications to insert checksum computations in the source codes are automated within a parallelizing compilation framework [7]. The increase in compilation time caused by our approach was about 19% when averaged over all the codes used in our evaluation. The largest compilation time increase was approximately 41%.

Table I gives the major simulation parameters used in this study. The parameter r specifies the magnitude of the leakage power consumption when a processor core/L1 cache is placed into the low-power mode. More specifically, when $r=r^*$, this means that the leakage energy consumed by the core/L1 pair in the low power mode is $r \times 100\%$ of the leakage energy consumption of an active processor core/L1 pair. We assume that the leakage energy per cycle for an 8KB SRAM (our L1) is equal to the dynamic energy consumed per access to a 32 byte data from that SRAM, similar to the assumption made by [11].

The seven benchmark codes used in this study and their important characteristics are given in Table II. These benchmark codes are extracted from the Perfect Club, Spec, Livermore,

TABLE II
BENCHMARK CODES USED IN OUR EXPERIMENTS AND THE STATISTICS COLLECTED WHEN NO POWER OPTIMIZATION AND NO DUPLICATION IS USED.

Benchmark Name	Number of Nests	Cycles (Million)	Dynamic Energy (uJ)	Leakage Energy (uJ)
3step-log	3	14688	20195.27	29390.57
adi	2	217	589.45	706.24
btrix	7	82336	92524.07	83096.42
eflux	2	1236	2018.51	2496.56
full-search	3	97640	137887.96	18937.01
n-real-updates	3	160	395.45	423.66
tsf	4	246	752.84	282.55

and DSPStone benchmark suites. The values listed in this table are obtained by executing the benchmarks in our simulation environment *without* any power management and *without* any computation duplication. All benchmark codes have been run to completion. In the rest of our discussion, we refer to this version of a benchmark as the *base version,* or the *original version.* The third column gives the number of cycles. The last two columns give the dynamic and leakage energy consumptions under the 70nm process technology. These values include the energies consumed in the processor cores, L1 and L2 caches, and off-chip main memory.

As stated earlier, our approach can work under any loop parallelization strategy. The particular strategy used in this work is based on locality of reference. In this strategy, a loop nest is parallelized such that each processor accesses data mostly with temporal or spatial reuse. However, each loop is parallelized using the minimum number of processors; that is, increasing the number of processors beyond this minimum number does *not* improve performance any further. Table III gives statistics on the execution of the base version under the default machine configuration. Each column (starting with the second one) in this table corresponds to a loop nest in the application and each cell shows the minimum processor count (as explained above) that gives the best performance for that nest. That is, using more processors for the nest does not improve its performance further. An observation that one can make from the values in this table is that, in almost all the loop nests, there exist some idle processors, which can potentially be placed into the low-power mode to save energy, or can be used for duplicating some computation to increase error detection capabilities. In fact, most of the nests use only 4 or fewer processor (of a total of 8 processors) to generate the best execution cycles. We need to point out that this low processor utilization is typical in parallel processing (i.e., it is not a particular characteristic of the applications used in this study), and can be explained as follows. As we increase the number of processors over which a loop nest is parallelized, at least two types of overheads increase. The first of these is the time/energy spent in spawning the additional threads and finalizing them when the parallel loop execution is over. The second one is the synchronization costs due to increased inter-processor communication. In addition, some data dependences across loop iterations generate the best results with a particular processor count. All these factors are effective in preventing us from using all processors in the chip multiprocessor. As a result, going beyond a given processor size increases only the overheads without bringing any performance benefits. In particular, using more processors than necessary can have devastating results from the energy consumption angle.

TABLE III

MINIMUM NUMBER OF PROCESSORS THAT GENERATE THE
OPTIMUM PERFORMANCE FOR EACH NEST OF EACH BENCHMARK
CODE. NOTE THAT, EACH APPLICATION HAS A DIFFERENT
NUMBER OF NESTS. NI REPRESENTS THE ITH NEST IN THE
CORRESPONDING APPLICATION.

Benchmark	N1	N2	N3	N4	N5	N6	N7
3step-log	1	1	5				
adi	4	5					
btrix	2	1	7	6	1	3	8
eflux	2	3					
full-search	2	2	6				
n-real-updates	4	4	4				
tsf	1	7	2	4			

As stated earlier, we use the energy-delay-fallibility product (denoted EDF henceforth) as our metric in this paper. In our context, energy corresponds to the sum of the energies consumed in the processor cores, L1 and L2 caches, and off-chip memory, and delay is the parallel execution time (measured in cycles). Fallibility is the opposite of reliability; the latter being defined as the percentage of loop iterations that are duplicated. Clearly, we want the value of the EDF metric to be as small as possible. As mentioned earlier, our approach can involve array duplication in certain cases, and this can in turn affect the energy consumption and execution cycles of the application (due to the degradation in cache performance). *All these overheads are included in the EDF values presented in the next subsection.* While, as mentioned earlier, we can duplicate a local iteration space twice (instead of just once) to correct errors (e.g., through a majority voting based scheme), in this paper we present the results with single duplication only; i.e., we focus primarily on the error detection problem. It should be mentioned that the impact of our approach on original execution cycles is not excessive. In fact, there are only three potential reasons why the performance can be affected because of our approach. The first of these is due to array duplication. As we mentioned earlier, it does not occur very frequently, and as a result, the overheads it incurs are not excessive. The second overhead is due to comparing the checksums (which also involves inter-processor synchronization), and its impact is not very high. The third one is the reactivation cost when a processor/L1 pair is powered down, which is not very frequent (as it is done only between nest boundaries). Overall, we observed that the performance degradation due to our approach was less than 2% when averaged over all the benchmark codes in our experimental suite. In any case, the EDF values presented below include this small degradation in performance as well. However, before presenting the benefits of our approach, we give in Figure 4 the percentage contribution of each type of overhead. We see that, while the overheads due to array duplication dominate the others, the other two factors also play a role. However, as mentioned above, their cumulative impact on the performance is less than 2% on the average.

B. Results

Our first set of EDF results are presented in Figure 5. Each point on the x-axis represents a percentage of idle processors used for computation duplication. The remaining processors are placed into the low-power mode along with their L1 caches. Note that, these percentages are valid for each nest of each application. That is, in each loop nest, the same percentage of

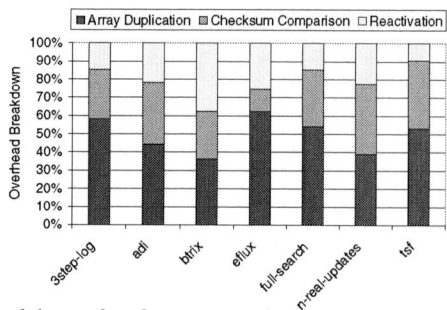

Fig. 4. Breakdown of performance overheads into three categories.

TABLE IV

STATISTICS WHEN 20% OF THE IDLE PROCESSORS ARE USED
FOR COMPUTATION DUPLICATION. AS COMPARED TO TABLE III,
WE HAVE MORE CYCLES AND MORE ENERGY CONSUMPTION
HERE DUE TO DUPLICATION.

Benchmark Name	Cycles (Million)	Dynamic Energy (uJ)	Leakage Energy (uJ)
3step-log	15623	25253.15	35313.51
adi	242	756.49	836.72
btrix	91089	127330.08	104521.98
eflux	1318	2816.13	2753.38
full-search	103233	165538.25	25913.14
n-real-updates	182	520.32	519.68
tsf	276	969.60	387.67

idle processors are used for duplication. In the results shown in Figure 5, the EDF values for each application are *normalized* with respect to the EDF values when 20% of the idle processors are used for computation duplication (the absolute cycles, dynamic and static energy consumption values for the case when 20% of the idle processors are used for duplication are given in Table IV). It can be observed from the results in Figure 5 that the trends in general are similar across the different applications. Most of the applications have a decrease at first in EDF as we increase the percentage of idle processors used for computation duplication, which can be attributed to the increased reliability we have as more iterations are duplicated. However, we also observe that, beyond a certain point, using more idle processors for duplication increases EDF since the increased energy consumption and execution time (to a lesser extent) starts to offset the benefits brought by more duplicated iterations. Therefore, *it is important to pick a suitable percentage of duplication for a given application to reach a good tradeoff point between performance, energy, and reliability.* Only two applications, namely full-search and tsf, exhibit slightly different trends. Their curves keep decreasing as the percentage of duplicates increases. This can be attributed to the relative lower leakage energy consumed by these two applications (as compared to their dynamic energy consumptions). Lower leakage energy consumption means less energy savings brought by putting the idle processors into the low-power operating mode. Therefore, as we use more processors for duplication, the benefits coming from the increased reliability (i.e., the decreased fallibility) can be offset by the increased energy consumption (as far as the EDF metric is concerned). We can also observe from Figure 5 that the different applications reach their optimum (minimum) EDF values at different points. For example, eflux reaches its optimum result at 30%, whereas 3-step-log achieves its optimum result at 50%. As has been discussed

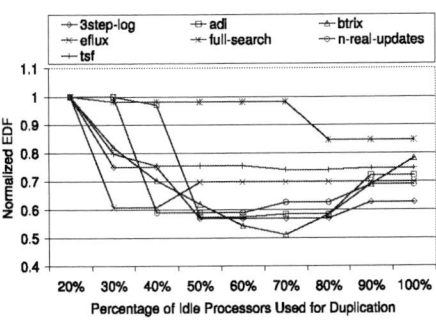

Fig. 5. The EDF values with the different percentage of idle processors being used for computation duplication. The EDF values for each application are normalized with respect to the EDF value when 20% of the idle processors are used for duplication.

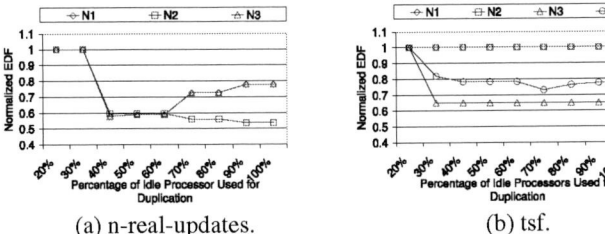

(a) n-real-updates.　　　　(b) tsf.

Fig. 6. Normalized EDF for the different loop nests in n-real-updates (left) and tsf (right). n-real-updates has three loop nests, denoted by N1, N2, N3, and tsf has four loop nests, denoted by N1, N2, N3, and N4.

above, leakage energy consumption behavior can affect the optimum point. In addition to this, the overheads incurred by computation duplication are also a factor for some applications. If the overhead brought by computation duplication is high, a small percentage in duplication will be favored. Otherwise, using more processors for duplication can be expected to be more beneficial as far as the EDF metric is concerned. Overall, these results suggest that the percentage of computation duplication used should be tuned for each application separately.

Figure 6 presents the EDF curves for individual loop nests in two benchmarks: n-real-updates and tsf. It can be seen from Figure 6(a) that the curves for N1 and N3 in n-real-updates are almost overlapped, whereas N2 exhibits a different pattern. Specifically, one would prefer to use a relatively small percentage value, say about 40%, for N1 and N3, but use a large percentage value for N2, as suggested by the trend exhibited by the N2's curve. The curves for loop nests of tsf, shown in Figure 6(b), are relatively flat. It can still be observed though that the different loop nests have different trends. For example, N4 reaches its optimum point at 70%, while the other three loop nests are not very sensitive to the percentage change. All of these observations indicate that, to obtain the best tradeoffs, one should consider adaptive computation duplication in the granularity of loop nests, in addition to adaptive computation duplication at application level. To sum up, the results presented in Figure 5 and Figure 6 motivate for application-level and loop nest-level adaptation, respectively, in utilizing the idle processors.

V. CONCLUSIONS

Single metric based compilation strategies are not sufficient for current complex embedded systems, where multiple constraints (e.g., power, performance, reliability) are important and need to be accounted for at the same time. In this context, maybe the most useful class of optimizations are those that optimize for one metric without impacting the others excessively. Motivated by this view, in this paper we evaluate a reliability oriented compilation strategy for embedded chip multiprocessors based on computation duplication. Using a new metric, called the energy-delay-fallibility product (EDF), we study the impact of the percentage of idle processors used for computation duplication. Our results suggest that an adaptive scheme in choosing the number of idle processors for computation duplication is needed to achieve the best tradeoffs between performance, energy efficiency, and reliability.

REFERENCES

[1] C. Bolchini. A Software Methodology for Detecting Hardware Faults in VLIW Datapaths. *IEEE Transactions on Reliability*, 52(4):458-468, December 2003.

[2] Chip Multiprocessing. http://industry.java.sun.com/javanews/stories/print/ 0,1797,32080,00.html

[3] Chip Multiprocessing. ITWorld.Com, http://www.itworld.com/Comp/ 1092/CW-STO54343/

[4] V. Degalahal, R. Rajaram, N. Vijaykrishanan, Y. Xie , and M. J Irwin. The Effect of Threshold Voltages on Soft Error Rate. In *Proc. ISQED*, San Jose, CA, 2004.

[5] A. Fraboulet, K. Kodary, and A. Mignotte. Loop Fusion for Memory Space Optimization. In *Proc. the International Symposium on System Synthesis*, Montreal, Canada, September 30-October 3, 2001.

[6] C. Gong, R. Melhem and R. Gupta. Compiler-Assisted Fault Detection for Distributed Memory Systems. In *Proc. the Scalable High Performance Computing Conference*, Knoxville, TN, 1994.

[7] M. W. Hall, J. M. Anderson, S. P. Amarasinghe, B. R. Murphy, S.-W. Liao, E. Bugnion, and M. S. Lam. Maximizing Multiprocessor Performance with the SUIF Compiler. *IEEE Computer*, Dec 1996.

[8] J. G. Holm and P. Banerjee. Low Cost Concurrent Error Detection in a VLIW Architecture Using Replicated Instructions. In *Proc. the International Conference on Parallel Processing*, pp. 192-195, 1992.

[9] C-H. Hsu and U. Kremer. Single Region vs. Multiple Regions: A Comparison of Different Compiler-Directed Dynamic Voltage Scheduling Approaches. In *Proc. PACS Workshop*, Cambridge, MA, February 2002.

[10] I. Kadayif, M. Kandemir, and M. Karakoy. An Energy Saving Strategy Based on Adaptive Loop Parallelization. In *Proc. Design Automation Conference*, June 10-14, 2002, New Orleans, Louisiana, USA.

[11] S. Kaxiras, Z. Hu, M. Martonosi. Cache Decay: Exploiting Generational Behavior to Reduce Cache Leakage Power. In *Proc. the 28th International Symposium on Computer Architecture*, Sweden, June 2001.

[12] S. Kim and A. Somani. Area Efficient Architectures for Information Integrity Checking in Cache Memories. In *Proc. ISCA*, May 1999.

[13] V. Krishnan and J. Torrellas. A Chip Multiprocessor Architecture with Speculative Multi-threading. *IEEE Transactions on Computers, Special Issue on Multi-threaded Architecture*, September 1999.

[14] M. Lorenz, L. Wehmeyer, and T. Drager. Energy-Aware Compilation for DSPs with SIMD Instructions. In *Proc. LCTES*, Berlin, Germany, 2002.

[15] K. Olukotun, B. A. Nayfeh, L. Hammond, K. Wilson, and K. Chang. The Case for a Single Chip Multiprocessor. In *Proc. ASPLOS*, 1996, pp. 2–11.

[16] K. Richter, M. Jersak, and R. Ernst. A Formal Approach to MpSoC Performance Verification. *IEEE Computer*, (Vol. 36, No. 4), April 2003.

[17] Simics Tool-set. http://www.simics.com.

[18] R. Sasanka et al. The Energy Efficiency of CMP vs. SMT for Multimedia Workloads. In *Proc. ICS*, June 2004.

[19] P. Stanley-Marbell and D. Marculescu. Dynamic Fault-Tolerance and Metrics for Battery Powered, Failure-Prone Systems. In *Proc. ICCAD*, San Jose, CA, 2003.

[20] M. Strout, L. Carter, J. Ferrante, and B. Simon. Schedule-Independent Storage Mapping in Loops. In *Proc. the International Conference on Architectural Support for Programming Languages and Operating Systems*, October, 1998.

[21] W. Thies et al. A Unified Framework for Schedule and Storage Optimization. In *Proc. PLDI*, Snowbird, UT, June, 2001.

[22] Y. Zhao and S. Malik. Exact Memory Size Estimation for Array Computations without Loop Unrolling. In *Proc. DAC*, June 1999.

Object Duplication for Improving Reliability

G. Chen, G. Chen, M. Kandemir, N. Vijaykrishnan, M. J. Irwin

Department of Computer Science and Engineering
The Pennsylvania State University
University Park, PA 16802, USA
e-mail: {guilchen,gchen,kandemir,vijay,mji}@cse.psu.edu

Abstract— Soft errors are becoming a common problem in current systems due to the scaling of technology that results in the use of smaller devices, lower voltages, and power-saving techniques. In this work, we focus on soft errors that can occur in the objects created in heap memory, and investigate techniques for enhancing the immunity to soft errors through various object duplication schemes. The idea is to access the duplicate object when the checksum associated with the primary object indicates an error. We implemented several duplication based schemes and conducted extensive experiments. Our results clearly show that this spectrum of schemes enable us to balance the tradeoffs between error rate and heap space consumption.

I. INTRODUCTION

A major reliability concern due to the increasing size of embedded memories is *soft errors*. Soft errors occur when a memory bit flips its value due to external radiation effects, thus corrupting the stored data. The need for reliable memory has become even more acute due to the use of power-savings techniques such as voltage scaling in current embedded systems.

A common approach to handling these memory errors is to use error detection and correction hardware [14, 13]. However, embedded systems are usually sold in huge quantities and thus tend to be more sensitive to the per device cost as compared to their high-performance counterparts. Consequently, a hardware approach, which increases the overall cost of the system, may not be attractive for low-cost embedded systems. Further, an embedded system may run a set of applications and not all of them may require fault-tolerance. Employing expensive hardware for just a few applications that need fault-tolerance may not be the best economic option. In comparison, a software scheme can take application specific requirements into account and tune the policy, considering the limited resources in the embedded device.

The focus of this work is on handling soft errors in object-oriented frameworks. We select an embedded Java Virtual Machine (JVM) as our target object-oriented environment, and inject errors into the heap memory that stores the objects in order to investigate techniques for enhancing immunity to soft errors through various object duplication schemes. While a lot of work has been done on the problem of reliable computation at the circuit, architectural, operating system, and application levels [1, 2, 10, 11, 15, 17, 18], our work focuses explicitly on the integrity of objects, and is complementary to model checking and verification based work [7, 12].

The rest of this paper is organized as follows. Section II discusses our error injection model. Section III

presents our object duplication schemes, including full duplication, compression-based duplication, and selective duplication schemes. Section IV presents an experimental evaluation of these schemes. Section V concludes the paper with a summary of our major observations.

II. THE ERROR INJECTION MODEL

We use Sun's KVM [5] to implement the object duplication-based error protection techniques proposed in this work. KVM is a compact, portable Java Virtual Machine specifically designed for small, resource-constrained devices. KVM uses a handle-free mark-sweep-compact collector. An error management module is added into KVM to store the error information for each object. For every bytecode executed, KVM invokes our error injection function to inject errors into the object instances in the heap. The error injection function scans the heap; every bit in the object instances has a fixed probability of incurring an error. When an object is accessed, we check the error management module to determine whether the accessed part has any error(s) in it. The default value for the error injection probability for our base experiments is 10^{-10}. While we perform experiments with different error injection rates, the rates used in our experiments are generally higher than those with the current technology. The main reason for this is that errors are more likely to happen only when an application executes for long durations of time or when it is executed repeatedly, and we need an accelerated testing environment. It should be noted that accelerated testing is meaningful because there are many embedded Java applications that need to be operational without errors for long durations ranging from several hours (e.g., cell phones) to months (e.g., sensors).

III. DUPLICATION SCHEMES

A. Motivation for Object Duplication

In this work, unless stated otherwise, we assume that each object is protected using a "checksum-based scheme" (called CHK). In this scheme, each object has a single checksum attached to it. The checksum calculations are performed in a similar fashion to that in [3]. Specifically, each object header is extended with one additional word to store the precomputed checksum. This checksum is checked upon a read request to a field, and updated upon a write request.

The Java applications used in this study are given in Table I. Calc, firstaid, jpeg, and mvideo are taken from the

TABLE I

THE JAVA BENCHMARK CODES AND THEIR CHARACTERISTICS.

Benchmark	Description	Execution Cycles	Errors Injected/ Consumed/Detected
auction	ticket auction	467.4	75 / 27 / 25
calc	calculator	338.5	71 / 14 / 14
firstaid	firstaid info	618.5	423 / 127 / 126
jpeg	jpeg viewer	1,052.9	1314 / 329 / 313
image	photo album	1,157.1	430 / 302 / 271
manyballs	bouncing balls	475.2	53 / 13 / 11
mvideo	video player	2,732.6	63 / 44 / 40
pushpuzzle	puzzle game	479.7	72 / 14 / 13

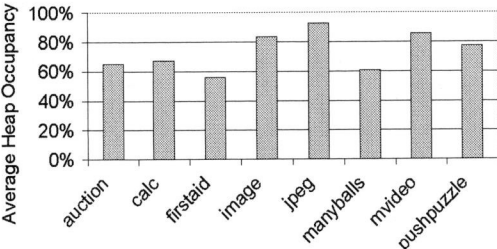

Fig. 1. Average heap occupancy of our applications.

http://www.microjava.com site, and auction, image, manyballs, and pushpuzzle come with the MIDP 1.0.3 reference implementation [8]. The third column gives the execution cycles (in millions) for the "base execution." In this base execution (denoted BASE in the rest of the paper), the objects are not protected. The last column shows statistics on the behavior of the CHK scheme. It gives the total number of errors injected into the memory , the number of errors that have been consumed by the application, and the number of errors detected by CHK. Note that many of the injected errors have not been consumed, meaning that the memory location with the error has not been accessed. We see that although CHK is successful in detecting most of the consumed errors (it detects 93.1% of the errors on the average) , it will not be able to correct any of them. One of our objectives in this work is to correct as many of these errors with as little performance overhead as possible.

We next look at the heap usage statistics of our applications. Figure 1 shows the average heap space used by each application over the time as a fraction of the peak heap space used by that application. For example, the first bar indicates that, on the average, the benchmark auction uses 65.2% of the peak heap space it needs for its objects. One can see from these results that, across all applications, only 74.1% of the peak heap usage is utilized on the average. The remaining heap space can be used for some other purpose. Moreover, the actual maximum heap space allocated for the objects of the application can even be larger than the peak space demanded by an application. That is, in reality, we may have a larger unused heap area that can be exploited for some other purpose.

B. Full Duplication

In this paper, we use this unused heap space for duplicating objects. In this scheme (called DUPL), each time a new object is created, we also create a duplicate object in the heap. Both the primary object and duplicate are protected using checksums. When accessing an object, we first access the primary object. If its checksum indicates no error, we continue with

execution as usual. On the other hand, if its checksum indicates an error, we access the duplicate and check its checksum. Note that, under realistic operating conditions, the chances that the checksum of the duplicate also shows an error will be very low. Therefore, we should be able to correct the error in the primary copy most of the time. An important advantage of this scheme is that as long as there is no error, we incur very small performance penalty over CHK. Our performance overheads are mainly due to creation of duplicate objects and updating them on writes. And, when an error occurs, correcting it using one more object access should be acceptable to most operating environments.

The primary and the duplicate objects are allocated in the primary area and duplicate area, respectively. The primary area starts from the lowest address of the heap, and grows toward the higher addresses. The duplicate area, on the other hand, starts from the highest address, and grows toward the lower addresses. When the boundaries of these two areas meet with each other, a mark/compact garbage collector [9] is invoked. To compact the heap, the primary objects are slided toward the lowest address, and the duplicate objects are slided toward the highest address. The header of each primary object contains a pointer to its duplicate, which is called the *forward pointer*. Similarly, each duplicate object contains a pointer to its primary object, which is referred to as the *backward pointer*. During the mark-compact garbage collection, if a primary object is moved, the backward pointer in its duplicate should be updated. Similarly, if a duplicate object is moved, we need to update the forward pointer in its primary copy. Forward-backward pointer pairs allow the primary and duplicate objects to find each other. Further, this mechanism enables us detect or repair the errors in the forward or backward pointers. Note that, though not evaluated in this work, it is possible to have more efficient strategies to connect primary and duplicate objects (e.g., having one object with duplicated fields and two checksums, or placing the duplicate at a fixed distance in memory from the primary object). The reason that we use the pointer-based scheme explained above is that it suits better for more sophisticated duplication strategies, such as the selective scheme as will be discussed later.

Discussion: A full duplication scheme doubles the memory requirement of heap objects, which might be undesirable for an embedded environment. There exist at least two ways of reducing the memory space and/or performance overheads associated with DUPL. First, since the duplicate objects are read only when there is an error in the primary copy, we can compress them so that their heap space occupancy can be reduced. The downside is that when we need to access a compressed duplicate, it first needs to be decompressed before the access can take place. Therefore, there is a tradeoff between performance and heap space saving. The second alternative for reducing the overheads is to use duplication selectively based on object lifetimes. If we are careful in identifying the objects that really need duplicates, we can reduce the overall overhead. The next two sections investigate these two approaches.

C. Compression-Based Full Duplication

Since it is known that, in most Java applications, the objects contain a lot of zero bytes [4], we use a "zero-removal" algorithm to compress the duplicate objects to reduce their heap space requirements. We modified KVM to implement object

31	16	15	0
Compressed Size		Original Size	
Backward Pointer			
Check Sum			
Bitmap			
Non-zero byte array			

Fig. 2. The format of a compressed duplicate.

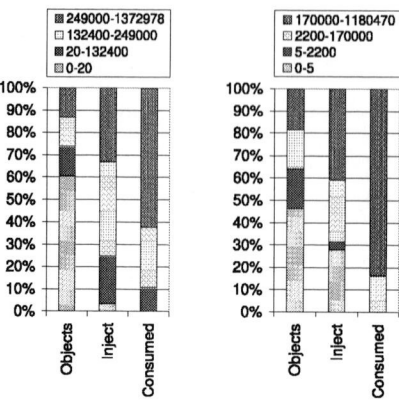

Fig. 3. The breakdown of objects, injected errors, and consumed errors into four life groups. Left: firstaid. Right: pushpuzzle. Each portion of the first bars indicates the percentage of objects that fall into that life group (in terms of a cycle range). Each portion of the second (third) bars represents the percentage of errors injected into (consumed by) objects within that life group.

compression. Figure 2 shows the format of a compressed duplicate object. The compressed object contains a bitmap and a non-zero byte array. Each bit in the bitmap corresponds to one byte of the object in the uncompressed format. A 0-bit indicates that the corresponding byte is zero and this byte is not stored; a 1-bit indicates that the corresponding byte is stored in the non-zero byte array. The details of the zero-removal compression/decompression algorithm are beyond the scope of this paper.

When an error is detected in the primary object during execution, we check if its compressed duplicate is corrupted. If the duplicate is not corrupted, we correct the error in the primary object by decompressing the duplicate into the address of the primary copy. Otherwise, we terminate the program and report the non-correctable error. When the contents of the primary object are updated, we check the bitmap in the compressed duplicate to determine the number of bytes that are currently used to store the accessed field. If this number is not smaller than the number of non-zero bytes in the updated value, we update the corresponding bitmap bits and non-zero bytes of the accessed field in the compressed duplicate. Otherwise, we need to discard the current duplicate, and create a new one since the new compressed duplicate cannot fit in the space reserved for the old one. It should be noted that, during garbage collection, the reference fields of the primary objects may be updated due to compaction. Therefore, we discard and collect all the duplicates whose primary objects contain reference fields. After the compaction phase, we re-create those discarded duplicates from the contents of their primary objects. This compression-based version of DUPL is referred to as COMPDUPL in the rest of this paper.

D. Selective Duplication

In this strategy, the main goal is to maintain as few duplicates as possible without significantly hurting the error correction rate achieved using full duplication. Recall that both DUPL and COMPDUPL maintain duplicates as long as the primary object is alive. However, since the duplicate is only needed when there is an error in the primary object, we can get rid of the duplicates under certain circumstances.

One example of how a duplicate can be eliminated is based on an analysis of object "drag times". The drag time of an object is the ratio between the time the object spends beyond its last use and the time since its creation to its death [16]. We found that the drag times for auction, calc, firstaid, image, jpeg, manyballs, mvideo, and pushpuzzle are 53%, 45%, 59%, 53%, 82%, 54%, 88%, and 48%, respectively. That is, the objects in our embedded applications spend a large percentage of their lifetimes in the heap beyond their last-use. Our first selective scheme is specifically designed to exploit this observation to reduce the lifetime of duplicates. Another way of cutting the number of duplicates is to consider the lifetime of the objects.

Figure 3 gives for two applications, firstaid and pushpuzzle, the breakdown of the objects created, of the errors injected, and of the errors consumed into four categories formed based on their lifetimes. We see that most of the objects created are small in size, and in comparison, the injected errors are more uniformly distributed across the different life groups. However, the last bars clearly indicate that the most of the consumed errors are in long-living objects. Consequently, one can expect a protection strategy that pays special attention to long-living objects to be effective in practice. Our second and third selective duplication strategies are designed to take advantage of this observation.

Early Termination of Duplicates: In the first selective scheme, when we create an object, we create its duplicate as usual. However, we also predict the *last-use* of the object. At each invocation of the garbage collector, we now collect not only the unreachable primary objects and their duplicates, but also the duplicates whose primary objects have become last-used. To implement this scheme, each duplicate is augmented with a time-stamp that records its allocation time. At each garbage collection, the duplicates that are older than a threshold are collected. The success of this strategy critically depends on the accuracy of the last-use prediction (i.e., the threshold value used). To determine good prediction values, we plotted the CDF (cumulative distribution function) for object last-uses. Each (x, y) point on the curve of a specific application in Figure 4 indicates that the last-use of $y\%$ of the total object words occurs within the first x cycles after its creation. From this graph, one can determine good estimates to use for predicting last-uses. For example, we see from the calc's curve that, if we create the duplicate at the same time when the primary object is created and then discard the duplicate about 3,000 cycles after the creation time, 90% of the total object words will be beyond their last-use point when their duplicates are discarded. Beyond its last-use point, an object will not be accessed by the application any more. Therefore, the errors occurring beyond the last-use point will not affect the correctness of the application. The downside of this scheme is that the objects that have not reached their last-uses will be vulnerable between their predicted last-uses and their actual last-uses. In addition, if the predicted last-use is longer than the actual last-use of an object, we increase heap occupancy unnecessarily. Therefore, an accurate last-use prediction is very important.

2006 Asia and South Pacific Design Automation Conference 2A-3

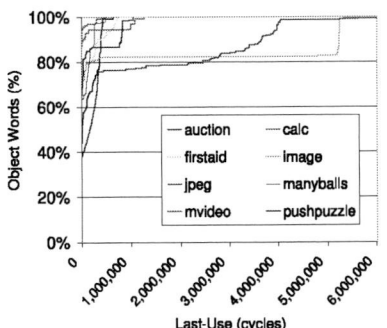

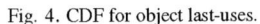

Fig. 4. CDF for object last-uses.

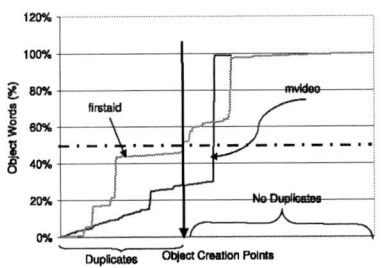

Fig. 5. CDF for object sizes created by each object creation point for firstaid and mvideo. The figure also shows how the object creation points are marked for firstaid.

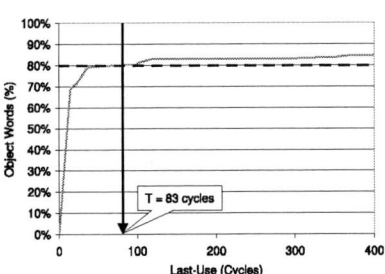

Fig. 6. CDF for last-uses of the objects in mvideo.

Allocation Site Based Selective Duplication: Our second selective scheme is based on profiling. We collected information on the number of accesses to the objects created by each object creation point for our applications. We observed that, for many object creation points, the objects created are not accessed very frequently. Consequently, one can choose not to provide duplicates for such objects. Figure 5 shows how we select the creation points for which we want to provide duplicates. The x-axis represents the object creation points sorted (from left to right) according to non-increasing number of accesses to the objects created by each point. The y-axis shows CDF for the sizes of the total objects created by each object creation point. For example, if we want to provide duplicates for only 50% of the frequently accessed objects, we need to draw a horizontal line from the y-axis and find the corresponding point on the x-axis. All the object creation points on the right of this point are then annotated. These annotations are recognized and handled by the JVM, and no duplicate is created for the objects allocated by these points.

Lazy Duplication: Our last selective scheme, referred to as DELAYED, defers the creation of the duplicate to a point, where we expect the object to be long-living and frequently-accessed once it reaches that point. In other words, this scheme is lazy in creating object duplicates. It is implemented as follows. Each object has a time-stamp, indicating its creation time. At each access to the object, we compare the current time against the time-stamp. If the difference between them is larger than a threshold (T), we create a duplicate for the object. This selective strategy does not create duplicates for the objects whose last accesses are shorter than T. A critical issue here is how to determine a suitable value for T. Note that, if we are late in creating the duplicate, we can increase the number of non-correctable errors since the object is vulnerable until a duplicate is created for it. On the other hand, if we create the duplicate too early, there will be very little improvement in heap space consumption over DUPL. To determine good threshold values, one can use the CDF curves presented in Figure 4. For example, Figure 6 zooms in the initial portion of the curve for mvideo. If we want to avoid the duplicates for 80% of the total allocated object words, we draw a horizontal line, determine the corresponding values on the x-axis, and use that value as T.

IV. EXPERIMENTAL RESULTS

In our experimental evaluation, we collect three types of statistics:
- *Heap space results:* These results indicate the memory space overhead due to object duplication. We are interested in

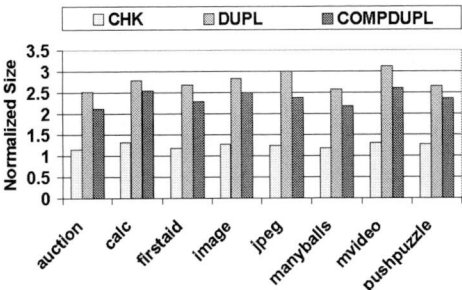

Fig. 7. Allocated heap space during the entire execution.

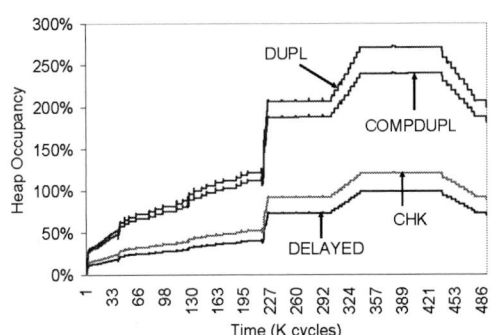

Fig. 8. Heap space occupancy by the application objects during execution of manyballs. The y-axis is normalized with respect to the peak heap space required by BASE.

two metrics: (1) the total amount of heap space allocated during execution, and (2) the variance in heap occupancy during the course of execution. We define "heap occupancy" in this context as the total size of the "live application objects" in the heap.
- *Error resilience results:* These results are collected by injecting errors into the heap memory and instrumenting the application code to collect error statistics. The main metric that we are interested in is the "non-correctable error rate", which is the percentage of soft errors that are detected by checksums but could not be corrected by a given protection scheme.
- *Performance results:* These results correspond to execution cycles, and are obtained using an enhanced version of the Shade tool-set [6].

A. Heap Space Results

The first two bars for each application in Figure 7 give the increase in the total size of the heap space allocated with CHK

143

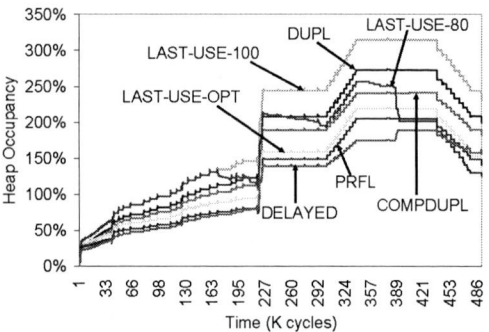

Fig. 9. Heap space occupancy for the application objects during the execution of manyballs for selective schemes.

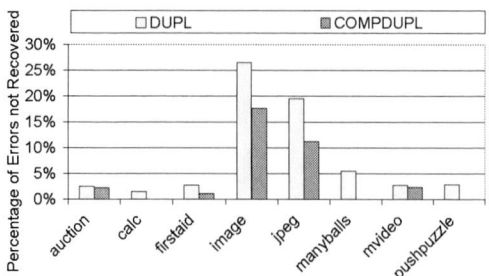

Fig. 10. Non-correctable error rates for the different schemes.

and DUPL during the entire execution. One can see that, while the allocated memory space increase due to CHK is within reasonable limits (around 24% increase over the base execution), DUPL increases the size of the allocated heap space by a factor of 2.77 on the average. We also see that the value for COMPDUPL is lower than the corresponding value for DUPL.

To see how the heap space requirements change during execution, we give in Figure 8 the variance in the heap space occupancy (y-axis) over the time (x-axis) with the different schemes for the manyballs benchmark. Recall that the peak point of a heap occupancy curve gives the minimum heap space required for the user objects. In a multiprogrammed environment, the entire shape of the curve can also be important as memory space saved at any point can be reused by some other application. When we look at the curves for CHK and DUPL, we see that at any given time DUPL occupies much more heap space than CHK. Also, its peak heap occupancy is 124% higher than that of the CHK scheme. We also observe that while object compression (COMPDUPL) brings some heap space savings, the results are not as good as someone might want. The main reason for this is the fact that most of the objects in our embedded applications are smaller than 16 bytes, which makes object compression less effective.

The heap occupancy of the selective duplication schemes is plotted in Figure 9 for manyballs. The curves marked as LAST-USE-OPT, LAST-USE-100, and LAST-USE-80 represents the selective schemes based on early termination of duplicates. LAST-USE-OPT represents an optimal version where we detect the last-uses of the objects by looking at the object traces we gleaned. In other words, in this post-processing based approach, we have a prediction accuracy of 100%, and we get rid of the useless duplicates without any delay. As a result, we do not incur any extra non-correctable errors over DUPL. In comparison, the curves marked as LAST-USE-100 and LAST-USE-80 show the results from our actual implementation. The difference between them is that, in the first one we predict, for each application, the last-use in such a way that it includes the last-uses of 100% of the total object words. In other words, we use the longest last-use time of all objects. In the LAST-USE-80 version, we want to cover the last-uses of 80% of the total object words. We see from Figure 9 that LAST-USE-OPT performs very well, and its heap occupancy is lower than that of the COMPDUPL scheme. Similarly, LAST-USE-80 also exhibits a good heap occupancy. The main reason for this is that it eliminates the duplicate objects aggressively. In contrast, LAST-USE-100 does not perform well from the heap occupancy perspective as it waits too much for removing the dupli-

cates. It should be mentioned that all these three selective versions incur an additional space overhead due to the time-stamp maintained. The line LAST-USE-100 is higher than DUPL due to such overheads. The curve marked PRFL in Figure 9 shows the heap occupancy of the allocation site based selective scheme. In this particular execution, we created duplicates for 60% of the frequently used objects. We observe from these results that the heap occupancy of this scheme is better than that of the first selective strategy in most of the cases. However, this behavior can change if we annotate the object creation points more or less aggressively We also observe from Figure 9 that the lazy duplication scheme, marked as DELAYED, generates better behavior than COMPDUPL.

B. Error Resilience Results

Having looked at the heap space occupancy, we next consider the error resilience of our duplication based schemes. The graph in Figure 10 shows the non-correctable error rate for the different schemes. We see that DUPL corrects more than 96% of the errors in five of our eight benchmarks (auction, calc, firstaid, mvideo, and pushpuzzle). It is not very successful with the image and jpeg benchmarks, mainly due to large number of errors consumed in these applications (see Table I). While the success of DUPL in correcting errors is significant when one considers the entire benchmark suite, there is still a large number of errors not corrected. One reason for this is the high error-injection rate we used (10^{-10}). When error-injection rate increases, the chance that both the original object and its duplicate being injected with errors will increase. To see how DUPL would behave under lower error rates, we also performed experiments with error rates 10^{-11} and 10^{-12}. The results given in Figure 11 indicate that the DUPL scheme is very successful in correcting errors with these rates. More specifically, with an error rate of 10^{-12}, it corrects all the errors detected by checksums. We also observe in Figure 10 that COMPDUPL performs better than DUPL. Specifically, it reduces the average non-correctable error rate from 8.2% (DUPL) to 4.3%. This is because of the reduction in the heap space allocated to the duplicates in this approach.

The non-correctable error rates for the three selective duplication schemes are shown in Figure 12. We see that the error correction behaviors of LAST-USE-OPT and LAST-USE-100 are the same as that of DUPL (given in Figure 10) since they destroy a duplicate only when they are certain that the primary object has reached its last-use. On the other hand, LAST-USE-80 incurs extra non-correctable errors over the DUPL scheme as it eliminates some duplicates while the primary copies are still alive. We also see that, as compared to DUPL, PRFL incurs significantly more non-correctable errors in some appli-

2006 Asia and South Pacific Design Automation Conference

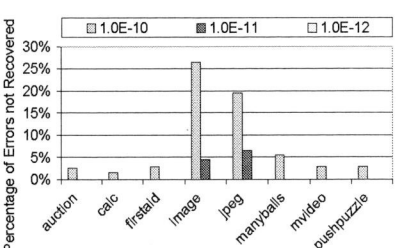

Fig. 11. Non-correctable error rates for DUPL with different error injection rates. Note that, with the last error rate, DUPL corrects all the errors.

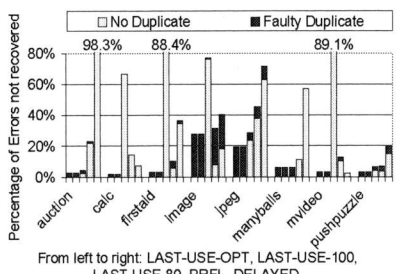

From left to right: LAST-USE-OPT, LAST-USE-100, LAST-USE-80, PRFL, DELAYED.

Fig. 12. Non-correctable error rates for the different selective schemes. Each bar is broken to show whether the non-correctable error is due to lack of the duplicate or due to faulty duplicate.

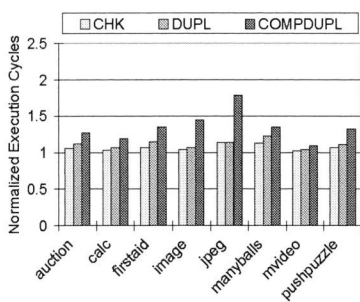

Fig. 13. Normalized execution cycles for the different schemes.

cations, since we create duplicates only for 60% of the frequently used objects. The error resilience can be improved if we increase the percentage of objects that we create duplicates. Finally, the non-correctable error rates for DELAYED is not excessive in some applications.

Overall, with the particular experimental parameters, used above, LAST-USE-OPT, LAST-USE-100, LAST-USE-80, PRFL, and DELAYED reduced the maximum heap occupancy of DUPL by 22.0%, -14.0%, 18.4%, 17.4%, 23.1%. The non-correctable error rates for these five schemes are 8.2%, 8.2%, 50.2%, 19.1%, and 41.6% in that order.

C. Performance Results

The bar-graph in Figure 13 gives the normalized execution cycles for the different schemes. Note that these results include the extra GC time due to duplication. We see that the performance penalty incurred by DUPL is not too high. Specifically, it increases the execution cycles of BASE by 11.3%, and 6.9% of this comes from the checksum overhead itself. This relatively small increase in execution cycles over CHK can be explained as follows. The only time we spend extra cycles with DUPL is when we create an object or write to an already existing object. Since both of these events are very rare as compared to object reads, DUPL does not bring an excessive performance overhead over CHK. Specifically, the number of object writes (creations) is less than 10% (1%) of the number of object reads in our applications. COMPDUPL increases the execution cycles of DUPL by around 20% across all applications. This is because of the compressions and decompressions that need to be performed during execution. While the decompression needs to be performed only when there is an error, the compression is done whenever an object creation or an object write takes place. Overall, the compression-based duplication helps reduce the non-correctable error rate and heap occupancy, but increases execution cycles of DUPL. Therefore, it is more suitable for embedded environments where performance degradation can be tolerated. Finally, we found that the performance of all the selective schemes we implemented is very close to that of DUPL since they eliminate only some writes for which they have already destroyed the duplicate.

V. CONCLUDING REMARKS

In this paper, we demonstrated how duplication can improve the data integrity of objects by recovering a significant percentage of the errors detected by a checksum-based scheme. Our baseline full duplication based scheme recovered 91.8%

of the errors at the expense of 11.6% degradation in performance. Compressing the duplicates brought up the error coverage to 95.7% and reduced average heap occupancy of the full duplication-based scheme by 18.6%. However, we found that it also increased the execution cycles significantly. We also presented results from three different implementations based on selective object duplication. Using the three selective schemes proposed in this paper, we can tradeoff different requirements of the application and help the designer to determine the best operating point considering both maximum heap space consumption and error rate.

ACKNOWLEDGEMENTS

This work was supported in part by GSRC and NSF Career Award #0093082.

REFERENCES

[1] A. Benso, S. Chiusano, P. Prinetto, and L. Tagliaferri, "A C/C++ source-to-source compiler for dependable applications," in *Proc. DSN'00*.

[2] C. Chen and A. K. Somani, "Fault containment in cache memories for TMR redundant processor systems," *IEEE Transactions on Computers*, 48(4):386–397, March 1999.

[3] D. Chen, A. Messer, P. Bernadat, G. Fu, Z. Dimitrijevic, D. Lie, D. Mannaru, A. Riska, and D. Milojicic, "JVM susceptibility to memory errors," in *Proc. JVM'01*.

[4] G. Chen, M. Kandemir, N. Vijaykrishnan, M. J. Irwin, B. Mathiske, and M. Wolczko, "Heap compression for memory-constrained Java environments," in *Proc. OOPSLA'03*.

[5] "CLDC and the K virtual machine (KVM)," http://java.sun.com/products/cldc/.

[6] B. Cmelik and D. Keppel, "Shade: a fast instruction-set simulator for execution profiling," in *Proc. ACM SIGMETRICS Conference on the Measurement and Modeling of Computer Systems*, 1994.

[7] B. Demsky and M. Rinard, "Automatic detection and repair of errors in data structures," in *Proc. OOPSLA'03*.

[8] "J2ME Mobile Information Device Profile," http://java.sun.com/j2me/.

[9] R. Jones and R. D. Lins, *Garbage Collection Algorithm for Automatic Dynamic Memory Management*, John Wiley & Sons, 1999.

[10] W. Kao, R. K. Iyer, and D. Tang, "FINE: A fault injection and monitoring environment for tracing the UNIX system behavior under faults," *IEEE Transactions on Software Engineering*, SE-19(11):1105–1118, November 1993.

[11] S. S. Mukherjee, C. Weaver, J. Emer, S. K. Reinhardt, and T. Austin, "A systematic methodology to compute the architectural vulnerability factors for a high-performance microprocessor," in *Proc. Micro'03*.

[12] C. S. Pasareanu, M. B. Dwyer, and W. Visser, "Finding feasible counter-examples when model checking Java programs," in *Proc. TACAS'01*.

[13] R. Phelan, "Addressing soft errors in ARM core-based designs," *White Paper*, ARM Limited, 2003.

[14] C. Pyyhtia, "Quality issues facing embedded memory," in *Proc. Sophia Antipolis Conference on Micro Electronics*, 2002.

[15] S. K. Reinhardt and S. S. Mukherjee, "Transient fault detection via simultaneous multithreading," in *Proc. ISCA'00*.

[16] R. Shaham, E. K. Kolodner, and S. Sagiv, "Heap profiling for space-efficient Java," in *Proc. PLDI'01*.

[17] P. P. Shirvani, N. R. Saxena, and E. J. McCluskey, "Software-implemented EDAC protection against SEUs," *IEEE Transactions on Reliablity*, 49(3):273–284, 2000.

[18] C. Weaver and T. Austin, "A fault tolerant approach to microprocessor design," in *Proc. DSN'00*.

Mapping and Configuration Methods for Multi-Use-Case Networks on Chips

Srinivasan Murali
CSL, Stanford University
Stanford, USA
smurali@stanford.edu

Martijn Coenen, Andrei Radulescu, Kees Goossens
Philips Research Laboratories
The Netherlands
{martijn.coenen,andrei.radulescu,kees.goossens}@philips.com

Giovanni De Micheli
LSI, EPFL
Switzerland
giovanni.demicheli@epfl.ch

ABSTRACT

To provide a scalable communication infrastructure for *Systems on Chips* (SoCs), *Networks on Chips* (NoCs), a communication centric design paradigm is needed. To be cost effective, SoCs are often programmable and integrate several different applications or use-cases on to the same chip. For the SoC platform to support the different use-cases, the NoC architecture should satisfy the performance constraints of each individual use-case. In this work we motivate the need to consider multiple use-cases during the NoC design process. We present a method to efficiently map the applications on to the NoC architecture, satisfying the design constraints of each individual use-case. We also present novel ways to dynamically reconfigure the network across the different use-cases and explore the possibility of integrating *Dynamic Voltage and Frequency Scaling (DVS/DFS)* techniques with the use-case centric NoC design methodology. We validate the performance of the design methodology on several SoC applications. The dynamic reconfiguration of the NoC integrated with DVS/DFS schemes results in large power savings for the resulting NoC systems.

Keywords: Systems on Chips, Networks on chips, Use-Cases, Multiple application platforms, Dynamic, Reconfiguration, Voltage Scaling, Frequency Scaling, Guaranteed Throughput, Best Effort.

I. INTRODUCTION

As the number of transistors on a chip increases with every technological generation, the number of processor, memory and hardware cores available on the chip also increases. Thus, functionalities that were carried out by several different chips are being integrated on to a single chip, forming a *Systems on Chip (SoC)*. This, coupled together with the increase in the operating speed of the transistors has created the availability of large computational power for such systems. The challenges facing the SoC designer are to efficiently tap the available computational power under tight power budgets and meet the tight time-to-market constraints.

As the computational loads on the SoC increases, so does the load on the communication architecture. To support the high communication needs of multi-application SoC platforms, scalable on-chip interconnection networks are needed. A communication centric design paradigm, *Networks on Chips (NoCs)*, has been presented to address the interconnect issues

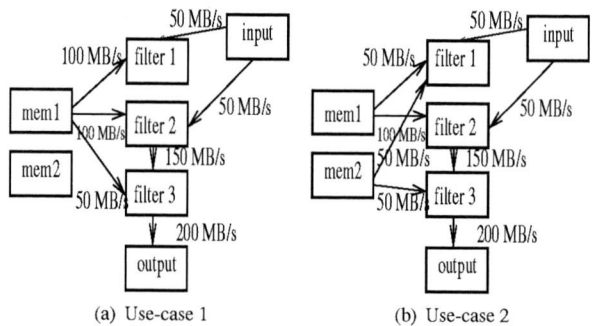

Fig. 1. A fragment of communication for two different use-cases of a set-top box SoC

of SoCs [2]-[6]. NoCs provide a scalable communication infrastructure with structured and modular wiring between the components. NoCs also help meet the tight time-to-market constraints, as the scalable architecture can be re-used across multiple platforms.

To be cost effective, SoCs are often programmable and integrate several different applications or use-cases on to the same chip. As an example, a SoC for a set-top box has multiple resolution video processing capabilities (like high definition, standard definition), multiple picture modes (like split-screen, picture-in-picture), video recording features, high speed internet access and file transfer services, etc. [9]. Such convergence of multiple use-cases on to the same device is being observed in other electronic devices as well, such as the cell-phone or the personal digital assistant.

The different use-cases run on the SoC, although share many of the hardware components, could have very different performance requirements and design constraints for the communication architecture. As an example, we consider a simplified version of a SoC used in television set-top boxes [9], with support for four different use-cases. The communication bandwidth requirements for some of the connections between the components of the SoC for two of the use-cases are shown in Figure 1. Although we want the NoC to support all the use-cases, a NoC that is designed to run exactly one use-case does not necessarily meet the design constraints of the other user cases. In many of the existing NoC design methods, the NoC is designed and optimized for a single use-case or for a single application-trace of the design [10]-[15]. Such a trace based approach captures the characteristics and constraints of a single use-case very well, but fails to capture the multiple use-case scenario. Such a method averages out the communication effects across all the use-cases, which may result in a design that is unacceptable for many use-cases. As an example, when such a method is applied to perform NoC mapping for the set-top box SoC, the resulting NoC violates the design constraints of all the four use-cases.

Today's high-end SoCs support several hundred use-cases and manually checking whether the design constraints of the individual use-cases are satisfied by the NoC is a tedious process. Moreover, if the NoC design for the use-cases is carried out individually, it is difficult to converge to a single NoC design that satisfies the design constraints of all the use cases. In this work we motivate the need to consider the design constraints of the individual use-cases during the NoC design process. We present a design method for mapping of cores on to the NoC, considering the NoC configuration (i.e. path selection and resource reservation in the NoC) as sub-problems during the mapping phase, such that the resulting design satisfies the constraints of all the use-cases of the SoC. We then present methods to decrease the required network resources by dynamically reconfiguring the network across different use-cases. We also explore the effect of DVS/DFS techniques for reducing the power consumption of the network across the different use-cases. The methods are validated by performing experiments on several SoC designs.

II. PREVIOUS WORK

Several researchers have proposed different architectures and design methodologies for the switches, links and Network Interfaces (NI), which are the major components of a NoC [18, 7, 20, 8]. Design flows that automate many of the steps of the design process have been presented in [17, 19]. In [8], the Æthereal architecture that supports Quality-of-Service (QoS) for applications by using Guaranteed Throughput (GT) connections for traffic streams that has bandwidth/latency constraints and by using Best Effort (BE) connections for the remaining traffic streams is presented.

The topology selection process and mapping of applications on to NoC architectures have been explored by many researchers. In [10, 11], branch-and-bound algorithms to map cores on to a mesh NoC topology for different routing functions are presented. In [12, 13], design methods and tools for mapping applications on to regular NoC topologies and automating the topology selection process has been presented. In [16], the methods are extended to consider the QoS constraints during the mapping phase. Building application specific buses and NoC topologies has been presented in [21, 14]. In [15], a tool that automates the combined mapping and NoC configuration steps for the Æthereal is presented. In all these NoC design works, the design methods assume a single set of communication constraints, which is obtained either for a single application or is obtained from a single trace for multiple applications.

In the RAW chip-multiprocessor, the interconnection network connectivity is reconfigured with the assistance of the compiler [23]. In the FLEXBUS architecture [22], the authors present methods to dynamically remove the overhead of bridges in multi-bus communication and provide methods

Multiple Use cases (design constraints,communication patterns)

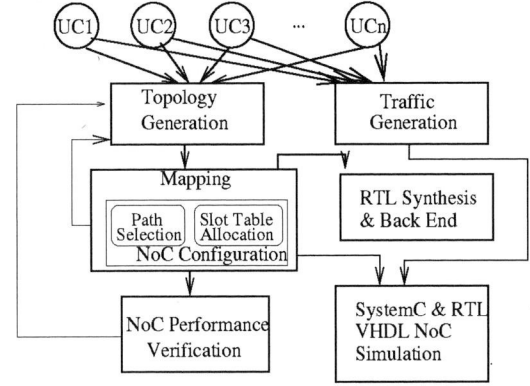

Fig. 2. Design Flow for NoCs

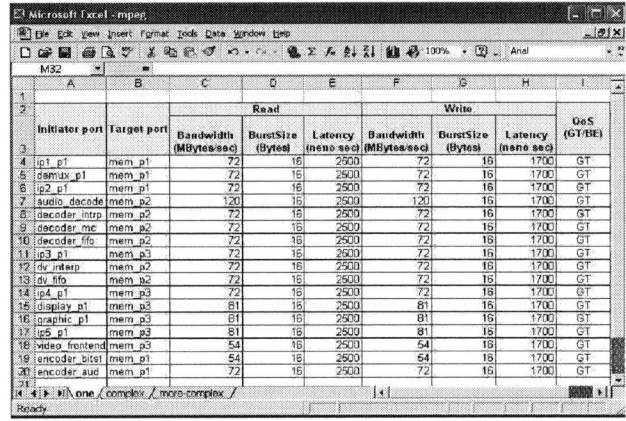

Fig. 3. Example input file with design constraints for an MPEG application

where a core can be connected to different buses dynamically.

III. THE USE-CASE CENTRIC DESIGN FLOW

In this section we present the NoC design flow with the support for multiple use-cases integrated in to the flow (Figure 2). The NoC design flow and the mapping algorithms for the NoC for a single use-case were presented in [15, 17]. In this work, we extend the tool chain to support the multiple use-case scenario that is commonly encountered in SoCs. The communication design constraints for the different use-cases of the SoC are input to the design flow in the excel and xml file formats. The communication design constraints for each use-case includes the required bandwidth for various connections between the cores in the use-case, the maximum latency allowed for the connection, the QoS level required for the connection (like GT or BE), etc. An example fragment of the input file is presented in Figure 3.

With the different use-cases as input, in the first two phases of the design flow, the topology exploration and mapping of the use-cases on to the NoC are performed. The NoC configuration phase in which path selection and TDMA slot-table allocation (required for the GT traffic) are performed, is integrated with the mapping phase. The RTL level VHDL and SystemC models for the resulting NoC configuration are then automatically generated, which can then be simulated. The performance of the NoC can also be verified in parallel by the automatic performance verifier, which analytically checks whether the design constraints are met. The extension of the tool chain to support multiple use-cases is performed in a modular fashion without affecting most of the existing flow. As the multi-use-case NoC design methods are integrated with the tool chain, performance validations of the methods can be easily carried out to analyze the efficiency of the design methods.

IV. THE MAPPING ALGORITHM

In this section, we first present the mapping algorithm for a single use-case and then present methods to extend the algorithm for multiple use-cases.

A. Mapping Algorithm for single use-case

The mapping algorithm for a single use-case is presented in detail in [15]. In this sub-section, we only present a brief version of the algorithm highlighting the major phases. As in general, graph mapping is a NP-Hard problem [10, 13], a heuristic algorithm is used to perform the mapping. The selection of paths for the different traffic flows and the reservation of TDMA slot-table entries for the GT traffic flows are unified with the mapping process. The mapping algorithm is presented in Algorithm 1. At the outermost level of the algorithm, a NoC topology is generated. In the outer loop, the size of the topology is increased until a feasible mapping is obtained in the subsequent phases. Initially, all the cores of the SoC are unmapped. In the first step of the mapping algorithm, the traffic flows between the communicating cores are sorted in a non-increasing order. Then for each flow in the order, the source and destination cores of the flow, if they are not already mapped, are mapped on to the NoC. When performing the mapping of these cores, the path with the least cost that satisfies the bandwidth and latency constraints for the flow is chosen and the cores are mapped to the NIs in the path. A path is assumed to originate from a NI, traverse one or more switches and terminate in a NI. The cost of the path is a combined metric that considers an affine combination of the latency and bandwidth requirements for the flow. The slot-table reservation for the flow is also carried out in this step. The procedure is repeated for all the flows in the SoC. The approach also takes in to account the possibility of multiple cores sharing a single NI for communication. Note that once the initial mapping step is performed, the solution space can be explored by considering swapping of vertices using simulated annealing or tabu search, as performed in [16].

Algorithm 1 Mapping Algorithm for a single use-case

OUTERLOOP: Generate a NoC topology.
  1. Sort the traffic flows between the cores in a non-increasing order of the bandwidth requirements.
  2. For each flow in order:
    a. Choose a least-cost path for the flow that satisfies the bandwidth, latency constraints.
    b. If the source or destination cores of the flow are not yet mapped, map them on to the NIs in the path.
    c. Reserve the required bandwidth across the ports and reserve the slot-table entries for the flow.

If the resulting mapping violates design constraints, increase the size/resources of the topology and go to OUTERLOOP.

We refer the interested reader to [15] for the time complexity of the algorithm, details of path selection, other optimizations carried out and for the performance evaluation of the algorithm for several SoC designs.

B. Mapping Design Approach for Multiple Use-cases

When the SoC has multiple use-cases, we assume that all of the use-cases utilize the same mapping of cores on to the NoC components. This is because, if each individual use-case has a different mapping, then each core potentially needs to be connected to several different NIs, which may not be feasible because of physical layout restrictions and wiring complexity.

A direct extension of the single use-case mapping algorithm to support multiple use-cases would be to perform the mapping for the most communication intensive use-case and reuse the mapping for the other use-cases. However, as the design constraints of the use-cases can be very different, such a method may result in a mapping that does not satisfy the performance constraints of many of the use-cases. As an example, when such an approach is applied to the SoC considered in section I, the resulting NoC design satisfies only 2 of the 4 use-cases.

We use the following design method to extend the mapping design procedure for multiple use-cases (Figure 4(a)). In order to obtain a mapping that satisfies all the use-cases, we construct a synthetic Worst-Case (WC) use-case from the given set of input use-cases. For the communication flow between every pair of cores, the maximum required bandwidth values and the minimum required latency values for the flow across all the use-cases are selected and used in the WC use-case. A small example is presented in Figure 4(b). Thus the design constraints of all the individual use-cases are subsumed in the WC use-case and any NoC design that satisfies the constraints in the WC use-case will satisfy the constraints of each individual use-case. The WC use-case is then used for the mapping process. Due to the manner in which the WC use-case is constructed, the selected paths and slot-table allocations from the mapping process will satisfy the design constraints of each individual use-case.

Once the mapping is obtained from the WC use-case, we perform an optimization step, where we fix the mapping that is obtained from the WC use-case, but re-run the path selection and slot-table reservation phases individually for each use-case. We perform this for two reasons. First, the WC use-case had the worst case constraints from each use-case and by re-running the path selection and slot-table reservation steps and choosing the maximum values from the individual use-cases we can reduce the NoC resources, while still satisfying the constraints of all the use-cases (experimental evidence presented in Section VI A). Second, when the configuration time between the use-cases is large, the frequency and voltage of operation of the NoC can be scaled to match each individual use-case, which can result in significant power savings for the system.

In general, when the paths and slot-tables used by the different use-cases are different, we need mechanisms to store them in memory and load them on to the network dynamically or compute them on the fly for the use-case. This is explored in the next section.

V. DYNAMIC RECONFIGURATION OF THE NOC

For most SoC designs, when the system switches between use-cases, some configuration time is needed for loading the

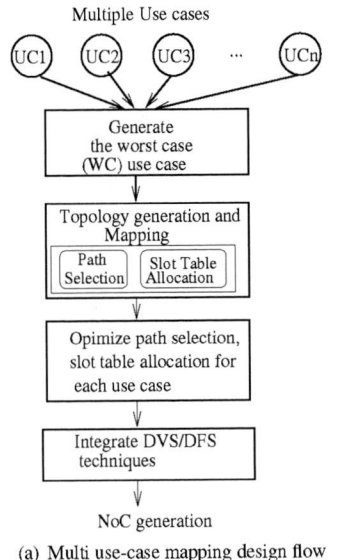

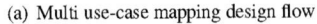

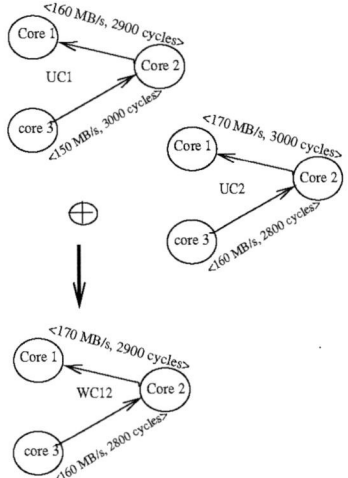

(a) Multi use-case mapping design flow

(b) Example of WC use-case constraints generation

Fig. 4. Multi use-case mapping flow and WC use-case generation

new use-case. This is mostly attributed for loading the use-case data and code, sending control signals to different parts of the design and for gracefully shutting down the already running use-case. In many designs, this use-case switching time is of the order of few milli-seconds. This configuration time can be utilized by the NoC for switching to a different path and slot-table allocation for the mapping. This time delay can also be utilized to vary the clock frequency/voltage of the NoC to match the use-case performance level.

In the *Æthereal* architecture [8], a static path routing scheme is used, where the paths are selected at the source NI of the traffic flow. Thus, the NIs maintain the path and slot-tables for the various connections. When the paths and slot-tables used by the NIs vary across different use-cases, the tables need to be stored in memory. As the on-chip memory available is mostly limited and as the use-case switching time is large, we use the off-chip memory to store the paths for the different use-cases.

We investigated the overhead for the reconfiguration mechanism for the set-top box SoC. The amount of data required to store the path and slot-table information for each use-case is around 560 Bytes. With 4 use-cases, the memory requirement for the reconfiguration mechanism is 2.24 KB. The time required to load the data from the memory and spread it around the NoC for an use-case is of the order of micro-seconds and the energy dissipation is of the order of micro-Joules. Using traditional mechanisms to scale the frequency and voltage of the system may require few milliseconds for configuration. Thus we can envision three different ways of NoC operation. First, when the use-cases that run on the SoC switch very frequently or when the initial configuration times are not acceptable (as in real-time use-cases), the different use-cases can use the WC use-case configuration. In this configuration, all the use-cases will use the same set of paths and slot-table allocations, thereby not requiring the NoC to be re-configured when the use-cases switch, resulting in seamless switching between

the use-cases. However, this leads to an over-design of the network when compared to the scenario where the NoC is reconfigured to suit the individual use-cases. Second, when the use-case switching is not that frequent, the NoC configuration (path and slot-tables) can be changed dynamically across use-cases, leading to a smaller NoC design (in terms of network components or frequency of operation). Third, when the use-cases are expected to run for a long time, the voltage or frequency of operation of the NoC can be varied to match the use-cases, resulting in large power savings for the system. The simulation results for these cases are presented in the next section.

VI. SIMULATION RESULTS

We present simulation results on applying the multi-use-case design procedure on to 4 different SoC designs: P1 (with 2 use-cases), P2 (2 use-cases), P3 (4 use-cases) and P4 (8 use-cases). The designs P1-P3 are simplified versions of set-top box SoCs [9] and the design P4 is a video processing SoC used in TVs. Each use-case has a large number of (50 to 150) communicating pairs of components. A fragment of two of the use-cases used in the P3 design was presented earlier in Figure 1. The set-top box SoCs and the TV processor have different functionalities and communication patterns. The designs P1-P3 use an external memory for storing and retrieving data and the amount of data communicated to the memory is very large when compared to the rest of the design. The P4 design uses a streaming architecture with local memories on the chip, there by distributing the communication load across several components. We apply our design method to these SoCs with different architectures to validate the generality of the methods.

A. Effect of Mapping on the NoC Frequency

To evaluate the mapped designs, we fixed the topology and the maximum slot-table size for each design and we found the

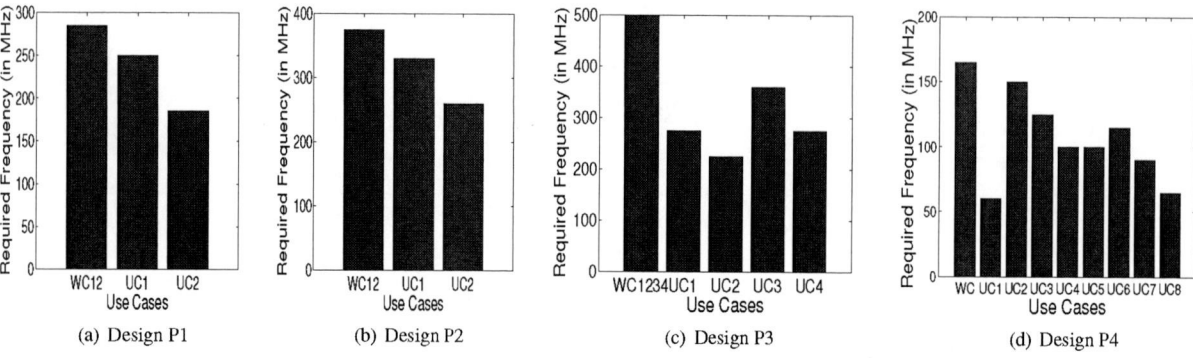

(a) Design P1 (b) Design P2 (c) Design P3 (d) Design P4

Fig. 5. The NoC operating frequencies required to support the different use-cases for the various designs. The WC use-case values are obtained when the path selection and the slot-table reservation are based on the WC use-case. The other values show the effect of re-applying the path selection and slot-table reservation for each use-case with the mapping obtained from the WC use-case.

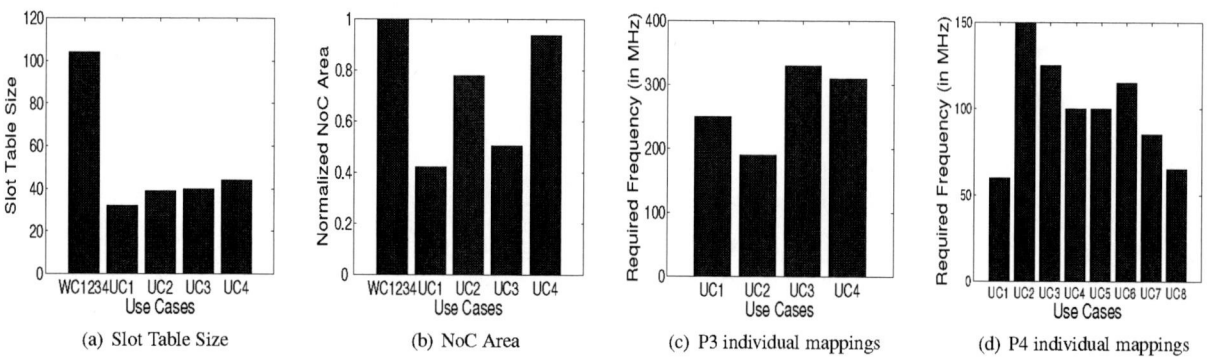

(a) Slot Table Size (b) NoC Area (c) P3 individual mappings (d) P4 individual mappings

Fig. 6. (a)-(b) The effect of the mapping design procedure on the slot-table size and NoC area, (c)-(d) NoC frequency requirements for individual mapping of use-cases in designs P3 and P4.

minimum frequency of operation required by the NoC to support the different use-cases. The results of the mapping procedure for the 4 SoC designs are presented in Figure 5.

The frequency of operation required for the WC use-case is obtained from applying the NoC configuration (i.e. the path selection and the slot-table reservation) procedure on the WC use-case. The frequency of operation of the NoC after re-applying the configuration procedure for each of the use-cases, fixing the mapping from the WC use-case is also presented in the figures. When DVS/DFS techniques are not used and a single frequency of operation is used for all the use-cases, we need to take the maximum of the frequencies of each of the individual use-cases as the operating frequency of each design. In this case, re-applying the NoC configuration step results in 9% to 38% reduction in the required NoC operating frequency across the different designs. A lower operating frequency implies lower power consumption and smaller impact of noise sources.

In the above analysis, we assumed that the slot-table size is fixed and the NoC frequency is varied to support the use-cases. We also explored the effect of fixing the NoC frequency (at 500 MHz) and varying the slot-table size. As similar results were observed for all the designs, we only present the results

for the P3 design (Figure 6(a)). We obtain 58% reduction in the slot-table size for the design by re-applying the NoC configuration step. A smaller slot-table size usually corresponds to a lower area for the NoC and lower packet latencies (as the traffic streams wait lesser to get the slots). The NoC area reduction due to the slot-table reduction (Figure 6(b)) is 10% for this design (the NoC area includes the area of the switches and the network interfaces). Trade-offs involving the frequency savings and area savings can also be explored.

B. Comparisons with the individual use-case mappings

To evaluate the optimality of the NoC design produced by the above method, we performed individual mappings for each of the use-cases in the P3 and P4 designs. The required NoC frequencies for the use-cases in the resulting designs are presented in Figures 6(c) and 6(d). These frequency values are the minimum possible values for the use-cases, as we have done the mappings individually and they provide a lower bound on the quality of solutions that can be obtained when all the use-cases share the same mapping[1]. When the results from Figures 6(c) and 6(d) are compared with the results from Figures 5(c)

---

[1]Note that the heuristic nature of the mapping algorithm can sometimes invalidate this general statement.

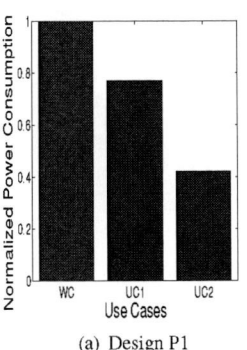 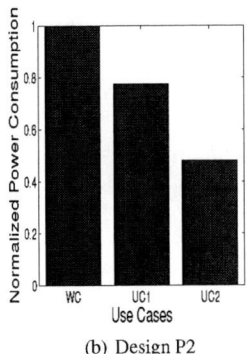 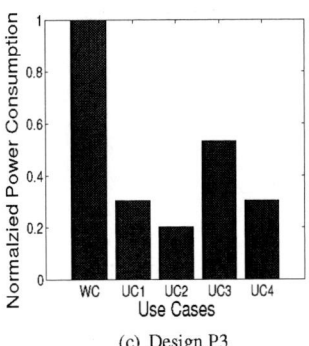 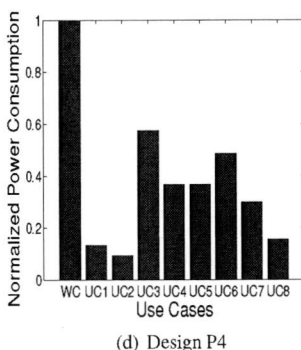

(a) Design P1   (b) Design P2   (c) Design P3   (d) Design P4

Fig. 7. Effects of DVS/DFS.

and 5(d), we find that the multi-use-case mapping design procedure results in mappings that require operating frequencies that are with in 10% of the minimum possible operating frequencies. We also performed experiments fixing the frequency of operation for the multi-use-case mappings to be the same as the individually mapped designs and varied the network resources needed to support all the use-cases in the designs. The multi-use-case mappings required slightly more resources (1%-10% increase in the NoC area) to support the same frequency of operation as the individual mappings.

C. Effect of DVS/DFS

When the frequency of the NoC is scaled to match the frequencies required for the individual use-cases, large power savings can be achieved. As the frequency of the network is scaled, the supply voltage required for operation can also be scaled to match the frequency. We use a conservative model for voltage scaling, where we assume that the square of the voltage scales linearly with the frequency [24]. The power savings achieved by the DVS/DFS techniques for the entire SoC platform depends on the amount of time each use-case is expected to run. Thus, in this experiment, we present the power savings achieved for each use-case of the platform separately. The power consumption of each of the use-cases, normalized with respect to the power consumption of the WC use-case is presented in Figure 7. On average, we obtain 59.21% power savings by using the DVS/DFS techniques across the different use-cases for the designs.

VII. CONCLUSIONS

As the number of applications or use-cases integrated on to a single SoC increases, the designer is faced with the challenge of building an interconnect structure that supports the design constraints of all the use-cases. In this paper we motivated the importance of the problem and presented use-case centric design methods to map applications on to NoC architectures. We also presented a way to dynamically configure the interconnect to support multiple use-cases and integrated Dynamic Voltage and Frequency (DVS/DFS) techniques with the reconfiguration mechanism. In future, we plan to extend the algorithms for supporting concurrent operation of use-cases and apply the use-case models for addressing other NoC design issues such as the application specific topology design.

REFERENCES

[1] R. Ho et al., "The Future of Wires", Proceedings of the IEEE, April 2001, pages 490-504.

[2] L. Benini and G.De Micheli, "Networks on Chips: A New SoC Paradigm", IEEE Computers, pp. 70-78, Jan. 2002.

[3] W. J. Dally, B. Towles, "Route packets, not wires: on-chip interconnection networks", Proc. DAC 2001.

[4] D.Wingard,"MicroNetwork-Based Integration for SoCs", Design Automation Conference DAC 2001, pp. 673-677, Jun 2001.

[5] F.Karim et al., "On-chip communication architecture for OC-768 network processors", Design Automation Conference, pp. 678-678, June 2001.

[6] M. Sgroi et al. , "Addressing the System-on-a-Chip Interconnect Woes Through Communication-Based Design", Proc. DAC 2001.

[7] P.Guerrier, A.Greiner,"A generic architecture for on-chip packet switched interconnections", DATE 2000, pp. 250-256, March 2000.

[8] E. Rijpkema et al., "Trade offs in the design of a router with both guaranteed and best-effort services for networks on chip", DATE 2003.

[9] S. Dutta et al., "Viper: A multiprocessor SOC for advanced set-top box and digital TV systems", IEEE Design and Test of Computers, pages 21-31, Sept-Oct 2001.

[10] J. Hu, R. Marculescu, "Energy-Aware Mapping for Tile-based NOC Architectures Under Performance Constraints", Proc. ASP-DAC 2003.

[11] J. Hu, R. Marculescu, "Exploiting the Routing Flexibility for Energy/Performance Aware Mapping of Regular NoC Architectures", Proc. DATE 2003.

[12] S. Murali, G. De Micheli, "Bandwidth-Constrained Mapping of Cores onto NoC Architectures", Proc. DATE 2004.

[13] S. Murali, G. De Micheli, "SUNMAP: a tool for automatic topology selection and generation for NoCs", Proc. DAC 2004.

[14] A. Pinto et al., "Efficient Synthesis of Networks On-Chip", Proc. ICCD, 2003.

[15] A. Hansson et al., "A unified approach to constrained mapping and routing on network-on-chip architectures", pp. 75-80, Proc. ISSS 2005.

[16] S. Murali et al., "Mapping and Physical Planning of Networks-on-Chip with Quality-of-Service Guarantees", Proc. ASPDAC 2005.

[17] K. Goossens et al., "A Design Flow for Application-Specific Networks on Chip with Guaranteed Performance to Accelerate SOC Design and Verification", pp. 1182-1187, DATE 2005.

[18] S.Kumar et al., "A network on chip architecture and design methodology", ISVLSI 2002, pp.105-112, Apr 2002.

[19] D. Bertozzi et al., "NoC Synthesis Flow for Customized Domain Specific Multi-Processor Systems-on-Chip", IEEE Transactions on Parallel and Distributed Systems, Feb 2005.

[20] M. Dall'Osso et. al, "xpipes: a Latency Insensitive Parameterized Network-on-chip Architecture For Multi-Processor SoCs", pp. 536-539, ICCD 2003.

[21] A. Pinto et al., "Constraint-Driven Communication Synthesis", Proc. DAC 2002.

[22] K. Sekar et al., "FLEXBUS: a high-performance system-on-chip communication architecture with a dynamically configurable topology", Proc. DAC 2005:

[23] M. B. Taylor et al., "Scalar Operand Networks: On-chip Interconnect for ILP in Partitioned Architectures", HPCA 2003.

[24] J. Rabaey et al., "Digital Integrated Circuits", Prentice Hall, 2002.

Conversion of Reference C Code to Dataflow Model:
H.264 Encoder Case Study

Hyeyoung Hwang   Taewook Oh   Hyunuk Jung*   Soonhoi Ha

The School of Electrical Engineering And Computer Science
Seoul National University KOREA
* Samsung Electronics
{hyhwang, twoh, jung, sha}@iris.snu.ac.kr

Abstract – Model-based design is widely accepted in developing complex embedded system under intense time-to-market pressure. While it promises improved design productivity, the main bottleneck lies not in the design methodology but in constructing the initial algorithm representation in the specified model. It is particularly true if a complicated multimedia application is given in the form of a sequential reference C code. In this paper we propose a systematic procedure for converting a sequential C code to a dataflow specification that has been widely used in many design environments for DSP systems. The proposed technique is successfully applied to H.264 encoder algorithm as a case study.

I. Introduction

In HW/SW co-design methodology of embedded systems, system level specification enables us to model and analyze the system behavior in high level of abstraction to cope with the ever-increasing complexity of system design under relentless time-to-market pressure. Especially model-based specification is widely accepted because mapping the system behavior to processing components can be easily performed. In a model-based approach, the system algorithm is specified using a block diagram or a composition of function blocks. Once functional blocks are built and completely tested, they can be reused in many other systems thereby saving time and costs compared to traditional design approaches [1].

In this paper we use an extended synchronous dataflow (SDF[2]) model for algorithm specification. In SDF model, a node, or a block, represents a coarse grain function block whose body is described in C language. An arc represents a channel that carries streams of data samples from the source node to the destination node. The number of samples produced (or consumed) per block execution is called the output (or the input) sample rate of the block. In case the number of samples consumed or produced on each arc is statically determined and can be any integer, the graph is called a synchronous dataflow graph (SDF). A block is executable only after it receives the specified number of samples at all input ports. These restrictions make the model formal and data-driven. The SDF model has been widely used in many system-level design environments especially for digital signal processing systems [2][3].

The inherent difficulty of model-based approach lies in constructing the initial algorithm representation in the specified model, dataflow model in this paper. In most cases the algorithm description is given in the form of a reference C code. To get the benefits of model-based design approach, we have to convert the sequential reference code somehow to a dataflow specification. It is not a simple task for a system designer who may not know the algorithm details in particular. For instance, the reference C code of H.264 encoder algorithm [6] is about 32000 lines long and is scattered into 55 files.

In this paper, we propose a systematic procedure for converting a sequential C code to a dataflow specification. The procedure has been established after numerous hands-on experiences and successfully applied to H.264 encoder algorithm. It consists of three phases. First, we transform the reference code into the same code structure as would be automatically generated from a dataflow specification. In the second phase we identify the functional blocks from the transformed code and analyze their dependencies. At last we draw a dataflow graph, synthesize the sequential C code from the dataflow specification, and check the correctness and the performance of the code by comparing it with the reference code.

The rest of the paper is organized as follows. In Section 2, some background information is reviewed on the dataflow specification and the H.264 encoder algorithm. Section 3 presents the problem definition and the overview of the proposed solution. The detailed description of the procedure is explained in Section 4 with the H.264 encoder case study. Clustering and scheduling of the data flow specification are explained in section 5. The experimental results are presented in Section 6. Section 7 concludes the paper.

II. Background

A. Software Synthesis from Dataflow Model

Fig. 1 depicts the process of software code synthesis from an SDF specification [3]. A simple SDF graph in Fig. 1(a) contains three nodes, labeled A, B, and C. Each arc is annotated with the number of samples consumed or produced per node execution.

To generate a code from the given SDF graph, the order of block executions is determined at compile time, which is called "scheduling". Since a dataflow graph specifies only partial orders between blocks, there are usually more than one valid schedule. Fig. 1(b) shows a valid schedule. The parenthesized terms in schedule are used to express repetitive invocation patterns. These are called schedule

0-7803-9451-8/06/$20.00 ©2006 IEEE.

loops. In Fig. 1(b), the term 2(B(2C)) represents the invocation sequence BCCBCC. Each schedule loop is implemented as a loop structure in the synthesized code.

A code template according to the schedule is shown in Fig. 1(c). The function body of each function block is placed in the scheduled position. The code structure of the synthesized code from dataflow specification has the following characteristics.
- It may have nested loops.
- All function blocks appear in the main loop. A function block may not be called in another function block.
- State variables of function blocks may not be shared.

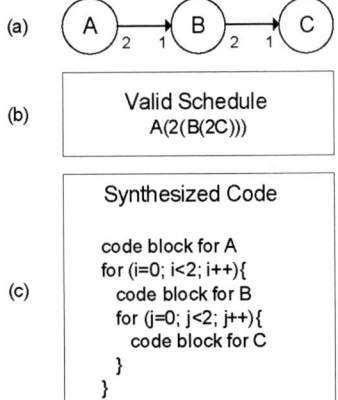

Fig. 1. An example of synthesized code from SDF

Fractional Rate Dataflow (FRDF) is an extension to the dataflow model in which fractional number of samples can be produced and consumed [4]. In the FRDF model, a constituent data type is considered as a fraction of the composite data type. For instance, a 16x16 macro-block is considered as 1/99 of a QCIF frame (176x144). Existing integer rate dataflow models can be easily extended to incorporate the fractional rates without loosing analytical properties. But it is reported that the FRDF model can reduce the buffer memory requirement in the synthesized code significantly, up to 70%, for some multimedia application.

B. H.264 Encoder Algorithm

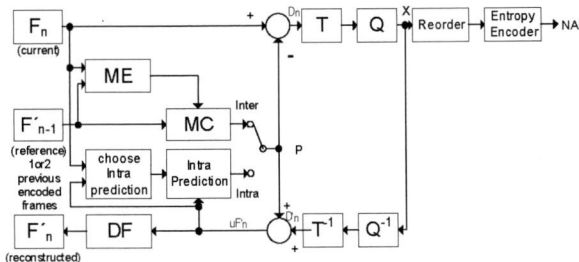

Fig. 2. H.264 encoder (baseline profile) block diagram

H.264 is a video coding standard (or Recommendation) made by the Video Coding Experts Group (VCEG) of the International Telecommunication Union Telecommunication Standardization Sector (ITU-T)[6]. While H.264 achieves bit rate saving ratio up to 50% compared to H.263+ and offers consistently good video quality at most bit rates, algorithm complexity grows significantly. More efficient compression and high quality video are attributed to the following features: Enhanced motion compensation, small blocks (4x4) for transform coding, improved in-loop deblocking filter, and enhanced entropy coding. The standard specification defines many options hence the reference code is very complex. In this work, we consider the baseline profile with a single slice mode for simplicity[7].

III. Problem Definition

Fig. 3 shows the skeleton of the H.264 reference code.

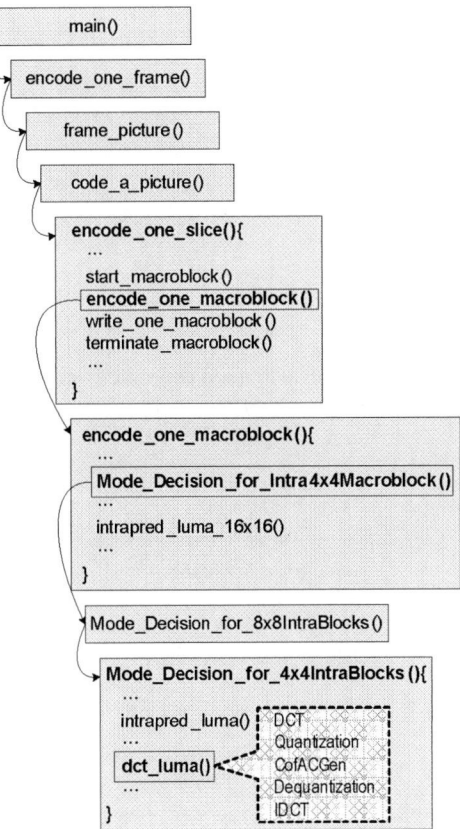

Fig. 3. Skeleton of H.264 reference code

As apparent from the Fig. 3 shown, the function call depth of the reference code is very deep and the key functions reside close to the bottom in the call graph. They should be elevated up to the top if they are to be defined as function blocks in the dataflow specification. While not shown in the Fig. 3, numerous global variables are defined and accessed in various places. Therefore, it is not easy to grasp the data flow dependency between functions because data is coupled very tightly.

On the contrary, Fig. 4 illustrates the code structure of the synthesized code. So in the first phase of the proposed conversion procedure we transform the reference code to this code structure. The proposed code transformation consists of three steps that are applied repeatedly until no more transformation is needed. They are "Function restructuring", "Variable classification", and "Data sample rate decision" as shown in Fig. 5.

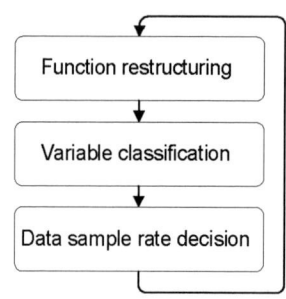

Fig. 4. Transformed code skeleton

Fig. 5. Code transformation overview

In the Function-restructuring step, basic function blocks are identified and moved up to the top level in the call graph. In the Variable-classification step, the scope of variables are classified and analyzed to remove the global variables as much as possible. After this step, the redundant dependency between function blocks are removed so that the total ordering of function blocks is converted to partial ordering between function blocks. In the Data-sample-rate-decision step, we determine the sample rates of all function blocks. We use fractional rates as much as possible to reduce the

memory requirements in the synthesized code. The process of transforming the reference code is repeated until the code structure of Fig. 4 is obtained.

IV. Code Transformation Techniques

In this section, we describe in detail the code transformation steps overviewed in the previous section. We also explain how the proposed technique is applied to the H.264 encoder example.

A. Function Restructuring

In this step we define 4 kinds of transformation. They are flattening, splitting, merging, and duplication, as demonstrated in Fig. 6.

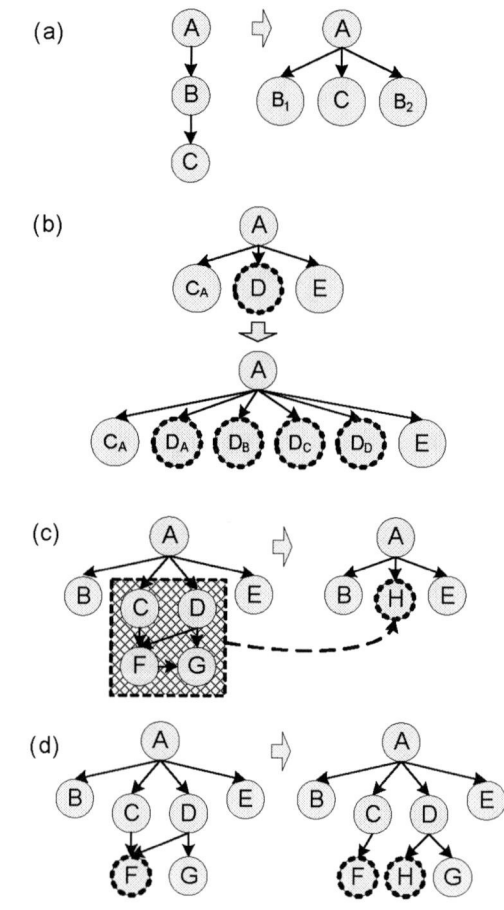

Fig. 6. Function restructuring transformations: (a) flattening, (b) splitting, (c) merging, (d) duplication.

Fig. 6(a) shows that flattening technique can be used to move a function up to one level higher. The caller function may need to be split into two parts that is called before and after the callee. In Fig. 3, the two functions *intrapred_luma()* and *dct_luma()* are called from inside *Mode_decision_for_4x4IntraBlocks()*. These two functions were first moved to same level as *intrapred_luma_16x16* as

154

shown in Fig. 4. Note that the function names of Fig. 4 are renamed to have a prefix, "intra4", for better readability.

Fig. 6(b) shows the case where a single function is split into multiple functions. The 5 different functionalities are included in function *dct_luma()*. The functionalities are dct, quantization, cofACGen, dequantization, and idct. So, as can be seen in Fig. 4, the *dct_luma()* is split and replaced with five functions.

Fig. 6(c) shows the opposite case where multiple functions are merged into a single function. When the call dependency is tightly coupled or data dependency between functions is tightly coupled, it is impossible to divide the functions into separate function blocks without re-programming. In this case, the tightly-coupled functions are merged into a single function to preserve the functionality of the code. In the H.264 encoder example, the functions for inter-prediction are tightly coupled so that we merge them into a single function block, called *inter()* in Fig. 4

Lastly, Fig. 6(d) shows the case where functions are duplicated. If the function F uses no shared variable or static variable inside, the function it can be duplicated without side-effect. Otherwise, two caller functions C and D should be merged by the merging transformation. When a function is duplicated, it should be renamed to avoid naming conflict. While function duplication has a drawback of increased code size, it increases the modularity and reusability of the function block.

B. Variable Classification

Variable classification is the most critical step to isolate the function blocks that are communicated with other function blocks only via port variables in dataflow specification. In the reference code, functions are tightly coupled with shared variables, so they are closely inter-dependent. The purpose of this step is to identify the true dependency between the functions by classifying the global and static variables.

> **Variable**
> ➢ Non-constant variable
>   ▪ Data-path
>   ▪ Block state
>   ▪ Block local
> ➢ Constant variable
>   ▪ Block parameters

Fig. 7. Variable classification

As shown in Fig. 7, variables are classified into non-constant variables and constant variables. Non-constant variables, updated at run-time, create the dependency between functions. The non-constant variables are further classified into *data-path, block state*, and *block local* variables. They can be classified using the life time chart as illustrated in Fig. 8. The life time of a variable is defined as a set of durations what starts with a write operation and ends with the last read operation. Integer variable, *int a*, in Fig. 8

has multiple life durations since it is reused several times.

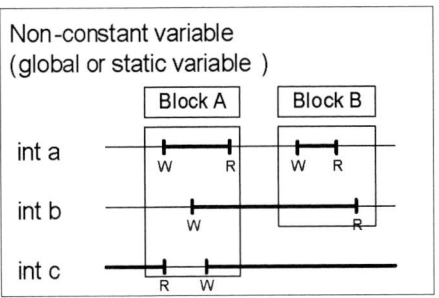

Fig. 8. Life time chart of non-constant variables

A data-path variable is a variable that is used as an interface between function blocks, especially at the top level, *i.e.* inside the main function. Integer variable *int b* in Fig. 8 is an example of a data-path variable. The life time of variable *b* starts at block A and ends at block B. A data path variable is translated to a port variable in the dataflow model.

Integer variable *int c* in Fig. 8 is not only read but also written during the execution of block A. The written value affects the next execution of block A. Therefore the variable *c* is classified as a block state variable and translated into a block state in the dataflow model.

Integer variable *int a* is a global variable that can be accessed from both block A and block B. But the value of *int a* inside one function does not affect the outcome of the other function. Hence, variable *a* can be classified as block local.

The constant variables are the variable whose value is not changed inside the main loop. They are translated into block parameters or global parameters without the risk of side-effect.

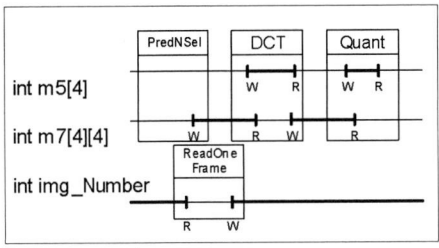

Fig. 9. Examples of variable classification

Fig. 9 shows the life time of some variables used in function *intra4x4()*. Variable *m5[4]*, which is shared between DCT and Quant functions, is classified to a block local variable. Variable *m7[4][4]* is a global variable used for sending the value written from the previous function to the next function – from PredNSel to DCT, and from DCT to Quant. Therefore, the variable is classified as a data-path variable. Lastly, variable *img_Number* is classified as a block state variable since the value modified at the previous instance of the ReadOneFrame function affects the next instance of the function.

C. Data Sample Rate Decision

After function restructuring and data classification is completed, the sample rates of each function block are determined by examining how many data samples are produced and consumed at each port per function invocation. It also determines the execution frequency of function blocks. Fig. 10(a) shows a part of the translated code , which highlights the imgY_org data-port variable that connects two function blocks. By analyzing the data producing and consumption rates of two blocks, it is identified that one invocation of the ReadOneFrame block triggers the intra4_prediction block 16x99 times. Then we have three possible assignments of sample rates between two blocks. Note that a sample rate can be a fractional number.

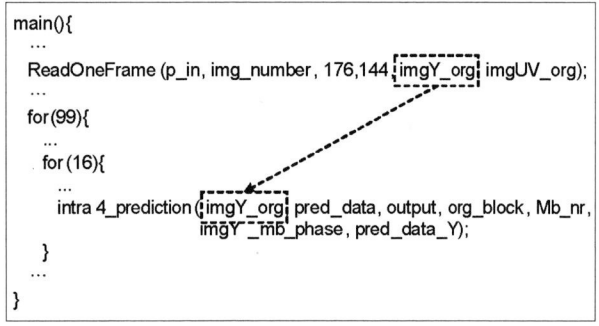

(a)

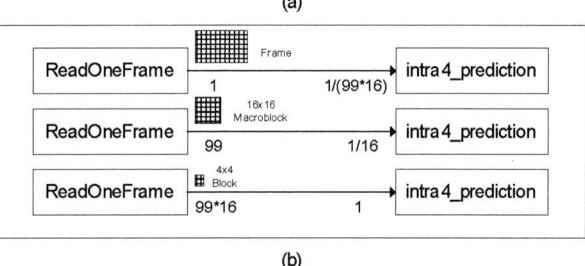

(b)

Fig. 10. Data sample rate decision between the ReadOneFrame and intra4_prediction blocks.

The first assignment in Fig. 10(b) indicates that the entire frame is passed from the *ReadOneFrame()* function to *intra4_prediction()*. In this case, the port buffer becomes *imgY_org[1]* which is a frame-type buffer whose size is 176x144 in the QCIF format. The data sample rate in this case is set to 1:1/(99x16). In the second assignment, the frame is broken down and passed in the unit of macroblock. In this case, the port buffer becomes *imgY_mb[99]* which is a macroblock array. In this case, the data sample rate becomes 99:1/16. Lastly, in the third example, the frame is passed in unit of 4x4 blocks. In this case, the buffer becomes *imgY_block[99*16]* and the data sample rate becomes 99*16:1.

Among three possible assignments we selected the second assignment to make a two-level nested loop as shown in Fig. 10(a). The corresponding dataflow subgraph of Intra4x4

prediction subsystem is shown in Fig. 11

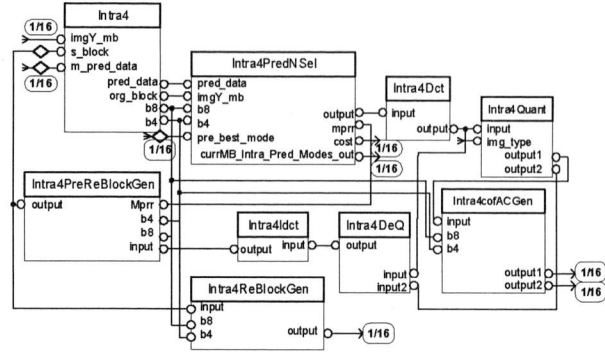

Fig. 11. Dataflow subgraph for Intra4x4 prediction subsystem

V. Clustering and Scheduling

After the conversion to a dataflow specification is completed, we have to find a valid schedule that generates the code structure as expected. Since obtaining an optimal schedule is beyond the scope of this paper, we outline how the valid schedule is constructed from the given dataflow specification. We make the clusters of function blocks that have the same sampling rates on the connected arc. Then the function blocks associated with intra4x4 prediction are merged together into a cluster. The Intra4x4 subsystem of Fig. 11 becomes a single cluster in the clustered graph. The cluster constructs a loop body in the final code as illustrated in Fig. 4.

The next step is the looping step of the clusters. If Intra4x4 subsystem is looped 16 times, the sample rates of this looped cluster become the same as the Intra16x16 subsystem. Then, we apply the merging step again to make it a hierarchically clustered graph. The second level loop of Fig. 4 is achieved in this way. We repeat the merging and looping steps until only one cluster remains at the top, which defines the main schedule body.

VI. Experimental Result

As a case study we convert the H.264 reference C code into an extended SDF model using our HW/SW codesign environment [8] that supports automatic C code generation from dataflow specification. Fig. 12 shows the schematic captured from the environment.

We first cut down the reference C code (JM original) to a light-weight version of the reference code, called Modified JM. The Modified JM code keeps the codes only for simple profile with single-slice mode. We then converts the Modified JM code to the dataflow graph applying the proposed methodology. From the dataflow specification, we obtain the synthesized code (Synthesized JM) from the design environment. We have compared these three codes in respect of code size, data size, and encoding time. All experiments are done on a Linux platform (kernel version 2.6.8), and each C code is compiled using GNU gcc version 3.3.5.

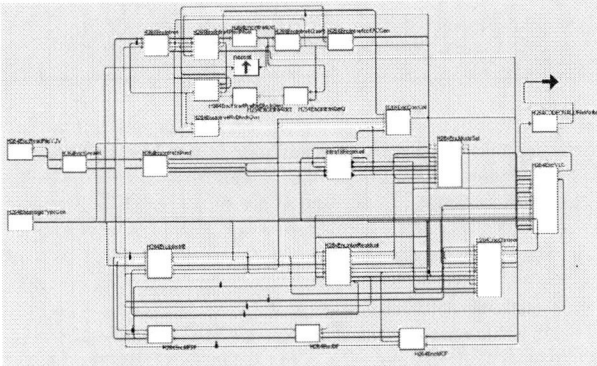

Fig. 12. SDF modeled H.264 encoder in PeaCE [8]

Table 1. Result on three different H.264 encoder codes

		JM original	Modified JM	Synthesized JM
Encoding time per Frame	I Frame	0.04 sec	0.04 sec	0.10 sec
	P Frame	0.70 sec	0.14 sec	0.31 sec
Code Size		369275 bytes	115863 bytes	148442 bytes
Data Size		7793536 bytes	2747900 bytes	2051908 bytes

Table 1 shows the experimental results on the three different H.264 encoder codes. As shown in the Table 1, the JM original code shows poor performance, and requires large code and data size, because it implements all options. So it is fair to compare the Modified JM code and the synthesized JM code. In this comparison, we do not perform any optimization on the dataflow specification.

As for the encoding performance, the Modified JM code shows about twice better performance than the synthesized code. This performance discrepancy is mainly due to data copying overhead between function blocks, which is the main target of optimization. In the current definition of function blocks, it includes several functions inside. Since these internal functions maintain the same variable sharing mechanism of the reference code, we explicitly copy the port buffers to the shared variables or vice versa. This copy overhead does not exist in the Modified JM code. It also causes code size expansion. It remains as a future work to remove this overhead since performance optimization is not a key objective of this paper.

On the other hand, the synthesized code shows better results than the Modified JM code in terms of the data size. This is because the Modified JM code allocates more frame –size variables than necessary. And the synthesized code allocates only a portion of frame for macro-block delivery between blocks and shares it through iteration, while the Modified JM code uses a frame-size variable. If we apply the buffer sharing optimization on the dataflow specification, the gap will be wider.

It took two man-months to transform the reference code

to the code structure as shown in Fig. 4. It is performed by an expert in dataflow modeling, but he has only the basic knowledge on the encoding algorithm. So, two man-month includes the learning time of the H.264 encoder algorithm. It took another two man-month to draw a dataflow graph from the transformed code, which includes function block definition and code synthesis. It is performed by two graduate students who have no experience in this task.

VII. Conclusion

In this paper, we presented a systematic procedure to convert a sequential C code to a dataflow specification. The key technique is to transform the C code into a well-structured form for dataflow model. We experimented with an H.264 encoding algorithm to demonstrate how the proposed methodology can be applied to a complicated real-life multimedia application. The proposed technique includes function restructuring, variable classification, and data sample rate decision. These transformation techniques are applied in turn and repeatedly until the well-structured form is obtained.

We successfully obtain the dataflow specification and compared the synthesized code from dataflow specification with the reference code in terms of encoding time, code size, and data size. While the automatically synthesized code shows worse performance and code size, it shows better result on the data size. Optimization of the dataflow specification is left as a future work.

ACKNOWLEDGEMENTS

This work was supported by National Research Laboratory Program(number M1-0104-00-0015), Brain Korea 21, SystemIC 2010 Project and IT Leading R&D Project funded by Korean MIC. ICT and ISRC at Seoul National University provided research facilities for this study.

References

[1] K. Jerry, "Model-Based Design and Beyond: Solution for Today's Embedded Systems Requirements," *American Technology International.*

[2] E. A. Lee, D. G.. Messerschmitt, "Synchronous Data Flow", *A Proceeding of the IEEE, Vol. 75, NO. 9, September 1987.*

[3] S. S. Bhattacharyya, P. K. Murthy, E. A. Lee, "Synthesis of Embedded Software from Synchronous Dataflow Specification," *Journal of VLSI Signal Processing 21, 151-166(1999).*

[4] H. Oh, S. Ha, "Fractional Rate Dataflow Model for Efficient Code Synthesis."

[5] S. Kwon, H. Jung, S. Ha, "H.264 Decoder Algorithm Specification and Simulation in Simulink and PeaCE."

[6] H.264/AVC Software Coordination http://bs.hhi.de/~suehring/tml/

[7] Iain E. G. Richardson, H.264 and MPEG-4 Video Compression, Willy, 2003

[8] PeaCE(Ptolemy extension as Codesign Environment) project homepage http://peace.snu.ac.kr/research/peace/

2B-1

2006 Asia and South Pacific Design Automation Conference

SAVS: A Self-Adaptive Variable Supply-Voltage Technique for Process- Tolerant and Power-Efficient Multi-issue Superscalar Processor Design

Hai Li	Yiran Chen	Kaushik Roy	Cheng-Kok Koh
Qualcomm Inc.	Synopsys Inc.	Purdue University	Purdue University
5775 Morehouse Dr.	700 East Middlefield Road	ECE Department	ECE Department
San Diego, CA, USA	Mountain View, CA, USA	West Lafayette, IN, USA	West Lafayette, IN, USA
Tel : +1-858-845-7393	Tel : +1-650-584-4885	Tel : +1-765-494-2361	Tel : +1-765-496-3683
e-mail: hail@qualcomm.com	e-mail: yiran.chen@synopsys.com	e-mail: kaushik@ecn.purdue.edu	e-mail: chengkok@ecn.purdue.edu

Abstract - *Technology scaling and sub-wavelength optical lithography is associated with significant process variations. We propose a self-adaptive variable supply-voltage scaling (SAVS) technique for multi-issue out-of-order pipeline to improve parametric yield with minimal power dissipation. Our error-correction circuitry and recovery mechanism allow the proposed fault-tolerant pipeline to work at a dynamically tuned supply voltage with a very low error rate. Experiments on an 8-issue, out-of-order superscalar processor show that SAVS can achieve 93.3% yield with 8.66% total power reduction under a scaled V_{DD}, compared to the same yield achieved by conventional microarchitecture. The increased execution time is negligible (0.014%).*

1. Introduction

With technology scaling, power dissipation has become a limiting factor in high-performance microprocessor design. Among existing power management techniques, supply voltage (V_{DD}) scaling has been proven to be effective for both dynamic and leakage power reduction: with V_{DD} scaling, dynamic power decreases quadratically [1] and leakage power decreases exponentially [2], associated with the increase of circuit delay.

The latency increases of different circuit styles due to V_{DD} scaling are different [3]. Fig. 1 shows the simulation results of the relative latency increases of a gate-dominant circuit and an interconnect-dominant circuit when V_{DD} is scaled, under BPTM 70nm technology. Compared to interconnect-dominant circuit (e.g., result bus), the latency of gate-dominant circuit (or logic circuit, e.g., ALU), is less sensitive to V_{DD} scaling. Obviously, the circuit whose latency increases slower with V_{DD} scaling may get a larger benefit from V_{DD} scaling.

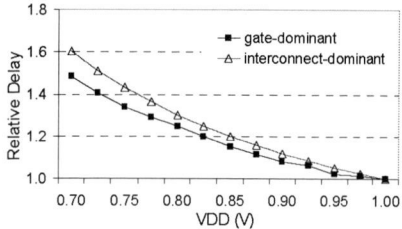

Fig. 1. V_{DD} scaling and relative delay

To ensure correct timing and consequently, acceptable chip yield at scaled V_{DD}, most of V_{DD} scaling techniques have to increase the clock period. The prolonged clock period leads to a longer execution time and consequently, the degradation of system performance. Hence, V_{DD} scaling is usually adopted in embedded systems that have relatively less stringent performance requirements [3][4]. For performance-oriented systems, V_{DD} scaling is applied only when the system has a low throughput. For example, processor can work at a scaled V_{DD} with a low clock frequency when pipeline idles during L2 cache misses [1]. The conflict between V_{DD} scaling and circuit delay (and consequently, the chip yield), severely limits the application of V_{DD} scaling technique in high-performance microprocessors.

In reality, due to the variability of device parameters (e.g. random doping, transistor dimension and threshold voltage variation) and

the fluctuation of environment factors (e.g. temperature shifting and power supply voltage noise), a margin has to be added to the supply voltage adopted in the design time. This ensures a certain chip yield, which is defined as the ratio of the number of chips that work properly over the total number of chips, under a certain V_{DD}. However, the traditional corner-based V_{DD} selection, which assumes that all worst-case conditions occur simultaneously, may heavily overestimate the actual needed V_{DD}.

Several studies show that for logic circuits, the worst-case V_{DD} requirement may seldom occur. For example, [5] shows that for random input vectors, the average carry propagation length of a carry look-ahead adder (CLA) is much less than 1/3 of the longest one. Therefore, power management techniques that scale V_{DD} *below* the worst-case V_{DD} requirement, have been recently investigated: In [6] and [7], ALU works under a scaled V_{DD} and timing-errors due to incomplete operation of ALU are detected and corrected by/from a result checker or a shadow latch.

We note that timing-error correction techniques can be also adopted to tolerate circuit delay variation due to the process parameter fluctuations and environment factor variations. By detecting and correcting the incomplete operations of circuit at the scaled V_{DD}, timing-error correction mechanism can improve chip yield as well as reduce the power dissipation. This is especially important to the application of V_{DD} scaling in high-performance processors and is the motivation of our work.

In this paper, we propose a self-adaptive variable supply-voltage scaling (SAVS) technique that targets the process-tolerance and the power-efficiency in multi-issue high-performance microprocessors. Timing-error correction mechanism is applied to selected pipeline stages for chip yield enhancement by correcting the errant timing due to the delay variation. The selected stages can work at a scaled V_{DD} with a tolerable timing error rate while the chip yield is still maintained.

Simulations on 23 SPEC2000 benchmarks show that on average, SAVS can reduce up to 8.66% of microprocessor power with negligible instruction per cycle-based (IPC-based) performance degradation (0.014%) while maintaining a required chip yield of 93.3% and same clock frequency.

2. Self-Adaptive Variable Supply-Voltage Scaling (SAVS)

2.1. SAVS Mechanism

Shadow latch has been proven to be effective and economic for data retention and checking. For example, in [8], data is stored in shadow latch when circuit switches to power-saving mode. When circuit switches back to active mode, system status is restored from the data stored in shadow latch. In [7], a shadow flip-flop-based technique (called Razor latch) is proposed for timing-error detection and recovery. The mechanism of shadow flip-flop can be summarized as follows:

At the end of each clock cycle, the output of pipeline stage L1 is latched by the main flip-flop (FF) (Fig. 2(a)). When an errant output occurs, i.e., when operation latency L_{op} exceeds the original clock period T_{clk}, the incomplete output is latched by main FF at the end of clock cycle $i+1$. After time $L_{op}-T_{clk}$, operation in L1 completes and the output of L1 switches to the correct data. Time ΔT after the end of cycle $i+1$, detection signal SHW triggers shadow latch to capture the correct data. If the data captured by shadow latch is different from the data stored in main FF, an 'ERROR' signal is

The work is sponsored in part by Marco Gigascale Systems Research Center (GSRC) and Semiconductor Research Corp.

0-7803-9451-8/06/$20.00 ©2006 IEEE.

158

2006 Asia and South Pacific Design Automation Conference

2B-1

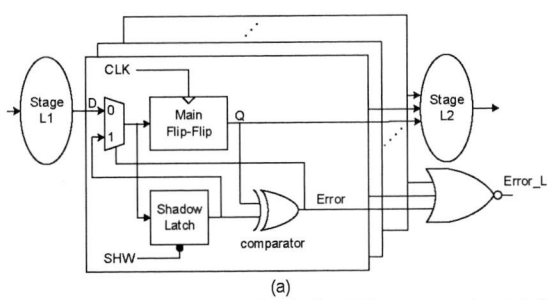

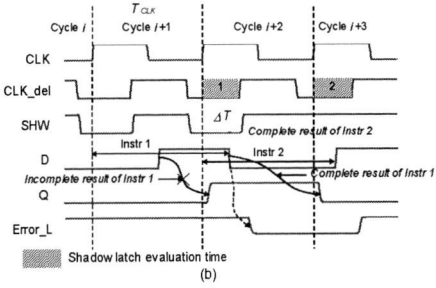

(a)

(b)

Fig. 2. Shadow flip-flop mechanism for timing failure correction (a) Schematic (b) Timing diagram (From HSPICE)

generated in the subsequent cycle $i+2$ and the correct data is restored to main FF. Obviously, any operation with the latency longer than $T_{clk}+\Delta T$ cannot be captured by shadow latch. In such a case, system may not recover from the timing-error. The corresponding timing diagram, which is extracted from HSPICE simulation, is shown in Fig. 2(b). After a timing-error has occurred at the end of cycle $i+1$, the errant output data of stage L1 is sent to the subsequent stage L2. Hence, the instruction executed in the subsequent stage L2 in cycle $i+2$ must be re-executed after getting the correct input from L1 at the beginning of cycle $i+3$. One cycle penalty is introduced in the procedure above since the execution of Instr 1 in stage L1 actually takes two clock cycles.

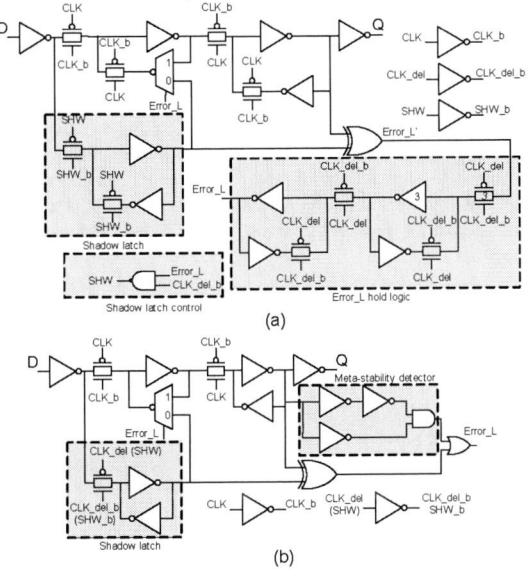

(a)

(b)

Fig. 3. Shadow flip-flop Design (a) Proposed Design (b) Original Design

In our SAVS technique, a modified robust shadow FF is developed for high-performance application and shown in Fig. 3(a). Compared to the shadow flip-flop design in [7] (shown in Fig. 3(b)), our shadow FF design has the following advantages:

1. Additional control transmission gate in the inverter loop of each latch and careful sizing up of inverters and transmission gates provide [9] more robust design to prevent the "drive fight" of two inverters and consequently, occurrence of meta-stability.

2. In the Error_L hold logic in Fig. 3(a), signal Error_L is triggered by a delayed clock signal CLK_del, based on the result of comparison of the values stored in the main slave latch and the shadow latch. This design avoids the false switching of Error_L due to any glitch at the circuit output D in shadow latch evaluation time 1 (see Fig. 2(b)).

3. When an errant output is detected, the FF structure of Error_L control logic keeps Error_L at logic ZERO until the next shadow latch evaluation time 2 (see Fig. 2(b)) completes. This mechanism masks the evaluation signal SHW of shadow latch in shadow latch evaluation time 2 and prevents the complete result of Instr 1 at Q from being corrupted by the result of Instr 2 (if Instr 2 is a long-latency instruction and switches in the shadow latch evaluation

time 2).

We note that only *one* such a flip-flop-based Error_L control logic is required by the whole pipeline stage L1 in Fig. 2(a): The Error_L signal in Fig. 3(a) is the Error_L signal in Fig. 2(a), which indicates any Error in any output bits of stage L1. The incurred area/power overhead is negligible.

For the non-critical pipeline stages whose latency is always short enough (less than one clock cycle at scaled V_{DD}), no error correction circuitry is required. Moreover, the output bits that are not located in the critical data paths also do not require error correction circuit. Here the critical data paths are defined as the data paths that may not complete execution within one clock cycle at the scaled V_{DD}, due to process variations or other environmental factors.

As pointed out in [7], in Fig. 2(b), if the execution time (T_S) of Instr 2 in the cycle $i+2$ is shorter than ΔT, the complete output of Instr 1 may be corrupted by the result of Instr 2. Hence, buffers need to be added at some inputs of stage L1 to ensure $T_S > \Delta T$ for the output bits with shadow latch. Such input buffers do not increase the critical path of stage L1.

The simulation of a 32-bit CLA under BPTM 70nm process [10] considering Vt variations shows that that only 7 output bits may generate errors when scaling the V_{DD} from 1.0V to 0.725V for a chip yield of 93.3%. More details of experiment setup are given in Section 4.1. Because the scaled V_{DD} applied to the stages with SAVS technique (SAVS stages) is different from the normal V_{DD} of other stages, FFs with level-conversion function (FFLC) [11] may be required at the output of SAVS stages. Our simulation shows the modification of the execution/bypass stage in SAVS technique results in about 6% area overhead, with respect to the conventional execution/bypass stage design in an 8-issue out-of-order superscalar microarchitecture [12][13].

2.2. Principle of SAVS

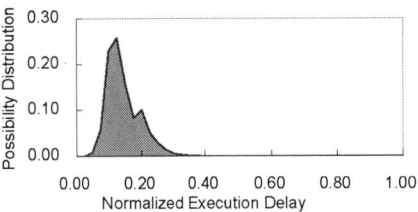

Fig. 4. Delay distribution of 32-bit CLA

Fig. 4 shows the delay distribution of the 32-bit CLA over 8192 random inputs. We can observe that at most of time, the delay of CLA is far shorter than the longest possible delay (normalized to 1 in Fig. 4). *Very few operations really go through the longest data path of CLA.* If we lower the V_{DD} and are able to correct all possible timing errors due to process variations or environmental factor fluctuations, power dissipation can be lowered without any degradation of chip yield while maintaining the same working frequency.

For some interconnect-dominant circuits, the delay also varies from case to case. Fig. 5 shows the layout of an 8-way bypass mechanism in execution/bypass stage of pipeline. Result bus bypasses the data between 8 different ALUs. To reduce the bypass delay, Register File (REG) is located at one end [12][13].

159

2B-1

2006 Asia and South Pacific Design Automation Conference

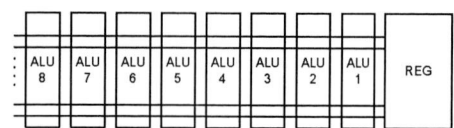

Fig. 5. Layouts for 8-way bypassing

Bypass delay includes bypass logic delay and interconnect delay. The bypass delays between different ALU pairs in Fig. 5 are shown in Fig. 6 (delays are normalized over the longest bypass delay occurring between ALU8 and ALU1). We use the dimensions of ALUs given in [12] and carefully scale them to 70nm technology. Repeaters are inserted in some result bypass buses to reduce the long RC delay of long metal interconnect.

It is known that for superscalar pipeline, the usage of ALUs is limited by the ILP (instruction level parallelism). Most of time, bypassing is constrained among first several ALUs that are in physical proximity in layout. The corresponding bypass delays are much shorter than the longest one (from ALU8 to ALU1).

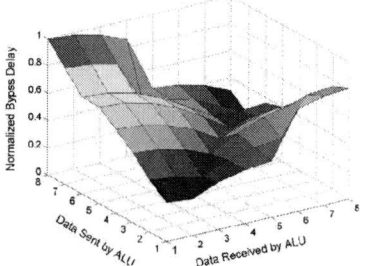

Fig. 6. Bypass delay between different ALU pairs

We note that the latency of circuit does *not* rely on the input vector when: 1) in gate-dominant circuit, every operation has to go through the longest data path; 2) in interconnect-dominant circuit, the data transmission is point-to-point, i.e., transmitter- receiver pair. In such cases, V_{DD} scaling may result in a high error rate with a fixed clock frequency. SAVS may not be applicable.

2.3. Overview of SAVS in High-Performance Microprocessors

Fig. 7 depicts the general pipeline model for a superscalar processor [13]. The delays of each stage in the 8-issue baseline superscalar pipeline are carefully analyzed in [7], [13] and [14]. The delays of every stage in terms of FO4 delay, which measures the delay of an inverter driving a-fanout-of-four, are shown in Table I for 70nm technology. It has been shown that in 0.18μm technology and beyond, for the superscalar microprocessor whose issue width is equal to or more than 6, execution/bypass stage becomes the critical stage (e.g., execution/bypass stage with a delay of 18+18=36 FO4 in Table I in an 8-issue pipeline) [13][14]. The corresponding simulation parameters of baseline pipeline will be shown in Section 4.3.1.

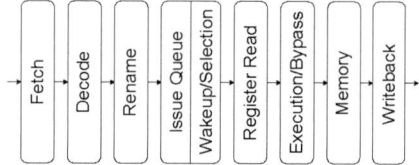

Fig. 7. Baseline superscalar model

Instruction cache (I-cache), data cache (D-cache) and register files involve RAM (random access memory) structure [14]. As the analysis in Section 2.2, due to different interconnect lengths from the active cell to sense amplifier, cache core access delay (along wordline and bitline) relies on the physical position of data cell in RAM array. However, the cache core access time usually occupies limited percentage (less than 35%) of total cache access time [15]. The delays of other peripheral circuitries, including decoder, sense amplifier and data bus, are insensitive to the data address in RAM.

Therefore, SAVS may *not* be applicable to fetch, register read, D-cache and writeback stages since the delays of those RAM-based stages (30 FO4 for Fetch (I-Cache) and D-Cache, 24 FO4 for register read and writeback) are close to the clock period (36 FO4).

The register rename logic is used to translate logical register designators into physical register designators. Its two functions – mapping and check-pointing – are also usually implemented by RAM structure [13][14]. However, the delay of rename stage (18 FO4) is only around half of the clock period (36 FO4). Therefore, SAVS can be applied to rename stage with error-free operations (at scaled V_{DD} of SAVS: 0.725V in our simulation, Section 4.1).

Table I
Stage Delays of 8-issue Superscalar Pipeline

	Delay (FO4)		Delay (FO4)
Fetch	30	Decode	18
Rename	18	Register Read	24
Issue Queue	12	Execution	18
Selection	12	Bypass	18
Load/Store. Queue	12	Writeback	24
I-Cache	30	D-Cache	30

The complexity and the latency of decode stage greatly depends on ISA (instruction set architecture) and circuit design. In many practical microprocessor design, decode stage latency is shorter than the execution time of ALU [7][16]. In the execution/bypass stage of multi-issue out-of-order pipelines, bypass logic delay is equal to or greater than ALU delay [12][13][14]. In such a case, the decode delay is around half of clock cycle. Consequently, SAVS *may* be applied to decode stage with error free operations. However, to be conservative, we did *not* apply SAVS to decode stage in our design.

The issue queue is a CAM (content addressable memory) structure [13][14]. In every entry of issue queue, the content is the decoded instruction and the tag is the sources of instruction's operands. Wakeup logic works as follows: The tags associated with the executed results of ALU are broadcasted to all entries in issue queue. If each tag in an entry matches one tag associated with the execution results, the corresponding instruction is ready to be issued in the next cycle. Wakeup logic latency is mostly determined by the length of issue queue and the physical position of stored instruction in issue queue.

Selection logic is composed of stacked arbiters to select instructions to be issued from the pool of ready instructions in issue queue. The number of instructions to be issued in the next cycle determines the latency of selection logic: more arbiters are required when more instructions are to be issued [17]. Although the latency of wakeup/selection stage is input vector (address) dependent, the extremely high-cost recovery mechanism makes the application of SAVS difficult: after an instruction is issued, it is popped out from the issue queue right away. Recovery from such errors may need to restore the issued instructions back to issue queue, and, introduce large power/area overhead and design complexity. To be conservative, we did *not* apply SAVS to wakeup/selection stage.

The load/store queue in memory stage is also a CAM structure. The delay of writing into or reading from memory queue (12 FO4) is around 1/3 of the delay of critical stage (36 FO4). Hence, SAVS can be applied to memory stage (not cache).

Because of the increasing gap between the delays of execution/bypass stage and other stages with technology scaling, the yield of pipeline is mainly determined by the timing error in execution/bypass stage. In execution/bypass stage, each of the ALU execution and data bypass takes about half of clock period [12][13][14]. Correcting the timing error in execution/bypass stage at the scaled V_{DD} can improve the yield of execution/bypass stage and consequently, the yield of whole pipeline, with minimal power dissipation. Carefully choosing the range of V_{DD} scaling ensures that the error rate of execution/bypass stage and the incurred performance penalty are low while other pipeline stages still work with error free.

We note that the discussion above may not be applicable to some deeper pipelines since the design objective of deep pipeline is to

160

uniform the delay of each stage of pipeline [14]. However, deep pipeline has been proven unsuitable to scaled technology because of the extremely high power dissipation and a large number of critical paths [18].

In summary, for a multi-issue out-of-order pipeline, SAVS can be easily applied to rename, execution/bypass stages and load/store queue. V_{DD} scaling causes the timing-errors only in execution/bypass stage. Hence a timing-error correction circuitry is required in execution/bypass stage. Although other stages keep working at the conservatively high V_{DD} (it is important if cache is another critical stage), our experimental results in Section 4 show a significant power reduction with negligible performance penalty and energy overhead, when SAVS is applied.

3. Implementation of SAVS

3.1. Pipeline recovery mechanism

In Fig. 2(b), the errant outputs of execution/bypass stage at the end of cycle $i+1$ may be bypassed back to the inputs of the execution/bypass stage and also sent to the memory stage in the subsequent cycle $i+2$. In such a case, the operations committed in execution/bypass stage and memory stage in the current cycle $i+2$ must be re-executed (or re-sent) in the following cycle $i+3$. The error recovery mechanism of the pipeline needs to ensure that: (1) no new instructions are issued out from the issue queue until the re-executions complete; (2) the incorrect execution results in the previous cycle is flushed out from the pipeline; and (3) no errant register or cache writing is committed.

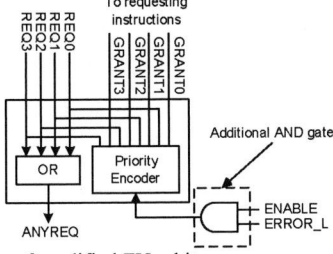

Fig. 8. Diagram of modified FU arbiter

Fig. 8 shows the modified FU arbiter for SAVS scheme. The ready instructions in issue queue can be issued out only when the ENABLE signal of the corresponding FU is raised to logic ONE [13][14][17]. In Fig. 2(b), the ERROR_L is pulled down to logic ZERO once an error in execution/bypass stage is detected in cycle $i+2$. We piggyback on this ERROR_L signal to block the ENABLE signal and prevent instructions from issuing out in cycle $i+3$. No extra performance penalty is introduced.

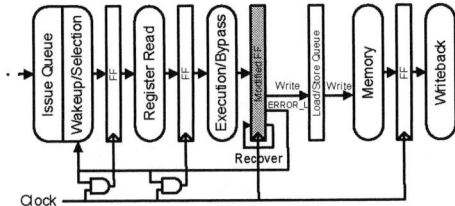

Fig. 9. Clock-gating-based stalling scheme

During the re-execution of instructions, the output latches of wakeup/selection stage and register read stage must be stalled at the end of cycle $i+2$. The data stored in the latches of register file stage and the correct data bypassed from execution/bypass stage are resent into execution/bypass stage for re-execution in cycle $i+3$. In modern high-performance microprocessors, the output of execution/bypass stage is sent to a load/store queue before it is written into the memory. Hence, the errant entry of load/store queue can easily recover by using the modified FFs of execution/bypass stage (see Section 2.1). A clock-gating-based stalling scheme is shown in Fig. 9. Compared to the modified in-order pipeline in [7], the additional pipeline stage between execution/bypass stage and memory stage (to prevent writing the errant data to the memory) is removed.

The whole pipeline is separated by issue queue into the front-end and the back-end. The stages in the front-end – fetch, decode and rename – are not stalled when the re-execution is committed. New instructions can still be fetched and sent to issue queue as long as issue queue is not full. Unlike the global clock-gating scheme implemented in an in-order pipeline [7], in our design, clock-gating is applied to the latches of only two back-end stages: wakeup/selection and register read. These two circuit blocks are located in physical proximity in the floorplan. The clock-gating control signal, which is the ERROR_L signal generated by timing-error correction circuit, can easily reach those latches by the end of current cycle.

Pipeline writes the results to register file at writeback stage. Since the information of date correctness has been available at the end of memory stage, errant writing to register file is avoided.

For pipelined multi-cycle ALU, e.g., multiplier and floating point ALU, the timing-error correction circuit is implemented on the output of each stage. Moreover, similar clock-gating scheme can be applied to every slice of multi-cycle ALU for the recovery from any possible timing error in the slice of multi-cycle ALU.

3.2. Supply-Voltage Scaling Control

3.2.1. Supply-voltage control scheme

Due to bypass mechanism, even if only one ALU operation is incorrect in the previous cycle, all the instructions being executed in the present cycle need to be re-executed. Hence, the error-induced performance and energy overheads are determined only by the error rate of program (ER_p). We also refer the error rate of an ALU as ER_a. If each ALU's error rate is independent of others', ER_p approximately equals $ER_a \cdot IPC$ when ER_a is small.

Because of the process variations and the time-varying environmental factors, ER_p changes from program to program or even from portion to portion within the same program. Hence, V_{DD} must be automatically adjusted to achieve an acceptable chip yield with acceptable performance and energy overheads.

In our design, we select two switching thresholds, ER_{high} and ER_{low}, and monitor the ER_p of the running program. V_{DD} does not change when ER_p is between ER_{high} and ER_{low}. Otherwise, we decrease V_{DD} to reduce power consumption if ER_p is lower than ER_{low}, or increase V_{DD} to avoid unacceptable performance/energy overhead if ER_p is higher than ER_{high}.

However, our V_{DD} adjustment mechanism *never* reduces V_{DD} under a critical voltage level V_{DD-min}, and hence, guarantees a desired chip yield. This constraint is *very crucial* when the chip yield is considered. More details will be given in Section 4.2.

3.2.2. Limitations of the V_{DD} scaling

V_{DD} ramping is limited by two factors: the response time of voltage regulator and the power supply noise tolerance. [7] described a typical commercial voltage regulator that takes 10's of microseconds to adjust V_{DD} by 100mV. Moreover, V_{DD} shifting results in the charging/discharging of the intrinsic capacitance of circuits and consequently, current surge in power supply network. V_{DD} ramping must be slow enough so that the induced power supply noise is below the allowed threshold. In our experiments, we conservatively set V_{DD} ramping rate to 5mV/μs.

4. Experimental results

4.1. Yield Analysis with Timing-Error Correction Mechanism

Our Monte-Carlo simulation on chip yield is conducted by HSPICE with BPTM 70nm Technology. Without loss of generality, only the variations of Vt and interconnect width are considered in our experiments. The STDs of both inter-die and intra-die Vt variations are set to 30mV [19]. The lumped RC model of interconnect [20] is used while the STD of interconnect width for each segment of lumped RC model is set to 10% of the nominal width. The spatial correlation coefficient between the intra-die Vt variations of any two transistors and the one between the widths of any two interconnect segments are both set to 0.4. Execution/bypass (E/B) stage is implemented with 32-bit CLA and the corresponding bypass logic. Fig. 10 shows the yield of E/B stage under different V_{DD}'s. Clock period is selected to ensure that when E/B stage operates at the normal V_{DD} (1.0V), chip yield is 93.3% ($\Phi(1.5)$, where Φ is the cumulative distribution function of a standard normal distribution).

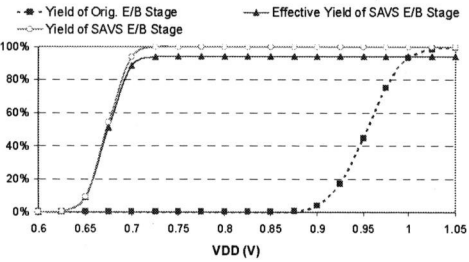

Fig. 10. Yield of ALU under different V_{DD}'s.

To take account of the area overheads of timing-error correction circuit, we introduce the effective yield, defined as:

$$Y_{chip}^{eff} = \frac{A_{orig}}{A_{orig} + A_{ovhd}} Y_{chip} \qquad (1)$$

Here A_{orig} and A_{ovhd} are the area of E/B stage and the area overhead incurred by timing-error correction circuit, respectively. Y_{chip} is the yield of whole E/B stage and Y_{chip}^{eff} is the effective yield. The effective yields of the E/B stage with SAVS under different V_{DD}'s are also shown in Fig. 10. Here the switching time of the error-detection signal ΔT is set at half clock cycle (see Fig. 2(b)). The whole pipeline's yield is largely determined by the yield of E/B stage, which is the critical stage for sub-0.18μm technology and beyond [13][14]. Hence, the yield of whole chip can be improved significantly by SAVS with the negligible area overhead (compared to the whole chip area of pipeline). *It should be noted that yield considered in this paper is only due to SAVS. In reality, the exact yield calculation for a system depends on multitude of factors and is a lot more involved.*

Increasing V_{DD} can also enhance the chip yield, albeit with significant power penalty. Although the highest possible effective yield of modified E/B stage is limited by the area overhead of timing-error correction circuit, when V_{DD} is low (less than the actual needed V_{DD}, say, 1.0V), the effective yield of modified E/B stage is much higher than the yield of non-modified ones. For an acceptable yield, 93.3% (say), the modified E/B stage requires only V_{DD}=0.725V while the non-modified one requires V_{DD}=1.0V. We select the minimal V_{DD} that ensure the desired chip yield in SAVS technique as the critical V_{DD} of SAVS (V_{DD-min}).

4.2. Error Rate with Aggressively Scaled V_{DD}

In our error rate simulation, if the latency of an operation at the scaled V_{DD} exceeds original clock cycle (considering flip-flop set up time and hold time), it is counted as an error. Obviously, the actual error rate of a circuit greatly depends on the process parameter and environmental factor variations. Fig. 11 shows the error rates of a CLA-based ALU with 8192 random input vectors at different V_{DD}'s: when the Vt's of all transistors in the 32-bit CLA: 1) equal the designed value (-designed). Here, the designed value is the Vt adopted in design time; 2) +42.4mV deviation from the designed value toward 0V (-slow); 3) -42.4mV deviation from the designed value toward 0V (-fast). We note that the error rate of ALU actually relies on the particular benchmark.

The error rate of bypass logic is different for various benchmarks. Based on the simulation results in Fig. 6, we use Wattch [21] to simulate the data bypass in an 8-way, out-of-order pipeline for 23 SPEC2000 benchmarks [22]. Results show that benchmark *art* has the highest error rate of bypass logic.

Fig. 11 also shows the error rate of E/B stage (including both ALUs and bypass logic): when the interconnect width of bypass

logic: 1) equals the designed value (-designed); 2) -10% deviation from the designed value toward 0 (-slow); 3) +10% deviation from the designed value toward 0 (-fast). Here, we assume that the longest latencies of ALU and data bypass take half clock period at V_{DD}=1.0V, respectively [12][13][14]. The behaviors of ALU and bypass logic are based on 8192 random input vectors for benchmark *art*.

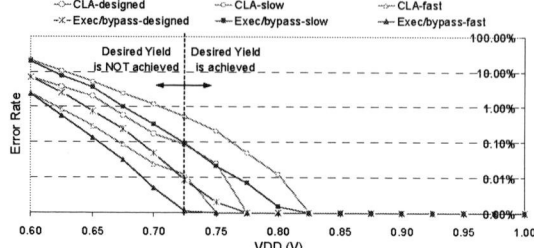

Fig. 11. Error rate of CLA and execution/bypass stage

We point out that, when V_{DD} is scaled under 0.725V, the error rate may still be at a low level: when V_{DD} is 0.7V, the corresponding error rate of E/B stage is only 0.05%. The incurred performance and power overheads is still small. However, the over scaled V_{DD}, i.e., 0.7V, cannot guarantee a desired chip yield (say, 93.3%): Some errors cannot be detected and recovered by the error correction circuitry (Section 2.1). Thus, when chip yield is considered, a pure error-rate-based (or performance/energy penalty-oriented) dynamic V_{DD} adjustment mechanism [7] is not sufficient: to maintain the desired chip yield, V_{DD} should *not* be scaled under the critical V_{DD} of SAVS (0.725V in our simulation).

In our simulations in Section 4.3, we assume all Vt's of transistor and the bypass interconnect width in microprocessors equal the designed (nominal) value. The corresponding error rates of E/B stage are shown as curve "Exec/bypass-designed" in Fig. 11. [7] shows that the random input-based simulation overestimates the error rate of an ALU in reality. Since we also adopted the error rate of bypass logic in benchmark *art* to come up with the curve "Exec/bypass-designed", in our simulation, the performance penalty and energy overheads due to the errant operations are actually overestimated. Consequently, for the error-rate-driven V_{DD} scaling, power reduction is underestimated.

[7] also shows that complex logic structures, such as multipliers, generally have lower error rate than adder-based ALU's. In our simulation, we assume that integer multiplier and floating-point ALU have the same error rate as that of CLA at the same V_{DD}. This assumption does not underestimate the error rate or overestimate the effectiveness of our methodology.

4.3. System-Level Simulations

4.3.1. Simulation environment

Table II
Baseline Processor Configuration

Processor	8-way issue, 128 RUU, 64 LSQ, 8 integer ALUs, 2 integer mul/div units, 4 FP ALUs, 4 FP mul/div units, uses clock gating (DCG) and s/w prefetching
Branch prediction	8K/8K/8K hybrid predictor; 32-entry RAS, 8192-entry 4-way BTB, 8 cycle misprediction penalty
Caches	64KB 2-way 2-cycle I/D L1, 2MB 8-way 12-cycle L2, both LRU
MSHR	IL1 - 32, DL1 – 32, L2 – 64
Memory	Infinite capacity, 400 cycle latency
Memory bus	32-byte wide, pipelined, split transaction, 4-cycle occupancy

We used a modified version of Wattch to simulate an 8-way, out-of-order SAVS processor, which is summarized in Table II, under BPTM 70nm process. A deterministic clock gating (DCG) technique [23] is involved in all simulations. The extra power dissipation of the timing-error correction circuitry and the additional energy overhead due to the re-execution of errant operation have been included in our power saving and overhead estimation.

Alpha-SPEC2000 binaries that are pre-compiled with SPEC *peak* setting are accepted in our simulation. We used *ref* inputs, fast-forwarded 2 billion instructions, and simulated 500 million instructions.

In our simulation, ER_{high} and ER_{low} are set to 0.1% and 0.05%, respectively. V_{DD} ramping rate is set to 5mV/μs. Under the assumption of the 2GHz clock frequency, it takes 2000 cycles to ramp the V_{DD} up/down by 5mV.

4.3.2. Effectiveness of SAVS

Fig. 12 depicts the simulation results of SAVS pipeline on 23 SPEC2000 benchmarks. The gray bar (with Y-axis on the left) represents the total microprocessor power savings ratio (including the cache power), with respect to the 8-issue baseline processor under V_{DD}=1.0V. The line (with Y-axis on the right) shows the percentage of the execution time increase of SAVS pipeline, with respect to the 8-issue baseline pipeline under V_{DD}=1.0V. The X-axis shows benchmarks.

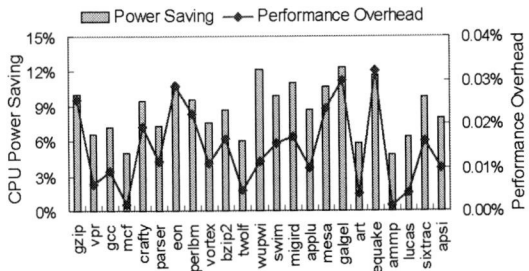

Fig. 12. Power savings and performance overhead of SAVS

The simulation results show that SAVS effectively reduces the power consumption of processor with negligible performance penalty while maintaining the same chip yield: on average, SAVS improves 8.66% of processor power consumption by paying only 0.014% performance penalty (the maximum performance overhead is within 0.032%). The average energy reduction is 8.64% while the chip yield is maintained at 93.3%.

The benchmarks with high ILP achieve higher power savings. For example *wupwise*, whose IPC is 4.10, achieves 12.1% reduction of processor power dissipation. The power savings for the benchmarks with low ILP, such as *ammp*, are insignificant. It can be explained as following: under clock gating scheme, for the program with higher ILP, ALUs are used more frequently and consume more power. Hence, more power saving can be achieved when SAVS is applied.

We point out that the application of SAVS in the critical stage of pipeline decreases the critical V_{DD} for whole system. Hence, the V_{DD} applied to other non-SAVS stages can be reduced to lower level that is determined by the new critical stage(s) of pipeline (for example, 0.85V in our simulation), without incurring additional performance, power and yield penalties. This fact provides an new methodology to design a high performance, power efficient, pipelined system under scaled technology.

5. Conclusion

In this paper, we proposed an error-tolerant self-adaptive variable supply-voltage scheme (SAVS) for multi-issue out-of-order superscalar microprocessors. Besides maintaining the chip yield, SAVS can effectively reduce power dissipation of high performance microprocessor with negligible performance degradation. SAVS is suitable for scaled technologies where process parameter fluctuations and environment factor variations are significant. SAVS also provides an alterative choice to simultaneously achieve the high throughput and low power in pipelined systems under scaled technologies.

Reference

[1] H. Li, *et. al.*, VSV: L2-Miss-Driven Variable Supply-Voltage Scaling for Low Power, *36th IEEE/ACM Int'l Symp. on Microarch.*, pp. 19-28, Dec. 2003.

[2] S. Mukhopadhyay, *et. al.*, Gate Leakage Reduction for Scaled Devices using Transistor Stacking, *IEEE Trans. on Very Large Integ. (VLSI) Sys.*, Vol. 11-4, pp. 716-730, Aug. 2003.

[3] T. D. Burd, *et. al.*, A dynamic Voltage Scaled Microprocessor System, *IEEE Jour. of Solid State Ckts.*, Vol. 35-11, pp. 1571-1580, Nov. 2000.

[4] J. Pouwelse, *et al.*, Dynamic Voltage Scaling on a Low-power Microprocessor, *Mobile Comp. Conf*, pp. 251-259, Jul. 2001.

[5] R. Ramachandran, *et al.*, Carry Logic, *Wiley Encyclopedia of Electr. and Electron. Engineering*, Edited by Joh G. Webster, 1999.

[6] T. Liu and S.-L. Lu, Performance Improvement with Circuit-level Speculation, *33rd IEEE/ACM Int'l Symp. on Microarch.*, pp. 348-355, Dec. 2000.

[7] D. Ernst, *et. al.*, Razor: A Low-Power Pipeline Based on Circuit-Level Timing Speculation, *36th IEEE/ACM Int'l Symp. on Microarch.*, pp. 7-18, Dec. 2003.

[8] D. Lammers, TI Moves Ahead with 65-nm Chips by Next Year, *EE Times*, Mar. 22, 2004.

[9] L. Kim and R. W. Dutton, Metastability of CMOS Latch/Flip-Flop, *IEEE Jour. of Solid State Ckts.*, Vol. 25-4, pp. 942-951, Aug. 1990.

[10] BPTM, http://www-device.eecs.berkeley.edu/~ptm

[11] K.Usami, *et al.*, Design Methodology of Ultra Low-power MPEG4 Codec Core Exploiting Voltage Scaling Techniques, *35th Des. Auto. Conf.*, pp. 483-488, June, 1998.

[12] Eric S. Fetzer, *et. al.*, A Fully Bypassed Six-Issue Integer Datapath and Register File on the Itanium-2 Microprocessor, *IEEE Journal of Solid State Circuits*, Vol. 37-11, pp. 1433-1440, Nov. 2002.

[13] S. Palacharla, *et al.*, Quantifying the Complexity of Superscalar Processors. *Technical report CS-TR-96-1038*, Dept. of CS., Univ. of Wisconsin, 1996.

[14] Z. Chishti, and T. N. Vijaykumar, Wire Delay Is Not a Problem for SMT (in the near future), *31st Annual Int'l Symp. on Comp. Arch.*, pp. 40-51, Jun. 2004.

[15] A. Agarwal, *et al.*, A Single-Vt Low-Leakage Gated-Ground Cache for Deep Submicron, *IEEE Jour. of Solid State Ckts.*, Vol.38-2, pp. 319-328, Feb. 2003.

[16] S. Virtanen and J. Lilius, The TACO Protocol Processor Simulation Environment, *9th Int'l. Symp. on Hardware/Software Codesign*, pp. 201–206, Apr. 2001.

[17] Y. Chen, *et al.*, Integrated Architectural/Physical Planning Approach for Minimization of Current Surge in High Performance Clock-gated Microprocessors, *Int'l Symp. on Low Power Electr. Des. 2003*, pp. 229-234, Aug. 2003.

[18] T. Karnik, Probabilistic and Variation-Tolerant Design: Key to Continued Moore's Law Scaling, *Invited talk in ACM/IEEE Int'l TAU Workshop on Timing Issues*, Feb. 2004.

[19] A. Bhavnagarwala, *et. al.*, The impact of Intrinsic Device Fluctuations on CMOS SRAM Cell Stability, *IEEE Jour. of Solid State Ckts.*, Vol. 36, No. 4, pp. 658-665, Apr. 2001.

[20] Y. Chen, *et. al.*, Model Reduction in the Time-domain Using Laguerre Polynomials and Krylov Methods, *2002 Des. Auto. and Test in Euro. Conf. and Exhi.*, pp. 931-935, Mar. 2002.

[21] D. Brooks, *et. al.*, Wattch: A Framework for Architectural-level Power Analysis and Optimizations, *27th Int'l Symp. on Comp. Arch.*, pp. 83-94, June 2000.

[22] http://www.eecs.umich.edu/~chriswea/benchmarks/spec2000.html

[23] H. Li, *et. al.*, Deterministic Clock Gating for Microprocessor Power Reduction, *9th Int'l Symp. on High-Perf. Comp. Arch.*, pp. 113-122, Feb. 2003.

The Design and Implementation of a Low-Latency On-Chip Network

Robert Mullins, Andrew West and Simon Moore
Computer Laboratory, University of Cambridge
Robert.Mullins@cl.cam.ac.uk

Abstract—Many of the issues that will be faced by the designers of multi-billion transistor chips may be alleviated by the presence of a flexible global communication infrastructure. In the short term, such a network will provide scalable chip-wide communication and ease the complexity of handling multi-cycle communications. In the long term, the network will become a primary tool for optimising power and data transfers and for scheduling computations. This paper details the design and implementation of a low-latency on-chip network. The network's speculative routers are in the best case able to route flits in a single clock cycle, helping to minimise on-chip communication latencies and maximise the effectiveness of buffering resources. Results from our 180nm test chip demonstrate an inter-router data transfer rate in excess of 16Gbit/s for each link. In the best case each router hop adds just 1 clock cycle to the final communication latency.

I. INTRODUCTION

Transistor switching speeds are continually improved through scaling. Unfortunately, the impact of scaling on long wires is a negative one. This forces an increase in communication latencies and the energy required to communicate each bit of information. The growing disparity between communication and switching times will soon make the provision of a chip-wide communication infrastructure a central problem in achieving performance and power dissipation goals. The resulting shift in design trade-offs will lead to an era of "communication-centric" system design.

While constant length global wires fail to scale well, the distance reachable in a single clock cycle in multiples of λ does remain essentially constant [7]. This allows the performance of designs of fixed complexity to scale when ported to the next technology node. This observation leads to the concept of a scalable architecture composed of a number of tiles or modules of fixed complexity. The performance of each tile scales as expected and additional performance is possible by adding tiles as scaling permits. Inter-tile communication is handled by an on-chip network which consumes only a few percent of the total chip area. The way in which such a system could scale is illustrated in Table I. In this example the die size is assumed to be fixed at $256mm^2$. Each tile contains around 11M transistors and the clock period is set to 16 FO4 delays. The table shows how the frequency, size (width) and number of tiles scale. The interconnect delay along one edge of a tile remains constant at around one clock cycle.

The calculation of cycle time in Table I assumes one FO4 delay may be calculated as $500ps * L_{gate}$ (where L_{gate} is the physical gate length as specified in [13]). The channel delays were estimated using results from [1], [7]. In all cases it should

Technology Node	No. of Tiles	Width of Tile	Tile Frequency
90nm	32	2.8mm	3.4GHz
65nm	64	2mm	5.0GHz
45nm	128	1.4mm	7.0GHz
32nm	256	1mm	13.0GHz

TABLE I

PREDICTED SCALING OF A GENERIC TILE-BASED SYSTEM

be possible to traverse the channel between two routers in less than one clock cycle. Of course, if a longer clock period is employed a smaller number of larger tiles may be used.

Such tile-based systems may implement arrays of homogeneous processor/cache tiles [9], [10], finer-grain computing fabrics [14] or networks of heterogeneous IP blocks. Such approaches provide highly reconfigurable platforms for a wide range of performance hungry applications. The provision of an efficient chip-wide dynamic on-chip network is fundamental in achieving performance goals, flexibility and mitigating complexity in such systems.

Packet-switched networks employing Virtual Channel (VC) flow control have recently been proposed as one approach to implementing a chip-wide interconnection network [3]. Figure 1 illustrates the major components of a generic virtual-channel router. Packets gain access to a physical channel by first obtaining a virtual-channel (*VC allocation*). Each of these virtual-channels has its own private input FIFO at the destination router allowing flits[1] from different packets to be sent in an interleaved manner. Access to a physical channel is now allocated on a cycle-by-cycle basis (*switch allocation*) amongst waiting flits from any of the buffered packets which have been assigned a VC. This scheme improves both throughput and latency when compared to a simple wormhole routed network by allowing blocked packets to be bypassed. Particular classes of traffic may be restricted to a subset of the available virtual-channels in order to provide QoS enhancements or circumvent message-dependent deadlocks.

II. SPECULATIVE ROUTER ARCHITECTURES

The description of virtual-channel flow control in Section I implies that VC allocation and switch allocation are performed sequentially. Peh and Dally [12] describe how this dependency may be relaxed if we speculate that a waiting packet will be successful in acquiring a VC. In this way both VC and switch allocation may be performed in parallel. In order to avoid a negative impact on performance, the

---

[1]A packet is composed of a number of flits (flow-control digits)

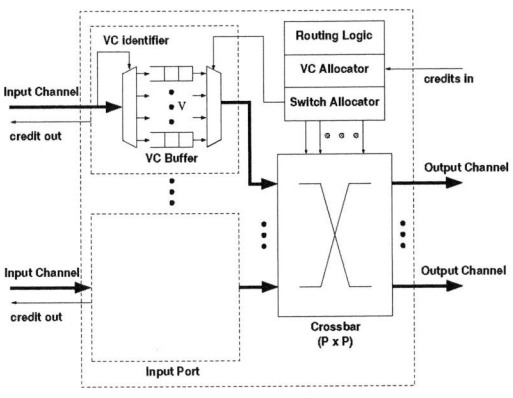

Fig. 1. A Virtual-Channel Router

switch allocator must prioritise non-speculative requests over speculative ones. This is achieved by implementing two switch allocators: one handling non-speculative requests from packets which have been allocated a VC and one for requests from packets awaiting VC allocation. We will refer to these as the *high* and *low priority switch allocators* respectively from this point onwards. Speculative requests are only granted for a particular output when no regular requests are present. In the case that a speculative request is granted we must ensure that the VC has in fact been allocated and buffer space exists downstream. Fortunately, such checks may be performed in parallel with crossbar traversal.

III. SINGLE-CYCLE ROUTERS

The introduction of further speculative optimisations to reduce the router pipeline depth to a single pipeline stage was proposed in [11]. These optimisations almost completely remove any control overhead from the critical path. Both VC and switch allocation are now performed concurrently with the transport of flits across the datapath and physical channel. The ability to make such optimisations is based on the following observations: if we assume that the network is heavily loaded it should be possible to make scheduling decisions accurately one clock cycle in advance. This is because all the information necessary to make such a decision is present when many packets are buffered. At the other extreme, when the network is very lightly loaded, we may assume that contention for a VC or physical channel is low. In this case it is also possible to schedule one cycle in advance by speculating that any new request for a VC or physical channel may be granted immediately. Simulation results predicted that for all intermediate throughputs the router sacrifices only a few percent of performance over a perfect single-cycle sequential scheme [11].

Figure 2 provides an outline of the single-cycle router architecture. In this scheme VC and switch allocation is effectively performed one cycle in advance and concurrently with the transport of flits. Each allocator's output is a set of grant-enable signals which are registered and used on the succeeding clock cycle to generate VC and switch allocation grant signals. The presence of buffered flits allow resources to be scheduled one cycle in advance, in this case the asserted grant-enable signals correspond to the subset of requests to be granted on the next clock cycle. If it is not possible to schedule a particular VC or output in advance, a prediction is made that there will be only one subsequent request for the resource. In this case multiple grant-enable signals are set. This allows any request on the next clock cycle, for the resource in question, to be successful. Cases where multiple requests are made on the following clock cycle are detected by the abort logic described in Section III-A.

Datapath control signals are produced early in the clock cycle by the *"fast"* logic blocks, simply by combining the output port requests from each buffered (or newly arrived) flit and the registered grant-enable signals. The output port required by each flit is known without the need to first evaluate a routing function by performing this task in the previous router (look-ahead routing [6]).

A. Abort Detection

One issue which must be considered carefully is the case when our prediction that requests on the next clock cycle will not contend is subsequently proven false. Fortunately, to detect these abort cases we only need consider newly arrived flits. If flits were buffered on the previous clock cycle, speculation would not have been necessary.

The abort logic associated with both VC and switch allocation consists mainly of a comparison between each of the output ports required by each new flit[2]. If we assume there are P-input ports this requires $P(P-1)/2$ comparisons. The abort logic detects cases where we are speculating and two or more flits requiring the same output port resource (physical link or VC) have arrived simultaneously. In these cases the allocation of the resource is blocked incurring a one clock cycle penalty. The correct scheduling of the resource takes place during this time and non-speculative grant-enable signals are generated for use on the following clock cycle.

In order to use the simple abort logic described above some additional logic is required to account for one remaining corner case. This is the scenario when a tail flit leaves an input buffer and exposes a new packet (buffered head flit). As this packet can now request any output port it is possible it will contend with another such packet or newly arrived flit. The solution adopted for such cases is to always stall such head flits for one cycle to ensure they are handled properly by the switch and VC allocation logic and need no further special treatment. This has a negligible impact on performance.

B. Calculating the next set of requests

To enable the VC and switch allocators to produce accurate grant-enable signals for the next clock cycle they may be fed a set of requests that we know will be present on the *next* clock cycle. These may be calculated by considering available

---

[2]At most one new flit may be received at each input port per clock cycle

2B-2

2006 Asia and South Pacific Design Automation Conference

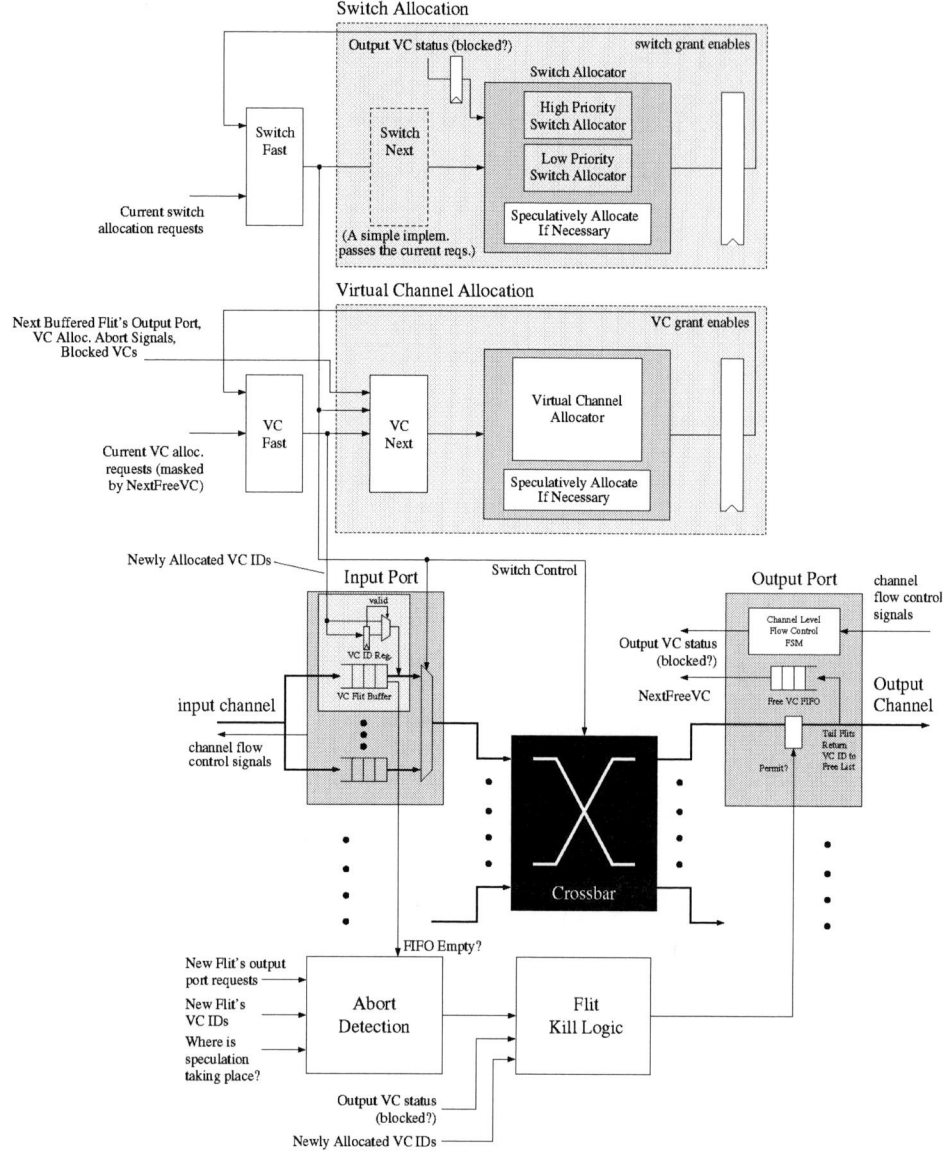

Fig. 2. A single-cycle speculative virtual-channel router architecture. When necessary the router is able to speculate that flits arriving on the next clock cycle may be routed without contention. During switch allocation the router is also able to speculate on the successful acquisition of VCs by new packets and on the availability of buffer space at the flit's destination.

information such as the current requests and those granted on the current cycle. Information about the next buffered flit in each VC buffer may also be exploited.

To ensure that the abort logic is the only place where we need to handle mispredictions, it is important that the set of requests output by the *next request logic* contains at least the requests from those flits already buffered. Presenting additional requests, e.g. those granted on the current cycle, may reduce performance but will not cause the router to malfunction. If requests that are to be made by buffered flits are not considered, grant-enable signals may be set speculatively enabling multiple buffered flits to gain access

to the same output (or VC). As only newly arrived flits are considered by the abort logic, this problem would go unchecked.

In the final router implementation we chose to accurately calculate VC next requests using all the information available. In the case of the switch scheduler we simply used the current set of requests to schedule the switch for the next cycle. This provided a significant improvement in cycle time with a small architectural performance penalty (see comparison between *spec-fast* and *spec-accurate* in Section V). The simplification is aided by the fact that those switch requests recently granted have a low arbitration priority.

C. Pipelining the use of VC state

The use of VC status information provides an example of how internal control paths may be pipelined with only minor changes to the architecture. In order to reduce cycle time it was advantageous to pipeline the VC status data used by the switch allocation logic. By adding a pipelining register, the information provided to the switch allocator about which VC is blocked becomes more out-of-date. In order to ensure the quality of the switch schedule does not suffer significantly the availability of both high- and low-priority switch allocators is exploited. If a request is associated with a VC that appears to be blocked it is steered to the low-priority allocator. Actual VC blocked status is checked when the flit is selected for transport (in parallel with its journey to its output port).

This sort of modification is simplified by the way in which the architecture decouples scheduling from the datapath. The allocator's task is simply to provide the best schedule it can for the next clock cycle with the information available. Final checks on the validity of the schedule are delayed until the schedule is applied. If advantageous, further pipelining of the control logic internally could be exploited without compromising the best case single cycle routing latency. This could involve further pipelining of the allocators themselves.

IV. IMPLEMENTATION

The Lochside test chip consists of 16 traffic generating tiles interconnected by a 4x4 mesh network. The chip is implemented in UMC's L180 logic process (1.8V core, 0.18μm) with all aluminium interconnect. Tiles and routers are interconnected as shown in Figure 3. Each router is connected to its neighbour using two unidirectional 80-bit channels (64-bits of data and 16-bits of control information). Each of the router's input ports support 4 virtual-channels and may buffer 4 flits on each virtual-channel.

The implementation is fully testable via traditional scan chain techniques. An on-chip PLL may be used to provide a clock source and is distributed to each tile using a simple hand-crafted H-tree. Alternatively, a Distributed Clock Generator (DCG) [5] may be selected as the global clock source. In both cases, tile level clock distribution was achieved by running a standard-cell clock tree synthesis tool.

The vast majority of the design is implemented in a standard cell style. Exceptions include the DCG nodes and latch-based virtual-channel buffers which benefited from a full-custom implementation. The final router design was generated from a highly parameterised network router model that allows a range of router designs to be synthesized.

The performance of the design is limited by our current PGA package (due to both thermal and bond-wire IR drop issues). This limits the performance when running all traffic generators to around 250MHz. If only two random packet sources are enabled the maximum clock rate may be increased to 300MHz.

Once a flit is received at a router's input port it may be allocated a virtual-channel and access to an output port, traverse the crossbar and arrive at the destination router in a single clock cycle (best case latency is simply one cycle per hop). At 250MHz each router is able to transfer data at a maximum rate of 16Gbits/s on each input and output link.

V. RESULTS

Each tile's traffic generator is able to produce a wide range of traffic patterns. Traffic destinations may be selected randomly or deterministically with control over packet length. Error detecting code and flit ordering checks are also performed at each tile. Each tile maintains statistics on the number of packets sent and received, together with the timing information necessary to calculate average latency and throughput. Each tile is able to inject traffic at a controlled rate into a tile output queue. Packets injected into this queue when it is full may be counted. The configuration system also provides the necessary control logic in order to synchronise the execution of commands at each tile, e.g. in order to start and stop all tiles simultaneously.

Figure 5 shows the recorded average packet latency versus measured throughput for our 4x4 mesh network. For each experiment 16K packets were sent to uniformly distributed destinations from each tile. Curves are plotted for a range of fixed packet lengths. Experiments which resulted in the tile output queue becoming full and generated packets being dropped are not plotted.

A. Performance

The router was synthesized to operate at 200MHz under worst case PVT operating conditions (around 35 FO4 including clocking overhead). If the speculative scheduling optimisations are removed but VC and switch allocation is still performed in parallel, the clock period is extended by a factor of 1.65. It may be noted that the optimised router does

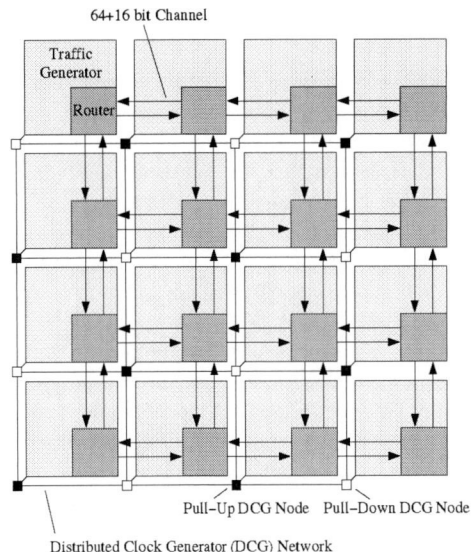

Fig. 3. Block Diagram of the Lochside Chip

2B-2

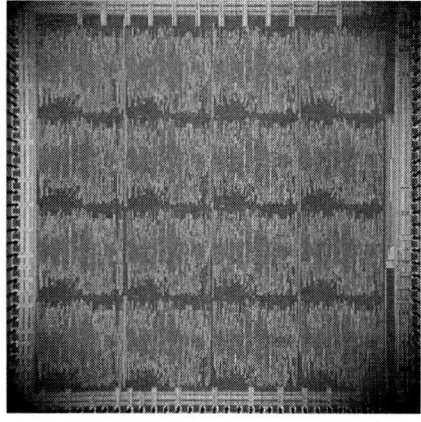

Fig. 4. Lochside Die Micrograph. Die size is 5mm x 5mm. The chip contains approximately 5 million transistors.

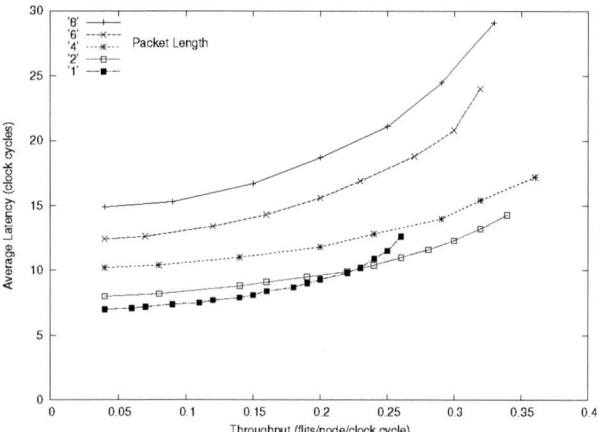

Fig. 5. Latency versus throughput measured from test-chip for a range of packet lengths.

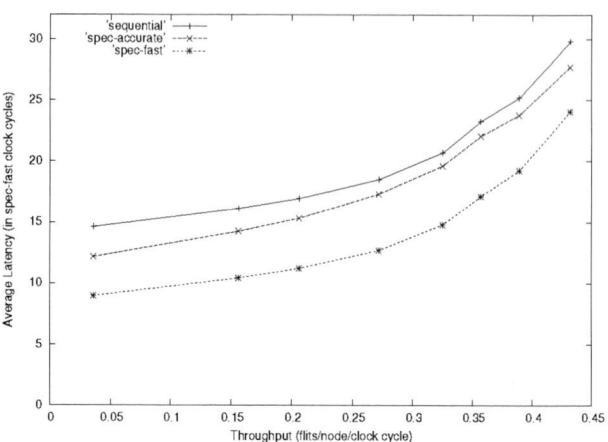

Fig. 6. Latency/Throughput comparison for three router architectures. Latency is scaled to account for differences in each router's cycle time.

not quite achieve a speedup equal to the reduction in clock cycle time. This is due to our router producing a slightly inferior routing and VC allocation schedule as a result of the approximations exploited to reduce cycle time. It is not, as may be expected, directly as a consequence of VC and switch aborts. The number of aborts is in fact consistently very low. Overall our speculative scheduling optimisations reduce average communication latency by a factor of 1.3 to 1.6.

Figure 6 plots latency against throughput for three router architectures: (*spec-fast*) full speculation with no switch next request logic, (*spec-accurate*) full speculation with accurate switch next request logic and (*sequential*) concurrent switch and VC allocation followed by switch traversal in the same clock cycle. Each architecture was synthesized to calculate its minimum clock period and the latency figures scaled to account for these differences. The clock period results were, 30, 41 and 50 FO4 delays respectively (including 2 FO4 of clock uncertainty). The packet length in these experiments was

256-bits (4 flits).

In our final design the switch allocation critical path is composed of the following delays: (32%) input register, steering and buffering switch request to allocator, (53%) switch allocation, (15%) selecting speculative or non-speculative switch allocator result and speculatively setting grant-enables if necessary.

The performance achieved by this implementation closely tracks that predicted by earlier router simulation models.

B. Area

A 4x4 mesh network is not really practical for the size of our test chip or its technology. The small tile size is dominated by the area of each router (more than two thirds of a tile's area is taken by the network). However, if we move to the next technology node (130nm) and imagine a larger chip (4x4 array of 3mm x 3mm tiles) synthesis results have shown that the network area overhead is reduced to only $5 - 6\%$. This overhead would drop even further if the scaling in Table I was adopted.

The area overhead of the speculative single-cycle architecture is small at around 8%. This compares the area of two single cycle routers one with our optimisations and one without. If the unoptimised case was pipelined, the difference in area would fall as registers would be required to buffer intermediate results.

VI. RELATED WORK

Our implementation compares favourably to other on-chip network designs and implementations published to date. Comparable networks which have been implemented include the Philips Æthereal network-on-chip [4] and the RAW processor's dynamic networks [14]. The Philips team report a similar peak link bandwidth of 16Gbit/s while operating at 500MHz in a 130nm process. Control decisions are actually taken at 166MHz. The router differs from ours in its ability to offer guaranteed services by reserving consecutive

168

routing slots in consecutive routers. Virtual-channels are not supported for improving the performance of best-effort traffic. RAW's dynamic networks operate at 225 MHz (worst-case PVT). In this case the whole network is duplicated in preference to exploiting virtual-channels. The RAW processor was implemented using IBM's 180nm 6LM ASIC copper process

Other research projects include Netchip [8] which aims to automatically generate application-specific on-chip networks. The authors emphasise the need to maintain a high switch operating frequency and adopt a deeply pipelined architecture (7-stage router). Unfortunately this significantly increases buffering requirements by extending round-trip time. It also incurs a significant overhead in terms of the additional pipelining registers required. Even if a very short clock period of less than 10 FO4 is possible, best case communication latencies would still be more than double that of our current single cycle design.

A study of virtual-channel router implementations undertaken by Peh and Dally [12] suggests that an on-chip network typically requires 3 pipeline stages operating at a clock frequency of 20 FO4. While our actual switch and VC allocator implementations offer improvements over the published delay models, clocking and test overheads and internal buffering delays extend our clock period to around 35 FO4 in the final implementation. Improvements to the input port logic to reduce this delay are ongoing. Even at 35 FO4 our network's best case latency would be nearly half that of their reported pipelined design.

A. Global Synchronisation

The speculative techniques at the heart of our router exploit the presence of a global clock. Global synchronisation offers regular snapshots of state and ensures the system proceeds in a deterministic fasion. This simplifies the implementation of the speculative scheduling mechanisms and ensures abort detection and handling mispredictions is relatively simple.

The cost of providing a low-skew high-frequency global clock is in both its complexity and the power it consumes. In many designs this cost may be considered to be too high. This has prompted asynchronous on-chip interconnect techniques to be investigated [2], [15]. While such approaches are promising, it is also possible to make similar trade-offs while retaining a synchronous router implementation. The first approach is to exploit known relationships between router clock signals while relaxing global synchronisation. Examples include source-synchronous communication and the use of clock predictive synchronisers. Global synchronisation may be relaxed further by generating clock pulses locally on demand or in a data-driven manner. This allows each router to operate at a rate dictated by the data it is transporting. This both reduces synchronisation overheads and provides a simple high-level approach to clock gating. Work in this area is ongoing. Techniques such as the DCG [5] may also be employed as previously discussed.

VII. CONCLUSION

This paper has detailed the design of an on-chip network which can provide an efficient global communications infrastructure for future gigascale ICs. A speculative architecture is able to accurately produce datapath control signals one cycle in advance of their use. This enables both datapath and control logic to operate concurrently providing significant latency improvements over previously published work. A number of trade-offs between cycle time and speculation accuracy have also been introduced and evaluated.

The optimisations proposed are orthogonal to other well known techniques for boosting performance such as adaptive routing and are independent of the network topology selected.

ACKNOWLEDGEMENTS

This work is supported by EPSRC (grant GR/L86326) and the Cambridge-MIT Institute.

REFERENCES

[1] V. Agarwal, M.S.Hrishikesh, S. W. Keckler, and D. Burger. Clock Rate versus IPC: The End of the Road for Conventional Microarchitectures. In *Proceedings of the 27th Annual International Symposium on Computer Architecture (ISCA)*, 2000.

[2] J. Bainbridge and S. B. Furber. Chain: A delay-insensitive chip area interconnect. *IEEE Micro*, 22(5):16–23, 2002.

[3] W. J. Dally and B. Towles. Route Packets, Not Wires: On-Chip Interconnection Networks. In *Proceedings of the 38th Design Automation Conference (DAC)*, June 2001.

[4] J. Dielissen, A. Radulescu, K. Goossens, and E. Rijpkema. Concepts and Implementation of the Philips Network-on-Chip. In *IP-Based SOC Design*, Grenoble, France, Nov 2003.

[5] S. Fairbanks and S. Moore. Self-timed circuitry for global clocking. In *Proceedings of the 11th International Symposium on Asynchronous Circuits and Systems*, 2005.

[6] M. Galles. Scalable Pipelined Interconnect for Distributed Endpoint Routing: The SGI SPIDER Chip. In *Proceedings of Hot Interconnects Symposium IV*, 1996.

[7] R. Ho. *On-Chip Wires: Scaling and Efficiency*. PhD thesis, Stanford University, 2003.

[8] A. Jalabert, S. Murali, L. Benini, and G. D. Micheli. xpipesCompiler: A tool for instantiating application specific Networks on Chip. In *Design, Automation and Test in Europe (DATE)*, Paris, France, Feb 2004.

[9] R. Krashinsky, C. Batten, M. Hampton, S. Gerding, B. Pharris, J. Casper, and K. Asanovic. The Vector-Thread Architecture. In *31st International Symposium on Computer Architecture (ISCA-31)*, Munich, Germany, June 2004.

[10] K. Mai, T. Paaske, N. Jayasena, R. Ho, W. Dally, and M. Horowitz. Smart Memories: A Modular Reconfigurable Architecture. In *27th International Symposium on Computer Architecture (ISCA-27)*, June 2000.

[11] R. D. Mullins, A. F. West, and S. W. Moore. Low-Latency Virtual-Channel Routers for On-Chip Networks. In *Proceedings of the 31st Annual International Symposium on Computer Architecture (ISCA)*, 2004.

[12] L.-S. Peh and W. J. Dally. A Delay Model and Speculative Architecture for Pipelined Routers. In *International Symposium on High-Performance Computer Architecture*, pages 255–266, Jan 2001.

[13] Semiconductor Industry Association. International technology roadmap for semiconductors (2004 update), 2004.

[14] M. B. Taylor et al. Evaluation of the Raw Microprocessor: An Exposed-Wire-Delay Architecture for ILP and Streams. In *The 31st Annual International Symposium on Computer Architecture (ISCA-31)*, Munich, Germany, June 2004.

[15] T.Felicijan and S.B.Furber. An Asynchronous On-Chip Network Router with Quality-of-Service (QoS) Support. In *Proceedings IEEE International SOC Conference*, pages 274–277, Santa Clara, CA, September 2004.

2B-3

2006 Asia and South Pacific Design Automation Conference

A Near Optimal Deblocking Filter for H.264 Advanced Video Coding

Shen-Yu Shih   Cheng-Ru Chang   Youn-Long Lin

Department of Computer Science
National Tsing Hua University
Hsin-Chu, Taiwan 300
Tel : +886-3-573-1072
e-mail: ylin@cs.nthu.edu.tw

Abstract - We propose a near optimal hardware architecture for deblocking filter in H.264/MPEG-4 AVC. We propose a novel filtering order and a data reuse strategy that result in significant saving in filtering time, local memory usage, and memory traffic. Every 16x16 macroblock requires 192 filtering operations. After a few initialization cycles, our 5-stage pipelined architecture is able to perform one filtering operation per cycle. Compared with some state-of-the-art designs, our architecture delivers the fastest level of performance while using much smaller gate count and memory. We have implemented and integrated the proposed deblocking filter into an H.264 main profile video decoder and verified it with an FPGA prototype.

I. Introduction

H.264/MPEG-4 AVC is an emerging video coding standard [1][2]. Compared with the most popular standard MPEG-2, it can save more than half of the bit-rate. The saving is gained from heterogeneous video coding algorithms, such as multi-mode intra-prediction, multi-frame variable-block-size quarter-pixel-accurate inter-prediction, integer discrete cosine transform (DCT), context adaptive binary arithmetic coding (CABAC), and deblocking filter. One of the most special features in H.264/MPEG-4 AVC is deblocking filter [3]. It is applied to reduce the blocking artifact generated by block-based motion compensated prediction, intra prediction, and integer discrete cosine transform. In H.264/MPEG-4 AVC, the filter for eliminating blocking artifact is embedded within the coding loop. Therefore, it is also called in-loop filter. According to some experiments, it is able to achieve up to 9% bit-rate saving [4] at the expense of large amount of computation. Even with the fastest CPU, it is hard to perform software-based real-time decoding or encoding of high quality video sequences. Consequently, a hardware accelerator is indeed required.

Fig. 1 shows an H.264 main profile decoder proposed by our research laboratory. The Variable-Length deCoding (VLC) module reads in encoded video stream and generates slice-level parameters for several other modules and macroblock-level bit-stream information for the CABAC module. The CABAC module generates syntax elements and stores them into *MBinfo mem* and *Coeff mem*. Then, according to the current slice type, one of either Motion Compensation (MC) or Intra Prediction (Ipred) module is

activated to perform compensation. Meanwhile, the Inverse Quantization and Inverse DCT (IQ/IDCT) module reads coefficient data from *Coeff mem* and transforms them back to residuals. The Picture Reconstruction (Pic Rec) module combines the compensated data with residuals. Finally, the Deblocking Filter (DF) module gets reconstructed data to perform filtering and outputs the filtered macroblock to *refMB mem* for reference and display. This paper presents our design and implementation of the DF module.

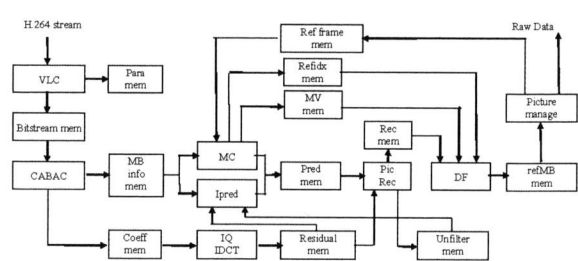

Fig. 1. An H.264 decoder that employees the proposed filter.

The rest of this paper is organized as following. Section II describes the deblocking filter algorithm. Section III presents our hardware architecture in detail.   In Section IV, we present our synthesis and FPGA-prototyping results and compare it with previous work. Finally, we draw some concluding marks and point to possible directions for future research in Section V.

II. Deblocking Filter Algorithm

A. Overview

The deblocking filter is used to eliminate blocking artifact and thus generate a smooth picture. The inter prediction module finds a block similar to the current block from reference frames. The found block usually cannot perfectly match with the current block resulting in prediction error. For coding efficiency, the error is DCT-transformed and quantized. After the decoding process, the reconstructed block is different from the original block. Especially, discontinuity is likely to appear at the block edge. To alleviate the degree of discontinuity, the deblocking filter process is applied.

0-7803-9451-8/06/$20.00 ©2006 IEEE.

170

Inputs to the deblocking filter include pixels, boundary strength, and threshold values as shown in Fig. 2. The pixels of a macroblock are filtered by an edge filter in a specific order, and each pixel may be filtered multiple times. After the whole picture is filtered, it is ready for display as well as being a reference picture.

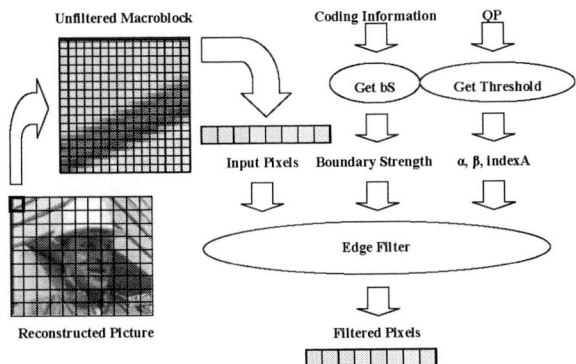

Fig. 2. Inputs to and outputs from the deblocking filter.

B. Filter Order

The deblocking filter process consists of a horizontal filtering across all vertical edges and a vertical filtering across all horizontal edges. Fig. 3(a) illustrates the filtering process for the 16x16 luma component of a macroblock. Each small box denotes a pixel, and a dotted one represents a pixel from neighboring macroblocks. The top part shows the horizontal filtering. Vertical edge 0 is filtered horizontally first from top to bottom, followed by edge 1, edge 2, and edge 3. For luma filtering, the edge filter takes as its inputs eight pixels, p3, p2, p1, p0, q0, q1, q2, and q3. At most 6 pixels will be modified by the filter as shown in the shadowed part of the figure. Because there are overlapping area between the filtering of two adjacent edges, some pixels (actually, half of them) may be filtered twice.

The vertical filtering shown in the bottom part of Fig. 3(a) is performed after horizontal filtering in a similar way. Edge 0 is vertically filtered from left to right, followed by edge 1, edge 2, and edge 3.

The filtering process of chroma components is similar to that of luma components as depicted in Fig. 3(b). It is first horizontally applied on edge 0 from top to bottom, followed by edge 1. After the vertical edges are filtered, the horizontal edges are then filtered from edge 0 to edge 1. Note that unlike a luma edge which is of length 16, a chroma edge is of length 8, and there are only 5 input pixels, p1, p0, q0, q1, and q2 with two possible pixel modifications per filtering.

C. Boundary Strength

The boundary strength (bS) is derived from the coding information [5] of the macroblock. Two adjacent 4x4 blocks share a bS value. Its value ranges from 4 to 0, 4 for the strongest filtering and 0 for no filtering. Fig. 4 gives a flowchart for calculating bS value. If any one of the two

adjacent 4x4 blocks is coded with intra prediction mode and they are on the macroblock edge, the bS is set to 4. If any one of them is intra-coded and they are not on the MB edge, the bS is 3. If any of them contains non-zero transform coefficients, the bS is set to 2. Finally, if different reference pictures are used or the difference between two motion vectors of the two blocks is greater than or equal to 4 in units of quarter pixels, the bS shall be equal to 1. For the remaining cases, the bS is set to 0.

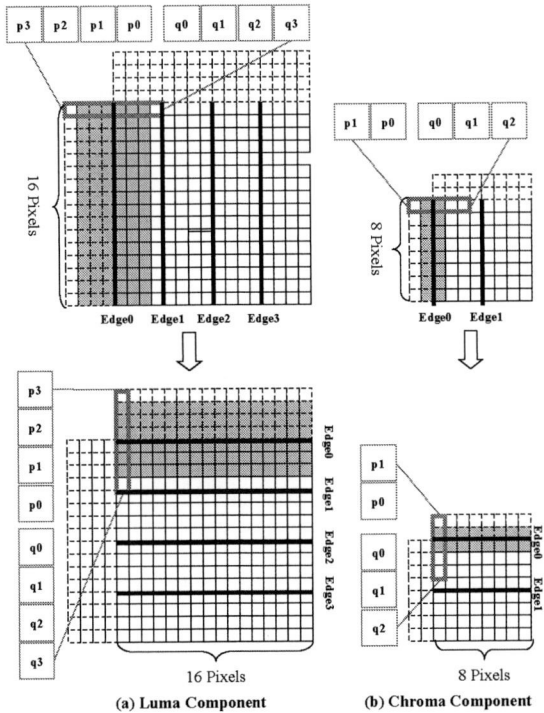

Fig. 3. Horizontal filtering and vertical filtering of luma component (a) and chroma component (b).

D. Threshold

Three threshold variables, α, β, and indexA, are used to prevent true edges from being filtered. Their values depend on the quantization parameters as described in [2]. The flag *filterSamplesFlag* is used to decide whether the filtering process should be carried out. It is set to true if (1) is true.

$$bS\,!=0\ \&\&\ |p0\text{-}q0|<\alpha\ \&\&\ |p1\text{-}p0|<\beta\ \&\&\ |q1\text{-}q0|<\beta \qquad (1)$$

E. Edge Filter

The edge filter starts to filter when the input pixels, boundary strength, and threshold variables are ready. First, if the flag *filterSamplesFlag* is equal to 1, the current edge is very likely to be a blocking artifact instead of a true edge. Thus, the filtering process should be applied. If the filtering process needs to be performed, there is a branch depending on the value of bS. If bS is smaller than 4, there are at most 4 pixels to be modified. Otherwise, there are at most 6 pixels to be modified. The detailed filtering operations are listed in [2].

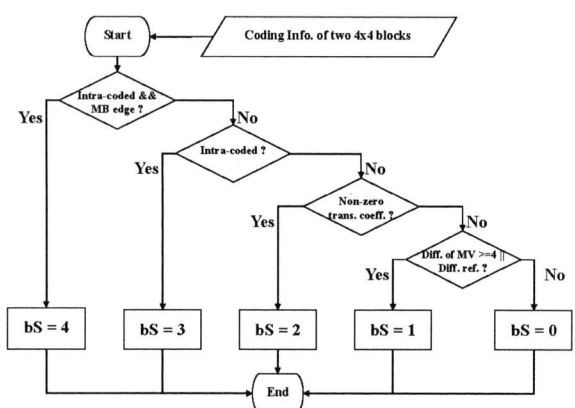

Fig. 4. Flowchart of boundary strength calculation.

III. Proposed Architecture

A. Deblocking Filter Architecture

Fig. **5** gives a top view of our deblocking filter architecture. The blocks outside the dotted box are local memories of our H.264 decoder. Through these memories or buffers, the deblocking filter gets data produced by other modules in different pipelined stages. For example, *reconstruct mem* stores the reconstructed pixels combining the data from the motion compensation unit and the IQ/IDCT unit. The coding information in such memories as *Ref idx mem, MV mem, Para mem, MBinfo mem* is used for calculating the boundary strength. After the filtering process completes, the output data is written back to *refMB mem*.

Inside the dotted box is our implementation of the deblocking filter. The module *Generate bS & Threshold* fetches data from external memories to calculate bS and threshold values. Two local memories *local mem 0* and *local mem 1* are used for storing pixels from neighboring macroblocks. Two transpose registers, *T0* and *T1*, are used for buffering and transposing pixels. In the center of the dotted box is the edge filter with 5 pipeline stages. After the pixels are filtered, the results will be written out via the *Write Back Unit*.

B. Local Memory Organization

Fig. **6** shows our memory organization. There are three local memory modules. The pixels of the currently under-filtered macroblock are stored in *reconstruct mem*. The two-port SRAM, *local mem 0*, stores the intermediate results of filtering process. For data reuse, we use a single-port SRAM, *local mem 1*, to buffer a frame-wide row of 4x4 blocks. Note that chroma filtering requires only half of 4x4 block. The size of *local mem 1* depends on the frame

width. For example, for CIF video, the memory is (1.5x352) x 32 bits.

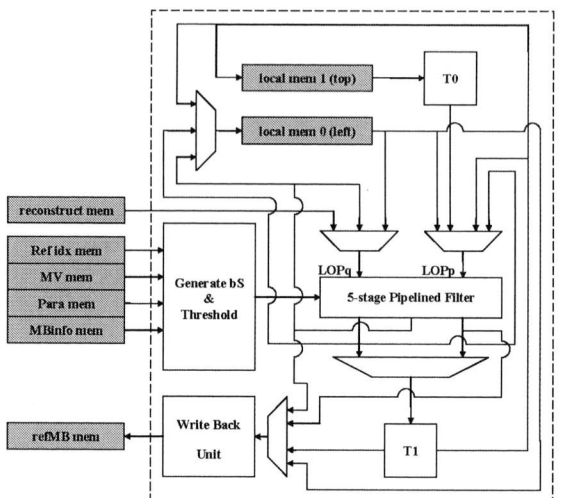

Fig. 5. Proposed deblocking filter architecture.

C. Filtering Order

Our filtering order is illustrated in Fig. **7**. Each circle stands for a step with 4 cycles. In order to preserve the rightmost column (i.e., B3, B7, B11 and B15 for luma component) of the current macroblock for the filtering of the next macroblock, we filter the edges from the left to right.

In Step 1, blocks L0 and B0 are read from *local mem 0* into the edge filter. In Step 2, Blocks L1 and B4 are filtered while Block B0 is stored back to *local mem 0*, and Block L0 is written out via *Write Back Unit*. In Step 3, Blocks L2 and B8 are filtered while Block B4 is stored back to *local mem 0*, and Block L1 is written out via *Write Back Unit*. In Step 4, Blocks B0, B4, B8, and B12 are filtered horizontally, and Block LT0 is loaded into transposed register *T0*. Note that Block B12 is still in the pipelined filter. In Step 5, blocks B0 and B1 are horizontally filtered. In Step 6, transposed blocks LT0 and B0 are vertically filtered. With the proposed filtering order, we can filter a macroblock in 192 cycles, which is optimal.

D. Pipelined Filter

Fig. **8** depicts our 5-stage pipelined filter architecture. Stage 1 reads pixels from various memories. Stage 2 calculates such parameters as *filterSamplesFlag* described in Section II. Stage 3 filters pixels with bS equal to 4. In Stage 4, pixels with bS equal to 3, 2, or 1 are filtered, and clipping performed. Finally, Stage 5 stores filtered pixels back to memory or transpose registers.

Multiplexers are added to resolve pipeline hazards. Let's take filtering Step 5 and Step 6 shown in Fig. 9 as an example. Filtering Step 6 requires the transposed pixels of Block B0 and LT0. However, pixels of Block B0 are still in

172

the pipeline. Therefore, we add some forwarding logic to get register transferring behavior as illustrated in Fig. 10.

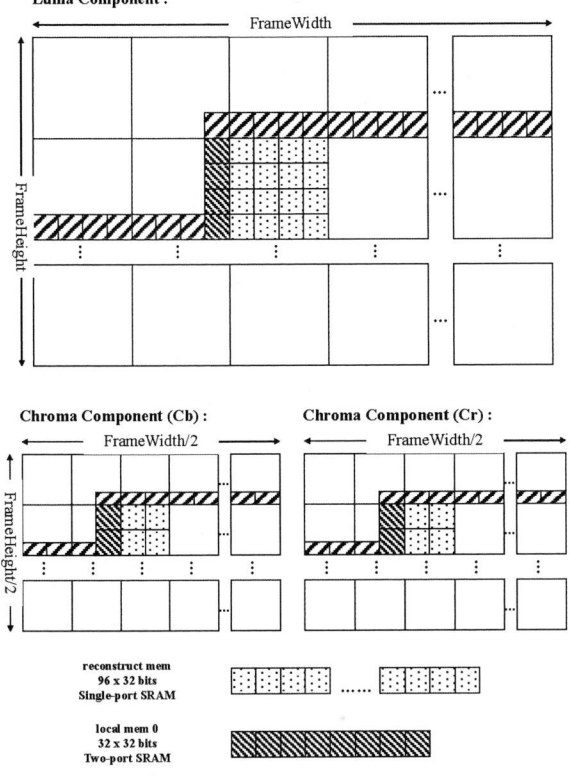

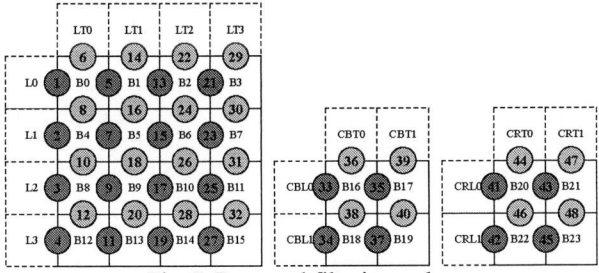

Fig. 6. Local memory organization.

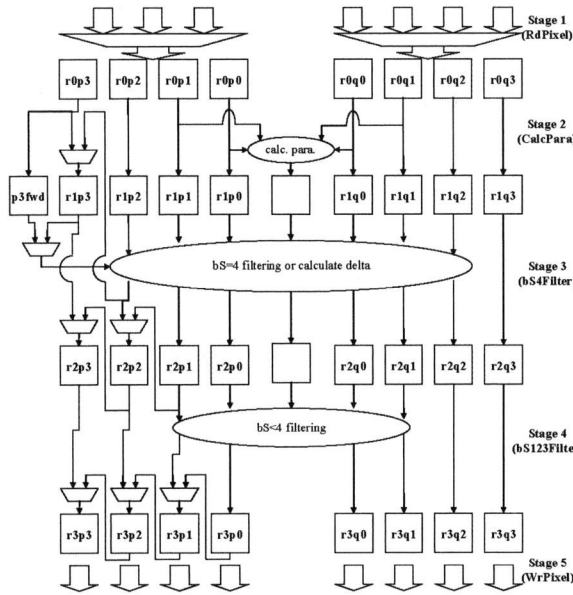

Fig. 8. Proposed 5-satge pipelined edge filter.

At the next cycle, parameters such as *filterSamplesFlag* have been calculated. Note that it requires 8 pixels, a30, a31, a32, a33, b30, b31, b32, and b33, to perform filtering for block B0 and B1 in Stage 3. The forwarding pixels are not fed back to the edge filter. Instead, it is directly output to the transpose register *T1* or *Write Back Unit* in some cases such as Step 7. We use the register *p3fwd* as shown in Fig. 8 to keep the required pixel, a30.

The multiplexers in the data path depicted in Fig. **8** denoted the forwarding paths.

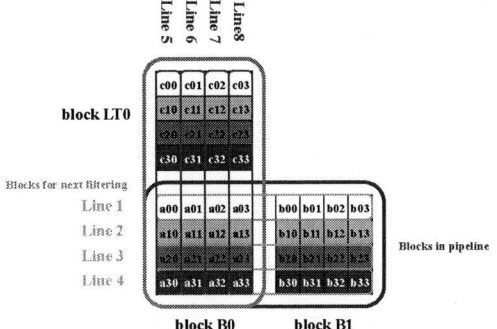

Fig. 9. Pipeline hazard illustration.

Fig. 7. Proposed filtering order.

When we begin to do the first filtering (Line 5) of Step 6, we need pixels of Line 1. As described in Sub-Section C, we have put Block LT0 into transpose register *T0* in Step 4, and thus each column of Block LT0 can be read. To get the lower part of Line 5, which is still inside the pipe stage, we insert a forwarding logic to select pixels marked with a00, a10, a20, a30 from different pipeline stages.

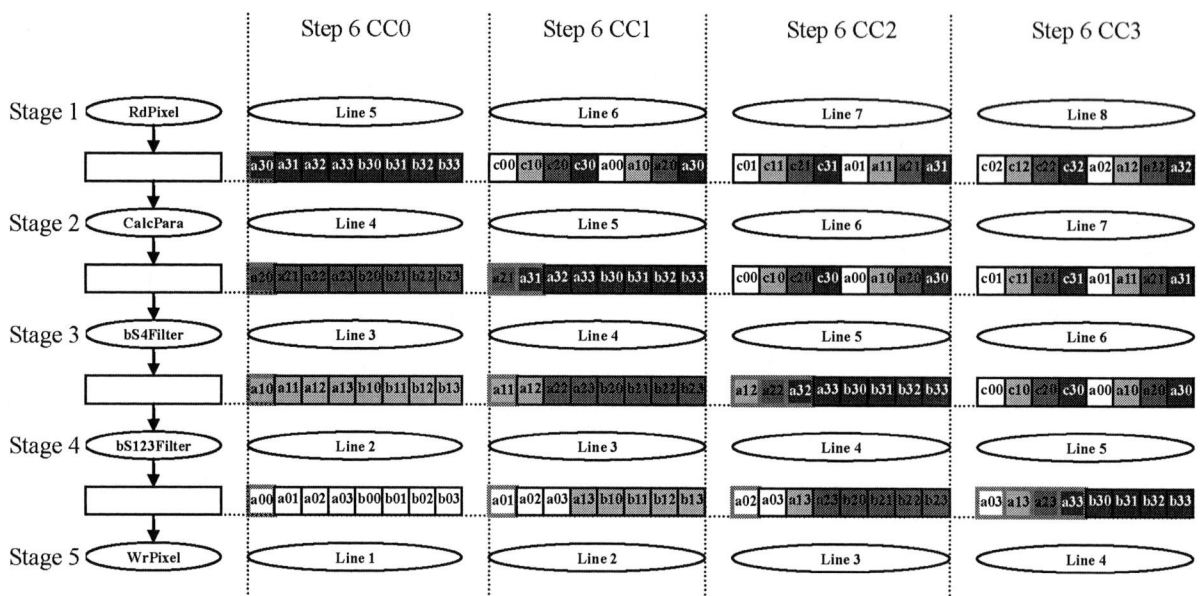

Fig. 10. Pipeline forward for filtering step 6.

IV. Experimental Results

We have implemented the proposed deblocking filter architecture in Verilog RTL and verified its integration with an H.264 decoder with FPGA prototyping. We synthesize our design using Synopsys Design Compiler targeted towards an Artisan UMC 0.18um cell library. The clock frequency is set to 100MHz.

Fig. 11 analyzes the number of required processing cycles of the proposed architecture. At the beginning, we take 14 cycles to read coding information [5] necessary for filtering the first pair of 4x4 blocks. It takes 192 cycles for both filtering and writing out processed pixels. This is optimal because there are exactly 192 filtering operations needed according to the following calculation. For the luma component, in each of the vertical and horizontal filtering process, there are 4 edges each requires 16 filtering operations. Therefore, we need 2x4x16 = 128 filtering operations. For the two chroma components, the number is 2x2x2x8 = 64. Therefore, the total is 128+64 = 192.

After initialization, the calculation of boundary strength is overlapped with filtering. After filtering, we need 8 extra cycles for flushing the pipeline. If the current macroblock is not the rightmost macroblock of a row of the picture, filtering one MB requires 14+192+8 = 214 cycles; otherwise, 32 additional cycles are needed to write out the rightmost column of 4x4 blocks to the external memory. Taking these 32 cycles into account, we need 246 cycles to filter one MB in the worst case. The average number of cycles per MB for a video sequence in CIF format is about 216 cycles.

Table I compares our work with some state-of-the-art designs. In terms of total number of cycles needed per MB, ours is better than every one. References [7][9][10] do not give the numbers while Reference [8]'s numbers varies due to skip mode implementation. In terms of the number of cycles spent in the kernel filter, 192 is the optimal. Reference [7] reduces it to 136 by employing two filters. From the table, we can conclude that most of the work can do very well in the kernel filter design. However, it is the data transportation that makes the difference. Most of the design spends more time in data transport than filter operation. By means of overlapping filtering and data transport, we are able to achieve near optimal performance. We use less local memory because only one half of chroma blocks are stored. Our hardware design can easily meet the requirements for real-time decoding of video sequences with 1280x720p and 30fps resolution.

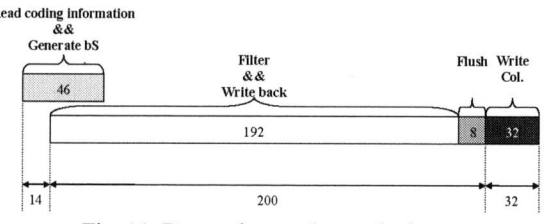

Fig. 11. Processing cycles analysis.

TABLE I Comparisons among various deblocking filters

	[5]	[6]	[7]	[8]	[9]	[10]	[11]	[12]	[13]	Proposed
Cycles/MB	614	566	N/A	Max:342 Min:50 Avg.:86-241[3]	N/A	N/A	446	238[6]	250	214 or 246[9]
Filtering Cycles/MB	240	192	136	0-342	214[5]	336	192	236	250	192
SRAM for Pixels[1]	2P 96x32 2P 64x32	8 DP 80x8	DP 88x32 DP 72x32 1P 32x32	1P 96x32	DP 16x32	1P 80x32	DP 64x32 2 2P 96x32	DP 96x32 DP 64x32 1P (2xFW[7])x32	2 1P 96x32 1P (2xFW[7])x32	1P 96x32 2P 32x32 1P (1.5xFW[7])x32
# of 4x4 Arrays	4	0	11	2	6	2	8	9	4	2
# of Edge Filters	1	1 Pipelined	2	1	1	1	1	1	1	1 Pipelined
Process(um)	.25	.35	N/A	.18	N/A	.18	.25	.18	.18	.18
Gate Count[2]	20.66K	9.35K	N/A	11.8K[4]	N/A	9.16K[4]	24K	14.5K[8]	19.64K	20.9K

1. DP : Dual-port SRAM with two R/W ports; 1P : Single-port SRAM with one R/W port; 2P : Two-port SRAM with one read and one write ports
2. Gate Count does not include SRAM
3. The performance is evaluated by QCIF video sequences with 1I+149P
4. The gate count does not include boundary strength calculation logic and coding information registers
5. The filtering cycles do not include filtering chroma components
6. The cycles do not include boundary strength calculation time
7. FW stands for Frame Width
8. The gate count does not include coding information registers
9. It takes 246 cycles to filter one MB at right picture boundary

V. Conclusions

We have proposed a near optimal architecture for deblocking filter in H.264/AVC. We implemented the design in synthesizable Verilog RTL and verified it with reference software [14]. The result shows that the performance of our design is near optimal but the usage of local memory is less compared with previous work. Besides, with a pipelined architecture, our design can achieve higher performance with increasing clock frequency. We have integrated the hardware accelerator into our H.264 decoder and verify it on a FPGA development board. The result shows that our design works correctly and the performance for decoding greatly increases compared with pure software solution or platform-based methodology. In the future, we will work on reducing the power consumption of our decoder. Meanwhile, we will use the deblocking filter in the development of both H.264 encoder and CODEC.

Acknowledgement

This research is supported in part by the National Science Council of Taiwan, the Ministry of Economic Affairs of Taiwan, and Taiwan Semiconductor Manufacturing Company under grant no. NTHU-0416.

References

[1] T. Wiegand, G. J. Sullivan, G. Bjntegaard, and A. Luthra, "Overview of the H.264/AVC video coding standard," *IEEE Trans. on Circuits and Systems for Video Technology*, vol. 13, pp. 560-576, 2003.

[2] "Draft ITU-T recommendation and final draft international standard of joint video specification (ITU-T Rec. H.264 | ISO/IEC 14496-10 AVC)," JVT G050, 2003.

[3] A. Luthra, G. J. Sullivan, and T. Wiegand, "Introduction to the special issue on the H.264/AVC video coding standard," *IEEE Trans. on Circuits and Systems for Video Technology*, vol. 13, pp. 557-559, 2003.

[4] P. List, A. Joch, J. Lainema, G. Bjntegaard, and M. Karczewicz, "Adaptive deblocking filter," *IEEE Trans. on Circuits and Systems for Video Technology*, vol. 13, pp. 614-619, 2003.

[5] Y. W. Huang, T. W. Chen, B. Y. Hsieh, T. C. Wang, T. H. Chang, and L. G. Chen, "Architecture design for deblocking filter in H.264/JVT/AVC," *IEEE Int'l Conf. on Multimedia and Expo*, 2003.

[6] L. Li, S. Goto, and T. Ikenaga, "An efficient deblocking filter architecture with 2-dimentional parallel memory for H.264/AVC," *Asia South Pacific Design Automation Conf.*, 2005.

[7] V. Venkatraman, S. Krishnan, and N. Ling, "Architecture for deblocking filter in H.264," *Picture Coding Symposium*, 2004.

[8] S. C. Chang, W. H. Peng, S. H. Wang, and T. Chiang, "A platform based bus-interleaved architecture for deblocking filter in H.264/MPEG-4 AVC," *IEEE Trans. on Consumer Electronics*, vol. 51, pp. 249-255, 2005.

[9] M. Sima, Y. Zhou, and W. Zhang, "An efficient architecture for adaptive deblocking filter of H.264/AVC video coding," *IEEE Trans. on Consumer Electronics*, vol. 50, pp. 292-296, 2004.

[10] C. C. Cheng, and T. S. Chang, "An hardware efficient deblocking filter for H.264/AVC," *IEEE Int'l Conf. on Consumer Electronics*, pp. 235–236, 2005.

[11] B. Sheng, W. Gao, and D. Wu, "An implemented architecture of deblocking filter for H.264/AVC," *IEEE Int'l Conf. on Image Processing*, vol. 1, pp. 665–668, 2004.

[12] G. Zheng, and L. Yu, "An efficient architecture design for deblocking loop filter," *Picture Coding Symposium*, 2004.

[13] T. M. Liu, W. P. Lee, T. A. Lin, and C. Y. Lee, "A memory-efficient deblocking filter for H.264/AVC video coding," *IEEE Int'l Symposium on Circuit and Systems*, 2005.

[14] JVT H.264/AVC Reference Software JM 8.3

Image Segmentation and Pattern Matching Based FPGA/ASIC Implementation Architecture of Real-Time Object Tracking

K. Yamaoka, T. Morimoto, H. Adachi, T. Koide, and H. J. Mattausch

Research Center for Nanodevices and Systems, Hiroshima University,

1-4-2 Kagamiyama, Higashi-Hiroshima, 739-8527, Japan

Phone:+81-82-424-6265, Fax:+81-82-424-3499

Email:yamaoka, morimoto, adachi, koide, hjm@sxsys.hiroshima-u.ac.jp

Abstract— A novel algorithm for object tracking in video pictures, based on image segmentation and pattern matching, as well as its FPGA/ASIC implementation architecture are presented. With image segmentation, we can detect all objects in the images no matter whether they are moving or not. Using image segmentation results of successive frames, we exploit pattern matching in a simple object feature space for tracking of objects. The proposed algorithm can be applied to multiple moving and still objects even in the case of a moving camera. The FPGA/ASIC implementation architecture is verified to enable real-time tracking of up to 220 objects, when realized with modern FPGA hardware [1].

I. INTRODUCTION

For intelligent information processing of visual data, the moving object tracking in video pictures has attracted recently a great deal of interest [2]. Scene surveillance or object recognition are typical examples, where object tracking is an indispensable technology. Many moving object tracking algorithm and architectures have already been proposed. Most of them are based on difference evaluation between the current image and a previous image or a background image [3, 4]. However, algorithms based on the difference of images have problems with following practical cases. (1) Still objects included in the tracking task exist. (2) Multiple moving objects are present in the same frame. (3) The camera is moving. (4) Occlusion of objects occurs. Our novel algorithm for object tracking [5], based on image segmentation and pattern matching (Fig. 1), aims at solving above problems. In this algorithm we extract all objects from an input image by image segmentation. Next we extract simple object features and use these features to form pattern representing the objects. Then we compare the features of extracted objects in the current frame and those of extracted objects in the preceding frame by pattern matching. The most similar objects (i.e. the objects which have the smallest distance) between successive frames are judged to be corresponding objects. In spite of the motion condition of objects, this algorithm is effective because each object's motion vector is determined and used as one of its features. Additionally, we can increase the number of extracted object features from segmentation results, so

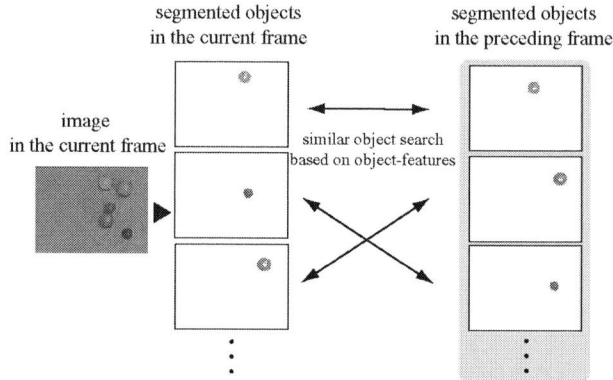

Fig. 1. Object-tracking based on image segmentation and similar object feature matching.

that detection and tracking accuracy can be improved to a suitable level. However, image segmentation processing needs a great deal of calculations, so that realizing real-time object tracking with the proposed algorithm by software implementation is difficult. That is why, we have developed an FPGA/ASIC architecture for realizing real-time object tracking. For the segmentation part we exploit a previously developed digital image segmentation architecture [6]. Furthermore, pipeline processing of segmented objects is applied in order to achieve tracking of more objects in real time.

II. OBJECT TRACKING ALGORITHM

A. Concept

In the proposed object tracking algorithm, we employ image segmentation of each frame and extraction a number of features for all segmented objects. Then pattern matching with the objects of the previous frame is carried out. A coarse flow chart of the proposed algorithm is shown in Fig. 2. The detailed processing consists of the following steps.

Step 1: With the image segmentation algorithm, we extract all objects in the input image.

Step 2: Then we extract coordinates of 4 object-pixels which are indicated in Fig. 3(a). P_{xmax} and P_{xmin} have the maximum and minimum x-component, while P_{ymax} and P_{ymin} have the maximum and minimum y-component, respectively.

Step 3: Afterwards we calculate characteristic features of

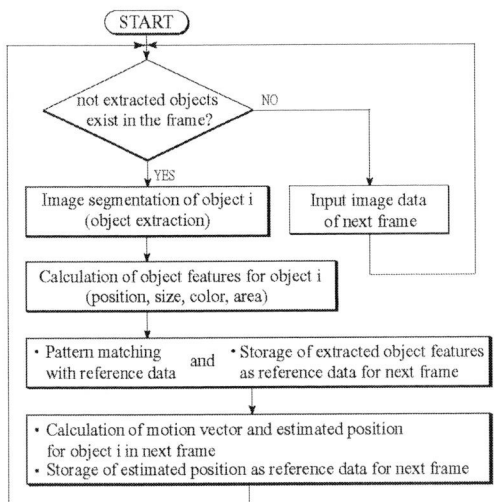

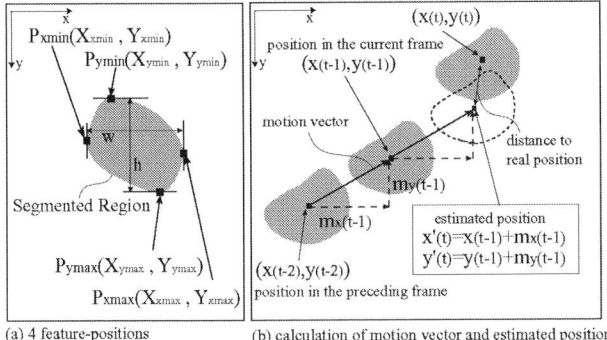

Fig. 2. Flowchart of the proposed algorithm based on image segmenation and pattern matching.

Fig. 3. Definition of four feature-positions, motion vector and estimated position.

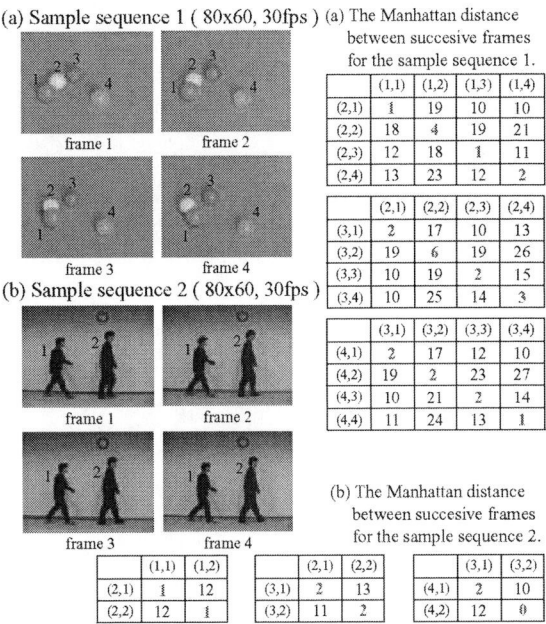

(a) Sample sequence 1 (80x60, 30fps)

(b) Sample sequence 2 (80x60, 30fps)

(a) The Manhattan distance between succesive frames for the sample sequence 1.

	(1,1)	(1,2)	(1,3)	(1,4)
(2,1)	1	19	10	10
(2,2)	18	4	19	21
(2,3)	12	18	1	11
(2,4)	13	23	12	2

	(2,1)	(2,2)	(2,3)	(2,4)
(3,1)	2	17	10	13
(3,2)	19	6	19	26
(3,3)	10	19	2	15
(3,4)	10	25	14	3

	(3,1)	(3,2)	(3,3)	(3,4)
(4,1)	2	17	12	10
(4,2)	19	2	23	27
(4,3)	10	21	2	14
(4,4)	11	24	13	1

(b) The Manhattan distance between succesive frames for the sample sequence 2.

	(1,1)	(1,2)
(2,1)	1	12
(2,2)	12	1

	(2,1)	(2,2)
(3,1)	2	13
(3,2)	11	2

	(3,1)	(3,2)
(4,1)	2	10
(4,2)	12	0

Fig. 4. Successive image sequence for algorithm simulation.

(Fig. 3(b)). This estimate position is exploited instead of the extracted position $(x(t-1), y(t-1))$ for pattern matching after a start-up phase from the 3rd frame onwards.

Step 6: By carrying out this matching procedure with all segments obtained for the current frame, we can identify all objects one by one and can maintain tracking of all objects between frames.

B. Verification

For verifying the effectiveness of the proposed object tracking algorithm, we tested sample picture sequences with 80×60 pixels per frame, consisting of four successive frames (30fps) as shown in Fig. 4. The sample sequence 1 is an example, which includes multiple moving objects and also the occlusion effect among objects. The sample sequence 2 is an example, which includes non-rigid objects. Note that the object labels are explicitly shown in the pictures. Tables (a) and (b) in Fig. 4 show Manhattan distances between the objects of successive frames. We employ the simple Manhattan distance as the distance measure, because matching quality is found to be approximately the same as with the more complicated Euclidean distance [5]. The distances express the similarity of objects, namely as the Manhattan distance approaches to 0 the similarity increases. In this table, the notation (t,i) stands for the objects i in the t-th frame. For making possible distance contributions of each object feature equal, we have normalized their numerical values. From the table(a) in Fig. 4, we can confirm correct matching between objects in successive frames and thus confirm the validity of the proposed algorithm. Furthermore, we also have confirmed the effectiveness of the proposed algorithm for many other difficult cases such as rapid direction of movement changes by object collision, rotating complex objects or non-rigid objects like walking humans (sample

the segmented object, that is, object position (x, y), object size (width, height), color information (R, G, B), and object area, respectively. Object position (x, y), width w and height h are calculated according to below equations.

$$w = X_{xmax} - X_{xmin}, \qquad h = Y_{ymax} - Y_{ymin},$$
$$x = \frac{X_{xmax} + X_{xmin}}{2}, \qquad y = \frac{Y_{ymax} + Y_{ymin}}{2}.$$

The object area is determined by counting the number of its constituting pixels. As object color information, average RGB data of the 4 pixels, P_{xmax}, P_{xmin}, P_{ymax} and P_{ymin}, are used.

Step 4: The minimum distance search in the feature space is performed between each object in the current frame and all objects in the preceding frame. Then we identify each object in the current frame with the object in the preceding frame which has the minimum distance or in other words which is the most similar object.

Step 5: Afterwards, we calculate the motion vector $(m_x(t-1), m_y(t-1))$ from the difference in position between the object in the current frame and matching object in the preceding frame (Fig. 3(b)). By adding the motion vector $(m_x(t-1), m_y(t-1))$ to the current position $(x(t-1), y(t-1))$ of the object, we determine an estimate for the object's position $(x'(t), y'(t))$ in the next frame

177

sequence 2 in Fig. 4). Since the object feature difference is also very small between successive frames for non-rigid objects, the proposed pattern matching based method can correctly track the same objects. By doing the verification based on the algorithm's software implementation, we could confirm that unless very small and very complex objects are tracked, sufficient tracking reliability is achieved with an image resolution of 80×60 pixels. For QVGA or VGA size images, we can shrink the original image to 80×60 pixel size image. Therefore, we design the FPGA/ASIC implementation architecture aiming at video pictures reduced to 80×60 pixels in size.

III. Proposed Architecture for FPGA/ASIC Implementation

This section presents the FPGA/ASIC architecture for implementing the proposed algorithm in more detail.

A. Architecture Overview

Fig. 5 shows the overall block diagram of the developed FPGA/ASIC implementation architecture. This architecture roughly consists of 4 blocks. The first block is the image segmentation cell-network in which all objects of the frame are extracted. The second block is the feature extraction block in which object features for each segmented object are calculated using the image segmentation results. The third block is the pattern matching block in which the most similar object is searched among the reference data from the previous frame. The fourth block is the estimated position calculation block in which the estimated position of each object in the next frame is calculated.

The image segmentation cell-network implements a region-growing algorithm and has the structure of a two-dimensional array of image segmentation cells corresponding to the pixels of an input image. By taking advantage of the cell-network, we can access the segmentation result of each cell in parallel in x-direction and in y-direction. This is done in the feature extraction block where the width of each segmented object is calculated from the cell-network data which is outputted in parallel into y-direction and where the height and the area of each object are calculated from the cell-network data which is outputted in parallel into x-direction. With this chosen approach it is not necessary to scan pixels belonging to a segmented region one by one, and thus high speed processing is achieved. Then the determined object-features are transmitted to pattern matching block so as to search the most similar object among the reference data in the previous frame. In the estimated position calculation block, the estimated position of the current input object in the next frame is calculated from the positions of the matched object and the input object. Then the estimated position is stored in the pattern matching block as one feature of the input object's reference pattern in the next frame.

Due to the sequential nature of the segmentation by region growing, we can apply pipeline processing, as shown in Fig. 6, to interleave the processing steps of image seg-

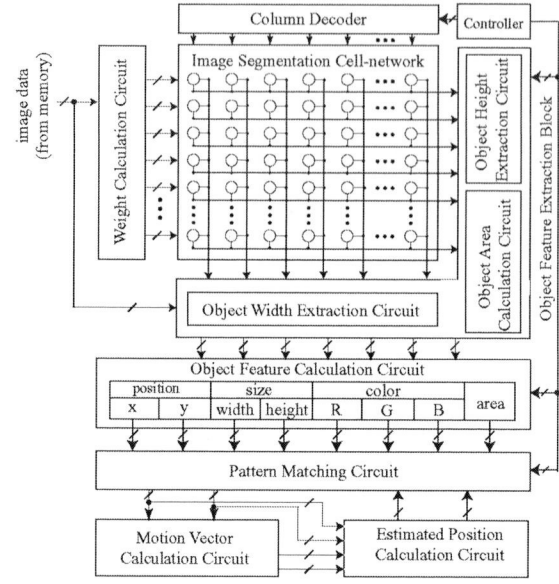

Fig. 5. Block diagram of proposed FPGA/ASIC implementation architecture.

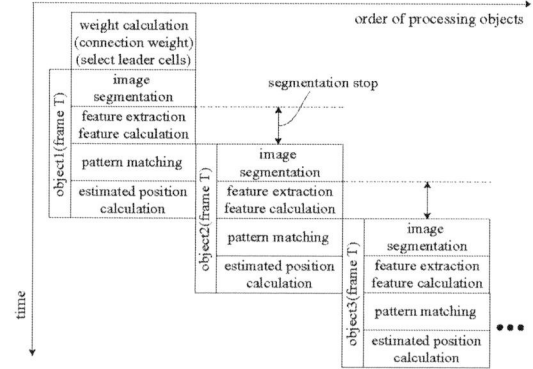

Fig. 6. Process flow of the FPGA/ASIC implementation architecture when using pipeline processing.

mentation, feature extraction and pattern matching. As compared to the case of completing the frame's segmentation before advancing to feature extraction, higher processing speed of the complete algorithm is achieved, so that more objects can be tracked in real time. This is evaluated quantitatively in SectionIV.

In following subsections, we explain the circuit structure and function of each of the 4 blocks.

B. Image Segmentation Cell-Network

The proposed implementation architecture uses a region-growing-type image segmentation cell-network [6, 7]. This cell-network realizes high-speed object segmentation from an input image. The cell-network is an array of image segmentation cells, where each cell corresponds a pixel of the input image. In a preprocessing step to the segmentation, connection weights between adjacent pixels are calculated from the pixel's RGB data in a weight calculation circuit. These calculated connection-weights are transmitted to the image segmentation cell-network,

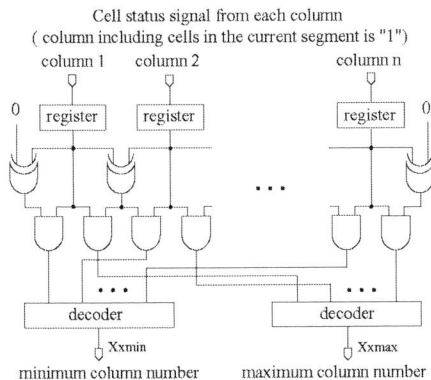

Fig. 7. Width extraction circuit for the current segment.

and image segmentation is done based on these connection weights. In the image segmentation cell-network, all segments are grown sequentially by a region-growing process starting from leader cells, which are also determined by the connection weights. After segmentation of one object, if a pixel belongs to the currently segmented object, a flag register of the corresponding cell has been set to 1 and a label number of the object has been stored in an internal register.

C. Feature Extraction and Calculation Block

C.1 Basic Feature Extraction Circuits

The feature extraction circuit is shown in Figs. 7 and 8. The four feature-positions shown in Fig. 3(a) are extracted in these circuits. The circuit shown in Fig. 7 extracts the maximum and the minimum x-coordinates, X_{xmin} and X_{xmax}, for calculating width of the segmented object. As shown in Fig. 5, status signals are transmitted from each column of the image segmentation cell-network to this circuit in parallel. The signal from each column becomes 1 if the column contains the pixel belonging to current segment. Otherwise it becomes 0. These input signals are stored in registers corresponding to each column in Fig. 7. The boundaries between 1 and 0 are detected with the shown combinational circuit consisting of EXOR and AND gates, and are transformed to the column index in the two decoders. The max-min detection unit (y-component) shown in Fig. 8 extracts the maximum and the minimum y-coordinates, Y_{ymin} and Y_{ymax}, for calculating height of the current segment and calculates the area at the same time. To extract Y_{ymin} and Y_{ymax}, the same circuit shown in Fig. 7 is used. After an image segmentation of an object finished, cell status signals (0 or 1) are inputted to registers which connected from rows of the cell-network in column parallel and Y_{ymax} and Y_{ymin} of each column are sequentially outputted from decoders. Then by inputting cell status signals of X_{xmax}-th and X_{xmin}-th column, Y_{xmax} and Y_{xmin} are calculated. To extract remaining coordinates, X_{ymin} and X_{ymax}, min and max position detection units (x-component) depicted in Fig. 8 are used. Input signals to these units are the output signal from each decoder and the current column number. When the output from each decoder becomes minimum or maximum, the current column number is

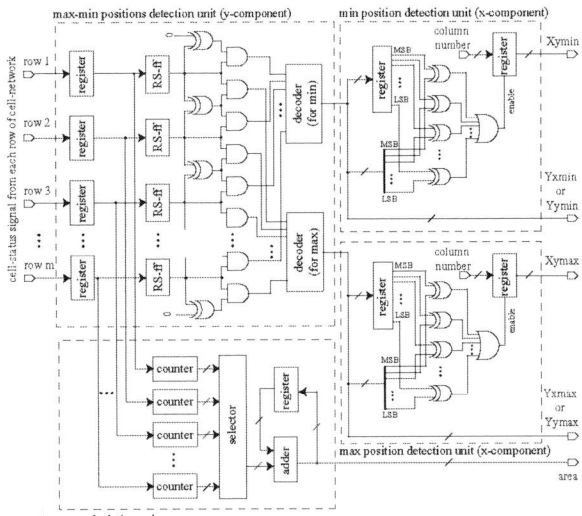

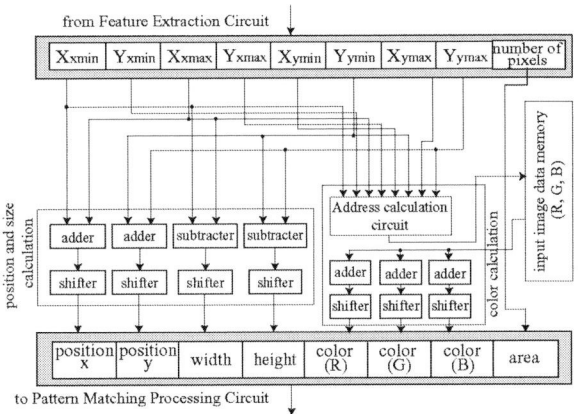

Fig. 8. Second basic feature extraction circuit which extracts height and area of the segmented region.

[Fig. 9 diagram]

Fig. 9. Dependent feature calculation circuit.

stored to each register as X_{ymin} or X_{ymax}. Concurrently, area of the current segment is calculated by counting the number of the cells in the current segment of each row, and all counter's values are finally summed up.

C.2 Dependent Feature Calculation Circuit

As shown in Fig. 9, the dependent feature calculation circuit consists of registers, adders, subtracters, and shifters. This circuit calculates the eight features, position(x, y), size(width, height), color(R, G, B) and area, by using 4 feature-positions detected with the basic feature extraction circuit and outputs these object features to pattern matching circuit. Color information is calculated by reading four boundary pixels data shown in Fig. 3(a) and taking the average of luminances (R, G, B) of four pixels.

D. Pattern Matching Processing Circuit

The structure of the pattern matching circuit is shown in Fig. 10. Two memories, memory-A and memory-B are used for pattern matching. One of the memories stores the object-features in the current frame, the other stores the object-features in the preceding frame as reference data.

179

2B-4

2006 Asia and South Pacific Design Automation Conference

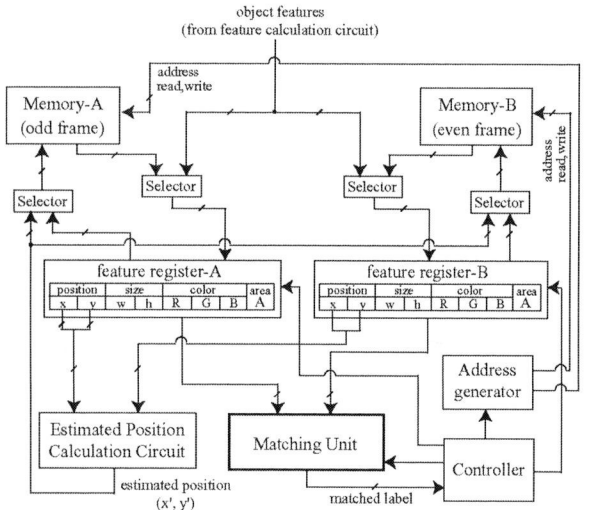

Fig. 10. Block diagram of the pattern matching circuit and estimated position calculation circuit.

For odd frames, an input data is written in memory-A and the reference data is read from memory-B. For even frames, an input data is written in memory-B and the reference data is read from memory-A. In this way, the functionarity of each memory is switched when processing frame is changed.

The pattern matching process is done in the matching unit. All combinations of the current object and all reference objects in the preceding frame are checked in the matching unit one by one. For that purpose, a feature-register-A and a feature-register-B are temporarily used as the buffer to storing the current object and a reference object. Then the matching unit compares the input data and the reference data which are set in both feature-registers.

The block diagram of the matching unit is shown in Fig. 11. This unit consists of a Manhattan distance calculation circuit which calculates the distance (i.e. similarity) between the input data and all reference data, and a Manhattan distance comparison circuit, which searches the most similar combination (i.e. the smallest distance). Each difference is calculated sequentially per object feature pair and is normarized by shifters. Afterward all distances components are added serially and the calculated Manhattan distance is transmitted to the Manhattan distance comparison circuit, and is stored in a comparison register (A or B) of the circuit with the label of the reference object. Then, this newly calculated distance is compared with up to now smallest distance between the input data and other reference data. And smaller distance is selected to the candidate of the most similar object. The selected distance is kept intact in its register and the other is deleted from its respective register. Repeating these processes for all reference data of the previous frame in memory, the most similar reference object is selected and the matched object-label is outputted from the matching unit.

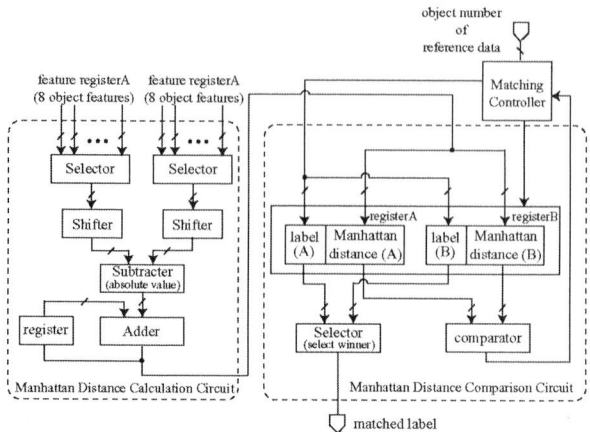

Fig. 11. Details of the matching unit.

E. Estimated Position Calculation Circuit

The estimated position calculation circuit consists of adder and subtracter and a motion vector of input object is calculated with distance between the matched object position in the preceding frame and the input object position in the current frame (Fig. 10). The estimated position of the input object for the next frame is calculated by adding the motion vector to the current position of the input object, and is stored to the reference memory to apply to the pattern matching of the next frame.

IV. PERFORMANCE ESTIMATION OF PROPOSED ARCHITECTURE

Here we estimate the processing speed of the proposed FPGA/ASIC architecture and judge the possibility of real time processing (30fps). The process flow of the proposed architecture is shown in Fig. 6 and we assumed that the achievable clock frequency with the FPGA is 20MHz in the worst case. First, when an image is inputted, the connection weights are calculated and leader cells are determined in the weight calculation circuit. These data are then transfered to image segmentation cell-network. With the assumed worst-case operating frequency of 20MHz, it takes 480μsec to finish calculating and transfering the connection-weight and leader cell data. Next, image segmentation is done, which requires about 200μsec for 80×60 image size at 20MHz. After image segmentation is finished, feature extraction processing is started. In the basic feature extraction circuits, calculating the area of segmented objects needs scanning of all columns belonging to the segmented object, so that area calculation becomes the critical path. In the worst-case it takes 80 clock cycles for counting the number of pixels belonging to the segmented region. Additionally, it takes 60 clock cycles to calculate the sum of the counted pixels of all columns. The delay of the dependent feature calculation circuit is only 5 clock cycles and thus quite small. Therefore it takes all together 145 clock cycles or 7.25μsec at 20MHz to extract the features of one object in the worst case.

After finishing the extraction of object features, pattern matching is done. With the described implementa-

180

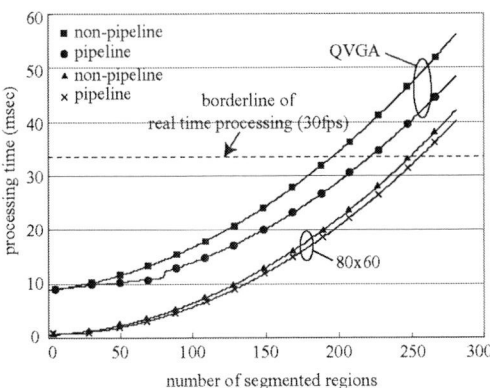

Fig. 12. Relationship between the number of segmented regions and the processing time

tion circuit it takes 10 clock cycles or 0.5μsec at 20MHz to finishing the processing for one reference pattern. In the case of N objects existing in the current frame and correspondingly N reference pattern from the previous frame, it takes therefore $0.5 \times N^2 \mu$sec to execute pattern matching for all objects. The final processing in the estimated position calculation circuit needs only a few clock cycles and can be neglected.

Fig. 12 shows the graph which plots the number of segmented regions vs. the processing time for two kinds of image size, QVGA and 80×60 pixels. The process of pattern matching with the sequential implementation of Figs. 10 and 11 requires much time. But as shown in Fig. 6 we can shorten the total amount of time by executing image segmentation, feature extraction and pattern matching in a processing pipeline. In this way, about 220 objects for QVGA image size and about 255 objects for 80×60 pixel image size can be handled in a real-time tracking application under the condition that the operating frequency is 20MHz. If the number of objects is less than about 50, then we can slow down the clock frequency to reduce the power dissipation. Futhermore Fig. 12 shows, that as the image size becomes larger, the effect of pipeline processing on processing time becomes more significant.

We have coded the proposed FPGA/ASIC architecture in Verilog-HDL and carried out logic synthesis with a standard cell library in 0.35μm CMOS technology. The image segmentation cell-network in the case of 80×60 pixel images is estimated 124mm^2, and therefore consumes the largest area. All other elements of the proposed object-tracking circuit can be implemented on an area of only 0.94mm^2 when memories are excluded. This gives a total of about 125mm^2 for an object tracking ASIC (80×60 pixel images) in 0.35μm CMOS with external memories. The area can be reduced to about 20mm^2 with a state-of-the-art 100nm CMOS technology, so that memories can also be integrated on the ASIC.

V. Conclusions

We have proposed an object tracking architecture for video pictures, based on image segmentation and pattern matching of the segmented objects between frames in a simple feature space. The suitability of the proposed algorithm was verified by simulation. It could be comfirmed that the algorithm overcomes all insufficiencies of the conventional approach, which is based on the difference between the current image and a previous image or a background image. In particular the cases of a moving camera or object occlusion, where objects are hard to track, are no problem for our proposed algorithm. Then we have proposed an FPGA/ASIC implementation architecture, realizing its algorithm for real time object tracking. The detailed circuitry of each processing block of this architecture was described. Furthermore, we estimated the processing speed of the proposed architecture for FPGA implementation with 80×60 pixel and QVGA size images. We comfirmed that even this FPGA implementation can track multiple moving objects in the same frame in real time. The estimated area of proposed architecture for 80×60 pixel with an ASIC in 0.35μm CMOS technology is about 125mm^2. By applying state-of-the-art CMOS technology at the 100nm node, the ASIC area can be reduced to about 20mm^2, so that larger image sizes for the tracking algorithm become possible. It will be also possible to implement the proposed architecture with state-of-the-art FPGA and we have already proposed a compact image segmentation architecture for FPGA implementation in Ref.[8].

Acknowledgements

The authors are grateful to O. Kiriyama, Y. Harada, from Hiroshima University, Japan, for fruitful advices and discussions. Part of this work was supported by the 21st Century COE program "Nanoelectronics for Tera-bit Information Processing", a Grant-in- Aid for Young Scientists (B) (No.16700184), Ministry of Education, Culture, Sports, Science and Technology, Japanese Government and a Grant-in-Aid for JSPS Fellows, 1650741, 2005.

References

[1] StratixII, Altera Corporation, 2005,
URL: http://www.altera.com/products/devices/stratix2/.

[2] See, for example, W. G. Kropatsch and H. Bischof, "Digital Image Analysis," Springer, 2001.

[3] S. W. Seol et al., "An automatic detection and tracking system of moving objects using double differential based motion estimation," *Proc. of International Technical Conference on Circuits/Systems, Computers and Communications*, pp. 260–263, 2003.

[4] H. Kimura and T. Shibata, "Simple-architecture motion-detection analog V-chip based on quasi-two-dimensional processing," *Extended Abstracts of the 2002 International Conference on Solid State Devices and Materials*, pp. 240-241, 2002.

[5] T. Morimoto et al., "Object tracking in video pictures based on image segmentation and pattern matching," *Proc. of the IEEE Int. Symp. on Circ. and Syst.(ISCAS2005)*, pp.3215-3218, 2005.

[6] T. Morimoto et al., "Efficient video-picture segmentation algorithm for cell-network-based digital CMOS implementation," *IEICE Transactions on Information and Systems*, Vol. E87-D, No. 2, pp. 500-503, 2004.

[7] T. Morimoto et al., "Pixel-parallel digital CMOS implementation of image segmentation by region growing," *IEE Proceedings Circuits, Devices & Systems*, in press.

[8] H. Adachi et al., "Image-scan architecture for efficient FPGA/ASIC implementation of video-segmentation by region growing," *Proc. of The International SoC Design Conference(ISOCC2005)*, pp. 301-304, 2005.

Prefetching-Aware Cache Line Turnoff for Saving Leakage Energy *

Ismail Kadayif	Mahmut Kandemir	Feihui Li
Dept. of Computer Engineering	Dept. of Computer Sci. & Eng.	Dept. of Computer Sci. & Eng.
Canakkale Onsekiz Mart University	Pennsylvania State University	Pennsylvania State University
Canakkale 17100, TR	University Park, PA 16802, USA	University Park, PA 16802, USA
kadayif@comu.edu.tr	kandemir@cse.psu.edu	feli@cse.psu.edu

Abstract— While numerous prior studies focused on performance and energy optimizations for caches, their interactions have received much less attention. This paper studies this interaction and demonstrates how performance and energy optimizations can affect each other. More importantly, we propose three optimization schemes that turn off cache lines in a prefetching-sensitive manner. These schemes treat prefetched cache lines differently from the lines brought to the cache in a normal way (i.e., through a load operation) in turning off the cache lines. Our experiments with applications from the SPEC2000 suite indicate that the proposed approaches save significant leakage energy with very small degradation on performance.

I. INTRODUCTION

Caches are critical components from both performance and energy viewpoints, and are being increasingly employed in mobile and embedded environments. From the performance angle, they sit on the critical path of execution, and their hit/miss characteristics usually determine the overall performance of an application. From the energy angle, they are responsible from up to 42% of overall on-chip energy consumption [11]. Therefore, optimizing their performance and energy characteristics is very important and is expected to be even more so in the future.

Due to importance of cache memories, they have been explicit target of many previous optimizations from both performance [12, 6] and power angles [3, 8, 7]. While these techniques have been evaluated thoroughly in isolation (from both performance and power perspectives in many cases), their interaction with each other, when they co-exist together in the same system, took relatively much less attention in the past. For example, it is not clear how data/instruction prefetching would interact with cache line turn-off, used for leakage reduction. Studying this interaction is critical since many embedded environments today demand both high-performance and low-power. Without capturing the power impact of performance optimizations and performance impact of power optimizations, one will not be able to perform the necessary tradeoffs in designing and optimizing an embedded system.

This paper has two major goals. First, focusing on a specific performance optimization (prefetching) and a specific power optimization (cache line turn-off), it presents energy and performance results, emphasizing on how these two optimizations interact with each other from both performance and power angles. Second, it proposes three novel cache line prefetching techniques that take prefetching employed by the hardware into account. The proposed approaches exploit the knowledge on prefetched lines by employing a different decay interval (time frame after which the cache line is turned off) for prefetched lines. In other words, in the proposed strategies, the prefetched lines and the normal lines (i.e., the lines brought into the cache through execution of load operations) use different thresholds. This in turn minimizes the potential negative impact of prefetching on energy consumption. In more detail, prefetching brings data/instructions from main memory to cache before they are actually needed. But, more importantly, when a cache line is placed into a low-leakage mode (using a cache line turnoff strategy), prefetching data/instructions into it forces it to be transitioned to the normal operation mode (active mode), thereby affecting leakage behavior as well. Our goal in this paper is to study these interactions using a set of benchmarks in a two-level cache hierarchy, and quantify the potential benefits of exploiting this interaction using different prefetching-aware cache line turn-off schemes. We present experimental data – using a simulation environment and the SPEC2000 benchmarks – that emphasize the importance of capturing the interactions between performance and energy optimizations, and show how the proposed optimization schemes improve power-performance tradeoff when prefetching and cache line turnoff co-exist in the same system. The experiments also indicate that the performance overheads caused by the proposed schemes are very low.

This paper is structured as follows. The next section summarizes the particular prefetching and cache line turnoff schemes studied in this paper, and explains their interaction qualitatively. Section III gives our experimental setup, and Section IV presents results from our implementation. In Section V, we propose three optimization strategies which treat the prefetched cache lines differently than those brought into cache in normal way to save further leakage energy in caches. We conclude the paper in Section VI with a summary of our major observations.

II. PRELIMINARIES: PREFETCHING AND CACHE LINE TURNOFF

There are a number of prefetching techniques in literature. For example Lai, et al. [10] use trace of memory references to predict when a block becomes dead, and exploits the ad-

---

*This work is supported in part by NSF Career Award 0093082 and a grant from GSRC.

dress correlation to predict which subsequent block to prefetch. In this study we used a hardware-based technique called the *tagged prefetch* [12], which is implemented in several commercial architectures including HP PA7200 [5]. It is based on one block lookahead, which initiates a prefetch for block b+1 when block b is accessed under two different scenarios. First, if there is a miss when b is accessed. Second, if b is brought into cache via prefetching and it is accessed for the first time. We used this approach in an L1-L2 cache hierarchy for both data and instruction prefetching.

While prefetching targets at improving performance, the leakage control mechanisms try to reduce the energy consumption. An important requirement to reduce leakage energy using either a state-destroying (cache line turnoff) or a state-preserving leakage control mechanism is the ability to identify unused resources (cache blocks).[1] Kaxiras et al. [8] present a state-destroying leakage energy reduction technique for cache memories. This technique, called *cache decay*, is based on the idea that a cache block (line) that is not used for a sufficiently long period of time can be considered dead. More specifically, with each cache block, they associate a small 4-state FSM (finite state machine). The FSM steps through these states as long as the cache block is not being accessed. When the last state is reached, the cache block is turned off. Li et al. [9] propose several architectural techniques that exploit data duplication across the different levels of cache hierarchy. They employ both state-preserving (data-retaining) and state-destroying leakage control mechanisms for the L2 sub-blocks when their data also exist in L1. Among their strategies, S-SP-Lazy (Speculative, State-Preserving, and Lazy) generates the best leakage energy savings. In this strategy, when a data is brought from L2 to L1, the corresponding L2 sub-block is put in a state preserving leakage control mode. In this study, the cache decay technique and the S-SP-Lazy technique are integrated to save leakage energy in the L1-L2 cache hierarchy. Specifically, we use cache decay for L1, but both cache decay and S-SP-Lazy for L2. The L2 cache is energy-managed at the sub-block granularity and the sub-block is put into state-preserving leakage mode immediately after it is brought into the L1 cache. Further, if the sub-block is not accessed for a sufficiently long period of time, it is transitioned to the state-destroying mode. If all the sub-blocks of an L2 block are in the state-destroying mode, the block is invalidated and subsequently becomes a candidate for LRU replacement.

From both performance and leakage energy viewpoints, it is important to ensure that the prefetched cache lines are used before being replaced. Therefore, we can divide prefetches in two groups, namely, *useful* and *non-useful*. If a prefetched cache line is discarded from the cache without being accessed, this means that it is prefetched unnecessarily. In this paper, we call this a non-useful prefetching, as opposed to useful prefetches whose data are used at least once before being displaced from the cache. Non-useful prefetches can cause performance degradation due to keeping the bus busy and potentially introducing extra cache misses. Further, they increase the both dynamic energy consumption (because of unnecessary cache access) and leakage energy consumption (if the

prefetched block is brought into a cache line which is in the leakage-saving mode).

A cache line can be in one of the three power modes (states): active (AC) mode (consuming full leakage power), state preserving (SP) low-power mode (we assume that, in this mode, the cache line consumes 10% of the leakage energy of active mode as in [9]), and state destroying (SD) power mode (no energy consumption at all).

TABLE I
BASE CONFIGURATION.

Processor Core	
Functional Units	4 integer and 4 FP ALUs
	1 integer multiplier/divider
	1 FP multiplier/divider
LSQ Size	64 Instructions
LRU Size	64 Instructions
Fetch/Decode/Issue/Commit Width	4 instructions/cycle
Fetch Queue Size	4 Instructions
Cache and Memory Hierarchy	
L1 Instruction/Data Cache	32KB, 32 byte blocks
	2-way, 1 cycle latency
L2 Cache	1MB unified, 128 byte blocks
	2-way, 10 cycle latency
Data/Instruction TLB	128 entries, full-associative,
	30 cycle miss latency
Memory	100 cycle latency
Energy Management	
Technology	0.07 micron
Supply Voltage	1.0V
Dynamic Energy per L1 Access	0.186nJ
Leakage Energy per L1 Block/	
L2 Sub-block per Active Cycle	0.182pJ
Leakage Energy per L2 Sub-block	
per Standby Cycle (state preserving)	0.018pJ
Leakage Energy per L1 block/	
L2 Sub-block per Standby Cycle	
(state-destroying)	0pJ

III. EXPERIMENTAL SETUP

We used SimpleScalar 3.0 [4] to implement our prefetching and leakage-saving optimization strategies, and study their interactions. SimpleScalar is a tool-set to simulate application programs on a range of processors and systems using fast execution-driven simulation. In this work, we used the sim-outorder component. Table I lists the simulation parameters used for our base configuration. Note that, embedded systems are increasingly using multiple issue processors. We used the 70nm technology and energy models from CACTI 3.2 [2] to get the dynamic energies of accessing L1 and L2 caches. We define the *leakage factor* represented with parameter k as follows: the ratio between the leakage energy per cycle of the entire L1 cache and the dynamic energy consumed per access. In our study we assume k=1. (Larger k values reflect the future designs, and although we only provide results for k=1 due to lack of space, for larger k values our schemes accomplish more savings in overall energy.) Further, we assume that the leakage energy of an L2 sub-block is equal to that of the L1 block. We used five randomly-selected benchmarks from the SPEC2000 suit [1] in our experiments. Since it takes a long time to simulate any benchmark from the SPEC2000 suit when it runs to completion, we fast forwarded the first 300 million instructions, and then simulated the next 200 million instructions.

---

[1] When there is no confusion, we use the term "cache line turnoff" to cover for both state-destroying and state-preserving mechanisms.

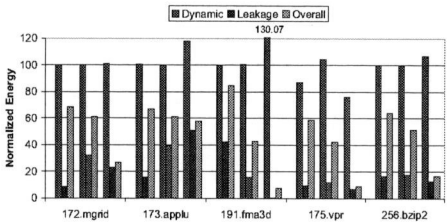

Fig. 1. The normalized energy consumption of optimized codes when k=1.

IV. BASE RESULTS

In the remainder of this paper, we refer to a benchmark optimized by prefetching and leakage control mechanisms as *optimized*. In this section we give energy and performance results for optimized codes.

A. Energy Savings

The normalized energy consumption of the optimized codes with respect to energy consumption of the original codes are given in Figure 1. In dynamic energy calculations, we conservatively assumed that each prefetch attempt (even for the blocks already in the cache) consumes some dynamic energy, amount of which is equal to the dynamic energy consumed per access. In the figure, for each benchmark, the first and the second group of bars are for the L1 instruction cache, and the L1 data cache, respectively, whereas the last group of bars is for the L2 cache. The bars in each group, from left to right, show the dynamic energy, leakage energy, and overall energy of the corresponding cache. As can be seen from this figure, for all benchmarks, the dynamic energy overhead introduced by the prefetching mechanism and the leakage control mechanism together is so small that can be omitted for L1 caches. This is not the case for the L2 cache though. Our optimizations increase the dynamic energy of the L2 cache 17.83%, and 30.07% for 173.applu, and 191.fma3d, respectively. Interestingly, the 175.vpr benchmark benefits, as far as the dynamic energy consumption in the L1 instruction and L2 caches are concerned, from the optimizations. The reason of this is that the prefetching in this benchmark reduces the cache misses for the caches in question dramatically, thereby lowering the number of total cache accesses. When we look at the leakage results, we see that, for each benchmark there is a considerable saving in all caches. An important observation from Figure 1 is that the leakage energy consumption dominates the dynamic energy consumption in the L2 cache even after cache line turnoff. This can be attributed to the fact that leakage energy consumption was extremely large for this cache in the original codes.

B. Execution Cycles

While prefetching potentially increases the performance by bringing cache blocks early in caches, a leakage control mechanism usually degrades performance since there is an overhead to be incurred when a cache line is transitioned from the state preserving/state destroying leakage mode to the full

TABLE II
NORMALIZED EXECUTION CYCLES.

172.mgrid	173.applu	191.fma3d	175.vpr	256.bzip2
65.07%	76.08%	118.22%	54.67%	67.83%

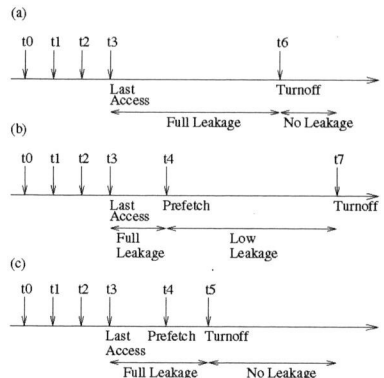

Fig. 2. (a) No prefetching. Full leakage power is consumed during the period [t3,t6]. (b) S-SP strategy. There is a prefetch at time t4, and full leakage power is consumed only during the period [t3,t4]. (c) L-SD strategy. Prefetched line is transitioned into SD leakage control mode at t5 (i.e., after a small decay period after being prefetched), so consuming full leakage power during the period [t3,t5]. However, there is no leakage consumption after t5.

power (i.e., active) mode before being accessed. In our experiments, we assumed that it takes one extra cycle to bring up a cache line into the active mode. Table II shows the execution cycles for the optimized codes, normalized with respect to the corresponding execution cycles of the original codes. On an average, we have 21.62% performance improvement for the optimized codes.

V. CUSTOMIZING CACHE LINE TURNOFF FOR PREFETCHED LINES

Until this point in our experimental evaluation, all the cache lines are treated exactly the same way, whether they have been brought into the cache through prefetching or through a normal load operation. In particular, a prefetched cache line is put in SD power mode if it is not touched during the decay interval. In this section, we discuss three different optimization strategies that treat the prefetched cache lines *differently* from the normal (non-prefetched) cache lines. The first strategy places the prefetched cache lines into the SP mode immediately after the prefetching is performed. In the rest of this paper, we refer to this scheme as the S-SP approach (which stands for Speculative and State Preserving). The second scheme, on the other hand, employs a new decay period for the prefetched cache lines to put them into the SD power mode; we call it L-SD (Lazy and State Destroying). As we will explain shortly, the decay period for the prefetched cache lines in L-SD is much shorter than that for the normal cache lines. The last strategy treats a prefetched cache line based on the characteristics of the previous prefetch into the same line. We refer to the third scheme as the P-H (Predictive and Hybrid). It is predictive in a sense that it tries to figure out whether the prefetching into the line in question would be useful or not. Furthermore, this strategy is hybrid as far as deciding the power status of the prefetched cache line (after prefetching) is concerned, since the power status of the cache line (after prefetching) can be either AC or SP (i.e., both are possible).

A. S-SP Scheme

This scheme places the prefetched cache line in the SP leakage control mode speculatively, immediately after the prefetching is complete. In this way, if the cache line is prefetched unnecessarily (i.e., the cache line would be discarded from the cache without being accessed at all), the line in question will be in SP leakage control mode until it is being replaced (or during the decay period), instead of being in the AC leakage control mode consuming full leakage power. Also, with this strategy, some non-useful prefetches are not bad at all in terms of leakage energy; on the contrary, they may contribute to the leakage energy savings, which can be explained as follows. Figure 2(a) shows the lifetime of a typical cache line. At time t0, a new data is brought into the line in question, and at time t3 the last access to the data occurs. Suppose that the cache line has not been accessed during the decay period, it is transitioned to the SD leakage control mode at time t6, so consuming no leakage power after t6. Figure 2(b) illustrates the cache line in question when the S-SP strategy is used. Assume that the data is brought into the cache line at t4 by prefetching, and at the same time the cache line is transitioned to the SP mode. Consequently, the cache line will consume full leakage power only during the period [t3,t4], instead of [t3,t6], and consume low leakage power during the period [t4,t7], resulting in leakage energy saving. If the original data in that cache line is not referenced later by the program, this prefetch (although it is not useful in itself) can translate to overall power savings.

The drawback of this strategy is that it puts not only the non-useful prefetched cache lines in SP leakage control mode but also the useful prefetched cache lines. So, when the useful prefetched cache line is accessed, the line needs one extra cycle to be waken up, causing performance degradation.

B. L-SD Scheme

In this scheme, we define a new decay period for the cache lines brought into the cache via prefetching. Because of the principle of locality, it is fair to assume that the useful prefetched cache lines, in general, would be accessed within a short period of time after they are brought into the cache. Therefore, it makes sense to put the prefetched cache line into SD leakage control mode if it is not accessed within this short decay period (anticipating that it will not be accessed at all). To explain this strategy better, let us consider the Figure 2(c). The cache line is brought into the cache via prefetching at time t4. If the cache line is not accessed within the period [t4,t5] (note that t5-t4 is equal to new cache decay period), it is placed in the SD leakage control mode at time t5. Consequently, the non-usefully prefetched cache line will consume full leakage power only from t4 to t5 instead of from t4 to t7 (t7-t4 is equal to the decay period of the normal cache lines). This results in reducing overhead of non-useful prefetched cache lines.

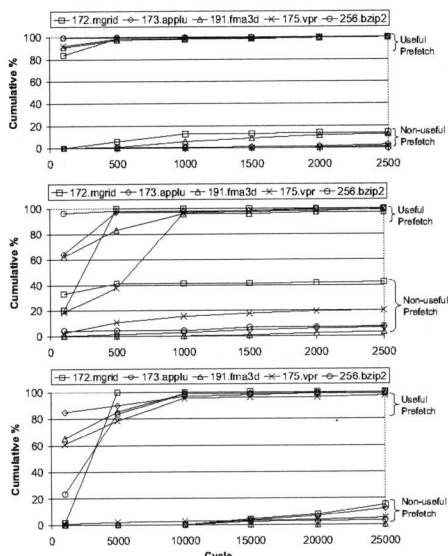

Fig. 3. Cumulative distribution of access interval of prefetched cache lines. The top and middle graphs correspond to the L1 instruction and data caches, respectively, and the bottom graph is for the L2 cache. The upper lines in each graph show the access intervals for the useful prefetched cache lines, while the lower lines indicate the access intervals for the non-useful prefetched lines.

B.1 Finding the Decay Period for Prefetched Lines

In L-SD scheme it is very important to choose a suitable decay period for the prefetched cache lines. If we choose a very short decay period, a usefully-prefetched line may be turned off too early before it is accessed, resulting in performance and energy degradation. On the other hand, if the decay period is chosen too long, a non-useful prefetched cache line will be in the AC leakage control mode unnecessarily long, increasing the leakage energy overhead. To decide a suitable cache decay period for prefetched cache lines, we conducted a set of experiments, which we explain below.

The cumulative distribution of the access intervals for the prefetched L1 instruction, L1 data, and L2 lines are shown in Figure 3. In all three graphs, the upper lines capture the access intervals for the useful prefetched cache lines, whereas the lower lines depict the access intervals for the non-useful prefetched cache lines. From this figure, we can make the following observations. First, in general, the useful prefetched lines in the L1-L2 hierarchy are touched within very short interval after being prefetched. For example, on average, 98.84% (97.65%) of the useful prefetched L1 instruction (L1 data) cache lines are touched within a 1K cycles interval after they are brought in the corresponding cache via prefetching. Furthermore, within an interval of 10K cycles after being brought into the L2 cache, 97.95% of the useful prefetched cache lines are touched. These results are compatible with the rule of spatial locality, that is, when an instruction/data is accessed it is very likely that the neighboring instruction/data will be accessed within a short period of time. Second, the non-useful prefetched lines spend considerable amount of time before being discarded from the cache.

Based on these two observations, one can reduce the decay

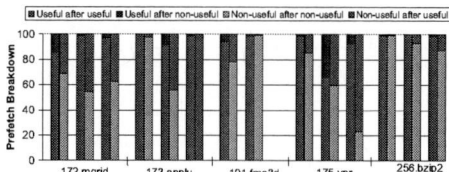

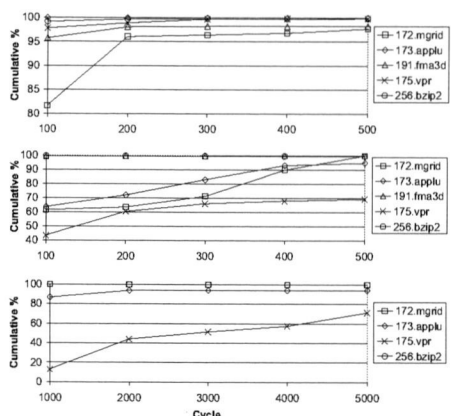

Fig. 4. Breakdown of the useful prefetches into useful after useful and useful after non-useful cases, and breakdown of non-useful prefetches into non-useful after non-useful and non-useful after useful cases.

Fig. 5. Cumulative distribution of the differences between the access intervals of the two successive useful prefetches into the same cache line. The top and middle graphs are for the L1 instruction and data caches, and the lower graph is for the L2 cache.

interval of prefetched cache lines to cut down the leakage energy overhead of the non-useful prefetched cache lines, with negligible degradation in performance, by allowing them to spend less time in the AC leakage control mode. For the rest of our experimental analysis, we selected the decay period of the prefetched cache lines as 1K cycles for the L1 instruction and data caches, and 10K cycles for the L2 cache.

C. P-H Scheme

Before going on to the details of this strategy, let us explain some experimental results consolidating the idea behind it. First, we tried to categorize the useful/non-useful prefetches based upon the previous prefetch into the corresponding cache line. If the prefetch is useful after a useful prefetch into the same line, we call it *useful after useful*. Similar definitions can be made for *useful after non-useful*, *non-useful after non-useful*, and *non-useful after useful* prefetches.

Figure 4 shows the breakdown of the useful/non-useful prefetches into categories. For each benchmark, there are three groups of bars; the first and second correspond to the L1 instruction and data caches, respectively, while the third one corresponds to the L2 unified cache. As can be seen from the figure, for all caches, if the prefetch is useful, the next prefetch into the same line will more likely be useful. Furthermore, except for L2 cache with the 173.applu and 175.vpr benchmarks, if the data is brought into a line via prefetching and replaced before being accessed, the next data brought into the same line via prefetching will likely be discarded before being accessed. Second, we tried to see to what extent we can predict the time when the prefetched data will be accessed by just looking at the access interval of the previous data brought into the same line via prefetching. What we mean by "access interval" is the difference between the time when the data is brought into cache via prefetching and the time when the data is actually accessed. Figure 5 shows the cumulative distribution of differences between the access intervals of two consecutive useful prefetches into the same lines. We can observe that, for all caches in our architecture, the access interval of a prefetched cache line can be predicted quite accurately within a range based on the access interval of the previous prefetch into the same line. For each prefetched line, the P-H strategy checks whether the previous prefetch into the same line was useful. If the previous prefetch was non-useful, the line is placed in SP leakage mode (as in the S-SP scheme) in case the current prefetch would be useful. Otherwise, the prefetch is assumed to be useful and its leakage mode is determined by considering the behavior of the previous useful prefetch into the same line. If the access

interval of the previous prefetch is smaller than a *threshold* value, the prefetched data is assumed to be accessed within very short interval after the prefetching is complete; so, it is placed in the AC leakage mode so that it will be ready when it is needed (as the threshold value, we used 200 cycles for both the L1 instruction and data caches, and 2000 cycles for the L2 cache). If this is not the case, the prefetched line is placed in SP leakage mode for a period, which is smaller than the access interval of the previous prefetched data into the same line by the threshold. Tuning the value of this threshold parameter is critical since it affects both power and performance. After the interval, if the line is still in the SP mode (i.e., the line is not accessed within the interval), it is placed into the AC leakage mode so that it would be ready when it is required. If the prefetched line is not accessed during the decay period used for the prefetched lines after it is prefetched (regardless of the outcome of the previous prefetch), it is placed into SD leakage mode as in the L-SD scheme to save further leakage energy. In our experiments, as the decay period for the prefetched cache lines, we adopted the same values we used in the L-SD strategy, i.e., 1K/10K cycles for L1/L2 caches.

D. Leakage Energy Savings

In this section, we first give the results regarding the leakage energy overhead caused by non-useful prefetches and the maximum leakage savings when an *oracle* predictor is used, and then give the leakage energy savings achieved through the L-SD, S-SP, and P-H schemes described above. For maximum leakage savings, we assume that we have an oracle that can predict precisely whether a prefetch is useful or not. The percentage of leakage overheads introduced by non-useful prefetches and the leakage savings achieved via the oracle in the L1-L2 cache hierarchy for the optimized codes are shown in Figure 6. For each benchmark, the first, second, and third groups of bars correspond to the L1 instruction cache, L1 data cache, and L2 cache, respectively. In each group, the first bar illustrates the overhead incurred by non-useful prefetches, the second bar depicts the leakage saving obtained using the oracle. One can see that, for any benchmark, in general, the

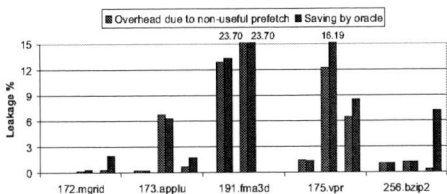

Fig. 6. Leakage energy overheads due to non-useful prefetches and the leakage savings by the oracle.

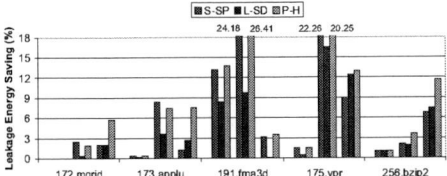

Fig. 7. Leakage energy savings with the L-SD, S-SP, and P-H schemes.

leakage energy saving for any component is larger than the overhead introduced by the non-useful prefetches. Moreover, the 191.fma3d benchmark has 23.70% leakage overhead and saving in the L1 data cache, the largest percentage of leakage overhead and saving in any component in the L1-L2 cache hierarchy among all our benchmark codes. On the other hand, the same benchmark does not incur any overhead in the L2 cache since all its prefetches into L2 are useful.

Figure 7 shows the leakage energy savings in the L1-L2 cache hierarchy when the L-SD, S-SP, and P-H schemes are applied. All the savings are given as percentages with respect to the leakage energy consumption of the optimized codes (i.e., the codes optimized by prefetching and leakage control mechanisms). For each benchmark, there are three groups of bars, where the first and second groups are for the leakage savings in the L1 instruction and L1 data caches, and the third group is for the leakage savings in the unified L2 cache. In each group, the first bar shows the leakage energy saving when the L-SD scheme is used, whereas the second and third bars indicate the leakage energy savings with the S-SP and P-H schemes, respectively. From this figure, we can make the following observations. First, the S-SP scheme outperforms the L-SD scheme in both the L1 instruction and L1 data cache leakage energy savings. This is due to fact that S-SP scheme puts the prefetched cache lines in SP leakage control mode immediately after the prefetch operation is complete, and these L1 lines spend at most 10K cycles (not a long duration at all) before being turned off. On the other hand, the L-SD scheme waits for 1K cycles to put the prefetched lines in the L1 caches in the SD mode, resulting in more leakage energy consumption. Second, as opposed to the case with the L1 caches, in general, the L-SD scheme does a better job than the S-SP scheme in saving the L2 leakage energy. This can be attributed to the fact that the L-SD scheme transitions the prefetched cache line to the SD leakage control mode in a relatively short period of time (10K cycles) after the prefetch is complete, thereby resulting in considerable leakage saving. On the other hand, with the S-SP scheme, a prefetched cache line may spend quite a long

time (about 1M cycle) in the SP leakage control mode if it is not accessed. This makes the S-SP scheme less efficient than the L-SD scheme as far as the savings in L2 are concerned. As shown in the figure, the behavior of the L-SD scheme in L2 energy saving is better than that of the S-SP scheme for all the benchmarks except 191.fma3d. The different behavior observed with this benchmark is due to the fact that there is no non-useful prefetches in L2 for this benchmark; so, the L-SD scheme can not take advantage of them. Our next observation is that, the P-H scheme outperforms the L-SD scheme in leakage savings for all caches in the hierarchy. Fourth, the performance of the L-SD scheme depends on the number of the non-useful prefetches. If the number of non-useful prefetches are low as in the 191.fma3d benchmark with the L2 cache, its performance will be very poor. However, this is not the case for the P-H scheme since it may place the usefully-prefetched cache lines into the SP leakage power mode. Due to space limitation we are not giving the performance details here. The average performance overhead is around 1% for each scheme, and their performance ranking is S-SP, P-H, and L-SD.

VI. Conclusions

This paper makes two important contributions. First, it presents a detailed quantification of the power-performance interactions between prefetching and cache line turnoff. Second, based on this quantification, it proposes and evaluates three different cache line turn off schemes that treat prefetched lines differently from the lines brought into the cache through a normal load operation.

References

[1] Spec cpu2000 benchmark. http://www.spec.org/.

[2] http://research.compaq.com/wrl/people/jouppi/CACTI.html

[3] B. Batson and T. N. Vijaykumar. Reactive associative caches. In Proc. *International Conference on Parallel Architectures and Compilation Techniques*, 2001.

[4] D.C Burger and T. M. Austin. The SimpleScalar toolset, version 2.0, Tech. Rep. 1342, *Dept. of Computer Science*, UW, June 1997.

[5] K. K. Chan. Design of the HP PA 7200 cpu. *Hewlett-packard J.*, vol. 47, no. 1, pp. 25-33.

[6] R. Cooksey, S. Jourdan, and D. Grunwald. A stateless contend-directed data prefetching mechanism. In Proc. *International Conference on Architectural Support for Programming Languages and Operating Systems*, October 2002.

[7] K. Flautner, N. S. Kim, S. Martin, D. Blaauw, and T. Mudge. Drowsy caches: simple techniques for reducing leakage power. In Proc. *International Symposium on Computer Architecture*, 2002.

[8] S. Kaxiras, Z. Hu, and M. Martonosi. Cache decay: exploiting general behavior to reduce cache leakage power. In Proc. *the 28th Annual International Symposium on Computer Architecture*, 2001.

[9] L. Li, I. Kadayif, Y. F. Tsai, N. Vijaykrishnan, M. Kandemir, M. J. Irwin, and A. Sivasubramaniam. Leakage energy management in cache hierarchies. In Proc. *International Conference on Parallel Architectures and Compilation Techniques*, September 2002.

[10] A. C. Lai, C. Fide, and B. Falsafi. Dead-block prediction and dead-block correlating prefetchers. In Proc. *International Symposium on Computer Architecture*, 2001.

[11] J. Montenaro et al. 160 mHz 32b 0.5w CMOS RISC Microprocessor. In Proc. *International Solid State Circuits Conference*, 1996.

[12] S. P. Vanderwiel and D. J. Lilja. Data prefetch mechanisms. *ACM Computing Surveys*, vol. 32, no. 2, June 2000.

A Robust Detailed Placement for Mixed-Size IC Designs

Jason Cong and Min Xie

Computer Science Department, University of California, Los Angeles
California 90095 USA
e-mail: {cong,xie}@cs.ucla.edu

Abstract— The rapid increase in IC design complexity and wide-spread use of intellectual-property (IP) blocks have made the so-called mixed-size placement a very important topic in recent years. Although several algorithms have been proposed for mixed-sized placements, most of them primarily focus on the global placement aspect. In this paper we propose a three-step approach, named XDP, for mixed-size detailed placement. First, a combination of constraint graph and linear programming is used to legalize macros. Then, an enhanced greedy method is used to legalize the standard cells. Finally, a sliding-window-based cell swapping is applied to further reduce wirelength. The impact of individual techniques is analyzed and quantified. Experiments show that when applied to the set of global placement results generated by APlace [1], XDP can produce wirelength comparable to the native detailed placement of APlace, and 3% shorter wirelength compared to Fengshui 5.0 [2]. When applied to the set of global placements generated by mPL6 [3], XDP is the only detailed placement that successfully produces legal placement for all the examples, while APlace and Fengshui fail for $4/9$ and $1/3$ of the examples. For cases where legal placements can be compared, the wirelength produced by XDP is shorter by 3% on average compared to APlace and Fengshui. Furthermore, XDP displays a higher robustness than the other tools by covering a broader spectrum of examples by different global placement tools.

I. INTRODUCTION

Placement is a critical step in VLSI circuit design because it determines the interconnect more than any other step in physical design. The rapid increase in IC design complexity, and the wide-spread use of intellectual-property (IP) blocks have made the so-called mixed-size placement a very important topic in recent years.

Formally, mixed-size placement solves the following problem: Given a rectangular region R and a netlist N, place standard cells and macros within the region without overlap. The optimization objective can be the minimization of total half-perimeter wirelength, routed wirelength, performance, power, etc.

A number of algorithms have been proposed for mixed-size placement, and they can be divided into two classes. The first class of algorithms removes the overlap between placeable objects during global placement, leaving detailed placement with only the task of further wirelength reduction. In this class is a two-pass approach that combines a recursive min-cut-based placer, Capo, and a fixed-outline floorplanner, Parquet [4]. Macros are first shredded into pieces and placed by the standard cell placer. The locations of macros are subse-

quently derived by reassembling the component pieces, and residual overlap is removed through the floorplanner. The second pass places standard cells with all macros fixed. A top-down "correct-by-construction" approach was proposed in [5] that may invoke Parquet many times in intermediate levels. Another algorithm in this class is mPG-ms [6], which uses simulated annealing to gradually legalize macros and fix them in the intermediate levels of the multilevel optimization. Dragon2005 [7] is a two-pass simulated annealing-based placer. Standard cells and macros are placed together in the first pass. In the second pass, the macros are held fixed, and the standard cells are placed again. To further reduce wirelength, it shifts cells when swapping cells from different rows during detailed placement. The most successful algorithm in this class is the recently published PolarBear [8], which combines recursive min-cut with an extra legalization step for every placement subproblem generated by the partitioning. With white space at 5%, PolarBear produces placement with wirelength 10% shorter than Capo 9.3, while Fengshui 5.1 often fails to find legal solutions.

The second class of algorithms may leave overlaps between macros and cells after global placement. Most analytical placers, including Kraftwerk [9], BonnPlace [10, 11, 12], Aplace [1], FDP [13], mPL5 [14], UPlace [15], and some min-cut-based placers, such as Fengshui [16, 2], belong to this category. In this case, the detailed placement is expected to remove the overlap, as well as reduce the wirelength. BonnPlace uses a quadratic programming-based approach coupled with quadri-section [10]. To legalize macros, a bottom-up branch and bound search with linear programming (LP) is proposed. Standard cells are evened out between placement regions with a min-cost-max-flow formulation. Further wirelength reduction is achieved by solving a LP formulation on each row. Fengshui [16, 2] uses a greedy scheme that considers simultaneously perturbation of macros and wirelength minimization for legalization. Windows spanning multiple rows for cell permutation are used for wirelength reduction. Domino [17] iteratively improves wirelength by shredding cells into uniform pieces and solving a min-cost-max-flow formulation. UPlace [15] applies zone refinement for both legalization and wirelength reduction purposes. The objective it considers combines wirelength and zone height.

Most macro legalization schemes used in the second class suffer from two limitations. First, they may not produce a legal placement in the end. Second, they may cause a large perturbation to the global placement during legalization, resulting in longer wirelength. Fig. 1 and Fig. 2 show an example of applying Fengshui's legalization scheme on a global placement generated by an analytical placer, mPL6 [3]. The legalized wirelength increases by more than 10% compared to global placement wirelength.

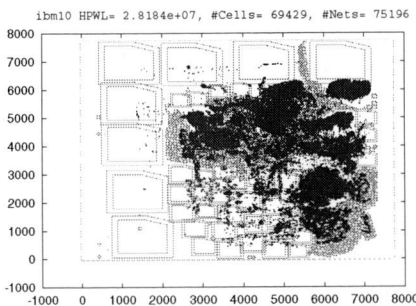

Fig. 1. An example global placement generated by mPL6 [3].

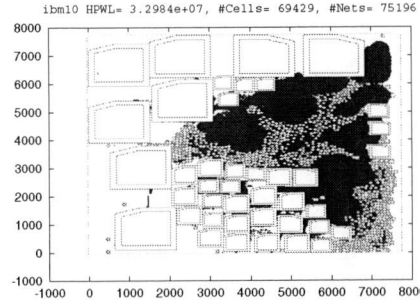

Fig. 2. Legalization by applying Fengshui's greedy method. The wirelength increases by more than 10%.

In this paper we present XDP, a three-step approach for mixed-size detailed placement. First, a combination of constraint graph and linear programming is used to legalize macros. Then, an enhanced greedy method is used to legalize the standard cells. Finally, a sliding-window-based cell swapping is applied to further reduce wirelength. The impact of individual techniques is analyzed and quantified. Experiments show that when applied to the set of global placement results generated by APlace [1], XDP can produce wirelength comparable to the native detailed placement of APlace, and 3% shorter wirelength when compared to Fengshui 5.0 [2].[1] When applied to the set of global placements generated by mPL6 [3], XDP is the only detailed placement that successfully produces legal placement for all the examples, while APlace and Fengshui fail for 4/9 and 1/3 of the examples. For cases where legal placements can be compared, the wirelength produced by XDP is shorter by 3% on average. Furthermore, XDP displays a higher robustness than the other tools by covering a broader spectrum of examples generated by different global placement tools.

The remainder of this paper is organized as follows: Section II presents each step of our detailed placement algorithm: Section II.A describes the macro legalization step, Section II.B describes the cell legalization step, and Section II.C presents the wirelength reduction step. Section III presents experiment results, and Section IV provides conclusions and future work.

---

[1]A number of other placement tools have mixed-size capability. Among them, only APlace and Fengshui exports their detailed placement capability.

II. MIXED-SIZE DETAILED PLACEMENT

II.A. Macro Legalization

The first step of our algorithm removes the overlap between macros in the global placement, which can be formulated as the following problem:

Given a set of rectangular blocks, $M = \{m_1, m_2...m_n\}$, pack the blocks within a rectangular region R without overlap. The objective is to minimize the perturbation, i.e., total movement of the blocks from their original locations.

Before a detailed description, we introduce some notations here.

Let m_i be the ith macro. Its center coordinate in global placement is (x_i, y_i). Its width and height is w_i and h_i respectively. The coordinate of m_i after macro legalization is denoted as (x'_i, y'_i).

Let the lower left corner of the placement region R be $(0, 0)$, the top right corner be (W, H).

Let G_h be a directed acyclic graph (DAG). For each macro m_i, v_{h_i} is the corresponding node in G_h. G_h has a source node v_{h_s} and a sink node v_{h_t}.

To represent the constraint that m_i should be on the left of m_j, a directed edge from v_{h_i} to v_{h_j} will be inserted into G_h. The edge weight is set to be $\frac{w_i + w_j}{2}$. Our graph definition is similar to those widely used in floorplaning [18].

For each node in G_h, we calculate two values, $L(v_{h_i})$ and $R(v_{h_i})$, using Equation 1.

$$
\begin{aligned}
L(v_{h_s}) &= 0 \\
L(v_{h_j}) &= \max(L(v_{h_i}) + weight(e_{ij})) \; \forall e_{ij} \in G_h \\
R(v_{h_t}) &= \max(L(v_{h_t}), W) \\
R(v_{h_i}) &= \min(R(v_{h_j}) - weight(e_{ij})) \; \forall e_{ij} \in G_h
\end{aligned}
\tag{1}
$$

For each edge e_{ij} in G_h, we calculate $slack(e_{ij})$ using Equation 2.

$$
slack(e_{ij}) = R(v_{h_j}) - L(v_{h_i}) - weight(e_{ij}) \; \forall e_{ij} \in G_h \tag{2}
$$

It can be seen that the definitions are analogous to those defined for timing analysis. For each node, we also calculate value $disp(v_{h_i})$ using Equation 3. This is to model the potential displacement for each macro.

$$
disp(v_{h_i}) = \begin{cases} L(v_{h_i}) - x_i & if\, L(v_{h_i}) \geq x_i \\ x_i - R(v_{h_i}) & if\, R(v_{h_i}) \leq x_i \\ 0 & otherwise \end{cases} \tag{3}
$$

In the end the total displacement of a constraint graph is defined using Equation 4.

$$
disp(G_h) = \sum_{v \in G_h} disp(v) \tag{4}
$$

Similarly, we can define G_v and the corresponding values for the vertical direction.

II.A.1. Initial Constraint Graph Generation

Given a global placement, we examine each pair of macros m_i and m_j, and create a constraint edge between them. The edge can be either horizontal or vertical, depending on the relative locations of m_i and m_j. Fig. 3 gives three relative locations that we consider. The type of the constraint edges is such that the macros are given the most flexibility in the constraint graphs. The edge weights are assigned accordingly.

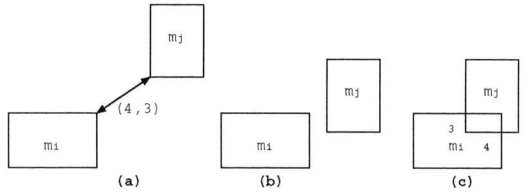

Fig. 3. Three types of relative macro locations which are used to determine the constraint type between each pair of macros. Constraint edge weight is assigned accordingly.

II.A.2. Constraint Graph Adjustment

After the constraint graph construction, we traverse each graph and calculate the longest path. [2] If the longest path exceeds the chip dimension, some of the edges need to be adjusted to reduce the longest path. By adjustment we mean change an edge's direction from horizontal to vertical while keeping its head and tail,[3] or vice versa. In the following discussion, we assume the longest path in G_h exceeds the chip width, while the path in G_v is within the chip height.[4]

Formally, we need to solve the following subproblem:

Given G_h with $L(v_{h_t})$ greater than r, select a subset of constraint edges in G_h to move the G_v, so that the $L(v_{h_t})$ after adjustment is reduced, subject to the constraint that $L(v_{v_t})$ after adjustment should not be greater than H. The objective is to minimize $disp(G_v)$ after the adjustment.

This problem needs to be addressed since identifying the right set of edges for adjustment may not be trivial under certain circumstances. Fig. 4 presents a global placement of macros with dimensions. The dimension of the placement region is 25×10. Fig. 5(a) presents the G_h corresponding to Fig. 4. Edges in the critical path are highlighted with weights. Since macro 2 and 3 have the same width, we have two converging paths with the same length. Fig. 5(b) gives the corresponding G_v. A straightforward method that examines one edge at a time will not pick e_{12}, e_{13}, e_{24} or e_{34} for adjustment, since the final longest path will not change. This leaves us with only the choice of e_{45} or e_{57}. However, adjusting either of them will make the longest path on the Y direction exceed the chip height.

To solve this problem, we extract a subgraph of G_h, consisting of edges with zero slack. This graph is similar to that used for timing optimization in logic synthesis [19, 20, 21, 22]. We name this subgraph the zero-slack network of G_h. According to this network, another DAG, G_c, will be constructed. Each

edge and node in the zero-slack network have a corresponding counterpart in G_c. For an edge e_{ij} in the network, if adjusting causes the longest path in the Y direction to exceed the chip height, the corresponding edge capacity in G_c will be set to $+\infty$. Otherwise, the capacity is set using Equation 5.

$$\max(y_i - R(v_j) + \tfrac{h_i+h_j}{2}, 0) + \max(L(v_i) + \tfrac{h_i+h_j}{2} - y_j, 0) \tag{5}$$

The first component is the potential perturbation on m_i's x coordinate because of the constraint edge under consideration. The second component is the potential perturbation on m_j's x coordinate because of the constraint edge adjustment. To reduce the complexity, we use $L(v_i)$ and $R(v_j)$ before the adjustment, rather than those values after the adjustment. All edges incident on v_{h_s} or v_{h_t} will be assigned a capacity $+\infty$. It can be seen that the definition of edge capacity is set to encourage choosing edges with potentially large slack on the orthogonal direction. A min-cut is then calculated on G_c. For each edge in the cut, the corresponding edge in G_h will be adjusted. Compared to [23], instead of permuting the sequence pair and evaluating the impact of the constraint graphs, we operate directly on the graphs, giving us more flexibility and finer granularity in the operations. Furthermore, our basic operations are more targeted to meeting the packing constraints.

Fig. 5(c) gives the G_c for G_h with edge capacity assigned. The solution for this instance is the min-cut formed by e_{12} and e_{13}. Adjusting this increases the longest path on the Y direction to 9, but is still within the chip height. Fig. 6 gives the final constraint graphs after the adjustment.

The adjustment process iterates until the longest paths in both graphs are shorter than the chip dimension, indicating we have found a set of non-overlapping constraints that can be satisfied. Empirically, it terminates after a few iterations. [5]

II.A.3. Macro Coordinate Determination

The constraint graphs and the subsequent adjustment are essentially used to find a set of non-overlapping constraints that can be satisfied. Our next stage is to determine the exact locations of the macros so that the total perturbation on macros is minimized. This can be formulated as the following linear programming problem:

$$\begin{aligned}
\min \sum_{i=1}^{n} & (w_{x_i} \times dx_i + w_{y_i} \times dy_i) \\
s.t. \ -dx_i & \leq x_i' - x_i \leq dx_i \\
-dy_i & \leq y_i' - y_i \leq dy_i \\
x_j' - x_i' & \geq \tfrac{w_i+w_j}{2} \quad \text{if } \exists e_{ij} \in G_h \\
y_j' - y_i' & \geq \tfrac{y_i+y_j}{2} \quad \text{if } \exists e_{ij} \in G_v \\
\tfrac{w_i}{2} & \leq x_i' \leq W - \tfrac{w_i}{2} \\
\tfrac{h_i}{2} & \leq y_i' \leq H - \tfrac{h_i}{2}
\end{aligned} \tag{6}$$

Here, the dx_i and dy_i are used to quantify the perturbation of m_i. w_{x_i} and w_{y_i} are positive weights that can be set to either one or the number of connections on each macro. The next two inequalities are derived from the edges in G_h and G_v. The last two constraints force the macros to stay with the chip

---

[2] In case the graph thus constructed has cycles, we first derive a sequence-pair representation of the macros, and construct the constraint graphs according to the representation.

[3] We also investigated the alternative of swapping the head and tail of the constraint edge, depending on the global placement. Overall, we do not observe significant improvement in the final quality.

[4] In case the longest path in G_v exceeds the chip height already, we temporarily lift the chip height to be the same as the longest path in G_v.

[5] It is possible that the iterations may not find a feasible solution. In reality, we have not observed any instance of failure on the example we tested, even with only 2 to 3% of white space.

2006 Asia and South Pacific Design Automation Conference   **2C-1**

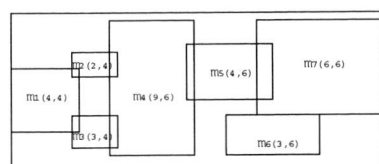

Fig. 4. An example of macros with overlap.

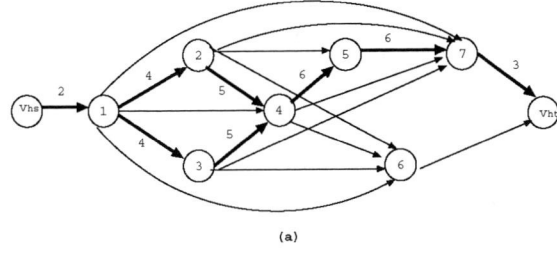

(a)

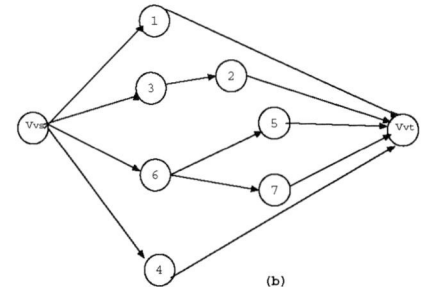

(b)

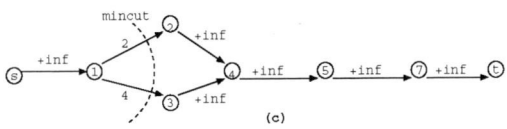

(c)

Fig. 5. (a) Constraint graph G_h. (b) Constraint graph G_v. (c) Corresponding G_c with edge capacity assigned. The min-cut identifies the set of edges which will be transformed from horizontal to vertical.

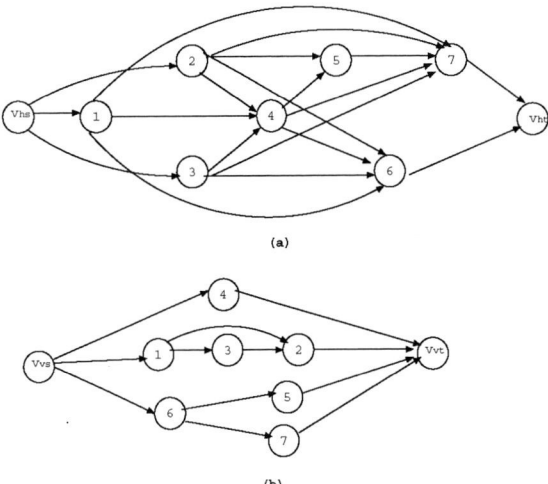

(a)

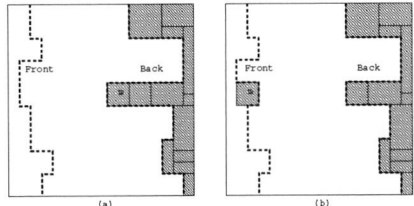

(b)

Fig. 6. (a) Constraint graph G_h after adjustment. (b) Constraint graph G_v after adjustment.

Fig. 7. (a) Front-end designates the leftmost site that can be occupied without overlap with already legalized objects. Back-end designates the rightmost site that can be occupied without overlapping with macros that have not been legalized yet. Back-end contour is initialized as the contour of macros if they are packed to the right boundary. (b) In addition to updating the front-end contour, the back-end contour of rows crossed by a macro will be updated after the macro is legalized.

region. Although the formulation is similar to that in [10, 24], we do not go through the bottom up branch and bound process, as proposed in [10]. Our constraint-graph-based method helps to prune the search space by following the relative order in the global placement. Furthermore, we only solve the LP after a legal packing of macros is guaranteed, while a LP may be tried for every possible combination of non-overlapping constraints [10] in the worst case. The objective can also be enhanced to consider wirelength by the formulation of Mongrel [25], as in [26]. To solve the LP, we used a public domain interior-point LP solver, BPMPD [27].

II.B. Cell Legalization

Following macro legalization, the second step removes the overlap between standard cells. This step is to solve the following problem:

Given a placement where overlap only exists between cells, or cells and macros, remove the overlap between all objects and obtain a legal placement. The objective is still minimization of wirelength.

A greedy heuristic has been proposed for this purpose in [16], as an extension of [28] for mixed-size placement. A

front-end contour designating the leftmost empty site on each row is maintained. Movable objects are traversed in ascending order of the x coordinate. The location of each object is determined by considering the combination of incident wirelength and displacement penalty. The front-end contour is updated after each object is placed. Although it gives a satisfactory result, this method can not guarantee that all the macros can fit within the chip boundary when the legalization finishes. To mitigate this drawback, the global placement of Fengshui takes a conservative approach, packing the macros and cells very tightly to increase the chance of success during legalization [16]. Another alternative by APlace is to iteratively "squeeze" the cell locations and restart cell legalization until a legal soltion is obtained [29]. However, as we will show in Section III, this strategy may not find a legal solution either.

We enhanced this method by introducing a back-end contour, which is initialized as the left contour of macros if they are packed to the right. Fig. 7(a) illustrates the initialization of a back-end contour.

Before legalization, all the movable objects are sorted in ascending order of their left boundary. The placeable objects are examined one at a time. If the object is a cell, we scan each row and pick the site between the two contours that gives the shortest wirelength for the nets connected with it. The front-

191

end on the target row is updated. If no site can be found for a cell, it will be temporarily put on its original location with its physical dimension ignored. This will result in cell area overflow in certain regions of the chip, which will be dealt with in the additional step that follows. If the object is a macro, it will only be considered for movement between the interval determined by the two contours. This restriction guarantees legality of macros obtained from II.A. An additional step after each macro legalization is to update the back-end contour of rows that the macro crosses, as shown in Fig. 7(b).

Depending on the global placement, if cell area overflow remains in certain part of the chip, we partition the chip into regions, and use the min-cost-max-flow formulation in [10, 12] to even out cells between different regions. Each region is represented as a node in a graph. A bi-directional edge is set up between each pair of adjacent regions, as illustrated by Fig 8. The node capacity is the difference between the region area and the total cell area in the region. The unit cost of an edge is the center-to-center distance between the two regions it connects. Since the edge cost is positive, the final solution has no cycles. A dynamic programming-based method is used to select the cells to move between regions. The occurrence of this situation depends partly on the global placement. Among the 18 examples we tested in Section III, five of them still need this adjustment. However, the wirelength usually increases after this adjustment.

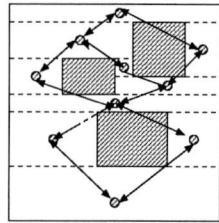

Fig. 8. Network flow based formulation to even out cells.

II.C. Further Wirelength Reduction

After a legal placement is obtained, the last step of our algorithm is to further reduce the wirelength. Here, we use a window spanning a single row or multiple rows and slide it across the chip. We enumerate all the possible configurations and pick the one giving the shortest wirelength of nets connected with the cells. After permutation is finished, the window is slid by half its width. This process is iterated until no further wirelength reduction is possible. This is the same process as that described in [16].

III. EXPERIMENT RESULTS

In this section we evaluate the effectiveness of our detailed placement, XDP, using ICCAD04-MS. These circuits were introduced in [5] to test the placer's capability to handle macros of different shapes and aspect ratios. Table I describes the characteristics of ICCAD04-MS. All the experiments were done on an Intel Xeon processor with 2.4GHz and 2GB memory.

TABLE I

CHARACTERISTICS OF ICCAD04-MS. ALL MACROS IN THE CIRCUITS ARE MOVABLE.

circuit	#cell	#macro	#pad	#net	#row	utilization
ibm01	12507	246	246	14111	144	80%
ibm02	19343	271	259	19584	203	80%
ibm03	22854	290	283	27401	219	80%
ibm04	27221	295	287	31970	213	80%
ibm05	28147	0	1201	28446	148	80%
ibm06	32333	178	166	34826	204	80%
ibm07	45640	291	287	48117	240	80%
ibm08	51024	301	286	50513	256	80%
ibm09	53111	253	285	60902	293	80%
ibm10	68686	786	744	75196	482	80%
ibm11	70153	373	406	81454	322	80%
ibm12	70440	651	637	77240	425	80%
ibm13	83710	424	490	99666	350	80%
ibm14	147089	614	517	152772	375	80%
ibm15	161188	393	383	186608	422	80%
ibm16	182981	458	504	190048	507	80%
ibm17	184753	760	743	189581	454	80%
ibm18	210342	285	272	201920	406	80%

III.A. The Effectiveness of XDP

First, we compared the effectiveness of XDP with the detailed placement of Fengshui 5.0 [2], and APlace [1]. We generated two sets of global placements using APlace and mPL6.[6]

Table II lists the overall comparison of the three algorithms on global placements generated by APlace. Column "FWL" gives the final wirelength. Column "RT(s)" gives the runtime.[7] XDP produces results comparable to the native detailed placement of APlace, and 3% shorter wirelength compared to Fengshui. Table III lists the overall comparison of the three algorithms on global placements generated by mPL6. XDP is the only detailed placer that produces legal placement for all the examples, while Fengshui and APlace can not produce meaningful results for 1/3 and 4/9 of the examples, respectively. Furthermore, for those examples where legal results can be compared, XDP produces wirelength 3% shorter than APlace and Fengshui.

III.B. Impact of Individual Techniques

Next, we analyze the impact of each heuristic used in XDP by experimenting with global placements generated by mPL6. We first evaluate the effectiveness of LP during macro coordinate assignment. For comparison purpose, we implemented a greedy method for macro assignment. The nodes in the constraint graph are traversed in topological order, and their coordinates are chosen as the closest location allowed by the constraints. Note that once a macro coordinate is assigned, it will have a ripple effect on its successor's leftmost/lowest allowable coordinate.

Table IV summarizes the overall impact of LP. The column "GPWL" gives the global placement wirelength. The column "Greedy" corresponds to the greedy method. Column "LP"

---

[6]We did not obtain meaningful global placements from Fengshui 5.0 due to orientation specification issues.

[7]N/A or Overlap means the placer either crashed or produced illegal placement.

TABLE II

COMPARISION OF DETAILED PLACEMENT ALGORITHMS USING GLOBAL PLACEMENTS GENERATED BY APLACE.

circuit	GPWL	Aplace		Fengshui		XDP	
		FWL	RT(s)	FWL	RT(s)	FWL	RT(s)
ibm01	2.16E+06	2.14E+06	24	Overlap		2.08E+06	24
ibm02	4.84E+06	4.65E+06	50	N/A		4.65E+06	53
ibm03	6.95E+06	6.71E+06	58	N/A		6.73E+06	52
ibm04	7.57E+06	7.57E+06	62	N/A		7.48E+06	64
ibm05	9.83E+06	9.69E+06	54	9.84E+06	59	9.59E+06	72
ibm06	6.38E+06	6.02E+06	76	Overlap		6.11E+06	73
ibm07	1.04E+07	1.00E+07	111	Overlap		9.94E+06	119
ibm08	1.29E+07	1.25E+07	131	Overlap		1.24E+07	152
ibm09	1.26E+07	1.21E+07	154	Overlap		1.21E+07	140
ibm10	3.11E+07	2.88E+07	296	Overlap		2.90E+07	324
ibm11	1.96E+07	1.87E+07	215	Overlap		1.87E+07	199
ibm12	3.50E+07	3.34E+07	279	Overlap		3.38E+07	260
ibm13	2.34E+07	2.28E+07	279	Overlap		2.30E+07	239
ibm14	3.74E+07	3.59E+07	445	Overlap		3.55E+07	413
ibm15	4.88E+07	4.68E+07	648	Overlap		4.67E+07	565
ibm16	5.83E+07	5.45E+07	798	Overlap		5.43E+07	666
ibm17	6.69E+07	6.57E+07	735	Overlap		6.53E+07	667
ibm18	4.48E+07	4.20E+07	706	Overlap		4.17E+07	672
Avg.		1.00	1.00	1.03	0.73	1.00	0.98

TABLE III

COMPARISON OF DETAILED PLACEMENT ALGORITHMS USING GLOBAL PLACEMENTS GENERATED BY MPL6.

circuit	GPWL	Aplace		Fengshui		XDP	
		FWL	RT(s)	FWL	RT(s)	FWL	RT(s)
ibm01	2.10E+06	N/A		2.23E+06	13	2.18E+06	37
ibm02	4.54E+06	5.12E+06	70	5.09E+06	22	4.74E+06	70
ibm03	6.94E+06	N/A		Overlap		6.64E+06	54
ibm04	7.37E+06	7.84E+06	98	N/A		7.53E+06	75
ibm05	9.36E+06	9.79E+06	80	9.80E+06	38	9.73E+06	66
ibm06	6.39E+06	N/A		N/A		5.97E+06	64
ibm07	1.00E+07	1.04E+07	150	Overlap		1.01E+07	120
ibm08	1.25E+07	1.25E+07	192	1.25E+07	68	1.19E+07	157
ibm09	1.37E+07	1.32E+07	233	1.38E+07	81	1.27E+07	146
ibm10	3.01E+07	N/A		Overlap	105	2.95E+07	321
ibm11	1.76E+07	N/A		1.91E+07	131	1.82E+07	206
ibm12	3.67E+07	N/A		N/A		3.44E+07	330
ibm13	2.26E+07	N/A		2.40E+07	145	2.36E+07	242
ibm14	3.62E+07	N/A		3.65E+07	540	3.53E+07	443
ibm15	5.57E+07	5.27E+07	1230	5.24E+07	420	5.00E+07	552
ibm16	5.73E+07	5.55E+07	1425	5.59E+07	447	5.30E+07	671
ibm17	6.67E+07	6.70E+07	1103	6.71E+07	493	6.53E+07	923
ibm18	4.41E+07	4.41E+07	1472	4.42E+07	550	4.31E+07	724
Avg.		1.03	1.52	1.03	0.60	1.00	1.00

TABLE IV IMPACT OF LP-BASED MACRO COORDINATE ASSIGNMENT.

circuit	GPWL	Greedy		LP	
		FWL	RT(s)	FWL	RT(s)
ibm01	2.10E+06	2.22E+06	30	2.18E+06	37
ibm02	4.54E+06	5.00E+06	87	4.77E+06	70
ibm03	6.94E+06	6.67E+06	58	6.68E+06	54
ibm04	7.37E+06	7.52E+06	66	7.59E+06	75
ibm05	9.36E+06	9.76E+06	66	9.76E+06	66
ibm06	6.39E+06	6.00E+06	66	6.06E+06	64
ibm07	1.00E+07	1.01E+07	123	1.02E+07	120
ibm08	1.25E+07	1.21E+07	152	1.19E+07	157
ibm09	1.37E+07	1.29E+07	145	1.28E+07	146
ibm10	3.01E+07	2.91E+07	340	2.90E+07	321
ibm11	1.76E+07	1.82E+07	195	1.80E+07	206
ibm12	3.67E+07	3.52E+07	380	3.48E+07	330
ibm13	2.26E+07	2.35E+07	242	2.34E+07	242
ibm14	3.62E+07	3.55E+07	452	3.54E+07	443
ibm15	5.57E+07	5.04E+07	650	5.03E+07	552
ibm16	5.73E+07	5.32E+07	665	5.31E+07	671
ibm17	6.67E+07	6.52E+07	948	6.52E+07	923
ibm18	4.41E+07	4.32E+07	715	4.31E+07	724
Avg.		1.01	1.02	1.00	1.00

corresponds to the LP-based method. Column "FWL" gives the final wirelength. Column "RT(s)" gives the runtime. The LP-based strategy helps to reduce the final wirelength by 1%, and the runtime by 2%.

In Table V we compare two alternative strategies in the cell legalization step. Column "Movable" corresponds to the strategy that allows macros to move when necessary, but only horizontally, as described in Section II.B. Column "Fixed" corresponds to the alternative which fixes macros after the macro legalization step. We list both the final wirelength and the runtime in columns labeled "FWL" and "RT(s)." It can be seen that giving flexibility to macros helps to reduce the final wirelength by 4%. The runtime for the movable strategy is longer by 11%. This suggests that it is useful to give flexibility to macros during legalization when there is plenty of white space in the placement examples.

Table VI shows the impact of back-end contour. Without it, only eight of the examples can be legalized successfully without adjusting the macros, because some macros are pushed outside the chip. With back-end contour for budgeting the legal sites on each row, we can legalize all examples.

IV. CONCLUSION AND FUTURE WORK

A three-step mixed-size detailed placement, named XDP, is described in this paper. A combination of constraint graph and linear programming is used to remove the overlap between macros. Standard cells are legalized by an enhanced greedy method. Sliding-window-based cell swapping is applied to further reduce wirelength. The impact of individual techniques is analyzed and quantified. When applied to global placement results generated by APlace, XDP can produce wirelength comparable to the native detailed placement of APlace, and 3% shorter wirelength compared to Fengshui 5.0. When applied to global placements generated by mPL6, XDP is the only detailed placement that successfully produce a legal placement for all the examples, while APlace and Fengshui fail for 4/9

and 1/3 of the examples. For cases where legal placements can be compared, the wirelength produced by XDP is shorter by 3% on average. Furthermore, XDP displayed a higher robustness by covering a broader spectrum of examples by different global placement tools. Future work includes extention to placement instances where both movable macros and fixed macros are present. Consideration for routability and performance may also be included.

ACKNOWLEDGEMENTS

This research is partially supported by Semiconductor Research Corporation under grant 2003-TJ-1091, supported by the National Science Foundation under grant CCF 0430077, and supported by Magma Design Automation under UC Micro Program 05-096. The authors would like to thank Prof. Andrew Kahng and Mr. Qinke Wang for providing APlace, Prof. Patrick Madden for providing Fengshui for use.

TABLE V IMPACT OF MOVABLE MACROS DURING LEGALIZATION.

circuit	GPWL	Fixed		Movable	
		FWL	RT(s)	FWL	RT(s)
ibm01	2.10E+06	2.25E+06	37	2.18E+06	37
ibm02	4.54E+06	4.83E+06	62	4.77E+06	70
ibm03	6.94E+06	6.93E+06	63	6.68E+06	54
ibm04	7.37E+06	7.97E+06	67	7.59E+06	75
ibm05	9.36E+06	9.75E+06	68	9.76E+06	66
ibm06	6.39E+06	6.21E+06	66	6.06E+06	64
ibm07	1.00E+07	1.09E+07	138	1.02E+07	120
ibm08	1.25E+07	1.18E+07	117	1.19E+07	157
ibm09	1.37E+07	1.31E+07	123	1.28E+07	146
ibm10	3.01E+07	3.10E+07	236	2.90E+07	321
ibm11	1.76E+07	1.90E+07	175	1.80E+07	206
ibm12	3.67E+07	3.90E+07	320	3.48E+07	330
ibm13	2.26E+07	2.53E+07	244	2.34E+07	242
ibm14	3.62E+07	3.62E+07	354	3.54E+07	443
ibm15	5.57E+07	5.13E+07	490	5.03E+07	552
ibm16	5.73E+07	5.34E+07	466	5.31E+07	671
ibm17	6.67E+07	6.65E+07	746	6.52E+07	923
ibm18	4.41E+07	4.45E+07	635	4.31E+07	724
Avg.		1.04	0.89	1.00	1.00

TABLE VI IMPACT OF BACKEND CONTOUR.

circuit	GPWL	w/o backend		w/ backend	
		FWL	RT(s)	FWL	RT(s)
ibm01	2.10E+06	N/A		2.18E+06	37
ibm02	4.54E+06	N/A		4.77E+06	70
ibm03	6.94E+06	N/A		6.68E+06	54
ibm04	7.37E+06	N/A		7.59E+06	75
ibm05	9.36E+06	9.76E+06	67	9.76E+06	66
ibm06	6.39E+06	6.06E+06	65	6.06E+06	64
ibm07	1.00E+07	N/A		1.02E+07	120
ibm08	1.25E+07	N/A		1.19E+07	157
ibm09	1.37E+07	N/A		1.28E+07	146
ibm10	3.01E+07	N/A		2.90E+07	321
ibm11	1.76E+07	1.80E+07	205	1.80E+07	206
ibm12	3.67E+07	N/A		3.48E+07	330
ibm13	2.26E+07	N/A		2.34E+07	242
ibm14	3.62E+07	3.54E+07	445	3.54E+07	443
ibm15	5.57E+07	5.03E+07	565	5.03E+07	552
ibm16	5.73E+07	5.31E+07	674	5.31E+07	671
ibm17	6.67E+07	6.52E+07	943	6.52E+07	923
ibm18	4.41E+07	4.31E+07	729	4.31E+07	724

REFERENCES

[1] A. Kahng and Q. Wang, "Implementation and extensibility of an analytic placer," in *Proc. Intl Symposium on Physical Design*, pp. 18–25, 2004.

[2] A. Agnihotri, S. Ono, and P. H. Madden, "Recursive bisection placement: Feng shui 5.0 implementation details," in *Proc. Intl Symposium on Physical Design*, pp. 230–232, April 2005.

[3] T. Chan, J. Cong, M. Romesis, J. Shinnerl, K. Sze, and M. Xie, "mpl6: A robust multilevel mixed-size placement engine.," in *Proc. Intl Symposium on Physical Design*, pp. 227–229, April 2005.

[4] S. N. Adya and I. L. Markov., "Consistent placement of macro-blocks using floorplanning and standard-cell placement," in *Proc. Intl Symposium on Physical Design*, pp. 12–17, April 2002.

[5] S. N. Adya, S. Chaturvedi, J. A. Roy, D. A. Papa, and I. L. Markov, "Unification of partitioning, placement and floorplanning," in *Proc. Int. Conf. on Computer Aided Design*, pp. 550–557, November 2004.

[6] C.-C. Chang, J. Cong, and X. Yuan, "Multilevel placement for large-scale mixed-size ic designs," in *Proc. Asia South Pacific Design Automation Conf.*, pp. 325–330, January 2003.

[7] T. Taghavi, X. Yang, and B.-K. Choi, "Dragon 2005: Large-scale mized-size placement tool," in *Proc. Intl Symposium on Physical Design*, April 2005.

[8] J. Cong, M. Romesis, and J. Shinnerl, "Robust mixed-size placement under tight white-space constraints.," in *Proc. Int. Conf. on Computer Aided Design*, pp. 165–173, November 2005.

[9] H. Eisenmann and F. M. Johannes, "Generic global placement and floor-planning," in *Proc. 35th ACM/IEEE Design Automation Conference*, pp. 269–274, 1998.

[10] J. Vygen, "Algorithms for large-scale flat placement," in *Proc. 34th ACM/IEEE Design Automation Conference*, pp. 746–751, 1997.

[11] J. Vygen, "Algorithms for detailed placement of standard cells," in *Proc. Conf. Design, Automation and Test in Europe*, pp. 321–324, 1998.

[12] U. Brenner, A. Pauli, and J. Vygen, "Almost optimum placement legalization by minimum cost flow and dynamic programming," in *Proc. Intl Symposium on Physical Design*, pp. 2–9, April 2004.

[13] K. P. Vorwerk, A. Kennings, and A. Vannelli, "Engineering details of a stable force-directed placer," in *Proc. Int. Conf. on Computer Aided Design*, pp. 573–580, November 2004.

[14] T. Chan, J. Cong, and K. Sze, "Multilevel generalized force-directed method for circuit placement," in *Proc. Intl Symposium on Physical Design*, April 2005.

[15] B. Yao, H. Chen, C.-K. Cheng, N.-C. Chou, L.-T. Liu, and P. Suaris, "Unified quadratic programming approach for mixed mode placement," in *Proc. Int'l Symposium on Physical Design*, April 2005.

[16] A. Khatkhate, C. Li, A. R. Agnihotri, M. C. Yildiz, S. Ono, C.-K. Koh, and P. H. Madden, "Recursive bisection based mixed block placement," in *Proc. Intl Symposium on Physical Design*, pp. 84–89, April 2004.

[17] K. Doll, F. Johannes, and K. Antreich, "Iterative placement improvement by network flow methods," *IEEE Transactions on Computer-Aided Design*, vol. 13, no. 10, October 1994.

[18] H. Murata, K. Fujiyoshi, S. Nakatake, and Y. Kajitani, "Rectangle-packing-based module placement," in *Proc. Int. Conf. on Computer Aided Design*, pp. 472–479, November 1995.

[19] G. D. Micheli, "Performance-oriented synthesis of large-scale domino cmos circuits," *IEEE Trans. on Computer-Aided Design of Integrated Circuits and Systems*, vol. 6, pp. 751–765, 1987.

[20] S. K, A. Wang, R. Brayton, and A. Sangiovanni-Vincentelli, "Timing optimization of combinatorial logic," in *Proc. Int. Conf. on Computer Aided Design*, pp. 282–285, November 1988.

[21] K. J. Singh, *Performance Optimization for Digital Circuits*. PhD thesis, Department of Computer Science, University of California Berkeley, 1992.

[22] S. Xu, *Synthesis for Hign-Density and High-Performance FPGA*. PhD thesis, Computer Science Department, University of California, Los Angeles, 2000.

[23] S. Nag and K. Chaudhary, "Post-placement residual-overlap removal with minimal movement," in *Proc. Conf. Design, Automation and Test in Europe*, pp. 581–586, 1999.

[24] R.Okuda, T. Sato, H. Onodera, and K. Tamaru, "An efficient algorithm for layout compaction problem with symmetry constraints," in *Proc. Int. Conf. on Computer Aided Design*, pp. 148–153, November 1989.

[25] S.-W. Hur and J. Lillis, "Mongrel: Hybrid techniques for standard-cell placement," in *Proc. IEEE International Conference on Computer Aided Design*, (San Jose, CA), pp. 165–170, Nov 2000.

[26] X. Tang, R. Tian, and M. D. F. Wong, "Optimal redistribution of white space for wire length minimization," in *Proc. Asia South Pacific Design Automation Conf.*, pp. 412–417, January 2005.

[27] M. Csaba, "Fast cholesky factorization for interior point methods of linear programming," *Computers and Mathematics with Applications*, vol. 31, pp. 49–51, 1996.

[28] D. Hill, "Method and system for high speed detailed placement of cells within an integrated circuit design," *US Patent No. 6,370,673*, 2002.

[29] A. Kahng, S. Reda, and Q. Wang, "Architecture and details of a high quality, large-scale analytical placer," in *Proc. Int. Conf. on Computer Aided Design*, pp. 891–899, November 2005.

FastPlace 2.0: An Efficient Analytical Placer for Mixed-Mode Designs *

Natarajan Viswanathan, Min Pan and Chris Chu
Department of Electrical and Computer Engineering
Iowa State University, Ames, IA 50011-3060, USA
email: {nataraj, panmin, cnchu} @iastate.edu

Abstract— In this paper, we present *FastPlace 2.0* – an extension to the efficient analytical standard-cell placer - *FastPlace* [15], to address the mixed-mode placement problem. The main contributions of our work are: (1) Extensions to the global placement framework of *FastPlace* to handle mixed-mode designs. (2) An efficient and optimal minimum perturbation macro legalization algorithm that is applied after global placement to resolve overlaps among the macros. (3) An efficient legalization scheme to legalize the standard cells among the placeable segments created after fixing the movable macros. On the ISPD 02 Mixed-Size placement benchmarks [3], our algorithm is $16.8X$ and $7.8X$ faster than state-of-the-art academic placers *Capo 9.1* and *Fengshui 5.0* respectively. Correspondingly, we are on average, 12% and 3% better in terms of wirelength over the respective placers.

I. INTRODUCTION

The explosive growth in the size of integrated circuits has imposed enormous challenges on placement algorithms. Placement tools have to produce good-quality results satisfying various design objectives, such as timing, congestion etc. Simultaneously, they have to be computationally efficient to deliver these solutions in a reasonable amount of runtime.

As the time to market for designs is constantly shrinking, there has been a steady increase in the re-use of pre-designed or generated macro blocks like IP cores, embedded memories, analog blocks etc. Designs today often contain a combination of a large number of macro blocks and millions of standard cells. This design style, known as mixed-mode design or mixed-size design complicates the placement step and imposes a lot of difficulty on placement tools due to the varied sizes of the placeable components.

Traditionally, the mixed-mode placement problem was divided into two stages namely, floorplanning or block/module placement and cell placement. Large macro blocks were handled during the floorplanning stage followed by cell placement wherein the macro blocks were treated as fixed. Current designs can have thousands of large and medium sized macros along with millions of standard cells. As a result, traditional floorplanning techniques cannot scale to this problem both in terms of runtime as well as solution quality. With an ever-increasing trend toward mixed-mode design, it is necessary to have efficient techniques that can simultaneously handle this combination of placeable objects.

Over the last few years, the mixed-mode placement problem has generated a lot of interest. Placement algorithms handling this problem employ various approaches including partitioning [2, 4, 11], clustering and simulated annealing [7] and analytical placement [6, 8, 10, 16–19]. Analytical placement techniques based on the force-directed method are promising for handling the mixed-mode placement problem. This is because force-directed methods can seamlessly handle the varied sizes of placeable objects without employing additional techniques like partitioning or clustering [8,18]. Secondly, they can be very efficient and scalable to handle large-scale placement problems [15].

In this paper we present *FastPlace 2.0*, an efficient analytical placer for mixed-mode designs. The main contributions of our work are:

- Extensions to the Cell Shifting technique of *FastPlace* [15] to handle mixed-mode designs.

- An efficient macro legalization algorithm that perturbs the macros by the minimum possible distance to resolve overlaps created during global placement. The macro legalization problem is solved by a floorplanning approach that uses the sequence pair to represent the relative positions of the macros. We prove that for a given sequence pair our algorithm is optimal. We then use simulated annealing to generate a good sequence pair and a non-overlapping placement of the macros with minimum perturbation from their global placement positions.

- An efficient legalization scheme that legalizes the standard cells among the placeable segments created after fixing the movable macros.

The rest of this paper is organized as follows: Section II gives an overview of *FastPlace* for standard-cell placement. Section III outlines the mixed-mode placement flow and describes the extensions to the global placement framework for handling mixed-mode designs. Section IV describes the legalization scheme for macros and standard cells. Section V describes the detailed placement technique. Experimental results are provided in Section VI followed by the conclusions in Section VII.

II. FASTPLACE: STANDARD-CELL PLACEMENT

In this section, we give an overview of the *FastPlace* analytical standard-cell placer described in [15]. *FastPlace* utilizes a quadratic wirelength objective function and is based on three

---

*This work was supported by the Semiconductor Research Corporation under Task ID: 1206.

key features: Cell Shifting, Iterative Local Refinement and a Hybrid Net Model.

The Cell Shifting technique is used to remove cell overlap and spread the cells over the core region. This technique roughly maintains the relative order of the cells as obtained by solving the quadratic program. During Cell Shifting the core region is binned and the utilization of each bin is computed. The cells are then spread depending on the utilization of their respective bins. The basic intuition behind Cell Shifting is to even out the utilization of adjacent bins. This is done by constructing an unequal bin structure from the regular bin structure. Cells are then mapped from the regular bin structure to the unequal bin structure. After each iteration of Cell Shifting, additional forces are added to the cells by way of pseudo nets connected to pseudo pins on the placement boundary. This prevents the cells from collapsing back to their previous positions during the next quadratic programming step.

The Iterative Local Refinement technique is used to reduce the wirelength of the placement based on the half-perimeter wirelength measure. This technique uses a greedy heuristic to move the cells based on a weighted score of the linear wirelength and the placement utilization.

FastPlace uses the pre-conditioned conjugate gradient method to minimize the quadratic objective function. The runtime of the solver is directly proportional to the number of non-zero entries in the connectivity matrix. To improve the speed of the solver, the algorithm uses a Hybrid Net Model to transform the circuit netlist for quadratic placement. [15] showed that the model results in a $2.95X$ reduction in the number of non-zero entries in the connectivity matrix and a $1.5X$ speed-up in the solver as compared to the clique model on the ISPD04 IBM Standard Cell Benchmarks [14].

For standard-cell placement, *FastPlace* achieves comparable placement solutions to other state-of-the-art academic placers, but in a significantly lesser runtime. We now build on this ultra-fast placement tool to handle mixed-mode designs.

III. MIXED-MODE PLACEMENT

Our mixed-mode placement flow is summarized in Figure 1. For the global placement stage, we employ the same top-level flow as [15]. During legalization, we first remove the overlaps among the macros and assign them to legal positions in the core region. Once legalized, the macro positions are fixed and they behave as placement blockages for all subsequent steps. These placement blockages fragment the rows in the core region into placeable segments. In the next step of legalization we move the standard cells among the placeable segments to satisfy their respective capacities. Finally, we legalize the standard cells within the segments. Following legalization we perform detailed placement on the standard cells to further reduce the wirelength of the placement.

A. Cell Shifting for Mixed-Mode Placement

As described in Section II, during Cell Shifting, the cells are spread over the core region by attempting to even out the utilization of adjacent bins in the regular bin structure. For standard-cell placement, the width of the bins in the regular

---

Algorithm *Mixed-Mode Placement*
Stage 1: Global Placement
  Step 1: Coarse Global Placement
    Repeat
      1. Solve the quadratic program
      2. Perform Cell Shifting on standard cells and macro blocks and Add Spreading Forces
    Until the placement is roughly even
  Step 2: Wirelength Improved Global Placement
    Repeat
      1. Solve the quadratic program
      2. Perform Iterative Local Refinement on standard cells and macro blocks
      3. Perform Cell Shifting on standard cells and macro blocks and Add Spreading Forces
    Until the placement is very even
Stage 2: Legalization
  1. Legalize Macro Blocks
  2. Fix Macros and move standard cells among placeable segments to satisfy segment capacity
  3. Legalize standard cells within segments
Stage 3: Detailed Placement

Fig. 1. The *Mixed-Mode* placement flow.

---

bin structure is greater than the average cell width. Hence, the movement of any cell has an influence on the utilization of only the adjacent bins. On the other hand, for mixed-mode placement, the movement of a macro will influence the utilization of all the bins spanned by the macro. Therefore, to move a macro during Cell Shifting we need to consider a larger region that is proportional to the size of the macro.

Shifting of the macros follows the same two-step process as the standard cells. We first construct an unequal bin structure from the regular bin structure. The macros are then linearly mapped from the regular bin structure to the unequal bin structure. The only difference between Cell Shifting for the macros and the cells is the construction of the unequal bin structure. Since Cell Shifting is independent and similiar in the vertical and horizontal directions, we describe the technique for the horizontal direction. Figure 2 illustrates the construction of the unequal bin structure for horizontal shifting. From Figure 2(a), for the regular bin structure, let,

- N: Total number of bins spanned by the macro.

- x_span: Total number of columns spanned by the macro.

- OB_L: x-coordinate of the left boundary of the leftmost bins spanned by the macro.

- OB_R: x-coordinate of the right boundary of the rightmost bins spanned by the macro.

- U_C: Sum of the utilizations of the N bins spanned by the macro (shaded region with lines to the right bottom).

- U_L: Sum of the utilizations of N bins to the left of the macro. (shaded region with lines to the left bottom).

- U_R: Sum of the utilizations of N bins to the right of macro. (shaded region with lines to the left bottom).

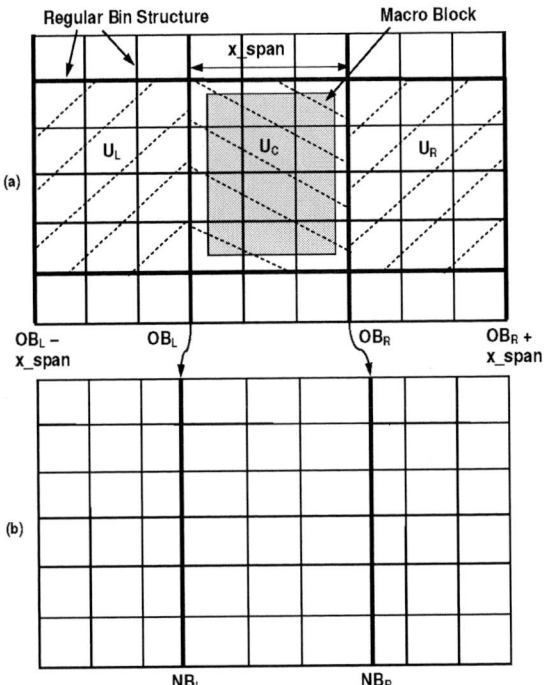

Fig. 2. (a) Regular bin structure (b) Unequal bin structure for macro block cell shifting.

From Figure 2(b), for the unequal bin structure, let,

- NB_L: x-coordinate of the left boundary of the leftmost bins spanned by the macro.

- NB_R: x-coordinate of the right boundary of the rightmost bins spanned by the macro.

Then,
$$NB_L = \frac{(OB_L - x\_span)(U_C + \delta) + OB_R(U_L + \delta)}{U_L + U_C + 2\delta}$$
and,
$$NB_R = \frac{OB_L(U_R + \delta) + (OB_R + x\_span)(U_C + \delta)}{U_R + U_C + 2\delta}$$

As in [15], the parameter δ is set to a value of 1.5 to prevent cross-over of bin boundaries in the unequal bin structure. For performing the linear mapping, if,

- x: x-coordinate of the macro before mapping.

- x': x-coordinate of the macro after mapping.

Then,
$$x' = \frac{NB_R(x - OB_L) + NB_L(OB_R - x)}{OB_R - OB_L}$$

Once the macro is moved, we add the spreading force to the macro and update the connectivity matrix for the next quadratic programming step in the same fashion as [15].

IV. LEGALIZATION

A key issue with analytical placement is that it generates overlaps among the cells or macros that need to be resolved.

We divide our legalization stage into two steps. First, we ignore all the standard cells and resolve overlaps among the macros and assign them to legal positions. In the next step, we fix the macros and legalize the standard cells. These steps are described in more detail below.

A. Macro Block Legalization

During legalization, we want to maintain the macro positions in the global placement solution as much as possible. If we denote the original position of a macro, determined by global placement, as its *target position*, then, the macro block legalization problem is to minimize the total perturbation of all the macros from their target positions such that there are no overlaps among them.

This problem is solved by using a fixed-outline floorplanning approach. We use the sequence pair [12] to represent the floorplan and enforce the non-overlapping constraints among the macros. We can also easily incorporate other floorplanning representations in our approach. We formally describe the problem of finding a minimum perturbation placement for a given sequence pair below.

Minimum Perturbation Floorplan Realization (MPFR) Problem:

Given: n macros with target coordinates (x_i^*, y_i^*) for $i = 1, \ldots, n$ and a sequence pair (p, q).

Determine: Legalized coordinates (x_i, y_i) s.t. $\sum_{i=1}^{n} |x_i - x_i^*| + |y_i - y_i^*|$ is minimized.

In the following sub-sections we first describe the *Iterative Clustering Algorithm* that is used to solve the *MPFR* problem. We then describe the top-level flow for macro legalization using simulated annealing. Since the horizontal and vertical non-overlapping constraints can be handled independently, we only discuss the horizontal problem.

A.1 Iterative Clustering Algorithm

The basic idea of the *Iterative Clustering Algorithm* is that if we know which macros abut with each other to form a cluster in the optimal solution, then the position of the cluster is easy to find. To determine which macros should be grouped in the same cluster, we always shift all clusters to their optimal positions. In doing so if there are any overlaps among some clusters, then we know that these clusters should be merged to form larger clusters. In Figure 3 we give the pseudo-code of the Iterative Clustering Algorithm.

From Figure 3, in step 1, immediate neighbours of macros are those that can potentially abut. They are associated with the non-transitive edges in the constraint graph. The immediate neighbours of all macros can be found in $O(n^2)$ time. In steps 3-4, the macros are placed one at a time from left to right (i.e., according to the sequence p). Then the clustering is updated according to steps 5-11. The condition in step 5 and the closest cluster in step 6 can be determined by considering constraints of the immediate left neighbours of modules in C. The shifting in step 8 is easy according to the following lemma.

Iterative Clustering Algorithm:
1. Find the immediate left and right neighbours of all macros
2. **for** $i = 1$ **to** n
3.   Place macro p_i in its target position
4.   Let C be a new cluster consisting of p_i
5.   **while** C overlaps with other clusters **do**
6.   Merge C with the closest cluster on its left
7.   Let C be the new cluster formed
8.   Shift C to its optimal position
9.   **if** macro m in C is at its target position **do**
10.   Detach m from C if necessary and goto step 8
11.   **endwhile**
12. **endfor**

Fig. 3. Iterative Clustering Algorithm.

Lemma 1 *For a cluster C, its position is optimal if the number of macros perturbed to the left from their target positions is equal to the number perturbed to the right.*

Note that since we add macros from left to right, macros will always be added to the right of a stationary cluster. So the clusters will always shift left. Therefore, it is very easy to find the correct shift amount of the newly formed clusters. In step 9, after shifting a cluster C, a macro $m \in C$ may potentially reach its target position. If m does not have any other macros in the same cluster to its right then it should be detached from the cluster. If not, m will move with the cluster during subsequent steps and hence its position will not be optimal in the final solution. The condition to detach m can be checked by looking at its immediate right neighbours.

Although the while loop in steps 5-11 looks complicated, we can show with careful implementation and analysis that the runtime complexity of the Iterative Clustering Algorithm is $O(n^2)$. We show in Section VI that its runtime is insignificant in practice.

A.2 Macro Legalization by Simulated Annealing

The aim of the macro legalization algorithm is to obtain a sequence pair such that the corresponding placement obtained from the *Iterative Clustering Algorithm* will resolve overlaps among the macros with minimum perturbation from the global placement solution. Another factor to be considered during placement is that the macros have to be placed in legal positions within the core region. Hence, the cost function is defined as a weighted sum of the total perturbation along with a penalty for being out of bound. We use simulated annealing to search for a sequence pair with low cost.

If (p, q) represents the sequence pair. Then, the initial sequence for p/q is generated by sorting the macros in ascending order according to the Manhattan distance from the upper left / lower left corner to their target positions. This sequence pair closely corresponds to the original placement and is usually quite good. Hence, a low-temperature annealing is sufficient to generate a good result. Besides, we restrict each annealing move to randomly exchange two adjacent macros in one of the

two sequences so as to not disturb the current solution significantly.

In Figures 4 and 5 we plot the placement of the macros before and after legalization for the circuit ibm01. From the two figures, we can see that the macros have moved by a very small amount from the global placement solution.

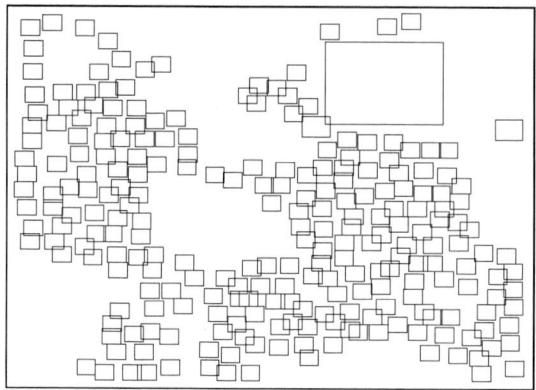

Fig. 4. Circuit ibm01 before legalization of movable macros.

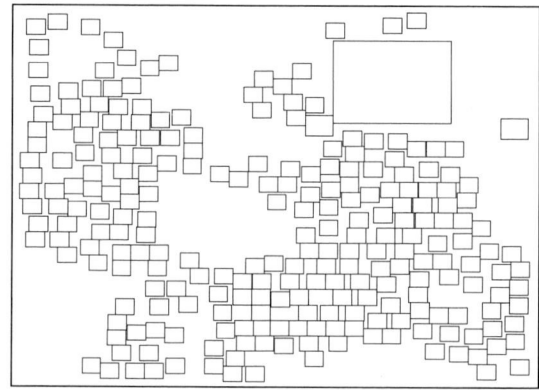

Fig. 5. Circuit ibm01 after legalization of movable macros.

B. Standard Cell Legalization

Once the overlaps among the movable macros have been resolved, we fix their positions for all subsequent steps and treat them as placement blockages. We then divide each row in the core region into placeable segments based on the overlap of the blockages with the row. A placeable segment is defined as the maximal part of a row that is not covered by a placement blockage. We then move the standard cells among the placeable segments to satisfy their respective capacities. Finally, we legalize the standard cells within the segments.

To move the cells among the placeable segments, we use a greedy heuristic similar to the Iterative Local Refinement technique of [15]. For every cell present in a segment, we compute 8 *scores* based on moving the cell to its nearest 8 neighboring segments. For calculating the score, we assume that a cell is moving from its current position in a *source* segment to the nearest possible position in the *target* segment. Each score is

TABLE I
PLACEMENT BENCHMARK STATISTICS.

Circuit	#Cells	#Macros	#Pads	#Nets	%Cell Area	%Macro Area
ibm01	12260	246	246	14111	37.23	42.76
ibm02	19071	271	259	19584	24.69	55.31
ibm03	22563	290	283	27401	30.04	49.96
ibm04	26925	295	287	31970	38.03	41.98
ibm05	28146	0	1201	28446	80.01	0.00
ibm06	32154	178	166	34826	34.60	45.41
ibm07	45348	291	287	48117	44.07	35.93
ibm08	50722	301	286	50513	38.79	41.20
ibm09	52857	253	285	60902	40.18	39.82
ibm10	67899	786	744	75196	20.34	59.66
ibm11	69779	373	406	81454	42.36	37.63
ibm12	69788	651	637	77240	28.35	51.65
ibm13	83285	424	490	99666	43.82	36.18
ibm14	146474	614	517	152772	60.36	19.64
ibm15	160794	393	383	186608	53.26	26.74
ibm16	182522	458	504	190048	42.11	37.89
ibm17	183992	760	743	189581	62.80	17.20
ibm18	210056	285	272	201920	71.31	8.69

TABLE II
BREAK-UP OF TOTAL RUNTIME (ALL VALUES IN SECONDS)

Ckt	Stage 1 Global Placement	Stage 2 Legalize Macros	Stage 2 Legalize Cells	Stage 3 Detailed Placement	Total Time
ibm01	6.73	0.85	0.93	1.97	10.48
ibm02	21.29	1.01	2.34	9.22	33.86
ibm03	17.96	1.00	2.23	6.16	27.35
ibm04	28.16	1.07	2.60	6.50	38.33
ibm05	17.54	0.00	2.34	14.15	34.03
ibm06	24.86	0.39	3.34	8.88	37.47
ibm07	84.49	1.21	5.37	14.10	105.17
ibm08	79.20	1.10	7.62	33.30	121.22
ibm09	66.80	0.68	10.21	15.91	93.60
ibm10	104.10	5.52	12.47	39.66	161.75
ibm11	96.55	1.96	10.07	23.32	131.90
ibm12	116.91	3.73	11.17	57.48	189.29
ibm13	116.87	2.30	19.22	31.49	169.88
ibm14	220.86	5.40	23.21	66.45	315.92
ibm15	305.02	2.03	31.08	76.83	414.96
ibm16	265.54	2.74	31.70	117.25	417.23
ibm17	425.82	7.94	46.30	143.81	623.87
ibm18	526.18	1.18	37.00	204.83	769.19

a weighted sum of two components: The first being the half-perimeter wirelength reduction for the move. The second being a function of the utilization of the source and target segments. Since the legalization technique is mainly used to even out the placement and bring all the segments within capacity, a higher weight is assigned to the second component. If all the scores are negative, the cell will remain in the current segment. Otherwise, it will move to the target segment with the highest score for the move. During one iteration, we traverse through all the segments in the core region and follow the above steps for cell movement. Subsequently, this iteration is repeated until all the segments are within their respective capacities. We then assign the cells to legal positions within each segment.

V. DETAILED PLACEMENT

The aim of the detailed placement stage is to further reduced the wirelength of the placement. We adopt the *FastDP* detailed placement algorithm described in [13] for the same. The detailed placement algorithm is based on four key techniques: global swap, vertical swap, local re-ordering and single-segment clustering. All the techniques act only on the standard cells and do not modify the positions of the macro blocks.

Briefly, the global swap uses the median idea of [9] to swap the standard cells for wirelength reduction. This technique operates on the entire core region. The vertical swap is similar to the global swap but it only considers cells in adjacent rows for swapping. The local re-ordering technique picks a subset of cells within a segment and tries out all possible left-right orderings of the cells to pick the one giving the best possible wirelength. Finally, retaining the order determined by local re-ordering, an optimal single-segment clustering algorithm is used to cluster the cells within a segment for further wirelength reduction.

VI. EXPERIMENTAL RESULTS

Our algorithm was tested on the ISPD02 IBM-MS Mixed-size Placement Benchmarks [3–5]. These designs are relatively large and contain many macro blocks and standard cells. All macro blocks are assumed to be hard blocks with fixed aspect ratios. Each design contains around 20% of whitespace. The circuit characteristics listed in Table I include the number of cells, macros, pads, nets and the area occupied by the cells and macro blocks as a percentage of the total placement area.

Table II gives the break-up of the runtime of *FastPlace 2.0* for the 18 benchmark circuits. From Column 3, it can be seen that on average, the macro block legalization algorithm takes only 1.9% of the total runtime over the 18 benchmark circuits. This demonstrates that the runtime of the algorithm is negligible compared to the other parts of the flow and it is highly efficient in resolving the overlaps among the macros.

In Table III, we compare *FastPlace 2.0* with various academic placers. Results for the *capo-parquet-capo* flow [4], *mPG-MS* [7], *Fengshui 2.4* [11], and *BonnPlace* [6] are as reported in the respective publications. We do not report runtimes for these placers as they were run on different machines. For runtime comparison we run *Capo 9.1* and *Fengshui 5.0*, which are updated versions of the tools published in [2] and [11] respectively. Both placers are run in their default mode. All experiments are run on an Intel Xeon, 3.06GHz CPU.

From Table III, we are on average, 12% and 3% better in terms of wirelength over *Capo 9.1* and *Fengshui 5.0* respectively. Correspondingly, we are $16.8X$ and $7.8X$ faster. We are on average 2% more in terms of wirelength as compared to *BonnPlace*. But accounting for the differences in the processors based on data obtained from [1], we are approximately $20X$ faster.

2006 Asia and South Pacific Design Automation Conference

TABLE III

COMPARISON OF OUR PLACEMENT RESULTS WITH VARIOUS ACADEMIC PLACERS.
CAPO-I, MPG-MS, FENGSHUI 2.4 (FS 2.4), CAPO 9.1, FENGSHUI 5.0 (FS 5.0) AND BONNPLACE (BP)

Ckt	Half Perimeter Wirelength							RunTime				
	Our	$\frac{Capo-I}{Our}$ [4]	$\frac{mPG-MS}{Our}$ [7]	$\frac{FS2.4}{Our}$ [11]	$\frac{Capo9.1}{Our}$ [2]	$\frac{FS5.0}{Our}$	$\frac{BP}{Our}$ [6]	Our (sec)	Capo 9.1 (sec)	$\frac{Capo9.1}{Our}$	FS 5.0 (sec)	$\frac{FS5.0}{Our}$
ibm01	2.45	1.62	1.23	0.98	1.05	1.01	0.92	10	219	20.90	142	13.55
ibm02	4.91	1.70	1.51	1.09	1.06	1.08	1.00	34	457	13.50	245	7.24
ibm03	7.32	1.66	1.53	1.03	1.20	1.16	0.96	27	735	26.87	284	10.38
ibm04	8.14	1.66	1.29	0.98	1.11	1.05	1.01	38	771	20.11	323	8.43
ibm05	10.24	1.12	1.06	0.99	1.00	0.96	0.98	34	684	20.10	372	10.93
ibm06	6.01	1.71	1.53	1.13	1.25	1.14	1.09	37	809	21.59	437	11.66
ibm07	10.99	1.43	1.25	1.07	1.11	1.05	0.95	105	1236	11.75	586	5.57
ibm08	12.38	1.71	1.32	1.10	1.13	1.04	1.02	121	1322	10.91	647	5.34
ibm09	13.79	1.42	1.35	1.00	1.11	1.00	0.96	94	1375	14.69	660	7.05
ibm10	31.65	1.92	1.38	1.18	1.18	1.11	1.04	162	2666	16.48	1085	6.71
ibm11	20.30	1.40	1.31	0.98	1.08	0.97	0.94	132	2172	16.47	891	6.76
ibm12	34.18	1.51	1.30	1.04	1.17	1.06	0.93	189	3413	18.03	1011	5.34
ibm13	25.21	1.56	1.50	0.99	1.16	0.98	0.96	170	4288	25.24	1189	7.00
ibm14	37.76	1.49	1.15	1.02	1.07	1.03	1.00	316	5091	16.11	2553	8.08
ibm15	52.56	1.34	1.25	0.99	1.13	0.97	0.94	415	6399	15.42	3171	7.64
ibm16	58.37	N/A	1.24	1.05	1.21	1.03	0.99	417	7211	17.28	3626	8.69
ibm17	69.89	1.32	1.12	1.01	1.08	0.99	0.95	624	6782	10.87	3935	6.31
ibm18	45.39	1.21	1.12	0.99	1.05	0.98	1.01	769	5163	6.71	3471	4.51
Average		**1.52**	**1.30**	**1.03**	**1.12**	**1.03**	**0.98**			**16.84**		**7.84**

VII. CONCLUSION AND FUTURE WORK

In this paper we extend the efficient analytical placement tool *FastPlace* to handle mixed-mode designs. The current implementation handles the wirelength minimization problem. It produces better results than state-of-the-art academic placers in a significantly lesser runtime.

Routability and timing are key concerns for industrial designs. Future extensions to our work would be in considering the problem of timing driven placement and routability driven placement. Also, the current implementation does not handle rotation and mirroring for the macro blocks. We will be working on handling these constraints for mixed mode placement in the future.

REFERENCES

[1] Standard performance evaluation corporation. http://www.spec.org/.

[2] S. N. Adya, S. Chaturvedi, J. A. Roy, D. Papa, and I. L. Markov. Unification of partitioning, floorplanning and placement. In *Proc. IEEE/ACM Intl. Conf. on Computer-Aided Design*, pages 550–557, 2004.

[3] S. N. Adya and I. L. Markov. ISPD02 IBM-MS Mixed-size Placement Benchmarks. http://vlsicad.eecs.umich.edu/BK/ISPD02bench/.

[4] S. N. Adya and I. L. Markov. Consistent placement of macro-blocks using floorplanning and standard-cell placement. In *Proc. Intl. Symp. on Physical Design*, pages 12–17, 2002.

[5] S. N. Adya and I. L. Markov. Combinatorial techniques for mixed-size placement. *ACM Trans. Design Automation of Electronics Systems*, 10(1):58–90, January 2005.

[6] U. Brenner and M. Struzyna. Faster and better global placement by a new transportation algorithm. In *Proc. ACM/IEEE Design Automation Conf.*, pages 591–596, 2005.

[7] C. C. Chang, J. Cong, and X. Yuan. Multi-level placement for large-scale mixed-size IC designs. In *Proc. Asia and South Pacific Design Automation Conf.*, pages 325–330, 2003.

[8] H. Eisenmann and F. Johannes. Generic global placement and floorplanning. In *Proc. ACM/IEEE Design Automation Conf.*, pages 269–274, 1998.

[9] S. Goto. An efficient algorithm for the two-dimensional placement problem in electrical circuit layout. *IEEE Trans. Circuits and Systems*, CAS-28(1):12–18, 1981.

[10] A. B. Kahng and Q. Wang. An analytical placer for mixed-size placement and timing-driven placement. In *Proc. IEEE/ACM Intl. Conf. on Computer-Aided Design*, pages 565–572, 2004.

[11] A. Khatkhate, C. Li, A. R. Agnihotri, M. C. Yildiz, S. Ono, C.-K. Koh, and P. H. Madden. Recursive bisection based mixed block placement. In *Proc. Intl. Symp. on Physical Design*, pages 84–89, 2004.

[12] H. Murata, K. Fujiyoshi, S. Nakatake, and Y. Kajitani. VLSI module placement based on rectangle-packing by the sequence pair. *IEEE Trans. Computer-Aided Design*, 15(12):1518–1524, December 1996.

[13] M. Pan, N. Viswanathan, and C. Chu. An efficient and effective detailed placement algorithm. In *Proc. IEEE/ACM Intl. Conf. on Computer-Aided Design*, pages 48–55, 2005.

[14] N. Viswanathan and C. C.-N. Chu. ISPD04 IBM Standard Cell Benchmarks with Pads. http://www.public.iastate.edu/~nataraj/ISPD04_Bench.html.

[15] N. Viswanathan and C. C.-N. Chu. FastPlace: Efficient analytical placement using cell shifting, iterative local refinement and a hybrid net model. In *Proc. Intl. Symp. on Physical Design*, pages 26–33, 2004.

[16] K. Vorwerk and A. Kennings. An improved multi-level framework for force-directed placement. In *Proc. Conf. on Design Automation and Test in Europe*, pages 902–907, 2005.

[17] K. Vorwerk, A. Kennings, and A. Vannelli. Engineering details of a stable force-directed placer. In *Proc. IEEE/ACM Intl. Conf. on Computer-Aided Design*, pages 573–580, 2004.

[18] B. Yao, H. Chen, C.-K. Cheng, N.-C. Chou, L.-T. Liu, and P. Suaris. Unified quadratic programming approach for mixed mode placement. In *Proc. Intl. Symp. on Physical Design*, pages 193–199, 2005.

[19] H. Yu, X. Hong, and Y. Cai. MMP: A novel placement algorithm for combined macro block and standard cell layout design. In *Proc. Asia and South Pacific Design Automation Conf.*, pages 271–276, 2000.

2006 Asia and South Pacific Design Automation Conference

2C-3

Timing-Driven Placement Based on Monotone Cell Ordering Constraints

Chanseok Hwang
Department of Electrical Engineering-Systems
Univ. of Southern California, Los Angeles, CA 90089
Tel : 1-213-740-4472   Fax : 1-213-740-9803
Email : chanseoh@usc.edu

Massoud Pedram
Department of Electrical Engineering-Systems
Univ. of Southern California, Los Angeles, CA 90089
Tel : 1-213-740-4458   Fax : 1-213-740-9803
Email : pedram@usc.edu

Abstract– In this paper, we present a new timing-driven placement algorithm, which attempts to minimize zigzags and crisscrosses on the timing-critical paths of a circuit. We observed that most of the paths that cause timing problems in the circuit meander outside the minimum bounding box of the start and end nodes of the path. To limit this undesirable behavior, we impose a physical constraint on the placement problem, i.e., we assign a preferred signal direction to each critical path in the circuit. Starting from an initial placement solution, by using a move-based optimization strategy, these preferred directions force cells to move in a direction that maximizes the monotonic behavior of the timing-critical paths in the new placement solution. To make the direction assignment tractable, we implicitly group all circuit paths into a set of input-output conduits and assign a unique preferred direction to each such conduit. We integrated this idea into a recursive bipartitioning-based placement framework with a min-cut objective function. Experimental results on a set of standard placement benchmarks show that this approach improves the result of a state-of-the-art industrial placement tool for all the benchmark circuits while increasing the wire length by a tolerable amount.

I. Introduction

Timing optimization during placement has been an active area of research and development. This is in part due to the increasing ratios of the interconnect delays to the gate delays in deep submicron designs and the huge impact of cell placement on wire lengths, and therefore, longest path delays in the circuit. In general, a "good" timing-aware cell placement tool can positively influence the timing closure of the circuit, and thus, greatly reduce the overall design turn-around-time. There is therefore a need for efficient timing-driven placement algorithms especially for the design of high-performance ASICs.

Many techniques have been developed to optimize circuit delay during placement. These techniques may be broadly classified into two categories depending on whether they modify the netlist or not. Circuit delay during placement can be optimized by using buffer insertion, logic replication, or retiming techniques [1-4]. On the other hand, many techniques [5-12] do not alter the circuit netlist. These techniques often give high weights to or specify physical length constraints for the edges that lie on the critical timing paths of the circuit. These methods therefore require an a priori classification of signal nets into critical and non-critical ones based on a static timing analysis of the circuit. Most of the reported works use slack values to identify critical nets, and decide the net weights or net length constraints. Since net weights do not bear a direct relation to the circuit delay, it has been quite difficult to stabilize the net weights in order to achieve good timing convergence [6]. Net length (or size of net bounding box) constraints have a more direct relation to the timing constraints. However, it has been difficult to effectively incorporate these constraints in a placement tool without creating "solution oscillation" problems whereby the constraints on the current set of critical nets are satisfied at the expense of making some other nets

timing-critical. In addition, these techniques tend to over-exert the current set of constraints by making the lengths of the critical nets much shorter than what they have to be in order to satisfy the current timing constraints. A number of researchers [7][8] have used the signal direction as an indicator of the *timing gain* function during the move-based partitioning process. Examples include "backward edges" [7] and "V-shaped nodes" [8]. These early results motivate the use of signal direction to guide the performance-driven placement process (see also the last paragraph of Section III(A))

In this paper, we introduce a novel approach to timing-driven placement, which employs a new type of physical constraint imposed on the circuit. More precisely, we impose constraints that specify preferred signal directions for the timing-critical *input-output conduits* in a circuit (see Section III for a formal definition of I/O conduits). These constraints then guide the cell placement so that timing-critical paths satisfy a type of monotonicity property in their cell ordering. Figure 1 depicts a critical path which has (a) non-monotone cell ordering and (b) monotone cell ordering. Clearly, the path with the monotone cell ordering will have a lower delay than the other path. This notion of monotonic path has also been used in logic synthesis to consider interconnect delay [13][20]. In [4], the logic replication was used to make such paths "straightened" for FPGA applications. Unlike their approach which uses logic replication, we employ the new physical constraint specifying preferred signal directions of the timing-critical *input-output conduits*. This idea has been integrated into a recursive bipartitioning-based placement framework with min-cut objective, which is a general top-down placement algorithm like that in [15]. The notion of the preferred signal directions of input-output conduits was described in [21]. The focus of that paper was however on circuit partitioning and does not consider two-dimensional placement in any form.

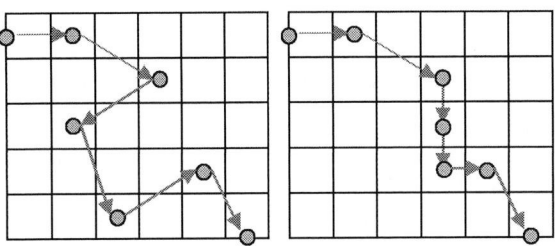

(a) Non-monotone cell ordering   (b) Monotone cell ordering

Figure 1. An example of a critical timing path.

II. Problem Statement

In this section, we describe our basic approach for timing optimization during a recursive partitioning-based placement.

Consider a sequential circuit, represented by a directed graph $G=(V, E)$. Each node $v_i \in V$ represents a combinational cell or flip-flop in the design. It has a weight $w(v_i)$ which specifies its

0-7803-9451-8/06/$20.00 ©2006 IEEE.

layout area. Let's denote the set of primary inputs of a circuit as PI, the set of primary outputs as PO, and the set of flip-flops as FF. We assume that the target chip area is known a priori and that PI and PO are placed at the boundary of the chip and remain fixed during placement. A path in the circuit is defined as the set of nodes and edges that connect a $p_i \in$ PI (or FF) to a $p_o \in$ PO (or FF). Path delay $d(\pi)$ can be calculated by the summation of delays of all the edges and nodes along path π. The minimum cycle time of graph G is denoted by Φ_G and is equal to $Max\ d(S)$ where S is a set of all paths in a circuit. The primary objective of a timing-driven placement tool is to minimize the cycle time of a circuit.

The timing optimization procedure in the context of recursive partitioning-based global placement engine typically consists of weighted wirelength-driven partitioning (WWP) and static timing analysis (STA.) More precisely, critical nets in the circuit are first identified based on STA and assigned higher weights. Next WWP decomposes the given placement instance into smaller instances by dividing the placement region into two sub-regions, and assigning cells to one or the other sub-region such that the weighted wire length is minimized and a balance condition on the total cell area of each sub-region is satisfied. This process continues until each region contains fewer then a certain number of cells.

III. Proposed Approach

A. Signal Direction Constraints

For completeness, we review here the notion of a signal direction of *input-output conduits* from [21], and the resulting constraint, which will be used to straighten critical paths in order to optimize circuit delay.

An input-output (I/O) conduit is defined as the set of all paths from some input node (in PI or FF) to some output node (in PO or FF). An I/O conduit, σ, is simply identified by the corresponding input ($p_i \in$ PI or FF) and output ($p_o \in$ PO or FF.) Notice that the maximum number of I/O conduits in a sequential circuit netlist is $(n_I+n_F).(n_O+n_F)$ where n_I, n_O and n_F denotes the cardinality of PI, PO and FF, respectively. An I/O conduit then belongs to one of the following types: PI→PO, PI→FF, FF→FF, or FF→PO.

In our approach a timing constraint is not explicitly specified for an individual path. Instead, it is defined for an I/O conduit (thereby it implicitly represents a constraint on a large number of paths.) We denote a timing constraint for an I/O conduit σ by $c(\sigma)$. The delay of a I/O conduit is $d(\sigma) = max\ d(\Pi)$ where Π denotes the set of all paths between p_i and p_o of the I/O conduit. Then *critical I/O conduits* are defined as the set of I/O conduits Γ, such that for every $\sigma \in \Gamma$, $d(\sigma) \geq c(\sigma)$.

Signal direction constraints for critical I/O conduits are illustrated in Figure 2. A critical I/O conduit σ_1 from pi_1 to po_1 comprises of two critical paths $pi_1 \rightarrow v_1 \rightarrow v_2 \rightarrow v_3 \rightarrow po_1$ and $pi_1 \rightarrow v_1 \rightarrow v_4 \rightarrow v_5 \rightarrow po_1$. To achieve a monotone cell ordering of these paths, the signal directions of edges of σ_1 should be from part M_0 to part M_1. Let $P(v_i)$ denote the part that node v_i is assigned to i.e., $P(v_i) = 0$ if v_i is put in M_0, otherwise, $P(v_i) = 1$. Notice that $P(v_i)$ of the source node v_i of an edge e of σ_1 should not be any larger than $P(v_j)$ of the target node v_j of that edge, and then both critical paths in σ_1 have a monotone cell ordering.

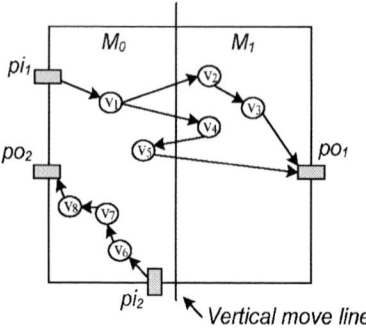

σ_1: $pi_1 \rightarrow v_1 \rightarrow v_2 \rightarrow v_3 \rightarrow po_1$, $e_1(pi_1,v_1)$, $e_2(v_1,v_2)$, $e_3(v_2,v_3)$, $e_4(v_3,po_1)$
$\quad pi_1 \rightarrow v_1 \rightarrow v_4 \rightarrow v_5 \rightarrow po_1$, $e_1(pi_1,v_1)$, $e_5(v_1,v_4)$, $e_6(v_4,v_5)$, $e_7(v_5,po_1)$
σ_2: $pi_2 \rightarrow v_6 \rightarrow v_7 \rightarrow v_8 \rightarrow po_2$, $e_8(pi_2,v_6)$, $e_9(v_6,v_7)$, $e_{10}(v_7,v_8)$, $e_{11}(v_8,po_2)$

Signal Direction Constraints:
$$P(s(e_i)) \leq P(t(e_i)),\ 1 \leq i \leq 7 \qquad \text{for } \sigma_1$$
$$P(s(e_i)) = P(t(e_i)) = 0,\ 8 \leq i \leq 11 \quad \text{for } \sigma_2$$
where $P(v_i)$ is a part number (0 or 1) of v_i, and $s(e_i)$ and $t(e_i)$ are a source and target nodes of edge e_i, respectively.

Figure 2. Signal direction constraints of critical I/O conduits.

This means that all critical paths in a critical I/O conduit σ have a monotone cell ordering if and only if all edges of such critical paths satisfy signal direction constraints for σ. Notice that in the Figure 2 edge e_6 violates the signal direction constraints for σ_1, resulting in a non-monotone cell ordering. In addition, for I/O conduit σ_2, comprising of a single path $pi_2 \rightarrow v_6 \rightarrow v_7 \rightarrow v_8 \rightarrow po_2$, both source and target nodes of edges on σ_2 should be put in M_0 in order to satisfy the signal direction constraint of σ_2.

Let L, R, B, and T denote left, right, bottom, and top, respectively. Based on the above discussion, we define *signal direction constraints* (SDC's) for a *vertical move line* as follows:

SDC^1: if $SD(\sigma)=LL$, $\forall\ e_i \in \sigma$, $\quad P(s(e_i)) = P(t(e_i)) = 0$

SDC^2: if $SD(\sigma)=RR$, $\forall\ e_i \in \sigma$, $\quad P(s(e_i)) = P(t(e_i)) = 1$

SDC^3: if $SD(\sigma)=LR$, $\forall\ e_i \in \sigma$, $\quad P(s(e_i)) \leq P(t(e_i))$

SDC^4: if $SD(\sigma)=RL$, $\forall\ e_i \in \sigma$, $\quad P(s(e_i)) \geq P(t(e_i))$

where $SD(\sigma)$ denotes the signal direction of σ, which is one of LL, RR, LR, or RL for a vertical move line. Clearly, LL (RR) implies that both start and end nodes of the conduit are located in M_0 (M_1), whereas LR (RL) means that the start node of the conduit is in M_0 (M_1) while the end node of the conduit is in M_1 (M_0). The SDC's for a *horizontal move line* are obtained similarly (by replacing LL with BB, RR with TT, LR with BT, and RL with TB in the above equations.) For the remainder of this paper, we will only refer to vertical move lines since the case of a horizontal move line is really the same.

Based on the above definitions, each edge of every path in an I/O conduit has the same preferred signal direction. Therefore, although many paths of a conduit can go through an edge, the edge will have only one signal direction constraint (SDC) for the conduit. However, a placement solution that satisfies all of the SDC's associated with the timing-critical I/O conduits seldom exists for any realistic netlist. This is because, in general, an edge may belong to several critical conduits in the circuit, each assigning a preferred signal direction to the edge. Therefore, we give up on the idea of trying to strictly impose SDC's. Instead we resort to minimizing a cost function which is proportional to the number of SDC violations.

We denote a violation of an SDC by SDV, which stands for a *signal direction violation*. To manage the circuit delay as a scalar objective function rather than as a set of signal direction constraints, we make use of the violation counts of signal directions as defined above. More precisely, in the framework of move-based local neighborhood search algorithm which is used during partition-based placement, we define a *timing gain, $TG(v_i)$*, to exactly quantify the desirability of moving v_i from M_0 to M_1. The timing gain for a node v_i is thus obtained by summing the number of SDV's of each edge e_i connected to node v_i as follows.

SDV^1:   if $v_i = s(e_i)$ and $P(s(e_i)) = P(t(e_i)) = 0$, then

$$TG(v_i) \; -= \; (SDC^1\text{-}cnt(e_i) + SDC^3\text{-}cnt(e_i))$$

SDV^2:   if $v_i = s(e_i)$ and $P(s(e_i)) = P(t(e_i)) = 1$, then

$$TG(v_i) \; -= \; (SDC^2\text{-}cnt(e_i) + SDC^4\text{-}cnt(e_i))$$

SDV^3:   if $v_i = s(e_i)$ and $P(s(e_i)) > P(t(e_i))$, then

$$TG(v_i) \; += \; (SDC^1\text{-}cnt(e_i) + SDC^3\text{-}cnt(e_i))$$

SDV^4:   if $v_i = s(e_i)$ and $P(s(e_i)) < P(t(e_i))$, then

$$TG(v_i) \; += \; (SDC^2\text{-}cnt(e_i) + SDC^4\text{-}cnt(e_i))$$

SDV^5:   if $v_i = t(e_i)$ and $P(s(e_i)) = P(t(e_i)) = 0$, then

$$TG(v_i) \; -= \; (SDC^1\text{-}cnt(e_i) + SDC^4\text{-}cnt(e_i))$$

SDV^6:   if $v_i = t(e_i)$ and $P(s(e_i)) = P(t(e_i)) = 1$, then

$$TG(v_i) \; -= \; (SDC^2\text{-}cnt(e_i) + SDC^3\text{-}cnt(e_i))$$

SDV^7:   if $v_i = t(e_i)$ and $P(s(e_i)) > P(t(e_i))$, then

$$TG(v_i) \; += \; (SDC^1\text{-}cnt(e_i) + SDC^4\text{-}cnt(e_i))$$

SDV^8:   if $v_i = t(e_i)$ and $P(s(e_i)) < P(t(e_i))$, then

$$TG(v_i) \; += \; (SDC^2\text{-}cnt(e_i) + SDC^3\text{-}cnt(e_i))$$

where $SDC^*\text{-}cnt(e_i)$ represents the number of signal direction constraints of type * (ranging from 1 to 4) for edge e_i, that is, the number of timing-critical I/O conduits with the corresponding signal direction that go through the edge. Notice that these counter values are pre-computed before we begin the cell movements for the purpose of timing optimization. The algorithm for setting the SDC-count is described in Section III(B).

Note that the early works [7][8] that use the signal direction to minimize cutsize cannot solve the problem globally. More precisely, in these references, the authors attempt to optimize local directions of edges without considering the parent path and its criticality. Unlike these methods, we aggregate preferred signal directions for all critical paths that pass through an edge, which in turn enables us to exactly calculate the global signal directions, resulting in maximization of the monotonic behavior of the critical paths.

B. Timing Optimization Process

In a recursive bipartitioning-based timing-driven placement, STA is performed at each level of the partitioning hierarchy in order to first identify the timing-critical nets, and then to assign them higher weights in order to prevent them from being cut at the subsequent partitioning step. From our experimentations, we have observed that timing analysis and optimization at early hierarchical levels are not helpful in reducing the circuit delay.

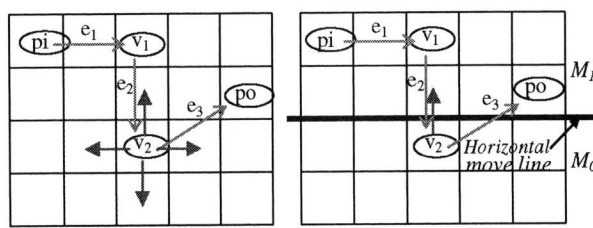

(a) Move directions   (b) v_2 is moved to the upper region

σ: $pi \rightarrow v_1 \rightarrow v_2 \rightarrow po$, edges: $e_1(pi,v_1)$, $e_2(v_1,v_2)$, $e_3(v_2,po)$

SDC^2: $SD(\sigma)=TT$, $\forall\, e_i \in \sigma$, $P(s(e_i)) = P(t(e_i)) = 1$

$SDC^2\text{-}count(e_2) = 1$, $SDC^2\text{-}count(e_3) = 1$

$VC_{(v_2:P(v_2)=0)} = 2$ // SDC violations before v_2-move

     $\Rightarrow SDC^2$ *violated for e_2 and e_3.*

$VC_{(v_2:P(v_2)=1)} = 0$ // SDC violations after v_2-move

     $\Rightarrow SDC^2$ *violations for e_2 and e_3 are eliminated.*

$\therefore TG(v_2) = VC_{(v_2:P(v_2)=0)} - VC_{(v_2:P(v_2)=1)} = 2$

(c) Computation of timing gain for v_2-moving to the upper region

Figure 3. An example of a cell move for timing optimization.

This is because the size of net bounding box, which is typically used for calculating the interconnect parasitics, is too rough at such levels where the chip area is divided into only a few sub-regions. Based on this observation, we start our timing optimization process after a few runs of the recursive partitioning with min-cut objective. The starting level of hierarchy for the timing optimization process is obviously a function of the circuit netlist size and the chip bounding box. In this way, we start with *an initial global placement* which has been optimized for minimum wire length objective.

We use an accurate internal STA engine, which uses the Elmore delay model and net-length estimation method proposed in [14] to calculate the wire delay, and a commercial timing library to obtain the gate delays. Based on the results of the timing analyzer, we identify the critical edges and nodes as follows: edges with negative slack values are marked as *critical edges* and nodes which have at least one critical incoming and/or outgoing edge are marked as *critical nodes*. Next, we find *critical I/O conduits* for each critical edge using a modified depth-first-search algorithm (MDFS), which visits only those successor nodes that are connected to their parents by critical edges. We add a source node and a sink node to the directed graph. Next we run a reverse-MDFS to find all transitive PI's and FF's for each node v_i and stored them as set S_i at that node. Similarly, all transitive PO's and FF's are searched for and stored at set T_i at the node during another MDFS. As a result, we can determine, C_{ij}, the set of all conduits that go thru any critical edge e_{ij} between nodes v_i and v_j in the circuit graph as the Cartesian product of the sets S_i and T_j. Now, we count the number of critical conduits of type *LL, LR, RL,* and *RR* in C_{ij} for a vertical move line, and thereby, initialize the corresponding SDC-count for all critical edges in the circuit.

We explain the timing gain calculation with the help of example in Figure 3. In the Section III(A), we described the timing gain calculation for the case of a move to a neighboring region over a vertical move line. The timing gain for a move across a horizontal move line can be calculated in a similar manner. Consider moving a critical node v_2 in one of four directions, calculating the timing gain for each direction of movement. v_2 will be moved in the direction with the highest gain. The timing gain calculation for v_2 moving to the top region is shown in Figure 3(c). Signal direction constraint of the critical I/O conduit σ passing v_2 is SDC^2 since both pi and po of this conduit are in the upper region over the horizontal move line. There are two edges connected to this node. Edges e_2 and e_3 do not satisfy SDC^2 of conduit σ. This is because $SD(\sigma)=TT$ but the source and target nodes of these two edges are not in M_1. The number of SDC violations is thus 2. After v_2 is moved to the upper region, SDC^2 can be satisfied for both e_2 and e_3. As a result, the total number of SDC violations are reduced by two, i.e., the timing gain for the v_2-move is two, $TG(v_2) = 2$. We calculate timing gains for other directions in the same way, resulting in $TG(v_2) = -2$ for v_2 moving to the left region, $TG(v_2) = 0$ for v_2 moving to the right region and $TG(v_2) = -2$ for v_2 moving to the bottom region. The maximum timing gain of v_2 is then 2.

After computing the timing gains for all critical nodes, we put them into a gain heap where nodes are sorted by their gain (highest gain move is root of the heap.) Next we extract the root node from the heap. Whenever a node v_i moves to its preferred region r_p, we update gains of nodes connected to v_i which are remaining in the heap and re-order it so that the root is the node with highest gain. If the remaining capacity of the region r_p is zero, then we choose a node v_j among non-critical nodes in that region based on the computation of wirelength gains for those nodes, and move it to the region where v_i is coming. This process continues until the timing gain heap is empty. The running sum of the total timing gain for the moves is constructed during this process in order to identify a sequence of moves that produces the maximum total gain. Moves that are not part of the accepted move sequence are reversed. We call these steps as a *pass*, which is similar to the mechanism used in a general FM partitioner[16]. We go through multiples passes until no further timing gain can be achieved. Figure 4 shows the pseudo-code for the proposed timing optimization flow.

C. Timing-driven Placement

In this section, we describe the flow of our proposed timing-driven placement. The placement framework is based on a recursive bipartitioning-based placement algorithm, which comprises of a hierarchical bipartitioning, terminal propagation and legalization. Our proposed timing optimization process is integrated into this framework. We used hMetis [17] as a bipartitioning algorithm, which consists of three phases: coarsening, initial partitioning and uncoarsening phases.

After each bipartitioning, those cells in a sub-region which are connected to external cells are propagated to the boundaries of the corresponding sub-regions. This terminal propagation is performed in a straight-forward manner based on shortest path (or a low-cost Steiner tree) connection of connected terminals. Finally, we allocate all cells that are contained in each sub-region into placement rows when the recursive bipartitioning reaches a certain pre-specified *end level*. This step is typically called *legalization*. We employed a simple technique whereby we divide each row to several equal-sized *bins*, and then, assigned cells in a sub-region to bins according to their coordinates. This assignment may cause unbalances in the total cell size of each bin. To reduce the unbalance, we move cells from "overfilled" bins to

"underfilled" bins, by a technique similar to that in [18]. Next, within each row, cell positions are adjusted to eliminate any cell overlaps. This whole procedure is described in Figure 5 and the layout hierarchy of bipartitioning-based placement is shown in Figure 6.

First, we calculate the start hierarchy level based on the target size of the smallest region before we start the timing optimization procedure. We obtained the target initial size of a region by experimentation, and from that size, calculated the start level. The end level is reached when the size of a sub-region of a hierarchy level becomes smaller than ten times the average cell size in the design. Next we ran a wirelength-driven bipartitioning-based placement algorithm until we reached the start level. This step resulted in the *initial global placement*.

Timing_Optimization_PSD (P,T)

P : An initial hierarchical placement solution with J regions
T : Timing constraints
  1. Perform static timing analysis;
  2. From T, find critical edges, nodes, and I/O conduits (initialize corresponding SDC-count for all critical edges);
  3. Compute initial timing gains for all critical nodes;
  4. Put all critical nodes into a timing gain heap;
  5. While (heap != empty)
  6.   Extract root node v_i from the heap and move it in its preferred direction to a neighbor region in P;
  7.   If the region capacity is violated, select a non-critical node in the region and move it back to the parent region of v_i;
  8.   Update timing gains and restructure the heap as needed;
  9. Find a sequence of moves that produces max_total_gain;
  10. Undo moves that are not in the selected sequence;
  11. If max_total_gain > 0 then goto step 3;
  12. Else exit;

Figure 4. Flow of the proposed algorithm for timing optimization with preferred signal directions.

PSD_Placement (G, T)

G : A directed graph representing a sequential circuit
T : Timing constraints
  1. Calculate the start and end levels of timing-driven global placement;
  2. Do initial wirelength-driven global placement from level one to start level;
  3. While (start_level $\leq i \leq$ end_level)
  4.   While ($j=0; j <$ number of sub_regions in level $i; j++$)
  5.   Generate a bipartitioning-based placement $P_{i,j}$ of subregion j;
  6.   Do Timing_Optimization_PSD(Pi ,T);
  7. Do the legalization;

Figure 5. Flow of the proposed preferred signal direction placement algorithm.

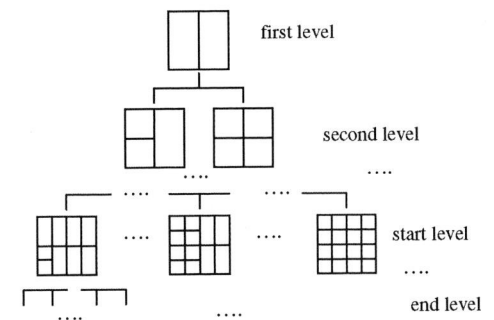

Figure 6. The layout hierarchy of bipartitioning-based placement with level descriptions.

Next we applied the *Timing_Optimization_PSD* to each level of the hierarchy between the start and end levels. Note that the timing optimization procedure is performed only once per hierarchical level on placement solution P_i, which itself comprises of $J=2^i$ sub-regions. In Figure 6, for example, at the start level, first eight bipartitionings are performed to divide the chip area into 16 equal-sized sub-regions. Next, the timing optimization is done on the global placement solution with 16 sub-regions. After reaching the end level, to allocate cells into placement rows without overlaps, a legalization step is performed.

IV. Experimental Results

We have implemented the proposed timing optimization algorithm and bipartitioning-based hierarchical placement flow in C++ on a Sun Ultra Sparc II machine, and tested it on six industry circuits. Four of them, matrix, vp2, mac1 and mac2, are among the ISPD 2001 Circuit Benchmarks that first appeared in [19]. These circuits are also used in [5]. The characteristics of the benchmark circuits are summarized in Table 1. We call our timing-driven placement approach as PSDP (stands for Preferred Signal Direction Placement). We compared PSDP with Capo-boost [22], which attempts to improve circuit delay by reducing the number of global interconnects, and an industrial placement tool, which we call QuadP[1]. We use a 0.18μm standard-cell library to report the delay results.

TABLE 1. The characteristics of benchmark circuits

Circuits	#Cells	#Nets	#IOs
indust1	5931	5969	179
indust2	20193	21699	351
matrix	3,083	3,200	117
vp2	8,714	8,789	321
mac1	8,902	9,115	211
mac2	25,616	26,017	415

---

[1] *QuadP* represents the virtual name of a commercial state-of-the-art placement tool.

Let *total negative slack*, TNS, denote the sum of the slacks of all paths with negative margins. Table 2 compares TNS between the non-timing mode and the timing-driven mode of PSDP. PSDP in non-timing mode (wirelength-driven) is the same as algorithm in Figure 5 with step 6 removed. To obtain the TNS values, we used our STA engine (which uses a commercial timing library to obtain the gate delays and relies on the Elmore delay calculation for interconnects) and assigned the clock cycle time of each circuit as the maximum of "no-wiring path delays [6]" in that circuit. The "no-wiring path delay" accounts for the delay of all gates on the path, but sets the corresponding wire delays to zero. We achieved an average of 44.5% improvement in TNS by using PSDP timing-driven mode.

TABLE 2. Comparison of TNS (total negative slack of all timing endpoints) between wirelength-driven and timing-driven mode of PSDP with the zero-loading delay clock cycle.

Benchmark circuits	Clock cycle	Wirelength-driven mode	Timing-driven mode	% Improvement
indust1	5.54	-38.2	-24.4	36.1%
indust2	8.75	-204.5	-93.1	54.5%
matrix	3.23	-5.8	-4.3	25.9%
vp2	3.67	-68.3	-25.1	63.3%
mac1	2.07	-21.4	-13.5	36.9%
mac2	2.35	-125.4	-62.7	50.2%
Average				**44.5%**

Table 3 compares PSDP with QuadP in wirelength-driven mode and in timing-driven mode, and Capo-boost in terms of the post placement wirelength (HPWL) and post routing wirelength (RWL), and the post-routing *worst negative slack* (WNS). We perform Cadence WarpRoute to route the placements obtained from each placer, extract RC values, and run Pearl to perform static timing analysis (STA). We use the values in [5] as the clock cycle for the corresponding four circuits. The other two circuits are available in complete LEF/DEF/GCF format. The wirelength and worst negative slack are represented in microns and in nanoseconds, respectively.

We observe that PSDP in timing-driven mode improved WNS for all circuits compared to QuadP in the wirelength-driven mode, on average, by 31%, while increasing the total wirelength of post placement and post routing, on average, by 5% and 4%, respectively. In addition, PSDP usually has a better result in terms of WNS compared to the other two placers, QuadP in timing mode and Capo-boost; our placer outperformed those placers for all benchmark circuits except one. PSDP runs on average 48% slower than QuadP in non-timing mode, but PSDP is on average 58% faster than QuadP in timing-driven mode.

V. Conclusions

The paper integrates wire planning into timing-driven min-cut placement. It formulates a new kind of constraint on cell locations based on preferred signal directions. These preferred directions are deduced by grouping all paths from one major source to one major sink into I/O conduits. All paths in the entire circuit are grouped into these conduits. Constraints are computed for all cells in this way, and they are then used to guide the optimization step by forcing the cells to move in a direction such that the timing-critical paths exhibit a monotonic behavior in their cell ordering. The advantage of the new methodology has been confirmed by experimental results; our placer achieves on average 31% improvement on WNS compared to a leading industry placer at the expense of wirelength increase, on average, by 5%.

Reference

[1] J. Cong and X. Yuan, "Multilevel Global Placement with Retiming", In *Proceedings of the ACM/IEEE DAC*, 208-213, 2003.

[2] M. Hrikic, J. Lillis and G. Beraudo, "An Approach to Placement-Coupled Logic Replication", In *Proceedings of the ACM/IEEE DAC*, 711-716, 2004.

[3] P. Saxena and B. Halpin, "Modeling Repeaters Explicitly Within Analytical Placement", In *Proceedings of the ACM/IEEE DAC*, 699-704, 2004.

[4] G. Beraudo and J. Lillis, "Timing Optimization of FPGA Placements by Logic Replication", In *Proceedings of the ACM/IEEE DAC*, 196-201, 2003.

[5] X. Yang, B. Choi and M. Sarrafzadeh, "Timing-Driven Placement using Design Hierarchy Guided Constraint Generation", In *Proceedings of the IEEE ICCAD*, 177-180, 2002.

[6] K. Rajagopal, T. Shaked, Y. Parasuram, T. Cao, A. Chowdlhary and B. Halpin, "Timing Driven Force Directed Placement with Physical Net Constraints", In *Proceedings of the ACM/IEEE ISPD*, 60-66, 2003.

[7] J. Cong and S.K. Lim, "Performance Driven Multiway Partitioning", In *Proceedings of the ACM/IEEE ASP-DAC*, 441-446, 2000.

[8] A. B. Kahng and X. Xu, "Local Unidirectional Bias for Smooth Cutsize-Delay Tradeoff in Performance-driven bipartitioning." In *ACM/IEEE ISPD*, 81-86, 2003.

[9] A. B. Kahng S. Mantik and I. L. Markov, "Min-Max Placement for Large-Scale Timing Optimization", In *Proceedings of the ACM/IEEE ISPD*, 143-148, 2002.

[10] W. Choi and K. Bazargan, "Incremental Placement for Timing Optimization", In *Proceedings of the IEEE ICCAD*, 463-466, 2003.

[11] B. Halpin, C. Y .Chen and N. Sehgal, "Timing Driven Placement using Physical Net Constraints", In *Proceedings of the ACM/IEEE DAC*, 780-783, 2001.

[12] S. Hur, T. Cao, K. Rajagopal, Y. Parasuram and B. Halpin, "Force Directed Mongrel with Physical Net Constraints", In *Proceedings of the ACM/IEEE DAC*, 214-219, 2003.

[13] W. Gosti, A. Narayan, R. K. Brayton and A. L. Sangivanni-Vincentelli, "Wireplanning in Logic Synthesis", In *Proceedings of the IEEE ICCAD*, 26-33, 1998.

[14] C. Ababei, N. Selvakkumaran, K. Bazargan, and G. Karypis, "Multi-objective Circuit Partitioning for Cutsize and Path-based Delay Minimization", In *Proceedings of the IEEE ICCAD*, 181-185, 2002.

[15] A. E. Caldwell, A. B. Kahng, and I. L. Markov, "Can recursive bisection produce routable placements", In *Proceedings of the ACM/IEEE DAC*, 477-482, 2000.

[16] C. Fiduccia and R. Mattheyses, "A Linear Time Heuristic for Improving Network Partitions", In *ACM/IEEE DAC*, 175-181, 1988.

[17] G. Karypis, R. Aggarwal, V. Kumar, and S. Shekhar, "Multilevel Hypergraph partitioning", In *Proceedings of the ACM/IEEE DAC*, 526-529, 1997.

[18] N. Viswanathan and C. C. Chu, "FastPlace: Efficient Analytical Placement using Cell Shifting, Iterative Local refinement and a Hybrid Net Model", In *Proceedings of the ACM/IEEE ISPD*, 26-33, 2004.

[19] Y. Chou and Y. Lin, "A Performance-driven Standard-Cell Placer Based on a Modified Force-Directed Algorithm", In *Proceedings of the ACM/IEEE ISPD*, 24-29, 2001.

[20] S. Iman, M. Pedram, C. Fabian, and J. Cong, "Finding uni-directional cuts based on physical partitioning and logic restructuring", *In Proceedings of the 4th ACM/IEEE* Physical Design Workshop, 187-198, 1993.

[21] C. Hwang and M. Pedram, "PMP: Performance-driven multilevel partitioning by aggregating the preferred signal directions of I/O conduits", In *Proceedings of the ACM/IEEE ASP-DAC*, 428-432, 2005.

[22] A. B. Kahng, I. L. Markov and S. Reda, "Boosting: Min-Cut Placement with Improved Signal Delay," In *Proceedings of the IEEE DATE*, 1098-1103, 2004.

TABLE 3. Timing-driven results of PSDP for six industry circuits with comparison to QuadP and Capo-boost.

Benchmark circuits	Clock cycle	QuadP (wirelength-driven mode)			QuadP (timing-driven mode)			Capo-boost			PSDP (timing-driven mode)		
		HPWL	RWL	WNS	HPWL	RWL	WNS	HPWL	RWL	WNS	HPWL	RWL	WNS
indust1	6.60	350134	461533	-1.23	358551	465394	-1.22	354437	472033	-1.85	357728	479551	-0.89
indust2	15.50	1573453	2754704	-4.31	1567428	2810432	-3.81	1638655	2866492	-3.52	1606993	2906574	-3.17
matrix	3.89	104695	116987	-2.2	107921	120481	-2.06	105133	115670	-2.04	111958	122867	-2.01
vp2	4.57	370677	450872	-3.02	377096	453074	-3.21	364578	482548	-3.21	381118	489366	-2.95
mac1	3.85	443460	506880	-0.56	444704	509136	-0.49	476643	523736	-0.41	481045	524894	-0.30
mac2	7.67	2247603	3244264	-14.46	2249426	3297112	-3.63	2354646	2948992	-1.01	2408205	3123254	-3.73
Ratio		1.00	1.00	1.00	1.01	1.01	0.83	1.03	1.01	0.85	1.05	1.04	0.69

2C-4

Constraint Driven I/O Planning and Placement for Chip-package Co-design *

Jinjun Xiong[1]   Yiu-Chung Wong[2]   Egino Sarto[2]   Lei He[1]

EE Department, University of California at Los Angeles[1], CA 90095, USA
Rio Design Automation, Inc.[2], Santa Clara, CA 95054, USA

ABSTRACT

System-on-chip and system-in-package result in increased number of I/O cells and complicated constraints for both chip designs and package designs. This renders the traditional manually tuned and chip-centered I/O designs suboptimal in terms of both turn around time and design quality. In this paper we formally introduce a set of design constraints suitable for chip-package co-design. We formulate a constraint-driven I/O planning and placement problem, and solve it by a multi-step algorithm based upon integer linear programming. Experiment results using real industry designs show that the proposed algorithm can effectively find a large scale I/O placement solution and satisfy all given design constraints in less than 10 minutes. In contrast, the state-of-the-art without considering those design constraints simply *cannot* meet all design constraints by relying solely upon the conventional iterative approach.

1. INTRODUCTION

I/O placement plays a key role as the interface between chip and package designs in a co-design flow. I/O placement not only significantly affects chip performance, but also determines the feasibility of package designs. Moreover, because the manufacturing cost is proportional to the number of routing layers used for both chips and packages, a good I/O placement not only helps to achieve design closure, but also reduces the number of layers. However, because of the ever-increasing requirement for functionality, the number of I/O cells in a single die keeps increasing, rendering traditional manually tuned I/O placement extremely difficult.

Recently, *flip-chip* packaging emerges as an increasingly popular alternative technology for many high performance IC designs [1]. Compared to wire-bonding packaging, flip-chip technology allows shorter connection between chips and packages and it permits more I/O cells to be implemented on the die. However, flip-chip packaging also brings many new design challenges for I/O placement. For example, instead of being restricted to chip peripherals, I/O cells now can be placed anywhere on the die, and the placement of I/O cells also needs to consider the bump locations on the package in order to minimize the number of extra die layers for connecting I/O cells to the bumps. Therefore, the traditional timing-driven I/O placement formulation [2, 3, 4, 5] without considering package design issues is no longer applicable. A more realistic I/O placement formulation is necessary to

---

*This work was partially supported by NSF award CCR-0093273/0401682.

support chip-package co-design. Moreover, I/O placement also needs to address many issues on timing closure, signal integrity (SI) and power integrity for chip-package co-design. To tackle these problems, complicated design constraints are generated in practice to guide the I/O placement. However, to the best of our knowledge, there is no study on I/O placement in the literature that has formally considered these real design constraints [2, 3, 6, 7] except [8], where only I/O standard compatibility constraints are considered for FPGA I/O placement. By making use of FPGA restricted but well-defined regular structures, [8] proposed to solve it via integer linear programming. However, for high speed ASIC designs, design constraints on I/O placement are more complicated than those in FPGA.

The major contributions of this work include: (1) a formal definition of a set of *design constraints* suitable for chip-package co-design; (2) a new formulation of constraint-driven I/O placement problem ($CIOP$); and (3) an effective multi-step algorithm to solve $CIOP$ for chip-package co-design. To the best of our knowledge, it is the first automatic I/O planning and placement algorithm available in industry for chip-package co-design.

2. PRELIMINARY

In traditional wire-bonding designs, I/O cells are placed on chip boundaries and I/O pads on these cells are then bonded to the substrate through wires. Because of the limited boundary area, the number of I/O cells is also limited. Moreover, high inductance and high crosstalk effects due to wire bonding also limit the use of this traditional packaging technique in today's high performance IC designs.

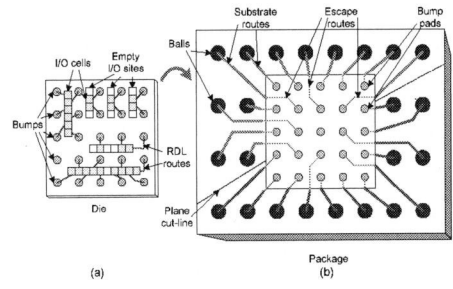

Figure 1: Area I/O Flip-chip design.

On the contrary, flip-chip technology eliminates wires for chip-package bonding. The bonding is achieved through *bumps* via the *surface mount technology* (SMT). As shown in Figure 1, I/O cells are first connected to bumps on the die via *redistribution layer* (RDL) routing, then the die is "flipped" and mounted on the surface of the substrate, where bumps are connected to *bump pads* on the substrate. Finally, package trace routing is performed to furnish the connection between bump pads to *balls* (or *package pins*). Because of the pitch mis-match between bumps (on the chip side) and balls (on the package side), package trace routing can be further divided into two parts. The first part is routing traces under the die, which is called *escape route* as its main goal is to *escape* traces from the die through an appropriate number of substrate layers. The second part is routing traces after escaping and we call it *substrate route*. Package trace routing is preferred to be *planar*, as it not only reduces the number of high cost *buried vias* on the package, but also makes timing and SI analysis more predictable as transmission line modeling can be used for package traces.

3. I/O PLACEMENT CONSTRAINTS

To tackle timing closure, signal integrity, and power integrity problems resulting from chip-package co-design, complicated design constraints are generated in practice to guide I/O placement. We discuss some of the common design constraints that we encountered in a number of real industrial designs in this section.

3.1 Power Integrity Constraints

Signal I/O cells' voltage specification describes the nominal voltage level as well as various voltage levels associated with the signal switching, like the allowable voltage overshoots, undershoots and ring-back values. All signal I/O cells that share the common power and ground nets fall into one *power domain*, and they are expected to be physically placed close to each other.

Moreover, in order to provide good reference planes for signal I/O traces in the package, the power/ground planes in the package have to be cut based upon I/O cells' power domain properties and their physical locations. For example, the power/ground planes in Figure 2 are cut into three parts by the *plane cut-lines*. How to define plane cut-lines heavily depends on the physical locations of I/O cells. If no attention were paid to I/O cells power domain constraints, I/O cells might scatter all over the die, which means that signal traces originating from I/O cells of the same power domain would also be scattered on the package and interleaved with signal traces from other power domains. This would either require many small zig-zag cuts on the package power/ground reference planes, or some I/O signal traces may not have the correct power/ground reference planes. In either cases, SI and power integrity are compromised. Therefore, a good I/O placement should not only consider the power domain constraints, but also should find a solution that would minimize the number of cuts on package's power/ground planes.

Power/ground (P/G) nets that provide power supply to signal I/O cells also require a set of corresponding *P/G driver cells* to be connected with the package power/ground planes. To ensure low voltage drop and minimize Ldi/dt noise, it is required that an enough number of corresponding P/G driver cells be interleaved with signal I/O cells in the final placement solution. We call P/G driver cells that provide primary power supply to signal I/O cells as *primary* P/G cells. In addition to primary P/G cells, there could be other set(s) of P/G driver cells that may supply power at levels different from those of the primary P/G cells, and are required for either SI or power integrity concerns. We call these P/G cells as *secondary* P/G cells. Examples of secondary P/G cells include pre-drive power cells, reference power cells, and core power cells. In practice, it is required that a *ratio* between the number of I/O cells to the number of neighboring P/G cells (the so-called *signal-power-ground ratio*, or *SPG ratio*) be maintained such that the design can have a reliable power supply. Different SPG ratios may be derived for different groups of signal I/O cells.

3.2 Timing Constraints

It is observed that substrate routes in a package vary significantly. For example, for a typical chip size, the substrate route length can span from $1mm$ up to $21mm$. One of the impacts of such variation is on timing measured from I/O cells to package pins. Through 3D EM simulation [9], we find that ignoring package substrate route length variation per se can impact the timing by more than $70ps$ for 2.5V SSTL_2 [10] I/O cells. In another words, different substrate route lengths result in significantly different delays. If power supply variation and package stackup variation are taken into account, more significant delay variation would exhibit, making timing closure extremely difficult to attain. Therefore, for I/O cells that have critical timing relations like differential pairs, we have to take this delay variation into account when we place them. A common practice is to place differential pairs close to each other so that the corresponding package routes will have similar route length.

3.3 I/O Standard Related Constraints

It is not unusual to see that a number of common I/O interfaces are implemented in the same chip in today's high-speed IC designs (e.g., DDR2, SSTL, PCI-express, Serdes). Each I/O interface has its own specification on the relative timing requirements for signals within that interface (like differential pairs). Moreover, because all signals belonging to the same I/O interface will be very likely routed to the same I/O interface in other chips on the PCB, it is desirable to have the I/O cells belonging to the same interface physically close to each other (or even in a preferred order), which reduces the delay and SI variations between signals of the same interface. In particular, differential signal pairs are usually required to escape and to be routed together on the package. This imposes not only a *closeness* constraint but also *bump assignment feasibility* constraints (e.g., bumps escaped on the same layers).

3.4 Floorplan Induced Region Constraints

Some I/O cells may have *region* preference or constraints that are imposed by either a chip floorplan or PCB floorplan. For example, in a top-down design hierarchy, the placement of I/O signals may have side preferences imposed by a board level floorplan. Or in a bottom-up design style, the placement of core dictates that some I/O cells be placed within certain regions. Without respecting these region constraints, it could result in significant wire length increase, hence degrading both routability and performance.

4. CIOP PROBLEM FORMULATION

We define the area where I/O cells can be placed as *I/O sites* and the area where bump pads locate as *bump arrays*.

I/O cells are connected to bump pads by RDL routing, and we treat direct bumping as a special case of RDL routing. We associate every I/O site with a set of bump pads in the bump array, so that I/O cells placed in one I/O site are allowed to connect to any one of these bump pads via RDL routes. Therefore, a general I/O placement problem for chip package co-design consists of three essential sub-problems, with each sub-problem fulfilling different design constraints: (1) the placement of bump arrays, (2) the placement of I/O sites, and (3) the placement of I/O cells. We propose a constraint-driven I/O placement ($CIOP$) problem as follows:

FORMULATION 1. *CIOP* **Problem**: *Given a fixed die size, a chip net-list with I/O cells to be implemented on the die, and a set of design constraints (arising from both package and chip aspects as discussed in Section 3) for I/O placement, determine (1) the placement of bump arrays, (2) the placement of I/O sites, and (3) the legal placement of I/O cells with respect to the defined I/O sites as well as assignment of these I/O cells to the bumps, such that the specified design constraints are satisfied.*

The $CIOP$ problem is more difficult than the conventional purely wire length minimization core placement or pure I/O placement problems. In addition to finding a legal I/O placement solution, $CIOP$ also needs to satisfy many complicated design constraints. It is not likely that a one-step algorithm can solve such a problem. Therefore, we propose to solve the $CIOP$ problem via the following two-step algorithm, *constraint driven I/O planning* ($CPPL$) and *constraint driven detailed I/O placement* ($CDPL$).

5. CONSTRAINT-DRIVEN I/O PLANNING

The purpose of $CPPL$ is to determine the placement of bump arrays, I/O sites, and the rough locations of I/O cells, subject to the given design constraints. Figure 2 illustrates one complete output of $CPPL$.

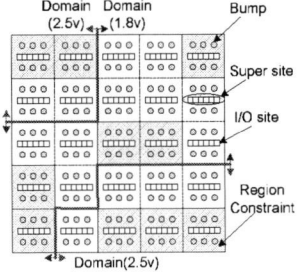

Figure 2: Output of constraint-driven I/O planning.

I/O placement and core placement are conventionally done in two separate steps. A recent study by [2] has shown that traditional global placement techniques (e.g. min-cut based or force-directed placement algorithms) can be extended to handle the co-placement of both core logic and I/O cells, and it has shown that co-placement of I/O and core could achieve better timing closure than the the conventional separated two-step approach. However, such a study is purely based upon wire length (timing-driven) minimization without considering real design constraints. Therefore, in the following we discuss the necessary changes we have made to

a conventional wire-length minimization analytic placer in order to handle the design constraints as discussed in section 3. Details of these conventional analytical placers is not repeated. Note that the focus of $CPPL$ is I/O planning, not the core placement.

To avoid congestion in the core area and to enable escapability and planar routability of bumps on the package, we use a bin-based density metric to measure the uniform distribution of cells for the analytical placer that spreads the core logic cells and I/O cells over the die evenly. Different from [11, 12] where only core logic cells are considered, however, we use different bin sizes for core logic cells and I/O cells during the placement procedure. Figure 3 illustrates the output of a sample run on a small benchmark we have tested, where the dark shapes are I/O cells and light shapes are core logic cells. The dashed lines represent the grid we used to distribute I/O cells.

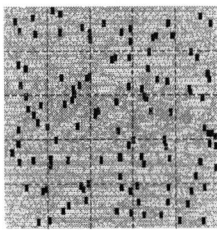

Figure 3: Global co-placement of core and I/O cells.

To consider region constraints, we model cell c_i's region constraint, also known as move bound, by a rectangle $R_i = (\overline{x_{l_i}}, \overline{y_{l_i}}, \overline{x_{h_i}}, \overline{y_{h_i}})$. The constraint is then formulated as a penalty function that is added to the objective function for the placer. Cells with region constraints contribute a penalty m_i to the placement objective as given by:

$$m_i(x_i, y_i) = m(x_i, \overline{x_{l_i}}, \overline{x_{h_i}}) + m(y_i, \overline{y_{l_i}}, \overline{y_{h_i}}), \qquad (1)$$

where function $m(x_i, \overline{x_{l_i}}, \overline{x_{h_i}})$ (similarly, $m(y_i, \overline{y_{l_i}}, \overline{y_{h_i}})$) is defined as:

$$m(x_i, \overline{x_{l_i}}, \overline{x_{h_i}}) = \begin{cases} (\overline{x_{l_i}} - x_i)^2, & x_i < \overline{x_{l_i}} \\ (x_i - \overline{x_{h_i}})^2, & x_i > \overline{x_{h_i}} \\ 0, & \overline{x_{l_i}} \leq x_i \leq \overline{x_{h_i}} \end{cases} \qquad (2)$$

In other words, cells placed outside their constrained regions will increase the objective function by a quadratic term that is proportional to the distance between cells and its desired region boundaries.

It is usually very hard, if not infeasible, to model *SI and escapability constraints* exactly during placement. However, a high level approximation is still beneficial. We divide the die into grids of suitable size, and a capacity limit is specified on each grid for special types of I/O cells (e.g. differential signal pair that has to be escaped on a certain layer). Also a SPG ratio is specified for each power domain so that the grid assigned to a domain has to reserve enough power/ground I/O cells for SI purpose. Such capacity requirements are translated into density constraints to the placer.

Another important goal in this step is to place I/O cells belonging to the same power domain close to each other and away from I/O cells belonging to a different power domain, thus minimizing the number of potential power plane cut-lines on the die[1]. To meet this goal, we add a virtual net to connect I/O cells belonging to the same domain and apply the placement algorithm to obtain an initial distribution of I/O cells on the die. We then subdivide the die into appropriately-sized bins. Each bin is assigned to at most one power domain based on the composition of the I/O cells residing within that bin. Adjacent bins assigned to the same domain are merged together. If one power domain is too fragmented (i.e., too many bins are not adjacent), the corresponding virtual net will be given a higher weight and we will re-run the placement algorithm. In order to make this "plane-cutting" process converge, I/O cells belonging to domains that are not severely fragmented may be artificially connected to some anchor points, so that they will be kept more or less in place during the placement iterations.

Having decided a rough location for each I/O cell from the global placement, we then proceed to synthesize bump arrays by combining algorithms from [13] and [14]. As I/O cells are roughly evenly distributed on the die, the bump arrays can also be evenly distributed. Moreover, we also need to reserve some extra bumps for power and ground (P/G) cells, which are needed to satisfy both the signal integrity and power integrity constraints. We then define the exact locations of I/O sites on the die where I/O cells will be finally placed as shown in Figure 2 based upon RDL routability estimation. As RDL routing is essentially a planar routing problem, we refer readers to the rich literature on this subject for details (e.g., [15]).

6. CONSTRAINT-DRIVEN I/O PLACEMENT

Finally, we need to assign I/O cells to the specific I/O sites such that no I/O cells overlap with each other. Since every I/O cell has already been assigned to one particular power domain after section 5, we can solve the I/O placement problem on a per domain basis, and we call it constraint-driven detailed I/O placement problem ($CDPL$).

6.1 CDPL Constraints

The design constraints discussed in Section 3 are refined with more detailed information after section 5 and can be formally described as follows. (1) One *primary SPG ratio* constraint ($N_{s,0}$:1:1). For every neighborhood consisting of at most $N_{s,0}$ number of signal I/O cells, there is at least one primary power cell and one primary ground cell. (2) A set of *secondary SPG ratio* constraints ($N_{s,i}$:1). For every $N_{s,i}$ constraint, it dictates that there is at least one corresponding secondary power cell for every neighborhood consisting of at most $N_{s,i}$ number of signal I/O cells. (3) A set of *region* constraints (R_i, C_i^R). A region R_i defines a rectangular area within the power domain, and a region constraint (R_i, C_i^R) specifies that the set of I/O cells C_i^R must be placed within the region. Some region constraints may come from board level or chip level floorplan; some may come from power domain definition; and some may come from $CPPL$ as we want to minimize the disturbance to the global optimal $CPPL$ solution. (4) A set of *clustering* constraints (L_i, C_i^L). L_i is a pair of length limits ($\overline{l_{i,x}}, \overline{l_{i,y}}$) and C_i^L is a cluster consisting of a set of I/O cells such that in the final placement the spreading of these cells in the x-axis (respectively y-

axis) is within a range given by $\overline{l_{i,x}}$ (respectively $\overline{l_{i,y}}$). (5) A set of *differential pair* constraints ($D_i=(c_{i0}, c_{i1})$). Cells that form a differential pair (c_{i0}, c_{i1}) should be connected to two bumps that have similar substrate route characteristics. (6) A set of *escape layer* constraints (or bump assignment feasibility constraints) (E_i, C_i^E). E_i are the escape layer properties of bumps as determined from section 5. C_i^E are sets of I/O cells that are required to be escaped and routed on layer E_i in package. Examples of C_i^E may come from I/O cells that form certain I/O standards or may be determined from SI or power integrity analysis.

6.2 CDPL Algorithm

We propose to solve the $CDPL$ problem via the following three-step algorithm. In the first step, we honor the power domain related constraints (Constraint 1 and 2). As the total number of signal cells N to be placed in the power domain is known, we can compute the total required number of primary power cells (and of ground cells) as $\lceil N/N_{s,0} \rceil$, and the total required number of secondary power cells as $\sum_i \lceil N/N_{s,i} \rceil$, respectively. We then insert the required number of primary and secondary power/ground cells into the design and distribute them evenly over the power domain.

In the second step, we formulate an ILP feasibility problem to satisfy Constraint 3, 4 and 5. A straight-forward formulation of ILP would make the problem size too large to be solved efficiently. Therefore, we reduce the ILP problem size by introducing the concept of *super site*, which is an abstraction of a cluster of physically continuous I/O sites. By properly defining the number of I/O sites in a super site (or super site's *granularity*) and formulating the ILP in terms of super site instead of I/O sites directly, we can reduce the ILP problem size greatly without sacrificing too much accuracy. The granularity of super sites is left as a tuning parameter for a particular design.

For a chosen granularity, we group all I/O sites in the power domain into a set of *super sites* $\mathcal{S} = \{S_i\}$ as shown in Figure 2. Each super site S_j is defined by its center location (a_j, b_j) on the die and a set of nearby bumps to which I/O cells assigned to this super site can connect. The number of bumps in S_j defines its bump capacity $\overline{p_j}$, which limits the total number of I/O cells that we can assign to S_j to find a feasible RDL routing solution after I/O placement. Among all bumps in S_j, the number of bumps that have similar substrate route characteristics further defines S_j's differential pair bump capacity $\overline{d_j}$. We define the binary integer variable $x_{i,j}$ for every pair of I/O cell c_i and super site S_j in the domain such that $x_{i,j}=1$ if c_i is assigned into S_j, and $x_{i,j}=0$ otherwise. Moreover, we denote the bounding box of the die by (u_L, u_B, u_T, u_R). Then we have the following ILP feasibility problem:

$$\sum x_{i,j} = 1, \quad \forall c_i \tag{3}$$

$$\sum x_{i,j} \leq \overline{p_j}, \quad \forall S_j \tag{4}$$

$$\sum (x_{i0,j} + x_{i1,j}) \leq \overline{d_j}, \quad \forall S_j \tag{5}$$

$$l_{i,x}^{min} \leq a_j \cdot x_{k,j} + u_R \cdot (1 - x_{k,j}), \quad \forall c_k \in C_i^L, \forall L_i, \forall S_j \tag{6}$$

$$a_j \cdot x_{k,j} + u_L \cdot (1 - x_{k,j}) \leq l_{i,x}^{max}, \quad \forall c_k \in C_i^L, \forall L_i, \forall S_j \tag{7}$$

---

[1]It is acceptable that I/O cells forming a large power domain may be separated into more than one physically disjoint power domain on the die as shown in section 7.

$$l_{i,x}^{max} - l_{i,x}^{min} \leq \overline{l_{i,x}}, \quad \forall L_i \qquad (8)$$

$$l_{i,y}^{min} \leq b_j \cdot x_{k,j} + u_T \cdot (1 - x_{k,j}), \quad \forall c_k \in C_i^L, \forall L_i, \forall S_j \quad (9)$$

$$b_j \cdot x_{k,j} + u_B \cdot (1 - x_{k,j}) \leq l_{i,y}^{max}, \quad \forall c_k \in C_i^L, \forall L_i, \forall S_j (10)$$

$$l_{i,y}^{max} - l_{i,y}^{min} \leq \overline{l_{i,y}}, \quad \forall L_i \qquad (11)$$

$$x_{k,j} = 0, \quad \forall c_k \in C_i^R, \forall S_j \notin R_i \qquad (12)$$

$$x_{i0,j} = x_{i1,j}, \quad \forall D_i, \forall S_j \qquad (13)$$

where (3) dictates that one cell can only be placed into one super site; (4) says that every super site cannot hold more I/O cells than its bump capacity; (5) specifies that every super site cannot hold more differential pairs than its differential bump capacity; (6) to (11) together enforce the clustering constraints; (12) captures the region constraints; and (13) requires us to put a differential pair into the same super site. Note that in the above ILP formulation, we do not consider wire length explicitly. Instead, we assume that there is a local region constraint for each I/O cell such that the I/O cell will be confined to a local neighborhood of its original location decided by $CPPL$. In this way the disturbance of $CDPL$ to $CPPL$ is controlled explicitly, and wire length minimization is achieved indirectly.

When the problem size of the above ILP formulation is small, we use the general branch and bound technique to solve the binary ILP problem optimally. When the above ILP problem size gets relatively large, instead of solving the ILP directly, we solve the corresponding linear programming (LP) problem by relaxing the binary variables $x_{i,j}$ to continuous ones, as $0 \leq x_{i,j} \leq 1$, and then round them back to integers afterward. Our experiment results show that such an approximation is very effective in practice and for some of the designs, we can even find the corresponding integer solutions directly.

After we assign I/O cells to super sites, in the final step of our $CDPL$ algorithm, we find a legal I/O placement within each super site with consideration of Constraint 5 and 6. We solve such a problem by formulating a min-cost-max-flow problem for each super site. We first build a bipartite network $G(V_1, V_2, E)$, where V_1 is the set of I/O cells assigned to the super site; V_2 is the set of I/O sites within the super site; E is the connection between V_1 and V_2 and are formed as follows. Because each super site contains a set of bumps and RDL routes that connect I/O sites and bumps after $CPPL$, by querying each bump's escape layer properties and substrate routing characteristics, we know whether it can be used by a given I/O cell without violating Constraint 5 and 6. Based upon such information we can build the edges of E between V_1 and V_2 in G. Moreover, we associate each edge with a cost that measures the preference of assigning one I/O cell to a particular I/O site. The cost is a weighted function of three components that we want to minimize: difference between an I/O cell's current location and its I/O site location, RDL routing length, and mismatch between I/O cells' requirements on substrate routing and bumps' package substrate routing characteristics. The final network flow problem can be obtained by adding one source and one sink into G, connecting the source to every node in V_1 with cost of zero, connecting every node in V_2 to the sink with cost of zero, and associate each edge in the network

with capacity as one. It is easy to see that the assignment of I/O cells to I/O sites within each super site can be determined optimally by solving a min-cost-max-flow problem.

7. EXPERIMENT RESULTS

Four test cases derived from real industrial custom designs are used to illustrate the effectiveness of our $CIOP$ problem formulation and solution. Table 1 shows the characteristics of the test cases.

Design	# Signal I/O	# Power Domain	# Constraints
d1	1221	4	1801
d2	504	6	814
d3	450	4	934
d4	641	25	1433

Table 1: Test case characteristics.

We report the experiment results in Table 2 including the number of bumps, the number of physically disjoint power domains, the number of inserted P/G cells (both primary and secondary), and constraint satisfaction ratio (CSR), which is defined as the ratio between the number of satisfied constraints and the total number of given constraints[2]. According to Table 2, we see that our $CIOP$ algorithm can effectively find an I/O cell placement solution and satisfy all design constraints simultaneously in the first run (first-time right).

Design	Bumps	Domains	P/G Cells	CSR	Runtime(s)
d1	1560	6	328	100%	538
d2	963	12	445	100%	177
d3	906	6	453	100%	132
d4	1187	71	459	100%	81

Table 2: Experiment results for $CIOP$.

In the following, we compare our results with the conventional two-step approach followed by iterative improvement. In the first experiment, we perform the conventional co-placement of core and I/O cells followed by detailed I/O placement targeting at wire length minimization [2], but without considering the above design constraints. We denoted it as $TIOP$. After that, we measure the quality of design in terms of CSR. Obviously, as design constraints are totally ignored, there may be many design constraint violations. Therefore, we perform a local refinement procedure to remedy this problem by iteratively swapping, shifting or relocating I/O cells to improve CSR. The local refinement's search region is increasingly expanded heuristically after each iteration. As shown in Figure 4(a), before local refinement procedure (zeroth iteration), the CSR is very low (no more than 50%). Local refinement procedure does improve the design, but it is only effective for the first iteration, after that the improvement is very slim. For all designs, the CSR after five iterations of local refinement range from 40% to 70% and start to plateau. This shows that I/O placement without considering design constraints is *impossible to meet all design constraints by relying solely upon the conventional iterative approach*.

In the second experiment, we perform our $CPPL$ followed by conventional detailed I/O placement. We denote it as $TCIOP$. As shown in Figure 4(b), $CPPL$ indeed helps to

---

[2]Region constraints generated *internally* after $CPPL$ are not included.

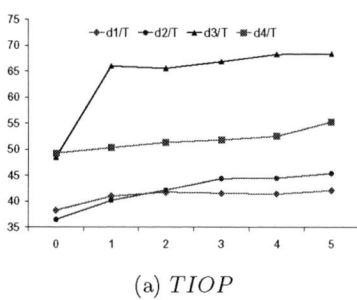

(a) *TIOP*

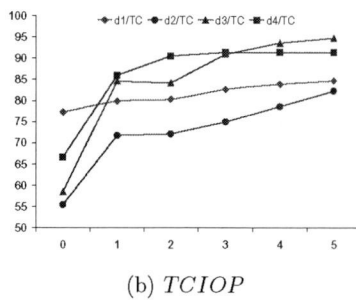

(b) *TCIOP*

Figure 4: Comparison of *CSR*. The *x*-axis is the iteration number and the *y*-axis is *CSR* in percentage, with *CIOP*'s *CSR* being 100%.

obtain better design quality. For the zeroth iteration, the *CSR* is more than 55% and can be up to 77% compared to *TIOP*. After five iterations of local refinement, more than 80% constraints are satisfied, but sill not 100% compared to *CIOP*. Therefore, we conclude that *I/O planning and placement should be constraint driven to be effective in achieving first-time right design closure*.

We further report the wire length as another metric to measure quality of design. We normalize the wire length with respect to the conventional wire length minimization approach (i.e., the zeroth iteration in Figure 4(a). The percentage of wire length increase is shown in Figure 5. It shows that our *CIOP* incurs the largest wire length increases among all designs for the first iteration. However, considering the fact that our *CIOP* satisfies all design constraints while others do not, we believe such a small amount of wire length increase (no more than 8%) is reasonable.

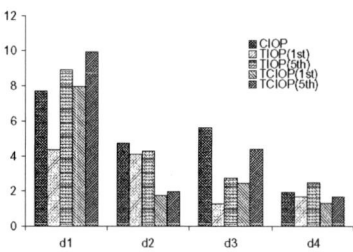

Figure 5: Wire length increase for *CIOP*, *TIOP* and *TCIOP*.

We report the runtime in second for *CIOP* in Table 2.

The machine has a 3.0GHz P4 CPU with 2G memory running Linux. It is observed that the total runtime is less than 10 minutes for the largest design. Therefore, our *CIOP* algorithm can provide a quick turn-around time for chip-package co-design by enabling the designer to explore more design alternatives (through different specification of the constraints) in a reasonable amount of time during the co-design process.

8. CONCLUSION AND DISCUSSION

In this paper, we have formally defined a set of common design constraints for chip-package co-design. Based upon these real design constraints, a detailed constraint-driven I/O placement problem (*CIOP*) has been formulated and solved effectively via a multi-step algorithms. Experiment results based on real industry designs have shown that the proposed algorithm can effectively find a large scale I/O placement solution while satisfying all design constraints in less than 10 minutes.

9. REFERENCES

[1] G. Pascariu, P. Cronin, and D. Crowley, "Next generation electronics packaging utilizing flip chip technology," in *Electronics Manufacturing Technology Symposium, IEEE 28th International*, pp. 423–426, July 2003.

[2] A. Caldwell, A. Kahng, S. Mantik, and I. Markov, "Implications of area-array i/o for row-based placement methodology," in *IC/Package Design Integration, IEEE Symposium on*, pp. 93 – 98, Feb. 1998.

[3] J. Wang, K. Muchherla, and J. Kumar, "A clustering based area I/O planning for flip-chip technology," in *Quality Electronic Design, 5th International Symposium on*, pp. 196 – 201, 2004.

[4] R. Farbarik, X. Liu, M. Rossman, P. Parakh, T. Basso, and R. Brown, "CAD tools for area-distributed I/O pad packaging," in *Multi-Chip Module Conference, IEEE*, pp. 125 – 129, Feb. 1997.

[5] P. Dehkordi, C. Tan, and D. Bouldin, "Intrinsic area array ICs: what, why, and how," in *Multi-Chip Module Conference, IEEE*, pp. 120 – 124, Feb 1997.

[6] M. Pedram, K. Chaudhary, and E. Kuh, "I/O pad assignment based on the circuit structure," in *Computer Design: VLSI in Computers and Processors, IEEE International Conference on*, pp. 314 – 318, Oct. 1991.

[7] B. Chen and M. Marek-Sadowska, "Timing driven placement of pads and latches," in *ASIC Conference and Exhibit, Fifth Annual IEEE International*, pp. 30–33, Sept. 1992.

[8] W.-K. Mak, "I/O placement for FPGAs with multiple I/O standards," *IEEE Trans. on Computer-Aided Design of Integrated Circuits and Systems*, vol. 23, pp. 315–320, February 2004.

[9] "SpeedXP user manual," in *http://www.sigrity.com/*.

[10] E. I. A. J. S. S. T. Division, "Stub series terminated logic for 2.5 volts (SSTL_2)," in *EIA/JEDEC Standard*, Sept. 1998.

[11] N. Viswanathan and C. C.-N. Chu, "Fastplace: efficient analytical placement using cell shifting, iterative local refinement and a hybrid net model," in *ISPD '04: Proceedings of the 2004 international symposium on Physical design*, pp. 26–33, 2004.

[12] K. Vorwerk, A. Kennings, and A. Vannelli, "Engineering details of a stable force-directed placer," in *ICCAD'04: Proceedings of the 2004 international conference on computer aided design*, pp. 573–580, 2004.

[13] C. Tan, D. Bouldin, and P. Dehkordi, "Design implementation of intrinsic area array ICs," in *Advanced Research in VLSI, Seventeenth Conference on*, pp. 82 – 93, Sept. 1997.

[14] M. M. Ozdal and M. D. Wong, "Simultaneous escape routing and layer assignment for dense PCBs," in *Proc. Int. Conf. on Computer Aided Design*, Nov. 2004.

[15] J. Cong, M. Hossain, and N. Sherwani, "A provably good multilayer topological planar routing algorithm in IC layout designs," *IEEE Trans. on Computer-Aided Design of Integrated Circuits and Systems*, vol. 12, pp. 70 – 78, Jan. 1993.

2006 Asia and South Pacific Design Automation Conference

Simultaneous Block and I/O Buffer Floorplanning for Flip-Chip Design*

Chih-Yang Peng[1], Wen-Chang Chao[1], Yao-Wen Chang[2], and Jyh-Herng Wang[3]

[1]Graduate Institute of Electronics Engineering, National Taiwan University, Taipei 106, Taiwan

[2]Department of Electrical Engineering & Graduate Institute of Electronics Engineering, National Taiwan University, Taipei 106, Taiwan

[3]Faraday Technology Corporation, Hsinchu 300, Taiwan

Abstract

The flip-chip package gives the highest chip density of any packaging method to support the pad-limited ASIC design. One of the most important characteristics of flip-chip designs is that the input/output buffers could be placed anywhere inside a chip. In this paper, we first introduce the floorplanning problem for the flip-chip design and formulate it as assigning the positions of input/output buffers and first-stage/last-stage blocks so that the path length between blocks and bump balls as well as the delay skew of the paths are simultaneously minimized. We then present a hierarchical method to solve the problem. We first cluster a block and its corresponding buffers to reduce the problem size. Then, we go into iterations of the alternating and interacting global optimization step and the partitioning step. The global optimization step places blocks based on simulated annealing using the B*-tree representation to minimize a given cost function. The partitioning step dissects the chip into two subregions, and the blocks are divided into two groups and are placed in respective subregions. The two steps repeat until each subregion contains at most a given number of blocks, defined by the ratio of the total block area to the chip area. At last, we refine the floorplan by perturbing blocks inside a subregion as well as in different subregions. Compared with the B*-tree based floorplanner alone, our method is more efficient and obtains significantly better results, with an average cost of only 51.8% of that obtained by using the B*-tree alone, based on a set of real industrial flip-chip designs provided by leading companies.

I. INTRODUCTION

A. Flip-chip Design

Flip-chip bonding gives the highest chip density of any packaging method to support the pad limited ASIC design. The important characteristics of flip-chip designs is that the signals or power could be imported from the signal bumps or power bumps distributed on the whole chip, and the input and output buffers could be placed anywhere inside a chip, like core cells. There exist a few different flip-chip architectures. See Figure 1 for an example layout of the flip-chip design available from UMC and ASE. We use the top metal or an extra metal layer, called *Re-Distributed Layer* (RDL), to connect input or output buffers to bump balls. Figure 2 illustrates the cross section of RDL. Bump balls are placed on RDL and use RDL to connect to IO buffers. Therefore, bump balls can overlap with input/output buffers and blocks.

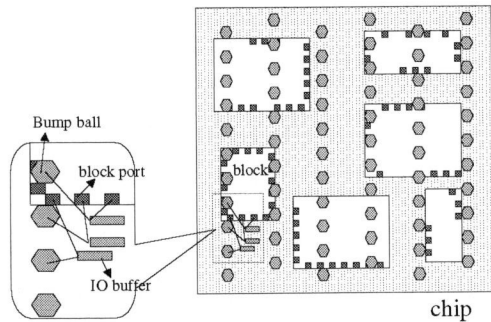

Fig. 1. Example layout of the flip chip.

*This work was partially supported by SpringSoft, Inc. and National Science Council of Taiwan under Grant No's. NSC 93-2215-E-002-009, NSC 93-2220-E-002-001, and NSC 93-2752-E-002-008-PAE.

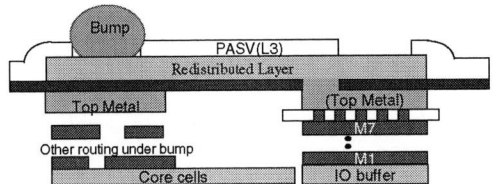

Fig. 2. Cross section of RDL.

In a wire-bond IC, in contrast, the circuit core is surrounded by the I/O pads on the perimeter of the chips. In general, the interconnection between an input/output gate and an I/O pad consists of two segments: the inner part and outer part. The inner-part segment is the portion of interconnection within the core while the outer part is between the core and the pads. Unlike the wire-bond I/O pads which are placed only on the perimeter of the chip, the flip-chip I/O pads are placed within the chip core. The inner-part routing can be minimized by placing the I/O pads. In the wire-bond layout, the area between the core and the pads is utilized for routing the interconnection of the outer-part segment. This area can be eliminated in a flip-chip IC by integrating the I/O pads within the core. This is a great saving in silicon area and generally occurs when there is a small core with a high number of I/O pins. The study conducted in [11] showed the reduction of die size and the increase in I/O count when the peripheral wire-bond technology was replaced by the flip-chip technology. Further, the bump balls in the flip chip have lower inductance than the bond-wires in the classical IC.

In this paper, we assume that all the bump balls are placed at pre-defined locations and their signals are determined, which is true for most real applications since they are often predefined by the packaging site. All core cells are partitioned or grouped into blocks. The input/output signals are connected to block ports through the input/output buffers to the bump balls. We need to place the input/output buffers and the blocks without overlapping with each other into a pre-defined chip area so that the path length between blocks and bump balls is minimized.

For most practical designs, like memory controllers, there are a large number of input/output pins being used as data buses. For such designs, we have to control the timing of the input/output signals. In other words, we have to make sure that the input signals arrive at the core simultaneously. In the same way, it also needs to make sure that the output signals arrive at bump balls simultaneously. This can be achieved through controlling the positions of bump balls, input/output buffers and blocks to minimize the signal skew.

B. Previous Work

The placement problem for the classical wire-bond IC's has been studied very extensively [1, 2, 5, 7, 8, 12, 13, 16, 17, 18, 22, 24, 25]. Nevertheless, most of these previous works target on standard cell designs, for which cells are of the same height and are placed in rows. For the floorplanning problem addressed here, it does not have such restrictions, making the previous works not flexible enough to the flip-chip floorplanning problem. (Note that although a few existing placers can handle the mixed-size placement problem, they usually focus more on the standard-cells of the same heights. Therefore, the placers cannot handle the floorplanning problem well. We have tried well-known publicly available placers such as Feng Shui 2.6/5.0 [13] and mGP [8, 22]. They all cannot obtain desirable floorplans for the flip-chip design directly.) Recently, Hsieh and Wang presented an analytical formulation for flip-chip placement in [15]. The work targets at an objective function of the sum of path delay and sum of skew between all input paths, which runs in quadratic time. Further, the sum of skew does not model the skew cost well.

0-7803-9451-8/06/$20.00 ©2006 IEEE.

In contrast, most floorplanning techniques can handle much more general objective functions by applying simulated annealing [9, 14, 21, 23, 26]. However, traditional floorplanning/placement algorithms do not scale well as the circuit size, complexity, and constraints increase. The B*-tree, in contrast, has been shown an efficient and effective data structure for floorplanning [6, 9, 20]. (In particular, the B*-tree tool is available on-line [4].) The B*-tree is particularly suitable for representing a floorplan/placement with mixed-size blocks, like the blocks and the I/O buffers for the flip-chip design; further, it does not have the cell height and the row placement constraints, imposed by the classical placement algorithms. Therefore, we shall take advantage of the nice properties of the B*-tree to develop our algorithm for flip-chip placement. Nevertheless, a key limitation of the B*-tree based floorplanner lies in its packing nature—the B*-tree based floorplanner always compacts blocks to the left and bottom as shown in Figure 3. If the total block area is much smaller than the chip area, then some blocks might not be placed at the desired positions to optimize the interconnection cost. As illustrated in Figure 3, the top and right sides of the chip are empty.

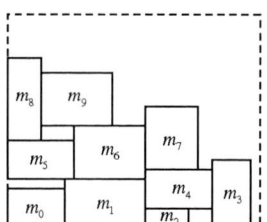

Fig. 3. The B*-tree based floorplanner always packs blocks to the left and to the bottom.

C. Our Contributions

In this paper, we first introduce the floorplanning problem for the flip-chip design and formulate it as assigning the positions of input/output buffers and first-stage/last-stage blocks so that the path length between blocks and bump balls as well as the delay skew of the paths are simultaneously minimized. In this formulation, we address practical issues in the industrial flip-chip design, such as the minimization of interconnection delay and data bus skew. To handle such objectives, the classical standard-cell/mixed-size placement techniques (like Aplace [16], Capo [1], GORDIAN [18], GORDIAN-L, mPG [8], mPL [7], FastPlace [24], Feng Shui [17], Dragon [25]) and the B*-tree floorplanning technique alone have their limitations. For example, the skew objective leads to a non-quadratic, non-convex term for which most analytic placers (such as Aplace, GORDIAN, mPL) rely on for the global optimization, the cell height and row placement constraints make the classical standard-cell placers not directly applicable to the flip-chip placement, and the compaction nature of the B*-tree limits the quality of interconnection optimization.

To remedy the limitations of the classical standard-cell/mixed-size placers and the B*-tree, we present a hierarchical top-down method to solve the problem based on a more accurate and efficient cost function. We first cluster a block and its corresponding buffers to reduce the problem size. Then, we go into iterations of the alternating and interacting global optimization step and the partitioning step. The global optimization step places blocks based on simulated annealing using the B*-tree representation to minimize a given cost function. The partitioning step dissects the chip into two subregions, and the blocks are divided into two groups and are placed in respective subregions. The two steps repeat until each subregion contains at most a given number of blocks, defined by the ratio of the total block area to the chip area. At last, we refine the floorplan by perturbing blocks inside a subregion as well as in different subregions. Compared with the B*-tree based floorplanner alone, our method obtains significantly better results, with an average cost of only 51.8% of that obtained by using the B*-tree alone, based on a set of real industrial flip-chip designs provided by leading companies. Further, our floorplanner is more efficient than the B*-tree alone.

The remainder of this paper is organized as follows. Section 2 formulates the problem of block and input/output buffer floorplanning for flip-chip design. Section 3 reviews the B*-tree representation. Section 4 presents our algorithm for handling the floorplanning problem. Section 5 reports the experimental results, and finally the conclusions and future work are given in Section 6.

II. PRELIMINARIES

We consider the block-based design, for which blocks and buffers are rectangular and can be placed anywhere in the given flip-chip to minimize the objective function. All the input/output signals are connected to block ports through the input/output buffers. We assume that all the bump balls are placed at pre-defined locations (typically defined by packaging sites) and their signals are determined. We intend to minimize the interconnection length among the bump balls, the I/O buffers, and blocks. For most practical designs, as mentioned earlier, there are a large number of input/output pins being used for data buses. Therefore, it is also desired to make sure that all input signals from bump balls via input buffers to blocks arrive simultaneously and output signals from blocks via output buffers to bump balls also arrive simultaneously. To achieve the goal, we define the objective function Γ as follows:

$$\Gamma \quad = \quad \alpha\phi_1 + \beta\phi_2, \tag{1}$$

where

$$\phi_1 \quad = \quad \sum_{j=1}^{n1} d_j^i + \sum_{j=1}^{n2} d_j^o$$

$$\phi_2 \quad = \quad \left(\max_{1 \leq j \leq n1} d_j^i - \min_{1 \leq j \leq n1} d_j^i \right)^2 + \left(\max_{1 \leq j \leq n2} d_j^o - \min_{1 \leq j \leq n2} d_j^o \right)^2.$$

In Γ, ϕ_1 gives the sum of path delays, and ϕ_2 gives the sum of the squares of the maximum (critical) input and output signal skews. (Note that we adopt the squares of the signal skews in order to match the magnitude of the path delay cost.) Here, α and β are the user-specified weighting factors, $n1$ and $n2$ are the numbers of input and output signals, respectively, and d_j^i and d_j^o are the respective path delays of the jth input signal and the jth output signal. The path delay of an input signal is the delay of a path from a bump ball via an input buffer to a block port; the path delay of an output signal is the delay of a path from a block port via an output buffer to a bump ball. The path delay is measured by the rectilinear path length between two circuit components (bump balls, buffers, or block ports), i.e., the Manhattan distance between the two points. Minimizing the above objective function means that it needs to minimize the critical skew of the path delays of all input signals, output signals, and the total path delay.

It should be noted that the above objective function does address the requirements needed for the recent real industrial flip-chip designs (e.g., from the leading foundry UMC and its design service company Faraday) to be reported in Section 5. Also, unlike the objective function used in [15] which cannot model the skew cost accurately and needs quadratic time for evaluation, the new objective function is more accurate and needs only linear time for evaluation.

III. THE B*-TREE REPRESENTATION

As mentioned earlier, we extend the B*-tree representation to handle the problem of block and I/O buffer floorplanning for flip-chip design. Thus, we shall give a review of the B*-tree representation.

Given a compacted placement P that can neither move down nor move left (called an *admissible placement* [14]), we can represent it by a unique B*-tree T [9]. (See Figure 4(b) for the B*-tree representing the placement shown in Figure 4(a).) A B*-tree is an ordered binary tree (a restriction of the O-tree [14] with faster and more flexible operations) whose root corresponds to the block on the bottom-left corner. Using the depth-first search (DFS) procedure, the B*-tree T for an admissible placement P can be constructed in a recursive fashion. Starting from the root, we first recursively construct the left subtree and then the right subtree. Let R_i denote the set of blocks located on the right-hand side and adjacent to m_i. The left child of the node n_i corresponds to the lowest block in R_i that is unvisited. The right child of n_i represents the lowest block located above m_i, with its x-coordinate equal to that of m_i.

Figure 4(b) illustrates the resulting B*-tree for the placement shown in Figure 4(a). The B*-tree keeps the geometric relationship between two blocks as follows. If node n_j is the left child of node n_i, block m_j must be located on the right-hand side and adjacent to block m_i in the admissible placement; i.e., $x_j = x_i + w_i$. Besides, if node n_j is the right child of n_i, block m_j must be located above block m_i, with the x-coordinate of m_j equal to that of m_i; i.e., $x_j = x_i$. Also, since the root of T represents the bottom-left block, the x- and y-coordinates of the block associated with the root $(x_{root}, y_{root}) = (0, 0)$. Therefore, given a B*-tree, the x-coordinates of all blocks can be determined by traversing the tree once. The y-coordinate can be computed based on the

2006 Asia and South Pacific Design Automation Conference

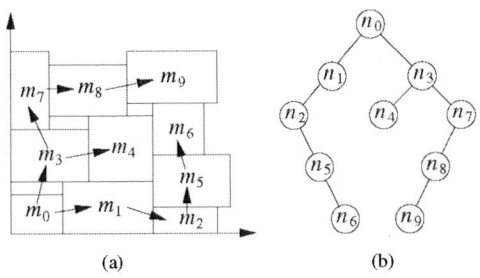

Fig. 4. (a) An admissible placement. (b) The corresponding B*-tree.

contour data structure presented in [14] in amortized $O(1)$ time for each node. Therefore, an n-node B*-tree can be evaluated very efficiently in amortized $O(n)$ time.

IV. OUR ALGORITHM

Our algorithm is illustrated in Figure 5. Given inputs of net list and the geometry of the chip, we first cluster a block and its corresponding I/O buffers into a clustered block. Then we go into the main steps of alternating and interacting global optimization and partitioning steps. The global optimization step places blocks based on simulated annealing using the B*-tree representation to minimize a given cost function. The partitioning step dissects the chip into two subregions, and the blocks are divided into two groups according to their coordinates and are placed in respective subregions. Until each region contains at most q clustered blocks, we decluster these clustered blocks. After the declustering step, the global optimization and the partitioning steps repeat until each region contains at most k blocks. At last the final floorplanning step starts. In the final floorplanning step, we refine the floorplan by perturbing blocks inside a subregion as well as in different subregions. Note that the values of q and k control the resulting number of subregions. The smaller the values, the more the resulting subregions. If the area utilization ratio (the total block area divided by the total chip area) is large, it is harder to place blocks into subregions if we cut the chip into too many small subregions, and thus we shall favor larger q and k for this situation. We shall explain each step of the algorithm and the choices of q and k in the following sections.

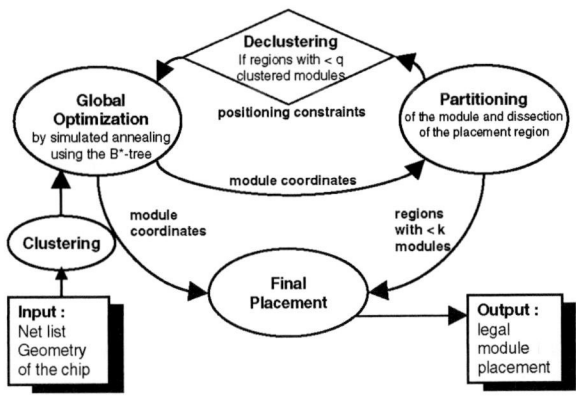

Fig. 5. Our algorithm.

A. Clustering

In this step, we apply simulated annealing using the B*-tree representation to group a block and its I/O buffers to a clustered block. The objective function is defined by area and the path delay between the input (output) port of a block and the output (input) port of an I/O buffer. By this process, I/O buffers will be clustered around its corresponding block. See Figure 6 for an example.

We introduce a node in the B*-tree based on the placement of a block and its clustered I/O buffers, called a *block node*. The width and height of the

block node are equal to the respective width and height of the floorplan of the block and its clustered I/O buffers. When we perform the global optimization and partitioning, each block node represents its block and corresponding I/O buffers until declustering. Therefore, the problem size can significantly be reduced by clustering.

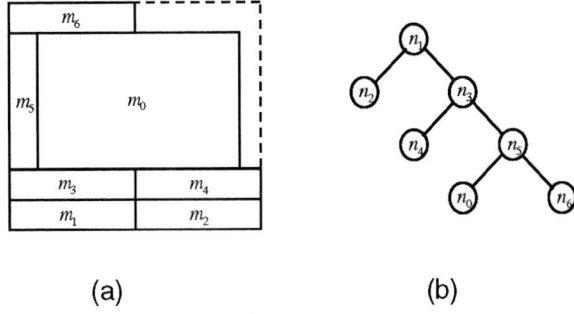

Fig. 6. (a) The geometry of a block and its clustered I/O buffers. (b) The B*-tree topology. m_0 is the block and $m_1, m_2, \cdots, m_n$ are the I/O buffers. The dotted line gives the boundary of the clustered block.

B. Global Optimization and Partitioning

The floorplanning procedure is composed of alternating and interacting global optimization and partitioning steps. In the global optimization step, we place blocks by simulated annealing using the B*-tree representation.

For the region ρ, the positions of all blocks in the region, denoted by M_ρ, are derived from simulated annealing using the B*-tree representation. We assume that W_ρ^r (H_ρ^r) is the width (height) of region ρ, and W_ρ^m (H_ρ^m) is the width (height) of the floorplan in region ρ. The width (W_ρ^r), height (H_ρ^r), and the coordinates of regions are determined in the partitioning step. The coordinate of each region is set to the bottom-left corner of the region, and the coordinates of blocks in M_ρ are relative to the coordinate of the region ρ.

There are two stages in the global optimization step, distinguished by the declustering step. For the global optimization before declustering, we place blocks only to minimize the objective function ϕ_1, and blocks might be placed out of the region at this stage. Here, ϕ_1 denotes the sum of wirelengths between the clustered blocks and bump balls. This process makes a clustered block closer to its corresponding bump balls. Although we do not consider the signal skew and the fixed outline of the flip chip at this stage, we can still fix/refine the solution at the final floorplanning stage or the global optimization step after declustering.

For the global optimization after declustering, we apply the objective function Γ. In order to place blocks into their region boundary (i.e., fixed-outline floorplanning), we shall also consider the width and height of the resulting floorplan individually so that neither dimension violates the outline constraint. To do so, we modify the objective function as follows:

$$\Gamma' = \Gamma + \gamma\Phi, \qquad (2)$$

where

$$\Phi = \max(0, W_\rho^m - W_\rho^r) + \max(0, H_\rho^m - H_\rho^r). \qquad (3)$$

Here, cost Φ is used to force blocks to be packed into the chip during simulated annealing. In order to satisfy the fixed-outline constraint, γ is set to a huge constant (say, 1000) to guarantee that the cost Φ is much bigger than any cost Γ if the fixed-outline constraint is violated.

After each global optimization process, a new partitioning step starts; we divide the region into two subregions and partition blocks into two groups depending on their positions. In the partitioning step, for each region ρ with $|M_\rho| > k$, if the region width (W_ρ^r) is larger than its height (H_ρ^r), the blocks in M_ρ are sorted according to the x-coordinates of the blocks, and the region is cut vertically. In contrast, if the region height is larger than its width, the blocks in M_ρ are sorted according to the y-coordinates, and the region is cut horizontally. Then, M_ρ is divided into $M_{\rho'}$ and $M_{\rho''}$ such that the summation of the block areas in $M_{\rho'}$ and $M_{\rho''}$ are approximately the same. The rectangular area of region ρ is dissected accordingly. See Figure 7 for an illustration of the processing.

215

2C-5

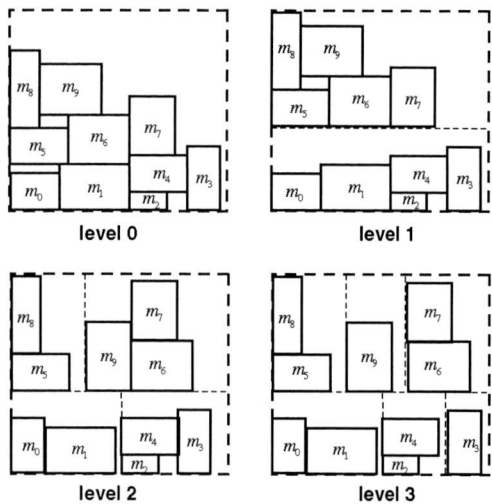

Fig. 7. Illustration of the interactive global optimization and partitioning steps ($q = 2$).

C. Declustering

If the number of blocks in any region is smaller than q (so the problem size is small enough), we shall ungroup each block which was formed by clustering a block and its corresponding I/O buffers previously in this region. Each node in the B*-tree is then expanded into a subtree representing the block's components which are constructed at the clustering step. The number of nodes after declustering is definitely larger. To cope with the increasing problem size, we process as follows to avoid a dramatic change in the resulting floorplan due to the declustering.

Suppose that the original tree has the nodes $n_0, n_1, \cdots, n_n$, denoting blocks $m_0, m_1, \cdots, m_n$, respectively; each block m_i corresponds to the subtree T_i constructed at the clustering step. There are two kinds of relations between two connected nodes in the original tree; a node is a left child or a right child of another node. Let n_l and n_r denote the left and right child of the node n_i; m_l and m_r represented by n_l and n_r are located right to or above m_i.

First we expand the parent node n_i into the subtree T_i and record the last contour when performing the B*-tree packing. The root node of the contour c_{root} represents the left- and top-most cell of the block m_i, and the tail node of the contour c_{tail} represents the right- and top-most cell. See Figure 8 for an illustration.

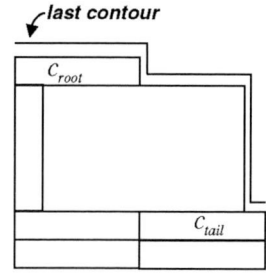

Fig. 8. The inter blocks and the last contour of the clustered block m_i.

Then, if the child of n_i is a left child n_l, we make the root node of the subtree T_l as the left child of the node c_{tail}. Thus the block represented by the root node of the subtree T_l is located adjacent to the right side of the right-most cell of block n_i, as illustrated in Figure 9. In contrast, if the child of n_i is a right child n_r, the root node of subtree T_l becomes the right child of the node c_{root}. The root cell will be located above the top-most cell of block n_i, as illustrated in Figure 10. By this process, the floorplan will not be changed dramatically after declustering.

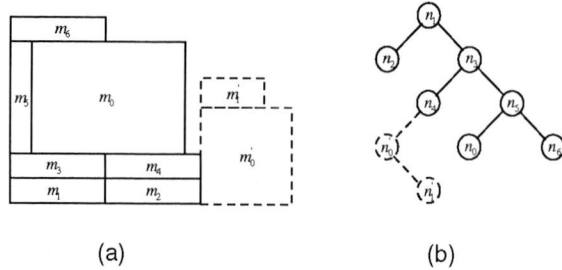

(a)   (b)

Fig. 9. The blocks of solid lines belong to a clustered block m_i, and the blocks of dotted lines belong to another clustered block m_l. m_l is the left child of m_i. (a) m_i and m_l after declustering. (b) The corresponding tree topology after declustering. (Here, m_4 is the tail bock on the last contour.)

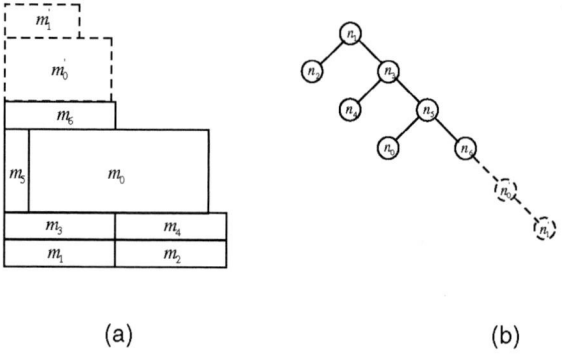

(a)   (b)

Fig. 10. The blocks of dotted lines belong to the clustered block m_r. m_r is the right child of m_i. (a) m_i and m_r after declustering. (b) The corresponding tree topology after declustering.

D. Final Floorplanning

In this step, the chip has been dissected into several subregions, and blocks have been divided into several groups and placed in respective subregions. The simulated annealing process starts again. We refine the floorplan by perturbing blocks inside a subregion as well as in different subregions. The objective function is the same as the function in the global optimization step, but the perturbation operations are different. We select two blocks randomly and swap them if the swap will not cause any outline violation. It gives a chance that a block can change its subregion. After changing blocks, we only re-compute the coordinates of blocks at the changed subregions. Doing so, we have a chance to further refine the floorplan solution.

E. Summary of Our Algorithm

The flow of our algorithm is shown in Figure 12 (the procedure is shown in Figure 5). The result of our floorplanning method may be influenced by the parameters q and k. The parameter q controls the degree of the partitioning step. A smaller q implies that more subregions will be partitioned. We suggest that if the total block area is much smaller than the chip area (i.e., the chip utilization ratio is small), the parameter q should be smaller to generate more subregions to prevent blocks from being packed together at the bottom-left corner of some subregion. (Note that this is an intrinsic behavior of a compacted floorplanner like the B*-tree.) Otherwise, we shall choose a larger q since it is harder to place blocks into subregions if we cut the chip into too many small subregions. The parameter k plays the same role as parameter q after declustering. The optimal q and k may be different for different test cases. Nevertheless, we propose a heuristic to define them based on the ratio of total block area to the chip area; the heuristic is given in Figure 11. It is clear that this heuristic leads to appropriate q and k for flip-chip design.

216

```
1    q = # clustered_blocks;
2    k = # blocks;
3    r = total_blocks_area/chip_area  /* utilization ratio */
4    if ( r < 0.75 )
5        q = 10 × r;
6        k = 10 × r × (#blocks/#clustered_blocks)
7    else
8        k = 10 × r × (# blocks / # clustered_blocks)
9    q = max{q, 3}; /* the smallest q = 3 */
10   k = max{k, 20} /* the smallest k = 20 */;
```

Fig. 11. A heuristic to define the parameters q and k.

Circuit	# blocks	# buffers	chip area	block area /chip area	α	β
fc1	6	25	1040x1040	0.4216	0.5	0.5
fc2	12	168	3440x3440	0.5598	0.5	0.5
fc3	23	320	4240x4240	0.6584	0.7	0.3
fc4	28	384	4440x4440	0.7276	0.7	0.3
fc5	28	384	4440x4440	0.7276	0.7	0.3
fc6	28	384	4040x4040	0.8788	0.7	0.3
fc7	28	384	4040x4040	0.8788	0.7	0.3

TABLE I
STATISTICS OF THE TEST CIRCUITS.

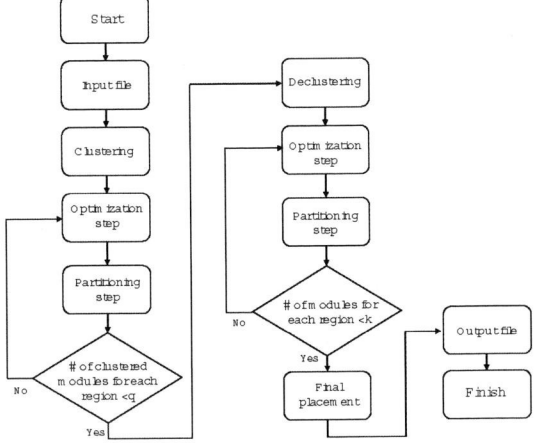

Fig. 12. The flow chart of our floorplanning method.

Ckt		B*-tree alone		TCG alone		Our Method	
fc1	Tot. path delay	23390	1.32	28430	1.60	17760	1.0
	Max. input skew	160	1.33	120	1.00	120	1.0
	Max. output skew	100	1.11	100	1.11	90	1.0
	Cost Γ	2.95e+06	1.46	2.641e+06	1.31	2.01e+06	1.0
	CPU Time	1 s	0.73	32 s	33.83	1 s	1.0
fc2	Tot. path delay	521030	1.44	750450	2.08	361650	1.0
	Max. input skew	1360	1.37	1390	1.38	1010	1.0
	Max. output skew	1890	1.36	1740	1.25	1390	1.0
	Cost Γ	2.97e+08	1.79	2.855e+08	1.72	1.66e+08	1.0
	CPU Time	20 s	1.29	9944 s	631.76	16 s	1.0
fc3	Tot. path delay	1033800	1.67	NR	-	619200	1.0
	Max. input skew	3320	2.00	NR	-	1660	1.0
	Max. output skew	2500	1.47	NR	-	1700	1.0
	Cost Γ	1.24e+09	3.00	NR	-	4.14e+08	1.0
	CPU Time	85 s	1.66	>10 hr	-	51 s	1.0
fc4	Tot. path delay	1153560	1.59	NR	-	726040	1.0
	Max. input skew	3380	1.54	NR	-	2190	1.0
	Max. output skew	2820	1.18	NR	-	2380	1.0
	Cost Γ	1.39e+09	1.84	NR	-	7.54e+08	1.0
	CPU Time	130 s	1.80	>10 hr	-	72 s	1.0
fc5	Tot. path delay	969140	1.37	NR	-	707430	1.0
	Max. input skew	3300	1.91	NR	-	1730	1.0
	Max. output skew	3200	1.48	NR	-	2160	1.0
	Cost Γ	1.51e+09	2.71	NR	-	5.57e+08	1.0
	CPU Time	130 s	1.66	>10 hr	-	78 s	1.0
fc6	Tot. path delay	1233720	1.65	NR	-	745880	1.0
	Max. input skew	3580	1.19	NR	-	3000	1.0
	Max. output skew	4360	1.39	NR	-	3140	1.0
	Cost Γ	2.26e+09	1.69	NR	-	1.34e+09	1.0
	CPU Time	108 s	0.68	>10 hr	-	160 s	1.0
fc7	Tot. path delay	1159560	1.59	NR	-	729180	1.0
	Max. input skew	3880	1.11	NR	-	3500	1.0
	Max. output skew	4720	1.65	NR	-	2860	1.0
	Cost Γ	2.65e+09	1.82	NR	-	1.45e+09	1.0
	CPU Time	251 s	1.11	>10 hr	-	226 s	1.0

TABLE II
EXPERIMENTAL RESULTS OF OUR FLOORPLANNING METHOD, THE
B*-TREE REPRESENTATION ALONE AND TCG REPRESENTATION ALONE.
*NR: NO RESULTS OBTAINED.

V. EXPERIMENTAL RESULTS

We implemented our algorithm in the C++ programming language on a 1.2GHz SUN Blade 2000 workstation with 8 GB memory. (We will make this tool available to the public after this work is published to facilitate future research along this direction.) The benchmark circuits fc1, fc2, ..., fc7 are real consumer designs (DVD players, MP3, etc) and were provided by the leading foundry UMC and its design service company Faraday. Table I lists the names of circuits, the number of blocks, the number of buffers, the chip areas, the ratio of the total blocks area (including blocks and I/O buffer blocks) to the chip area, and the parameters α and β (also defined by the company). The parameter α is the weighting factor of the path delay part ϕ_1 of the objective function Γ, and the β is that of the skew part ϕ_2 of Γ. The test cases fc4, fc5, fc6, and fc7 are for the same design with different assignments of block ports and wire connections; therefore, their chip sizes and the block sizes are all the same. (So the problem sizes range from 31 blocks + I/O buffers to 412 blocks + I/O buffers, representing the typical problem sizes for recent applications.)

We compared our algorithm with the state-of-the-art B*-tree floorplanner and the TCG [21] one using the same cost function Γ. It should be noted that we do not compare with the classical standard-cell placers. As mentioned earlier, the classical standard-cell and/or mixed-size placers (such as the famous Aplace, Capo, GORDIAN, GORDIAN-L, mPG, mPL, FastPlace, Feng Shui, Dragon) cannot directly apply to the flip-chip floorplanning problem well because of the cell height and row placement constraints and the non-quadratic, non-convex term in the problem formulation. (As mentioned earlier, we have tried well-known publicly available placers such as Feng Shui 2.6/5.0 [13] and mGP [8, 22]. They all cannot obtain desirable floorplans for the flip-chip design directly.) We shall also note that the B*-tree floorplanner is considered a leading tool in block floorplanning [6, 10, 20].

The results are listed in Table II. The B*-tree package that we used here is the state-of-the-art version used in [20], which has been shown to be able to handle up to thousands of blocks. The source code of the B*-tree package is available to the general public on-line [4]. We implemented our algorithm based on the same simulated annealing scheme as that used by the B*-tree package. Unlike the B*-tree based floorplanner which always compacts blocks to the left and bottom, the TCG based floorplanner results in general floorplans, which well addresses the layout requirement for the flip-chip design. Nevertheless, TCG has a higher complexity of $O(n^2)$ time for its operations and packing with n blocks. This limits the applicability and quality of TCG for large-scale designs.

As shown in Tables II and III, our method obtains significantly better results in total path delays and the input/output signal skews; the B*-tree based algorithm (the TCG based algorithm) results in the overall cost of 2.04 times (1.52 times) of that of our algorithm. Note that because of the higher complexity in operations and packing, the TCG based floorplanner alone is only feasible for the first two cases. Further, our method is more efficient than the B*-tree and the TCG-based floorplanners. The results justify the effectiveness and efficiency of our method; the B*-tree based algorithm (the TCG based algorithm) needs 1.28 times (more than 332 times) of our CPU time. The results show the effectiveness and efficiency of our algorithm. The resulting layout fc3 is shown in Figures 13.

It should be noted that the reason why our method is even more efficient than the B*-tree based floorplanner mainly lies in the hierarchical framework and the clustering and declustering schemes adopted in our work. By using the framework and the schemes, we can control the problem sizes well at each

	Total path delays	Max. input skew	Max. output skew	Cost Γ	CPU Time
Our method	1.00	1.00	1.00	1.00	1.00
B*-tree alone	1.52	1.49	1.38	2.04	1.28
TCG alone	1.84	1.19	1.18	1.52	332.80

TABLE III

THE AVERAGE COST AND CPU TIME RATIOS FOR THE B*-TREE AND THE TCG BASED ALGORITHMS VS. OUR ALGORITHM FOR ALL TEST CIRCUITS.

stage. Therefore, our method has better scalability than the B*-tree and TCG-based floorplanners for handling the flip-chip floorplanning of various problem sizes.

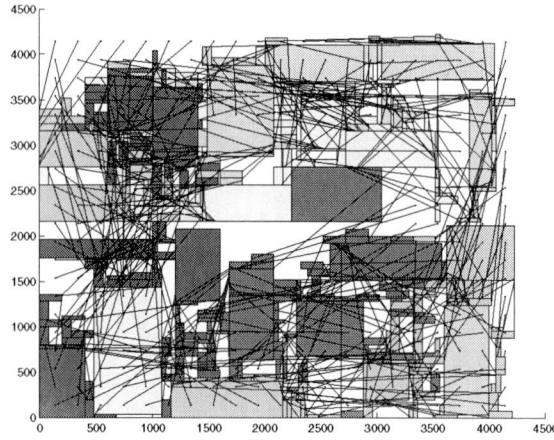

Fig. 13. The floorplanning result of fc3.

VI. CONCLUSION

We have presented a B*-tree based hierarchical top-down method for the block and input/output buffer floorplanning for flip-chip design. This method not only remedies the limitations of the classical standard-cell placers and the B*-tree, but also speeds up the running time by applying the hierarchical top-down scheme. Experimental results based on real industrial flip-chip designs provided by leading companies have shown the effectiveness and efficiency of our algorithm. Future work lies in developing other heuristics to slice the chip to further improve the results. Also, the routing and tighter integration of layout and packaging co-synthesis for the flip-chip design are on-going.

REFERENCES

[1] S. N. Adya, I. L. Markov, and P. G. Villarrubia, "On whitespace in mixed-size placement and physical synthesis," *Proc. of IEEE/ACM Int. Conf. on Computer-Aided Design*, pp. 311–318, 2003.

[2] A. R. Agnihotri, M. C. Yildiz, A. Khatkhate, A. Mathur, S. Ono, and P. H. Madden, "Fractical cut: improved recursive bisection placement," *Proc. of IEEE/ACM Int. Conf. on Computer-Aided Design*, pp. 307–310, 2003.

[3] P.H. Buffet, J. Natonio, R.A. Proctor, Yu H. Sun, and G. Yasar, "Methodology for I/O cell placement and checking in ASIC designs using area-array power grid," *Proc. of IEEE Custom Integrated Circuits Conf.*, pp. 125–128, 2000.

[4] B*-tree: http://cc.ee.ntu.edu.tw/~ywchang/research.html.

[5] A. E. Caldwell, A. B. Kahng, and I. L. Markov, "Can recursive bisection alone produce routable placement?," *Proc. of ACM/IEEE Design Automation Conf.*, pp. 477–482, 2000.

[6] H. H. Chan, S. N. Adya, and I. L. Markov, "Are floorplan representations important in digital design?" *Proc. of ACM International Symposium on Physical Design*, pp. 129–136, 2005.

[7] T. Chan, J. Cong, and K. Sze, "Multilevel generalized force-directed method for circuit placement," *Proc. of ACM International Symposium on Physical Design*, 2005.

[8] C. C. Chang, J. Cong, and X. Yuan, "Multilevel placement for large-scale mixed-size IC designs," *Proc. of ACM/IEEE Asia and South Pacific Design Automation Conf.*, pp. 325–330, 2003.

[9] Y.-C. Chang, Y.-W. Chang, G.-M. Wu, and S.-W. Wu, "B*-Trees: a new representation for non-slicing floorplans," *Proc. of ACM/IEEE Design Automation Conf.*, pp. 458–463, 2000.

[10] J. Con, G. Nataneli, M. Romesis, and J. R. Shinnerl, "An area-optimality study of floorplanning," *Proc. of ACM International Symposium on Physical Design*, pp. 78–83, April 2004.

[11] P. Dehkordi and D. Bouldin, "Design for packageability: the impact of bonding technology on the size and layout of VLSI dies," *Proc. Multichip Module Conf.*, pp. 153–159, 1993.

[12] H. Eisenmann and F. M. Johannes, "Generic global placement and floorplanning," *Proc. of ACM/IEEE Design Automation Conf.*, pp. 269–274, 1998.

[13] *FengShui Placer.* http://vlsicad.cs.binghamton.edu/software.html.

[14] P.-N. Guo, C.-K. Cheng, and T. Yoshimura, "An O-tree representation of non-slicing floorplan and its applications," *Proc. of ACM/IEEE Design Automation Conf.*, pp. 268–273, 1999.

[15] H.-Y. Hsieh and T.-C. Wang, Simple yet effective algorithms for block and I/O buffer placement in flip-chip designs, *Proc. of IEEE International Symposium on Circuits and Systems*, pp. 1879–1882, May 2005.

[16] A. B. Kahng and Q. Wang, "Implementation and extensibility of an analytic placer," *Proc. of ACM International Symposium on Physical Design*, pp. 18–25, April 2004.

[17] A. Khatkhate, C. Li, A. R. Agnihotri, M. C. Yildiz, S. Ono, C.-K. Koh, and P. H. Madden, "Recursive bisection based mixed block placement," *Proc. of ACM International Symposium on Physical Design*, pp. 84–89, April 2004.

[18] J.M. Kleinhans, G. Sigl, F.M. Johannes, K.J. Antreich, "GORDIAN: VLSI placement by quadratic programming and slicing optimization," *IEEE Trans. Computer-Aided Design*, pp. 356–365, 1991.

[19] J.N. Kozhaya, S.R. Nassif, F.N. Najm, "I/O buffer placement methodology for ASICs," *Proc. of IEEE International Conference on Electronics, Circuits and System*, pp. 245–248, 2001.

[20] H.-C. Lee, Y.-W. Chang, J.-M. Hsu, and H. Yang, "Multilevel floorplanning/placement for large-scale modules using B*-trees," *Proc. of ACM/IEEE Design Automation Conf.*, Anaheim, CA, June 2003.

[21] J.-M. Lin and Y.-W. Chang, "TCG: A transitive closure graph based representation for non-slicing floorplans," *Proc. of ACM/IEEE Design Automation Conf.*, pp. 764–769, Las Vegas, NV, June 2001.

[22] *mGP: Multilevel Global Placement.* http://ballade.cs.ucla.edu/mGP/.

[23] Parquet: Fixed-Outline Floorplanner, http://vlsicad.eecs.umich.edu/BK/parquet/

[24] N. Viswanathan and C. C.-N. Chu, "FastPlace: efficient analytical placement using cell shifting, iterative local refinement and a hybrid net model," *Proc. of ACM International Symposium on Physical Design*, pp. 26–33, April 2004.

[25] M. Wang, X. Yang, and M. Sarrafzadeh, "Dragon2000: standard-cell placement tool for large industry circuits," *Proc. of IEEE/ACM Int. Conf. on Computer-Aided Design*, pp. 260–263, June 2000.

[26] H. Zhou and J. Wang, "ACG–Adjacent constraint graph for general floorplans," *Proc. of IEEE Int. Conf. on Computer Design*, pp. 572–575, October 2004.

Electrothermal Analysis and Optimization Techniques for Nanoscale Integrated Circuits

Yong Zhan, Brent Goplen, and Sachin S. Sapatnekar
Department of Electrical and Computer Engineering
University of Minnesota
Minneapolis, MN 55455, USA.

Abstract— With technology scaling, on-chip power densities are growing steadily, leading to the point where temperature has become an important consideration in the design of electrical circuits. This paper overviews several methods for the analysis and optimization of thermal effects in integrated circuits. Thermal analysis may be carried out efficiently through the use of finite difference methods, finite element methods, or Green function based methods, each of which provides different accuracy-computation tradeoffs, and the paper begins by surveying these. Next, we overview a restricted set of thermal optimization methods, specifically, placement techniques for thermal heat-spreading, and then we conclude by summarizing a set of future directions in electrothermal design.

I. INTRODUCTION

Thermal considerations are playing an increasingly important role in high-performance integrated circuits, necessitating a greater role for the thermal analysis and optimization in the design cycle. Temperature effects are important for several reasons. First, they can cause the transistor delay to change: elevated temperatures cause the threshold voltage to drop and degrade the mobility: depending on which effect wins out, the delay can either increase or decrease. Second, wire resistances increase with temperature, leading to larger interconnect delays. Third, leakage power is particularly sensitive to thermal variations, and varies exponentially with temperature. Worse yet, such an increase in leakage power could, in turn, lead to an increase in the temperature, leading to the potential for positive feedback and an effect known as thermal runaway, in which the chip can physically burn out. Fourth, high temperatures can lead to a variety of reliability problems, including negative temperature bias instability (NBTI), oxide breakdown, and electromigration.

As a result, there has been a great deal of research in the area of thermal analysis and optimization in recent years. This paper overviews some of these issues. Section II presents methods for thermal analysis, and is followed by Section III, which outlines methods for thermally-driven optimization. We conclude with a description of future directions in Section IV.

II. THERMAL ANALYSIS

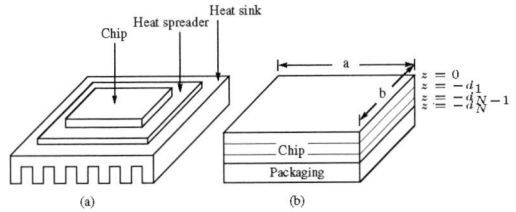

Fig. 1. Schematic of a VLSI chip with packaging (a) integrated circuit and the packaging structure (b) simplified model of the chip and packaging.

Fig. 1(a) shows an integrated circuit chip with the associated packaging, and Fig. 1(b) shows a schematic of the structure in Fig. 1(a) where the packaging including the heat spreader and the heat sink has been simplified, but the multilayered structure of the

This work was supported in part by DARPA under N66001-04-1-8909, SRC under 2003-TJ-1092, and NSF under CCR-0205227.

chip is explicitly shown. The steady state temperature distribution inside the chip is governed by Poisson's equation

$$\nabla^2 T(\mathbf{r}) = -\frac{g(\mathbf{r})}{k_{l(\mathbf{r})}} \tag{1}$$

where $\mathbf{r} = (x, y, z)$, $T(\mathbf{r})$ is the temperature (°C) distribution inside the chip, $g(\mathbf{r})$ is the volume power density (W/m³), and $k_{l(\mathbf{r})}$ is the thermal conductivity (W/(m·°C)) of the layer where point $\mathbf{r}$ is located [1]. The vertical surfaces and the top surface of the chip are assumed to be adiabatic [2], and the bottom surface of the chip is assumed to be convective, with an effective heat transfer coefficient h (W/(m²·°C)) [3]. In mathematical form, these boundary conditions can be expressed as

$$\left.\frac{\partial T(\mathbf{r})}{\partial x}\right|_{x=0,a} = \left.\frac{\partial T(\mathbf{r})}{\partial y}\right|_{y=0,b} = 0 \tag{2}$$

$$\left.\frac{\partial T(\mathbf{r})}{\partial z}\right|_{z=0} = 0 \tag{3}$$

$$\left.k_N \frac{\partial T(\mathbf{r})}{\partial z}\right|_{z=-d_N} = h(T(\mathbf{r})|_{z=-d_N} - T_a) \tag{4}$$

where T_a is the ambient temperature, and k_N is the thermal conductivity of the bottom layer of the chip. In addition, we enforce the continuity conditions at the interface between adjacent layers within the multilayered chip, i.e.,

$$T(\mathbf{r})|_{z=-d_i+\epsilon} = T(\mathbf{r})|_{z=-d_i-\epsilon} \tag{5}$$

$$\left.k_i \frac{\partial T(\mathbf{r})}{\partial z}\right|_{z=-d_i+\epsilon} = \left.k_{i+1} \frac{\partial T(\mathbf{r})}{\partial z}\right|_{z=-d_i-\epsilon} \tag{6}$$

where ϵ is an infinitesimally small quantity and k_i is the thermal conductivity of the ith material layer in the multilayered chip structure. Transient analysis is not explicitly treated here, but extensions of these methods as well as thermal-ADI techniques [4] may be used.

The most widely used numerical techniques for full-chip thermal analysis fall into three broad categories: finite difference, finite element and boundary element methods using Green functions. These are discussed in the following sections.

A. Finite Difference Analysis

The finite difference method (FDM) meshes the chip into (typically) rectangular cuboidal regions, assuming that the temperature of the region is represented by the temperature at its center. Writing the partial derivative as a finite difference between the temperature in neighboring regions leads to a well-established *thermal-electrical analogy*, where the thermal conductivity of the material is mapped to thermal resistors, and power sources (which correspond to the power drawn by logic cells) are mapped to current sources. This leads to a formulation of the type

$$G\mathbf{T} = \mathbf{P} \tag{7}$$

where G is the conductance matrix of thermal resistances, $\mathbf{T}$ is the vector of unknown temperatures at the center of each region, and $\mathbf{P}$ is the vector of power sources.

These equations closely resemble the equations for a power grid, and therefore, can be solved by a variety of techniques. Interestingly, although the substrate is also gridded into regions, it is often the case that the only points of interest are in the topmost device layer, and therefore, the bulk can be macromodeled using Schur decomposition methods as proposed in [5], and improved upon in [6].

B. Finite Element Analysis

As in the case of the FDM, finite element analysis (FEA) also meshes the design space into elements. Various element shapes can be used such as tetrahedra and hexahedra, but a rectangular prism is ideal in that it makes the meshing of the rectangular-shaped substrate easier, and it can simulate heat conduction in lateral directions without aberrations in the prime directions.

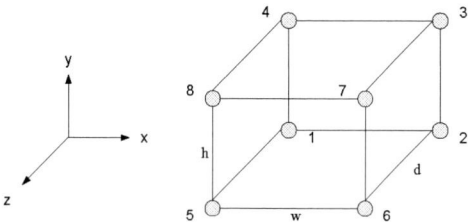

Fig. 2. An eight-node rectangular prism element for FEA.

In FEA, the temperatures are calculated at discrete points, the nodes of the elements, and the temperatures elsewhere within the elements are interpolated using a weighted average of the temperatures at the nodes. In deriving the finite element equations, the differential equation describing heat conduction is approximated within the elements using this interpolation. For an 8-node hexahedral element shown in Fig. 2, a trilinear interpolation function is used to describe the temperature within each element based on the nodal temperatures:

$$T(x, y, z) = \mathbf{N}^T \mathbf{T} \tag{8}$$

where $\mathbf{N} = [N_1 N_2 \cdots N_8]^T$ and $\mathbf{T} = [T_1 T_2 \cdots T_8]^T$, where T_i is the temperature at the ith of eight vertices of the rectangular prism, and N_i is the shape function for the ith vertex. The shape functions are determined by the width, w, height, h, and depth, d, of the element.

Similar to circuit simulation using the modified nodal analysis (MNA) method [7], stamps are created for each element and added to the global system of equations. In FEA, these stamps can be derived from the element interpolation functions using the variational method [8]. For an eight-node rectangular prism, a heat conduction stamp is produced as an 8×8 matrix, whose rows and columns correspond to the element's nodes. Stamps are also produced for surfaces exposed to convective boundary conditions. These element matrices are combined into a global matrix, K, by adding the matrix components that correspond to the same node in the global mesh together, and this produces a global system of equations

$$K\mathbf{T} = \mathbf{P} \tag{9}$$

where $\mathbf{T}$ is the vector of nodal temperatures and $\mathbf{P}$ is the vector of nodal powers. If isothermic boundary conditions are present, the resulting fixed temperatures and their corresponding rows and columns are removed from the system of equations with the right hand side modified accordingly.

C. Enhanced solution techniques for FDM and FEM

Several computational techniques may be employed to improve the solution speed of FDM and FEM. We overview two such methods in this section.

1) Random walk methods: Random walk methods have been used very successfully for the analysis of large RC networks, in the context of power grids [9], [10]. Such methods can easily be applied to the large resistive networks that appear in steady-state thermal analysis, and can be extended to transient thermal analysis as well. They can perform incremental analysis rapidly and efficiently, so that they are excellent candidates for incremental placement.

Consider the solution of a resistive network for the voltages; the thermal analog, of course, is that the voltages in the resistive network

are the temperatures, and the current sources are the power values. It has been shown that the solution to these networks can be obtained through a set of random walks on a graph. The nodes and edges of this graph are the nodes and connections in the initial circuit, respectively, and the transition probabilities from a node along each edge correspond to the ratio of the edge conductance to the total conductance connected to the node. A random walk begins at any node and along the way, at each node, it pays a levy that is related to the ratio of the current source connected to that node to the total conductance at the node. When the walk reaches a node whose voltage is known, it terminates and receives a reward that equals the voltage at the node. For each node, the expected value of earnings over all random walks beginning at that node can be shown to be the voltage of the node. A fuller exposition of this method is provided in [11], and it is shown to be efficient at finding the solution to a large network. For a more exact solution and better numerical efficiency, the method in [12] uses the approximate solution from the random walk solver to efficiently create a preconditioner for an iterative solver.

2) Multigrid Methods: An alternative method for solving systems of FDM equations may use the multigrid method, a multi-level iterative scheme. The essential idea of a multigrid scheme is that at every level, it coarsens the grid and then refines it: the coarsened grid is smaller and hence easier to solve, and captures the low-spatial-frequency component of thermal variations, while the refined grid capture high-spatial-frequency variations. Moving between a coarsened solution and a refined solution requires the use of interpolation and restriction operators, whose definitions are a key step in using multigrid methods. Depending on the coarsening/ refinement sequence, this may result in schemes such as the V cycle or the W cycle, or alternatives. Conventional multigrid schemes are classified as either geometric multigrid methods, which use some knowledge of the problem structure to perform the coarsening and refinement operations, or algebraic multigrid methods, which perform these steps automatically. The work in [13] uses a geometric multigrid method for solving FDM equations. A grid hierarchy is defined, with coarse grid operators

D. Green Function Based Methods

An alternative to the FEM and FDM methods, which mesh up the entire substrate, is a boundary element method using Green functions. The partial differential equation to be solved for thermal analysis is linear when the material properties are region-wise uniform, and therefore, conceptually, the problem can be solved by superposition, considering one source at a time. A Green function enables such a computation: it is the response in a field region to a power source in a source region; in the presence of multiple power sources, superposition can be used to sum up the responses at a field point due to each of the sources.

Let $G(\mathbf{r}, \mathbf{r}')$, with $\mathbf{r} = (x, y, z)$ and $\mathbf{r}' = (x', y', z')$, be the distribution of temperature above T_a in the multilayer chip structure when a unit point power source of 1W is placed at position $\mathbf{r}'$. Then $G(\mathbf{r}, \mathbf{r}')$ satisfies the equation

$$\nabla^2 G(\mathbf{r}, \mathbf{r}') = -\frac{\delta(\mathbf{r} - \mathbf{r}')}{k_{l(\mathbf{r})}} \tag{10}$$

and the boundary conditions

$$\left.\frac{\partial G(\mathbf{r}, \mathbf{r}')}{\partial x}\right|_{x=0,a} = \left.\frac{\partial G(\mathbf{r}, \mathbf{r}')}{\partial y}\right|_{y=0,b} = 0 \tag{11}$$

$$\left.\frac{\partial G(\mathbf{r}, \mathbf{r}')}{\partial z}\right|_{z=0} = 0 \tag{12}$$

$$\left.k_N \frac{\partial G(\mathbf{r}, \mathbf{r}')}{\partial z}\right|_{z=-d_N} = hG(\mathbf{r}, \mathbf{r}')|_{z=-d_N} \tag{13}$$

$$G(\mathbf{r}, \mathbf{r}')|_{z=-d_i+\epsilon} = G(\mathbf{r}, \mathbf{r}')|_{z=-d_i-\epsilon} \tag{14}$$

$$\left.k_i \frac{\partial G(\mathbf{r}, \mathbf{r}')}{\partial z}\right|_{z=-d_i+\epsilon} = \left.k_{i+1} \frac{\partial G(\mathbf{r}, \mathbf{r}')}{\partial z}\right|_{z=-d_i-\epsilon} \tag{15}$$

where $\delta(\mathbf{r}, \mathbf{r}') = \delta(x-x')\delta(y-y')\delta(z-z')$ is the three-dimensional Dirac delta function, and $G(\mathbf{r}, \mathbf{r}')$ is the Green function. The temperature field under an arbitrary power density distribution can be obtained easily as

$$T(\mathbf{r}) = T_a + \int_0^a dx' \int_0^b dy' \int_{-d_N}^0 dz' G(\mathbf{r}, \mathbf{r}') g(\mathbf{r}') \qquad (16)$$

For thermal problems encountered in chip design, both the source regions, where powers are generated, and the field regions, whose temperatures are to be computed, are located on discrete planes. Thus, in the following analysis, we will focus on a single source plane and a single field plane, i.e., a particular z and z'. For these planes, it can be shown [14] that the Green function is given by

$$G'(x, y, x', y') \triangleq G(\mathbf{r}, \mathbf{r}')|_{z,z'} =$$
$$\sum_{m=0}^\infty \sum_{n=0}^\infty C_{mn} \cos\left(\frac{m\pi x}{a}\right) \cos\left(\frac{n\pi y}{b}\right) \cos\left(\frac{m\pi x'}{a}\right) \cos\left(\frac{n\pi y'}{b}\right) \qquad (17)$$

where the coefficient C_{mn} only depends on z and z'.

The above expression is complicated, both visually and computationally, and it involves a double summation to infinity. Fortunately, several methods are available for managing the computation. The work in [14], based on the substrate analysis methods in [15], uses the discrete cosine transform (DCT) and table lookups to accelerate the Green function based thermal analysis; for the analysis of a single layer, the computational complexity is $O(N_g^2)$, where N_g is the number of regions. An improved method in [16] reduces the complexity from quadratic to $O(N_g \log N_g)$. The essential idea is to recognize that the bottleneck corresponds to a convolution operation, and this can be performed efficiently in the frequency domain: the primary cost here is in the transform from the space domain to the frequency domain.

III. THERMAL PLACEMENT

While several methods can be used for thermal optimization, we will focus here on thermal placement.

At the placement stage, as far as the thermal constraints are concerned, the goals are to minimize the maximal on-chip temperature gradient and obtain an even temperature distribution, while controlling conventional metrics such as the wire length. Thermal analysis involves the solution of a system of linear equations of the type

$$G\mathbf{T} = \mathbf{P} \qquad (18)$$

where G is the thermal conductance matrix, $\mathbf{T}$ is the vector of temperatures, and $\mathbf{P}$ is the vector of power dissipations. For the finite difference method, G refers to the matrix associated with the thermal conductances, while for the finite element method, this corresponds to the stiffness matrix. The value of each P_i is not constant, but may change depending on which cell is located in a particular grid of the layout. Additionally, the power consumption of a cell varies with the interconnect capacitance that it drives, i.e., the length of the nets that it drives. During placement, these values are liable to change. The total power is actually dissipated by both the switching transistors and the interconnecting wires. Except for long global wires, the driver resistance is typically much larger than the metal resistance, and therefore most of the power is dissipated in the cells, and it is reasonable to ignore the part consumed by the metal wires. Even though self-heating of wires plays a very important role in the electromigration lifetime of the metal wires [17], during the placement stage, it may be ignored. It is potentially possible to take this into account during a later stage of placement using the congestion information.

A. Partitioning-based Placement

Partitioning-based approaches to placement are based on the idea of recursively dividing the layout into regions and assigning cells to each region. The key issue in partitioning-based placement, tackled in [6], is to simplify the thermal model at each level of partitioning to achieve the goal of placing the cells so that T is evenly distributed across the chip. The equation $G\mathbf{T} = \mathbf{P}$ cannot be directly used, since at each partition level, the only location information that is available is identity of the partitioning blocks that the cell belongs to. Assuming that all the cells belonging to a block are located at

its center, the corresponding analysis will correspond to an incorrect thermal analysis of the partition. For example, it is easy to see that this will result in an exaggerated thermal gradient within the partition, since the temperature at the center of the partition will be much higher than that at its periphery.

For simplicity, consider the thermal model for a top-down bipartitioning process; this process can be easily extended to a top-down k-way partitioning process. At any one particular partition stage, a single block is being partitioned into two sub-blocks so that the cuts between the boundary of sub-blocks are minimized. For now, the actual temperature profile in the sub-blocks is not of direct concern, since cells inside the blocks will be further partitioned later, and this can be considered at that time. However, it is important to minimize the temperature discrepancy between the blocks. If some high-power cells are accumulated into one of the blocks, then at a later stage it will not be possible to move these cells out of the block, due to the divide-and-conquer nature of top-down partitioning. It is reasonable to assume that the temperature inside each of the sub-blocks are uniform, since if such an objective were to be enforced at every step of the partitioning, then a uniform temperature distribution would indeed result. Under this assumption, a simplified thermal model is obtained, and assumptions about the precise cell locations inside the sub-block need not be made.

In the first step in top down partitioning, the chip is partitioned into two blocks, the left block and the right block. For simplicity, assume that the number of thermal cells in each region is the same, although this assumption can easily be discarded. The thermal cells $i = 1 .. m/2$ will be said to be in the left block and cells $i = m/2 + 1 .. m$ in the right block. Now assume, as stated above, that the block on the left has an even temperature of T_l and the temperature of the block to the right is T_r. The thermal equation can now be simplified to:

$$\begin{bmatrix} \sum_{i=1}^{m/2} \sum_{j=1}^{m/2} G_{ij} & \sum_{i=1}^{m/2} \sum_{j=\frac{m}{2}+1}^{m} G_{ij} \\ \sum_{i=\frac{m}{2}+1}^{m} \sum_{j=1}^{m/2} G_{ij} & \sum_{i=\frac{m}{2}+1}^{m} \sum_{j=\frac{m}{2}+1}^{m} G_{ij} \end{bmatrix} \begin{bmatrix} T_l \\ T_r \end{bmatrix} = \begin{bmatrix} \sum_{i=1}^{m/2} P_i \\ \sum_{i=\frac{m}{2}+1}^{m} P_i \end{bmatrix} \qquad (19)$$

This reasoning can be extended to a general case where the chip is partitioned into k regions, each with possibly a different number of cells, where a positive definite system of k equations in k variables must be solved.

The approach in [6] uses a top-down two-way partitioner based on the Fiduccia-Mattheyses algorithm [18], but can be extended naturally to incorporate a state-of-the-art multi-level partitioner. To reduce the computational cost for incremental temperature updates, the notion of an "effective thermal influence region" of a block is introduced. For a unit heat source on a block, this corresponds to the area outside of which the temperature induced by the unit heat source is less than a certain percentage of the maximum temperature induced by the unit heat source: this can be easily computed once the thermal resistance matrix is known. At each partitioning step, the cells are moved to provide a near-uniform thermal profile, while attempting to control the wire length.

B. Force-directed Placement

In force-directed methods, an analogy to Hooke's law is used by representing nets as springs and finding the placement corresponding to the system's minimum energy state. Attractive forces are created between interconnected cells and are made proportional to the separation distance and interconnectivity. Other design criteria such as cell overlap, timing, and congestion are used to derive the repulsive forces. After repulsive forces are added, the system is solved for the minimum energy state, i.e., the equilibrium location. Ideally, this minimizes the wire lengths while at the same time satisfying the other design criteria.

The work in [19] presents a force-directed approach to thermal placement. The application domain is in the design of 3D integrated circuits, where chips have multiple levels of active devices. Therefore, placement must be carried out in not just the xy-plane, but the entire xyz-space in three dimensions. In current technologies, in the z dimension, the number of layers is restricted to a small number. The work in [19] uses a force-directed framework with FEA-based thermal analysis, using repulsive forces to avoid hot spots. The thermal forces are calculated using the temperature gradient, which itself can be related to the stiffness matrix and its derivative. The temperature gradient determines both the direction and relative magnitude of the thermal forces, thereby moving cells away from areas with high temperature.

Fundamentally, force-directed methodologies involve minimizing an objective function corresponding to a summation of cost components from each net. For 3D layouts, this takes the form

$$c_{ij} \left[(x_i - x_j)^2 + (y_i - y_j)^2 + (z_i - z_j)^2 \right] \qquad (20)$$

where c_{ij} is the weight of the connection between the two nodes. If the c_{ij} coefficients are combined into a global C matrix, an objective function can be written for the entire system:

$$\frac{1}{2}\mathbf{x}^T C \mathbf{x} + \frac{1}{2}\mathbf{y}^T C \mathbf{y} + \frac{1}{2}\mathbf{z}^T C \mathbf{z} \qquad (21)$$

where $\mathbf{x}$, $\mathbf{y}$, and $\mathbf{z}$ are the x, y, and z coordinates of all cells and points of interest. This objective function can be minimized by solving the following three systems of equations:

$$C\mathbf{x} = \mathbf{f}_x, C\mathbf{y} = \mathbf{f}_y, C\mathbf{z} = \mathbf{f}_z \qquad (22)$$

In the absence of external repulsive forces, the total force vectors, $\mathbf{f}_x$, $\mathbf{f}_y$, and $\mathbf{f}_z$, would be zero. The net stiffness matrix, C, describes the entire net connectivity. Fixed coordinate values, created by physical constraints such as I/O pads, can be used to reduce and solve the system of equations, much like the isothermic boundary conditions in FEA.

Generally, an iterative force-directed approach follows the following steps in the main loop. Initially, forces are updated based on the previous placement. Using these new forces, the cell positions are then calculated. These two steps of calculating forces and finding cell positions are repeated until the exit criteria are satisfied.

IV. Future Directions

Several issues remain to be solved in the area of electrothermal design:

- Leakage power is becoming a major contributor to the total power: leakage depends exponentially on temperature, and a rise in this component of power can itself cause the temperature to increase. Therefore, electrothermal analysis must be self-consistent in taking this feedback effect into account in the high-leakage regime. Since leakage power also varies significantly with process variations, even a circuit that is designed to meet its power and thermal requirements may fail these specifications after manufacturing. Adaptive methods such as the application of body biases may be used to allow a circuit to recover from the increased leakage power due to process and temperature variations [20], [21]. Moreover, temperature causes circuit delays to change, both by changing transistor characteristics and by altering interconnect delays, resulting in altered power-delay tradeoffs.

- A number of reliability issues come into play as temperatures rise. Negative temperature bias instability (NBTI) [22] causes the threshold voltages of devices to shift with time, and is an aging process that is accentuated under thermal stress. Electromigration is also hastened under high temperatures [23], [24], and effects such as these must be controlled.

- Traditional solution separate the problems of power reduction from the problem of heat removal. More efficient techniques can develop spot cooling methods for heat removal. One example of such a method utilizes thermal vias to remove heat from 3D integrated circuits [25].

Many of these problems have seen little research in the past, and are fertile grounds for future work.

References

[1] M.N. Ozisik. *Boundary Value Problems of Heat Conduction.* Dover, New York, NY, 1968.

[2] A. G. Kokkas. Thermal analysis of multi-layer structures. *IEEE Transactions on Electron Devices,* 21(11):674–681, November 1974.

[3] Y. K. Cheng, P. Raha, C. C. Teng, E. Rosenbaum, and S.-M. Kang. ILLIADS-T: An electrothermal timing simulator for temperature-sensitive reliability diagnosis of CMOS VLSI chips. *IEEE Transactions on Computer-Aided Design,* 17(8):668–681, August 1998.

[4] T.-Y. Wang and C. C.-P. Chen. 3-D thermal-ADI: A linear-time chip level transient thermal simulator. *IEEE Transactions on Computer-Aided Design,* 21(12):1434–1445, December 2002.

[5] C. Tsai and S. Kang. Cell-level placement for improving substrate thermal distribution. *IEEE Transactions on Computer-Aided Design,* 19(2):253–266, February 2000.

[6] G. Chen and S. S. Sapatnekar. Partition-driven standard cell thermal placement. In *Proceedings of the International Symposium on Physical Design,* pages 75–80, 2003.

[7] C. Ho, A. E. Ruehli, and P. Brennan. The modified nodal approach to network analysis. *IEEE Transactions on Circuits and Systems,* 22(6):504–509, June 1975.

[8] D. L. Logan. *A First Course in the Finite Element Method.* Brooks/Cole Pub. Co., 3 edition, 2002.

[9] H. Qian, S. R. Nassif, and S. S. Sapatnekar. Random walks in a supply network. In *Proceedings of the ACM/IEEE Design Automation Conference,* pages 93–98, 2003.

[10] H. Qian and S. S. Sapatnekar. Hierarchical random-walk algorithms for power grid analysis. In *Proceedings of the Asia/South Pacific Design Automation Conference,* pages 499–504, 2003.

[11] H. Qian, S. R. Nassif, and S. S. Sapatnekar. Power grid analysis using random walks. *IEEE Transactions on Computer-Aided Design,* 24(8):1204–1224, August 2005.

[12] H. Qian and S. S. Sapatnekar. A hybrid linear equation solver and its application in quadratic placement. In *Proceedings of the IEEE/ACM International Conference on Computer-Aided Design,* 2005.

[13] P. Li, L. T. Pileggi, M. Asheghi, and R. Chandra. Efficient full-chip thermal modeling and analysis. In *Proceedings of the IEEE/ACM International Conference on Computer-Aided Design,* pages 319–326, 2004.

[14] Y. Zhan and S. S. Sapatnekar. Fast computation of the temperature distribution in vlsi chips using the discrete cosine transform and table look-up. In *Proceedings of the Asia/South Pacific Design Automation Conference,* pages 87–92, 2005.

[15] R. Gharpurey and R. G. Meyer. Modeling and analysis of substrate coupling in integrated circuits. *IEEE Journal of Solid-State Circuits,* 17(4):305–315, March 1996.

[16] Y. Zhan and S. S. Sapatnekar. A high efficiency full-chip thermal simulation algorithm. In *Proceedings of the IEEE/ACM International Conference on Computer-Aided Design,* 2005.

[17] K. Banerjee, A. Mehrotra, A. Sangiovanni-Vincentelli, and C. Hu. On thermal effects in deep sub-micron vlsi interconnects. In *Proceedings of the ACM/IEEE Design Automation Conference,* pages 885–891, 1999.

[18] C. M. Fiduccia and R. M. Mattheyses. A linear time heuristic for improving network partitions. In *Proceedings of the ACM/IEEE Design Automation Conference,* pages 175–181, 1982.

[19] B. Goplen and S. S. Sapatnekar. Efficient thermal placement of standard cells in 3d ics using a force directed approach. In *Proceedings of the IEEE/ACM International Conference on Computer-Aided Design,* pages 86–89, 2003.

[20] J. W. Tschanz, J.T. Kao, S. G. Narendra, R. Nair, D.A. Antoniadis, A.P. Chandrakasan, and V. De. *Adaptive Body Bias for reducing impacts of die-to-die and within-die parameter variations on microprocessor frequency and leakage. IEEE Journal of Solid-State Circuits,* 37(11):1396–1402, November 2002.

[21] S. Kumar, C. H. Kim, and S. S. Sapatnekar. Mathematically-assisted adaptive body bias (abb) for temperature compensation in gigascale lsi systems. In *Proceedings of the Asia/South Pacific Design Automation Conference,* 2005.

[22] D. K. Schroder and J. A. Babcock. Negative bias temperature instability: Road to cross in deep submicron semiconductor manufacturing. *Journal of Applied Physics,* 94:1–18, 2003.

[23] J. R. Black. Mass transport of aluminum by momentum exchange with conducting electrons. In *Proceedings of the IEEE International Reliability Physics Symposium,* pages 148–159, 1969.

[24] Z. Lu, W. Huang, J. Lach, M. Stan, and K. Skadron. Interconnect lifetime prediction under dynamic stress for reliability-aware design. In *Proceedings of the IEEE/ACM International Conference on Computer-Aided Design,* pages 327–334, 2004.

[25] B. Goplen and S. S. Sapatnekar. Thermal via placement in 3d ics. In *Proceedings of the International Symposium on Physical Design,* pages 167–174, 2005.

Electrothermal Engineering in the Nanometer Era:
From Devices and Interconnects to Circuits and Systems

Kaustav Banerjee, Sheng-Chih Lin and Navin Srivastava

Department of Electrical and Computer Engineering, University of California, Santa Barbara, CA 93106
{kaustav, sclin, navins}@ece.ucsb.edu

Abstract—*Management of electrothermal (ET) issues arising due to power dissipation both at the micro- and macro- scale is central to the development of future generation microprocessors, integrated networks, and other highly integrated circuits and systems. This paper will provide a broad overview of various ET effects in nanoscale VLSI and highlight both technology and design choices that are thermally-aware. First, effects at the micro scale--in interconnects and devices and their implications for performance, reliability and design are discussed. Next, macro scale--circuit and system level issues including substrate temperature gradients as well as strong ET couplings between supply voltage, frequency, power dissipation and junction temperature in leakage dominant technologies are outlined. A recently developed system level ET analysis methodology and tool that comprehends ET couplings in a self-consistent manner and can generate accurate thermal profile of the substrate is summarized. The application of the ET-tool is demonstrated in a number of areas from power-performance-cooling cost tradeoff analysis to circuit optimization, full-chip leakage estimation, and temperature/reliability aware design space generation. Implications of chip cooling for nanometer scale bulk and SOI based CMOS technologies are also discussed. The ET analysis tool is also shown to be useful for hot-spot management. The paper ends with a brief discussion of electrothermal issues in emerging 3-D ICs and highlights the advantages of employing hybrid Carbon Nanotube-Cu interconnects in both 2-D and 3-D designs.*

I. Introduction

During the past two decades CMOS integrated circuits (ICs) have witnessed unprecedented improvements in their functionality and performance. This was primarily achieved by aggressive technology scaling, which resulted in device density and performance doubling roughly every 18 months as per Moore's law while achieving a remarkable 25% per year improvement in cost per chip function. As CMOS scaled from generation to generation, power dissipation increased proportionately to increasing transistor density and switching speeds. However, with the minimum feature size of the transistor entering the nanometer regime (< 100 nm), power dissipation is increasing ominously especially due to a substantial increase in the leakage power. Leakage power used to be insignificant for earlier generations of ICs but is becoming an increasing fraction of the total power [1]. Moreover, most leakage mechanisms are strongly temperature dependent. This strong coupling between temperature and leakage can cause further increase in total power dissipation. In fact, the International Roadmap for Semiconductors (ITRS) [2] forecasts that high-performance ICs will dissipate around 200 W within a few years. Hence, in spite of the ongoing efforts to reduce power through clever design and technological innovations in the form of new materials and device structures, management of thermal issues will be central to the development of nanoscale VLSI and future generation microprocessors, integrated networks, and other highly integrated systems [1, 3-4]. Since power consumed by the ICs is converted into heat, the corresponding heat densities are rising exponentially. These *electrothermal effects* within the chip are leading to issues and

challenges in the design and analysis of high-performance ICs that previous generations did not exhibit [1, 3, 5-7]. For example, since power dissipation in high-performance ICs is spatially non-uniform across the chip and localized heating can occur much faster than chip-scale heating, *hot-spots* and *temperature gradients* are formed that can cause timing errors and reliability problems [8-10]. In short, problems over the heat generated by semiconductor chips are becoming so severe that they threaten to slow, or even limit the development of the entire IC industry. Since, higher temperature slows down the speeds of transistors and interconnects, it is no longer prudent to target highest packing densities in the designs and instead, if the design can be made in a *thermally aware* manner, higher performance and reliability can be achieved for any given technology generation.

In general, these electrothermal effects may impact device, circuit, and system level metrics that are strongly interlinked with one another. For instance, in addition to increasing the leakage, temperature is also known to degrade *device* (due to reduced mobility of electrons and holes) and *interconnect* delay (due to increased resistivity of metal wires arising from increased scattering of electrons with the lattice) and also degrade their reliability (since, mean time-to-failure for reliability mechanisms, including electromigration (EM) and time-dependent gate-oxide breakdown (TDDB), have an inverse exponential dependence on temperature). This in turn leads to degradation of *circuit* level parameters including timing, which eventually degrades *system* level performance. Moreover, several circuit architectures such as dual-V_{th}/V_{dd} designs, and physical design issues including P/G integrity and placement and routing are strongly affected by temperature [7, 11-12]. Furthermore, increased device leakage eventually manifests itself as increased chip power and junction temperature, which in turn, will affect *system level* thermal management (packaging and cooling) solutions and eventual cost per chip function [3, 6]. Therefore, the design process must comprehend thermal effects to ensure improved circuit performance and reliability. Also, in order to formulate comprehensive solutions to the thermal problems, it is necessary to comprehend the correlation between the problems at various levels of the design process, the reliability and performance issues at all levels, and the implications arising due to technology and process conditions. Hence, an integrated holistic approach towards understanding these electrothermal effects at various levels is absolutely vital. **Fig. 1** illustrates the *Electrothermal Engineering* vision to convey its broad scope. Electrothermal Engineering involves development of methods and tools to facilitate the incorporation of temperature and temperature-dependent issues (including reliability) at various stages in the VLSI design process. It also includes exploration of novel thermal management techniques, either through new materials or *thermally-aware* device/interconnect/circuit/system design. We feel that such efforts would be necessary for correctly designing future generations of integrated circuits in a cost effective manner. Furthermore, the lessons learned under electrothermal engineering of nanoscale CMOS based designs are directly applicable to most emerging technologies such as 3-D ICs due to the common issues arising from increased power densities and/or increased temperature sensitivity of performance metrics.

The paper has been organized as follows. Section II discusses various ET issues at the micro-scale: in devices and interconnects. Section III illustrates macro-scale ET issues in circuits and systems. Electrothermal issues in emerging technologies such as 3-D ICs are discussed briefly in Section IV and the positive implications of employing hybrid carbon nanotube-Cu interconnects in both 2-D and 3-D ICs are highlighted. At this point, it is instructive to rigorously define micro- and macro-scale electrothermal effects. As shown in **Fig. 2**, micro- and macro- scale ET effects differ mainly in the length and time scales involved. Micro-scale ET effects involve length/time scale of 10nm-10µm/0.1ns-10µs, while macro-scale issues involve much larger length/time scales.

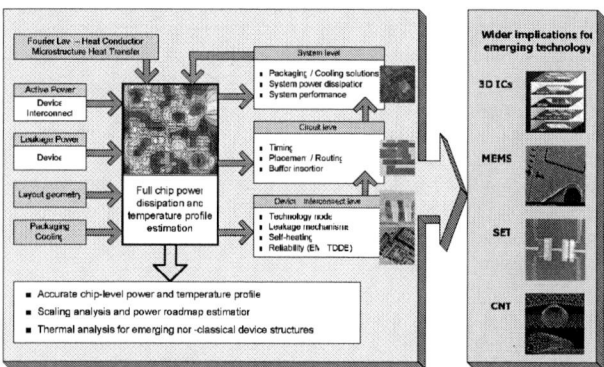

Fig. 1: Schematic overview of the scope of electrothermal engineering research in nanometer scale VLSI and wider implications for emerging technologies such as 3-dimensional ICs, micro-electromechanical systems (MEMS), single electron transistors (SET) and carbon nanotubes (CNT).

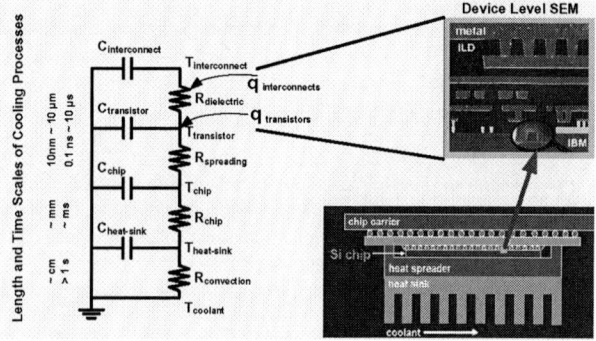

Fig. 2: Global view of heat transfer in integrated circuits: illustrating the different length and time scales of cooling processes at the micro-scale: devices and interconnects and the macro-scale: circuits and system.

II. Electrothermal Engineering at the Micro-Scale: Transistors and Interconnects

II.A. Electrothermal Engineering of Nanoscale Transistors

Conventionally, thermal transport in semiconductors has been modeled through classical diffusion theory [13-14]. However, as device dimensions scale to the orders of tens of nanometer or comparable to the mean free path of phonons (~300nm for bulk Si at 300K), the physics of thermal transport gets affected resulting in localization of heat. At these length scales, the transfer of energy from the charge carriers to the lattice occurs at a faster rate than the relaxation of thermal energy in the lattice, leading to local non-equilibrium in the lattice. Consequently the local hot spot temperature rises beyond the diffusion theory predictions. This is predicted to be a potential obstacle to the continued scaling of devices. This effect has been illustrated through simulations involving the solution of the Boltzmann Transport Equation (BTE) in MOSFETs (**Fig. 3**) [15]. This also has significant implications for

device reliability. For example, it has been shown that under electrostatic discharge (ESD) conditions, which involves very high current (> few Amps) for a very short time (<100 ns), the classical diffusion theory will not be able to predict the failure temperature accurately leading to poor correlation between measurements and simulations [16]. This effect is expected to become even more important for sub-90 nm technologies, where it can influence other parameters such as device delay and leakage.

Furthermore, it has been predicted that scaling CMOS to and beyond the 22-nm technology node will probably require the introduction of several new material and structural changes to the MOSFET to sustain performance increases as per the ITRS projections and to manage short channel effects as shown in **Fig. 4** [17-18]. These new technologies have been shown to alleviate short channel effects, reduce leakage, increase drain saturation current at even lower operating voltage. Material changes will include strained silicon n- and p-channels, high-k gate dielectric and metal gate electrode. Structural changes could include fully depleted Ultra Thin Body (UTB) SOI single- gate MOSFETs, followed by fully depleted UTB double-gate structures. Strained-Si increases drive current due to strain induced enhanced mobility and average carrier velocity. The UTB SOI MOSFET consists of a very thin ($t_{si} \leq 10$ nm), fully depleted (FD) transistor body to ensure good electrostatic control of the channel by the gate in the off state. The use of a lightly doped or undoped body enhances carrier mobility for higher transistor drive current.

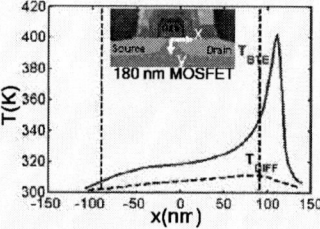

Fig. 3: Temperature distribution along the channel region [15].

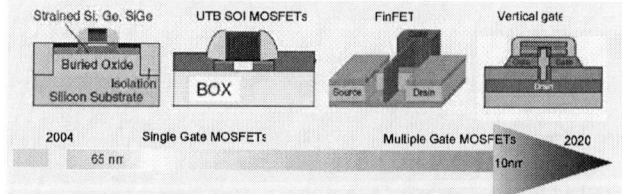

Fig. 4: Evolution of CMOS technology towards transport-enhanced (Strained-Si, SiGe) and UTB devices. These devices exhibit poorer thermal properties due to confined geometry and poor thermal conductivity materials.

However, the use of buried oxide in SOI type devices or SiGe graded layer in Strained-Si devices increases the thermal resistance of the device due to low thermal conductivity of these materials [19, 20]. It has been shown that drain current can reduce by as much as 15% due to self-heating in a 100 nm strained-Si device [19]. Also, the thermal conductivity of thin Si layer degrades due to increased phonon boundary scattering [21], which in turn, results in poor thermal properties of the channel layer in an UTB device [21-23]. Furthermore, in highly confined geometries such as double-gate and tri-gate structures, thermal resistance is further exacerbated due to multiple interfaces. The poor thermal properties of these non-classical devices results in higher temperature rise and subsequently lower drive current and higher leakage than predicted.

Consequently, it is extremely important to analyze and optimize these electrothermal effects in non-classical devices for adequate use of these in nanoscale circuits and systems. For example, ongoing work in our group indicates that in FinFET devices while reducing the fin thickness gives improved channel control, self-heating increases substantially due to increased phonon-boundary scattering. Hence, there is a trade-off between channel control and junction temperature. Similar electrical-thermal trade off exists for other FinFET device parameters such as fin height, BOX thickness and fin pitch for a multi-fin structure. Thus, the scaling of silicon transistors will, in general, lead to increased temperature rise. However, understanding the physics of heat conduction in the device may provide an additional way of controlling the temperature rise and allow devices to be operated closer to their peak performance levels. Existing diffusion based thermal transport models underestimate the peak temperature rise in these non-classical devices where non-equilibrium thermal transport can lead to dramatically high temperatures.

Micro-scale electro-thermal modeling in the literature can be widely separated into two categories. The first method considers heat generation and transport as two separate problems and solves them rigorously, resulting in non self-consistent and inaccurate results. The second method develops a fully coupled model making simplistic assumptions [24]. However, with device scaling these coupled solutions become more and more inaccurate due to assumptions invalid at nanoscale. We are developing a fully coupled method while removing these assumptions by employing more sophisticated methods for heat generation via electron Monte Carlo simulations and for heat conduction via phonon BTE [25-26]. Once the electron MC and the phonon BTE methods are developed, they will be coupled together for electrothermal simulations as shown in **Fig 5**. Electron scattering mechanisms are temperature dependent, therefore the temperature distributions along the device will be fed back to the MC simulations. The resulting heat generation profile will be fed back into the phonon BTE for self-consistent solutions.

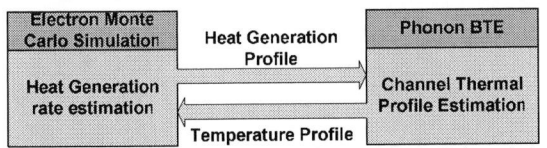

Fig. 5: Schematic illustrating the coupled electron Monte-Carlo and phonon BTE approach.

II.B. Electrothermal Engineering of Nanoscale Interconnects

With the aggressive scaling of VLSI technology, cross-sectional dimensions of on-chip interconnects in current technologies [2] are of the order of the mean free path of electrons in copper (40 nm at room temperature). At such dimensions, the increase in Cu interconnect resistivity due to the presence of a highly resistive barrier layer (which occupies a significant fraction of the drawn wire width) is further exacerbated by the increased scattering of electrons at the surface and grain boundaries, as shown in **Fig. 6(a)** [27]. The low-k dielectric materials used for intra- and inter- metal layer dielectric (ILD) in current VLSI technologies have much lower thermal conductivities than silicon dioxide (**Fig. 6(b)**) [27]. Alongside the lower thermal conductivity of inter-layer dielectrics, the scaling of technology also results in higher current density demands on interconnects [28].

The combined effect of increasing copper interconnect resistivity, decreasing thermal conductivity of ILD materials and rising current densities in on-chip wires results in significant rise in interconnect temperatures, especially at the global metal layers which are furthest away from the heat sink (**Fig. 7(a)**) [27]. These high temperatures will become a major concern for interconnect reliability as the mean time to failure due to electromigration depends exponentially on metal temperature. The maximum current density that can be supported by these interconnects will thus be severely limited due to reliability constraints (**Fig. 7(b)**) [28].

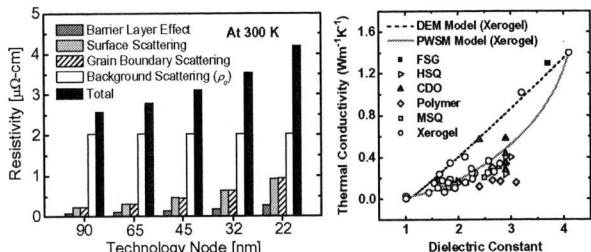

Fig. 6: **(a)** Scaling of Cu resistivity for the ITRS intermediate wires (at 300K). The total resistivity is the sum of all resistivity components. Parameter values are: $\rho_o=2.04$ $\mu\Omega$-cm (300K), $\lambda=37.3$ nm (300K), $p=0.41$, and $R=0.22$. **(b)** Correlation of the thermal conductivity and dielectric constant showing the simultaneous decrease of thermal conductivity and dielectric constants of several materials along with physical models that predict this behavior [27].

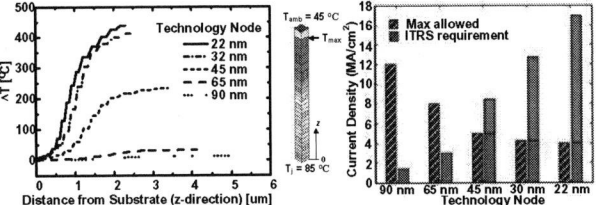

Fig. 7: **(a)** Temperature rise (ΔT) along the vertical distance from the substrate at different technology nodes [27]. The temperature contour plot of 45 nm technology node is also shown as an example. **(b)** Maximum allowed current density (duty ratio=0.001) in local vias from self-consistent electromigration lifetime estimation vs. the ITRS requirement for current density [28].

III. Electrothermal Engineering at the Macro-Scale: Circuits and Systems

III.A. Circuits

III.A.1. Interconnect Power Dissipation

While typical local interconnect delay is expected to decrease with technology scaling (mainly as a result of the higher packing density of devices), global interconnect delay increases. In order to keep the delay of global interconnects under control, repeaters (buffers) are inserted at regular intervals to drive signals faster. As technology scales, an increasing number of buffers is required for the global interconnection system on a chip (**Fig. 8**). These repeaters can contribute significantly to total chip power dissipation, which is a critical problem for high-performance ICs.

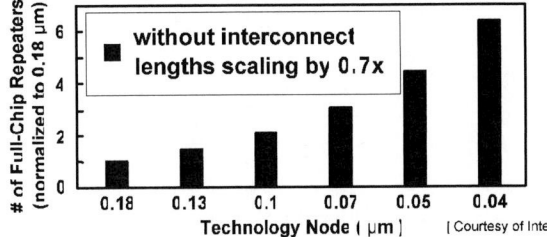

Fig. 8: Number of repeaters increases rapidly with technology scaling.

Increase in effective wire resistivity will further increase wire delay resulting in more numerous repeaters and larger chip power dissipation. However, it has been shown that when delay is not of critical importance "power-optimal" repeater insertion [29] can be used to achieve large power savings. This methodology assumes tremendous importance in the light of power-limited technologies of current and future IC generations as the power savings for a given amount of delay penalty increases as technology scales (**Fig. 9(a)**). This is mainly due to the increasing leakage power dissipation in

CMOS devices (**Fig. 9(b)**). In the presence of variations in devices as well as interconnects, the power optimal repeater insertion can be significantly impacted as shown in **Fig. 10** [30]. Also, as shown in **Fig. 11**, for a given delay penalty, power savings are greater under higher percentage of variations, mainly because of the increase in leakage power under variations [30]. As a result the impact of the power-optimal repeater insertion methodology becomes more significant in aggressively scaled technologies which are more susceptible to variations.

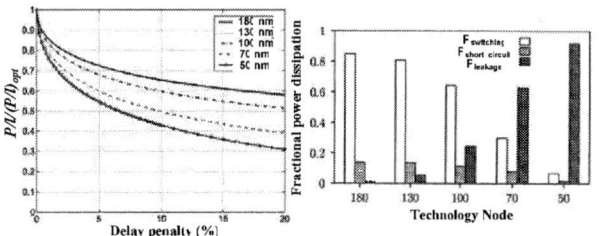

Fig. 9: (a) Normalized power per unit length for a buffered global interconnect (normalized to power per unit length for optimally buffered case) as a function of delay penalty, at different technology nodes. **(b)** Relative contributions of three components of overall power dissipation for 5% delay penalty at different technology nodes [29].

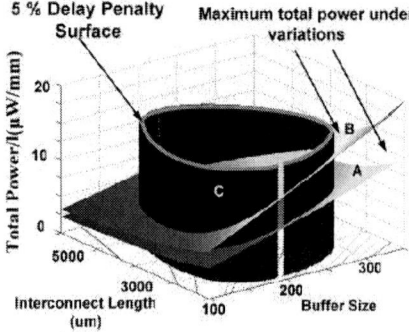

Fig. 10: Maximum value of total power per unit length as a function of buffer size (s) and interconnect length (l) under two different cases (amount of parameter variations is much higher in Case B as compared to Case A) [30]. Also drawn is a delay penalty surface, along which $(\tau/l) = 1.05\ (\tau/l)_{opt}$. The 2-D contour that gives 5% delay penalty has been extended in 3-D plane (surface C) for better visualization.

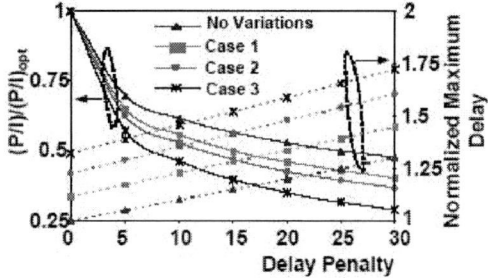

Fig. 11: Power per unit length normalized to optimal power per unit length, for different delay penalties under different conditions [30]. Normalized maximum delay curves are also drawn. Parameter variations increase from Case 1 to Case 3, as 10%, 20% and 30% of mean value.

III.A.2. Impact of On-Chip Thermal Gradients

Recent work indicates that, in high performance ICs, the peak chip temperature can rise up to 140°C in 90 nm technology node and is expected to rise even further for future technology nodes [31]. Since, these peak temperatures always occur at the top of the chip, they can significantly increase the interconnect resistance, which would in turn increase the signal propagation delay in the

interconnect line. In nanometer scale high performance ICs, large temperature gradients can also occur in the substrate. These gradients, for example, may be created due to "spotty" gate-level switching activity and/or because various functional blocks are put in different operational modes, e.g., active, standby, or sleep modes [32]. Dynamic power management (DPM) [33] and clock gating can also be major contributors to a non-uniform substrate temperature. In fact, thermal gradients as large as 50°C can exist across high-performance microprocessor substrates (**Fig. 12**). **Fig. 13** shows the error in interconnect delay that can occur as a result of assuming a uniform average temperature for two different temperature gradients existing along the line. **Fig. 14** depicts the percentage skew between wires 1 and 2 (which are of identical length = 2000 μm) as a function of the position x where the thermal gradient occurs, while wire 1 is at a uniform temperature of $T_1 = 100$°C. **Fig 15** shows that by neglecting the thermal effects of hot spots on the resistivity of the global layers, worst-case voltage drop can not be predicted correctly. Finally, **Fig 16** shows the difference resulting from using a temperature-aware buffer insertion technique and the increasing improvement in performance that can be obtained as the thermal gradient increases.

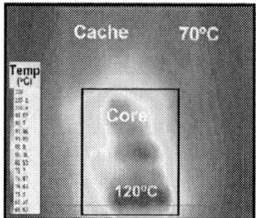

Fig. 12: Varying temperature profile on a microprocessor [courtesy S. Borkar, Intel].

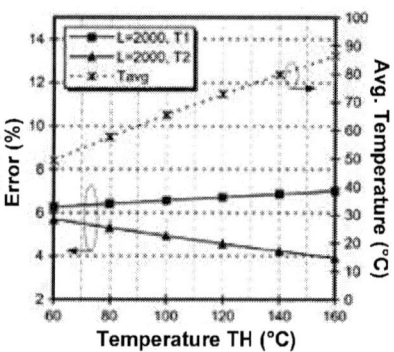

Fig. 13: Percentage delay differences between the non-uniform temperature-dependent interconnect delay and the delay at a uniform line temperature T_{avg}. T_1 and T_2 denote positive and negative exponential temperature gradients between a low temperature T_L of 40°C and a high temperature T_H shown on the x-axis [10].

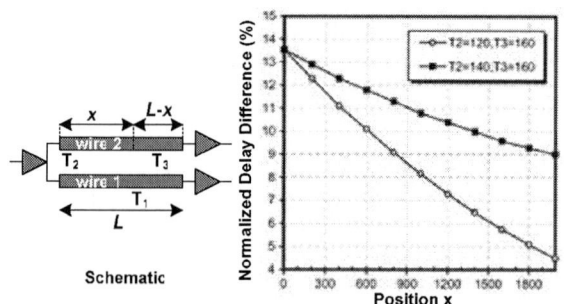

Fig. 14: Percentage of normalized delay difference between wire segment 1 and wire segment 2 as a function of break point x, the point at which the thermal gradient occurs in wire segment 2 [10].

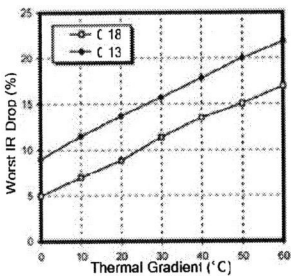

Fig. 15: Worst-case voltage drop (V_{IR}/V_{dd}) increase in presence of hot spots modeled by constant-peak Gaussian distribution as a function of thermal gradient magnitudes (°C), shown for two technology feature sizes [34].

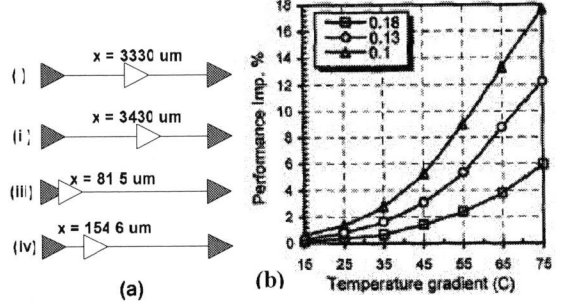

Fig. 16: (a) Location of an inserted buffer in a 6660 um line (180 nm node): (i) standard technique, (ii-iv) temperature aware technique with only variable interconnect resistance (ii), only variable driver resistance (iii) and variable interconnect and driver resistances (iv). **(b)** Delay improvement due to the thermally aware buffer insertion for one buffer as a function of different thermal gradients between the two ends of the line in comparison with the standard buffer insertion techniques for different technologies [35].

III.B. Systems

III.B.1. Electrothermal (ET) Couplings in Nanoscale ICs

Fig. 17(a) illustrates the interdependencies between various design parameters including performance (frequency), power dissipation, supply voltage, threshold voltage, and substrate (junction) temperature [36]. The total power dissipation has two major components: switching and leakage power dissipation. The switching power increases as the chip frequency (performance) and supply voltage increase. Moreover, the performance itself is dependent on temperature due to the dependence of the transistor on-current on substrate temperature. Continuous increase of integration density and power consumption elevates on-chip temperature. Higher power dissipation and temperature of the chip, which further increases the subthreshold leakage, thereby creates a strong feedback loop leading to various ET couplings that used to be inconspicuous for earlier generation of ICs. By taking into account these couplings, a system level ET analysis methodology and tool has been developed [36], using which design tradeoffs between power-performance-cooling cost can be evaluated as shown in **Fig. 17(b)**.

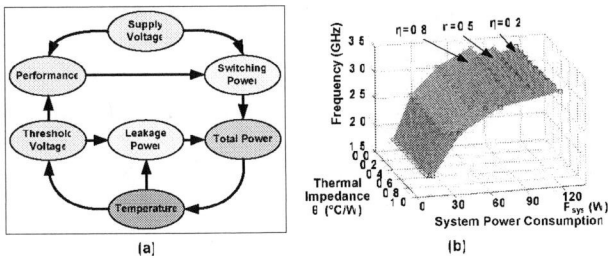

Fig.17: (a) Schematic view of ET couplings between different design parameters. **(b)** Power-performance-cooling cost tradeoff analysis [36].

III.B.2. Implications for Circuit Optimization

For power-constrained applications, lowering supply voltage (V_{dd}) offers the biggest potential to decrease the active power consumption, since CMOS switching power has a quadratic dependence on supply voltage. On the other hand, lowering supply voltage degrades the performance of circuits. It is, however, possible to maintain the performance by decreasing the threshold voltage (V_{th}) at the same time, but then the subthreshold leakage power increases exponentially. Consequently, the need for low power and high performance circuit applications motivates the search for an optimal set of supply and threshold voltages to tradeoff performance and power consumption.

Traditionally, circuit designers use energy-delay product (EDP) as the design metric to minimize both power and delay of a circuit [37]. **Fig. 18(a)** has been generated simply by direct numerical evaluation of energy and delay for a specific design. However, as pointed out in the preceding discussion, it is crucial to incorporate electrothermal couplings when evaluating the power and delay. Recently, an electrothermally coupled EDP methodology has been developed [38]. This methodology takes these electrothermal couplings into consideration and incorporates both analytical models and results from the circuit simulator based on an integrated device, circuit, and system level modeling approach. In comparison with **Fig. 18(a)**, it can be observed from **Fig. 18(b)** that not only the EDP contours and iso-performance curves shift but also the design space gets restricted by thermal constraint that cannot be predicted by traditional evaluation. Consequently, if electrothermal couplings are not considered, power dissipation and delay evaluations will be inaccurate and mislead the design optimization process.

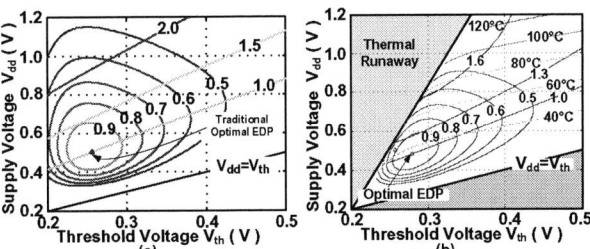

Fig. 18: (a) Traditional EDP evaluation. The EDP contours can be found by normalizing with respect to the value of the EDP at the optimal point as indicated by 'Δ'. For instance, any point on the curve labeled *0.5* has an EDP value twice that of optimal (EDP = $2 \cdot EDP_{opt}$), i.e., minimum value. The numbers on the iso-performance curves indicate the normalized value of the frequency with respect to the frequency at EDP_{opt}. **(b)** Electrothermally coupled EDP evaluation. Note that the design space gets restricted by thermal constraint (thermal runaway) that is determined by a passive cooling model [36], assuming junction-to-ambient thermal resistance $\theta_j = 0.85$ °C/W.

Most reliability mechanisms are highly temperature sensitive. The electrothermally coupled EDP methodology also provides a reliability and thermally aware "design space" that can be used to optimize and compare various designs and also evaluate the relative impact of various reliability constraints on a given design space.

Different reliability and thermal constraints can be applied at the design stage by using the iso-temperature curves shown in **Fig. 18(b)**. It can be observed that the junction temperature must be maintained below 40°C to achieve optimal EDP and if the junction temperature is required to be 100°C due to some cost constraints then the EDP will be higher by a factor of 2.5.

In order to highlight the impact of technology scaling on this methodology, **Fig. 19(a)** shows the reduction of design space due to increasing I_{off}. On the other hand, lowering of the junction temperature by employing advanced packaging and cooling techniques with lower thermal impedance (θ_j) will expand the design space as shown in **Fig. 19(b)**. Thus, tradeoffs between cooling cost and the benefit from design space relaxation also can be evaluated by the proposed methodology. Moreover, the significance of applying this methodology is expected to increase when parameter variations such as process, supply voltage and temperature variations are also taken into account since they are known to increase subthreshold leakage significantly.

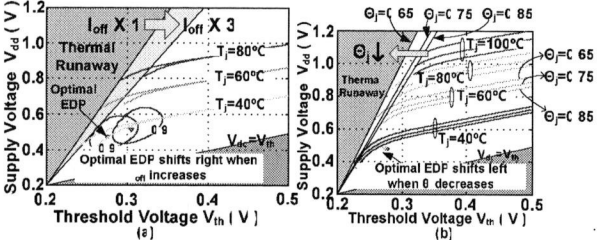

Fig. 19: (a) Impact of technology scaling on design space. While the leakage increases due to technology scaling or process variations, the operation region prohibited by thermal runaway expands. (b) Relaxation of design space by lowering of thermal impedance (θ_j). The iso-temperature curves move upwards by employing advanced packaging and cooling techniques with lower θ_j and the operation region prohibited by thermal runaway reduces. Therefore, the design space expands.

Different optimization metrics result in different design choices. Besides energy-delay product (EDP), other design metrics are also used for different applications such as power-delay product (PDP) and power-energy product (PEP). Moreover, in [39], Pénzes and Martin showed that the Et^n metric characterizes any feasible tradeoff. Hofstee [40] conclude that optimal metric is not unique for all designs but depends on the desired level of performance.

In order to choose an appropriate design metric in a systematic manner, an electrothermally-aware methodology has been developed [38]. The method captures the relative importance of power dissipation and performance to achieve design-specific targets and also provides a more meaningful basis to optimize supply and threshold voltages as they change from one technology generation to the next. As shown in **Fig. 20**, the optimal operating locus obtained for different optimization metrics based on a generalized metric (PT^μ) shifts when technology scales from 100nm to 70nm nodes where μ represents the ratio of exponent of delay to that of power.

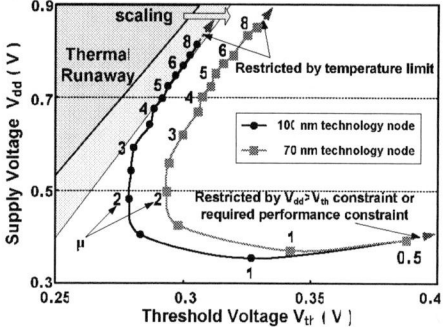

Fig. 20: Electrothermally-aware optimal operation locus for different μ. Note that the region (thermal runaway) expands due to technology scaling.

III.B.3. Implications for Full-Chip Subthreshold Leakage Estimation

For nanometer scale CMOS technologies, leakage power forms a significant component of the total power dissipation, especially due to within-die and die-to-die variations in process (P), temperature (T) and supply voltage (V). As a result of these variations, leakage power has been reported to have 20X variations for a 180 nm CMOS technology [41]. Thus, designing with the

worst-case leakage values may result in excessive guard-banding while underestimating the leakage might result in highly optimistic designs. Additionally, due to a 5X increase of total leakage power every generation [42], the design constraint based on leakage power may soon limit the yield. Under this scenario, a probabilistic framework for accurately estimating full-chip subthreshold leakage power distribution under P-T-V variations has been developed, which can be subsequently used to accurately estimate the yield [43].

Fig. 21 shows that apart from within-die variations, die-to-die parameter variations such as channel length and temperature can strongly impact the leakage power. It can be clearly observed from the figure that die-to-die temperature variations significantly increase the leakage due to the electrothermal couplings between subthreshold leakage power dissipation and die temperature, especially at higher operating temperatures.

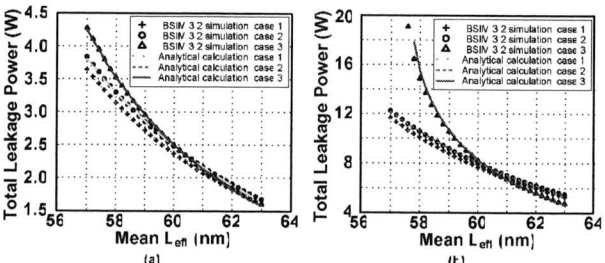

Fig. 21: Total subthreshold leakage power vs. mean die-channel length. **(a)** Average junction temperature = 300K **(b)** Average junction temperature = 320K. In case 1, only die-to-die channel length variations are considered. In case 2, besides die-to-die channel length variations, within-die channel length variations are also considered, while case 3 considers die-to-die temperature variations together with all the variations considered in case 2.

III.B.4. Implications for IC Cooling and Hot-Spot Management

Cooled chip operation is being seriously evaluated as a practical technique for boosting the performance of high-end microprocessors. While the drive current increases at lower temperature, it is observed that SOI type transistors show greater sensitivity to temperature due to the less increase of body to source voltage (V_{BS}) of PD-SOI at low temperature [44] which causes a smaller increase in the saturated threshold voltage of PD-SOI type transistors, as shown in **Fig. 22(a)** [45]. Thus, the enhancement of drive current of PD-SOI transistors at low temperature is higher than that in bulk transistors.

It has been shown that cooling leads to benefit at the device and circuit levels [45]. However, in order to understand the real benefit from cooling, system level considerations need to be taken into account. **Fig. 22(b)** shows the leakage power dissipation for two identical test microprocessor designs at different technology nodes under the application of active cooling. As expected, the leakage power decreases at lower temperature and the reduction of leakage becomes greater as technology scales. **Fig. 23** shows that the chip power (including active and leakage power dissipation) decreases as more cooling is applied mainly as a result of decreasing leakage power which is shown in **Fig. 22(b)**. However, the total system power (including chip power and cooling power consumption), determines the practical limit beyond which further cooling does not lead to any overall power savings and the limit occurs at an increasingly lower temperature as technology scales. Thus, as technology scales, the benefit that can be derived from cooling increases.

Highly integrated circuits with different functional blocks and finite thermal conductivity of silicon and packaging materials will create a non-uniform power density across the chip surface. Those regions with higher heat flux densities (hot-spots) affect performance and reliability and lead to a general over-design in the microprocessor packaging and cooling solutions. In order to comprehend the impact of cooling on thermal gradients and hot-spots, a self-consistent electrothermal methodology for estimating substrate temperature profile has been developed (**Fig. 24**) [45].

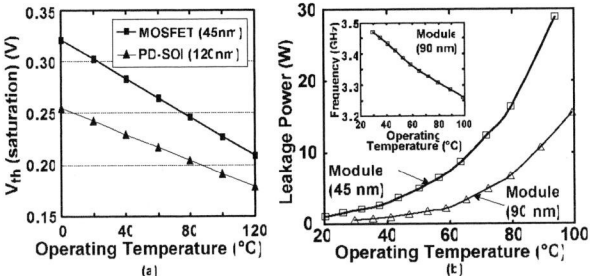

Fig. 22: (a) The increasing rate of saturated threshold voltage for bulk MOSFET (0.9mV/°C) is larger than PD-SOI type transistor (0.6mV/°C). (b) Leakage power dissipation as a function of operating temperature. The inset shows chip performance increases as operating temperature decreases [45].

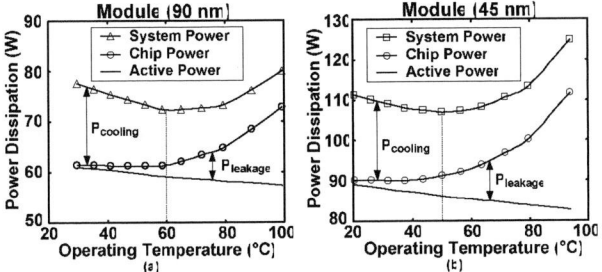

Fig. 23: Electrothermally-aware system level evaluation of power dissipation [45]. A minimum system power determines the practical limit beyond which further cooling does not lead to any power saving. (a) 90 nm test module (b) 45 nm test module.

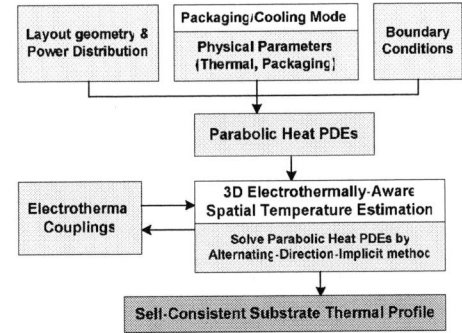

Fig. 24: Overview of the electrothermally-aware substrate thermal profile generator [45].

An example chip design (die size: 10 mm × 10 mm) with power densities per functional block is shown in **Fig. 25(a)**. The spatial substrate temperature profile, **Fig. 25(b)**, shows several hot-spots and the highest junction temperature is around 133°C. Although the results shown here are specific to the above mentioned IC, the conclusions drawn are more generic. **Fig. 26** shows the effect of applying global and localized cooling strategies on hot-spot management. As shown in **Fig. 26(a)**, a lower junction-to-ambient thermal resistance (θ_j) reduces the maximum junction temperature by applying global cooling (through better interface material, higher cooling efficiency, etc.). However, on-chip hot-spots and thermal gradients still remain. On the other hand, localized cooling solutions such as local spray cooling, thin-film thermoelectric coolers, can be applied to electronic application to eliminate the hot-spots. For

example, if two thin-film thermoelectric coolers can be placed on the backside of the wafer below the locations of hot-spots, as shown in **Fig. 26(b)**, it can effectively eliminate the targeted hot-spots.

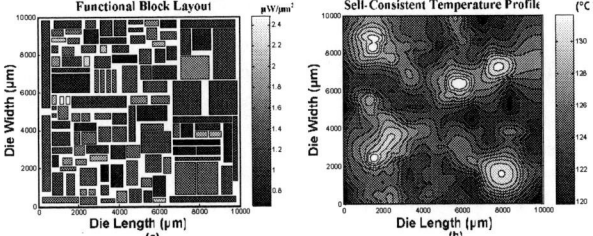

Fig. 25: (a) Functional block layout of a test chip showing power density associated with each block. Nominal total power consumption is 75 W. (b) Spatial substrate temperature profile generated using methodology shown in Fig. 24 (θ_j=1.1 °C/W). Five hot-spots can be observed [45].

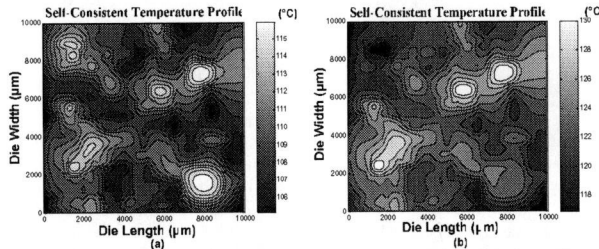

Fig. 26: Spatial substrate temperature profile: (a) Applying global cooling (θ_j reduces 20%). Although highest temperature decreases, the hot-spots remain. (b) Integrating two thin-film thermoelectric coolers at the top-left and bottom-right hot-spots. Only three hotspots can be observed [45].

IV. Electrothermal Engineering in Emerging Technologies

IV.A. Three-Dimensional (3-D) ICs

The challenges from on-chip interconnects in nanometer scale VLSI have led researchers to seek innovative design solutions, circuit or interconnect optimization techniques and material solutions so that the chip's wires do not offset the benefits of continued device scaling. Three-dimensional integration to create multi-layer ICs [46] is a concept that reduces the number and the average lengths of the longest global wires seen in traditional 2-D chips by providing shorter "vertical" paths for connection and also allows integration of disparate technologies and substrates. However, this technology still needs to overcome difficult challenges such as the development of new system architectures and tools. A critical problem in 3-D ICs is the thermal management of internal (stacked) active layers [31] (**Fig. 27(a)**). Hence, the electrothermal couplings described in Section III.B.1 also have a significant impact on the temperature and power dissipation of 3-D ICs (**Fig. 27(b)**).

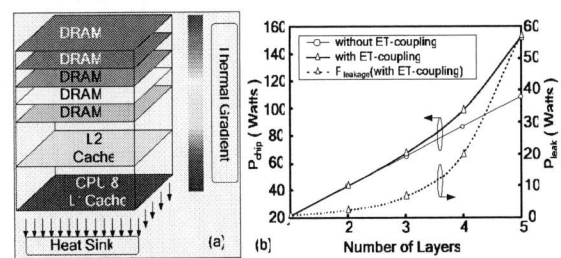

Fig. 27: (a) The prevalent high temperatures on stacked layers in a 3-D integrated circuit. Performance evaluation of such chips must account for the negative impact of these high temperatures. (b) Chip power and leakage power dissipation vs. number of layers. As the number of layers is increased the total power dissipation becomes a strict nonlinear function of the number of layers as opposed to a linear function observed by non-self-consistent treatment of the problem.

IV.B. Hybrid Carbon Nanotube-Cu Technologies

Carbon nanotube (CNT) bundle interconnects are possibly the least disruptive of all alternatives to copper interconnects that have been suggested so far. Although there are some technological issues that must be resolved before CNT interconnects can be used in practice [28], they have the potential to meet interconnect challenges without the need for paradigm changes in VLSI circuit design techniques and tools or extra circuitry. Besides the performance benefits of CNT bundle interconnects [47], their high thermal conductivity makes them very effective in controlling the large backend temperature rise expected with metallic interconnects (**Fig. 28**), and hence in improving overall interconnect performance and lifetime [48]. The thermal advantage of CNT bundle vias will also have significant implications for 3-D ICs where thermal management is a big concern.

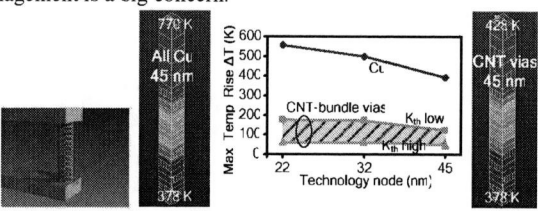

Fig. 28: Maximum interconnect temperature rise for Cu interconnect stack with Cu vias compared to CNT bundle vias integrated with Cu interconnects (see schematic on left [courtesy Infineon]). For CNT bundles, the shaded region shows the range of thermal conductivity 1750 W/mK < Kth < 5800 W/mK [48]. Reference (substrate) temperature = 378K.

Acknowledgements

This work was supported by Intel Corp., Fujitsu Labs. of America, Mentor Graphics Corp., NIST, SRC and the University of California-MICRO program.

References

[1] V. De and S. Borkar, "Technology and design challenges for low power and high performance microprocessors," *ISLPED*, 1999, pp. 163 –168.

[2] International Technology Roadmap for Semiconductors (ITRS), 2004

[3] R. Viswanath et al., "Thermal performance challenges from silicon to systems," *Intel Technology Journal 3rd quarter*, 2000.

[4] I. Aller et al., "CMOS circuit technology for sub-ambient temperature operation," *ISSCC*, 2000, pp. 214-215.

[5] S. Borkar et al., "Parameter variations and impact on circuits and microarchitecture," *DAC*, 2003, pp. 338-342.

[6] P. E. Ross, "Beat the heat," *Spectrum, IEEE*, Vol. 41, pp. 38-43, 2004.

[7] C. X. Zhang, "Timing-, heat- and area-driven placement using self-organizing semantic maps," *ISCAS*, 1993, pp. 2067-2070.

[8] Y-K Cheng et al., "ILLIADS-T: An electrothermal timing simulator for temperature-sensitive reliability diagnosis of CMOS VLSI chips," *IEEE TCAD*, Vol. 17, pp.668-681, 1998.

[9] C. C. Teng et al., "iTEM: A Temperature-dependent electromigration reliability diagnosis tool," *IEEE TCAD*, Vol. 16, pp. 882-893, 1997.

[10] A. H. Ajami, K. Banerjee and M. Pedram, "Modeling and analysis of non-uniform substrate temperature effects on global ULSI interconnects", *IEEE TCAD*, Vol. 24, pp. 849-861, 2005.

[11] J. Lee, "Thermal placement algorithm based on heat conduction analogy," *IEEE Trans. on Components and Packaging Tech.*, Vol. 26, pp.473-482, 2003.

[12] C. H. Tsai and S. M. Kang, "Cell-level placement for improving substrate thermal distribution," *IEEE TCAD*, Vol. 19, pp. 253-266, 2000.

[13] L.T. Su, et al., "Measurement and modeling of self-heating in SOI nMOSFETs", *IEEE TED*, Vol. 41, pp. 69-75, 1994.

[14] Y-K. Leung, et al., "Heating mechanisms of LDMOS and LIGBT in ultra-thin SOI", *IEEE EDL*, Vol. 18, pp. 414-416, 1997.

[15] E. Pop et al., "Localized heating effects and scaling of sub-0.18 micron CMOS devices," *IEDM*, 2001, pp. 677-680.

[16] P. G. Sverdrup et al., "Sub-continuum thermal simulations of deep sub-micron devices under ESD conditions," *SISPAD*, 2000, pp. 54-57.

[17] A. M. Ionescu and K. Banerjee, *Emerging Nanoelectronics: Life With and After CMOS*, Springer, 2005.

[18] T. Skotnicki et al., "End of CMOS scaling", *IEEE Circuits and Devices Magazine*, Vol. 21, pp. 16-26, 2005.

[19] K. A. Jenkins and K. Rim, "Measurement of the effect of self-heating in strained-silicon MOSFETS", *IEEE EDL*, Vol. 23, pp. 360-362, 2002.

[20] W. Liu and M. Asheghi, "Thermal modeling of self-heating in strained-silicon MOSFETs", *IEEE Inter society conference on thermal phenomena*, Vol. 2, pp. 605-609, 2004.

[21] W. Liu and M. Asheghi, "Thermal conductivity of ultra-thin single crystal silicon layers, part I - experimental measurements at room and cryogenic temperatures," *J. Heat Transfer*, 2004.

[22] E. Pop et al, "Thermal analysis of ultra-thin body device scaling", *IEDM*, 2003, pp. 883-886.

[23] M. Asheghi et al., "Thermal conduction in doped single-crystal silicon films," *JAP*, Vol. 91, pp. 5079-5088, 2002.

[24] J. Lai and A. Majumdar, "Concurrent thermal and electrical modeling of sub-micrometer silicon devices," *JAP*, Vol. 79, pp. 7353–7363, 1996.

[25] E. Pop et al., "Analytic band Monte Carlo model for electron transport in Si including acoustic and optical phonon dispersion," *JAP*, 96, 4998, 2004.

[26] S Sinha et al., "A split-flux model for phonon transport near hotspots", *J. Heat Transfer*, 2005.

[27] S. Im et al., "Scaling analysis of multilevel interconnect temperatures for high performance ICs", *IEEE TED*, 2005 (in press).

[28] N. Srivastava and K. Banerjee, "A comparative scaling analysis of metallic and carbon nanotube interconnections for nanometer scale VLSI technologies," *VMIC*, 2004, pp. 393-398.

[29] K. Banerjee and A. Mehrotra, "A power-optimal repeater insertion methodology for global interconnects in nanometer designs," *IEEE TED*, Vol. 49, pp. 2001-2007, 2002.

[30] V. Wason and K. Banerjee, "A probabilistic framework for power-optimal repeater insertion for global interconnects under parameter variations," *ISLPED*, 2005, pp. 131-136.

[31] S. Im and K. Banerjee, "Full chip thermal analysis of planar (2-D) and vertically integrated (3-D) high performance ICs," *IEDM*, 2000, pp. 727-730.

[32] Z. Yu et al., "Full chip thermal simulation," *ISQED*, 2000, pp.145-149.

[33] Q. Wu, Q. Qiu, and M. Pedram, "Dynamic power management of complex systems using generalized stochastic Petri nets," *DAC*, 2000, pp. 352-356.

[34] A. H. Ajami, K. Banerjee and M. Pedram, "Scaling analysis of on-chip power grid voltage variations in nanometer scale ULSI", *J. of Analog Integrated Circuits and Signal Processing*, Vol. 42, pp. 277-290, 2005.

[35] A. H. Ajami, K. Banerjee and M. Pedram, "Analysis of substrate thermal gradient effects on optimal buffer insertion," *ICCAD*, 2001, pp. 44-48.

[36] K. Banerjee et al., "A self-consistent junction temperature estimation methodology for nanometer scale ICs with implications for performance and thermal management," *IEDM*, 2003, pp. 893-896.

[37] M. Horowitz et al., "Low power digital design," *ISLPED*, 1994, pp. 8-11.

[38] S-C. Lin et al., "A thermally aware methodology for design-specific optimization of supply and threshold voltages in nanometer scale ICs," *ICCD*, 2005, pp. 411-416.

[39] P. I. Pénzes and A. J. Martin, "Energy-delay efficiency of VLSI computations," *GLSVLSI*, 2002, pp. 104–111.

[40] H. P. Hofstee, "Power-constrained microprocessor design," *ICCD*, 2002, pp. 14–16.

[41] S. Borkar et al., "Parameter variations and impact on circuits and microarchitecture," *DAC*, 2003, pp. 338-342.

[42] S. Borkar, "Design challenges of technology scaling," *IEEE Micro*, Vol. 19, pp. 23-29, 1999.

[43] S. Zhang et al., "A probabilistic framework to estimate full-chip subthreshold leakage power distribution considering within-die and die-to-die P-T-V variations," *ISLPED*, 2004, pp. 156-161.

[44] M.M. Pelella, J.G. Fossum, and S. Krishnan, "Control of off-state current in scaled PD/SOI CMOS digital circuits," *International SOI conference*, 1998, pp. 147-148.

[45] S-C. Lin et al., "Analysis and implications of IC cooling for deep nanometer scale CMOS technologies", *IEDM*, 2005 (to appear).

[46] K. Banerjee, et al, "3-D ICs: A novel chip design for improving deep submicron interconnect performance and systems-on-chip integration," *Proceedings of the IEEE*, Vol. 89, pp. 602-633, 2001.

[47] N. Srivastava and K. Banerjee, "Performance analysis of carbon nanotube interconnects for VLSI applications", *ICCAD*, 2005, pp. 383-390.

[48] N. Srivastava, R. V. Joshi and K. Banerjee, "Carbon nanotube interconnects: implications for performance, power dissipation and thermal management", *IEDM*, 2005 (to appear).

Area Optimization for Leakage Reduction and Thermal Stability in Nanometer Scale Technologies

Ja Chun Ku and Yehea Ismail

EECS Department
Northwestern University
Evanston, IL 60208
{jck273, ismail}@ece.northwestern.edu

Abstract - Traditionally, minimum possible area of a VLSI layout is considered the best for delay and power minimization due to decreased interconnect capacitance. This paper shows however that the use of minimum area does not result in the minimum power and/or delay in nanometer scale technologies due to thermal effects, and in some cases, may result in thermal runaway. A methodology using area as a design parameter to reduce the leakage power, and prevent thermal runaway is presented. A 16-bit adder example in a 70nm technology shows a total power savings of 17% with 15% increase in area, and no increase in delay. The power savings using this technique are expected to increase in future technologies.

I. Introduction

As CMOS devices continue to scale down, the decrease in transistor threshold voltage in order to maintain the performance has resulted in an exponential increase in the subthreshold leakage current [9]. The leakage power that mainly comes from the subthreshold current has already become comparable to the dynamic power in many applications, and it is projected to dominate the total power in sub-100nm technologies, especially at high operating temperatures [10, 11, 17, 20]. Thus, much effort has been put in order to minimize the power consumption through suppressing the subthreshold current [12, 18, 20].

Traditionally, minimum possible area is considered the best for a VLSI layout as it minimized both delay and power consumption due to the decreased interconnect capacitance. However, the use of minimum area has also resulted in an increase in the power density of circuit modules, and hence, the junction temperature increased, which has an exponential impact on the subthreshold current. In other words, minimum area may no longer be the optimum point for power due to the leakage. Furthermore, the use of minimum area may cause thermal runaway for some designs as the leakage power and the temperature are locked into an increasing positive feedback loop. Therefore, controlling power density has to be a crucial part of the design in nanometer scale technologies where leakage is the dominant power dissipation source.

In this paper, area is used as a design parameter to reduce the power density (junction-to-ambient thermal resistance), and it is shown that minimum area does not correspond to the optimum point for power minimization in nanometer scale technologies. By using a larger area for a hot module, its junction temperature is lowered, which reduces the leakage power significantly. However, on the other hand, the increase in interconnect ground capacitance makes the dynamic power increase. The delay traditionally decreases as the temperature drops, but in nanometer scale technologies where the supply voltage is low ($V_{dd} \approx 1$V), the improvement in the delay is very small due to the negative temperature dependence of the threshold voltage. Thus overall, an increase in the area results in a slight increase in the delay. In addition, area can also be used in the design to guarantee thermal stability of a system. This paper provides an analysis of thermal stability condition, and use of the area to prevent thermal runaway.

The next section presents the models used in this work for power and delay as functions of temperature. The thermal model relating power, temperature, thermal resistance, and area is also presented along with the derivation of an analytical expression for steady-state temperature calculation. In the third section, the impact of area optimization is illustrated with a 16-bit adder example in a 70nm technology using the models presented. The fourth section discusses the condition for thermal stability, and shows how the area can be used to prevent thermal runaway using an analytical methodology. Finally, the last section concludes the paper with a summary.

II. Power, Delay, and Thermal Models

In order to evaluate the effect of area on power, delay, and temperature, the first two subsections develop temperature-dependent power and delay models, and compare them to SPICE BSIM3v3 models. Then, the relationship between temperature, power, thermal resistance, and area is modeled in the third subsection with a closed-form expression for steady-state temperature that includes electrothermal coupling.

A. Power Model

Power dissipation in CMOS circuits can be divided into two major components

$$P = P_{dynamic} + P_{leakage} \qquad (1)$$

There is a third component, short-circuit power that results from a direct-path current when both the pull up and pull down networks are simultaneously on while inputs switch. However, short-circuit power is usually small compared to the other two components of total power, and also is expected to become even less significant as technology scales down [3]. Thus, short-circuit power is ignored to simplify the power models [6, 12].

Dynamic power, $P_{dynamic}$ is given by

$$P_{dynamic} = Na\left(C_g + MCF_p C_c\right)V_{dd}^2 f \qquad (2)$$

where N is the number of gates in the design, a is the average switching activity factor, V_{dd} is the supply voltage, and f is the operating frequency. The average load capacitance driven by a gate can be subdivided into the ground capacitance, C_g and the coupling capacitance, C_c. The coupling capacitance is multiplied by MCF_p, the average Miller coupling factor for average power consumption that takes into account the relative switching patterns of the interconnect wires. In nanometer scale CMOS circuits, the coupling capacitance is more dominant compared to the ground capacitance due to the increased interconnect density and smaller feature sizes. To a very good approximation, the capacitance is temperature-independent, and hence, the dynamic component of the power can be assumed to be temperature-independent unless the operating frequency is indirectly affected by the temperature. In this paper, area of a small unit such as a module is considered for optimization, and thus, the chip operating frequency is assumed to

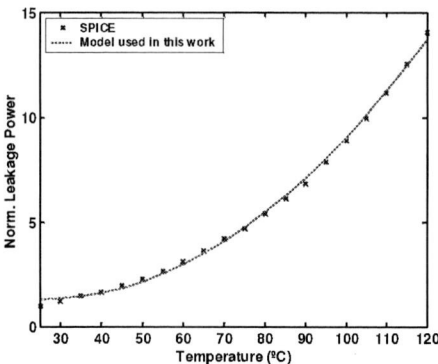

Figure 1. Temperature model of the leakage power compared to SPICE.

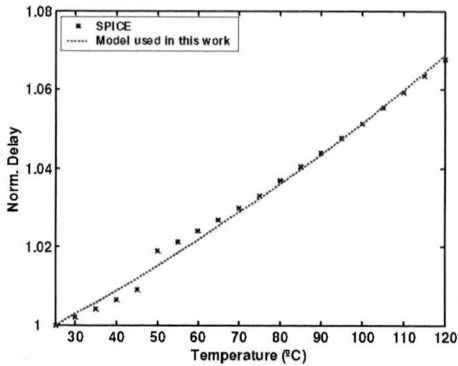

Figure 2. Temperature model of the delay compared to SPICE.

be kept constant.

The other major component is the leakage power, $P_{leakage}$, which is dominated by the subthreshold current, and thus it is expressed as

$$P_{leakage} = NI_{sub}(T)V_{dd} \qquad (3)$$

The temperature dependence of the average subthreshold current, I_{sub} for a gate can be written as a quadratic function of the temperature as similarly done by Su et al. in [26].

$$I_{sub}(T) = WI_{sub0}\left\{c_1(T_j - T_0)^2 + c_2(T_j - T_0) + c_3\right\} \qquad (4)$$

W and I_{sub0} are the average gate width and subthreshold current at nominal temperature, T_0 (which is usually 25°C), respectively. T_j is the junction temperature, and c_1, c_2, c_3 are constants. The reason that a quadratic function is used rather than an exponential one is because it makes derivation of the expression for steady-state temperature easier in the following subsection. Figure 1 compares the temperature model of the leakage power with a SPICE result in the BPTM [1] 70nm technology. Note that the leakage power more than doubles for every 30°C rise in the temperature.

The gate oxide leakage current which is ignored in this paper is also on the trend of becoming comparable to the subthreshold current as the thickness of the gate oxide layer shrinks with technology scaling. Unfortunately, SPICE BSIM3v3 (level 49) does not model gate oxide leakage. However, gate oxide leakage is expected to be small compared to the subthreshold leakage in the BPTM 70nm technology used in this paper as can be seen from the data in [16]. Furthermore, gate oxide leakage is relatively temperature-independent, which means that it will be even smaller compared to the subthreshold leakage at high temperatures.

The total power consumption per gate is therefore given by

$$P/N = a\left(C_g + MCF_p C_c\right)V_{dd}^2 f + WI_{sub0}\left\{c_1(T_j - T_0)^2 + c_2(T_j - T_0) + c_3\right\}V_{dd} \quad (5)$$

B. Delay Model

As for the delay model, an expression based on the alpha-power law is used [12, 21]

$$D = N_c \frac{\left(C_g + MCF_d C_c\right)V_{dd}}{2I_{Di}(T)} \qquad (6)$$

where N_c is the number of gates on the critical path, and MCF_d is the average Miller coupling factor for worst-case delay. I_D is the average drain current at saturation region for each gate, and it can be expressed as [9]

$$I_D(T) = KWv_{sat}(T)(V_{dd} - V_t(T))^\alpha \qquad (7)$$

K is a constant specific to a given technology. v_{sat} is the saturation velocity, and α is the velocity saturation index whose value is between 1 and 2 in the deep submicron region [21]. The saturation velocity is given by

$$v_{sat} = \mu E_c \qquad (8)$$

where μ and E_c are mobility and the critical electric field for velocity saturation, respectively. Although both saturation velocity and mobility have a negative temperature dependence, its magnitude is weaker for saturation velocity because the value of E_c also becomes higher as the temperature is raised. Note that the drain current stays in the saturation region for almost the entire duration of the transition. Thus, the temperature dependence of the drain current follows that of the saturation velocity rather than the mobility. The temperature dependence of saturation velocity is almost linear [7, 19], and can be written as

$$v_{sat}(T) = v_{sat}(T_0) - \eta(T - T_0) \qquad (9)$$

η is the saturation velocity temperature coefficient whose value obtained from SPICE simulations is around 140 ms^{-1}/°C for a 70nm technology. The threshold voltage also decreases as the temperature is raised, and is given by

$$V_t(T) = V_t(T_0) - \kappa(T - T_0) \qquad (10)$$

κ is the threshold voltage temperature coefficient whose value is about 0.7mV/°C in deep submicron technologies [27]. When the supply voltage is high, the effect of the saturation velocity dominates the overall temperature dependence of the drain current since the change in $(V_{dd} - V_t(T))^\alpha$ is relatively insignificant. However, when the supply voltage is low (close to 1V), the change in $(V_{dd} - V_t(T))^\alpha$ becomes more important, and cancels the temperature dependence of the drain current due to the saturation velocity. Figure 2 compares the temperature model of delay with a SPICE result in BPTM [1] 70nm technology operating at $V_{dd} = 1.1$V. The delay increases only by 6-7% after a temperature rise of 80-90°C.

C. Thermal Model

The heat generated from a chip is dissipated through the silicon substrate, and the cooling system in the package. The heat flow in the package is a function of many parameters such as geometry, flux source and placement, package orientation, next-level package attachment, heat sink efficiency, and method of chip connection [15]. In this paper, we consider a typical flip-chip C4 package adapted from models by Kromann in [15] as shown in Figure 3. Most of the heat is dissipated through the heat sink that is usually attached to the back-side of the silicon substrate. This constitutes the primary heat transfer path where the heat generated is conducted upwards through the silicon to the thermal paste,

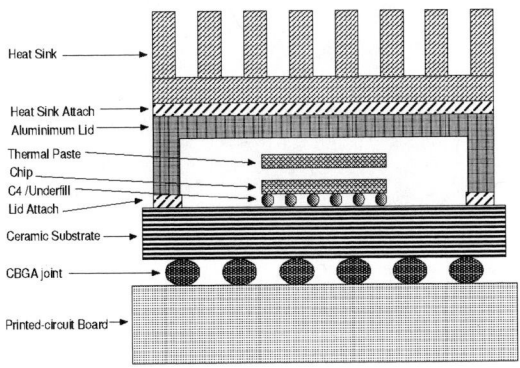

Figure 3. A typical flip-chip C4 package.

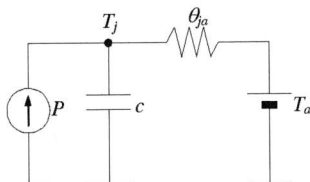

Figure 4. One-dimensional chip thermal model.

aluminum cap, heat sink attach, and heat sink, then convectively removed to the ambient air.

Since different parts of a chip have different activities and power densities, there is a variation in the thermal profile across a chip [4]. Accurate thermal modeling of a whole chip would thus require a 3-D analysis. However, the target of the area optimization technique proposed in this paper is not the entire chip, but only a small unit of a chip such as a module that is hot and leaking much (hot spot), and there is not much variation in temperature within such units. Hence, the following 1-D model is a valid approximation for its thermal analysis.

$$\theta_{ja}c\frac{dT_j}{dt} + T_j = P(T_j)\theta_{ja} + T_a \qquad (11)$$

where θ_{ja} is the junction-to-ambient thermal resistance of the silicon substrate and the package, c is the thermal capacitance of the system, T_j is the junction temperature, P is the power dissipation, and T_a is the ambient air temperature which is usually assumed to be 45°C. Figure 4 shows an equivalent electrical circuit for the thermal model [14]. Note that power is a function of the junction temperature with a positive dependence as it can be seen in (5). A rise in the temperature results in an increase in the power, which in turn, raises the temperature even higher, thus creating a positive feedback loop until the system reaches a steady-state (electrothermal coupling).

The thermal resistances of the silicon, the aluminum cap and the heat sink attach are small, and their contribution to the temperature drop can be omitted for a first-order analysis [15]. Hence, the junction-to-ambient thermal resistance can be expressed as

$$\theta_{ja} = \theta_{thermalpaste} + \theta_{heat\,sin\,k} \qquad (12)$$

It is shown by Kromann in [15] that the thermal paste resistance is reduced as the chip area increases. This is because the thermal resistance can be written as [5]

$$\theta = R_{th} \big/ A_c \qquad (13)$$

where R_{th} is the unit thermal resistance, and A_c is the cross-sectional area. An increase in the chip area directly increases the area of the thermal paste placed above it, thus assuming the chip area equals to the thermal paste area, the steady-state temperature (when the time derivative of the junction temperature equals to zero) can be written as

$$T_j = \left(P(T_j) \big/ A_{chip} \right) R_{thermalpaste} + P(T_j)\theta_{heat\,sin\,k} + T_a \qquad (14)$$

where $P(T_j)/A_{chip}$ represents the power density of the chip, and $R_{thermalpaste}$ is the unit thermal resistance of the heat sink. Convective thermal resistance of the heat sink, $\theta_{heatsink}$ is affected less by the chip area since the heat is usually spread out more uniformly (using a heat spreader) before it reaches the heat sink. However, in case of adapting an advanced fan heat sink as it is commonly done in today's technology, the heat sink resistance becomes small enough that the thermal paste resistance takes up the majority of the total junction-to-ambient thermal resistance (more than 60%) [15]. Therefore, increasing an area in the chip can significantly lower the junction temperature.

In order to derive an expression for the steady-state temperature after the electrothermal coupling, the temperature dependence of the power in (5) is substituted in (11). After some rearrangement, the following equation can be obtained.

$$\theta_{ja}c\frac{dT_j}{dt} = AT_j^2 + BT_j + C$$

where $\qquad\qquad (15)$

$A = c_1\theta_{ja}NWI_{sub0}V_{dd}$

$B = \theta_{ja}NWI_{sub0}V_{dd}(c_2 - 2c_1T_0) - 1$

$C = T_a + \theta_{ja}NV_{dd}\left\{ a\left(C_g + MCF_pC_c\right)V_{dd}f + WI_{sub0}\left(c\,T_0^2 - c_2T_0 + c_3\right)\right\}$

For steady-state, the time derivative of the junction temperature is equal to zero. Then, the quadratic equation on the right-hand side in (15) is solved to obtain the steady-state temperature value.

$$T_j = \frac{-B \pm \sqrt{B^2 - 4AC}}{2A} \qquad (16)$$

Since the equation is quadratic, there are two solutions. The smaller one of the two corresponds to a stable point while the other corresponds to a metastable point. The stability of the two solutions is illustrated in Figure 5 which plots the time derivative of temperature, T_j' against the temperature, T_j. Positive T_j' indicates that T_j is increasing while a negative T_j' indicates that T_j is decreasing. Hence, if the curve crosses $T_j' = 0$ with a negative slope, a small perturbation to either direction would bring T_j back to the point where $T_j' = 0$. On the other hand, if the curve crosses $T_j' = 0$ with a positive slope, any small perturbation would make T_j run

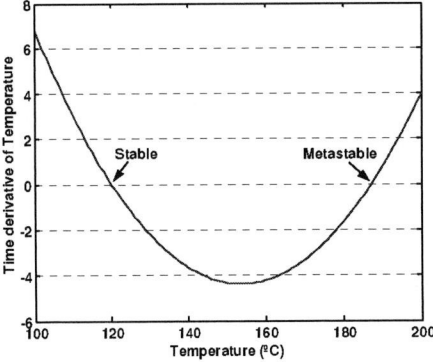

Figure 5. Solutions for the steady-state temperature.

233

away from the point. Thus, the relevant solution for steady-state temperature is only the one that corresponds to a stable point. After calculating the steady-state temperature, the steady-state value for the power consumption can be easily obtained as well by using (5).

III. Area Optimization for Leakage Reduction

In this paper, when an area increase of a design is considered, the length of the interconnect wires and the space between the wires are increased by the same ratio, while the height, thickness, and width of the wires are kept unchanged. Increasing the gate size along with the layout area was also initially considered in the optimization process. However, it was found that the increase in the load capacitance (which affect both delay and dynamic power) and the leakage power by increasing the gate size makes the initial gate size the optimum in terms of power-delay product. This is true also because of the fact that the interconnect capacitance does not change significantly with area due to the cancellation between ground and coupling components in the capacitance (which will be further explained in the following paragraphs). Thus, the sizes of the devices are kept unchanged.

The area scaling factor is called x in this paper. Thus, both the length of the wires and the space between the wires are increased by a factor of $x^{1/2}$. Figure 6 shows the effects of an area scaling by a factor if x on the interconnect wires. Increasing the area has two major impacts on the power and delay. The first impact is due to a decrease in the thermal resistance as shown in (13). The decrease in the thermal resistance results in a lower steady-state temperature as explained in the previous section. The subthreshold current will decrease exponentially, which in turn, results in a lower junction temperature. This means that the steady-state value for the power consumption becomes significantly lower. The drain current from (9) will also increase as the temperature drops, resulting in a slightly better delay. However, this improvement in the delay is not significant as discussed earlier.

The second impact is caused by changes in the load capacitances as the interconnect wires become longer. The coupling capacitance of the wires increases linearly as the length of the wires increases by $x^{1/2}$. However at the same time, the coupling capacitance of the wires also decreases superlinearly with an exponent around 1.34 as the space between the wires widens according to the work by Sakurai in [22]. Since the space between the wires is increased by $x^{1/2}$, the coupling capacitance decreases by $(x^{1/2})^{1.34}$. This superlinear decrease in the coupling capacitance outweighs the linear increase due to increasing wire length, resulting in an overall decrease in the coupling capacitance by $x^{0.17}$. On the other hand, the ground capacitance increases by $x^{1/2}$ as the length of the wires increases, countering the effects of the decreasing coupling capacitance. Hence, the total interconnect capacitance after an area scaling by x can be described by

$$x^{0.5}C_g + MCFx^{-0.17}C_c \qquad (17)$$

where C_g and C_c are the ground and coupling capacitance of the

Figure 7. Layout of the 16-bit adder.

wires before area scaling, respectively. The total interconnect capacitance may initially decrease slightly for small x since coupling capacitance is more dominant compared to ground capacitance in nanometer scale technologies. However, the total interconnect capacitance will start to increase for larger x because the magnitude of the positive exponent of the ground capacitance is greater than that of the negative exponent of the coupling capacitance. The increase in the interconnect capacitance results in an increase in both power and delay. The observations above indicate that there is an optimum area that designers can choose to reduce leakage.

In order to evaluate the effectiveness of the area optimization technique proposed in this paper, a 16-bit adder is laid out for 70mn BPTM [1] technology operating at $V_{dd} = 1.1V$ (Figure 7). Then, the models presented in the previous section are used to simulate the impact of area scaling on junction temperature, power, and delay. The value of the junction-to-ambient thermal resistance, which depends on the packaging technology, was chosen such that the junction temperature is around 120°C. Figure 8 illustrates the relationship between the junction temperature and the area. Notice that the temperature initially drops significantly when the area increases, then starts to saturate as x becomes larger. The initial big drop in the temperature is due to the exponential reduction in the leakage power. However, as more leakage power is eliminated, the dynamic power becomes more dominant, and increases slightly for larger x due to the increase in the interconnect capacitance, countering the decrease in the leakage power. This trend implies that the effect of the temperature drop on power and delay will be strongest during the initial increase in the area, and start to decay for larger x. Figure 9 shows how power and delay are affected as the area is increased. As expected, there is a big drop in power in the beginning, then it soon starts to saturate, following the behavior of the temperature. It can be seen from Figure 8 that the delay improves slightly in the beginning due to the initial decrease in the interconnect capacitance as discussed above (plus the temperature drop), however it soon starts to increase as x becomes larger. The reason the delay keeps increasing without saturating is that the

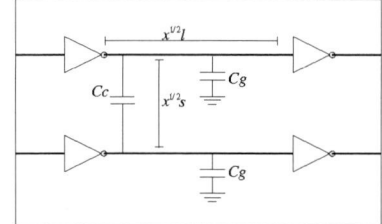

Figure 6. Changes in the wire dimensions after an area scaling by a factor of x.

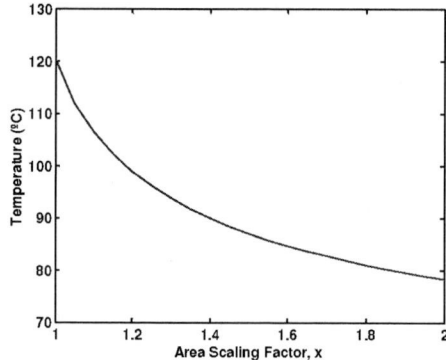

Figure 8. Area dependence of the junction temperature.

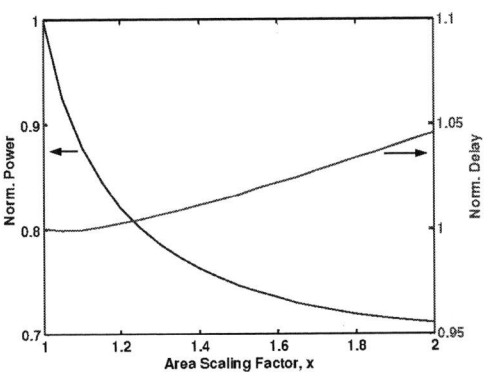

Figure 9. Area dependence of the power and delay.

temperature effect is not significant enough to counter the increase in the interconnect capacitance. However, the increase in the delay due to an area increase is still negligible (*increases only about 4-5% even when the area is doubled*). Note that the power consumption is reduced about 17% when the area is increased by 15% with almost no change in the delay.

The choice of the area would differ between each design depending on the design goals and constraints. It also strongly depends on how much weight the area cost carries. Although it depends on the purpose of the chip, area cost is becoming less significant as technology scales down especially in the case of high-performance ICs. In future technologies, delay will have a higher sensitivity to area due to smaller supply voltage applied. However at the same time, leakage power will dominate the total power even more, and since the leakage power increases exponentially with technology scaling, the reduction in the power through an area increase will easily exceed the increase in the delay. Furthermore, coupling capacitance will take up a larger fraction of the total interconnect capacitance, which will counter the effect of increased ground capacitance. Therefore, the proposed area optimization technique will prove to be even more useful in future technologies.

IV. Area Optimization for Thermal Stability

The possibility of thermal runaway in current and future technologies has been discussed in some literatures [14, 24]. The main cause for thermal runaway is the rapid increase in leakage current as the technology scales down. A large leakage power drives a high junction temperature as described by (12), and the new temperature pushes the leakage power up to an even higher value creating a positive feedback loop. If the gain of this loop is higher than a certain value, the feedback results in thermal runaway, and the system fails.

Traditionally, solutions to the thermal management problems have been mostly focused on the package cooling technology. However, there is usually a disparity between the maximum power and typical power consumed, and this gap is increasing with technology scaling. The package designed for worst-case (maximum power) is too expensive since the cooling cost increases nonlinearly with thermal dissipation [13]. Thus, packages today are typically designed for the *typical* power consumption to significantly reduce the package cost, and other dynamic alternatives such as frequency scaling, voltage scaling, and global clock gating are used in order to maintain the temperature in the permissible range when the power consumption increases beyond the typical case [13, 24]. In nanometer scale technologies where leakage dominates the power consumption, an increase in the power

will not just raise the temperature, but also may result in thermal runaway due to the electrothermal coupling effect.

The stability condition can actually be derived from (16). The term inside the square root has to be positive or zero for real solutions to exist for the temperature. A negative value inside the square root corresponds to thermal runaway since that means there is no real stable solution. Therefore, the following condition has to be satisfied for the system to be stable.

$$B^2 - 4AC \geq 0 \qquad (18)$$

It can be seen from (18) that parameters such as the supply voltage, operating frequency, junction-to-ambient thermal resistsance, and ambient temperature can cause thermal runaway when their values increase above certain critical values. In other words, such parameters can also be used to control and prevent thermal runaway.

As mentioned above, there are existing dynamic techniques that prevent thermal runaway. Typical examples are frequency scaling, voltage scaling, and global clock gating [24]. When the power of a system exceeds a certain threshold value (that may cause failure of the system) the operating frequency and/or the supply voltage of the chip are immediately dropped to a level where the power would be reduced enough to maintain the stability of the system. As for global clock gating, the global clock is stopped until the temperature falls down to a safe value. The common drawback of these techniques is that the whole chip has to be slowed down even though a hot small area (hot spot) may be the only cause of thermal runaway, which is often the case. In addition, all these techniques require sensors on chip for real-time temperature sensing which are sensitive to lithographic variations and supply-current variations causing imprecision [24].

An alternative way to prevent thermal runaway proposed in this paper is to increase the local area of hot spots (reducing its junction-to-ambient thermal resistance). This way, the performance of the chip is likely to be unaffected since there is no change in the delay for a small increase in area as it was shown in the previous sections. Hence, the operating frequency of the chip does not need to be dropped. From (18), the critical value of the junction-to-ambient thermal resistance, θ_c for the system to be stable can be derived by rearranging (18) as a quadratic function of θ_c.

$$X\theta_c^2 + Y\theta_c + Z = 0$$
$$where \qquad (19)$$
$$X = N^2 WI_{sub0} V_{dd}^2 \{WI_{sub0}(c_2^2 - 3c_1^2 T_0^2 - 4c_1 c_3) - 4c_1 a(C_g + MCF_p C_c)V_{dd} f\}$$
$$Y = -NWI_{sub0} V_{dd}(2c_2 - 4c_1 T_0 + 4c_1 T_a)$$
$$Z = 1$$

The solution of the quadratic equation in (19) is

$$\theta_c = \frac{-Y - \sqrt{Y^2 - 4X}}{2X} \qquad (20)$$

As it can be seen from (20), there is only one solution for (19) because the other solution leads to a lower junction temperature as the thermal resistance increases, which physically does not make sense. Thus, the value of junction-to-ambient thermal resistance has to be equal or smaller than θ_c in order to maintain thermal stability of the system. A junction-to-ambient thermal resistance that is greater than θ_c leads to thermal runaway. Figure 10 shows the transient response of temperature for two different values of thermal resistance. The minimum area, A_{min} that is required for thermal stability can be therefore derived as well by relating (20) to (12) and (13).

$$A_{\min} = \frac{R_{thermalpase}}{\theta_c - \theta_{heat \sin k}} \qquad (21)$$

235

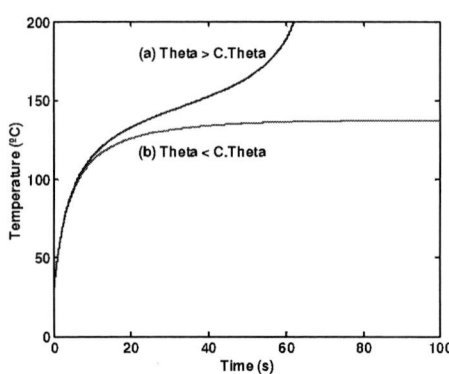

Figure 10. Transient response of temperature for (a) theta > critical theta (b) theta < critical theta.

V. Conclusion

It was shown in this paper that area can be used as a degree of freedom to reduce leakage, and prevent thermal runaway in nanometer scale technologies. Increasing the area reduces the thermal resistance, thus lowering the junction temperature. The power is initially reduced significantly as the area is increased because the drop in the temperature results in an exponential decrease in the leakage power. However, the reduction in the power saturates as the area is increased further since the dynamic power rises slowly, and the leakage power becomes smaller. Delay worsens slightly as the area is increased due to the increasing interconnect capacitance. However, the change in the delay is relatively insignificant. An analytical model was presented for the calculation of steady-state junction temperature after the electrothermal coupling. The area optimization technique for leakage reduction was evaluated using the power, delay, and thermal models presented with on a 16-bit adder example in a 70nm technology. It was shown that by increasing the area of a hot spot a significant power reduction is obtained. This trend is expected to continue as technology scales down because the decrease in the leakage power will easily exceed the increase in the delay. The possibility and condition for thermal runaway were also discussed. It was shown that the method of increasing a local area where it may cause thermal runaway can be used to maintain thermal stability without slowing down the whole chip. An analytical design methodology has been used to derive the required area for thermal stability.

References

[1] http://www-devices.eecs/berkeley.edu/~ptm/mosfet.html

[2] K. Banerjee, S-C. Lin, A. Keshavarzi, S. Narendra and V. De, "A self-consistent junction temperature estimation methodology for nanometer scale ICs with implications for performance and thermal management," in Proc. IEDM, pp. 887-890, 2003

[3] K. Banergee and A. Mehrotra, "A power-optimal repeater insertion methodology for global interconnects in nanometer designs," in IEEE Trans. Electron Devices, vol. 49, pp. 2001-2007, Nov. 2002

[4] S. Borkar et al, "Parameter variations and impact on circuits and microarchitecture," in Proc. DAC, pp. 338-342, 2003

[5] A. Chapman, Fundamentals of heat transfer. Macmillan Press, 1987

[6] A. Chatterjee, M. Nandakumar, and I. Chen, "An investigation of the impact of technology scaling on power wasted as short current in low voltage CMOS," in Proc. ISLPED, pp. 145-150, Aug. 1996

[7] Y. Cheng, K. Imai, M. Jeng, Z. Liu, K. Chen, and C. Hu, "Modeling temperature effects of quarter micrometer MOSFETs in BSIM3v3 for circuit simulation," in Semicond. Sci. Tech., pp. 1349-1354, 1997

[8] J. Daga, E. Ottaviano, D. Auvergne, "Temperature effect on delay for low voltage applications," in Proc. DATE, pp. 680-685, 1998

[9] V. De and S. Borkar, "Technology and design challenges for low power and high performance," in Proc. ISLPED, pp. 163-168, 1999

[10] V. De and S. Borkar, "Low power and high performance design challenges in future technologies," in Proc. GLSVLSI, pp. 1-6, 2000

[11] D. Genossar and Y. Shemir, "Intel Pentium M processor power estimation, budgeting, optimization, and validation," in Intel Tech. J., vol. 7, May 2003

[12] R. Gonzalez, B. Gordon, M. Horowitz, "Supply and threshold voltage scaling for low power CMOS," in IEEE J. Solid-State Circuits, vol. 32, pp.1210-1216, Aug. 1997

[13] S. Gunther, F. Binns, D. Carmean, and J. Hall, "Managing the impact of increasing microprocessor power consumption," in Intel Tech. J. Q1, 2001

[14] K. Kanda, K. Nose, H. Kawaguchi, and T. Sakurai, "Design impact of positive temperature dependence on drain current in sub-1-V CMOS VLSIs," in IEEE J. Solid-State Circuits, vol. 36, pp1559-1564, Oct. 2001

[15] G. Kromann, "Thermal modeling and experimental characterization of the C4/surface-mount-array interconnect technology," in IEEE Trans. Component, Packaging, and Manufacturing Technology, vol. 18, no. 1, pp. 87-93, Mar. 1995

[16] D. Lee, D. Blaauw, and D. Sylvester, "Gate oxide leakage current analysis and reduction for VLSI circuits," in IEEE Trans. VLSI, vol. 12, pp. 155-166, Feb. 2004

[17] S. Lin, A. Basu, A. Keshavarzi, V. De, A. Mehrotra, K. Banerjee, "Impact of off-state leakage current on electromigration design rules for nanometer scale CMOS technologies," in Proc. IRPS, pp. 74-78, 2004

[18] K. Nose and T. Sakurai, "Optimization of V_{dd} and V_{th} for low-power and high-performance applications," in Proc. ASP-DAC, pp. 469-474, 2000

[19] R. Quay, C. Moglestue, V. Palankovski, S. Selberherr, "A temperature dependent model for the saturation velocity in semiconductor materials," in Material Science in Semiconductor Processing 3, pp. 149-155, 2000

[20] K. Roy, S. Mukhopadhyay, H. Mahmoodi-Meimand, "Leakage current mechanisms and leakage reduction techniques in deep-submicrometer CMOS circuits," in Proc. IEEE, vol. 91, pp. 305-327, Feb. 2003

[21] T. Sakurai and A. Newton, "Alpha-power law MOSFET model and its application to CMOS inverter delay and other formulas," in IEEE J. Solid-State Circuits, vol. 25, pp. 584-593, Apr. 1990

[22] T. Sakurai, "Closed-form expression for interconnection delay, coupling, and crosstalk in VLSI's," in IEEE Trans. Electron Devices, vol. 40, no. 1, pp. 118-124, Jan. 1993

[23] O. Semenov, A. Vassighi, M. Sachdev, A. Keshavarzi, and C. Hawkins, "Effect of CMOS technology scaling on thermal management during burn-in," in IEEE Trans. Semiconductor Manufacturing, vol. 16, no. 4, pp. 686-695, Nov. 2003

[24] K. Skadron, M. Stan, W. Huang, S. Velusamy, K. Sankaranarayanan, and D. Tarjan, "Temperature-aware microarchitecture," in Proc. ISCA, 2003

[25] S. Strogatz, Nonlinear dynamics and chaos. Westview Press, 1994

[26] H. Su, F. Liu, A. Devgan, E. Acar, S. Nassif, "Full chip leakage estimation considering power supply and temperature variations," in Proc. ISLPED, pp. 78-83, 2003

[27] Y. Taur and T. Ning, Fundamentals of modern VLSI devices. Cambridge University Press, 1998

2006 Asia and South Pacific Design Automation Conference

Compact Thermal Models for Estimation of Temperature-dependent Power/Performance in FinFET Technology

Aditya Bansal, Mesut Meterelliyoz, Siddharth Singh[*], Jung Hwan Choi, Jayathi Murthy[§], Kaushik Roy

School of Electrical and Computer Engineering, Purdue University,
[§]Department of Mechanical Engineering, Purdue University
West Lafayette, IN 47907
[*]Department of Electrical Engineering, Osmania University, India
Email: {bansal, mesut, choi56, jmurthy, kaushik}@purdue.edu

Abstract: With technology scaling, elevated temperatures caused by increased power density create a critical bottleneck modulating the circuit operation. With the advent of FinFET technologies, cooling of a circuit is becoming a bigger challenge because of the thick buried oxide inhibiting the heat flow to the heat sink and confined ultra-thin channel increasing the thermal resistivity. In this work, we propose compact thermal models to predict the temperature rise in FinFET structures. We develop cell-level compact thermal models for standard INV, NAND and NOR gates accounting for the heat transfer across the six faces of a cell. Temperature maps of benchmark circuits exhibit close correspondence with dynamic power maps because of confined regions of heat generation separated by low thermal conductivity material. It is illustrated that temperature-aware timing analysis is imperative, because of high inter-cell temperature gradient. Accurate prediction of temperature in the early phase of design cycle will give valuable estimation of power/performance/reliability of a circuit block and will guide in the design of more robust circuits.

I. Introduction

The need for higher performance in smaller area has always been the driving force for the semiconductor industry. Aggressive technology scaling has led to increase in transistor density on a chip and innovative device structures (UTB-SOI, FinFET, Tri-gate etc.) to achieve the desired performance. Increasing packing density has led to power density to become a critical bottleneck in the design of microelectronics. The local temperature rise can result in circuit malfunction and can also impact performance, power and reliability. For every 10°C increase in temperature, a MOSFET's drive current decreases approximately 4% and interconnect (Elmore) delay increases approximately 5% [1]. To avoid the increase in temperature (or to dissipate heat), a heat sink is integrated on the chip package. However, since the heat sink is away from the device layer, it is not very efficient in taking the heat directly away from the transistors.

Several architecture and circuit level techniques have been proposed to distribute the temperature uniformly over the chip and reduce the hot spots. At architecture level, several dynamic thermal management (DTM) schemes have been proposed including dynamic voltage scaling (DVS) [2], discrete frequency scaling (DFS), migrating computation (MC) [3] and dynamic clock throttling (DCT) [4] etc. These techniques depend on the accurate thermal modeling of circuit blocks. Conventionally, circuit blocks are modeled as heat sources with uniform temperature distribution inside the block. There are several thermal models available in the

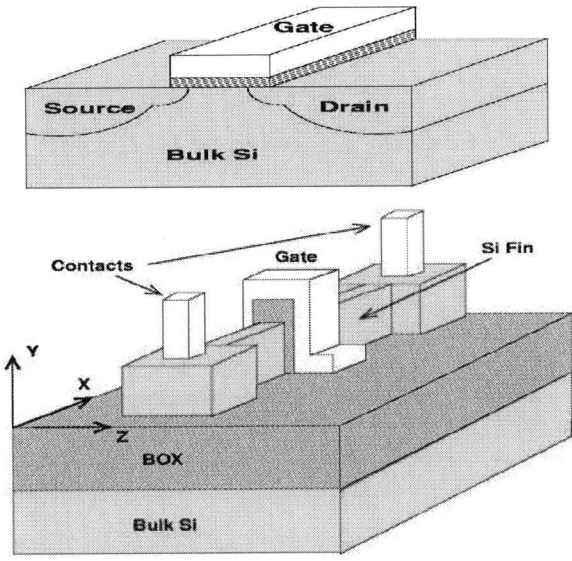

Fig. 1: Schematics of bulk-Si MOSFET and FinFET. In FinFET, gate surrounds the channel which is separated from the bulk-Si by thick BOX.

literature at different stages of design cycle of VLSI circuits. For example, a dynamic compact thermal model at microarchitecture level is presented in [3]. Grid-like thermal models for temperature-aware-design at arbitrary granularities are presented in [5]. Weiping et al. [6] modeled the dependence of power and performance on the temperature at microarchitecture level. However, thermal modeling at finer granularity level i.e., transistor level or logic gate level is required for more accurate estimation of local hot spots.

The increasing thermal problems are more aggravated in next generation transistor technologies such as Ultra-thin-body (UTB) Silicon-On-Insulator (SOI) MOSFETs and FinFETs. FinFETs have gate insulator all around the channel where heat is generated. This oxide layer results in less efficient dissipation of heat compared to bulk-MOSFETs where heat mainly dissipates through the substrate [7]. Eric et al. [8] [9] discuss the impact of confined dimensions and complicated geometries on self-heating in these devices. Ultra-thin silicon body improves the device scalability by reducing short-channel-effect, higher on-current and near ideal sub-threshold slope. However, the thermal resistivity of the thin active region is higher compared to thick active region in bulk-MOSFETs. Also, because of increased surface

0-7803-9451-8/06/$20.00 ©2006 IEEE.

2D-4 2006 Asia and South Pacific Design Automation Conference

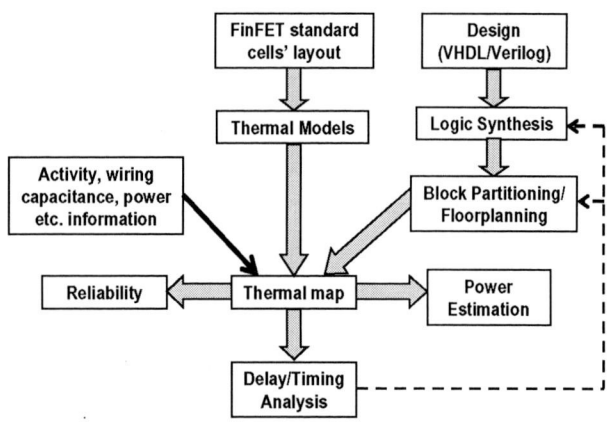

Fig. 2: Design flow for temperature based power/ performance/reliability estimation.

to volume ratio in UTB-SOI devices, thermal boundary resistances of various materials' further impede the heat flow [9]. Fig.1 shows the schematics of bulk- MOSFET and FinFET devices. In bulk-MOSFETs, heat generated in device layer as well as interconnect layers is mainly dissipated through the heat sink integrated with bulk substrate silicon. However, in FinFETs, device layer is separated from bulk-Si substrate by thick buried oxide layer (which has two orders of magnitude lower thermal conductivity than silicon [10]). Therefore, heat dissipation is mainly through the contact holes and interconnects. This elevates the local temperature rise in FinFET circuits. Because of temperature dependence of critical electrical parameters, it is becoming increasingly clear that electro-thermal co-design of transistors and circuits is necessary to arrive at optimal power and performance. However, a typical chip may have hundreds of millions of transistors, and a direct thermal simulation of such a collection of devices is all but impossible with current computing power. Instead, in this paper, we develop compact thermal models for circuit components used in cell-level electrical models in circuit design. Detailed computational models for three cell-level components – NAND, NOR and INV – are created based on Fourier theory.

Fig. 2 shows the design flow adopted in this work for the estimation of temperature-dependent power/performance/ reliability in FinFET based circuits. Compact thermal models are generated for standard cell layouts. The temperature information for each standard cell is associated with its heat generation and heat flow across the boundaries to the neighboring cells. Note that temperature of a cell is dependent on the heat generation in the transistors and temperatures of neighboring cells. Hence, real temperature can not be determined till a floorplan is generated. These models are used in generating the temperature profile of a circuit for a given floorplan. Wiring capacitance is accounted for as lumped capacitive load at the output of driving gate. The temperature map is used to estimate the increase in power dissipation, deterioration in performance and lifetime reliability because of self-heating and local temperature rise. This estimation can be primarily used in two ways for robust

circuit design: (1) It can be used to define an appropriate cost function to develop thermally-aware placement and optimization strategies for VLSI circuits, (2) Temperature map can be used for better delay analysis of the critical paths in early phase of the design cycle.

In this work, we are targeting the problem of heat dissipation in predictive 28nm ITRS [11] specified technology node for FinFET circuits. In section 2, we discuss the impact of local temperature rise on the power and performance of the standard cells. In section 3, we develop the compact thermal models for standard logic cells designed using NMOS and PMOS FinFETs. The joule heating in devices is simulated for the worst case input pattern (each transistor switching at every clock cycle) using Taurus Device Simulator [12]. Switching activity dependent temperature rise in a standard cell is modeled based on three-dimensional thermal resistances of the cell. In section 4, we generate the temperature maps of benchmark circuits and discuss the local temperature rise in FinFET circuits followed by conclusions in section 5.

II. Performance and Power Estimation

The temperature rise in a transistor affects its electrical characteristics. Mobility and sub-threshold slope degrade with the increase in temperature resulting in reduced on-current and hence, increased delay. Also, static leakage increases with temperature because of increase in sub-threshold leakage. In scaled technology generations, the static power is becoming a significant component of the total power dissipation, especially in the low activity sections of a chip like memory. Moreover, reliability is strongly dependent on the temperature. Increasing the temperature exponentially decreases the lifetime of a circuit. A first order model of mean-time-to-failure (MTF) can be given by Arrhenius equation:

$$MTF = MTF_0\, exp(E_a/k_B T) \qquad (1)$$

exhibiting the exponential deterioration in reliability with temperature. We quantitatively analyze the temperature dependence of delay and power in a 2-input NOR gate. Similar analysis has been done for INV and 2-input NAND gates, however, results are omitted for brevity.

A. Delay dependence on temperature

With the increase in temperature, mobility degrades in transistors. This effect is more dominant in bulk MOSFETs because of heavily doped body used to reduce short-channel-effect. However, in fully-depleted double gate SOI devices, body is left undoped resulting in less deterioration of mobility with temperature. Mobility degradation reduces on-current, however, with increase in temperature, threshold voltage decreases resulting in improved on-current. These two effects counter each other and the net variation in on-current is dependent on the relative sensitivities of mobility and threshold voltage to the temperature. To analyze the impact of temperature on the intrinsic delay, we simulated a three NOR stage ring oscillator with inputs of each NOR gate tied together. Fig. 3 shows the intrinsic delay increase with temperature. It can be seen that intrinsic delay increases

238

2006 Asia and South Pacific Design Automation Conference

2D-4

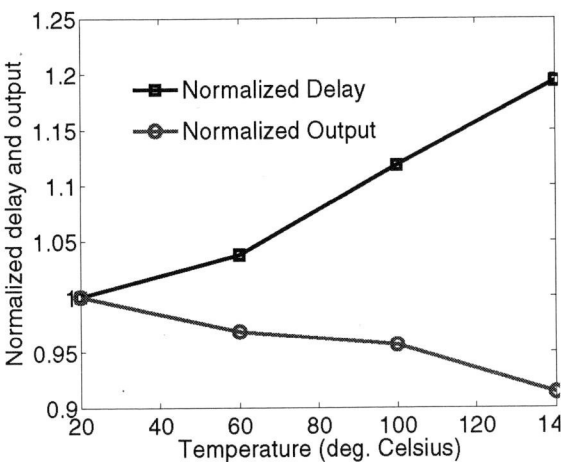

Fig. 3: Normalized delay and output swing of 3 NOR stage ring oscillator with temperature.

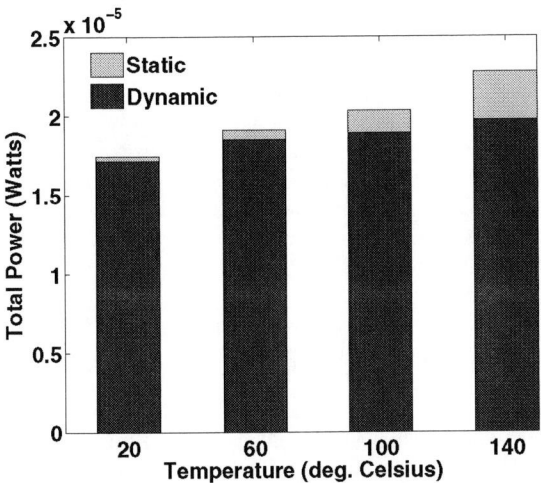

Fig. 4: Total power dissipation in NOR cell with temperature.

approximately 20% for 120°C rise in cell temperature. This increase in intrinsic delay coupled with increase in Elmore delay of interconnects because of increased electrical resistivity can pose serious challenges to circuit designers demanding more rigorous timing analysis. One more observation is the decrease in output swing (Fig. 3) because of increase in sub-threshold slope and reduced I_{on}/I_{off} ratio. Output swing decreases 9% for 120°C rise in temperature. This will result in reduced noise margins in cascaded logic gates at high temperatures.

B. Power dependence on temperature

Power dissipation in a circuit depends on its operating mode and can be given by

$$P_{total} = P_{static} + P_{dynamic} \quad (2)$$

Static power dissipation (P_{static}) is due to the leakage currents – sub-threshold leakage, gate direct tunneling leakage and reverse-biased junction band-to-band tunneling leakage – in off-state of a transistor [13]. Sub-threshold

leakage increases exponentially with the decrease in threshold voltage and is given by,

$$I_{sub} = I_o e^{q(V_{gs}-V_t)/mk_BT} (1 - e^{-qV_{ds}/k_BT}) \quad (3)$$

This results in 12X increase in static leakage for the temperature rise of 120°C in a NOR gate (Fig. 4).
Dynamic power dissipation ($P_{dynamic}$) is mainly due to:

(1) charging/discharging of the capacitances at the output and internal nodes of a cell ($P_{cap,dyn}$).

(2) short-circuit power due to direct current path between the power supply and ground ($P_{sc,dyn}$).

The switching of the capacitances at the output and internal nodes of a cell depends on the input pattern. Average dynamic power dissipation can be obtained by considering the average number of transitions per clock cycle. The dynamic power can be given by,

$$P_{cap,dyn} = 0.5 \times V_{DD}^2 \times f \times \left[(C_{load} \times \alpha_{out}) + \sum (C_i \times \alpha_i) \right] \quad (4)$$

where, V_{DD} is the supply voltage and f is the clock frequency. C_{load} includes the output load capacitance and the device capacitances of transistors connected at the output node and α_{out} is the average number of output transitions in a clock cycle. C_i is the internal node capacitance of the ith node and α_i is the average number of switching transitions per clock cycle at the ith node.

Short-circuit power is due to direct path of current flow between the power supply and ground during output transition. Because of the non-zero rise/fall times of an input signal, pull-down and pull-up networks can be simultaneously conducting in a CMOS circuit resulting in a direct current path between power supply and ground. With the increase in temperature, sub-threshold slope of the transistors increases and I_{on}/I_{off} ratio decreases. This results in increased short circuit power. Fig. 4 shows the dynamic and static power dissipation in a NOR gate. Dynamic power dissipation is calculated by integrating the currents in a 3-stage ring oscillator. It can be seen that dynamic power dissipation slightly increases with temperature mainly because of increase in short-circuit power.

III. Compact Thermal Models

A. Detailed Component Model

Detailed computational models for three cell-level components – NAND, NOR and INV are created based on Fourier theory. A typical NOR gate structure is shown in Fig. 5, including two PMOS and two NMOS FinFETs at the 28 nm node. The gates and fins are also shown. The simulation includes portions of the metallic contacts, but not all the metallization layers. These structures are enclosed in a domain of size 504 nm x 300 nm x 539 nm and are located on the y=150 nm plane. The rest of the interior volume is filled with SiO₂. In this orientation, the metallization layers would lie above the y=300 nm boundary, while the heat sink would lie below the y=0 boundary. Other details for the three components are shown in Table 1. The gate material is assumed to be aluminum, as are the metallic contacts, and

239

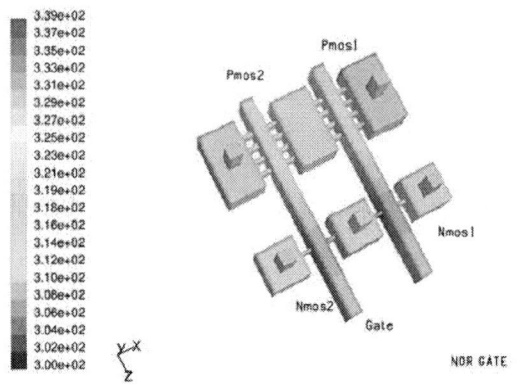

Fig. 5: Temperature field in NOR gate on the y=150 nm plane. Two PMOS and two NMOS FinFETs at the 28 nm technology node are shown. Heat generation in the fingers causes a temperature rise of 39°C above ambient.

standard temperature-independent thermal properties for bulk Si, SiO$_2$ and aluminum are assumed.

The volumetric heat generation due to self-heating is computed using the TAURUS device simulator [12]. The electrical simulation is 2D, in the x-z plane in Fig. 5, and yields the heat source **J.E** (x, z) for a typical fin in the PMOS or NMOS device. Thermal transport in the geometry is necessarily 3D with the primary direction for heat transfer being perpendicular to the x-z plane. The spatial (x, z) distribution of the heat generation is assumed to hold throughout the y-depth of the fins. The FLUENT CFD software [14] is used to compute the temperature field using the unstructured mesh technique described in [15].

B. Compact Model Generation

The generation of the compact model exploits the fact that the Fourier conduction equation with constant thermal properties is a linear elliptic boundary value problem. Thus, the temperature at any location (x_1, y_1, z_1), or indeed, the

average temperature of the domain, can be uniquely written as:

$$T\left(x_1,y_1,z_1\right)=\sum_{f=1}^{6}a_f\left(x_1,y_1,z_1\right)T_f+a_0\left(x_1,y_1,z_1\right)\alpha q \quad (5)$$

where the coefficients a_f are unique to the spatial location (x_1, y_1, z_1), and determine the influence of the six boundary temperatures of the cuboidal domain on $T(x_1, y_1, z_1)$. The coefficient a_0 quantifies the influence of the heat generation rate q for a given activity α. The coefficients a_f have the property:

$$\sum_{f=1}^{6}a_f=1 \quad (6)$$

By the same token, the heat transfer rates q_{bj} out of each of the six boundaries of the domain may also be written as:

$$q_{bj}=\sum_{f=1}^{6}b_{fj}T_f+b_{0j}\alpha q \quad j=1,2,...6 \quad (7)$$

For a given geometry, materials properties, and spatial pattern of heat generation, the coefficients a_f, a_0, b_{fj} and b_{0j} are uniquely determined, and constitute the compact model of the NOR, NAND or INV component. Seven temperature calculations are done using seven different sets of boundary temperatures to obtain a total of seven coefficient values (six a_f's and a_0). The same seven runs are also sufficient to determine seven values (b_{fj} and b_{0j}) for each of the six boundary face heat transfer rates q_{bj}, j=1,2,...6. A typical compact model for a NOR gate is present in Table 2. Here, the average component temperature is related to the six boundary temperatures of the computational domain and the heat generation rate; similarly, the heat transfer rates at the six boundary faces. This compact model will recover *exactly the same* average temperature and boundary heat transfer as the detailed model for any set of Dirichlet boundary conditions on the boundaries.

Table I: Parameters for Thermal Simulations

Comp-onent	Devices	Domain Size	Mesh Size (Hexah-edra)	Number of Fins	Heat Generation Per Fin (W)	Net Heat Generation in Device (W)
NOR	2 PMOS, 2 NMOS	504 nm x 300 nm x 539 nm	267,376	PMOS1 : 4 PMOS2: 4 NMOS1: 1 NMOS2: 1	PMOS1 : 4.4399x10^{-6} PMOS2: 5.211x10^{-6} NMOS1: 8.2332x10^{-5} NMOS2: 8.2332x10^{-5}	2.0326x10^{-4}
NAND	2 NMOS 2 PMOS	504 nm x 300 nm x 406 nm	146,260	PMOS1 : 2 PMOS2: 2 NMOS1: 2 NMOS2: 2	PMOS1 : 3.1178x10^{-5} PMOS2: 2.1156x10^{-5} NMOS1: 8.4242x10^{-6} NMOS2: 8.4242x10^{-6}	1.3836x10^{-4}
INV	1 PMOS 1 NMOS	336 nm x 300 nm x 399 nm	228,214	PMOS1 : 2 NMOS1: 1	PMOS1 : 9.496x10^{-7} NMOS1: 3.4595x10^{-6}	5.3589x10^{-6}

Table II: Compact thermal model for NOR gate

$T_{avg} = 0.1258 T_{left} + 0.1256 T_{right} + 0.2326 T_{bottom} + 0.2747 T_{top} + 0.1228 T_{back} + 0.1185 T_{front} + 3.5039 \times 10^4 \alpha q$
$q_{left} = 7.2206 \times 10^{-6} T_{left} - 6.8573 \times 10^{-8} T_{right} - 2.2603 \times 10^{-6} T_{bottom} - 2.5307 \times 10^{-6} T_{top} - 1.1725 \times 10^{-6} T_{back}$ $-1.1885 \times 10^{-6} T_{front} - 0.1141 \, \alpha q$
$q_{right} = -6.8573 \times 10^{-8} T_{left} + 7.2188 \times 10^{-6} T_{right} - 2.2603 \times 10^{-6} T_{bottom} - 2.5301 \times 10^{-6} T_{top} - 1.1719 \times 10^{-6} T_{back}$ $-1.1879 \times 10^{-6} T_{front} - 0.1119 \, \alpha q$
$q_{bottom} = -2.2603 \times 10^{-6} T_{left} - 2.2603 \times 10^{-6} T_{right} + 9.2783 \times 10^{-6} T_{bottom} - 5.0752 \times 10^{-7} T_{top} - 2.0967 \times 10^{-6} T_{back}$ $-2.1534 \times 10^{-6} T_{front} - 0.1168 \, \alpha q$
$q_{top} = -2.5307 \times 10^{-6} T_{left} - 2.5301 \times 10^{-6} T_{right} - 5.0752 \times 10^{-7} T_{bottom} + 1.0289 \times 10^{-5} T_{top} - 2.3629 \times 10^{-6} T_{back}$ $-2.3574 \times 10^{-6} T_{front} - 0.3877 \, \alpha q$
$q_{back} = -1.1725 \times 10^{-6} T_{left} - 1.1719 \times 10^{-6} T_{right} - 2.0967 \times 10^{-6} T_{bottom} - 2.3629 \times 10^{-6} T_{top} + 6.9145 \times 10^{-6} T_{back}$ $-1.1047 \times 10^{-7} T_{front} - 0.1264 \, \alpha q$
$q_{front} = -1.1885 \times 10^{-6} T_{left} - 1.1879 \times 10^{-6} T_{right} - 2.1534 \times 10^{-6} T_{bottom} - 2.3574 \times 10^{-6} T_{top} - 1.1047 \times 10^{-7} T_{back}$ $+6.9977 \times 10^{-6} T_{front} - 0.1430 \, \alpha q$
Left (x=0); right (x=x_{max}); bottom(y=0); top (y=y_{max}); back (z=0); front (z=z_{max})

IV. Results and Discussions

Once the average temperature and boundary heat transfer dependence of a logic cell on the boundary temperatures of the cuboid are known, the next step is cell placement and the generation of the thermal map of a circuit block. Though the primary heat flow direction is perpendicular to the plane in which the logic cells are placed, there is sufficient in-plane thermal non-uniformity that the thermal transport in all three coordinate directions must be considered. The logic cells are arranged in planar mesh of cuboidal cells, with those cells not containing logic elements assumed to contain SiO_2. Convective heat transfer boundary conditions are posed on all external boundaries, with the heat transfer coefficients being chosen to correctly model the thermal resistance due to metallization layers, the wafer, as well as lateral losses to other circuit blocks. By enforcing continuity of heat transfer rate at logic cell cuboid faces, equations for the face temperatures are found. These, in turn, are used to evaluate the average logic cell temperature. Heat generation and hence local temperature rise in a cell depends on the input data pattern. Therefore, to obtain the worst cast temperature map, an exhaustive set of test patterns need to be applied. However, to reduce the computational complexity, we obtain the average switching activity of each cell in a circuit block for a large set of random input patterns.

Fig. 6 shows the dynamic power and temperature distribution in the layouts of *alu4* and *x3* (MCNC'91 [16]) benchmark circuits. Circuits have been modified to use standard cells. Floorplans are generated to arrange cells in a planar grid of 30x30. Each cell volume is either occupied by a standard cell – NAND, NOR, INV – or filled with SiO_2 insulator. The maximum temperature difference inside the circuit blocks is 20°C. It can be seen that in FinFET based circuits, the temperature distribution closely corresponds to the dynamic power map (Fig.6), unlike bulk MOSFETs where hot spot region distributes over several cells. This is attributed to the confined silicon channels and lack of common high conductivity bulk silicon under the device layer.

In FinFET circuits, lateral heat flow is mainly through interconnects because of the lack of common low thermal resistivity substrate. This can result in large temperature differences between the neighboring cells. Power and ground lines shared by the adjacent cells help in heat flow and temperature distribution to some extent. Because of high inter-cell temperature gradients, it's imperative to perform temperature-dependent timing analysis at the granularity of gate level in FinFET circuits. With this estimation, some circuit level techniques, such as sizing, can be employed to prevent the failures.

V. Conclusions

In this work, we propose gate level compact thermal models for estimating temperature rise in FinFET circuits. The effect of low thermal conductivity buried oxide, confined ultra-thin channel and thermal boundary resistances of different materials' on temperature rise is taken into account. The floorplans of benchmark circuits show close correspondence between the temperature maps and dynamic power maps. This can be attributed to the confined channels where heat is generated and lack of high thermal conductivity material (bulk silicon) under the device layer impeding the lateral heat distribution. Results show that 120°C rise in temperature can result in 20% increase in intrinsic delay of a 2-input NOR gate and 12X increase in leakage power dissipation. The proposed thermal models can be used in estimating temperature rise in early phase of design cycle for proper timing analysis. The models can also be integrated in floorplanning algorithms to remove hot spots.

References

[1] C.-H. Tsai et al., "Standard Cell Placement for Even On-Chip Thermal Distribution," *Proc. of Int. Sym. on Phys. Design*, 1999, pp. 179-184.

[2] D. Brooks et al., "Dynamic Thermal Management for High-Performance Microprocessors," *Proc. 7th Int. Sym. High-Perf. Comp. Arch.*, 2001, pp. 171-182.

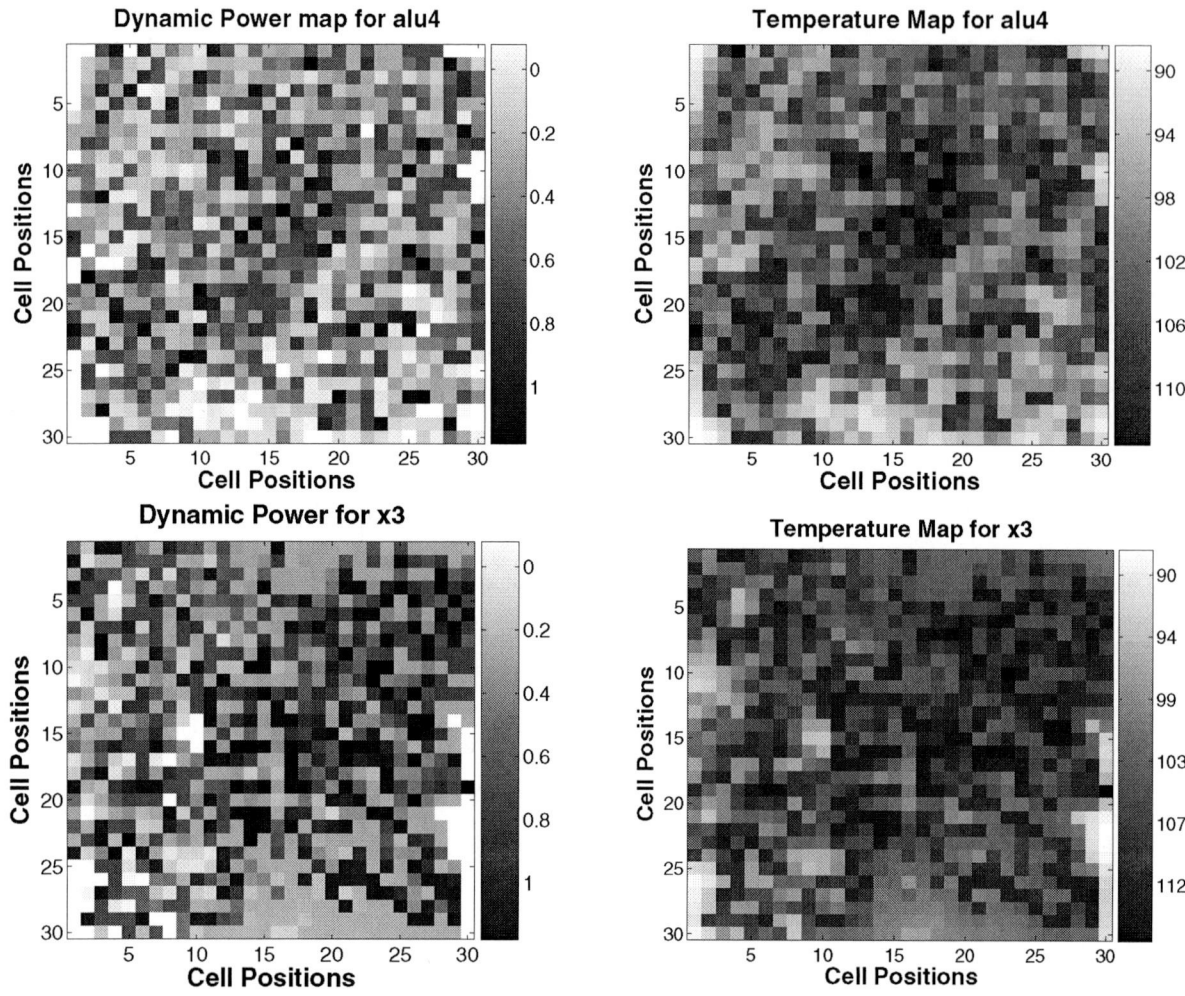

Fig. 6 Dynamic power and temperature maps for *alu4* and *x3* benchmark circuits. x-y axes represent the positions of standard cells. Dynamic power is normalized to maximum, where zero value corresponds to empty cells filled with oxide.

[3] K. Skadron et al., "Temperature-Aware Microarchitecture: Modeling and Implementation" *ACM Transactions on Architecture and Code Optimization*, Vol. 1, No. 1, March 2004, pp. 94-125.

[4] W. R. Daasch et al., "Design of VLSI CMOS Circuits Under Thermal Constraint," *IEEE Trans. on Ckts. and Sys. – II*, Aug. 2002, pp. 589-593.

[5] W. Huang et al., "Compact Thermal Modeling for Temperature-Aware Design," *DAC*, 2004, pp. 878-883.

[6] W. Liao et al., "Temperature and Supply Voltage Aware Performance and Power Modeling at Microarchitecture Level," *IEEE Trans. on Computer-Aided Design of Integrated Circuit and Systems*, Vol. 24, No. 7, July 2005.

[7] M. Berger et al., "Estimation of Heat Transfer in SOI MOSFET's," *IEEE Trans. on Elec. Dev.*, 1991, pp. 871-875.

[8] E. Pop et al., "Thermal Analysis of Ultra-thin Body Device Scaling," *IEDM*, 2003, pp. 883-886.

[9] E. Pop et al., "Thermal Phenomena in Nanoscale Transistors," *ITherm*, 2004, pp. 1-7

[10] L. T. Su et al., "Measurement and Modeling of Self-Heating in SOI NMOSFETs," *IEEE Trans. on Elec. Dev*, 1994, pp. 69-75.

[11] The International Technology Roadmap for Semiconductors, *Semiconductor Industry Assoc*, 2004update.

[12] Taurus Device Simulator v2004.09, *Synopsys Inc.*

[13] K. Roy et al., "Leakage Current Mechanisms and Leakage Reduction Techniques in Deep-Submicrometer CMOS Circuits," *Proc. of IEEE*, 2003, pp. 305-327.

[14] Fluent User Manual, Fluent Inc., Lebanon, NH03766, 2005.

[15] S. R. Mathur and J. Y. Murthy, *Advances in Numerical Heat Transfer*. Taylor and Francis, 2000, vol. 2, pp. 37-67.

[16] S. Yang, *Logic Synthesis and Optimization Benchmarks User Guide Version 3.0*. MCNC, Research Triangle Park, NC, Jan. 1991.

An Anytime Symmetry Detection Algorithm for ROBDDs

Neil Kettle

University of Kent, UK
e-mail: njk4@kent.ac.uk

Andy King

University of Kent, UK
e-mail: a.m.king@kent.ac.uk

Abstract— Detecting symmetries is crucial to logic synthesis, technology mapping, detecting function equivalence under unknown input correspondence, and ROBDD minimization. State-of-the-art is represented by Mishchenko's algorithm. In this paper we present an efficient anytime algorithm for detecting symmetries in Boolean functions represented as ROBDDs, that output pairs of symmetric variables until a prescribed time bound is exceeded. The algorithm is complete in that given sufficient time it is guaranteed to find all symmetric pairs. The complexity of this algorithm is in $O(n^4 + n|G| + |G|^3)$ where n is the number of variables and $|G|$ the number of nodes in the ROBDD, and it is thus competitive with Mishchenko's $O(|G|^3)$ algorithm in the worst-case since $n \ll |G|$. However, our algorithm performs significantly better because the anytime approach only requires lightweight data structure support and it offers unique opportunities for optimization.

I. INTRODUCTION

Symmetry detection has been important since the days of Shannon [1] who observed that symmetric functions have particularly efficient switch network implementations. Symmetry detection is no less important these days and knowledge of symmetric variables has many applications in logic synthesis [2,3], technology mapping [4,5], ROBDD minimization [6,7] and detecting equivalence of Boolean functions for which input correspondence is unknown [8,9].

The challenge in symmetry detection is to find efficient algorithms for detecting all symmetric variables pairs (x_i, x_j) of a given Boolean function $f(x_1 \ldots x_n)$, that is, find all pairs (x_i, x_j) such that $f(\ldots, x_i, \ldots, x_j, \ldots) = f(\ldots, x_j, \ldots, x_i, \ldots)$. The intuition being that f remains unchanged under the switching of the variables x_i and x_j. This symmetry is formally known as the first-order classical symmetry, or the non-skew non-equivalence symmetry [10]. It can be shown from Boole's expansion theorem [11] this is equivalent to checking equality of the co-factor pair $f|_{x_i \leftarrow 0, x_j \leftarrow 1} = f|_{x_i \leftarrow 1, x_j \leftarrow 0}$. This formulation shows that it is possible to find the set of all symmetric pairs by calling the co-factoring operation no more than $n^2 - n$ times, where n is the number of variables. Early work on detecting symmetric variables in Boolean functions has focussed on the computation of these co-factor pairs and symmetry detected by checking their equivalence [12]. The use of ROBDDs to represent Boolean functions enables co-factor equivalence to be checked in constant time, however, repeated co-factoring involves the creation and deletion of many intermediate ROBDD nodes and for very large ROBDDs this overhead can be prohibitive. This method is often referred to as the naïve method [12]. Möller, Mohnke and Weber [12] thus advocate the use of two preprocessing algorithms — two sieves — that detect pairs of asymmetric variables. These linear-time sieves significantly reduce the number of co-factor pairs that need to be computed. In general, however, the method still requires naïve co-factor computation, that is, calls to the standard co-factoring algorithm the complexity of which is in $O(|G| \lg |G|)$ [13]. Methods that rely on asymmetry sieves, such as those proposed in [7,12], are said to be based upon the so-called *negative-thinking* paradigm [14]. That is, they obtain the symmetric variable pairs from the set of all variables pairs by systematically removing all asymmetric variable pairs.

Because of the cost of repeated co-factoring, many symmetry detection methods endeavor to avoid naïve co-factor computation. Möller *et al.* [12] and Panda *el al.* [6] detect all symmetries between variables adjacent in the variable order with an algorithm in $O(|G|)$. Rudell's dynamic variable reordering algorithm [15] has also been used to detect symmetries, although the aim is not symmetry detection *per se*, but ROBDD minimization. Rudell's algorithm considers each variable in turn moving it up and down in the variable ordering (subject to complexity limits) so as to minimize the ROBDD. Panda *et al.* [6] modify Rudell's algorithm to detect symmetries between variables that become adjacent when one of the variables is repositioned in the ROBDD variable ordering. Symmetric variables are then grouped, and any subsequent reordering that is applied is required to preserve a contiguous variable ordering within each group. This approach to symmetry detection does not require naïve co-factor computation, but there is no guarantee that all symmetries will be found. State-of-the-art is represented by Mishchenko's algorithm [14] that detects all symmetric variable pairs in a ROBDD in $O(|G|^3)$. (Note that this algorithm is parameterized by the underlying set representation that is used to store the variable pairs, and therefore this complexity result does not consider the complexity of the set operations themselves. Most conservatively, assuming all set operations are linear, the overall running time is at least $O(n^2|G|^3)$ since each set contains potentially $O(n^2)$ elements). Algorithms such as those of Mishchenko [14] and Panda *et al.* [6] are based on the so-called *positive-thinking* paradigm [14]. That is they compute variable pairs that are symmetric, and in the case of Mishchenko's algorithm, because of its completeness, those pairs not found to be symmetric are then known to be asymmetric.

The problem with existing symmetry detection methods is that they are either monolithic, inefficient, or incomplete. A monolithic algorithm has to be run to completion before it can return any answer; the value of such an algorithm is compromised if the running time is prohibitive. Mishchenko's [14] algorithm falls into this class. Practically all engineering tasks (and logic synthesis is no exception) require an acceptable answer to be found in a reasonable amount of time rather than the optimal answer in an exorbitant amount of time. This is relevant in the context of symmetry detection because the running time of the state-of-the-art algorithm [14] can exceed 12 hours on some ROBDDs of less than a million nodes (actually this was benchmark `simpl2`). This motivates the need for a so-called anytime algorithm that will incrementally detect pairs of symmetric variables until some given time bound is exceeded. Symmetry detection algorithms [7,12] based on naïve co-factor computation can be considered to be incremental but, alas, this approach is inefficient. The algorithm of Panda *et al.* [6] is an interesting example of an incremental algorithm that does not require co-factor computation but, unfortunately, the algorithm is incomplete for the purposes of symmetry detection.

In this paper we present a novel anytime algorithm for symmetry detection based on the negative-thinking paradigm, whose efficiency compares very favorably against that of Mishchenko in the case when all the pairs require to be enumerated. The algorithm demonstrates that, with careful construction, it is possible to detect symmetries incrementally without compromising efficiency. Our anytime algorithm is inspired by that of Mishchenko, but the correctness of our algorithm is surprisingly subtle in that it depends on paths not passing through given nodes in the ROBDD. For pedagogical purposes, two versions of the algorithm are presented: a simple version that contains a minimal number of components to ensure correctness; and a refined version that demonstrates how an incremental algorithm has computational advantages over a comparable monolithic algorithm. These two algorithms are respectively presented in Sections III and IV. An intriguing aspect of the anytime approach is that it permits transitivity to be fully exploited. It is well-known that if $(x_i, x_j), (x_j, x_k)$ are symmetric then so is (x_i, x_k) [16, 17], but this observation has had scant consideration in the symmetry detection literature. Möller *et al.* [12, p 681] state that "we also use the fact that if $\{x_i, x_j\}$ and $\{x_j, x_k\}$ are pairs of symmetric variables, then $\{x_i, x_k\}$ is a pair of symmetric variables as well", seemingly missing the fact that if (x_i, x_j) are symmetric and (x_i, x_k) are asymmetric then (x_j, x_k) are asymmetric. Due to the way our anytime algorithm decomposes symmetry detection into a series of passes, one for each variable, we are free to apply asymmetry/symmetry propagation between each of these passes to reduce the expected cost of each pass. This is discussed in Section IV-C. Sections IV-A and IV-B show how the algorithm can be accelerated using more well-known techniques that relate to adjacent symmetries [12] and positive satisfy counts [8]. Extensive experimental results that are given in Section V demonstrate the value of these refinements, compare the algorithm against that Mishchenko and demonstrate the anytime nature of the algorithm. The remainder of this paper is organized as follows: Section II presents definitions used within the paper and Section VI presents the concluding discussion. For clarity, we summarize our contributions as follows:

- The paper presents a novel incremental, anytime algorithm for symmetry detection based on the negative-thinking paradigm.

- In theory, the algorithm is in $O(n^4 + n|G| + |G|^3)$ where $n \ll |G|$ (even considering the complexity of all set operations) which compares favorably against state-of-the-art [14].

- The paper shows that an anytime algorithm can put low computational demands on the underlying data-structures that represent pairs of symmetric variables. Thus anytime generality does not have to sacrifice efficiency, indeed the converse is true.

- The paper explains how an incremental anytime approach offers special opportunities for optimization, in that classical assymetry/symmetry sieves can precede the algorithm and assymetry/symmetry propagation techniques can be inserted into the main loop of the algorithm.

- The paper also reports a hitherto overlooked subtlety of symmetry detection: it seems that at least $O(n|G|)$ preprocessing steps must be performed before incremental symmetry detection may commence. Rather surprisingly, the correctness of our algorithm critically depends upon an $O(n|G|)$ asymmetry sieve [12], that relates to paths that can arise within an ROBDD in the presence of symmetries. (As a consequence, we conjecture that there is no way to construct an incremental, complete symmetry detection algorithm without first applying preprocessing).

II. PRELIMINARIES

In this paper we consider completely specified Boolean functions $f : \{0, 1\}^n \to \{0, 1\}$ that are conventionally written as Boolean formulae defined over a variable set $X = \{x_1, \ldots, x_n\}$. The satisfy-count of an n-ary Boolean function f is defined as $\|f\| = |\{(b_1, \ldots, b_n) \mid f(b_1, \ldots, b_n) = 1\}|$ [13]. The (Shannon) co-factor of a function f w.r.t a variable x_i and a Boolean constant b is defined by $f|_{x_i \leftarrow b} = f(x_1, \ldots, x_{i-1}, b, x_{i+1}, \ldots, x_n)$. Multiple variable co-factors can be defined inductively as $f_0 = f$, $f_i = f_{i-1}|_{x_i \leftarrow b_i}$ and $f|_{x_1 \leftarrow b_1, \ldots, x_n \leftarrow b_n} = f_n$. A function f over X is symmetric in a pair of variables (x_i, x_j) iff $f|_{x_i \leftarrow 0, x_j \leftarrow 1} = f|_{x_i \leftarrow 1, x_j \leftarrow 0}$, otherwise it is asymmetric in (x_i, x_j).

ROBDDs are obtained by inducing a total-order on X. A BDD is a rooted directed acyclic graph where each internal node is labeled with a Boolean variable. Each internal node has one successor node connected via an edge labeled 0, and another successor connected via an edge labeled 1. Each external (leaf) node is either 0 or 1. The Boolean function represented by a BDD can be evaluated for a given variable assignment by traversing the graph from the root, taking the 1 edge at a node when the variable is assigned to 1 and the 0 edge when the variable is assigned to 0. The external node reached in this traversal indicates the value of the Boolean function for the assignment. An OBDD is a BDD with the restriction that the label of a node is always less than the label of any internal node reachable via its successors. An ROBDD is an OBDD with the additional constraint that the successors of any internal node do not represent the same Boolean function. Note that any internal node of an ROBDD is itself the root of an ROBDD. An ROBDD f is symmetric in a pair of variables (x_i, x_j) iff the Boolean function it represents is symmetric in (x_i, x_j). Finally, let $|G|$ denote the number of internal nodes in a ROBDD G.

III. ANYTIME SYMMETRY DETECTION ALGORITHM

In this section we propose a novel, anytime approach to symmetry detection. The algorithm presented in Algorithm 1 contains the minimum number of components required so as to ensure correctness. The algorithm takes as input an ROBDD f and returns the set S of symmetric variable pairs. The algorithm is composed of two distinct procedures. ComputeAsymmetry(f) performs two depth-first search (dfs) traversals over the ROBDD f, to detect pairs of variables that are asymmetric (in the particular sense that is described in Section III-A). RemoveAsymmetry(f, i, C) filters a set of variables C whose symmetry relationship with variable x_i is unknown to return the set $C' \subseteq C$ of variables that are symmetric with x_i (this procedure is detailed in Section III-B).

Algorithm 1 ComputeSymmetricPairs(f)

$A \leftarrow$ ComputeAsymmetry(f)
$S \leftarrow \emptyset$
for $i = 1$ **to** $n - 1$ **do**
$\quad C \leftarrow \{ j \mid (i, j) \notin (S \cup A) \wedge i < j \}$
$\quad D \leftarrow$ RemoveAsymmetry(f, i, C)
$\quad S \leftarrow S \cup \{(i, k), (k, i) \mid k \in D\}$
$\quad A \leftarrow A \cup \{(i, l), (l, i) \mid l \in C \setminus D\}$
return S

The call to ComputeAsymmetry initializes the set of asymmetric variable pairs A; S is initially empty. The remainder of the algorithm considers each of the n variables in turn. Firstly, a set C is constructed that contains all variables whose symmetry relation with x_i has not yet been ascertained. Secondly, the set of symmetric variables D returned from RemoveAsymmetry is used to extend S and A. Observe that the sets S and A can be augmented in $O(n)$ time when C and D are represented as arrays. Furthermore, observe that C can be constructed in $O(n)$ time when the sets of pairs S and A are represented as adjacency matrices. Finally, observe that actually only $n - 1$ iterations of the loop are required because of the structure of C. Further details of these two procedures are given in Sections III-A and III-B.

A. Computing Asymmetries

The algorithm that initializes A is constructed from lemmas that detail how symmetric variables place structural constraints on ROBDDs [12]. For completeness, we state these lemmas below:

Lemma 1. *If an ROBDD f over a set of variables $X = \{x_1, \ldots, x_n\}$ is symmetric in the pair (x_i, x_j) and $i < j$, then every ROBDD rooted at a node labeled x_i must contain a node labeled x_j.*

Lemma 2. *If an ROBDD f over a set of variables $X = \{x_1, \ldots, x_n\}$ is symmetric in the pair (x_i, x_j) and $i < j$, then every path from the root of f to a node labeled x_j must visit a node labeled x_i.*

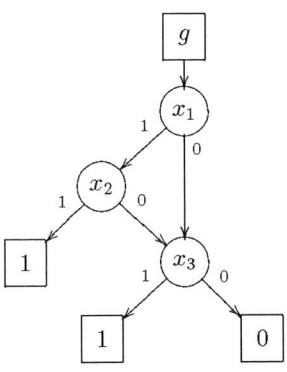

Fig. 1. The ROBDD g for the propositional formula $(x_1 \wedge x_2) \vee x_3$

Lemma 1 and Lemma 2 provide two conditions under which asymmetry can be observed. For any given node labeled x_i we can compute the set of all variables x_j that appear in a ROBDD that is rooted at that node, and any variable not appearing in this set is necessarily asymmetric with x_i. Furthermore, for any given node labeled x_j, we can compute the set of all variables x_i that appear on *all* paths from the root of the ROBDD to the node, and any variable not appearing in this set is asymmetric with x_j. The asymmetry conditions of Lemma 1 and Lemma 2 can be checked in two dfs traversals of the ROBDD, each traversal taking $O(n|G|)$ time.

Each iteration of the loop in Algorithm 1 considers a variable x_i and forms the set C from those variables whose symmetry relationship with variable x_i is not yet known. The validity of this decomposition into multiple passes, is justified by the proposition which itself is a consequence of the following lemma [12]:

Lemma 3. *A Boolean function f over a set of variables $X = \{x_1, \ldots, x_n\}$ is symmetric in the pair (x_i, x_j) iff both cofactors $f|_{x_k \leftarrow 0}$ and $f|_{x_k \leftarrow 1}$ are symmetric in the pair (x_i, x_j).*

Proposition 1. *If an ROBDD f over a set of variables $X = \{x_1, \ldots, x_n\}$ is symmetric in the pair (x_i, x_j) and $i < j$ iff*

- *every ROBDD rooted at a node labeled x_i is symmetric in (x_i, x_j) and,*
- *every path from the root to a node labeled x_j passes through a node labeled x_i.*

Proof. The proposition follows by applying the lemma inductively on the variables $x_{i-1}, \ldots, x_1$, though for brevity we consider only the first inductive step. Consider an ROBDD g whose root node is labeled with x_{i-1}. There are four cases to consider. First, the roots of both co-factors $g|_{x_{i-1} \leftarrow 0}$ and $g|_{x_{i-1} \leftarrow 1}$ are labeled x_i. By Lemma 3, g is symmetric in (x_i, x_j) iff $g|_{x_{i-1} \leftarrow 0}$ and $g|_{x_{i-1} \leftarrow 1}$ are symmetric in (x_i, x_j). Observe that every path from the root of g to x_j passes through a node labeled x_i. Second, the root of $g|_{x_{i-1} \leftarrow 0}$ is labeled with x_i whereas $g|_{x_{i-1} \leftarrow 1}$ is not. Again, g is symmetric in (x_i, x_j) iff $g|_{x_{i-1} \leftarrow 0}$ and $g|_{x_{i-1} \leftarrow 1}$ are symmetric in (x_i, x_j). Observe $g|_{x_{i-1} \leftarrow 1}$ is symmetric in (x_i, x_j) iff $g|_{x_{i-1} \leftarrow 1}$ contains no node labeled x_j, or equivalently, every path from the root of g to x_j passes through a node labeled x_i. The third and fourth cases are respectively analogous and similar to the second. □

The proposition allows exhaustive checking to be decomposed into a series of passes; one pass for each variable x_i. The crucial point is that when the loop is entered, we have already removed all pairs of variables (x_i, x_j) such that there exists a path from the root to a node labeled x_j which does not pass through a node labeled x_i. Hence, for correctness, the body of the loop in Algorithm 2, must only check the first condition of the proposition. The counterexample given in Figure 1 illustrates the necessity of the second condition in the proposition, or put another way, it shows that correctness

is compromised if the preprocessing is omitted from the algorithm. Observe that in Figure 1 that the variable pair (x_2, x_3) is symmetric in the ROBDD rooted at x_2, however (x_2, x_3) are asymmetric in the ROBDD g since there exists a path from the root of g (x_1) to the node x_3 that does not visit a node labeled x_2. In fact, disabling the preprocessing gives the following asymmetry and symmetry sets A_i and S_i after i iterations of the loop: $A_0 = S_0 = \emptyset$, $A_1 = \{(x_1, x_3), (x_3, x_1)\}, S_1 = \{(x_1, x_2), (x_2, x_1)\}$, $A_2 = A_1, S_2 = S_1 \cup \{(x_2, x_3), (x_3, x_2)\}$. Observe the erroneous pair (x_2, x_3) contained within S_2.

B. Removing Asymmetries

After the initial preprocessing, incremental symmetry detection can commence. The procedure given below takes as input an ROBDD f, a variable index i, and a set C of variable indices corresponding to those variables whose symmetry relation with variable i is unknown.

Algorithm 2 RemoveAsymmetry(f, i, C)

---

if $C = \emptyset$ **then**
    return $\emptyset$
$j \leftarrow$ index(f)
if $j > i \vee f = $ **true** $\vee f = $ **false then**
    return C
if $j = i$ **then**
    return RemoveAsymmetryVar($f|_0, f|_1, C$)
else
    $C \leftarrow$ RemoveAsymmetry($f|_0, i, C$)
    return RemoveAsymmetry($f|_1, i, C$)

The function index(f) returns the index of the root node of f, that is, i if the root is labeled x_i. The test $j > i$ implements a form of early termination: if the test is satisfied then the ROBDD f can contain no node labeled x_i. The external nodes **true** and **false** also trigger early termination. At the heart of RemoveAsymmetry is a call to RemoveAsymmetryVar which encapsulates the logic to cofactor f_0 and f_1 so as to perform the symmetry check. The pseudocode for this procedure is given in Algorithm 3. Whenever the call RemoveAsymmetryVar is reached, it examines the co-factors of f to remove variables from C that are asymmetric w.r.t x_i. First, consider the case when both root nodes of f_0 and f_1 are labeled with the same variable x_j. In this case we compute $f_0|_{x_j \leftarrow 1}$ and $f_1|_{x_j \leftarrow 0}$ and check for equivalence. Second, when f_0 is labeled with x_j and f_1 is labeled with x_k where $j < k$, the check reduces to $f_0|_{x_j \leftarrow 1} = f_1$. Third, the $k < j$ case is analogous to the second. The recursive calls follow the co-factor check because it is necessary to check symmetry across all variable assignments. Note that both RemoveAsymmetry and RemoveAsymmetryVar terminate as soon as $C = \emptyset$.

When C is implemented as an array, the complexity of a single call to RemoveAsymmetryVar is $O(|G|^2)$. This follows since co-factor comparison and $C \setminus \{l\}$ are in $O(1)$, as is the test $C = \emptyset$ when C is augmented with a counter to record $|C|$. Overall, RemoveAsymmetryVar can only be invoked a total of $|G|$ times from within Algorithm 1, thus RemoveAsymmetryVar contributes $O(|G|^3)$ to the overall running time. The $n - 1$ calls to RemoveAsymmetry cumulatively cost $O(n|G|)$.

IV. Optimized Anytime Symmetry Detection Algorithm

In this section we propose a series of optimizations for Algorithm 1. This refined algorithm retains the incremental nature of the original algorithm, and in fact shows how this can be exploited by several optimizations. These optimizations seek to reduce the size of the set C, and hence the running time of the call

3A-1

Algorithm 3 RemoveAsymmetryVar(f_0, f_1, C)

> **if** $C = \emptyset$ **then**
>> **return** $\emptyset$
> **if** $(f_0 = \mathtt{true} \vee f_0 = \mathtt{false}) \wedge (f_1 = \mathtt{true} \vee f_1 = \mathtt{false})$ **then**
>> **return** C
> $j \leftarrow \mathtt{index}(f_0)$
> $k \leftarrow \mathtt{index}(f_1)$
> **if** $j = k$ **then**
>> $(l, f_{00}, f_{01}, f_{10}, f_{11}) \leftarrow (j, f_0|_{j\leftarrow 0}, f_0|_{j\leftarrow 1}, f_1|_{k\leftarrow 0}, f_1|_{k\leftarrow 1})$
> **else if** $j < k$ **then**
>> $(l, f_{00}, f_{01}, f_{10}, f_{11}) \leftarrow (j, f_0|_{j\leftarrow 0}, f_0|_{j\leftarrow 1}, f_1, f_1)$
> **else**
>> $(l, f_{00}, f_{01}, f_{10}, f_{11}) \leftarrow (k, f_0, f_0, f_1|_{k\leftarrow 0}, f_1|_{k\leftarrow 1})$
> **if** $f_{01} \neq f_{10}$ **then**
>> $C \leftarrow C \setminus \{l\}$
> $C \leftarrow \mathtt{RemoveAsymmetryVar}(f_{00}, f_{10}, C)$
> **return** $\mathtt{RemoveAsymmetryVar}(f_{01}, f_{11}, C)$

RemoveAsymmetry(f, i, C), by enriching the sets A and S on-the-fly before, and between, iterations of the main loop. The symmetry sieve algorithms presented by [7, 12] give a way to refine the sets A and S before the loop is entered. When the loop is entered, it is possible to take advantage of the transitivity of the symmetry relation to add further pairs to A and S. The optimized symmetry detection algorithm presented in Algorithm 4 takes as input an ROBDD f and returns the set S of symmetric variable pairs. The new algorithm includes three additional procedures, namely, ComputeSatisfyCounts(f), ComputeAdjSymmetry(f) and SymmetryClosure(A, S) which are detailed in Sections IV-A and IV-B, IV-C respectively.

Algorithm 4 OptimizedSymmetricPairs(f)

> $A \leftarrow \mathtt{ComputeAsymmetry}(f)$
> $M \leftarrow \mathtt{ComputeSatisfyCounts}(f)$
> **for** $i = 1$ **to** n **do**
>> **for** $j = i + 1$ **to** n **do**
>>> **if** $M(i) \neq M(j)$ **then**
>>>> $A \leftarrow A \cup \{(i, j), (j, i)\}$
> $S \leftarrow \mathtt{ComputeAdjSymmetry}(f)$
> **for** $i = 1$ **to** $n - 2$ **do**
>> $(A, S) \leftarrow \mathtt{SymmetryClosure}(A, S)$
>> $C \leftarrow \{j \mid (i, j) \notin (S \cup A) \wedge i + 1 < j\}$
>> $D \leftarrow \mathtt{RemoveAsymmetry}(f, i, C)$
>> $S \leftarrow S \cup \{(i, k), (k, i) \mid k \in D\}$
>> $A \leftarrow A \cup \{(i, l), (l, i) \mid l \in C \setminus D\}$
> **return** S

ComputeSatisfyCounts(f) returns a mapping M from variable indices to a natural number that can be used to distinguish pairs of asymmetric variables, that is, if $M(i) \neq M(j)$ then (x_i, x_j) are asymmetric. ComputeAdjSymmetry(f) returns the set of symmetric variable pairs for those pairs that are adjacent in the ROBDD ordering (which permits the number of loop iterations to be relaxed to $n - 2$). Finally, SymmetryClosure(A, S) takes as input two sets A and S of variable pairs known to be asymmetric and symmetric respectively. Two new sets $A' \supseteq A$ and $S' \supseteq S$ are output that are derived by exploiting the transitivity of symmetry.

A. Positive Satisfy-Counts

A consequence of symmetry, which can also be used to detect asymmetry [8], relates to the satisfy count of one positive co-factor of a variable to the satisfy count of another:

Lemma 4. *If a Boolean function f over a set of variables $X = \{x_1, \ldots, x_n\}$ is symmetric in the pair (x_i, x_j), then $\|f|_{x_i \leftarrow 1}\| = \|f|_{x_j \leftarrow 1}\|$.*

Computing the satisfy counts of all co-factors can be realized using a single dfs traversal of the ROBDD in $O(n|G|)$ time [8]. Finding the resultant asymmetries requires n^2 comparisons in Algorithm 4, and thus the overall complexity of this phase is $O(n^2 + n|G|)$.

B. Adjacent Symmetries

The following lemma details a special case of symmetry, which relates to variables that are adjacent in the ROBDD ordering:

Lemma 5. *If a ROBDD f over a set of variables $X = \{x_1, \ldots, x_n\}$ is symmetric in the pair (x_i, x_{i+1}) iff $g|_{x_i \leftarrow 0, x_{i+1} \leftarrow 1} = g|_{x_i \leftarrow 1, x_{i+1} \leftarrow 0}$ holds for each ROBDD g that is rooted at a node labeled x_i*

This lemma leads to an $O(|G|)$ time algorithm that can detect all symmetry and asymmetry relationships between adjacent variables [12]. (In fact the algorithm of Möller *et al.* can be improved to detect asymmetry for a pair of non-adjacent variables, that is, a pair (x_i, x_k) is asymmetric if there exists a node g labeled x_i with successor nodes labeled x_k and x_l where $i + 1 < k \leq l$ and $g|_{x_i \leftarrow 0, x_k \leftarrow 1} \neq g|_{x_i \leftarrow 1, x_k \leftarrow 0}$.)

C. Symmetry Propagation

The final lemma can be obtained by recalling that a function f remains unchanged under the switching of any symmetric variables:

Lemma 6. *If a Boolean function f over a set of variables $X = \{x_1, \ldots, x_n\}$ is symmetric in the pairs (x_i, x_j) and (x_j, x_k) then f is also symmetric in the pair (x_i, x_k).*

This transitivity result provides a way of enriching the set S, that is, if $(x_i, x_j), (x_j, x_k) \in S$ then it follows that (x_i, x_k) is also a symmetric pair. Further, given $(x_i, x_j) \in S, (x_i, x_k) \in A$ then it follows that the pair (x_j, x_k) is asymmetric, that is, A can possibly be enriched too. This follows since if (x_j, x_k) is symmetric then by the lemma it follows that (x_i, x_k) is symmetric, which is a contradiction. Adding those variable pairs to A and S which can be inferred through transitivity is not dissimilar to computing the transitive closure of a binary relation. This motivates adapting the Floyd-Warshall [18, 19] all-pairs-shortest-path algorithm to this task by representing the sets of pairs A and S as an adjacency matrix of n^2 size. The pseudo-code for this algorithm is given in Algorithm 5.

Algorithm 5 SymmetryClosure(A, S)

> **for** $i = 1$ **to** n **do**
>> **for** $j = i + 1$ **to** n **do**
>>> **for** $k = 1$ **to** n **do**
>>>> **if** $(k, i) \in S \wedge (k, j) \in S$ **then**
>>>>> $S \leftarrow S \cup \{(j, i), (i, j)\}$
>>>> **else if** $(k, i) \in A \wedge (k, j) \in S$ **then**
>>>>> $A \leftarrow A \cup \{(j, i), (i, j)\}$
>>>> **else if** $(k, i) \in S \wedge (k, j) \in A$ **then**
>>>>> $A \leftarrow A \cup \{(j, i), (i, j)\}$
> **return** (A, S)

The complexity of Algorithm 5 is in $O(n^3)$ since membership check and single element insertion can be performed in $O(1)$ time for an adjacency matrix representation. Note that although the worst-case running time is not dependent on the number of symmetries present, larger symmetry sets induce more propagation which reduces the overall running time.

TABLE I
EXPERIMENTAL RESULTS

Circuit	# In	# Out	Σ|G|	|S|	read	naïve	[14]	§ III	A	A+B	A+B+C
pair	173	137	118066	1910	0.20	132.46	6.62	2.37	2.18	2.16	2.08
s4863	153	104	126988	547	2.63	20.60	5.30	1.41	1.08	1.01	0.82
s9234.1	247	250	4434504	3454	20.14	>7200	1407.20	183.84	158.36	145.94	141.26
s38584.1	1464	1730	150554	15629	3.70	337.59	16.70	3.12	3.04	3.01	2.80
C880	60	26	600998	262	8.29	704.54	13.90	7.75	6.84	5.63	5.20
C3540	50	22	4618194	81	21.80	>7200	132.72	71.64	68.23	66.08	65.04
simp10	105	1	722074	19	58.45	>7200	661.70	65.28	47.53	43.90	40.88
simp12	117	1	758330	23	76.23	>7200	>7200	105.67	61.94	59.87	57.59
simp14	120	1	562326	36	70.38	>7200	1114.29	75.75	38.48	36.17	30.63
hom06	104	1	1176845	20	65.22	>7200	274.90	115.66	91.70	88.31	81.50
hom08	95	1	893312	16	56.48	>7200	135.79	67.79	54.99	50.89	49.00
hom10	130	1	309221	29	29.98	>7200	1510.32	35.85	33.39	31.61	31.21
ca004	53	1	782640	2	5.40	>7200	147.97	31.35	12.33	12.33	12.10
ca008	96	1	682617	16	20.40	>7200	326.92	53.54	44.69	43.05	42.78
ca016	107	1	861209	26	60.10	>7200	305.11	72.68	59.96	50.90	50.80
urquhart2_25	48	1	722657	5	3.06	>7200	70.50	26.22	20.23	20.21	17.95
urquhart3_25	62	1	1771025	24	6.22	>7200	>7200	82.98	81.14	76.97	72.80
urquhart4_25	68	1	1736705	27	5.96	>7200	>7200	83.44	81.84	76.48	72.02
rope_0002	54	1	634914	3	3.06	>7200	192.77	22.48	18.53	18.47	18.50
rope_0004	62	1	1052214	10	4.73	>7200	487.26	41.71	39.70	37.90	37.82
rope_0006	61	1	759039	13	3.14	>7200	657.74	35.78	30.76	30.64	30.68
ferry8	111	1	290127	30	78.35	>7200	95.15	30.10	29.56	23.21	22.99
ferry10	116	1	539419	38	88.08	>7200	1866.62	70.34	69.84	54.19	53.42
ferry12	123	1	277291	36	47.96	>7200	142.10	37.63	37.50	30.98	30.95
gripper10	125	1	393485	28	69.08	>7200	261.32	52.97	50.53	45.38	44.74
gripper12	129	1	667877	43	50.95	>7200	368.50	106.32	102.87	85.43	84.90
gripper14	118	1	767735	40	47.29	>7200	415.57	111.49	110.40	73.48	71.34

V. EXPERIMENTAL RESULTS

To assess the efficiency of the anytime approach, the algorithm, complete with all its refinements was implemented using the CUDD [20] Decision Diagram package. The rationale for this choice of library was that the Extra DD library [21], which implements Mishchenko's algorithm, also uses CUDD. Experiments were performed on an UltraSPARC IIIi 900MHz based system with 16GB RAM under the Solaris 9 Operating System. All programs — the CUDD package, the Extra library, and our algorithm — were compiled with the GNU C Compiler version 3.3.0 with -O3 enabled. The algorithms were run against a range of MCNC and ISCAS benchmark circuits of varying size [22], as well as several other benchmarks derived from the SAT literature. All timings were averaged over four runs and are given in seconds. Table I presents the results of these tests, the first four columns of Table I give the circuit name, number of inputs, outputs and the total number of nodes over all outputs respectively. Column five indicates the total number of all symmetric pairs found over each of the outputs of the circuit. Column six gives the time in seconds to read in the benchmark circuit and construct the ROBDD applying variable sifting. The remaining six columns give the runtimes required to compute all symmetric and asymmetric pairs. The first of these is the naïve method for computing all co-factor pairs. The second is Mishchenko's implementation of his own algorithm [21]. The third column is the unoptimized algorithm presented in Section III. The remaining three columns relate to the refinements presented in Section IV, that is, with the optimizations of Sections IV-A, IV-B and IV-C cumulatively enabled. The rationale for implementing the naïve method was to verify the implementation of our algorithm and Mishchenko's; the performance numbers are included to quantify the value of Mishchenko's algorithm. Note, that these figures present Mishchenko's algorithm in best light since when garbage collection is enabled the performance of Mishchenko's implementation can degrade, presumably because of its extensive use of ZDDs [23] to represent sets. For example, the circuit pair requires 33.40s compared to 6.62s with garbage collection disabled. Enabling garbage collection has no perceivable impact on our algorithm.

Figure 2 illustrates the outcome of some experiments designed to explore the anytime nature of the algorithm. In these experiments, the optimized algorithm was stopped after progressively larger timeouts were exceeded. The graphs display the number of symmetries found against these timeouts. Future work will investigate whether reordering the iterations in the main loop, for example, choosing i with the largest number of unknowns, increases the proportion of symmetries found early in the search.

VI. DISCUSSION

This paper presents a novel anytime symmetry detection algorithm, that is capable of detecting all symmetric variable pairs. The startling speed-ups over Mishchenko's algorithm stem from our use of a single static adjacency matrix rather than sets of pairs that are repeatedly generated. It is important to appreciate that there is no obvious way to re-engineer Mishchenko's algorithm to use a static adjacency matrix. This is because Mishchenko's algorithm is a bottom-up, divide and conquer algorithm that derives the solution to a problem by obtaining, and combining, the solutions to several sub-problems. Mishchenko [14, p 1590] points out that caching of the answers to these sub-problems is required to reduce the computational complexity from exponential to polynomial yet this requires multiple data structures to be maintained. By contrast, the anytime approach merely has to mark nodes as visited in any of the ROBDD traversals. Moreover, the only set operations that the anytime algorithm require are atomic $O(1)$ insertions and deletions, which finesses the otherwise $O(n^2)$ overhead of set intersection and union. This partly explains the speed of the anytime approach.

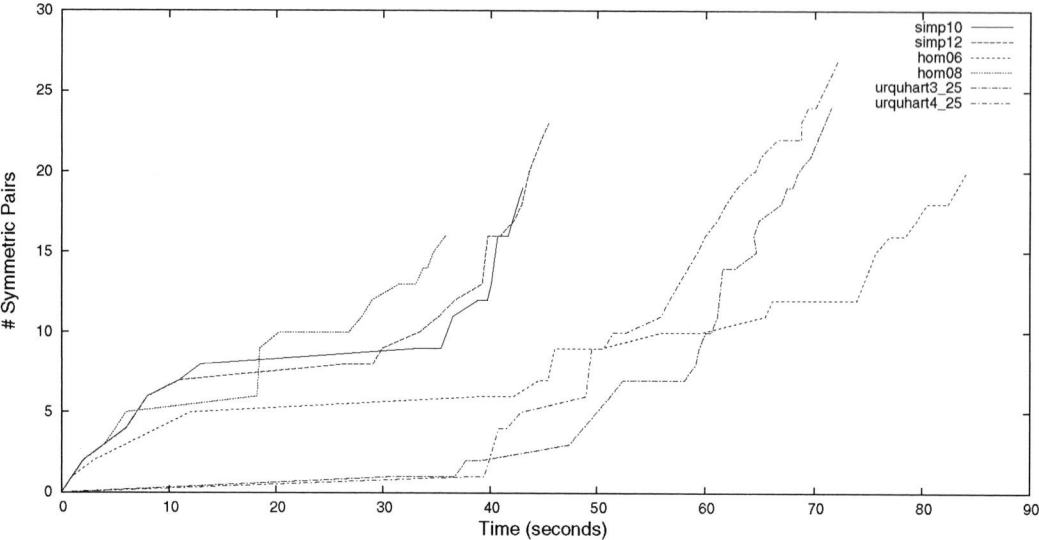

Fig. 2. Time against number of symmetries

Another source of speedup, in the anytime approach, is its amenability to optimization by enriching the A and S sets on-the-fly. One would think that computing the transitive closure is prohibitively expensive, but close inspection of the SPARC assembler revealed that the GNU compiler was able to generate very tight code from the regular structure of the closure algorithm.

Finally, Mishchenko's algorithm [14] is capable of detecting all four basic types of symmetry, namely, non-skew non-equivalence symmetry — the notion of symmetry considered in this paper — (NE), non-skew equivalence symmetry (E), skew non-equivalence symmetry (!NE) and the skew equivalence symmetry (!E). In this more general setting, a pair of variables are asymmetric if they do not satisfy any of these four symmetry types. A key component of our optimized algorithm, SymmetryClosure, can be straightforwardly generalized to infer these transitive symmetries by using a 4-bit encoding to indicate which symmetry types apply. A lookup-table of $(4^2) \times (4^2) = 256$ entries can then be used to obtain the transitive symmetry types without any impact on the asymptotic running time.

Acknowledgments We thank Arnaud Gotlieb and Peter Schachte for useful discussions. This work was supported, in part, by EPSRC Grant-EP/C015517 and British Council Grant-PN 05.021.

REFERENCES

[1] C. E. Shannon, "A Symbolic Analysis of Relay and Switching Circuits," *AIEE Trans.*, vol. 57, pp. 713–723, 1938.

[2] B. G. Kim and D. L. Dietmeyer, "Multilevel Logic Synthesis of Symmetric Switching Functions," *IEEE Trans. Computer-Aided Design*, vol. 10, no. 4, pp. 436–446, 1991.

[3] C. R. Edward and S. L. Hurst, "A Digital Synthesis Procedure Under Function Symmetries and Mapping Methods," *IEEE Trans. Comput.*, vol. C-27, no. 11, pp. 985–997, 1978.

[4] F. Mailhot and G. De Micheli, "Technology Mapping Using Boolean Matching and Don't Care Sets," in *European Design Automation Conference*, 1990, pp. 212–216.

[5] Y. T. Lai, S. Sastry, and M. Pedram, "Boolean Matching Using Binary Decision Diagrams with Applications to Logic Synthesis and Verification," in *International Conference on Computer-Aided Design*, 1992, pp. 452–458.

[6] S. Panda, F. Somenzi, and B. F. Plessier, "Symmetry Detection and Dynamic Variable Ordering of Decision Diagrams," in *International Conference on Computer-Aided Design*, 1994, pp. 628–631.

[7] C. Scholl, D. Möller, P. Molitor, and R. Drechsler, "BDD Minimization Using Symmetries," *IEEE Trans. Computer-Aided Design*, vol. 18, no. 2, pp. 81–100, 1999.

[8] J. Mohnke and S. Malik, "Permutation and Phase Independent Boolean Comparison," *INTEGRATION, The VLSI Journal*, vol. 16, pp. 109–129, 1993.

[9] D. I. Cheng and M. Marek Sadowska, "Verifying Equivalence of Functions with Unknown Input Correspondence," in *European Design Automation Conference*, 1993, pp. 272–277.

[10] V. N. Kravets and K. A. Sakallah, "Generalized Symmetries in Boolean Functions," in *International Conference on Computer-Aided Design*, 2000, pp. 526–532.

[11] G. D. Hachtel and F. Somenzi, *Logic Synthesis and Verification Algorithms.* Kluwer Academic Publishers, 1996.

[12] D. Möller, J. Mohnke, and M. Weber, "Detection of Symmetry of Boolean functions Represented by ROBDDs," in *International Conference on Computer-Aided Design*, 1993, pp. 680–684.

[13] R. E. Bryant, "Graph-based Algorithms for Boolean Function Manipulation," *IEEE Trans. Comput.*, vol. 35, no. 8, pp. 677–691, 1986.

[14] A. Mishchenko, "Fast Computation of Symmetries in Boolean Functions," *IEEE Trans. Computer-Aided Design*, vol. 22, no. 11, pp. 1588–1593, 2003.

[15] R. Rudell, "Dynamic Variable Ordering for Ordered Binary Decision Diagrams," in *International Conference on Computer-Aided Design*, 1993, pp. 42–47.

[16] D. Bochmann and B. Steinbach, *Logikenwurf mit XBOOLE: Algorithmen und Programme.* Berlin: Verlag Technik GmBH, 1996.

[17] C. C. Tsai and M. Marek Sadowska, "Generalized Reed-Muller Forms as a Tool to Detect Symmetries," *IEEE Trans. Comput.*, vol. 45, no. 1, pp. 33–40, 1996.

[18] R. W. Floyd, "Algorithm 97: Shortest Path," *Commun. ACM*, vol. 5, no. 6, p. 345, 1962.

[19] S. Warshall, "A Theorem on Boolean Matrices," *Journal of the ACM*, vol. 9, no. 1, pp. 11–12, 1962.

[20] F. Somenzi, "CUDD Package, Release 2.4.0." [Online]. Available: http://vlsi.colorado.edu/~fabio/CUDD/cuddIntro.html

[21] A. Mishchenko, "Extra Library of DD Procedures." [Online]. Available: http://www.ee.pdx.edu/~alanmi/research/extra.htm

[22] "Lgsynth93 Benchmark Set." [Online]. Available: http://www.bdd-portal.org/benchmarks/

[23] S. Minato, "Zero-Suppressed BDDs for Set Manipulation in Combinatorial Problems," in *Design Automation Conference*, 1993, pp. 272–277.

2006 Asia and South Pacific Design Automation Conference

High Level Equivalence Symmetric Input Identification

Ming-Hong Su   Chun-Yao Wang

Department of Computer Science
National Tsing Hua University, HsinChu, Taiwan, R.O.C

{harrysu, wcyao}@cs.nthu.edu.tw

Abstract — *Symmetric input identification is an important technique in logic synthesis. Previous approaches deal with this problem by building BDDs and developing algorithms to determine symmetric inputs. For the design whose corresponding BDDs cannot be built, BDD-based approaches cannot be applied on this problem. To avoid the limitations of BDD-based approaches, simulation-based methods have been proposed. It is applicable to designs described in arbitrary level, especially to high-level and black box designs. Previous simulation-based approaches focus on determining the inputs of nonequivalence symmetry. In this paper, we propose a simulation-based approach to identify equivalence symmetric inputs. The experimental results on a set of ISCAS-85 and MCNC benchmarks are also presented.*

I. INTRODUCTION

Symmetry input identification is to find the symmetric relation of all inputs. Symmetric input sets are the subsets of inputs. Grouping symmetric inputs to form symmetric input sets and thus any permutations of the inputs within a subset leave the function invariant. The problem to find the maximal symmetric inputs sets has been formulated in [3] [4]: Given a function $f(x)$, find maximal subsets of inputs $x_1, x_2,..., x_n \subseteq X$, such that $x_1 \cup x_2 \cup...\cup x_n = X$ and the inputs in every x_i can be permuted in any fashion without changing the functionality.

Method to finding the maximal symmetric input sets is based on finding all pairs of symmetric inputs and then take the unions of all the pairs having nonempty intersection. Previous approaches [2][3][6] are based on checking the equality of cofactors of the function. The maximal symmetric input sets could be computed after checking all symmetric pairs. However, those approaches are very time-consuming and not feasible for large functions. While the number of inputs is large, representing functions using BDDs will improve the efficiency of cofactor computation. A simple symmetry test is to check whether the BDD representations of two cofactor functions are isomorphic or not. This can be seen in Figure 1. However, computing multiple cofactor pairs is still expensive for large functions. This is because the repeated computation leads to create

----

This work was supported in part by the National Science Council of R.O.C. under Grant NSC94-2220-E-007-041.

and delete a large number of intermediate BDD nodes. Previous approaches based on BDDs and Boolean functions are summarized as follows. [6] avoids redundant cofactor computation by some criterions and thus speed up the computation process. An efficient algorithm without computing cofactors is proposed in [2]. [10] uses a K-Disjointness Paradigm which can compute disjoint situations with Hamming distance K between Boolean function to find the maximum symmetries. [11] formulates symmetry identification as an equation without using cofactor computation and equivalence checking.

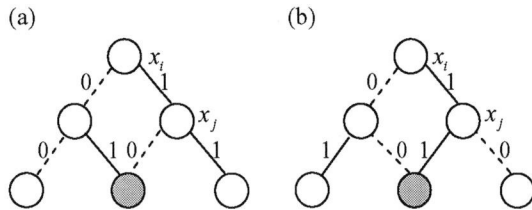

Figure 1: a) Nonequivalence Symmetry
b) Equivalence Symmetry

In addition to BDD-based and Boolean function-based methods, simulation-based approaches were applied to circuits without having compact BDDs or Boolean functions representation. [5] establishes two stages to accomplish the symmetric input identification and using a simulation-based method as the first stage of its two-stage algorithm.

Most of previous works focus on determining *nonequivalence* (NE) symmetry. For NE symmetry, symmetric pair is an equivalence relation. However, this is not true for *equivalence* (E) symmetry. For example, three inputs { x_i, x_j, x_k } could be distinguished as NE symmetric inputs by any two symmetric pairs are held because of transitivity. But for E symmetry, three pairs of inputs have to be symmetric simultaneously. Consequently, determining E symmetry needs more efforts than determining NE symmetry. This paper proposes a simulation-based approach to determining E symmetry.

The remainder of the paper is organized as follows. In the next section we briefly overview different types of symmetries and introduce the representation of maximal symmetric input sets. Our algorithm will be presented in Section 3. The experimental results of E symmetry are

0-7803-9451-8/06/$20.00 ©2006 IEEE.

249

presented in Section 4. Finally, Section 5 concludes the paper.

II. PRELIMINARIES

This section reviews different types of symmetries at first. Then, we introduce a representation for maximal symmetric input sets. Finally, we show the naïve approach to identifying symmetric inputs.

A. *Overview of Symmetries*

A *cofactor* of a function $f(x)$ with respect to variables x_i and x_j is the function resulting from the substitution into $f(x)$ of specific values for x_i and x_j. For example, the cofactor of $f(x)$ with respect to $x_i = 0$ and $x_j = 1$ is $f(x_1, ..., 0, ..., 1, ..., x_n)$, which is denoted as $f_{\bar{x}_i x_j}$ [8].

For any pairs of variables x_i and x_j, there are four cofactors, which are $f_{\bar{x}_i \bar{x}_j}$, $f_{\bar{x}_i x_j}$, $f_{x_i \bar{x}_j}$, and $f_{x_i x_j}$. Different categories of symmetries can be defined according to the equality of two cofactors among them. A function $f(x)$ exhibits a nonequivalence (NE) symmetry in inputs x_i and x_j, if $f_{x_i \bar{x}_j} = f_{\bar{x}_i x_j}$. When $f_{\bar{x}_i \bar{x}_j} = f_{x_i x_j}$, the function is said to exhibit equivalence (E) symmetry with respect to x_i and x_j. The illustrations of NE and E symmetries are shown in Figure 2 and Figure 3, respectively.

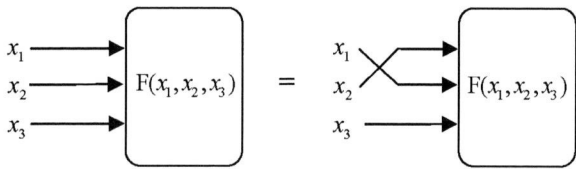

Figure 2: Illustration of nonequivalence symmetry

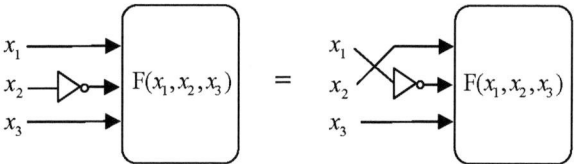

Figure 3: Illustration of equivalence symmetry

B. *Symmetric-ASymmetric Inputs (SASIs) Representation*

For an N-input circuit, there are C_2^N symmetric pairs of all inputs. The maximal symmetric input sets could be computed after checking all symmetric pairs. A naïve way to presenting the result is to construct an $N \times N$ triangular matrix for an N-input circuit. Each entry in the triangular matrix shows the symmetry of the corresponding inputs x_i and x_j except the diagonal entries. This representation is very simple but hard to understand globally. Therefore, we use *Symmetric -ASymmetric Inputs* (SASIs) [9][1], which is an *implicit* representation to present the maximal symmetric input sets. By the SASIs, if any two inputs are not in the same group, then they are asymmetric inputs. Otherwise they are "possibly" symmetric. For an N-input circuit, we number the inputs from 1 to N. Initially, we can assume that all inputs are possibly symmetric, so the SASIs representation is (1 2 3 .... N). If we can confirm that input i is asymmetric to the other inputs, the SASIs representation is (i) (1 2 .... i-1 i+1 N). The following example demonstrates the details of the SASIs representation.

Example 2.1: Given a 10-input circuit, the inputs are numbered from 1 to 10. Initially, we assume all inputs are possibly symmetric and thus the corresponding SASIs representation is (1 2 3 4 5 6 7 8 9 10). While the SASIs representation is (1 2 3 4 5) (6 7 8) (9) (10), it indicates that input 9 and input 10 are asymmetric to the other inputs. If SASIs representation could be divided into ten groups, then we claim that all inputs are asymmetric inputs.

C. *Naïve Approach*

A pair of patterns whose assignments are identical except on inputs x_i and x_j, and $x_i = x_j$ (00 or 11) in each pattern, is capable of distinguishing whether x_i and x_j are E symmetric or not. These pairs of patterns are called *legal pattern pairs*.

There are 2^{N-2} legal pattern pairs for any two inputs in an N-input circuit. Two inputs are E symmetric while the outputs of each legal pattern pair are identical. Otherwise they are E asymmetric inputs. We illustrate it using the following example.

Example 2.2: For a 5-input circuit, if we want to recognize whether input 1 and input 2 are E symmetric inputs or not. We have to exhaustively simulate $2^{5-2} = 8$ legal pattern pairs, which are {(00000, 11000), (00100, 11100), (00010, 11010), (00001, 110001), (00110, 11110), (00101, 11101), (00011, 11011), (00111, 11111)}. If the outputs of each legal pattern pairs are identical, input 1 and input 2 are symmetric. Otherwise they are asymmetric.

It is clear that to identify two inputs are E asymmetric is easier than to identify they are E symmetric. Thus, our approach will target at the identification of asymmetric inputs.

Definition 2.1: A pair of inputs x_i and x_j is denoted as *VP* (x_i, x_j) if they have not been recognized as symmetric or asymmetric.

The naïve approach is an exhaustive approach. It simulates all legal pattern pairs to recognize whether the targeted VP is asymmetric or not. If there exits one legal pattern pair with different outputs, then the process to recognizing the targeted VP would cease.

III. EQUIVALENCE SYMMETRY IDENTIFICATION ALGORITHM

Since the naïve approach needs a great number of

patterns and comparisons to identify symmetric inputs. A heuristic which applies two sets of patterns to recognize all VPs simultaneously is investigated.

Definition 3.1 [1]: For an N-input combination circuit, the set consists of all patterns with m 1s and $(N\text{-}m)$ 0s is denoted as θ_m^N, where $m \in [0, 1, 2, ..., N\text{-}1, N]$. The size of θ_m^N is the number of patterns in θ_m^N and is denoted as $|\theta_m^N|$ and $|\theta_m^N|$ equals C_m^N, where

$$C_m^N = \frac{N!}{(N-m)! \times m!}$$

Following equations represent the relations of θ_m^N for different m and N:

$$|\theta_m^N| \leq |\theta_{m+1}^N| \; for \; m \in \{1, 2,...., \lfloor (N\text{-}1)/2 \rfloor\}$$

$$|\theta_m^N| \leq |\theta_{m-1}^N| \; for \; m \in \{\lceil (N+1)/2 \rceil ,...., N\text{-}1\}$$

Theorem 3.1: For any two pattern sets { θ_i^N , θ_{i+2}^N }, those two pattern sets can be used to recognize all VPs.

However, this heuristic is infeasible while i increases. For example, considering two pattern sets { θ_1^{100} , θ_3^{100} }, there are C_1^{100} =100 patterns in θ_1^{100}, and C_3^{100} =161,700 patterns in θ_3^{100}. It conducts $100 \times 161,700 = 16,170,000$ comparisons for identifying E symmetry.

The heuristic approach fails due to a great number of patterns and comparisons have to be generated and conducted. The number of patterns in a pattern set depends on two factors, one is the length of a pattern, i.e., the number of inputs. The other is the number of 1s in a pattern of the pattern set. Since the pattern set is determined by the number of 1s, it seems difficult in reducing this factor. Therefore, we attempt to divide inputs into as many groups as possible. Therefore, an improved approach is introduced.

Definition 3.2[1]: A *multiple element group* (MEG) is a group that contains more than one element in the SASIs representation. A *single element group* (SEG) is a group that contains only one element.

The improved approach aims at each MEG and generates the corresponding pattern sets for each MEG. While the size of MEG is reduced, the number of pattern in a pattern set could also be reduced.

Example 3.1: For a 10-input circuit and assume the SASIs representation is (1 2 3 4 5) (6 7 8 9 10) after generating { θ_1^{10} , θ_3^{10} }. The second step is to generate { θ_2^{10} , θ_4^{10} } and we have to generate ($C_2^{10} + C_4^{10}$) = 255 patterns and conduct ($C_2^{10} \times C_4^{10}$) = 9,450 comparisons by using the heuristic approach. But in considering the improved approach, for the MEG (1 2 3 4 5), it only generates { θ_2^5 , θ_4^5 }. It is the same to the MEG (6 7 8 9 10). The total number of patterns and comparisons by using the improved approach are ($C_2^5 + C_4^5$) × 2 = 30 and ($C_2^5 \times C_4^5$)×2 = 100, respectively. As compared with the heuristic approach, the improved approach is effective in diminishing the total number of patterns and comparisons.

If the size of MEG is large, the number of patterns to be generated is still large. Thus, next we will propose an algorithm that systematically generates smaller number of patterns to distinguish as many E-asymmetric inputs as possible.

Definition 3.3: The *distance* of VP(x_i , x_j) in an MEG is the difference of relative position of x_i and x_j.

Theorem 3.2: For an MEG with K elements, the number of VPs with distance i is $(K - i)$ and the maximal distance among all VPs is $(K - 1)$.

Example 3.2: For an MEG (2 3 5 6 7 8 9), we number the position from left to right as 1 to 7. Please note the distance of a VP is the difference of relative position. While the initial position (position number is 1) or the allocation of elements in an MEG is changed, the distance of VPs would also be changed. All VPs in the MEG are listed by their distances in Table I.

Table I The distance of VPs

Distance	VP	|VP|
1	(2,3),(3,5),(5,6),6,7),(7,8),(8,9)	6
2	(2,5),(3,6),(5,7),(6,8), (7,9)	5
3	(2,6),(3,7),(5,8),(6,9)	4
4	(2,7),(3,8),(5,9)	3
5	(2,8),(3,9)	2
6	(2,9)	1

Definition 3.4: For an N-input circuit, *circular pattern set* for an MEG with K elements is the set that consists of all patterns which satisfy following conditions in θ_m^N and is denoted as $\alpha_{m,i}^K$.

1). Initial position is circularly set in each element of the MEG.

2). The distance of elements assigned value 1 is 1 except on the last two elements. The distance of the last two elements assigned value 1 is i.

3). If m=1, then i is 1.

Example 3.3: For a 10-input circuit, there is an MEG with 7 elements and assume the SASIs representation is (2 3 5 6 7 8 9). The circular pattern set $\alpha_{3,1}^7$ is {0110100000, 0010110000, 0000111000, 0000011100, 0000001110, 0100000110, 0110000010}. To represent the patterns concisely, a simple representation that indicates which inputs are assigned 1 is used. Hence the simple representation of $\alpha_{3,1}^7$ is {(2,3,5), (3,5,6), (5,6,7), (6,7,8), (7,8,9), (8,9,2), (9,2,3)}. The circular pattern set $\alpha_{3,2}^7$ is {0110010000, 0010101000, 0000110100, 0000011010, 0100001100, 0010000110, 0100100010} and the corresponding simple representation is {(2,3,6), (3,5,7), (5,6,8), (6,7,9), (7,8,2), (8,9,3), (9,2,5)}.

Theorem 3.3: For an MEG with K elements, a couple of circular pattern sets { $\alpha_{m,1}^K$, $\alpha_{m+2,i}^K$ } can be used to recognize VPs with distance i and $(K - i)$.

Proof: The distances of patterns in $\alpha_{m,1}^N$ is 1, hence the distance sequence is $(x_1 =1, x_2 =1,..., x_{i-1} =1)$ for all patterns in $\alpha_{m,1}^N$. The patterns in $\alpha_{m+2,i}^N$ with the distance sequence $(x_1 =1, x_2 =1, ..., x_{i-1} =1, x_i =1, x_{i+1} =i)$ and the patterns in $\alpha_{m,1}^N$ could be used to recognize VPs with distance i. Since i represent the distance of last two critical 1s and we regards the distance as circular, VPs with distance $(N - i)$ could be treated as i. Therefore, $\{ \alpha_{m,1}^N, \alpha_{m+2,i}^N \}$ could be used to recognize VPs with distance i and $(N - i)$. ∎

Now, we will explain how to utilize circular pattern set to recognize symmetric inputs. Considering an MEG with K elements, there are C_2^K VPs and could be divided into $(K-1)$ sets by the distance. Since a couple of $\{ \alpha_{m,1}^K \ \alpha_{m+2,i}^K \}$ could be used to recognize VPs with distance i and $(K - i)$, and the possible distance of all VPs is from 1 to $(K - 1)$, we apply the following rules to recognize all VPs.

Rule 1: Choosing circular pattern set $\alpha_{m,1}^K$ in θ_m^N.

Rule 2: Choosing circular pattern sets $\alpha_{m+2,i}^K$ in θ_{m+2}^N

where, $i = 1, 2, ..., \lceil (N-1)/2 \rceil$ in θ_{m+2}^N.

Example 3.4: For a 10-input circuit and we assume the SASIs representation is (2 3 5 6 7 8 9)(1)(4)(10) after generating and comparing $\{ \theta_0^{10}, \theta_2^{10} \}$. Next we choose circular pattern set $\alpha_{1,1}^7$ in θ_1^{10} and $\alpha_{3,1}^7$ in θ_3^{10} for the MEG (2 3 5 6 7 8 9). Those two circular pattern sets can be used to recognize VPs with distance 1 and 6. This can be seen in Figure 4. Similarly, circular pattern set $\alpha_{3,2}^7$ and $\alpha_{3,3}^7$ could be used to recognize VPs with distance 2 and 5 as well as 3 and 4 as comparing $\alpha_{1,1}^7$, respectively. Those two illustrations could be seen in Figure 5 and Figure 6. It is obvious that those three circular pattern sets can cover all distances of all VPs.

$\alpha_{1,1}^7$	$\alpha_{3,1}^7$		VPs	Distance
2	2 3 5		(3, 5)	1
3	3 5 6		(5, 6)	1
5	5 6 7		(6, 7)	1
6	6 7 8	⟹	(7, 8)	1
7	7 8 9		(8, 9)	1
8	8 9 2		(9, 2)	6
9	9 2 3		(2, 3)	1

Figure 4: Comparing ($\alpha_{1,1}^7$, $\alpha_{3,1}^7$) covers VPs with distance 1 and 6

$\alpha_{1,1}^7$	$\alpha_{3,2}^7$		VPs	Distance
2	2 3 6		(3, 6)	2
3	3 5 7		(5, 7)	2
5	5 6 8		(6, 8)	2
6	6 7 9	⟹	(7, 9)	2
7	7 8 2		(8, 2)	5
8	8 9 3		(9, 3)	5
9	9 2 5		(2, 5)	2

Figure 5: Comparing ($\alpha_{1,1}^7$, $\alpha_{3,2}^7$) covers VPs with distance 2 and 5

$\alpha_{1,1}^7$	$\alpha_{3,3}^7$		VPs	Distance
2	2 3 7		(3, 7)	3
3	3 5 8		(5, 8)	3
5	5 6 9		(6, 9)	3
6	6 7 2	⟹	(7, 2)	4
7	7 8 3		(8, 3)	4
8	8 9 5		(9, 5)	4
9	9 2 6		(2, 6)	3

Figure 6: Comparing ($\alpha_{1,1}^7$, $\alpha_{3,3}^7$) covers VPs with distance 3 and 4

Figure 7 shows the flow chart that we proposed for finding maximal symmetric inputs sets. Our approach reads a design with arbitrary levels and generates patterns. The results of patterns provide information to the remaining VPs. Grouping all remaining VPs to from the updated SASIs. Then further heuristic patterns are generated and simulated again by the updated SASIs in the next iteration. If all inputs are recognized as asymmetric or the iterations are over the bound, our approach will be terminated and the maximal symmetric input sets will be returned.

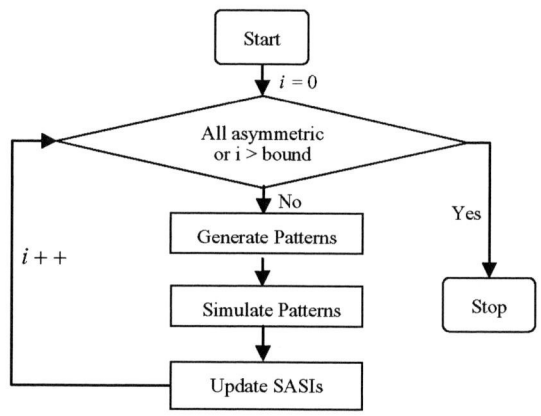

Figure 7: The flow chart of our approach

IV. EXPERIMENTAL RESULTS

We have implemented the proposed algorithm in Verilog HDL. Experiments are conducted over a set of ISCAS-85 and MCNC benchmarks which are described in Verilog HDL.

We compare the experimental results with [10]. [10] is an BDD-based approach for E symmetry identification. It claims that all VPs can be identified exactly.

Table II summarizes the experimental results of [10] and ours. The first column shows the name of each benchmark and the following two columns **#in** and **#out** represent the number of inputs and outputs. The following columns show the CPU time measured in second and the results. In [10], the time for building BDDs was not listed. We construct the BDDs for each benchmark by CUDD package [7] without using any reordering technique and its time is shown in the "**reading**" column on a "SUN SPARC II" workstation measured in second. The last column shows the number of variable pairs that cannot be recognized as asymmetric inputs by [10] and our approach. According to Table II, our CPU time is less than that of [10] with including the time of BDD construction, and our results are the same with [10] for most benchmarks. In c2670 and c7552, however, our approach returns more VPs that cannot be recognized as asymmetric than [10], but the CPU time is less than that of [10]. Note that our approach is applicable to the designs whose compact BDDs cannot be built. For example, c6288 is a multiplier design, one cannot have an efficient BDD representation for it. Hence, the proposed approach is a robust approach to some degree.

V. CONCLUSIONS

Random simulation could also find some asymmetric VPs. But it may generate redundant patterns for some recognized asymmetric VPs. Thus, simulation with randomly generated patterns is inefficient. In this paper, we propose a systematic patterns search algorithm for computing maximal symmetric inputs sets. It is applicable to designs described in arbitrary level, especially to high-level and black box designs. Experimental results on ISCAS-85 and MCNC benchmarks demonstrate the effectiveness and efficiency of our approach.

REFERENCES

[1] C.-L Chou, C.-Y Wang, G.-W Lee, and J.-Y Jou, "Graph automorphism-based algorithm for determining symmetric inputs," *in Proceedings of IEEE International Conference on Computer Design*, pp. 417-419, 2004.

[2] A. Mishchenko, "Fast computation of symmetries in Boolean functions," *IEEE Transactions on Computer-Aided Design of Integrated Circuits and Systems*, vol. 22, pp. 1588-1593, Nov. 2003.

[3] E. J. McCluskey, "Detection of group invariance or total symmetry of a Boolean function," *Bell System Technology Journal*, pp. 1445-1453, Nov. 1956.

[4] E. J. McCluskey, *Logic Design Principles with Emphasis on Testable Semicustom Circuits, Prentice-Hall*, 1986.

[5] I. Pomeranz and S. M. Reddy, "On determining symmetries in inputs of logic circuits," in *Proceedings of IEEE International Conference on Computer-Aided Design*, pp. 500-507, 1993.

[6] C. Scholl, D. Moller, and P. Molitor, "BDD minimization using symmetries," *IEEE Transactions on Computer-Aided Design of Integrated Circuits and Systems*, vol. 18, pp. 81-100, Feb. 1999.

[7] F. Somenzi, "CUDD: CU Decision Diagram Package, Release 2.3.1," University of Colorado at Boulder, 2001.

[8] C.-C. Tsai and M. Marek-Sadowska, "Generalized Reed-Muller forms as a tool to detect symmetries," *IEEE Transactions of Computers*, vol. 45, pp. 33-40, Jan. 1996.

[9] C.-Y Wang, S.-W Tung, and J.-Y Jou, "On Automatic Verification Pattern Generation for SoC with Port Order Fault Model", *IEEE Transactions on Computer-Aided Design of Integrated Circuits and Systems*, vol. 21, pp. 466-479, Apr. 2002.

[10] K.-H. Wang and J.-H. Chen, "Symmetry Detection for Incompletely Specified Functions with K-Disjointness Paradigm," *in Proceedings of the IEEE Asia and South Pacific Design Automation Conference*, vol. 2, pp. 994-997, 2005.

[11] K.-H. Wang and J.-H. Chen, "Symmetry detection for incompletely specified functions," *in Proceedings of IEEE Design Automation Conference*, pp. 434-437, 2004.

Table II
Experimental results of the equivalence symmetric input identification

circuit	#in	#out	time (s)			symmetry pair	
			reading	[10]	ours	[10]	ours
c880	60	26	11.57	0.03	1.75	0	0
c1355	41	32	1.30	0.05	0.68	0	0
c1908	33	25	--	--	0.28	--	0
c432	36	7	--	--	0.19	--	0
c499	41	32	1.17	0.05	0.66	0	0
c3540	50	22	18.96	0.08	2.42	0	0
c5315	178	123	>1hr	0.02	49.38	0	0
c2670	233	140	>1hr	0.08	593.04	28	227
c7552	207	108	>1hr	0.17	633.61	6	160
c6288	32	32	--	--	0.25	--	0
des	256	245	3.42	0.03	14.36	0	0
rot	135	107	3.08	0.07	26.31	1	1
9sym	9	1	--	--	0.46	--	0
alu4	14	8	--	--	0.71	--	0
cordic	23	2	--	--	2.89	--	0
t481	16	1	--	--	0.98	--	0

Fast Multi-Domain Clock Skew Scheduling For Peak Current Reduction

Shih-Hsu Huang, Chia-Ming Chang, and Yow-Tyng Nieh

Department of Electronic Engineering
Chung Yuan Christian University
Chung Li, Taiwan, R.O.C.

Abstract - Given several specific clocking domains, the peak current minimization problem can be formulated as a 0-1 integer linear program. However, if the number of binary variables is large, the run time is unacceptable. In this paper, we study the reduction of this high computational expense. Our approach includes the following two aspects. First, we derive the ASAP schedule and the ALAP schedule to prune the redundancies without sacrificing the exactness (optimality) of the solution. Second, we propose a zone-based scheduling algorithm to solve a large circuit heuristically.

I. Introduction

It is well known that the clock skew can be utilized to shorten the clock period [1] or reduce the peak current [2]. However, it is very difficult to implement a wide spectrum of dedicated clock delays.

The architecture of multiple clocking domains [3] provides an alternative to unconstrained clock skew scheduling. Fishburn [4] presented a polynomial time complexity algorithm to derive a multi-domain clock skew schedule for a target clock period. Ravindran, Keuhlmann, and Sentovich [5] also studied the clock period minimization problem using multiple clocking domains combined with small within-domain latency.

Vittal, Ha, Brewer and Marek-Sadowska [6] used the 0-1 integer linear programming (ILP) formulations to minimize the peak current under the constraint of multi-domain clock skew scheduling. Although their approach guarantees the optimality (in terms of the specified cost function), the run time of the 0-1 ILP solver grows dramatically with the increase of binary variables. The high computational expense has limited the use of their approach.

This paper presents an approach to overcoming the limitation. Our approach includes the following two aspects:
(1) We derive the ASAP (as-soon-as-possible) schedule and the ALAP (as-late-as-possible) schedule of each register. As a result, we can prune all the redundancies without sacrificing the exactness (optimality) of the solution.

(2) For a large circuit, the number of binary variables and the number of constraints are still very large even though all the redundancies are pruned. Thus, we propose a zone-based approach to solve a large circuit heuristically.

Note that the proposed algorithms are *independent* of the used peak current model. Therefore, in fact, our approach is a *general methodology* for the reduction of peak current under the constraint of multi-domain clock skew scheduling.

II. Preliminaries

Multi-domain clock skew scheduling only uses n discrete clocking domains: d_1, d_2, ...and d_n. A clocking domain d_k, where k = 1, 2, ... and n, corresponds to a clock arrival time. Without loss of generality, we assume that $d_1 \leq d_2 \leq ... \leq d_n$.

The clock arrival time of each register must be one of the n clocking domains. Thus, the clock arrival time of register R_i can be represented by n binary variables: $S_{i,1}$, $S_{i,2}$, ..., $S_{i,n}$, where $S_{i,k}$ is 1 if and only if T_{Ci} is equal to d_k. For each register R_i, we have

$$\sum_{k=1}^{n} S_{i,k} = 1.$$

For each data path $R_i \rightarrow R_j$, the double clocking constraint can be rewritten as below:

$$\sum_{k=1}^{n} d_k \cdot S_{j,k} - \sum_{k=1}^{n} d_k \cdot S_{i,k} \leq T_{PDi,j(min)} ;$$

For each data path $R_i \rightarrow R_j$, the zero clocking constraint can be rewritten as below:

$$\sum_{k=1}^{n} d_k \cdot S_{i,k} - \sum_{k=1}^{n} d_k \cdot S_{j,k} \leq P - T_{PDi,j(max)} .$$

The objective is to minimize the value *peak_current*. Without loss of generality and for the convenience of illustration, in the following, we assume that the total current derivative at the clocking domain d_k is $\sum_{i=1}^{r} S_{i,k} \cdot I_i$, where r denotes the number of registers and I_i denotes the maximum possible current of register R_i caused by the clock edge. As a

0-7803-9451-8/06/$20.00 ©2006 IEEE.

result, for each clocking domain d_k, the peak current constraint is as below:

$$\sum_{i=1}^{r} S_{i,k} \cdot I_i \le peak\_current.$$

Thus, the peak current minimization problem formulated by [6] has $r+2s+n$ constraints, where s is the number of data paths.

Using Fig. 1 as an example, the circuit graph has 6 registers and 10 data paths. Suppose that clock period P = 6 tu (time units) and $I_1 = I_2 = I_3 = I_4 = I_5 = I_6 = c$. If the circuit is fully synchronous, a huge peak current $6*c$ is observed at the clock edge. Assume that we are given 3 clocking domains: $d_1 = -2$ tu, $d_2 = 0$ tu and $d_3 = 2$ tu. If we directly use the 0-1 ILP formulations as described in [6], the number of binary variables is $6*3 = 18$ and the number of constraints is $6+2*10+3 = 29$. After solving the 0-1 ILP formulations, we find that the peak current is reduced to $2*c$ under the multi-domain clock skew schedule in which $T_{C1} = d_1 = -2$ tu, $T_{C2} = d_2 = 0$ tu, $T_{C3} = d_1 = -2$ tu, $T_{C4} = d_3 = 2$ tu, $T_{C5} = d_2 = 0$ tu, and $T_{C6} = d_3 = 2$ tu.

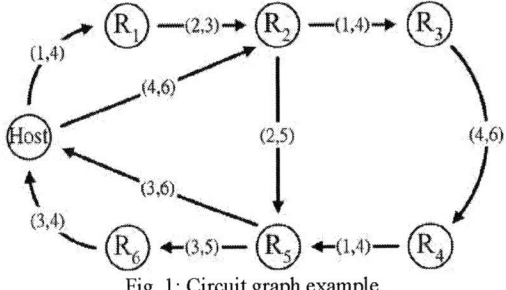

Fig. 1: Circuit graph example.

III. The Approach

Section 3.1 derives an ASAP schedule and an ALAP schedule to prune the redundancies. Section 3.2 presents a zone-based scheduling algorithm for large circuits.

A. ASAP and ALAP

In fact, due to the zero clocking constraints and the double clocking constraints, many binary variables used in the 0-1 ILP formulations are definitely to be 0. Therefore, a large fraction of the binary variables are redundant. Moreover, a large fraction of the constraints used in the 0-1 ILP formulations are also redundant as they are implied by some of the other constraints. If these redundant binary variables and redundant constraints can be pruned, the problem size can be significantly reduced. Thus, our approach is to find a tight bound on the clock arrival time of each register without sacrificing the exactness (optimality) of the solution.

Given the clock arrival time x of a register, we define the following two functions: *floor* and *ceil*. The function floor(x) gives the latest clocking domain before the clock arrival time x; the function ceil(x) gives the earliest clocking domain after the clock arrival time x. The pseudo codes of the two functions are shown in Fig. 2 and Fig. 3, respectively, where n is the number of clocking domains.

Given a circuit graph G and a target clock period P, the procedure *OBTAIN_ASAP_ALAP* is to find a tight bound on the clock arrival time of each register. Let $ASAP_i$ and $ALAP_i$ denote the allowable earliest clocking domain and the allowable latest clocking domain of register R_i, respectively. Thus, we say that $[ASAP_i, ALAP_i]$ is the allowable clocking range of register R_i. Initially, we let the allowable clocking range $[ASAP_i, ALAP_i]$ be $[d_1, d_n]$ for each register R_i. Then, by iteratively applying the clocking constraints under the target clock period P, we narrow the allowable clocking range of each register. The repeat-until-loop iteration repeats until the allowable clocking range of each register is stable. Thus, the number of repeat-until-loop iterations is O(r*n). Fig. 4 gives the pseudo code

```
Function floor(x)
begin
for i = n downto 1 do
    if (x ≥ d_i) then return(d_i);
return(d_1);
end.
```

Fig. 2: Function floor.

```
Function ceil(x)
begin
for i = 1 to n do
    if (x ≤ d_i) then return(d_i);
return(d_n);
end.
```

Fig. 3: Function ceil.

```
Procedure OBTAIN_ASAP_ALAP(G,P)
begin
  let [ASAP_host, ALAP_host] be [0,0];
    for each register R_i do
      [ASAP_i, ALAP_i] = [d_1, d_n];
    repeat
      update = 0;
    for each data path R_i→R_j do
      begin
        apply clocking constraints in the forward direction to narrow
        the allowable clocking range [ASAP_j, ALAP_j];
        if [ASAP_j, ALAP_j] is narrowed then update = 1;
      end;
    for each data path R_i→R_j do
      begin
        apply clocking constraints in the backward direction to narrow
        the allowable clocking range [ASAP_i, ALAP_i];
        if [ASAP_i, ALAP_i] is narrowed then update = 1;
      end
    until (update == 0);
end.
```

Fig. 4: Procedure OBTAIN_ASAP_ALAP.

There are two directions to apply the clocking constraints of data paths: *forward direction* and *backward direction*. The first for-loop applies clocking constraints in the forward direction (i.e., in the direction of data path); and the second for-loop applies clocking constraints in the backward direction (i.e., against the direction of data path). Note that the sequence of data paths in each for-loop, which applies the clocking constraints, can be arbitrarily specified without affecting the calculation of ASAP schedule and ALAP schedule. The details are described as below.

(1) If the clocking constraint of data path $R_i \rightarrow R_j$ is applied in the forward direction, the allowable clocking range $[ASAP_j, ALAP_j]$ of register R_j is narrowed under the assumption that the allowable clocking range $[ASAP_i, ALAP_i]$ of register R_i is fixed. As a result, we have:

$ASAP_j = maximum(ASAP_j, ceil(ASAP_i + T_{PDi,j(max)} - P))$;
$ALAP_j = minimum(ALAP_j, floor(ALAP_i + T_{PDi,j(min)}))$.

(2) If the clocking constraint of data path $R_i \rightarrow R_j$ is applied in the backward direction, the allowable clocking range $[ASAP_i, ALAP_i]$ of register R_i is narrowed under the assumption that the allowable clocking range $[ASAP_j, ALAP_j]$ of register R_j is fixed. As a result, we have:

$ASAP_i = maximum(ASAP_i, ceil(ASAP_j - T_{PDi,j(min)}))$;
$ALAP_i = minimum(ALAP_i, floor(ALAP_j + P - T_{PDi,j(max)}))$.

Let's use the circuit graph shown in Fig. 1 as an example to demonstrate the procedure OBTAIN_ASAP_ALAP. Suppose that clock period $P = 6$ tu, $I_1 = I_2 = I_3 = I_4 = I_5 = I_6 = c$, and we are given 3 clocking domains: $d_1 = -2$ tu, $d_2 = 0$ tu and $d_3 = 2$ tu. At the beginning, we have $[ASAP_i, ALAP_i] = [-2,2]$ for each register R_i where $i = 1, 2, 3, 4, 5$ and 6. Then, we enter the repeat-until-loop. We assume that the first for-loop (i.e., applying clocking constraints in forward direction) tackles the data paths in the sequence of host$\rightarrow R_1$, host$\rightarrow R_2$, $R_1 \rightarrow R_2$, $R_2 \rightarrow R_3$, $R_2 \rightarrow R_5$, $R_3 \rightarrow R_4$, $R_4 \rightarrow R_5$, and $R_5 \rightarrow R_6$, and the second for-loop (i.e., applying clocking constraints in backward direction) tackles the data paths in the sequence of $R_6 \rightarrow$host, $R_5 \rightarrow$host, $R_5 \rightarrow R_6$, $R_4 \rightarrow R_5$, $R_3 \rightarrow R_4$, $R_2 \rightarrow R_5$, $R_2 \rightarrow R_3$, and $R_1 \rightarrow R_2$. Table I gives the snapshot. The repeat-until-loop iteration takes two times. The column *forward* denotes the allowable clocking range after the first for-loop is completed, and the column *backward* denotes the allowable clocking range after the second for-loop is completed.

Table I. The snapshots of procedure OBTAIN_ASAP_ALAP.

	First repeat-until-loop		Second repeat-until-loop	
	forward	backward	forward	backward
R_1	[-2,0]	[-2,0]	[-2,0]	[-2,0]
R_2	[0,2]	[0,0]	[0,0]	[0,0]
R_3	[-2,2]	[-2,0]	[-2,0]	[-2,0]
R_4	[-2,2]	[-2,2]	[-2,2]	[0,2]
R_5	[-2,2]	[-2,0]	[0,0]	[0,0]
R_6	[-2,2]	[-2,2]	[0,2]	[0,2]

After the procedure OBTAIN_ASAP_ALAP is completed, we have $[ASAP_1, ALAP_1] = [-2,0]$, $[ASAP_2, ALAP_2] = [0,0]$, $[ASAP_3, ALAP_3] = [-2,0]$, $[ASAP_4, ALAP_4] = [0,2]$, $[ASAP_5, ALAP_5] = [0,0]$ and $[ASAP_6, ALAP_6] = [0,2]$. Since $ASAP_2 = ALAP_2 = 0$ tu and $ASAP_5 = ALAP_5 = 0$ tu, we have $T_{C2} = d_2 = 0$ tu and $T_{C5} = d_2 = 0$ tu. Thus, we only need to use 8 binary variables: $S_{1,1}$, $S_{1,2}$, $S_{3,1}$, $S_{3,2}$, $S_{4,2}$, $S_{4,3}$, $S_{6,2}$ and $S_{6,3}$. Moreover, we can use both the ASAP schedule and the ALAP schedule to simplify the zero clocking constraints and the double clocking constraints. For example, the double clocking constraint of the data path $R_1 \rightarrow R_2$ can be simplified to $2*S_{1,1} \leq T_{PD1,2(min)} = 2$ and thus can be easily justified as a redundant constraint. The justification of redundant constraints is in polynomial time complexity. By exploiting both the ASAP schedule and the ALAP schedule, we find that all zero clocking constraints and all the double clocking constraints in this circuit graph become redundant. Consequently, the number of constraints is reduced from 29 to 7. As a result, we can rewrite the 0-1 ILP formulations as below:

minimize *peak_current*
subject to
$S_{1,1} + S_{1,2} = 1$;
$S_{3,1} + S_{3,2} = 1$;
$S_{4,2} + S_{4,3} = 1$;
$S_{6,2} + S_{6,3} = 1$;
$S_{1,1}*I_1 + S_{3,1}*I_3 \leq peak\_current$;
$S_{1,2}*I_1 + 1*I_2 + S_{3,2}*I_3 + S_{4,2}*I_4 + 1*I_5 + S_{6,2}*I_6 \leq peak\_current$;
$S_{4,3}*I_4 + S_{6,3}*I_6 \leq peak\_current$;

By applying the 0-1 ILP solver, we have the results that $S_{1,1} = 1$, $S_{1,2} = 0$, $S_{3,1} = 1$, $S_{3,2} = 0$, $S_{4,2} = 0$, $S_{4,3} = 1$, $S_{6,2} = 0$ and $S_{6,3} = 1$. In other words, the peak current is $2*c$ under the multi-domain clock skew schedule in which $T_{C1} = d_1$, $T_{C2} = d_2$, $T_{C3} = d_1$, $T_{C4} = d_3$, $T_{C5} = d_2$ and $T_{C6} = d_3$. Clearly, both the number of binary variables and the number of constraints are dramatically reduced without scarifying the exactness of the solution.

B. Zone-Based Scheduling

For a large circuit, the number of binary variables is still very large even though all the redundant binary variables are pruned. Given a circuit graph G and a target clock period P, Fig. 5 gives the procedure ZONE_BASED_SCHEDULING to deal with a large circuit. At the beginning, in addition to prune all the redundant binary variables, we also use the algorithm proposed in [4] to derive a feasible multi-domain clock skew schedule for the circuit graph G to work with clock period P. We let the feasible schedule be the initial solution. Note that the algorithm proposed in [4] does not attempt to minimize the peak current. The kernel of zone-based scheduling is an iteration process of the procedure ZONE_PARTITION_SCHEDULING.

The procedure ZONE_PARTITION_SCHEDULING is to reduce the peak current by re-scheduling the registers in the set RS. Fig. 6 gives the pseudo code. By specifying the constraint on the maximum number of binary variables in the ILP formulations, we partition the set of registers RS into some sub-sets. Each sub-set is called a zone. The philosophy of zone partitioning is as below. Intuitively, if a register is associated with few binary variables, the flexibility in assigning its clock arrival time is limited. On the other hand, if a register is associated with many binary variables, the flexibility in assigning its clock arrival time is very high. Therefore, in order to evenly spread the currents over the clocking domains, a reasonable heuristic is to first re-schedule the registers that are associated fewer binary variables. Based on this philosophy, a zone is created to accommodate as many registers as possible. The fewer associated binary variables, the higher priority (to be assigned into the zone) the register has. If the current zone cannot accommodate any more register (due to the limitation on the maximum number of binary variables), we create another new zone. The iteration of zone creation repeats until all the registers are assigned. Note that the earlier created zone has a higher priority to be re-scheduled. When a zone is re-scheduled, the clock arrival times of other registers are assumed to be fixed. Since the maximum number of binary variables within a zone is limited, the 0-1 ILP solver can re-schedule each zone efficiently.

Procedure ZONE_BASED_SCHEDULING(G,P)
begin
  prune all the redundant binary variables according to the results of the procedure OBTAIN_ASAP_ALAP(G,P);
  apply [4] to derive a feasible multi-domain clock skew schedule for the circuit graph G to work with clock period P;
  phase = 1;
  RS = all the registers in the circuit graph G;
  call ZONE_PARTITION_SCHEDULING(phase,RS, G,P);
  phase = 2;
  repeat
    RS=all the registers scheduled into hottest clocking domains;
    call ZONE_PARTITION_SCHEDULING(phase,RS,G,P);
  until (the peak current cannot be further reduce);
end.

Fig. 5: Procedure ZONE_BASED_SCHEDULING.

In fact, the zone-based scheduling uses two phases to apply the procedure ZONE_PARTITION_SCHEDULING as below.

(1) The first phase is to derive a multi-domain clock skew schedule that attempts to minimize the peak current. We use the procedure ZONE_PARTITION_SCHEDULING to re-schedule all the registers in the circuit graph G. When a zone is re-scheduled, our objective is to minimize the peak current produced by the registers that have been re-scheduled. In the pseudo code of Fig. 6, we use the notation RP to denote the set of all the registers that have been re-scheduled.

(2) The second phase attempts to further reduce the peak current. A clocking domain is called the hottest clocking domain, if and only if the total current derivative at this clocking domain is exactly the same as the peak current. The peak current cannot be further reduced, unless some registers scheduled in hottest clocking domains can be re-scheduled into other clocking domains. Thus, we iteratively apply the procedure ZONE_PARTITION_SCHEDULING to re-schedule the registers scheduled in hottest clocking domains. The iteration process repeats until the peak current cannot be further reduced. Note that, in the second phase, we let the set RP be all the registers in the circuit graph G.

Procedure ZONE_PARTITION_SCHEDULING(phase,RS,G,P)
begin
  partition all the registers in the set RS into zones (under the constraint on the maximum number of binary variables);
  if (phase == 1) **then** RP = ∅
  else RP = all the registers in the circuit graph G;
  for each zone **do** (in the sequence of their priorities)
  begin
    if (phase == 1) **then** add all the registers in this zone into RP;
    re-schedule this zone under all the clocking constraints to reduce the peak current produced by the registers in the set RP;
  end
end.

Fig. 6: Procedure ZONE_PARTITION_SCHEDULING.

Let's use Fig. 1 as an example. First, we derive both the ASAP schedule and ALAP schedule as shown in Table I. Also, we derive a feasible multi-domain clock skew schedule in which $T_{C1} = d_2$, $T_{C2} = d_2$, $T_{C3} = d_2$, $T_{C4} = d_2$, $T_{C5} = d_2$ and $T_{C6} = d_2$ to work with clock period 6 tu as the initial solution. Therefore, initially, the peak current is 6*c. Suppose that the maximum number of binary variables involved in a zone is only 4. The process of zone-based scheduling is as below.

First, we enter the first phase of zone-based scheduling. The first phase re-schedules all the registers in the circuit graph. From both the ASAP schedule and the ALAP schedule, we know that R_1, R_2, R_3, R_4, R_5 and R_6 are associated with 2, 0, 2, 2, 0 and 2 binary variables. Thus, all the registers in the circuit graph are partitioned into two zones: zone $Z_1 = \{R_1, R_2, R_3, R_5\}$ and zone $Z_2 = \{R_4, R_6\}$, and zone Z_1 has a higher priority to be re-scheduled. In the following, we reduce the peak current in the sequence of zone Z_1 and zone Z_2.

We reduce the peak current by re-scheduling zone Z_1 under $T_{C4} = d_2$ and $T_{C6} = d_2$. The 0-1 ILP formulations are as below:
minimize *peak_current_Z_1*

subject to
$S_{1,1} + S_{1,2} = 1$;
$S_{3,1} + S_{3,2} = 1$;
$S_{1,1}*I_1 + S_{3,1}*I_3 \leq peak\_current\_Z_1$;
$S_{1,2}*I_1 + 1*I_2 + S_{3,2}*I_3 + 1*I_5 \leq peak\_current\_Z_1$;

We have the results that $peak\_current\_Z_1$ = 2*c, $S_{1,1}$ = 1, $S_{1,2}$ = 0, $S_{3,1}$ = 1 and $S_{3,2}$ = 0. As a result, we have $T_{C1} = d_1$, $T_{C2} = d_2$, $T_{C3} = d_1$, $T_{C4} = d_2$, $T_{C5} = d_2$, $T_{C6} = d_2$ and the peak current is 4*c. Thus, the peak current is reduced from 6*c to 4*c.

We further reduce the peak current by re-scheduling zone Z_2 under $T_{C1} = d_1$, $T_{C2} = d_2$, $T_{C3} = d_1$ and $T_{C5} = d_2$. The 0-1 ILP formulations are as below:
minimize $peak\_current\_Z_1\_Z_2$
subject to
$S_{4,2} + S_{4,3} = 1$;
$S_{6,2} + S_{6,3} = 1$;
$1*I_1 + 1*I_3 \leq peak\_current\_Z_1\_Z_2$;
$1*I_2 + S_{4,2}*I_4 + 1*I_5 + S_{6,2}*I_6 \leq peak\_current\_Z_1\_Z_2$;
$S_{4,3}*I_4 + S_{6,3}*I_6 \leq peak\_current\_Z_1\_Z_2$;

We have the results that $peak\_current\_Z_1\_Z_2$ = 2*c, $S_{4,2}$ = 0, $S_{4,3}$ = 1, $S_{6,2}$ = 0 and $S_{6,3}$ = 1. Thus, the peak current is reduced from 4*c to 2*c.

Next, we enter the second phase of zone-based scheduling. The peak current is 2*c. Since the total current derivative at clocking domains d_1, d_2 and d_3 are 2*c, 2*c, 2*c, respectively, they are hottest clocking domains. We find that R_1 and R_3 are scheduled to d_1, R_2 and R_5 are scheduled to d_2, and R_4 and R_6 are scheduled to d_3. Thus, we partition these six registers into two zones: zone Z_3 = {R_1, R_2, R_3, R_5} and zone Z_4 = {R_4, R_6}, and zone Z_3 has a higher priority to be re-scheduled.

We re-schedule zone Z_3 under $T_{C4} = d_3$ and $T_{C6} = d_3$. The 0-1 ILP formulations are as below:
minimize $peak\_current$
subject to
$S_{1,1} + S_{1,2} = 1$;
$S_{3,1} + S_{3,2} = 1$;
$S_{1,1}*I_1 + S_{3,1}*I_3 \leq peak\_current$;
$S_{1,2}*I_1 + 1*I_2 + S_{3,2}*I_3 + 1*I_5 \leq peak\_current$;
$1*I_4 + 1*I_6 \leq peak\_current$;

We have the results that $peak\_current$ = 2*c, $S_{1,1}$ = 1, $S_{1,2}$ = 0, $S_{3,1}$ = 1 and $S_{3,2}$ = 0. The peak current is still 2*c.

We re-schedule zone Z_4 under $T_{C1} = d_1$, $T_{C2} = d_2$, $T_{C3} = d_1$ and $T_{C5} = d_2$. The 0-1 ILP formulations are as below:
minimize $peak\_current$
subject to
$S_{4,2} + S_{4,3} = 1$;
$S_{6,2} + S_{6,3} = 1$;
$1*I_1 + 1*I_3 \leq peak\_current$;
$1*I_2 + S_{4,2}*I_4 + 1*I_5 + S_{6,2}*I_6 \leq peak\_current$;
$S_{4,3}*I_4 + S_{6,3}*I_6 \leq peak\_current$;

The second phase of zone-based scheduling is finished with the results that $peak\_current$ = 2*c, $S_{4,2}$ = 0, $S_{4,3}$ = 1, $S_{6,2}$ = 0

and $S_{6,3}$ = 1. After the zone-based scheduling is completed, the peak current is 2*c under the multi-domain clock skew schedule in which $T_{C1} = d_1$, $T_{C2} = d_2$, $T_{C3} = d_1$, $T_{C4} = d_3$, $T_{C5} = d_2$ and $T_{C6} = d_3$.

IV. Experimental Results

Our approach has been implemented in a C++ program running on a personal computer with AMD K8-3GHz CPU and 1G Bytes RAM. We use Extended LINGO Release 8.0 as the 0-1 ILP solver. The circuits from ISCAS'89 benchmark suites are targeted to a 0.35μm cell library for the experiments. Without loss of generality, in each benchmark circuit, we assume that: (1) the clock period P is the longest path delay; (2) we are given $2 \times \left\lfloor \dfrac{P}{0.3} \right\rfloor + 1$ clocking domains, where 0.3 ns denotes the clock skew between two consecutive clocking domains; (3) each clocking domain d_k = $0.3 \times (k - \left\lfloor \dfrac{P}{0.3} \right\rfloor - 1)$ ns, where k = 1, ..., $2 \times \left\lfloor \dfrac{P}{0.3} \right\rfloor + 1$; (4) the maximum possible current of each register is c (at the clock arrival time); and (5) if the circuit is fully synchronous, the peak current occurs when all the registers are switching simultaneously.

Table II gives the characteristics of benchmark circuits. The column *zero skew* denotes the peak current of fully synchronous implementation, while the column [6] denotes the peak current solved by the 0-1 ILP formulations as described in [6]. We use the notation -- to denote that the problem complexity cannot be solved within 12 hours.

Table II. Characteristics of benchmark circuits.

circuit	registers	gates	data paths	clock period	peak currrent	
					zero skew	[6]
S444	21	119	175	4.24	21*c	2*c
S499	22	120	528	3.99	22*c	6*c
S1269	37	437	455	9.89	37*c	4*c
S1512	57	413	686	5.50	57*c	6*c
S3271	116	1035	1137	8.50	116*c	4*c
S3384	183	1070	1831	19.86	183*c	--
S5378	179	1001	2313	5.16	179*c	11*c
S6669	239	2155	2179	28.42	239*c	--
S9234	228	2680	2830	9.42	228*c	--
S13207	669	2573	4660	10.92	669*c	--
S15850	597	3448	16863	14.23	597*c	--
S35932	1728	12204	5159	5.14	1728*c	288*c
S38584	1452	11448	15329	11.85	1452*c	--

Table III demonstrates the advantage of exploiting both the ASAP schedule and the ALAP schedule to prune the

258

redundancies. Note that, for each benchmark circuit, both the ASAP schedule and the ALAP scheduled can be derived within 1 second. The column [6] describes the original 0-1 ILP formulations. The column *Reduced formulations* describes the reduced 0-1 ILP formulations, in which all the redundancies are pruned. The column *#vars* denotes the number of binary variables, the column *#cons* denotes the number of constraints, and the column *CPU time* gives the CPU time in seconds.

Table III. The advantage of pruning redundancies.

Circuit	[6]			Reduced Formulations		
	#vars	#cons	CPU time (s)	#vars	#cons	CPU time (s)
S444	609	400	4	239	355	1
S499	594	1105	6	80	524	1
S1269	2405	1012	9	312	411	3
S1512	2109	1466	5	326	825	1
S3271	6612	2447	30087	3342	2027	88
S3384	24339	4316	--	4917	3531	24
S5378	6265	4860	16046	2088	342	59
S6669	45171	5476	--	3951	3265	122
S9234	14364	6803	--	6554	6042	232
S13207	48837	10062	--	20057	8416	362
S15850	56715	34418	--	18124	28314	--
S35932	60480	12081	714	17516	3714	271
S38584	114708	32189	--	40018	29569	--

Table IV demonstrates the results of zone-based scheduling. The benchmark circuits that have more than 2000 irredundant binary variables are used to test the effectiveness of zone-based scheduling. Without loss of generality, here we use the fully synchronous solution as the initial solution. The column *tackled zones* denotes the number of times to use the 0-1 ILP solver to re-schedule a zone. The CPU time includes the time spent in both our C++ program and the 0-1 ILP solver. For the convenience of comparisons, we also report the results obtained by *whole-circuit scheduling*, which solves the whole circuit as a single zone. We find that the whole-circuit scheduling cannot solve benchmark circuits S15850 and S38584 within 12 hours. However, on the other hand, the

zone-based scheduling can solve each benchmark circuit within 712 seconds no matter the constraint on the maximum number of binary variables involved in a zone is 500, 1000 or 2000.

V. Conclusions

This paper investigates a fast multi-domain clock skew scheduling for peak current reduction. The main contribution of our work is that it overcomes the high computational expense of previous work. Benchmark data consistently show that our approach achieves very good results within an acceptable run time.

Acknowledgements

This work was supported in part by the National Science Council of R.O.C. under grant number of NSC 93-2215-E-033-004.

References

[1] J.P. Fishburn, "Clock Skew Optimization", IEEE Trans. on Computers, Vol. 39, No. 7, pp. 945—951, 1990.

[2] L. Benini, P. Vuillod, A. Bogliolo, and G. De Micheli, "Clock Skew Optimization for Peak Current Reduction", Journal of VLSI Signal Processing, vol. 16, pp. 117—130, 1997.

[3] A. Vittal and M. Marek-Sadowska, "Power-Optimal Buffered Clock Tree Design", in the Proc. of IEEE/ACM Design Automation Conference, pp. 497—502, 1995.

[4] J.P. Fishburn, "Solving a System of Difference Constraints with Variables Restricted to a Finite Set", Information Processing Letters, vol. 82, no. 3, pp. 143—144, 2002.

[5] K. Ravindran, A, Keuhlmann, and E. Sentovich, "Multi-Domain Clock Skew Scheduling", in the Proc. of IEEE/ACM International Conference on Computer Aided Design, pp. 801—808, 2003.

[6] A. Vittal, H. Ha, F. Brewer, and M. Marek-Sadowska, "Clock Skew Optimization for Ground Bounce Control", in the Proc. of IEEE/ACM International Conference on Computer Aided Design, pp. 395—399, 1996.

Table IV. The results of zone-based scheduling.

circuit	500 variables in a zone			1000 variables in a zone			2000 variables in a zone			whole-circuit	
	peak current	tackled zones	CPU time (s)	peak current	tackled zones	CPU Time (s)	peak current	tackled zones	CPU time (s)	peak current	CPU time (s)
S3271	4*c	8	10	4*c	6	14	4*c	4	88	4*c	88
S3384	5*c	10	5	5*c	6	8	5*c	4	14	5*c	24
S5378	11*c	7	7	11*c	5	28	11*c	2	54	11*c	59
S6669	11*c	10	11	11*c	6	20	11*c	5	26	11*c	122
S9234	7*c	8	23	7*c	7	25	7*c	5	37	7*c	232
S13207	16*c	28	205	16*c	13	238	16*c	11	285	16*c	362
S15850	15*c	29	501	15*c	19	557	15*c	12	712	--	--
S35932	288*c	35	133	288*c	18	169	288*c	9	240	288*c	271
S38584	38*c	54	334	37*c	32	420	36*c	21	513	--	--

Low Area Pipelined Circuits by Multi-clock Cycle Paths and Clock Scheduling

Bakhtiar Affendi Rosdi, Atsushi Takahashi

Department of Communications and Integrated Systems, Tokyo Institute of Technology
2-12-1-S3-58 Ookayama, Meguro-ku, Tokyo 152-8552 Japan
Tel:+81-3-5734-2665 Fax:+81-3-5734-2902
E-mail:{fendi, atushi}@lab.ss.titech.ac.jp

Abstract— A new algorithm is proposed to reduce the number of intermediate registers of a pipelined circuit using a combination of multi-clock cycle paths and clock scheduling. The algorithm analyzes the pipelined circuit and determines the intermediate registers that can be removed. An efficient subsidiary algorithm is presented that computes the minimum feasible clock period of a circuit containing multi-clock cycle paths. Experiments with a pipelined adder and multiplier verify that the proposed algorithm can reduce the number of intermediate registers without degrading performance, even when delay variations exist.

I. Introduction

The sustained progress of VLSI technology has led to increasing wire delays, shrinking clock period and growing chip size. Circuit pipelining is one technique that has been used in order to shrink the clock period. Pipelining is a method in which a circuit is divided into a small number of stages and intermediate registers are inserted between stages to store the intermediate data. With this method, extra circuit area is required to situate the additional intermediate registers and the size of the clock tree is also increased.

Recently, to overcome this problem, several studies have been carried out on wave pipelining [1], which is a method of speeding up the circuit without the insertion of intermediate registers. However, wave pipelining requires tighter timing constraints. In wave pipelining, there may exist a number of 'waves' of data in a circuit at any given time. Therefore, to avoid data collisions, delay balancing is required, which increases the circuit area.

This paper presents a new algorithm to reduce the number of intermediate registers of a pipelined circuit by using a combination of multi-clock cycle paths and clock scheduling. A multi-clock cycle path is a path from register to register where data transmission takes more than one clock period. Note that in wave pipelining, all paths are multi-clock cycle paths. Introducing a multi-clock cycle path into a pipelined circuit allows some intermediate registers to be removed. However, as mentioned above, certain timing constraints must be satisfied. Therefore, there is a tradeoff between area reduction from register removal and area increase from delay balancing. Clock scheduling is a technique in which the clock skew of a register is intentionally introduced to improve circuit performance by relaxing the timing constraints. Using clock scheduling, more intermediate registers can be removed,

without the need for delay balancing.

The minimum feasible clock period in terms of clock scheduling is obtained by linear programming [2], by graph-theoretic approaches with binary search [3, 4], or by graph-theoretic approaches without binary search [5]. Graph-theoretic approaches are based on construction of a constraint graph that represents various constraints and which can handle a circuit of practical size. The constraints are feasible if and only if the constraint graph contains no negative cycle. In graph-theoretic approaches with binary search [4], the Bellman-Ford shortest path algorithm is used to decide whether the graph contains a negative cycle and a simple negative cycle detection method is employed to increase speed. The algorithm proposed in [4] is for a circuit that contains single-clock cycle paths only. However, when the algorithm [4] is applied to a circuit containing multi-clock cycle paths, there are some cases for which the minimum feasible clock period cannot be determined. The clock period for such a circuit is bounded above, unlike the situation for a circuit containing only single-clock cycle paths. This range of feasible clock periods has to be taken into account in clock-schedule design.

In this paper, we initially discuss the constraints on a circuit containing multi-clock cycle paths. These constraints take into account the range of feasible clock periods required to make the circuit tolerant of clock jitter. Using the constraints, we enhance the algorithm in [4] to find the minimum feasible clock period of a circuit that contains multi-clock cycle paths. The enhanced minimum clock-period algorithm uses the existence of a negative cycle to narrow the binary search interval efficiently. A negative cycle exists whenever the constraints are infeasible. Then, we propose an algorithm to reduce the number of intermediate registers of a pipelined circuit by introducing multi-clock cycle paths with clock scheduling. In the proposed algorithm, all intermediate registers of the pipelined circuit are initially removed. Then the minimum feasible clock period of the resulting circuit is computed by the proposed minimum clock-period algorithm. If this value is too high, i.e. greater than the target minimum clock period, then intermediate registers are repeatedly inserted into the multi-clock cycle paths until the minimum feasible clock period has been sufficiently reduced.

Experiments with a pipelined adder and multiplier verify that, given a particular target clock-period range, the proposed algorithm can reduce the number of intermediate registers, even when delay variations are present.

II. Preliminaries

We consider a circuit with a single clock consisting of registers linked by combinatorial circuits. The clock timing $s(v)$ of register v is the difference in clock signal arrival time between v and an arbitrarily chosen (perhaps hypothetical) reference register. The set of clock timings is called a clock-schedule.

We make the basic assumption that a circuit works correctly if the following two types of constraint are satisfied for each register pair with signal propagation [2, 6]:

Setup Constraint

$$s(u) - s(v) \leq \beta_{u,v} T - d_{\max}(u, v) \qquad (1)$$

Hold Constraint

$$s(v) - s(u) \leq d_{\min}(u, v) - \alpha_{u,v} T \qquad (2)$$

where T is the clock period, $d_{\max}(u, v)$ ($d_{\min}(u, v)$) is the maximum (minimum) propagation delay from register u to register v along a combinatorial circuit, and $\beta_{u,v}$ and $\alpha_{u,v}$ are given constants ($\beta_{u,v} > \alpha_{u,v} \geq 0$). Note that for a pair of registers with a single-clock cycle path, $\beta_{u,v}$ and $\alpha_{u,v}$ are given by 1 and 0, respectively. This formulation is sufficiently general to deal with multi-clock cycle paths, a mixture of positive-edge and negative-edge registers, latch based circuitry, and multi-clocks that have different periods.

If $\alpha_{u,v}$ is 0 for every pair, the feasible clock period has no upper bound, i. e. if the clock period T is feasible then any T' (where $T' \geq T$) is feasible. However, the feasible clock period is bounded above if $\alpha_{u,v}$ is not 0 for some pair (u, v).

From the above constraints, when the clock schedule and the signal propagation delay are known, the minimum and maximum feasible clock period, $T_{\min}$ and $T_{\max}$, can be determined from the setup and hold constraints, respectively.

If the clock timing is not fixed, then $T_{\min}$ and $T_{\max}$ depend on each other. $T_{\min}$ has to be minimized under the constraint that the circuit works correctly throughout a certain clock-period range, in order for the circuit to tolerate clock jitter. The above constraints become:

Setup Constraint

$$s(u) - s(v) \leq \beta_{u,v} T_{\min} - d_{\max}(u, v) \qquad (3)$$

Hold Constraint

$$s(v) - s(u) \leq d_{\min}(u, v) - \alpha_{u,v} T_{\min} - \alpha_{u,v} \delta \qquad (4)$$

where δ is the clock-period range, i.e. $\delta = T_{\max} - T_{\min}$. Therefore if δ is given, then, by using the above constraints, clock timings can be determined so that the circuit works correctly for a clock period between $T_{\min}$ and $T_{\min} + \delta$. In the following, our target is to minimize $T_{\min}$ under the constraint that the circuit is feasible throughout the given clock-period range, i.e. that constraints (3) and (4) hold.

These constraints are represented by the constraint graph $G(V, E)$ of the circuit, which is defined as follows: a vertex $v \in V$ corresponds to a register; a directed edge $(u, v) \in E$ corresponds to either type of constraint; an edge (u, v) corresponding to the setup (hold) constraint

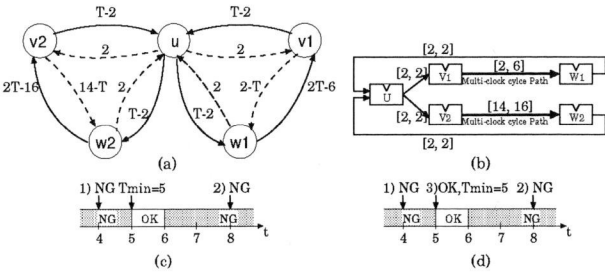

Fig. 1. (a) Constraint graph G^1. (b) Circuit graph CG^1 (containing multi-clock cycle paths). (c) Min clock-period computation by the algorithm shown in [4]. (d) Min clock-period computation by proposed algorithm.

is called a Z-edge (D-edge), and the weight $w(u, v)$ of (u, v) is $\beta_{u,v} T - d_{\max}(u, v) (d_{\min}(u, v) - \alpha_{u,v} \delta - \alpha_{u,v} T)$. The constraint graph G corresponding to clock period t is denoted by G_t.

In a constraint graph G, for any cycle C, the cycle weight $w(C)$ is the sum of edge weights over the cycle. The cycle weight can be expressed as $kT + w$, where T is the clock period and k and w are constants.

Definition 1 *In the constraint graph G, a cycle C for which $w(C) = kT + w$ is said to be of **type 0**, **type P**, or **type M**, as $k = 0$, $k > 0$, or $k < 0$, respectively.*

Theorem 1 *Let C be a negative cycle in the constraint graph G_t. If C is of **type 0**, then for any t', there exists a negative cycle in the constraint graph $G_{t'}$.*

Theorem 2 *Let C be a cycle in the constraint graph G such that $w(C) = kT + w$. If C is of **type P**, then for any $t < w/k$, there exists a negative cycle in the constraint graph G_t, whereas, if C is of **type M**, then for any $t > -w/k$, there exists a negative cycle in the constraint graph G_t.*

Definition 2 *Let C be a cycle in the constraint graph G such that $w(C) = kT + w$. Then $Bound(C) = w/k$, $-w/k$, or ∞, according to whether C is of **type P**, **type M**, or **type 0**, respectively.*

Example: For the constraint graph G^1 shown in Fig. 1 (a), the cycle $C_1 = (u, w2, v2, u)$ with $w(C_1) = 4T - 20$ is of **type P** and $Bound(C_1) = 5$. The cycle $C_2 = (u, v1, w1, u)$ with $w(C_2) = -T + 6$ is of **type M** and $Bound(C_2) = 6$.

Note that, in a constraint graph of a circuit that contains just single-clock cycle paths, only **type P** and **type 0** cycles can exist, whereas in a constraint graph of a circuit that contains multi-clock cycle paths, all three cycle types can be present.

III. Minimum Feasible Clock Period

A. A circuit that contains just single-clock cycle paths

The minimum feasible clock period of a circuit that contains just single-clock cycle paths can be determined by graph-theoretic approach with binary search [4]. The maximum signal propagation delay from a register to the same register gives a lower bound of feasible clock period.

3A-4
2006 Asia and South Pacific Design Automation Conference

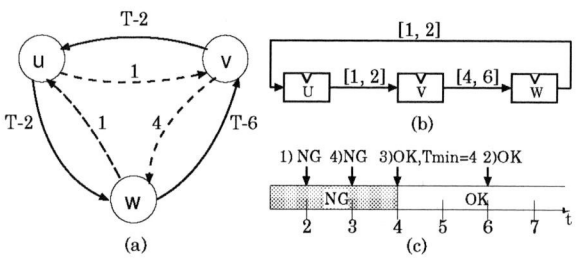

Fig. 2. (a) Constraint graph G^2. (b) Circuit graph CG^2 (containing just single-clock cycle paths). (c) Min clock-period computation by the algorithm shown in [4].

The difference of the maximum and minimum signal propagation delay from a register to another register gives also a lower bound of feasible clock period. They adopt the larger of these two lower bounds as an initial lower bound L of the binary search. They adopt the maximum signal propagation delay between registers as an initial upper bound U since it gives a feasible clock period even in zero clock-skew framework.

In the algorithm [4], the initial lower bound L and upper bound U are initially checked. If L is feasible the algorithm is stopped and output L as the minimum feasible clock period. While, if L is infeasible, U is checked to confirm there is no negative cycle of **type 0**. If there exists a negative cycle of **type 0**, the circuit is infeasible and the algorithm is stopped. If U is feasible, the algorithm does binary search by adjusting the lower and upper bounds to determine the minimum feasible clock period.

Using the algorithm shown in [4] let us determine the minimum feasible clock period of the circuit shown in Fig. 2 (b). Note that, throughout this paper, the precision used is 1. In this example, initial lower bound $L = 2$ and initial upper bound $U = 6$. So, the algorithm does binary search between 2 and 6 as follows:

Check $L(L=2)$: G_2^2 is infeasible, cycle $C_1 = (u, w, v, u)$ with $w(C_1) = 3T - 10$ is negative, so next step is check U.

Check $U(U=6)$: G_6^2 is feasible, so next step is check $M = (U + L/2)$.

Check $M(M = 4 = (6 + 2/2))$: G_4^2 is feasible, so 4 becomes new upper bound U.

Check $M(M = 3 = (4 + 2/2))$: G_3^2 is infeasible, cycle $C_1 = (u, w, v, u)$ with $w(C_1) = 3T - 10$ is negative. Since $U - L = 1$, output 4 as the minimum feasible clock period T.

The flow when we apply the algorithm is shown in Fig. 2 (c).

B. A circuit that contains multi-clock cycle paths

For a circuit that contains just single-clock cycle paths, if the circuit is feasible then the circuit is feasible at the initial upper bound U, otherwise the circuit is infeasible. However, for a circuit that contains multi-clock cycle paths, even if the circuit is infeasible at initial upper bound U, there are some possibilities that the circuit is feasible at clock period t ($t < U$ or $t > U$).

Input: circuit graph CG, target clock-period range δ.
Output: minimum feasible clock period T.

1. Construct constraint graph G by CG and δ.
2. $L_{\text{self}} := \max_{(u,u) \in E_{hold}} d_{\max}(u, u)$.
   $L_{\text{diff}} := \max_{(u,v) \in E_{hold}} (d_{\max}(u, v) - d_{\min}(u, v))$.
   $L := \max\{L_{\text{self}}, L_{\text{diff}}\}$.
   $U := \max_{(u,v) \in E_{hold}} \{(d_{\max}(u, v) + s(u) - s(v))/\beta_{u,v}\}$.
3. Check whether G_L is feasible.
   if there is no negative cycle in G_L return L.
   else if there exists a negative cycle C of type 0 or type M return ∞.
4. Check whether G_U is feasible.
   if there exists a negative cycle C
     case C is of type 0 return ∞.
     case C is of type P then repeat the following
       $L := Bound(C)$.
       if there is no negative cycle in G_L return L.
       else if there exists a negative cycle C'
         if C' is of type 0 or type M return ∞.
         else $C \leftarrow C'$.
     case C is of type M then $U := Bound(C)$ and if $U < L$ return ∞.
5. While $(U - L > \varepsilon)$ do
     $M := (U + L)/2$.
     check whether G_M is feasible.
     if there is no negative cycle in G_M then $U := M$.
     else let C be the negative cycle.
       case C is of type 0 return ∞.
       case C is of type P then
         $L := Bound(C)$.
         if there is no negative cycle in G_L return L.
         else if there exists a negative cycle C' of type 0 or type M return ∞.
       case C is of type M then $U := Bound(C)$.
   endwhile.
6. $T := U$. return T.

Fig. 3. Minimum feasible clock period algorithm of the circuit that contains multi-clock cycle paths.

When the initial upper bound U is infeasible, the algorithm [4] concludes that the circuit is infeasible at any clock period and stop. For example, let us determine the minimum feasible clock period of the circuit shown in Fig. 1 (b). Initial lower bound $L = 4$ and initial upper bound $U = 8$. So, the algorithm does binary search between 4 and 8 as follows:

Check $L(L=4)$: G_4^1 is infeasible, cycle $C_1 = (u, w2, v2, u)$ with $w(C_1) = 4T - 20$ is negative, so next step is check U.

Check $U(U=8)$: G_8^1 is infeasible, cycle $C_2 = (u, v1, w1, u)$ with $w(C_2) = -T + 6$ is negative and the algorithm is stopped.

The flow when we apply the algorithm is shown in Fig. 1 (c). As you can see from the above example, the algorithm [4] is stopped after checking the initial upper bound U and concludes that the circuit is infeasible at any clock period.

However as we mentioned earlier, the conclusion is correct for a circuit that contains just single-clock cycle paths, while for the circuit that contains multi-clock cycle paths the conclusion might be incorrect. Therefore, the above approach might miss the minimum feasible clock period. In fact, in this case, the algorithm cannot determine the minimum feasible clock period which is 5.

We enhance the algorithm that has been introduced in [4] to determine the minimum feasible clock period of a circuit that contains multi-clock cycle paths. The algorithm does binary search between lower and upper bounds

262

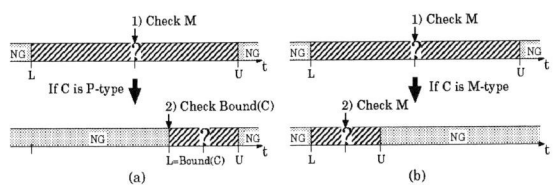

Fig. 4. (a) If C is of **type P** then $L = Bound(C)$ and check L. (b) If C is of **type M** then $U = Bound(C)$.

same as in the algorithm shown in [4]. We extend the algorithm [4] by introducing checking the type of cycle when there exists a negative cycle in the constraint graph. If the circuit is infeasible at given clock period, there exists a negative cycle. The lower and upper bounds are adjusted based on the type of negative cycle and the *Bound* value.

The new algorithm to determine the minimum feasible clock period of a circuit that contains multi-clock cycle paths is shown in Fig. 3.

For the initial value of lower bound L and upper bound U of the binary search, we adopt the same approach as in the algorithm shown in [4]. Initial lower bound L will be checked whether it is feasible or not, if L is feasible, then output L as the minimum feasible clock period. Otherwise, there exists a negative cycle C. If C is of **type 0** or **type M**, the circuit is infeasible and the algorithm is stopped. While, if C is of **type P** then an initial upper bound U will be checked whether it is feasible or not.

If the initial upper bound U is feasible, then the algorithm does binary search to determine the minimum feasible clock period. Otherwise, there exists a negative cycle C. In case C is of **type 0**, the circuit is infeasible and the algorithm is stopped. In case C is of **type P**, $Bound(C)$ is our new lower bound L and L will be checked whether it is feasible or not. If our new lower bound L is feasible then output L as the minimum feasible clock period. Otherwise, repeat until L is feasible or C is of **type 0** or **type M**, where the circuit is infeasible and the algorithm is stopped. In case C is of **type M**, $Bound(C)$ is our new upper bound U and the algorithm will check whether $U < L$ or not. If $U < L$, then the circuit is infeasible and the algorithm is stopped. Otherwise, our algorithm does binary search by adjusting the lower and upper bounds to determine the minimum feasible clock period.

In binary search, the algorithm will check whether the constraint graph $G_M(M = (U + L)/2)$ containing any negative cycle or not. If there are no negative cycles in G_M, then M is our new upper bound U and continue do binary search. Otherwise, if there exists a negative cycle in G_M then the algorithm will check the type of it. In case C is of **type 0**, the circuit is infeasible and the algorithm is stopped. From Theorem 2, in case C is of **type P** then $Bound(C)$ is our new lower bound L and L will be checked whether it is feasible or not (Refer Fig. 4 (a)). If our new lower bound L is feasible then output L as the minimum feasible clock period, otherwise, continue do binary search. In case C is of **type M** then $Bound(C)$ is our new upper bound U (Refer Fig. 4 (b)), and continue do binary search.

Using the proposed algorithm, let us find the minimum feasible clock period of the circuit shown in Fig. 1 (b). Our target clock-period range δ is 0. Initial lower bound $L = 4$ and initial upper bound $U = 8$. So, the algorithm

does binary search between 4 and 8 as follows:

Check $L(L = 4)$: G_4^1 is infeasible, cycle $C_1 = (u, w2, v2, u)$ with $w(C_1) = 4T - 20$ is negative and of **type P**. So next step is check U.

Check $U(U = 8)$: G_8^1 is infeasible, cycle $C_2 = (u, v1, w1, u)$ with $w(C_2) = -T + 6$ is negative and of **type M** and $Bound(C_2) = 6$, therefore 6 is our new U.

Check $M(M = 5 = (6 + 4/2))$: G_5^1 is feasible. Since $U - L = 1$, output 5 as the minimum feasible clock period T.

The flow when we apply the algorithm is shown in Fig. 1 (d). The algorithm can determine the minimum feasible clock period of the circuit which is 5.

IV. REDUCTION ON THE NUMBER OF INTERMEDIATE REGISTERS

In this paper we consider a problem on how to reduce the number of intermediate registers of a pipelined circuit, subject to the minimum feasible clock period is lower than or equal to the original circuit and works at target clock-period range

In the proposed algorithm, all intermediate registers of the pipelined circuit are initially removed. Then the minimum feasible clock period is computed using the algorithm shown in Fig. 3. When the intermediate registers are removed, the minimum delay is reduced compared with the original circuit. Therefore, the hold constraints which correspond to the D-edges in the cycles become tight. If the circuit is infeasible at any clock period, there exists a negative cycle in the constraint graph. As we mentioned earlier, the hold constraints are the reason why the timing constraints become tight. Therefore to make the circuit feasible at target clock-period range, an intermediate register which corresponds to a D-edge contained in the negative cycle is inserted to the multi-clock cycle path.

Our target is to get a circuit, where the minimum feasible clock period is lower than or equal to the original circuit. If the minimum feasible clock period of the obtained circuit is higher than the original circuit, the minimum feasible clock period needs to be reduced. Our proposed algorithm find a critical cycle in the constraint graph that decide the minimum feasible clock period of the obtained circuit. Then, in order to reduce the minimum feasible clock period, an intermediate register which corresponds to a D-edge contained in the critical cycle is inserted to the multi-clock cycle path. The details of our proposed algorithm is omitted here due to the space constraint.

To explain the behavior of the algorithm, we implemented the algorithm to the pipelined circuit shown in Fig. 5. In this example, the timing of each I/O pin is fixed at 0, while the timing of each register is scheduled. We also assume that Setup and Hold Time for registers are 0 and the minimum and maximum delay of the intermediate registers are 1 and 2, respectively. Our target clock-period range δ is 4. Note that in the constraint graph, vertices In and Out are the same vertex because we fix the timings of I/O pins to 0. In the figure, vertices In and Out are draw in different vertex to make it easy

3A-4   2006 Asia and South Pacific Design Automation Conference

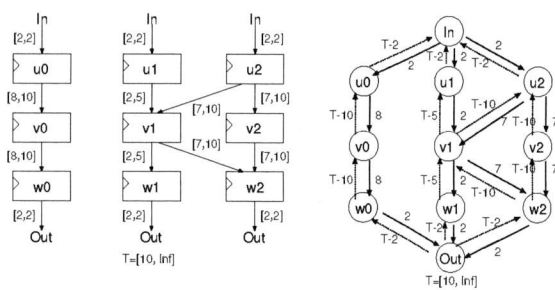

Fig. 5. Pipelined circuit with intermediate registers and the corresponding constraint graph G^{in}. $T_{\min}(G^{in}) = 10$.

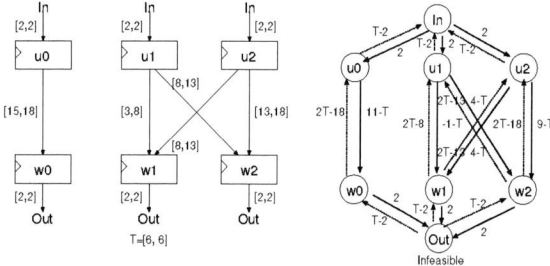

Fig. 6. Pipelined circuit after removing all intermediate registers and the corresponding constraint graph G^0 (D-edge weight of the multi-clock cycle paths in G^0 is reduced 4 compared with the corresponds minimum delay in the circuit).

to understand. For the original circuit with zero clock-skew, the minimum feasible clock period $T_{\min}(G^{in})$ is 10. The circuit after removing the intermediate registers v_0, v_1, v_2 is shown in Fig. 6. Note that the D-edge weight of the multi-clock cycle paths in the constraint graph is reduced 4 compared with the corresponds minimum delay in the circuit because δ is 4. When the minimum clock period of constraint graph G^0 is computed, there exists a negative cycle $C^0 = (in, u_1, w_1, out)$ which is of **type M** in the constraint graph G^0_9. The minimum clock period algorithm concludes that it is infeasible. This means that the circuit works at clock period 6 but δ is not secured. Since path (u_1, w_1) is a D-edge which is multi-clock cycle path, intermediate register v_1 is inserted. The circuit after inserting the intermediate register v_1 is shown in Fig. 7. $T_{\min}(G^1) = 11 > T_{\min}(G^{in})$, therefore the minimum fea-

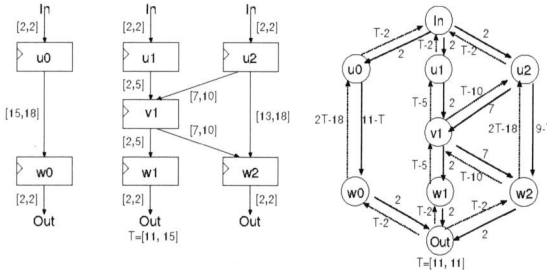

Fig. 7. Pipelined circuit after inserting the intermediate register v_1 and the corresponding constraint graph G^1 (D-edge weight of the multi-clock cycle paths in G^1 is reduced 4 compared with the corresponds minimum delay in the circuit). $T_{\min}(G^1) = 11 > (T_{\min}(G^{in}) = 10$.

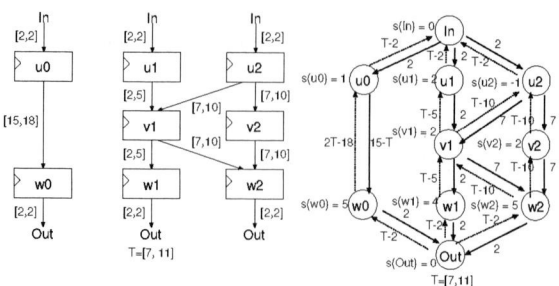

Fig. 8. Pipelined circuit after inserting the intermediate registers v_1 and v_2, and the corresponding constraint graph G^2. $T_{\min}(G^2) = 7$.

TABLE I
STATISTICS OF ADDER AND MULTIPLIER

circuit	# FF	Circuit delay [ps]			
		1st stage		2nd stage	
		min	max	min	max
4bitadd	32	588	2454	588	2840
8bitadd	60	598	4079	598	4474
16bitadd	116	598	7239	598	7634
16bitmul	120	757	5075	373	4050

sible clock period of the obtained circuit need to be reduced. The critical cycle $C^1 = (u2, w2, v1, u2)$ is found in the constraint graph G^1_{11}. Since path (u_2, w_2) is a D-edge which is multi-clock cycle path, intermediate register v_2 is inserted. The circuit after inserting the intermediate register v_2 is shown in Fig. 8. $T_{\min}(G^2) = 7 < T_{\min}(G^{in})$, so the algorithm stop and output the circuit and clock timings as shown in Fig. 8.

V. EXPERIMENTS

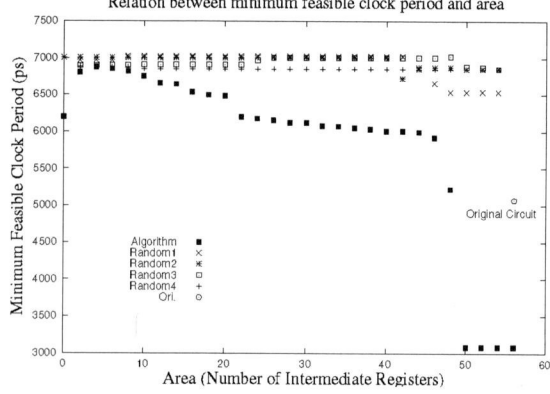

Fig. 9. Relation between $T_{\min}(ps)$ and # Int. FF of a 16 bit multiplier. $\delta = 500ps$. Delay variation = 0%.

The proposed algorithms were written in C++ and implemented on a Pentium 4 (CPU 3GHz, memory 513764kb). Since there are no benchmark examples of pipelined circuits, two simple examples, briefly described below, were constructed for our experiments.

264

TABLE II
Experimental Results

Circuit	δ (ps)	Delay variation = 0%				Delay variation = 20%				
		T_{min} (ps)	(%)[a]	Int. FF #	(%)	T_{min} (ps)	(%)[a]	(%)[b]	Int. FF #	(%)
4bit add	Ori.	2840	(100)	10	(100)	3093	(110)	(100)	10	(100)
	0	1422	(50)	0	(0)	2047	(72)	(66)	0	(0)
	100	1522	(54)	0	(0)	2147	(76)	(69)	0	(0)
	200	1622	(57)	0	(0)	2247	(79)	(73)	0	(0)
	300	1722	(61)	0	(0)	2347	(83)	(76)	0	(0)
	400	1822	(64)	0	(0)	2447	(86)	(79)	0	(0)
	500	1922	(68)	0	(0)	2547	(90)	(82)	0	(0)
8bit add	Ori.	4474	(100)	18	(100)	4890	(110)	(100)	18	(100)
	0	1702	(38)	0	(0)	2629	(59)	(54)	0	(0)
	100	1802	(40)	0	(0)	2729	(61)	(56)	0	(0)
	200	1902	(43)	0	(0)	2829	(63)	(58)	0	(0)
	300	2002	(45)	0	(0)	2929	(66)	(60)	0	(0)
	400	2102	(47)	0	(0)	3029	(68)	(62)	0	(0)
	500	2202	(49)	0	(0)	3129	(70)	(64)	0	(0)
16bit add	Ori.	7634	(100)	34	(100)	8366	(110)	(100)	34	(100)
	0	3024	(40)	0	(0)	4533	(59)	(54)	0	(0)
	100	3124	(41)	0	(0)	4633	(61)	(55)	0	(0)
	200	3224	(42)	0	(0)	4733	(62)	(57)	0	(0)
	300	3324	(44)	0	(0)	4833	(63)	(58)	0	(0)
	400	3424	(45)	0	(0)	4933	(65)	(59)	0	(0)
	500	3524	(46)	0	(0)	5033	(66)	(60)	0	(0)
16bit mul	Ori.	5075	(100)	56	(100)	5551	(110)	(100)	56	(100)
	0	4722	(93)	48	(86)	3590	(71)	(65)	49	(88)
	100	4822	(95)	48	(86)	3590	(71)	(65)	49	(88)
	200	4922	(97)	48	(86)	3590	(71)	(65)	49	(88)
	300	5022	(99)	48	(86)	3590	(71)	(65)	49	(88)
	400	3086	(61)	49	(88)	3670	(72)	(66)	49	(88)
	500	3086	(61)	49	(88)	3770	(74)	(68)	49	(88)

[b]Compared with original circuit with zero clock-skew and delay variation = 20%.

[a]Compared with original circuit with zero clock-skew and delay variation = 0%.

- n-bit ($n = 4, 8, 16$) add: A 2-stage adder that added four n-bit numbers (A, B, C and D) [7]. The first stage computed the partial sum A + B and C + D and the second stage computed the final sum. Each adder was of ripple-carry type.

- 16-bit mul: A 2-stage multiplier that multiplied two 16-bit numbers. The first stage used a carry-save adder with Wallace tree structure [8] and the second stage used a carry-look-ahead adder.

The statistics of the circuits are shown in Table I. The ROHM 0.35 process library was used for these experiments. The timing of each I/O pin was scheduled as well as the timing for each register.

Table II shows the results when the algorithm shown in section IV was implemented. Ori. is the original circuit containing the intermediate registers and with the clock timing of all registers fixed at 0 (zero clock-skew). "δ" and "$T_{min}(ps)$" are the target clock-period range and output

minimum feasible clock period, respectively. "Int. FF (#)" is the number of intermediate registers, and "Int. FF (%)' is the percentage of the number of intermediate registers present compared with the total in the original circuit. Delay variation was 20%, i.e. the delay variation for each gate and register was set at ±10%.

The results show that by a combination of multi-clock cycles and clock scheduling, the number of intermediate registers and the minimum feasible clock period can be reduced, even in the presence of delay variations in gates and registers.

The relation between the minimum feasible clock period and the number of intermediate registers of the 16 bit multiplier is shown in Fig. 9. In the graph, the label "Algorithm" indicates results using the proposed algorithm for insertion of the intermediate registers, while "Random1-4" labels indicate results when the intermediate registers are inserted randomly. The graph shows that the proposed algorithm can construct an equivalent circuit using fewer registers and with a smaller minimum feasible clock period.

VI. Conclusion

It has been shown that the number of intermediate registers of a pipelined circuit can be reduced by implementing a multi-clock cycle path technique together with clock scheduling. The proposed algorithm inserts intermediate registers without considering delay balancing in order to make the circuit work correctly throughout the target clock-period range. We believe that by using delay balancing together with intermediate register insertion, circuit area can be further reduced. This is a topic for future investigation.

Acknowledgements

This work is supported by VLSI Design and Education Center (VDEC), the University of Tokyo in collaboration with Synopsys, Inc., Cadence Design Systems, Inc., Rohm Corporation and Toppan Printing Corporation.

References

[1] W. J. Kim and Y. Kim, "Clocking for correct functionality on wave pipelined circuits," in *Proc. IEEE International ASIC/SOC Conference*, 2003, pp. 161–164.

[2] J. P. Fishburn, "Clock skew optimization," *IEEE Trans. on Computers*, vol. 39, no. 7, pp. 945–951, 1990.

[3] R. B. Deokar and S. S. Sapatnekar, "A graph-theoretic approach to clock skew optimization," in *Proc. International Symposium on Circuits and Systems (ISCAS)*, vol. 1, 1994, pp. 407–410.

[4] A. Takahashi, "Practical fast clock schedule design algorithms," in *Proc. 18th Karuizawa Workshop*, 2005, pp. 515–520.

[5] A. Takahashi and Y. Kajitani, "Performance and reliability driven clock scheduling of sequential logic circuits," in *Proc. Asia and South Pacific Design Automation Conference (ASP-DAC)*, 1997, pp. 37–42.

[6] B. A. Rosdi and A. Takahashi, "Reduction on the usage of intermediate registers for pipelined circuits," in *Proc. the Workshop on Synthesis and System Integration of Mixed Technologies (SASIMI 2004)*, 2004, pp. 333–338.

[7] S.Malik, K.J.Singh, R.K.Brayton, and A.Sangiovanni-Vincentelli, "Performance optimization of pipelined circuits," in *Proc. IEEE/ACM International Conference on Computer Aided Design (ICCAD)*, 1990, pp. 410–413.

[8] C. Wallace, "A suggestion for fast multiplier," *IEEE Trans. on Electronic Computers*, vol. 13, no. 2, pp. 14–17, 1964.

A Transduction-based Framework to Synthesize RSFQ Circuits

Shigeru Yamashita

Nara Institute of Sci. and Tech.
8916-5 Takayama
Ikoma 630-0192, Japan
Tel: +81-743-72-5301
Fax: +81-743-72-5309
e-mail: ger@is.naist.jp

Katsunori Tanaka

NEC Corporation
1753 Shimonumabe, Nakahara-ku,
Kawasaki 211-8666, Japan
Tel: +81-44-431-7541
Fax: +81-44-431-7589
e-mail k-tanaka@jm.jp.nec.com

Hideyuki Takada

Kyoto University
Yoshida-Honmachi, Sakyo
Kyoto 606-8501, Japan
Tel: +81-75-753-5375
Fax: +81-75-753-4970
e-mail: htakada@db.soc.i.kyoto-u.ac.jp

Koji Obata   Kazuyoshi Takagi

Nagoya University
Furo-cho, Chikusa-ku, Nagoya 464-8603, Japan
Tel: +81-52-789-4597
Fax: +81-52-789-3798
{obata, ktakagi}@takagi.nuie.nagoya-u.ac.jp

Abstract— In this paper, we propose a new framework to synthesize rapid single flux quantum (RSFQ) logic circuits. In our framework, we construct a virtual cell, which we call "2-AND/XOR," from the RSFQ logic primitives. By using 2-AND/XOR cells, we can successfully adopt the conventional logic design techniques into our framework, and thus we can successfully generate RSFQ circuits in reasonable time even for large benchmark circuits that have not been reported in the existing researches.

I. INTRODUCTION

Rapid single flux quantum (RSFQ) integrated circuits composed of Josephson-junction devices have been intensively studied because of their potentially high performance with high clock frequency and extremely low power consumption [7]. RSFQ technology has the following features [7].

- Ultrafast digital signals can be passed along the chips ballistically with a propagation speed approaching that of light.

- Intrinsic switching time of the Josephson junction is also very short, typically a few picoseconds.

- The power dissipated by a Josephson junction is typically below one microwatt. Hence, the problem of removal of heat is quite solvable. Currently this is not essentially true since we need some cooling system for the whole RSFQ circuits themselves. Although special cooling systems cannot be used for a consumer PCs, we may be able to afford special cooling devices for high-end computing servers. Also, at this

moment, we cannot deny the possibility that we can construct ultra low power systems by RSFQ technology in the future.

- The Josephson junction fabrication technologies are considerably simpler than those of the conventional semiconductor (both Si and GaAs) transistors with similar design rules.

Although it currently requires a refrigeration technique, such as liquid-helium cooling, the benefits of RSFQ technology would become huge in the near future.

As the current CMOS technology is approaching "Red Brick Wall" [1], the RSFQ technology is considered as one of the promising next generation technologies. Indeed, the RSFQ device is listed as one of the most promising ones in the "Emerging Research Devices" list by ITRS [1]. Actually RSFQ digital circuits containing several thousands of Josephson junctions have been successfully implemented and their high performance has been confirmed [12].

Although it still appears quite challenging to realize primitive RSFQ logic cells efficiently, it is also a challenge to establish systematic logic design methods for large RSFQ circuits. The reason is that the logic primitives in RSFQ circuits are very much different from those in the conventional CMOS technology, and thus we cannot directly utilize the conventional logic design techniques. Therefore, we also need to do researches for the logic design methodologies for RSFQ logic circuits.

Among the researches for logic design methods for RSFQ circuits, the cell-based methods [6, 16, 17] have been studied like conventional CMOS technology. In their methods, a logic primitive called "RSFQ D_2 flip-flop" is used to replace a node of BDD (Binary Decision Diagram) [2]. Recently a systematic design method [9]

0-7803-9451-8/06/$20.00 ©2006 IEEE.

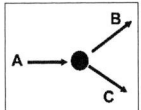 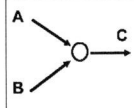 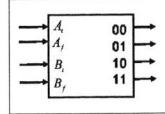

Fig. 1. SPL.   Fig. 2. CB.   Fig. 3. 2x2-Join.

TABLE I
2X2-JOIN PULSE OPERATION

Inputs				Outputs			
A_t	A_f	B_t	B_f	00	01	10	11
Pulse	No	Pulse	No	No	No	No	Pulse
Pulse	No	No	Pulse	No	No	Pulse	No
No	Pulse	Pulse	No	No	Pulse	No	No
No	Pulse	No	Pulse	Pulse	No	No	No

has been proposed to use another primitive called "2x2-Join" [5]. The method is also based on BDDs. For small circuits, BDD-based methods may generate good circuits. However, it is obvious that BDD-based methods such as [9] cannot be applied directly to large circuits since their manipulation of BDDs essentially takes a lot of time for the large functions. Thus it is desirable to have another approach that can be applied to large circuits.

For that purpose we propose a framework that is not based on the BDD manipulation. Our framework consists of two phases which are very much alike to the conventional logic design methodologies for AND/OR/NOT gates. At first phase, we generate initial circuits consisting of two-input nodes by using any conventional techniques. A two-input node is transformed to a virtual logic cell called *2-AND/XOR* cell which is introduced in this paper. Then, at the second phase, we optimize the circuit by using a transformation-based heuristic method like the conventional logic design. For that, we modify the original Transduction Method [8] to be suitable for the 2-AND/XOR cells.

A 2-AND/XOR cell exploits the property of a 2x2-Join fully, i.e., it can represent all the possible functions realized by a 2x2-Join. More importantly, by considering logic circuits consisting of 2-AND/XOR cells, we are able to adopt the conventional logic design techniques for RSFQ circuit design. Indeed our method can handle large circuits that have not been reported in existing researches [9]. Accordingly, our framework can be complementarity to the existing BDD-based methods. Also it would be possible to construct an efficient logic design system from the combination of our method and the above-mentioned BDD-based methods [6, 16, 17, 9].

This paper is organized as follows. In Sec. II, we provide minimum information to understand the contents of this paper. Then, Sec. III is devoted to explain our proposed logic design framework. We show some experimental results to demonstrate the effectiveness of our method in Sec. IV, and discuss the comparison between our method and related work in Sec. V. Finally, we conclude the paper in Sec. VI.

II. PRELIMINARIES

A. RSFQ Logic Primitives

In this paper, we consider a design framework of RSFQ logic circuits. In an RSFQ circuit, single flux quantum (SFQ) pulses are used for representing logic values. Pulses are generated and propagated by the combination of super-conducting rings with Josephson junctions. There has been proposed many logic primitives to manipulate pulses in RSFQ circuit in logic level. Among

them, in this paper, we use the following three logic primitives [5, 7].

SPL (Splitter) This logic primitive generates two pulses from a single pulse. We express this primitive as a black circle as shown in Fig. 1.

CB (Confluence Buffer) This logic primitive generates a single pulse when two input pulses arrive. (The two input pulses are not allowed to arrive at the very same time.) We express this primitive as a white circle as shown in Fig. 2.

2x2-Join 2x2-Join has four inputs and four outputs as shown in Fig. 3, and it generates one pulse at one of the four outputs depending on the combination of input pulses. The relationship between inputs and outputs are described in Table I. For example, if it receives two pulses at A_t and B_t, then it generates a pulse at the output 11. Note that the output pulse is not dependent on the arrival order of the input pulses.

B. Dual-Rail Logic Design

At the early times when researches for RSFQ circuit started, a clock signal was used to translate a pulse on a data line in a "clock window" as logic "1" and no pulse as logic "0" like conventional synchronous circuit design. The reason is that we need to specify the exact time when a pulse comes, or otherwise we cannot distinguish between logic "0" and logic "1" while pulse has not arrived yet. Therefore, unlike the conventional technologies, for RSFQ circuit logic design, careful delay estimation and clock design are required [4] since an improper arrival order of data and clock pulses leads to erroneous data transfer. Facing the above-mentioned timing problems, the RSFQ technology has been paying an attention to an asynchronous approach. More precisely, dual-rail data encoding is used to enables clock free data transfer [3]. In the dual-rail scheme, a pair of (true- and false-) data lines carries 1-bit binary information. The propagation of a pulse on the true-line or the false-line represents logic "1" and "0," respectively. No race occurs because only one pulse propagates either true- or false-line during 1-bit data transfer. Therefore, for a large circuit, it is indispensable to adopt dual-rail scheme, and thus, in our framework we use dual-rail scheme; we use two lines (true- line and false- line) to propagate 1-bit information of an intermediate logic function as will be mentioned in the next section.

3A-5

C. Transformation-based Optimization Methods

Since it is totally impossible to generate an optimal logic circuit by one method (especially for a large problem), we iteratively apply optimization methods to transform an initial circuit generated by some methods into a smaller circuit. This strategy should also be important for RSFQ circuit design. Thus, we utilize the Transduction Method [8], which is one of such optimization methods, in our framework for the same purpose. the Transduction Method is based on the concept of **permissible functions (PFs)**. Intuitively, a permissible function at a gate (or a connection) is a set of functions any one of which can be used at the gate (or the connection) without changing the functionality of the circuit. A compatible set of permissible functions (CSPFs) is defined as a set of PFs that can be used independently on a set of gates (and/or connections) without changing functionality. Originally CSPFs were defined for NOR gates [8]. Later, they were generalized to arbitrary Boolean nodes and called compatible observability don't cares (CODCs) [10]. CODCs differ also in that they are expressed in terms of intermediate signals. CODCs are used in SIS [13], and transformation-based optimization methods are widely used in many conventional logic synthesis tools. We refer readers to [8] for the details of the calculation of CSPFs and the transformation procedures.

III. THE PROPOSED FRAMEWORK TO DESIGN RSFQ LOGIC CIRCUITS

A. Overview of the Proposed Framework

As mentioned in the previous section, the logic primitives for RSFQ logic circuits are different from the conventional logic gates. Therefore, it is not possible to directly apply the conventional logic design techniques for designing RSFQ logic circuits. The difficulties are summarized as follows.

- A CB works in a very similar way to a conventional OR gate, but the two input pulses are not allowed to arrive at the same time for a CB. Thus two input functions, h_1 and h_2, to a CB should satisfy the condition $h_1 \cdot h_2 = 0$.

- A 2x2-Join works as generating a logic combination with respect to its input pulses as shown in Table I. However, the behavior of a 2x2-Join is not known for the combination of input pulses that are not described in the table (e.g., when a 2x2-Join receives pulses at both A_t and A_f). Thus, A_t and A_f (B_t and B_f also) should be exactly the negation to each other; we should always generate $\overline{f}$ as well if we want to use f as an intermediate function.

By considering the above issues, we restrict ourselves to design a logic circuit by using a virtual logic cell called 2-AND/XOR which will be mentioned below. As we will see soon, using 2-AND/XOR cells solves the above issues very naturally, and moreover, it utilizes almost all the abilities of 2x2-Joins.

B. 2-AND/XOR Cell

Before introducing our 2-AND/XOR cells, let us consider how to use 2x2-Joins to make logic functions. Suppose we want to construct f with respect to two intermediate functions h_1 and h_2. As we take dual-rail scheme, we also assume that there are $\overline{h_1}$ and $\overline{h_2}$. Then, we connect h_1, $\overline{h_1}$, h_2 and $\overline{h_2}$ to A_t, A_f, B_t and B_f, respectively, of a 2x2-Join as shown in Fig. 4. Then the outputs of the 2x2-Join, 00, 01, 10 and 11 generate pulses corresponding to the logic functions of $(\overline{h_1} \cdot \overline{h_2})$, $(\overline{h_1} \cdot h_2)$, $(h_1 \cdot \overline{h_2})$ and $(h_1 \cdot h_2)$, respectively. They are the four minterms of h_1 and h_2, and thus, any function with respect to h_1 and h_2 can be constructed by merging some of four outputs, 00, 01, 10 and 11, with CBs. We can also construct $\overline{f}$ by merging the outputs of the 2x2-Join that are not used for f. An example where $f = h_1 \cdot h_2$ is shown in Fig. 4. To sum up, by using a 2x2-Join with two CBs, we can construct any dual-rail logic function f (both f and $\overline{f}$) of two intermediate functions. Therefore, from the conventional logic point of view, we can consider this 2-input sub-circuit as a 2-input LUT (look-up table).

There is another usage of a 2x2-Join, i.e, we consider making multiple functions from a single 2x2-Join. For example, we can make AND and XOR functions of two inputs by using a single 2x2-Join, three CBs and three SPLs as shown in Fig. 5. Indeed, we can make all the possible sixteen 2-input functions at the same time by a single 2x2-Joins with many CBs and SPLs. However, since we use dual-rail logic, we do not need to consider the polarity of the inputs and the outputs of the functions, and therefore, it is enough to consider the following five functions with respect to inputs h_1 and h_2: $f_1 = \overline{h_1} \cdot \overline{h_2}$, $f_2 = \overline{h_1} \cdot h_2$, $f_3 = h_1 \cdot \overline{h_2}$, $f_4 = h_1 \cdot h_2$ and $f_5 = h_1 \oplus h_2$. (Note that we may want to implement some of these five functions *at the same time*; we need to have four functions for AND type functions at the same time.)

With the above discussions in mind, we introduce a virtual logic cell called 2-AND/XOR that has two inputs, h_1 and h_2, and five outputs, $f_1 = \overline{h_1} \cdot \overline{h_2}$, $f_2 = \overline{h_1} \cdot h_2$, $f_3 = h_1 \cdot \overline{h_2}$, $f_4 = h_1 \cdot h_2$ and $f_5 = h_1 \oplus h_2$ as shown in Fig. 6. Obviously we can design a 2-AND/XOR cell by a single 2x2-Join, CBs and SPLs. This cell corresponds to all the possible output functions that can be implemented by a single 2x2-Join. Also we can always generate negations of f_1 to f_5. It should also be noted that the usage of CBs in a 2-AND/XOR cell satisfies the condition of CBs, i.e., two input pulses do not arrive at the same time since an output pulse is generated at only one of four outputs of a 2x2-Join. Thus the problems mentioned in Sec. III-A are naturally solved if we use only 2-AND/XOR cells to design logic circuits. Moreover, a 2-AND/XOR cell also naturally corresponds to a circuit consisting of the conventional AND, XOR and NOT gates; we can utilize the conventional logic design techniques.

C. Initial Circuits Synthesis

As mentioned in the previous section, a sub-circuit realizing any 2-input logic function can be naturally mapped

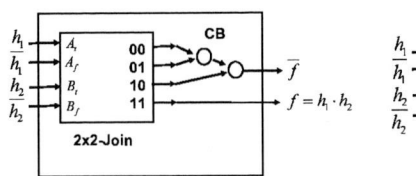

Fig. 4. 2x2-Join Usage: Single Function.

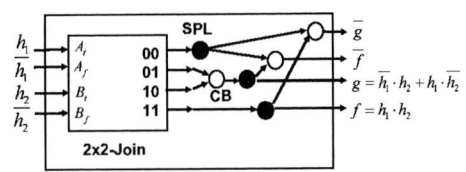

Fig. 5. 2x2-Join Usage: Multiple Functions.

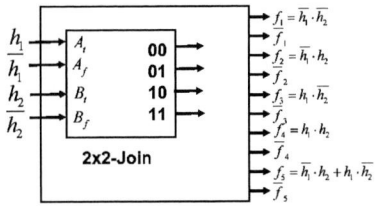

Fig. 6. 2-AND/XOR Cell.

to a 2-AND/XOR cell. Therefore, any logic circuit consisting of only 2-input nodes can be mapped to a circuit consisting of only 2-AND/XOR cells. Thus, our logic synthesis starts with a circuit consisting of only 2-input nodes. This initial circuit can be generated by conventional logic design tools, such as SIS (A System for Sequential Circuit Synthesis) [13]. For example, by using a standard script of SIS, we can obtain a circuit consisting of 2-input ANDs, 2-input XORs, and NOTs. Then, we can map this circuit naturally into one consisting of 2-AND/XOR cells. Note that four of five outputs of each 2-AND/XOR cell is not used (i.e., redundant) at this moment, however, the redundant outputs are very useful in our optimization procedure mentioned in the next section.

D. Transduction Method for 2-AND/XOR Circuits

After obtaining the initial circuit consisting of 2-AND/XOR cells, we apply the following procedure where we consider the whole circuit as just a conventional circuit consisting of 2-input XORs, 2-input ANDs and NOTs. However, while each AND or XOR gate is considered as a single gate in the conventional Transduction Method, the four AND gate and the XOR gate in a 2-AND/XOR cell should be handled together as we will mention.

Step 1. Calculate CSPFs of all connections and gates in the circuit.

Step 2. For an XOR gate, remove an input connection of the gate if the CSPF at the connection contains the constant-0 function. For an AND gate, remove an input connection of the gate if the CSPF at the connection contains the constant-1 function.

Step 3. If the function at a gate is included in the CSPF at a connection, replace the connection with a new connection from the gate.

We repeat the above transformation until there is no change.

This scheme is exactly the same as that of the conventional Transduction Method (See [8] for more details). However, note that we can remove a 2-AND/XOR cell only when all the four AND and the XOR outputs are removed. Therefore, unlike the case of conventional circuit optimization methods, it is not important to remove only one of XOR and AND gates in a 2-AND/XOR cell. Thus, we modify the conventional Transduction Method as follows. The modification is considered to be very natural for 2-AND/XOR cells.

- If there are multiple candidates for the alternative connection at Step 3, we choose the output of 2-AND/XOR whose total number of fanouts (i.e., the number of fanouts of all the four AND and the XOR outputs) is the largest. In contrast, in the conventional Transduction Method, we consider only the number of fanouts of the gate that is chosen as a replacement. This means that if we take the conventional strategy, we always consider only one of AND or XOR outputs in 2-AND/XOR cells, and therefore, we may miss a chance to remove a whole 2-AND/XOR cell even though we can remove one of its output gate.

- At Step 3, to select a candidate function, if there is a gate whose output function is f, we can also use $\overline{f}$ since we use dual-rail logic.

- At Step 2, we remove an AND (or XOR) gate only when all the four AND and the XOR gates in the 2-AND/XOR cell become redundant. By this modification, we can continue to have a possibility to use AND (or XOR) function to replace another connection even though it is currently not used.

It is very interesting to note that any functional redundancy cannot be obtained if we apply the CSPF calculation directly to circuits consisting of 2x2-Joins and CBs because of their difference from the conventional gates described in Sec. III-A. In other words, we can successfully utilize the Transduction Method by introducing 2-AND/XOR cells and considering the whole circuit as just a conventional circuit consisting of 2-input XORs, 2-input ANDs and NOTs. This enable us to optimize RSFQ circuits more as we will see in our experimental results in Sec. IV.

As we mentioned in Sec. III-B, we can consider the logic primitives as 2-input LUTs that can be constructed as shown in Fig. 4. For a logic circuit consisting of LUTs, there is an efficient optimization method that utilizes SPFD [14]. (SPFD is a generalization of CSPF to the case of LUTs where we can utilize the flexibility of LUTs, i.e., we can change the internal functions of LUTs.) Therefore, one might wonder if the following strategy is more natural and better than our strategy.

- Consider the logic primitives as 2-input LUTs which can be constructed as shown in Fig. 4.

- Apply SPFD-based optimization method [14].

It should be noted that the above strategy is essentially the same as the ours since it is sufficient to consider AND and XOR for 2-input functions when we use dual-rail logic. In other words, our strategy to use 2-AND/XOR cell essentially has the same power as the one that uses SPFDs. Therefore, for our problem, we do not need to calculate and manipulate SPFDs that are more complicated than CSPFs.

E. Optimality of Our Framework

Our framework successfully utilizes the conventional logic design techniques by considering virtual cells called 2-AND/XOR cells in stead of considering directly 2x2-Joins, CBs and SPLs. Then, one may wonder how much we may miss the possibility to have a smaller circuit compared to the case where we design a circuit directly from 2x2-Joins, CBs and SPLs. As mentioned, as far as we consider a function realized by only a single 2x2-Join with SPLs and CBs, we do not miss the opportunity to have a small circuit, i.e., our 2-AND/XOR cell essentially represents all the possible functions realized by ORing any combination of the four outputs of a single 2x2-Join, i.e., any function with respect to two inputs of the cell.

However, we cannot generate a logic function realized by ORing two functions from two different 2x2-Joins. Suppose h_1 and h_2 be the functions realized at the outputs of two different 2x2-Joins. Then, if $h_1 \cdot h_2 = 0$ is satisfied we can make $f = h_1 + h_2$ by only a single CB. Obviously our framework cannot deal with the above usage of CBs. It should be noted that such a situation does not happen so frequently by the following reason. Since we take dual-rail scheme, if we generate $f = h_1 + h_2$, we also need to generate $\overline{f}$, i.e., we need to have h_3 and h_4 such that $h_3 \cdot h_4 = 0$ and $h_3 + h_4 = \overline{f}$. Thus, we need to find h_3 and h_4 with the above difficult conditions when we want to generate OR functions of h_1 and h_2 with a CB; there are not so many such situations. Therefore, missing the above usage of CBs may be allowed if we consider the advantage of our framework, i.e., we can adopt conventional logic design techniques.

One may also consider that our usage of 2x2-Joins seems to increase the number of CBs and SPLs since we use a lot of CBs and SPLs to generate multiple outputs from a single 2x2-Join in a 2-AND/XOR cell. (See Fig. 5 again.) However, we would like to stress that the number of CBs and SPLs are not increased, or even decreased in many cases. The reason is as follows. Consider a case when we want to generate the two functions, f and g, which are shown in Fig. 5. If we do not use our 2-AND/XOR cells, i.e., just use two 2x2-Joins to implement two functions separately, the situation can be expressed as shown in Fig. 7. Note that we need four SPLs before two 2x2-Joins. Thus, by comparing Fig. 5 and Fig. 7, one can easily see that our usage of 2x2-Joins decreases the number of not only 2x2-Joins but also CBs and SPLs. (In this example, we can decrease the number of CBs since $\overline{f}$ and g share a common minterm.)

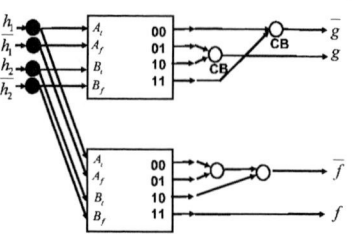

Fig. 7. Two 2x2-Joins without Sharing.

IV. Experimental Results

We have implemented the methods presented in the previous sections and performed preliminary experiments on MCNC [15] benchmark circuits. To synthesize initial circuits we used the following recommended script of SIS [13].
(1) eliminate 2,   (2) gkx -ac,   (3) simplify -d,
(4) xl_part_coll -m -g 2   (5) xl_coll_ck,
(6) xl_partition -m,   (7) simplify.

Table II shows the results of our optimization procedure described in Sec. III-D. In Table II, "Join," "Conn." and "Lev." show the number of 2x2-Joins, the number of connections between them and the level (depth) of the circuits, respectively, and "Time" shows optimization time our method took in these experiments. In the lowest row, with respect to the number of 2x2-Joins, the number of connections and the circuit level, we show the ratio to the initial circuits, and with respect to the optimization time, we show the average of those for the benchmark circuits. Here we consider only the number of 2x2-Joins like the existing researches by observing that the implementation cost of 2x2-Joins will be much larger than those of the other primitives although it is a little bit early to discuss the real implementation costs. It should also be noted that the numbers of CBs and SPLs are also decreased if we can decrease the number of 2x2-Joins as mentioned in Sec. III-E. In Table II, we can observe that our optimization method reduced the number of 2x2-Joins, the number of connections and the circuit level by 32.0%, 31.8% and 17.3%, respectively, while our optimization procedure took 6.08 seconds on average.

We consider that this large reduction is due to the change from the single-output LUT to the multi-output 2-AND/XOR composed of a 2x2-Join, SPLs and CBs. It is apparent that two 2-AND/XOR cells can be merged into one if their inputs are from the same 2-AND/XOR cells. Otherwise, we cannot determine whether they can be merged or not by only observing the circuit configuration. Our method can find more cases where we can merge 2-AND/XOR cells by using CSPFs. This feature makes our optimization method more powerful. The number of such mergers are reported in "Share" in Table II, and the ratio to the total number of the 2x2-Joins is in (%).

As the case of conventional logic synthesis, the logic optimization should be also very important for the RSFQ

TABLE II
EXPERIMENTAL RESULTS

Circuits	PI	PO	Initial Circuits			Trans.					
			Join	Conn.	Lev.	Join	Conn.	Lev.	Time (s)	Shared	(%)
C1355	41	32	234	468	17	174	348	14	1.17	6	3.4
C7552	207	108	1504	3059	32	994	2049	32	122.14	136	13.7
alu2	10	6	386	773	42	236	473	28	3.35	41	17.3
alu4	14	8	667	1334	43	479	958	36	11.26	84	17.5
cmb	16	4	44	88	6	25	51	5	0.01	4	16.0
dalu	75	16	1172	2344	36	809	1618	18	36.59	84	10.4
f51m	8	8	113	227	10	64	129	9	0.11	10	15.6
i8	133	81	1260	2520	19	971	1942	15	155.37	127	13.1
lal	26	19	88	177	8	61	123	8	0.06	7	11.5
my_adder	33	17	96	192	48	64	128	48	0.08	16	25.0
t481	16	1	1690	3380	20	1073	2146	19	110.66	57	5.3
term1	34	10	259	519	16	116	234	11	0.68	15	12.9
ttt2	24	21	182	364	10	127	254	10	0.41	19	15.0
x3	135	99	724	1448	14	548	1096	12	14.19	41	7.5
z4ml	7	4	44	88	9	12	25	8	0.01	6	50.0
Average			100	100	100	68.0	68.2	82.7	6.08		11.4

logic circuit design since it is difficult to design optimum RSFQ circuits directly from the specifications. From this viewpoint, our optimization method might be useful in the second step of any circuit synthesis method for RSFQ logic circuits. Moreover, although the conventional mapping tools mainly produce single-output LUT circuits, our method can transform them into multi-output 2-AND/XOR circuits. Therefore, by taking the advantage of the multi-output feature, our method can optimize the 2-input circuit mapped by SIS as the experimental results show.

V. COMPARISON WITH RELATED WORKS

To the best of our knowledge, there are only two researches [16] and [9] for systematic methodology to design RSFQ logic circuits. Both of them are based on BDDs. Unlike our framework, they directly consider RSFQ logic primitives, Bina [16] or 2x2-Joins [9]. Although the paper [16] does not provide any results of benchmark circuits, it seems that the method proposed in [9] is more promising as they claim in [9].

Although we do not have comparison data, we can observe the disadvantages of the BDD-based methods as follows:

- BDD-based methods need relatively large time (compared with conventional logic synthesis techniques used in our framework) to manipulate the BDDs in a special manner.

- BDD-based methods need to consider BDD variable ordering and/or the division of functions if necessary, and the variable ordering should have a great influence on the resultant circuits.

- The level of the generated circuits is essentially the

same as the number of inputs of the functions to be synthesized if we use the BDD-based methods.

Note that it is well-known that, for some functions, it is much better to use functional decomposition based on BDDs [11] than to use conventional logic design techniques, such as SIS. Thus, it might be true that the BDD-based methods should produce better circuits than our method especially for small circuits. However, it is also obvious that for some cases our method may be better. It is also true that our method can be applied for larger circuits to which BDD-based methods are not feasible. In conclusion, our method is complementary to the existing BDD-based methods.

The obvious advantages of the method [9] over our method is the following. Our method cannot find the special usage of CBs as mentioned in Sec. III-E. On the other hand, the method [9] sometimes extracts the special usage of CBs from their BDD representation since the method constructs circuits directly from RSFQ logic primitives. As we mentioned, by ignoring this special usage of CBs, we can successfully utilize the conventional logic techniques in our framework. Note also that the optimization phase in our framework can handle this special CBs by treating them as OR gates with some special conditions. Thus, it is easy to modify our Transduction Method so that it can be applied to the circuit obtained by the method [9]. Such modifications and further comparison with the BDD-based methods should be done as our future work.

VI. CONCLUSIONS

In this paper, we have presented a new framework to synthesize RSFQ logic circuits from 2x2-Joins. Our method can utilize conventional logic design techniques.

Indeed we show how to use logic synthesis tools such as SIS [13] and a transformation-based heuristic method based on the Transduction Method [8] in our framework. Our contributions are summarized as follows.

- We propose a framework to use a 2x2-Join as a 2-AND/XOR cell for RSFQ logic circuit synthesis. By this we can utilize the conventional logic synthesis techniques including initial circuit design and optimization methods.

- We propose an optimization method by modifying the original Transduction Method so that it can utilize the property of 2-AND/XOR cells.

- By using our framework, we can successfully generate RSFQ circuits in reasonable time for large benchmark circuits that have not been reported in the existing researches.

The experimental results show that our Transduction Method reduces the number of 2x2-Joins by 32.0% on average.

There are other logic primitives for RSFQ circuits, such as RSFQ D_2 flip-flop. Our future work is to treat other logic primitives in our framework. Then, we would also like to combine our method with the existing methods [6, 16, 17] that use RSFQ D_2 flip-flops for their primitives.

Acknowledgment

We are deeply grateful to the late Professor Yahiko Kambayashi for his initiation and encouragement of our research philosophy through the study of the Transduction Method. The first author is supported by MEXT.KAKENHI ((B) 16700067) and the Okawa Foundation Research Grant.

References

[1] International Technology Roadmap for Semiconductors. Technical Report http://public.itrs.net/, 2004.

[2] R. E. Bryant. Graph-based algorithm for Boolean function manipulation. *IEEE Transactions on Computers*, C-35(8):667–691, August 1986.

[3] J. Deng, S. Whiteley, and T. Van Duzer. Data-Driven Self-Timing of RSFQ Digital Integrated Circuits. In *Extended Abstracts of ISEC'95*, pages 189–191, September 1995.

[4] K. Gaj, E. G. Friedman, and M. J. Feldman. Timing of multi-gigahertz rapid single flux quantum digital circuits. *IEEE Journal of VLSI Signal Processing*, 16(2-3):247–276, 1997.

[5] Y. Kameda, S.V. Polonsky, M. Maezawa, and T. Nanya. Self-timed Parallel Adders based on DI RSFQ Primitives. *IEEE Trans. Appl. Superconductivity*, 9(2):4040–4045, June 1999.

[6] J. Koshiyama and N. Yoshikawa. A Cell-Based Design Approach for RSFQ Circuits Based on Binary Decision Diagram. *IEEE Trans. Appl. Superconductivity*, 11(1):263–266, March 2001.

[7] K. K. Likharev and V. K. Semenov. RSFQ logic/memory family: A new Josephson-junction technology for sub-terahertz-clock frequency digital systems. *IEEE Trans. Appl. Superconductivity*, 1(1):3–28, March 1991.

[8] S. Muroga, Y. Kambayashi, H. C. Lai, and J. N. Culliney. The Transduction Method - Design of Logic Networks Based on Permissible Functions. *IEEE Transactions on Computers*, 38(10):1404–1424, October 1989.

[9] K. Obata, K. Takagi, and N. Takagi. Design Method of Dual-Rail RSFQ Logic Circuits Using 2x2-Join. *IEICE Trans.*, J88-C(3):202–209, March 2005. (In Japanese)

[10] H. Savoj and R. K. Brayton. The Use of Observability and External Don't Cares for Simplification of Multi-Level Networks. pages 297–301, June 1990.

[11] H. Sawada, T. Suyama, and A. Nagoya. Logic Synthesis for Look-up Table Based FPGAs Using Functional Decomposition and Support Minimization. pages 353–358, November 1995.

[12] V. K. Semenov, Yu. A. Polyakov, and D. Schneider. Implementation of Oversampling Analog-to-Digital Converter Based on RSFQ Logic. In *Extended Abstracts of ISEC'97*, volume 1, pages 41–43, June 1997.

[13] E. M. Sentovich, K. J. Singh, L. Lavagno, C. Moon, R. Murgai, A. Saldanha, H. Savoj, P. R. Stephan, R. K. Brayton, and A. Sangiovanni-Vincentelli. SIS: A System for Sequential Circuit Synthesis. Technical Report UCB/ERL M92/41, Univ. of California, Berkeley, May 1992.

[14] S. Yamashita, H. Sawada, and A. Nagoya. SPFD: A New Method to Express Functional Permissibilities. *IEEE Transactions on Computer-Aided Design of Integrated Circuits*, 19(8):840–849, August 2000.

[15] S. Yang. Logic synthesis and optimization benchmarks user guide version 3.0. *MCNC*, January 1991.

[16] N. Yoshikawa and J. Koshiyama. Top-Down RSFQ Logic Design Based on a Binary Decision Diagram. *IEEE Trans. Appl. Superconductivity*, 11(1):1098–1101, March 2001.

[17] N. Yoshikawa, H. Tago, and K. Yoneyama. A New Design Approach for RSFQ Logic Circuits Based on the Binary Decision Diagram. *IEEE Trans. Appl. Superconductivity*, 9(2):3161–3164, June 1999.

Fast Simulation of Large Networks of Nanotechnological and Biochemical Oscillators for Investigating Self-Organization Phenomena

Xiaolue Lai and Jaijeet Roychowdhury

Department of Electrical and Computer Engineering, University of Minnesota

Email: {laixl, jr}@ece.umn.edu

Abstract—We address the problem of fast and accurate computational analysis of large networks of coupled oscillators arising in nanotechnological and biochemical systems. Such systems are computationally and analytically challenging because of their very large sizes and the complex nonlinear dynamics they exhibit. We develop and apply a nonlinear oscillator macromodel that generalizes the well-known Kuramoto model for interacting oscillators, and demonstrate that using our macromodel provides important qualitative and quantitive advantages, especially for predicting self-organization phenomena such as spontaneous pattern formation. Our approach extends and applies recently-developed computational methods for macromodelling electrical oscillators, and features both phase and amplitude components that are extracted automatically (using numerical algorithms) from more complex differential-equation oscillator models available in the literature. We apply our approach to networks of Tunneling Phase Logic (TPL) and Brusselator biochemical oscillators, predicting a variety of spontaneous pattern generation phenomena. Comparing our results with published measurements of spiral, circular and other pattern formation, we show that we can predict these phenomena correctly, and also demonstrate that prior models (like Kuramoto's) cannot do so. Our approach is more than 3 orders of magnitude faster than techniques that are comparable in accuracy.

I. INTRODUCTION

Coupled self-oscillating systems appear in diverse natural and physical systems. For example, in nanoelectronics, the tunneling phase logic (TPL) [1], [2] device, which makes use of the bistability of single-electron tunneling oscillation to realize logic in phase are proposed for large scale circuits, due to its extremely high gate density and ultra low power dissipation. This concept has been applied [3] to implement cellular nonlinear networks (CNN) [4], [5], with ultra-high integration levels far beyond even DSM CMOS. Such CNN systems, consisting of large populations of interacting TPL oscillators, constitute a promising approach for implementing future large-scale high performance image processing systems.

It has long been known empirically that in systems of coupled oscillating entities, self-organizing collective behavior begins to occur when coupling exceeds a certain threshold; some entities start to synchronize spontaneously while others remain incoherent. Such "co-operation" and "competition" engendered by oscillatory nonlinear dynamics can produce complicated and beautiful spatio-temporal patterns of collective behavior. In particular, pattern formation arising from the nonlinear interaction of many individual cells in biochemical systems is a fascinating and extensively noted phenomenon (*e.g.*, [6]–[8]) — for example, an important issue in developmental biology is understanding the formation of spatial patterns in the embryo. These patterns can be explained by reaction diffusion theory [9], which shows that a system of reacting and diffusing chemicals can evolve spontaneously, from an initially uniform or random state, into spatial patterns known as Turing structures [9]. Such a process can be modeled as a network of biochemical oscillators interacting with each other. A typical example is the Brusselator reaction-diffusion system [10], whose patterns have recently been studied by experiments and simulation, especially under periodic perturbations [11], [12].

Although the dynamics of coupled oscillating systems have been well studied experimentally, analytical understanding of the details of pattern formation remains a challenge. A fundamental difficulty is that while it is often possible to understand specific systems of a few coupled oscillators at an analytical level, collective self-organization typically manifests itself only when *large numbers* of oscillators are networked. Many important characteristics of such large networks have been essentially impossible to understand fully using hand analysis so far, in part because pattern generation is often very sensitive to small details of the nature of and coupling between individual network entities [13], [14]. Thus, analytical simplifications

that may be justified for small systems are potentially inapplicable to much larger ones and can lead to egregious mispredictions. As a result, accurate and detailed computational techniques are particularly critical for developing "understanding" of such systems. Because of the huge sizes and complex dynamics of these oscillatory networks, however, the computational problem is very challenging.

A straightforward computational method for nonlinear systems is to use numerical differential equation solvers to simulate waveforms in the time domain. But while such "transient" methods work well for non-oscillatory systems, they are far less suitable for simulating oscillators, especially coupled oscillators, due to inherent error buildup in phase. Very small timesteps have to be taken within each oscillation cycle and a complex integration methods need to be used to provide acceptable accuracy over long simulations. These problems are very familiar to circuit designers using simulation programs like SPICE [15] on oscillators, even though conventional electronic oscillator systems are usually very small relative to networks of nano and biochemical oscillators. Because envelopes and phase transition in oscillatory systems can evolve very slowly over thousands or millions of cycles, the computational challenge is dramatically exacerbated.

An extensive and deep literature is available on analytical approaches for understanding coupled oscillator systems; here, we provide a very brief synopsis of the main approaches in order to better place our contributions. One of the earliest studies was made by the legendary Norbert Wiener [16], [17], who studied collective synchronization phenomena using a Fourier-integral-based method. Later, Winfree formulated an equation governing phase transitions in populations of coupled limit-cycle oscillators and presented the concept of a phase sensitivity function [18]. Winfree's approach was abstracted by Kuramoto and applied to systems of identical oscillators with equally weighted, all-to-all, purely sinusoidal couplings, resulting in a well-known phase-based nonlinear differential equation model for coupled oscillator systems, the *Kuramoto model* [19], [20].

Using his model, Kuramoto developed a steady-state 'locking' (or 'drifting') condition which has been shown to successfully predict bifurcation of phase transitions in coupled oscillator systems. Kuramoto's model has been extremely influential because it has provided a relatively simple means of understanding self-organizing phenomena in systems of coupled oscillators. For predicting detailed pattern formation in a variety of real systems, however, Kuramoto's approach has accuracy limitations that can compound, in large networks, to the extent that wrong patterns can emerge, as we show in this paper.

The main contribution of this work is a much more powerful model that alleviates the lack of accuracy and general applicability of Kuramoto's model, while retaining its advantages of relative simplicity and computational efficiency; and indeed, significantly enhancing the convenience and ease with which the model can be specialized for any specific system of interest. Our model consists of nonlinear phase macromodel, together with amplitude macromodel that capture dominant amplitude components. The nonlinear phase macromodel is a scalar, nonlinear differential equation for phase deviations. A fundamental mechanism in collective synchronization is phase pulling/locking between oscillators; our phase macromodel captures these phenomena very effectively [21], [24].

In some situations (*e.g.*, strong loading effects due to coupling, unlocked mutually pulling oscillators, *etc.*), amplitude variations are large and couple significantly with phase effects, hence it is necessary to take them into account. To do so, we incorporate an amplitude macromodel [21] together with our phase equations. Thus, we are able to predict a much broader range of coupled oscillatory phenomena, with better accuracy than when using the phase macromodels alone. We develop models of common coupling mechanisms between

0-7803-9451-8/06/$20.00 ©2006 IEEE.

oscillators that capture both phase and amplitude implications.

Importantly, both phase and amplitude macromodels can be extracted from the oscillators' differential equations automatically via numerical algorithms, making this method very easily applicable to a diverse variety of large and complex oscillator systems.

We apply and validate our technique on networks of biochemical and nanoelectronical oscillators: a large coupled Brusselator chemical dynamics system [10] and TPL-based cellular nonlinear network (CNN) [3]. We show that our methods are able to reproduce the formation of two important patterns families in biochemical systems: target patterns and spiral wave patterns, showing that they closely match measurements reported in the literature [6]–[8]. Furthermore, we show that Kuramoto's model is unable to predict the correct patterns. For the TPL network, we validate bistability behavior and demonstrate image processing ability (*e.g.*, edge detection) in the TPL-CNN system, predicting patterns very close to measurements reported in [3]. Our methods provide speedups of more than 3 orders of magnitude over direct time-stepping simulation of the original oscillator network.

The remainder of the paper is organized as follows: in Section II, we review the Winfree and Kuramoto models for the analysis of collective synchronization in coupled systems. In Section III, we summarize the nonlinear oscillator macromodel that we employ in this work. In Section IV, we describe our nonlinear macromodel-based technique for simulating coupled oscillating systems. Finally, in Section V, we present applications to the TPL-CNN and biochemical Brusselator systems.

II. REVIEW OF PREVIOUS WORK

In this section, we summarize relevant previous work in the analysis of coupled oscillator systems.

A. Winfree's Equation For Coupled Systems

For the study of phase dynamics in coupled systems, a fruitful approach, pioneered by Winfree in 1967 [18], is based on the intuition that if an oscillator is perturbed by an external input at a given moment, its phase change should equal to the product of the perturbation strength and a "phase sensitivity" of the oscillator at that moment. Based on this intuition, Winfree formulated the governing equation for a coupled oscillating system with the assumptions of weak coupling and nearly identical oscillators. His phase model is expressed by

$$\dot{\theta}_i = \omega_i + (\sum_{j=1}^{n} X(\theta_j))Z(\theta_i), \qquad i = 1, ..., n, \qquad (1)$$

where n is the number of oscillators in the system, θ_i is the phase of oscillator i, θ_j is the phase of oscillator j and ω_i is the free running frequency of oscillator i. $X(\theta_j)$ denotes the influence on oscillator i exerted by oscillator j, and $Z(\theta_i)$ is the phase sensitivity function of oscillator i.

The chief difficulty with Winfree's model is that no clear or convenient means is provided for obtaining the phase sensitivity functions, which is not easily obtained. As we show later in this paper, our approach completely resolves this issue.

B. Kuramoto's Model For Coupled Systems

Kuramoto extended and simplified Winfree's approach [19] by showing (using asymptotic expansion theory [22] and averaging) that the long-term phase dynamics of weakly-coupled simple-harmonic-type oscillators can be predicted using

$$\dot{\theta}_i = \omega_i + \sum_{j=1}^{n} \Gamma_{ij}(\theta_j - \theta_i), \qquad i = 1, ..., n, \qquad (2)$$

where n is the system size and Γ_{ij} are "interaction functions".

Obtaining the interaction functions faces the same difficulties as Winfree's phase sensitivity function. Kuramoto made simplifications using the assumption that the coupled system was made up of identical oscillators with equally weighted, all-to-all, and purely sinusoidal coupling, resulting in

$$\Gamma_{ij}(\theta_j - \theta_i) = \frac{K}{n} \sin(\theta_j - \theta_i), \qquad i = 1, ..., n. \qquad (3)$$

K is a coupling strength. The phase equation is thus simplified to

$$\dot{\theta}_i = \omega_i + \frac{K}{n} \sum_{j=1}^{n} \sin(\theta_j - \theta_i), \qquad i = 1, ..., n. \qquad (4)$$

To visualize the phase dynamics, Kuramoto introduced the complex order parameter

$$re^{i\psi} = \frac{1}{n} \sum_{j=1}^{n} e^{i\theta_j}, \qquad (5)$$

where $r(t)$ is the radius which measures the phase coherence, and $\psi(t)$ is the average phase. Equating the imaginary part of (5), the right hand side of (4) can be rewritten as

$$\frac{K}{n} \sum_{j=1}^{n} \sin(\theta_j - \theta_i) = Kr\sin(\psi - \theta_i). \qquad (6)$$

Thus, (4) becomes

$$\dot{\theta}_i = \omega_i + Kr\sin(\psi - \theta_i) \qquad (7)$$

Kuramoto showed that the locking condition of oscillator i in a coupled system is

$$|\Delta\omega_i| \leq Kr, \qquad (8)$$

where $\Delta\omega_i$ is the difference between the free-running frequency of oscillator i and the mean frequency of the coupled system.

As noted in the introduction, the simplifications inherent in Kuramoto's model makes its predictions questionable when applied to large systems of self-organizing oscillators.

III. NONLINEAR OSCILLATOR MACROMODEL

In this paper, we present a generally applicable approach for predicting nonlinear dynamics in coupled oscillating systems. Our method is based upon a nonlinear oscillator macromodel originally developed for predicting noise, injection locking and other phenomena in electrical oscillators [23], [24]. In this section, we briefly review the nonlinear oscillator macromodel we employ for the simulation of coupled oscillating systems.

A. Oscillator Phase Macromodel - Intuition

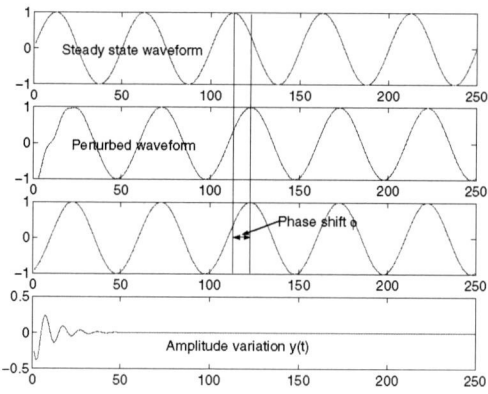

Fig. 1. Decomposition of phase and amplitude variation of oscillator under impulse perturbation.

Figure 1 depicts the decomposition of phase and amplitude variation of oscillators under impulse perturbation. The first figure is the steady state waveform of the oscillator without perturbation. The second figure depicts the waveform of the oscillator if an impulse perturbation is applied to the oscillator at $t = 0$: the oscillator features some transient amplitude variation which vanishes as time goes on, until finally, it converges to its steady state again, but with a permanent phase shift of ϕ. Intuitively, we know this perturbed waveform can be decomposed into a steady state waveform with the time shift ϕ, and the amplitude response $y(t)$, as shown in the third and fourth figure respectively. Since the amplitude response of a perturbed oscillator is stable, this implies that the impulse response $y(t)$ should die out as time goes to infinity. As a result, the decomposition of the phase deviation ϕ and the amplitude response $y(t)$ should be unique, and the response of the perturbed oscillator can be represented as

$$x_p(t) = x_s(t + \phi) + y(t), \qquad (9)$$

where $x_s(t)$ is the steady state of the unperturbed system and $x_p(t)$ represents the waveforms of the perturbed system.

Since oscillators are periodic systems, phase shift ϕ is dependent on the time when the impulse injection is applied to the oscillator. If we sweep the impulse injection from $t = 0$ to $t = T$, we obtain a periodic waveform of phase shift $\phi(t)$, which we call the phase sensitivity waveform, and which is very important for predicting the oscillator's jitter performance. Since it is not practical to obtain the phase sensitivity waveform using impulse injection, a method for calculating it without performing the full simulation is of great value.

B. Nonlinear Oscillator Phase Macromodel

In [23], Demir *et al* put this intuition on a solid mathematical foundation. Using Floquet theory [25], [26], Demir presented the method for obtaining phase sensitivity waveforms from an oscillator's linearized periodically time varying (LPTV) systems, and formulated a nonlinear scalar differential equation to capture the oscillator's phase deviation due to perturbations.

A general oscillator under perturbation can be described by

$$\dot{x} + f(x) = b(t), \qquad (10)$$

where $b(t)$ is a vector of perturbation signals applied to the free running oscillator. The corresponding LPTV system can be obtained by linearizing this oscillator about its steady state orbit:

$$\begin{aligned} \dot{w}(t) &\approx -\frac{\partial f(x)}{\partial x}\Big|_{x_s(t)} w(t) + b(t) \\ &= A(t)w(t) + b(t), \end{aligned} \qquad (11)$$

where $x_s(t)$ is the steady state orbit of the oscillator. Via Floquet decomposition of the homogeneous part of this LPTV system, a series of Floquet exponents and corresponding eigenvectors can be obtained. Since we know, from Figure 1, that phase deviation ϕ never vanishes, it should correspond to the Floquet exponent with the value of 0. [23] showed that the phase sensitivity waveform of the oscillator can be extracted from the eigenvector associated with the Floquet exponent 0, and that the phase deviation $\alpha(t)$ is governed by a simple one-dimensional nonlinear differential equation [23]

$$\dot{\alpha}(t) = V_1^T(t + \alpha(t)) \cdot b(t), \qquad (12)$$

where $V_1(t)$ is the perturbation projection vector (PPV). $V_1(t)$ is a vector with the size of system size n; Each element in $V_1(t)$ represents the oscillator's phase sensitivity to the perturbation applied to the corresponding node. The PPV, or the phase sensitivity vector, has periodic waveforms that have the same frequency as that of the oscillator. Various methods [23], [27]–[29], both in the time domain and the frequency domain, have been presented for calculating the PPV from SPICE-level circuit descriptions of oscillators. In (12), the phase deviation $\alpha(t)$ has units of time. To obtain the phase deviation in radians, we need to multiply $\alpha(t)$ by the oscillator's free-running frequency ω_0.

C. Amplitude Macromodel

Once the phase deviation $\alpha(t)$ is obtained by solving (12), a macromodel for dominant amplitude components can be built as well, by linearizing the oscillator over its perturbed time-shifted orbits $x_s(t + \alpha(t))$. In [21], a method is presented to construct amplitude macromodels of oscillators. The oscillator is first linearized on $x_s(t + \alpha(t))$:

$$\begin{aligned} \dot{y}(t) &\approx -\frac{\partial f}{\partial x}\Big|_{x_s(t+\alpha(t))} y(t) + b(t) \\ &= A(x_s(t + \alpha(t)))y(t) + b(t), \end{aligned} \qquad (13)$$

where $x_s(t)$ is the oscillator's steady-state orbit, $\alpha(t)$ is the phase deviation due to perturbation $b(t)$, and $y(t)$ is a small amplitude deviation from the phase-shifted orbit, due to the perturbation $b(t)$. By introducing a new variable $\hat{t} = t + \alpha(t)$ and defining $\hat{y}(\hat{t}) = y(t)$

and $\hat{b}(\hat{t}) = b(t)$, we obtain a linear periodic time-varying (LPTV) system

$$\dot{\hat{y}}(\hat{t}) = A(x_s(\hat{t}))\hat{y}(\hat{t}) + \hat{b}(\hat{t}). \qquad (14)$$

Applying Floquet decomposition, the LPTV system can be decomposed into a diagonalized LTI system with periodic input/output vectors:

$$\hat{y}(\hat{t}) = \sum_{i=1}^{n} u_i(\hat{t}) \int_0^{\hat{t}} \exp(\mu_i(\hat{t} - \tau)) v_i^T(\tau)\hat{b}(\tau)d\tau, \qquad (15)$$

where μ_i are Floquet exponents, and $v_i(t)$ and $u_i(t)$ are periodic input/output vectors. By dropping the Floquet exponent corresponding to phase and other less important Floquet exponents, we obtain a reduced amplitude macromodel.

When both the phase shift $\alpha(t)$ and amplitude variations $y(t)$ are available, the oscillator's orbit under perturbation can be obtained by the equation

$$x_p(t) = x_s(t + \alpha(t)) + y(t), \qquad (16)$$

where $x_s(t)$ is the steady state orbit of the oscillator, and $x_p(t)$ is the orbit of the oscillator under perturbation.

IV. SIMULATING COUPLED OSCILLATORS USING THE NONLINEAR OSCILLATOR MACROMODEL

Once oscillator macromodels are obtained using the methods in Section III, the complex oscillator equations in the coupled system can be replaced with the nonlinear scalar phase equation (12), and the coupling between oscillators can be modeled as the inputs applied to (12). The resulting reduced system can be simulated using any transient simulator, with great speedups compared to the full system. Moreover, since the system is simulated in the phase domain directly, the simulation efficiency can be improved by using larger timesteps and simpler integration methods, without appreciable loss of accuracy.

A. Nonlinear Phase Equation For Coupled Systems

Since in (12) $V_1(t)$ is a vector in which each element represents the phase sensitivity of the corresponding node in the oscillator, and $b(t)$ is also a vector of size n that models the perturbation on each oscillator circuit node, our method can handle a system consisting of oscillators with different characteristics, coupled with very complex topology. For purposes of illustration of simplicity, we assume that the coupled system consists of identical oscillators, and coupling only occurs on one node with the phase sensitivity function $v(t)$ and the steady state waveform $x(t)$. This leaves the following governing equation of the coupled system:

$$\dot{\alpha}_i(t) = v(t + \alpha_i(t)) \cdot \gamma_i(t), \qquad i = 1, ..., N, \qquad (17)$$

where N is the network size or number of oscillators in the coupled system, $\alpha_i(t)$ is the phase shift of oscillator i due to coupling, $v(t)$ is the phase sensitivity of the node on which coupling occurs and $\gamma_i(t)$ is the coupling function that models the coupling force applied to oscillator i. If the coupling $\gamma_i(t)$ and phase sensitivity $v(t)$ are purely sinusoidal waveforms, it is easy to show that (17) is equivalent to Kuramoto's model. However, when the coupling and the phase sensitivity functions are not purely sinusoidal, (17) is far more accurate than Kuramoto's model, since it considers all harmonics.

B. Modeling Coupling In Coupled Systems

To solve (17), we need to formulate the coupling function $\gamma_i(t)$, which models the coupling force applied to oscillator i from other oscillators in the coupled system. Here, we model three typical couplings: resistive coupling, capacitive coupling and idealized coupling, shown in Figure 2.

1) Resistively-loaded coupling: Figure 2(a) depicts a system coupled by resistors; the resistance between oscillator i and oscillator j is $R_{i,j}$ ($R_{i,j} = \infty$ if there has no coupling between oscillator i and oscillator j). Such coupling adds a load to the oscillator and this can lead to significant amplitude effects if the coupling is strong. We call this the *loading effect*. For such a system, the coupling function can be written as

$$\gamma_i(t) = \sum_{j=1}^{n} (x_j(t) - x_i(t))/R_{i,j} \qquad (18)$$

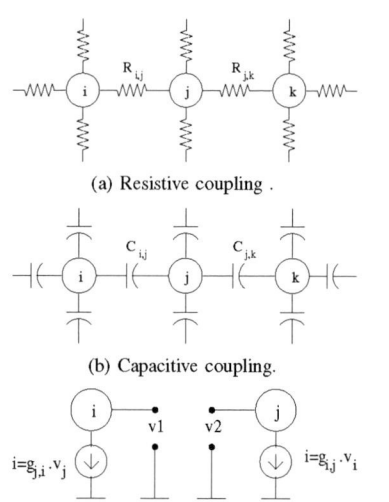

(a) Resistive coupling .

(b) Capacitive coupling.

(c) Ideal coupling.

Fig. 2. Typical coupling in coupled oscillating systems.

and models the perturbation current injected into oscillator i, where $x_j(t)$ and $x_i(t)$ are the voltage waveforms of oscillator j and oscillator i respectively. According to (16), the oscillator waveform has the form

$$x_i(t) = x(t + \alpha_i(t)) + y_i(t), \qquad i = 1,..,N, \qquad (19)$$

where $x(t)$ is the oscillator's steady state waveform, $\alpha_i(t)$ is the oscillator's phase shift due to coupling, and $y_i(t)$ is the oscillator's amplitude variations due to coupling. These can be calculated using the method in [21]. In the case that the coupling is weak and the loading effect is negligible, we can ignore the amplitude variations and obtain a simpler form:

$$x_i(t) = x(t + \alpha_i(t)), \qquad i = 1,..,N. \qquad (20)$$

If the coupling is not weak, the loading effect may change the free-running characteristics of the oscillator (*e.g.*, as in relaxation and ring oscillators). In such cases, Kuramoto's model cannot apply, since it assumes oscillators converge to a mean frequency that can be changed due to the loading effect.

2) Capacitively-loaded coupling: If the system is coupled by capacitors, as shown in Figure 2(b), and the capacitance between oscillator i and oscillator j is $C_{i,j}$ ($C_{i,j} = 0$ if there has no coupling between oscillator i and oscillator j), the coupling function can be written as

$$\gamma_i(t) = \sum_{j=1}^{n} C_{i,j} \cdot (\dot{x}_j(t) - \dot{x}_i(t)), \qquad (21)$$

where $\dot{x}_j(t)$ and $\dot{x}_i(t)$ are derivatives of waveforms on oscillator j and oscillator i. If the oscillator waveforms have the form of (20), their derivatives can be written as

$$\dot{x}_i(t) = \dot{x}(t + \alpha_i(t)) \cdot (1 + \dot{\alpha}_i(t)), \qquad i = 1,...,N. \qquad (22)$$

Considering that $\dot{\alpha}(t) \ll 1$ if coupling in the system is weak, we can simplify (22) as

$$\dot{x}_i(t) = \dot{x}(t + \alpha_i(t)), \qquad i = 1,...,N \qquad (23)$$

for the weak coupling case.

If the coupling is not weak, the loading effect needs to be taken into consideration, and we need to incorporate the amplitude macromodel. Furthermore, capacitive loading may change the free-running frequency of LC type oscillators, since it changes the equivalent tank capacitance. For such a system, the oscillator will not converge to a mean frequency, and Kuramoto's model fails for this case as well, since it assumes that oscillators converge to a mean frequency.

3) Idealized coupling (without loading): Figure 2(c) depicts the idealized coupling case. The coupling from oscillator j to oscillator i is modeled by coupling factor $g_{j,i}$. Such systems have no loading effect, and is thus the only case in which Kuramoto's model can

apply. The coupling function can be written as

$$\gamma_i(t) = \sum_{j=1}^{n} g_{j,i} \cdot x_j(t), \qquad i = 1,...,N. \qquad (24)$$

Incorporating the coupling function $\gamma(t)$ into the governing phase equation (17), we can solve the phase dynamics of complex coupled systems in this simplified form using the traditional transient integration method. Since the system size is reduced and the phase is simulated directly, we obtain speedups without appreciably sacrificing accuracy, especially in the case of large coupled systems with complex oscillator models.

V. APPLICATION TO LARGE OSCILLATOR NETWORKS

In this section, we apply and evaluate our technique. We apply our method to simulate the collective behavior of different coupled oscillating systems, including nanoelectronic systems and biochemical systems. We first simulate a Brusselator system to show that our model is able to reproduce complex pattern formation processes in biochemical systems. We then apply our technique to a TPL-based CNN system to demonstrate its ability in simulating next-generation nano-scale systems. All simulations are performed using **MATLAB** on a Linux system. We construct oscillator macromodels using the method described in Section III, and simulate the behavior of different coupled systems using the method described in Section IV. Numerical results show our method is able to capture the phase and amplitude dynamics in coupled systems with good accuracy. We obtain speedups of about $3000\times$ in our simulations.

A. Pattern formation in a Brusselator biochemical network

Patterns widely exist in many biological systems, such as animal furs and human fingerprints. The pattern formation process, which is important for the understanding of biological mechanisms in biological systems, can be modeled as a reaction-diffusion system. In such systems, chemicals interact with each other, forming patterns from an initially uniform state. The simulation of these chemical interaction systems present a challenge for direct simulation methods, as the system sizes are very large. In this section, we simulate a large Brusselator biochemical system using our macromodel-based method.

A Brusselator system is a oscillating chemical system with two chemical species,

$$\begin{aligned} \frac{\partial u}{\partial t} &= A - (B+1)u + (1 + \gamma \cdot \sin(2\pi f t))u^2 v + D_u \nabla^2 u \\ \frac{\partial v}{\partial t} &= Bu - u^2 v + D_v \nabla^2 v, \end{aligned} \qquad (25)$$

where u and v are two species, A and B are constant parameters corresponding to feed concentrations, $\gamma \sin(2\pi f t)$ is the external force applied to the oscillator, and D_u and D_v are the diffusion coefficients. We chose $A = 0.5$ and $B = 1.5$ in our simulations.

We simulate a network of Brusselator oscillators with size of 400×400 (about 160000 oscillators). After discretizing the diffusion between oscillators, we obtain a coupled oscillating system coupled by resistive coupling. The coupling resistance R is dependent on diffusion coefficients D_u and D_v. We show that the coupled system forms different patterns when the coupling resistance R is varied.

We first choose the coupling resistance $R = 15$, and simulate the system using our method for 150 cycles. The simulation results are shown in Figure 3, which clearly depicts the pattern formation process in this coupled oscillating system. In this figure, we use colors to represent the phase of oscillators, *i.e.*, different colors indicate oscillators with different phases. At the beginning ($t = 0$), all oscillators are given a random phase: hence we cannot see any pattern. After 5 oscillating cycles ($t = 5T$), the collective synchronization phenomenon is clearly seen: oscillators synchronize their phase with the phase of their neighbors. As a result, we can see many color spots in the figure. After 20 oscillating cycles, some small target patterns appear, and a spiral wave pattern forms on the right side of the figure. From $t = 40T$ onwards, we can see those patterns grow, and merge together. Finally, after 150 cycles, we obtain a complex figure which combines both target pattern and spiral pattern. This pattern is very close to the experimental measurements reported in [8], [11].

Now we investigate the patterns of this biological system under different coupling strength, and plot the patterns in Figure 4. In the

276

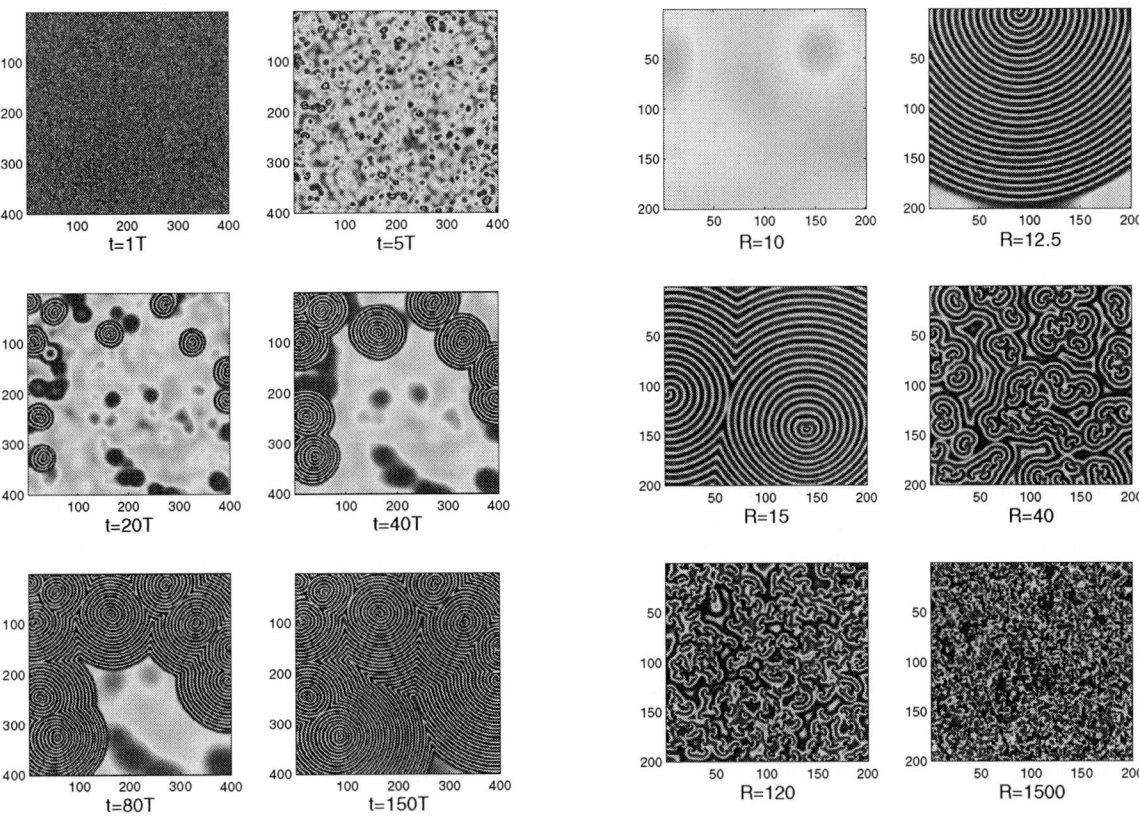

Fig. 3. Pattern formation in an unforced biochemical system.

Fig. 4. Biological patterns under different coupling strength.

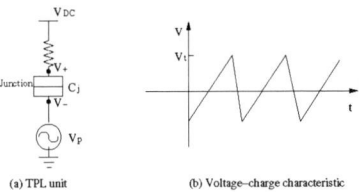

(a) TPL unit   (b) Voltage–charge characteristic

Fig. 5. TPL unit and its voltage-charge characteristic.

first figure, coupling is strong ($R = 10$), all oscillators lock to same phase, so we cannot see any pattern. In the second figure, we increase the coupling resistance to $R = 12.5$, a spiral wave pattern forms. We keep increasing the coupling resistance and obtain different kinds of patterns, as shown in Figure 4.

Such pattern formation processes are of great importance in biochemical systems, suggesting new insights in the understanding of biological processes. Our macromodel based technique offers a feasible approach for simulating the pattern formation in large bio-chemical systems. We obtain about $3000\times$ speedup in this simulation. If the system size is larger and oscillator model is more complex, speedups can be greater.

B. Nanoelectronic System – Tunneling Phase Logic Based Cellular Nonlinear Network

Figure 5 depicts a basic tunneling phase logic unit and its oscil-lating waveform. A basic TPL unit consists of an ultra-small SET junction with capacitance C_j, a DC bias V_{DC} and a pump voltage V_p. The SET junction has the property that when its voltage increases to a threshold V_T, single-electron tunneling occurs and the capacitor C_j is discharged. With the DC bias V_{DC} providing a bias current, the SET junction behaves as shown in Figure 5(b). The AC pump provides a sinusoidal voltage with amplitude V_p, which runs two times faster than the SET frequency. Therefore, if the SET is super-harmonically locked by the pump voltage, it has two steady states, with the phase difference π. If the phase of the SET oscillator is set to represent the logical values 0 and 1, we can realize logic in phase, instead of voltage as in traditional CMOS circuits.

We first use the nonlinear phase macromodel to verify bistability in the TPL. We simulate multiple TPL units with random initial phases, using the nonlinear phase equation (12). If the TPL has a bistable nature, the final phase of those TPL units will converge to two phases with the interval of π. We plot the simulation results in Figure 6. Four TPL units have different phases at $t = 0$. After about 20 cycles, the phases of TPL unit 1 and unit 4 converge to about $\frac{3\pi}{2}$; and the phases of TPL unit 2 and unit 3 converge to about $\frac{\pi}{2}$. Hence, our simulations verify that the TPL does indeed feature bistability.

In [3], a TPL based CNN implementation is proposed. Here, we use our macromodel based technique to simulate the TPL-CNN system. We adopt the near-neighbor coupling topology described in [3], and define the nonlinear output of the TPL unit as

$$f_{i,j}(\phi_{i,j}) = \begin{cases} 1, & \pi < \phi_{i,j} < 2\pi \\ -1, & 0 < \phi_{i,j} < \pi, \end{cases} \quad (26)$$

where $f_{i,j}$ is the output of the TPL cell on row i and column j, and

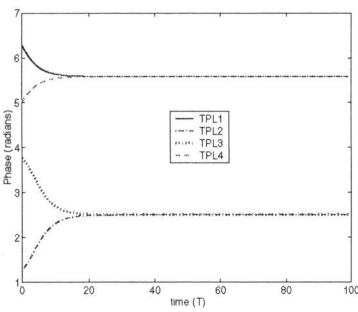

Fig. 6. The bistability of the TPL.

$\phi_{i,j}$ is the cell's phase.

A major potential application of CNNs is in image processing [30]. Here we present an example to show the image processing ability of TPL-CNNs: we use a TPL-CNN to detect an image edge. The original images are shown in Figure 7(a) and Figure 7(b). We transfer these images into two-color mode and apply them as inputs to the CNN network. In Figure 7(c) and Figure 7(d), we can see tunneling occurs and the inputs are replicated at the outputs of the CNN after 4 oscillating cycles. At $t = 8T$, the edges of the images are detected, as shown in Figure 7(e) and Figure 7(f).

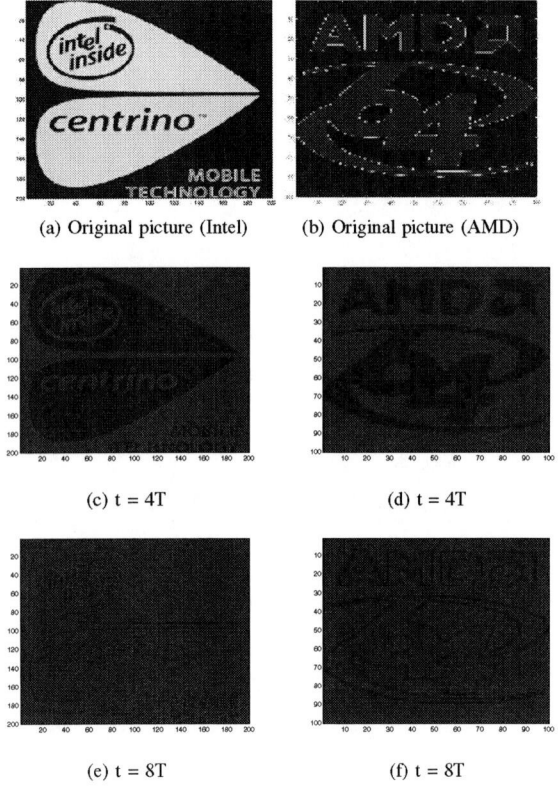

(a) Original picture (Intel)

(b) Original picture (AMD)

(c) t = 4T

(d) t = 4T

(e) t = 8T

(f) t = 8T

Fig. 7. Image edge detection performed by a TPL-CNN.

VI. CONCLUSION

We extend a nonlinear oscillator macromodelling technique, usually applied to the prediction of phase noise in electrical systems, to nano- and biochemical-systems, and demonstrate its ability in solving some difficult problems in these areas. Experimental results show that our technique is able to predict the behavior of very large-scale coupled oscillating systems, with great speedups while preserving the simulation accuracy. Our future work includes investigating large-scale oscillating systems under external forces.

REFERENCES

[1] T. Ohshima and R. A. Kiehl. Operation of bistable phase-locked single-electron tunneling logic elements. J. Appl. Phys., April 1996.

[2] R. A. Kiehl and T. Ohshima. Bistable locking of single-electron tunneling elements for digital circuitry. Appl. Phys. Lett., 67(17):2494–2496, Oct 1995.

[3] Tao Yang, R. A. Kiehl, and L. O. Chua. Tunneling phase logic cellular nonlinear networks. International Journal of Bifurcation and chaos, 11:2895–2911, 2001.

[4] L. O. Chua and L. Yang. Cellular neural networks: Theory. IEEE Trans. Circuit and Systems, pages 1257–1272, 1988.

[5] L. O. Chua and L. Yang. Cellular neural networks: Applications. IEEE Trans. Circuit and Systems, pages 1273–1290, 1988.

[6] A. J. Koch and H. Meinhardt. Biological pattern formation: from basic mechanisms to complex structures. Rev. Mod. Phys., 66:1481–1507, OCT 1994.

[7] D. A. Kessler and H. Levine. Pattern formation in dictyostelium via the dynamics of cooperative biological entities. Phys. Rev. E, 48:4801–4804, Dec 1993.

[8] P. K. Maini, K. J. Paintera, and H. N. P. Chaub. Spatial pattern formation in chemical and biological systems. J. Chem. Soc., Faraday T rans., pages 3601–3610, 1997.

[9] A. M. Turing. Philos. T rans. R. Soc. London, 1952.

[10] I. Prigogine and R. Lefever. J. Chem. Phys., 48:1695, 1968.

[11] A. L. Lin, A. Hagberg, A. Ardelea, M. Bertram, H. L. Swinney, and E. Meron. Four-phase patterns in forced oscillator systems. Physical Review E, 62:3790–3798, 2000.

[12] A. L. Lin, M. Bertram, K. Martinez, and H. L. Swinney. Resonant phase patterns in a reaction-diffusion system. Physical Review Letters, 84:4240–4243, May 2000.

[13] R.L. Schiek and E.M. May. Examining tissue differentiation stability through large scale, multi-cellular pathway modeling. In Nanotech 2005, May 2005.

[14] Tim Poston. Catastrophe Theory And Its Applications. Dover Publications, 1997.

[15] L. Nagel. SPICE2: A Computer Program to Simulate Semiconductor Circuits. Electron. Res. Lab., Univ. Calif., Berkeley, 1975.

[16] N. Wiener. Nonlinear Problems in Random Theory. MIT press, Cambridge, MA, 1958.

[17] N. Wiener. Cybernetics. MIT press, Cambridge, MA, 1961.

[18] A. T. Winfree. Biological rhythms and the behavior of populations of coupled oscillators. J. Theor. Biol., 1967.

[19] Y. Kuramoto. Chemical Oscillations, Waves, and turbulence. Springer, Berlin, 1984.

[20] Y. Kuramoto and I. Nishikawa. Cooperative Dynamics in Complex Physical systems. Springer, Berlin, 1989.

[21] X. Lai and J. Roychowdhury. Automated oscillator macromodelling techniques for capturing amplitude variations and injection locking. In IEEE/ACM International Conference on Computer Aided Design, November 2004.

[22] Tosiya Taniuti and Nobuo Yajima. Perturbation method for a nonlinear wave modulation. Journal of Mathematical Physics, 10:1369–1372, Aug 1969.

[23] A. Demir, A. Mehrotra, and J. Roychowdhury. Phase noise in oscillators: a unifying theory and numerical methods for characterization. IEEE Trans. on Circuits and Systems-I:Fundamental Theory and Applications, 47(5):655–674, May 2000.

[24] X. Lai and J. Roychowdhury. Capturing Oscillator Injection Locking via Nonlinear Phase-Domain Macromodels. IEEE Trans. Microwave Theory Tech., 52(9):2251–2261, September 2004.

[25] R. Grimshaw. Nonlinear Ordinary Differential Equations. Blackwell Scientific, New York, 1990.

[26] M. Farkas. Periodic Motions. Springer-Verlag, New York, 1994.

[27] A. Demir. Phase noise in oscillators: Daes and colored noise sources. In IEEE/ACM International Conference on Computer-Aided Design, November 1998.

[28] A. Demir, D. Long, and J. Roychowdhury. Computing phase noise eigenfunctions directly from steady-state jacobian matrices. In IEEE/ACM International Conference on Computer Aided Design, pages 283–288, November 2000.

[29] A. Demir and J. Roychowdhury. A reliable and efficient procedure for oscillator ppv computation, with phase noise macromodelling applications. IEEE Trans. on Computer-Aided Design of Integrated Circuits and Systems, 22(2):188–197, February 2003.

[30] T. Roska and J. Vandewalle. Cellular Neural Networks. Wiley, New York, 1993.

Newton: A Library-Based Analytical Synthesis Tool for RF-MEMS Resonators

Michael S. McCorquodale

Mobius Microsystems, Inc.
Detroit, MI 48226-1686
Tel: +1-313-420-5400x114
Fax: +1-313-420-5404
email: mccorquodale@mobiusmicro.com

James L. McCann

Carnegie Mellon University
Pittsburgh, PA 15213-3891
Tel: +1-412-268-8525
Fax: +1-412-268-5576
email: jmccann@cs.cmu.edu

Richard B. Brown

University of Utah
Salt Lake City, UT 48112
Tel: +1-801-585-7498
Fax: +1-801-581-8692
email: brown@coe.utah.edu

Abstract—*Newton* is a library-based CAD tool with an analytical synthesis engine which has been developed to support the direct synthesis of the physical design and an electromechanically equivalent model of RF-MEMS resonators based on process parameters and performance metrics. *Newton* provides accuracy comparable to finite element analysis while requiring a fraction of the computation and design time. A comparison of results from synthesis with *Newton*, design with FEA, and test results from fabricated devices is presented.

I. INTRODUCTION

Since its inception, the general trend in the field of design automation has been to consider the design activity from an increasingly abstract level, the advantages of which are many. First, design productivity can be increased substantially. Second, design effort can be focussed on functional, logical, and performance verification as opposed to physical design. Lastly, with the use of an APR tool, the entire physical design process can be automated. Of course, significant physical design verification effort is still required, but design iteration can be achieved in a fraction of the time that is required for full-custom physical design.

Certainly, a similar approach could be applied to the synthesis of microelectromechanical systems (MEMS) and the advantages would be similar. However, MEMS synthesis presents certain implementation challenges. Consider that synthesis of MEMS is not as straight-forward as analogous problems in the field such as digital logic synthesis. Further consider that MEMS devices must typically be modeled as continuous systems while logic can be described by a finite set of states. Also, a variety of MEMS topologies and manufacturing process technologies exist. Consequently, the standard approach to MEMS design has been with finite element analysis (FEA). With FEA, arbitrary structures can be designed in any material and simulated for a given set of boundary conditions. Though these tools are clearly indispensable for MEMS design, their use requires significant knowledge of MEMS device physics and manufacturing. Circumventing FEA would substantially increase design productivity. Similarly, some type of direct silicon compilation would reduce physical design time.

In this work, an alternative library-based analytical synthesis tool, *Newton*, is presented for use in the design of radio frequency (RF)-MEMS resonators. *Newton* supports the direct synthesis of the physical design and an electromechanically equivalent model for RF-MEMS resonators from process parameters and performance metrics. Moreover, *Newton* provides performance accuracy comparable to FEA while requiring a fraction of the computation time and design effort.

In the sections that follow, current MEMS simulation and synthesis approaches are described in greater detail where design effort and time are considered. The approach utilized in *Newton* is then presented and a sample derivation of the analytical expression for one key performance metric for one supported device is shown. The user interface is illustrated as a sample synthesis problem is solved, thus further describing the tool framework and its use. Experimental results from devices that have been fabricated from synthesis with *Newton* are presented along with FEA results. Good agreement is found between the specification and the measured results.

II. MEMS SIMULATION APPROACHES

A. Finite Element Analysis

FEA is an analysis technique by which an object is dissected into a finite number of elements and analyzed based on mechanical constraints, or boundary conditions, that are placed upon specific elements. The technique requires significant computation time for both finite element generation, or meshing, and solution convergence. Additionally, significant design effort is required to generate the 3D model of the structure to be simulated. Mesh and simulation time are largely contingent upon the mesh size and type. As the number of nodes in the mesh is increased, the accuracy is increased, but so is the computational complexity and simulation time. Each simulation typically requires several hours and a typical design effort can span days for a single device.

B. Nodal Analysis

In [1], a nodal analysis approach, similar to that implemented in SPICE, was developed in *Matlab* for MEMS simulation. Here, basic MEMS elements including beams, gaps, and anchors are parameterized based on geometry and can be coupled together. The law of static equilibrium is applied to each node and DC, steady-state, and transient results can be found by solving the resulting set of coupled ordinary differential equations using nodal analysis. With this approach, simulation time can be reduced dramatically as compared to FEA, though design iteration is still required.

III. MEMS SYNTHESIS APPROACHES

A. Automated Design Synthesis

Recent work has leveraged the speed at which MEMS can be simulated with the tool presented in [1] and lead to the exploration of automated synthesis approaches. A comparison of such approaches has been presented in [2] and includes the evolutionary approach presented in [3] where multi-objective genetic algorithms have been employed. Such an approach is general-purpose and allows the design space to be explored rapidly, but is suited to the design of arbitrary devices and not targeted at specific topologies of specific utility.

0-7803-9451-8/06/$20.00 ©2006 IEEE.

B. Library-Based Analytical Synthesis

A parameterized layout generator for low frequency lateral MEMS resonators was presented in [4]. Here, parameters such as the device resonant frequency were used to solve simple mechanical relationships from which the device geometry could be determined, thus determining the physical design. Additionally, an electromechanical analogy was employed to build a SPICE model for system-level simulation and optimization. Though this work demonstrated reasonably good agreement between measurements from fabricated devices and the specified resonant frequency, system-level design iteration was required and extension of the framework was not addressed. Moreover, it would likely be difficult to achieve high specification accuracy due to the simplicity of the mechanical relationships that were employed.

In contrast to this previous work, *Newton* has been developed with an extensible component library framework. The reality of MEMS design and development is that although an infinite number and variety of devices can be conceived and developed with FEA or evolutionary synthesis, only a finite number possess any practical use. For example, RF-MEMS resonators have recently been demonstrated successfully in oscillator circuits [5]. Moreover, the numerous MEMS foundries that emerged throughout the mid and late 1990's are now being consolidated and standardized, thus reducing the variety of available devices and process technologies. Considering these trends, *Newton's* library contains previously and analytically designed components that can be selected and synthesized based on performance parameters. This finite component library is certainly not a limitation of the tool as *Newton* is an extensible software framework into which more components can be integrated easily. Components currently supported include RF-MEMS varactors and RF-MEMS resonators of two topologies: clamped-clamped beam (CCB) and free-free beam (FFB).

The back-end of *Newton* solves analytical expressions which have been derived for the synthesis of the supported MEMS components. Thus solutions can be obtained directly from the performance specification and process parameters. With this approach, the costly simulation and design time associated with FEA is reduced substantially. As compared to other analytical approaches, MEMS-specific design knowledge is not required for synthesis, and convergence to a specification is achieved by *Newton*, not the designer. Lastly, in this approach, lumped-parameter equivalent circuits have been derived, which provide significant utility for co-simulation of MEMS and transistor circuitry. Moreover, the total time required for physical synthesis and model generation is mere seconds.

In the following section, an example of one of the derived performance-driven analytical expressions utilized in the back-end of *Newton* is presented for illustrative purposes. How such expressions are utilized within the framework of *Newton* is presented following the derivation.

IV. EXAMPLE ANALYTICAL EXPRESSION AND COMPUTATIONAL ALGORITHM

Here, the derivation of the analytical expression for synthesis for one performance metric (the resonant frequency) of one supported component (a CCB resonator) is shown.

A. CCB Resonator Design Overview

A CCB resonator is illustrated in Fig. 1 and has been presented previously in [6], though without complete analysis for design synthesis purposes. It is comprised of a prismatic beam that is fixed at both ends and suspended over an electrode with an air gap between the electrode and the beam. The device is typically utilized as a frequency reference for oscillator and filter applications. Thus the most significant performance metric is the mechanical resonant frequency of the device. Two general physics-based harmonic analysis approaches can be applied in an effort to calculate the resonant frequency of this device and the corresponding lumped-parameter electrical model. These approaches include the Euler-Bernoulli Method [7] and Timoshenko's method [8], the former of which is presented next. The results from this analysis provide the general purpose analytical synthesis expression for synthesizing CCB resonators based on the resonant frequency.

B. Euler-Bernoulli Method and Solution

For transverse vibrations of a simple beam, the mode shape equation takes the following form [9],

$$u(x) = C_1(\cos kx + \cosh kx) + C_2(\cos kx - \cosh kx) \quad (1)$$
$$-C_3(-\sin kx - \sinh kx) + C_4(\sin kx - \sinh kx)$$

where C_1, C_2, C_3, and C_4 are constants; k is the wave number; and x is distance along the beam with length L. For the clamped-clamped case, the boundary conditions are:

$$u(0) = 0, \left.\frac{du}{dx}\right|_{x=0} = 0, u(L) = 0, \text{ and } \left.\frac{du}{dx}\right|_{x=L} = 0 \quad (2)$$

The first two conditions are satisfied if $C_1 = C_3 = 0$. The remaining two conditions determine a set of equations which can be solved by evaluating the determinant. After some algebra, the determinant becomes,

$$\frac{1}{\cosh kL} = \cos kL \quad (3)$$

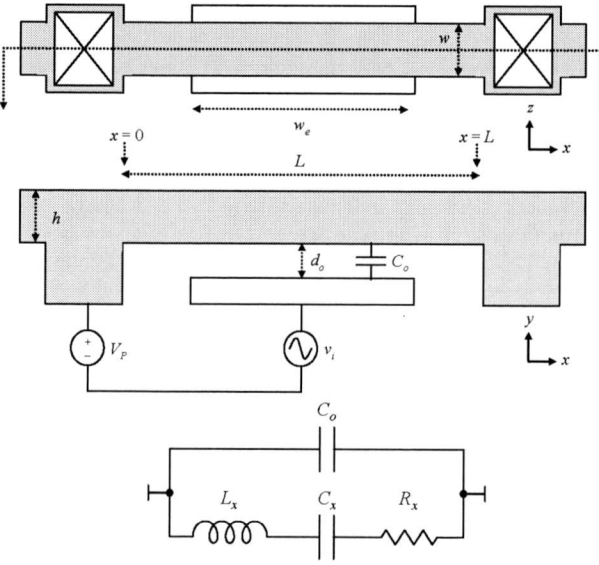

Fig. 1. Simple top and cross-sectional illustration of a CCB resonator along with equivalent lumped-parameter electromechanically equivalent circuit model at resonance.

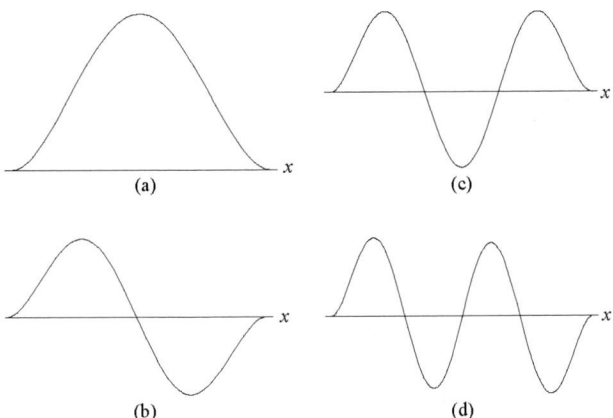

Fig. 2. First 4 mode shapes of a CCB resonator with length L: (a) Fundamental mode. (b) 2nd mode. (c) 3rd mode. (d) 4th mode.

There are multiple kL roots to (3). The first nontrivial root represents the fundamental mode shape of the device. This root, and others, are pictured in Fig. 2. Because L does not change for any of these mode shapes, it is useful to denote the wave number by k_n where n denotes the mode number. For each mode, C_1 and C_3 can be solved uniquely, thus $u(x)$ can be determined.

The fundamental resonant frequency, f_o, for any mechanically resonant device can be determined by solving the differential equations of simple harmonic motion. Without damping, the solution is [9],

$$f_o = \frac{1}{2\pi}\sqrt{\frac{k_m}{m}} \qquad (4)$$

where k_m is the mechanical stiffness of the resonating body and m is its mass. In [6], it has been shown how (3) and (4) can be evaluated for the case of the CCB and an arbitrary mode,

$$f_n = \frac{(k_nL)^2}{2\pi L^2}\sqrt{\frac{EI}{\rho A}} \qquad (5)$$

where f_n is the nth resonant mode, E is Young's modulus, I is the moment of inertia, ρ is the density of the material, and A is the cross-sectional area where $A = W_r h_r$. The moment of inertia for a simple prismatic beam is given by, $I = (1/12)W_r h_r^3$, and substituting yields,

$$f_n = \frac{(k_nL)^2}{2\pi\sqrt{12}}\sqrt{\frac{E}{\rho}}\frac{h_r}{L^2} \qquad (6)$$

At first glance, this expression appears to be sufficient for analytical evaluation. However, consider that the CCB actually experiences two spring effects: mechanical (k_m) and electrical (k_e). The mechanical spring constant is determined by the stiffness of the material and the anchors. The electrical spring constant is associated with the DC voltage that is applied across the beam which effectively "softens" the spring constant associated with the system because this force is subtractive. Considering this effect, a modified version of (6) can be written as,

$$f_n' = \frac{1}{2\pi}\sqrt{\frac{k_m - k_e}{m}} = f_n\sqrt{1 - \frac{k_e}{k_m}} = \frac{(k_nL)^2}{2\pi\sqrt{12}}\sqrt{\frac{E}{\rho}}\frac{h_r}{L^2}\left(1 - \frac{k_e}{k_m}\right)^{1/2} \qquad (7)$$

The force on the beam can be calculated by considering the energy stored between the beam and the electrode. The differential of this force, with respect to displacement, is the electrical spring constant. Referring to Fig. 1, displacement is in the y direction.

$$F_e = \frac{\partial E}{\partial y} = \frac{1}{2}\frac{\partial C}{\partial y}V_p^2 = \frac{1}{2}\frac{C_o V_p^2}{d_o} \qquad (8)$$

$$k_e = \left|\frac{\partial F_e}{\partial y}\right| = \frac{C_o V_p^2}{d_o^2} = \frac{\varepsilon A V_p^2}{d_o^3} \qquad (9)$$

Consider that both k_m and m are functions of x, the position along the beam. Return to the general expression for the mechanical resonant frequency in (6) and consider the following relationship,

$$f_n = \frac{(k_nL)^2}{2\pi\sqrt{12}}\sqrt{\frac{E}{\rho}}\frac{h_r}{L^2} = \frac{1}{2\pi}\sqrt{\frac{k_m}{m}} \qquad (10)$$

and solve for the position-dependent mechanical stiffness, $k_m(x)$.

$$k_m(x) = \left[\frac{(k_nL)^2 h_r}{\sqrt{12}L^2}\right]^2 \frac{E}{\rho}m(x) \qquad (11)$$

Now using (9) and (11) derive an expression for k_e/k_m over a differential length of the beam, dx, for a beam of length L. Such an expression becomes,

$$\frac{k_e}{k_m}(L)dx = \frac{V_p^2\varepsilon W_r}{d_o^3}\frac{\rho L^4}{\frac{(k_nL)^4}{12}Eh_r^2 m(x)}dx \qquad (12)$$

The expression in (12) should be integrated over the region in which there exists a beam-electrode overlap because this is the region in which a DC bias exists across the device, thus softening the system spring constant. Assuming that the beam is positioned in x between $x = 0$ and $x = L$ and that the electrode is centered under the beam, as illustrated in Fig. 1, (12) becomes,

$$\frac{k_e}{k_m}(L) = \frac{V_p^2\varepsilon W_r\rho L^4}{\frac{(k_nL)^4}{12}Eh_r^2 d_o^3}\int_{\frac{L-W_e}{2}}^{\frac{L+W_e}{2}}\frac{1}{m(x)}dx \qquad (13)$$

At this point (13) could be evaluated by integration, except that no expression for $m(x)$ has been determined yet. This expression is determined next using a generalized equivalent mass technique [10].

The kinetic energy, KE, for a body in motion is given by,

$$KE = \frac{1}{2}mv^2 \qquad (14)$$

where v is the velocity of the mass. If the equivalent mass of the body varies along the position of the body, then so must the equivalent velocity, such that energy is conserved. Thus, (14) can be rewritten as an integral in the form,

TABLE I
CCB RESONATOR PROCESS AND PERFORMANCE VARIABLES

Design variable	Type	Description
ρ	Process	Density
E	Process	Young's Modulus
h	Process	Beam height
d_o	Process	Beam-electrode gap
k_n	Performance	Determined by mode
V_p	Performance	Bias voltage
W_r	Performance	Beam width
W_e	Performance	Electrode width
f_o	Performance	Resonant frequency

TABLE II
CCB RESONATOR CONSTANTS AND DERIVED VARIABLES

Design variable	Value/Expression	Description
ε	8.85×10^{-12} F/m	Permittivity of free space
A	$A = W_r h_r$	Beam cross-sectional area
I	$I = (1/12) W_r h_r^3$	Moment of inertia
$u(x)$	Defined by (1)	Mode shape function
L	From (17) in (7)	Beam length

correct root of (7) with (17) substituted. The remaining parameters are simply calculated from the respective definitions. This analysis demonstrates one example of the typical back-end computations performed by *Newton* for a specific device.

V. TOOL FRAMEWORK

Newton is partitioned into a graphical user interface (GUI) and a synthesis engine in order to facilitate the development and addition of new devices to the library. The tool has a uniform GUI shared by all components and individual synthesis scripts for each component. The interaction between the GUI and the synthesis engine is illustrated in Fig. 3. To add a new component, one need only add a new component script to the engine. This modularity greatly facilitates the ability to expand the capabilities of the tool. The synthesis engine is simply comprised of the synthesis scripts which have been coded based upon performance analyses such as the analysis presented in the previous section.

Fig. 4 and Fig. 5 outline the synthesis procedure for a CCB resonator where the user interaction with the GUI can be described in two phases. In the first phase the user selects a device from the library component browser and specifies the performance and process-dependent parameters within the component parameter interface. Then synthesis is initiated.

$$m_{eq}(x) = \frac{KE}{\frac{1}{2}v^2(x)} = \frac{\frac{1}{2}\rho A}{\frac{1}{2}v^2(x)} \int_0^L v^2(x)dx \qquad (15)$$

Velocity is the differential of position. Thus for an object at resonance, the velocity can be described in phasor form by, $v(x) = j\omega u(x)$, where ω is the radian resonant frequency and $u(x)$ is the displacement of the beam in the y direction at position x. Now (15) can be rewritten as:

$$m_{eq}(x) = \frac{\frac{1}{2}\rho A \int_0^L [j\omega u(x)]^2 dx}{\frac{1}{2}[j\omega u(x)]^2} = \frac{\rho A}{u^2(x)} \int_0^L u^2(x)dx \qquad (16)$$

Now with an expression for $m_{eq}(x)$, an expression for $k_e/k_m(L)$ can be determined,

$$\frac{k_e}{k_m}(L) = \frac{V_p^2 \varepsilon W_r \rho L^4}{\frac{(k_n L)^4}{12} E h_r^2 d_o^3} \int_{\frac{L-W_e}{2}}^{\frac{L+W_e}{2}} \left(\frac{\rho A}{u^2(x)} \int_0^L u^2(x)dx \right)^{-1} dx \qquad (17)$$

and finally (7) can be determined with the substitution of (17). Thus, an accurate physics-based analytical expression for the resonant frequency of the device has been determined. Moreover, the effect of spring softening on the resonant frequency of the device has also been included.

In a typical application, ρ, E, h_r, and d_o are determined by the fabrication process. Performance parameters k_n, V_p, W_r, W_e, and f_o are determined by design and ε and π are constants. Thus L can be synthesized from the design parameters using a numeric integration technique for solving (7) in L with (17) substituted.

Table I summarizes these process and performance parameters, while a summary of constants and derived variables is presented in Table II. The solutions to these expressions are determined in *Newton* via scripts written in *Mathematica*, a symbolic mathematics package. *Mathematica* was selected due to the fact that it supports symbolic integration, which appears often in this analysis. The computation algorithm involves using the design and process parameters in order to synthesize the device length, L, based upon convergence to the

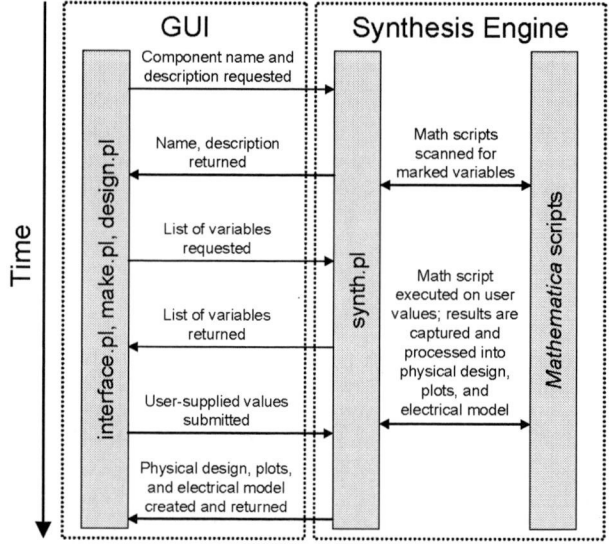

Fig. 3. The *Newton* framework partitioned into a GUI and a synthesis engine. The chronology of steps involved with a typical synthesis session is indicated by the time axis.

2006 Asia and South Pacific Design Automation Conference

3B-2

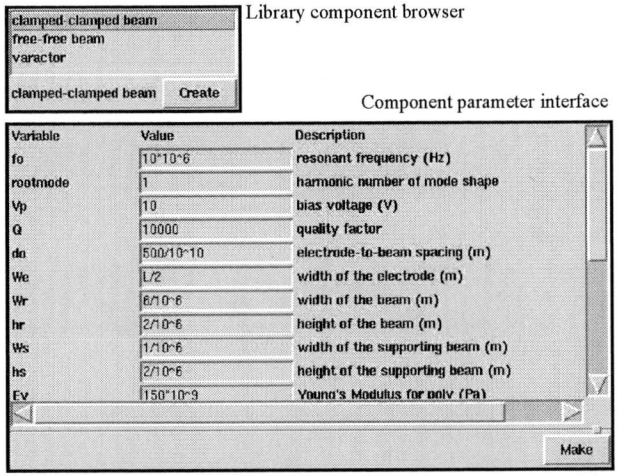

Library component browser

Component parameter interface

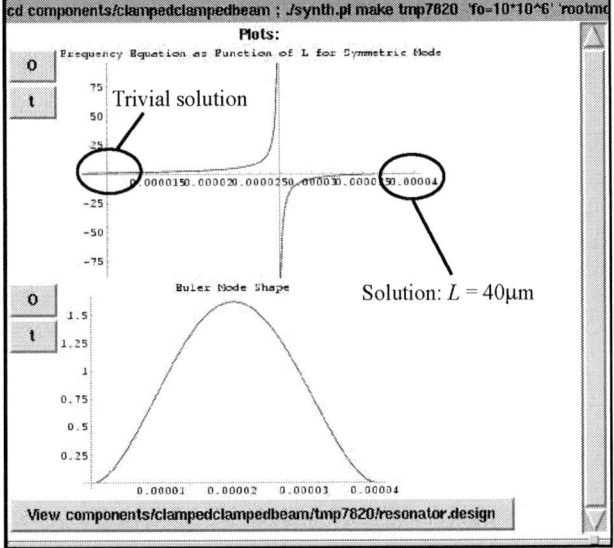

Fig. 4. Screenshot of *Newton*'s GUI and framework. Post synthesis results are shown for a 10MHz CCB resonator.

Results can be verified by examining the frequency equation graph and the mode shape as shown in Fig. 4. In the second phase, the user is presented with a graphical display of the mask set, shown in Fig. 5, for the component and allowed to adjust those parameters that are not fixed by the desired performance characteristics. These include options such as interconnect position and size. Once the design is complete, the user may export both the physical design in CIF for fabrication and the electromechanically equivalent model of the component in *SPICE* format for simulation with other MEMS devices or transistors.

A. Graphical User Interface

Perl and the GUI package Perl/Tk were utilized to develop *Newton*'s GUI since these packages are easily portable across operating environments and allow for rapid software development. The partition of *Newton*, however, allows new MEMS library component synthesis scripts to be written in any programming language. In this work, *Newton* was compiled for the *SUN Solaris* operating environment.

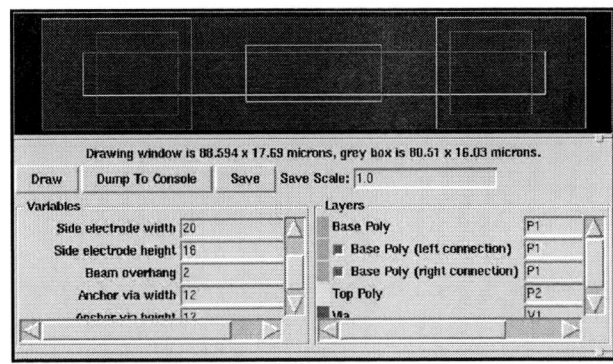

Fig. 5. *Newton*'s physical design viewpoint for a 10MHz CCB resonator, which is shown fabricated in Fig. 6.

B. Synthesis Engine

As described previously, synthesis scripts have been developed with *Mathematica* but could be eventually coded directly into the design framework with the aid of a math package. This would eliminate the need for a license for *Mathematica*. However, the synthesis scripts were developed with these tools due to the symbolic nature of the derived analytical expressions. Some challenges reside in porting the code for these analytical expressions to the code supported by a numeric math package, particularly in instances where intermediate symbolic solutions must be replaced with numeric solutions. For these reasons, this activity was not pursued in this work.

VI. A SYNTHESIS EXAMPLE

As a benchmark and synthesis example, a 10MHz polysilicon (poly-Si) CCB resonator was synthesized with *Newton* using the Euler-Bernoulli Method. The procedure and data from this design example are exactly those shown in Fig. 4 and Fig. 5. First, the CCB library part was selected from the library component browser. The process and performance parameters are entered into the synthesis form; the values for this example are listed for the reader's convenience in Tables III and IV. Synthesis was initiated by selecting the *Make* button. The frequency equation was plotted against L and displayed along with the mode shape equation. In Fig. 4, it can be seen that indeed the synthesis script has converged to the first nontrivial solution of the frequency equation and that the mode shape corresponds to the first resonant mode. The solution was $L = 40\mu m$ as shown. Once design convergence was

TABLE III
PERFORMANCE PARAMETERS FOR THE SYNTHESIS OF A 10MHz CCB RESONATOR.

Performance-Driven	Value
Resonant frequency, f_o	10MHz
Resonant mode number, n	1
Resonator width, W_r	6μm
Bias voltage, V_p	10V
Electrode width, W_e	$L/2$

TABLE IV
PROCESS PARAMETERS FOR THE SYNTHESIS OF A 10MHz CCB RESONATOR

Process-Dependent	Value
Density, ρ	2330kg/m^3
Young's modulus, E	150GPa
Resonator height, h_r	$2\mu m$
Resonator-electrode gap, d_o	500Å

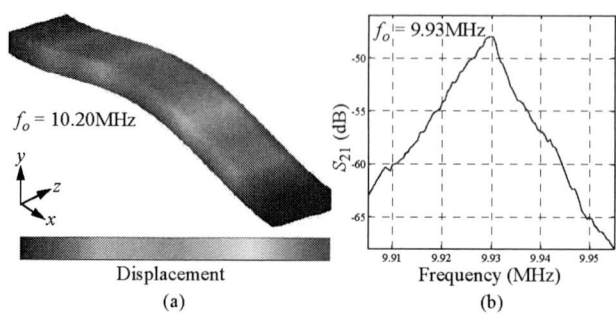

Fig. 7. (a) FEA simulation results. (b) Measured S_{21} response for the fabricated 10MHz CCB poly-Si resonator shown in Fig. 6.

verified, the physical design was edited in real time as shown in Fig. 5. The physical design was generated automatically and displayed within a form where attributes that do not affect performance can be modified. These parameters were adjusted appropriately and the design was exported to CIF for mask generation. The device was fabricated using a custom polysilicon surface micromachining process in the University of Michigan's Solid State Electronics Laboratory. An electron micrograph of fabricated device is shown in Fig. 6.

VII. EXPERIMENTAL RESULTS

Prior to testing the fabricated device, the physical design from *Newton* was imported into the FEA tool *Coventorware*. An accurate 3D model was generated from process parameters that corresponded to the fabrication process for the device. A manhattan brick mesh of the 3D device was created and harmonic analysis was performed. The simulated fundamental mode resonant frequency was 10.20MHz as shown in Fig. 7.

The resonant frequency of the fabricated CCB device was determined by measuring the transmission scattering parameter, S_{21}, with an HP4195 network analyzer. The device under test was placed under a vacuum pressure of approximately 100mTorr. The measured spectrum in shown in Fig. 7, where the resonant frequency peak is at 9.93MHz and the quality factor is approximately 1,500.

Results from both *Newton* and *Coventorware* are close to the measured resonant frequency for the device. The simulated resonant frequency within *Coventorware* was 2.7% in error as compared to fabricated devices while the *Newton* design was in error by only 0.70%. Accuracy differences between the two packages likely arise from the fact that the synthesis scripts in *Newton* account for electrical frequency pulling, while the modal analysis in *Coventorware* is a mechanical analysis that does account for this phenomenon.

In addition to the CCB resonator, all of the remaining devices in *Newton*'s component library have been synthesized, fabricated, and tested. Results are not presented here in the

interest of space, but performance is similar to that for the CCB resonator shown here.

VIII. CONCLUSION

Newton is the first complete CAD tool that supports library-based analytical synthesis of RF-MEMS resonators from a performance specification directly to physical design, while also generating an equivalent lumped-parameter electrical model for simulation with other devices. *Newton* has been shown to be a fast and accurate tool for the design of RF-MEMS resonators. As a benchmark, a 10MHz poly-Si CCB resonator was synthesized with *Newton* and the measured resonant frequency was in error by only 0.70%.

Newton could certainly be developed into a much more substantial tool through expansion of the component library. Each device would require analysis and derivation of critical performance parameters. The developed analytical expressions would then need to be coded into a math package, after which synthesis accuracy could be determined via fabrication and test. Beyond expanding the component library, the tool could be developed to increase automation. For example, process parameters could be set based on a selected foundry. The authors intend to explore such opportunities in future work.

REFERENCES

[1] N. Zhou, J. V. Clark, and K. S. J. Pister, "Nodal analysis for MEMS design using SUGAR v0.5," *Proc. of Modeling and Simulation of Microsystems, Semiconductors, Sensors, and Actuators*, 1998.

[2] R. Kamalian, N. Zhou, and A. M. Agogino, "A comparison of MEMS synthesis techniques," *Proc. of 1st Pacific Rim Workshop on Transducers and Micro/Nano Technologies*, 2002.

[3] R. H. Kamalian, H. Takagi, and A. M. Agogino, "Optimized design of MEMS by evolutionary multi-objective optimization with interactive evolutionary computation," *Proc. of the Genetic and Evolutionary Computation Conference*, 2004.

[4] N. R. Lo, *et al.*, "Parameterized layout synthesis, extraction, and SPICE simulation for MEMS," *Proc. of IEEE International Symposium on Circuits and Systems*, 1996.

[5] Y.-W. Lin, S. Lee, S.-S. Li, Y. Xie, Z. Ren, and C. T.-C. Nguyen, "60-MHz wine-glass micromechanical-disk reference oscillator," *Proc. of IEEE Int. Solid-State Circuits Conf. Dig. of Tech. Papers*, 2004.

[6] C. T.-C. Nguyen, *Micromechanical Signal Processors*, Ph.D. Dissertation, University of California-Berkeley, 1994.

[7] K. Wang, *Microelectromechanical Resonators and Filters for Communications Applications*, Ph.D. Dissertation, University of Michigan, 1999.

[8] R. A. Anderson, "Flexural vibrations in uniform beams according to the Timoshenko theory," *J. of Applied Mechanics*, 1953.

[9] W. Weaver, *et al.*, *Vibration Problems in Engineering*, 5th ed., New York: John Wiley & Sons, 1990.

[10] M. E. Frerking, *Crystal Oscillator Design & Temperature Compensation*, New York: Van Nostrand Reinhold Company, 1978.

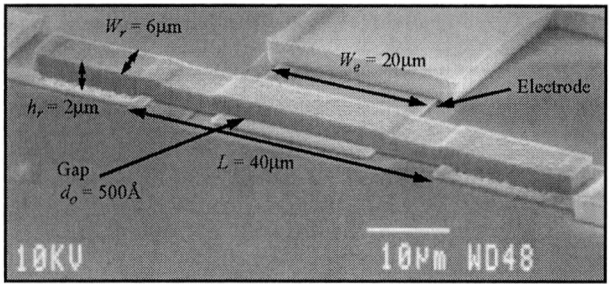

Fig. 6. Electron micrograph of a fabricated surface micromachined 10MHz CCB poly-Si resonator.

2006 Asia and South Pacific Design Automation Conference

Jitter Decomposition in Ring Oscillators *

Qingqi Dou

Computer Engineering Research Center
The University of Texas at Austin
Austin, TX 78712
e-mail: qdou@cerc.utexas.edu

Jacob A. Abraham

Computer Engineering Research Center
The University of Texas at Austin
Austin, TX 78712
e-mail: jaa@cerc.utexas.edu

Abstract— It is important to separate random jitter from deterministic jitter to quantify their contributions to the total jitter. This paper identifies the limitations of the existing methodologies for jitter decomposition, and develops a new and efficient approach using time lag correlation functions to decompose different jitter components. The theory of the approach is developed and it is applied to a ring oscillator simulated in a 0.6-um AMI CMOS process. Results show good agreement between the theory and hspice simulation.

I. INTRODUCTION

Ring oscillators have been integrated into many digital and communication systems due to their low area, low power and fast response time. They are used in serial links, disk-drive read channels in data communications, *Clock and Data Recovery* (CDR) circuitry in optical communication networks, and frequency synthesizers in wireless communications. In such applications, jitter is one of the most important parameters affecting the *Bit-Error Rate* (BER). BER is determined by the signal integrity, which is quantified by *Signal-to-Noise Ratio* (SNR) and *Total Jitter* (TJ) [3, 4]. Therefore, both signal amplitude noise and jitter magnitude must comply with specifications in the communication standards for a certain BER level.

One of the greatest challenges in integrating all circuitry on the same chip is in the placement of sensitive analog circuits (e.g., ring oscillators) and large digital blocks to meet the signal integrity requirements. With the global supply and ground busses, ring oscillators may experience performance degradation due to the substrate and supply noise. As demonstrated in [2, 1], the noise manifests itself as jitter at the output of ring oscillators. Though the jitter magnitude induced by the supply and substrate noise far exceeds that introduced by the device electronic noise, its contribution to the TJ is weighted approximately ten times less [4]. Therefore, jitter decomposition is necessary for accurately deriving TJ and BER.

This paper presents a new methodology for jitter decomposition using *Time Lag Correlation* (TLC) functions. TLC treats jitter in its original form, as a time series, resulting in good accuracy in the decomposition.

In section ii, we define and analyze the problems of jitter

decomposition in ring oscillators. The limitations of conventional methods are discussed in section iii. In section iv, the basic theory of TLC functions is presented and equations using them for jitter decomposition are derived. The generality of the theory is verified with different assumptions in section v. In section vi, the developed approach is applied to seperate *Random Jitter* (RJ) from *Sinusoid Jitter* (SJ) injected by the thermal and supply noise respectively, in a ring oscillator and the simulation results are presented.

II. PROBLEM DESCRIPTION

A. Jitter Definitions

Jitter is a general term; it can include absolute jitter, cycle jitter or cycle-to-cycle jitter (Figure 1). Cycle jitter is defined as the deviation of a signals nth timing event, t_n, from its intended (ideal) occurrence in time. The nth period is then defined as $T_n = t_{n+1}-t_n$, while the ideal point is normally the minus-to-plus zero crossing of the signal, which is determined by the period, T. Then, the cycle jitter is defined as $j_n = T_n\text{-}\overline{T}$, where $\overline{T}$ is the mean period, the average value of the measured periods.

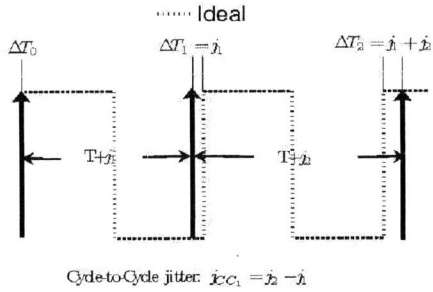

Fig. 1. Absolute jitter, cycle jitter, and cycle-to-cycle jitter

Absolute jitter defined [9] as

$$\Delta T_N = \sum_{n=1}^{N} j_n \qquad (1)$$

and is often used to quantify the jitter of Phase-Locked Loops (PLLs) or to derive the BER. This is, in fact, an integral of the

---

*This work was supported by the National Science Foundation under Grant No. CCR-0325371 and in part by Subcontract No. SA3271JB from UC Berkeley under prime Contract 2003-DT-660 from Microelectronic Advanced Research Corporation (MARCO).

0-7803-9451-8/06/$20.00 ©2006 IEEE.

cycle jitter; thus, the variance of ΔT_N diverges with time [2]. Therefore, this is not a good indicator for jitter analysis.

Cycle-to-cycle jitter is given by

$$j_{CC_n} \;=\; j_{n+1} - j_n \qquad (2)$$

The difference between the cycle jitter and the cycle-to-cycle jitter is that the former compares the measured period with the mean period while the latter compares with the preceding period. Hence, cycle-to-cycle jitter is the first difference of the cycle jitter, representing the short-term dynamics of the period, while cycle jitter contains no information about the dynamics. Though cycle jitter is used here, both ΔT_N and j_{CC_n} can be derived similarly.

In general, the jitter distributions encountered contain both bounded and unbounded random components as well as deterministic components. RJ is assumed to be a Gaussian distribution with a *Root-Mean-Square* (RMS) value while *Deterministic Jitter* (DJ) is represented by its *Peak-to-Peak* (PP) value. DJ can be further categorized according to the source of the jitter. For example, power switching noise leads to SJ; mismatch in the differential circuits gives *Duty Cycle Distortion* (DCD); while bandwidth limited channel introduces *Intersymbol Symbol Interference* (ISI), etc.

Provided that RJ and DJ are independent variables, the sum function in the time domain is equal to the convolution of the RJ and the DJ distribution [7]. There are several reasons to separate the different jitter components. A simple case of TJ consisting of an RJ with an RMS value of 0.5 ps and an SJ with a PP value of $5\sqrt{2}$ will be used to illustrate the problem. The problem encountered is that neither the RMS nor the PP value of the PDF represents either the RMS of the RJ or the PP value of the SJ (Tab. I). The PP value of the convoluted histogram is no longer constant because the RJ is included. Also, the RMS value of the convoluted histogram is not that of the RJ component.

Another reason to separate the RJ and the DJ is that they accumulate differently in a serial link. TJ is actually the sum of PP_{RJ} and PP_{DJ}, where $PP_{RJ}=K_\sigma*RMS_{RJ}$. K_σ denotes a weight factor associated with different BER boundaries [8] (Tab. II).

TABLE I
RMS AND PP OF A CONVOLUTED JITTER (PS)

Sample Size	RMS	PP
1000	5.02	8.21
11000	5.02	8.45
101000	5.03	8.85

B. Jitter Injection in a Ring Oscillator

Fig. 2 illustrates a three-stage ring oscillator, where all the unit delay cells share the power supply. SJ is introduced by coupling a sinusoid noise to the power supply voltage (Fig. 3) while RJ is injected through the thermal noise [10]. The jitter resulting from the simultaneous injection of the RJ and SJ is demonstrated in Fig. 4.

TABLE II
WEIGHT FACTOR K_σ VS. BER

BER	K_σ	BER	K_σ
10^{-4}	7.44	10^{-11}	13.41
10^{-5}	8.53	10^{-12}	14.07
10^{-6}	9.51	10^{-13}	14.70
10^{-7}	10.40	10^{-14}	15.30
10^{-8}	11.22	10^{-15}	15.80
10^{-9}	12.00	10^{-16}	16.44
10^{-10}	12.72	10^{-17}	16.93

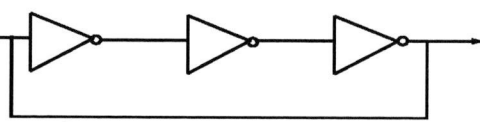

Fig. 2. Three-stage ring oscillator

III. LIMITATIONS OF CONVENTIONAL METHODS

Histogram-based analysis specifies only the TJ through *Peak-to-Peak* (PP) or *Root-Mean-Square* (RMS) value, ignoring the parameters of the deterministic jitter, such as the frequency of the SJ. Since SJ is a function of time, the PDF, a simple counting of events, contains no knowledge of how the events evolve in time. The events could have happened in ascending order, descending order, or in a completely random fashion. Figure 5 compares two TJs both consisting of a RJ with a RMS value of 5ps and an SJ with a PP value of $5\sqrt{2}$ but different frequencies. The two histograms differ little even with a large sample size.

The TailFit algorithm [5], extracts the RMS value of the RJ from the tail of the PDF and defines the difference between the two peaks in the histogram as the PP value of the DJ (Figure 6). However, it fails when the RMS value is comparable to the PP value. Figure 6 demonstrates that the value of the DJ cannot be extracted as there is only one peak. Even more samples do not help.

Generally, the time evolution of a process can be quantified by the power spectrum, providing an estimation of the contri-

Fig. 3. SJ is injected by the supply noise

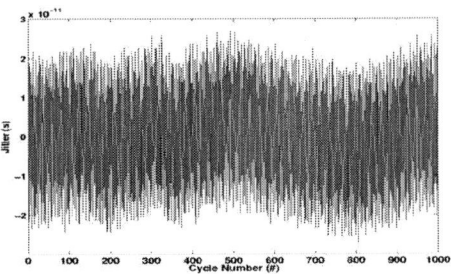

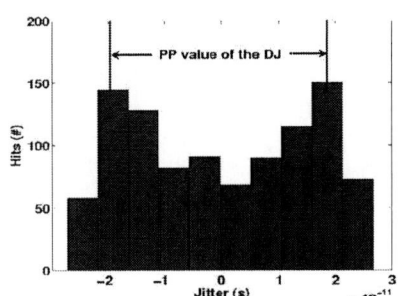

Fig. 4. Jitter induced by thermal and supply noise

Fig. 6. Histogram-based method

bution of each frequency to the total variance. Spectral analysis normally gives good precision in estimating the frequency of the SJs [6]. However, it is not straightforward to deal with the time series and there still exists several issues, such as, how to separate jitter with SJ, DCD, and other deterministic components; how to detect high-frequency jitter; and how to overcome the aliasing problem.

Another way to view the time evolution is in terms of TLC. This is efficient for our purposes because each component of jitter evolves differently with time. For example, RJ, as a random variable, has little correlation with its lagged time series while SJ has strong correlation with its lagged version. It will be illustrated in Section IV that it is possible to separate RJ, multiple SJs and DCD by their different functions of the TLC.

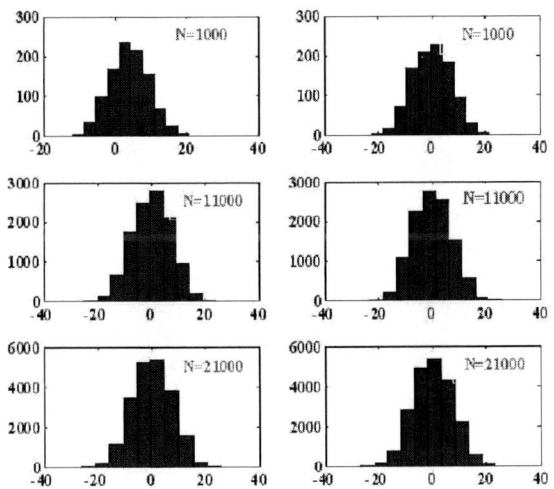

Fig. 5. Dependence of PDF on the sample size and its inability to extract the frequency of SJ, f_m

IV. THE PROPOSED METHODS

This section illustrates how the TLC functions of different jitter components differ. and how the TLC function of the time series for jitter helps to separate SJ, RJ, and DCD. The basic theory of TLC and its efficiency in separating RJ and DJ will be demonstrated in the following. Jitter as a time series can be

represented by TLC, which is defined as

$$C_j(m) = lim_{N \to \infty} \frac{1}{N} \sum_{n=1}^{N} (j_{n+m} j_n), \qquad (3)$$

where $C_j(m)$ is the TLC of the cycle jitter, defined as the correlation of a variable with itself over m successive time lags with a sample size of N. Ideally, a random variable only has autocorrelation when time lag is zero

$$C_j(0) = \sigma_{RJ}^2 \qquad (4)$$

which means that the TLC with a zero argument is the variance of the cycle jitter. The TLC drops rapidly as it departs from zero, finally approaching zero when the time lags approaches infinity (Fig. 7).

The RMS of the cycle-to-cycle jitter for a RJ can also be expressed by the TLC of the cycle jitter.

$$\sigma_{CC}^2 = lim_{N \to \infty} \frac{1}{N} \sum_{n=1}^{N} (j_{n+m} - j_n)^2 \qquad (5)$$

This indicates that the variance will double if two Gaussian variables are subtracted with identical means and standard deviations as $C_j(1)$ is nearly zero here.

Similarly, the cycle jitter and the cycle-to-cycle jitter of an SJ can also be expressed by the TLC, with all values being the PP values. Assume the SJ to be

$$j(m) = PP * \cos(\frac{m \varpi_m}{f_0}), \qquad (6)$$

where ϖ_m is the angular frequency of the SJ and f_0 is the signal frequency. The time lag number is represented by m, which corresponds to a time lag of $\frac{m}{f_0}$. Then the TLC of the SJ is

$$\begin{aligned} C_j(m) &= lim_{N \to \infty} \frac{1}{N} \sum_{n=1}^{N} (j_{n+m} j_n) \\ &= \frac{PP^2}{2} cos(\frac{m \varpi_m}{f_0}). \end{aligned} \qquad (7)$$

When m=0,

$$C_j(0) = \frac{PP^2}{2} = \sigma_{SJ}^2 \qquad (8)$$

287

where σ_{SJ} is the RMS of the SJ. Equation 8 represents the fact that the PP value of the SJ is $\sqrt{2}\sigma_{SJ}$, while Equation 7 indicates that the TLC of SJ is dependent upon the PP value and the frequency of SJ (Fig. 8).

If the SJ and the RJ coexist in the cycle jitter, the TLCs will filter out the RJ when the number of time lag is nonzero.

$$C_j(0) \quad = \quad \frac{PP^2}{2} + \sigma_{RJ}^2 \tag{9}$$

$$C_j(m) \quad = \quad \frac{PP^2}{2}\cos(\frac{m\varpi_m}{f_0}) \tag{10}$$

, where m $\neq$ 0. Equations 9 and 10 indicate that only three autocorrelation values $C_j(0), C_j(m_1)$, and $C_j(m_2)$ are necessary to derive the RMS value of the RJ, PP value of the SJ and the frequency of the SJ, f_m.

If multiple SJs are considered, for example, two SJ, j_{s_1} and j_{s_2}, then the TLC will be

$$C_j(0) \quad = \quad \frac{PP_1{}^2}{2} + \frac{PP_2{}^2}{2} + \sigma_{RJ}^2 \tag{11}$$

and

$$C_j(m) \quad = \quad \frac{PP_1{}^2}{2}\cos(\frac{m\varpi_{m_1}}{f_0})$$
$$+ \quad \frac{PP_2{}^2}{2}\cos(\frac{m\varpi_{m_2}}{f_0}) \tag{12}$$

, where m $\neq$ 0. This indicates that five $C_j(m)$ values are necessary to derive PP_1, PP_2, ϖ_{m1} and ϖ_{m2} of the SJs and the RMS value of the RJ.

It is still possible to separate the different components of jitter if it consists of RJ, SJ, and DCD as they differ in their TLCs, for example, for a single SJ, DCD and RJ,

$$C_j(0) \quad = \quad \frac{PP^2}{2} + \sigma_{RJ}^2 + \frac{DCD^2}{4}, \tag{13}$$

$$C_j(2k+2) \quad = \quad \frac{PP^2}{2}\cos(\frac{2k\varpi_m}{2f_0}) + \frac{DCD^2}{4}, \tag{14}$$

and

$$C_j(2k+1) \quad = \quad \frac{PP^2}{2}\cos(\frac{(2k+1)\varpi_m}{f_0})$$
$$- \quad \frac{DCD^2}{4}. \tag{15}$$

, where k = 0, 1, 2, $\cdots$.

V. VERIFICATION OF THE THEORY

This section presents a case of jitter composed of an SJ and an RJ to verify the proposed method. The key factors, the sample size and the error are studied with SJ frequency, f_m; system central frequency, f_0; the ratio of the RMS value of RJ to the PP value of SJ, α; and the ratio of the TJ to a bit period, *i.e.*, *Unit Interval* (UI).

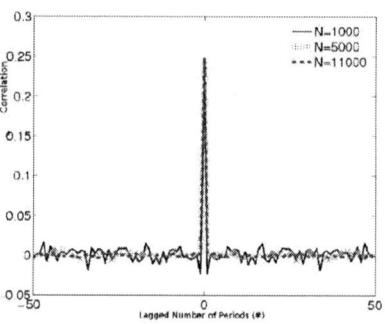

Fig. 7. TLC of RJ

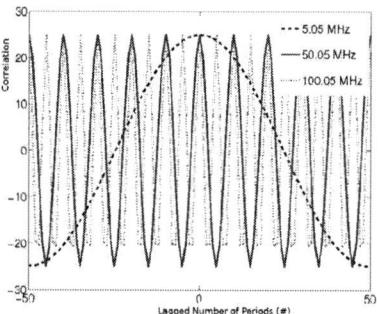

Fig. 8. TLC of SJ with same PP at different frequencies

A. Total Jitter versus Unit Interval

The increasing importance of jitter in high data rate systems derives from the increasing weight of its ratio to the bit period. This section investigates whether the ratio of the TJ to UI will affect the sample size and the error in the algorithm. Specifically, for a system with a central frequency of 1 GHz, four cases are studied: 0.05 UI, 0.1 UI, 0.15 UI and 0.2 UI while the other two factors, f_m and α, are assumed constant. For this, we make f_m = 420 MHz and α = 10. The results in Tab. III show good accuracy, even when the sample size is only 1000.

B. Root Mean Square Value of Random Jitter versus Peak-to-Peak Value of Sinusoid Jitter

Typically, the RMS of RJ is approximately ten times less than the PP of SJ [2], thus, ignored by the designers. However, as Tab. II illustrates, the weight factor, K_σ, distinguishes RJ in the TJ. Therefore, it is necessary to discuss whether the method developed here is robust to different values of α. Four cases will be studied: α=1,5,10 and 20 while assuming other parameters are held constant. For this, we make f_0 = 1 GHz, f_m = 420 MHz, K_σ = 14, and TJ = 0.1UI.

Results illustrate that the error of extracting the PP value of the SJ is inversely proportional to α (Fig. 9), while it is the opposite for the RJ parameters.

288

TABLE III
f_m PERCENTAGE ERROR (%) VS. TJ

Sample Size	0.05UI	0.1UI	0.15UI	0.2UI
1000	0.024	0.014	0.013	0.01
5000	0.003	0.006	0.005	0.005

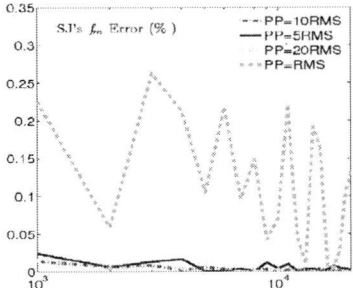

Fig. 9. f_m Error (%) vs. Sample Size

C. f_m of Sinusoid Jitter versus System Central Frequency f_0

For the frequency case, as the SJ is sampled by f_0, it is necessary to specify the frequency of the SJ, f_m within a frequency range, such as $0 \sim \frac{f_0}{2}, \frac{f_0}{2} \sim f_0, \ldots, \frac{kf_0}{2} \sim \frac{(k+1)f_0}{2}$ to overcome the aliasing problem. It is also required to investigate whether this method has problem dealing with high or low f_m or f_0.

Tab. IV (with sample size =3000) shows that the error is independent of the f_m frequency ranges and f_0, even when f_0 increases to 10 GHz.

VI. SIMULATION RESULTS IN RING OSCILLATOR

A current starved three-stage ring oscillator built with the unit delay cell shown in Fig. 10 is simulated in the 0.6-um AMI CMOS process. The oscillation frequency and the voltage swing are controlled by the pMOS current source and the diode connected nMOS device, respectively. This circuit oscillates at 341 MHz without perturbation under a supply voltage of 3 V and a control voltage of 1.5 V.

TABLE IV
PERCENTAGE ERROR (%) VS. f_0 (HZ)

f_0	$f_m=0.0112f_0$			$f_m=0.112f_0$		
	f_m	PP	RMS	f_m	PP	RMS
10M	2.48	-0.02	1.23	0.004	-0.03	1.42
10G	2.43	-0.05	1.08	0.006	-0.03	1.32

f_0	$f_m=1.012f_0$			$f_m=10.012f_0$		
	f_m	PP	RMS	f_m	PP	RMS
10M	0.01	0.02	1.05	0.002	0.04	1.34
10G	0.02	-0.04	0.57	0.002	-0.01	1.89

Three runs are required to verify the accuracy of the proposed method. The first run is a single injection of thermal noise while the second is a single injection of supply noise. The third run injects both the noise sources. The golden values of the SJ and RJ are extracted from the first two runs and are referred to as the actual values. The linear superposition of the thermal and supply noise sources is verified through the *Discrete Fourier Transform* (DFT) of the output jitter signal. The plot in Fig. 11 indicates that the modulation between the thermal noise and the supply sinusoid noise is negligible. The results also show that the white noise floor of the SJ is around 15 dB below that of the RJ. Thus, it is assumed that the thermal noise causes RJ while the supply sinusoid noise gives rise to SJ. Therefore, the total induced jitter can be treated simply as a single SJ superimposed with RJ.

To investigate the effects of f_m and the α ratio on the accuracy of the proposed method, two levels of the frequencies of the SJ (f_m = 100 MHz and 200 MHz) are combined with two different amplitudes of the sinusoid noise, 2.5 mV and 1.25 mV as the test cases. The results in Tab. V demonstrate that the accuracy of the proposed method is not highly dependent on the f_m while a smaller α gives a more accurate RMS value of the RJ. These observations are consistent with the conclusions drawn in section V.

Fig. 12 illustrates that the PP value of the SJ in cycle jitter is almost linearly dependent on the noise amplitude, and mostly independent of f_m, as demonstrated in [2]. The results at the upper and lower parts of the plot correspond to the cases in which the supply sinusoid noise of 2.5 mV and 1.25 mV are injected, respectively. The results show that the proposed method extracts the PP values of the SJ with a maximum error of 8% (sample size=1000) while the histogram-based method fails as the two peaks for predicting the PP value of the SJ converge. Fig. 13 indicates that the histogram-based method is only applicable to the cases in which the PP value of the SJ far exceeds the RMS value of the RJ.

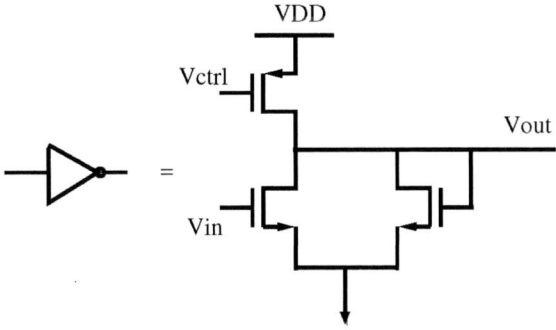

Fig. 10. Unit delay cell

VII. CONCLUSIONS AND FUTURE WORK

We have demonstrated an efficient technique for jitter decomposition using time lag correlation functions. This method extracts the frequency of the SJ (f_m)and is applicable to any α ratio, overcoming the limitations of the histogram-based method. This approach was applied to a ring oscillator, and

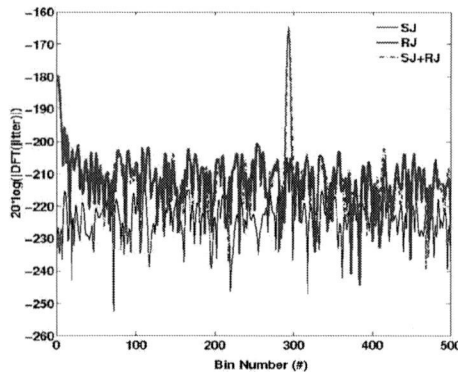

Fig. 11. DFT of the output jitter signal

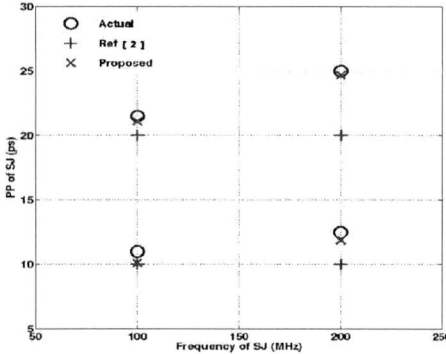

Fig. 12. Comparision of the PP values of the SJ

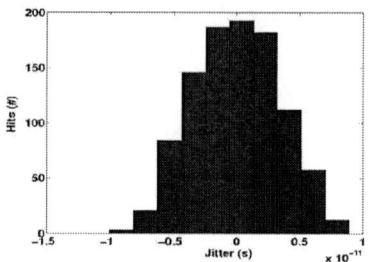

Fig. 13. Supply sinusoid noise =1 mV

TABLE V
COMPARISION BETWEEN ACTUAL VALUES AND THE RESULTS FROM THE PROPOSED METHOD

V_n (mV)	Actual Values			Proposed Method		
	f_m (MHz)	PP (ps)	RMS (ps)	f_m (MHz)	PP (ps)	RMS (ps)
2.5	100	21.5	2.5	99.4	21.2	3.5
1.25	100	11	2.5	97.2	10.1	3.4
2.5	200	25	2.5	164.8	24.7	3.5
1.25	200	12.5	2.5	161.8	11.9	3.4

good agreement between theory and hspice simulation results was obtained. In future work we plan to apply this technique to estimate the BER. This requires separation of the DJ and RJ components in the absolute jitter.

REFERENCES

[1] A. Demir. Phase Noise and Timing Jitter in Oscillators with Colored-noise Sources. *IEEE Trans. on Circuits and Systems I: Fundamental Theory and Applications*, 49(12):1782–1791, Dec. 2002.

[2] F. Herzel and B. Razavi. A Study of Oscillator Jitter Due to Supply and Substrate Noise. *IEEE Trans. on Circuits and Systems II: Analog and Digital Signal Processing*, 46(1):56–62, Jan. 1999.

[3] D. Hong, C.-K. Ong, and K.-T. Cheng. BER Estimation for Serial Links Based on Jitter Spectrum and Clock Recovery Characteristics. In *Proc. International Test Conference*, pages 1138–1147. IEEE, Oct. 2004.

[4] M. Kossel and M. L. Schmatz. Jitter Measurements of High-Speed Serial Links. *Design & Test of Computers*, 21(6):536–543, Nov. - Dec. 2004.

[5] M. Li, J. Wilstrup, R. Jessen, and D. Petrich. A New Method for Jitter Decomposition Through Its Distribution Tail Fitting. In *Proc. International Test Conference*, pages 788–794. IEEE, Sept. 1999.

[6] C.-K. Ong, K.-T. C. Dongwoo Hong, and L.-C. Wang. A Scalable On-chip Jitter Extraction Technique. In *Proc. VLSI Test Symposium*, pages 267–272. IEEE, Apr. 2004.

[7] A. Papoulis. *Probability, Random Variables and Stochastic Probability Processes*. McGraw Hill, New York, 1984.

[8] K. Shanmugan and A. Breipohl. *Random Signals: Detection, Estimation and Data Analysis*. John Wiley & Sons, Ontario, Canada, 1988.

[9] T. Yamaguchi, M. Ishida, M. Soma, K. Ichiyama, K. Christian, K. Ohsawa, and M. Sugai. A Real-time Jitter Measurement Board for High-performance Computer and Communication Systems. pages 77–84, Nov. 2004.

[10] C. Zhang, X. Wang, and L. Forbes. Simulation Technique for Noise and Timing Jitter in Electronic Oscillators. In *Proc. Circuits, Devices and Systems*, pages 184–189. IEEE, Apr. 2004.

A Fast Methodology for First-Time-Correct Design of PLLs using Nonlinear Phase-Domain VCO Macromodels

Prashant Goyal
Indian Institute of Technology, Kanpur, India

Xiaolue Lai, Jaijeet Roychowdhury
University of Minnesota, Twin Cities, USA

Abstract— **We present a novel methodology suitable for fast, correct design of modern PLLs. The central feature of the methodology is its use of accurate, nonlinear behavioral models for the VCO within the PLL, thus removing the need for many time-consuming SPICE-level simulations during the design process. We apply the new methodology to design a novel injection-aided PLL that acquires lock $3\times$ faster than prior designs, without trading off other design metrics such as jitter. We demonstrate how existing design methodologies based on behavioral simulation are incapable of leading to our new PLL design. The nonlinear behavioral simulations employed in our methodology are about 2 orders of magnitude faster than transistor-level ones, resulting in an overall design productivity gain of an order of magnitude.**

I. INTRODUCTION

Phase locked loops or PLLs are important in virtually all mixed-signal and digital systems. For example, PLL synthesizers are employed frequently in mobile communications and wireless communication transceivers. In high-speed data communications systems such as Ethernet transceivers, disk drive read/write channels, digital mobile receivers, high-speed memory interfaces and so forth, PLLs are widely used as clock generators. Other uses include clock and data recovery (CDR) direct FM-demodulation in RF systems (see, *eg.*, [1]).

The design of PLLs constitutes one of the most challenging problems in mixed-signal design today. Because of complex nonlinear dynamics in their transient operation, achieving the right balance between various PLL design metrics — such as settling time, phase noise or jitter performance, lock and capture ranges, *etc.*— for a given application is far from simple. It is not uncommon, therefore, for many months to be required to finalize the design of today's advanced PLLs. Employing effective design methodologies, supported heavily by simulation at different abstraction levels, is crucial in PLL design. Unfortunately, existing methodologies for PLL design are often inefficient or ineffective, with the result that it is not uncommon for 5 or more re-spins to be required before the PLL functions correctly.

In existing methodologies, a fresh PLL design often starts from a simple first-principles block structure such as that shown in Figure 1, or from an existing PLL design. Rough hand calculations, based on simple classical linearized analysis of a PLL feedback loop in lock, are first performed by the designer to estimate lock range, jitter, *etc.*. During the course of the design, behavioral simulation using *phase-domain macromodels* is extensively applied for greater accuracy. When the design is finalized at the transistor level, full SPICE-level simulation is heavily used for final verification.

Important steps in this flow break down in today's methodologies. It is for this reason that, as mentioned above, PLL design tends to be extremely time consuming and error prone. Problems exist at each level of the above flow that contribute to the breakdown:
• hand calculation level: Existing hand-analysis techniques [1], [2] for dynamics, noise, jitter, *etc.*, in PLLs are all based on linear analysis of the PLL around a locked steady state. The few nonlinear analyses that are amenable to hand calculation (*eg.*, for estimating lock range [3]) are overly simplistic for most practical designs; for example, they do not take dynamics, which are very important in determining

PLL responses, into account. Therefore, the rôle of simulation in PLL design assumes much greater importance than for the design of simpler systems like op-amps.
• system-level simulation with behavioral models (or *macromodels*): Behavioral simulation using phase-domain macromodels is extremely important in PLL design [4], [5] because of the great speedups it offers over transistor-level full simulation. Existing behavioral simulation of PLLs relies largely on using *linear* models for most components, especially for the VCO phase macromodel. The main issue with VCO behavioral models is unacceptable loss of accuracy and predictive power. Although it has generally been assumed that linear VCO macromodels[1] [4], [6] are adequate for behavioral simulation of PLLs, it has recently been demonstrated that using them can lead to very serious prediction errors [7], especially in the presence of nonlinear transient effects such as those involved in the capture, lock acquisition, and slipping processes in PLLs. The predictive power of linear VCO models is particularly poor for advanced PLL designs that use feed-forward or injection-aided mechanisms to enhance performance [8]–[10], as we investigate in detail in this paper. (Section III explains these mechanisms and design techniques in more detail.)
• transistor-level circuit simulation: In view of the significant accuracy problems in hand- and behavioral-level analysis of PLLs, designers rely heavily on transistor-level circuit simulation in existing PLL design methodologies. Such full simulation has the great advantage that it is able to predict non-ideal and nonlinear effects accurately. Unfortunately, as practitioners are well aware, full simulation of PLLs is extraordinarily time consuming. For example, a single jitter simulation for an industrial PLL can take days. The reason for the inefficiency of full SPICE-level simulation of PLLs stems from the fact that loop dynamics are typically orders of magnitude slower than the oscillation frequency of the VCO, resulting in a classic fast/slow timescale situation, where very small simulation time-steps need to be taken over a very long total simulation period. Because the only option for accurate PLL simulation in today's methodologies is so slow, designers are often forced to ignore large parts of the design space or to skip important verification steps simply due to time pressure. It is mainly for this reason that PLL design tends to be particularly error prone.

In this paper, we present a fast, accurate and extremely effective methodology for designing any kind of PLL. Our methodology involves extensive use of behavioral simulation using *nonlinear VCO phase-domain macromodels* that are automatically generated via algorithm from transistor-level VCO circuits. Our use of nonlinear phase macromodels is motivated by recent work [7], [11] which has established their suitability for predicting a variety of advanced or non-ideal effects, such as injection locking, capture and acquisition transients, jitter due to power supply variations, *etc.*. The most important benefit of the proposed methodology is that it dispenses with the need for time-consuming transistor-level simulations to a much greater extent than previously possible. The nonlinear behavioral

---
[1]*ie.*, the VCO inside the PLL is modeled as a linear integrator [6].

0-7803-9451-8/06/$20.00 ©2006 IEEE.

simulations employed instead produce results virtually identical in accuracy, while being orders of magnitude faster.

We demonstrate this methodology by applying it to the complete design of an advanced new injection-aided PLL. Using our new methodology, we are able to design the new PLL to lock three times faster than similar conventional PLLs, without having to make tradeoffs that sacrifice other performance metrics such as noise/jitter.

We show in this paper how our new design methodology results in significantly improved design creativity and productivity. Using the proposed PLL design methodology (implemented in MATLAB) we are able to accurately simulate transient capture/locking within approximately one minute. In contrast, transistor-level simulation in the same simulation framework takes about fifty minutes. This speedup also has a great impact on the overall work flow of a designer, because the flow of ideas and design decisions is significantly improved by fast simulation turnaround times. With accurate simulations completing in a few seconds or minutes instead of in hours or days, it becomes possible and convenient to investigate many more different design scenarios or parameter sets. If time for thinking and design decisions (based on information from prior simulation runs) is included, we estimate that a typical designer can run approximately 5–6 PLL simulations per hour using our behavioral methodology, as opposed to an entire day for the same level of productivity.

Crucial to the effectiveness of our methodology is the fact that the fast behavioral simulations we employ *do not appreciably sacrifice accuracy* relative to full SPICE-level simulation. To validate the methodology, we compare against full transistor level simulations and always achieve excellent match, implying that far fewer full transistor-level simulations are needed when our methodology is employed. We also explore conventional behavioral methodologies [6] and demonstrate although they are equally fast, they completely fail to predict correct results.

The remainder of the paper is organized as follows. In Section II, we briefly describe conventional PLL design methodologies. In Section III we provide background on advanced PLL design concepts such as injection locking. In Section IV, we describe our new PLL design methodology and in Section V, we apply it to an injection-aided PLL design and describe its benefits.

II. CONVENTIONAL PLL DESIGN METHODOLOGIES AND LIMITATIONS

Conventional PLL design methodologies for behavioral simulation of PLLs are typically based on linearized analysis around a locked state. Figure 1 depicts the structure of a simple PLL and its linearized phase-domain model when in lock. The phase/frequency detector (PFD) is modeled as a multiplier with a gain K_d, low-pass filter (LPF) with a transfer function of $F(s)$ and a voltage-controlled oscillator (VCO) as a linear integrator. When locked, the negative feedback

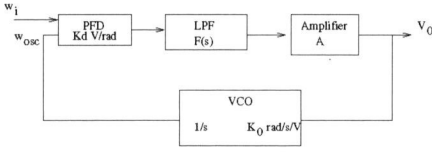

Fig. 1. Linear PLL Model

loop ensures that the frequency ω_{osc} from the VCO is identical to the input reference frequency ω_i. Using classical linear feedback control theory, the closed loop transfer function of the system can be derived to be

$$\frac{V_0}{\omega_i} = \frac{1}{K_0}\left(\frac{K_v F(s)}{s + K_v F(s)}\right), \qquad (1)$$

where $K_v = K_d K_0 A$ is the loop gain, which determines the lock-in range. For a first-order loop filter

$$F(s) = \frac{1}{1 + \frac{s}{\omega_1}}, \qquad (2)$$

the poles of the transfer function are

$$s = -\frac{\omega_1}{2}\left(1 \pm \sqrt{1 - \frac{4K_v}{\omega_1}}\right). \qquad (3)$$

On comparing with a standard two-pole transfer function [12], the natural frequency ω_n and damping ratio ζ can be found to be

$$\omega_n = \sqrt{K_v \omega_1}, \qquad \zeta = \frac{1}{2}\sqrt{\frac{\omega_1}{K_v}}.$$

Similar results are also easily obtained for third and higher order loop filters [13]. Thus we see that the poles of the system function (which determine transient settling behavior for linear systems) have their real parts directly dependent on the loop filter bandwidth ω_1. ζ, which determines peaking in frequency response, is also dependent on ω_1. The loop filter bandwidth ω_1 also governs noise performance, while K_v determines the lock-in range and other performance metrics. Thus, it is apparent that even when only a linearized methodology is used, it can be a non-trivial optimization problem to achieve the right balance between different performance metrics.

But as demonstrated shortly, linearizations are at best a simplification valid in a narrow region around lock, so optimization of the linearized PLL is of limited value in any case. In particular, as has been demonstrated [7], PLL capture phenomena are inherently strongly nonlinear. In the methodology developed in this paper, we provide an effective means to take nonlinearities into account during PLL design.

III. NONLINEAR EFFECTS IN ADVANCED PLL DESIGNS

Possibly the most important functional component inside any PLL is the voltage controlled oscillator (VCO). As is well known, amplitude-stable operation of oscillators is a fundamentally nonlinear process (*eg.* [14]). This leads to many important and fascinating effects in oscillators that are impossible to understand using linear concepts only.

A. Injection locking

One such effect, important for advanced PLL design as shown later, is *injection locking*. As the term implies, when an external weak signal is injected into an oscillator, then, under certain conditions, the oscillator's frequency changes to become identical to that of perturbing signal. Even if perfect lock to the external signal is not achieved, interesting and useful "frequency pulling" phenomena typically occur. Injection locking effects have been extensively studied (*eg.*, Adler [15], Kurokawa [16] and others [17]). However, several prior approaches to predict injection locking have relied on approximations like simplifying nonlinearities and neglecting higher order harmonics. Therefore these approaches are not able to predict results accurately when the circuit deviates from these assumptions significantly. In the methodology presented in this paper, we emphasize full consideration of nonlinearities to capture such phenomena accurately.

B. Injection-aided PLL design

There has been growing awareness in recent years that injection locking can be used to advantage in circuits that rely on phase synchronization. For example, injection locking has been used for quadrature generation in mixers [18]. In PLLs, injection locking has been applied to improve locking range, phase noise and jitter performance [8]–[10]. The increasing importance of new injection-aided PLL architectures has placed existing methodologies for PLL design [4], [5] under even greater stress, since injection locking is a fundamentally nonlinear phenomenon that linearized approaches are completely incapable of predicting.

In this work, we also present new PLL design concepts enabled by our methodology. We design an injection-aided PLL prototype that significantly reduces capture and lock acquisition time. Further, we employ a soft switching approach to remove the injection locking path once the PLL has achieved lock, so as to enable complete freedom in optimizing other performance metrics (such as jitter). We believe that investigation and refinement of these ideas would have been impractical without our design methodology.

IV. NONLINEAR PHASE MACROMODEL BASED PLL DESIGN METHODOLOGY

Our new methodology is based on nonlinear phase domain VCO macromodels. In our methodology, the VCO inside the PLL loop is modeled as a *nonlinear* element as compared to previous linear integrator models.

Existing linear VCO phase macromodels have the form

$$\dot{\alpha}(t) = K_{vco}b(t), \quad \text{or} \quad \alpha(t) = K_{vco}\int b(\tau)\,d\tau, \qquad (4)$$

where α is the phase deviation of the VCO caused by an external input (or perturbation) $b(t)$. In contrast, our nonlinear phase equation [14], [19] has the form

$$\dot{\alpha}(t) = v_1^T(t + \alpha(t)) \cdot b(t). \qquad (5)$$

In this equation, $v_1(t)$ — called the perturbation projection vector (PPV) — is a vector of highly nonlinear, periodic, waveforms. Each node of the VCO has an associated PPV waveform component. These PPV components determine the effect of perturbations at the node on the output phase of VCO. It is this relationship that is captured by the nonlinear differential equation (5).

From a methodological viewpoint, the PPV waveforms for any oscillator can be easily extracted from a SPICE-level circuit of the oscillator, using numerical algorithms [19], [20]. It has already been established [7] that use of such macromodels leads to excellent prediction of capture/lock transients, cycle-slipping, static phase offsets and other effects in which injection locking/pulling plays a key rôle.

A. Design intuition from VCO PPV waveforms

A major benefit of our methodology is that examination of the PPV waveforms $v_1(t)$ can used to obtain direct design intuition and insight. Two PPV waveforms of an LC VCO are shown in Figure 2.

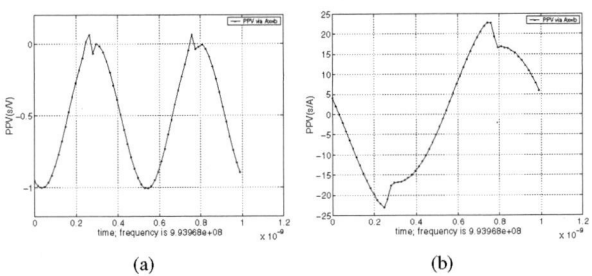

(a)   (b)

Fig. 2. The PPV of control node and capacitor voltage node of the LC VCO

If a VCO is initially locked at ω_0 and an input frequency signal of frequency ω_1 disturbs the initially locked loop, then the output phase ϕ_{out} of the VCO can be expressed as

$$\phi_{out}(t) = \omega_0(t + \alpha(t)). \qquad (6)$$

Thus, if the VCO locks to ω_1, we have

$$\omega_1 t + \theta = \omega_0(t + \alpha(t)), \qquad (7)$$

or

$$\alpha(t) = \frac{\omega_1 - \omega_0}{\omega_0}t + \frac{\theta}{\omega_0}. \qquad (8)$$

Therefore, for locking to occur, $\alpha(t)$ should change with time linearly, ie., $\dot{\alpha}(t)$ *must have a constant DC value.*

From (5), we see that $\dot{\alpha}(t)$ is a multiplication of two waveforms: the PPV and the external input. If the PPV waveform is an AC waveform, then an AC input signal is required to obtain a DC component (thereby changing the VCO's frequency for injection lock); while if the PPV contains both AC and DC terms, then a DC signal can also make the VCO locked. The *frequency control node of VCOs is typically designed to have predominantly DC terms in its PPV component.* By injecting signals into other VCO nodes with strong AC terms in their PPVs, injection locking to aid lock acquisition can be usefully induced, as we describe further in Section V.

B. Nonlinear phase equation based design methodology

Before demonstrating our PLL design methodology by applying it to design a PLL in Section V, we first summarize its main steps:

1) Use existing phase domain behavioral models for the PLL's phase detector (PD) and frequency divider.
2) Model the loop filter, which is typically small, at the voltage level as a circuit or behavioral block.
3) Model the VCO using (5). Obtain the PPV $v_1(t)$ from the *full SPICE-level VCO circuit* via numerical algorithms [19], [20], thus setting up the nonlinear macromodel correctly. Crucially, identify all relevant inputs to the VCO, including the traditional control, auxiliary inputs like injection locking feeds, power supply and ground nodes (for jitter), *etc.*. Examine the PPV components of these nodes with a view to exploiting them during design.
4) Compose the PLL behavioral model using the above blocks and use it for simulations. Re-extract the VCO phase macromodel if changes are made to the internal circuitry of the VCO during design.
5) To regenerate voltage-domain waveforms from phase-domain ones, retain the VCO steady-state obtained during PPV extraction. Regenerate voltage-domain outputs using the steady state waveforms as described in [7].

V. FIRST-TIME-CORRECT DESIGN OF INJECTION-AIDED PLL FOR FAST LOCK ACQUISITION

In this section, we use our methodology to design a novel injection-aided PLL with enhanced lock acquisition properties. We first apply behavioral simulation to the simple PLL shown in Figure 1. Then, applying the new design methodology and leveraging design intuition gained from each step, we improve the design in steps. After finalizing the design, we compare predictions from our behavioral simulations against full simulation to confirm correctness of our design.

A. Simulation of a Basic PLL

First, we simulate transient responses in the simple PLL loop. The (initially locked) VCO frequency f_0 is chosen to be 1Ghz and the loop filter bandwidth is taken to be 15 Mhz. We compare the step response of the PLL obtained using our methodology with that from a linear methodology, as well as against full SPICE-level simulation.

We inject a reference frequency signal of $1.05 f_0$ and simulate the capture/lock transients of PLL until it reaches steady state. Figure 5 and Figure 3 depict lock acquisition transients, as simulated by linear

293

3B-4

2006 Asia and South Pacific Design Automation Conference

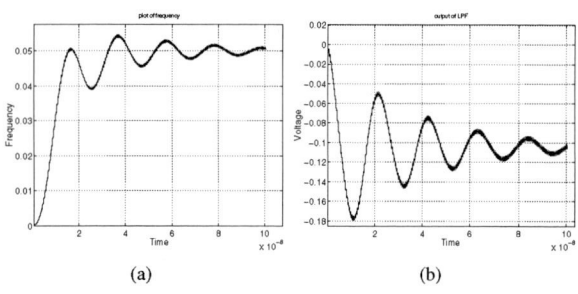

(a)　　　　　　　　(b)

Fig. 3. VCO frequency shift and control node voltage waveform using nonlinear macromodel

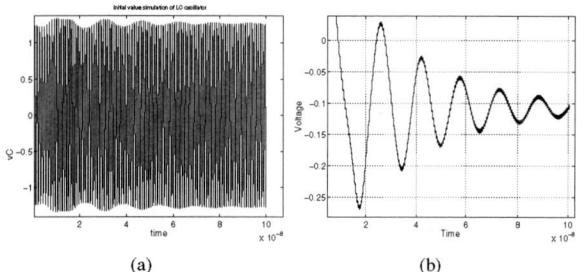

(a)　　　　　　　　(b)

Fig. 4. VCO frequency shift and control node voltage waveform using full simulation

and nonlinear macromodels. As is apparent from the figures, the VCO frequency tracks the reference frequency after approximately 100 T (where T is the time period of the oscillator), when transients die out and the frequency shift settles to a constant factor of 0.05. Full simulation verifies that results from both macromodel simulations are approximately correct. The control node voltage waveforms also show good matches with that of full simulation – settling finally to -0.1 volt, close to estimates from hand calculations.

Thus, we see that for the simple PLL loop perturbed from lock, linear as well as nonlinear VCO macromodels work well. This is not surprising, since linearization is relatively valid for small perturbations from lock. The use of such macromodels provides approximately a 50 times speed up over full simulation.

B. Improving settling time response using injection locking

Using standard PLL design techniques, it is difficult to reduce the PLL's settling time without sacrificing aspects of noise performance. Improving loop settling time requires increasing loop filter bandwidth ω_1 (as described in Section II), thus resulting in more mixer noise propagating to the VCO input.

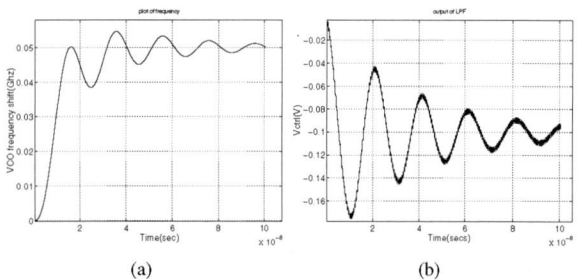

(a)　　　　　　　　(b)

Fig. 5. VCO frequency shift and control node voltage waveform using linear macromodel

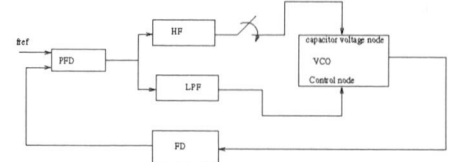

Fig. 6. New PLL Design

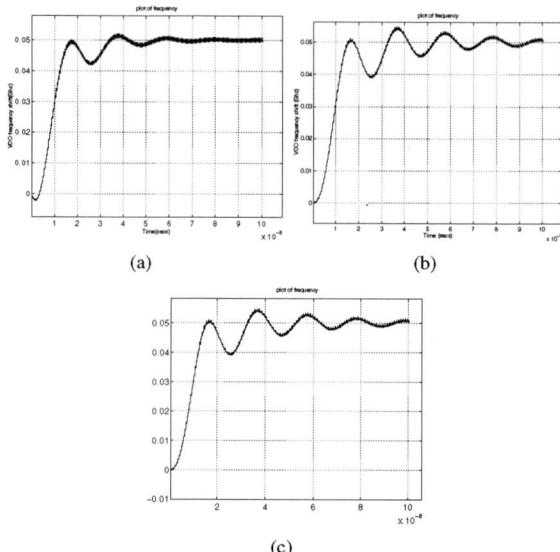

(a)　　　　　　　　(b)

(c)

Fig. 7. VCO frequency shift using different injection locking paths

To avoid having to make this undesirable tradeoff, we next investigate how an additional path that excites injection locking can be used to improve settling response, without changing the loop filter. This technique is motivated by observing strong AC terms in the PPV components of internal VCO nodes, eg., the PPVs shown in Figure 2. The extra HF injection locking path is shown in Figure 6, immediately following the mixer. The rôle of the switch (which is normally closed during lock acquisition) will be explained shortly.

C. Injection to VCO capacitor node with different injection levels

In search of an effective injection locking path, we try three different injection locking signals to the capacitor voltage node of VCO: injection of the reference signal itself, injection of the full phase detector output signal, and a high-pass filtered version of the phase detector's output signal. We first attenuate the injection to make it one tenth of oscillator's free-running amplitude. We apply this injection to the capacitor voltage node of VCO. Results from behavioral simulations of this setup are shown in Figure 7.

From Figure 7, we see that for the case of injection of the reference frequency signal, the settling response of the PLL loop is improved, while in the other two cases injection has little effect on capture/lock transients of PLL loop. Keeping in mind that different injection levels lead to different levels of locking and pulling (as described in [11]), we try other injection signal levels. We increase injection signal levels to be comparable of that of oscillator signal and inject them into the capacitor voltage node of the VCO.

As we see from the simulation results in Figure 8, injection of the high-pass filtered phase detector output leads to considerable improvement in the settling time of the PLL loop. Injection of the reference frequency also speeds up the PLL's settling response. Although direct injection of the phase detector output also improves

294

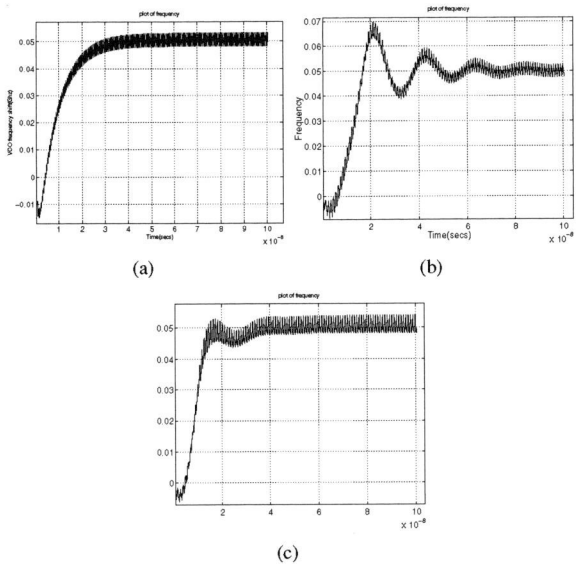

Fig. 8. VCO frequency shift using different injection locking paths

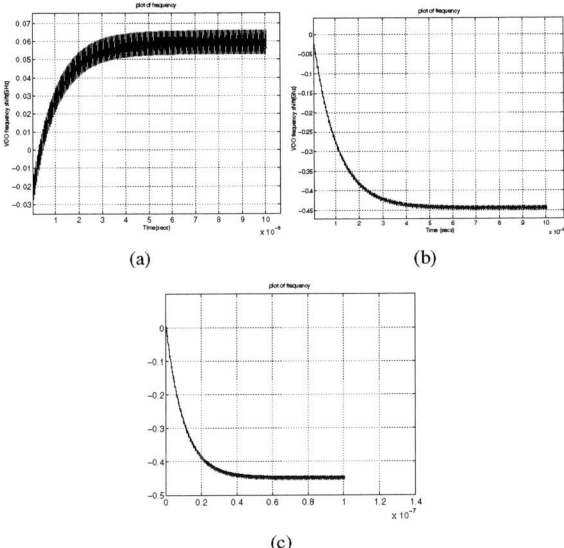

Fig. 9. VCO frequency shift using different injection paths

the PLL's settling response, the effect is less than by the other two paths.

Encouraged by these observations, we further increase the injection signal level, to five times that of the oscillation amplitude. However, we now observe from Figure 9 that this results in the oscillator's locking to a completely different frequency (a subharmonic of the reference frequency).

After experimenting with several other injection levels, we are able to find optimal injection paths and an optimal injection signal level, which leads to a speed up of three in settling-time response, compared to the PLL without injection-aided locking. These optimal paths are injection of reference frequency signal and injection of high pass filtered phase detector output with injection signal level comparable to oscillator signal.

It is worth mentioning here that conducting one such experiment with full SPICE level simulation takes approximately 50 minutes, as

compared to about 1 minute for nonlinear behavioral simulation. As mentioned earlier, this speedup is crucial for enabling new design insights and ideas. We also emphasize that the above experiments, involving PLL design space exploration, could not have been carried out using a traditional linearized PLL design methodology [6]. Erroneous simulation results obtained by using linear macromodels (for the case of Figure 8) are shown in Figure 10 below. As already noted, linear models cannot account for effects such as injection locking.

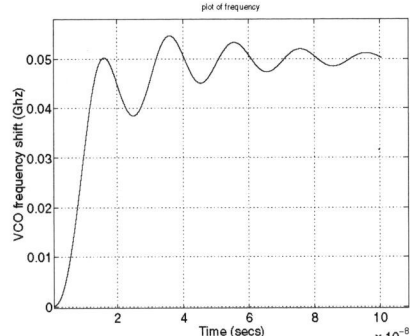

Fig. 10. The Plot of VCO phase shift using linear macromodel simulation for new PLL

D. Switching to disable injection aided operation after lock

Once the PLL has acquired lock, it can be desirable to remove the injection locking path from the loop for complete freedom in optimizing other performance metrics (such as jitter). This enables very easy augmentation of existing PLL designs to employ our injection-aided lock acquisition technique as described above.

To remove the injection, we cut the injection path using the switch shown in Figure 6. We first employ hard (abrupt) switching once the PLL has acquired lock. As we see from Figure 11, as soon

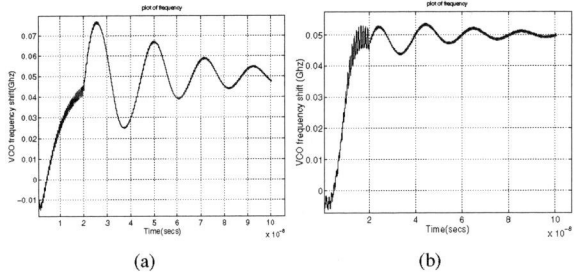

Fig. 11. Hard switching simulation for new PLL

as the switch is turned off, the PLL response shifts to that of the simple PLL with no injection path. This scheme fails as it increases the settling time. The reason stems from that the static phase offset for the injection-aided PLL is very different from that for the PLL without injection; abruptly removing the injection results in the PLL's losing lock again.

In order to remove the injection while ensuring that the "regular" PLL loop always remains in lock, we next try a soft switching approach, ie., taking the injection locking path out "slowly". As seen in Figure 12(a), the injection locking mechanism gradually relinquishes control of the VCO's frequency to the normal VCO control node and the static phase offset changes smoothly, without loss of lock at any point, to the value it has in the absence of injection locking.

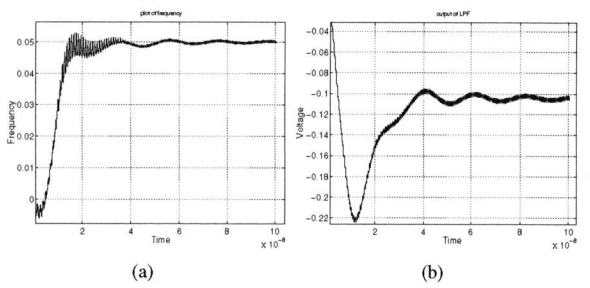

(a)   (b)

Fig. 12.   Soft switching simulation results: frequency offset and LPF output

Thus we see that injection of the high-pass filtered phase detector output improves PLL settling time to approximately one-third that possible via linear design techniques. Furthermore, using soft switching, this is achieved without affecting any other performance metrics.

E. Final verification against full SPICE-level simulation

To verify that the design indeed functions correctly, we compare the results of behavioral simulation of the final design against full SPICE-level simulation. The full SPICE level simulation voltage waveform,

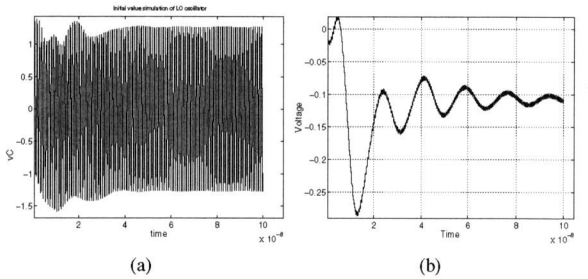

(a)   (b)

Fig. 13.   Full simulation results: VCO output voltage, LPF output

as shown in Figure 13(a), shows a constant envelope after a time period of about 40 cycles, as predicted by the macromodel-simulated VCO phase shift in Figure 12(a). Comparing this with the voltage waveform of the simple PLL loop in Figure 4(a) (which features a constant envelope in about 100 cycles), we confirm that the new PLL design settles approximately 3 times faster.

For further comparison, we also run behavioral simulations of the new PLL with linear VCO behavioral models used in existing PLL design methodologies. As shown in Figure 14, we find that they still represent the same transient behaviour as that of the simple PLL with no injection. These results confirm that linear macromodel based methodologies completely fail for such designs.

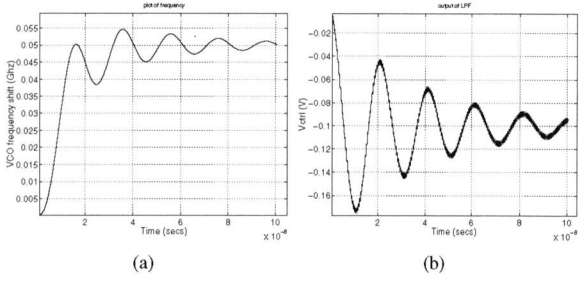

(a)   (b)

Fig. 14.   Simulations with linear macromodel for new PLL

VI. CONCLUSIONS

We have presented a new PLL design methodology based on nonlinear VCO macromodels. Using the new methodology, we have successfully designed a new type of PLL which exploits injection locking to speed lock acquisition by a factor of 3. The behavioral simulations used in our methodology run about two orders of magnitude than SPICE-level simulations, while retaining excellent quantitative and quantitive accuracy. This leads to design productivity improvements of an order of magnitude or greater. We have also shown how existing methodologies for PLL design, which do not account for VCO nonlinearities, would have been inadequate for this design. We anticipate that adoption of our methodology in industrial PLL design will significantly cut the time and cost of obtaining correctly functioning silicon.

REFERENCES

[1] J.L. Stensby. Phase-locked loops: Theory and applications. CRC Press, New York, 1997.

[2] U.L.Rohde. Microwave and wireless synthesizers: Theory and design. Wiley,Hardcover, 1997.

[3] S.H.Lewis P.R.Gray, Paul J.Hurst and R.G.Meyer. Analysis and design of analog integrated circuits. John Wiley and Sons,Inc., 2001.

[4] K. Kundert. Predicting the Phase Noise and Jitter of PLL-Based Frequency Synthesizers. www.designers-guide.com, 2002.

[5] A. Demir, E. Liu, A.L. Sangiovanni-Vincentelli, and I. Vassiliou. Behavioral simulation techniques for phase/delay-locked systems. In Proceedings of the Custom Integrated Circuits Conference 1994, pages 453–456, May 1994.

[6] M. Takahashi, K. Ogawa, and K.S. Kundert. VCO jitter simulation and its comparison with measurement. In Proceedings of Design Automation Conference 1999, pages 85–88, June 1999.

[7] X.Lai,Y.Wan and J.Roychowdhury. Fast PLL simulation using nonlinear VCO macromodels for accurate prediction of jitter and cycle-slipping due to loop-nonidealities and supply noise. In Proc. Asia South-Pacific Design Automation Conference, January 2005.

[8] Peter K.Runge. Phase-locked loops with signal injection for increased pull-in range and reduced output phase jitter. Communications, IEEE Transactions on, 24(6), June 1976.

[9] Kwing F.Lee B.Razavi and Ran H. Yan. Design of high speed, low-power frequency dividers and phase-locked loops in deep submicron cmos. Solid-State Circuits, IEEE Journal of, 30(2), February 1995.

[10] T.Berceli K.Kudszus, M.Neumann and W.H.Haydl. fully integrated 94-ghz subharmonic injection-locked pll circuit. IEEE Trans. Microwave Theory Tech., 10(2), February 2000.

[11] X.Lai and J.Roychowdhury. Capturing injection locking via non-linear phase domain macromodels. IEEE Trans. Microwave Theory Tech., 52(9), September 2004.

[12] J.D.Powell G.F.Franklin and A.Naeini. Feedback control of dynamic systems. Pearson Education, 2002.

[13] I.V.Thompson and P.V.Brennan. Fourth-order pll loop filter design technique with invariant natural frequency and phase margin. IEE Proc. Circuits Devices Syst., 152(2), April 2005.

[14] X.Lai and J.Roychowdhury. Automated Oscillator macromodelling techniques for capturing amplitude variations and injection locking. In Proc.IEEE International Conference on Computer-Aided Design, November 2004.

[15] R. Adler. A study of locking phenomena in oscillators. Proceedings of the I.R.E. and Waves and Electrons, 34:351–357, June 1946.

[16] K. Kurokawa. Injection locking of microwave solid state oscillators. Proc. IEEE, 61:1386–1410, October 1973.

[17] M. Armand. On the output spectrum of unlocked driven oscillators. Proc. IEEE, 57:798–799, May 1969.

[18] P. Kinget, R. Melville, D. Long, and V. Gopinathan. An injection-locking scheme for precision quadrature generation. IEEE Journal of Solid-State Circuits, 37(7):845–851, July 2002.

[19] A. Demir, A. Mehrotra, and J. Roychowdhury. Phase noise in oscillators: a unifying theory and numerical methods for characterization. IEEE Trans. on Circuits and Systems-I:Fundamental Theory and Applications, 47(5):655–674, May 2000.

[20] A. Demir and J. Roychowdhury. A reliable and efficient procedure for oscillator ppv computation, with phase noise macromodelling applications. IEEE Trans. on Computer-Aided Design of Integrated Circuits and Systems, 22(2):188–197, February 2003.

DOUBLE EDGE TRIGGERED FEEDBACK FLIP-FLOP IN SUB 100NM TECHNOLOGY

S. H. Rasouli
Nanoelectronics Center of
Excellence
ECE Dept.
University of Tehran
Tehran , Iran
Tel: 9821-8209-4359
Email: s_hrasouli@yahoo.com

A. Amirabadi
Nanoelectronics Center of
Excellence
ECE Dept.
University of Tehran
Tehran , Iran
Tel: 9821-8209-4359
Email: A.amirabadi@ece.ut.ac.ir

A. Seyedi
Nanoelectronics Center of
Excellence
ECE Dept.
University of Tehran
Tehran , Iran
Tel: 9821-8209-4359
Email: A.Seyedi@ece.ut.ac.ir

A. Afzali-Kusha
Nanoelectronics Center of
Excellence
ECE Dept.
University of Tehran
Tehran , Iran
Tel: 9821-8209-4359
Fax: 9821-8877-8690
Email: afzali@ut.ac.ir

ABSTRACT

In this paper, a new flip-flop called Double-edge triggered Feedback Flip-Flop (DFFF) is proposed. The dynamic power consumption of DFFF is reduced by avoiding unnecessary internal node transition. The subthreshold current in the flip-flops is very low compared to other structures. Reducing the number of transistor in the stack and increasing the number of charge path leads to higher operational speed compared to others flip-flops. The simulation results show an improvement of 44% in the speed and 45% in the static leakage power.

1. INTRODUCTION

The power consumption of the systems is a critically important parameter in modern VLSI circuits especially for low power applications and, hence, the power optimization techniques should be applied at different levels of the digital design. One of these techniques is to use low power logic styles which should be used in design of latches and flip-flops (FF's) which are among the components widely used in digital systems [1][2]. There are other concerns in the design of DFF's such as T_{clk-q} (delay from clk to output of FF) and C_{clk} (the load capacitance of the clock) which are also should be minimized to maximize the FF performance. Among these

parameters, reducing the C_{clk} or the frequency of clock has a great impact on the power consumptions of clock tree and the logic [3].

In addition to the dynamic power consumption, the high leakage current in deep sub-micron regimes is a significant contributor to the power dissipation of CMOS circuits as the CMOS technology scales down [4]. The subthreshold leakage power is expected to become a significant fraction of the total power in the sub-100 nm CMOS technology where reducing the subthreshold leakage power of the circuit is crucial.

Several flip-flops have been proposed in the literature for improving the speed and/or reducing the power consumption (see, e.g. [3], [5], [7], [9]). A static single edge-triggered flip-flop called Hybrid Latch Flip-Flop (HLFF) has been proposed in [5]. It is based on generating an explicit transparency window for the time that the transition is allowed. Its idea is similar to a latch because it can provide a soft clock edge which allows for slack passing and minimizes the effect of clock skew on the cycle time [6]. However, the existence of redundant transitions in the internal nodes of HLFF leads to more power consumption. Semi-Dynamic Flip-Flop (SDFF) which is a single edge-triggered FF and faster than HLFF has been proposed in [7]. The existence of 1-1 glitch leads to an

0-7803-9451-8/06/$20.00 ©2006 IEEE.

undesired power dissipation. The number of transistors in this logic is greater than that of HLFF. Conditional Capture Flip-flop (CCFF) has been proposed to reduce redundant transitions at internal nodes [3]. The conditional capture technique needs many additional transitions for certain flip-flops which themselves cause an extra power consumption.

The dynamic power consumption in the clock tree depends on the frequency, the voltage swing, and the load of clock tree [8]. If the sampling of the input is performed in both rising and falling edge of clock (double-edge triggered), then for same applications and operational speeds, the frequency of the clock can be half of the clock frequency of the single edge triggered FF. This has been the motivation for proposing double-edge triggered flip-flops. In [9], Low-Swing clock Double-edge triggered Flip-Flop (LSDFF) has been described. In their work, the power consumption in the clock tree is reduced using a low swing clock and low-V_{th} transistors in the FF. The subthreshold current of low-V_{th} transistors in the main logic is controlled by high-V_{th} transistors. However, the subthreshold current of low-V_{th} transistors in the inverters used in the clock tree incur more power consumption especially in very deep submicron technology. Furthermore, the number of transistors in this logic is much greater than previous works.

In this paper, a Double-edge triggered Feedback Flip-Flop (DFFF) is proposed which has less dynamic power consumption, static power, and delay compared to the previous flip-flops. This paper is organized as follows. In Section 2, the structures of single-edge and double-edge triggered FF's are described and compared. The subthreshold leakage currents of the flip-flops are discussed in Section 3 while section 4 contains the simulation results. The paper ends with summary and conclusions in Section 5.

2. FLIP-FLOP STRUCTURES

A. Single-edge triggered Flip-Flops

The structure of Hybrid Latch Flip-flop (HLFF) is shown in Figure 1 [6]. While HLFF has a very simple circuit, its unnecessary internal transitions increase the total power consumption of the flip-flop. In each clock cycle, when the

input is high, regardless of previous state of the output a glitch is generated [3]. Furthermore, the transistors in the stack degrade the performance of the logic. These disadvantages make HLFF not suitable for low power applications.

In Figure 2, the circuit diagram of Semi-Dynamic Flip-Flop (SDFF) is illustrated [7]. This logic is faster than HLFF due to its lower number of transistor in the stack. However, the total number of transistors is greater than HLFF and, similar to HLFF, unnecessary internal node transitions exist in SDFF.

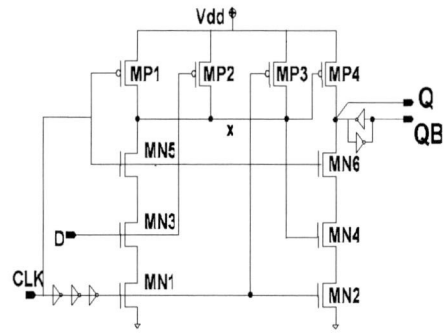

Figure 1. Circuit diagram of HLFF [6].

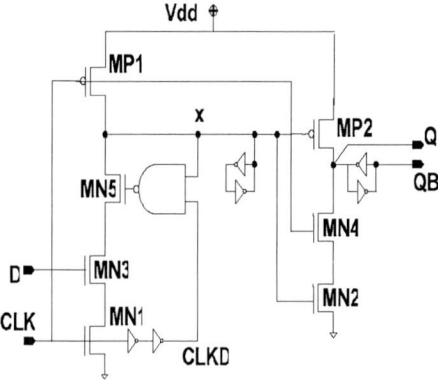

Figure 2. Circuit diagram of SDFF [7].

To see the first drawback of this FF more clearly, suppose that input is high in two successive clock cycles. Before the rising edge of the second clock, the node Q is high while the node X is pre-charged to V_{dd}. At rising edge of the second clock cycle, there is a short circuit path from Q to ground until the node X is discharged. This leads to a 1-1 glitch which consumes unnecessary power.

298

B. Double-edge triggered flip-flops

The circuit diagram of Low Swing clock Double edge Flip-Flop (LSDFF) is depicted in Figure 3 [9]. The input of the flip-flop is transferred to the output at the rising and falling edges of the clock. To reduce the power consumption of the clock tree, a low swing clock is used in this logic. To have a proper functioning, some of high-V_{th} transistors are replaced with low-V_{th} transistors whose subthreshold currents are controlled by high-V_{th} transistors. For the same throughput, the frequency of the clock in LSDFF could be half of the

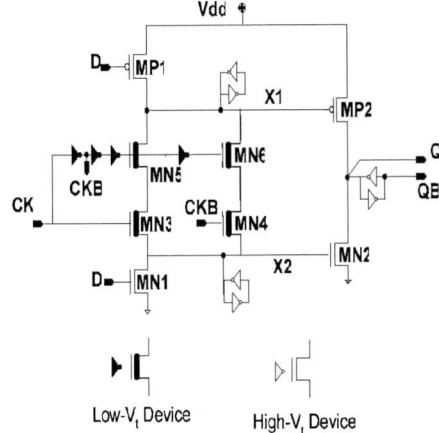

Figure 3. Circuit diagram of LSDFF [9].

frequency of the clock in HLFF or SDFF.

The power consumption of the clock tree is proportional to the clock load, frequency and the swing of clock. Since compared to the previous FF's, the swing and the frequency of the clock is lower, the power consumption of LSDFF clock tree could be lower than those of others. However, uncontrolled subthreshold current low-V_{th} transistors in the clock tree leads to a more power consumption. In addition, since the charging (discharging) the internal node X2 (X1) is done through three transistors, the speed of the circuit is reduced.

To avoid unnecessary transitions in the previous flip-flops, we propose a Double edge-triggered feedback flip-flop (DFFF) whose circuit is shown in Fig. 4. In this flip-flop, the node transitions occur only when the inputs are different in two successive clocks. The operational principle of this work

is explained here. When the clock (CLK) makes a transition from low to high, CLKBD remains high for a period equal to the delay of the three inverters creating a transparency window. In this period, C1 is high turning on MN1 and MN3. In this window, if D is low and Q is high (D was high in the previous clock), MP2 becomes on turning on MN2 which forces the output to low. If both D and Q are low, MP1 and MN2 are on before the beginning of the transparency window making the delay zero (similar to previous flip-flops). If D is high and Q is low, node X becomes low turning on MP3 which forces the output to high.

Note that, as MP1 is a weak transistor, the fighting problem during the output change is alleviated. If D is high and Q is high, node X will not change and, therefore, contrary to the other flip-flops discussed here, redundant transitions are avoided. As another advantage of this logic compared to the other flip-flops, note that there is no delay whenever D is high in two successive clock cycles. Additionally, the charging of the node X is done through two paths where one path consists of MP1 and MP2 (similar to others) and the other consists of MN1 at rising edge of the clock and MN2 at the falling edge. This increases the speed of the FF compared to the previous ones.

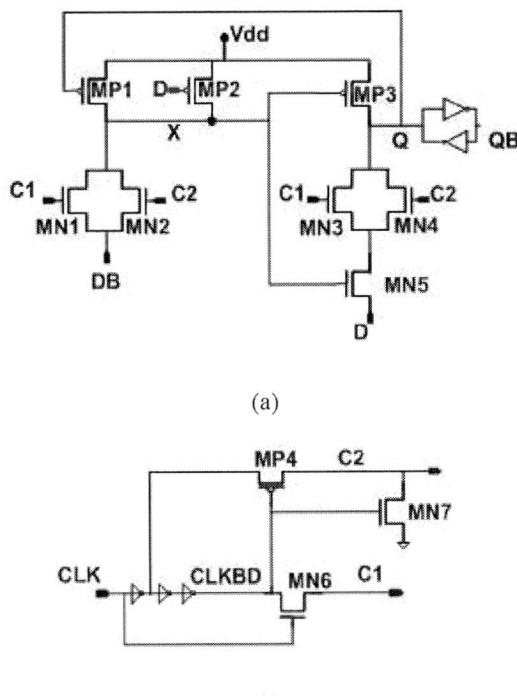

(a)

(b)

Figure 4: Structure of (a) DFFF, (b) clock-tree.

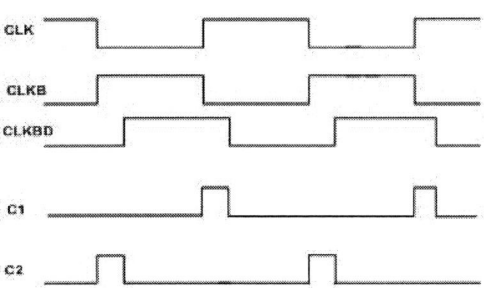

Figure 5: The timing diagrams of C1 and C2 in DFFF.

Also, it should be noted that the charging of node X is needed when DB is high and discharging of node X occurs when it is low. As another advantage of this logic is that the node X is discharged through only one transistor (MN1 or MN2) that again leads to the reduction of the DFFF delay. Finally, we should mention that the node Q also can be charged through MN3 and MN5 at the rising edge of clock and MN4 and MN5 at the falling edge of the clock whenever needed (*i.e.*, when D is high). Contrary to previous logic, there is no unnecessary transition in X and, hence, no extra power consumption occurs. Choosing MP1 as a small pull-up device, a weak fighting might exist during an input state change in two successive clock cycles.

The operation of the logic at the falling edge of the clock is similar to its operation at the rising edge except that C2 is high rather than C1 (Fig. 5) and MN2 and MN4 play the role of MN1 and MN3, respectively. The waveform of C1, C2, and the output using HSPICE is depicted in Figure 6.

3. SUBTHRESHOLD CURRENT

Subthreshold or weak inversion conduction current between the source and drain in an MOS transistor occurs when the gate voltage is below V_{th} [4]. Weak inversion typically dominates modern device off-state leakage due to the low-V_{th} [4]. The weak inversion current can be expressed as [10]

$$I_{ds} = \mu_0 C_{ox} \frac{W}{L}(m-1)(v_T)^2 \times \exp[\frac{(V_g - V_{th})}{mv_T}]$$

$$\times (1 - \exp[\frac{-V_{DS}}{v_T}]) \qquad (1)$$

where

$$m = 1 + \frac{3t_{ox}}{W_{dm}} \qquad (2)$$

where V_{th} is the threshold voltage, and $v_T = KT/q$ is the thermal voltage, C_{ox} is the gate oxide capacitance, μ_0 is the zero bias mobility; and m is the subthreshold swing coefficient (also called body effect coefficient). W_{dm} is the maximum depletion layer width, and t_{ox} is the gate oxide thickness [4]. As it is obvious from (1), if $V_{DS} = 0$, then subthreshold current will be zero.

Based on the above discussion, here we present a brief description of the previous flip-flop structures. In HLFF (Fig. 1) and SDFF (Fig. 2), when the node X is high, a voltage equal to V_{dd} is applied across the first branch in the pull down network (consisting of MN1, MN3 and MN5). On the other hand, when the node X is low then Q (output) will be high and output pull down tree sustains a voltage equal to V_{dd}. This high V_{DS} voltage drop causes large leakage currents and hence high leakage powers. The situation is even worse in the case of SDFF where this voltage exists across two transistors compared to the case of HLFF where three transistors exist in the output pull down network. Let's explain the situation in LSDFF (Fig. 3). Suppose that D is low, and then the voltage of node X2 as well as V_{DS} of MN1 is equal to V_{dd}. In the case that D is high, the V_{DS} of MN2 will be equal to V_{dd} and, hence, only one transistor has a high V_{DS} drop. As a result of this, the leakage current will be higher than the previous flip-flops. With the same argument, it can be observed that LSDFF would have more leakage current due to low-V_{th} transistors in its clock tree.

The subthreshold current in DFFF is very low which is due to the fact that the V_{DS} of each transistor in the pull-down network will be zero. Assuming D is high (DB is low), node X will be high, and, hence, both the drain and the source of MN1 and MN2 have high logic values leading to an approximately zero V_{DS} for these transistors. When D as well as Q is high, the voltage drop across the output pull-down tree will be approximately zero too. Compared to other flip flops, subthreshold current in DMHLFF is very low. These very low V_{DS} minimize the subthreshold leakage current of the flip-flop.

300

4. SIMULATION RESULTS

To evaluate the performance of the proposed flip-flop compared to other flip-flop circuits, all the discussed flip-flops have been simulated in a 70 *nm* CMOS process [Ref]. The HSPICE simulation results for V_{dd} = 0.7V are given in Table 1. The clock frequencies were 100 and 50 MHz for single-edge and double-edge triggered FF's, respectively. The load capacitance for flip-flops was assumed to be 10 fF. As is observed from the table, DFFF has lower power consumption, delay, and area (transistor count) compared to other flip-flop structures. The simulation result shows that the power-delay product of DFFF is 73% better than LSDFF. These improvements are 82% and 83% compared to SDFF and HLFF, respectively. The leakage powers of different flip-flops are given in Table 2 which shows the smallest leakage for DFFF as was expected.

5. SUMMARY AND CONCLUSION

In this work, we proposed a new Double edge triggered Feedback Flip-flop (DFFF) which had a better performance compared to previous logic. By a proper circuit design, unnecessary internal node transitions were

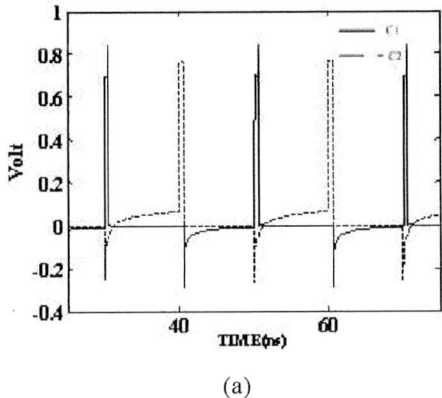

(a)

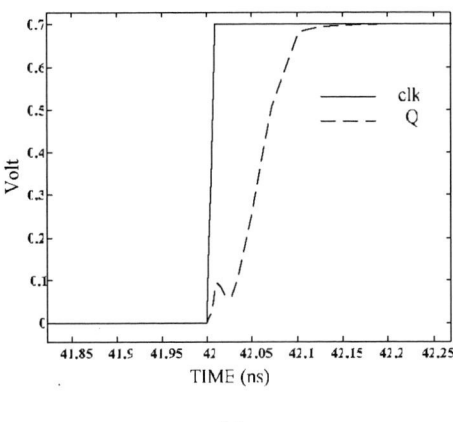

(b)

Figure 6: The waveform of (a) the controlling signal (*i.e.*, C1 and C2) (b) the output of DFFF.

avoided in this logic. Since the flip-flop is double-edge triggered, this logic may work with a lower clock frequency compared to single edge triggered flip-flops. These two reduced the power consumption of the flip-flop compared to other flip-flops. Furthermore reducing the number of transistor in the stack for both the internal and the output nodes and increasing the number of charging and discharging paths decreased the delay of the logic. The simulation results indicate that the improvement in the performance of DFFF is approximately between 70% and 84% compared to previous works.

REFERENCES

[1] M. Hamada, T. Terazawa, T. Higashi, S. Kitabayashi, S. Mita, Y. Watanabe, M. Ashino, and H. Hara, T. Kuroda, "Flip-Flop Selection Technique for Power-Delay Trade-off," in IEEE Int. Solid-State Circuits Conf., pp. 270–271, Feb. 1999.

[2] B. Nicolick, V. Stojanovic, V. G. Oklobdzija, W. Jia, J. chiu, M. Leung . "Sense Amplifier-Based Flip-Flop" in IEEE Int. Solid-State Circuits Conf. , pp. 282-283, Feb. 1999.

[3] B. Kong, S.-S. Kim, and Y.-H. Jun, "Conditional-Capture Flip-Flop Technique for Statistical Power Reduction," in IEEE Int. Solid-State Circuits Conf., pp. 290–291, Feb. 2000.

[4] K. Roy, S. Mukhopadhyay, and H. Mahmoodi-Meimand, "Leakage current mechanisms and leakage reduction techniques in deep-submicrometer CMOS circuits," in

Proceedings of IEEE, vol. 91, issue 2, pp. 305 – 327, Feb. 2003.

[5] N. Nedovic, V.G. Oklobdzija, "Hybrid latch flip-flop with improved power efficiency," in *Proceedings of IEEE Symp. on Integrated Circuits and Systems Design*, pp. 211-215, Sep. 2000.

[6] E. Partovi, R. Burd, U. Salim, F. Weber, L. DiGregorio, and D. Draper, "Flow-Through Latch and Edge-Triggered Flip-Flop Hybrid Elements," in *Proceedings of IEEE Int. Solid-State Circuits Conf.*, Feb. 1996.

[7] F. Klass, "Semi-Dynamic and Dynamic Flip-Flops with Embedded Logic," in *Symp. VLSI Circuits Dig. Tech. Papers*, pp. 108–109, June 1998.

[8] H. Kojima, S. Tanaka, and K. Sasaki, "Half-Swing Clocking Scheme for 75% Power Saving in Clocking Circuitry," *IEEE J. Solid-State Circuits*, vol. 30, pp. 432–435, Apr. 1995.

[9] C. Kim, S.-M. (Steve) Kang, "A Low-Swing Clock Double-Edge Triggered Flip-Flop" *IEEE J. of Solid-State Circuits*, vol. 37, no. 5, May 2002.

[10] Y. Taur and T. H. Ning, *Fundamentals of Modern VLSI Devices*, New York, Cambridge Univ. Press, 1998, ch. 3, pp. 120–128.

[11] V. Stojanovic and V.G. Oklobdzija, "Comparative Analysis of Master–Slave Latches and Flip-Flops for High-Performance and Low-Power Systems," *IEEE J. Solid-State Circuits*, vol. 34, pp. 536–548, Apr. 1999.

Table 1: Comparing various structures of DFF

Style	No. of Tr.	No. of Clked Tr.	Clk-Q (ps)	Power (μW)	P.D. (fj)	Improvement
SDFF	23	5	132	2.12	0.280	82%
HLFF	20	4	145	2.06	0.299	83%
LSDFF	28	3	106	1.8	0.191	73%
DFFF	21	3	59	0.88	0.051	-

Table 2: Comparison between the leakage powers of different flip-flops.

Leakage Power (nW)	SDFF	HLFF	LSDFF	DFFF
	86	49	82	27

2006 Asia and South Pacific Design Automation Conference

Post-Routing Redundant Via Insertion for Yield/Reliability Improvement*

Kuang-Yao Lee and Ting-Chi Wang
Department of Computer Science, National Tsing Hua University, Hsinchu, Taiwan
d924347@oz.nthu.edu.tw, tcwang@cs.nthu.edu.tw

Abstract - Reducing the yield loss due to via failure is one of the important problems in design for manufacturability. A well known and highly recommended method to improve via yield/reliability is to add redundant vias. In this paper we study the problem of post-routing redundant via insertion and formulate it as a maximum independent set (MIS) problem. We present an efficient graph construction algorithm to model the problem, and an effective MIS heuristic to solve the problem. The experimental results show that our MIS heuristic inserts more redundant vias and distributes them more uniformly among via layers than a commercial tool and an existing method. The number of inserted redundant vias can be increased by up to 21.24%. Besides, since redundant vias can be classified into on-track and off-track ones, and on-track ones have better electrical properties, we also present two methods (one is modified from the MIS heuristic, and the other is applied as a post processor) to increase the amount of on-track redundant vias. The experimental results indicate that both methods perform very well.

I. Introduction

With the advent of the very deep submicron (VDSM) technologies, the process variations become more and more serious, and thus achieving high yield rates on semiconductor chips will be more difficult. In order to reduce the burden of manufacturers to maintain the manufacturability and high yield rates, a new design methodology, design for manufacturability (DFM), is suggested. This design methodology proposes that in order to improve the manufacturability and yield of a design, the manufacturability issues could be considered during the physical design stage [1].

In an IC layout, a via provides a connection between two net segments from adjacent metal layers. Due to the growing of the design scale and/or the jumper-based solution to avoid the antenna effect [11], the number of vias could become very large. However, due to various reasons such as cut misalignment in a manufacturing process, electromigration and thermal stress, a via may fail partially or completely. For a partially failed via, the contact resistance and the parasitic capacitance will increase and may induce timing problems. On the other hand, a complete via failure will leave an open net on the circuit. These may heavily impact the functionality and yield of a design. Therefore, reducing the yield loss due to via failure is one of the most important problems in DFM.

A well known and highly recommended method to improve via yield is to add a redundant via adjacent to a single via [2,3], enabling a single via failure to be tolerated. Therefore, redundant vias will improve the reliability of a design.

Although major EDA vendors have already added the redundant via insertion feature to their routers, their results still have space to improve. (The details will be discussed in section VI.) The tools EYE/PEYE [4] reported in the literature are designed specially to insert redundant vias in the post layout stage but the details of how they do redundant via addition are not given. Besides, according to [4] and the results of the commercial tool used in our experiments, redundant vias are not evenly added on via layers.

[5] is the first work to consider redundant via insertion during the routing stage, but it will overcount the number of alive vias when all alive vias are critical, and cannot estimate the number of free neighbors of alive vias accurately in the general case. (In [5],

the number of free neighbors of a via is the number of redundant vias that can be inserted adjacent to the via without inducing any design rule violation; a via with at least one free neighbor is called an alive via.) [6] simultaneously considers redundant via insertion and via minimization during routing. However, in order to reduce the number of vias, the routed wire segments could become longer and violate the antenna rules, and thus need to introduce more vias to fix antenna problems in the post-routing stage. Besides, post-routing ECO operations might also change the routing result and introduce extra vias into the design. Therefore, no matter whether the router considers the redundant via insertion issue or not, it is usually necessary to consider redundant via insertion after detailed routing to improve yield and reliability.

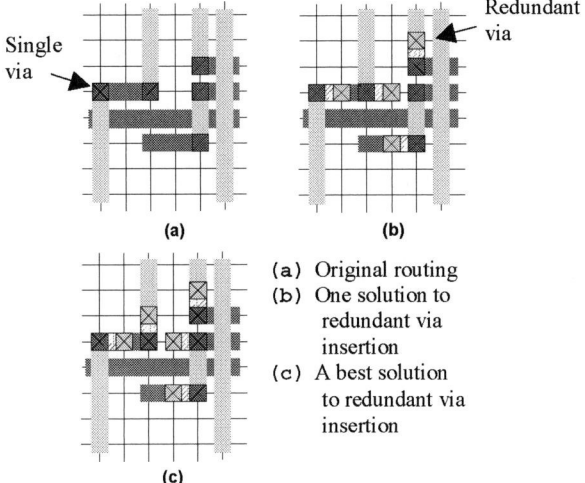

Fig. 1. Illustration of redundant via insertion.

(a) Original routing
(b) One solution to redundant via insertion
(c) A best solution to redundant via insertion

Given a detailed routing solution, because the positions of inserted redundant vias will affect the number of redundant vias that can be inserted into the design, how to decide the position of each inserted redundant via after detail routing is an important problem. As shown in Fig. 1, we can see that there are only four redundant vias inserted in (b), but as illustrated in (c), all of five single vias can be inserted with redundant vias.

Therefore, in this paper we study the post-routing redundant via insertion problem, and our contributions are threefold. First, we reduce the problem into the maximum independent set problem. All the vias of a circuit are considered simultaneously, and we believe that doing this can get better results than considering redundant via insertion layer by layer. Second, we present an efficient algorithm to construct the conflict graph (to model the problem) from a given detailed routing solution, and an effective heuristic to find a maximal independent set of the graph. The experimental results show that our MIS heuristic not only can insert more redundant vias but also can make the inserted redundant vias more evenly distributed among via layers, as compared to a commercial tool and a method based on [6]. Third, since redundant vias can be classified into on-track and off-track ones, and on-track ones have better electrical properties, we also propose two methods (one is modified from the MIS heuristic, and the other is applied as a post processor) to increase the amount of on-track redundant vias. The experimental results indicate that both methods perform very well.

The rest of this paper is organized as follows. In section II we show that the redundant via insertion problem can be transformed into the maximum independent set problem. In section III, an

---

* The work was partially supported by the National Science Council of Taiwan under Grant No. NSC-94-2220-E-007-015, and by the Ministry of Economic Affairs of Taiwan under Grant No. MOEA-94-EC-17-A-01-S1-031.

0-7803-9451-8/06/$20.00 ©2006 IEEE.

3C-1

algorithm for constructing the conflict graph from a given detailed routing solution is presented, and then we describe a heuristic method for solving the maximum independent set problem on the conflict graph in section IV. In section V, the methods for increasing the ratio of on-track redundant vias are presented. Section VI gives experimental results, and we conclude the paper in section VII.

II. Problem Formulation

A. Technology

We assume that the manufacturing technology used in this paper consists of $2m+1$ layers denoted by ME_1, VIA_1, ME_2, VIA_2, ..., ME_m, VIA_m, ME_{m+1}, where for all i and j, $1 \leq i \leq m+1$ and $1 \leq j \leq m$, ME_i and VIA_j represent the ith metal layer and the jth via layer, respectively. A via on VIA_i involves the layers ME_i, VIA_i, and ME_{i+1}. We also assume that a set of design rules is given, and SP is the spacing between two metals or cuts[1].

B. Double vias

The redundant via insertion process is to add a redundant via adjacent to a single via without violating any design rule. For simplicity we name the single via and the inserted redundant via adjacent to it as a *double via*. According to the position of a redundant via, we can categorize a double via into four types, as shown in Fig. 2; a single via is illustrated in (a) and its position is defined at its center; (b), (c), (d) and (e) are the illustrations of the four different double vias, and their types are named *DVU*, *DVD*, *DVL*, and *DVR*, respectively. Given a single via i, its double via of type j ($j \in \{DVU, DVD, DVL, DVR\}$) is denoted by $dv(i,j)$. For each single via, it has four choices to insert a redundant via if they do not violate any design rule.

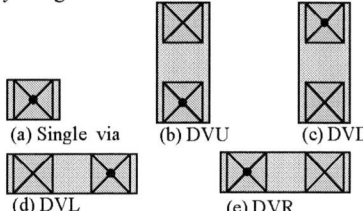

Fig. 2. Double via types.

Definition 1. (Feasible double via)
A double via of a single via is said to be feasible if replacing the single via with the double via will not violate any design rule, assuming none of the other single vias has a redundant via inserted in the design; otherwise the double via is defined as an infeasible one.

C. Post-routing redundant via insertion

With the definition of a double via, the post-routing redundant via insertion problem is defined as follows.

Problem 1. *Given a detailed routing solution, without re-routing any signal net, the problem asks to replace single vias on signal nets with double vias as many as possible subject to the following conditions: First, each single via either remains unchanged or is replaced by a double via. Second, after double via replacement, no design rule is violated.*

In the next two subsections, we will discuss two possible formulations, maximum bipartite matching and maximum independent set, to model Problem 1, and explain why the formulation of maximum bipartite matching might not work.

C1. Maximum bipartite matching formulation

---

[1] Depending on the technology, the spacing between metals could be different from the spacing between cuts. Also these space rules could vary on different layers. Nevertheless, our redundant via insertion methods presented in this paper can be easily modified to handle all these cases.

[6] reports that Problem 1 can be easily formulated as a maximum bipartite matching problem but without giving any further details. However, we find that either the formulation cannot capture optimal solutions, or some maximum bipartite matchings do not satisfy design rules. We use Fig. 3 to explain it.

Fig. 3(a) gives two different nets, and Fig. 3(b) is their 3D illustrations. Fig. 3(c), (d) and (e) show feasible double vias D_1, D_3 and D_2 for single vias V_1, V_3 and V_2, respectively. Assume that the double vias D_1 and D_2 will introduce some design rule violations if they both exist in the design, and so do the double vias D_3 and D_2. However, because D_1 and D_3 belong to the same net, they can both exist in the design, as shown in Fig. 3(f).

We now describe how to formulate this example as a maximum bipartite matching problem. We construct the bipartite graph $G = (V, E)$ as follows, where $V = X \cup Y$ and there is no edge between any two vertices in X (or between any two vertices in Y). Each single via corresponds to a vertex in X. Each feasible double via corresponds to at least one edge in E. For two feasible double vias originating from different single vias, if their existence in the design will violate design rules, their corresponding edges in E will be incident to the same vertex in Y. Fig. 3(g), (h) and (i) are three possible bipartite graphs obtained from this formulation. In graph Fig. 3(g) or (h), the set of bold edges is a maximum bipartite matching solution. However, neither of them is a legal solution to this example. On the other hand, the bipartite graph shown in Fig. 3(i) does not include the optimal solution to this example. We are not aware of any other way to construct the bipartite graph, but at least the three ones shown in Fig. 3(g), (h) and (i) cannot model Problem 1 correctly.

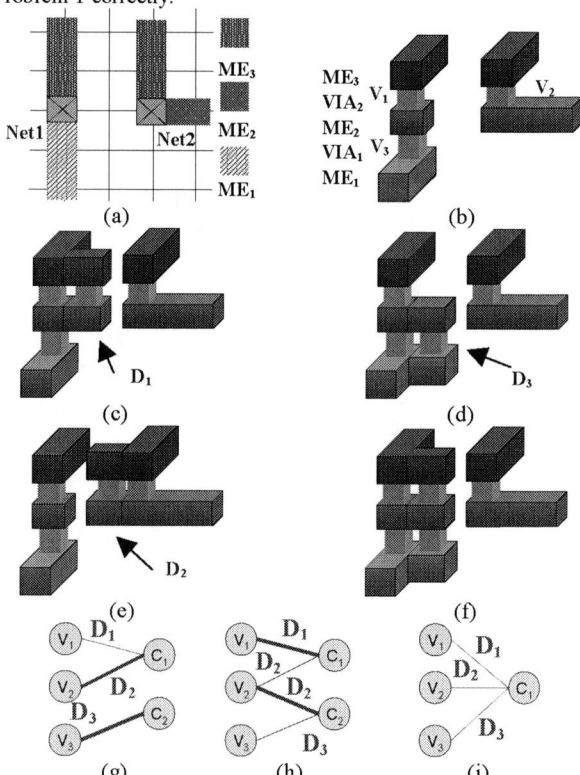

Fig. 3. Limitations with maximum bipartite matching.

C2. Maximum independent set formulation

Before introducing the maximum independent set formulation, we need to define what a conflict graph is first.

Definition 2. (Conflict graph)
A conflict graph $G(V,E)$ is an undirected graph constructed from a detailed routing solution. For each single via i on a signal net, if its double via of type j (i.e., $dv(i,j)$) is feasible, there exists a vertex $v_{i,j}$ in V. An edge $(v_{i,j}, v_{i'j'}) \in E$ if and only if $i = i'$, or $dv(i,j)$ and

304

dv(i',j') will cause design rule violations when both exist in the design.

Lemma 1. *Problem 1 can be reduced into the maximum independent set problem.*

Proof. Consider the conflict graph $G(V,E)$ constructed from a routed design. A maximum independent set MV of G is a maximum vertex set such that, $\forall\ v_{i,j}, v_{i',j'} \in MV$, $(v_{i,j}, v_{i',j'}) \notin E$. A vertex of G represents a feasible double via, and if two vertices are the endpoints of an edge, the corresponding double vias will violate design rules or they come from the same single via. Hence a maximum independent set of G is a set having the maximum number of double vias that can be inserted into the design. ∎

With Lemma 1, Problem 1 can be reduced to the following problem.

Problem 2. *Given a detailed routing solution, the problem asks to first construct a conflict graph from the design, then find a maximum independent set of the conflict graph, and finally for each vertex $v_{i,j}$ in the maximum independent set, replace the single via i with the double via dv(i, j).*

In the following two sections, we will describe how to efficiently construct a conflict graph and find a maximal independent set of the conflict graph.

III. Conflict Graph Construction

The construction of a conflict graph can be briefly divided into the vertex construction step and the edge construction step.

For the vertex construction step, we have to identify the feasible double vias of each single via. First, under the consideration of time complexity, we construct an R-tree [7,8,9] for each metal layer instead of constructing a single R-tree for all metal layers. An R-tree and its variants are data structures that are similar to a B-tree, but are used for indexing multi-dimensional information. In this paper, we use an R-tree for indexing 2-dimensional information. Typical queries on an R-tree specify a window of interest and retrieve all data intersecting or contained in the specified query window.

For a metal layer, the corresponding R-tree consists of the bounding box[2] of each object such as a wire segment, pin, or obstacle on the layer; besides, the bounding box of the vias on adjacent via layers are also included in the R-tree.

Definition 3. *(DVE)*
Suppose the bounding box of a single via i is $R_i=[x_{11},x_{12}]\times[y_{11},y_{12}]$, where (x_{11},y_{11}) and (x_{12},y_{12}) are the coordinates of the lower left corner and the upper right corner of the bounding box, respectively (see Fig. 4(a)); suppose the bounding box of a double via dv(i,j) is $R_{dv(i,j)}=[x_{21},x_{22}]\times[y_{21},y_{22}]$ (see Fig. 4(b)). The reduced bounding box of dv(i,j), denoted by DVE(i,j), is defined as $R_{dv(i,j)}-R_i=[x_{e1},x_{e2}]\times[y_{e1},y_{e2}]$ (see Fig. 4(c) for the illustration of DVE(i,DVU)).

Definition 4. *(DRW)*
Given a double via dv(i,j), suppose the bounding box of the redundant via contained in dv(i, j) is $R_{rv}=[x_{r1},x_{r2}]\times[y_{r1},y_{r2}]$. Then, the reduced design rule window of dv(i,j) is defined to be $DRW(i,j)=[x_{r1}-SP,\ x_{r2}+SP]\times[y_{r1}-SP,\ y_{r2}+SP]$. (See Fig. 4(d) for the illustration of DRW(i,DVU) which is the region with oblique lines.)

Definition 5. *(DRWSET and DVESET)*
The DRW set and DVE set of a single via i, denoted by DRWSET(i) and DVESET(i), are defined to be {DRW(i,j) | dv(i,j) is a feasible double via} and {DVE(i,j) | dv(i,j) is a feasible double via}, respectively.

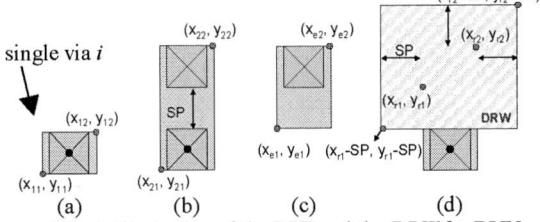

Fig. 4. Illustration of the *DVE* and the *DRW* for DVU.

For the vertex construction step, since a vertex in the conflict graph corresponds to a feasible double via, we need to check each double via and decide if it is feasible. In the following, we describe the details of the vertex construction step.

A. Vertex set construction

For each double via *dv(i,j)* originating from a single via *i* on layer VIA_k, we construct *DRW(i,j)* and use it as a query window to perform the range query on R-trees of ME_k and ME_{k+1}.

If there are any objects intersecting with *DRW(i,j)*, we cannot replace single via *i* with *dv(i,j)*, because it will induce design rule violations. Hence, there will never be a vertex on the conflict graph for *dv(i,j)*. On the other hand, if there is no object intersecting with *DRW(i,j)*, we add a vertex $v_{i,j}$ to the conflict graph.

After constructing the vertex set of a conflict graph, we should start the edge construction step. However, if we construct the edges of the conflict graph after completing the vertex construction step, the time complexity will be $O(n^2)$, where *n* is the number of vertices in the conflict graph. In fact, we can construct the edges more efficiently by constructing the vertex and edge sets simultaneously, as detailed in following subsection.

B. Graph construction algorithm

If there is an edge connecting two vertices in a conflict graph, the two double vias corresponding to the ends of the edge will belong to the same single via, or induce some design rule violations if they both exist in the design. Furthermore, a double via may introduce design rule violations to another one only if their corresponding single vias locate in nearby grids. Therefore, we first sort all single vias by their x-coordinates in the non-decreasing order. We then construct an R-tree for each metal layer, and finally according to the sorted order of vias (denoted *1, 2, ..., n*,), we perform the following four steps (i.e., Step 1 through Step 4) for each single via to get the conflict graph.

Before stating the details of the graph construction algorithm (called *GCA*), we introduce another R-tree named *VNC* first. *VNC* consists of the *DVEs* of feasible double vias, and initially it is empty. Once a single via *i* has been processed, each element of *DVESET(i)* will be inserted into *VNC*. For each element of *VNC*, if it will never intersect with any element of *DRWSET(j)*, for those sigle via *j*'s that have not been processed, it will be deleted from *VNC*. With *VNC*, we can construct edges efficiently.

Step 1. Suppose *i* is the single via being under consideration and x_{ll} is the x-coordinate of the lower left corner of the bounding box of *i*. If *i* is located in (x_i,y_i) and none of the x-coordinates of single vias *1, 2, ..., i-1* is equal to x_i, we retrieve the elements of *VNC* contained in the range $[-\infty, x_{ll}-SP]\times[-\infty,+\infty]$ and delete them from *VNC*, since these elements will never overlap with any element of *DRWSET(j)* with $j\geq i$.

Step 2. We start the vertex construction step for *i*. For each *dv(i,j)*, we use *DRW(i,j)* as the query window to do the range query on the R-trees of adjacent metal layers (Details are as described in the previous subsection.). Suppose the set of added vertices for *i* is called *FV(i)*.

Step 3. First, we add an edge for each vertex pair of *FV(i)* to the conflict graph. Then, we use each element of *DRWSET(i)* as the query window to do the range query on *VNC*. For each vertex $v_{i,j}\in FV(i)$, we can get a vertex set *V'*, where for each $v_{i',j'}\in V'$, the corresponding element in *VNC*, i.e., *DVE(i',j')*, intersects with *DRW(i,j)*. However, we cannot

---

[2] The bounding box of an object in the design is the contour of its 2-dimensional structure.

directly add an edge $(v_{i,j}, v_{i',j'})$ to the conflict graph, because $v_{i,j}$, $v_{i',j'}$ may belong to the same net. Therefore, we need to check each pair $(v_{i,j}, v_{i',j'})$ to see if they really introduce any design rule violation.

Step 4. We insert each element of $DVESET(i)$ to VNC.

Note that in Step 3, we need to check each pair $(v_{i,j}, v_{i',j'})$ to see if they really introduce any design rule violation, because there are cases where even if $DVE(i',j')$ intersects with $DRW(i,j)$, inserting both double vias $dv(i,j)$ and $dv(i',j')$ into the design still will not violate any design rule. A possible case is depicted in Fig. 5, where single vias V_1 and V_2 belong to the same net. In Fig. 5(b), $DVE(V_1, DVR)$ ($DVE(V_2, DVL)$, respectively) intersects with $DRW(V_2, DVL)$ ($DRW(V_1, DVR)$, respectively). However, they will not violate any design rule because they belong to the same net.

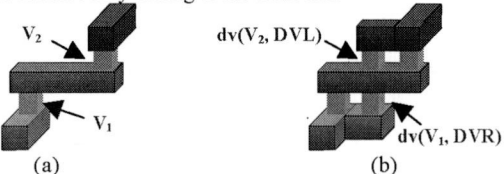

Fig. 5. A case where no design rule is violated.

Fig. 6 illustrates how the graph construction algorithm GCA works. In Fig. 6(a), there are four single vias in the design, and they are numbered to form the sorted sequence. After processing via 2, the elements of VNC and the conflict graph are shown in Fig. 6(b). When processing via 3, suppose $DVE(1, DVU)$ and $DVE(1, DVR)$ are contained in $[-\infty, x_{3,ll} - SP] \times [-\infty, +\infty]$, where $x_{3,ll}$ is the x-coordinate of the lower left corner of the bounding box of via 3. Therefore, after Step 1, $DVE(1, DVU)$ and $DVE(1, DVR)$ are deleted from VNC and the remaining elements in VNC are shown in Fig. 6(c). Then, after Step 2, the conflict graph gets updated as shown in Fig. 6(c). Finally, after Steps 3 and 4, VNC and the conflict graph become those shown in Fig. 6(d). Via 4 will be processed similarly.

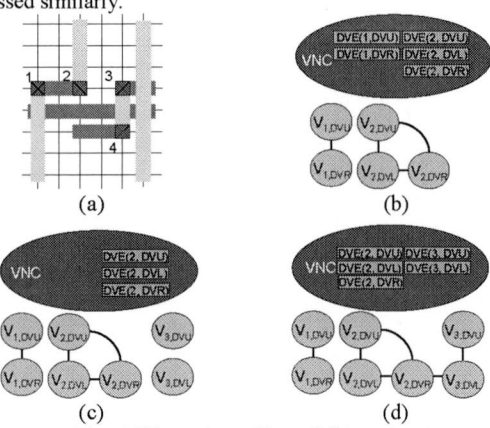

Fig. 6. Illustration of how GCA works.

IV. Heuristic for Solving the MIS Problem

Now we present a heuristic that solves the maximum independent set (MIS) problem on the conflict graph.

It is well known that the MIS problem is an NP-hard problem, so it is unlikely that we can get an optimal solution in polynomial time. Besides, the time complexities of MIS solvers are usually growing very fast as the numbers of vertices and edges in the graph increase. Therefore, our heuristic (called $H2K$) will solve the MIS problem in an iterative manner. In each iteration, a subgraph of size k (which specifies the maximum number of vertices in the subgraph and is a user-specified constant) is extracted from the conflict graph, a maximal independent set solution to the subgraph is sought and added to the final solution, and the conflict graph is updated. $H2K$ will terminate when the conflict graph has no remaining vertices.

Before describing the details of $H2K$, we define the "feasible number" for each vertex. The *feasible number* of each vertex $v_{i,j}$ in the conflict graph is equal to the number of vertices $v_{i',j'}$'s

(excluding $v_{i,j}$ itself) in the conflict graph such that $i=i'$ (i.e., the number of the other feasible double vias originating from the same single via). Initially, the feasible number of each $v_{i,j}$ is equal to $|DVESET(i)|-1$, where $|DVESET(i)|$ is the cardinality of $DVESET(i)$. The feasible number and degree of each vertex will decrease during the execution of $H2K$. The detailed steps of $H2K$ are given as follows.

Step 1. For the conflict graph $G(V,E)$, we construct a priority queue Q of V by using the feasible number and degree of a vertex as the first and second keys. We give a vertex a higher priority if it has smaller feasible number and degree. In addition, we define a vertex set V_{sol} to be the maximal independent set solution to G. Initially V_{sol} is an empty set.

Step 2. We extract the set $V_{sub} = \{v_1, v_2, ..., v_k\}$ of the first k vertices from Q, and construct the graph $G' = (V_{sub}, E')$, where $\forall v_i, v_j \in V_{sub}, (v_i, v_j) \in E'$ if $(v_i, v_j) \in E$.

Step 3. Solve the MIS problem on G' and get the solution denoted V_{tsol}.

Step 4. We set $V_{sol} = V_{sol} \cup V_{tsol}$ and then delete the vertices of V_{tsol} and their adjacent vertices from G and Q. Moreover, each edge incident to any deleted vertex is also removed from G. Finally, we update the feasible number and degree of each remaining vertex which is originally adjacent to some deleted vertex. In addition, Q is also updated.

Step 5. If V is empty, the vertex set V_{sol} is our final solution; otherwise we go back to Step 2.

The rationale behind subgraph extraction (i.e., Step 2) is that if a vertex with smaller feasible number and degree appears in the maximal independent set solution to G', less vertices will be deleted from the conflict graph in Step 4. Therefore, we prefer solving the MIS problem on a subgraph containing vertices with smaller feasible numbers and degrees.

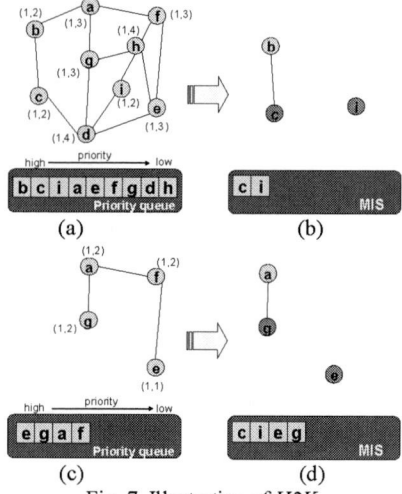

Fig. 7. Illustration of $H2K$.

Fig. 7 illustrates how our $H2K$ works, where each vertex is attached with a pair of numbers; the first number is the feasible number, and the second number is the degree. To simplify the example, we assume the feasible number of each vertex is equal to one. In the beginning, the conflict graph G and the priority queue Q are shown in Fig. 7(a). In Step 2, suppose k is set to 3, and the extracted subgraph G' has the vertex set $\{b, c, i\}$ as shown in Fig. 7 (b). Suppose the maximal independent set solution to G' found in Step 3 is $\{c, i\}$. Then in Step 4, G and Q are updated by deleting vertices c, i, and their adjacent vertices; each edge incident to any deleted vertex is also removed from G. The resultant G and Q are shown in Fig. 7 (c). At the second iteration, G' will be the one shown in Fig. 7 (d) and the maximal independent set solution to G' is assumed to be $\{g, e\}$. After Step 4 is done, G is empty, and hence the final solution found by $H2K$ will be $\{c, e, i, g\}$.

306

V. On- and Off-track Redundant Vias

As shown in Fig. 8, a redundant via rv of a single via v is called an *on-track* redundant via if rv is inserted on a wire segment connecting to v; otherwise, rv is called an *off-track* redundant via. Since an on-track redundant via takes less routing resource and has better electrical properties than an off-track redundant via, on-track redundant vias are more preferable. Therefore, if two solutions contain the same number of redundant vias, we prefer the one with more on-track redundant vias.

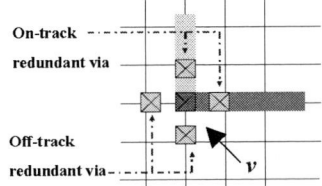

Fig. 8. Illustration of on- and off-track redundant vias.

A double via is said to be *on-track* if its associated redundant via is an on-track redundant via; otherwise it is an *off-track* double via. We now modify Problem 1 to consider the preference of on-track redundant vias as well.

Problem 3. *Given a detailed routing solution, without re-routing any signal net, the problem asks to replace single vias on signal nets with double vias as many as possible, and the ratio of on-track double vias should be also as high as possible. In addition, two conditions should be satisfied. First, each single via either remains unchanged or is replaced by a double via. Second, after double via replacement, no design rule is violated.*

We present two methods to solve Problem 3. The first one is to modify *H2K* by adding the third key to each vertex in the priority queue. If a vertex corresponds to an on-track double via, it will have a higher priority on this key. With this modification, for vertices having the same feasible number and degree, on-track ones will be extracted first, and hence have higher chances to be included in the maximal independent set solution than off-track ones. We call this method as *H3K*.

In addition, we also present a post processing heuristic (called *PPH*). Given a redundant via insertion solution, *PPH* will increase the amount of on-track double vias as many as possible while at the same time without decreasing the total number of double vias. *PPH* works as follows. It takes a conflict graph $G(V,E)$ and a redundant via insertion solution $RVIS_{org}$ as the input, and will generate another vertex set $RVIS_{mod}$ as the output. Initially $RVIS_{mod}$ is an empty set. In addition, a Boolean flag IS_DEL is used in *PPH*. Without loss of generality, $RVIS_{org}$ is assumed to be a set of vertices, and we will interchangeably use vertices and double vias. Each vertex v of $RVIS_{org}$ will be processed by the following four steps in a random order.

Step 1. Set IS_DEL to *FALSE*.

Step 2. If v is an on-track double via, go to Step 4. Otherwise, go to Step 3.

Step 3. Check each adjacent vertex v' of v in G. If v' is an on-track double via and each adjacent vertex of v' (excluding v) is not in $RVIS_{org} \cup RVIS_{mod}$, add v' to $RVIS_{mod}$ and set IS_DEL to *TRUE*.

Step 4. If IS_DEL is *FALSE*, v will be moved from $RVIS_{org}$ to $RVIS_{mod}$. Otherwise, v will be deleted from $RVIS_{org}$.

VI. Experimental Results

The technology used in our experiment has 5 metal layers. For simplicity we directly used the R*-tree package [9] for indexing 2-dimensional information of each metal layer. Moreover, we used the qualex-ms [10] as our MIS solver; we tried many different sizes when extracting a subgraph, and found that if we limited the subgraph to consist of 1500 vertices at most, it could get the best performance in terms of the number of inserted redundant vias.

Table 1: The experimental results on test cases

C1 Statistics	Via1	Via2	Via3	Via4	Total	CPU(s)
Original	11979	11111	1462	42	24594	
Upper	5218	10819	1443	42	17522	
CT	2125	10797	1438	42	14402	19
RatC(%)	17.74	97.17	98.36	100	58.56	
FNF	5165	10788	1438	42	17433	34
RatF(%)	43.12	97.09	98.36	100	70.88	
ImpF(%)	143.06	-0.08	00.00	00.00	21.05	
H2K	5175	10803	1441	42	17461	32
Rat2K(%)	43.20	97.23	98.56	100	71.00	
Imp2K(%)	143.53	00.06	00.21	00.00	21.24	
C2 Statistics	Via1	Via2	Via3	Via4	Total	CPU(s)
Original	17208	18086	4745	1118	41157	
Upper	6078	17066	4359	1088	28591	
CT	3476	17005	4351	1086	25918	28
RatC(%)	20.20	94.02	91.70	97.14	62.97	
FNF	6059	16982	4325	1085	28451	45
RatF(%)	35.21	93.90	91.15	97.05	69.13	
ImpF(%)	74.31	-0.14	-0.60	-0.90	09.77	
H2K	6069	17011	4341	1086	28507	43
Rat2K(%)	35.27	94.06	91.49	97.14	69.26	
Imp2K(%)	74.60	00.04	-0.23	00.00	09.99	
C3 Statistics	Via1	Via2	Via3	Via4	Total	CPU(s)
Original	55878	55252	13066	2863	127059	
Upper	23755	52780	12407	2785	91727	
CT	13179	52506	12365	2777	80827	101
RatC(%)	23.59	95.03	94.63	97.00	63.61	
FNF	23634	52539	12358	2784	91315	190
RatF(%)	42.30	95.09	94.58	97.24	71.84	
ImpF(%)	79.33	00.06	-0.06	00.25	12.98	
H2K	23687	52615	12375	2784	91461	192
Rat2K(%)	42.39	95.23	94.71	97.24	71.98	
Imp2K(%)	79.73	00.21	00.08	00.25	13.16	
C4 Statistics	Via1	Via2	Via3	Via4	Total	CPU(s)
Original	57216	64879	20864	8953	151912	
Upper	14917	61300	17950	8180	102347	
CT	4677	60978	17777	8142	91574	120
RatC(%)	08.17	93.99	85.20	90.94	60.28	
FNF	14750	60848	17711	8148	101457	201
RatF(%)	25.78	93.79	84.89	91.01	66.79	
ImpF(%)	215.37	-0.21	-0.37	00.07	10.79	
H2K	14805	61008	17791	8161	101765	203
Rat2K(%)	25.88	94.03	85.27	91.15	66.99	
Imp2K(%)	216.55	00.05	00.08	00.23	11.13	
C5 Statistics	Via1	Via2	Via3	Via4	Total	CPU(s)
Original	148661	158862	40726	9137	357386	
Upper	62312	148592	35729	8668	255301	
CT	33216	147781	35505	8640	225142	311
RatC(%)	22.34	93.02	87.18	94.56	63.00	
FNF	62033	147757	35453	8656	253899	697
RatF(%)	41.73	93.01	87.05	94.74	71.04	
ImpF(%)	86.76	-0.02	-0.15	00.19	12.77	
H2K	62174	148063	35535	8656	254428	710
Rat2K(%)	41.82	93.20	87.25	94.74	71.19	
Imp2K(%)	87.18	00.19	00.08	00.19	13.01	

[6] points out a simple heuristic for redundant via insertion and its idea is that if there is only one feasible redundant via for a single via, it adds the redundant via first. However, [6] does not provide any further details. We also based on the above idea and implemented a heuristic called *FNF* for comparative studies. Its details are as follows. *FNF* takes a conflict graph as the input, and creates a priority queue for vertices such that a vertex with smaller feasible number has a higher priority. *FNF* iteratively extracts the vertex with the smallest feasible number from the priority queue, adds it into the final solution, and updates the priority queue and

the conflict graph. When the conflict graph or priority queue is empty, *FNF* terminates.

We first compared our approach *H2K* with a commercial tool and *FNF* on five real circuits C1-C5. Our experimental flow is as follows. We used the commercial tool to generate the routed circuit, and then inserted redundant vias by its redundant via insertion feature. Each conflict graph used by *H2K* and *FNF* was generated by our *GCA* algorithm that took the routed design as the input. Then, *H2K* and *FNF* generated the circuits with inserted redundant vias. Finally, the results obtained by the commercial tool, *H2K* and *FNF* were verified with the built-in DRC and LVS verifier of the commercial tool.

The results are shown in Table 1. "Original" gives the number of single vias on each via layer before performing redundant via insertion. "Upper" denotes the number of single vias that have at least one feasible double via. "CT", "FNF" and "H2K" are the numbers of redundant vias inserted by the commercial tool, *FNF* and *H2K*, respectively. "RatC(%)", "RatF(%)" and "Rat2K(%)" are the ratios of "CT", "FNF" and "H2K" to "Original", respectively. "ImpF(%)" and "Imp2K(%)" represent the improvement rates of *FNF* and *H2K* over the commercial tool, respectively. "CPU(s)" gives the CPU time in seconds of different approaches. The commercial tool was executed on a Sun Fire V440 machine with four CPUs and 8GB memory; *H2K*, *GCA* and *FNF* were implemented in C++ language running on a Linux based machine with 2.4G processor and 2GB memory. Because *H2K*, *GCA* and *FNF* used some Linux based packages, they could not be executed on a Sun based platform. It should be noted that the CPU times for the commercial tool only record the redundant via insertion step, and before this step the design has been loaded into memory. The CPU times of *H2K* and *FNF* include the time spent by *GCA*.

From Table 1, we can see that our approach *H2K* can insert 9.99%-21.24% more redundant via than the commercial tool. Besides, the number of redundant vias inserted on each layer by *H2K* is very close to the upper bound in all test cases, but the number of redundant via inserted on Via1 by the commercial tool is much smaller than the upper bound. Hence, the redundant vias inserted by *H2K* are distributed more uniformly among via layers. Moreover, the experimental results show that although *FNF* also inserts more redundant vias than the commercial tool, its improvement rate is still less than our approach *H2K* for each test case. *H2K* can insert up to 529 more redundant vias than *FNF* with comparable CPU time. In every test case, there is at least one via layer on which *FNF* inserts less redundant vias than the commercial tool. Nevertheless, our approach *H2K* can always insert more or the same number of redundant vias among each via layer than *FNF* and the commercial tool.

Table 2 shows the results of our approaches *H3K* and *PPH* when considering on-track redundant vias. "FNF+PPH", "H2K+PPH" and "H3K+PPH" indicate that *PPH* was applied after *FNF*, *H2K*, and *H3K*, respectively. It should be mentioned that although *H3K* is design to consider on-track redundant vias directly, we would like to see if its result still has room to improve, and therefore we also applied *PPH* after *H3K*.

The columns "MISo" and "ONo" show the numbers of double vias and the on-track double vias from each original solution, respectively. After running *PPH*, the numbers of inserted double vias and on-track double vias are shown in the columns "MISm" and "ONm", respectively. The column "Imp(%)" denotes the improvement rate on the number of on-track double vias achieved by *PPH*. "CPU(s)" gives the CPU time of *PPH*, but for "H3K", it represents the total CUP time of *H3K*.

From Table 2, we can see that even if we prefer on-track redundant vias, the total number of inserted redundant vias can still remain the same or even larger while the CPU time spent by *PPH* is no more than 3 seconds. Compared to *H2K*, *H3K* can increase the number of on-track double vias by up to 65.31% while almost having the same number of inserted redundant vias and spending the same or less CPU time. As for *PPH*, it helps to increase the amount of on-track double vias by 19.99%-21.90% and 18.58%-

20.54% for *FNF* and *H2K*, respectively. Besides, for some test cases, *PPH* can also slightly increase the total number of redundant vias. Finally, we observe that running *H3K* alone is always good enough to beat both "FNF+PHP" and "H2K+PHP" on the number of on-track redundant vias, although its result can still be improved by *PHP* for more than half of the test cases.

Table 2: The experimental results for *H3K* and *PPH*.

C1 Statistics					
	MISo	ONo	MISm	ONm	Imp(%)	CPU(s)
FNF+PPH	17433	7128	17433	8553	19.99	<1
H2K+PPH	17461	7167	17461	8552	19.32	<1
H3K	17461	11848	-	-	-	32
H3K+PPH	17461	11848	17461	11878	00.25	<1
C2 Statistics					
	MISo	ONo	MISm	ONm	Imp(%)	CPU(s)
FNF+PPH	28451	13132	28451	15986	21.73	1
H2K+PPH	28507	13406	28507	16047	19.70	<1
H3K	28506	20508	-	-	-	43
H3K+PPH	28506	20508	28506	20519	00.05	<1
C3 Statistics					
	MISo	ONo	MISm	ONm	Imp(%)	CPU(s)
FNF+PPH	91315	42084	91318	50551	20.12	1
H2K+PPH	91461	42397	91461	50275	18.58	1
H3K	91461	66205	-	-	-	190
H3K+PPH	91461	66205	91461	66212	00.01	1
C4 Statistics					
	MISo	ONo	MISm	ONm	Imp(%)	CPU(s)
FNF+PPH	101457	47649	101459	58084	21.90	1
H2K+PPH	101765	48073	101765	57946	20.54	<1
H3K	101765	70696	-	-	-	201
H3K+PPH	101765	70696	101765	70696	00.00	1
C5 Statistics					
	MISo	ONo	MISm	ONm	Imp(%)	CPU(s)
FNF+PPH	253899	117432	253903	142331	21.20	3
H2K+PPH	254428	118557	254428	142251	19.99	2
H3K	254428	180512	-	-	-	680
H3K+PPH	254428	180512	254428	180513	00.00	1

VII.   Conclusions

In this paper we consider the post-routing redundant via insertion problem which is formulated as the maximum independent set problem. We present an efficient graph construction algorithm to model the problem, and an effective heuristic to solve the maximum independent set problem. Besides, we also describe how to modify the MIS heuristic and give a post-processing method to increase the amount of on-track redundant vias. Promising experimental results are shown to support all our methods.

VIII.   References

[1] L. K. Scheffer, "Physical CAD Changes to Incorporate Design for Lithography and Manufacturability", Proc. of ASPDAC, 2004.
[2] TSMC Reference Flow 5.0.
[3] Y. Zorian, D. Gizopoulos, C. Vandenberg and P. Magarshack, "Guest Editors' Introduction: Design for Yield and Reliability", IEEE Trans on Design & Test of Computers, vol. 21, May 2004.
[4] G. A. Allan, "Targeted Layout Modifications for Semiconductor Yield/Reliability Enhancement", IEEE Trans on Semiconductor Manufacturing, vol. 17, Nov. 2004.
[5] G. Xu, Li-Da Huang, D. Z. Pan and M. D. F. Wong, "Redundant-Via Enhanced Maze Routing for Yield Improvement", Proc. of ASPDAC, 2005.
[6] H. Yao, Y. Cai, X. Hong and Q. Zhou, "Improved Multilevel Routing with Redundant Via Placement for Yield and Reliability", Proc. of GLSVLSI, 2005.
[7] A. Guttman, "R-Trees: A Dynamic Index Structure for Spatial Searching", Proc of SIGMOD, 1984.
[8] M. de Berg, J. Gudmundsson, M. Hammar and M. H. Overmars, "On R-trees with Low Stabbing Number", Proc. European Symposium on Algorithms, 2000.
[9] N. Beckmann, H.-P. Kriegel, R. Schneider and B. Seeger, "The R*-Tree: An Efficient and Robust Access Method for Points and Rectangles", Proc. of SIGMOD, 1990.
[10] http://www.busygin.dp.ua/npc.html
[11] P. H. Chen, S. Malkani, C.-M. Peng and J. Lin, "Fixing Antenna Problem by Dynamic Diode Dropping and Jumper Insertion", Proc. of ISQED, 2000.

Temperature-Aware Routing in 3D ICs *

Tianpei Zhang, Yong Zhan, and Sachin S. Sapatnekar
Department of Electrical and Computer Engineering, University of Minnesota
zhangt, yongzhan, sachin@ece.umn.edu

Abstract

Three-dimensional integrated circuits (3D ICs) provide an attractive solution for improving circuit performance. Such solutions must be embedded in an electrothermally-conscious design methodology, since 3D ICs generate a significant amount of heat per unit volume. In this paper, we propose a temperature-aware 3D global routing algorithm with insertion of "thermal vias" and "thermal wires" to lower the effective thermal resistance of the material, thereby reducing chip temperature. Since thermal vias and thermal wires take up lateral routing space, our algorithm utilizes sensitivity analysis to judiciously allocate their usage, and iteratively resolve contention between routing and thermal vias and thermal wires. Experimental results show that our routing algorithm can effectively reduce the peak temperature and alleviate routing congestion.

1 Introduction

By stacking multiple device layers into a monolithic structure, three-dimensional integrated circuits (3D ICs) can achieve higher transistor densities and shorter interconnect lengths than two-dimensional (2D) ICs, and work as a platform to integrate different components and provide good substrate isolation among them [1]. Despite the advantages of 3D ICs over 2D ICs, thermal effects are expected to be significantly exacerbated in 3D ICs due to higher power density and greater thermal resistance of the insulating dielectric, and this can cause greater degradation in device performance and chip reliability which have already plagued 2D ICs [1]. Thus, it is essential to develop 3D-specific design tools that take a thermal co-design approach so as to address the thermal effects and generate reliable and high performance designs. This work considers the problem of developing routing solutions for 3D ICs.

To address thermal issues in global routing, an efficient and accurate thermal computation method is needed. Previous work on on-chip thermal analysis falls into the following categories: the finite difference method (FDM) and the finite element method (FEM) as in [3–5], and Green function method as in [6, 7]. In this paper, we utilize the finite difference based thermal circuit model presented in [3] to obtain fast temperature and sensitivity analysis. Several 3D routing algorithms have addressed the problems of 3D channel routing [8], 3D maze routing [9], 3D Field-Programmable Gate Array (FPGA) routing [10], and 3D hierarchical routing [11]. However, none of these approaches considers the thermal problem associated with 3D ICs in the routing phase. A first approach to this problem is presented as a thermally-driven 3D routing algorithm in [12] with the planning of thermal vias to reduce temperature. However, it does not fully address the contention issues between thermal vias and routing resources.

In this paper, we propose a novel 3D routing algorithm which can effectively reduce on-chip temperatures by appropriate insertion of "thermal vias" and a new construct that we call "thermal wires," and generate a routing solution free of thermal and routing capacity violations. Thermal vias correspond to vertical interlayer vias that do not have any electrical function, but are explicitly added as thermal conduits. Thermal wires perform a similar function, but conduct heat laterally within the same layer. Thermal vias perform the bulk of the conduction to the heat sink, while thermal wires help distribute the heat paths over multiple thermal vias. The routing scheme begins with routing congestion estimation and signal interlayer via assignment, followed by thermally-driven maze routing. Sensitivity analysis is employed and linear programming (LP) based thermal via/wire insertion is performed to reduce temperature. The above process is iteratively repeated until temperature and routing capacity violations are resolved. Experimental results show

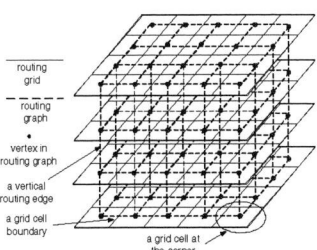

Figure 1: Routing grid and routing graph for a four-layer 3D circuit.

that the scheme can effectively resolve the contentions between thermal via/wire and routing, generating a solution satisfying both congestion and temperature requirements.

2 3D IC Global Routing Model

2.1 3D Interconnects and Routing Model

In 3D ICs, devices are fabricated on a number of active layers, which are separated by silicon dioxide and joined by an adhesive material. Within each device layer, interconnections among devices can be achieved with traditional interconnect wires and vias. Connections between active layers are facilitated by vertical interconnect vias that span through multiple layers, providing a means for electrically connecting wires in those layers. This type of via is different from a regular 2D via: in particular, it is significantly taller than conventional vias, and has a larger landing pad to maintain a viable aspect ratio. We refer to such vias as *interlayer vias*.

The multiple-layer structure of 3D IC complicates the global routing problem, but we may extend some basic constructs from 2D routing to 3D. As shown in Figure 1, our global routing model tessellates each layer in the 3D circuit into a two-dimensional array, referred to as the *routing grid*. A net N consists of a set of electrically equivalent pins, $\{n_0, n_1, n_2, ..., n_k\}$, distributed in different routing grid cells (possibly in different active layers), of which n_0 is the source and $n_1, n_2, ..., n_k$ are sinks. If a net has pins distributed on different layers, it is referred to as an *interlayer net*; otherwise it is called a *2D net*.

The dual graph of a 2D routing grid tessellation is a *2D routing graph*. By stacking 2D routing graphs of all layers and connecting each pair of vertically neighboring nodes with a vertical routing edge, we build a *3D routing graph*, denoted as G, which is shown by the dashed lines in Figure 1, and the routes for each net are to be determined along the edges of this 3D routing graph. There are two classes of edges in the graph: (i) Each lateral edge in the 3D routing graph corresponds to a boundary e_{ij} in the routing grid that connects grid cells i and j. Due to geometrical limitations on each lateral boundary, we require $W_e \leq C_e$, in which W_e is the total width (including wire spacing) used by all wires, and C_e is the geometrical width of boundary e, or the *boundary capacity*. Any violation of this requirement results in a *boundary overflow*. (ii) Each vertical edge in the 3D routing graph corresponds to the interlayer vias connecting two vertically neighboring grid cell positions. Since interlayer vias go through the active device layer(s), and can only use the white space that is not occupied by devices, there is an upper limit on the number of interlayer vias for each vertical edge in the routing graph, referred to as *interlayer via capacity* U_i, and it is determined by the white area of grid cell i through which the interlayer vias pass, and the size of the landing pad of the interlayer via. We require $V_i \leq U_i$, in which V_i is the actual number of interlayer vias through grid cell i.

---

*This work was supported in part by DARPA under grant N66001-04-1-8909, and NSF under award CCR-0205227.

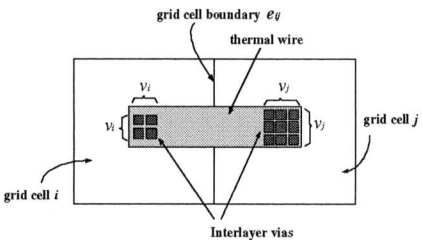

Figure 2: Reduction of lateral routing capacity due to the interlayer vias in neighboring grid; thermal wires are lumped, and together with thermal vias, form a thermal dissipation network.

2.2 Thermal Vias and Thermal Wires

The silicon dioxide layer acts as a thermal insulator that strongly inhibits heat flow, potentially leading to elevated temperatures. Interlayer vias connecting different active layers serve to to conduct heat and alleviate hot spots in 3D ICs [13,14]. Straightforward extensions of traditional routing scheme to 3D are inadequate, since they leave out several new 3D-specific complications. The locations of the interlayer vias that carry signals from one layer to another should not be determined purely by the signal wire routing process, but should also incorporate thermal considerations, determining the most desirable interlayer via positions from the point of view of heat conduction. Moreover, a sufficient number of interlayer vias must be used so that the temperature can be reduced to below the target level.

Our work uses two types of vias: (i) thermal vias are deliberately introduced interlayer vias that serve no electrical function, but are dedicated to the purpose of temperature reduction [15], and (ii) signal interlayer vias, which carry electrical signal, and thus perform signal and heat conduction simultaneously. Like signal interlayer vias, thermal vias can be planned in each grid cell position i between two layers. We require that V_i, the total number of signal interlayer vias and thermal vias going through grid cell i, be smaller than its capacity U_i, defined above. Here we assume that the interlayer vias for power grid are predetermined and there are dedicated resources for them, which are excluded from U_i.

Although interlayer vias can effectively reduce the on-chip temperature, their large size also acts as a significant routing blockage; as more heat removal is achieved, less routing space is available, and this problem has been identified in the packaging of multi-chip modules (MCMs) [16]. Figure 2 shows how interlayer vias can reduce the routing capacity of a neighboring lateral routing edge. If $v_i \times v_i$ interlayer vias pass through grid cell i, and $v_j \times v_j$ interlayer vias pass through the adjacent grid cell j, the signal routing capacity of boundary e_{ij} will be reduced from the original capacity, C_e, to a reduced capacity, $C_{e,red.}$, and we require that signal wire usage W_e should satisfy:

$$C_{e,red.} = \min \left(C_e - v_i \cdot w, C_e - v_j \cdot w \right)$$
$$W_e \leq C_{e,red.} \tag{1}$$

where w is the geometrical width of an interlayer via. Here we define the smaller of the two reduced routing widths as the reduced edge capacity, so that there can be a feasible translation from the global routing result to a detailed routing solution. On the other hand, given the actual signal wire usage W_e of a routing edge, we can also use equation (1) to determine how many interlayer vias can go through the neighboring grid cells so that there is no overflow at the routing edge. Since temperature reduction requires insertion of a large number of thermal vias, careful planning is necessary to meet both the temperature and routability requirements.

Just as thermal vias enhance vertical heat conduction, we can also introduce *thermal wires* to improve lateral heat conduction. The lateral routing tracks in the circuits are used by power grid wires and signal wires. For simplicity of analysis, we assume a predetermined power grid architecture as well as dedicated routing tracks for power wires, and they are excluded from the routing capacity calculation. Along a routing edge, signal wires may not utilize all of the routing tracks, and we employ the remaining tracks to connect the thermal vias in adjoining grid cells with *thermal wires*. Like thermal vias, thermal wires do not carry signal, and therefore, they can be connected directly to thermal vias to form an efficient heat dissipation network as shown in Figure 2. Thermal wires enable the conduction of heat in lateral direction, and can thus help vertical thermal vias to reduce hot spots temperature efficiently: for those

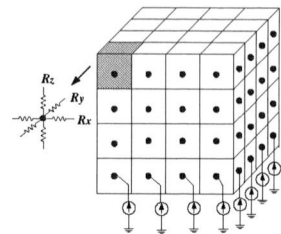

Figure 3: Equivalent thermal circuit model of a 3D IC.

hot spots where only a restricted number of thermal vias can be added, we can use thermal wires to conduct heat laterally, and then remove heat through thermal vias in adjoining grids. Note that although signal wires can also conduct heat laterally, they must be separated from the thermal vias by an oxide that has high thermal resistivity, and are therefore not efficient in lateral heat removal. Moreover, unlike signal routes that only use a small number of interlayer signal vias, the thermal vias form a global net and are therefore very effective in heat removal. Another advantage of thermal wires is improved design for manufacturing. Filling the remaining tracks with thermal wires can create a dense metal fill, which improves manufacturability and performance predictability which may be deteriorated by the Chemical-Mechanical Planarization (CMP) step. Since thermal wires contend for lateral routing resources with signal wires, they should be well planned to satisfy the temperature and routability requirements. Along a routing edge e with total routing capacity C_e, the number of signal wires s_e and the number of thermal wires m_e should satisfy

$$(s_e + m_e) \cdot w_t \leq C_e \tag{2}$$

where w_t is the width of a routing track.

2.3 Temperature-Aware 3D Global Routing Problem

Our formulation of the temperature-aware 3D global routing problem is as follows. The inputs to the problem include:

1. The result of placement for a 3D IC
2. 3D technology parameters
3. A specified value of the maximum temperature, T_{spec}

Our global routing algorithm will route the circuit netlist and efficiently insert thermal vias and thermal wires at appropriate locations so that the solution satisfies the requirements on the routing capacities and the interlayer via capacities. At the same time, the peak temperature will be lowered towards T_{spec}. Moreover, for reasons related to reliability and to the limited availability of interlayer via resources, we require that the number of signal interlayer vias used in routing should be minimized.

3 Thermal Analysis Model

Equivalent thermal circuit models have been used extensively in the simulation of thermal effects because of their clear physical origins, ease of implementation, and flexibility in handling different kinds of boundary conditions and non-uniform thermal conductivity. Our thermal analysis uses a grid of the same size as that of global routing. We model a 3D chip as a resistive thermal network and solve the corresponding linear equation to obtain the temperature of each node. As shown in Figure 3, we use a node to represent all of the heat dissipating devices in a global routing grid cell, using an equivalent current source that represents the total power generated in the cell. Adjacent nodes are connected by a thermal resistance, forming a thermal resistive grid. In this work, we only consider power dissipation from active devices, since heat generated by interconnect wires is small because of their relatively small resistance; however, it is possible to extend the approach to incorporate this.

The numerical value of thermal resistance connecting to a node in x, y, and z directions can be computed as:

$$R_x = R_{oxide,x} || R_{int.,x} || R_{mix,x}$$
$$R_y = R_{oxide,y} || R_{int.,y} || R_{mix,y}$$
$$R_z = R_{oxide,z} || R_{int. via,z} || R_{mix,z} \tag{3}$$

where the symbol "||" refers to a parallel connection between the resistors. The thermal resistance in each direction is determined by the materials contained in the cuboid volume extending to the neighboring

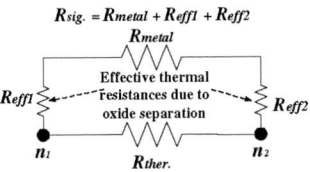

Figure 4: Thermal circuit model of signal wire and thermal wire.

grid cell and can be calculated as parallel connections of three parts. $R_{oxide,x}$, $R_{oxide,y}$ and $R_{oxide,z}$ are the oxide thermal resistance of pure oxide extending to neighboring node in x, y and z directions respectively. The terms $R_{int., x}$ and $R_{int., y}$ are the thermal resistance of interconnect wires running along x and y directions, respectively. $R_{int. via, z}$ is the thermal resistance of the vertical interlayer vias which act as one of the major heat conduction paths, and it is proportional to the number of interlayer vias connecting the vertically neighboring grid cells. Along each direction, besides pure oxide and metal wires or vias extending in that direction, the remaining volume is a mixture of metal wires/vias extending in the orthogonal directions and the oxide separating them; along the direction we are considering, alternate metal and oxide form a "sandwich" structure. We denote the expected thermal resistance of such a mixture as $R_{mix,x}$, $R_{mix,y}$ and $R_{mix,z}$ for x, y and z direction respectively, and their values are calculated as series thermal resistance of the metal wires/vias (running in orthogonal directions) and the oxide which separates the metal.

To differentiate between the the heat conducting abilities of signal wires and thermal wires, we use the model depicted in Figure 4. We model a signal wire of thermal resistance $R_{sig.}$, connecting neighboring grid cells n_1 and n_2 as series connection of metal resistance R_{metal} and two additional resistances R_{eff1} and R_{eff2} connecting to the thermal via nodes at the center of n_1 and n_2. R_{eff1} and R_{eff2} correspond to the effective thermal resistance of the oxide which separates signal wires from thermal vias. The effective oxide thickness is estimated by a probabilistic approach which assumes a uniform distribution of signal wires crossing a grid cell boundary. By comparison, the thermal resistance of a thermal wire connecting n_1 and n_2 only consists of the metal resistance $R_{ther.}$.

The vertical edges of the 3D chip are modeled to be adiabatic to the ambient, and the bottom of the chip is connected to an isothermal heat sink by a bulk substrate. With the thermal circuit established, the nodal analysis (NA) equations for the thermal grid can be set up as

$$M\mathbf{T} = \mathbf{P} \qquad (4)$$

where M, $\mathbf{T}$ and $\mathbf{P}$ are the NA coefficient matrix, the temperature profile vector and power dissipation vector for the grid cells, respectively. This equation is sparse and symmetric positive definite (SPD), and we use the LASPACK package [17] as the linear solver to obtain the temperature profile efficiently.

To effectively remove temperature violations in circuit under constrained resources, it is very important that we insert thermal vias and thermal wires in the *right* places; otherwise even a large number of them can not effectively reduce the high temperatures. Thus, we perform adjoint sensitivity analysis which can help determine the potential thermal via and thermal wire locations where insertion of a thermal via/wire can reduce hot spot temperature effectively. Adjoint sensitivity analysis has the advantage of obtaining the sensitivities of one output value to many parameters simultaneously and efficiently. We now briefly describe the calculation of the sensitivity of the temperature at a hot spot to the thermal via density as an example; sensitivity analysis for hot spot temperature to thermal wire density can be performed similarly.

The temperature profile of a 3D circuit is determined by equation $M\mathbf{T} = \mathbf{P}$ as described above. For a hot grid cell in the circuit with temperature T_i at the ith position of vector $\mathbf{T}$, the sensitivity of T_i to matrix element M_{kl} is expressed as follows [18]:

$$\frac{\partial T_i}{\partial M_{kl}} = -\xi_{ik}T_l \qquad (5)$$

in which ξ_i is the solution of equation $M^T\xi_i = \mathbf{e}_i$, and $\mathbf{e}_i$ is a column vector with all zeros except for a one in its ith row. We can then obtain the sensitivity s_{ij} of the temperature T_i to the number of interlayer vias

at location j, V_j, by applying the chain rule:

$$s_{ij} = \frac{\partial T_i}{\partial V_j} = \sum_{M_{kl} \in S_j} \frac{\partial T_i}{\partial M_{kl}} \frac{\partial M_{kl}}{\partial V_j} \qquad (6)$$

where set S_j contains all of the entries, M_{kl}, that depend on V_j. There is a small number of such entries in each S_j, since the V_j term only appears in the vertical and lateral thermal conductance related to this location (thus, there are at most 12 entries in each S_j since each conductance appears four times in matrix M). This sensitivity calculation indicates how effective thermal via insertion can be in reducing high temperatures, and we can use it to guide thermal via insertion. Similarly, we can perform sensitivity analysis for hot spot temperature to thermal wire density and identify the most effective location for thermal wire insertion. Due to the advantage of adjoint sensitivity and fast iterative solution of LASPACK, practically the complexity of obtaining sensitivity information is only $O(RG)$, in which R is the number of temperature violation spots, and G is the number of nodes in the 3D routing graph.

4 3D Temperature-Aware Global Routing

In addition to the traditional 2D-like challenges of wire length and congestion, 3D global routing must deal with the complexity introduced by higher dimensional routing and by temperature constraints. Our global routing approach resolves these challenges in two phases. Phase I first performs signal interlayer via assignment, and then utilizes a thermally-driven 2D maze routing algorithm in each layer to generate an initial routing solution. Based on that, phase II resolves temperature and congestion violations iteratively: it identifies hot spots and sensitivity information, and then judiciously inserts thermal vias and thermal wires in sensitive locations; the insertions, together with rip-up-and-reroute are iteratively performed until both congestion and temperature violations are resolved. In the above rip-up-and-reroute, we process one net at a time and maintain a fixed order of all nets. We have experimentally found that, under different randomly chosen net orderings, the results change very little as long as we maintain the same fixed net order through all of the iterations. This is due to the fact that early iterations are seen to create good estimates of resource utilization, and this reduces the order dependence. This is consistent with the observation in [20]. Since thermal effects pose a significant problem in 3D ICs and the thermal via and thermal wire insertion for temperature reduction is a core part of our algorithm, we will first describe routing phase II in our paper; the initial routing procedures of phase I and the overall flow are discussed following that.

4.1 Thermal Via and Thermal Wire Insertion for Temperature Reduction

In an initial global routing solution generated from phase I, the high power density and low thermal conductivity of silicon dioxide can result in hot spots in 3D ICs, and we can employ thermal vias and thermal wires to build effective heat conduction paths. Assuming a predetermined distribution of wires and interlayer vias for power lines, we can analyze the heat conduction for a 3D IC. The role of the thermal via and thermal wire insertion algorithm is to reduce the peak temperature towards T_{spec}, and at the same time, generate a routable design. This task must take into account the routing blockages and resource utilization (via capacity and wire capacity) as these are added. Our algorithm iteratively inserts thermal vias and thermal wires with a linear programming (LP) approach, and performs a rip-up-and-reroute step to obtain a solution free from temperature and congestion violations.

4.1.1 Thermal Via and Thermal Wire Insertion Algorithm

Figure 5 shows the flow of the algorithm. The LP-based thermal via and thermal wire insertion and rip-up-and-reroute are iteratively performed until there is no violation or no further improvement. The latter is easily identified if it is detected that the congestion map changes insignificantly, or the peak temperature reduces trivially. In each iteration, we relax the specified temperature T_{spec} to a target temperature $T_{target} = T_{max}^\mu \cdot T_{spec}^{1-\mu}$ in identifying temperature violation, where T_{max} is the current highest temperature, and μ is a constant with value $0 < \mu < 1$ (we take $\mu = 0.2$ in practice); also if this calculated value of T_{target} is smaller than $T_{spec} + \Delta T_{min}$, we assign $T_{target} = T_{spec}$, where ΔT_{min} is a positive constant. By introducing ΔT_{min} we can avoid infinite number of temperature decreases before T_{target} reaches T_{spec}; in practice, we set $\Delta T_{min} = 5°C$. As the final step we greedily insert thermal vias and thermal wires using all of the remaining space.

```
Algorithm: Thermal_via/wire_insertion_for_temperature_reduction
Input: Initial 3D global routing result and specified peak temperature requirement T_spec.
Output: A routing solution with thermal via and thermal wire insertion that is free from
congestion and temperature violation.
begin
  DO
  {
      Update temperature profile, get current highest temperature T_max;
      Update target temperature T_target according to T_max as described in Section 4.1.1;
      Perform sensitivity analysis for temperature violating spots with T > T_target;
      Construct and solve LP formulation: equations (7), (8), (9), (10) and (11);
      Insert thermal vias and thermal wires and update congestion information;
      Rip-up-and-reroute for resolving overflow;
  } UNTIL (free from temperature and congestion violation or no further improvement)

  Post insertion of thermal vias and thermal wires using all available space;
end
```

Figure 5: Thermal via and thermal wire insertion algorithm for temperature and congestion reduction.

4.1.2 Linear Programming Based Thermal Via and Thermal Wire Insertion

In the thermal via and thermal wire insertion algorithm of Figure 5, we first perform thermal analysis to update the temperature profile and identify hot spots that violate the temperature specifications. The thermal via and thermal wire planning procedure is then formulated as a LP problem to reduce the temperature of hot spots and minimize total thermal via and thermal wire usage, while leaving ample space for lateral routing.

For each of n temperature-violating hot spots with temperature $T_i > T_{target}$, $i = 1, 2, ..., n$, we perform sensitivity analysis for T_i with respect to the number of thermal vias at each thermal via location as in Section 3, and this can be performed fast due to the advantage of adjoint sensitivity analysis. However, to reduce the complexity of the LP problem, we only record those non-trivial sensitivity values $s_{v,ij} \geq s_{th}$, in which s_{th} is a constant threshold of sensitivity and in practice we take $s_{th} = 0.01°C$ per via; the associated location j is then defined as a candidate thermal via location, and $j = 1, 2..., p$, where p is the total number of candidate thermal via locations. Similarly, we can define candidate thermal wire location and obtain the sensitivities $s_{w,ik}$ for T_i with respect to the number of thermal wires at each candidate location k, and $k = 1, 2, ..., q$, in which q is the total number of such locations. Based on the sensitivity analysis results, we can economically plan the number of inserted thermal vias, $N_{v,j}$, at each candidate thermal via location j, and the number of inserted thermal wires, $N_{w,k}$, at each candidate thermal wire location k, so that the temperature at each hot spot i can be reduced by ΔT_i. The LP problem is formulated as follows:

$$\text{minimize} \sum_{j=1}^{p} N_{v,j} + \sum_{k=1}^{q} N_{w,k} + \Gamma \sum_{i=1}^{n} \delta_i \quad (7)$$

$$\text{subject to:} \sum_{j=1}^{p} -s_{v,ij} N_{v,j} + \sum_{k=1}^{q} -s_{w,ik} N_{w,k} + \delta_i \geq \Delta T_i,$$

$$i = 1, 2, ..., n, \ \Delta T_i = T_i - T_{target} \quad (8)$$

$$N_{v,j} \leq \min((1+\beta)R_{v,j}, \ U_j - V_j), \ j = 1, 2, ..., p \quad (9)$$

$$N_{w,k} \leq (1+\beta)R_{w,k}, \ k = 1, 2, ..., q \quad (10)$$

$$\delta_i \geq 0, \ i = 1, 2, ..., n; \ N_{v,j} \geq 0, \ j = 1, 2, ..., p;$$
$$N_{w,k} \geq 0, \ k = 1, 2, ..., q \quad (11)$$

The objective function that minimizes the total usage of thermal vias and wires is consistent with the goal of routing congestion reduction. To guarantee the problem is feasible, we introduce relaxation variables δ_i, $i = 1, 2...n$. Γ is a constant which remains the same over all iterations. It is chosen to be a value that is large enough to suppress the value of δ_i to be 0 when the thermal via and thermal wire resources in constraint (9) and (10) are enough to reduce temperature as desired.

Constraint (8) requires the temperature reduction at hot spot i, plus a relaxation variable δ_i, to be at least ΔT_i during the current iteration, where ΔT_i is the difference between current temperature T_i and target temperature T_{target}, and δ_i is introduced to ensure that the problem is feasible. Constraints (9) and (10) are related to capacity constraints on the thermal vias and thermal wires, respectively, based on lateral bound-

ary capacity overflows in the same layer, and interlayer via capacity overflows across layers. Constraint (9) sets the upper limit for the number of thermal via insertions $N_{v,j}$ with two limiting factors. $R_{v,j}$ is the maximum number of additional thermal vias that can be inserted at location j without incurring lateral routing overflow on neighboring edge, and it is calculated as $R_{v,j} = v_j - v_{cur.,j}$, in which $v_{cur.,j}$ is the current interlayer via usage at location j, and v_j is the maximum number of interlayer vias that can be inserted at location j without incurring lateral overflow which can be calculated from equation (1). Adding more interlayer vias in the most sensitive locations can be very influential in temperature reduction, and therefore, we intentionally amplify the constraint by a factor β to temporarily permit a violation of this constraint, but which will allow better temperature reduction. This can potentially result in lateral routing overflow after the thermal via assignment, but this overflow can be resolved in the iterative rip-up-and-rerouting phase; in practice, we find $\beta = 0.1 \sim 0.2$ works well. A second limiting factor for $N_{v,j}$ is that the total interlayer via usage can not exceed U_j, which is the interlayer via capacity at position j, and we take the minimum of the two limiting factors in the constraint formulation. Similarly, constraint (10) sets a limit on the number of thermal wire insertions with the consideration of lateral routing overflow. $R_{w,k}$ is the maximum number of additional thermal wires that can be inserted at location k without incurring lateral routing overflow, and it is calculated as $R_{w,k} = m_k - m_{cur.,k}$, where $m_{cur.,k}$ is the current thermal wire usage at location k, and m_k is the maximum number of thermal wires at location k without incurring lateral overflow, which can be calculated from equation (2). In the same spirit of encouraging temperature reduction, we relax $R_{w,k}$ by factor of β, and any potential overflow will be resolved in the rip-up-and-rerouting phase.

The LP problem above can be efficiently solved with the *lp_solve* package [25]. The solution to this LP formulation are rounded to the ceiling integer to guarantee the optimality, and the corresponding number of thermal vias and thermal wires are then inserted into the 3D circuit; the resulting overflow can be resolved in the following rip-up-and-reroute. This LP based insertion and rip-up-and-reroute are iteratively performed until a feasible solution is reached.

4.2 Temperature-Aware 3D Global Routing Flow

We now describe the steps involved in our 3D routing algorithm. Briefly, these include a Minimum Spanning Tree (MST) generation and routing congestion estimation step, a signal interlayer via assignment step, and a thermally-driven 2D maze routing in each active layer; these steps constitute phase I of our routing algorithm. The phase II step is iterative routing involving rip-up-and-reroute and LP-based thermal via/wire insertion which has been described above in Section 4.1.

4.2.1 3D MST and Routing Congestion Estimation

A 3D net $N = \{n_0, n_1, n_2, ..., n_k\}$ can have more than two pins distributed on different layers (if not, as described earlier, we refer to it as a 2D net). We first build a minimum spanning tree, using Prim's algorithm [19], for each multiple-pin net, so as to decompose it into a set of two-pin nets. The cost function for two pins of a net in Prim's algorithm is essentially the Manhattan distance but with large weighting factor assigned before the vertical separation distance, and this large weight factor will discourage the generation of interlayer two-pin net. Therefore, the usage of this cost function can reduce total wire length and minimize the use of signal interlayer vias. With a set of two-pin nets from MST decomposition, we can statistically estimate the routing congestion over each lateral routing edge using the L-Z shape statistical routing model proposed in [21]. To extend the model in [21] for 3D routing, we assume that a two-pin net with pins on different layers has an equal probability of utilizing any interlayer via position within the bounding box defined by the pins. The estimated congestion map provides important information for the later signal interlayer via assignment stage.

4.2.2 Signal Interlayer Via Assignment

Based on the two-pin interlayer decompositions generated from 3D MST step, our approach uses a hierarchical min-cost network flow heuristic to simultaneously assign interlayer vias for all interlayer nets and minimize a cost function. This heuristic builds a group hierarchy by recursively dividing all device layers into two neighboring groups of equal size. Signal interlayer via assignment is then performed at the boundaries of group pairs at each level in a top-down way following the hierarchy: the assignment is conducted for the topmost level group boundary first, and then at the boundaries of group pairs of lower levels in the hierarchy. Figure 6(a) shows an example of signal interlayer via assignment

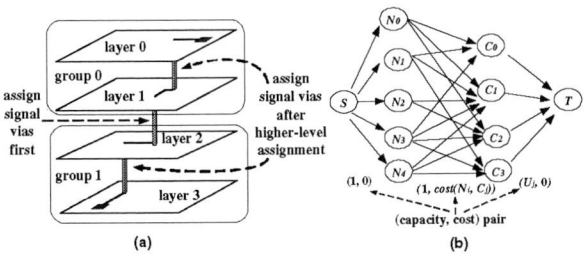
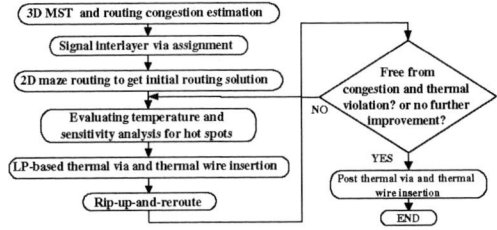

Figure 6: (a) Example of hierarchical signal via assignment for a four-layer circuit. (b) Example of min-cost network flow heuristics to solve signal via assignment problem at each level of hierarchy.

for a decomposed 2-pin signal net in a four-layer circuit with two levels of hierarchy. The signal interlayer via assignment is first performed at the boundary of group 0 and group 1 at topmost level, and then it is processed for layer boundary within each group.

At each level of the hierarchy, the problem of signal interlayer via assignment is formulated as a min-cost network flow. Figure 6(b) shows the network flow graph for assigning signal interlayer vias of five interlayer nets to four possible interlayer via positions. Each interlayer net is represented by a node N_i in the network flow graph; each possible interlayer via position is indicated by a node C_j. If C_j is within the bounding box of the two-pin interlayer net N_i, we build a directed edge from N_i to C_j, and set the capacity to be 1, the cost of the edge to be $cost(N_i, C_j)$. The $cost(N_i, C_j)$ is evaluated as the shortest path cost for assigning interlayer via position C_j to net N_i when both pins of N_i are on the two neighboring layers; otherwise it is evaluated as the average shortest path cost over all possible unassigned signal interlayer via positions in lower levels of the hierarchy. The shortest path cost is obtained with Dijkstra's algorithm [19] in the 2D congestion map generated from the previous estimation step, and the cost function for crossing a lateral routing edge is a combination of edge length and a overflow cost function which is similar to that in [22]. The incorporation of overflow into cost function can guide signal interlayer via assignment to find the best location for minimizing routing congestion in each layer. The network flow graph in Figure 6(b) also includes a source node S and a target node T. There is an edge from S to every N_i with capacity 1 and cost 0; an edge is built from every C_j to target T with capacity U_j (the interlayer via capacity at C_j) and cost 0. There is a supply of N for source node S and a demand of N for target node T, where N is the number of interlayer net nodes in the flow graph. For the network flow graph, the min-cost network flow problem is optimally solved in polynomial time with the cs2 package [23], and we can thus obtain the signal interlayer via assignment for all interlayer nets.

4.2.3 Thermally-driven 2D Maze Routing

After signal interlayer via assignment, we can decompose all interlayer nets into a set of 2D nets by adding a pseudo-pin at each signal interlayer via landing position. Moreover, with a known signal interlayer via distribution, we can perform temperature analysis as discussed in Section 3 and obtain an initial temperature map to guide 2D routing. 2D routing is performed on all nets using the maze routing algorithm from [24], The cost function is similar to that in signal interlayer via assignment, but with an additional temperature term in the cost, so that the resulting route will have an incentive to choose a low temperature path. This strategy not only decreases the temperature influence on the delay of signal nets [2], but also reduces congestion in hot regions, which is beneficial, since heuristically more thermal vias will later be inserted in these regions to reduce temperature. The overflow will be resolved by iterative rip-up and reroute. After 2D routing, we can further connect the pins at different layers for those interlayer nets by employing the previously assigned signal interlayer vias, thus build an initial 3D routing solution.

4.2.4 Overall Flow of the 3D Global Routing Algorithm

We summarize the individual pieces of the algorithm and outline the overall flow in Figure 7. The input is a tessellated 3D circuit with given power distribution. An initial 3D routing solution is found by generating a 3D MST, and then performing a network flow based interlayer via assignment, followed by thermally-driven 2D maze routing. Based on the thermal profile and sensitivity information, we then perform a LP-based thermal via and thermal wire insertion procedure. This insertion is per-

Figure 7: Overall flow for the temperature-aware 3D global routing algorithm.

formed iteratively; each time after the insertion, a rip-up-and-reroute step is employed to resolve the lateral routing congestion and overflow. This continues until there is no temperature and congestion violation or no further improvement is possible. The majority of the algorithm run time is on the iterative thermal via and thermal wire insertion and rip-up-and-reroute, and each such iteration has a complexity of $O(NGlogG + G^3)$, in which N is the number of nets in circuit, and G is the number of grid cells in the routing graph. Practically, it is seen that the optimized solution can be reached in a small number of iterations.

5 Experimental Results

The temperature-aware 3D router was implemented in C++ and tested on an Intel Pentium 4 2.8GHz Linux machine, on benchmarks from the MCNC [26] and IBM placement benchmark [27] suites. Table 1 lists parameters of the benchmark circuits and routing grid size for each layer. The benchmarks are placed with the placer from [5]. Four layers are used, the chip size is fixed at 5mm × 5mm; the oxide layer separation is 7μm, and the silicon substrate is 500μm thick. The power of each circuit is randomly generated with a power density between 10W/cm^2 to 800W/cm^2. In practice, the power density at each cell position can be estimated with two components: dynamic power and static power. In evaluating dynamic power, the effects of interconnect wire load capacitance on switching power should be considered, and the interconnect length can be estimated with half-perimeter wire length from placement results. The white spaces of grid cells is calculated and the average is about 20% of the total chip area. The interlayer via has a size of 5μm × 5μm. The thermal conductivity of silicon, oxide and metal are 119W/(m°C), 1W/(m°C) and 396W/(m°C), respectively. The bottom of the chip was made isothermic with the ambient temperature to represent the heat sink, and the top and sides of the chip are assumed to be adiabatic to the ambient. The temperature of the heat sink is set to a reference value of 0°C for convenience; if the temperature is nonzero, it is well known that the calculated values can be obtained by a simple translation.

Circuit	# cells	# nets	Grid	Circuit	# cells	# nets	Grid
biomed	6417	5743	28 ×28	ibm02	19321	18429	34×34
industry2	12149	12696	31×31	ibm03	22207	21905	37×37
industry3	15059	21939	35×35	ibm04	26633	26451	36×36
ibm01	12282	11754	31×31	ibm06	32185	33521	40×40

Table 1: Benchmark circuit parameters.

The experimental results are listed in Table 2. We compare our algorithm of temperature-aware 3D global routing (denoted as TA) with three other 3D global routing schemes. The first comparison scheme performs initial 3D global routing in the same way as our algorithm, but after that, both thermal vias and thermal wires are post-inserted into the 3D circuit in a greedy way, under the constraint of existing routing resource usage and white space allowance, and we denote it as P in Table 2. The second comparison scheme follows the same routing procedure as our approach, but only thermal vias are inserted and optimized to improve heat conduction; thermal wires are not used, and we denote this scheme as V. The third comparison scheme uses the same number of thermal vias and thermal wires as our routing algorithm, but they are distributed uniformly across the chip. The level of thermal via distribution is equal to the average number of thermal vias per grid cell position from our algorithm, and the thermal wire level is assigned in the similar uniform way. We denote this comparison scheme as U in the table. For all four approaches, we set up our specified peak temperature T_{spec} to be 80°C to guide the temperature reduction process.

Circuit	T_{init} (°C)	Peak temperature (°C)				Ave. top 1% temperatures (°C)				Wire length ($\times 10^5$)				run time (seconds)			
		TA	P	V	U	TA	P	V	U	TA	P	V	U	TA	P	V	U
biomed	237.1	81.9	105.6	115.3	87.5	73.7	97.8	80.0	76.6	1.82	1.77	1.78	1.76	255	137	188	131
industry2	207.5	82.4	106.6	116.3	98.0	64.0	75.2	77.9	68.3	6.04	5.92	6.01	5.90	855	473	591	519
industry3	202.0	79.2	99.1	112.5	89.8	68.0	78.0	74.9	71.9	9.85	9.75	9.71	9.93	1807	1405	1686	1193
ibm01	264.8	79.1	109.9	99.3	108.4	61.7	86.9	68.7	74.4	2.63	2.46	2.63	2.56	238	87	224	87
ibm02	257.5	80.5	105.6	111.6	97.5	57.3	65.2	63.3	60.8	7.73	7.57	7.68	7.60	769	400	717	469
ibm03	218.5	82.9	106.9	123.4	85.4	68.1	78.1	77.8	67.3	10.27	10.30	10.26	10.00	1645	1490	1344	864
ibm04	218.1	80.0	96.0	107.8	84.6	68.6	73.3	78.1	68.7	8.07	8.15	8.21	8.25	886	597	974	581
ibm06	236.4	81.2	99.2	131.4	89.2	63.0	69.3	66.9	63.8	18.08	18.19	18.12	18.08	2956	1585	2274	1581

Table 2: Comparison of temperature and routing performance results among four approaches: 3D global routing using our thermal-aware (TA) method; using post (P) insertion of thermal vias and thermal wires; using thermal via (V) insertion only; and using uniform (U) thermal via and thermal wire insertion.

Routing results and temperature performance for each approach are shown in Table 2, for a set of circuits listed in the first column. The second column T_{init} lists the peak temperature after simply performing initial 3D routing but without any thermal via and thermal wire insertion. As we can see, the peak temperature is well above our expected value without insertion of thermal vias and thermal wires, which shows the necessity of these insertions in 3D ICs. The following four columns list the peak temperature for all routing algorithms. Experimental results show that our temperature-aware (TA) algorithm can successfully bring the peak temperatures of circuits down to specified temperature T_{spec} and performs much better than other routing schemes. The post-insertion of thermal vias and thermal wires in approach P does not optimize resource usage during routing phase and results in inferior performance; the peak temperature from P can be as high as 30.8°C more than that from TA algorithm (ibm01). Approach V does not employ thermal wires as an effective lateral heat conduction resource, thus for some spots where thermal via insertion is restricted from routing congestion, hot spots are generated due to the lack of lateral thermal conduction path; experimental results show a maximum peak temperature difference of 50.2°C (ibm06) between V and TA. The uniform insertion approach U applies thermal vias and wires uniformly instead of distributing them according to sensitivity analysis, therefore its performance is inferior to our TA routing algorithm. Moreover, it creates great routing overflow as discussed later. The next four columns listed the average temperature of top 1% hot spots for all four approaches. The results show the same tendency as peak temperature results, our temperature-aware algorithm can reduce the temperature of hot spots better than other approaches from an iterative routing and temperature reduction approach.

Circuit	# routing overflows				Circuit	# routing overflows			
	TA	P	V	U		TA	P	V	U
biomed	11	0	8	6750	ibm02	34	0	15	4262
industry2	1	0	4	6201	ibm03	11	0	1	16873
industry3	14	0	1	10021	ibm04	2	0	1	5596
ibm01	34	0	15	4262	ibm06	7	0	5	5404

Table 3: Comparison of routing overflow of four approaches.

Table 2 also lists the wirelength and run time results. All four approaches report similar wirelength results. Averagely, our TA algorithm generates 0.6% longer wirelength than approach P, because during iterations TA algorithm will detour to leave space for thermal via and thermal wire insertion for temperature reduction and thus increase wirelength; however, this increase is trivial. TA and V algorithms need longer run time, which is due to the iterations to resolve temperature violation and routing congestion. Table 3 reports the lateral routing overflow of four algorithms. The TA algorithm can successfully resolve routing congestion and temperature violations by employing LP-based thermal via and thermal wire insertion as well as iterative approach, and reports trivial routing overflow. However, the uniform insertion algorithm U distributes thermal vias and thermal wires uniformly across the chip without addressing the routing capacity reduction from them, thus approach U generates huge routing overflow, which makes the routing infeasible. Overall, the experimental results show that our temperature-aware algorithm can effectively reduce hot spot temperature in 3D ICs to specified value, and maintain a good wirelength and overflow property through an iterative LP-based thermal via and thermal wire insertion as well rip-up-and-rerouting.

References

[1] K. Banerjee, S. J. Souri, P. Kapur and K. C. Saraswat, "3-D ICs: A Novel Chip Design for Improving Deep Submicrometer Interconnect Performance and System-on-Chip Integration," *Proc. of IEEE*, 89(5), pp. 602-633, May 2001.

[2] K. Banerjee, M. Pedram and A. Ajami, "Analysis and Optimization of Thermal Issues in High-Performance VLSI," *Proc. ISPD*, 2001, pp. 230-237.

[3] C. H. Tsai and S. M. Kang, "Cell-Level Placement for Improving Substrate Thermal Distribution," *IEEE Trans. CAD*, 19(2), pp. 253-266, Feb. 2000.

[4] P. Li, T. Pileggi, M. Asheghi and R. Chandra, "Efficient Full-Chip Thermal Modeling and Analysis," *Proc. ICCAD*, 2004, pp. 319-326.

[5] B. Goplen and S. S. Sapatnekar, "Efficient Thermal Placement of Standard Cells in 3D ICs Using a Force Directed Approach," *Proc. ICCAD*, 2003, pp. 86-89.

[6] Y. K. Cheng and S. M. Kang, "An Efficient Method for Hot-Spot Identification in ULSI Circuits," *Proc. ICCAD*, 1999, pp. 124-127.

[7] Y. Zhan and S. S. Sapatnekar, "Fast Computation of the Temperature Distribution in VLSI Chips Using the Discrete Cosine Transform and Table Look-up," *Proc. ASP-DAC*, 2005, pp. 87-92.

[8] R. J. Enbody and K. H. Tan, "Routing the 3-D Chip," *Proc. DAC*, 1992, pp. 132-137.

[9] Y. Deng and W. Maly, "Physical Design of the "2.5D" stacked System," *Proc. ICCD*, 2003, pp. 211-217.

[10] C. Ababei, H. Morgal and K. Bazargan, "Three-dimensional Place and Route for FPGAs," *Proc. ASP-DAC*, 2005, pp. 773-778.

[11] S. Das, A. Chandrakasan and R. Reif, "Design Tools for 3-D Integrated Circuits," *Proc. ASP-DAC*, 2003, pp. 53-56.

[12] J. Cong and Y. Zhang, "Thermal-Driven Multilevel Routing for 3-D ICs," *Proc. ASP-DAC*, 2005, pp. 121-126.

[13] T-Y Chiang, K. Banerjee and K. C. Saraswat, "Effect of Via Separation and Low-k Dielectric Materials on the Thermal Characteristics of Cu Interconnects," *Tech. Dig. IEDM*, 2000, pp. 261-264.

[14] T-Y Chiang, K. Banerjee and K. C. Saraswat, "Compact Modeling and SPICE-Based Simulation for Electrothermal Analysis of Multilevel ULSI Interconnects," *Proc. ICCAD*, 2001, pp. 165-172.

[15] B. Goplen and S. S. Sapatnekar, "Thermal Via Placement in 3D ICs," *Proc. ISPD*, 2005, pp. 167-174.

[16] S. Lee, T. F. Lemczyk and M. M. Yovanovich, "Analysis of Thermal Vias in High Density Interconnect Technology," *Proc. IEEE Semi-Therm Symp.*, 1992, pp. 55-61.

[17] http://www.tu-dresden.de/mwism/skalicky/laspack/laspack.html

[18] L. Pillage, R. Rohrer and C. Visweswarish, "Electronic Circuit and System Simulation Methods," McGraw-Hill, New York, NY, 1995.

[19] T. H. Corman, C. E. Leiseron, R. L. Rivest and C. Stein, "Introduction to Algorithms," 2nd edition, MIT Press, Cambridge, MA, 2001.

[20] R. Nair, "A Simple Yet Effective Technique for Global Wiring," *IEEE Trans. CAD*, 6(2), pp. 165-172, March 1987.

[21] J. Westra, C. Bartels and P. Groeneveld, "Probabilistic Congestion Prediction," *Proc. ISPD*, 2004, pp. 204-209.

[22] R. T. Hadsell and P. H. Madden, "Improved Global Routing through Congestion Estimation," *Proc. DAC*, 2003, pp. 28-34.

[23] A. V. Goldberg, "An Efficient Implementation of a Scaling Minimum-Cost low Algorithm",*Journal of Algorithms*, 22, pp. 1-29, 1997.

[24] C.J. Alpert, T.C. Hu, J.H. Huang, A.B. Kahng and D. Karger, "Prim-Dijkstra Tradeoffs for Improved Performance-Driven Routing Tree Design," *IEEE Trans. CAD*, 14(7), pp. 890-896, July 1995.

[25] http://groups.yahoo.com/group/lp_solve/

[26] http://www.cbl.ncsu.edu/pub/Benchmark_dirs/LayoutSynth92

[27] http://er.cs.ucla.edu/benchmarks/ibm-place/

Closed Form Solution for Optimal Buffer Sizing Using The Weierstrass Elliptic Function

Sebastian Vogel

Institute of Microelectronic Systems
Darmstadt University of Technology
D-64283 Darmstadt
e-mail: sebastian.vogel@stud.tu-darmstadt.de

Martin D. F. Wong

Department of Electrical and Computer Engineering
University of Illinois at Urbana-Champaign
Urbana, IL 61801
e-mail: mdfwong@uiuc.edu

Abstract— **This paper presents a fundamental result on buffer sizing. Given an interconnection wire with n buffers evenly spaced along the wire, we would like to size all buffers such that the Elmore delay is minimized. It is well known that the problem can be solved by an iterative algorithm which sizes one buffer at a time. However, no closed form solution has ever been reported. In this paper, we derive a closed form buffer sizing function $f(x)$ where $f(x)$ gives the optimal buffer size for the buffer at position x. We show that $f(x)$ can be expressed in terms of the Weierstrass elliptic function $\wp(x)$ and its derivative $\wp'(x)$.**

I. INTRODUCTION

Buffer insertion, buffer sizing and wire sizing have been shown to be effective techniques in reducing interconnect delay [1]. This paper focuses on the buffer sizing problem under the Elmore delay model [2]. Suppose we only have one buffer size and we want to insert a fixed number of buffers into an interconnection wire for delay minimization, it is well known that the buffers will be uniformly spaced in the optimal solution [3, 4]. Provided that only buffers of identical size are allowed, closed-form solutions for the optimal number of buffers [5, 3] and their uniform size [5] have already been reported. Clearly if we allow to size the buffers, the overall delay will be further reduced. Therefore, this paper considers the following buffer sizing problem:

Given an interconnection wire with n buffers evenly spaced along the wire, we would like to size all buffers such that the Elmore delay is minimized. It is well known that the problem can be solved by an iterative algorithm which sizes one buffer at a time [6]. Chu and Wong [7] even considered a more general problem than ours and described an algorithm running in polynomial time. However, no closed form solution has ever been reported.

In this paper, we present a closed form solution to the buffer sizing problem. Without loss of generality, we may assume that the interconnection wire is of unit length represented by the interval $[0, 1]$ with source (driver) at 0 and sink (load) at 1. Let $x_i = \frac{i}{n+1}$ be the position of buffer i, for $1 \leq i \leq n$. We derive a continuous function $f(x)$ such that $f(x_i)$ gives the optimal buffer size for buffer i,

for all i. The buffer sizing function $f(x)$ can be expressed in terms of the Weierstrass elliptic function $\wp(x)$ and its derivative $\wp'(x)$ as follows:

$$f(x) = a + \frac{b\wp'(x) + c\wp(x) + d}{2(\wp(x) - e)^2}, \tag{1}$$

where a, b, c, d, and e are constants.

The reminder of the paper is organized as follows. Section II presents our circuit model and derives a recurrence relation for optimal buffer sizing. Section III shows that the recurrence relation for optimal buffer sizing implies an ordinary differential equation. In Section IV, we give a brief overview of the Weierstrass $\wp(x)$ function and its fundamental difference equation that will be used in Section V to give our closed form expression (1). Section VI briefly discusses an integration constant that arose during our derivations and in Section VII, we show our experimental results. Finally we conclude the paper in Section VIII.

II. THE RECURRENCE RELATION

This section introduces our circuit model of an interconnect with several equally spaced buffers. We derive a recurrence relation for the buffer sizes and show that it is both a necessary and sufficient condition to minimize the Elmore delay on the wire.

Consider an interconnect of length L that has a total resistance of R and capacitance of C. The interconnect has a driver resistance R_D at the source and a load capacitance C_L at the sink. To minimize the propagation delay on the interconnect, we want to insert n buffers at equally spaced locations, thus we split the interconnect into $n + 1$ segments; each segment has a length of $\frac{L}{n+1}$.

We have $R_S = \frac{R}{n+1}$ and $C_S = \frac{C}{n+1}$ being the segment resistance and capacitance, respectively. (Since we do not perform wire sizing, we can include the fringing capacitance into the unit capacitance of our wire model and can thus avoid limitations that occurred, for example, in [8]). Each wire segment is modeled as a π-type RC-circuit.

A buffer of size b is represented by a switch-level RC-circuit as shown in Fig. 1, where R_B/b denotes its output resistance, $C_B b$ denotes its input capacitance and $C_D b$

3C-3

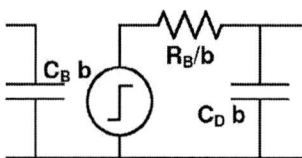

Fig. 1. RC switch-level model of a buffer of size b.

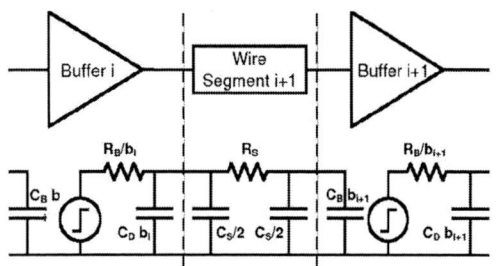

Fig. 2. RC switch level interconnect between buffers i and $i+1$.

denotes its output capacitance. For the ease of presentation, we also model the driver and the load as buffers of fixed sizes $b_0 = \lambda$ and $b_{n+1} = \mu$, respectively, so that $R_D = R_B/b_0$ and $C_L = C_B b_{n+1}$. The Elmore delay between buffer i and buffer $i+1$ is, as shown in Fig. 2,

$$ED_i = \frac{R_B}{b_i}(C_D b_i + C_S + C_B b_{i+1}) + R_S(\frac{C_S}{2} + C_B b_{i+1}) \quad (2)$$

and the total delay from the driver to the load is

$$ED = \sum_{i=0}^{n} ED_i. \quad (3)$$

Clearly, (3) is only a function of $b_1, b_2, ..., b_n$ since all other parameters depend on the physical device characteristics. As it was stated above, $b_0 = \lambda$ and $b_{n+1} = \mu$ define boundary conditions because the driver and the load are fixed. The problem therefore consists in finding positive $b_1^*, b_2^*, ..., b_n^*$ that minimize (3).

A neccessary condition is that the partial derivatives of ED with respect to all b_i are equal to zero, i. e.,

$$\frac{\partial}{\partial b_i} ED = 0 \quad \forall i = 1...n. \quad (4)$$

Since b_i only appears in the expressions of ED_{i-1} and ED_i, we get

$$R_S C_B + R_B C_B \frac{1}{b_{i-1}} - R_B C_S \frac{1}{b_i^2} - R_B C_B \frac{b_{i+1}}{b_i^2} = 0. \quad (5)$$

Solving for b_i yields our basic recurrence relation for optimal buffer sizing,

$$b_i^2 = \frac{b_{i+1} + C_S/C_B}{b_{i-1}^{-1} + R_S/R_B} \quad \forall i = 1, ..., n. \quad (6)$$

To simplify the notation in the following sections, we define

$$\frac{\alpha}{n+1} = \frac{C_S}{C_B} = \frac{C}{(n+1)C_B}, \frac{\beta}{n+1} = \frac{R_S}{R_B} = \frac{R}{(n+1)R_B}. \quad (7)$$

Then (6) becomes

$$b_i^2 = \frac{b_{i+1} + \alpha/(n+1)}{b_{i-1}^{-1} + \beta/(n+1)}. \quad (8)$$

Note that (8) gives a recurrence relation for the buffer sizes with b_i expressed as a function of b_{i-1} and b_{i+1} with boundary values $b_0 = \lambda$ and $b_{n+1} = \mu$. Furthermore, it is both a necessary and sufficient condition to minimize the Elmore delay expression (3). To see this, we note that (3) is of the form

$$ED(b_1, ..., b_n) = \sum_{i=1}^{T} a_i \prod_{j=1}^{n} b_j^{c_{ij}} \quad (9)$$

where the a_i are non-negative, b_j are positive and c_{ij} are real numbers. Equation (9) is a posynomial with T terms and n variables. Under a change of variables $b_i = e^{d_i}$, $ED(d_1, ..., d_n)$ is a convex function of $(d_1, ..., d_n)$. Therefore, the local optimum determined by (8) is simultaneously a global optimum.

III. ODE Formulation

In the following sections, we derive our closed form expression $f(x)$. First, we show that the recurrence relation for optimal buffer sizing (8) implies a second order ordinary differential equation. We will use this ODE then in Section V to find the buffer sizing function $f(x)$.

In a first step, we replace b_i in (8) by $f(x_i)$ and have

$$f(x)^2 = \frac{f(x_{i+1}) + \alpha/(n+1)}{f(x_{i-1})^{-1} + \beta/(n+1)} \quad (10)$$

Next we set $x_i = x$ and $\Delta x = \frac{1}{n+1}$. We have $x_{i+1} = x + \Delta x$, $x_{i-1} = x - \Delta x$, and $\frac{1}{n+1} \cdot \frac{1}{n} \approx \Delta x^2$. Equation (10) becomes

$$f(x)^2 = \frac{f(x + \Delta x) + \Delta x^2 n \alpha}{f(x - \Delta x)^{-1} + \Delta x^2 n \beta}. \quad (11)$$

To show that the RHS of (11) effectively defines a second order ordinary differential equation, it is useful to substitute $f(x) = e^{g(x)}$ and to perform a Taylor expansion of the resulting exponential terms on the RHS, so we have

$$e^{g(x+\Delta x)} =$$
$$e^{g(x)} + \Delta x g' e^g + \frac{1}{2} \Delta x^2 e^g (g'' + g'^2) + O(\Delta x^3) \quad (12)$$

for the numerator and a similar expression for the denominator. Collecting terms, (11) becomes the quotient of two

316

polynomials in Δx,

$$f(x)^2 = e^{2g(x)} =$$
$$\frac{e^g(1 + \Delta x g' + \Delta x^2(\frac{1}{2}g'^2 + \frac{1}{2}g'' + \alpha n e^{-g})) + O(\Delta x^3)}{e^{-g}(1 + \Delta x g' + \Delta x^2(\frac{1}{2}g'^2 - \frac{1}{2}g'' + \beta n e^g)) + O(\Delta x^3)}$$
$$= \frac{P(\Delta x)}{Q(\Delta x)} = A(\Delta x). \quad (13)$$

It is now our goal to find $A(\Delta x) = a_0 + a_1\Delta x + a_2\Delta x^2 + O(\Delta x^3)$ so that $f(x)^2 = A(\Delta x)$. The a_i are determined by

$$(a_0 + a_1\Delta x + a_2\Delta x^2 + ...)(q_0 + q_1\Delta x + q_2\Delta x^2 + ...)$$
$$= p_0 + p_1\Delta x + p_2\Delta x^2 + O(\Delta x^3). \quad (14)$$

and we simply have to compare coefficients to get

$$a_0 = \frac{p_0}{q_0} = e^{2g} \quad (15)$$

$$a_1 = \frac{1}{q_0}(p_1 - a_0 q_1) = e^{2g}g' - e^{2g}g' = 0 \quad (16)$$

$$a_2 = \frac{1}{q_0}(p_2 - a_0 q_2 - a_1 q_1)$$
$$= g'' e^{2g} + \alpha n e^g - \beta n e^{3g}$$
$$= e^{2g}(g'' + \alpha n e^{-g} - \beta n e^g). \quad (17)$$

Hence, (13) becomes

$$A(\Delta x) = e^{2g}(1 + \Delta x^2(g'' + \alpha n e^{-g} - \beta n e^g)) + O(\Delta x^3)$$
$$= e^{2g(x)} = f(x)^2 \quad (18)$$

For the last step, we drop the terms of third and higher order and require that $g(x)$ satisfies the differential equation $g'' = \beta n e^g - \alpha n e^{-g}$.

An intermediate result is that the buffer sizing function $f(x)$ must satisfy

$$f(x) = e^{g(x)}, \quad x = \frac{i}{n+1} \quad \forall i = 0, ..., n+1 \quad (19)$$

where $g(x)$ is a solution to the second-order ordinary differential equation

$$g'' = \beta n e^g - \alpha n e^{-g} \quad (20)$$

with the boundary conditions

$$g(0) = \ln(\lambda) \text{ and } g(1) = \ln(\mu). \quad (21)$$

IV. WEIERSTRASS ELLIPTIC FUNCTION

Our buffer sizing function $f(x)$ will be obtained by solving an elliptic integral and has the form of a rational function in terms of the Weierstrass $\wp$-function and its derivative $\wp'$. Before we solve the ODE from the last section, we briefly state important properties of the Weierstrass elliptic function. Due to space constraints, we refer to the excellent presentation in [9] for further details.

Apart from being a prototype for all elliptic functions, the Weierstrass $\wp$-function has the fundamental property that it satisfies the differential equation

$$(\wp'(x))^2 = 4\wp^3(x) - g_2\wp(x) - g_3. \quad (22)$$

$\wp(x)$ and its derivative $\wp'(x)$ have two parameters, g_2 and g_3, which are called *invariants*. We will use the shorter notation $\wp(x)$ and $\wp'(x)$ instead of $\wp(x; g_2, g_3)$ and $\wp'(x; g_2, g_3)$ in this paper.

The elliptic integral

$$x = \int_y^\infty (4t^3 - g_2 t - g_3)^{-\frac{1}{2}} dt \quad (23)$$

implicitly defines y as a function of x and cannot be solved using elementary functions. However, differentiating leads to

$$(\frac{dy}{dx})^2 = 4y^3 - g_2 y - g_3 \quad (24)$$

which is of the same form as (22). The solution is then given by $y = \wp(x)$ with g_2, g_3 as parameters [9]. This relationship can be used to solve elliptic integrals of the form

$$x = \int_{y_0}^y \frac{dt}{\sqrt{h(t)}}, \quad (25)$$

where $h(t)$ is a cubic or quartic polynomial. Provided that y_0 is a root of $h(t)$, the solution to (25) is given by [9]

$$y = y_0 + \frac{\frac{1}{4}h'(y_0)}{\wp(x) - \frac{1}{24}h''(y_0)}, \quad (26)$$

where $\wp(x)$ is dependent on the invariants g_2 and g_3 of the polynomial $h(t)$. However, for our purposes, the more general formula given by Weierstrass [9] allows the lower bound y_0 of the integral (25) to be any constant a, not necessarily a root of $h(t)$. The solution is then slightly more complex, but it is still only a rational function involving $\wp(x)$, $\wp'(x)$ and some constants,

$$y = a + \frac{\sqrt{h(a)}\wp'(x)}{2(\wp(x) - \frac{1}{24}h''(a))^2 - \frac{1}{48}h(a)h^{iv}(a)}$$
$$+ \frac{\frac{1}{2}h'(a)(\wp(x) - \frac{1}{24}h''(a)) + \frac{1}{24}h(a)h'''(a)}{2(\wp(x) - \frac{1}{24}h''(a))^2 - \frac{1}{48}h(a)h^{iv}(a)}. \quad (27)$$

Given the invariants g_2 and g_3 of $h(t)$ as parameters, $\wp$ and $\wp'$ can be calculated using standard mathematical software such as Mathematica. The derivations of (26) and (27) are rather lengthy and therefore omitted from this paper.

In the rest of this paper, we derive an elliptic integral similar to (25) from the set of recurrence relations (8). Its explicit solution $f(x)$ is our optimal buffer sizing expression and has the same form as (27).

V. CLOSED FORM SOLUTION

In this section, we give the closed form expression $f(x)$ for optimal buffer sizing which has the form of a rational function in terms of the Weierstrass functions $\wp(x)$ and $\wp'(x)$. We have found so far that $f(x) = e^{g(x)}$, where $g(x)$ satisfies the ODE (20). We will now derive an ODE in f which finally can be solved using the Weierstrass $\wp$-function.

As a first step, we multiply both sides of (20) by $2g'$ and apply the chain rule of differentiation backwards. Hence,

$$2g'g'' = 2g'(n\beta e^g - n\alpha e^{-g}) \tag{28}$$

$$\frac{d}{dx}(g'^2) = 2n\frac{d}{dx}(\beta e^g - \alpha e^{-g}) \tag{29}$$

$$g'^2 = 2n(\beta e^g - \alpha e^{-g}) + c, \tag{30}$$

where c is a constant of integration that has to be determined numerically. We briefly deal with this problem in Section VI.

We now undo the substitution made in (13) so that $g(x) = \ln f(x)$ and therefore $g' = f'(x) \cdot \frac{1}{f(x)}$. The desired buffer sizing function $f(x)$ is now given as the solution to the differential equation

$$f'^2 = \left(\frac{df}{dx}\right)^2 = 2n\beta f^3 + cf^2 + 2n\alpha f. \tag{31}$$

Moving all terms with f to one side and integrating gives an elliptic integral

$$\int dx = x = \int_{f(0)}^{f(x)} \frac{dt}{\pm\sqrt{2n\beta t^3 + ct^2 + 2n\alpha t}} \tag{32}$$

which essentially has the same form as (25). It only remains to find the invariants g_2 and g_3 in order to apply (27) to get a solution for $f(x)$. This can be done by a simple transform of variables

$$t' = \frac{2}{n\beta} \cdot t - \frac{c}{6n\beta} \tag{33}$$

in (32). This changes the expression under the radical sign $2\beta n t^3 + ct^2 + 2\alpha n t$ into

$$4t'^3 - \left(\frac{c^2}{12} - \alpha\beta n^2\right)t' - \left(\frac{1}{12}\alpha\beta n^2 c - \frac{c^3}{216}\right)$$

so that the invariants become

$$g_2 = \frac{c^2}{12} - \alpha\beta n^2, \quad g_3 = \frac{1}{12}\alpha\beta n^2 c - \frac{c^3}{216}. \tag{34}$$

This leads to our central result. The buffer sizing function $f(x)$ for optimal buffer sizing at equally spaced points $x = x_i = \frac{i}{n+1}$, $i = 0...n+1$, is given by

$$f(x) = \lambda + \frac{\sqrt{h(\lambda)}\wp'(x)}{2(\wp(x) - \frac{1}{24}h''(\lambda))^2}$$

$$+ \frac{\frac{1}{2}h'(\lambda)(\wp(x) - \frac{1}{24}h''(\lambda)) + \frac{1}{24}h(\lambda)h'''(\lambda)}{2(\wp(x) - \frac{1}{24}h''(\lambda))^2}, \quad (35)$$

where $h(t) = 2\beta n t^3 + ct^2 + 2\alpha n t$ and the Weierstrass function has the invariants (34) as parameters. $f(x)$ fulfills the boundary values $f(0) = \lambda$ and $f(1) = \mu$.

VI. INTEGRATION CONSTANT

In the last section, we did not give a description of the integration constant c. For completeness, this section shows that one can distinguish four non-trivial cases for the behavior of $f(x)$. Each case has its different expression that determines c.

1. From (20), g'' can be either purely non-negative, then $g(x)$ and consequently $f(x) = e^{g(x)}$ are convex functions and $f(x)$ has the shape of a U, limited by the boundary values. Setting $g'' = 0$, we find that all b_i, including $b_0 = \lambda$ and $b_{n+1} = \lambda$, must be larger than $e^{\ln\sqrt{\frac{\alpha}{\beta}}} = \sqrt{\frac{\alpha}{\beta}}$. The slope f' from (31) must consist of a decreasing part from λ to a minimum buffer and an increasing part from the minimum to μ. It is not hard to see that the value of this minimum is given by the zero of (31) that lies to the right of $\sqrt{\frac{\alpha}{\beta}}$ which is

$$b_{min}(c) = \frac{-c + \sqrt{c^2 - 16\alpha\beta n^2}}{4\beta n} \tag{36}$$

We then have a simple condition that the constant c must fulfill: integrating and summing the two parts of the slope must be equal to one,

$$\begin{aligned} 1 &= \int_{b_{min}(c)}^{\lambda} \frac{dy}{\sqrt{2n\beta y^3 + cy^2 + 2n\alpha y}} \\ &+ \int_{b_{min}(c)}^{\mu} \frac{dy}{\sqrt{2n\beta y^3 + cy^2 + 2n\alpha y}}. \end{aligned} \tag{37}$$

2. Conversely, if g'' is strictly non-positive, $f(x)$ is concave and all b_i are below $\sqrt{\frac{\alpha}{\beta}}$. This case is analog to the first one and leads to a similar condition.

3. Also, f' can have only one monotonic part and is strictly increasing on the whole interval $[\lambda; \mu]$, then $\lambda \leq \sqrt{\frac{\alpha}{\beta}}$ and $\mu \geq \sqrt{\frac{\alpha}{\beta}}$. We integrate the (positive) slope f' from λ to μ and have a simpler condition for c,

$$1 = \int_{\lambda}^{\mu} \frac{dy}{\sqrt{2n\beta y^3 + cy^2 + 2n\alpha y}}. \tag{38}$$

4. The case that f' is strictly decreasing and thus $\lambda \geq \sqrt{\frac{\alpha}{\beta}}$ and $\mu \leq \sqrt{\frac{\alpha}{\beta}}$ can be handled in a similar manner as the previous one.

5. Note that if f' is equal to zero, $f(x) = \sqrt{\frac{\alpha}{\beta}}$ follows as a trivial case.

For the cases 1-4, one can construct simple procedures to find c. For case 3, for example, one can interpret (38) as a function

$$d(c) = \int_{\lambda}^{\mu} \frac{dy}{\sqrt{2n\beta y^3 + cy^2 + 2n\alpha y}} - 1. \tag{39}$$

It is not hard to show that $d(c)$ is monotonic with respect to c and has a unique root which can be determined efficiently by bisection, for example.

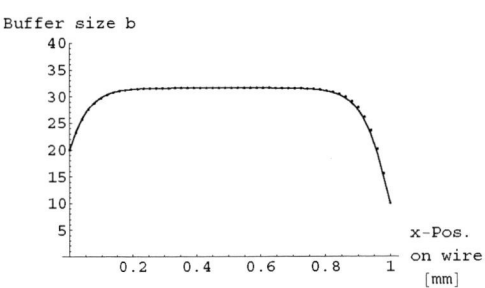

Fig. 3. $n = 50$, $b_0 = 20$, $b_{51} = 10$.

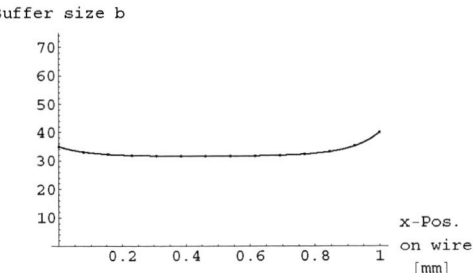

Fig. 4. $n = 12$, $b_0 = 35$, $b_{13} = 40$.

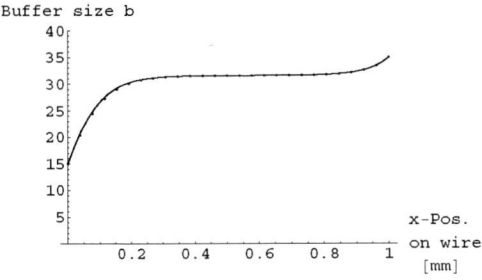

Fig. 5. $n = 25$, $b_0 = 15$, $b_{26} = 35$.

TABLE I

PHYSICAL PARAMETERS FOR UNIT BUFFER AND WIRE

Buffer		Wire	
Output res. [kΩ]	4.5	resistance $[\frac{\Omega}{mm}]$	620
Input cap. [fF]	0.425	capacitance $[\frac{fF}{mm}]$	58.5
intrinsic delay [ps]	45.8	length [mm]	1

VII. EXPERIMENTAL RESULTS

In this section, we show that our solution performs very well in practice. We compare buffer sizes obtained by an iterative method [7, 6] with values calculated by evaluating (35) at equally spaced points on the interval $[0, 1]$.

For the physical parameters, we rely on values given in [1]. More specifically, we consider a copper wire ($\rho = 2.2\mu\Omega \cdot cm$) with a width of 130 nm, a length of 1 mm

and an aspect ratio of 2.1:1. We add unit-length area and fringing capacitance. These values are summarized in table I. Applying (7), we have $\alpha = 137.65$ and $\beta = 0.1378$. The following Fig. 3 - 5 plot iteratively obtained buffer sizes (dots) and those calculated by applying (35) (straight curve) for several combinations of b_0, b_{n+1} and n. We observed that our closed form expression shows a very good agreement with iteratively calculated buffer sizes.

VIII. CONCLUSION

In this paper we addressed the problem of sizing n uniformly spaced buffers on an interconnection wire to minimize Elmore delay.

Previously there was no known closed form solution to this problem. We presented a closed-form buffer sizing function $f(x)$, expressed in terms of the Weierstrass elliptic function $\wp(x)$ and its derivative $\wp'(x)$, such that $f(x_i)$ gives the optimal buffer size for buffer i (at position x_i), $1 \le i \le n$.

We showed that the buffer sizes obtained by our closed form expression matched very well with those obtained by the iterative method [7, 6].

Clearly, our approach has some limitations, some of which we would like to address in the future: First, our solution is restricted to non-branch lines, however most interconnects have a line topology. Second, the derivation in Section III relies on the assumption of equally spaced buffers with continuous size. In practice, obstacles and area constraints require a different setup. Currently, one has to resort to iterative procedures such as in [7].

REFERENCES

[1] J. Cong, "An interconnect-centric design flow for nanometer technologies," *IEEE Proc.*, Vol. 89, No. 4, pp. 505-528, 2001.

[2] W. C. Elmore, "The transient response of damped linear network with particular regard to wideband amplifiers," *Journal of Applied Physics*, Vol. 19, pp. 55-63, 1948.

[3] C. Alpert and A. Devgan, "Wire segmenting for improved buffer insertion," *Proc. ACM/IEEE Design Automation Conference*, pp. 588-593, 1997.

[4] S. Dhar and M. A. Franklin, "Optimum buffer circuits for driving long uniform lines," *IEEE J. Solid-State Circuits*, Vol. 26, No. 1, pp. 32-40, 1991.

[5] H. B. Bakoglu, *Circuits, Interconnections, and Packaging for VLSI*, Addison-Wesley, USA, 1990.

[6] C. N. Chu and D. F. Wong, "Greedy wire-sizing is linear time," *IEEE Transactions on CAD*, Vol. 18, No.4, pp. 398-405, 1999.

[7] C. N. Chu and D. F. Wong, "An efficient and optimal algorithm for simultaneous buffer and wire sizing," *IEEE Transactions on CAD*, Vol. 18, No. 9, pp. 1297-1304, 1999.

[8] C. N. Chu and D. F. Wong, "Closed form solution to simultaneous buffer insertion/sizing and wire sizing," *Proc. Intl. Symp. on Physical Design*, pp. 192-197, 1997.

[9] E. T. Whittaker and G. N. Watson, *A Course of Modern Analysis*, Academic Press, USA, 1962.

An $O(mn)$ Time Algorithm for Optimal Buffer Insertion of Nets with m Sinks

Zhuo (Robert) Li and Weiping Shi
Department of Electrical and Computer Engineering
Texas A&M University, College Station, TX 77843

Abstract—**Buffer insertion is an effective technique to reduce interconnect delay. In this paper, we give a simple $O(mn)$ time algorithm for optimal buffer insertion, where m is the number of sinks and n is the number of buffer positions. This is the first linear time buffer insertion algorithm for nets with constant number of sinks. When m is small, it is a significant improvement over our recent $O(n \log^2 n)$ time algorithm, and the $O(n^2)$ time algorithm of van Ginneken. For b buffer types, the new algorithm runs in $O(b^2 n + bmn)$ time, an improvement of our recent $O(bn^2)$ algorithm. The improvement is made possible by a clever bookkeeping method and an innovative linked list data structure that can perform addition of a wire, and addition of a buffer in amortized $O(1)$ time. On industrial test cases, the new algorithm is faster than previous best algorithms by an order of magnitude.**

I. INTRODUCTION

Delay optimization techniques for interconnect are increasingly important for achieving timing closure of high performance designs. One popular technique for reducing interconnect delay is buffer insertion. A recent study by Saxena *et al* [1] projects that 35% of all cells will be intra-block repeaters for the 45 nm node. Consequently, algorithms that can efficiently insert buffers are essential for the design automation tools.

This paper studies buffer insertion in interconnect with a set of possible buffer positions and a discrete buffer library. In 1990, van Ginneken [2] proposed an $O(n^2)$ time dynamic programming algorithm for buffer insertion with one buffer type, where n is the number of possible buffer positions. His algorithm finds a buffer insertion solution that maximizes the slack at the source. In 1996, Lillis, Cheng and Lin [3] extended van Ginneken's algorithm to allow b buffer types in time $O(b^2 n^2)$. In 2003, Shi and Li [4] used a number of techniques to improve the time complexity to $O(b^2 n \log n)$ for 2-pin nets, and $O(b^2 n \log^2 n)$ for multi-pin nets. To reduce the quadratic effect of b, Li and Shi [5] recently proposed an algorithm with time complexity $O(bn^2)$. However, all these algorithms do not utilize the fact that in real applications most nets have small numbers of pins and large number of buffer positions. As a result, the running time is still long for large nets, especially when other constraints such as slew and cost are considered.

In this paper, we first propose a new algorithm that performs optimal buffer insertion for 2-pin nets in time $O(b^2 n)$. The speedup is achieved by an observation that the best candidate to be associated with any buffer must lie on the convex hull of the (Q, C) plane, a clever bookkeeping method and an innovative linked list that allow $O(1)$ time update for adding a wire or a candidate. The new data structure, which is a linked list, is much simpler than the candidate tree used in [4] and the skip list used in [6]. We then extend the algorithm to m-pin nets in time $O(b^2 n + bmn)$. Experimental results show that our algorithm is faster than previous best algorithms by an order of magnitude. Note that all previous research assumed m and n are of the same order. But in fact, m is often much less than n. For example, in one group of nets reported in [13], which is extracted from industrial ASIC chips with 300k+ gates and consists the 5000 most run time consuming nets for buffer insertion, over 90% of nets has less than 50 sinks. For another group with 1000 nets, all of nets has less than 20 sinks. On the other hand, each net has several hundreds to thousands of buffer positions. Note that even if $m > n$, we can merge sinks in a branch that contains no buffer position, without changing the problem. Therefore in this paper we assume $m \leq n$.

Many extensions have been made based on van Ginneken style algorithms, include wire sizing [3], simultaneous tree construction [7], [8], noise constraints [9] and resource minimization [3], [10]. Our new algorithms are fundamental improvements, and are therefore applicable to some of these extensions.

Finally, we note that for 2-pin nets, when there is no restriction on where a buffer is allowed, Dhar and Franklin [11] gave a closed form solution, assuming buffers can be continuously sized. Chu and Wong [12] proposed a convex quadratic programming method to find the optimal buffer insertion location and buffer sizing with discrete set of buffers. However in real applications, buffer blockage is always a serious restriction. Such information should be considered as early as possible to reduce the design cycle. Therefore these algorithms are often used in the very early stage of design planning when buffer blockage information is not available, not in actual physical synthesis. Also for multi-pin nets, no simple closed form solution is available.

The paper is organized as follows. Section II formulates the problem. Section III describes the new algorithm for 2-pin nets. Section IV extends the algorithm to multi-pin nets. Simulation results are given in Section V, and conclusions are drawn in Section VI.

II. PRELIMINARY

A net is a tree $T = (V, E)$, where $V = \{v_0\} \cup V_s \cup V_n$, and $E \subset V \times V$. Vertex v_0 is the *source* vertex and also

---

This research was supported by the NSF grants CCR-0098329, CCR-0113668, EIA-0223785, ATP grant 512-0266-2001.

0-7803-9451-8/06/$20.00 ©2006 IEEE.

the root of T, V_s is the set of *sink* vertices, and V_n is the set of *internal* vertices. Each sink vertex $s \in V_s$ is associated with sink capacitance $C(s)$ and required arrival time $RAT(s)$. A buffer library $\boldsymbol{B}$ contains b types of buffers. For each buffer type $B \in \boldsymbol{B}$, the intrinsic delay is $K(B)$, driving resistance is $R(B)$, and input capacitance is $C(B)$. A function $f : V_n \rightarrow 2^{\boldsymbol{B}}$ specifies the types of buffers allowed at each buffer position. Each edge $e \in E$ is associated with lumped resistance $R(e)$ and capacitance $C(e)$.

Following previous researchers [2], [3], [4], [7], we use Elmore delay for the interconnect and the linear delay for buffers due to their high fidelity in the synthesis stage. For each edge $e = (v_i, v_j)$, signals travel from v_i to v_j. The Elmore delay of e is

$$D(e) = R(e) \left(\frac{C(e)}{2} + C(v_j) \right),$$

where $C(v_j)$ is the downstream capacitance at v_j. If v_i is inserted a buffer of type B_k, then the buffer delay is

$$D(v_i) = R(B_k) \cdot C(v_i) + K(B_k),$$

where $C(v_i)$ is the downstream capacitance at v_i, and the capacitance viewed from the upper stream is $C(B_k)$.

For any vertex v, let $T(v)$ be the subtree downstream from v, and with v being the root. Once we decide where to insert buffers in $T(v)$, we have a *candidate* α for $T(v)$. The delay from v to sink $s_i \in T(v)$ under α is

$$D(v, s_i, \alpha) = \sum_{e=(v_j, v_k)} (D(v_j) + D(e)),$$

where the sum is over all edges e in the path from v to s_i. If a buffer is inserted at v_j in α, then $D(v_j)$ is the buffer delay. Otherwise, $D(v_j) = 0$. The slack of v under α is

$$Q(v, \alpha) = \min_{s_i \in T(v)} \{RAT(s_i) - D(v, s_i, \alpha)\}.$$

Max-Slack Buffer Insertion Problem: Given a net $\boldsymbol{T} = (V, E)$, capacitance and RAT for all sinks, capacitance and resistance for all edges, possible buffer position function f, and buffer library $\boldsymbol{B}$, find a candidate α for $\boldsymbol{T}$ that maximizes $Q(v_0, \alpha)$.

The effect of a candidate α for tree $T(v)$ at v to the upstream is traditionally described by slack $Q(v, \alpha)$ and downstream capacitance $C(v, \alpha)$ [2]. For any two candidates α_1 and α_2 of $T(v)$, we say α_1 *dominates* α_2, if $Q(v, \alpha_1) \geq Q(v, \alpha_2)$ and $C(v, \alpha_1) \leq C(v, \alpha_2)$. The set of *nonredundant candidates* of $T(v)$, which we denote as $N(v)$, is the set of candidates such that no candidate in $N(v)$ dominates any other candidate in $N(v)$, and every candidate of $T(v)$ is dominated by some candidates in $N(v)$. Once we have $N(v_0)$, the candidate that gives the maximum $Q(v_0, \alpha)$ can be found easily.

III. TWO-PIN NETS

In this section, we show how to compute optimal buffer insertion for 2-pin nets in $O(b^2 n)$ time. We use van Ginneken style dynamic programming paradigm, enhanced with two techniques 1) convex pruning to find the best candidate and delete redundancy, and 2) a simple implicit data structure to store and update (Q, C) values. Our data structure is inspired by the candidate tree of Shi and Li [4], but much simpler.

A. Convex Pruning

The concept of convex pruning was first proposed by Li and Shi [5]:

Definition 1: Let α_1, α_2 and α_3 be three nonredundant candidates of $T(v)$ such that $C(\alpha_1) < C(\alpha_2) < C(\alpha_3)$ and $Q(\alpha_1) < Q(\alpha_2) < Q(\alpha_3)$. If

$$\frac{Q(\alpha_2) - Q(\alpha_1)}{C(\alpha_2) - C(\alpha_1)} < \frac{Q(\alpha_3) - Q(\alpha_2)}{C(\alpha_3) - C(\alpha_2)}, \tag{1}$$

then we call α_2 *non-convex*, and prune it.

Convex pruning can be explained by Figure 1. Consider Q as the Y-axis and C as the X-axis. Then candidates are points in the two-dimensional plane. It is easy to see that the set of nonredundant candidates $N(v)$ is a monotonically increasing sequence. Candidate $\alpha_2 = (Q_2, C_2)$ in the above definition is shown in Figure 1(a), and is pruned in Figure 1(b). The set of nonredundant candidates after convex pruning $M(v)$ is a convex hull.

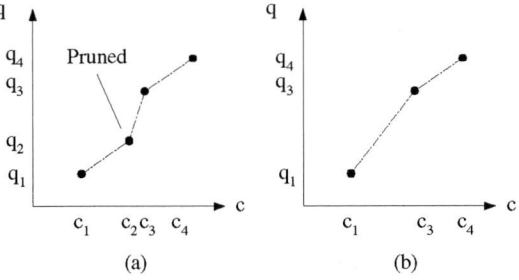

Fig. 1.   (a) Nonredundant candidates $N(v)$. (b) Nonredundant candidates $M(v)$ after convex pruning.

Lemma 1: For 2-pin nets, convex pruning preserves optimality.

Proof: Let α_1, α_2 and α_3 be candidates of $T(v)$ that satisfy the condition in Definition 1. In a 2-pin net, every candidate will be connected to some wires, which could be empty, before reaches an upstream buffer or the source driver. Let v' be the upstream buffer or driver, D be the total sum of the delay of wires from v' to v and the delay of the buffer or driver at v' driving wires from v' to v, and R be the sum of the resistance of wires from v' to v and the resistance of the buffer or driver at v'. Then

$$Q(v', \alpha_i) = Q(v, \alpha_i) - R \cdot C(v, \alpha_i) - D,$$

where $i = 1, 2$ or 3. Therefore when

$$R < \frac{Q(v, \alpha_3) - Q(v, \alpha_2)}{C(v, \alpha_3) - C(v, \alpha_2)},$$

we have

$$R(C(v, \alpha_3) - C(v, \alpha_2)) < Q(v, \alpha_3) - Q(v, \alpha_2),$$

and

$$Q(v, \alpha_2) - R \cdot C(v, \alpha_2)) < Q(v, \alpha_3) - R \cdot C(v, \alpha_3).$$

Therefore

$$Q(v', \alpha_2) \quad < \quad Q(v', \alpha_3).$$

On the other hand when

$$R \geq \frac{Q(v, \alpha_3) - Q(v, \alpha_2)}{C(v, \alpha_3) - C(v, \alpha_2)},$$

condition (1) implies

$$R > \frac{Q(v, \alpha_2) - Q(v, \alpha_1)}{C(v, \alpha_2) - C(v, \alpha_1)}.$$

Therefore

$$R(C(v, \alpha_2) - C(v, \alpha_1)) > Q(v, \alpha_2) - Q(v, \alpha_1),$$

which implies

$$Q(v', \alpha_1) > Q(v', \alpha_2).$$

This shows α_2 gives a slack that is worse than either α_1 or α_3 when the source or an upstream buffer is reached. When a buffer is attached, the input capacitance of that buffer will reset $C(\alpha_i)$. Therefore α_2 is redundant. ■

We note that this lemma only applies to 2-pin nets. For multi-pin nets when the upstream could be a merging vertex, nonredundant candidates that are pruned by convex pruning could still be useful.

Convex pruning of a list of non-redundant candidates sorted in increasing (Q, C) order can be performed in linear time [5]. Furthermore, when a new candidate is inserted to the list, we only need to check its neighbors to decide if any candidate should be pruned under convex pruning. The time is $O(1)$, amortized over all candidates.

B. Best Candidates

Assume v is a buffer position, and we have computed the set of nonredundant candidates $N'(v)$ for $T(v)$, where $N'(v)$ does not include candidates with buffers inserted at v. Now we want to add buffers at v and compute $N(v)$. Define $P_i(v, \alpha)$ as the slack at v if we add a buffer of type B_i for any candidate α:

$$P_i(v, \alpha) = Q(v, \alpha) - R(B_i) \cdot C(v, \alpha) - K(B_i). \quad (2)$$

If we do not insert any buffer, then every candidate in $N'(v)$ is a candidate in $N(v)$. If we insert a buffer, then for every buffer type B_i, $i = 1, 2, \ldots, b$, there will be a new candidate β_i:

$$\begin{aligned} Q(v, \beta_i) &= \max_{\alpha \in N'(v)} \{P_i(v, \alpha)\}, \\ C(v, \beta_i) &= C(B_i). \end{aligned}$$

Define the *best candidate* for B_i as the candidate $\alpha \in N'(v)$ such that α maximizes $P_i(v, \alpha)$ among all candidates in $N'(v)$. If there are multiple α's that maximize $P_i(v, \alpha)$, choose the one with minimum C. From Lemma 1, it is easy to see that all best candidates are on the convex hull.

The following lemma says that if we sort candidates in increasing Q and C order from left to right, then as we add wires to the candidates, we always move to the left to find the best candidates.

Lemma 2: For any $T(v)$, let nonredundant candidates after convex pruning be $\alpha_1, \alpha_2, \ldots, \alpha_k$, in increasing Q and C order. Now add wire e to each candidate α_j and denote it

as $\alpha_j + e$. For any buffer type B_i, if α_j gives the maximum $P_i(\alpha_j)$ and α_k gives the maximum $P_i(\alpha_k + e)$, then $k \leq j$.

Proof: From the definition,

$$\begin{aligned} P_i(\alpha_j + e) &= Q(v, \alpha_j + e) - R(B_i)C(v, \alpha_j) \\ &\quad - R(B_i)C(e) - K(B_i) \\ &= P_i(\alpha_j) - R(e)C(\alpha_j) \\ &\quad - R(e)C(e)/2 - R(B_i)C(e). \end{aligned}$$

Since $P_i(\alpha_j + e) \leq P_i(\alpha_k + e)$, we have

$$P_i(\alpha_j) - R(e)C(\alpha_j) \leq P_i(\alpha_k) - R(e)C(\alpha_k),$$

which is equivalent to

$$P_i(\alpha_j) - P_i(\alpha_k) \leq R(e)(C(\alpha_j) - C(\alpha_k)).$$

On the other hand, $P_i(\alpha_j) \geq P_i(\alpha_k)$ and $R(e) > 0$, therefore

$$C(\alpha_j) - C(\alpha_k) \geq 0.$$

This implies $k \leq j$. ■

The following lemma says the best candidate can be found by local search, if all candidates are convex.

Lemma 3: For any $T(v)$, let nonredundant candidates after convex pruning be $\alpha_1, \alpha_2, \ldots, \alpha_k$, in increasing Q and C order. If $P_i(\alpha_{j-1}) \leq P_i(\alpha_j)$, $P_i(\alpha_j) \geq P_i(\alpha_{j+1})$, then α_j is the best candidate for buffer type B_i and

$$\begin{aligned} P_i(\alpha_1) &\leq \cdots \leq P_i(\alpha_{j-1}) \leq P_i(\alpha_j), \\ P_i(\alpha_j) &\geq P_i(\alpha_{j+1}) \geq \cdots \geq P_i(\alpha_k). \end{aligned}$$

Proof: From $P_i(\alpha_{j-1}) \leq P_i(\alpha_j)$, we have

$$Q(\alpha_{j-1}) - R(B_i)C(\alpha_{j-1}) \leq Q(\alpha_j) - R(B_i)C(\alpha_j).$$

Therefore,

$$R(B_i) \leq \frac{Q(\alpha_j) - Q(\alpha_{j-1})}{C(\alpha_j) - C(\alpha_{j-1})}.$$

Since all candidates are convex, (1) is false. Hence

$$R(B_i) \leq \frac{Q(\alpha_{j-1}) - Q(\alpha_{j-2})}{C(\alpha_{j-1}) - C(\alpha_{j-2})},$$

which implies $P_i(\alpha_{j-2}) \leq P_i(\alpha_{j-1})$. Then, we can easily get

$$P_i(\alpha_1) \leq \cdots \leq P_i(\alpha_{j-1}) \leq P_i(\alpha_j).$$

The other direction is similar. From $P_i(\alpha_j) \geq P_i(\alpha_{j+1})$, we have

$$Q(\alpha_j) - R(B_i)C(\alpha_j) \geq Q(\alpha_{j+1}) - R(B_i)C(\alpha_{j+1}).$$

Therefore,

$$R(B_i) \geq \frac{Q(\alpha_{j+1}) - Q(\alpha_j)}{C(\alpha_{j+1}) - C(\alpha_j)}.$$

Since all candidates are convex, (1) is false. Hence

$$R(B_i) \geq \frac{Q(\alpha_{j+2}) - Q(\alpha_{j+1})}{C(\alpha_{j+2}) - C(\alpha_{j+1})},$$

which implies $P_i(\alpha_{j+1}) \geq P_i(\alpha_{j+2})$. We can also easily get

$$P_i(\alpha_j) \geq P_i(\alpha_{j+1}) \geq \cdots \geq P_i(\alpha_k).$$

Since $P_i(\alpha_j)$ is the maximum $P_i(\alpha)$ among all candidates, α_j is the best candidates for buffer type B_i. ■

C. Data Structure

We store all nonredundant candidates of $T(v)$ in a linked list $L(v)$ of the following data structure:

```
typedef struct Candidate {
    double q, c;
    Candidate *next, *prev;
} Candidate;
```

We also have three global variables:

```
double Qa, Ca, Ra;
```

$L(v)$ is organized in increasing C and Q order, and pruned by convex pruning. The value of Q and C of each candidate α, pointed by a, are given by fields a->q and a->c, as well as global variables Qa, Ca and Ra:

$$
\begin{aligned}
Q(\alpha) &= (\text{a->q}) - \text{Qa} - \text{Ra} \cdot (\text{a->c}), \\
C(\alpha) &= (\text{a->c}) + \text{Ca}.
\end{aligned} \tag{3}
$$

To facilitate the search for best candidates and the insertion of new candidates, we have two arrays of pointers:

```
Candidate *best[b], *new[b];
```

where best[i] points to the most recent best candidate for B_i, and new[i] points to the most recent new candidate for B_i.

D. Algorithm

When we reach an edge e with resistance e->R and capacitance e->C, we update Qa, Ca and Qa to reflect the new values of Q and C of all candidate in L in $O(1)$ time, without actually touching any candidate:

```
void AddWire (e)
{
    Qa = Qa + e->R*e->C/2 + e->R*Ca;
    Ca = Ca + e->C;
    Ra = Ra + e->R;
}
```

This is similar to Shi and Li's algorithm [4], but much simpler.

When we reach a buffer position, we may generate a new candidate for each buffer type B_i. But first, we have to find the best candidate for B_i. This is done by pointer best[i]:

```
void AddBuffer (i)
{
    Candidate *a;
    while (P(i, best[i]->prev) >
      P(i, best[i]))
        best[i] = best[i]->prev;
    ...
```

Function P(i, ...) computes P_i of a candidate defined in (2). From Lemma 2, the best candidate is always to the left of where we found the best candidate last time. From Lemma 3, we can find the best candidate by local search. Therefore the while loop can find the best candidate that gives the maximum P_i. Now form the new candidate:

```
    ...
    a = new Candidate;
    a->c = B[i]->C - Ca;
```

```
    a->q = P(i, best[i]) + Qa + Ra*a->c;
    ...
```

With Eqn. (3), it is easy to verify that the above transformation of q and c fields will make the new candidate consistent with every other candidate in L. Now insert the new candidate into L:

```
    while (a->c < new[i]->c)
        new[i] = new[i]->prev;
    a->next = new[i]->next;
    new[i]->next->prev = a;
    a->prev = new[i];
    new[i]->next = a;
    ...
```

The location to insert new candidates also moves to the left in L, because the capacitances of all candidates increase when wires are added. Finally, we perform convex pruning around the new candidate:

```
    if (! Convex(a->prev, a, a->next)) {
        a->prev->next = a->next;
        a->next->prev = a->prev;
        Delete(a);
        return;
    }
    while (! Convex(a, a->next,
      a->next->next)) {
        a->next = a->next->next;
        a->next->next->prev = a;
        Delete(a->next);
    }
    while (! Convex(a->prev->prev,
      a->prev, a)) {
        a->prev = a->prev->prev;
        a->prev->prev->next = a;
        Delete(a->prev);
    }
}
```

Function Convex(...) checks if the middle candidate is convex. Function Delete(...) deletes a candidate, and moves best and new pointers to the right by one if the pointer points to the candidate to be deleted. Now we describe the entire algorithm:

Algorithm 2-Pin

Input Routing tree $T(v_1)$ consists of path $v_1, \ldots, v_{n+1}$, where v_{n+1} is the sink.

Output Nonredundant candidates of $T(v_1)$ stored in linked list L.

Begin

1: Let Qa=0, Ca=0, Ra=0;
2: Let L contain one candidate (Q, C), where $Q = RAT(v_{n+1})$ and $C = C(v_{n+1})$;
3: Let all best and new pointers point to the only candidate in L;
4: **For** $i = n$ **to** 1 **do**
5:   AddWire(e), where $e = (v_i, v_{i+1})$;
6:   **For** each buffer type B_j allowed at v_i **do**
7:     AddBuffer(j);
8: Return L;

End.

Theorem 1: Algorithm 2-Pin finds the optimal buffer insertion of any 2-pin nets in worst-case time $O(b^2 n)$.

Proof: The only difference between our algorithm and previous algorithms, other than speedup, is convex pruning. Lemma 1 guarantees convex pruning does not lose the optimality. Therefore our algorithm is correct.

Now consider the time complexity. The outer loop between lines 4 and 7 is executed n times. The inner loop between lines 6 and 7 is executed b times. This requires $O(bn)$ time. In addition, the number of times that any pointers best[i] and new[i] move equals the total number of candidates, which is bn. Since there are b best pointers and b new pointers, the total time to move these pointers is $O(b^2 n)$. The total deletion time is the same as the number of candidates, which is $O(bn)$. Therefore, the overall time complexity of our algorithm is $O(b^2 n)$. ∎

Some properties can be used to speed up the implementation, but it does not change the asymptotic time complexity. If buffers are sorted in decreasing driving resistance $R(B_1) \geq R(B_2) \geq \cdots \geq R(B_b)$, and let α_i be the best candidate for B_i. Then it is easy to see that $C(\alpha_1) \geq C(\alpha_2) \geq \cdots \geq C(\alpha_b)$. This helps to reduce the search time for best pointers. A similar order can be explored to reduce the search time for new pointers.

IV. MULTI-PIN NETS

We now extend the 2-pin algorithm to multi-pin nets. In a multi-pin net, a candidate for a 2-pin segment may be merged with a candidate of a different branch, before associated with a buffer. In this case, optimal solution could come from a non-convex candidate. Therefore we need all nonredundant candidates of every 2-pin segment, not only the convex ones.

This is done by a subroutine 2PinSubroutine(...) for 2-pin segments. The subroutine is similar to Algorithm 2-Pin, but in addition to list $L(v)$, maintains a second list $A(v)$. $A(v)$ contains ALL nonredundant candidates of $T(v)$, including non-convex ones. So $A(v)$ is a superset of $L(v)$. Best candidates are still found through L, yet new candidates are inserted to both L and A. Note that to facilitate the insertion of new candidates into A, another array of pointers newA[b] is used and the operation is similar to new[b]. For any 2-pin segment $u_1, u_2, \ldots, u_k$, the subroutine takes as input $A(u_k)$, prunes non-convex ones to get $L(u_k)$, and computes each $L(u_i)$ and $A(u_i)$ as it moves to u_1.

Algorithm M-Pin

Input Routing tree $T(v)$ with root v.

Output List $A(v)$ that contains all nonredundant candidates of $T(v)$.

Begin

1: **If** $T(v)$ consists of path v to v_1 where v_1 is a branch vertex **then**
2:     Recursively compute $A(v_1)$ for $T(v_1)$;
3:     $A(v) = \text{2PinSubroutine}(A(v_1))$;
4: **Else** $T(v)$ consists of subtrees $T(v_1)$ and $T(v_2)$
5:     Recursively compute $A(v_1)$ and $A(v_2)$;
6:     Merge $A(v_1)$ and $A(v_2)$ to form $A(v)$;
7: Return $A(v)$;

End.

Theorem 2: Algorithm M-Pin computes the optimal buffer insertion of an m-pin net in time $O(b^2 n + bmn)$.

Proof: We compute the same set of all nonredundant candidates as previous algorithms. Therefore the algorithm is correct.

For all 2-pin segments, the total time is bounded by $O(b^2 n)$. At each branch vertex, the time is $O(bn)$. Therefore the total time is $O(b^2 n + bmn)$. ∎

Our new algorithm can be easily integrated with predictive pruning [10], [4], and inverting buffer types [3].

V. SIMULATION

All algorithms are implemented in C and run on a Sun SPARC workstations with 400 MHz clock and 2 GB memory. The device and interconnect parameters are based on TSMC 180 nm technology and are same as those used in [5] and [4]. We have 4 different buffer libraries, of size 1, 4, 8, and 16 respectively. The value of $R(B_i)$ is from 180 Ω to 7000 Ω, $C(B_i)$ is from 0.7 fF to 23 fF, and $K(B_i)$ is from 29 ps to 36.4 ps. The sink capacitances range from 2 fF to 41 fF. The wire resistance is 0.076 $\Omega/\mu m$ and the wire capacitance is 0.118 fF/μm.

Table I shows for a 2mm long two-pin net with different possible buffer insertion locations, the new algorithm is up to 20 times faster than previous best algorithms. Table II shows for large industrial multi-pin nets where m is as high as 337, the new algorithm is still faster than previous best algorithms. All algorithms generate same slacks.

VI. CONCLUSION

We presented a new $O(mn)$ algorithm for optimal buffer insertion on nets with m sinks. When m is small, the new algorithm is a significant improvement over the recent $O(n \log^2 n)$ time algorithm [4], and the $O(n^2)$ time algorithm of van Ginneken. Also, the new algorithm is much simpler than the $O(n \log^2 n)$ algorithm. Simulation results show the new algorithm is faster than these algorithms by an order of magnitude. In addition, for large buffer libraries, the new algorithm is faster than recent $O(bn^2)$ algorithm [5]. In the journal version of this paper, we will apply our algorithm to resource minimization to show significant speedup. Since the new algorithm could run for large number of buffer positions and large buffer libraries in just few seconds, synthesis tool can use very refined buffer positions to select the best quality solutions with small amount of run time overhead.

REFERENCES

[1] P. Saxena, N. Menezes, P. Cocchini, and D. A. Kirkpatrick, "Repeater scaling and its impact on CAD," *IEEE Trans. Computer-Aided Design*, vol. 23, no. 4, pp. 451–463, 2004.

[2] L. P. P. P. van Ginneken, "Buffer placement in distributed RC-tree network for minimal Elmore delay," in *Proc. IEEE Int. Symp. Circuits Syst.* 1990, pp. 865–868.

[3] J. Lillis, C. K. Cheng, and T.-T. Y. Lin, "Optimal wire sizing and buffer insertion for low power and a generalized delay model," *IEEE J. Solid-State Circuits*, vol. 31, no. 3, pp. 437–447, 1996.

2006 Asia and South Pacific Design Automation Conference

3C-4

TABLE I

SIMULATION RESULTS FOR A 2 MM TWO-PIN NET.

Buffer pos. n	Library size b	CPU Time (sec)			
		Lillis-Cheng-Lin [3] $O(b^2 n^2)$	Shi-Li [4] $O(b^2 n \log n)$	Li-Shi [5] $O(bn^2)$	New $O(b^2 n)$
404	1	0.02	0.01	0.03	0.001
	4	0.04	0.11	0.04	0.01
	8	0.08	0.41	0.04	0.02
	16	0.14	1.64	0.06	0.04
2044	1	0.51	0.10	0.80	0.01
	4	1.08	0.70	0.84	0.04
	8	1.78	2.50	0.92	0.10
	16	3.28	9.09	1.01	0.21
10404	1	13.70	0.56	21.85	0.05
	4	28.11	4.33	23.01	0.23
	8	46.71	16.18	23.26	0.49
	16	83.97	59.64	23.75	1.10

TABLE II

SIMULATION RESULTS FOR INDUSTRIAL MULTI-PIN TEST CASES.

Sinks m	Buffer pos. n	Library size b	CPU Time (sec)			
			Lillis-Cheng-Lin [3] $O(b^2 n^2)$	Shi-Li [4] $O(b^2 n \log^2 n)$	Li-Shi [5] $O(bn^2)$	New $O(b^2 n + bmn)$
25	107	1	0.01	0.002	0.002	0.002
		4	0.01	0.03	0.01	0.01
		8	0.02	0.16	0.01	0.01
		16	0.05	0.67	0.02	0.03
	1337	1	0.24	0.04	0.14	0.02
		4	1.06	0.48	0.44	0.11
		8	1.95	2.06	0.60	0.20
		16	3.32	8.62	0.78	0.33
	2567	1	0.75	0.08	0.50	0.05
		4	4.08	1.04	1.47	0.19
		8	7.07	4.30	2.07	0.36
		16	12.12	17.94	2.58	0.64
337	337	1	0.02	0.02	0.02	0.03
		4	0.05	0.04	0.04	0.06
		8	0.09	0.75	0.08	0.12
		16	0.19	3.23	0.14	0.20
	5647	1	0.89	0.17	0.41	0.22
		4	2.51	2.03	0.98	0.59
		8	4.46	8.34	1.51	0.98
		16	7.34	31.55	2.03	1.73
	10957	1	3.40	0.34	1.24	0.42
		4	9.29	4.10	2.95	1.16
		8	16.03	16.88	4.44	1.93
		16	26.96	64.59	5.85	3.26

[4] W. Shi and Z. Li, "A fast algorithm for opitmal buffer insertion," *IEEE Trans. Computer-Aided Design*, vol. 24, no. 6, pp. 879–891, 2005.

[5] Z. Li and W. Shi, "An $O(bn^2)$ time algorithm for buffer insertion with b buffer types," in *Proc. Design, Automation and Test in Europe* 2005, pp. 1324–1329.

[6] R. Chen and H. Zhou, "A flexible data structure for efficient buffer insertion, " in *Proc. IEEE Int. Conf. Computer Design* 2004, pp. 216–221.

[7] T. Okamoto and J. Cong, "Buffered steiner tree construction with wire sizing for interconnect layout optimization," in *Proc. IEEE/ACM Int. Conf. Computer-Aided Design* 1996, pp. 44–49.

[8] M. Hrkic and J. Lillis, "S-tree: a technique for buffered routing tree synthesis," in *Proc. ACM/IEEE Design Automation Conf.* 2002, pp. 578–583.

[9] C. J. Alpert, A. Devgan, and S. T. Quay, "Buffer insertion for noise and delay optimization," in *Proc. ACM/IEEE Design Automation Conf.* 1998, pp. 362–367.

[10] W. Shi, Z. Li, and C. J. Alpert, "Complexity analysis and speedup techniques for optimal buffer insertion with minimum cost," in *Proc. Asia South Pacific Design Automation Conf.* 2004, pp. 609–614.

[11] S. Dhar and M. A. Franklin, "Optimum buffer circuits for driving long uniform lines," *IEEE J. Solid-State Circuits*, vol. 26, no. 1, pp. 32-40, 1991.

[12] C. C. N. Chu and D. F. Wong, "A quadratic programming approach to simultaneous buffer insertion/sizing and wire sizing," *IEEE Trans. Computer-Aided Design*, vol. 18, no. 6, pp. 787-798, 1999.

[13] Z. Li, C. N. Sze, C. J. Alpert, J. Hu and W. Shi, "Making fast buffer insertion even faster via approximation techniques," in *Proc. Asia South Pacific Design Automation Conf.*, 2005, pp. 13–18.

Spec-based flip-flop and latch repeater planning

Man Chung Hon
Intel Corportation
Santa Clara, CA 95054, USA
manch.c.hon@intel.com

Abstract— Shrinking process geometries and frequency scaling give rise to an increasing number of interconnects that require multiple clock cycles. This paper explores efficient techniques to insert flip-flops and latches to meet pre-determined latency and margin constraints at the receivers. Previous approaches push timing margins to either ends of interconnect. We present an $O(n \log n)$-time algorithm to insert flip-flops that evens out timing margins across the entire interconnect, resulting in more robust designs and faster design convergence. An $O(n \log n)$-time extension to handle symmetric, two-phases latches is also presented. Experimental results verify the correctness and practicality of our approach.

I. INTRODUCTION

Deep submicron process has created a number of new design challenges. Primary among them is the increasing dominance of interconnects. Current scaling trends show that the frequency of high-performance ICs approximately doubles and the die size grows by 25% with every process generation. Interconnect optimization techniques such as repeater insertion and gate sizing have proven effective in reducing interconnect delay. However, with such short clock cycles, the delay on global signals may be longer than one clock cycle even *after* being optimized with these techniques. Insertion of clocked repeaters such as flip-flops and latches become necessary on these signals.

In a typical design flow, microarchitects determine the target latencies for each driver-receiver pair in the design during the early phase of design cycle. Traditionally, flip-flops and latches are coded into RTL manually. It is difficult for the circuit designers to try out different configurations and placements. A recent study on the effect of process scaling on repeaters [9] shows that, not only is a clock cycle's worth of metal length shrinking faster than the scaling of geometries, it is also decreasing at a much faster rate than the repeater-to-repeater distance. It is predicted that at the 45-nm process node, as many as one in every four repeaters is clocked [9].

There has been growing interests among researchers on the problem of multi-cycle global interconnects. Lu, Zhong, Koh and Chao [8] proposed an analytical formulation for the simultaneous insertion of flip-flops and

buffers to minimize latencies. Hassoun, Alpert and Thiagarajan [3] combined routing, buffer and flip-flop insertion in multiple clock domain systems. Both [8] and [3] only work with two-pin nets, and do not respect latency constraints specified *a priori*. Cocchini [2] extended van Ginneken's classical dynamic programming structure [12] to simultaneously insert flip-flops and buffers, and can be directed to either minimize latencies or to meet specified latency constraints. This algorithm was in turn extended by Seth, Zhao and Hu [10] to incorporate single phase level-sensitive latches. As pointed out by Akkiraju and Mohan [1], Cocchini's approach puts all margins, both positive and negative, to the first segment right after the driver. (On a 2-pin net bounded by two flip-flops, margin is defined to be the difference between a clock cycle and the arrival time at the receiving flip-flop. On a multifanout net, we take the minimum from all the receivers.) Instead, [1] proposed another extension of van Ginneken to insert flip-flops in such a way that margins are evenly distributed in the driver and the receiver ends, while holding the middle segments timing-tight. Researchers have also investigated techniques other than clocked repeaters. Zhang, Hu and Chen [14] suggested a new global interconnect architecture using the wave-pipelining technique. Lin and Zhou [7] applied retiming to improve the clock speed on an initial flip-flop placement.

In this paper, we present an algorithm to insert flip-flops to even out margins throughout the entire interconnect. This way of margin distribution is often what a circuit designer wants intuitively. In the case of *positive* margin, putting all at the extremities of the interconnect, as is done in Cocchini [2] and Akkiraju and Mohan [1], makes the middle segments timing-critical. A large number of tight segments could pose challenges to downstream design stages. This is especially problematic when designers seek to improve the chip's speed in minor revisions of a taped-out design. The middle segments, which barely pass the original frequency goal, may now come in with negative margins. In the case of *negative* margin, putting all at the ends of the interconnect makes the timing problem more difficult to fix. As an example, imagine a 2-pin net with 9 flip-flops in between, and a negative margin of -100 pico-seconds. Fixing 10 nets each with a -10 ps margin is arguably easier than fixing one net with a -100 ps margin (Cochinni), or two nets with -50 ps margin

0-7803-9451-8/06/$20.00 ©2006 IEEE.

(Akkiraju and Mohan)

The remainder of the paper is organized as follows. Section II sets up the flip-flop and latch insertion problems in terms of equivalent retiming problems. Section III presents an efficient $O(n \log n)$-time flip-flop insertion algorithm that evens out margins. Section IV extends the flop algorithm to handle 2-phase symmetric latches, which also runs in $O(n \log n)$ time. Both algorithms are asymtotically faster and easier to implement than previous work. Section V presents experimental results, and we conclude in Section VI.

II. PRELIMINARIES

In this paper, we model the topology of an interconnect as a tree $T(V, E)$. $V = \{v_d \cup S_N \cup S_{TN}\}$ is a set of n nodes, with root v_d and leaves S_N of T being the interconnect driver and receivers respectively. The intermediate Steiner nodes S_{TN} contain candidate locations for flop and latch insertion. E is a set of $|V| - 1$ edges corresponding to wires. A *stage* is an ordered subset of edges $(u, u_1), \ldots, (u_k, v)$ in E such that u is either v_d or contains a flop (latch), v is either in S_N or contains a flop (latch), and $u_1, \ldots, u_k$ are free of flops and latches. Each receiver $v_r \in S_N$ has a non-negative, integeral latency requirement $l(v_r)$, and a non-negative required margin $m(v_r)$. $l(v_r)$ specifies the number of flops, or in the case of latches, half the number of latches, on the unique path from v_d to v_r. The margin requirement states that the signal must arrive at v_r no later than $m(v_r)$ time-units before the falling clock edge. We model the delay inside a flop or a latch by a single, fixed real number $d_{cell} \geq 0$.

Our modeling of wire delay follows [1]. We assume the delay of a wire to be proportional to the square root of its RC-content. The maximum unrepeated distance for a wire was computed based on performance and reliability requirements. Buffers were inserted based on this distance and simulations were done to compute the delay across the wire. The delays were then curve-fitted to a linear equation with respect to the square root of the wire's total RC-content. This allows us to quickly estimate the wire delay $d(e)$, under the assumption that the wire is part of a buffered interconnect path.

We now describe our notion of proper latch timing. We follow closely the definition in [4]. In particular, we require the latches to hold the same values as in an identical circuit in which all elements, including both gates and wires, have *zero* propagation delay. The notion of proper latch timing is *structural* in nature, in the sense that the circuit should operate correctly regardless of the functions it computes.

Formally, we seek to solve the following two problems:

1. **Flip-Flop Repeater Insertion Problem** Assign flops to S_{TN} such that: (i) margins are equally distributed among all stages; and (ii) latency and margin requirements at the receivers are satisfied.

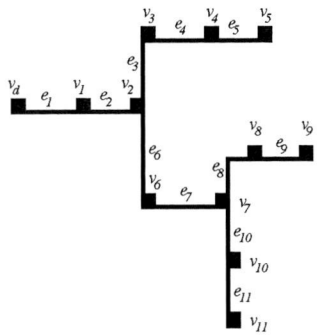

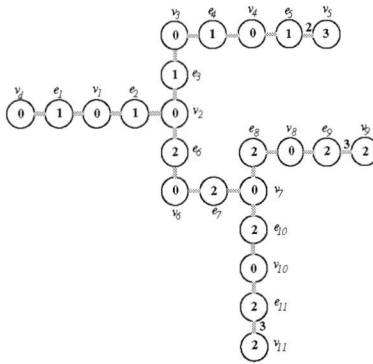

Fig. 1. Top: Original graph model $T(V, E)$. $d(e_1), \cdots, d(e_5) = 1$, $d(e_6), \cdots, d(e_{11}) = 2$. Receivers have required margins $m(v_5) = 3, m(v_9) = m(v_{11}) = 2$, and required latencies $l(v_5) = 2, l(v_9) = l(v_{11}) = 3$. Bottom: Transformed graph model $T'(V', E')$. Edges have implicit weight 0, unless otherwise specified. Numbers on nodes are associated delays.

2. **Latch Repeater Insertion Problem** Assign two-phase, symmetric level-sensitive latches to S_{TN} such that: (i) latency and margin requirements at the receivers are satisfied; and (ii) timing is structurally correct.

Let us break the circuit T into $|v_r|$ sub-circuits T_r, each consisting of the unique 2-pin sub-net from the driver v_d to a single receiver v_r. It is clear that distributing margins evenly on the 2-pin net T_r amounts to minimizing the clock period of T_r. The clock period of T is given by the relation $\text{period}(T) \geq \max_{v_r} \text{period}(T_r) \geq \text{period}(T_r)$ for all v_r. Minimizing clock period of T evens out margins by proxy.

Retiming [5] is a technique that relocates registers to reduce cycle time while preserving circuit functionality. As a first step, we convert each edge $e \in E$ into a vertex with propagation delay $d(e)$. Formally, we transform $T(V, E)$ into another tree $T'(V', E')$, with $V' = V \cup E$ and each edge $u \xrightarrow{e} v$ in the original edge set E spawns two edges in E': $u \to e$ and $e \to v$.

There are three possibilities for $d(v')$, $v' \in V'$. If v' comes from a receiver v_r in the original vertex set V,

$d(v') = m(v_r)$. If v' comes from an edge $e \in E$, $d(v') = d(e)$. For all other vertices, $d(v') = 0$.

As in [5], sequential elements are represented implicitly by a non-negative integral *weight* $w(e')$ on $e' \in E'$. In our case, $w(e') = 0$ except for the edges that are adjacent to a receiver v_r, in which case $w(e') = l(v_r)$ ($2l(v_r)$ for latches) (Fig. 1.) Note that the original flop or latch placement, if there is any, is ignored. Our algorithms operate on the transformed tree model T'. For ease of exposition, we will refer to both as T in the rest of the paper. Note that the asymptotic bounds derived for T' also hold for T.

A retiming solution can be viewed as an integral labeling r of vertices. $r(v)$ represents the number of sequentials moved from v's outputs toward its inputs. In particular, if we require that $r(v_d)$ at the driver and $r(v_r)$ at all receivers to be *equal*, it is guaranteed that latencies at the receivers remain the same after retiming. Since we model delays across a clocked repeater as a constant d_{cell}, it does not factor in the selection of one retiming solution over another. For the rest of the paper we assume $d_{cell} = 0$.

III. FLIP-FLOP INSERTION

A general framework to calculate the minimum period is to perform a binary search over a range of possible clock periods until a user-defined accuracy is reached. At each step in the binary search, a new clock period is tried. Retiming is performed to recalculate $r(v)$ at each node in an attempt to make the circuit work with the current clock period. The smallest period for which retiming succeeds is returned as the best clock period.

The efficiency of the binary search depends strongly on the retiming step in its inner loop. One algorithm for the retiming step is the relaxation method, modelled after the Bellman-Ford algorithm for the single source shortest path problem [5, 11]. Shown in Fig. 2, relaxation works on general graphs. We denote by $\delta(v)$ the latest arrival time at v along any *combinational* path that terminates at v. It is obvious that the clock period c is given by $c = max_{v \in V}\delta(v)$. For a given set of $r(v)$, we calculate $\delta(v)$ for all v, which takes $O(E)$ time. The relaxation method thus runs in $O(VE)$ time.

An analysis by Yen [13, 6] explains why the $V - 1$ loop in lines 2-7 of Graph_Flop_Clk_Feas (Fig. 2) works. The latest arrival time $\delta(v)$ converges to its final value if the edges along the longest path (in terms of propagation time) from the source node v_d to v are relaxed in order. The sequence of edges relaxed in the loop consists of $V - 1$ copies of some ordering of E, and therefore contains every vertex-disjoint path as a subsequence. Since T is a tree, there is a unique path from v_d to v. There is thus no need for the $V - 1$ loop. Instead, we combine the calculation of $\delta(v)$ with the relaxation of r in a single topological tour of the nodes. The new linear-time relaxation algorithm is shown in Fig. 2.

We bound the binary search in Fig. 2 from above with

```
Graph_Flop_Clk_Feas(G,r,c)
/* Input:  Graph G(V,E), label r, period c
 * Output:  Feasibility of c */
1   for each v ∈ V  r(v) ← 0
2   for |V| - 1 {
3       Compute retimed edge weights
4       Compute δ(v) for all v ∈ V
5       for all v ∈ V such that δ(v) > c
6           r(v) ← r(v) + 1
7   }
8   Compute retimed edge weights
9   if max_{v∈V} δ(v) > c
10      then no feasible retiming
11      else the current r is legal
```

```
Tree_Flop_Clk_Feas(T,r,c)
/* Input:  Tree T(V,E), label r, period c
 * Output:  Feasibility of c */
1   for each v ∈ V  r(v) ← 0
2   for v ∈ V in topological order {
3       Compute δ(v)
4       if δ(v) > c
5           then r(v) ← r(parent(v)) + 1
6                δ(v) ← d(v)
7           else r(v) ← r(parent(v))
8   }
9   Compute retimed edge weights
10  if max_{v∈V} δ(v) > c
11      then no feasible retiming
12      else the current r is legal
```

```
Tree_Flop_Insert(T,ε)
/* Tree Flip-Flop Insertion
 * Input:  Tree T(V,E), rel error ε
 * Output:  Retiming label r*/
1   d_max ← max_{v∈V} d(v)
2   c_hi ← max_{v_r} d(v_d ↝ v_r)
3   c_lo ← max_{v_r}{d(v_d ↝ v_r)/(l(v_r) + 1)}
4   r ← 0
5   while c_hi - c_lo > ε * d_max {
6       c ← (c_hi + c_lo)/2
7       if Tree_Flop_Clk_Feas(T,r,c) = true
8       then c_hi ← c
9       else c_lo ← c
10  }
11  return r
```

Fig. 2. Clock feasibility test in FF retiming for general graphs, clock feasibility test for trees and FF insertion for trees

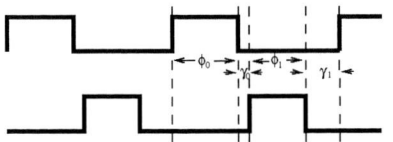

Fig. 3. Two phase clocking scheme, reproduced from [4].

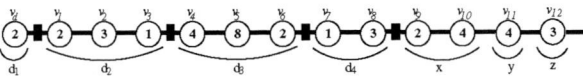

Fig. 4. The path $v_4 \rightsquigarrow v_8$ satisfying Condition 1 implies $v_5 \rightsquigarrow v_7$ meets the same condition. d_i denotes the sum of vertex delays in the i-th stage.

$c_{hi} = max_{v_r} d(v_d \rightsquigarrow v_r)$. This corresponds to the circuit being entirely combinational. The search is bounded from below by $c_{lo} = max_{v_r}\{d(v_d \rightsquigarrow v_r)/(l(v_r)+1)\}$, which corresponds to margins being distributed perfectly evenly along every driver to receiver path. Note that for $c >= c_{lo}$, it is guaranteed that Tree_Flop_Clk_Feas leaves $r(v_r) = 0 = r(v_d)$ for all receivers v_r: the latency requirement is satisfied throughout the binary search. Tree_Flop_Insert solves Problem 1 in $O(V \log V)$ time for any fixed ϵ.

IV. Latch Insertion

In a two-phase clocking scheme $\langle \phi_0, \gamma_0, \phi_1, \gamma_1 \rangle$, two clocking phases are employed (Fig. 3). ϕ_0, ϕ_1 denote the duty cycle of the first and second phases, and γ_0, γ_1 denote the gaps. The period is given by $\phi_0 + \gamma_0 + \phi_1 + \gamma_1$. A two-phase symmetric clocking scheme is characterized by equal duty cycles and gaps: $\phi_0 = \phi_1 = \phi$ and $\gamma_0 = \gamma_1 = \gamma$.

The conditions for proper timing of T are based on considering the operation of T when all propagation delays are 0. It can be shown that T is properly timed if and only if for every path $u \overset{p}{\rightsquigarrow} v$, the following condition is satisfied [4]:

$$d(p) \le c \left(\frac{1 + w(p)}{2} \right) + \phi. \tag{1}$$

Furthermore, we only need to make sure (1) is satisfied for *maximal* paths. These are paths that start right after a latch and ends right before another. To see this, note that the right hand side of (1) stays the same as we take in more nodes into the path, as long as the number of latches does not change. On the other hand, more nodes could potentially increase $d(p)$, leading to a tighter inequality. To reduce clutter on the formulas, we set $\phi = 0$ for the rest of the paper.

(1) leads to a greedy algorithm. Nodes are visited in topological order staring with the driver. In Fig. 4, suppose all the maximal paths up until v_{10} satisfy (1), and

we are considering whether a latch *must* be inserted between v_{10} and v_{11} to maintain timing feasibility. Applying (1) to the maximal paths that end with v_{11}, we consider whether the following inequalities are satisfied:

$$
\begin{aligned}
y &\le \frac{c}{2} - x \\
y &\le \frac{c}{2} * 2 - d_4 - x \\
&\;\;\vdots \\
y &\le \frac{c}{2} * 5 - d_1 - d_2 - d_3 - d_4 - x \quad (2)
\end{aligned}
$$

Case 1 Not satisfied. In this case we must insert a latch. x becomes the new d_5, the right hand sides of (2) adjusted, and a new inequality is added to the set:

$$
\begin{aligned}
z &\le \frac{c}{2} - y \\
&\;\;\vdots \\
z &\le \frac{c}{2} * 6 - d_1 - d_2 - d_3 - d_4 - d_5 - y \quad (3)
\end{aligned}
$$

Case 2 Satisfied. We can avoid adding a latch:

$$
\begin{aligned}
z &\le \frac{c}{2} - x - y \\
&\;\;\vdots \\
z &\le \frac{c}{2} * 5 - d_1 - d_2 - d_3 - d_4 - x - y \quad (4)
\end{aligned}
$$

The inequality satisfaction tests can be done more efficiently by replacing (2) with a single inequality $y \le d_{min}$, where d_{min} is the minimum of the right hand sides in (2):

$$d_{min} = \min(\frac{c}{2} - x, \ldots, \frac{c}{2} * 5 - d_1 - d_2 - d_3 - d_4 - x).$$

Comparing the updated inequalities in (3) and (4) with (2), it is evident that d_{min} can be computed efficiently. Starting with $d_{min}(v_d) = \frac{c}{2}$ at the driver, $d_{min}(v)$ is computed recursively from $d_{min}(u)$, where $u \rightarrow v$ is an edge in E, as follows:

$$
d_{min}(v) = \begin{cases}
\frac{c}{2} + \min(0, d_{min}(u) - d(u)) \\
\qquad \text{if } u \rightarrow v \text{ is latched} \\
d_{min}(u) - d(u) \quad \text{otherwise}
\end{cases} \tag{5}
$$

With (5), feasibility of any period c can be determined in linear time (Fig. 5). Using this as a subroutine, a binary search over the range of possible periods is performed. Let $D = max_{v \in V} d(v)$. The search is bounded by $|V| * D + 2 * \phi$ from above, and $2 * \phi$ from below. This results in an $O(n \log n)$-time algorithm (Fig. 5).

V. Experimental Results

We implemented three different spec-based sequential insertion algorithms in C++, and compared them on a block from an Intel Xeon^{TM} microprocessor designed on 90-nm process technology.

TABLE I

RESULTS OF VARIOUS SEQUENTIAL INSERTION ALGORITHMS. THE FLOP NUMBER OF *LATCH* IS HALF THE NUMBER OF LATCHES.

Algorithm	Flops	RunTime (seconds)
SBFIA	14989	3104.38
RFLOP	14907	11.54
LATCH	15028.5	11.92

```
Tree_Latch_Clk_Feas(T,r,c)
/* Input:  Tree T(V,E), label r, period c
 * Output:  Feasibility of c*/
1    for each v ∈ V  r(v) ← 0
2    for v ∈ V in topological order {
3        Compute d_min(v) according to Eq 5
4        if v is source or sink
5            then r(v) ← 0
6            elseif d(v) > d_min(v)
7                then r(v) ← r(parent(v)) + 1
8                else r(v) ← r(parent(v))
9    }
10   for e = u → v ∈ E {
11       if w(e) − r(u) + r(v) < 0
12           then no feasible retiming
13   }
14   The current r is legal
```

```
Tree_Latch_Insert(T,ε)
/* Input:  Tree T(V,E), rel error ε
 * Output:  Retiming label r*/
1    d_max ← max_{v∈V} d(v)
2    c_hi ← |V| * d_max + 2 * φ
3    c_lo ← 2 * φ
4    r ← 0
5    while c_hi − c_lo > ε * d_max {
6        c ← (c_hi + c_lo)/2
7        if Tree_Latch_Clk_Feas(T,r,c) = true
8        then c_hi ← c
9        else c_lo ← c
10   }
11   return r
```

Fig. 5. Latch clock feasibility test for trees and latch insertion for trees

- SBFIA - Spec-based FF insertion from [1].

- RFLOP - FF insertion algorithm from Section III.

- LATCH - Latch insertion algorithm from Section IV.

The design has 1769 multi-cycle nets, with fanouts ranging from 1 to 34. The nets' topologies are discretized by breaking the wires with a candidate insertion node every 100 microns. The relative error ϵ is set to 0.001. All experiments were run on a workstation with four Intel Xeon^{TM}2.8GHz/512KB processors and 8GB of memory, although none of the three algorithms exploit the parallelism afforded by the machine.

As shown in Table I, the three algorithms used similar amount of sequential resources. For accounting purpose a latch is counted as half a flop. Note that even though minimizing flop usage is not an explicit goal of *RFLOP*, it used slightly fewer flops than *SFBIA*, while *LATCH* used 0.26% more. A more detailed study of the result shows that *RFLOP* was more flop-efficient than *SBFIA* in every fanout bucket.

The comparison on runtime is more pronounced. Both *RFLOP* and *LATCH* finished under 12 seconds, while the $O(n^2 \log n)$-time *SFBIA* took over 50 minutes.

We observe that *stage spread* – the difference between maximum and minimum flop-to-flop propagation delay – is smaller for *RFLOP* than *SBFIA* (Table II), validating our design goal of a more even margin distribution. Errors from discretization and imbalance among multiple receivers keep the spread positive. Perfectly balanced margins would have given a spread of *zero*.

Both *RFLOP* and *LATCH* improved negative slacks compared to *SFBIA* (Table III). In the case of *LATCH*, all the negative slacks were wiped out. The median gain on frequency over *SFBIA* ranges from 15.96% to 35.11% for *RFLOP*, and 22.42% to 44.27% for *LATCH*.

VI. CONCLUSIONS

We have presented fast and practical algorithms for flop and latch insertion to facilitate design of pipelined interconnects within the latency constraints at the receivers. We argued that evening out timing margins across the entire interconnect is more beneficial than pushing them to either ends, and made it an explicit goal of our algorithms to distribute margins evenly. Experimental results

TABLE II
STAGE SPREAD IS MEASURED IN PICOSECONDS. THE NUMBER OF
NETS IS GIVEN IN PARENTHESES NEXT TO THE FANOUTS.

Algorithm	No. Flops	Stage Spread	
		Median	Average
1-fanout (941)			
SBFIA	5766	162.12	175.96
RFLOP	5766	58.23	80.31
2-fanout (456)			
SBFIA	4968	249.78	253.46
RFLOP	4956	146.73	152.98
3-fanout (187)			
SBFIA	1877	269.94	256.91
RFLOP	1876	183.30	181.92
4-fanout to 6-fanout (145)			
SBFIA	1703	281.67	281.46
RFLOP	1671	205.34	200.73
7-fanout+ (40)			
SBFIA	675	333.72	319.72
RFLOP	638	230.98	224.37

TABLE III
NEGATIVE SLACK IS MEASURED IN PICOSECONDS. THE NUMBER OF
NETS IS GIVEN IN PARENTHESES NEXT TO THE FANOUTS.

	Negative Slack			Freq Gain	
	Total	Worst	Nets	Median	Average
1-fanout (941)					
SBFIA	-38242.2	-213.4	707	0	0
RFLOP	-253.2	-16.6	192	17.51%	19.36%
LATCH	0.0	0.0	0	24.35%	29.67%
2-fanout (456)					
SBFIA	-17859.8	-220.0	413	0	0
RFLOP	-69.1	-6.3	59	17.47%	17.13%
LATCH	0.0	0.0	0	23.47%	25.53%
3-fanout (187)					
SBFIA	-7704.4	-180.8	168	0	0
RFLOP	-278.6	-15.0	44	15.96%	17.45%
LATCH	0.0	0.0	0	22.42%	26.48%
4-fanout to 6-fanout (145)					
SBFIA	-7756.8	-189.9	142	0	0
RFLOP	-335.8	-15.8	60	16.24%	21.66%
LATCH	0.0	0.0	0	24.05%	29.27%
7-fanout+ (40)					
SBFIA	-3312.1	-169.0	40	0	0
RFLOP	-119.4	-13.2	22	35.11%	32.36%
LATCH	0.0	0.0	0	44.27%	39.58%

demonstrate the efficiency and efficacy of the algorithms on industrial test cases. unsrt

REFERENCES

[1] N. Akkiraju and M. Mohan. Spec based flip-flop and buffer insertion. In *Proceedings of the 21st International Conference on Computer Design*, pages 270–275. IEEE Computer Society, 2003.

[2] P. Cocchini. A methodology for optimal repeater insertion in pipelined interconnects. *Computer-Aided Design of Integrated Circuits and Systems, IEEE Transactions on*, 22(12):1613–1624, 2003.

[3] S. Hassoun, C. J. Alpert, and M. Thiagarajan. Optimal buffered routing path constructions for single and multiple clock domain systems. In *Proceedings of the 2002 IEEE/ACM international conference on Computer-aided design*, pages 247–253. ACM Press, 2002.

[4] A. T. Ishii, C. E. Leiserson, and M. C. Papaefthymiou. Optimizing two-phase, level-clocked circuitry. *J. ACM*, 44(1):148–199, 1997.

[5] C. Leiserson and J. Saxe. Retiming synchronous circuitry. *Algorithmica*, 6(1):5–35, 1991.

[6] C. E. Leiserson and J. B. Saxe. A mixed-integer linear programming problem which is efficiently solvable. *J. Algorithms*, 9(1):114–128, 1988.

[7] C. Lin and H. Zhou. Retiming for wire pipelining in system-on-chip. In *Proceedings of the 2003 international conference on on Computer-aided design*, pages 215–220. IEEE Computer Society, 2003.

[8] R. Lu, G. Zhong, C. Koh, and K. Chao. Flip-flop and repeater insertion for early interconnect planning. In *Proceedings of the conference on Design, automation and test in Europe*, page 690. IEEE Computer Society, 2002.

[9] P. Saxena, N. Menezes, P. Cocchini, and D. Kirkpatrick. Repeater scaling and its impact on cad. *Computer-Aided Design of Integrated Circuits and Systems, IEEE Transactions on*, 23(4):451–463, 2004.

[10] V. Seth, M. Zhao, and J. Hu. Exploiting level sensitive latches in wire pipelining. In *Proceedings of the 2004 international conference on on Computer-aided design*, pages 283–290. IEEE Computer Society, 2004.

[11] N. Shenoy. Retiming: theory and practice. *Integr. VLSI J.*, 22(1-2):1–21, 1997.

[12] L. van Ginneken. Buffer placement in distributed rc-tree networks for minimal elmore delay. In *Proceedings of the IEEE International Symposium on Circuits and Systems*, volume 2, pages 865–868. IEEE Computer Society, 1990.

[13] J. Yen. An algorithm for finding shortest routes from all source nodes to a given destination in general networks. *Quarterly of Applied Mathematics*, 27(4):526–530, 1970.

[14] L. Zhang, Y. Hu, and C. Chen. Wave-pipelined on-chip global interconnect. In *Proceedings of the Design Automation Conference Asia and South Pacific*, page to appear. IEEE Computer Society, 2005.

3D-1

2006 Asia and South Pacific Design Automation Conference

Current Trends in Flash Memory Technology
(Invited Paper)

Sang Lyul Min   Eyee Hyun Nam

School of Computer Science and Engineering
Seoul National University
Seoul 151-742, KOREA
Tel : +82-2-880-7047
Fax : +82-2-886-7589
e-mail : symin@snu.ac.kr

Abstract - In this paper, we describe the basics of flash memory technology in general and flash memory drive in particular, and explain the current trends of major components of a flash memory drive including flash memory chips, host interface and flash memory controller.

I. Introduction

This paper explains the current trends of flash memory technology in general and flash memory drive in particular. The paper is organized as follows. Section II surveys the current trends of the two components that interface a flash memory drive: flash memory chips and host interface. Then, in Section III we explain various techniques to enhance the performance of a flash memory drive. Section IV briefly explains a recent storage system called a hybrid hard disk drive that combines the advantages of flash memory drive and hard disk drive. Finally, we offer conclusions in Section V.

II. Flash memory chips and host interface

As we can see in Fig. 1 that shows the basic architecture of a flash memory drive, there are two components that interface to the flash memory controller in a flash memory drive: flash memory chips and host interface.

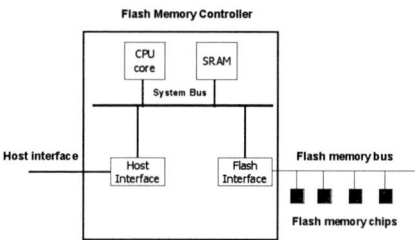

Fig. 1. Basic architecture of flash memory drive.

Fig. 2 gives a high-level interface of a NAND flash memory chip, the type of flash memory used for bulk storage, from a software perspective. The NAND flash memory chip consists of a set of blocks that in turn consist of a set of pages where each page has the data part that stores the user data and the spare part that stores meta data associated with user data such as ECC. Although different sizes may be used, currently the most popular block size is 128 Kbytes consisting of 64 pages of 2 Kbytes. There are three possible operations to a flash memory chip: **read page**,

program page, and **erase block**. The read page operation, given the chip number, the block number, and the page number returns the contents of the addressed page, which takes about 20 us. Likewise, the program page operation writes the supplied contents to the target page and takes about 200 us. Unlike a write operation to other types of storage medium, the program operation can change the stored bits from 1 to 0 only. Therefore, the write operation is implemented by selectively changing bits from 1 to 0 to match the supplied contents assuming all bits in the target page are 1's before the program operation. In flash memory, the only way to change a bit in a page from 0 to 1 is to erase the block that contains the page. The erase operation sets all bits in the block to 1. The erase block operation takes about 2 ms.

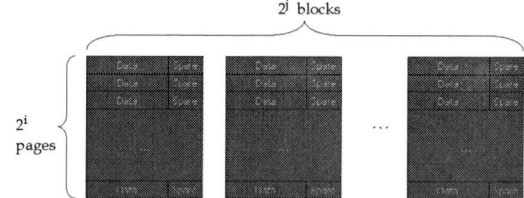

Fig. 2. NAND flash memory chip.

Fig. 3 shows the density and price trends of NAND flash memory chips. One notable point in the figure is the capacity of NAND flash memory has been doubled every year over the past five years. Another interesting point is that a flash drive whose capacity is comparable to that of hard disk drive used in a notebook these days (> 20 Gbytes) will become affordable (< $500) for high-end users within two or three years.

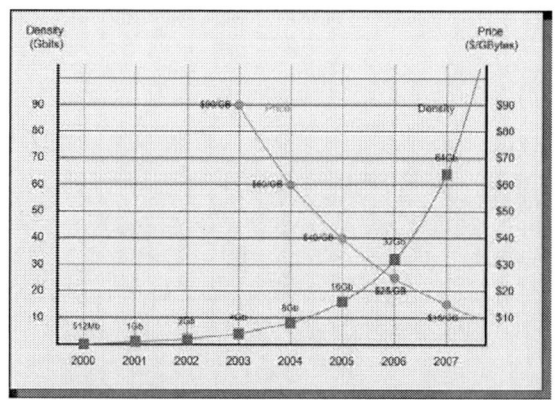

Fig. 3. NAND flash memory density and price trends.

0-7803-9451-8/06/$20.00 ©2006 IEEE.
332

On the host interface side, the two most notable trends are (1) the maximum host interface bandwidth has increased exponentially and (2) most of recently proposed host interface standards are serial. We expect that these trends will continue for the foreseeable future and will assure a scalable performance for future flash memory drives.

III. Flash memory controller architecture

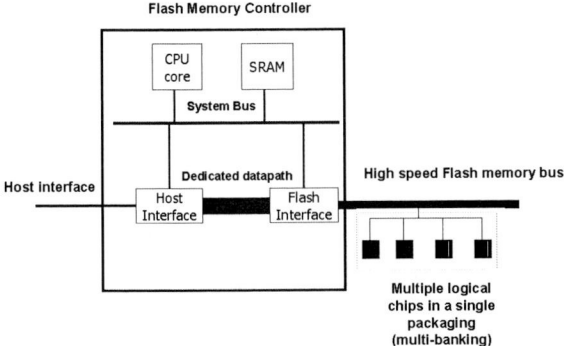

Fig. 4. Advanced flash memory controller architecture.

Fig. 4 shows various enhancements that can be made to the basic architecture of a flash memory drive in Fig.1. First of all, there is a lot of room for improvement in the architecture of flash memory chip itself. Currently, the maximum write bandwidth that can be provided by a single flash memory chip is 10 Mbytes/s (= 2 Kbytes/200us) that is certainly not sufficient for high-performance flash memory drives. Higher write bandwidth can be obtained by using interleaving within a chip or across multiple chips. This interleaving will give benefits to the read bandwidth and erase bandwidth as well.

Another place for improvement is flash memory bus. Currently the maximum flash memory bus bandwidth is 33 Mbytes/s with a 8-bit data bus. This means that the maximum performance of any flash memory drive that uses a single flash memory bus is limited to 33 Mbytes/s, which is also not sufficient for high-performance flash memory drives. Of course, multiple flash memory buses can be used for higher bus bandwidth but this approach will require a higher pin count and a bus interleaving logic within the flash memory controller. A preferred approach would be to use a single high-performance flash memory bus, an approach currently taken by the OneNAND bus architecture that provides up to 108 Mbytes/s with a 16-bit bus [1].

The third place for improvement is the datapath between host interface and flash memory. Traditionally, the datapath was implemented by a cascaded DMA transfers using SRAM within the controller but this approach not only increases the latency but also consumes a significant amount of system bus bandwidth making the system bus an overall system bottleneck that limits the maximum performance. An alternative approach is to use a dedicated datapath between host interface and flash memory as shown in Fig. 4 that not only reduces the latency but also has an advantage in terms

of scalability.

IV. Hybrid hard disk drive

One of recent developments in storage systems is a so-called hybrid hard disk drive shown in Fig. 5. It adds a flash memory chip to a hard disk drive for reduced power consumption and faster start-up. In a hybrid hard disk drive, write requests from host are buffered in flash memory with the spindle motor spun down. When the flash write buffer is full, the spindle motor spins up and buffered requests are flushed to disk storage. This writing buffering using a 128 Mbytes of flash memory is shown to save up to 80% of power consumption in a 2.5in hard disk drive [2]. Moreover, by caching in flash memory those contents that will be used in the next boot/resume and using them while the hard disk drive is being spun up, the boot/resume time can be drastically reduced.

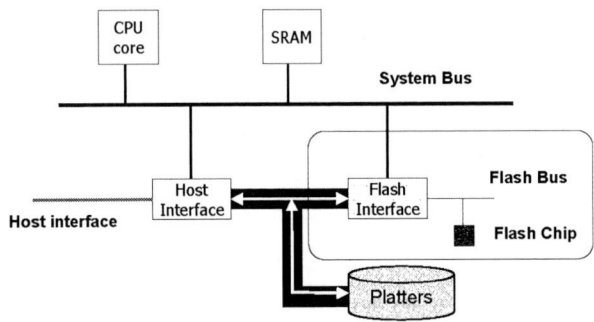

Fig. 5. Hybrid hard disk drive

V. Conclusions

This paper gave basics of flash memory drive and explained the current trends of its components including flash memory chips, host interface, and flash memory controller. The market of storage devices based on flash memory is expanding rapidly and this expansion will pose many challenges not only in flash memory technology but also in related areas as in the case of hybrid hard disk drives.

Acknowledgements

This work was supported in part by the IDEC (IC Design Education Center), by the IT-SoC project, and by Brain Korea 21 project. ICT at Seoul National University provided research facilities for this study.

References

[1] B. Kim, S. Cho, Y. Choi, and Y. Choi, "OneNAND(TM): A High Performance and Low Power Memory Solution for Code and Data Storage," Proc. 20th Non-Volatile Semiconductor Workshop, 2004.

[2] J. Creasey, "Hybrid Hard Drives with Non-Volatile Flash and Longhorn," WinHEC 2005.

Configurability of Performance and Overheads in Flash Management*

Tei-Wei Kuo	Jen-Wei Hsieh	Li-Pin Chang	Yuan-Hao Chang
Dept. of Computer Science and Information Engineering National Taiwan University Taipei, Taiwan 106, R.O.C. Tel: +886-2-23625336 ext 205 Fax: +886-2-23628167 ktw@csie.ntu.edu.tw	Dept. of Computer Science and Information Engineering National Taiwan University Taipei, Taiwan 106, R.O.C. Tel: +886-2-23625336 ext 436 Fax: +886-2-23628167 d90002@csie.ntu.edu.tw	Dept. of Computer Science National Chiao-Tung University Hsinchu, Taiwan 300, R.O.C. Tel: +886-3-5712121 ext 56685 Fax: +886-3-5721490 lpchang@cis.nctu.edu.tw	Graduate Institute of Networking and Multimedia National Taiwan University Taipei, Taiwan 106, R.O.C. Tel: +886-2-23625336 ext 436 Fax: +886-2-23628167 d93944006@csie.ntu.edu.tw

Abstract— Flash memory has been widely considered as a good alternative for storage system implementations because it offers superior vibration tolerance and power efficiency, compared to hard-disks. Because of its unique characteristics, direct applications of disk management methods over flash memory might result in performance degradation and even the reducing of the lifetime. The management issues become even more challenging, especially when the capacity of flash memory increases significantly in the past few years. In this paper, we summarize our work on several important issues in flash memory management, where system performance and management overheads are considered. The capability of the proposed methodology was evaluated by a series of experiments to provide more insights in system designs.

I. INTRODUCTION

Flash memory is widely adopted as an alternative for storage system designs because of its nature in non-volatility, shock-resistance, and low power consumption. It is also considered as being low cost (compared to SRAM or DRAM) and having good performance (compared to disks) in storage system implementations. Due to its unique characteristics in manipulations and market definitions, different challenging issues are raised, compared to those based on disks. There are two critical and inter-dependent issues that must be addressed by most vendors and researchers: Performance and overheads. While good system performance is a must for many applications, most vendors would only give restricted budgets on various system overheads, such as the memory space size for flash management. How to provide a good design with reasonable performance under given overheads constraints is always a question faced by many vendors and researchers.

The management of flash memory is carried out by either software on a host system (as a raw medium) or hard-

---

*Supported in part by a research grant from Genesys Logic, Inc. and the Taiwan, ROC National Science Council under Grants NSC94-2752-E-002-008, NSC94-2219-E-002-013, and NSC94-2213-E-002-007.

ware/firmware of the device for flash memory. In particular, Kawaguchi, Nishioka, and Motoda [13] proposed a flash-memory translation layer to provide a transparent way to access flash memory through the emulating of a block device. Wu and Zwaenepoel [19] proposed to integrate a virtual memory mechanism with a non-volatile storage system based on flash memory. Native flash-memory file systems were also presented without imposing any disk-aware structures on the management of flash memory [9, 16]. Kuo and Chang explored performance issues of flash-memory storage systems by considering new system architectures [4], an energy-aware scheduler [7], and a deterministic garbage collection mechanism [6]. In [18], Wu, Kuo, and Chang provided efficient roll back and quick mounting for flash-memory file systems. How to efficiently handle fine-grained updates due to index access of spatial data over flash memory is also discussed [17]. In addition to the work from the academics, many implementation designs and specifications were proposed in the industry, e.g., [1, 2, 14, 11].

This paper summarizes our work in several design issues that are involved with system performance and overheads considerations [4, 5, 12]. We shall first present our work on how to efficiently identify hot data in data access over flash memory under a very restricted memory-space constraint, where the identification of hot data is important in the improvement of system performance and the reducing of overheads in garbage collection. We shall then summarize our work on performance improvement with multiple banks, where the relationship among performance, capacity utilization, and garbage collection overheads is considered. We will then present an efficient space-management scheme with variable granularities for large-scale flash memory, in which memory-space overheads will become overwhelming in careless designs. Experimental results are presented to demonstrate the capability of the proposed methodology.

The rest of this paper is organized as follows: In Section II, the designs of flash-memory storage systems and the motivation of this work are presented. Section III summarizes our

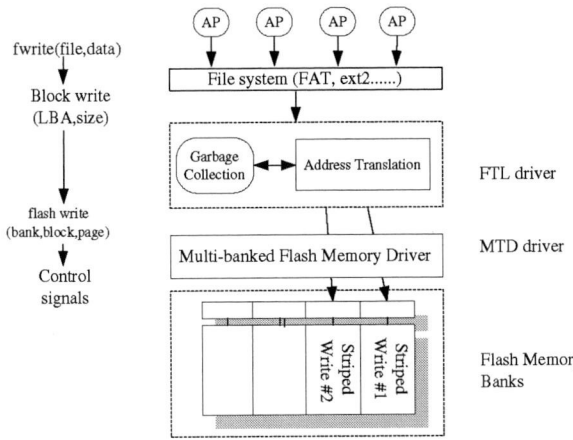

Fig. 1. System Architecture.

work that is involved with system performance and overheads considerations [4, 5, 12]. Some experimental results are shown in Section IV. Section V is the conclusion.

II. DESIGNS OF FLASH-MEMORY STORAGE SYSTEMS

Layered designs are usually adopted for the implementations of flash-memory storage systems, regardless of hardware or software implementations of certain layers. The Memory Technology Device (MTD) driver and the Flash Translation Layer (FTL) driver are the two major layers for flash-memory management, as shown in Fig. 1. Each flash-memory bank can operate independently and is composed of flash-memory chips. The MTD driver provides lower-level functionalities of a storage medium, such as read, write, and erase. Based on these services, higher-level management algorithms, such as wear-leveling, garbage collection, and physical/logical address translation, are implemented in the FTL driver. The objective of the FTL driver is to provide transparent services for user applications and file systems to access flash memory as a block-oriented device.

A flash memory chip is partitioned into blocks, where each block has a fixed number of pages, and each page is of a fixed size, e.g., 512B. Due to hardware architecture, pages are basic write-operation units while blocks are basic erase-operation units. Initially, all pages in flash memory are considered as "free." After a page has been written, it is no longer available unless an erase operation is performed. When a piece of data over a page needs to be modified, out-place update is usually adopted for performance consideration since erase operations take time. The pages stored the old versions of the data are considered as "dead," while the page stored the newest version of data is considered as "live." After sustained write operations, the number of free pages would be low, and the system must reclaim free pages (referred to as *garbage collection*) for

further writes.

The operation model of flash memory, in general, consists of two phases: setup phase and busy phase. For example, the first phase (setup phase) of a write operation is for command setup and data transfer. The command, the address, and the data are written to proper registers of flash memory in order. The second phase (busy phase) is for busy-waiting of the data being flushed into flash memory. The operation of reads is similar to that of writes, except that the sequence of data transfer and busy-waiting is inverted. The phases of an erase is as the same as those of a write, except that no data transfer is needed in the setup phase. The control sequence of read, write, and erase are illustrated in

The implementation of the FTL driver could consist of an *allocator* and a *cleaner*. The allocator is responsible to the finding of proper pages on flash memory to dispatch writes, and the cleaner is responsible to the reclaiming of pages with invalidated data, where space reclaiming is referred to as garbage collection. Since the unit of erase operations is block, live pages over the selected block (if any) must be copied to some free pages of other blocks before the erasure. With a proper garbage-collection policy, the number of overall live-page copying could be much reduced, from which the free pages can be utilized efficiently. On the other hand, a block might be worn out after about 10^6 erasures under the current technology. When a block is worn out, its reliability can no longer be guaranteed. A poor garbage collection policy could quickly wear out some blocks and, thus, a flash memory chip. A strategy called "wear-leveling" with the intention to erase all blocks as evenly as possible is widely adopted to achieve durability.

III. A CONFIGURABLE FLASH-MEMORY MANAGEMENT SYSTEM

The configurability of performance and overheads in flash management is explored from three different perspectives. In Section A, a hash-based hot-data identification mechanism with scalability considerations on precision and memory-space overheads is presented to provide a highly efficient on-line spatial-locality analysis. In Section B, an adaptive striping mechanism with consideration of garbage collection is presented. The goal is to boost the system performance with better parallelism in executing operations, where issues of the capacity utilization and the wear leveling of each bank become important. An efficient scheme with variable granularity is presented in Section C. It aims at the reduction in the main-memory footprint and the improvement on system performance for large-scale flash-memory management. The results of this section are based on the work in [4, 5, 12].

A. Efficient On-Line Hot-Data Identification

The identification of hot data could significantly affect the performance of garbage collection and wear-leveling, because any recycling of a block with lots of live-and-hot data would be relatively inefficient, and hot data could wear blocks out faster than non-hot data do. When large-scale flash memory is considered, many previous approaches introduce either considerable memory/processor overheads or poor accuracy in identifying hot data. In this section, an on-line hot-data identification mechanism is presented to efficiently and accurately capture run-time spatial locality with reduced requirements of memory-space and processor time [12].

A.1 A Multi-Hash-Function Framework

The proposed framework adopts K independent hash functions to hash a given LBA into multiple entries of a M-entry hash table to track the write number of the LBA, where each entry is associated with a counter of C bits. Whenever a write is issued to the FTL, the corresponding LBA is hashed simultaneously by K given hash functions. Each counter corresponding to the K hashed values (in the hash table) is incremented by one to reflect the fact that the LBA is written again. Note that we do not increase any counter for a read because there is no invalidation of any page for a read. Whenever an LBA needs to be verified to see if it is associated with hot data, the LBA is hashed simultaneously and in the same way by the K hash functions. The data addressed by the given LBA is considered as hot data if the H most significant bits of every counter of the K hashed values contain a non-zero bit value.

Fig. 2.(a) shows the increment of the counters that correspond to the hashed values of K hash functions for a given LBA, where there are four given independent hash functions, and each counter is of four bits. Fig. 2.(b) shows the hot-data identification of an LBA, where only the first two most significant bits of each counter is considered to verify whether the LBA corresponds to hot data. The rationale behind the adopting of K independent hash functions is to reduce the chance for the false identification of hot data. Because hashing tends to randomly maps a large address space into a small one, it is possible to falsely identify a given LBA as a location for hot data. With multiple hash functions adopted in the proposed framework, the chance of false identification might be reduced. In addition to this idea, the adopting of multiple independent hash functions also helps in the reducing of the hash table space, as indicated by Bloom [3].

For every given number of sectors that have been written, called the "decay period" of the write numbers, the values of all counters are divided by 2 in terms of a right shifting of their bits. It is an aging mechanism to exponentially decay the values of all write numbers as time goes on.

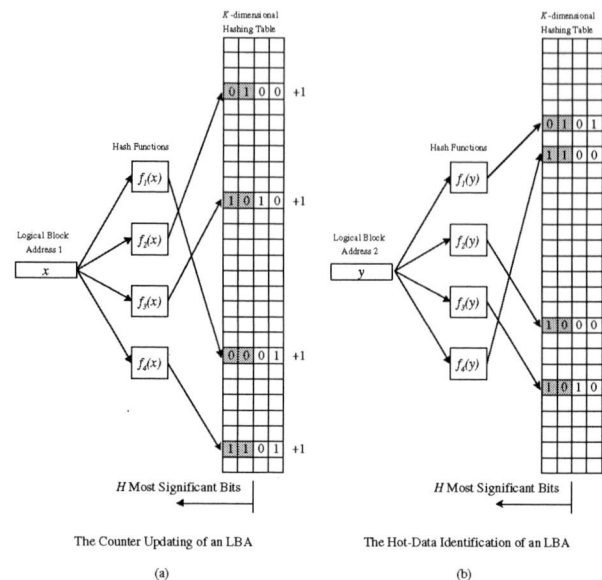

Fig. 2. The Counter Updating and the Hot-Data Identification of an LBA, where $C = 4$, $K = 4$, and $H = 2$.

A.2 Implementation Strategies

Instead of enlarging the hash table to improve false identification, it is proposed to increase only counters of the K hashed values that have the minimum value to improve false identification. The rationale behind the counter-increment policy is as follows: The reason for false identification is because counters of the K hash values of a non-hot LBA are also increased by other data writes, due to hashing collision. If an LBA is for hot data, then the policy in the increasing of small counters for its writes would still let all of the K counters corresponding to the LBA go over $2^{(C-H)}$ (because other writes would make up the loss in counter increasing). However, if an LBA is for non-hot data, then the policy would reduce the chance of false identification because a less number of counters will be falsely increased due to collision. The revised policy in counter increasing would introduce extra time complexity in the hot-data verification of each LBA because of the locating of counters with the minimum value. The revised policy would certainly increase the implementation difficulty of the algorithm with a certain degree, regardless of whether this algorithm is implemented in software, firmware, or even hardware. The performance improvement, compared to the basic framework proposed in Section A.1, will later be shown in the experiments.

B. An Adaptive Striping Architecture

In this section, we present a striping architecture to introduce I/O parallelism to flash-memory storage systems based on the work in [4]. An adaptive bank assignment method is

3D-2

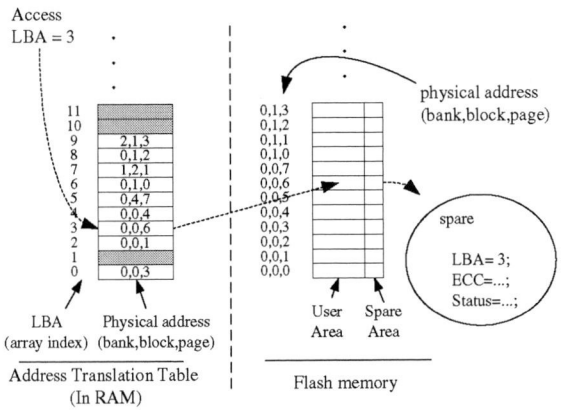

Fig. 3. Address Translation Mechanism.

presented to improve the system overheads on garbage collection. We must point out that although people always believe performance improvement in the application of the striping technology, little work has been done in the exploring of system behaviors on striping for flash-memory management, such as bank capacity utilization and garbage collection. There does exist a tradeoff between the striping level (and performance) and the garbage-collection overheads in system designs in many cases.

B.1 Multi-Bank Address Translation

In order to provide transparent data access, a dynamic address translation mechanism is usually adopted in the FTL driver. The dynamic address translation is accomplished by using an address translation table in main memory, e.g., [8, 11, 13, 19]. In a typical multi-bank storage system, each entry in the table could be a triple ($bank\_num$, $block\_num$, $page\_num$), indexed by LBA. Such a triple indicates that the corresponding logical block resides at Page $page\_num$ of Block $block\_num$ of Bank $bank\_num$. In a typical NAND flash memory, a page consists of a user area and a spare area, where the user area is for the storage of a logical block, and the spare area stores the corresponding LBA, ECC, and other information. The status of a page can be either "live", "dead", or "free" (i.e., "available"). Whenever a write is processed, the FTL driver first finds a free page and then writes the written data and the corresponding LBA to the user area and the spare area, respectively. The address translation table is updated accordingly. Whenever a system is powered up, the address translation table is re-built by scanning the spare area of all pages. As an example shown in Fig. 3, when a logical block with LBA 3 is accessed, the corresponding table entry (0,0,6) shows that the logical block resides at the 7th page (i.e., (6+1)th page) of the first block on the first bank.

B.2 Bank Assignment Policies

Three kinds of operations are supported on flash memory: read, write, and erase. Reads and erases do not need any bank assignment because they are already stored in specific locations. Writes would need a proper bank assignment policy to utilize the parallelism of multiple banks. When the FTL driver receives a write request, it will break the write into a number of page writes. There are two types of striping for write requests: static and dynamic striping.

Under the static striping, the bank number of each page write is derived based on the corresponding LBA of the page as follows, where the definition of RAID-0 is adopted as an example design: **Bank address = LBA % (number of banks)**. Each sizable write is striped across banks "evenly". However, we must point out that a static bank assignment policy could not provide even usages of banks in many real cases. A system that adopts a static bank assignment policy might suffer from a large number of data copyings (and thus a degraded performance level) and different wearing-out time for banks because of the characteristics of flash. The phenomenon is caused by two reasons: (1) the locality of write requests, and (2) an uneven capacity utilization distribution over banks.

Note that an uneven utilization distribution would result in not only different capacity utilizations of banks but also significant degradation of the wear-leveling effects over banks. As a result, multi-bank flash memory might be worn out much faster than single-bank flash memory. It is because a static bank assignment policy always dispatches write requests to their statically assigned banks. Some banks might have more hot data than others do. Since hot data are invalidated very often and result in many dead pages on their residing banks, their residing banks must do garbage collecting frequently. On the other hand, the banks which have more non-hot data might have a high capacity utilization, since non-hot data would stay on their banks for a longer period of time. Due to the uneven capacity distribution among banks, the performance of garbage collection on each bank might vary widely since the system performance also highly depends on the capacity utilization [10, 15].

To resolve the above issue, a dynamic bank assignment policy is presented: When a write request is received by the FTL driver, we propose to scatter page writes of the write request over banks which are idle and have free pages. The parallelism of multiple banks is achieved by switching over banks without having to wait for the completion of issued page-writes. The general mechanism is to choose an idle bank that has free pages to store the written data. One important guideline is to further achieve the "fairness" of bank usages by analyzing the attributes (hot or non-hot) of the written data:

Before a page write is assigned a bank address, the attributes of the written data must be identified. We propose to write hot data to the bank that has the smallest erase-count (which is the number of erases ever performed on the bank) for the consider-

337

ation of wear-leveling, since hot data will contribute more live page copyings and erases to the bank. The strategy in writing hot data prevents hot data from clustering on some particular banks. On the other hand, non-hot data are written to the bank that has the lowest capacity utilization to achieve a more even capacity utilization over banks. The strategy in writing non-hot data intends to achieve a more even capacity utilization distribution since non-hot data will reside at their written locations for a longer period of time. Because flash memory management already adopts a dynamic address translation scheme, it is intuitive to implement a dynamic bank assignment policy in FTL. In Section 4, we shall present some experimental results on the performance improvement based on striping and its relationship to garbage collection/capacity utilization.

C. A Management Scheme for Large-Scale Flash

The purpose of this research is on the minimization of the main-memory footprint and the amount of house-keeping data written for flash-memory management. The objective is to design a highly efficient large-scale flash-memory storage system with small overheads on main-memory usages. While many previously proposed flash-memory management schemes adopt one or few fixed granularity sizes for both space management and address translation, this section presents a buddy-system-based tree structure and an extendable-hash-based table for available and used space management of flash-memory, respectively [5].

C.1 Space Management

A *physical cluster* (PC) is defined as a set of contiguous pages on flash memory. The corresponding data structure for each PC is stored in the main memory. The status of a PC could be a combination of (free/live) and (clean/dirty). A free PC simply means that the PC is available for allocation, and a live PC is occupied by valid data. A dirty PC is a PC that might be involved in garbage collection for block recycling, where a clean PC does not. In other words, An LCPC, an FCPC, and an FDPC are a set of contiguous live pages, free pages, and dead pages, respectively. Similar to LCPC's, an LDPC is a set of contiguous live pages, but it could be involved in garbage collection.

The handling of PC's is close to the manipulation of memory chunks in a buddy system, where each PC is considered as a leaf node of a buddy tree. PC's in different levels of a buddy tree correspond to PC's with different sizes (in a power of 2). A tree structure of PC's is maintained in the main memory. The initial tree structure is a hierarchical structure of FCPC's based on their LBA's. In the tree structure all internal nodes are initially marked with CLEAN_MARK. On the splitting of an FCPC and a live PC (LCPC/LDPC) the internal nodes generated are marked with CLEAN_MARK and DIRTY_MARK, respectively. When a write request arrives, the system will locate an FCPC with a sufficiently large size. If the allocated

FCPC is larger than the requested size, then the FCPC will be split until an FCPC with the requested size is acquired. New data will be written to the resulted FCPC (i.e., the one with the requested size), and the FCPC becomes an LCPC. Because of the data updates, the old version of the data should be invalidated.

Garbage collection could be done based on the concept of PC: Consider the results of a partial invalidation on an 128KB LCPC (in the shadowed region) in Fig. 4. Let the partial invalidation generate internal nodes marked with DIRTY_MARK. Note that the statuses of pages covered by the subtree with a DIRTY_MARK root have not been updated on flash memory. A subtree is considered *dirty* if its root is marked with DIRTY_MARK. The subtree in the shadowed region in Fig. 4 is a proper dirty subtree, and the flash-memory address space covered by the proper dirty subtree is 128KB. The *proper dirty subtree* of an FDPC is the largest dirty subtree that covers all pages of the FDPC.

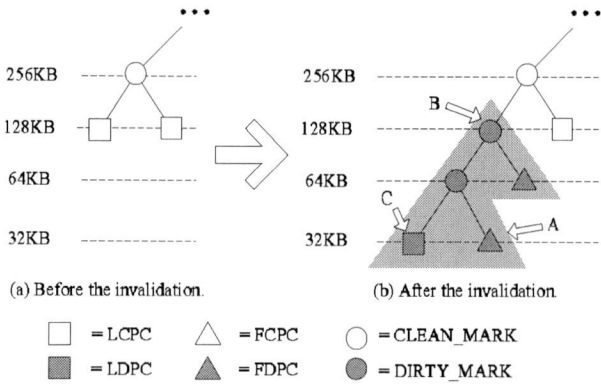

(a) Before the invalidation.   (b) After the invalidation.

☐ = LCPC   △ = FCPC   ○ = CLEAN_MARK
▨ = LDPC   ▲ = FDPC   ◉ = DIRTY_MARK

Fig. 4. A Proper Dirty Subtree with Two FDPC's and One LDPC.

There are three cases for space allocation when a new request arrives: The priority for allocation is on Case 1 and then Case 2. Case 3 will be the last choice.

Case 1: *There exists an FCPC that can accommodate the request.* The searching of such an FCPC could be done by a best-fit algorithm. That is to find an FCPC with a size closest to the requested size. Note that an FCPC consists of 2^i pages, where $0 \leq i$. If the selected FCPC is much larger than the request size, then the FCFC could be split according to the mechanism just presented in this Section.

Case 2: *There exists an FDPC that can accommodate the request.* The searching of a proper FDPC is based on the weight function value of PC's. We shall choose the FDPC with the largest function value, where any tie-breaking could be done arbitrarily.

Case 3: *Otherwise. (That is no single type of PC's that could accommodate the request.)* To handle such a situation, we should "merge" FCPC's and FDPC's repeatedly until an FCPC that can accommodate the request size appears.

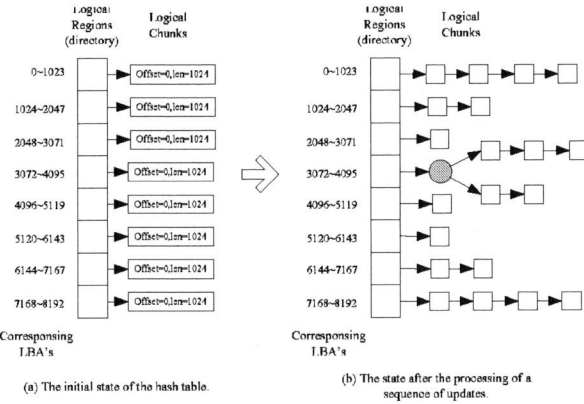

(a) The initial state of the hash table.

(b) The state after the processing of a sequence of updates.

Fig. 5. Example Layout of a Hash Table for Logical-to-Physical Address Translation.

C.2 Logical-to-Physical Address Translation

The space management unit for the proposed approach is a physical cluster (PC), instead of a page. A main-memory-resident hash table is proposed, where each hash entry is a chain of tuples for collision resolution. Each tuple (starting logical address, starting physical address, the number of pages) represents a *logical chunk* (LC) of pages in consecutive locations, and the number of pages in an LC does not need to be a power of 2.

The logical address space of flash memory is first exclusively partitioned into equal-sized regions referred to as *logical regions (LR's)*. Suppose that the total logical address space is from page number 0 to page number $2^n - 1$, and each LR is of 2^m pages. A dynamic-hashing-based method could be used as follows: Initially we have a directory which is a static array with 2^{n-m} entries, where one entry points to one *bucket*. Each LC is an LR in the beginning, and all LC's are hashed into the hash table, as shown in Fig. 5.(a). The hash function is defined as the first $(n - m)$ bits of a given logical address. When a bucket is overflowed, it is split into two buckets, and all of the LC's in the old bucket are distributed among the two bucket based on their corresponding logical addresses, as shown in Fig. 5.(b). Note that LC's in the hash table could also be split and merged to reflect the new logical addresses of a piece of data when invalidations and/or garbage collection occur.

IV. PERFORMANCE EVALUATION

A series of simulations was conducted to evaluate the capability of the proposed flash-memory management schemes. The trace of data access for performance evaluation was collected over a mobile PC with a 20GB hard disk, 384MB RAM, and an Intel Pentium-III 800MHz processor. The operating system was Windows XP, and the hard disk was formatted as NTFS.

A. Hot-Data Identification

Fig. 6 shows the ratio of false hot-data identification for the multi-hash-function framework (denoted as *basic* in the figure) and the framework with an enhanced counter update policy (denoted as *enhanced* in the figure), compared to the direct address method (that denoted an optimal method). Let X be the number of LBA's being identified as non-hot data by the direct address method but being identified as hot data by the (basic/enhanced) multi-hash-function framework for every 5117 writes. Y was 5117. The ratio of false hot-data identification for the (basic/enhanced) multi-hash-function framework was defined as (X/Y). As shown in Fig. 6, the enhanced multi-hash-function framework outperformed the basic multi-hash-function framework. When the hash table size reached 2KB, the performance of the (basic/enhanced) multi-hash-function framework was very close to that of the direct address method.

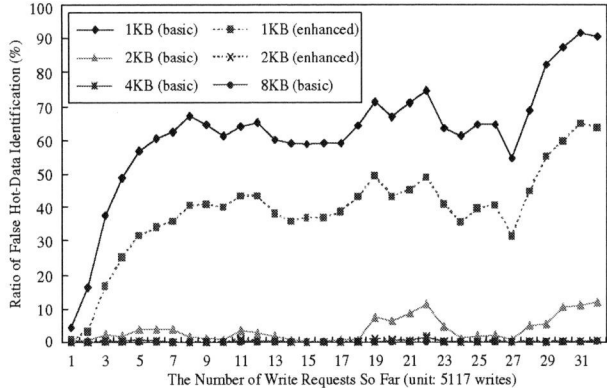

Fig. 6. Ratio of False Identification for Various Hash-Table Sizes.

Fig. 7 shows the performance gap achieved by the framework and the direct address method, when the decay period ranged from twice of the original setup to a quarter of the original setup. It was shown that the performance of the multi-hash-function framework was close to that of the direct address method when the decay period was about 1.25 of the original setup, i.e., a decay per 6396 writes. We should also point out that when the decay period was too large, the chance of false hot-data identification might increase more than expected because the results of "incorrect" counter increments would be accumulated. If we had to set the decay period as a unreasonably large number, then we should have a large hash table!

B. An Adaptive Striping Architecture

Fig. 8 shows the average response time of writes under various numbers of banks of an 8MB-flash storage system. The X-axis reflects the number of write requests processed so far in the experiments. The system performance was substantially improved when the number of banks increased from one to two

339

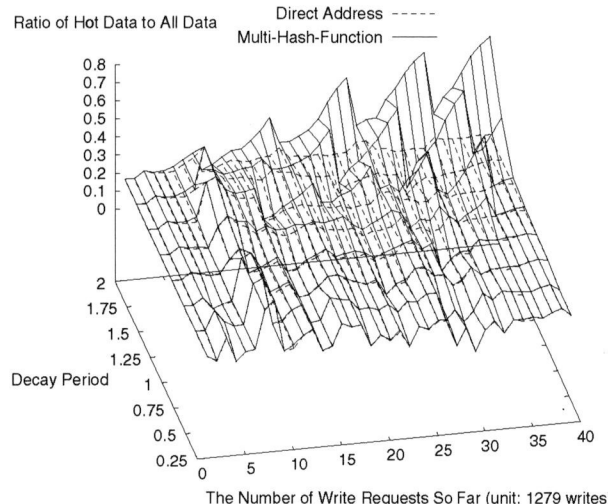

Fig. 7. The performance gap achieved by the multi-hash-function framework and the direct address method

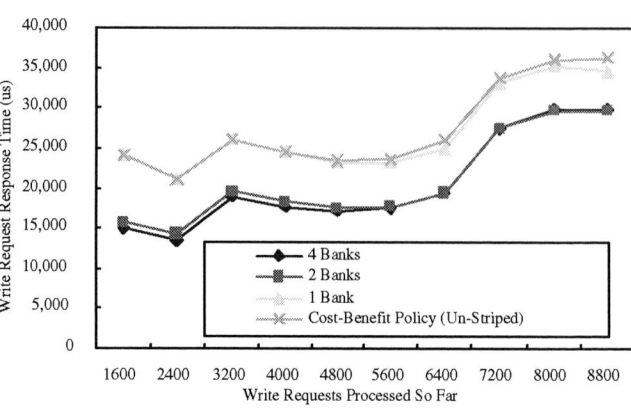

Fig. 8. Average Response Time of Writes under Different Bank Configurations.

because of more parallelism in processing operations. However, when the number of banks increased from two to four, the improvement was not so significant. One reason behind the observation was because W_{setup} was larger than W_{busy}. Another major reason behind the observation was because a larger number of banks mean a smaller capacity for each bank, where the total flash memory capacity was assumed being 8MB. Since the page size was fixed for all configurations in the experiment, a smaller capacity for a bank mean a relatively larger access unit, i.e., the page size, to a bank. As a result, garbage collection cost would be higher for a small bank, compared to a large bank. When the number of banks increased from two to four, the improvement due to striping was offset by the increased garbage collection overheads. Note that the garbage collection started happening after 3,200 write requests were processed in the experiments (due to the exhaustion of free pages). As a result, there was a significant performance degradation after the garbage collection activities began.

C. A Management Scheme for Large-Scale Flash

With a 20GB flash-memory storage system, a fixed-granularity scheme and the proposed variable-granularity scheme were evaluated over the multimedia data access patterns, as shown in TABLE I. The **Fixed Scheme** denotes the fixed-granularity scheme (with a granularity size provided), and the **Flexible Scheme** denotes the proposed variable-granularity scheme. The total number of pages actually written by the clients of the storage system was 41,943,168 pages. The proposed variable-granularity scheme reduces both the memory usage and the number of pages written in the experiments, and was proven being significantly better than the fixed-

granularity scheme.

Scheme	Footprint Size	Pages Written
Fixed Scheme, (1 page)	321MB	41,943,168
Fixed Scheme, (1 block)	10MB	52,106,912
Flexible Scheme	**3.18MB**	**41,943,168**

TABLE I
RESULTS OF EVALUATED SCHEMES UNDER THE MULTIMEDIA DATA ACCESS PATTERNS.

Figure 9 evaluates overheads imposed on the fixed scheme and the flexible scheme due to power-up initialization, where the X-axis denoted the granularity sizes in pages (for the fixed scheme), and the Y-axis denoted the time needed to completely initialize RAM-resident data structures. The space utilization was fixed at 96%. Under the fixed scheme, intuitively, when the granularity size was a 512B, a very long initialization time was observed (i.e., 1,342 seconds or 33,554,432 spare area reads). That was because every spare area of the entire flash memory needed to be scanned. On the other hand, with a 16KB granularity, it became significantly faster (i.e., 41 seconds or 1,048,576 spare area reads) because only the space area of the first page of every block needed to be scanned. Regarding the flexible scheme, since a significant portion of the entire flash-memory space could be managed by large PC's and only the spare area of the first page of a PC needed to be fetched, the flexible scheme took only 17 seconds (i.e., 434,111 reads of spare areas) for its initialization.

V. CONCLUSION

As high-capacity flash memory becomes much more affordable than ever, many existing flash-memory management

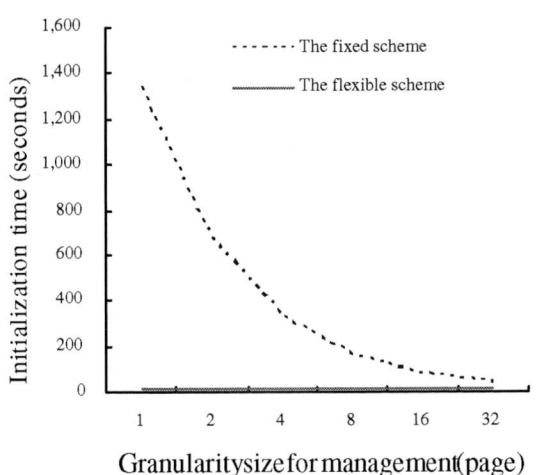

Fig. 9. Initialization Overhead of the Fixed Scheme and the Flexible Scheme.

methods might face serious overhead problems. In this paper, novel frameworks are presented to offer the flexibility in tuning up the system performance with overheads. The configurability in flash-memory management is considered from three aspects: (1) We present a highly efficient hash-based method to identify hot data effectively with limited memory-space overheads. The effectiveness of hot-data identification was shown with different memory-space overheads. (2) An adaptive striping mechanism is presented for multi-bank flash memory. Striping issues were explored with respect to system performance, bank utilization and garbage collection overheads. (3) A flash memory scheme with variable granularities is presented to minimize the number of pages written to the flash memory and the memory-space overheads. A series of experiments was conduced to demonstrate the trade-off between performance and overheads. For future research, we shall further address the reliability issues in flash management, with respect to the system performance.

REFERENCES

[1] Understanding the Flash Translation Layer (FTL) Specification. Technical report, Intel Corporation, Dec 1998.

[2] Compact Flash Association. $CompactFlash^{TM}$ 1.4 Specification, 1998.

[3] Burton H. Bloom. Space/Time Trade-offs in Hash Coding with Allowable Errors. *Communications of the ACM*, 13(7):422–426, July 1970.

[4] Li-Pin Chang and Tei-Wei Kuo. An Adaptive Striping Architecture for Flash Memory Storage Systems of Embedded Systems. In *IEEE Real-Time and Embedded Technology and Applications Symposium*, pages 187–196, 2002.

[5] Li-Pin Chang and Tei-Wei Kuo. An Efficient Management Scheme for Large-Scale Flash-Memory Storage Systems. In *ACM Symposium on Applied Computing (SAC)*, pages 862–868, Mar 2004.

[6] Li-Pin Chang and Tei-Wei Kuo. Real-time Garbage Collection for Flash Memory Storage Systems of Real-Time Embedded Systems. *ACM Transactions on Embedded Computing Systems (TECS)*, 3:837–863, Nov 2004.

[7] Li-Pin Chang, Tei-Wei Kuo, and Shi-Wu Lo. A Dynamic-Voltage-Adjustment Mechanism in Reducing the Power Consumption of Flash Memory for Portable Devices. In *IEEE Conference on Consumer Electronic (ICCE 2001)*, pages 218–219, Jun 2001.

[8] M. L. Chiang, Paul C. H. Lee, and R. C. Chang. Managing Flash Memory in Personal Communication Devices. In *Proceedings of the 1997 International Symposium on Consumer Electronics (ISCE'97)*, pages 177–182, Dec 1997.

[9] Aleph One Company. Yet Another Flash Filing System.

[10] F. Douglis, R. Caceres, F. Kaashoek, K. Li, B. Marsh, and J.A. Tauber. Storage Alternatives for Mobile Computers. In *Proceedings of the USENIX Operating System Design and Implementation*, pages 25–37, 1994.

[11] SSFDC Forum. $SmartMedia^{TM}$ Specification, 1999.

[12] Jen-Wei Hsieh, Li-Pin Chang, and Tei-Wei Kuo. Efficient On-Line Identification of Hot Data for Flash-Memory Management. In *Proceedings of the 2005 ACM symposium on Applied computing*, pages 838–842, Mar 2005.

[13] Atsuo Kawaguchi, Shingo Nishioka, and Hiroshi Motoda. A Flash-Memory Based File System. In *Proceedings of the 1995 USENIX Technical Conference*, pages 155–164, Jan 1995.

[14] M-Systems. Flash-memory Translation Layer for NAND flash (NFTL), 1998.

[15] M. Rosenblum and J. K. Ousterhout. The Design and Implementation of a Log-Structured File System. *ACM Transactions on Computer Systems*, 10(1), 1992.

[16] David Woodhouse. JFFS: The Journalling Flash File System. In *Ottawa Linux Symposium*, 2001.

[17] Chin-Hsien Wu, Li-Pin Chang, and Tei-Wei Kuo. An Efficient R-Tree Implementation over Flash-Memory Storage Systems. In *ACM 11th International Symposium on Advances on Geographic Information Systems (ACM-GIS)*, Nov 2003.

[18] Chin Hsien Wu, Tei-Wei Kuo, and Li-Pin Chang. Efficient Initialization and Crash Recovery for Log-based File Systems over Flash Memory. In *Proceedings of the ACM Symposium on Applied Computing (SAC'06)*, April 2006.

[19] Michael Wu and Willy Zwaenepoel. eNVy: A Non-Volatile Main Memory Storage System. In *Proceedings of the Sixth International Conference on Architectural Support for Programming Languages and Operating Systems*, pages 86–97, 1994.

Delay Defect Screening for a 2.16GHz SPARC64 Microprocessor

Noriyuki Ito, Akira Kanuma, Daisuke Maruyama, Hitoshi Yamanaka, Tsuyoshi Mochizuki, Osamu Sugawara, Chihiro Endoh, Masahiro Yanagida, Takeshi Kono, Yutaka Isoda, Kazunobu Adachi, Takahisa Hiraide[1], Shigeru Nagasawa, Yaroku Sugiyama, Eizo Ninoi

Fujitsu Limited, [1]Fujitsu laboratory
4-1-1 Kamikodanaka, Nakahara-ku,
Kawasaki, 211-8588, Japan
ito.noriyuki@jp.fujitsu.com

ABSTRACT

This paper presents a case-study of delay defect screening applied to Fujitsu 2.16GHz SPARC64 microprocessor. A non-robust delay test is used while each test vector is compacted to detect multiple transition faults in a standard scan-based design targeting a stuck-at fault test. Our test technique applied to a microprocessor designed with 6M gate logic, 4MB level 2 cache, and 239K latches, achieves 90% coverage using 3,103 test vectors. We estimate the distribution of the delay of paths covered by our delay test. We also show the effectiveness of our method by discussing the correlation between the screening result and the actual number of delay defects.

Keywords: Microprocessor, screening, delay defect, at-speed

1. Microprocessor Overview

Figure 1. Fujitsu SPARC64 Microprocessor

As the successor to the 1.3GHz SPARC64 microprocessor [1,2] designed in 130 nm technology, we have developed a 2.16GHz SPARC64 microprocessor [4] in 90 nm technology. The chip die image is shown in Figure 1. The right-hand area is the level 2 cache, and the left-hand area is the logic. We used a hierarchical incremental custom design methodology supported mainly by in-house CAD tools [3] which have been customized for high performance microprocessor design. Our SPARC64 microprocessors have the same reliability, availability, and serviceability (RAS) functions as the mainframe; they are employed in the UNIX servers used for a wide range of applications including the mission critical ones. The specification of our 2.16GHz SPARC64 microprocessor is as follows:

Process: 90nm, Cu metallization, 10 metal layers
Frequency: 2.16GHz
Die size: 18.46mm x 15.94mm

Transistor count: 400M
Level 2 on-chip cache: 4MB
I/O signals count: 279
Power dissipation: less than 65W

For testability purposes, scan chains are designed to scan data latches. Scan chains are selected and controlled by a test port controlling macro (TPCM). For the memory, a high speed test based on BIST is used in addition to the test through scan chains. For the delay test performed by two consecutive at-speed clock pulses, a 2-pulse generator is built in the chip. These design-for-testability circuits are verified based on logic simulation by a CAD tool to check testability design rules. The features of these testability circuits are as follows:

Scan clock: 2-phased clocks
Scan chain count: 16
Scan latch count: 238,620
Additional circuits: TPCM, 2-pulse generator, RBIST

The rest of this paper is organized as follows. In Section 2, the test flow from chip manufacturing to shipment of server products in which the microprocessor is used is overviewed. Section 3 presents the general concept of a delay test and our algorithm of test vector generation applied to a 2.16GHz SPARC64 microprocessor design. In Section 4, we show vector generation results, and compare the results with the results of using a robust test method. Finally, we conclude the paper in Section 5.

2. Delay Defect and Chip Screening

Delay defect is a defect that does not influence the functionality of a circuit but changes the speed of the circuit. The cause of this type of defect is mainly an excessive delay due to a high resistance by an open wire or interconnect via, or due to a large capacitance caused by bridging short circuits between wires [5]. Since this kind of delay defect is fatal in high performance microprocessors, it has to be detected as early as possible after manufacturing. Figure 2 shows our test flow from chip manufacturing to shipment of server products. At the end of wafer fabrication, a go/no-go test is performed for each chip using a set of functional test vectors. For good chips, a delay defect screening is performed. Chips that pass all these tests are shipped to a server product factory. In the server product factory, the chips are packaged after an acceptance test such as external inspection is performed. Then, a burn-in test is applied to induce time and stress dependent failures by applying thermal and electrical stresses. Next in a *speed binning* test, which determines the maximum functional operating speed of each chip, a test program is loaded into the memory of the chip. Then, speed binning at several frequencies is performed by executing the test program on the chip. After the maximum speed is determined for each chip, they are used inside units. Then, a unit test is performed under a high temperature and supply voltage. After the unit test, each system is configured according to a customer's requirement and the server is shipped to the customer after running a test for several days. In this paper, we refer to the test performed after the burn-in test as the system test.

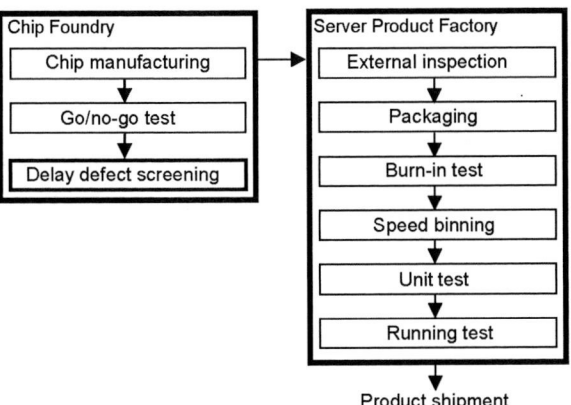

Figure 2. Test Flow of Server Products

In our test flow, we can reduce delay defects found after packaging by screening which is done through performing a delay test at wafer-level. Since the packaging cost is wasted if a defect is found after packaging, this screening is very important to reduce the manufacturing cost. In the delay test at the wafer-level, a set of two consecutive at-speed pulses is applied for each pair of test vectors at the same or a slower speed than the production specified speed. The important thing before adding a delay test step to the test flow is to confirm that a go/no-go result by a delay test correlates well with the delay defect found in a system test. To the best of our knowledge, few analyses on the correlation between delay test results and actual delay defects have been reported [6]. In [6,7] to confirm the correlation, chips failing the delay test are tested again at a lower speed. In our experiment, we confirmed it using the data of chips that failed the speed binning. If there is a good correlation and there is a possibility of reducing the manufacturing cost, the screening by the delay test is applied. Table 1 shows four possible cases depending on the results of the delay test at chip-level and the system test performed after packaging the chip. Case A is when a chip passes both delay and system tests. Case B is when a chip is *under-killed* by the delay test. The chip passes the delay test but fails a system test. Since the delay defect is found in a system test after packaging, the packaging cost is wasted. Case C is when a chip is *over-killed* by the delay test. The chip fails the delay test but passes the system test. In this case, screening regards a good chip that passes a system test as a defective chip. This means that good chips are wasted. Case D is when a chip fails both delay and system tests. By screening out chips with delay defects, there is no waste of manufacturing costs for packaging and the system test. When the delay test is used for screening, ideally cases B and C should not happen. The reason some chips are under-killed (Case B) can be

the low coverage achieved for critical paths and the low frequency of the clock applied for the delay test. The reason some chips are over-killed (Case C) is the excessive usage of test vectors that test functionally untestable paths [6,7]. A functionally untestable path is a path not activated by any combination of instructions in a microprocessor.

Table 1. Correlation between Delay Test and System Test

		Delay test	
		Pass	Fail
System test after packaging	Pass	Case A: Real pass No loss	Case C: Over-kill Loss
	Fail	Case B: Under-kill Loss	Case D: Real fail No loss

When the delay test is applied, the clock is set to the same frequency or a slower frequency than the target frequency of the chip. According to the frequency of the clock in the delay test, the percentages of chips categorized into cases A, B, C, and D change. In general, when the frequency increases, the percentages of cases A and B decrease and the percentages of cases C and D increase. Here, let N_A, N_B, N_C, and N_D be the numbers of chips categorized into cases A, B, C, and D, respectively. Further, let *UP*, *PC*, *STC*, and DTC be the unit price, the packaging cost, the system test cost, and the delay test cost respectively. When comparing the case the delay test is applied with the case the delay test is not applied, the loss of the manufacturing cost ΔLMC is expressed as,

$$\Delta LMC = N_D \cdot (PC + STC) - N_C \cdot UP - \sum_{i=A,B,C,D} N_i \cdot DTC \quad (1)$$

Applying the delay test is beneficial only if ΔLMC is positive. The value of ΔLMC depends on the ability of the delay test to detect actual delay defects and the frequency of the clock used in the test. If the ability of the delay test is fixed, it is possible to use a frequency which maximizes ΔLMC.

3. Delay Test

In a standard scan design, all latches are connected into a single or multiple scan chains through which the values of latches can be loaded and unloaded. In the delay test for a standard scan design, the test vector is loaded and the result is unloaded through scan chains. To generate test vectors for the delay test, *transition fault model* [8] or *path delay fault model* [9] is used. In the delay test based on the path delay model, paths to be tested are selected from critical paths. A path which causes an incorrect operation of a chip due to a small excessive delay can be tested with a high probability. However, the test coverage for all delay faults in the entire chip is typically low because only a small number of paths are tested; the reason is the limitations on test generation and application times. On the other hand, the coverage for all delay faults in the entire chip is higher if a delay test based on the transition fault model is used. Unfortunately the delay test may not test critical paths with a high probability. Since the coverage for all delay faults in the entire chip is very important if the delay test is used to screen out actual delay defects, we select the transition fault model instead of the path delay fault model to achieve a high coverage. In the transition fault model, *slow-to-rise* and *slow-to-fall* delay faults are assumed at all input and output pins of each gate in a circuit. A slow-to-rise fault is a fault that makes a rising transition slow, while a slow-to-fall fault is a fault that makes a falling transition slow. In this fault model, it is assumed the delay increase due to a defect is large enough to make a chip operate incorrectly at the desired clock speed. The delay test for detecting a transition fault is done using a pair of test vectors $\langle V_1, V_2 \rangle$. The delay test is performed using this vector pair and two consecutive at-speed clock pulses. The first test vector V_1 is loaded into launching latches through scan chains. Then, the first clock pulse is issued so that the launching latches capture outputs of the combinational circuit. The outputs of launching latches become inputs to the combinational circuit to generate a transition. The generated transition is rising for a slow-to-rise fault and falling for a slow-to-fall fault. The second at-speed clock pulse is issued to capture the generated transition at capturing latches. When a value captured by a capturing latch is the same as the value after a transition, there is no actual delay defect at the assumed location. When the captured value is the same as the value before a transition, there is an actual delay defect at the assumed location. Figure 3 shows the sequence of the abovementioned delay test. In our delay test, the second test vector is loaded into launching latches through the combinational circuit by issuing the first clock to latches. This technique is called *functional justification*. There are two other techniques to load the second test vector, *enhanced scan* and *skewed load*. [10] compares these three techniques. We use functional justification because it needs no additional hardware. Functional justification is less likely to test functionally untestable paths than other two techniques [10,11]. Thus, the possibility of over-kill in the functional justification is least among the three techniques.

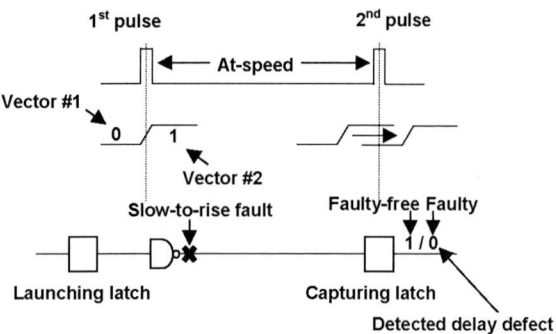

Figure 3. Delay Test by Two At-speed Clock Pulses

3.1 Design for Testability Circuits

Our delay test based on a standard scan design needs no additional hardware except for a circuit to generate two at-speed clock pulses. In this generator, a 2-pulse extractor and a selector are added after a PLL circuit as shown in Figure 4. The 2-pulse extractor is a circuit to extract two consecutive pulses from the PLL output and the selector selects the PLL output or the 2-pulse extractor output according to the test mode signal that controls the

selector. The selected clock pulses are supplied to the system clock distribution circuit.

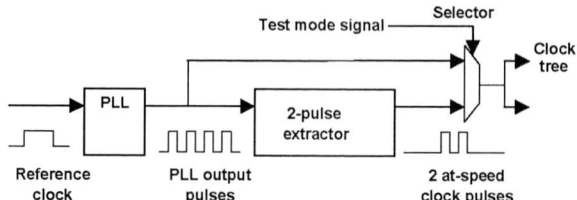

Figure 4. The Circuit Used to Generate Two At-speed Clock Pulses

3.2 Test Vector Generation

To generate test vectors for the delay test, our ATPG tool for stuck-at faults is enhanced in two areas. The first is to handle two consecutive time-frames corresponding to time frames before and after the first at-speed clock pulse. By this enhancement, the ATPG tool can generate a set of two test vectors by processing two time frames in the reverse order (i.e., t and t-1). In the original ATPG, expression (2) is applied only as a test generation condition to detect a stuck-at fault on line p. In the enhanced ATPG, expression (3) is added to generate a transition. Expression (2) corresponds to time frame t and expression (3) corresponds to time frame t-1. Our enhanced ATPG processes expression (2) before expression (3).

$$\text{state}(t,p) = T/F \tag{2}$$
$$\text{state}(t\text{-}1,p) = F/F \tag{3}$$

Here, the term state(t,p) in the left hand side of expression (2) denotes the state of line p at time frame t. The T/F in the right hand side represents value T in the fault-free circuit and value F in the faulty circuit. A slow-to-rise fault takes T=1 and F=0, and a slow-to-fall fault takes T=0 and F=1. The term state(t-1,p) in the left hand side of expression (3) denotes the state of line p at time frame t-1. The notation F/F in the right side represents value F in both the fault-free and faulty circuits. A slow-to-rise fault takes F=0 and a slow-to-fall fault takes F=1. In the second enhancement, a feature to generate multiple transitions is added to our ATPG. This is done by applying a dynamic compaction technique at time frame t-1 to detect as many faults as possible using a single test vector pair. Table 2 shows the comparison of off-path values in robust, non-robust, and our tests. A delay test is called *robust* if it can detect the target delay fault independent of other delay faults. On the other hand, it is called *non-robust* if the delay test for the target delay fault is invalidated by other delay faults. In a robust test, off-path values at time frame t-1 are set to the values they had at time frame t so that other delay faults do not invalidate the delay test. Since invalidation is permitted in a non-robust test, off-path values at time frame t-1 can take any arbitrary value. Our method [12] tries to generate as many transitions as possible by selecting different values for off-path signals at time frames t-1 and t. We denote these off-path values by A0 and A1 (Table 2) to distinguish them from don't care (X) used in the non-robust test. A0 (A1) means the value 0 (1) is selected if possible. When a rising transition on an on-path input of an AND or a NAND gate is propagated, the off-path input can be set to A0 (see Figure5). When

344

a falling transition on an on-path input of an OR or a NOR gate is propagated, the off-path input can take A1. Since in other two possible cases no transition can propagate on an off-path if a transition propagates on an on-path, off-paths can be set to neither A0 nor A1. For A0 (A1), a test vector compaction algorithm assigns 0 (1) value to off-path inputs. More details on this can be found in [12]. In the non-robust delay test described before, it is not possible to analyze a specific path because other delay faults may invalidate the test for that specific path. On the other hand in our method, delay faults that invalidate a test are also detectable. For example, consider an on-path and an off-path input of an AND gate. A slow-to-rise delay fault on an on-path input may invalidate a slow-to-rise delay test of the other input and vice versa. If only one of the inputs has a slow-to-rise delay fault, then the defect is detectable. Therefore, our method can detect delay faults using a small number of test vectors.

Table 2 Off-path conditions in Delay Test

	On-path		Off-path					
			Our Method		Non-Robust		Robust	
Time frame	t-1	t	t-1	t	t-1	t	t-1	t
AND/NAND	0	1	A0	1	X	1	1	1
	1	0	X	1	X	1	1	1
OR/NOR	0	1	X	0	X	0	0	0
	1	0	A1	0	X	0	0	0

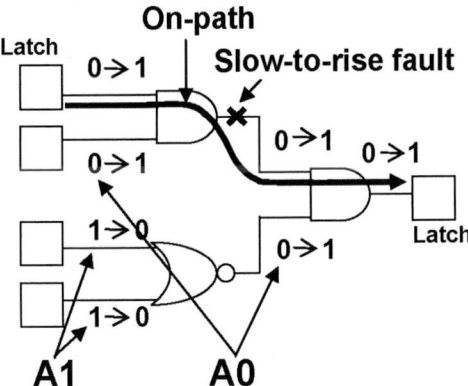

Figure 5. Generation of Multiple Transitions

4. Result

In this section, we summarize statistics such as the number of test vectors for each test item and the test vector generation time. Then, we present the followings: 1) a comparison of the number of test vectors and the test vector generation time for the robust test method and our method, 2) the delay distribution for paths covered by our delay test method, 3) the correlation between screening results and the actual delay defects.

4.1 Test Items and Generated Test Vectors

We have developed an in-house ATPG tool at Fujitsu and have been using it since the ECL mainframe era. The reason of developing an in-house tool is that an ATPG program needs to be optimized for design-for-testability circuits. All test vectors including the ones used for the delay test are generated by our ATPG program. Table 3 shows test vector generation results for a 2.16GHz microprocessor design described in Section 1. SCAN is a test for scan chains, FUNCTION is a functional test based on a stuck-at fault model, RBIST is a Built-In-Self-Test for memories, and DELAY is the delay test to screen out delay defects. We verify all test vectors from functional and timing viewpoints before they are applied to chips. The verification is done by an in-house test vector verification tool named VERIFIER. Using the delay data generated by a static timing analyzer in SDF format, VERIFIER verifies the expected values and timing by performing timing simulation. By this, it is possible to verify not only the expected values for input vectors but also the timing information reported by the static timing analyzer. Table 4 shows the CPU time needed for the verification.

Table 3. Test Items

Test	# Faults	# Vectors	Coverage	Time (Hours)
SCAN	9,059,216	14	99.9%	0.22
FUNCTION	21,803,669	2,014		14.15
RBIST	N/A	N/A	N/A	0.12
DELAY	9,750,387	3,103	90.0%	31.11

Table 4. Time Required for Verification and Test

Test	Verification (Hours)	Relative Verification Time
SCAN	101.02	2.9%
FUNCTION	16.80	0.5%
RBIST	3,346.04	96.1%
DELAY	16.24	0.5%

We generated test vectors using a Fujitsu 1.3GHz PRIMEPOWER, and we verified test vectors on IA servers with an Intel Pentium4 2.4GHz CPU. The verification time in Table 4 corresponds to the case when one CPU is used; we accelerated the verification by parallel execution on 12 CPUs. Overall, 2.9%, 0.5%, 96.1%, and 0.5% of the total time was used for the scan chain test, the function test, the memory test, and the delay test, respectively.

4.2 Comparison with Robust Delay Test

First we used the robust test to screen out delay defects, but it was not practical because of large generation time and large number of test vectors. Therefore, we compare our test with the robust test to show the effectiveness of our method. The left graph in Figure 6 shows the change of coverage with the number of test vectors. In our method, the coverage reaches 90% using only 3,103 test vectors. The robust test requires twice the number of test vectors to achieve the same coverage. The right graph in Figure 6 shows the change of coverage as a function of the execution time. In our method, the coverage reaches 90% in 31 hours, but it takes the robust test three times longer to reach the same coverage. Therefore, to screen out delay defects of microprocessors, using the robust test is not practical. Table 5 summarizes the above results. When performing our delay test, the maximum memory usage was about 5 GB.

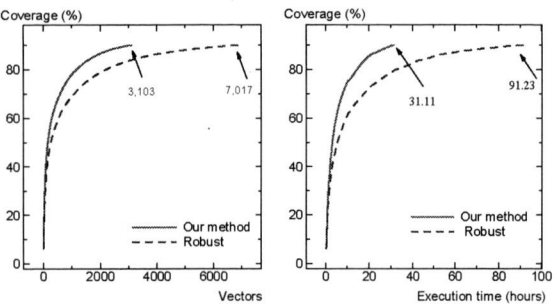

Figure 6. Change of Coverage of Delay Test

Table 5. Summary of delay test results

Test	# Faults	# Vectors	Coverage	Time (Hours)
Robust	9,749,914	7,017	90.0%	91.3
Our Method	9,750,387	3,103	90.0%	31.1

4.3 Delay of Paths Covered by Delay Test

The transition delay fault assumes an infinite excessive delay due to a defect. More than one path may pass location x where a delay fault is assumed. In a delay test based on a transition fault, the path with the maximum delay out of all paths passing location x is not necessarily tested by a test vector generated for this purpose. We estimate the distribution of the delay of paths covered by our delay test by VERIFIER described in Section 4.1. VERIFIER performs timing simulation using input vectors and delay information. Then simulated output values are compared against the expected values. In our verification tool, we can specify the time to compare the values (i.e., the strobe time). We performed the verification several times using different values for the strobe times; we used frequencies both lower and higher than the target frequency.

If the simulated values of latches are different than the expected values, the applied vector tests the delay of a path that is slower than the frequency corresponding to the strobe time. Therefore, we can get the delay distribution of paths tested by the applied test vectors from the number of latches in which simulated values are different than the expected ones. However, the exact number of paths is not known from the number of those latches. If the simulated value of a latch is different than its expected value, then some of the paths terminating at the latch are tested. For each test vector we count the number of latches whose simulated values are different than the expected values. Then, we calculate the sum of the number of latches for all test vectors. We plot the results in a

graph (see Figure 7). The graph *Dist_DT* shows the delay distribution of paths covered by the delay test. For *Dist_DT* the vertical axis corresponds to the numbers of latches. Figure 7 also shows *Dist_STA*, the actual delay distribution of paths reported by the static timing analyzer. The vertical axis in this case corresponds to the number of paths.

Since the scale of the vertical axis is different for *Dist_DT* and *Dist_STA*, we cannot compare the two graphs directly to find out how many paths are covered by the delay test. Therefore, we enlarged *Dist_DT* graph vertically so that two graphs intersect at frequency A.

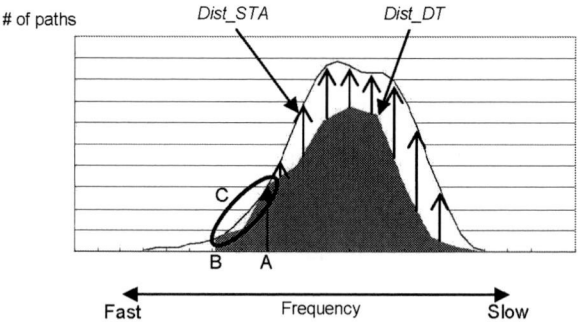

Figure 7. Delay Distribution of the Paths Covered by Delay Test

Since the fastest frequency at which VERIFIER can run is B, we collect data for frequencies less than it. Assume most paths with a speed between frequencies A and B in *Dist_STA* are almost fully covered by the delay test (see the circled area marked with C). On the other hand, *Dist_STA* is not fully covered by *Dist_DT* for frequencies faster than A. Therefore, we speculate paths with a speed faster than frequency A in *Dist_STA* are not fully covered by the delay test. However, we can see that the delay test can to some extent cover paths in each frequency according to the delay distribution *Dist_STA* although the coverage is smaller in low frequencies. Therefore, we can expect that screening by a delay test based on a transition fault model can also cover to some extent slower paths including critical paths. We report experimental results in Section 4.4.

4.4 Screening Results

We perform the screening for delay defects at 1.5GHz which is about 70% of the target frequency of the chip and at the normal operating voltage. The screening ratio for the total actual defective chips is 5.0%. In general, it is effective to detect delay defects at a lower voltage [13]. However, there is a risk to over-kill chips in a delay test at a lower voltage. Therefore, the important thing to do before actually applying a delay test to screening is to ensure that go/no-go results by a delay test correlate well with defective chips found by the system test. If the correlation between the delay test and the system test is confirmed, the screening for delay defects is applied if the value of ΔLMC in expression (1) is positive.

We apply the delay test at a lower voltage for chips that pass all functional tests as well as the delay test at the normal voltage. We set the clock frequency to 1.5GHz. Then, regardless of the result, speed binning by the system test is performed on the chips. We used about 4,000 chips in our experiment. Table 6 shows the ratios of pass and fail in the system test for both outcomes of the delay test. Since the defects found in the system test are not only delay defects but also memory defects, we exclude chips that have a memory defect. From the table, we can see that 1.7% of the chips that passed the system test failed the delay test at a lower voltage. Further, 46.7% of the chips that failed the system test also failed the delay test at a lower voltage. This means the delay test may waste 1.7% of good chips that pass the system test. Furthermore, 46.7% of packaged chips that fail the system test can be eliminated through screening by a delay test. At the time of this paper submission, we do not have delay test results for frequencies different than 1.5GHz due to the limited tester resource. If the frequency applied in the delay test is changed, we believe it is possible to achieve a detection ratio higher than 46.7%. In any case, the decision whether to apply it to an actual screening must be done based on the value of ΔLMC in expression (1). Otherwise, there is a risk of wasting good chips.

Table 6. Screening Result

		Delay test		Total
		Pass	Fail	
System test after packaging	Pass	98.3%	1.7%	100%
	Fail	53.3%	46.7%	100%

5. Conclusions

We presented a screening technique based on the non-robust delay test model that can detect multiple transition faults simultaneously using a single pair of vectors. We showed the results of applying the delay test to screen 2.16GHz microprocessors. Our delay test based on the standard scan design can generate vectors within practical limitations on the number of test vectors and generation time and it can achieve 90% coverage. By applying this delay test to delay defect screening, we can screen out chips with defects at the normal voltage. The screening ratio at the normal voltage is 5.0% for the total actual defective chips. This screening can reduce the manufacturing cost by detecting delay defects before chips are packaged even if the applied frequency is slower than the target frequency. By applying the delay test at a lower voltage for chips that pass the delay test at a normal voltage, it is shown that 46.7% of chips that fail at the system test may be screened out before packaging.

Since our delay test described in this paper is based on the transition fault model, it does not need any delay information of gates or wires when the test vector is generated. Therefore, the coverage of critical paths of the circuit may not be enough. To increase the coverage, it is necessary to test more critical paths. For this, some test vectors generated based on the path delay model should be added to our test vectors. We are planning to enhance

our ATPG tool to generate these test vectors using the delay information of gates and wires.

6. ACKNOWLEDGMENTS

We would like to thank Kaoru Kawamura and other members of CAD group at Fujitsu Labs. Ltd. as well as the members of Advanced CAD Technology group at Fujitsu Labs. of America for their helpful suggestions. We would also like to thank the members of Server CAD group and Technology group for supporting this development. Special thanks go to Yuji Oinaga for encouraging us to write this paper.

7. REFERENCES

[1] H. Ando, Y. Yoshida, et al, "A 1.3GHz Fifth Generation SPARC64 Microprocessor," Proceedings International Solid-State Circuits Conference, pp. 246-247, 2003.

[2] H. Ando, Y. Yoshida, et al, "A 1.3GHz Fifth Generation SPARC64 Microprocessor," Proceedings Design Automation Conference, pp. 702-705, 2003.

[3] N. Ito, H. Komatsu, et al, "A Physical Design Methodology for 1.3GHz SPARC64 Microprocessor," Proceedings International Conference on Computer Design, pp. 204-210, 2003.

[4] A. Inoue, "SPARC64^{TM} V/VI for Mission-Critical Servers," presented at Fall Processor Forum, 2004.

[5] K. Baker, G. Gronthoud, et al, "Defect-Based Delay Testing of Resistive Vias-Contacts A Critical Evaluation," Proceedings International Test Conference, pp. 467-476, 1999.

[6] G. Vandling, "Modeling and Testing the GEKKO Microprocessor, an IBM PowerPC Derivative for Nintendo," Proc. International Test Conference, pp.593-599, 2001.

[7] K. S. Kim, S. Mitra, et al, "Delay Defect Characteristics and Testing Strategies," IEEE Design & Test of Computers, vol.20, issue 5, pp.8-16, 2003.

[8] J. A. Waicukauski, et al, "Transition Fault Simulation by Parallel Pattern Single Fault Propagation," Proceedings International Test Conference, pp. 542-549, 1986.

[9] G. L. Smith, "Model for Delay Faults Based on Paths," Proceedings International Test Conference, pp. 342-349, 1985.

[10] J. Rearick, "Too Much Delay Fault coverage Is a Bad Thing," Proceedings International Test Conference, pp. 624-633, 2001.

[11] W. C. Lai, A. Krstic, et al, "On Testing the Path Delay Faults of a Microprocessor Using its Instruction Set," Proceedings VLSI Test Symposium, pp.15-20, 2000.

[12] D. Maruyama, A. Kanuma, N. Ito, et al, "Detection of Multiple Transitions in Delay Fault Test of SPARC64 Microprocessor", Proceedings International Conference on Computer Aided Design, pp. 893-898, 2004.

[13] P. Gillis, K. McCauley, et al, "Low Overhead Delay Testing of ASICs", Proceedings International Test Conference, pp. 534-542, 2004.

A Dynamic Test Compaction Procedure for High-quality Path Delay Testing

Masayasu Fukunaga[1], Seiji Kajihara[2], Xiaoqing Wen[2],
Toshiyuki Maeda[3], Shuji Hamada[3], and Yasuo Sato[3]

[1]Fujitsu Ltd., Kawasaki, Japan
fukunaga-m@jp.fujitsu.com
[2]Kyushu Institute of Technology, Iizuka, Japan
{kajihara, wen}@cse.kyutech.ac.jp,
[3]Semiconductor Technology Academic Research Center, Yokohama, Japan
{maeda,hamada.shuji,satoh.y}@starc.or.jp

Abstract - We propose a dynamic test compaction procedure to generate high-quality test patterns for path delay faults. While the proposed procedure generates a compact two-pattern test set for paths selected by a path selection criterion, the generated test set would detect not only faults on the selected paths but also faults on many unselected paths. Hence both high test quality by detecting untargeted faults and test cost reduction by reducing test patterns can be achieved. Experimental results show that the proposed procedure could generate a compact test set that detect many untargeted path delay faults certainly, compared with the static test compaction method previously proposed in [15].

I. Introduction

Path delay fault model [1] is known as a powerful delay fault model to detect defects which lead to the timing violation. Since the path delay fault model models localized as well as distributed excessive delays, test patterns generated for a path delay fault can detect most of other types of delay faults such as gate delay faults [2] on the path. However it is practically impossible to generate test patterns for all paths because there are a huge number of paths in a logic circuit. Therefore we need to select only a subset of paths and target it in test generation.

In path selection for test generation, it is important to select paths which are likely to be faulty, i.e. longer paths. Path selection criteria are categorized into two approaches, which are based on static timing analysis and statistical or dynamic timing analysis. The former selects structurally longest paths in the circuit, and are categorized into two approaches further. One approach is to select N longest paths in order of the path length. The length of any selected path is longer than the length of any unselected path. However since the selected paths may not be distributed all over the circuit and may be locally concentrated in a part of the circuit. The other approach is to select a set of paths which contains at least one of the longest paths through each line [3-8]. If we select paths based on this approach, the selected paths would be distributed all over the circuit. However the structurally longest paths may not be actual longest paths in a manufactured circuit due to process variation and/or circuit noise [9,10]. On the other hand, [11,12] tried to select actual

longest paths by using statistical or dynamic timing analysis. However it is difficult to know exact delay distribution of manufactured circuits. In addition, the longest paths may be different for each manufactured circuit. Hence actual longest paths cannot be selected necessarily.

A test generation method proposed in [13,14] selects two subsets of paths. For paths in the primary set consisting of structurally longest paths, test patterns are guaranteed to be generated. For paths in the secondary set consisting of next longest paths, fault detection is not guaranteed, but it is considered so as to maximize accidental detection by the test patterns for paths in the primary set.

Recently an idea of test generation for a given set of path delay faults was proposed [15]. A test generation procedure based on the idea can bring an effective solution for two major problems in test generation for path delay faults, namely reducing the number of test patterns and achieving high fault coverage against process variation and noise. In test compaction, while each two-pattern test is generated for more than one fault in the given fault list as well as ordinary test compaction methods, the faults simultaneously detected are selected such that paths with the faults have many cross points. When crossing paths on which there is a common gate are tested simultaneously, non-target paths consisting of partial paths of the paths can be accidentally sensitized and faults on the non-targeted paths can be detected simultaneously too. Note that the accidentally detected faults may not be included in the fault list. Hence even if longer paths in a manufactured circuit are not structurally long paths and not included in the target fault list, the compact test set generated by the method would detect the longer untargeted faults. Although a test compaction procedure based on this idea was given in [15], it is a simple static test compaction [16] and hence the advantage of the idea has not been derived very much.

In this paper we propose a dynamic test compaction procedure based on the concept described in [15]. Dynamic compaction [16] is a well-known compaction procedure with a higher ability of compaction than static compaction. While the proposed procedure selects a primary target fault and secondary target faults from the fault list, the secondary targeted faults are selected such that there are many cross points with the primary fault or other secondary faults processed in the test pattern. Experimental results showed that the size of test sets generated by the proposed dynamic compaction procedure is about six times smaller than that of uncompacted test sets, and 1.5 times smaller than that generated by the static test compaction procedure described in [15]. Though final fault coverage of the test sets by the proposed procedure is lower a little due to the much smaller size of the test sets, certainly detected faults by specified inputs at test generation are increased.

This paper is organized as follows. In Section 2, we explain the method of test compaction for high-quality path delay testing described in [15]. In Section 3, we propose a dynamic test compaction procedure to detect many untargeted path delay faults. In Section 4, experimental results and discussions are given. Finally, we conclude this paper in Section 5.

II. Path Delay Tests with Process Variation Tolerance

In [15] an idea of test generation for path delay faults has been proposed. The compaction method tries to generate not only compact test patterns for a given set of path delay faults but also detect many path delay faults not included in the fault set. The basic idea of this method is to test path delay faults on crossing paths simultaneously. Fig. 1 illustrates a simple example. Assume that there are two paths, $PI_1\text{-}g\text{-}PO_2$, and $PI_2\text{-}g\text{-}PO_1$ which cross at a gate g. If test patterns are generated for $PI_1\text{-}g\text{-}PO_2$ and $PI_2\text{-}g\text{-}PO_1$ separately, path delay faults on $PI_1\text{-}g\text{-}PO_1$ and $PI_2\text{-}g\text{-}PO_2$ may not be tested by the generated test patterns. On the other hand, if paths $PI_1\text{-}g\text{-}PO_2$ and $PI_2\text{-}g\text{-}PO_1$ are tested by a same test pattern t, then t can test both $PI_1\text{-}g\text{-}PO_1$ and $PI_2\text{-}g\text{-}PO_2$ simultaneously. When faults on the two paths, $PI_1\text{-}g\text{-}PO_1$ and $PI_2\text{-}g\text{-}PO_2$ are included in the fault list, we have no need to generate additional test patterns for them. This situation would lead to efficient generation of a compact test set. Even when two paths, $PI_1\text{-}g\text{-}PO_1$ and $PI_2\text{-}g\text{-}PO_2$ are not included in the fault list, t would enhance fault coverage.

In order to simultaneously test crossing paths on which there is a common gate, the paths must be satisfied with following conditions:

(1) The crossing paths have same transition at the common gate each other.
(2) The transition at the common gate is from the controlling value [22] of the gate to the non-controlling value.

Since paths that are likely to be faulty should be tested, longer paths are selected according to a criterion. Test patterns generated would detect path delay faults on the

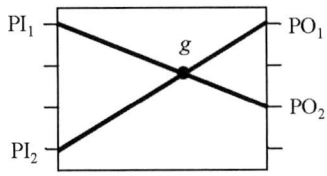

Fig. 1: Two paths with one cross point.

selected paths certainly if they are testable. However, it is difficult to predict the delay size of a path in manufactured circuits because of process variation or noise. As a result, there remain paths that are more likely to be faulty than the selected ones and the generated test patterns might miss a fault on the paths. Test patterns generated by this method, however, would detect not only faults on the selected paths but also some faults on unselected paths. If the unselected paths whose faults are accidentally detected consist of parts of the selected paths, the length of the unselected paths is relatively long because the selected paths are long. Therefore the test patterns potentially compensate the detection of untargeted faults.

III. The proposed dynamic compaction

A. Outline of the Proposed Procedure

In this section we propose a dynamic compaction procedure to generate test patterns for a given set of path delay faults which are selected with a criterion of path selection.

First we pick up an undetected fault from the fault list and generate an initial test pattern with unspecified bits for the fault. We call the fault a primary fault. Next, another

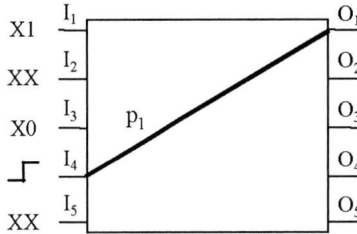

(a) Initial path selection and test generation

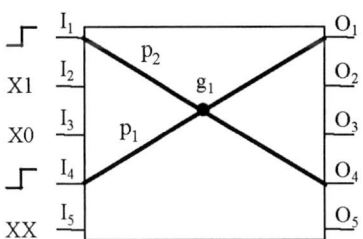

(b) Dynamic compaction for $p2$

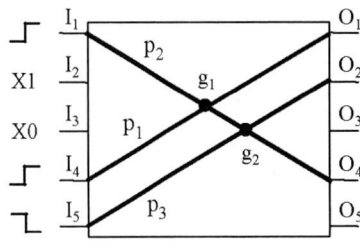

(c) Dynamic compaction for $p3$

Fig. 2: Outline of the proposed procedure

undetected fault, which is called a secondary fault, is picked up from the fault list. When the secondary fault is selected, we care whether the path of the secondary fault crosses the path of the primary fault or not. Test generation for the secondary fault tries assigning logic values to unspecified bits in the initial test pattern for the primary fault.

Fig. 2 illustrates the outline of the proposed dynamic compaction procedure. Suppose that path p_1, p_2 and p_3 is included in the given fault list. First, we pick up an undetected fault and generates an initial test pattern like Fig. 2 (a), where the path of the primary fault is path p_1, and the generated initial test pattern is $(I_1, I_2, I_3, I_4, I_5) = (X, X, X, 0, X)$, $(1, X, 0, 1, X)$. Next we search an undetected fault in the fault list that satisfies the following conditions:

(1) The fault is detected by filling unspecified bits of the initial test pattern.
(2) The path of the fault produces more sensitized paths additionally by crossing the path of the primary fault than any other undetected fault in the fault list.

In Fig. 2(b), assume that the path p_2 is satisfied with above conditions, and that the new test pattern, $(I_1, I_2, I_3, I_4, I_5) = (0, X, X, 0, X)$, $(1, 1, 0, 1, X)$, is generated by dynamic compaction. Two paths p_1 and p_2 are tested by the test pattern simultaneously, and more two additional paths, $I_1\text{-}g_1\text{-}O_1$, $I_4\text{-}g_1\text{-}O_4$ can be tested by the test pattern. After that we repeat to select secondary faults and to assign values for the detection of the secondary faults as long as there is a candidate of the secondary fault. Fig. 2(c) illustrates the repetition of the dynamic compaction. Path p_3 is processed after the dynamic compaction for p_2. In this case, more three additional paths can be tested by the test pattern, $(I_1, I_2, I_3, I_4, I_5) = (0, X, X, 0, 1)$, $(1, 1, 0, 1, 0)$. Finally, five paths in the given fault list can be tested by one test pattern, and five paths are tested by the test pattern additionally.

B. Selection of a primary fault

In selecting a primary fault, it is important to select one which gives a chance to cross with other paths to be selected as secondary faults. Such a path as a primary fault would satisfy the conditions below.

(1) The path has more off-inputs on each gate having the transition from the controlling value of the gate to the non-controlling value.
(2) The path has more fan-out branches associated with

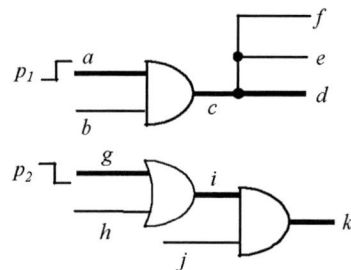

Fig. 3: Deference of the first selected path.

the path.

Fig. 3 shows two examples. Assume that two sub-circuits in Fig. 3 are a part of a logic circuit, and the path p_1, (a-c-d with rising transition at a), and the path p_2, (g-i-k with falling transition at g) are included in the fault list. The path p_1 can cross a path including b-c-e with rising transition at b and a path b-c-f with rising transition at b. If the three paths can be tested by the same test pattern, six paths can be tested additionally. On the other hand, the path p_2 can cross only a path h-i-k with falling transition at b. No other paths can be crossed with the two paths. Therefore when we select a primary fault to generate an initial test pattern, we had better care the above two conditions.

C. Detailed procedure

We implemented a dynamic test compaction procedure shown below. This procedure generates test patterns for faults in a given fault list with heuristics to detect as many untargeted path delay faults as possible by crossing paths in the fault list. Note that, in this procedure, P_{init} is the given fault list, and T_{fin} is the final test pattern set.

Step1: Select a primary fault p from P_{init}, which is undetected and satisfied with two conditions described in the previous section. If P_{init} is empty, go to Step 8. Otherwise go to Step 2.

Step2: Remove p form P_{init}, and go to Step 3.

Step3: If p is untestable, go to Step 1. Otherwise generates the initial test pattern t for p, which has unspecified bits, and go to Step4.

Step4: Search a fault q in P_{init} which is satisfied with below two conditions.

(1) The test pattern for q can be generated by filling unspecified bits of t.

(2) The path of q produces most additional paths by crossing the path p than any other paths in P_{init}.

If there is a path in P_{init}, which are satisfied with these conditions, go to Step 5. Otherwise go to Step 7.

Step5: Generate a new test pattern t' for q by filling unspecified bits of t. Remove q from P_{init}, define a new test pattern t' as new t, and go to step 6.

Step6: If P_{init} is not empty, return to Step 4. Otherwise go to Step 7.

Step7: Output the generated test pattern t to T_{fin}, and go to Step 8.

Step8: If P_{init} is not empty, return to Step1. Otherwise this procedure is finished.

IV. Experimental Results and Discussions

We implemented the proposed procedure of dynamic test compaction using C programming language on a PC (Pentium III Xeon 2GHz, 4GB memory) and applied it to full scan version of ISCAS'89 benchmark circuits. We assume single path delay fault, and refer to untestable paths as non-robust untestable paths in the rest of this paper. However, discussions in this paper are valid for any other sensitizing conditions of path delay faults. We constructed a given fault list such that one of the longest potentially testable paths through each line of the circuit are included. In the construction of a given fault list, we use the partial path sensitization method to identify untestable paths. Note that the length of a path is determined by the number of logic gates on the path. In our experiments, we compared with uncompacted test sets, compacted test sets generated by static compaction procedure described in [15] and compacted test sets generated by the proposed dynamic compaction procedure.

Table 1 shows statistics of each circuit in terms of testable paths and selected paths. The columns of Table 1 give the circuit name, the total number of logical paths i.e. path delay faults, the number of testable paths which can be calculated by ATPG for all logical paths, and the number of selected paths and the number of testable paths out of the selected paths. The given fault list consists of the selected paths. In the selected paths some untestable paths existed except for s35932 because of the incompleteness of the partial path sensitization method in path selection. However most of potentially testable paths were testable.

Table 2 gives the number of generated test patterns. The first column shows circuit names. The second column shows the results of test generation without test compaction where each test pattern is generated for an undetected fault in the fault list and fault simulation is performed for the generated test pattern after random-filling for unspecified bits. The third column shows the result of static test compaction by using the procedure introduced in [15]. The last column shows the results of the proposed test compaction procedure. Table 2 shows that the dynamic test compaction could generate about 6 times smaller test pattern sets than uncompacted test pattern sets on the average. In addition, compared with the test pattern sets generated by static test

compaction, the size of test pattern sets generated by dynamic test compaction is about 2/3. This result strongly suggests that the proposed dynamic test compaction procedure generates very small test sets.

Table 3 shows the number of path delay faults which are certainly detected by ATPG. In this result, accidentally detected path delay faults by random filling for remaining unspecified bits are not included. In Table 3, the second column and third column and fifth column show the number of certainly detected path delay faults by ATPG when using generated uncompacted or compacted test patterns. The forth column and the last column show the percentage of increased path delay faults that detected by the compact test patterns generated compared with the uncompacted test patterns. Table 3 shows that the test patterns generated by static compaction can certainly detect about 20% of more path delay faults than that by uncompacted test patterns, and the test patterns generated by dynamic compaction can certainly detect about 26% of more path delay faults than uncompacted test patterns. Therefore the generated test patterns can detect many untargeted path delay faults. In addition, the generated test pattern by dynamic compaction can detect more path delay faults than that by static compaction. Hence we should generate test patterns using dynamic compaction method.

Table 4 shows the coverage of untargeted path delay faults in 10000 longest testable paths. This experiment brings out how generated test sets cover structurally longest paths. In this experiment, we calculated the number of certainly detected paths by ATPG for an uncompacted test set, a compacted test set generated by static compaction and a compacted test set generated by dynamic compaction. Accidentally detected path delay faults by random filling for unspecified bits are not included in these results. In Table 4, the second column shows the number of untargeted path delay fault in 10000 longest testable paths. The third, fifth and seventh columns show the number of detected path delay faults in the untargeted path delay faults for each test pattern. The forth, sixth and eighth columns show the percentage of detected path delay faults for the number of untargeted paths delay faults in 10000 longest testable paths. From Table 4, we can observe that the generated compacted test patterns by dynamic compaction could certainly detect about 10% more path delay faults compared with the

Table 1: The number of selected paths and testable paths.

circuit	#total paths	#testable paths	#selected paths	#testable paths in selected paths
s5378	27,084	21,928	4,170	4,133
s9234	489,708	59,854	5,193	5,159
s13207	2,690,738	476,145	8,792	8,723
s15850	329,476,092	10,782,994	10,027	9,950
s35932	394,282	58,657	28,549	28,549
s38417	2,783,158	1,138,194	28,713	27,496
s38584	2,161,446	334,927	30,891	30,730

Table 2: The size of test pattern sets.

circuit	uncompacted tests	compacted tests (static compation)	compacted tests (dynamic compation)
s5378	920	254	173
s9234	1287	429	296
s13207	1441	511	384
s15850	1855	922	456
s35932	257	32	24
s38417	6799	833	596
s38584	2803	522	347

uncompacted test patterns. Although the generated compacted test patterns by static compaction could certainly detect many untargeted path delay faults, the number of detected paths are less than that by dynamic compaction From this result, since the compacted test patterns can detect many longer paths, even if critical paths are distributed by process variation or electrical noise, real critical path delay faults would detect by the compact test patterns. Therefore, to generate the compacted test patterns by the proposed method is to generate high quality compact test patterns for recent DSM circuits.

Table 5 shows the number of detected faults and fault efficiency of each test pattern set. In this experiment, unspecified bits of each test pattern are filled with 0 or 1 at random. Therefore detected faults include in faults accidentally detected by random-filling. From Table 5, we can observe that uncompacted test pattern sets have higher fault efficiency than the other compacted test sets. Since the number of uncompacted test pattern sets is larger than that of compacted test sets and uncompacted test sets would have many unspecified bits, by filling unspecified bits with 0 or 1 at random, accidentally detected faults would be increase.

Table 3: Certainly detected path delay faults.

| circuit | uncompacted tests | compated tests | | | |
| | | static compaction | | dynamic compaction | |
	#detected faults	#detected faults	%increase	#detected faults	%increase
s5378	7252	9345	28.86%	10110	39.41%
s9234	10804	14209	31.52%	15136	40.10%
s13207	80001	81961	2.45%	82068	2.58%
s15850	365360	450920	23.42%	483689	32.39%
s35932	34719	37027	6.65%	36307	4.57%
s38417	104912	140404	33.83%	154017	46.81%
s38584	82144	91158	10.97%	92989	13.20%
average			19.67%		25.58%

Table 4: Coverage for 10000 longest testable paths by certainly detected path delay faults.

| circuit | untergeted path delay faults | uncompacted tests | | compacted tests | | | |
| | | | | static compaction | | dynamic compaction | |
		#detected untergeted	%coverage	#detected untergeted	%coverage	#detected untergeted	%coverage
s5378	7672	663	8.64%	1736	22.63%	2043	26.63%
s9234	9463	1217	12.86%	1499	15.84%	1890	19.97%
s13207	9205	4199	45.62%	4209	45.73%	4186	45.48%
s15850	9734	2752	28.27%	3470	35.65%	3684	37.85%
s35932	4324	1522	35.20%	2163	50.02%	2222	51.39%
s38417	8413	3079	36.60%	4122	49.00%	4356	51.78%
s38584	9563	2717	28.41%	2847	29.77%	2669	27.91%
average			27.94%		35.52%		37.29%

Table 5: Fault efficiency of each test pattern set.

| circuit | uncompacted tests | | compated tests | | | |
| | | | static compaction | | dynamic compaction | |
	#detected faults	fault efficiency	#detected faults	fault efficiency	#detected faults	fault efficiency
s5378	17240	78.62%	15992	72.93%	15436	70.39%
s9234	31174	52.08%	27509	45.96%	27118	45.31%
s13207	167612	35.20%	146771	30.82%	138209	29.03%
s15850	1537904	14.26%	1411797	13.09%	1293744	12.00%
s35932	57904	98.72%	46005	78.43%	44172	75.31%
s38417	405046	35.59%	346901	30.48%	332759	29.24%
s38584	180556	53.91%	167841	50.11%	152199	45.44%
average		52.63%		45.98%		43.82%

On the other hand, compacted test pattern sets would have a small number of unspecified bits in itself. Even if certainly detected faults are increased by the proposed dynamic compaction method, the total number of detected faults is not increased so much. However these accidentally detected faults cannot guarantee the quality of generated test pattern, since they may not be detected by another random-filling for unspecified bits. To guarantee the quality of generated test pattern set, many faults should be detected certainly. Therefore generated test pattern sets by the proposed dynamic compaction have high quality to guarantee the circuit operation.

V. Conclusions

In this paper, we proposed a dynamic compaction procedure to test paths with cross points simultaneously so as to accidentally detect many faults which may not be included in the target fault list. The proposed procedure for path delay faults brought improvement of test quality in spite of reduction of test patterns. Experimental results showed that the proposed procedure could generate a compact two-pattern test set and it could detect many untargeted path delay faults efficiently. Our future work is to improve the heuristics algorism to detect more certainly detected paths.

Acknowledgements

This work was supported by the New Energy and Industrial Technology Development Organization (NEDO).

References

[1] G. L. Smith, "Model for delay faults based upon paths," International Test Conf., pp.342-349, 1985.

[2] Z.Barzilai and B.K.Rosen, "Comparison of AC Self-testing Procedures," International Test Conf., pp.89-91, 1983.

[3] W.-N. Li, S. M. Reddy, S. K. Sahni, "On Path Selection in Combinational Logic Circuits," IEEE Trans. on CAD., vol.8, no.1, pp.56-63, 1989.

[4] A. Murakami, S. Kajihara, T. Sasao, I. Pomeranz, and S. M. Reddy, "Selection of Potentially Testable Path Delay Faults for Test Generation," International Test Conf., pp. 376-384, 2000.

[5] M. Sharma and J. H. Patel, "Finding a Small Set of Longest Testable Paths that Cover Every Gate," International Test Conf., pp.974-982, Oct. 2002.

[6] Y. Shao, S. M. Reddy, I. Pomeranz, S. Kajihara, "On Selecting Paths to Test in Scan Designs," Journal of Electronic Testing Theory and Applications, volume 19, pp.447-456, August 2003.

[7] W. Qiu and D. M. H. Walker, "An Efficient Algorithm for Finding the K Longest Testable Paths Through Each Gate in a Combinational Circuit," International Test Conf., pp.592-601, Sept. 2003.

[8] S. Tragoudas, S. Padmanaban, "A Critical Path Selection

Method for Delay Testing," International Test Conf., pp. 232-241, Oct. 2004.

[9] L.-C. Chen, S. K. Gupta and M. A. Breuer, "High Quality Robust Tests for Path Delay Faults", in Proc. VLSI Test Symp., pp.88-93, April 1997.

[10] K.-T. Cheng, S. Dey, M. Rodgers, K. Roy. "Test Challenges for Deep Sub-Micron Technologies," Design Automation Conf., pp.142-149, June 2000.

[11] J.-J. Liou, A. Krstic, Y.-M. Jiang and K.-T. Cheng, "Path Selection and Pattern Generation for Dynamic Timing Analysis Considering Power Supply Noise Effects", International Conf. on Computer-Aided Design, pp.493-496, Nov. 2000.

[12] J.-J. Liou, A. Krstic, L.-C. Wang, K.-T. Cheng. "False-Path-Aware Statistical Timing Analysis and Efficient Path Selection for Delay Testing and Timing Validation. Design Automation Conf., pp.566-569, 2002.

[13] I. Pomeranz and S. M. Reddy, "Test Enrichment for Path Delay Faults Using Multiple Sets of Target Faults", Conf. on Design Automation and Test in Europe, pp.722-729, March 2002.

[14] I. Pomeranz and S. M. Reddy, "A Postprocessing Procedure of Test Enrichment for Path Delay Faults", Asian Test Symposium, pp.448-453, Nov. 2004.

[15] S. Kajihara, M. Fukunaga, X. Wen, T. Maeda, S. Hamada, Y. Sato,"Path Delay Test Compaction with Process Variation Tolerance," Design Automation Conference, pp. 845-850, June 2005.

[16] P. Goel and B. C. Rosales, "Test Generation & Dynamic Compaction of Tests", in Digest of Papers 1979 Test Conf. , pp.189-192, Oct. 1979.

[17] I. Hamzaoglu, J.H. Patel, "Compact two-pattern test set generation for combinational and full scan circuits," International Test Conf., pp.944-953, Oct. 1998.

[18] S. Kajihara, I. Pomeranz, K. Kinoshita and S. M. Reddy, "Cost-Effective Generation of Minimal Test Sets for Stuck-at Faults in Combinational Logic Circuits," IEEE Trans. Computer-Aided Design of Integrated Circuits and Systems, vol.14, no.12, pp.1496-1504, Dec. 1995.

[19] I. Hamzaoglu and J. H. Patel, "Test Set Compaction Algorithms for Combinational Circuits," Intl. Conf. on Computer-Aided Design, pp. 283-289, Oct. 1998.

[20] S. Bose, P. Agrawal, V. Agrawal, "Generation of compact delay tests by multiple path activation," International Test Conf., pp.714-723, Oct. 1993.

[21] J. Saxena; D.K.Pradhan, "A method to derive compact test sets for path delay faults in combinational circuits," International Test Conf., pp. 724-733, Oct. 1993.

[22] M. Abramovici, M. A. Breuer, A. D. Friedman, Digital Systems Testing and Testable Design, Piscataway, NewJersey: IEEE Press, 1990.

Delay Variation Tolerance for Domino Circuits

Kai-Chiang Wu, Cheng-Tao Hsieh, Shih-Chieh Chang

Department of CS, National Tsing Hua University, Hsinchu, Taiwan

Alexe@nthucad.cs.nthu.edu.tw, jdshieh@nthucad.cs.nthu.edu.tw, scchang@cs.nthu.edu.tw

ABSTRACT

Factors of delay variation, such as process variation and noise effects, may cause a manufactured chip to violate the pre-specified timing constraint. In this paper, we propose a novel re-synthesis technique to tolerate delay variation for domino circuits. Note that the slacks of nodes along critical paths are zero; any delay addition to those zero-slack nodes will worsen the final performance of a circuit. Our basic idea is to increase the slacks of nodes in the critical region by appending a redundant auxiliary sub-circuit to the original circuit. The auxiliary sub-circuit can cause critical paths to become false paths or imperceptible paths [7] so as to improve the capability of delay variation tolerance. Experimental results are very encouraging.

1. Introduction

Circuit delay in advanced technologies becomes increasingly sensitive to process variation and noise [1][2]. Those factors would cause a circuit's performance to fluctuate and, in the worst case, a timing violation may occur. Especially for high performance designs, timing critical regions are often implemented in domino circuit style. As a result, the delay variation problem becomes a critical issue for domino circuits. In this paper, we propose a re-synthesis method to tolerate delay variation for a domino circuit.

The slacks of nodes along critical paths are zero; any delay addition to those zero-slack nodes will worsen the final performance of a circuit. The degree of delay variation tolerance in [3] is formulated by the concept of slacks. The authors proposed a re-synthesis technique which incorporates Triple Module Redundancy (TMR) like structure into a circuit so that the circuit can tolerate a given range of delay uncertainty. Their results show that by adding 40% area overhead, a certain degree of variation tolerance can be achieved for static CMOS circuits.

Unlike static circuits, dynamic domino circuits operate in two phases: the pre-charge phase and the evaluation phase. To avoid the racing problem, domino circuits require all

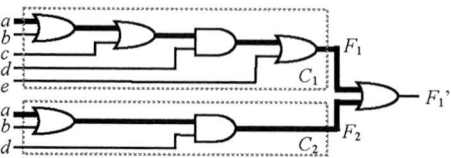

Figure 1: An example demonstrating delay variation tolerance for a domino circuit.

signals, except primary inputs, to have only rising transitions during the evaluation phase. Due to the rising-transition-only property, delay variation tolerance in domino circuits is much easier to accomplish than that in static circuits. In other words, directly applying the same method as [3] to domino circuits may have an unnecessarily large area.

Our basic idea of delay variation tolerance is to combine a (target) domino circuit with a redundant auxiliary sub-circuit, which is illustrated in the following example. A domino circuit C_1 in Figure 1 implements logic function $F_1 = (a+b+c)d+e = ad+bd+cd+e$. For delay tolerance on C_1, we construct an auxiliary circuit C_2 implementing $F_2 = (a+b)d = ad+bd$ and generate a new output $F_1' = F_1+F_2$. Because the on-set of F_1 apparently covers that of F_2 (i.e., $F_1 \supseteq F_2$), the new function $F_1' = F_1+F_2$ is identical to the original function F_1. We will show that appending circuit C_2 does not change the functionality of the original circuit but has the effect of delay tolerance. Consider an input pattern $(a, b, c, d, e) = (1, 0, 0, 1, 0)$ which induces transitions propagating along critical (highlighted) paths in both C_1 and C_2. Since both F_1 and F_2 have only rising transitions and feed into an OR gate, whichever rising signal arrives earlier will dominate the OR gate's output. In other words, the earlier arrival of either F_1 or F_2 determines the output value instead of the late one. Delay variation tolerance is consequently achieved because late arriving signals will not influence the whole circuit's delay.

In addition to proposing a new structure for domino circuits, this paper also derives novel theorems which allow us to carry out a smaller C_2 than those from [3]. We also show that delay variation tolerance can be applied to both internal nodes and primary outputs to achieve better results.

0-7803-9451-8/06/$20.00 ©2006 IEEE.

354

The experiments show that the area overhead of our method is less than a half of that from [3]. Also, we have performed Monte-Carlo experiments. The delay of a gate is given as a probability density function similar to [6]. The results show that about 77% of samples of the original circuit $C6288$ can meet a certain delay requirement; however, about 99% of samples of the corresponding re-synthesized circuit can satisfy the same requirement.

We would like to mention that there have been several studies [4][5][7] attempting to optimize the statistical timing results by gate sizing techniques. These methods require statistical models to be given for all gates. We think that if statistical distributions of gates and statistical correlations among gates are given precisely, methods [4][5][7] can be more efficient than redundant structures as described by this paper. On the other hand, this paper does not assume any statistical timing models. Therefore, our method will be better when accurate statistical models are not available. For example, very few chips are manufactured for new technologies, such as 90nm and 65nm, so it is difficult to gather or to verify statistical models for new technologies. In addition to process variation, delay variation due to noise issues, such as IR drop, is not easy to model.

2. Delay Variation Tolerance in Duplex Domino Systems

Let a domino circuit C_1 be under consideration for delay variation tolerance. A duplex system is shown in Figure 2, which consists of the original circuit C_1, its duplication C_2, and an OR gate taking the outputs of C_1 and C_2 as its inputs. There are two properties for a duplex domino system. First, it performs the same function as the original circuit. Secondly, a transition traveling along a path in C_1 also travels along the mirror path in C_2. Since a domino circuit has only 0-to-1 transitions, the OR gate's output transits from 0 to 1 when the earlier transition arrives. Any delay increase which is postponing either C_1 or C_2 will not affect the eventual timing result. Therefore, each node (except the OR gate) in the

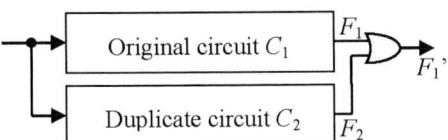

Figure 2: A duplex system.

duplex system has an infinite slack. Still, there is more than 100% area overhead for a duplex system, making the duplex system impractical. In the remainder of this paper, we will explain how to reduce C_2 while maintaining a given degree of delay variation tolerance.

Let us first define the degree of delay variation tolerance. The degree of delay variation tolerance can be quantified with the smallest slack of gates/wires in a circuit. A circuit is defined as having d_t *delay tolerance* [3] if the slack of each gate/wire is at least d_t. Given a delay tolerance value d_t and a circuit, our objective is to re-synthesize the circuit so that every gate/wire in the new circuit can tolerate at least delay variation d_t. The slack of each gate/wire in a duplex domino system is infinite (i.e, $d_t = \infty$), which is over-protective for the delay variation problems. In general, delay tolerance of 10%~20% of the original circuit delay is sufficient for our consideration of process variation and noise effects.

3. Re-synthesis for Delay Variation Tolerance

To accomplish a given delay tolerance value d_t without adding too much area overhead, a practical scheme originated from a duplex system is proposed. Our re-synthesis steps are as follows. (1) Begin with a duplex system in Figure 2. (2) Remove and modify some wires in C_2 for area reduction while maintaining the tolerance value. We now discuss how to perform wire removal and modification in C_2.

Assume the original circuit C_1 implements the logic function F_1, the redundant auxiliary circuit C_2 implements F_2, and the combinative output is $F_1' = F_1 + F_2$. If the on-set of F_2 is a sub-set of that of F_1 (i.e., $F_2 \subseteq F_1$), we can preserve the original functionality of F_1' ($= F_1 + F_2 = F_1$). While there are many possible Boolean functions which satisfy $F_2 \subseteq F_1$, we only consider wire removal in C_2 to reduce the on-set of F_2. Before presenting our theorems, we describe what wires in C_2 can be removed to maintain $F_2 \subseteq F_1$ in the following lemma.

Lemma 1: All direct input wires to OR gates in C_2 are redundant and can be **simultaneously** removed.

Proof: Omitted.

Let the delay tolerance value be d_t. A path is called a d_t-*critical path* if its delay is greater than the timing requirement minus d_t. In other words, d_t delay increment on a d_t-critical path will cause the path's delay to exceed the

4A-3

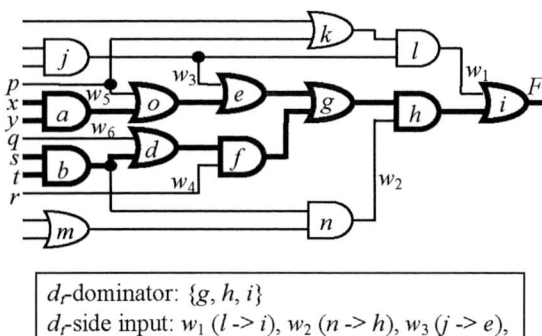

d_r-dominator: $\{g, h, i\}$
d_r-side input: w_1 ($l \rightarrow i$), w_2 ($n \rightarrow h$), w_3 ($j \rightarrow e$),
$\quad\quad\quad\quad\quad w_4$ ($r \rightarrow f$), w_5 ($p \rightarrow o$), w_6 ($q \rightarrow d$)

Figure 3: The original circuit.

timing requirement. In addition, we say that a node is a d_r-*critical node* if it is along a d_r-critical path. One can also find that the slack of a d_r-critical node is smaller than d_t. Consider the example in Figure 3, where the delay of each gate is 1. Suppose the delay tolerance d_t is 1 and the timing requirement d_r is 6. In this example, the timing requirement is equal to the length of the longest path. Path $\{s\text{-}b\text{-}d\text{-}f\text{-}g\text{-}h\text{-}i\}$ is a d_r-critical path because its delay is 6, greater than d_r-d_t (6-1=5). In fact, all highlighted paths in Figure 3 are d_r-critical paths. Node g is a d_r-critical node because it is along a d_r-critical path. Besides, all highlighted nodes $\{a, b, o, d, e, f, g, h, i\}$ are d_r-critical nodes. Note that the slacks of these d_r-critical nodes are 0, smaller than d_t.

A node is defined to be a d_r-*dominator* if it is a d_r-critical node and all d_r-critical paths to a primary output must pass through the node. A wire $n_1 \rightarrow n_2$ is a d_r-*side input* if node n_1 is not a d_r-critical node but node n_2 is a d_r-critical node. In the same example, nodes $\{g, h, i\}$ are d_r-dominators because all d_r-critical paths to the primary output must pass through these nodes. Wire w_1 ($l \rightarrow i$) is a d_r-side input because node l is not a d_r-critical node but node i is. Similarly, wires $\{w_2, w_3, w_4, w_5, w_6\}$ are also d_r-side inputs.

Theorem 1 [3]: A d_r-side input wire w to an OR d_r-dominator in C_2 can be removed (replaced by a non-controlling value, i.e. a logic 0) without violating the requirement of d_t delay tolerance.

Proof: Omitted.

Take the original circuit in Figure 3 as an example. A duplex system can be constructed by two duplicates (C_1 and C_2) of the original circuit. According to Theorem 1, we can remove wire w_1 in C_2 by replacing it with a logic 0 since it is a d_r-side input to OR d_r-dominator i.

We say that a node is a *transitive fanout* of wire w if

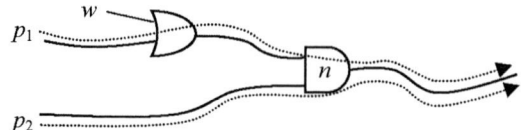

Figure 4: Node n is the AND-converging node of d_r-critical paths p_1 and p_2.

there is a path from wire w to the node. In addition, in Figure 4, two paths may "converge" on a node, which is called the *converging* node of the two paths. We define node n to be the AND-*converging node* of two d_r-critical paths if these two paths converge on node n and node n is an AND gate. For example in Figure 4, d_r-critical paths p_1 and p_2 converge on an AND gate n so node n is the AND-converging node of p_1 and p_2. We have the following theorem.

Theorem 2: Let wire w be a d_r-side input to an OR gate in C_2. If there is no AND-converging node of d_r-critical paths in wire w's transitive fanout, wire w can be removed without violating d_t delay tolerance.

Proof: Omitted.

For example, wire w_3 ($j \rightarrow e$) can be removed according to Theorem 2. First, wire w_3 is a d_r-side input to OR gate e. The transitive fanout nodes of w_3 consist of $\{e, g, h, i\}$, among which OR gate g is the only converging node of d_r-critical paths. Since there is no AND-converging node in w_3's transitive fanout, wire w_3 in C_2 can be removed. In fact, wires $\{w_3, w_5, w_6\}$ all satisfy the condition in Theorem 2 so wires $\{w_3, w_5, w_6\}$ in C_2 are removable. The resultant circuit after removing wires w_1 (by Theorem 1), w_3, w_5, and w_6 in C_2 is shown in Figure 5.

We can also adopt signal sharing as in [3] to further reduce the area overhead. Two signals which implement the

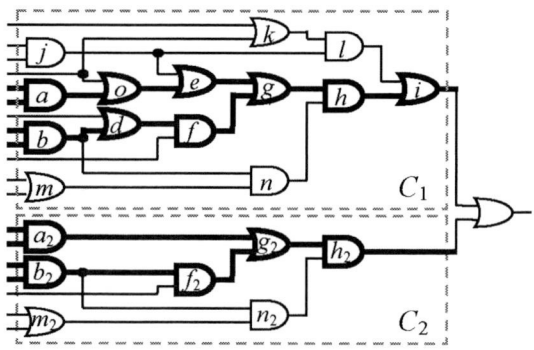

Figure 5: The re-synthesized circuit after wire removal.

356

equivalent Boolean function but do not belong to d_t-critical paths can be shared. For example in Figure 5, the output of node n in C_1 and that of node n_2 in C_2 have the same functionality. We can share the output signals of n and n_2, as demonstrated in Figure 6. Suppose all equivalent signals are allowed to be shared without violating the requirement of d_t delay tolerance. The final circuit is shown in Figure 6.

A path is said to be an *imperceptible* path if any change (increase or decrease) on the path's delay can never affect the circuit delay [7].

Theorem 3: After wire removal according to Theorem 1 and Theorem 2, all d_t-critical paths are either false paths or imperceptible paths [7].

Proof: Omitted.

The intuition for this theorem is as follows. Our wire removal theorems guarantee that after wire removal, circuits C_1 and C_2 still have the same output value whenever d_t-critical paths in C_1 or C_2 are activated. Therefore, any delay increment on a d_t-critical path will not affect a circuit's delay. In Figure 6, all highlighted paths are either false paths or imperceptible paths. We now discuss slacks of nodes after re-synthesis. Consider node e in Figure 6. The longest true path passing through node e is $\{j\text{-}e\text{-}g\text{-}h\text{-}i\}$ whose path length is 5. Assume the timing requirement is 6. Therefore, node e has the slack of 1 in the re-synthesized circuit in Figure 6 while it has the slack of 0 in the original circuit in Figure 3. In another example, since all paths passing through node a are false paths, the slack of node a is infinite. Generally, after re-synthesis, paths whose delays are greater than 5 become either false paths or imperceptible paths; that is, the longest true path in the re-synthesized circuit has the delay of 5. As a result, the slack of each node in the re-synthesized circuit is at least 1 ($=d_t$).

4. Delay Tolerance on Internal Signals

The re-synthesis methodology described previously is applied to primary outputs whose arrival time is susceptible to delay variation. We can also employ an identical approach to protect the arrival time of internal signals from delay variation. Delay tolerance on internal signals can have the advantage of area reduction.

In Figure 7, consider the circuit which is almost the same as that in Figure 3 except node g in Figure 7 is an AND gate. Bold lines in Figure 7 represent d_t-critical nodes and d_t-critical paths. If the re-synthesis technique is applied to only the primary output, only d_t-side input wire w_1 in C_2 can be removed and the resultant circuit is shown in Figure 8. In this example, the area overhead is large because there is only one removable d_t-side input wire. We will show that by employing delay tolerance on internal signal w_f, the area penalty can be reduced. The result after performing re-synthesis on w_f is shown in the gray blocks in Figure 9. According to Theorem 3, path segment $\{s\text{-}b\text{-}d\text{-}f\}$ becomes either a false path (segment) or an imperceptible path (segment). Thus, path $\{s\text{-}b\text{-}d\text{-}f\text{-}x\text{-}g\text{-}h\text{-}i\}$ passing through a false path (segment) $\{s\text{-}b\text{-}d\text{-}f\}$ is also a false path. Similarly, all other paths passing through $\{t\text{-}b\text{-}d\text{-}f\}$, $\{s_2\text{-}b_2\text{-}f_2\}$, and $\{t_2\text{-}b_2\text{-}f_2\}$ are either false paths or imperceptible paths. Those

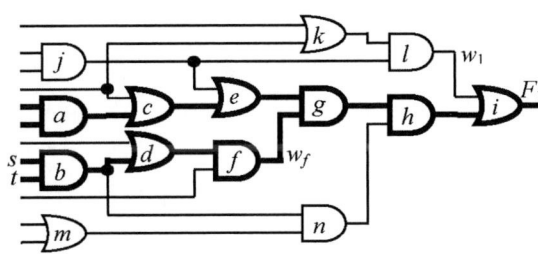

Figure 7: An example demonstrating delay tolerance on internal signals.

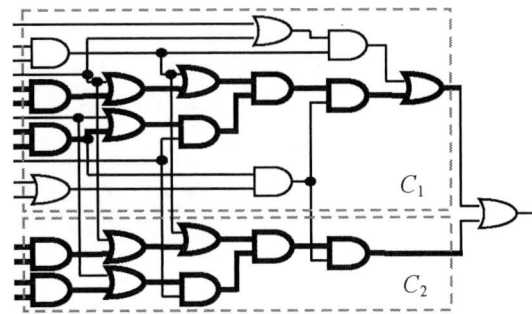

Figure 8: The re-synthesized circuit by applying delay tolerance on the primary output in Figure 7.

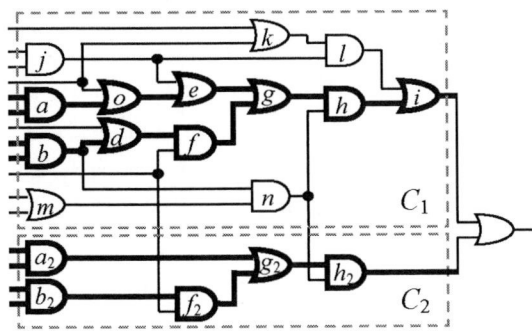

Figure 6: The re-synthesized circuit after signal sharing.

4A-3

2006 Asia and South Pacific Design Automation Conference

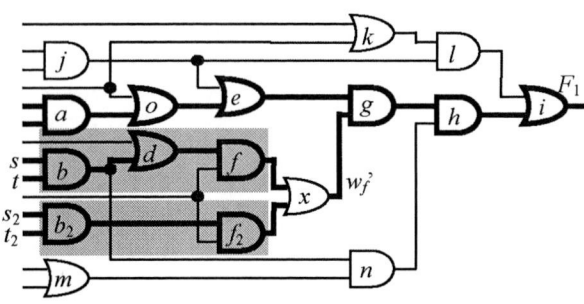

Figure 9: The re-synthesized circuit by applying delay tolerance on internal signal w_f in Figure 7.

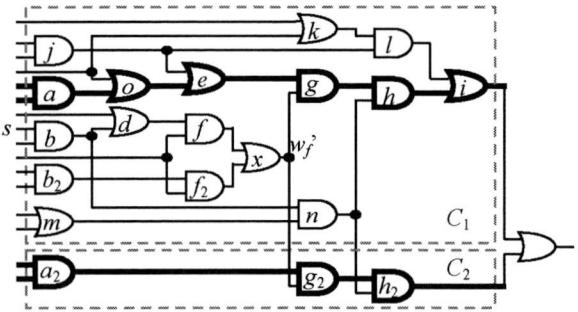

Figure 10: The re-synthesized circuit by applying delay tolerance on the primary output in Figure 9.

false paths or imperceptible paths no longer belong to d_r-critical paths. Consequently, nodes $\{a, o, e, g, h, i\}$ in Figure 9 become d_r-dominators. When we continue to perform delay tolerance on the primary output, d_r-side input wires to three OR d_r-dominators $\{o, e, i\}$ in C_2 become removable. The re-synthesized circuit is shown in Figure 10. The total area overhead of the re-synthesized circuit in Figure 10 is 7 (gates), whereas that of the re-synthesized circuit in Figure 8 without delay tolerance on internal signals is 9 (gates).

5. Experimental Results

We have implemented the re-synthesis method [3] and ours in SIS environment, and also experimented on a set of MCNC and ISCAS benchmarks. Table 1 provides the comparison of the area overhead between the method in [3] and ours. Table 2 demonstrates the advantages of delay variation tolerance for domino circuits. We first optimized a circuit with "*script.delay*," and then used delay tolerance values of 10% and 15% of the original circuit delay to re-synthesize a circuit. The timing requirement for each

re-synthesized circuit is set to be the delay of the corresponding original circuit.

In Table 1, column one lists the name of the original circuit. Columns two and three provide the area and delay of the original circuit, respectively. Columns four and five show the results for 10% delay tolerance and columns six to seven for 15% delay tolerance. For a given delay tolerance requirement, we re-synthesized the original circuit by the method in [3] and ours individually. Then we compared the area overhead between the re-synthesized circuits. For example, circuit *C880* has the area of 3569 and the delay of 33.6. When the delay tolerance is 3.36 (= 10%*33.6), the re-synthesized circuit by [3] has the area overhead of 31.2% of the original circuit's area. On the other hand, the re-synthesized circuit by our method has the area overhead of 14.9% of the original circuit's area. When the delay tolerance is 5.04(=15%*33.6), the re-synthesized circuit by [3] has the area overhead of 53.1% while by our method has the area overhead of 25.4%. On the average, for 10% (15%) delay tolerance, the area overhead from [3] and ours is 29% (48%) and 12% (21%), respectively. Both the methods usually spend few seconds for each circuit.

We have also performed Monte-Carlo experiments to demonstrate the effect of delay variation tolerance in Table 2. In the experiment, the delay of a gate is given as a probability density function similar to [6]. We obtained 1000 samples of an original circuit and 1000 samples of its re-synthesized circuit with 10% delay tolerance from Monte-Carlo experiments. After that, the delay of each sample is calculated. In Table 2, column one lists the name of the original circuit. Let d_r be the delay of the circuit. Column two shows the number of circuit samples whose delays are smaller than $\{0.8*d_r\}$ for the original circuit, and column three shows the number for its re-synthesized circuit. Similarly, columns four to eleven report the numbers of samples whose delays are smaller than $\{0.9*d_r\}$, $\{1.0*d_r\}$, $\{1.1*d_r\}$, and $\{1.2*d_r\}$ for both the original circuit and its re-synthesized circuit. Take circuit *t481* as an example. Let the timing requirement be $1.1*d_r$ ($1.1*34.1 = 37.51$), 815 samples of the original circuit meet the timing requirement while 963 samples of the re-synthesized circuit satisfy the same requirement.

In Figure 11, we drew the distribution curves of all Monte-Carlo samples of circuit *t481* and its re-synthesized circuit with 10% delay tolerance. The standard deviation of

358

Table 1: Comparison between [3] and our method

Circuit	Original circuit Area	Original circuit Delay	10% delay tolerance [3] Area overhead (%)	10% delay tolerance Ours Area overhead (%)	15% delay tolerance [3] Area overhead (%)	15% delay tolerance Ours Area overhead (%)
C432	1568	35.9	6.0	2.5	6.0	2.5
C880	3569	33.6	31.2	14.9	53.1	25.4
C1355	5854	35.8	29.3	10.4	47.1	20.4
C1908	6080	48.2	35.0	16.7	54.5	26.3
C2670	11281	62.2	23.2	11.4	34.2	16.8
C3540	19170	85.4	30.2	15.0	45.0	22.4
C5315	29937	90.0	38.1	18.6	59.5	29.4
C6288	46379	161.7	39.2	19.3	59.2	29.2
pair	12941	28.6	33.1	14.0	51.3	22.3
rot	5915	35.7	30.8	11.8	50.9	19.2
t481	4927	34.1	6.9	1.1	11.4	2.4
9symml	1317	21.4	45.3	10.9	74.1	21.9
alu2	3855	50.7	39.8	19.0	57.0	27.2
apex6	5220	56.2	18.5	7.2	63.8	24.4
apex7	1836	26.9	32.9	12.5	48.0	17.4
Avg.			29.3	12.3	47.7	20.5

Table 2: Statistical results of Monte-Carlo experiments

Circuit	$0.8*d_r$ O	R	$0.9*d_r$ O	R	$1.0*d_r$ O	R	$1.1*d_r$ O	R	$1.2*d_r$ O	R
C432	0	0	1	0	125	280	663	852	977	1000
C880	0	0	2	117	265	819	907	996	999	1000
C1355	0	0	0	0	0	7	68	445	583	862
C1908	0	0	0	0	1	0	248	389	964	976
C2670	0	0	8	14	150	186	515	680	893	965
C3540	0	0	5	2	66	50	361	386	855	910
C5315	0	0	7	1	96	62	346	570	796	992
C6288	0	0	0	0	35	553	769	999	1000	1000
pair	0	0	0	0	1	13	162	386	753	912
rot	0	0	0	17	75	299	486	706	877	976
t481	1	0	29	19	373	463	815	963	984	1000
9symml	0	0	0	0	3	3	138	249	639	926
alu2	0	0	0	5	75	259	551	825	940	1000
apex6	248	398	368	504	470	601	564	718	693	836
apex7	4	1	71	47	264	298	622	711	780	890
Avg.	17	27	33	48	133	260	481	658	849	950

O: Original circuit   R: Re-synthesized circuit

the original circuit is 2.39 while that of the re-synthesized circuit is 1.92. The curves also reveal that the re-synthesized circuit has more stable timing behavior than the original circuit.

6. Conclusions

We have proposed a framework to re-synthesize a given domino circuit for d_t delay tolerance. Our method begins with a duplex system; we then adopt wire removal and signal sharing to reduce the area overhead of the delay tolerance structure. Two novel theorems for further area reduction are presented. Experimental results demonstrate the advantages of delay variation tolerance for a re-synthesized domino circuit.

References

[1] K. Baker, G. Gronthoud, M. Lousberg, I. Schanstra, and C. Hawkins, "Defect-based delay testing of resistive vias-contacts, a critical evaluation," *Proc. of International Test Conference*, pp. 467-476, Sept. 1999.

[2] M. A. Breuer, C. Gleason, and S. Gupta, "New validation and test problems for high performance deep sub-micron VLSI circuits," *Tutorial Notes, VLSI Test Symposium*, April 1997.

[3] Shih-Chieh Chang, Cheng-Tao Hsieh, and Kai-Chiang Wu, "Re-synthesis for delay variation tolerance," *Proc. of Design Automation Conference*, pp. 814-819, June 2004.

[4] Seung Hoon Choi, Bipul C. Paul, and Kaushik Roy, "Novel sizing algorithm for yield improvement under process variation in nanometer technology," *Proc. of Design Automation Conf.*, pp. 454-459, June 7-11, 2004.

[5] E.T.A.F. Jacobs and M.R.C.M. Berkelaar, "Gate sizing using a statistical delay model," *Proc. of DATE*, pp. 27-30, 2000.

[6] Jing-Jia Liou, A. Krstic, Li-C. Wang, and Kwang-Ting Cheng, "False-path-aware statistical timing analysis and efficient path selection for delay testing and timing validation," *Proc. of Design Automation Conference*, pp. 566-569, June 2002.

[7] Sreeja Raj, Sarma B. K. Vrudhula, Janet Wang, "A methodology to improve timing yield in the presence of process variations," *Proc. of Design Automation Conf.*, pp. 448-453, June 7-11, 2004.

[8] Alexander Saldanha, "Functional timing optimization," *Proc. of International Conference on Computer-Aided Design*, pp. 539-543, Nov. 1999.

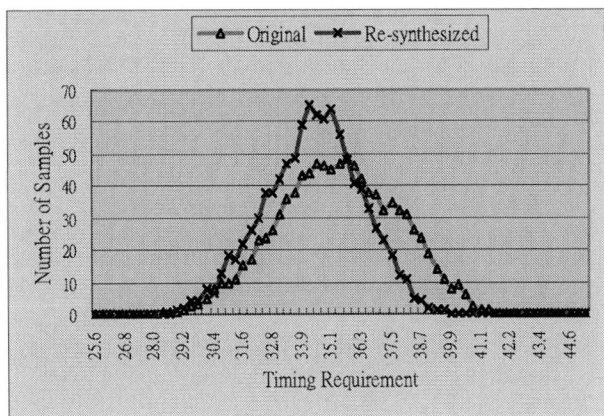

Figure 11: The distribution curves of Monte-Carlo samples of circuit *t481* and its re-synthesized circuit.

Efficient Identification of Multi-Cycle False Path

Kai Yang, Kwang-Ting Cheng
University of California, Santa Barbara
{kyang, timcheng}@ece.ucsb.edu

Abstract

Due to false paths and multi-cycle paths in a circuit, using only topological delay to determine the clock period could be too conservative. In this paper, we address the timing analysis problem by considering both single-cycle and multi-cycle operations. We give a precise definition of multi-cycle false paths and provide the necessary conditions for multi-cycle sensitizable paths. We then propose an efficient algorithm to identify multi-cycle false paths. By considering both single-cycle and multi-cycle false paths, we could derive a shorter clock period than that determined by existing methods. Finally, we propose an algorithm to compute the valid clock period and demonstrate the improvement in clock frequency by taking multi-cycle false paths into account.

1 Introduction

The clock period of a circuit is determined by the delay of the longest path in the circuit. Static timing analysis, which computes path delay primarily based on topological delay, is an efficient approach for determining the valid clock period.

However, previous research shows that due to false paths [1–3] and multi-cycle paths [4–9] , it might be too conservative to use the topological delay for calculating the path delay. By taking into account both false paths and multi-cycle paths, we find that the timing constraints of the circuit could be relaxed.

The actual delay of a circuit is determined by the delay of its longest sensitizable paths. A false path is a path that cannot be sensitized by any input vector. The definition of false paths and algorithms for identifying all or a subset of them have been studied for years [1–3,10–12]. In [10], Cheng and Chen proposed the notion of functional unsensitizable paths and demonstrated that those paths will not affect the circuit timing under *any* delay configuration.

A multi-cycle path in a sequential circuit is a combinational path which does not have to complete the propagation of the signals along the path within one clock cycle. A k-cycle path would have up to k clock cycles to propagate the transition from the source to the destination. The clock period can therefore be relaxed if the longest paths are multi-cycle paths. Figure 1 shows a circuit containing multi-cycle paths.

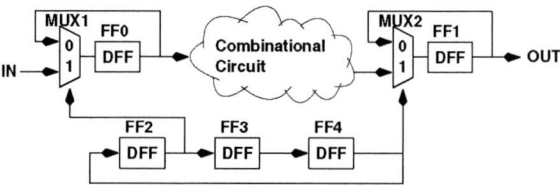

Figure 1: Example of Multi-Cycle Paths [5]

The initial values in flip-flops $\{FF_2, FF_3, FF_4\}$, which form an autonomous circular shift register, are $\{1, 0, 0\}$ respectively. Therefore, the sequence of states in these three flip-flops would be: $\{1, 0, 0\} \rightarrow \{0, 1, 0\} \rightarrow \{0, 0, 1\} \rightarrow \{1, 0, 0\}$. Multiplexer MUX1 selects primary input data IN when $\{FF_2, FF_3, FF_4\}$ are in state $\{1, 0, 0\}$. Then FF_0 latches the value IN and starts launching the new value toward the combinational circuit while $\{FF_2, FF_3, FF_4\}$ are switched to state $\{0, 1, 0\}$. Meanwhile, MUX2 selects the output of the combinational block when $\{FF_2, FF_3, FF_4\}$ are in state $\{0, 0, 1\}$. Then FF_1 latches the new value when $\{FF_2, FF_3, FF_4\}$ are switched to state $\{1, 0, 0\}$. Since it requires two cycles for $\{FF_2, FF_3, FF_4\}$ from $\{0, 1, 0\}$ to $\{1, 0, 0\}$. Therefore, all the paths from FF_0 to FF_1 are 2-cycle paths. That is, the circuit would allow 2 clock cycles to complete the propagation of transitions along these paths.

Multi-cycle paths have been studied for several years [4–9]. For microprocessor designs, the method proposed in [8] utilizes functional information such as instruction-set architecture (ISA), to extract multi-cycle paths. The proposed procedure, manually extracting the functional constraints, may not be scalable for larger and complex designs. The methods proposed in [5–7] are based on the stable-state analysis to identify multi-cycle flip-flop pairs. All paths between the multi-cycle flip-flop pairs are multi-cycle paths. Such stable-state analysis could be done by BDD [5], SAT [6], or ATPG [7] techniques.

However, considering signals' stable states only, which is the assumption made in all previous methods for multi-cycle path identification [5–7], may result in optimistic (and, thus, invalid) prediction of the clock period. In [7], the problem introduced by the presence of static hazards was described but no clear analysis was provided to address this problem.

In this paper, we first define the multi-cycle false paths and multi-cycle sensitizable paths. We then provide some necessary conditions for multi-cycle sensitizable paths. We use the functional sensitization criterion, introduced in [10], to check path sensitizability, so the static-hazards problem can be implicitly considered. Then, we address the problem of determining valid clock periods for circuits containing multi-cycle paths. We propose an efficient method for computing the valid clock period, which takes into account both single-cycle and multi-cycle false paths.

The rest of the paper is organized as follows. In Section 2, we review some representative approaches on this topic. In Section 3, we review the definitions to be used throughout this paper. In Section 4, we give the definition of multi-cycle paths. The necessary conditions for single-cycle and multi-cycle sensitizable paths are derived in Section 5. An efficient algorithm to identify the multi-cycle false path is introduced in Section 6. In Section 7, we address the problem of computing valid clock period. An iterative method to calculate the valid clock period is proposed. In Section 8, we show some experimental results, followed by the conclusions.

2 Background

A multi-cycle path in a sequential circuit is a combinational path which does not have to complete the propagation of the signal transition from the source to the destination of the path in single clock cycle. A k-cycle path is a path that is allowed to use k clock cycles to propagate the transition. By considering multi-cycle paths, the timing constraints could be relaxed. That is, the minimal clock period could be shorter than the delay of the multi-cycle path.

To analyze multi-cycle operations, the method proposed in [6, 7] is based on checking the stable states in flip-flop pairs. A flip-flop pair (FF_i, FF_j) is classified as a multi-cycle flip-flop pair, if there exist input vectors to satisfy Equation 1, where $FF_i(t)$ denotes the stable state in flip-flop i at cycle t:

$$FF_i(t) \neq FF_i(t+1) \Rightarrow FF_j(t+1) = FF_j(t+2) \tag{1}$$

All paths between flip-flop pairs (FF_i, FF_j) are then declared as multi-cycle paths. However, because the stable-state checking only checks the necessary conditions for multi-cycle paths, it might not result in correct classification of multi-cycle flip-flop pairs due to the presence of static-hazards [7]. The problem was pointed out in [7] but no solution was provided. Moreover, the stable-state checking could not determine the multiplicity, k, of the identified multi-cycle paths. None of the previous work has yet addressed the problem of determining the valid clock period for circuits with multi-cycle paths.

In the following section, we first give the notation and definition of path sensitization used throughout the paper.

3 Definition

A path $P^x = (G_0, f_0, G_1, f_1, ..., f_{m-1}, G_m)$ is a sequence of gates and signals associated with a transition $x \in \{rising, falling\}$ at the source of the path. Gate G_0 is either a primary input or the output of an FF, G_m is either a primary output or the input of an FF. Signal $f_i, 0 \leq i \leq m - 1$ is an **on-input** of P^x which connects gate G_i to G_{i+1}. A signal is called a **side-input** of P^x associated with f_i if it is connected to G_i, but not originated from G_{i-1}. The delay of a path P^x is denoted as $d(P^x)$. A path set η contains all paths in a circuit.

A state of a circuit, denoted as S, indicates the values in all or a subset of flip-flops at certain cycle. The controlling (non-controlling) value of gate G is denoted by $cv(G)$ ($ncv(G)$).

Let $v = < v_1, v_2 >$ be an input vector pair and assume v_2 becomes stable at time $t = 0$. The logic value stabilized at a signal f or the output of a gate G is called its stable value under v_2. The time when the value at a signal f or the output of a gate G becomes stable is called its stable time under v_2.

The latest arrival time of signal f for value x, as $ar_{max}^x(f)$, is the delay of the longest path from any primary input or the output of an FF to signal f for a corresponding transition. Similarly, the earliest arrival time is denoted as $ar_{min}^x(f)$.

3.1 Path Sensitization

Signal f, which is connected to G, is considered to **dominate** G if the stable value and the stable time at G are determined by those at f. A path is considered to be sensitized, under a delay configuration, by a vector pair if each on-input of the path dominates its connected gate. Given a delay configuration, a path is a **true** (**sensitizable**) path if there exists at least one vector pair which sensitizes the path. Otherwise, it's a **false** path.

Under a delay configuration, a vector pair sensitizes a path *iff* each on-input of the path is either the earliest controlling value or the latest non-controlling input with all its side-inputs being non-controlling inputs. This criterion is called the **exact sensitization criterion** [2].

To efficiently check the sensitizability of target paths, a delay-independent method, using the **functional sensitization criterion**, is proposed in [10]. If there exists an input vector v such that all the side-inputs of f_i along P^x settle to non-controlling values on v when the on-input f_i has a non-controlling value, then P^x is functional sensitizable. Otherwise, P^x is functional unsensitizable.

Because functional sensitization, a delay-independent criterion, is only a necessary condition of exact sensitization, an identified functional sensitizable path might not be sensitizable under certain delay configurations. On the other hand, the identified functional unsensitizable paths must be false under any arbitrary delay configuration.

4 Multi-Cycle Path

In this section, we give a precise definition of multi-cycle path and introduce a model for illustration and analysis.

A k-cycle path P^x could complete the propagation of the signal transition from the source to the destination in k cycles. This implies that the clock period clk could be shorter than the delay of P^x. The transition, along P^x, travels through a segment seg_i in cycle i ,and then continues propagating through segment seg_{i+1} in the next cycle. A k-cycle path can therefore be divided into several non-overlapping segments seg_i where $P^x = \cup(seg_i)$ for $1 \leq i \leq k$. The delay $d(seg_j)$ of each segment seg_j, for $1 \leq j \leq k-1$, is equal to the clock period clk. The delay of the last segment seg_k, $d(seg_k)$ is equal to $modulo(d(P^x), clk)$. Each segment seg_i represents the segment through which the transition travels in cycle i. Since every segment needs to consist of integral number of stages (gates), $d(seg_j)$ might not perfectly match clk. In this case, we ignore the gates which are overlapped for sensitization checking.

We use the **timeframe expanded model** of the sequential circuit to illustrate how a transition propagates through a multi-cycle path. Segment seg_i in timeframe-i (TF-i) represents the transition propagation path in cycle i. Figure 2 shows an example of a 2-cycle path. The same path appears in both TF-1 and TF-2. The transition is launched at the source of the path in TF-1. The transition travels through segment seg_1 in TF-1 (bold line shown inside TF-1) and then continues to travel through seg_2 to reach the destination in TF-2 (bold line shown inside TF-2).

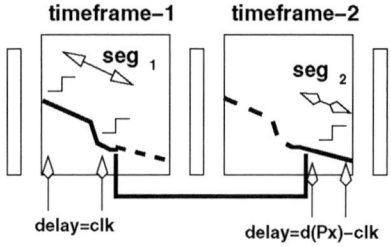

Figure 2: Example of 2-Cycle Path

5 Necessary Conditions for Path Sensitization

Due to the delay variation, the sensitizability of a path might be different from chip to chip. To identify paths which are false under any delay configuration, we derive necessary conditions , which are delay-configuration independent, for both single-cycle and multi-cycle path sensitization. Paths which cannot satisfy these necessary conditions are false paths.

5.1 Single-Cycle Paths

The following is a necessary condition for a single-cycle sensitizable path: A single-cycle sensitizable path P^x must satisfy the functional sensitization criterion [10]. Otherwise, P^x is false.

The set of single-cycle false paths, under a delay configuration M and a state S, is denoted as $\zeta_1^{M,S}$. The set of other paths which do not belong to $\zeta_1^{M,S}$ is denoted as $\omega_1^{M,S}$. That is, $\omega_1^{M,S} \cup \zeta_1^{M,S} = \eta$ and $\omega_1^{M,S} \cap \zeta_1^{M,S} = \phi$.

5.2 Multi-Cycle Paths

The following is a necessary condition for a multi-cycle sensitizable path: Under the timeframe expanded model, each segment seg_i of a multi-cycle sensitizable path P^x must satisfy the functional sensitization criterion in its corresponding timeframe. Otherwise, P^x is false.

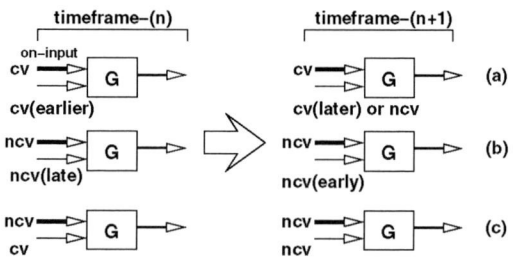

Figure 3: Side-input dominates gate G in TF-(n)

We prove the necessary condition for a multi-cycle sensitizable path by contradiction. We show that if the propagated transition is blocked in any segment (i.e. the functional sensitization criterion is violated), then the transition cannot continue propagating through the multi-cycle path (i.e. it's a multi-cycle false path). The left hand side of Figure 3 enumerates all possible conditions that a transition, propagating through multi-cycle path P^x and arriving at gate G in timeframe-(n), would be blocked by the side-input. The right hand side shows the values of on-input and side-input of gate G in timeframe-$(n + 1)$. In the first case, as shown in Figure 3 (a), the on-input propagates a controlling value which arrives at gate G in timeframe-(n). If the side-input has a controlling value which arrives earlier than the on-input, then the side-input dominates gate G in timeframe-(n). In this case, regardless of the side-input value in timeframe-$(n + 1)$, either non-controlling or controlling, the output of gate G will not change its value in timeframe-$(n + 1)$ (i.e. no transition occurs at gate G). Therefore any transition propagated through this gate must be dominated by the side-input. For cases shown in Figure 3 (b)(c), the on-input propagates a non-controlling value which arrives at gate G in timeframe-(n). If the side-input has a controlling value (case (c)), or a non-controlling value which arrives later than the on-input (case (b)), then the side-input dominates gate G in timeframe-(n). In this case, even if the side-input becomes a non-controlling value in timeframe-$(n+1)$, the output of gate G either will not change its value or the new value will be dominated by the side-input in timeframe-$(n+1)$. Therefore, we can conclude that, in order to propagate a transition along a multi-cycle path P^x, each segment needs to be sensitizable in its corresponding timeframe.

The set of k-cycle false paths, under a delay configuration M and an initial state S, is denoted as $\zeta_k^{M,S}$. The initial state S indicates the state at which the transition is launched from the source of P^x. The set of other paths which do not belong to $\zeta_k^{M,S}$ is denoted as $\omega_k^{M,S}$. That is, $\omega_k^{M,S} \cup \zeta_k^{M,S} = \eta$ and $\omega_k^{M,S} \cap \zeta_k^{M,S} = \phi$.

6 Identification of Multi-Cycle False Paths

In this section, we introduce a segment-based checking algorithm to identify multi-cycle false paths. The algorithm checks the necessary condition described in Section 5.2. Under the timeframe expanded model, the algorithm, summarized in Algorithm 1, checks the sensitizability of each segment of the multi-cycle path at each timeframe. The inputs of the algorithm are a path or a partial path P^x, the multiplicity k, and the clock period clk.

Algorithm 1 Segment-Based Checking(P^x, k, clk)

For each f, calculate $ar_{max}^x(f)$, for $x = \{0, 1\}$
Construct timeframe expanded model $comb^k$ for P^x
for $i = 1, i \leq k, i = i + 1$ **do**
  // check the sensitization of all side-inputs in seg_i
  for each side-input f connected to G in seg_i **do**
    if $clk \geq ar_{max}^{ncv(G)}(f)$ **then**
      if on-input of G propagates an $ncv(G)$ **then**
        impose $ncv(G)$ to f at timeframe-i of $comb^k$
      end if
    end if
  end for
end for
Perform logic implication. If conflict, P^x is false.
Otherwise P^x is true

We first calculate the latest arrival time for value $x = 0$ and $x = 1$ for each signal f, which is denoted as $ar_{max}^x(f)$. We then construct the time-frame expanded model $comb^k$ and divide P^x into non-overlapping segments. The delay of each segment seg_j for $1 \leq j \leq k-1$ is clk. The delay of the last segment seg_k is $modulo(d(P^x), clk)$.

We check the sensitizability of P^x segment by segment through $comb^k$. For each segment seg_i, the sensitization constraints are imposed, followed by implication, on the expanded circuit $comb^k$ in its corresponding timeframe i. Note that the side-inputs of seg_i may not reach its stable value in timeframe i because a side-input may also be part of a multi-cycle path. Therefore, instead of imposing all logic constraints required for functional sensitization of seg_i, we impose constraints only at side-inputs which can reach stable value in timeframe i. For example, if value $ncv(G)$ is needed at side-input f sensitizing segment seg_i in timeframe-i, and the latest arrival time of f for value $ncv(G)$, $ar_{max}^{ncv(G)}(f)$, is smaller than clk (which means signal f would become stable at $ncv(G)$ within this cycle), then we impose value $ncv(G)$ at f in timeframe-i of $comb^k$. Otherwise, no side-input constraint is imposed. Figure 4 shows an example of the segment-based checking for a 2-cycle path. To sensitize the multi-cycle path P^x, signals a and b need to be at logic "1" in timeframe-1. Similarly, to sensitize the segment in timeframe-2, signals c and d need to be at logic "1".

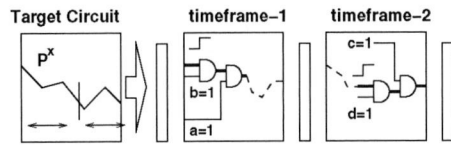

Figure 4: Example of Segment-Based Checking

After imposing the sensitization constraints, we perform logic implication on $comb^k$ to see whether there is any conflict. If a conflict occurs, P^x is false. Otherwise it classified as true. Please note that the identified false paths must be false. However, because we only check a necessary condition, the rest may or may not be true.

7 Valid Clock Period

Traditionally, the valid clock period is determined by the delay of the longest single-cycle sensitizable path. However, due to the presence of multi-cycle paths, the clock period might be relaxed. Equation (2) summarizes the constraints for the valid clock period in presence of multi-cycle operations.

$$clk \geq (\max(d(P^x))/m, \forall P^x \in \omega_m^{M,S}), 1 \leq m \leq k \tag{2}$$

Based on Equation (2), we propose a method to calculate the valid clock period for circuits with multi-cycle paths.

7.1 Calculation of Valid Clock Period

Using an exemplar circuit architecture, shown in Figure 5, as the driver, illustrate the proposed iterative method for determining the valid clock period. In Figure 5, flip-flops $\{FF_{c0}, FF_{c1}, ..., FF_{cn}\}$ forms an autonomous circular shift register. The transitions at the inputs of the combinational sub-circuit are launched when the state in $\{FF_{c0}, FF_{c1}, ..., FF_{cn}\}$ switches from $\{1, 0, 0, ..., 0\}$ to $\{0, 1, 0, ..., 0\}$.

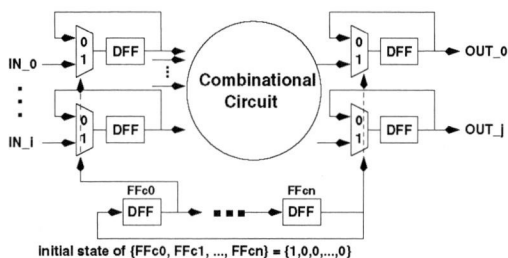

Initial state of {FFc0, FFc1, ..., FFcn} = {1,0,0,...,0}

Figure 5: Exemplar Circuit Architecture

We make the following observations regarding the output-MUX in Figure 5, whose gate-level implementation is shown in Figure 6. The select-line (sel) of the output-MUX always has a controlling value of gate G_1 when $\{FF_{c0}, FF_{c1}, ..., FF_{cn}\}$ are not in state $\{0, 0, 0, ..., 1\}$. In addition, the latest arrival time of the select-line is always earlier than the earliest arrival time of In_1. Therefore, based on the exact sensitization criterion, transitions traveling through In_1 to gate G_1 will always be blocked by the select-line when $\{FF_{c0}, FF_{c1}, ..., FF_{cn}\}$ are not in state $\{0, 0, 0, ..., 1\}$.

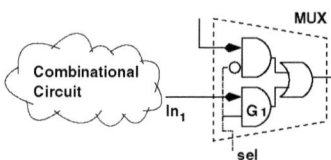

Figure 6: Output MUX

To calculate the valid clock period for a multi-cycle circuit similar to the one in Figure 5, we propose an iterative method which is shown in Figure 7. In each iteration, we identify the longest k-cycle true path and update the circuit maximum delay MAX. The valid clock period will then be determined by the final MAX.

In the preprocess step, we first identify all signals connected to the output-MUXs (such as signal In_1 in Figure 6) and group these signals as FS. The iterative procedure then identifies the longest single-cycle ($k = 1$) true path, which starts from the output of an FF and a PI. To efficiently identify the longest true path, we utilize the Best-First-Search (BFS) method [11] to identify the longest path. We then check

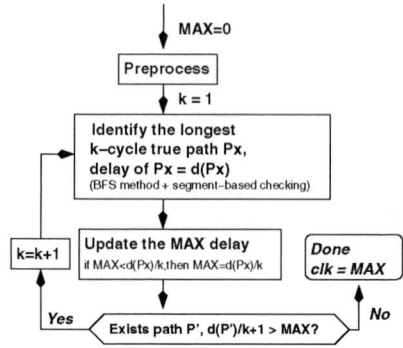

Figure 7: Calculation of Valid Clock Period

the sensitizability using the segment-based checking algorithm. The inputs to the checking procedure are the path/partial-path identified by the BFS, k, and the delay of the path/partial path. In addition to checking functional sensitizability, we also utilize FS (the set of signals that we group in the preprocessing phase) to help identify the false paths. Assume the propagated transition of a path P^x arrives at signal f in cycle-i, at which $\{FF_{c0}, FF_{c1}, ..., FF_{cn}\}$ are not in state $\{0, 0, 0, ..., 1\}$. If the f belongs to FS, then P^x is false. Once the longest single-cycle sensitizable path is identified, its delay is then assigned to MAX.

In the next iteration, we first identify paths P whose delay divided by $k + 1$ are longer than MAX. If no such path exists, the process stops and MAX is the minimal valid clock period. Otherwise, we check $k+1$-cycle sensitizability of each path in P. If a path in P is identified as a $k + 1$-cycle true path and its delay divided by $k + 1$ is longer than MAX, then we update MAX. After all paths in P are processed, we continue on to the next iteration.

8 Experimental Result

For the experiments, we constructed 2-cycle operation circuits by replacing the combinational sub-circuit in Figure 5 with ISCAS-85 circuits and constructed the autonomous circular shift register with three registers $\{FF_{c0}, FF_{c1}, FF_{c2}\}$. The initial values in flip-flops $\{FF_{c0}, FF_{c1}, FF_{c2}\}$ are $\{1, 0, 0\}$ respectively.

For the given circuits, we compare the resulting clock periods for three different methodologies: static timing analysis, single-cycle-false-path-aware timing analysis [12], and the proposed method which considers both single-cycle and multi-cycle false paths. All the methods are implemented by C++ and run on Linux workstations. Table 1 summarizes the experimental results.

The first column shows the name of the circuit used as the combinational sub-circuit in Figure 5. The second column shows the delay of the longest path of the circuit which is calculated by static timing analysis, denoted as clk_{sta}. The third column shows the valid clock period derived by the longest single-cycle sensitizable path, denoted as clk_{sc}. The fourth column shows the resulting clock clk_{mc} calculated by the proposed method, and the CPU time for computing the valid clock period is listed in the fifth column. As indicated, the clock period derived by the proposed method is shorter than those derived by traditional methods.

Table 1: Valid Clock Period

Replace-ckt	$clk_{sta}(ns)$	clk_{sc}	clk_{mc}	$CPU(s)$
c17	13.4	13.4	6.7	0.0
c1355	135.2	135.2	67.6	0.11
c1908	184.8	184.8	92.4	0.07
c2670	125.7	125.7	62.8	0.11
c3540	226.1	224.8	112.4	0.48
c432	110.5	107.4	53.7	0.6
c449	135.4	135.4	67.7	0.55
c5315	216.6	202.0	101.0	1.92
c7552	188.6	181.7	90.8	1.77
c880	109.6	109.6	54.8	0.04

The reason for the conservative result of clk_{sc} is that the traditional methods do not take the circuit states into account. It's assumed that all primary inputs and outputs of FFs could have any value. So some paths classified as false by the proposed method are classified true in traditional methods.

Previous work [7] addressed the problem of sampling error in presence of static hazards. Such a problem will not occur if the clock period is determined by the proposed method. If a static-hazard is propagated, then it must go through at least one true path. In the proposed method, the clock period is derived from the longest true path of the circuit, so the static-hazard will disappear, and the signals should become stable before a FF latches its data.

9 Conclusion

In this paper, we first define the multi-cycle false paths and multi-cycle sensitizable paths. We then provide the necessary conditions for multi-cycle sensitizable paths. We use the functional sensitization criterion, introduced in [10], to check the path sensitizability, so the static-hazards problem, ignored in previous approaches, can be implicitly considered. Then, we provide a thorough analysis for the problem of determining valid clock period for circuits containing multi-cycle paths. We have proposed an algorithm to compute the valid clock period and demonstrate the improvement to clock frequency by considering multi-cycle false paths.

References

[1] P. McGeer and R. Brayton, *Efficient Algorithms for computing the Longest Viable Path in a Com-*

binational Network. Proc. IEEE/ACM Int. Conf. Computer-Aided Design (ICCAD), 1989.

[2] H.-C. Chen, D. H.-C. Du, and L.-R. Liu, "Critical path selection for performance optimization", *IEEE Trans. Computer-Aided Design of Integrated Circuits and Systems*, vol. 12, no. 2, pp. 185–195, Feb. 1993.

[3] S. Davadas, K. Keutzer, S. Malik, and A. Wang, "Certified timing verification and the transition delay of a logic circuit", *IEEE Trans. VLSI Systems*, vol. 2, no. 3, pp. 333–342, Sept. 1994.

[4] A. P. Gupta and D. P. Siewiorek, *Automated Multi-Cycle Symbolic Timing Verification of Microprocessor-based Designs*. Proc. IEEE/ACM Design Automation Conf. (DAC), 1994.

[5] K. Nakamura, K. Takagi, S. Kimura, and K. Watanabe, *Waiting False Path Analysis of Sequential Logic Circuits for Performance Optimization*. Proc. IEEE/ACM Int. Conf. Computer-Aided Design (ICCAD), Nov. 1997.

[6] K. Nakamura, S. Maruoka, S. Kimura, and K. Watanabe, *Multi-Cycle Path Detection based on Propositional Satisfiability with CNF Simplification using Adaptive Variable Insertion*. IEICE Trans. on Fundamentals, Dec. 2000.

[7] H. Higuchi, *An Implication-based Method to Detect Multi-Cycle Paths in Large Sequential Circuits*. Proc. IEEE/ACM Design Automation Conf. (DAC), June 2002.

[8] W.-C. Lai, A. Krstic, and K.-T. Cheng, "Functionally testable path delay faults on a microprocessor", *IEEE Design & Test of Computers*, vol. 15, 2000.

[9] P. Ashar, S. Dey, and S. Malik, *Exploiting multicycle false paths in the performance optimization of sequential circuits*. Proc. IEEE/ACM Int. Conf. Computer-Aided Design (ICCAD), 1992.

[10] K.-T. Cheng and H.-C. Chen, "Classification and identification of nonrobust untestable path delay faults", *IEEE Trans. Computer-Aided Design of Integrated Circuits and Systems*, vol. 15, no. 8, pp. 845–853, Aug. 1996.

[11] W. Qiu and Walker D.M.H., *An efficient algorithm for finding the k longest testable paths through each gate in a combinational circuit*. Proc. Int. Test Conf. (ITC), 2003.

[12] J.-J. Liou, A. Krstic, L.-C. Wang, and K.-T. Cheng, *False-path-aware statistical timing analysis and efficient path selection for delay testing and timing validation*. Proc. IEEE/ACM Design Automation Conf. (DAC), 2002.

IEEE Standard 1500 Compatible Interconnect Diagnosis for Delay and Crosstalk Faults

Katherine Shu-Min Li[1], Yao-Wen Chang[2], Chauchin Su[3], Chung-Len Lee[1], Jwu E Chen[4]

[1]Department of Electronics Engineering, National Chiao Tung University, Hsichu, Taiwan
[2]Department of Electrical Engineering & Graduate Institute of Electronics Engineering, National Taiwan University, Taipei, Taiwan
[3] Department of Department of Electronic Control, National Chiao Tung University, Hsichu, Taiwan
[4]Department of Electrical Engineering, National Central University, Chungli, Taiwan

Abstract – We propose an interconnect diagnosis scheme based on Oscillation Ring test methodology for SOC design with heterogeneous cores. The target fault models are delay faults and crosstalk glitches. We analyze the diagnosability of an interconnect structure and propose a fast diagnosability checking algorithm and an efficient diagnosis ring generation algorithm which achieves the *optimal diagnosability*. Two optimization techniques improve the efficiency and effectiveness of interconnect diagnosis. In all experiments, our method achieves 100% fault coverage and the optimal diagnosis resolution.

I. Introduction

Interconnect delays, rather than gate delays, dominate overall circuit performance in the nanometer era [1-2], especially for System-on-chip (SOC) ICs. *Interconnect diagnosis*, including the detection and location of faulty nets, plays a key role in enhancing circuit reliability and yield. It is not easy to directly apply those existing interconnect diagnosis techniques to SOC designs, and the diagnosis costs greatly increase for manufacturing and yield enhancement. Therefore, it is desired to develop an effective test scheme to reduce the costs of interconnect diagnosis. Interconnect diagnosis for various applications, such as printed circuit board (PCB) and multi-chip module (MCM) has been studied extensively in the literature [3-8]. However, their target fault models are mainly traditional stuck-at and bridging faults. The diagnosis algorithms include counting sequence, walking-0 and walking-1 sequence, maximal independent test set, and focus mainly on special interconnect structures, especially for bus-oriented or FPGA designs. On the other hand, in this paper we consider delay and crosstalk glitch faults, which are important in nanotechnology.

Oscillation ring based test is an efficient and effective method to detect faults in a circuit or a device [9-10]. An oscillation ring is a closed loop with an odd number of signal inversions. Once the ring is constructed, oscillation signal appears on the ring. For a circuit with faults, some rings will not oscillate correctly. Once a set of oscillation tests have been conducted, we can locate some or all of the faults according to the test outcome [11]. Whether each fault can be correctly identified, or *diagnosed*, depends on the interconnect structure and the test rings applied.

The advantage of oscillation ring based diagnosis for the interconnect structure is that, in addition to functional faults like stuck-at and open faults, it is also capable of identifying delay faults and crosstalk glitch faults, the main sources for the loss of signal integrity [2]. Therefore, the *oscillation ring* based technique is an ideal approach to interconnect diagnosis.

In this paper, we propose an oscillation ring based scheme to diagnose interconnect faults to reduce the test time for SOC interconnect diagnosis. This approach is compatible with the P1500 standard [12], providing structural support for core testing as well as interconnect testing in SOC. We analyze the diagnosability of an interconnect structure and propose a fast diagnosability checking algorithm and an efficient ring generation algorithm. We prove that the generation algorithm can find the optimal diagnosability for any interconnect structure. The predetermined diagnosis method achieves the optimal diagnosability (i.e. the maximum diagnosis resolution). We also propose two optimization techniques for test time reduction with no hardware overhead. The first one is an adaptive diagnosis method, which reduces test time by 1.54X-2.67X. The other is a concurrent diagnosis method, which improves test effectiveness by up to 9.66%. Experiments on the MCNC benchmark circuits show our methods achieve 100% fault coverage and the optimal diagnosis resolution. (Here, the diagnosis resolution is defined as the cardinality of the largest set of indistinguishable faults, and the maximum diagnosis resolution or the optimal diagnosability implies that the cardinality is 1.)

The proposed approach provides many advantages. First, it is applicable to arbitrary global interconnect. In contrast, previous diagnosis methods are more concentrated on special structures. Second, our approach can deal with faults that cause signal integrity problems, while it is difficult to handle such faults under traditional methods. We provide ring generation algorithms that achieve 100% fault coverage and the optimal diagnosis resolution for the modeled faults. A fast diagnosability checking methodoloty is given in this paper, which greatly reduce the execution time.

II. Oscillation Ring Test Scheme for Interconnect Detection and Diagnosis

A. The OR Test Architecture

We discuss the *interconnect oscillation ring test* (*IORT* for short) for SOC interconnects [11]. Figure 1 illustrates a counter-based test architecture for both delay and crosstalk glitch detection for SOC ICs with the compatible IEEE P1500 core test standard. In P1500, each input/output pin of a core is attached with a *wrapper cell*, and a centralized *test access mechanism* (*TAM*) is provided to coordinate all test processes. In addition to the normal input/output connections, all wrapper cells in a core can also be connected with a shift register, usually referred to as a scan path, to facilitate test access. A modified wrapper cell design has been proposed to provide extra connections and inversion control so that the oscillation rings can be constructed through the wires and the boundary scan paths in cores [11]. For example, the oscillation ring test architecture in Figure 1 consists of one oscillation ring and two neighboring nets.

The target fault models of this test architecture are stuck-at, open, delay and crosstalk glitch faults. In addition to fault detection, measuring the delay fault can also be achieved. If an oscillation ring fails to oscillate, there exists stuck-at or open fault(s) in the components of the oscillation ring. The period of the oscillation

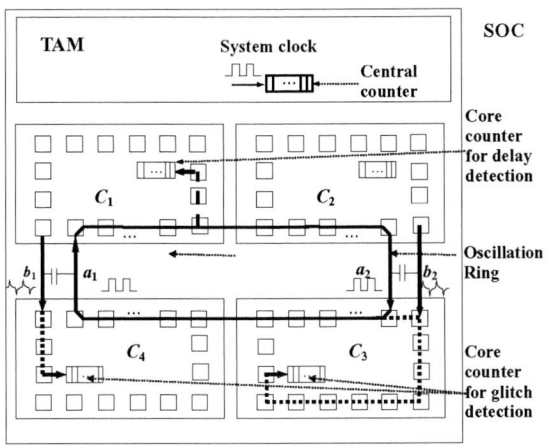

Figure 1. Test architecture for delay and crosstalk detection.

signal is measured by using a delay counter in a core to test delay faults, and a similar scheme is also applied for crosstalk glitch detection.

A local counter is included in each core, and a central counter is in the TAM of an SOC. The central counter in the TAM is enabled by the signal *OscTest*, and triggered by the system clock. A local counter is connected to one wrapper cell in each core; however, it can be accessed by every wrapper cell through the wrapper cell chain. When an oscillation ring passes a core, an internal scan path is formed to connect the oscillation signal to the local counter. For example, core C_1 is passed by the oscillation ring in Figure 1. The oscillation signal is fed to the local counter through a series of modified wrapper cells. When an oscillation test session starts ($OscTest = 1$), the TAM enables its own central counter as well as all local counters in cores. After the central counter in the TAM counts to a specific number n, the oscillation test session terminates and all local counters are disabled ($OscTest = 0$). Then all the local counter contents can then be scanned out to ATE for inspection.

Assume that m oscillation rings are tested. Let the frequency of the system clock be f, and the delay counter contents of the rings be n_1, n_2, ..., n_m, respectively. An estimation of the i-th ring's oscillation frequency f_i can be approximated by

$$f_i = f \times n_i / n \qquad (1)$$

Since the frequency of each ring is predetermined during the design phase, a delay fault is detected and measured by inspecting the contents of the delay counters. Let the oscillation frequency of the rings, according to the timing specification, be $f_{min} \leq f_i \leq f_{max}$, with the unit of measuring T_0 (= n/f). Thus, we have $n_{min} \leq n_i \leq n_{max}$, where $n_{min} = f_{min} \times T_0$ and $n_{max} = f_{max} \times T_0$. Let ξ be the resolution of delay measurement, and ε be the maximum measurement error. Since a counter's maximum measurement error is ± 1, the requirement for ε should be the reciprocal of f_{min} and T_0.

$$\varepsilon = \frac{1}{f_{min} \times T_0} \leq \zeta \qquad (2)$$

Let the frequency specification of the oscillation rings be 4 MHz to 400 MHz and ξ be 0.001, implying the counter content n_{min} is at least 1000. From (2), we have the required T_0 to be 250μs. This example illustrates the feasibility of the oscillation test scheme from a measurement prospect, and this frequency specification is actually compliant with ATE specifications.

B. P1500 Wrapper Cell Design

An oscillation ring consists of interconnect wires and part of the scan path in each core where the ring passes. Thus, a P1500 wrapper cell must provide necessary paths between input/output ports and scan in/scan out ports. If an oscillation test is used to test wires connected to pads, the boundary scan cells also have to be modified in a similar way. In order to facilitate the scheme, the P1500 boundary wrapper cells need to be modified.

A normal wrapper cell provides two types of paths: a scan path connecting all wrapper cells into a shift register, and an interface buffering between internal core and the wire connected to the pin. Whenever an oscillation test is applied, a third combination path must be provided. For an input pin, the wrapper cell must connect the pin input (IN) to the scan output (SO); while for an output pin, it should connect scan in (SI) to pin output (OUT) during an oscillation test session.

The modified wrapper cell designs are shown in Figure 2. In each cell, two MUXs are added for path selection. For an input wrapper cell, the extra paths are SI→SO and IN→SO; while for an output cell, the extra paths are SI→SO and SI→OUT. The added inverting and non-inverting buffers are used to generate oscillation signals for the OR test; however, in an input wrapper cell, only one type of buffer is provided due to the limited control signals. Two control signals are needed in each modified wrapper: signal *OscTest* is a global control signal; while the signal *sel* is only used in the input wrapper cell, and the signal *inv* is only used in the output wrapper cell to ensure the odd parity of each ring. Signals *sel* and *inv* are set individually and scanned into the wrapper cells before an oscillation ring test session starts.

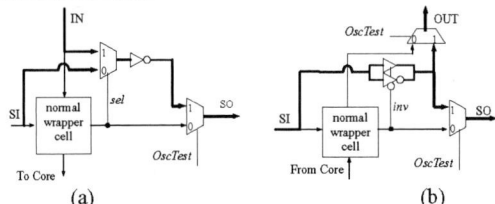

Figure 2. Modified wrapper cells with forced inversion; (a) input (b) output.

C. Interconnect Diagnosis for SOC

A circuit consisting of three cores (C_1, C_2, and C_3) and three nets (n_1, n_2, and n_3) is shown in Figure 3. The first ring consists of nets n_1 (and its right-hand side branch), n_2, and n_3, and it passes all three cores. The second ring consists of n_1 (and its left-hand side branch) and n_3, and scan paths in C_1 and C_3. The oscillation ring test scheme detects which line is faulty (n_1, n_2 or n_3), and the oscillation ring diagnosis scheme diagnoses which segment is faulty (n_{11}, n_{12}, n_{13}, n_2 or n_3).

In order to simplify the interconnect diagnosis problem, we model the SOC circuit in Figure 3 by a *hypergraph*, and model interconnects by a *hypernet* as shown in Figure 4.

Definition 1: A *hypergraph* $H = (V, L)$ consists of a vertex set V and an edge set L. A multi-terminal edge connects a set of vertices $V_i \subseteq V$, $|V_i| \geq 2$, and it is referred to as a *hypernet*.

This hypergraph model is not good enough for diagnosis, since different parts of the same net (i.e. different net segments) affect different rings. Consider the 5-terminal hypernet with seven edge segments e_1 to e_7 as shown in Figure 4(a). If edge e_1 is faulty, all four rings will not oscillate correctly. A faulty e_2 affects rings 1 and 2, a faulty e_3 affects rings 3 and 4, and faults on edges e_4, e_5, e_6 and

e_7 affect rings 1, 2, 3, and 4, respectively. For diagnosis purpose, all these seven segments are different.

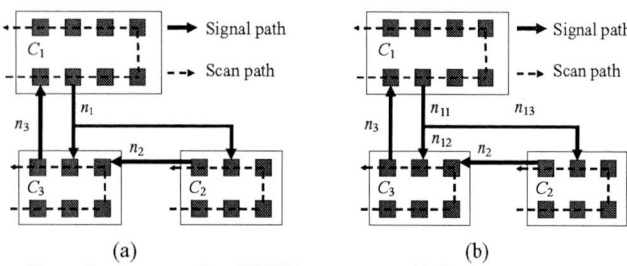

(a)   (b)

Figure 3. An example of SOC interconnect (a) interconnect detection for each net (b) interconnect diagnosis for each net segment.

Definition 2: A directed *graph* $G = (V, E)$ consists of a vertex set V and an edge set E, and each edge in E is an ordered pair (u, v), where $u, v \in V$.

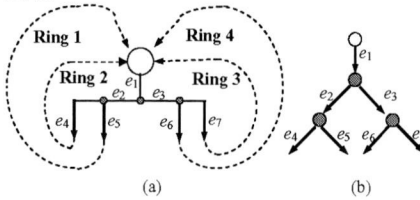

(a)   (b)

Figure 4. (a) a hypernet, and (b) the graph model for diagnosis.

III. Interconnect Diagnosability

A. Diagnosability Analysis

Given a circuit consisting of n edges $E = \{e_1, e_2, ..., e_n\}$ and a set of m oscillation rings $R = \{r_1, r_2, ..., r_m\}$. Once a ring is constructed, the test outcome is either "*pass*" (P) or "*fail*" (F). When an edge e_i is faulty, the test outcome of applying the m rings is said to be the *syndrome* of faulty e_i.

Definition 3: A fault on edge e_i and a fault on edge e_j are *distinguishable* under the test set R if the syndrome of faulty e_i and faulty e_j are different.

Definition 4: An edge is said to be *single-fault diagnosable* under the test set R if a faulty edge can be correctly identified, given that there is at most one fault in the interconnect structure.

Lemma 1: A fault on edge e_i and a fault on edge e_j are *distinguishable* under the test set $R \Leftrightarrow R_i \neq R_j$.

Proof: $\Leftarrow$ The fact $R_i \neq R_j$ implies that there exists a ring r such that either (1) $r \in R_i \land r \notin R_j$, or (2) $r \in R_j \land r \notin R_i$. Thus, the syndromes of faulty e_i and faulty e_j are different.

$\Rightarrow$ When $R_i = R_j$, both faulty e_i and faulty e_j fail the same set of rings, and thus they have the same syndrome. $\square$

Theorem 2: Edge e_i is *single-fault diagnosable* $\Leftrightarrow R_i \neq R_j$ for all $1 \leq j \leq n$ and $j \neq i$.

The correctness of Theorem 2 follows the result of Lemma 1. It takes $O(n^2 m)$ time to verify Theorem 2, since each pair of edges have to be compared. In order to reduce the complexity for diagnosability check, the following theorems can be used.

Theorem 3: Edge e_i is single-fault diagnosable if $|E_i| = 1$.

Proof: Assume that edge e_i is not single-fault diagnosable. From Theorem 2, there must exist an edge e_j such that $j \neq i$ and $R_i = R_j$. Therefore, both e_i and e_j belong to E_i and thus $|E_i| > 1$.

Theorem 4: Let R_i' be any non-empty subset of R_i for an edge e_i, and $E_i' = \bigcap_{r \in R_i} r$. Edge e_i is single-fault diagnosable $\Leftrightarrow \forall\, e_k \in E_i' - \{e_i\}$, e_i and e_k are distinguishable.

Proof: $\Leftarrow$ When at least one ring in R_i' oscillates correctly, e_i must be fault-free. On the other hand, when no rings in R_i' oscillate correctly, at least one edge in E_i' is faulty. Since all edges in $E_i' - \{e_i\}$ are distinguishable from e_i, we know whether e_i is faulty. Therefore, e_i is also single-fault diagnosable.

$\Rightarrow$ Assume that there is an $e_k \in E_i' - \{e_i\}$ and e_k is not distinguishable from e_i. When every ring in R_i' fails, it may be attributed to either e_k or e_i. Thus, e_i is not single-fault diagnosable. $\square$

Theorem 4 shows that not all rings in R_i are necessary to diagnose e_i, and a subset R_i' is informative enough if and only if e_i is distinguishable.

Corollary 5: Let R_i' be any non-empty subset of R_i for an edge e_i, and $E_i' = \bigcap_{r \in R_i'} r$. If for each $e_k \in E_i' - \{e_i\}$, e_k is single-fault diagnosable, then edge e_i is also single-fault diagnosable.

An example for the above definitions, theorems and corollaries is shown in Figure 5. Let the edge under consideration be e_i, then $R_i = \{r_1, r_2, r_3, r_4\}$, and $E_i = \{e_i, e_j, e_k\}$. Since R_i' can be any non-empty subset of R_i, we may choose $R_i' = \{r_2, r_3\}$, and thus $E_i' = \{e_i, e_j, e_k\}$. It is not necessary to have both e_j and e_k diagnosable to make e_i diagnosable. For example, let faults on e_j and e_k be indistinguishable; if a fault on e_i is distinguishable with $\{e_j, e_k\}$, then e_i is diagnosable according to Theorem 4.

Note that the above analysis applies to all types of faults except crosstalk glitches since they can be located directly from the test results of each ring.

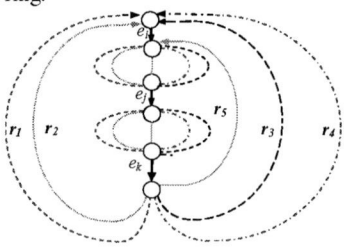

Figure 5. An interconnect diagnosis graph example.

B. Heuristic Diagnosability Check

In order to accelerate the process of diagnosability analysis, we propose a diagnosability check heuristic. Consider two edges e_i and e_j. According to Lemma 1, faults on these two edges are distinguishable if $|R_i| \neq |R_j|$. Thus, as the first step, we sort and partition all edges according to the number of rings passing them (i.e., $|R_i|$ for edge e_i). For example, in Figure 5, e_j and e_k are in the same group as $|R_j| = |R_k| = 5$, distinguishable from $|R_i| = 4$.

The second heuristic is to apply Theorem 3 first to check the diagnosability of an edge. Since the condition of Theorem 3, $|E_i| = 1$, is only sufficient but not necessary to guarantee that e_i be single-fault diagnosable, it is still possible that e_i is single-fault diagnosable when $|E_i| \neq 1$. In this case, we need to compare R_i with R_j for each e_j in the same group as e_i. To avoid the aforementioned problem, a third heuristic is used. The most likely reason for diagnosable e_i with $|E_i| \neq 1$ is that there exists an e_j such that $R_j \supset R_i$. When the edge e_j has been checked and removed from the check list before edge e_i is processed, we shall not run into this problem by Corollary 5. The flowchart of the diagnosis checking heuristic is shown in Figure 6.

Finally, when two faults are indistinguishable, they are put into the same *equivalent class* so as not to be compared twice.

The interconnect diagnosis heuristic algorithm is illustrated as follows. Consider the graph shown in Figure 7. There are three rings in the figure: $r_1 = \{e_1, e_4\}$, $r_2 = \{e_2, e_5\}$, and $r_3 = \{e_1, e_2, e_3\}$.

368

The diagnosis matrix representation for Figure 7 is illustrated in Figure 8(a), where each column represents an edge and each row represents a ring. A "1" is put in cell (i,j) if ring i passes edge j. Note that the edges are sorted and partitioned into two groups that are separated by the broken line. The first group consists of edges e_1 and e_2, and each of them is passed by two rings (i.e., $|R_1|=|R_2|=2$). The second group consists of three edges, and each of them is passed by one ring only (i.e., $|R_3|=|R_4|=|R_5|=1$).

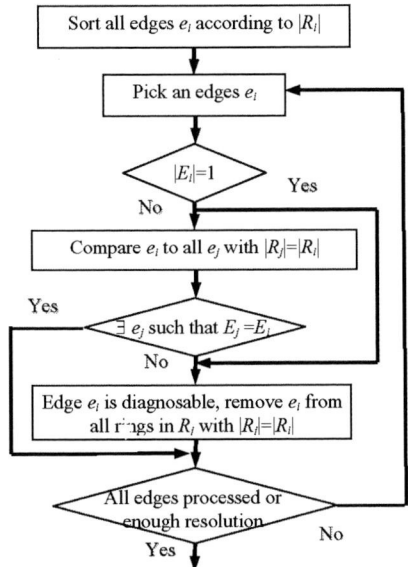

Figure 6. Flow chart of the heuristic for diagnosability checking.

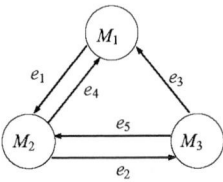

Figure 7. An illustrative diagnosability example.

The diagnosability checking process works as follows. First, apply Theorem 3 to edge e_1. We see that it is passed by rings r_1 and r_3, and the intersection of these two rings is $\{e_1\}$ (i.e., $|E_1|=1$). Thus, edge e_1 is single-fault diagnosable. Similarly, edge e_2 is also diagnosable as shown in Figure 8(b).

Syndrome of $e_1 = \{101\}$ indicates that the test results of r_1 and r_3 are incorrect and r_2 is correct when e_1 is faulty; syndrome of $e_2 = \{011\}$ indicates that r_2 and r_3 are incorrect and r_1 is correct when e_2 is faulty. Since the diagnosability analysis starts with the group with the highest $|R_i|$, we start with group $|R_i|=2$, including e_1 and e_2 and then group $|R_i|=1$, consisting of edges e_3, e_4, and e_5. Then, edges e_1 and e_2 are then marked and removed from the rings, as shown in Figure 8(b). There is only one edge remained in each ring, thus edges e_3, e_4 and e_5 are single-fault diagnosable due to Corollary 5.

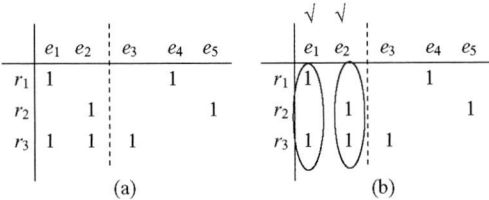

Figure 8. Matrices for the heuristic diagnosability checking.

IV. Interconnect Diagnosis Algorithm

In order to uniquely identify the faulty net segment, we need to ensure the optimal diagnosability or the maximum diagnosis resolution. The diagnosis resolution is defined as the largest number of nets with the same syndrome under a given set of test rings. Our goal is to diagnose every fault on every net segment, defined as the optimal diagnosability or the maximum diagnosis resolution.

We propose a heuristic to find a small set of rings for single fault diagnosis. The algorithm is a modified depth-first search. The SOC under test is modeled as a hypergraph H. This graph is then transformed into graph $G = (V, E)$ as outlined in Section 3.1. The vertex set V consists of cores and fanout points (intermediate nodes). The edge set E consists of edge segments partitioned from the original hypernets as explained in Figure 4(b). Our goal is to generate a predetermined set of rings to diagnose all edges in E. Since we need to detect the interconnect structure before diagnosis, the set of fault-detection test rings R_t should be applied first. A heuristic to find R_t is outlined below in Figure 9.

Algorithm: IORT (Interconnect Oscillation Ring Generation for Fault Detection)

Input: A hypergraph $H = (V, L)$ representing a circuit

Output: A list of rings R_t

1.   Transform hypergraph H into a new graph $G = (V', E)$ with equivalent *2-pin nets*;
2. $R_t := \varnothing$;
3.   **for every** $e = (u, v) \in E$ and e is not visited
4. $R_t = R_t \cup$ find_ring(G, e);
5.   reverse-order simulation for rings in R_t.

function find_ring(G, e)
1.   Let $e = (u, v)$ and v is an input pin in core C;
2.   **if** v is a pin in the starting core
3.   **return** the ring and mark all nets as visited;
4.   **for every** output pin w in C
5.   **if** there is an unvisited edge (w, x)
6.   find_ring$(G, (w,x))$;
7.   **else if** there is an untried output net (w, x)
8.   find_ring$(G, (w,x))$;
9.   **else**
10.   **return** $\varnothing$;
11. **end function**

Figure 9. The ring generation for fault detection algorithm.

For interconnect detection in the IORT scheme, in order to find R_t, we propose a heuristic algorithm to find a minimum set of rings that cover all 2-pin nets under test. We generate a ring containing a 2-pin net $(u, v) \in E$ by starting from vertex v, an input pin. Then we find an output pin w that locates in the same core as v, and w is connected to a 2-pin net that is not yet covered by any other ring. Each new ring may cover as many other uncovered nets as possible. After all rings having been generated, a simple reverse order simulation is conducted to remove redundant rings. A net is oscillation ring testable if there exists at least one ring containing this net.

Our goal for the interconnect diagnosis in the IORD scheme is to find a small set of rings R_d that can uniquely identify the faulty *edge* or *net segment*. The hypernet graph model for interconnect diagnosis (R_d) is the *2-pin net segment* model shown in Figure 4(b), different from the *2-pin net* model for interconnect detection (R_t). The set R_d is obtained by augmenting R_t as follows. We first apply the diagnosability checking techniques discussed in Section 3 to R_t to find out the net segments that are not diagnosable. For an edge e that is not single fault diagnosable, we try to find a new ring passing it without going through the edges that are indistinguishable to e. If such a ring exists, it will be included in R_d. The diagnosability checking should be conducted for each added ring so that other edges that become diagnosable with the new ring will be found.

This algorithm can be adjusted to the required diagnosis resolution to reduce the number of diagnostic rings in Figure 10.

Algorithm: IORD (<u>I</u>nterconnect <u>O</u>scillation <u>R</u>ing Generation for Fault <u>D</u>iagnosis)

Input: A hypergraph $H = (V, L)$ representing a circuit
Output: A set of rings R_d

1. Transform hypergraph H into a new graph $G = (V'', E)$ with equivalent *2-pin net segments*;
2. Generate a set of rings R_t for fault detection;
3. $R_d = R_t$;
4. Conduct diagnosability check;
5. **for every** $e \in E$ {
6.   **if** (e is not single-fault diagnosable)
7.   Find a ring r to make e diagnosable;
8. $R_d = R_d \cup \{r\}$;
9.   Modify the diagnosability of all edges in E;
       }
10. return R_d;

Figure 10. The ring generation for fault diagnosis algorithm.

The flowchart illustrating the process of diagnosis ring generation is given in Figure 11.

V. Optimization Techniques for Interconnect Diagnosis

Multiple oscillation rings cannot be applied simultaneously if they share some net segment (common edge constraint), or they go through the same scan path in a core (scan path conflict). In order to achieve the maximum concurrency (i.e., parallel test), we model all the constraints by a *conflict graph*, in which each ring is represented by a node, and two nodes are connected by an edge if they interfere with each other. The problem of finding the maximum concurrency tests can thus be reduced to the well-known *graph coloring problem*.

The number of test patterns can be greatly reduced whenever adaptive diagnosis is possible. In the adaptive diagnosis, a test pattern is selected according to the result of previous tests. An adaptive diagnosis tree, typically a binary tree, can be constructed according to the test patterns. For example, the adaptive diagnosis tree for the diagnosis example given in Figures 7 and 8 is illustrated in Figure 12.

For an *n*-net system, initially there are $n+1$ possible diagnosis results, namely fault-free ($\varnothing$) and a single fault on net e_i (f_{ei}) for $1 \leq i \leq n$. Each node in the tree represents a test pattern (ring), and the test outcome can be either pass (P) or fail (F). If the tree is balanced, the minimum number of diagnosis patterns required is $\lceil \log_2(n+1) \rceil$.

In order to construct a balanced adaptive diagnosis tree, in each internal tree node we need to select the test pattern (i.e. test ring) that evenly partitions the possible outcomes into two groups: Fail (F) and Pass (P). For example, in Figure 15, we choose the test pattern r_3 as the first test, since it evenly partitions the six possible outcomes into Fail (f_{e1}, f_{e2}, f_{e3}) and Pass ($\varnothing$, f_{e4}, f_{e5}). It can be seen that, in Figure 15, each test partitions possible outcomes into two groups whose cardinalities differ by at most 1.

The upper bound on the number of adaptive diagnosis test sessions needed in our method can be computed as follows. Let the number of test rings (without diagnosis) be $|R_t|$, and the length of the longest test ring be L_h. In the worst case, we need to apply $|R_t|$ rings to find out that there is a faulty net, and the last ring contains L_h net segments that are all passed by the ring only. It takes up to L_h-1 rings to distinguish these L_h possible faults, and thus the maximum number of diagnosis rings is $|R_t|+ L_h-1$.

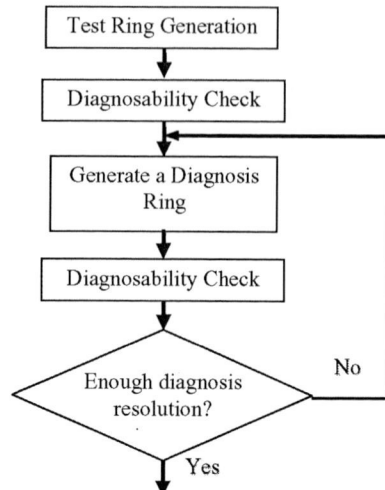

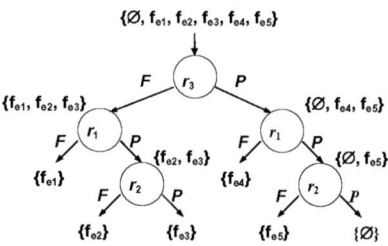

Figure 11. Diagnosis ring generation procedure.

Figure 12. An adaptive diagnosis tree.

VI. Experimental Results

We tested the diagnosis algorithm based on six benchmark circuits. In Table I, where the first column gives the circuit names, and the next four columns give the circuit statistics ("*Statistics*"), including the number of cores (*#core*), the number of pads (*#pads*), the number of hypernets (*#hyp*), and the number of net segments (*#net_segment*). The 5th column, *#net_segment*, lists the number of net segments to be diagnosed in each benchmark. The next three columns ("*Predetermined*") give the experimental results for predetermined diagnosis, including the number of rings required to detect all *2-pin nets* ($|R_t|$) and to diagnose all single faults ($|R_d|$). The last column, $|R_d|/|R_t|$, gives the ratio of rings from 1.25X to 2.81X for the maximum diagnosis resolution vs. for fault detection. This ratio means that we need extra test time of 1.25X to 2.81X to diagnose the single fault in each net segment under the predetermined diagnosis method, compared to the IORT scheme. In each case, we also give the estimated testing time (given in parenthesis), obtained by assuming only 4 MHz measuring period as discussed in Section 2.1 to estimate the *longest* test application time for each ring. The time needed to set up the rings should be roughly proportional to the testing time.

The next four columns ("*analysis*") give the diagnosis related information after applying R_t rings. The column *#OneRing* gives the number of nets passed by only one ring. Since the purpose of R_t is to detect faults with the minimum number of rings, it is not surprising that most nets are passed by one ring only. Most nets that are not diagnosable at this stage fall into this set. Columns "*#NoDiag*" and "*#EquClass*" give the number of nets that are <u>not</u> <u>diag</u>nosable and the number of <u>equ</u>ivalence <u>class</u>es after applying R_t, respectively. Two faults are in the same equivalence class if their syndromes for the tests are identical. The last column in this group ("$|R_d|-|R_t|$") gives the number of extra diagnosis rings required in each case to make all nets single-fault diagnosable. Assume that there are m equivalence classes whose sizes are s_1, s_2, ..., s_m, respectively. The upper bound on the number of additional diagnosis rings "$|R_d|-|R_t|$" can be expressed as follows:

$$\sum_{i=1}^{m}(S_i-1)=\sum_{i=1}^{m}S_i-m=\#NoDiag-\#EquClass \quad (3)$$

The upper bound on the required number of extra rings ($|R_d|-|R_t|$) is "($\#NoDiag$)–($\#EquClass$)". The empirical results "$|R_d|-|R_t|$" differs from the theoretical results "($\#NoDiag$)–($\#EquClass$)" given in Equation (3) by small differences of only up to 6.64%.

The last three columns ("*adaptive*") compare the number of rings required in both predetermined ($|R_d|$) and adaptive diagnosis ($|R_a|$). After applying R_t rings, the size of the largest equivalence class for each benchmark is given in the column "*max. EC*". In the worst case, the adaptive diagnosis needs to apply $|R_t|$ rings, and then *(max. EC)*–1 rings for diagnosis. The number of the worst-case adaptive diagnosis rings is given in column "$|R_a|$". The last column ($|R_d|/|R_a|$) shows the ratio of rings for the predetermined vs. adaptive diagnosis schemes. For the results shown in the column, the adaptive algorithm obtains 1.23X to 2.67X improvements over the predetermined diagnosis scheme. Also, from the normalized $|R_a|$ and $|R_t|$, the test time of adaptive diagnosis is approximately equal to that for detection alone and this reveals the effectiveness of adaptive diagnosis.

The experimental results for the concurrent test are given in Table II. The 3rd column ($|R_c|$) lists the number of test sessions after applying the concurrency test under the assumed worst-case scenario of net directions, core lists, scan paths and boundary scan paths. When a set of rings are applied concurrently, we refer to these rings as a test session. The 4th column ($|R_d|-|R_c|$) gives the percentage of improvements. The improvement can be even better for general interconnect structures. The reduction in test time due to the concurrent test ranges from 0.27% to 9.66% *with no hardware overhead*.

VII. Concluding Remarks

We have presented an IORD scheme for interconnect faults in SOC. In addition to the *100% fault detection coverage for each net* achieved by the IORT scheme, we have shown that fault location or fault diagnosis can also be done by including some extra test rings to achieve the *optimal diagnosability (or the maximal diagnosis resolution) for each net segment*. We have also presented two heuristics, diagnosability check and diagnosis ring generation, with theoretical study and integrated them into the IORD algorithm. Finally, two optimization techniques for improving interconnect diagnosability are proposed and showed to be effective. We have further compared the predetermined, adaptive and concurrent diagnosis schemes. Experimental results have justified the efficiency and effectiveness of the proposed methods.

References

[1] M. Tehranipour, N. Ahmed, M. Nourani, "Testing SoC Interconnects for signal integrity using boundary scan", in *Proc VTS*, 2003.

[2] Semiconductor Industry Association (SIA), *International Technology Roadmap for Semiconductors 2003 Edition* (ITRS), 2003.

[3] W. K. Kautz, "Testing of faults in wiring interconnects," *IEEE Trans. Computers*, vol. C-23, no. 4, pp. 358-363, Apr. 1974.

[4] X.-T. Chen, F. J. Meyer, and F. Lombardi, "Structural diagnosis of interconnects by coloring," *ACM Trans. Design Automation Electronic Systems*, vol. 3, no. 2, pp. 249-271, Apr. 1998.

[5] Y. Kim, H.-D. Kim, and S. Kang, "A new maximal diagnosis algorithm for interconnect test," *IEEE Trans. VLSI*, vol. 12, no. 5, pp. 532-537, May 2004.

[6] J.-C. Lien and M. A. Breuer, "Maximal diagnosis for wiring networks," in *Proc. ITC*, pp. 71-77, 1991.

[7] W.-T. Chen, J.-L. Lewandowski, and E. Wu, "Optima diagnostic methods for wiring interconnects," *IEEE Trans. Computer-Aided Design*, vol. 11, no. 9, pp. 1161-1166, Sep. 1992.

[8] E.J. Marinissen, B. Vermeulen, H. Hollmann, and R.G. Bennetts, "Minimizing pattern count for interconnect test under a ground bounce constraint," *IEEE Design &. Computers*, Vol. 20, No. 2, pp. 8-18, Mar-April, 2003.

[9] M. Kaneko and K. Sakaguchi, "Oscillation fault diagnosis for analog circuits based on boundary search with perturbation model," in *Proc. ISCAS*, pp93-96, 1994.

[10] K. Arabi and B. Kaminska, "Oscillation-based test strategy for analog and mixed-signal integrated circuits," in *Proc. VTS*, 1996.

[11] K. S.-M. Li, C.-L. Lee, C. Su, J.E. Chen, "Oscillation ring based interconnect test for SOC" in *Proc. ASPDAC*, pp. 184-187, 2005.

[12] F. DaSilva, Y. Zorian, L. Whetsel, K. Arabi, R. Kapur, "Overview of the IEEE P1500 standard," in *Proc. ITC.*, pp. 988-997, 2003.

Table I: Experimental results for Interconnect Diagnosis both for Predetermined and Adaptive Methods.

Circuit	Statistics				Predetermined			Analysis				Adaptive		
	#core	#pad	#hyp	#net_segment	$|R_t|$	$|R_d|$	$|R_d|/|R_a|$	#One Ring	#No Diag	#Equ Class	$|R_d|-|R_t|$	max. EC	$|R_a|$	$|R_d|/|R_a|$
ac3	27	75	211	416	133(33.3ms)	374(93.5ms)	2.81	389	323	68	241	8	140(35ms)	2.67
ami33	33	42	117	343	242(60.5ms)	303(75.8ms)	1.25	309	126	59	61	5	246(61.5ms)	1.23
ami49	49	22	361	475	156(39ms)	386(96.5ms)	2.47	406	337	88	230	9	162(40.5ms)	2.38
apte	9	73	92	136	73(18.3ms)	122(30.5ms)	1.67	127	94	40	49	4	76(19ms)	1.61
hp	11	45	72	195	81(20.3ms)	164(41ms)	2.02	176	145	51	82	7	87(21.8ms)	1.89
xerox	10	2	161	356	218(54.5ms)	342(85.5ms)	1.57	346	214	86	124	5	222(55.5ms)	1.54
Comp.					0.9679								1	

Table II: Concurrent Test Sessions.

Circuit	$|R_d|$	$|R_c|$ (worst case)	$|R_d|-|R_c|$
ac3	374	373	1 (0.27%)
ami33	303	290	17 (5.86%)
ami49	386	352	34 (9.66%)
apte	122	119	3 (2.52%)
hp	164	160	4 (2.50%)
xerox	342	327	15 (4.59%)
Comparison		1	4.57%

Table III: Comparison between Theoretical Bounds and Experimental Results.

Circuit	#NoDiag	#EquClass	Eq (3) (#NoDiag-#EquClass)	Extra Rings ($|R_d|$-$|R_t|$)	(#NoDiag-#EquClass) and (R_d-R_t)
ac3	323	68	255	241	14 (5.49%)
ami33	126	59	67	61	6 (8.96%)
ami49	337	88	249	230	19 (7.63%)
apte	94	40	50	49	1 (2.00%)
hp	145	51	94	82	12 (12.77%)
xerox	214	86	128	124	4 (3.13%)
Comparison				1	6.64%

High-Level Architecture Exploration for MPEG4 Encoder with Custom Parameters

Marius Bonaciu, Aimen Bouchhima, Wassim Youssef, Xi Chen, Wander Cesario*, Ahmed Jerraya

TIMA Laboratory, Grenoble, FRANCE, +33(0) 4 76 57 43 34, {firstname.surname@imag.fr}
*MND, Paris, FRANCE, +33 (0) 1 30 57 61 90, wcesario@mnd.fr

Abstract - this paper proposes the use of a high-level architecture exploration method for different MPEG4 video encoders using different customization parameters. The targeted architecture is a heterogeneous MP-SoC which may include up 2 coarse grain SIMD (task level SIMD) subsystems to perform the computations. The customization parameters are related to video resolution, frame rate, Communication Network, level of parallelism and CPU types. These parameters are determined during the high-level architecture exploration, by estimating the architecture performances at early stages of the design flow. Experiments shows that the error factor of these high-level performances estimations are less than 10% compared to those obtained with final manually implemented RTL architecture. This method was used successfully for exploration of different MPEG4 architecture configurations with different customization parameters. We consider these experiments a breakthrough because they show how a complex design can be mastered through a set of pragmatic choices.

Keywords – Multiprocessors SOC architecture, Video encoder, MPEG4, Architecture exploration, Customization

I. Introduction

Video encoding is widely included in most of consumer, multimedia, mobile and telecommunication applications [1], and becomes a key technology for many future applications. These different applications impose different constraints on the encoding parameters (i.e. video resolution) and on the resulting design (cost, speed and power). Even if MPEG4 is an accepted common standard for most embedded systems domains, a plethora of MPEG4 architectures exist today to comply with different applications [2].

MPEG4 requires a huge amount of computations, and thus needs parallel computations and hardware accelerations. Encoding for digital cinema system (HDTV 1920x1080 video resolution) using full motion search, it requires 32TIPS (Tera Instructions Per Second, 10^{12}) as computation power. This corresponds to a generic platform with 32000 RISC processors running at 1 GHz in parallel. Current technology doesn't allow such integration, and such design is difficult to program and debug [3]. This implies an expensive design process, and it's out of reach of many applications domains.

Implementing an MP-SoC until the RTL level, starting from a wrong set of ad-hoc parameters (i.e. CPUs number or communication topology) might turn out to be very costly. Each modification of parameters, will lead to the need of expensive modifications (which might also lead to a deteriorated final result, because new bugs may result after these "forced" modifications). In the worst case, it might require a complete redesign of the architecture. Very few products may justify such design budget, and the only working solution to get video encoding for low cost products (such as consumers) is to reduce the design cost of the product. The key solution to reduce the design cost is to explore the architecture at high-level, before the low-level architecture is implemented.

This paper proposes an efficient high-level architecture exploration method for different MPEG4 video encoders, using different customization parameters. This work concentrates only on performance estimations in term of speed.

A. Solution space for MPEG4 encoder on MP-SoC

Implementations of MPEG4 video encoder on MP-SoC can be applied in multiple domains: video surveillance, camera recorders, mobile telecommunications, home entertainment, etc. Each of them requires specific architecture configurations, and imposes their own constraints in term of speed, power and chip surface. Finding the final implementation solution requires adjusting a large number of parameters. These parameters can be split into two categories:

1) *Standard MPEG4 Algorithm parameters* are related specifically to the algorithm functionality: video resolution, frame rate, bitrate, quantization range, quantization type, motion estimation precision, motion search area, progressive/interlaced encoding, key frame rate, scene change detection, etc [4]. As it will be shown later in this paper, these algorithm parameters are not sufficient for implementing the MPEG4 video encoder on MP-SoC. To be able to implement the MPEG4 video encoder on a parallel architecture, the algorithm should be able to be easily parallelized / pipelined, by adding parameters for parallelism/pipelining support.

2) *Architecture parameters* are related to the targeted MP-SoC architecture: number of CPUs to be used, type of CPUs, HW-SW partitioning, communication topology, blocking/non-blocking protocol, arbitration type, message sizes, data width, maximum allowed data transfer latency, transfer initialization latency, etc.

B. Classical exploration flow

In classical exploration flows [5][6][7][8] (Fig.1a) , the designer implements the *Algorithm Specifications* starting from a set of already chosen *Algorithm Configurations*. After that, the *Architecture Specifications* is implemented, which should match with the *Algorithm Specifications*. In the end, the *Algorithm Specifications* and *Architecture Specifications* are combined, to obtain an *Algorithm/Architecture Executable Model*. This model simulates the algorithm and architecture running together, and it's used for *Performance Estimations*. If these estimations are not satisfying the requirements, the designer has to modify/redesign the algorithm and/or architecture specifications. This flow has some weak points:

a) The exploration space is highly reduced. The reason is that when having to change the algorithm and/or architecture specifications, the only things which can be changed is related to the parallelism/pipelining functionality of the algorithm on the architecture, some mapping decisions [8], data organizations and communications [5][6]. Any change leads to the need of completely redesigning the specifications, which means "restarting" the project.

b) Building the algorithm/ architecture executable model has to be done manually, which is a fastidious work, requires long design time, and might induces many errors[7]. Also, this model has to be re-designed every time the algorithm and/or architecture specifications are changed. The simulation speed of this model depends on the used abstraction level. If the abstraction level is too low [7], the simulation speed becomes unacceptable long.

c) The performance estimation precision depends on the used abstraction level. No matter the level of abstraction, the estimation precision represents a key issue. In [5][6], the performance estimations are covering only the communication. In [8], the performance estimations are covering precisely the computations, but the communication performances are estimated using an "always available" shared memory. This is insufficient if other communication topology is required. In [7], the estimations are precise, but the lack of any abstractions makes the simulations very long.

C. Contribution

The key contribution of this paper is a working solution for architecture exploration, used for the implementation of the MPEG4 video encoder on MP-SoC. We use a flexible target architecture and a flexible modeling strategy, that allow for both algorithm/architecture exploration. Compared with the previously presented classical exploration flow, our proposed exploration flow (Fig.1b) is able to cover multiple requirements:

a) The need to explore a large solution space is solved by automatically generating the *Executable Algorithm / Architecture Model*. Different customized *Executable Algorithm / Architecture Models* can be obtained from a unique *Flexible Algorithm/Architecture Model for MPEG4*. This model provides the possibility of automati-

0-7803-9451-8/06/$20.00 ©2006 IEEE.

cally customize the algorithm and build the abstract architecture, based on a set of *Algorithm/Architecture Configurations*.

b) The need of obtaining a fast simulation is covered by doing the architecture exploration at high-level. As result, by ignoring many low-level architecture details in *the Algorithm/Architecture Executable Model*, the simulation becomes fast.

c) The need of precise simulation results is solved by using a High-Level Architecture Exploration which provides estimation results with high precision. This is done by using precise estimations for the computations and communication times, by time annotating both computations and communications. Additionally, the exploration is capturing the computations and communications running together, to estimate the performances of the entire system.

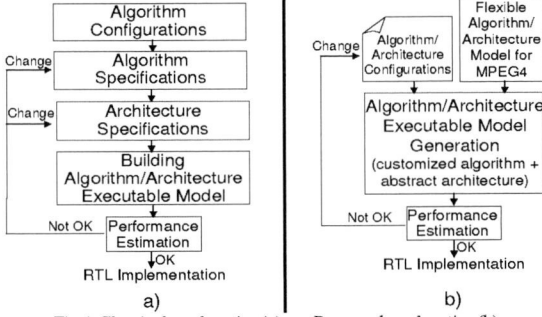

Fig.1 Classical exploration(a) vs. Proposed exploration(b)

By using such high-level architecture exploration, the time required to obtain efficient architectures of MPEG4 video encoder into MP-SoC is decreased drastically. This architecture exploration is achieved at High-Level using the *Flexible Algorithm / Architecture Model for MPEG4*. This decreases the time needed to obtain and test multiple architecture configurations. Thus, the time required to find acceptable architecture configurations is reduced. The proposed approach was applied successfully for the generation of several configurations of MPEG4 encoders.

The rest of the paper is organized as following. Section 2 presents the *Flexible Algorithm/Architecture Model for MPEG4*. Section 3 details the models used during the proposed high-level architecture exploration. Section 4 shows the architecture exploration flow for MPEG4 encoder with custom parameters. Section 5 presents the experiments and results, followed by conclusions in Section 6.

II. Flexible Algorithm/Architecture Model for MPEG4

This section presents the MPEG4 encoder algorithm and the *Flexible Algorithm/Architecture Model for MPEG4*, which will be used to generate different models required during the exploration.

A. MPEG4 video encoder algorithm

In this work we used the DivX specifications. The DivX is a popular algorithm implementation of the MPEG-4 video compression technology (ISO/IEC 14496-2). The idea of this technology is to compress and store only the *spatio-temporal* differences between consecutive frames. A block diagram of the DivX algorithm is shown in Fig.2. Describing each block is out of this paper's scope, and more details can be found in [4][9].

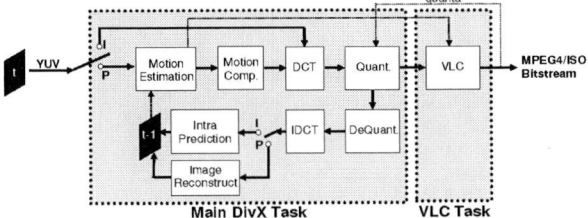

Fig.2 Block diagram of the DivX algorithm

The initial MPEG4 encoder algorithm is a sequential algorithm. The *Standard MPEG4 Algorithm parameters* don't provide any automatic parallelism/pipeline support, which makes difficult to

implement the MPEG4 encoder on multi-processors. This drastically reduces the exploration space.

So, the need of inserting parallelism and pipeline support is required. The parallelism/pipeline shouldn't change the algorithm specifications, only the implementation will be different. Our goal is to build a *Flexible MPEG4 Encoder algorithm*, which supports the *Standard MPEG4 Algorithm parameters* plus parameters for the *level of Parallelism/Pipeline*.

For this, the MPEG4 algorithm was grouped into 2 pipelined tasks: *MainDivX task* and the *VLC (Variable Length Coding) task*. The *MainDivX* is processing the current image relative to the previous one, and its results are motion vectors, quantized DCT *Macro-Blocks* (image zones of 16x16 pixels). The *VLC* is compressing these results using a Zigzagging and Huffman compression. The final output respects the MPEG4/ISO standard.

Several approaches for parallelization can be found in [10][11]. In [10] and [11], the image is split into smaller areas to be able to achieve parallel computations. In their approach, the *MainDivXs* and *VLCs* were not separated, thus splitting the image into areas required an equal number of *MainDivXs* and *VLCs*. Also, a *VLC* has to wait for the corresponding *MainDivX* to finish its computations for all the *Macro-Blocks*, which means that the *VLC* is 90% into idle mode.

In our work the algorithm is divided into two pipeline stages: the first contains the *MainDivXs*, and the second contains the *VLCs*. We use SIMD architectures for each of the pipeline stage, to handle heavy computations. Since the two stages require different computation powers, the structure of both SIMD may be different in term of number and type of CPU. To adapt the algorithm to this architecture, the image is split into areas, and the computations are done in parallel for each of them.

We've exploited the fact that the *VLC task* doesn't have to wait for the *MainDivX task* to finish processing the entire image. By adapting the *VLC* task to work at *Macro-Block* level, once the *MainDivX* task finished processing a *Macro-Block*, the *VLC* task can start to compress it, while the *MainDivX* task continues to process the next *Macro-Block*. Also, the *VLC* task requires very few computations but large memory (because the need to store its standard Huffman tables). To reduce the memory for the *VLC tasks*, we use a number of *VLC tasks* much smaller than the number of *MainDivX tasks*, but just enough not to become a computational bottleneck. As results, the processing and compression are executed in parallel, and the application's memory is reduced.

Along with these tasks, the use of 4 other smaller tasks is required: *Video*, *Splitter* (for image splitting), *Combiner* (for final reordering) and *Storage* (Fig.3).

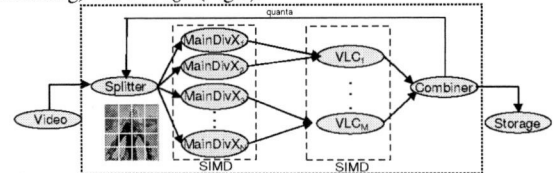

Fig.3 Flexible MPEG4 Encoder algorithm

The *Video* task doesn't belong in the final design. It's a test-bench task that simulates a video source. It sends the video under the form of a pixels stream, compatible with YUV420 standard to the *Splitter*. The *Splitter* divides the image and routes the pixels to the corresponding *MainDivX*, which processes the image. Once the *MainDivXs* processed a *Macro-Block*, its results are sent to a corresponding *VLC*, which will compress the *Macro-Blocks* one by one.

The compressed *Macro-Blocks* are then sent to the *Combiner*, which reorders all the *VLC* results, in order to obtain an MPEG4/ISO bitstream. Also it adjusts the quantization value to be used for the encoding of the future images. The bitstream is sent to the output *Storage* task, which is another test-bench task simulating a storage support. As result, the architecture's behavior is composed of 2 pipelines. One pipeline at frame level between the *Splitter* and the rest of the architecture using a lock step synchronization at frame level, and a second local pipeline at *MacroBlock* level between the *MainDivXs*, *VLCs* and *Combiner*.

Coarse grain parallelization was chosen instead of fine grain one (i.e. every basic function of the *MainDivX task* to be a different task) for simplicity and efficiency. Instead of achieving the parallelism by dividing the algorithm into multiple tasks, resulting in computation distribution on the architecture, the parallelism is done using data distribution. If one of the tasks becomes a bottleneck, the amount of data associated to that task is reduced [12]. Also, using fine grained partitioning for highly called tasks induces serious performance degradation, because of the required big number of context switches.

B. Configuration parameters

To explore the architecture for the MPEG4 video encoder, we use 2 categories of configurations parameters, shown in Table 1.

Algorithm parameters	Architecture parameters
Level of Parallelism/Pipeline	Number of CPUs
Video resolution	Type of CPUs
Frame rate	HW-SW partitioning
Bitrate	Communication topology
Key frame	Blocking/Non-blocking comm.
MotionEstimation precision	Arbitration type
MotionEstimation search area	Message size
Progressive/Interlaced mode	Data width
Scene change detection	Data transfer latency
Quantization range	Transfer initialization latency
Quantization type (H263,MPEG4)	Transfer close latency

Table 1. Algorithm and Architecture parameters to be explored

C. Flexible Algorithm/Architecture Model

The *Flexible Algorithm/Architecture Model for MPEG4* (Fig.4) is composed of *Modules* and an *Abstract interconnect execution model*. Each *Module* contains one task, which can be of 2 types:
a) *flexible tasks*, which has at least one of following characteristics:
- flexible computations – tasks which are belonging to a SIMD: *MainDivX* and *VLC*
- flexible input – tasks which are receiving data from a SIMD: *VLC* and *Combiner*
- flexible output – tasks which are sending data to a SIMD: *Splitter* and *MainDivX*
b) *fixed tasks* – none of the above: *Video* and *Storage*

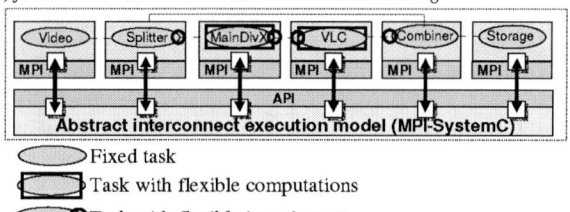

Fixed task

Task with flexible computations

Task with flexible input/output

Fig.4 Flexible Algorithm/Architecture Model for MPEG4

Tasks are communicating via *API* calls. The interconnections are done through an *Abstract interconnect execution model*, which in [13] is called *High-Level Parallel Programming Model* (HLPPM). A HLPPM hides completely the low-level architecture details: Communication Network, HW/SW & HW/HW Interfaces.

The targeted architecture model is also flexible as shown in Fig.5. It features 2 SIMD and an interconnect structure. The *Splitter* and *Combiner* may be HW or SW. For example, in case of Fig.5, it was freely chosen to map the *Splitter* and *Combiner* on HW, to avoid them to become I/O bottlenecks.

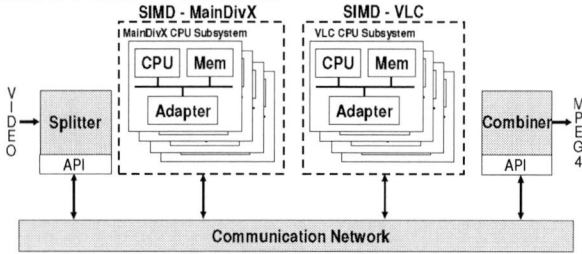

Fig.5 Abstract architecture model with 2 SIMD

III. Algorithm and architecture representation

During the architecture exploration, a unique representation will be used to combine both the architecture and the MPEG4 algorithm. This combined architecture/algorithm format will be used to model different customized algorithm/architecture instances at different levels of abstraction, starting from the pure algorithm model down to the architecture. All these models are executable SystemC models that can be simulated and used for performance estimations, debug, and as entry for design. This section describes in details all the models used during the high-level explorations.

A. Flexible Algorithm/Architecture Model for MPEG4

This model is a macro-code made of a set of generic SystemC modules containing each of them a single task written in C/C++. Tasks are communicating through message passing by calling a set of MPI primitives (see Fig.6)

```
MP_Init(*this,argc,argv);
MP_Finalize(*this);

MP_[I]Send(*this,buf,count,datatype,dest,tag,comm);
MP_[I]Recv(*this,buf,count,datatype,source,tag,comm,status);

MP_[I]BSend(*this,buf,count,datatype,dest,tag,comm);
MP_[I]BRecv(*this,buf,count,datatype,source,tag,comm,status);

MP_[I]SSend(*this,buf,count,datatype,dest,tag,comm);
MP_[I]SRecv(*this,buf,count,datatype,source,tag,comm,status);

MPI_Wait(*this,request,status);
MPI_Test(*this,request,flag,status);
```

Fig.6 MPI communication primitives subset

An example of a task using MPI primitives is illustrated in Fig.7.

```
//---------------- MainDivX"N" task ----------------
void MainDivX"N"_MAIN(*image_memory"N", height"N", length"N", top_border"N",
                      left_border"N", bottom_border"N", right_border"N",&result)
{
  //initialization of computations
  MainDivX"N"_INIT (&image_memory"N", height"N", length"N");

  //data_receive_communication from the Splitter
  MPI_"PROTOCOL"Recv(this,&image_memory"N", sizeof(image_memory"N"),
              "DATA_WIDTH",SPLITTER_ID,22,MPI_COMM_WORLD);

  //calls the function with flexible computations
  MainDivX"N"_COMPUTE (&image_memory"N",height"N", length"N",
                      top_border"N", left_border"N",
                      bottom_border"N", right_border"N",&result);

  //send_results communication to the coresponding VLC
  MPI_"PROTOCOL"Send(this,&result,sizeof(result),"DATA_WIDTH",
              VLC["target_vlc"]_ID,22, MPI_COMM_WORLD);
}
```

Fig.7 Example of task description using MPI primitives

MPI_"PROTOCOL"Recv(this,&image_memory"N",sizeof(image _memory"N"),"DATA_WIDTH",SPLITTER_ID,22,MPI_COMM _WORLD) receives data from other task. This primitive specifies the pointer were data will be stored, the amount of data, the communication data width and the unique ID assigned to the source task (*Splitter*). It can be noticed that most parameters are not yet fixed.

Flexible tasks with flexible computations are parameterized to be duplicated and work on different data. For example, the *MainDivX* task uses a set of parameters that specify the address where the image is stored, the image size, the border characteristics, the results storing address, and it uses them to call its computations.

Flexible I/Os are coded using Generate like loops. Fig.8 shows a part of the code of the *Splitter*.

```
void Splitter()
{
  for (target=0; target<"N"; target++)
  {
    MPI_"PROTOCOL"Send(&this,data[target],"BURST_SIZE",
              "DATA_WIDTH",MainDivX[target]_ID,
              22,MPI_COMM_WORLD);
  }
}
```

Fig.8 Describing the flexible I/O in the Splitter

The loop executes *"N"* MPI_Send to split the image among the *MainDivX* modules performing the encoding. Even the protocol can be parameterized. In Fig.8, the MPI_*"PROTOCOL"*Send can be expanded into MPI_*I*Send, MPI_*B*Send or MPI_*S*Send, according to the *"PROTOCOL"* parameter of communication. [13] details the differences between these protocols. The *"BURST_SIZE"* sets the communication message size, and the *"DATA_WIDTH"* defines the communication data width. This flexible model is used to generate an executable SystemC Model.

The *Flexible Algorithm/Architecture Model for MPEG4* is made of a set of *Modules* interconnected through *MPI-SystemC HLPPM* (Fig.4). The *MPI_SystemC HLPPM* is a runtime execution environment for message passing communication using the subset of MPI primitives presented in Fig.6. It's similar to MPICH [14] (supports the same MPI primitives) but with the possibility of including configurable timing annotations for the communication, using SystemC libraries. Fig.9 shows that the communication between 2 tasks are done using <u>C</u>ommunication <u>U</u>nits (CU) (one CU for each task), which manages the MPI requests from the tasks, the communication with other CUs, and inserts the timing annotations. Since a CU can be connected to many other CUs, the MPI-SystemC HLPPM can support point to point and bus topologies.

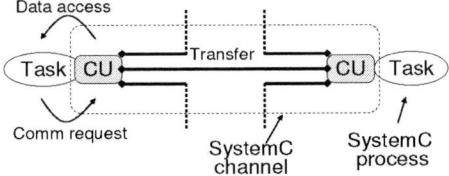

Fig.9 Task to Task communication using MPI-SystemC

Using the *Flexible Architecture/Algorithm Model for MPEG4*, many customized models can be macro-generated. This is done by expanding the flexible model with the desired algorithm / architecture configuration parameters, with the approach presented at [15].

Fig.10 shows an example of macro-generated SystemC Model with 2 SIMD subsystems (*MainDivX* and *VLC*), and the data dependencies between the tasks (the dotted arrows). This model is called *Executable SystemC Model of Combined Architecture/ Algorithm*. It is an un-timed model, and it captures both the architecture and the algorithm. Fig.11 (the WAITs will be explained later) shows the C/C++ code of the resulted *MainDivX1* task after the macro-expansion. It can be seen that all the algorithm/architecture parameters are now fixed. For different configuration parameters, different Executable SystemC Models are obtained. The key advantage of such model is its suitability for performances analysis, algorithm debug, syncronization debug, etc.

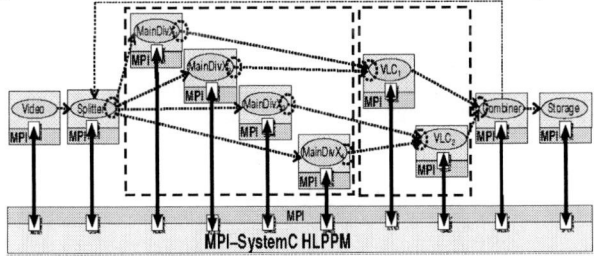

Fig.10 Executable SystemC Model of Combined Architecture/Algorithm

For performance estimations, a *Timed Executable SystemC Model* is used. This model is obtained by inserting time annotations for computations and communications, into the tasks code of the previously untimed Executable SystemC Model. The time annotations for computations are done by inserting WAIT calls to simulate the computations delays [14]. The time annotations for communications are embedded within the MPI-SystemC.

Fig.11 shows the code of the resulted Timed Model for the *MainDivX* task, which contains the time annotations for the computations, and the time annotations for the communications (integrated into MPI-SystemC HLPPM). The values for these delays are captured in tables and depend on the configurations chosen for the computations (i.e. CPU model, CPU clock frequencies) and communication primitives (i.e. data width, message sizes, latencies). This *Timed Model* allows performance estimations at High-Level for different algorithm/architecture configurations.

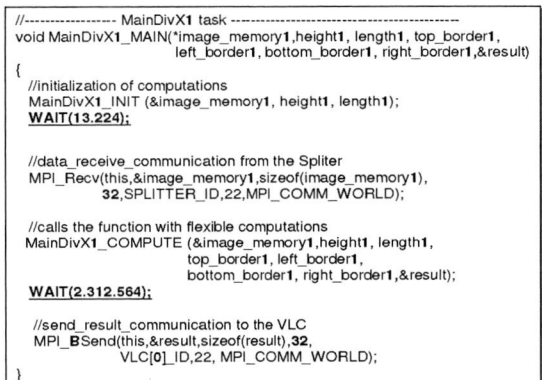

```
//------------- MainDivX1 task ------------------------
void MainDivX1_MAIN(*image_memory1,height1, length1, top_border1,
                left_border1, bottom_border1, right_border1,&result)
{
    //initialization of computations
    MainDivX1_INIT (&image_memory1, height1, length1);
    WAIT(13.224);

    //data_receive_communication from the Spliter
    MPI_Recv(this,&image_memory1,sizeof(image_memory1),
        32,SPLITTER_ID,22,MPI_COMM_WORLD);

    //calls the function with flexible computations
    MainDivX1_COMPUTE (&image_memory1,height1, length1,
                top_border1, left_border1,
                bottom_border1, right_border1,&result);
    WAIT(2.312.564);

    //send_result_communication to the VLC
    MPI_BSend(this,&result,sizeof(result),32,
        VLC[0]_ID,22, MPI_COMM_WORLD);
}
```

Fig.11 Timed Model for the MainDivX task

IV. High-level architecture exploration flow for MPEG4

This section describes in details the high-level architecture exploration flow (Fig.12) used for the MPEG4 video encoder.

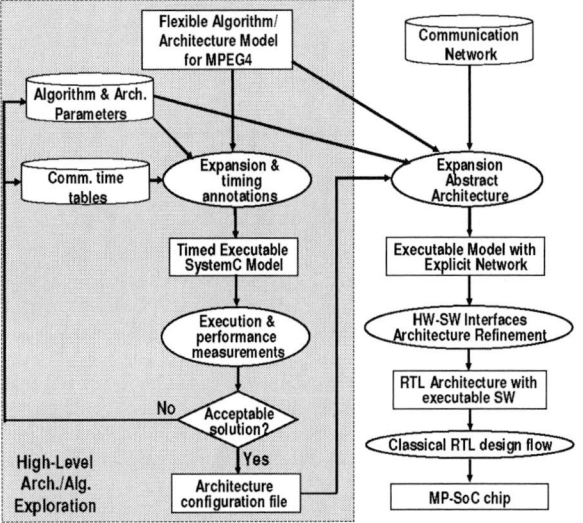

Fig.12 Detailed representation of the design flow

This flow is composed of 3 major phases: (1) obtaining the *Timed Executable SystemC Model*; (2) performance estimation and reconfiguration; (3) building the final RTL architecture. Only the first 2 phases are part of the high-level architecture exploration and will be detailed. Presenting the phase of building the final RTL architecture is outside the scope of this paper [21].

A. Obtaining the Timed Executable SystemC Model

The *Timed Executable Model* is obtained in two steps. First the *Flexible Algorithm/Architecture Model for MPEG4* is macro-expanded to obtain the *Executable SystemC Model of Combined Algorithm/Architecture*. This initial model is used to compute the delays. Afterwards, the delays are inserted in the executable model.

Delays are obtained using a classical approach consisting of executing the code on an Instruction Set Simulator (ISS) of the targeted CPU. This gives approximate number of clock cycles required by the different tasks, independently of the communications. The obtained times aren't 100% accurate, because the scheduling effect isn't captured with this approach. However, the experiments show that the precision is enough for our architecture exploration, as will be shown later in this paper.

Communication times are given for different communication configurations (message size, data width, protocol, transfer latencies). In this work, these times are given as parameterized delay functions associated to each MPI primitive. The execution of each primitive is broken into 3 steps: initialization (initial synchronizations), transfer (for each data) and close (communication release). An execution time is associated to each of these steps, allowing a detailed viewing/analyzing of the communication behavior.

B. Performance estimations and architecture exploration

By compiling and executing the *Timed Executable SystemC Model*, performances can be measured using the function *sc_simulation_time()* after encoding every frame. The execution of this timed model gives an estimation of performances. The obtained performances can be represented using performances diagrams (graphic tables), and they include the time annotated computations and communications running together. Also, in the same graphic can be displayed for comparison the performances measured for multiple different algorithm/architecture configurations, to help the designer to take the next decisions. Fig.13 gives the estimated performance for the execution of MPEG4 for 25 frames of QCIF (176x144) video resolution movie, using 1,2,4,8,16 and 32 CPUs ARM7[17] at 60MHz for the MainDivX tasks and 1 CPU for VLC.

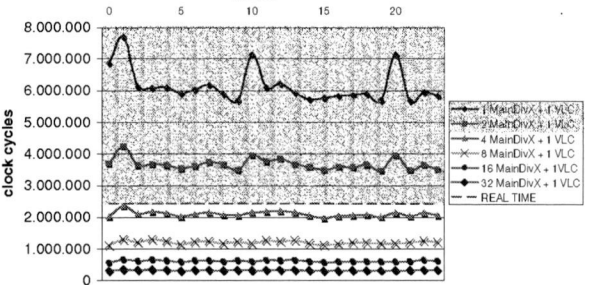

Fig.13 Performance estimated for QCIF, using ARM7, 60MHz

As benchmark movie we used one second (25 frames) of „snow-show" movie (similar to what the TV receivers show when there is no signal on the antenna). This represents the worst case scenario for the MPEG4 application. Consequently, is assured the real-time encoding for any other input cases. Also, the used search area for the Motion Estimation is 16x16. The reason is that previous research experiments showed that for QCIF (176x144) and CIF (352x288) resolutions, the full area search can be discarded, because the compression gain doesn't pay for the performance loss. However, this isn't true for higher video resolutions.

Fig.14 shows the estimated performances using ARM946E-S, 4kI$, 4kD$ CPUs at 60MHhz [17]. In order to achieve real-time, maximum 2.400.000 cycles are allowed for compressing 1 frame. From this figures, we can determine that minimum 5 ARM7 or 2 ARM946E-S processors are required to achieve real-time.

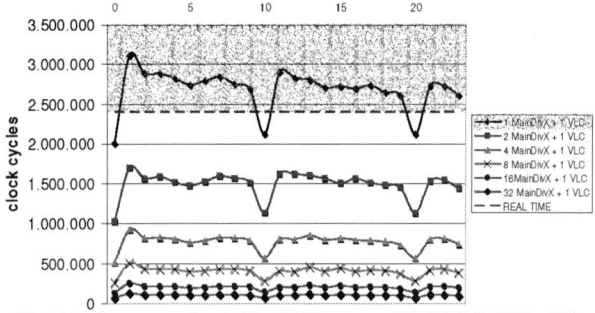

Fig.14 Performance estimated for QCIF, using ARM946E-S CPUs, 4kI$, 4kD$, 60 MHz

Different curves of these simulations are obtained doing a new macro-expansion of the initial model with different parameters. Besides number and types of CPU, several other parameters may be explored. For example, the communication may be explored via message size, data width, protocols and latencies.

C. Validation of the High-Level simulation results

The architecture exploration allows fixing a set of parameters that will define the number of required CPUs, models of CPUs, commu-

nication protocols, message sizes, maximum latencies, etc. These parameters will be followed during the architecture implementation. Fig.15 shows an example of obtained configurations.

```
CPUs Number      → 5 (4 MainDivX + 1 VLC)
IPs Number       → 2
Splitter         → IP
MainDivX₁        → CPU (ARM7)
MainDivX₁        → 60MHz
MainDivX₁        → no cache
... the same for the other 3 MainDivX
VLC₁             → CPU (ARM7)
VLC₁             → 60MHz
VLC₁             → no cache
Combiner         → IP
Splitter-MainDivX₁ send protocol      → Blocking
MainDivX₁-Splitter recv. protocol     → Non-Blocking
Splitter-MainDivX₁ burst size         → 128 bytes
Splitter-MainDivX₁ data_width         → 32 bits
Splitter-MainDivX₁ init_latency       → 2 cycles
Splitter-MainDivX₁ data_latency       → 3 cycles
MainDivX₁-VLC₁ send protocol          → Blocking (FIFO 810bytes)
VLC₁-MainDivX₁ recv. protocol         → Blocking
VLC₁ recv. arbitration                → AnySource
MainDivX₁-VLC₁ burst size             → 810 bytes
MainDivX₁-VLC₁ data_width             → 32 bits
.... similar for the other modules
```

Fig.15 Example of architecture configuration file

V. Experiments and results analysis

This section presents the experiments results obtained for the architecture exploration of the MPEG4 application for QCIF (176x144) and CIF (352x288) video resolution at 25 frames/sec, using ARM7 and ARM946E-S processors running at 60 MHz.

For QCIF video resolution using only ARM7 processors running at 60MHz, the resulted architecture required 5 processors: 4 processors for 4 *MainDivX* tasks, and 1 processor for 1 *VLC* task. Simulation for 1, 2, 4, 8, 16 and 32 CPUs for *MainDivX* and 1 for VLC was shown in Fig.13. The obtained architecture configurations are shown in Fig.15.

For CIF (352x288) video resolution, the same experiments were conducted. In case of using only ARM7 processors, the architecture required 23 processors: 20 for *MainDivX* task and 3 for *VLC* task. Initially, 16 processors were sufficient for the *MainDivX* tasks, but the communication degradation made this impossible. So it was opted for more processors, instead of choosing a very "super" communication. Fig.16 shows the performance diagram using ARM7 CPUs at 60 MHz.

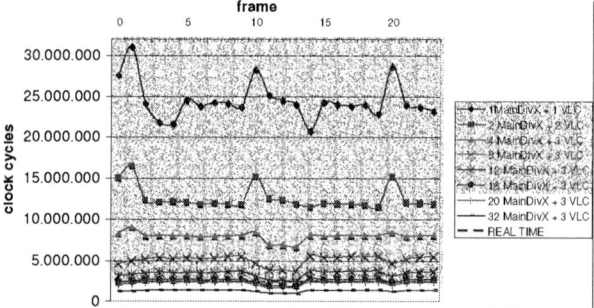

Fig.16 Performance estimated for CIF, using ARM7, 60MHz

When ARM946E-S processors were chosen, things got simpler because of the higher provided computation power. Fig.17 shows the obtained performance diagram for CIF video resolution, using ARM946E-S processors. Fig.17 shows that 10 ARM946E-S processors at 60 MHz were required to obtain a real-time functionality: 8 for *MainDivX* tasks and 2 for *VLC* tasks.

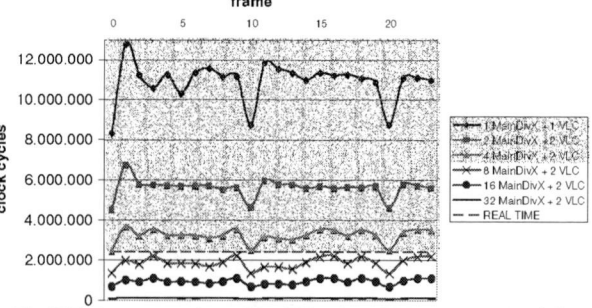

Fig.17 Performance estimated for CIF, using ARM946E-S CPUs, 4kI$, 4kD$ at 60 MHz

376

For higher resolutions of MPEG4, ARM7 and ARM9 are not enough. Additionally the amount of embedded memory gets higher than what the current technology may allow. For memory, an off chip memory may be required which might change the required interconnect. For computations, powerful DSP or VLIW processors are needed, or HW instructions [18][19].

To validate the precision of the high-level architecture exploration, an RTL architecture was built more or less manually, using one of the architecture configurations obtained during the high-level architecture exploration. Fig.18 shows that the precision of the performance estimations, obtained during the high-level architecture exploration, are very close with the one measured at RTL level. The communication infrastructure used in case of the RTL architecture, is a highly performant and customizable data transfer architecture [20] that can be easily configured with the communication parameters obtained by architecture explorations.

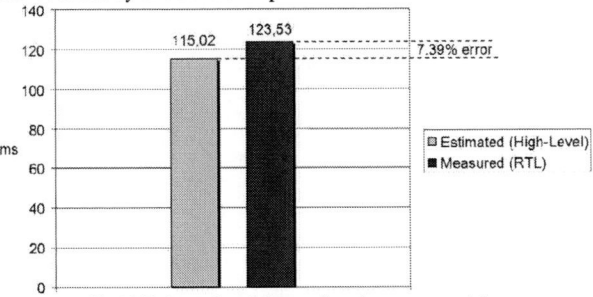

Fig.18 Estimated vs. Measured performance precision
[QCIF, 1 frame, 2 ARM7 CPUs (1 MainDivX + 1 VLC), 60MHz]

To compress 1 frame at QCIF resolution, using 2 ARM7 CPUs (1 *MainDivX* + 1 *VLC*) running at 60MHz, the high-level estimations predicted that 115.02 ms are required (in Fig.13, 6.82 million cycles is equivalent with 115.02 ms). The performance measured for the obtained RTL architecture proved that 123.53 ms were required to compress 1 frame. The 7.39% precision error comes from the impossibility to capture with our proposed high-level estimations, the performance degradations of the:
- OS (scheduling, service calls latencies induced by the API calls)
- Interconnect between the CPU buses and communication infrastructure (the conflicts for local bus grant between the CPUs and the Network Interfaces).
- HW/SW Wrappers

By using the proposed high-level architecture exploration, different and already validated architecture configurations were explored very quickly, even for a big number of CPUs and complex communications. This process dramatically shortened the time required to do the architecture exploration. As an example, in case of 25 frames of QCIF resolution video and using ARM7 processors running at 60MHz, approximately 15 minutes were required to generate the Timed Model. The simulation for 25 frames took approximately 2 minutes. Exploring one architecture solution takes less than one hour. This is the time required just to simulate one frame at RTL level. Approximately 25 hours were required to simulate 25 frames using the RTL model. So, the high-level performance estimations error of less than 10% compared with the low-level performance measurements is more then acceptable, considering the gain of design time. In these experiments we've used a Pentium4, 3GHz, 1Gbytes RAM, using Linux Mandrake 9.2.

Currently, in order to adapt this approach to other applications, the Initial Specifications Model must be adapted manually. Also, if the communication network is changed (uses different topologies with the ones already supported), the MPI-SystemC HLPPM model needs to be adapted to the new communication constrains. Automating these tasks or finding a method to reduce the effort needed for applying the proposed design paradigm to different applications are, in our opinion, open research subjects, and we have chosen to leave this point for future works.

VI. Conclusions

The architecture exploration of the MPEG4 video encoder into MP-SoC raises many challenges, because its complexity in term of computation, communications and memory requirements. Architecture exploration at low-level is a very long time consuming process. This paper proposed the use of a high-level architecture exploration method, since early stages of the design flow. The design effort gain is more important when lots of processors and complex communication networks are used, as showed by our case study. Because the architecture validation is done at High-Level, finding the optimal architecture configurations becomes possible in a much shorter time, compared with the validation at Low-Level. The proposed approach was successfully used during the architecture exploration of MPEG4 video encoders. The time required to explore different architectures was reduced from days to approximately 1 hour. This method can be extended for different other application, by adapting the Flexible Algorithm/Architecture Model for this new application.

References

[1] AV140 Video Recorder, http://www.archos.com/products/prw_500431.html

[2] eXpressDSP compliant MPEG4 Simple Profile Video Encoder for TI TMS320C64x DSPs, http://focus.ti.com/catalog/docs/thirdpartysoftwaref older.tsp?softwareId=193

[3] M.W.Youssef et al, "*Debugging HW/SW Interface for Multiprocessor SoC: Video Encoder System Design Case*", 41st DAC, San Diego, CA, June 2004

[4] I. Richardson, "*H.264 and MPEG-4 Video compression*", white paper

[5] Pieter van der Wolf et al, „*Design and Programming of Embedded Multiprocessors: An Interface-Centric Approach*", CODES-ISSS, Stockholm, Sweden, September 2004

[6] Satya Kiran M.N.V et al, "*A Complexity Effective Communication Model for Behavioral Modeling of Signal Processing Applications*", 40th DAC, Anaheim, CA, June 2003

[7] L. Formaggio et al, "*A Timing-Accurate HW/SW Co-Simulation of an ISS with SystemC*", CODES-ISSS 2004, Stockholm, Sweden, September 2004

[8] Xtensa LX Xplorer 1.0.1, www.tensilica.com

[9] V. Bhaskaran et al, "*Image and Video Compression Standards: Algorithms and Architecture*", Norwell,MA:Kluwer 1995

[10] Y. He et al: "*A Software Based MPEG-4 Video Encoder Using Parallel Processing*", IEEE Trans. on Circuits and Systems for Video, Nov. 1998

[11] K-K. Leung et al, "*Parallelization Methodology for Video Coding – An Implementation on the TMS320C80*", IEEE Transaction on Circuits and Systems for Video Technology, Dec.2000

[12] M.Raulet et al, "*Automatic coarse-grain partitioning and automatic code generation for heterogeneous architectures*", in proceedings of IEEE Workshop on Signal Processing Systems 2003, Seoul, Korea, Aug.2003

[13] C-C. Chiang, "*High-Level Heterogenous Distributed Parallel Programming*", Proceedings of the 2004 International Symposium on Information and Communication Tech., Las Vegas, Nevada, June 2004.

[14] MPICH – A Portable MPI Implementation, http://www-unix.mcs.anl.gov/mpi/mpich/

[15] M4 - a GNU implementation of the UNIX macro processor, http://www.seindal.dk/rene/gnu/whatis.htm

[16] Luciano Lavagno et al, "*Specification, Modeling and Design Tools for System-on-Chip*", ASP-DAC'02/VLSI Design, Jan. 2002

[17] ARM7 & ARM946E-S, http://www.arm.com/products/CPUs/index.html

[18] Pierre Paulin et al , "*Parallel Programming Models for a Multi-Processor SoC Platform Applied to High-Speed Traffic Management*", Best paper, ISSS/CODES 2004, September 2004, Stockholm, Sweden,

[19] Tensilica XTENSA-LX, http://www.tensilica.com

[20] S.-Il Han et al "*An Efficient Scalable and Flexible Data Transfer Architecture for Multiprocessor SoC with Massive Distributed Memory*", DAC'04, San Diego, USA, June 2004

[21] A. A. Jerraya, W. Wolf, "*Multiprocessor Systems-on-Chips*", Morgan Kaufmann Publishers, ISBN 0-12-385251-X, September 2004

Programmable Numerical Function Generators Based on Quadratic Approximation: Architecture and Synthesis Method

Shinobu Nagayama

Dept. of CE
Hiroshima City Univ.
Hiroshima 731-3194, Japan

Tsutomu Sasao

Dept. of CSE
Kyushu Inst. of Tech.
Iizuka 820-8502, Japan

Jon T. Butler

Dept. of ECE
Naval Postgraduate School
CA 93943-5121, USA

Abstract— This paper presents an architecture and a synthesis method for programmable numerical function generators (NFGs) for trigonometric, logarithmic, square root, and reciprocal functions. Our NFG partitions a given domain of the function into non-uniform segments using an LUT cascade, and approximates the given function by a quadratic polynomial for each segment. Thus, we can implement fast and compact NFGs for a wide range of functions. Implementation results on an FPGA show that: 1) our NFGs require only 4% of the memory needed by NFGs based on the linear approximation with non-uniform segmentation; and 2) our NFGs require only 22% of the memory needed by NFGs based on the 5th-order approximation with uniform segmentation. Our automatic synthesis system generates such compact NFGs quickly.

I. INTRODUCTION

Numerical function generators (NFGs) are often used in computer graphics, digital signal processing, communication systems, robotics, astrophysics, fluid physics, etc. The functions realized include trigonometric, logarithmic, square root, and reciprocal functions. High-performance CPUs usually have numerical coprocessors. However, embedded CPUs and CPUs on FPGAs do not have such coprocessors. Thus, FPGA implementation of numerical functions $f(x)$ is needed. Implementation by a single lookup table for $f(x)$ is simple and fast. For low-precision computations of $f(x)$ (e.g. x and $f(x)$ have 8 bits), this implementation is straightforward. For high-precision computations, however, the single lookup table implementation is impractical due to the huge table size. For such applications, the CORDIC (COordinate Rotation DIgital Computer) algorithm [1, 21] has been often used. Although CORDIC is implemented with compact hardware, it is iterative and therefore slow. For numerically intensive applications, faster evaluation of numerical function is required.

For fast evaluation of numerical functions, polynomial approximations have been used [9, 10, 19, 20]. These methods approximate the given numerical functions by piecewise polynomials, and realize the polynomials with hardware. Linear or quadratic approximations offer fast and relatively high-precision evaluation of numerical functions. However, the methods proposed so far are ad-hoc and not systematic. This paper proposes an architecture and a systematic synthesis method for NFGs based on quadratic approximation. By using the LUT cascade [8], many numerical functions are efficiently

approximated by piecewise quadratic functions. Our synthesis method can be automated, so that fast and compact NFGs can be produced by non-experts. Fig. 1 shows the synthesis flow for the NFG. It converts the Design Specification described by Scilab [18], a MATLAB-like software, into HDL code. The Design Specification consists of a function $f(x)$, a domain for x, and an acceptable error. This system first partitions the domain into segments, and then approximates $f(x)$ by a quadratic function for each segment. Next, it analyzes the errors, and derives the necessary precision for computing units in the NFG. Then, it generates HDL code to be mapped into an FPGA using an FPGA vendor tool. Due to the page limitation, the error analysis for our NFGs is omitted here, but it is available in [14]. This paper extends [17] to quadratic approximations.

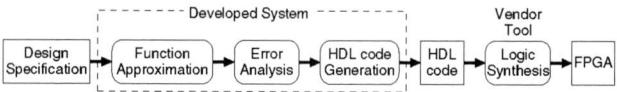

Fig. 1. Synthesis flow for NFGs.

II. PRELIMINARIES

Definition 2.1 *The* **binary fixed-point representation** *of a value r has the form*

$$d_{n\_int-1} d_{n\_int-2} \ldots d_1 d_0 . d_{-1} d_{-2} \ldots d_{-n\_frac}, \quad (1)$$

where $d_i \in \{0,1\}$, n_int is the number of bits for the integer part, and n_frac is the number of bits for the fractional part of r. The representation in (1) is two's complement, and so

$$r = -2^{n\_int-1} d_{n\_int-1} + \sum_{i=-n\_frac}^{n\_int-2} 2^i d_i.$$

Definition 2.2 *Error is the absolute difference between the original value and the approximated value.* **Approximation error** *is the error caused by a function approximation, and* **rounding error** *is the error caused by a binary fixed-point representation.* **Acceptable error** *is the maximum error that an NFG may assume.* **Acceptable approximation error (AAE)** *is the maximum approximation error that a function approximation may assume.*

0-7803-9451-8/06/$20.00 ©2006 IEEE.

Definition 2.3 *Precision is the total number of bits for a binary fixed-point representation. Specially, **n-bit precision** specifies that n bits are used to represent the number; that is, $n = n\_int + n\_frac$. An **n-bit precision NFG** has an n-bit input.*

Definition 2.4 *Accuracy is the number of bits in the fractional part of a binary fixed-point representation. Specially, **m-bit accuracy** specifies that m bits are used to represent the fractional part of the number; that is, $m = n\_frac$. An **m-bit accuracy NFG** is an NFG with m-bit fractional part of the input, m-bit fractional part of the output, and a 2^{-m} acceptable error.*

Input:	Numerical function $f(x)$, Domain $[a, b]$ for x, Acceptable approximation error ε.
Output:	Segments $[s_0, e_0], [s_1, e_1], \ldots, [s_{t-1}, e_{t-1}]$.

Process:
1.   Let $s_0 = a$ and $i = 0$.
2.   Find a value $p (\geq s_i)$ where $\varepsilon_2(s_i, p) = \varepsilon$.
3.   If $p > b$, then let $p = b$.
4.   Let $e_i = p$ and $i = i + 1$.
5.   If $p = b$, then let $t = i$, and stop the process.
6.   Else, let $s_i = p$, and go to step 2.

Fig. 2. Non-uniform segmentation algorithm for the domain.

III. QUADRATIC APPROXIMATION ALGORITHM

To approximate the numerical function $f(x)$ using quadratic functions, first, we partition the domain for x into segments. For each segment, we approximate $f(x)$ using a quadratic function $g(x) = c_2 x^2 + c_1 x + c_0$. In this case, the approximation error depends on the segmentation method and the values of coefficients c_2, c_1, and c_0 in the approximation polynomial.

For piecewise polynomial approximations, in many cases, the domain is partitioned into uniform segments [2, 6, 19]. Such methods are simple and fast, but for some kinds of numerical functions, too many segments are required, resulting in large memory.

For a given error, non-uniform segmentation of the domain uses fewer segments than the uniform segmentation [9, 17]. However, a non-uniform segmentation often requires a complicated segment index encoder (see Section IV), and results in larger and slower NFGs. To overcome this problem, a special non-uniform segmentation has been proposed [9]. This method produces a simple segment index encoder by restricting the segmentation points, and results in fewer segments as well as faster and more compact NFGs than produced by uniform segmentation. However, it is ad-hoc and non-optimum for the given function. Our NFG can implement any non-uniform segmentation with a fast and compact segment index encoder by using an LUT cascade [17] with a synthesis method that can be automated.

Selection of the approximation polynomial influences the number of non-uniform segments as well as the approximation error. In this paper, we use the 2nd-order Chebyshev approximation to approximate $f(x)$ with fewer non-uniform segments, and compute the approximated value. Since coefficients of the Chebyshev approximation polynomial are easily computed, it is suitable for automatic synthesis.

A. Segmentation Algorithm

For a segment $[s, e]$ of $f(x)$, the maximum approximation error $\varepsilon_2(s, e)$ of the 2nd-order Chebyshev approximation [11] is given by

$$\varepsilon_2(s, e) = \frac{(e - s)^3}{192} \max_{s \leq x \leq e} |f^{(3)}(x)|, \tag{2}$$

where $f^{(3)}$ is the 3rd-order derivative of f. From (2), $\varepsilon_2(s, e)$ is a monotone increasing function of segment width $e - s$. Using this property, we partition a domain into as wide segments as possible such that the approximation error is less

than the specified error. Fig. 2 shows the non-uniform segmentation algorithm. The inputs for this algorithm are a numerical function $f(x)$, a domain $[a, b]$ for x, and an acceptable approximation error ε. Then, this algorithm approximates $f(x)$ with the acceptable approximation error ε, and produces t segments $[s_0, e_0], [s_1, e_1], \ldots, [s_{t-1}, e_{t-1}]$. For step 2 in Fig. 2, the accurate computation of the value p where $\varepsilon_2(s_i, p) = \varepsilon$ is difficult. Thus, we obtain the maximum value p' satisfying $\varepsilon_2(s_i, p') \leq \varepsilon$. Such p' can be found by scanning values of n-bit input x. However, it requires $O(2^n)$ search, and is time-consuming. Therefore, we compute the maximum value p' by setting 0 or 1 from MSB to LSB of x such that $\varepsilon_2(s_i, p') \leq \varepsilon$. This requires $O(n)$ search. In the computation of $\varepsilon_2(s_i, p')$, the value of $\max_{s_i \leq x \leq p'} |f^{(3)}(x)|$ is computed by the nonlinear programming algorithm, which is one of the most efficient [7].

B. Computation of Approximate value

For each $[s_i, e_i]$, $f(x)$ is approximated by the corresponding quadratic function $g_i(x)$. That is, the approximated value y of $f(x)$ is computed as follows:

$$y = g_i(x) = c_{2i} x^2 + c_{1i} x + c_{0i}, \tag{3}$$

where the coefficients c_{2i}, c_{1i}, and c_{0i} are derived from the 2nd-order Chebyshev approximation polynomial [11]. Substituting $x - q_i + q_i$ for x in (3) yields the transformation

$$\begin{aligned}
g_i(x) &= c_{2i}(x - q_i)^2 + (c_{1i} + 2c_{2i}q_i)(x - q_i) \\
&\quad + c_{0i} + c_{1i}q_i + c_{2i}q_i^2.
\end{aligned} \tag{4}$$

In (4), let $c'_{1i} = c_{1i} + 2c_{2i}q_i$ and $c'_{0i} = c_{0i} + c_{1i}q_i + c_{2i}q_i^2$. Then, we have

$$g_i(x) = c_{2i}(x - q_i)^2 + c'_{1i}(x - q_i) + c'_{0i}. \tag{5}$$

This transformation reduces the multiplier size.

IV. ARCHITECTURE FOR NFGs

Fig. 3 shows the architecture that realizes (5). It uses 7 units: the segment index encoder that computes the index i for segment $[s_i, e_i]$ including the input value x; the coefficients table for $-q_i, c_{2i}, c'_{1i}$, and c'_{0i}; the adder for $x + (-q_i)$; the squaring unit; two multipliers; and the final adder.

4B-2

2006 Asia and South Pacific Design Automation Conference

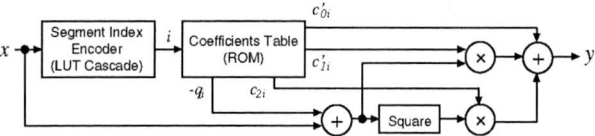

Fig. 3. Architecture for NFGs.

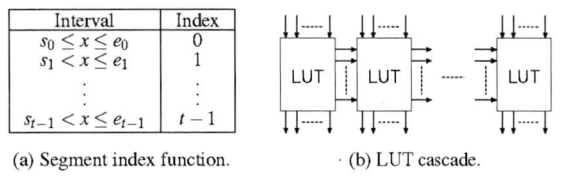

(a) Segment index function.   (b) LUT cascade.

Fig. 4. Segment index encoder.

A *segment index encoder* converts x into a segment index i. It realizes the segment index function $seg\_func(x)$: $B^n \rightarrow \{0, 1, \ldots, t-1\}$ shown in Fig. 4 (a), where x has n bits, $B = \{0, 1\}$, and t denotes the number of segments. In [9], to simplify the segment index encoder, the values of s_i and e_i are restricted to what can be produced by a simple combinational logic circuit. Such a segmentation method results in many segments since it does not adapt to the given function. Our synthesis system uses the *LUT cascade* [8, 15, 16] shown in Fig. 4 (b) to realize arbitrary $seg\_func(x)$. It can be designed by functional decomposition using BDDs (Binary Decision Diagrams) representing $seg\_func(x)$. Our synthesis system uses a nonrestrictive segmentation. It is suitable for automatic synthesis. In LUT cascades, the interconnecting lines between adjacent LUTs are called *rails*. The size of an LUT cascade depends on the number of rails. The next theorem shows that the segment index functions are realized by compact LUT cascades.

Theorem 4.1 *[16] Let $seg\_func(x)$ be a segment index function with t segments. Then, there exists an LUT cascade for $seg\_func(x)$ with at most $\lceil \log_2 t \rceil$ rails.*

Our synthesis system uses heterogeneous MDDs (Multi-valued Decision Diagrams) [13] to find compact LUT cascades. Since the LUT cascade is suitable for the pipeline processing, it offers a fast and compact circuit. In Section VI, we will show that our architecture produces fast and compact NFGs for various numerical functions.

V. IMPLEMENTATION WITH FPGA

Modern FPGAs consist of logic elements (LEs) or configurable logic blocks (CLBs), synchronous memory blocks, multipliers (DSP units), etc. Our synthesis system efficiently generates NFGs using these components. Each unit for the NFG shown in Fig. 3 is implemented by the following components in an FPGA: 1) Segment index encoder (LUT cascade) and coefficients table: by synchronous memory blocks; 2) Squaring unit: by logic elements; 3) Multiplier: by DSP units; and

4) Adder: by logic elements. Our synthesis system derives the appropriate bit-width for each component by automatic error analysis.

A. Size Reduction of Multiplier

Although modern FPGAs have dedicated multipliers, large multipliers are slow. In our architecture, the multiplier often has the longest delay time among all the units. Thus, to implement a fast NFG, reducing multiplier size is important. Since the size of multipliers depends on the number of bits for c_{2i}, c'_{1i}, and $x - q_i$, it is important to reduce the number of bits to represent these values.

First, we consider the case where the absolute values of c_{2i} and c'_{1i} are large. Our synthesis method uses a *scaling method* [9]. We represent c_{2i} and c'_{1i} as $c_{2i} = c_{2i} \times 2^{-l_{2i}} \times 2^{l_{2i}}$ and $c'_{1i} = c'_{1i} \times 2^{-l_{1i}} \times 2^{l_{1i}}$, respectively. That is, instead of the original values of c_{2i} and c'_{1i}, we store the values of $c_{2i} \times 2^{-l_{2i}}$, l_{2i}, $c'_{1i} \times 2^{-l_{1i}}$, and l_{1i} in the coefficients table. In this case, the products $c_{2i}(x - q_i)^2$ and $c'_{1i}(x - q_i)$ are computed using multipliers and shifters. The use of l_{2i} and l_{1i} reduces the number of bits to represent the values of $c_{2i} \times 2^{-l_{2i}}$ and $c'_{1i} \times 2^{-l_{1i}}$, but increases the rounding errors. Our synthesis method finds optimum values of l_{2i} and l_{1i} for each segment such that an acceptable error is achieved. When l_{2i} and l_{1i} are 0 for all the segments, no shifter is implemented, that is, $c_{2i}(x - q_i)^2$ and $c'_{1i}(x - q_i)$ are directly implemented with multipliers.

Next, we consider the value of $x - q_i$. The number of bits for $x - q_i$ influences the sizes of the squaring unit and multipliers. Thus, reducing the value of $x - q_i$ reduces the sizes of the squaring unit and multipliers, and also the error. From (5), we can choose any value for q_i. To reduce the value of $x - q_i$, for a segment $[s_i, e_i]$, we set $q_i = (s_i + e_i)/2$. Then, we have $|x - q_i| \leq (e_i - s_i)/2$. Thus, reducing the segment width $e_i - s_i$ reduces the value for $x - q_i$. However, this also increases the number of segments, and results in increased memory size. The rest of this section shows a reduction method of segment width without increasing the memory size.

The coefficients table in Fig. 3 has 2^k words, where $k = \lceil \log_2 t \rceil$ and t is the number of segments. Therefore, we can increase the number of segments up to $t = 2^k$ without increasing the memory size. From Theorem 4.1, the size of LUT cascade also depends on the value of k. However, increasing the number of segments to $t = 2^k$ seldom increases the size of the LUT cascade. We reduce the size of segments by dividing the largest segment into two equal sized segments up to $t = 2^k$. This method reduces both the number of bits for $x - q_i$ and the error without increasing the memory size.

B. Pipeline Processing

To implement a high-throughput NFG in an FPGA, our synthesis system inserts pipeline registers between all units in the architecture. Since all units operate in parallel, and each unit has a short delay time, our NFGs achieves high throughput. Table I shows the units and the number of pipeline stages for them. Our NFGs have $n\_cas + (5 \text{ or } 6)$ pipeline stages, where n_cas is the number of LUTs for the LUT cascade.

380

TABLE II
NUMBER OF SEGMENTS FOR VARIOUS APPROXIMATION METHODS.

Function $f(x)$	Domain	AAE $= 2^{-17}$				AAE $= 2^{-25}$			
		Linear Non	2nd-Chebyshev Uniform	Non	Time [msec]	Linear Non	2nd-Chebyshev Uniform	Non	Time [msec]
2^x	[0, 1]	128	9	7	0.1	2048	65	44	70
$1/x$	[1, 2)	124	16	11	0.1	1982	128	64	60
$\sqrt{x}$	[1/32, 2)	193	252	24	10	3082	2016	138	150
$1/\sqrt{x}$	[1, 2)	46	16	8	0.1	1024	128	46	50
$\log_2(x)$	[1, 2)	128	16	10	10	2048	128	56	70
$\ln(x)$	[1, 2)	89	16	9	10	1437	128	50	50
$\sin(\pi x)$	[0, 1/2)	127	17	12	10	2027	129	74	90
$\cos(\pi x)$	[0, 1/2)	127	17	12	10	2027	129	74	90
$\tan(\pi x)$	[0, 1/4)	112	33	12	10	1787	129	73	110
$\sqrt{-\ln(x)}$	[1/32, 1)	354	31744	52	70	5933	8126464	331	720
$\tan^2(\pi x) + 1$	[0, 1/4)	256	33	17	20	4096	257	101	170
Entropy	[1/256, 255/256]	520	509	40	30	8320	4065	234	300
Sigmoid	[0, 1]	127	33	13	20	2020	129	76	160
Gaussian	[0, 1/2]	32	5	4	0.1	512	33	18	30
Average		· 170	2337	17	20	2739	580995	99	100

AAE: Acceptable Approximation Error.
Linear: Linear approximation.
Uniform: Uniform segmentation.
Experiment environment:
OS: Redhat (Linux 7.3)

Time: CPU time for our non-uniform segmentation algorithm.
2nd-Chebyshev: 2nd-order Chebyshev approximation.
Non: Non-uniform segmentation.
CPU: Pentium4 Xeon 2.8GHz   Memory: 4GB
C compiler: gcc -O2

TABLE I
NUMBER OF PIPELINE STAGES FOR NFGS.

	Name of units	Pipeline stages
1.	Segment index encoder	n_cas
2.	Coefficients table	1
3.	Adder for $x + (-q_i)$	1
4.	Squaring unit	1
5.	Multipliers (parallel)	1
6.	Shifter (optional)	0 or 1
7.	Final adder	1
	Total pipeline stages	$n\_cas + (5 \text{ or } 6)$

n_cas: Number of LUTs for LUT cascade.

VI. EXPERIMENTAL RESULTS

A. Number of Segments and Computation Time of Algorithm

Table II compares the number of segments for various approximation methods for the functions in [16]. In this table, *Entropy*, *Sigmoid*, and *Gaussian* are

$$Entropy = -x\log_2 x - (1-x)\log_2(1-x),$$

$$Sigmoid = \frac{1}{1+e^{-4x}}, \quad \text{and} \quad Gaussian = \frac{1}{\sqrt{2\pi}}e^{-\frac{x^2}{2}}$$

In Table II, the columns "Linear Non" show the number of non-uniform segments for linear approximation in [17], and the columns "2nd-Chebyshev Uniform" and "2nd-Chebyshev Non" show the number of uniform segments and non-uniform segments for 2nd-order Chebyshev approximation, respectively. The columns "Time" show the CPU time for our non-uniform segmentation algorithm applied to functions, in milliseconds.

Table II shows that, for many functions, the 2nd-order Chebyshev approximations require many fewer segments than the linear approximation. However, for some functions, such as $\sqrt{-\ln(x)}$, the 2nd-order Chebyshev approximation based on uniform segmentation requires many more segments than the linear and 2nd-order Chebyshev approximations based on non-uniform segmentations. Many existing polynomial approximation methods are based on uniform segmentation. For trigonometric and exponential functions, approximation methods based on uniform segmentation require relatively few segments. However, for some kinds of functions such as $\sqrt{-\ln(x)}$, the uniform 2nd-order approximation method requires excessively many segments. On the other hand, our quadratic approximation based on non-uniform segmentation requires fewer segments for a wide range of functions. Also, Table II shows that the CPU time is strongly correlated to the number of segments. Smaller acceptable approximation error (AAE) requires more segments and longer computation time. However, Table II shows that, for all functions in the table, the CPU times are shorter than 1 second when the acceptable approximation error is 2^{-25}.

These results show that, for various functions, our segmentation algorithm partitions a domain into fewer non-uniform segments quickly, and it is useful for automatic synthesis.

B. Memory Sizes of Various NFGs

This section compares the memory sizes of our NFGs with three existing NFGs [17, 3, 4]. Table III compares NFGs using linear approximation shown in [17]. This linear approximation is based on non-uniform segmentation. In Table III, the columns "R" show the following values:

$$R = \frac{\text{memory size of quadratic approximation}}{\text{memory size of linear approximation}} \times 100.$$

Table III shows that NFGs using quadratic approximation require much smaller memory than ones using linear approximation. Especially, 24-bit precision NFGs using quadratic approximation can be implemented with only 4% of the memory size needed for a linear approximation. From the relation between precision and memory size shown in Table III, we can see that increasing the precision decreases the ratio of memory sizes in NFGs.

TABLE III

COMPARISON WITH LINEAR APPROXIMATION BASED ON NON-UNIFORM SEGMENTATION.

Function $f(x)$	16-bit precision			24-bit precision		
	Memory [bits]		R	Memory [bits]		R
	Linear	Quad.	[%]	Linear	Quad.	[%]
2^x	20992	1112	5	696320	19072	3
$1/x$	21248	2432	11	700416	19136	3
$\sqrt{x}$	43776	5536	13	1425408	86784	6
$1/\sqrt{x}$	10176	1104	11	343040	19008	6
$\log_2(x)$	20864	2464	12	694272	19072	3
$\ln(x)$	20096	2448	12	700416	19136	3
$\sin(\pi x)$	19456	2336	12	661504	38656	6
$\cos(\pi x)$	19584	2336	12	663552	38784	6
$\tan(\pi x)$	19712	2304	12	667648	38272	6
$\sqrt{-\ln(x)}$	74240	11264	15	2662400	173056	7
$\tan^2(\pi x) + 1$	37632	4960	13	1290240	39040	3
Entropy	106496	10688	10	3768320	83968	2
Sigmoid	21120	2432	12	702464	40320	6
Gaussian	4416	444	10	156672	8384	5
Average	31415	3704	11	1080905	45906	4

Memory: Memory size.   Linear: Linear approximation [17].
Quad.: 2nd-order Chebyshev approximation. R: Ratio.

TABLE IV

COMPARISON WITH 5TH-ORDER APPROXIMATION BASED ON UNIFORM SEGMENTATION.

Func. $f(x)$	Domain	Acc.	Memory size [bits]		Ratio [%]
			5th-order (Uniform)	Quad. (Non)	
$\sin(\pi x)$	[0, 1/4]	2^{-23}	70528	18048	26
$\exp(x)$	[0, 1]	2^{-24}	82432	43136	52
$2^x - 1$	[0, 1]	2^{-24}	89600	19968	22

Acc.: Accuracy.
5th-order: 5th-order approximation [3].
Quad.: 2nd-order Chebyshev approximation.

Table IV and Table V compare our NFGs with NFGs using 5th-order Taylor expansion [3] and NFGs using 2nd-order minimax approximation by the Remez algorithm [4], respectively. Both approximations in [3, 4] are based on uniform segmentation. Thus, their NFGs require no segment index encoder. On the other hand, since our approximation is based on non-uniform segmentation, the memory size is obtained by the sum of the coefficients table and the segment index encoder. As shown in [17] and Table II, for trigonometric and exponential functions, the difference of the number of uniform segments and non-uniform segments is not so large under the same approximation polynomial. For such functions, NFGs based on uniform segmentation (needing no segment index encoder) often require smaller memory than non-uniform segmentations. Although our NFGs require the segment index encoder and use approximation polynomials with larger approximation error than approximation polynomials in [3, 4], our NFGs for such functions are implemented with only 22% to 52% of the

TABLE V

COMPARISON WITH QUADRATIC APPROXIMATION BASED ON UNIFORM SEGMENTATION.

Func. $f(x)$	Domain	Acc.	Memory size [bits]		Ratio [%]
			Minimax (Uniform)	Cheb. (Non)	
$\sin(\pi x/4)$	[0, 1)	2^{-24}	16288	19200	118
$2^x - 1$	[0, 1)	2^{-16}	2208	2512	114

Minimax: 2nd-order minimax approximation [4].
Cheb.: 2nd-order Chebyshev approximation.

memory sizes of NFGs in [3], and with memory size comparable to [4]. In [3, 4], memory sizes of NFGs for $\sqrt{x}$ and $\sqrt{-\ln(x)}$ are unavailable. However, from Table II, we can see that the memory size of their NFGs for $\sqrt{x}$ and $\sqrt{-\ln(x)}$ is excessively large. On the other hand, our NFGs can realize a wide range of functions with small memory size.

C. FPGA Implementation Results

Table VI compares the FPGA implementation results of our NFGs with NFGs using linear approximation [17].

Since the architecture of linear NFG is simpler than quadratic NFG, linear NFGs are faster, and require fewer logic elements and DSP units than quadratic NFGs. However, linear approximation requires more segments and larger memory than quadratic approximation, as shown in Table II and Table III. Table VI shows that 24-bit precision linear NFGs cannot realize any function except *Gaussian* with the FPGA (the smallest device in the Stratix family) due to the excessive memory size although many logic elements and DSP units are unused. The most crucial issue in the FPGA implementation is the constraints on these hardware resources. For 24-bit precision, the linear approximation requires a larger FPGA due to the excessive memory size. However, in the larger FPGA, more logic elements and DSP units are left unused and wasted. On the other hand, the quadratic NFGs can be implemented with a smaller FPGA since they require much less memory size than the linear NFGs and reasonable sizes of logic elements and DSP units. In fact, 24-bit precision quadratic NFGs can be implemented with lower cost and more compact FPGAs (Cyclone II).

VII. CONCLUSION AND COMMENTS

We have demonstrated an architecture and a synthesis method for programmable NFGs for trigonometric functions, logarithm functions, square root, reciprocal, etc. Our architecture can efficiently realize any non-uniform segmentation using a compact LUT cascade, and approximate many numerical functions by quadratic polynomials. Therefore, our architecture is suitable for automatic synthesis of fast and compact NFGs. Implementation results on an FPGA show that our synthesis method can approximate a wide range of functions with a small number of non-uniform segments, and generate NFGs with small memory size. For 24-bit precision, our NFGs can be implemented with only 4% of the memory size of NFGs based on the linear approximation with non-uniform segmentation, and with only 22% of the memory size of NFGs based on the 5th-order approximation with uniform segmentation. NFGs based on the linear approximation are faster than the quadratic ones, but for high-precision, they require a large FPGA due to the excessive memory size. On the other hand, our quadratic NFGs can be implemented with more compact and low-cost FPGA by using hardware resources on the FPGA efficiently.

ACKNOWLEDGMENTS

This research is partly supported by the Grant in Aid for Scientific Research of the Japan Society for the Promotion of

TABLE VI

FPGA IMPLEMENTATION OF NFGS FOR LINEAR AND QUADRATIC APPROXIMATIONS.

FPGA device:	Altera Stratix (EP1S10F484C5: 10570 logic elements, 48 DSP units)											
Logic synthesis tool:	Altera QuartusII 5.0											
Synthesis options:	speed optimization, timing requirement: 200MHz											

Function $f(x)$	16-bit precision						24-bit precision					
	Logic elements		DSP units		Freq. [MHz]		Logic elements		DSP units		Freq. [MHz]	
	Linear	Quad.	Linear	Quad.	Linear	Quad.	Linear	Quad.	Linear	Quad.	Linear	Quad.
2^x	167	482	2	4	195	185	604	758	2	10	–	131
$1/x$	204	376	2	4	234	186	636	859	2	10	–	134
$\sqrt{x}$	270	496	2	4	237	179	1211	822	2	16	–	124
$1/\sqrt{x}$	186	475	2	4	237	186	402	753	2	10	–	131
$\log_2(x)$	163	381	2	4	194	186	597	757	2	10	–	131
$\ln(x)$	170	379	2	4	197	185	416	863	2	10	–	131
$\sin(\pi x)$	154	424	2	4	197	192	480	646	8	10	–	134
$\cos(\pi x)$	172	354	2	4	237	179	412	647	8	10	–	131
$\tan(\pi x)$	234	382	2	4	237	178	655	604	2	10	–	131
$\sqrt{-\ln(x)}$	304	623	2	10	215	135	854	942	8	16	–	130
$\tan^2(\pi x)+1$	132	282	2	4	194	215	991	720	2	10	–	135
Entropy	141	403	2	4	235	206	1370	914	2	16	–	128
Sigmoid	167	430	2	4	194	191	627	706	2	10	–	131
Gaussian	181	419	2	4	237	186	303	747	2	10	216	129
Average	189	422	2	4	217	185	683	767	3	11	–	131

Linear: Linear approximation [17].   Quad.: 2nd-order Chebyshev approximation.   Freq.: Operating frequency.
–: NFGs cannot be mapped into the FPGA due to the excessive memory size.
Memory sizes are omitted in this table (see Table III).

Science (JSPS), funds from Ministry of Education, Culture, Sports, Science, and Technology (MEXT) via Kitakyushu innovative cluster project, and NSA Contract RM A-54.

REFERENCES

[1] R. Andraka, "A survey of CORDIC algorithms for FPGA based computers," *Proc. of the 1998 ACM/SIGDA Sixth Inter. Symp. on Field Programmable Gate Array (FPGA'98)*, pp. 191–200, Monterey, CA, Feb. 1998.

[2] J. Cao, B. W. Y. Wei, and J. Cheng, "High-performance architectures for elementary function generation," *Proc. of the 15th IEEE Symp. on Computer Arithmetic (ARITH'01)*, Vail, Co, pp. 136–144, June 2001.

[3] D. Defour, F. de Dinechin, and J.-M. Muller, "A new scheme for table-based evaluation of functions," *36th Asilomar Conference on Signals, Systems, and Computers,*, Pacific Grove, California, pp. 1608–1613, Nov. 2002.

[4] J. Detrey and F. de Dinechin, "Second order function approximation using a single multiplication on FPGAs," *Proc. Inter. Conf. on Field Programmable Logic and Applications (FPL'04)*, pp. 221–230, 2004.

[5] N. Doi, T. Horiyama, M. Nakanishi, and S. Kimura, "Minimization of fractional wordlength on fixed-point conversion for high-level synthesis," *Proc. of Asia and South Pacific Design Automation Conference (ASPDAC'04)*, pp. 80–85, 2004.

[6] H. Hassler and N. Takagi, "Function evaluation by table look-up and addition," *Proc. of the 12th IEEE Symp. on Computer Arithmetic (ARITH'95)*, Bath, England, pp. 10–16, July 1995.

[7] T. Ibaraki and M. Fukushima, *FORTRAN 77 Optimization Programming*, Iwanami, 1991 (in Japanese).

[8] Y. Iguchi, T. Sasao, and M. Matsuura, "Realization of multiple-output functions by reconfigurable cascades," *International Conference on Computer Design: VLSI in Computers and Processors (ICCD'01)*, Austin, TX, pp. 388–393, Sept. 23–26, 2001.

[9] D.-U. Lee, W. Luk, J. Villasenor, and P. Y.K. Cheung, "Non-uniform segmentation for hardware function evaluation," *Proc. Inter. Conf. on Field Programmable Logic and Applications*, pp. 796–807, Lisbon, Portugal, Sept. 2003.

[10] D.-U. Lee, W. Luk, J. Villasenor, and P. Y.K. Cheung, "A hardware Gaussian noise generator for channel code evaluation," *Proc. of the 11th Annual IEEE Symp. on Field-Programmable Custom Computing Machines (FCCM'03)*, Napa, CA, pp. 69–78, April 2003.

[11] J. H. Mathews, *Numerical Methods for Computer Science, Engineering and Methematics*, Prentice-Hall, Inc., Englewood Cliffs, NJ, 1987.

[12] J.-M. Muller, *Elementary Function: Algorithms and Implementation*, Birkhauser Boston, Inc., Secaucus, NJ, 1997.

[13] S. Nagayama and T. Sasao, "Compact representations of logic functions using heterogeneous MDDs," *IEICE Trans. on fundamentals*, Vol. E86-A, No. 12, pp. 3168–3175, Dec. 2003.

[14] S. Nagayama, T. Sasao, and J. T. Butler, "Error analysis for programmable numerical function generators based on quadratic approximation," http://www.lsi-cad.com/Error-QNFG/.

[15] T. Sasao, M. Matsuura, and Y. Iguchi, "A cascade realization of multiple-output function for reconfigurable hardware," *Inter. Workshop on Logic Synthesis (IWLS'01)*, Lake Tahoe, CA, pp. 225–230, June 12–15, 2001.

[16] T. Sasao, J. T. Butler, and M. D. Riedel, "Application of LUT cascades to numerical function generators," *Proc. the 12th workshop on Synthesis And System Integration of Mixed Information technologies (SASIMI'04)*, Kanazawa, Japan, pp. 422–429, Oct. 2004.

[17] T. Sasao, S. Nagayama, and J. T. Butler, "Programmable numerical function generators: architectures and synthesis method," *Proc. Inter. Conf. on Field Programmable Logic and Applications (FPL'05)*, Tampare, Finland, pp. 118–123, Aug. 2005.

[18] Scilab 3.0, INRIA-ENPC, France, http://scilabsoft.inria.fr/

[19] M. J. Schulte and J. E. Stine, "Approximating elementary functions with symmetric bipartite tables," *IEEE Trans. on Comp.*, Vol. 48, No. 8, pp. 842–847, Aug. 1999.

[20] J. E. Stine and M. J. Schulte, "The symmetric table addition method for accurate function approximation," *Jour. of VLSI Signal Processing*, Vol. 21, No. 2, pp. 167–177, June 1999.

[21] J. E. Volder, "The CORDIC trigonometric computing technique," *IRE Trans. Electronic Comput.*, Vol. EC-820, No. 3, pp. 330–334, Sept. 1959.

An Automated Design Flow for 3D Microarchitecture Evaluation*

Jason Cong[1]   Ashok Jagannathan[1]   Yuchun Ma[1,2]   Glenn Reinman[1]   Jie Wei[1]   Yan Zhang[1]

[1]University of California, Los Angeles, California 90095, U. S. A.
[2]Tsinghua University, Beijing, P.R.China

Abstract - Although the emerging three-dimensional integration technology can significantly reduce interconnect delay, chip area, and power dissipation in nanometer technologies, its impact on overall system performance is still poorly understood due to the lack of tools and systematic flows to evaluate 3D microarchitectural designs. The contribution of this paper is the development of MEVA-3D, an automated physical design and architecture performance estimation flow for 3D architectural evaluation which includes 3D floorplanning, routing, interconnect pipelining and automated thermal via insertion, and associated die size, performance, and thermal modeling capabilities. We apply this flow to a simple, out-of-order superscalar microprocessor to evaluate the performance and thermal behavior in 2D and 3D designs, and demonstrate the value of MEVA-3D in providing quantitative evaluation results to guide 3D architecture designs. In particular, we show that it is feasible to manage thermal challenges with a combination of thermal vias and double-sided heat sinks, and report modest system performance gains in 3D designs for these simple test examples.

1. Introduction

Due to the aggressive scaling of semiconductor technology, interconnect delays have become the dominant factor limiting system performance. Three-dimensional integration (3D ICs) has been proposed, such as [1, 4, 5, 13, 25, 28], to address this problem by reducing the wirelength, and therefore the interconnect delay, in future technologies. Moreover, reduced interconnection lengths in 3D ICs will help reduce power dissipation since the number of repeaters required for the wires can be reduced and the wire capacity load is reduced. However, thermal dissipation is one of the biggest concerns in 3D designs. The increased power density due to the decreased die area and the low conductivity inter-layer dielectrics (ILD) between the device layers can lead to thermal problems. For this reason, we need to evaluate 3D integration systematically, with regard to both performance and temperature.

Although a significant reduction of wirelength has been reported using 3D technology [13, 14, 15], its impact on the microarchitecture is still poorly understood due to the lack of tools that enable such an evaluation. While earlier work studied the impact of 3D IC on the memory subsystem [19] of the architecture or a specific implementation of an IA32 microprocessor [5], there is no existing flow that allows us to evaluate 3D implementations of architectures systematically and study them from both a performance and thermal perspective. We present an automated 3D microarchitecture evaluation flow (MEVA-3D) that can be used to compare and contrast 2D and 3D implementations of a given architecture. Specifically, we make the following contributions in this work:

This research is supported by the National Science Foundation under Grant CCR-0096383 and DARPA under Prime Contract DAAH01-03-C-R193 through CFDRC Corporation.

First, we develop a floorplanner to optimize a microarchitecture for area, wirelength, performance, and temperature both for 2D and 3D integration. The performance is optimized by reducing the latency along critical loops in the architecture through interconnect pipelining at a given target frequency. Thermal evaluation is done using a resistive network model considering whitespace and thermal via insertion. Finally, we use our 3D router to do a detailed physical implementation.

Based on our automated evaluation flow, we study the impact of 3D integration on the performance and temperature of a processor model at 70nm technology and target frequencies ranging from 3GHz to 8GHz. For this simple architecture design driver, we can improve performance by around 11% at several frequencies by eliminating most of the wire latencies in 2D. We also show that the on-chip temperature can vary by up to 4.78x depending on what 3D IC fabrication technology is used, and we show that it is feasible to manage temperature challenges through a combination of thermal vias and double-sided heat sinks.

The remainder of this paper is organized as follows: Section 2 gives a brief overview of the different 3D fabrication technologies and the technology assumption we use in this paper; Section 3 introduces our architectural evaluation framework, including thermal evaluation and optimization, performance estimation, and the microarchitectural floorplanner with wire pipelining; Section 4 presents the baseline architecture which is evaluated by our framework and the performance and the temperature results of the baseline architecture. We conclude the paper and discuss the possible future research directions in Section 5.

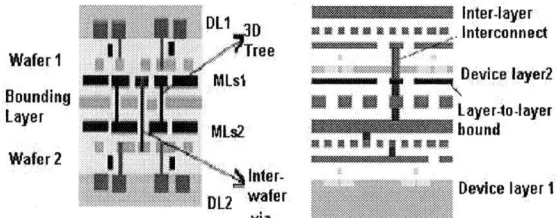

(a) With flipping the top layer   (b) Without flipping the top layer

Fig. 1 A 3D IC example with two device layers

2. Technology Background

3-D IC fabrication technologies include multi-chip module (MCM) packaging [1, 25], wafer bonding [4, 5, 13], solid-phase recrystallization [4], etc. Different fabrication technologies will greatly affect the circuit performance, manufacturing cost, on-chip temperature, etc. In this work we will use wafer bonding 3D IC technology, where each device layer is fabricated separately and then put together into one design. We will evaluate the same microprocessor architecture in two different kinds of wafer bonding strategies described in [5, 13], as shown in Fig. 1. Both technologies contain two device layers: in case (a), the top device layer is flipped upside down, and in case (b), the top device layer is oriented in the same direction as the bottom device layer. In case (a), the interlayer connect vias do not go through the device layers and therefore do not increase the chip size. Also in case (a), both silicon device layers are close to the boundary of the circuit, which will help the heat dissipation. However, case (a) cannot handle more than two device layers.

3. 3D Microarchitecture Evaluation Flow

Evaluating 3D architectures is a complex process that involves optimizing several parameters such as frequency, die area, performance, whitespace for thermal via insertion, and the temperature itself. An overview of our 3D Microarchitecture Evaluation Flow (MEVA-3D) with its components is shown in Fig. 2. The upper half of Fig. 2 shows the estimation part of our framework and the bottom half shows the validation part, which is used to verify the results generated by the estimation process.

The MEVA-3D flow takes as input: (1) a microarchitecture description in terms of the parameters of the blocks, such as the sizes of the structures, number of functional units etc.; (2) a target frequency for evaluating the microarchitecture; (3) a set of architecture-level critical paths whose latency will impact the performance of the microarchitecture, and a notion of performance sensitivity for each critical path; and (4) estimated power density numbers for the blocks in the processor. Notice that (3) and (4) are estimates of performance and power for the microarchitecture components that can drive a subsequent physical planning process.

In MEVA-3D, we propose an automated floorplanner that can be configured to optimize the floorplan for die area, performance, and temperature with consideration of interconnect pipelining. The performance estimation models are used to evaluate the impact of wire pipelining, and the power density estimates are used for the thermal calculation during floorplanning. The floorplanner is flexible enough to consider 2D and 3D placement of blocks, and allows one to configure the number of device layers for 3D integration. The output of the floorplanning engine consists of the locations of the blocks and their shapes, the total latency along the critical architectural paths including both blocks and wires, and the maximum on-chip temperature. Thermal via insertion and global routing can be employed to get the optimized thermal and routing profile. Once the exact block

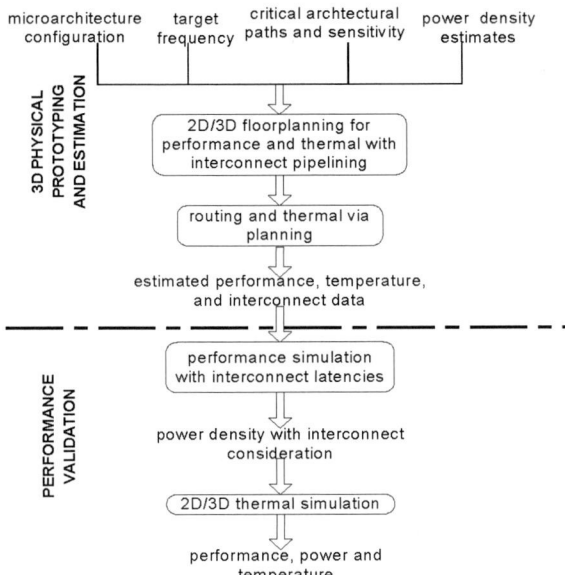

Fig. 2 Overview of MEVA-3D

positions and wire latencies are known, this information is fed to our validation flow. A detailed cycle-accurate simulator that considers wire latency and power, and is coupled with power and thermal models, is used to validate the results from the MEVA-3D flow. In the sections that follow, we explain the components of our flow, mainly the performance optimization and the thermal evaluation methodology.

3.1 Thermal Evaluation and Optimization

High on-chip temperatures will degrade timing, including both the gate and the interconnect delay, increase leakage power, and even cause logic failure. Therefore, temperature should be considered in any realistic 3-D microprocessor design. This section will describe the thermal evaluation tool and the automatic thermal via insertion tool that we use in the MEVA-3D flow.

3.1.1 Resistive thermal network model

Considering both accuracy and runtime, we chose to make use of a compact, thermal resistive model proposed by Wilkerson [26], which explicitly models the 3-D circuit structure and the thermal vias. In [26], the circuit stack is first divided into tiles, each tile being the size of a thermal via pitch, as shown in Fig. 3(a). A tile stack is a vertical array of tiles, as shown in Fig. 3(b). A tile either contains one thermal via at the center, or no thermal via at all. A tile stack is modeled as a resistive network, as shown in Fig. 3(c). Fig. 3(c) shows the cases with two heat sinks. In cases with only one heat sink, one of the voltage sources can be taken away. The isothermal bases of room temperature are modeled as a voltage source. A current source is present at every node in the network to represent the heat sources. The tile stacks are connected by lateral resistances. The values of the resistances in the network are determined by a commercial FEM- based thermal simulation tool [3, 24] and it shows that our model has a small error within 5% [20].

3.1.2 Thermal via insertion

Chip cooling techniques can be divided into two groups. One of these is heat sink optimization, which tries to cool down the heat sinks through packaging level techniques such as fans and micro-channels. However, in 3-D designs, the poor thermal conducting ILD layers impede the internal heat dissipation from the heat sources to the heat sink. Even with the perfect heat sinks, the maximum on-chip temperature can still be very high. The other group of cooling technologies, which optimizes the internal heat dissipation paths, includes temperature-aware physical design tools [10, 11], thermal via insertion [11], and 3-D IC micro-channel techniques [15].

In this work, the temperature-aware floorplanning, thermal-aware routing and thermal via insertion techniques are applied to cool down the 3-D circuits. The temperature-aware floorplanning tool discussed in Section 3.3 tries to put the "hot" blocks close to the heat sinks to avoid hot spots. After floorplanning, we can further reduce the temperature by thermal via insertion. In [11], a simultaneous routing and thermal via planning method is proposed. Since the detailed netlist information is not available at the architectural level, we will concentrate more on the thermal via planning in the evaluation framework. A through-the-silicon via is a via that goes through a device layer. Under the current technology, through-the-silicon vias (pitch ≈ 5μm × 5μm) [16] are usually much larger than the metal wires. Through-the-silicon vias are very effective for heat dissipation and can decrease the maximum temperature over 50% [9]. When the signal vias are not sufficient to bring the chip temperature down to a satisfactory level, additional thermal vias can be inserted into the chip. The only purpose for thermal via insertion is for temperature reduction purposes —— the thermal vias have no wire connection. Experiments show that the temperature reduction will increase with the increase of the thermal via numbers, however, the temperature reduction will also gradually saturate with the increase of thermal via density. Therefore, we want to insert more thermal vias at hot places and spread the thermal vias whenever possible. The details of the thermal via planning algorithm can be found in [11].

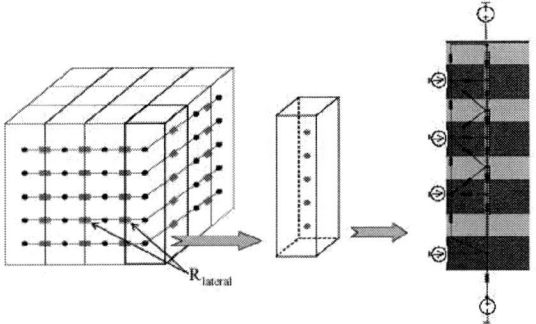

(a) Tiles Stack Array   (b) Single Tile Stack   (c) Tile Stack Analysis

Fig. 3 Resistive thermal model for a 3D IC

3.2 Performance Estimation

Our approach to calculate the performance of the processor during the floorplanning process in Fig. 2 is similar to the method used in [18]. As the target frequency during floorplanning is fixed, we want to calculate the IPC degradation caused by the extra latency introduced by the interconnects in the layout. The work by Borch et al. [6] showed that the IPC of a microarchitecture depends on a set of critical processor loops, and extra latency along these loops can cause the IPC to degrade.

For each critical path, we develop IPC performance sensitivity models similar to the work by Sprangle and Carmean [23] which provides information about IPC degradation due to extra latency along the path. During floorplanning, we calculate the total latency of each critical path including the blocks and the wires, and determine the total number of cycles at the target frequency required to cover this path latency. Extra latency from the wires is used to compute the new IPC, and hence the performance of the processor for that floorplan.

3.3 Microarchitecture Floorplanning with Wire Pipelining

Traditional floorplanning optimizes the area and wirelength of the packing, but a minimal wirelength is not enough for performance optimization. Instead, the floorplanner should be used to effectively explore a large space of architectural configurations efficiently and accurately, measuring the trade-offs on each circuit path. This is particularly true for superscalar microarchitectures which can feature critical timing paths of varying importance —— minimizing overall wirelength may not be as important as minimizing critical loop wirelength.

The floorplanning problem that we investigate here considers several components in its objective function that are important tradeoffs in 3D architectures. Specifically, we consider the die area (footprint), the performance of the microarchitecture in *BIPS*, the maximum on-chip temperature, and the wirelength so that the power from the interconnects can be reduced. Formally, we define the problem as follows:

Given:
(1) target cycle time T_{cycle}
(2) clocking overhead $T_{overhead}$
(3) list of blocks in the microarchitecture with their area, dimensions and total logic delay
(4) set of critical microarchitectural paths with performance sensitivity models for the paths
(5) average power density estimates for the blocks.

Objective: Generate a floorplan which optimizes for the die area, performance, and maximum on-chip temperature.

The 3D thermal-driven floorplanner [10] used in this work is based on a simulated annealing framework with a 3D extension of the TCG representation [20]. We extended the floorplanner to consider the performance of the design instead of a wirelength optimization objective. Our cost function uses a weighted combination of area, performance, and temperature, and can be represented by

$$\cos t = w1 * \frac{1}{BIPS} + w2 * Area + w3 * Temp + w4 * Wire$$

where *BIPS* corresponds to the performance of the microarchitecture with that floorplan of the blocks, *Area* is the total area of the floorplan, and *Temp* corresponds to the maximum on-chip temperature calculated by the thermal model described in Section 3.1. The performance (*BIPS*) is calculated through the method presented in Section 3.2. The coefficients of *w1, w2, w3* and *w4* are used to control the different weight for each component. In our test evaluation, the performance component is given a high weight and will be optimized when the simulated annealing engine tries to minimize the cost function.

3.4 Performance Validation

Once we have finished the physical planning stage, we will have selected a 3D floorplan optimized for our performance and thermal estimation. Once we have the critical loop latencies and cycle time from this stage, we can input these, along with the architectural configuration, into our cycle-accurate simulation framework. We adapted the SimpleScalar 3.0 tool set [8], a suite of functional and timing simulation tools for the Alpha AXP ISA, for our simulation framework. This simulator allows execution-based simulation (including simulation down mispredicted branch paths) of a microarchitecture, with a wide range of customizable parameters. This framework gives performance statistics in instructions per cycle (IPC) that can be combined with the cycle time from the floorplanning stage to give a result in BIPS. This validation phase can provide feedback to the performance estimators in the floorplanning stage if the performance does not match the expected result.

4. Case Study for a Design Driver

In this section we present detailed evaluation results obtained for our design driver microarchitecture. This architecture, illustrated in Fig. 4, is an out-of-order microprocessor with detailed parameters shown in Table 1. We modified SimpleScalar to model this architecture and to parameterize the different critical path latencies found through the floorplanning process. To perform our evaluation, results were collected for 24 SPEC2000 [2] benchmarks.

As we consider automated floorplanning with performance and thermal optimization, it is necessary to estimate the area, wirelength, delay, and power associated with each block. We model

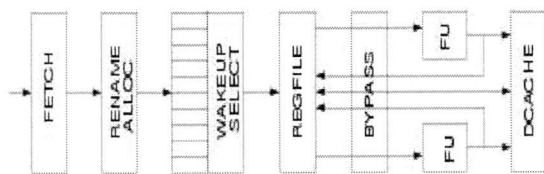

Fig. 4 Superscalar pipeline explored in this paper

the area and delay of the blocks using assumptions similar to Wilton and Jouppi [27] and Palacharla et al. [22] in which the authors analyzed the delay of many of the microprocessor logic blocks. The power numbers are modeled based on the models proposed by Brooks [7]. The delay of each block was derived by adding the delays along the critical path of the block. The aspect ratio for the blocks is also derived under these array layout assumptions. The power is calculated by adding the capacitances from the transistors as well as the wires inside these logic blocks.

Table 1. Baseline processor parameters

Instruction Cache	32KB, 32B/block, 2-way
Decode Width	8
ROB Size	128 entries
Issue Queue	32 entries
Issue Width	8
Register File	70 INT and 70 FP
Functional Units	Units 4 IntALU, 1 FPALU, 2 IntMult, 1 FPMult
Load/Store Queue	32 entries
L1Data Cache	16KB, 32B/block, 4-way, 2RW ports
Unified L2 cache	1MB, 64B/block, 8-way

4.1 Performance Impact of 3D Integration

In order to consider the impact of pipelining based on interconnect delays, we use the critical paths in Table 2 for this study. The area and delay of the blocks were derived based on [22, 27] for a futuristic 70nm process technology. Based on [17], we assume that the clock cycle overhead is 46ps, which corresponds to roughly 1.8FO4 (fan-out-of-four) for 70nm technology. Thus, for a 4GHz target cycle time, we set the useful time for computation as 204ps and use this to calculate the number of pipeline stages required to cover a given path delay. As we study the impact of semi-global interconnects in the microarchitecture, we assume that all of the signals are routed in the routing layers meant for semi-global wires. The delay of interconnects is derived using the IPEM models [12] which consider several optimizations such as wire sizing, buffer insertion and buffer sizing, etc.

To facilitate the insertion of repeaters, flip-flops, vias, etc., which are inevitable to achieve the required interconnect performance, we assume that 10% of each block's area is reserved around the block in the floorplan. Moreover, as the L2 cache occupies more than 50% of our die size, we allow the four L2 cache banks to be placed separately so that the floorplanner has more flexibility in packing the blocks. To study the impact of 3D integration on the performance of the microarchitecture, we generated the best performance results for the 2D and 3D cases by

Table 2. Different critical paths considered during layout optimization

Wakeup Latency	Latency to wake up the dependent instruction
ALU Bypass	Latency of the bypass wires between the ALUs
DL1 Latency	Load latency though the L1 data cache
L2 Latency	Latency for access to L2 cache
MPLAT	Latency through the branch resolution path

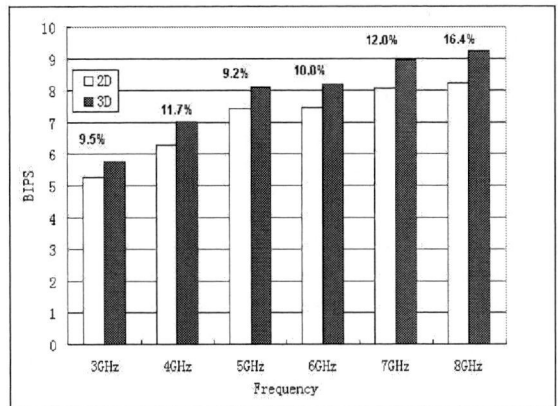

Fig. 5 Performance numbers for the microarchitecture with 2D and 3D layout at different target frequencies

running the floorplanning engine 20 times and picking the best solution for each case. Fig. 5 shows that the impact of wire latency reduction due to 3D integration frequencies. Overall, we can see that a performance improvement of about 11% can be obtained by using a 3D design instead of a 2D design. Table 3 shows the detailed information for these test results. By using the 3D architecture, the die size can be shrunk to 53% of the die size in 2D design, and the wirelength can be reduced by about 46% at the same time. With the elimination of the extra latencies generated by the long wires along the critical paths, the performance measured in BIPS improves by around 11%. Table 4 shows the detailed information for the number of extra cycles caused by the wire delay in both 2D and 3D design. In 3D design, the long wires along the critical paths can be reduced, therefore, some of the extra cycles from wire can be eliminated. In Table 4, all the extra interconnect latencies are removed in 3D design for 3GHz frequency. Therefore we can see about 9.5% improvement in terms of performance. In 4GHz case, the critical path of ALU bypass has one extra cycle because the blocks along this path are spread out because of the limitation of 2D layout. But in 3D design, since the ALU units can be put on different layers to reduce the interconnection between them, the extra cycle for this loop is removed. Similarly, the extra cycles along the other loops are also reduced and the performance improvement from 2D design to 3D design is about 11.7%.

Table 3. The results for different frequencies

Frequency		Area mm²	Wire mm	BIPS	T(C) 1 sink	Via inserted T(C)	Via Area %	T(C) 2 sinks
3G	2D	33.7	149	5.266	31.59	--	--	--
	3D	19.4	71.8	5.769	132.3	52.4	4.93	32.6
4G	2D	33.6	151	6.301	33.08	--	--	--
	3D	18.3	60.9	7.038	150.6	56.9	4.94	33.0
5G	2D	34.0	153	7.425	34.84	--	--	--
	3D	17.8	69.4	8.110	166.2	61	6.16	34.9
6G	2D	33.5	127	7.455	35.76	--	--	--
	3D	17.4	62.6	8.202	185.7	65.2	7.21	36.2
7G	2D	33.7	162	8.06	38.5	--	--	--
	3D	16.6	72.9	8.96	196.7	75.2	8.91	38.6
8G	2D	33.5	144	8.23	41.7	--	--	--
	3D	18.0	89.8	9.22	203.6	82.2	9.3	40.6
Average 3D/2D ratio		53%	48%	115%	478%	181%	--	100%

387

To provide further insight, we show the performance simulation results for several benchmarks with the packing generated by our framework at a 6GHz target (at 6GHz frequency, the performance is the best in our test). In Fig. 6, we show ten test cases from the SPEC2000 [2] benchmarks. We can observe that the automatic floorplanner used in our work produces good quality floorplans, where blocks whose communication latency is critical for performance are placed close to one another in 2D as well as in 3D. The 3D layout eliminated the extra cycle from wires in the wakeup loop, and also removed some of the extra cycles on the branch misprediction resolution loop, the DL1 access loops, and the L2 access loops. Thus, the total improvement in performance comes to around 10.0% as shown in Fig. 5. Even though the 3D architecture could not remove all the extra cycles from the wire latency, it helped to reduce the latency along the most critical loops to gain good performance improvement. For the test cases shown in Fig. 6, each benchmark sees performance improvements in the range of 7% to 24%. On average, the performance improvement in Fig. 5 is about 10% —— a relatively large improvement given the simplicity of the microarchitecture, the limited die size, the small number of critical loops exposed in the floorplan optimization and simple re-use of 2D architecture components. More complex designs should see even more improvement from 3D technologies, especially when we consider the re-design of 3D components.

Table 4. The number extra cycles for critical paths group

	3G		4G		5G		6G		7G		8G	
	2D	3D	2D	3D	2D	3D	2D	3D	2D	3D	2D	3D
Wakeup	1	0	1	1	1	1	1	0	2	1	2	1
ALU	0	0	1	0	0	0	0	0	1	0	1	0
DL1	0	0	0	0	1	0	2	1	1	1	2	1
L2	2	0	2	1	3	1	5	1	5	2	5	1
MPLAT	1	0	1	0	2	1	2	1	2	1	5	2

4.2 Thermal Impact of 3D Integration

We also present thermal results corresponding to the best performance solutions presented in the previous section for both 2D and 3D. In order to calculate the power density of the blocks for thermal calculations, we use Wattch [7] models integrated with the simulator to calculate the dynamic power associated with each block at the specified target frequency. We add 30% of dynamic logic power as the additional power from the interconnects and associated repeaters. We chose this fraction based on observed interconnect power trends in recent studies [21]. We then assume that the leakage power is a fraction of the dynamic power, and add this fraction to obtain the total power dissipated by each block including the logic, wires, and leakage. For our experiments, we set the leakage to be 80% of the total dynamic power, as this corresponds to leakage contributing to 45% of the overall chip power. For reference, the leakage power of a Pentium 4 processor in 130nm technology is around 40% [21]. [*] Based on the above method, we calculated the maximum power density at any point on the chip to be less than $2W/cm^2$ for the 2D chip. While 3D integration usually helps to reduce interconnect power as well as the associated leakage power from the repeaters, we assume the same total power values for both the 2D and the 3D case in our thermal evaluation for easier comparison. Fig. 7 shows the maximum on-chip temperature obtained using the thermal

simulator discussed in Section 3.1. The results are shown for the following four cases: (1) a 2D layout with a heat sink at the bottom (2D-HS1); (2) 3D integration in the case (b) technology in Fig. 1, with one heat sink at the bottom and without inserting any thermal vias (3D-HS1-no_via); (3) 3D integration in the case (b) technology in Fig. 1, with one heat sink at the bottom and thermal via insertion to reduce temperature (3D-HS1-via); and (4) 3D integration in the case (a) technology in Fig. 1, with two heat sinks, one at the top and one at the bottom (3D-HS2). As we expected, adding another device layer will generally cause a drastic temperature increase, and case (2) shows a temperature increase of over 4.78× on average. After thermal via insertion (case (3)), we can reduce the maximum on-chip temperature by an average of about 62%. And the area occupied by the thermal vias (Via Area %) is about 5% to 9% of the total chip area. However, we can see that the temperature is still quite high compared to case (4), under the wafer bonding technology with two heat sinks and the flipped top device layer. In case (4), both device layers are located very close to the heat sinks, which provides very good heat dissipation paths to the "hot" devices.

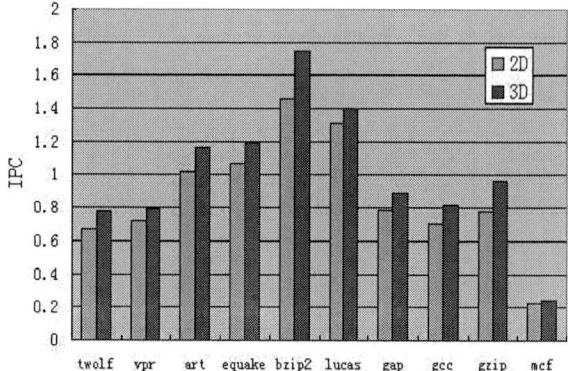

Fig. 6: Performance improvement for a 6GHz processor from 2D to 3D

5. Conclusions and Future Works

In this paper we presented an automatic evaluation flow named MEVA-3D for 3D architectures from 3D microarchitecture floorplanning, performance evaluation to thermal evaluation and thermal via planning. By using this flow, we can systematically evaluate the 3D architecture from both the performance side and the thermal side. Through one specific design driver, we see that 3D integration can help improve the performance by 10% with comparable maximum on-chip temperature. Since our automatic evaluation flow has the capability to handle different kinds of 3D architectures, there is great potential for an even further benefit from novel designs that can leverage 3D IC to improve individual blocks. To simulate the thermal effect dynamically, we are currently working to refine the power model. And the architecture components can be re-designed in 3D technologies, which can reduce the power consumptions and optimize the performance of the design.

---

[*] In this evaluation, since the design driver as both 2D and 3D designs have a similar temperature profile after thermal via insertion, we assume the leakage to be a fixed percentage of dynamic power.

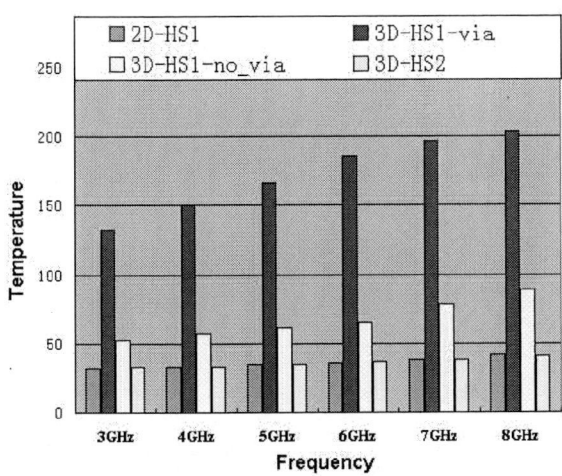

Fig. 7: Maximum on-chip temperature for different 2D and 3D integration for the best performance solution for the micro-architecture. HS denotes a heat sink, and the 3D integration allows the insertion of thermal vias to reduce temperature.

6. References

[1] *http://www.irvine-sensors.com/r_and_d.html#high*

[2] The Standard Performance Evaluation Corporation, 2000. http://www.spec.org.

[3] *CFD-ACE+ Module Manual*, 2002.

[4] K. Banerjee, S. Souri, P. Kapur, and K. Saraswat. 3-D ICs: A Novel Chip Design for Improving Deep-Submicrometer Interconnect Performance and Systems-on-Chip Integration, *Proc. of the IEEE*, 89(5):602–633, May 2001.

[5] B. Black, D. W. Nelson, C. Webb, and N. Samra. 3D Processing Technology and its Impact on IA32 Microprocessors. *Proc. Of ICCD*, pp.316-318,2004.

[6] Eric Borch, Eric Tune, Srilatha Manne, and Joel Emer. Loose Loops Sink Chips, *Proc. of 8th Internatonal Symposium on High-Performance Computer Architecture*, January 2002.

[7] D. Brooks, V. Tiwari, and M. Martonosi. Wattch: A Framework for Architectural-level Power Analysis and Optimizations. *Proc. of International Symposium on Computer Architecture*, June 2000.

[8] D. C. Burger and T. M. Austin. The SimpleScalar Tool Set, Version 2.0, Technical Report CS-TR-97-1342, University of Wisconsin, Madison, June 1997.

[9] T.-Y. Chiang, S. J. Souri, C. O. Chui, and K. C. Saraswat. Thermal Analysis of Heterogeneous 3-D ICs with Various Integration Scenarios, *IEEE International Electron Devices Meeting (IEDM) Technical Digest*, pp. 681–684, Dec. 2001.

[10] J. Cong, J.Wei, and Y. Zhang. A Thermal-Driven Floorplanning Algorithm for 3D ICs, *Proc. IEEE International Conference on Computer-Aided Design*, 2004.

[11] J. Cong and Y. Zhang. Thermal-Driven Multilevel Routing for 3-D ICs, *Proc. of the Asia South Pacific Design Automation Conference*, pp. 121–126, January 2005.

[12] Jason Cong and David Zhigang Pan. Interconnect Estimation And Planning For Deep Submicron Designs. *Proceedings of the 36th ACM/IEEE Conference on Design Automation*, pp. 507–510, 1999.

[13] S. Das, A. Chandrakasan, and R. Reif. Design Tools for 3-D Integrated Circuits, *Proc. Asia and South Pacific Design Automation Conf.*, pp. 53–56, January 2003.

[14] S. Das, A. Chandrakasan, and R. Reif. Timing, Energy, and Thermal Performance of Three-Dimensional Integrated Circuits, *Proc. of GLSVLSI*, 2004.

[15] Shamik Das. Design Automation and Analysis of Three-Dimentional Integrated Circuits, PhD thesis, Massachusetts Institute of Technology, May 2004.

[16] S. B. Horn. Vertically Integrated Sensor Arrays VISA (Invited). *Defense and Security Symposium*, 2004.

[17] M. S. Hrishikesh, K. Farkas, N. P. Jouppi, D. C. Burger, S. W. Keckler, and P. Sivakumar. The Optimal Logic Depth Per Pipeline Stage is 6 to 8 FO4 Inverter Delays. *Proc. of 29th International Symposium on Computer Architecture*, May 2002.

[18] Ashok Jagannathan, Hannah Honghua Yang, Kris Konigsfeld, Dan Milliron, Mosur Mohan, Michail Romesis, Glenn Reinman, and Jason Cong. Microarchitecture Evaluation with Floorplanning and Interconnect Pipelining, *Proc. of the Asia Pacific Design Automation Conference*, 2005.

[19] M. B. Kleiner, S. A. Kuhn, P. Ramm, and W. Weber. Performance and Improvement of the Memory Hierarchy of Risc-Systems by Application of 3-D Technology, *IEEE Trans. Comp. Packag, Manufact. Technol. B*, 19, 1996.

[20] J.-M. Lin and Y.-W. Chang. TCG: A Transitive Closure Graph Based Representation for Non-Slicing Floorplans, *Proc. of Design Automation Conference*, pp. 764–769, June. 2001.

[21] Nir Magen, Avinoam Kolodny, Uri Weiser, and Nachum Shamir. Interconnect-power Dissipation in a Microprocessor, *Proceedings of the 2004 International Workshop on System Level Interconnect Prediction*, pp. 7–13, 2004.

[22] Subbarao Palacharla, Norm Jouppi, and J. E. Smith. Complexity Effective Superscalar Processors, *Proc. International Symposium on Computer Architecture*, pp. 206–218, June 1997.

[23] Eric Sprangle and Doug Carmean. Increasing Processor Performance by Implementing Deeper Pipelines, *ISCA '02: Proceedings of the 29th Annual International Symposium on Computer Architecture*, pp. 25–34, 2002.

[24] Z. Q. Tan, M. Furmanczyk, M. Turowski, and A. Przekwas. CFD-Micromesh: A Fast Geometrical Modeling and Mesh Generation Tool for 3D Microsystem Simulations, *Int. Conf. MSM 2000*, pp. 712–715, March. 2000.

[25] Y. K. Tsui, S. W. R. Lee, J. S. Wu, J. K. Kim, and M. M. F Yuen. Three-Dimensional Packaging for Multi-chip Module with Through-the-Silicon Via Hole, *Electronics Packaging Technology, 2003 5th Conference*, pp. 1–7, 2003.

[26] P. Wilkerson, M. Furmanczyk, and M. Turowski. Compact Thermal Modeling Analysis for 3D Integrated Circuits. *11th International Conference Mixed Design of Integrated Circuits and Systems, Szczecin, Poland*, June. 2004.

[27] S. Wilton and N. Jouppi. CACTI: An Enhanced Cache Access and Cycle Time Model, *IEEE Journal of Solid-State Circuits*, May 1996.

[28] Y. Deng and W. Maly. Interconnect Characteristics of 2.5-D System Integration Scheme, *Proc. Int. Symp. On Physical Design*: pp.171-175, 2001.

Optimal Topology Exploration for Application-Specific 3D Architectures [*]

Ozcan Ozturk, Feng Wang, Mahmut Kandemir, and Yuan Xie
Computer Science and Engineering Department
Pennsylvania State University
e-mail: {ozturk,fenwang,kandemir,yuanxie}@cse.psu.edu

Abstract— As technology scales, increasing interconnect costs make it necessary to consider alternate ways of building integrated circuits. One promising option along this direction is 3D architectures where a stack of multiple device layers, with direct vertical tunneling through them, are put together on the same chip. In this paper, we explore how processor cores and storage blocks can be placed in a 3D architecture to minimize data access costs under temperature constraints. This process is referred to as the topology exploration. Using integer linear programming, we compare the best 2D placement with the best 3D placement, and show through experiments with both single-core and multi-core systems that the 3D placement generates much better results (in terms of data access costs) under the same temperature bounds. We also discuss the tradeoffs between temperature constraint and data access costs.

I. INTRODUCTION

As technology scales, the International Technology Roadmap for Semiconductors projects that on-chip communications will require new design approaches to achieve system level performance targets [15]. Three-dimensional integrated circuits (3D ICs) [6, 10, 11, 12] are attractive options for overcoming the barriers in interconnect scaling, offering an opportunity to continue the CMOS performance trend. In a three-dimensional (3D) chip, multiple device layers are stacked together with direct vertical interconnects tunneling through them. Consequently, one of the most important benefits of a 3D chip over a traditional two-dimensional (2D) design is the reduction on global interconnect. Other benefits of 3D ICs include: (i) higher packing density and smaller footprint due to the addition of a third dimension to the conventional two-dimensional layout; (ii) higher performance due to reduced average interconnect length; (iii) lower interconnect power consumption due to the reduction in total wiring length; and (iv) support for realization of mixed-technology chips.

As heat is generated by the power dissipated on the chip, the on-chip junction temperatures also increase, resulting in higher cooling/packaging costs, acceleration of failure mechanisms, and degradation of performance. For 3D ICs, the thermal issues are even more pronounced than the 2D designs due to higher packing density, especially for the inner layer of the die. It is often considered as a major hindrance for 3D integration [6]. For example, temperature increases force the design to slow down to cool the chip, such that the actual performance benefits from reducing global interconnect could be

offset. Therefore, thermal-aware design is very critical to extract the maximum benefits from the 3D integration.

As technology moves towards 3D designs, one of the challenging problems in the context of multi-core systems is the placement of processor cores and storage blocks across the multiple layers available. This is a critical problem as both power and performance behavior of a design are significantly influenced by the data communication (data access) distances between the processor cores and storage blocks. In particular, if a computation block (e.g., a processor core) frequently accesses data from certain storage blocks, these storage blocks should be placed into positions close (in vertical or horizontal sense) to that processor core. Similarly, in a multi-core design, if two cores frequently share certain data residing in a given storage block, that storage block should be put close to both these cores to minimize the data communication distances, thereby, improving potentially both performance and power consumption.

The important point to note, however, is that each embedded application can require a different placement of processor cores and storage blocks for achieving the minimum data communication distances. Therefore, in this paper, our focus is on *application-specific placement* of processor cores and storage blocks in a 3D design space. For this purpose, we propose an integer linear programming (ILP) based processor core/storage block placement for single-core and multi-core embedded designs. This placement problem is also referred to as topology exploration. While ILP based solutions are known to take large solution times, this issue does not seem to be very pressing in our case because (i) we design a customized placement for a given embedded application, and thus, can afford large solution times as design quality is of utmost importance, and (ii) since a given design typically has only a small number of processor cores and storage blocks, the solution times can be kept under control.

We implemented our ILP based solution within a commercial solver [23], and performed experiments with different types of applications. The first group of applications are a set of six sequential Spec2000 codes and used for single-core designs in this work. The second group of applications, on the other hand, are for multi-core designs and include four array-based codes parallelized through an optimizing compiler. Our experiments that consider a set of cache lines as a storage block reveal two important results: (i) the best 3D designs significantly outperform the best 2D designs (also obtained using ILP) for a given application under the same temperature constraint for both single-core and multi-core applications, and

---

[*]This work is supported in part by NSF Career Award #0093082 and by a grant from GSRC.

(ii) optimized placement of blocks is very important in the 3D domain as the difference between the optimized and random core/storage block placements (in 3D) is very significant in all the cases tested.

The rest of this paper is organized as follows. The next section discusses the related work on 3D and customized memory design. Section III explains the thermal model used in this work. Section IV presents the details of our ILP formulation, and Section V shows the working of our approach through an example. Section VI gives an experimental evaluation of our approach. Finally, Section VII presents our concluding remarks.

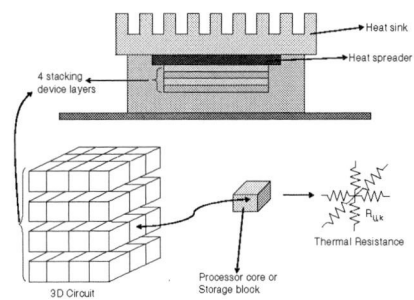

Fig. 1. 3D resistor mesh model.

II. RELATED WORK

We discuss the related efforts in two categories: 3D design and customized memory design for embedded systems. Design techniques and methodologies for 3D architectures have been investigated to efficiently exploit the benefits of 3D technologies. Recent efforts have focused on developing tools for supporting custom 3D layouts and placement tools [11]. In Deng et al [12], the technology and testing issues are surveyed and a 3D integration framework is presented. However, the investigation of the benefits of 3D design at the architectural level is still in its infancy. Das et al [10] study the energy and thermal behavior of 3D designs under a supplied time constraint, and their tool is based on a standard-cell circuit layout. A recent paper provides an overview of the potential benefits of designing an Itanium processor in the 3D technology [6]. However, it does not provide details of the design of the individual components. Several recent efforts also study employment of 3D designs in reconfigurable fabrics [1, 3]. Our work is different from all these prior efforts as we are interested in application-specific placement of processor cores and storage blocks in 3D.

Memory system design and optimization has been a popular area of research. The prior work mostly focus on single processor based systems [5, 8, 9, 18, 19, 20]. An embedded system consisting of a VLIW processor, instruction cache, data cache, and second-level unified cache has been investigated by Abraham and Mahlke [2]. A hierarchical approach has been used to partition the system into components. Meftali et al [16] attack the memory space allocation (partitioning) problem. In their work, an integer linear programming model has been applied to obtain an optimal distributed shared memory architecture. The objective is to minimize the global cost to access the shared data in the application and the memory cost. A packet routing switch example has been used to test the effectiveness of the proposed approach. Embedded memory design for application-specific multi-core system-on-chips has been investigated by Gharsalli et al [13]. In this methodology, they try to integrate the standard memory components. Our work is different from these prior efforts discussed above since we focus on a 3D architecture and study optimal block placement under temperature constraints.

III. 3D THERMAL MODEL

In order to facilitate the thermal-aware processor core/storage block placement process, a compact thermal model is needed to provide the temperature profile. Numerical computing methods (such as finite difference method (FDM) [22]) are very accurate but computationally intensive, while the simplified close-form formula [14] is very fast but inaccurate. Skadron et al proposed a thermal model called Hotspot [21], which is based on lumped thermal resistances and thermal capacitances. It is more efficient than the prior low-level approaches since the variances at temperature are tracked at a granularity of functional block level. In our research, we use a simplified analytical model called the 3D resistor mesh (shown in Figure 1), which is similar to the approach taken by Hotspot, to facilitate the thermal analysis. The model employs the principle of thermal-electrical duality to enable efficient computation of the thermal effects at the block level. The transfer thermal resistance $R_{i,j}$ of block i with respect to block j can be defined as the temperature rise at the block i due to one unit of power dissipated at block j:

$$R_{i,j} = \frac{\Delta T_{i,j}}{\Delta P_{i,j}}.$$

In this simplified model, the device layers are stacked together with the heat sink as the bottom layer, and each device layer consists of the blocks, which are the storage (cache memory) blocks or the processor core(s). These blocks define the thermal nodes of the thermal resistor mesh. (For more details, the reader is referred to [21].) A 3D resistor mesh consists of vertical and lateral thermal resistors, which model the heat flow from wafer to wafer, wafer to heatsink, and heat transferring between the blocks. These vertical and lateral resistances are calculated based on Hotspot [21].

After the 3D thermal resistance network has been determined, the temperature rise at each block can be estimated by solving the following equation:

$$T = R \times P,$$

where T is the temperature of each block, R is a matrix of thermal resistances, and P is the power consumption of each block. One of the nice properties of this high-level abstraction of temperature behavior is that it can easily be embedded within an ILP-based opimization framework (since it is linear), as will be discussed in the next section.

IV. ILP FORMULATION

Our goal in this section is to present an ILP formulation of the problem of minimizing data communication cost of a given

TABLE I

THE CONSTANT TERMS USED IN OUR ILP FORMULATION. THESE ARE EITHER ARCHITECTURE SPECIFIC OR PROGRAM SPECIFIC. NOTE THAT C_Z CAPTURES THE NUMBER OF LAYERS IN THE 3D DESIGN. BY SETTING C_Z TO 1, WE CAN MODEL A CONVENTIONAL 2D (SINGLE LAYER) DESIGN AS WELL. THE VALUES OF $FREQ_{p,m}$ ARE OBTAINED BY COLLECTING STATISTICS THROUGH SIMULATING THE CODE AND CAPTURING ACCESSES TO EACH STORAGE BLOCK.

Constant	Definition
P	Number of processor cores
M	Number of storage blocks
C_X, C_Y, C_Z	Dimensions of the chip
P_X, P_Y	Dimensions of a processor core
$SIZE_m$	Size of a storage block m
$FREQ_{p,m}$	Number of accesses to storage block m by processor p
$R_{l,v}$	Thermal resistance network
T_B	Temperature bound

application by determining the optimal placement of storage blocks and processor cores under a given temperature bound. A storage block in this paper corresponds to a set of consecutive cache lines. The data cache is assumed to be divided into *storage blocks* of equal size. In this paper, our focus is on the data cache only; however, the proposed approach can be applied to the instruction cache as well.

ILP provides a set of techniques that solve those optimization problems in which both the objective function and constraints are linear functions and the solution variables are restricted to be integers. The 0-1 ILP is an ILP problem in which each (solution) variable is restricted to be either 0 or 1 [17]. Table I gives the constant terms used in our ILP formulation. We used *Xpress-MP* [23], a commercial tool, to formulate and solve our ILP problem, though its choice is orthogonal to the focus of this paper. In our ILP formulation, we view the chip area as a 3D grid, and assign storage blocks and cores into this grid.

Assuming that P denotes the total number of processor cores, M the total number of storage blocks, (C_X, C_Y, C_Z) the dimensions of the chip, (P_X, P_Y) the dimensions of the processor core, our approach uses 0-1 variables to specify the coordinates of each storage block and processor core.

We use PC to identify the coordinates of a processor core. More specifically,

- $PC_{p,x,y,z}$: indicates whether processor core p is in (x, y, z).

Similarly, MC is used in our formulation to identify the coordinates of a storage block.

- $MC_{m,x,y,z}$: indicates whether storage block m is in (x, y, z).

Although the size of a storage block is given, its dimensions may vary. We use MD to capture the dimensions of a storage block. Specifically, we have:

- $MD_{m,x,y}$: indicates whether storage block m has dimensions of (x, y).

The mapping between the coordinates and the blocks is ensured by variable $MMap$ for the storage blocks, and variable $PMap$ for the processor cores. That is,

- $MMap_{x,y,z,m}$: indicates whether coordinate (x,y,z) is assigned to storage block m.
- $PMap_{x,y,z,p}$: indicates whether coordinate (x,y,z) is assigned to processor core p.

The distances between a processor core and a storage block on each axis (x,y and z) are captured by $Xdist_{p,m,x}$, $Ydist_{p,m,y}$, and $Zdist_{p,m,z}$. Specifically, we have:

- $Xdist_{p,m,x}$: indicates whether the distance between processor core p and storage block m is equal to x on the x-axis.
- $Ydist_{p,m,y}$: indicates whether the distance between processor core p and storage block m is equal to y on the y-axis.
- $Zdist_{p,m,z}$: indicates whether the distance between processor core p and storage block m is equal to z on the z-axis.

In order to facilitate the thermal-aware core/storage block placement, power and temperature values need to be calculated. Temperature of each block (T_i) is obtained using the resistance vector (R) and the corresponding power consumption values (obtained through a Wattch-based simulation [7]).

- $Power_m$: is the power consumption for block m.
- $Temp_m$: is the temperature of block m. A processor core needs to be assigned to a single coordinate:

$$\sum_{i=0}^{C_X-1} \sum_{j=0}^{C_Y-1} \sum_{k=0}^{C_Z-1} PC_{p,i,j,k} = 1, \quad \forall p. \tag{1}$$

In the above equation, i, j and k correspond to the x, y and z coordinates, respectively. A storage block also needs to be assigned to a unique coordinate:

$$\sum_{i=0}^{C_X-1} \sum_{j=0}^{C_Y-1} \sum_{k=0}^{C_Z-1} MC_{m,i,j,k} = 1, \quad \forall m. \tag{2}$$

A storage block needs to have unique dimensions:

$$\sum_{i=1}^{C_X} \sum_{j=1}^{C_Y} MD_{m,i,j} = 1, \quad \forall m. \tag{3}$$

Each storage block should have dimensions in such a way that its size, $SIZE_m$ (given as input), will fit into the allocated space. That is, $SIZE_m = width \times height$:

$$SIZE_m = \sum_{i=1}^{C_X} \sum_{j=1}^{C_Y} MD_{m,i,j} \times i \times j, \quad \forall m. \tag{4}$$

Storage blocks should be mapped to the chip based on the coordinate and dimensions of the corresponding storage block. This requirement can be captured as follows:

$$MMap_{x,y,z,m} \geq MC_{m,x_1,y_1,z-1} + MD_{m,dx,dy} - 1,$$
$$\forall m, x, x_1, dx, y, y_1, dy, z, z_1, dz \text{ such that}$$
$$x_1 + dx \geq x > x_1, \text{ and } y_1 + dy \geq y > y_1. \tag{5}$$

In this expression, x_1, y_1 and z_1 denotes the x, y, and z coordinates of a storage block. Similarly, dx and dy denote the dimensions of the storage block. Based on these values, $MMap$ assigns the corresponding coordinates to the storage block m. Similarly, processor cores should be mapped to the chip, which can be expressed as follows:

$$PMap_{x,y,z,p} \geq PC_{p,x_1,y_1,z-1}, \forall p, x, x_1, y, y_1, z, z_1$$
such that $x_1 + P_X \geq x > x_1$, and $y_1 + P_Y \geq y > y_1$. (6)

In order to prevent multiple mappings of a coordinate in our grid, we force a coordinate to belong a single processor core or

TABLE II
THE ACCESS PERCENTAGE OF EACH BLOCK BY DIFFERENT PROCESSORS.

Processor	B_1	B_2	B_3	B_4	B_5	B_6	B_7	B_8	B_9	B_{10}
1	0.20%	0.18%	61.76%	0.19%	0.17%	0.20%	3.48%	2.86%	0.20%	0.20%
2	0.20%	0.19%	61.86%	0.19%	0.22%	4.07%	2.30%	0.18%	0.19%	0.18%
3	0.18%	0.18%	61.83%	0.18%	0.18%	4.05%	2.34%	0.19%	0.21%	0.19%
4	0.18%	0.22%	61.76%	0.19%	0.18%	4.05%	2.32%	0.18%	0.20%	0.18%
Processor	B_{11}	B_{12}	B_{13}	B_{14}	B_{15}	B_{16}	B_{17}	B_{18}	B_{19}	B_{20}
1	22.83%	0.22%	0.20%	0.18%	0.20%	0.19%	0.18%	0.19%	1.83%	4.55%
2	22.64%	0.20%	0.18%	0.18%	0.17%	0.19%	0.18%	1.91%	4.49%	0.28%
3	22.65%	0.20%	0.20%	0.22%	0.19%	0.20%	0.18%	1.89%	4.47%	0.29%
4	22.59%	0.17%	0.18%	0.18%	0.19%	0.21%	0.18%	1.89%	4.49%	0.31%

a single storage block:

$$\sum_{i=1}^{M} MMap_{i,x,y,z} + \sum_{i=1}^{P} PMap_{i,x,y,z} = 1, \forall x, y, z. \quad (7)$$

Manhattan Distance is assumed to be the cost of the *data communication* between a storage block and a processor core. This is also referred to as the *data access cost* in this paper, and is the metric whose value we want to minimize. Note that in our architecture processor cores communicate by sharing data in storage blocks. To capture the Manhattan Distance, we use two variables, namely, $Xdist_{p,m,x}$ and $Ydist_{p,m,y}$, and employ the following constraints:

$$Xdist_{p,m,x} \geq PC_{p,x_1,y_1,z_1} + MC_{m,x_2,y_2,z_2} - 1,$$
$$\forall p, m, x, x_1, x_2, y_1, y_2, z_1, z_2 \text{ such that } x = |x_1 - x_2|. \quad (8)$$

$$Ydist_{p,m,y} \geq PC_{p,x_1,y_1,z_1} + MC_{m,x_2,y_2,z_2} - 1,$$
$$\forall p, m, x_1, x_2, y, y_1, y_2, z_1, z_2 \text{ such that } y = |y_1 - y_2|. \quad (9)$$

In addition to the x and y dimensions, there is a communication cost due to z dimension, which captures the vertical dimension. Data communication across the layers is not the same as the communication within a given layer, because it only depends on the wafer thickness, which does not change with the block size, so it needs to be captured separately as follows:

$$Zdist_{p,m,z} \geq PC_{p,x_1,y_1,z_1} + MC_{m,x_2,y_2,z_2} - 1,$$
$$\forall p, m, x_1, x_2, y_1, y_2, z, z_1, z_2 \text{ such that } z = |z_1 - z_2|. \quad (10)$$

In wafer-bonding 3D technology, the dimensions of the vertical through-wafer interconnect are not expected to scale at the same rate as feature size, because wafer-to-wafer alignment tolerances during bonding pose limitations on the scaling of the through-wafer interconect. Current dimensions of through-wafer via sizes vary from 1μm-by-1μm to 10μm-by-10μm [6, 11, 12]. The relatively large size of via makes the interconnect delay going through wafer to be relatively much smaller.

We next calculate the temperature of each block using the $Temp = R \times Power$ equation. More specifically,

$$Temp_m = \sum_{j=1}^{P+M+1} R_{m,j} \times Power_j, \quad \forall m. \quad (11)$$

Finally, the temperature constraint is enforced using the following expression:

$$Temp_m \leq T_B, \quad \forall m. \quad (12)$$

TABLE III
SINGLE-CORE BENCHMARK CODES USED IN THIS STUDY.

Benchmark Name	Source	Description	Number of Data Accesses
ammp	Spec	Computational Chemistry	86967895
equake	Spec	Seismic Wave Propagation Simulation	83758249
mcf	Spec	Combinatorial Optimization	114662229
mesa	Spec	3-D Graphics Library	134791940
vortex	Spec	Object-oriented Database	163495955
vpr	Spec	FPGA Circuit Placement and Routing	117239027

TABLE IV
MULTI-CORE BENCHMARK CODES USED IN THIS STUDY.

Benchmark Name	Source	Description	Number of Data Accesses
3step-log	DSPstone	Motion Estimation	90646252
adi	Livermore	Alternate Direction Integration	71021085
btrix	Spec	Block Tridiagonal Matrix Solution	50055611
tsf	Perfect Club	Nearest Neighbor Computation	54917732

Having specified the necessary constraints in our ILP formulation, we next give our objective function. We define our cost function as the sum of the data communication distances in all 3 dimensions. X_{Cost}, Y_{Cost}, and Z_{Cost} denote the total data communication distances traversed along dimensions x,y, and z, respectively. The communication cost due to dimension x is:

$$X_{Cost} = \sum_{i=1}^{P} \sum_{j=1}^{M} \sum_{k=1}^{C_X-1} FREQ_{i,j} \times Xdist_{i,j,k} \times k. \quad (13)$$

Similarly, the cost due to dimension y can be expressed as:

$$Y_{Cost} = \sum_{i=1}^{P} \sum_{j=1}^{M} \sum_{k=1}^{C_Y-1} FREQ_{i,j} \times Ydist_{i,j,k} \times k. \quad (14)$$

Finally, the cost due to dimension z can be written as:

$$Z_{Cost} = \sum_{i=1}^{P} \sum_{j=1}^{M} \sum_{k=1}^{C_Z-1} FREQ_{i,j} \times Zdist_{i,j,k} \times k. \quad (15)$$

Consequently, our objective function can be expressed as:

$$\min \quad (\alpha \times (X_{Cost} + Y_{Cost}) + \beta \times Z_{Cost}). \quad (16)$$

To summarize, our topology exploration problem can be formulated as "minimize $\alpha \times (X_{Cost} + Y_{Cost}) + \beta \times Z_{Cost}$ under constraints (1) through (15)." It is important to note that this ILP formulation is very flexible as it can accomodate different number of processor cores, storage blocks, and layers.

V. EXAMPLE

In this section, we give an example demonstrating the effectiveness of our approach. As our example, we use 3step-log,

4B-4

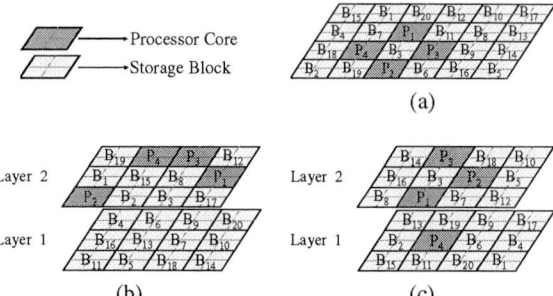

Fig. 2. The topologies generated with 4 processor cores (3step-log). (a) Optimal 2D. (b) Random 3D. (c) Optimal 3D. $P_1 \ldots P_4$ are processor cores and $B_1 \ldots B_{20}$ are storage blocks.

one of our benchmarks, with 4 processor cores and 20 storage blocks. In the 2D case, we assume the dimensions of the chip (C_X, C_Y and C_Z) as $12 \times 8 \times 1$, and in the 3D case, we assume the dimensions to be $8 \times 6 \times 2$. In both the cases, the total number of coordinates (total chip area) is equal to 96. Both the storage blocks and the processor cores are assumed to be 2×2, though they can be set to any value in our formulation. Table II shows the percentage of accesses to each block by different processor cores.

In Figure 2, the topologies generated by three different schemes are depicted. Figure 2(a) illustrates the best 2D placement (determined using ILP). On the other hand, a randomly-generated storage block/processor core placement is shown in Figure 2(b) for 3D. Finally, Figure 2(c) shows the best placement for 3D (determined using ILP). Note that, each processor core and each storage block is mapped to 4 coordinates since the size of a storage block/processor core is assumed to be 4. To explain these mappings, let us consider Figure 2(c). In this topology, B_{15} is mapped to coordinates $(0,0,0), (0,1,0), (1,0,0)$, and $(1,1,0)$. As it can be seen from Table II, data block B_3 is the most frequently accessed block by all the processor cores, and that is why it is put into a very close position to all 4 processor cores in optimal 2D and optimal 3D. We also see that, B_{11} is the next most frequently accessed block by all the processor cores in this example.

VI. EXPERIMENTAL EVALUATION

A. Setup

We performed several experiments with two different set of benchmarks. The first group of applications are sequential Spec2000 codes and used for single-core designs in this work. This benchmark set consists of six applications randomly-selected from the Spec2000 benchmark suite. Table III lists these codes and their important characteristics. In collecting the statistics on accesses to storage blocks, for each benchmark, we fast-forwarded the first 2 billion instructions, and simulated the next 300 million instructions. The fourth column of Table III gives the number of data accesses for each application. The second group of applications, on the other hand, are for multi-core designs and include several array-based codes parallelized through an optimizing compiler built upon SUIF [4]. The exact mechanism used in parallelizing the code is orthogonal to the focus of this paper. What we mean

by parallelization in this context is distributing loop iterations across the processors; each processor is typically assigned a subset of loop iterations. Table IV lists the important characteristics of this second group of benchmarks. As before, the third column gives the description of the benchmarks, and the last column gives the number of data accesses. The ILP solution times on an Intel Pentium III Xeon Processor of 549MHz with 1GB of RAM varied between 144 seconds and 3.5 hours, averaging on about 25 minutes. The default simulation parameters used in our experiments are presented in Figure 3. As our base configuration, we assumed a stack of two device layers connected to each other. We also assumed that the chip is composed of 24 blocks, each of which can be a storage block or a processor core. We conservatively assumed that the block-to-block distance is ten times costlier than that of the layer-to-layer distance. This is denoted as the ratio of $\frac{\alpha}{\beta}$ in this paper.

We performed experiments with four different execution models for each benchmark code in our experimental suite:

- **2D-Random:** This is in a sense a conventional topology which uses a single wafer and the storage blocks and the processor cores are placed randomly.

- **2D-Opt:** This is an integer linear programming based strategy, wherein the storage blocks and processor cores are distributed on the die in a way that minimizes the data communication cost of the whole system. Note that, this is an optimal core/storage block placement scheme for 2D.

- **3D-Random:** This is same as the 2D-Random case except that there are possibly multiple device layers.

- **3D-Opt:** This is the integer linear programming based placement strategy for 3D proposed in this paper, wherein the storage blocks and processor cores are placed on several wafers optimally. This scheme represents the optimal placement for 3D.

B. Results

Figure 4 gives our results based on two layers. Note that, these results are obtained using the values given in Figure 3. All results are normalized with respect to those of the 2D-Random scheme. Figure 4(a) shows the improvement brought by our approach for the single-core designs, whereas Figure 4(b) gives the similar results for the multi-core designs. We see that the overall average reduction in data access costs with 2D-Opt is around 63% and 58% for the single-core case and the multi-core case, respectively. On the other hand, the 3D-Opt scheme reduces the costs by about 82% and 69% on average for the single-core case and the multi-core case, respectively. In other words, the best 3D design generates much better results than the best 2D design.

To see the impact of the optimal placement of processor cores/storage blocks, we next compare the optimized and random designs in 3D. In Figure 5, we compare our 3D approach against the randomly-generated placements obtained through 3D-Random. As before, the results are normalized with respect to the 2D-Random scheme. The average data communication cost reductions are 42% and 51% for the single-core case and for the multi-core case, respectively. Overall, the results presented in Figure 4 and Figure 5 clearly show that employing a 3D design with optimal placement is critical for the best results.

Parameter	Value
Number of processor cores (in multi-core designs)	4
Number of blocks	24
Number of layers	2
$\frac{\alpha}{\beta}$	10
Total storage capacity	128KB
Set associativity	2 way
Line size	32 Bytes
Number of lines per block	90
Temperature bound	110°C

Fig. 3. The default simulation parameters.

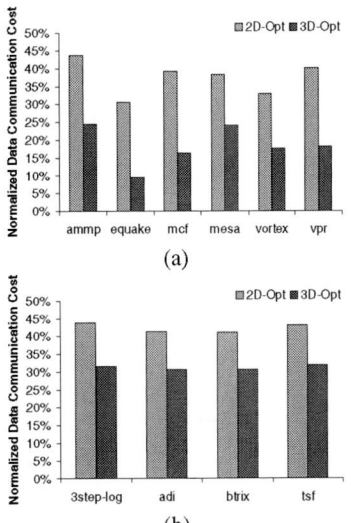

Fig. 4. Data communication costs for 2D-Opt and 3D-Opt normalized with respect to the 2D-Random scheme. (a) Single-core design. (b) Multi-core design (with 4 processor cores).

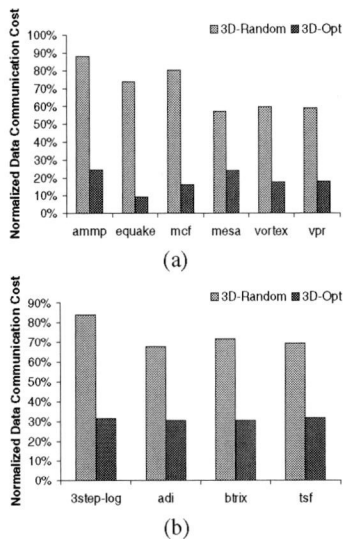

Fig. 5. Data communication costs for 3D-Random and 3D-Opt normalized with respect to the 2D-Random scheme. (a) Single-core design. (b) Multi-core design (with 4 processor cores). The results of 3D-Random are obtained by taking the average over five different experiments with random placement.

VII. CONCLUDING REMARKS

Ever-shrinking process technology coupled with increasing data communication requirements of embedded applications make on-chip interconnects an increasing bottleneck in embedded system design. One of the promising solutions to this global interconnect problem is to move towards 3D designs, where direct verical tunneling allows multiple layers to be stacked, one on top of the other. The work described in this paper studies application-specific placement of processor cores and storage blocks in a customized 3D design. We formulated this problem using ILP and solved it using a commercial solver. Our experiments with both the single-core and multi-core designs and with both the 2D and 3D designs indicate that the optimal placement of storage blocks and processor cores (i.e., optimal topology discovery) is very important in a 3D design (i.e., it makes a huge difference in terms of data communication costs) and that the best 3D designs consistently generate better results than the best 2D designs.

REFERENCES

[1] C. Ababei, H. Mogal, and K. Bazargan, *Three-dimensional Place and Route for FPGAs*, In Proc. of Asia South-Pacific Design Automation Conference (ASP-DAC), 2005.

[2] S. G. Abraham and S. A. Mahlke, *Automatic and Efficient Evaluation of Memory Hierarchies for Embedded Systems*, In Proc. of the 32nd Annual International Symposium on Microarchitecture, Haifa, Israel, November 1999.

[3] A. J. Alexander, et al, *Three-Dimensional Field-Programmable Gate Arrays*, In Proc. of The Eighth Annual IEEE International Application Specific Integrated Circuits Conference, pp. 253-256, 1995.

[4] S. P. Amarasinghe, J. M. Anderson, M. S. Lam, and C. W. Tseng. *The SUIF compiler for scalable parallel machines*, In Proc. SIAM Conference on Parallel Processing for Scientific Computing, February, 1995.

[5] F. Angiolini, L. Benini, and A. Caprara, *Polynomial-Time Algorithm for On-Chip Scratch-Pad Memory Partitioning*, In Proc. of the International Conference on Compilers, Architectures and Synthesis for Embedded Systems, San Jose, CA, 2003.

[6] B. Black, et al, *3D Processing technology and Its Impact on IA32 Microprocessors*, In Proc. of ICCD, 2004.

[7] D. Brooks, V. Tiwari, and M. Martonosi, *Wattch: A framework for architectural-level power analysis and optimizations*, In Proc. of the 27th Annual International Symposium on Computer Architecture, June 2000.

[8] Y. Cao, H. Tomiyama, T. Okuma, and H. Yasuura, *Data Memory Design Considering Effective Bit-width for Low-Energy Embedded Systems*, In Proc. of the 15th International Symposium on System Synthesis, Kyoto, Japan, October 2002.

[9] F. Catthoor, et al, *Custom Memory Management Methodology – Exploration of Memory Organization for Embedded Multimedia System Design*, Kluwer Academic Publishers, 1998.

[10] S. Das, *Timing, energy, and Thermal performance of Three-Dimensional Integrated Circuits*, In Proc. of GLSVLSI, 2004.

[11] S. Das, et al, *Technology, Performance, and computer-aided Design of Three-Dimensional integrated Circuits*, In Proc. of ISPD, 2004.

[12] Y. Deng, et al, *2.5D System Integration: A Design Driven System Implementation Schema*, In Proc. of ASP-DAC, 2004.

[13] F. Gharsalli, S. Meftali, F. Rousseau, and A. A. Jerraya, *Automatic Generation of Embedded Memory Wrapper for Multiprocessor SoC*, In Proc. of the 39th Design Automation Conference, New Orleans, Louisiana, 1999.

[14] S. Im and K. Banerjee, *Full Chip Thermal Analysis of Planar (2D) and Vertically Integrated (3D) High Performance ICs*, Tech. Digest IEDM 2000, pp.727-730.

[15] International Technology Roadmap for Semiconductors, http://www.itrs.net/Common/2004Update/2004Update.htm.

[16] S. Meftali, F. Gharsalli, F. Rousseau, and A. A. Jerraya. *An Optimal Memory Allocation for Application-Specific Multiprocessor System-on-Chip*, In Proc. of the International Symposium on Systems Synthesis, Montreal, Canada, 2001.

[17] G. Nemhauser, L. Wolsey. *Integer and Combinatorial Optimization*, Wiley-Interscience Publications, 1988.

[18] P. R. Panda and L. Chitturi, *An Energy-Conscious Algorithm for Memory Port Allocation*, In Proc. of the 2002 IEEE/ACM International Conference on Computer-Aided Design, San Jose, November 2002.

[19] A. Ramachandran and M. F. Jacome, *Xtream-Fit: An Energy-Delay Efficient Data Memory Subsystem for Embedded Media Processing*, In Proc. of the 40th Design Automation Conference, Anaheim, CA, June 2003.

[20] W.-T. Shiue and C. Chakrabarti, *Memory Exploration for Low-Power Embedded Systems*, In Proc. of the 36th Design Automation Conferences, New Orleans, LA, 1999.

[21] K. Skadron, et al, *HotSpot Thermal Modeling Simulator*, http://lava.cs.virginia.edu/HotSpot/

[22] C. Tsai and S. Kang, *Cell-Level Placement for Improving Substrate Thermal Distribution*, IEEE Transactions On Computer-Aided Design of Integrated Circuits and Systems, vol. 19, pp. 253-266, Feb. 2000.

[23] Xpress-MP, http://www.dashoptimization.com/pdf/Mosel1.pdf, 2002.

Task Placement Heuristic Based on 3D-Adjacency and Look-Ahead in Reconfigurable Systems

Jesús Tabero

Dept. Programas Espaciales
Instituto Nacional de Técnica Aeroespacial
Madrid 28850, Spain
Tel : +34-91-520-1693
Fax : +34-91-520-1492
taberogj@inta.es

Julio Septién, Hortensia Mecha, Daniel Mozos

Dept. Arquitectura de Computadores y Automatica
Universidad Complutense de Madrid
Madrid 28040, Spain
Tel : +34-91-394-7617/18/19
Fax : +34-91-394-7527
{jseptien,horten,mozos}@dacya.ucm.es

Abstract- To get efficient HW management in 2D Reconfigurable Systems, heuristics are needed to select the best place to locate each arriving task. We propose a technique that locates the task next to the borders of the free area for as many cycles as possible, trying to minimize the area fragmentation. Moreover, we combine it with a look-ahead heuristic that allows delaying the scheduling of a task to the next event, increasing the solution search space.

I. Introduction

Current multimedia systems have a very dynamic behaviour. Moreover, sometimes the user wants to upgrade their functionality by loading new applications. Traditional platforms based on processors are not able to provide the needed performance, whereas ASICs lack flexibility to deal with such a changing environment.

A technology that lies between processors and ASICs is reconfigurable hardware. It participates of the flexibility of processors and, at the same time, it is able to offer very good results in terms of performance. In recent years this flexibility has increased with the possibility of partially reconfiguring the hardware at run time [1]. This allows changing the functionality of a digital system under user demand, as multimedia applications require. It also allows true hardware multitasking through space multiplexing.

In order to use all these capabilities, the functionality of the operating system must be increased to manage this kind of resources. In [2] some of the main problems of designing such an operating system are outlined.

One of the most interesting problems is to decide where to locate the bitmap of a new task in the FPGA when it must be run. A data structure is needed that maintains information about the available free area, and the algorithm must choose the best place to locate the arriving task, trying to use the reconfigurable area as efficiently as possible.

Our algorithm chooses the best location for an arriving task trying to minimize the fragmentation that this mapping will produce. The fragmentation of the free area is an indirect measure of the probability of finding a suitable location for a new task in the FPGA in the near future.

The rest of the paper is organized as follows. Section II reviews related work in this field. Section III presents the algorithm for area management. Section IV describes the heuristics used to select a location for an arriving task. And finally, in sections V and VI some experimental results and conclusions are presented.

II. Related work

Task allocation, free space management and area fragmentation are fundamental problems of managing 2D reconfigurable resources. They have been dealt with recently by several research teams.

Diessel et al. [3] have developed a quad-tree structure to store the information of the available FPGA area. Such structure can be travelled and updated quite fast, but it does not guarantee that an adequate place is found, even if there is enough area to store the task, but split among different branches of the tree. This solution does not take into account the resultant free area fragmentation to select the position where the task is mapped to. On the contrary, it deals with fragmentation by proposing several high-cost defragmentation processes.

Bazargan et al. [4] deal with the area allocation problem by using a bin-packing approach and applying some of the classical algorithms for such theoretical problem. They propose several strategies for on-line 2D bin-packing of the arriving rectangular tasks. These strategies differ mainly in the way the free area is managed. One of them keeps track of all the maximum empty rectangles (MER) where an arriving task could be placed. Such approach guarantees that, if an adequate place exists, it can be found, but at the cost of a very high complexity. A second approach tries to use heuristics in order to reduce the number of rectangles considered when updating the rectangle list. When a free rectangle is selected to store the arriving task, the excess area

0-7803-9451-8/06/$20.00 ©2006 IEEE.

is divided in only two, non-overlapping, new rectangles. Bazargan offers several criteria to do this splitting, but does not decide clearly for one of them. Anyway, by selecting some of the possible rectangles, situations can arise where existing room cannot be used to store a task, because it is split among several rectangles.

Walder et al. propose in [5] an enhanced version of Bazagan's partitioner with the same efficiency but improved placement quality. This enhanced method delays the basic vertical/horizontal split decision and manages overlapping rectangles in a restricted form. They also present a hash matrix approach to find a placement in constant time, but the updating of this structure is very time consuming.

Ahmadinia et al. [6] present another version of Bazagan´s partitioner, managing the occupied area instead of the free area, so in most cases the number of rectangles is much lower, though the complexity order is the same. The heuristic used to allocate a task tries to minimize the distance to the tasks it communicates with. But it does not take into account the area fragmentation, nor time or data constraints during scheduling.

Handa et al. [7] present a fast algorithm for finding empty space in a FPGA which uses a staircase data structure to report the empty area in the form of a list of MER. The search of such structure has a complexity order of $O(m*n)$ where m and n are the number of columns and rows respectively, but there is no details of the MER selection criterion, and the complexity of the task placement algorithm is not considered.

III. Vertex-List Based Area Management

Our approach to reconfigurable HW management keeps track of the available FPGA free area with a vertex-list structure that has been described in detail in [8]. Such structure can be travelled with different heuristics in order to choose the vertex where the arriving tasks will be placed.

As fig.1 shows, we use a 2D FPGA model, an homogeneous two dimensional grid formed by W*H basic reconfigurable blocks, that we will use as "area units" all along. We suppose that each basic block includes processing elements as well as I/O resources. The FPGA has also some dedicated resources to manage the task I/O. A task can be made of an arbitrary number of such basic blocks, but always with a rectangular shape. The tasks are relocatable and can be inserted at arbitrary row and column offsets. The tasks are independent, with no data constrains between them, but there can be real-time constrains that must be satisfied.

Each task is defined by the following parameters: $T_i = \{ w_i, h_i, t\_ex_i, t\_arr_i, t\_max_i \}$, where w_i and h_i indicate the task size, t_ex_i is the task execution time, t_arr_i the task arrival time and t_max_i the maximum time allowed for the task to finish execution. Therefore, the task T_i can't be scheduled later than $t\_max_i - t\_ex_i$.

Fig. 1 summarizes the operation of our HW manager, that consists of three main modules, the Task Scheduler, the Vertex Selector and the Vertex List Updater, and uses three

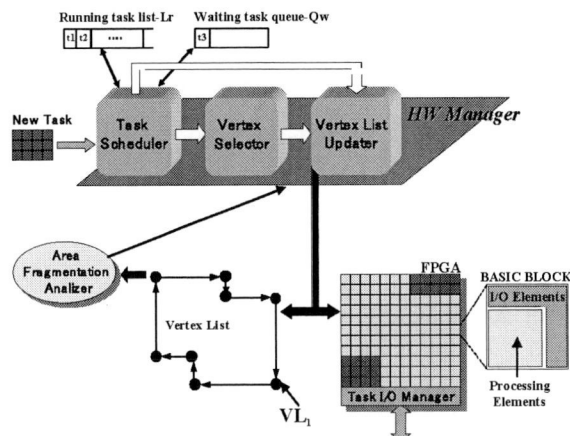

Fig. 1. HW manager and vertex-list structure

important data structures, the Running Task List, L_r, the Waiting Task Queue, Q_w, and the Vertex List Set, VLS, that describes all the available FPGA free space, with a different Vertex-List VL_i for each FPGA free hole.

The Free Area Fragmentation Analizer is a module that computes the fragmentation of the FPGA free area for a given FPGA status, by using a new fragmentation metrics described in detail in section IV. This value can be used either by the HW Manager to perform a defragmentation process, or by the Vertex Selector Module if a fragmentation-based heuristic is used.

When a new task (an arriving or waiting task) is considered, the Task Scheduler calls the Vertex Selector to check whether a feasible position exists where the task could be mapped to. The vertex is chosen among the candidates according to the selected heuristic. The candidates can be of several types: bottom-left (BL), bottom-right (BR), top-left (TL) or top-right (TR). Then the task is inserted and the VLS is updated by the Vertex List Updater accordingly. When a task ends, the VLS is also updated, and different special situations such as hole merging or island managing, that are described in detail in [8], must be dealt with. If there is no place for an arriving task, the task is temporally stored at the queue Qw, and if timeout happens it is discarded.

IV. Heuristics for Location Selection

An heuristic is used to choose a given vertex among all the feasible candidates to locate the task. A simple approach based on a First-Fit criterion was proposed in [8]. A more efficient alternative was presented in [9], with a Best-Fit approach based on a cost function computed with a task-adjacency criteria. We also presented a fragmentation metric that was used to evaluate the quality of the FPGA status at any given time. A 2D-Adjacency example is shown in Fig. 2a.

4B-5

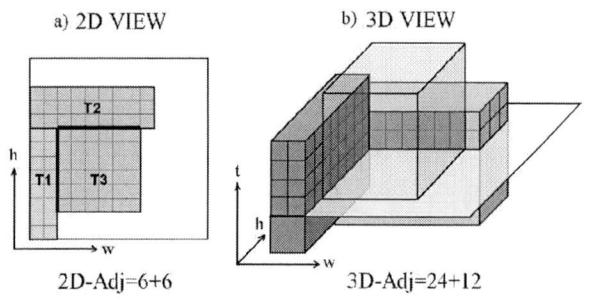

a) 2D VIEW   b) 3D VIEW

2D-Adj=6+6   3D-Adj=24+12

Fig. 2. 2D-Adjacency (a) and 3D-Adjacency (b)

What we are proposing now is a new alternative, where the cost function considers not only 2D-spatial adjacency, but also temporal adjacency. Then we will develop a look-ahead heuristic based on this 3D-adjacency cost function that will give even better experimental results.

A. 2D-Adjacency Heuristic

This heuristic inserts the task at the vertex position where the arriving task achieves the higher contact level between the new task borders and the envelope defined by the VL_i. The adjacency is computed in terms of block-length units. This heuristic was presented in [9], and an example is shown in Fig. 2a, where the adjacency value obtained when T3 is placed next to T1 and T2 is of 12 units.

B. 3D-Adjacency Heuristic

This heuristic is an evolution of the 2D-Adjacency heuristic described above, but extended to the time axis. Considering the FPGA as a bin and the tasks as smaller boxes, with time as vertical axis, the algorithm tries to pile up the new box as close as possible to the rest of boxes already in the FPGA or to the FPGA borders. This heuristic is similar to the one we would use in the real world to place the boxes inside the bin. This approach is shown in Fig. 2b, where the 3D-adjacency value obtained when T3 is placed next to T1 and T2 is now of 36 units (6x4+6x2).

To take this into account, it is necessary to store a new value for each edge of the vertex list: the remaining execution time of the task the edge belongs to, a kind of "edge lifetime". In order to simplify, if an edge belongs to several adjacent tasks, the one with the shortest remaining time is chosen as lifetime for the whole edge.

The 3D adjacency is then computed as the sum of the length of the new task in contact with each adjacent edge, multiplied by the temporal adjacency between the new task and the edge. This temporal adjacency is the minimum between the edge's lifetime and the t_ex of the arriving task. When the arriving task is in touch with any edge of the FPGA perimeter, then t_ex of the arriving task is considered. Thus, when it is possible at the current time, the task will be preferably placed next to FPGA borders, or to other long-life tasks.

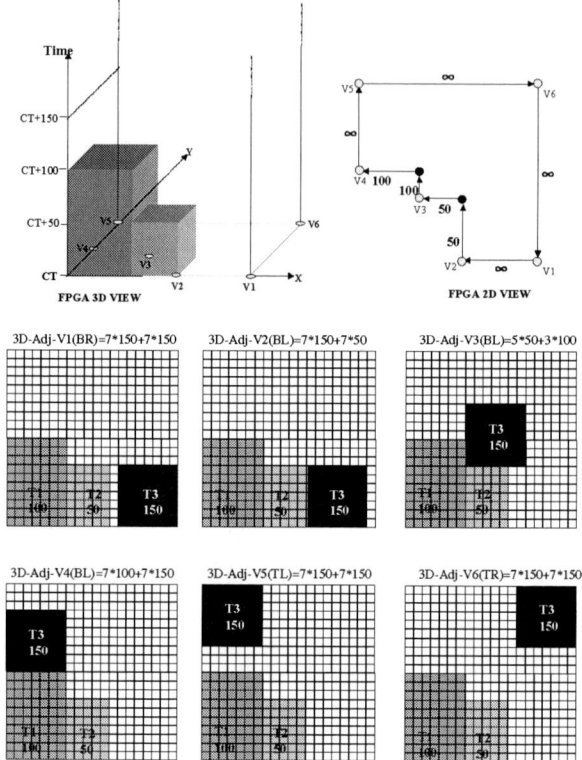

Fig. 3. 3D-Adjacency Heuristic computation

Fig. 3 shows this concept using an FPGA with two running tasks T1 and T2. When T3 arrives, with a size of 7*7 basic cells and a t_ex value of 150 time units, the HW Manager tries to schedule it at CT (Current Time). The 3D-adjacency value for all the candidates vertex is calculated as it is shown in the figure. Therefore the new task T3 will be inserted on the candidate vertex V1, the first one where the 3D-adjacency value is maximum.

Fig. 4 shows a simple example to illustrate the behavior of the heuristic. The FPGA initial status is shown on top. There are four currently running tasks, with their remaining execution time shown inside, and the corresponding VL. When a new task, of 3*4 basic cells and a t_ex value of 150 time units arrives, this approach computes the 3D-adjacency value for each feasible candidate position. As it can be seen, this heuristic would place the task at candidate A (B is equivalent).

We can analyse the quality of the decisions taken by the 3D-Adjacency heuristic by using the fragmentation metric proposed in [8], where the fragmentation level of the free area for a given FPGA status was estimated as follows:

$$F = 1 - \Pi_i \left[(4/V_i) * (A_i/A_F) \right]$$

where the term between brackets represents a kind of suitability for each free hole i, with area A_i and V_i vertices, to accommodate future tasks, while A_F stands for the whole free area in the FPGA.

398

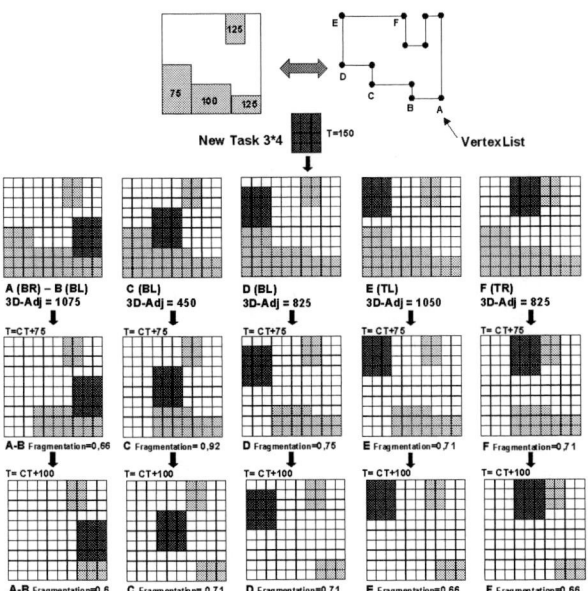

Fig. 4. 3D Adyacency heuristic example

The task insertion at candidate A, the one selected by the 3D-Adjacency heuristic, leads to future FPGA states with lowest fragmentation values, according to our metric, at different simulation steps. When currently running tasks finish execution at T=CT+75 and at T=CT+100, the fragmentation obtained for candidate A is the lowest, as it can be seen in fig. 4 where this fragmentation value is labelled at the bottom of each feasible candidate position.

C. Look-ahead 3D Heuristic

This heuristic uses the 3D-adjacency value, computed as it was described earlier, but in this approach this value is calculated for all feasible candidates at both the current simulation time and at the next event time (when the next task-end happens), performing thus a one-level look-ahead scheduling. The idea is that we could get a better 3D-adjacency value for a location only available in the near future.

However, sometimes it is not possible to perform this look-ahead calculation, if the new task must be scheduled before any running task can exit the FPGA, due to its timeout t_max_i.

Fig.5 shows a comparative between 3D-Adjacency (3D-Ad) and Look-Ahead-3D (LA-3D) heuristics behavior, when two new tasks, T3 and T4, arrive simultaneously.

We have considered a 20*20 FPGA, with two currently running tasks defined by the tuple of parameters (described in section III): T1={12,12,6,98,110} and T2={2,13,15,98,118}, placed at bottom-left and top-right FPGA corners respectively.

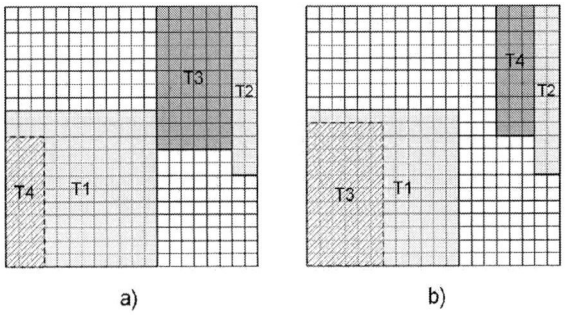

a)   b)

Fig. 5. a) 3D-Ad Heuristic   b) LA-3D Heuristic

Let us suppose T3={6,11,8,100,120} is processed then. The 3D-Ad heuristic would give at current simulation time, CT=100, a maximum value for TR candidate vertex at coordinates X=18, Y=20. In the other hand, the LA-3D heuristic would also compute the 3D-adjacency value at the next event time, and would place T3 at the BL candidate at coordinates X=0, Y=0, at Time=104 when T1 finishes execution (fig. 5b).

When T4={3,10,21,100,123} is processed, at CT=100 also, the 3D-Ad heuristic can not find a feasible candidate and T4 would be rejected at Time =103 because of its t_max value. On the contrary, LA-3D heuristic would find a feasible candidate for T4 at X=18, Y=20 which satisfies its real-time constrains.

V. Experimental Results

To evaluate the quality of our new approaches, we have made experiments using four different algorithms for area management:

a. *Classical FF with BL heuristic(FF_BL)*. When a new task arrives to the FPGA, it performs an exhaustive search through the block matrix, from left to right and from bottom to top, in order to find a feasible location for the arriving task.

b. *Vertex List with BF_2D-Adjacency heuristic (2D-Ad)*. Presented in [9].

c. *Vertex List with BF-3D-Adjacency heuristic (3D-Ad)*.

d. *Vertex List with BF_LA-3D-Adjacency heuristic (LA-3D)*.

These four algorithms have been tested with a simulated FPGA of 100*100 basic blocks, and different data sets of 100 tasks each. They have been randomly generated with a task size range and ratio, similar to others found in many multitasking environments. Table 1 shows these data sets which have been classified in three classes, considered as representative scenarios, depending on the task size ranges and their temporal features.

4B-5

TABLE I
Data set classes

Data Sets	Min. task area	Max. task area	Data Set features
D1, D2	5*5	40*40	Low arrival frequency Low execution time Low task size
D3, D4	10*10	50*50	Medium arrival frequency Medium execution time Medium task size
D5, D6	10*10	60*60	High arrival frequency High execution time High task size

We have used two different parameters to evaluate the results obtained. Table 2 includes for each data set, the percentage of the computing volume rejected for each algorithm. This volume represents all the tasks that were rejected because the manager was not able to find a proper location in time to meet the task's time constraint. For each task, the volume is the product of the task area multiplied by its execution time. The other parameter shown in table 2 is the average FPGA occupation maintained by each algorithm when processing the different data sets.

Figure 6 shows the detailed and the average computing volume rejected by the four algorithms, respectively. 3D-Adjacency and LA-3D-Adjacency show better performance than the others. Moreover, LA-3D is able to execute all the tasks, giving a zero value for these parameters in all data sets. Also LA-3D produces the best average occupation level as it is shown in fig. 7.

TABLE II
Experimental results

Data Set	% Computing Volume Rejected				% Average Occupation Level			
	FF-BL	2D-Adj	3D-Adj	LA-3D-Adj	FF-BL	2D-Adj	3D-Adj	LA-3D-Adj
D1	8,4	1,5	0	0	22,6	24,5	24,9	24,9
D2	8,5	5,8	2,4	0	28,8	29,6	30,7	31,4
D3	4,5	7,8	4,3	0	35,2	32,9	35,2	36,6
D4	9,5	2,2	5,6	0	32,9	35,05	34,2	36,1
D5	3,32	11,5	0	0	38,7	34,2	40	40
D6	14,9	14,5	1,9	0	41,4	42,1	48	48,7

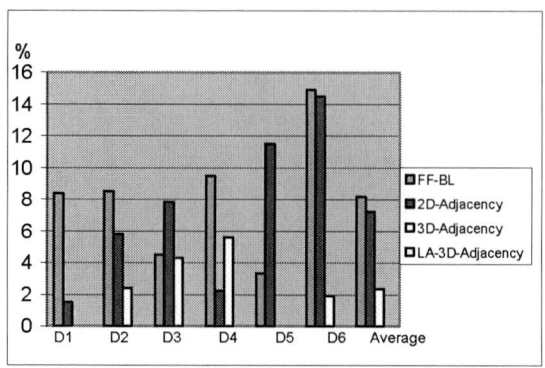

Fig. 6. Detailed and Average Computing Volume Rejected
(Percentage)

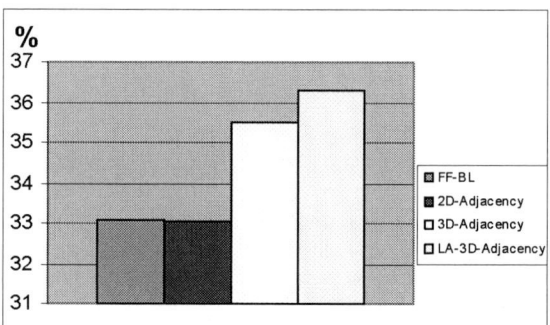

Fig. 7. Average FPGA Occupation Level (Percentage)

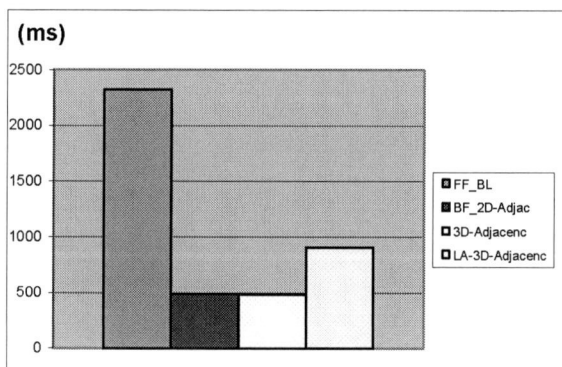

Fig. 8. Relative Average Execution Times

Finally, fig. 8 shows a comparative of the average execution time of the algorithms for the different heuristics. Logically 2D and 3D-Adjacency present the same values because they have the same complexity, and LA-3D takes almost twice this time, because it has to perform the same computation at current time and at the next event time. Finally FF_BL, being the algorithm with a higher complexity order, spends much more time than the others.

400

VI. Conclusions

We have presented a new approach to HW multitasking allocation, with an area manager that maintains the free area with a vertex-list structure. As an adequate management of the fragmentation problem has revealed itself crucial, our heuristics tries to keep it as low as possible. This is done by allocating the tasks next to the borders of the free area for as many cycles as possible. This heuristic has produced very good results in terms of FPGA area occupation and number of task executed. Finally, we have proposed a look-ahead heuristic that allows delaying the allocation of a task to the next event in order to increase the solution search space. It has clearly shown the best behavior.

Acknowledgements

This work has been supported by Spanish Government research grant TIC2002-00160.

References

[1] K. Compton, S. Hauck, "Reconfigurable computing: a survey of systems and software", ACM Computing Surveys, Vol. 34, No. 2, pp 171-210. June 2002.

[2] O. F. Diessel, G. Wigley, "Opportunities for Operating Systems Research in Reconfigurable Computing", Technical Report ACRC-99-018. Advanced Computing Research Centre, School of Computer and Information Science, University of South Australia, 1999.

[3] O. F. Diessel, H. Elgindy, "On dynamic task scheduling for FPGA-based systems", International Journal of Foundations of Computer Science, IJFCS'01, Vol. 12, No. 5, 2001.

[4] K. Bazargan, R. Kastner, M. Sarrafzadeh, "Fast template placement for reconfigurable computing systems", IEEE Design and Test of Computers, Vol. 17, pp 68–83, 2000.

[5] H. Walder, C. Steiger, M. Platzner, "Fast online task placement on FPGAs: free space partitioning and 2D-Hashing", IPDPS-2003 (RAW'03), vol. 00, p. 178b, Munich, Germany, April 2003.

[6] A. Ahmadinia, C. Bobda, M. Bednara, J. Teich, "A new approach for on-line placement on reconfigurable devices", IPDPS-2004 (RAW'04), vol. 04, n°4, p. 134a, 2004.

[7] Handa and R. Vemuri, "An efficient algorithm for finding empty space for online FPGA placement," Design Automation Conference, p. 960-965, San Diego, CA, June 2004.

[8] J. Tabero J. Septién, H. Mecha, D. Mozos, "A vertex-list approach to 2D HW multitasking management in RTR FPGAs", DCIS 2003, Ciudad Real, Spain, pp. 545-550, Nov 2003.

[9]. J. Tabero J. Septién, H. Mecha, D. Mozos, "A low fragmentation heuristic for task placement in 2D RTR HW management", FPL 2004, Antwerp, Belgium, pp. 241-250, Sep 2004

4C-1

A Quasi-Newton Preconditioned Newton-Krylov Method for Robust and Efficient Time-Domain Simulation of Integrated Circuits with Strong Parasitic Couplings*

Zhao Li[†] and C.-J. Richard Shi

Department of Electrical Engineering, University of Washington, Seattle, WA 98195, USA
{lz2000, cjshi}@ee.washington.edu

Abstract: In this paper, the Newton-Krylov method is explored for robust and efficient time-domain VLSI circuit simulation. Different from the LU-factorization based direct method, the Newton-Krylov method uses a preconditioned Krylov-subspace iterative method for linear system solving. Our key contribution is to introduce an effective quasi-Newton preconditioning scheme for Krylov-subspace methods to reduce the number and cost of LU factorizations during time-domain circuit simulation. Experimental results on a collection of digital, analog and RF circuits have shown that the quasi-Newton preconditioned Krylov-subspace method is as robust and accurate as SPICE3. The proposed Newton-Krylov method is especially attractive for simulating circuits with a large amount of parasitic RLC elements for post-layout verification.

1. Introduction

In modern deep-submicron meter very-large-scale integrated (VLSI) circuits, parasitic effects are no longer ignorable with the higher operation frequency, lower supply voltage and smaller device feature size. It is desirable, especially for sensitive analog and mixed-signal circuit design, to perform full-chip time-domain simulation of massively parasitic-coupled systems [14] before tape-out using SPICE-like circuit simulators [12]. This is, however, an extremely time-consuming task, since SPICE-like circuit simulators use LU factorization based *direct* methods for solving systems of nonlinear differential equations. For massively parasitic coupled systems, the complexity of LU factorization is approaching its worst case $O(n^3)$, where n is the size of a circuit, instead of its average case $O(n^{1.1-1.5})$ [12].

Existing work to reduce the cost of LU factorization for circuit simulation can be classified into three categories. The first category is so-called relaxation-based methods [15], including relaxation methods for nonlinear iteration, such as Gauss-Jacobi, Gauss-Seidel, Successive Over-Relaxation (SOR), etc. [17]. In [2], the preconditioner for the Gauss-Seidel method has been carefully studied to improve its robustness, which leads to the partial Gauss-Seidel (PGS) method. For time-domain simulation, semi-implicit integration methods [20], including alternating-direction-implicit methods [17], have been applied to substrate analysis [14] and power/ground network analysis [10]. The drawback of relaxation-based methods is that their stability and convergence properties strongly depend on circuit structures.

The second category, which has been explored extensively since the inception of SPICE, belongs to a class of methods known as quasi-Newton methods [5]. The purpose of quasi-Newton methods is to reduce the number of LU factorizations for circuit simulation. For numerical integration, several fixed leading coefficient integration methods [4][7][11] have been proposed for variable time step-size integration to keep the circuit matrix constant – thus to have a less number of LU factorizations. For nonlinear iteration, virtually all SPICE-like circuit simulators explore quasi-Newton methods to reduce the number of LU factorizations [1][11], also

known as Jacobian matrix bypass. However, it is well known that the convergence rate of quasi-Newton methods for nonlinear iteration can degrade to linear.

The third category is to apply Krylov-subspace based iterative methods [17] to solve the system of linearized equations in the inner Newton-Raphson iteration, so called the Newton-Krylov method [9]. The Newton-Krylov method can have the same stability and convergence properties as the classical Newton-Raphson method provided that the linearized system can be solved accurately by Krylov-subspace methods. The key issue is how to construct an effective preconditioner for iterative methods. For special applications such as in the shooting-Newton method and harmonic balance analysis for RF circuit simulation [18], where a circuit matrix is generally block-diagonally dominant, a preconditioner can be well constructed. Similarly, Krylov-subspace methods are applied for model order reduction of interconnect lines [13], substrates [8], and power/ground networks and direct analysis of well-structured large-scale linear circuits, such as substrate [14] and power/ground networks [3].

In this paper, we explore Krylov-subspace methods for time-domain simulation of general nonlinear circuits. The basic idea of our approach is to reuse the LU factorized matrices from the previous time point/nonlinear iteration as the preconditioner for Krylov-subspace methods. With this, the number and cost of LU factorizations can be reduced dramatically. Additionally, the stability and convergence properties will not be compromised, thanks to Krylov-subspace methods. The Texas Instrument research team explored a similar idea using a conjugate gradient squared method with a partial LU preconditioner, called PLUCGS [2]. However, it was concluded in [2] that the PLUCGS method was even less efficient than the PGS method, a variant of the Gauss-Seidel method.

The main contribution of our work is a systematic method to construct effective preconditioners based on quasi-Newton principles to perform as few LU factorizations as possible during the whole time domain nonlinear circuit simulation. Our implementation is as robust as LU factorization based direct methods, and can be orders of magnitude faster than SPICE for time-domain simulation of VLSI circuits with a large amount of linear parasitic elements.

Specifically, the efficiency comes from the following four ideas:
1. The preconditioner is kept constant when time step-sizes vary within a predefined range.
2. We propose a generalized way to partition the entire operating region of nonlinear devices into piecewise weakly nonlinear (PWNL) regions. With this, the preconditioner is kept constant if all nonlinear devices reside in their present operating PWNL regions.
3. When nonlinear devices switch their operating PWNL regions during nonlinear iteration, the low-rank update technique [6] is applied to update the preconditioner efficiently rather than performing new LU factorization.
4. We further explore incomplete LU preconditioners derived from factorized full L and U matrices for the best efficiency.

This paper is organized as follows. In Section 2, we provide an overview of quasi-Newton methods and Krylov-subspace methods. Section 3 introduces Newton-Krylov based time-domain nonlinear circuit simulation with the quasi-Newton preconditioning scheme.

---

* This research was supported by the DARPA NeoCAD Program under Grant No. N66001-01-8920, an NSF CAREER Award under Grant No. 9985507, and the SRC/NSF Mixed-Signal Initiative under Contract No. CCR0120371.

[†] Zhao Li is now with Cadence Design Systems, Inc. (zhaoli@cadence.com)

0-7803-9451-8/06/$20.00 ©2006 IEEE.

Section 4 describes the proposed time-domain nonlinear circuit simulation flow. Experimental results on general nonlinear circuits and power/ground network examples are reported in Section 5. Section 6 concludes the paper.

2. Iterative Methods for Time-Domain Circuit Simulation

LU factorization based direct methods are efficient for small-scale to medium-scale circuit simulation. However, the cost of LU factorization is becoming the dominant per-iteration cost for large-scale circuit simulation incorporating parasitic effects [11][14]. To tackle this problem and to continue exploiting the robustness of LU factorization, a key idea is to reuse the previous LU factorization to solve circuit matrix equations $Ax=b$ for as many time points and/or nonlinear iteration steps as possible. This leads to two categories of iterative methods – quasi-Newton methods and Krylov-subspace methods.

2.1 Quasi-Newton methods

Suppose that we have a LU factorized matrix M, which is considered to be close enough to the circuit matrix A, circuit matrix equations $Ax=b$ can be solved by Eq. (1) derived from the first-order Taylor expansion with the matrix M as the approximate Jacobian matrix (representing the first-order derivatives).

$$x^{(k)} = x^{(k-1)} + M^{-1}(b - Ax^{(k-1)}) \qquad (1)$$

For nonlinear circuits, Eq. (1) is further written as follows,

$$x^{(k)} = x^{(k-1)} + M^{-1}(b^{(k-1)} - A^{(k-1)}x^{(k-1)}) = x^{(k-1)} + M^{-1}(-f^{(k-1)}) \quad (2)$$

where f is the vector contributed by input sources, nonlinear devices, and numerical integration of charge/flux storage devices. It should be noted that Eq. (2) will reduce to the Newton-Raphson method if $M=A$.

It should be noted that the search direction with quasi-Newton methods is $M^{-1}(b - Ax^{(k-1)})$ for each nonlinear iteration step.

2.2 Krylov-subspace methods

Given an initial guess $x^{(0)}$ to the circuit matrix equation $Ax=b$, Krylov-subspace methods seek an approximate solution $x^{(m)}$ from the subspace of $x^{(0)}+K_m(A, x^{(0)})$ by imposing the Petrov-Galerkin condition [17],

$$b - Ax^{(m)} \perp L_m(A, x^{(0)}) \qquad (3)$$

where $K_m(A, x^{(0)})=span\{r^{(0)}, Ar^{(0)}, A^2r^{(0)}, ..., A^{m-1}r^{(0)}\}$, $r^{(0)} = b - Ax^{(0)}$, and $L_m(A, x^{(0)})$ is a subspace of dimension m.

It is well known that a preconditioner [17] (or a preconditioning matrix) M is the key to the fast convergence of Krylov-subspace methods. The purpose of a preconditioner is to make the preconditioned matrix $M^{-1}A$ as close to the identity matrix as possible. With left-preconditioned Krylov-subspace methods, circuit matrix equations to be solved become $M^{-1}Ax = M^{-1}b$ and the Krylov subspace K_m is defined as follows,

$$K_m=span\{r^{(0)}, M^{-1}Ar^{(0)}, (M^{-1}A)^2r^{(0)}, ..., (M^{-1}A)^{m-1}r^{(0)}\} \qquad (4)$$

where $r^{(0)} = M^{-1}(b - Ax^{(0)})$. It is not surprising that the effect of the preconditioner M on preconditioned Krylov-subspace methods is similar to that of the approximate Jacobian matrix M on quasi-Newton methods.

The advantage of preconditioned Krylov-subspace methods over quasi-Newton methods is that, for each nonlinear iteration step, an orthogonal Krylov subspace K_m is used for constructing the search direction, rather than only one single search direction $M^{-1}(b - Ax^{(k-1)})$ as in quasi-Newton methods. As the result, the search direction of the Newton-Raphson method could be well approximated with the Newton-Krylov method.

We choose to use the flexible GMRES (FGMRES) method [16], an extension of the original right-preconditioned GMRES method [17], to solve the linearized circuit equations $(AM^{-1})(Mx)=b$.

3. Quasi-Newton Preconditioner Construction

In this section, we present a systematic method to construct effective preconditioners based on quasi-Newton methods. Section 3.1 presents adaptive time-step size control for preconditioner computation. Section 3.2 describes the generation of generalized PWNL definition of nonlinear devices.

3.1 Quasi-Newton preconditioners by adaptive time step-size control

Suppose that h is the base time-step size, and h_n is the current time-step size. To develop a guideline for adaptive time step-size control for preconditioner computation, let us write the system of linearized circuit equations as:

$$Gx + C\dot{x} = b \qquad (5)$$

where G and C represent the conductance and susceptance (capacitance) matrices, and b is the vector due to input sources and nonlinear devices. Replace time derivatives by the standard trapezoid formula, we have

$$\left(G + \frac{2C}{\alpha h}\right)x_n^{(k)} = \frac{2C}{\alpha h}x_{n-1} + C\dot{x}_{n-1} + b \qquad (6)$$

where $h_n = \alpha h$. To solve the above equation with preconditioned Krylov-subspace methods, the preconditioner we use is chosen to be $\left(G + \frac{2C}{h}\right)$, which should be as close to $\left(G + \frac{2C}{\alpha h}\right)$ as possible. Therefore, we introduce a parameter $0 < \eta < 1$ so that the preconditioner should satisfy the following inequality

$$\left\|\left(G + \frac{2C}{h}\right)^{-1}\left(G + \frac{2C}{\alpha h}\right) - I\right\| = \left\|\left(G + \frac{2C}{h}\right)^{-1}\left(1 - \frac{1}{\alpha}\right)\frac{2C}{h}\right\| < \eta < 1 \quad (7)$$

where $\|\bullet\|$ represents the spectral radius of the matrix. The above inequality can be re-written as

$$\left|\frac{1 - 1/\alpha}{1 - z}\right| < \eta < 1 \qquad (8)$$

where $z = -h/(2\tau)$ and τ is an eigenvalue of the matrix $G^{-1}C$. Let us refer to the region defined by the above inequality as the effective preconditioner region. To ensure the effective preconditioner region to include the left half of the complex-z plane for covering all poles of a decaying system, we always choose $\eta = |1 - 1/\alpha| < 1$ so that the effective preconditioner region is $|z - 1| > \frac{|1 - 1/\alpha|}{\eta} = 1$. Then we can draw the effective preconditioner region in the complex-z plane as in Fig. 1, in which the black region represents the effective preconditioner region.

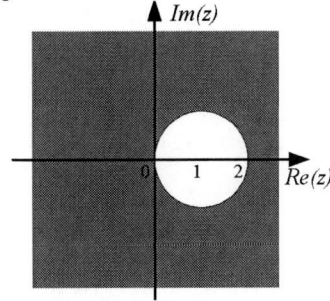

Figure 1. The effective preconditioner region.

A large range of α is helpful to reduce the number of LU factorizations when time steps vary. However, the preconditioner M^{-1} will diverge from A^{-1} when α is far away from 1. This will unfortunately increase the cost of Krylov-subspace methods. In our implementation, $0.625 < \alpha < 2.5$ is used for a tradeoff. We note that the effective preconditioner region is similar to the convergence region for the iterative integration formulae in SILCA [11].

3.2 Quasi-Newton preconditioners by piecewise weakly nonlinear definition of nonlinear devices

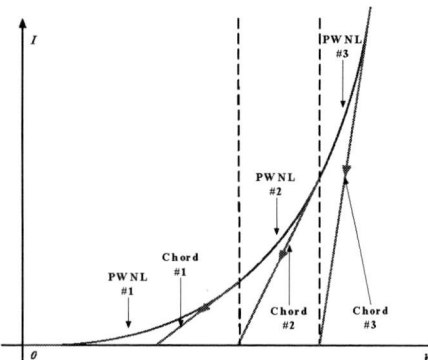

Figure 2. The PWNL definition of a nonlinear function.

In this sub-section, we present a systematic method for quasi-Newton preconditioner computation during nonlinear iteration. The central idea is to partition any nonlinear device function into a collection of regions, where in each region the nonlinear function is a weakly nonlinear function, i.e., its derivatives can be approximated by a constant (called a chord). As an example, Figure 2 shows an example of the PWNL definition of a nonlinear function, where three PWNL regions are defined with three different chords (fixed first-order derivatives).

Now consider how to generate PWNL regions automatically for arbitrary nonlinear functions. Suppose that nonlinear iteration is performed within a PWNL region of a nonlinear function $f(x)$ to solve $f(x)=0$, the nonlinear iteration equation can be expressed by,

$$x_{i+1} = x_i - \frac{f(x_i)}{g} \tag{9}$$

where g is the chord for this PWNL region. Let the exact solution be $x^* = x_i + \varepsilon_i = x_{i+1} + \varepsilon_{i+1}$. Subtracting x^* from the both sides of Eq. (9) gives

$$\varepsilon_{i+1} = \varepsilon_i + \frac{f(x_i)}{g} \tag{10}$$

After applying the Taylor expansion on $f(x)$ at x_i, we obtain the following error estimation,

$$\varepsilon_{i+1} \approx \varepsilon_i(1 - \frac{f'(x_i)}{g}) - \varepsilon_i^2 \frac{f''(x_i)}{2g} \tag{11}$$

From Eq. (11), if g is always equal to $f'(x_i)$, as in the Newton-Raphson method, the convergence rate is quadratic. The smaller the $|1 - f'(x_i)/g|$ is, the closer to the quadratic convergence rate Eq. (11) is. On the other hand, the larger the $|1 - f'(x_i)/g|$ is, the larger the range of a PWNL region could be. In practice, we define the following condition with a parameter $0 < \delta < 1$,

$$\left| 1 - \frac{f'(x_i)}{g} \right| < \delta \tag{12}$$

In our implementation, for simplicity, the chord is chosen to be the maximum first-order derivative in each PWNL region. Note that PWNL regions for a nonlinear function are equivalent to piecewise constant (PWC) regions for first-order derivatives of the same nonlinear function.

Now that the chord is chosen to be the maximum first-order derivative in each PWNL region, PWNL regions for MOSFETs can be generated automatically with the following rules:

1. The maximum voltages of V_{ds} and V_{gs} are predefined. In our experiments, we use V_{dd} as the maximum voltage for both of

them. Given model parameters, the maximum g_{ds} and g_m for all operating regions can then be calculated.

2. With a predefined $0 < \delta < 1$, PWC region values for g_{ds} and g_m can be calculated as follows,

$$g_n = g_{max}$$
$$g_{i-1} = (1-\delta)g_i, \quad i = n, n-1, ..., 2$$

3. A lower bound of g_{ds} and g_m is predefined, so that rule (2) will stop whenever g_{ds} and g_m are less than the predefined lower bound. This is necessary to avoid a PWC region for g_{ds} and g_m to be too narrow.

4. A voltage step-size is chosen so that at least one PWC region exists for each of the calculated PWC region values of g_{ds} and g_m in the V_{ds}–$(V_{gs}-V_{th})$ plane. A uniform voltage step-size has been used in our implementation for simplicity. Once the voltage step-size is finalized, g_{ds} and g_m at each grid point of the V_{ds}–$(V_{gs}-V_{th})$ plane can be evaluated, so that each patch of the V_{ds}–$(V_{gs}-V_{th})$ plane will be allocated to a PWC region of g_{ds} and g_m.

As an example, the PWC regions for g_{ds} of the MOSFET level 1 model are shown in Fig. 3, where δ is set to 1/3. It can be seen that there are a total of six PWC region values for g_{ds} (including the cutoff region #0).

The proposed method can be easily incorporated into model compilers [19] for automatic generation of piece-wise weakly nonlinear regions for any nonlinear device with any device model. It should be noted that rules (1)~(4) are applicable to multi-dimensional PWNL region generation. However, one thing to keep in mind is that the PWNL definition of nonlinear devices is only used for the preconditioner, rather than the circuit matrix equation $Ax=b$. Therefore, reasonable approximation can be made to ease the PWNL region generation.

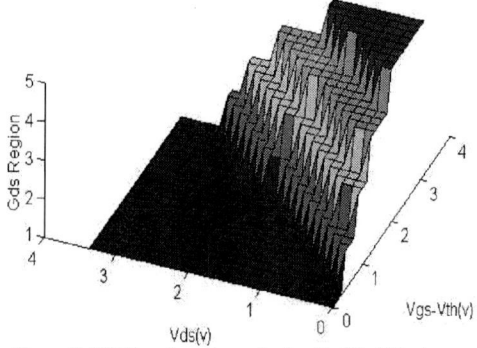

Figure 3. PWC regions for g_{ds} in the V_{ds}–$(V_{gs}-V_{th})$ plane.

The PWNL definition of nonlinear devices will introduce discontinuous first-order derivatives of a device model equation. It is well known that the continuity of a device model equation and its first-order derivatives is important for the successful convergence of the Newton-Raphson method. However, this requirement is not necessary for the Newton-Krylov method. The merit of the Newton-Krylov method in our framework is that the exact Newton-Raphson direction, which is determined by the exact first-order derivatives of a device model equation, can be well represented by the Krylov subspace constructed based on the approximate first-order derivatives (i.e., PWC first-order derivatives).

3.3 Low-Rank Updating

If only a few nonlinear devices change their operating PWNL regions during nonlinear iteration, the low-rank update technique [6] can be used to efficiently update the previously factorized L and U matrices rather than LU factorization. Therefore, in practice, the ratio of the number of nonlinear devices switching their operating PWNL regions vs. the total number of nonlinear and linear devices could be used as a guide for choosing either the low-rank update technique or LU factorization. To utilize the low-rank update technique, it is required that the new circuit matrix A_{new} be derived from the old circuit matrix A_{old} as follows,

$$A_{new} = A_{old} + cr^T \tag{13}$$

where A_{new} and A_{old} are both $n \times n$ matrices, c and r are both $n \times m$ matrices, and $m \ll n$.

When a MOSFET switches its operating PWNL region within either the normal operating mode or the reverse operating mode (i.e., the drain and source terminals are flipped), its (g_{ds}, g_m, g_{mbs}) stamp on a circuit matrix will change. The stamp change of this MOSFET on the circuit matrix can be represented as follows,

$$
\begin{array}{c}
 \\
D \\
S
\end{array}
\begin{array}{cccc}
D & G & S & B \\
\left[\begin{array}{cccc}
\Delta g_{ds} & \Delta g_m & -\Delta g_{ds} - \Delta g_m - \Delta g_{mbs} & \Delta g_{mbs} \\
-\Delta g_{ds} & -\Delta g_m & \Delta g_{ds} + \Delta g_m + \Delta g_{mbs} & -\Delta g_{mbs}
\end{array}\right]
\end{array}
$$

where D, G, S, and B represent the drain, gate, source and bulk terminal indexes in the circuit matrix, respectively. It can be further represented in the following rank-one update ($m=1$) format,

$$
\left[\begin{array}{c} \sqrt{a} \\ -\sqrt{a} \end{array}\right]
\left[\begin{array}{cccc}
\dfrac{\Delta g_{ds}}{\sqrt{a}} & \dfrac{\Delta g_m}{\sqrt{a}} & -\dfrac{\Delta g_{ds} + \Delta g_m + \Delta g_{mbs}}{\sqrt{a}} & \dfrac{\Delta g_{mbs}}{\sqrt{a}}
\end{array}\right],
$$

$$a = \max(|\Delta g_{ds} + \Delta g_m + \Delta g_{mbs}|, |\Delta g_{ds}|, |\Delta g_m|, |\Delta g_{mbs}|)$$

When a MOSFET switches its operating PWNL region from the normal mode to the reverse mode, the contribution of this MOSFET to the circuit matrix is changed similarly.

The low-rank update is efficient when the number of switching nonlinear devices is much less than the total number of nonlinear devices and linear elements. This is generally true in our framework, due to the two reasons. First, the PWNL definition of nonlinear devices are used for preconditioner construction, which requires a fairly coarse region partition than in the traditional piece-wise linear device modeling for device model evaluation. Second, our focused application is parasitic-coupled VLSI systems, where the number of linear parasitic elements is generally dominant. Therefore, we always use low-rank update during nonlinear iteration.

4. FGMRES-based Transient Simulation Flow
Algorithm I. Transient simulation flow.

DC operating point analysis
Choose an initial step size h_0, the basis step size $h = h_0$, $t = 0$
WHILE ($t < T_{final}$){
    OUTER LOOP: do{
      $\alpha = h_n/h$, iter_no = 0
      INNER LOOP: do{
        IF($0.625 < \alpha < 2.5$){
          IF(PWNL region is changed)
            Apply low-rank update on L/U matrices
        }ELSE{
          IF(iter_no==0)
            Apply LU factorization
          ELSE{
            IF(PWNL region is changed)
               Apply low-rank update on L/U matrices
          }
        }
        Apply the preconditioned FGMRES method
        iter_no = iter_no + 1
      } while (not converged)
      Choose a new h_n based on LTE requirement
    } while (LTE greater than predefined error limit)
    $t = t + h_n$
}

The flow of the proposed method for time-domian nonlinear circuit simulation is shown in Algorithm I. It can be seen that LU factorization is only performed when time step-sizes vary out of the predefined h_n/h range ($0.625 < h_n/h < 2.5$ has been set to make comparison with SILCA [11]). In other cases, L and U matrices are either kept unchanged or updated by the low-rank update technique when nonlinear devices change their operating PWNL regions. During the whole process, L and U matrices are used for preconditioning the FGMRES method.

Two types of preconditioners have been tested in our experiments:

1) A LU preconditioner composed of factorized full L and U matrices.

2) An incomplete LU preconditioner composed of matrices approximated from the factorized full L and U matrices – a matrix element $l(i,j)$ is removed if $|l(i,j)| < c \cdot \max(|l(*,j)|)$ in L or $|u(i,j)| < c \cdot \max(|u(i,*)|)$ in U. c is a coefficient for the incomplete LU factorization, 0.001 is mainly used in our experiments. Since the incomplete LU preconditioner we use is derived from the already factorized L and U matrices, it is more robust and effective than general incomplete LU preconditioners [17] at the cost of more memory resources. The incomplete LU preconditioner is helpful to reduce the cost of forward/backward substitution during the preconditioning process.

5. Experimental Results
5.1 General nonlinear circuits

To verify the robustness of the proposed method for the simulation of general nonlinear circuits, several digital, analog and RF circuits have been tested. The simulation results are summarized in Table I. During the test, the full LU preconditioner has been used for the FGMRES method. Our implementation is based on SPICE3. For simplicity, the MOSFET level 1 model is used in our test. Parameter δ is set to 1/3 to automatically generate PWNL regions for MOSFETs.

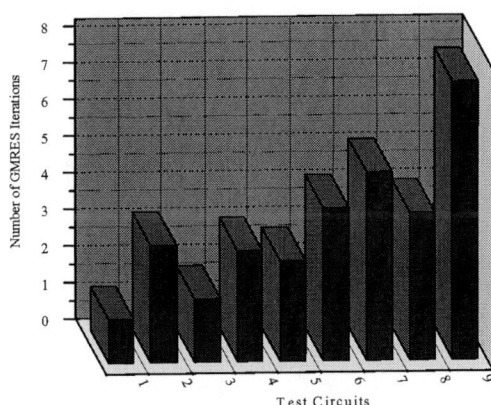

Figure 4. The average number of FGMRES iterations for general nonlinear circuits.

It can be seen from Table I that, with the preconditioned FGMRES method, the number of LU factorizations (#Tran LU) during transient simulation is reduced dramatically compared to that with SPICE3 (#Tran LU = #Tran iteration). Further, the number of total simulated time points (#Total points), the number of accepted time points (#Accepted points), and the number of transient iterations (#Tran iteration) with the preconditioned FGMRES method are kept almost the same as those with SPICE3, and are generally less than those with quasi-Newton base SILCA [11]. Figure 4 shows the average number of FGMRES iterations in each FGMRES solving process for test circuits (the dimension m of the Krylov subspace K_m). The average number of FGMRES iterations is below 5 for most of test circuits.

405

4C-1

5.2 Power/ground network examples

To test the efficiency of the proposed method for the simulation of VLSI circuits with a large amount of parasitic elements, a power/ground network example as that used in [11] is simulated. The power and ground supply networks are modeled as two RCL mesh layers. In our example, between these two layers is a 20-stage inverter chain representing nonlinear circuits, different inverters of which are connected to different power/ground nodes. Furthermore, RCL loads are added for each inverter to model interconnect lines between adjacent stages. The size of two RCL meshes is changed to vary the number of linear parasitic elements (#Elemts in Table II and III). Parameter δ is set to 1/3 to generate PWNL regions for MOSFETs.

Table II summarizes the simulation results for the power/ground network example using SPICE3 and the FGMRES method with the full LU preconditioner. The error tolerance ε is set to 1e-10 for the preconditioned FGMRES method. The speedup over SPICE3 is over 10X for the largest example we test. It can be expected that more speedup could be achieved for larger power/ground networks. The average number of FGMRES iterations in each FGMRES solving process (#FGMRES Iter / #Tran Iter) is about 5. It is worthy noting that the number of LU factorization (#Tran LU) is reduced greatly with the preconditioned FGMRES method compared to SPICE3 (#Tran LU = #Tran Iter). As shown in Table II, the number of transient iterations with the preconditioned FGMRES method has been kept almost the same as that with SPICE3, which shows that the SPICE-like convergence property has been preserved.

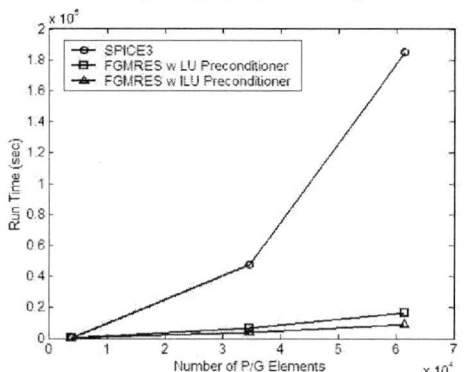

Figure 5. The run time vs. the number of elements in the power/ground network.

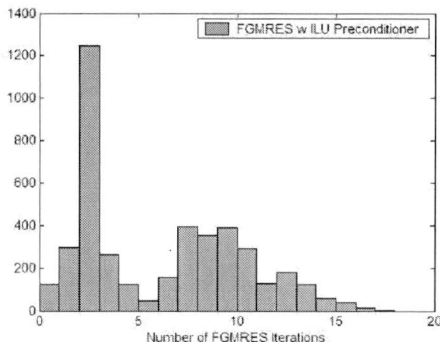

Figure 6. The histogram of the number of FGMRES iterations with the ILU preconditioner.

The simulation results of the FGMRES method with the incomplete LU (ILU) preconditioner are shown in Table III. It is seen that the FGMRES method with the ILU preconditioner achieves the best speedup over SPICE3 for the largest power/ground network – 20.68X, which is about 2X speedup over the FGMRES method with the LU preconditioner (The run time comparison is shown in Fig. 5.). The reason is that the number of matrix elements

in the ILU preconditioner is much less than those in the LU preconditioner. For the power/ground network example with 61602 elements, the histogram of the number of FGMRES iterations per FGMRES solving process is shown in Fig. 6. The average number of FGMRES iterations is about 6 to 7.

Finally, to test how the efficiency of the proposed method is affected by the percentages of linear parasitic elements in a circuit, we vary the amount of transistors in the logic circuit of the power-ground example. The results are shown in Table IV. Clearly the larger amount of linear parasitic elements, the more speed up.

From Table IV, we can see that for the P/G example (1600/4482), the cost of LU factorization has been reduced greatly, and the cost of FGMRES is comparable to that of LU factorization. However, the cost of low-rank update is becoming dominant. Shown in Fig. 7 is the histogram of the number of MOSFETs switching PWNL regions during one low-rank update. It can be seen that the number of MOSFETS switching their PWNL regions can be more than 200 during one low-rank update, and it is more expensive to compute LU factors by low-rank update than by full LU factorization. Therefore, in practice, whether to use low-rank update or full LU factorization should be determined based on the pre-set ratio of the number of switching nonlinear devices over that of total nonlinear devices. It is worthy noting that from Fig. 7, for most low-rank updates, only a few MOSFETS switch their PWNL regions.

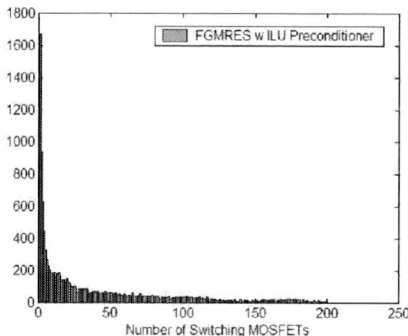

Figure 7. The histogram of the number of FGMRES iterations with the ILU preconditioner.

6. Conclusions and Future Research

In this paper, a quasi-Newton preconditioned Newton-Krylov method has been presented and implemented for efficient, accurate, and robust time-domain simulation of VLSI circuits with large amount of parasitic elements. Systematic methods of adaptive time-step size control and piece-wise weakly nonlinearity (PWNL) partitioning of any nonlinear device, combined with low-rank updating, have been proposed to minimize the number and cost of LU factorizations during the entire variable time step-size transient simulation. Orders of magnitude speedup has been achieved on power/ground network examples with the SPICE-like accuracy and robustness.

References

[1] E. Acar, F. Dartu, and L. T. Pileggi, "TETA: Transistor-level waveform evaluation for timing analysis", IEEE Trans. on Computer-Aided Design, vol. 21, no. 5, pp. 605-616, May 2002.

[2] R. Burch, P. Yang, P. Cox, and K. Mayaram, "A new matrix solution technique for general circuit simulation", IEEE Trans. on Computer-Aided Design, vol. 12, no. 2, pp. 225-241, Feb. 1993.

[3] T.-H. Chen and C. C.-P. Chen, "Efficient large-scale power grid analysis based on preconditioned Krylov-subspace iterative methods", Proc. IEEE/ACM Design Automation Conference, pp. 559-562, June 2001.

[4] P. F. Cox, R. G. Burch, P. Yang, and D. E. Hocevar, "New implicit integration method for efficient latency exploration in circuit

simulation", IEEE Trans. on Computer-Aided Design, vol. 8, no. 10, pp. 1051-1064, Oct. 1989.

[5] J. E. Dennis and J. J. Moré, "Quasi-Newton methods, motivation and theory", SIAM Review, vol. 19, no. 1, pp. 46-89, Jan. 1977.

[6] T. Fujisawa, E. S. Kuh, and T. Ohtsuki, "A sparse matrix method for analysis of piecewise-linear resistive networks", IEEE Trans. on Circuit Theory, vol. CT-19, no. 6, pp. 571-584, Nov. 1972.

[7] K. R. Jackson and R. Sacks-Davis, "An alternative implementation of variable step-size multistep formulas for stiff ODEs", ACM Trans. on Math. Soft., vol. 6, no. 3, pp.295-318, Sept. 1980.

[8] K. J. Kerns, I. L. Wemple, and A. T. Yang, "Stable and efficient reduction of substrate model networks using congruence transforms", Proc. IEEE/ACM Int. Conf. on Computer-Aided Design, pp. 207-214, Nov. 1995.

[9] D. A. Knoll and D. E. Keyes, "Jacobian-free Newton-Krylov methods: A survey of approaches and applications", Journal of Computational Physics, vol. 193, pp. 357-297, 2004.

[10] Y.-M. Lee and C. C.-P. Chen, "Power grid transient simulation in linear time based on transmission-line-modeling alternating-direction-implicit method", Proc. IEEE/ACM Int. Conf. on Computer-Aided Design, pp. 75-80, Nov. 2001.

[11] Z. Li and C.-J. R. Shi, "SILCA: Fast-yet-accurate time-domain simulation of VLSI circuits with strong parasitic coupling effects", Proc. IEEE/ACM Int. Conf. on Computer-Aided Design, pp. 793-799, Nov. 2003.

[12] L. W. Nagel, SPICE: A Computer Program to Simulate Semiconductor Circuits, University of California, Berkeley, Tech. Rep., UCB/ERL M520, May 1975.

[13] A. Odabasioglu, M. Celik, and L. T. Pillegi, "PRIMA: Passive reduced-order interconnect macromodeling algorithm", IEEE Trans. on Computer-Aided Design, vol. 17, no. 8, pp. 645-654, Aug. 1998.

[14] J. R. Phillips and L. M. Silveira, "Simulation approaches for strongly coupled interconnect systems", Proc. IEEE/ACM Int. Conf. on Computer-Aided Design, pp. 430-437, Nov. 2001.

[15] A. R. Newton and A. Sangiovanni-Vincentelli, "Relaxation-based electrical simulation", IEEE Trans. Computer-Aided Design, vol. 3, no. 4., pp. 308-331, 1984.

[16] Y. Saad, "A flexible inner-outer preconditioned GMRES algorithm", SIAM J. Scientific Computing, vol. 14, no. 2, pp. 461-469, Mar. 1993.

[17] Y. Saad, Iterative Methods for Sparse Linear Systems, 2nd Edition, SIAM, 2003.

[18] R. Telichevesky, K. Kundert, and J. White, "Fast simulation algorithms for RF circuits", Proc. IEEE Custom Integrated Circuit Conf, pp. 437-444, May 1996.

[19] B. Wan, B. Hu, L. Zhou and C.-J. R. Shi, ``MCAST: An abstract-syntax-tree based model compiler for circuit simulation'', Proc. IEEE Custom Integrated Circuits Conf., pp. 249-252, San Jose, CA, Sept. 2003.

[20] J. K. White and A. Sangiovanni-Vincentelli, Relaxation Techniques for the Simulation of VLSI Circuits, Kluwer Academic Publishers, 1987.

Table I. Simulation results on test circuits.

	Test Circuits	SPICE3			FGMRES w LU Preconditioner				
		#Total points	#Accepted points	#Tran iteration	#Total points	#Accepted points	#Tran iteration	#Tran LU	#Low-rank update
1	Inverter	142	127	340	141	127	338	63	64
2	20-stage inverter chain	356	260	1159	369	266	1185	69	643
3	Nand2	132	123	305	132	123	305	64	51
4	One-shot trigger	431	371	1360	431	371	1348	169	639
5	Comparator	140	126	410	140	126	403	47	163
6	Opamp follower	220	148	814	221	149	819	18	44
7	Ring oscillator	243	173	1020	256	176	1060	21	746
8	VCO	1281	887	4529	1280	887	4524	30	2251
9	Power amplifier	873	587	3451	840	581	3331	43	1761

Table II. Simulation results for the power/ground network example (ε=1e-10, LU preconditioner).

	SPICE3			Preconditioned FGMRES						
#Elems	#Tran Iter	Tran LU (sec)	Tot Tran (sec)	#Tran Iter	#Tran LU	#FGMRES Iter	Tran LU (sec)	FGMRES (sec)	Tot Tran (sec)	Speedup
4002	4023	371.20	403.99	4106	54	19982	4.89	92.70	110.00	3.67
34802	4006	4.549e4	4.760e4	4087	55	18879	730.97	5919.82	6944.40	6.85
61602	4377	1.797e5	1.848e5	4253	53	20341	2279.88	13647.74	16556.52	11.16

Table III. Simulation results for the power/ground network example (ε=1e-10, ILU preconditioner).

#Elems	#Tran Iter	#Tran LU	#FGMRES Iter	Tran LU (sec)	FGMRES (sec)	Tot Tran (sec)	Speedup
4002	4241	52	24208	4.83	67.20	84.01	4.81
34802	4199	53	28311	688.40	3081.03	4066.19	11.71
61602	4254	56	28901	2323.74	5995.01	8938.69	20.68

Table IV. Simulation results for the power/ground network example (ε=1e-10, ILU preconditioner).

#MOSFETs vs. #RCL	SPICE3 (sec)				FGMRES w ILU Preconditioner (sec)					
	LU	FBS	Load	Total	LU	Low-rank	FGMRES	Load	Total	Speed-up
400/4122	2742.41	90.24	158.62	2991.27	59.74	184.09	234.84	55.21	533.88	5.60
800/4242	5269.12	297.40	544.76	6111.28	158.69	679.79	803.22	224.80	1866.50	3.27
1600/4482	35998.68	750.74	1235.39	37984.81	1723.75	10254.65	1533.65	472.36	13984.41	2.72

An Efficient and Globally Convergent Homotopy Method for Finding DC Operating Points of Nonlinear Circuits

Kiyotaka Yamamura and Wataru Kuroki
Faculty of Science and Engineering, Chuo University
Tokyo, 112-8551 Japan
Email: yamamura@elect.chuo-u.ac.jp

Abstract— Finding DC operating points of nonlinear circuits is an important problem in circuit simulation. The Newton-Raphson method employed in SPICE-like simulators often fails to converge to a solution. To overcome this convergence problem, homotopy methods have been studied from various viewpoints. There are several types of homotopy methods, one of which succeeded in solving bipolar analog circuits with more than 20000 elements with the theoretical guarantee of global convergence. In this paper, an improved version of the homotopy method is proposed that can find DC operating points of practical nonlinear circuits smoothly and efficiently. It is also shown that the proposed method can be easily implemented on SPICE without programming.

I. INTRODUCTION

Finding DC operating points of nonlinear circuits is an important problem in circuit simulation. SPICE-like circuit simulators, which are widely used in LSI design, employ the Newton-Raphson (NR) method for solving modified nodal (MN) equations. However, the NR method or its variants often fails to converge to a solution unless the initial point is sufficiently close to the solution. Therefore, many circuit designers experience difficulties in finding DC operating points, especially for bipolar analog integrated circuits.

To overcome this convergence problem, globally convergent homotopy methods have been studied by many researchers from various viewpoints. By these studies, the application of the homotopy methods in practical circuit simulation has been remarkably developed, and the homotopy method termed the Newton homotopy (NH) method succeeded in solving bipolar analog circuits with more than 20000 elements (that belong to a class of the largest-scale circuits available with the current bipolar analog LSI technology) with the theoretical guarantee of global convergence [1]–[5]. However, since the NH method is globally convergent only when we choose an initial point on which the uniform passivity holds, we cannot choose a *good* initial point (e.g., point in the forward active operation region of transistors).

The Newton-fixed-point homotopy (NFPH) method is an improved version of the NH method [6],[7]. In this method, we can trace a solution curve from a good initial point, which often makes the solution curve short and the algorithm efficient. However, the auxiliary equation of this method contains a linear function that has no relation to the original nonlinear function, which sometimes causes complicated movement of solution curves, especially in the neighbourhood of $\lambda = 1$.

As another efficient approach of the homotopy method, the variable gain homotopy (VGH) method is well-known [8], which is an extension of the homotopy method termed the fixed-point homotopy (FPH) method [9]. Since this method includes the excellent idea of variable gain, solution curves often become smooth. However, in this method, we sometimes have to trace a solution curve from an initial point far from the solution; namely, the initial state is sometimes far from the normal operation of transistor circuits. In order to solve large-scale circuits more efficiently, it is necessary to develop a more efficient homotopy method.

In this paper, an efficient homotopy method termed the *variable gain Newton homotopy (VGNH) method* is proposed that is based on the idea of the NFPH method and that of the VGH method. The proposed method has the following advantages: i) The auxiliary equation is closely related to the original nonlinear equation. ii) Since this method is globally convergent for any initial point, we can choose a good initial point. iii) The idea of variable gain is introduced. Therefore, we can trace solution curves smoothly and efficiently. By numerical examples, it is shown that the proposed method finds DC operating points of practical transistor circuits more efficiently than the conventional methods. It is also shown that the proposed method can be easily implemented on SPICE without programming.

II. HOMOTOPY METHODS FOR MN EQUATIONS

We first review the homotopy methods for solving systems of nonlinear equations of the form:

$$f(x) = 0, \quad f : R^n \to R^n. \quad (1)$$

In the MN equation, (1) is written as follows [2],[4]:

$$\boldsymbol{f}_g(\boldsymbol{v}, \boldsymbol{i}) \triangleq \boldsymbol{D}_g g(\boldsymbol{D}_g^T \boldsymbol{v}) + \boldsymbol{D}_E \boldsymbol{i} + \boldsymbol{J} = 0$$
$$\boldsymbol{f}_E(\boldsymbol{v}, \boldsymbol{i}) \triangleq \boldsymbol{D}_E^T \boldsymbol{v} - \boldsymbol{E} = 0, \qquad (2)$$

where $\boldsymbol{v} \in \boldsymbol{R}^N$ is the variable vector denoting the node voltages to the datum node, $\boldsymbol{i} \in \boldsymbol{R}^M$ is the variable vector denoting the branch currents of the independent voltage sources, $\boldsymbol{g} : \boldsymbol{R}^K \to \boldsymbol{R}^K$ is a VCCS type continuous function, $\boldsymbol{D}_g$ and $\boldsymbol{D}_E$ are $N \times K$ and $N \times M$ (resp.) reduced incidence matrices, $\boldsymbol{J} \in \boldsymbol{R}^N$ and $\boldsymbol{E} \in \boldsymbol{R}^M$ are source vectors, $\boldsymbol{f} = (\boldsymbol{f}_g, \boldsymbol{f}_E)^T : \boldsymbol{R}^n \to \boldsymbol{R}^n$, $\boldsymbol{x} = (\boldsymbol{v}, \boldsymbol{i})^T \in \boldsymbol{R}^n$, and $n = N + M$.

In transistor circuits, the branch $\boldsymbol{g}$ is composed of transistors, diodes, resistors, etc. The relationship between the branch voltage vector $\boldsymbol{v}_q = (v_{\text{be}}, v_{\text{bc}})^T$ and the branch current vector $\boldsymbol{i}_q = (i_e, i_c)^T$ of a bipolar junction transistor (BJT), for instance, is described by the Ebers-Moll model as follows:

$$\boldsymbol{i}_q(\boldsymbol{v}_q) = \boldsymbol{T}\boldsymbol{q}(\boldsymbol{v}_q), \qquad (3)$$

where

$$\boldsymbol{T} = \begin{bmatrix} 1 & -\alpha_r \\ -\alpha_f & 1 \end{bmatrix} \qquad (4)$$

and

$$\boldsymbol{q}(\boldsymbol{v}_q) = \begin{bmatrix} m_e(\exp(n_e v_{\text{be}}) - 1) \\ m_c(\exp(n_c v_{\text{bc}}) - 1) \end{bmatrix}. \qquad (5)$$

Also, i_e (i_c) denotes the emitter (collector) current and v_{be} (v_{bc}) denotes the base to emitter (collector) voltage, respectively. The model parameters $\alpha_f, \alpha_r, m_e, m_c, n_e,$ and n_c are required to satisfy the passivity, no-gain, and reciprocity conditions [10].

In the homotopy methods [9], we consider an auxiliary equation $\boldsymbol{f}^0(\boldsymbol{x}) = 0$ with a known solution $\boldsymbol{x}^0$ (or a solution easily obtained) and define a homotopy function:

$$\boldsymbol{h}(\boldsymbol{x}, \lambda) = \lambda \boldsymbol{f}(\boldsymbol{x}) + (1 - \lambda)\boldsymbol{f}^0(\boldsymbol{x}), \qquad (6)$$

where $\lambda \in [0, 1]$ is the homotopy parameter. Then, the solution curve (often called the path) of the homotopy equation:

$$\boldsymbol{h}(\boldsymbol{x}, \lambda) = 0 \qquad (7)$$

is traced from the initial point $(\boldsymbol{x}^0, 0)$ at $\lambda = 0$. If the solution curve reaches the $\lambda = 1$ hyperplane at $(\boldsymbol{x}^*, 1)$, then a solution $\boldsymbol{x}^*$ of (1) is obtained.

There are several types of homotopy methods for solving MN equations. The NFPH method [6],[7] uses the homotopy function:

$$\boldsymbol{h}(\boldsymbol{x}, \lambda) = \boldsymbol{f}(\boldsymbol{x}) - (1 - \lambda)\boldsymbol{f}(\boldsymbol{x}^0) + (1 - \lambda)\boldsymbol{A}(\boldsymbol{x} - \tilde{\boldsymbol{x}}^0), \quad (8)$$

where $\boldsymbol{A}$ is an $n \times n$ matrix represented as follows:

$$\boldsymbol{A} = \begin{bmatrix} \boldsymbol{D}_g \boldsymbol{G}_{FP} \boldsymbol{D}_g^T & 0 \\ 0 & -R_{FP} \boldsymbol{1}_M \end{bmatrix}. \qquad (9)$$

In (9), $\boldsymbol{G}_{FP}$ is a positive semi-definite diagonal matrix whose diagonal elements are positive and others are zero. Also, R_{FP} is a scalar positive value and $\boldsymbol{1}_M$ denotes an $M \times M$ identity matrix. Note that the auxiliary function $\boldsymbol{f}^0(\boldsymbol{x})$ at $\lambda = 0$ contains a linear function that has no relation to the original nonlinear function $\boldsymbol{f}(\boldsymbol{x})$.

The VGH method [8] uses the homotopy function:

$$\boldsymbol{h}(\boldsymbol{x}, \lambda) = \boldsymbol{f}(\boldsymbol{x}, \lambda\boldsymbol{\alpha}) + (1 - \lambda)\boldsymbol{G}(\boldsymbol{x} - \boldsymbol{a}), \qquad (10)$$

where $\boldsymbol{\alpha}$ is a vector consisting of forward current gains α_f and reverse ones α_r of transistors, $\boldsymbol{a}$ is a random vector, and $\boldsymbol{G}$ is an $N \times N$ diagonal matrix. The VGH method is a two-stage procedure. In Phase 1, the initial point $\boldsymbol{x}^0$ that satisfies $\boldsymbol{h}(\boldsymbol{x}, 0) = 0$ is computed by the modified NR method. In Phase 2, the solution curve of $\boldsymbol{h}(\boldsymbol{x}, \lambda) = 0$ is traced from $(\boldsymbol{x}^0, 0)$. In Phase 1, the circuit described by $\boldsymbol{h}(\boldsymbol{x}, 0) = 0$ contains diodes as only nonlinear elements, hence it has a unique solution.

Under some regularity assumptions of $\boldsymbol{h}$, the solution curve of $\boldsymbol{h}(\boldsymbol{x}, \lambda) = 0$ is guaranteed to reach the $\lambda = 1$ hyperplane if the uniqueness condition at $\lambda = 0$ and the boundary free condition hold [4],[9]. Several homotopy methods including the NFPH method are proven to be globally convergent for MN equations [1],[4],[6],[7]; namely, the solution curve of $\boldsymbol{h}(\boldsymbol{x}, \lambda) = 0$ is guaranteed to reach the $\lambda = 1$ hyperplane.

III. Proposed Method

In this section, we propose a new homotopy method that is not only globally convergent but also very efficient. We first consider the following homotopy function:

$$\boldsymbol{h}(\boldsymbol{x}, \lambda) = \boldsymbol{f}(\boldsymbol{x}, \lambda\boldsymbol{\alpha}) - (1 - \lambda)\boldsymbol{f}(\boldsymbol{x}^0, 0 \cdot \boldsymbol{\alpha}), \qquad (11)$$

where $0 \cdot \boldsymbol{\alpha}$ implies the product of 0 and $\boldsymbol{\alpha}$. Note that this homotopy function includes the concept of variable gain. If we consider a circuit described by $\boldsymbol{h}(\boldsymbol{x}, \lambda) = 0$, then each transistor of the circuit can be described by (3) with $\boldsymbol{T}$ replaced by

$$\boldsymbol{T}_\lambda = \begin{bmatrix} 1 & -\lambda\alpha_r \\ -\lambda\alpha_f & 1 \end{bmatrix}. \qquad (12)$$

If we put

$$\tilde{\boldsymbol{T}} = \begin{bmatrix} 0 & \alpha_r \\ \alpha_f & 0 \end{bmatrix}, \qquad (13)$$

then

$$\boldsymbol{T}_\lambda = \boldsymbol{T} + (1 - \lambda)\tilde{\boldsymbol{T}} \qquad (14)$$

holds.

Next, consider the following function:

$$\tilde{\boldsymbol{f}}(\boldsymbol{x}) \triangleq \begin{bmatrix} \boldsymbol{D}_g \tilde{g}(\boldsymbol{D}_g^T \boldsymbol{v}) \\ 0 \end{bmatrix}, \qquad (15)$$

where the components $\tilde{g}_i$ ($i = 1, 2, \cdots, K$) of $\tilde{\boldsymbol{g}} = (\tilde{g}_1, \tilde{g}_2, \cdots, \tilde{g}_K)^T$ are defined as follows:

1. If g_i and g_{i+1} are a pair of transistor branches, that is,

$$\begin{bmatrix} g_i \\ g_{i+1} \end{bmatrix} = \boldsymbol{T}\boldsymbol{q}(\boldsymbol{v}_q), \qquad (16)$$

then the corresponding function $\tilde{\boldsymbol{g}}_q = (\tilde{g}_i, \tilde{g}_{i+1})^T$ is

$$\begin{bmatrix} \tilde{g}_i \\ \tilde{g}_{i+1} \end{bmatrix} = \tilde{\boldsymbol{T}}\boldsymbol{q}(\boldsymbol{v}_q). \qquad (17)$$

2. If g_i is not a transistor branch, then

$$\tilde{g}_i = 0. \qquad (18)$$

Then, from (14)–(18), it is easily seen that $\boldsymbol{f}(\boldsymbol{x}, \lambda\boldsymbol{\alpha}) = \boldsymbol{f}(\boldsymbol{x}) + (1 - \lambda)\tilde{\boldsymbol{f}}(\boldsymbol{x})$ and $\boldsymbol{f}(\boldsymbol{x}^0, 0 \cdot \boldsymbol{\alpha}) = \boldsymbol{f}(\boldsymbol{x}^0) + \tilde{\boldsymbol{f}}(\boldsymbol{x}^0)$ hold. Hence, (11) can be rewritten as:

$$\boldsymbol{h}(\boldsymbol{x}, \lambda) = \boldsymbol{f}(\boldsymbol{x}) + (1 - \lambda)\tilde{\boldsymbol{f}}(\boldsymbol{x}) - (1 - \lambda)(\boldsymbol{f}(\boldsymbol{x}^0) + \tilde{\boldsymbol{f}}(\boldsymbol{x}^0))$$
$$(19)$$

and we have

$$\boldsymbol{h}(\boldsymbol{x}, \lambda) = \boldsymbol{f}(\boldsymbol{x}) - (1-\lambda)\boldsymbol{f}(\boldsymbol{x}^0) + (1-\lambda)(\tilde{\boldsymbol{f}}(\boldsymbol{x}) - \tilde{\boldsymbol{f}}(\boldsymbol{x}^0)). \quad (20)$$

Note that (20) is equivalent to (11). In this paper, we propose a homotopy method using the homotopy function (11) or (20). From the form of (11), the proposed method may be called the variable gain Newton homotopy (VGNH) method.

The basic VGH method requires some modifications to the model subroutines such as (12), but as seen from (20), the proposed method requires only additional subroutines of $\tilde{\boldsymbol{T}}$. Thus, the proposed method can be implemented on the SPICE-like simulators with no modification to the existing model subroutines.

For the global convergence property of the proposed method, the following theorem holds.

Theorem 1 Consider the homotopy function defined by (11) or (20). Assume that $\boldsymbol{g}$ is uniformly passive [4] on certain points. Then, for any initial point $\boldsymbol{x}^0 \in \boldsymbol{R}^n$, the solution curve of $\boldsymbol{h}(\boldsymbol{x}, \lambda) = 0$ starting from $(\boldsymbol{x}^0, 0)$ reaches $\lambda = 1$. □

Proof : To prove the theorem, it is sufficient to show that i) $\boldsymbol{x}^0$ is the unique solution of $\boldsymbol{h}(\boldsymbol{x}, 0) = 0$, and ii) $\boldsymbol{h}$ is boundary free [4],[9]. i) Consider the circuit described by $\boldsymbol{h}(\boldsymbol{x}, 0) = 0$. Since $\boldsymbol{T}_\lambda = \begin{bmatrix} 1 & 0 \\ 0 & 1 \end{bmatrix}$ holds at $\lambda = 0$, this circuit has diodes as only nonlinear elements as well as in Phase 1 of the VGH method. Hence, this circuit has a unique solution, which implies that the solution of $\boldsymbol{h}(\boldsymbol{x}, 0) = 0$ is unique. ii) It is evident that the branch $\boldsymbol{g} + (1-\lambda)\tilde{\boldsymbol{g}}$ satisfies the uniform passivity on certain points for all $\lambda \in [0, 1]$. Hence, following the proofs discussed in [1] and [4], it is trivial to show that $\boldsymbol{h}$ is boundary free. Thus, the global convergence of the proposed method is guaranteed for any initial point $\boldsymbol{x}^0 \in \boldsymbol{R}^n$. □

Note that a fairly general class of resistive elements including BJTs, diodes, tunnel diodes, and positive linear resistors are known to be uniformly passive on certain points [4]. Thus, the uniform passivity is a very mild condition.

Next, we discuss the computational efficiency of the proposed method, considering the factors that degrade the efficiency in the conventional methods. In the VGH method, the initial point $\boldsymbol{x}^0$ in Phase 2 is obtained by solving a circuit that contains diodes as only nonlinear elements. However, since the structure of such a circuit often makes the operation of some transistors not be the normal (foward active) operation of transistor circuits, the initial point obtained in Phase 1 is sometimes far from the solution. In the NFPH method, a good initial point can be used as discussed in [7]. However, as stated before, the auxiliary function contains a linear function that has no relation to the original nonlinear function. Moreover, this method requires relatively large values of some elements of $\boldsymbol{G}_{FP}$ in (9) to guarantee the uniqueness condition at $\lambda = 0$ [7]. Such linear function with large $\boldsymbol{G}_{FP}$ sometimes causes complicated movement of solution curves, especially in the neighbourhood of $\lambda = 1$ [6],[7].

We show here that the proposed method is free from the difficulties of the VGH method and the NFPH method. First, since $\boldsymbol{h}(\boldsymbol{x}^0, 0) = 0$ holds for any $\boldsymbol{x}^0$, we can choose a good initial point as discussed in [7]. Secondly, the homotopy function (11) or (20) contains no linear function, and the auxiliary equation $\boldsymbol{h}(\boldsymbol{x}, 0) = 0$ is closely related to the original nonlinear equation $\boldsymbol{f}(\boldsymbol{x}) = 0$. Hence, the proposed method is free from the problem of the NFPH method mentioned above. Moreover, since the proposed method includes the concept of variable gain, it is expected that the solution curves become smooth and short.

We now propose an efficient variation of the proposed method. As has been discussed, $\boldsymbol{x}^0$ is the unique solution of $\boldsymbol{h}(\boldsymbol{x}, 0) = 0$ defined by (20), independent of the circuit parameters or topologies. However, it is well-known among circuit designers that many practical transistor circuits cannot have multiple solutions under the condition $\alpha_f \leq 0.5$ $(\beta_f = \alpha_f/(1 - \alpha_f) \leq 1)$ [10]. Considering this property, we can propose a more practical homotopy function:

$$\boldsymbol{h}(\boldsymbol{x}, \lambda) = \boldsymbol{f}(\boldsymbol{x}) - (1-\lambda)\boldsymbol{f}(\boldsymbol{x}^0) + (1-\lambda)\frac{(\tilde{\boldsymbol{f}}(\boldsymbol{x}) - \tilde{\boldsymbol{f}}(\boldsymbol{x}^0))}{2},$$
$$(21)$$

which is obtained by replacing $\boldsymbol{T}_\lambda$ in (14) with

$$\boldsymbol{T}_\lambda = \boldsymbol{T} + (1 - \lambda)\frac{\tilde{\boldsymbol{T}}}{2}$$
$$= \begin{bmatrix} 1 & -\alpha_r \\ -\alpha_f & 1 \end{bmatrix} + (1 - \lambda)\begin{bmatrix} 0 & \dfrac{\alpha_r}{2} \\ \dfrac{\alpha_f}{2} & 0 \end{bmatrix}. (22)$$

This is the second homotopy function proposed in this paper, where the current gains change from $\boldsymbol{\alpha}/2$ to $\boldsymbol{\alpha}$,

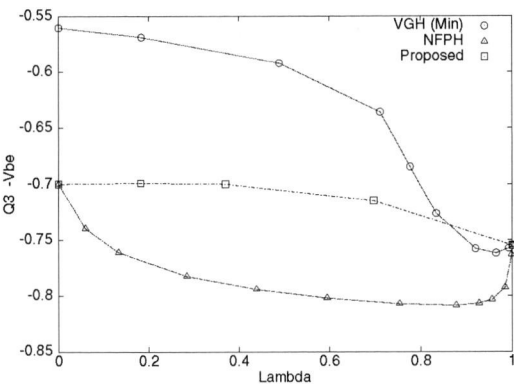

Fig. 1. Solution curves for HVRef.

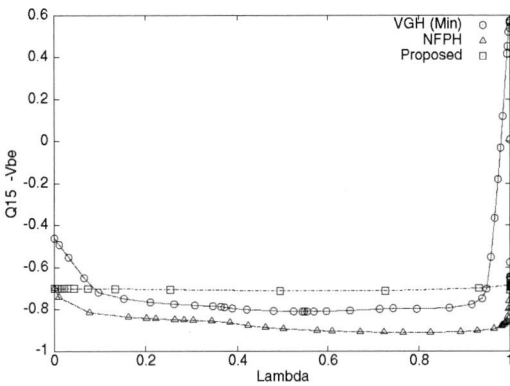

Fig. 2. Solution curves for μA741.

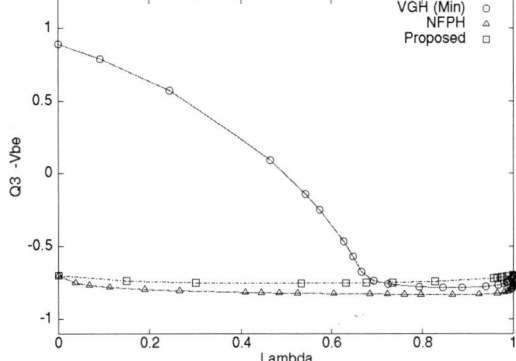

Fig. 3. Solution curves for RegCkt.

not from 0 to $\boldsymbol{\alpha}$. Hence, it is expected that $\boldsymbol{h}(\boldsymbol{x}, 0) = 0$ defined by (21) is closer to the original nonlinear equation than that defined by (20). Thus, from the practical viewpoint, we recommend the following algorithm.

1. We first apply the homotopy method using the homotopy function (21).

2. If the above method happens to fail, then we apply the homotopy method using the homotopy function (20), whose global convergence is theoretically guaranteed.

IV. Numerical Examples

We implemented the proposed method on a Sun Blade 2000 (UltraSPARC-III Cu 1.2GHz) and have confirmed the effectiveness of the proposed method using many practical transistor circuits. In all of the numerical experiments, the proposed method was the most efficient. In this section, we show the results applied to five types of practical transistor circuits widely used in analog LSIs; namely, the hybrid voltage reference circuit (HVRef) [6]–[8], a basic two-stage operational amplifier (2sOA) [7], a six-stage limiting amplifier (6sLA) [7], a high-gain operational amplifier μA741 that consists of 29 elements including 22 BJTs, and a regulator circuit (RegCkt) with an output voltage of 4.2 V that is used in bipolar LSIs and consists of 41 elements including 24 BJTs.

In the numerical experiments, we used the typical set of model parameters as those used in [7] for BJT models. We chose the initial points in the forward active operation region for all transistors [7]. It is natural to use $\boldsymbol{v}_q^0 = (-0.7, 0)^T$ as a typical forward active state for silicon npn transistors and $\boldsymbol{v}_q^0 = (0.7, 0)^T$ for pnp. We also used the spherical method [1],[2] for tracing solution curves.

Figs. 1–3 show the solution curves for HVRef, μA741, and RegCkt. In these figures, the emitter to base voltage $-v_{\mathrm{be}}$ of a certain BJT is plotted, where marks indicate the steps. In each step, a system of $n+1$ nonlinear equations

is solved by the NR method. From these figures, it is seen that the proposed method traces solution curves more smoothly and efficiently than the conventional methods (VGH and NFPH).

For the comparison of computational efficiency, we summarize in Table I the number of steps and the total number of Newton iterations of the above three methods. For the VGH method, the minimum (Min) and maximum (Max) numbers are shown in the hundred trials, and the left-hand numbers in the parentheses indicate the modified NR iterations in Phase 1. From the viewpoint of Newton iterations, the proposed method is several times more efficient than the VGH (Min) method and the NFPH method.

V. Extension to Circuits Containing Tunnel Diodes

In this section, we extend the proposed method to circuits containing tunnel diodes.

In general, the v-i characteristic of a tunnel diode, which will be written by $i = g(v)$, is represented by a polynomial:

$$g(v) = av^3 - bv^2 + cv, \qquad a, b, c > 0. \qquad (23)$$

411

TABLE I
COMPARISON OF COMPUTATIONAL EFFICIENCY.

Method	VGH (Min)		VGH (Max)		NFPH		Proposed	
Circiut	n	Number of steps	(Number of Newton iterations)					
HVRef	41	9	(11+25)	35	(13+99)	11	(44)	4	(16)
2sOA	42	4	(11+17)	21	(9+71)	3	(13)	3	(8)
6sLA	80	14	(10+46)	48	(10+152)	3	(12)	3	(9)
μA741	95	72	(16+177)	213	(17+494)	47	(173)	12	(33)
RegCkt	95	36	(16+99)	60	(16+161)	46	(128)	22	(67)

TABLE II
COMPARISON OF CONVERGENCE RATE (%).

n	NH	Proposed
10	29	100
20	15	100
50	1	100
100	0	100

TABLE III
COMPARISON OF COMPUTATINAL EFFICIENCY.

	NH method			Proposed method		
n	S	L	T (s)	S	L	T (s)
500	5 040	1 570	87	22	83	0.5
1000	16 040	4 480	1 591	52	153	8
1500	57 838	8 026	15 037	88	215	51
2000	83 941	12 352	55 410	120	282	190
2500	–	–	–	151	339	416
3000	–	–	–	145	389	658
3500	–	–	–	166	443	1 059
4000	–	–	–	176	495	2 977
4500	–	–	–	251	544	3 278
5000	–	–	–	236	600	3 934

This function is not monotone because of the existence of the second term $-bv^2$. Hence, we consider the following function:

$$g(v, \lambda) = av^3 - \lambda bv^2 + cv, \quad a, b, c > 0. \quad (24)$$

At $\lambda = 0$, $g(v, 0)$ is a monotone function and satisfies the uniform passivity on any point. Hence, the uniqueness condition at $\lambda = 0$ holds for any initial point. Moreover, the function defined by (24) changes continuously from a monotone function $g(v, 0)$ to the original function $g(v)$ as λ changes from 0 to 1, which often makes the solution curve smooth and short. This idea may be considered as an extension of the VGH method to tunnel diode circuits. Note that (24) can be realized by using $\tilde{g}(v) = -bv$ in (15) and (20).

Now we show some numerical examples. We consider the circuit containing n tunnel diodes discussed in [5].

We first compare the global convergence property of the NH method and the proposed method. Table II compares the convergence rate when we applied the two methods from randomly chosen one hundred initial points for $n = 10, 20, 50,$ and 100. As seen from the table, the NH method often fails to converge; more precisely, the solution curve of the NH method often returned back to $\lambda = 0$. This is because the global convergence of the NH method is guaranteed only when we choose an initial point on which the uniform passivity hold [1],[4]. However, the proposed method always converged to a solution from any initial point as guaranteed in Theorem 1.

We next compare the computational efficiency of the NH method[1] and the proposed method in Table III, where

---

[1] In the comparison, we chose the NH method because the VGH method cannot be applied to this circuit, and the NFPH method seems to be less efficient than the NH method for this circuit when we used $x^0 = 0$.

S denotes the total number of steps, L denotes the arc-length of solution curves, T (s) denotes the computation time, and "–" denotes that it could not be computed in one day. We used the initial point $x^0 = 0$, on which $g(v)$ is uniformly passive. Both methods found the same solution for all n. It is seen that the proposed method is much more efficient than the NH method, and could solve this circuit for $n = 5000$ in about one hour. It is also seen that the arc-length of solution curves is much smaller in the proposed method. Typical examples are shown in Figs. 4 and 5, where Fig. 4 shows the solution curve of the NH method for $n = 200$ and Fig. 5 shows the solution curve of the proposed method for $n = 5000$. In both figures, the vertical line denotes the arc-length of the solution curve. In Fig. 4, we can see a complicated solution curve with many sharp turning points. Such solution curves have often been observed when we applied the conventional homotopy methods to large-scale circuits [1]. However, In Fig. 5, we can see a smooth and short solution curves although n is large.

VI. IMPLEMENTATION OF THE VGNH METHOD ON SPICE WITHOUT PROGRAMMING

Thus, the proposed method is not only globally convergent for any initial point but also efficient because we can use good initial points and the solution curves tend to become smooth and short. In a sense, the proposed method has all the advantages of the NH, NFPH, and VGH methods, and is free from the difficulties of these

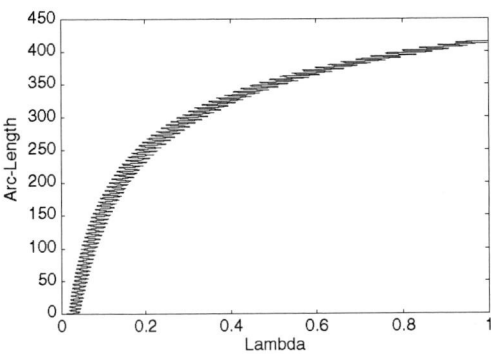

Fig. 4. Solution curve of the NH method ($n = 200$).

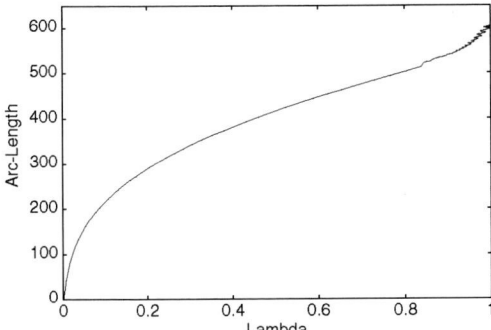

Fig. 5. Solution curve of the proposed method ($n = 5000$).

methods.

However, the programming of sophisticated homotopy methods is often difficult for non-experts or beginners. In this section, we propose an effective method for implementing the VGNH method on SPICE using the idea of *SPICE-oriented numerical methods* [1],[11]–[13].

In the VGNH method, we trace the solution curve of $\boldsymbol{h}(\boldsymbol{x}, \lambda) = 0$ where $\boldsymbol{h}$ is defined by (19). The solution curve can be traced by integrating a system of the algebraic-differential equations:

$$\boldsymbol{f}(\boldsymbol{x}) + (1 - \lambda)\tilde{\boldsymbol{f}}(\boldsymbol{x}) - (1 - \lambda)(\boldsymbol{f}(\boldsymbol{x}^0) + \tilde{\boldsymbol{f}}(\boldsymbol{x}^0)) = 0 \quad (25a)$$

$$\sum_{i=1}^{m} \left(\frac{dv_{\text{be}i}}{ds}\right)^2 + \left(\frac{d\lambda}{ds}\right)^2 = 1 \quad (25b)$$

starting from $(\boldsymbol{x}^0, 0)$ [1],[11]–[13], where m denotes the number of transistors contained in the circuit, s denotes the arc-length of the solution curve starting from $(\boldsymbol{x}^0, 0)$, and $v_{\text{be}i}$ denotes the base-emitter voltage of the ith transistor. Note that the points on the solution curve are considered as functions of s, namely, they can be written as $(\boldsymbol{x}(s), \lambda(s))$.

In the approach of the SPICE-oriented numerical methods, we consider a circuit described by (25). Then, we perform the transient analysis of SPICE to the circuit

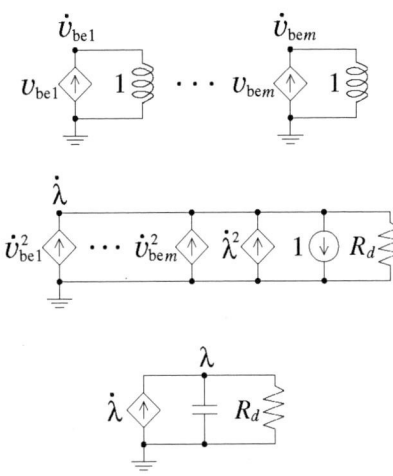

Fig. 6. Circuits that describe (25b).

starting from $(\boldsymbol{x}^0, 0)$, by which numerical integration is applied to (25) and the solution curve of (25a) is traced.

As discussed in [1] and [11]–[13], (25b) is described by the circuits shown in Fig. 6. In this figure, $\dot{v}_{\text{be}i}$ or $\dot{\lambda}$ denotes a node voltage that is independent of $v_{\text{be}i}$ or λ but is equal to $dv_{\text{be}i}/ds$ or $d\lambda/ds$ as a result, respectively. Such circuits are called *path following circuits*.

Next, we consider a circuit that is described by (25a). However, this is not an easy task because of the following reasons.

1. In the VGNH method, we first determine a good initial point $\boldsymbol{x}^0$. It is natural to choose $\boldsymbol{x}^0$ so that $\boldsymbol{v}_q = (v_{\text{be}}, v_{\text{bc}})^T$ becomes a point in the forward active operation region for all transistors. However, since $\boldsymbol{v}_q$ is a vector consisting of branch voltages but $\boldsymbol{x}$ is a vector consisting of node voltages and branch currents of the independent voltage sources, we have to calculate the initial point $\boldsymbol{x}^0$ such that $\boldsymbol{v}_q$ becomes a point in the forward active operation region.

2. We have to determinte the constant term $\boldsymbol{f}(\boldsymbol{x}^0) + \tilde{\boldsymbol{f}}(\boldsymbol{x}^0)$ in (25a), which cannot be obtained by substituting $\boldsymbol{x}^0$ to $\boldsymbol{f}(\boldsymbol{x})$ or $\tilde{\boldsymbol{f}}(\boldsymbol{x})$ because the formulas of $\boldsymbol{f}(\boldsymbol{x})$ do not appear explicitly in SPICE.

Therefore, the proposed implementation method consists of two phases.

A. Determination of the initial point $\boldsymbol{x}^0$ and the constant term $\boldsymbol{f}(\boldsymbol{x}^0) + \tilde{\boldsymbol{f}}(\boldsymbol{x}^0)$.

In Phase 1 of the proposed method, we first set $\boldsymbol{v}_q$ of all transistors in the forward active operation region [e.g., $\boldsymbol{v}_q = (0.7, 0)^T$]. Let such a point be $\boldsymbol{v}_q^0 = (V_{\text{be}}^0, V_{\text{bc}}^0)^T$. Then, we connect the independent voltage sources V_{be}^0 and V_{bc}^0 to each transistor as shown in Fig. 7.

413

4C-2   2006 Asia and South Pacific Design Automation Conference

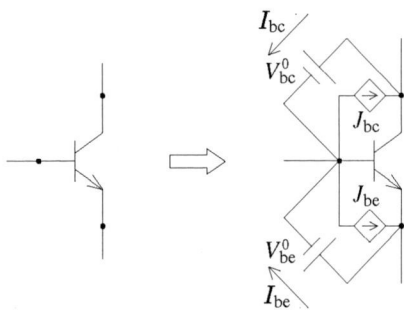

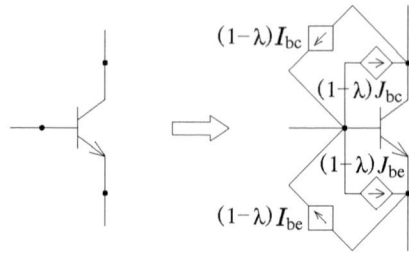

Fig. 8. The circuit that describes (25a).

Fig. 7. The initial circuit for determining the initial point $\boldsymbol{x}^0$ and the constant term $\boldsymbol{f}(\boldsymbol{x}^0) + \tilde{\boldsymbol{f}}(\boldsymbol{x}^0)$.

TABLE IV
COMPARISON OF COMPUTATION TIME.

Circuit	n	Program		SPICE	
		S	T (s)	S	T (s)
HVRef	41	21	0.117	21	0.060
2sOA	42	19	0.133	19	0.020
6sLA	80	19	0.500	19	0.120
μA741	95	28	1.517	28	0.320
RegCkt	95	48	2.600	48	0.400

We next connect the controlled current sources J_{be} and J_{bc} to each transistor as shown in Fig. 7, where currents of J_{be} and J_{bc} are described by $\tilde{g}_i$ and $\tilde{g}_{i+1}$ in (17), respectively. By the definition of $\tilde{\boldsymbol{f}}(\boldsymbol{x})$, it is easily seen that connecting these controlled sources is equivalent to adding $\tilde{\boldsymbol{f}}(\boldsymbol{x})$ to the left-hand side of the original MN equations $\boldsymbol{f}(\boldsymbol{x}) = 0$. Such a circuit where two independent voltage sources and two controlled current sources are connected to each transistor of the original circuit is called the *initial circuit*.

Then, we solve the initial circuit by the DC analysis of SPICE. Since the initial circuit is essentially a linear circuit, it can be solved by the DC analysis of SPICE. Let the solution of the initial circuit be $\boldsymbol{x}^0$. Since $\boldsymbol{v}_q = (V_{\text{be}}^0, V_{\text{bc}}^0)^T$ holds in $\boldsymbol{x}^0$, it can be used as a good initial point of the VGNH method. Moreover, since the original circuit is described by $\boldsymbol{f}(\boldsymbol{x})$ and the controlled sources J_{be} and J_{bc} are described by $\tilde{\boldsymbol{f}}(\boldsymbol{x})$, considering the Kirchhoff's current law, it is easily seen that the currents of the independent voltage sources V_{be}^0 and V_{bc}^0 (that are denoted by I_{be} and I_{bc} in Fig. 7) give the linear term $\boldsymbol{f}(\boldsymbol{x}^0) + \tilde{\boldsymbol{f}}(\boldsymbol{x}^0)$. Note that it is sufficient to consider the constant term only at the nodes where transistors are connected, because at the node n_j where transistors are not connected, $f_j(\boldsymbol{x}^0) = 0$ and $\tilde{f}_j(\boldsymbol{x}^0) = 0$ hold.

Thus, by solving the initial circuit, the initial point $\boldsymbol{x}^0$ and the constant term $\boldsymbol{f}(\boldsymbol{x}^0) + \tilde{\boldsymbol{f}}(\boldsymbol{x}^0)$ are obtained.

B. Solving circuits that describe (25).

Now it is clear that (25a) is described by a circuit as shown in Fig. 8, where four controlled current sources are connected to each transistor. Namely, by connecting $(1 - \lambda)J_{\text{be}}$ and $(1 - \lambda)J_{\text{bc}}$, $(1 - \lambda)\tilde{\boldsymbol{f}}(\boldsymbol{x})$ is described, and by connecting $(1 - \lambda)I_{\text{be}}$ and $(1 - \lambda)I_{\text{bc}}$, $-(1 - \lambda)(\boldsymbol{f}(\boldsymbol{x}^0) + \tilde{\boldsymbol{f}}(\boldsymbol{x}^0))$ is described. In Phase 2 of the proposed method, we perform the transient analysis of SPICE to this circuit together with the path following circuits shown in Fig. 6, and trace the solution curve of (25a).

C. Proposed method.

Thus, the proposed implementation method is summarized as follows.

1. We solve the initial circuit as shown in Fig. 7 by the DC analysis of SPICE, and obtain the initial point $\boldsymbol{x}^0$ and the constant term $\boldsymbol{f}(\boldsymbol{x}^0) + \tilde{\boldsymbol{f}}(\boldsymbol{x}^0)$ of the VGNH method. (Since the initial circuit is essentially a linear circuit, it can be solved by the DC analysis of SPICE.)

2. We perform the transient analysis of SPICE to the circuits shown in Figs. 6 and 8 starting from $(\boldsymbol{x}^0, 0)$ and trace the solution curve of (25a). If the solution curve reaches the $\lambda = 1$ hyperplane at $(\boldsymbol{x}^*, 1)$, then a solution $\boldsymbol{x}^*$ of (2) is obtained.

Since SPICE contains various efficient techniques such as sparse matrix techniques, variable-step variable-order implicit integration methods, and time-step control algorithms, a high-level VGNH method can be realized by the proposed method. Moreover, programming is not necessary and making the netlist of Figs. 6–8 is quite easy in the proposed method.

D. Examples.

We have applied the proposed implementation method to many practical circuits and have obtained good results. In this subsection, we show some examples. We used SPICE3f5 and the Sun Blade 2000.

Table IV shows the result of computation when we applied the VGNH method realized by i) our own program

414

2006 Asia and South Pacific Design Automation Conference

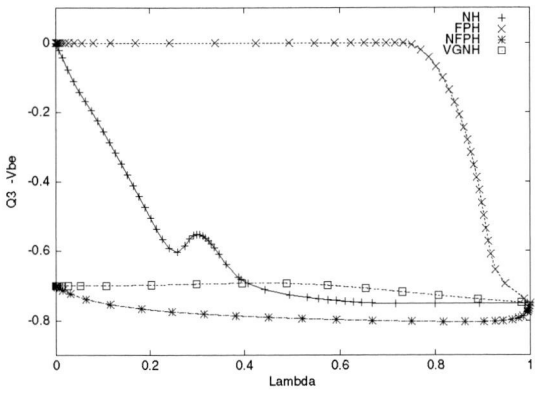

Fig. 9. Solution curves for HVRef (obtained by SPICE).

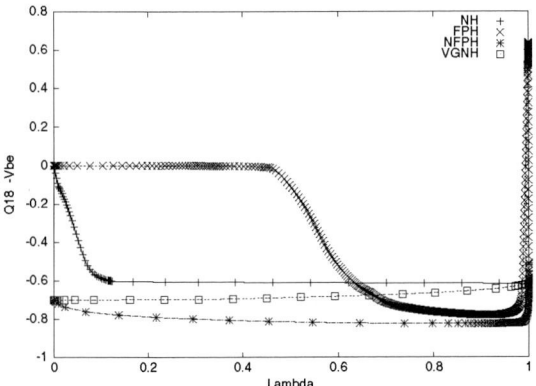

Fig. 10. Solution curves for μA741 (obtained by SPICE).

that was used in Section IV and ii) the proposed implementation method to the five transistor circuits discussed in Section IV. Since the number of steps changes by the parameters such as the initial step size and the maximum step size, we chose the parameters so that S becomes the same in the two approaches. From the table, it is seen that the proposed method is more efficient if the number of steps is the same. This is because SPICE contains various efficient techniques such as the sparse matrix techniques. Thus, we can implement an efficient VGNH method by using SPICE.

We also implemented the NH, FPH, and NFPH methods on SPICE using the similar idea. Figs. 9 and 10 show the solution curves when we applied these three methods and the VGNH method to HVRef and μA741, respectively. From these figures, it is seen that the number of marks of the VGNH method is the smallest, which implies that the solution curves are traced smoothly and efficiently in the proposed method.

VII. Conclusion

In this paper, an efficient and globally convergent homotopy method has been proposed for finding DC operating points of nonlinear circuits. Since the proposed method can use good initial points, and since it includes the concept of variable gain and does not include linear auxiliary functions, we can trace solution curves smoothly and efficiently. Furthermore, by using the method proposed in Section VI, we can implement a "sophisticated VGNH method with various efficient techniques" "easily" "without programming," "although we do not know the homotopy method well."

References

[1] Y. Inoue and K. Yamamura, "Practical algorithms for dc operating-point analysis of large-scale circuits," in *Proc. 1995 Int. Symp. Nonlinear Theory and its Applications*, Las Vegas, Nevada, pp. 1153–1158, Dec. 1995. (Invited)

[2] K. Yamamura, "Spherical methods for tracing solution curves," in *Proc. 1995 Int. Symp. Nonlinear Theory and its Applications*, Las Vegas, Nevada, pp. 1177–1182, Dec. 1995. (Invited)

[3] K. Yamamura, "Pathways from theory to practice," *J. IEICE*, vol. 81, no. 1, pp. 33–36, Jan. 1998. (Invited)

[4] K. Yamamura, T. Sekiguchi, and Y. Inoue, "A fixed-point homotopy method for solving modified nodal equations," *IEEE Trans. Circuits Syst.-I*, vol. 46, no. 6, pp. 654–665, June 1999.

[5] K. Yamamura, "Research topics and results on globally convergent algorithms for finding DC operating points of nonlinear circuits," in *Proc. 17th Workshop on Circuits and Systems in Karuizawa*, Japan, pp. 345–350, April 2004. (Invited)

[6] Y. Inoue, K Yamamura, T. Takahashi, and S. Kusanobu, "A globally convergent algorithm using the Newton-fixed-point homotopy for finding DC operating points of nonlinear circuits," *J. Japan Society for Simulation Technology*, vol. 22, no. 1, pp. 47–54, March 2003.

[7] Y. Inoue, S. Kusanobu, K. Yamamura, and M. Ando, "An initial solution algorithm for globally convergent homotopy methods," *IEICE Trans. Fundamentals*, vol. E87-A, no. 4, pp. 780–786, April 2004.

[8] R. C. Melville, L. Trajković, S. C. Fang, and L. T. Watson, "Artificial parameter homotopy methods for the DC operating point problem," *IEEE Trans. Computer-Aided Design*, vol. 12, no. 6, pp. 861–877, June 1993.

[9] C. B. Garcia and W. I. Zangwill, *Pathways to Solutions, Fixed Points, and Equilibria*. Englewood Cliffs, NJ: Prentice-Hall, 1975.

[10] L. Trajković and A. N. Wilson, Jr., "Theory of DC operating points of transistor networks," *Int. J. Electorn. Commun.*, vol. 46, no. 4, pp. 228–241, 1992.

[11] A. Ushida, Y. Yamagami, Y. Nishio, I. Kinouchi, and Y. Inoue, "An efficient algorithm for finding multiple DC solutions based on SPICE-oriented Newton homotopy method," *IEEE Trans. Computer-Aided Design*, vol. 21, no. 3, pp. 337–348, 2002.

[12] K. Yamamura, W. Kuroki, and Y. Inoue, "Path following circuits —SPICE-oriented numerical methods where formulas are described by circuits—," *IEICE Trans. Fundamentals*, vol. E88-A, no. 4, pp. 825–831, April 2005.

[13] K. Yamamura, "SPICE-oriented numerical methods for solving nonlinear problems where formulas are described by circuits," *J. IEICE*, vol. 88, no. 12, Dec. 2005. (Invited)

415

Optimization of Circuit Trajectories: an Auxiliary Network Approach

Baohua Wang
Dept. of EECS, University of Michigan, Ann Arbor
E-mail: baohuaw@eecs.umich.edu

Pinaki Mazumder
Dept. of EECS, University of Michigan, Ann Arbor
E-mail: mazum@eecs.umich.edu

Abstract—On optimizing circuit trajectories, i.e. continuous paths of circuit parameters, the paper presents an auxiliary network approach, which utilizes Pontryagin's Minimum Principle. Based on a set of circuit element correspondence rules, the introduced approach establishes an auxiliary network for a given circuit to be optimized, then circuit trajectories are optimized in a process of simulating the given circuit and the auxiliary network. The auxiliary network approach facilitates establishing analytic models in designing high-performance circuits that require fine tuning circuit trajectories. The paper details the theoretical framework of auxiliary network, and provides practical examples of its application in adiabatic circuit design.

I. INTRODUCTION

In designing high-performance circuits, multiple design objectives as diverse as circuit area, timing, power, noise immunity, yield, etc., need be accommodated, which need be aided by the automatic circuit optimization techniques. Traditional optimization concentrated on tuning sets of circuit parameters to maximize circuit performance. The typically considered parameter optimization problem can be described by

$$\underset{p}{\text{minimize}} \int_0^T \Phi(x(t), p, t)\, dt, \text{ s.t. } \dot{x}(t) = f(x(t), p, t). \quad (1)$$

Here Φ is a cost function, e.g. the square error between the desired transient output of a circuit and its actual output, x is a vector of circuit state variables, T specifies a time duration, and p is a vector of parameters to be tuned, e.g. transistor size, interconnect geometry, etc. The set of circuit state equations have been set as the optimization constraints.

Instead of optimizing circuit parameters, tuning paths of circuit parameters can also significantly improve circuit performance. For example, the voltage waveform that turns on the sense amplifier in a DRAM circuit can be optimized to reduce the sensing time [1]; an adiabatic circuit tunes the power clock waveform to reduce the energy dissipation in charging its loads [2], [3], [4]. This type of optimizations involve tuning different paths or trajectories of circuit parameters, to achieve high performance. Mathematically, a circuit trajectory optimization problem is solved:

$$\underset{u(t) \in U}{\text{minimize}} \phi(x(T), T) + \int_0^T \Phi(x(t), u(t), t)\, dt \quad (2)$$
$$\text{s.t. } \dot{x}(t) = f(x(t), u(t), t)$$

Compared to (1), an optimal vector of time-varying trajectories $u(t)$, for $t \in [0, T]$, instead of a vector of scalar parameters p,

are chosen from the set of admissible time-varying trajectories U, to minimize the integral cost Φ, plus a cost $\phi(x(T), T)$.

Solving a circuit parameter optimization problem like (1) often employs circuit sensitivity, which can be computed via adjoint network or sensitivity circuit [5, 6]. An example of parameter optimization of VLSI circuits is transistor resizing, which was reported could improve microprocessor performance by as much as 15% [7], [8]. Compared to parameter optimization, optimization of circuit trajectories can improve performance in innovative ways. However circuit trajectory optimization has seldom been investigated.

To solve a circuit trajectory optimization problem as (2), this paper introduces an auxiliary network approach, by utilizing Pontryagin's Minimum Principle [9]. In the approach, a set of circuit element correspondence rules are used to establish an auxiliary network for the given circuit to be optimized; then the optimal circuit trajectories are solved by simulating the given circuit and the auxiliary network, thus making it possible to employ the state of the art circuit simulation techniques. Auxiliary network also facilitates deriving analytic models in designing high-performance circuits that employ trajectory optimization strategies.

The rest of the paper is organized as follows. Section II introduces a circuit trajectory optimization example, and a solution method of directly employing Pontryagin's Minimum Principle. Section III presents the theoretical framework of auxiliary network. Section IV gives the applications of auxiliary network in adiabatic circuit design. Section V presents the experimental results.

II. CIRCUIT TRAJECTORY OPTIMIZATION: DIRECT METHOD

This section introduces a circuit trajectory optimization example: power clock optimization, and a solution method of directly using Pontryagin's Minimum Principle.

A. A Circuit Trajectory Optimization Example

Fig.1 shows two adiabatic circuits: an adiabatic buffer and a stepwise driver [10], [11], and gives a General Adiabatic Multi-driver Model (GAMM). A given adiabatic circuit can be described into the GAMM, by modeling its loads with the GAMM's Π blocks, and its nonlinear devices with those nonlinear dependent sources $G_1, \ldots, G_n$. In the shown GAMM, node voltages are denoted by capitalized letters, e.g. V_ks and $\bar{V}_k$s, while branch voltages are denoted by small letters, e.g. v_s, v_ks and v_{rk}s. The same convention is used in the rest.

An adiabatic circuit curtails energy dissipation during charging its load by using a timing-varying power clock V_{PC}. Based

2006 Asia and South Pacific Design Automation Conference

4C-3

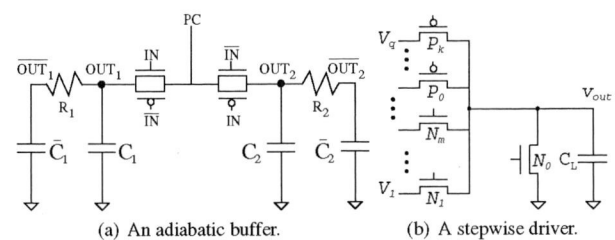

(a) An adiabatic buffer.   (b) A stepwise driver.

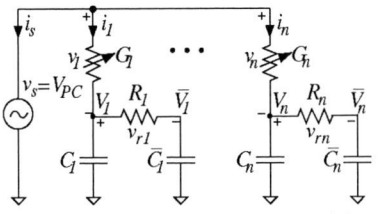

(c) A general adiabatic multi-driver model.

Fig. 1. Two adiabatic circuit examples and a General Adiabatic Multi-driver Model (GAMM).

on the GAMM, the issue of optimizing power clock to minimize the total energy dissipation can be described into the form of (2): replace its cost function Φ by the transient power P:

$$P(V,\bar{V},V_{PC},t) = \sum_{k=1}^{n} \left[(V_{PC}-V_k)\, i_k\,(V_{PC},V_k,t) + \frac{(V_k-\bar{V}_k)^2}{R_k} \right];$$

replace its constraints by circuit equations:

$$\dot{V}_k = f_k \text{ and } \dot{\bar{V}}_k = \bar{f}_k, \text{ for } k = 1,2,\ldots,n, \qquad (3)$$

where $f_k = \frac{i_k(V_{PC},V_k,t)}{C_k} - \frac{(V_k-\bar{V}_k)}{C_k R_k}$ and $\bar{f}_k = \frac{V_k-\bar{V}_k}{R_k\bar{C}_k}$; set time T to the final charging time; set $\phi(\cdot)$ to zero; the trajectory vector $u(t)$ to be optimized becomes the power clock voltage $V_{PC}(t)$. In other words, one is to shape the power clock waveform, to minimize the total energy dissipation during charging each capacitor from a specified initial voltage to a specified final voltage within a time period T. To solve this power clock optimization problem, having been formulated into (2), the paper employs Pontryagin's Minimum Principle.

B. Pontryagin's Minimum Principle

In using Pontryagin's Minimum Principle (PMP) to solve (2), Hamiltonian H at time t is defined:

$$H(x(t),u(t),\lambda(t),t) = \Phi(x(t),u(t),t) + \lambda(t)^T f(x(t),u(t),t), \quad (4)$$

where λ is a vector of adjoint variables, and $\lambda(T)$ equals to $\partial\phi(x(T),T)/\partial x$, if those partial derivatives exist, otherwise $\lambda(T)$ is unspecified. Then PMP is stated as follows.

Theorem 1 *Let $u^*(t) \in U$ be the optimal trajectory, i.e. the solution to (2); $x^*(t)$ the resultant transient circuit waveforms; $\lambda(t)$ the solution to the adjoint equations defined by*

$$\dot{\lambda}(t) = -\frac{\partial H(x^*(t),u^*(t),\lambda(t),t)}{\partial x}. \qquad (5)$$

Then $u^*(t)$ minimizes H at each time $t \in [0,T]$, that is

$$u^*(t) = \arg \min_{u(t)\in U} H(x^*(t),u(t),\lambda(t),t), \text{ for } t \in [0,T]. \qquad (6)$$

C. Solving the Power Clock Optimization Problem by PMP

Introduce two adjoint variable vectors λ and $\bar{\lambda}$, then from (4), Hamiltonian of the power clock optimization problem is

$$H(V,\bar{V},V_{PC},\lambda,\bar{\lambda},t) = P(V,\bar{V},V_{PC},t) + \lambda^T f + \bar{\lambda}^T \bar{f}.$$

According to (5), the adjoint equations are: for $k = 1,2,\ldots,n$,

$$\frac{d\bar{\lambda}_k}{dt} = -\frac{\partial H}{\partial \bar{V}_k} = \beta_k = \frac{1}{R_k}\left[2(V_k-\bar{V}_k) - \frac{\lambda_k}{C_k} + \frac{\bar{\lambda}_k}{\bar{C}_k} \right], \quad (7)$$

$$\frac{d\lambda_k}{dt} = -\frac{\partial H}{\partial V_k} = i_k(V_{PC},V_k,t) - \left(V_{PC}-V_k + \frac{\lambda_k}{C_k} \right)\frac{\partial i_k}{\partial V_k} - \beta_k.$$

Relation (6) implies that the optimal power clock $V_{PC}(t)$ minimizes H. Therefore $\partial H/\partial V_{PC}$ is set to zero, i.e. for each time $t \in [0,T]$,

$$\frac{\partial H}{\partial V_{PC}} = \sum_{k=1}^{n}\left[i_k(V_{PC},V_k,t) + \left(V_{PC}-V_k+\frac{\lambda_k}{C_k} \right)\frac{\partial i_k}{\partial V_{PC}} \right] = 0. \quad (8)$$

The previous procedure has set up a system of differential-algebraic equations (DAEs), consisting of (3), (7) and (8). Under a given initial value of λ, or $\lambda(0)$, the power clock solution to the DAEs optimally drives the capacitors from the specified initial voltages to some final voltages. To reach the specified final capacitor voltages, the value of $\lambda(0)$ need be determined in an iterative procedure, e.g. using the Newton-Raphson method.

The previous procedure by directly employing PMP is less convenient, as in trajectory optimization of even a small circuit, all the previously exemplified steps need be conducted to build the DAEs. The other issue is that as several methods exist in formulating circuit equations, e.g. sparse tableau approach and modified nodal analysis, the built DAEs can have different forms, which can hardly spur insight into the underlying trajectory optimization issue. Hence this paper presents an alternative approach of using auxiliary network, which is established for a given circuit from a set of circuit element correspondence rules. Then a circuit trajectory optimization problem is solved by simulating the given circuit and its auxiliary network.

III. THE AUXILIARY NETWORK

This section presents the theoretic framework of auxiliary network for solving the circuit trajectory optimization issue.

A. Trajectory Optimization Template

Problem (2) is extended to a template problem that includes equality constraints:

$$\begin{aligned} \underset{u(t)\in U}{\text{minimize}} \ & \phi(x(T),T) + \int_0^T \Phi(x(t),y(t),u(t),t)\,dt \\ \text{s.t. } & \dot{x}(t) = f(x(t),y(t),u(t),t) \\ & g(x(t),y(t),u(t),t) = 0 \end{aligned} \qquad (9)$$

Here g consists of equality constraints, x consists of the regular state variables, and y consists of the other variables.

By Lagrange multiplier theory [9], eliminate g from (9)'s constraints via a Lagrange multiplier vector μ. Then the

417

template problem is transformed to the same form as (2), with the objective function of (2) replaced by $\phi(x(T),T) + \int_0^T \left[\Phi(x,y,u,t) + \mu^T g(x,y,u,t)\right] dt$, and its constraints replaced by $\dot{x}(t) = f(x(t),y(t),u(t),t)$. Then using PMP leads to the optimal conditions for the template problem (9):

$$-\frac{\partial H(x,y,u,\mu,\lambda,t)}{\partial x} = \dot{\lambda}(t) \qquad (10a)$$

$$\frac{\partial H(x,y,u,\mu,\lambda,t)}{\partial y} = 0 \qquad (10b)$$

$$u^* = \arg\min_{u \in U} H(x,y,u,\mu,\lambda,t) \qquad (10c)$$

where H is the Hamiltonian:

$$H(x,y,u,\mu,\lambda,t) = \Phi(x,y,u,t) + \mu^T g(x,y,u,t) + \lambda^T f(x,y,u,t).$$

The paper names (10c) the minimization condition. From the optimal conditions in (10), the paper derives an auxiliary network for a given circuit trajectory optimization problem, by exploiting the regularity of circuit equations.

B. The Set of Circuit Constraints

For a given circuit, Kirchhoff's current and voltage laws are described by $A^T i_b = 0$ and $AV = v_b$. Here A is the branch versus nodal incidence matrix, with each row corresponding to one branch in the circuit, and each column one node in the circuit, excluding the datum node. In a row, $+1$ corresponds to the source node, and -1 the destination node. i_b and v_b are the vectors of branch currents and voltages, and V the vector of node voltages. For a branch x, i_x and v_x indicate its branch current and voltage.

The branch constitutive relations (BCR's) for the capacitive and inductive branches in the circuit are given by $i_c = C\dot{v}_c$ and $v_l = L\dot{i}_l$, where C and L are the capacitance and inductance matrices. The two sets of BCR's correspond to the f part in (9). BCR's for the other branches are given in the form of $g(i_q,v_q,u,t) = 0$, which correspond to the g part in (9). Here $i_c, i_l, i_q \subset i_b$ and $v_c, v_l, v_q \subset v_b$. Note that a BCR can have the trajectory vector u and time t as its dependent variables.

Incorporate the mentioned circuit constraints consisting of

$$\dot{v}_c = C^{-1}i_c, \dot{i}_l = L^{-1}v_l, A^T i_b = 0, v_b - AV = 0, g(i_q,v_q,u,t) = 0$$

into the template (9), then Hamiltonian H, named the circuit Hamiltonian, results:

$$H = \Phi(i_\Phi, v_\Phi, u, t) +$$
$$[\lambda_c^T, \lambda_l^T]\begin{bmatrix} C^{-1}i_c \\ L^{-1}v_l \end{bmatrix} + [\mu_i^T, \mu_v^T, \mu_q^T]\begin{bmatrix} A^T i_b \\ v_b - AV \\ g(i_q,v_q,u,t) \end{bmatrix}, \quad (11)$$

where $i_c, i_q, i_\Phi \subset i_b$, $v_l, v_q, v_\Phi \subset v_b$ and $i_c \cap i_q, v_l \cap v_q = \emptyset$. i_Φ and v_Φ are branch currents and voltages dependent by cost function Φ. u is the trajectory vector to be optimized, μ the vector of Lagrange multipliers, and λ the vector of adjoint variables.

The following derives the auxiliary network.

C. The Topology of the Auxiliary Network

Firstly determine the topology of the auxiliary network. From the circuit Hamiltonian H in (11), one has $\frac{\partial H}{\partial V} = -A^T \mu_v$. Then applying (10b) for node voltage vector V leads to the optimal condition $\frac{\partial H}{\partial V} = -A^T \mu_v = 0$. Rename μ_v to $\hat{i}_b$, and also define a branch voltage vector $\hat{v}_b$ by $\hat{v}_b = A\hat{V}$, i.e.

$$A^T \hat{i}_b = 0 \quad \text{and} \quad \hat{v}_b = A\hat{V}, \qquad (12)$$

where $\hat{V}$ is the renaming of μ_i to manifest a vector of node voltages, and renaming μ_v to $\hat{i}_b$ is to manifest a vector of branch currents.

Conclusion 1 *Vectors $\hat{V}$, $\hat{v}_b$ and $\hat{i}_b$, by satisfying (12), correspond to the vectors of node voltages, branch voltages and currents for an auxiliary network that has the same topology as the circuit to be optimized.*

Since a circuit and its auxiliary network have the same topology, the *same branch names* are used in the two circuits. Next step is to determine the BCR's of the constructed auxiliary network, from the optimal conditions in (10).

D. The Capacitive and Resistive Branches

For the capacitive branches c in the original circuit, assume $v_c \cap v_\Phi, i_c \cap i_\Phi = \emptyset$ (these can always be satisfied with some simple circuit transformations). Differentiate the circuit Hamiltonian H in (11) with respect to i_c and v_c:

$$\frac{\partial H}{\partial v_c} = \frac{\partial \mu_v^T v_b}{\partial v_c} = \hat{i}_c \text{ and } \frac{\partial H}{\partial i_c} = C^{-1}\lambda_c + \frac{\partial \mu_i^T A^T i_b}{\partial i_c}.$$

$\frac{\partial H}{\partial i_c}$ can then be simplified as $\frac{\partial H}{\partial i_c} = C^{-1}\lambda_c + \hat{v}_c$, since $\frac{\partial \mu_i^T A^T i_b}{\partial i_c} = \frac{\partial (A\mu_i)^T i_b}{\partial i_c} = \frac{\partial \hat{v}_b^T i_b}{\partial i_c} = \hat{v}_c$. Optimal condition (10a) implies $\dot{\lambda}_c = -\frac{\partial H}{\partial v_c}$, and (10b) implies $\frac{\partial H}{\partial i_c} = 0$, hence, $\dot{\lambda}_c = -\hat{i}_c$ and $C^{-1}\lambda_c + \hat{v}_c = 0$. Then it follows that $\hat{i}_c = C\dot{\hat{v}}_c$.

Conclusion 2 *For the capacitive branches c in the original circuit, branches c in the auxiliary network remain as capacitive with BCR's given by $\hat{i}_c = C\frac{d\hat{v}_c}{dt}$.*

Consider a resistive branch r in the original circuit, whose BCR can generally be given by $g_r(v_r,u,t) - i_r = 0$. Denote the Lagrange multiplier for this BCR as μ_{qr}, a component of μ_q in (11). Differentiate the circuit Hamiltonian H with respect to i_r and v_r:

$$\frac{\partial H}{\partial i_r} = \hat{v}_r - \mu_{qr} + \frac{\partial \Phi}{\partial i_r} \text{ and } \frac{\partial H}{\partial v_r} = \hat{i}_r + \mu_{qr}\frac{\partial g_r}{\partial v_r} + \frac{\partial \Phi}{\partial v_r}.$$

Since optimal condition (10b) implies $\frac{\partial H}{\partial i_r} = 0$ and $\frac{\partial H}{\partial v_r} = 0$,

Conclusion 3 *Corresponding to a resistive branch r in the original circuit, of BCR given by $g_r(v_r,u,t) - i_r = 0$, branch r in the auxiliary network has its BCR specified with $\hat{i}_r = -\left(\hat{v}_r + \frac{\partial \Phi}{\partial i_r}\right)\frac{\partial g_r}{\partial v_r} - \frac{\partial \Phi}{\partial v_r}$.*

If the cost function Φ does not depend on branch current i_r and branch voltage v_r, branch r in the auxiliary network is simply a linear resistor of conductance $-\frac{\partial g_r}{\partial v_r}$.

E. The Dependent Source Branches

Consider a dependent source example: the voltage controlled voltage source (VCVS), with controlling branch 1 and controlled branch 2. The BCR's of the two branches are given by $i_1 = 0$ and $g_2(v_1, X) - v_2 = 0$, with X denoting any other dependent variables. The Lagrange multipliers for the two BCR's are μ_{q1} and μ_{q2}, which are components of μ_q in (11). Assume $i_1 \cap i_\Phi$, $i_2 \cap i_\Phi$, $v_1 \cap v_\Phi$ and $v_2 \cap v_\Phi$ are empty set $\emptyset$, otherwise use simple circuit transformations to make the assumptions true. Optimal condition (10b) implies the partial derivatives of circuit Hamiltonian H with respect to i_1, v_1, i_2 and v_2 are zero, hence,

$$\frac{\partial H}{\partial i_1} = \hat{v}_1 + \mu_{q1} = 0, \quad \frac{\partial H}{\partial v_1} = \hat{i}_1 + \mu_{q2}\frac{\partial g_2}{\partial v_1} = 0,$$

$$\frac{\partial H}{\partial i_2} = \hat{v}_2 = 0, \quad \frac{\partial H}{\partial v_2} = \hat{i}_2 - \mu_{q2} = 0.$$

Conclusion 4 *a VCVS element in the original circuit corresponds to a CCCS element, or a current controlled current source, in the auxiliary network, with the roles of "controlling" and "controlled" exchanged. For a branch controlled by multiple sources in the original circuit, each controlling branch in the original circuit will correspond to a controlled source branch in the auxiliary network.*

For nonlinear storage elements in the original circuit, the corresponding BCR's in the auxiliary network can be derived by using the charge or flux representations of BCR's. Table I shows the derived circuit element correspondence rules, the Lagrange multipliers and the adjoint variables. In the table, a Lagrange multiplier with subscript qx indicates it associates to branch x's BCR.

The auxiliary network has incorporated the first two set of optimal conditions in (10). It can be conveniently established from the derived circuit element correspondence rules. In principle, the auxiliary network customizes the optimal conditions in (10), by exploiting the regularity of circuit equations.

F. Discussions on Auxiliary Network

A circuit trajectory optimization problem can be solved by simulating the original circuit and its auxiliary network, meantime imposing minimization condition (10c). For a typical circuit trajectory optimization problem that has specified the initial and final voltages/currents of the storage elements, the optimal trajectories can be found by an iterative procedure shown in Fig.2.

In certain trajectory optimization cases that the terminating state for a storage element can be in a specified interval, instead of a fixed value, Lagrange multiplier theory [9] can be employed to formulate a template problem. Usually changing terminating conditions only requires modifying some sources in the auxiliary network, and does not affect its circuit element types.

In using auxiliary network for circuit trajectory optimization, timing-varying circuits need be simulated. Simulating timing-varying circuits is also required in circuit parameter optimization problems, when circuit sensitivities are computed using adjoint network or sensitivity circuit [5, 6].

1. The initial voltages/currents for the storage elements in the auxiliary network are guessed first.

2. Simulate the original circuit and its auxiliary network under the given initial states for the storage elements in both circuits.

3. At every simulation time step t, the optimization trajectory vector $u(t)$ is determined by the minimization condition (10c).

4. At the final simulation time T, if the reached voltages or currents for those storage elements in the original circuit match their specified values, the optimal trajectories are found. Otherwise, modify the initial voltage/current values using the Newton-Raphson method, and go to Step 2 to begin the next run of simulation.

Fig. 2. The auxiliary network approach for circuit trajectory optimization.

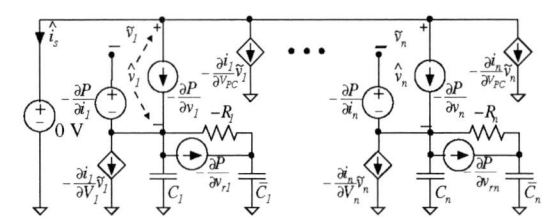

Fig. 3. The auxiliary network for the GAMM.

IV. THE AUXILIARY NETWORK APPLICATIONS

This section demonstrates the applications of auxiliary network in adiabatic circuit optimization.

A. Power Clock Optimization

The power clock optimization problem in Section II is again solved, however, by utilizing the auxiliary network of the GAMM in Fig.1(c). The cost function Φ in the template problem (9) becomes the transient power P, represented with branch voltages and currents by $P = \sum_{k=1}^{n} \left(v_k i_k + v_{rk}^2/R_k \right)$.

Fig.3 shows the auxiliary network of the GAMM, established from the circuit element correspondence rules in Table I. In the auxiliary network, the linear resistors and nonlinear dependent sources in the original circuit become negative linear resistors and dependent sources, the power clock source in the original circuit changes to a voltage source of magnitude $-\frac{\partial P}{\partial i_s}$, i.e. the zero voltage source input in Fig.3, and the capacitors remain. When $\frac{\partial i_k}{\partial V_{PC}} = -\frac{\partial i_k}{\partial V_k}$, the auxiliary network will be simpler.

According to the minimization condition (10c), V_{PC} minimizes the circuit Hamiltonian H, defined in (11). Therefore $\frac{\partial H}{\partial V_{PC}} = 0$. Since only item $\mu_{qs}(v_s - V_{PC})$, which corresponds to the power clock source branch, relates to V_{PC} in the circuit Hamiltonian H, a simplified minimization condition results

$$\frac{\partial H}{\partial V_{PC}} = -\mu_{qs} = \hat{i}_s = 0. \tag{13}$$

The auxiliary network in Fig.3 and the minimization condition (13) can be verified are simplified forms of the optimal conditions given by (7) and (8). In general, the optimal

TABLE I

CIRCUIT ELEMENT CORRESPONDENCE RULES IN ESTABLISHING A CIRCUIT'S AUXILIARY NETWORK.

Elements & BCR's in Original Circuit			Elements & BCR's in Auxiliary Network			λ or μ	
C	$i_c = C\frac{dv_c}{dt}$		C	$\hat{i}_c = C\frac{d\hat{v}_c}{dt}$		$\lambda_c = -C\hat{v}_c$	
L	$v_l = L\frac{di_l}{dt}$		L	$\hat{v}_l = L\frac{d\hat{i}_l}{dt}$		$\lambda_l = -L\hat{i}_l$	
R	$i_r = g_r(v_r,u,t)$		-	$\hat{i}_r = -\left(\hat{v}_r + \frac{\partial\Phi}{\partial i_r}\right)\frac{\partial g_r}{\partial v_r} - \frac{\partial\Phi}{\partial v_r}$		$\mu_{qr} = \hat{v}_r + \frac{\partial\Phi}{\partial i_r}$	
VS	$v_s = vs(u,t)$		VS	$\hat{v}_s = -\frac{\partial\Phi}{\partial i_s}$		$\mu_{qs} = -\hat{i}_s$	
CS	$i_s = cs(u,t)$		CS	$\hat{i}_s = -\frac{\partial\Phi}{\partial v_s}$		$\mu_{qs} = -\hat{v}_s$	
VCVS	$i_1=0$	$v_2 = g_2(v_1,X)$	CCCS	$\hat{v}_2=0$	$\hat{i}_1 = -\hat{i}_2\frac{\partial g_2}{\partial v_1}$	$\mu_{q1}=-\hat{v}_1$	$\mu_{q2}=\hat{i}_2$
VCCS	$i_1=0$	$i_2=g_2(v_1,X)$	VCCS	$\hat{i}_2=0$	$\hat{i}_1 = -\hat{v}_2\frac{\partial g_2}{\partial v_1}$	$\mu_{q1}=-\hat{v}_1$	$\mu_{q2}=\hat{v}_2$
CCVS	$v_1=0$	$v_2=g_2(i_1,X)$	CCVS	$\hat{v}_2=0$	$\hat{v}_1=-\hat{i}_2\frac{\partial g_2}{\partial i_1}$	$\mu_{q1}=-\hat{i}_1$	$\mu_{q2}=\hat{i}_2$
CCCS	$v_1=0$	$i_2=g_2(i_1,X)$	VCVS	$\hat{i}_2=0$	$\hat{v}_1=-\hat{v}_2\frac{\partial g_2}{\partial i_1}$	$\mu_{q1}=-\hat{i}_1$	$\mu_{q2}=\hat{v}_2$

conditions for a circuit trajectory optimization problem can have various forms, depending on how the circuit equations for the original circuit are established, e.g. by sparse tableau approach, nodal analysis, modified nodal analysis, etc. Auxiliary network, however, physically implements the optimal conditions for a circuit trajectory optimization problem, and has only one configuration. The auxiliary network can lead to more simplified optimal conditions, compared to the previous direct method.

B. Transistor Sizing for Stepwise Driver

For the stepwise driver in Fig.1(b), constant power supplies $V_1,\ldots,V_q$ continuously deliver charges to its load, by sequentially turning on transistors $N_1\ldots P_k$. At any time, only one transistor is made on, by applying a short voltage pulse on its gate. Fig.4(a) shows the stepwise driver model, which uses power clock $V_{PC}(t)$ to describe the effective supply voltage produced by $V_1,\ldots,V_q$, and G_r to model the on-transistor.

Under fixed supply voltages $V_1,\ldots,V_q$, each transistor in the stepwise driver can be sized to minimize the total energy dissipation during charging the load to voltage V_{DD} within time period T. This sizing issue can be formulated as a circuit parameter optimization problem. However the paper formulates it as a trajectory optimization problem, to be solved by the auxiliary network approach.

The width, drain-source voltage drop and driving current of the transistor that charges the loading capacitor at time t are denoted by continuous functions $w(t)$, $v_r(t)$ and $i_r(t)$, respectively. As shown in Fig.4(a), the supply voltage at time t is denoted by a continuous function $V_{PC}(t)$, which closely approximates the effective supply voltage produced by $V_1,\ldots,V_q$. Similarly $w(t)$ is a close approximation to the time-varying widths of the on-transistors in $N_0\ldots P_k$.

Without loss of generality, assume that the transistor driving current is proportional to its width, i.e. G_r's BCR is given by $w(t)g_r(V_{out}(t),t) - i_r(t) = 0$, where $g_r(\cdot)$ denotes the driving current of the reference transistor. The energy consumed in turning on all the driving transistors is modeled as $\frac{N}{T}\int_0^T \alpha V_{DD}w(t)dt$, where α is a factor relating gate width to the energy consumed in turning on it, and N is the number of transistors to be sized. The objective function of transistor sizing is the total energy dissipation, given by

$$\int_0^T \Phi(\cdot)dt = \int_0^T \left[v_r(t)i_r(t) + \frac{\alpha N V_{DD}}{T}w(t)\right]dt \qquad (14)$$

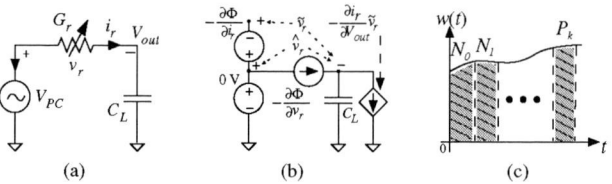

Fig. 4. Transistor sizing of the stepwise driver: (a).Continuous time-domain circuit model; (b).The auxiliary network; (c).The implementation of transistor sizing.

Fig.4(b) shows the auxiliary network, established from the circuit element correspondence rules in Table I. Using the minimization condition (10c) leads to $\frac{\partial H}{\partial w} = \frac{\partial\Phi}{\partial w} + \mu_{qr}g_r(V_{out},t) = 0$, where μ_{qr} is the Lagrange multiplier associated to G_r's BCR, and $\mu_{qr} = \tilde{v}_r$. The minimization condition is simplified to

$$\tilde{v}_r g_r(V_{out},t) = -\frac{\alpha N V_{DD}}{T}. \qquad (15)$$

The approach in Fig.2 can then be employed to obtain the optimal trajectory of transistor width $w(t)$. In employing the approach, the initial voltage for C_L in the auxiliary network need be iteratively updated, to satisfy the two voltage conditions $V_{out}(0) = 0$ and $V_{out}(T) = V_{DD}$ for C_L in the original circuit. In actually sizing the transistors, $w(t)$ is broken into N pieces, and each transistor in the stepwise-driver has its width sized to the average value of $w(t)$ in the corresponding piece, as illustrated by Fig.4(c). In [11], rules of thumb are given for sizing transistor in the stepwise driver. These rules, however, assume a uniform power supply voltage distribution, i.e. $V_{PC}(t)$ is a ramp voltage source, as the voltage increment in each step is a constant.

V. EXPERIMENTAL RESULTS

First the paper gives an example of power clock optimization of an adiabatic buffer. The tested adiabatic buffer is modeled by the GAMM in Fig.1(c), with only one driving block included, i.e. using G_1 to denote one of the transmission gates, and using C_1, R_1 and $\bar{C}_1$ to model the active driving block of the adiabatic buffer. Consequentially the auxiliary network in Fig.3 need include only one block. Here $R_1 = 200\ \Omega$ and $C_1 = \bar{C}_1 = 30$ fF. The paper used 0.13 μm MOSFET models.

2006 Asia and South Pacific Design Automation Conference

4C-3

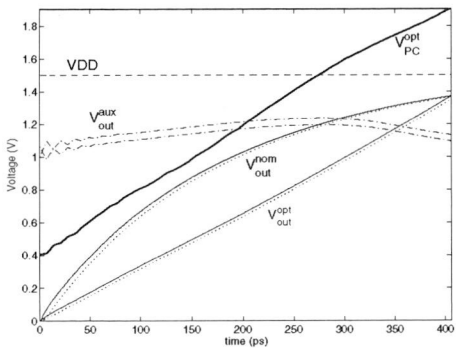

Fig. 5. An example of power clock optimization for an adiabatic buffer.

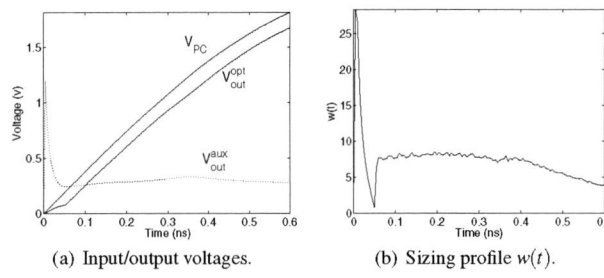

(a) Input/output voltages.

(b) Sizing profile $w(t)$.

Fig. 6. An example of stepwise driver transistor sizing.

TABLE II
MULTI-DRIVER POWER CLOCK OPTIMIZATION EXAMPLES.

Name	$\hat{C}$ (fF)	$\hat{R}$ (Ω)	$\hat{\tilde{C}}$ (fF)	P (mW)	OP(mW)	*Improv.*
Cir1	136.9	70.7	254.2	0.429	0.200	53.2%
Cir2	185.7	169.8	188.8	0.411	0.208	49.3%
Cir3	151.8	60.8	258.9	1.035	0.507	50.9%
Cir4	163.4	46.6	230.6	2.034	1.102	45.8%
Cir5	125.2	76.6	206.6	0.547	0.392	28.3%
Cir6	185.3	181.0	199.1	0.969	0.493	49.1%
Cir7	126.8	120.8	154.1	0.710	0.383	46.1%
Cir8	225.3	63.2	320.0	0.895	0.529	40.9%

Fig.5 shows the simulation results. V_{PC}^{opt} denotes the optimal power clock obtained using the approach in Fig.2. V_{out}^{aux} indicates the output voltages of the two capacitors in the auxiliary network. V_{out}^{opt} indicates the output voltages of the two capacitors in the original circuit when driven by V_{PC}^{opt}, while V_{out}^{nom} indicates the output voltages, when the circuit is driven by a 1.5 V VDD. The chosen time duration T is 406 ps. Using the optimal power clock leads to an average power consumption of 106.6 μW, while using the constant supply leads to a value of 164.6 μW, which is 54.4% more than the optimal solution.

Table II shows the power clock optimization results for the GAMM with more than two driving blocks. In the table, $\hat{C}$, $\hat{R}$ and $\hat{\tilde{C}}$ are the average values of the Π-load parameters for each tested circuit. P denotes the average power consumption in charing the loads, by using a constant power supply to drive the tested circuit, while OP denotes the value, by using the optimal power clock. As indicated by the last column, around 25% to 50% power consumption reductions were observed by using the optimal power clocks.

Fig.6 shows the results of using auxiliary network in transistor sizing for a stepwise driver. The driver has a 150 fF load that need be charged to 1.5 V in 0.6ps. Instead of using the ramp source in [11], the paper uses a 2 V 300 MHz sinusoidal power clock source. The driver has five stages, therefore 5 driving blocks need be sized. V_{out}^{opt} is the output voltage of the driver under the optimal transistor sizing. V_{out}^{aux} is the output voltage of the capacitor in the auxiliary network, which is shown in Fig.4(b). $w(t)$ in the rightmost diagram is the optimal transistor sizing in the time domain. It uses a unit driving block as the reference. This continuous sizing curve need be transformed into 5 pieces for the 5 blocks as illustrated in Fig.4(c).

VI. CONCLUSIONS

On circuit trajectory optimization issue, the paper introduced an auxiliary network approach, by utilizing Pontryagin's Minimum Principle. By the approach, a given circuit's auxiliary network is established from a set of circuit element corresponding rules; then the optimal trajectories are found by a process of simulating the given circuit and its auxiliary network. The approach was demonstrated more convenient than the method that directly employs Pontryagin's Minimum Principle to establish DAEs. It can lead to more simplified optimal conditions. Auxiliary network facilitates the establishment of analytical models in designing high-performance circuits that require optimizing circuit trajectories. The paper demonstrated the application of auxiliary network in optimizing adiabatic circuits.

REFERENCES

[1] N. Wang, "On the design of mos dynamic sense amplifiers," *IEEE Trans. Circuits Syst.*, vol. 29, no. 7, pp. 467–477, Jul. 1982.

[2] W. Athas, L. Svensson, J. Koller, N. Tzartzanis, and E. Ying-Chin Chou, "Low-power digital systems based on adiabatic-switching principles," *IEEE Trans. VLSI Syst.*, vol. 2, no. 4, pp. 398–407, Dec. 1994.

[3] A. Dickinson and J. Denker, "Adiabatic dynamic logic," *IEEE Journal of Solid-State Circuits*, vol. 30, no. 3, pp. 311–315, Mar. 1995.

[4] B. Wang and P. Mazumder, "On optimality of adiabatic switching in mos energy-recovery circuit," in *Proc. IEEE Int. Symp. Low-Power Electronics and Design*, Aug. 2004, pp. 236–239.

[5] S. W. Director and R. A. Rohrer, "The generalized adjoint network and network sensitivities," *IEEE Trans. Circuit Theory*, vol. 16, no. 3, pp. 318–323, Aug. 1969.

[6] D. Hocevar, P. Yang, T. Trick, and B. Epler, "Transient sensitivity computation for mosfet circuits," *IEEE Trans. Computer-Aided Design*, vol. 4, no. 4, pp. 609–620, Oct. 1985.

[7] C. Visweswariah and A. Conn, "Formulation of static circuit optimization with reduced size, degeneracy and redundancy by timing graph manipulation," in *Proc. IEEE Int. Conf. Computer-Aided Design*, Nov. 1999, pp. 244–251.

[8] G. Northrop and P.-F. Lu, "A semi-custom design flow in high-performance microprocessor design," in *Proc. Design Automation Conference*, June 2001, pp. 426–431.

[9] D. P. Bertsekas, *Dynamic Programming and Optimal Control*, 2nd ed. Athena Scientific, 2001.

[10] M. Alioto and G. Palumbo, "Power estimation in adiabatic circuits: a simple and accurate model," *IEEE Trans. VLSI Syst.*, vol. 9, no. 5, pp. 608–615, Oct. 2001.

[11] L. Svensson and J. Koller, "Driving a capacitive load without dissipating fcv^2," in *Proc. IEEE Int. Symp. Low-Power Electronics and Design*, Oct. 1994, pp. 100–101.

4C-4

SASIMI: Sparsity-Aware Simulation of Interconnect-Dominated Circuits with Non-Linear Devices

Jitesh Jain, Stephen Cauley, Cheng-Kok Koh, and Venkataramanan Balakrishnan
School of Electrical and Computer Engineering
Purdue University, West Lafayette, IN 47907-1285
{jjain,stcauley,chengkok,ragu}@ecn.purdue.edu

Abstract

We present a technique for the fast and accurate simulation of large-scale VLSI interconnects with nonlinear devices, called SASIMI. The numerical efficiency of this technique is realized through linear-algebraic techniques that exploit the sparsity and structure of the matrices that are encountered in VLSI structures. Numerical results show that SASIMI is up to 1400 times as fast as commercial-grade SPICE, for moderate-size circuits, with little sacrifice in simulation accuracy.

1 Introduction

With aggressive technology scaling, the accurate and efficient modeling and simulation of interconnect effects has become (and continues to be) a problem of central importance. For accurate modeling of the distributive effects of interconnects, it is necessary to model a long wire using many segments of lumped RLC elements. Owing to the inductive and capacitive coupling between these elements, direct simulation of the resulting models comes at a very high (and often unacceptable) simulation cost. It has been highlighted in the past that SPICE [11] is not amenable to the simulation of interconnect-dominated structures, such as power/ground networks, clock networks, and busses.

There have been many studies to eliminate the bottleneck due to SPICE in the past several years. A representative selection of these advances in the simulation of large-scale interconnects, which exploit the *locality* of capacitive and inductive coupling effects, can be classified into techniques at the modeling and simulation levels as follows:

Modeling level techniques: In [5], the authors propose a new circuit element K to capture the inductive coupling using the inverse of the inductance matrix. Window-based extraction of K elements are proposed in [2], [15], and [16]. In [8], the authors use controlled voltage and current sources to construct a SPICE-compatible circuit model for K elements. A similar concept, called Vector Potential Equivalent Circuit (VPEC), is introduced in [12] to obtain a localized circuit model for inductive interconnects. These techniques involve the inversion of the inductance matrix. To avoid matrix inversion, the authors of [16] propose a wire duplication-based interconnect model, in which the authors construct a sparse equivalent circuit by windowing the inductance matrix. However, it should be emphasized that all these techniques ultimately rely on SPICE (and its variants) for the simulation of dynamic responses, and thus are constrained by the limitations of SPICE.

Simulation level techniques: The underlying solver in a simulation tool essentially addresses the problem of solving $Ax = b$ fast. In [14], the authors propose a hierarchical analysis of power distribution networks. In this work the authors partition the power grid and model each partition as a macro-model with a sparsified port admittance matrix. In [3], INDUCTWISE, an efficient simulation tool for cir-

cuits modeled using conductance (G), capacitance (C), and K elements, is proposed. In [14], and [3], it is shown that the Cholesky factorization of the A matrix, which is composed of G, C, and/or K, can be very efficient due to the inherent sparsity of A. However, the inverse of A is dense, which in turn implies that each simulation step involves dense matrix-vector multiplications. In contrast, the authors of [7] perform linear circuit simulation using a new formulation, called RLP, to model circuits by resistance (R), inductance (L), and the inverse of capacitance (P). Although the resulting A matrix in the RLP formulation is dense, its inverse is sparse, which enables fast sparse matrix inversion and sparse matrix-vector multiplication in each simulation step. The above methods however, are unable to handle non-linear devices. This problem is addressed in [4], where the approach in [14] is extended to handle non-linear devices. However, it also inherits the limitation that the sparsity of matrix A is exploited only for fast Cholesky factorization, and not in each simulation step.

In this paper, we propose SASIMI which uses *sparsity-aware simulation techniques for interconnect-dominated circuits with non-linear devices.* Our contribution in this paper is two fold. First, we extend the RLP formulation as described in [7] to include non-linear devices, without sacrificing the computational benefits achieved due to sparsity of the linear system. It should be noted that the A matrix involved in the solution of the linear system is constant throughout the simulation. In contrast, the A matrix involved in solving the non-linear system changes in each simulation step. However, the A matrix is sparse. Due to the sparse and time varying nature of the problem at hand Krylov subspace based iterative methods could be used for efficient simulation. Our second contribution is to introduce a novel preconditioner constructed based on the sparsity structure of the non-linear system. The inverse of the preconditioner has a compact representation in the form of the Hadamard product [10], which facilitates not only the fast computation of the inverse, but also the fast dense matrix-vector product. Experimental results show that SASIMI is up to 1400 times faster than commercial grade SPICE, even for moderate-size circuits.

2 Mathematical Preliminaries

VLSI interconnect structures, with non linear devices can be analyzed using the Modified Nodal Analysis (MNA) formulation which yields equations of the following form

$$\tilde{G}x + \tilde{C}\dot{x} = b, \tag{1}$$

where

$$\tilde{G} = \begin{bmatrix} \mathcal{G} & A_l^T \\ -A_l & 0 \end{bmatrix}, \ \tilde{C} = \begin{bmatrix} C & 0 \\ 0 & L \end{bmatrix}, \ x = \begin{bmatrix} v_n \\ i_l \end{bmatrix},$$

0-7803-9451-8/06/$20.00 ©2006 IEEE.   422

$$b = \begin{bmatrix} A_i^T I_s + I_{\text{nl}} \\ 0 \end{bmatrix}, \quad G = A_g^T R^{-1} A_g, \text{ and } C = A_c^T C A_c.$$

R denotes the resistance matrix. The matrices G, L and C are the conductance, inductance and capacitance matrices respectively, with corresponding adjacency matrices A_g, A_l and A_c. I_s is the current source vector with adjacency matrix A_i, and v_n and i_l are the node voltages and inductor currents respectively.

Vector, I_{nl} captures the effect of non-linear loads and depends on the node voltages as $I_{\text{nl}} = f(v_n)$. f is a function which varies depending on the load characteristics and in general can be a non-linear function.

With N denoting the number of inductors, we note that

$$L, C, R \in \mathbf{R}^{N \times N}, \quad C, G \in \mathbf{R}^{2N \times 2N}.$$

Differential equations such as (1) can be numerically solved using standard algorithms like the trapezoidal method [1]. Considering a uniform discretization of the time axis with resolution h, $x^k = x(kh)$. Using the approximations

$$\left. \frac{d}{dt} x(t) \right|_{t=kh} \approx \frac{x^{k+1} - x^k}{h} \text{ and } x^k \approx \frac{x^{k+1} + x^k}{2}$$

over the interval $[kh, (k+1)h]$, the determination of x^{k+1} from x^k requires the solution of a set of linear and nonlinear equations:

$$\left(\frac{\tilde{G}}{2} + \frac{\tilde{C}}{h} \right) x^{k+1} = -\left(\frac{\tilde{G}}{2} - \frac{\tilde{C}}{h} \right) x^k + \frac{b^{k+1} + b^k}{2} \tag{2}$$

and

$$I_{\text{nl}}^{k+1} = f\left(v_n^{k+1} \right). \tag{3}$$

The nonlinearity in the above set of equations can be handled by the standard Newton-Raphson technique of linearizing (3) and iterating until convergence: Equation (2) is a linear equation of the form $\mathcal{L}(x) = 0$, where we have omitted the iteration index k for simplicity. Equation (3) is a nonlinear equation of the form $g(x) = 0$. Let $\tilde{g}(x) \approx g(x)$ be a linear approximation of $g(x)$, linearized around some $x = x_0$. Then, simultaneously solving $\mathcal{L}(x) = 0$ and $\tilde{g}(x) = 0$ yields numerical values for x and hence v_n. These values are then used to obtain a new linear approximation $g(x) \approx \tilde{g}_{\text{new}}(x)$, and the process is repeated until convergence. A good choice of the point x_0 for the initial linearization at the kth time-step is given by the value of v_n from the previous time-step.

A direct implementation of this algorithm requires $O(pqn_1^3)$ operations, where p is the number of time steps, q is the maximum number of Newton-Raphson iterations in each time step, and $n_1 = 3N$.

3 The RLP formulation

The mathematical framework that underlies our approach is an alternative formulation of the MNA equations that uses the resistance, inductance and the inverse of the capacitance matrix. This is the so-called "RLP formulation", first proposed in [7].

We begin by decomposing C, A, and A_i as:

$$C = \begin{bmatrix} C_{cc} & C_{cv} \\ C_{vc} & C_{vv} \end{bmatrix}, \quad A^T = \begin{bmatrix} A_1^T \\ A_2^T \end{bmatrix} \quad A_i^T = \begin{bmatrix} A_{i1}^T \\ A_{i2}^T \end{bmatrix}$$

$$I_{\text{nl}} = \begin{bmatrix} 0 \\ I_v \end{bmatrix}. \quad v_n = \begin{bmatrix} v_c \\ v_v \end{bmatrix}. \text{ Here } C_{vv} \text{ denotes the sub-matrix}$$

of the capacitance matrix that changes amid the simulation, while all other sub-matrices remain constant. The matrix C_{vv} captures the drain, gate and bulk capacitances of all devices, which are voltage-dependent, while C_{cc}, and C_{cv} are the capacitance matrices that arise from interconnects and are hence constant.

For typical interconnect structures, the above decomposition allows us to manipulate the MNA equations (2) and (3):

$$\underbrace{\left(\frac{L}{h} + \frac{R}{2} + \frac{h}{4} A_1 P_{cc} A_1^T \right)}_{X} i_l^{k+1}$$
$$= \underbrace{\left(\frac{L}{h} - \frac{R}{2} - \frac{h}{4} A_1 P_{cc} A_1^T \right)}_{Y} i_l^k$$
$$+ \quad A_1 v_c^k + \frac{h}{4} A_1 P_{cc} A_{i1}^T \left(I_s^{k+1} + I_s^k \right)$$
$$- \quad A_1 P_{cc} C_{cv} \left(v_v^{k+1} - v_v^k \right) + \frac{A_2}{2} \left(v_v^{k+1} + v_v^k \right), \tag{4}$$

$$\begin{aligned} v_c^{k+1} &= v_c^k - \frac{h}{2} P_{cc} A_2^T \left(i_l^{k+1} + i_l^k \right) + \frac{h}{2} P_{cc} A_{i1}^T \left(I_s^{k+1} + I_s^k \right) \\ &- P_{cc} C_{cv} \left(v_v^{k+1} - v_v^k \right), \end{aligned} \tag{5}$$

$$\begin{aligned} C_{vv} v_v^{k+1} &= C_{vv} v_v^k - \frac{h}{2} A_2^T \left(i_l^{k+1} + i_l^k \right) + \frac{h}{2} A_{i2}^T \left(I_s^{k+1} + I_s^k \right) \\ &- C_{vc} \left(v_c^{k+1} - v_c^k \right) + \frac{h}{2} \left(I_v^{k+1} + I_v^k \right), \end{aligned} \tag{6}$$

$$I_v^{k+1} = f\left(v_v^{k+1} \right). \tag{7}$$

Here r denotes the size of interconnect structure connected directly to non linear circuit, and given $l = N - r$ we note that

$$C_{cc} \in \mathbf{R}^{l \times l}, \quad C_{vv} \in \mathbf{R}^{r \times r}.$$

$P_{cc} = C_{cc}^{-1}$ is the inverse capacitance matrix, and A is the adjacency matrix of the circuit. A is obtained by first adding A_g and A_l and then removing zero columns (these correspond to intermediate nodes, representing the connection of a resistance to an inductance).

The development thus far is similar to that in [7], with the major difference being the addition of (6) and (7), which account for the nonlinear elements. The main contribution in [7] was the fast solution of (4) and (5), where all matrices are constant over the simulation period. We will show in §4.2 that the techniques in [7] can be extended to handle the case when nonlinear elements are present.

For future reference, we will call the technique of directly solving (4), (5), (6), and (7) as the "Exact-RLP" algorithm. It can be shown that the computational complexity of the Exact-RLP algorithm is $O\left(l^3 + pq\left(l^2 + r^3 \right) \right)$. For large VLSI interconnect structures we have $l >> r$, reducing the complexity to $O\left(l^3 + pq\left(l^2 \right) \right)$.

4 Computationally efficient implementation

We now turn to the fast solution of equations (4) through (7). Recall that the nonlinear equation (7) is handled via the Newton-Raphson technique. This requires, at each time step, linearizing (7) and sub-

4C-4

2006 Asia and South Pacific Design Automation Conference

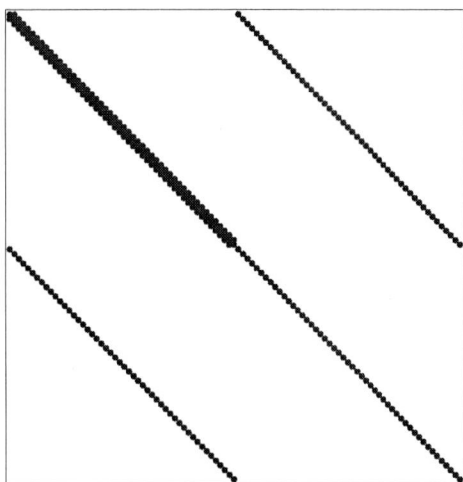

Figure 1: Sparsity structure of A. The nonzero entries are shown darker.

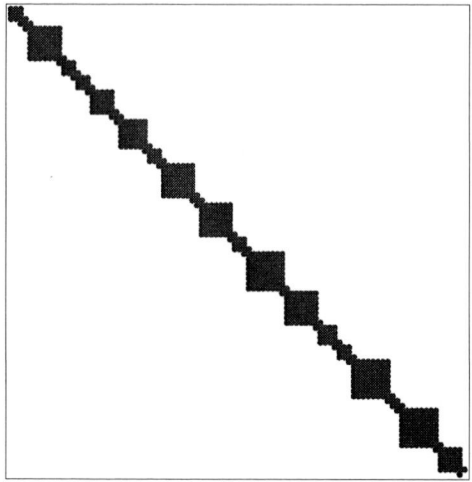

Figure 2: Sparsity structure of A. The non-zero entries are shown darker.

stituting it into (6). The resulting set of linear equations have very specific structure:

- Equations (4) and (5) are of the form $Ax = b$ where A is fixed (does not change with the time-step). Moreover, A^{-1} *is typically approximately sparse* (For details, see §4.2).

- Equation (6) (after the substitution of the linearized (7)) is again of the form $Ax = b$, where the matrix A is obtained by adding C_W and the coefficient of the first-order terms in the linearized equation (7). Recall that the matrix C_W captures the drain, gate and bulk capacitances of all devices. It also contains the interconnect coupling capacitances between gates and drains of different non-linear devices in the circuit. As each non-linear device is connected to only a few nodes and the capacitive effects of interconnects are localized, *the A matrix is observed to be sparse in practice* (For details, see §4.1). Note that A changes with each Newton-Raphson iteration and with the time-step.

Thus the key computational problem is the solution of a sparse time-varying set of linear equations, coupled with a large fixed system of linear equations $Ax = b$ with A^{-1} being sparse.

4.1 Solving sparse time-varying linear equations

Krylov subspace methods have been shown to work extremely well for sparse time-varying linear equations [6]. Specifically, the GMRES (Generalized Minimum Residual) method of Saad and Schultz [13] allows the efficient solution of a sparse, possibly non-symmetric, linear system to within a pre-specified tolerance. This method performs a directional search along the orthogonal Arnoldi vectors which span the Krylov subspace of A. That is, given an initial guess x_0 and corresponding residual $r_0 = b - Ax_0$, orthogonal vectors $\{q_1, q_2..., q_m\}$ are generated with the property that they span S_m, the solution search space at iteration m.

$$
\begin{aligned}
S_m &= x_0 + span\{r_0, Ar_0, ..., A^m r_0\} \\
&= x_0 + \kappa(A, r_0, m) \\
&\subseteq span\{q_1, q_2..., q_m\}.
\end{aligned}
\tag{8}
$$

These vectors are chosen according to the Arnoldi iteration: $AQ_m = Q_{m+1}H_m$ where $Q_m = \{q_1, q_2..., q_m\}$ is orthogonal and $H_m \in \mathbf{R}^{m+1 \times m}$

is an upper Heisenberg matrix.

For these methods the choice of a preconditioner matrix M, which is an approximation of A, can greatly affect the convergence. A good preconditioner should have the following two properties:

- $M^{-1}A \approx I$.

- It must accommodate a fast solution to an equation of the form $Mz = c$ for a general c.

Figure 1 depicts the sparsity structure of the A matrix for a circuit example of parallel wires driving a bank of inverters. For such a sparsity structure, an appropriate choice of the preconditioner could be of the form as shown in Figure 3. Although we have chosen a circuit with only inverters for simplicity, a more complicated circuit structure would simply distribute the entries around the diagonal and off-diagonal bands and lead to possibly more off diagonal bands. To see this, consider an extreme case where the circuit under consideration has only non-linear devices and does not comprise of interconnects. In this case the sparsity pattern of the A matrix is as shown in Figure 2. Therefore, the chosen preconditioner would encompass not only the sparsity structure shown in Figure 1 but also other sparsity patterns that might arise with the analysis of more complicated non-linear devices. Correspondingly the structure of the preconditioner (see Figure 3) would have additional bands.

Matrices of the form shown in Figure 3 have the following two properties which make them an ideal choice for preconditioner.

- The inverses of the preconditioner matrix can be computed efficiently in linear time, $O(r)$ (r denotes the size of interconnect structure directly connected to non-linear devices), by exploiting the Hadamard product formulation as shown in [10].

- It can also be shown that this formulation facilitates the fast matrix-vector products, again in linear time ($O(r)$), which arise while solving linear systems of equations with the preconditioner matrix.

A simple example which best illustrates these advantages is a

424

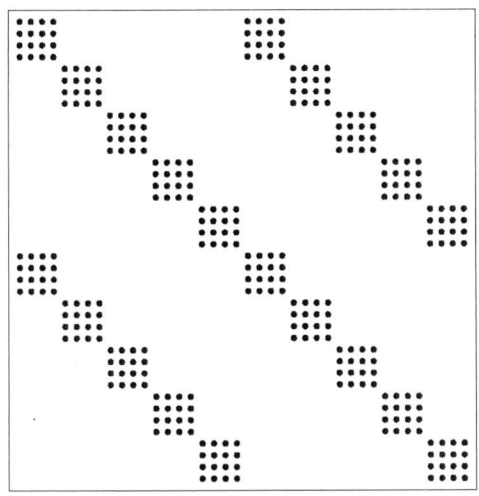

Figure 3: Preconditioner matrix.

symmetric tridiagonal matrix.

$$
B = \begin{pmatrix}
a_1 & -b_1 & & & \\
-b_1 & a_2 & -b_2 & & \\
& \ddots & \ddots & \ddots & \\
& & -b_{n-2} & a_{n-1} & -b_{n-1} \\
& & & -b_{n-1} & a_n
\end{pmatrix} \tag{9}
$$

The inverse of B can be represented compactly as a Hadamard product of two matrices, which are defined as follows:

$$
B^{-1} = \underbrace{\begin{pmatrix}
u_1 & u_1 & \cdots & u_1 \\
u_1 & u_2 & \cdots & u_2 \\
\vdots & \vdots & \ddots & \vdots \\
u_1 & u_2 & \cdots & u_n
\end{pmatrix}}_{U} \circ \underbrace{\begin{pmatrix}
v_1 & v_2 & \cdots & v_n \\
v_2 & v_2 & \cdots & v_n \\
\vdots & \vdots & \ddots & \vdots \\
v_n & v_n & \cdots & v_n
\end{pmatrix}}_{V}. \tag{10}
$$

There exists an explicit formula to compute the sequences $\{u\}, \{v\}$ efficiently in $O(n)$ operations which is detailed in [10]. In this case, if we are interested in solving a linear system of equations $By = c$, we only need to concern ourselves with the matrix-vector product $B^{-1}c = y$. This computation can also be performed efficiently in $O(n)$ computations as outlined below:

$$
P_{u_i} = \sum_{j=1}^{i} u_j c_j, \quad P_{v_i} = \sum_{j=i}^{n} v_j c_j, \quad i = 1, \ldots, n,
$$

$$
y_1 = u_1 P_{v_1},
$$

$$
y_i = v_i P_{u_{i-1}} + u_i P_{v_i}, \quad i = 2, \ldots, n. \tag{11}
$$

The above formulation for a tridiagonal matrix could be easily extended to handle the more general case when the preconditioner matrix is a zero padded block tridiagonal matrix (matrix with zero diagonals inserted between the main diagonal and the non-zero super-diagonal and sub-diagonal of tridiagonal matrix) as in Figure 3. Elementary row and column block permutations could be performed on such a matrix to reduce it into a block tridiagonal matrix. This has

been shown with a small example as below.

$$
B = \begin{pmatrix}
a_1 & 0 & -b_1 & 0 \\
0 & a_2 & 0 & -b_2 \\
-b_1 & 0 & a_3 & 0 \\
0 & -b_2 & 0 & a_4
\end{pmatrix} \tag{12}
$$

$$
= P \underbrace{\begin{pmatrix}
a_1 & -b_1 & 0 & 0 \\
-b_1 & a_2 & 0 & 0 \\
0 & 0 & a_3 & -b_2 \\
0 & 0 & -b_2 & a_4
\end{pmatrix}}_{X} P^T, \tag{13}
$$

where

$$
P = \begin{pmatrix}
1 & 0 & 0 & 0 \\
0 & 0 & 1 & 0 \\
0 & 1 & 0 & 0 \\
0 & 0 & 0 & 1
\end{pmatrix}.
$$

Hence

$$
B^{-1} = P X^{-1} P^T
$$

$$
= \underbrace{\begin{pmatrix}
u_1 & 0 & u_1 & 0 \\
0 & u_2 & 0 & u_2 \\
u_1 & 0 & u_3 & 0 \\
0 & u_2 & 0 & u_4
\end{pmatrix}}_{U} \circ \underbrace{\begin{pmatrix}
v_1 & 0 & v_3 & 0 \\
0 & v_2 & 0 & v_4 \\
v_3 & 0 & v_3 & 0 \\
0 & v_4 & 0 & v_4
\end{pmatrix}}_{V}. \tag{14}
$$

We have not included block matrices for simplicity of presentation, however the zero padded block tridiagonal case is a natural extension of the above example. All the entries in U, V matrices have to be now replaced by blocks and accordingly the row and column permutations would be replaced by their block counterparts with an identity matrix of appropriate size replacing the 'ones' in the P matrix. Table 1 gives the comparison for Incomplete-LU preconditioner and the zero-padded (Z-Pad) preconditioner. Simulations were done on circuits consisting of busses with parallel conductors driving bank of inverters. 'Size' denotes the number of non-linear devices. All the results are reported as a ratio of run-time and iteration-count (number of iterations for the solution to converge to within a tolerance of 1e-10) of Z-Pad to the Incomplete-LU preconditioner. As can be seen from Table 1, Z-Pad offers a substantial improvement in run time as compared to the Incomplete-LU preconditioner.

Size	400	800	1600	3200
Runtime	.44	.42	.42	.43
Iterations	5/10	5/10	5/10	5/10

Table 1: Preconditioner comparison.

4.2 Solving $Ax = b$ with a constant, approximately sparse A^{-1}

We now turn to the solution of equations (4) and (5). As mentioned earlier, these equations reduce to the form $Ax = b$ with a constant, approximately sparse A^{-1}. A (corresponding to X in (4)) is composed of L, R and P. Each of these matrices has a sparse inverse for typical VLSI interconnects which then leads to a approximately sparse A^{-1} (Note that this argument is used for motivating the sparsity inherent in A^{-1} and cannot be used as a theoretical proof for the same). In addition this sparsity has a regular pattern which can be explained on the basis of how inductance and capacitance matrices are extracted. The distributed *RLC* effects of VLSI interconnects can be modeled by dividing conductors into small subsets of segments,

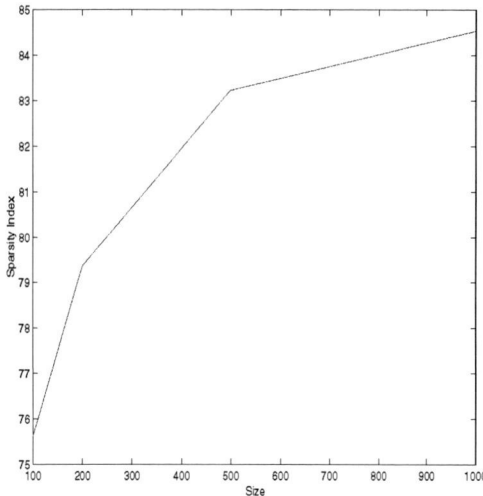

Figure 4: Average sparsity versus circuit size.

σ	ρ=5	ρ=20	ρ=50
100	.0054	.0053	.0088
200	.0078	.0052	.0071
500	.0006	.0022	.0001
1000	.0003	.0005	.0004
2000	.0003	.0004	.0004

Table 2: RMSE comparison.

5 Numerical results and conclusions

We implemented the Exact-RLP and RLP (SASIMI) algorithms in C++. A commercially available version of SPICE with significant speed-up over the public-domain SPICE has been used for reporting all results with SPICE. Simulations were done on circuits consisting of busses with parallel conductors driving bank of inverters, with wires of length 1mm, cross section $1\mu m \times 1\mu m$, and with a wire separation of $1\mu m$. A periodic 1V square wave with rise and fall times of 6ps each was applied to the first signal with a time period of 240ps. All the other lines were assumed to be quiet. For each wire, the drive resistance was 10Ω. A time step of 0.15ps was taken and the simulation was performed over 30 ps (or 200 time steps). For the inverters the W/L ratio of NMOS and PMOS were taken to be $.42\mu m/.25\mu m$ and $1.26\mu m/.25\mu m$ respectively.

In order to explore the effect of the number of non-linear elements relative to the total, three cases were considered. With ρ denoting the ratio of the number of linear elements to that of non-linear elements, the experiments were performed for ρ equaling 5, 20 and 50. The number of linear elements in the following results is denoted by σ.

We first present results comparing the accuracy in simulating the voltage waveforms at the far end of the first line (after the inverter load). The metric for comparing the simulations is the relative mean square error (RMSE) defined as

$$\frac{\sum_i (v_i - \widetilde{v}_i)^2}{\sum_i v_i^2}$$

where v and $\widetilde{v}$ denote the waveforms obtained from Exact-RLP and SASIMI respectively.

Table 2 presents a summary of the results from the study of simulation accuracy. It can be seen that the simulation accuracy of the Exact-RLP algorithm is almost identical to that of SPICE, while the SASIMI has a marginally inferior performance as measured by the RMSE. The error values for SASIMI are compared simply with the Exact-RLP as it had the same accuracy as SPICE results for all the experiments run. A plot of the voltage waveforms at the far end of the active line, obtained from SPICE, Exact-RLP and SASIMI algorithms, is shown in Figure 5. (The number of conductors in this simulation example is 200.) There is almost no detectable simulation error between the SASIMI, Exact-RLP and SPICE waveforms over 200 time steps. To give a better picture, the accuracy results reported are for a larger simulation time of 2200 time steps.

We now turn to a comparison of the computational requirements between Exact-RLP, SASIMI and SPICE. Table 3 summarizes the findings. For a fair comparison our total simulation time is compared against the transient simulation time for SPICE(i.e we have not included any of the error check or set up time for SPICE). As can be seen from the table, SASIMI outperforms the Exact-RLP algorithm and SPICE. For the case of 500 conductors with $\rho = 50$, the Exact-RLP algorithm is 390 times as fast compared to SPICE. SASIMI is about 1400 times faster as compared to SPICE, and more than three times faster than Exact-RLP. As can be seen, the computational savings increase as the ratio of linear to non-linear elements is

each of which are aligned [9, 3]. Each of these subsets leads to a sparsity pattern (corresponding to a band in A^{-1}). All the effects when summed up lead to a A^{-1} matrix that has a regular sparsity pattern. Window selection algorithm as described in [16, 3] could then be employed to find out the sparsity pattern in A^{-1}. It has been recognized in earlier work that this property (sparsity) yields enormous computational savings; it has been shown in [7] that an approximate implementation of the Exact-RLP algorithm, referred to simply as the "RLP algorithm" provides an order-of-magnitude in computational savings with little sacrifice in simulation accuracy.

To proceed, we rewrite (4) and (5) as

$$i_l^{k+1} = X^{-1}Yi_l^k + X^{-1}A_1v_c^k + \frac{h}{4}X^{-1}A_1P_{cc}A_{i1}^T\left(I_s^{k+1} + I_s^k\right)$$
$$- X^{-1}A_1P_{cc}C_{cv}\left(v_v^{k+1} - v_v^k\right) + \frac{A_2}{2}\left(v_v^{k+1} + v_v^k\right), \quad (15)$$

$$X^{-1}A_1v_c^{k+1} = X^{-1}A_1v_c^k - X^{-1}A_1\frac{h}{2}P_{cc}A_2^T\left(i_l^{k+1} + i_l^k\right)$$
$$+ \frac{h}{2}X^{-1}A_1P_{cc}A_{i1}^T\left(I_s^{k+1} + I_s^k\right) - X^{-1}A_1P_{cc}C_{cv}\left(v_v^{k+1} - v_v^k\right). \quad (16)$$

Although X is a dense matrix, X^{-1} turns out to be an approximately sparse matrix. Moreover the matrices $X^{-1}Y$, $X^{-1}A_1$, $X^{-1}A_1P_{cc}A_{i1}^T$, $X^{-1}A_1P_{cc}C_{cv}$ are also approximately sparse [7]. This information can be used to reduce the computation significantly by noting that each step of trapezoidal integration now requires only sparse vector multiplications. Solving sparse (15) and (16) along with (6) and (7) is termed as the RLP algorithm (SASIMI). To analyze the computational saving of the approximate algorithm over the Exact-RLP algorithm, we denote "sparsity index" of a matrix A as ratio of the number of entries of A with absolute value less than ε to the total number of entries. The computation required for each iteration of (15) and (16) is then $O\left((1-v)l^2\right)$, where v is the minimum of the sparsity indices the matrices $X^{-1}Y$, $X^{-1}A_1$, $X^{-1}A_1P_{cc}A_{i1}^T$, $X^{-1}A_1P_{cc}C_{cv}$. Figure 4 provides the average sparsity for the matrices for a system with parallel conductors driving a bank of inverters. The sizes in consideration are 100, 200, 500 and 1000. On top of this the computation time of X^{-1} can be reduced to $O(l)$ by using the windowing techniques (details in [16]). Hence the computational complexity of RLP is $O\left(pq(1-v)l^2\right)$ as compared to $O\left(pqn_1^3\right)$ for the MNA approach.

σ	$\rho=5$			$\rho=20$			$\rho=50$		
	SPICE	Exact-RLP	SASIMI	SPICE	Exact-RLP	SASIMI	SPICE	Exact-RLP	SASIMI
100	11.96	1.34	1.26	13.73	.27	.21	13.54	.15	.12
200	100.25	3.28	2.68	68.72	.64	.28	67.68	.55	.22
500	3590.12	17.13	4.872	1919.21	13.47	3.01	1790.67	4.58	1.30
1000	>12hrs	87.75	22.71	>10hrs	79.07	16.49	>10hrs	77.56	15.20
2000	> 1day	545.6	78.06	> 1day	526.23	59.33	> 1day	408.54	56.05

Table 3: Run time (in seconds) comparisons.

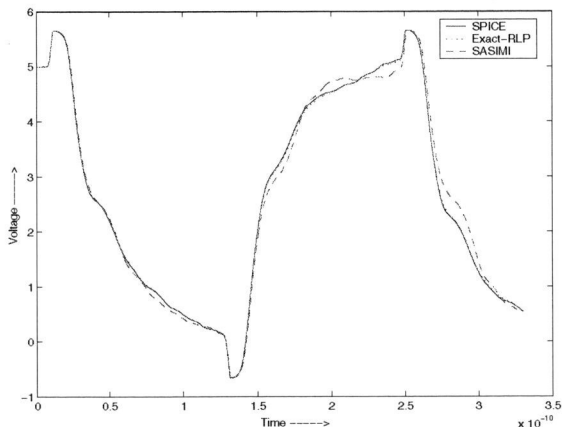

Figure 5: The voltage waveforms obtained through SPICE, Exact-RLP and SASIMI.

increased from 5 to 50. The savings also increase with increase in the size of the problem considered. The computational efficiency of the SASIMI can be explained on the use of sparsity-aware algorithms for both the linear and non-linear parts of the problem.

6   Acknowledgments

This material is based on work supported by the NASA, under Award NCC 2-1363, and by National Science Foundation under Award CCR-9984553 and CCR-0203362.

References

[1] U. M. Ascher and L. R. Petzold. *Computer Methods for Ordinary Differential Equations and Differential-Algebraic Equations*. SIAM, 1998.

[2] M. Beattie and L. Pileggi. Modeling magnetic coupling for on-chip interconnect. In *Proc. Design Automation Conf*, pages 335–340, 2001.

[3] T. H. Chen, C. Luk, H. Kim, and C. C.-P. Chen. IN-DUCTWISE: Inductance-wise interconnect simulator and extractor. In *Proc. Int. Conf. on Computer Aided Design*, pages 215–220, 2002.

[4] T.-H. Chen, J.-L. Tsai, C. C.-P. Chen, and T. Karnik. HISIM: Hierarchical interconnect-centric circuit simulator. In *Proc. Int. Conf. on Computer Aided Design*, pages 489–496, 2004.

[5] A. Devgan, H. Ji, and W. Dai. How to efficiently capture on-chip inductance effects: Introducing a new circuit element K. In *Proc. Int. Conf. on Computer Aided Design*, pages 150–155, 2000.

[6] G. H. Golub and C. F. Van Loan. *Matrix Computations*. John Hopkins University Press, 1996.

[7] J. Jain, C.-K. Koh, and V. Balakrishnan. Fast simulation of VLSI interconnects. In *Proc. Int. Conf. on Computer Aided Design*, pages 93–98, 2004.

[8] H. Ji, Q. Yu, and W. Dai. SPICE compatible circuit models for partial reluctance K. In *Proc. Asia South Pacific Design Automation Conf.*, pages 786–791, 2004.

[9] T. Lin, M. W. Beaftie, and L. T. Pileggi. On the efficacy of simplified 2D on-chip inductance. In *Proc. Design Automation Conf*, pages 757–762, 2002.

[10] R. Nabben. Decay rates of the inverses of nonsymmetric tridiagonal and band matrices. *SIAM Journal on Matrix Analasis and Applications*, 20(3):820–837, May 1999.

[11] L. W. Nagel. SPICE2: A computer program to simulate semiconductor circuits. Technical report, U.C. Berkeley, ERL Memo ERL-M520, 1975.

[12] A Pacelli. A local circuit topology for inductive parasitics. In *Proc. Int. Conf. on Computer Aided Design*, pages 208–214, 2002.

[13] Y. Saad and M. Schultz. GMRES: A generalized minimal residual algorithm for solving non-symmetric linear systems. *SIAM Journal on Scientific Computing*, pages 856–869, 1986.

[14] M. Zhao, R. V. Panda, S. S. Sapatnekar, and D. Blaauw. Hierarchical analysis of power distribution networks. *IEEE Trans. on Computer-Aided Design of Integrated Circuits and Systems*, pages 159–168, February 2002.

[15] H. Zheng, B. Krauter, M. Beattie, and L. Pileggi. Window-based susceptance models for large scale RLC circuit analyses. In *Proc. Design Automation and Test in Europe Conf.*, pages 628–633, 2002.

[16] G. Zhong, C.-K. Koh, and K. Roy. On-chip interconnect modeling by wire duplication. In *Proc. Int. Conf. on Computer Aided Design*, pages 341–346, 2002.

An Unconditional Stable General Operator Splitting Method for Transistor Level Transient Analysis

Zhengyong Zhu, Rui Shi, Chung-Kuan Cheng
Department of Computer Science and Engineering
University of California, San Diego
La Jolla, CA 92093
Email: {zzhu,rshi,kuan}@cs.ucsd.edu

Ernest S. Kuh
Department of Electrical Engineering and Computer Sciences
University of California, Berkeley
Berkeley, CA 94720
Email: kuh@eecs.berkeley.edu

Abstract—In this paper, we introduce a general operator splitting method for transient simulation of VLSI circuits. The proposed approach generates special partitions of the circuits and alternates the explicit and implicit integrations between the partitions. We prove that the method is unconditionally stable independent of the step size. The splitting scheme greatly reduces the nonzero fill-ins generated in direct methods like LU decomposition. Orders of magnitude speedup over Berkeley SPICE3 is observed for sets of circuits.

I. INTRODUCTION

With increasing design complexity, huge size of extracted interconnect data is pushing the capacity of transistor level simulation tools to the limits. Direct methods like Gaussian Elimination used in Berkeley SPICE and its variations is prohibitive because of the above linear complexity $O(n^{1.5})$ where n is the number of circuit nodes.

In last decade, there is rich literature in the field of circuit analysis to improve the performance of simulation under the rising demand from advanced technologies. Among them, Conjugate Gradient Method and Multigrid Method were used on linear networks [3]–[5]. Model order reduction methods [1], [2] reduced the circuit size by generating stable and passive macromodels in time domain simulation. For timing analysis, Acar et al. [6] introduced a waveform evaluation engine using Successive Chord and macromodeling approach.

At transistor level, Sakallah and Directors [7] saved unnecessary computation by applying different integration method (explicit or implicit) on subcircuits according to their activities. Li and Shi [11] tried to reduce the number and cost of LU decompositions by using low cost integration approximation and Successive Chord Method with approximated device model.

Since direct methods such as LU decomposition remain to be efficient for small circuits with up to tens of thousands of nodes, partition-based simulation methods are widely used in commercial tools [8]–[10]. However, the convergence of those methods cannot be guaranteed and is sensitive to the partition algorithm and propagation order.

The operator splitting method has been adopted to partition the system based on the geometry of the physical adjacency and the locality of the processes. In 1999, Namiki and Ito [16] adopted its special form, the Alternating Direction Implicit (ADI), to simulate a two dimensional electromagnetic wave. They demonstrated the unconditional stability of the finite difference time domain analysis independent of the time step size under the proposed geometric structure. Later, Zheng et al extended the structure to three dimensions [13]. Since then, the method has been applied to tackle huge problems for finite difference time domain analysis [18]. In 2001 and 2003, Lee and Chen proposed TLM-ADI approach [14], [15] for power grid analysis, based on implicit FDTD methods. In 2003, Guo and Tan applied the ADI method to circuit-level power grid analysis [20], which splits power mesh along horizontal and vertical directions and iterates between partitions at each time point till converge with fixed time step size.

In this paper, we present a generalized operator splitting method and demonstrate that the generalized method is unconditionally stable. Following the generalized approach, we partition the circuits using a network splitting algorithm with guaranteed DC paths and alternate the explicit and implicit integrations between the partitions. The splitting algorithm partitions the circuit into structures that produces much fewer nonzero fill-ins during LU factorization. Thus direct methods can remain efficient for large-scale circuits. Unlike [20], the proposed approach has no geometrical constrains and can handle general circuits. We can also prove that there is no iteration needed between partitions at each time point since the operator splitting approach is actually an A-stable numerical integration method and the local truncation error can be controlled by dynamic time step estimation.

The rest of this paper is organized as follows. General operator splitting method and its application on linear and nonlinear circuits are discussed in section II. Section III proves the unconditional stability of the proposed method. Section IV discusses the local truncation error (LTE) estimation and dynamic time step control. Experimental results are then demonstrated in section V. The paper is wrapped up with conclusion and future directions.

II. GENERAL OPERATOR SPLITTING METHOD

The operator splitting method [17] was first introduced as a technique for solving partial differential equations. The basic idea of operator splitting can be explained with the following initial value problem (IVP) of a simple ordinary differential equation (ODE) [19],

$$\frac{\delta u}{\delta t} = Lu \qquad (1)$$

0-7803-9451-8/06/$20.00 ©2006 IEEE.

where L is a linear or nonlinear operator and can be written as a linear sum of m suboperators of u,

$$Lu = L_1u + L_2u + \cdots + L_mu \qquad (2)$$

Suppose $U_1, U_2, \cdots, U_m$ are updating operators on u with respect to $L_1, L_2, \cdots, L_m$ from time step n to time step $n+1$, the operator splitting approach has the form of:

$$
\begin{aligned}
u^{n+(1/m)} &= U_1(u^n, h/m) \\
u^{n+(2/m)} &= U_2(u^{n+(1/m)}, h/m) \\
&\cdots \\
u^{n+1} &= U_m(u^{n+(m-1)/m}, h/m)
\end{aligned}
\qquad (3)
$$

where each partial operation acts with all the terms of the original operator.

Our invention introduced in the next subsection generalizes the operator splitting method to graph based modeling. The generalization frees us from the geometry or locality constraints. We prove that the method is unconditionally stable.

A. Formulation

We use a general circuit system to describe our operator splitting method. The circuit contains resistors, capacitors, and inductors with mutual couplings. For linear circuits, the modified nodal analysis using Backward Euler Integration can be expressed as below:

$$
\begin{bmatrix} \frac{C}{h} + G & -A^T \\ A & \frac{L}{h} + R \end{bmatrix} \begin{bmatrix} V(t+h) \\ I(t+h) \end{bmatrix} = \begin{bmatrix} \frac{C}{h} & 0 \\ 0 & \frac{L}{h} \end{bmatrix} \begin{bmatrix} V(t) \\ I(t) \end{bmatrix} + U(t+h)
\qquad (4)
$$

where C, L, R, G are the matrices of capacitances, inductances, resistances, and conductances. Matrix A is an incidence matrix linking between the topology of capacitance nodes and inductance branches. Vectors V, I, and U describes the voltages of capacitance nodes, currents of inductance branches, and system inputs. Scalar h is the time step from time t to $t+h$. Note that the four matrices, C, L, R, and G, are symmetric by construction and are positive semidefinite because the elements: capacitances, inductances, resistances, and conductances, are non-active. In addition, we can assume that matrices C and L are positive definite for a nondegenerated case.

The generalized operator splitting formulation allows us to make arbitrary partitions of the circuit. Thus, we have corresponding partitions of matrices A, R, and G, i.e. $A = A_1 + A_2$, $R = R_1 + R_2$ and $G = G_1 + G_2$. By construction, matrices R_i and G_i for $i \in \{1, 2\}$ remain to be symmetric and positive semidefinite. Following the circuit partition, we divide the integration into two half steps and alternates the forward and backward integrations between the partitions as shown in formulation (5). In the first half step, we use forward integration for the subcircuit with matrices A_2, G_2 and R_2. Then, in the second half step, we use forward integration for the subcircuit with matrices A_1, G_1 and R_1. In both half steps, the other partition is integrated by backward implicit integration.

$$
\left\{
\begin{aligned}
&\begin{bmatrix} \frac{2C}{h} + G_1 & -A_1^T \\ A_1 & \frac{2L}{h} + R_1 \end{bmatrix} \begin{bmatrix} V(t+\frac{h}{2}) \\ I(t+\frac{h}{2}) \end{bmatrix} = \\
&\qquad \begin{bmatrix} \frac{2C}{h} - G_2 & A_2^T \\ -A_2 & \frac{2L}{h} - R_2 \end{bmatrix} \begin{bmatrix} V(t) \\ I(t) \end{bmatrix} + U(t+\frac{h}{2}) \\
&\begin{bmatrix} \frac{2C}{h} + G_2 & -A_2^T \\ A_2 & \frac{2L}{h} + R_2 \end{bmatrix} \begin{bmatrix} V(t+h) \\ I(t+h) \end{bmatrix} = \\
&\qquad \begin{bmatrix} \frac{2C}{h} - G_1 & A_1^T \\ -A_1 & \frac{2L}{h} - R_1 \end{bmatrix} \begin{bmatrix} V(t+\frac{h}{2}) \\ I(t+\frac{h}{2}) \end{bmatrix} + U(t+h)
\end{aligned}
\right.
\qquad (5)
$$

If the two left-hand-side matrices correspond to trees or forest structures, a direct matrix inversion will be very efficient to solve those two equations because there is no nonzero fill-ins and the computational cost is linearly proportional to the number of elements.

Let $P_1 = \begin{bmatrix} G_1 & -A_1^T \\ A_1 & R_1 \end{bmatrix}$, $P_2 = \begin{bmatrix} G_2 & -A_2^T \\ A_2 & R_2 \end{bmatrix}$, $S = \begin{bmatrix} \frac{2C}{h} & 0 \\ 0 & \frac{2L}{h} \end{bmatrix}$, and $X = \begin{bmatrix} V \\ I \end{bmatrix}$ then the notation of the two half step of operator splitting formulation (5) can be simplified as:

$$
\left\{
\begin{aligned}
(P_1 + S)X(t+\tfrac{h}{2}) &= -(P_2 - S)X(t) + U(t+\tfrac{h}{2}) \\
(P_2 + S)X(t+h) &= -(P_1 - S)X(t+\tfrac{h}{2}) + U(t+h)
\end{aligned}
\right.
\qquad (6)
$$

B. Splitting Operation

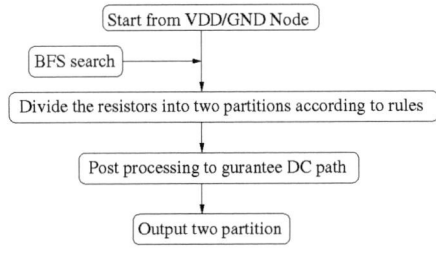

Fig. 1. Splitting Algorithm Flow

1) Network Splitting with Guaranteed DC Paths: The performance of direct methods such as LU decomposition can still beat those of iterative methods for small circuits with up to tens of thousands of nodes. Direct methods become prohibitive for large circuits because of the significant amount of nonzero fill-ins generated during factorization. However, it can be proved that the LU decomposition method does not create nonzero fill-ins for circuits in tree/forest structure if nodes elimination always starts from leaves. The elimination order can be captured by ordering algorithms based on minimum degrees. Following this observation, the proposed operator splitting algorithm tries to split the circuits into two partitions in structures close to tree or forests such that the number of nonzero fill-ins is minimized. Even though general circuits may not be able to get optimized partitions in terms of the number of nonzero fill-ins because of its structure limitation or the restriction of DC paths (Splitting algorithm should not generate floating nodes at DC stage), the number of overall nonzero fill-ins is significantly reduced for most circuits.

4C-5

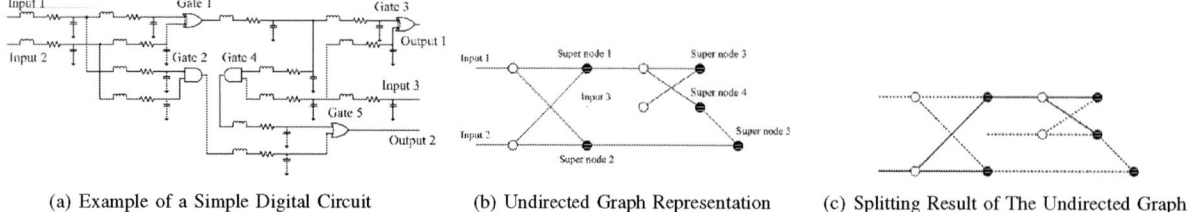

(a) Example of a Simple Digital Circuit   (b) Undirected Graph Representation   (c) Splitting Result of The Undirected Graph

Fig. 2. Example for Undirected Graph Representation and Splitting

2) Applications on Linear Circuits: Only resistive connections are considered during partition. Capacitors and inductors are duplicated into both partitions. Resistors are divided into two partitions using graph theory algorithms. In order to obtain DC convergence, some adjustments are needed to ensure that every node in both partitions has a DC path to some voltage source. When solving each partition, the rest of circuit is modelled as equivalent current sources, following the operator splitting formulation (5). Detailed algorithm is discussed in section II-B.1.

3) Applications on Circuits with transistors: We extend the algorithm to handle circuits with transistors. In typical digital circuits, transistors are grouped as various gates. Taking into consideration the nonlinear property of transistor devices and gates, the proposed approach does not split a single transistor or gate into different partitions; instead, each partition has a full-version of all transistor devices. In other words, transistor devices are duplicated in both partitions and solved at every half time point.

The splitting algorithm actually regards each gate as a super node. The details inside each gate are invisible to the splitting algorithm. The same splitting operation for linear circuits is applied on the super node structure instead of the original circuits.

Moreover, the introduction of super node brings some advantages during LU decomposition if all the internal nodes in one gate are eliminated together. Even though each gate may have many internal nodes, it usually has only a few input/output nodes connecting to the outside, which implies that the nonzero fill-ins are confined inside each gate and will not propagate to outside if all the internal nodes in the same gate are eliminated together during LU decomposition. However, the ordinary minimum degree algorithm or the Markowitz Product method used in SPICE3 does not have a global view of the circuit structure and introduce many unnecessary nonzero fill-ins. We modified the ordering algorithm in SPICE3 to incorporate the super node concept.

We define an undirected graph $G = (V, E)$ to represent the circuit structure. There are two kinds of nodes in this graph: super node and branch node. Super node denotes end point of resistors in large linear networks or a single gate. Branch node represents end point of resistors on signal wires connecting gates in the circuit. The edge denotes the resistor branch in the circuit since only resistors are divided into partitions.

Figure 2 gives the undirected graph representation of a simple digital circuit. The original circuit shown in Figure 2(a) is mapped to an undirected graph in Figure 2(b). Figure 2(c) demonstrates one possible splitting result of the graph, where dashed lines and solid lines denote two different partitions.

The splitting algorithm divides the graph into two partitions. The objective is to minimize the total number of non-zero fill-ins of both partitions generated in LU decomposition. The basic idea of our rule-based splitting algorithm is to distribute the edges of every node into different partitions and avoid loops as much as possible, because sparse and tree-like structures produce much fewer nonzero fill-ins.

Figure 1 summarizes the splitting algorithm flow. The input information includes the undirected graph, super node identification and VDD/GND nodes set. There are two main steps, breath first search (BFS) partition and DC path post-processing. In the BFS partition step, we start from VDD/GND nodes simultaneously, go through all the nodes in the graph using BFS and divide the edges of every node into two partitions according to the partition rules defined below. Nodes and edges are associated with labels to record the partition and DC path connection status. Based on the labelling information, the partition rules are defined to benefit DC path available for all nodes. In the post-processing step, we adjust the partition for nodes without a DC path to VDD/GND.

We define seven types of node labels: CONNECTED, UNCONNECTED, ZERO_UNCONNECTED, ONE_UNCONNECTED, ZERO_ONE_UNCONNECTED, ZERO_CONNECTED, ONE_CONNECTED, where "ZERO" and "ONE" represent different partitions. The label describes the partition and DC path status of the node. For example, label "ZERO_UNCONNECTED" means the node has an edge in partition ZERO without a DC path to VDD/GND. Similarly, we define five types of edge labels: UNCONNECTED, ZERO_UNCONNECTED, ONE_UNCONNECTED, ZERO_CONNECTED, ONE_CONNECTED.

There are four partition rules for the splitting process.

1) Branch rule: the edges in one branch belong to the same partition. In the undirected graph, a branch is consisted of edges connected by branch nodes. Branch rule assigns nodes on the same signal wires into one partition, which will accelerate the nonlinear convergence.

2) Degree rule: the edges of node with degree two should be assigned to the same partition. This is because the line structure would not cause many non-zero fill-ins and will be propitious to provide DC path in both partitions.

3) Loop rule: the loop will be avoided in both partitions if possible. Edge loops will potentially introduce certain number of non-zero fill-ins.

4) Balance rule: the edges for each node in the graph will be evenly divided into two partitions. Thus, each partition will be much simpler than the original graph.

As an example, Figure 3 illustrates the step by step splitting process on a 6x6 mesh shown in Figure 3(b). The legend used to represent different types of nodes and edges is given in Figure 3(a). Figure 3(c) - 3(g) reveal the stepwise changes of splitting status for all the nodes and edges in BFS partition stage. Figure 3(h) gives the final splitting result after the post-processing step. Both partitions in the final result have a tree/forest structure, which will greatly benefits the LU decomposition.

III. UNCONDITIONAL STABILITY ANALYSIS

For the analysis of the error propagation, we can ignore the inputs in the operator splitting formulation (5). We then combine the two half steps and reduce the two equations (5) to a recursive formula

$$X_{(k+1)} = \Lambda X_{(k)} \tag{7}$$

where $\Lambda = (P_2 + S)^{-1}(P_1 - S)(P_1 + S)^{-1}(P_2 - S)$.

In the proof of the convergence, we use the norm $\|x\|_{S^{-1}} = (x^T S^{-1} x)^{1/2}$. Note that matrix S^{-1} is positive definite because matrix S is positive definite and the inverse of a positive definite matrix remains to be positive definite. We first state the theorem of the unconditional stability. The proof of the statement is assisted by the lemmas which follow the theorem.

Theorem 3.1: The operator splitting formula (5) is stable independent of the step size h

Proof: Let $\rho(\Lambda) = max(|\lambda_i(\Lambda)|)$, where $\lambda_i(\Lambda)$ is the ith eigenvalue of matrix Λ. The proposed operator splitting approach is stable if $\rho(\Lambda) <= 1$.

From Lemma 3.4, we have

$\|(P_1 - S)(P_1 + S)^{-1}x\|_{S^{-1}} \leq \|x\|_{S^{-1}}$ and

$\|(P_2 - S)(P_2 + S)^{-1}x\|_{S^{-1}} \leq \|x\|_{S^{-1}}$.

Let $\rho(\bar{\Lambda}) = (P_1 - S)(P_1 + S)^{-1}(P_2 - S)(P_2 + S)^{-1}$

From Lemma 3.2 and 3.3, we can deduce: $\rho(\Lambda) = \rho(\bar{\Lambda}) \leq 1$.

Lemma 3.2: $\rho((P_2 + S)^{-1}(P_1 - S)(P_1 + S)^{-1}(P_2 - S)) = \rho((P_1 - S)(P_1 + S)^{-1}(P_2 - S)(P_2 + S)^{-1})$

Proof: We can derive that $\rho(AB) = \rho(BA)$ if matrix A or B is nonsingular. Thus, we prove the lemma by setting $A = (P_2 + S)^{-1}$ and $B = (P_1 - S)(P_1 + S)^{-1}(P_2 - S)$.

Lemma 3.3: Given a real matrix M, if $\|Mx\|_{S^{-1}} \leq \gamma \|x\|_{S^{-1}}$ for all real x, then $\rho(M) \leq \gamma$.
The proof can be found in [12].

Lemma 3.4: $\|(P_i - S)(P_i + S)^{-1}x\|_{S^{-1}}^2 \leq \|x\|_{S^{-1}}^2$ for $i \in \{1, 2\}$ and every real vector x.

Proof:

$\|(P_i - S)(P_i + S)^{-1}x\|_{S^{-1}}^2 \leq \|x\|_{S^{-1}}^2$ is equivalent to $\|(P_i - S)y\|_{S^{-1}}^2 \leq \|(P_i + S)y\|_{S^{-1}}^2$ where $y = (P_i + S)^{-1}x$

We expand the inequality expression according to the definition of the norm.

$$y^T(P_i^T - S^T)S^{-1}(P_i - S)y \leq y^T(P_i^T + S^T)S^{-1}(P_i + S)y \tag{8}$$

We expand the product terms and cancel the common items on the two sides of the inequality. The expression is reduced to:

$$y(P_i + P_i^T)y^T \geq 0 \tag{9}$$

which is true since $P_i + P_i^T$ is positive semidefinite for $i \in \{1, 2\}$.

IV. LOCAL TRUNCATION ERROR AND TIME STEP CONTROL

Though the general operator splitting approach is A-stable, the local truncation error still need to be controlled below the error tolerance in order to ensure the accuracy. By estimating the local truncation error at each time point, we can dynamically adjust the time step to control the local truncation error.

Given the system equation before numerical integration (10),

$$\begin{bmatrix} C & 0 \\ 0 & L \end{bmatrix} \begin{bmatrix} \dot{V}(t) \\ \dot{I}(t) \end{bmatrix} = \begin{bmatrix} -G & A^T \\ -A & -R \end{bmatrix} \begin{bmatrix} V(t) \\ I(t) \end{bmatrix} + U(t) \tag{10}$$

Let $M = \begin{bmatrix} C & 0 \\ 0 & L \end{bmatrix}$, $N = \begin{bmatrix} -G & A^T \\ -A & -R \end{bmatrix}$ and ignore the input vector U, Equation (10) can be simplified as:

$$\begin{aligned} M\dot{X} &= NX \\ \dot{X} &= M^{-1}NX \end{aligned} \tag{11}$$

Here, we regard the system equations above as circuit state equations for the sake of clarity, which implies that matrix M is non-singular in the following derivation.

The exact analytic solution X with time step h can be derived as below:

$$\begin{aligned} X_{n+1} &= e^{M^{-1}Nh}X_n \\ &= (1 + M^{-1}Nh + \frac{h^2(M^{-1}N)^2}{2} + \frac{h^3(M^{-1}N)^3}{6} + O(h^4))X_n \end{aligned} \tag{12}$$

The general operator splitting approach can also be formulated as:

$$\frac{M}{h}(\hat{X}_{n+1} - X_n) = N_1\hat{X}_{n+1} + N_2X_n \tag{13}$$

where $N = N_1 + N_2$, N_1 represents the the partition applied Backward Euler and N_1 denotes the partition applied forward Euler integration method.

The analytic solution of operator splitting approach is derived as below:

$$(\frac{M}{h} - N_1)\hat{X}_{n+1} = (\frac{M}{h} + N_2)X_n \tag{14}$$

$$\hat{X}_{n+1} = [I + hM^{-1}N + h^2M^{-1}N_1M^{-1}N + O(h^3)]X_n \tag{15}$$

The local truncation error is the difference of operator splitting solution $\hat{X}$ and exact solution X:

$$LTE = \|h^2M^{-1}(\frac{N}{2} - N_1)\dot{X}_n + O(h^3)\| \tag{16}$$

The local truncation error at each time step should not exceed the error tolerance. If we ignore the high order terms of local truncation error, the time step when forward Euler integration is applied to partition corresponding to N_1 is estimated as:

$$h_1 < \sqrt{\frac{\text{Error Tolerance}}{\|M^{-1}(\frac{N}{2} - N_1)\dot{X}_n\|}} \tag{17}$$

Similarly, the time step when forward Euler integration is applied to partition corresponding to N_2 is estimated as:

$$h_2 < \sqrt{\frac{\text{Error Tolerance}}{\|M^{-1}(\frac{N}{2} - N_2)\dot{X}_n\|}} \tag{18}$$

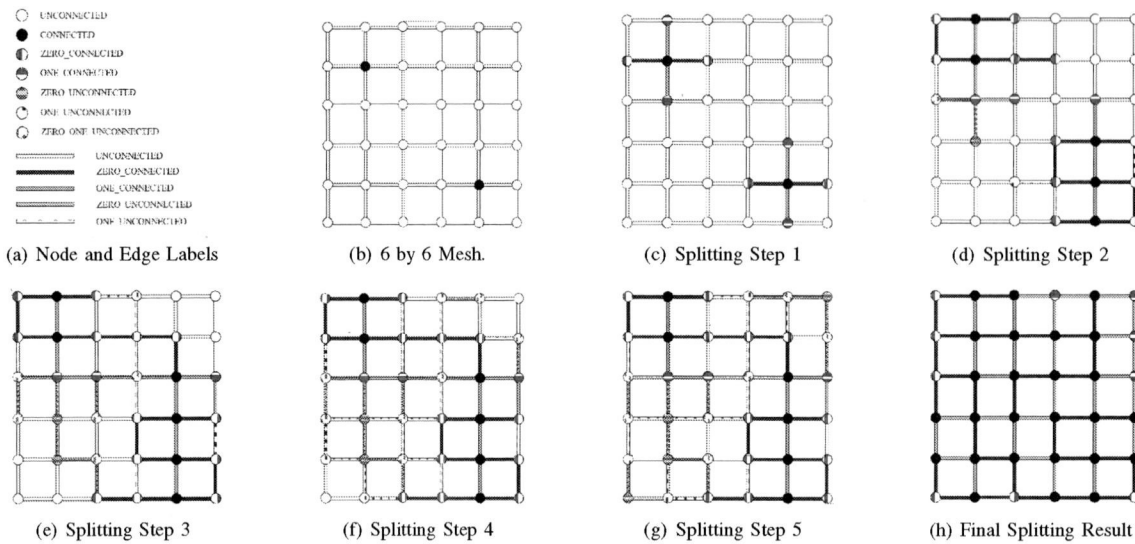

(a) Node and Edge Labels (b) 6 by 6 Mesh. (c) Splitting Step 1 (d) Splitting Step 2

(e) Splitting Step 3 (f) Splitting Step 4 (g) Splitting Step 5 (h) Final Splitting Result

Fig. 3.   6x6 mesh splitting

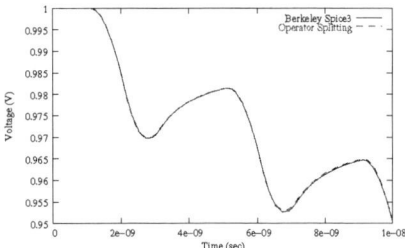

Fig. 4.   Transient Response of Circuit3

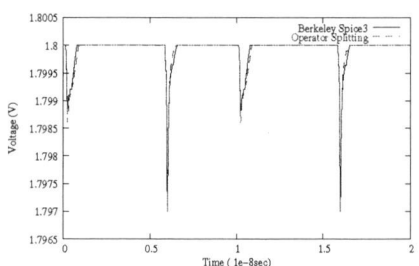

Fig. 5.   Voltage Drop of RLC Power and Clock Network Example

And the new time step h is twice of the minimum time step of each partition:

$$h = 2min(h_1, h_2) \qquad (19)$$

Our experiment shows that the operator splitting time step size is about 25% of the original SPICE time step size.

V. EXPERIMENTAL RESULTS

The proposed approach is implemented in C programming language. For all circuits presented in this section we compare the proposed approach with Berkeley SPICE3 using BSIM3 models for transistor devices. We did not compare the result with commercial fast simulators because we do not trade any accuracy for speed. Convergence and accuracy are guaranteed. Examples are tested on a Linux machine with 2.6 GHz CPU and 4 Gigabytes memory.

A. Power Network with Nonlinear Current Sinks

We test a number of RLC power networks with size ranging from 11k nodes to 160k nodes. Various transistor gates draw current from the power networks. Those power networks are approximately in mesh structures. The splitting algorithm results in very limited nonzero fill-ins and we observe linear runtime of the proposed method. The CPU runtime is given in

Table I. Orders of magnitude speedup (8.1x to 58.2x) against SPICE3 is obtained. The transient waveform circuit3 is given in Figure 4.Since we only replace the LU decomposition procedure inside the SPICE3, other overhead such as device evaluation and dynamic time step control take more than 30% of the total runtime, thus limits the overall speedup.

B. Power and Clock Network

The Power and clock network case contains an RLC power ground network and a two-level H-tree clock. Figure 5 shows the voltage drop at one node of the power network. Transient simulation of 10ns is completed in 649.5 seconds, which is 18.5 times faster than SPICE3 as shown in Table I.

C. Large Power Network Example

This example contains a huge RC power network (0.6 million nodes) with very irregular structure (some nodes have thousands of neighbors). The switching activities that draw current from the power network are modeled as piecewise linear current waveform. Berkeley SPICE3 fails to execute due to the capacity limit. The operator splitting approach finished the transient analysis of 10ns in just 4083 seconds. Figure 6 illustrates the voltage drop of a node on on the power network.

TABLE I
TRANSIENT SIMULATION RUNTIME

Examples	circuit1	circuit2	circuit3	circuit4	Power/Clock	Power Network	1K-cell	10K-cell
#Nodes	11,203	41,321	92,360	160,657	29,100	615,446	10,200	123,600
#Transistors	74	512	1,108	2,130	720	0	6,500	69,000
Simulation Period	10ns	10ns	10ns	10ns	10ns	10ns	20ns	20ns
SPICE3(sec)	602.44	8268.92	39612.32	N/A	12015	N/A	2121	44293
Operator Splitting(sec)	74.64	305.38	681.18	1356.21	649.5	4083.7	415.9	3954.7
Speedup	8.1x	27.1x	58.2x	N/A	18.5x	N/A	5.1x	11.2x

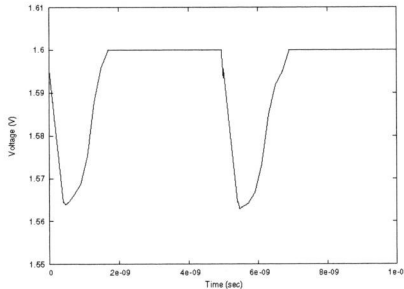

Fig. 6. Voltage drop on the power network

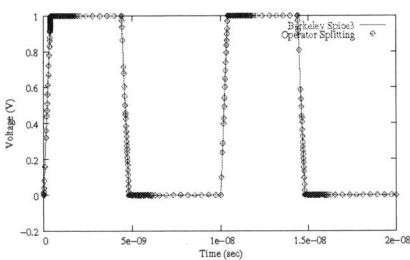

Fig. 7. Transient Waveform of 1K cell Design

D. ASIC designs

Two 1K and 10K cells ASIC designs are tested to demonstrate the proposed approach's ability of handling transistor dominated circuits. The 1k cell circuit has 10,200 nodes and 6,500 transistors. The 10k cell circuit has 123,600 nodes and 69,000 transistors. We assume that ideal power and ground supply is provided in those RC examples. The proposed approach takes 415.9 seconds for 1K cell circuit and 3954.7 seconds for 10K cell circuit to finish 20ns transient simulations. The speedup over SPICE3 is 5.1x and 11.2x for these two examples (Table I). We observe accurate waveform match for both examples. Figure 7 shows the transient waveform of a gate output in the 1K cell design.

VI. CONCLUSION

In this paper, we introduce an general unconditional stable operator splitting method for transistor level transient simulation. Orders of magnitude speedup over Berkeley SPICE3 is observed. Experimental results demonstrate accurate waveform match with SPICE3.

REFERENCES

[1] P. Feldmann, R. W. Freund, "Reduced-Order Modeling of Large Linear Subcircuits via a Block Lanczos Algorithm," DAC, pp. 376–80, 1995.

[2] A. Odabasioglu, M.Celik, and L.T. Pileggi, "PRIMA: Passive Reduced-Order Interconnect Macromodeling Algorithm," ICCAD, 1997.

[3] Z. Zhu, B. Yao, and C.K. Cheng, "Power Network Analysis Using an Adaptive Algebraic Multigrid Approach," DAC, pp. 105-108, 2003

[4] J. N. Kozhaya, S. R. Nassif, F. N. Najm, "Multigrid-like Technique for Power Grid Analysis," ICCAD, pp. 480-487, 2001

[5] T. Chen and C. Chen, "Efficient Large-Scale Power Grid Analysis Based on Preconditioned Krylov-Subspace Iterative Methods," DAC, pp.559-562, 2001.

[6] E. Acar, F. Dartu and L. T. Pileggi, "TETA: Transistor level Waveform Evaluation for Timing Analysis," IEEE Trans. on Computer-Aided Design, Vol. 21, No. 5, May 2002

[7] K.A.Sakallah and S.W.Director,"SAMSON2: An Event Driven VLSI Circuit Simulator," IEEE Trans. on Computer-Aided Design of ICs and Sytems, vol. 4(4), pp. 668-684, October 1985.

[8] www.nassda.com/hsim.html

[9] www.synopsys.com/products/mixedsignal/nanosim

[10] www.cadence.com/products/custom_ic/ultrasim/

[11] Z. Li, C. J. Shi, "SILCA: Fast-Yet-Accurate Time-Domain Simulation of VLSI Circuits with Strong Parasitic Coupling Effects," ICCAD, pp.793-799, 2003

[12] E. L. Wachspress and G. J. Habetler, "An alternating-direction-implicit iteration technique," J. Soc. Ind. and Appl. Math. 8, 403-424(1960)

[13] F. Zheng, Z. Chen, J. Zhang, "Toward the development of a three-dimensional unconditionally stable finite-difference time-domain method," IEEE Tran. Microwave Theory and Techniques, vol 48, No. 9, Sep 2000

[14] Y.-M Lee and C.P. Chen. "Power grid transient simulation in linear time based on transmission-line-modeling alternating-direction-implicit," ICCAD 75-80, 2001.

[15] Y.-M Lee and C.P. Chen. "The power grid transient simulation in linear time on 3D alternating-direction-implicit," Date 2003

[16] T. Namiki and K. Ito, "New FDTD algorithm free from the CFL condition restraint for a 2D-TE wave," IEEE Antennas Propagat. Symp. Dig., pp, 192-195, July 1999.

[17] W. F. Ames, "Numerical Methods for Partial Differential Equations," 2nd ed. New York Academic Press, 1977

[18] W. Sui, "Time-Domain Computer Analysis of Nonlinear Hybrid Systems," CRC Press, 2002.

[19] W. H. Press, S. A. Teukolsky,W. T. Vetterling, "Numerical Recipe in C," 2nd ed. Cambridge University Press, 1992

[20] W. Guo and S. X.-D. Tan, "Circuit level alternating-direction-implicit approach to transient analysis of power distribution networks," Proc. 5th International Conference on ASIC, Oct.2003. pp.246-249.

4D-1

2006 Asia and South Pacific Design Automation Conference

An Introduction to OpenAccess

An Open Source Data Model and API for IC Design

Michaela Guiney

Co-Chief Architect
OpenAccess Change Team
Cadence Design Systems
San Jose, CA 95134
Tel : 408-428-5923
Fax : 408-428-5380
e-mail:mguiney@cadence.com

Eric Leavitt

IC Framework Technology
Cadence Design Systems
San Jose, CA 95134
Tel : 408-894-2247
Fax : 408-428-5380
e-mail:eric@cadence.com

Abstract - The OpenAccess database provides a comprehensive open standard data model and robust implementation for IC design flows. This paper describes how it improves interoperability among applications in an EDA flow. It details how OA benefits developers of both EDA tools and flows. Finally, it outlines how OA is being used in the industry, at semiconductor design companies, EDA tool vendors, and universities.

I. Introduction

The purpose of this paper is to provide an introduction to OpenAccess, and how it differs from previous standards efforts in EDA data models. The improvements in interoperability and efficiency for design flows are described. The capabilities provided by the API for application developers are also discussed. Finally, the paper lists some of the companies and universities that are using OpenAccess, and where to get additional information.

II. What is OpenAccess?

OpenAccess is an advanced EDA database designed to enable interoperability among IC design tools through an open standard data access interface (API) and Reference Implementation of that API. The OpenAccess data model spans the EDA design space. It can be used to represent designs from post-synthesis netlists to tapeout.

Standardization of EDA data models have been attempted in the past without significant success. OpenAccess took a unique approach. In addition to defining a standard API and data model, OpenAccess provides a Reference Implementation. The implementation was donated by Cadence Design Systems, who contributed their leading edge IC design database in 2002. Cadence continues to develop and maintain OpenAccess as the Integrator, and has invested over eighty person years in the project. The availability of a high-quality implementation eliminates a barrier to using OpenAccess in design flows.

Changes to the OpenAccess API are managed by the

OpenAccess Coalition, a group of over 30 companies who are leaders in the EDA, electronics, and semiconductor business. Twelve members of the OAC are elected to the change team, which champions, refines, and approves changes to the API. Another key differentiator is that changes, once approved, must undergo user-level testing before they are integrated and distributed. This approach ensures the usability and quality of proposed changes. The OpenAccess effort is managed by Si2, the Silicon Integration Initiative.

The first release of OpenAccess was version 2.0, in January 2003. OA 2.1 followed in June 2003. In October 2004, OpenAccess 2.2 release was made available. OA 2.2 is the release targeted for production usage by most companies.

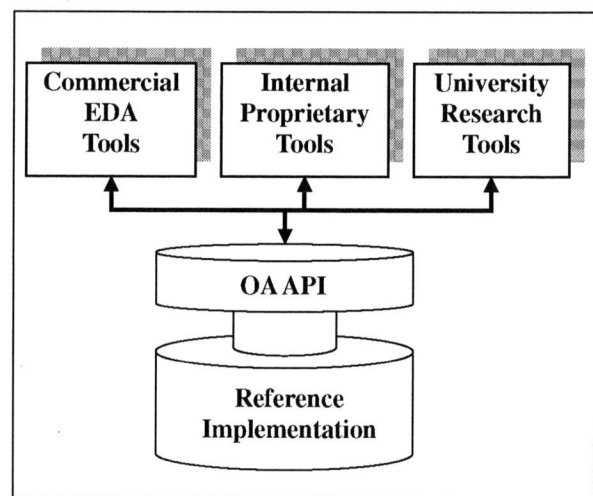

Fig. 1. The OpenAccess API and Reference Implementation : Available to All Tools

III. Why Use OpenAccess?

OpenAccess provides advantages to developers of both design flows and EDA tools. We will discuss both.

0-7803-9451-8/06/$20.00 ©2006 IEEE.

434

A. Advantages of OpenAccess in Design Flows

Today, many design flows use common file formats such as Verilog, DEF, GDSII, and SPEF to exchange information between tools. There are two main issues with this approach. First, the various data files for a design are usually incomplete and inconsistent. Second, each application which needs design information must parse one or more of these files and translate it into an internal data structure. This translation is inefficient, and can often misinterpret or lose information, due to ambiguities in the format specification. OpenAccess can solve both of these issues. First, the OpenAccess information model is more complete, unambiguous and consistent than the collection of data formats typically used. In addition, an OpenAccess database for a design can be read by applications through the API much more efficiently than these file formats can be translated, leading to greater efficiency in the overall flow. OpenAccess provides much tighter integration than was previously possible for a design flow with tools from multiple sources. The most efficient approach is for applications to operate directly on the OpenAccess data model. However, some applications may have algorithms that need specialized data structures. In such cases, applications can use OpenAccess data to build the specialized data structures, calculate their results, and store them back into OpenAccess.

Although a well-specified data model reduces the need to modify data in order to ensure it is correctly interpreted by the next tool in a flow, sometimes customer-specific needs require manipulating data and building customized tools.

Efficient access to design data via a C++ API enables more efficient flow customization than is possible when limited to accessing data through extension languages. OpenAccess users have seen dramatic reductions in runtime of applications ported to OpenAccess from extension languages and scripts. [1]

The ease of integration opens the door to new possibilities. University research developed on OpenAccess can be quickly integrated into a design environment. Exchange of design data and development tools with business partners can be achieved more easily. The effort required to evaluate new internal or commercial tools based on OpenAccess can be greatly reduced. Overall, OpenAccess can increase the ability to evolve flows to respond to new design challenges.

B. Advantages of OpenAccess for EDA tool Developers

OpenAccess is an object-oriented API written in C++. It was built from the start for open community use. Using C++ ensures a strongly typed interface, preventing many programming errors. Consistency was emphasized during the design of the API, in order to make it easier to understand and use. It also has extensive online documentation.

The OpenAccess information model covers a large portion of the EDA domain. It represents both logical and physical hierarchy and connectivity, as well as an occurrence model which relates the two. It includes custom geometry, routing topology and floor planning information, parameterized cells, scan chains, and technology information. It also provides a

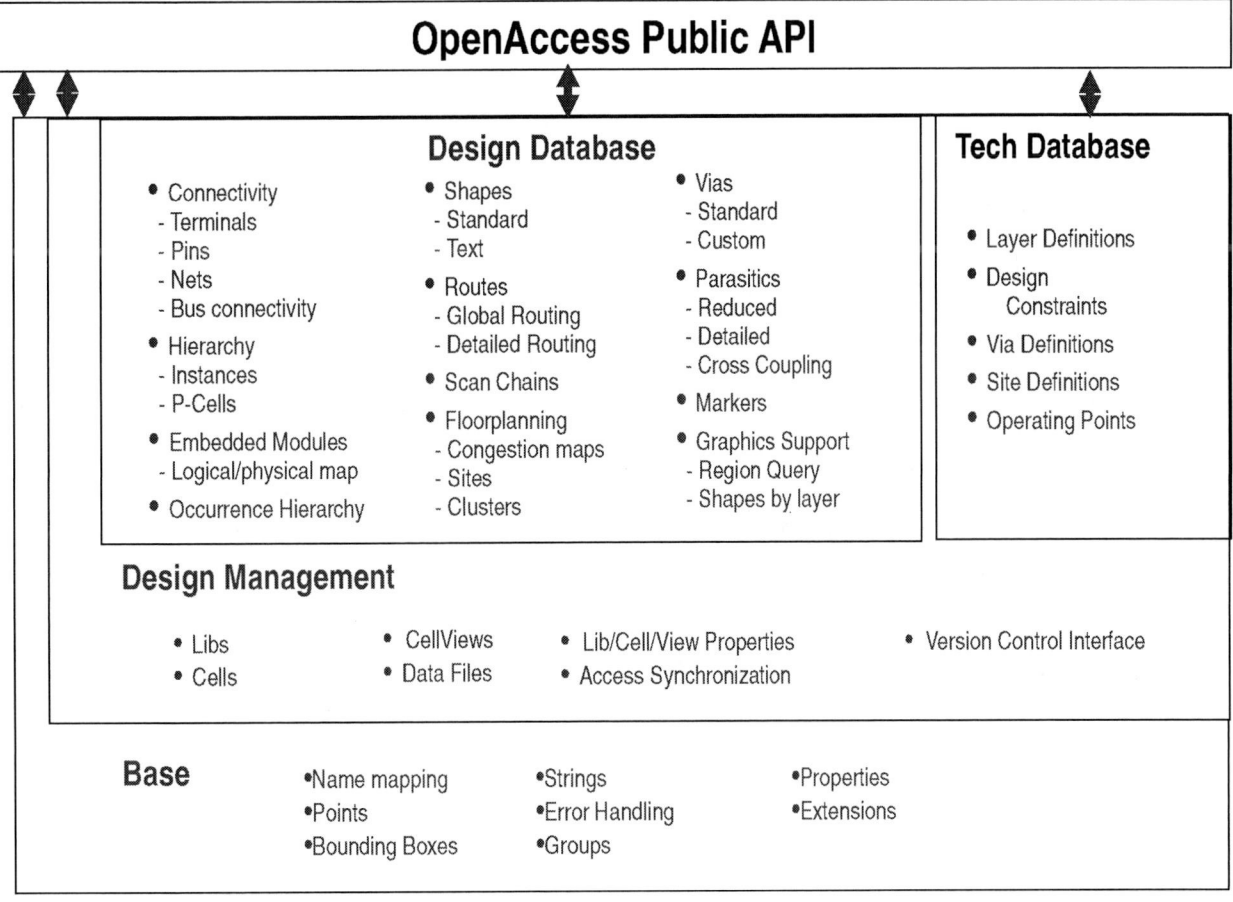

Fig. 2: Information modeled by OpenAccess

parasitic network API which enables access to parasitics on a per-net or even per-subnet basis, to minimize memory usage. An observer mechanism can be used by applications to take action when database objects are modified. Finally, the API supports defining extensions to most built-in objects as well as new kinds of objects; those extensions can optionally be saved. The extension mechanism is highly efficient, and can be used by developers to extend the database to support their application's needs. [2]

The API has been designed to support typical application access to EDA data. Various collections are supported, with filters for different types of usage. The API supports efficient searches, utilities such as Region Query, and name mapping capabilities. Multiple Design Management (DM) implementations are possible via a plug-in mechanism. The Reference Implementation has been tuned for improved performance and memory efficiency. Information and relationships are cached on the fly, depending on application access. Tcl, Python, and Skill extension languages are also available for use.

The breadth of the information model is one reason why many EDA tool developers have chosen OpenAccess as their development platform. The availability, quality and performance of the implementation provided by the Integrator (Cadence), and used in production by Cadence and other major EDA vendors, is another key factor. There are other factors as well. The availability of the source gives developers the best of both worlds. They can rely on the Integrator to provide maintenance and enhancements, but for time-critical bug fixes, they have the source available to them to fix issues themselves if required. Also, the quality and extent of the documentation surpasses that of any other EDA API available today, and certainly surpasses that of most internally deployed databases. Finally, the momentum behind OpenAccess ensures that their OA-based applications can interoperate efficiently and correctly with a growing number of tools.

IV. Where is OpenAccess Being Used Today?

Many companies are actively working with OpenAccess today. Freescale, IBM, AMD, HP, AMI, Renesas, and LSI Logic are just a few. LSI Logic is using OpenAccess in its RapidWorx design flow [3]. IBM is using some OA 2.2 based tools in their custom design flow [4]. Renesas is basing their integrated EDA system on OpenAccess [5] [6]. AMI Semiconductor is using OpenAccess as the basis for their flow to convert ASIC or FPGA designs to their environment [7]. Some companies are developing their own OpenAccess-based tools. Others are using tools from Cadence and other vendors based on OpenAccess 2.2. Cadence's custom design tools, digital implementation tools, extraction and physical verification tools all work with OpenAccess 2.2. In addition, a number of other EDA vendors have products that use OA 2.2, including MatrixOne, Mentor Graphics, Silicon Navigator, Cira Nova, Atrenta, and Gradient.

Universities such as UC Berkeley, Carnegie Mellon University, and the University of Michigan are using OA to build tools and toolkits to further research. They have put a set of tools together into a package called OA Gear. The package includes a viewer, a static timing engine, a placement interface, a functional/logic representation package and an and-inverter graph package. [8] In addition, North Carolina State University is independently utilizing OpenAccess to write tools for 3D thermal analysis and visualization. [9]

IV. Where To Get More Information

For more information on the OpenAccess Coalition, see www.si2.org/?page=69. Materials describing companies' experience with OpenAccess can be found through links to various conference and panel presentations at the Si2 "News" website, http://www.si2.org/?page=12. To download OpenAccess code and documentation, go to www.openeda.si2.org. For more information on the OA Gear project coordinated by Cadence Berkeley Labs, go to http://openedatools.si2.org/projects/oagear.

V. Summary and Conclusions

OpenAccess provides a comprehensive data model, API, and Reference Implementation for IC design. It is being adopted as the foundation by a number of EDA vendors, and is seeing more and more usage at IC design companies. The time is right for companies seeking efficient integration of tools and data in their design flows to investigate OpenAccess.

Acknowledgements

Thanks to the OpenAccess Development Team, the OpenAccess Coalition, the Change Team, and Si2.

References
[1] T. Heiter and J. Glas, "How Infineon Implemented OpenAccess", EETimes, November 11, 2004, http://www.eetimes.com/news/design/features/showArticle.jhtml;jsessionid=NHADQETGW5UI4QSNDBOCKHSCJUMEKJVN?articleID=52601068
[2] D. Mallis, OpenAccess 2.2: Standard API Tutorial, 2005. Available at www.si2.org
[3] R. Goering, "OpenAccess adoption challenging but worth it", EETimes, November 22, 2004, http://www.eedesign.com/article/showArticle.jhtml?articleId=53701169
[4] K. Barkley, "IBM: OpenAccess Adoption from End User's Perspective", 7th OpenAccess Conference, November 2005, http://www.si2.org/?page=613
[5] Y. Inoue, "OpenAccess: Reducing Handling Mistakes and Smoothing Logic Design and Layout Design", DesignWave Magazine, August 2005, page 99.
[6] I. Kojima, "Design View from Japan: Renesas Leverages OpenAccess Database to Pursue EDA Integration", NE Asia Online, http://neasia.nikkeibp.com/archive_magazine/nea/200405/columns_305267.php
[7] K. Reynolds, "AMI: OpenAccess Adoption from End User's Perspective", 7th OpenAccess Conference, November 2005, http://www.si2.org/?page=613.
[8] R. Davis, "OpenAccess Tools for 3D Integration", 7th OpenAccess Conference, November 2005, http://www.si2.org/?page=613.
[9] A. Hurst et al, "OpenAccess Gear Functionality: A Platform for Functional Representation, Synthesis and Verification", 7th OpenAccess Conference, November 2005, http://www.si2.org/?page=613.

OPEN ACCESS OVERVIEW "Industrial Experience"

Yoshio Inoue
Renesas Technology Corp
4-1, Mizuhara, Itami-shi,
Hyogo, 664-0005, Japan
inoue.yoshio@renesas.com

Abstract - Renesas Technology Corp. designers turned to OpenAccess to address the major design challenges with systems on chip for the automotive, wireless, digital consumer and industrial markets. OpenAccess provides Renesas with an industry standard database that has the capacity and performance needed for today's largest designs. The C++ API (C++ Application Programming Interface) facilitates fast access to a unified data model for both logical and physical design. It enables an efficient level of access to the data model to integrate tools developed in-house with commercially available tools for translation free interoperability.

I.Introduction

Renesas is advancing the development of design system (named REAP) for SoC that uses OpenAccess as the design database. In this session, the performance data of the thought of the REAP system development in Renesas, System configuration, Function, and a part of flow, function is described as "Industrial Experience".
Also, and it introduces the design rule checker that uses C++API and it Renesas originally has as an example of using more positive OpenAccess and the outline of the composition of the prototyping system

II.Renesas' Design Environment "REAP"

In design system REAP v1.X of RENESA, conversion, the generation at each format even if the data format that the tool supported by a general and EDA vender leaving and the management of the data of a variety of design road inside in the step of each design when SoC (System on a Chip) is designed and it develops handles is various and is the data of the same content as straightening, and leaving are current states.
The problem of often happening only has to notice the mistake at once by using old data by forgetting to generate the mistake of the data format that should be converted or data, and it will be discovered mostly as verification

trouble in the design step the next from now on. There is a thing that hangs for about 1-3 weeks though the problem is corrected and a necessary design step is done over again in the worst case, too.
The adoption of OpenAccess was decided to REAP v2.X as a design system that managed the data under the design and had various, new features.

It comes the problem of disappearing of the necessity that generates the format of the data of each EDA tool with the introduction of OpenAccess and manages on the site of each SoC design, and originating in a simple mistake that makes a mistake in the data format that should be converted is able to be able not only to be solved but also to connect EDA vender tool and an in-house development tool easily. The EDA design system can be smoothly constructed for that.
In RENESAS, it is prepared with OpenAccess.
 ・ Abundant C++ API and manual
 ・ LEF/DEF interface
 ・ Verilog interface
 ・ SPEF interface
 ・ Stream interface
 ・ Milkyway translator
 ・ API access environment of Tcl language base
is used, and the development of REAP v2.X is advanced.

The feature of REAP v2.X is dual scripts environment that values the interface of the TCL base that values continuance from REAP v1.X and the extendibility of REAP v2.X and builds in Python.
Other features are thing to have used the EMH function that OpenAccess offers positively. Because the correction generated on the layout system chip implementation side is reflected in the logical circuit correction dynamically with a logical hierarchy maintained, efficiently doing the static, logical agreement verification with the formal verification tool becomes possible. .

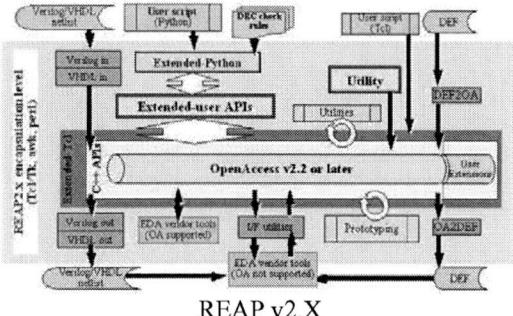

REAP v2.X

III. Summary

RENESAS developed the SoC design environment that used OpenAccess. REAP continuously is improved, enhanced, and comes to be offered to SoC designer outside the company.

EDA Vendor Adoption

Hillel Ofek
Sagantec

As stated in the introduction to the Open Access session, rapid IC technology advances towards deep sub-micron technologies produce ever growing pressure on EDA Vendors. This is evident when examining the requirements to support 65nm and 45nm IC technologies and beyond. The reason is simple. EDA design systems and complex design flows require close cooperation of various analysis and optimization tools that originate from multiple vendors. Also, sharing design data in memory is a performance and ease of use requirement. The current version of Open Access (OA) has evolved from a single vendor (Cadence) to a standard in which contributions from many member companies have been incorporated. It is now used to integrate tools from EDA vendors, internal development groups at semiconductor companies, and universities.

This paper will provide a view of Open Access from the vantage point of an EDA Vendor. It will address the positives and the challenges facing both the EDA Vendors and the EDA Users. Hopefully, the paper will contribute to the understanding of OA adoption, its benefits, and by way of example illustrate current state and show the way to broad adoption of this essential standard.

One can say that a lot has already been achieved by Open Access. Still, many things are needed for a real seamless integration of EDA software, particularly, a complete integration of technology data within OA (design-rules, process models)

Also, Interoperability benefits can only be realized when demanded by EDA customers (and incorporated in their design flows), not just by the EDA vendors.

The following talking points will illustrate the discussion delivered by this paper:

1. Today, OA promise of interoperability is not delivered for full custom design flows. For example, P-cell evaluation in custom IC design requires a plug-in (mechanism exists), which at the very least should be licensed by Cadence to all segments of the OA community.

2. Sagantec vision to "migrate" from proprietary layout design structure and File format to native OA is essential. When implemented in full, Sagantec tools will read the same database as other tools within the design flow. This will no longer require format translations and avoid format conversion issues. It will also allow us to build better solutions than before. An example of a better solution is the P-cell parameter selection. To accomplish that, OA needs a richer set of technology design-rules and it needs to be extended for design-manufacturing data-exchange (DTMC).

3. From an EDA vendor perspective, Sagantec is almost ready with OA implementation and usage. We are now waiting for customers to start adopting OA in their design flows so that our tools and other vendors' tools can be used smoothly.

Another topic of interest is the engineering benefit and cost of Open Access to an EDA vendor's engineering organization. This can be summarized by the following points:

1. Less interface effort

2. Influence on content

3. Less maintenance effort

4. Easier problem solving

5. Incorporation cost

6. Additional overhead

Additionally, we have had the experience of adding new technology to one of our products to create a new integrated product. An effort that usually takes a long time required just 2 months, thanks to the use of OA as a platform and an infrastructure. Essentially, OA served as a backplane to integrate technologies from different sources and different disciplines, by plugging in various old and the new code packages.

In summary, OA is an excellent vehicle to help in enabling users and developers of EDA tools to rapidly and simply construct complex design and DFM flows with components originating from many sources.

Utility of the OpenAccess Database in Academic Research

David. A. Papa
The University of Michigan
Department of EECS
Ann Arbor, MI, 48109-2122
iamyou@eecs.umich.edu

Igor L. Markov
The University of Michigan
Department of EECS
Ann Arbor, MI, 48109-2122
imarkov@eecs.umich.edu

Philip Chong
Cadence Berkeley Labs
Cadence Design Systems
Berkeley, CA, 94704
pchong@cadence.com

ABSTRACT

The proliferation of OpenAccess is opening promising new research opportunities to academic communities. The benefits of adopting an OpenAccess based approach to EDA research are growing, and we review a number of them. Among them are the ability to learn about a domain while writing software for it, increased ease of code reuse, new high-quality benchmarks, and enhanced industry adoption.

1. INTRODUCTION

Twenty years ago, there was a common design tool which was free and open-source and satisfied the needs of many academic circuit designers. The Magic VLSI Layout System introduced at DAC in 1984 [7] was an effective solution that enabled academic designers to comply with design rules of that time. However, the size of designs and complexity of design rules has increased drastically since then, and the strain created a need for newer more scalable tools. To enable collaboration one needs a common database, but maintaining software to handle the increased complexity in one academic group is hopeless. Given the amount of investment necessary to develop a single integrated approach, one can only hope to accomplish this task with the help of industry. CAD Framework Initiative, Inc. was formed in 1988 to organize these efforts [3]. Unfortunately, this work never took off, and until recently all major EDA companies had their own proprietary data models that were developed to support their specific flows. OpenAccess is a new attempt at the same goal and aims to meet the needs of everyone. CAD Framework Initiative is now known as SI2 [8] – an organization of industry-leading silicon systems and tool companies that maintain the OpenAccess standard. The collaborative technology developed by SI2 benefits academic researchers in a number of ways.

2. EDUCATIONAL PROTOTYPING

OpenAccess is a featureful data model which has been designed to efficiently perform the operations of typical VLSI design tasks. This should be leveraged when developing new algorithms, but is particularly helpful for prototyping.

Often in research, the easiest way to evaluate an optimization is to implement it and measure its impact empirically. When these optimizations are not helpful or cost-effective (e.g. too slow) they are abandoned in favor of a more promising direction. Therefore it is desirable to quickly develop prototypes that can be used to extrapolate if an optimization is viable.

To support academic prototyping efforts, the free, open-source OpenAccess Gear toolkit provides several useful EDA components, such as an incremental static timing analysis tool and an interface to the Capo placement engine [6]. Such tools allow users to use OpenAccess to build complex flows on top of OpenAccess which would otherwise be difficult or costly to implement in an academic environment, such as timing-driven placement flows. Recently several new features have been added to OpenAccess Gear, including a logical representation layer, which allows academics to build physically-driven synthesis tools on top of OpenAccess. Such an ability to inexpensively and rapidly prototype interesting tools in universities is made possible only due to the open nature of OpenAccess and OpenAccess Gear.

3. CODE REUSE

One of the greatest motivations for development in OpenAccess is the increased ease of code reuse. Once a *service* or *transformation* of a VLSI design is defined for the OpenAccess data model, it can be used as a part of larger OpenAccess-based applications without costly conversions from one data representation to another. Such straightforward reuse is particularly attractive to academia because it avoids the burden of extending home-grown data models to efficiently handle the operations of new algorithmic components brought in from 3rd party sources.

One prominent case of this type of reuse is in the creation of a timing-driven version of the Warp placer [9]. Xiu et al. estimate that the development effort was reduced from 3 months to 1 month by the ability to reuse the OpenAccess Gear timer. Because of the incremental nature of timing queries in the process of placement, using a standalone timer would not be adequate. Rather, calls to the timer must be integrated into placement. Since Xiu et al. were able to reuse timer code, they saved themselves the effort of writing a timing analyzer required for timing-driven placement.

4. HIGH-QUALITY BENCHMARKS

The reported improvements in many physical design papers are so small that it is necessary to check improvements against additional realistic designs. To this end, the importance of a diverse set of high-quality benchmarks has been demonstrated for VLSICAD tools [1]. That work notes several problems with open benchmark suites such as designs that are too small, artificially constructed, or missing important information.

Many of these problems are addressed in a new open suite of designs in the OpenAccess and Verilog formats [5]. This suite contains 84 real designs with up to 185k registers and 900k gates. All designs are mapped with Cadence RTL Compiler to a 180nm library which is included in the suite. Using a tool that comes free with OpenAccess Gear [6], physical information is generated for these designs including core area, core rows and routing tracks. In addition, much of the information which has been lost in previous benchmark suites is preserved for these designs, such as signal directions of nets and logic information of cells. The presence of this information enables new experiments of interest to the academic community. For example, signal directions allows one to examine and optimize signal paths, and logic information allows for detecting equivalence of signals – also useful in circuit optimization. This information is available in some small benchmarks, but not on large realistic designs.

5. INDUSTRY ADOPTION

The International Technology Roadmap for Semiconductors [2] calls for a shift in the architecture of VLSI design systems into an

Copyright is held by the author/owner.
Copyright 200X ACM X-XXXXX-XX-X/XX/XX ...$5.00.

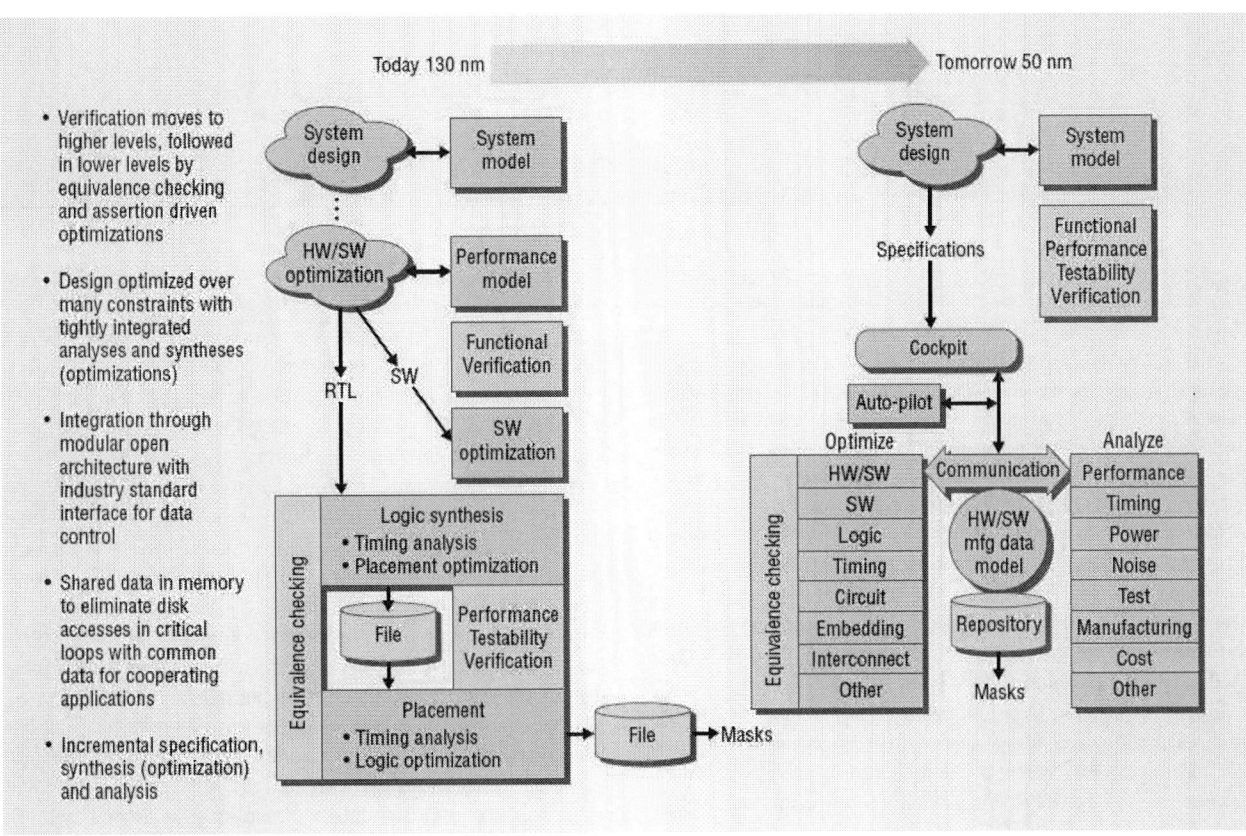

Figure 1: This figure from the 2001 ITRS shows that as tools become more interoperable, design system architecture will change toward a more integrated approach. OpenAccess is one example of the Repository in the architecture at 50nm.

integrated system of interoperable tools. Central to this theme is a single data model used by all tools, with analysis tools used to drive optimization steps. Figure 1 shows how design system architecture will evolve with increased tool interoperability. OpenAccess is likely to lead the way in this worthwhile paradigm shift. Until recently this functionality was unavailable to academia due to lack of infrastructure. With OpenAccess, a flow similar to the right flowchart of Figure 1 is possible. This allows research on reordering of optimization steps of traditional VLSI design flows, e.g., applying logic optimization after placement [4].

Industry partners will be much more likely to adopt research software that works with the same data model, and is built using the same architecture as their commercial tools. This is mutually beneficial because it allows industry to reap immediate benefit from new research, and thus benefits academic research by encouraging industry to support it.

6. CONCLUSIONS

OpenAccess brings many benefits to academic researchers including improved open benchmarks, easier code reuse, faster prototoying and enhanced industry adoption. However, there are some downsides associated with adopting an approach using OpenAccess. Converting established software to run natively on OpenAccess can be a huge endevour. Additionally, since many academic groups distribute open-source software, licensing conflicts can arise with OpenAccess. This is because many open-source licenses are less restrictive than the current OpenAccess license which can restrict users of open-source software. This means that supporting Open-

Access in open-source software is likely to be an addition to, rather than as a substitution for existing data models.

7. ACKNOWLEDGEMENTS

The authors thank the OpenAccess Coalition and the OpenAccess development team for providing the OpenAccess database to the academic community under an open-source license, allowing academic research to move closer to industry needs. The authors also acknowledge the contributions of the OpenAccess Gear team, especially Zhong Xiu, Christoph Albrecht and Andreas Kuehlmann, who made the initial release of OpenAccess Gear possible.

8. REFERENCES

[1] S. N. Adya et al., "Benchmarking for Large-Scale Placement and Beyond," *IEEE Trans. on CAD*, vol. 23, no. 4, 2004, pp. 472–487.

[2] A. Allan et al., "The International Technology Roadmap for Semiconductors (ITRS), 2001 edition," *International Sematech*, Austin Texas, 2001.

[3] "CAD Framework Initiative, Inc.," http://www.si2.org/?page=42/

[4] K.-H. Chang, I. L. Markov and V. Bertacco, "Post-Placement Rewiring and Rebuffering by Exhaustive Search For Functional Symmetries," to appear, *ICCAD* 2005.

[5] "IWLS 2005 benchmarks," http://iwls2005/benchmarks.html

[6] "OpenAccess Gear," http://openedatools.si2.org/projects/oagear/

[7] J. K. Ousterhout et al., "Magic: A VLSI Layout System," *DAC* 1984, pp. 152–159.

[8] "SI2," http://www.si2.org/

[9] Z. Xiu et al., "Early Research Experience with OpenAccess Gear: An Open Source Development Environment for Physical Design," *ISPD* 2005, pp. 94–100.

Depth-Driven Verification of Simultaneous Interfaces

Ilya Wagner   Valeria Bertacco   Todd Austin

Advanced Computer Architecture Lab

University of Michigan

Ann-Arbor, MI 48109

e-mail: {iwagner, valeria, austin}@umich.edu

Abstract— The verification of modern computing systems has grown to dominate the cost of system design, often with limited success as designs continue to be released with latent bugs. This trend is accelerated with the advent of highly integrated system-on-a-chip (SoC) designs, which feature multiple complex subcomponents connected by simultaneously active interfaces.

In this paper, we introduce a closed-loop feedback technique targeting the verification of multiple components connected by parallel interfaces. We utilize an environment with hierarchical Markov models, where top-level submodels specify overarching simulation goals of the system, while lower-level submodels specify the detailed component-level input generation. Test accuracy is improved through the use of depth-driven random test generation. The approach allows users to specify correctness properties and key activity nodes in the design to be exercises. We examine three non-trivial designs, two microprocessors and a chip-multiprocessor router switch, and we demonstrate that our technique finds many more bugs than constrained-random test generation technique and reduces the simulation effort in half, compared to previous Markov-model based solutions.

I. INTRODUCTION

Systems-on-a-chip (SoC) are becoming predominant in application-specific domains, such as portable media devices, network processors, and so on. The development of an SoC entails the integration of multiple heterogeneous integrated circuit (IC) blocks, with a mix of components developed in-house and acquired through third-party IP, producing extremely complex IC systems in a relative short time. Unfortunately, the verification of SoC has become a bottleneck in the design process, as it commonly absorbs 70% of the design costs. In fact, the International Technology Roadmap for Semiconductors (ITRS) has determined that the most important productivity challenge to be overcome, in order to maintain the current growth trends in SoC development, is precisely verification [4].

Two aspects of SoC make their functional verification particularly challenging: first, the integration density offered by state-of-the-art fabrication technologies is impressive: a typical SOC could contain hundreds of millions of transistors. The result is a large complex stateful design that does not lend itself to existing verification techniques, either because of the small fraction of state space that simulation-based approaches can explore, or because the sheer complexity is beyond the grasp of current formal and semi-formal technology. Second, SOCs feature many heterogeneous autonomous components connected by multiple simultaneously active interfaces, often abiding to complex communication protocols. The inability of exploring all the possible interactions among these components further underscores the challenge of SoC verification.

A variety of techniques have been explored to assist the designer in locating design bugs on systems such as this. Simulation-based random test generation is a long-standing approach used to locate design errors [5, 9, 10, 8, 12]. However, due to the huge state spaces of even simple devices, it is impossible to achieve sufficient confidence in the correctness of a design by just using random generation techniques. Formal and semi-formal solutions can help by providing powerful mechanisms to achieve high coverage in verification, but can only be deployed on the smallest components of a design. The need for solutions that can address both the high-coverage and design size requirements has been recently voiced by Intel, which predicted that by 2007 designs will deploy 100M transistors and require 2000 person-years of effort for verification [13]. The work presented here attempts a step in this direction by proposing a novel solution which targets both high-coverage and scalability in the verification of complex systems with multiple parallel interfaces.

A. Contributions of This Work

In this paper, we introduce a novel closed-loop test generation technique targeting the construction of high-quality tests for large complex designs with multiple, simultaneously active interfaces. Our solution generates constrained, automatically-biased tests by modeling the interfaces' communication protocols through abstract Markov models, and by sampling the system's reactions through signal sensors placed at relevant internal nodes of the designs. Specifically, the work makes two novel contributions compared to previous efforts in this field:

1. The complications of modeling the input protocols and the interactions among multiple simultaneous interfaces are mitigated through the use of a **hierarchical modeling environment**. The environment dictates the specification of a high-level Markov model (through the use of simple templates) concerned with describing the overarching testing goals. Subsequently, this high-level model spawns off lower-level Markov models – threads – that specify component-specific input generation.

2. To address complexity challenges, our constrained random test generation technique introduces a **depth-driven test quality evaluation** technique. In this context, the

user hints to the system the critical regions to cover through the specification of key signals within the design. The test generator then tunes its focus on input stimuli that exercise the logic closer to these key signals (where "closeness" is defined in terms of logic depth from the key signal, hence the name).

In addition, we implement our proposed test generation infrastructure in a tool called **IQTest** (Interface Quality Tester), and demonstrate that it is capable of finding more bugs than constrained random testing techniques, and with less simulation effort than previous Markov-model based solutions.

The remainder of this paper is organized as follows. Section II highlights previous related work. Section III overviews the high-level architecture of *IQTest*, while section IV provides details on the novel aspects of the work. Rationale and details of the implementation of *IQTest* are also discussed. Section V presents experiments that compare IQTest to previous solutions and naive random testing. We examine three non-trivial designs, two microprocessors and a chip-multiprocessor network router switch. Finally, Section VI gives conclusions and suggests future directions.

II. PRIOR WORK

Random test generation has long been a focus of industry and the academic research community [6]. The key aspects of systems that differentiate the verification solutions are whether or not simulation is pure random or directed, how constraints on the random tests are specified, and the mechanics of the underlying random test generation engine.

Recent advances in random test generation have focused on generating feedback on previous tests to influence the generation of future tests. Tools such Specman Elite [2] and Vera [1] provide on-the-fly data assertion and checking and methods for validation of generated tests. In both cases the generation process is directed by dynamically changing constraints based on functional coverage analysis. Although these tools simplify the work of the end user with GUIs and powerful verification languages, most of the test set up and decision process is still left to the verification engineer, who must specify functional test plans [2] or implement constraint adjustment policies [1].

A number of tools have been developed to enable verification engineers to have more control over the generation of random tests. In particular, some of these techniques involve the use of program templates that define the structure of the desired test, along with primitives to control the randomization of the related data, such as opcodes, register operands, and memory addresses [3]. Most tools employ a coverage-directed generation process, but use sophisticated techniques for representing relationships between coverage and input generation through Markov-models (as in this work) [11, 14] or with Bayesian networks and computer learning [7].

StressTest was recently proposed as a technique to implement closed-loop feedback directed random testing [14]. The tool is based on an abstract representation of the input model, specified using a template based specification language. In ad-

dition, users specify "activity monitors" which represent signals in the design that correspond to coverage concerns, so that the underlying simulation engine directs simulation towards the excitation of the activity monitors. Our solution, implemented in *IQTest*, borrows the specification language and Markov-model test generation from [14]. However, we make a number of significant advances over that work. In particular, we extend the template language to include support for hierarchical specifications, which simplifies the modeling of simultaneous interfaces. Moreover, we provide a highly responsive and accurate link between activity monitors and model response, through the use of a depth-driven activity analyzer.

III. AN OVERVIEW OF IQTEST

Figure 1 illustrates the high-level architecture of IQTest's random testing infrastructure. There are four primary inputs to the system (colored in gray): 1) the design under test (DUT), 2) a known-correct golden model, 3) the input template specification, and 4) the activity signals. Based on the input template, IQTest generates an input sequence which is fed to both the DUT and the golden model. The outputs of the two models are compared and, if any discrepancy is noted, a bug is detected.

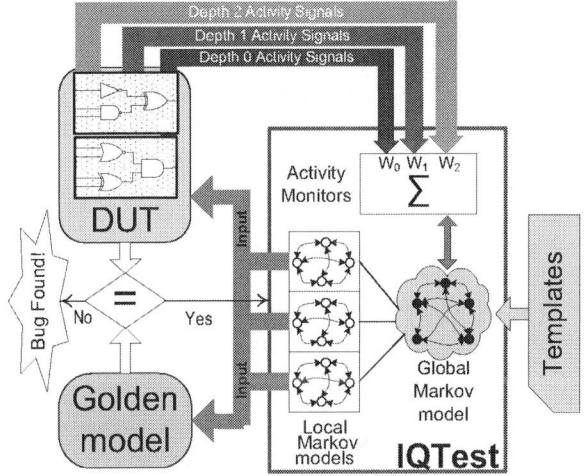

Fig. 1. IQTest Structure. A template describing the legal DUT stimuli is used by IQTest to generate input sequences. In addition, an activity analyzer closes the feedback loop by scoring the quality of the stimuli. Bugs are detected by validation against a golden model.

Markov models and Templates. The Markov models are in charge of the stimuli generation in IQTest. In this context, a Markov model is a graph where each vertex describes a legal input sequence for the design. For instance, if the input interface was a processor's instruction bus, a vertex could describe a single instruction, or a program segment. Or, if the input interface was a bus protocol, than a vertex could describe a possible transaction. The vertices in the graph are connected by edges labeled with the probability of performing a graph transition through them: at the beginning of the test generation, all the outgoing edges from a vertex have equal probability. Probabilities are adjusted over time so as to bias the stimuli towards the interesting "activity", as specified by the "activity signals".

Moreover, the input sequence described by each vertex incorporates random aspects, based on the template specification. As discussed in more detail in Section IV, templates are text files providing a model for the input sequences at each vertex of the Markov graph. For more details on the basics of templates and the related Markov model construction see [14]. To address the specific needs of SoC verification, we devised a novel technique to structure the Markov models hierarchically, so that a system with multiple, simultaneously active interfaces could be stimulated by IQTest through all its input channels asynchronously and in parallel.

Activity Signals and Analyzer. The activity signals are internal nodes of a design under test selected by a user because they are representative of critical activity in a component. Examples of such signals are collision indicators between different ports of a network switch (which can be used to check the correctness of the switch at high utilization), or a branch misprediction signal (used to verify a pipeline recovery mechanism). The activity analyzer gathers the switching activity of these signals and steers the test generation by indicating to IQTest which transactions in the Markov models generate the highest activity, and thus are most relevant in exercising critical circuitry. During simulation, the edges of the Markov models are adjusted continuously so that high-activity transactions are associated to higher probabilities.

Activity signals may be any design's internal node deemed relevant by the user, or checkers (that is, properties that we are trying to falsify). In the latter case, IQTest focuses on activating the output signal of the checker: the detection of switching activity at such a node corresponds to having triggered the checker. Note that, particularly in the case of checkers, the activity observed at the node would be non-existent during the whole simulation, until when the checker is fired, thus reducing IQTest to a mere constrained random test generation with no adaptive biasing. To solve this problem, IQTest introduces a novel depth-driven activity feedback solution. The approach works as follows: from the signals selected as activity sensors, we derive an additional set of auxiliary signals, whose values closely affect the value of the activity sensors. The activity monitor then estimates the change to apply to the Markov models' edges based on a weighted sum of the activity in the extended set of signals. The weights are heavier for the activity sensors and lighter for the auxiliary signals. We call the approach "depth-driven" because the weights are inversely proportional to the logical depth of a given auxiliary signal from the sensor. We find that this approach reaches significantly more bugs than simple constrained random simulation.

IV. ARCHITECTURAL INSIGHTS

In this section we focus on the two main contributions of this paper, namely:

- The ability to specify the communication protocols of the DUT hierarchically. The architecture of *IQTest* and the template language we defined facilitates the simulation and verification of simultaneous interfaces.

- The activity analyzer, our technique to produce an accurate analysis of the design response to a stimulus transaction. At each simulation step we consider the signal transitions at multiple nodes in the design and evaluate the quality of the last transaction by weighing this data based on the circuit depth from the critical node

A. Hierarchical Specification of the Stimulus Generator

For the specification of the input protocol at each interface of the design, we deploy a hierarchical Markov model that generates valid legal input sequences based on the template specifications. In previous work proposing random input generation based on a Markov model [14], the model was used to partition the set of inputs of a microprocessor core and generate instruction sequences based on activity feedback obtained from the design. However, it is often the case that a design has multiple parallel input interfaces, unlike pipelines which can be viewed as having only one stimuli entry point. A good example of such a design is a network switch or a crossbar that has multiple ports. In such situations, it is critical to be able to generate stimuli at each individual port that are time- and data-independent. Therefore, it is often more desirable to generate multiple input streams in parallel and observe the interactions between them. Since often the hardest bugs in an SoC design are found when multiple input requests are competing for the same hardware resources, we deemed crucial to allow the designer to create sequences of input stimuli where the traffic at each interface is independent, so that it becomes easier to produce relevant input sequences. Finally, a methodology that allows for hierarchical specification of the stimuli leads to a description of the input transactions that is simpler and easier to understand.

To cope with this problem we devised a new approach that employs a multi-level Markov model. Each of the parallel inputs of the circuit is assigned to an individual Markov model for generating valid input stimuli according to the interface specifications. However, some information between these models can be shared to increase the competition for the resources and intensify the pressure on the design. To allow this information passing between the models, we use a global variable space that is accessible from any model. Note that the models can still generate valid input sequences independently of their peers through local variables.

In addition to the individual Markov models assigned to input ports of the design, a global model is used to encode possible scenarios of simultaneous stimuli generation. For example, the model can determine the number of ports of the design activated simultaneously and values of the controlling inputs to the design. The objective of the global Markov model is to coordinate local models assigned to individual design ports. Note that individual models supply sequences of stimuli with dependencies between them, while the global model orchestrates them to exert simultaneous pressure on the design. The activity feedback from individual inputs of the design in this framework is used to reinforce transitions in local models. Moreover, combined activity measures from different points in the circuit are used for adjusting edges in the global Markov model.

```
/ Global variable space /
global {
   dest(probCache=0.7,cacheSize=8,lambda=.5,
                            minVal=1,maxVal=15);
   srcW(probCache=1,cacheSize=20,lambda=3)={'b0000, 'b1111};
}
/ Global Markov model/
none : TopModel(global) {
   rand-send-one(probCache=1)={'b1000,'b0100,'b0010,'b0001};
   command[3:0] : { portD, portC, portB, portA };
   vertex(send_one_pkt) {  command = 'bCCCC;
                           field(C)=$rand-send-one.read(); }
   vertex(send_all) { command = 'b1111; } };
/ Local Markov models descriptions/
switchPort {
   using global::dest;
   using global::srcW;
   vertex (message) {  input='bDDDDSSSS;
                       field(D)=$dest.read();
                       field(S)=$srcW.read();
                       $dest.write(field(D)); } };
burstingPort {
   Bsrc  (probCache=0,minVal=0,maxVal=15);
   vertex (message){  input='b0100SSSS;
                      field(S)=$Bsrc.read();
                      $Bsrc.write(field(S)); } };
/ Local Markov models binding /
dut.port_a : switchPort(portA) ;
dut.port_b : burstingPort(portB) ;
dut.port_c : switchPort(portC) ;
dut.port_d : switchPort(portD) ;
```

Fig. 2. Example Template file. Shown are a global Markov model along with two local models bound to each port of a network switch.

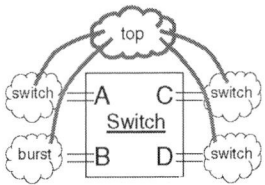

Fig. 3. Switch and hierarchical Markov model for the template example.

An example of a hierarchical template file shown in Figure 2 indicates how hierarchical Markov models can be used to model different interfaces of a DUT. See [14] for details of Markov model and variable specification. In the case shown in the example, the DUT is a network switch with four simultaneous interfaces *port_a* to *port_d*. The two input-generating models *switchPort* and *burstingPort* produce different kinds of data packets, while the global top level Markov model creates different scenarios and orchestrates the work of the packet generators. The messages described here consist of a packet with 8 bits of data.

The top portion of the template contains the global variable space with two random variables, *dest* and *srcW*. They are visible to any of the local models that declare using them via a *using* clause, as shown in model *switchPort*. The global Markov model either runs a scenario where input is sent to only one port or where it is sent to all ports simultaneously, by signaling to the low level models with command bits. Model *switchPort* produces packets with source and destination fields generated by accessing the global variables, while model *burstingPort* produces all messages to destination 0100.

B. Depth-Driven Activity Monitoring

One important assumption that was made in prior work related to Markov-model based testing is that a handful of key activity signals is sufficient to guide the simulation towards areas of interest and expose hard-to-find bugs. However, we found this approach to be somewhat coarse. In other words, the scores reported by the activity monitors manifest bipolar behavior. Therefore the system was likely, after just a handful of cycles, to strongly reinforce input sequences leading to high activities and almost eradicate the possibility of generating inputs that lead to low activity rates.

In our improved test generation methodology, a different approach was taken. We have created a tool that traverses the hierarchy of the design and extracts additional signals that influence the behavior of the selected activity points. Preference was given to control signals that were relatively easy to identify from a register-transition level design description. The signals that directly influence the primary activity points were assigned depth one, the signals that influence them were assigned depth two, and so on. The activity analyzer incorporates these secondary signals using a weight proportional to depth of the signal exercised (Figure 4-left). Note that the selection is done automatically and does not require any additional user effort.

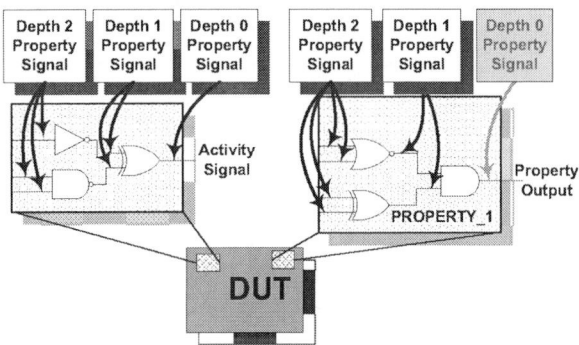

Fig. 4. Depth-driven activity monitoring selects auxiliary signals to monitor based on their logic depth from the property output or activity signal

In the context of a verification methodology that attempts to falsify properties embedded in a design, we deploy a specialized adapted technique for the evaluation of the activity analyzers. Often, a design includes several checker modules that track signals vital to system operation, and are triggered when a property is violated. Typically, these checkers are derived from the design specification or embedded by the developer. The analysis of the outputs of these checkers is the critical aspect of validating a design's correctness, however, since these properties are never asserted until the corresponding bug is exposed, the guidance provided to the input generator is poor. The input signals to the property expression, on the other hand, change frequently and can be selected automatically from the description of the checker (Figure 4-right). As it is shown in our results, this depth-directed activity analysis technique is able to achieve greater bug coverage with less effort, compared to approaches that observe only a handful of key points.

445

V. EXPERIMENTAL EVALUATION

In this section, we introduce our experimental evaluation framework and the designs we tested. We then compare the performance of our proposed technique against an open-loop random instruction generator and a recently proposed Markov-model based test generator, comparing both coverage of bugs and number of simulated instructions required to expose them.

A. Experimental Framework

Evaluation of the random test generators was performed on three designs: two microprocessors and a chip-multiprocessor router switch. The two microprocessor designs that we tested were the DLX pipeline (MIPS-Lite ISA) and a DEC Alpha processor core. For both designs we wrote a behavioral golden model, against which the correctness of the design was compared. Also we created 30 cores for the DLX and 10 for Alpha, each containing one bug. The simplest bugs included incorrect opcode interpretation and erroneous ALU operations, while the hardest ones could only be exposed through complex interactions between instructions. The templates for these experiments were derived directly from ISA of the processors, with one vertex corresponding to one instruction type. The amount of effort needed to create the templates was under 8 man-hours. The chip-multiprocessor router switch was included because it features simultaneous parallel interfaces. The switch was initially designed for testing performance and traffic patterns of different routing algorithms. In all experiments, we used a version of the switch utilizing an adaptive cut-through minimal-path routing algorithm for two-dimensional mesh networks. The design consists of five input ports with three virtual channels each, five output ports, and crossbar logic.

For the switch experiments, the generator was a hierarchical Markov model with the top level model specifying the number of packets to send simultaneously. The top model also specified the state of the back pressure signals in the network surrounding the switch. The local Markov models were used to generate valid packets that were likely to have similar destinations, thereby exerting high pressure on the switch. The amount of effort to write templates for both global and local Markov models for the switch was around 16 man-hours. To guide the test generation during switch verification, we utilized checkers written in Verilog. We derived ten distinct checkers from the high level description of the switch routing algorithm and buffer functionality. Each property module had a single-bit output, which depended only on particular signals in the design. The *Depth-0* experiment only monitored the output bits of the properties, while IQTest monitored the inputs to the properties, *i.e. Depth-1* signals, as well.

Incidentally, we were able to find three actual design bugs during verification of the switch: one in the buffer control logic and two in the routing logic of the crossbar. All of these bugs were hard corner cases of the switch's behavior, for example, in one bug several internal counters were incorrectly handled during a buffer-overflow situation, however from the error could be seen several tens of cycles later.

B. Results and Analysis

Since our verification mechanism uses a random generator in its kernel, each buggy design in both experiments was run 25 times with different random seeds and afterwards we calculated average effort and coverage. In the first test, the maximum allowed time to search for a bug was limited to 75000 cycles for both DLX and Alpha. For performance evaluation each bug was checked by an open-loop constrained random generator (*Random*). We also compared IQTest against *StressTest*, which is based on a closed-loop Markov model structure but it observes only a handful of crucial control signals in the pipeline. *IQTest* itself was implemented with different depths of the observed signals, labeled correspondingly (*Depth-1, Depth-2, and Depth-3*).

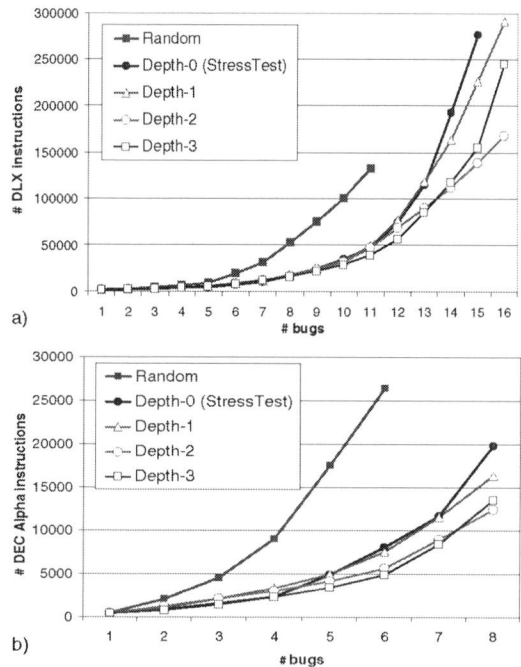

Fig. 5. Effort vs. Bug coverage for DLX (a) and Alpha (b) cores

Figure 5 shows the effort in terms of number of instructions produced, of these three techniques vs. the number of bugs discovered. Here we present the results for the 16 hard bugs in DLX and 8 hard bugs in the Alpha design. Since the easy bugs were found by all techniques in just several hundred instructions, for these bugs there is no clear differences between the approaches. As can be seen in both graphs, the curve for *Random* stops short of the rightmost edge, that is, the hardest bugs, while both *StressTest* and *IQTest* were able to find all the bugs in the Alpha pipeline. Note that *StressTest* was unable to find the most complex bug in the DLX pipeline. In general, we note that a deeper activity analysis is better for harder bugs.

The results of the switch verification experiment are presented in Figure 6. The effort in this case is measured in number of packet transactions required to find a bug. Again only the hard to find bugs are reported, since all three systems performed equally well in discovering easy bugs. The three bugs

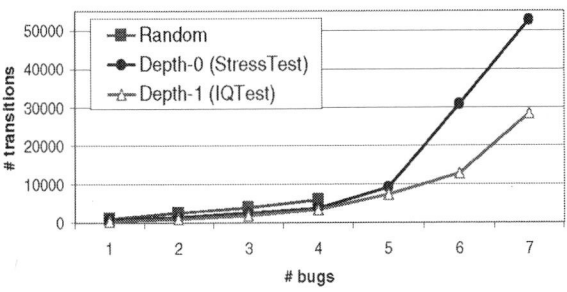

Fig. 6. Effort vs. Bug coverage for the switch design

TABLE I - TIME COMPARISON OF RANDOM AND IQTEST

	DLX	Alpha	Switch
	up to bug 25	up to bug 8	up to bug 7
Random	135.5s	35.7s	39.1s
Depth-0	N/A	N/A	32.1s
Depth-1	70.1s	21.3s	31.6s
Depth-2	75.7s	21.0s	N/A
Depth-3	79.1s	21.5s	N/A

on the far right of the graph are the three original bugs found during the verification of the switch. Once again, the *Random* configuration was unable to discover the last three bugs and requires significant effort for covering easier bugs. That explains why the bugs were not discovered during the design of the switch where only random testing was used. As can be seen in the graph, the approach of monitoring just the output bits of the property modules performs relatively well for intermediate bugs. We believe that is partially because our approach enabled information sharing between local Markov models. Moreover, this system explores more distinct input sequences than *Random*. This is because observing property outputs produces low activity scores that stimulate negative reinforcement in the Markov model, thus continuously steering the generation away from past sequences. The third approach produces positive and negative reinforcement and therefore significantly reduce the effort needed to uncover the bugs.

Finally, in Table I, we compare the wall-clock performance of *IQTest* vs *Random* (we run on a 1 GHz UltraSPARC IIIi machines with 1GB of RAM). Since *Random* discovers only a fraction of the bugs that IQTest finds, we used the best achievements of *Random* as a reference. As shown in the table, IQTest performs always better than Random. And after Random runs out of steam, IQTest keeps finding more complex bugs.

VI. CONCLUSIONS

In this paper, we introduced a random simulation technique targeting SoC devices, with stateful components connected by parallel interfaces. The approach is based on a Markov-model driven random simulation, with two novel enhancements: To produce effective random vectors for multiple components with parallel interfaces, we utilized a hierarchical modeling environment where "global" models specify overarching simulation goals of the system, while the "local" models specify the specific component-level input generation. To address the statefulness of SoC devices, we utilize a depth-driven random simulation search engine. The search engine allows users to specify correctness properties and key activity points in the design that are of particular concern. A closed-loop feedback system then tailors the hierarchical Markov-models into one that stresses the points of concern in the design.

We found that the hierarchical specification and depth-driven random test generation are quite effective in reducing the amount of effort required to expose bugs in complex designs, both in terms of the amount of human effort to craft the verification scripts and the amount of simulation required to reach the bugs. We examined three non-trivial designs: two microprocessors and a chip-multiprocessor router switch. We demonstrate that hierarchical specification of input models yields a compact precise representation. Additionally, we showed that depth-driven random simulation finds more bugs more quickly than simple constrained random simulation. Moreover, we found that depth-driven simulation cuts the simulation effort in half, compared to a recently published Markov-model based verification technique.

REFERENCES

[1] Constrained-random test generation and functional coverage with Vera. Technical report, Synopsys, Inc, Feb. 2003.

[2] Specman elite - testbench automation, 2004. http://www.verisity.com/products/specman.html.

[3] A. Adir et al. Genesys-pro: Innovations in test program generation for functional processor verification. *IEEE Design & Test of Computers*, 21(2):84–93, 2004.

[4] A. Allan et al. 2001 technology roadmap for semiconductors. *IEEE Computer*, pages 42–53, Jan. 2002.

[5] B. Bentley. Validating the Intel Pentium 4 microprocessor. In *DAC, Proceedings of Design Automation Conference*, pages 224–228, 2001.

[6] E.A.Poe. Introduction to random test generation for processor verification. Technical report, Obsidian Software, 2002.

[7] S. Fine and A. Ziv. Coverage directed test generation for functional verification using bayesian networks. In *DAC, Proceedings of Design Automation Conference*, 2003.

[8] I.Silas et al. System-level validation of the Intel Pentium M processor. *Intel Technology Journal*, 07:38–43, May 2003.

[9] J. M. Ludden et.al. Functional verification of the POWER4 microprocessor and POWER4 multiprocessor systems. *IBM Journal of Research and Development*, 46:53–76, Jan. 2002.

[10] Y. Levhari. Verification of the PalmDSPCore using pseudo random techniques. Technical report, VeriSure Consulting, Ltd.

[11] S. Tasiran et al. A functional validation technique: Biased-random simulation guided by observability-based coverage. *ICCD, Proceedings of the International Conference on Computer Design*, pages 82–88, 2001.

[12] S. Taylor et al. Functional verification of a multiple-issue, out-of-order, superscalar Alpha processor: The DEC Alpha 21264 microprocessor. In *DAC, Proceedings of Design Automation Conference*, pages 638–644, 1998.

[13] G. Spirakis. Opportunities and challenges in building silicon products in 65nm and beyond. In *Design and Test in Europe (DATE-2004)*, 2004.

[14] I. Wagner, V. Bertacco, and T. Austin. Stresstest: An automatic approach to test generation via activity monitors. In *DAC, Proceedings of Design Automation Conference*, 2005.

FSM-Based Transaction-Level Functional Coverage
for Interface Compliance Verification

Man-Yun Su, Che-Hua Shih, Juinn-Dar Huang, and Jing-Yang Jou

Department of Electronics Engineering
National Chiao Tung University
Hsinchu, Taiwan, R.O.C.
e-mail:{powmei, matar}@eda.ee.nctu.edu.tw, jdhuang@mail.nctu.edu.tw, jyjou@faculty.nctu.edu.tw

Abstract – Interface compliance verification plays a very important role in modern SoC designs. In order to perform a quantitative analysis of simulation completeness, adequate coverage metrics are mandatory. In this paper, we propose a finite state machine (FSM) based transaction-level functional coverage methodology for interface compliance verification. A language, State-Oriented Language (SOL), is developed to specify functional transactions mainly at the higher FSM level instead of lower logic or signal level. By utilizing SOL, it is simple and rigorous to specify interesting transactions from the specification FSM of the target interface protocol. Experimental results show that the proposed methodology can effectively improve the verification quality as well as increase the efficiency of regression verification.

1. Introduction

In designing a modern system-on-a-chip (SoC), the platform-based design methodology with reusable intellectual property (IP) cores is usually adopted to accelerate the design and verification process [1]. Each pre-verified IP core is wrapped with certain interface logic and integrated into a system platform which is based on that interface protocol. In order to ensure that each component can concordantly communicate with others within the system, it is very important to guarantee that the interface logic of each utilized IP core conforms to the protocol. Hence, interface compliance verification becomes an essential part of the SoC verification flow.

Though there are numerous existing functional verification methods, simulation is still the most commonly used technique. During simulation, coverage metrics are usually adopted to perform a quantitative analysis of simulation completeness. Coverage metrics can not only measure how well a design is verified objectively but also help improve the quality of verification patterns. That is, they are capable of guiding either direct (deterministic) or random patterns to target those unverified design corners. Therefore, exploring adequate coverage metrics is a very crucial issue in today's functional verification.

In general, there are two major categories of coverage metrics [2]: code coverage and functional coverage. Code coverage methods concentrate on identifying which part of the hardware description language (HDL) code has been

executed in the design under verification (DUV). That is, they measure how much of the HDL implementation has been exercised [3-6]. For example, statement coverage, branch coverage, and condition coverage are well-known code coverage metrics. However, the fundamental issue of all code coverage metrics is that they can only measure how well the structural HDL code has been exercised. They are not sufficient to represent the whole functionality of the design specification. Namely, the verification quality is generally considered not enough for modern complex SoC designs even if a high code coverage is achieved. Thus, the functional coverage is usually applied to further boost the verification quality.

Functional coverage, as its name implies, focuses on the design functionality. It measures how much of the original design specification has been verified. That is, the coverage is independent of the details of HDL implementation, and thus is considerably hard to measure. Many methods are proposed to address this issue. In [7], a user-defined cross-product coverage measurement tool is developed. In [8-9], the cross-product functional coverage is further improved either in quality or efficiency. In [10-11], the specification must be first given as a proprietary graph. Then the functional coverage analyzer can be automatically generated by traversing the graph. The methods mentioned above do really help interface compliance verification. However, these techniques generally require users to specify what they want to cover in proprietary input formats or languages uncommon to typical designers.

In this paper, we propose a transaction-level functional coverage methodology and provide a means to specify functional transactions at a higher FSM level, which is popular and familiar to most designers. First, the interface protocol is given as a specification FSM (spec FSM) by using the concepts in [12-13]. Then *a transaction can be defined as a specific sequence of state transitions within the spec FSM*. Meanwhile, we develop a transaction description language, State-Oriented Language (SOL), which is capable of modeling diverse state transition sequences precisely and rigorously. The transactions can then be specified in an easier and more readable way even by common designers. Moreover, the specified transactions with the spec FSM can be further translated into the corresponding functional coverage analyzer automatically.

0-7803-9451-8/06/$20.00 ©2006 IEEE.

The rest of this paper is organized as follows. Section 2 introduces the basic concepts and the related works of transaction-level functional coverage. In Section 3, the state-based transaction description language SOL and the details of our verification methodology are presented. Section 4 demonstrates the proposed methodology with the AMBA AHB slave protocol and shows the experimental results. Finally, the conclusions are given in Section 5.

2. Transaction-level functional coverage

As mentioned, functional coverage is favorable to improve the verification quality. Transaction-level functional coverage is one of the commonly used methods to measure the functional coverage for an interface design [13-16]. An interface specification usually defines a set of different transaction types. A transaction can be considered as the transfer of data and control over an interface to perform certain basic operation. For example, a transaction can be a 4-beat burst or an 8-beat burst, or a 4-beat burst followed by an 8-beat one. Transaction-level functional coverage is generally measured by how many types of transactions are exercised. However, two designs may have different sets of interesting transactions even if they comply with the same interface protocol. Therefore, the interesting transactions of a given design are usually derived manually.

Several approaches are proposed for the transaction-level functional coverage. For M-path coverage [13], the protocol is first modeled as a spec FSM. Then an M-path is defined as a path of state transitions which can form a complete bus transfer in the FSM model. In other words, an M-path, which is a finite sequence of state transitions, is actually a simple transaction. M-paths are used as the targets for coverage measurement.

In [14], Component Wrapper Language (CWL) is used to describe signal sequences based on regular expressions. In CWL, the input and output signals must be declared first. Then signal values at each cycle are defined as signal sets. Next, each simple transaction is modeled by utilizing the defined signal sets. Finally, a more complex transaction can be built up by assembling simple ones. In this approach, values of individual signals are required when describing thorough transactions. If the interesting transactions are getting more complex, it might be troublesome and time-consuming to author the corresponding CWL descriptions.

In general, it is tedious and error-prone for human to specify transactions if the detailed signal values are required. To cope with this issue, it is a better idea to provide a simple, human-friendly, rigorous, and systematic way to specify transactions at a higher level of abstraction instead of at the signal level. In our work, the interface protocol is specified as a spec FSM by using the methods in [12-13]. *A transaction can then be defined as a specific sequence of state transitions. This enables the use of states in the spec FSM as basic elements to describe transactions.*

The proposed method can raise the transaction description to the FSM level which is well understood by most designers. It facilitates the encapsulation of the details of low-level signals so that the detailed signal values at each cycle are no longer required. Hence, one can put more emphasis on the functionality at the familiar FSM level.

3. Proposed approach

3.1. Our methodology

In this paper, we propose an FSM-based transaction-level functional coverage methodology. In order to provide a means to specify transactions at the FSM level, we develop a transaction description language, State-Oriented Language (SOL), mainly based on the Property Specification Language (PSL) [17]. Because PSL provides a richer set of expressive and readable language constructs than typical regular-expression-based approaches do, SOL adopts most PSL constructs used to describe temporal sequences. In SOL, the PSL-like syntax is used to represent *a sequence of state transitions*. Though SOL is similar to PSL, the fundamental conceptual difference between them is that *SOL uses states as the atomic elements when defining a transaction*. Hence, it is easier for designers to author complex state-based transactions by using SOL.

The flow of our methodology is illustrated in Figure 1. The interface protocol needs to be specified as a spec FSM first. Note that the spec FSM can be translated into an interface protocol checker [13]. Meanwhile, the interesting transactions are manually specified by using SOL. These transactions with the spec FSM are further translated into a functional coverage analyzer automatically. Next, we simulate the whole system, including the DUV, verification patterns, checker, and coverage analyzer. According to the outcome of the checker, we can know if the DUV conforms to the interface protocol. From the coverage analyzer, the report tells how many interesting transactions have been verified or not. Moreover, the coverage information can guide the development of either direct or random patterns to hit those unverified corner cases.

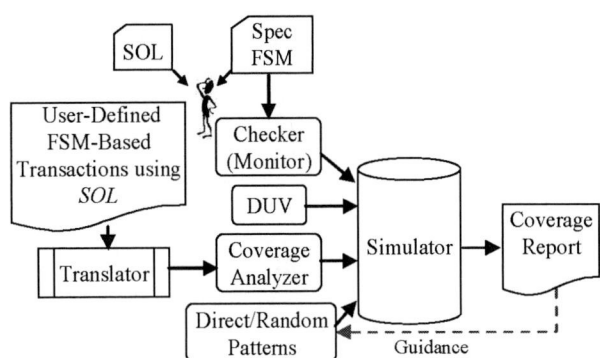

Figure 1. The flow of our methodology.

3.2. The transaction description language SOL

The syntax of SOL is based on the following principles:
- Since a transaction is defined as a specific sequence of state transitions in the spec FSM. *States* are used as basic elements to describe sequences.
- In order to keep the spec FSM as simple as possible, extra signals can be included in additional to the states while defining a transaction.
- A sequence can be defined once as a *named sequence* and then be reused later. The *assignment* operator is used to define a named sequence. The left-hand-side of the assignment operator becomes a synonym for the sequence on the right-hand-side.
- Sequence name is enclosed in *braces* when referred.
- A *sequence set* comprises one or more sequences. Sequences are enclosed in *angle brackets* and separated by *commas*.

The syntax of SOL is briefly introduced below (shown in shaded area). The FSM shown in Figure 2 is taken as an example to introduce operators in SOL.

3.2.1. Extra signal qualification ("").
Extra signals can be qualified while making a state transition. The *Boolean expression* built from the extra signals should be enclosed in *double quotes*.

3.2.2. Concatenation (;).
Two sequences can be concatenated into one by the concatenation operator.

Example 1 In Figure 2(a), T1 is a transaction with the state transitions that starts from S1, then moves through S3, S4, and ends at S1.

T1: S1 → S3 → S4 → S1

T1 = { S1 ; S3 ; S4 ; S1 };

Example 2 In Figure 2(b), T2 is another transaction with the same state transitions sequence as T1 while the extra signal V must be true when moving from S1 to S3.

T2 : S1 $\xrightarrow{V=1}$ S3 → S4 → S1

T2 = { S1 "V == 1" ; S3 ; S4 ; S1 };

3.2.3. Repetition ([ ]).
The repetition operators are used to describe repeated concatenations of a sequence. There are three types of the repetition operators: consecutive repetition ([*]), non-consecutive repetition ([=]), and goto repetition ([→]).

(a) consecutive repetition ([*]).

Example 3 In Figure 2(a), T3 is a transaction with the state transitions that starts from S1, moves to S2, and stays at S2 for three consecutive cycles, then ends at S1.

T3 : S1 → S2 → S2 → S2 → S1

T3= { S1 ; S2[*3] ; S1 };

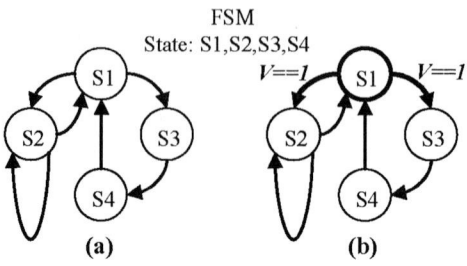

Figure 2. An example FSM.

Example 4 In Figure 2(a), T4 is a transaction with the state transitions that starts from S1, moves to S2, and stays at S2 for one to five consecutive cycles, then ends at S1.

T4 : S1 → S2 (1~5 cycles) → S1

T4 = { S1 ; S2[*1:5] ; S1};

(b) non-consecutive repetition ([=]).

Example 5 In Figure 2(a), T5 is a transaction with the state transitions that starts from S1, and then visits S2 three times. The visits of S2 need not to be in consecutive cycles. In addition, T5 holds after the 3rd S2 is visited and still holds before the 4th S2 appears.

T5 : S1 →...→ S2 →...→ S2 →...→ S2 →...→ S2 →...

T5 = { S1 ; S2[=3] };

(c) goto repetition ([→]).

Example 6 In Figure 2(a), similar to T5, T6 is also a transaction with the state transitions that starts from S1, and then moves to S2 three times (can be non-consecutive). In addition, T6 holds only at the cycle in which the 3rd S2 is visited.

T6 : S1 →...→ S2 →...→ S2 →...→ S2 →...→ S2 →...

T6 = { S1 ; S2[→3] };

3.2.4. Sequence AND (&&).
The transaction comprising two sequences using the sequence AND operator holds only if both sequences hold and complete at the same cycle.

Example 7 In Figure 2(a), similar to T6, T7 is also a transaction with the state transitions that starts from S1, and then visits S2 three times (can be non-consecutive). However, S3 is strictly not allowed showing up in the sequence T7.

T7: S1 →...(!S3) → S2 →...(!S3) → S2 →...(!S3) → S2

T7 = { S1 ; {S3[=0]} && S2[→3] };

3.2.5. Sequence OR (|).
The transaction comprising two sequences using the sequence OR operator holds if one of two alternative sequences holds.

Example 8 In Figure 2(a), T8 is a transaction shown below,

T8 : S1→S3→S4→S1 OR S1→S2→S2→S2→S1

T8 = { {S1;S3;S4;S1} | {S1;S2[*3];S1} };

Note that above two sequences are previously defined as T1 and T3. Hence, T8 can also be defined in terms of these named sequences.

T8 = { {T1} | {T3} };

3.2.6. Sequence fusion (:). Similar to the concatenation operator, a sequence fusion operator concatenates two sequences overlapping by one cycle.

Example 9 In Figure 2(a), T9 is a transaction shown below,

T9 : S1→S3→S4→S1→S2→S2→S2→S1

T9 = { S1;S3;S4;S1;S2[*3];S1 };

T9 can also be treated as two sequences that overlap each other for one cycle as shown below:

T9 : S1→S3→S4→S1 : S1→S2→S2→S2→S1

T9 = { {S1;S3;S4;S1} : {S1;S2[*3];S1} };

Again, T9 can also be defined in terms of T1 and T3.

T9 = { {T1} : {T3} };

3.2.7. Sequence set cross (**). A sequence set cross operator is used to represent a set of back-to-back consecutive transactions.

Example 10 Assume the following 8 transactions are interesting.

{{T1}:{T3}:{T8}}, {{T1}:{T4}:{T8}}, {{T1}:{T3}:{T9}}, {{T1}:{T4}:{T9}}, {{T2}:{T3}:{T8}}, {{T2}:{T4}:{T8}}, {{T2}:{T3}:{T9}}, {{T2}:{T4}:{T9}};

The following expression utilizing the sequence set cross operator provides a much more elegant but equivalent representation for the set of 8 interesting transactions.

<{T1},{T2}> ** <{T3},{T4}> ** <{T8},{T9}>;

3.3. SOL examples

To apply our methodology, the interface protocol should be given as a spec FSM first. The details about how to construct a spec FSM can be found in [12-13]. The AMBA AHB slave interface protocol [18] is adopted here to demonstrate how to define transactions in SOL. The spec FSM of the simplified AMBA AHB slave protocol is given in Figure 3.

Example 1 1-beat burst transaction.

A 1-beat burst transaction basically means the given design moves to the state NSEQ/SEQ (S1) one time and can not move to the state ERROR (S4), i.e.,

{{S4[=0]} && {S1[→1]}}

In addition, a 1-beat burst transaction consists of two cases. One starts from the state ORIG (S0), which indicates the slave is just selected and going to do the first transaction. The other starts from the state NSEQ/SEQ (S1), which implies the slave is already selected and going to do another transaction. Besides, the signal HBURST must be set to 0 for a 1-beat burst transaction.

(1) starting from the state ORIG (S0) :

One_S0 = {S0 "HBURST==0";{S4[=0]}&&{S1[→1]}};

(2) starting from the state NSEQ/SEQ (S1) :

One_S1 = {S1 "HBURST==0";{S4[=0]}&&{S1[→1]}};

The 1-beat burst transaction is composed of the sequence One_S0 and the sequence One_S1 by using a sequence OR operator. That is,

One = {{One_S0} | {One_S1}};

Example 2 4-beat burst transaction.

Similar to a 1-beat burst transaction, a 4-beat burst one also consists of two cases. But the design must visit the state NSEQ/SEQ (S1) four times. The signal HBURST should also be set to 2 or 3 for a 4-beat transfer.

(1) starting from the state ORIG (S0) :

Four_S0 = {S0 "HBURST==2 || HBURST==3"; {S4[=0]} && {S1[→4]}};

(2) starting from the state NSEQ/SEQ (S1) :

Four_S1 = {S1 "HBURST==2 || HBURST==3"; {S4[=0]} && {S1[→4]}};

The 4-beat burst transaction can then be written as,

Four = {{Four_S0} | {Four_S1}};

Example 3 A 4-beat burst transaction instantly followed by an 8-beat write burst transaction.

A 4-beat burst transaction (i.e., Four) is defined before, and an 8-beat write burst transaction (i.e., EightWrite) can also be specified in the similar way. Since the required transaction can be defined by fusing these two transactions, it can be written as {{Four}:{EightWrite}}; .

4. Experiments

4.1. Experimental environment

To demonstrate our methodology, we choose the AMBA AHB slave interface protocol [18] as an example. The spec FSM of the simplified AHB slave protocol is given in Figure 3. Figure 4 illustrates the experimental environment used in this work. It consists of three parts: a DUV, a constraint-driven random pattern generator, and the proposed verification framework.

(1) The experiments are conducted over three real AHB slave designs. The basic information of these designs is shown in Table 1. The design RGB2YCrCB is an RGB-to-YCrCB color space converter. The design MAC is a multiply-accumulator. The design Convolution is a convolution calculator to be used in discrete wavelet transfer.

(2) The constraint-driven random pattern generator is an AHB master which generates verification patterns based on

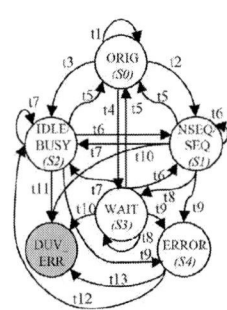

t1:~HSEL
t2:HSEL•(HTRANS=NSEQ)•HRADY
t3:HSEL•(HTRANS=IDLE)•HREADY
t4:HSEL•~HREADY
t5:~HSEL•(HRESP=OKAY)
t6:HSEL•(HTRANS=NSEQ||SEQ)•
   HREADY•(HRESP=OKAY)
t7:HSEL•(HTRANS=IDLE||BUSY)•
   HREADY•(HRESP=OKAY)
t8:HSEL•~HREADY•(HRESP=OKAY)
t9:HSEL•~HREADY•(HRESP=ERROR)
t10:HSEL•HREADY•(HRESP≠OKAY)
t11:HSEL•(~HREADY+HRESP≠OKAY)
t12:HSEL•(HTRANS=IDLE)•HREADY•
   (HRESP=ERROR)
t13:HSEL•(~HREADY+HRESP≠ERROR)

Figure 3. The spec FSM of the simplified AMBA AHB slave protocol.

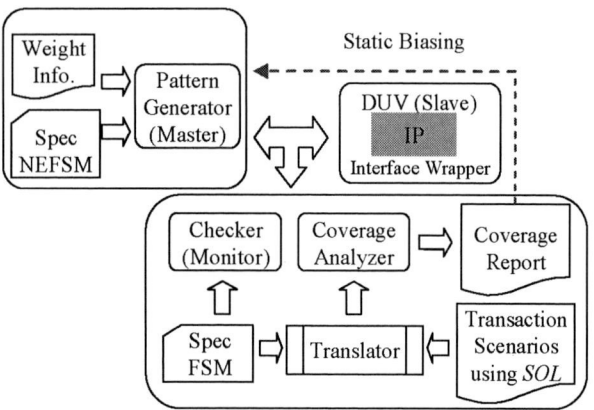

Figure 4. Experimental environment.

an NEFSM (Non-deterministic Extended FSM) with the weighted state transitions. The weight of each transition is configurable. The transitions are assigned with an *equal weight* initially.

(3) We develop a translator which accepts the spec FSM and user-defined SOL transactions then produces the corresponding coverage analyzer. The reported coverage is used to help statically bias the random pattern generator to create more effective verification patterns.

4.2. Experimental results

Two experiments are conducted: coverage comparison and efficiency improvement. In the first experiment, four coverage results (state, state transition, M-path, and our transaction coverage) are compared for three designs, respectively. In the second experiment, the coverage information is sent back to bias the random pattern generator to produce more effective patterns.

4.2.1. Coverage comparison

Case 1. The interesting transactions are defined as 10 basic read and write transactions, e.g., {OneRead};, {OneWrite};, {FourRead};, etc.

The comparison results are shown in Table 2. For the design RGB2YCrCb, it takes 4/16/82/492 cycles to reach 100% state/transition/M-path/transaction coverage. As the state/transition/M-path coverage reach 100%, the transaction coverage is only 0/10/20%. For the other two designs, the results are similar. It is observed that the transaction coverage is very low while the other three coverage metrics reach 100%.

Table 1. Design information.

Design	Supported AHB responses	# of state/transition/M-path		
RGB2YCrCb	OKAY	3	8	14
MAC	OKAY, ERROR	4	10	12
Convolution	OKAY (wait)	4	10	16

Table 2. Coverage comparison for Case 1.

Design	Coverage	# of cycles to reach 100%	Transaction coverage (%)
RGB2YCrCb	State	4	0 (0/10)
	Transition	16	10 (1/10)
	M-path	82	20 (2/10)
	Transaction	492	100 (10/10)
Design	**Coverage**	**# of cycles to reach 100%**	**Transaction coverage (%)**
MAC	State	61	30 (3/10)
	Transition	61	30 (3/10)
	M-path	33	10 (1/10)
	Transaction	9644	100 (10/10)
Design	**Coverage**	**# of cycles to reach 100%**	**Transaction coverage (%)**
Convolution	State	12	10 (1/10)
	Transition	47	20 (2/10)
	M-path	102	30 (3/10)
	Transaction	787	100 (10/10)

Case 2. Make the interesting transactions more complex by adding 15 more transactions with BUSY/WAIT (e.g., {OneWithWAIT};,{FourWithBUSY};,etc.) and 25 consecutive transactions (e.g., <{Incr},{One},{Four},{Eight},{Sixteen}>** <{Incr},{One},{Four},{Eight},{Sixteen}>;).

The comparison results are shown in Table 3. For the design Convolution, it still takes 12/47/102 cycles to reach 100% state/transition/M-path coverage. But it takes 11135 cycles to reach 100% transaction coverage. As the state/transition/M-path coverage reach 100%, the transaction coverage is only 4/8/12%. It is shown that the transaction coverage is even lower than that in **Case 1** as the other three coverage metrics reach 100%.

We get some conclusions from the above 2 cases. While the set of interesting transactions becomes larger and more complex, it needs a significantly (non-linearly) longer simulation time to reach 100% transaction coverage. Moreover, even the state/transition/M-path coverage reach 100%, the transaction coverage can still be extremely low. The situation is getting worse when more complicated transactions are concerned. It means that even a pattern set developed to reach 100% state/transition/M-path coverage may not provide a satisfied functional coverage. Experimental results exactly show that the classical coverage metrics are not capable of providing enough verification quality.

4.2.2. Efficiency improvement

After analyzing the coverage report of **4.2.1 Case 2**, we find the major reason why so many cycles are required to reach 100% transaction coverage is the seldom occurrence of BUSY transactions. Hence, it is possible to reduce the simulation time by statically biasing the pattern generator. The biasing information is shown in Table 4.

In *bias1*, we intuitively increase the weights of transitions that can generate BUSY transactions. This biasing indeed decreases the simulation time to 1864 cycles, which is only 16.7% of the original one. In *bias2*, the weights of INCR burst, 1-beat burst, 4-beat burst, 8-beat

Table 3. Coverage comparison for Case 2.

Design	Coverage	# of cycles to reach 100%	Transaction coverage (%)
Convolution	State	12	4 (2/50)
	Transition	47	8 (4/50)
	M-path	102	12 (6/50)
	Transaction	11135	100 (50/50)

Table 4. Efficiency improvement.

Design	Bias	# of cycles to reach 100%	Factor
Convolution	equal weight	11135	1
	bias1	1864	0.167
	bias1 + bias2	981	0.088

burst, and 16-beat burst are given in *decreasing order* because the BUSY transaction takes place more frequently in long-beat transfers. Combining *bias1* with *bias2*, the simulation time can be further decreased to 981 cycles, which is only 8.8% of the original one.

The results show that the coverage information can help bias the random pattern generator to create more effective patterns and help verify the DUV in a shorter time. This technique is extremely useful while developing a regression verification environment in which the compact and effective pattern suites are crucial to minimize the required simulation time. That is, the proposed methodology can increase the efficiency of the regression verification process.

5. Conclusions

In the paper, we propose an FSM-based transaction-level functional coverage methodology for interface compliance verification. To provide a familiar, user-friendly, but still rigorous, and systematic way to specify transactions at a higher FSM level, we develop a PSL-like transaction description language SOL. The expressive power of SOL is generally stronger than that of previous regular-expression-based approaches. It is shown that SOL is capable of modeling very complicated functional transactions. Meanwhile, a translator is also developed to automatically convert a set of SOL-based transactions with the spec FSM into the corresponding functional coverage analyzer. The experimental results demonstrate that the proposed methodology can indeed improve the verification quality as well as increase the efficiency of regression verification. In a near future, we plan to develop a technique that can automatically and dynamically bias the pattern generator by instantly analyzing the functional coverage on-the-fly and then integrate this technique into our methodology.

References

[1] M. Keating and P. Bricaud, "Reuse Methodology Manual for System-On-A-Chip Designs, 3rd Edition," *Kluwer Academic Publishers*, July 2002.

[2] J. Bergeron, "Writing Testbenches: Functional Verification of HDL Models, 2nd Edition," *Kluwer Academic Publishers*, February 2003.

[3] D. Drako and P. Cohen, "HDL Verification Coverage," *Integrated System Design Magazine*, pp. 46-52, June 1998.

[4] F. Fallah, S. Devadas, and K. Keutzer, "OCCOM: Efficient Computation of Observability-Based Code Coverage Metrics for Functional Verification," *Proceedings of the Design Automation Conference*, pp. 152-157, June 1998.

[5] P. A. Thaker, V. D. Agrawal, and M. E. Zaghloul, "Validation Vector Grade (VVG): A New Coverage Metric for Validation and Test," *Proceedings of the IEEE VLSI Test Symposium*, pp. 182-188, April 1998.

[6] B. Min and G. Choi, "ECC: Extended Condition Coverage for Design Verification Using Excitation and Observation," *Proceedings of the Pacific Rim International Symposium on Dependable Computing*, pp. 183-190, December 2001.

[7] R. Grinwald, E. Harel, M. Orgad, S. Ur, and A. Ziv, "User Defined Coverage - A Tool Supported Methodology for Design Verification," *Proceedings of the Design Automation Conference*, pp. 158-163, June 1998.

[8] S. Asaf, E. Marcus, and A. Ziv, "Defining Coverage Views to Improve Functional Coverage Analysis," *Proceedings of the Design Automation Conference*, pp. 41-44, June 2004.

[9] A. Ziv, "Cross-product Functional Coverage Measurement with Temporal Properties-based Assertions," *Proceedings of the Design, Automation and Test in Europe Conference and Exhibition*, pp. 834-839, March 2003.

[10] Y.-S. Kwon, Y.-I. Kim, and C.-M. Kyung, "Systematic Functional Coverage Metric Synthesis from Hierarchical Temporal Event Relation Graph," *Proceedings of the Design Automation Conference*, pp. 45-48, June 2004.

[11] Y.-S. Kwon and C.-M. Kyung, "Functional Coverage Metric Generation from Temporal Event Relation Graph," *Proceedings of the Design, Automation and Test in Europe Conference and Exhibition*, pp. 670-671, February 2004.

[12] Y.-C. Yang, J.-D. Huang, C.-C. Yen, C.-H. Shih, and J.-Y. Jou, "Formal Compliance Verification of Interface Protocols," *Proceedings of the IEEE International Symposium on VLSI Design, Automation, and Test*, pp. 12-15, April 2005.

[13] H.-M. Lin, C.-C. Yen, C.-H. Shih, and J.-Y. Jou, "On Compliance Test of On-Chip Bus for SOC," *Proceedings of the Asia and South Pacific Design Automation Conference*, pp. 328-333, January 2004.

[14] K. Ara and K. Suzuki, "A Proposal for Transaction-Level Verification with Component Wrapper Language," *Proceedings of the Design, Automation and Test in Europe Conference and Exhibition*, pp. 82-87, March 2003.

[15] C. Browy, "Comparing TestWizard and Specman for Transaction-level Verification," white paper, available at http://www.avery-design.com/twwp.html.

[16] H.-J. Schlebusch, G. Smith, D. Sciuto, D. Gajski, C. Mielenz, C. K. Lennard, F. Ghenassia, S. Swan, and J. Kunkel, "Transaction based design: Another Buzzword or the Solution to a Design Problem?," *Proceedings of the Design, Automation and Test in Europe Conference and Exhibition*, pp. 876-877, March 2003.

[17] Property Specification Language – Language Reference Manual, Ver. 1.1, http://www.eda.org/vfv/docs/PSL-v1.1.pdf.

[18] ARM Limited, *AMBA Specification (Rev 2.0)*, May 1999.

5A-3

Hardware Debugging Method Based on Signal Transitions and Transactions

Nobuyuki Ohba   Kohji Takano

IBM Research, Tokyo Research Laboratory, IBM Japan Ltd.
Yamato city, Kanagawa, Japan 242-8502
Tel: +81-46-215-4547
Fax: +81-46-273-7413
e-mail: {ooba, chano}@jp.ibm.com

Abstract - This paper proposes a hardware design debugging method, Transition and Transaction Tracer (TTT), which probes and records the signals of interest for a long time, hours, days, or even weeks, without a break. It compresses the captured data in real time and stores it in a state transition format in memory. It can be programmed to generate a trigger for a logic analyzer when it detects certain transitions. The visualizer, which shows the captured data in the matrix, timing-chart, and state-transition diagram formats, helps the engineer effectively find bugs.

I. Introduction

System On Chip (SoC) design is widely used to boost the performance, lower the power consumption, and reduce the overall system costs by integrating many resources. In ASIC/SoC design, however, growing design complexity has forced engineers to tackle deeper bugs, and thus the development requires more work in the test and debugging. Hardware prototyping is widely used [1, 2] for accelerating the work. It sure is a powerful tool giving outstanding test speed, which is usually 100 to 100,000 times faster than software simulation. It even allows the engineer to run real firmware, an operating system, and applications [3, 4].

As the size and complexity of ASIC/SoC increases, many more test cases are required to achieve sufficient test coverage in the verification. In addition, long-running tests taking hours or even days are needed to remove bugs from the product. In such testing and debugging, engineers often face difficulties in identifying what causes a bug. This is especially the case if the error is intermittent and not reproducible, as when the test program sometimes does and sometimes does not generate the error.

To trace the behavior of signals in a hardware prototype, designers normally use a logic analyzer and FPGA built-in signal trace tools, such as the Xilinx ChipScope [5] or Altera SignalTap [6]. The logic analyzer is a powerful tool for debugging since it allows the engineer to see what is happening in the target hardware in real time. However, the logic analyzer also has weak points:

- A logic analyzer has a limited amount of memory, so that it records signal behavior for only part of a test run, as shown in .
- It is not always obvious to the engineer as to which trigger conditions will pinpoint the source of the bug.

- Human designers find it difficult to fully understand the large amount of collected data in the timing-chart format.

During hardware debugging, we occasionally came across problems, which are hard to solve by using conventional approaches. Let us show three typical problems:

1) I ran a test program on the prototype board and got an error. Running the test program again, there was no error. I ran it ten more times, but still had no error. Where is the error gone?
2) I connected my new core to Design X from Company Y. I ran a test program and got an error. We only have a minimal data sheet for Design X. According to the specifications, my core should work. Why not?
3) I am using a logic analyzer to trace an error, but I have no idea as to what kind of trigger condition I have to set.

To address these problems, we have been developing a hardware debugging tool named Transition and Transaction Tracer (TTT). Our experiences in hardware development show that the target ASIC/SoC is well verified for the transactions that occur frequently. However, the ASIC/SoC tends to have potential flaws in processing transactions that rarely happen. For this reason, TTT captures the time varying signals as a series of vectors, and records them for a long time, such as for hours, days, or even weeks without a break. TTT constantly monitors the transition counts between states to help the engineer effectively find the problem. When it detects a new or unexpected transition, it calls the attention of the engineer by generating a trigger for the logic analyzer.

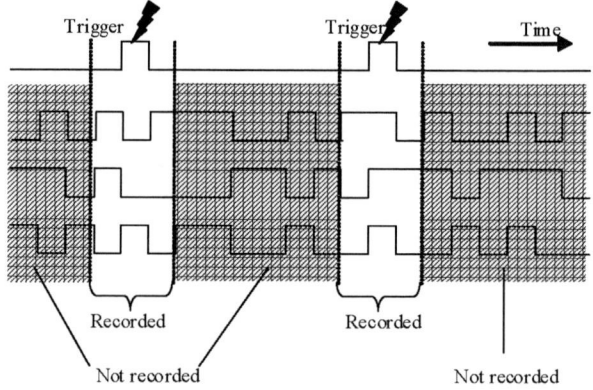

Fig. 1. Signal transitions recorded by a logic analyzer

0-7803-9451-8/06/$20.00 ©2006 IEEE.   454

2006 Asia and South Pacific Design Automation Conference

5A-3

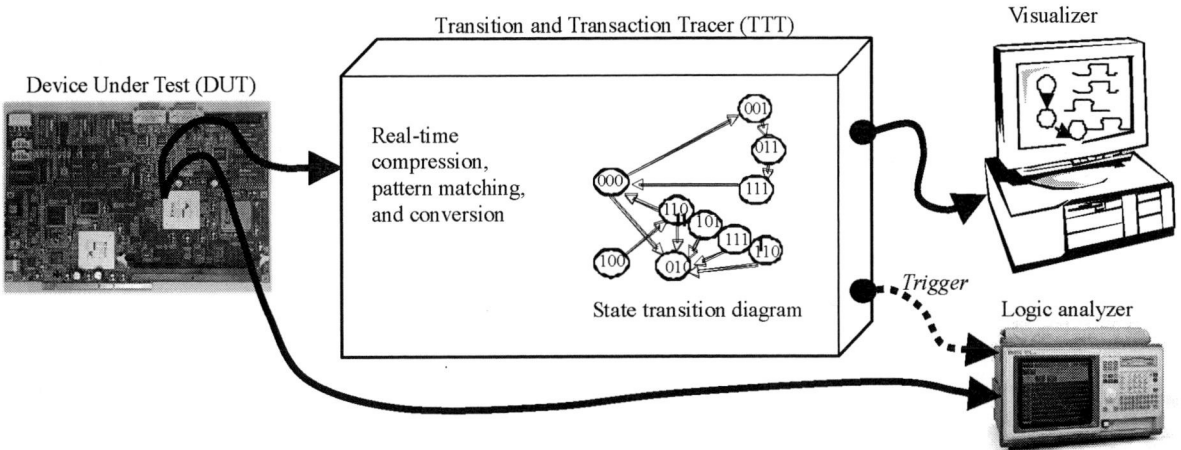

Fig. 1. Debugging System Overview

To help engineers easily perceive what happens in the target hardware, TTT shows the signal behavior not only in the timing-chart format but also in the state transition diagram. To automatically identify and extract a transaction from the signal transition sequence, we developed an idle state detector, which splits two adjacent transactions.

This paper is organized as follows: Section 2 describes the hardware architecture of TTT; Section 3 shows an implementation of TTT using an FPGA prototyping board; Section 4 shows the results obtained by running TTT on a PCI bus; and Section 5 offers concluding remarks and discusses future work.

II. Hardware Architecture of Transition and Transaction Tracer

A. Overview of the debugging system

TTT captures the transitions of the target signals and records them in a state transition form for a long time, for hours, days, or even weeks, without a break. This greatly decreases the possibility of missing the signal behavior associated with an error. In contrast to a conventional logic analyzer, TTT generates a trigger for signal capture when any new transition is detected.

Fig. 1 is the overview of the typical debugging system, which consists of a Device Under Test (DUT), Transition and Transaction Tracer (TTT), a visualizer of TTT, and a logic analyzer. TTT probes the target signals of the DUT, captures the signal transitions, compresses the data in real time, performs matching against the state transitions already stored, and sends the analyzed data to the visualizer. It also has a trigger generator connected to the logic analyzer.

B. Internal Structure of Transition and Transaction Tracer

Fig. 2 shows the internal structure of TTT. It is composed of a transition recorder and transaction tracer. The transition recorder focuses on the transitions between pairs of adjacent states. The transaction tracer, on the other hand, has a higher view of completed transactions, where the transactions are delimited by idle states.

The transition recorder consists of a transition memory (U1) and a counter memory (U2). It captures the state transition between adjacent states, which is the vectors at Clock N-1 (previous state) and Clock N (current state). The vector pair is stored in the transition memory. The counter memory records the number of each transition between adjacent states. If the transition has not yet been stored in the transition memory, the transition recorder generates a trigger. If the transition has already been stored, the transition recorder increments the counter that is associated with the captured transition.

Fig. 3 is a user interface example, which shows the state transitions in a two-dimension matrix format. In the figure the number 43,123 (marked with *) is the transition count from state A to state B. The cell is automatically shaded because the transition count exceeds the user specified threshold, 10,000 in this case. In like manner, the number 3 (marked with **) shows the transition count from states D to B. The cell color is reversed because the transition count is lower than the user-specified threshold of 10, helping the user easily spot the rare transitions.

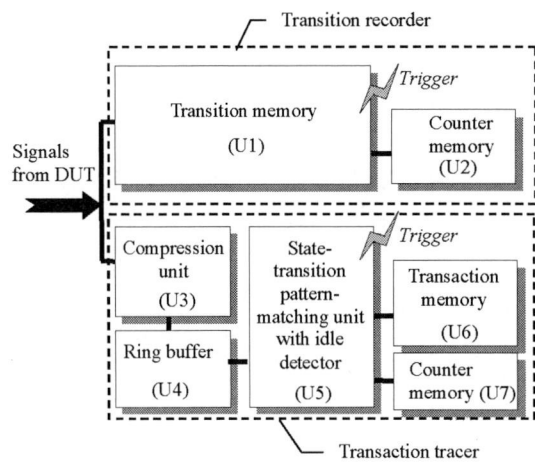

Fig. 2. Internal Structure of Transition and Transaction Tracer

455

		State at Clock N			
		A	B	C	...
State at Clock N-1	A	121	43,123*	0	...
	B	0	324	0	...
	C	815	0	132	...
	D	0	3**	43	...
	...	...	...	...	...

High threshold = 10,000 Low threshold = 10	(defined by the user)

Fig. 3. Visualization example in the matrix format

The transaction tracer captures the signal behavior as a transaction, which is a series of states between idle states. The idle states are defined by the user or by the idle state detector. The transaction tracer consists of a compression unit (U3), a ring buffer (U4), a state-transition pattern-matching unit with an idle state detector (U5), a transaction memory (U6), and a counter memory (U7). The compression unit carries out run length coding to compress the captured data. The state-transition pattern-matching unit searches for the captured transaction pattern in the transaction memory. If it finds the transaction to be new, it stores the data in the transaction memory. The counter memory stores the number of each specific transactions observed. The state transition diagram drawn in the visualizer is constructed by using the information stored in the transaction memory.

Table 1 shows an example of the compression procedure. The transaction is defined as a series of states starting with the transition from the idle state to a non-idle state and ending with the transition from a non-idle state to the idle state. The run-length expression in the table shows how the input states are compressed in the run-length form. This is based on blocks of repeated states and their repetition counts.

The compressed data is stored in the transaction memory. The transaction tracer uses hashing for the search and store operations. Since transactions vary in length, the compressed data are stored in a linked list form.

C. Idle State Detector

In the first version of TTT, the idle state was defined by the user. By experiences of debugging work, we have learned that the idle state can easily be detected in many cases. The bus is not busy all the time; rather it frequently "hovers" over the

idle state, in which the signals tend to keep the constant value. From a power saving perspective, it is a good idea to keep the signals unchanged in the idle state, and many hardware designs adhere to this design scheme.

If the input signals are unchanged more than C cycles, the idle state detector registers the state as IDLE, where C is a user defined integer (C ≥ 2). By definition, one or more states can be registered as IDLE. The transaction tracer uses this state or these states to find a series of signal transitions between the idle states and records them as a transaction.

III. Hardware Implementation

We implemented the transition and transaction tracer using a custom-made FPGA prototyping board. Fig. 4 is a photograph of the board. It has two Xilinx VirtexII XC2V4000 FPGAs and a card-edge PCI interface, so that it can be installed in a PCI slot.

We made the design in VHDL and synthesized it for the FPGA configuration data using Xilinx ISE version 6.3. The transition memory (U1) in Fig. 2 is a fully-associative memory. The other memories and counters (U2, U6, and U7) use a two-port 18 Kb block SRAM provided in the FPGA [7]. All the counters are 36 bits wide. Table 2 shows the number of FPGA slices and 18 Kb SRAM modules used for the implementation. In the table, the capacity is the maximum number of transitions that can be stored in the transition memory (U1). The transition recorder uses many more slices than the transaction tracer. This is because the transition memory is a fully-associative CAM running as fast as the DUT signals. The CAM is implemented by using the primitive latches and comparators of the FPGA. The depth is the maximum recordable block length of a repetition. In Table 1, for example, the block length is two for (BC).

The amount of hardware resources for implementing the transition and transaction tracer depends on the capacity and depth, both of which are related to the complexity of the target signal behavior, but not to the usage duration.

Table 1: An example of a compression procedure

Sampling time	Input state	Direct expression	Run-length expression
+0	<idle>		
+1	A	A	A
+2	B	AB	AB
+3	C	ABC	ABC
+4	B	ABCB	ABCB
+5	C	ABCBC	A(BC)1
+6	B	ABCBCB	A(BC)1B
+7	C	ABCBCBC	A(BC)2
+8	D	ABCBCBCD	A(BC)2D
+9	<idle>		

Fig. 4. FPGA prototyping board

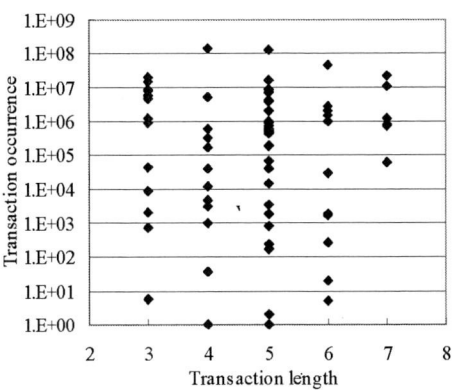

Fig. 5. Occurrence of PCI transactions

Table 2: Hardware resources

(1) Transition recorder		
Capacity	Slices	SRAM modules
64	2,720	2
128	3,065	2
256	4,118	2
512	12,871	2
1,024	26,705	4

(3) Transaction memory		
Capacity	Slices	SRAM modules
512	315	2
1,024	315	3
2,048	315	5
4,096	315	9
8,192	315	17
16,384	315	33

(2) Transaction tracer excluding transaction memory		
Depth	Slices	SRAM modules
2	84	0
3	137	0
4	207	0
5	288	0
6	382	0
7	492	0
8	621	0
9	767	0
10	928	0

(4) Idle state detector		
C	Slices	SRAM modules
31	674	3

IV. Evaluation

To see the effectiveness of the proposed method, we captured the behavior of a 32-bit 33 MHz PCI bus. The capacity of the transition recorder we chose for this application was 64. The depth and capacity of the transaction tracer were 4 and 1,024, respectively. By using the speed optimized option in the synthesizer, the Virtex-II XC2V4000-6 FPGA can run at up to 100 MHz.

In addition to the FPGA board, a network card and an IDE control card were resident in the PCI bus. Ten PCI control signals, FRAME#, IRDY#, TRDY#, DEVSEL#, STOP#, GNT#, C/BE#[3,2,1,0], were monitored.

Firstly, the bus state was manually defined as IDLE if FRAME#, IRDY#, and TRDY# were all inactive. All of the PCI transactions were monitored for about an hour. During the test run, 78 types of transactions were captured. Fig. 5 shows the occurrence counts of the recorded transactions. The transaction length is defined as the number of the states between the idle states. The figure shows that the occurrence count for the most frequently captured transaction was more than 10^8, and a transaction that happened only once was also recorded.

Without any background knowledge, it is very hard for conventional methods to capture such a rare transaction. The method proposed in this paper records all of the transactions and display them in the state transition diagram, which helps the engineer clearly understand what is happening in the target hardware.

Secondly, we engaged the idle state detector. Value 31 was assigned to C, and therefore the states that stay unchanged for more than 30 cycles were automatically defined as IDLE. After two hour test run, two states shown in Table 3 were registered as IDLE.

This result conforms to the PCI specification [8]. FRAME#, IRDY#, TRDY#, DEVSEL#, and STOP# are all sustained tri-state[1] signals, and therefore they must be in H state for the idle cycle. C/BE#[3,2,1,0], on the other hand, are tri-state signals and can be left H or L after a transaction is completed.

To help the engineer understand the target behavior, we developed a visualizer, which runs on Windows and Linux. It shows the signal transitions in three formats: 1) transition matrix, 2) timing chart, and 3) state transition diagram. The transition matrix, as illustrated in Fig. 3, is a two dimensional matrix, which shows the transition counts for adjacent states. Fig. 6 is a screen shot of the visualizer, which shows a state transition diagram and a timing chart. While TTT is monitoring the signals, the state transition diagram is generated on the fly. Newly recorded transitions are colored red to be easily identified.

To increase the bus bandwidth, address pipelining is used in several bus protocols, such as IBM CoreConnect and ARM AMBA. By definition, the address phase begins before the previous data phase is completed. In such a case, a traced transaction path delimited by an IDLE sate will contain two or more transactions. Although the state transition diagram becomes bigger, it still gives important information on how two or more transactions are overlapped. To make it easier for the user to distinguish address and data phases, the visualizer can color the states in accordance with the specified signal status.

Table 3. Idle states

Signal	IDLE1	IDLE2
FRAME#	H	H
IRDY#	H	H
TRDY#	H	H
DEVSEL#	H	H
STOP#	H	H
GNT#	H	H
C/BE#[3]	H	L
C/BE#[2]	H	L
C/BE#[1]	H	L
C/BE#[0]	H	L

(H: high voltage, L: low voltage)

---

[1] Sustained Tri-State is an active low tri-state signal owned and driven by one and only one agent at a time. The agent that drives a sustained tri-state pin low must drive it high for at least one clock before letting it float.

5A-3

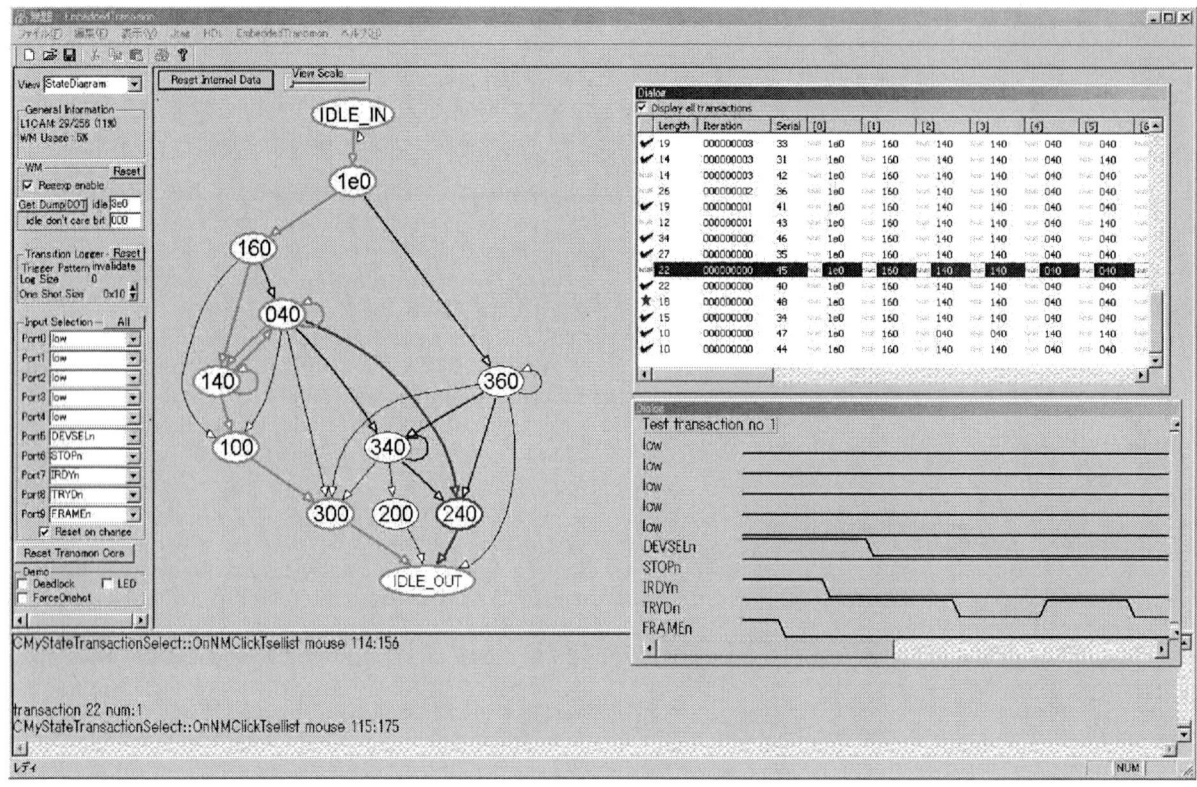

Fig. 6. Screen shot of the visualizer

V. Concluding Remarks

This paper proposes a new hardware debugging method, a transition and transaction tracer, which focuses on the signal transitions and transactions to accelerate ASIC/SoC debugging. We implemented it using a prototyping FPGA board and evaluated the hardware resources and operating speed. A state transition diagram for PCI transactions was shown as an example.

To review, here are the problems listed in the introduction and how we tackled them:

1) I ran a test program on the prototype board and got an error. Running the test program again, there was no error. I ran it ten more times, but still had no error. Where is the error gone?

A1) With TTT, I probed the crucial signals and recorded all the transitions in each test run. The state transition diagram from the test that caused an error has a transition that does not appear in the error-free state transition diagrams. It turned out that the transition occurred only when three devices simultaneously requested a bus. TTT provided a hint to solve the problem.

2) I connected my new core to Design X from Company Y. I ran a test program and got an error. We only have a simple data sheet for Design X. According to the specifications, my core should work. Why not?

A2) TTT showed that the error occurred in a specific situation that is not described in the data sheet of Design

X. My core was never expected to encounter this behavior. I contacted an engineer from Company Y and we sorted out the problem.

3) I am using a logic analyzer to trace an error, but I have no idea as to what kind of trigger condition I have to set.

A3) I configured TTT to trigger the logic analyzer to record data for each new transition. By analyzing the captured data in the logic analyzer, I found that the error was associated with a very rare transition event, which appeared roughly once an hour.

Design and verification engineers often delude themselves that a certain input or transition is impossible. Looking at the state transition diagram created with TTT, we sometimes discover the reality is different from what we had believed. We have been using TTT for practical ASIC development and succeeded in accelerating our debugging work.

We are now developing a new function that incorporates TTT with an assertion-based verification method [9]. The assertion specifies state transitions that must not happen (illegal transitions) and those that must happen (transitions that must be tested) in the test run. We define these transitions in the TTT prior to the test run. TTT gives warning to the engineer if it detects an illegal transition. It also records the number of captured states and transitions in each test run for measuring the test coverage.

We are also studying how TTT gives more useful information to the user in pipelined and split-transaction buses.

References

[1] N. Ohba and K. Takano, "An SoC Design Methodology Using FPGAs and Embedded Microprocessors," *Proceedings of Design Automation Conference*, pp.747-752, 2004.

[2] J. O. Hamblen, "Rapid Prototyping Using Field-Programmable Logic Devices," *IEEE Micro*, pp.29-37, 2000.

[3] T. Matsumura, N. Yamanaka, R. Yamaguchi and K Ishikawa, "Real-time Emulation Method for ATM Switching Systems in Broadband ISDN," *Proceedings of IEEE International Workshop on Rapid System Prototyping*, pp.19-21, 1996.

[4] M. Courtoy, "Rapid System Prototyping for Real-Time Design Verification," *Proceedings of Ninth International Workshop on Rapid System Prototyping*, pp.108-112, 1998.

[5] Xilinx Inc., "Chipscope Pro Software and Cores User Guide," October, 2004.

[6] Altera Inc., "Design Debugging Using the SignalTap II Embedded Logic Analyzer," December, 2004.

[7] Xilinx Inc., "Virtex-II Platform FPGA Handbook," December, 2001.

[8] PCI Special Interest Group, "PCI Local Bus Specification – Revision 2.2," December, 1998.

[9] Accellera, "Property Specification Language – Reference Manual," http://www.eda.org/ieee-1850/, version 1.1, 2004.

Cycle Error Correction in Asynchronous Clock Modeling for Cycle-Based Simulation

Junghee Lee and Joonhwan Yi

Telecommunication R&D Center
Samsung Electronics
{junghee77.lee, joonhwan.yi} @ samsung.com

Abstract— **As the complexity of SoCs is increasing, hardware/software co-verification becomes an important part of system verification. C-level cycle-based simulation could be an efficient methodology for system verification because of its fast simulation speed. The cycle-based simulation has a limitation in using asynchronous clocks that causes inherent cycle errors. In order to reuse the output of a C-level cycle-based simulation for the verification of a lower level model, the C-level model should be cycle-accurate with respect to the lower level model. In this paper a cycle error correction technique is presented for two asynchronous clock models. An example design is devised to show the effectiveness of the proposed method. Our experimental results show that the fast speed of cycle-based simulation can be fully exploited without sacrificing the cycle accuracy.**

I. Introduction

Ever increasing complexity of systems-on-a-chip (SoCs) makes the verification harder and harder [1]. Co-simulation/verification of hardware and software is now widely recognized as an important and viable verification approach [1].

Traditional event-driven simulator has limitations on run-time performance [1]. Cycle-based simulation can be an alternative solution for system level co-verification because of its fast simulation speed. System-level simulation usually does not require detail signal transition information but requires fast simulation speed due to huge amount of simulation data. A cycle-based simulator evaluates signal values once at an active clock edge instead of evaluating whenever a signal transition occurs, which can significantly accelerate the simulation speed.

One of the challenging problems in cycle-based simulation is to handle multiple asynchronous clocks [7] because the simulation is synchronized by a single clock. To our knowledge, all of cycle based simulation researches including [2][3][4] constrain their applications to synchronous circuits. Modern SoCs, however, often have multiple asynchronous clocks. In order to support multiple asynchronous clocks in cycle-based simulation, asynchronous clocks can be approximated as synchronous clocks —, two approximations, namely early- and late-edge models, are introduced in Section III. In this case, inherent clock cycle errors compared to RTL simulation are inevitable.

More details are discussed in Section III

The accuracy of asynchronous clock modeling is especially important in verification vector reuse between C-level and register transfer level (RTL) as shown in Fig. 1. Note that C-level modeling usually sacrifices timing accuracy in some degree to achieve the faster simulation speed. Murali [5] proposed a verification framework by using a C simulator that drives stimulus for an RTL circuit. Whereas Murali deals with the signal conversion from C-level to RTL, we are focusing on the clock cycle accuracy of the C-level model compared to RTL model. It is difficult to automate the verification reuse if the C-level model is not cycle-accurate. Kirk [6] suggested a technique to reuse C simulation outputs as a testbench for RTL verification. To overcome the cycle errors between C-level and RTL, both C and RTL models need to have extra handshaking signals that, of course, cause area and pin count overhead. If the model executed on the C simulator is developed as a cycle-accurate level model, the output of the simulator can be directly used without any modification.

A circuit with single clock can be easily modeled cycle accurately in C-level. On the contrary, a circuit with multiple (asynchronous) clocks needs to be modeled carefully to prevent cycle errors at the point of clock domain crossing even though each block is modeled cycle accurately. Cycle errors between C-level and RTL circuit models come from either the inaccurate clock models or simulation mechanisms. In this paper, we address the conditions that cycle errors occur, and propose a technique to correct the cycle errors. Note that synchronizers are assumed to be at the signal paths of clock domain crossing, which is very reasonable and practical assumption. When a signal is transferred across clock domains, there exists a metastability problem [8]. To overcome the metastability, the synchronizers are widely used [11].

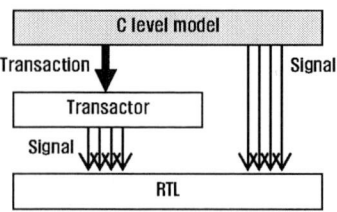

Fig. 1 Using a C level model as testbench of RTL

0-7803-9451-8/06/$20.00 ©2006 IEEE.

We first introduce basic concepts of cycle based simulation and clock in Section II. Then, we anayze the conditions that cycle errors occur and propose a correction method in Section III. In Section IV, the cycle errors due to simulation mechanisms are discussed. Experimental results are shown in Section V and we conclude the paper in Section VI

II. Preliminary

In cycle-based simulation all signals are evaluated only at an active edge of a reference clock that is called a *simulation clock*. Note that the simulation clock is the only built-in clock in cycle-based simulation. In order to simulate concurrency of hardware, the signal evaluation at an active clock edge is divided into a few phases. For cycle-based simulation, mainly two simulation mechanisms are used.

First, there is a message-passing mechanism [9] where an evaluation is achieved through 3 phases: a phase to communicate messages as scheduled (P1), a phase to update the signal values of resources according to the messages received at P1 (P2), and a phase to communicate messages generated at P2 (P3). Next, there is a communication and update mechanism [10] where an evaluation is achieved through 2 phases: communicate phase and update phase. Comparing to the message-passing mechanism, the *communicate phase* corresponds to P1 and the *update phase* corresponds to P2.

Consider the C-level circuit model C^C that models the RTL circuit C^{RTL}. Assume that signal (or net) S^{RTL} in C^{RTL} corresponds to signal S^C in C^C. When S^{RTL} is triggered by a clock CK^{RTL}, S^C is triggered by a clock CK^C modeling CK^{RTL}. Let $CK^{RTL}(n)$ denote the n-th active edge of the CK^{RTL} and $CK^C(m)$ denote the m-th active edge of CK^C. When $CK^C(m)$ corresponds to $CK^{RTL}(n)$, the relationship between m and n is determined by the clock model as described in Section III.

We take an RTL circuit model as a reference for calculating the cycle error of its C-level model because the physical behavior of a signal crossing clock domains may have non-deterministic cases like metastability [8]. As addressed before, in order to reuse a C-level testbench for RTL verification without any modification, the C-level model should have the same behavior of its corresponding RTL model and should have no cycle error. In order to exploit the fast simulation speed of cycle-based simulation methods, it is desirable to abstract the description in C-level model as much as possible.

Because the only built-in clock in cycle-based simulation is the simulation clock CK^C_{Sim}, a clock generator is needed to generate various clocks using the simulation clock. For example, consider a clock CK^C_k that is n times slower than CK^C_{Sim}. Then, a clock generator counts the number of active edges of CK^C_{Sim}, and generates an active edge of CK^C_k at every k active edges of CK^C_{Sim}. An asynchronous clock whose period is not a multiple of the period of the simulation clock cannot be modeled such a simple way. In the next Section, we

explain how to model an asynchronous clock for cycle-based simulation and the inevitable cycle errors due to the clock models. We also propose a technique to correct the cycle errors.

III. Cycle Error due to Asynchronous Clock Models

Asynchronous clocks can be modeled in various ways. Among them, three simplest models are considered here: greatest common divisor (GCD) model, early edge model, and late edge model, see Fig. 2. The number n in the right hand side of an active edge means that the edge is n-th active edge. In the *GCD (clock) model*, the period of the simulation clock (CK^C_{SimGCD}) becomes the GCD of periods of all clocks in a system. The GCD model is accurate and easy to implement but impedes the simulation speed. Let us assume that there are three clocks CK^{RTL}_{125}, CK^{RTL}_{1024}, and CK^{RTL}_{20000} with periods of 125ns, 1024ns, and 20ms. Then the GCD of the 3 clocks is 1(ns) and thus the period of the simulation clock CK^C_{SimGCD} becomes 1ns. Now, $CK^{RTL}_{125}(n) = CK^C_{SimGCD}(125n)$, $CK^{RTL}_{1024}(n) = CK^C_{SimGCD}(1024n)$, and $CK^{RTL}_{20000}(n) = CK^C_{SimGCD}(20000n)$. To generate CK^{RTL}_{125}, the clock generator makes an active clock edge at every 125 edges of CK^C_{SimGCD}. That is, 124 cycles are used just for counting to know the period of the 125ns clock not for proceeding simulation. Similarly, for CK^{RTL}_{1024}, and CK^{RTL}_{20000}, 1023 and 19999 cycles are wasted in vain.

The overhead caused by the wasted cycles is imposed only on the clock generator. Although the portion of the clock generator in a circuit system is small, the performance degradation due to the GCD model is not because the clock generator is computed at every active edge of the simulation clock. In the example above, even the fastest clock of 125ns period in the system uses only 1/125 of the simulation clock edges. More than 99% cycles are wasted. Furthermore, in real system, the periods of clocks are often not an integer in nanosecond unit. As a result, the period of the simulation clock could be scaled down to the pico second unit if the GCD model is used. This time precision is nearly the same time resolution of the event-driven simulation. Therefore, we cannot exploit the fast simulation speed of cycle-based simulation.

To overcome inefficiency of the GCD model, *early edge* and *late edge models* can be used. In the early (late) edge model, the fastest clock in the system becomes the simulation

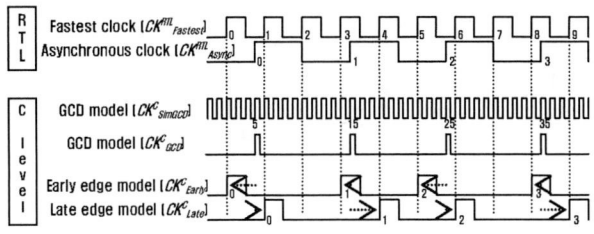

Fig. 2 Candidate asynchronous clock models

461

clock denoted by CK^C_{Early} (CK^C_{Late}). Other clocks are synchronized with the simulation clock by forcefully moving their edges to the preceding (following) edges of the simulation clock. The early and the late edge models are illustrated in Fig. 2. Consider the active edge $CK^{RTL}_{Async}(0)$ of an asynchronous clock between $CK^{RTL}_{Fastest}(0)$ and $CK^{RTL}_{Fastest}(1)$ of the fastest clock. If the early (late) edge model is used, $CK^{RTL}_{Async}(0)$ is moved to $CK^{RTL}_{Fastest}(0)$ ($CK^{RTL}_{Fastest}(1)$). The early and late edge models can maximize the simulation performance because they have less idle cycles than the GCD model. In the previous example of CK^{RTL}_{125}, CK^{RTL}_{1024}, and CK^{RTL}_{20000}, the period of the simulation clock becomes 125ns when early or late edge model is used. Then CK^{RTL}_{125} has an active edge every simulation clock cycle. For CK^{RTL}_{1024}, an active cycle is made every 8 or 9 simulation clock cycle thus only 7 or 8 cycles are wasted. Similarly, to generate CK^{RTL}_{20000}, an active edge is made at every 160 simulation clock cycle, thus only 159 cycles are wasted. The number of wasted cycles is drastically reduced. Note that there exists one-to-one correspondence between an RTL clock edge and a C-level clock edge in early and late edge models.

It is favorable to use either the early edge or the late edge model for clocks in order to maximize the simulation performance. But using those models causes inherent cycle errors at the point of clock domain crossing. Cycle errors caused by those asynchronous clock models occur regardless of the simulation mechanism.

The *cycle error* of CK^C compared to CK^{RTL} which is denoted by $CE(CK^C)$ is α if the behavior of an RTL signal S^{RTL} at the edge of $CK^{RTL}(m)$ corresponds to that of a C-level signal S^C at $CK^C(m\pm\alpha)$. The definition of cycle error can be extended for the GCD model. But the extension is not necessary because there is no cycle error when the GCD model is used.

RTL event-driven simulation traces and their corresponding C-level cycle-based simulation traces with the late edge clock model are depicted in Figures 3,4 and 5. The C-level late edge model of the RTL asynchronous clock CK^{RTL}_{Async} is CK^C_{Async}. In Fig. 3, a signal is transferred from the domain of the simulation clock CK^{RTL}_{Sim} to the domain of CK^{RTL}_{Async}. The signal is received at $CK^{RTL}_{Async}(1)$ in RTL, and at $CK^C_{Async}(1)$ in C-level. Thus, there is no cycle error. Since $CK^{RTL}_{Async}(n)$ will be mapped to $CK^C_{Async}(n)$ that will be always behind the sending clock edge by the late edge model, the cycle error $CE(CK^C_{Async})$ is always zero. When a signal is transferred from the domain of CK^C_{Async} to the domain of the simulation clock as shown in Fig. 4, the signal is received at $CK^{RTL}_{Sim}(4)$ in RTL and at $CK^C_{Sim}(5)$ in C-level. Thus the cycle error $CE(CK^C_{Sim})$ is one. But if the sending edge of CK^{RTL}_{Async} is synchronized with an edge $CK^{RTL}_{Sim}(n)$ of the simulation clock, the receiving edge will be $CK^{RTL}_{Sim}(n+1)$ in both RTL and C-level. In this case, $CE(CK^C_{Sim})$ is zero. Thus, $CE(CK^C_{Sim})$ can be one or zero. Fig. 5 depicts the case that a

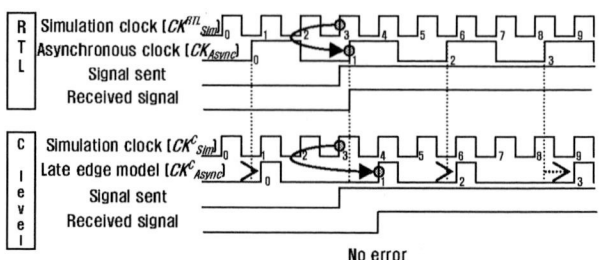

Fig. 3 The late edge model: simulation to asynchronous

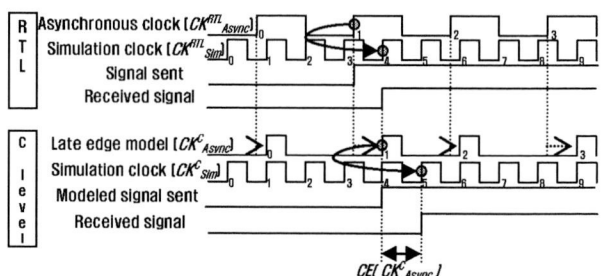

Fig. 4 The late edge model: asynchronous to simulation

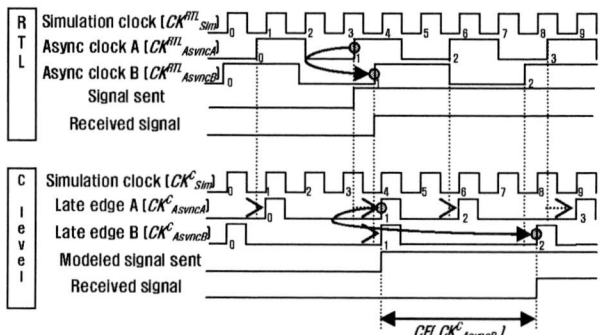

Fig. 5 The late edge model: between asynchronous

signal is transferred between asynchronous clocks when the late edge model is applied. In this case $CE(CK^C_{AsyncB})$ is one at most.

Similarly, the cycle errors due to the early edge model for various conditions can be computed. Note that the cycle errors can be formulated by the clock models and the clock domains that send and receive the signals. This is summarized in TABLE I. The cycle error is advent when the edges of the simulation and the asynchronous clocks become identical – that is, they happen at the same time – in C-level but their corresponding RTL edges are not. Now, a cycle error correction method for the late edge clock model is presented. Although only the late edge clock model is considered in this paper, it is easy to extend this discussion for the early edge clock model. As mentioned before, every active clock edge in RTL has corresponding active clock edge in C-level. To correct the cycle errors, the principle is to make signals be transferred at the corresponding edge as illustrated in Fig. 6.

TABLE I. The cycle errors due to clock models

Clock model	Clock domain From	To	Cycle error	Illustrated in
Late edge model	Sim	Async	$CE(CK^C_{Async})$ = zero	Fig. 3
	Async	Sim	$CE(CK^C_{Sim})$ = one or zero	Fig. 4
	Async A	Async B	$CE(CK^C_{AsyncB})$ = one or zero	Fig. 5
Early edge model	Sim	Async	$CE(CK^C_{Async})$ = one or zero	
	Async	Sim	$CE(CK^C_{Sim})$ = zero	
	Async A	Async B	$CE(CK^C_{AsyncB})$ = one or zero	

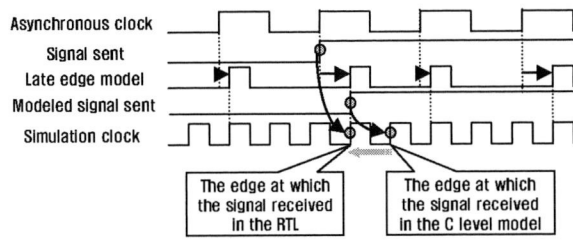

Fig. 6 Basic idea of cycle correction

We assume that synchronizers are inserted in every path crossing clock domains to prevent the malfunction of a circuit due to metastability [8]. Generally synchronizers consisting of two D-type flip-flops (DFFs) are used as shown in Fig. 7. Two DFFs cause two cycle delay at the receiving clock domain. The cycle error can be corrected if the synchronizer is operated as it has only one DFF when an one-cycle error occurs. To do so, a synchronizer like Fig. 8 is needed. The select signal should be zero when the cycle error occurs. The select signal should be provided by a controller. A way to generate the select signal is described in Section IV.

IV. Cycle Error in Communicate & Update Mechanism

The communicate and update mechanism is computationally more efficient than the message-passing mechanism because it utilizes less number of update and communication operations for a signal evaluation. Thus it is desirable to use the communicate and update mechanism to achieve the faster simulation speed in cycle-based simulation. However, unlike the message-passing mechanism, a cycle error may occur unless the modeled circuit has a single clock domain. In this section cycle errors by using the communicate and update mechanism and a correction technique of them are presented.

When the communicate and update mechanism is used for a cycle-based simulation, the communicate phase of all components is executed first to exchange signals among components. Then the update phase is executed to update shared resources. A clock CK^C_{Gen} is called a generated clock of a clock CK^C_{Sim} if CK^C_{Gen} is created by a logic circuit including a set of flipflops activiated by CK^C_{Sim}. CK^C_{Sim} is

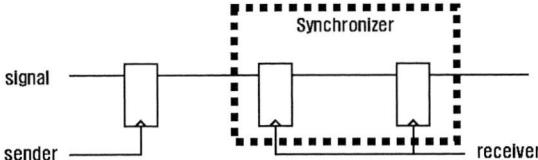

Fig. 7 A normal synchronizer

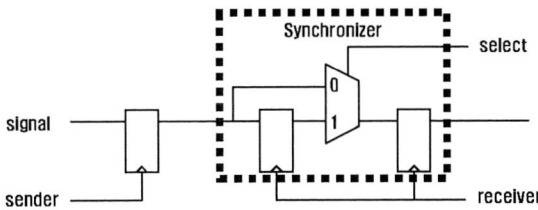

Fig. 8 Synchronizer model for correcting cycle error

called the source clock of CK^C_{Gen}. The active edge of CK^C_{Gen} is synchronized with either the rising or the falling edge of CK^C_{Sim}. Fig. 9 shows a usual logic design that implements CK^C_{Gen}, a generated clock of CK^C_{Sim} by the factor of two. Fig. 9 illustrates an example that a signal is transferred from the simulation clock CK^C_{Sim} to CK^C_{Gen}. D0, D1 and D2 are DFFs. $\alpha0$ and $\alpha1$ are inputs and $\beta0$ and $\beta1$ are outputs of D0 and D1, respectively. $\alpha2$ is the input of another component receiving sig_c.

Assume that $\alpha0$ is changing from 0 to 1 at $CK^C_{Sim}(0)$ during the communicate phase. A general event-driven simulator will evaluate $\alpha0$ to 0 due to the delta delay. But a cycle-based simulator evaluates signals only at an active edge. This is an implementation issue. Regardless of the implementation, the cycle errors occur when the communicate and update mechanism is used.

The values are transferred to $\alpha0$ and $\alpha1$ during the communication phase of CK^C_{Sim} (0). Now, $\alpha0$ and $\alpha1$ become 1 and 0, respectively. During the update phase of CK^C_{Sim} (0), $\beta0$ and $\beta1$ are updated to 1 and 0, respectively. In the

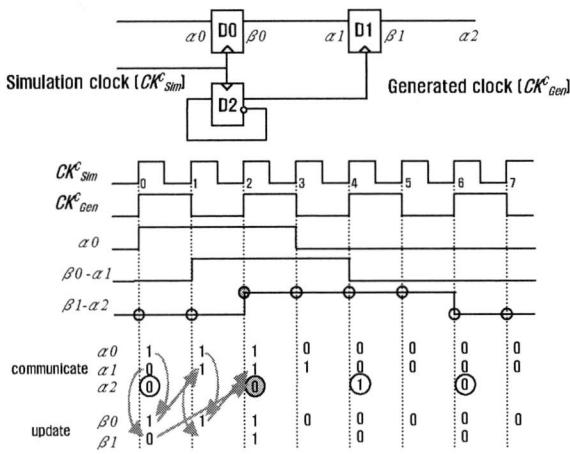

Fig. 9 Communicate and update example

communication phase of $CK^C_{Sim}(1)$, $\alpha 0$ and $\alpha 1$ become 1 and 1, respectively. Then, in the update phase of $CK^C_{Sim}(1)$, only $\beta 0$ is updated to 1 because D1 is not active at that time. Note that $\beta 0$ and $\beta 1$ are 1 and 0, respectively. During the communication phase of $CK^C_{Sim}(2)$, $\alpha 0$, $\alpha 1$, and $\alpha 2$ become 1, 1, and 0, respectively. In the update phase, both $\beta 0$ and $\beta 1$ are updated to 1. Note that $\alpha 2$ will remain as 0 although $\alpha 2$ is supposed to be 1 unless the output value 1 of D1 is transferred to $\alpha 2$ by a communication phase. This errorneous value of $\alpha 2$ will last till the next active edge of CK^C_{Gen}. This is because the updated value is transferred to $\alpha 2$ during the communicate phase of the next active edge. In the message-passing mechanism [9], this error is eliminated by the last communication phase after the update phase during an evaluation.

This erroneous behavior due to the communicate and update mechanism occurs when signals are transferred between asynchronous clocks. If the receiving clock is synchronous with the sending clock and they have different periods, one extra communicate phase should be enforced after the update phase. However, if the receiving clock is asynchronous with the sending clock, the method using a synchronizer proposed in Section III can be used to correct the cycle errors.

Fig. 10 shows a flowchart of the function *GenFlag* which generates the enable flags, `Enable_A` and `Enable_B`, for asynchronous clocks and the select signal of the mux shown in Fig. 8, `Select_A,B` and `Select_B,A`, for synchronizers. The communicate and the update phases will be executed only when the enable flag is set. This function is included in the clock generator and called every edge of the simulation clock.

When *GenFlag* function is called, it first calculates the next active edge time `NextActiveEdge_Clk` of every clock in the system except for the simulation clock. Each clock *Clk* has its own active edge counter `CntEdge_Clk`. For example, CK^{RTL}_{Clk}(`CntEdge_clk`) denotes the `CntEdge_clk`-th active edge of *Clk*. `NextActiveEdge_Clk` is the time of the next active edge of *Clk* relative to the simulation clock. Reference of `NextActiveEdge_Clk` is the period of the simulation clock. If the ceiling of the next active edge time equals to that of the simulation clock, the enable flag `Enable_Clk` is set. If we use the flooring instead of the ceiling, we can implement the early edge model instead of the late edge model. For every pair of clock *A* and clock *B* there are two select signals of the mux shown in Fig. 8. `Select_A,B` is for the synchronizers from clock *A* domain to clock *B* domain. `Select_B,A` is for those from clock *B* domain to clock *A* domain. Fig. 11 shows an example illustrating function *GenFlag*.

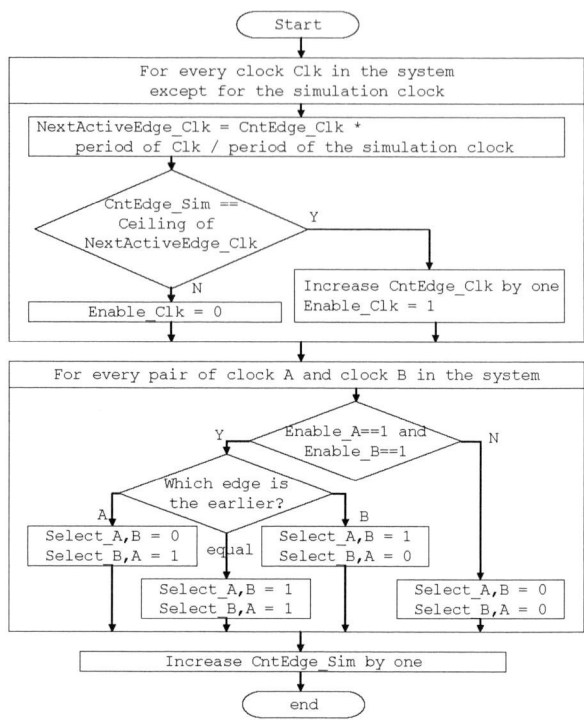

Fig. 10 Flowchart of the *GenFlag*

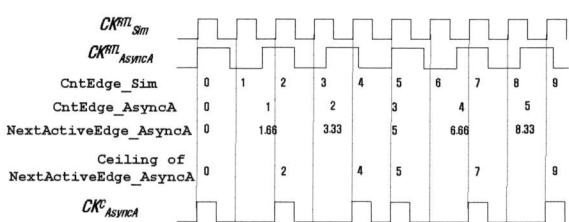

Fig. 11 An example of implementing the late edge model

After the next edge time is caculated, then it generates select signals. For every pair of clocks in the system, *GenFlag* determines which edge is earlier if both clock edges are active. To determine which edge is the earlier, it uses difference between `NextActiveEdge_Clk` and ceiling of it. The larger the difference is, the earlier the edge is.

V. Experiment

An example circuit shown in Fig. 12 is implemented to demonstrate the effectiveness of the proposed method.

The example system consists of three clock domains: CK^C_{Sim}, CK^C_{AsyncA}, and CK^C_{AsyncB}. Clock CK^C_{Sim} is the fastest clock and thus the simulation clock. Clocks CK^C_{AsyncA} and CK^C_{AsyncB} are asynchronous clocks. Each clock domain receives signals from other clock domains through a two-DFF synchronizer. Signals to be transferred to other clock domains are stable for at least one period of the receiving clock so that it is assured that the signal is transferred to the other clock domain.

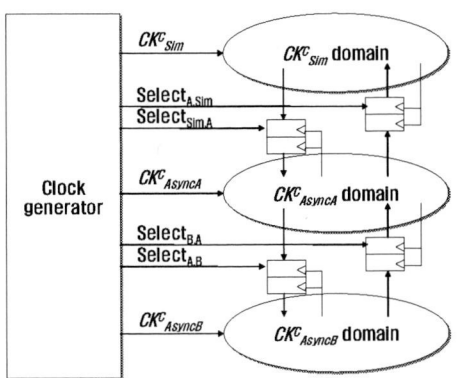

Fig. 12 An example system

When a request from CK^C_{AsyncA} domain goes to the CK^C_{Sim} domain, the CK^C_{Sim} domain returns a data to the CK^C_{AsyncA} domain with a certain delay. Then the received data is passed to the CK^C_{AsyncB} domain. The CK^C_{AsyncB} domain accumulates the received data and passed it to the CK^C_{AsyncA} domain with a certain delay. That operation is executed 1000 times in this example. The periods of CK^C_{Sim}, CK^C_{AsyncA}, and CK^C_{AsyncB} are 40ns, 110ns, and 250ns, respectively.

We simulated the sample circuit on a cycle-based simulator MaxSim [10]. In MaxSim the communicate phase and the update phase are implemented as function calls. Each component implements the communicate function and the update function and the clock generator calls the communicate functions and the update functions of components at every active clock edge. To correct cycle errors the clock generator generates select signals to let synchronizers know whether they should operate as one DFF or two DFFs.

We performed experiments to verify the cycle error correction technique and measure the accuracy and the performance. TABLE II shows the cycle accuracy of various models: RTL, the GCD model, the late edge model

TABLE II Cycle accuracy of each model

	Total clock cycle used for the complete simulation	Cycle difference from RTL	Cycle accuracy
RTL	93654		
GCD	93654	0	100.0 %
Late edge model with correction	93654	0	100.0 %
Late edge model without correction	105669	12015	87.2 %

TABLE III Simulation speed of each model

	Cycle/sec	Ratio
GCD	410702	100.0 %
Late edge model with correction	588638	143.3 %
Late edge model without correction	688227	167.6 %

with cycle error correction, and the late edge model without cycle error correction. The accuracy of the late edge model with cycle error correction is 100% while the one without cycle error correction shows only 87.2% accuracy.

TABLE III shows the simulation speed of each model on a PC with a 2.53 GHz Pentium 4 processor with 512 MB memories. As can be seen, the late edge model can be implemented 1.4 times faster then the GCD model without sacrificing the cycle accuracy. Although the late edge model without cycle error correction can be 1.6 times faster than GCD, it has drawback due to the cycle inaccuracy.

VI. Conclusion

A cycle error correction technique for a cycle-based simulation with asynchronous clocks is proposed. A two-flipflop synchronizer is assumed to be inserted on every clock domain crossing path. Three clock models for cycle-based simulation are introduced: greatest common divisor (GCD), early edge, and late edge models. Although the early and the late edge clock models are more efficient for faster simulation speed, they pose inherent cycle errors with respect to register-transfer level. It is demonstrated that a 100% cycle accurate cycle-based simulation is possible with the early and late edge clock models by using the proposed technique.

References

[1] Lisa Guerra et al., "Cycle and Phase Accurate DSP Modeling and Integration for HW/SW Co-Verification," *Proc. of Design Automation Conference*, pp.964-969, 1999

[2] G. Cabodi et al., "Exploiting timed transition relations in sequential cycle-based simulation of embedded systems," *Proc. of Computers and Digital Techniques*, pp.305-312, 2000

[3] B.H. Yaran, B.H., D. Rahmati, and A.S. Zebardast, "Applying cycle-based simulation technique to VITAL as a VHDL gate level standard," *Proc. of Canadian Conference on Electrical and Computer Engineering*, pp.1076-1084, 2001

[4] L. Ghasemzadeh and Z. Navabi, "A fast cycle-based approach for synthesizable RT level VHDL simulation," *Proc. of International Conference on Microelectronics*, pp.281-284, 2000

[5] Murali Kudlugi, Soha Hassoun, Charles Selvidge, and Duaine Pryor, "A Transaction-Based Unified Simulation/Emulation Architecture for Functional Verification," *Proc. of Design Automation Conference*, pp.623-628, 2001

[6] Kirk Ober, "Doing Behavioral Design the Right Way Minimizes Verification," *Proc. of Design and Verification Conference and Exhibition*, 2004

[7] K. Olukotun, M. Heinrich, and D. Ofelt, "Digital system simulation: methodologies and examples," *Proc. of Design Automation Conference*, pp.658-663, 1998

[8] T.J. Gabara, G.J. Cyr, and C.E. Stroud, "Metastability of CMOS master/slave flip-flops," *Proc. of Custom Integrated Circuits Conference*, pp.29.4.1-29.4.6, 1991

[9] G. Maturana et al., "Incas: a cycle accurate model of UltraSPARC," *Proc. of IEEE International Conference on Computer Design: VLSI in Computers and Processors*, pp.103-135, 1995

[10] MaxSim designer's guide, AXYS design automation, Inc., 2004

[11] J. Walker and A. Cantoni, "A new synchronizer design," *Proc. of IEEE Transactions on Computers*, pp 1308-1311, 1996

5A-5

A Fast Logic Simulator Using a Look Up Table Cascade Emulator

Hiroki Nakahara   Tsutomu Sasao   Munehiro Matsuura

Depertment of Computer Science and Electronics
Kyushu Institute of Technology, Iizuka 820-8502, Japan

Abstract— This paper shows a new type of a cycle-based logic simulation method using a Look-Up Table (LUT) cascade emulator. The method first transforms a given circuit into LUT cascades through BDD (Binary Decision Diagram). Then, it stores LUT data to the memory of an LUT cascade emulator. Next, it generates the C code representing the control circuit of the LUT cascade emulator. And, finally, it converts the C code into the execution code. This method is compared with a Levelized Compiled Code (LCC) simulator with respect to the simulation time and setup time. Although we used standard PC to simulate the circuit, experimental results show that this method is 12-64 times faster than the LCC.

I. INTRODUCTION

With the increase of the integration of LSIs, the time for the verification of the design increases. Thus, high-speed logic simulators are needed.

Logic simulators can be roughly divided into two types: event-driven simulators and cycle-based simulators. In an event-driven simulator, only the logic gates whose input signal change are evaluated. On the other hand, in a cycle-based logic simulator, the operation order of gates are determined statically beforehand, and all the logic values of the gates are evaluated for each clock cycle. Although the cycle-based logic simulator does not perform the timing verification, it is often faster than the event-driven simulator.

An **LCC** [1] is a kind of a cycle-based logic simulator using a general-purpose CPU. An LCC generates a program code for each gate of a logic circuit, and evaluates the circuit in a topological order from the inputs towards the outputs. An event-driven simulator that emulates only logic gates whose outputs change has been developed [16]. It is at most two times faster than the LCC. In this paper, we will present a cycle-based logic simulator using an LUT cascade emulator. An LUT cascade emulator [2] consists of a control part, memories, and registers. Each register is connected to a programmable interconnection circuit, and the LUT cascade emulator evaluates the logic circuit stored in the memory. Murgai-Hirose-Fujita [11] also developed a logic simulator using large memories. Their method first converts a given circuit into a random logic network of single-output LUTs, then stores them in the memory, and finally evaluates the circuit by an event-driven logic simulator implemented by a hardware accelerator. In our method, we first convert the given circuit into a cascade rather than ran-

dom logic, so the control part is simpler than Murgai-Hirose-Fujita's method. Also, our method uses multiple-output LUTs rather than single-output LUTs. In this paper, we consider the software-based logic simulation system where the LUT cascade emulator is simulated on a PC. Compared with the hardware-based logic simulator, logic simulator using a standard PC is much cheaper, and can be enhanced with the improvement of the performance of PCs.

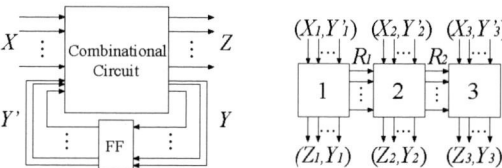

Fig. 1. A model for a sequential circuit. Fig. 2. LUT cascade.

II. LUT CASCADE EMULATOR

A. LUT Cascade

Fig. 1 shows a model for a sequential circuit, where X denote inputs, Z denote outputs, Y denotes the inputs to flip-flops, Y' denotes the outputs of flip-flops, and $|Y|$ denotes the number of state variables. We first consider an **LUT cascade** [3] that realizes the combinational part of the sequential circuit, then consider the LUT cascade emulator that emulates the LUT cascade.

An LUT cascade is shown in Fig. 2, where multiple-output LUTs (**cells**) are connected in series to realize a multiple-output function. The wires connecting adjacent cells are called **rails**. Also, each cells may have external outputs in addition to the rail outputs. In this paper, X_i denotes the **external inputs** to the i-th cell; Y_i' denotes the **state inputs** to the i-th cell; Z_i denotes the **external outputs** of the i-th cell; Y_i denotes the **state outputs** of the i-th cell; R_{i-1} denotes the **rail inputs** to the i-th cell; and R_i denotes the **rail outputs** from the i-th cell. We can obtain the LUT cascades by applying **functional decomposition** repeatedly to the BDD that represents the multiple-output function [4].

Definition 2.1 *Let $\vec{X} = (x_1, x_2, \ldots, x_n)$ be the input variables, $\vec{Y} = (y_1, y_2, \ldots, y_m)$ be the output variables, and $\vec{f} = (f_1(\vec{X}), f_2(\vec{X}), \ldots, f_m(\vec{X}))$ be the corresponding out-*

0-7803-9451-8/06/$20.00 ©2006 IEEE.

466

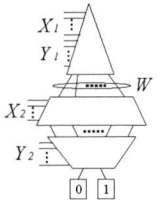

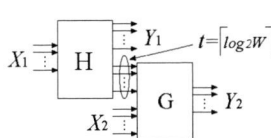

Fig. 3. BDD_for_CF. Fig. 4. Functional decomposition.

put functions. **The characteristic function of the multiple-output function** *is* $\vec{\chi}(\vec{X}, \vec{Y}) = \bigwedge_{i=1}^{m} (y_i \equiv f_i(\vec{X}))$.

The characteristic function of an n-input m-output function is a two-valued logic function with $(n + m)$ inputs. It has input variables x_i ($i = 1, 2, \ldots, n$), and output variables y_j for output f_j. Let $B = \{0,1\}$, $\vec{a} \in B^n$, $\vec{F} = (f_1(\vec{a}), f_2(\vec{a}), \ldots, f_m(\vec{a})) \in B^m$, and $\vec{b} \in B^m$. Then, the characteristic function satisfies the relation

$$\vec{\chi}(\vec{a}, \vec{b}) = \begin{cases} 1 & (\text{when } \vec{b} = \vec{F}(\vec{a})) \\ 0 & (\text{otherwise}) \end{cases}$$

Definition 2.2 **A support variable** *of a function* f *is a variable on which* f *actually depends.*

Definition 2.3 *[5]* **The BDD_for_CF** *of a multiple-output function* $\vec{f} = (f_1, f_2, \ldots, f_m)$ *is the ROBDD [10] for the characteristic function* $\vec{\chi}$. *In this case, we assume that the root node is in the top of the BDD, and the variable* y_i *is below the support variable of* f_i, *where* y_i *is the variable representing* f_i.

Definition 2.4 **The width of the BDD_for_CF** *at height* k *is the number of edges crossing the section of the graph between* x_k *and* x_{k+1}, *where the edges incident to the same nodes are counted as one. Also, in counting the width of the BDD_for_CF, we ignore the edges that incident to the constant 0 node.*

Let X_1 and X_2 be sets of input variables, Y_1 and Y_2 be sets of output variables, (X_1, Y_1, X_2, Y_2) be the variable ordering of a BDD_for_CF for the multiple-output function $\vec{f} = (f_1, f_2, \ldots, f_m)$, and W be the width of the BDD_for_CF at the height (X_1, Y_1) in Fig. 3. By applying functional decomposition to $\vec{f}$, we obtain the network in Fig. 4, where the number of lines connecting two blocks is $t = \lceil \log_2 W \rceil$ [4].

Theorem 2.1 *[5] Let* μ_{max} *be the maximum width of the BDD_for_CF that represents an* n-input logic function $\vec{f}$. *If* $u = \lceil \log_2 \mu_{max} \rceil \leq k - 1$, *then* $\vec{f}$ *can be realized by a circuit shown in Fig. 4, where* $|X_1| = k$. *By applying functional decompositions* $s - 1$ *times, we have the cascade having the structure of Fig. 2.*

B. LUT Cascade Emulator

Fig. 5 shows an **LUT cascade emulator** for a sequential circuit.

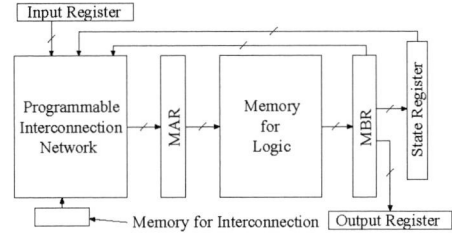

Fig. 5. LUT cascade emulator.

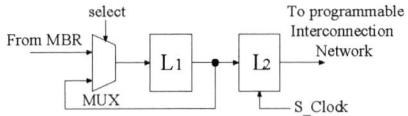

Fig. 6. Double rank flip-flop.

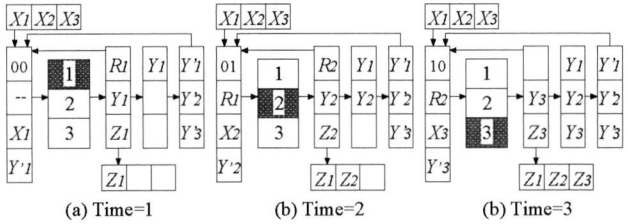

(a) Time=1 (b) Time=2 (b) Time=3

Fig. 7. Emulation of a sequential circuit

An LUT cascade emulator stores the cell data of an LUT cascade in the **memory for logic**. The address of cell data is calculated from inputs, state variables, and rail outputs of the preceding cell. The LUT cascade emulator reads the cell outputs from the memory for logic, and send them to the **State Register** and the **Output Register**. The **Input Register** stores the values of the primary inputs; the **MAR** (Memory Address Register) stores the address of the memory; the **MBR** (Memory Buffer Register) stores the outputs of the memory; the **Programmable Interconnection Network** connects the input register, the state register, and the MAR, also it connects the MBR and the MAR; the **Memory for Interconnection** stores data for the interconnections; and the **Control Network** generates necessary signals to obtain functional values.

To emulate a sequential circuit, the LUT cascade emulator stores state variables and output variables in the registers. Fig. 6 shows the **Double-Rank Flip-Flop** for the state register and the output register, where L_1 and L_2 are D-latches. Set the select signals to high when all the cells in a cascade are evaluated, and send the values into L_1 latches. When all the cascades are evaluated, the values of the state variables are sent to L_2 latches. This can be done by adding a pulse to S_Clock.

Example 2.1 *Fig. 7 illustrates the emulation of the sequential circuit whose combinational part realizes the LUT cascade in Fig. 2.*

At $Time = 1$, to evaluate Cell 1, the two most significant bits of the address are set to (0,0) to specify Page 1. Also, the inputs to Cell 1 X_1 and Y_1' are set to the lower address bits through the programmable interconnection network. By reading Page 1, the outputs of Cell 1 are sent to MBR. For the outputs that become the primary output Z_1, store it in the output register, while for the output that becomes the state output Y_1, store it in the state register (Fig. 7(a)).

At $Time = 2$, to evaluate Cell 2, the two most significant bits of the address are set to $(0,1)$ to specify Page 2. Also, the outputs of Cell 1, R_1 are connected to the middle address bits, and the input variables of Cell 2 X_2 and Y_2' are set to the lower address bits through the programmable interconnection network. By reading Page 2, the outputs of Cell 2 are sent to MBR. For the output that becomes the primary output Z_2, store it in the output register, while for the output that becomes the state output Y_2, store it in the state register (Fig. 7(b)).

At $Time = 3$, to evaluate Cell 3, the two most significant bits of the address are set to $(1,0)$ to specify Page 3. Also, the outputs of Cell 2, R_2 are connected to the middle address bits, and the input variables of Cell 3 X_3 and Y_3' are set to the lower address bits through the programmable interconnection network. By reading Page 3, the outputs of Cell 3 are sent to MBR. For the output that becomes the primary output Z_3, store it in the output register, while for the output that becomes state output Y_3, store it in the state register (Fig. 7(c)).

When all the cells of cascades are evaluated, Control Network sends a pulse to S_Clock of the state register and the output register, and the values of state outputs Y_1, Y_2, Y_3 are sent to the L_2 latches. Also, the values of the output register Z_1, Z_2, Z_3 are sent to the primary outputs. (End of Example)

III. Synthesis of the LUT Cascade Emulator

A. Partition of the outputs

When the number of outputs is large, we partition the outputs into groups, and realize a cascade for each group independently. Usually, the BDD_for_CF for all the outputs are too large to construct. Even if the BDD_for_CF is constructed, it can be too large to be realized by an LUT cascade. Also, constructing a single BDD_for_CF for all the outputs is inefficient, since the optimization of a large BDD_for_CF is time consuming.

In order to construct as small BDD_for_CFs as possible, we partition the outputs so that each group has a small number of support variables.

Definition 3.5 *Let* $F = \{f_1, f_2, \ldots, f_m\}$ *be the set of the outputs functions,* $G \subseteq F$, *and* $f_i \in F - G$. *Then, the **similarity** of the output* f_i *with* G *is defined as follows:*

$$Similarity(i, G, F) = |Sup(f_i) \cap Sup(G)|, \quad (1)$$

where $Sup(F)$ *denotes the set of support variables of* F.

Algorithm 3.1 *(Partition the Outputs and Construction of BDD_for_CF)*
Let $F = \{f_1, f_2, \ldots, f_m\}$ *be the set of* m *logic functions,*

$\mathcal{Z} = \{G_1, G_2, \ldots, G_r\}$ *be the set of subset of output functions after partitioning,* r *be the number of partitions, and* Th_node *be the threshold of the number of nodes.*

1. $\mathcal{Z} \leftarrow \phi, r \leftarrow 0$.
2. While $F \neq \phi$, do Steps (a)-(d).

   (a) $Node \leftarrow 0, G_r \leftarrow \phi$.
   (b) While $Node \leq Th\_node$, $F \neq \phi$, and G_r is cascade realizable, do Steps i-iii.

       i. Select f_i with the maximum $Similarity(i, G_r, F)$. If $G_r = \phi$, then select f_i that has the largest support.
       ii. $G_r \leftarrow G_r \cup \{f_i\}, F \leftarrow F - \{f_i\}$.
       iii. Construct BDD_for_CF that realizes G_r, and $Node \leftarrow$ *(the number of nodes)*.

   (c) If G_r is not cascade realizable, then $G_r \leftarrow G_r - \{f_i\}, F \leftarrow F \cup \{f_i\}$.
   (d) $\mathcal{Z} \leftarrow \mathcal{Z} \cup \{G_r\}, r \leftarrow r + 1$.

3. Terminate.

This method merges outputs into a group while a cascade is realizable and the number of nodes in the BDD is equal to or less than the threshold. Algorithm 3.1 partitions the outputs into the groups so that the resultant BDDs are small enough.

B. Memory Packing

By Alogrithm 3.1, a given multiple-output function is represented by a set of BDD_for_CFs. Then, the LUT cascades are constructed, and LUT data is allocated into the memory of the LUT cascade emulator.

Example 3.2 *Fig. 8 shows the LUT cascade consisting of 4-input cells. Fig. 9(a) illustrate the memory map of cell data, where the memory has 6-bit address inputs, and each word consists of four bits. The dark parts in the figure are unused, and* P_i *denotes the page number.* (End of Example)

In Example 3.2, each cell data are stored in a separate page of the memory. The data of a cell must be stored in the same page, and must be read simultaneously. However, if there are any extra space in the same page, multiple cell data can be stored in the same page. This method to reduce the memory area is called **memory-packing** [6].

Example 3.3 *In Fig. 9(a), by storing the cell data* r_5 *and* z_1 *to Page 1, we have the memory map in Fig. 9(b), where a half of the memory is enough to store all the data.* (End of Example)

IV. Logic Simulation using an LUT Cascade Emulator

A. Generation of the execution code for the simulation

Fig. 10 shows the data flow of the logic simulation system using LUT cascade emulator.

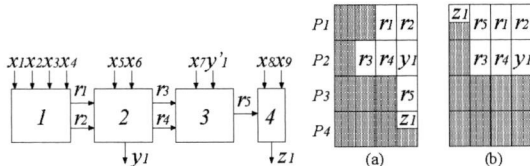

Fig. 8. Example of LUT cascade.   Fig. 9. Example of memory-packing.

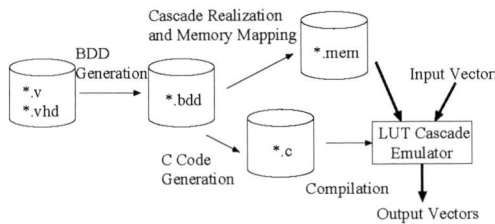

Fig. 10. Flow of data for the logic simulation system using LUT cascade emulator.

First, it converts the Verilog-VHDL code describing the given circuit into shared-BDDs [9]. Then, it reduces the number of nodes of BDDs by changing variable orders [7]. Next, it generates the LUT cascades from BDDs using the functional decomposition described in Chapter 2, and it maps them to the memory of the LUT cascade emulator. Also, it generates the C code that simulates the control circuit of the LUT cascade emulator. Next, it complies the C code into the execution code for simulation of the LUT cascade emulator. And, finally the simulator evaluates the output of the given circuit, by using the memory map of the LUT cascade emulator.

B. Program code that simulates the LUT cascade emulator

This system generates the program code that describes the following operations.

Step 1 Set the input register, and initialize the state register.
An input value is set to the input register. Also, the value of the state register is initialized.

Step 2 Evaluate of each cell.

Step 2.1 Simulate the programmable interconnection network.
The address of the memory for logic is generated from the values of the input register, the state input register, the MBR, and a page address.

Step 2.2 Read the memory for logic.
The content of the memory for logic is read using the address generated by Step 2.1.

Step 2.3 Distribute output values of the memory for logic.
The values read by Step 2.2 are sent to the output register and the state register.

Step 3 Perform the state transitions.
The output values of the state register is updated by S_Clock.

Assigning each memory output to each register consumes CPU time. Fortunately, the memory outputs are stored in the order of primary outputs, state outputs, and rail outputs. For a 32-bit processor, we can evaluate up to 32 outputs at a time. To obtain required outputs, we shift the memory outputs covered by a mask, and assign to a 32-bit variable. In this way, we can evaluate the multiple outputs simultaneously. Also, there is an additional merit for performing the state transition. Let $|Y|$ be the number of state variables for given logic function, then the number of evaluations for the state transition is $\left\lceil \frac{|Y|}{32} \right\rceil$ for a 32-bit machine.

Since cascades have much fewer signal lines than the original circuit, the compilation time for cascades are much shorter than that of the conventional LCC method.

C. Analysis of Simulation Time

When the LUT cascade emulator is implemented on a dedicated hardware [2], the evaluation time is proportional to the number of cells. However, when the LUT cascade emulator is implemented on a standard PC, we need extra time since the inputs and outputs of a cell are evaluated sequentially.

To do high-speed simulation for the LUT cascade emulator on a PC, we have to consider two different objects:

a. Reduction of the number of cells.
This can be done by increasing the number of inputs of each cell. However, the increase of the number of inputs of each cell also increases the evaluation time per cell.

b. Reduction of the number of cell inputs.
This decreases the evaluation time per cells, but increases the number of cells.

To find the best strategy, we did following experiments. We implemented 10 MCNC benchmark functions [8] on the LUT cascade emulator. By changing the maximum number of inputs for cells, we obtained the average number of cell inputs, the number of cells, and the execution time of the LUT cascade emulator. Fig. 11 shows the experimental results, where the horizontal axis denotes the maximum number of cell inputs; 0 denotes the lower bound on the maximum number of inputs of cells, that is $\lceil \log_2 \mu_{max} \rceil + 1$; the vertical axis denotes the ratios of the number of cells, the number of the average cell inputs, and simulation time. We set 1.00 to the ratios when the number of cell inputs is $\lceil \log_2 \mu_{max} \rceil + 1$.

Fig. 11 shows that the simulation time increases with the number of cell inputs. To compute the address of the memory for logic, we need CPU time. This CPU time increase with the number of inputs of a cell. Therefore, our strategy is to reduce the number of cell inputs in the LUT cascade emulator.

V. EXPERIMENTAL RESULTS

A. Comparison with LCC

We simulated selected MCNC benchmark functions by LUT cascade emulator and LCC on a same PC. Table I shows the

5A-5

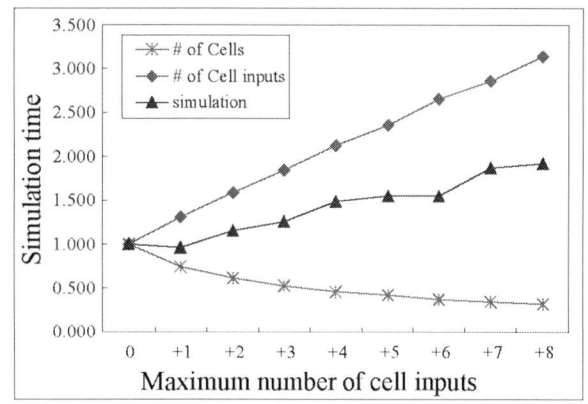

Fig. 11. Relation between the maximum number of cell inputs and simulation time.

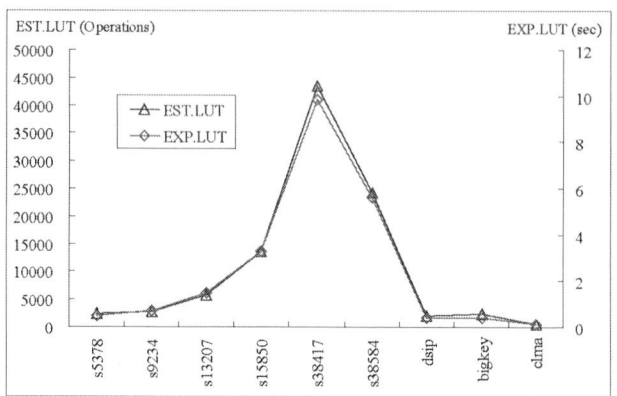

Fig. 12. Simulation time for LUT cascade emulator.

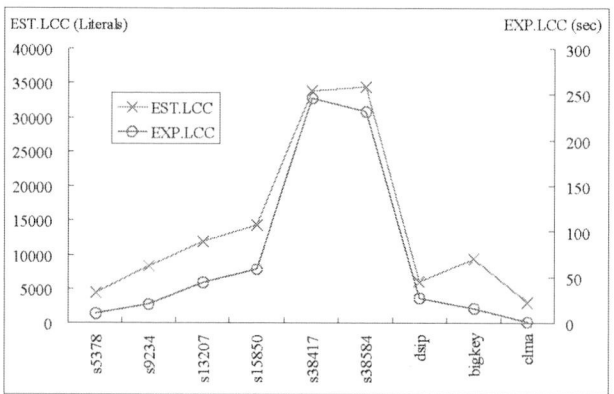

Fig. 13. Simulation time for LCC.

experimental results. *Name* denotes the name of benchmark function; *In* denotes the number of inputs; *Out* denotes the number of outputs; *State* denotes the number of state variables; *Cas* denotes the number of LUT cascades; *Cell* denotes the total number of cells; and *Mem* denotes the amount of memory (Mega Bits). Also, *EXT.in* denotes the average number of external inputs to cells; *P.out* denotes the total number of cells with external output(s); and *S.out* denotes the total number of cells with state output(s). *Sim* denotes the evaluation time (sec). In order to obtain the raw evaluation time for the simulation, we generated the one million random test vectors, and evaluated the time excluding the time for reading and writing vectors. *Setup* denotes the setup time (sec) for a simulation. *Setup* of LCC is the time for the C code generation and the compilation, while *Setup* of the LUT cascade emulator is the time for BDD generation, LUT cascades synthesis, memory mapping, C code generation, and the compilation. *Literals* denotes the total number of literals in expressions of lines of the C code generated by the LCC. *Ratio* denotes that of the simulation setup time or that of the simulation execution time (LCC/LUT cascade emulator). To produce the executable code for LCC, we used gcc compiler with optimization option -O3. Also, we generated program codes for LUT cascade emulator, and compiled them with the same conditions as LCC. In the experiments, we used an IBM PC/AT compatible machine, Pentium4 Xeon 2.8GHz, L1 Cache: 8KB, L2 Cache: 512KB, Memory: 4GByte, and OS: Redhad (Linux 7.3). For the benchmark function *clma*, it has many redundant primary inputs, primary outputs, and logic gates. For pre-processing, we simplified the Verilog-VHDL descriptions of benchmark functions by using Quartus II version 5.0 [13]. Note that, the time for the pre-processing is not included in the setup time.

Table I shows that the LUT cascade emulator is 12 - 64 times faster than the LCC. Also, the setup speed for the LUT cascade emulator is 1.20 - 5.22 times faster than the LCC, except for s5378 and *clma*.

The number of operations in the LUT cascade emulator is

estimated as follows:

$$EST.LUT = EXT.in \times Cell \\ + Cell + P.out + S.out + Rail, \quad (2)$$

where $Rail = Cell - Cas$. The first term of expression (2) corresponds to the setup time of all the external inputs of the cells; the second term corresponds to the access time to the memory for logic; the third term corresponds to the setup time for the output register; the fourth term corresponds to the setup time for the state register; and the last term corresponds to the setup time for the rail inputs.

In Fig. 12, the right vertical axis denotes the experimental value EXP.LUT (sec), and the left vertical axis denotes the estimated number of operations EST.LUT. Also, we conjecture that *Literals* is proportional to the simulation time for LCC. In Fig. 13, the right vertical axis denotes the experimental value EXP.LCC (sec), and the left vertical axis denotes the estimated number of literals EST.LCC = *Literals*. Figs. 12 and 13 show that the EXP.LUT and EXP.LCC can be estimated from EST.LUT and EST.LCC, respectively.

In Figs. 12 and 13, note that EXP.LUT is smaller than EXP.LCC, while EST.LUT (number of operations) is larger than EST.LCC (number of literals) for some functions. This may look strange. To see the reason, we converted the C codes

470

TABLE I
RESULTS OF REALIZATION OF BENCHMARK FUNCTIONS.

Name	In	Out	State	LUT cascade emulator									LCC			Ratio	
				Cas	Cell	Mem [Mbit]	EXT.in	P.out	S.out	Setup [sec]	Sim [sec]	Literals	Setup [sec]	Sim [sec]	Setup	Sim
s5378	35	49	164	40	543	0.82	2.39	105	28	14.59	0.49	4424	10.26	10.46	0.70	21.35
s9234	36	39	211	44	599	0.91	2.63	26	120	14.68	0.71	8220	41.01	20.64	2.79	29.07
s13207	62	152	638	93	1245	3.28	2.27	109	390	31.09	1.46	11954	83.04	43.76	2.67	29.97
s15850	77	150	534	105	3370	8.95	1.95	115	338	79.25	3.25	14328	115.72	58.71	1.46	18.06
s38417	28	106	1636	389	9411	45.36	2.55	60	964	763.07	9.87	33769	917.40	245.63	1.20	24.89
s38584	38	304	1426	270	5118	16.90	2.55	232	956	159.09	7.61	34485	830.27	230.76	5.22	30.33
dsip	229	197	224	45	473	6.60	1.96	108	115	10.90	0.42	5959	26.39	27.08	2.42	64.48
bigkey	263	197	224	48	541	3.76	1.97	171	121	11.74	0.42	9262	27.58	16.26	2.35	38.71
clma	101	81	33	9	45	0.16	3.28	27	21	3.59	0.11	2994	2.87	1.37	0.79	12.45

of the LUT cascade emulator and the LCC into the assembly-codes, and analyzed them. The size of the assembly-code for the LCC is several times larger than EST.LCC. This is because the LCC compiler generates extra codes to evaluate the negative literals and the logic gates, and to produce the output signals. On the other hand, the size of the assembly-code for LUT cascade emulator does not depend on EST.LUT, and is much smaller than EST.LUT. In the LCC, it's operands frequently move between register and memory. For the gate with fan-outs, the LCC stores the output values of the gate into a variable temporarily, and uses it as the input of two or more gates. On the other hand, the LUT cascade emulator uses only the rail values in the single register variable. Therefore, only the input and output register and the memory for logic requires memory references. Experimental results show that the simulator based on an LUT cascade emulator is 12-64 times faster than the LCC. One reason for this is the difference of the representations: the cascade has many fewer signals than the random logic network. Another reason is due to the CPU architecture of the PC. The access time of the data in the main memory is about 200 times longer than one in the L1 cache. So, the CPU time heavily depends on the frequency of cache miss. In the case of the LCC simulator, the circuit data and control are mixed, and the instruction data is too large to be stored in the data cache. On the other hand, in the case of an LUT cascade emulator, the cascade data and control are separated. Control data is in the instruction cache, while the cascade data is in the data cache. Thus, we can expect fewer cache miss in the LUT cascade emulator.

B. Comparison with Commercial Tools

We compare the our method with two commercial simulators: Super-FinSim [14] version 6.2.9 and ModelSim [15] Altera-Edition (AE) 6.0c. ModelSim (AE) is an event-driven logic simulator, bundled with Quartus II. It supports functional simulation involving the 'zero-delay'. To obtain the evaluation time for the functional simulation, first, we generate a verilog code for a testbench including 10^4 test vectors. Then, we compiled the verilog codes, representing benchmark circuit and testbench, using a command 'vlog [benchmark name.v]' with option '+notimingcheck' and 'no_notifier'. Next, we evaluate the simulated time using commands which are 'vsim -c' and

TABLE II
RESULTS OF COMPARISON WITH COMMERCIAL TOOLS.

Name	Simulation time			Ratio	
	FinSim	ModelSim	LUT Cascade Emulator	v.s.FinSim	v.s.ModelSim
s5378	1.45	35.77	0.09	16.11	397.44
s9234	1.73	44.72	0.11	15.73	406.55
s13207	6.59	46.50	0.17	38.76	273.53
s15850	14.84	68.63	0.58	25.59	118.33
s38417	17.93	72.70	3.28	5.47	22.16
s38584	58.21	155.84	1.76	33.07	88.55
dsip	14.31	30.42	0.16	89.44	190.13
bigkey	2.99	37.28	0.08	37.38	466.00
clma	41.13	88.86	0.09	457.00	987.33

'run 10000'. Super-FinSim supports both the enhanced cycle simulation (ECS) and the event-driven simulation at the same time. To compare the our method with the ECS, we generated the testbench similar to ModelSim, and we complied verilog codes using a command 'finvc' with option '+delay_mode_zero', '-dsm com', '-ol 1', '-acc', '-fastgate', '+no_timingcheck', and '+no_notifier'. Then, we built an execution file using the command 'finbuild', and evaluated the simulation time. We used Microsoft Visual C++ version 6.0 for Super-FinSim. In the experiments, we used an IBM PC/AT compatible machine, Pentium4 3.06GHz, L1 Cache: 8KB, L2 Cache 512KB, Memory 2GByte, and OS: Windows XP Professional SP2. To realize the LUT cascade emulator for Linux on Windows, we used gcc compiler version 3.2 on the cygwin.dll version 1.3.22 which emulates Linux API using Windows API.

Table II compares the results. *Simulation time* denotes the actual simulation time including the time for reading and writing 10^4 vectors; and *Ratio* denotes that of the simulation execution time (Super-FinSim / LUT cascade emulator) and (ModelSim / LUT cascade emulator), respectively.

Table II shows that the LUT cascade emulator is 22.16-987.33 times faster than ModelSim, and 5.47-457.00 times faster than FinSim.

Especially, for benchmark *clma*, our method is hundreds times faster than commercial tools. *clma* is a special benchmark circuit with redundant input-and-output signals and redundant gates. In order to perform a high-speed simulation, our method converted the circuit into BDDs to remove these

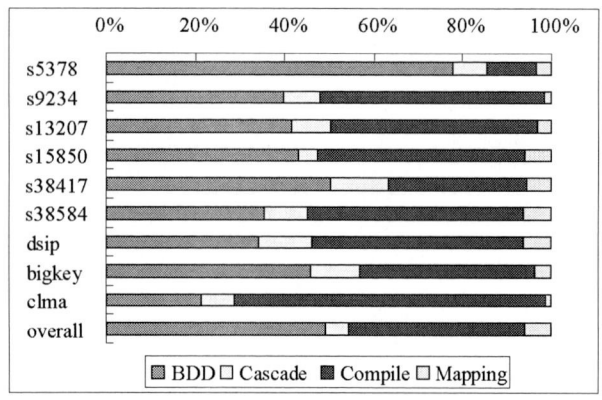

Fig. 14. Percentage of each process for setup time.

redundancy. However, ModelSim and FinSim used original circuits, so the redundancy of circuits affected the simulation time.

C. Setup Time for the LUT Cascade Emulator

Fig. 14 shows the percentage of each process of the setup time for the LUT cascade emulator. The labels of the vertical axis show function names. The horizontal axis shows the percentage of the processing time for BDD for CF generation (*BDD*); LUT cascade synthesis (*cascade*); C code generation and compilation (*compile*); and memory mapping (*mapping*). *overall* denotes the average percentage of each process. Fig. 14 shows that BDD generation and compilation consume most of the CPU time. For BDD generation, optimization of variable orders for BDDs consumes most of the CPU time. Reduction of the code size, e.g., reduction of the number of cells, is effective to reduce the compilation time. Increase of cells increases the simulation setup time. The time for memory mapping increases with the number of cell inputs. The amount of memory for each cell is $2^k \cdot u$, where k is the number of inputs of a cell, and u is the number of outputs of a cell. Therefore, the number of cell inputs influences the total amount of data. In our experiment, since the average number of cell inputs is small, the memory mapping time did not influence the setup time.

VI. CONCLUSION AND COMMENTS

In this paper, we showed a cycle-based logic simulator using the LUT cascade emulator running on a standard PC. This method converts the circuit into LUT cascades. Then, it stores the LUT data in the memory of the LUT cascade emulator. Next, it generates the program code for the control circuit of the LUT cascade emulator. The program code is suitable for logic simulation on a standard PC due to the better memory reference patterns. Experimental results using benchmark functions show that this method is 12-64 times faster than LCC on a standard PC.

Our method converts the circuit into BDDs, where the sizes of BDDs representing the circuits are limited due to the avail-

able memory. To avoid this limitation, we partition the output functions into groups. However, when the BDD representing a single output is excessively large, our method fails to perform the emulation.

One of the future projects is to derive a partition method for the circuit and to represent the circuits by smaller BDDs.

VII. ACKNOWLEDGMENTS

This research is partly supported by Japan Society for the Promotion of Science (JSPS), MEXT, and Kitakyushu Innovative Cluster.

REFERENCES

[1] M. Abramovici, M. A. Breuer, and A. D. Friedman, *Digital Systems Testing and Testable Design,* Wiley-IEEE Press; Rev. Print edition, Sept. 1994.

[2] T. Sasao, Y. Iguchi, and M. Matsuura, "LUT cascades and emulators for realizations of logic functions," *RM2005*, Tokyo, Japan, Sept. 5-6, 2005, pp.63-70.

[3] T. Sasao, M. Matsuura, and Y. Iguchi, "A cascade realization of multiple-output function for reconfigurable hardware," *IWLS-2001*, Lake Tahoe, CA, June 12-15,2001, pp.225-300.

[4] T. Sasao, and M. Matsuura, "A method to decompose multiple-output logic functions," *Proc. Design Automation Conference*, San Diego, CA, USA, June 2-6, 2004, pp.428-433.

[5] P. Ashar, and S. Malik, "Fast functional simulation using branching programs," *ICCAD'95*, Nov.1995, pp.408-412.

[6] T. Sasao, M. Kusano, and M. Matsuura, "Optimization methods in look-up table rings," *IWLS-2004*, June 2-4, 2004, Temecula, California, USA, pp.431-437.

[7] R. Rudell, "Dynamic variable ordering for ordered binary decision diagrams," *ICCAD'93*, pp. 42-47, 1993.

[8] S. Yang, Logic synthesis and optimization benchmark user guide version 3.0, MCNC, Jan. 1991.

[9] S. Minato, N. Ishiura, and S. Yajima, "Shared binary decision diagram with attributed edges for efficient boolean function manipulation," *Proc. Design Automation Conference*, June 1991, pp.52-57.

[10] R. E. Bryant, "Graph-based algorithms for boolean function manipulation," *IEEE Transact. Comput.*, C-35, Aug.1986, pp.677-691.

[11] R. Murgai, F. Hirose, and M. Fujita, "Logic synthesis for a single large look-up table," *ICCD1995*, pp.415-424, Oct. 1995.

[12] P. C. McGeer, K. L. McMillan, A. Saldanha, A. L. Sangiovanni-Vincentelli, and P. Scaglia, "Fast discrete function evaluation using decision diagrams," *ICCAD'95*, pp.402-407, Nov. 1995.

[13] Altera: http://www.altera.com/

[14] Fintronic USA, Inc.: http://www.fintronic.com/

[15] Mentor Graphics Corporation: http://www.model.com/

[16] P. M. Maurer, "The Inversion algorithm for digital simulation," *ICCAD'94*, pp.258-261, Nov. 1994.

Power-Aware Scheduling and Dynamic Voltage Setting for Tasks Running on a Hard Real-Time System[1]

Peng Rong

Dept. of Electrical Engineering
University of Southern California
Los Angeles, CA 90089
e-mail : prong@usc.edu

Massoud Pedram

Dept. of Electrical Engineering
University of Southern California
Los Angeles, CA 90089
e-mail : pedram@ceng.usc.edu

Abstract - **This paper addresses the problem of minimizing energy consumption of a computer system performing periodic hard real-time tasks with precedence constraints. In the proposed approach, dynamic power management and voltage scaling techniques are combined to reduce the energy consumption of the CPU and devices. The optimization problem is first formulated as an integer programming problem. Next, a three-phase solution framework, which integrates power management scheduling and task voltage assignment, is proposed. Experimental results show that the proposed approach outperforms existing methods by an average of 18% in terms of the system-wide energy savings.**

I. INTRODUCTION

Reducing power consumption is a key requirement for extending the battery service lifetime of portable devices. Even in high-end computer systems, expensive cooling and packaging cost and declined reliability associated with high levels of power dissipation, make low power design a critical design consideration. Dynamic power management (DPM) and dynamic voltage scaling (DVS) have both proven to be highly effective techniques for reducing power dissipation in such systems. DPM refers to a selective shut-off of idle system components, while DVS slows down underutilized resources and decreases their operating voltages. A detailed survey of DPM techniques can be found in [1].

Most researches on low-power task scheduling focus only on reducing the CPU power by using DVS techniques. However, in reality, executing a useful task on a computer system requires cooperation between the CPU and many other system components, e.g., memory, disk drives, wireless devices, etc., which can also consume significant amounts of power. These components generally have their own voltage levels and may or may not support DVS, which makes it difficult to apply DVS techniques to the CPU only and achieve total system power savings. In fact, DVS when applied to CPU only may even increase the overall system energy consumption for executing a given set of tasks. At the same time, DPM is known to be an effective approach for reducing the power consumption of the various peripheral

components and I/O devices. Thus DVS combined with DPM has the potential to achieve power savings, not possible by either DPM or DVS.

This paper addresses the problem of power optimization of a real-time system having heterogeneous components and performing periodic hard real-time tasks. The dependencies between the tasks are described by a directed acyclic graph (DAG), sometimes referred to as a *task graph*.

Most related work on low power scheduling for dependent tasks concentrate on DVS techniques. Some authors have considered voltage assignment on distributed embedded systems. The approach proposed in [2] first schedules tasks based on a list-scheduling algorithm by using the reciprocal of the slack time as the task priority, and next tries to evenly distribute the available positive slack time among tasks on each critical path and thereby reduce the operating voltages and save energy. Reference [3] assumes a given task schedule and assignment and proposes an extended list-scheduling algorithm. At each time step, the energy saving of a task is calculated as the difference between the expected energies given the task is scheduled at this step or at the next step. A task with a higher energy saving and less slack time has a higher priority. The authors of [4] present a two-phase framework. In the first phase, a version of the early-deadline-first scheduling is used to assign a task to a best-fit processor in terms of the task ready time and the processor free time. In the second phase, an ILP optimization problem is formulated and solved in order to determine the voltage level of the processor used to run each task.

Several works on DPM-based task scheduling have also been proposed in the literature. An online scheduling algorithm for independent tasks is presented in [5]. This algorithm attempts to reduce the number of device on/off transitions by greedily extending the pattern for current device usage so as to reduce average power consumption in the near future. Reference [6] proposes an offline branch-and-bound algorithm to search for the energy optimal task scheduling. In [7], the authors prove that solving energy optimal task scheduling for DPM on multiple devices is an NP hard problem even for a simple case where no timing dependency is considered. References [8] and [9] start with a

---

[1] This project was funded in part by the NSF CNS grant no. 0509564.

0-7803-9451-8/06/$20.00 ©2006 IEEE.

given timing-fixed task sequence and propose algorithms to determine an energy-minimal state transition sequence for devices while satisfying hard time constraints.

More recently, a number of researchers have reported DVS algorithms taking into account energy consumption of the system components. In [10] the authors present a DVS heuristic based on the *critical speed* of each task, which is defined as the CPU speed at which the execution of a task consumes the least total system energy. Reference [11] proposes a DVS technique based on a precise energy model considering both the active power and standby component of the system power.

In the literature, several works have been proposed on combining DVS and DPM. Reference [12] present a Markovian decision processes based DPM model which is a uniform modeling framework for both DVS and DPM. In [13], the authors combine DVS with their previously proposed renewal theory based DPM approach. These two stochastic approaches are unable to handle tasks with hard-time constraints or dependency. The problem of combining DVS and DPM for hard real-time tasks is studied in reference [14], where a scheduling algorithm for a single processor with a sleep state is presented which is proved having a competitive ratio of 3. Task dependency is not considered in this work either.

To the best of our knowledge, no proposed research work is conducted to combine DVS and DPM techniques for hard real-time dependent tasks running on multiple devices. This is specifically the contribution of the present paper. In particular, an integer programming based formulation is first provided to exactly state the optimization problem to be addressed. Next, a three-phase algorithm is proposed to solve the power-aware task scheduling and voltage-to-task assignment problems with the objective of minimizing the total system energy consumption. The three steps are power-aware task scheduling, task-level voltage assignment, and task rescheduling and voltage level refinement.

The remainder of the paper is organized as follows: The problem formulation is presented in section 2. The three steps of the proposed algorithm are described in sections 3, 4 and 5, respectively. Experimental results and conclusions are given in sections 6 and 7.

II. PROBLEM FORMULATION

This paper targets a real-time system which has a single CPU and κ system devices (e.g., various I/O devices, main memory.) The CPU is considered to be device number 0 whereas other devices are numbered from 1 to κ. The CPU has a discrete number of *performance states* corresponding to different supply voltage levels and clock frequencies and one *sleep* state. All other devices have a *functional* state during which they provide service and a low power *sleep* mode during which they cannot provide any services. [2]

---

[2] It is straight-forward to extend the mathematical formulation to handle I/O devices with multiple low-power states (e.g., standby, drowsy, and sleep.)

Furthermore, a device which is in the performance/functional state can be in one of two sub-states: 1) actively performing services; 2) waiting for service requests to arrive. We will refer to sub-state 1 as the *active* state and sub-state 2 as the *idle* state. We assume that each device k consumes the same amount of power when they are in active or idle mode (denoted by $funcpow_k$), but significantly less power when it is in the sleep mode ($sleepow_k$.)

A set of n non-preemptive dependent tasks periodically run on the system with a time period T_d. The data dependency (precedence) constraints between the tasks are described by a directed acyclic task graph, called a *task graph*, $G(V, E)$, where each node v denotes a task and a directed edge $e(u, v)$ represents a data flow between task u and v and implies that task v can be executed only after task u finishes. Every task has to be performed on the CPU, and may require support (services) from some (or all) of the system devices. It is assumed that during the run time of a task, all devices whose services are required by the task in question will stay in their active modes. The problem is to solve the optimal task scheduling and task-level voltage assignment with the objective of minimizing the total system energy consumption during period T_d.

Let V_i, $i = 1, \ldots, m$, denote the m operating voltages for the CPU and f_i the clock frequency of the CPU at voltage V_i. We define the workload of task u as the number of CPU cycles without considering memory and IO device access delay. Let $N_{u,i}$ denote the actual number of CPU cycles required to complete task u at operating voltage V_i. We define variable $x(u, i)$ to represent the percentage of the workload of task u which is performed at voltage V_i. Note that there are $m \cdot n$ such variables. The execution time (*duration*) of task u is calculated as

$$dur_u = \sum_{i=1}^{m} \frac{x(u,i) \cdot N_{u,i}}{f_i}, \tag{2-1}$$

where $\sum_{i=1}^{m} x(u,i) = 1 \cdot \tag{2-2}$

We introduce $n \cdot \kappa$ 0-1 integer variables, $Z_k(u)$, as follows: $Z_k(u) = 1$ exactly if task u requires service from device k. The energy consumption due to execution of task u is equal to

$$ene_u = c_u \sum_{i=1}^{m} x(u,i) \cdot N_{u,i} \cdot V_i^2 + \sum_{k=1}^{K} Z_k(u) \cdot P_k \cdot dur_u, \tag{2-3}$$

where c_u is the effective switched capacitance per CPU cycle; and P_k is the power consumption of device k in the active mode.

Let $s(u)$ denote the start time of task u. Thus the precedence constraint is expressed as

$$s(u) + dur_u \leq s(v) \quad \forall e(u,v) \in E \tag{2-4}$$

To formulate the energy consumed by the CPU and devices during idle time, we need to introduce two virtual (dummy) tasks: task 0 of duration zero which is placed at exactly the start of period T_d and task $n+1$ of duration zero which is placed at the end of period T_d. We define tasks 0 and n+1 so as to require all devices, i.e.,

$Z_k(0) = 1$ and $Z_k(n+1) = 1$, $\forall k$. Notice that the interval between task 0 and the first task executed on device k denotes the first idle period. Similarly, the last idle period is defined as the interval between the last task executed on device k and task $n+1$. Also notice that.

We introduce $(n+2)^2 k$ 0-1 integer *scheduling variables* $Y_k(u,v)$ as follows: $Y_k(u,v) = 1$ exactly if task u is executed on device k immediately before task v is executed on the same device. Since on each device, every task has only one immediate successor, the following constraint on $Y_k(u,v)$ should be respected

$$\sum_{v=1}^{n+1} Y_k(u,v) = \begin{cases} 1, & Z_k(u) = 1 \\ 0, & \text{otherwise} \end{cases}, \quad u = 0,1,\ldots,n \qquad (2\text{-}5)$$

Similarly, every task has only one immediate predecessor; i.e.,

$$\sum_{u=0}^{n} Y_k(u,v) = \begin{cases} 1, & Z_k(v) = 1 \\ 0, & \text{otherwise} \end{cases}, \quad v = 1,2,\ldots,n+1 \qquad (2\text{-}6)$$

There is also a precedence constraint between task v and its immediate successor, both of which are executed on device k, as follows

$$\sum_{u=0}^{n} (s(u) + dur_u) \cdot Y_k(u,v) \le s(v) \quad \forall v \in V, k \in devs_v \qquad (2\text{-}7)$$

With variable $Y_k(u,v)$, we can express the duration of the idle time of device k just before it provides service to task v, $it_{k,v}$, as

$$it_{k,v} = s(v) - \sum_{u=0}^{n} (s(u) + dur_u) \cdot Y_k(u,v) \cdot \qquad (2\text{-}8)$$

Let function $idlene_k(it)$ return the energy consumed by device k during idle time of length it. Note that the device may be placed in a low-power state during its long idle times, as suggested, for instance, in [8][9]. For the illustration purpose, assume that device k has two power states: active and sleep. Let p_a and p_s denote the power consumptions of the device in the active and sleep states, respectively. Let ε_{tr} and τ_{tr} denote the summation of energy overheads and latency overheads associated with the two transitions into and out of the sleep state, respectively. Recall that the breakeven time is equal to $\tau_{BE} = \varepsilon_{tr}/(p_a - p_s)$. Then,

$$idlene_k(it) = \begin{cases} \varepsilon_{tr} + p_s \cdot it, & it \ge \max(\tau_{BE}, \tau_{tr}) \\ p_a \cdot it, & \text{otherwise} \end{cases}$$

Thus, the total energy consumed by the system during time period T_d is calculated as

$$E_{sys} = \sum_{u=1}^{n} ene_u + \sum_{k=0}^{K} \sum_{v=1}^{n} idlene_k(Z_k(v) \cdot [1 - Y_k(0,v)] \cdot it_{k,v}) \\ + \sum_{k=0}^{K} idlene_k(it_{k,n+1} + \sum_{v=1}^{n} Y_k(0,v) \cdot it_{k,v}) \qquad (2\text{-}9)$$

Notice that when executing a periodical task set, for a device, the idle time before the first task starts and the idle time after the last task finishes actually constitute a single idle period. The third term on the right-hand side of equation (2-9) calculates the device energy consumption for such an idle period. The second term on the RHS handles all the other idle times.

The optimization problem is to minimize E_{sys} with respect to constraints (2-1) to (2-8). Note that in this formulation, we ignore the energy and timing overhead associated with the voltage changes because switching of the CPU voltage normally takes between 10-100 microseconds depending on the hardware support for the DVS function. This is negligible compared to the device on/off transition times, which tend to be in the range of a few tenths of a second. The corresponding energy overhead is also small.

This problem is a nonlinear non-convex integer program over variables $s(u)$, $x(u,i)$ and $Y_k(u,v)$; the worst-case computational complexity of exactly solving this problem is expected to be exponential. So we propose a three-step heuristic approach to solve the problem as follows:

1. **Task Ordering**: Derive a linear ordering of tasks (i.e., calculate $Y_k(u,v)$ values) by performing an interactive minimum-cost matching on some appropriately constructed graph (cf. section 3.)

2. **Voltage Assignment**: Given the task ordering implied by the schedule obtained in step 1, assign voltages and task durations (i.e., calculate $x(u,i)$ values) and exact start times (i.e., calculate $s(u)$ values) to each task so as to meet a target cycle time, T_d (cf. section 4.)

3. **Refinement**: Improve the task scheduling and voltage assignment of steps 1 and 2 to increase the energy efficiency of the resulting solutions (cf. section 5.)

III. TASK ORDERING

In this step, we assume that the CPU voltage level is set to the maximum possible value and that the task execution times (durations) are calculated on this basis.[3] The goal is to take the task graph with known task execution times and schedule it on the CPU (device 0) so as to minimize the total energy dissipation due to I/O devices $(1,\ldots,\kappa)$ staying in the idle mode and that caused by transitioning the devices from their high-power functional state to the low-power sleep state. Notice that the summation of energy dissipation in all devices $(0,\ldots,\kappa)$ when these devices are in active states is fixed and independent of the scheduling. The scheduling only changes the duration of the idle times and the number of on to off transitions for the I/O devices.

Let $tasks_k$ denote the set of tasks running on device k and dev_u denote the set of devices that are needed by task u. A lower bound on the total system energy dissipation, $totene_{LB}$ can be obtained by assuming that there is no energy overhead for the transitions between idle and sleep states of any device

$$totene_{LB} = \sum_{u \in V} \sum_{k \in dev_u} funcpow_k \cdot dur_u + \sum_{k=0}^{\kappa} sleepow_k \cdot (T_d - \sum_{u \in tasks_k} dur_u)$$

Now, the actual total energy includes the energy consumed by various devices when they stay in their idle modes and

---

[3] This is a simple heuristic used to assign task durations for this step. Other heuristic assignments are possible. Note, however, that we are only interested in the ordering of tasks after the completion of this step and will in fact calculate the exact task schedule and execution times after voltage assignment.

when they transition in and out of the sleep modes. Let's denote the schedule, Λ, by the start times of all tasks in the given task graph. Based on this information, one can linearly order the set of tasks and represent the active times of each device as a set of closed intervals. More precisely, device k will be represented by a segment set, $S_k = \{s_{k,1}, \ldots, s_{k,z}\}$ $(z \leq n)$ corresponding to the time intervals during which the device is in its active state.

$$t_{nonactive}(s_{k,i}, s_{k,i+1}) \triangleq \begin{cases} start(s_{k,i+1}) - end(s_{k,i}) & \text{if } i < |S_k| \\ T_d + start(s_{k,1}) - end(s_{k,|S_k|}) & \text{otherwise} \end{cases}$$

$$F_{k,i} \triangleq F_k(s_{k,i}, s_{k,i+1}) = \begin{cases} 1 & \text{if } t_{nonactive}(s_{k,i}, s_{k,i+1}) \geq t_{BE,k} \\ 0 & \text{otherwise} \end{cases}$$

$$totene(\Lambda) = totene_{LB} +$$
$$\sum_{k=1}^{K} \sum_{i=1}^{|S_k|-1} (F_{k,i} \cdot transene_k + (1 - F_{k,i}) \cdot funcpow_k \cdot t_{nonactive}(s_{k,i}))$$

Here $start(s)$ and $end(s)$ denote the start time and end time of segment s while $transene_k$ denotes the total transition energy cost of device k to go from idle mode to the sleep mode *and* to return to the active mode.

Next we construct an *augmented task graph* (ATG) $A(V,C)$ from the given task graph $G(V,E)$ by copying $G(V,E)$ and subsequently adding/deleting some edges to/from E. More precisely, the new edge set, C, does not contain any directed edge uv such that there exists another directed path from u to v in C. In addition, C contains undirected edges qr if tasks associated with q and r can be scheduled next to each other in some order. In addition, each node, q, in V (task) has three attributes: task execution time, dur_q, task energy consumption, ene_q, and the list of devices that are required by the task, dev_q. Finally, each directed edge qr in C, has an associated energy cost, $extraene_{qr}$, calculated as follows

$$s_{k,q} \triangleq [start(r) - dur_q, start(r)]; \quad s_{k,succ(r)} \triangleq [start(r) + dur_r, T_d]$$
$$F_{k,q,r} \triangleq F_k(s_{k,q}, s_{k,succ(r)})$$
$$extraene_{qr} =$$
$$\sum_{k \in dev_q - dev_r} \sum_{i=1}^{|S_k|-1} (F_{k,q,r} \cdot transene_k + (1 - F_{k,q,r}) \cdot funcpow_k \cdot t_{nonactive}(s_{k,i}))$$

Each undirected edge between nodes q and r will have two such energy costs corresponding to directed edges qr and rq. Note however that at most one of the two directed edges may be chosen as part of the scheduling solution.

The basic flow of the proposed scheduling algorithm is to iteratively find the edge with the *least* extra energy value and merge its two end nodes, implying that the corresponding tasks will be scheduled to run in immediate succession. For a directed edge, the ordering is fixed a priori whereas for the undirected edge, the algorithm will choose one of the two possible orderings and fix it. After each merge, the ATG is updated by removing all edges that become invalid and calculating the attributes for the newly generated node. The process continues until a single node is left in the ATG, which corresponds to a total ordering (scheduling) of all the tasks. The process continues until exactly one node remains in the modified ATG (i.e., a complete schedule is obtained.)

If at any step of the algorithm, there is a tie between the extra energy costs of two candidate edges qr and uw, then we will choose the edge that would result in the minimum *total extra energy cost*, $extraene_{tot}$, of the resulting graph if the merge was performed. Now, $extraene_{tot}$ is calculated as the summation of the node weights of the resulting graph where the node weight is itself calculated as the average of the extra edge costs of outgoing edges from that node.

Example 1: Consider a task graph depicted in Figure 1(a). Assume that there are four devices $\{0,1,2,3\}$ with the following device utilization sets:

$dev(u1) = \{0\}$

$dev(u2) = dev(u4) = \{0,1\}$

$dev(u3) = dev(u5) = dev(u6) = \{0,2,3\}$

For the sake of simplicity, we assume that each task has a unit time duration (which is longer than its breakeven time) and that the idle power consumption of all devices is the same. In addition, each device consumes 1 unit of energy for each transition to and from the sleep state. The ATG graph of this task set is given in Figure 1(b). The directed edge $u1u2$ exists in ATG, because there is a precedence constraint between nodes $u1$ and $u2$ and $u2$ can be scheduled immediately after $u1$. The presence of undirected edge $u2u3$ implies that $u2$ and $u3$ can be scheduled next to the other without any ordering constraint. The edge labels denote the energy consumption if the start and end nodes of the edge are scheduled one after the other. For simplicity, assume all node energies are 0.

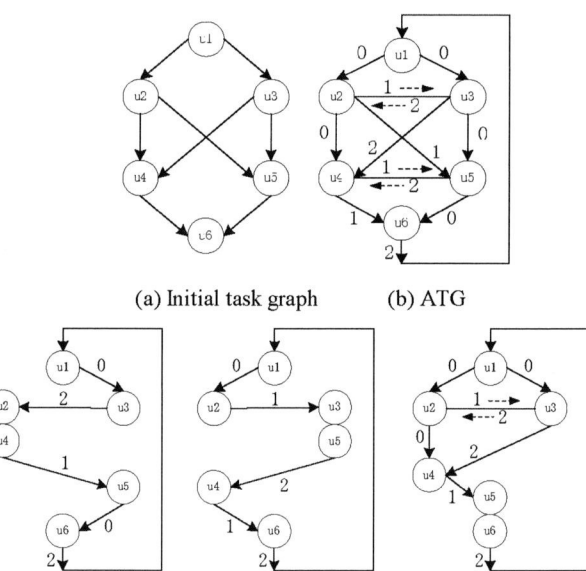

(a) Initial task graph   (b) ATG

(c) ATG after merging a pair of node

Figure 1. Illustrative example for power-aware task schedule.

With this ATG, we can start the task scheduling for power management. There are five edges with minimal edge energy equal to 0. That is, we can merge the pair of nodes associated with each of these edges without incurring additional energy cost. In Figure 1(c), three ATGs are presented, each corresponding to the merging of the node pair for one of the edges. Let us consider the left-most ATG which is generated after merging $u2$ and $u4$. Since there is a directed edge from $u2$ to $u4$, $u4$ must be scheduled after $u2$. After the $u2$-$u4$ merge, the edge from $u3$ to $u2$ becomes a directed edge, because originally $u3$ had to be executed before $u4$ which has now been merged with $u2$ into a single node. The directed edge from $u1$ to $u2$ in the initial ATG disappears because after the $u2$-$u4$ merge, $u3$ stands between $u1$ and $u2$ in the precedence chain.

The left-most ATG in Figure 1(c) has the minimal E_{ATG} value equal to 5. So the merge of $u2$ and $u4$ is selected for the first step.

IV. VOLTAGE ASSIGNMENT

Having generated the task schedule, we fix the ordering of tasks, but otherwise, ignore the task execution times and start times, which were heuristically set as explained at the beginning of section 3, we can easily calculate the $Y_k(u,v)$ values. Thus by substituting the value of $Y_k(u,v)$ into the optimization problem formulated by (2-1) through (2-9), all constraints becomes linear constraints and the only unknowns become $s(u)$ and $x(u,i)$ variables. However, the optimization problem cannot be solved exactly and efficiently, because the objective function remains a non-convex function of idle times, it. We thus propose two approaches to get around this non-convexity issue.

The first one simply ignores the energy components introduced by $idlene_k$ in equation (2-9). The optimization problem thus becomes a linear programming problem over continuous variable $x(u,i)$, which can be solved in polynomial time. It is worth pointing out that, strictly speaking, $x(u,i)$ should take discrete values instead of continuous ones, because the number of CPU cycles executed at each operating voltage is an integer. However, when we consider a task executed in hundreds of thousands of CPU cycles, the effect introduced by rounding up to one cycle can be safely ignored.

The second approach introduces new 0-1 integer variables $W_k(v,h)$ to approximate the idle time $it_{k,v}$ as follows: $W_k(v,h) = 1$ exactly if $t_{k,h} \le it_{k,v} < t_{k,h+1}$. $t_{k,h}$ and $t_{k,h+1}$ take values from a discrete set $\{t_{k,1}, t_{k,2}, ..., t_{k,H}\}$. The value of $idlene_k(it)$ is approximated by

$$idlene_k(it_{k,v}) = \sum_{h=1}^{H} idlene_k(t_{k,h}) W_k(v,h), \text{ with } \sum_{h=1}^{H-1} W_k(v,h) = 1.$$

Consequently, constraint (2-8) is thus rewritten as

$$\sum_{h=1}^{H-1} t_{k,h} W_k(v,h) \le s(v) - \sum_{u=0}^{n} (s(u) + dur_u) \cdot Y_k(u,v)$$
$$\le \sum_{h=1}^{H-1} t_{k,h+1} W_k(v,h).$$

And equation (2-9) becomes an integer linear cost function. As a result, the original optimization problem is approximated with a mixed integer linear program. The value of parameter H can be adjusted to trade-off the computational complexity and the approximation accuracy.

V. REFINEMENT

In this section, we provide a top-level overview of an algorithm that we have developed to improve the results

obtained by the first two steps. Starting from the solution obtained from steps 1 and 2, we shift the tasks together to remove redundant positive slack times. Next, we apply a greedy refinement algorithm on this solution to improve the total energy cost while meeting the timing constraint. In particular, we identify the set of critical tasks whose duration has a large impact on the system energy dissipation, e.g. a small change of the duration could enable device transitions to low power states, and change their voltage assignments accordingly. Detailed are omitted.

VI. EXPERIMENTAL RESULTS

This experiment is conducted on a system comprising of a single CPU and three other devices. The CPU has three operating voltage/frequency levels: 1V/200MHz, 1.1V/300MHz and 1.3V/400MHz [11]. The CPU and all devices support only one low-power sleep state. The power consumptions in different states, energy and timing overheads of state transitions for both the CPU and the three devices are reported in Table 1.

TABLE 1

Power and transition parameters

Device	Active Power	Sleep Power	Energy Overhead	Timing Overhead
SDRAM	0.3W	~0	~0	~0
HDD	2.1W	0.85W	0.6J	400ms
WLAN	0.7W	0.05W	0.04J	100ms
CPU	1.0W (200MHz)	0.05W	0.3J	400ms

To evaluate the effectiveness of our proposed approach, we generated five task graphs by using software package, TGFF [15], which is a randomized task graph generator widely used in the literature to evaluate the performance of scheduling algorithms. Each task graph consists of 20 to 200 tasks. All tasks require supports from SDRAM. The dependency of tasks on HDD and WLAN were randomly generated and fixed before optimization. The characteristics of different task graphs are given in the following table. For example, for task graph G1, when the CPU has its highest frequency setting, the cpu is used during 61% of the total execution whereas the SDRAM, HDD and WLAN are used for 61%, 29% and 42% of the total time.

TABLE 2

Characteristics of task graphs

Task Graph	No. of Tasks	CPU and device utilization factors at max speed for the CPU			
		CPU	SDRAM	HDD	WLAN
G1	28	0.61	0.61	0.29	0.42
G2	65	0.72	0.72	0.51	0.39
G3	110	0.34	0.34	0.12	0.23
G4	159	0.48	0.48	0.30	0.25
G5	204	0.55	0.55	0.28	0.36

In this experiment, we compare the total system energy consumptions of the following methods:

M1: No DVS, no DPM. The CPU always operates at the highest voltage level and devices are kept active during the whole execution time. This provides the baseline compare against.

M2: DPM without any task scheduling. Tasks are executed on the CPU (which has assumed its highest frequency and voltage setting) in an un-optimized order based on their ID numbers after they become available. A method similar to the approach in [8] is used to determine the state transition sequences of all devices and the CPU.

M3: DPM with task scheduling. This method is similar to M2, except that our proposed power-aware task scheduling algorithm is used to determine the task execution sequence.

M4: Conventional cpu-driven DVS plus DPM. Similar to M2, except that the task operating voltage is assigned to minimize the CPU power consumption. More specifically, the operating voltage setting for each task is obtained by solving the optimization problem defined in section 2 without considering the energy consumption of devices.

M5: Proposed system-aware DVS plus DPM (which have called, Power-aware Scheduling and Voltage Setting or PSVS for short.) Task scheduling and operating voltage settings are determined through the proposed three-phase framework.

TABLE 3
Normalized energy consumption results for different techniques

Task Graph	M2	M3	M4	M5
G1	0.54	0.50	0.58	0.47
G2	0.67	0.59	0.63	0.53
G3	0.28	0.26	0.32	0.25
G4	0.40	0.35	0.42	0.34
G5	0.43	0.37	0.39	0.33

The energy consumptions of different techniques are compared in Table 3. These values have been normalized with respect to the baseline energy consumption of M1, e.g., for G1, M2 results in total system energy consumption which is 54% of the baseline energy consumption. From this table, it is seen that compared to DPM technique without task scheduling, our proposed DPM with task scheduling can reduce energy consumption by an average of 11%. Furthermore, when this method is combined with our proposed voltage assignment technique (resulting in M5 or PSVS), an additional 9% energy saving is achieved.

VII. CONCLUSIONS

This paper addresses the problem of minimizing energy consumption of a computer system performing periodic hard real-time tasks with precedence constraints. In the proposed approach, dynamic power management and voltage scaling techniques are combined to reduce the energy consumption of the CPU and devices. The optimization problem is first formulated as an integer programming problem. Next, a three-phase solution framework, which integrates power management scheduling and task voltage assignment, is proposed. Experimental results demonstrate efficiency of the proposed approach.

REFERENCES

[1] L. Benini, A. Bogliolo and G. De Micheli, "A survey of design techniques for system-level dynamic power management," *IEEE Trans on VLSI*, vol.8 iss.3, pp.299-316, 2000.

[2] J. Luo and N. Jha, "Static and dynamic variable voltage scheduling algorithms for real-time heterogeneous distributed embedded systems," *ASP-DAC*, pp. 719-26, 2002.

[3] F. Gruian and K. Kuchchinski, "LEneS: task scheduling for low-energy systems using variable supply voltage processors," *ASP-DAC*, pp. 449-55, 2001.

[4] Y. Zhang, X. Hu, and D.Z. Chen, "Task scheduling and voltage selection for energy minimization," *DAC*, pp. 183-8, 2002.

[5] Y-H Lu, L. Benini and G. De Micheli, "Low-power task scheduling for multiple devices," *CODES*, pp. 39-43, 2000.

[6] V. Swaminathan and K. Chakrabarty, "Pruning-based energy-optimal device scheduling for hard real-time systems," *CODES*, pp.175-80, 2002.

[7] Y-H Lu, L. Benini and G. De Micheli, "Power-aware operating systems for interactive systems," *IEEE Trans. on VLSI*, vol.10 iss.2, pp. 119-34, 2002.

[8] V. Swaminathan and K. Chakrabarty, "Energy-conscious, deterministic I/O device scheduling in hard real-time systems," *IEEE Trans. on CAD*, vol.22 iss.7, pp.847-58, 2003.

[9] J. Liu and P.H. Chou, "Optimizing mode transition sequences in idle intervals for component-level and system-level energy minimization," *ICCAD*, pp. 21-28, 2004.

[10] R. Jejurikar and R. Gupta, "Dynamic voltage scaling for system-wide energy minimization in real-time embedded systems," *ISLPED*, pp. 78-81, 2001.

[11] K. Choi, W. Lee, R. Soma and M. Pedram, "Dynamic voltage and frequency scaling under a precise energy model considering variable and fixed components of the system power dissipation," *ICCAD*, pp. 29-34, 2004.

[12] Q. Qiu and M. Pedram, "Dynamic power management based on continuous-time Markov decision processes," *DAC*, pp. 555-561, 1999.

[13] T. Simunic, L. Benini, A. Acquaviva, P. Glynn and G. De Micheli, "Dynamic voltage scaling and power management for portable systems," *DAC*, pp.524-529, 2001.

[14] S. Irani, S. Shukla and R. Gupta, "Algorithms for power savings," *SODA*, pp. 37 – 46, 2003.

[15] http://ziyang.ece.northwestern.edu/tgff.

Optimal TDMA Time Slot and Cycle Length Allocation for Hard Real-Time Systems

Ernesto Wandeler   Lothar Thiele
Computer Engineering and Networks Laboratory
Swiss Federal Institute of Technology (ETH) Zurich, Switzerland
E-mail: {wandeler,thiele}@tik.ee.ethz.ch

Abstract— We present an analytic method to determine the provably smallest possible slot length that must be allocated in a TDMA resource, to serve an event-triggered hard real-time load with arbitrary deterministic timing behavior. Based on this method, we then present constructive methods to find all feasible as well as the optimal cycle length in a TDMA resource, and we show how to determine the minimum required bandwidth of a TDMA resource. We demonstrate the applicability and computational efficiency of the presented methods in a case study of a large distributed embedded system with a TDMA bus, where we will find the optimal parameter set for the TDMA bus.

I. INTRODUCTION

In large distributed embedded systems, TDMA scheduling policies play an increasingly important role. TDMA is often employed in such systems on backbone communication resources that typically interconnect a large number of the present embedded computing units (ECU's), as shown in Fig. 1. This trend can best be observed in the area of safety-critical automotive and avionic systems, where TDMA-based communication protocols such as TTP, or more recently the mixed TDMA/FTDMA-based FlexRay replace more and more the formerly omnipresent CAN protocol. But also for communication on MpSoC's [7], as well as to provide QoS guarantees in network on chips [6], TDMA gets increasingly important.

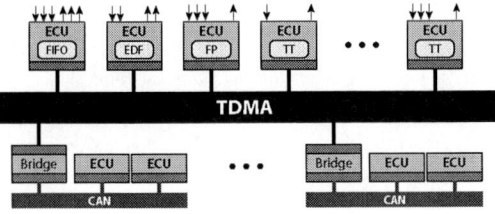

Fig. 1. A distributed system with TDMA communication.

TDMA protocols possess a number of advantages compared to event-triggered communication protocols such as CAN, that make them interesting for the use in such systems. First of all, TDMA resources support temporal composability, by clearly separating resource access of different subsystems that therefore do not interfere with each other. Moreover, TDMA resources have a very deterministic timing behavior, can be made fault tolerant, and support error detection, as well as error contention, i.e. a faulty subsystem does not affect the correct behavior of the remaining system. Note, that due to these properties, TDMA is also often applied for single processor scheduling, for example to enable composable and hierarchical scheduling, see e.g. [13].

A major difficulty that arises however during the design process of systems with TDMA-scheduled resources is parameter selection for the TDMA resources. Customizable parameters are typically the total bandwidth B of the resource, the cycle length c of the TDMA round, as well as the individual slot lengths s_i for the different service consumers of the TDMA resource.

In purely time-triggered systems, an optimal communication schedule that defines slot and cycle lengths can be constructed at design-time [8], but in reality heuristics are often used to find a valid communication schedule due to the computational complexity of finding the optimal schedule.

Many large distributed embedded systems are however not anymore designed as purely time-triggered systems, but contain instead mixed time- and event-triggered components. Be it because of the co-existence of time- and event-triggered subsystems (clusters) that are connected with each other by bridges, as considered in [12], or be it because of the existence of some event-triggered ECU's, as considered for example in [10]. Both is shown in Fig. 1.

When we get to such mixed time- and event-triggered systems, parameter selection for TDMA resources gets even more challenging. In [11], slot lengths are chosen as a fraction of a fixed cycle length, such that every service consumer with event-triggered load receives an individual total bandwidth from the TDMA resource. While the method presented in [11] can be used for systems with non-real-time event-triggered loads, it is only applicable by trial-and-error for systems with real-time event-triggered loads, i.e. loads with deadline constraints. For such systems, [12] presents a heuristic to assign slot lengths of a TDMA resource with fixed bandwidth and cycle length. [12] however only deals with strictly periodic loads and does therefore not consider buffering effects that may occur when serving loads with jitter or bursts.

Very recently, a method for slot as well as cycle length optimization based on evolutionary search techniques was presented in [7]. This method can be used to parameterize slot and cycle lengths of TDMA resources with fixed bandwidth, and it can handle real-time event-triggered loads with jitter and bursts. Since the method is based on evolutionary algorithms, it is however computationally expensive, and cannot guarantee a global optimal solution for a predetermined optimality criterion.

Contributions of this work:

- We present an analytic method to determine the provably smallest possible slot length that must be allocated in a TDMA resource with fixed cycle length and bandwidth, to serve a hard real-time load with arbitrary deterministic timing behavior.

- We present constructive methods to find the optimal cycle length and minimum required bandwidth of a TDMA resource.

- We show the applicability and computational efficiency of the presented methods in a case study of a large distributed embedded system, where we will find the optimal parameter set for a

TDMA bus that interconnects 21 ECU's that send a total of 30 different hard real-time message streams with jitter and bursts over this shared communication resource.

- The presented work is based on an existing theoretical framework for modular system level performance analysis of hard real-time systems. We present a TDMA component that will extend this framework to enable performance analysis and interface-based design of distributed real-time systems with TDMA.

II. MODULAR PERFORMANCE ANALYSIS

In the domain of communication networks, powerful abstractions have been developed to model flow of data through a network. In particular Network Calculus [9] provides means to deterministically reason about timing properties of data flows in queuing networks. Real-Time Calculus [14] extends the basic concepts of Network Calculus to the domain of real-time embedded systems, and in [5] a unifying approach to Modular Performance Analysis with Real-Time Calculus has been proposed. It is based on a general event and resource model, allows for hierarchical scheduling and arbitration, and can take computation and communication resources into account.

The following sections introduce some concepts of Network and Real-Time Calculus, that build the foundation of this work. While we introduce these concepts from a communications point of view, all results can also be applied directly to the analysis and design of hardware/software components in a system. Then, messages correspond to tasks, and the message size to a task's execution time.

A. From Components to Abstract Components

In this work, we consider distributed embedded systems, consisting of a number of embedded computing units and a shared communication network. Every embedded computing unit (ECU) is connected to the communication network (CN) via a communication network interface (CNI), that uses the services of the communication network via a communication controller (CC). The ECU processes incoming event streams, and generates message streams that must be sent in real-time to other ECU's via the communication network. To initiate the sending of a message, the ECU places the message into one of possibly several FIFO buffers of the CNI. For real-time communication, every message stream has an associated maximum communication delay, that denotes the maximum time interval between the time a message of the stream is placed into the buffer, and the required delivery time of the message. The CNI uses the services of the CC to send the buffered messages over the network, and it applies an arbitration policy to share the available communication resource amongst the queued messages. Figure 2 shows on the left side a block diagram of the connection of an ECU to a communication network.

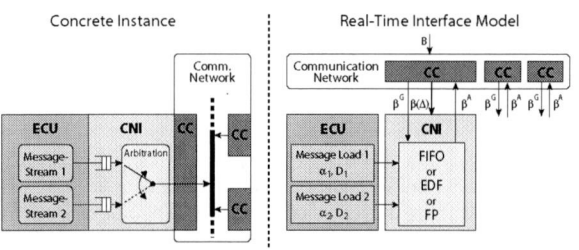

Fig. 2. A concrete ECU connected to a communication system (left), and the corresponding Real-Time Interface model (right).

We will use Real-Time Calculus [14, 5] to abstractly describe the real-time properties of such a CNI component. In comparison to the concrete CNI, an abstract CNI sends an abstract message stream over an abstract communication resource. In Real-Time Calculus, the timing and resource demand properties of the concrete message streams that enter the CNI are abstracted by Variability Characterization Curves (VCC) that are called arrival curves $\alpha(\Delta)$, following [9]. Together, $\alpha(\Delta)$ and the maximum allowable delay D of every message stream describe the properties of a message stream that are essential for real-time analysis, see [14]. The communication resource that enables the sending of the messages is modeled by a service curve $\beta(\Delta)$, see also [9]. $\beta(\Delta)$ is also a VCC, and describes the essential properties of the communication resources that are available to an abstract CNI component.

B. Variability Characterization Curves

An arrival curve $\alpha(\Delta) \in \mathbb{R}^{\geq 0}, \Delta \in \mathbb{R}^{\geq 0}$ provides an upper bound on the communication resource demand that messages arriving from an ECU in *any* time interval of length Δ create on a CNI, i.e. the messages that arrive on a stream within a time interval $[t, t+\Delta)$ may create a resource demand of at most $\alpha(\Delta)$ on an CNI, for all $t \geq 0$.

Arrival curves substantially generalize the classical representations of standard event arrival patterns such as sporadic, periodic, periodic with jitter or others. For example a message stream with messages of size e, that arrive with a period p, a jitter j, and a minimum inter arrival distance d, can be modeled by an arrival curve as follows:

$$\{p, j, d, e\} \Rightarrow \alpha(\Delta) = \min\left\{\left\lceil\frac{\Delta+j}{p}\right\rceil e, \left\lceil\frac{\Delta}{d}\right\rceil e\right\} \quad (1)$$

Besides being able to represent any message stream with known timing behavior, it is also possible to determine arrival curves corresponding to any finite length message trace, obtained for example from observation or simulation. Some examples of arrival curves are shown in Fig. 3.

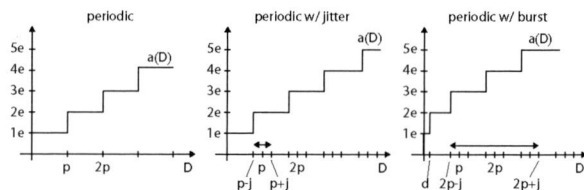

Fig. 3. Arrival curves for standard event arrival patterns.

Similarly, the resource availability of an abstract communication resource is modeled by a VCC called service curve. A service curve $\beta(\Delta) \in \mathbb{R}^{\geq 0}, \Delta \in \mathbb{R}^{\geq 0}$ provides a lower bound on the available resources, e.g. bus cycles, in any time interval Δ, i.e. in a time interval $[t, t+\Delta)$ at least $\beta(\Delta)$ bus cycles are available.

C. Real-Time Calculus

Network and Real-Time Calculus provide powerful methods to analyze different performance properties of systems, as for example detailed throughput or resource consumption analysis, see [9], [14] and [5]. Here, we however only consider delay and backlog analysis.

In Real-Time Calculus, the maximum *delay* Del_{max} experienced by messages of a message stream with arrival curve $\alpha(\Delta)$ that is sent over an abstract communication resource with service curve $\beta(\Delta)$ is bounded by the maximum horizontal distance between $\alpha(\Delta)$ and $\beta(\Delta)$:

$$Del_{max} \leq \sup_{\Delta \geq 0}\{\inf\{\tau \geq 0 : \alpha(\Delta) \leq \beta(\Delta+\tau)\}\} \quad (2)$$

Analogously, the maximum *backlog* Buf_{max} of a message stream in the buffer of a CNI is bounded by the maximum vertical distance between $\alpha(\Delta)$ and $\beta(\Delta)$:

$$Buf_{max} \leq \sup_{\Delta \geq 0} \{\alpha(\Delta) - \beta(\Delta)\} \qquad (3)$$

If a CNI serves more than one message stream, i.e. if it has more than one arrival curve as input, the total required buffer can be computed by replacing $\alpha(\Delta)$ with the sum of all arrival curves $\sum \alpha_i(\Delta)$ in (3). It is also possible to compute the different maximum delays, experienced by the message streams. For this, the arbitration policy of the CNI must be considered, see [5].

D. Real-Time Interfaces

Real-Time Interfaces were first introduced in [2], and they connect the principles of interface-based design [?] and Real-Time Calculus. The central idea of interface-based design is to describe components by a component interface that, if well-designed, provides enough information to decide whether a component can work properly in a system together with other components. Real-Time Interfaces can thereby be considered as a special instance of assume/guarantee interfaces, see also [?]. Components with an assume/guarantee interface make assumptions on the values of their input variables and give in return guarantees on the values of their output variables.

When we look at the connection of an ECU to a communication network in Fig. 2, then we see that the communication network provides a communication resource supply, represented as a service curve $\beta(\Delta)$, to a CNI. $\beta(\Delta)$ is therefore the output of an abstract communication network, and the input to an abstract CNI.

A communication network can therefore be modeled as an abstract component with a service output variable $\beta(\Delta)$, and its real-time interface would provide the output guarantee $\beta(\Delta) \geq \beta^G(\Delta)$ on this output variable. Through this output guarantee, the network component expresses that the service $\beta(\Delta)$, that is provided by the communication network is larger or equal $\beta^G(\Delta)$. $\beta^G(\Delta)$ is sometimes also referred to as supply bound function $\mathbf{sbf}(\Delta)$, see e.g. [4] or [13].

An abstract CNI on the other hand can be modeled as a component with a service input variable $\beta(\Delta)$, and its real-time interface would make the input assumption $\beta(\Delta) \geq \beta^A(\Delta)$ on this input variable. Through this input assumption, the abstract CNI expresses that the service $\beta(\Delta)$, that is provided on its service input must be larger or equal $\beta^A(\Delta)$. In return, the abstract CNI then guarantees to send all messages within the required maximum delay. $\beta^A(\Delta)$ is sometimes also referred to as demand bound function $\mathbf{dbf}(\Delta)$.

Figure 2 shows on the right side the Real-Time Interface model of an abstract communication network component with service guarantee $\beta^G(\Delta)$ that is connected to an abstract CNI component with service assumption $\beta^A(\Delta)$. In order to determine, whether the communication network provides enough service to the CNI, we only need to check that the following predicate is true:

$$\beta^G(\Delta) \geq \beta^A(\Delta) \quad \forall \Delta \geq 0 \qquad (4)$$

The difficulty is then to find appropriate values for $\beta^G(\Delta)$ and $\beta^A(\Delta)$. The service assumption $\beta^A(\Delta)$ of an abstract CNI that serves a message stream with arrival curve $\alpha(\Delta)$ and maximum delay D is given by (see also Example 1):

$$\beta^A(\Delta) = \alpha(\Delta - D) \qquad (5)$$

If on the other hand the CNI serves several message streams with an EDF arbitration policy, the service assumption can be computed as $\beta_{EDF}^A(\Delta) = \sum \alpha_i(\Delta - D_i)$, and for FIFO it can be computed as $\beta_{FIFO}^A(\Delta) = \sum \alpha_i(\Delta - D_{min})$. To compute the service assumption under a fixed priority arbitration policy is slightly more involved and is described in detail in [2], while [1] describes how to

obtain the service assumption of components with mixed static and dynamic arbitration policies. $\beta^G(\Delta)$ on the other hand can be given as $\beta^G(\Delta) = B \cdot \Delta$ for a fully available communication network with bandwidth B, and in the next section, we will establish $\beta^G(\Delta)$ for a communication network with TDMA.

III. Performance Analysis of TDMA Systems

In this work, we analyze a real-time communication system consisting of n communication network interfaces that access via communication controllers a bus with bandwidth B that implements a TDMA protocol. The TDMA cycle length is denoted as $\bar{c}$ and can only take on values that are multiples of the cycle length quantum q_c. In every TDMA cycle, one single communication slot of length s_i is assigned to every CNI that is connected to the bus. In a realistic system, the slot lengths s_i can only take on values that are multiples of a slot length quantum q_s. We denote a quantized slot length as $\bar{s}_i = \lceil s_i / q_s \rceil \cdot q_s$. Further, every slot typically involves a slot overhead o_s, while the cycle itself involves a cycle overhead o_c. These overheads account for example for required network idle times between consecutive slots and cycles, CRC codes for channel fault detection, time synchronization data, or any other protocol related overhead. Depending on the different timing specifications, some bandwidth may remain unused in every communication cycle. Fig. 4 depicts the timing specifications in this TDMA protocol.

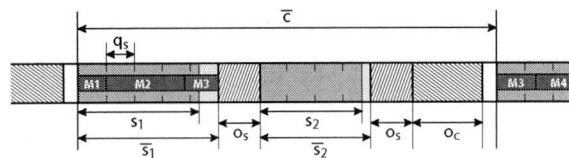

Fig. 4. TDMA protocol timing specifications.

In the CNI, a message service layer provides a packet service that fragments outgoing messages into packets and that reassembles incoming packets into messages. This message service layer guarantees that a CNI can always use the complete time slot assigned to it, as long as there are messages waiting to be sent. The overhead for the message fragmentation is accounted for in the slot overhead o_s. Since this fragmentation overhead depends on the message sizes and the arbitration policy of a CNI, one could introduce individual slot overheads $o_{s,i}$ for every CNI. The methods in this work could easily be extended accordingly.

A. TDMA Resource Component Interface

An abstract communication (or computation) resource with TDMA can be modeled by a real-time interface as depicted in Fig. 5. The component has an input B, that determines the total bandwidth of the underlying resource, and the cycle length of the TDMA protocol is specified by the parameter $\bar{c}$. The component further has n service outputs, that provide a service with the output guarantees $\beta_i(\Delta) \geq \beta_i^G(\Delta)$ to n connected CNI components with input assumptions $\beta_i^A(\Delta)$.

Internally, the service output guarantee $\beta_i^G(\Delta)$ can be determined by the slot length guarantee s_i^G that is assigned to the ith CC. Analogously, we show in section IV, how the service input assumption $\beta_i^A(\Delta)$ of a connected CNI component can be transformed into a minimum slot length assumption s_i^A. The fulfillment of the communication service demand of a connected abstract CNI component can then be guaranteed directly by guaranteeing $s_i^G \geq s_i^A$.

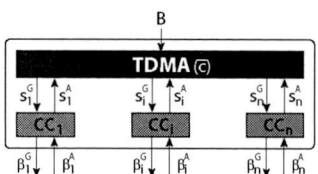

Fig. 5. Interface of an abstract resource component with TDMA.

B. Service Supply, Schedulability, Feasibility and Utilization

In a communication resource with TDMA, the ith communication controller may not have access to the resource during a time interval that is limited by $\Delta = \bar{c} - s_i^G$. After this interval, the CC has exclusive access to the resource during a time interval of length s_i^G. A CC can therefore not guarantee any service to a connected CNI during any time interval $0 \le \Delta < \bar{c} - s_i^G$, but it can guarantee a service of $B(\Delta - (\bar{c} - s_i^G))$ in any time interval $\bar{c} - s_i^G \le \Delta < \bar{c}$. This service guarantee can be expressed as

$$\beta_i^G(\Delta) = B \max\left(\left\lfloor \frac{\Delta}{\bar{c}} \right\rfloor s_i^G, \Delta - \left\lceil \frac{\Delta}{\bar{c}} \right\rceil (\bar{c} - s_i^G)\right) \quad (6)$$

or more compactly as

$$\beta_i^G(\Delta) = B \sup_{0 \le \lambda \le \Delta}\left\{\lambda - \left\lceil \frac{\lambda}{\bar{c}} \right\rceil (\bar{c} - s_i^G)\right\} \quad (7)$$

We define that a real-time communication system is said to be *schedulable*, if the connected CNI's can fulfill the real-time requirements of all message streams, i.e. if all messages in the system can be delivered within an time interval that is limited by their maximum allowable delay D. According to the theory of Real-Time Interfaces, a real-time communication system is therefore schedulable if

$$\beta_i^G(\Delta) \ge \beta_i^A(\Delta) \quad \forall i, \forall \Delta \ge 0 \quad (8)$$

or equivalently if $s_i^G \ge s_i^A \quad \forall i$.

We further define that a real-time communication system with TDMA is *feasible*, if the sum of the required slot lengths and the protocol overhead is less or equal the cycle length, i.e. if

$$\bar{c} \ge \sum_{\forall i} s_i^A + o_c + no_s \quad (9)$$

Following this definition of feasibility, the slot length guarantee s_i^G to the ith CC can then be computed as

$$s_i^G = \bar{c} - \left(\sum_{\forall j \ne i} s_j^A + o_c + no_s\right) \quad (10)$$

We define the utilization σ_i^A as the quotient of the slot length s_i^A divided by the cycle length $\bar{c}$. Analogously to (9), a TDMA system is then feasible if the total utilization is less or equal one:

$$1 \ge \sum_{\forall i} \sigma_i^A + \frac{o_c + no_s}{\bar{c}} \overset{def}{=} \sigma_{tot}(\bar{c}) \quad (11)$$

C. Delay and Backlog

For delay and backlog analysis in a communication system with TDMA as defined above, we may now use (2) and (3). We only need to replace $\beta(\Delta)$ with $\beta_i^G(\Delta)$ in the corresponding formulas.

EXAMPLE 1. *Suppose we have a message stream M_0 with $p_0 = 198ms$, $j_0 = 387ms$ and $d_0 = 48ms$. The single messages have a size of 12 units, and a maximum allowable delivery delay of $D_0 =*

$110ms$. *The messages are sent over a communication resource with TDMA, that has a total bandwidth of $B = 1'000units/s$. The TDMA cycle has a length of $\bar{c} = 80ms$, and a slot of length $s_0^G = 20ms$ is assigned to the CNI that sends the message stream M_0.*

Figure 6 shows the arrival curve α_0, the service assumption β_0^A, as well as the service guarantee β_0^G of the above system, computed according to (1), (5) and (7), respectively. Since $\beta_0^G \ge \beta_0^A$, we know that the real-time requirements of the message stream are fulfilled, following (8). And using (2) and (3), we can compute the maximum delay experienced by a message as $Del_{max,0} = 96ms$, and the maximum backlog in the CNI as $Buf_{max,0} = 24units$.

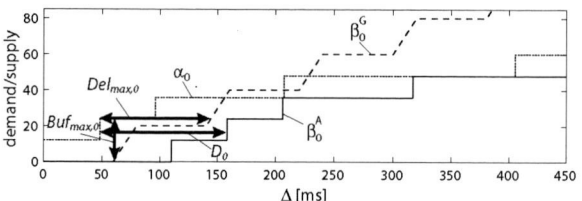

Fig. 6. Arrival and service curves of Example 1.

IV. MINIMUM SLOT TIME ALLOCATION

Based on the powerful abstractions of service guarantee and service assumption curves, and the explicit schedulability requirement (8) coming from the theory of Real-Time Interfaces, it is possible to determine the exact minimum time slot s_i^A that must be assigned to a CNI with service assumption $\beta_i^A(\Delta)$ to be schedulable on a TDMA communication network with bandwidth B and cycle length $\bar{c}$. For this, we need to construct the inverse of (6) with respect to s_i^G as the smallest s_i^G that leads to a service supply that fulfills the schedulability requirement (8):

$$s_i^A = \sup_{\Delta \ge 0}\left\{\min\left(\frac{\beta_i^A}{B\left\lfloor \frac{\Delta}{\bar{c}} \right\rfloor}, \frac{\beta_i^A - B\Delta + B\left\lceil \frac{\Delta}{\bar{c}} \right\rceil \bar{c}}{B\left\lceil \frac{\Delta}{\bar{c}} \right\rceil}\right)\right\} \quad (12)$$

This minimum time slot s_i^A is provably the smallest possible time slot allocation that guarantees a service supply $\beta_i^G(\Delta) \ge \beta_i^A(\Delta)$ on a TDMA resource with bandwidth B and slot length $\bar{c}$.

EXAMPLE 2. *Figure 7 shows the minimum slot length $s_0^A(\bar{c})$ of message stream M_0 from Example 1, as a function of the TDMA cycle length $\bar{c}$. We see that the minimum required slot length increases with increasing cycle length, and it is lower bounded by $s_0^A(\bar{c}) \ge \bar{c} - (D_0 - e_0)$, because the gap between two consecutive slots must never be greater than $D_0 - e_0$.*

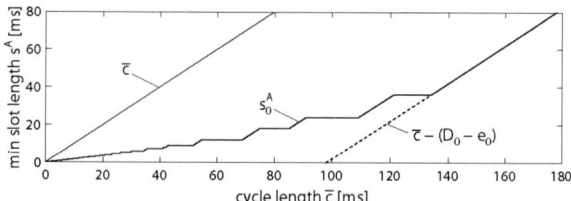

Fig. 7. Minimum slot lengths for message stream M_1.

Figure 8 shows the minimum service guarantees $\beta_0^G(\Delta)$ for three different cycle lengths, computed by setting $s_0^G(\bar{c}) = s_0^A(\bar{c})$.

Using (12), we can now compute the minimum slot lengths for all CNI's in a communication network with bandwidth B and TDMA

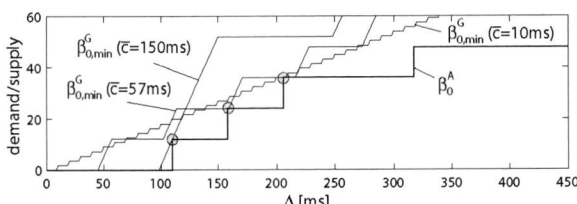

Fig. 8. Minimum service guarantees for different cycle lengths.

cycle length $\bar{c}$. If a slot allocation with these minimum slot lengths leads to a feasible communication system according to (9), i.e. if the sum of the minimum slot lengths plus the protocol overhead is smaller than the cycle length, then we can use $\bar{c}$ and $s_i = s_i^A(\bar{c})$ as TDMA settings. Otherwise we are guaranteed that no feasible slot allocation exists for the cycle length $\bar{c}$.

EXAMPLE 3. *Consider a communication network with 10 CNI's, each one serving one of the message streams M_0 to M_9 specified in Table I. This specification equals the specification of System 3 in [7]. According to [7], this was the most difficult system to optimize in [7].*

TABLE I– EXAMPLE SYSTEM WITH 10 MESSAGE STREAMS.

	M0	M1	M2	M3	M4	M5	M6	M7	M8	M9
p	198	102	283	354	239	194	148	114	313	119
j	387	70	269	387	222	260	91	13	302	187
d	48	45	58	17	65	32	78	-	86	89
e	12	7	7	11	8	5	13	14	5	6
D	110	140	115	145	180	140	200	120	140	100

To find feasible TDMA parameters for this system, we computed $\sigma_{tot}(\bar{c})$ for $\bar{c} \in [0.1ms \ldots 600ms]$ with a cycle quantum $q_c = 100\mu s$. In [7] no slot length quantization or protocol overhead was considered, Fig. 9 depicts the corresponding results. Additionally, we computed $\bar{\sigma}_{tot}(\bar{c})$ with a slot length quantum $q_s = 10\mu s$, but still without protocol overhead. The results are also shown in Fig. 9.

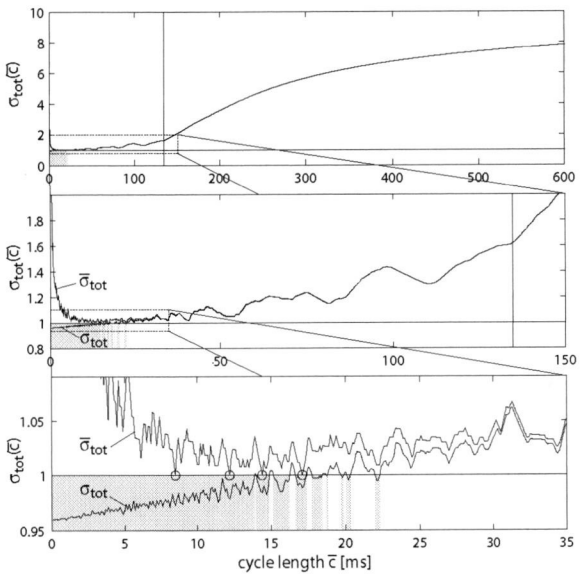

Fig. 9. Total utilization as a function of the TDMA cycle length.

Remember, that feasible TDMA parameters exists for a cycle length $\bar{c}$ if, and only if $\sigma_{tot}(\bar{c}) \leq 1$. Without considering quantization, this is the case for all values in the grey shaded areas in Fig. 9. When we

consider a small slot quantization however, only the four encircled values of $\bar{c}$ still lead to feasible TDMA parameters.

In general, if we do not consider quantization effects and protocol overhead, then the smallest possible $\bar{c}$ will always lead to feasible TDMA parameters, if the total bandwidth B is large enough. As soon as we consider quantization effects and protocol overhead however, arbitrary small values for $\bar{c}$ are not feasible anymore.

But also arbitrary large values for $\bar{c}$ will not lead to feasible TDMA parameters, because the slot lengths are lower bounded by $s_0^A(\bar{c}) \geq \bar{c} - (D_0 - e_0)$. $\sigma_{tot}(\bar{c})$ will therefore strive towards the number of CNI's in the system for large $\bar{c}$, i.e. towards 10 in Example 3.

Feasible TDMA parameters will therefore only exist for cycle lengths $\bar{c}$ that are not too small and not to large. And as we can be seen in Fig. 9, the total utilization $\sigma_{tot}(\bar{c})$ as a function of the TDMA cycle length has a very complex and nonlinear behavior. Often, intervals of feasible cycle lengths are even non-contiguous.

V. OPTIMAL CYCLE LENGTH

In a typical TDMA system, not only the slot lengths s_i, but also the cycle length $\bar{c}$ is a customizable parameter. To find an optimal cycle length for a TDMA system, we first need to define an optimality criterion. One possible optimality criterion that we will use in this work is the average remaining bandwidth $\sigma_r = 1 - \sigma_{tot}$. This remaining bandwidth could be distributed additionally to the existing CNI's, or it could be used to admit additional future load in a dynamic system.

Note, that if we want to account for up to m dynamically added CNI's in a dynamic system, we need to consider the slot overheads for these dynamically added CNI's. Further, for systems with a slot quantum, we need to consider that only multiples of full slot quantums can be assigned to existing or future CNI's:

$$\bar{\sigma}_r(\bar{c}) = \left\lfloor \left(1 - \sum_{\forall i} \bar{\sigma}_i(\bar{c}) - \frac{(n+m)o_s + o_c}{\bar{c}}\right) \frac{\bar{c}}{q_s} \right\rfloor \frac{q_s}{\bar{c}} \quad (13)$$

In Fig. 9, we have seen that the total utilization $\sigma_{tot}(\bar{c})$ as a function of the TDMA cycle length has a very complex and nonlinear behavior. To find the optimal cycle length with the maximum remaining bandwidth σ_r, we therefore have no choice but to compute σ_r for all possible values of $\bar{c}$.

However, in Example 2 and in Fig. 7, we have seen that the required slot lengths are lower bounded by $s_1^A(\bar{c}) \geq \bar{c} - (D_1 - e_1)$. It is therefore possible to find an upper bound to feasible cycle lengths:

$$\bar{c}_{max} = \sup_{\bar{c} \geq 0} \left\{ \bar{c} \ : \ \bar{c} \geq \sum_{\forall i} \max(0, \bar{c} - (D_i - e_i)) \right\} \quad (14)$$

Due to this upper bound, σ_r needs only to be computed for $\bar{c}_{max}/q_c$ different values. The upper bound for the system in Example 3 is $\bar{c}_{max} = 134, 7ms$ and is indicated by the vertical bar in Fig. 9.

VI. MINIMUM TOTAL SERVICE BANDWIDTH

At design time, the service bandwidth B is often also a customizable parameter. The minimum total service bandwidth B_{min} is the smallest possible service bandwidth B of a TDMA system with service assumptions β_i^A, for which feasible slot allocations s_i exists:

$$B_{min} = \inf_{B \geq 0} \{B \ : \ \sigma_{tot,min}(B) \leq 1\} \quad (15)$$

with

$$\sigma_{tot,min}(B) = \inf_{\bar{c} \leq \bar{c}_{max}} \{\sigma_{tot}(\bar{c}, B)\} \quad (16)$$

483

From (12), it can be seen that the minimum slot length s_i^A is monotonically decreasing with increasing service bandwidth B, i.e. $s_i^A(B + dB) \leq s_i^A(B)$. Because of this, the total utilization σ_{tot} is also monotonically decreasing, and as a consequence also the minimum total utilization $\sigma_{tot,min}$ is monotonically decreasing with increasing service bandwidth B, i.e. $\sigma_{tot,min}(B+dB) \leq \sigma_{tot,min}(B)$. We can therefore find the minimum total service bandwidth B_{min} (or B^A in Real-Time Interface notation) by using binary search.

VII. CASE STUDY

The case study systems consists of 21 ECU's that are interconnected with a TDMA bus, and that send a total of 30 different hard real-time message streams with jitter and bursts that are specified in Table II. The TDMA bus has a cycle quantum $q_c = 1ms$, a slot quantum $q_s = 0.5ms$, a slot overhead $o_s = 0.5ms$, and a cycle overhead $o_c = 2.5ms$. On ECU_0 and ECU_1, EDF is used to arbitrate the sending of $M_0 - M_3$ and $M_4 - M_6$, FIFO is used on ECU_2 to arbitrate $M_7 - M_9$, and FP is used on ECU_3 to arbitrate $M_{10} - M_{12}$. The remaining 17 ECU's send only a single message stream each.

TABLE II– CASE STUDY SYSTEM WITH 30 MESSAGE STREAMS.

	M0	M1	M2	M3	M4	M5	M6	M7	M8	M9
p	196	245	105	147	231	308	275	234	273	182
j	387	70	269	387	222	260	91	387	70	269
d	48	-	58	17	65	-	-	48	-	58
e	7	4	6	1	8	2	3	5	5	3
D	176	237	115	488	206	311	275	207	178	198
	M10	M11	M12	M13	M14	M15	M16	M17	M18	M19
p	153	357	476	302	258	424	287	451	539	309
j	80	70	177	967	719	257	38	159	11	153
d	-	-	-	27	89	-	-	-	-	-
e	2	2	3	2	2	4	7	6	13	11
D	153	423	556	511	371	315	210	245	196	413
	M20	M21	M22	M23	M24	M25	M26	M27	M28	M29
p	506	357	304	510	298	243	457	502	247	226
j	250	393	278	296	184	400	300	312	365	278
d	-	3	40	-	-	18	-	-	83	85
e	4	3	4	5	4	5	7	3	8	7
D	245	336	378	126	161	469	574	560	133	301

We first search the minimum required total service bandwidth for the TDMA bus in this system. From this we learn, that a minimum bandwidth of $B_{min} = 1.27Mbit/s$ is required. With this bandwidth, feasible TDMA settings exist for a cycle length $\bar{c} = 92ms$, and lead to a total utilization of $\bar{\sigma}_{tot} = 1$.

In a next step, we choose a TDMA bus with total bandwidth of $B = 1.5Mbit/s \geq B_{min}$. We want to optimize slot and cycle lengths on this bus according to section V, such that 5 additional ECU's could be added at a later point of time. For this we compute (13) up to $\bar{c}_{max} = 169ms$. The results are shown in Fig. 10 and suggest to use a cycle length of again $\bar{c} = 92ms$. This leads to a maximum remaining average bandwidth of $\bar{\sigma}_r = 0.11$.

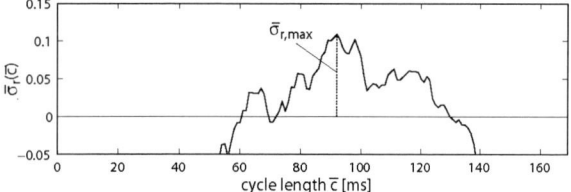

Fig. 10. Remaining average bandwidth in the case study system.

VIII. COMPUTATIONAL COMPLEXITY

To analyze the examples and case study in this paper, we used a prototype implementation of Real-Time Calculus and Real-Time Interfaces that is implemented in Java and uses Matlab as user frontend. We run this prototype tool on a Pentium Mobile 1.6 GHz.

Computing Fig. 9 took $7.5s$, while computing the same results up to the required upper bound $\bar{c}_{max}$ took only $1.1s$. In the case study, finding the minimum total bandwidth and the corresponding cycle and slot lengths took $0.31s$ per iteration, and optimizing the cycle and slot lengths for the TDMA bus with fixed bandwidth took also $0.31s$.

When computation needs to be faster (e.g. for exploration), linear approximated VCC's could be used, that trade off computational complexity with the tightness of the results, see e.g. [3].

IX. CONCLUSIONS

We presented an analytic method to determine the provably smallest possible slot length that must be allocated in a TDMA resource, to serve an event-triggered hard real-time load. We further presented constructive methods to find the optimal cycle length as well as the minimum required bandwidth of a TDMA resource. Using these new methods, it is now possible to determine the minimum required bandwidth, as well as the optimal slot and cycle parameters for a TDMA resource, when we are initially only given a set of real-time load specifications that must be served by the TDMA resource. The applicability and computational efficiency of the presented methods was shown in a case study, where finding the optimum TDMA parameter set for a large distributed embedded system with 21 ECU's that send a total of 30 different hard real-time message streams took less than a second. Finally, we extended an existing theoretical framework for modular system level performance analysis of hard real-time systems by introducing a component that models a TDMA resource. This component enables to use the framework for performance analysis and interface-based design of complete distributed real-time systems with TDMA.

REFERENCES

[1] Interface-Based Design of Real-Time Systems with Hierarchical Static and Dynamic Scheduling. Authors omitted for blind review.

[2] Real-Time Interfaces for Interface-Based Design of Real-Time Systems with Fixed Priority Scheduling. Authors omitted for blind review.

[3] K. Albers and F. Slomka. An event stream driven approximation for the analysis of real-time systems. In *Proceedings of the 16th Euromicro Conference on Real-Time Systems (ECRTS'04)*. IEEE Press, 2004.

[4] S. K. Baruah. Dynamic- and static-priority scheduling of recurring real-time tasks. *Real-Time Systems*, 24(1):93–128, 2003.

[5] S. Chakraborty, S. Künzli, and L. Thiele. A general framework for analysing system properties in platform-based embedded system designs. In *Proc. 6th Design, Automation and Test in Europe (DATE)*, pages 190–195, March 2003.

[6] K. Goossens, J. Dielissen, J. van Meerbergen, P. Poplavko, A. Rădulescu, E. Rijpkema, E. Waterlander, and P. Wielage. Guaranteeing the quality of services in networks on chip. In *Networks on chip*, pages 61–82. Kluwer Academic Publishers, Hingham, MA, USA, 2003.

[7] A. Hamann and R. Ernst. TDMA Time Slot and Turn Optimization with Evolutionary Search Techniques. In *Design, Automation and Test in Europe (DATE 2005)*, 2005.

[8] H. Kopetz. *Real-Time Systems - Design Principles for Distributed Embedded Applications*. Kluwer Academic Publishers, 1997.

[9] J. Le Boudec and P. Thiran. *Network Calculus - A Theory of Deterministic Queuing Systems for the Internet*. LNCS 2050, Springer Verlag, 2001.

[10] R. Obermaisser. CAN Emulation in a Time-Triggered Environment. In *Proceedings of the 2002 IEEE International Symposium on Industrial Electronics (ISIE)*. IEEE, 2002.

[11] R. Obermaisser. *Event-Triggered and Time-Triggered Control Paradigms*, volume 22 of *Real-Time Systems Series*. Springer, 2005.

[12] P. Pop, P. Eles, Z. Peng, V. Izosimov, M. Hellring, and O. Bridal. Design Optimization of Multi-Cluster Embedded Systems for Real-Time Applications. In *Design, Automation and Test in Europe (DATE 2004)*, pages 1028–1033, 2004.

[13] I. Shin and I. Lee. Periodic resource model for compositional real-time guarantees. In *Proceedings of the Real-Time Systems Symposium (RTSS)*, pages 2–13. IEEE Press, 2003.

[14] L. Thiele, S. Chakraborty, and M. Naedele. Real-time calculus for scheduling hard real-time systems. In *Proc. IEEE International Symposium on Circuits and Systems (ISCAS)*, volume 4, pages 101–104, 2000.

Author Index

A

Abbaspour, Soroush	p. 947	(9C-2)
Abdi, Samar	p. 126	(1D-19)
Abraham, Jacob	p. 285	(3B-3)
Absar, Javed	p. 820	(8B-5)
Adachi, Hidekazu	p. 176	(2B-4)
Adachi, Hiroaki	p. 624	(6C-2)
Adachi, Kazunobu	p. 342	(4A-1)
Adamez, Jesus	p. 485	(5B-3)
Afzali-Kusha, Ali	p. 297	(3B-5)
Agarwal, Vineet	p. 718	(7C-3)
Agarwal, Vineet	p. 724	(7C-4)
Al-Hashimi, Bashir M.	p. 36	(1B-2)
Al-Hashimi, Bashir M.	p. 923	(9B-3)
Alizadeh, Bijan	p. 7	(1A-2)
Amirabadi, Amir	p. 297	(3B-5)
Anderson, Hans-Werner	p. 871	(8D-4)
Andraus, Zaher Semon	p. 19	(1A-4)
Asai, Hideki	p. 832	(8C-2)
Austin, Todd	p. 442	(5A-1)

B

Balakrishnan, Venkataramanan	p. 422	(4C-4)
Balakrishnan, Venkataramanan	p. 509	(5C-2)
Balasa, Florin	p. 802	(8B-2)
Banerjee, Kaustav	p. 223	(2D-2)
Banerjee, Sudarshan	p. 491	(5B-4)
Bansal, Aditya	p. 237	(2D-4)
Basu, Prasenjit	p. 13	(1A-3)
Behnen, Erwin	p. 871	(8D-4)
Ben-Romdhane, Mohamed	p. 30	(1B-1)
Bertacco, Valeria	p. 442	(5A-1)
Bhardwaj, Sarvesh	p. 953	(9C-3)
Bhunia, Swarup	p. 665	(7A-4)
Bhunia, Swarup	p. 712	(7C-2)
Blasco, Francisco	p. 485	(5B-3)
Boerstler, David	p. 856	(8D-1)
Bolliger, Mark	p. 871	(8D-4)
Bonaciu, Marius	p. 372	(4B-1)
Bouchhima, Aimen	p. 372	(4B-1)
Bozorgzadeh, Elaheh	p. 491	(5B-4)
Brockmeyer, Erik	p. 606	(6B-4)

Brokenshire, Daniel	p. 860	(8D-2)
Brown, Richard B.	p. 84	(1C-5)
Brown, Richard B.	p. 94	(1D-3)
Brown, Richard B.	p. 279	(3B-2)
Butler, Jon	p. 378	(4B-2)

C

Cai, Yici	p. 582	(6A-5)
Cai, Yici	p. 826	(8C-1)
Cai, Yuan	p. 923	(9B-3)
Cao, Yu	p. 953	(9C-3)
Cao, Zhen	p. 618	(6C-1)
Catthoor, Francky	p. 42	(1B-3)
Catthoor, Francky	p. 606	(6B-4)
Catthoor, Francky	p. 820	(8B-5)
Cauley, Stephen F	p. 422	(4C-4)
Cesario, Wander	p. 372	(4B-1)
Chae, Soo-Ik	p. 588	(6B-1)
Chae, Soo-Ik	p. 935	(9B-5)
Chakrabarti, Partha P	p. 13	(1A-3)
Chang, Cheng-Ru	p. 170	(2B-3)
Chang, Chia-Ming	p. 254	(3A-3)
Chang, Li-Pin	p. 334	(3D-2)
Chang, Shih-Chieh	p. 354	(4A-3)
Chang, Tian-Sheuan	p. 742	(7D-2)
Chang, Yao-Wen	p. 213	(2C-5)
Chang, Yao-Wen	p. 366	(4A-5)
Chang, Yao-Wen	p. 636	(6C-4)
Chang, Yen-Jen	p. 917	(9B-2)
Chang, Yuan-Hao	p. 334	(3D-2)
Chao, Chie-Min	p. 118	(1D-15)
Chao, Wen-Chang	p. 213	(2C-5)
Chaudhry, Rajat	p. 867	(8D-3)
Cheema, Muhammad Omer	p. 54	(1B-5)
Chen, Charlie Chungping	p. 941	(9C-1)
Chen, Chien-Chang	p. 777	(8A-3)
Chen, Chien-Hua	p. 600	(6B-3)
Chen, Guangyu	p. 128	(2A-1)
Chen, Guangyu	p. 140	(2A-3)
Chen, Guangyu	p. 808	(8B-3)
Chen, Guangyu	p. 814	(8B-4)
Chen, Guilin	p. 134	(2A-2)
Chen, Guilin	p. 140	(2A-3)

Chen, Hongyu	p. 73	(1C-3)
Chen, Jian-Wen	p. 736	(7D-1)
Chen, Jwu E	p. 366	(4A-5)
Chen, Liang-Gee	p. 750	(7D-3)
Chen, Po-Yuan	p. 576	(6A-4)
Chen, Tai-Chen	p. 636	(6C-4)
Chen, Tung-Chien	p. 750	(7D-3)
Chen, Wei-Nien	p. 742	(7D-2)
Chen, Weijen	p. 941	(9C-1)
Chen, Xi	p. 372	(4B-1)
Chen, Yiran	p. 158	(2B-1)
Cheng, Chung-Kuan	p. 73	(1C-3)
Cheng, Chung-Kuan	p. 428	(4C-5)
Cheng, Kai-Ting	p. 742	(7D-2)
Cheng, Kwang-Ting	p. 25	(1A-5)
Cheng, Tim	p. 360	(4A-4)
Cho, Hansu	p. 126	(1D-19)
Cho, Minsik	p. 765	(8A-1)
Choi, Jung Hwan	p. 237	(2D-4)
Choi, Jung Hwan	p. 712	(7C-2)
Choi, Kiyoung	p. 905	(9A-5)
Choi, Kyu-Myung	p. 551	(5D-4)
Choi, Kyu-Myung	p. 905	(9A-5)
Chong, Philip	p. 440	(4D-4)
Chu, Chris	p. 195	(2C-2)
Chu, Fangqing	p. 100	(1D-6)
Chung, Eui-Young	p. 551	(5D-4)
Chung, Eui-Young	p. 905	(9A-5)
Coenen, Martijn	p. 146	(2A-4)
Cong, Jason	p. 188	(2C-1)
Cong, Jason	p. 384	(4B-3)
Cordone, Roberto	p. 898	(9A-4)

D

Dai, Wayne Wei-Ming	p. 730	(7C-5)
Das, Sayantan	p. 13	(1A-3)
Dasgupta, Pallab	p. 13	(1A-3)
Datta, Animesh	p. 712	(7C-2)
De Micheli, Giovanni	p. 146	(2A-4)
DeHon, Andre M.	p. 789	(8A-5)
Deng, Liang	p. 965	(9C-5)
Devgan, Anirudh	p. 61	(1C-1)
Dhong, Sang	p. 867	(8D-3)
Dick, Robert P.	p. 879	(9A-1)
Dong, Sheqin	p. 694	(7B-4)

Dou, Qingqi	p. 285	(3B-3)
Du, Yu	p. 521	(5C-4)
Dutt, Nikil	p. 30	(1B-1)
Dutt, Nikil	p. 491	(5B-4)
Dutt, Nikil	p. 497	(5B-5)

E

Eda, Tsutomu	p. 124	(1D-18)
Ejlali, Alireza	p. 923	(9B-3)
Ekpanyapong, Mongkol	p. 959	(9C-4)
Endoh, Chihiro	p. 342	(4A-1)
Enomoto, Tadayoshi	p. 90	(1D-1)
Eo, Soo-Kwan	p. 551	(5D-4)

F

Fan, Yibo	p. 122	(1D-17)
Fatemi, Hanif	p. 947	(9C-2)
Feng, Xingguang	p. 905	(9A-5)
Feng, Yan	p. 771	(8A-2)
Feng, Zhe	p. 630	(6C-3)
Ferrandi, Fabrizio	p. 898	(9A-4)
Fujiwara, Hideo	p. 671	(7A-5)
Fukazawa, Mitsuya	p. 106	(1D-9)
Fukunaga, Masayasu	p. 348	(4A-2)

G

Gajski, Daniel	p. 116	(1D-14)
Gajski, Daniel	p. 126	(1D-19)
Goel, Ashish	p. 665	(7A-4)
Goossens, Kees	p. 146	(2A-4)
Goplen, Brent	p. 219	(2D-1)
Gorjiara, Bita	p. 116	(1D-14)
Goto, Satoshi	p. 112	(1D-12)
Goto, Yuichi	p. 856	(8D-1)
Goyal, Prashant	p. 291	(3B-4)
Gu, Zhenyu (Peter)	p. 879	(9A-1)
Guiney, Michaela	p. 434	(4D-1)
Gulati, Kanupriya	p. 850	(8C-5)
Guo, Jin	p. 42	(1B-3)
Gupta, Sanjay	p. 871	(8D-4)
Guthaus, Matthew R.	p. 84	(1C-5)

H

Ha, Soonhoi	p. 152	(2A-5)

Ha, Soonhoi	p. 497	(5B-5)
Ha, Yajun	p. 886	(9A-2)
Hailu, Eskinder	p. 856	(8D-1)
Hamada, Shuji	p. 348	(4A-2)
Hammami, Omar	p. 54	(1B-5)
Han, Jun	p. 122	(1D-17)
Han, Sang-Il	p. 935	(9B-5)
Hang, Hsueh-Ming	p. 742	(7D-2)
Hanna, Ziyad	p. 25	(1A-5)
Hara, Hiroyuki	p. 533	(5D-1)
Harvey, Paul	p. 871	(8D-4)
Hashimoto, Masanori	p. 515	(5C-3)
He, Lei	p. 207	(2C-4)
He, Lei	p. 630	(6C-3)
Hedrich, Lars	p. 700	(7B-5)
Higami, Yoshinobu	p. 659	(7A-3)
Hiraide, Takahisa	p. 342	(4A-1)
Hiraoka, Daisuke	p. 860	(8D-2)
Hofstee, Peter	p. 871	(8D-4)
Hon, Man Chung	p. 326	(3C-5)
Hong, Xianlong	p. 582	(6A-5)
Hong, Xianlong	p. 618	(6C-1)
Hong, Xianlong	p. 630	(6C-3)
Hong, Xianlong	p. 694	(7B-4)
Hong, Xianlong	p. 826	(8C-1)
Horikawa, Kazunari	p. 1	(1A-1)
Hsieh, Cheng-Tao	p. 354	(4A-3)
Hsieh, Jen-Wei	p. 334	(3D-2)
Hu, Bo	p. 92	(1D-2)
Hu, Qubo	p. 606	(6B-4)
Hu, Xiaodong	p. 618	(6C-1)
Hu, Yu	p. 618	(6C-1)
Hu, Yuhen	p. 941	(9C-1)
Huang, Chao-Wei	p. 118	(1D-15)
Huang, Jiun-Lang	p. 648	(7A-1)
Huang, Juinn-Dar	p. 448	(5A-2)
Huang, Juinn-Dar	p. 600	(6B-3)
Huang, Shih-Hsu	p. 254	(3A-3)
Huang, Xuan-Lun	p. 648	(7A-1)
Huang, Yu-Hui	p. 576	(6A-4)
Hutton, Mike	p. 73	(1C-3)
Hwang, Chanseok	p. 201	(2C-3)
Hwang, Hyeyoung	p. 152	(2A-5)
Hwang, TingTing	p. 576	(6A-4)
Hyun, Chung Jin	p. 758	(7D-4)

I

Ignjatovic, Aleksandar	p. 612	(6B-5)
Ignjatovic, Aleksandar	p. 796	(8B-1)
Ikeda, Hiroshi	p. 624	(6C-2)
Ikenaga, Takeshi	p. 112	(1D-12)
Imai, Satoshi	p. 541	(5D-2)
Inoue, Atsuki	p. 541	(5D-2)
Inoue, Yoshio	p. 437	(4D-2)
Irwin, Mary Jane	p. 140	(2A-3)
Ishida, Koichi	p. 98	(1D-5)
Ishikawa, Tatsuyuki	p. 112	(1D-12)
Ishizaka, Kinya	p. 624	(6C-2)
Ismail, Yehea	p. 231	(2D-3)
Isoda, Yutaka	p. 342	(4A-1)
Isoda, Yutaka	p. 624	(6C-2)
Ito, Noriyuki	p. 342	(4A-1)
Ito, Noriyuki	p. 624	(6C-2)

J

Jagannathan, Ashok	p. 384	(4B-3)
Jain, Jitesh	p. 422	(4C-4)
Janapsatya, Andhi	p. 612	(6B-5)
Janapsatya, Andhi	p. 796	(8B-1)
Jangkrajarng, Nuttorn	p. 92	(1D-2)
Jen, Chein-Wei	p. 118	(1D-15)
Jerraya, Ahmed	p. 372	(4B-1)
Jerraya, Ahmed Amine	p. 935	(9B-5)
Jia, Yongbo	p. 730	(7C-5)
Jing, Tong	p. 618	(6C-1)
Jing, Tong	p. 630	(6C-3)
Johns, Charles	p. 871	(8D-4)
Jou, Jing-Yang	p. 448	(5A-2)
Jou, Jing-Yang	p. 600	(6B-3)
Jung, Hyunuk	p. 152	(2A-5)

K

Kadayif, Ismail	p. 182	(2B-5)
Kahle, Jim	p. 871	(8D-4)
Kajihara, Seiji	p. 348	(4A-2)
Kameyama, Atsushi	p. 871	(8D-4)
Kandemir, Mahmut	p. 128	(2A-1)
Kandemir, Mahmut	p. 134	(2A-2)
Kandemir, Mahmut	p. 140	(2A-3)
Kandemir, Mahmut	p. 182	(2B-5)

Kandemir, Mahmut . . . . p. 390 (4B-4)
Kandemir, Mahmut . . . . p. 808 (8B-3)
Kandemir, Mahmut . . . . p. 814 (8B-4)
Kandemir, Mahmut . . . . p. 929 (9B-4)
Kankani, Navneeth . . . . p. 724 (7C-4)
Kanuma, Akira . . . . p. 342 (4A-1)
Kao, Chao-Yang . . . . p. 736 (7D-1)
Katagiri, Hideaki . . . . p. 624 (6C-2)
Kato, Toshiyuki . . . . p. 124 (1D-18)
Katsuki, Kazuya . . . . p. 110 (1D-11)
Kawasaki, Kenichi . . . . p. 541 (5D-2)
Keaty, John . . . . p. 871 (8D-4)
Kettle, Neil . . . . p. 243 (3A-1)
Khatri, Sunil . . . . p. 850 (8C-5)
Ki, Wing-Hung . . . . p. 96 (1D-4)
Ki, Wing-Hung . . . . p. 102 (1D-7)
Ki, Wing-Hung . . . . p. 104 (1D-8)
Kim, Chris H . . . . p. 559 (6A-1)
Kim, Hyung-Ock . . . . p. 565 (6A-2)
Kim, Hyunsuk . . . . p. 551 (5D-4)
Kim, Seon Wook . . . . p. 120 (1D-16)
Kim, Suki . . . . p. 120 (1D-16)
Kim, Sung Dae . . . . p. 758 (7D-4)
Kimura, Shinji . . . . p. 1 (1A-1)
Kimura, Shinji . . . . p. 653 (7A-2)
King, Andy . . . . p. 243 (3A-1)
Kitahara, Takeshi . . . . p. 533 (5D-1)
Kjeldsberg, Per Gunnar . . . . p. 606 (6B-4)
Ko, Sung Jea . . . . p. 120 (1D-16)
Kobayashi, Fuminori . . . . p. 108 (1D-10)
Kobayashi, Kazutoshi . . . . p. 110 (1D-11)
Kobayashi, Masatsugu . . . . p. 124 (1D-18)
Kobayashi, Nobuaki . . . . p. 90 (1D-1)
Kobayashi, Shin-ya . . . . p. 659 (7A-3)
Kodakara, Sreekumar V. . . . . p. 61 (1C-1)
Koh, Cheng-Kok . . . . p. 158 (2B-1)
Koh, Cheng-Kok . . . . p. 422 (4C-4)
Koh, Cheng-Kok . . . . p. 509 (5C-2)
Kohara, Shunitsu . . . . p. 594 (6B-2)
Koide, Tetsushi . . . . p. 176 (2B-4)
Kolcu, Ibrahim . . . . p. 814 (8B-4)
Komatsu, Hiroaki . . . . p. 624 (6C-2)
Kong, Jeong-Taek . . . . p. 551 (5D-4)
Kono, Takeshi . . . . p. 342 (4A-1)
Koon, Suet-Chui . . . . p. 102 (1D-7)
Kosaka, Daisuke . . . . p. 677 (7B-1)

Kotani, Manabu . . . . p. 110 (1D-11)
Ku, Ja Chun . . . . p. 231 (2D-3)
Kubota, Hidemasa . . . . p. 832 (8C-2)
Kuh, Ernest S. . . . . p. 428 (4C-5)
Kulkarni, Sarvesh Hemchandra . . . . p. 838 (8C-3)
Kumar, Sanjay V . . . . p. 559 (6A-1)
Kuo, Ming-Hsine . . . . p. 783 (8A-4)
Kuo, Tei-Wei . . . . p. 334 (3D-2)
Kuo, Yu-Ting . . . . p. 118 (1D-15)
Kuroki, Wataru . . . . p. 408 (4C-2)

L

Lai, Xiaolue . . . . p. 273 (3B-1)
Lai, Xiaolue . . . . p. 291 (3B-4)
Lai, Xiaolue . . . . p. 527 (5C-5)
Lam, Yat-Hei . . . . p. 102 (1D-7)
Lam, Yat-Hei . . . . p. 104 (1D-8)
LaMeres, Brock . . . . p. 850 (8C-5)
Le, Bob . . . . p. 871 (8D-4)
Leavitt, Eric . . . . p. 434 (4D-1)
Lee, Chung-Len . . . . p. 366 (4A-5)
Lee, Geeng-Wei . . . . p. 600 (6B-3)
Lee, Ikhwan . . . . p. 551 (5D-4)
Lee, Jeong Hoo . . . . p. 758 (7D-4)
Lee, Junghee . . . . p. 460 (5A-4)
Lee, Kuang-Yao . . . . p. 303 (3C-1)
Lee, Sang . . . . p. 871 (8D-4)
Lee, Suh Ho . . . . p. 120 (1D-16)
Li, Fehui . . . . p. 128 (2A-1)
Li, Feihui . . . . p. 134 (2A-2)
Li, Feihui . . . . p. 182 (2B-5)
Li, Feihui . . . . p. 808 (8B-3)
Li, Feihui . . . . p. 929 (9B-4)
Li, Hai . . . . p. 158 (2B-1)
Li, Katherine Shu-Min . . . . p. 366 (4A-5)
Li, Wei . . . . p. 100 (1D-6)
Li, Zhao . . . . p. 402 (4C-1)
Li, Zhuo Robert . . . . p. 320 (3C-4)
Lian, Chung-Jr . . . . p. 750 (7D-3)
Liffiton, Mark Hammond . . . . . . p. 19 (1A-4)
Lim, Sung Kyu . . . . p. 959 (9C-4)
Lin, Hung-Chih . . . . p. 742 (7D-2)
Lin, Rung-Bin . . . . p. 783 (8A-4)
Lin, Sheng-Chih . . . . p. 223 (2D-2)
Lin, Shyh-Chang . . . . p. 636 (6C-4)

Lin, Tao	p. 67	(1C-2)
Lin, Tay-Jyi	p. 118	(1D-15)
Lin, Tsai-Ying	p. 783	(8A-4)
Lin, Youn-Long	p. 170	(2B-3)
Lin, Youn-Long	p. 736	(7D-1)
Liu, Bin	p. 582	(6A-5)
Liu, Chih-Wei	p. 118	(1D-15)
Long, Di	p. 694	(7B-4)
Luican, Ilie I.	p. 802	(8B-2)
Luo, Fangyi	p. 730	(7C-5)

M

Ma, Yuchun	p. 384	(4B-3)
Maeda, Toshiyuki	p. 348	(4A-2)
Mahmoodi, Hamid	p. 665	(7A-4)
Mak, Wai-Kei	p. 777	(8A-3)
Malani, Parth	p. 911	(9B-1)
Marchal, Pol	p. 42	(1B-3)
Markov, Igor	p. 440	(4D-4)
Marsman, Eric D.	p. 94	(1D-3)
Maruyama, Daisuke	p. 342	(4A-1)
Matsumoto, Yuki	p. 124	(1D-18)
Matsumura, Motoaki	p. 541	(5D-2)
Matsuura, Munehiro	p. 466	(5A-5)
Mattausch, Hans Juergen	p. 176	(2B-4)
Mazumder, Pinaki	p. 416	(4C-3)
McCann, James L.	p. 279	(3B-2)
McCorquodale, Michael S.	p. 94	(1D-3)
McCorquodale, Michael S.	p. 279	(3B-2)
Mecha, Hortensia	p. 396	(4B-5)
Meil, Gavin	p. 860	(8D-2)
Meterelliyoz, Mesut	p. 237	(2D-4)
Michael, Brad	p. 860	(8D-2)
Miki, Kazuhiko	p. 856	(8D-1)
Miki, Takuji	p. 124	(1D-18)
Min, Sang Lyul	p. 332	(3D-1)
Minami, Fumihiro	p. 533	(5D-1)
Miyaoka, Yuichiro	p. 594	(6B-2)
Miyazaki, Masahide	p. 671	(7A-5)
Mochizuki, Isamu	p. 547	(5D-3)
Mochizuki, Tsuyoshi	p. 342	(4A-1)
Moondanos, John	p. 25	(1A-5)
Moore, Simon	p. 164	(2B-2)
Mori, Yutaka	p. 624	(6C-2)
Morikawa, Kimihiro	p. 547	(5D-3)

Morimoto, Takashi	p. 176	(2B-4)
Mozos, Daniel	p. 396	(4B-5)
Mukhopadhyay, Saibal	p. 712	(7C-2)
Mullins, Robert	p. 164	(2B-2)
Murali, Srinivasan	p. 146	(2A-4)
Murthy, Jayathi	p. 237	(2D-4)

N

Nagasawa, Shigeru	p. 342	(4A-1)
Nagata, Makoto	p. 106	(1D-9)
Nagata, Makoto	p. 677	(7B-1)
Nagayama, Shinobu	p. 378	(4B-2)
Nakahara, Hiroki	p. 466	(5A-5)
Nam, Eyee Hyun	p. 332	(3D-1)
Nazarian, Shahin	p. 67	(1C-2)
Neto, Horácio	p. 48	(1B-4)
Nguyen, Tuyen	p. 871	(8D-4)
Nieh, Yow-Tyng	p. 254	(3A-3)
Ninoi, Eizo	p. 342	(4A-1)
Nishimura, Tsutomu	p. 124	(1D-18)
Noguchi, Koichiro	p. 106	(1D-9)
Nonomura, Kazuhiro	p. 624	(6C-2)
Nugroho, Arif	p. 114	(1D-13)

O

Obata, Koji	p. 266	(3A-5)
Oda, Noriaki	p. 706	(7C-1)
Ofek, Hillel	p. 439	(4D-3)
Oh, Hyunok	p. 497	(5B-5)
Oh, Taewook	p. 152	(2A-5)
Ohba, Nobuyuki	p. 454	(5A-3)
Ohkubo, Naoaki	p. 570	(6A-3)
Ohtsuki, Tatsuo	p. 594	(6B-2)
Ohtsuki, Tatsuo	p. 653	(7A-2)
Onodera, Hidetoshi	p. 110	(1D-11)
Onodera, Hidetoshi	p. 515	(5C-3)
Onouchi, Masafumi	p. 547	(5D-3)
Ou, Shih-Hao	p. 118	(1D-15)
Ozturk, Ozcan	p. 390	(4B-4)
Ozturk, Ozcan	p. 814	(8B-4)

P

Palermo, Gianluca	p. 898	(9A-4)
Palkovic, Martin	p. 606	(6B-4)

Pan, David Z. . . . . p. 61 (1C-1)
Pan, David Z. . . . . p. 503 (5C-1)
Pan, David Z. . . . . p. 765 (8A-1)
Pan, Min . . . . p. 195 (2C-2)
Pan, Sung-Jui . . . . p. 25 (1A-5)
Panda, Preeti Ranjan . . . . p. 892 (9A-3)
Papa, David . . . . p. 440 (4D-4)
Papanikolaou, Antonis . . . . p. 42 (1B-3)
Parameswaran, Harindranath . . . . . p. 79 (1C-4)
Parameswaran, Sri . . . . p. 612 (6B-5)
Parameswaran, Sri . . . . p. 796 (8B-1)
Park, Ji Hwan . . . . p. 120 (1D-16)
Park, Sanggyu . . . . p. 588 (6B-1)
Pasricha, Sudeep . . . . p. 30 (1B-1)
Pedram, Massoud . . . . p. 67 (1C-2)
Pedram, Massoud . . . . p. 201 (2C-3)
Pedram, Massoud . . . . p. 473 (5B-1)
Pedram, Massoud . . . . p. 947 (9C-2)
Peng, Chih-Yang . . . . p. 213 (2C-5)
Petrovick, John . . . . p. 871 (8D-4)
Pettengill, Sarah . . . . p. 856 (8D-1)
Pham, Dac . . . . p. 871 (8D-4)
Pham, Mydung . . . . p. 871 (8D-4)
Pille, Juergen . . . . p. 871 (8D-4)
Posadas, Hector . . . . p. 485 (5B-3)
Posluszny, Stephen . . . . p. 867 (8D-3)
Posluszny, Stephen . . . . p. 871 (8D-4)
Pratap, Rajendra . . . . p. 79 (1C-4)
Pu, Yu . . . . p. 886 (9A-2)

Q

Qi, Jieming . . . . p. 856 (8D-1)
Qiu, Qinru . . . . p. 911 (9B-1)

R

Radulescu, Andrei . . . . p. 146 (2A-4)
Rahmatullah, Nursani . . . . p. 114 (1D-13)
Ramalingam, Anand . . . . p. 61 (1C-1)
Rasouli, Seid Hadi . . . . p. 297 (3B-5)
Reddy, Sudhakar M. . . . . p. 923 (9B-3)
Reinman, Glenn . . . . p. 384 (4B-3)
Ren, Junyan . . . . p. 100 (1D-6)
Reshadi, Mehrdad . . . . p. 116 (1D-14)
Riley, Mack . . . . p. 871 (8D-4)
Rong, Peng . . . . p. 473 (5B-1)

Rosdi, Bakhtiar Affendi . . . . p. 260 (3A-4)
Roy, Kaushik . . . . p. 158 (2B-1)
Roy, Kaushik . . . . p. 237 (2D-4)
Roy, Kaushik . . . . p. 665 (7A-4)
Roy, Kaushik . . . . p. 712 (7C-2)
Roychowdhury, Jaijeet . . . . p. 273 (3B-1)
Roychowdhury, Jaijeet . . . . p. 291 (3B-4)
Roychowdhury, Jaijeet . . . . p. 527 (5C-5)

S

Sakallah, Karem Ahmad . . . . p. 19 (1A-4)
Sakurai, Takayasu . . . . p. 98 (1D-5)
Sallay, Balazs . . . . p. 860 (8D-2)
Saluja, Kewal K. . . . . p. 659 (7A-3)
Sanchez, Pablo . . . . p. 485 (5B-3)
Santambrogio, Marco Domenico . . p. 898 (9A-4)
Sapatnekar, Sachin S. . . . . p. 559 (6A-1)
Sapatnekar, Sachin S. . . . . p. 219 (2D-1)
Sapatnekar, Sachin S. . . . . p. 309 (3C-2)
Sapatnekar, Sachin S. . . . . p. 771 (8A-2)
Sarto, Egino . . . . p. 207 (2C-4)
Sasao, Tsutomu . . . . p. 378 (4B-2)
Sasao, Tsutomu . . . . p. 466 (5A-5)
Sato, Yasuo . . . . p. 348 (4A-2)
Schmitz, Marcus T. . . . . p. 36 (1B-2)
Schmitz, Marcus T. . . . . p. 923 (9B-3)
Sciuto, Donatella . . . . p. 898 (9A-4)
Sekine, Hidetoshi . . . . p. 547 (5D-3)
Senger, Robert M. . . . . p. 94 (1D-3)
Septien, Julio . . . . p. 396 (4B-5)
Seyedi, Azam . . . . p. 297 (3B-5)
Shafi, Hazim . . . . p. 860 (8D-2)
Shang, Li . . . . p. 879 (9A-1)
Shao, Hui . . . . p. 96 (1D-4)
Shi, Jin . . . . p. 826 (8C-1)
Shi, Richard . . . . p. 92 (1D-2)
Shi, Richard . . . . p. 402 (4C-1)
Shi, Rui . . . . p. 428 (4C-5)
Shi, Sean X. . . . . p. 503 (5C-1)
Shi, Weiping . . . . p. 320 (3C-4)
Shi, Yiyu . . . . p. 618 (6C-1)
Shi, Yiyu . . . . p. 630 (6C-3)
Shi, Youhua . . . . p. 653 (7A-2)
Shih, Che-Hua . . . . p. 448 (5A-2)
Shih, Shen-Yu . . . . p. 170 (2B-3)

A-6

Shimizu, Kazunori............p. 112 (1D-12)
Shin, Hongjoong ....p. 765 (8A-1)
Shin, Youngsoo ....p. 565 (6A-2)
Shiratake, Shinichiro..........p. 533 (5D-1)
Shrivastava, Sachin ....p. 79 (1C-4)
Singh, Siddharth ....p. 237 (2D-4)
Son, Seung Woo.............p. 128 (2A-1)
Srivastava, Navin ....p. 223 (2D-2)
Stasiak, Daniel ....p. 867 (8D-3)
Su, Chauchin ....p. 366 (4A-5)
Su, Feng....................p. 96 (1D-4)
Su, Man-Yun.................p. 448 (5A-2)
Su, Ming-Hong ....p. 249 (3A-2)
Suga, Atsuhiro ....p. 541 (5D-2)
Sugawara, Osamu ....p. 342 (4A-1)
Sugiura, Hiroaki..............p. 124 (1D-18)
Sugiyama, Hiroyuki ....p. 624 (6C-2)
Sugiyama, Yaroku............p. 342 (4A-1)
Sugiyama, Yaroku............p. 624 (6C-2)
Sunwoo, Myung Hoon ....p. 758 (7D-4)
Sylvester, Dennis.............p. 84 (1C-5)
Sylvester, Dennis.............p. 838 (8C-3)

T

Tabero, Jesus.................p. 396 (4B-5)
Takada, Hideyuki.............p. 266 (3A-5)
Takagi, Kazuyoshi............p. 266 (3A-5)
Takahashi, Atsushi ....p. 260 (3A-4)
Takahashi, Atsushi ....p. 642 (6C-5)
Takahashi, Hiroshi............p. 659 (7A-3)
Takamatsu, Yuzo ....p. 659 (7A-3)
Takano, Kohji................p. 454 (5A-3)
Taki, Kazuo..................p. 106 (1D-9)
Tamtrakarn, Atit..............p. 98 (1D-5)
Tan, Shelton X-D.............p. 826 (8C-1)
Tan, Ying....................p. 911 (9B-1)
Tanaka, Katsunori ....p. 266 (3A-5)
Tanamura, Yoshiyasu ....p. 624 (6C-2)
Tanji, Yuichi ....p. 832 (8C-2)
Thiele, Lothar................p. 479 (5B-2)
Togawa, Nozomu.............p. 594 (6B-2)
Togawa, Nozomu.............p. 653 (7A-2)
Tomioka, Yoichi.............p. 642 (6C-5)
Tomono, Naoki ....p. 594 (6B-2)
Tsai, Chia-Yang..............p. 742 (7D-2)

Tsai, Jengliang ....p. 941 (9C-1)
Tsai, Shr-Cheng..............p. 783 (8A-4)
Tseng, Wei-Chiu ....p. 783 (8A-4)
Tsuchiya, Akira ....p. 515 (5C-3)
Tsuchiya, Takehiko............p. 1 (1A-1)
Tsui, Chi-Ying ....p. 96 (1D-4)
Tsui, Chi-Ying ....p. 102 (1D-7)
Tsui, Chi-Ying ....p. 104 (1D-8)
Tsukiboshi, Yoshiki...........p. 533 (5D-1)
Tuncer, Emre ....p. 67 (1C-2)

U

Uchida, Jumpei ....p. 594 (6B-2)
Unnikrishnan, Priya ....p. 929 (9B-4)
Usami, Kimiyoshi ....p. 570 (6A-3)
Utsumi, Tetsuaki ....p. 533 (5D-1)

V

Vandecappelle, Arnout........p. 606 (6B-4)
Verma, Manuj................p. 79 (1C-4)
Verock, Joseph ....p. 871 (8D-4)
Véstias, Mário Pereira ....p. 48 (1B-4)
Vijaykrishnan, Narayanan ....p. 140 (2A-3)
Villar, Eugenio...............p. 485 (5B-3)
Viswanathan, Natarajan ....p. 195 (2C-2)
Vogel, Sebastian..............p. 315 (3C-3)
Vrudhula, Sarma ....p. 953 (9C-3)

W

Wagner, Ilya ....p. 442 (5A-1)
Wakayama, Cherry ....p. 92 (1D-2)
Wandeler, Ernesto ....p. 479 (5B-2)
Wang, Baohua ....p. 416 (4C-3)
Wang, Chun-Yao ....p. 249 (3A-2)
Wang, Feng..................p. 390 (4B-4)
Wang, J.-H...................p. 213 (2C-5)
Wang, Janet ....p. 718 (7C-3)
Wang, Janet M ....p. 724 (7C-4)
Wang, Jia....................p. 879 (9A-1)
Wang, Ting-Chi ....p. 303 (3C-1)
Wang, Xiaoying..............p. 700 (7B-5)
Wang, Xiren ....p. 683 (7B-2)
Wang, Yu-Jen ....p. 742 (7D-2)
Wang, Zeyi ....p. 521 (5C-4)

Wang, Zeyi	p. 683	(7B-2)
Wang, Zeyi	p. 844	(8C-4)
Warnock, James	p. 871	(8D-4)
Watanabe, Minoru	p. 108	(1D-10)
Watanabe, Takayuki	p. 832	(8C-2)
Watanabe, Yukio	p. 860	(8D-2)
Watewai, Thaisiri	p. 959	(9C-4)
Wei, Jie	p. 384	(4B-3)
Weitzel, Steve	p. 871	(8D-4)
Wen, Xiaoqing	p. 348	(4A-2)
Wendel, Dieter	p. 871	(8D-4)
West, Andrew	p. 164	(2B-2)
Wong, Martin D. F.	p. 965	(9C-5)
Wong, Martin D.F.	p. 315	(3C-3)
Wong, Yiu-Chung	p. 207	(2C-4)
Wrighton, Michael G.	p. 789	(8A-5)
Wu, Dong	p. 36	(1B-2)
Wu, Kai-Chiang	p. 354	(4A-3)
Wu, Meng-Chiou	p. 783	(8A-4)
Wu, Min	p. 122	(1D-17)
Wu, Qing	p. 911	(9B-1)
Wu, Yongyi	p. 122	(1D-17)

X

Xie, Min	p. 188	(2C-1)
Xie, Yuan	p. 390	(4B-4)
Xiong, Jinjun	p. 207	(2C-4)
Xu, Xingwen	p. 1	(1A-1)

Y

Yamada, Kenta	p. 706	(7C-1)
Yamada, Tetsuya	p. 547	(5D-3)
Yamamura, Kiyotaka	p. 408	(4C-2)
Yamanaka, Hitoshi	p. 342	(4A-1)
Yamaoka, Kousuke	p. 176	(2B-4)
Yamashita, Ryoichi	p. 624	(6C-2)
Yamashita, Shigeru	p. 266	(3A-5)
Yamauchi, Hironori	p. 124	(1D-18)
Yan, Changhao	p. 844	(8C-4)

Yan, Guiying	p. 618	(6C-1)
Yanagida, Masahiro	p. 342	(4A-1)
Yanagisawa, Masao	p. 594	(6B-2)
Yanagisawa, Masao	p. 653	(7A-2)
Yang, Kai	p. 360	(4A-4)
Yang, Peng	p. 551	(5D-4)
Yang, Ya-Chi	p. 509	(5C-2)
Yang, Yonghong	p. 879	(9A-1)
Yao, Bo	p. 73	(1C-3)
Ye, Zuochang	p. 689	(7B-3)
Yeh, Shang-Yu	p. 742	(7D-2)
Yi, Joonhwan	p. 460	(5A-4)
Yoda, Tomoyuki	p. 533	(5D-1)
Yoneda, Tomokazu	p. 671	(7A-5)
Yoo, Jun-hee	p. 905	(9A-5)
Yoo, Sungjoo	p. 551	(5D-4)
Yoon, Sang-Yong	p. 588	(6B-1)
Yoshitake, Akihiko	p. 624	(6C-2)
Youssef, Wassim	p. 372	(4B-1)
Yu, Wenjian	p. 521	(5C-4)
Yu, Wenjian	p. 683	(7B-2)
Yu, Wenjian	p. 844	(8C-4)
Yu, Zhiping	p. 689	(7B-3)

Z

Zeng, Xiaoyang	p. 122	(1D-17)
Zhan, Yong	p. 219	(2D-1)
Zhan, Yong	p. 309	(3C-2)
Zhan, Yong	p. 771	(8A-2)
Zhang, Lizheng	p. 941	(9C-1)
Zhang, Mengsheng	p. 521	(5C-4)
Zhang, Tianpei	p. 309	(3C-2)
Zhang, Yan	p. 384	(4B-3)
Zhou, Lili	p. 92	(1D-2)
Zhou, Qiang	p. 582	(6A-5)
Zhou, Shuo	p. 73	(1C-3)
Zhu, Hongwei	p. 802	(8B-2)
Zhu, Yi	p. 73	(1C-3)
Zhu, Zhengyong	p. 428	(4C-5)

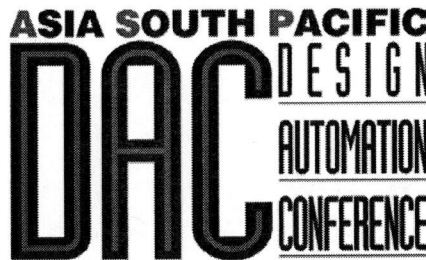

Call for Papers
ASP-DAC 2007
Asia and South Pacific Design Automation Conference 2007
http://www.aspdac.com/aspdac2007/
January 23-26, 2007
Pacifico Yokohama, Yokohama, JAPAN

Aims of the Conference:

ASP-DAC 2007 is the twelfth in a series of annual international conferences on VLSI design automation. Asia and South Pacific region is one of the most active regions of design and fabrication of silicon chips in the world. The conference aims are providing the Asian and South Pacific CAD/DA and Design community with opportunities of interchanging ideas and collaboratively discussing the directions of the technologies related to all of Electronic Design Automation (EDA). The goal of the conference is to provide a forum for presentation, discussion, and observation of the state-of-the-art of EDA technologies and design methodologies of electronic systems. The format of the meeting intends to cultivate and promote an instructive and productive interchange among EDA researchers/developers, and system/circuit/device designers. A wide variety of those scientists, engineers, and students who are interested in theoretical issues on EDA are also welcome.

Areas of Interest:
Original papers on, but not limited to, the following areas are invited.

[1] System Level Design:
System VLSI and SOC design methods, System specification, Specification languages, Design languages, Design reuse and IPs, Tools/methods for low power system design, Platform-based design, Network on chip design

[2] Embedded and Real-Time Systems:
Hardware-software co-design, Co-simulation, Co-verification, Real-time OS and middleware, Design language for embedded systems, Compilation techniques, ASIP synthesis

[3] Behavioral/Logic Synthesis and Optimization:
Behavioral/RTL synthesis, Technology independent optimization, Technology mapping, Interaction between logic design and layout, Sequential and asynchronous logic synthesis

[4] Validation and Verification for Behavioral/Logic Design:
Logic simulation, Symbolic simulation, Formal verification, Equivalence checking, Transaction-level/RTL and gate level modeling and validation

[5] Physical Design (Routing):
Routing, Repeater issues, Interconnect optimization, Interconnect planning, Module generation, Layout verification

[6] Physical Design (Placement):
Placement, Floorplanning, Partitioning, Hierarchical design

[7] Timing, Power, Signal/Power Integrity Analysis and Optimization:
Timing analysis, Power analysis, Signal/power integrity, Clock and global signal design

[8] Interconnect, Device and Circuit Modeling and Simulation:
Interconnect modeling, Interconnect extraction, Package modeling, Circuit simulation, Device modeling/simulation, Library design, Design fabrics, Design for manufacturability, Yield optimization, Reliability analysis, Emerging technologies

[9] Test and Design for Testability:
Test design, Fault modeling, ATPG, BIST and DFT, Memory, core and system test

[10] Analog, RF and Mixed Signal Design and CAD:
Analog/RF synthesis, Analog layout, Verification, Simulation techniques, Noise analysis, Analog circuit testing, Mixed signal design considerations

[11] Leading Edge Design Methodology for SOCs and SIPs:
Design methodology for Microprocessors, DSP, IP-core, multimedia processors, wireless communication systems, A/D mixed circuits, Memories, Sensors, MEMS chips, FPGAs, Novel reconfigurable systems, Rapid prototyping
Please note that ASP-DAC 2007 University LSI Design Contest encourages original papers on LSI design and implementation at universities and other educational organizations.

Submission of Papers:
Deadline for submission:　**5 pm JST, July　　10 (Mon), 2006**
Notification of acceptance:　　　**September 29 (Fri), 2006**
Deadline for final version:　**5 pm JST, November 17 (Fri), 2006**

Specification of the paper submission format will be available at our WEB site:
`http://www.aspdac.com/aspdac2007/`

Panels, Special Sessions and Tutorials:
Suggestions and proposals are welcome and have to be addressed to the Conference Secretariat (e-mail:aspdac2007@aspdac.com) no later than 5 pm JST, June 9 (Fri.), 2006.

Prospective Sponsors:
ACM SIGDA, IEEE Circuits and Systems Society, IEICE ESS (Institute of Electronics, Information and Communication Engineers, Engineering Sciences Society), IPSJ SIG-SLDM (Information Processing Society of Japan, SIG System LSI Design Methodology)

ASP-DAC2007 Chairs:
General Chair: Hidetoshi Onodera (Kyoto Univ.)
Technical Program Chair: Yusuke Matsunaga (Kyushu Univ.)
Technical Program Vice Chair: Kiyoung Choi (Seoul National Univ.)

Conference Secretariat:
Please contact Conference Secretariat (e-mail:aspdac2007@aspdac.com), if you have questions or comments.

ASIA SOUTH PACIFIC

DAC
DESIGN AUTOMATIOM CONFERENCE

University LSI Design Contest

Call for Designs
ASP-DAC 2007
University LSI Design Contest

http://www.aspdac.com/aspdac2007/
January 23-26, 2007
Pacifico Yokohama, Yokohama, JAPAN

Aims of the Contest:

As a unique feature of ASP-DAC 2007, the University LSI Design Contest will be held. The aim of the Contest is to encourage education and research on VLSI design at universities and other educational organizations. We solicit designs that fit in one or more of the following categories:

(1) Designed, and actually implemented on chips in universities or other educational organizations during the last two years.
(2) Designs that report actual measurements from implementations.
(3) Innovative design prototypes.

Interesting or excellent designs selected will be honored by providing the opportunities for presentation in a special session at the conference. Awards will be given to a few number of outstanding designs, selected from those presented at the conference.

Areas of Design:

Application areas, or types of circuits, of the original LSI circuit designs include (but are not limited to):
  (1)Analog, RF and Mixed-Signal Circuits, (2) Digital Signal Processing, (3) Microprocessors, (4) Custom ASIC.
Methods, or technology, used for implementation include:
  (a) Full Custom and Cell-Based LSIs, (b) Gate Arrays, (c) FPGA/PLDs

Submission of Design Descriptions:

A camera-ready summary is requested to be prepared within 2 pages including figures, tables, and references. It is strongly recommended that measured experimental results and a chip micrograph are included in the original LSI circuit design. If the experimental results and the chip micrograph have not been prepared before the deadline of submission, the authors can send the revised paper including them later. Please do not submit the same paper as a regular paper.

Specification of the submission format and the predetermined design tasks are available at
http://www.aspdac.com/aspdac2007/

Deadline for summary:	July	10(Mon),	2006
Notification of acceptance:	September	29(Fri),	2006
Deadline for camera-ready:	November	17(Fri),	2006

Review:

Submitted designs will be reviewed by the Design Contest Committee in a process similar to the review process for the technical papers. The following criteria will be applied in the selection of designs:
(1) Reliability of design and implementation, (2) Quality of implementation, (3) Performance of the design, (4) Novelty of application, algorithm, architecture, (5) Others.
Interesting or excellent designs selected will be presented at a special session of the conference.

Presentation:

An author of each selected design will be required to make a short presentation at a special session of ASP-DAC 2007. A digest of each design to be presented will be included in the conference proceedings.

ASP-DAC 2007 Chairs
  General Chair: Hidetoshi Onodera (Kyoto Univ.)
  Technical Program Chair: Yusuke Matsunaga (Kyushu Univ.)
  Design Contest Chair: Makoto Nagata (Kobe Univ.)

Proceedings of the

ASP-DAC 2006

Asia and South Pacific Design Automation Conference
2006

January 24-January 27, 2006
Pacifico Yokohama
Yokohama, Japan

Sponsored by:
IEEE Circuits and Systems Society
ACM SIGDA
IEICE (Institute of Electronics, Information and
Communication Engineers)
IPSJ (Information Processing Society of Japan)
Supported by:
JEITA (Japan Electronics and Information Technologies
Industry Association)
STARC (Semiconductor Technology Academic
Research Center)
SCAT (Support Center for Advanced Telecommunications
Technology Research, Foundation)
In Cooperation with:
JNTO (Japan National Tourist Organization)
City of Yokohama

Additional Copies may be ordered from:

IEEE Order Dept.
Hoes Lane
P.O. Box 1331
Piscataway, NJ 08854, U.S.A.

Copyright and Reprint Permission: Abstracting is permitted with credit to the source. Libraries are permitted to photocopy beyond the limit of U.S. copyright law for private use of patrons those articles in this volume that carry a code at the bottom of the first page, provided the per-copy fee indicated in the code is paid through Copyright Clearance Center, 222 Rosewood Drive, Danvers, MA 01923. For other copying, reprint or republication permission, write to IEEE Copyrights Manager, IEEE Operations Center, 445 Hoes Lane, P.O. Box 1331, Piscataway, NJ 08855-1331. All rights reserved. Copyright ©2006 by the Institute of Electrical and Electronics Engineers, Inc.

IEEE Catalog Number: 06EX1199 (CD-ROM Version: 06EX1199C)
ISBN: 0-7803-9451-8 (CD-ROM Version: 0-7803-9452-6)
Library of Congress: 2005932176

ASP-DAC 2006 General Chair's Message

On behalf of the Organizing Committee, I would like to welcome you to the Asia and South Pacific Design Automation Conference 2006 (ASP-DAC 2006) being held here at Pacifico Yokohama jointly with the Electronic Design and Solution Fair 2006. ASP-DAC is a sister conference of DAC, DATE and ICCAD, and it is the 11th event of this conference series.

ASP-DAC is the meeting place where researchers and engineers come together to learn and discuss state of the art technologies of system/SoC design, EDA and design methodologies. This year, we put special effort into attracting designers and design industries to produce Designers' Forum.

We have three keynote speakers from academia, the semiconductor industry and the systems industry. Professor Alberto Sangiovanni-Vincentelli, University of California at Berkeley, will explore future system design perspectives in automotive electronics. Satoru Ito, President & CEO of RENESAS Technology Corp. will discuss challenges of device innovation. Yukichi Niwa, Senior Advisory Director of CANON INC. will present the company's key concept of architecting platform based design.

The technical program was selected from 424 papers from 27 countries. The 64 members of the Technical Program Committee chaired by Professor Onodera and helped by over 250 reviewers had to make difficult choices to carefully select 135 papers. It is an outstanding program that covers a variety of key topics from system level design to physical design. Designers' Forum is a new program that shares design experience and solutions of real product designs of the industries whose topics include the CELL and mobile designs, panels of top 10 design issues and system verification. The University Design Contest is also an important event of ASP-DAC, which focuses on a real chip design in academia. On Tuesday, one full-day and six half-day tutorials are scheduled to provide introductions to hot topics like DFM, lowpower, packaging/interconnect, system level design and verification.

An event like ASP-DAC can happen only with the efforts of many people. We wish to express our appreciation to all authors, speakers, reviewers, session organizers, moderators, panelists, session chairs, keynote speakers and committee members. Also my sincere thanks to the dedicated members of the Organizing Committee, the Technical Program Committee, the University LSI Design Contest Committee, the Industry Liaison and the Steering Committee.

Finally, special thanks to all ASP-DAC attendees. I hope you will have a productive and exciting experience at ASP-DAC 2006.

Fumiyasu Hirose
General Chair
ASP-DAC 2006

Message from Technical Program Committee

Hidetoshi Onodera   Yusuke Matsunaga

On behalf of the Technical Program Committee for the Asia and South Pacific Design Automation Conference 2006, we would like to welcome all of you to the conference held from January 24 through 27, 2006 at Pacifico Yokohama Conference Center in Yokohama, Japan.

This year, ASPDAC received 424 paper submissions, which is the second highest number in our history. This is a 46% increase in submissions over the last ASPDAC held in Japan two years ago. The submissions span 27 countries/regions in Asia, North America, South America, Europe, Oceania, and Africa.

The Technical Program Committee was composed of 64 professionals who are experts on EDA, LSI design, and embedded system design, and was organized into 11 sub-committees. The number of committee members is relatively small compared to the previous conferences. However, all the members were committed to make in-depth reviews for all the papers assigned to each sub-committee and physically attend the TPC meeting for paper selection. Based on the result of a rigorous and thorough review followed by a full day face-to-face discussion, we have selected 135 papers and compiled them into a 3-day program of 27 technical sessions in three parallel tracks. The technical program is further enriched by multiple special sessions and panels in one more track, resulting in four parallel tracks of exciting presentations and discussions.

Each day, the technical program starts with a keynote address that is organized through the leadership of Fumiyasu Hirose, General Chair, followed by regular and special sessions. We have 7 special sessions in total. On Wednesday, selected submissions to the University Design Contest will be presented at Session 1D. Session 2D focuses on electro-thermal design and Session 3D discusses flash memory in embedded systems. On Thursday, Session 4D addresses an overview of Open Access and Session 5D focuses on low power design challenges for mobile applications. On Friday, Session 7D discusses H.264 design issues and Session 8D addresses design methodology for "Cell" processor development. It should be mentioned that Sessions 5D and 8D are organized as Designers' Forum that is a brand new event for encouraging mutual exchange between/within designers in industry and EDA researchers. Besides two special sessions, Designers' Forum will organize two panel discussions on functional verification (Session 6D, Thursday) and top-10 design issues (Session 9D, Friday). These special sessions will provide you with a wide variety of hot and exciting topics from system level down to physical level, and from industrial design issues to theoretical fundamentals.

On behalf of the Technical Program Committee, we would like to thank all people involved in the 2006 event. In particular, we would like to thank the members of the Organizing and Technical Program Committees, the members of Industry Liaison, special session organizers, and everyone at JESA for conference management. Also, we would like to express our sincere thanks to all the authors who submitted papers with valuable results, since their contributions form the basis of our technical excellence.

We would be more than happy if you could attend the conference and find something new in the directions of EDA and technologies during ASP-DAC 2006.

Technical Program Chair
Hidetoshi Onodera
Kyoto University

Technical Program Vice Chair
Yusuke Matsunaga
Kyushu University

University LSI Design Contest

The University LSI Design Contest was conceived as a unique program of ASP-DAC Conference. The purpose of the Contest is to encourage education and research in LSI design, and its realization on chips at universities, and other educational organizations by providing opportunities to present and discuss innovative and state-of-the-art designs at the conference. Application areas and types of circuits include (1) Analog and Mixed-Signal Circuits, (2) Digital Signal processing, (3) Microprocessors, and (4) Custom Application Specific Circuits. Methods or technology used for implementation include (a) Full Custom and Cell-Based LSIs, (b) Gate Arrays, and (c) Field Programmable Devices, including FPGA/PLDs.

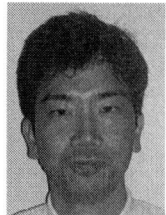 

Kazutoshi Kobayashi   Takahiko Arakawa

This year, nineteen selected designs from seven countries/areas will be disclosed in Session 1D with a short presentations followed by live discussions in front of posters with light meals. Submitted designs were reviewed by the members of the University Design Contest Committee based on the following criteria: Reliability of design and implementation, Quality of implementation, Performance of the design, Novelty, and Additional special features. In the selection process, emphasis was placed more on reliability, quality, and performance. As a result, the nineteen designs were selected. Also, we have instituted one outstanding design award and two special feature awards.

It is with great pleasure that we acknowledge the contributions to the Design Contest, and it is our earnest belief that it will promote and enhance research and education in LSI design in academic organizations. It is also our hope that many people not only in academia but in industry will attend the contest and enjoy the stimulating discussions.

Co-Chairs, University LSI Design Contest Committee

Co-Chair
Kazutoshi Kobayashi
Kyoto University

Co-Chair
Takahiko Arakawa
RENESAS Technology Corp.

Designers' Forum

This brandnew event, Designers' Formum, is conceived as a unique program for ASP-DAC to encourage mutual exchange both between and within designers in industry, researchers in the area of EDAs, and EDA developers. Here, designs will be presented focusing on design styles, design issues, and ways to tackle design issues. Panel discussions will also be held for the latest design issues.

Haruyuki Tago   Makoto Ikeda

This year, we will have 2 special sessions in ASP-DAC 2006 as follows. 4 presentations related to low power designs, power models, and power estimation frameworks for the real SoC designs will be given by Toshiba, Fujitsu, Hitachi and Renesas, and Samsung in session 5D, "Low Power Design." In session 8D, "Cell processor", 4 presentations will be given fucusing on simulations, tests, verifications, power estimation, and design methodology of Cell; and PLL design employed in the Cell processor.

In addition to the special sessions, we will have 2 panel discussions. Session 6D, "Functional Verification -now and future-" will be modareted by Dr. Y. Masubuchi of Toshiba, who has been deeply engaged in the architecture design and verifications of the Cell processor. Three panelists are LSI designers, working for functional verifications in LSI design, and one panelist is from an EDA vendor, developing functional verification tools. Session 9D, "Top 10 design issues seen by LSI designers versus EDA developers", on the other hand, will be moderated by Dr. Y. Hagihara of Sony. Three panelists are managers of SoC and system designs, will focus on the top 10 design issues seen by LSI designers; and 3 panelists are techology officers from 3 major EDA vendors, and they will focus on top 10 design issues seen by EDA developers. Discussions will be led toward the perspectives of the future SoC design issues, comparing with the two top 10 issues seen by both LSI designers and EDA developers.

This Designers' Forum is planned by the Industry Laison Members of ASP-DAC2006. It is with great pleasure that we acknowledge the contributions to the Designers' Forum, and it is our earnest belief that this forum will promote mutual exchange of designers and EDA researchers and developers, toward nano-meter SoC design issues.

Industry Liaison Chair
Haruyuki Tago
TOSHIBA CORPORATION

Designers' Forum Chair
Makoto Ikeda
University of Tokyo

Keynote Address I

Automotive Electronics: Steady Growth for Years to Come!

Alberto Sangiovanni-Vincentelli

The Edgar L. and Harold H. Buttner Chair of Electrical
Engineering and Computer Science, University of California,
Berkeley, and Chief Technology Advisor, Member of the Board
and Co-founder, Cadence Design Systems, United States

The world of electronics is witnessing a revolution in the way products are conceived, designed and implemented. The ever growing importance of the web, the advent of microprocessors of great computational power, the explosion of wireless communication, the development of new generations of integrated sensors and actuators are changing the world in which we live and work. The new key words are:

- Disappearing electronics, i.e., electronics has to be invisible to the user, it has to help unobtrusively.

- Pervasive computing, i.e., electronics is everywhere, all common use objects will have an electronic dimension.

- Ambient intelligence, i.e., the environment will react to us with the use of electronic components. They will recognize who we are and what we like.

- Wearable computing, i.e., the new devices will be worn as a watch or a hat. They will become part of our clothes. Some of these devices will be tags that will contain all important information about us.

- Know more, carry less, i.e., the environment will know more about us so that we will not need to carry all the paraphernalia of keys, credit cards, personal I.D.s, access cards, access codes.

The car as a self-contained microcosm is experiencing a similar revolution: all the key words listed above are going to have a great impact on the automotive world. We need to rethink what a car really is and the role of electronics in it. Electronics is now essential to control the movements of a car, of the chemical and electrical processes taking place in it, to entertain the passengers, to establish connectivity with the rest of the world, to ensure safety. What will an automobile manufacturer's core competence become in the next few years? Will electronics be the essential element in car manufacturing and design? The challenges and opportunities are related to

- how to integrate the mechanical and the electronics worlds, i.e., how to make mechatronics a reality in the automotive world,

- how to integrate the different motion control and power-train control functions so that important synergies can be exploited,

viii

- how to combine entertainment, communication and navigation subsystems,

- how to couple the world of electronics where the life time of a product is around 2 years and shrinking, with the automotive world, where the product life time is 10 years and possibly growing,

- how to develop new services based on electronics technology,

- how to exploit communication among cars and between cars and infrastructure such as Global Positioning Systems and cellular networks,

- how are the markets evolving (for example, what will be the size of the after-market sales for automotive electronics, if any?).

We will pose these questions while reviewing some of the most important technology and product developments of the past few years. We will also present new trends on how the design of electronics of the car should be carried out. We will finally analyze the dynamics of the automotive electronics industry that is bound to produce a major shake-up in the structure of the design chain with particular emphasis on the AUTOSAR consortium.

Keynote Address II

Challenging Device Innovation

Satoru Ito

President & CEO

RENESAS Technology Corp., Japan

The semiconductor industry has continuously transformed our way of life, through a number of underlying technology breakthroughs and innovations over the past years. There are currently two challenges that this industry faces: a limitation of miniaturization technology and a difficulty in maintaining an economy of scale. To cope with these challenges, there is a growing need to work closely with partners and customers who have business related to semiconductors, in addition to semiconductor manufacturers. Especially in the area of semiconductor design, we see a need to create a new EDA methodology that broadens the definition of traditional EDA and re-defines the connection among system designers, SoC designers and development tool designers. As we move closer to the realm driven by the convergence of applications and advancements in miniaturization technology, I'd like to discuss the associated technological challenges as well as economical challenges, and present to you our strategy to overcome these issues.

Keynote Address III

Effective Platform-based Development for Large-scale Systems Design

Yukichi Niwa

Senior Advisory Director,
Group Executive of Platform Technology Development
Headquarters
CANON INC., Japan

Platform-based development (PBD) aims to continuously add new value in both cases of incremental development and product planning based development. By adding new technology to previously existing technology and by storing the technologies as reusable assets, PBD enables high quality, low cost, and short turnaround time development. Furthermore, PBD allows target-oriented development where we can select and concentrate technology to eliminate unnecessary development.

In order to execute effective PBD, it is important to introduce the firm layer structuring of digital/analog technology so that individual professionals in independent layer can maximize their efficiency without any restraint. The act of layer structuring is nothing but the architectural design of the development methodology. Thus, it's no exaggeration to say that success in business profitability management directly depends on the presence of the good architect.

The important thing in the next stage is to optimize the design process by investing in computer resources. For example, it is necessary to thoroughly adapt simulation technology to the development of high quality imaging technology, embedded system (hardware/software) technology, or communication technology. The quantitative evaluation from the early design phase and the workflow based on the accumulated design know-how (IP, methodology) will accelerate technology innovation and strengthen the platform even further. Eventually, management can directly obtain absolute advantage of large-scale system design effectiveness.

ASP-DAC 2005 Best Papers

Best Paper Award

Speed and Voltage Selection for GALS Systems based on Voltage/Frequency Islands
Koushik Niyogi, Diana Marculescu (Carnegie Mellon University)

The Polygonal Contraction Heuristic for Rectilinear Steiner Tree Construction
Yin Wang, Xianlong Hong, Tong Jing, Yang Yang (Tsinghua University, Beijing), Xiaodong Hu, Guiying Yan (Institute of Applied Mathematics, Chinese Academy of Sciences)

Fast PLL Simulation Using Nonlinear VCO Macromodels for Accurate Prediction of Jitter and Cycle-Slipping due to Loop Non-idealities and Supply Noise
Xiaolue Lai, Yayun Wan, Jaijeet Roychowdhury (University of Minnesota)

Design Contest Award

A Bandwidth Efficient Subsampling-based Block Matching Architecture for Motion Estimation
Hao-Yun Chin, Chao-Chung Cheng, Yu-Kun Lin, Tian-Sheuan Chang (Chiao Tung University)

A Low-Power Video Segmentation LSI with Boundary-Active-Only Architecture
Takashi Morimoto, Osamu Kiriyama, Hidekazu Adachi, Zhaomin Zhu, Tetsushi Koide, Hans Jürgen Mattausch (Hiroshima University)

ASP-DAC 2006 Best Papers

Best Paper Award

1B-1 **Constraint-Driven Bus Matrix Synthesis for MPSoC**
Sudeep Pasricha, Nikil Dutt (Univ. of California, Irvine, United States), Mohamed Ben-Romdhane (Conexant, United States)

3C-1 **Post-Routing Redundant Via Insertion for Yield/Reliability Improvement**
Kuang-Yao Lee, Ting-Chi Wang (National Tsing Hua Univ., Taiwan)

Best Paper Candidates

1B-1 **Constraint-Driven Bus Matrix Synthesis for MPSoC**
Sudeep Pasricha, Nikil Dutt (Univ. of California, Irvine, United States), Mohamed Ben-Romdhane (Conexant, United States)

3C-1 **Post-Routing Redundant Via Insertion for Yield/Reliability Improvement**
Kuang-Yao Lee, Ting-Chi Wang (National Tsing Hua Univ., Taiwan)

7B-1 **Equivalent Circuit Modeling of Guard Ring Structures for Evaluation of Substrate Crosstalk Isolation**
Daisuke Kosaka, Makoto Nagata (Kobe Univ., Japan)

7C-2 **Speed Binning Aware Design Methodology to Improve Profit under Parameter Variations**
Animesh Datta (Purdue Univ., United States), Swarup Bhunia (Case Western Reserve Univ., United States), Jung Hwan Choi, Saibal Mukhopadhyay, Kaushik Roy (Purdue Univ., United States)

8A-1 **Fast Substrate Noise-Aware Floorplanning with Preference Directed Graph for Mixed-Signal SOCs**
Minsik Cho, Hongjoong Shin, David Z. Pan (Univ. of Texas, Austin, United States)

8B-1 **Finding Optimal L1 Cache Configuration for Embedded Systems**
Andhi Janapsatya, Aleksandar Ignjatovic, Sri Parameswaran (Univ. of New South Wales, Australia)

9A-1 **TAPHS: Thermal-Aware Unified Physical-Level and High-Level Synthesis**
Zhenyu (Peter) Gu (Northwestern Univ., United States), Yonghong Yang (Queen's Univ., Canada), Jia Wang, Robert P. Dick (Northwestern Univ., United States), Li Shang (Queen's Univ., Canada)

9C-3 **Statistical Leakage Minimization through Joint Selection of Gate Sizes, Gate Lengths and Threshold Voltage**
Sarvesh Bhardwaj, Yu Cao, Sarma Vrudhula (Arizona State Univ., United States)

Design Contest Award

Outstanding Design Award

1D-15 **A 52mW 1200MIPS Compact DSP for Multi-Core Media SoC**
Shih-Hao Ou, Tay-Jyi Lin, Chao-Wei Huang, Yu-Ting Kuo, Chie-Min Chao, Chih-Wei Liu (National Chiao Tung Univ., Taiwan), Chein-Wei Jen (STC, ITRI, Taiwan)

Special Feature Award

1D-1 **A Low Dynamic Power and Low Leakage Power 90-nm CMOS Square-Root Circuit**
Tadayoshi Enomoto, Nobuaki Kobayashi (Chuo Univ., Japan)

1D-8 **Adaptively-Biased Capacitor-Less CMOS Low Dropout Regulator with Direct Current Feedback**
Yat-Hei Lam, Wing-Hung Ki, Chi-Ying Tsui (Hong Kong Univ. of Science and Tech., Hong Kong)

ASP-DAC 2006 Organizing Committee

General Chair

Fumiyasu Hirose
Cadence Design Systems, Japan

SC Chair

Hiroto Yasuura
Kyushu University

Secretary

Atsushi Takahashi
Tokyo Institute of Technology

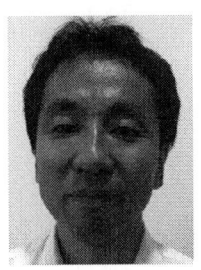

SC Vice Chair

Takeshi Yoshimura
Waseda University

Secretary

Yoshinori Takeuchi
Osaka University

Technical Program Chair

Hidetoshi Onodera
Kyoto University

Assitant Secretary

Asako Kaneko
Cadence Design Systems, Japan

Technical Program Vice Chair

Yusuke Matsunaga
Kyushu University

Past Chair

Masaharu Imai
Osaka University

TPC Secretary

Masanori Hashimoto
Osaka University

Design Contest Co-Chair

Kazutoshi Kobayashi
Kyoto University

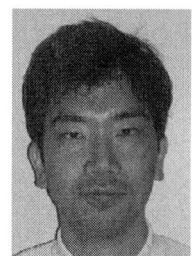

Tutorial Vice Chair

Hiroyuki Higuchi
FUJITSU LABORATORIES LTD.

Design Contest Co-Chair

Takahiko Arakawa
RENESAS Technology Corp.

Finance Co-Chair

Naoya Tohyama
System Fabrication Technologies, Inc

Designers' Forum Chair

Makoto Ikeda
University of Tokyo

Finance Co-Chair

Shinji Kimura
Waseda University

Tutorial Co-Chair

Kazutoshi Wakabayashi
NEC Corporation

Publicity Chair

Masahiro Fukui
Ritsumeikan University

Tutorial Co-Chair

Chung-Kuan Cheng
University of California, San Diego

Publication Chair

Nozomu Togawa
Waseda University

Web Publicity

Bakhtiar Affendi
Tokyo Institute of Technology

ASP-DAC Rep. at DAC

Yusuke Matsunaga
Kyushu University

Industry Liaison Chair

Haruyuki Tago
TOSHIBA CORPORATION

ASP-DAC Rep. at DATE

Masaharu Imai
Osaka University

EDSF Chair

Mitsuru Nadaoka
Oki Electric Industry Co., Ltd.

ASP-DAC Rep. at ICCAD

Kazutoshi Wakabayashi
NEC Corporation

ASP-DAC Japan Council Rep.

Tokinori Kozawa

IEICE/CAS Rep.

Takao Nishitani
Kochi University of Technology

ASP-DAC Japan Council Rep.

Kenji Yoshida
Cadence Design Systems, Japan

IEICE/ICD Rep.

Masao Nakaya
RENESAS Technology Corp.

IEICE/VLD Rep.

Shinji Kimura
Waseda University

Secretariat

Yoshinori Ishizaki
Japan Electronics Show Association

IPSJ/SLDM Rep.

Takashi Kambe
Kinki University

Secretariat

Mieko Mori
Japan Electronics Show Association

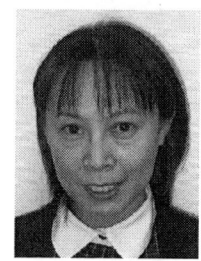

JEITA/EDA TC Rep.

Yoshio Okamura
RENESAS Technology Corp.

Secretariat

Fumiaki Yoshinaga
Japan Electronics Show Association

IWCM Rep.

Shigetaka Kumashiro
STARC

Secretariat

Kayoko Oda
Japan Electronics Show Association

Secretariat

Jiro Irie
Japan Electronics Show Association

ASP-DAC Steering Committee

Chair

Hiroto Yasuura
Kyusyu University
yasuura@c.csce.kyushu-u.ac.jp

ASP-DAC 2006 General Chair

Fumiyasu Hirose
Cadence Design Systems, Japan

Vice Chair

Takeshi Yoshimura
Waseda University
t-yoshimura@waseda.jp

ASP-DAC 2005 General Chair

TingAo Tang
Fudan University

Secretary

Toshihiro Hattori
RENESAS Technology Corp.
hattori.toshihiro@renesas.com

ACM SIGDA Rep.

Nikil Dutt
University of California at Irvine

Secretary

Kazutoshi Wakabayashi
NEC Corporation
wakaba@bl.jp.nec.com

IEEE CAS Rep.

Georges Gielen
The Katholieke Universiteit Leuven

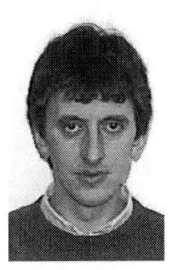

PAST Chair

Tatsuo Ohtsuki
Waseda University

DAC Representative

Steven P. Levitan
University of Pittsburgh

xix

DATE Representative

Georges Gielen
The Katholieke Universiteit Leuven

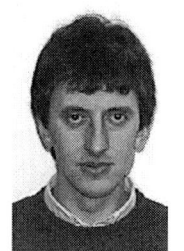

IEICE TGVLD Chair

Shinji Kimura
Waseda University

ICCAD Representative

Kazutoshi Wakabayashi
NEC Corporation

IEICE TGICD Chair

Masao Nakaya
RENESAS Technology Corp.

JEITA Representative

Yoshio Okamura
RENESAS Technology Corp.

IPSJ SIG SLDM Chair

Takashi Kambe
Kinki University

EDSF Chair

Mitsuru Nadaoka
Oki Electric Industry Co., Ltd.

STARC Representative

Takeshi Imamura
Semiconductor Technology Academic Research Center

IEICE TGCAS Chair

Takao Nishitani
Kochi University of Technology

JIEP Representative

Yoichi Shiraishi
Gunma University

International Members

Sunil D. Sherlekar
Tata Consultancy Services

Youn-Long Steve Lin
Tsing Hua University, Hsin-Chu

Richard M M Chen
City University of Hong Kong

Alexander Stempkovsky
Russin Academy of Sciences

Sri Parameswaran
The University of New South Wales

Qianling Zhang
Fudan University

Xian-Long Hong
Tsinghua University, Beijing

Advisory Members

Masaharu Imai
Osaka University

Chong-Min Kyung
Korea Advanced Institute of Science and
Technology

Basant R. Chawla
Genentech

Hideo Fujiwara
Nara Institute of Science and Technology

Hidetoshi Onodera
Kyoto University

Satoshi Goto
Waseda University

Isao Shirakawa
Professor Emeritus of Osaka University

Fumiyasu Hirose
Cadence Design Systems, Japan

Kenji Yoshida
Cadence Design Systems, Japan

Takashi Kambe
Kinki University

Tokinori Kozawa

Hiroaki Kunieda
Tokyo Institute of Technology

ASP-DAC 2006 Technical Program Committee

Chair

Hidetoshi Onodera
Kyoto University
onodera@i.kyoto-u.ac.jp

Vice Chair

Yusuke Matsunaga
Kyushu University
matsunaga@c.csce.kyushu-u.ac.jp

Secretary

Masanori Hashimoto
Osaka University
hasimoto@ist.osaka-u.ac.jp

Subcommittees (∗ indicates the subcommitte chair.)
[1] System Level Design Methodology

∗**Youn-Long Lin**	**Ahmed Jerraya**	**Yoshinori Takeuchi**
National Tsing Hua University	TIMA	Osaka University
Soonhoi Ha	**Tsuneo Nakata**	
Seoul National University	Fujitsu Lab.	

[2] Embedded and Real-Time Systems

∗**Hiroyuki Tomiyama**	**Tei-Wei Kuo**	**Sungjoo Yoo**
Nagoya University	National Taiwan University	Samsung
Pai Chou	**Sri Parameswaran**	
University of California, Irvine	University of New South Wales	

[3] Behavioral/Logic Synthesis and Optimization

∗**Kiyoung Choi**	**Shinji Kimura**	**Shigeru Yamashita**
Seoul National University	Waseda University	AIST Nara
Shih-Chieh Chang	**Diana Marculescu**	
National Tsing Hua University	Carnegie Mellon University	

[4] Validation and Verification for Behavioral/Logic Design

∗**Kiyoharu Hamaguchi**	**Shin'ichi Minato**	**Farn Wang**
Osaka University	Hokkaido University	National Taiwan University
Jin-Young Choi	**Karem Sakallah**	
Korea University	University of Michigan	

xxiii

[5] Physical Design (Routing)

*Martin D. F. Wong
University of Illinois, Urbana Champaign
Tong Jing
Tsinghua University

Youichi Shiraishi
Gunma University
Atsushi Takahashi
Tokyo Institute of Technology

Ting-Chi Wang
National Tsing Hua University

[6] Physical Design (Placement)

*Shin'ichi Wakabayashi
Hiroshima City University
Yao-Wen Chang
National Taiwan University

Jason Cong
University of California, Los Angeles
Shigetoshi Nakatake
University of Kitakyushu

Evangeline F. Y. Young
Chinese University of Hong Kong

[7] Timing, Power, Signal/Power Integrity Analysis and Optimization

*Sachin Sapatnekar
University of Minnesota
Shabbir Batterywala
Synopsys (India)
Jin-Jia Liou
National Taiwan University

Takashi Sato
Renesas
Weiping Shi
Texas A&M University
Youngsoo Shin
KAIST

Sheldon Tan
University of California, Riverside
Ryuichi Yamaguchi
Matsushita

[8] Interconnect, Device and Circuit Modeling and Simulation

*Hideki Asai
Shizuoka University
Arun Chandrasekhar
Intel (India)
Charlie Chung-Ping Chen
National Taiwan University

Eli Chiprout
Intel
Yungseon Eo
Hanyang University

Hiroo Masuda
STARC
Jae-Kyung Wee
Soongsil University

[9] Test and Design for Testability

*Seiji Kajihara
Kyushu Institute of Technology
Masaki Hashizume
Tokushima University

Sungho Kang
Yonsei University
XiaoWei Li
China Academy of Sciences

Prab Varma
Veritable

[10] Analog, RF and Mixed Signal Design and CAD

*Makoto Nagata
Kobe University
Seijiro Moriyama
PDF Solutions

Hong-June Park
POSTECH
Jaijeet Roychowdhury
University of Minnesota

Chau-Chin Su
National Chao-Tung University

[11] Leading Edge Design Methodology for SOCs and SIPs

∗Hideharu Amano
Keio University
Ing-Jer Huang
National Sun-Yat-Sen University

Satoshi Matsushita
NEC
Borivoje Nikolic
University of California, Berkeley

In-Cheol Park
KAIST
Yulu Yang
Nankai University

University LSI Design Contest Committee

Co-Chairs

Kazutoshi Kobayashi
Kyoto University
kobayasi@kuee.kyoto-u.ac.jp

Takahiko Arakawa
RENESAS Technology Corp.
arakawa.takahiko@renesas.com

Chih-Wei Liu
National Chiao Tung University

Hideo Ohwada
FUJITSU LABORATORIES LTD.

Takao Onoye
Osaka University

Mengtian Rong
Shanghai Jiao Tong University

Hao San
Gunma University

Makoto Takamiya
University of Tokyo

Chi-ying Tsui
The Hong Kong University of Science
and Technology

Tomohisa Wada
University of the Ryukyus

Hideki Yamauchi
SANYO Electric Co., Ltd.

Xiaooyang Zeng
Fudan University

Industry Liaison

Chair
Haruyuki Tago
TOSHIBA CORPORATION
haruyuki.tago@toshiba.co.jp

Design

Yoshiaki Hagihara
Sony Corporation

Sunao Torii
NEC Corporation

Kunio Uchiyama
Hitachi, Ltd.

Shigeru Watari
Matsushita Electric Industrial Co., Ltd.

Takeshi Yamamura
FUJITSU LABORATORIES LTD.

JEITA

Yoshitada Fujinami
NEC Electronics Corporation

Shigemi Saito
Sony Corporation

EDA

Hiromitsu Fujii
Nihon Synopsys Co.,Ltd.

Fumiyasu Hirose
Cadence Design Systems, Japan

Satoshi Kojima
Mentor Graphics Japan Co.,Ltd.

List of Reviewers

Rohit A
Bakhtiar Affendi
Yong-Jin Ahn
Hideharu Amano
Chirayu Amin
Zaher Andraus
Tetsuya Aoyama
Hideki Asai
Shabbir Batterywala
Valeria Bertacco
Sandeep Bhatia
Sambuddha Bhattacharya
Srinivas Bodapati
Zhen Cao
Sourav Chakravarty
Jeremy Chan
Arun Chandrasekhar
Kai-hui Chang
Shih-Chieh Chang
Soon-Jyh Chang
Tsin-Yuan Chang
Yao-Wen Chang
Yuan-Hao Chang
Yun-Nan Chang
Kai-Yuan Chao
Charlie Chung-Ping Chen
De-Sheng Chen
Hsin-Chen Chen
Huang-Yu Chen
Hung-Ming Chen
Jian-Jia Chen
Liang-Bi Chen
Sao-Jie Chen
Tai-Chen Chen
Tung-Chieh Chen
Ya-Shu Chen
Yan-Bin Chen
Yi-An Chen
Yu-Cheng Chen
Yu-Zhi Chen
C.K. Cheng
Li Chia
Charles Chiang
Mei-Fang Chiang
Eli Chiprout
Jih-Ching Chiu

Youngchul Cho
Jin-Young Choi
Kiyoung Choi
Michael Yee-Jern Chong
Pai Chou
Szu-Jui Chou
Chris C.N. Chu
Jason Cong
Florentin Dartu
Liang Deng
Lan-Rong Dung
Yungseon Eo
Jeffrey Fan
Jia-Wei Fang
Yu-Luen Fang
Zhe Feng
Ryuichi Fujimoto
Kunihiro Fujiyoshi
Youxin Gao
Eric Grimme
Ruei-Ting Gu
Hi Annie Guo
Zheng Guo
Soonhoi Ha
Kiyoharu Hamaguchi
Kee-Sung Han
Masaki Hashizume
Atsushi Hatabu
Hao-Chiao Hong
Jin-Hua Hong
Jen-Wei Hsieh
Wang-Jui Hsieh
Pi-Cheng Hsiu
Chin-Hsiung Hsu
Heng-Ruey Hsu
Ren-Chien Hsu
Tien-Chang Hsu
Ching-Chi Hu
Jiang Hu
Wui Hu
Xiaodong Hu
Yu Hu
Chih-Yuan Huang
Chung-Yang (Ric) Huang
Geng-Dian Huang
Ing-Jer Huang

Jiun-Lang Huang
Jun-Dar Huang
Po-Chiun Huang
Shih-Hsu Huang
Szu-Wei Huang
Wen-Kai Huang
Wen-Pin Huang
Chia-Mei Hung
Ting-Ting Hwang
Hiroyuki Igura
Masato Inagi
Shigeto Inui
Takeshi Inuo
Ryosuke Isotani
Tsuyoshi Isshiki
Hiroaki Iwashita
Andhi Janapsatya
Ahmed Jerraya
Hui-Ru Jiang
Jie-Hong Roland Jiang
Rong Jiang
Zhe-Wei Jiang
Tong Jing
Jinyong Jung
Nobuki Kajihara
Seiji Kajihara
Sungho Kang
Hiroshi Kawaguchi
Shoji Kawahara
Hideyuki Kawakita
Masahiro Kawakita
Mahesh Ketkar
Sunil Khatri
Mary Kiemb
Tomohiro Kikuma
Chris H. Kim
Jin-Hyun Kim
Sun-Gyeum Kim
Shinji Kimura
Tomohisa Kimura
Lih Wen Koh
Yukihide Kohira
Tetsushi Koide
Satoshi Komatsu
Michio Komoda
Shiann-Rong Kuang

Yukiko Kubo
Shuichi Kunie
Ko-Chi Kuo
Tei-Wei Kuo
Shin-ya Kuwamura
Chi-Ping Lai
Chun-Hung Lai
Gang-Hee Lee
Imyong Lee
Katherine S.M. Lee
Kuang-Yao Lee
Sung-Hyun Lee
Wan-Ping Lee
Yu-Min Lee
Melinda Ler
Guo-Liang Li
Hang Li
XiaoWei Li
Yih-Lang Li
Guang-Wan Liao
Cheng-Min Lien
John Lillis
Chi-Hung Lin
Chung-Wei Lin
I-Jye Lin
I-Ting Lin
Jiun-Ren Lin
Rung-Bin Lin
Youn-Long Lin
Tsui-Yee Ling
Jin-Jia Liou
Bin Liu
Bo Liu
Chien-Nan Liu
Hung-Yi Liu
Meng-Chung Liu
Pu Liu
Hong-Wen Lu
Yung-Feng Lu
Danny W.S. Luk
Jianfeng Luo
Yuchun Ma
Patrick H. Madden
Amit Majumdar
Arthur W.K. Mak
Wai-Kei Mak
Usama Malik
Junichi Mano
Diana Marculescu

Dejan Markovic
Hiroo Masuda
Osamu Matsumoto
Hidetoshi Matsuoka
Satoshi Matsushita
Hiroki Matsutani
Yasuyuki Matsuya
Noel Menezes
Shin'ichi Minato
Natasa Miskov-Zivanov
Hiroshi Miyashita
Maher Mneimneh
Seijiro Moriyama
Tatsuji Moriyoshi
Rafael Kazumichi Morizawa
Hiroshi Murata
Kouhei Nadehara
Makoto Nagata
Masaki Nakanishi
Koichi Nakashiro
Tsuneo Nakata
Shigetoshi Nakatake
Shogo Nakaya
Takashi Nakayama
Myung-Jin Nam
Borivoje Nikolic
Kengo Nishino
Dongkeun Oh
Michiroh Ohmura
Kenichi Okada
Makiko Okumura
Mustafa Ozdal
Liang-Teck Pang
Sanjay Pant
Sri Parameswaran
Hong-June Park
In-Cheol Park
Priyadarsan Patra
Jorgen Peddersen
Nei-Chiung Perng
Frederic Petrot
Stephen Plaza
Jan Poland
Swarna Radhakrishnan
Roshan Ragel
Subramanian Rajagopalan
Frederic Rousseau
Jaijeet Roychowdhury
Saowanee Saewong

Karem Sakallah
Roy Sanghamitra
Sachin Sapatnekar
Takashi Sato
Prashant Saxena
Muzhou Shao
Seng Lin Shee
Farhana Sheikh
Hossein Sheini
Weiping Shi
Atsufumi Shibayama
Youngsoo Shin
Joseph Shinnerl
Youichi Shiraishi
Alan Su
Chau-Chin Su
Qing Su
Dong-kwan Suh
Pei-Lun Sui
Masaya Sumita
C. N. Sze
Kenton Sze
Akira Tada
Masamichi Takagi
Atsushi Takahashi
Yasuhiro Takashima
Koichiro Takayama
Yoshinori Takeuchi
Emil Talpes
Yutaka Tamiya
Sheldon Tan
Genichi Tanaka
Katsunori Tanaka
Song-Jian Tang
Hiroshi Tanimoto
Yuichi Tanji
Yoshinori Tomita
Hiroyuki Tomiyama
Masahiko Toyonaga
Ming-Chao Tsai
Ren-Song Tsay
Hsueh-Wen Tseng
Shuji Tsukiyama
Yojiro Uchimura
Prab Varma
Ilya Wagner
Shin'ichi Wakabayashi
Chun-Yao Wang
D. H. Wang

Farn Wang
Kuo-Hua Wang
Ting-Chi Wang
Xiaoyi Wang
Takayuki Watanabe
Jae-Kyung Wee
Xinjie Wei
Martin D. F. Wong
Y.C. Wong
Chin-Hsien Wu
David Wu
Min Xie
Jingyu Xu
Ryuichi Yamaguchi

Tatsuya Yamamoto
Shigeru Yamashita
Chuan-Yue Yang
Fu-Ching Yang
Yang Yang
Yulu Yang
Zijiang Yang
Akira Yasuda
Tse-Chen Yeh
Shoko Yonezawa
Jun-hee Yoo
Seung-Mok Yoo
Sungjoo Yoo
Atsushi Yoshikawa

Evangeline F.Y. Young
Ping-Hung Yuh
Nacer eddine Zergainoh
Lizheng Zhang
Yan Zhang
Zhengya Zhang
Xin Zhao
Hai Zhou
Yi Zhou
Haikun Zhu
Qiang Zhu
Radu Zlatanovici

Contents

ASP-DAC 2006 General Chair's Message	iii
Message from Technical Program Committee	iv
University LSI Design Contest	vi
Designers' Forum	vii
Keynote Addresses	viii
ASP-DAC 2005 Best Papers	xii
ASP-DAC 2006 Best Papers	xiii
ASP-DAC 2006 Organizing Committee	xv
ASP-DAC Steering Committee	xix
ASP-DAC 2006 Technical Program Committee	xxiii
University LSI Design Contest Committee	xxvi
Industry Liaison	xxvii
List of Reviewers	xxviii

CONTENTS

Session 1A
Formal Methods for Coverage and Scalable Verification

Chair(s): Kiyoharu Hamaguchi, Valeria Bertacco

1A-1 **Transition-Based Coverage Estimation for Symbolic Model Checking**
Xingwen Xu, Shinji Kimura, Kazunari Horikawa, Takehiko Tsuchiya 1

1A-2 **Word Level Functional Coverage Computation**
Bijan Alizadeh 7

1A-3 **Discovering the Input Assumptions in Specification Refinement Coverage**
Prasenjit Basu, Sayantan Das, Pallab Dasgupta, Partha P Chakrabarti 13

1A-4 **Refinement Strategies for Verification Methods Based on Datapath Abstraction**
Zaher Semon Andraus, Mark Hammond Liffiton, Karem Ahmad Sakallah 19

1A-5 **Generation of Shorter Sequences for High Resolution Error Diagnosis Using Sequential SAT**
Sung-Jui Pan, Kwang-Ting Cheng, John Moondanos, Ziyad Hanna 25

Session 1B
Interconnect for High-End SoC

Chair(s): Yoshinori Takeuchi, Juinn-Dar Huang

1B-1 **Constraint-Driven Bus Matrix Synthesis for MPSoC**
Sudeep Pasricha, Nikil Dutt, Mohamed Ben-Romdhane 30

1B-2 **Improving Routing Efficiency for Network-on-Chip through Contention-Aware Input Selection**
Dong Wu, Bashir M. Al-Hashimi, Marcus T. Schmitz 36

1B-3 **Physical Design Implementation of Segmented Buses to Reduce Communication Energy**
Jin Guo, Antonis Papanikolaou, Pol Marchal, Francky Catthoor 42

1B-4 **Co-Synthesis of a Configurable SoC Platform based on a Network on Chip Architecture**
Mário Pereira Véstias, Horácio Neto 48

1B-5 **Customized SIMD Unit Synthesis for System on Programmable Chip - A Foundation for HW/SW Partitioning with Vectorization**
Muhammad Omer Cheema, Omar Hammami 54

CONTENTS

Session 1C
Timing Analysis and Optimization

Chair(s): Ryuichi Yamaguchi, Atsushi Kurokawa

1C-1 **Robust Analytical Gate Delay Modeling for Low Voltage Circuits**
Anand Ramalingam, Sreekumar V. Kodakara, Anirudh Devgan, David Z. Pan 61

1C-2 **CGTA: Current Gain-based Timing Analysis for Logic Cells**
Shahin Nazarian, Massoud Pedram, Tao Lin, Emre Tuncer 67

1C-3 **Efficient Static Timing Analysis Using a Unified Framework for False Paths and Multi-Cycle Paths**
Shuo Zhou, Bo Yao, Hongyu Chen, Yi Zhu, Chung-Kuan Cheng, Mike Hutton . 73

1C-4 **Crosstalk Analysis using Reconvergence Correlation**
Sachin Shrivastava, Rajendra Pratap, Harindranath Parameswaran, Manuj Verma 79

1C-5 **Process-Induced Skew Reduction in Nominal Zero-Skew Clock Trees**
Matthew R. Guthaus, Dennis Sylvester, Richard B. Brown 84

Session 1D
University Design Contest

Chair(s): Kazutoshi Kobayashi, Takahiko Arakawa

1D-1 **A Low Dynamic Power and Low Leakage Power 90-nm CMOS Square-Root Circuit**
Tadayoshi Enomoto, Nobuaki Kobayashi 90

1D-2 **A High-Throughput Low-Power Fully Parallel 1024-bit 1/2-Rate Low Density Parity Check Code Decoder in 3-Dimensional Integrated Circuits**
Lili Zhou, Cherry Wakayama, Nuttorn Jangkrajarng, Bo Hu, Richard Shi 92

1D-3 **A 16-Bit, Low-Power Microsystem with Monolithic MEMS-LC Clocking**
Robert M. Senger, Eric D. Marsman, Michael S. McCorquodale, Richard B. Brown 94

1D-4 **Ultra-Low Voltage Power Management Circuit and Computation Methodology for Energy Harvesting Applications**
Chi-Ying Tsui, Hui Shao, Wing-Hung Ki, Feng Su 96

1D-5 **A 0.5-V Sigma-Delta Modulator Using Analog T-Switch Scheme for the Subthreshold Leakage Suppression**
Koichi Ishida, Atit Tamtrakarn, Takayasu Sakurai 98

1D-6 **An Implementation of a CMOS Down-Conversion Mixer for GSM1900 Receiver**
Fangqing Chu, Wei Li, Junyan Ren 100

xxxiii

CONTENTS

1D-7 Integrated Direct Output Current Control Switching Converter using Symmetrically-Matched Self-Biased Current Sensors
Yat-Hei Lam, Suet-Chui Koon, Wing-Hung Ki, Chi-Ying Tsui 102

1D-8 Adaptively-Biased Capacitor-Less CMOS Low Dropout Regulator with Direct Current Feedback
Yat-Hei Lam, Wing-Hung Ki, Chi-Ying Tsui 104

1D-9 A Built-in Power Supply Noise Probe for Digital LSIs
Mitsuya Fukazawa, Koichiro Noguchi, Makoto Nagata, Kazuo Taki 106

1D-10 A 476-gate-count Dynamic Optically Reconfigurable Gate Array VLSI chip in a standard 0.35um CMOS Technology
Minoru Watanabe, Fuminori Kobayashi 108

1D-11 Measurement Results of Within-Die Variations on a 90nm LUT Array for Speed and Yield Enhancement of Reconfigurable Devices
Kazuya Katsuki, Manabu Kotani, Kazutoshi Kobayashi, Hidetoshi Onodera .. 110

1D-12 High-Throughput Decoder for Low-Density Parity-Check Code
Tatsuyuki Ishikawa, Kazunori Shimizu, Takeshi Ikenaga, Satoshi Goto 112

1D-13 Hardware Implementation of Super Minimum All Digital FM Demodulator
Nursani Rahmatullah, Arif Nugroho 114

1D-14 Designing a Custom Architecture for DCT Using NISC Technology
Bita Gorjiara, Mehrdad Reshadi, Daniel Gajski 116

1D-15 A 52mW 1200MIPS Compact DSP for Multi-Core Media SoC
Shih-Hao Ou, Tay-Jyi Lin, Chao-Wei Huang, Yu-Ting Kuo, Chie-Min Chao, Chih-Wei Liu, Chein-Wei Jen 118

1D-16 Implementation of H.264/AVC Decoder for Mobile Video Applications
Suh Ho Lee, Ji Hwan Park, Seon Wook Kim, Sung Jea Ko, Suki Kim 120

1D-17 A High-Performance Platform-Based SoC for Information Security
Min Wu, Xiaoyang Zeng, Jun Han, Yongyi Wu, Yibo Fan 122

1D-18 Configurable Multi-Processor Architecture and its Processor Element Design
Tsutomu Nishimura, Takuji Miki, Hiroaki Sugiura, Yuki Matsumoto, Masatsugu Kobayashi, Toshiyuki Kato, Tsutomu Eda, Hironori Yamauchi 124

1D-19 Design and Implementation of Transducer for ARM-TMS Communication
Hansu Cho, Samar Abdi, Daniel Gajski 126

Session 2A
Software Techniques for Efficient SoC Design

Chair(s): Qiang Zhu, Ahmed Jerraya

2A-1 Energy Savings through Embedded Processing on Disk System
Seung Woo Son, Guangyu Chen, Mahmut Kandemir, Fehui Li 128

CONTENTS

2A-2 **Energy-Aware Computation Duplication for Improving Reliability in Embedded Chip Multiprocessors**
Guilin Chen, Mahmut Kandemir, Feihui Li 134

2A-3 **Object Duplication for Improving Reliability**
Guilin Chen, Guangyu Chen, Mahmut Kandemir, Narayanan Vijaykrishnan, Mary Jane Irwin 140

2A-4 **Mapping and Configuration Methods for Multi-Use-Case Networks on Chips**
Srinivasan Murali, Martijn Coenen, Andrei Radulescu, Kees Goossens, Giovanni De Micheli 146

2A-5 **Conversion of Reference C Code to Dataflow Model: H.264 Encoder Case Study**
Hyeyoung Hwang, Taewook Oh, Hyunuk Jung, Soonhoi Ha 152

Session 2B
Application Examples with Leading Edge Design Methodology

Chair(s): In-Cheol Park, Hideharu Amano

2B-1 **SAVS: A Self-Adaptive Variable Supply-Voltage Technique for Process - Tolerant and Power-Efficient Multi-issue Superscalar Processor Design**
Hai Li, Yiran Chen, Kaushik Roy, Cheng-Kok Koh 158

2B-2 **The Design and Implementation of a Low-Latency On-Chip Network**
Robert Mullins, Andrew West, Simon Moore 164

2B-3 **A Near Optimal Deblocking Filter for H.264 Advanced Video Coding**
Shen-Yu Shih, Cheng-Ru Chang, Youn-Long Lin 170

2B-4 **Image Segmentation and Pattern Matching Based FPGA/ASIC Implementation Architecture of Real-Time Object Tracking**
Kousuke Yamaoka, Takashi Morimoto, Hidekazu Adachi, Tetsushi Koide, Hans Juergen Mattausch 176

2B-5 **Prefetching-Aware Cache Line Turnoff for Saving Leakage Energy**
Ismail Kadayif, Mahmut Kandemir, Feihui Li 182

Session 2C
Placement

Chair(s): Evangeline F.Y. Young, Shin'ichi Wakabayashi

2C-1 **A Robust Detailed Placement for Mixed-Size IC Designs**
Jason Cong, Min Xie 188

2C-2 **FastPlace 2.0: An Efficient Analytical Placer for Mixed-Mode Designs**
Natarajan Viswanathan, Min Pan, Chris Chu 195

xxxv

CONTENTS

2C-3 Timing-Driven Placement Based on Monotone Cell Ordering Constraints
Chanseok Hwang, Massoud Pedram 201

2C-4 Constraint Driven I/O Planning and Placement for Chip-package Co-design
Jinjun Xiong, Yiu-Chung Wong, Egino Sarto, Lei He 207

2C-5 Simultaneous Block and I/O Buffer Floorplanning for Flip-Chip Design
Chih-Yang Peng, Wen-Chang Chao, Yao-Wen Chang, J.-H. Wang 213

Session 2D
Special Session: Electrothermal Design of Nanoscale Integrated Circuits

Chair(s): Dennis Sylvester, Mongkol Ekpanyapong

2D-1 Electrothermal Analysis and Optimization Techniques for Nanoscale Integrated Circuits
Yong Zhan, Brent Goplen, Sachin S. Sapatnekar 219

2D-2 Electrothermal Engineering in the Nanometer Era: From Devices and Interconnects to Circuits and Systems
Kaustav Banerjee, Sheng-Chih Lin, Navin Srivastava 223

2D-3 Area Optimization for Leakage Reduction and Thermal Stability in Nanometer Scale Technologies
Ja Chun Ku, Yehea Ismail 231

2D-4 Compact Thermal Models for Estimation of Temperature-dependent Power/Performance in FinFET Technology
Aditya Bansal, Mesut Meterelliyoz, Siddharth Singh, Jung Hwan Choi, Jayathi Murthy, Kaushik Roy 237

Session 3A
Logic Synthesis

Chair(s): Shinji Kimura, Shih-Chieh Chang

3A-1 An Anytime Symmetry Detection Algorithm for ROBDDs
Neil Kettle, Andy King 243

3A-2 High Level Equivalence Symmetric Input Identification
Ming-Hong Su, Chun-Yao Wang 249

3A-3 Fast Multi-Domain Clock Skew Scheduling for Peak Current Reduction
Shih-Hsu Huang, Chia-Ming Chang, Yow-Tyng Nieh 254

3A-4 Low Area Pipelined Circuits by Multi-clock Cycle Paths and Clock Scheduling
Bakhtiar Affendi Rosdi, Atsushi Takahashi 260

CONTENTS

3A-5 **A Transduction-based Framework to Synthesize RSFQ Circuits**
Shigeru Yamashita, Katsunori Tanaka, Hideyuki Takada, Koji Obata, Kazuyoshi Takagi 266

Session 3B
Future Technical Directions for Design Automation

Chair(s): Makoto Nagata, Ryuichi Fujimoto

3B-1 **Fast Simulation of Large Networks of Nanotechnological and Biochemical Oscillators for Investigating Self-Organization Phenomena**
Xiaolue Lai, Jaijeet Roychowdhury 273

3B-2 **Newton: A Library-Based Analytical Synthesis Tool for RF-MEMS Resonators**
Michael S. McCorquodale, James L. McCann, Richard B. Brown 279

3B-3 **Jitter Decomposition in Ring Oscillators**
Qingqi Dou, Jacob Abraham 285

3B-4 **A Fast Methodology for First-Time-Correct Design of PLLs Using Nonlinear Phase-Domain VCO Macromodels**
Prashant Goyal, Xiaolue Lai, Jaijeet Roychowdhury 291

3B-5 **Double Edge Triggered Feedback Flip-Flop in Sub 100nm Technology**
Seid Hadi Rasouli, Amir Amirabadi, Azam Seyedi, Ali Afzali-Kusha 297

Session 3C
Routing and Interconnect Optimization

Chair(s): Youichi Shiraishi, Tong Jing

3C-1 **Post-Routing Redundant Via Insertion for Yield/Reliability Improvement**
Kuang-Yao Lee, Ting-Chi Wang 303

3C-2 **Temperature-Aware Routing in 3D ICs**
Tianpei Zhang, Yong Zhan, Sachin S. Sapatnekar 309

3C-3 **Closed Form Solution for Optimal Buffer Sizing Using The Weierstrass Elliptic Function**
Sebastian Vogel, Martin D.F. Wong 315

3C-4 **An O(mn) Time Algorithm for Optimal Buffer Insertion of Nets with m Sinks**
Zhuo Robert Li, Weiping Shi 320

3C-5 **Spec-based Flip-Flop and Latch Repeater Planning**
Man Chung Hon 326

xxxvii

CONTENTS

Session 3D
Special Session: Flash Memory in Embedded Systems

Chair(s): Tohru Ishihara, Hiroyuki Tomiyama

3D-1 **Current Trends in Flash Memory Technology**
Sang Lyul Min, Eyee Hyun Nam 332

3D-2 **Configurability of Performance and Overheads in Flash Management**
Tei-Wei Kuo, Jen-Wei Hsieh, Li-Pin Chang, Yuan-Hao Chang 334

Session 4A
Resolving Timing Issues: Design and Test

Chair(s): Masaki Hashizume, Kazumi Hatayama

4A-1 **Delay Defect Screening for a 2.16GHz SPARC64 Microprocessor**
Noriyuki Ito, Akira Kanuma, Daisuke Maruyama, Hitoshi Yamanaka, Tsuyoshi Mochizuki, Osamu Sugawara, Chihiro Endoh, Masahiro Yanagida, Takeshi Kono, Yutaka Isoda, Kazunobu Adachi, Takahisa Hiraide, Shigeru Nagasawa, Yaroku Sugiyama, Eizo Ninoi 342

4A-2 **A Dynamic Test Compaction Procedure for High-quality Path Delay Testing**
Masayasu Fukunaga, Seiji Kajihara, Xiaoqing Wen, Toshiyuki Maeda, Shuji Hamada, Yasuo Sato 348

4A-3 **Delay Variation Tolerance for Domino Circuits**
Kai-Chiang Wu, Cheng-Tao Hsieh, Shih-Chieh Chang 354

4A-4 **Efficient Identification of Multi-Cycle False Path**
Kai Yang, Tim Cheng 360

4A-5 **IEEE Standard 1500 Compatible Interconnect Diagnosis for Delay and Crosstalk Faults**
Katherine Shu-Min Li, Yao-Wen Chang, Chauchin Su, Chung-Len Lee, Jwu E Chen 366

Session 4B
Leading Edge Design Methodology for SoCs and SiPs

Chair(s): Satoshi Matsushita

4B-1 **High-Level Architecture Exploration for MPEG4 Encoder with Custom Parameters**
Marius Bonaciu, Aimen Bouchhima, Wassim Youssef, Xi Chen, Wander Cesario, Ahmed Jerraya 372

4B-2 **Programmable Numerical Function Generators Based on Quadratic Approximation: Architecture and Synthesis Method**
Shinobu Nagayama, Tsutomu Sasao, Jon Butler 378

xxxviii

CONTENTS

4B-3 **An Automated Design Flow for 3D Microarchitecture Evaluation**
Jason Cong, Ashok Jagannathan, Yuchun Ma, Glenn Reinman, Jie Wei, Yan Zhang 384

4B-4 **Optimal Topology Exploration for Application-Specific 3D Architectures**
Ozcan Ozturk, Feng Wang, Mahmut Kandemir, Yuan Xie 390

4B-5 **Task Placement Heuristic Based on 3D-Adjacency and Look-Ahead in Reconfigurable Systems**
Jesus Tabero, Julio Septien, Hortensia Mecha, Daniel Mozos 396

Session 4C
Advanced Circuit Simulation

Chair(s): Hideki Asai, C.J. Richard Shi

4C-1 **A Quasi-Newton Preconditioned Newton-Krylov Method for Robust and Efficient Time-Domain Simulation of Integrated Circuits with Strong Parasitic Couplings**
Zhao Li, Richard Shi 402

4C-2 **An Efficient and Globally Convergent Homotopy Method for Finding DC Operating Points of Nonlinear Circuits**
Kiyotaka Yamamura, Wataru Kuroki 408

4C-3 **Optimization of Circuit Trajectories: An Auxiliary Network Approach**
Baohua Wang, Pinaki Mazumder 416

4C-4 **SASIMI: Sparsity-Aware Simulation of Interconnect-Dominated Circuits with Non-Linear Devices**
Jitesh Jain, Stephen F Cauley, Cheng-Kok Koh, Venkataramanan Balakrishnan 422

4C-5 **An Unconditional Stable General Operator Splitting Method for Transistor Level Transient Analysis**
Zhengyong Zhu, Rui Shi, Chung-Kuan Cheng, Ernest S. Kuh 428

Session 4D
Special Session: Open Access Overview

Chair(s): John Darringer

4D-1 **An Introduction to OpenAccess -An Open Source Data Model and API for IC Design-**
Michaela Guiney, Eric Leavitt 434

4D-2 **Open Access Overview "Industrial Experience"**
Yoshio Inoue 437

4D-3 **EDA Vendor Adoption**
Hillel Ofek 439

CONTENTS

4D-4 Utility of the OpenAccess Database in Academic Research
David Papa, Igor Markov, Philip Chong 440

Session 5A
Advances in Simulation Technologies

Chair(s): Shin'ichi Minato, Karem Sakallah

5A-1 Depth-Driven Verification of Simultaneous Interfaces
Ilya Wagner, Valeria Bertacco, Todd Austin 442

5A-2 FSM-Based Transaction-Level Functional Coverage for Interface Compliance Verification
Man-Yun Su, Che-Hua Shih, Juinn-Dar Huang, Jing-Yang Jou 448

5A-3 Hardware Debugging Method Based on Signal Transitions and Transactions
Nobuyuki Ohba, Kohji Takano 454

5A-4 Cycle Error Correction in Asynchronous Clock Modeling for Cycle-Based Simulation
Junghee Lee, Joonhwan Yi 460

5A-5 A Fast Logic Simulator Using a Look Up Table Cascade Emulator
Hiroki Nakahara, Tsutomu Sasao, Munehiro Matsuura 466

Session 5B
Scheduling for Embedded Systems

Chair(s): Sri Parameswaran, Sang Lyul Min

5B-1 Power-Aware Scheduling and Dynamic Voltage Setting for Tasks Running on a Hard Real-Time System
Peng Rong, Massoud Pedram 473

5B-2 Optimal TDMA Time Slot and Cycle Length Allocation for Hard Real-Time Systems
Ernesto Wandeler, Lothar Thiele 479

5B-3 POSIX modeling in SystemC
Hector Posadas, Jesus Adamez, Pablo Sanchez, Eugenio Villar, Francisco Blasco 485

5B-4 PARLGRAN: Parallelism Granularity Selection for Scheduling Task Chains on Dynamically Reconfigurable Architectures
Sudarshan Banerjee, Elaheh Bozorgzadeh, Nikil Dutt 491

5B-5 Memory Optimal Single Appearance Schedule with Dynamic Loop Count for Synchronous Dataflow Graphs
Hyunok Oh, Nikil Dutt, Soonhoi Ha 497

xl

CONTENTS

Session 5C
High Frequency Interconnect Effects in Nanometer Technology

Chair(s): Charlie Chung-Ping Chen, Noel Menezes

5C-1 **Wire Sizing with Scattering Effect for Nanoscale Interconnection**
Sean X. Shi, David Z. Pan 503

5C-2 **Adaptive Admittance-based Conductor Meshing for Interconnect Analysis**
Ya-Chi Yang, Cheng-Kok Koh, Venkataramanan Balakrishnan 509

5C-3 **Interconnect RL Extraction at a Single Representative Frequency**
Akira Tsuchiya, Masanori Hashimoto, Hidetoshi Onodera 515

5C-4 **An Efficient Algorithm for 3-D Reluctance Extraction Considering High Frequency Effect**
Mengsheng Zhang, Wenjian Yu, Yu Du, Zeyi Wang 521

5C-5 **Macromodelling Oscillators Using Krylov-Subspace Methods**
Xiaolue Lai, Jaijeet Roychowdhury 527

Session 5D
Designers' Forum: Low Power Design

Chair(s): Haruyuki Tago, Makoto Ikeda

5D-1 **Low-Power Design Methodology for Module-wise Dynamic Voltage and Frequency Scaling with Dynamic De-skewing Systems**
Takeshi Kitahara, Hiroyuki Hara, Shinichiro Shiratake, Yoshiki Tsukiboshi, Tomoyuki Yoda, Tetsuaki Utsumi, Fumihiro Minami 533

5D-2 **Single-Chip Multi-Processor Integrating Quadruple 8-Way VLIW Processors with Interface Timing Analysis Considering Power Supply Noise**
Satoshi Imai, Atsuki Inoue, Motoaki Matsumura, Kenichi Kawasaki, Atsuhiro Suga 541

5D-3 **A System-level Power-estimation Methodology based on IP-level Modeling, Power-level Adjustment, and Power Accumulation**
Masafumi Onouchi, Tetsuya Yamada, Kimihiro Morikawa, Isamu Mochizuki, Hidetoshi Sekine 547

5D-4 **PowerViP: SoC Power Estimation Framework at Transaction Level**
Ikhwan Lee, Hyunsuk Kim, Peng Yang, Sungjoo Yoo, Eui-Young Chung, Kyu-Myung Choi, Jeong-Taek Kong, Soo-Kwan Eo 551

xli

CONTENTS

Session 6A
Power Optimization of Large-Scale Circuits

Chair(s): Sheldon Tan, David Z. Pan

6A-1 **Mathematically Assisted Adaptive Body Bias (ABB) for Temperature Compensation in Gigascale LSI Systems**
Sanjay V Kumar, Chris H Kim, Sachin S Sapatnekar 559

6A-2 **Analysis and Optimization of Gate Leakage Current of Power Gating Circuits**
Hyung-Ock Kim, Youngsoo Shin 565

6A-3 **Delay Modeling and Static Timing Analysis for MTCMOS Circuits**
Naoaki Ohkubo, Kimiyoshi Usami 570

6A-4 **Switching-Activity Driven Gate Sizing and Vth Assignment for Low Power Design**
Yu-Hui Huang, Po-Yuan Chen, TingTing Hwang 576

6A-5 **Power Driven Placement with Layout Aware Supply Voltage Assignment for Voltage Island Generation in Dual-Vdd Designs**
Bin Liu, Yici Cai, Qiang Zhou, Xianlong Hong 582

Session 6B
Advanced Memory and Processor Architectures for MPSoC

Chair(s): Soonhoi Ha, Youn-Long Lin

6B-1 **Reusable Component IP Design using Refinement-based Design Environment**
Sanggyu Park, Sang-Yong Yoon, Soo-Ik Chae 588

6B-2 **An Interface-Circuit Synthesis Method with Configurable Processor Core in IP-Based SoC Designs**
Shunitsu Kohara, Naoki Tomono, Jumpei Uchida, Yuichiro Miyaoka, Nozomu Togawa, Masao Yanagisawa, Tatsuo Ohtsuki 594

6B-3 **A Real-Time and Bandwidth Guaranteed Arbitration Algorithm for SoC Bus Communication**
Chien-Hua Chen, Geeng-Wei Lee, Juinn-Dar Huang, Jing-Yang Jou 600

6B-4 **Hierarchical Memory Size Estimation for Loop Fusion and Loop Shifting in Data-Dominated Applications**
Qubo Hu, Arnout Vandecappelle, Martin Palkovic, Per Gunnar Kjeldsberg, Erik Brockmeyer, Francky Catthoor 606

6B-5 **A Novel Instruction Scratchpad Memory Optimization Method based on Concomitance Metric**
Andhi Janapsatya, Aleksandar Ignjatovic, Sri Parameswaran 612

xlii

CONTENTS

Session 6C
New Routing Techniques

Chair(s): Ting-Chi Wang, Vijay Pitchumani

6C-1 **DraXRouter: Global Routing in X-Architecture with Dynamic Resource Assignment**
Zhen Cao, Tong Jing, Yu Hu, Yiyu Shi, Xianlong Hong, Xiaodong Hu, Guiying Yan 618

6C-2 **Diagonal Routing in High Performance Microprocessor Design**
Noriyuki Ito, Hideaki Katagiri, Ryoichi Yamashita, Hiroshi Ikeda, Hiroyuki Sugiyama, Hiroaki Komatsu, Yoshiyasu Tanamura, Akihiko Yoshitake, Kazuhiro Nonomura, Kinya Ishizaka, Hiroaki Adachi, Yutaka Mori, Yutaka Isoda, Yaroku Sugiyama 624

6C-3 **CDCTree: Novel Obstacle-Avoiding Routing Tree Construction based on Current Driven Circuit Model**
Yiyu Shi, Tong Jing, Lei He, Zhe Feng, Xianlong Hong 630

6C-4 **A Novel Framework for Multilevel Full-Chip Gridless Routing**
Tai-Chen Chen, Yao-Wen Chang, Shyh-Chang Lin 636

6C-5 **Monotonic Parallel and Orthogonal Routing for Single-Layer Ball Grid Array Packages**
Yoichi Tomioka, Atsushi Takahashi 642

Session 7A
Minimization of Test Cost and Power

Chair(s): Seiji Kajihara, Satoshi Ohtake

7A-1 **A Routability Constrained Scan Chain Ordering Technique for Test Power Reduction**
Xuan-Lun Huang, Jiun-Lang Huang 648

7A-2 **FCSCAN: An Efficient Multiscan-based Test Compression Technique for Test Cost Reduction**
Youhua Shi, Nozomu Togawa, Shinji Kimura, Masao Yanagisawa, Tatsuo Ohtsuki 653

7A-3 **Compaction of Pass/Fail-based Diagnostic Test Vectors for Combinational and Sequential Circuits**
Yoshinobu Higami, Kewal K. Saluja, Hiroshi Takahashi, Shin-ya Kobayashi, Yuzo Takamatsu 659

7A-4 **Low-Overhead Design of Soft-Error-Tolerant Scan Flip-Flops with Enhanced-Scan Capability**
Ashish Goel, Swarup Bhunia, Hamid Mahmoodi, Kaushik Roy 665

xliii

CONTENTS

7A-5 **A Memory Grouping Method for Sharing Memory BIST Logic**

Masahide Miyazaki, Tomokazu Yoneda, Hideo Fujiwara 671

Session 7B
Substrate Coupling and Analog Synthesis

Chair(s): Jaijeet Roychowdhury, Tomohisa Kimura

7B-1 **Equivalent Circuit Modeling of Guard Ring Structures for Evaluation of Substrate Crosstalk Isolation**

Daisuke Kosaka, Makoto Nagata 677

7B-2 **A New Boundary Element Method for Accurate Modeling of Lossy Substrates with Arbitrary Doping Profiles**

Xiren Wang, Wenjian Yu, Zeyi Wang 683

7B-3 **Parasitics Extraction Involving 3-D Conductors based on Multi-layered Green's Function**

Zuochang Ye, Zhiping Yu 689

7B-4 **Signal-Path Driven Partition and Placement for Analog Circuit**

Di Long, Xianlong Hong, Sheqin Dong 694

7B-5 **An Approach to Topology Synthesis of Analog Circuits Using Hierarchical Blocks and Symbolic Analysis**

Xiaoying Wang, Lars Hedrich 700

Session 7C
Statistical and Yield Analysis

Chair(s): Hiroo Masuda, Seijiro Moriyama

7C-1 **Statistical Corner Conditions of Interconnect Delay (Corner LPE Specifications)**

Kenta Yamada, Noriaki Oda 706

7C-2 **Speed Binning Aware Design Methodology to Improve Profit under Parameter Variations**

Animesh Datta, Swarup Bhunia, Jung Hwan Choi, Saibal Mukhopadhyay, Kaushik Roy 712

7C-3 **Yield-Area Optimizations of Digital Circuits Using Non-dominated Sorting Genetic Algorithm (YOGA)**

Vineet Agarwal, Janet Wang 718

7C-4 **A Probabilistic Analysis of Pipelined Global Interconnect Under Process Variations**

Navneeth Kankani, Vineet Agarwal, Janet M Wang 724

7C-5 **Yield-Preferred Via Insertion Based on Novel Geotopological Technology**

Fangyi Luo, Yongbo Jia, Wayne Wei-Ming Dai 730

xliv

CONTENTS

Session 7D
Special Session: H.264/AVC Design Challenges and Solutions

Chair(s): Wayne Wolf

7D-1 **Introduction to H.264 Advanced Video Coding**
Jian-Wen Chen, Chao-Yang Kao, Youn-Long Lin 736

7D-2 **Algorithms and DSP Implementation of H.264/AVC**
Hung-Chih Lin, Yu-Jen Wang, Kai-Ting Cheng, Shang-Yu Yeh, Wei-Nien Chen,
Chia-Yang Tsai, Tian-Sheuan Chang, Hsueh-Ming Hang 742

7D-3 **Hardware Architecture Design of an H.264/AVC Video Codec**
Tung-Chien Chen, Chung-Jr Lian, Liang-Gee Chen 750

7D-4 **ASIP Approach for Implementation of H.264/AVC**
Sung Dae Kim, Jeong Hoo Lee, Chung Jin Hyun, Myung Hoon Sunwoo 758

Session 8A
Floorplanning

Chair(s): Yao-Wen Chang, Shigetoshi Nakatake

8A-1 **Fast Substrate Noise-Aware Floorplanning with Preference Directed Graph**
for Mixed-Signal SOCs
Minsik Cho, Hongjoong Shin, David Z. Pan 765

8A-2 **A Fixed-die Floorplanning Algorithm Using an Analytical Approach**
Yong Zhan, Yan Feng, Sachin S. Sapatnekar 771

8A-3 **A Multi-Technology-Process Reticle Floorplanner and Wafer Dicing Plan-**
ner for Multi-Project Wafers
Chien-Chang Chen, Wai-Kei Mak 777

8A-4 **Design Space Exploration for Minimizing Multi-Project Wafer Production**
Cost
Rung-Bin Lin, Meng-Chiou Wu, Wei-Chiu Tseng, Ming-Hsine Kuo, Tsai-Ying
Lin, Shr-Cheng Tsai 783

8A-5 **SAT-Based Optimal Hypergraph Partitioning with Replication**
Michael G. Wrighton, Andre M. DeHon 789

Session 8B
Memory Optimization for Embedded Systems

Chair(s): Hiroyuki Tomiyama, Preeti Ranjan Panda

8B-1 **Finding Optimal L1 Cache Configuration for Embedded Systems**
Andhi Janapsatya, Aleksandar Ignjatovic, Sri Parameswaran 796

CONTENTS

8B-2   Memory Size Computation for Multimedia Processing Applications
Hongwei Zhu, Ilie I. Luican, Florin Balasa 802

8B-3   Maximizing Data Reuse for Minimizing Memory Space Requirements and Execution Cycles
Mahmut Kandemir, Guangyu Chen, Feihui Li 808

8B-4   Compiler-Guided Data Compression for Reducing Memory Consumption of Embedded Applications
Ozcan Ozturk, Guangyu Chen, Mahmut Kandemir, Ibrahim Kolcu 814

8B-5   Analysis of Scratch-Pad and Data-Cache Performance Using Statistical Methods
Javed Absar, Francky Catthoor 820

Session 8C
Inductive Issues in Power Grids and Packages

Chair(s): Takashi Sato, Yehea Ismail

8C-1   Efficient Early Stage Resonance Estimation Techniques for C4 Package
Jin Shi, Yici Cai, Shelton X-D Tan, Xianlong Hong 826

8C-2   Parallel-Distributed Time-Domain Circuit Simulation of Power Distribution Networks with Frequency-Dependent Parameters
Takayuki Watanabe, Yuichi Tanji, Hidemasa Kubota, Hideki Asai 832

8C-3   Power Distribution Techniques for Dual VDD Circuits
Sarvesh Hemchandra Kulkarni, Dennis Sylvester 838

8C-4   Calculating Frequency-Dependent Inductance of VLSI Interconnect by Complete Multiple Reciprocity Boundary Element Method
Changhao Yan, Wenjian Yu, Zeyi Wang 844

8C-5   Controlling Inductive Cross-talk and Power in Off-chip Buses using CODECs
Brock LaMeres, Kanupriya Gulati, Sunil Khatri 850

Session 8D
Designers' Forum: "Cell" Processor

Chair(s): Haruyuki Tago, Makoto Ikeda

8D-1   A New Test and Characterization Scheme for 10+ GHz Low Jitter Wide Band PLL
Kazuhiko Miki, David Boerstler, Eskinder Hailu, Jieming Qi, Sarah Pettengill, Yuichi Goto 856

8D-2   An SPU Reference Model for Simulation, Random Test Generation and Verification
Yukio Watanabe, Balazs Sallay, Brad Michael, Daniel Brokenshire, Gavin Meil, Hazim Shafi, Daisuke Hiraoka 860

xlvi

CONTENTS

8D-3 **A Cycle Accurate Power Estimation Tool**
Rajat Chaudhry, Daniel Stasiak, Stephen Posluszny, Sang Dhong 867

8D-4 **Key Features of the Design Methodology Enabling a Multi-Core SoC Implementation of a First-Generation CELL Processor**
Dac Pham, Hans-Werner Anderson, Erwin Behnen, Mark Bolliger, Sanjay Gupta, Peter Hofstee, Paul Harvey, Charles Johns, Jim Kahle, Atsushi Kameyama, John Keaty, Bob Le, Sang Lee, Tuyen Nguyen, John Petrovick, Mydung Pham, Juergen Pille, Stephen Posluszny, Mack Riley, Joseph Verock, James Warnock, Steve Weitzel, Dieter Wendel 871

Session 9A
High-Level Synthesis

Chair(s): Shigeru Yamashita, Youngsoo Shin

9A-1 **TAPHS: Thermal-Aware Unified Physical-Level and High-Level Synthesis**
Zhenyu (Peter) Gu, Yonghong Yang, Jia Wang, Robert P. Dick, Li Shang 879

9A-2 **An Automated, Efficient and Static Bit-width Optimization Methodology Towards Maximum Bit-width-to-Error Tradeoff With Affine Arithmetic Model**
Yu Pu, Yajun Ha 886

9A-3 **Abridged Addressing: A Low Power Memory Addressing Strategy**
Preeti Ranjan Panda 892

9A-4 **Using Speculative Computation and Parallelizing Techniques to Improve Scheduling of Control based Designs**
Roberto Cordone, Fabrizio Ferrandi, Gianluca Palermo, Marco Domenico Santambrogio, Donatella Sciuto 898

9A-5 **Worst Case Execution Time Analysis for Synthesized Hardware**
Jun-hee Yoo, Xingguang Feng, Kiyoung Choi, Eui-Young Chung, Kyu-Myung Choi 905

Session 9B
Modeling, Compilation and Optimization of Embedded Architectures

Chair(s): Hiroyuki Tomiyama, Lovic Gauthier

9B-1 **Workload Prediction and Dynamic Voltage Scaling for MPEG Decoding**
Ying Tan, Parth Malani, Qinru Qiu, Qing Wu 911

9B-2 **Lazy BTB: Reduce BTB Energy Consumption Using Dynamic Profiling**
Yen-Jen Chang 917

9B-3 **Cache Size Selection for Performance, Energy and Reliability of Time-Constrained Systems**
Yuan Cai, Marcus T. Schmitz, Alireza Ejlali, Bashir M. Al-Hashimi, Sudhakar M. Reddy 923

xlvii

CONTENTS

9B-4 **Reducing Dynamic Compilation Overhead by Overlapping Compilation and Execution**
Priya Unnikrishnan, Mahmut Kandemir, Feihui Li 929

9B-5 **Functional Modeling Techniques for Efficient Sw Code Generation of Video Codec Application**
Sang-Il Han, Soo-Ik Chae, Ahmed Amine Jerraya 935

Session 9C
Statistical Design

Chair(s): Sachin Sapatnekar, Sunil Khatri

9C-1 **Convergence-Provable Statistical Timing Analysis with Level-Sensitive Latches and Feedback Loops**
Lizheng Zhang, Jengliang Tsai, Weijen Chen, Yuhen Hu, Charlie Chungping Chen 941

9C-2 **Parameterized Block-Based Non-Gaussian Statistical Gate Timing Analysis**
Soroush Abbaspour, Hanif Fatemi, Massoud Pedram 947

9C-3 **Statistical Leakage Minimization through Joint Selection of Gate Sizes, Gate Lengths and Threshold Voltage**
Sarvesh Bhardwaj, Yu Cao, Sarma Vrudhula 953

9C-4 **Statistical Bellman-Ford Algorithm With An Application to Retiming**
Mongkol Ekpanyapong, Thaisiri Watewai, Sung Kyu Lim 959

9C-5 **An Exact Algorithm for the Statistical Shortest Path Problem**
Liang Deng, Martin D. F. Wong 965

Author Index A-1

POSIX modeling in SystemC

Hector Posadas, Jesús Ádamez, Pablo Sánchez, Eugenio Villar
Microelectronics Engineering Group, University of Cantabria, Santander, Spain
www.teisa.unican.es/gim

Francisco Blasco
DS2, Valencia, Spain

Abstract - Early estimation of the execution time of Real-Time embedded SW is an essential task in complex, HW/SW embedded system design. Application SW execution time estimation requires taking into account the impact of the underlying RTOS. As a consequence, RTOS modeling is becoming an active research area. SystemC provides a framework for multiprocessing, HW/SW co-simulation at several abstraction levels. In this paper, a SystemC library for POSIX modeling and simulation is presented. By using the library, the SystemC specification using POSIX functions is converted automatically into a timed simulation estimating the execution time of the application SW running on the POSIX platform. The library works directly on the source code. Therefore, it provides an early and fast estimation of the performance of the system as a consequence of the architectural mapping decisions. Although accuracy is lower than when using lower-level techniques, it supports high-level design-space exploration as simulation time is significantly less than RT (ISS) simulation[1].

I. Introduction

Cost of design has been identified as the greatest threat to the continuation of microelectronic technology improvement towards larger integration scales. Among the different costs, embedded software development represents the major part of the SoC development cost [1]. In this context, there is a clear need for new methodologies supporting efficient design of Real-Time, Embedded (RT/E) systems on complex platforms [2].

Performance analysis is an increasingly important and challenging task in embedded system design since performance parameters (time, size, consumption, cost, etc.) can be as important as functional requirements [3]. Complex, HW/SW embedded systems demand accurate estimation of their timing characteristics before implementation. Such system timing analysis requires system modeling taking into account the close interaction between the application SW and hardware-dependent SW running on the different processors and the application-specific HW through the platform communication resources [2-3].

Time execution estimation has been a traditional problem in real-time embedded SW engineering [4]. Execution time figures are necessary to develop timed SW simulation models [5].

Accurate SW simulation requires taking into account the effect of the RTOS [6-12]. In order to be efficient and, therefore, applicable to complex systems, the SW code can be directly executed using an abstract model of the RTOS. System specification languages like SpecC [6-7] or SystemC [10][12-13] have proven to provide a useful HW/SW co-simulation platform.

SystemC can be used to effectively create accurate models of complex, HW/SW embedded systems. SystemC allows the hardware and software design team to develop an executable specification of the system which can be used to quickly simulate and explore various algorithms, and validate and optimize the design [14]. SystemC supports efficient generation of the embedded software including interface drivers and the RTOS [15]. Nevertheless, certain RTOS characteristics such as priority-based preemption are very difficult to be adequately modeled in SystemC.

In this paper, a high-level, POSIX simulation library in SystemC is presented. The library allows the designer a fast, sufficiently accurate, timed simulation of the application SW running on top of POSIX [16]. As most current RTOSs support this standard, the library is portable to different development frameworks. Moreover, SystemC provides a flexible infrastructure for multiprocessing, HW/SW co-simulation at different abstraction levels. As a consequence, the POSIX simulation library can be used in any SystemC HW/SW co-simulation environment.

The structure of the paper is the following. In the next section, the contribution of the paper is described in the context of related work. In Section 3, the POSIX modeling and simulation methodology is presented. Experimental results are provided in Section 4. Finally, the main conclusions of the work are presented.

II. Related work

Traditionally, simulation capability has not been a main criterion in selecting a RTOS [17]. Nevertheless, most RTOS vendors provide a simulator of their OS to help software designers to develop and emulate application code on the host before having developed the actual hardware prototypes [18]. These simulators usually take the application code, compile it with a native compiler, link it with the OS Kernel libraries and produce a host executable. The simulator is completed with the libraries of the additional RTOS modules (TCP/IP stack, file system, link handler, etc.) in the simulated

---

[1]This work has been partially supported by the ITEA IP 03002 Medea project and the TIC-2002-00660 project.

environment. Simulation cannot be timing accurate because it relies upon the host operating system scheduling. Although it can guarantee the order of the events, the simulation is untimed. Very few commercial RTOS simulators are timed simulators [19]. They are based on proprietary languages and do not support HW/SW co-simulation.

Architectural exploration of complex, HW/SW embedded systems require accurate profiling of their timing characteristics. Precise co-simulation at the RT level is unfeasible for system-level profiling due to its excessive host execution time. SW simulation using cycle-accurate Instruction-Set Simulators (ISS) only alleviates the problem. Moreover, it requires setting up a co-simulation infrastructure [9]. Faster co-simulation can be achieved by directly executing the code in a SW-HDL co-simulation environment [20]. With this approach, the SW execution time is very difficult to model. A rough approximation of the SW execution time can be achieved using the native system clock [11].

Fast, sufficiently accurate, HW/SW co-simulation requires adequate modeling of the RTOS in a system-level language like SpecC [6-7] or SystemC [10][12]. The RTOS model is abstract, based on APIs including the most common RTOS functions [7], wrappers [8] or an implementation-specific model of each channel [10]. An advantage of these techniques, in addition to the improvement in simulation time, is that they can be applied early in the design process avoiding costly design iterations. As these techniques use an abstract RTOS model, they cannot simulate embedded code using actual functions of any specific RTOS. As a consequence, the simulation methodology does not support refinement of the original specification apart from the inclusion of the RTOS modeling itself [7]. Moreover, the application code has to be modified with specific RTOS modeling functions [7-8]. Only by direct modeling of a standard RTOS it is possible to use the same code for both simulation and implementation [11-12].

Execution time of both the application code and the RTOS is taken into account during simulation by introducing 'wait' statements [5-12]. More the number of 'wait' statements introduced, higher the accuracy. Nevertheless, the 'wait' statements are introduced at certain static, predetermined points. As a consequence, low-level, dynamic timing characteristics of the RTOS like time-slicing, preemptive scheduling, interrupts and exceptions are very difficult to model.

A SystemC library called PERFidy was developed for system-level, timed simulation and performance analysis [10]. The library is able to obtain timing cost estimations from a set of values that characterize the target platform in which the embedded systems will be implemented. This is made possible by redefining all the C++ operators with new ones with the same functionality but estimating the corresponding execution time on the chosen platform resource. This detailed estimation of the SW execution time allows deciding dynamically the place of the application code where the task (process or thread) has to be preempted. As a consequence, the low-level, dynamic characteristics of the RTOS can be efficiently modeled.

In this paper, an extension of the timed-simulation methodology used in PERFidy is proposed in order to allow the timed-simulation of RTOS functions in SystemC. Although the SystemC 2.0 simulation kernel executes processes following a non-preemptive scheduling policy without priorities, the proposed simulation methodology is able to model preemption as well as different scheduling policies based on priorities. In order to ensure independence from any specific RTOS, the simulation methodology supports POSIX, thus ensuring wide portability. In this way, the proposed extension, called PERFidiX supports the early performance analysis of the system specification as well as the impact of the decisions taken during design refinement. Using SystemC ensures a flexible, portable framework for multiprocessing, HW/SW co-simulation.

III. POSIX modeling in SystemC

A. Analysis of the POSIX standard

The POSIX standard covers a wide range of system facilities. As the scope of this work is embedded systems, the focus of interest of the paper is the Real-Time extension. Concurrency is supported both by processes and threads. Three standard scheduling policies are defined: FiFo (FF), Round Robin (RR) and Sporadic Server (SS). Furthermore, other policies can be offered but they are implementation specific. The standard also defines several mechanisms for communication and synchronization. Apart from shared variables, they are mutexes, conditional variables, message queues, semaphores, sockets, streams, signals, etc. Timing facilities are based on timers and real-time clocks. Traces are supported for system analysis and verification. Additionally, file systems and memory management are also included.

Real-time requirements define functions to support source portability of applications. The presence of many of these functions is dependent on support for implementation options described in the standard. The specific functional areas included in this section include the following:

- Priority Scheduling
- Semaphores
- Timers
- Real-Time Signal Extension
- Synchronized Input and Output
- Inter-process Communication
- Process Memory Locking
- Memory Mapped Files and Shared Memory Objects

The first six functional areas require an appropriate modeling in order to ensure a correct timed and functional simulation of the code. As the simulation runs over one process of the host OS, the modeling of the last two features using the host memory is not straightforward. The influence of these facilities in the execution time will be included in the simulation, but the development of SystemC functions that implement this functionality is currently under study.

The model assumes the correct memory access of the code.

B. POSIX simulation library

The correct simulation of all the POSIX functions described above requires extending the SystemC simulation kernel with additional functionality not covered by PERFidy. This modification does not affect the original SystemC simulation kernel. This means that the standard kernel has not been modified; the required functionality is emulated over the original one by using its facilities. This increases the portability of the library, as it does not require the installation of a completely new version of SystemC; the standard one can be used. Moreover, the library is external, so the use of future SystemC versions will produce no problems since compatibility with previous versions will be ensured.

The PERFidiX library including all facilities described above contains three different parts. The first one contains the execution support that is provided over the original SystemC kernel. It implements process and thread management, including scheduling capabilities, and timing facilities. The second part implements the POSIX API by using the facilities provided by the new execution support kernel. The last part carries out the performance analysis task by using the timing and traces facilities, and reusing the technology of the original PERFidy library [10].

C. Concurrency modeling

Parallelism is modeled by using the SC_THREAD process of SystemC. Therefore, both POSIX processes and threads are modeled in the same way. Thus, the library implements the required actions that give each element its own characteristics. The characteristics of processes and threads are loaded in a list when they are created and these parameters can be modified during simulation using the methods the POSIX standard defines. However, modeling the capabilities derived by the use of separate memory spaces in SystemC is not straightforward[2].

SystemC does not allow dynamic thread creation. In order to support dynamic thread creation, a thread-pool is initialized with the simulation. This pool has a predefined number of SC_THREADS (the number can be modified in the source code) starting in a blocked state. Each time a new thread is declared, a thread in the pool is resumed. This means that there are a maximum number of dynamic threads that can be executed at the same time.

D. Modeling the scheduler

The SystemC underlying kernel activates in each δ-cycle all the threads that are not blocked without any considerations about priorities. However, the scheduler model running over the SystemC kernel has to ensure that only one thread is executed in each processor each time. To solve the problem it is necessary to ensure that all threads remain blocked, except the one with the greatest priority, which is awoken. It is also necessary to manage preemption and priorities. As many schedulers are modeled as processors are defined, and each process is assigned only to one scheduler.

Following the POSIX standard, there is one thread list for each priority. A runnable thread is placed on the thread list for that thread's priority when it is ready to execute. When a thread reaches a blocking point or its time slot is consumed, it releases the processor, and in the second case, it is added to the corresponding list. Then, the first thread of the first non-empty list is deleted from there and resumed.

However this approach does not model preemption as a SC_THREAD is executed until a wait statement is reached. POSIX defines that a thread must be preempted when a thread with a higher priority is awoken. In order to model preemption adequately, a new list is instantiated. This list is similar to the previous one, but it does not contain the threads ready to execute, but the threads that are under execution. As a consequence, the execution state has two parts. The first one is the functional execution, and the second one is the temporal execution. That is, the code is executed in zero time (in the simulation) and then the thread is slept to take up the corresponding time in the processor. This placement in time is produced before a data exchange is made. If during the time the thread is slept, another thread with higher priority is awoken, it is executed, and it indicates to the other thread the time it has been preempted. After checking if preemption has taken place, and it is slept again during this time. As a result, when the data exchange is made, the execution state is correctly placed in time. This implementation needs another list including the processes in the execution state, because preemptions can be chained. Considering that global variables and signals are generated only in the same processor where they are delivered, it is possible to support preemption and delivering of signals just by placing in time the thread execution.

As commented above, only software timers, channel accesses (e.g. a mutex unlock) or signals could resume threads. Moreover, only a thread, a timer or an external interruption can generate a signal. Then, the time when the events will be produced is known except for external interruptions or channel accesses. The events generated by the software are known in advance. Thus, the way to model preemption is to place in time not only channel accesses but also these temporal points, and then resume the adequate thread. This means that all accesses to global variables can be done in the right order and no errors are made.

Correct timed simulation of the SW code including the RTOS requires stopping the running thread each tick-time interval. In this way, any expected or external event can be taken into account. In Figure 1, an example is used to show the result when using the proposed solution. The system is composed of three threads:

Thread 1 presents the lowest priority and a Round Robin policy with a time limit of 20µs. The thread is ready in T=0.

Thread 2 has a medium priority and a FIFO policy. The

---

[2]This feature should be taken into account in embedded processors with MMU. Certain specific functions are still under study.

thread is blocked with a timeout that expires at T=30μs.

Thread 3 has the highest priority and FIFO scheduling policy. It is blocked awaiting a hardware interruption (a POSIX signal).

Although PERFidiX takes into account the execution time of the OS, it will not be considered here for the sake of simplicity. At T=0 only thread 1 is ready and it is moved to execution state. Two time events have to be considered, the time limit (T=20μs) and the timeout expiration (T=30μs). Thus, the thread can be executed for 20μs (or until another event is detected such as a channel access, a new timer, etc.). Then, the thread code is executed until PERFidiX estimates that the executed code will take 20μs in the target platform. This is done in zero time in the SystemC simulation. To make the simulation take into account the estimated execution time, the thread is slept until T=20μs and then the process is moved to the ready state and the scheduler is awoken. In this way, the actual behavior of the processor is closely modeled.

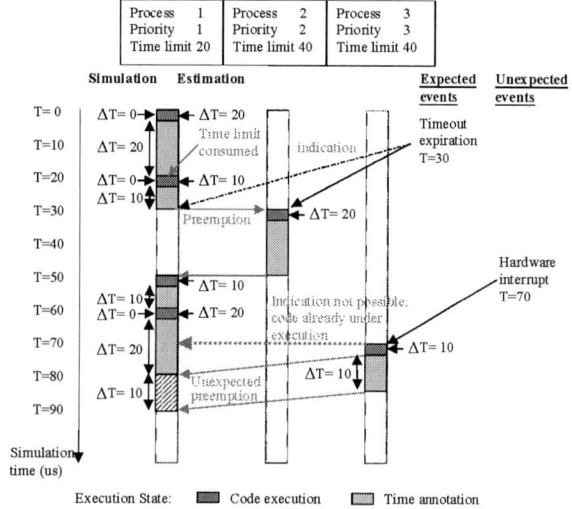

Figure 1: Preemption modeling in PERFidiX.

At T=20μs only thread 1 is ready so it is executed. However, this time it can be executed during 10μs because at T=30μs the timeout expires. Then the same procedure described previously is followed. At T=30μs, thread 1 has used 10μs of the time interval and another 10μs remains. However, thread 2 is ready and has a higher priority, so thread 1 has to be preempted, and thread 2 resumed. Thread 2 execution takes 20μs until it is slept (no events are expected) at T=50μs. At that time, thread 1 is resumed and it finishes the 10 μs of the time limit. At T=60μs it is moved to the ready state and the scheduler is awoken. However, as there are no other threads in this state, it is resumed with a new time limit, until T=80μs. Nevertheless, at T=70μs a hardware interruption is produced and thread 3 is ready with highest priority, so thread 1 should be preempted. Thread 1 has scheduled its finishing event at T=80μs while, as a consequence of the execution of thread 3, its actual finishing time is T=90μs. To correctly model this behavior, thread 3 is

executed at T=80μs with a finishing event scheduled at T=90μs and then, thread 1 variable where preemption is indicated, is incremented in 10μs. As a consequence, at T=80μs, thread 1 is slept again during the time the preemption variable indicates and it is set to 0. Results (as well as the implementation) are deterministic if no global variables are used from T=70μs to T=90μs in both thread 1 and thread 3. As explained above, threads that are awoken by unexpected signals should not make use of global variables without including additional synchronization mechanisms.

As explained before, this entire model does not modify the SystemC kernel. It is based on the use of "wait()" and "notify" SystemC primitives. The scheduler also provides the functions to model the blocks produced by communication and synchronization POSIX facilities.

F. Modeling Signals

Once the scheduler is implemented, signals can be modeled as defined by the POSIX standard. The signal manager can access the scheduler to allow all blocking communications to implement signals that mean that a thread can be stopped or unblocked independently of the cause that produced this blockage. There is a SystemC thread used only by the signal manager that will execute the actions related to signals that have to be delivered, since no other process can execute them.

G. Clocks and timers

The POSIX real-time standard requires the implementation of clocks for each process and thread, and for the whole simulation. Timers, sleep facilities and alarms are defined by using these clocks. The values of the clocks are updated and the execution time estimated by PERFidiX for each code segment. The actions of the elements declared over them, are executed by adding the time each event will take to the events list of the scheduler.

The elements that depend on the real-time clock of the system have been implemented in a different way. With this purpose, a SC_THREAD has been defined that is slept until the next event of that clock is required.

H. POSIX Interface modeling

POSIX services are provided in three different ways. Some of them use the underlying host functions, others are completely new, and those that depend strongly on the hardware platform have to be adapted to model correctly its platform-dependent functionality.

If the OS of the host computer is POSIX based, such as UNIX or Linux platforms, some of the host POSIX functions can be reused. These functions are basically those that are platform independent. Mathematical functions, string management, etc., maintain their functionality in every platform and they do not interfere with the scheduler or the parallelism capabilities of the system. Thus, they can be used to model, at least, the platform functionality. To include the timing cost, these functions will be wrapped into new functions that will take into account the time the function will take in the final processor.

The second group of the API functions is composed of those facilities that allow the designer to interact with the elements that have been implemented in the software execution support described below. Parallelism, scheduling, communication, synchronization and timing features are completely platform dependent, so new implementations are required. They have been developed in two different ways. Some of them provide access to the software support but they have no functionality implemented inside. These are the functions used to indicate to the kernel the characteristics of threads and processes, to activate the timers, to obtain data from the system clocks, to indicate to the signal manager the generation of a new signal, etc. The other functions implement their own functionality using the facilities provided by the simulation kernel. Mutexes, semaphores or message queues use the kernel facilities to block or resume threads, generate signals or implement timeouts.

The last group of POSIX API functions is composed of those functions whose implementation is strongly dependent on the hardware platform. Thus, a general platform execution support model is not possible. Some examples are the I/O functions, which strongly depend on the system drivers, so the implementation cannot be reusable on different platforms. Therefore, PERFidiX cannot provide accurate implementations that model all platform implementations. Instead of that, models that allow the designer to simulate the functionality are provided. In some cases, the host equivalent functions can be used to simulate it by adding some parameters that model their timing characteristics. In other cases new functions have to be developed because the host versions can interfere with the parallelism and scheduling capabilities of the platform model. In this context, files and sockets are in the same situation.

I. Traces

The original PERFidy library implements a set of functions to allow the designer to indicate and get the data required in order to obtain an adequate performance analysis. Now this set has to be extended to implement the functionality described in the POSIX standard to achieve this goal. First, the events are traced in POSIX standard to two classes: User trace events that are generated by the application of instrumentation functions and system trace events, which are generated by the operating system. Each trace event of the latter group may be an implementation-defined action such as a context switch, or an application-programmed action such as a call to a specific operating system service (for example, *fork*()) or a call to *posix_trace_event*().

IV. Experimental results

PERFidiX has been evaluated in a sufficiently complex case study, the EN 301 245 vocoder for GSM applications

standard of ETSI [21]. The original, sequential, standard golden model was structured in 13.500 lines of SystemC code. The SystemC specification was refined into a POSIX version, closer to the final implementation.

Experimental results are shown in Table I. The first column shows the estimated execution times for each thread and the RTOS obtained with PERFidy from the system-level, SystemC specification (SCS). The second column shows the corresponding estimated execution times obtained by PERFidiX from the POSIX code (PP). The third column shows the actual execution times obtained from the implementation of the code on an OpenRisk 1500 platform [22]. Execution times have been obtained directly from the prototype board inserting counters for each thread directly in the RTOS (eCos) code.

TABLE I
Estimated and actual execution times of the experiment

Thread	SCS (ms)	PP (ms)	IP (ms)
pre_filtering	1,085.10	1,035.60	1,061.16
homing_frame_test	84.72	73.76	73.12
frame_lsp_func	12,774.98	11,955.65	11,901.95
frame_int_tol_fun	12,308.99	11,646.93	11,246.62
subframe_coder_fun	40,004.07	37,585.17	37,498.38
serializer_fun	158.24	184.44	173.41
vad_comp_fun	10.89	11.82	11.44
CN_encoder_fun	54.74	61.85	58.72
sid_encoding_fun	223.32	249.53	269.15
RTOS	7,021.70	7,721.35	7,581.67

The corresponding error percentages are shown in Table II. As expected, the error obtained on the refined, POSIX code is smaller than the error at the original SystemC code. This indicates the accuracy of the technique in estimating the time slicing and the context changes. In any case, the error is kept smaller than 8%.

TABLE II
Estimation accuracy

Thread	SCS (%)	PP (%)
pre_filtering	3.45	2.41
homing_frame_test	13.67	0.88
frame_lsp_func	4.51	0.45
frame_int_tol_fun	0.17	3.56
subframe_coder_fun	3.04	0.23
serializer_fun	6.72	6.36
vad_comp_fun	9.36	3.34
CN_encoder_fun	7.45	5.33
sid_encoding_fun	12.18	7.29
RTOS	3.94	1.84

As shown in Table III, the increase in simulation time

implied by PERFidiX is negligible with respect to PERFidy. In any case, the gain with respect to a cycle accurate ISS is large (77 times faster).

TABLE III

Simulation times

	SCS (ms)	PP (ms)	ISS (ms)
Simulation time	1,070	1,560	124,000

V. Conclusions

Like other recent contributions in the field, the paper shows that a high-level modeling and timed simulation of application SW is possible at the source code level including the chosen RTOS. The main advantage of the proposed underlying technology over other similar techniques is that it avoids the complex, three-step process of estimating the execution times, annotating them in the appropriate points of the code and simulating. This technology was proven effective in PERFidy, a SystemC performance analysis library. In the paper, this technology is applied to the modeling and simulation of POSIX. The new library models the concurrency, communication and synchronization functionality required by the standard without modifying the original SystemC simulation kernel. This ensures full portability.

The new library, called PERFidiX, supports refinement of the original SystemC specification by optimizing the code including POSIX functions. Reusability is also improved as now it is possible to include legacy code and COST components making use of POSIX functions.

SystemC has been shown to be a flexible framework for system specification, refinement and simulation at different abstraction layers.

Experimental results assess the execution time estimation capability of the library. Although some error is unavoidable, accuracy is enough to allow the designer a fast and early performance estimation of the system taking into account the architectural mapping decisions, but avoiding the need for HW/SW synthesis.

References

[1] ITRS. International Technology Roadmap for Semiconductors: 2003 Edition. http://public.itrs.net.

[2] A.A. Jerraya, S. Yoo, D. Verkest and N. When: "Embedded Software for SoC", *Springer*, 2003.

[3] A. Sangiovanni-Vincentelli and G. Martin: "Platform-based design and software design methodology for embedded systems", *IEEE Design and Test of Computers*. November-December, 2001, 23-33.

[4] P. Puschner and C. Koza: "Calculating the maximum execution time of real-time programs", *The Journal of Real-Time Systems, 1*, 1989, 159-176.

[5] S. Yoo, I. Bacivarov, A. Bouchima, Y. Paviot and A. Jerraya: "Building fast and accurate SW simulation models based on hardware abstraction layer and simulation environment abstraction layer", in *Proceedings of the Design, Automation and Test Conference*, IEEE, 2003, 550-555.

[6] H. Tomiyama, Y. Cao and K. Murakami: "Modeling fixed-priority preemptive multi-task systems in SpecC", *Proceedings of the 10th Workshop on System And System Integration of Mixed Technologies (SASIMI '01)*, IEEE, 2001.

[7] A. Gerstlauer, H. Yu and D. Gajski: "RTOS modeling for system-level design", in Embedded Software for SoC, A.A. Jerraya, S. Yoo, D. Verkest and N. When (Eds.), *Springer*, 2003.

[8] S. Yoo, G. Nicolescu, L. Gauthier and A. Jerraya: "Automatic generation of fast timed simulation models for operating systems in SoC design", in *Proceedings of the Design, Automation and Test Conference*, IEEE, 2002, 620-625.

[9] Y. Yi, D. Kim and S. Ha: "Fast and time-accurate cosimulation with OS scheduler modeling", *Design Automation of Embedded Systems, 8*, 2003, 211-228.

[10] H. Posadas, F. Herrera, P. Sánchez, E. Villar and F. Blasco: "System-level performance analysis in SystemC", in *Proceedings of the Design, Automation and Test Conference*, IEEE, 2004, 378-383.

[11] S. Honda, T. Wakabayashi, H. Tomiyama and H. Takada: "RTOS-centric HW/SW cosimulator for embedded system design", *Proceedings of CoDes-ISSS'04*, ACM, 2004.

[12] M.A. Hassan, K. Sakanushi, Y. Takeuchi and M. Imai: "RTK-Spec TRON: A simulation model of an ITRON based RTOS kernel in SystemC", *Proceedings of the Design, Automation and Test Conference*, IEEE, 2005.

[13] L. Benini, D. Bertozzi, D. Bruni, N. Drago, F. Fummi and M. Ponzino: "SystemC cosimulation and emulation of multiprocessor SoC design", *IEEE Computer,* April, 2003.

[14] W. Müller, W. Rosenstiel and J. Ruf: "SystemC: Methodologies and Applications", *Springer*, 2003.

[15] F. Herrera, V. Fernández, P. Sánchez and E. Villar: "Embedded software generation from SystemC for platform-based design", in SystemC: Methodologies and Applications, W. Müller, W. Rosenstiel and J. Ruf (Eds.), *Springer*, 2003.

[16] IEEE: "Information technology-Portable Operating System Interface", IEEE Std 1003.1, 2004.

[17] G. Hawley: "Selecting a RTOS", *Embedded Systems Programming Europe*, May, 1999.

[18] ENEA: "OSE Soft Kernel Environment", available in http://www.ose.com/products.

[19] AXLOG, information available in http://www.axlog.fr.

[20] C. Liem, F. Naçabal, C. Valderrama, P. Paulin and A. Jerraya: "System-on-a-Chip cosimulation and compilation", *IEEE Design & Test of Computers*, April-June 1997.

[21] EN 301.245, ETSI, December, 1997.

[22] M. Bolado, H. Posadas, J. Castillo, P. Huerta, P., Sánchez, C. Sánchez, H. Fouren. and F. Blasco: "Platform based on Open-Source Cores for industrial applications", in *Proceedings of the Design, Automation and Test Conference*, IEEE, 2004, 1014-1019.

PARLGRAN: Parallelism granularity selection for scheduling task chains on dynamically reconfigurable architectures [*]

Sudarshan Banerjee   Elaheh Bozorgzadeh   Nikil Dutt
Center for Embedded Computer Systems
University of California, Irvine, CA, USA
{banerjee,eli,dutt}@ics.uci.edu

ABSTRACT

Partial dynamic reconfiguration, often called RTR (run-time reconfiguration) is a key feature in modern reconfigurable platforms. While partial RTR enables additional application performance, it imposes physical constraints necessitating simultaneous scheduling and placement while mapping application task graphs onto such architectures. In this paper we present PARLGRAN, an approach that maximizes performance of application *task chains* by selecting a suitable granularity of data-parallelism for individual *data parallel* tasks. Our approach focusses on reconfiguration delay overhead and placement-related issues (such as fragmentation) while selecting individual data-parallelism granularity as an integral part of simultaneous scheduling and placement. We demonstrate that our heuristic generates high-quality schedules on an extensive set of over a 1000 synthetic experiments by comparing the results with an approach that tries to statically maximize data-parallelism, i.e., does not consider the overheads and constraints associated with partial RTR. A detailed case-study on JPEG encoding additionally confirms that blindly maximizing data-parallelism can result in schedules even worse than that generated by a simple (but RTR-aware) approach oblivious to data-parallelism.

Keywords: Partial dynamic reconfiguration, data-parallelism, granularity selection, linear placement, scheduling

1. INTRODUCTION

Reconfigurable architectures are popular for applications with intensive computation such as image processing, since a limited amount of logic can be customized to set up deep pipelines, and/or exploit more coarse-grain parallelism, etc. Partial dynamic reconfiguration, or, run-time reconfiguration (RTR) allows additional customization during application execution, making it possible to obtain increased performance [11]. Our overall goal is to maximize performance of applications represented as precedence-constrained task DAGs (directed acyclic graphs) on *single-context* architectures with partial RTR (Xilinx Virtex-II is a commercial instance of such architectures). Some key issues in mapping applications onto such devices are the significant reconfiguration delay overhead, physical (placement) constraints, etc.

In this paper, we focus on precedence-constrained *task chains*, common in image-processing applications [7], [4]. In such applications, area-execution time characteristics of key tasks such as IDCT, Quantize, etc, are predictable because of complete pipelining. Additionally, key tasks such as DCT are completely *data-parallel*, i.e., results of task execution on a block of data are completely independent of results when the same task is executed on a disjoint block of data. On an architecture with partial RTR, it is possible to improve application execution time by *dynamically* adjusting the parallelism granularity of such tasks, i.e., reconfig-

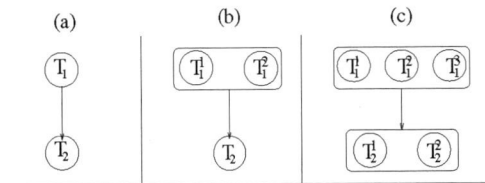

Figure 1: Granularity of individual data-parallel tasks

uring the architecture to instantiate multiple copies of such tasks *during application execution* – each copy (instance) uses an identical amount of HW resources, but processes only part of the data. Due to complete pipelining, execution time of such tasks is directly proportional to the volume of data processed, and thus, reducing the data volume proportionally improves (reduces) the application execution time. Note that on architectures with no partial RTR, the scope of exploiting such data-parallelism is much more limited – partial RTR enables resource reuse, significantly expanding the potential of exploiting data-parallelism.

As an example, we consider a simple chain with two tasks, as shown in Figure 1. Assuming that there are enough resources to simultaneously execute 3 copies of task T_1 *or* 2 copies of task T_2, (b) and (c) show some possible task graph configurations after such a transformation. However, such a transformation can be quite costly on architectures with partial RTR– each new task instance (copy) adds a significant reconfiguration overhead. Therefore the transformations need to be guided by selecting the right granularity of parallelism that masks the reconfiguration overhead and maximizes performance. One important issue is that because of the reconfiguration penalty, multiple copies of a task may not be able to start at the same time– therefore, individual execution time (workloads) of the multiple copies may vary.

We propose an approach, PARLGRAN, that attempts to maximize application performance on architectures with partial RTR by choosing the right parallelism granularity for each individual data-parallel task. By granularity we mean both the **number of instances** (copies) of that task, and, the **workload** (execution time) of each copy. Our approach considers physical (placement) constraints, and utilizes configuration prefetch [10] to reduce the latency. The key constraints of such architectures necessitate joint scheduling and placement [8], [1]. Our approach therefore, incorporates granularity selection as an integral part of simultaneous scheduling and placement. To the best of our knowledge, ours is the first effort to solve this problem.

We validate the quality of our proposed approach against an approach that tries to statically maximize performance gain from data parallelism without considering the constraints and overheads due to partial RTR. A large set of over a thousand synthetic experiments demonstrates that the average improvement in schedule length by using our approach is over 15%. A detailed case study of JPEG encoding additionally confirms that the static parallelization approach

---

[*]This work was partially supported by NSF Grants CCR-0203813 and CCR-0205712

5B-4

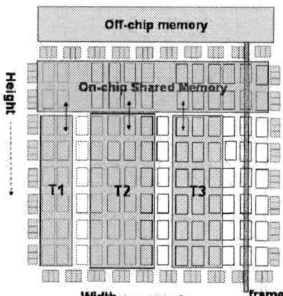

Figure 2: Target dynamic architecture

can end up generating schedules much worse than a simple (but RTR-aware) approach oblivious to data-parallelism.

2. RELATED WORK

There exists a large body of work for mapping task chains typical in image processing to reconfigurable architectures. A significant amount of the work such as [7] does not consider partial run-time reconfiguration (RTR). Recent work that does consider partial RTR such as [4], often focuses on multicontext architectures [9], where the reconfiguration overhead is negligible- unfortunately there is a very significant area overhead in such architectures. Other recent work such as [2] focusses on the key problem of task reuse as another technique for reducing the reconfiguration overhead, an aspect we do not address in this work. Our work focuses on reducing reconfiguration overhead by considering configuration prefetch as an integral aspect of joint scheduling and placement.

While there is a growing body of work in joint scheduling and placement on such architectures [3], [8], they typically ignore key architectural constraints such as the resource contention due to a single reconfiguration controller, prefetch to reduce the latency, etc. Ignoring these key issues makes the problem closer to the rectangle packing problem [13] and does not realistically exploit RTR.

Of course, there is a vast body of knowledge in the compiler domain about extracting parallelism from programs at different levels of granularity [14]. However, detailed consideration of placement and other architectural aspects related to partial RTR make our work significantly different.

3. PROBLEM OVERVIEW

Target architecture: Our target dynamically reconfigurable device as shown in Figure 2 consists of a set of configurable logic blocks (CLB) arranged in a two-dimensional matrix. The basic unit of configuration for such a device is a frame spanning the height of the device. A column of resources consists of multiple frames. A task occupies a contiguous set of columns. The reconfiguration time of a task is directly proportional to the number of columns (frames) occupied by the task implementation. One key constraint is that only one task reconfiguration can be active at any time instant. An example of our target device is the Xilinx Virtex-II series where constraints such as dynamic tasks occupying a contiguous set of columns are critical for realization of partial run-time reconfiguration.

Application specification: A task T_i executing on such a system can be represented as a 3-tuple (c_i, t_i, r_i) where c_i is the number of resource columns occupied by the task, t_i and r_i are the execution time and reconfiguration overhead respectively. Each task needs to be reconfigured before its execution is scheduled. The physical constraints on such a device necessitates joint scheduling and placement [8], [1].

In image processing applications, we often find chains (linear sequences) of such tasks. For a chain of n tasks, $(T_1..T_n)$, each task in the chain has exactly one predecessor and one successor. Of course, the first task, T_1, has no predecessor, and the last task, T_n, has no successor. A predecessor task utilizes a shared memory mechanism to communicate necessary data to its successor– this shared memory can be physically mapped to local on-chip memory and/or off-chip memory depending upon memory requirements of the application.

Our overall goal is to maximize performance (minimize schedule length) under physical and architectural constraints, given a resource constraint of C columns available for the application, $C < \sum_{i=1}^{n}(c_i)$. An additional goal is that our approach should have a low computational overhead.

4. MOTIVATION

Ideally, the degree of parallelism for a data-parallel task is limited only by the availability of HW resources. Let us consider a chain with only a single task T_1 that executes in time t_1 using c_1 columns. Given a resource constraint of C columns, performance is maximized when this task is instantiated $\lfloor C/c_1 \rfloor$ times, as shown in Figure 3. In this figure, the X-axis represents the columnar area constraint, C, and, the Y-axis represents the schedule length. For sequential tasks (0 degree of data-parallelism), the execution of task T_i is represented as E_i and the reconfiguration of task T_i is represented as R_i, as in Figure 3 (a). For data-parallel tasks, we additionally denote the execution of j-th instance (copy) of the task as E_i^j and the reconfiguration for this instance (copy) as R_i^j, as shown in Figure 3 (b). For uniform treatment, we assume that the compilation cost includes reconfiguration overhead for the first task in the chain, T_1– the schedule length is always computed from beginning of execution of T_1. However, this *ideal* performance gain is typically not achievable while considering realistic issues on such architectures, as discussed next.

Reconfiguration overhead: For modern *single-context* architectures that support partial RTR, the large reconfiguration delay is a key bottleneck in achieving *ideal* parallelism. To illustrate this, we consider Figure 4. We assume that the reconfiguration controller is available at the beginning of the execution of the first copy of the task. Next we instantiate two copies of this task with the intention of equally distributing the workload (execution time). However, execution of the second copy E_1^2 can start only after the reconfiguration overhead, r_1. Thus, instead of the ideal workload of $t_1/2$, the workload of the second task is only: $(t_1 - r_1)/2$, leading to less performance improvement than expected.

For a single task, a simple equation suffices to compute the best performance improvement, leading to the following lemma:

LEMMA 1. *For parallelizing a task into j instances, and given that the reconfiguration controller is available at the beginning of execution of the first instance, the best performance (least execution time) is obtained when the workload (execution time) of the j-th instance is:* $((t_1 - r_1 * (j * (j-1)/2))/j$.

Due to lack of space, detailed proof is in [16]. However, it is important to note the following key ideas, evident from Figure 4.
• Even if sufficient HW resources are available, the large reconfiguration delay may prevent further performance improvement if more than a few copies of a task are instantiated. In Figure 4, the 4'th copy does not improve performance (shorten schedule length).
• Maximizing performance involves *unequal* workload (execution time) distribution between multiple copies of a task, to compensate for the reconfiguration overhead.

Next, we consider the additional complications introduced by precedence constraints.

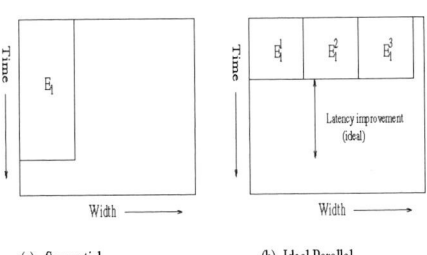

(a) Sequential   (b) Ideal Parallel

Figure 3: "Ideal" parallelism

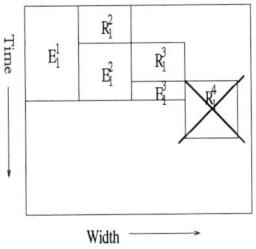

Figure 4: Parallelism degree determined by overhead

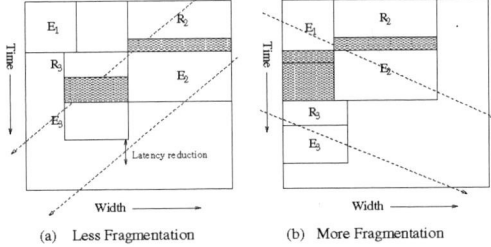

Figure 5: Precedence constraints- choices

Precedence constraints: For precedence-constrained tasks, simple equations such as Lemma 1 do not work any longer – we now need to consider the interaction of the resource demands of the tasks, as shown in Figure 5 for a simple chain with two tasks T_1 and T_2. The HW resource constraint allows three copies (instances) of T_1, **or**, two copies of T_2 to be executing simultaneously. We show some of the possible schedules and their transformed task graphs in Figure 5 (b) and (c). Note that in our execution model, *all copies of a parent task must finish execution before any copy of a child task starts execution*. Also, Figure 5 (b) shows how the high reconfiguration delay necessitates configuration prefetch to improve the latency.

We can now formulate our problem as:

For a precedence-constrained chain with some data-parallel tasks, we want to compute both the **number** of copies for each data-parallel task, and, the **workload** (execution time), t_i^j, of each (j-th) copy of such tasks, $\sum_j (t_i^j) = t_i$. Our goal is to maximize performance (minimize schedule length) given that only a fixed number of contiguous columns are available for mapping the task chain. In this work, we make an additional assumption that sufficient communication bandwidth is available for satisfying the data requirements of the multiple copies of a task.

5. PROPOSED APPROACH

In this section, we first present MFF, a heuristic for scheduling simple task chains. While MFF is oblivious to data-parallelism, it provides the core concepts underlying PARLGRAN, our proposed approach for chains with data-parallel tasks.

5.1 MFF (modified first fit)

For architectures with partial RTR, the physical (placement) constraints and, the architectural constraint of the single reconfiguration mechanism, make it difficult to achieve the ideal schedule length $L_{ideal} = \sum_{i=1}^{n} (t_i)$. In fact, this *simple* problem of minimizing schedule length for a chain, under constraints related to partial RTR, is actually NP-complete, as proved in [15]. MFF, our proposed heuristic to solve this problem, essentially tries to satisfy task resource constraints, and, attempts simple local optimizations to *reduce fragmentation*, and, hence, the schedule length.

Approach: **MFF (modified first-fit)**

Place task T_1 starting from leftmost column

for each task $(T_i, i > 1)$

   F_i^S = earliest time-slot enough space is available (last-fit)

   F_i^R = earliest time-slot reconfiguration controller is available

   $R_i^{start} = \text{MAX} (F_i^S, F_i^R)$

   $E_i^{start} = \text{MAX} (R_i^{start} + r_i, E_{i-1}^{end})$

   if (T_i aligned with rightmost column)

     local optimization: Adjust immediate ancestor placement

     (and start time) if possible to improve start time of T_i

endfor

MFF is based on a first-fit approach. To get an idea why a first-fit approach works well in practical scenarios, we take a look at Figure 6 (a). The tasks are essentially laid out in the form of diagonals running from the top-right of the placed schedule towards the bottom-left. As long as a task does not "fall off" the diagonal, it is possible to overlap at least part of the reconfiguration overhead with the execution of its immediate ancestor. Once a task "falls off" the diagonal and is placed at the rightmost column C, it is essentially trying to reuse the area of ancestor tasks higher up in the chain. Given that for tasks in a chain the execution components have to be in sequence, a more distant ancestor is guaranteed to finish earlier than a closer ancestor. This increases significantly the possibility of being able to overlap reconfiguration of this task with the execution of ancestors that are closer to it in the chain. Effectively the chain property causes a "window" of tasks: tasks within a window affect each other much more strongly than tasks outside the window.

Simple fragmentation reduction: One minor modification for reducing fragmention in MFF compared to pure first-fit is shown in Figure 6. Our observations indicate that in tightly-constrained scenarios (few columns available for task mapping), placing the second task T_2 adjacent to task T_1, as in Figure 6 (b), often leads to immediate fragmentation– though enough area is available to reconfigure task T_3 in parallel with execution of task T_2, this area is not contiguous, and thus task T_3 gets delayed. MFF takes care of this by placing T_2 at the right-hand corner. Of course, this simple modification is not applicable to all scenarios.

Local optimization: Exploiting slack in reconfiguration controller: A more interesting local optimization to reduce fragmentation is shown in Figure 7 (a). While scheduling task T_4, we notice that it is possible to exploit slack in the reconfiguration mechanism to *postpone* the reconfiguration R_3 of task T_3 without delaying the actual execution E_3 of task T_3. We can thus make better use of the available area (HW resources) to reschedule (and change placement of) task T_3 – as a result, reconfiguration R_4 of task T_4 can now execute in parallel with E_3, leading to a reduction in schedule length, as shown in Figure 7 (b).

Before proceeding to PARLGRAN, it is important to understand that the fragmentation problems we try to address in MFF (and

Figure 6: Simple chain- right placement of task 2

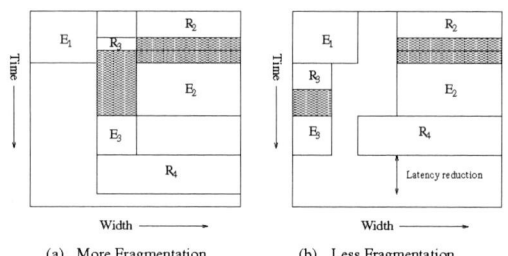

(a) More Fragmentation (b) Less Fragmentation

Figure 7: Exploiting slack in reconfiguration controller

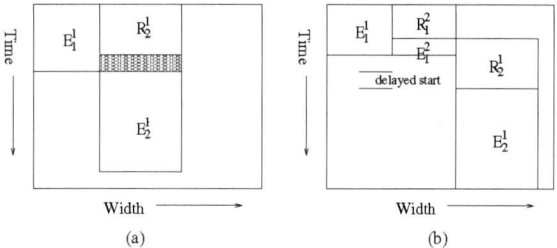

(a) (b)

Figure 8: Static pruning based on timing

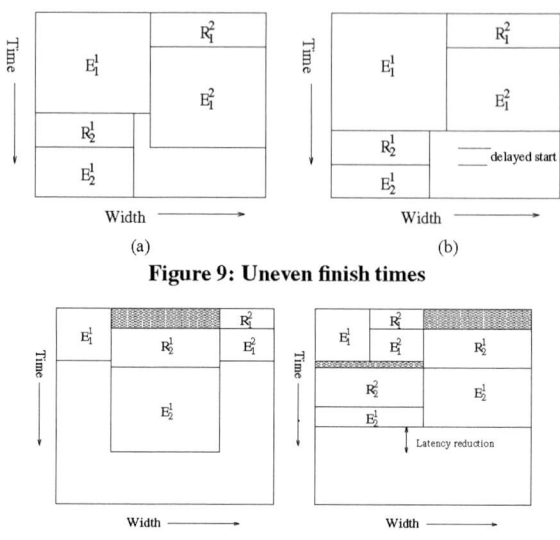

(a) (b)

Figure 9: Uneven finish times

(a) More Fragmentation (b) Less Fragmentation

Figure 10: Left placement for copies of first task

PARLGRAN) are because we are trying to jointly schedule and place while satisfying a host of other constraints– thus, other free space coalescing techniques for partially reconfigurable architectures, such as [6], are not directly applicable.

5.2 PARLGRAN

We use the insights obtained from the chain-scheduling problem as the basis for our granularity selection approach. Detailed analysis of chain-scheduling shows that applying local optimizations can improve the performance. We additionally want to design an approach such that the algorithm execution time is comparable to the execution time of the tasks. So, our proposed algorithm is simple and greedy, but, uses specific problem properties to try and improve the solution quality.

Our approach consists of two steps:
- Static pruning
- Dynamic granularity selection

5.2.1 Static pruning

First, we utilize some simple facts to statically prune regions of the search space. As an example of pruning, consider Figure 8. If we schedule exactly one copy each for tasks T_1 and T_2, then task T_2 can start as soon as T_1 ends, i.e., at t_1, as in Figure 8 (a). If we schedule another copy of task T_1, the execution time of T_1 improves. However, now the reconfiguration controller becomes the bottleneck, as shown in Figure 8 (b). Now, task T_2 can start only at $(r_1 + r_2)$, which is greater than t_1. In general, the number of copies of a task is limited by the impact of its reconfiguration overhead on its successors.

5.2.2 Dynamic granularity selection

We next consider work distribution (load balancing) issues for the multiple task copies.

Uneven finish times: From our initial discussion on data-parallelism (as shown earlier in Figure 4), it seems that it is a good idea to always generate as many copies as possible subject to performance improvement and get them to finish at the same time instant. However, with the introduction of task dependencies, it is possible to modify this approach in certain cases to improve performance, as

shown in Figure 9. In Figure 9 (a) let FT_1^1 denote the time instant the earlier copy of task T_1, that is E_1^1 ends. Task T_2 can start at: $ST_2^1 = FT_1^1 + r_2$. However, if both copies of T_1 end at the same time instant as shown in Figure 9 (b), this time-instant is given by:
$$FT_1^{equal} = FT_1^1 + r_2/2$$
As a result, reconfiguration R_2 for task T_2 gets delayed and execution E_2 for task T_2 can only start at
$$FT_1^{equal} + r_2 = FT_1^1 + 3 * r_2/2$$
Of course, if the area of task T_2 is greater than the area of task T_1, letting both copies of T_1 end at the same time instant would lead to a shorter schedule.

One other minor observation to improve MFF specifically for parallelism granularity selection is shown in Figure 10. Placing multiple copies of a task adjacent to each other intuitively helps reduce fragmentation.

PARLGRAN is an adaptation of MFF that essentially tries to greedily add multiple copies of data parallel tasks as long as it estimates that adding a new copy is beneficial for performance (shorter schedule length). The concepts of dynamically adjusting the workload combined with local optimizations makes it effective. We summarize our PARLGRAN approach below.

Approach: **PARLGRAN (Parallelism Granularity Selection)**
Place first copy of task T_1 starting from leftmost column
for each task (T_i, i > 1)
    Compute earliest execution start of task (space search by last-fit)
    if (parent task is data-parallel)
        while (no degradation in start time of T_i)
            add new copy of parent (assign start time, physical location)
            adjust workload of existing scheduled copies of parent
    Schedule (and place) T_i
    apply local optimizations from MFF for improving schedule
endfor

While this approach appears to be simplistic, experimental results in the following section show it typically does better than statically deciding to parallelize each task to its maximum degree. For applications like JPEG encoding, blind parallelization can lead to *significantly inferior* results, even worse than RTR-aware first-fit, because of the reconfiguration overhead and the physical (placement) constraints.

6. EXPERIMENTS

We conducted a large set of experiments with over a 1000 datapoints to demonstrate the quality of schedules generated by our heuristics. In this section we present the key results from our experiments. It is important to remember that our goal is to maximize the performance (minimize schedule length) of a chain of tasks given a hard constraint on the available area. Therefore, while it is possible to fit our applications onto suitably sized target devices, we assume for experimental purposes that the resource constraint is less than the aggregate size of all tasks. This makes our approach suitable for a scenario where multiple processes are executing simultaneously on the reconfigurable device and the area available for mapping the task graph is not known beforehand.

Experimental setup

We assumed a target device organized as a CLB matrix of 56 rows, 48 columns, similar to Xilinx XC2V2000. From the XC2V2000 data sheet, we estimate that the reconfiguration overhead for the smallest task occupying one column on our architecture is 0.19 ms at the maximum suggested reconfiguration frequency of 66 MHz. We obtained area and timing data for tasks like Huffman, DCT, etc., by synthesizing tasks under columnar placement and routing constraints on the XC2V2000, similar to the Xilinx methodology suggested for "reconfigurable modules".

We explored a large set of possible scenarios by generating task chains which varied the following parameters: (1) varying chain length, in the range of 4 to 20 tasks in the chain. (2) varying task parameters for each task in a chain of given length (3) varying area constraints.

That is, for each chain length in the given range (4 to 20), we had a set of testcases where the tasks had different parameters. For each such testcase, we varied the area constraint across a wide range to represent loose as well as tight constraints, thus generating data for over a thousand individual experiments.

The task parameters (execution time, reconfiguration delay, number of columns) were randomly selected from our database of synthesized tasks. The database consisted of information corresponding to images of various sizes- since each task is completely pipelined, the reconfiguration delay and number of columns occupied by the task is independent of the image size, but, the execution time is directly proportional to the image size.

We measure schedule quality of our proposed heuristic by comparing against two approaches: FF (first-fit) and MAXPARL (maximum parallelization). MAXPARL attempts to maximize parallelization by statically generating the maximum number of copies possible for each task subject to resource constraints only, and assigns equal workload to each copy. In subsequent discussions, the following notation denotes schedule lengths generated by the various approaches on an individual experiment:

L_{ff}: corresponds to first-fit approach

L_{max}: corresponds to MAXPARL (maximum parallelization) approach

L_{mff}: corresponds to our proposed MFF approach

L_{gran}: corresponds to our proposed PARLGRAN (parallelism granularity selection) approach

Schedule quality of MFF

Our first set of experiments consisted of comparing schedule lengths generated by MFF with that of first-fit, on the set of experiments as described above.

The experimental data confirmed that schedules generated by MFF were almost always equal to or better than first-fit. The sched-

Chain length	PARLGRAN Vs FF	PARLGRAN Vs MAXPARL		
	Avg	Avg	Best	Worst
4-7	46.3%	9.8%	142.5%	-49.6%
8-11	51.7%	15.8%	109.6%	-30.9%
12-15	55.0%	18.5%	82.3%	-15.5%
16-20	58.3%	33.8%	151%	-17.5%
Avg gain	>50%	> 15%		

Table 1: Reduction in schedule length for completely data parallel chains with PARLGRAN

ule lengths generated by MFF were better in 243 out of 1140 tests, i.e., approximately 20% of the tests, worse in 5 out of 1140 tests. In 113 tests, around 10% of the total, MFF was better by at least 3%. In the worst experiment for MFF, first-fit generated a schedule longer by 0.44%. Overall, on longer chains (more tasks) and looser constraints (more columns), both algorithms were almost equally able to hide the reconfiguration overhead. However, on more constrained problems with shorter chains and tighter area constraints, MFF tends to generate better schedules.

Schedule quality of PRLGRAN

Next, in Table 1, we present a summary of results conducted on our PARLGRAN (parallelism granularity selection) approach. The data in each row of the table corresponds to experiments on chains of corresponding length– as an example, data in row 2 (chain length 8-11) was obtained from experiments on chains with at least 8 tasks and at most 11 tasks. Note that this set of experiments is identical to that we used to validate MFF– the difference is that we now assume each task in the chain is completely data-parallel. For comparison with MAXPARL and FF, our quality measure is simply the percentage increase in schedule length generated by the other approach compared to PARLGRAN.. As an example, for comparison with MAXPARL, the quality measure is simply:

$$((L_{max} - L_{gran})/L_{gran}) * 100$$

The first column in Table 1 represents the *Average* percentage improvement of PARLGRAN as compared to FF. Each entry in the first column is an average of a large number of experiments conducted on chains of corresponding length. The second, third and fourth columns respectively represent the *Average*, the *Best* and the *Worst* performance of our approach compared to MAXPARL. As an example, the data in row 2, column 3, states that on a large number of chains with chain length between 8 and 11 tasks, the best result generated by our approach corresponds to an experiment where MAXPARL generated a schedule 109% longer.

The table clearly shows that our 'granularity selection' heuristic, PARLGRAN, generates increasingly better results compared to MAXPARL when more space is available. Intuitively, with more space, it is possible to make more instances of the data-parallel tasks. However, with each additional instance, the workload (execution time) decreases per instance, making the execution time comparable to the reconfiguration overhead – PARLGRAN is much better capable of deciding when to stop instantiating multiple copies, as opposed to MAXPARL. The local optimizations in PARLGRAN play an active role in such circumstances to help improve the schedule length. One key experimental aspect we would like to mention is that for smaller chains, our presented results cover a very large range of varying area constraints– for the longer chains, the presented results cover the scenarios where the available HW area is at most around 40% of the aggregate HW area of the tasks. More detailed results in [16] confirm that the benefit of our approach over MAXPARL increases significantly as chain length gets longer and available area increases.

Case	C	L_{mff} (ms)	L_{max} (ms)	L_{gran} (ms)
256X256 JPG	5	12.71	12.73	**12.36**
	6	11.24	12.52	**10.81**
	7	11.24	11.38	**10.05**
	8	11.24	12.11	**9.08**
	9	10.10	12.79	**9.08**
512X512 JPG	5	42.86	40.68	40.30
	6	41.34	35.32	35.13
	7	41.34	34.18	34.37
	8	41.34	29.08	28.60
	9	40.20	28.38	27.71

Table 2: Case study of JPEG encoding: Schedule Length with different image size and area constraints

Case study of JPEG encoding

After conducting a wide range of experiments on synthetic graphs, we conducted a case study on the JPEG encoding algorithm, represented as a chain of four key tasks: RGB2YCbCr, DCT, Quantize, Huffman. Table 2 presents the consolidated results from the case study. Entries in the first column, CASE, denote the image size – 256X256 denotes experiments on a 256X256 colour image. For each case, we varied the number of columns and observed the resulting schedule lengths. The second column C represents the area constraint in columns. The third, fourth and fifth columns correspond to schedule lengths (in ms) generated by MFF, MAXPARL, and PARLGRAN respectively.

For the 256X256 image, the reconfiguration overheads are comparable to the task execution times. Our approach frequently does much better than statically parallelizing everything, as in MAXPARL– the results confirm that such blind parallelization can often lead to results worse than a simple sequential scheduling approach. For an area constraint of 8 columns, schedule length of FF is longer than PARLGRAN by (11.24-9.08)/9.08 = 23.5%. Blind (static) parallelization leads to significantly worse schedule longer by (12.11-9.08)/9.08 = 33.3%. This is in spite of the fact that the effective transformed graph from MAXPARL consisted of 9 tasks with apparently more parallelism, while the transformed graph from PARLGRAN consisted of 7 tasks only.

For the 512X512 image, each task execution time is significantly greater than the reconfiguration overhead. In such a scenario, where, additionally, the chain length is short, MAXPARL generates good results – of course, PARLGRAN typically does somewhat better. But, both parallelizing approaches result in significant speedups.

Estimated run-time: Preliminary estimates indicate that the run-time of our approach on a PowerPC processor at 400 MHz (available on the Virtex-II Pro platform from Xilinx) is around 3-4 ms for a reasonably large experiment with 12 tasks and 20 columns. Considering the fact that the run-time of the DCT task for 512X512 colour image is around 11 ms on the target architecture, our approach is suitable for *semi-online* scenarios where the task precedence relations, and the task area-timing characteristics are available at compile-time, while the available HW area for for mapping a DAG is known only at run-time. Task management under such dynamic resource availability is a key issue in modern operating systems for reconfigurable architectures [5].

Our wide range of experiments and case studies confirm that PARLGRAN generates high-quality schedule in all situations– tightly constrained problems with shorter chains, fewer columns, as well as problems with more degrees of freedom, i.e., longer chains, more available columns. Additionally, the estimated run-time of our approach on a typical embedded processor is comparable to the HW task execution times.

7. CONCLUSION

In this paper, we proposed PARLGRAN, an approach that selects granularity of data-parallelism to maximize performance of application *task chains* executing on an architecture with partial RTR (run-time reconfiguration). Our approach selects both the number of instances of a data-parallel task, and, the execution time of each such instance – it is integrated in a joint scheduling and placement formulation, necessitated by the underlying physical and architectural constraints imposed by partial RTR. Experimental results on a significantly large space of over a 1000 synthetic experiments confirm that our approach generates schedules that are on an average better by 15% over an approach that tries to *statically* maximize data-parallelism. A detailed case study on JPEG encoding confirms that in realistic scenarios, an approach that simply tries to maximize data parallelism without accounting for the underlying constraints can end up generating schedules *much worse* than even a data-parallelism-oblivious (but RTR-aware) approach. Initial estimates indicate PARLGRAN is fast enough to be suitable for integration in a *semi-online* scheduling methodology where the goal is to maximize performance of an application given an area constraint known only at run-time.

While our approach demonstrates the potential for significant performance improvement, there are some key aspects that we want to address in our future work. First, we have assumed in this work that we are not constrained by memory/communication bandwidth. As we increase the task granularity (make more instances to exploit more data-parallelism), the data transfer to and from memory, both on-chip, and, off-chip, has the potential to become a bottleneck and will be considered in future work. We also plan to study the performance versus energy characteristics of such implementations on reconfigurable architectures with partial RTR.

8. REFERENCES

[1] S Banerjee, E Bozorgzadeh, N Dutt, "Physically-aware HW-SW Partitioning for reconfigurable architectures with partial dynamic reconfiguration", DAC, 2005

[2] J Harkin, T M Mcginnity, L P Maguire, "Modeling and Optimizing Run-Time reconfiguration using evolutionary computation", ACM TECS, V-3, Nov 2004

[3] P-H Yuh, C-L Yang, Y-W Chang, H-L Chen, "Temporal floorplanning using the T-tree formulation", ICCAD, 2004

[4] J Noguera, R M Badia, "Power-Performance trade-offs for reconfigurable computing", CODES+ISSS, 2004

[5] C Steiger, H Walder, M Platzner, "Operating systems for reconfigurable embedded platforms: Online Scheduling of Real-Time Tasks", IEEE Trans on Computers, V-53, 11, Nov 2004

[6] M Handa, R Vemuri, "An efficient algorithm for finding empty space for online FPGA placement", DAC, 2004

[7] H Quinn, L A Smith King, M Leeser, W Meleis, "Runtime Assignment of Reconfigurable Hardware Components for Image Processing Pipelines", FCCM, 2003

[8] S P Fekete, E Kohler, J Teich, "Optimal FPGA module placement with temporal precedence constraints", DATE, 2001

[9] H Singh, G Lu, E M C Filho, R Maestre, M-H Lee, F J Kurdahi, N Bagherzadeh, "MorphoSys: case study of a reconfigurable computing system targeting multimedia applications", DAC, 2000

[10] S Hauck, "Configuration pre-fetch for single context reconfigurable processors", FPGA, 1998

[11] M J Wirthlin, "Improving functional density through Run-time Circuit Reconfiguration", PhD Thesis, Electrical and Computer Engineering Dept, Brigham Young Univesity, 1997

[12] G Brebner, "A virtual hardware operating system for the Xilinx XC6200", FPL, 1996

[13] H Murata, K Fujiyoshi, S Nakatake, Y Kajitani, "Rectangle-packing based module placement", ICCAD, 1995

[14] S Muchnick, "Advanced Compiler design and implementation", Morgan Kaufmann, 1997

[15] J Augustine, Personal communication

[16] S Banerjee, E Bozorgzadeh, N Dutt, "Selecting granularity of parallelism for tasks executing on dynamically reconfigurable architectures", CECS Technical Report, UC Irvine.

Memory Optimal Single Appearance Schedule with Dynamic Loop Count for Synchronous Dataflow Graphs

Hyunok Oh, Nikil Dutt
Center for Embedded Computer Systems
University of California, Irvine, CA
{hoh,dutt}@ics.uci.edu

Soonhoi Ha
School of EECS
Seoul National University, Seoul, Korea
sha@iris.snu.ac.kr

Abstract— In this paper, we propose a new single appearance schedule for synchronous dataflow programs to minimize data memory and code memory size simultaneously. While a single appearance schedule promises only one appearance of each node definition in the generated code, it requires significant amount of data memory overhead compared with a buffer optimal schedule allowing multiple appearance. The key idea of the proposed technique is to make a dynamic decision of loop count to make a schedule *quasi-static*. The proposed quasi-static schedule produces a single appearance schedule code with minimum data memory requirement. We prove that every buffer optimal schedule can be transformed to our single appearance schedule which requires optimal buffer size for arbitrary synchronous dataflow graphs. The only penalty for the proposed technique is slight performance overhead of computing loop counts dynamically. In order to minimize the overhead we propose optimization techniques. Experimental results show that the proposed algorithm reduces 20% total memory with less than 1% performance overhead compared with the previous single appearance schedule algorithms.

I. INTRODUCTION

As system complexity increases and fast design turn-around time becomes important, high level software design methodologies become critical. In the context of DSP applications, there have been several approaches to automatic code generation from block diagram specification including COSSAP [1], GRAPE [6], and Ptolemy [4]. It is also the main concern of this paper.

In a hierarchical dataflow program graph, a node, called a actor or a block, represents a function that transforms input data streams into output streams. The functionality of an atomic node is described in a high-level language such as C or VHDL. An arc represents a channel that carries streams of data samples from the source node to the destination node. The number of samples produced (or consumed) per node firing is called the output (or the input) sample rate of the node. In case the number of samples consumed or produced on each arc is statically determined and can be any integer, the graph is called a synchronous dataflow graph (SDF) [7] which is widely adopted in aforementioned design environments. We illustrate an example of SDF graph in Figure 1(a). Each arc is

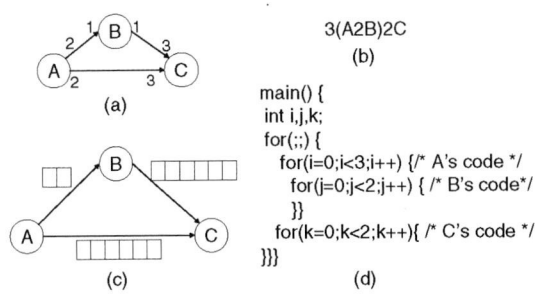

Fig. 1. (a) SDF graph example, (b) a scheduling result, (c) a code template, and (d) buffer allocation

annotated with the number of samples consumed or produced per node execution.

To generate a code from the given SDF graph, the order of node executions is determined at compile time by static *scheduling* of the graph. Since a dataflow graph specifies only partial orders between nodes, there are usually several valid schedules that satisfy the partial ordering. Figure 1(b) shows one of many possible scheduling results in a list form, where 2C means that node C is executed twice. The schedule will be repeated with the streams of input samples to the application. A code template according to the schedule of Figure 1(b) is shown in Figure 1(d). The block definition is inlined in the generated code, it is called inline-style. When a software code is automatically synthesized from an SDF graph, buffer space is allocated to each arc to store the data samples between the source and the destination blocks as shown in Figure 1 (c). The total buffer size becomes 14 in this example. The number of allocated buffer entries should be no less than the maximum number of samples accumulated on the arc at run-time.

If a schedule contains only one lexical appearance of each node, this schedule is called a single appearance schedule(SAS) (e.g. as 3(A2B)2C in Figure 1(b)). A single appearance schedule minimizes the code memory size since each block has a single definition in a generated code. Consider another schedule that is a non single appearance schedule, 2(A2B)C(A2B). Then, the generated code has two instances for nodes A and B while it reduces the data buffer size from 14 to 10 in Figure 1. In general while an SAS is preferable to minimize the code memory size, it requires larger buffer memory than a non SAS. The buffer size on each arc in a SAS is no less than the least common multiplier of the producing sample rate and consuming sample rate for the arc.

In this paper we propose a novel single appearance scheduling technique whose key idea is introducing a dynamic decision of loop count to make a schedule *quasi-static*. The proposed quasi-static schedule produces a single appearance schedule code with minimum data memory requirement. Section II defines some notations and section III reviews the related works. In section IV, we introduce motivational examples. The proposed technique is explained in section V. We will show experimental results in section VI and make a conclusion in section VII.

II. TERMINOLOGY

We use the following notation to represent the parameters of arc a and node v in SDF graphs.

$src(a)$: the source node of a that produces samples on the arc

$sink(a)$: the sink node of a that consumes samples from the arc

$p(a)$: the number of samples produced by an invocation of $src(a)$

$c(a)$: the number of samples consumed by an invocation of $sink(a)$

$d(a)$: the number of initial delay samples on arc a

$inputArc(v)$: a set of incoming arcs to node v.

$outputArc(v)$: a set of outgoing arcs from node v.

For arc AB in Figure 1, $src(AB) = A$, $sink(AB)=B$, $p(AB)=2$, $c(AB)=1$, $d(AB)=0$, $outputArc(A) = \{AB, AC\}$, $inputArc(B) = \{AB\}$,$outputArc(B) = \{BC\}$,and $inputArc(C) = \{AC, BC\}$.

III. RELATED WORKS

Since minimization of memory requirements in embedded system is crucial, many researches have been performed to find a schedule to minimize data memory and/or code memory.

Ade et al. [2] have developed the formula on the upper bounds on the minimum buffer memory requirement for a number of restricted subclasses of delayless, acyclic graphs, including arbitrary-length chain-structured graphs. Some of these bounds have been generalized to handle delays in [9] which has shown that the problem of constructing a schedule that minimizes the buffer requirement is NP-complete.

Ritz et al. [12] have proposed a buffer sharing optimization among a subset of single appearance schedules, called flat single appearance schedule. Since the flat SAS does not allow nested loops, it usually requires large buffer memory even though it shares buffers allocated on each arc. Murthy et al. have developed several heuristics that produce SAS with nested loop: APGAN, RPMC, and GDPPO [9]. These algorithms have an inherent limitation that they require at least buffer memory of $LCM(p(a), c(a))$ for each arc a.

To overcome the limitation of SAS, some techniques have been developed, which give up the single appearance constraint for overall memory saving [13, 5]. These approaches observe the tradeoff of code and data memory size and try to minimize the code memory overhead by generating function-style codes instead of inline-style code. By defining each block as a function call, a generated code from a non SAS has only one definition of each block but paying the extra overhead of function calls.

Buffer sharing algorithms [8, 11] have been proposed to minimize data memory. These sharing algorithms analyze buffer life time and share buffers of which life-times are not overlapped with each other.

Dynamic loop count schedule for a chained structure graph has been developed, which requires optimal buffer size [10]. However the schedule is not optimal for general graphs with delay samples and feedback arcs. In this paper, we extend the previous schedule to memory optimal schedule for arbitrary SDF graphs.

IV. MOTIVATION

As discussed earlier, single appearance scheduling algorithms pay huge penalty of data memory for a graph with large sample rate changes. Moreover, no SAS exists for cyclic graphs in general. The following two examples show these limitations of SAS. With those examples we will introduce the proposed scheduling technique.

The first example is shown in Figure 2(a). The previous SAS algorithms produce 2A3B5C as the schedule result, which requires 6 and 15 data buffers on arc AB and arc BC respectively. If a buffer optimal non SAS algorithm is applied, the schedule becomes ABCABCCBCC(=2(ABC)CB2C) requiring 4 and 7 data buffers, which is minimum buffer size while additional code memory is necessary to represent multiple appearances of node B and C in a code.

To avoid the multiple lexical appearances of nodes in buffer optimal non SAS, we propose a dynamic loop count single appearance scheduling called dlcSAS which converts a buffer optimal non SAS to a single appearance schedule while preserving the minimum buffer size. Examine the non SAS in Figure 2 (b). In the buffer optimal schedule, whenever a sink node has enough samples on its input arc it should be executed. Hence, node B can be executed twice after the second invocation of node A while node B can be executed only once after the first invocation of node A. In the proposed dlcSAS, we notate this varying loop count of node B as 2(A{1,2}B) meaning that the loop count values of node B are 1 and 2 alternatively every invocation of node A. Similarly, node C can be executed twice after the second and the third invocations of node B while it can be executed only once after the first invocation of node B. The schedule is represented as 3(B{1,2,2}C) in the proposed dlcSAS. By combining the two schedules, we obtain the final dlcSAS, 2(A{1,2}(B{1,2,2}C)). The generated code template from this dlcSAS is shown in Figure 2 (c). Note that the generated code has a single appearance of each block while preserving the minimum buffer memory as the buffer optimal non SAS.

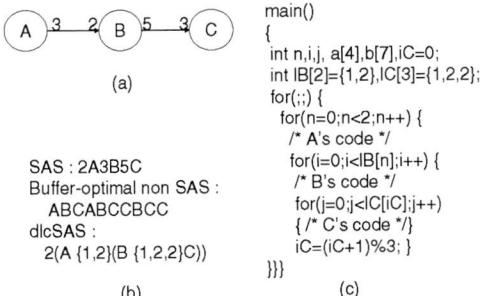

SAS : 2A3B5C
Buffer-optimal non SAS :
  ABCABCCBCC
dlcSAS :
  2(A {1,2}(B {1,2,2}C))

(b)

```
main()
{
  int n,i,j, a[4],b[7],iC=0;
  int IB[2]={1,2},IC[3]={1,2,2};
  for(;;) {
    for(n=0;n<2;n++) {
      /* A's code */
      for(i=0;i<IB[n];i++) {
        /* B's code */
        for(j=0;j<IC[iC];j++)
        {/* C's code */}
        iC=(iC+1)%3; }
}}}
```

(c)

Fig. 2. (a) An SDF graph (b) schedule results and (c) generated code by dlcSAS

The second example illustrates a cyclic graph that has no valid SAS as shown in Figure 3 where there are 4 initial delay samples on arc BA. 2ABAB is the only valid schedule and it is a buffer optimal non SAS. We can translate it as a dlcSAS that is 2({2,1}A B). It means that the first loop count of node A is 2 and the second is 1.

For the simple examples discussed above, dlcSAS may be regarded as a different representation of non SAS. In order to transform non SAS to dlcSAS, we first choose the appearance order of each node by applying topological sort. And then, we determine the loop count of each node by comparing the non SAS with the appearance order.

For instance, assume that ABACABBD schedule is given. We choose the appearance order as "ABCD". By comparing AB with ABCD, we make a schedule of {1}A{1}B{0}C{0}D. By comparison of AC with ABCD, we build a schedule of {1,1}A{1,0}B{0,1}C{0,0}D. Finally, the schedule becomes {1,1,1}A{1,0,2}B{0,1,0}C{0,0,1}D by comparing ABBD with ABCD.

Even though we can build dlcSAS equivalent to any schedule, we are interested in a code with simple expression of loop count computation to minimize code memory and performance overhead. In the following section, we will discuss how to compute the loop count with simple computation.

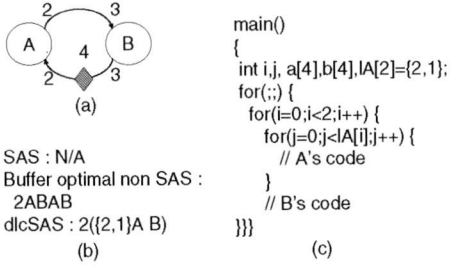

SAS : N/A
Buffer optimal non SAS :
  2ABAB
dlcSAS : 2({2,1}A B)

(b)

```
main()
{
  int i,j, a[4],b[4],IA[2]={2,1};
  for(;;) {
    for(i=0;i<2;i++) {
      for(j=0;j<IA[i];j++) {
        // A's code
      }
      // B's code
}}}
```

(c)

Fig. 3. (a) An SDF graph with delay samples (b) schedule results and (c) generated code by dlcSAS

V. DYNAMIC LOOP COUNT SINGLE APPEARANCE SCHEDULING ALGORITHM

A. Dynamic Loop Count for a Chained Structure Graph

In this section, we briefly explain dlcSAS for chained structure in the previous paper [10].

When we change the loop count value of a source node, the source node should be executed multiple times to produce the sufficient number of samples for the sink node. Let r be the accumulated number of samples on the arc. After the source node is executed h times, $h * p(a) + r$ samples are available. Since $h * p(a) + r \geq c(a)$ in order to execute the sink node, h becomes $\lceil \frac{c(a)-r}{p(a)} \rceil$. After executing both nodes, the accumulated number r is updated as $r + h * p(a) - c(a)$.

When we change the loop count value of a sink node, the sink node can be executed until the accumulated samples are exhausted. Let k be the loop count of the sink node. Since there are $r + p(a)$ samples after execution of the source node and $r + p(a) - k * c(a) \geq 0$, k becomes $\lfloor \frac{r+p(a)}{c(a)} \rfloor$. After executing both nodes, there are $r + p(a) - k * c(a)$ samples on the arc. Note that we use $_{p(a)}h_{c(a)}$ and $_{p(a)}k_{c(a)}$ to represent loop counts.

Equation 1 summarizes the formulation of dynamic loop count in both cases.

Equation 1. *For each arc a, $r = d(a)$ initially and*

(i) if a schedule is $(h\ src(a))(sink(a))$, $h = \lceil \frac{c(a)-r}{p(a)} \rceil$ and $r = r + h * p(a) - c(a)$.

(ii) if a schedule is $(src(a))(k\ sink(a))$, $k = \lfloor \frac{p(a)+r}{c(a)} \rfloor$ and $r = r + p(a) - k * c(a)$.

B. Dynamic Loop Count for General Graphs

Now we explain the main dlcSAS algorithm for general graphs. It is complex to minimize data buffer memory size for a arbitrary graph. While for an acyclic graph without delay samples an algorithm of which time complexity is $O(e^3)$ have been developed [2] where e is the number of arcs, the buffer minimum scheduling becomes NP [9] for a graph with delay samples. Therefore it is practically impossible to build a schedule with simple expression of loop count computation without knowing buffer size.

Fortunately, when we know buffer size on each arc at compile time we can compute loop count for each node at run time with slight performance overhead. Therefore we assume that the buffer size for each arc is already computed at compile time by using existing heuristics in this paper.

Since a node consumes samples on input arcs and produces samples onto output arcs, it can be executed while there are sufficient samples on all input arcs and the number of samples on every output arc does not exceed the given buffer size.

Since a node can be executed while there are enough samples on all of its input arcs, the maximum loop count k for the node becomes minimum value among the number of live samples ($r(e_i)$) over the consuming rate ($c(e_i)$) on arc e_i that is an input arc. Since it consumes $k * c(e_i)$ samples that should be no greater than the number of samples $r(e_i)$ on arc e_i when the node is executed k times, $k * c(e_i) \leq r(e_i)$ and $k \leq \frac{r(e_i)}{c(e_i)}$. Moreover the node can produce samples while the number of samples does not exceed buffer size on every arc. When the node is executed h times, the number of produced samples is $h * p(e_j)$. Since there are $r(e_j)$ samples, the total number of samples is $h * p(e_j) + r(e_j)$ that should be no greater than buffer size $bs(e_j)$ on arc e_j. So $h * p(e_j) + r(e_j) \leq bs(e_j)$ and $h \leq \frac{bs(e_j) - r(e_j)}{p(e_j)}$.

A loop count computation for a node A is summarized as following:

Equation 2.

$$LoopCount = \min(k, h)$$

$$k = \min_{e_i \in inputArc(A)} \lfloor \frac{r(e_i)}{c(e_i)} \rfloor$$

$$h = \min_{e_j \in outputArc(A)} \lfloor \frac{bs(e_j) - r(e_j)}{p(e_j)} \rfloor$$

$$For\ e_i \in inputArc(A), e_i -= LoopCount * c(e_i)$$

$$For\ e_j \in iutputArc(A), e_j += LoopCount * p(e_j)$$

where $r(e_i)$ indicates the number of remained samples on arc e_i.

In Equation 2, k denotes loop count by considering input arcs and h considering output arcs. Since the final loop count is constrained by the number of samples on the input arcs and the remained buffer size on the output arcs, the loop count becomes minimum number between k and h. After determining the loop count, we update the number of samples on each arc connected with the node. By using Equation 2, we can build memory optimal dlcSAS for arbitrary SDF graphs.

Theorem 1. *Every buffer optimal schedule for synchronous dataflow graphs can be transformed to an equivalent dynamic loop count single appearance schedule that requires same buffer size.*

Since we consider all arcs to compute loop count with preserving the number of samples, the run time overhead of the proposed scheduling is proportional to the number of arcs in the given graph.

Consider Figure 4(a) in [2], in which *bs* on each arc indicates optimal buffer size. Since the subgraph of node A and B is chained structure, the algorithm for chained structure is applied. Therefore we build $(_4h_6\text{A})\text{B}$ schedule. For the remained nodes, we need to apply the scheduling algorithm for general graphs. First we determine lexical appearance order by applying topological sort. In this example, the order becomes (AB) C D E F G. Note that we use topological ordering to minimize performance overhead by minimizing zero loop count even though any ordering is applicable. Since the loop count of node B is dependent on arc BC and BD, the loop count denoted l_B is the minimum between (10-rBC)/4 and (5-rBD) where rBC and rBD indicate the number of samples on arc BC and BD respectively. Similarly, we can compute loop counts

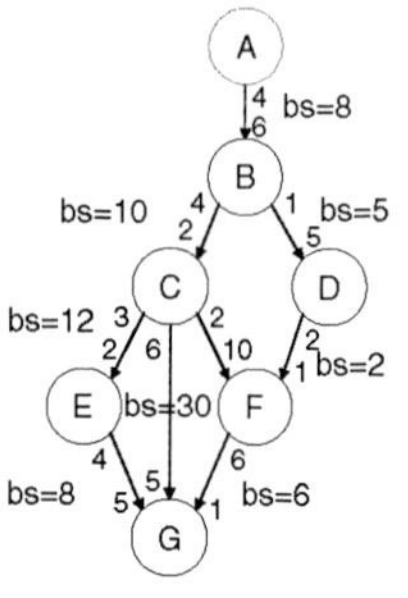

(a)

$l_B((_4h_6A)B)\ l_C(C)\ l_D(D)\ l_E(E)\ l_F(F)\ l_G(G)$

$l_B=\min(h_{BC},h_{BD}), l_C=\min(k_{BC},h_{CE},h_{CF},h_{CG}),$
$l_D=\min(k_{BD},h_{DF}), l_E=\min(k_{CE},h_{EG}),$
$l_F=\min(k_{CF},k_{DF},h_{FG}), l_G=\min(k_{CG},k_{EG},k_{FG})$

(b)

```
main() {
int I,j,IA,IB,IC,ID,IE,IF,IG;
int rA=0;rBC=0,rBD=0,rCE=0;
Int rCF=0,rCG=0,rDF=0,rEG=0,rFG=0;
for(;;) {
 IB = min((10-rBC)/4,(5-rBD));
 rBC+=4*IB; rBD+=IB;
 for(i=0;i<IB;i++) {
  if(rA <2) {IA=2;rA+=2;} else {IA=1;rA-=2;}
  for(j=0;j<IA;j++) /* A */
  /* B */ }
 IC = min(BC/2,(12-rCE)/3,(10-rCF)/2,(30-rCG)/6);
 rBC-=2*IC;rCE+=3*IC;rCF+=2*IC;rCG+=6*IC;
 for(i=0;i<IC;i++) /* C */
 ID = min(rBD/5,(2-rDF)/2);
 rBD-=5*ID; rDF+=2*ID;
 for(i=0;i<ID;i++) /* D */ }
 IE = min(rCE/2,(8-rEG)/4);
 rCE-=2*IE;rEG+=4*IE;
 for(i=0;i<IE;i++) /* E */
 IF=min(rCF/10,rDF,(6-rFG)/6);
 rCF-=10*IF;rDF-=IF;rFG+=6*IF;
 for(i=0;i<IF;i++) /* F */
 IG=min(rCG/5,rEG/8,rFG);
 for(i=0;i=IG;i++) /* G */
}}
```

(c)

Fig. 4. (a) A general SDF graph, (b) schedule the graph and (c) generated code by dlcSAS

for the other nodes by applying Equation 2 as shown in Figure 4(b). Finally we can generate a code as shown in Figure 4(c).

C. Optimization of Schedule for General Graphs

Since the loop count of a node is constrainted by the accumulated samples and buffer size on each connected arc, $2 * e$ computations are required where e denotes the number of arcs in the graph. Some constraints on arcs, however, can be eliminated when the constraints have been never used. Hence optimization techniques eliminate unnecessary constraints.

First we examine the loop count computation dependency by running the unoptimized schedule for an iteration period. If the number of samples on an arc is not used for computation of a loop count then the expression referring to the arc can be eliminated. Furthermore, when no node refers to an arc, variables on the arc are removed.

If a loop count is computed by an expression on an arc, dlcSAS for chained structure graphs is applied. If the loop count of a node is dependent on its output arc then Equation 1(i) is applied. It is, however, not applicable when the loop count is not zero even if the loop count of the sink node is zero. For instance, the schedule of ABCAC = {1,1}A {1,0}B {1,1}C cannot be represented by $l_B((h\ A)B)l_C C$ since $(h\ A)B$ cannot express {1}A {0}B schedule. Similarly, Equation 1(ii) is applicable if the loop count of the sink node is only dependent on its input arc and its loop count becomes zero whenever its source node loop count is zero.

In addition, if the loop count only has 0 or 1 then more compact code can be generated.

We can summarize the optimization techniques as follows:

Algorithm

1: Run unoptimized dlcSAS for an iteration period.

2: Eliminate a loop count computation if the computation is not used to compute minimum loop count value.

3: Eliminate the updating code of the number of samples which is not referred to in a loop count computation.

4: For arc a, if the loop count $l_{src(a)}$ of source node $src(a)$ is only dependent on arc a and $l_{src(a)}=0$ whenever $l_{sink(a)}=0$ then Equation 1(i) is applied. Similarly if $l_{sink(a)}$ is only dependent on arc a and $l_{sink(a)}=0$ whenever $l_{src(a)}=0$ then Equation 1(ii) is used.

5: If loop count l_A of node A has only 0 or 1 then "if-statement" code is generated instead of "for-loop" as following:

$$\text{if}(\bigwedge_{e_i \in outputArc(A)} (r(e_i) \leq bs(e_i) - p(e_i))$$
$$\bigwedge_{e_j \in inputArc(A)} (r(e_j) \geq c(e_j)))$$

```
{
For all e_i ∈ outputArc(A), r(e_i)+ = p(e_i);
For all e_j ∈ inputArc(A), r(e_j)- = c(p_j);
/* A's code */
}
```

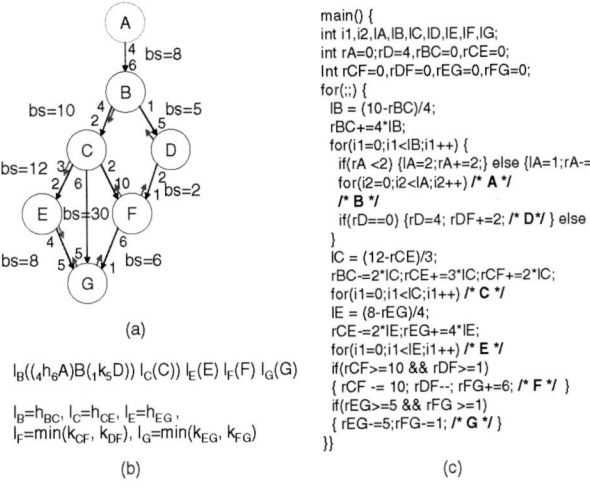

Fig. 5. (a) Examine whether each arc contributes loop count computation or not. Short arrows indicate the loop count of the node is dependent on the arc. (b) Optimized schedule and (c) optimized code by dlcSAS

Now, we apply the optimization techniques to Figure 4 (a) graph. During simulating the code of Figure 4 (c), we inspect which arcs each loop count is dependent on. In Figure 5 (a), short arrows represent the dependency of loop count on arcs. The loop count of node B is $(10-rBC)/4$ ignoring $(5-rBD)$ since it is dependent on arc BC only. Node C just requires rCE value to compute its loop count and $l_C = h_{CE} = (12-rCE)/3$ although it has three arcs BC, CE and EF. Similarly, we know that loop count of node D is dependent on rBD, node E is on rEG, node F on rCF and rDF, and node G on rEG and rFG. Since no node refers to arc CG, rCG is not necessary to be maintained.

Figure 5(b) represents optimized schedule. By the optimization, we can reduces loop count computations from 16 expressions to 7 expressions. Furthermore, we can eliminate 8 condi-

tional expressions to find minimum values from 10 conditions in the unoptimized schedule.

Since the loop counts of node F and node G have only two values of 0 and 1, the generated code is more compact as shown in Figure 5 (c).

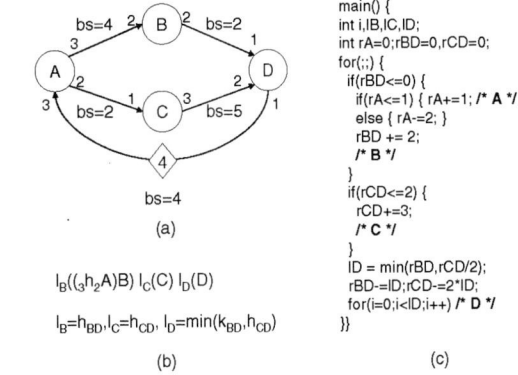

Fig. 6. (a) cyclic SDF graph (b) Optimized schedule and (c) optimized code by dlcSAS

Figure 6(a) indicates a graph with a cycle. First, it acquires buffer size on each arc by using existent heuristics. And then we determine the lexical appearance order by applying topological sort. In this example, assume that the order is ABCD. By running a graph with $(l_A A)(l_B Br)(l_C C)(l_D D)$ schedule, we examine the dependency of loop count. This example shows that node A can be clustered into node B since the loop count of node A is only dependent on arc AB and the loop count of node A is always 0 when that of node B is 0. Figure 6(b) represents the optimized schedule. The loop counts of node B and C rely on arc BD and CD respectively and the loop count of node D on both arc BD and CD. Figure 6(c) shows a generated code with minimal performance overhead for computation of loop counts.

VI. EXPERIMENTS

We have experimented several examples to demonstrate effectiveness of our approach. Table I represents minimum buffer size for various examples between previous SAS(APGAN) and proposed dlcSAS. The last column indicates the buffer size reduction by dlcSAS compared with the previous SAS, which is computed by (previousSAS-dlcSAS)/previousSAS. For Figure 6 that contains feedback cycle, the previous SAS is not applicable since there are not enough delay samples on arc DA. Note that 6 delay samples are required on arc DA for the previous SAS to produce a schedule. Since sample rate is stable in the modem application [3], both the previous SAS and dlc SAS require same size buffer.

In order to measure memory size and performance overhead on real platform, we used the arm compiler and armulator for ARM920T processor.

Figure 7 represents the SDF graph of a 4-channel non-uniform filterbank. The sample rates are shown on each arch

TABLE I
COMPARISON OF BUFFER SIZE

application	SAS(APGAN)	dlc SAS	reduction(%)
Figure 1	14	9	36
Figure 4	194	81	58
Figure 6	N/A	17	N/A
modem [3]	38	38	0
Figure 5 in [3]	120	28	77

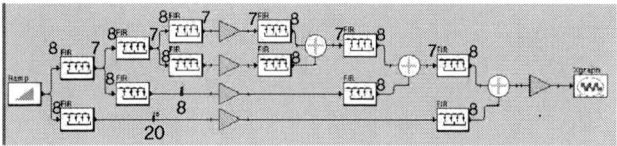

Fig. 7. SDF graph for a non-uniform filterbank. The highpass channel retains 1/8 of the spectrum and the lowpass channel retains 7/8 of the spectrum

whenever they are different from unity. In the 4-channel non-uniform filterbank, the lowpass filters retain 7/8 of the spectrum while the highpass filters retain 1/8. We can also save more than 20% total memory with less than 1% performance overhead in this example.

VII. CONCLUSION

In this paper, we presented a new single appearance scheduling algorithm to minimize data memory and code memory jointly for synchronous dataflow graphs. Our algorithm is different from previous algorithms in terms of determining loop counts at run time even though the SDF graphs can be scheduled at compile time. Therefore while it introduces performance overhead to compute loop counts(which is much lower than function call approaches), it reduces buffer memory requirement to buffer lower bounds of non single appearance schedule for arbitrary graphs. Therefore we can argue that the proposed schedule is memory optimal. For non uniform filter bank application, we can reduce more than 20% of total memory size with less than 1% performance overhead compared with the previous single appearance schedules.

In the future, we will extend the schedule to consider buffer sharing.

TABLE II
COMPARISON FOR NON-UNIFORM FILTER BANK EXAMPLE

	previous SAS	dlcSAS	ratio(%)
code memory	13128 bytes	13540 bytes	3.14
data memory	15720 bytes	9664 bytes	-38.52
total memory	28848 bytes	23204 bytes	-19.56
cycles	71060K cycles	71363K cycles	0.43

VIII. ACKNOWLEDGMENTS

This work was partially supported by NSF grants CCR-0203813, ACI-0204028, National Research Laboratory Program (Grant No. M1-0104-00-0015), and IT leading R&D Support Project funded by Korean MIC.

REFERENCES

[1] *COSSAP User's Manual.* Synopsys Inc. 700 E. Middlefield Rd. Mountain View,CA94043, USA.

[2] M. Ade, R. Lauwereins, and J. A. Peperstraete. Data memory minimization for synchronous data flow graphs emulated on dsp-fpga targets. In *DAC*, June 1997.

[3] S. S. Bhattacharyya, P. K. Murthy, and E. A. Lee. Synthesis of embedded software from synchronous dataflow specifications. In *Journal of VLSI Signal Processing*, volume 21, pages 151–166, June 1999.

[4] J. T. Buck, S. Ha, E. A. Lee, and D. G. Messerschimitt. Ptolemy: A framework for simulating and prototyping heterogeneous systems. In *Int. Journal of Computer Simulation, special issue on Simulation Software Development*, volume 4, pages 155–182, April 1994.

[5] M. Ko, P. K. Murthy, and S. S. Bhattacharyya. Compact procedural implementation in DSP software synthesis through recursive graph decomposition. In *Proceedings of the International Workshop on Software and Compilers for Embedded Processors*, pages 47–61, Amsterdam, The Netherlands, September 2004.

[6] R. Lauwereins, M. Engels, J. A. Peperstraete, E. Steegmans, and J. V. Ginderdeuren. Grape: A case tool for digital signal parallel processing. In *IEEE ASSP Magazine*, volume 7, pages 32–43, April 1990.

[7] E. A. Lee and D. G. Messerschmitt. Static scheduling of synchronous dataflow programs for digital signal processing. In *IEEE Transaction on Computer*, volume C-36, pages 24–35, January 1987.

[8] P. K. Murthy and S. S. Bhattacharyya. Shared buffer implementations of signal processing systems using lifetime analysis techniques. In *IEEE Transactions on Computer-Aided Design of Integrated Circuits and Systems*, volume 20, pages 177–198, February 2001.

[9] P. K. Murthy, S. S. Bhattachayya, and E. A. Lee. Joint minimization of code and data for synchronous dataflow programs. In *Journal of Formal Methods in Systems Design*, volume 11, pages 41–70, July 1997.

[10] H. Oh, N. Dutt, and S. Ha. Single appearance schedlue with dyanmic loop count for sy. In *CASES2005*, volume 2005, pages 514–529, Sept 2005.

[11] H. Oh and S. Ha. Memory-optimized software synthesis from dataflow program graphs with large size data samples. In *EURASIP Journal on Applied Signal Processing*, volume 2003, pages 514–529, May 2003.

[12] S. Ritz, M. Willems, and H. Meyr. Scheduling for optimum data memory compaction in block diagram oriented software synthesis. In *Proceedings of the ICASSP 95*, May 1995.

[13] W. Sung and S. Ha. Memory efficient software synthesis using mixed coding style from dataflow graph. In *IEEE Transaction on VLSI Systems*, volume 8, pages 522–526, October 2000.

Wire Sizing with Scattering Effect for Nanoscale Interconnection

Sean X. Shi and David Z. Pan
Department of Electrical and Computer Engineering
University of Texas at Austin
Austin, TX 78712, USA
Email: sean.shi@mail.utexas.edu, dpan@ece.utexasedu

Abstract—For nanoscale interconnection, the scattering effect will soon become prominent due to scaling. It will increase the effective resistivity and thus interconnection delay significantly. Existing works on scattering effect are mostly performed using very complicated physics-based models, while the scattering impact on nanoscale VLSI interconnect and optimization have not been studied. In this paper, we first present a simple, closed-form scattering effect resistivity model based on extensive empirical studies on measurement data. Then we apply the proposed scattering model to revisit several classic wire sizing/shaping problems. Our experimental results show that if the scattering effect is ignored or characterized inaccurately beyond 65nm, the resulting interconnect optimization might be way off from the real optimal solution, e.g., up to 70% underestimation of the delay, or 20x oversizing. We also obtain the new closed-form wiresizing functions with consideration of scattering effects.

I Introduction

AS the feature size continues to shrink, the lateral dimension of conductors will be approaching the mesoscopic regime in which the diameter of the wire is in the range of or smaller than the mean free path of the electrons (λ, about $40nm$ for copper at room temperature), and the electrical resistivity of metallic conductors is increased compared to the resistivity of bulky metal. The earliest work on this dates back to 1938, when an expression for the resistivity of metal thin films (1-dimensional) was derived by Fuchs [1]. Later on it was extended to 2-dimensional by the FS model [2] for thin/narrow wires. Basically, the FS model accounts for the *surface* scattering. Later on, the MS model [3] was developed to incorporate the *grain boundary* scattering, which also increases the wire resistivity. For both surface and grain boundary scattering effects, very complicated quantum mechanical effects can be applied to obtain the empirical parameters in FS or MS model [4, 5].

While MS and FS models have been tested with measurement data for polycrystalline nanowires [6-8], very few experimental results on copper (Cu) film or interconnect have been reported until recently, e.g., the size effect of copper thin film was studied in [9], and the

resistivity of copper wires with widths under 50nm was reported in [10-13]. The key observation is that the resistivity of copper wires will increase significantly as the wires width decreases [9-14]. While exploratory structures such as carbon nanotubes are being studied as possible replacement of copper interconnect [15-17], at least by 22nm technology, it is not likely that copper will be replaced by carbon nanotube or other materials [15]. There are also efforts in manufacturing improvement to partially reduce the scattering effect of copper wires [18, 19], but even so the scattering effect can no longer be ignored as the technology continues to scale down.

A popular method to reduce the interconnection delay is wire sizing. In fact, there have been a lot of works on it, e.g., the wire sizing and shaping without fringing capacitance [20, 21], with fringing capacitance [22, 23], under transmission line model [24], or with single-width sizing (1-WS) or two-width sizing (2-WS) [28]. Wire sizing for multi terminal nets have also been extensively studied [25-27]. However, all these previous wire sizing algorithms are based on the constant resistivity model without considering the scattering effects.

In this paper, we systematically study the impact of scattering effect on copper interconnection and wire sizing/shaping. Our key contributions include:

- Since existing physics-based scattering models are too complicated for nanoscale interconnect optimization, we propose a simple yet high-fidelity width-dependent resistivitiy model based on the experimental data for copper wire [10], thin film [9], and from ITSR 2004 [14]. Our proposed model fits well compared to these data with the regression coefficient R≥0.999.

- Based on our simplified closed-form model, we revisit various wire sizing issues with consideration of the scattering effect. As wire width shrinks, wire sizing will be more effective and necessary to reduce the interconnection delay because of scattering effect. Our experimental results also show that if the scattering effect is ignored or characterized inaccurately, the resulting interconnect optimization might be totally off from the real optimal solution, e.g., up to 70% delay underestimation, or 20x over sizing.

- We obtain the new closed-form wire sizing functions with consideration of scattering effects, for both discrete wire sizing (with one sizable width), and continuous wire sizing (with infinite number of widths). The differences compared to those without scattering effects are pointed out.

The rest of the paper is organized as follows. In section II, we present the preliminaries and parameters used for our study [14, 28]. Section III presents our scattering effect modeling for resistivity of nanoscale Cu interconnection, and shows the delay impact caused by the scattering effect. The new wire sizing formulae and results are presented in Sections IV & V, followed by the conclusion in Section VI.

II PRELIMINARIES

This section presents the preliminaries and key parameters used in this paper. The driver of the interconnection is modeled as an effective resistance R_d connected to an ideal voltage source and a sink as a load capacitance C_L. The Elmore delay model [29, 30] is used to compute the interconnection delay. The notations for the key interconnect and device parameters include:

w_{min}, minimum wire width, nm;

c_a, unit area capacitance, fF/μm;

c_f, unit effective-fringing capacitance, fF/μm;

c_g, input capacitance of a minimum device, fF;

r_g, output resistance of a minimum device, kΩ;

ρ_0, resistivity of Cu, assume no scattering, μΩ-cm;

ρ_{max}, Max Cu resistivity of the minimum width, μΩ-cm;

AR, aspect ratio;

$t = w_{min} \cdot AR$, metal thickness;

$R_d = r_g/100$, driver resistance;

$C_L = 100 \cdot c_g$, load capacitance.

The basic parameters used in this paper are shown in Table 1. Note that these parameters are used mainly to illustrate the impact of scattering effect. The values are extracted according to [10, 14, 28, 31]. If necessary, more complete and specific parameters can be used.

Table 1 Basic Parameters

Year	2004	2007	2010	2013	2016
w_{min}	90	65	45	32	22
c_g	0.0625	0.0573	0.0375	0.0246	0.0161
r_g	24.09	22.75	24.82	27.07	29.53
c_a	0.056	0.056	0.056	0.056	0.056
c_f	0.04	0.04	0.04	0.04	0.04
ρ_0	2.2	2.2	2.2	2.2	2.2
ρ_{max}	3.35	3.79	4.49	5.42	6.88
AR	1.7	1.8	1.8	1.9	2

III SCATTERING EFFECT MODELING & IMPACT ON INTERCONNECTION DELAY

In this section, an empirical closed-form scattering effect model for nanoscale copper interconnect is proposed and validated, then it impact on interconnect delay is shown

A Model of Scattering Effect

Equation (1) (at the bottom of this page) shows the size-dependent part of the scattering effect model for metal wires, where only the grain boundary scattering (MS model [3]) is considered [8]. Obviously, such model is too complicated to be used for interconnection delay calculation and wire sizing. A simplified analytical model will be desirable for VLSI physical design applications.

B Simplified Closed-Form Model of Scattering Effect

Based on our empirical study on both MS and FS models and curve fitting, we obtain the following simple closed-form width-dependent resistivity model with scattering effect.

$$\rho(w) = \rho_B + \frac{K_\rho}{w} \quad (2)$$

Fig. 1 shows that the simple resistivity model fits well with the measured experimental data [10]. We have also verified this model based on the measurement data from [9] and the complicated model used in ITRS 2004 [14].

In the rest of this paper, we use the fitting parameters ρ_B=2.202 μΩ·cm, and K_ρ=1.030×10^{-15} Ω·m^2 based on [10]. Note that ρ_B is almost the same as ρ_0, as when the dimension of Cu wire is large enough, the scattering effect can be ignored.

$$\frac{\rho_0}{\rho} = \frac{3}{4\pi hw} \int_{-h/2}^{h/2} dy \int_{-w/2}^{w/2} dx \int_{-\pi+\arctan(w/h)}^{\arctan(-w/h)} d\varphi \int_0^\pi \sin(\theta)\cos^2(\theta) \left[1-(1-p)\frac{\exp\left(-\frac{w}{2\lambda\cos(\theta)\cos(\varphi)}\right)}{1-p\cdot\exp\left(-\frac{w}{2\lambda\cos(\theta)\cos(\varphi)}\right)} \right] d\theta \quad (1)$$
$$+ \frac{3}{4\pi hw} \int_{-h/2}^{h/2} dy \int_{-w/2}^{w/2} dx \int_{\arctan(-w/h)}^{\arctan(w/h)} d\varphi \int_0^\pi \sin(\theta)\cos^2(\theta) \left[1-(1-p)\frac{\exp\left(-\frac{w}{2\lambda\cos(\theta)\cos(\varphi)}\right)}{1-p\cdot\exp\left(-\frac{w}{2\lambda\cos(\theta)\cos(\varphi)}\right)} \right] d\theta$$

2006 Asia and South Pacific Design Automation Conference

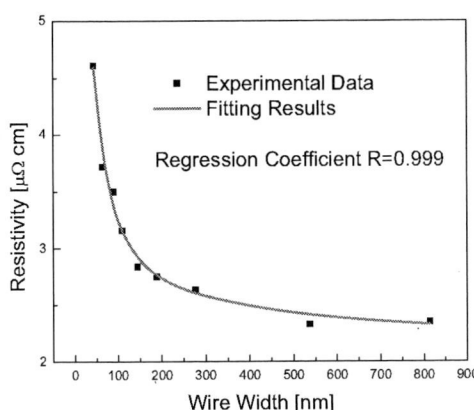

Fig.1 Resistivity of nanoscale wire. Observe that the resistivity *increases dramatically* as we decrease the wire width

C Impact of Scattering Effect on Interconnection Delay

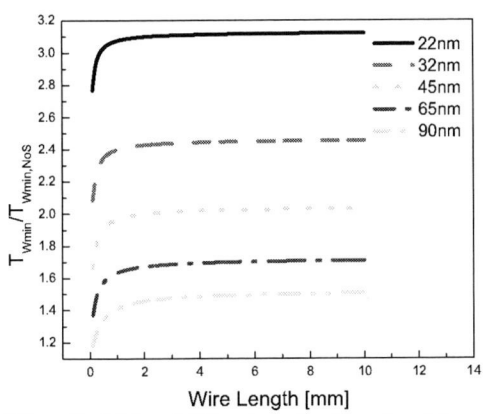

Fig.2 Normalized delay of different wirelengths under minimum wire width. The normalized delay is the ratio of delay with scattering effect to delay without scattering effect. Observe that the ratio is *always greater than 1* and it *worsens* with decreasing feature size.

The delay of single width wire including scattering effect can be written as follows

$$T_{1\ WS} = R_d \cdot \left[c_a \cdot l \cdot w + c_f \cdot l + C_L \right]$$
$$+ \left[\frac{c_a \cdot l \cdot w}{2} + \frac{c_f \cdot l}{2} + C_L \right] \cdot \left[\rho_B + \frac{K_\rho}{w} \right] \cdot \frac{l}{w \cdot t} \quad (3)$$

where l is the wire length. In Fig. 2, the delay of the minimum width wires is calculated, and the ratio shows the delay with scattering effect over that without scattering consideration. We can see that in nanoscale manufacturing, scattering effect should be considered. Otherwise, interconnect delay may be underestimated significantly, and cause timing error.

IV WIRE SIZING FOR INTERCONNECT DELAY AND AREA ESTIMATION BASED ON SCATTERING EFFECT

In this section, we will derive the new wire sizing

formulae based on the new width-dependent resistivity model with scattering effect Eqn.(1). Our experimental results show that scattering effect will have major impact on wire sizing in nanoscale interconnections.

A Efficiency of Wire Sizing

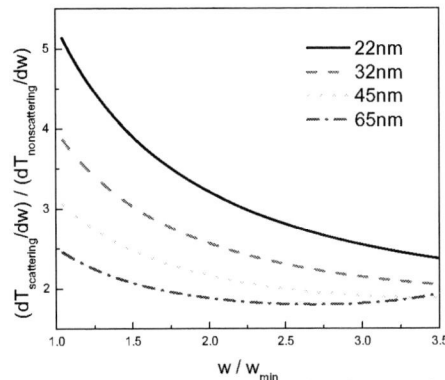

Fig.3 Compare the efficiency of wire sizing based on scattering and non-scattering. The efficiency defined as differential of delay over width. Wire length is 0.01 mm. Because of *scattering effect*, wire sizing beomes more effecient to reduce interconnection delay.

As mentioned in section III, in the nanoscale interconnection, scattering effect induces bigger resistivity, which in turn causes bigger delay. Thus, wider wires become more effective than before to compensate the more rapidly increase of wire resistance of narrow wires (additional effect due to increasing resistivity). In other words, wire sizing considering scattering effect becomes more efficient and more necessary. The efficiency of wire sizing can be defined as $\partial T/\partial w$, the sensitivity of delay reduction due to wire sizing. Eqn. (4) is the traditional wire sizing efficiency with a constant resistivity (bigger value can be used to mimic the impact of scattering effect). And (5) is the wire sizing efficiency with consideration of width-dependent scattering effect.

$$\frac{\partial T_{NonScattering}(w,l,S)}{\partial w} = R_d \cdot c_a \cdot l - \frac{1}{w^2} \cdot \left[\frac{1}{2} \cdot c_f \cdot \rho \cdot \frac{l^2}{t} + C_L \cdot \rho \cdot \frac{l}{t} \right] \quad (4)$$

$$\frac{\partial T(w,l,S)}{\partial w} = R_d \cdot c_a \cdot l - \frac{1}{w^2} \cdot \left[\frac{1}{2} \cdot c_f \cdot \rho_B \cdot \frac{l^2}{t} + C_L \cdot \rho_B \cdot \frac{l}{t} \right] \quad (5)$$
$$- \frac{1}{w^2} \cdot \frac{1}{2} \cdot c_a \cdot K_\rho \cdot \frac{l^2}{t} - \frac{1}{w^3} \cdot \left[c_f \cdot \frac{l^2}{t} + 2 \cdot C_L \cdot \frac{l}{t} \right] \cdot K_\rho$$

In Fig. 3, wire sizing efficiencies with considering scattering and non-scattering are compared. It shows that wire sizing became more efficient than before where no scattering effect needs to be considered.

B 1-WS Model

In previous wire sizing and planning works [28], resistivity is assumed to be constant, and the optimal single width sizing (1-WS) can be calculated by:

505

$$w_{optimal,no-scattering} = \sqrt{\frac{c_f \cdot l + 2 \cdot C_L}{2 \cdot R_d \cdot c_a \cdot t} \cdot \rho} \qquad (6)$$

In nanoscale interconnection, the scattering effect is prominent. To calculated nanoscale interconnection delay, the resistivity of bulky Cu should be replaced by a bigger value to avoid underestimation of delay (see section III-C). But if we put (6) into the use of wire sizing of nanoscale interconnection just by replacing old resistivity with a bigger value (such as ρ_{max}, which is the resistivity of the minimum width wire of some specific technology node) and regarding this value as a constant, the optimal width of nanoscale Cu wire will be overestimated for the same reason that the resistance will decrease faster during wire width widening because of scattering effect.

According to (3), to minimize the interconnection delay, $\partial T_{1\,WS}/\partial w = 0$. It becomes a cubic equation of wire width. Calculation shows that, from 130nm to 18 nm technology, only one real solution is bigger than the minimum wire width. The analytical function is shown as below:

$$w_{Optimal} = w_1 = 2\sqrt{\frac{a_1}{3}} \cos\left(\frac{\vartheta}{3}\right) \ge w_{min} \qquad (7)$$

where, $\vartheta = \cos^{-1}\left(\dfrac{K_\rho (c_f l + 2C_L)\sqrt{54 R_d c_a t}}{\left(c_a K_\rho l + c_f \rho_B l + 2 C_L \rho_B\right)^{3/2}}\right)$

$$a_1 = \frac{c_a K_\rho l + c_f \rho_B l + 2 C_L \rho_B}{2 \cdot R_d \cdot c_a \cdot t}$$

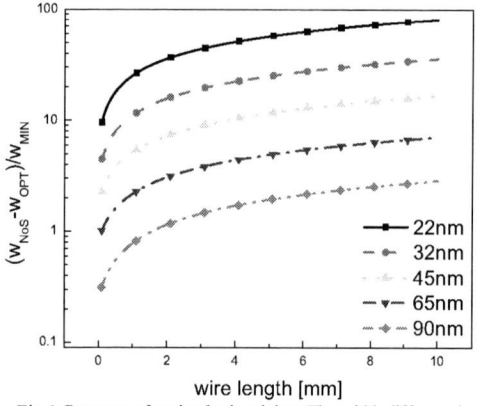

Fig.4 Compare of optimal wire sizing. The width diffrence is normalized in the way of being divided by minimum width. Observe that it is *alway bigger than 1* after 65nm node and it *worsens to more than 10x* after 22nm node.

Fig.4 compares the optimal wire sizes calculated without and with the width-dependent scattering effect (3). The optimal width of 1-WS with the constant resistivity ρ_{max} can be over 10 or 20x bigger than that of 1-WS with the width-dependent resistivity from scattering effect. This would cause excessive area waste and routability problem. On the other hand, if we use a smaller constant resistivity ρ_0 (in Table 1) for 1-WS, the resulting "optimal" width will be less, but such "optimal" delay can be much off from the real optimal obtained based on the width-dependent resistivity caused by the scattering effect. Using (7), the optimal wire width can be calculated, and the result of wire sizing is shown in Fig. 5. The delay can be reduced by 50-90% when using 1-WS optimization, compared to the minimum width.

Thus, it is very important to consider accurate scattering effect during the interconnect delay optimization and interconnect planning.

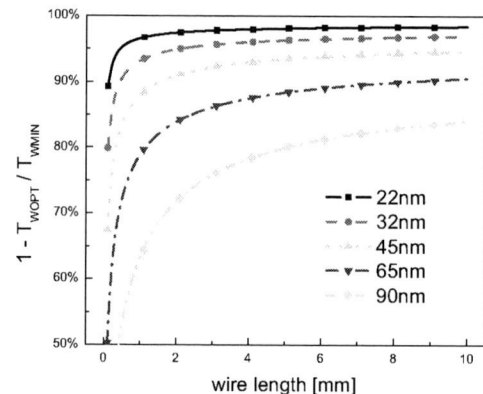

Fig.5 Delay reduction by wire sizing in nanoscale. Because of *scattering effect*, interconnection delay can be efficiently reduced by wire sizing (compared with the delay of minum width wire).

V   CONTINUOUS WIRE SIZING

In this section, continuous wire sizing/shaping is discussed. Different from classic wire shaping function [22], with the consideration of scattering effect, not only the theoretical solution but also the approximate solution will be more complicated than before.

A Euler's Differential Equation (math background)

$$I = \int_{x_0}^{x_1} F\left(x, u(x), u'(x)\right) dx$$

If *u(x)* is a function which can produce minimum *I*, according to [32], *u(x)* should satisfy Euler's Differential Equation:

$$F_u\left(x, u(x), u'(x)\right) = \frac{d}{dx} F_{u'}\left(x, u(x), u'(x)\right) \qquad (8)$$

B Wire Shaping to Minimize Elmore Delay

The wire shape can be defined as *f(x)*, and we define the terminal to contact driver R_d as *x=0*, the terminal to contact load C_L is *x=L* (*L* is the wire length). As the scattering effect is considered, the function definition for Euler's equation is different from the previous work [22].

$$T = \int_0^L R_d \cdot \left[c_a \cdot dx \cdot f(x) + c_f \cdot dx + C_L \right]$$

$$+ \int_0^L \left[C_L + \int_0^x \left(c_a f(l) + c_f \right) dl \right] \left[\frac{\rho_B}{f(x)} + \frac{K_\rho}{f^2(x)} \right] \frac{dx}{t} \qquad (9)$$

$$= R_d \cdot C_L + \int_0^L F(x, u, u') \cdot dx$$

Where, $u(x) = \int_0^x f(l) \cdot dl$, $u'(x) = f(x)$

To minimize (9), Euler's equation (8) can be rewritten as

$$c_a (u')^2 \left[2\rho_B u' + 3K_\rho \right] + c_f u' \left[\rho_B u' + 2K_\rho \right] \qquad (10)$$

$$= \left[C_L + c_a u + c_f x \right] \left[2\rho_B \cdot u' + 6K_\rho \right] u''$$

If

$$u(x) = \sum_{n=0}^{\infty} a_n \cdot x^n \qquad (11)$$

For $a_0 = 0$ and a_1 is wire width at $x=0$. According to (10), $a_2, a_3, \cdots$ can be calculated one by one, for example, a_2 is:

$$a_2 = \frac{c_a a_1^2 \left(2\rho_B a_1 + 3K_\rho \right) + c_f a_1 \left(\rho_B a_1 + 2K_\rho \right)}{4C_L \left(\rho_B a_1 + 3K_\rho \right)}$$

C Approximate Model

The continuous wire sizing/shaping function f(x) can be approximated by $g_n(x)$ with different orders of accuracy.

$$g_1(x) = a_1; \quad g_2(x) = a_1 + \left(a_2 x + a_3 x^2 \right)/2; \quad \cdots$$

$$g_n(x) = \sum_{i=1}^{2n-3} a_i x^{i-1} + \frac{1}{2} \left(a_{2i-2} x^{2i-3} + a_{2i-1} x^{2i-2} \right)$$

Fig. 6 shows the difference from $g_1(x)$ to $g_5(x)$ of 45nm technology. In most cases, the approximation using $g_3(x)$ is good enough to get the optimal wire shaping.

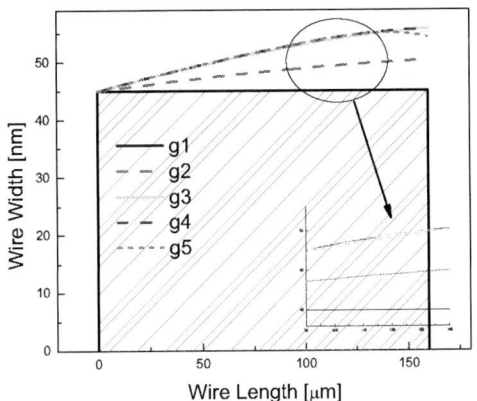

Fig. 6 Wire shaping with different order polynomial approximation

It shall be noted that $g_n(x)$ is different from [22], e.g., $g_3(x)$ is a quintic function while in [22] it is a cubic function. The scattering effect also increases the complexity of the approximate solutions.

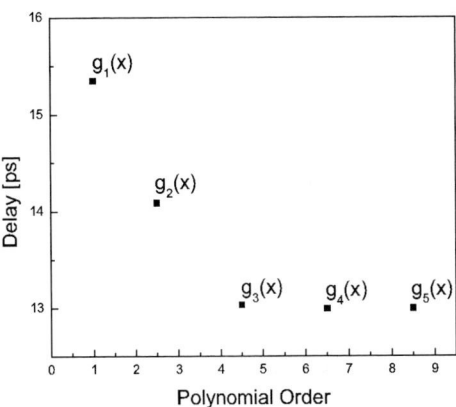

Fig. 7 Delay estimation of wire shaping (45nm node),
compared of different orders of polynomial approximation.
Observe that difference among g3, g4 and g5 is less than 0.5%

Fig.7 gives the delay comparison of different orders of polynomial approximation of wire shaping function. In this case, we can succeed reduce the interconnection delay by *18%* when using closed form $f(x)=g_3(x)$ of wire shaping instead of using minimum width $f(x)= g_3(x)=a_1$. From Fig.6, the area of wire just increase by *16%* for wire shaping. And closed form $f(x)= g_3(x)$ is accurate enough for wire shaping because the deference between this closed form function and more accurate ones are no more than 1%.

VI CONCLUSION

This paper studies an emerging topic for nanoscale interconnection, the scattering effect, as its impact on delay will soon become very important due to further technology scaling. We obtain a simple, width-dependent resistivity model due to scattering effect based on extensive empirical studies on measurement data. We then apply it to the classic wire sizing problems and show the importance of considering scattering effects for future interconnect optimizations.

To our best knowledge, this is the first work that addresses scattering effects on the nanoscale interconnect sizing. We plan to extend the model to consider other nanoscale effects such as the surface roughness.

ACKNOWLEDGMENT

This work is supported in part by IBM Faculty Award, Fujitsu, and equipment donations from Intel. The authors would like to thank Gang Xu, Anand Ramalingam and Tao Luo at UT Austin for their helpful discussions.

REFERENCES

[1] K. Fuchs, "The conductivity of thin metallic films according to the electron theory of metals," *Proc. Cambridge Philosophical Society*, 1938, pp. 100-108.

[2] E. H. Sondheimer, "The Mean Free Path of Electrons in Metals," *Adv. Phys.*, vol. 1, pp. 1-42, 1952.

[3] A. F. Mayadas and M. Shatzkes, "Electrical-Resistivity Model for Polycrystalline Films: the Case of Arbitrary Reflection at External Surfaces," *Phys. Rev*, vol. B 1, pp. 1382–1389, 1970.

[4] Z. Tesanovic, M. V. Jaric, and S. Maekawa, "Quantum Transport and Surface Scattering," *Phys. Rev. Lett.*, vol. 57, pp. 2760-2763, 1986.

[5] R. Dannenberg and A. H. King, "Behavior of grain boundary resistivity in metals predicted by a two-dimensional model," *Journal of Applied Physics*, vol. 88, pp. 2623, 2000.

[6] J. Vancea, "Unconventional Features of Free Electrons in Polycrystalline Metal Films," *International Journal of Modern Physics B*, vol. 3, pp. 1455-1501, 1989.

[7] J. Vancea, G. Reiss, and H. Hoffmann, "Mean-free-path concept in polycrystalline metals," *Phys. Rev. B*, vol. 35, pp. 6435-6437, 1987.

[8] C. Durkan and M. E. Welland, "Size effects in the electrical resistivity of polycrystalline nanowires," *Phys. Rev. B*, vol. 61, pp. 14215–14218, 2000.

[9] S. M. Rossnagel and T. S. Kuan, "Alteration of Cu conductivity in the size effect regime," *Journal of Vacuum Science & Technology B: Microelectronics and Nanometer Structures*, vol. 22, pp. 240, 2004.

[10] W. Steinhoegl, G. Schindler, G. Steinlesberger, and M. Engelhardt, "Size-dependent resistivity of metallic wires in the mesoscopic range," *Physical Review B (Condensed Matter and Materials Physics)*, vol. 66, pp. 075414, 2002.

[11] G. Steinlesberger, M. Engelhardt, G. Schindler, W. Steinhogl, A. v. Glasow, K. Mosig, and E. Bertagnolli, "Electrical assessment of copper damascene interconnects down to sub-50 nm feature sizes," *Microelectron. Eng.*, vol. 64, pp. 409-416, 2002.

[12] W. Wen and K. Maex, "Studies on size effect of copper interconnect lines," *Proc. International Conference on Solid-State and Integrated-Circuit Technology (ICSICT)*, 2001, pp. 416.

[13] C.-U. Kim, J. Park, N. Michael, P. Gillespie, and R. Augur, "Study of Electron-Scattering Mechanism in Nanoscale Cu Interconnects," *Journal of Electronic Materials*, vol. 32, pp. 982-987(6), 2003.

[14] "International Technology Roadmap for Semiconductors (ITRS)," 2004.

[15] A. Naeemi, R. Sarvari, and J. D. Meindl, "Performance comparison between carbon nanotube and copper interconnects for gigascale integration (GSI)," *Electron Device Letters*, vol. 26, pp. 84, 2005.

[16] A. Raychowdhury and K. Roy, "A circuit model for carbon nanotube interconnects: comparative study with Cu interconnects for scaled technologies," *Proc. ICCAD*, 2004, pp. 237.

[17] N. Srivastava and K. Banerjee, "A Comparative Scaling Analysis of Metallic and Carbon Nanotube Interconnections for Nanometer Scale VLSI Technologies," *Proc. the 21st International VLSI*

Multilevel Interconnect Conference (VMIC), Waikoloa, HI, 2004, pp. 393-398.

[18] P. Kapur, J. P. McVittie, and K. C. Saraswat, "Technology and reliability constrained future copper interconnects. I. Resistance modeling," *TED*, vol. 49, pp. 590, 2002.

[19] S. H. Brongersma, K. Vanstreels, W. Wu, W. Zhang, D. Ernur, J. D'Haen, V. Terzieva, M. Van Hove, T. Clarysse, L. Carbonell, W. Vandervorst, W. De Ceuninck, and K. Maex, "Copper grain growth in reduced dimensions," *Proc. Interconnect Technology Conference*, 2004, pp. 48-50.

[20] J. P. Fishburn and C. A. Schevon, "Shaping a distributed-RC line to minimize Elmore delay," *IEEE Transactions on Circuits and Systems I: Fundamental Theory and Applications*, vol. 42, pp. 1020, 1995.

[21] C.-P. Chen, Y.-P. Chen, and D. F. Wong, "Optimal wire-sizing formula under the Elmore delay model," *Proc. DAC*, 1996, pp. 487.

[22] J. P. Fishburn, "Shaping a VLSI wire to minimize Elmore delay," *Proc. European Design and Test Conference*, 1997, pp. 244 - 251.

[23] C.-P. Chen and D. F. Wong, "Optimal Wire-sizing Function With Fringing Capacitance Consideration," *Proc. DAC*, 1997, pp. 604.

[24] Y. Gao and D. F. Wong, "Shaping a VLSI wire to minimize delay using transmission line model," *Proc. ICCAD*, 1998, pp. 611.

[25] J. J. Cong and K.-S. Leung, "Optimal wiresizing under Elmore delay model," *TCAD*, vol. 14, pp. 321, 1995.

[26] C. C. N. Chu and D. F. Wong, "An efficient and optimal algorithm for simultaneous buffer and wire sizing," *TCAD*, vol. 18, pp. 1297, 1999.

[27] J. Cong and Z. Pan, "Interconnect performance estimation models for design planning," *TCAD*, vol. 20, pp. 739, 2001.

[28] J. Cong and Z. Pan, "Wire width planning for interconnect performance optimization," *TCAD*, vol. 21, pp. 319, 2002.

[29] W. C. Elmore, "The Transient Response of Damped Linear Networks with Particular Regard to Wideband Amplifiers," *Journal of Applied Physics*, vol. 19, pp. 55, 1948.

[30] J. Rubinstein, P. Penfield, and M. A. Horowitz, "Signal Delay in RC Tree Networks," *TCAD*, vol. 2, pp. 202, 1983.

[31] S. Borkar, "Design challenges of technology scaling," *Micro*, vol. 19, pp. 23, 1999.

[32] L. Euler, *Methodus inveniendi lineas curvas maximi minimive proprietate gaudentes sive solutio pro blematis isoperimetrici lattissimo sensu accepti.* Lausanne, Geneva, 1744.

2006 Asia and South Pacific Design Automation Conference

Adaptive Admittance-based Conductor Meshing for Interconnect Analysis

Ya-Chi Yang, Cheng-Kok Koh, Venkataramanan Balakrishnan
School of Electrical and Computer Engineering
Purdue University, West Lafayette, IN 47907-2035
e-mail: {yang36, chengkok, ragu}@purdue.edu

Abstract— We present a new algorithm for discretizing interconnects, a step that is typically performed to account for the non-uniformity of current flow at high frequencies. The algorithm is based on an easily-computable measure that correlates well with the model accuracy. This measure is used to refine the discretization of interconnects in an adaptive scheme so as to systematically trade off computation against model accuracy. We apply the proposed discretization technique on two classes of problems in the analysis of VLSI interconnects: simulation and frequency-dependent inductance extraction. Numerical results establish that with the interconnect discretizations generated by our algorithm, a reduction in simulation and extraction times by a factor between three and seven can be realized with negligible sacrifice in model accuracy ($< 1\%$ error).

I. INTRODUCTION

The current distribution across modern interconnects is far from uniform owing to two main reasons. The first is the *skin effect* that occurs due to high signal frequencies. For at least the case of a single conductor carrying a signal at a single frequency, the current density $J(x)$ drops off exponentially as a function of the distance x from the conductor surface into the interior: $J(x) = J(0)e^{-x/\delta}$, where δ is the skin depth [1]. The second reason for the non-uniformity of current flow is the *proximity effect*, caused by the presence of nearby conductors. The usual technique to account for skin and proximity effects is to discretize the conductor into filaments, which results in a larger lumped circuit model. The finer the discretization, the more accurate the model; however, this accuracy comes with the high computational cost that is required to manipulate and simulate the high-order model.

Perhaps the simplest discretization scheme is the *uniform meshing* (UM) scheme, where a conductor is discretized so that the filaments are all of the same cross-sectional area, with the width and the height of each filament being no larger than the skin depth δ. However, the skin and proximity effects cause the current flow to crowd at the edges and at the corners. This motivates the *exponential meshing* (EM) where the width and height of filaments increase exponentially from outside to inside [1], with the width and the height of the smallest filament being no larger than the skin depth. The ratio of the width of a filament to a smaller adjacent filament is typically taken as two. Fig. 1 shows the cross-section of conductor with these two meshing schemes.

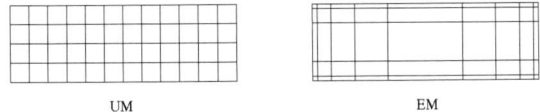

UM   EM

Fig. 1. Meshing of conductor.

A fundamental issue with the UM and EM schemes is where they fit in the context of the accuracy versus computation trade-off. Numerical experiments reveal that both these schemes perform excellently in terms of model accuracy. Moreover, the EM scheme, as expected, is slightly inferior to the UM scheme, but with a much reduced computational demand. We will establish in this paper that considerable further reduction in computation can be realized with negligible sacrifice to modeling accuracy.

A simple exploration of reducing the computational demand with meshing schemes would proceed as follows: Beginning with any scheme whose accuracy is acceptable, coarsen the discretization until the loss of accuracy is no longer acceptable. The term "accuracy" here refers to the model accuracy, the determination of which typically involves simulation, and is therefore computationally demanding. We will demonstrate in §II the existence of a quantity that is (i) easily computable given any meshing scheme and (ii) correlates very well with the model accuracy. This quantity can thus be used to quickly estimate the model accuracy at a given level of discretization, replacing the costly simulation. This idea forms the heart of the adaptive schemes that we propose in §III and §IV to determine a discretization that maximizes the reduction in computation, yet yielding an acceptable model accuracy. Through an analysis of the complexity of the proposed adaptive meshing schemes, we will establish that the computational overhead is indeed minor. Numerical results in §V demonstrate the superior performance of our technique in the problems of simulation and frequency-dependent inductance extraction.

II. A QUANTITY THAT TRACKS SIMULATION ERROR

To begin, consider a single conductor carrying a signal at a single angular frequency ω. Let $\mathcal{G}_1$, $\mathcal{G}_2$, ... describe a sequence of discretizations with the property that $\mathcal{G}_{i+1}$ is "strictly finer" than $\mathcal{G}_i$, i.e., every filament in $\mathcal{G}_{i+1}$ is contained in some filament in $\mathcal{G}_i$, and at least one filament in $\mathcal{G}_i$ is not contained in any filament in $\mathcal{G}_{i+1}$ (see Fig. 2). Moreover, let the largest dimension (width or height) of any filament go to zero with i. Let L_i and R_i be the inductance and resistance matrices with discretization $\mathcal{G}_i$. Then, with a voltage difference of one Volt between the two terminals, the vector of filament currents I_i at discretization $\mathcal{G}_i$ is given by

$$I_i = (R_i + j\omega L_i)^{-1}\mathbf{1}, \tag{1}$$

where $\mathbf{1}$ denotes a vector of ones. The total current in the conductor $\mathbf{1}^T I_i = \mathbf{1}^T(R_i + j\omega L_i)^{-1}\mathbf{1}$, thus is the conductor-level complex admittance Y_i at discretization $\mathcal{G}_i$.

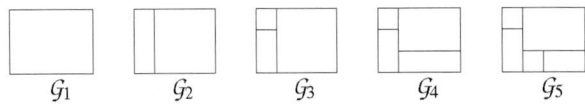

Fig. 2. An example of the sequence of meshing: ($\mathcal{G}_1, \mathcal{G}_2, \mathcal{G}_3, \mathcal{G}_4, \mathcal{G}_5$).

We first note that $|Y_i|$ is an *increasing* sequence. This follows from physical reasoning, as with increasing i, an increasing number of current paths are included in the model: The currents of the conductor with discretization G_i contain all the currents of the conductor with discretization G_{i-1}, thereby increasing the admittance magnitude. We next note that the sequence $|Y_i|$ converges to the true value of the conductor-level admittance, denoted $|Y_\infty|$. Indeed, the latter observation holds true for *any* sequence of meshing schemes where the discretizations become finer, i.e., when the width and height of the largest filament go to zero. Combining these two observations, we can conclude that the discretization with the higher admittance magnitude is closer to the ideal, and is hence more accurate. Moreover, the admittance magnitude serves as a measure to compare the accuracy of any two discretizations: The admittance error $|Y_\infty| - |Y_i|$ is equal to the steady-state error in the current amplitude when a unit sinusoidal voltage at frequency ω is applied between the terminals of the conductor. These observations provide a straightforward method for determining an acceptable discretization: Begin with a coarse grid, and continue with a sequence of finer grids until $|Y_\infty| - |Y_i|$ is small. However, the quantity $|Y_\infty|$ is seldom available, and even its approximate computation requires manipulation of very large matrices of doubtful conditioning. We now present a simple variant of this idea that circumvents this difficulty.

Instead of the current error (which equals the admittance error $|Y_\infty| - |Y_i|$), consider instead the *change in the current error*, whose magnitude equals $|Y_{i+1}| - |Y_i|$. Intuitively, as the discretization becomes finer, the change in the magnitude of the current error should go to zero; thus, we will use this easily computable quantity to define the stopping criterion of the meshing schemes. Therefore, the basic idea behind much of the development in this paper can be summarized as: "Begin with a coarse grid, and continue with a sequence of finer grids until $|Y_{i+1}| - |Y_i|$ is small."

The preceding discussion applies to steady-state analysis at a single frequency. We now establish via a simple numerical example that the principles outlined above apply to analysis with more realistic signals as well. Consider a single conductor with cross-section $3 \times 3\mu m$ and length $1000\mu m$, with a voltage signal that is a 1V ramp with a rise-time of 10ps. Let $\{G_i\}$ denote a sequence of increasingly fine discretizations with the UM scheme. Figure 3 compares two quantities: The first is the change in the admittance magnitude $|Y_{i+1}| - |Y_i|$ computed at a frequency of 100GHz. The second is the change in the simulation error, defined as $\max(|I(t) - I_i(t)|)$, where $I(t)$ is the "true" current in the conductor (computed with a very fine UM discretization, size 29×29) and $I_i(t)$ the current obtained using SPICE simulations of the UM discretization G_i. It is evident that the change in the admittance magnitude indeed tracks the change in the simulation error, even for the more commonly ramp waveforms.

III. ADAPTIVE MESHING SCHEMES

Recalling that the current flow typically decreases exponentially into the interior of the conductor, we next consider meshing schemes that retain the principle behind the conventional exponential meshing, yet adaptively try to obtain the coarsest meshing scheme of acceptable accuracy. For easy reference, we briefly describe two exponential meshing (EM) schemes which we call EM-1 and EM-2.

EM-1 is a popular meshing scheme, described in [1]. We describe the meshing along the width (the meshing along the height proceeds similarly). Given a wire of width w, let N_1 and N_2 be the smallest integers satisfying

$$\frac{w}{2(1+2+2^2+\cdots+2^{N_1-1})} \le \delta,$$

and

$$\frac{w}{2(1+2+2^2+\cdots+2^{N_2-2})+2^{N_2-1}} \le \delta.$$

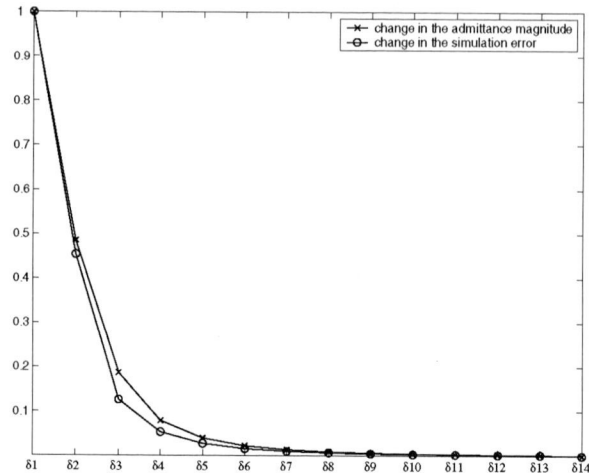

Fig. 3. The change in the admittance magnitude closely tracks the change in the simulation error ("δi" denotes the quantity of $|Y_{i+1}| - |Y_i|$).

Then, the number of filaments along the width with EM-1 is $\min\{2N_1, 2N_2 - 1\}$. An example of a conductor discretized along its width using the EM-1 scheme is shown in Fig. 4 where $N_2 = 3$ and $\delta' = \dfrac{w}{2(1+2)+2^2}$.

Fig. 4. An example of EM-1.

The feature characterizing EM-1 is that the smallest filament is of width less than or equal to the skin depth, with the filament widths doubling towards the interior. The number of filaments is determined automatically. In contrast, with EM-2, a second exponential meshing scheme [4], the number of filaments is specified a priori. This meshing scheme then holds the smallest filament width at the skin depth, and adjusts the ratio between adjacent filaments so as to fit the filaments inside the given width w. More precisely, given a conductor of width w and N, the number of desired filaments, the ratio r between adjacent filaments satisfies

$$w = \begin{cases} 2\delta\left(\dfrac{r^{N/2}-1}{r-1}\right), & \text{even } N, \\[2ex] 2\delta\left(\dfrac{r^{(N-1)/2}-1}{r-1}\right) + \delta r^{(N-1)/2}, & \text{odd } N. \end{cases} \quad (2)$$

While EM-1 and EM-2 perform excellently in terms of model and simulation accuracy and are widely used in practice, we will show that *significant* savings in computation can be realized with alternate meshing schemes, at *negligible* loss in accuracy. In other words, conventional EM schemes are an over-kill. Instead, we propose two meshing schemes, where the number of filaments is increased until it can be established that additional filaments yield no discernible improvement in accuracy.

The first such scheme that we consider is an *adaptive exponential meshing* scheme, denoted AEM-1. This scheme is inspired by the conventional exponential meshing scheme EM-1. The basic idea is to hold the outermost (smallest) filaments at a width of δ, and begin adding filaments whose widths double towards the interior, until the admittance magnitude does not increase appreciably or when there is no room for adding any more filaments. We note that if the latter happens, AEM-1 is very similar to EM-1; however, we will demonstrate through numerical examples that this is seldom the case; the AEM-1 meshing scheme typically terminates well before the EM-1 limit is reached, accruing considerable computational savings and indicating that the EM-1 scheme over-meshes.

An example of the AEM-1 discretization sequence is shown in Fig. 5. Note that the number of filaments with AEM-1 is almost always odd. Discretization G_1 has only one filament (i.e., no discretization). Discretization G_2 has three filaments, with the outermost filaments of width δ, the skin-depth (this assumes that the dimension that is being discretized is of width at least 2δ; otherwise the meshing scheme terminates with G_1). If the width is at least 6δ, there exists discretization G_3 with five filaments, with the outermost filaments of width δ, the next pair of inner filaments of width 2δ, with the middle filament spanning the rest of the width.

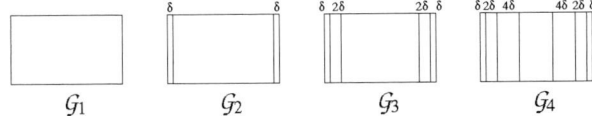

Fig. 5.   The discretization sequence of AEM-1.

The second adaptive exponential meshing scheme, denoted AEM-2, is very similar to the conventional exponential meshing scheme EM-2. Here, holding the width of the outermost filaments constant at the skin depth, the number of filaments N is incremented with the ratio r between adjacent filaments satisfying (2), until the admittance magnitude does not increase appreciably. Note that AEM-2 is identical to EM-2 when the number of filaments match; the crucial difference is that with EM-2, the number of filaments is to be specified a priori, whereas with AEM-2, the the number of filaments is adaptively determined.

Fig. 6 illustrates an example of an AEM-2 discretization sequence. As with AEM-1, the number of filaments with AEM-2 is almost always odd. The discretization pattern of G_1 and G_2 match those of AEM-1 (see Fig. 5). With discretization G_3, there are five filaments, with the outermost filaments of width δ, the next pair of inner filaments of width $r\delta$, and with the middle filament of width $r^2\delta$.

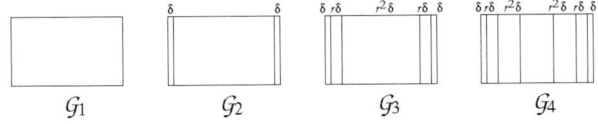

Fig. 6.   The discretization sequence of AEM-2.

Recall that in §II (see Fig. 3 in particular), we demonstrated that the change in the admittance magnitude tracks the true simulation error fairly well. We now show that this observation can be used in conjunction with the proposed meshing schemes AEM-1 and AEM-2 to achieve an acceptable modeling and simulation accuracy with far less computation. We will present detailed simulations backing this claim in §V; we give a brief preview here, considering a conductor with the cross-section of $12.6 \times 1.05 \mu m$ and the length of $1000 \mu m$. The different meshing schemes compared are: UM (60×5 grid; this will be taken as the "standard"); EM-1 (10×4); EM-2, also with an 10×4 grid; AEM-1 (5×3 grid); and AEM-2 (5×3 grid)

(see Fig. 7). The relative simulation error of the various meshing schemes are presented in Fig. 8 (SPICE was used to perform the numerical simulations). Here, the simulation error is defined as $\max(|I(t) - \hat{I}(t)|)$, where $I(t)$ is the current in the conductor obtained with UM, and $\hat{I}(t)$ the current obtained from one of the four (non-adaptive and adaptive) exponential meshing schemes. The results confirm that the resulting loss in accuracy obtained with the coarser meshes in the AEM-1 and AEM-2 schemes is insignificant. We shall show in §V that the computation required by the AEM schemes is considerably less than that required by the conventional EM schemes.

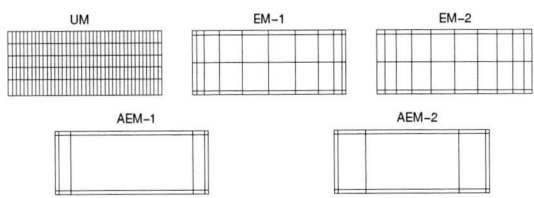

Fig. 7.   Various meshing schemes used in the comparative study.

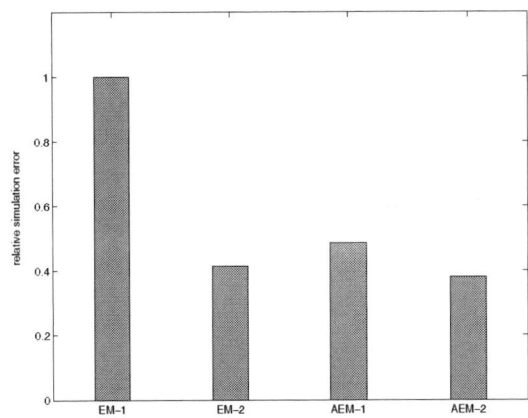

Fig. 8.   The relative simulation error of the various meshing schemes.

IV. IMPLEMENTATIONS OF ADAPTIVE MESHING SCHEMES

While we have established that it is possible to achieve sufficient modeling accuracy with meshing schemes that are coarser than conventional EM schemes, the question of systematically determining such a coarse mesh remains. The approach that we present uses the following simple steps: Given a discretization that needs refinement, we consider two candidate discretizations, one with two more filaments along the width, and the other with two more filaments along the height. We then compute the admittance magnitude at the working frequency ω, and choose the discretization that yields a higher admittance magnitude. Beginning with a 1×1 grid, the refinement proceeds along a "binary" decision diagram, shown in Fig. 9, to define a sequence of finer grids. The process stops when the change in the admittance magnitude between two successive discretizations is below a suitably chosen threshold. For example, the admittance values corresponding to the different discretizations in Fig. 9 for a single wire of length 1mm with cross section $10 \mu m \times 3 \mu m$ are shown in Fig. 10. This figure also shows the sequence of discretizations traversed by our algorithm.

The discussion so far has focused on a single conductor, thus addressing the non-uniformity in current flow arising from the skin

5C-2 2006 Asia and South Pacific Design Automation Conference

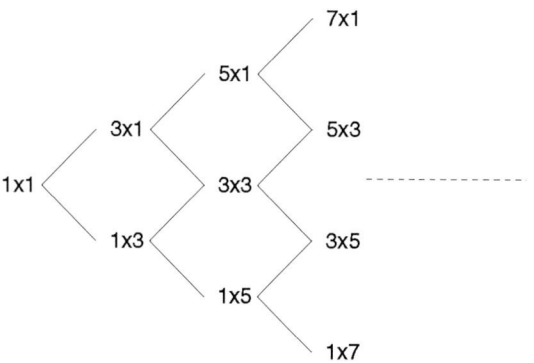

Fig. 9. "Binary" decision diagram of meshing.

effect. A natural question is whether proximity effects (i.e., non-uniformity of current flow attributable to the presence of nearby conductors) should be separately accounted for. We now present an intuitive argument that answers this in the negative. (We will also present empirical evidence that backs this claim.) The argument is simply that discretizing a conductor effectively treats a conductor as a bundle of filaments, whose "proximity" effect is indeed the skin effect. As the filaments in bundle are much closer to each other than the filaments from any other conductors, the interaction of the former overwhelm any effects from the latter.

To illustrate the above arguments, consider three conductors all of length 1mm and cross section $10\mu m \times 3\mu m$, with the middle one being the victim that is sandwiched between two "aggressors." We begin with a discretization of the victim (5×3) that is based on an analysis of its "standalone" admittance magnitude. We then consider its admittance in the presence of the aggressors (which are also discretized using a standalone analysis), and consider the change in this true admittance when the victim is discretized with finer and finer meshing schemes (holding the aggressor discretization constant). Table I shows the current error (obtained via a SPICE variant when the aggressors are triggered by a 1V ramp with a rise-time of 10ps) in the victim for a number of discretizations for the victim, and for a number of values of inter-conductor spacing. It is evident that the current error does not change with finer discretizations of the victim beyond the 5×3 mesh.

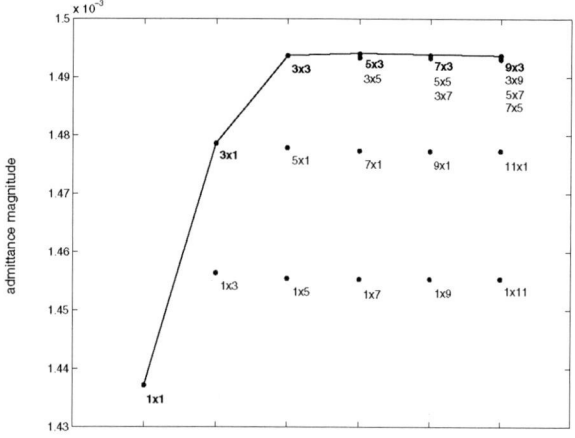

Fig. 10. Admittance values of a wire corresponding to the different discretizations in Fig. 9.

ADAPTIVE-MESHING($\mathcal{G}_{\text{initial}}$, ε)

1 $Y_{\text{initial}} \leftarrow$ COMPUTE-ADMITTANCE($\mathcal{G}_{\text{initial}}$)
2 $Y_{\text{parent}} \leftarrow Y_{\text{initial}}, Y_{\text{change}} \leftarrow Y_{\text{parent}}$
3 **while** $Y_{\text{change}} > \varepsilon$
4   **do** $Y_{\text{right}} \leftarrow$ COMPUTE-ADMITTANCE($\mathcal{G}_{\text{right}}$)
5        $Y_{\text{left}} \leftarrow$ COMPUTE-ADMITTANCE($\mathcal{G}_{\text{left}}$)
6   **if** $Y_{\text{right}} \geq Y_{\text{left}}$
7   **then** $Y_{\text{change}} \leftarrow |Y_{\text{right}} - Y_{\text{parent}}|$
8   parent $\leftarrow$ right child
9   **else** $Y_{\text{change}} \leftarrow |Y_{\text{left}} - Y_{\text{parent}}|$
10   parent $\leftarrow$ left child

Fig. 11. Implementation of Adaptive Meshing Schemes.

Summarizing the above discussions, Fig. 11 illustrates the overall implementation of our technique. Inputs are any single conductor with an initial discretization $\mathcal{G}_{\text{initial}}$ and the user-defined threshold ε. The operation COMPUTE-ADMITTANCE in line 4-5 is performed via (1), and requires $O(l^3)$ computation, where l is the number of filaments. Assume that k iterations are needed for the **while** loop, which is dominated by the operation COMPUTE-ADMITTANCE, the total complexity of ADAPTIVE-MESHING is

$$O(\sum_{i=1}^{k} (i^2)^3) = O(k^7).$$

With i denoting the iteration index, the number of filaments are $(i+1)(i-1)$ and i^2 if i is even and odd, respectively. At termination, the number of filaments n is equal to k^2 in the worst case, and therefore the complexity of ADAPTIVE-MESHING is $O(n^{3.5})$. Assuming that there are m conductors, the overall complexity of our technique is $O(mn^{3.5})$ where n is small because the procedure ADAPTIVE-MESHING terminates quickly in general. However, the complexity can be further reduced in several ways. First, we note that the discretization of a conductor depends only on its cross-section, and thus the run-time of the proposed adaptive meshing schemes is dependent only on the number of distinct cross-sections $\hat{m}$ (and not the number of conductors m). Second, we can sort the distinct cross-sections in an increasing order before running the procedure ADAPTIVE-MESHING. Once the discretization for a smaller cross-section is determined, we can use it to initialize the algorithm for the discretization of the next (larger) cross-section. As a result, the number of iterations in ADAPTIVE-MESHING can be substantially reduced.

V. EXPERIMENTAL RESULTS

We present application of AEM schemes on two classes of problems in the analysis of VLSI interconnects. The first is one of simulation, and the second is on the frequency-dependent inductance extraction. Note that the run-times of AEM schemes throughout this section include not only the simulation and extraction times but also the pre-processing time for determining the discretizations.

A. Simulation of wires of uniform cross-section

Consider an example of five parallel conductors of size of $3 \times 1 \times 1000\mu m$ each, with an inter-conductor separation of $1\mu m$. Both AEM-1 and AEM-2 applied with a threshold of 10^{-6} for the change in the admittance magnitude yield a discretization of 5×3 for the conductors. The center conductor is taken as the victim, with the four aggressors (two on each side) triggered by a ramp signal with a rise-time of 10ps. Owing to the enormous computational demands of SPICE, the simulation of only one input pattern is shown. Table II

512

TABLE I
AN EXAMPLE OF THREE CONDUCTORS SHOWING THE NON-EFFECT OF PROXIMITY.

Mesh size	spacing (μm)									
	0.1		0.5		1.0		1.5		2.0	
	$\|Y\|(\times 10^{-2})$	error	$\|Y\|(\times 10^{-2})$	error	$\|Y\|(\times 10^{-2})$	error	$\|Y\|(\times 10^{-2})$	error	$\|Y\|(\times 10^{-2})$	error
5×3	2.5094	0.9793%	1.5437	0.8933%	1.1063	0.8442%	0.8973	0.8155%	0.7724	0.8188%
7×3	2.5166	0.8609%	1.5465	0.7788%	1.1078	0.7296%	0.8984	0.6995%	0.7732	0.8188%
5×5	2.5077	0.9375%	1.5421	0.8622%	1.1050	0.8196%	0.8962	0.7946%	0.7715	0.8188%
9×3	2.5176	0.8143%	1.5470	0.7425%	1.1082	0.6992%	0.8987	0.6720%	0.7734	0.8188%
7×5	2.5135	0.9400%	1.5440	0.8639%	1.1060	0.8209%	0.8970	0.7957%	0.7720	0.8188%

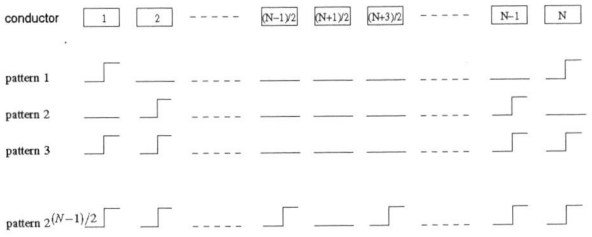

Fig. 12. Input patterns of simulation.

lists the discretizations, relative simulation errors and run-times for the various meshing schemes. The simulation arising from the UM (15×5 grid) is taken as the standard. It is evident from Table II that the runtime arising from discretizations suggested by the AEM schemes is only 0.066 times that of EM, with a negligible loss in accuracy.

Second, we consider N conductors in parallel, each with a cross-section of $5 \times 1\mu$m and length 1000μm, with an inter-conductor spacing of 1μm. As SPICE simulation is impractical, a fast simulator (from [2]) is used for the simulations. The center conductor is taken as the victim, and the aggressor inputs are ramp signals with a rise-time of 10ps. For simplicity, the aggressor switching patterns are assumed to be symmetric with respect to the victim, and a total of $2^{(N-1)/2}$ switching patterns are simulated (see Fig 12). The simulation arising from the UM discretization (25×5 grid) is taken as the standard, and the relative simulation errors, averaged over all conductors and switching patterns, are listed in Table III. The table also lists the run-times arising from discretizations suggested by different meshing schemes, normalized with respect to those of the EM-1 scheme. The conclusion is that compared to the simulation result of EM, the error of AEM is within 1% of UM and the speed-up is as high as 3×.

B. Simulation of wires of non-uniform cross-sections

Typical VLSI interconnects have non-uniform cross-sections. Here, we consider an example of five parallel conductors and an example of seven parallel conductors. In both examples, the length and height of each conductor are 1mm and 3μm, respectively. The widths are randomly chosen from the set $\{3\mu m, 5\mu m, 7\mu m\}$. As before, the

TABLE II
A SMALL EXAMPLE : 5 PARALLEL CONDUCTORS

meshing scheme	EM-1	EM-2	AEM-1	AEM-2	UM
discretization	7×4	7×4	5×3	5×3	15×5
simulation error	0.327%	0.174%	0.374%	0.542%	0.000%
Runtime (hour)	3.32	3.25	0.22	0.21	298.39

center conductor is taken as the victim, the aggressor inputs are ramp signals with a rise-time of 10ps, and the aggressor switching patterns are assumed to be symmetric with respect to the victim. The simulation arising from the UM discretization (15×15, 25×15, and 35×15 grids for 3μm $\times 3\mu$m, 5μm $\times 3\mu$m, and 7μm $\times 3\mu$m cross-sections, respectively) is taken as the standard, and the average relative simulation errors and normalized run-times are listed in Table IV. Compared to the simulation result of EM, the error of AEM is within 0.8% of UM, and the speed-up is as high as 7×.

Note that in general, signals in real circuits have multiple frequency components. For the purpose of determining a discretization that is appropriate for a conductor over a frequency range, we use the highest frequency of signals for determining the discretization.

C. Extraction of frequency-dependent inductance

We have also successfully applied the AEM schemes to speed-up the extraction of frequency-dependent inductance (or in general, frequency-dependent impedance) with little loss in accuracy. In this experiment, we consider a *spiral inductor* composed of segments with a uniform cross-section of 3μm $\times 1\mu$m. The longest segment is 20μm long and the separation between any pair of adjacent parallel segments is 1μm. The underlying algorithm used for the extraction of frequency-dependent inductance is FastHenry [3]. The discretizations determined using the UM, EM, and AEM schemes over a set of frequency points taken from the 10GHz–100GHz range are used by FastHenry for the frequency-dependent inductance extraction. For EM-1, AEM-1, and AEM-2, the respective discretizations at each frequency point are determined as outlined in §II and §IV. For EM-2, the discretization at each frequency point is determined based on the number of filaments used in a corresponding EM-1 discretization. For UM, which is again taken to be the standard, we use the coarsest uniform discretization where the width and the height of each filament are no larger than the skin depth δ. Table V lists the relative errors, averaged over all frequency points, and the total run-times arising from different discretization schemes. Compared to the result of EM, the error of AEM is within 0.6% of UM and the speed-up is as high as 3×.

TABLE V
PERFORMANCE OF INDUCTANCE EXTRACTION FOR SPIRAL INDUCTOR.

meshing scheme	EM-1	EM-2	AEM-1	AEM-2
error	0.0913%	0.0885%	0.6063%	0.1726%
runtime(second)	12.60	12.33	4.01	7.60
speedup	1.00	1.02	3.14	1.66

VI. CONCLUSION

We have presented a new adaptive algorithm for discretizing interconnects to account for high-frequency effects. The algorithm uses

TABLE III

PERFORMANCE OF CIRCUIT SIMULATION FOR INTERCONNECTS WITH UNIFORM CROSS-SECTION.

number of conductors	discretization							
	EM-1		EM-2		AEM-1		AEM-2	
	8×4		8×4		5×3		5×3	
N	error	runtime	error	runtime	error	runtime	error	runtime
5	0.1492%	1.00	0.1101%	1.07	0.5936%	0.67	0.5346%	0.67
7	0.1571%	1.00	0.1160%	1.07	0.6835%	0.58	0.5742%	0.60
9	0.1581%	1.00	0.1183%	1.03	0.7202%	0.44	0.5841%	0.44
11	0.1576%	1.00	0.1184%	1.01	0.8279%	0.29	0.5865%	0.29
13	0.1635%	1.00	0.1231%	0.97	1.0085%	0.35	0.5993%	0.35

TABLE IV

PERFORMANCE OF CIRCUIT SIMULATION FOR INTERCONNECTS WITH VARIOUS CROSS-SECTIONS.

number of conductors	EM-1		EM-2		AEM-1		AEM-2	
N	error	runtime	error	runtime	error	runtime	error	runtime
5	0.2121%	1.00	0.1263%	1.04	0.5682%	0.42	0.8481%	0.44
7	0.2133%	1.00	0.1311%	1.07	0.5521%	0.14	0.7421%	0.14

an easily-computable quantity that correlates well with simulation errors that arise from discretizations; this quantity is used in the adaptive meshing schemes to determine the coarsest discretization of conductors that provides an acceptable model accuracy, thus obtaining the maximum possible reduction in computation time. Experimental results demonstrate that the proposed adaptive meshing schemes yield a significant speed-up in inductance extraction and interconnect simulation with little sacrifice in accuracy.

VII. ACKNOWLEDGMENTS

This material is based on work supported by National Science Foundation under Award CCR-0203362 and ECS-0200320.

REFERENCES

[1] C. Cheng, J. Lillis, S. Lin, and N. Chang. *Interconnect Analysis and Synthesis*. John-Wiley, 2000.

[2] J. Jain, C.-K. Koh, and V. Balakrishnan. Fast simulation of VLSI interconnects. In *Proc. Int. Conf. on Computer Aided Design*, pages 93–98, 2004.

[3] M. Kamon, M. J. Tsuk, and J. K. White. FASTHENRY: A multipole-accelerated 3-D inductance extraction program. *IEEE Journal on Microwave Theory and Techniques*, 42(9):1750–1758, Sept. 1994.

[4] T. Makkonen, V. P. Plessky, S. Kondratiev, and M. M. Salomaa. Electromagnetic modeling of package parasitics in SAW-duplexer. In *Proc. IEEE Ultrasonics Symposium*, pages 29–32, 1996.

Interconnect RL Extraction at a Single Representative Frequency

Akira Tsuchiya[1] Masanori Hashimoto[2] Hidetoshi Onodera[1]

[1]Dept. CCE, Kyoto University, [2]Dept. ISE, Osaka University

e-mail: {tsuchiya,onodera}@vlsi.kuee.kyoto-u.ac.jp, hasimoto@ist.osaka-u.ac.jp

Abstract— This paper proposes a method to determine a single frequency for interconnect RL extraction. Resistance and inductance of interconnects depend on frequency, and hence the extraction frequency strongly affects the modeling accuracy of interconnects. The proposed method determines an extraction frequency based on the transfer characteristic of interconnects. By choosing the frequency where the transfer characteristic becomes maximum, the extracted RL values achieve the accurate modeling of the waveform. We experimentally verify that the proposed method provides accurate transition waveforms over various interconnect topologies.

I. Introduction

According to advancements in LSI fabrication technology, performance of LSI chips is predicted to improve continuously [1]. As improving the chip performance, on-chip interconnects become important and accurate modeling of interconnects is crucial for circuit design. One difficulty of interconnect modeling is frequency dependency of the characteristics. Resistance and inductance depend on frequency because of skin- and proximity-effect [2]. In digital circuits, pulse waveforms are commonly used. The frequency spectrum of pulse waveforms widely spreads from DC to frequency several times as high as clock frequency. Therefore to model the behavior of interconnects precisely, designers have to take the frequency characteristics into consideration. To treat frequency dependent interconnects, several modeling methods are proposed [3–5]. These frequency-dependent models improve simulation accuracy. However, in circuit design, the conventional frequency-independent model has an advantage that there are a number of techniques and methods developed so far. Inductance-aware circuit design techniques, such as analytical performance estimation, circuit reduction, buffer insertion and timing analysis, have been widely studied [2, 6–8]. Most of these techniques assume that interconnects can be modeled as a frequency-independent RLC ladder circuit. However, how to cope with frequency-dependency in modeling interconnects as a RLC ladder has not been studied enough, though its modeling accuracy affects design quality. Therefore developing an accurate modeling technique by frquency-independent model is indispensable for designing high-performance circuits, and hence we focus on RL extraction at a single frequency in this work.

In this paper, the extraction frequency based on the transfer characteristic of interconnects is proposed. It is commonly adopted to determine the extraction frequency from the shape of an input signal waveform, especially from the rise time, focusing on the spectrum of the input signal [2]. This is natural and reasonable when we analyze the incident waveform to the near-end (driver output) of the interconnects. On the other hand, our main interest is the analysis of the waveform at the far-end (receiver input). As signals are propagating through an interconnect, high-frequency components easily attenuate. The dominant frequency components that determine the far-end waveform are different from those for the near-end waveform. We observe that the transfer characteristic of interconnects is playing an important role in the waveforms at the far-end of interconnects. Therefore we focus on the transfer characteristic of interconnects and select the frequency where the transfer characteristic becomes maximum as the frequency to use for interconnect RL extraction. A preliminary work is presented in Ref. [9], however, the frequency determination method in Ref. [9] is not practical because it can treat only open-ended uniform transmission-lines without branches. If the interconnect is branching or nonuniform, the transfer characteristic of each segment is different. The proposed method in this paper gives a respective extraction frequency to every segment of an interconnect instead of enforcing a single extraction frequency on the entire interconnect. The proposed method systematically determines the extraction frequencies successively from the sinks to the source by replacing the downstream interconnect with the equivalent load impedance. We experimentally verify that the equivalent circuit of interconnects extracted at the proposed frequency can achieve the most accurate waveform modeling compared with the conventional extraction frequencies. Experimental results show the maximum errors are below 10% in signal delay and signal transition time. The contribution of this paper is that our method realizes accurate transient analysis using frequency-independent interconnect model. The proposed method is effective when the topology and the length of interconnects are fixed, for example, post-layout extraction.

In Section II, the problems in interconnect modeling are described. Section III explains the detail of the proposed method. We then show experimental results in Section IV. Section V concludes the discussion.

II. Problem description

This section describes the problem discussed in this paper. We first show frequency-dependence of interconnect characteristics and demonstrate its impact on transient analysis.

A. Frequency-dependence of interconnect characteristics

Frequency-dependence of interconnect characteristics is mainly caused by skin-effect and proximity effect. The characteristics variation is strongly related with the interconnect structure as well as the frequency. Skin- and proximity-effects are remarkable on wide and thick interconnects because skin

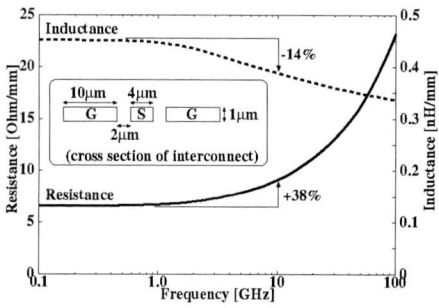

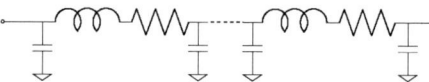

Fig. 1. Frequency-dependence of resistance and inductance. (co-planar structure, signal line width 4μm, ground line width 10μm, spacing 2μm)

Fig. 2. RLC ladder circuit model.

depth becomes comparable to the interconnect size in relatively lower frequency.

Figure 1 shows an example of resistance and inductance characteristics. The resistance and inductance values are calculated by a field-solver [10]. The assumed interconnect structure is co-planar, and the width of the signal line is 4μm, the width of the ground line is 10μm and their spacing is 2μm. In this case, the resistance increases by 38% from DC to 10GHz, and the inductance decreases by 14% from DC to 10GHz. The resistance and the inductance start changing at relatively low frequency of 1 to 2GHz, and thus frequency-dependence is not negligible to model interconnects in current high-performance circuits any longer.

B. Conventional extraction frequencies

In digital circuits, a trapezoidal pulse that contains multiple frequency components is a common waveform. To model long interconnects that have transmission-line characteristics, an RLC ladder circuit as Fig. 2 is used. This ladder model cannot consider the frequency-dependence of interconnect RL values. In order to derive frequency-independent model of Fig. 2, we have to choose a single extraction frequency.

There are several representative frequencies of periodic pulse waveform. One of them is significant frequency [2]. The significant frequency f_{sig} is expressed by signal transition time t_r and defined such that the signal energy from DC to f_{sig} becomes 85% of all signal energy. In the range $7 \le T_w/t_r \le 13$ where T_w is the pulse width, f_{sig} is given by $0.34/t_r$ [2]. On the other hand, DC is often used for extraction. Reference [11] concludes that the extraction at DC is accurate enough to estimate signal delay and overshoot/undershoot. DC extraction is acceptable when frequency-dependence is weak, e.g. for narrow interconnects or in low frequency. However as shown in the next section, interconnect modeling at these frequencies causes error in evaluating the propagation waveform.

C. Interconnect models and their impact on waveform

Generally, interconnects in VLSIs are expressed by lumped RLC for circuit design. As explained in the previous section, the frequency-independent RLC ladder circuit in Fig. 2 is used to model on-chip interconnects. A number of frequency-dependent models are proposed [3–5]. In this paper, we use the model of Ref. [5] as a golden frequency-dependent model. It is implemented in HSPICE [12] as w-element model. Although frequency-dependent models such as Ref. [5] can provide accurate waveforms, the frequency-dependent model does not tell the designers which frequency component is important in circuit design. Conversely, if we know which frequency component dominantly forms and affects the waveforms at the far-end, we do not necessarily have to use the frequency-dependent model, and we can use many design methods and techniques based on conventional frequency-independent model extracted at the dominant frequency. However the frequency spectrum spreads widely and depends on circuit behavior and interconnect characteristics, and hence it is difficult to specify the most representative frequency from the frequency spectrum. The goal of this research is to determine the representative frequency for modeling interconnects at a single frequency.

Figure 3 shows the impact of frequency-dependence on transient analysis. The simulated circuit is shown in Fig. 3. The interconnect shown in Fig. 1 is driven by a voltage source and a resistor R_d that correspond to a CMOS driver whose output impedance is 150Ω. The solid line labeled "FD" shows the voltage waveform at the far-end by the frequency-dependent model. In this paper, we use "FD" as the abbreviation of "Frequency-Dependent model". The dashed lines labeled "DC" and "f_{sig}" are the results of frequency-independent models shown in Fig. 2. "DC" means the RLC ladder model extracted at DC, and "f_{sig}" corresponds to RLC extraction at the significant frequency. The number of ladder is 51. As you see, both waveforms of the conventional frequency-independent models ("DC" and "f_{sig}") are far from that of frequency-dependent model ("FD"). In the signal propagation delay time and the signal transition time (from 0.2V to 0.8V), the errors of "DC" are -28% in delay and -13% in transition time. The errors of "f_{sig}" are 19% in delay and 10% in transition time. When R and L are extracted at DC, the extracted resistance is too low, and, the resistance extracted at significant frequency is too high. From the above observations, we can expect that a certain frequency between DC and significant frequency provides the waveform that is close to the waveform of the frequency-dependent model. If the representative frequency can be determined systematically, we can model interconnects by a single frequency. In the following section, we discuss the way to determine the representative frequency to model interconnects at a single frequency.

III. A METHOD TO DETERMINE AN EXTRACTION FREQUENCY

In this section, we propose a method to determine the representative frequency for interconnect RL extraction. The pro-

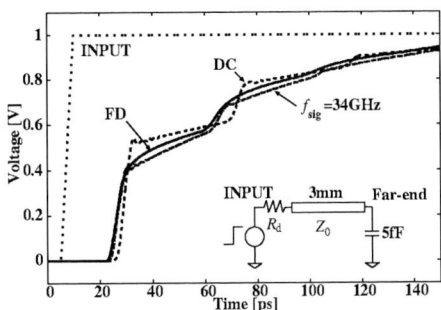

Fig. 3. The impact of frequency-dependence on transition waveform. (interconnect structure is shown in Figure 1, $R_{\rm d} = 150\Omega$)

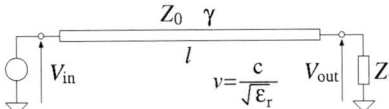

Fig. 4. Transmission-line with impedance load.

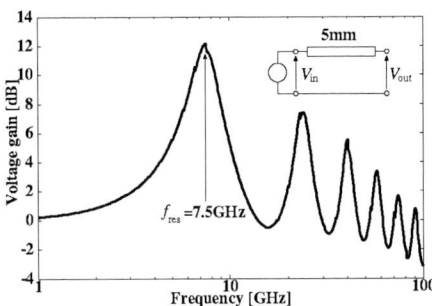

Fig. 5. transfer characteristic of a transmission-line with 5mm long.

posed method determines the extraction frequency focusing on the transfer characteristic of interconnects. We first explain the transfer characteristic of transmission-lines. We then show the detail of the proposed method.

A. Transfer characteristic of transmission-lines

The basic idea of our method is to choose the frequency where the transfer characteristic becomes maximum. We here explain the nature of the transfer characteristic of interconnects based on the transmission-line theory. We here discuss a simple transmission-line as shown in Fig. 4. The characteristic impedance is Z_0, the propagation constant is γ, the length of the transmission-line is l and the velocity of electromagnetic wave is v. The velocity v is equal to $c/\sqrt{\varepsilon_{\rm r}}$, where c is the velocity of light in vacuum and $\varepsilon_{\rm r}$ is the relative dielectric constant of insulator. The load impedance $Z_{\rm t}$ is connected to the far-end of transmission-lines. We write the voltage at the near-end as $V_{\rm in}$ and the voltage at the far-end as $V_{\rm out}$.

According to the transmission-line theory, the voltage transfer characteristic $V_{\rm out}/V_{\rm in}$ is expressed as

$$\frac{V_{\rm out}}{V_{\rm in}} = \frac{1}{\cosh \gamma l + \frac{Z_0}{Z_{\rm t}} \sinh \gamma l}. \tag{1}$$

Figure 5 shows an example of the transfer characteristic of an open-ended transmission-line. We use the transmission-line shown in Fig. 1 and the relative dielectric constant $\varepsilon_{\rm r}$ is 4. We extract the frequency characteristics by a 2D field solver and model the interconnect by the frequency-dependent model [5]. When the transmission-line is open-ended, the transfer characteristic $V_{\rm out}/V_{\rm in}$ becomes maximum where the quarter wavelength $\lambda/4$ is equal to the line length l. This nature is used for quarter wavelength transmission-line resonators. In this case, the resonance frequency $f_{\rm res}$ where $V_{\rm out}/V_{\rm in}$ becomes maximum is $v/4l = 1.5 \times 10^8/(4 \times 5 \times 10^{-3}) = 7.5{\rm GHz}$.

This transfer characteristic affects the waveform at the far-end of transmission-lines. The frequency components near the resonance frequency tend to appear at the far-end. On the other hand, the frequency components near the antiresonance frequency hardly affect the waveform at the far-end. Therefore the frequency spectrum at the far-end depends on the input pulse and the transfer characteristic of the interconnect. If the frequency spectrum of the input pulse spreads over the resonance frequency, the frequency components around the resonance frequency are expected to affect the waveform at the far-end. If the resonance frequency is higher than the significant frequency, the frequency components around the resonance frequency are small because almost all of the frequency components concentrate in the range from DC to the significant frequency.

Figure 6 shows the frequency spectrum of the waveform at the far-end in the case that trapezoidal pulses with various transition time $t_{\rm r}$ are injected to the near-end of the interconnect shown in Fig. 5. The transition time $t_{\rm r}$ is varied from 10ps to 50ps, and hence the significant frequency changes from 34GHz to 6.8GHz. The resonance frequency is 7.5GHz. In this case, the significant frequency is nearly equal to or higher than the resonance frequency. The frequency spectrum has is a unique peak at the resonance frequency even if the signal transition time $t_{\rm r}$ is changed. This result indicates the frequency component at the resonance frequency is an important factor which determines the waveform at the far-end of the interconnect. Reference [9] reports that the resonance frequency is suitable for the extraction frequency of open-ended uniform transmission-lines without branches. However, on-chip interconnects is not uniform and have branches. If the interconnect is branching, the transfer characteristic from the driver to one receiver is not the same as that from the driver to the other receiver. Therefore we cannot apply the method in Ref. [9] to branching wire directly.

B. Flow of the proposed method

This section proposes a method to determine the extraction frequency based on the transfer characteristic. We explain the flow of the proposed method. Figure 7 is the conceptual diagram of the proposed method. We divide an interconnect into segments at the branch points or discontinuous points and consider as a tree such that the root node is the output of the driver

5C-3

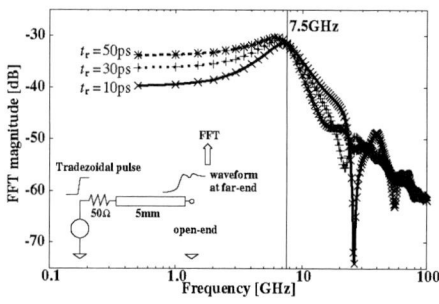

Fig. 6. Frequency spectrum of waveform at the far-end.

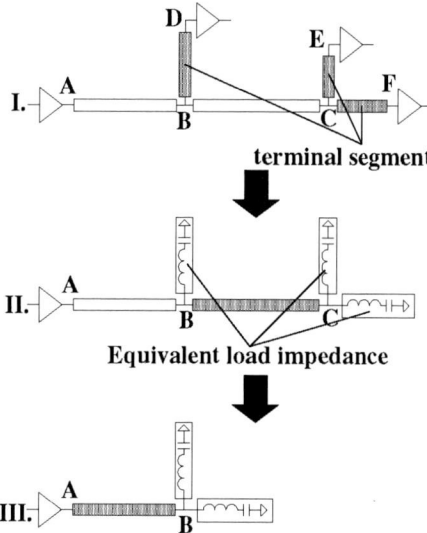

Fig. 7. Conceptual diagram of the proposed method.

and the input of the receiver is the leaf node. The proposed method determines the extraction frequencies for each segment from leaf to root by replacing the downstream branches with the equivalent load impedances.

Assumptions of the proposed method

The proposed method determines the extraction frequency from the topology of interconnects and the length of each segment. We assume that the velocity of electromagnetic wave v is known and a constant value. In LSIs, this assumption is valid because the velocity v depends on the relative dielectric constant ε_r and it is constant in the same fabrication process. We also assume that the significant frequency f_{sig} is known. As described in following step, we use the significant frequency as the upper limit of the extraction frequency. Additionally, we assume that the characteristic impedances of each segment are the same. Strictly speaking, this assumption is not correct. However in LSIs, the characteristic impedance of on-chip interconnects does not vary drastically even if the interconnect structure changes. The typical value of co-planar structures is from 50Ω to 100Ω. In the following section, we experimentally verify that these assumptions are reasonable.

Step 1. Determine the extraction frequency for terminal segments

If the length of the segment and the load impedance are known, the frequency f_{res} at which the transfer characteristic of the segment becomes maximum can be determined from Eq.(1). The resonance frequency f_{res} depends on the length of the segment and the load impedance. Generally, as short the segment is, the extraction frequency becomes higher. The frequency f_{res} can be too high frequency to use as an extraction frequency. If the frequency component of the input is too small at the frequency, extracting at the frequency at which the transfer characteristic becomes maximum is meaningless. We therefore set the upper limit of the extraction frequency. As mentioned so far, the significant frequency f_{sig} is defined as the frequency that the 85% of all the energy is included from DC to f_{sig}. In the case that the resonance frequency is higher than significant frequency, the frequency components around the resonance frequency are small and hardly affect the waveform at the far-end. Therefore it is reasonable to use the significant frequency as the upper limit of the extraction frequency. Extraction frequency $f_{proposed}$ is expressed as

$$f_{proposed} = \min\left(f_{res}, f_{sig}\right). \qquad (2)$$

Reference [9] reports the case of open-ended uniform transmission-lines. In CMOS circuits, the input capacitance of the gates is small and we can assume the segments connected to the receiver are open-ended transmission-line. The extraction frequencies of these open-ended branches are the frequency where the quarter wavelength is equal to the interconnect length. When the length of a segment is l, the resonance frequency f_{res} is $v/4l$. We cannot apply the method in Ref. [9] to interconnects which have a large capacitive load or a resistive termination because Ref. [9] assumes open-ended transmission-lines. By using Eq. (1), the proposed method can be applied to the interconnects that we cannot be regarded as open-ended transmission-lines.

Step 2. Replacing terminal segments with equivalent load impedances

At step 1, the extraction frequencies of terminal segments are decided. To decide the extraction frequencies of the preceding segments, we replace the segments whose extraction frequency is already decided with equivalent load impedances. This step corresponds to Fig. 7. II. By replacing with the equivalent load impedance, we can calculate the extraction frequency by Eq. (1). For example, in Fig. 7, we can calculate the extraction frequency for the segment B-C by replacing the segment C-E and C-F with the equivalent load impedances.

The load impedance of a certain segment is the input impedance of the downstream branches. For example, the load impedance of the segment B-C in Fig. 7 is the input impedance of the segment C-E and the segment C-F. As shown in Fig 7. II, the input impedance of transmission-lines can be modeled by a RLC series resonator circuit whose resonance frequency is equal to f_{res}. We can ignore the resistance because the proposed method needs only the resonance frequency. The input

518

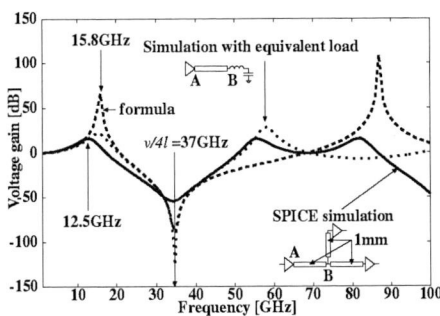

Fig. 8. Voltage gain estimated by the equivalent load impedance.

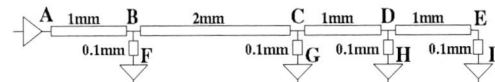

Fig. 9. Stub-bus topology.

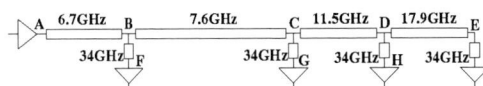

Fig. 10. Extraction frequencies by the proposed method.

impedance of a certain segment is expressed as

$$Z_{\mathrm{in}} = \sqrt{\frac{L}{C}} \frac{1-\omega^2 LC}{\mathrm{j}\omega\sqrt{LC}} \simeq \frac{Z_0}{\mathrm{j}\tan\left(\frac{\omega l'}{v}\right)} \qquad (3)$$

where l' is equivalent length defined as $v/4f_{\mathrm{res}}$. We assume that the characteristic impedances Z_0 of each segment are the same. Therefore the value of the inductance L and the capacitance C are determined from the characteristic impedance Z_0 and the resonance frequency f_{res}. This means that once the resonance frequency f_{res} is calculated, the equivalent load impedance is uniquely determined.

Figure 8 shows an example of transfer characteristic estimated by Eq. (1) and the equivalent load impedance defined by Eq. (3). Figure 8 is the voltage gain between node A and node B. The interconnect topology is a branching wire as shown in Fig. 8. The solid line is the transfer characteristic by SPICE AC analysis. The dashed line labeled "Simulation with equivalent load" is the result of SPICE simulation using the equivalent load, and the dashed line labeled "formula" is that by Eq. (1) with the equivalent load impedance. The resonance frequency f_{open} when we assume the segment A-B as open-ended is 37.5GHz. As shown in Fig. 8, the transfer characteristic estimated by Eq. (1) with equivalent load impedance is valid to estimate the peak of the transfer characteristic. On the other hand, the frequency f_{open} becomes antiresonance frequency on the segment A-B and the transfer characteristic at f_{open} becomes minimum. This result shows that we have to consider the load impedance to estimate the first peak frequencies of the transfer characteristic. From the above discussion, the transfer characteristic can be estimated by Eq. (1) with equivalent load impedance by Eq. (3).

By replacing the terminal segments with equivalent load impedance, the terminal segments are eliminated and other segments become terminal segments. We can return to step 1 and determine the extraction frequency for new terminal segments. The proposed method determines extraction frequencies for each segment by iterating the step 1 and step 2.

IV. EXPERIMENTAL RESULTS

This section demonstrates experimental results. We show the experimental results of a major interconnect topology, stub-

bus. We then show statistics of experimental results in various situations.

A. Stub-Bus topology

We show the experimental result of stub-bus structure. Figure 9 shows the interconnect topology. The bus line A-B-C-D-E is a fat wire that the signal width $W_s = 10\mu\mathrm{m}$, the ground width $W_g = 10\mu\mathrm{m}$ and the spacing $S = 2\mu\mathrm{m}$. The stubs are short and thin wires, that $W_s = 1\mu\mathrm{m}$, $W_g = 1\mu\mathrm{m}$ and $S = 2\mu\mathrm{m}$. The driver output impedance is 50Ω and the transition time of input is 10ps.

We here show the frequency determination process step by step. First, the stubs (B-F, C-G, D-H and E-I) are the terminal segment. The resonance frequency of the stubs is 375GHz from the length. The resonance frequency is higher than the significant frequency, 34GHz. Therefore the extraction frequency of stubs is determined to 34GHz. Then we can replace the stubs by the equivalent load from Eq. (3). By replacing the stub E-I, the resonance frequency of the segment D-E is determined from Eq. (1). By iteration of step 1 and step 2, we can obtain the extraction frequency of each segment in the order of D-E, C-D, B-C, A-B. Finally the extraction frequencies by the proposed method are determined as shown in Fig. 10. Figure 11 shows the waveform at the node I and Table I shows the errors in delay and transition time. The ladder model extracted at DC or significant frequency causes serious error especially in signal transition time. As shown in Fig. 11, DC extraction underestimates the attenuation and f_{sig} overestimates. DC extraction also causes 8% error in delay because the extraction at DC causes estimation error in phase velocity. On the other hand, the RLC ladder by the proposed method provides accurate modeling of frequency-dependent interconnect.

B. Results of overall experiments

We evaluate our method under various conditions for verification. In this section, we show the statistical summary of all experiments. We vary the topology of net, lengths of each segment, interconnect structure, driver size and transition time of input. The number of segments in one net is varied from 1 to 5. The length of segment is $200\mu\mathrm{m}$–5mm. We use co-planar structure, whose signal width W_s and ground width are $1\mu\mathrm{m}$–$10\mu\mathrm{m}$ and the spacing S is $2\mu\mathrm{m}$–$8\mu\mathrm{m}$. The output impedance of the driver is 25Ω–100Ω. The transition time of input pulse is 10ps–100ps. By changing those parameters, we evaluate 9,545

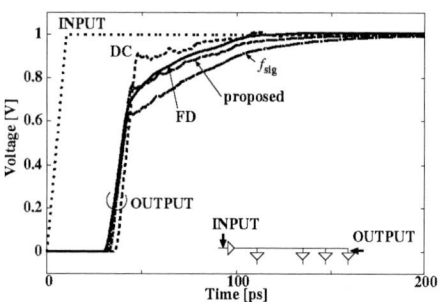

Fig. 11. Waveform at the node I of stub-bus.

TABLE I
ERRORS IN THE DELAY TIME AND THE TRANSITION TIME AT THE NODE I
OF THE STUB-BUS.

Extraction	delay		transition time	
Frequency	[ps]	error [%]	[ps]	error [%]
FD	34.5	—	19.5	—
DC	37.3	8.1	6.8	−65.4
Proposed	35.3	2.3	21.2	8.7
f_{sig}	35.3	2.3	40.6	108.2

patterns of net and observe the waveforms at 43,199 nodes in total.

Table II shows the summary of overall experiments. Table II contains the maximum error in delay time and transition time (rows of "Max. error"), and the ratio of nodes where the error is over 5% (rows of "> 5%"). The ladder extracted at DC tends to underestimate the delay and transition time, and the ladder extracted at significant frequency f_{sig} overestimates. In the case of the ladder extracted at DC, the error in delay exceeds 5% at the 12% of all nodes and the maximum error is −88.1%. In transition time, the error at 28% of nodes exceeds 5% and the maximum error is −71.9%. In the case of significant frequency, the error in delay at about 12% of all nodes is over 5% and the maximum is 110.0%, and the error in transition time at about 35% of all nodes is over 5% and the maximum is over 160%. Those errors are serious problem for evaluating the circuit behavior, such as timing analysis. On the other hand, the proposed method achieves the error less than 10% in both delay and transition time. The results above confirm the RLC ladder extracted at the proposed frequency provides accurate modeling of frequency-dependent interconnects.

V. CONCLUSION

The frequency that should be used to extract RL values is discussed. When we use frequency-independent equivalent circuits for circuit design, the extraction frequency must be carefully determined to maximize the fidelity in interconnect characteristics. We propose an RL extraction scheme that uses the frequency determined by interconnect length. We experimentally verify that the proposed frequency achieves the most

TABLE II
STATISTICAL SUMMARY OF OVERALL EXPERIMENTS.

Extraction	delay		transition time	
Frequency	Max. error	> 5%*	Max. error	> 5%*
DC	−88.1 %	11.5 %	−71.9 %	27.8 %
proposed	−9.9 %	5.4 %	−9.8 %	12.5 %
f_{sig}	110.0 %	12.2 %	160.3 %	35.2 %

(∗ : The ratio of the experiments that the error is over 5%.)

accurate estimation in signal propagation delay and transition time. The maximum error is within 10% in delay and in transition time in our experiments. With the proposed representative frequency, RL extraction at a single frequency becomes accurate enough to model interconnect characteristics, and hence we can exploit many effective design and analysis techniques developed ignoring frequency-dependence.

ACKNOWLEDGEMENT

This work is supported in part by the 21st Century COE Program (Grant No. 14213201).

REFERENCES

[1] Semiconductor Industry Association, "International Technology Roadmap for Semiconductors", 2003 ed., 2003.

[2] C.-K Cheng, J. Lillis, S. Lin, and N. H. Chang, "Interconnect Analysis and Synthesis," A Wiley-Interscience Publication., 2000.

[3] H. A. Wheeler, "Formulas for the Skin-Effect," *Proc. Institute of Radio Engineers*, vol.30, pp.412–424, Sept 1942.

[4] B. Krauter and S. Mehrotra, "Layout Based Frequency Dependent Inductance and Resistance Extraction for On-Chip Interconnect Timing Analysis," *Proc. DAC*, pp.303–308, 1998.

[5] D. B. Kuznetsov and J. E. Schutt-Ainé, "Optimal Transient Simulation of Transmission Lines," *IEEE Trans. Circuits and Systems*, vol.43, no.2, pp.110–121, Feb 1996.

[6] J. A. Davis and J. D. Meindl, "Compact Distributed RLC Interconnect Models — Part I: Single Line Transient, Time Delay, and Overshoot Expressions," *IEEE Trans. Electron Devices*, vol.47, no.11, pp.2068–2077, Nov 2000.

[7] H. B. Bakoglu, "Circuits, Interconnections, and Packaging for VLSI," Addison-Wesley Publishing Company, Inc., 1990.

[8] Y. I. Ismail and E. G. Friedman, "Effects of Inductance on the Propagation Delay and Repeater Insertion in VLSI Circuits," *Proc. DAC*, pp.721–724, 1999.

[9] A. Tsuchiya, M. Hashimoto, and H. Onodera, "Representative Frequency for Interconnect R(f)L(f)C Extraction," *Proc. ASP-DAC2004*, pp.691–696, Jan 2004.

[10] Avant! Corporation, "Raphael Reference Manual, Release 4.2", May 1998.

[11] Y. Cao, X. Huang, D. Sylvester, T.-J. King, and C. Hu, "Impact of On-Chip Interconnect Frequency-Dependent R(f)L(f) on Digital Design," *Proc. International ASIC/SOC Conference*, pp.438–442, Sep 2002.

[12] Synopsys Corp., "Star-Hspice Manual", 2004.

An Efficient Algorithm for 3-D Reluctance Extraction
Considering High Frequency Effect

Mengsheng Zhang   Wenjian Yu   Yu Du   Zeyi Wang

Dept. Comuter Science & Tech.,
Tsinghua University,
Beijing 100084, China
mszhang00@mails.tsinghua.edu.cn

Dept. Comuter Science & Tech.,
Tsinghua University,
Beijing 100084, China
yu-wj@tsinghua.edu.cn

Synopsys Inc.,
CA 94043, USA,
duyu@synopsys.com

Dept. Comuter Science & Tech.,
Tsinghua University,
Beijing 100084, China
wangzy@tsinghua.edu.cn

Abstract – As shown in literatures, partial reluctance based circuit analysis is efficient in capturing on-chip inductance effect, because the partial reluctance exhibits much better locality than partial inductance. However, most previous works on reluctance extraction did not take high frequency effect into account and were not efficient enough for 3-D complex structure. In this paper, a new reluctance extraction algorithm is proposed considering the high frequency effect. Numerical experiments demonstrate that our algorithm can handle complex 3-D interconnect structures while exhibiting high accuracy and a speed-up ratio of several tens to hundreds over FastHenry.

I. Introduction

The industry of modern VLSI circuits is advancing toward ultra deep sub-micron technology or even nano technology, and the operating frequency of circuit has reached multiple giga-hertz (GHz). Currently, the interconnect delay and coupling has become critical for the design of high performance circuits, such that the conventional RC model of interconnect is not enough for accurate circuit analysis. Therefore, accurate and efficient modeling of on-chip inductance effect becomes very important. Moreover, the reduction of resistance by copper and capacitance by low-κ dielectric, denser geometries and growing complexity of interconnect design also make on-chip inductance indispensable [1][2].

Modeling the on-chip inductance effect is difficult because of the unknown circuit return path prior to the extraction and simulation. Thus, on-chip inductance extraction and the circuit return path becomes a "chicken-egg" paradox. With the partial element equivalent circuit (PEEC) model [3], the return path problem is solved by defining the concept of partial inductance whose return path is at infinity. However, since the coupling of partial inductance is among all the conductor segments, the resulting partial inductance matrix is a dense one. Although faraway terms in the partial inductance matrix maybe small, simply truncating them may lead to the system unstable [4].

In order to overcome this problem, a new concept 'partial reluctance' was first proposed in [5], it was also called K element. The partial reluctance matrix is denoted by

K, and its definition is the inversion of partial inductance matrix, i.e.:

$$K = L^{-1} \qquad (1)$$

where L is the partial inductance matrix. The partial reluctance has locality similar to capacitance, therefore circuit analysis based on it is much more efficient than that based on partial inductance. Experiments in later papers like [6] showed that circuit simulation based on partial reluctance also has great advantage over partial inductance based simulation both in speed and accuracy. Ref. [7] proved that by ignoring small elements in matrix, the sparsified partial reluctance matrix is positive definite and the circuit simulation based on it is stable.

As mentioned before, the operating frequency of current VLSI circuit has reached several GHz, so high frequency effect becomes more and more important. However, among the existing literatures about partial reluctance extraction, only [8] and [11] considered the high frequency effect. Ref. [8] demonstrated the necessity of considering high frequency effect through simulation experiments, and presented an algorithm for extracting frequency-dependent reluctance. But the inversion of matrices in the algorithm prohibits its performance especially when the structure of input circuit is complex. Besides, [11] proposed an impedance extraction algorithm based on the idea of K element. However, the algorithm in [11] extended the locality of reluctance to the admittance (reciprocal of impedance), which actually only holds for very high frequency. In this paper, we propose a new algorithm for frequency-dependent reluctance extraction. This algorithm considers the physical meaning of partial reluctance, and extract the reluctance directly without any matrix inversion. This is superior to the algorithms in [8] and [11], because there is no expensive operation of matrix inversion in the proposed algorithm. Numerical experiments are performed on interconnect structure generated randomly and some practical structures, and their results demonstrate the efficiency of our new algorithm.

The rest of this paper is organized as follows. Section II discusses how to extract the partial reluctance considering high frequency effect. In Section III, the details of our new algorithm which is called direct reluctance extraction is introduced. The experiment results are given in Section IV, where both interconnect structure generated randomly and practical structures from industry are tested. Finally, Section V is the conclusion.

This work is supported by the China National Science Foundation under Grant 90407004 and 60401010. And, it is partly supported by National High Technology Research and Development Program of China (No. 2004AA1Z1050).

II. Partial Reluctance Extraction Considering High Frequency Effect

A. Capturing High Frequency Effect

It is well known that as the frequency goes high, the current on a conductor is no long evenly distributed. In order to capture such high frequency effect, the conductors need to meshed into filaments along the current direction. With the magneto quasi static (MQS) assumption, the current can be considered to flow parallel to conductor surface, and the current is distributed evenly in each filament.

Since the current density near the surface of the conductor is larger than that far from the surface because of the skin effect, we can mesh the cross section of a conductor segment non-uniformly. How many filaments should be meshed for a conductor mainly depends on the operating frequency and the size of the conductor. Fig. 1 shows an example of two parallel conductors which are mashed into 5×3 filaments each.

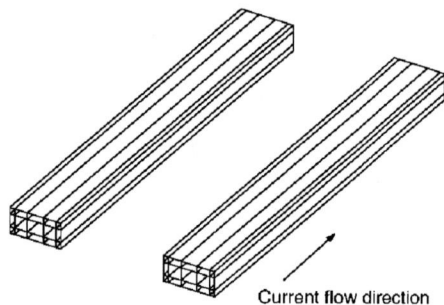

Current flow direction

Fig. 1. An example of two parallel conductors which are mashed into 5×3 filaments non-uniformly.

B. Frequency-Dependent Reluctance Extraction

The general process of frequency-dependent reluctance extraction can be divided into the following steps. First, for each conductor, a proper window is selected for it using some window selection strategy. We will discuss it in detail in next subsection. Second, in order to capture the high frequency effect, each conductor is meshed into some filaments as described in last subsection. After filaments meshing, the local reluctance matrix for each window is calculated using some methods. Finally, the local reluctance matrices are combined into a big one, which is the whole reluctance matrix of the circuit.

The differences between different methods of frequency dependent reluctance extraction are mainly focused on the second step, i.e. how to extract the local reluctance matrix for each window. The second step of the algorithm presented in [8] can be summarized as follows. For each window, It first calculate the impedance matrix of all filaments in it. Then it sets a conductor as the aggressor, i.e. the voltage on it is set to 1 while others are 0 and the filaments current distribution is calculated. Adding the currents in the same conductor together, the result is equal to one column of the local admittance matrix. The above process is repeated again and again until all conductors in the window are set as the aggressor one by one. Now, it gets the local admittance matrix, by inversing it, the local impedance matrix of conductors is obtained. Select the imaginary part of the local impedance matrix and inverse it again, the resulting matrix is the local reluctance matrix.

From the above summary, we know the algorithm in [8] needs two times matrix inversion for each window. It is clear that the matrix inversion work is costly especially when the structure is complex and it has a lot of windows. So developing an extraction method for directly extracting partial reluctance without the matrix inversion would be valuable. We propose such a novel algorithm in this paper. Actually, in our algorithm, we only need to slove an equation for each window, the details will be discussed in Section III.

C. Window Selection Strategy

How to select the coupling window is the key point to assure the accuracy of reluctance extraction. Too large window size will bring high accuracy as well as long running time, and too small window size may result in less accuracy. A window selection algorithm has been proposed based on the shield effect [10], which however was applied to 2-D simple structures. Based on some definitions and tips in it, a more complicated window selection method dealing with 3-D complex structures was proposed in [11]. This method considers both shield effect and distance among conductors, and is used in our frequency-dependent reluctance extraction.

To select the coupling window is actually to choose some nearby conductors for the specified aggressor conductor (the rest are usually called victim conductors). In [11], a coupling level is defined for each victim according to its position relation with the aggressor. Then all victims whose coupling level is less than a specified "max coupling level" form the window.

For a general 3-D structure, the procedure of determining the coupling levels of the victims is performed on the conductor projections in XOY, ZOX, and YOZ planes, respectively. In each coordinate plane, the projections are sorted in two directions at first to get two queues of projections. For each queue, the projection is set as aggressor in turn, then the coupling level of other projection is calculated according to the shield level and their distance. After implementing the above procedure for all the six projection queues, the minimum of the levels of conductor i and j is used as the final result. For more detail, please refer to Section III of [11].

III. Direct Extraction of Frequency-Dependent Reluctance

In this section, we propose a novel algorithm which extract K matrix directly without any matrix inversion work.

A. Details of Direct Partial Reluctance Extraction

With the sinusoidal steady-assumption, for a system with n conductors at angular frequency ω, the relation of current vector $I \in \mathbb{C}^n$ and voltage vector $V \in \mathbb{C}^n$ on the conductors can be expressed as:

$$(R + j\omega L)I = V \qquad (2)$$

where $R, L \in \mathbb{R}^{n \times n}$ are the resistance matrix and the partial inductance matrix of the system respectively. Transforming Eq. (2) and combining it with the definition of partial reluctance matrix K in Eq. (1), we have:

$$j\omega I = K(V - RI) \qquad (3)$$

Notice that $V\text{-}RI$ is an $n \times 1$ vector, if we set entry i of $V\text{-}RI$ to $j\omega$ and other entries to zero, then the resulting current distribution I would be equal to the ith column of K.

Now we talk about the physical meaning of it. According to the definition of magnetic vector potential A and Faraday's law, we have:

$$E = -j\omega A - \nabla\Phi \qquad (4)$$

where E is the electric field, Φ is the scalar potential. Integrating both side of Eq. (4) along the length direction of the conductor from one end a to the other b, it becomes:

$$V_{ab} = -(\Phi_b - \Phi_a) = j\omega \int_b^a A dl + El_{ab} \qquad (5)$$

As E contributes to the resistance potential drop from $J = \sigma E$, so:

$$El_{ab} = RI \qquad (6)$$

Combining Eq. (5) and Eq. (6), we have:

$$V_{ab} - RI = j\omega \int_b^a A dl \qquad (7)$$

Eq.(7) tells us that setting entry i of $V\text{-}RI$ to $j\omega$ equals to setting entry i of the magnetic vector potential drop $\int A dl$ to 1. Now, the physical meaning of partial reluctance is clear. Actually, the ith column of partial reluctance matrix K is the current distribution on the conductors when we set the ith entry (corresponding to the aggressor) of the vector potential drop to 1 and other entries (corresponding to the victims) to 0. The physical meaning of partial reluctance makes it possible to be extracted directly instead of inverse the partial inductance L.

In order to capturing the high frequency effect, we have already meshed each conductor into several filaments as discussed in Section II. Assuming each filament is thin enough that the current can be approximately uniformly distributed in it. We note the current and voltage of filament i as $\hat{I}_i$ and $\hat{V}_i$ respectively. The relation between them can be expressed as:

$$\hat{R}_{ii}\hat{I}_i + j\omega \sum_{j=1}^{m} \hat{L}_{ij}\hat{I}_j = \hat{V}_i \qquad (8)$$

where m is the number of filaments. The filament's DC resistance $\hat{R}_{ii}$ can be obtained by:

$$\hat{R}_{ii} = l_i / \sigma \hat{a}_i \qquad (9)$$

where σ is the conductivity of the conductor, l_i is the length of it and $\hat{a}_i$ is the cross-section area of the filament. The mutual partial inductance between the filaments can be obtained by [3]:

$$\hat{L}_{ij} = \frac{\mu_0}{4\pi \hat{a}_i \hat{a}_j} [\int_{\hat{a}_i} \int_{\hat{a}_j} \int_{l_i} \int_{l_j} \frac{\vec{dl_i}\vec{dl_j}}{\|\vec{r_i} - \vec{r_j}\|} d\hat{a}_i d\hat{a}_j] \qquad (10)$$

where μ_0 is the magnetic permeability. Put Eq. (8) of each filaments together, we get the matrix form of them like:

$$(\hat{R} + j\omega\hat{L})\hat{I} = \hat{V} \qquad (11)$$

where $\hat{R}$ is an $m \times m$ diagonal matrix and $\hat{L}$ is an $m \times m$ matrix. They can be easily obtained according to Eq. (9) and Eq. (10). $\hat{I} \in \mathbb{C}^m$ is the vector of current on the m filaments, which is what we want to know and $\hat{V} \in \mathbb{C}^m$ is the vector of potential drop on the m filaments.

Because there is no transverse current between filaments, the potential drop on different filaments in the same conductor are equal, and the current of a conductor is equal to the sum of the filaments current in the same conductor. For convenience, we define a mesh incidence matrix $M \in \mathbb{R}^{n \times m}$ as:

$$M_{ij} = \begin{cases} 1 & \text{when filament } j \text{ is in coductor } i \\ 0 & \text{otherwise} \end{cases} \qquad (12)$$

Then the above relation can be expressed as:

$$\hat{V} = M'V, \quad I = M\hat{I} \qquad (13)$$

Since the partial reluctance matrix K is real, the resulting current I must be real when we set the vector potential drop $\int A dl$ along the aggressive conductor to 1 and others to 0, i.e.,

$$I_{im} = 0 \qquad (14)$$

Rewrite Eq. (11) as:

$$\begin{cases} \hat{R}\hat{I}_{re} - \omega\hat{L}\hat{I}_{im} = \hat{V}_{re} = M'V_{re} \\ \omega\hat{L}\hat{I}_{re} + \hat{R}\hat{I}_{im} = \hat{V}_{im} = \omega M' \int A dl \end{cases} \qquad (15)$$

Put Eq. (14) and Eq. (15) together into a matrix form, we have:

$$\begin{bmatrix} \hat{R} & -\omega\hat{L} & -M^t \\ \omega\hat{L} & \hat{R} & 0 \\ 0 & M & 0 \end{bmatrix} \begin{bmatrix} \hat{I}_{re} \\ \hat{I}_{im} \\ V_{re} \end{bmatrix} = \begin{bmatrix} 0 \\ \omega M^t \int Adl \\ 0 \end{bmatrix} \quad (16)$$

By solving this linear equation, we can get $\hat{I}$. Then, the current on conductors I can be got by $I = M\hat{I}$. Suppose we set conductor i to be the aggressor, then I is equal to the ith column of partial reluctance matrix K. In this way, we can obtain the whole partial reluctance matrix K, by setting each conductor to be the aggressor one by one and repeat this process.

B. Condensing the Coefficient Matrix

The dimension of linear system (16) is $2m+n$, where m is the number of filaments and n is the number of conductors in a window. Actually, we do not need to solve linear system (16), just solving a linear system with dimension $m+n$ is enough. Since m is much larger than n, reducing the problem dimension from $2m+n$ to $m+n$ would cut the equation solving time by nearly 7/8. Notice Eq. (14) shows that when we set the vector potential drop along the aggressor to 1 and others to 0, the resulting current vector is real. That means we only need to know $\hat{I}_{re}$ without caring $\hat{I}_{im}$. In the following, we will show how to condense the coefficient matrix.

From the second equation of Eq. (16) we have:

$$\hat{I}_{im} = \omega\hat{R}^{-1}M^t \int Adl - \omega\hat{R}^{-1}\hat{L}\hat{I}_{re} \quad (17)$$

Replace $\hat{I}_{im}$ in the first and the third equation of Eq. (16) with Eq. (17), the resulting two equations are:

$$(I + (\omega\hat{R}^{-1}\hat{L})^2)\hat{I}_{re} - \hat{R}^{-1}M^tV_{re} = \omega^2\hat{R}^{-1}\hat{L}\hat{R}^{-1}M^t \int Adl \quad (18)$$

$$M\hat{R}^{-1}\hat{L}\hat{I}_{re} = M\hat{R}^{-1}M^t \int Adl \quad (19)$$

where I is the indentity matrix. Combining Eq. (18) and Eq. (19), we get the condensed form of the linear system:

$$\begin{bmatrix} I + (\omega\hat{R}^{-1}\hat{L})^2 & -\hat{R}^{-1}M^t \\ M\hat{R}^{-1}\hat{L} & 0 \end{bmatrix} \begin{bmatrix} \hat{I}_{re} \\ V_{re} \end{bmatrix} = \begin{bmatrix} \omega^2\hat{R}^{-1}\hat{L}\hat{R}^{-1}M^t \int Adl \\ M\hat{R}^{-1}M^t \int Adl \end{bmatrix} \quad (20)$$

After the above process, the dimension of equation is reduced to $m+n$.

It should be pointed out that the computational expense of this condensing procedure is very little, because the $\hat{R}$ in Eq. (20) is a diagonal matrix whose inverse is trivial. Besides, the M matrix is very sparse, taking it into account also saves computing time.

C. Algorithm Description

Given a circuit, for each conductor in it, our partial reluctance extraction algorithm first selects a window for the conductor using the window selection strategy presented in Section II. When extracting the partial reluctance, we only consider the conductors in the same window, this is according to the excellent locality of partial reluctance. For each conductor, we set it as the only aggressor in its window, and extract the column of partial reluctance of the whole partial reluctance matrix corresponding to it. The proposed reluctance extraction algorithm considering high frequency effect is summarized as follows.

Direct partial reluctance extraction algorithm

1. For each conductor in the given circuit, select a proper window for it using the window selection strategy in Section II.

2. Mesh each conductor into several filaments non-uniformly, the number of filaments mainly depends on the frequency and the size of the conductor.

3. Suppose there are N conductors in the circuit, for conductor i (i from 1 to N) do:
   (a) Pick out all the conductors in conductor i's window, suppose there are n_i conductors and m_i filaments.
   (b) Form the resistance matrix R, mutual inductance matrix L of filaments and the mesh incidence matrix M of the conductors in this window using Eq. (9) and Eq. (10).
   (c) Set the vector potential drop along the aggressor to 1 and others to 0.
   (d) Form the condensed form of the linear system for this window, whose dimension is m_i+n_i.
   (e) Solve the linear system to get $\hat{I}_{re}$. Since the dimension is not very large, we use direct method based on LU decomposition to solve it.
   (f) Calculate the current of the conductors in the window by $I = M\hat{I}_{re}$.
   (g) Put the items of vector I into the corresponding positions of the whole partial reluctance matrix K_{asym}. These items form the column of K_{asym} that corresponding to conductor i. Other positions in this column are filled by 0.

4. Make the K symmetric by $K = (K_{asym} + K^t_{asym})/2$.

IV. Numerical Results

The proposed algorithm has been implemented in C language as a software prototype for 3-D interconnect partial reluctance extraction. With it, we did a series of experiments and compared our results with famous frequency-dependent inductance extraction tool: FastHenry [9]. All the experiments are run on our Sun Fire V880 server with a 750MHz CPU. FastHenry is used to extract the impedance matrix and the imaginary parts of it elements are the partial inductances. The result of our algorithm is partial reluctance matrix. In order to compare the result with FastHenry, it needs to be inversed to get the partial inductance matrix. Finally, we compute the loop inductance of each pair of conductors, and compare it with that from the results of FastHenry.

The first example is a structure generated randomly, including 300 conductors. The conductors in the circuit have unequal length and are placed parallel, with width and height set to 1.0 μm. The conductivity is used copper conductivity as default. The second and the third examples are real package structure from industry, obtained from the webpage of FastHenry. The second example is a 30 pins structure, it contains 260 conductor segments, as shown in Fig. 2. The third one is a 35 pins structure and contains 175 conductor segments, as shown in Fig. 3. In our experiment, the operating frequency of all the three examples are set to 10 GHz. For the three examples, each conductor is partitioned into 2×2, 3×3 and 3×5 filaments, respectively. And, the same filament meshing scheme is used in our algirithm and FastHenry.

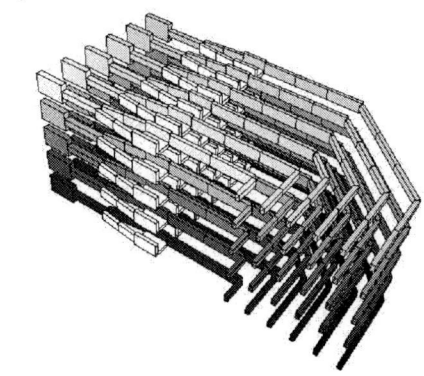

Fig. 2. A real industry package structure with 30 pins.

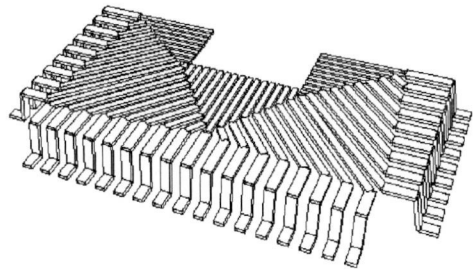

Fig. 3. A real industry package structure with 35 pins.

The results of our experiments are shown in Table I and Table II. Table I shows the error distribution of loop inductance for the three examples while using our algorithm. Table II shows the CPU time of the proposed algorithm and FastHenry. From the results, it is clear that the proposed algorithm has several tens to several hundreds speedup over FastHenry, while maintaining good accuracy. For the random interconnect structure, the errors of loop inductance are mainly less than 6%. Actually, the speed and the accuracy is a tradeoff, we can decrease the speed to increase the accuracy or vice versa according to the requirement.

In this paper, a direct extraction algorithm for partial reluctance matrix is proposed. In our method, we take the high frequency effect into account by meshing the conductor segments into filaments non-uniformly. According to the excellent locality of partial reluctance matrix, we use an efficient window selection strategy to select a window for each conductor. In each window, the partial reluctance is directly extracted without the expensive matrix inversion usually employed in previous methods. Numerical results demonstrate that our new algorithm has several tens to hundreds speedup ratio to FastHenry while preserving high accuracy.

The future work may be the combination of our algorithm to the reluctance based circuit simulator, and the improvement of equation solution for each window.

References

[1] K. Gala, V. Zolotov, R. Panda, et al., "On-chip inductance modeling and analysis," in *Proc. Design Automation Conference*, pp. 63-68, June 2000.

[2] M. W. Beattie and L. T. Pileggi, "Inductance 101: modeling and extraction," in *Proc. Design Automation Conference*, pp. 323-328, June 2001.

[3] A. E. Ruehil, "Inductance calculation in a complex integrated circuit environment," *IBM Journal of Research and Development*, vol. 16, no. 5, pp. 470-481, September 1972.

[4] Z. He, M. Celik, and L. T. Pileggi, "Spie: sparse partial inductance extraction," in *Proc. Design Automation Conference*, pp. 137-140, June 1997.

[5] A. Devgan, H. Ji, and W. Dai, "How to efficiently capture on-chip inductance effect: introducing a new circuit element K," in *Proc. IEEE International Conference on Computer Aided Design*, pp. 150-155, Nov 2000.

[6] H. Ji, A. Devgan, and W. Dai, "KSim: a stable and efficient *RKC* simulator for capturing on-chip inductance effect," in *Proc. Asia and South-Pacific Design Automation Conference*, pp. 379-384, Jan 2001.

[7] Y. Du and W. Dai, "Partial reluctance based circuit simulation is efficient and stable," in *Proc. Asia and South-Pacific Design Automation Conference*, pp. 483-488, Jan 2005.

[8] C. Luk, T. H. Chen, C. P. Chen, "Frequency-dependent reluctance extraction," in *Proc. Asia and South-Pacific Design Automation Conference*, pp. 793-798, Jan 2004.

[9] M. Kamon, M. J. Tsuk, and J. K. White, "Fasthenry: a multipole-accelerated 3-D inductance extraction program," *IEEE Trans. on MTT*, pp. 216-230, Sep 1994.

[10] T. Chen, C. Luk, H. Kim, and C. Chen, "Inductwise: inductance-wise interconnect simulator and extractor," in *Proc. IEEE International Conference on Computer Aided Design*, pp. 215–220, Nov. 2002.

[11] H. Wei, W. Yu, L. Yang and Z. Wang, "Fast 3-D impedance extraction of VLSI interconnects based on the K element," *International Conference on Communications, Circuits and Systems(ICCCAS)*, pp. 1201-1205, May 2005.

V. Conclusion

TABLE I
Error Distribution of Loop Inductance for the Examples (Compared with FastHenry)

Error of loop inductance(%)	<3%	3-6%	6-9%	9-12%	12-15%	>15%
example 1	95.5%	4.2%	0.3%	0.0%	0	0
example 2	46.4%	26.4%	12.3%	6.8%	2.5%	5.1%
example 3	72.7%	20.7%	5.7%	0.8%	0.1%	0.0%

TABLE II
Computational Time for the Examples (Unit in Second)

	FastHenry	Our algorithm	Speed-up ratio
example 1	855.2	9.12	**94**
example 2	5650.2	9.83	**575**
example 3	5796.3	122.51	**47**

Macromodelling Oscillators Using Krylov-Subspace Methods

Xiaolue Lai and Jaijeet Roychowdhury
Department of Electrical and Computer Engineering
University of Minnesota
Email: {laixl, jr}@ece.umn.edu

Abstract— We present an *efficient* method for automatically extracting unified amplitude/phase macromodels of arbitrary oscillators from their SPICE-level circuit descriptions. Such comprehensive oscillator macromodels are necessary for accuracy when speeding up simulation of higher-level circuits/systems, such as PLLs, in which oscillators are embedded. Standard MOR techniques for linear time invariant (LTI) and varying (LTV) systems are not applicable to oscillators on account of their fundamentally nonlinear phase behavior. By employing a cancellation technique to deflate out the phase component, we restore the validity and efficacy of Krylov-subspace-based LTV MOR techniques for macromodelling oscillator amplitude responses. The nonlinear phase response is re-incorporated into the macromodel after the amplitude components have been reduced. The resulting unified macromodels predict oscillator waveforms, in the presence of any kind of input or interference, at far lower computational cost than full SPICE-level simulation, and with far greater accuracy compared to existing macromodels. We demonstrate the proposed techniques on LC and ring oscillators, obtaining speedups of 30-120× with no appreciable loss of accuracy, even for small circuits.

I. INTRODUCTION

Oscillators are important building blocks in electronic and optical systems. For example, they are often used for frequency-translation of information signals in communication systems. Voltage-controlled oscillators (VCOs) are key components of phase-locked loops (PLLs), which are widely used in both digital and analog circuits for clock generation and recovery, frequency synthesis, *etc*. Despite their widespread use, the simulation of oscillators and oscillator-based systems still poses significant challenges.

Traditional circuit simulators, such as SPICE [1], are far from ideally suited for simulating oscillators. One key problem is that transient simulation accumulates numerical phase errors without limit; furthermore, it is also difficult to extract phase information from time-domain voltage/current waveforms accurately. To improve phase accuracy, many timesteps need to be taken in each oscillation cycle, with transient simulations of high-Q oscillators requiring many thousands of cycles, hence suffering from great inefficiency.

Since phase responses are of major concern for oscillators, various specialized and approximate techniques (*e.g.*, [2]–[7]) have been developed for predicting phase information directly without relying on transient simulation of the full circuit. Simulation in the phase domain can result in great speedups. However, phase macromodels do not capture amplitude variations at the output of the oscillator, which can be important in many situations. For example, in pico-radio systems, "radio nodes" must be ultra-low power, leading to novel, very simple PLL systems where VCO outputs are in essence directly fed to analog mixers, with no intervening amplitude stabilization or clipping. VCO amplitude variations in such systems change gains of PLL loops dynamically, thereby affecting important phenomena such as jitter, lock/capture behavior, *etc*. Similar design philosophies are emerging for a variety of low power systems in wireless communication and mobile systems.

In [8], a technique was presented to macromodel both phase and amplitude in oscillators under perturbation. The phase deviation was calculated via a nonlinear phase equation [9], while the amplitude macromodel was extracted by full Floquet decomposition [10] of the

linearized, time-varying oscillator. The method applies to any kind of oscillator, including LC, ring, *etc*. However, it has a drawback: Floquet decomposition becomes very computationally expensive, and can also suffer from numerical errors, as system sizes increase. An oscillator macromodelling technique that scales gracefully with circuit size is of great practical interest, given that on-chip RF oscillators today can have many thousands of nodes.

In this paper, we present a novel method to circumvent this issue. Instead of performing full Floquet decomposition on the linearized oscillator equations, we apply the time-varying Padé (TVP) method [11], to reduce the oscillator system to a smaller LPTV system which captures the important amplitude-variation components of the original oscillator accurately. The transfer function of the LPTV oscillator system is expanded into matrix forms using time- or frequency-domain methods (such as FDTD or harmonic balance), and then reduced using Krylov-subspace methods [12]–[14]. A major issue faced for oscillators, that we solve in this paper, is that the oscillator's LPTV system incorporates both phase and amplitude information, leading to inaccurate amplitude macromodels if the phase component is not separated out correctly. Moreover, because oscillators are fundamentally phase-unstable [9] and sustained small perturbations can change the phase of the oscillator unboundedly, additional issues are faced if the phase component is not dealt with specially.

Instead of using full Floquet decomposition as in [8], we use a novel alternative technique in this paper, based on canceling out components of the input that excite phase responses, that enables application of Krylov-subspace methods to oscillators. The perturbation input to the oscillator is decomposed into two parts with one contributing to phase and the other to amplitude. We change the input vector of the LPTV system to remove the input corresponding to phase and apply Krylov-subspace methods to reduce this modified system. The resulting reduced system contains no phase information since the pole corresponding to phase has a very small residual, thus having been effectively eliminated by the Krylov reduction process.

Once the reduced amplitude macromodel, without any phase information, has been obtained, we re-incorporate the phase component by coupling the amplitude macromodel with the nonlinear, scalar phase macromodel presented in [9]. The resulting unified oscillator macromodel is able to predict both phase and amplitude variations accurately at far lower computational cost than that of full SPICE-level simulation. Compared to the method described in [8], the key difference is that the method in [8] requires full decomposition of the whole LPTV system, while our method relies on Krylov-subspace methods, with far lower computational cost. Hence, our method scales well to large oscillator circuits. The generated macromodels can be easily encapsulated into other circuit simulation tools (*e.g.*, MATLAB/Simulink, Verilog-A, *etc*.) to predict the comprehensive behavior of oscillators in a variety of system-level situations.

We verify our macromodelling technique on ring and LC oscillators. Comparing simulation results between our macromodels and SPICE-level full circuit simulation, we show that our macromodels are able to reproduce the waveforms of oscillators under various perturbations accurately, while obtaining impressive speedups. Even with the relatively small oscillators we have used for testing purposes, we obtain speedups in the range of 1-2 orders of magnitude; we expect much greater speedups for larger circuits with complex device models, which we are currently in the process of incorporating into our simulation infrastructure.

The remainder of the paper is organized as follows. In Section II, we show that linear perturbation analysis is not valid for oscillators. In Section III, we review nonlinear perturbation analysis for oscillators and the nonlinear oscillator macromodel we employ in this work. In Section IV, we summarize the time-varying Padé technique for reducing LPTV systems. In Section V, we describe our phase-component deflation technique for macromodelling amplitude variations of oscillators. In Section VI, we present simulation results on ring and LC oscillators.

II. LINEAR PERTURBATION ANALYSIS IS NOT SUITABLE FOR OSCILLATORS

The traditional method to analyze perturbed nonlinear system is to linearize the nonlinear equations on its unperturbed orbit. In this section, we will show that this method is not suitable for oscillators.

A general oscillator under perturbations can be described by

$$\dot{x} + f(x) = Bb(t), \qquad (1)$$

where $b(t)$ is perturbation signal applied to the free running oscillator. Since perturbation signal has small amplitude, we can linearize (1) on its unperturbed steady-state orbit and get a linearized periodic time-varying system

$$\dot{z}(t) + G(t)z(t) \approx Bb(t), \qquad (2)$$

where $G(t) = \frac{\partial f(x)}{\partial x}|_{x_s(t)}$ is the linearized system on oscillator's steady-state orbit $x_s(t)$.

According to Floquet theory [10], the state transition matrix of the homogeneous part of (2) can be given by

$$\Phi(t,\tau) = U(t)e^{D(t-\tau)}V^T(\tau) \qquad (3)$$

where $U(t) = [u_1(t), u_2(t), ..., u_n(t)]$ and $V(t) = [v_1(t), v_2(t), ..., v_n(t)]$ are T-periodic nonsingular matrix, satisfying biorthogonality conditions $v_i^T(t)u_j(t) = \delta_{ij}$, and $D = diag[\mu_1, ..., \mu_n]$, where μ_i are Floquet exponents. The particular solution of (2) under perturbation $b(t)$ is given by

$$z(t) = \sum_{i=1}^{n} u_i(t) \int_0^t e^{\mu_i(t-\tau)} v_i^T(\tau) Bb(\tau) d\tau. \qquad (4)$$

For an oscillator system, one of the Floquet exponents must be 0 [9]. Without loss of generality, we assume $\mu_1 = 0$, and $e^{\mu_1(t-\tau)}$ term vanishes. Thus, we can always choose a perturbation $b(t)$ to satisfy $v_1^T(t)Bb(t)$ has nonzero average value, then $z(t)$ will grow unboundedly on t even though $b(t)$ have very small amplitude. This contradicts the assumption that $z(t)$ is small variation, thus the linear perturbation analysis is inconsistent.

III. PREVIOUS WOEK: NONLINEAR OSCILLATOR MACROMODEL

In [9], authors show that the Floquet exponent $\mu_1 = 0$ in (4) is corresponding to the oscillator's phase deviation, which means the oscillator's phase deviation grows unboundedly even though the perturbation signal is very small. Due to oscillator's neutral phase stability, linearizing the oscillator on its steady state is meaningless since a small perturbation will change the phase, or the steady-state orbit of the oscillator dramatically.

An idea to overcome this limitation is to linearize the oscillator circuits over its perturbed time-shifted orbit. Based on this idea, in [8], a novel approach is presented to construct the oscillator macromodels.

The macromodel produced is a combination of a scalar nonlinear differential equation [9] that solves the phase deviation and a reduced linear time-varying system for predicting amplitude variation, which is computationally simpler than the original oscillator system. Here, we summarize this method.

A. Nonlinear Oscillator Phase Macromodel

According to [9], the solution of the oscillator under perturbation can be expressed as

$$x(t) = x_s(t + \alpha(t)) + y(t), \qquad (5)$$

where $x(t)$ is the orbit of the oscillator under perturbation, $x_s(t)$ is the unperturbed steady-state orbit of the oscillator, $\alpha(t)$ is the phase deviation, and $y(t)$ is the amplitude variation. If we can calculate the phase deviation $\alpha(t)$ and amplitude variation $y(t)$, we can rebuild the waveforms of the oscillator under perturbations using (5).

The phase deviation $\alpha(t)$ is governed by the nonlinear differential equation [9]

$$\dot{\alpha}(t) = v_1^T(t + \alpha(t)) \cdot Bb(t), \qquad (6)$$

where $v_1(t)$ is the perturbation projection vector (PPV) that corresponds to oscillator's phase sensitivity to perturbations, and $b(t)$ is the perturbation applied to the oscillator. The PPV has periodic waveforms that have the same period as the oscillator. Various methods [9], [15]–[17], both in time domain and frequency domain, have been presented for calculating the PPV from the oscillator circuit equations. When the PPV is available, the phase deviation $\alpha(t)$ can be efficiently calculated by solving this simple one-dimension differential equation.

B. Amplitude Macromodel

In [8], a method is presented to to construct the amplitude macromodel of the oscillator. the oscillator is first linearized on its perturbed time-shifted orbit

$$\begin{aligned}
\dot{o}(t) &\approx -\frac{\partial f}{\partial x}|_{x_s(t+\alpha(t))} o(t) + Bb(t) \\
&= A(x_s(t+\alpha(t)))o(t) + Bb(t), \qquad (7)
\end{aligned}$$

where $x_s(t)$ is oscillator's steady-state orbit, $o(t)$ is small variations due to perturbation $b(t)$ and $\alpha(t)$ is phase deviation. Since $A(x_s(t + \alpha(t)))$ is not periodic, we introduce a new variable $\hat{t} = t + \alpha(t)$ and define $\hat{o}(\hat{t}) = o(t)$ and $\hat{b}(\hat{t}) = b(t)$. After dropping a quadratic term, we can obtain a linear periodic time-varying system

$$\dot{\hat{o}}(\hat{t}) = A(x_s(\hat{t}))\hat{o}(\hat{t}) + B\hat{b}(\hat{t}). \qquad (8)$$

Applying Floquet decomposition to this LPTV system, the solution of this system can be expressed as

$$\hat{o}(\hat{t}) = \sum_{i=1}^{n} u_i(\hat{t}) \int_0^{\hat{t}} \exp(\mu_i(\hat{t} - \tau)) v_i^T(\tau) B\hat{b}(\tau) d\tau. \qquad (9)$$

where $u_1 = 0$ is corresponding to oscillator's phase deviation, we need to drop it. The resulting system can be reduced by dropping some less important Floquet exponents. The limitation of this method is that Floquet decomposition is very slow and numerically unstable when the system size is large.

IV. REDUCED-ORDER MODELING FOR LPTV SYSTEM

In [11], the time-varying Padé (TVP) method is presented to reduce the LPTV system.

A nonlinear system can be expressed as a differential-algebraic equation (DAE)

$$\frac{\partial q(x(t))}{\partial t} + f(x(t)) = b_L(t) + Bb(t)$$

$$y(t) = d^T x(t), \tag{10}$$

where $b_L(t)$ is a large signal, which determines the system's steady-state orbit, and $Bb(t)$ is small perturbations applied to the system. B and d are input/output vectors. For small perturbation analysis, we linearize this system over the orbit generated by the large signal. The time-varying Padé (TVP) method can be applied to reduce this linearized time-varying system. Here, we summarize this method.

The transfer function of the linearized system can be written as

$$H(t_1, s) = d^T \left(\frac{D}{dt_1}[\,] + sC(t_1) + G(t_1) \right)^{-1} [B], \tag{11}$$

where $\frac{D}{dt_1}[\,]$ is a differential operator [11]. Assume $C(t_1)$ and $G(t_1)$ to be periodic with angular frequency w_0, and define $W(t_1, s)$ to be the operator-inverse in (11), we have

$$W(t_1, s) = \left(\frac{D}{dt_1}[\,] + sC(t_1) + G(t_1) \right)^{-1} [B] \tag{12}$$

$$\Rightarrow \left(\frac{D}{dt_1}[\,] + sC(t_1) + G(t_1) \right) W(t_1, s) = B. \tag{13}$$

Assume $W(t_1, s)$ is also a periodic function, we can express (13) in frequency domain

$$[sC_{FD} + J_{FD}]\vec{W}_{FD} = \vec{B}_{FD}, \tag{14}$$

where

$$J_{FD} = (G_{FD} + \Omega C_{FD}), \tag{15}$$

$$C_{FD} = \begin{pmatrix} \vdots & \vdots & \vdots \\ \dots C_0 & C_{-1} & C_{-2} \dots \\ \dots C_1 & C_0 & C_{-1} \dots \\ \dots C_2 & C_1 & C_0 \dots \\ \vdots & \vdots & \vdots \end{pmatrix} \tag{16}$$

$$G_{FD} = \begin{pmatrix} \vdots & \vdots & \vdots \\ \dots G_0 & G_{-1} & G_{-2} \dots \\ \dots G_1 & C_0 & G_{-1} \dots \\ \dots G_2 & G_1 & G_0 \dots \\ \vdots & \vdots & \vdots \end{pmatrix} \tag{17}$$

$$\Omega = j\omega_0 \begin{pmatrix} \ddots & & & & & \\ & -2I & & & & \\ & & -I & & & \\ & & & 0I & & \\ & & & & I & \\ & & & & & iI \\ & & & & & & \ddots \end{pmatrix} \tag{18}$$

$$\vec{B}_{FD} = [\dots, 0, 0, B^T, 0, 0, \dots]^T. \tag{19}$$

Then the transfer function in frequency domain can be written as

$$\vec{H}_{FD}(s) = D^T [sC_{FD} + J_{FD}]^{-1} \vec{B}_{FD}, \tag{20}$$

where

$$D = \begin{pmatrix} \ddots & & & & \\ & d & & & \\ & & d & & \\ & & & d & \\ & & & & d \\ & & & & & \ddots \end{pmatrix}. \tag{21}$$

We can use any Krylov-subspace method to reduce (20) and obtain a smaller LPTV system.

$$\vec{H}_q(s) = L_q^T [I_{q \times q} + sT_q]^{-1} R_q, \tag{22}$$

and the reduced system equation is

$$T_q \dot{\tilde{x}}(t) + \tilde{x}(t) = R_q u(t), \tag{23}$$

where $\tilde{x}(t)$ is a vector of size q, which is much smaller than that of the original system.

V. Reducing Oscillator System Using TVP method

Reduce-order modeling of oscillators is slightly different than that of other systems, such as mixers and converters. Oscillator systems are neutral stable, traditional TVP methods may not be able to produce stable reduced system. In this section, we present our method for modeling amplitude variations of oscillators.

A. Traditional TVP Method Is Not Suitable For Oscillators

TVP methods [11] are very useful for reducing LPTV systems, such as mixers and convertors, *etc.* However, the method is not suitable for oscillators, as oscillator systems are neutral stable. We have shown in Section II that the Floquet decomposition of oscillator system has a Floquet exponent of 0, which contributes to phase response. If we expand the oscillator transfer function to frequency domain using the method described in Section IV, the resulting frequency domain transfer function (20) has many poles on the imaginary axis, with interval $j2\pi\omega_0$. Figure 1 depicts the pole distribution of an LC oscillator. In this figure, the oscillator has poles on $\pm jn\omega_0$, some poles even have positive real part due to numerical integration error. Performing MOR methods on such a system will produce unstable reduced system, especially when the expansion point is close to the imaginary axis. Moreover, since the phase and amplitude information are mixed together in the oscillator LPTV system, the reduced system is not able to correctly represent the amplitude variation of the oscillator, even though we can obtain a stable reduced system using the TVP method.

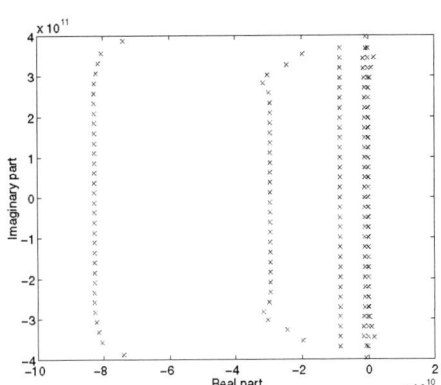

Fig. 1. Pole distribution of an LC oscillator.

B. Oscillation Pole Cancellation Method

The instability of the reduced system is due to oscillation poles in (20), or the Floquet exponent $\mu_1 = 0$ in the LPTV system (2). In [8], Floquet decomposition is used to eliminate the Floquet exponent $\mu_1 = 0$; however, this method is expensive and numerically unstable for large systems. In this work, we present a novel method to cancel the effect of μ_1 without performing the Floquet decomposition.

According to [8], we can obtain amplitude macromodel of the oscillator by linearizing the oscillator over its perturbed time-shifted orbit $x_s(\hat{t})$, where $\hat{t} = t + \alpha(t)$ and $\alpha(t)$ is the phase deviation due to perturbation. The resulting LPTV system is

$$\dot{\hat{o}}(\hat{t}) = A(x_s(\hat{t}))\hat{o}(\hat{t}) + B\hat{b}(\hat{t}). \tag{24}$$

The solution of above system can be expressed as

$$\hat{o}(\hat{t}) = \sum_{i=1}^{n} u_i(\hat{t}) \int_0^{\hat{t}} \exp(\mu_i(\hat{t} - \tau)) v_i^T(\tau) B\hat{b}(\tau) d\tau, \tag{25}$$

where μ_i are Floquet exponents, μ_1 has a value of 0 [9], $u_i(t)$ and $v_i(t)$ are T-periodic vectors, satisfying biorthogonality conditions $v_i^T(t)u_j(t) = \delta_{ij}$. In [8], the μ_1 term is dropped after the Floquet decomposition since it contributes to phase deviation.

To avoid the Floquet decomposition, we cannot use this method to eliminate the μ_1 term; instead, we eliminate the input that contributes to phase deviation. From (25), we know that the input for the μ_1 term is $v_1^T(\tau)Bb(\tau)$. Since $u_i(t)$ and $v_i(t)$ are biorthogonal, we can eliminate this input using the projection method. We define a new input vector

$$\tilde{B} = B - v_1(\tau)Bu_1(\tau). \tag{26}$$

Using this new input vector, the input to the μ_1 term is

$$\begin{aligned}
v_1^T(\tau)\tilde{B}b(\tau) &= v_1^T(\tau)Bb(\tau) - v_1(\tau)v_1(\tau)Bu_1(\tau)b(\tau) \\
&= v_1^T(\tau)Bb(\tau) - v_1(\tau)Bv_1(\tau)u_1(\tau)b(\tau) \\
&= v_1^T(\tau)Bb(\tau) - v_1^T(\tau)Bb(\tau) = 0,
\end{aligned} \tag{27}$$

and inputs for other μ_is $(i > 1)$ are

$$\begin{aligned}
v_i^T(\tau)\tilde{B}b(\tau) &= v_i^T(\tau)Bb(\tau) - v_1(\tau)v_i(\tau)Bu_1(\tau)b(\tau) \\
&= v_i^T(\tau)Bb(\tau) - v_i(\tau)Bv_i(\tau)u_1(\tau)b(\tau) \\
&= v_i^T(\tau)Bb(\tau) - 0 = v_i^T(\tau)Bb(\tau).
\end{aligned} \tag{28}$$

Hence, the new input vector $\tilde{B}$ only eliminates the effect of the Floquet exponent $\mu_1 = 0$, but preserves inputs of all other Floquet exponents in the system. We can replace the original input vector B with $\tilde{B}$ without changing the amplitude response of the LPTV system. (24) can be rewritten as

$$\dot{\hat{o}}(\hat{t}) = A(x_s(\hat{t}))\hat{o}(\hat{t}) + \tilde{B}\hat{b}(\hat{t}). \tag{29}$$

(29) is stable, because we have minimized the residuals of oscillation poles in (20), even though we do not really eliminate them. We can apply the TVP method described in [11] to reduce this modified system and obtain a stable reduced system for amplitude variations.

C. Detailed Procedure Of Oscillator MOR

In this subsection, we summarize the procedures for constructing and using the oscillator macromodel.

1) Constructing Oscillator Macromodel: Following procedure will construct the oscillator macromodel.

1) Obtain oscillator steady-state $x_s(t)$ using time domain or frequency domain methods.
2) Calculate $u_1(t) = \dot{x}_s(t)$.
3) Calculate the PPV $v_1(t)$ using numerical methods [9], [15]–[17].
4) Construct new input vector $\tilde{B} = B - v_1(t)^T Bu_1(t)$, where $\tilde{B}$ is a T-periodic vector.
5) Perform the TVP method on the new LPTV system

$$\dot{o}(t) = A(x_s(t))o(t) + \tilde{B}b(t) \tag{30}$$

and obtain the reduced system

$$T_q\dot{\tilde{o}}(t) = \tilde{o}(t) + R_qb(t), \tag{31}$$

where $\tilde{o}(t)$ has size of q, which has size smaller than that of the original system.

2) Using Oscillator Macromodel: To reproduce oscillator waveforms under perturbations using our macromodel, we need to integrate both the phase and amplitude equations. The phase equation (6) is solved on time t; however, the amplitude equation (29) is integrated on the shifted time $\hat{t} = t + \alpha(t)$. Below is the pseudocode to rebuild the oscillator' waveform using our macromodel.

```
1   t=0; α(0)=0; i=0
2   t=t+Δt; i=i+1
3   Calculate α(i) by solving (6) on t
4   ĩ = t+α(i)
5   Calculate amplitude variation o(t)=ô(ĩ) by solving (31) on ĩ
6   Rebuild oscillator's waveforms using (5)
7   goto 2
```

VI. NUMERICAL RESULTS

In this section, we evaluate the technique presented above using ring and LC oscillators. All simulations were performed using **MATLAB** on an Intel-architecture machine, running Linux. We constructed oscillator macromodels using the method described in Section V, simulated oscillator waveforms using the constructed macromodels, and compared the results with SPICE-like simulations of the full oscillator circuits in the same **MATLAB** environment. Experiment results show that our macromodels are able to capture the amplitude variations of oscillators accurately. The rebuilt waveforms using our macromodel match the results from full SPICE-level simulation perfectly, with about 30–120 times speedups.

A. 3-stage Single-Ended Ring Oscillator

Our first example is a 3-stage single-ended ring oscillator, as shown in Figure 2. This oscillator has a system size of 8 and an oscillation frequency of $f_0 = 1GHz$.

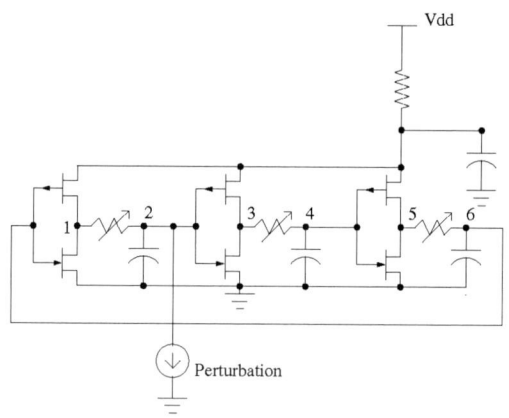

Fig. 2. A 3-stage single-ended ring oscillator.

We first calculate the steady-state of the oscillator using the harmonic balance method, and extract the PPV of the oscillator using the Monodromy method [9], [15]. We then linearize the oscillator on its steady-state orbit and reduce the resulting LPTV system using the method we presented in Section V. Since we are interested in perturbations whose frequency is close to the oscillator's oscillation frequency, we choose the Arnoldi expansion point $s_0 = 2\pi f_0$, where f_0 is the oscillator's free-running frequency. The original nonlinear system has the size of 8, using our method, we can reduce the system size to 3. The resulting small LPTV system can be simulated much faster than the original nonlinear system.

To verify that our macromodel can capture the oscillator amplitude variations correctly, we apply a periodic perturbation current $b(t) = 0.0001sin(\omega_1 t)$ to the node 2 of the oscillator, as shown in Figure 2. We apply different perturbation frequencies ω_1, rebuild the oscillator's output voltage on node 2 using our macromodel and compare with SPICE-level full simulation. The simulation results are shown in Figure 3 and Figure 4. Our macromodel is able to match the full simulation perfectly, with great speedups. The SPICE-level full simulation takes about 6 minutes for a simulation time of 100 cycles; however, it takes only 4 seconds to simulate the same number of cycles using our macromodel. This gives us approximately 120 times speedup on this small oscillator circuit. For larger oscillator circuits with more nonlinear components and complex device model, we expect more significant speedups.

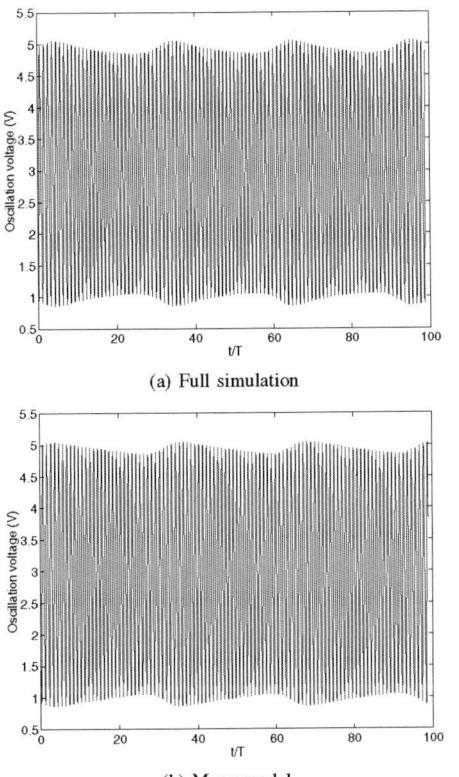

(a) Full simulation

(b) Macromodel

Fig. 3. Output waveform of the 3 stage ring oscillator under perturbation $b(t) = 0.0001sin(1.04\omega_0 t)$.

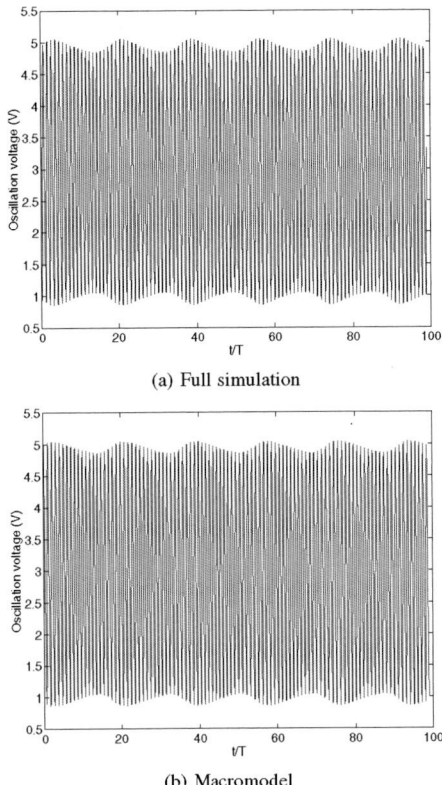

(a) Full simulation

(b) Macromodel

Fig. 4. Output waveform of the 3 stage ring oscillator under perturbation $b(t) = 0.0001sin(1.06\omega_0 t)$.

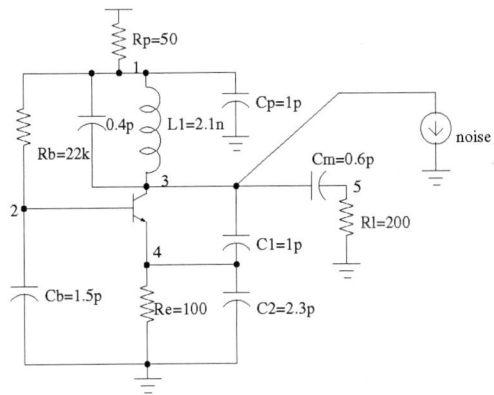

Fig. 5. A 4GHz Colpitts LC oscillator.

B. 4GHz Colpitts LC Oscillator

Our second test circuit is a 4GHz Colpitts LC oscillator. The circuit and parameters of this oscillator are shown in Figure 5. The oscillator has a system size of 6 and a free-running frequency of $f_0 = 4GHz$.

We apply similar method as we describe in the ring oscillator case to calculate the PPV and construct the amplitude macromodel. We evaluate our macromodel with a current perturbation injected into the node 3 of the oscillator, as shown in Figure 5. The perturbation current has an amplitude of $0.1mA$. We simulate the oscillator's output voltage on the node 3 under different perturbation frequencies using our macromodel and compare with SPICE-level full simulation. The simulation results are shown in Figure 6 and Figure 7. Our macromodel is able to match the full simulation with acceptable accuracy. In this case, we obtain about 30 times speedup. This

speedup is not as good as the ring oscillator case, because this LC oscillator has only one nonlinear component, hence, its full SPICE-level simulation is very fast.

VII. CONCLUSIONS

We have presented a novel technique to extract simple amplitude macromodels from SPICE-level circuit descriptions of any oscillator circuit. Combining it with the oscillator phase macromodel presented in [9], we can rebuild the output waveforms of any perturbed oscillator. We have tested our macromodel using LC and ring oscillators and provided detailed comparisons against full SPICE-level circuit simulation. Numerical results show our macromodels are able to predict oscillator amplitude and phase variations well in the presence

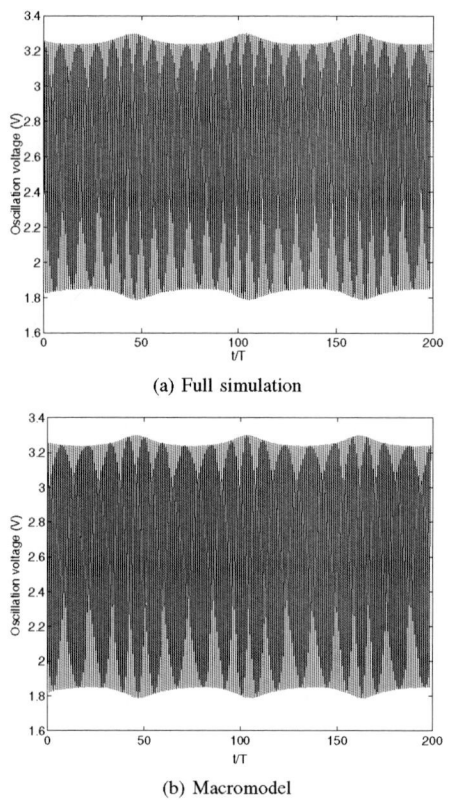

(a) Full simulation

(b) Macromodel

Fig. 6. Output waveform of the Colpitts LC oscillator under perturbation $b(t) = 0.0001 sin(1.02\omega_0 t)$.

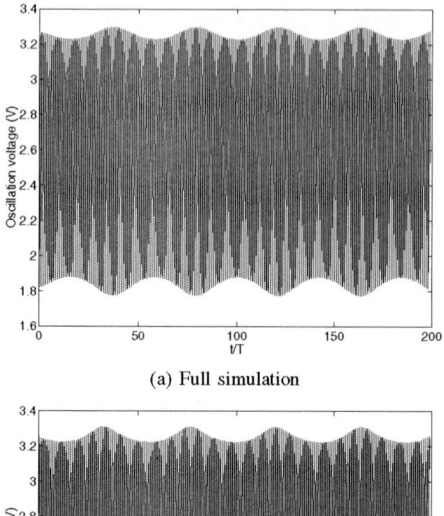

(a) Full simulation

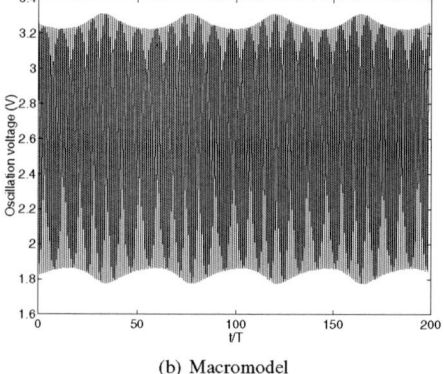

(b) Macromodel

Fig. 7. Output waveform of the Colpitts LC oscillator under perturbation $b(t) = 0.0001 sin(1.025\omega_0 t)$.

of perturbations, with great speedups over SPICE-level simulation. Currently, we are working on validating our technique on larger circuits, for which we expect speedups of 3-4 orders of magnitude.

REFERENCES

[1] L. Nagel. SPICE2: A Computer Program to Simulate Semiconductor Circuits. Electron. Res. Lab., Univ. Calif., Berkeley, 1975.

[2] M. Gardner. Phase-Lock Techniques. Wiley, New York, 1966.

[3] A. Hajimiri and T.H. Lee. A general theory of phase noise in electrical oscillators. IEEE Journal of Solid-State Circuits, 33(2), February 1998.

[4] A. Demir, E. Liu, A.L. Sangiovanni-Vincentelli, and I. Vassiliou. Behavioral simulation techniques for phase/delay-locked systems. In Proceedings of the Custom Integrated Circuits Conference 1994, pages 453–456, May 1994.

[5] A. Costantini, C. Florian, and G. Vannini. Vco behavioral modeling based on the nonlinear integral approach. IEEE International Symposium on Circuits and Systems, 2:137–140, May 2002.

[6] L. Wu, H.W. Jin, and W.C. Black. Nonlinear behavioral modeling and simulation of phase-locked and delay-locked systems. In Proceedings of IEEE CICC, 2000, pages 447–450, May 2000.

[7] M. F. Mar. An event-driven pll behavioral model with applications to design driven noise modeling. In Proc. Behav. Model and Simul.(BMAS), 1999.

[8] X. Lai and J. Roychowdhury. Automated oscillator macromodelling techniques for capturing amplitude variations and injection locking. In IEEE/ACM International Conference on Computer Aided Design, November 2004.

[9] A. Demir, A. Mehrotra, and J. Roychowdhury. Phase noise in oscillators: a unifying theory and numerical methods for characterization. IEEE Trans. on Circuits and Systems-I:Fundamental Theory and Applications, 47(5):655–674, May 2000.

[10] R. Grimshaw. Nonlinear Ordinary Differential Equations. Blackwell Scientific, New York, 1990.

[11] J. Roychowdhury. Reduced-order modelling of time-varying systems. IEEE Trans. Ckts. Syst. – II: Sig. Proc., 46(10), November 1999.

[12] P. Feldmann and R.W. Freund. Efficient linear circuit analysis by pade approximation via the lanczos process. IEEE Trans. on Computer-Aided Design, 14:639–649, May 1995.

[13] A. Odabasioglu, M. Celik, and L.T. Pileggi. PRIMA: passive reduced-order interconnect macromodelling algorithm. In Proceedings of IEEE ICCAD 1997, pages 58–65, November 1997.

[14] R.W. Freund. Reduced-order modeling techniques based on Krylov subspaces and their use in circuit simulation. Technical Report 11273-980217-02TM, Bell Laboratories, 1998.

[15] A. Demir. Phase noise in oscillators: Daes and colored noise sources. In IEEE/ACM International Conference on Computer-Aided Design, November 1998.

[16] A. Demir, D. Long, and J. Roychowdhury. Computing phase noise eigenfunctions directly from steady-state jacobian matrices. In IEEE/ACM International Conference on Computer Aided Design, pages 283–288, November 2000.

[17] A. Demir and J. Roychowdhury. A reliable and efficient procedure for oscillator ppv computation, with phase noise macromodelling applications. IEEE Trans. on Computer-Aided Design of Integrated Circuits and Systems, 22(2):188–197, February 2003.

2006 Asia and South Pacific Design Automation Conference

5D-1

Low-Power Design Methodology for Module-wise Dynamic Voltage and Frequency Scaling with Dynamic De-skewing Systems

Takeshi Kitahara, Hiroyuki Hara, Shinichiro Shiratake, Yoshiki Tsukiboshi[*],
Tomoyuki Yoda, Tetsuaki Utsumi, and Fumihiro Minami

TOSHIBA Corporation Semiconductor Company   *TOSHIBA Microelectronics Corporation
580-1, Horikawa-cho, Saiwai-ku, Kawasaki, 212-8520, JAPAN
kitahara@dad.eec.toshiba.co.jp

Abstract - This paper discusses design methodology for a module-wise dynamic voltage and frequency scaling(DVFS) technique which adjusts the supply voltage for a module appropriately to reduce the power dissipation. A circuit is able to work even when the supply voltage is in transition, by using our dynamic de-skewing system(DDS). We propose a novel clock design methodology to minimize the inter-module clock skew for solving one of the major design issues in the module-wise DVFS. We also describe a method of determining the minimum supply voltage value for a module. We lead the issue to a problem of solving simultaneous polynomial inequalities. Our experimental results show that the module-wise DVFS can reduce 53% power compared with the chip-wise DVFS, and 17% more reduction was achieved by applying the minimum supply voltage proposed.

I. Introduction

As the clock frequency has become higher, reducing the power dissipation has become a more serious issue. Chips for portable electric appliances are required to reduce the power in order to make the battery life longer, and chips for high-end digital systems should be cared about issues of the heat and the packaging cost. Recently, the low power design is also an important topic from the ecological point of view.

The DVFS technique controls the supply voltage and the clock frequency dynamically, according to the performance requirement. DVFS makes a deep impact on the dynamic power reduction, because the energy consumption of a chip is nearly proportional to the square of the supply voltage. In many cases, DVFS is employed for a whole chip, that is the supply voltage and the clock frequency are changed for all of the digital parts on a chip, such as Intel XScale[R] technology and Transmeta LongRun[TM] technology[1]. Recently, however, the technique to apply DVFS to the module on a chip has been proposed[2-4]. It is effective for reducing the power dissipation, as the possibility of lowering the supply voltage increases, compared with DVFS which is applied to a whole chip. When controlling a whole chip, the supply voltage and the clock frequency cannot be lowered, if there exists only one module which requires high performance[2].

Lately, the leakage power dissipation has been increasing. Its reduction is a major subject to be tackled under the 90nm technology era and beyond, as lowering the supply voltage makes the threshold voltage lower. The dynamic power dissipation decreases according to the supply voltage with the quadratic order when lowering the supply voltage; however, the sub-threshold leakage power increases with the exponential order when lowering the threshold voltage [5]. So, the ratio of the leakage power to the dynamic power in the total power dissipation is increasing, nowadays. Some papers proposed the technique of reducing the sum of the dynamic power dissipation and the leakage power dissipation by DVFS [6-8]. Multi-threshold CMOS (MT-CMOS), selective multi-threshold CMOS (Selective-MT) and variable threshold CMOS (VT-CMOS)[9-12] are the effective techniques to reduce the leakage power dissipation. The techniques can be applied, when a circuit or a module is on standby. The technique to apply DVFS to the module is also effective to reduce the leakage power, because the standby mode can be defined for each module, in accordance with its functionality. The standby time assigned to the module is longer than the one assigned to the whole chip. The module-wise DVFS has an effect not only on the dynamic power reduction, but also on the leakage power reduction.

The H.264/MPEG-4 codec LSI with the module-wise DVFS was developed, and its effectiveness was verified[2]. One feature of this LSI is that it operates even when the supply voltage is in transition. Fig.1 shows the way how the supply voltage and the clock frequency change. This mechanism is realized by using DDS[2]. In [13], the voltage transition overhead is considered. The optimal scheduler is proposed in a discretely variable voltage environment. It is assumed that no instructions in the task are executed during the transition interval; however, instructions can be executed during the interval by using DDS. Neither voltage transition overhead nor performance degradation happens.

The clock delay in a DVFS module changes continuously, when the supply voltage is in transition. On the other hand, the clock delay does not change for a module whose supply voltage is fixed. In this situation, the clock skew between a DVFS module and a non-DVFS module may become larger,

0-7803-9451-8/06/$20.00 ©2006 IEEE.

533

5D-1

and a larger clock skew causes the timing violation. DDS can minimize the clock skew even during the transition interval, and novel methodology of generating the clock tree is required to minimize the clock skew all the time. In this paper, we propose design methodology for the module-wise DVFS. Novel strategy of the clock tree synthesis is presented for a circuit with DDS, and a method of determining the minimum supply voltage is also introduced. The remainder of the paper is organized as follows. Section II explains the module-wise DVFS using DDS. A delay control circuit(DCC), which controls the clock-delay for a DVFS module, is also described. Clock design methodology for a circuit using DDS, is given in Section III. Section IV presents conditions of the timing closure. Experimental results are shown in Section V.

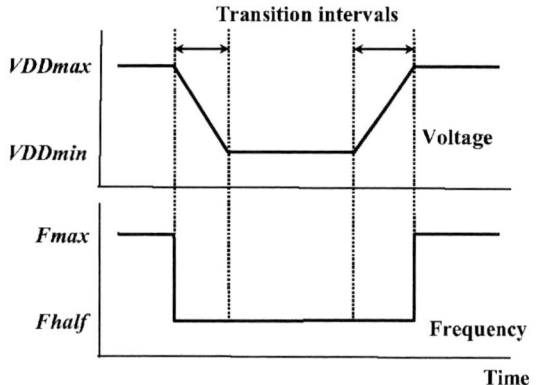

Fig.1. Behavior of voltage and frequency

II. Module-wise DVFS

The module-wise DVFS is a low power technique operating a module at an appropriate frequency and an adequate supply voltage, by using a frequency gear and a supply voltage controller. The frequency change is performed instantly; however the supply voltage transition with high accuracy and high speed is difficult. Dynamic skew control for the DVFS module, such as DDS, is necessary, so as not to interrupt the processing for the DVFS module during the supply voltage transition.

A. Dynamic De-skewing System(DDS)

DDS is introduced to realize the module-wise DVFS. Fig.2 shows the structure of DDS with the DVFS module and the non-DVFS module. In Fig.2, a variable voltage value is supplied for the DVFS module, and a fixed voltage value is supplied for the non-DVFS module. The clock signal is connected directly to the non-DVFS module, and it is connected to the DVFS module via DCC. The function of DCC is to control the clock delay to minimize the clock skew between the DVFS module and the non-DVFS module. Details of DCC are explained in the following subsection. The clock signal of the DVFS module and that of the non-DVFS module have paths which reach DCC. They are shown as the return signal A and B in Fig.2. The delay of the return signal A is adjusted to that of B. DDS measures the skew between the two return signals, and adjusts the clock delay for the DVFS module in each clock cycle. According to the measured skew, an adequate delay is added to the clock signal for the DVFS module in the next cycle.

B. Delay Control Circuit(DCC)

DCC consists of one skew measure unit and two delay generation units, as shown in Fig.2. Fig.3(a) shows the schematic of the delay generation unit A. The port CKI is connected to the root clock signal ROOTCLK, and the adjusted clock signal is supplied to the DVFS module from the port CKO. The delay of each component Ai is designed to be same for $i \in \{1,2,...,n\}$. (In Fig.3(a), the case of $n=5$ is shown.) The signal CKI is connected to the n switches, and

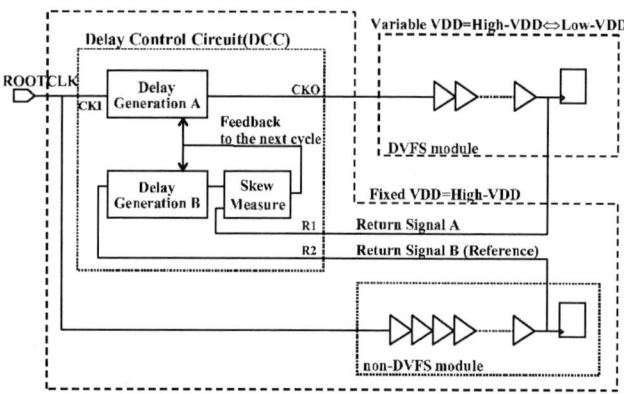

Fig.2. Dynamic de-skewing system structure

534

they are controlled by skew indication signals, which are described in the following explanation. Only one of the skew indication signals is set to 'H', so only one path from the signal CKI to CKO is activated when CKI changes from 'L' to 'H'. As the delay of the component Ai can be defined as a unit delay, the delay of the signal from CKI to CKO can be controlled by the step of the unit delay. For example, 3 unit delays are added in Fig.3(a).

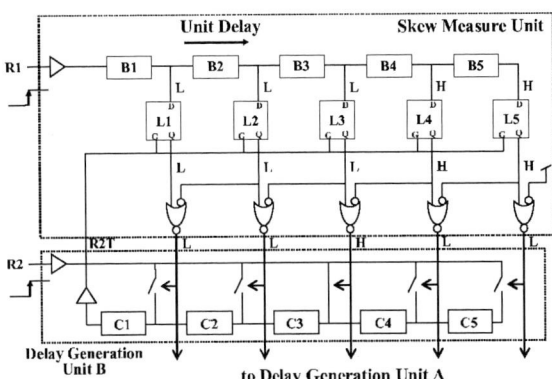

Fig.3(b). Skew measure unit and delay generation unit B

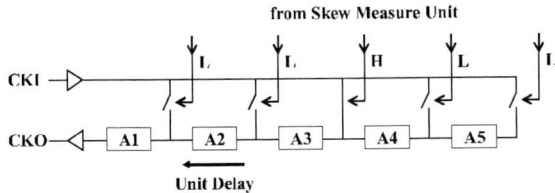

Fig.3(a). Delay generation unit A

Fig.3(b) shows the schematic of the skew measure unit and the delay generation unit B. It has two input ports. The port R1 is connected to the return signal A, and the port R2 is connected to the return signal B. It measures the time between the rise edge of the signal R1 and that of R2, as follows. In Fig.3(b), the delay of each component Bi and Ci is same as that of Ai in Fig.3(a) for $i \in \{1,2,...,n\}$. The case of $n=5$ is shown in Figure 3(b). If the signal R1 changes from 'L' to 'H', the output value of B1 changes from 'H' to 'L' after 1 unit delay, and that of $Bi(i=2,3,...,n)$ changes from 'H' to 'L' one after another. When the signal R2 changes from 'L' to 'H', each output value of $Bi(i=1,2,...,n)$ is captured by each transparent latch $Li(i=1,2,...,n)$ after some unit delays which is determined by the delay generation unit B. There exists some m such that the value of $Li(1 \leq i \leq m)$ is equal to 'L' and that of $Li(m < i \leq n)$ is equal to 'H'. For example, $m=3$ in Fig.3(b). The skew measure unit has the skew indication signals. They are generated by the NOR gates whose inputs are $Li(i=1,2,...,n)$, and only one of them is set to 'H', according to the skew between the delay of the input R1 and that of the signal R2T in Fig.3(b). According to the skew between the two, the clock delay from CKI to CKO is adjusted in the next cycle. When the supply voltage of the DVFS module and that of the non-DVFS module are same, the clock delay inside DCC is the largest. As the supply voltage of the DVFS module goes down, the clock delay becomes smaller.

III. Clock Design Methodology

This section shows the clock design methodology for a circuit including DDS. We assume that the clock tree structure is employed for the clock design.

A. Problem Formulation

When designing the clock, the Clock Tree Synthesis(CTS) technique is commonly employed. Since CTS can systematically minimize the clock skew, it is helpful in making the clock design task easier and reducing the design turn around time. Clock design of DDS is complicated, because there are feedback clock signals to DCC as shown in Fig.2. The conventional CTS technique is not constructed on the assumption with feedback loops on clocks. So, a novel CTS technique is required to design a chip with the module-wise DVFS effectively. Determining the supply voltage value for the lower clock frequency is another thing to be considered. We want to let the supply voltage value as lower as possible, for reducing the power dissipation of a chip.

Our overall goal is the following. (1)Constructing novel clock design methodology which employs the CTS technique for the module-wise DVFS. (2)Clarifying the minimum clock skew the methodology can achieve. The setup and the hold timing constraints are examined with the clock skew taken into consideration. (3)Determining the lowest supply voltage which guarantees a chip to work correctly.

B. Clock Design Strategy

Fig.4 shows the clock network of DDS, and we present the clock design strategy to generate a clock tree structure for it. Let *Fmax* be the clock frequency of the clock signal ROOTCLK. The clock signal having a half of the frequency (*Fhalf*) is generated via D-type flip-flop(F/F). We assume that the frequency of the clock signal which is connected to DCC, switches between *Fmax* and *Fhalf* according to the performance requirement as explained in Fig.1. The frequency of the clock signal which is connected to the non-DVFS module is always *Fmax*. This design style is commonly employed in a practical design. For example, in the audiovisual chip design[2], the audio module works with

5D-1

2006 Asia and South Pacific Design Automation Conference

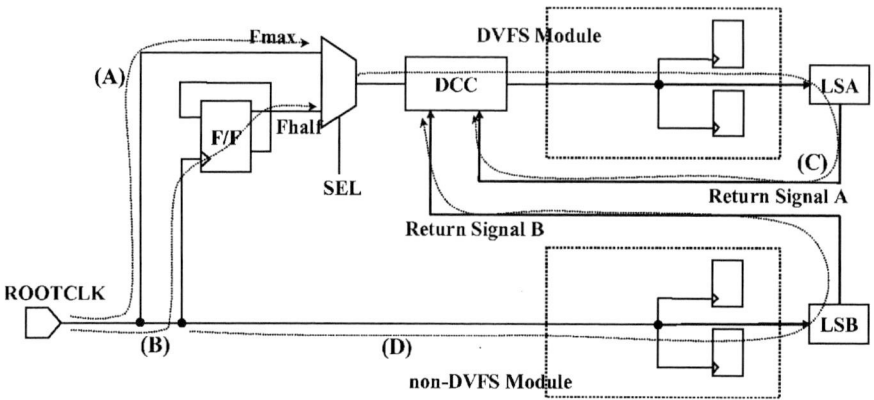

Fig.4. Clock network of Dynamic de-skewing system(DDS)

180/90MHz, because the workload of it tends to be small in many applications, while the video modules always work with 180MHz. With the exception of the clock which generates a half of frequency, Fig.4 is almost same as Fig.2. The following is the flow of generating a clock tree structure for the clock network shown in Fig.4.

III. Clock Design Methodology

This section shows the clock design methodology for a circuit including DDS. We assume that the clock tree structure is employed for the clock design.

A. Problem Formulation

When designing the clock, the Clock Tree Synthesis(CTS) technique is commonly employed. Since CTS can systematically minimize the clock skew, it is helpful in making the clock design task easier and reducing the design turn around time. Clock design of DDS is complicated, because there are feedback clock signals to DCC as shown in Fig.2. The conventional CTS technique is not constructed on the assumption with feedback loops on clocks. So, a novel CTS technique is required to design a chip with the module-wise DVFS effectively. Determining the supply voltage value for the lower clock frequency is another thing to be considered. We want to let the supply voltage value as lower as possible, for reducing the power dissipation of a chip.

Our overall goal is the following. (1)Constructing novel clock design methodology which employs the CTS technique for the module-wise DVFS. (2)Clarifying the minimum clock skew the methodology can achieve. The setup and the hold timing constraints are examined with the clock skew taken into consideration. (3)Determining the

lowest supply voltage which guarantees a chip to work correctly.

B. Clock Design Strategy

Fig.4 shows the clock network of DDS, and we present the clock design strategy to generate a clock tree structure for it. Let *Fmax* be the clock frequency of the clock signal ROOTCLK. The clock signal having a half of the frequency (*Fhalf*) is generated via D-type flip-flop(*F/F*). We assume that the frequency of the clock signal which is connected to DCC, switches between *Fmax* and *Fhalf* according to the performance requirement as explained in Fig.1. The frequency of the clock signal which is connected to the non-DVFS module is always *Fmax*. This design style is commonly employed in a practical design. For example, in the audiovisual chip design[2], the audio module works with 180/90MHz, because the workload of it tends to be small in many applications, while the video modules always work with 180MHz. With the exception of the clock which generates a half of frequency, Fig.4 is almost same as Fig.2. The following is the flow of generating a clock tree structure for the clock network shown in Fig.4.

Step 1: To minimize the skew between the delay of the following two clock paths, delay buffers are inserted into the clock paths:
(A) the direct path from ROOTCLK to the multiplexer,
(B) the path from ROOTCLK to the multiplexer via D-type flip-flop.
Let $Delay_{Fmax}$ and $Delay_{Fhalf}$ be the delay of the path (A) and (B), respectively, and α be the minimum delay of a delay buffer among all delay buffers in a cell library. Then, we have:

$$| Delay_{Fmax} - Delay_{Fhalf} | < \alpha . \qquad (1)$$

Step 2: The conventional CTS technique is applied to all

536

flip-flops in both the DVFS module and the non-DVFS module with assigning the root as ROOTCLK in Fig.4. Clock buffers are added to the clock paths in the DVFS module and the non-DVFS module, in order to construct a clock tree structure. Let β be the maximum clock skew among all of the flip-flops, and $Delay_{FFi}$ be the clock delay between ROOTCLK and each flip-flop FFi. Then we have the following inequality, for any pair (i, j) such that FFi and FFj are flip-flops in the DVFS module or the non-DVFS module, under a fixed supply voltage:

$$| Delay_{FFi} - Delay_{FFj} | \leq \beta . \qquad (2)$$

Remark 1: When the supply voltage value for the DVFS module is equal to the scaling upper limit(i.e. $VDDmax$ in Figure 1), the path delay from CKI to CKO inside the delay control unit A shown in Figure 3(a) is the maximum, that is the path $An \rightarrow An-1... \rightarrow A1$ is activated. The clock delay inside the DVFS module is the smallest when the supply voltage value is the upper limit, so the largest path delay is required for the delay control unit A, to minimize the clock skew between a flip-flop in the DVFS module and one in the non-DVFS module.

Remark2: The supply voltage of the DVFS module is lower than that of DCC in some periods. So, it is necessary to add a level shifter to the outputs of the DVFS module [14]. The function of the level shifter is to lift the output voltage, when the output signal value is equal to 'H'. The level shifter is shown as LSA in Fig.4. The level shifter LSB is also added to the output of the non-DVFS module. These level shifters are treated as the leaves when the CTS is applied. So, the skew among these two and all the flip-flops is minimized, and the difference between any two of them is smaller than β.

Let γ be the unit delay in DCC, which is the delay of Ai. Let $VDDmax$ and $VDDmin$ be the maximum and the minimum voltage scaling value for the DVFS module. When the supply voltage value changes between $VDDmax$ and $VDDmin$, the maximum clock skew caused by DCC is γ. So, we have the following inequality, for any pair (i, j) such that FFi and FFj are flip-flops in the DVFS module or the non-DVFS module, under any supply voltage value V such that $VDDmin \leq V \leq VDDmax$ for the DVFS module:

$$| Delay_{FFi} - Delay_{FFj} | \leq \beta + \gamma . \qquad (3)$$

But, to be precise, only the clock skew between a flip-flop in the DVFS module and one in the non-DVFS module is affected by the delay control unit. If FFi and FFj are in the same module, there happens no clock skew penalty caused by DCC. So, we have the following inequality, for any pair (i, j) such that FFi and FFj are flip-flops in the DVFS module, under any supply voltage value V such that $VDDmin \leq V \leq VDDmax$ for the DVFS module:

$$| Delay_{FFi} - Delay_{FFj} | \leq \beta . \qquad (4)$$

We also have (4) for any pair (i, j) such that FFi and FFj are flip-flops in the non-DVFS module. When comparing the rising edge of R1 and that of R2, the chip was designed so that the former is always earlier. Supposing that the

difference of the above two is $n * \gamma + \gamma'$, where $n \in \mathbf{N}$ ($\mathbf{N}$ is the set of natural numbers) and $0 \leq \gamma' < \gamma$, γ' is rounded off by $n * \gamma$. So, we have the following two asymmetrical inequalities, for any pair (i, j) such that FFi is a flip-flop in the DVFS module and FFj is one in the non-DVFS module, under any supply voltage value V such that $VDDmin \leq V \leq VDDmax$ for the DVFS module:

$$Delay_{FFi} - Delay_{FFj} \leq \beta , \qquad (5)$$

$$Delay_{FFj} - Delay_{FFi} \leq \beta + \gamma . \qquad (6)$$

Step 3: To equalize the delay of the return signal A with that of B, delay buffers are inserted into one of the above two. Let $Delay_{RA}$ and $Delay_{RB}$ be the delay of the return signal A and B, respectively, then we have:

$$| Delay_{RA} - Delay_{RB} | < \alpha , \qquad (7)$$

as α is the minimum delay of a delay buffer among all delay buffers in a cell library. Here, we observe the delay difference between the following two clock paths.
(C)ROOTCLK $\rightarrow$ multiplexer $\rightarrow$ DCC
$\quad\quad \rightarrow$ DVFS module $\rightarrow$ LSA $\rightarrow$ return signal A $\rightarrow$ DCC,
(D)ROOTCLK $\rightarrow$ non-DVFS module $\rightarrow$ LSB
$\quad\quad \rightarrow$ return signal B $\rightarrow$ DCC.
Let $Delay_C$ and $Delay_D$ be the delay of the paths (C) and (D), respectively. We can reach the following inequality by (1) and (7):

$$| Delay_C - Delay_D | < 2 * \alpha . \qquad (8)$$

Let us remember that the clock delay difference inside the DVFS module and the non- DVFS module can be ignored, as its effect is already considered in (5) and (6). The difference expressed in (8) can be reduced by the following Step 4, and we have the following inequality, under the condition that the clock frequency value F is equal to $Fmax$ or $Fhalf$:

$$| Delay_C - Delay_D | < \alpha . \qquad (9)$$

Step 4: When the following two inequalities are both satisfied, the delay buffer whose delay is equal to α is inserted into the return signal A:

$$0 < Delay_{Fhalf} - Delay_{Fmax} < \alpha , \qquad (10)$$

$$0 < Delay_D - Delay_C < \alpha . \qquad (11)$$

It can be also realized by removing the one from the return signal B if exists, instead of the insertion. When the following two inequalities are both satisfied, the delay buffer is inserted into the return signal B:

$$-\alpha < Delay_{Fhalf} - Delay_{Fmax} < 0, \qquad (12)$$

$$-\alpha < Delay_D - Delay_C < 0. \qquad (13)$$

In other cases, (9) is satisfied without inserting or removing delay buffers.

IV. Timing Closure

A. Timing Constraints

Based on the observation in the previous section, we express the setup and the hold timing constraints for the flip-flops in the DVFS module and the non-DVFS module. Let $Min\_delay(FFi(ClkIn) \rightarrow FFj(DataIn))$ be the minimum delay in the set of a path whose start point is the clock-input port of a flip-flop FFi and whose end point is the data-input port of FFj. To be precise, Min_delay is a mapping of $\mathbf{P^2} \rightarrow \mathbf{R}$, where $\mathbf{P}$ is the set of ports of flip-flops and $\mathbf{R}$ is the set of real numbers. The function Max_delay is defined, similarly. $Clock\_delay(FFi)$ denotes the clock delay between the ROOTCLK and the clock-input port of the flip-flop FFi shown in Fig.4.

(I)The following is the hold timing constraint when $V=VDDmax$ and $F=Fmax$, for any pair (i, j) such that FFi is a flip-flop in the DVFS module and FFj is one in the non-DVFS module, or vice versa;

$$Min\_delay(FFi(ClkIn) \rightarrow FFj(DataIn))$$
$$> Clock\_delay(FFj) - Clock\_delay(FFi) + \alpha + \beta . \quad (15)$$

(II)The following is the setup timing constraint when $V=VDDmax$ and $F=Fmax$, for any pair (i, j) such that FFi is a flip-flop in the DVFS module and FFj is one in the non-DVFS module, or vice versa;

$$Max\_delay(FFi(ClkIn) \rightarrow FFj(DataIn))$$
$$< Clock\_delay(FFj) - Clock\_delay(FFi)$$
$$+ (1/Fmax) - \alpha - \beta . \quad (16)$$

(III)Inequality (15) is the hold timing constraint when $V \neq VDDmax$ and $F=Fhalf$, for any pair (i, j) such that FFi is a flip-flop in the non-DVFS module and FFj is one in the DVFS module. When FFi is a flip-flop in the DVFS module and FFj is one in the non-DVFS module, the hold timing constraint is as follows;

$$Min\_delay(FFi(ClkIn) \rightarrow FFj(DataIn))$$
$$> Clock\_delay(FFj) - Clock\_delay(FFi)$$
$$+ \alpha + \beta + \gamma . \quad (17)$$

(IV)Inequality (16) is the setup timing constraint when $V \neq VDDmax$ and $F=Fhalf$, for any pair (i, j) such that FFi is a flip-flop in the DVFS module and FFj is one in the non-DVFS module. When FFi is a flip-flop in the non-DVFS module and FFj is one in the DVFS module, the setup timing constraint is as follows;

$$Max\_delay(FFi(ClkIn) \rightarrow FFj(DataIn))$$
$$< Clock\_delay(FFj) - Clock\_delay(FFi)$$
$$+ (1/Fhalf) - \alpha - \beta - \gamma . \quad (18)$$

Remark: When FFi and FFj are flip-flops in the same module, the timing constaints are relaxed. The inequalities by setting $\alpha = 0$ to (15) and (16), are the setup and the hold timing constraints, respectively.

B. Method for Timing Closure

Next, we discuss a flow of the timing closure. The main focus is the way how to generate a circuit which satisfies the inequalities from (15) to (18) above. We use the scalable polynomial delay model (SPDM)[15] which can be generated by the curve-fitting method. By using this model, the cell delay can be expressed by the variables c, s and v

which denote the output capacitance, the input slew and the supply voltage, respectively:

$$Delay = Delay(c,s,v) = \sum_{i,j,k=0}^{l,m,n} a_{ijk} * c^l * s^j * v^k . \quad (19)$$

The following is the flow for the timing closure.

Step 1: A logic circuit is synthesized and optimized under the condition of the supply voltage $V = VDDmax$ for the DVFS module. Every path in a logic circuit is adjusted to satisfy the inequality both (15) and (16). The non linear delay model (NLDM) can be used at this point, instead of SPDM.

Step 2: The minimum supply voltage $VDDmin$ is determined. As the path delay can be given like (19) by using SPDM, the functions of Max_delay and $Clock\_delay$ are expressed by polynomials of the variable v. So, we can obtain the minimum value $v = VDDmin$ such that (18) is satisfied, by solving the n-th degree inequality.

Step 3: If (17) is satisfied for any v such that $VDDmin \leq v \leq VDDmax$, the timing constraints are satisfied and the timing closure is done. If not, it is necessary to add some delay buffers to fix the hold violation. To determine which and how many delay buffers are to be added, a polynomial is added to the left side of (17). It expresses the delay of an adequate delay buffer. If (17) is satisfied for any v such that $VDDmin \leq v \leq VDDmax$ by this addition, the timing closure can be done by adding such delay buffer. If not, it is necessary to add more delay buffers.

V. Experimental Results

We applied the clock design methodology described in Section III to the H.264/MPEG-4 LSI [2]. This LSI was implemented using our TC300C(90nm) technology[16]. Fig.5(a) shows the clock delay distribution map without using our novel clock design methodology. The horizontal axis denotes the clock delay between ROOTCLK and a flip-flop, and the vertical axis denotes the number of flip-flops whose clock delay is the value of the horizontal. The graph "DVFS(VDDmax)" indicates the clock delay distribution of the flip-flops in the DVFS-module when the supply voltage value is equal to $VDDmax$. "DVFS (VDDmin)" indicates the one when the supply voltage value is equal to $VDDmin$. The graph "non-DVFS" indicates the number of the flip-flops in the non-DVFS module. Here, we picked up flip-flops in the non-DVFS module which are interactive with flip-flops in the DVFS module, because the clock delay variation in the DVFS module gives a timing impact on only these flip-flops. The maximum clock skew is nearly 1.0ns. One remark is that the shape of the graph "DVFS(VDDmax)" and "DVFS (VDDmin)" are different. It indicates that the ratio of the clock delay under $VDDmax$ and the one under $VDDmin$ is different for each clock path. One clock path may become slower than another under $VDDmin$, even if the path delay of the two is same under $VDDmax$. Fig.5(b) shows the clock delay distribution map with using our novel clock design methodology. The three graphs gather around the clock delay 4.3ns, and the maximum clock skew is reduced to nearly 0.35ns.

Next, we applied our timing closure technique which is presented in Section IV. In [2], 0.90V was put in the place of *VDDmin*. We developed an SPDM cell library for the TC300C technology, and analyzed the timing of the chip. We set the supply voltage value from 1.20V to 0.70V, every 0.025V. Fig.6 shows the result of the static timing analysis. The horizontal axis shows the supply voltage, and the vertical axis denotes the timing slack of the chip. The slack for the frequency *Fhalf*(=90MHz) is shown in Fig.6. If the timing slack is positive, the timing constraints are satisfied. When the DVFS module works with the frequency *Fmax*(=180MHz), the supply voltage value is equal to 1.20V. So, the slack value of nearly 5.6ns is necessary under *VDDmax*(=1.20V). We can find that the *VDDmin* is equal to nearly 0.82V by Fig.6, as the slack value is nearly 0ns when the supply voltage value is around 0.82V.

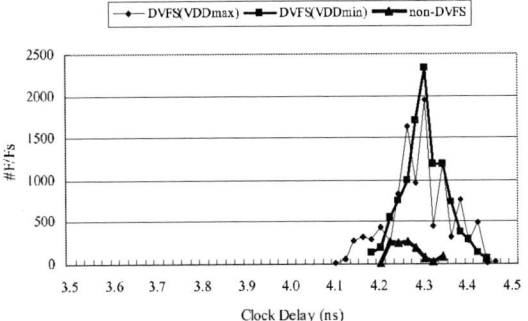

Fig.5(b). Clock delay distribution with the clock design methodology

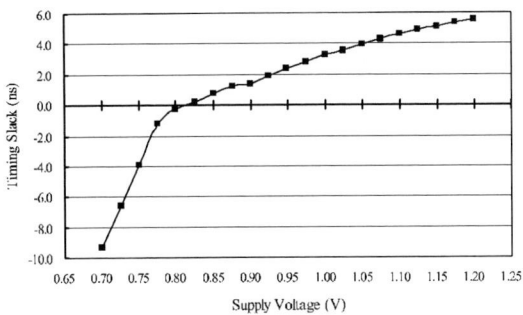

Fig.6. Relation between supply voltage and timing slack

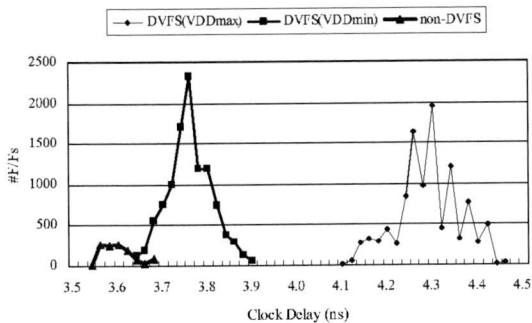

Fig.5(a). Clock delay distribution without the clock design methodology

Table 1 shows the power dissipation data of the chip. We measured the power dissipation of the LSI by two functions. One is the video data decoding and the other is the data encoding. The column "Chip-wise" denotes the power dissipation data, when the DVFS technique is employed for the whole chip. "Module-wise(0.90V)" and "Module-wise (0.82V)" denote the data, when the module-wise DVFS technique is employed for *VDDmin*=0.90V and 0.82V, respectively. The upper table shows the power dissipation data of the DVFS module. About 60% power reduction has been achieved by the module-wise DVFS, compared with the chip-wise DVFS. As for the chip-wise DVFS, it was unable to lower the voltage so much because often one of the modules in a chip requires high throughput. By setting the proposed derived *VDDmin*, about 17% power reduction has been achieved by the module-wise DVFS. The lower table shows the power dissipation data of the sum of the DVFS module and the non-DVFS modules. Nearly 9% power reduction can be achieved by the module-wise DVFS, but the effectiveness is less than what is shown in the upper table. To improve the effectiveness, it is necessary to adopt the module-wise DVFS to more than two modules.

Table 1. Power dissipation of codec LSI

DVFS module	Chip-wise	Module-wise (0.90V)	Module-wise (0.82V)
decoding	5.3mW (100.0%)	2.5mW (47.1%)	2.1mW (39.6%)
encoding	6.8mW (100.0%)	3.3mW (48.5%)	2.7mW (39.7%)

DVFS module + non-DVFS modules	Chip-wise	Module-wise (0.90V)	Module-wise (0.82V)
decoding	41.3mW (100.0%)	38.5mW (93.2%)	38.1mW (92.3%)
encoding	50.9mW (100.0%)	47.4mW (93.1%)	46.8mW (91.9%)

VI. Conclusion

We have proposed the clock design methodology for the

module-wise DVFS technique. We can achieve more power reduction by the technique, as the possibility of lowering the supply voltage increases, compared with the chip-wise DVFS technique. By using DDS, instructions can be executed, even when the supply voltage is in transition. It gets rid of the performance loss. We have presented the clock design strategy to minimize the clock skew among inter-modules, even when the clock delay in each module changes, according to its supply voltage value. We have also described the condition of the timing closure which consists of the setup and the hold constraints, and shown the way of finding the minimum supply voltage value, by solving simultaneous polynomial inequalities. We have demonstrated that the 17% additional power reduction can be achieved by using the minimum supply voltage value. Adopting the module-wise DVFS technique to plural modules is a future work, in order to achieve more power reduction.

Acknowledgements

The authors would like to thank T.Yoshimori, T.Furuyama, K.Ochii, S.Imai, M.Murakata, H.Ohta, Y.Oowaki, M.Takahashi, M.Yamada, and A.Watanabe for their support. They also would like to thank T.Sakamoto and K.Miyabe for providing the SPDM environment, and thank T.Nishikawa and Y.Kikuchi for helping the static timing analysis.

References

[1]M.Fleischmann, "LongRun™ Power Management – Dynamic Power Management for Crusoe Processors", *http://www.transmeta.com/corporate/pressroom/whitepapers .html*, January, 2001.

[2]T.Fujiyoshi et al., "An H.264/MPEG-4 Audio/Visual Codec LSI with Module-Wise Dynamic Voltage/ Frequency Scaling", *Proc. ISSCC*, pp.132–133, 2005.

[3]S.Lee et al., "Reducing pipeline Energy Demands with Local DVS and Dynamic Retiming", *Proc. ISLPED*, pp.319–324, 2004.

[4]B.Gorjiara, N.Bagherzadeh and P.Chou, "An Efficient Voltage Scaling Algorithm for Complex SoCs with Few Number of Voltage Modes", *Proc. ISLPED*, pp.381–386, 2004.

[5]A.Chandrakasan and R.Brodersen, *Low Power Digital CMOS Design*, Kluwer Academic Publishers, 1995

[6]R.Jejurikar, C.Pereira and R.Gupta, "Leakage Aware Dynamic Voltage Scaling for Real-Time Embedded Systems", *Proc. DAC*, pp.275–280, 2004.

[7]K.Choi, W.Lee, R.Soma and M.Pedram, "Dynamic Voltage and Frequency Scaling Under a Precise Energy Model Considering Variable and Fixed Components of the System Power Dissipation", *Proc. ICCAD*, pp.29–34, 2004.

[8]A.Andrei, M.Schmitz, P.Eles, Z.Peng and B.Al-Hashimi, "Simultaneous Communication and Processor Voltage Scaling for Dynamic and Leakage Energy Reduction in Time-Constrained Systems", *Proc. ICCAD*, pp.362–369, 2004.

[9]M.Anis, S.Areibi, M.Mahmoud and M.Elmasry, "Dynamic and Leakage Power Reduction in MTCMOS Circuits Using an Automated Efficient Gate Clustering Technique", *Proc. DAC*, pp.480–485, 2002.

[10]K.Usami, N.Kawabe, M.Koizumi, K.Seta and T.Furusawa, "Automated Selective Multi-Threshold Design For Ultra-Low Standby Applications", *Proc. ISLPED*, pp.202-206, 2002.

[11]T.Kitahara, N.Kawabe, F.Minami, K.Seta and T.Furusawa, "Area-efficient Selective Multi-Threshold CMOS Design Methodology for Standby Leakage Power Reduction", *Proc. DATE*, pp.646-647, 2005.

[12]T.Kuroda et al., "A 0.9V 150MHz 10mW 4mm2 2-D Discrete Cosine Transform Core Processor with Variable-Threshold-Voltage Scheme", *IEEE JSSC*, vol.31, pp.1770 –1779, November, 1996

[13]J.Seo, T.Kim and K.Chung, "Profile-based Optimal Intra-task Voltage Scheduling for Hard Real-Time Applications", *Proc. DAC*, pp.87–92, 2004.

[14]K.Usami et al., "Design Methodology of Ultra Low-power MPEG4 Codec Core Exploiting Voltage Scaling Techniques", *Proc. DAC*, pp.483–488, 1998.

[15]*http://www.scd.magma-da.com/articles/SiliconSPDM.pd f.*

[16]*http://www.semicon.toshiba.co.jp/eng/prd/asic/doc/pdf/b ce0012a.pdf.*

Single-Chip Multi-Processor Integrating Quadruple 8-Way VLIW Processors with interface timing analysis considering power supply noise

Satoshi Imai   Atsuki Inoue   Motoaki Matsumura   Kenichi Kawasaki   Atsuhiro Suga

System LSI development laboratories
Fujitsu Laboratories Ltd.
4-1-1 Kamikodanaka, Nakhara-ku, Kawasaki, Japan
Tel : +81-44-754-2783
Fax : +81-44-754-2691
e-mail : {imai.satoshi-02,inoue.atsuki,matsumura.motoa,k.kawasaki,suga.atsuhiro}@jp.fujitsu.com

Abstract - This paper introduces a 51.2Gops, 1.0GB/s-DMA single-chip multi-processor integrating quadruple cores and proposes a new power integrity analysis. Our multi-processor is designed to decode MP@HL streams without any dedicated circuits. To achieve such high performance, data throughput as well as processing capability is important, requiring a large number of high speed I/Os. However, this makes for a high level of power supply noise. We then applied an interface timing margin analysis tool that took power supply noise into account, and succeeded in putting reasonable restrictions on LSI design, as well as that for the printed circuit board. As a result, we succeeded in operating the processor at 533MHz with the 2ch 64bit main memory IF at 266MHz and 64bit system bus at 178MHz.

I. Introduction

There are two major methodologies for increasing the processing performance of processor: namely increasing the frequency and increasing the total number of processing elements in parallel. Processors for commercial products must satisfy three requirements: high performance, low pricing and low power consumption. For example, a processor for a personal computer achieves high performance; however, it requires a forced air cooling device or an expensive sealing package for the chip to work within a 100W power consumption limit. Use of such expensive devices for the highly competitive commercial supercomputer market is not possible. Generally, the price level for a built-in processor is set much lower than general purpose processors. Thereby, we have to use a less expensive package and cool it without using a fanning device. To meet this condition, we have to make a low priced LSI with low power consumption: approximately one fiftieth that of a PC microprocessor. Hence, we decided to achieve high performance and low power consumption by keeping the fundamental idea of FR-V and increase the total number of processing elements in parallel. We decided not to depend on increasing the frequency. The FR550[5] used

VLIW and SIMD with eight instruction level paralleled processing and four data level paralleled processing (Fig. 1). Moreover, without increasing the frequency, it was necessary to run multiple chunks of instructions larger than those at the instruction level in parallel to boost the level of performance by a few times. As semiconductor process technology entered the 90nm age, the multi-core processor FR1000 with VLIW processor that we had planned at the beginning was finally realizable [6].

Our target was the realization of media processing, such as images and audio, exclusively by software. This kind of media processing was previously performed by configuring the control CPU with dedicated logic (ex. ASIC)[7]. Until now, a supercomputer or computing server divides the task into plural parallel tasks, mainly utilizing the loop parallelism in source code. However, a media processing program such as MPEG decoding includes in itself parallel tasks such as IDCT functions and a large chunk of slice level tasks. So we decided to equip the multiple core VLIW processor with our processor. A thread level (which is a larger chunk than instruction level) parallel processing was performed on multiple processor cores, and the instruction level parallel processing was performed in each processor core using VLIW architecture as well as the existing FR-V processor had done. As a result, we achieved higher performance and lower power consumption. For example, the decoding of MPEG2 MP@HL, which was six times the processing volume of MPEG2 MP@ML, consumed only about twice as much as power as the decoding of MP@ML streams by FR550.

Recently, power supply noise has become a critical design issue in LSI design. In particular, the problem of simultaneous switching noise (SSN) by I/O buffers is becoming serious. This is caused by the following two factors.

The first is an increase of SSN. Higher I/O bandwidth of system interface between LSIs causes an increase of dI/dt (I is the drive current of I/O buffers) as well as an increase of the current itself. Because SSN is proportional to dI/dt and total inductance of power supply line, an increase of dI/dt leads an increase of SSN.

The second factor is a decrease in the noise tolerance of the circuits. The more advanced process technology becomes, the lower power supply voltage becomes in order to maintain transistor reliability. Lower voltage brings lower power consumption. However the noise margin of the circuits decreases significantly at the same time.

SSN not only causes false operation of logic circuits but also adds a delay penalty to the timing of the system interface between LSIs, such as high speed memory interface. So it is essential to evaluate the power supply noise and the delay penalty accurately in order to feed these results back to PCB designers.

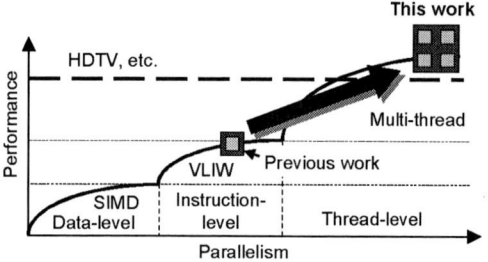

Fig. 1. Effects of processor-core parallelism on the limits of VLIW architecture

II. Hardware Architecture

The FR1000 processor is a multi-core processor loaded on one chip with four VLIW processor cores inside. Each VLIW processor core is able to execute up to eight parallel processes at one time.

As stated above, the processor core on the FR1000 is a FR550 compatible processor. The instructions executed on the FR550 processor core consist of integer arithmetic instructions, floating point arithmetic instructions, and the media instructions with 16 bit fixed point arithmetic operations. Media instruction can process four or eight operations in parallel with the SIMD method. A processor core which can execute eight parallel instructions at the same time can execute twenty eight operations at the same time in one cycle. Consequently, the FR1000 processor is capable of executing 112 operations at the same time in one cycle.

A block diagram of this processor is shown in Fig. 2. In aggregate view, the chip consists of four processor cores, a main memory controller with two channels, a DMA controller (DMAC) for transfer between local memory units, a DMA controller (DMAC) for transfer to outside, and a 64 bit system bus interface. If the chip is configured with a

system that requires the processing of a huge volume of image data such as HDTV, a high speed/high precision printing system or a graphic system, it is very important to provide high speed capability to transfer data including I/O accesses, and to transfer data between memory units. High performance arithmetic processing capability is also very important. For example, in an arithmetic instruction, if the data transfer time from memory to arithmetic unit takes 90 cycles and the operation in the processor core is executed in 10 cycles, then ninety percent of total processor capability is controlled by the data transfer. The bandwidth of memory is not necessarily cause for concern with a single processor, however it becomes a major factor in a multi-core processor.

Consequently, we adopted the configuration below to avoid data transfer becoming a bottleneck in performance capability.

1) Four processor cores, two 64 bit channels of main memory interface at 266MHz, and with one of the system bus interfaces at 178MHz. Connections were made using cross bar methodology

2) To reduce external memory accesses, each processor core is equipped with 128KB SRAM (local memory unit) as local storage.

3) The DMA controller is functionally divided into DMAC for internal data transfer (internal DMAC) and DMAC for external data transfer (external DMAC), and each DMAC runs independently at the same time. The internal DMAC is used to control data transfer between processor cores, processor core and external memory, and memory and memory. The external DMAC is used to control the data transfer between memory and system bus.

4) In addition to the cross bar for data transfer described above, a communication control mechanism is installed for dedicating the instruction data transfer between processors.

The local memory unit, built into each processor core, is connected with a dedicated cross bar. All cores can access all local memory units. The internal DMAC and external DMAC described above are equipped with 16 channels respectively. With these bus architectures, the FR1000 processor can simultaneously process data transfer between the built-in local memory units, data transfer between areas on the memory, and data transfer between the memory and an external device. As shown Fig. 3, the data transfer speed between memory and external device reaches up to 1GB/s thanks to this architecture.

542

2006 Asia and South Pacific Design Automation Conference

5D-2

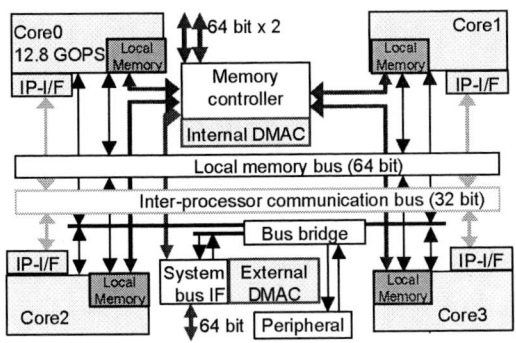

Fig. 2. Block Diagram of FR1000

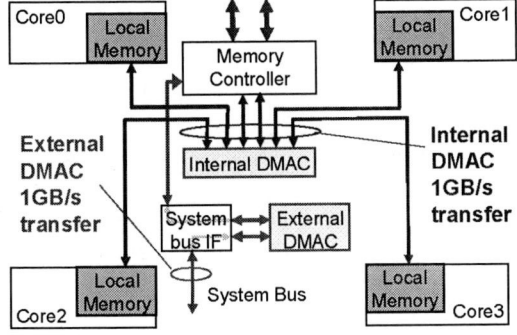

Fig. 3. Data Transfer performance

III. Performance Evaluation

Fig. 5 shows the increase in application performance when optimizing the MPEG2 MP@HL decoding software. The vertical axis represents the operating frequency required for decoding the MPEG2 MP@HL. First, we applied MPEG2 MP@ML decoding software for single core to MP@HL decode and found that more than 1GHz of operating frequency is required. Next, we simply divided the decoding software into four cores using the method shown in Fig. 4. Large streaming data was divided into slice level data and processing for each slice level was assigned to each core. However, we could not increase the performance at all if we simply modified the program to fit the multi-core processor. We still needed over 1GHz of frequency (Single HL to 4PE HL in the Fig. 5). After we analyzed the memory access profile, we found that heavy memory access came from plural cores to the same main memory channel at the same time, which brought about a bottle neck in performance. This situation easily came about because many functions were accessing just a single memory unit at the

same time in a multi-core environment. Accordingly, to reduce the load of memory accesses, we changed the memory map for MPEG2 decode processing and equalized the load of data accesses to memory as well. We thereby succeeded in increasing the performance to twice that of before this adjustment, requiring operating frequency for the MP@HL decoding to be reduced to approximately 500MHz.

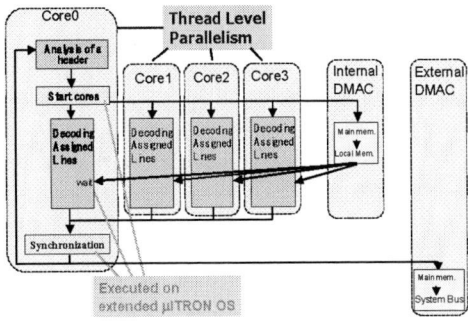

Fig. 4. Parallel processing of MPEG@HL streaming data on four cores

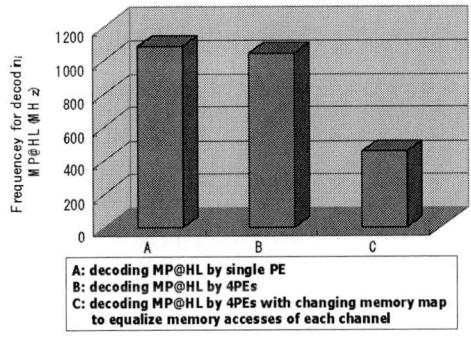

Fig. 5. Performance optimization result on decoding MPEG-2 MP@HL

IV. Power integrity analysis

A. Power supply noise problem

Power supply noise inside LSI is becoming a critical design issue in realizing a high speed LSI with a large number of high speed I/Os, such as the FR1000 described in the previous chapter. We need to evaluate the power supply noise generated from LSI properly and its influence on the timing margin of high speed I/Os to provide reasonable design guidelines to PCB designers. The major concerns for LSI designers are as follows.

1) The SSN of I/O circuits is becoming large enough to interfere with other circuit operations and/or increase timing fluctuations.

2) The timing margin of system interfaces between LSIs is

543

decreasing as demand for higher I/O bandwidth is increasing.

In the case of FR1000, there are 230 SSTL I/O buffers for the two channels of memory interface which operate at a 266Mbps data rate, and 160 conventional CMOS I/O buffers for the system bus interface toggling at 178MHz.

The major factors reducing the timing margin are clock jitter and skew, the difference of the signal trace length on the package (PKG) and the printed circuit board (PCB), x-talk noise on PKG and PCB, and the I/O delay penalty caused by power supply noise. It is necessary to take all these factors into consideration to achieve robust design for processing and supply voltage variations. In particular, budgeting the interface timing, including the delay penalty by the SSN, is essential because the influence of the noise on timing is too large to be ignored.

In this chapter, we explain our method for analyzing power supply noise and delay penalty as well as their analysis models. Then, we discuss the analyzed results of the delay penalty for the FR1000's memory interface caused by power supply noise [1] [2].

B. Power supply noise analysis model

The power supply noise analysis model is a combination of three components: the DIE, PKG, and PCB models. To analyze the power supply noise accurately, not only PKG and PCB but also the DIE model should be generated properly, because the nature of the noise source is important to this analysis.

We proposed the LSI power model based on the power unit abstraction for the DIE, as shown in Fig. 6 [3]. The size of each power unit is typically chosen as 100 or 200 um squared. The model is composed of four parts: the power supply line network, capacitors, current sources, and I/O buffers. The power supply network is modeled as series LR circuit in each power unit. The capacitors represent the capacitance between power and ground such as decoupling capacitor, gate capacitance, and junction capacitance. The current sources represent the switching behavior of the logic gate and RAM inside each power unit. The I/O buffers are modeled separately in each pin as a transistor level netlist. We developed an automatic extraction tool from the layout design.

The PKG and PCB model are composed of power and ground supply network, signal traces, receivers, and decoupling capacitors. The signal traces are extracted as w-element or LRC ladder circuit, and the receivers are modeled as capacitor element.

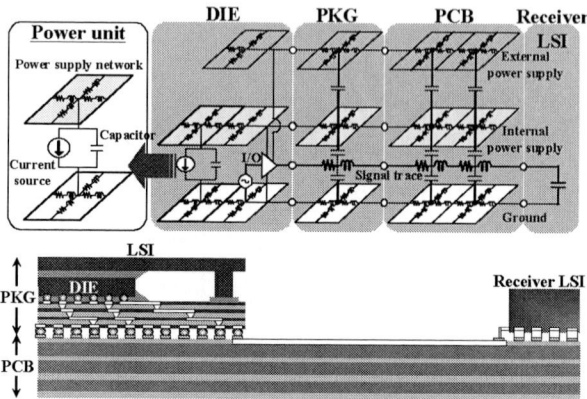

Fig. 6. Power supply noise analysis model

C. Timing Analysis method

The analysis method of power supply noise is separated into two phases as shown in Fig. 7.

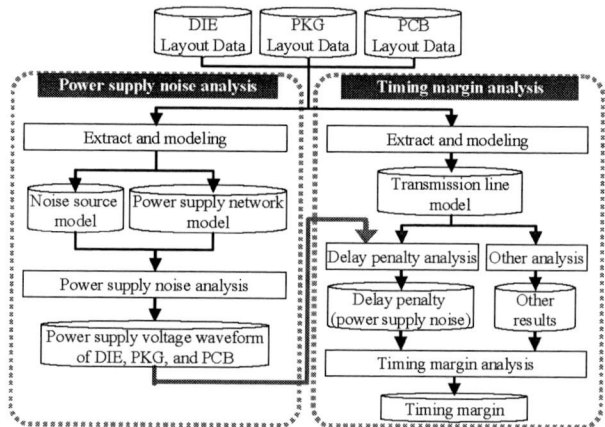

Fig. 7. Timing analysis method

In the power supply noise analysis phase, noise source and power supply network models are extracted from each set of layout data of DIE, PKG, and PCB. The power supply voltage waveforms of DIE, PKG, and PCB can be simulated by combining these models with appropriate stimulus.

In the timing analysis phase, the transmission line model is extracted from each set of layout data of DIE, PKG, and PCB. Then, the power supply voltage waveforms obtained in the previous noise simulation are applied to this model and the delay penalty of each I/O is evaluated. (Fig. 7, Fig. 8) In order to obtain the worst case delay penalty value, the generating timing of power supply noise is swept within possible timing window, as shown in Fig. 9. The signal arrives at the receiver either earlier or later depending on the applied supply noise timing. The largest timing fluctuation in the receiver is defined as the worst case delay penalty.

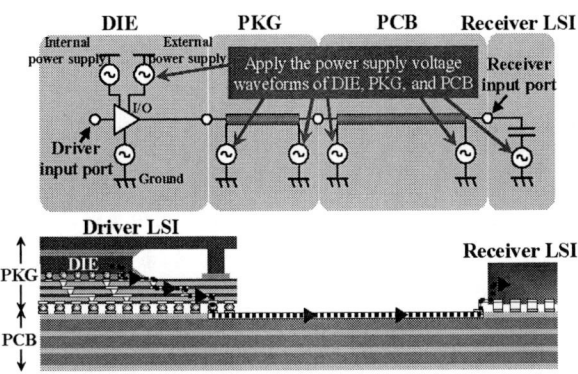

Fig. 8. Delay penalty analysis model

Finally, interface timing analysis is performed to check the timing margin considering this penalty, as well as others such as x-talk penalty, clock jitter, clock skew and so on.

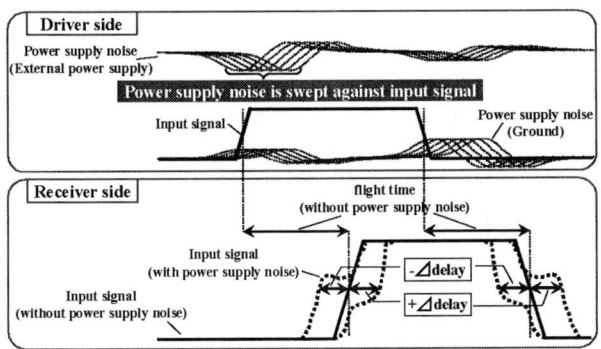

Fig. 9. Delay penalty by power supply noise analysis method

D. Timing analysis of memory interface

We analyzed the delay penalty for the memory interface of FR1000 using the method described in the previous section. Because each signal operates at a different frequency and has different load capacitance and allowable timing margin, we classified the signals into four groups: address (Address), clock (Clock), data (DQ) and data strobe (DQS), and analyzed the I/O delay penalty for each group at first.

TABLE I shows the worst case delay penalty analysis results at the worst case process condition, supply voltage and temperature condition. The delay penalty of Address and Clock is larger than those of DQ and DQS. This is because the load capacitance of Address and Clock is larger than those of DQ and DQS, which causes the slew rate of Address and Clock to be larger than those of others. Generally, signals with a large slew rate are easily affected by power supply noise.

TABLE I
Worst case delay penalty results of memory interface

Signal	Delay penalty [ns]	
	+$\triangle$Delay	-$\triangle$Delay
DQ	0.284	0.220
DQS	0.285	0.220
Address	0.447	0.303
Clock	0. 440	0.380

Next, we discuss the timing analysis results of our memory interface. Figure 10 shows the write timing diagram of DQS and DQ at a 266Mbps data rate. Ideally, 1.88ns, which is the DQS-DQ phase difference on the driver buffer, is also kept at the memory input pins. The ideal setup margin of DQ against DQS becomes 1.38ns, assuming a 0.5ns setup time (tDS) taken from the main memory data specification sheet. Actually, this margin is reduced by various factors such as clock jitter and skew, a difference in the signal trace between DQS and DQ, and the I/O delay penalty caused by power supply noise. In particular, the delay penalty of DQS and DQ reduces the setup margin to 0.504ns (the sum of 284ps and 220ps) at the worst case process condition. In a word, the delay penalty reduces the setup margin by as much as 36.5%.

In TABLE II, other timing restrictions are summarized as well as the result of DQ-DQS timing analysis. From these results, we found that the delay penalty by power supply noise significantly influences the timing margin of the memory interface in the worst case. It is important to take it into account when providing appropriate design guidelines to PCB designers.

In a physical design of FR1000, we budgeted ideal timing margin to clock jitter and skew, flight time difference on PKG and PCB, x-talk noise delay penalty on PKG and PCB, and the I/O delay penalty by power supply noise for all of the interface signals. Though the I/O delay penalty was large, we succeeded in designing the FR1000, securing a sufficient interface timing margin by optimizing other timing budgets.

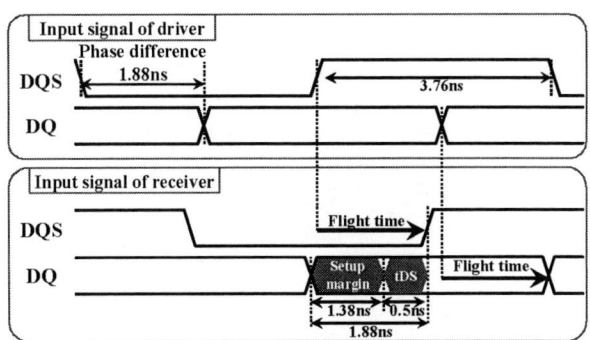

Fig. 10. Timing diagram of DQS and DQ for both driver and receiver

TABLE II
Delay penalty of each signal group in memory interface

Timing restriction	Ideal margin [ns] ※1	Total delay penalty [ns]	Ratio [%] ※2
Clock-Address setup timing	2.86	0.827	28.9
Clock-Address hold timing	2.86	0.743	26.0
DQS-DQ setup timing	1.38	0.504	36.5
DQS-DQ hold timing	1.38	0.505	36.6
Clock-DQS skew	1.88	0.665	35.4

※1. Ideal margin is the margin without considering any delay penalty.
※2. Ratio [%] = (Total delay penalty / Ideal margin) × 100

V. Chip implementation

TABLE III provides the specifications of FR1000. This processor was fabricated using a 90nm, nine-metal layer CMOS process technology. The number of transistors is 28M for logic and 55M for RAM. Transistors totalling 83M are integrated on a 10.3mm x 11.9mm die as shown in Fig. 11. The bus logic part, which operates at half core clock frequency, is arranged in center of the die.

The typical power supply voltage is 1.2V, and the power consumption measured to be 3.0W at MP@HL stream decode.

TABLE III
Chip Specifications

Core	4 cores with 8-way VLIW architecture
Memory	32 KB+32 KB/core (D-cache, I-cache) 128 KB/core (Local memory)
DMA controller	16 ch (Internal) , 16 ch (External)
Interface	Main mem IF 266 MHz 64 bit x 2ch System Bus 178 MHz 64 bit
Technology	90-nm CMOS, 9-metal layers
Transistor count	28M (Logic), 55M (Memory)
Operating frequency	533 MHz @1.2 V
Power consumption	3.0 W @1.2 V, 533 MHz
Package	900-pin FCBGA

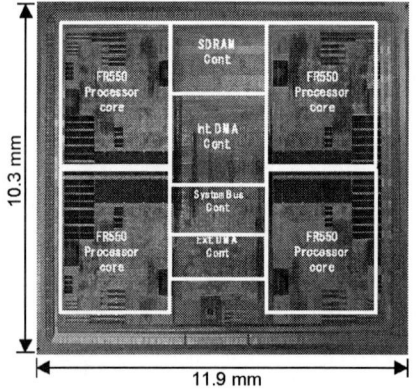

Fig. 11. Chip Die micrograph

VI. Conclusion

The FR1000 is designed to achieve 51.2Gops, 1.0GB/s-DMA and have four processor cores, internal DMAC, external DMAC, two 64bit channels of main memory interfaces at 266MHz, and a system bus interface at 178MHz. We demonstrated MPEG2 MP@HL stream decoding without any dedicated circuits with only a 3.0W power dissipation.

We investigated the influence of power supply noise and found the simultaneous switching noise by large numbers of high speed I/O buffers is enough large to reduce the timing margin of a high speed interface. We prepared reasonable PCB design guidelines taking this effect into account. Thus, we succeeded in operating the processor without any LSI re-spin.

Acknowledgements

The authors would like to thank Hiromasa Takahashi, Yukihito Kawabe, Wataru Shibamoto, Atsushi Sato, Tetsutaro Hashimoto, Hideo Miyake, Yasuki Nakamura, Hiroshi Okano, Fumihiko Hayakawa, Shinichiro Tago, Teruhiko Kamigata, Atsushi Tanaka and Takahisa Suzuki for their contributions to this work.

References

[1] M.Matsumura et al., "Analysis of the Effects of Simultaneous Switching Noise on System Interface Timing Between LSIs," Collected papers of the 18th workshop on circuits and systems in Karuizawa, pp.43-48, 2005
[2] K.Kawasaki et al., "Single-Chip Multi-Processor integrating Quadruple Processors on 90nm CMOS Process," IEICE Technical Report, Vol.105, No.95, pp.7-12, May.2005
[3] T.Sato et al., "LSI Noise Model for Power Integrity Analysis," FUJITSU, Vol.55, No.6, pp.608-613, Nov 2004
[4] Suga et al., "Introducing The FR500 Embedded microprocessor," IEEE Micro, pp. 21-27, July/Aug 2000.
[5] H. Okano et al., "An 8-way VLIW Embedded Multimedia Processor Build in 7-layer Metal 0.11µm CMOS Technology," ISSCC Dig. Tech. Papers, pp. 374-375, Feb. 2002.
[6] Shiota et al., "A 51.2GOPS, 1.0GB/s-DMA Single-Chip Multi-Processor Integrating Quadruple 8-Way VLIW Processors," ISSCC Dig. Tech. Papers, pp.18-19, 2005.
[7] Irvin M et al., "A new generation of MPEG-2 video encoder ASIC and its application to new technology markets," Broadcasting Convention, International (Conf. Publ. No. 428)12-16 Sept. 1996 Page(s):391 - 396.

2006 Asia and South Pacific Design Automation Conference

A System-level Power-estimation Methodology based on IP-level Modeling, Power-level Adjustment, and Power Accumulation

Masafumi Onouchi†, Tetsuya Yamada†, Kimihiro Morikawa‡, Isamu Mochizuki‡ and Hidetoshi Sekine‡

†Hitachi Ltd., 1-280, Higashi-Koigakubo, Kokubunji-shi, Tokyo, 185-8601, Japan
‡Renesas Technology Corp., 5-20-1, Josuihon-cho, Kodaira-shi, Tokyo, 187-8588, Japan

Abstract— We have developed a specialized rapid power-estimation methodology for multimedia applications. This methodology has adequate accuracy for the first design of a complicated SoC. For a multimedia application, we developed three new methodologies: an IP-level modeling, a power-level adjustment methodology, and a power accumulation methodology. With these methodologies, the system-level power estimation on a SoC executing a practical application becomes so precise and easy that we can revise the SoC design to reduce its power. According to a comparison of the system-level power estimated with these methodologies to board-measured power, the error between the two powers is less than 5.6%.

I. INTRODUCTION

Recently, more and more intellectual properties (IPs) are being integrated on a single chip. These large-scale integration (LSI) chips are known as system-on-a-chips (SoCs). SoCs are mainly applied in digital appliances, such as cellular phones and digital cameras, because of their compactness. Executing applications with the limited power resources of mobile appliances, however, means that SoCs face a severe power restriction. To reduce SoC power dissipation, it is thus important to know which IP dissipates the largest power when executing practical applications. We call power dissipated in practical applications, system-level power. Accordingly, revising this IP's design to reduce system-level power by means of careful analysis of the estimated power, will lead to a power-aware SoC.

Power dissipation can be classified into two categories: dynamic power and leakage power. Leakage power can simply be estimated from the total gate width of the transistors on a SoC, because it is always dissipated through a transistor's gates, sources, and drains. On the other hand, it is difficult to estimate dynamic power because it depends on the switching rates of the transistor's gates and its connected wires.

Present methodologies using abstract models on a SoC

can rapidly estimate the dynamic power dissipation in the early stages of design [1–5]. However, these methodologies do not include the gate-level behaviors of a SoC, therefore the precise estimation of the system-level power is not obtained. Moreover, the practical application, such as a digital broadcast TV service or a TV phone service, is too large to simulate in reasonable time. To improve the accuracy of the estimation, the methodology using gate-level simulation on an IP was developed [6]. In this methodology the gate-level simulation, which supplies more precise estimations, was enabled by splitting both the gate-level description and the benchmark into small units. However, its so expensive to split both of them that it is impossible to simulate the whole application.

In this study, with the above issues in mind, we present rapid and precise system-level-power-estimation methodologies specialized on multimedia applications. Firstly, we estimate an IP's precise dynamic power with a simple benchmark by a gate-level simulation. Secondly, we abstract one-frame processes from the application and split these processes into individual IP processes. This is how we make an IP-level model. What makes this power estimation very simple is that the simple repetition of this model approximately equals the whole application. In calculating system-level power by adjusting the IP's precise dynamic power according to the IP-level model, we developed a power-level adjustment methodology and a power accumulation methodology. These methodologies enable us to design a power-aware SoC in the early stages of design.

II. SYSTEM-LEVEL POWER-ESTIMATION METHODOLOGIES

We developed specialized power-estimation methodologies for multimedia applications on a SoC (Fig. 1). Traditional methodologies for the estimation of dynamic power are based on an abstract model of a SoC's architecture. These methodologies, however, do not include the gate-level behavior of a SoC, therefore it is impossible to precisely estimate the system-level power. In contrast to

0-7803-9451-8/06/$20.00 ©2006 IEEE.

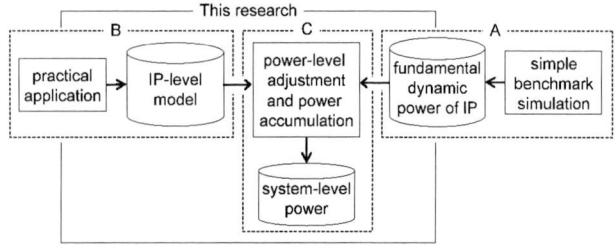

Fig. 1. SoCs system-level power-estimation methodologies. The rectangular region is this work.

these methodologies, we firstly calculate the fundamental dynamic power of an IP using a simple benchmark simulation with a gate-level net list (Fig. 1, region A). This simulation is executed with such a simple but typical benchmark for each IP that the estimated power has adequate accuracy. Secondly, we developed an IP-level model abstracted from the practical application processes (Fig. 1, region B). Thirdly, to adjust the IP's fundamental dynamic power according to the IP-level model, we developed both the power-level adjustment and power accumulation methodologies (Fig. 1, region C).

As a target of this estimation, a low-power embedded-application SoC for cellular phones, i.e., an SH-Mobile3AS [7], was used (Fig. 2). The SH-Mobile3AS was designed by the Renesas 90-nm low-power process and consists of many IPs, such as a CPU core (SH-X2), a H.264 IP, and a camera IP. The CPU core is designed with a fine-grained clock gating to reduce the power consumption of the flip-flops and a clock-tree [8]. The CPU core is a master IP, which can handle all slave IPs (H.264, camera, and so on). Slave IPs perform boot-ups and are stopped by the master IP.

CPG : Clock Pulse Generator

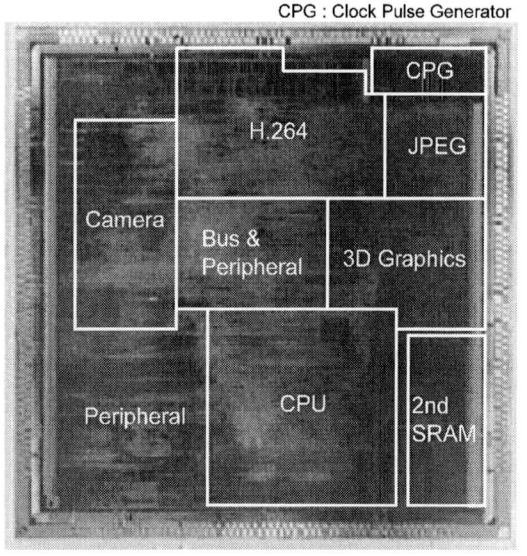

Fig. 2. Photograph of sample chip integrating CPU core (SH-X2) and many other IPs.

A. Calculation of fundamental dynamic power of an IP by simple benchmark simulation

We can use two environments for calculating dynamic-power dissipation. One is a Register-Transfer Level (RTL) evaluation for calculating the indirect dynamic power dissipation. In the RTL evaluation, it does not take such a long time because only the toggle information on the flip-flops' wires is extracted by the simulation; thus, it is suitable for short TAT analysis. The toggle information indicates how many times each wire is charged or discharged, which causes power dissipation.

The other environment is a gate-level evaluation for calculating the absolute dynamic power dissipation. In the gate-level evaluation, an IP's estimated power has adequate accuracy because all gates' and wires' toggle information is obtained and dissipated power on each cell and wire is calculated. The gate-level evaluation has two versions: pre-layout and post-layout evaluations. The post-layout version has high accuracy with a synthesized clock-tree and a back-annotated net load. The pre-layout version has no clock-tree or net load. The power dissipated on the clock-tree has a large uncertainty, because the clock-tree structure has not yet been synthesized. Therefore, we use pseudo-CTS (clock-tree synthesis) for the pre-layout version [8]. In pseudo-CTS, clock-buffer trees are inserted into a clock-tree net, whose fan-out and structure are similar to those of the post-layout.

The gate-level evaluation consists of the two stages shown in Fig. 3. Firstly, a simple benchmark is simulated with a gate-level net list to obtain the toggle information. The simulation generally takes a long time to get the toggle information of all wires. Even the gate-level evaluation can be executed in hours, because a simple benchmark, such as a one-frame H.264, decodes a small picture. Secondly, the dynamic power of IPs is calculated with power libraries and the toggle information obtained in the first stage. We named this calculated power, fundamental dynamic power. This power is classified into circuit components, such as clock-trees, random logic, flip-flops, and SRAM.

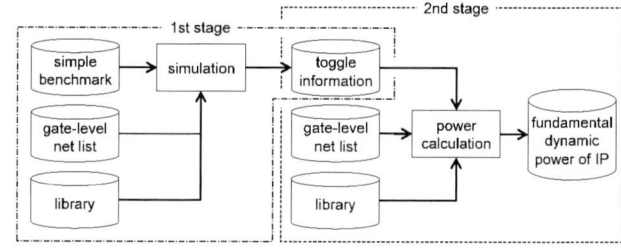

Fig. 3. Evaluation flow of fundamental dynamic power of IPs with gate-level net list.

The fundamental dynamic power on the SH-Mobile3AS estimated by the pre-layout gate-level simulation is shown in Fig. 4. This figure shows the IP's name, dynamic power, and the number of operating cycles for several benchmarks. The processes, such as the rotation, format transformation, and display, are less cost effective than H.264 decoding. Several of these processes can be simultaneously executed on the camera IP.

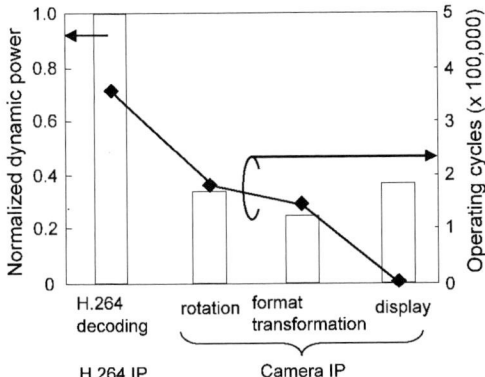

Fig. 4. Fundamental dynamic power and number of operating cycles for simple benchmarks.

B. IP-Level model of a practical application

For abstracting a practical application we developed an IP-level model of it. A digital broadcast TV service, which is a new multimedia application on third-generation cellular phones, was assumed. The digital broadcast TV, which can supply high-definition movies, applies H.264 decoding to animation and AAC decoding for sound.

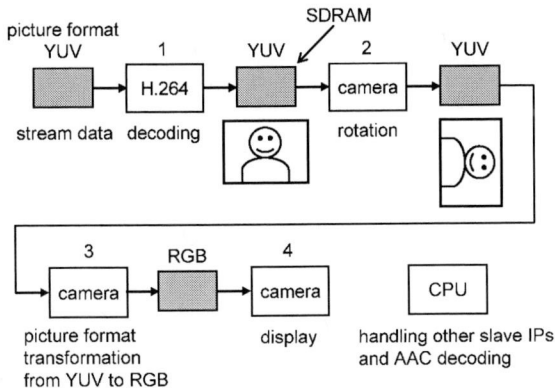

Fig. 5. IP-level model in digital broadcast TV service.

Figure 5 illustrates the IP-level model of the digital broadcast TV service. In this model, the application is split into each IP process in executing a one-frame picture. For example, the CPU, as a master IP, handles some slave IPs and executes AAC decoding for sound. A slave IP (H.264 IP and Camera IP) executes the "H.264 decoding," "rotation," "picture-format transformation," and "display" in sequence. Its data are transferred by a bus, peripheral IP, and SDRAMs. The picture size is

QVGA (320×240 pixels) for all processing, and the picture format is converted from a YUV format to an RGB format in process 3 in Fig. 5.

Power estimation becomes very simple if we focus on the fact that the repetition of this model approximately equals to the whole practical application, namely the averaged power dissipated in one-frame-picture processing approximately equals that of in the whole multimedia applications. Therefore system-level power can easily be estimated by summing up only the IPs power contained in this model.

C. Power-level adjustment methodology and power accumulation methodology with fundamental dynamic power

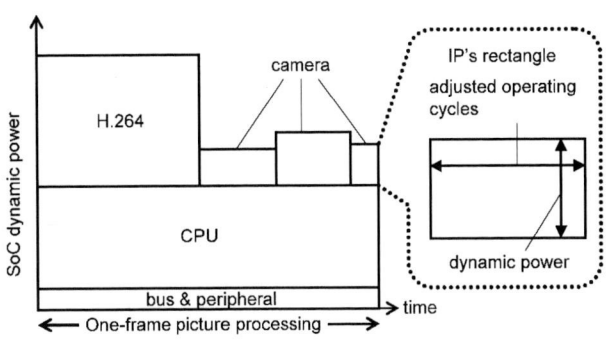

Fig. 6. Power-level adjustment methodology using each fundamental dynamic power and operating cycles.

To adjust the fundamental dynamic power according to the IP-level model, the power-level adjustment methodology was developed (Fig. 6). In this case, we also used the digital broadcast TV service. In Fig. 6, the vertical length of the rectangle denotes the fundamental dynamic power, and the horizontal length of the rectangle denotes the IP's adjusted operating cycles. In the slave IPs, the operating cycles are mainly determined by the picture size. For example, in the H.264 IP, the number of macro blocks, which are fundamental units in encoding/decoding picture, determines its operating cycles. The number of macro blocks doubles, so twice the number of cycles is required. In this model, we assume that the clock signal for the slave IP is running only when the IP is executing the processes. In the master IP (CPU), we estimate the operating cycles using an instruction-set simulator (ISS). Lastly, in the bus and peripheral IPs, we use the averaged IP power executing the benchmark. Because these IPs are almost always active when the other IPs are executing the processes, and power is mainly dissipated in the clock-tree and flip-flops, that the contents of the transferred data do not affect the power dissipated in the random logic circuits. Therefore, we can obtain each IP's actual operating cycle by adjusting it with this methodology.

549

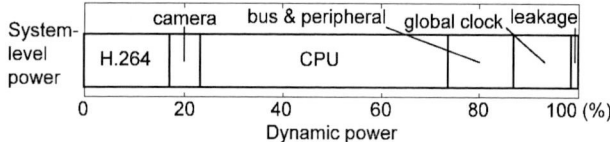

Fig. 7. Accumulated system-level power.

To obtain actual system-level power, we developed the power accumulation methodology. Firstly we estimate each IP's actual dynamic power by calculating the area of the rectangle(Fig. 6). Secondly we accumulate these powers and other components such as the "global clock" and "leakage" power. The "global clock" includes the clock generation and clock distribution from a phase locked loop (PLL) to each IP. The leakage power is mainly from the transistor's gate and subthreshold leakage.

III. EXPERIMENTAL RESULTS

To compare the estimated system-level power and an actual power, we measured the SH-Mobile3AS board under almost the same conditions as the estimation in section II.

In this estimation, because we use the gate-level net list of the pre-layout version, the value estimated in section II needs to be revised for the post-layout version. In the post-layout version, many extra buffers are inserted to fix the hold-time violations, and a number of transistor's gate sizes enlarge to meet the set-up-time constraints [8].

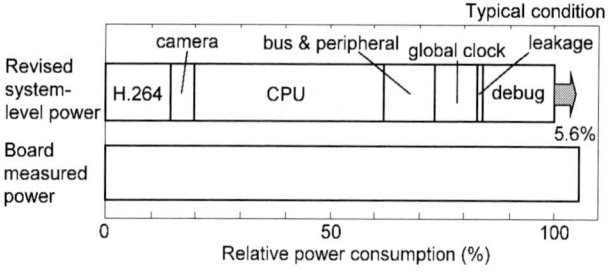

Fig. 8. Comparison between revised system-level power and board measured power for sample chip.

The system-level power and board-measured power are shown in Fig. 8. This figure includes "debug" power, which is dissipated in the executing software for the debugging, and estimated from the board-measured power when the debugging software is executed on the CPU.

According to this comparison, the revised system-level power is 5.6% smaller than the board-measured power. The accuracy of this estimation is adequate enough for the first design of a SoC. There are mainly two reasons for this error. Firstly, because the software used in the board measurement is not fully optimized, the clock signal for the slave IPs is not completely stopped when these IPs

do not execute any processes. The second reason is the lack of accuracy in estimating each IP's load. With more mature software and more accurate estimations of IP's loads, the accuracy of the estimation will improve.

IV. CONCLUSION

We developed a IP-level modeling, a power-level adjustment methodology, and a power accumulation methodology for the system-level power estimation on a complicated SoC. Firstly, an IP's fundamental dynamic power is estimated with a gate-level simulation. Secondly, we make an IP-level model, abstracting a practical multimedia application. We then can obtain the system-level power only by calculating the averaged power dissipated in this model, because the repetition of this model approximately equals the whole practical application. To adjust the fundamental dynamic power according to the IP-level model, a power-level adjustment methodology and a power accumulation methodology were developed. Therefore, we can quickly estimate the system-level power without executing a time-consuming simulation. A comparison of the estimated system-level power with these methodologies to a board-measured power shows that the error between the simulations is less than 5.6%.

V. ACKNOWLEDGEMENTS

The authors would like to thank Kunio Uchiyama, Naohiko Irie, Kenichi Osada, Hiromi Watanabe, Hiroshi Hatae and Hiroshi Ueda for their continuing support.

REFERENCES

[1] D. Brooks et al., "Wattch: a framework for architectual-level power analysis and optimizations," *Proc. of the 27th ISCA*, pp. 83-94, 2000.

[2] A. Dhodapkar et al., "TEMPEST: A thermal enabled multi-model power/performance estimator," *Proc. of the Workshop on Power-Aware Computer Systems*, pp. 112-125, 2000.

[3] S. Gunther et al., "Managing the Impact of Increasing Microprocessor Power Consumption," *Intel Tech. Journal*, Q1, 2001.

[4] D. Brooks et al., "New methodology for early-stage, microarchitecture-level power-performance analysis of microprocessors," *IBM Journal of Research and Development*, Vol. 47, Num. 5/6, pp. 653-670, 2003.

[5] K. M. Büyükşahin et al., "Early Power Estimation for VLSI Circuits," *IEEE Trans. on Computer-aided design of integrated circuits and systems*, Vol. 24, pp. 1076-1088, 2005.

[6] Y. Nakamura et al., "A Fast Chip-Scale Power Estimation Method for Large and Complex LSIs Based on Hierarchical Analysis," *ISCAS*, pp. 628-631, 2005.

[7] M. Saen et al., "Elastic Shared Resource Scheduling SOC Interconnect Architecture for Real-time System," *CICC Dig. Tech. Papers*, pp. 787-790, 2005.

[8] T. Yamada et al., "Low-power design of 90-nm SuperH^{TM} processor core," *ICCD Dig. Tech. Papers*, pp. 258-263, 2005.

PowerV*i*P: SoC Power Estimation Framework at Transaction Level

Ikhwan Lee[1], Hyunsuk Kim[1], Peng Yang[1], Sungjoo Yoo[1], Eui-Young Chung[2],
Kyu-Myung Choi[1], Jeong-Taek Kong[1], and Soo-Kwan Eo[1]

[1]CAE center, System LSI division, Semiconductor Business, Samsung Electronics, Co. Ltd.
San 24 Nongseo-Dong, Giheung-Gu, Youngin, Gyeonggo-Do, 449-711, Korea
e-mail: {ikhwan.lee, hyunsuk71.kim, peng.yang, sungjoo.yoo, kmchoi, jkong, sookwan.eo}@samsung.com

[2]School of Electrical and Electronic Engineering, Yonsei University
134 Sinchon-Dong, Seodaemun-Gu, Seoul, 120-749, Korea
e-mail: eychung@yonsei.ac.kr

Abstract - In this work, we propose a SoC power estimation framework built on our system-level[1] simulation environment. Our framework provides designers with the system-level power profile in a cycle-accurate manner. We target the framework to run fast and accurately, which is enabled by adopting different modeling techniques depending on the power characteristics of various IP blocks. The framework can be applied to any target SoC design.

I. Introduction

System-level design paradigm has been widely adopted to cope with the ever-increasing complexity of System-on-Chip (SoC) design. The high simulation speed at the system level allows designers to explore the huge design space of modern SoC designs.

Design space exploration of modern SoC devices usually deals with three design constraints: performance, area and power consumption. The former two constraints have been relatively well understood in the traditional design flow. However, the market needs for low power devices have introduced the third constraint, power consumption. Since typical SoC devices have many components heavily interacting with each other, it is essential to examine the power consumption of each component in the system context [1]. The power profile generated by independent simulation of each component may mislead designers to a local power optimal design. This means that power estimation should also be performed at the system level.

In order to perform system-level power estimation, we need to build power models of all components. However, a salient feature of SoC is that it has many heterogeneous components with varying power characteristics, ranging from very regular structures such as on-chip SRAM to irregular custom IP blocks such as video codec. This makes it extremely difficult to derive a single modeling methodology that can cover every component constituting a SoC device. Thus, different approaches are adopted for different components. In any cases, however, we need to consider the relationship among the following three factors:

simulation speed, estimation accuracy and modeling effort.

Estimation accuracy is often compromised for simulation speed and modeling effort. By exploiting the heterogeneity of SoC, we can make a good trade-off among them. For example, components with little power variation can employ simple power models to reduce the modeling effort while boosting the simulation speed. For custom IP blocks, we also need to take into account the effort to build the system-level simulation models. Unlike processor cores and bus fabrics whose models are provided by the vendors, legacy custom IP blocks that exist in the form of RTL usually do not have system-level models.

In this paper, we present a system-level power estimation framework, PowerV*i*P, built on our system-level simulation framework, V*i*P [2]. In PowerV*i*P, different power modeling techniques are employed for each component: processor cores, bus fabrics, custom IP blocks and memories. Moreover, for custom IP blocks, an RTL to V*i*P model translation technique is adopted to reduce the modeling effort. PowerV*i*P provides designers with useful power information fast and accurately as well as easy modeling capability.

The paper is organized as follows. Section II describes related work and Section III presents our contributions. The details on power modeling of the processor cores, bus fabrics, and custom IP blocks are presented in Section IV, V, and VI, respectively. The memory power model used in this work is briefly described in Section VII. Section VIII presents the integrated framework based on the separately modeled and validated components. Section IX concludes the paper.

II. Related Work

Extensive studies dealing with the problem of power estimation have been proposed, ranging from circuit-level modeling to behavioral modeling approaches [1, 3]. While highest accuracy is achieved at the lowest level, estimation speed degrades significantly as we move down to lower levels. Therefore, it is crucial to derive a method that performs the best trade-off between estimation accuracy and speed.

Co-simulation based approach is one way to achieve a

---

[1] In this paper, we will use the term transaction level and system level interchangeably to represent their union.

good trade-off between accuracy and speed. In [1], multiple power simulation engines work in a concurrent and synchronous manner. In an effort to minimize the speed degradation caused by co-simulation, they propose several speed-up techniques. In our work, we adopt a model translation technique to completely eliminate the overhead of co-simulation.

A dynamic power model selection scheme at the system level is proposed in [3], where computation effort among different SoC components is allocated at run-time for the best estimation time and accuracy trade-off.

Most of the work on power modeling has focused on power modeling of individual components such as processors, bus fabrics, memories, and custom IP blocks.

The first work on processor power modeling is reported in [4]. Their model quantifies instruction base energy and inter-instruction energy effects to enable fast software energy estimation. Wattch [5] and SimplePower [6] are two well-known power estimation tools in academia. A power model tailored for the Intel XScale processor is proposed in [7]. Their power model is based on module activities, where each module has its power equation embedded in Sim-XScale simulator. The power equations are constructed using transistor level schematics of functional units and a high-level view of transistor gate and drain capacitances. A software power estimation tool, JouleTrack, is presented in [8]. They propose a power characterization methodology that avoids explicit power characterization for each differentiated instruction class.

Bus system power modeling and estimation has been addressed in many different flavors, from the simple analytic model to the detailed gate-level switching activity based model [9 - 12]. Several papers are published to address the problem at a higher abstraction level, the system level [13 - 15]. In practice, most current commercial design flows utilize RTL and gate-level power estimation tools. However, due to their poor efficiency, it is impractical to apply them at the early stage of design, when many different architecture options have to be explored. Most of the work considers only the global wire, which is comparatively easy to model, but not the communication architecture components. This is incomplete because, as pointed out in [16], for complex communication network, the global wire only contributes a small potion of total power consumption.

Most of the existing work on IP power modeling takes RTL level approach. A few suggest behavioral-level methods; however, their accuracy is too low because of the mismatch between the behavioral description and the real implementation. The estimation accuracy becomes even lower when we employ an analytical method. Therefore, significant amount of work has been done on RTL macro modeling of IP power consumption [17]. Although the procedure to build an RTL power macro model is clear, it is still difficult to automate the process.

III. Contributions

System-level simulation at the early design phase has

become essential to search for optimal system architecture and also to enable early software development. We have developed a cycle-based simulation framework, called V*i*P, which can perform concurrent, cycle-accurate and synchronous system-level simulation [2].

On top of V*i*P, we build power models for each major SoC component to estimate the power consumption as well as the performance in a synchronous manner. Note that since the V*i*P framework provides synchronous activity information, the power models can provide more accurate power numbers in the system context [1].

Our goal in PowerV*i*P development is three-fold:

- maintain the target accuracy level
- keep the simulation overhead incurred by power estimation low
- make the power models easy to be customized to an arbitrary target design, since it is used at the system-level design phase where architecture exploration is performed

To achieve the conflicting goal of building a fast, accurate and easy-to-build power model, we take a component-based approach. As shown in Fig. 1, heterogeneous components of a SoC can be categorized into processors, bus fabrics, custom IP blocks, and memories. To achieve the best trade-off between simulation speed and accuracy, we apply different modeling techniques for each component depending on the power characteristics.

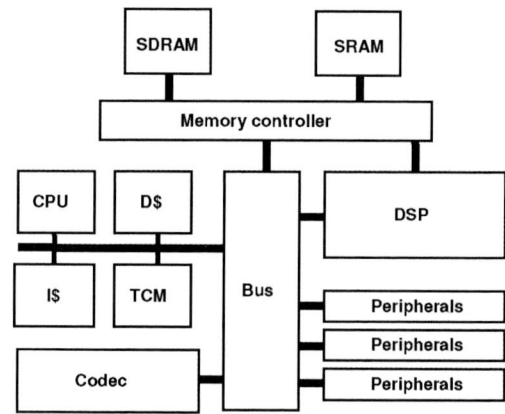

Fig. 1. A System-on-Chip design.

The procedure to build PowerV*i*P follows three steps in general. First, we set up a gate-level or RTL power analysis environment per IP component to extract (characterize) its power values. Next, we build a power model with the extracted power numbers. Finally, we annotate the power model into the system-level model of the IP component to generate power numbers during system-level simulation. For seamless adaptation of the power model to technology transitions, e.g., 130nm to 90nm or high speed to low power process, we automate this process by provisioning scripts that perform the power characterization and model parameter extraction steps.

In the following sections, we propose new approaches

applied to a SoC platform in detail in the following order: processor cores, bus fabrics, custom IP blocks, and memories. Power model development procedure and its validation result are separately presented in each section.

IV. ARM926EJ-S Processor Power Model Development

The ARM926EJ-S processor is widely adopted in SoC as a controller as well as a small data processing engine; thus, we first embark on power modeling of the processor. Currently, power modeling of the ARM1176 processor is being conducted using the methodology described herein.

A. ARM926EJ-S architecture

The ARM926EJ-S processor has a five stage pipelined data path and a Harvard cache architecture. The size of the caches can be from 4KB to 128KB. The ARM926EJ-S processor also has a fill buffer (FB) that keeps the most recently fetched cache line.

In the ARM926EJ-S processor, any instruction that modifies the program counter (such as a branch, or 'MOV pc, r0') causes a non-sequential instruction accesses on the next cycle. An instruction access by 'PC increment by 4' that crosses the cache line boundary also causes a non-sequential access. In Fig. 2 (a), a non-sequential (NS) access causes all four cache tag memories and data memories to be accessed along with the fill buffer. Whereas a sequential access (SEQ) causes only the data memory where the data is located is accessed as in Fig. 2 (b). In Fig. 2 (c), if the data is accessed from the fill buffer, there in no access to the cache.

For data caches, load multiple (LDM) and store multiple (STM) instructions support sequential accesses. LDR and STR instructions incur non-sequential accesses.

B. ARM926EJ-S power states

We separate the processor power model into two parts: Processor core model and cache model. This separation comes from two observations. One is that caches can be configured differently (in terms of size, associativity, etc.) for various applications. Thus, one single model will not give an accurate estimation. The other observation is that the power consumption of caches gives a large variation. In the ARM926EJ-S processor, the cache power consumption ranges from 3% up to 60% of the total power. Therefore, we decide to model the core logic block and cache memory separately.

Processor core: two simple power states

We observe that the core logic can be in one of the two states: busy state and idle state (stalled by interlocks). There are numerous studies on processor power modeling, where more complex instruction level power states are identified [4, 7, 8]. However, in our work, we find that the two-state core power model gives more than 95% of the core power

estimation accuracy for all of our benchmarks. On the other hand, one state model performs very poorly with its accuracy level of less than 70% for some benchmarks. Thus, we adopt the two-state power model for the processor core.

Activity-based coarse-grain cache power model

Most of the previous work on cache power modeling has exploited circuit-level information such as bit line and word line capacitive loads to generate flexible cache power models [5-7]. In industry, cache memories use memory compiler-generated SRAMs, where power values for each module are also provided for each type of read and write access. Thus, our cache power is modeled as a sum of power values for all accessed SRAM modules. For SRAM modules not accessed during the cycle, their static power values are added.

The ARM926EJ-S cache access behavior can be categorized into three different types as shown in Fig. 2. In power perspective, a non-sequential access consumes more than four times of power than a sequential access, since the cache power is the sum of dynamic power of all activated modules (tag memories and data memories) and static power of inactive modules. It dictates that in the ARM926EJ-S caches, non-sequential accesses and sequential accesses should be differentiated for accurate power estimation.

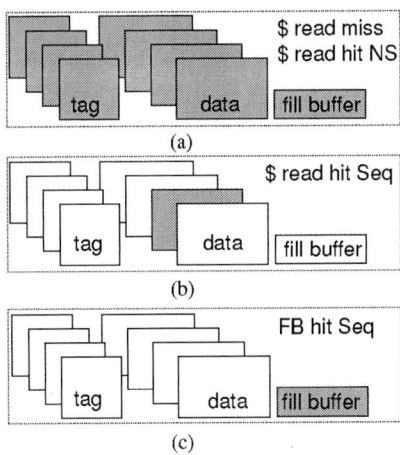

Fig. 2. ARM926EJ-S cache activity patterns.

Table I lists our identified data cache states and their corresponding module activities and power equations. In the table, Tr (Tw) and Dr (Dw) represent module power numbers for $Tag\ read$ ($Tag\ write$) and $Data\ read$ ($Data\ write$), respectively, obtained from our in-house memory compiler. The states are identical for the instruction cache except that there is no cache write hit or miss states. In this work, we ignore the power consumed by fill buffers.

Instructions and data are accessed from the fill buffer until it is evicted to the cache in two cycles (as shown 1st write-back and 2nd write-back in Table I) by the following cache line fetched in from the bus. Instruction fill buffer (I-FB) hit counts accounts for approximately 10% of the instruction cache hit counts in our *dhrystone* benchmark. If an instruction fill buffer hit is encountered and the PC increments by 4, then it is I-FB sequential read, where negligible amount of power is consumed by the fill buffer. Therefore, it should be distinguished if the data is read from the cache or fill buffer to estimate power accurately.

TABLE I. Activity-based cache power model.

Cache states	Module activity	Power Equation
sequential (cache) read	1 data read	Dr
non-sequential (cache) read / read miss	4 tag reads and 4 data reads	Tr*4+ Dr*4
Data cache write hit	4 tag reads, 1 tag write and 1 data write	Tr*4+Tw +Dw
Data cache write miss	4 tag reads	Tr*4
FB -> cache write (1st write-back)	1 tag write and 4 data writes	Tw+Dw* 4
FB -> cache write (2nd write-back)	4 data writes	Dw*4
sequential FB read	-	-

C. Power (re-)characterization flow

Power consumption is a complex function of many parameters. Depending on the quality of implementation, the same RTL can result in very different power values at the gate-level netlist. For example, two of our sample designs of the ARM926EJ-S show as much as twice power difference at the same frequency level, even though they are implemented with the same technology library. This implies that 'characterize once' approach might not hold true in real applications.

In general, power characterization at the gate level proceeds as follows: (1) obtain the signal toggle information from gate-level simulation, (2) estimate the gate-level power from the toggle information using power libraries, and (3) calculate per-state power values using the estimated power information. If the power characterization is performed manually for each different gate-level netlist, it will be long, tedious, and error-prone task.

To reduce the characterization efforts, we set up an automated characterization flow as shown in Fig. 3, where designers can characterize power values repeatedly without investing much effort. The characterized power values are simply read by our simulator annotated with the power model to produce software power profiles. Note that the power model itself does not need any modification. We find that the power model itself is valid for different implementations of the same RTL.

Fig. 3 shows our power characterization flow. We first build a gate-level and RTL co-simulation template, where an RTL testbench with a simple bus and memory module drives the simulation with the ARM926EJ-S gate-level netlist of interest to generate the cycle-by-cycle signal toggle information as well as signal traces to infer the power states, using *dhrystone* benchmark. The toggle information is then fed into our in-house gate-level power estimation tool to generate cycle-by-cycle power values. The per-state power value is obtained by averaging the estimated cycle-by-cycle power values. All the aforementioned steps are performed automatically without any user intervention. The obtained per-state power value is finally annotated into our power simulator. We use the characterization flow to obtain the core power states in our power model. Note that the cache power model is activity-based and its SRAM module power value is provided by our memory compiler.

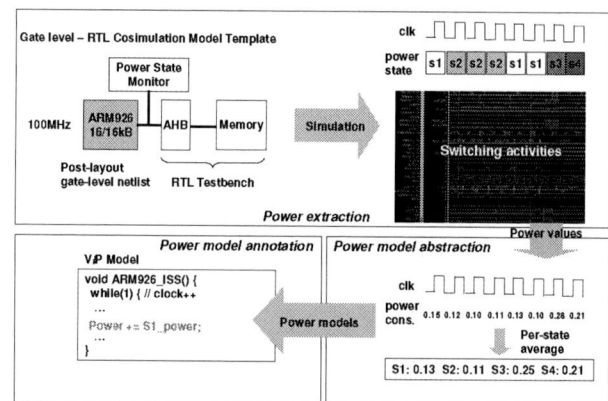

Fig. 3. Power characterization flow.

D. Power model validation

Our simulation with five benchmarks shows 93%~98% of average power estimation accuracy. Fig. 4 shows cycle-by-cycle estimation result for a short code segment. It can be seen that the estimated power values closely track the power values measured at the gate-level. Regarding the power estimation speed of our simulator, it performs approximately 1600 times faster than gate-level estimation.

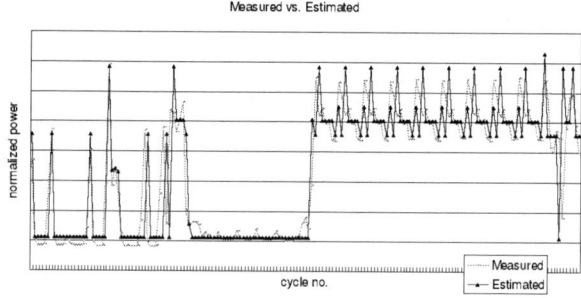

Fig. 4. Cycle-by-cycle estimation accuracy.

V. Component-based Transaction-level Power Modeling for ARM AXI Bus System

The development of today's semiconductor technology provides unprecedented computing speed that is shifting the IC design bottleneck from computation capacity to communication bandwidth and flexibility. The global communication becomes so difficult that more and more designs turn to SoC architecture where a set of local blocks are connected with a communication network. Recent research [19] shows that the on-chip interconnect architecture not only has significant impact on the system performance and energy efficiency, it is also a significant source of power consumption, which is still increasing with the complexity of the system. Managing and optimizing power of this important SoC component require a detailed understanding of its characteristics.

A. AXI Bus architecture

The AMBA AXI 3.0 protocol is targeted at high-performance, high-frequency system designs and includes a number of features that make it suitable for a high-speed sub-micron interconnects.

AXI 3.0 supports simultaneous read and write accesses. Both READ and WRITE transactions have their independent address and data channel. It also supports split transaction, which means the address/command phase and the data phase are separated. In addition, if configured to use ID, the AXI 3.0 compliant masters will assign an ID number to each outgoing transaction and check the ID of each in-coming response. This enables the implementation of multiple outstanding transactions and out-of-order transaction completion, which provide efficient communication for a wider variety of slaves.

The AXI 3.0 protocol only defines the interface between a master and a slave. To connect multiple masters and slaves, ARM suggests the new bus architecture, ARM Configurable Interconnect (ACI). Different from its predecessor AMBA AHB, ACI uses crossbar architecture to provide much higher communication bandwidth. Whenever there is no conflict, e.g., two masters sending commands to one slave at the same cycle, the transactions issued by one master will not be interfered or obstructed by other masters.

B. Power modeling

What ACI differs from other bus system, e.g., AHB, is its crossbar architecture, which allows parallel bus accesses from multiple masters and introduces complex competition and coupling effects. For instance, the power consumption of one master at one specific cycle is not only decided by the communication between this master and the bus, but also depends on the communications conducted by other masters. To make it even worse, each master can do READ and WRITE in parallel. All these result in a huge number of states which are needed to represent the behavior of the bus.

To counter the problem, we have developed our transaction level bus power estimation method using a component based approach. The crossbar consists of components such as decoders, routers and arbitrators. The basic idea is to identify, at each basic state, which component is active and how much power it consumes. At run time, the bus state can be looked as a combination of these basic states, e.g., master 1 is decoding the address while master 2 is sending write data. Hence, to model all the bus states, we need only to characterize the limited number of basic states.

Power model extraction

The model extraction environment is shown in Fig. 5. We use the ARM ACI toolset to configure the bus interconnect, generate the RTL code and synthesize the circuit. Then the gate-level RTL code is simulated and an in-house tool is used to collect the gate-level switching activity information. Based on this information, the gate-level power number could be collected. The Cadence eVC environment is used to feed in random test sequences into the calibration flow.

Extraction of the power model of each component:

Given an ACI bus configuration, i.e., the number of masters, the number of slaves and other parameters such as data width, we synthesize the gate-level RTL code for our flow. During this step, we consider only the case when one master is communicating with one slave and configure the eVC to generate 8 random test sequences:

- READ with burst length of 1, 4, 8 and 16 words, respectively
- WRITE with burst length of 1, 4, 8 and 16 words, respectively

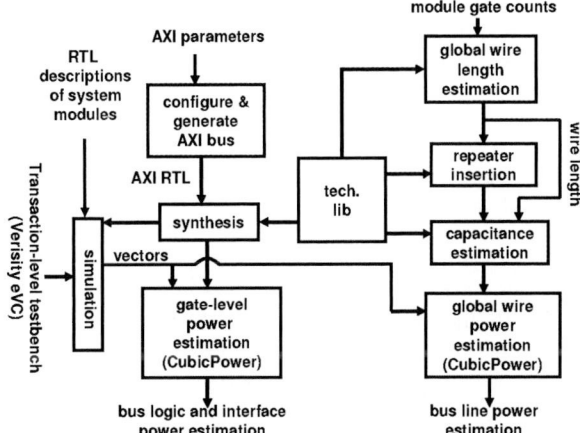

Fig. 5. The power model extraction environment.

This will rule out the coupling between multiple parallel transactions and allow us to focus on the behavior of each ACI component under all the basic transaction states.

Coupling effect:

To account for the coupling effect caused by multiple masters and slaves, another set of coefficients are introduced. We separate the coupling contribution into two items, one is related to the number of bursts the bus transferred during a period, and the other is related to the number of active cycles in that period.

With our eVC automatic test sequence generation environment, we systematically vary the number of active masters and slaves, and calibrate the gate-level bus power consumption under each situation. From these results, linear regression is applied to extract the coupling coefficients.

5D-4

Power estimation

After extracting the transaction level bus power model, we apply the model to a sequence of bus transactions to estimate the bus power consumption of the bus during that period. The input can be at both the transaction level and the signal level.

For each bus cycle, we count the number of READ related transactions and the number of WRITE related transactions being performed by the bus. From them, we could know the specific state of each component and their basic power number accordingly. Thus we obtain the basic power estimation, without considering the coupling effect. To compensate the coupling effect when there are more than one active master and slave, a simple linear equation is applied. By accumulating the power estimation of each bus cycle for a period of time that we are interested in, we could finally obtain the total power consumption during that period.

C. Model validation

In this section, we validate our model by presenting the estimation result of our transaction level power model with various bus configurations. Our experiment set-up is as follows. Given a configuration, such as frequency, data bandwidth, the number of masters, and the number of slaves, we use Synopsys Design Compiler to synthesize the bus architecture. Then our model extraction flow is applied to extract all the coefficients, based on the gate-level simulation result.

Fig. 6 summarizes our result for a 4x4 ACI bus, i.e., a configuration with four masters and four slaves. Due to the large amount of data, we present only the result when the burst length is 8 words. The other results with different burst length show similar characteristics.

In Fig. 6, the horizontal axis shows the number of active masters and slaves, which represents how many masters are really communicating with how many slaves. The vertical axis gives the estimation error of our transaction-level power model from the gate-level simulation. The results of the estimation with coupling effect and without coupling effect are separately plotted as well as the type (read/write) of the transactions. We can see clearly that the estimation error is significantly reduced if the coupling effect is taken into account. The maximum error drops from 16.24% to 5.57%, and the average error drops from 6.23% to 1.87%.

We have applied our power model to four different bus configurations. For all our experiments, the maximum error against the gate-level estimation is less than 10%, and on the average, it is below 5%.

The power model integrated into the V*i*P runs more than 100 times faster, compared with gate-level power estimation.

VI. Automatic Generation of Power-Annotated V*i*P Models for Custom IP Blocks

Power modeling and estimation of custom IP blocks encounter two major challenges. First, a large amount of

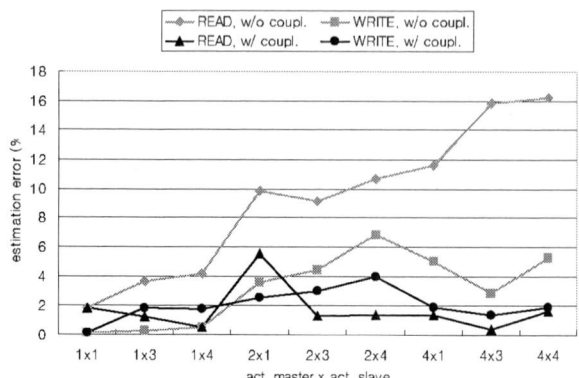

Fig. 6. Power estimation result, burst length 8.

work is required to build the high-level simulation model. This becomes critical when we consider frequently updated IP blocks with low reusability. Second, IP blocks with irregular power characteristics make it difficult to characterize their power consumption.

We propose a practical technique to meet the challenges for custom IP blocks. Our target is to perform a good trade-off between estimation accuracy and speed while minimizing the effort spent on power and V*i*P modeling. High accuracy is achieved by employing RTL power macro modeling method, and the speed overhead of co-simulation is completely eliminated by using automatic RTL to V*i*P model translation technique. The automatic RTL to V*i*P model translation technique additionally enables designers to minimize the V*i*P modeling effort. The overall flow of our technique is illustrated in Fig. 7. Three rectangular boxes represent the state-of-the-art EDA technologies that we employ: RTL to V*i*P translator, transaction-level simulation platform, and simulation based RTL power estimator. We will elaborate them in the following sections.

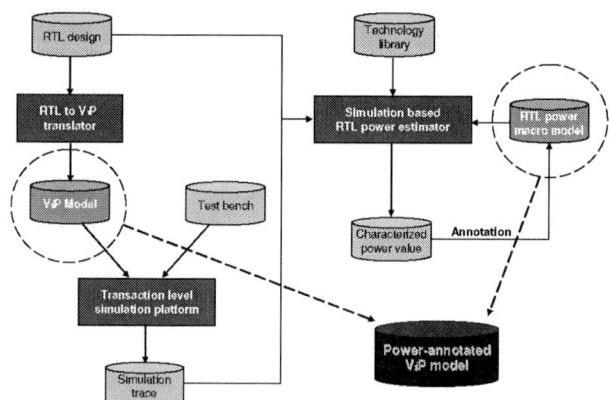

Fig. 7. Overall custom IP power modeling flow.

*A. V*i*P model and RTL power macro model generation*

Two intermediate outputs of our flow are circled in dotted circles. The V*i*P model is generated from RTL description

556

by using an RTL to V*i*P translator, and the RTL power macro model is built by extracting representative state machine variables from the RTL source code. At the end of the flow, two intermediate outputs are merged to make the power back-annotated V*i*P model.

B. System-level simulation

System-level simulation is performed to obtain simulation traces that are required for the power characterization step. Since the generated V*i*P model contains all the structural information of the original RTL design, it is possible to dump RTL switching activities to output VCD files. This enables the simulation based power estimation. The inputs to this step are the generated V*i*P model and testbench. Since most of the IP blocks have a general interface such as a standard bus protocol, the testbench can usually be reused.

C. Simulation based RTL power estimation

As mentioned previously, we use an RTL power estimator for characterization. The inputs to this step are as follows: the simulation trace in VCD format, the original RTL design, the technology library, and the initial RTL power macro model. The RTL power estimator reads in the technology library file and quickly synthesizes the original RTL design into a gate-level netlist in the target technology. The simulation trace is applied to this gate-level netlist to perform gate-level power estimation. As a result, the initial power macro model is back-annotated with characterized power numbers. The initial RTL power macro model must be evaluated at this point to confirm that it satisfies the accuracy constraint. If the accuracy of the initial model is too low, the RTL power macro model should be refined and this step must be repeated.

This step can be easily substituted with a gate-level estimation tool if the gate-level netlist and testbenches are available. That is, we can come back to this point after the RTL freeze for more accurate power estimation.

D. Integration of the power model into the ViP model

The last step of our flow is the integration of the power macro model with the V*i*P model. Since the power macro model is built by extracting representative state machine variables, it is easy to automatically embed the power macro model into the V*i*P model.

We implement a monitor inside the V*i*P model, which tracks the list of signals representing the power macro model. Then the power consumption of the IP is reported according to the signal values during simulation. That is, the power back-annotated V*i*P model can estimate the power consumption on a cycle-by-cycle basis by monitoring its internal switching activities.

E. Validation of the flow

The validation of our custom IP power modeling technique is in progress. For an 80K-gate IP block that we have tested so far, the modeling effort in terms of man-month is reduced by an order of magnitude, and the accuracy is 80% as compared to the gate-level estimation.

VII. Memory Power Models

Various types of memories are used for different purposes. SRAM and SDRAM are two of the most commonly used memories in a SoC system. In our framework, SRAM is easily modeled with energy per read, write, or idle operations. In most cases, power variation caused by the data dependency of each operation can be ignored. For SDRAM, we adopt the widely accepted approach, which is well-described in [18]. Read, write, activate, precharge, and refresh power are separately calculated using the DC characteristic numbers provided in the SDRAM datasheets. Our experience shows that these models are sufficient to explore the power consumption in the system context.

VIII. PowerV*i*P: A SoC Power Estimation Framework

The separately prepared power models are integrated in the V*i*P framework. In this study, we only present the results for the processor and AXI bus fabrics since the validation of the custom IP power modeling flow is in progress We build a system composed of an ARM926 processor, an AHB-to-AXI converter, AXI bus fabric, a memory controller, and an external SDRAM. A set of test bench suite is run on the ARM processor; we believe that this configuration conforms to our goal of providing a feasibility of our study.

PowerV*i*P produces profile information in HTML format for all the functions run on the ARM926EJ-S processor. As shown in Fig. 8, power consumption and other statistics for each function are reported, e.g., total cycles and bus utilization. Note that the power numbers in Fig. 8 are normalized with respect to the highest power consuming function. It is sorted in decreasing order of total cycles as shown in the second column. From the result, designers can find which function is most power-hungry and start cycle-by-cycle power and performance simulation at the beginning of the function.

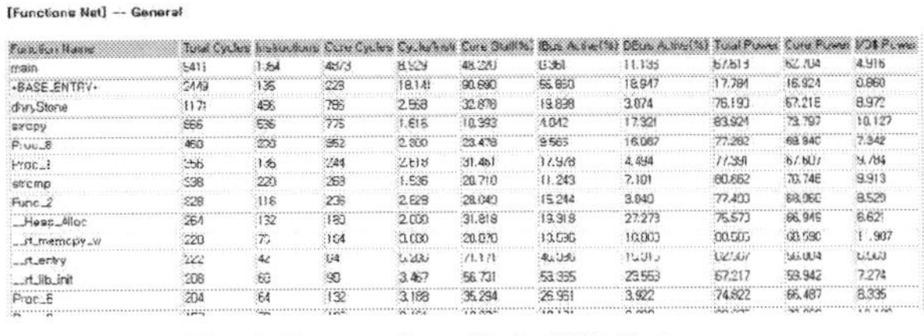

Fig. 8. Reported profile in HTML format.

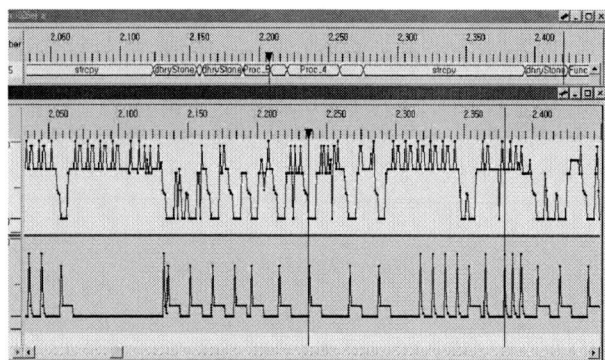

Fig. 9. Graphical power profile in time axis.

Fig. 9 shows the power consumption of the ARM926EJ-S processor and the AXI bus fabrics graphically (on the lower window) as a series of functions such as *strcpy* and *dhryStone* are executed (on the upper window). The power graph shows distinct pattern for each function. In this way, the exact cycle when a power surge occurs can be pin-pointed. Then the debugger can trace the source code line-by-line from the pin-pointed cycle to identify the cause of the power surge. For example, if high instruction cache power consumption is attributed to the power surge, it can be monitored if the instruction cache misses keep occurring at the time, or if the tight loop body incurs frequent branches, causing power-hungry non-sequential instruction cache accesses. All the necessary information is reported in the simulator in a synchronized fashion.

IX. Conclusions

We tackle the heterogeneity of SoC by adopting different power modeling techniques for different components. Simple power models are used for components that have simple power characteristics, i.e., SRAM and processor core. We build an activity-based coarse-grain power model for the cache memory and develop a component-based approach for bus fabrics. For custom IP blocks, we use RTL power macro modeling combined with the RTL to ViP model translation technique. On the average, our power models for ARM926EJ-S and AXI bus fabrics show 93% and 95% of estimation accuracy respectively as compared to gate-level estimation.

Although our custom IP power modeling strategy still needs to be validated, we are confident that with the help of PowerViP, system designers can explore various architectural choices at the early design phase to find a power-optimal design before the RTL freeze. Currently, our design teams are adopting the framework at the early design exploration stage. We plan to enrich the power model database by providing power models for other commonly used processor core families, bus fabrics, custom IP blocks, and memory devices.

References

[1] M. Lajolo, A. Raghunathan, S. Dey, and L. Lavagno, "Cosimulation-based power estimation for system-on-chip design," *IEEE Trans. on VLSI Systems*, vol. 10, no. 3, pp. 253-266, 2002.

[2] Samsung Electronics, "Samsung's ViP design methodology reduces SoC design time up to 40 percent," http://www.samsung.com/Products/Semiconductor/News/SystemLSI/SystemLSI_20040914_0000069677.htm.

[3] N. Bansal, K. Lahiri, A. Raghunathan, and S. T. Chakradhar, "Power Monitors: a framework for system-level power estimation using heterogeneous power models," in *Proc. Int. Conf. on VLSI Design*, 2005, pp. 579-585.

[4] V. Tiwari, S. Malik, and A. Wolfe, "Power analysis of embedded software: a first step towards software power minimization," *IEEE Trans. on VLSI systems*, vol. 2, no. 4, pp. 437-445, 1994.

[5] D. Brooks, V. Tiwari, and M. Martonosi, "Wattch: a framework for architectural-level power analysis and optimizations," in *Proc. ISCA*, 2000, pp. 83-94.

[6] W. Ye, N. Vijaykrishnan, M. Kandemir, and M. J. Irwin, "The design and use of SimplePower: a cycle-accurate energy estimation tool," in *Proc. DAC*, 2000, pp. 340-345.

[7] G. Gontreras, M. Martonosi, J. Peng, R. Ju, and G.-Y. Lueh, "XTREM: a power simulator for the Intel XScale® core," in *Proc. LCTES*, 2004, pp. 115-125.

[8] A. Sinha and A. Chandrakasan, "JouleTrack: a web based tool for software energy profiling," in *Proc. DAC*, 2001, pp. 220-225.

[9] J. Y. Chen, W. B. Jone, J. S. Wang, H.-I. Lu, and T. F. Chen, "Segmented bus design for low power," *IEEE Trans. on VLSI systems*, vol. 7, no. 1, pp. 25-29, 1999.

[10] C.-T. Hsieh and M. Pedram, "Architectural power optimization by bus splitting," *IEEE Trans. Computer Aided Design*, vol. 21, no. 4, pp. 408-414, 2002.

[11] P. P. Sotiriadis and A. P. Chandrakasan, "A bus energy model for deep submicron technology," *IEEE Trans. on VLSI systems*, vol. 10, no. 3, pp. 341-350, 2002.

[12] L. Benini, A. Macii, M. Poncino, and R. Scarsi, "Architecture and synthesis algorithm for power-efficient bus interfaces," *IEEE Trans. Computer Aided Design*, vol. 19, no. 9, pp. 969-980, 2000.

[13] M. Caldari et al., "System-level power analysis methodology applied to the AMBA bus," in *Proc. DATE*, 2003, pp. 32-37.

[14] U. Neffe et al., "Energy estimation based on hierarchical bus models for power-aware smart card," in *Proc. DATE*, 2004, pp. 300-305.

[15] A. Bona, V. Zaccaria, and R. Zafalon, "System level power modeling and simulation of high-end industrial network-on-chip," in *Proc. DATE*, 2004, pp. 318-323.

[16] K. Lahiri and A. Raghunathan, "Power analysis of system-level on-chip communication architectures," in *Proc. CODES+ISSS*, 2004, pp. 236-241.

[17] S. Ravi, A. Raghunathan, and S. Chakradhar, "Efficient RTL power estimation for large designs," in *Proc. Int. Conf. on VLSI Design*, 2003, pp. 431-439.

[18] Micron Technology, "Calculating DDR memory system power" http://www.micron.com/products/dram/ddrsdram/technote.html,

[19] R. Kumar, V. Zyuban, and D. Tullsen, "Interconnections in multi-core architectures: understanding mechanisms, overheads and scaling", in *Proc. ISCA*, 2005, pp. 408-419.

2006 Asia and South Pacific Design Automation Conference

6A-1

Mathematically Assisted Adaptive Body Bias (ABB) for Temperature Compensation in Gigascale LSI Systems

Sanjay V. Kumar, Chris H. Kim, and Sachin S. Sapatnekar
Department of Electrical and Computer Engineering, University of Minnesota, Minneapolis, MN 55455
sanjay,chriskim,sachin@ece.umn.edu

Abstract—Process variations and temperature variations can cause both the frequency and the leakage of the chip to vary significantly from their expected values, thereby decreasing the yield. Adaptive Body Bias (ABB) can be used to pull back the chip to the nominal operational region. We propose the use of this technique to counter temperature variations along with process variations. We present a CAD perspective for achieving process and temperature compensation using bidirectional ABB. Mathematical models are used to determine the exact amount of body bias required to optimize the delay and leakage, and an algorithmic flow that can be adopted for gigascale LSI systems is provided.

Index Terms : Delay, Leakage, Adaptive Body Bias (ABB), Process Variations, Temperature Variations, Nonlinear Programming Problem (NLPP), Enumeration

I. INTRODUCTION

With technology scaling, the effects of process parameter variations and on-chip temperature variations have caused the delay and leakage of modern-day processors to vary significantly from their desired values. Some of the dies may satisfy the delay constraint but leak too much, while others may leak nominally but fail to meet the target frequency. Thus, a significant fraction of the total number of acceptable dies may fail to achieve the performance goals. This has led to the evolution of methodologies to perform post-silicon tuning for yield improvement. Adaptive Body Bias (ABB) provides a viable control technique that can counter the effects of on-chip variations.

Two of the significant contributors to on-chip variability arise from changes in process parameters and in the operating temperature. Process variations lead to fluctuations in parameters such as transistor channel lengths, oxide thicknesses, and dopant concentrations. These cause variations in the delay and leakage of the circuit, thereby affecting performance. On-chip temperature variations, on the other hand, change the mobilities of electrons and holes. An increase in the operating temperature causes the mobilities to decrease, thereby decreasing the I_{on} current, which, in turn, reduces the speed of the circuit. Further, elevated temperatures also lead to an increase in the leakage current. On-chip variations can be categorized as lot-to-lot (L2L), wafer-to-wafer (W2W), die-to-die (D2D), and within-die (WID) variations [1].

Adaptive Body Bias (ABB) is a dynamic technique that helps tighten the distribution of the *maximum operational frequency* and the *maximum leakage power* in the presence of WID variations, and thereby helps improve the yield significantly. It was first proposed by Wann *et al.* in [2] and was further explored by Kuroda [3] during the design of a DSP Processor. Bidirectional Adaptive Body Bias has been shown to reduce the impact of D2D and WID parameter variations on microprocessor frequency and leakage in [4], [5], [6] and [1]. Typically, devices that are slow but do not leak too much can be Forward Body Biased (FBB) to improve the speed, whereas devices are fast and leaky can be Reverse Body Biased (RBB) to meet the leakage budget. The work in [4], [7] performs process variation-based ABB, and divides the die into a set of WID variational regions. In each region, test structures, which are replicas of the critical path, are built. The delay and leakage of these test structures are measured, and used to determine the exact body bias values that are required to counter process variations at room temperature. The application of a WID-ABB technique for one-time compensation during the test-phase, in [4], shows that 100% of the dies can be salvaged, while 99% of them operate at the highest frequency bin.

Traditionally, ABB has been used only to compensate for process variations [4–6]. However, on-chip temperature changes can also significantly vary the delay and leakage of nanometer-scale devices, thereby necessitating the need to mitigate the effects of these thermal variations as well. Only a limited amount of work so far has addressed this problem, such as [8], which focuses purely on temperature effects. In this work, we apply a combination of temperature-based ABB (TABB) and process-based ABB (PABB) to permit the circuit to recover from changes due to both temperature and process variations. In order to be able to adaptively body bias all of our dies at all operating temperatures, we utilize an efficient self-adjusting mechanism that can sense the operating temperature, and thereby dynamically regulate the voltages that must be applied to the body of the devices to meet the performance constraints. We propose a general architecture and an implementation scheme to achieve this.

The contribution of this paper is to provide a strategy for determining the exact amount of bias required to achieve process and temperature compensation through a combination of simulation, probabilistic design and post-silicon tuning in order to maximize the yield subject to frequency and leakage constraints. This method is aptly termed PTABB (process and temperature-based adaptive body bias). The final set of PTABB voltages that can counter process and temperature variations at all operating conditions is thus a combination of PABB and TABB. We propose two methods to compute the TABB values, namely, an enumeration based method and a mathematical model based method. Enumeration based TABB involves simulating the circuit at discrete points in the solution space and finding the best solution. In contrast, mathematically assisted TABB assumes a continuous search space and provides an exact solution using a model that captures the effect of body bias on delay and leakage and a simple nonlinear programming problem (NLPP) formulation. PABB can be performed by building test structures with critical path replicas on each WID-variational region [4]. The exact amount of body bias to counter the effects of process variations at room temperature is determined by measuring the delay and leakage of the circuit, and choosing the optimal solution.

The concept of using mathematical models to formulate expressions for delay and leakage, and thereby to obtain exact solutions for the ABB voltages, is in itself a new and attractive approach. Compared to prior approaches that determine the exact body bias required during run-time by monitoring the delay and leakage (listed in [9]), our scheme uses a simple look-up table (similar in concept to that used in [8]), that stores these pre-computed values, and hence, only requires a temperature sensor to monitor the variations in on-chip temperature. This eliminates the need for circuits like leakage current monitor, substrate charge injector, self substrate bias, *etc.*, since the determination of the TABB voltages is carried out at the design stage. Further, the idea of *one-time compensation* for process variations and *run-time compensation* for temperature variations is effectively combined. The generation of these additional body bias voltages and their distribution on chip is not considered to be within the scope of our work. We present the algorithm, implementation and results of this novel scheme in the subsequent sections.

0-7803-9451-8/06/$20.00 ©2006 IEEE.

559

II. CENTRAL IDEA

In this section, we present an overall picture of the proposed implementation. The die is partitioned into several WID variational regions, and each of these regions is separately compensated. Our target technology in this work uses a triple well process although the idea can be generalized to any other process. The central body bias generator consists of a PVT invariant voltage reference source, and is capable of generating one pair of voltages applicable to the NWELL and PWELL of each WID variational region separately. (Alternatively, the body bias generators can be replicated in each WID zone to locally generate and distribute the required voltages, but a central PVT invariant voltage reference is still required). A temperature sensor is placed in each of these regions in order to detect on-chip temperature variations. A small ROM is fabricated in every region, whose look up table consists of a pair of voltages (v_{bn}, v_{bp}) for each temperature being compensated.

The TABB values $(v_{bn}, v_{bp})_{TABB}$, that can compensate for temperature variations only, assuming ideal process conditions are determined during the design stage. Similar values $(v_{bn}, v_{bp})_{PABB}$, to compensate for process variations at room temperature, are calculated through post-silicon tuning. These values can be combined as shown in Fig. 1 to get one pair of bias values for every block (WID variational region) at each operating temperature for all the dies. The final bias pair, denoted by $(v_{bn}, v_{bp})_{PTABB}$, is programmed into the ROM.

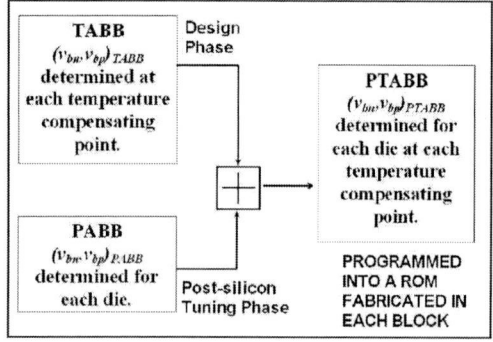

Fig. 1. **Generation of PTABB values for every block (WID variational region).**

When the circuit is in operation, the temperature sensor detects changes in the on-chip temperature. The corresponding values of v_{bn} and v_{bp} are read from the ROM and fed as inputs to the central bias generator. These voltages are generated by the central bias generator and distributed to NWELL and PWELL through the bias distribution network. The overall architecture is shown in Fig. 2.

III. PTABB ALGORITHM

In this section, we present the algorithm that determines the body bias required for process and temperature variation compensation. Since we assume the existence of a triple well process, the bodies of both NMOS and PMOS devices can be biased independently. However, the algorithm can be easily modified for a twin well process. We present SPICE-calibrated models that express the delay, and leakage in terms of the bias voltages and determine the optimal bias voltages based on operational constraints.

The effects of process and temperature variations on the delay of a combinational circuit can be represented as:

$$D = f(\mathbf{x}, T) \qquad (1)$$

where D is the delay of the circuitry, $\mathbf{x}$ is a vector of process variables and T is the operating temperature of the chip. Let $\mathbf{x_0}$ and T_0 denote the values of the process and temperature variables

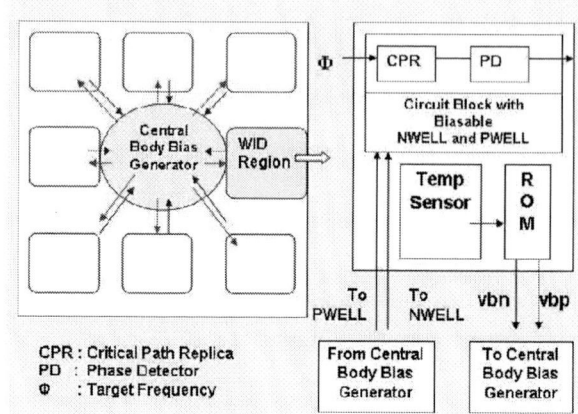

Fig. 2. **Block diagram for PTABB implementation.**

under ideal conditions where there is no variation. The increase in delay at any point $(\mathbf{x}, T)$ can be written as:

$$\Delta D = f(\mathbf{x}, T) - f(\mathbf{x_0}, T_0) \qquad (2)$$

where $\mathbf{x}$ is the vector of process variables of a particular die, while T is the operating temperature of the die. If $\mathbf{x}$ and T are independent variables, the effect of simultaneously varying $\mathbf{x}$ and T, from $(\mathbf{x_0}, \mathbf{T_0})$ to $(\mathbf{x_1}, \mathbf{T_1})$ can be approximated as varying $\mathbf{x}$ and T individually from their original values and adding their effects, i.e.,

$$\Delta D \approx [f(\mathbf{x_0}, \mathbf{T_1}) - f(\mathbf{x_0}, \mathbf{T_0})] + [f(\mathbf{x_1}, \mathbf{T_0}) - f(\mathbf{x_0}, \mathbf{T_0})] \quad (3)$$

where $f(\mathbf{x_0}, \mathbf{T_1})$ is the delay with temperature variations only, while $f(\mathbf{x_1}, \mathbf{T_0})$ is the delay considering the effect of process variations only.

The above assumption of independence is justified since process and temperature variations have different device level effects, and hence their impacts on the delay can be treated as independent of one another. Process variations affect parameters such as channel length, oxide thickness, and dopant concentration, thereby altering the delay, while temperature variations affect the mobilities of electrons and holes, which influences the on-current, and hence, the delay of the circuit. Further, the results shown in Table I indicate the validity of the assumption. The delay of a ring oscillator is measured through simulations performed using BPTM [10] 100nm model files at $T = 50°C$ and $T = 75°C$ at the two extreme process corners:

1) *Low-Vt corner* which is the case where process variations cause the threshold voltages of both NMOS and PMOS devices to decrease by 10%.
2) *High-Vt corner* which is the case where process variations cause an increase in both V_{tn0} and V_{tp0} by 10%.

The column labeled **Nom-Delay** in Table I indicates the delay at $T = 25°C$ under ideal process conditions. The delay considering the effect of both process and temperature variations is shown in the column labeled **Delay$_{PT}$** and the variation in delay calculated directly, using (2) is shown in column 7. Columns labeled **Delay$_T$** and **Delay$_P$** list the delay considering temperature variations and process variations respectively. The change in delay, expressed as a sum of the change in delays due to process and thermal effects using (3) is listed in column 8. It can be seen from the last column in the table that the difference in delay between the two measurements are negligible compared to the actual circuit delay values. Thus, we can indeed decompose the delay expressions into a temperature-dependent term and a process-dependent term. We use the above findings to perform temperature compensation and process compensation independently of each other.

TABLE I

DELAY DECOMPOSITION INTO PROCESS DEPENDENT AND TEMPERATURE DEPENDENT TERMS FOR A RING OSCILLATOR

Temp °C	Process Corner	Nom-Delay $f(\mathbf{x_0}, \mathbf{T_0})$ (ps)	Delay$_{PT}$ $f(\mathbf{x_1}, \mathbf{T_1})$ (ps)	Delay$_P$ $f(\mathbf{x_1}, \mathbf{T_0})$ (ps)	Delay$_T$ $f(\mathbf{x_0}, \mathbf{T_1})$ (ps)	ΔD from Eqn (2) $f(\mathbf{x_1}, T_1) - f(\mathbf{x_0}, T_0)$ (ps)	ΔD from Eqn (3) $[f(\mathbf{x_0}, \mathbf{T_1}) - f(\mathbf{x_0}, \mathbf{T_0})]+$ $[f(\mathbf{x_1}, \mathbf{T_0}) - f(\mathbf{x_0}, \mathbf{T_0})]$(ps)	Difference (ps)
50	Low V_t	151.0	145.6	141.8	154.9	-5.4	-5.2	-0.2
50	High V_t	151.0	165.3	161.2	154.9	14.3	14.2	0.1
75	Low V_t	151.0	149.2	141.8	158.6	-1.8	-1.5	-0.3
75	High V_t	151.0	169.3	161.2	158.6	18.3	17.8	0.5

A. Temperature Compensation

Generally, the delay of a circuit exhibits negative temperature dependence, *i.e.* the delay increases with an increase in temperature due to a reduction in the mobility of electrons and holes. Hence, we need to forward body bias the devices to reduce the delay at higher operating temperatures, at the expense of leakage. However, at low-V_{dd} operations, the reduction in V_t has a higher impact than the reduction in mobility and an increase in temperature allows the circuits to operate at a higher speed. This effect, described as *positive temperature dependence*, can be used to achieve TABB as described in [11]. In such cases, the devices may be less forward biased (or relatively reverse body biased) at higher temperatures to achieve leakage savings. We hereby present two methodologies to determine the amount of FBB needed to meet the delay constraint, thereby minimizing leakage, for the general case of negative temperature dependence.

B. Enumeration based TABB

The task of ABB compensation is to determine the optimal value of the biases for the NWELL and PWELL, that brings the delay back to specifications, with a minimal leakage overhead. The basic idea of enumeration is to traverse through the entire search space and find this solution. However, since it is infeasible to find the delay and leakage over all possible values of v_{bn} and v_{bp}, we discretize the voltage levels and perform the enumeration over a limited set of values. The maximum amount of FBB that can be applied is restricted by the diode turn on voltage of the source-substrate junction and is process-dependent. The minimum resolution of voltage that can be applied is set by the designer and is constrained by the bias generation network. A method for determining the optimal values is shown in Algorithm 1. We wish to operate the circuit at the highest possible frequency, and the target delay of the circuitry (D^*) is determined by a simulation at the nominal temperature. Since we have assumed negative temperature dependence, the delay of the circuit at a higher operating temperature is greater than D^*, hence requiring FBB. The circuit is simulated with the upper bound of the search space (v_{bnmax}, v_{bpmax}[1]) to determine if maximum FBB can pull the circuit delay back to D^*. If the maximum applicable bias fails to meet the target delay, the operational frequency of the circuit block needs to be reduced. Otherwise, we set this as our initial solution and seek better solutions than (v_{bnmax}, v_{bpmax}) within the search space since (v_{bnmax}, v_{bpmax}) is overkill in terms of leakage. The circuit is simulated at each of the bias pair points and the solution that has the minimum leakage is chosen. If the final leakage of the block is still greater than the allocated budget, then the operational frequency is reduced, D^* thereby increased, and the process of enumeration is repeated.

C. Mathematically assisted TABB

Enumeration over the entire two dimensional search space to determine the optimum bias ordered pair is a costly process for large circuits since it requires simulations at each bias value (worst case)

Algorithm 1 Enumeration Algorithm for TABB

1: Simulate circuit with zero body bias at $T = 25°C$ to obtain D^*
2: Leakage budget for the circuit is denoted by L_{max}
3: **for** each temperature T being compensated **do**
4:    Measure best-case delay $\tilde{D}(v_{bnmax}, v_{bpmax})$
5:    **if** $D(v_{bnmax}, v_{bpmax}) \geq D^*$ **then**
6:      Reduce operational frequency
7:      {Maximum ABB cannot meet delay requirement; decrease target delay D^*.}
8:    **else**
9:      Initial Solution = (v_{bnmax}, v_{bpmax})
10:     $L_{min} = \infty$
11:   **end if**
12:   **for** ($v_{bn} = v_{bnmax}$ to v_{bnmin}) **do**
13:     **for** ($v_{bp} = v_{bpmax}$ to v_{bpmin}) **do**
14:       Apply (v_{bn}, v_{bp}) and simulate at temperature T
15:       **if** $D(v_{bn}, v_{bp}) \leq D^*$ **then**
16:         {Likely solution since it meets delay.}
17:         **if** $L(v_{bn}, v_{bp}) \leq L_{min}$ **then**
18:           New solution = (v_{bn}, v_{bp})
19:           $L_{min} = L(v_{bn}, v_{bp})$
20:         **end if**
21:       **else**
22:         break inner for loop
23:         {Lower values of vbp do not meet delay.} [2]
24:       **end if**
25:       $v_{bp} = v_{bp} - v_{step}$
26:       {v_{step} is the minimum resolution of bias that can be applied.}
27:     **end for**
28:     **if** $D(v_{bn}, v_{bpmax}) \geq D^*$ **then**
29:       break outer for loop
30:       {Lower values of vbn do not meet delay.} [2]
31:     **end if**
32:     $v_{bn} = v_{bn} - v_{step}$
33:   **end for**
34:   **if** $L_{min} \geq L_{max}$ **then**
35:     Reduce Operational Frequency and go to Line 5
36:   **end if**
37: **end for**

---

and has a cost of $O(n^2)$, where n is the number of bias voltages available. Hence, we propose an efficient algorithm based on a simple nonlinear programming problem (NLPP) that requires the simulation of the circuit for delay and leakage at a few points only, to determine the exact body bias pair required. The crux of this method is as follows.

The delay and leakage of a circuitry can be altered by applying a bias voltage v_{bn} to the body of the NMOS transistors and $(V_{dd} - v_{bp})$ to that of the PMOS transistors. Since analytical expressions that can quantize the effect of body bias on delay and leakage at the circuit level do not exist, we use polynomial best fit curves to realize these models. Simulation results show that second order polynomials in both v_{bn} and v_{bp} provide a reasonably accurate model of both delay and leakage. Thus we have the expressions:

$$D(v_{bn}, v_{bp}) = D_0[\sum_{i=0}^{2}(\sum_{j=0}^{2} a_{ij} v_{bn}^j) v_{bp}^i] \quad (4)$$

$$L(v_{bn}, v_{bp}) = L_0[\sum_{i=0}^{2}(\sum_{j=0}^{2} b_{ij} v_{bn}^j) v_{bp}^i] \quad (5)$$

where D_0 and L_0 are the delay and leakage values at the given

---

[1]The actual voltage applied to the body of the PMOS transistors is ($V_{dd} - v_{bp}$). For simplicity, we refer to this as v_{bp}.

[2]If a bias pair (v_{bn1}, v_{bp1}) does not satisfy the delay requirement, all bias pairs with ($v_{bn} \leq v_{bn1}$) and ($v_{bp} \leq v_{bp1}$) fail to meet the delay requirement and hence can be directly eliminated.

operating temperature under the condition where process variation effects are ignored. Since we have two variables v_{bn} and v_{bp}, it is desirable to model the effects of these individually, independently of each other and finally superpose their effects. In other words, we wish to re-write (4) as:

$$D(v_{bn}, v_{bp}) = D_0 f(v_{bn}) g(v_{bp}) \qquad (6)$$

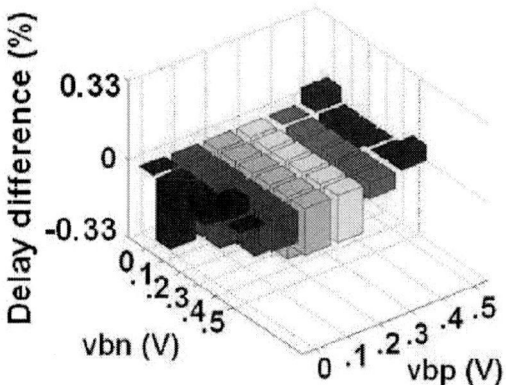

Fig. 3. **Accuracy of polynomial curve fits compared with HSPICE simulations for a ring oscillator at** $T = 25°C$ **and** $V_{dd} = 1.0V$. **Reported values are the difference in delays at each simulation point.**

We verified the possibility of this decomposition on the delay of a Ring Oscillator (RO) and the results are shown in Fig. 3. The reference delays of the RO following the application of body bias are measured through HSPICE simulations performed using BPTM 100nm model files. The delay due to varying v_{bn} only (measured at $v_{bp} = V_{dd}$) is approximated using a second order best fit curve as,

$$f(v_{bn}) = (1 + x_1 v_{bn} + x_2 v_{bn}^2) \qquad (7)$$

Similarly, the delay due to varying v_{bp} only (measured at $v_{bn} = 0$) is approximated using the polynomial $g(v_{bp})$ as,

$$g(v_{bp}) = (1 + y_1 v_{bp} + y_2 v_{bp}^2) \qquad (8)$$

The new delay of the ring oscillator at any point (v_{bn}, v_{bp}) is calculated as a product of the polynomials $f(v_{bn})$ and $g(v_{bp})$. Finally, the difference between the reference values and the new delay values, calculated at each point, is shown in Fig. 3. It can be seen that this difference is negligible, thereby conforming the predicted trend. Hence (4) can be re-written as,

$$D(v_{bn}, v_{bp}) = D_0(1 + x_1 v_{bn} + x_2 v_{bn}^2)(1 + y_1 v_{bp} + y_2 v_{bp}^2) \quad (9)$$

However, leakage does not show a similar trend. The coefficients in (5) and (9) can be determined by simulating the circuit at well-spaced sample points. The desired accuracy for these curve-fitted expressions determines the number of points chosen to obtain the best-fit curve. The Nonlinear Programming Problem can be formulated as:

$$\text{minimize } L(v_{bn}, v_{bp}) = L_0 \left[\sum_{i=0}^{2} \left(\sum_{j=0}^{2} b_{ij} v_{bn}^j \right) v_{bp}^i \right] \qquad (10)$$

subject to

$$D(v_{bn}, v_{bp}) = D_0(1 + x_1 v_{bn} + x_2 v_{bn}^2)(1 + y_1 v_{bp} + y_2 v_{bp}^2) \le D^*$$
$$0 \le v_{bn} \le v_{bnmax}$$
$$0 \le v_{bp} \le v_{bpmax} \qquad (11)$$

Practically, if $D(v_{bn}, v_{bp})$ exceeds D^*, then the minimum leakage solution corresponds to the case where (11) is an equality. Therefore,

the above problem can be solved by eliminating one of the variables v_{bn} or v_{bp} in (11) and finding the minimum value of L in (10) with respect to the other variable using the Newton-Raphson method to obtain the optimum solution, $(v_{bn}, v_{bp})_{TABB}$.

D. Process Compensation

In order to perform process compensation using ABB, a test structure consisting of the critical path replica is built in each of the WID variational regions. PABB is performed in [4] by applying an NMOS bias (v_{bn}) from an off-chip source and automatically adapting the PMOS bias to meet the target frequency. The process is repeated for all possible values of v_{bn} and the bias pair which results in lowest leakage is chosen as the final solution. This scheme requires a 5 bit counter and a DAC (Digital to Analog Converter) in the test structures, to automatically determine the PMOS bias for each NMOS bias applied.

This methodology can be simplified with the use of external voltage sources for biasing both the NWELL and PWELL and an NLPP formulation to determine the exact PABB values. The test structure now consists of a critical path replica and a phase detector only as shown in Fig. 2. The NLPP formulation outlined in the previous sub-section is employed to determine the exact PABB values. The coefficients in (5) and (9) are now determined by actual measurements on chip, instead of circuit simulations for the TABB case. Off-chip sources are used to bias the wells, and the delay and leakage values are measured at some points. The NLPP is formulated in an identical manner as that in (10) and (11), with D_0 and L_0 being the measured delay and leakage values of the *WID-variational region* at *nominal temperature*. The NLPP is solved to obtain the optimal bias values $(v_{bn}, v_{bp})_{PABB}$. The final value that can counter both process and temperature variations for each WID-variational region is calculated by summing the values obtained individually through PABB and TABB:

$$(v_{bn}, v_{bp})_{PTABB} = (v_{bn}, v_{bp})_{PABB} + (v_{bn}, v_{bp})_{TABB} \qquad (12)$$

However if the final values are greater than v_{bnmax} or v_{bpmax}, the solution must be legalized by considering the minimum of the sum of the leakage due to PABB and TABB. The NLPP must be re-formulated as:

$$\text{minimize } L(v_{bn}, v_{bp}) = L(v_{bn}, v_{bp})_{PABB} + L(v_{bn}, v_{bp})_{TABB} \qquad (13)$$

subject to

$$
\begin{aligned}
D(v_{bn}, v_{bp})_{TABB} &\le D^* \\
D(v_{bn}, v_{bp})_{PABB} &\le D^* \\
v_{bnmin} &\le v_{bnTABB} \le v_{bnmax} \\
v_{bnmin} &\le v_{bnPABB} \le v_{bnmax} \\
v_{bn} &= v_{bnTABB} + v_{bnPABB} \\
v_{bnmin} &\le v_{bn} \le v_{bnmax} \\
v_{bpmin} &\le v_{bpTABB} \le v_{bpmax} \\
v_{bpmin} &\le v_{bpPABB} \le v_{bpmax} \\
v_{bp} &= v_{bpTABB} + v_{bpPABB} \\
v_{bpmin} &\le v_{bp} \le v_{bpmax}
\end{aligned}
$$

where $D(v_{bn}, v_{bp})_{TABB}$ and $L(v_{bn}, v_{bp})_{TABB}$ are the delay and leakage values from (5) and (9) considering temperature variations only while $D(v_{bn}, v_{bp})_{PABB}$ and $L(v_{bn}, v_{bp})_{PABB}$ are the delay and leakage values from (5) and (9) with process variations only. The limits v_{bnmin}, v_{bnmax}, v_{bpmin} and v_{bpmax} are determined by the process-technology used. The final values (v_{bn}, v_{bp}) are programmed into the ROM, as described in Fig. 1. When the circuit is in operation, these values are referenced from the ROM, based on the output of the temperature sensor and the corresponding bias values are applied to recover performance.

562

TABLE II
TABB COMPENSATION VALUES FOR ISCAS BENCHMARKS

| Bench mark | No Body Bias | | | | Enumeration based TABB | | | | Mathematically assisted TABB | | | | | | | | Run-time ratio |
| | | | | | | | | | NLPP Solution | | | | Solution after snapping[3] | | | | |
	D^* (ns)	Temp (C)	Delay (ns)	Lkg (uW)	Delay (ns)	Lkg (uW)	v_{bn} (V)	v_{bp} (V)	Delay (ns)	Lkg (uW)	v_{bn} (V)	v_{bp} (V)	Delay (ns)	Lkg (uW)	v_{bn} (V)	v_{bp} (V)	
C17	0.067	50	0.070	0.067	0.067	0.167	0.1	0.3	0.067	0.159	0.11	0.27	0.067	0.167	0.1	0.3	1.50
C17	0.067	75	0.073	0.158	0.067	0.759	0.4	0.5	0.067	0.811	0.44	0.50	0.067	0.89	0.5	0.5	4.00
C432	0.902	50	0.941	4.78	0.897	11.5	0.2	0.2	0.902	8.58	0.13	0.13	0.907	9.53	0.1	0.2	0.51
C432	0.902	75	0.986	11.2	0.897	47.8	0.4	0.4	0.902	46.1	0.36	0.42	0.902	47.8	0.4	0.4	1.63
C880	0.763	50	0.801	2.90	0.757	8.09	0.2	0.3	0.763	6.83	0.16	0.24	0.757	8.09	0.2	0.3	0.52
C880	0.763	75	0.838	6.85	0.763	37.5	0.5	0.5	0.763	37.6	0.49	0.44	0.763	37.5	0.5	0.5	3.11
C1355	0.83	50	0.841	5.06	0.825	15.7	0.2	0.3	0.83	12.8	0.17	0.24	0.825	15.7	0.2	0.3	0.55
C1355	0.83	75	0.879	11.9	0.825	72.2	0.5	0.5	0.83	69.2	0.50	0.50	0.825	72.2	0.5	0.5	3.10
C3540	1.33	50	1.39	16.0	1.32	31.30	0.2	0.1	1.33	28.4	0.19	0.08	1.32	31.3	0.2	0.1	0.41
C3540	1.33	75	1.45	37.50	1.33	135	0.3	0.4	1.33	136	0.37	0.32	1.33	136	0.4	0.3	0.89
C5315	1.20	50	1.25	14.9	1.19	35.5	0.2	0.2	1.20	30.6	0.13	0.19	1.19	35.5	0.2	0.2	0.42
C5315	1.20	75	1.30	35.0	1.19	144	0.3	0.5	1.20	147	0.40	0.38	1.19	148	0.4	0.4	1.17
C6288	3.64	50	3.82	24.7	3.63	58.7	0.2	0.2	3.64	55.8	0.17	0.19	3.63	58.7	0.2	0.2	0.47
C6288	3.64	75	3.99	57.7	3.61	276	0.4	0.5	3.64	256	0.37	0.46	3.61	276	0.4	0.5	1.75

IV. RESULTS FOR ISCAS BENCHMARK CIRCUITS

In this section, we apply the above described design flow on 7 ISCAS combinational benchmark circuits and present the results obtained. A static timing analyzer (STA) is implemented to determine the delay and leakage of the benchmark circuits. The library consists of 26 gates (10 NOT gates, 5 NAND2 gates, 5 NOR2 gates, 3 NAND3 gates and 3 NOR3 gates) of different sizes, and has been characterized using HSPICE simulations performed using the BPTM 100nm technology [10] with $V_{dd} = 1.0V$. The benchmark circuits have been synthesized based on this library using SIS [12]. Since each ISCAS benchmark is rather small, we consider a test case where all of the benchmarks are placed in different regions of the same chip. Specifically, we assume that each of these benchmarks is in a different WID variational zone, and can be compensated independently of each other.

A. Results of TABB

To determine the optimal amount of TABB required, we assume that there are no process variations, and that the on-chip temperature varies from $25°C$ to $75°C$. We also choose $T = 50°C$ and $T = 75°C$ as the points at which we will determine the optimal bias required to maintain the delay. The results obtained through enumeration as well as mathematically assisted methods are explained below:

1) *Enumeration based TABB:* We assume that the devices can be body biased between the range of $[0V, 0.5V]$ with a step of $0.1V$. A step of $0.1V$ is chosen assuming that this is the lowest resolution of voltages that can be generated by the central body bias generator. Thus, 6 possible voltage levels exist for both v_{bn} and v_{bp}, leading to 36 candidate solutions. The benchmarks are simulated at these points, and the solution that satisfies the delay and has the minimum leakage is chosen as the final optimal solution, based on Algorithm 1. The results are tabulated in Table II.

2) *Mathematically assisted TABB:* At each of the temperature points, the delay of the benchmarks are measured at $v_{bn} = [0V, 0.1V, 0.2V, 0.3V, 0.4V, 0.5V]$ with $v_{bp} = V_{dd}$, and at $v_{bp} = [0V, 0.1V, 0.2V, 0.3V, 0.4V, 0.5V]$ with $v_{bn} = 0$. Leakage values are measured [4] at $v_{bn}=[0V, 0.25V, 0.5V]$ and $v_{bp}=[0V, 0.25V, 0.5V]$. Second order best fit expressions for delay and leakage are obtained as outlined in Section III-A. The NLPP is formulated, as explained in (10) and (11), and the solution obtained for different temperatures is tabulated in Table II. When the NLPP solutions are snapped to points in the discrete solution space, three options exist namely:

a) Snap both v_{bn} and v_{bp} to the next higher voltage.

b) Snap v_{bn} to the next higher voltage while v_{bp} to the nearest lower voltage.

c) Snap v_{bp} to the next higher voltage while v_{bn} to the nearest lower voltage.

The delay and leakage of these three points are compared and the best solution is chosen. The results after snapping are also shown in Table II.

It can be seen from the table that all benchmarks require FBB at higher operating temperatures to compensate for the increase in delay due to reduction in mobilities. Further, most of the NLPP solutions when snapped to the nearest discrete voltage levels give solutions which are identical to that obtained by enumeration. However, for C17 at $T = 75°C$, mathematically assisted TABB returns a solution which is one grid higher for v_{bn} as compared with enumeration due to slight inaccuracies in the delay-leakage model. Due to the same reason, for C432 at $T = 50°C$, mathematically assisted TABB returns a solution which is better than enumeration (but does not meet the delay requirement when back-annotated using STA). Similarly, solutions for C3540 and C5315 at $T = 75°C$ are slightly inferior than the corresponding values obtained through enumeration. The final column in Table II compares the run-time for the two implementations measured on a Linux workstation with a 2.8GHz Pentium CPU. While it can be seen that mathematically assisted TABB is approximately two times faster than enumeration based TABB at $T = 50°C$ (with the exception of the smallest benchmark C17), enumeration outperforms the former for most benchmarks at $T = 75°C$. This is due to the fact that fewer bias pairs at $T = 75°C$ satisfy the delay requirement, and hence the number of candidate solutions for enumerating is quite low. (At $T = 75°C$, only $(v_{bn}, v_{bp}) = (0.5, 0.5)$ satisfies the delay requirement for C17, C880 and C1355, and hence enumeration is more than three times faster than mathematically assisted TABB.) However, at $T = 50°C$, the search space for enumeration based TABB increases, and significant speed-up is obtained by the other method.

B. Results of PABB

While PABB is actually performed through post-silicon tuning, we perform the same using statistical simulations to get an overview of the nature of results obtainable by our method. The test structures to compensate for process variations in each WID variational zone are assumed to consist of a simple ring oscillator (RO) circuit. Simulations are performed on the ring oscillator using BPTM 100nm. model files [10]. The delay of the RO simulated at $V_{dd} = 1.0V$ and $T = 25°C$ with nominal threshold voltage values ($V_{tn0} = 0.2607V$ and $V_{tp0} = -0.303V$) is $151ps$, while the leakage power

---

[3] The delay and leakage numbers reported are the STA values obtained after back-annotating the bias voltages.

[4] A minimum of 9 points is required for the leakage interpolation.

is $5.253nW$. We wish to maintain the delay of the RO at this value, denoted by D^*, despite process variations.

The effects of process variations on transistor threshold voltages are quantized using Gaussian distributions for V_{tn0} and V_{tp0}. For simplicity, it is assumed that the statistical distribution of transistor threshold voltages in each WID variational region is the same. This simplification helps us to perform Monte Carlo simulations with one set of Gaussian distribution parameters for transistor threshold voltages, and use the results over all benchmarks. In order to obtain an estimate of the yield without adaptive body bias, Monte Carlo simulations are performed on this ring oscillator with 50 runs at each temperature. If the delay of the RO does not meet the target value, it is assumed that the die fails to meet the delay requirement. The number of dies that satisfy the delay requirement at each temperature is shown in Fig. 4. It can be seen from the figure that only about 50% of the dies are acceptable at room temperature, and this number steadily decreases with increase in temperature. This is attributed to changes in threshold voltages caused by process variations, thereby necessitating compensation using PABB.

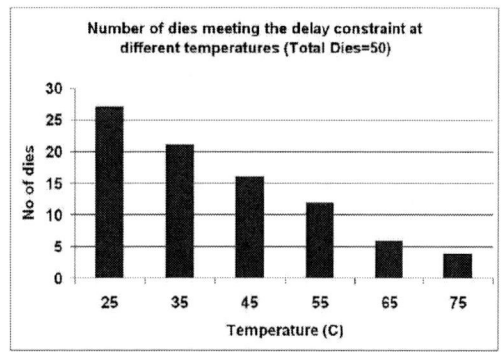

Fig. 4. **Yield at different temperatures for the die**

In order to determine the PABB voltages for each die, the delay and leakage distributions of the test-structure are characterized based on the method described in Section III D. The delay and leakage values with body biasing are measured through simulations, and second order polynomials, as indicated in (10) and (11) are obtained. The NLPP is formulated and solved for each die to determine its optimal bias. All 50 dies have been successfully biased. 42 dies require RBB for PMOS and FBB for NMOS while 6 dies require RBB for both NMOS and PMOS and the remaining 2 require FBB for both NMOS and PMOS. Most dies need FBB for PMOS to increase the speed and RBB for NMOS to minimize the leakage. This is consistent with the observation made by the authors in [1].

The PABB values can be combined with the TABB data obtained from the previous sub-section to determine the PTABB values required for each benchmark at each operating temperature, according to (12). The amount of dies which meet the delay requirement at $T = 50°C$ and $T = 75°C$ for the benchmark circuits and the nature of bias required is shown in Table III. Although 100% of the dies cannot be recovered at $T = 75°C$, the yield can be significantly improved.

V. CONCLUSION

Temperature variations and process variations in nanometer-scale devices can cause the delay and leakage of dies to vary significantly. Bidirectional Adaptive Body Bias can be used to improve the yield of dies for reasonable ranges of operating temperatures. We propose an algorithm to compute the exact amount of body bias required to perform run-time compensation to counter thermal variations. We determine these bias values during the design stage using mathematical models and thereby eliminate the need for complex on-chip circuitry to monitor delay and leakage. We also present a unique methodology

TABLE III
PTABB COMPENSATION FOR ISCAS BENCHMARKS

Bench mark	Temp (C)	Accepted Dies	P-FBB N-RBB	P-RBB N-FBB	P-FBB N-FBB
C17	50	50	27	0	23
C17	75	30	0	0	30
C432	50	50	35	0	15
C432	75	38	0	0	38
C880	50	50	24	0	26
C880	75	27	0	0	27
C1355	50	50	24	0	26
C1355	75	24	0	0	24
C3540	50	50	0	4	46
C3540	75	46	0	0	46
C5315	50	50	29	0	21
C5315	75	38	0	0	38
C6288	50	50	27	0	23
C6288	75	39	0	0	39

to decouple the effects of process and temperature variations. We use ABB to counter these individually, and finally combine the values effectively to achieve compensation under all operating conditions. The results indicate that ABB can be used as a successful means to combat both process and temperature variations and improve the performance of gigascale LSI systems.

REFERENCES

[1] T. Chen and S. Naffziger. "Comparison of adaptive body bias (ABB) and adaptive supply voltage (ASV) for improving delay and leakage under the presence of process variation". In *IEEE Transactions on Very Large Scale Integration (VLSI) Systems*, pages 888–899, October 2003.

[2] C. H. Wann, H. Chenming, K. Noda, D. Sinitsky, F. Assaderaghi, and J. Bokor. "Channel doping engineering of MOSFET with adaptable threshold voltage using body effect for low voltage and low power applications". In *International Symposium of VLSI Technology*, pages 159–163, June 1995.

[3] T. Kuroda, T. Fujita, S. Mita, T. Nagamatu, S. Yoshioka, F. Sano, M. Norishima, M. Murota, M. Kako, M. Kinugawa, M. Kakumu, and T. Sakurai. "A 0.9 V 150 MHz 10 mW 2-D discrete cosine transform core processor with variable-threshold-voltage scheme". In *IEEE International Solid-State Circuits Conference*, pages 166–167, August 1996.

[4] J. W. Tschanz, J.T. Kao, S. G. Narendra, R. Nair, D.A. Antoniadis, A.P. Chandrakasan, and V. De. "Adaptive body bias for reducing impacts of die-to-die and within-die parameter variations on microprocessor frequency and leakage". *IEEE Journal of Solid-State Circuits*, 37(11):1396–1402, November 2002.

[5] J. W. Tschanz, S. G. Narendra, Y. Ye, B. A. Bloechel, S. Borkar, and V. De. "Dynamic sleep transistor and body bias for active leakage power control of microprocessors". *IEEE Journal of Solid-State Circuits*, 38(11):1838–1845, November 2003.

[6] J. W. Tschanz, S. G. Narendra, R. Nair, and V. De. "Effectiveness of adaptive supply voltage and body bias for reducing impact of parameter variations in low power and high performance microprocessors". *IEEE Journal of Solid-State Circuits*, 38(5):826–829, May 2003.

[7] S. Narendra, A. Keshavarzi, B. A. Bloechel, S. Borkar, and V. De. "Forward body bias for microprocessors in 130-nm technology generation and beyond". *IEEE Journal of Solid-State Circuits*, 38(5):696–701, May 2003.

[8] G. Ono, M. Miyazaki, H. Tanaka, N. Ohkubo, and T. Kawahara. "Temperature referenced supply voltage and forward-body-bias control (TSFC) architecture for minimum power consumption". In *European Solid State Circuits Conference*, pages 391–394, September 2004.

[9] T. Kuroda. "Low power CMOS digital design for multimedia processors". In *International Conference for VLSI and CAD*, pages 359–367, October 1999.

[10] Device Group at UC Berkeley. "Berkeley Predictive Technology Model", 2002. Available at http://www-device.eecs.berkeley.edu/~ptm/.

[11] H. Ananthan, C. H. Kim, and K. Roy. "Larger-than-Vdd forward body bias in sub-0.5V nanoscale CMOS". In *International Symposium on Low Power Electronic Design*, pages 8–13, 2004.

[12] E. M. Sentovich, K. J. Singh, L. Lavagno, C. Moon, R. Murgai, A. Saldanha, H. Savoj, P. R. Stephan, R. K. Brayton, and A. Sangiovanni-Vincentelli. "SIS: A system for sequential circuit synthesis". Technical Report UCB/ERL M92/41, 1992. Available at http://www-cad.eecs.berkeley.edu/research/sis.

2006 Asia and South Pacific Design Automation Conference

6A-2

Analysis and Optimization of Gate Leakage Current of Power Gating Circuits*

Hyung-Ock Kim   Youngsoo Shin

Dept. of Electrical Engineering
Korea Advanced Institute of Science and Technology (KAIST)
Daejeon 305-701, Korea
ppc750@kaist.ac.kr, youngsoo@ee.kaist.ac.kr

Abstract— Power gating is widely accepted as an efficient way to suppress subthreshold leakage current. Yet, it suffers from gate leakage current, which grows very fast with scaling down of gate oxide. We try to understand the sources of leakage current in power gating circuits and show that input MOSFETs play a crucial role in determining total gate leakage current. It is also shown that the choice of a current switch in terms of polarity, threshold voltage, and size has a significant impact on total leakage current. From the observation of the importance of input MOSFETs, we propose the power optimization of power gating circuits through input control.

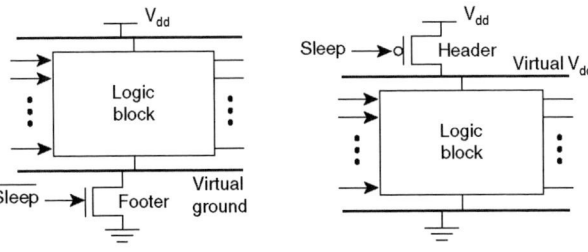

Fig. 1. Power gating circuits.

I. INTRODUCTION

With CMOS technology scaling, subthreshold leakage current experiences an exponential growth mainly due to the scaling down of the threshold voltage. Many circuit level approaches have been proposed including dual threshold CMOS [1], variable threshold CMOS [2], input vector control [3], power gating [4], and so on. Power gating is shown to be especially efficient and conceived as a main circuit technique by many semiconductor companies [5], [6].

Power gating is realized by using a current switch, which can be either nMOSFET (footer) or pMOSFET (header), as shown in Fig. 1. When the power management unit (not shown in Fig. 1) detects the sufficiently long period of idle time, a sleep signal is applied to the gate of the current switch to disconnect V_{dd} or ground from a logic block. When it detects the logic block to be requested, the sleep signal is de-asserted to connect the logic block to the power rail. The logic block usually employs low V_t to sustain its performance, while high V_t is used for the current switch to suppress its subthreshold leakage current. Nevertheless, the same low V_t can be used for the current switch with the advantage of using single V_t on a chip and of using the switch of smaller size, but at the cost of increased subthreshold leakage current [5].

While being efficient to suppress subthreshold leakage current, power gating circuits suffer from *gate leakage current*, which grows very fast with scaling down of gate oxide. In fact, gate leakage current is expected to exceed subthreshold leakage current when channel length is reduced to 60nm [7]. There are two kinds of gate leakage current depending on the bias condition of MOSFET: forward biased one and reverse biased one [8]. The former mainly consists of gate-to-channel tunneling current, which flows from gate to source/drain through channel. The latter is mainly composed of edge-direct-tunneling current, which flows from source/drain to gate through source-drain extension. The reverse biased gate leakage current is less than the forward biased one: e.g. about 30% less in our experiment with 45nm predictive technology model [9] [10] when the source and drain voltage levels are equal, and the gate has opposite voltage level.

Once in sleep mode, i.e. the current switch is turned off, the drain voltage of the current switch becomes close to either V_{dd} (footer) or ground (header). Since all internal node voltages of the logic block are close to V_{dd} or ground as well, the overall leakage current is determined by subthreshold and gate leakage current of the current switch [11]. Gate leakage current is dominant when high V_t is used for the current switch to suppress its own subthreshold leakage current. Since gate leakage current of pMOSFET is an order of magnitude smaller than that of nMOSFET, header is preferred in view of total leakage current [11].

However, this is not true in practice, since the primary inputs of most power gating circuits maintain their logic states even when sleep. This is because logic blocks that drive them may

---
*This work was supported by Samsung Electronics.

0-7803-9451-8/06/$20.00 ©2006 IEEE.

565

6A-2

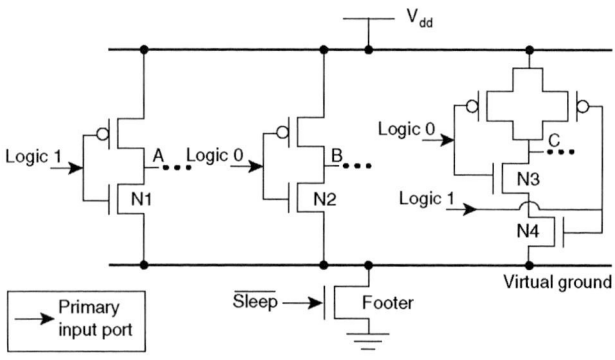

Fig. 2. Power gating circuit with a footer.

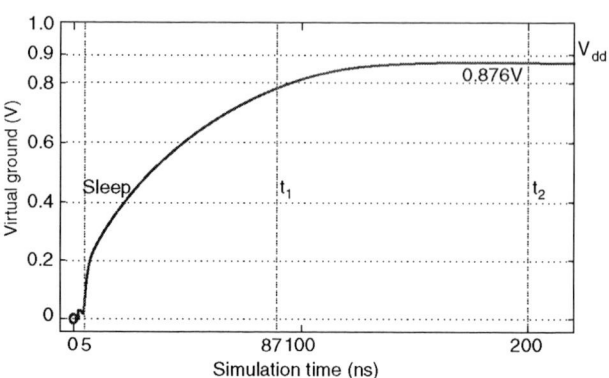

Fig. 3. The trace of virtual ground with a footer turned-off.

not exploit power gating (thus, always maintain logic) or may be in active. Since the sources and drains of the input transistors, which are internal nodes of the logic block, are biased to potential close to either V_{dd} or ground, there are reverse biased or forward biased gate leakage current depending on the bias condition of gate inputs.

In this paper, we study the sources of leakage current in power gating circuits and show that the input transistors play a crucial role in determining overall leakage current. It is also shown the choice of a current switch in terms of V_t voltage and size for optimization of total leakage current of a power gating circuit in idle state. From the observation of the importance of the input transistors, we propose the power optimization of power gating circuits through input control.

The remainder of the paper is organized as follows. In the next section, we study the sources of leakage current in power gating circuits, and address the problem of current switch design. In section III, we present power optimization of power gating circuits through input control and draw conclusions in section IV.

II. GATE LEAKAGE CURRENT AND CURRENT SWITCH DESIGN OF POWER GATING CIRCUITS

A. Mechanism of Gate Leakage Current in Power Gating Circuits

Fig. 2 shows a power gating circuit with a footer. When the circuit is in active (the footer is turned on), the voltage of virtual ground is close to ground, which is determined by the size of the footer and the current that flows into the footer. Once in sleep (the footer is turned off), virtual ground slowly goes up until it reaches a steady state potential, which is close to V_{dd}. The steady state potential and the time it takes to reach are determined by the amount of current flows in the logic block and in the footer. Fig. 3 shows how virtual ground reaches a steady state for one of ISCAS benchmark circuit, called C5315. It takes 82ns ($t_1 - 5$) after the footer is turned off for virtual ground to reach 90% of the steady state (t_2) voltage of 876mV, which is 24mV less than V_{dd}.

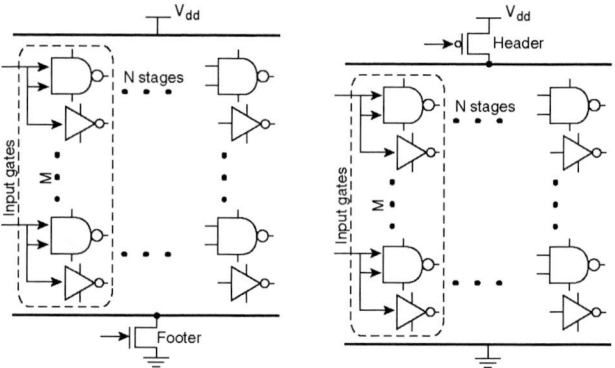

Fig. 4. Simulation circuits for measuring input gate leakage current.

Once in steady state, the footer induces reverse biased gate leakage current due to its bias condition. Also, nMOSFETs connected with primary inputs are another sources of gate leakage current (input gate leakage current). Assuming that the primary inputs are driven by logic blocks that are in active, N2 induces large reverse biased gate leakage current (70% of forward biased one) due to the large voltage difference between its gate and source/drain. Note that reverse biased gate leakage current of pMOSFET is negligible. N1 induces gate leakage current as well, but it is negligible compared to that of N2, since the voltage difference between the gate and source/drain is small. In the case of nMOSFET stack (N3 and N4), N3 induces reverse biased gate leakage current while the gate leakage current in N4 is negligible.

B. Impact of Input Gate Leakage Current

The impact of input gate leakage current is determined by the number of transistors that are connected to primary inputs. Fig. 4 shows the simulation circuits to study the influence of input transistors on total leakage currents. The logic block consists of two types of gate chains: inverter chain and 2-input nand chain. [1] We vary the number of each type of gate chains, denoted as M, and the number of stages of gate chains, denoted

566

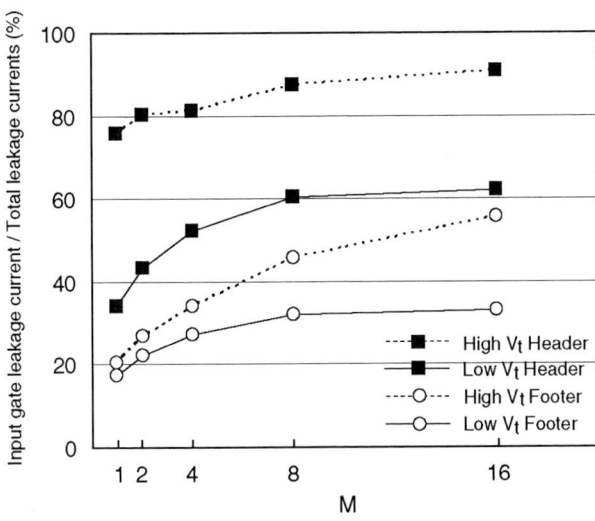

Fig. 5. Input gate leakage current (T=27°C, low V_t = 220mV, high V_t = 400mV).

TABLE I
POWER EFFICIENT HIGH V_t SWITCH

Tech. (nm)	Delay Penalty	M			
		2	4	8	16
45	3%	Header	Header	Footer	Footer
	6%	Header	Header	Footer	Footer
65	3%	Header	Header	Header	Footer
	6%	Header	Header	Footer	Footer

C. Design of Power Optimal Current Switch

If we neglect input gate leakage currents, a header is preferred since the gate leakage current of pMOSFET is an order of magnitude smaller than that of nMOSFET [11]. However, as we discussed in the previous subsection, input gate leakage current can take up large portion of total leakage currents, and this depends on many factors: polarity (footer or header) and V_t of a current switch, delay penalty that affects the size of a current switch, the percentage of input transistors in a logic block, logic states of primary inputs when sleep, and temperature. The choice of a current switch (polarity, V_t, and size) thus lends itself to a non-obvious problem.

Specifically, in power-gated circuits with a footer, leakage current consists of subthreshold and reverse biased gate leakage current of a footer and reverse biased input gate leakage currents (for nMOSFETs whose gates are driven by primary inputs of logic 0). For power-gated circuits with a header, main leakage components are subthreshold leakage current of a header and forward biased input gate leakage currents (for nMOSFETs whose gates are driven by primary inputs of logic 1). Thus, in terms of input gate leakage current, a header is inferior to a footer since forward biased gate leakage current is larger than reverse biased one. Note that, though, this depends on input patterns, and gate leakage current of a header is negligible compared to that of a footer.

We use the same circuits in Fig. 4 to gain an understanding of the design of power optimal current switch. We change the number of gate chains, get a suitable size of a current switch depending on its polarity and delay penalty that can be tolerated, apply the randomly generated input patterns such that 1's and 0's are equally probable, and see which polarity is better for a current switch. TABLE I shows the results for 45nm and 65nm predictive technology models.

It can be seen that a header is preferred when the number of primary inputs are small. This can be understood because a footer is superior to a header in terms of input gate leakage current (reverse biased vs. forward biased), but a header is superior in terms of total leakage current of a switch itself. Thus, the benefit of a largely sized header outweighs the disadvantages of input gate leakage current. We repeat the same experiment with 65nm technology and observe the same result. We also perform the experiment with one of ISCAS benchmark circuits, called C5315, whose input transistors occupy 9.2% of total transistors. A header is recommended when a switch is

as N, while keeping the number of inverters and nand gates to 48 respectively (i.e. there are M-inverter chains and M-nand chains with each chain consisting of N gates while keeping $M \times N = 48$). We assign logic 0 to all primary inputs when a footer is used as a current switch, since it maximizes input gate leakage current (recall the previous subsection). Similarly we assign logic 1 to all primary inputs in the case of a header. The size of the switch is determined such that the delay of the logic block increases less than 3% compared to non-power-gated circuit.

Fig. 5 is the simulation result with 45nm predictive technology model, and shows that input gate leakage current can be a large portion of total leakage currents when power gating circuits are in sleep. As an example, input gate leakage current is about 34% of total leakage currents when a high V_t footer is used and the input transistors are 8.3%[2] ($M = 4$) of total transistors. This goes up to 45% when the input transistors get twice ($M = 8$). When we use low V_t for a footer (e.g. when dual V_t is not available or when we want to reduce the size of a footer while keeping the same performance), subthreshold leakage current increases substantially due to its exponential dependency on threshold voltage. This increases the total leakage currents, while input gate leakage current remains almost constant. Thus, the portion of input gate leakage current gets smaller as is evident in Fig. 5. If we use a header instead of a footer, input gate leakage current is even more important since gate leakage current of a pMOSFET switch (header) is an order of magnitude lower than that of an nMOSFET switch (footer).

---

[1]Gate leakage current is different for different types of gates, and we use inverter and 2-input nand gate to reflect gate-wise variation.

[2]Only nMOSFETs effectively contribute to input gate leakage current. Thus, if we consider total width of only input nMOSFETs, the percentage goes down (e.g. 2.9% for $N = 4$).

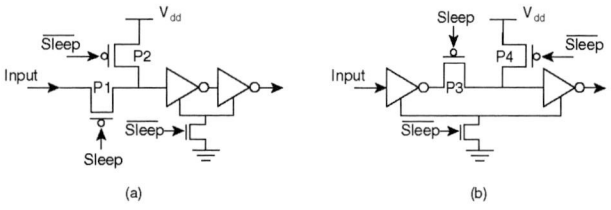

Fig. 6. Proposed circuits for input control.

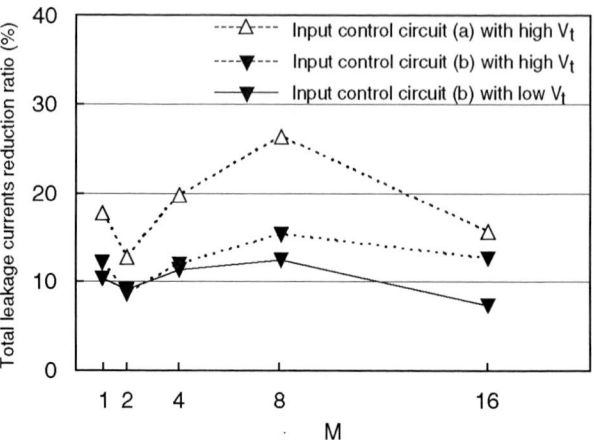

Fig. 7. Leakage current reduction with input control circuit.

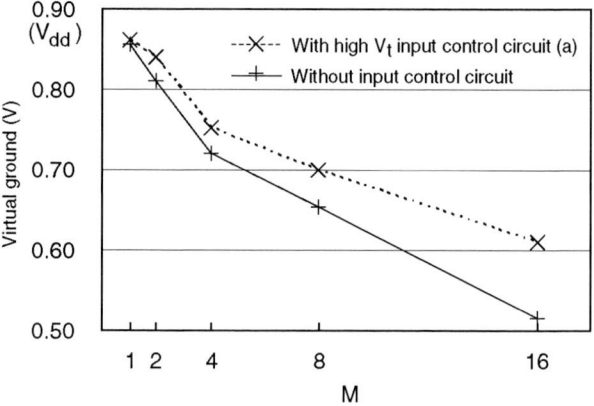

Fig. 8. Virtual ground voltage.

sized such that 3% delay penalty can be tolerated, but a footer is superior when delay penalty exceeds 6%.

When low V_t is used for a switch, a footer is always better than a header. This is because the subthreshold leakage current of a switch now takes up large portion of total leakage currents due to its exponential dependency on the threshold voltage, and the subthreshold leakage current is smaller in a footer than in a header due to the smaller size of a footer with the same delay penalty. Thus, a footer is superior to a header in terms of its own leakage current as well as in terms of input gate leakage current.

III. POWER OPTIMIZATION THROUGH INPUT CONTROL

Once the current switch is designed based on the method presented in the previous section, the power consumption of power gating circuits when sleep can be further reduced by controlling primary inputs. By providing logic 1 to *all* primary inputs of a power-gated circuit with a footer (similarly logic 0 in the case of a header), input gate leakage currents can be virtually eliminated.

Since the extra logic for input control can be an overhead in terms of power consumption, area, and delay, we propose two types of efficient circuits as shown in Fig. 6. In the circuit of Fig. 6(a), the input is transferred through P1 and two buffers in normal operation (i.e. when sleep=0). When sleep, the input is de-coupled to the buffers and the gate of the first buffer is driven by logic 1, which almost eliminates the gate current of the buffer. The original input transistors that are now connected to the second buffer do not induce gate currents any more, since their inputs are not directly driven by primary inputs. The main leakage component of the circuit in the sleep mode is the subthreshold leakage current of P1 when the primary input is 0, which can outweigh the advantage of input control especially when the width of P1 is large. However, this can be further reduced if high V_t is available. If we move the first buffer in front of P3 as shown in Fig. 6(b), the gate leakage current of the first buffer is now the main component of leakage current, which is in general less than the low V_t subthreshold leakage current of Fig. 6(a). The first buffer, minimum-size inverter, reduces total width of input transistors connected to primary inputs. If we replace P2 and P4 with nMOSFETs and connect their sources to ground instead of V_{dd}, we have input control circuits for a power gating circuit with a header.

We use the same circuits in Fig. 4 to see the efficiency of the proposed input control circuits. We insert the input control circuit at each primary input, simulate the entire circuit (thus, the overhead of extra logic is included) to obtain the total leakage currents, and then compare them to the total leakage currents of the original circuit. Fig. 7 shows the result in the case of a footer. When high V_t is available, input control circuit in Fig. 6(a) is more efficient than that in Fig. 6(b). However, if we use only low V_t, the opposite is true.

Input gate leakage currents can be virtually eliminated by the proposed circuits. However, gate leakage current of a footer goes up due to elevated voltage of a virtual ground. This is because the extra MOSFETs in input control circuits induce current that flows through a footer in addition to the current from the logic block. Fig. 8 shows the virtual ground voltage in steady state, which clearly shows the elevated voltage with input control circuits.

We perform another experiment with a set of circuits extracted from ISCAS benchmarks and 64-b carry look-ahead adder (CLA). The current switches are sized with average cur-

568

TABLE II
LEAKAGE CURRENT REDUCTION WITH INPUT CONTROL CIRCUIT

	Low V_t footer	High V_t footer	Low V_t header	High V_t header
C1908	2.0%	11.4%	-3.3%	35.3%
C2670	4.7%	22.6%	-17.1%	45.9%
C3540	7.1%	27.1%	1.1%	61.0%
C5315	8.3%	30.8%	-5.7%	52.0%
C7552	2.1%	11.9%	-8.3%	33.5%
64-b CLA	-10.2%	1.6%	-7.8%	43.7%
avg.	2.8%	19.1%	-7.4%	47.0%

rent methods [12]. For each circuit, we insert input control circuit in Fig. 6(a) when high V_t current switch is used (thus, high V_t is available), and the circuit in Fig. 6(b) otherwise. We apply randomly generated input patterns such that 1's and 0's are equally probable, and see how much leakage current can be reduced by the proposed method. TABLE II shows the result. Significant leakage reduction is observed when high V_t header or footer is used in conjunction with input control circuits in Fig. 6(a). Total leakage current increases rather than decreases in the case of low V_t header due to the overhead of input control circuits.

We measure the delay and the area of the proposed circuits. In the case of Fig. 6(a) with low V_t, the propagation delay for rising and falling inputs are 0.25ns and 0.34ns respectively. The rising and falling delays are not balanced due to the presence of pass gate (P1). The circuit of Fig. 6(b) has delays of 0.84ns and 0.65ns for rising and falling inputs respectively when all MOSFETs are in low V_t. Total width of MOSFETs in each circuit is 1.2μm. The presence of input control circuits may or may not be an overhead. As an example, 64-b CLA in TABLE II exhibits the delay of 1.02ns and the MOSFET total width of 654.7μm, which can be comparable to those of input control circuit. Larger circuits (e.g. C3540) may accommodate the proposed input control circuits, especially when they do not include timing critical paths inside.

IV. CONCLUSION

We study the influence of input gate leakage current of power gating circuits, which are widely used as a solution to suppress standby currents. It is shown that the input gate leakage current is indeed an important factor in determining the total leakage currents and varies widely depending on the polarity, threshold voltage, and size of current switch. This implies the importance of current switch design, which we also address in the paper. Finally, we propose the method of input control to reduce input gate leakage current, and propose new circuits to realize the proposed method.

REFERENCES

[1] L. Wei, Z. Chen, M. Johnson, K. Roy, and V. De, "Design and optimization of low voltage high performance dual threshold CMOS circuits," *Proc. Design Automat. Conf.*, June 1998, pp. 489-494.

[2] T. Kuroda, T. Fujita, S. Mita, T. Nagamatu, S. Yoshioka, F. Sano, M. Norishima, M. Murota, M. Kako, M. Kinugawa, M. Kakumu, and T. Sakurai, "A 0.9V 150MHz 10mW 4mm^2 2-D discrete cosine transform core processor with variable-threshold-voltage scheme," *Proc. Int'l Solid-State Circuits Conf.*, Feb. 1996, pp. 166-167.

[3] M. C. Johnson, D. Somasekhar, and K. Roy, "Models and algorithms for bounds on leakage in CMOS circuits," *IEEE Trans. on Computer-Aided Design*, vol. 18, no. 6, pp. 714-725, June 1999.

[4] S. Mutoh, T. Douseki, Y. Matsuya, T. Aoki, S. Shigematsu, and J. Yamada, "A 1-V power supply high-speed digital circuit technology with multithreshold-voltage CMOS," *IEEE Journal of Solid-State Circuits*, vol. 30, no. 8, pp. 847-854, Aug. 1995.

[5] S. V. Kosonocky, M. Immediato, P. Cottrell, and T. Hook, "Enhanced multi-threshold (MTCMOS) circuits using variable well bias," *Proc. Int'l Symp. on Low Power Electronics and Design*, Aug. 2001, pp. 165-169.

[6] H.-S. Won, K.-S. Kim, K.-O. Jeong, K.-T. Park, K.-M. Choi, and J.-T. Kong, "An MTCMOS design methodology and its application to mobile computing," *Proc. Int'l Symp. on Low Power Electronics and Design*, Aug. 2003, pp. 110-115.

[7] N. Sirisantana and K. Roy, "Low-power design using multiple channel lengths and oxide thicknesses," *IEEE Design & Test of Computers*, vol. 21, no. 1, pp. 56-63, Jan.-Feb. 2004.

[8] C.-H. Choi, K.-Y. Nam, Z. Yu, and R. W. Dutton, "Impact of gate direct tunneling current on circuit performance: a simulation study," *IEEE Trans. on Electron Devices*, vol. 48, no. 12, pp. 2823-2329, Dec. 2001.

[9] Nanoscale Integration and Modeling Group, "45nm BSIM4 model cards," *http://www.eas.asu.edu/~ptm/*.

[10] Y. Cao, T. Sato, D. Sylvester, M. Orshansky, and C. Hu, "New paradigm of predictive MOSFET and interconnect modeling for early circuit design," *Proc. Custom Integrated Circuits Conf.*, Sep. 2000, pp. 201-204.

[11] F. Hamzaoglu and M. R. Stan, "Circuit-level techniques to control gate leakage for sub-100nm CMOS," *Proc. Int'l Symp. on Low Power Electronics and Design*, Aug. 2002, pp. 60-63.

[12] S. Mutoh, S. Shigematsu, Y. Gotoh, and S. Konaka, "Design method of MTCMOS power switch for low-voltage high-speed LSIs," *Proc. Asia South Pacific Design Automat. Conf.*, Jan. 1999, pp. 113-116.

Delay Modeling and Static Timing Analysis for MTCMOS Circuits

Naoaki Ohkubo   Kimiyoshi Usami

Graduate School of Engineering, Shibaura Institute of Technology

307 Fukasaku, Munuma-ku, Saitama, 337-8570 Japan

E-mail: {m105021, usami}@sic.shibaura-it.ac.jp

Abstract - One of the critical issues in MTCMOS design is how to estimate a circuit delay quickly. In this paper, we propose a delay modeling and static timing analysis (STA) methodology targeting at MTCMOS circuits. In the proposed method, we prepare a delay look-up table (LUT) consisting of the input slew, the output load capacitance, the virtual ground length, and a power-switch size. Using this LUT, we compute a circuit delay for each logic cell by applying the linear interpolation. Experimental results show that the proposed methodology enables to estimate the critical path delay in a good accuracy.

Key words: MTCMOS, Selective-MT, Delay, Static timing analysis, Leakage power, Interpolation.

I. Background

As the transistor technology gets advanced, low-power design techniques become significantly important. In particular, leakage power reduction is strongly required in the LSI for portable information devices to prolong the battery life. In addition, high performance is also needed in the recent cell phones with multimedia capabilities.

Papers have been reported so far describing techniques to reduce standby leakage current while maintaining high performance. A Multiple-Threshold CMOS (MTCMOS) [1] is a well-known technique that efficiently reduces the standby leakage power. Figure 1 shows the circuit structure of an MTCMOS circuit. MTCMOS has low-Vth logic gates and hign-Vth power switches. These elements are connected each other by the wire named virtual ground (VGND) line. The low-Vth logic gates operate at high speed by turning on the power switch in the active mode, while in the standby mode high-Vth transistor cuts the leakage current by turning off the power switch.

As an improved MTCMOS methodology, the selective MT technique is presented in [2]. Selective MT applies MTCMOS technique only to the critical path and does not share the virtual ground line, as shown in Figure 2. Low-Vth gates are only applied in the critical path and each low-Vth gate is connected to the power switch individually. Since high-Vth gates are applied to the non-critical paths, it is also effective in the reduction of active leakage power. To downsize the

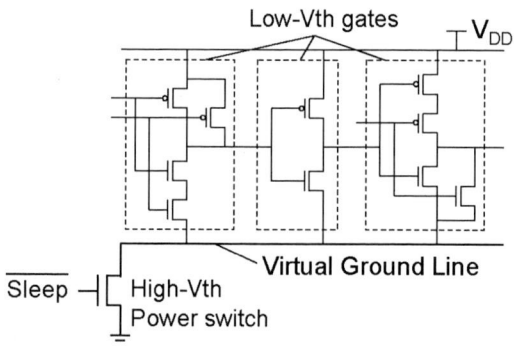

Figure 1: MTCMOS circuit

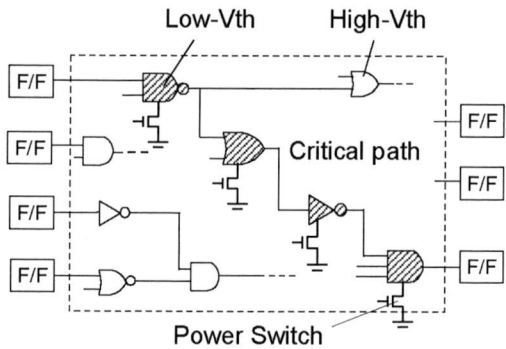

Figure 2: Selective MT technique

power switch area overhead in the selective MT design, an approach that shares the virtual ground line has been also proposed in [3]. This methodology can improve an area overhead and efficiency of leakage power reduction compared with the conventional selective MT technique.

Next, we discuss the problems in MTCMOS. One of the critical issues is the delay increase due to the voltage fluctuation of the virtual ground line [4]. When logic cells and flip-flops switch and discharge its output load capacitance, flowing the discharge current gets difficult because of the wire resistance of virtual ground and the resistance of power switches. These factors trigger the fluctuation of virtual ground and influence the circuit delay. The impact affected by this event is different if the virtual ground is shared or not. In the shared design, circuit delay increases depending on the discharge pattern of the logic gates. In general, it is necessary to resize the power switch to avoid this delay increase. Therefore, in this case, the power-switch size needs to be optimized considering the gate discharge patterns and the delay of the entire gates sharing the same virtual ground. Meanwhile, the unshared design is able to avoid the delay increase caused by the discharge current of the other logic gates. Although this will enable to optimize the power-switch size independently, area overhead is larger than the shared design.

Due to the fact presented above, timing verification gets difficult in MTCMOS design. To guarantee the circuit timing, timing analysis is essential considering the delay increase due to the resistance of the virtual ground and the power switch. Conventionally, the delay look-up table (LUT) consisting of the input slew and the output load capacitance is generally used in the STA methodology.

However, in the design applying the MTCMOS technique, we are not able to estimate delay increase by using the conventional STA methodology because the virtual ground length and the power-switch size affect the fluctuation of the virtual ground. This makes MTCMOS circuit design more difficult.

In this paper, we propose a delay modeling and static timing analysis (STA) methodology targeting at MTCMOS circuits. The proposed method enables to estimate the critical path delay by simply extending the conventional STA methodology.

The rest of this paper is organized as follows: Section 2 presents the layout model and delay modeling in MTCMOS circuits. Section 3 presents the proposed STA methodology. Section 4 presents experimental results and Section 5 concludes the paper.

II. Delay modeling for MTCMOS Circuits

A. Physical Implementation of Cell-based Selective MT

This section describes the layout architecture of the selective MT. In this architecture, we assume a cell-based methodology that combines Low-Vth MTCMOS logic cells and high-Vth cells. The MTCMOS logic cell has an extra virtual ground pin in addition to the conventional input and output pins. The virtual ground pin of the MTCMOS logic cell and the power-switch cell are connected by the virtual ground line. The layout architecture based on this implementation is shown in Figure 3. The virtual ground length depends on where the power switch is placed. Since the virtual ground line has wire resistance and capacitance, it is necessary to model the virtual ground. Based on this RC model, we analyze the impact on the circuit delay.

B. Delay Modeling of MTCMOS Circuits

We discuss factors that affect the MTCMOS circuit delay. Generally, a cell delay and an output slew are determined by an input slew and output load capacitance. However, MTCMOS has a virtual ground line and a power switch in addition to logic transistors in the conventional CMOS circuits. A wire resistance and a capacitance exist in virtual ground line and a channel resistance also exists in power switch. Those factors influence a circuit delay and an output slew especially when the output of the gate make a transition to "low". To analyze the impact of these factors, we model the MTCMOS circuit as shown in Figure 4. We investigate how the input slew, the output capacitance, the power-switch size, and the virtual ground length impact on the cell delay and output slew. We conducted analysis by using HSPICE simulations and the Toshiba 90nm device models. The input and the output slew are defined as transition times from 10% to 90% of the VDD. The cell delay is defined as the time from 50% of input to 50% of output. We assume the virtual ground line has the same line width as other inter-cell wires, and model it using the π-type lumped RC circuit. As the number of π-ladders, we chose three. This number is based on the result of the pre-analysis showing that no significant difference was observed between 3 and 4 as the number of π-ladders [5]. We assume that the virtual ground is drawn by using two routing layers. We also assume the wire resistance and capacitance are the same between the two layers. As the

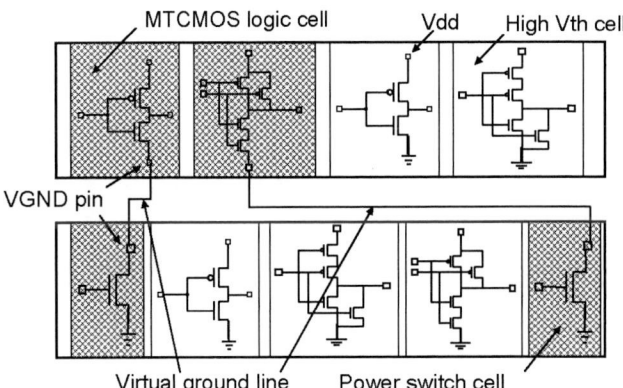

Figure 3: Layout architecture of cell-based selective MT

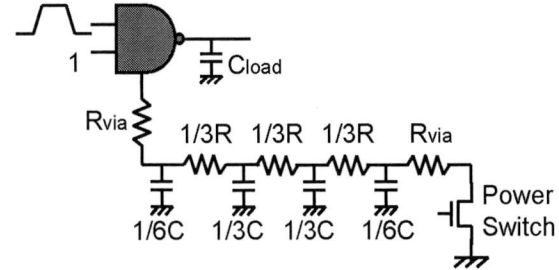

Figure 4: Circuit model for HSPICE simulation (NAND)

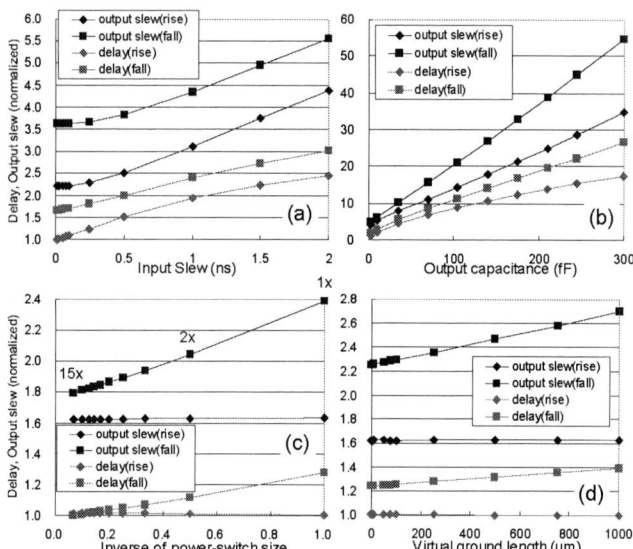

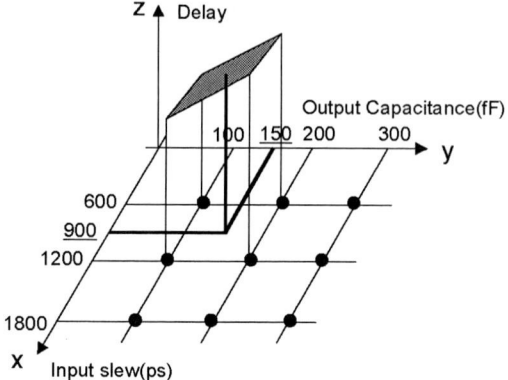

Figure 7: Two-dimensional interpolation example using LUT

negligible. We investigated this impact on 8x size AND cell. The circuit delay increase with the virtual ground length and the inverse of power-switch size, as shown in Figure 6. The additional delay should be taken into consideration in the case of rise transition.

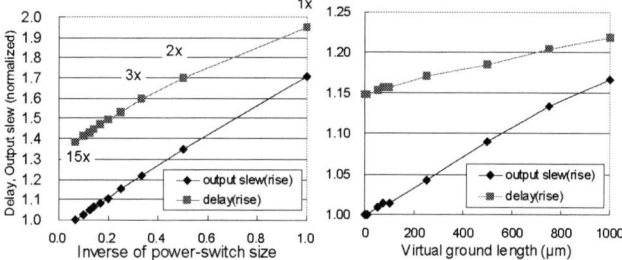

Figure 5: Influence of the delay and the output slew by the parameters (NAND 1x)

Figure 6: Delay increase due to the discharge within the cell (AND 8x)

default values, we chose the input slew of 1000ps, the load capacitance of 70fF, the virtual ground length of 50μm, and the x1 power-switch size. For example, when we investigate how the delay is affected by the input slew, we kept other three parameters as the default values. For the output slew, we analyzed in the same manner as the delay. Figure 5 shows the results of the analyses.

Results show that the delay and the output slew increase almost linearly with the parameters such as input slew, output capacitance and virtual ground. In addition, the cell delay and output slew increase linearly with inverse of power-switch size. In contrast, since the rise transition delay is the charging time for output capacitance, the influence from the virtual ground length and power-switch size is not significant.

However, the situation becomes different in gates such as AND or OR gate consisting of a NAND (or NOR) gate and an inverter inside the cell. For AND gates, when the output goes "high", the output of the internal NAND circuit goes "low". This means the discharge of capacitance occurs when the output of the gate is "high". Especially, since the large AND cell tends to have a large parasitic capacitance at the internal NAND circuit, the influence of this case is not

III. Proposed STA Methodology for MTCMOS

A. Proposed Delay Modeling

In the previous section, we proposed that the virtual ground length and the power-switch size influence the circuit delay, and that delay increases linearly with the virtual ground and the inverse of power-switch size. Based on this observation, we propose a novel delay modeling and STA methodology for MTCMOS design. The proposed method mainly targets at MTCMOS circuit that does not share the virtual ground line.

First, we present the conventional delay modeling. In the conventional delay modeling, the cell delay and the output slew are computed using LUT that consists of the input slew and the output capacitance. LUT has delay values corresponding to the discrete input slew and output capacitance. If given input slew and output capacitance are not in LUT, the circuit delay is estimated by interpolation using the values in LUT. For example, as shown in Figure 7, when the input slew is 900ps and the output capacitance is 150fF, the delay is computed from delay values corresponding to the two nearest input slew (i.e. 600ps and 1200ps) and two nearest output capacitance (i.e. 100fF and 200fF) through interpolation. The interpolation is performed by using the equation (1) below. For given input slew and output capacitance, two nearest values for them are picked up from the LUT and substituted for x and y in the equation (1). The corresponding cell delay is substituted for Z.

$$Z = A + Bx + Cy + Dxy \qquad (1)$$

Since we obtain simultaneous equations for A, B, C, and D, we solve them by applying the Gaussian elimination to obtain the coefficients A, B, C, and D. The same procedure is done for the interpolation of the output slew as well.

In the proposed methodology, the virtual ground length

and the inverse of power-switch size are added as the parameters. This is based on the fact that the cell delay and the output slew linearly increases with the virtual ground length and the inverse of power-switch size as previously presented in Section 2. This trend is well suited to the linear interpolation. In the proposed methodology, we employ LUT consisting of following parameters: the input slew, the output capacitance, the virtual ground length and the power-switch size. By using this LUT, we compute the cell delay and the output slew through the linear interpolation. Extending the equation (1) into the four-dimensional interpolation gives the equa- tion (2) as follows:

$$Z = A + Bv + Cw + Dx + Ey + Fvw + Gvx +$$
$$Hvy + Iwx + Jwy + Kxy + Lvwx + Mvwy + \quad (2)$$
$$Nvxy + Ovwy + Pvwxy$$

For given input slew, output capacitance, virtual ground length and power-switch size, two nearest values are picked up from LUT and substituted for v, w, x and y in the equation (2). Then, we solve simultaneous equations to obtain the coefficients A to P.

Although we extend to the four dimensions in the proposed methodology, there is the case that interpolation to only three dimensions is enough. In contrast to the parameters such as the input slew, the output capacitance, or virtual ground length, the power-switch size prepared in the cell library is discrete. If we prepare frequently used sizes in LUT, it is likely that we find the exact value for the power-switch size in the LUT. In this case, we do not need to interpolate the delay by using two nearest values for the power-switch size. In other words, the interpolation is performed by using three parameters of the input slew, the output capacitance and the virtual ground length. This results in reducing the dimension of the interpolation to three. It leads to making the simultaneous equations simpler.

In addition, preparing the LUT consists of four parameters for fall transition and preparing the LUT consists of the input slew and the output capacitance for rise transition is possible in which the gates does not discharge the parasitic capacitance of the internal circuit when the output goes "high". Except for AND or OR gates presented previously, the delay in the rise transition do not need the information of the virtual ground length and the power-switch size. This technique allows us to suppress the increase of the LUT size.

B. Design Flow

We propose a design flow that uses STA based on the delay modeling presented in the previous section. STA is generally applied after the logic synthesis. In MTCMOS design, it is difficult to apply the STA including the information of virtual ground length and power-switch size after the logic synthesis stage. This is because these parameters are optimized after the place. We perform STA after the place using the information on the power-switch size and the virtual ground length. This STA enables us to optimize the power-switch size and location in the cell-based MTCMOS design.

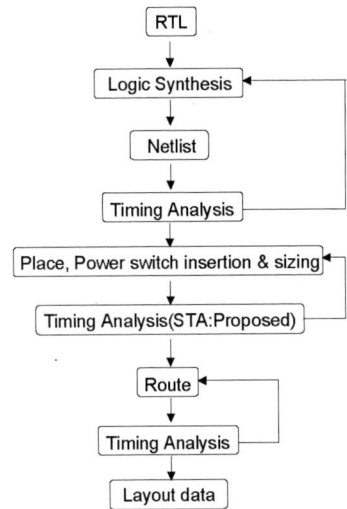

Figure 8: MTCMOS design flow

IV. Experimental results

A. Application to a NAND Gate

We describe the application of the proposed STA methodology to the NAND gate. We prepared the LUT1 and LUT2 for the interpolation, as shown in Table 1. In LUT1, we provide the delay for equal step of parameter values. For example, delay values for every 600ps of input slew are provided. In contrast, in LUT2, for smaller input slew we prepare the delay at fine steps, while for large input slew we prepare the delay at coarse steps. In this experiment, we examined the accuracy of this interpolation by using NAND circuits with various parameter values shown in Table 2. These parameter values were randomly generated within the compass of values that actually used in MTCMOS design. The objective is to examine if the proposed methodology enables to interpolate in a good accuracy under any con-

	LUT 1				LUT 2			
Input Slew(ps)	10	600	1200	1800	10	400	900	1800
Output capacitance(fF)	5	100	200	300	5	50	120	300
Virtual ground length(μm)	1	150	300	450	1	150	300	450
Power switch size	x1	x2	x4	x8	x1	x2	x4	x8

Table 1: Look up table

	Ex1	Ex2	Ex3	Ex4	Ex5	Ex6	Ex7	Ex8	Ex9	Ex10	Ex11	Ex12
Input slew (ps)	50	750	1600	230	610	1720	20	1010	1340	490	880	1130
Output capacitance (fF)	110	20	220	4	85	100	55	290	35	230	135	185
VGND length (μm)	380	22	75	320	120	440	95	160	310	200	140	190
Power switch size	x3	x1	x7	x9	x5	x1	x2	x3	x4	x2	x6	x8

Table 2: Circuit parameter examples for the interpolation (NAND)

dition. The comparison of interpolation results about the cell delay and the output slew with the SPICE simulation is shown in Figure 9.

Experimental results show that LUT2 allows us to interpolate at slightly better accuracy than LUT1, especially when the cell delay is small as shown in Figure 9 (c) and (d). This is because the delay does not increase linearly for small input slew and output capacitance values, as shown in Figure 5 (a) and (b). Since the delay is small for the small input slew and output capacitance, the relative error tends to become large. For these reasons, preparing delay values at fine steps for smaller input slew and output capacitance is preferable to obtain good accuracy.

Next, we describe the application of the proposed methodology to the critical path delay calculation. Using Synopsys Design Compiler, we synthesized a couple of modules with high-Vth cells of the Toshiba 90nm technology. As the modules, we used a circuit "SAND" in the MCNC Benchmark [6] and a CPU control unit ("CONTROL") of the SH3-compatible processor IP [7]. The number of gates of these circuits is 392 and 1766, respectively. We extracted the critical paths for these two circuits, as shown in Figure 10 and Figure 12. We replaced the high-Vth cells in the critical path with low-Vth MTCMOS logic cells. In addition, we connected them with each power-switch cell individually in the SPICE netlist. We modeled the virtual ground as π-type lumped circuit as previously illustrated in Figure 4. The values for output capacitance, virtual ground length and power-switch size were randomly generated, as the previously presented NAND gate. The input slew of the first gate is defined as 50ps, while the input slew for the gates at later stages is interpolated by using the output slew of previous gate computed by this methodology. The result is shown in Figure 11 and Figure 13. The result is shown as the error between the SPICE simulation and the interpolation

using LUT2.

To examine the effectiveness of this methodology, we also interpolate by using the real length for the virtual ground at the critical-path delay calculation of CONTROL. Using Synopsys Astro, we placed the CONTROL with the Toshiba 90nm technology. We estimated the real virtual ground length using Manhattan distance from the placement result. Real virtual ground length is shown in Figure 12. Experimental result is shown in Figure 14. Since the power switch is placed near the MTCMOS logic cells, the virtual ground length tends to be short in unshared design.

In this experiment, we found that the proposed methodology allows to interpolate with approximately 8% error in a cell and 1% error in the critical path. This result is almost the same even in the case of giving the real virtual ground length. The big relative error would appear when the output capacitance is small. More detailed analysis will be needed about how fine we should prepare the LUT for small output capacitance.

In this experiment, we use the same LUT even when the cell size is different. Large cell tends to have large output capacitance. When the parameter values do not exist in the range of LUT, we compute the cell delay by applying the extrapolation. However, this technique is difficult to obtain good accuracy. Hence, preparing wide range for larger cell size whereas preparing the narrow range for smaller cell size will lead to improving the accuracy.

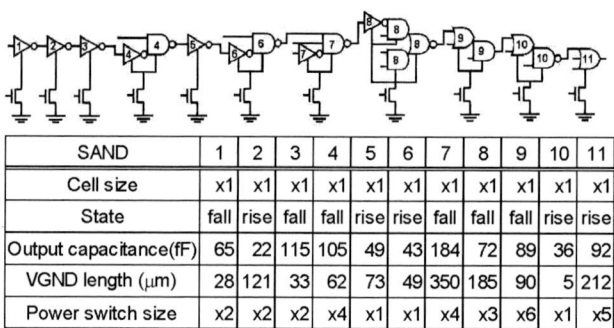

SAND	1	2	3	4	5	6	7	8	9	10	11
Cell size	x1	x1	x1	x1	x1	x1	x1	x1	x1	x1	x1
State	fall	rise	fall	fall	rise	rise	fall	fall	fall	rise	rise
Output capacitance(fF)	65	22	115	105	49	43	184	72	89	36	92
VGND length (μm)	28	121	33	62	73	49	350	185	90	5	212
Power switch size	x2	x2	x2	x4	x1	x1	x4	x3	x6	x1	x5

Figure 10: Critical path and parameters (SAND)

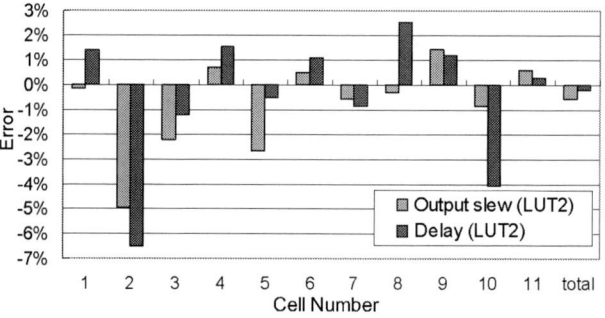

Figure 11: Interpolation result in critical path (SAND)

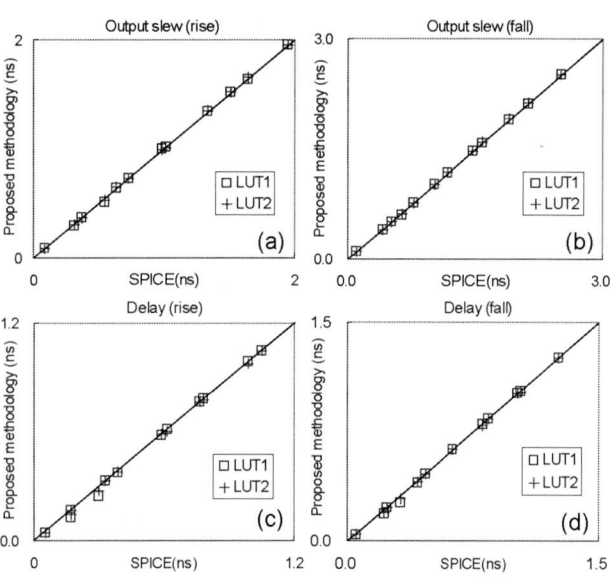

Figure 9: Interpolation result in NAND cell

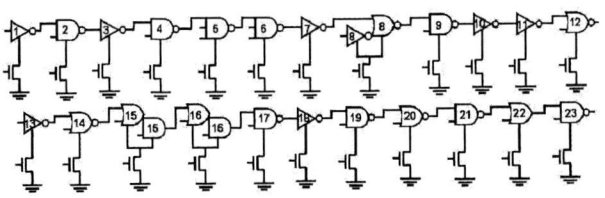

Control	1	2	3	4	5	6	7	8	9	10	11	12
Cell size	x1	x1	x1	x1	x1	x1	x1	x1	x1	x2	x2	x2
State	fall	rise	fall	rise	fall	rise	fall	rise	fall	rise	fall	rise
Output capacitance(fF)	73	48	38	69	55	32	112	22	49	90	27	241
VGND length : random(μm)	45	88	78	59	138	43	341	5	121	95	213	10
VGND length : estimated(μm)	2.2	11	1.4	1.8	2.2	12	2.5	1.6	9.8	9.9	5.8	4.8
Power switch size	x2	x1	x4	x3	x1	x1	x4	x1	x6	x6	x5	x3
Control	13	14	15	16	17	18	19	20	21	22	23	
Cell size	x1	x2	x1	x1	x1	x1	x2	x2	x1	x1	x1	
State	fall	rise	rise	rise	fall	rise	fall	rise	fall	fall	rise	
Output capacitance(fF)	91	193	70	78	68	11	189	279	138	76	41	
VGND length : random(μm)	32	30	21	443	79	132	52	21	62	16	76	
VGND length : estimated(μm)	3.4	15	2.3	7.9	7.7	13	3	4.8	4.1	5.4	6.5	
Power switch size	x2	x2	x3	x2	x4	x1	x7	x3	x8	x4	x1	

Figure 12: Critical path and parameters (CONTROL)

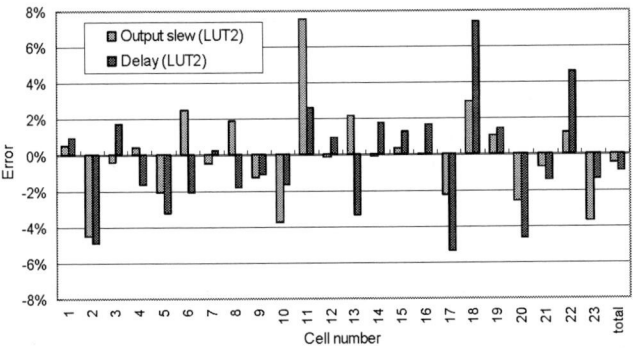

Figure 13: Interpolation result in critical path (CONTROL)

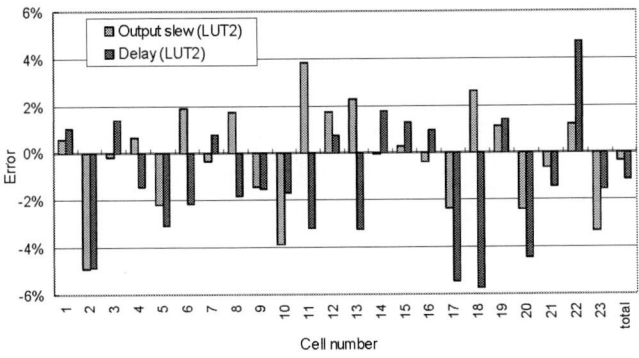

Figure 14: Interpolation result applying the measured virtual ground length (CONTROL)

V. Conclusion and Future Work

In this paper, we have proposed the STA methodology targeting at MTCMOS. We described that the cell delay and the output slew linearly increase with the virtual ground length and the inverse of power-switch size. Based on this observation, we proposed a delay computation scheme to use the four-dimensional linear interpolation. We found that this scheme allows us to estimate the delay in a good accuracy. We showed that STA could be applied to the MTCMOS circuits by simply extending the parameters and the interpolation equations.

In the MTCMOS design, the design to share the virtual ground is also done to reduce the area overhead. In case of the shared design, the factors affecting the delay become more complex. This is because the discharge current of the cell overlaps the current of the other cells. How to further extend the proposed methodology to shared virtual ground cases is the future work.

Acknowledgements

This work was supported by Toshiba Corporation. The authors would like to thank Mr. Masami Murakata and Mr. Takeshi Kitahara for their support and suggestions.

This work is supported by VLSI Design and Education Center (VDEC), the University of Tokyo in collaboration with Synopsys, Inc.

References

[1] S. Mutoh, T. Douseki, Y. Matsuya, T. Aoki, S. Shigematsu, and J. Yamada, "1-V Power Supply High-Speed Digital Circuit Technology with Multithreshold-Voltage CMOS," *J. of Solid State Circuits*, vol.30(8), pp.847-854, 1995.

[2] K. Usami, N. Kawabe, M. Koizumi, K. Seta, and T. Furusawa, "Automated Selective Multi-Threshold Design for Ultra-Low Standby Applications," *in Proc. ISLPED*, pp.202-206, August 2002.

[3] T. Kitahara, N. Kawabe, F. Minami, K. Seta, and T. Furusawa, "Area-efficient Selective Multi-Threshold CMOS Design Methodology for Standby Leakage Power Reduction," *in Proc. DATE*, pp.646-647, 2005.

[4] J. Kao, A. Chandrakasan, and D. Antoniadis, "Transistor Sizing Issues and Tools for Multi-Threshold CMOS Technology," *in Proc. DAC*, pp.409-414, 1997.

[5] K. Usami, N. Ohkubo, and M. Shirakawa, "Analysis on MTCMOS Circuit based on Lamped RC Model for Virtual Ground Line," *in Proc. ISOCC*, pp.116-119, 2005.

[6] "MCNC Benchmark" http://www.cbl.ncsu.edu/

[7] Y. Mitani, H. Uchida, T. Hironaka, J. MattauschHans, and T. Koide, "The Processor IP for Research with Software Development Environment," *Technical Report of IEICE*, VLD2001-109, pp.121-126, 2001. (in Japanese)

Switching-Activity Driven Gate Sizing and Vth Assignment for Low Power Design

Yu-Hui Huang

Department of Computer Science
National Tsing Hua University
HsinChu, Taiwan 300
yhhuang@nthucad.cs.nthu.edu.tw

Po-Yuan Chen

Department of Computer Science
National Tsing Hua University
HsinChu, Taiwan 300
pychen@cs.nthu.edu.tw

TingTing Hwang

Department of Computer Science
National Tsing Hua University
HsinChu, Taiwan 300
tingting@cs.nthu.edu.tw

Abstract

Power consumption has gained much saliency in circuit design recently. One design problem is modelled as "Under a timing constraint, to minimize power as much as possible". Previous research regarding this problem focused on either minimizing dynamic power by gate sizing, or reducing leakage power by dual threshold voltage assignment on non-critical path. However, given a timing constraint, an optimization algorithm must be able to utilize gate sizing and threshold-voltage assignment interchangeably, in order to minimize total power consumption including dynamic and leakage power in active mode and leakage power in idle mode. We find that switching-activity of a gate plays an important role in making decision as to choosing gate sizing or threshold-voltage assignment for performance improvement. For high switching-activity gates, threshold-voltage assignment should be used while for low switching-activity gates, gate sizing should be utilized. We develop an algorithm to perform gate sizing and threshold-voltage assignment simultaneously taking switching activity into consideration. The results show that under the same timing constraint, our circuits have 16.26%, and 18.53%, improvement of total power as compared to the original circuits for the cases where the percentage of active time are 100%, and 50%, respectively.

1 Introduction

Power consumption has gained much saliency in circuit design recently. In general, power sources are classified as dynamic and leakage power. Literature on low power design is abundant [3, 4, 5, 6] in which various techniques have been proposed to minimize either dynamic power or leakage power.

One design problem is modelled as "Under a timing constraint, to minimize power as much as possible". To solve this problem, one direction of research is focused on minimizing dynamic power. The minimization is based on the dynamic power consumption model where to maintain the performance, sizes of gates on critical paths remain unchanged while sizes of gates on non-critical paths are sized down to utilize their timing slack. By doing so, the gate capacitance is reduced, so is the dynamic power consumption.

The other direction is focused on reducing leakage power consumption by dual threshold assignment. It uses low-Vth transistors for gates on critical-path and high-Vth transistors on non-critical-path. This strategy is used in [3, 4, 5, 6, 8] and has shown that it has significant saving of leakage power and dose not mitigate the performance of circuit.

Gate sizing-up will increase gate capacitance hence large dynamic power and minor leakage. On the other hand, low threshold-voltage assignment will cause large leakage power increase. However, given a timing constraint, in order to minimize total power consumption including dynamic and leakage power in active mode and leakage power in idle mode, an optimization algorithm must be able to utilize sizing and threshold-voltage assignment interchangeably.

We find that switching activity of a gate plays an important role in making decision as to choosing gate sizing or threshold-voltage assignment. For high switching activity gates, re-assigning Vth should be used while for low switching activity gates, gate sizing should be utilized. Moreover, gates on both critical path and non-critical path should be taken into consideration. By utilizing these two techniques on both critical path and non-critical path, we can minimize the total power of circuit.

The rest of the paper is organized as follows. Section 2 will present our motivation. An algorithm taking switching activity into consideration for sizing and threshold-voltage assignment will be shown in Section 3. Benchmark results are presented in Section 4. Finally, conclusion remarks are drawn in Section 5.

2 Motivation

To enhance the performance of a circuit, we can size-up gate or change the Vth of gate from high to low. The former method will increase dynamic power and small leakage power while the latter method increase leakage power.

We observe that the switching activity of a gate will determine whether to use gate sizing or threshold-voltage assignment technique, in order to improve the timing of a circuit and to achieve lower power. If a gate has high switching activity, low Vth-assignment technique should be used while low switching activity, gate sizing-up is a better selection.

To understand if our observation is correct, we perform an experiment on two invertors, A and B. Invertor A has higher Vth and larger size than invertor B. The size and Vth are tuned so that the two invertors have the same

0-7803-9451-8/06/$20.00 ©2006 IEEE.

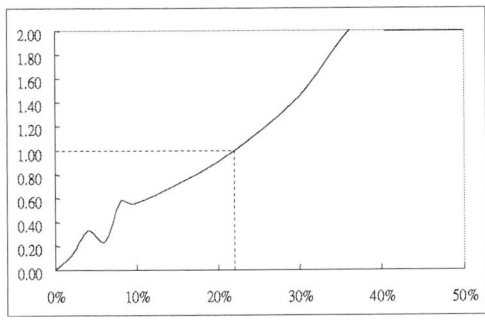

Figure 1: Relation between switching activity and $\frac{dyn(A)-dyn(B)}{lea(B)-lea(A)}$

Table 1: Switching activity(α) distribution of cells

α	Ratio(%)						
	TOP	MAC	AVG	GCC	RSA	AES	Avg.
$\alpha \leq 22\%$	71.0	48.9	70.9	55.3	84.5	60.8	65.3
$\alpha > 22\%$	29.0	51.1	29.1	44.7	15.5	39.2	34.7

delay under the same driving capability and output loading. The experiment is performed using *45nm* model from *PTM*[9] by *Hspice*. The result is shown in Figure 1. In this figure, X axis represents the switching activity, and Y axis is the result calculated by $\frac{dyn(A)-dyn(B)}{lea(B)-lea(A)}$, where $dyn(A)$ and $dyn(B)$ represent the dynamic power of A and B, and $lea(A)$ and $lea(B)$ represent the leakage power of A and B. Notice that invertor B is a better choice when the value of Y is larger than 1.

This figure shows that when the switching activity is lower than 22%, using invertor A (high Vth and larger cell) will produce less amount of total power. However, when the switching activity is more than 22%, invertor B (low Vth and smaller cell) will produce less power consumption. Based on the result in Figure 1, next, we would like to know if the switching activity of gates are indeed distributed in these two groups, with the switching activity less than and more than 22%. Table 1 shows the switching-activity distribution of gates in a set of circuits. In this table, we do see that on the average, 65% of the total gates are with switching activity between 0% to 22% and 34% with switching activity greater than 22%.

Moreover, previous work regarding minimizing power focused only on non-critical path. Here, we will minimize power on both critical and non-critical paths. On critical path, we will assign Vth to high and up-size gates which has small switching activity. On non-critical path, slack will be used to down-size gate or assign Vth to high.

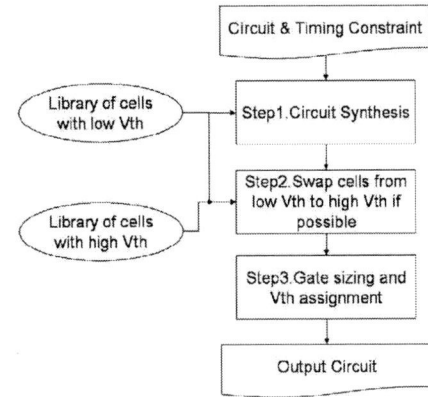

Figure 2: Design flow of the algorithm

3 Design Flow

Based on our motivation in Section 2, we propose a design flow to determine how to perform gate sizing and threshold voltage assignment. In Section 3.1, we first define the problem and show the design flow. In Section 3.2, the detailed algorithm will be presented.

3.1 Problem Definition and Design Flow

The problem can be defined as follows. Given a circuit, timing requirement, time profile of active and idle modes (i.e., the percentage of the total time that the circuit is in active mode and in idle mode) and cell library, to minimize the power consumption (including leakage in idle mode and dynamic and leakage in active mode) by gate sizing and threshold-voltage assignment.

To solve this problem, a design flow shown in Figure 2 is proposed. First, with the timing constraint, the circuit is synthesized using only cells with low Vth by synthesis tools. The reason to use only cell with low Vth is that a circuit synthesized using cells with the best timing (low threshold voltage) but minimal size allows gates to be sized up in the later optimization steps. Then, in the second step, all gates with low Vth are swapped to their corresponding high Vth cells. After this step, the timing constraint is no longer satisfied but the leakage is maximally reduced (the gate with low Vth has larger delay and less leakage). Finally, the last step is to perform gate sizing and threshold voltage re-assignment to restore the original timing performance. For nodes on critical path , our decision to choose gate for up-sizing or low threshold voltage assignment, will take the minimal increase in power consumption and area into consideration. For nodes on non-critical path, the algorithm utilize the slack of nodes to save more power consumption.

The first and the second steps are well understood. The detailed algorithm of the third step, **gate sizing and Vth assignment**, will be explained in the next subsection.

6A-4

2006 Asia and South Pacific Design Automation Conference

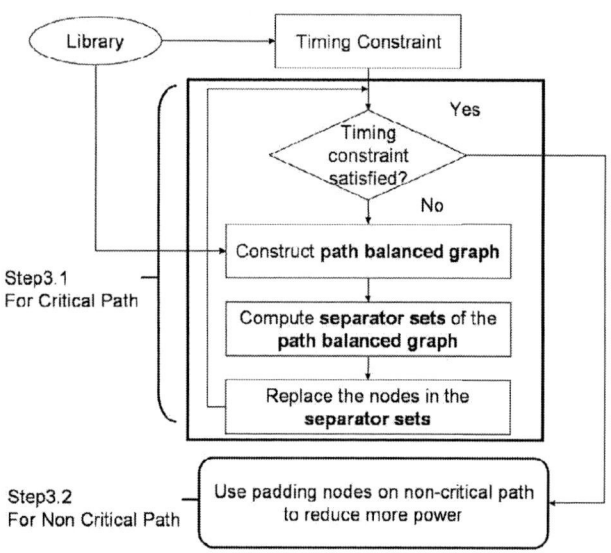

Figure 3: Step of **gate sizing** and **Vth** assignment

3.2 Gate Sizing and Threshold Voltage Assignment

Our third step, **gate sizing and Vth assignment**, is conducted in two phases. In the first phase, gate sizing or Vth voltage re-assigning is performed on nodes on critical path and in the second phase, on nodes on non-critical path. First, timing analysis on the circuit is performed. The arrival time, the required time and the slack of each gate are computed. Then, based on this timing information, a path balanced graph $G = (V, E)$ for the circuit is constructed. Next, separator sets of the graph are computed. The nodes in the separator sets are candidates for sizing or threshold voltage re-assignment. This step continues until the timing constraint is satisfied. Once the timing constraint is met, we continue to minimize the power consumption of circuit in the second phase by utilizing the remaining timing slack on non-critical paths.

The algorithm is depicted in Figure 3. The details are described in the following.

3.2.1 Optimization on Critical Path

A circuit can be viewed as a directed graph $G = (V, E)$, as shown in Figure 4, where x, y, z in (x, y, z) denote *slack, delay-reduction*, and *cost*, of the nodes, respectively (*delay reduction*, and *cost* will be defined later). After timing analysis, the arrival, required times and slack of nodes are computed. Based on this timing information, to improve the timing performance, a set of nodes can be selected to speed up. Since there is usually more than one critical path, the selection step requires a lot of attention. The objective is usually to select a set of nodes with minimum cost.

One way to select nodes which guarantee the circuit timing improvement is to select a separator set. However, simply selecting a separator set will not produce low cost result because slack on short path may not be fully utilized. Instead, we will select a separator set based on a *path*

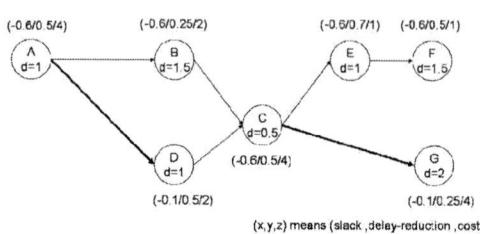

Figure 4: A circuit graph

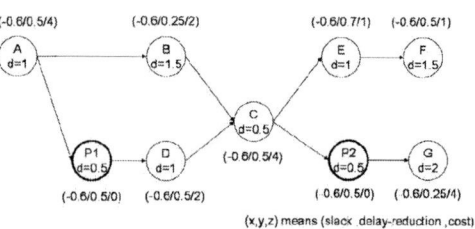

Figure 5: A *path balanced graph*

balanced circuit graph [7] in which slack of short paths can be fully exploited.

A *path balanced graph* is defined as follows [7]. First, for all edges e, $ds(e)$ is computed. $ds(e)$ is defined as the slack difference of nodes at the two ends of an edge e. It is computed as

$$ds(e) = slack(tail\_node(e)) - slack(head\_node(e)) \quad (1)$$

If $ds(e) > 0$, it means that input is from a short path. A *padding node* is inserted at e whose delay is $ds(e)$ and the cost of this node is *0*. By doing so, slack of all nodes become equal. To select a separator set with least cost, padding nodes are more likely to be selected. When the padding nodes are indeed selected in a separator for timing improvement, the cost of replacing *padding nodes* is 0. Figure 5 shows a *path balanced graph*, $G_{balanced} = (V, E)$ with padding nodes constructed from circuit graph, $G = (V, E)$ of Figure 4. In this figure, padding node $P1$ is added between node A and node D because the slack difference of the edge $A \rightarrow D$ is *0.5* (*slack(node D) - slack(node A)*). Similarly padding node $P2$ is added because the slack difference of the edge $C \rightarrow G$ is *0.5* (*slack(node G) - slack(node C)*).

Once the *path balanced graph* is constructed, we need to set the cost of a node. The cost is to be defined so that the less the cost of a node, the more likely the node is to be replaced for performance improvement. Before we present how to set the cost of a node, we need first to decide what the next candidate-change is for a node, in order to solve the timing violation problem. The objective of our algorithm is to select nodes with minimal power

578

and area increase for replacement. There are two ways to solve the timing violation: up-sizing gate or changing the gate with high Vth to its corresponding cell with low Vth. However, either way will increase the total power (dynamic and leakage) in which up-sizing gate increases the load capacitance of fan-ins and hence dynamic power while assignment of low threshold voltage increases leakage power. To make a choice between these two options, based on the observation in Section 2, we should take the switching activity of gates into consideration. For gates with fan-ins of low switching activity, up-sizing should be selected because the increase in dynamic power of fan-ins may be very small. On the other hand, for gates with fan-ins of high switching activity, assignment of low threshold voltage should be considered.

Based on this observation, we define a power penalty function, *penalty(g)* which is the penalty for the gate *g* when up-sizing or low Vth assignment is selected to replace the current gate *g*. It is calculated as .

$$penalty(g) = \alpha \cdot p\_penalty(g) + \beta \cdot a\_penalty(g) \quad (2)$$

In this equation, *p_penalty* and *a_penalty* are the power and area increase overhead, respectively, and α and β are parameters to control the weights of power and area penalty. The *p_penalty(g)* is further defined as

$$p\_penalty(g) = per \cdot (\sum_{j \in fanin(g)} E(j) \cdot C_{inc}(g)V^2 + leak_{inc}(g))$$

$$+ (1 - per) \cdot leak_{inc}(g) \quad (3)$$

where *per* is the percentage of the total time that the circuit is in the active mode, $E(j)$ is the switching activity of signal j, $C_{inc}(g)$ is the increased capacitance, V is the supply voltage, $leak_{inc}(g)$ is the increased leakage. The first term represents the power increase when the circuit runs in active mode and the second term the power increase when the circuit is in idle.

We compute the *penalty(g)* for the gate *g* for both up-sizing and low Vth assignment options. The option that has less penalty is selected as a candidate for replacement of the current gate *g*. Then, it is used to compute the *cost* of nodes in the *path-balanced graph*, $G_{balanced} = (V, E)$. Moreover, the delay reduction of the selected replacement is modelled as *delay-reduction* in the *path-balanced graph* .

Now, we show how to compute the *cost* of the *path-balanced graph* . The cost of a *padding node* is set to *0* and all other nodes *g* are computed as,

$$cost(g) = \gamma \cdot penalty(g) + \delta \cdot delay\_reduction(g) \quad (4)$$

where γ and δ are control parameters. Once the cost of the *path balanced graph* is computed, we will find a separator set of the graph. The nodes in the separator set are selected for replacement. Note that the *delay improvement* of this separator set, which is defined as the delay improvement of the circuit after the nodes of the separator set are replaced, is the minimum *delay-reduction* among the nodes in the separator set.

The next iteration will start with the timing analysis. If the timing constraint is not satisfied, the procedure continues.

Take the example shown in Figure 6 to demonstrate our selection algorithm. Figure 6(a) shows a separator

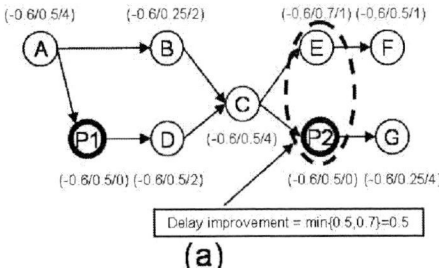

Delay improvement = min{0.5,0.7}=0.5

(a)

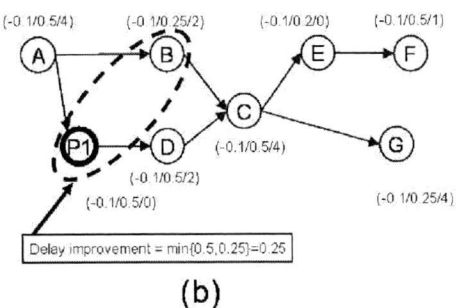

Delay improvement = min{0.5,0.25}=0.25

(b)

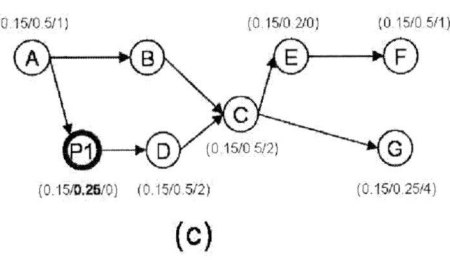

(c)

Figure 6: An example of running step 3.1 (a) iteration one (b) iteration two (c)a final *path balanced graph* after replacement

set of *path balanced graph* of the original graph shown in Figure 5. {E,P2} is selected to be the separator set with the minimal cost. The *delay improvement* of this set is *0.5*. The *delay-reduction* of node *E* and *P2* are decreased by *0.5* as shown in Figure 6(b). Since the *delay-reduction* of node *P2* equals to *0*, it is removed from the graph in Figure 6(b). In the next iteration, the separator {B,P1} is selected as shown in Figure 6(b) and it results in *0.25* delay reduction. We continue finding separator sets and updating *path balanced graph* until timing constraint is met. The final *path balanced graph* is shown in Figure 6(c).

3.2.2 Optimization on Non-Critical Path

After sizing and Vth re-assignment are performed on critical path, the timing constraint of circuit is met now. The objective of the next step is to utilize the remaining timing slack on *padding nodes* to reduce more power consumption. There are two ways to save the power consumption of a gate: down-sizing gate or changing the gate to high threshold voltage.

Node A can be delayed ϵ time if every path going out from A has ϵ slack. Recall that adding a *padding node* between node A and node B means that there is slack on the edge $A \rightarrow B$. In other words, if and only if all the *fan-out nodes* of node A are *padding nodes* with slack at least ϵ, node A can be delayed ϵ without affecting the timing of circuit. The maximum ϵ time of node A is computed as the minimum *delay reduction* of its fan-out *padding nodes*.

Therefore, for a node Ni in the graph, its available slack is computed by checking if the fan-outs of the output are all padding nodes. If they are, the available slack of Ni is computed as the minimum *delay-reduction* of Nj, where Nj is the fan-out node of Ni. Otherwise, the available slack of Ni is 0.

After we compute the available slack of each node, the next step is to utilize the available slack to reduce power consumption. First, we compute the delay penalty which will be caused by down-sizing nodes or re-assigning Vth to high. The *delay-penalty* of Ni is computed as $Delay(new\_Ni) - Delay(Ni)$. $Delay(Ni)$ is the delay of Ni with the current size and Vth, and $Delay(new\_Ni)$ is the delay of Ni after Ni is down-sized or re-assigned Vth to high.

If the delay penalty of both options are less than the available slack, we have a choice to select either down-sizing or re-assigning Vth to high. The choice will be based on the power saving of these two options. From our observation presented in Section 2, we know that if the switching activity is high, smaller gate with lower Vth should be selected while if the switching activity is low, high Vth with larger size gate is more power efficient. Therefore, we define a cost function, $p\_saving(g)$ for a gate g, to determine which option is better. For both down-sizing and re-assigning Vth, we compute

$$p\_saving(g) = p\_penalty(g) - p\_penalty(new\_g)$$

where new_g is a new implementation of gate g by either down sizing or high Vth voltage assignment and $p\_penalty$ is the cost function defined in equation (3).

The implementation which has more p_saving is selected for replacement. The procedure repeats until no more slack can be used or no more cell can be replaced.

Take the circuit in Figure 7 as an example. Let us consider nodes A and E. For node A, its fan-outs, $P1$ and $P2$, are both *padding nodes*. Since the minimum *delay reduction* of $P1$ and $P2$ is 0.4, the available slack of node A is 0.4. Suppose the delay penalty of node A be 0.3, which is less than the available slack 0.4. The size of node A can be reduced. On the contrary, for node E, it has two fan-outs but only one of them is *padding node*. Hence, the available slack of node E is 0, hence node E cannot be delayed without affecting the timing of circuit.

4 Experimental Results

Our power reduction algorithm is implemented in C language. The experimental process proceeds as follows. First, circuits are synthesized with low Vth library by *DesignCompiler*. Next, we obtain the critical paths of each synthesized circuit using *PrimeTime*. This critical timing plus ε will be used as timing constraint, where ε is set to be the timing-variance tolerance of the optimization process. In this experimental, ε is set to 2% of the critical timing

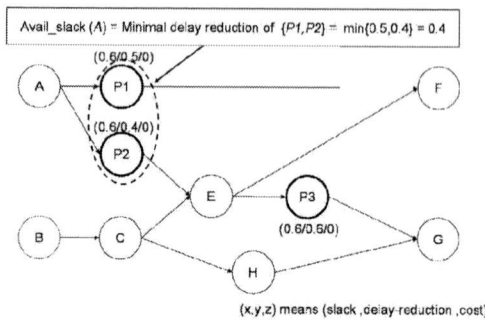

Figure 7: A updated *path balanced graph*

Table 2: Circuit descriptions

Cir.	CN	Characteristics
TOP	463	An Alarm Clock
MAC	2425	Multiplier and Accumulator
AVG	6361	Average Number Calculator
GCC	8204	Gravity Center Calculator
RSA	14815	Asymmetric Crypto-Processor
AES	16824	Advanced Encryption Core

of the synthesized circuit. Then, we change Vth from low to high and form a new circuit. Next, we use *PrimeTime* again to get the output capacitance of cells. Besides, we randomly generate some input patterns which are stored in *VCD* file. By reading this *VCD* file, *PrimePower* is used to report the toggle rate of every cell.

With the information mentioned above, we execute our program and output a *Verilog* file using cells of high Vth and low Vth cells from *TSMC 0.13µm* library. Finally, *PrimeTime* and *PrimePower* are used to report timing and total power consumption of the circuit, respectively.

Six circuits are used to examine the effectiveness of our algorithm. Table 2 shows the characteristic of our benchmark set. The columns labelled **CN**, and **Characteristics** are the cell number of the design, and the characteristics of the design, respectively.

The following tables show the power consumption and reduction after performing our algorithm for different cases where the percentages of active time and idle time are different. Let the percentage of active time and idle time be α. α is computed by $\alpha = \frac{active\_time}{active\_time + idle\_time} * 100\%$

Table 3, and Table 4 are the results for the cases where α is 100%, and 50%, respectively. The columns labelled P_O, P_A and P_B are the original power consumption, power consumption after executing **Steps 2 + 3.1** and power consumption after executing **Step 3.2**, respectively.

The original power is the result of the circuit with low Vth cells mapped by **DesignCompiler**. The columns labelled **A (Steps 2 + 3.1)** are the results that we change the Vth of cell from low to high and repair the timing violation problem on critical path. The columns labelled **B**

Table 3: The percentage of active time (α) is 100%

Cir.	P_O (mW)	A (Steps 2+3.1)		B (Step3.2)		
		P_A(mW)	Red_A	P_B(mW)	Red_B	Red_{total}
TOP	0.413	0.383	7.10%	0.363	4.85%	11.95%
MAC	0.970	0.843	13.06%	0.790	5.50%	18.56%
AVG	1.750	1.70	2.86%	1.65	3.07%	5.75%
GCC	0.806	0.764	5.19%	0.753	1.29%	6.48%
RSA	3.490	2.97	14.82%	2.12	24.38%	39.20%
AES	15.90	14.3	10.18%	13.4	5.42%	15.60%
Avg.	-	-	8.84%	-	7.42 %	16.26%

Table 4: The percentage of active time (α) is 50%

Cir.	P_O (mW)	A (Steps 2+3.1)		B (Step3.2)		
		P_A(mW)	Red_A	P_B(mW)	Red_B	Red_{total}
TOP	0.209	0.191	8.39%	0.179	5.85%	14.24%
MAC	0.504	0.424	15.89%	0.397	5.19%	21.09%
AVG	0.916	0.862	5.87%	0.835	2.92%	8.79%
GCC	0.451	0.418	7.39%	0.412	1.30%	8.69%
RSA	1.84	1.50	18.16%	1.08	23.24%	41.40%
AES	8.07	7.16	11.19%	6.70	5.79%	16.99%
Avg.	-	-	11.15%	-	7.38%	18.53%

(Step 3.2) are the results that we utilize the remaining slack on non-critical path to minimize the power consumption of the circuit.

The columns labelled Red_A are the power reduction rate computed as $Red_A = \frac{(P_O - P_A)}{P_O} * 100\%$ and the columns labelled Red_B and Red_{total} are similarly defined.

On the average, the power reduction rates of **A (Steps 2 + 3.1)**, **B (Step 3.2)**, and **total** of Table 3 are 8.84%, 7.42%, and 16.26%, respectively. Similar results are also produced in Table 4 where reduction rates are 11.15%, 7.38%, and 18.53%.

Table 5 shows the circuit timing information. The

Table 5: Timing comparisons

Cir.	Original T	α is 100%		α is 50%		α is 10%	
		N_T	TP	N_T	TP	N_T	TP
TOP	1.43	1.37	-4.2%	1.39	-2.8%	1.40	-2.1%
MAC	3.30	3.33	0.8%	3.33	0.8%	3.33	0.8%
AVG	23.78	23.13	-2.7%	23.46	-1.3%	23.54	-1.0%
GCC	26.30	26.65	1.3%	26.73	1.6%	26.34	0.2%
RSA	10.00	10.08	0.8%	10.03	0.3%	10.10	1.0%
AES	2.29	2.21	-3.5%	2.27	-0.9%	2.27	-0.9%

columns labelled **N_T** represent the timing of the new circuit. The columns labelled **TP** represent the timing penalty after performing our algorithm, which is calculated as $TP = 1 - \frac{(N\_T)}{Original\_T}$.

If **TP** is less than *0*, it means that timing becomes better than the original circuit. From the table, we can see that our algorithm do not cause serious timing violation problem. The largest timing penalty is 1.6% (circuit GCC when $\alpha = 50\%$), which is within our timing-variance tolerance, 2%.

From the table above, we found that our algorithm can effectively decrease the power consumption with very small timing penalty.

5 Conclusions

We have studied the problem that under a timing constraint, to minimize total power consumption. We have found that switching activity of a gate plays an important role in making decision as to choosing gate sizing or threshold assignment to improve timing performance. For high switching activity gates, Vth assignment should be used while for low switching activity gates, gate sizing should be utilized. The results showed that under the timing constraint, our circuits have 16.26%, and 18.53% improvement as compared the original circuits for cases where the percentage of active time are 100%, and 50%, respectively.

References

[1] Shekhar Borkar, "Low Power Challenges for the Decade", *Proceedings of ASP-DAC*, pp. 293-296, 2001.

[2] A. P. Chandrakasan and R. W. Brodersen, "Minimizing Power Consumption in Digital CMOS Circuits", *Procceddings of the IEEE*, pp. 498-523, 1995.

[3] L. Wei, Z. Chen, M. Johnson and K. Roy, "Design and Optimization of Low Voltage High Performance Dual Threshold CMOS Circuits", pp. 489-494, *Proceedings of the 35th DAC*, 1998.

[4] Vijay Sundararajan and Keshab K. Parhi, "Low Power Synthesis of Dual Threshold Voltage CMOS VLSI Circuits", *Proceedings of 1999 ISLPED*, pp. 139- 144, 1999.

[5] N. Tripathi, A. Bhosle, D. Samanta and A. Pal, "Optimal Assignment of High Threshold Voltage for Synthesizing Dual Threshold CMOS Circuits", *The 14th International Conference on VLSI Design*, pp. 227-232, 2001.

[6] Yen-Te, and TingTing Hwang, "Low Power Design Using Dual Threshold Voltages", *Proceedings of ASP-DAC 2004*, pp. 205-208, Japan, Jan. 2004.

[7] Yutaka Tamiya, "Performance Optimization Using Separator Sets", *Proceedings of ICCAD 1999*, pp. 191-194, 1999.

[8] D. Nguyen, A. Davare, M. Orshansky, D. Chinnery, B. Thompson, and K. Keutzer, "Minimization of Dynamic and Static Power Through Joint Assignment of Threshold Voltages and Sizing Optimization", *Proceedings of ISLPED* , pp. 156-163, 2003.

[9] Predictive Technology Model "http://www-device.eecs.berkeley.edu/p̃tm/".

Power Driven Placement with Layout Aware Supply Voltage Assignment for Voltage Island Generation in Dual-Vdd Designs

Bin Liu, Yici Cai, Qiang Zhou, Xianlong Hong

EDA Lab, Department of Computer Science and Technology, Tsinghua University, Beijing, China

Abstract— In this paper we propose a method for standard cell placement with support for dual supply voltages, aiming to reduce total power under timing constraints and to implement voltage islands with minimal overheads. The method begins with timing and power driven coarse placement, followed by a few iterations between voltage assignment and placement refinement to generate voltage islands. Several techniques, including timing and power driven net weighting, seed growth based voltage assignment, and soft clustering strategy for placement refinements are employed in our implementation. Experimental results on a set of MCNC benchmarks show that our approach is able to produce feasible placement for dual-Vdd designs and significantly reduce total power with a wirelength increase within 14% compared to a power and timing driven placer without voltage islands.

I. INTRODUCTION

Due to rapidly increasing on-chip power density with technology evolution and the growing market for battery powered devices, power dissipation has become one of the most critical concerns in modern chip designs. Researchers have proposed many low power design styles in recent years, among which multiple supply voltage (MSV) is a promising scheme to achieve significant reduction on both dynamic and static power while maintaining performance [1–4]. It is reported in [4] that dynamic power can be reduce by about 30% on average if an additional supply voltage is available. MSV can be combined with other low power techniques [5] and has been successfully implemented in some commercial chips [6, 7]. The design of MSV circuits can work at either macro module level during design planning [8] or cell level after logic synthesis [1, 2, 7]. In this paper, we focus on cell based designs with two supply voltages.

The basic idea behind MSV is to trade timing slacks for power reduction by using high voltage on cells with negative (or little) slack to maintain performance and using low voltage on others to save power. Previous works have demonstrated that the number of gates on critical paths accounts for only a small portion of total gates, while the majority of other gates have relatively large slacks, leaving much room for potential power reduction using MSV [2].

Despite its effectiveness and flexibility, MSV introduces some particular electrical and physical constraints. Level converters are needed whenever the supply voltage of a driver cell is lower than that of the receiver [9]. Moreover, cells under different voltages should be carefully placed so as to facilitate power network design and to reduce chip complexity. To save power while minimizing overheads induced by these effects, efforts must be made in two major operations in MSV design, voltage assignment and placement.

Most published algorithms for voltage assignment begin by setting all cells to VddH, and try to lower the supply voltages for some cells based on static timing analysis. Clustered voltage scaling (CVS) tries to reduce the supply voltages of cells from primary outputs to primary inputs in reverse topological order and does not allow voltage converters in the middle of a path [1]. A more flexible approach allowing level converters is called enhanced clustered voltage scaling (ECVS) [2], which proves to provide appreciably larger power reduction compared with CVS [4] . Other techniques for timing and power optimization, including gate sizing, multiple threshold voltages, can be combined with MSV for better timing-power tradeoff [5].

Cell placement is a critical step for MSV designs in that it greatly influences power grid complexity, as well as path delay, which can challenge timing closure especially when most slacks are traded for power saving after voltage assignment. There are two kinds of placement schemes for MSV designs: one is *row based* [10], where there are interleaving rows or half-rows for VddH cells and VddL cells; the other is *region based*, where the cells in each region (called voltage island) operate under the same supply voltage. Recently voltage island approach has been widely recognized as the state of the art in MSV design for its structural flexibility [3, 8, 11]. To generate voltage islands, cells with the same supply voltage must be physically clustered during placement, which is a new requirement for the placer. Although there are industrial efforts on tool development supporting MSV layout [7, 12, 13], detailed algorithms focusing on the physical implementation of fine-grained voltage islands are not seen in open literature.

As we examine the methodology for voltage island design, it is interesting to notice that voltage assignment is often performed prior to layout (e.g., during design planning or logic synthesis). Known voltage assignment algorithms tend to work without specific physical information [2, 4, 5], which may result in at least the following problems.

1. Interconnect delay, which dominates total path delay in deep submicron technology, can hardly be estimated accurately before placement. Thus, there can be either too many VddL cells, causing trouble to timing closure, or too few VddL cells, wasting slack that can be otherwise traded for power saving.

2. With final locations of cells unknown, the predefined assignment of voltages is likely to cause large wirelength and delay penalty even if flexible clustering strategies are employed in the placer.

The penalties caused by pre-placement voltage assignment give rise to the idea of layout aware voltage assignment. A concept of voltage assignment exploiting both logical and physical adjacency is mentioned in [7], but no detailed algorithm is described. Another work attempting to combine voltage assignment with placement is now in progress [14].

The purpose of this work is to develop a practical method for fine-grained voltage island generation in dual-Vdd designs. Specifically, we focus on two major aspects in design and implementation of chips with voltage islands: reducing total power under timing constraints, and reducing electrical and physical overheads. We propose a practical design flow for voltage island generation(outlined in Fig. 1). Our method begins with timing and power driven coarse placement, followed by a few iterations between layout aware voltage assignment and placement refinement. This flow is general enough to embrace many practical techniques and considerations. Preliminary algorithms are developed to support the flow. Experimental results on a set of MCNC benchmarks have demonstrated the effectiveness of our approach.

II. BACKGROUND

A. Timing and Power Analysis

In order to support flexible voltage assignment in dual-Vdd designs, at least some cells in the library should be designed to work with both supply voltages, possibly with different implementations. Gate level dual-Vdd design style allows replacing VddH cells with VddL cells along uncritical paths. Thus, exhaustive static timing analysis is a prerequisite to voltage assignment [1, 2, 4, 5].

In static timing analysis, the combinatorial part of a circuit is modelled as a weighted directed acyclic graph (DAG) called timing graph, where every node represents a signal pin in the netlist, and the weight of every directed edge represents either gate delay or wire delay between

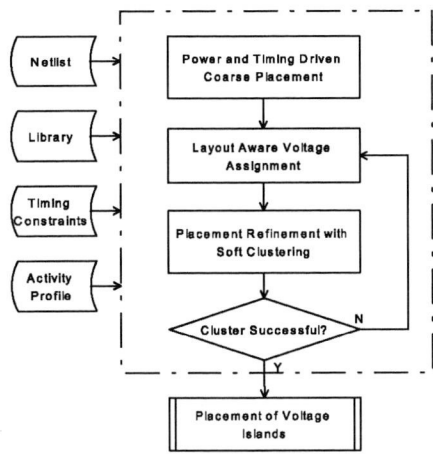

Fig. 1. Flow for voltage island generation.

two pins. Usually delay constraints are imposed by specifying the maximum delay between primary inputs and primary outputs. After a graph traversal in topological order (forward pass) and another in reverse topological order (backward pass), the arrival time, required time and slack at every node can be calculated. This process takes $O(|N| + |E|)$ time, where N is the set of nodes and E is the set of edges.

In this work, we use table-based models for gate delay, as well as leakage and short-circuit power for every cell. We use a Elmore-star model to computer wire delays [15]. Switching power is calculated with the Eqn. 1.

$$P_{sw} = \alpha f V_{dd}^2 C_{load} \qquad (1)$$

Where α is the activity factor, indicating the probability that the signal changes in one clock cycle; f is clock frequency; C_{load} is load capacitance of the gate, including gate capacitance and wire capacitance. Given the quadratic dependency of switching power on supply voltage, the efficacy of lowering supply voltage in power reduction is beyond controversy. Eqn. 1 also indicates other ways to save power by reducing the length of high activity wires, or using VddL cells to drive these nets, both of which are considered in our approach.

B. Cut-based Placement Paradigm

In this subsection, we review the framework of cut-based placement algorithms, and focus on the outline of Capo [16], which forms the base for our placement algorithm. A min-cut placement instance contains: 1) a rectangular region (referred to as *bin*) where cells are to be placed; 2) a hypergraph, with each node representing a cell and each hyperedge representing a signal net connecting two or more cells. A min-cut placer recursively partitions each bin and its associated hypergraph at current level, and assigns the subhypergraphs to subbins, minimizing total (weighted) net cuts for total (weighted) wirelength reduction. The cut direction usually alternates between horizontal and vertical cuts. For wirelength estimation during the placement process, the nodes in each bin can be considered as being placed at the geometric center of the bin.

Capo is an elegant cut-based placer, with several techniques to improve total wirelength including placement feedback, weighted terminal propagation, etc. The idea of feedback in placement [17], i.e., merging adjacent bins and perform repartition under the guidance of information from last iteration is an inspiration of our idea on placement refinement.

III. Voltage Island Generation Methodology

As illustrated in Fig. 1, our voltage island design methodology does not require pre-layout voltage assignment. Instead, layout aware voltage assignment is performed after a coarse placement result is available, so that delay due to interconnect effects can be more accurately captured in timing analysis, and physical adjacency information is available to guide the assignment of supply voltages.

It is natural to cluster cells with the same supply voltage in a region to form a voltage island. However, aggressively clustering cells is likely to result in wirelength increase and timing violation, especially when the region is relatively large and the number of VddH cells and VddL cells are close. We hereby further exploit the flexibility of voltage island generation by iteratively updating the voltage assignment and the layout. The placement of cells can be adjusted by merging adjacent bins and repartitioning the netlist, aiming to physically cluster cells with the same voltage. In our implementation, we do not require that the island be generated after adjusting placement. Instead, repartition is done after with some additional hyperedges on cells to be clustered(this is called *soft clustering*). After repartition, the assignment of voltages is also adjusted to fit changes in layout. Convergence of the iterations can be guaranteed with increasing weights of additional hyperedges and a decreasing threshold parameter used in incremental voltage assignment.

IV. Coarse Placement with Timing and Power Driven Net Weighting

Coarse placement is like the first several passes of a timing and power driven global placement with all cells operating under VddH. While the final objective of this work is to minimize power, it is helpful to optimize timing, instead of merely meeting timing constraints or optimizing path delay on a few critical paths during coarse placement, because additional slacks can provide more opportunities to use VddL cells. It is particularly advantageous that even the most critical path has some slack, which

makes the exploitation of physical adjacency possible for cells along critical paths.

Net weighting is a popular technique for large scale placement due to its flexibility and efficiency. There have been extensive works on net weighting strategies for timing optimization in placement [18, 19]; a few other works employ switching activity based net weighting to minimize total switching power [20]. Different from strategies that focus either timing or power, we seek a weighting scheme that improves both. The empirical formula in Eqn. 2 is used for net weighting taking account of both slack and switching power.

$$W = \begin{cases} (1 + c \times \alpha) \times (1 + \frac{T_0}{T_0/N + slack}), & slack \geq 0 \\ (1 + c \times \alpha) \times (1 + N), & else \end{cases}$$
$$(2)$$

Here α is the switching probability; *slack* is the minimum slack at the input of downstream cells; T_0 is typically several times larger than the gate delay; N and c are constant parameters. Other net weighting methods can also be used in coarse placement. It should be emphasized that coarse placement not only try to reduce delay along a few most critical paths, but also attempt to create more slacks along even moderately critical paths, because slacks can probably be traded for power saving afterward.

Net weights are incrementally updated in the feedback procedure at every partition level. The weight of an edge is updated considering both its previous value and new value.

$$W = \beta W_{new} + (1 - \beta)W_{orig} \qquad (3)$$

Our experience shows that β should be kept below 0.3 to maintain consistency and avoid oscillation of path delays.

V. Initial Voltage Assignment

Initial voltage assignment largely defines the final pattern of voltage islands. The proposed algorithm for initial assignment works in a seed-growth manner, exploiting both physical adjacency and logical adjacency [7]. The algorithm flow is described in Algorithm 1.

We associate a tendency value to every VddH cell. Similar to [4], the tendency is defined as follows.

$$tendency = \begin{cases} \frac{(G_{self} + \gamma G_{LC}) \times slack}{\Delta delay}, & slack > \Delta delay \\ 0, & else \end{cases}$$
$$(4)$$

Here G_{self} is the power reduction due to the use of VddL; G_{LC} is the reduction of needed level converters; *slack* is the minimum slack at the output pins; $\Delta delay$ is the increase of gate delay (measured with the maximum increase on pin-to-pin delay).

Note that the tendency of a cell depends on the status of its logical neighbors as well as its slack, both of which can change dynamically in the assignment process. Thus, incremental updates on the timing graph and cell tendencies are required. For efficiency consideration, these updates are performed lazily, and only the tendencies of relevant unprocessed cells are updated.

The seed growth process begins by selecting *seed bins*. We calculate a priority value for every bin by investigating the tendencies of cells in it. Bins with priority larger than a threshold are selected as seed bins. Priority of a bin is defined as the average tendency of cells in the bin, as shown in Eqn. 5, where N is the number of cells in the bin.

$$priority = \frac{1}{N} \sum_{i=1}^{N} tendency(cell_i) \quad (5)$$

After seed bins are selected, the algorithm enters a procedure of selecting cells to work under VddL. This is done in two phases. The first phase tries to assign VddL to every cell in seed bins if timing constraints are not violated, while the second phase tries to assign VddL to more cells across the chip. The purpose of the first phase is to generate some physical clusters of VddL cells with no wirelength penalty, which form the bases of VddL islands. The second phase can be viewed as a logical expansion of VddL cells according to on existing VddL cells, aiming to reduce power dissipation without adding level converters.

Algorithm 1 Initial Voltage Assignment
___
Require: *threshold*
  sort bins and cells according to priority and tendency;
  while there are unprocessed bins **do**
   $currentBin \leftarrow$ the unprocessed bin with highest priority;
   if $currentBin.priority < threshold$ **then**
     break;
   else
     try to use VddL for every cell in $currentBin$ without timing violation;
   end if
   update priorities of unprocessed bins;
  end while
  while there are unprocessed cells **do**
   $c \leftarrow$ the unprocessed cell with highest tendency;
   try to use VddL for c if timing budgets are met;
  end while
___

VI. Iterative Voltage Assignment and Placement Refinement

A. Placement Refinement

Placement refinement is a procedure that locally adjusts the locations of cells, so that cells with the same supply voltage get closer to each other. The refinement procedure is much the same as the feedback mechanism used in Capo [17]. After voltage assignment, it is probably that most bins contain both VddH cells and VddL cells. In placement refinement stage, neighboring bins are merged to generate a new larger bin, containing all cells in original bins. The new bin is repartitioned, taking into account both total wirelength and requirement for clustering. Instead of modifying the hypergraph partitioning algorithm, a simple method is adopted to incorporate clustering consideration into the partitioning problem. Pseudo hyperedges connecting cells with the same supply voltage are added to the hypergraph to be partitioned. Since min-cut placement minimizes total weighted wirelength naturally, cells connected with the pseudo hyperedges tend to get close to one another and clustering can be realized. This approach is referred to as *soft clustering*, because it differs from a strong clustering method (hard clustering) that combines all cells to be clustered into a soft macro-module and performs mixed-size placement afterward. Empirically, although hard clustering can be easily implemented with existing tools, it tends to increase total wirelength significantly; similar conclusion has been validated in the research of integrated floorplanning and placement [21]. The weights of pseudo edges reflect the desire of clustering. In order to reduce wirelength overheads, the total weights of the added edges should be kept small, at least in the first a few iterations, when complete isolation of VddH cells and VddL cells is not required. In order to accelerate convergence, these weights are increased iteration by iteration.

B. Incremental Update of Voltage Assignment

In the proposed flow, a voltage island is generated if a bin contains purely VddH cells or VddL cells. However, there can be many bins containing both kinds of cells even after placement refinement due to the inadequacy of soft clustering in some regions. Intuitively, if most part of a region is filled with VddH cells, it is desired that the supply voltages for all VddL cells in the region are raised to VddH. If only a small portion of cells are powered by VddH in a region, it is probable that the supply voltages of VddH cells in this region can be lowered after replacing some some VddL cells with VddH cells in some other bins. By increasing and decreasing the supply voltages region by region, timing slacks can be concentrated into some regions to form VddL islands, and other regions that can not work at VddL without timing violations eventually become VddH islands. This is done incrementally with the iterations.

The algorithm is illustrated in Algorithm 2. A queue containing all bins with both VddH cells and VddL cells is constructed, and the portion of VddH cells in every bin is monitored. At each iteration, the algorithm tries to boost existing dominance of VddH or VddL cells in some bins. For every bin with with the portion of VddH cells larger than a threshold, all other VddL cells in the bin are replaced with VddH cells, and the bin is removed from the queue. Then every other bin in the queue is examined in reverse order of VddH portion to see if all cells in it can be powered by VddL while meeting the constraints.

TABLE I
EXPERIMENTAL RESULTS OF THREE ALGORITHMS.

name	period (ns)	Capo 9.2				Capo 9.2+net weighting				Proposed algorithm			
		power	slack	time	HPWL	power	slack	time	HPWL	power	slack	time	HPWL
c880	2.4	0.022	-0.2	2.06	1.15E6	0.022	-0.12	5.74	1.23E6	0.019	-0.134	7.46	1.26E6
c1355	3.05	0.022	0.06	3.95	1.33E6	0.021	0.19	7.36	1.32E6	0.016	0.080	11.38	1.31E6
c1908	4.0	0.024	-0.24	8.71	2.08E6	0.025	0.02	14.78	2.26E6	0.018	0.108	22.30	2.29E6
c2670	4.0	0.046	-1.02	14.14	4.62E6	0.049	-0.64	33.89	5.02E6	0.038	-0.398	57.56	4.97E6
c3540	5.5	0.039	-0.39	15.12	5.29E6	0.040	0.19	26.13	5.41E6	0.021	-0.213	41.62	5.25E6
c5315	4.8	0.067	-1.001	28.70	7.58E6	0.072	0.25	53.54	8.59E6	0.045	-0.064	367.22	8.80E6
c6288	9.5	0.028	-2.31	23.76	5.21E6	0.029	-2.16	34.47	5.39E6	0.027	-1.965	86.25	5.30E6
c7552	5.0	0.090	-0.21	46.31	10.61E6	0.097	-0.07	91.95	11.21E6	0.065	-0.046	165.93	11.60E6
s1488	3.9	0.022	-0.05	5.51	1.84E6	0.022	0.15	7.46	1.84E6	0.015	0.125	14.83	1.93E6
s15850	8.6	0.123	-3.88	167.25	24.94E6	0.143	-1.78	325.18	32.03E6	0.097	-1.270	591.04	32.49E6
s35932	20.6	0.107	0.44	291.23	48.74E6	0.114	0.613	514.27	54.98E6	0.066	0.388	824.02	62.11E6
s38417	5.6	0.433	-3.30	402.51	55.49E6	0.471	-1.46	722.20	64.36E6	0.250	-0.942	1041.88	63.98E6
s38584	10.5	0.259	-1.40	376.92	68.15E6	0.267	0.743	666.47	71.23E6	0.174	0.796	975.89	72.53E6

period: clock period(ns); slack: worst slack(ns); time:running time(s).

Algorithm 2 Incremental Voltage Reassignment

Require: *threshold*

  construct a queue Q, containing all bins with both VddH cells and VddL cells, order by the portion of VddH cells;

  while Q is nonempty **do**

    $currentBin \leftarrow$ the bin with highest portion of VddH cells;

    if the portion of VddH cells is lower than *threshold* **then**

      return;

    end if

    assign VddH to all cells in *currentBin*;

    update timing graph;

    remove *currentBin* from Q;

    for all *bin* in Q in reverse order **do**

      try to use VddL for *currentBin*;

    end for

  end while

The threshold parameter is initially close to 1, and decreases toward 0 with iterations between placement refinement and incremental voltage assignment, which guarantees the convergence of the voltage island generation algorithm.

VII. EXPERIMENTAL RESULTS

The proposed algorithms have been implemented with C++ based on Capo 9.2. We create our dual-Vdd library based on a industrial 0.18um library with VddH=1.8V. We add a VddL(1.2V) alternative for every cell in the original library and compute its delay based on alpha-power law model.

Experiments are performed on a set of ISCAS85 and IS-CAS89 benchmarks with specified timing constraints and activity profiles. In order to evaluate the effectiveness of the proposed method, we examine the results of three algorithms: 1) original Capo(version 9.2), which aims at minimizing total half perimeter wirelength; 2) Capo with the net weighting strategy described in Section IV (referred to as CapoW), aiming at improving delay and power by minimizing the weighted total wirelength; 3) the proposed algorithm with support for dual-Vdd designs (referred to as CapoV). We measure total power, maximum slack, total half perimeter wirelength and running time for all the three algorithms (Table I).

When comparing CapoW with Capo, it can be noticed that timing is consistently improved in all the benchmarks, and the increase of total wirelength is within 12% except C5315, C6288, C15850 and C38417, which all have tight timing constraints. While it is evident that our net weighting strategy is effective in promoting timing closure, the total power is not remarkably reduced; actually it is increased in many designs (mostly by within 10%). This is probably because the timing constraints in many benchmarks are rather tight and the values of net weighting parameters in our implementation are tuned mainly for timing optimization. A lot of slacks are created on both critical paths and some uncritical paths, thus enlarging the room for power optimization in voltage assignment.

As expected, the results by CapoV show significant power saving with the use of dual-supply voltage, which corroborates previous works. What we are especially concerned is the implementation penalty on wirelength. Since CapoV and CapoW are based on the same coarse placement algorithm with the same net weighting parameters, and CapoV works with some additional clustering constraints while no further constraints are imposed on CapoW, the total wirelength produced by CapoW can be viewed as an "upper bound" for that by CapoV. (Our experiments show that for some benchmarks CapoV produces slightly shorter wirelength than CapoW, which is probably because the additional merge-and-repartition iterations are helpful to optimize wirelength (as well as tim-

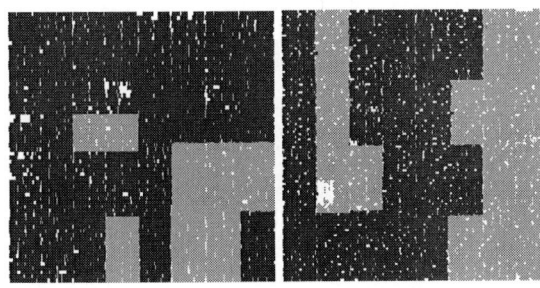

Fig. 2. Typical placement results with voltage islands. Left is s15850, and right is s38417. VddH areas are marked with dark purple and VddL areas with light yellow.

ing) at the cost of more running time.) Table I shows that CapoV produces results with wirelength increase of -2.96% to 13.7% compared with CapoW. We think it is not fair to compared CapoV and original Capo on wirelength, because CapoV works with net weights due to timing and power considerations, which inherently is inconsistent with wirelength optimization.

Fig. 2 illustrates the placement results of two benchmarks with voltage islands. Power grid design and verification for these designs will not be much complicated because the number of voltage islands are not large.

VIII. CONCLUSION

In this paper we present an effective methodology together with algorithms to reduce power dissipation under timing constraints in placement and to provide physical-level support for voltage island designs. The proposed layout aware voltage assignment and iterative adjustment on placement and voltage assignment have shown great effectiveness in the implementation of voltage islands with minimal wirelength penalty. Our results probably indicate that it is necessary to perform voltage assignment, or at least adjustment on voltage assignment, during placement in order to reduce physical overhead. In the current implementation, level converters are not dynamically inserted and deleted during voltage assignment. ECO placement should be performed afterward taking level converters into account. Future works can make efforts to accelerate design convergence and better consider level converter issues.

ACKNOWLEDGEMENT

The authors would like to thank Prof. Igor Markov, Prof. Patrick Madden and Prof. David Pan for their kind help. This work is supported by the Hi-Tech Research & Development (863) Program of China (No. 2005AA1Z1230) and the National Natural Science Foundation of China (NSFC) (No. 60476014).

REFERENCES

[1] K. Usami and M. Horowitz, "Clustered voltage scaling technique for low-power design, " in *Proc. ISLPED'95*, pp. 3-8.

[2] C. Chen, A. Srivastava and M. Sarrafzadeh, "On gate level power optimization using dual-supply voltages," *IEEE Trans. on VLSI Syst.*, vol. 9, pp. 616-629, Oct. 2001.

[3] D.E. Lackey, P.S. Zuchowski, T.R. Bednar, D.W. Stout, S.W. Gould, and J.M. Cohn, "Managing power and performance for system-on-chip designs using voltage islands," in *ICCAD'02*, pp.195-202, Nov. 2002.

[4] S.H. Kulkarni, A.N. Srivastava, and D. Sylvester, "A new algorithm for improved VDD assignment in low power dual VDD systems," in *Proc. ISLPED'04*, pp.200-205.

[5] A. Srivastava, D. Sylvester and D. Blaauw, "Power minimization using simultaneous gate sizing, dual-Vdd and dual-Vth assignment," in *Proc. DAC'04*, pp.783-787.

[6] S.K. Mathew, M.A. Anders, B. Bloechel, T. Nguyen, R.K. Krishnamurthy and S. Borkar, "A 4-GHz 300-mW 64-bit integer execution ALU with dual supply voltages in 90-nm CMOS," *IEEE J. Solid-State Circuits*, vol. 40, pp.44-51, Jan. 2005.

[7] R. Puri, L. Stok, J. Cohn, D. Kung, D. Pan, D. Sylvester, A. Srivastava and S. Kulkarni "Pushing ASIC performance in a power envelope," in *Proc. DAC'03*, pp.788-793.

[8] J. Hu, Y. Shin, N. Dhanwada and Radu Marculescu, "Architecting voltage islands in core-based system-on-a-chip designs," in *Proc. ISLPED'04*, pp.180-185.

[9] F. Ishihara, F. Sheikh and B. Nikolic, "Level conversion for dual-supply systems," *IEEE Trans. VLSI Syst.*, vol. 12, pp.185-195, Feb. 2004.

[10] C. Yeh, Y. Kang, S. Shieh and J. Wang, "Layout techniques supporting the use of dual supply voltages for cell-based designs," in *Proc. DAC'99*, pp.62-67.

[11] R. Puri, L. Stok, S. Bhattacharya, "Keeping hot chips cool," in *Proc. DAC'05*, pp.285-288.

[12] Synopsys inc., Galaxy design platform multi-voltage Design, available online, http://www.synopsys.com/products/power/multivoltage_bkgrd.pdf.

[13] Cadence inc., Cadence/TSMC Reference Flow 6.0.

[14] P.H. Madden, private communication.

[15] A.B. Kahng, S. Mantik and I.L. Markov, "Min-max placement for large-scale timing optimization," in *Proc. ISPD'02*, pp.143-148.

[16] A.E. Caldwell, A.B. Kahng, and I.L. Markov, "Can recursive bisection alone produce routable placements?," in *Proc. DAC'00*, pp.477-482.

[17] A.B. Kahng and S. Reda, "Placement feedback: a concept and method for better min-cut placements," in *Proc. DAC'04*, pp.357-362.

[18] T. Kong, "A novel net weighting algorithm for timing-driven placement," in *Proc. ICCAD'02*, pp.172-176.

[19] A.B. Kahng and Q. Wang, "Implementation and extensibility of an analytic placer," *IEEE Trans. Computer-Aided Design*, vol. 24, pp.734-747, May 2005.

[20] Y. Cheon, P.H. Ho, A.B. Kahng, S. Reda and Q. Wang, "Power aware placement," in *Proc. DAC'05*, pp.795-800.

[21] J.A. Roy, S.N. Adya, D.A. Papa and I.L. Markov, "Min-cut floorplacement,", *IEEE Trans. on Computer-Aided Design*, to appear.

Reusable Component IP Design using Refinement-based Design Environment

Sanggyu Park, Sangyong Yoon, and Soo-Ik Chae

Center for SoC Design Technology and
School of Electrical Engineering and Computer Science
Seoul National University,
Seoul 151-742, KOREA
email : {sanggyu, syyoon, chae}@sdgroup.snu.ac.kr

Abstract - We propose a method of enhancing the reusability of the component IPs by separating communication and computation for a system function. In this approach, we assume that the component designers describe mainly the computation part of the component, and the system designer can construct the communication part by using our refinement-based design environment. Moreover, we introduced a concept of the Communication Architecture Template Tree (CATree), which helps IP designers to effectively separate computation and communication for a system function. We confirmed that this approach is effective by applying it to a H.264 decoder design.

I. Introduction

Reuse-centric design methodologies including IP-based design and platform-based design have been widely accepted in the SoC industry . Its effectiveness is determined mainly with the richness of the component IPs and their reusability. We can enhance the component reusability substantially by using the standard bus interfaces [1] such as AMBA and Core Connect and generic memory interfaces such as an on-chip SRAM interface.

Although integrating the component IPs with standard interfaces is easier, there are still several limitations. First, connecting components with different protocols incurs considerable overheads. Second, standard interfaces limit the internal architecture of the component. Furthermore, the standard interfaces limit the system-level communication architecture. Consequently, there have been strong attentions on using flexible communication interfaces [2,4-6].

In this paper, we propose a method to enhance the reusability of component IPs by exploiting the concept of orthogonalization. We re-defined the roles of the component and system designers. The component designers should capture mainly the computation part of a function and its test model with only essential architectural hints for the communication refinement and provide a component IP package, which is described in details in Section IV, to the system designers. The system designers should configure only the communication part of the component IP to make it best fit to the system with our refinement-based design environment where we introduced a concept of the Communication Architecture Template Tree (CATtree).

The CATtree provides communication architecture templates to the system designers so that they can refine the communication part of the component for a system function before integrating it to the system. Therefore, the component designers just model the computation part of the component IP without worrying about the refinement of its communication part.

The rest of this paper is organized as follows. In Section II we review the previous works on reusable component designs and refinement-based design methodologies. We introduce the refinement-based design environment and explain the proposed component IP package in Sections III and IV, respectively. After describing a H.264 VLD component design example in Section V. we summarize our contributions and future works in Section VI.

II. Related Works

Several methods that provide more flexibility in the communication interface have been proposed. These component design methods can be summarized into two approaches: standard interface-based and abstract interface-based.

A. Standard Interface-based Component Design Approach

There are several popular standard interfaces for on-chip buses and memories, which are main primitives for the system integration. Among the various on-chip bus interfaces, AMBA is the most popular bus interface, which defines transaction functions, protocols and RT-level signals.

The example shown in Figure 1 is a design environment for a standard interface-based component that has three functions F_A, F_B and F_C, which communicate with peer functions A_F, B_F, and C_F, respectively, in a system with standard interfaces. Although its intended functions were initially only three, the final component IP contains eight functions including two bus interface, two buffer, one on-chip memory interface. Although integrating a component IP to the system is relatively easy, designing the component IP itself is more complicated. Especially, it is difficult and time-consuming to design and verify the bus interface logic. Moreover, in designing a component IP with a standard interface, the component designer should make several architectural decisions, for example,

+ Bus interface standard: AMBA or CoreConnect
+ Memory type: On-chip memory, External memory
+ # of bus interfaces and on-chip memory interfaces
+ Bus interface types: Bus master or Bus slave
+ Existence of internal buffers and those sizes

2006 Asia and South Pacific Design Automation Conference

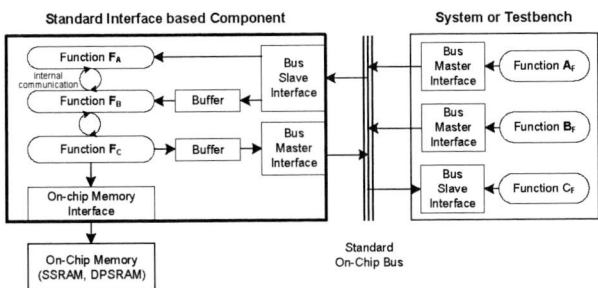

Figure 1. Standard Interface-based Component Design

To build an optimized system, these architectural decisions should be made not by the component designer, but by the system designers. Although this approach eases system-level hardware integration, it limits the system level design space.

B. Abstract Interface-based Component Design Approach

Exploiting the concept of orthogonalitzation can enhance the reusability of the component IPs. There are two types of orthogonalization: the separation of a function and its architecture and the separation of computation and communication for a function. Many researchers have made good contribution, which can be summarized in three stages: (1) finding a good abstraction of various interfaces, (2) generating wrappers efficiently and automatically, and (3) refining the architecture of system functional models

In the stage (1), VCI [4] and OCP-IP [5] defined generic protocol between the internal core part of a component IP and its bus wrapper. In figure 1, VCI and OCP-IP can be the interface between F_A and its bus slave wrapper. The bus interface controller, internal buffers and on-chip memory controllers are the wrappers in this category.

In the stage (2), methods for wrapper generation or synthesis were studied. Y. Hwang et. al. [2] proposed a method for generating communication wrappers from the timing diagram. They showed that synthesized wrappers are more efficient in terms of delay and area comparing to the generic wrappers and bridges.

In the stage (3), the refinement-based system design is an important issue in the design automation. F. Gharsalli, A. Jerraya et. al. [7,8] proposed an MPSoC design methodology. The computation part of the function is captured as a virtual component (VC), which is mapped onto architectural components: processors, memories and ASIC IP cores. Then, the architectural components are integrated with generic wrappers. S. Abdi, D. Shin, D. Gajski [9] proposed a communication synthesis tool, which is based on their own refinement-base design environment where they capture the communication function as a channel and refine it using a protocol library, which is a template set for the channel implementation.

III. Refinement-based Design Environment

The refinement-based design approach is a top-down system-level design methodology, in which the system level

functions are first captured at a higher abstraction level. A system function should be divided into computation and communication functions. Hereafter, a communication function is called as a channel. After each system-level function is captured, each of its computation functions or channels is refined into a more concrete implementation step by step by making a certain architectural decision such as HW-SW partitioning, types of processors, types and sizes of on-chip memories, and the number of ASIC cores or embedded FPGAs.

Computation functions and channels of a system-level function should be captured separately with the hints from the refinement-based design environment. Because the channels such as FIFOs, arrays, and buses are commonly used in the system modeling, we provide them as a primitive library. The functional model shown in Figure 2 is for the component design example described in Section II, which includes six computation functions, three FIFO channels and a memory channel. Each computation function and channel is connected to the others through abstract interfaces regardless of their architectures or implementations.

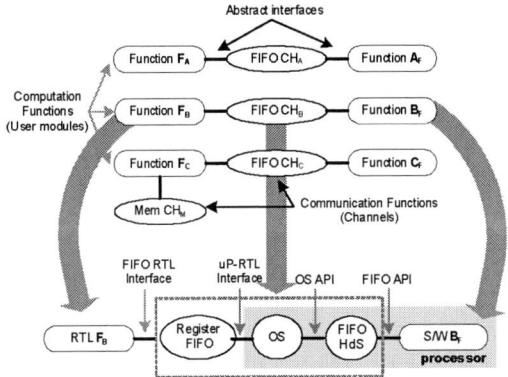

Figure 2. System Function Model and Partial Refinement

In our refinement-based design flow, a system-level function model is integration of user-defined computation functions with selected channels. A system implementation process can be seen as channel template replacement and high-level synthesis (or manual re-description) of the computation functions to a lower abstraction level. Our refinement-based design environment provides the following three types of hints to make modeling and refinement easy:
(1) hints for primitive channels that are the most frequently used for communication.
(2) hints for abstract interfaces for function modeling, RTL and S/W implementation.
(3) hints for architecture templates for refining the primitive channels.

These hints are represented with communication architecture template trees (CATtrees). Next, we explain primitive channels, abstract interfaces and channel templates before explaining the CATtrees

A. Primitive Channels

Models of computation like Kahn Process Network

589

(KPN) and Synchronous Dataflow Network (SDF) provide a well-defined set of abstract communication functions or channels. Although an abstract FIFO, which is the fundamental communication function of KPN and SDF, can model any point-to-point communication, it is too abstract to model an application function with the abstract FIFO. It is not easy to refine a MoC-modeled system function with it. Therefore, we replace it into the following four types of primitive channels, which are more concrete and widely used in the real application function modeling and system implementation.

(1) FIFO channels

A FIFO channel has two primitive functions: a blocking read and a blocking write, which are sufficient for capturing a high-level FIFO function. For the step-by-step architecture refinement, however, we additionally defined four additional functions such as peek, clear, more and sync. A peek function just checks the data without data retrieval, which is useful for H/W RTL implementation. A clear function initializes the FIFO channels. A more function returns true if there is more data to be written, which is useful to determine when to destroy a dynamically allocated channel. A sync function returns true if the written data is transferred to the reading counterpart, which is relevant when the refined architecture of a FIFO channel has intermediate buffers. The FIFO channel provides two abstract interfaces: abs_fifo_write interface and abs_fifo_read interface.

(2) Variable channels

A variable channel provides read and write functions, which do not have data synchronization. This channel is very useful in modeling memories updated infrequently. This channel provides abs_var_read and abs_var_write interfaces.

(3) Array channels

An array channel is a set of variable channels that is pointed by an index. Its primitive functions include index read and write functions. Although the array and variable channels are not required in the most MoCs, they are very useful in modeling and refining architecture. An array channel is an abstraction of memories including on-chip and external memories. Therefore, it is not easy to write a RTL model without using them in many cases. This channel provides abs_array_read and abs_array_write interfaces.

(4) Bus channels

A bus channel can perform point-to-point data transfers through a shared medium. In the top-down approach, a bus channel is not necessary in the function modeling. In the bottom-up approach, however, the bus channels are the most important communication pattern to connect components. In our refinement-based design environment, the bus channel is not used in function modeling but many partially refined (PR) channels, which will be explained later, have bus interfaces. For example, the PR channels and adapters related to the S/W have bus interfaces because RISC processors have a bus master interface.

B. Abstract Interfaces

In the system function model, an abstract interface is a boundary between a computation function and a channel. For computation of a function, its abstract interfaces mean primitive functions that it can use. For communication of a function, its abstract interfaces mean primitive functions that it must implement. Therefore, the abstract interfaces are a key to separate computation and communication for a function. An abstract interface is active for computation while it is passive for communication. An active interface of a computation model is connected to a passive interface of the channel. An abstract interface can be refined into three concrete interfaces: a TLM one for function capture and transaction level simulation and a RTL one for RT-level implementation, and a SW API one for S/W implementation.

C. Partially Refined Channels (PR channels)

Partially refined (PR) channels are used to refine primitive channels such as FIFOs, variables and arrays by making certain architectural decisions.

(1) Bus-FIFO Channels

A bus-FIFO is two FIFO channels that are connected using a bus channel, which can be refined to a Bus Master Write FIFO (top) and a Bus Master Read FIFO (bottom) as depicted in Figure 3(a). Although the internal architecture of a bus-FIFO is complex, its function is the same with that of the abstract FIFO channel.

(2) Cached Array Channels

A cached array channel is an array channel that contains a cache. An array channel is often refined into an external memory. However, an external memory has long latency and limited bandwidth. Therefore, an array channel can first be refined to a cached array channel that is connected to an external memory as depicted in Figure 3(b).

(3) Bus-Memory Channels

Some memories are shared with many computation functions. A bus-memory is an abstraction of shared memories that has a bus slave interface. A bus-memory channel can be refined to either an on-chip memory (left) or external SDRAM memory (right) as depicted in Figure 3(c).

(4) Channel Adapters

There are two types of channel adapters, which are interface and abstraction adapters. An interface adapter is used to connect two different types of channels. For example, a Bus Master Read FIFO channel shown in Figure 3(a) contains a bus_fifo_sender channel and a bus_fifo_receiver channel. These two PR channels are interface adapters. They adapt a FIFO channel to a bus channel. An abstraction adapter connects two channels in different abstraction levels. For example, a TLM-to-RTL adapter and a RTL-to-TLM adapter can connect a transaction-level model of computation to a RTL model of channel for a function or vice versa.

D. Channel Templates

Channel templates are configurable implementations of the primitive and PR channels, which are actually parameterized source files or generators. There are three templates for each channel: TLM, RTL, and S/W ones. A channel template has template parameters. To instantiate a channel instance, the value of the template parameters must be determined.

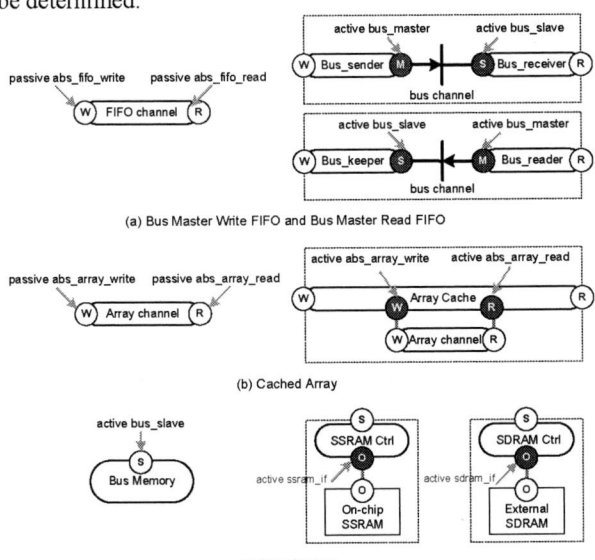

Figure 3. Partially refined channels

E. Communication Architecture Template Trees

A communication architecture template tree (CATtree) for an abstract channel is the collection of a primitive channel, an abstract interface and a set of templates for the abstract channel. We named CATtree because the architectures are expanded like a tree as shown in Figure 4.

Representing the architectural information for an abstract channel with its CATtree is an effective and integrated way of realizing the orthogonalization of function and architecture for communication. Providing a set of CATtrees can clearly expose what are communication functions and what are not for the designer. Once a communication function for a system function is captured with the CATtree for an abstract channel, refinement of that channel is guaranteed by its corresponding templates.

Computations for a system function are modeled with the TLM APIs in the CATtree and they are integrated to build its system function model. This system function model is the starting point of further implementation. Channels in the system function model are refined by replacing it with templates of the CATtree. Computations can be refined to RTL or S/W by high-level synthesis tools or by hand. Each computation and communication can be refined independently exploiting the abstract interfaces and adapters. Note that we do not cover the refinement of computation in this paper.

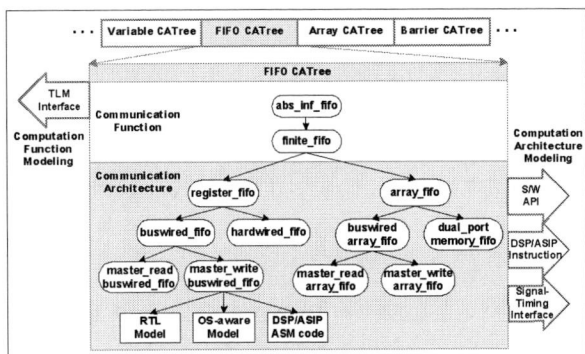

Figure 4. Communication Architecture Template Tree

IV. Reusable Component IP Design

Refinement-based design is an alternative to the reuse-centric design methodology. However, the platform-based design methodology is still attractive to the design teams who have systems that have been pre-integrated and verified. Therefore, reusable components are still important matters for them. However, conventional component IPs, which are designed to have standard interfaces, have limited reusability. Therefore, we introduce a new method of component IP design and delivery that utilizes our refinement-based design environment.

In this method, a component designer captures the component function using the TLM interfaces in the CATtrees. At the same time, he or she should model its test drivers. The test drivers generate stimulus to the captured component function and validate its response from the captured function model. By connecting the component function and its test-drivers with the TLM templates in the CATtrees, the component design obtains a testbench for validation, verification and interface refinement of the component IP. CATtree-based function capture of the component IP is relatively easier than bus-aware function modeling because he or she can concentrate on the computation and ignore communication related details.

The component designer can describe a RTL model of the computation function with the RTL interfaces. An abstraction adapter is inserted when a RTL computation function is connected with transaction-level channels and test drivers. Most of the designers have difficulty in designing a component because of the bugs related to its interfaces. In our approach, the designer can ignore the details about communication, which greatly reduces the complexity of the RTL design. For a computation function, we define the set of its transaction-level model, RTL model, and testbenches as the augmented deliverable package of its reusable component IP.

System designers can configure the communication part of the component IPs with our refinement-based design environment. The system designers replace an abstract channel with a more concrete one by using the CATtrees. Because they replace the channels instead of adapting the interfaces, the communication parts of the test drivers are also refined together. Therefore, without any manual

6B-1

modification of the testbenches, the system designers can execute them to estimate the performance of the component IP and to verify the refined model at each refinement step.

V. An H.264 VLD Component Design

We designed a new H.264 decoder system by reusing a system design shown in Figure 5. We decided to design and integrate a new dedicated hardware IP for the VLD operation. We followed the platform-based design methodology, utilizing the proposed method to design a more reusable VLD component IP as follows.

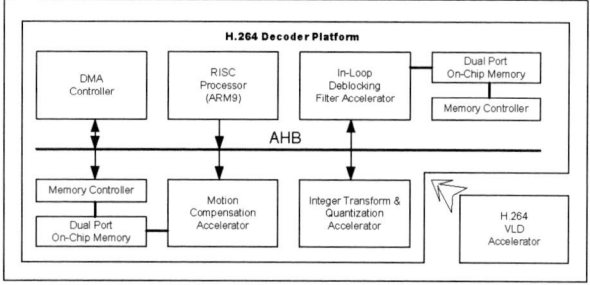

Figure 5. H.264 decoder SoC Platform

A. Reusable H.264 VLD Component IP design

A component designer designed an H.264 VLD component IP with the design flow we described. First, he captured its computation function and developed its test drivers, which took about a. week (Figure 6). The computation consists of a VLD core, an nC calculation block and a NAL decoder, which has 4 FIFO interfaces and 1 array interface. The VLD core decodes Exp-Golomb and CAVLC codes. The nC calculation block determines which mapping table is used in the VLD core to decode a CAVLC code. The NAL decoder eliminates the emulation prevention codes in the bit-stream.

After modeling the VLD function, he manually described its RTL model in HDL with the RTL interfaces of the CATtrees. The RTL model can decode an Exp-Golomb code in 1 cycle, a 2x2 chroma DC in 10 cycles (average) and 4x4 residual decoding in 30 cycles (average). We synthesized the RTL model using the Synopsys DesignCompiler™ and the estimated gate count was 8277 gates at 5 ns delay in 0.18 um process technology. He finished both description and verification of the VLD RTL model in two days. It was relatively faster than its function modeling because of the two reasons. First, we described only a computation part in details. In general, the communication part is more complex and error-prone than the computation part. In our approach, however, potential communication errors are eliminated by exploiting the simple abstract interfaces of the CATtrees. Second, in the verification of RTL description we reused all the test models used for function modeling.

B. Communication Refinement and System Integration

A system designer configures the communication part of the H.264 VLD computation and integrates it to an existing platform. Here we present two implementations of the VLD

IP in the H.264 decoder for QCIF (176*144) and HDV (1280*768) images, respectively.

(1) VLD decoding for QCIF images

In decoding QCIF images of 15fps, both computation and communication loads are low. The system designer decided to configure the VLD computation to have only an AHB slave interface for easier system integration. Because Flexible Macroblock Ordering (FMO) in the H.264 baseline profile, he also decided that the array channel connected to nC calculation sub-block has only 44 indices. In this configuration, the memory size is 220 bits and he finally decided to refine the array channel into a register array. Figure 7(a) shows refinement steps. The total gate count of the communication part was 12,150 gates

(2) VLD decoding for HDV images

In decoding HDV images of 30 fps, both computation and communication loads are very high. Additionally the H.264 decoder must support the FMO feature. Because the FMO feature requires all nC values of a frame, $ARRAY_{nC}$ must have at least 92160 indices, which is too big to be implemented with registers or on-chip memories. Thus, he decided to configure the H.264 VLD computation with the following five refinements:

- $FIFO_{CMD}$ to a 8 depth Bus-Master Write channel and refine $FIFO_{VLD}$ to a 8 depth Bus-Master Read channel
- a bus slave interface is shared by $FIFO_{CMD}$ and $FIFO_{VLD}$.
- $FIFO_{ITQ}$ to a dedicated register FIFO.
- $FIFO_{STRM}$ to a memory FIFO that have a cache with 16 registers of 8-bit width
- $ARRAY_{nC}$ to an external memory with cache.

Figure 7(b) shows the refinement steps according to the decisions listed above. The total gate count of the communication part was 10,559 gates.

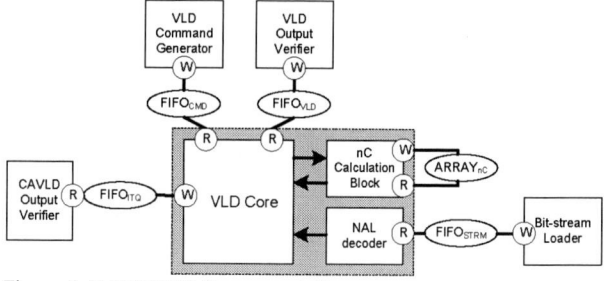

Figure 6. H.264 VLD Component IP

VI. Conclusions

We proposed an effective design approach to enhancing the reusability of component IPs. In this approach, we re-defined the roles of component and system designers. In designing a component IP, a component designer captures its computation and its communication in an abstract way while a system designer refines its communication part with our refinement-based design approach. We also proposed an augmented component IP deliverable package according to the redefined roles of the component and system designers.

592

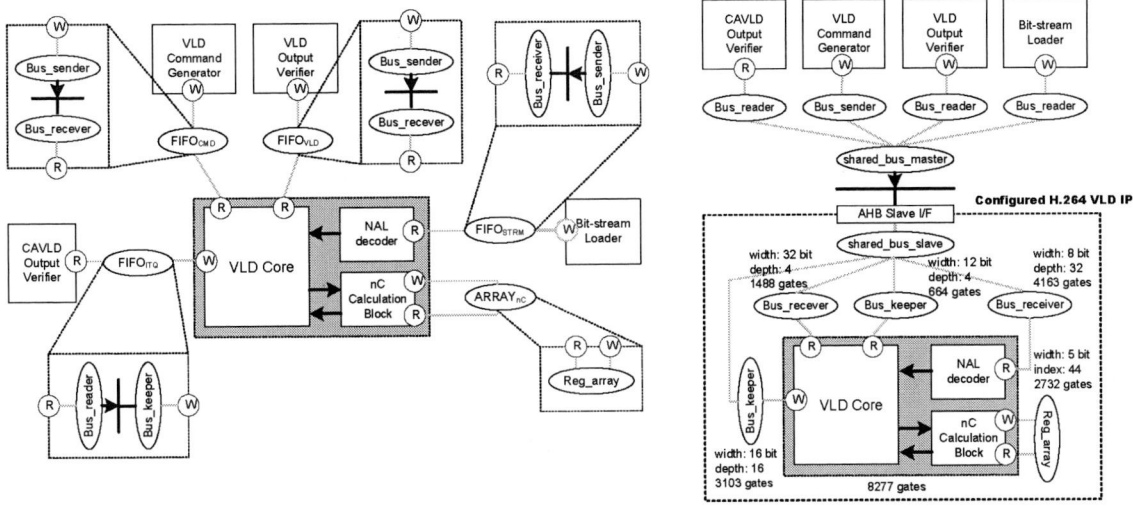

(a) H.264 VLD Component IP configuration for QCIF Decoding

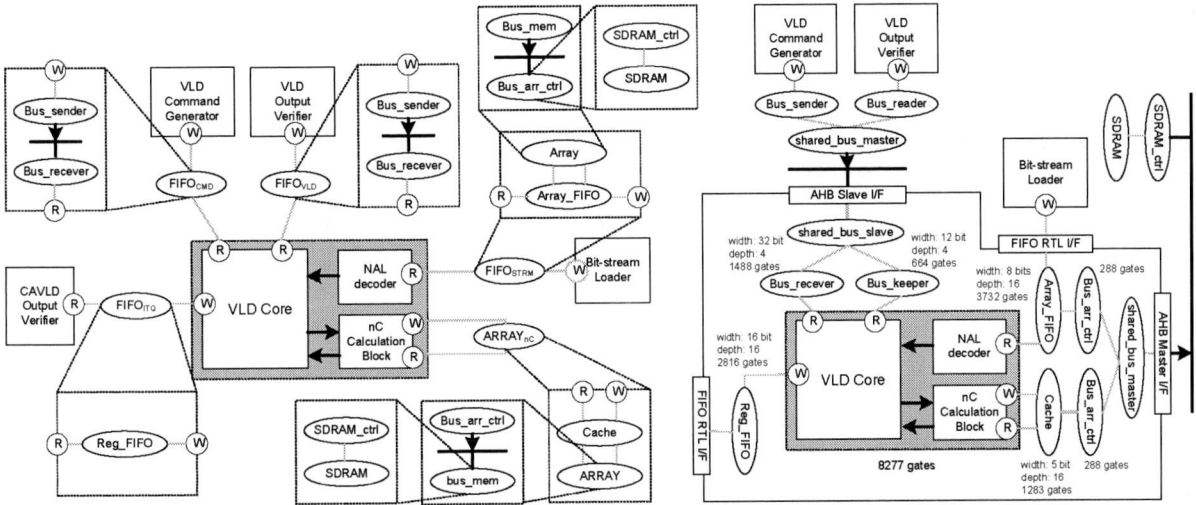

(b) H.264 VLD Component IP configuration for HDV Decoding

Figure 7. H.264 VLD Component IP Configuration Example

Our refinement-based design environment introduced a concept of the Communication Architecture Template Tree (CATtree), which can clearly separate computation function and communication architecture. Moreover, it provides generic interfaces to model both of them. After designing a VLD component for H.264, we found that the concept of the CATtree is effective for the communication refinement.

With this proposed approach, the component designers can design computation functions easily and the system designer can explore a larger design space with the help of the CATtree. We are now developing an H.264 decoder system with our design environment to exploit the full power of the CATtree. We will cover the refinement of computation later in another paper.

References

[1] Michael Keating, Pierre Bricaud, "Reuse methodology manual", Kluwer academic publisher, 2002

[2] Y.-T. Hwang and S.-C. Lin, "Automatic protocol translation and template based interface synthesis for IP reuse in SoC", *Proc of 10th ASP-DAC*, pp. 565-568, Dec. 2004

[3] K. Keutzer, S. Malik, et. al., "System-level design: Orthogonalization of concerns and platform-based design", *Trans. on Computer-Aided Design of Integrated Circuits and Systems*, Vol19, No. 12, pp. 1523-1542, Dec. 2000.

[4] VSI Alliance On-Chip Bus Development Working Group,"OCB 2.0", Apr. 2001

[5] Sonics Inc. "Open core protocol specification 1.0", 2000

[6] W. Cesario,D. Lyonnard,G. Nicolescu,Y. Paviot,S. Yoo,A. Jerraya, "Multiprocessor SoC Platforms : A Component Based Design Approach", *IEEE Design & Test of Computers*, pp. 52-63, Nov. 2002,

[7] F. Gharalli, D. Lyonnard, S. Meftali, F. Rousseau, A. A. Jerraya, "Unifying memory and processor wrapper architecture in multiprocessor SoC design", *proc of ISSS'02*, pp. 26-31, Oct. 2002.

[8] S. Abdi, D. Gajski, "Automatic generation of equivalent architecture model from functional specification", *Proc. of42th DAC*, pp. 608-613, June, 2004

An Interface-Circuit Synthesis Method with Configurable Processor Core in IP-Based SoC Designs

Shunitsu Kohara[†]   Naoki Tomono[†,‡]   Jumpei Uchida[†]   Yuichiro Miyaoka[†,*]
Nozomu Togawa[‡]   Masao Yanagisawa[†]   Tatsuo Ohtsuki[†]

[†] Department of Computer Science, Waseda University
[‡] Presently, the author is with the Toyota
[*] Presently, the author is with the Toshiba
E-mail: kohara@yanagi.comm.waseda.ac.jp

Abstract— In SoC designs, efficient communication between the hardware IPs and the on-chip processor becomes very important, however the interface is usually affacted by the processor core specification. Thus in this paper, we focus on developing an efficient interface circuit architecture for the communications between the on-chip processor and embedded hardware IP cores. we also propose a method to synthesize it. Experimental results show that our method could obtain optimal interface circuits and works well through designing a MPEG-4 encode application.

I. Introduction

The growing demand for hardware/software systems, together with the ability to put the entire system on a single chip using deep sub-micron technologies, has led to the evolution of complex hardware/software system-on-chips (SoCs). While the complexity of SoCs increases, so does the demand to reduce their time-to-market. Typically, IP-based SoC design contains the following steps such as application specification , hardware/software partition and hardware/software integration. Though the design time of SoCs can be greatly reduced by efficient re-use of intellectual property (IP) cores, how to develop an efficient interface circuit between the hardware IPs and the on-chip processor becomes an important task.

One of the solutions is to generate the one automatically [8]. Works in this approach include [10, 7, 5, 15]. In [10], an arbiter consists of protocol conversion FSM and FIFOs to regulate transfers and mismatched protocols are mapped into a standard communication scheme. In [7], regular expression is used to describe protocol and the interface is generated as a product machine from automata from both of two IPs using formal approach. In [5], communication protocol of IP is described as FSM, and a protocol translation algorithm is proposed, which derives an interface FSM between two IPs. In [15], for saving the complexity in design space exploration, parameterized templates are used to synthesize a hardware.

On the other hand, to integate a processor core into SoC, the practical method is not to use a processor core, but use a configurable one so as to satisfy the preformance requirements and the area constraints. When SoC application is implemented on a configurable processor core, such as [12, 13], and several hardware IPs, generating the interface between the processor core and the hardware IPs requires: (1) to communicate with the hardware IP (2) to communicate with the configurable processor core.

In the literature, most of the previous works assumes that both of connected IPs have same model for interface description. However most of the used IPs are not standard IPs, if one side of them is a configurable processor core, the interface is affected by the processor core specification, such as instruction set, pipeline stages.

In this paper, we propose an architecture of the interface circuit to communicate with a configurable processor core and a hardware IP. We also propose a method for synthesizing the one. The models of the interface in previous works are based on FSM and so on. We use architecture templates of the interface circuit. Our proposed architecture and method enable us to obtain optimal interface circuits.

This paper is organized as follows. Section II describes IP-based SoC design method and target architecture. Section III proposes an architecuter of an interface circuite (IFC) and a method for synthesizing the one (*IFC_Synthesizer*). Section IV shows the experimental results with the proposed method through a MPEG-4 encoder application. Section V gives the concluding remarks.

II. IP-Based SoC Design Method

In this section we describe IP-based SoC design method and define target architecture.

A. Design Method

Figure 1 shows a design method with an interface circuit synthesizer "*IFC_Synthesizer*". A designer searches hardware IPs for hardware parts from a hardware IP database

0-7803-9451-8/06/$20.00 ©2006 IEEE.

2006 Asia and South Pacific Design Automation Conference

6B-2

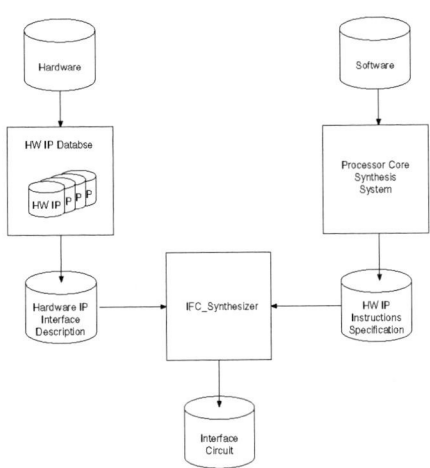

Fig. 1. Design method with *IFC_Synthesizer*

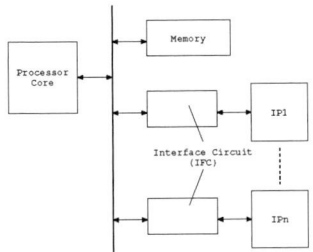

Fig. 2. Architecture model of the target SoC.

and uses a processor core synthesis system for software parts. The interface description of using hardware IP and instruction set for hardware IP generated by the processor core synthesis system are inputs for *IFC_Synthesizer*. The output of *IFC_Synthesizer* is an interface circuit (IFC), which communicates with the processor core and the hardware IP.

B. Architecture Model

Figure 2 shows the architecture model of the target SoC. The architecture consists of a processor core, a memory and several hardware IPs which are connected with each other via a shared bus. In our approach, first, the input application is partitioned into hardware/software parts, then the hardware parts are implemented with hardware IPs, and the software parts are implemented on a processor core.

B.1 Processor Core

The processor core is configurable. The configurable parameters include an instruction set, pipeline stages, hardware units such as ALU, multiplier, register files and so on.

The instruction set of the processor core includes hardware-IP-instructions. The hardware-IP-instructions

are described in Subsect. A.2.

B.2 Hardware IPs

Hardware IPs distributed in the market have various architecture and interface. In the target architecture, software on the processor core controls the hardware IPs with hardware-IP-instructions. To avoid bus conflicts, hardware IPs should not have data transfer unit but data-path for processing data.

We make a premise that in our work all the target hardware-IPs in Hardware-IP Database have an interface description in CWL [4]. *IFC_Synthesizer* synthesizes IFC from CWL description of the target hardware IP.

B.3 Memory

The memory in the target architecture is a simple model like SRAM.

III. IFC_SYNTHESIZER

In this section we propose an IFC architecture and a method for synthesizing it. In our work, the synthesizer is called as *IFC_Synthesizer*. In this section, we first illustrate the interface of processor core and hardware IPs, and then we propose an IFC architecture and an algorithm of *IFC_Synthesizer*, where *IFC_Synthesizer* is the name of the synthesizer developed in our work. Details will be explained in the followings.

A. Interface

The interface between processor core and hardware IP is based on ARM Coprocessor Interface [3]. The ARM Coprocessor Interface defines a signal interface and an instruction interface.

A.1 Signal Interface

Figure 3 shows a connection of the processor core and hardware IPs. The processor core can connect up to 16 hardware IPs. The processor core communicates with hardware IPs with three handshake signals as follows:

nCPI *not CoProcessor Instruction* (Processor Core → Hardware IPs): A processor core wants to execute hardware-IP-instruction.

CPA *CoProcessor Absent* (Hardware IP→Processor Core): There are no hardware IPs which can execute the hardware-IP-instruction.

CPB *CoProcessor Busy* (Hardware IP→Processor Core): Hardware IP can not execute the hardware-IP-instruction immediately since it is executing another hardware-IP-instruction.

595

6B-2

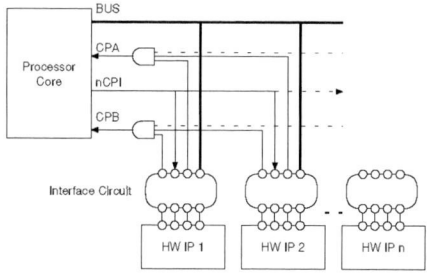

Fig. 3. Connection between a processor core and hardware IPs.

A.2 Instruction Interface

Processor core sends three type of hardware-IP-instructions: (a) CDP (processing data operations), (b) LDC/STC (transfer data operations from / to a shared memory) and (c) MCR/MRC (transfer data operations from / to a register in a processor core).

The format of hardware-IP-instruction is as follows:

```
CDP HW#, OP#
LDC HW#, N, Rd, Rn, offset
STC HW#, N, Rd, Rn, offset
MRC HW#, Rd1, Rd2
MCR HW#, Rd1, Rd2
```

CDP performs processing operation with a hardware IP. Each hardware IP is numbered. HW# is the number. A processor core operates a hardware IP to use HW#. When a hardware has several functions. Each function is numbered. OP# is the number. The processor core use OP# to select the function.

LDC/STC transfer data between a hardware IP and a shared memory.

MCR/MRC transfer data between a processor core register and a hardware IP register.

B. IFC

In the model of the target SoC, the processor core controls hardware IPs with the interface described in Subsect. A. Since hardware IPs distributed in the market might be provided by different vendors, they do not always have the standard interface. So it will cause many problems, such as:

1. they can not communicate on handshake communication with the signal interface.

2. hardware IP can not decode hardware-IP-instructions by the processor core.

3. therefore processor core can not control hardware IPs.

IFC synthesized by *IFC_Synthesizer* communicates with the processor core at the proxy of the hardware IP. Figure 4 shows the architecture of IFC. Each of units is as follows:

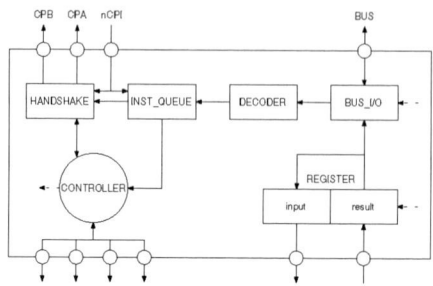

Fig. 4. Architecture of IFC

IFC defines the mapping of the external and internal ports of IFC.

BUS_I/O controls input/output data flow via shared bus. It (1) inputs data from shared bus to REGISTER (2) or inputs hardware-IP-instructions from shared bus to DECODER (3) or outputs data from REGISTER to a shared bus.

DECODER decodes instructions from the processor core. If the instruction is a kind of hardware-IP-instructions and HW# field and OP# field are validate for target hardware IP, DECODER decodes the instruction and queues it into INST_QUEUE.

INST_QUEUE preserves bit vectors decoded by DECODER. If the target hardware IP is not busy, it dequeue first bit vector to CONTROLLER.

HANDSHAKE deals with handshake protocol with the signal interface. It is controlled with nCPI signal from processor core and control signals from CONTROLLER, and output CPA and CPB signals for the handshake communication.

REGISTER saves data from / to a shared memory and the target hardware IP. input REGISTER saves data before hardware IP processing, and result REGISTER saves data after hardware IP processing.

CONTROLLER controls all units in IFC with control signals and controls hardware IP for processing data. It consists of counter and state machine. The input is given from INST_QUEUE and HANDSHAKE. It is described in detail in Sect. C.3.

C. IFC_Synthesizer

IFC_Synthesizer synthesizes IFC HDL from the interface description of a hardware IP (Fig. 5). The interface description is written in CWL. To communicate with both of a processor core and a hardware IP, IFC must have the interface of them. The interface to a processor core described in Sect. A has been defined. Since the interface to a hardware IP depends on its own specification, IFC is synthesized for each of the using hardware IPs.

596

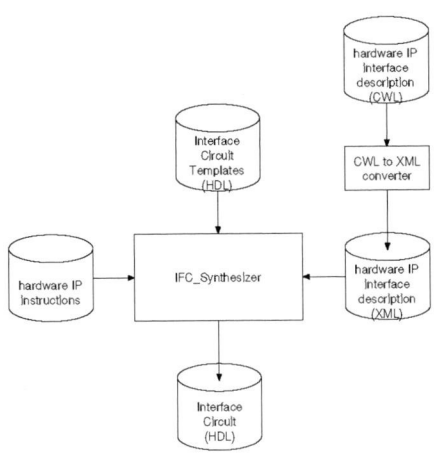

Fig. 5. IFC_Synthesizer

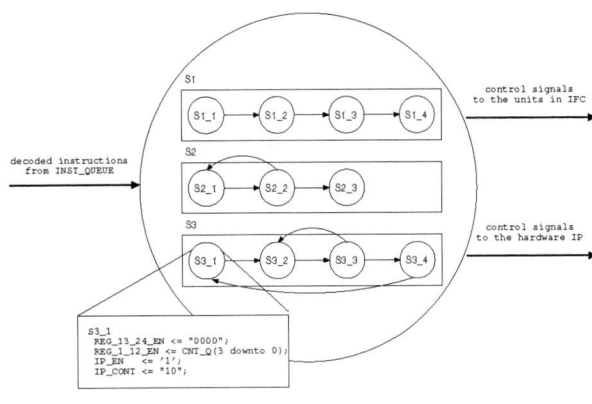

Fig. 6. state and sub-states in CONTROLLER.

One of the input for *IFC_Synthesizer* is HDL template description of IFC. Not all the units are synthesized for each of the using hardware IPs. The units which need not vary are template description as they are.

In Fig. 4, BUS_I/O, HANDSHAKE are interfaces to a processor core. Since the interface to a processor core has been defined, they are independent of using hardware IP.

DECODER and INST_QUEUE are also interfaces to a processor core. In the hardware-IP-instructions, HW# indicates which hardware IP is a target one, and OP# indicates which function are a target one. Since HW# and OP# depend on hardware IPs, the HDL description of DECODER varies for them. The queue length of INST_QUEUE is correspond with pipeline stages of a processor core.

CONTROLLER is the interface to a hardware IP. Since the interface to a hardware IP depends on its own specification, the HDL description of CONTROLLER varies for the hardware IP.

REGISTER is the interface to a hardware IP. The register size is decided by the specification of the hardware IP.

Methods of synthesizing the units which varies for using hardware IPs are described following sections.

C.1   DECODER

IFC_Synthesizer refers hardware-IP-instructions generated by compiler to synthesize DECODER. It preserves values of the HW# and OP# correspond with a target hardware IP. DECODER decodes hardware-IP-instructions correspond with the target hardware IP and ignores the others.

C.2   INST_QUEUE

IFC_Synthesizer refers hardware-IP-instructions generated by compiler to synthesize INST_QUEUE. INST_QUEUE depends on the decoded bits by DECODER and pipeline stages of a target processor core.

C.3   CONTROLLER

CONTROLLER sends control signals to all the units in IFC to execute hardware-IP-instructions. CONTROLLER has state machines to control all the units in IFC and the target hardware IP. The state machine has states correspond with hardware IP functions, and each of the states has several sub-states (Fig. 6) Control signals are defined every sub-state to execute hardware IP functions.

Figure 6 is an example to execute STC, which is one of the hardware-IP-instruction. When IFC received STC from a processor core, state S3, which is correspond with STC, starts and CONTROLLER sends control signals defined in the sub-state S3_1. Then sub-state is transitted to S3_2, and alike.

Control signals from/to CONTROLLER are classified into three groups:

- input signals from INST_QUEUE and output signals to BUS_I/O and HANDSHAKE.

- input/output signals from/to hardware IP. They are defined at the number of ports of the target hardware IP. The name of them begins "IP_".

- output signal to REGISTER. The bit width depends on the register size.

The algorithm of synthesizing CONTROLLER is as follows.

1. Ports Decision
   external and internal ports in IFC are decided.
2. States Decision
   states for processing and transferring data are decided.
3. Sub-states Decision
   sub-states, which define control signals to all the units, are decided.
4. Sub-state Transitions Decision
   transitions among sub-states are decided.

We illustrate them with an example CWL description in Fig. 7.

597

```
port:
 input.en          EN;
 input.control[1:0] CONT;
 input.data[7:0]    ADR;
 input.data[31:0]   DATA;
endport
alphabet;
 signalset a = {CLK, EN, CONT,  ADR, DATA};
         I : {R,    1,   2'b01, x,   Z   };
         N : {R,    1,   2'b00, x,   Z   };
     R(Xa) : {R,    1,   2'b10, Xa,  Z   };
     O(Xd) : {R,    1,   2'b11, x,   Xd  };
 endsignalset
endalphabet
word;
 proc(Xa,Xd):(R(Xa) N[2])[1,2] O(Xd)[3];
endword
```

Fig. 7. CWL description example.

```
BUS_IO_S: out std_logic;
HANDSHAKE_RUN: out std_logic;
HANDSHAKE_TR: out std_logic;
REG_EN: out std_logic_vector(3 downto 0);
IP_EN: out std_logic;
IP_CONT: out std_logic_vector(1 downto 0);
```

Fig. 8. Ports decision

Ports Decision *IFC_Synthesizer* decides control signals to BUS_I/O and HANDSHAKE. Control signals to a hardware IP are correspond with input.control in CWL description. output ports of a hardware IP are input ports of CONTROLLER. The bit width of CWL description is equal the one of HDL description.

output.data signals connects REGISTER. Control signals to REGISTER depend on the register size (Sect.C.4 in detail).

Figure 8 shows output signals of CONTROLLER in VHDL at the example in Fig. 7.

States Decision *IFC_Synthesizer* decides states correspond with hardware-IP-instructions from a processor core. For CDP instructions, the number of states are equal to the number of functions of the target hardware IP. In case of Fig. 7, if the target hardware IP is numbered as "1" and proc is numbered as "2", which means HW# = 1 and OP# = 2, the processor core sends CDP 1, 2. *IFC_Synthesizer* defines the state S_CDP_2 correspond with it.

Sub-states Decision *IFC_Synthesizer* decides the values of each sub-states.

When a target hardware-IP-instruction is LDC/STC or MCR/MRC, the control signals to hardware IP are "don't care". In case of receiving data instructions such as LDC and MRC, BUS_I/O behaves as a data receiver from the bus to the input REGISTER. On the contrary, in case of sending data instructions such as STC and MCR, BUS_I/O behaves as a data sender to the bus from the result REGISTER. In case of data transferring instructions such as LDC and STC, HANDSHAKE_TR is set to "1", which means transferring data.

```
if CURRENT_STATE = S_CDP_2_1 then
   BUS_IO_S    <= '0';
   HANDSHAKE_RUN <= '1';
   HANDSHAKE_TR  <= '0';
   REG_EN       <= CNT_Q(3 downto 0);
   IP_EN        <= '1';
   IP_CONT      <= "10";
```

Fig. 9. Control signals decision at the sub-state S_CDP_2_1

```
elsif CURRENT_STATE = S_CDP_2_2 then
  if CNT_Q = 3 then
     NEXT_STATE <= S_CDP_2_1;
  elsif CNT_Q = 6 then
     NEXT_STATE <= S_CDP_2_3;
  else
     NEXT_STATE <= S_CDP_2_2;
  end if;
```

Fig. 10. Sub-state S_CDP_2_2 transitions

When a target hardware-IP-instruction is CDP, the control signals to hardware IP are required. In case of Fig. 7, proc operation is defined at *word* section. Since proc consists of *alphabets* R(Xa), N, O(Xd), sub-states S_CDP_2_1, S_CDP_2_2, S_CDP_2_3, correspond with *alphabets* R(Xa),N, O(Xd), are defined. The values of the control signals every sub-state are defined as the value at *alphabets* section in CWL description. HANDSHAKE_RUN is set to "1", which means busy for processing data.

Figure 9 shows control signal decision at sub-state S_CDP_2_1 correspond with *alphabet* R(Xa).

Sub-state Transitions Decision *IFC_Synthesizer* decides a sequence of sub-states to execute operations. In CWL description, a sequence of *word* is expressed as regular expression of *alphabet*. In case of proc operation in Fig. 7, the sequence of *alphabet* is R, N, N, R, N, N, O, O, O. Figure 10 shows sub-state transitions of S_CDP_2_2 correspond with N.

C.4 REGISTER

When *IFC_Synthesizer* decides REGISTER, we must know the resister size required. The size is given by the processor core synthesis system. Hardware-IP-instructions used in the application, include the length of transferring data,therefore we can decide the size of required registers.

C.5 IFC

IFC is the top layer of all the units in IFC. *IFC_Synthesizer* decides the mapping of all the units and external ports in IFC. Mapping is independent of the target hardware IP. Though external ports to the bus and the processor core is also independent, external ports to the hardware IP vary for a target hardware IP. We can decide the number and bit width of them from *port* section in CWL description of the target hardware.

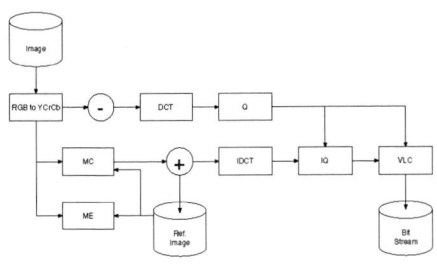

Fig. 11. MPEG4 encode algorithms.

TABLE I
HARDWARE IP INFORMATION.

function	area $[mm^2]$	response time $[cycles/\text{MB}]$
RGBtoYCrCb[1]	0.3904	489
DCT/IDCT[6]	2.4480	1192
ME/MC[2]	3.6000	1032

IV. EXPERIMENTAL RESULTS

We implement *IFC_Synthesizer* in Ruby Language [9]. We design MPEG-4 encoder as a SoC application in SystemC under the design framework[14]. The design environment is as follows: OS: Linux 2.4, CPU: Intel Pentium III 500MHz, RAM: 192MB.

Figure 11 shows MPEG-4 encode algorithms. We partition them into hardware parts and software parts to estimate the performance as a representative. Color space convert (RGBtoYCRCb), motion estimation (ME), motion compensation (MC) and discrete cosine transform / inverse discrete cosine transform (DCT/IDCT) are implemented by hardware IPs, quantization / inverse quantization (DCT/IDCT), variable length coding (VLC) are implemented by software.

Table I shows using hardware IPs information: the area of hardware IP, cycles during processing data.

Table II shows the results of synthesized processor cores[13]. The processor kernel is (1) RISC-type or (2) DSP-type. RISC-type kernel has five pipeline stages composed of IF , ID , EXE , MEM and WB stages. DSP-type kernel has three pipeline stages composed of IF, ID and EXE stages. The optional hardware units are functional units (ALU, multiplier), register files, and addressing units. They can be added to the processor kernel.

Table III shows the results of synthesizing IFC with synthesized processor cores in Tab. II by *IFC_Synthesizer*. *IFC_Synthesizer* synthesizes optimal IFCs correspond with a target hardware IP and the target processor core. The main reason the area of processor core B is larger than A is that the number of pipeline stage of B is more than the that of A.

TABLE II
CONFIGURATION OF SYNTHESIZED PROCESSOR CORES.

Name	Processor Core Area $[mm^2]$	Frequency [MHz]	Hardware configuration			
			Kernel	Issue	#ALUs	#Regs
A	5.9723	81.300	RISC	4	ALU*2,Mult*2	47
B	1.7554	70.225	DSP	2	ALU*1,Mult*1	8

TABLE III
SYNTHESIZED IFC INFORMATION.

function	Processor Core Name	IFC area $[mm^2]$
RGBtoYCrCb	A	0.1080
RGBtoYCrCb	B	0.1148
DCT/IDCT	A	0.1028
DCT/IDCT	B	0.1108
ME/MC	A	0.1547
ME/MC	B	0.1638

The maximum of the execution time of *IFC_Synthesizer* is 9.4436 [sec], the minimum is 4.3475 [sec], the average is 6.5534 [sec]. However, in case of designing manually, the design of IFC requires about three days. IFC_Synthesizer reduces the cost of designing IFC.

V. CONCLUSION

In this paper, we presented an architecture of interface circuit to communicate with a processor core and a hardware IP, and a method for synthesizing it. Using the synthesis system "IFC_Synthesizer", we can reduce the interface circuit development cost to less than 10 [min], while it would cost about three days by manual design. IFC_Synthesizer generates a HDL description of the interface circuit to communicate with the processor core and the hardware IP.

REFERENCES

[1] Amphion, "CS6400:Color Space Converter Datasheet," http://www.amphion.com/.

[2] Amphion,"CS6710:Motion Estimation Controller Accelerator Datasheet," http://www.amphion.com/.

[3] ARM. http://www.arm.com/.

[4] Hitachi, Ltd., CWL: Component Wrapper Language, http://www.labs.fujitsu.com/jp/techinfo/cwl/.

[5] V. D'silva, S. Ramesh, Arcot Sowmya, "Bridge Over Troubled Wrappers : Automated Interface Synthesis," Proc. 17th Int. Conf. VLSI Design, 2004.

[6] Ocean Logic Pty, "Discrete Cosine Transform Rev. 1.1 Datasheet," http://www.ocean-logic.com/.

[7] R. Passerone, J. A. Rowson, and A. Sangiovanni-Vincentelli, "Automatic Synthesis of Interfaces between Incompatible Protocols," Proc. 35th DAC, pp. 8–13, 1998.

[8] A. Rajawat, M. Balakrishnan, and A. Kumar, "Interface synthesis: Issues add approaches." Proc. 13th International Conference on VLSI Design, pp. 92–97, 2000.

[9] Ruby Language, http://www.ruby-lang.org/.

[10] J. Smith and G. D. Micheli, "Automated composition of hardware components," Proc. 35th DAC, 1998.

[11] SystemC, http://www.systemc.org/.

[12] Tensilica, Xtensa, http://www.tensilica.com/.

[13] N. Togawa, M. Yanagisawa, and T. Ohtsuki, "A hardware/software cosynthesis system for digital signal processor cores," IEICE Trans. Fundamentals, vol. E82–A, no.11, 1999.

[14] N. Tomono, S. Kohara, J. Uchida, Y. Miyaoka, N. Togawa, M. Yanagisawa, T. Ohtsuki, "A Processor Core Synthesis System in IP-Based SoC," Proc. 33rd ASP-DAC, pp. 527–532, 2004.

[15] Y. Hwang and S. Lin, "Automatic Protocol Translation and Template Based Interface Synthesis for IP Reuse in SoC," Proc. APCCAS 2004, 2004.

A Real-Time and Bandwidth Guaranteed
Arbitration Algorithm for SoC Bus Communication

Chien-Hua Chen, Geeng-Wei Lee, Juinn-Dar Huang, and Jing-Yang Jou
Department of Electronics Engineering
National Chiao Tung University
Hsinchu, Taiwan
e-mail: {tony, gwlee, jyjou}@eda.ee.nctu.edu.tw, jdhuang@mail.nctu.edu.tw

Abstract – In shared SoC bus systems, arbiters are usually adopted to solve bus contentions with various kinds of arbitration algorithms. We propose an arbitration algorithm, RT_lottery, which is designed to meet both hard real-time and bandwidth requirements. For fast evaluation and exploration, we use high abstract-level models in our system simulation environment to generate parameters for our configurable arbiter. The experimental results show that RT_lottery can meet all hard real-time requirements and perform very well in bandwidth allocation. The results also show that RT_lottery outperforms several commonly-used arbitration algorithms today.

1. Introduction

Although there are many possible communication architectures for inter-module communications in SoC systems, shared buses are still very popular among these architectures because of their simplicity and area efficiency. The masters on an SoC bus may issue requests simultaneously and hence an arbiter is required to decide which master is granted for bus access. In many applications, masters may have real-time and/or bandwidth requirements on requests. A master with a real-time requirement demands its transactions accomplished within a fixed number of clock cycles. On the other hand, a master with a bandwidth requirement must occupy a fixed fraction of total bandwidth of a bus. If designers find that the implemented arbitration algorithm cannot fulfill some requirements at late design stages, they have to return to a very early design stage to modify the original arbitration algorithm. This would result in a significant schedule delay.

Arbitration algorithms commonly used for shared buses include Static Priority, Time Division Multiplexing (TDM), and Round-Robin [1-4]. Lottery is the arbitration algorithm proposed recently [5] with the advantages of (i) providing designers with good control over bandwidth allocation for each master, and (ii) providing a high-priority master with quite low transaction latency. However, all arbitration algorithms mentioned above cannot well handle bandwidth and hard real-time requirements concurrently.

In this paper, we propose a two-level arbitration algorithm, RT_lottery, which is expected to meet hard real-time and bandwidth requirements of each master at the same time. At the 1st level, we use a Real-Time Handler to satisfy all hard real-time requirements. At the 2nd level, a Lottery-based algorithm with tuned weight is adopted for proper bandwidth allocation.

We compare RT_lottery with other three arbitration algorithms, Static Priority, Lottery, and TDM+Lottery (1st level: TDM, 2nd level: Lottery). The experimental results show that RT_lottery with parameters generated by our weight tuning flow can handle real-time and bandwidth requirements of each master better than the other arbitration algorithms.

The rest of this paper is organized as follows: previous work including the introduction to some common arbitration algorithms (Static Priority, TDM, and Lottery) is presented in Section 2. Section 3 describes the proposed arbitration algorithm (RT_lottery) and the flow for generating appropriate parameters of RT_lottery to meet bandwidth requirements. The experimental environment and results are shown in Section 4. Section 5 concludes this paper.

2. Preliminaries

2.1 Previous work

In this section, we briefly introduce several previous arbitration schemes [1-4].

1) Static Priority:

Each master is statically assigned a unique priority value. When multiple masters issue requests simultaneously, the master with the highest priority gets granted. The advantage of this arbitration scheme is its simple implementation and small area cost. However, if masters with higher priority issue requests excessively and frequently, other masters with lower priority may rarely be granted. This could introduce severe starvation of low-priority masters and result in extremely unfair bandwidth allocation.

2) TDM:

Time Division Multiplexing (TDM) algorithm divides access time on a bus into time slots and then allocates these slots to masters in certain way. If a master possessing the current time slot does not issue request, the time slot would be wasted. To mend this inefficiency, a 2nd level arbitration algorithm is usually adopted to reallocate the current slot to other requesting masters. Fig. 1 is an example architecture of two-level TDM.

For a two-level TDM arbitration algorithm, the 1st level uses a time wheel where each slot is statically reserved for a unique master and the 2nd level can adopt any arbitration algorithm depending on the target application. For example, if the bandwidth allocation among masters is important, 2nd level can use an arbitration algorithm with better ability of bandwidth allocation. Also note that Round-Robin is actually one kind of TDM algorithm.

3) Lottery [5-6]:

An arbiter implementing the Lottery arbitration algorithm is like a lottery manager deciding which lucky one wins a prize. Each mater on the bus is statically assigned a number of "lottery tickets". The lottery manager generates a pseudo random number, and the master having the ticket matched to this number is granted for access. Obviously, the master having more tickets is more likely granted.

Let the masters be M_1, M_2, ..., M_n and the number of tickets held by each master be t_1, t_2, ..., t_n. At any cycle, the set of pending requests is represented by a set of Boolean variables r_1, r_2, ..., r_n, where r_i =1 means that M_i has a pending request, and r_i =0 otherwise. The master to be granted is chosen with the probability given by the equation:

$$P(M_i) = \frac{r_i \cdot t_i}{\sum_{j=1}^{n} r_j \cdot t_j}$$

The number of tickets of each master can be regarded as its weight. A master with higher weight has higher probability to be granted. We represent the number of tickets possessed by a master as its weight in the following sections. In summary, the Lottery arbitration algorithm is (i) capable of providing designers with good control over bandwidth allocation for each master, and (ii) quite good at providing high priority master with low transaction latency.

2.2 Observations on Lottery arbitration algorithm

A real-time requirement in the previous work [5] is represented in terms of the average transaction latency. However, such a requirement can only be regarded as a loose real-time requirement since there may exist some extremely long-latency transactions. For hard real-time requirements, all transaction latencies (not the average transaction latency) must be smaller than the given requirement all the time.

Meanwhile, to meet the bandwidth requirements, masters are assigned weights according to the ratio of their required bandwidth [5]. Nevertheless, if the bus access behaviors are very diverse among masters, the actual bandwidth ratio would not conform to the weight ratio. The reason may be that the actual traffic load generated by some master is much less or much more than it requests. For example, a master asks for a large fraction of bus bandwidth but rarely issues requests. For this reason, we propose a weight tuning method for better bandwidth allocation.

To meet the real-time requirements, the weight of the master with the minimum latency requirement should be much larger than the others. However, it is really hard to assign a proper weight to each master if there are multiple masters with diverse real-time and bandwidth requirements. For instance, how to assign a proper weight to a master that has a tight real-time requirement but requires only a small fraction of bus bandwidth. Furthermore, if there are masters having hard real-time requirements, a probabilistic arbitration algorithm like Lottery is obviously not appropriate for such applications.

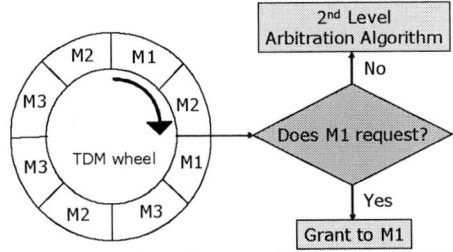

Fig. 1. An example architecture of the two-level TDM

3. Proposed Approach

3.1 Proposed arbiter architecture

Since probabilistic arbitration algorithms cannot handle hard real-time requirements, we propose a two-level arbitration algorithm, RT_lottery (R for **R**eal-time, T for **T**uned weight) to solve the problem. The proposed arbiter architecture is shown in Fig. 2. The 1ˢᵗ level, *Real-Time Handler*, is designed to handle real-time requirements. The 2ⁿᵈ level, *Lottery with tuned weight*, is designed to handle bandwidth requirements. The weight of each master is fine tuned by our weight tuning algorithm based on the evaluation results obtained from system simulation. The details of RT_lottery will be described in later sections.

3.2 Simulation model

In our model, it is assumed that once a master possesses the bus, other masters cannot access the bus until the possessing master releases the bus, i.e., each transaction is non-preemptive. An example of a system architecture containing four masters is shown in Fig. 3. Each master has a traffic generator. The behavior of each traffic generator is given by designers. The arbiter receives requests from all masters then decides which master should be granted.

There are four types of traffic behaviors that can be given for a master:

(1) R_{cycles}:

It is the real-time requirement (in clock cycles) of a master. For those masters without real-time requirements, this information should be left undefined.

(2) Beat number and probabilities:

It defines the probabilities of burst sizes possibly issued by a master. Take Table 2 for example, M3 issues requests of which 50% requests are 8-beat burst and the other 50% requests are 16-beat burst.

(3) Interval cycles and probabilities:

It determines the interval time between two successive requests issued by a master. However, the rule of deciding the interval time varies with different master types (explained later). For example, in Table 2, 10% of the request interval of M1 is 6 clock cycles while 20% is 7 clock cycles and so on.

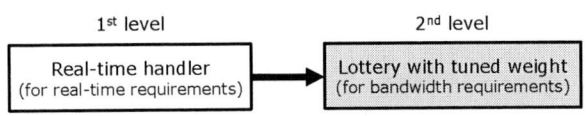

Fig. 2. Proposed arbiter architecture

(4) Type:

In our work, masters are classified into three types based on their traffic behaviors:

1. **D type** (D for Dependency):

D type masters have no real-time requirements and the next request is issued at the time depending on the finish time of the current request. For D type masters, the interval time between two successive requests is the time from the issued time of the former to the finish time of the latter. Fig. 4(a) shows an example. At cycle 2, assume the traffic generator generates a 4-beat burst. The request is not granted until cycle 5 and is finished at cycle 9 (4-beat burst). If the interval time is 10, then the next request is issued at cycle 19 (The issued time of the latter request is the finish time of the former plus 10 cycles).

2. **D_R type** (D for Dependency, R for Real-time):

D_R type masters are the same as D type masters except that they have extra real-time requirements. Fig. 4(b) is an example with the same parameters used in Fig. 4(a). In this example, the master has a real-time requirement, R_{cycle}, which is set to 10 cycles. Thus the request issued at cycle 2 must be finished before cycle 12 ($2 + R_{cycles} = 12$), which is shown as the dotted line in the figure. If the request is not finished before cycle 12, a real-time violation occurs.

3. **ND_R type** (ND for No Dependency, R for Real-time):

The issued time of a request from an ND_R type master is independent of the finish time of its previous request, and the interval time is the clock cycles between two successive requests. In Fig. 4(c), assume that the interval time is 15. The second request is issued at cycle 17, which directly depends on the issued time (at cycle 2) of the first request but not its finish time (at cycle 9). Since the current request must be finished before the next request, the reasonable value of R_{cycles} is supposed to be smaller than the minimum possible interval time. That is, designers can also assign a tighter real-time requirement. To ensure a reasonable R_{cycles}, we define $R_{cycles} = \min(t_{min\_interval}, t_{user\_given})$, where $t_{min\_interval}$ is the minimum possible interval time and t_{user_given} is the real-time requirement given by designers.

3.3 Proposed arbitration algorithm

In this section, the algorithms of the Real-Time Handler and the weight tuning process for Lottery are described in detail.

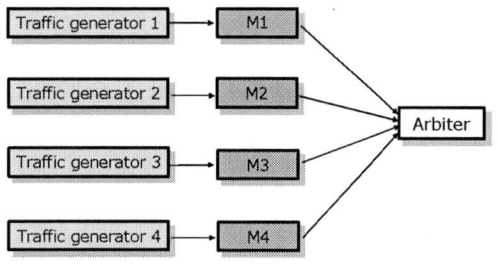

Fig. 3. An example architecture

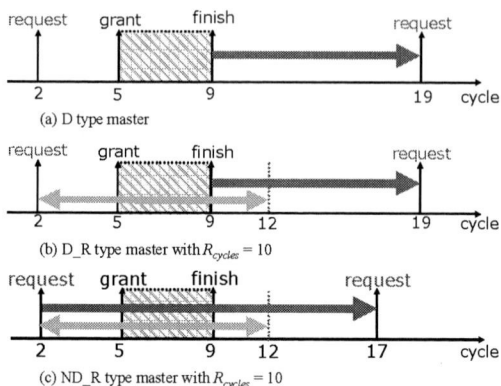

Fig. 4. The example of three types of masters

3.3.1 Real-Time Handler

The Real-Time Handler sets a real-time counter for each master according to their real-time requirements. When a master issues a request, the corresponding real-time counter is set to this master's R_{cycles}. The real-time counter is decremented by 1 every cycle until the master is granted. *warning_line* is a global constant value used to remind the arbiter to grant the most urgent master. The master would have higher priority if its corresponding real-time counter value is below the *warning_line*. When two or more real-time counters are below *warning_line*, the master with the smallest real-time counter value (more urgent) gets granted. Fig. 5 shows an example of Real-Time Handler's operation. We assume that M1 has $R_{cycles} = 30$ and the whole system has *warning_line* = 25.

Let us focus on cycle 3 and cycle 11:

(1) Cycle 3 (the left table in Fig. 5):

As M1 issues a request at this cycle, the real-time counter of M1 is set to its R_{cycles}, 30. All other masters also issue requests at this time, but only M2's real-time counter value is below *warning_line* and thus it is granted first.

(2) Cycle 11 (the right table in Fig. 5):

M2's request is a 8-beat burst, and therefore the request is finished at cycle 11. At this time, all real-time counters of the pending masters (M1 and M3) are decremented by 8. The values of real-time counters of M1 and M3 are both below *warning_line*. Since the value of M3's real-time counter is smaller, M3 is granted at this cycle.

To meet all real-time requirements in any circumstances, *warning_line* must be carefully set according to the worst contending case. That is,

warning_line = ∑(*maximum possible beat of D_R and ND_R masters*) + *maximum possible beat number of D masters*

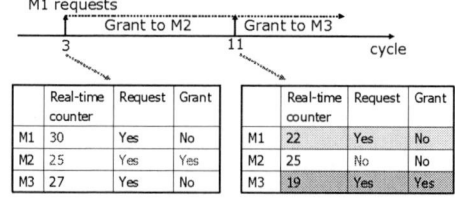

Fig. 5. R_{cycles} of M1 = 30, *warning_line* = 25

The idea behind *warning_line* is as follows: in the worst circumstance, the D type master with the maximum possible beat number issues a request of its maximum possible beat number and gets granted. At the next cycle, all other masters with real-time requirements all issue requests. It must be guaranteed that all the real-time requirements of these masters can still be met after the request of this D type master is finished.

Take Table 1 as an example and the worst contending case is shown in Fig. 6:

warning_line =

$\max(5,6,7,4,5,6)+\max(2,3,4)+\max(3,4,5)+\max(5,6,7) = 23$

If there is no master with R_{cycles} smaller than *warning_line*, the proposed arbiter is guaranteed to meet all hard real-time requirements.

3.3.2 Weight tuning flow for Lottery

In this section, we present the 2nd level of RT_lottery, Lottery with tuned weight. Fig. 7 shows the weight tuning flow.

First, we read in the traffic information of each master given by designers. Each master's required bandwidth must be smaller than its maximum bandwidth. The maximum bandwidth of a master is calculated by assuming there is only one master on the bus, i.e., all requests from the master are granted immediately. To screen out unreasonable bandwidth requirements, we evaluate the maximum bandwidth of each master first. Initial weight assignment is based on each master's maximum and required bandwidth.

Second, the weight tuning process tries to move bandwidth share from a master whose allocated bandwidth is more than its required bandwidth to another master whose allocated bandwidth is less than its required bandwidth. We say that a master has extra bandwidth if its allocated bandwidth is more than its required bandwidth. If there are no masters having extra bandwidth, the weight tuning process stops.

3.3.3 Algorithm of weight tuning

In this section, the greedy algorithm of the block named weight tuning in Fig. 7 is presented. First, we introduce some definitions:

- M_i: Each master in the system is marked as M_i, $i = 1 \sim n$, where n is the total number of masters in the system.
- S_{more}: If (M_i's simulated bandwidth – M_i's required bandwidth > 2%), $M_i \in S_{more}$.
- S_{less}: If (M_i's required bandwidth – M_i's simulated bandwidth > 2%), $M_i \in S_{less}$.
- S_{met}: If ($|M_i$'s required bandwidth – M_i's simulated bandwidth| < 2%), $M_i \in S_{met}$.
- m_{most}: The master with the most extra bandwidth in S_{more}.
- m_{least}: The master lacking the most bandwidth in S_{less}.
- t_m: The number of tickets m_{most} has.
- t_l: The number of tickets m_{least} has.

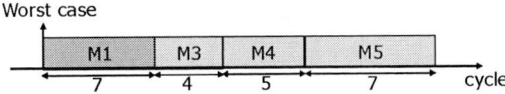

Table 1. A traffic pattern for the explanation of *warning_line*

	type	R_{cycles}	beat/prob.			interval/ prob.	
M1	D		5/20	6/40	7/40	40/50	50/50
M2	D		4/50	5/20	6/30	60/20	70/80
M3	D_R	200	2/30	3/30	4/40	40/50	60/50
M4	D_R	100	3/20	4/50	5/30	80/10	90/90
M5	ND_R	120	5/30	6/50	7/20	14/50	16/50

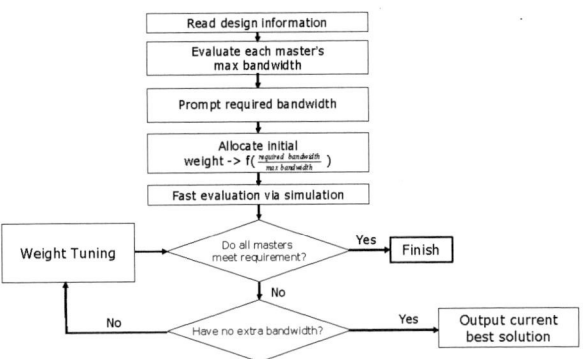

Fig. 6. The worst contending case in Table 1 for real-time

Fig. 7. The weight tuning flow for Lottery

- t_d: The number of tickets that we try to move from t_m to t_l each time.
- B: The bound used for deciding t_d.

The pseudo code of the weigh tuning algorithm is shown in Fig. 8. First, masters are classified into three exclusive sets, S_{more}, S_{less}, and S_{met} (*line 1*), and then B is initialized (*line 2*). The *while* loop (*line 5-19*) decides t_d (*line 6*) for new t_m and t_l (*line 12 and 13*) in each iteration. It stops on two conditions: (i) $t_d = 0$, the weight cannot be tuned any more (*line 7*); (ii) the new t_m and t_l do not result in moving masters whose bandwidth requirements are met originally (S_{more} and S_{met}) into S_{less} (*line 15*). Otherwise, another iteration proceeds and B is reduced to re-calculate a new t_d, t_m and t_l.

4. Experimental Results

4.1 Experimental environment setup

We compare RT_lottery with other three arbitration algorithms, Lottery, Static Priority, and TDM+Lottery. We use a system containing six masters for evaluation. The parameters of these arbitration algorithms are set as follows:
(1) Lottery:

The weight of each master is assigned according to its required bandwidth (weight ratio = required bandwidth ratio).
(2) Static Priority:

Each master is assigned a priority according to its required bandwidth. The master with higher required bandwidth has a higher priority.

```
1:  Classify masters;
2:  Initialize B = 1, finish = 0;
3:  t_m_old = t_m;        // Record the old value of t_m and t_l
4:  t_l_old = t_l;
5:  while ( finish == 0 ) {
6:      t_d = B * t_m / 2;
7:      if (t_d == 0) { // Loop breaks if it is not a meaningful action
8:          t_m = t_m_old;
9:          t_l = t_l_old;
10:         break;
11:     }
12:     t_m = t_m_old - t_d;
13:     t_l = t_l_old + t_d;
14:     simulate();
15:     if(requirements of the masters in S_more and S_met are still met)
16:         finish = 1;
17:     else
18:         B = B / 2;
19:}
```

Fig. 8. The pseudo code of weight tuning

(3) TDM+Lottery:

1st level - TDM: Masters with real-time requirements are allocated with time slots accordingly.

2nd level - Lottery: The weight of each master is assigned according to its required bandwidth (weight ratio = required bandwidth ratio).

4.2 Experiment 1

In this experiment, 6 masters are put on a bus with the traffic behaviors shown in Table 2 [3,7]. For each type of master, we design a heavy-traffic master and a light-traffic master. For example, both M1 and M2 are D type masters, and the requests issued by M1 have larger beat numbers and shorter average interval than those issued by M2. That means M1 generates a heavier traffic load to the bus than M2 does.

The difficulty to meet both real-time and bandwidth requirements generally depends on the total required bandwidth in a system. In the following, we conduct two experiment cases for observations. We consider a given total required bandwidth in one case, and consider a set of 100 different total required bandwidth cases randomly generated in the other case.

First, we evaluate the case that the total required bandwidth utilizes 94% of the entire bus bandwidth, as shown in Table 3. The evaluated maximum bandwidth and the given required bandwidth of masters are also shown in Table 3. From the table, we observe that the maximum bandwidth of each master is very different from each other because there are masters with heavy- and light-traffic loads.

All the experiments are conducted on a PC with a Intel Pentium 4 2.8G processor and 512MB DRAM. Following statistics are recorded during simulation for evaluation:

(1) *bw_miss_num*:

This value represents the number of masters whose bandwidth requirements are missed.

(2) *rt_vio_time*:

This value is calculated by: ∑*(the number of real-time violations of all masters' requests)*. If a request of M_i

with real-time requirements is not finished within M_i's R_{cycles} cycles, a real-time violation occurs on this request.

(3) *max_latency*:

During a simulation run, we record the latencies of all requests and pick the maximum latency among them as the *max_latency*.

The experimental results are shown in Table 4. On the ability of bandwidth allocation, Static Priority is poor as expected, but Lottery is surprisingly poor as well. This fact indicates that Lottery still needs a good weight tuning strategy for better bandwidth allocation. On the aspect of real-time handling ability, Lottery and Static Priority are failed to meet real-time requirements since they do not take real-time requirements into consideration. Note that Static Priority is even worse than Lottery because its *max_latency* is much longer than that of Lottery (7060 vs. 954). Though TDM+Lottery can handle real-time and bandwidth requirements better, it still fails in bandwidth allocation (*bw_miss_num* = 1).

In general, it is usually harder to meet requirements with higher total required bandwidth summed from all the masters with bandwidth requirements, i.e., the bus utilization is supposed higher. In the second experiment case, we use a generator that can randomly generate the required bandwidth for each master. And let R_{sum} represent the total required bandwidth in terms of the percentage of entire bus bandwidth. Here, we evaluate seven different values of R_{sum}, ranging from 65% to 95%. For each R_{sum}, 100 random cases are conducted to compare four arbitration algorithms. R_{sum_i} represents the i_{th} case ($i = 1 \sim 100$) of simulation for R_{sum}. The simulation time for each R_{sum_i} is less than one minute on our equipment. Following statistics are recorded during simulation for evaluation:

(1) *rt_vio_time_sum*:

∑(*rt_vio_time* in each R_{sum_i})

(2) *rt_fail_case_sum*:

The number of cases which contain one or more real-time violations among all 100 cases (R_{sum_i} is a case failed to meeting real-time requirements if *rt_vio_time* > 0 in R_{sum_i}).

(3) *bw_fail_case_sum*:

The number of cases which fail to meet bandwidth requirements among all 100 cases (R_{sum_i} is a case failed to meeting bandwidth requirements if *bw_miss_num* > 0 in R_{sum_i}).

(4) *fail_case_sum*:

The number of cases which fail to meet real-time or bandwidth requirements among all 100 cases, (R_{sum_i} is a failed case if *rt_vio_time* > 0 **or** *bw_miss_num* > 0 in R_{sum_i}).

The experimental results are shown in Table 5. We can see that it is harder to meet the requirements with larger R_{sum}. The value of *fail_case_sum* decreases as R_{sum} goes low. The summary of experimental results is shown in Table 6. RT_lottery can not only meet real-time requirements but also be good at bandwidth allocation for the masters.

4.3 Experiment 2

The objective of experiment 2 is to observe the impact of different burst beat numbers on the arbitration algorithms. The traffic patterns are given that all masters send the same beat numbers of 8, 16, and 32, respectively. Similar to the experiment 1, we run 100 random cases for each R_{sum}.

The experimental results are shown in Fig. 9. RT_lottery is the best among the four algorithms for fixed 8, 16, and 32-beat. RT_lottery and TDM+Lottery, which are capable of handling both bandwidth and real-time requirements, perform much better than the other two algorithms. Nevertheless, it is harder to meet requirements with larger fixed beat number for RT_lottery and TDM+Lottery, since the numbers of failed cases arise with larger beat numbers. The reason is that with larger beat number, the granularities of weight (ticket number) for RT_lottery and TDM+Lottery get coarser. Each time a fixed amount of weight is transferred from M_i to M_j, the influence of weight transfer on cases of 8 or 16 fixed beat number is smaller than that on the case of 32 fixed beat number.

5. Conclusions

The two-level arbitration algorithm, RT_lottery, is proposed in this paper. We use high abstract-level models and a fast simulation-based evaluation environment to generate appropriate parameters for RT_lottery. RT_lottery is guaranteed to meet all hard real-time requirements and perform very well in bandwidth allocation. Three existing arbitration algorithms, Static Priority, Lottery, and TDM+Lottery are compared with RT_lottery. The experimental results show that RT_lottery is the best among these four algorithms in the ability to handle real-time and bandwidth requirements.

Hence, the RT_lottery-based arbiter can be a better choice for those SoC systems containing masters with hard real-time and diverse bandwidth requirements.

Table 2. The traffic pattern for the experiment 1

	type	R_{cycles}	beat/prob.		interval/prob.				
M1	D		8/50	16/50	6/10	7/20	8/40	9/20	10/10
M2	D		1/50	4/50	10/10	11/20	12/40	13/20	14/10
M3	D_R	65	8/50	16/50	6/10	7/20	8/40	9/20	10/10
M4	D_R	85	1/50	4/50	10/10	11/20	12/40	13/20	14/10
M5	ND_R	65	8/50	16/50	65/10	66/20	67/40	68/20	69/10
M6	ND_R	85	1/50	4/50	85/10	86/20	87/40	88/20	89/10

Heavy traffic / Light traffic

Table 3. An example bandwidth requirement

	M1	M2	M3	M4	M5	M6
Maximum Bandwidth(%)	63	18	63	19	17	2
Required Bandwidth(%)	20	5	40	10	17	2	=> 94 % in total

Table 4. The experimental results of the experiment 1

	bw_miss_num	max_latency (cycle)	rt_vio_time
Static Fixed Priority	3 (50%)	7060	244
Lottery	3 (50%)	954	160
TDM+Lottery	1 (17%)	314	0
RT_lottery	0 (0%)	170	0

Table 5. The results of 100 random cases

RT_lottery R_{sum}	rt_v	bw_f	rt_f	fail		Lottery R_{sum}	rt_v	bw_f	rt_f	fail
95	0	87	0	87		95	12915	99	100	100
90	0	80	0	80		90	12150	97	100	100
85	0	79	0	79		85	11159	98	100	100
80	0	68	0	68		80	10535	86	100	100
75	0	66	0	66		75	9007	73	100	100
70	0	57	0	57		70	9022	58	100	100
65	0	38	0	38		65	8274	45	100	100

TDM+ Lottery R_{sum}	rt_v	bw_f	rt_f	fail		Static Priority R_{sum}	rt_v	bw_f	rt_f	fail
95	1	99	1	99		95	18577	100	100	100
90	8	96	8	96		90	17396	100	100	100
85	8	95	8	96		85	13739	100	99	100
80	6	91	6	91		80	14235	98	100	100
75	6	83	6	84		75	11200	88	99	100
70	3	75	3	75		70	11076	83	97	97
65	2	58	2	58		65	10345	82	96	98

rt_v : rt_vio_time_sum   rt_f : rt_fail_case_sum
bw_f : bw_fail_case_sum   fail : fail_case_sum

Table 6. The summery of the experimental results

Arbitration algorithm	Real-time capability	Bandwidth allocation capability
RT_lottery	Always holds	Best
TDM + Lottery	Only fails for critical cases	Good but requiring weight tuning
Lottery	No consideration	Good but requiring weight tuning
Static Fixed Priority	No consideration	Poor

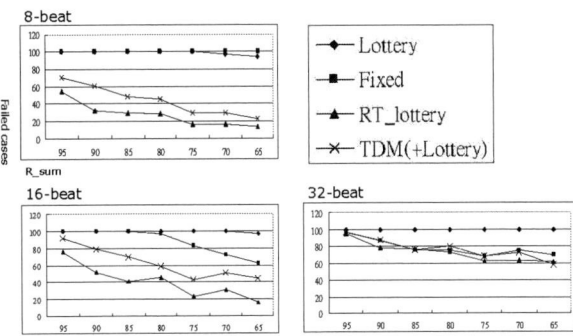

Fig. 9. Number of failed cases for 100 random cases

References

[1] C. H. Pyoun, C. H. Lin, H. S. Kim, and J. W. Chong, "The Efficient Bus Arbitration Scheme In Soc Environment," International Workshop on *System-on-Chip for Real-Time Applications*, 2003, Page(s):311 – 315.

[2] M. Yang, S. Q. Zheng, Bhagyavati, and S. Kurkovsky, "Programmable Weighted Arbiters for Constructing Switch Schedulers," Workshop on *High Performance Switching and Routing*, 2004, Page(s):203 – 206.

[3] M. Conti, M. Caldari, G. B. Vece, S. Orcioni, and C. Turchetti, "Performance Analysis of Different Arbitration Algorithms of the AMBA AHB Bus," *Design Automation Conference*, 2004, Page(s):618 – 621.

[4] F. Poletti, D. Bertozzi, L. Benini, and A. Bogliolo, "Performance Analysis of Arbitration Policies for SoC Communication Architectures," Journal of *Design Automation for Embedded Systems*, 2003, Page(s):618 – 621.

[5] K. Lahiri, A. Raghunathan, and G. Lakshiminarayan, "LOTTERYBUS: A New High-Performance Communication Architecture for System-on-Chip Designs," *Design Automation Conference*, 2001, Page(s):15 – 20.

[6] A. C. Waldspurger and W. E. Weih., "Lottery Scheduling: Flexible Proportional Share Resource Management," Symp. on *Operating Systems Design and Implementation*, 1994.

[7] K. Lahiri, A. Raghunathan, and S. Dey, "Evaluation of the Traffic Performance Characterization of System-on-Chip Communication Architectures," International Conference on *VLSI Design*, 2001, Page(s):29 – 35.

Hierarchical Memory Size Estimation for Loop Fusion and Loop Shifting in Data-Dominated Applications

Qubo Hu[*]   Arnout Vandecappelle[†]   Martin Palkovic[†]

Per Gunnar Kjeldsberg[*]   Erik Brockmeyer[†]   Francky Catthoor[†‡]

[*] Norwegian University of Science and Technology, Trondheim, Norway   {qubo.hu,pgk}@iet.ntnu.no

[†] IMEC vzw, Leuven, Belgium   {vdcappel,palkovic,brockmey,catthoor}@imec.be

[‡] also professor at Katholieke Universiteit Leuven, Belgium

Abstract — Loop fusion and loop shifting are important transformations for improving data locality to reduce the number of costly accesses to off-chip memories. Since exploring the exact platform mapping for all the loop transformation alternatives is a time consuming process, heuristics steered by improved data locality are generally used. However, pure locality estimates do not sufficiently take into account the hierarchy of the memory platform. This paper presents a fast, incremental technique for hierarchical memory size requirement estimation for loop fusion and loop shifting at the early loop transformations design stage. As the exact memory platform is often not yet defined at this stage, we propose a platform-independent approach which reports the Pareto-optimal trade-off points for scratch-pad memory size and off-chip memory accesses. The estimation comes very close to the actual platform mapping. Experiments on realistic test-vehicles confirm that. It helps the designer or a tool to find the interesting loop transformations that should then be investigated in more depth afterward.

I. INTRODUCTION

In most advanced embedded real-time communication and multimedia processing applications, the manipulation of large data sets has a major effect on both energy consumption and performance of the system. This is due to the huge amount of data transfers to/from large, energy consuming off-chip data memories. Globally optimizing the memory accesses of data-dominated applications is therefore critical for system performance and energy consumption. Loop transformations are important techniques for improving parallelism, performance and to reduce memory energy consumption [2, 19, 5]. They are usually performed on a Geometrical Model (GM) [18]. Loop fusion, or combining with loop shifting to satisfy dependency, is the basic transformation for improving data locality [8, 17]. It was shown [6] that the search for optimal loop fusion for global array contraction in general is an NP-complete problem. Heuristics based on data locality are hence used in the existing work. Still, data locality is a very abstract measure, so several techniques have been developed to estimate the real size requirements for the large data structures [1, 20, 9, 14, 10].

However, the size as such does not directly represent how the accesses to the costly off-chip memories can be reduced. Indeed, the minimal size for some array may still be larger than

the local memory. In addition, if sufficient locality between read accesses is present, a local copy of part of the array may already remove most of the off-chip accesses [7], making the actual size of that array less relevant.

To select the interesting loop transformation candidates, it is therefore necessary to estimate not just the size of each array, but also their mapping on the hierarchical memory architecture. The number of costly off-chip memory accesses depends on which arrays and which copies can fit in the local memory. In addition, the platform and hence the exact size of the memories are often not yet known at this early design stage, so the estimation must take this unknown parameters into account.

In this paper we present a technique and tool support for hierarchical memory size requirement estimation for loop fusion and shifting. The basic idea has previously been introduced in [11] but it is significantly extended here. Our focus is on the Scratch-Pad Memory (SPM) based memory hierarchy, which is a more energy-efficient alternative to caches. The global view taken during data mapping replaces the area and energy consuming hardware used in caches. The SPM is filled with not only arrays [15], but also copies of the currently used part of the arrays [4].

Our hierarchical memory size estimation is performed in two main phases: a data reuse analysis phase and a Memory Hierarchy Layer Assignment (MHLA) estimation phase. Data reuse analysis is performed on the geometrical model by determining, for every loop nest, the set of data which has reuse. The different ways to copy data across the memory hierarchy is represented in data reuse trees. Earlier work using exact data reuse analysis techniques [16, 3, 12, 13] is either too limited, or too slow to be used for the exploration of a huge number of loop transformations. Instead of doing the actual geometrical computations required for this analysis, we use a bounding box approximation of the domains. This can be an over-estimate, but in practical cases it turns out to be as good as an exact analysis. To further save computation time when loop shifting and fusion are applied, we propose to incrementally compute the data reuse trees based on the previous ones.

The MHLA estimation phase selects which arrays and copies are stored in the SPM, such that the number of off-chip memory accesses is minimized. The existing technique for MHLA [4] finds the optimal selection for a given memory hierarchy (SPM size) using backtracking. Their approach

0-7803-9451-8/06/$20.00 ©2006 IEEE.

```
for (y=0; y<=399; ++y)
  for (x=0; x<=639; ++x)
    image[x][y] = ...;            // S1
for (y=0; y<=399; ++y)
  for (x=1; x<=638; ++x)
    for (z=-1; z<=1; ++z)
      ... = g(image[x+z][y]);     // S2
```

Fig. 1.: Code example before loop transformation

is not feasible for our estimation purpose as the memory platform instance is usually not defined at the loop transformation stage: it is not realistic to perform an estimate for each possible memory hierarchy instance. Their heuristic has high complexity and is hence too slow for large applications, making it unfeasible to be used during the exploration of loop transformations. Instead, a platform-independent heuristic is used, which is fast but usually comes very close to their result. It outputs Pareto curves (SPM size vs. off-chip accesses) for different loop transformations and helps to find the possibly good loop transformation alternatives. The Pareto curve furthermore allows an early energy estimate of any two-layer memory hierarchy instance.

The rest of this paper is organized as follows. Section II reviews the basic concepts in the geometrical model on which the loop transformations are performed. Section III presents our algorithm for doing fast hierarchical memory size requirement estimation. Experiments on real-life applications are demonstrated in Section IV. Conclusions and future work are drawn in Section V.

II. OVERVIEW OF THE GEOMETRICAL MODEL

The targeted data-dominated applications are at the system level characterized by deep loop nests and multidimensional arrays as shown in Fig. 1. Loop transformations are usually performed on a geometrical model which uses multidimensional iteration domains and access mappings to represent all necessary information. The concepts needed to understand how our techniques use the geometrical model are presented below. Further details can be found in [18].

The iteration domain of a statement is a set of integer points where each point represents exactly one execution of this statement. Its description is derived from the constraints corresponding to the boundaries of the surrounding loops and conditions that restrict the execution of the statement. For example, the iteration domain of statement S2 in Fig. 1 is described as:

$$ID_{S2} = \{[y,x,z]|0 \le y \le 399 \wedge 1 \le x \le 638 \wedge -1 \le z \le 1\}$$

Note that we leave out the constraint of integer points, $[y,x,z] \in \mathbb{Z}^3$, to simplify the formulas.

Each statement has a number of accesses to variables. For our purpose, only the array accesses are important, as scalars are assumed to be mapped to local memory anyway. Each array reference (read or write) in the statement has an access mapping: a function mapping the iterators to the array indices. The access mapping for array `image` referenced in statement S2 is described as:

$$AM_{\texttt{image},S2} = \{[y,x,z] \mapsto [a_1,a_2]|a_1 = x+z \wedge a_2 = y$$
$$\wedge [y,x,z] \in ID_{S2}\}$$

The data domain of an array in a certain statement represents which elements are accessed in that statement. It is found by projecting the iteration dimensions from the access mapping:

$$DD_{\texttt{image},S2} = \{[a_1,a_2]|\exists y,x,z : a_1 = x+z \wedge a_2 = y$$
$$\wedge [y,x,z] \in ID_{S2}\}$$
$$= \{[a_1,a_2]|0 \le a_1 \le 639 \wedge 0 \le a_2 \le 399\}$$

Usually, geometrical models use polytopes to represent the domains. A large set of operations can be applied on polytopes, and they are sufficient to represent many practical applications. However, polytope operations are still rather computationally expensive, especially counting the number of integer points. Therefore, we use a simplified geometrical model which uses only bounding boxes; computations on it are extremely fast. The idea was first introduced in [11] (where it is called a hyperplane). A bounding box is specified by the lower and upper bounds of the corresponding domain in each dimension. In the examples given earlier, the bounding boxes (denoted by $\overleftrightarrow{D}$) are exact: $\overleftrightarrow{ID}_{S2} = \{[0,1,-1] \to [399,638,1]\}$ and $\overleftrightarrow{DD}_{\texttt{image},S2} = \{[0,0] \to [639,399]\}$.

III. HIERARCHICAL MEMORY SIZE ESTIMATION

This section explains how to do the platform-independent hierarchical memory size estimation at the early loop transformations design stage. Our approach consists of four steps. Each of them is explained in the following subsections.

A. Initial data reuse analysis

Our data reuse analysis is performed on the geometrical model and identifies the data (arrays or parts of arrays) that are most frequently accessed at each loop nest. It can potentially save energy and improve performance when the heavily accessed data is copied from the main memory to the smaller on-chip SPM from where it is accessed multiple times. The frequently accessed data to be copied are called copy candidates. The data reuse analysis is done for each array individually. Initially, all the array references for one array are considered together resulting in the declared array (root). This is represented geometrically with the union of the data domains of all array references. Then, the analysis is proceeded at each loop dimension, starting from the outermost dimension. The analysis is performed both for individual array references and between different references. The recursive analysis at all loop dimensions results a tree set of copy candidates, as shown in Fig. 3 (it was called copy candidate graph in [7]).

At a certain loop dimension, the data domain at that level is calculated by assuming that the current and all outermost dimensions remain constant. The descendant inner loop dimensions projected as for the total data domain. This leads to the following formulation for the data domain of array `image` in statement S2 at the level of the x-loop:

$$DD_{\texttt{image},S2|y=0,x=1} = \{[a_1,a_2]|\exists z : a_1 = 1+z \wedge a_2 = 0$$
$$\wedge -1 \le z \le 1\}$$
$$\overleftrightarrow{DD}_{\texttt{image},S2|y=0,x=1} = \{[0,0] \to [2,0]\}$$

6B-4

2006 Asia and South Pacific Design Automation Conference

```
for (y=0; y<=399; ++y)
  for (x=0; x<=639; ++x) {
    image[x][y] = ...;           // S1
    if (x>=2)
      for (z=-1; z<=1; ++z)
        ... = g(image[x+z][y]);  // S2
  }
```

Fig. 2.: Code example after loop fusion and shifting

The number of points in the data domain determines the size of the copy candidate: 3 in the example. The bounding box approximation allows the use of a constant instead of a symbolic for the data domain: the size is 3 independent of x. In addition, it allows a very simple formula for the size.

$$\#size_{CC} = \prod_{i=1}^{n}(UB_{DD_i} - LB_{DD_i} + 1)$$

where n is the number of array dimensions, UB and LB are the bounding box boundaries per dimension.

Data reuse is present at a certain loop dimension when the data domain at that level overlaps with the data domain at the same level in the next iteration. For instance, $\overrightarrow{DD}_{image,S2|y=0,x=2} = \{[1,0] \rightarrow [3,0]\}$ is overlapping with $\overrightarrow{DD}_{image,S2|y=0,x=1}$. In that case, a copy candidate is created at that dimension. The overlapping part is called the *reuse part*. The data in the *reuse part* does not need to be read from off-chip memory but can be accessed in the SPM. To keep the SPM up-to-date, in every iteration of the loop some data is copied from the off-chip memory to the SPM. This is called the *update part*. It is the difference between the data domain and the *reuse part*. In the example, the *reuse part* has 2 elements and the *update part* has 1 element. Note we only need to know the size, which makes the computations very fast again (computing set difference is rather complex, even with the bounding box approximation). The copy candidate also has to be initialized with the *reuse part* in the first iteration of the loop.

To evaluate how useful a copy candidate is, we need to know two figures: the number of accesses and the number of misses.

$$\#accesses_{CC} = \#iter_{all}$$

where $\#iter_{all}$ means the total number of the iterations for all loop dimensions surrounding the statement at where the array is referenced. In the example, $\#accesses$ is 768000.

$$\#misses_{CC} = \#iter_{outer} \cdot (\#size_{reuse\ part} \\ + \#size_{update\ part} \cdot \#iter_{cur})$$

where $\#iter_{cur}$ and $\#iter_{outer}$ are respectively the number of iterations at the analyzing dimension and at all ancestor loop dimensions. $\#misses$ at the x-dimension is hence 256000, calculated as $400 \cdot (2 + 1 \cdot 640)$.

Fig. 3.a shows the data reuse tree for the example code in Fig. 1. At the root, all array references are considered together. The analysis then continues at dimensions y, x and z individually. Copy candidates with reuse are detected at the y and x-dimensions. For the loop fused code shown in Fig. 2, the same procedure is repeated. This time, interesting copy candidates

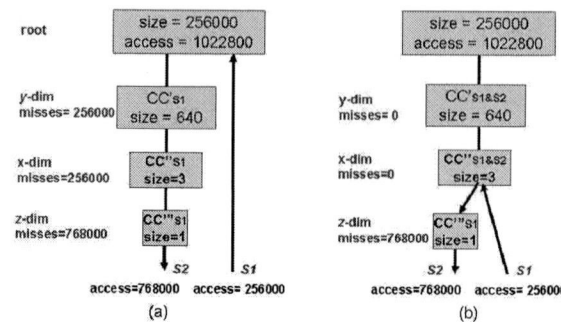

Fig. 3.: Data reuse trees for the example codes (a) before any transformation and (b) after loop shifting and fusion

are detected at y-dimension and at the x-dimension when the two array references are analyzed together. The analysis between multiple references is performed only if they are in the same loop nest till the current dimension and they have identical index coefficients till the current dimension. Note that the loop shifting and fusion has resulted in copy candidates with no misses to the original array. The array does not even have to be written to the main memory and can be kept completely in 3 SPM locations. This results in a significant energy reduction.

The data reuse trees show which copies are potentially interesting to put in the SPM. However, they do not yet show their impact on a hierarchical memory organization. That is analyzed in the next step.

B. Platform-independent MHLA estimation

The MHLA estimation is performed based on the data reuse trees. It maps the copy candidates together with the original arrays (root) onto a memory hierarchy in order to minimize the energy consumption. As there is usually no memory platform defined at the loop transformations stage, we propose a platform-independent MHLA estimation based on a two-layer memory hierarchy template. The size of the main memory is assumed to be unlimited, while the on-chip SPM layer has a varying size. The reason behind is that this memory hierarchy template enables us to simulate any two layer memory platform instances with an early power estimate as explained afterward.

As a starting point, the SPM is empty and all accesses from the processor go to the main memory. Then at each iteration, the candidate giving the biggest potential benefit, as explained below, is assigned to the SPM (replacing its children if they were present). This procedure is repeated until all copy candidates and arrays are assigned. At each iteration, the SPM size increases and the accesses to the main memory decreases.

The potential benefit of a copy candidate is quantified by its *gain_factor*. The one having the highest *gain_factor* among all the unconsidered candidates is selected for assignment.

$$gain\_factor_{CC} = \frac{\#accesses_{CC} - \#misses_{CC}}{\#size_{CC}}$$

The rationale behind this selection criterion is that the candidate with the highest *gain_factor* replaces, per size unit increase of SPM, the largest number of off-chip accesses with

608

accesses to the SPM. For example, the *gain_factor* for copy candidate $CC''_{S1\&S2}$ shown in Fig. 3.b is $\frac{1022800-0}{3} = 340933$.

Each iteration results in a Pareto trade-off point between the size of the SPM (denoted by *#SPM_size*) and the number of accesses to the main memory (denoted by *#MM_acc*).

$$\#SPM\_size = \sum \#size_{CC}$$

$$\#MM\_acc = \#accesses_{total} + \sum (\#misses_{CC} - \#accesses_{CC})$$

in which $\#accesses_{total}$ is the total number of accesses from the processor core, and the sum is over the currently selected copy candidates. In the end, all arrays are assigned to the SPM and there are no accesses to main memory. Different loop transformation alternatives will result in their own Pareto curves.

Because of the incremental assignment, where each array and copy candidate are only considered for assignment once, our algorithm is very fast. It has a complexity of $O(n \log n)$ where n is the total number of copy candidates and arrays considered. For comparison, The algorithm used in[4] has a complexity of $O(2^n n^2 \log n)$ for a predefined two layer memory hierarchy instance. Our platform-independent algorithm, on the other hand, can be used for a whole range of possible two-layer platform instantiations (e.g. with different SPM sizes or different SPM bank activations on a configurable organization). It gives a quick MHLA estimate with reasonable result, which is acceptable for the estimation purposes. This is substantiated on real-life applications in Section IV.

C. Data reuse analysis for incremental loop transformations

Previous techniques have no direct coupling between loop transformations and data reuse analysis when incremental transformations are performed. That is, the changed geometrical model after a loop transformation must be dumped to C-code which is then parsed back to the geometrical model for repeated data reuse analysis. This dumping and parsing procedure is time consuming and is redundant as we can simply update the geometrical model with the loop transformation performed and then rebuild data reuse trees directly based on the updated geometrical model.

Additionally, the data reuse tree rebuilding procedure can be sped up by just rebuilding the trees for the transformed arrays if not all arrays are affected at a time, as is typically the case when loop transformations are performed incrementally. If an array has only been transformed starting from a certain inner loop dimension, it is for sure no changes happened on its data reuse tree at the outer loop dimensions and those dimensions do not need to be updated. Fig. 4 shows our algorithm with three alternatives for incremental data reuse analysis. The choice of alternative is based on an evaluation of the loop transformation effects. As data reuse analysis is the most time consuming step in our estimation framework, this incremental data reuse analysis can significantly reduce the execution time especially when the incremental loop transformations only affect inner loop dimensions. Local data reuse tree update at the transformed loop dimensions only is also possible, but this cannot be proceeded without analysis of which transformations is exactly performed and it is considered for future work.

```
OMD = the outermost dimension
find all the transformed arrays in GM
update GM based on transformations
if ( all arrays are transformed at OMD ) then
    compute the trees for all arrays based on updated GM
else
    for each transformed array
        OMD_tra = its transformed outermost dimension
        if (OMD_tra == OMD)
            recompute tree for this array based on updated GM
        else
            update tree starting from OMD_tra down
```

Fig. 4.: Incremental data reuse analysis algorithm

D. Comparison between different Pareto curves

Based on the Pareto curves generated for different loop transformations, we determine the potentially good loop transformation alternatives. All Pareto curves are combined into a global Pareto curve, and any alternative contributing to the global curve is good for a certain platform instance. A point belongs to the global Pareto curve if it is not dominated in both *#SPM_size* and *#MM_acc* by any other point. This means less accesses to the main memory then any others. Since accessing on-chip SPM memory is more energy efficient and faster than accessing main memory, the global Pareto point will result in the most energy efficient solution at least for this memory platform instance. Hence, that loop transformation alternative results in minimal energy for certain memory platform instances.

This also demonstrates how we can simulate any two-layer memory platform instances based on the Pareto curve for specific loop transformation. The Pareto point selected for simulating the on-chip SPM layer should be the one having a size as close as possible to, but not larger than, the SPM size of the selected platform. The chosen Pareto point defines which data that should be mapped on the on-chip SPM layer. The off-chip memory of the selected platform should be large enough to store the remaining data. The energy can hence be estimated based on the number of accesses to each layer together with an abstract energy-per-access model, which depends on the SPM size. Energy estimation for a number of realistic two layer memory platform instances are demonstrated on the real life applications in the next section.

IV. EXPERIMENTAL RESULTS

Two realistic demonstrators are selected to present our automated estimation method. The first one is the cavity detection algorithm used for detection of cavity perimeters in medical imaging. The second one is the video compression algorithm Quadtree Structured Difference Pulse Code Modulation (QS-DPCM).

A. Cavity Detection

In the original cavity detection code, different intermediate arrays are produced and consumed in different loop nests. Data locality can be improved with loop fusion (combined with shifting to satisfy dependencies). Fig. 5(a) shows our estimation results with Pareto curves for four selected incremental loop fusion and shifting alternatives. The horizontal axis shows the SPM size required and the vertical axis shows

6B-4

2006 Asia and South Pacific Design Automation Conference

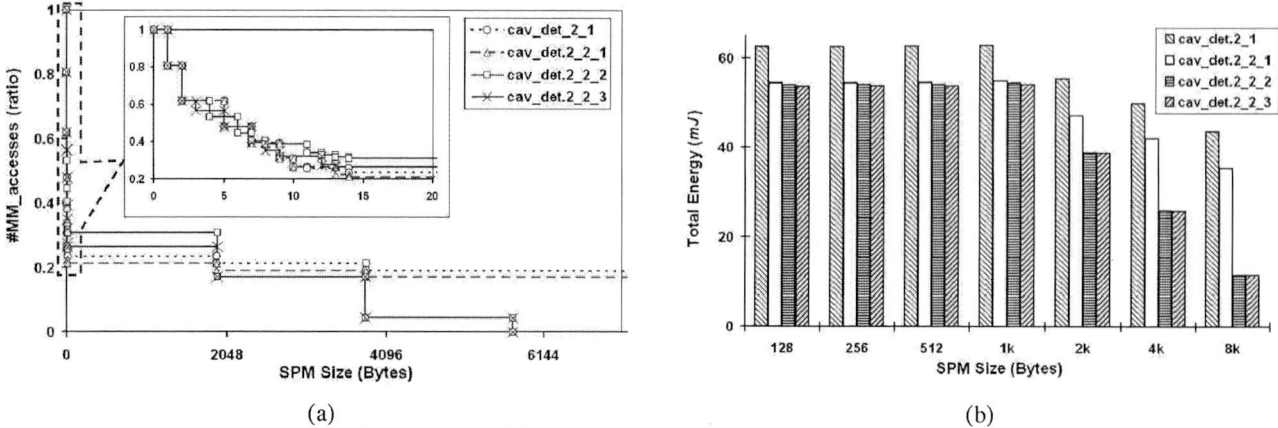

(a)   (b)

Fig. 5.: (a) Pareto curves and (b) energy estimate comparison for Cavity Detection

#MM_acc normalized over the total number of accesses from the processor core (29 million). When the SPM size increases, the main memory accesses decrease. As shown, the four Pareto curves all contribute to global Pareto points when the SPM size is 128 or smaller. This indicates that each loop transformation instance may result in low power memory hierarchy exploration with very small SPM size. cav_det.2_2_2 and cav_det.2_2_3 both contribute to the global Pareto points when the SPM size is larger than 1927. When SPM size is larger than 5748, there is no off-chip memory accesses for these two transformation alternatives, indicating that all data can be accessed on-chip.

As mentioned, any transformation alternatives that contribute to the global Pareto points can potentially result in minimal energy consumption. This is substantiated with the estimated energy for a number of two layer memory hierarchy instances shown in Fig. 5(b). The energy is calculated based on the number of accesses to each memory layer and an abstract energy-per-access model. For this experiment our energy estimate always has over-estimate with maximal 5% margin, compared to the detailed data reuse analysis and MHLA [4]. As shown in the energy estimate, significant energy reduction can be achieved when a suitable memory hierarchy instance, together with the right version of codes, is chosen. For example, the version of code cav_det.2_2_2 or cav_det.2_2_3 is selected for two layer memory hierarchy having 4K or 8K SPM size.

Fig. 5(a) also shows why it is important for the memory size estimation to take into account the memory hierarchy. The total memory size requirement is 5745 for cav_det.2_2_3 and cav_det.2_2_3, and 2536880 for cav_det.2_1 and cav_det.2_2_1. Without taking into account the memory hierarchy exploration, the conclusion would therefore be that cav_det.2_1 and cav_det.2_2_1 are not interesting at all. However, when the hierarchical memory size estimation is performed, it turns out that for SPM sizes up to 1927, cav_det.2_2_1 is a viable alternative. Since the original code cav_det.2_1 has lower complexity (the loop shifting adds `if`-clauses), this alternative is actually preferred for small SPM sizes. Analysis of the code complexity as a third trade-off axis is hence required for future work.

B. QSDPCM

The QSDPCM algorithm is an inter-frame compression technique for video images, which involves hierarchical motion estimation and a quadtree-based encoding of the motion compensated frame-to-frame differences. Fig. 6(a) shows the estimation outputs with four Pareto curves corresponding to the four selected incremental loop transformations alternatives. All these four Pareto curves contribute to global Pareto points. As shown, there are significant differences in the off-chip memory accesses between the first two and the last two transformation instances, especially when SPM size is between 1312 and 2496. Fig. 6(b) shows that choosing the right loop transformations among the 4 choices can give 25% reduction in total energy consumption for the memory platform having 2k SPM size.

As mentioned, speed is critical for the estimation among the larger number of loop transformation possibilities. It is also our motivation to do a fast estimation in order to help the designer find the right loop transformation alternatives while trading off a suitable memory hierarchy instance. Experiments show the usefulness of our techniques. For the cavity detection algorithm, the approach in [4] takes 1.54 seconds of CPU time for a single SPM size, compared to 0.30 seconds for all sizes in our approach. For the QSDPCM algorithm, theirs takes between 2 to 5 minutes for a single SPM size, compared to 3.0 seconds for ours. In particular, our approach further reduces the execution time during estimation for incremental loop transformations. The time varies between close to 0 and the time required for the first round estimate, depending on the incremental loop transformations' effects on data reuse trees and the choices chosen to rebuild the new data reuse trees. In contrast, the time is constant for each round analysis of the approach in [4]. The time difference will be very significant considering the exploration among a large number of loop transformation possibilities. Note that [4] approach can only estimate for one specific memory hierarchy instance at one time and our implementation is in python which can be a factor 10 slower.

V. CONCLUSION AND FUTURE WORK

This paper presents a technique for hierarchical memory size requirement estimation for loop fusion and loop shifting

610

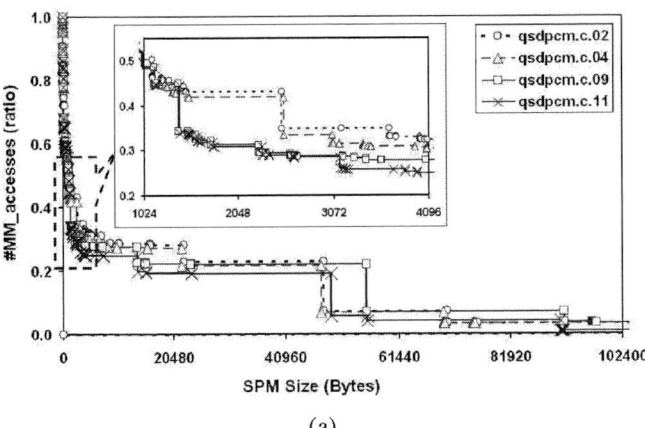 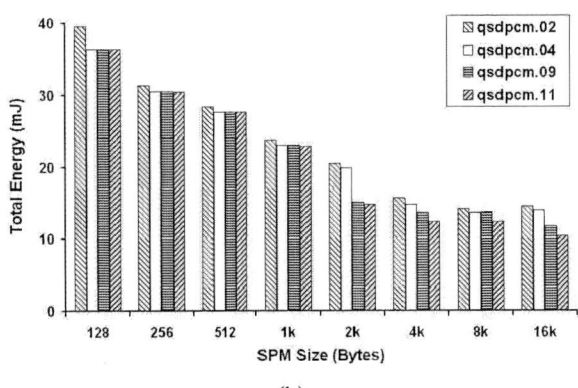

(a) (b)

Fig. 6.: (a) Pareto curves and (b) energy estimate comparison for QSDPCM

at the early loop transformations stage. As a large number of loop transformation possibilities exist and usually no memory platform is defined at this stage, we have proposed a fast platform-independent hierarchical memory size estimation algorithm. It outputs Pareto curves enabling to select the possibly good loop transformations. The Pareto curve also permits energy estimate for any two-layer memory hierarchy instances. This helps the designer to select a memory hierarchy instance, together with the right loop transformation alternatives. Experiments show the satisfactory estimation result with fast speed, which is critical for the exploration of the large number of loop transformation possibilities.

Loop transformations improving data locality also have a direct impact on array data lifetimes, which can potentially affect the memory size requirement both for individual arrays and between different arrays. This array lifetime analysis is not yet considered in our method. As taking it into account can potentially lead to better memory exploration, integrating it in our method is considered for future work. As our estimation method are in principle also applicable to any affine loop transformations, the automation of incremental estimation for general affine loop transformations is also left for future work.

REFERENCES

[1] F. Balasa, F. Catthoor, and H. De Man. Background memory area estimation for multi-dimensional signal processing systems. 3(2):157–172, June 1995.

[2] U. K. Banerjee. *Loop Transformations for Restructuring Compilers: The Foundations.* Kluwer Academic Publ., Norwell, MA, 1993.

[3] K. Beyls et al. Reuse distance-based cache hint selection. In *International Euro-Par Conference, 2002.*

[4] E. Brockmeyer, M. Miranda, H. Corporaal, and F. Catthoor. Layer assignment techniques for low energy in multi-layered memory organisations. In *Proc. 6th ACM/IEEE Design and Test in Europe Conf.*, pages 1070–1075, Munich, Germany, Mar. 2003.

[5] F. Catthoor, S. Wuytack, E. De Greef, F. Balasa, L. Nachtergaele, and A. Vandecappelle. *Custom Memory Management Methodology, Exploration of memory organization for embedded multimedia system design.* Boston, MA, 1998.

[6] A. Darte. On the complexity of loop fusion. *Parallel Computing*, 26(9):1175–1193, 2000.

[7] J.-P. Diguet, S. Wuytack, F. Catthoor, and H. De Man. Formalized methodology for data reuse exploration in hierarchical memory mappings. In *Proc. IEEE Int. Symp. on Low Power Design*, pages 30–35, Monterey CA, Aug. 1997. IEEE.

[8] A. Fraboulet, G. Huard, and A. Mignotte. Loop alignment for memory access optimization. In *Proc. 12th ACM/IEEE Int. Symp. on System Synthesis.*

[9] P. Grun, F. Balasa, and N. Dutt. Memory size estimation for multimedia applications. In *Proc. ACM/IEEE Wsh. on Hardware/Software Co-Design (CODES)*, pages 145–149, Seattle, WA, Mar. 1998.

[10] Q. Hu, M. Palkovic, and P. Kjeldsberg. Memory requirement optimization with loop fusion and loop shifting. In *Euromicro Symp. on Digital System Design (DSD'04)*, pages 272–278, Aug. 2004.

[11] Q. Hu, M. Palkovic, and P. Kjeldsberg. Memory requirement optimization with loop fusion and loop shifting. In *Euromicro Symp. on Digital System Design (DSD'04)*, pages 272–278, Aug. 2004.

[12] I. Issenin, E. Brockmeyer, M. Miranda, and N. Dutt. Data reuse analysis technique for software-controlled memory hierarchies. In *3rd ACM/IEEE Design and Test in Europe Conf.*, pages 202–207, Paris, France, Feb. 2004.

[13] M. Kandemir and A. Choudhary. Compiler-directed scratch pad memory hierarchy design and management. In *Proc. 39th ACM/IEEE Design and Test in Europe Conf.*, pages 690–695, Las Vegas, NV, June 2002.

[14] P. Kjeldsberg, F. Catthoor, and E. J. Aas. Data dependency size estimation for use in memory optimization. 22(7):908–921, July 2003.

[15] P. R. Panda, N. D. Dutt, and A. Nicolau. Efficient utilization of scratch-pad memory in embedded processor applications. In *Proc. 5th ACM/IEEE Europ. Design and Test Conf.*, pages 7–11, Paris, France, Mar. 1997.

[16] T. Van Achteren, F. Catthoor, R. Lauwereins, and H. De Man. Search space definition and exploration for nonuniform data reuse opportunities in data-dominant applications. 8(1):125–139, 2003.

[17] S. Verdoolaege, M. Bruynooghe, G. Janssens, and F. Catthoor. Multi-dimensional incremental loop fusion for data locality. In *Proc. Int. Conf. on Application-specific Systems, Architectures and Processors (ASAP)*, pages 17–27, Leiden, The Netherlands, June 2003.

[18] D. K. Wilde. A library for doing polyhedral operations. Master's thesis, Oregon State University, Corvallis, OR, Dec. 1993. also Technical Report PI-785, IRISA, Rennes, France.

[19] M. E. Wolf and M. S. Lam. A data locality optimizing algorithm. In *Proc. ACM SIGPLAN '91 Conf.*, pages 30–44, Toronto, Canada, June 26–28 1991.

[20] Y. Zhao and S. Malik. Exact memory size estimation for array computations without loop unrolling. In *Proc. 36th ACM/IEEE Design Automation Conf.*, pages 811–816, New Orleans, LA, June 1999.

A Novel Instruction Scratchpad Memory Optimization Method based on Concomitance Metric

Andhi Janapsatya†, Aleksandar Ignjatović†‡, Sri Parameswaran†‡

†School of Computer Science and Engineering, The University of New South Wales
Sydney, NSW 2052, Australia
‡NICTA, The University of New South Wales
Sydney, NSW 2052, Australia
{andhij,sridevan,ignjat}@cse.unsw.edu.au

ABSTRACT

Scratchpad memory has been introduced as a replacement for cache memory as it improves the performance of certain embedded systems. Additionally, it has also been demonstrated that scratchpad memory can significantly reduce the energy consumption of the memory hierarchy of embedded systems. This is significant, as the memory hierarchy consumes a substantial proportion of the total energy of an embedded system. This paper deals with optimization of the instruction memory scratchpad based on a novel methodology that uses a metric which we call the *concomitance*. This metric is used to find basic blocks which are executed frequently and in close proximity in time. Once such blocks are found, they are copied into the scratchpad memory at appropriate times; this is achieved using a special instruction inserted into the code at appropriate places. For a set of benchmarks taken from Mediabench, our scratchpad system consumed just 59% (avg) of the energy of the cache system, and 73% (avg) of the energy of the state of the art scratchpad system, while improving the overall performance. Compared to the state of the art method, the number of instructions copied into the scratchpad memory from the main memory is reduced by 88%.

1. Introduction

A designer looks beyond mere functionality of the required embedded system and optimizes for performance, energy consumption, and cost. Performance optimization allows for greater functionality, or the utilization of a lesser processor for the same task. Low energy consumption allows longer battery life, lower heat dissipation, and superior reliability. Reduced cost makes the system more competitive in the marketplace.

A customary method of improving performance and reducing energy consumption of a system is to use a cache. Despite the overall reduction in system energy, cache memory is known to consume up to half the total system energy. Instruction cache alone has been shown to consume up to 27% of total processor energy [1]. Thus, careful optimization in this area can reap rewards in terms of reduced energy consumption.

Prior research on optimization of instruction memory hierarchy for embedded systems saw the replacement of the instruction cache by a scratchpad memory (SPM) [2]. Even though cache and scratchpad are both made of SRAM cells, they operate differently. Cache is typically designed to improve performance of general-purpose processors, and is divided into tag RAM and data RAM. The tag RAM is compared against the instruction address requested, and if the instruction exists in the cache, it is sent to the processor, avoiding an access to the main memory. SPM, on the other hand, does not contain tag RAM, and forms a part of the main memory address map. By accessing the most frequently executed parts of the program from the SPM, one can reduce both the total execution time and the power

consumption, by avoiding tag RAM storage and comparison. As embedded systems typically execute a restricted number of applications, there are opportunities for improving their performance that are not available in the case of general purpose systems. For example, patterns of execution of the blocks of the code can be understood by the use of profiling. Once these patterns are known, they can be used to fill the SPM with appropriate segments of code at appropriate moments [11]. This reduces both the total access time and the total energy consumption of the system. Additionally, energy consumption is reduced due to the fact that the processor requires fewer wait cycles.

Selecting the correct code segments for placement in the SPM requires a careful analysis of the way such code segments are executed. All prior work in this area has focused upon loop analysis of the trace of a program as the method for finding the appropriate segments. Loop analysis has several drawbacks: (i) the structure and relationship of loops can be very complex; (ii) this structure can significantly vary for different inputs; (iii) the precise structure of loops is irrelevant for the placement of instructions in the SPM because only relative (temporal) proximity of executions matters, rather than the precise order of these instructions (as provided by the loop analysis) [12, 13, 14, 15].

In this paper we present, for the first time, an optimization method for utilizing instruction SPM that is based on an analysis of temporal correlation of instruction executions, rather than on loop analysis.

We introduce a class of metrics for estimating temporal proximity of consecutive executions of the same block of the code, and for estimating temporal proximity of interleaved executions of two different blocks of the code. The former is used to decide if a block should be executed from the SPM or from the main memory; the latter is used to decide if two blocks should be placed in SPM together or not. Such temporal information is gathered using a very efficient and adaptive algorithm whose parameters can easily be changed (various metrics for distance estimation). These metrics are used to estimate how correlated in time the execution of various blocks of code are, using an informative quantity that we call the *concomitance*. Our methods have a signal-processing flavor, because the trace is seen as a "signal" on which we perform a statistical, rather than structural analysis. Such analysis of the trace has proven to yield an algorithm for SPM placement with performance results that are not only superior to the previous state of the art, but that is also much simpler, more efficient, and adaptive to different types of applications. Recently we found out that a related, but somewhat cruder and less general idea has been used for cache management [16].

The benefits of using temporal proximity information is illustrated in the following example. Consider the "if-clause" shown in Figure 1(a) with the Control Flow Graph (CFG) shown in Figure 1(b). Looking at this code segment, for $K = 50$ and $M = 100$, a profiling

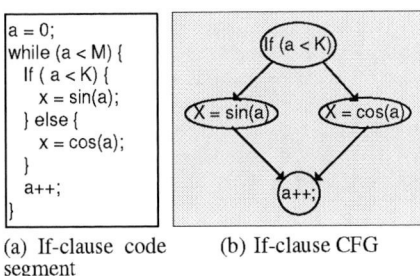

(a) If-clause code segment	(b) If-clause CFG

Figure 1: Motivational Example.

that takes into consideration temporal proximity of executions would find that the first 50 execution of the "if-clause" take the path containing $x = sin(a)$, and the next 50 execution of the "if-clause" take the path containing $x = cos(a)$. Thus, such algorithm will deduce that, since they are not interleaved in time, the block of instructions used for computing $x = sin(a)$ and the block of instructions used for computing $x = cos(a)$ can be placed onto SPM with overlap without degrading the performance. However, in a similar program, executions of $x = sin(a)$ and $x = cos(a)$ may be temporally interleaved. For example, if the "if-clause" is of the form $if(a = \text{"even number"})$, then the path with $x = sin(a)$ will be executed whenever a is an even number and the path with $x = cos(a)$ will be executed whenever a is an odd number. In such case the profiling will deduce that the executions of blocks used to compute these two functions alternate and consequently such temporally interleaved blocks must not be placed in the SPM with an overlap. Our concomitance metric is designed to distinguish between pairs of blocks whose executions are *likely* to be interleaved and blocks whose executions are *likely* to be separate in time, thus enabling proper placement on the scratchpad. Clearly, if only frequency of execution was determined as in the state of the art [11], these two patterns of execution would be indistinguishable and CFG would simply inform the algorithm that each path was executed 50 times. This shows the importance of measuring appropriately the temporal proximity of executions of blocks of code and making such information available to the SPM placement algorithm.

The rest of this paper is structured as follows: section 2 summarizes the related work and our contribution; section 3 presents the SPM system architecture; section 4 defines the *concomitance* and presents techniques for building the *concomitance table*; section 5 presents the optimization method; Section 6 explains the experimental setup and presents the results; finally, section 7 states the conclusions reached in this paper.

2. Related Work and Contributions

Existing work on the utilization of SPM can be categorized into three areas: (i) use of SPM for data memory only; (ii) use for instruction memory only; or (iii) both. The SPM utilization can be further divided into two classes: statically managed use and dynamically managed use. With static management, the SPM is filled at load time and its content does not change during the execution; in dynamic management, the contents of the SPM is changed during the program run-time.

Existing works on cache optimization techniques rely on careful placement of instructions and/or data within the memory to ensure low cache miss rates. Cache optimization methods generally increase the program memory size [21, 22, 23, 24, 25, 26, 27, 28]. The use of SPM to replace cache memory has been shown to improve the performance and reduce energy consumption [2].

SPM optimization methods for data memory are presented in [3, 4, 5, 8, 6, 17, 18, 19, 20]. In optimizing for data memory, data access patterns were analyzed to find frequently accessed variables and constants. A dynamic management scheme of data SPM was

first presented by Kandemir [5].

For static SPM, the most frequently executed basic blocks (e.g., loops) are kept within the SPM [2, 7]. Static SPM is simple, because it does not change its contents during execution. However, static SPM can be limiting, because it has to either keep all loops in the program by providing a large SPM to accommodate all the loops, or decide upon the most executed loops to be stored in a smaller SPM.

To overcome this limitation of static SPMs, dynamic SPMs update their contents during runtime. Steinke [9] used loop analysis to find loops within the program to dynamically allocate into SPM. They added a series of load and store instructions within the program to perform copying into SPM, resulting in greatly increased code size. In [11], Janapsatya et al. modified the processor architecture and inserted a custom instruction to perform copying into SPM. They used a graph partitioning procedure to perform loop analysis to find suitable loops within the program to place in the SPM. This reduced the number of instruction inserted, but was still dependent upon loop analysis. Their graph partitioning procedure was inefficient, since it used a global heuristic which performs badly for loops with basic blocks which are far apart in the program structure.

2.1 Our Contribution

In this paper, we define the notion of *concomitance* of blocks of code and use it to automatically identify blocks that are executed in clusters, each cluster consisting of many executions in close temporal succession. Such blocks are found to have high *concomitance* and will be placed in the SPM first. Further, pairs of basic blocks whose executions are interleaved are found to have high pairwise *concomitance* are thus simultaneously allocated to SPM. Hence, without the need to perform difficult loop analysis, we can identify which basic blocks are to be executed from SPM and which groups of instructions should be placed in the SPM simultaneously.

Our contribution includes replacement of difficult structural loop analysis of the program by an essentially statistical method for automated decision making regarding which basic blocks should be executed from SPM and, among them, which groups of basic blocks should be placed in the SPM simultaneously. Our method is very intuitive, flexible, and perspicuous, and is more akin to signal processing techniques rather than the existing algorithms for SPM management.

3. SPM System Architecture

3.1 Assumptions

The work presented in this paper was performed under the following assumptions:

- Size of scratchpad memory or cache memory is only available in powers of two and the size is known a priori. This is to enable better optimization.

- The applications and their execution patterns are known a priori.

3.2 Dynamic Instruction Scratchpad Memory

The architecture of the embedded system described in this work is presented in Figure 2. Our proposed SPM architecture uses a dynamic management scheme. The copying process can either use a series of load and store instructions as describe in [10] or use the special instruction method described in [11].

In our experimental setup, we decided to adopt the copying method described in [11] because it is the state of the art. To copy instructions into the SPM, the energy costs are incurred due to instructions fetch from DRAM, copying instructions into the SPM, and the energy cost of the memory copying hardware. Copying is performed by the custom copy instruction inserted within the program to inform the copy mechanism hardware when to start moving instruction into the SPM.

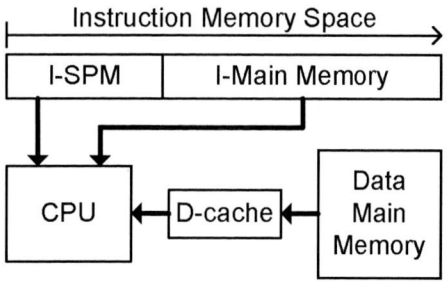

Instruction Memory Space

Figure 2: Embedded System Architecture.

Total energy cost of the system is given by summing the energy cost of all its components. For our dynamic scratchpad system, this is given by the following equation,

$$E_{system} = E_{CPU} + E_{SPM,insn} + E_{DRAM,insn} + E_{copy}$$
$$+ E_{cache,data} + E_{DRAM,data}$$

The parameters shown in the above equation are defined as follows:
E_{CPU} is the energy cost of the processor.
$E_{SPM,insn}$ is the sum of the energy costs of all instruction executed from scratchpad.
$E_{DRAM,insn}$ is the sum of the energy costs of all instruction executed from DRAM.
E_{copy} is the energy cost of all instructions copied into the SPM.
$E_{cache,data}$ is the energy cost of all data cache accesses.
$E_{DRAM,data}$ is the energy cost of all data cache misses.

The energy equation for calculating E_{copy} is given by,

$$E_{copy} = C \times E_{controller} + T \times (E_{SPM,write} + E_{DRAM,Burstread})$$

where C is the number of copy instructions executed, $E_{controller}$ is the energy cost of the copy mechanism, T is the total number of instructions copied into the SPM, $E_{SPM,write}$ is the cost of writing into the SPM, and $E_{DRAM,Burstread}$ is the cost for reading a sequential block of data from DRAM.

The heuristics behind our method can be explained as follows. If we are to place on to the SPM, a block of code a whose consecutive executions are often separated by execution of group g consisting of a large number of other distinct instructions, then either the part of the SPM occupied by a would not be utilized during the execution of g, or a would be overwritten by g and thus it would have to be reloaded. Consequently, we preferentially put in the SPM those blocks a that are executed frequently with a relatively small number of other distinct instructions executed between consecutive executions of a. Similarly, if two blocks of code a and b occur frequently in the trace in sequences of "sandwich forms" a,x,b,y,a or b,x,a,y,b such that x and y are some groups consisting in total of a relatively small number of distinct instructions of the program, then a and b should be allocated to the SPM in a non overlapping way. To formalize this heuristic, in section 4 we introduce the notion of *distance* between consecutive executions $e(b)$ and $e'(b)$ of a basic block b, as well as a notion of *concomitance* between basic blocks.

4. Concomitance

A basic block, by definition, is the largest chain of consecutive instructions that has the properties: (i), if the first instruction of the block is executed, then all instructions in the basic block will also be executed consecutively; and (ii), any instruction of the basic block is executed only as a part of the consecutive execution of the whole block.

The distance between two consecutive executions $e(b)$ and $e'(b)$ of a basic block b in the trace T of a run of a program is defined as follows. If between the executions $e(b)$ and $e'(b)$ of b there are

no other occurrences of b in T, we count the number of **distinct** instruction steps executed between $e(b)$ and $e'(b)$, including b. We call this value *the distance between $e(b)$ and $e'(b)$* and denote it by $d[e(b),\ e'(b)]$. For example, assume that "$bxyxyxyzxyxyb$" is a sequence of consecutive executions in a trace T, and that each of the basic blocks b, x, y and z contains ten distinct instructions; then the distance between $e(b)$, $e'(b)$ is 40, because only x,y,z appear between the two executions of the basic block b (and we include b itself in the count).

The weight function is used to give a decreasing significance to the two consecutive executions of the same block that are further apart in the sense of the above notion of distance. Thus, it is a non-negative real function $W(z)$ that is decreasing, i.e. $u \leq v$ implies $W(u) \geq W(v)$.

The trace *concomitance* $\tau(a,b,T)$ gives information about how tightly interleaved the executions of two distinct basic blocks a and b in the trace T are. Thus, for a basic block a we consider all of its consecutive executions $e(a)$, $e'(a)$ in the trace T, for which there exists at least one execution of the block b between the executions $e(a)$ and $e'(a)$; we denote such fact by $b \in [e(a), e'(a)]$. We now also reverse the roles of a and b, and define $\tau(a,b,T)$ by

$$\tau(a,b,T) = \sum_{\substack{b \in [e(a),e'(a)] \\ e(a) \in T}} W(d[e(a),e'(a)]) + \sum_{\substack{a \in [e(b),e'(b)] \\ e(b) \in T}} W(d[e(b),e'(b)])$$

Here $e(a) \in T$ in the sum means that $e(a)$ ranges over all executions of the basic block a that appear in the trace T. Note that for two distinct basic blocks a, b the *concomitance* of these two blocks will be large just in case b is often executed between two consecutive executions of a that are a short distance apart, and/or if a is often executed between two consecutive executions of b that are also a short distance apart, in the sense of distance defined above.

The trace self-*concomitance* $\sigma(b,T)$ of a basic block b is a measure of how clustered consecutive executions of the block b are, and is defined as:

$$\sigma(b,T) = \sum_{e(b) \in T} W(d[e(b),e'(b)])$$

Thus, trace self-*concomitance* $\sigma(b,T)$ has a large value for those basic blocks b whose executions appear in clusters, with all successive pairs of executions within each cluster separated by short distances. Note that even if b is executed relatively frequently, but such executions of b are dispersed in the trace T rather than clustered, then self-*concomitance* $\sigma(b,T)$ will still be low. On the other hand, if for certain input a particular loop is frequently executed, then the trace self-*concomitance* $\sigma(a,T)$ of each basic block a from this loop will be large. Thus, the loop structure of a program is reflected in the *statistics* of the *concomitance* values if such statistics is taken over runs with sufficient number of inputs reasonably representing what is expected in practice. This is the motivation for the following definitions.

The *concomitance* $\overline{\tau}(a,b)$ of a pair of basic blocks a,b for a given probability distribution of inputs is the corresponding **expected value** of trace *concomitance* $\tau(a,b,T)$.

The self-*concomitance* $\overline{\sigma}(b)$ of a basic block b for a given a probability distribution of inputs is the corresponding **expected value** of trace self-*concomitance* $\sigma(b,T)$.

To conveniently use the *concomitance* and self-*concomitance* in our scratchpad placement algorithms, we construct the *concomitance* table by the following profiling procedure.

• *Chose a suitable weight function $W(d)$.* In our experiments so far we have studied two types of weight functions: $W(d) = \frac{M}{d}$ and $W(d) = e^{\frac{-d^2}{M}}$, where M is a constant depending on the size of the scratchpad.

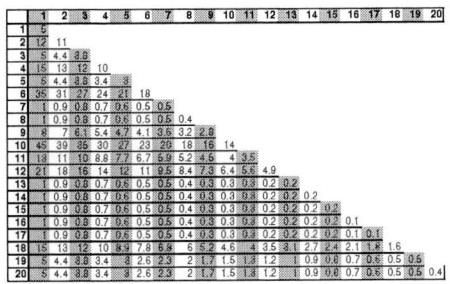

Figure 3: *Concomitance* **Table.**

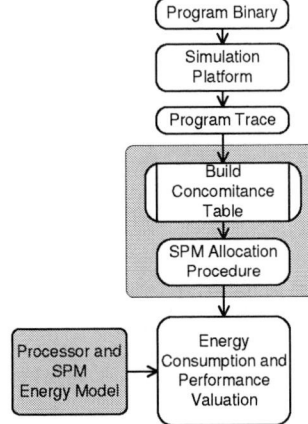

Figure 4: SPM Optimization Procedure.

- *Run the program* with inputs that reasonably represent the probability distribution of inputs expected in practice.
- *Calculate the average value of the trace self*-concomitance obtained from such runs, thus obtaining the self-*concomitance* value $\overline{\sigma}(b)$.
- *Set a threshold of significance* for the value of self-*concomitance* of basic blocks. The set S of all blocks with significant self-*concomitance* (i.e., larger than the threshold) is formed.
- *Calculate the* concomitance for all pairs of basic blocks from such set S, by finding the average of all trace *concomitances* obtained from the runs of the program and then form the corresponding table. Since the *concomitance* is commutative, such a table is symmetric and an example is shown in Figure 3; thus, we record only its lower left triangle. The self-*concomitance* $\overline{\sigma}(b)$ is conveniently placed on the diagonal of the table.

The size of the table is given by

$$Size = \frac{N*(N+1)}{2}$$

where N is the total number of basic blocks in the set S. Time complexity of the construction of the *concomitance* table is bounded by $2NT$, where T is the size of the trace.

5. SPM Optimization Procedure

The methodology for allocating basic blocks into the SPM is shown in Figure 4. A program binary is simulated along with its represented data to obtain the program trace. We then use our methodology to find segments in the program which are to be executed from the SPM. The methodology utilizes a SPM allocation procedure which uses the *concomitance* metric describe in section 4. This algorithm is shown in Figure 5.

The algorithm starts by building the *concomitance* table and the control flow graph (CFG) of the application. Each vertex in the CFG

Build Control Flow Graph.
Build *Concomitance* table of elements in S.
Perform Edge Cut procedure based on *concomitance* value.
Evaluate the need for insertion of copy instruction on all cut edges.
Calculate energy cost of the resulting sub-graphs.
Calculate performance result of the sub-graphs.

Figure 5: SPM Allocation Procedure.

Construct one group containing all the vertices in S;
Sort the *concomitance* values of all pairs $\{a,b\}$ of basic blocks
in ascending order;
Start from lowest *concomitance* value;
For each pair of basic blocks $\{a,b\}$ {
    if the resulting group size is larger than SPM size
        Cut the edge connecting a and b
}

Figure 6: Edge-cut Procedure.

represents a basic block. The weight of each edge a represents the *concomitance* value $\overline{\tau}(a,b)$ of a and b, obtained from the *concomitance* table. Edges of this graph are sorted in increasing order with respect to *concomitance* value. We then select from the lowest *concomitance* value and evaluate whether two vertices connected by the edge belong to a sub-graph with size larger than SPM. We cut all edges which connect two vertices with total size larger than the SPM. The edge cut procedure is shown in Figure 6.

For each subgraph, one or more copy instructions are to be added for copying the instructions within the subgraph into the SPM. Possible locations for adding copy instructions are the edges connecting subgraphs. To minimize the amount of copy instructions to be added, each edge is traversed up and a copy instruction is only added if another subgraph was possibly loaded into the SPM along the same execution path.

Time complexity of the SPM allocation procedure is bounded by N^3, where N is the total number of basic blocks in the set S.

6. Experimental Results

6.1 Setup

We simulated a number of benchmarks using the simplescalar/PISA 3.0d simulation environment [29] combined with DineroIV [30], to obtain memory access statistics. Power figures for the CPU were calculated using Wattch [31] ($0.18\mu m$). CACTI 3.2 [32] was used as the energy model for the cache memory. The energy model for the scratchpad memory was extracted from CACTI as in [2]. The DRAM power figures were taken from IBM embedded DRAM SA-27E [33]. We adopt the same Simplescalar CPU configuration and memory delay figures as described in [11] for ease of comparison of the results.

All benchmarks were obtained from the mediabench suite [34]. The total number of instructions executed in each benchmark is tabulated in Table 1. The data cache memory is fixed at 4K bytes giving a large enough data cache to ensure all benchmark applications cause less than 1% data cache misses.

Table 2 shows the SPM access time, SPM access energy, cache ac-

App.	Prog. size	Total no. of insn. Exec.	Copy insn. inserted	Avg. no. of insn. copied into SPM	SMI added [11]	Avg. no. of Insn. copied into SPM[11]
rawcaudio	9182	6689768	4.6	1247	30.4	741972
rawdaudio	9384	12414463	4.6	1263	29.7	2704463
g721enc	11052	314594475	6.2	33751706	63	130722589
g721dec	11066	302967631	4.8	12844416	58	87833725
mpeg2enc	26808	1134231679	23.4	4385032	272.86	197787388

Table 1: Cost of adding and executing copy instructions.

6B-5

Size (bytes)	Cache acc. time(ns)	SPM acc. time(ns)	ratio	Cache acc. energy(nJ)	SPM acc. energy(nJ)	ratio
512	1.19	0.74	1.61	1.37	0.18	7.61
1024	1.24	0.78	1.59	1.37	0.19	7.21
2048	1.30	0.83	1.57	1.39	0.20	6.95
4096	1.31	0.88	1.49	1.42	0.23	6.17
8192	1.34	1.05	1.28	1.49	0.29	5.14
16384	1.64	1.21	1.36	1.55	0.36	4.31

Table 2: Access time and energy consumption of static memory.

App.	SRAM size	Total cache misses assoc = 1	2	4	8	16	Total DRAM acc.[11]	Total DRAM acc.	% imp. cache	% imp. [11]
rawcaudio	1024	7408	6653	7818	8666	6807	3883	3004	59.4	22.6
	2048	7052	4629	2899	2846	2852	3427	2396	32.7	30.1
	4096	3931	4076	2334	2275	2240	3400	2092	24.1	38.5
	8192	2154	1981	1868	1847	1830	3400	2092	-8.5	38.5
	16384	2007	1841	1810	1799	1799	3400	1799	2.7	47.1
rawdaudio	1024	20899	26033	29336	30530	31148	31527	12488	53.7	60.4
	2048	14996	10389	4806	2914	2920	3470	2634	44.3	24.1
	4096	5465	5900	2398	2352	2322	3535	2160	29.8	38.9
	8192	2208	2035	1932	1915	1899	3535	2160	-8.5	38.9
	16384	2049	1898	1878	1867	1867	3535	1867	2.2	47.2
g721enc	1024	1.3E+8	1.3E+8	1.1E+8	1.1E+8	1.1E+8	1.5E+8	5.6E+7	52.6	62.8
	2048	1.1E+8	1.1E+8	1.1E+8	1.1E+8	1.1E+8	2.1E+8	5.6E+7	47.9	73.1
	4096	7.8E+7	8.3E+7	9.0E+7	1.0E+8	1.1E+8	3.3E+8	1.2E+8	-38.1	62.6
	8192	5.1E+7	2.1E+7	1.4E+7	1.1E+7	4.3E+6	7.5E+5	3.0E+5	97.2	59.5
	16384	3.9E+6	2.4E+6	1.1E+4	2.7E+3	2.7E+3	8.6E+3	2.7E+3	55.5	68.6
g721dec	1024	1.2E+8	1.2E+8	1.1E+8	1.1E+8	1.1E+8	1.3E+8	3.0E+7	73.5	77.1
	2048	1.0E+8	1.0E+8	1.0E+8	1.0E+8	1.0E+8	1.8E+8	3.9E+7	62.4	78.6
	4096	7.4E+7	8.0E+7	8.5E+7	9.2E+7	1.0E+8	1.1E+8	4.1E+7	51.8	62.5
	8192	4.1E+6	2.6E+7	8.6E+6	3.6E+6	3.7E+6	1.9E+5	4.7E+4	99.3	75.1
	16384	6.6E+4	3.5E+4	1.3E+4	3.6E+3	2.8E+3	6.8E+3	2.7E+3	58.8	60.0
mpeg2enc	1024	1.1E+8	1.4E+8	1.7E+8	1.8E+8	1.9E+8	2.4E+8	8.6E+7	43.6	64.2
	2048	2.7E+7	1.1E+7	1.1E+7	1.1E+7	1.1E+7	2.5E+7	5.2E+6	58.8	79.4
	4096	4.8E+6	4.1E+6	3.0E+6	2.8E+6	2.7E+6	6.7E+6	1.9E+6	43.4	71.9
	8192	1.9E+6	1.0E+6	7.6E+5	7.9E+5	8.4E+5	4.5E+6	7.7E+5	19.0	82.9
	16384	4.8E+5	3.7E+5	1.4E+5	1.1E+5	1.1E+5	2.8E+6	1.3E+5	21.1	95.4

Table 3: Total memory access Comparison.(Cache result shows the total cache misses. SPM total DRAM access is the total number of instructions executed from DRAM plus the number of instructions copied from DRAM to the SPM.)

cess time, and cache access energy [32] for an 8-way set associative cache (8-way is only shown here as an example). It can be seen that accessing SPM is approximately 1.5 times faster than a cache access and uses approximately 6 times less energy compared to an 8-way set associative cache. Instruction copying hardware cost is taken from [11] and stated to be at 2.94mW.

Experimental setup is shown in Figure 7. Comparison is made between results obtained in this work, work described in [11], and a conventional cache memory system.

6.2 Results

Table 3 showed a comparison of the total number of memory accesses. Performance and energy results measured from the experiments are shown in Table 4 and Table 5, and cost of adding copy locations within the program is shown in Table 1.

In Table 1, column 1 shows the application name, column 2 gives the size of the program, column 3 the number of instructions executed, column 4 gives the average number of copy locations to be inserted (average is taken from varying SPM sizes ranging from 1K bytes to 16K bytes), column 5 shows the average number of instructions that need to be copied into the SPM, column 6 presents the results from [11] showing the number of copy instruction (SMI) inserted into the program, and column 7 shows the total number of instruction copied into the SPM from [11]. Comparing figures in Table 1 column 5 and column 7 shows that our method significantly reduces the number of instructions to be copied into the SPM.

Table 3 compares the total number of memory accesses of a cache

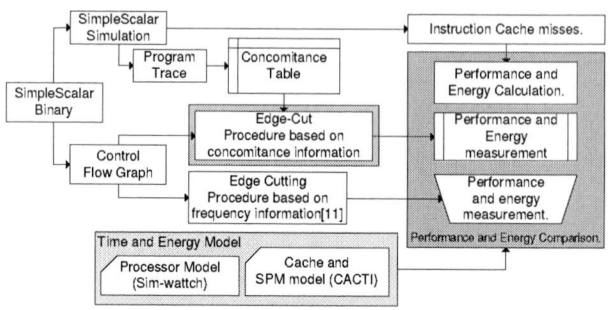

Figure 7: Experimental Setup.

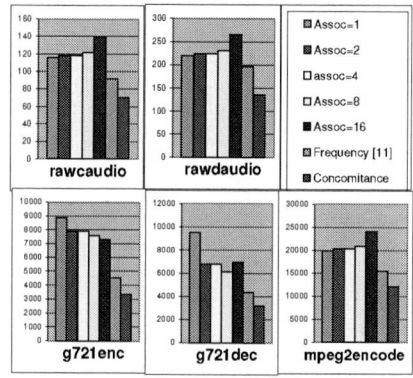

Figure 8: Energy Comparison for 8K bytes SRAM. (y-axis are energy in mJ)

system, the system presented in [11], and the system presented in this paper. Column 1 in Table 3 gives the application name; column 2 shows the cache or SPM size; column 3 to column 7 show the total number of cache misses for different cache associativities; column 8 gives the total DRAM accesses obtained from [11]; column 9 shows the total DRAM accesses obtained from our optimization method; column 10 shows the average percentage improvement of our method over the cache system; and column 11 the percentage improvement over the method described in [11]. In column 11, it is shown that our *concomitance* method reduces the total number of DRAM accesses by an average of 52.94% compared to the method described in [11]. When comparison is made with a cache based system (column 10), it is shown that the total number of DRAM accesses for our SPM system is not always less than the total number of cache misses. Despite the higher total DRAM accesses in a SPM system compared to the cache system, it does not always translate to worse energy consumption and worse performance compared to a cache system. This is because the energy and time cost per SPM access is far less compared to the energy and time cost per cache access (especially when compared to the energy and time cost of accessing a 16-way set associative cache.). It can also be noted that total number of DRAM accesses for a SPM system is comprised of both the number of instruction to be executed from DRAM and the total number of instruction copied from DRAM to SPM. Copying instructions from DRAM to SPM causes a sequential DRAM access which consumes less power and time compared to a random DRAM access that happens on each cache miss.

Table 4 shows energy comparison of our method with: (i) a cache system; and (ii) with the system presented in [11]. The table structure is identical to Table 3 except the comparison is now for energy. Figure 8 shows the energy improvement comparison between the cache system, our SPM allocation method, and the SPM allocation procedure described in [11] for all the benchmarks. The energy compari-

App.	SRAM size	Cache System's Energy (mJ) assoc =					SPM (mJ) [11]	SPM (mJ)	% imp. cache	% imp. [11]
		1	2	4	8	16				
rawcaudio	1024	101	105	105	113	130	69	59	45.9	13.6
	2048	103	106	106	117	132	73	64	43.2	12.6
	4096	109	111	111	118	138	77	66	43.2	14.3
	8192	116	118	118	122	140	92	70	42.5	23.3
	16384	140	139	139	147	165	99	75	48.5	24.2
rawdaudio	1024	194	201	201	217	249	157	116	45.1	26.1
	2048	197	200	200	222	249	161	122	42.4	24.2
	4096	207	211	211	223	260	170	127	42.3	25.0
	8192	219	223	223	230	264	196	135	41.7	31.4
	16384	265	263	263	277	310	210	143	47.7	31.6
g721enc	1024	25763	23507	23507	23753	24470	18397	10082	58.3	45.2
	2048	22782	22856	22856	23473	24123	22071	8920	61.6	59.6
	4096	18919	20190	20190	22273	24056	32151	14548	30.6	54.8
	8192	8880	7906	7906	7595	7314	4557	3356	57.5	26.4
	16384	7016	6568	6568	6932	7778	4832	3552	48.9	26.5
g721dec	1024	24830	22514	22514	22552	23315	16092	7185	68.9	55.4
	2048	21745	22026	22026	22589	23340	19071	6843	69.4	64.1
	4096	18141	19137	19137	20672	22841	13214	6968	64.9	47.3
	8192	9562	6779	6779	6104	6969	4357	3215	54.5	26.2
	16384	6382	6327	6327	6677	7493	4664	3422	48.3	26.6
mpeg2enc	1024	39931	44684	44684	49043	52774	34447	22183	51.6	35.6
	2048	19399	19849	19849	21853	24341	14118	11612	44.5	17.8
	4096	19292	19502	19502	20636	23970	13218	11636	43.1	12.0
	8192	19971	20294	20294	20941	24068	15421	12174	42.1	21.1
	16384	24032	23793	23793	25099	28151	16479	12907	48.1	21.7

Table 4: System's Energy Comparison.

App.	SRAM size	Cache System Execution Time (ms) assoc =					SPM (ms) [11]	SPM (ms)	% imp. cache	% imp. [11]
		1	2	4	8	16				
rawcaudio	1024	7.9	8.0	8.0	8.4	9.1	5.2	4.7	43.3	10.8
	2048	8.1	8.1	8.1	8.7	9.2	5.6	5.0	40.5	9.9
	4096	8.5	8.5	8.5	8.8	9.7	5.9	5.2	40.5	11.8
	8192	9.0	9.0	9.0	9.0	9.9	7.0	5.5	39.7	21.3
	16384	11.0	10.7	10.7	11.0	11.8	7.6	5.9	46.4	22.1
rawdaudio	1024	14.8	15.1	15.1	15.7	17.1	9.9	8.8	43.5	11.7
	2048	15.0	15.0	15.0	16.1	17.1	10.3	9.3	40.6	9.8
	4096	15.8	15.8	15.8	16.2	18.0	11.0	9.7	40.5	11.7
	8192	16.7	16.8	16.8	16.7	18.3	13.0	10.3	39.7	21.3
	16384	20.3	19.8	19.8	20.4	21.9	14.0	10.9	46.4	22.1
g721enc	1024	2016.8	1832.5	1832.5	1837.3	1864.5	1443.5	798.4	57.4	44.7
	2048	1782.8	1781.3	1781.3	1815.6	1837.2	1747.1	703.8	60.9	59.7
	4096	1479.5	1571.8	1571.8	1720.7	1831.8	2563.4	1147.7	29.4	55.2
	8192	690.9	607.3	607.3	567.7	517.3	335.0	261.8	55.8	21.9
	16384	545.1	502.2	502.2	516.9	554.1	355.4	277.1	47.0	22.0
g721dec	1024	1943.6	1754.9	1754.9	1744.0	1775.8	1260.1	568.2	68.9	55.4
	2048	1701.5	1716.5	1716.5	1747.1	1777.6	1507.9	538.6	68.9	64.3
	4096	1418.5	1489.5	1489.5	1596.0	1738.3	1034.2	548.0	64.4	47.0
	8192	744.5	519.2	519.2	451.5	492.2	319.1	256.0	52.7	21.5
	16384	495.5	483.6	483.6	497.8	533.6	342.2	266.8	46.4	22.0
mpeg2enc	1024	3142.5	3495.2	3495.2	3789.4	3978.2	2762.6	1742.9	51.0	36.9
	2048	1500.8	1509.6	1509.6	1616.0	1705.0	1123.1	899.1	42.5	19.9
	4096	1490.0	1479.5	1479.5	1516.2	1673.6	1048.2	900.3	40.9	14.1
	8192	1539.2	1538.7	1538.7	1534.4	1678.0	1221.2	942.9	39.7	22.8
	16384	1859.7	1812.6	1812.6	1865.2	1999.5	1300.3	1000.2	46.4	23.1

Table 5: System's Performance Comparison.

son shows that our method almost always performs better compared to the cache system, and superior results are seen when compared with results obtained from [11]. On average, the energy consumption by utilizing our method is 41.9% better than cache system energy, and 27.1% better than the method described in [11]. In particular our method is superior in cases where negative improvements over cache were shown in results from [11].

Performance result is shown in Table 5. Structure of the table is identical to Table 4; column 3 to column 7 shows the execution time of a cache based system; column 8 shows the performance results obtained from [11]; and column 9 shows our performance measurement results; column 10 shows the average performance improvement of our method over cache system; and column 11 shows performance improvement over method described in [11]. Our method improves the execution time by 40.0% compared to cache and 23.6% compared to the method described in [11].

Thus it is clear from the results that the method described here is a feasible method for dynamic SPM allocation. We show that the number of copy instructions inserted are far fewer than the state of the art. In addition, for the applications shown here, we show performance improvement and energy savings.

7. Conclusions

In this paper we have proposed a method to reduce energy and improve performance of an embedded system, containing an instruction scratchpad. Our system relies on a new metric called *concomitance*, which is used to identify basic blocks that should be placed together in the scratchpad. This method results in embedded systems with lower energy and higher performance, compared to a standard cache system and state of the art scratchpad instruction partitioning algorithm.

8. References

[1] J. Montanaro et al., "A 160MHz, 32b, 0.5W CMOS RISC microprocessor," *JSSC*, vol.31(11), pp. 1703-1712, 1996.
[2] R. Banakar et.al., "Scratchpad Memory: A Design Alternative for Cache On-chip Memory in Embedded Systems," *CODES*, 2002.
[3] O. Avissar and R Barua, "An Optimal Memory Allocation Scheme for Scratch-Pad-Based Embedded Systems," *ACM Trans. on Embedded Computing Systems*, vol. 1, pp. 6-26, 2002.
[4] P.R. Panda, "Efficient Utilization of Scratch-Pad Memory in Embedded Processor Applications," *European Design and Test Conference, Proceedings of*, 1997.
[5] M. Kandemir et.al., "Dynamic Management of Scratch-Pad Memory Space," *DAC*, 2001.
[6] M. Kandemir and A. Choudhary, "Compiler-Directed Scratch Pad Memory Hierarchy Design and Management," *DAC*, 2002.

[7] F. Angiolini et.al., "Polynomial-Time Algorithm for On-Chip Scratchpad Memory Partitioning," *CASES*, 2003.
[8] S. Udayakumaran and R. Barua, "Compiler-Decided Dynamic Memory Allocation for Scratch-Pad Based Embedded Systems," *CASES*, 2003.
[9] S. Steinke et.al., "Assigning Program and Data Objects to Scratchpad for Energy Reduction," *DATE*, 2002.
[10] S. Steinke et.al., "Reducing Energy Consumption by Dynamic Copying of Instructions onto Onchip Memory," *ISSS*, 2002.
[11] A. Janapsatya et.al., "Hardware/Software Managed Scratchpad Memory for Embedded System," *ICCAD*, 2004.
[12] G. Ramalingam, "On Loops, Dominators, and Dominance Frontier," *PLDI*, 2000.
[13] P. Havlak, "Nesting of Reducible and Irreducible Loops," *ACM Transactions on Programming Languages and Systems*, 1997.
[14] V. C. Sreedhar et. al., "Identifying Loops using DJ Graphs," *ACM Transactions on Programming Languages and Systems*, 1996.
[15] B. Steensgaard, "Sequentializing Program Dependence Graphs for Irreducible Programs," *Technical Report MSR_TR-93-14*, Microsoft Research, Redmond, Washington, October 1993.
[16] N. Gloy and M. D. Smith, "Procedure Placement Using Temporal-Ordering Information," *Programming Languages and Systems. ACM Transactions on*, Vol. 32, No. 5, Pages 977-1027, September 1999.
[17] P. Grun et. al., "Access Pattern-Based Memory and Connectivity Architecture Exploration," *Embedded Computing Systems. ACM Transactions on*, Vol. 2, No. 1, Pages 33.73, February 2003.
[18] A. Ramachandran and M. F. Jacome, "Xtream-Fit: An Energy-Delay Efficient Data Memory Subsystem for Embedded Media Processing," *DAC*, 2003.
[19] M. Verma et. al., "Dynamic Overlay of Scratchpad Memory for Energy Minimization," *CODES+ISSS*, 2004.
[20] M. Kandemir et. al., "Exploiting Scratch-Pad Memory Using Presburger Formulas," *ISSS*, 2001.
[21] S. Parameswaran and J. Henkel, "I-CoPES: fast instruction code placement for embedded systems to improve performance and energy efficiency," *ICCAD*, 2001.
[22] P. P. Chang and et.al., "IMPACT: an architectural framework for multiple-instruction-issue processors," *Computer Architecture News*, vol. 19, no. 3, 1991.
[23] S. McFarling, "Program optimization for instruction caches," *ASPLOS*, 1989.
[24] S.McFarling, "Procedure merging with instruction caches," *SIGPLAN Notices*, vol. 26, no. 6, 1991.
[25] P. Panda and et.al., "Memory Organization for Improved Data Cache Performance in Embedded Processors," *ISSS*, 1996.
[26] P. Panda and et.al., "A Data Alignment Technique for Improving Cache Performance," *ICCD*, 1997.
[27] H. Tomiyama and H. Yasuura, "Optimal code Placement of Embedded Software for Instruction Cache," *EDAC*, 1996.
[28] S. Bartolini and C. A. Prete, "A cache-aware program transformation technique suitable for embedded systems," *Information and Software Technology 44(13)*, 2002.
[29] D. Burger and T. M. Austin, "The SimpleScalar Tool Set, Version 2.0," *TR-CS-1342*, University of Wisconsin-madison, June 1997.
[30] J. Edler and M. D. Hill, "Dinero IV Trace-Driven Uniprocessor Cache Simulator," *http://www.cs.wisc.edu/ markhill/DineroIV/*.
[31] D. Brooks et.al., "Wattch: A Framework for Architectural-Level Power Analysis and Optimizations," *ISCA*, 2000.
[32] P. Shivakumar and N. P. Jouppi, "Cacti 3.0: An Integrated Cache Timing, Power, and Area Model," *Technical Report 2001/2*, Compaq Computer Corporation, August, 2001. 2001.
[33] IBM Microelectronics Division, "Embedded DRAM SA-27E," *http://ibm.com/chips*, 2002.
[34] C. Lee et.al., "MediaBench: A Tool for Evaluating Multimedia and Communications Systems," *IEEE MICRO 30*, 1997.

DraXRouter: Global Routing in X-Architecture with Dynamic Resource Assignment*

Zhen Cao[1], Tong Jing[1], Yu Hu[2], Yiyu Shi[2], Xianlong Hong[1], Xiaodong Hu[3], Guiying Yan[3]

[1] Computer Science & Technology Department
Tsinghua University
Beijing 100084, China
Phone: +86-10-62785564
Fax: +86-10-62781489
e-mail: caoz@mails.tsinghua.edu.cn

[2] Electrical Engineering Department
UCLA
Los Angeles, CA 90095, USA
Phone: (310) 267-5407
Fax: (310) 267-5407
e-mail: {hu, yshi}@ee.ucla.edu

[3] Institute of Applied Mathematics
Chinese Academy of Sciences
Beijing 100080, China
Phone: +86-10-62639192
Fax: +86-10-62574529
e-mail: {xdhu, yangy}@amss.ac.cn

Abstract – In recent years, the X-Architecture is introduced to obtain better performance for integrated circuit physical design. This paper reformulates the global routing problem in X-Architecture under the liquid routing model. Then, a dynamic resource assignment (Dra) method is presented to reduce potential vias. At last, a global router called DraXRouter, is designed, in which we adopt a dynamic-tabulist-based tree construction algorithm and a stochastic optimization strategy to gain high quality routing solution. Tested on ISPD'98 benchmarks, DraXRouter achieves better routing performance compared with two recent global routers.

I. Introduction

With advance in fabrication technology of integrated circuit (IC), the interconnect delay has become a significant factor affecting IC performance. The optimization capability of algorithms for interconnect performance optimization is limited since they were designed based on rectilinear interconnect architecture. Then, researchers begin to focus on other on-chip interconnect architectures to obtain better optimization results and higher performance. [4] indicates that non-Manhattan routing/interconnect optimization is now championed by the X-Architecture.

Compared with traditional Manhattan architecture, the X-Architecture reduces wire length by 20%, as well as via number by 30%. This results in a chip performance improvement of 10% and a power reduction of 20% [1]. However, the X-Architecture is not widespread today due to the lack of X-Architecture-based placement and routing algorithms.

In the traditional Manhattan architecture, some recent progresses focus on congestion reduction. [3] presented a global router, called labyrinth, which evaluates placement results in terms of congestion and wire length. By integrating amplified congestion estimates with the routing, [16] improved over-congestion and routing length. [12] presented an efficient congestion reduction algorithm for global routing based on search space traversing technology (SSTT), by which large circuits can be routed in a short running time while keeping good performance.

The routing problem in traditional architecture is well

studied while there is not much literature focusing on the X-Architecture-based routing problem. [4] proposed a new global router to explore modern interconnect problems in Manhattan and non-Manhattan architectures, which indicated that non-Manhattan architectures require more vias than Manhattan architecture under preferred direction routing. [15] introduced a layer balance approach for congestion reduction in both Manhattan and non-Manhattan architectures. [2] presented a routing algorithm with high complexity under liquid routing to reduce both total wire length and the number of vias, while keeping high routability. [20] proposed the first multilevel routing framework for the X-Architecture. Compared with router in Manhattan Architecture, this router can reduce wire length by 18.7% and average delay by 8.8% with similar routing completion rates. Via number was not reduced.

Liquid routing model [1], [18] is an effective way to reduce vias in X-Architecture. [10] considers the liquid routing model in global routing phase, and studies on the routing resource estimation method for liquid routing model in X-Architecture, based on which a global routing algorithm is presented. Compared with labyrinth [3] and SSTT [12] in Manhattan architecture, this algorithm can reduce the total wire length and total overflow more than 10% and 80% on average, respectively.

To explore the more benefit can be obtained from the X-Architecture routing, this paper employs liquid routing model in global routing. Our work aims to find a more accurate and effective way to estimate and utilize the routing resources in global routing stage and provide a better guidance for liquid routing after global routing. The main contributions of this paper are as follows.

(1) We present a dynamic resource assignment (Dra) method to guide liquid routing in X-Architecture, in order to utilize the routing resources more reasonably under liquid routing model.

(2) We design a global router, called DraXRouter, in which we adopt a dynamic-tabulist-based (DTB) tree construction algorithm and the stochastic optimization strategy to gain better routing performance.

Tested on ISPD'98 benchmarks, our router, DraXRouter, has gained good routing performance. Compared with x-labyrinth[1], DraXRouter improves utilization of routing resource with a shorter wire length. At the same time, the

---

* This work was supported in part by the NSFC under Grant No.60373012 and the SRFDP of China under Grant No.20050003099.

---

[1] We designed labyrinth algorithm [7] in X-Architecture, called x-labyrinth, for more comparison with our DraXRouter.

runtime of DraXRouter is 33% shorter. Compared with COCO algorithm [10], DraXRouter gains a better routing result with an acceptable runtime. Meanwhile, we compared the wire crossover in the global routing results produced by DraXRouter with the traditional resources assignment (RA) methods and that with the Dra method, respectively. The results indicated the Dra method can reduce the crossover number by 5.09% with even a shorter wire length.

The remainder of this paper is organized as follows. In Section II, we formulate the global routing problem in X-Architecture. Section III presents the Dra method. Section IV describes the details of the global routing algorithm and the implementation of our global router, DraXRouter. Experimental results are given in Section V. We conclude this paper in Section VI.

II. Preliminaries

A. GRG Generation in X-Architecture

In global routing problem people often partition the routing area into several global tiles or global routing grid (GRG) and map all physical pins into each tile, and then find a routing to connect the center of corresponding tiles. There exist several ways to partition routing area in non-Manhattan architectures. For X-Architecture, [8] presented the octagonal tiling. Since our global router was supposed to guide a gridless liquid detail routing, it's hard to use this GRG generation. As the placement tool in X-Architecture is not mature [1], we utilize the placement result of a traditional standard cell placer as the input of our global router. We adopt the GRG partition method in [10]. As Fig.1 shows, we add some diagonal edges to connect each two diagonal adjacent vertices.

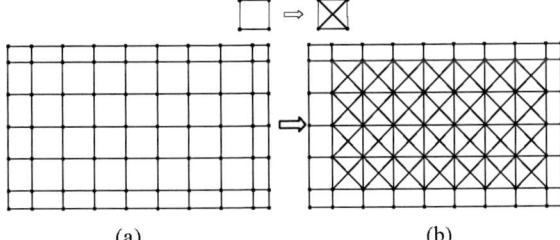

(a)   (b)

Fig. 1. GRG generation (a) GRG in Manhattan architecture, (b) GRG in X-Architecture.

B. Problem Formulation

Given c_e as the capacity of a GRG edge, d_e as the routing demand for edge e, the overflow of edge e is defined as follows.

$$overflow_e = \begin{cases} d_e - c_e, & if\ d_e > c_e \\ 0, & otherwise \end{cases}$$

The total overflow of the entire routing area is as follows.

$$tof = \sum_{e \in E} overflow_e \qquad (1)$$

The total wire length can be calculated as follows.

$$twl = \sum_{n=1}^{N_{net}} length_{net_n} = \sum_{n=1}^{N_{net}} \sum_{t=1}^{N_{edge}} l(e_t) \qquad (2)$$

Then, the global routing problem can be formulated as find routing solution for each edge e, such that minimize:

$$f(tof, twl) \qquad (3)$$

This function considers both wire length and total overflow, will be introduced in detail in Section IV.B.

C. Design Flow Under Liquid Routing Model

As routing with preferred direction in X-Architecture may cause an increase of via number, people adopt liquid routing [18], which utilize the gridless octilinear routing technology, to finish the detailed routing.

Our routing algorithm is designed to guide liquid detailed routing, so a layer assignment process should be employed to assign the intersectant segments into different layers.

III. Dynamic Resource Assignment

A. Capacity Upper Bound C_{eub}

Researches adopted liquid routing [1] to reduce via number in X-Architecture. However, in liquid routing model, the routing resource is not static. Routing in one edge will influence its adjacent edges. Considering the GRC (global routing cell) shown in Fig.2(a), K is the center of GRC CEGI, also a vertex in GRG. Routing in the direction KG will affect the routing resources of KF and KH. Routing in direction of KF will affect the routing resources of KE and KG.

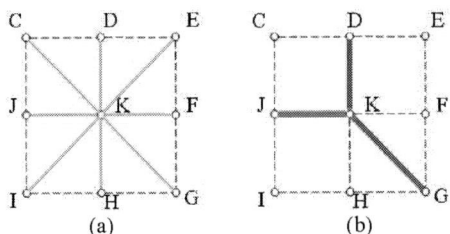

(a)   (b)

Fig.2. Routing resources interaction in liquid routing model (a) GRC in X-Architecture, (b) an extreme routing demand.

Fig.2(b) presents a case with an extreme routing demand. That is, the routing demand of KG is too much and the routing resources of KF and KH are all utilized by KG. Meanwhile, the routing resources of KE, KC, and KI are all utilized by KD and KJ (three blue segments in Fig.2(b)). In traditional static RA, such requirements can not be met due to the fixed pre-assignment. I.e. the pre-assignment resources for each edge can not be changed during the whole routing process.

In order to assign the routing resources more reasonably, we present a dynamic resource assignment (Dra) method to take on characters of liquid routing model, which assigns the routing resources in the routing process. In Dra method, a capacity upper bound C_{eub} (NOT a fixed pre-assignment capacity value) is given to each routing edge, which indicates the potential routing resources utilized by this edge. First, the C_{eub} of vertical and horizontal edges are calculated based on wire widths. Then, the C_{eub} of diagonal edges is estimated based on an area distribution. In the routing process, the C_{eub} of an edge is influenced by both routing over this edge and routing over its adjacent edges. Then, the routing resources will be assigned dynamically in the routing process to take on characters of liquid routing model.

B. Calculation of C_{eub}

The capacities of vertical and horizontal edges are calculated based on wire widths in traditional method. Then, we take half of the routing area of vertical and horizontal edges as the routing area for diagonal edges. As Fig.3 shows, assume the capacity of vertical edges is 20. As the capacity is taken by half, 10 left for vertical edges. And the capacity of diagonal edges is $10\sqrt{2}$ as the width of track is not changed.

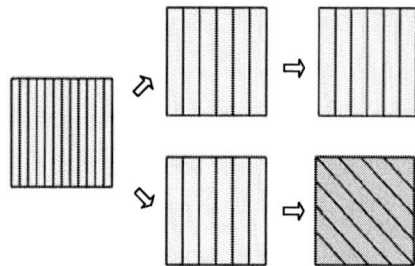

Fig.3. Calculation of capacity.

The C_{eub} of an edge indicates the potential routing resources utilized by this edge, then the initial C_{eub} should be twice of the capacity due to our calculation method. Meanwhile, Dra should guarantee the real resources utilized is the same as that of traditional method. Suppose the C_{eub} of a vertical edge is c_v and the C_{eub} of a horizontal edge is c_h, so the C_{eub} of a diagonal edge is $\sqrt{c_v^2 + c_h^2}$.

C. Dynamic Resource Assignment (Dra)

The assignment of Dra considers the influence of routing on one edge to its adjacent edges. See Fig.4(a) as a reference, if a net is routed over edge AR, the C_{eub} of AR will be cost one. As an influence of AR, the C_{eub} of AS and RQ will be taken a cost of $c_h / 4\sqrt{c_v^2 + c_h^2}$. And the C_{eub} of AQ and SR will be taken a cost of $c_v / 4\sqrt{c_v^2 + c_h^2}$. In Fig.4(b), if a net is routed over AS, the influence of AS to AR, SQ, SB, and AC will be $\sqrt{c_v^2 + c_h^2} / (4* c_h)$. Similarly, the influence of AQ to AR, QS, AT, and QU will be $\sqrt{c_v^2 + c_h^2} / (4* c_v)$.

The proportion of Dra method comes from the prove process as follows. Assume that the routing resources of a GRG are just utilized by all horizontal and vertical routing edges or all by diagonal edges. That is, c_v vertical edges

and c_h horizontal edges or $2\sqrt{c_v^2 + c_h^2}$ diagonal edges. This indicates these routing edges share the same routing resources. Then, the proportion of interaction can be calculated. This proportion guarantees the real resources utilized are the same as the traditional methods.

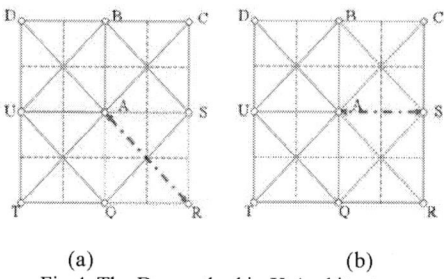

(a)   (b)

Fig.4. The Dra method in X-Architecture.

IV. DraXRouter Routing Algorithm

In this section, we design a new X-architecture-based global router, called DraXRouter. In DraXRouter, we construct initial trees firstly, and then rip-up and reroute congested nets to reduce congestion with several iterations. We first present our tree construction heuristic.

A. Octilinear Steiner Tree Construction Heuristic

Steiner minimal tree problem has been proved to be NP-complete in [19]. A random sub-trees growing heuristic (RSG) is introduced in [10] as a tree construction algorithm for global routing. This heuristic is efficient and can find Steiner points for the terminals to be connected. But it just searches the next one step of a growing point, which may cause an improper detour. As shown in Fig.5(a), the pins to be connected is A and B. The gray area indicates congestion. If the heuristic only searches one step, the detour will be started too late. The accurate detour is shown in Fig.5(b). In X-Architecture, routing can explore more directions than Manhattan cases, which indicates that we may get much worse results without a good control of the sub tree growing.

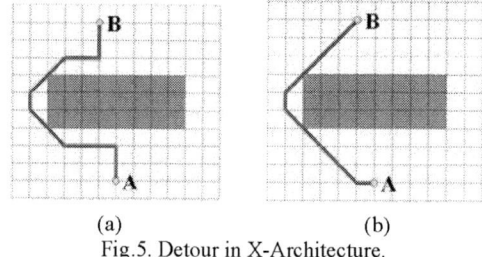

(a)   (b)

Fig.5. Detour in X-Architecture.

We design a dynamic-tabulist-based algorithm (DTB) to solve this problem, in which every pin of the net is initially regarded as a growing point (GP). And each GP will grow toward another GP according to the others' tabulist. When two GPs meet, they will merge into a new GP. This process repeats until there is one GP left and the tree is constructed.

2006 Asia and South Pacific Design Automation Conference

6C-1

Algorithm 1 DTB Heuristic
Input: Graph G and terminal set N
Output: An octilinear Steiner minimal tree in G

1:     Put a GP on every terminal.
2:     **while** the number of GPs > 1 **do**
3:         $t \leftarrow$ select a GP p according to estimate function (4), t is its current position;
4:         p grows from t;
5:         find the new current position v for p;
6:         **if** $v \in$ tabulist of another GP p' **then**
7:             $p_{new} \leftarrow$ merging p and p';
8:             Compute the location of P_{new};
9:         **else**
10:           update the tabulist of p;
11:         **end if**
12: **end while**
13: **return** *tree*;

Fig.6. The pseudo-code of DTB.

When growing, the current position of a GP and its tabulist is selected dynamically to find out the best topology for the Steiner tree. The tabulist of a GP is changed in construction process, and the influence of this GP to others is changed too. All GPs interact until the tree is constructed. The pseudo-code of DTB is shown in Fig.6.

The estimate function is defined as follows.

$$f_p(v) = w(p,v) + dis(s,v) + g(p,v) \qquad (4)$$

Where $w(p,v)$ presents the congestion information of the tabulist of p from source to v, $dis(s,v)$ presents the routing distance from source to v, $g(p,v)$ is the heuristic function, it allows every GP to grow toward other GPs.

The construction process of an 11-pin net is shown in Fig.7. Every GP is given a color. The color of a wire is the same as color of the active GP on this wire. The gray area indicates congestion. The tabulist of the orange GP affects tabulist of the green GP in the routing process, as shown in Fig.7(b) and Fig.7(c). Final topology of the tree is shown in Fig.7(d), in which the best topology is obtained. This shows DTB algorithm has an advantage of avoiding congestion and finding proper Steiner point.

B. Stochastic Optimization Approaches

Our routing algorithm uses DTB to construct initial trees. Then, several iterations are required for the rip-up and reroute process to refine the solution. For every iteration, we first find out all nets that go over a congested edge and pick up the nets to be rerouted randomly according to an proportion p calculated in Eq.(6).

$$p = P * cnum / tof \qquad (6)$$

where *cnum* is the number of congested nets, *tof* is the total overflow by Eq.(1), P is a constant.

Then, these nets are rerouted using DTB heuristic with the new routing demand estimation. The new *tof* and *twl* are calculated in Eq.(1) and (2). After the new routing solution of these nets is obtained, a judgment of whether to accept this routing solution is employed. This judgment is formulated as follows.

$$\begin{cases} d_{tof} \geq 0 \ and \ d_{twl} \geq 0 & p = 0 \\ d_{tof} < 0 \ and \ d_{twl} \leq 0 & p = 1 \\ d_{tof} < 0 \ and \ d_{twl} > 0 & p = e^{\left(d_{twl}/-d_{tof} - T\right)} \\ d_{tof} > 0 \ and \ d_{twl} < 0 & p = e^{\left(-d_{twl}/d_{tof} - T\right)} \end{cases}$$

where p is the probability of acceptance, T is defined as T = (N / K + 1) * D, N is the number of iterations, D is a constant, dtof and dtwf are the difference of the total overflow and total wire length between the previous iteration and the current iteration. If this solution is accepted, the new topologies will be stored and the new routing demand will be utilized by the next iteration. This judgment can avoid trapping into a local minimum.

V. Experimental Results

A. Experiments Setup

We test our global router with the ISPD'98 benchmarks. TABLE I shows the characteristics of all benchmarks.

As there are few global routers in X-Architecture, we studied the routing algorithm of labyrinth [3] and implemented it in the X-Architecture, called x-labyrinth, for comparison.

To demonstrate the optimization capability of our router, we compare our router with x-labyrinth and COCO [10]. For a fare comparison, we assign the same routing resources to all the three routers. For these three routers all have strong capabilities of congestion reduction, the experimental results of initial routing resource provided by the benchmarks show

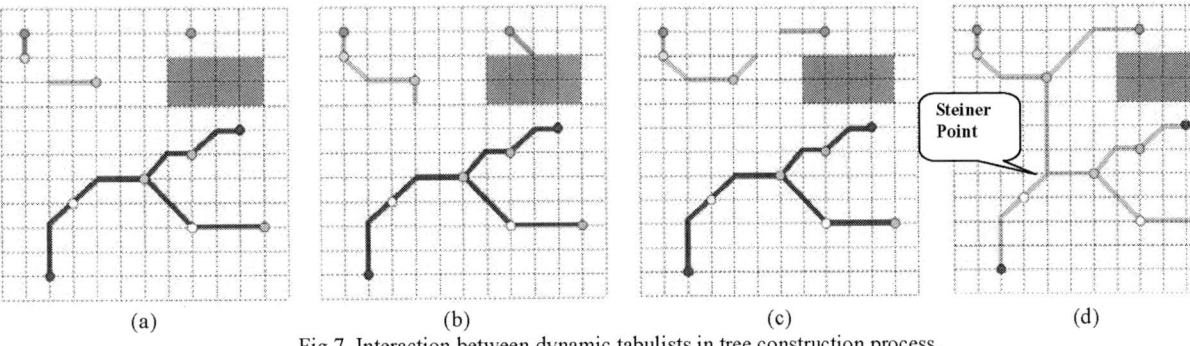

Fig.7. Interaction between dynamic tabulists in tree construction process.

that there exist no congestion, then it is hard to compare optimization capability of these routers, so we shrink the routing resources until overflow appears. The total overflow (*tof*), total wire length (*twl*) and runtime (CPU) of the routing results are compared and shown in TABLE II to TABLE V. Experiments in this paper are performed in 1.6GHz CPU / 680MB memory Linux.

TABLE I
Benchmark Data

Circuits	Net#	Grids	Circuits	Net#	Grids
ibm01	13k	64×64	ibm06	34k	128×64
ibm02	19k	80×64	ibm07	46k	192×64
ibm03	26k	80×64	ibm08	49k	192×64
ibm04	31k	96×64	ibm09	59k	256×64
ibm05	30k	128×64	ibm10	66k	256×64

To show the effectiveness of Dra method, we run our router under the static RA method [10] and Dra method, respectively. After the running, the total crossed segment numbers of both strategies are compared. The number of the crossovers indicates the number of potential vias, so the reduction of crossover brings reduction of via number in the gridless liquid detailed routing.

B. Wire length and Congestion Comparisons

The comparison of test results is shown in TABLE II to TABLE V. Compared with x-labyrinth, our algorithm can achieve 100% completion for global routing for almost all circuits (8 out of 10 testcases). On average, DraXRouter reduces total overflow by 99.49%, total wire length by 3.21%. Compared with COCO, our algorithm reduces total overflow by 99.93% and total wire length by 18.15% with an acceptable runtime.

TABLE II
Comparison between DraXrouter & x-laby on *twl*

Circuits	DraX	x-laby	Imp.(%)
ibm01	61011.9	65989.6	7.54
ibm02	168908	169388	0.28
ibm03	148970	163081	8.65
ibm04	163009	174081	6.36
ibm05	408902	421328	2.95
ibm06	275598	280265	1.67
ibm07	349383	355865	1.82
ibm08	389485	394140	1.18
ibm09	412195	416077	0.93
ibm10	565999	570143	0.73
Average	——	——	3.21

C. Crossover Number Reduction

In liquid routing, [2] indicates that any two crossed segments introduce a least one via. To show the effectiveness of

TABLE III
Comparison between DraXrouter & x-laby on *tof* and Running Time

Circuits	DraX		x-labyrinth		*Tof*	CPU
	Tof	CPU(s)	*Tof*	CPU(s)	Imp.(%)	Imp.(%)
ibm01	4	36	87	89	95.40	59.55
ibm02	0	78	185	119	100.00	34.45
ibm03	1	73	217	148	99.54	50.68
ibm04	0	137	199	179	100.00	23.46
ibm05	0	438	109	587	100.00	25.38
ibm06	0	331	63	410	100.00	19.27
ibm07	0	349	80	548	100.00	36.31
ibm08	0	429	144	701	100.00	38.80
ibm09	0	734	33	987	100.00	25.63
ibm10	0	1132	96	1309	100.00	13.52
Average		——	—	——	99.49	32.71

TABLE IV
Comparison between DraXrouter & COCO on *twl*

Circuits	DraX	COCO	Imp.(%)
ibm01	61011.9	74256.8	17.84
ibm02	168908	201484	16.17
ibm03	148970	182767	18.49
ibm04	163009	197607	17.51
ibm05	408902	502694	18.66
ibm06	275598	329378	16.33
ibm07	349383	434956	19.67
ibm08	389485	475549	18.10
ibm09	412195	510459	19.25
ibm10	565999	702568	19.44
Average	——	——	18.15

TABLE V
Comparison between DraXrouter & COCO on *tof* and Running Time

Circuits	DraX		COCO		*Tof* Imp.(%)
	Tof	CPU(s)	*Tof*	CPU(s)	
ibm01	4	36	786	18	99.49
ibm02	0	78	299	46	100
ibm03	1	73	494	43	99.79
ibm04	0	137	1022	47	100
ibm05	0	438	900	236	100
ibm06	0	331	748	176	100
ibm07	0	349	503	159	100
ibm08	0	429	425	247	100
ibm09	0	734	1043	312	100
ibm10	0	1132	1183	350	100
Average		——		——	99.93

the Dra method for via reduction in liquid routing, we compare the number crossed segments in the routing results produced by our router with and without the Dra Method. The crossed segments number is the sum of the crossover between nets. [2] adopts a similar method to calculate the crossover number but it takes area as a reference because it is considered in detailed routing. Obviously, the crossover number gives a good prediction for the number of vias produced by liquid routing.

The result is shown in TABLE VI, in which the router with the Dra method produces 5% lesser the crossed segments than the one with the traditional static RA method, and the wire length is even shorter. It indicates that the Dra method utilizes the routing resources more reasonably, so it provides a better guidance for liquid routing. Note that, both RA methods assign the same resource in the edges of a global routing graph to keep a fare comparison.

TABLE VI
Comparison between DraXrouter with old RA Method & DraXrouter with Dra Method

Circuits	Old		Dra		Cross
	Twl	Cross#	*Twl*	Cross#	Imp.(%)
ibm01	59804	234002	58031	223526	4.48
ibm02	162486	1521016	157341	1436766	5.54
ibm03	138126	1078131	134866	1012122	6.12
ibm04	156425	1081236	153320	1053306	2.58
ibm05	383450	5152323	381340	5041213	2.16
ibm06	268433	2289791	259542	2096781	8.43
ibm07	343957	2692672	337639	2541374	5.62
ibm08	382818	3213644	374226	3051586	5.04
ibm09	396096	2372184	385792	2215325	6.61
ibm10	548445	5377452	537612	5146802	4.29
Average	——	——	——	——	5.09

VI. Conclusions and Future Work

This paper reformulates the global routing problem in X-Architecture with the liquid routing model. A dynamic re-source assignment (Dra) method is presented to reduce potential vias. Based on our formulation, a global router, DraXRouter, is designed. In this router, we adopt a dynamic-tabulist-based tree construction algorithm and the stochastic optimization strategy to obtain better routing performance. Tested on ISPD'98 benchmarks, DraXRouter increases the routability substantially compared with two recent global routers, as well as reduces total wire length. The comparison with the traditional resource assignment method shows the Dra method can be a better guidance for liquid routing to reduce vias in detailed routing.

As future work, we will integrate the function considering timing and coupling issues into our router.

References

[1] Steven L Teig. The x architecture: Not your father's diagonal wiring. In Proceedings of the international work-shop on System-level interconnect prediction, pages 33–37, San Diego, California, USA., 2002.

[2] Martin Paluszewski, Pawel Winter, Martin Zachariasen. A New Paradigm for General Architecture Routing. GLSVLSI '04, April 26–28, 2004, Boston, Massachusetts, USA.

[3] Labyrinth. http://www.ece.ucsb.edu/~kastner/labyrinth/

[4] Cheng-Kok Koh and P. H. Madden. Manhattan or non-Manhattan? a study of alternative VLSI routing architectures. In Proceedings of the 10th ACM GLSVLSI, pages 47–52, Chicago, IL, USA, 2000.

[5] Qi Zhu, Hai Zhou, Tong Jing, Xianlong Hong, Yang Yang. Spanning Graph Based Non-Rectilinear Steiner Tree Algorithms. IEEE Trans. on CAD. 2005, 24(7).

[6] Majid Sarrafzadeh, C. K. Wong, Hierarchical Steiner Tree Construction in Uniform Orientations. IEEE Trans. on CAD, 11(9): pp.1095-1103, 1992.

[7] H. Chen, F Zhou and C. K. Cheng. The Y-architecture: yet another on-chip interconnect solution. In Proceedings of the ASP-DAC, pages 840-846, Kitakyushu, Japan, 2003.

[8] Yu Hu, Tong Jing, Xianlong Hong, Zhe Feng, Xiaodong Hu, Guiying Yan. An Efficient Rectilinear Steiner Minimum Tree Algorithm Based on Ant Colony Optimization. In: Proceedings of IEEE ICCCAS, 2004, Chengdu, China, 1276-1280.

[9] T. Mitsuhashi, K. Someha: Performance-oriented layout design, pervasive use of diagonal interconnects reduces wire-length. Design Wave Magazine (2001) 59–64

[10] Yu Hu, Tong Jing, Xianlong Hong, Xiaodong Hu, and Guiying Yan. A Routing Paradigm with Novel Resources Estimation and Routability Models for X-Architecture Based Physical Design. SAMOS V, Samos, Greece, 2005.

[11] S. Mand and W. C. K. An Introduction to VLSI Physical Design. McGraw Hill, USA, 1996.

[12] T. Jing, X. Hong, H. Bao, J. Xu, and J. Gu. SSTT: Efficient local search for GSI global routing. Journal of Compute Science and Technology, 18(5):632–639, September 2003.

[13] H. Chen, B. Yao, F. Zhou and C. K. Cheng, "Physical Planning of On-Chip Interconnect Architecture, In Proceedings of IEEE Int. Conference on Computer Design, pp. 30-35, Sept. 2002.

[14] The x initiative. http://www.xinitiative.org.

[15] A. R. Agnihotri and P. H. Madden. Congestion reduction in traditional and new routing architectures. In Proceedings of the 13th ACM GLSVLSI, pages 28–29, Washington, DC, USA., 2003.

[16] R. T. Hadsell and P. H. Madden. Improved global routing through congestion estimation. In Proceedings of the 40th conference on Design automation, Anaheim, CA, USA, 2003.

[17] H. Chen, B. Yao, F. Zhou and C. K. Cheng. Physical Planning of On-Chip Interconnect Architecture, In Proceedings of IEEE Int. Conference on Computer Design, pp.30-35, 2002.

[18] Takashi Mitsuhashi and Kenji Someha. Performance- Oriented Layout Design, Pervasive Use of Diagonal Interconnects Reduces Wire- Length, Design Wave Magazine, pages 59-64, September, 2001.

[19] M. R. Garey, R. L. Graham, and D. S. Johnson. The complexity of computing Steiner minimal trees. SIAM Journal on Applied Mathematics, 32(4):835–859, 1977.

[20] Tsung-Yi Ho, Chen-Feng Chang, Yao-Wen Chang and Sao-Jie Chen. Multilevel Full-Chip Routing for the X-Based Architecture. DAC 2005, pages597-602, June 13–17, 2005, Anaheim, CA, USA.

Diagonal Routing in High Performance Microprocessor Design

Noriyuki Ito, Hideaki Katagiri, Ryoichi Yamashita, Hiroshi Ikeda, Hiroyuki Sugiyama, Hiroaki Komatsu, Yoshiyasu Tanamura, Akihiko Yoshitake, Kazuhiro Nonomura, Kinya Ishizaka, Hiroaki Adachi, Yutaka Mori, Yutaka Isoda, Yaroku Sugiyama

Fujitsu Limited
4-1-1 Kamikodanaka, Nakahara-ku,
Kawasaki, 211-8588, Japan
ito.noriyuki@jp.fujitsu.com

ABSTRACT

This paper presents a diagonal routing method which is applied to an actual microprocessor prototype chip. While including the layout functions for the conventional Manhattan routing with horizontal and vertical directions, a new diagonal routing capability is added as one of the routing functions. With this enhancement, diagonal routing becomes an additional strategy for improving delays of critical paths in the microprocessor design. This method was applied to the prototype chip of the Fujitsu SPARC64 microprocessor with two CPU cores using 90nm process technology. By applying the diagonal routing to long distance nets, net length is reduced by 36% per net on average. When the diagonal routing is applied to a critical path, path delay is improved by as much as about 14 pico-seconds per net on a path. This improvement is more than the delay of a gate with no load. This prototype chip proved that our method was effective in reducing the total net length and improving path delays.

Keywords: Microprocessor, diagonal routing, Manhattan routing

1. Introduction

For the performance improvement in microprocessor design, the continuous improvement of circuit design techniques and process technology is necessary. In this improvement, the CAD system also plays an important role. Designers achieve the targeted performance by tuning logical and physical designs to the last pico-second with various techniques [1]. For path delay optimization, there are several conventional techniques such as reduction of logical stages, custom macro design, buffer insertion, gate sizing, placement change, fan-out splitting, etc. Since interconnect delay is dominant in the deep sub-micron technology, some new techniques are used in routing to improve the delay. Timing-driven routing, which minimizes the delay from the driver pin of a net to the most critical receiver pin, is an effective routing technique. Another technique is using wider wires. As for the wire resistance, it is smaller for upper metal layers. This is because in general, a wire is wider and taller on an upper metal layer. Therefore, routing a critical net on an upper metal layer is another useful technique for improving the delay. After all these techniques are utilized, designers still improve delay by other techniques to achieve the targeted performance with the assistance of CAD tools. Although the delay improvement is small, pin permutation, which swaps nets connected to symmetric input pins of a cell so that a critical net is connected to a pin with

smaller gate delay, is applied. Also, a limited number of low Vth cells are used to improve delay on critical paths, even though the power dissipation may increase due to leakage current. These various techniques are used according to the progress of design in each stage. The technique that is used in the final phase of a design is very important even if the absolute delay improvement is small (e.g., a few pico-seconds). Therefore, it is very meaningful to provide designers of high performance microprocessors with new methods to improve delay.

This paper presents diagonal routing as a technique to improve delay of critical paths in microprocessor design. In order to add the new diagonal routing function to the current design methodology, several modifications are added to the existing in-house CAD tools. We made these modifications with a target completion deadline of at most six months. The rest of this paper is organized as follows. In Section 2, several diagonal routing techniques are surveyed, and the difficulty in the use of diagonal routing in LSI design is addressed. In Section 3, we estimate the contribution of improvement by our proposed diagonal routing scheme. Sections 4 and 5 describe the routing and capacitance extraction methods respectively. Experimental results are shown in Section 6, and Section 7 concludes the paper with directions for future work.

2. Related Work

Diagonal routing is not a new technique; it is routinely used in the printed circuit board design. In one of the Fujitsu mainframes, FACOM M-780 [2], one CPU is composed of 336 LSI's. These LSI's were mounted on both sides of a printed circuit board in a high density configuration [2]. In this printed circuit board design, diagonal routing was used to reduce delay between LSI's on the board [3]. In the context of LSI designs, several diagonal routing techniques have been proposed to reduce wire length and to improve density [4,5,6,7,8,9]. However, the use of diagonal routing has been very limited in LSI design. The reason is the difficulty to support diagonal routing in the related CAD tools such as routing, timing analysis, noise analysis, layout verification and manufacturing data generation. Recently, the pervasive use of diagonal routing with X-architecture was reported [10]. In the X-architecture, diagonal routing in 45 and 135 degree directions is used in conjunction with conventional Manhattan routing in horizontal and vertical directions. This architecture was applied to a RISC processor core. It was reported that the path delay improved by 19.8% and the area reduced by about 10%. As another routing model, Y-architecture based on 0, 60 and 120 degree directions is proposed [11]. With this

0-7803-9451-8/06/$20.00 ©2006 IEEE.

architecture, the wire length reduced by 13.4%. On average, the wire length was only 4.3% more than that of X-architecture. Since these routing architectures are different from the conventional Manhattan architecture, the development of CAD tools to support them is not easy. Even if we try to find some commercially available tools, it is very limited or none for LSI design. Although there are several recent research works on congestion reduction and Steiner tree generation assuming the use of diagonal routing [12, 13], actual use of diagonal routing in an industrial chip is not popular yet. We believe that to develop a high performance microprocessor, it is very important to enhance the existing CAD system with the new capability of diagonal routing. In this paper, we propose a new routing technique that uses diagonal wiring partially on a critical net. Such limited usage of diagonal wires minimizes the enhancement needed for the existing CAD system.

3. Preliminaries

Given two pins A and B, let L_{xy} denote the length of the net when it is routed with conventional Manhattan routing. Let $L_{angle}(\alpha)$ denote the net length when the same net is routed diagonally in α degree direction. As shown in Figure 1, let the elevation from A to B be θ degree. Then, L_{xy} and $L_{angle}(\alpha)$ are expressed by (1) and (2) respectively when θ is within α and when θ is within 90 degree.

$$L_{xy} = L(1 + \tan\theta) \tag{1}$$

$$L_{angle}(\alpha) = L\,\tan\theta/\sin\alpha + L(1 - \tan\theta/\tan\alpha) \tag{2}$$

When diagonal routing is used in a chip, α is fixed. The length of a net routed with a diagonal wire is the minimum when θ is equal to α. This is because a diagonal wire can connect two points directly with no Manhattan wire. Assuming that θ is equal to α and that α varies from 0 to 90 degrees, the reduction ratio $(L_{xy} - L_{angle}(\alpha))/L_{xy}$ is the maximum when α is equal to 45 degrees. As shown in Figure 2, the maximum reduction ratio is about 29.3%. Therefore, 45 degree direction is the most effective choice for diagonal routing to reduce the length of a net.

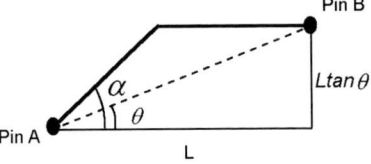

Figure 1. α-degree diagonal routing

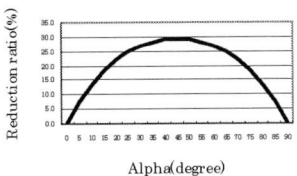

Figure 2. Length reduction ratio for α-degree diagonal

Before actually applying our diagonal routing to the microprocessor design, we evaluated the net-length reduction when diagonal routing is applied to the actual design. When net length with a diagonal wire is estimated, the center location of receiver pins is calculated first. Then, for the connection between a driver pin and the center location, net length is estimated by both Manhattan and diagonal routing, as shown in Figure 3. The two lengths are compared for each net. We performed this evaluation for 6,000 long distance nets. The results are shown in Table 1. We see that the net length is reduced by about 12% on average by diagonal routing.

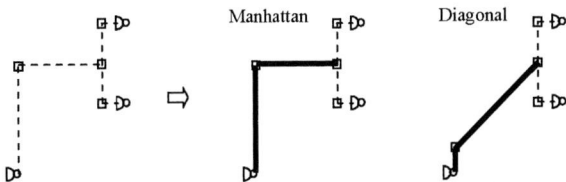

Figure 3. Manhattan routing and diagonal routing

Table 1. Length reduction by diagonal routing

Unit: grid

	Manhattan	Diagonal	difference
Total length	25,835,198	22,569,954	3,265,244
Average Net length	4,090	3,573	517

We also evaluated the improvement in delay for the typical model shown in Figure 4. In this model, the length of the net estimated using Manhattan routing is $3L$, and using diagonal routing is $(1 + \sqrt{2})L$. Diagonal routing can reduce the net length by about 20% in this case. For this model, delay is measured by HSPICE, and L is varied from 0 to 1500 grids. Figure 5 shows delay from the driver to a receiver for Manhattan and diagonal schemes. When L is 1500, the improvement of delay is about 7 pico-seconds. As it is clear from the graph, the improvement is larger when diagonal routing is applied to a longer net.

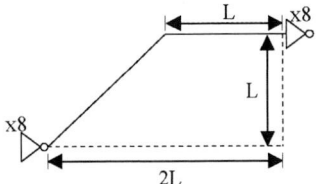

Figure 4. Model of a net with a diagonal wire

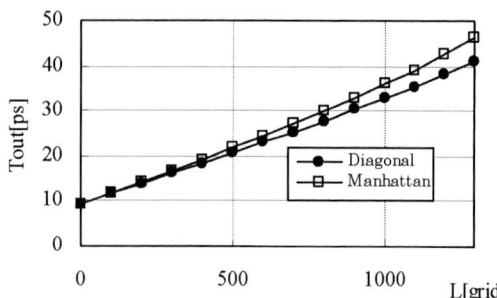

Figure 5. Reduction of delay by diagonal routing

4. 45, 135 Degree Diagonal Routing

In order to apply our diagonal routing technique to a microprocessor prototype, it is desirable that we enhance our existing CAD tools with minimum effort. Therefore, we limited the use of diagonal wire only to critical nets. To meet these requirements, a diagonal wire is routed first as a straight-line trunk between driver and receivers. Then, the connection between the driver and the diagonal wire is routed using conventional Manhattan routing. Finally, the connection between the diagonal wire and receivers is routed using Manhattan routing. Figure 6 shows this idea.

An algorithm *Diagonal_routing* to route a given net with diagonal wire is shown in Figure 7. A net is passed to *Diagonal_routing* with a set of pins P of the net. *Routable_check* checks whether diagonal routing can be applied or not. When a driver pin lies within the minimum bounding box of the receiver pins, the given net is routed by a conventional Manhattan router and the procedure *Diagonal_routing* terminates. When the drive pin lies outside the bounding box of receiver pins, a further check is performed. A Steiner tree S is created for P by *Steiner_generation*. S is passed to *Make_receiver_group*. This procedure finds a sub-tree S' of S that connects only receiver pins, and it returns a connection point R that connects S' and the rest of S. *Select_driver* returns a driver pin D from P. Then, both R and D are passed to *Diagonal_search*, which searches a diagonal route with no bend between R and D and returns the result as W_{diag}. The direction 45 or 135 degree is determined depending on the relative positions of driver and receivers. If no diagonal route is found, value *NONE* is returned. By passing W_{diag} and P to *Manhattan_routing*, the connection between the diagonal wire W_{diag} and the driver pin D, and connection between W_{diag} and receiver pins are routed by the conventional Manhattan routing function. In *Diagonal_search*, diagonal route search is not performed when the estimated length of the diagonal wire is less than some specific length. This is to apply diagonal routing only when its effect will be appreciable. When diagonal routing is applied between two points, the vertical and horizontal distances between these points for a diagonal route search are compared. When the horizontal distance is greater than the vertical distance, a diagonal route is searched in the horizontal direction by *Horizontal_search*. Otherwise, a diagonal route is searched in the vertical direction by *Verical_search*. In *Horizontal_search* or *Vertical_search*, the angle of diagonal routing (45 or 135 degree) is determined based on the relative locations of D and R. Figure 8 shows these two types of diagonal route search. The length of the

diagonal wire is constant in the range from (b) to (c). Since the length of the diagonal wire varies in the range from (a) to (b) or from (c) to (d), search is performed within the range such that the length is larger than some specific value.

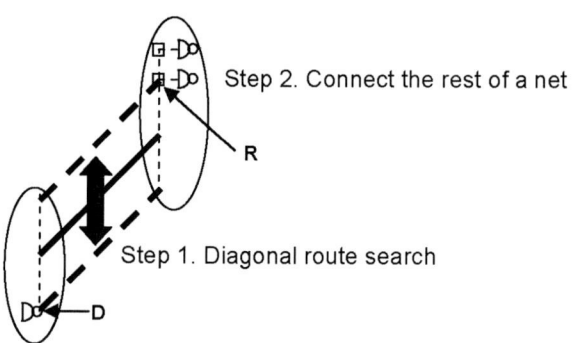

Figure 6. Basic procedure in diagonal routing

```
Diagonal_routing(P)
  flag = Routable_check(P)
  if flag == FALSE {
    W,V = Manhattan_routing(P)
  }
  else {
    S = Steiner_generation(P)
    R = Make_receiver_group(S)
    D = Select_driver(P)
    Wdiag = Diagonal_search(D,R)
    Wires,Vias = Manhattan_routing(P,Wdiag)
  }
  return(Wires,Vias)
end


Diagonal_search(D,R)
  L = Diagonal_length(D,R)
  if ( L < α ) then
    return(NONE)
  }
  else {
    x = Horizontal_distance(D,R)
    y = Vertical_distance(D,R)
    if ( x > y ) {
      diag_wire = Horizontal_search(D,R)
    }
    else {
      diag_wire = Vertical_search(D,R)
    }
    return(diag_wire)
  }
}
```

Figure 7. Diagonal routing algorithm

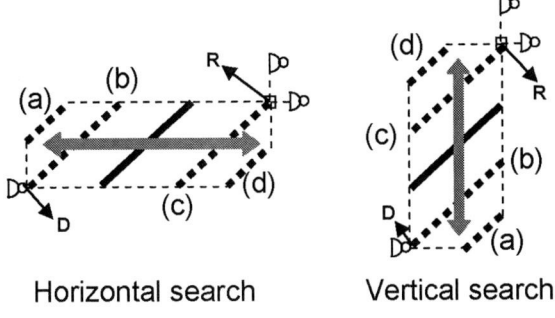

Figure 8. Search direction of diagonal route

Since diagonal routing is applied only to critical nets in our method, 45 and 135 degree layers are assigned to upper layers n and $n+1$, which have smaller wire resistance. After all nets with diagonal wiring are routed, some channels may be left unused on those layers. These available channels are used to route regular nets with the conventional Manhattan routing, as shown in Figure 9.

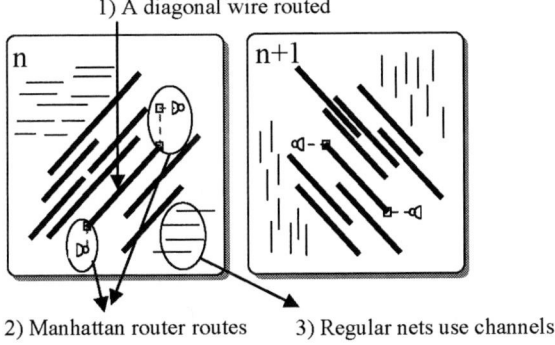

Figure 9. Manhattan and diagonal mixed layers

5. Capacitance Extraction

In order to run timing analysis for design data with diagonal wires, it is necessary to enhance the capacitance extraction tool to handle diagonal wires. Here, we will give an overview of the enhancement we made to our in-house capacitance extraction tool for microprocessor design. In this tool, capacitance for each wire is extracted grid by grid from a pre-computed table using the wire location and its adjacency with other wires. For a horizontal or vertical wire, a capacitance table CAPTBL_M, which assumes the interval between two wires as N (1, 2, 3, ...) times of a regular grid, is used. For a diagonal wire with 45 or 135 degree, an additional grid with the $1/\sqrt{2}$ grid length is introduced as shown in Figure 10. For the capacitance extraction of a diagonal wire, another capacitance table CAPTBL_D, which assumes the interval between two wires as N (1, 2, 3, ...) times of a $1/\sqrt{2}$ grid, is added.

With the addition of CAPTBL_D, a procedure to search wires adjacent to a diagonal wire in the orthogonal direction is added to

the extraction tool. Figure 11 shows an example in which a wire adjacent to a given 45 degree diagonal wire is searched for in the orthogonal direction. As shown in this figure, whether the other adjacent wire exists or not in the $1/\sqrt{2}*N$ grid is checked on the $1/\sqrt{2}$ grid by $1/\sqrt{2}$ grid. If N is even, the search point with the distance $1/\sqrt{2}*N$ lies on the regular grid. If N is odd, the search point in the orthogonal direction does not lies on the regular grid. In this case, two regular grid points nearest to the search point are checked. According to this search procedure, regular grids are checked in the order of 1, 2, 3, 4, 5 which are the labels on a regular grid as shown in Figure 11. In this example, an adjacent wire is found when the search reaches the point with distance of $1/\sqrt{2}*5$ grids from the given wire.

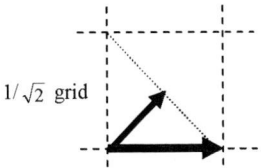

Figure 10. Basic grid for capacitance extraction

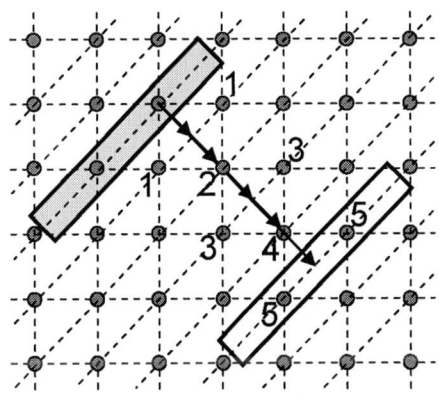

Figure 11. Capacitance extraction for a diagonal wire

6. Experimental Results

In order to apply the diagonal routing technique described in this paper, we enhanced the existing in-house CAD tools and database. The enhanced tools are automatic router, interactive placement and routing editor, RC extraction, timing analysis and layout verification. Calibre from Mentor Graphics was used for DRC and LVS sign-off. An actual prototype chip, which is called MTP (Multi Threaded Processor), is developed [14]. Figure 12 shows the micrograph with two CPU cores and six mega-byte second caches. This prototype chip is developed to verify MTP architecture, 90 nm process technology, diagonal routing, etc. In this chip, diagonal routing is applied to the most critical nets between an instruction unit and RAM's. The total number of nets to which diagonal routing is applied is about 1,500 nets per CPU core. The left figure in Figure 13 shows the 8th layer of an instruction unit, and the right figure zooms in on the layout part containing diagonal wires.

We evaluated the improvement in net length and path delay with diagonal routing vis-à-vis the conventional Manhattan routing. Figure 14 shows this comparison for net length for about 1,500 nets to which diagonal routing is applied. Diagonal routing reduces the net length by about 36%. Since diagonal routing is applied only to a trunk of a net with no vias, the total detour of the net with diagonal routing is smaller than that with the Manhattan routing. Therefore, the maximum ratio of length reduction exceeds 29.3%, which was estimated in Section 3.

Figure 15 shows how the path delay is improved when diagonal routing is applied to critical nets. The number of nets with diagonal routing is 1,500, and the number of paths that pass them is about 27,000. In our prototype chip, one net on each path is routed by diagonal routing. The maximum improvement in path delay is about 14 pico-seconds, the average improvement being about 6 pico-seconds. There are some paths whose path delay increases by 2 pico-seconds. This is because the coupling capacitance with diagonal routing happens to be larger than that with the Manhattan routing. In general, coupling capacitance that can cause cross-talk noise is less when diagonal routing is applied. This is due to the associated reduction of net length. The graph in Figure 16 shows the ratio of the coupling capacitance per net with diagonal routing to that with the Manhattan routing. Although the coupling capacitance increases in some cases, overall coupling capacitance decreases by about 15%. This implies, in turn, that diagonal routing is effective from the point of view of reducing cross-talk noise.

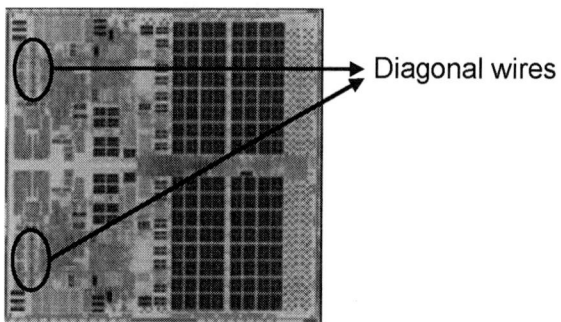

Figure 12. Prototype chip [14]

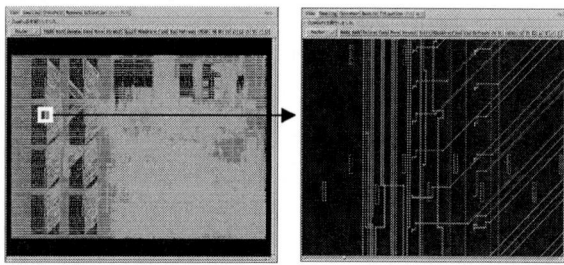

Figure 13. Actual diagonal wires

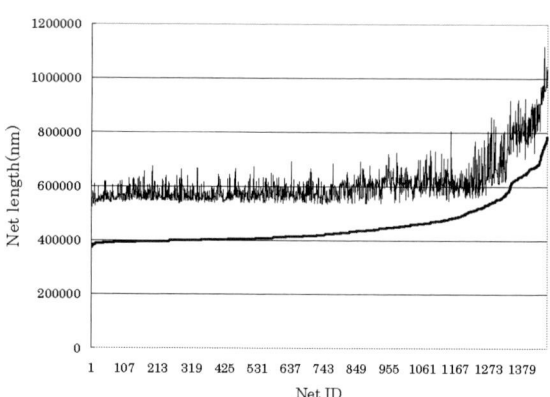

Figure 14. Reduction of net length

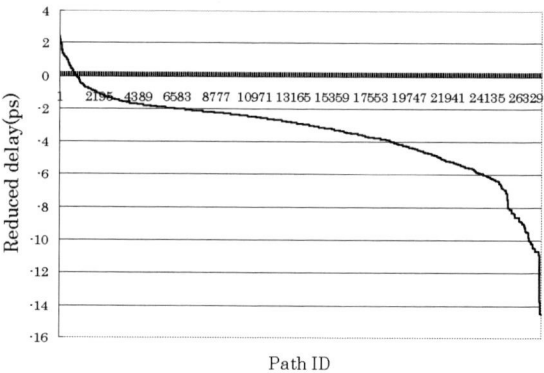

Figure 15. Reduction of path delay

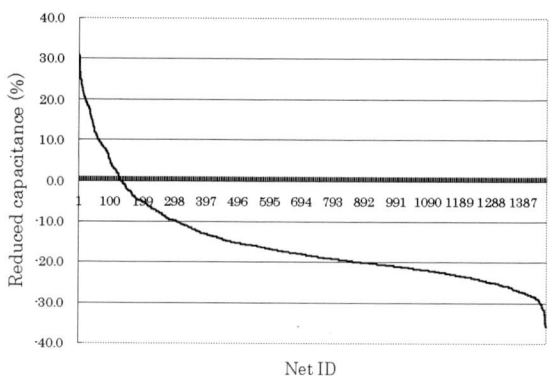

Figure 16. Reduction of net capacitance

7. Conclusion

With this work, we were able to provide microprocessor designers with diagonal routing capability as a new means to improve path delay. This capability is built into our in-house CAD system for microprocessor design. We applied it to an actual prototype chip of a microprocessor with two CPU cores. We were able to reduce the net length by 36% per net (on average), and path delay by up to 14 pico-seconds. This delay improvement is more than the delay of a gate with no load. These results prove that the use of diagonal wires can improve the delay of critical paths in microprocessor design.

In the prototype processor design, diagonal routing was applied to about 3,000 nets in the chip. To increase the number of nets to which we can apply diagonal routing, further enhancements are needed, such as automatic net selection and floorplanning & cell placement that make diagonal routing more effective. We plan to address these issues in the near future and hopefully make diagonal routing an effective strategy for improving path delays in microprocessor designs.

8. ACKNOWLEDGMENTS

We would like to thank Toshiyuki Shibuya and Kaoru Kawamura and other researchers in the CAD group of Fujitsu Laboratories for their enhancement of the automatic routing engine. We also wish to thank Eizo Ninoi and his technology group for the support they provide for this work.

9. REFERENCES

[1] N. Ito, H. Komatsu, et al, "A Physical Design Methodology for 1.3GHz SPARC64 Microprocessor", Proc. ICCD, pp. 204-210, 2003.

[2] M. Nishihara, T. Murase, et al, "Single-Board CPU Packaging for the FACOM M-780", Fujitsu Scientific and Technical Journal, 23, 4, pp.226-235, 1987.

[3] M. Oda, T. Hamaguchi, "Oblique Routing System for FACOM M-780 SSC", Fujitsu Scientific and Technical Journal, 23, 4, pp.236-242, 1987.

[4] H. H. Chen, "Routing L-Shaped Channels in Nonslicing-Structure Placement", Proc. DAC, pp. 152-158, 1987.

[5] D. Marple, M. Smulders, et al, "An Efficient Compactor for 45° Layout", Proc. DAC, pp. 396-402, 1988.

[6] C. Chiang, M. Sarrafzadeh, "Wirability of Knock-Knee Layouts with 45-degree wires", IEEE trans. On Circuits & Systems, vol. 38, No. 6, June 1991, pp. 613-624.

[7] D. T. Lee, C. F. Shen, et at, "On Steiner Tree Problem with 45° Routing", Proc. ISCAS, pp. 1680-1682a, 1995.

[8] S. Das, B. B. Bhattacharya, "Channel Routing in Manhattan-Diagonal Model", Proc. VLSI Design, pp. 43-48, 1996.

[9] C. K. Koh, P. H. Madden, "Manhattan or Non-Manhattan? A Study of Alternative VLSI Routing Architectures", Proc. GLSVLSI, pp. 47-52, 2000.

[10] I. Mutsunori, T. Mitsuhashi, et al, "A Diagonal-Interconnect Architecture and Its Application to RISC Core Deisgn", Proc. ISSCC, pp. 684-689, 2002.

[11] H. Chen, B. Yao, et al, "The Y-Architecture: Yet Another On-Chip Interconnect Solution", Proc. ASPDAC, pp. 840-846, 2003.

[12] A. R. Agnihotri, P. H. Madden, "Congestion Reduction in Traditional and New Routing Architecture", Proc. GLSVLSI, pp. 211-214, 2003.

[13] C. Chiang, Q. Su, "Wirelength Reduction by Using Diagonal Wire", Proc. GLSVLSI, pp. 104-107, 2003.

[14] Takumi Maruyama, "SPARC64 VI: Fujitsu's Next Generation Processor", presented at the microprocessor forum, 2003..

CDCTree: Novel Obstacle-Avoiding Routing Tree Construction based on Current Driven Circuit Model

Yiyu Shi[1], Tong Jing[2], Lei He[1],

Electrical Engineering Department[1]
UCLA[1]
Los Angeles, California,   90095-1594, USA
Tel: 310-206-2037
Fax: 310-206-4685
e-mail: {yshi, lhe}@ee.ucla.edu[1]

Zhe Feng[2] and Xianlong Hong[2]

Computer Science & Technology Department[2]
Tsinghua University[2]
Beijing, P.R.China   10084
Tel: +86-10-62785564
Fax: +86-10-62781489
e-mail: {jingtong, hxl-dcs}@tsinghua.edu.cn[2]

Abstract— Routing tree construction is a fundamental problem in modern VLSI design. In this paper we propose CDC-Tree, an Obstacle-Avoiding Rectilinear Steiner Minimum Tree (OARSMT) heuristic algorithm to construct an OARSMT. CDC-Tree is based on the current driven circuit (CDC) model mapped from an escape graph. The circuit structure comes from the topology of the escape graph, with each edge replaced by a resistor indicating the wirelength of that edge. By performing DC analysis on the circuit and selecting the edges according to the current distribution to construct an OARSMT, the wirelength of the resulting tree is short. The algorithm has been implemented and tested on cases of different scales and with different shapes of obstacles. Experiments show that CDCTree can achieve shorter wirelength than the existing best algorithm, An-OARSMan, when the terminal number of a net is less than 50.

I. INTRODUCTION

Routing tree construction is one of the major tasks in the routing phase. When wirelength is of interest, it is in essence an RSMT (*Rectilinear Steiner Minimum Tree*) construction problem on a given terminal set. In practical full-chip routing or detailed routing applications, we consider macro cells, IP blocks, and pre-routed nets as obstacles. Therefore, powerful algorithms of OARSMT (*Obstacle-Avoiding Rectilinear Steiner Minimum Tree*) construction are required to get short wirelength.

The RMST problem has been proved to be NP-complete [1]. However, the OARSMT problem is even more complicated and no polynomial-time algorithms have been proposed to solve it precisely. Maze algorithm was proposed in [2] and

The work at UCLA was partially supported by NSF award CCR-0093273/0401682. The work at Tsinghua University was partially supported by the NSFC under Grant No.60373012, the SRFDP of China under Grant No.20050003099.

several improvements on searching efficiency were made in [3]-[6] later.

The method of line search routing was introduced in [7] and [8]. However, these algorithms are only suitable for small-scale problems. Most existing OARSMT algorithms use the multi-terminal variant of the maze algorithm, which incurs the same space demand as that of the two-terminal variant with a result far from the optimal. [9] proposed an algorithm to construct optimal three-terminal or four-terminal OARSMT. Then, G3S, G4S, and B3S heuristics were proposed for the cases with less than twenty terminals. [10] provided an exact algorithm to find an obstacle-avoiding Euclidean Steiner tree with less than 150 terminals. [11] introduced an $O(mn)$ two-step heuristic of OARSMT, in which m is the number of obstacles and n is the number of terminals. The two-step heuristic works well when the terminal number is less than seven and the obstacles are convex. The most recent works are FORst [12] and An-OARSMan [13]. FORst can tackle large scale problems efficiently. An-OARSMan, based on the track graph put forward in [14], can achieve shorter wirelength than FORst when the terminal number is less than 100.

In practice, most nets in circuits are in common scale. If we research on powerful OARSMT algorithms to get short wirelength for such nets, we can have more opportunities to get better routing results. Meanwhile, there is still much room to improve the wirelength performance.

The main contribution of this paper is CDCTree, a heuristic algorithm of OARSMT construction based on a current driven circuit (CDC) model. The idea of CDCTree is different from those of existing algorithms. It maps the edges of escape graph into resistors, and adds a current source at each terminal. Then it makes use of Coulomb's Law, which indicates the repellency of currents, to construct an RSMT. Experimental results show that CDCTree can achieve shorter wirelength than An-OARSMan when the terminal number is less than 50. In addition, CDCTree can route among both convex and concave

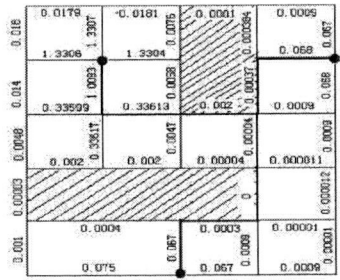

Fig. 1. Current distribution for a three-terminal current driven circuit model

Fig. 2. An example of topology mapping from (a) escape graph to (b) circuit structure

polygon obstacles.

The rest of the paper is organized as follows: In Section II, we discuss how *Coulomb's Law* can be applied to the OARSMT construction. In Section III, the detailed procedure of tree construction based on CDC model is described in detail. Section IV shows the experimental results and some discussions. Conclusions and remarks are given in Section V.

II. COULOMB'S LAW IN OARSMT CONSTRUCTION

The foundation of our CDCTree algorithm is different from the existing algorithms. In order to clearly describe the idea, we first describe how Coulomb's Law can be applied to the OARSMT problem.

Coulomb's Law indicates that charges of the same kind (negative or positive) are repellent. [15] first employed this law to solve planning problems. But we notice that this law can also be used in optimization problems. Basically speaking, electrical currents are composed of electrons, all of which are negatively charged. According to Coulomb's Law, the currents repel from each other. Here, we introduce this idea into the tree construction problem.

If we map an escape graph [9] into a circuit and inject current at each terminal, the current distribution of this current driven circuit can be obtained by performing DC analysis. Then the edges with the minimum currents are selected to construct the Steiner tree. The total length of the tree thus constructed should be short because the currents along the edges which are close to all the terminals are significantly repelled by current injections at those terminals. We observe that RSMT is usually constructed by the edges close to all the terminals. Edges far from some of the terminals are seldomly selected. Therefore, the tree constructed by the edges with minimum current has a short wirelength.

Figure 1 illustrates the current distribution in a three-terminal current driven circuit model. Currents are injected at the three vertices marked in black. Each edge represents a resistor with a given value. The RMST is composed by the edges in bold black lines. It can be easily verified that the currents flowing along the edges of the RSMT are small. Note that not all the edges with minimum currents are selected because of the *path selection* rules, which is discussed in the next section.

III. THE CDCTREE ALGORITHM

CDCTree algorithm is composed of four steps: *topology mapping, resistor selection, circuit simulation, and path selection*. The four steps are discussed in detail below.

A. Topology Mapping

CDCTree algorithm is based upon the escape graph [9], which enables us to get short wirelength. This step deals with the mapping from escape graph to circuit structure.

We first place a resistor at each edge of the escape graph, the value of which is to be decided in the next step, i.e. *resistor selection*. Next, we add a current source at each terminal. The value of the current source can be arbitrarily chosen as it does not influence the relative distribution of the currents. The edge with larger current always has larger current regardless of the current source value. In our experiments we set the current sources to be *5A*.

Finally, to let the circuit function correctly, we have to decide how to connect the circuit to the ground. One method is to extend the escape graph to infinite size. Then the nodes at infinite are automatically connected to ground. However, this method requires a large memory as well as high computation cost because we have to calculate the currents at each edge while the total edge number is infinite. Therefore, this method is impractical.

An alternative way is to connect the periphery nodes of the circuit to ground. Though much simplified, this method brings about another problem. Unlike the infinite structure, the finite one influences the current distribution of the circuit significantly. If we let the periphery nodes connect to the ground via a resistor, the side effects can be alleviated to some extent. In the fourth step (*path selection*), further techniques are employed to compensate for the side effects.

An example of topology mapping from the escape graph to the circuit structure is given in Figure 2. The terminals and edges in (a) are mapped into current sources and resistors in (b), respectively.

B. Resistor Selection

There are three different types of resistors in our circuit model. The first type, GND resistors, are used to connect the periphery nodes of the circuit to the ground. The second type, EDGE resistors, refer to the resistors on the edges of the graph except for the obstacle boundaries. The last type, OBST resistors, are the resistors on the edges of the obstacle boundaries.

1) GND resistors: GND resistors are used to provide paths for currents to flow to the ground. In our experiments, we set GND resistors to be a fixed small value, 0.1.

2) EDGE resistors: According to Ohm's Law, the larger the resistor, the smaller the current flowing through it. Since we choose edges based on minimum currents, we want less current to flow through paths with larger wirelength. Therefore, the longer the edge, the smaller the EDGE resistor should be. A mapping function $f(x)$ should thus be decided by which EDGE resistor R can be calculated as

$$R = f(L) \qquad (1)$$

where L is the length of the edge. The slope of $f(x)$ cannot be too large, otherwise it may significantly influence the current distribution. This causes the resulting tree to be constructed simply by the shortest edges without considering the topology of the tree. On the other hand, the slope can not be too small, which leads to neglecting the impact of the edge length. After experiments, we select the following function.

$$f(x) = K - ln(x) \qquad (2)$$

where K must be large enough to guarantee $f(x)$ to be positive in the range of x, i.e. edge lengths. In our experiments, the edge lengths are set to be between 100 and 10,000, so we choose K to be 10. The mapping function turns out to be

$$R = 10 - ln(L) \qquad (3)$$

In general cases, K should be larger than the logarithm of the maximum wirelength to keep the resistor value positive.

3) OBST Resistors: The existence of obstacles to some extent destroys the characteristics of the original circuit model. From our experiments, we find out that currents tend to gather at the edges of obstacles, which can be called as *Current Crowding Effect (CCE)*. To compensate for CCE, the resistors on obstacle edges should be set larger. In our experiments, we simply add a fixed value 10 to the value calculated from (3).

Following the four steps, we construct a current driven circuit model.

C. Circuit Simulation

The current through each edge can be solved by using NA (*Nodal Analysis*) method [16]. The main idea of this method is based upon KCL (*Kirchhoff's Current Law*), which states that the algebraic sum of the currents flowing into each node of the circuit must be zero. Set the voltages at the nodes as variables and the current through each edge can be expressed by those variables according to *Ohm's Law*. Then, apply KCL to get a set of linear equations. These equations can be written in matrix form as $Ax=b$, where A is a positive definite diagonally dominant sparse matrix and b is the port incidence matrix [16]. There are numerous efficient algorithms to solve it, such as *ICCG* (Incomplete Cholesky-conjugate Gradient method) [17], etc. For convenience's sake, we simply use Gaussian-Seidel iteration method in our experiments to verify the theory of CDCTree.

D. Path Selection

This step is of great importance as it directly decides the quality of the tree. The step is composed of three sub-functions: *preprocess*, *growth*, and *reduction*. The overall procedure of the *path selection* is shown in Algorithm 1. It first moves inside the terminals at periphery to alleviate the deficiency brought by GND resistors (in sub-function *preprocess*). Then, it decides what to do in each iteration according to the situation of surrounding terminals, whether *reduction* or *growth*.

The key point for this algorithm is to select the edges with the minimum currents. This is also the goal of sub-function *growth*. However, as we mentioned above, there are other factors that influence the current distribution. Some strategies must be employed to make up for the disturbance. The first and second sub-functions, *preprocess* and *reduction*, are put forward for this purpose.

This algorithm cannot guarantee that all the terminals can be connected. Therefore, a maximum iteration number must be set. If the above algorithm does not converge successfully after a certain number of iterations, it is forced to exit. The remaining unconnected terminals are connected using Maze Algorithm. This situation happens when there are minimum current loops in the circuit. In our experiments, we observe the unconnected terminal number is always less than three.

Algorithm 1 Overall Procedure of *Path Selection*

INPUT: terminal_number, terminal_list and current distribution of the escape graph;
OUTPUT: OARSMT;
initialization: active_number = terminal_number;
initialization: active_list = terminal_list;
initialization: dead_list = Φ
call: preprocess;
while *active_number* $\geq$ *2* **do**
  i = 0;
  Δ*active_number = 0;*
  while *i < active_number* **do**
    current_active_vertex = active_list(i);
    if *Any neighboring vertices of current_active_vertex has already been selected* **then**
      call: reduction;
    else
      call: growth;
    end if
    i ++;
  end while
  active_number = active_number - Δactive_number;
end while
if *dead_list* $\neq$ Φ **then**
  Use maze algorithm to connect the vertices in the dead_list sequentially;
end if

The sub-function *preprocess* moves the periphery terminals inside. The periphery vertices are connected to ground via a small resistor. So the current distribution at periphery is

2006 Asia and South Pacific Design Automation Conference

6C-3

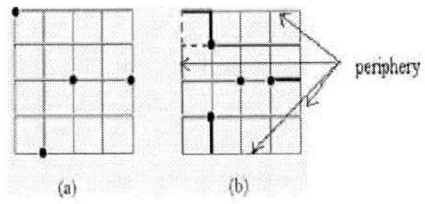

Fig. 3. Terminal distribution on a 5×5 grid: (a) Before preprocess and (b) After preprocess

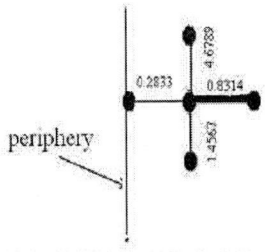

Fig. 4. An illustration of subfunction *growth at a vertex near periphery*

influenced. To avoid this, we move the periphery terminals inside to a new position, as shown in Figure 3. The terminals are denoted by black dots and the edges selected in the *preprocess* stage are denoted by bold lines. Note that the terminal on the upper-left corner is moved right and then down along the edges in black. It can also be moved inside along the dashed edges as the total wirelength remains the same.

A vertex is "active" when it is newly added to the Steiner tree during tree *growth*. For each active vertex, the subfunction *growth* selects the available neighboring edge with minimum current, adds the vertex at the other end of the edge to the current active vertex list, and removes itself from the active vertex list. An edge is not available when it connects the current active vertex with a vertex on the periphery. This is because the selection of this edge will cause the new active vertex to be moved to the periphery, where the current distribution is significantly influenced by the GND resistors.

Algorithm 2 provides the procedure of sub-function *growth*. Figure 4 shows an example of *growth* where the vertex in the center is the current active vertex and the value beside each edge is the current flowing through it. Although the edge heading left has the minimum current (0.2833), it can not be selected because it is not available. Therefore, the edge heading right is selected as it has the minimum current among the available edges.

Algorithm 2 Procedure of the sub-function *growth*

INPUT: current_active_vertex and the current distribution of the escape graph;
OUTPUT: updated dead_list, updated active_list, updated current_active_vertex and updated Δactive_number;
search for the edge e with the minimum current among all the available edges residing on the current_active_vertex;
if no such edge exists **then**
    remove current_active_vertex from active_list;
    add current_active_vertex to dead_list;
    Δactive_num - -;
else
    update current_active_vertex along edge e ;
    mark edge e as selected;
    mark current_active_vertex(i) as selected;
end if

The sub-function *reduction* is used to alleviate the influence of GND resistors. The periphery vertices are connected to ground via a small resistor, thus influencing the current

distribution. To compensate for this, *reduction* changes the tree structure when necessary. It functions when any of the neighboring vertices has already been selected. When one of the neighboring vertices has already been selected but has not been connected with the current active vertex yet, add the edge connecting them to the tree. Otherwise if it has already been connected with the current active vertex, add the edge connecting them to the tree and delete the edges on the path connecting these two edges until the connectivity of the tree is destroyed.

The detailed procedure of *reduction* is shown in Algorithm 3. Figure 5 is an example of *reduction*. The original tree is $A - B - C - D - E - F - G - H - I - J$. The improved tree after *reduction* is $A - B - C - F - G - H - I - J$, showing edges $C - D - E - F$ being replaced by edge $C - F$.

Algorithm 3 Procedure of the sub-function *reduction*

INPUT: current_active_vertex and the current distribution of the escape graph
OUTPUT: updated dead_list, updated active_list, updated current_active_vertex and updated Δactive_number;
if any neighboring vertex is selected but it is not connected with current_active_vertex **then**
    select the edge e connecting them;
    mark edge e as selected;
    delete current_active_vertex from active_list;
    Δactive_number - -;
end if
if any neighboring vertex is selected and it is already connected with current_active_vertex **then**
    select the edge e connecting them;
    delete the edges on the path connecting these two vertices until the connectivity of the tree is destroyed;
end if

IV. EXPERIMENTAL RESULTS AND DISCUSSIONS

We have implemented the CDCTree algorithm in C language. It was reported that the An-OARSMan algorithm produces the shortest wirelength among existing algorithms when routing less than 100 terminals [13]. Therefore, we compare our results with those of the An-OARSMan on an 800MHz Sun V880 fire workstation with Unix operating system. We

633

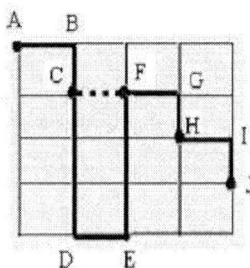

Fig. 5. An illustration of sub-function *reduction on a 5×5 grid*

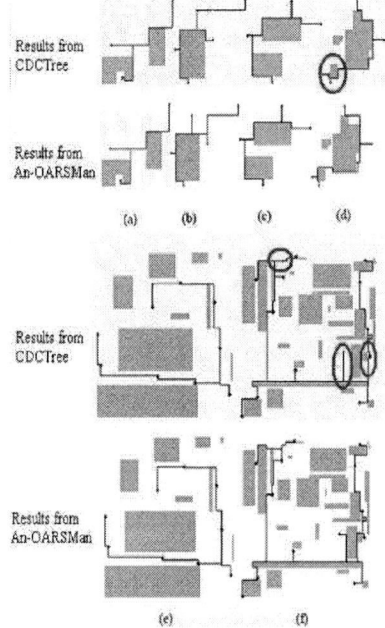

Fig. 6. Results comparison between CDCTree and An-OARSMan

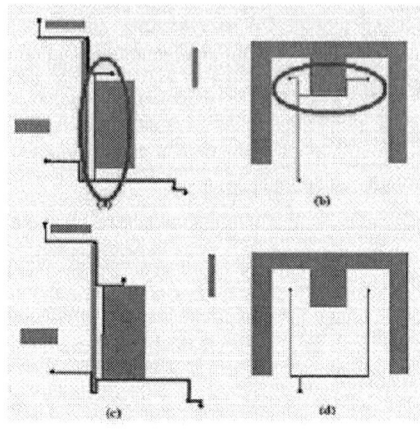

Fig. 7. Further results comparison between CDCTree and An-OARSMan

TABLE I

COMPARISON BETWEEN AN-OARSMAN AND CDCTREE

Terminal #	Obstacle #	An-OARSMan		CDCTree		
		Length	Runtime (s)	Length	Runtime (s)	
					I	II
3	3	2350	<0.01	2350	<0.01	<0.01
5	3	4380	0.01	4350	0.01	<0.01
7	5	9610	0.04	9610	0.02	0.01
10	5	11340	0.06	10980	0.08	0.02
20	7	14790	0.28	13110	0.84	0.13
30	10	21220	0.77	19970	1.38	0.25
40	10	28600	1.98	27300	4.32	0.77
50	15	31330	3.22	31310	10.25	1.02

test the cases presented in [13] using the CDCTree algorithm and the comparison between their results are shown in Figure 6. Both algorithms achieve the same optimal results in the case (a), (b), (c), and (e). In the cases (d) and (f), the results of CDCTree are better than those of An-OARSMan as shown in red circles.

Figure 7 provides more examples to show the advantages of CDCTree over An-OARSMan. In Figure 7, (a) and (b) are the results of CDCTree, while (c) and (d) are their respective counterparts produced by An-OARSMan. Improvements are shown in red circles. It is obvious that CDCTree algorithm produces shorter wirelength than An-OARSMan does.

Table 1 shows the comparison results upon test cases with a different number of terminals and obstacles. For each scale, we randomly select several different cases and represent the average results. The fi rst two columns are the terminal number and obstacle number of the test cases. The next four columns are the wirelength and runtime of An-OARSMan and CDCTree, respectively. From Table 1, we can see that when the terminal number is less than 50, CDCTree can achieve better result than An-OARSMan.

In physical design, most nets have less than 50 terminals, with the exception of clock trees and power grids. In addition, the terminals of the same net are usually placed close to each other in the placement stage. Therefore, CDCTree algorithm is practical in routing applications.

The time consumed by CDCTree can be divided into two parts. Part I is the time used to solve linear equations and Part II is the time used to construct OARSMT. CDCTree consumes a majority of the time in Part I. This is because we use the Gaussian-Seidel algorithm to solve the equations, which is extremely ineffi cient. As the coeffi cient matrix for the equations are sparse, positive-defi nite and symmetric, ICCG or other effi cient algorithms can be employed to dramatically reduce the run time. It is reported that ICCG is about $8,000X$ faster than the point Gaussian-Seidel method in solving linear equations [17]. In addition, more advanced algorithms can be employed to further reduce the runtime. For example, an improved ICCG method was put forward in [18], which can reduce the computation time by $30\% - 50\%$ compared with the ICCG method. Therefore, by using these algorithms, the runtime for equation-solving can be reduced to be small enough to be neglected.

V. SUMMARY AND CONCLUSIONS

This paper focuses on OARSMT routing tree construction. An heuristic algorithm named CDCTree is presented based on a current driven circuit model. The algorithm maps the escape graph into a circuit structure and replaces each edge with one resistor. By performing DC analysis on the circuit and selecting among the edges with the minimum currents to construct the OARSMT, the resulting tree has a short wirelength. Experimental results show that CDCTree achieves better results than the existing best algorithm An-OARSMan when the terminal number of a net is less than 50.

ACKNOWLEDGEMENT

The paper describes the research work performed cooperatively at University of California, Los Angeles (UCLA), USA and Tsinghua University, Beijing, P.R.China. The authors wish to thank the anonymous reviewers for their helpful comments.

REFERENCES

[1] M.R.Garey and D.S.Johnson, "The rectilinear Steiner tree problem is NP-complete", *SIAM Journal on Applied Mathematics*, 1977(32): pp. 826-834.

[2] C.Y.Lee, "An algorithm for path connections and its applications", *IRE Trans. on Electronic Computers*, 1961, 10: pp. 345-346.

[3] S.B.Akers, "A Modification of Lee's Path Connection Algorithm", *IEEE Trans. on Electronic Computer*, 1967, 16 (4): pp. 97-98.

[4] J.Soukup, "Fast Maze Router", In *Proc. of the 15th IEEE/ACM Design Automation Conferece*, 1978: pp. 100-102.

[5] Hadlock, "A Shortest Path Algorithm for Grid Graphs", *Networks*, 1977(7): pp. 323-334.

[6] F.Rubin, "The Lee Connection Algorithm", *IEEE Trans. on Computer*, 1974(23): pp. 907-914.

[7] D.W.Hightower, "A Solution to the Line Routing Problem on the Continous Plane", in *Proc. of the 6th Design Automation Conference*, 1969: pp. 1-24.

[8] K.Mikami and K.Tabuchi, "A Computer Program for Optimal Routing of Printed Circuit Connectors", in *Proc. of IFIPS*, 1968, H47: pp. 1475-1478.

[9] J.L.Ganley and J.P.Cohoon, "Routing a multi-terminal critical net: Steiner tree construction in the presence of obstacles", in *Proc. of IEEE ISCAS*, London, UK, 1994: pp. 113-116.

[10] M.Zachariasen and P.Winter, "Obstacle-avoiding Euclidean Steiner trees in the plane: an exact algorithm", extended abstract presented at *the Workshop on Algorithm Engineering and Experimentation (ALENEX)*, 1999.

[11] Y.Yang, Q.Zhu, T.Jing, X.L.Hong, and Y.Wang, "Rectilinear Steiner Minimal Tree among Obstacles", in *Proc. of IEEE ASICON*, Beijing, China, 2003: pp. 348-351.

[12] Yu Hu, Zhe Feng, Tong Jing, Xianlong Hong, Yang Yang, Ge Yu, Xiaodong Hu, Guiying Yan, "FORst: A 3-Step Heuristic for Obstacle-Avoiding Rectilinear Steiner Minimal Tree Construction", *Journal of Information and Computational Science*, 2004, 1(3): 107-116.

[13] Yu Hu, Tong Jing, Xianlong Hong, Zhe Feng, Xiaodong Hu, and Guiying Yan, "An-OARSMan: Obstacle-Avoiding Routing Tree Construction with Good Length Performance", in *Proc. of IEEE/ACM Asia and South Pacific Design Automation Conference*, 2005, Shanghai, China, pp. 7-12.

[14] Y.F.Wu, P.Widmayer, M.D.F.Schlag, and C.K.Wong, "Rectilinear Shortest Paths and Minimum Spanning Trees in the Presence of Rectilinear Obstacles", *IEEE Trans. on Computers*, 1987, 36(3): pp. 321-331.

[15] Yiyu Shi, Bike Xie and Yanjie Mao, "Circuit Model in Mathematical Modeling", *Journal of Mathematical Engineering*, 2004, 21(7): pp. 43-48.

[16] Ho C, Ruehli, Brennan P. "The Modified Nodal Approach to Network Analysis", *IEEE Trans. Circuits and Systems*, 1975, 39(22): pp. 504-509.

[17] D. S. Kershaw, "The incomplete Cholesky-conjugate gradient method for the iterative solution of systems of linear equations", *Journal of Computational Physics*, 26 I (Jan.) (1978), 43-65.

[18] J. Wang, D. Xie and Y. Yao, "The Modified Solutoin for Large Sparse Symmetric Linear Systems in Electromagnetic Field Analysis", *Transactions of China Electrotechnical Society*, 2001 16(2): pp. 26-29.

6C-4

A Novel Framework for Multilevel Full-Chip Gridless Routing *

Tai-Chen Chen
Graduate Institute of Electronics Engineering
National Taiwan University
Taipei 106, Taiwan
tcchen@eda.ee.ntu.edu.tw

Yao-Wen Chang
Graduate Institute of Electronics Engineering
and Department of Electrical Engineering
National Taiwan University
Taipei 106, Taiwan
ywchang@cc.ee.ntu.edu.tw

Shyh-Chang Lin
SpringSoft, Inc.
Hsinchu 300, Taiwan
chris@springsoft.com.tw

Abstract— Due to its great flexibility, gridless routing is desirable for nanometer circuit designs that use variable wire widths and spacings. Nevertheless, it is much more difficult than grid-based routing because of its larger solution space. In this paper, we present a novel *"V-shaped"* multilevel framework (called VMF) for full-chip gridless routing. Unlike the traditional *"Λ-shaped"* multilevel framework (inaccurately called the "V-cycle" framework in the literature), our VMF works in the V-shaped manner: top-down uncoarsening followed by bottom-up coarsening. Based on the novel framework, we develop a multilevel full-chip gridless router (called VMGR) for large-scale circuit designs. The top-down uncoarsening stage of VMGR starts from the coarsest regions and then processes down to finest ones level by level; at each level, it performs global pattern routing and detailed routing for local nets and then estimate the routing resource for the next level. Then, the bottom-up coarsening stage performs global maze routing and detailed routing to reroute failed connections and refine the solution level by level from the finest level to the coarsest one. We employ a dynamic congestion map to guide the global routing at all stages and propose a new cost function for congestion control. Experimental results show that VMGR achieves the best routability among all published gridless routers based on a set of commonly used MCNC benchmarks. Besides, VMGR can obtain significantly less wirelength, smaller critical path delay, and smaller average net delay than the previous works. In particular, VMF is general and thus can readily apply to other problems.

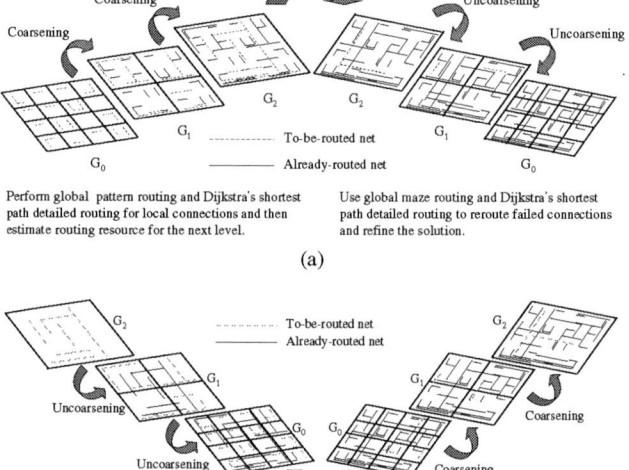

Perform global pattern routing and Dijkstra's shortest path detailed routing for local connections and then estimate routing resource for the next level.

Use global maze routing and Dijkstra's shortest path detailed routing to reroute failed connections and refine the solution.

(a)

Perform global pattern routing and Dijkstra's shortest path detailed routing for local nets and then estimate routing resource for the next level. (Longer nets have higher priority than shorter nets as far as timing is concerned.)

Use global maze routing and Dijkstra's shortest path detailed routing to reroute failed connections and refine the solution. (Shorter connections have higher priority than longer connections as far as routability is concerned.)

(b)

Fig. 1. (a) The Λ-shaped multilevel framework flow; (b) The V-shaped multilevel framework flow.

I. INTRODUCTION

Routing complexity is an important problem for modern routers. To cope with the increasing complexity, the multilevel framework is proposed to solve the routing problems (e.g., MRS [8], MARS [9, 10], MR [3, 22], CMR [12, 14], MGR [4], XMR [15]) as well as graph/circuit partitioning (e.g., Chaco [11], ML [1], hMETIS [19], HPM [7]), floorplanning (e.g., MB*-tree [21], MLGFA [16]), and placement (e.g., mPL [2] and APlace [17, 18]). All of the existing multilevel frameworks adopt a two-stage technique, bottom-up coarsening followed by top-down uncoarsening, which is known as the "Λ-shaped" framework. See Figure 1(a) for an illustration of the "Λ-shaped" multilevel routing framework. (Note that this framework is often called the "V-cycle" framework in the literature. However, we think that it is more appropriate to name it the "Λ-shaped" framework as it works bottom-up and then top-down.) These frameworks handle the target problems first bottom-up from local configurations to global ones and then refine the solutions top-down from global to local. It is obvious that there are significant limitations for the Λ-shaped framework to handle the global circuit effect, such as interconnection optimization, since only local information is available at the begin-

ning stages. A wrong choice made in such early stages may make the solution very hard to be refined during the top-down stage.

Most of the previous routing algorithms are grid-based, assuming uniform wire/via sizes. However, the grid-based approach is not effective to handle modern routing problems with nanometer electrical effects, such as optical proximity correction (OPC) and phase-shift mask (PSM). To cope with these nanometer electrical effects, we need to consider designs of variable wire/via widths and spacings, for which gridless routers are desirable due to their great flexibility. The gridless routing, however, is much more difficult than the grid-based routing because the solution space of gridless routing is significantly larger than that of grid-based routing. Cong et al. in [6] proposed a three-level routing scheme with a wire-planning phase between the global routing and the detailed routing. However, for large-scale designs, even with the three-level routing system, the problem size at each level may still be very large. Therefore, as the designs grow, more levels of routing are needed [10]. Recently, we proposed an OPC-aware multilevel gridless router based on the Λ-shaped framework [4], which integrates gridless global and detailed routing at each level. The router can handle non-uniform wire widths and reduce OPC pattern feature requirements.

In this paper, we present a new "V-shaped" multilevel routing framework (called VMF). Unlike the traditional Λ-shaped multilevel frameworks (called LMF) that apply bottom-up coarsening followed

---

*This work was partially supported by SpringSoft, Inc. and National Science Council of Taiwan under Grant No's. NSC 94-2215-E-002-005 and NSC 94-2215-E-002-030.

0-7803-9451-8/06/$20.00 ©2006 IEEE.

Work	Category of routing	Framework	Characteristics
Ours	• Multilevel gridless global and detailed routing	• Use V-shaped multilevel framework. • Before uncoarsening: channel density initialization. • Uncoarsening: GR+DR+RE. • Coarsening: global and detailed maze refinement.	• Perform global and detailed routing at each level. • Handle longer nets first and thus the wirelength and the critical path are reduced.
Chang et al. in [3, 22]	• Multilevel grid-based global and detailed routing	• Use Λ-shaped multilevel framework. • Coarsening: GR+DR+RE. • Uncoarsening: global and detailed maze refinement.	• Perform global and detailed routing at each level. • Lack initial global routing.
Chen et al. in [4]	• Multilevel gridless global and detailed routing	• Use Λ-shaped multilevel framework. • Coarsening: GR+DR+RE. • Uncoarsening: global and detailed maze refinement.	• Perform global and detailed routing at each level. • Lack initial global routing.
Cong et al. in [8, 9, 10]	• Multilevel gridless global routing + flat gridless detailed routing	• Use Λ-shaped multilevel framework. • Coarsening: RE. • Intermediate stage: multicommodity flow. • Uncoarsening: global maze refinement.	• Perform global and detailed routing separately.
Ho et al. in [12, 13, 14]	• Multilevel grid-based global and detailed routing	• Use Λ-shaped multilevel framework. • Coarsening: GR+RE. • Intermediate stage: track/layer assignment. • Uncoarsening: global and detailed maze refinement.	• Perform global and detailed routing separately.

TABLE I

MULTILEVEL FRAMEWORK COMPARISONS AMONG [3, 22], [4], [8, 9, 10], [12, 13, 14], AND VMGR. GR, DR, AND RE DENOTE GLOBAL ROUTING, DETAILED ROUTING, AND RESOURCE ESTIMATION, RESPECTIVELY.

by top-down uncoarsening, VMF adopts the two-stage technique of top-down uncoarsening followed by bottom-up coarsening. See Figure 1(b) for an illustration of VMF. The V-shaped multilevel framework was first introduced for interconnect-driven floorplanning [5]; it outperforms the Λ-shaped one in optimizing global circuit effects (such as wirelength, timing, and crosstalk optimization), since the V-shaped framework first considers the global configuration and then processes down to local ones level by level and thus the global effects can be handled at earlier stages.

Based on VMF, we develop a V-shaped multilevel full-chip gridless router (called VMGR) for large-scale circuit designs. The top-down uncoarsening stage of VMGR starts from the coarsest regions and then processes down to finest ones level by level; at each level, it performs global pattern routing and detailed routing for local nets and then estimate the routing resource for the next level. Then, the bottom-up coarsening stage performs global maze routing and detailed routing to reroute failed connections and refine the solution level by level from the finest level to the coarsest one.

In addition to the aforementioned characteristics, our VMF-based VMGR has the following distinguished features:

- The previous works [3, 12, 13, 14, 22] are *grid-based* multilevel router, which cannot handle designs of variable wire/via widths and spacings. Thus, they cannot effectively handle modern routing problems with nanometer electrical effects such as OPC.

- VMF considers the global longer nets first at the earlier uncoarsening stage, leading to better control on critical path delay and global interconnect effects.

- The previous works [3, 4, 22] perform *greedy global routing*, which determines the global path of the current net without considering the routing resource of succeeding nets. In contrast, VMGR employs a congestion map to guide the global routing at all stage. Initially, the map keeps the preliminary estimation of routing congestion based on the pin distribution. After routing a net, the map is updated dynamically based on the real route, previously routed nets, and estimated unrouted nets. As routing proceeds, we keep more and more accurate congestion information in the map. Therefore, we have better congestion control throughout the whole routing process.

- We use a new cost function based on *both* the total path congestion and the maximum channel congestion for global routing.

The cost function obtains better solutions than those consider only total path congestion or the maximum channel congestion.

- VMGR has higher flexibility and keeps more global views, and thus more routing objectives (such as crosstalk and OPC) can be more easily considered in VMGR since exact track and wiring information at each level after detailed routing is known.

Table I compares the existing multilevel routing frameworks among [3, 22], [4], [8, 9, 10], [12, 13, 14], and VMF.

Experimental results show that our VMGR achieves the best routability among all published gridless routers [4, 10] based on a set of commonly used MCNC benchmarks with non-uniform and uniform wire widths.

The rest of this paper is organized as follows. Section II presents the global, detailed, and V-shaped multilevel routing models. Section III presents our V-shaped multilevel routing framework. Experimental results are reported in Section IV. Finally, we give concluding remarks in Section V.

II. PRELIMINARIES

Routing in modern IC's is a very complex process, and we can hardly obtain high-quality solutions directly. Therefore, the routing problem is usually solved using the two-stage approach of global routing followed by detailed routing. Global routing first partitions the routing area into tiles and decides tile-to-tile paths for all nets while detailed routing assigns actual tracks and vias for nets.

A. Modeling of Global Routing

Our global routing algorithm is based on a graph search technique guided by the congestion associated with routing regions and topologies. The router assigns higher costs to route nets through congested areas to balance the net distribution among routing regions.

Before we can apply the graph search technique to multilevel routing, we first need to model the routing resource as a graph such that the graph topology can represent the chip structure. Fig. 2 illustrates the graph modeling. For the modeling, we first partition a chip into an array of rectangular subregions. These subregions are called *global routing cells (GRCs)*. A node in the routing graph represents a *GRC*

6C-4

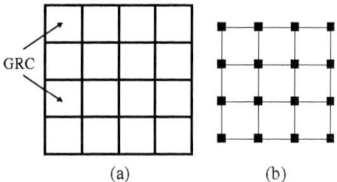

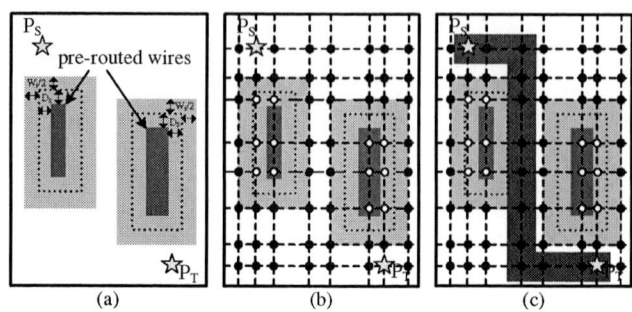

Fig. 2. Modeling of global routing: (a) Partitioned layout; (b) Routing graph.

Fig. 3. (a) A routing example. The gray areas denote the obstacle zones which are constructed by expanding a range which is the sum of the wire spacing and the half width of the routing wire. D_S and W_i are wire/via spacing and the width of the routing wire that satisfy the design rules, respectively. P_S and P_T are the source and target of the routing wire, respectively. (b) The implicit connection graph constructed by our detailed model. (c) A design-rule-correct path found through the eleven routable nodes.

in the chip, and an edge denotes the boundary between two adjacent *GRCs*. Each edge is assigned a capacity according to the width/height of a *GRC*. The routing graph is used to represent the routing area and is called a *multilevel routing graph*, denotes by G_k, where k is the level ID. A global router finds *GRC*-to-*GRC* paths for all nets on G_0 to guide the detailed router. The goal of global routing is to route as many nets as possible while meeting the capacity constraint of each edge and any other constraint, if specified. Note that, because of the gridless nature of our routing problem, the cost of routing a net is associated with the wire width and spacing.

B. Modeling of Detailed Routing

In the detailed routing stage, seeking high-quality and design-rule-correct paths in the routing region are two major concern. A suitable detailed routing model greatly affects these concerns. At first, for each obstacle, its obstacle zone is constructed by expanding the obstacle for a range which is the sum of the obstacle spacing and the half width of the routing wire. As shown in Fig. 3(a), the expanded range (gray area) is the sum of D_S and $W_i/2$, where D_S and W_i are the obstacle spacing to satisfy the design rules and the width of the routing wire, respectively. With the boundaries of each all extended regions and the center of the obstacle zone, three x-coordinates (the left boundary, the right boundary, and the center) and three y-coordinates (the top boundary, the bottom boundary, and the center) are obtained. The x-coordinates and y-coordinates of all obstacle zones and the source P_S and target P_T of the routing wire are stored into two sets, ICG_x and ICG_y, separately. Based on ICG_x and ICG_y, an implicit connection graph is constructed, as shown in Fig. 3(b). A vertical (horizontal) dashed lines in the implicit connection graph is generated through each x-coordinate (y-coordinate) in ICG_x (ICG_y). A node in the implicit connection graph denotes an intersection of a horizontal and a vertical dashed lines. There are two types of nodes, *routable nodes* and *unroutable nodes*. A routable node allows a routing path to pass through it without violating the design rules; it is unroutable, otherwise. As shown in Fig. 3(b), the respective black and white circles are the routable and unroutable nodes. To seek a design-rule-correct path from the source P_S to the target P_T, therefore, we only need to check if there exists a feasible path along which all nodes are routable. As shown in Fig. 3(c), a design-rule-correct path from P_S to P_T is found through the eleven routable nodes.

C. Modeling of V-Shaped Multilevel Routing

As illustrated in Figure 1(b), G_0 corresponds to the routing graph of the level 0 of the multilevel uncoarsening stage. Before the uncoarsening stage is performed, we need to determine the number of levels and build *GRCs* for each level. For each level i, we merge four GRC_i of G_i into a larger GRC_{i+1}. The process continues until the number of *GRCs* at level k is equal to one. Note that this process is just a pre-processing for determining k and no any global routing, detailed routing, or resource estimation is involved. Therefore, the pre-processing is different from the coarsening. After determining the number of levels, we start with the uncoarsening stage from the

k-th level. At each level i, our global router just finds routing paths for the *local nets* (a local net at level i denotes that all pins of the net can be included entirely by a GRC_i and cannot be included totally by a GRC_{i-1}), and then the detailed router is used to determine the exact wiring. After the global and detailed routing are performed, we expand each GRC_i to four finer GRC_{i-1} and at the same time perform resource estimation. The uncoarsening stage continues until the 0-th level is arrived. After finishing the uncoarsening stage, the coarsening stage tries to refine the routing solution starting from the level 0. During the coarsening stage, the unroutable connections during the uncoarsening stage are considered, and point-to-path maze routing and rip-up and re-route are performed to refine the routing solution. Then we proceed to the next level (i.e., level 1 here) of the coarsening stage by merging four adjacent GRC_0 into a larger GRC_1. The process continues until we go back to level k when the final routing solution is obtained.

III. V-SHAPED MULTILEVEL ROUTING FRAMEWORK

VMGR tends to route wider nets first since a wider net consumes more routing resource. Besides, VMGR tends to route longer nets first at the uncoarsening stage. It is obvious that the local nets at the higher level (say, level k) are usually longer than those at a lower level (say, level 0). Usually, a longer net has larger path delay. Thus, this observation implicitly suggests that a longer net has a higher priority than a shorter net as far as timing is concerned. Though this net ordering scheme may not be the optimal solution for some routing problems (for example, when routability is considered, routing shorter nets first often leads to a better completion rate), it is still a better alternative to the optimization of global interconnect effects.

A. Channel Density Initialization and Update

If global routing, detailed routing, and resource estimation are performed separately, the re-routing process conducted at the global routing stage may be in vain since it does not know if the re-routing is useful for the detailed routing. Also, the detailed router may fail to find a path because of the low flexibility induced from the separated global routing. Therefore, making the three tasks interact with each other can significantly improve routing quality [3, 22]. However, the concept can only guide the latter nets passing through the area with lower congestion and cannot avoid a wrong decision made by *greedy global routing* which determines the global path of an early routed net

638

without considering the routing resource of succeeding nets. Therefore, we initialize the routing congestion information based on the pin distribution and the global-path prediction of all nets, and then keep a congestion map that is updated dynamically based on both the already routed nets and the estimated unrouted nets. As routing proceeds, we keep more and more accurate congestion information in the map. Therefore, we have better congestion control throughout the whole routing process.

For a 2-pin connection c, we use L- and Z-shaped routes to determine the number of possible global routes n_c. We evenly distribute the wire density of the connection c, w_c, among all possible global routes. Therefore, the wire density of each possible global route is w_c/n_c. For each possible global route, we add the wire density of the possible global route to the channel density in the routing graph. After all 2-pin connections finish the process, we get an initial channel density. Note that the aforementioned approach is a natural way to estimate routing congestion, commonly used for interconnect-driven floorplanning.

At first, the channel density is totally estimated by the approach. After a connection has been routed successfully, the estimated cost induced by the connection will be removed from the channel density, and the wire density of the real path will be updated to the channel density (congestion map) dynamically. Therefore, our congestion control is based on congestion information induced by both the already routed nets and the estimated unrouted nets. As routing proceeds, we have more and more accurate congestion information for routing succeeding nets.

B. Cost Function for Global Routing

Let the multilevel routing graph be $G_0 = (V_0, E_0)$. Let $R_e = \{ e \in E_0 \mid e$ is the edge chosen for routing$\}$. We apply the cost function $\alpha : E_0 \to \Re$ to guide the global routing:

$$\alpha(R_e) = \max_{e \in R_e} c_e + \frac{1}{|R_e|} \sum_{e \in R_e} c_e, \qquad (1)$$

where c_e is the congestion of edge e and is defined by

$$c_e = \frac{d_e}{p_e},$$

where p_e and d_e are the capacity and channel density associated with e, respectively. We measure the routing congestion based on the *channel density* defined by the sum of wire spacing and wire width for gridless routing. (Note that the definition is different from the case in grid-based routing, for which channel density is defined as the maximum number of parallel nets passing through a routing channel.)

There are two advantages by using this cost function for global routing. First, this cost function can avoid that we select a path which has lower total path congestion with a higher channel congestion. Second, this cost function can prevent us from choosing a worse global path with the higher overall path congestion when two global paths have the same maximum channel congestion.

C. V-shaped Multilevel Gridless Routing

In the following, we present our framework for VMGR and summarize it in Figure 4.

Given a netlist, we first run a minimum spanning tree (MST) algorithm to construct the topology for each net, and then decompose each net into 2-pin connections, with each connection corresponding to an edge of the MST. According to those 2-pin connections, we use the heuristic in Section A to initialize the channel density in the routing graph by predicting the global paths of all nets in advance.

Algorithm: V-shaped-Multilevel-Gridless-Routing(G, N)
  Input: G - partitioned layout;
          N - netlist of multi-terminal nets;
  Output: routing solutions for N on G
begin
1 Partition layout;
2 For each net $n \in N$
3   Construct an MST;
4   Decompose the MST into 2-pin connections;
5   For each 2-pin connection
6   Initialize channel density;
7 **// Uncoarsening Stage**
8 For each level at the uncoarsening stage
9   Choose a local net n;
10   For each connection $c \in n$;
11   Perform global pattern routing;
12   Perform detailed routing;
13   Update channel density;
14 **// Coarsening Stage**
15 For each level at the coarsening stage
16   Choose a failed connection at the uncoarsening stage
17   Perform global maze routing;
18   Perform detailed routing;
19   Update channel density;
20 Analyze timing for all nets;
21 **return** the routing layout;
end

Fig. 4. Algorithm for V-shaped multilevel gridless routing.

VMGR starts from uncoarsening the coarsenest tile of level k. At each level, tiles are processed one by one, and only local nets are routed. At each level, the two-stage routing approach of global routing followed by detailed routing is applied. The global routing is based on the approach used in the Pattern Router [20] and first routes local nets on the tiles of level k. Let the multilevel routing graph of level i be $G_i = (V_i, E_i)$. Let $R_e = \{ e \in E_i \mid e$ is the edge chosen for routing$\}$. We apply the cost function in Section B to guide the routing.

After the global routing is completed, VMGR performs detailed routing with the guidance of the global-routing results and finds a real path in the chip. Our detailed router is based on the Dijkstra's shortest path algorithm and supports the *local refinement*. If detailed routing of a connection fails, it will be reconsidered (refined) at the coarsening stage. After a connection has been routed successfully, the estimated cost induced by the connection which calculated by the approach in Section A will be removed from the channel density, and the wire density of the real path will be updated to the channel density (congestion map) dynamically. This is called *resource estimation*. There are at least two advantages by using this approach. First, routing resource estimation is more accurate than that performing global routing alone since we can precisely evaluate the routing region. Second, we can obtain a good initial solution for the following refinement very effectively since pattern routing enjoys very low time complexity and uses fewer routing resources due to its simple L- and Z-shaped routing patterns.

The coarsening stage starts to refine each local failed connection, left from the uncoarsening stage. The global router is now changed to the maze router with the same cost function in the uncoarsening stage. Coarsening continues until the first level k is reached and the final solution is found. Note that the global maze routing here serves as an elaborate rip-up and re-route processor, in contrast to the simple L- and Z-shaped routing during uncoarsening. (For rip-up and re-route in VMGR, we mean the maze routing at the coarsening stage. It is only applied to global routing for better efficiency and quality trade-off.) This two-stage approach of global and local refinement of detailed routing gives our overall refinement scheme.

IV. EXPERIMENTAL RESULTS

We implemented VMGR in the C++ language on a 1 GHz SUN Blade-2000 workstation with 8 GB memory. We compared our results with the gridless routers presented in [4, 10] based on the 11 benchmark circuits provided by the authors. (Note that since the results of [10] is better than those of [8, 9], we just compare our results with [10].) The design rules for wire/via widths and wire/via spacings for detailed routing are the same as those used in [10].

Table II lists the set of benchmark circuits. In the table, "Circuit" gives the names of the circuits, "Size (μm^2)" gives the layout dimensions in μm^2, "#Layers" denotes the number of routing layers used, "#Nets" gives the number of two-pin connections after net decomposition, and "#Pins" gives the number of pins. For delay computation, we use the Elmore delay model. All the parameters are the same as those used in [4]. A via is modeled as the Π-model circuit, with its resistance and capacitance being twice of those of a wire segment. As pointed out in [3, 22], Mcc1, Mcc2, Struct, Primary1, and Primary2 do not have the information of net sources. Therefore, we cannot calculate the path delay for those benchmark circuits. In the following experiments, we represent the critical path and average net delays of these 5 benchmark circuits by the notation, –.

Circuit	Size (μm^2)	#Layers	#Nets	#Pins
Mcc1	45000×39000	4	1693	3101
Mcc2	152400×152400	4	7541	25024
Struct	4903×4904	3	3551	5471
Primary1	7522×4988	3	2037	2941
Primary2	10438×6488	3	8197	11226
S5378	435×239	3	3124	4818
S9234	404×225	3	2774	4260
S13207	660×365	3	6995	10776
S15850	705×389	3	8321	12793
S38417	1144×619	3	21035	32344
S38584	1295×672	3	28177	42931

TABLE II
THE BENCHMARK CIRCUITS.

A. Multilevel Gridless Routing with Uniform Nets

Table III lists the experimental results obtained by the Λ-shaped multilevel gridless routing in [4] (called LMGR), the Λ-shaped multilevel gridless routing system (multilevel global routing + flat gridless detailed routing) in [10] (called MARS), and VMGR. In the table, "WL (μm)" represents the wirelength in μm, "D_{max} (psec)" represents the critical path delay in pico-second, "D_{avg} (fsec)" represents the average net delay in femto-second, "Comp. Rates" gives the routing completion rates, and "Time (sec)" represents the runtime in second.

Compared with LMGR, the experimental results show that VMGR achieves a 5.19X runtime speedup while LMGR results in longer wirelength, larger critical path delay, and larger average net delay (1.02X wirelength, 1.21X critical path delay, and 1.00X average net delay). Compared with MARS, the experimental results show that VMGR achieves a 1.97X runtime speedup. (Note that it is hard to make a fair comparison between MARS and VMGR, because MARS and VMGR ran on different machines. Nevertheless, they both ran on

SUN workstations. Therefore, we try our best to make a fair comparison by normalizing the runtime based their clock rates.) Since MARS did not report their wirelength, critical path delay, and average net delay in their paper, we cannot compare those results in MARS with VMGR.

B. Multilevel Gridless Routing with Non-Uniform Nets

We also performed experiments on the benchmark circuits of non-uniform wire widths. We modify the original circuits of uniform wire sizes to generate a set of circuits of non-uniform wire sizes by using the following rules, which was proposed by [10]. The longest 10% nets are widened to twice the original width, while the next 10% are widened to 150% the original width. However, because the benchmark circuits S5378–S38584 are standard-cell designs, widening any pin violates the design rules for via spacing. Therefore, it is unreasonable and incorrect to test these six modified benchmark circuits.

In Table IV, "#Total Sub-nets" denotes the total number of 2-pin nets seen by the detailed router of MARS, since the detailed router of MARS segments long two-pin nets into short subnets. As shown in the table, VMGR still achieves 100% routing completion for all of the 5 circuits with 1.91X (1.19X) runtime speedup while [4] ([10]) completes routing for only 4 circuits. Note that VMGR is the first router to complete the routing for this set of benchmarks of non-uniform wire sizes. In particular, we expect that the difference will be much more significant for larger and difficult designs such as vd_Mcc2. Figures 5 and 6 show the full-chip and partial routing solutions for "vd_Mcc2" obtained from VMGR, respectively. The bounding box in Figure 5 is the boundary of this benchmark circuit. We can see in Figure 6 that the three left-most vertical lines have different widths.

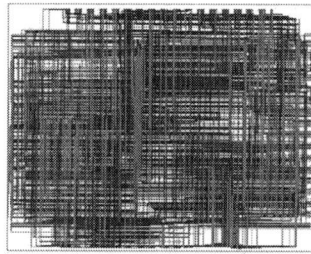

Fig. 5. The full-chip routing solution for "vd_Mcc2" obtained from VMGR. The bounding box is the boundary of this benchmark circuit.

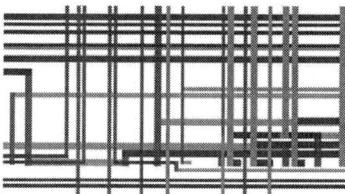

Fig. 6. A partial routing solution for "vd_Mcc2" obtained from VMGR. We can see that the three left-most vertical lines have different widths.

The two experimental results reveal the effectiveness of VMF for multilevel routing. Since VMF considers the global longer nets first at the earlier uncoarsening stage, it can have better control on the wirelength and the critical path delay. Besides, the runtime and solution quality are improved simultaneously. Also, compared with [4] that was based on LMF, the experimental results have shown that LMF leads to significantly better wirelength, critical path delay, and average net delay, and 100% routing completion rates.

640

Circuit	(A) Results of [4]					(B) Results of [10]					(C) Our Results				
	WL (μm)	D_{max} (psec)	D_{avg} (fsec)	Comp. Rates	Time (sec)	WL (μm)	D_{max} (psec)	D_{avg} (fsec)	Comp. Rates	Time (sec)	WL (μm)	D_{max} (psec)	D_{avg} (fsec)	Comp. Rates	Time (sec)
Mcc1	2.8e7	–	–	100%	190.2	NA	–	–	100%	105.1	2.7e7	–	–	100%	56.4
Mcc2	4.1e8	–	–	100%	3711.0	NA	–	–	100%	1916.9	4.0e8	–	–	100%	1353.8
Struct	8.5e5	–	–	100%	6.5	NA	–	–	100%	31.6	8.4e5	–	–	100%	4.4
Primary1	1.0e6	–	–	100%	5.1	NA	–	–	100%	33.5	1.0e6	–	–	100%	4.7
Primary2	4.2e6	–	–	100%	46.7	NA	–	–	100%	162.7	4.1e6	–	–	100%	27.5
S5378	7.6e4	21	780	100%	45.6	NA	NA	NA	100%	30.0	7.4e4	11	777	100%	5.7
S9234	5.5e4	18	681	100%	25.1	NA	NA	NA	100%	22.8	5.4e4	17	678	100%	4.3
S13207	1.8e5	37	828	100%	136.2	NA	NA	NA	100%	85.2	1.8e5	33	812	100%	17.9
S15850	2.2e5	87	855	100%	362.2	NA	NA	NA	100%	107.1	2.2e5	84	866	100%	22.7
S38417	4.8e5	183	759	100%	403.1	NA	NA	NA	100%	250.9	4.7e5	174	763	100%	70.7
S38584	6.7e5	1086	835	100%	765.1	NA	NA	NA	100%	466.1	6.6e5	1026	828	100%	209.0
Comp.	1.02	1.21	1.00	1	5.19				1	1.97*	1	1	1	1	1

TABLE III

COMPARISON AMONG (A) THE Λ-SHAPED MULTILEVEL GRIDLESS ROUTING [4], (B) THE Λ-SHAPED MULTILEVEL GRIDLESS GLOBAL ROUTING + FLAT GRIDLESS DETAILED ROUTING [10], AND (C) VMGR. NOTE: (A) AND (C) RAN ON A 1 GHz SUN BLADE-2000 WITH 8 GB MEMORY; (B) RAN ON A 440 MHz SUN ULTRA-10 WITH 384 MB MEMORY. (–: BECAUSE THOSE BENCHMARK CIRCUITS DID NOT HAVE THE INFORMATION OF NET SOURCES, WE CANNOT CALCULATE THE PATH DELAY FOR THEM.) (NA: [10] DID NOT REPORT THEIR WIRELENGTH, CRITICAL PATH DELAY, AND AVERAGE NET DELAY IN THEIR PAPER.) (*: FOR FAIR COMPARISONS, WE NORMALIZE THE RUNTIME OF [10] BY THE FACTOR 440/1000.)

Circuit	(A) Results of [4]				(B) Results of [10]				(C) Our Results			
	WL (μm)	#Failed Nets	Comp. Rates	Time (sec)	WL (μm)	#Failed Nets (#Total Sub-nets)	Comp. Rates	Time (sec)	WL (μm)	#Failed Nets	Comp. Rates	Time (sec)
vd_Mcc1	2.8e7	0	100%	199.6	NA	0	100%	148.1	2.7e7	0	100%	65.4
vd_Mcc2	4.1e8	383	98.5%	36581.5	NA	27(99715)	99.97%	3388.8	4.1e8	0	100%	23383.3
vd_Struct	8.5e5	0	100%	15.3	NA	0	100%	36.3	8.4e5	0	100%	10.3
vd_Primary1	1.0e6	0	100%	19.2	NA	0	100%	47.4	1.0e6	0	100%	12.2
vd_Primary2	4.2e6	0	100%	150.8	NA	0	100%	296.7	4.1e6	0	100%	80.0
Comp.	1.02		99.70%	1.91			99.99%	1.19*	1		1	1

TABLE IV

COMPARISON AMONG (A) THE Λ-SHAPED MULTILEVEL GRIDLESS ROUTING [4], (B) THE Λ-SHAPED MULTILEVEL GRIDLESS GLOBAL ROUTING + FLAT GRIDLESS DETAILED ROUTING [10], AND (C) VMGR. NOTE: (A) AND (C) RAN ON A 1 GHz SUN BLADE-2000 WITH 8 GB MEMORY; (B) RAN ON A 440 MHz SUN ULTRA-10 WITH 384 MB MEMORY. (NOTE THAT BECAUSE THE BENCHMARK CIRCUITS S5378–S38584 VIOLATE THE DESIGN RULES OF VIA SPACING, WE DID NOT LIST THESE CASES IN THIS TABLE.) (NA: [10] DID NOT REPORT THEIR WIRELENGTH IN THEIR PAPER.) (*: FOR FAIR COMPARISONS, WE NORMALIZE THE RUNTIME OF [10] BY THE FACTOR 440/1000.)

V. CONCLUSION

In this paper, we have proposed a novel V-shaped framework for multilevel, full-chip gridless routing. The V-shaped multilevel framework adopts a two-stage technique, top-down uncoarsening followed by bottom-up coarsening. Experimental results have shown that our V-shaped multilevel gridless router can obtain 100% routing completion rates with less wirelength, smaller critical path delay, and smaller average net delay than previous works. Besides, it can handle designs with non-uniform wire widths well and obtained better routing solutions than previous works. In particular, our gridless router is the first to complete the routing for the set of commonly used benchmarks of non-uniform wire sizes listed in the preceding section.

REFERENCES

[1] C. J. Alpert, J.-H. Huang, and A. B. Kahng, "Multilevel circuit partitioning," *IEEE Trans. CAD*, vol. 17, no. 8, pp. 655–667, August 1998.

[2] T. Chan, J. Cong, T. Kong, and J. Shinnerl, "Multilevel optimization for large-scale circuit placement," *Proc. ICCAD*, pp. 171–176, Nov. 2000.

[3] Y.-W. Chang and S.-P. Lin, "MR: A new framework for multilevel full-chip routing," *IEEE Trans. CAD*, vol. 23, no. 5, pp. 793–800, May 2004.

[4] T.-C. Chen and Y.-W. Chang, "Multilevel gridless routing considering optical proximity correction," *Proc. ASP-DAC*, pp. 1160–1163, Jan. 2005.

[5] T.-C. Chen, Y.-W. Chang, and S.-C. Lin, "IMF: Interconnect-driven multilevel floorplanning for large-scale building-module designs ," *Proc. ICCAD*, pp. 159–164, Nov. 2005.

[6] J. Cong, J. Fang, and K. Khoo, "DUNE: A multi-layer gridless routing system with wire planning," *Proc. ISPD*, pp. 12–18, April 2000.

[7] J. Cong, S. Lim, and C. Wu, "Performance driven multilevel and multiway partitioning with retiming," *Proc. DAC*, pp. 274–279, June 2000.

[8] J. Cong, J. Fang, and Y. Zhang, "Multilevel approach to full-chip gridless routing," *Proc. ICCAD*, pp. 396–403, Nov. 2001.

[9] J. Cong, M. Xie, and Y. Zhang, "An enhanced multilevel routing system," *Proc. ICCAD*, pp. 51–58, Nov. 2002.

[10] J. Cong, J. Fang, M. Xie, and Y Zhang, "MARS–A multilevel full-chip gridless routing system," *IEEE Trans. CAD*, vol. 24, no. 3, pp. 382–394, March 2005.

[11] B. Hendrickson and R. Leland, "A multilevel algorithm for partitioning graph," *Proc. Supercomputing*, pp. 1–24, July 1995.

[12] T.-Y. Ho, Y.-W. Chang, S.-J. Chen, and D.-T. Lee, "A fast crosstalk- and performance-driven multilevel routing system," *Proc. ICCAD*, pp. 382–387, Nov. 2003.

[13] T.-Y. Ho, Y.-W. Chang, and S.-J. Chen, "Multilevel routing with antenna avoidance," *Proc. ISPD*, pp. 34–40, April 2004.

[14] T.-Y. Ho, Y.-W. Chang, S.-J. Chen, and D.-T. Lee, " Crosstalk- and performance-driven multilevel full-chip routing," *IEEE Trans. CAD*, vol. 24, no. 6, pp. 869–878, June 2005.

[15] T.-Y. Ho, C.-F. Chang, Y.-W. Chang, and S.-J. Chen, " Multilevel full-chip routing for the X-based architecture," *Proc. DAC*, pp. 597–602, June 2003.

[16] C.-C. Hu, D.-S. Chen, and Y.-W. Wang, " Fast multilevel floorplanning for large scale modules," *Proc. ISCAS*, pp. 205–208, May 2004.

[17] A.B. Kahng and Q. Wang, Implementation and extensibility of an analytic placer," *Proc. ISPD*, pp. 18–25, April 2004.

[18] A.B. Kahng, S. Reda, and Q. Wang, APlace: A general analytic placement framework," *Proc. ISPD*, pp. 233–235, April 2005.

[19] G. Karypis, R. Aggarwal, V. Kumar, and S. shekhar, "Multilevel hypergraph partitioning: Application in VLSI domain," *IEEE Trans. VLSI Systems*, vol. 7, pp. 69–79, March 1999.

[20] R. Kastner, E. Bozorgzadeh and M. Sarrafzadeh, "Predictable routing," *Proc. ICCAD*, pp. 110–114, Nov. 2000.

[21] H.-C. Lee, Y.-W. Chang, J.-M. Hsu, and H. Yang, "Multilevel floorplanning/placement for large-scale modules using B*-trees," *Proc. DAC*, pp. 812–817, June 2003.

[22] S.-P. Lin and Y.-W. Chang, "A novel framework for multilevel routing considering routability and performance," *Proc. ICCAD*, pp. 44–50, Nov. 2002.

Monotonic Parallel and Orthogonal Routing for Single-Layer Ball Grid Array Packages

Yoichi Tomioka   Atsushi Takahashi

Department of Communications and Integrated Systems, Tokyo Institute of Technology
2–12–1–S3–58 Ookayama, Meguro-ku, Tokyo, 152–8550 Japan
Tel: +81-3-5734-2665
Fax: +81-3-5734-2902
e-mail: {yoichi,atsushi}@lab.ss.titech.ac.jp

Abstract— In this paper, we give the necessary and sufficient condition that all nets can be connected by monotonic routes when a net consists of a finger and a ball and fingers are on the two parallel boundaries of the Ball Grid Array package, and propose a monotonic routing method based on this condition. Moreover, we give a necessary condition and a sufficient condition when fingers are on the two orthogonal boundaries, and propose a monotonic routing method based on the necessary condition.

I. Introduction

Ball Grid Array (BGA) packages as shown in Fig. 1, in which I/O pins are placed in a grid array pattern, realize a number of connections between chips and the printed circuit board (PCB). Bonding fingers are connected to chips, and solder balls are I/O pins of the package in a grid array pattern. Since the structure of BGA packages is simple, many routes can be realized in few layers in the packages if connection requirements and routing patterns are suitable for the structure. In current package routing design, the designer generates satisfactory routing patterns by using the properties of connection requirements effectively. But it takes much time for large packages since the huge number of routes needs to be realized. So, the demand for automation of package routing is increasing. In this paper, we consider routing for a single-layer BGA package as the first step for BGA packages routing.

In the literature on for planar routing, there are a lot of problem formulations and approaches. For example, problem formulations for single-row and double-row routing, where terminals are placed on single-row and double-row, are proposed in [1] and [2], respectively. Though these problem formulations are similar to problems for single-layer BGA packages, approaches for them are not enough to obtain satisfactory routes for BGA packages. Actually, many parts of the routing process for BGA packages are realized manually with support tools.

In order to obtain a satisfactory routing pattern, the analysis of manual routing patterns is necessary. In the routing pattern by manual, though routes may snake, most of them do not go back. The route which do not go back are said to be monotonic. In monotonic routing patterns, it is expected that the total wire length tends to be small, and it is easy to decide the route of each net. But there exists a netlist that cannot be realized by monotonic routes in one layer with any design rule. There also exists a netlist in which a design rule may be satisfied if non-monotonic routes are allowed. In these cases, non-monotonic routes are needed. Though we aim to realize nets in one layer under a certain design rule, in this paper we propose an approach in which all nets are realized by monotonic routes. The obtained monotonic routes will be an initial solution in iterative improvement to satisfy the design rule.

In literatures for BGA package, several approaches focusing on monotonic routes were proposed. The first approach for single-layer BGA packages was proposed in [3] and it was improved in [4]. Their approach generates optimal uniform distribution of wire by generating connection requirements. An approach for 2-layer BGA packages was proposed in [5]. It is given connection requirements, and optimizes the total wire length and the wire congestion by improving via assignment.

Also, several approaches considering non-monotonic routes were proposed. The approach for multilayer Pin Grid Array (PGA) and Ball Grid Array packages were proposed in [6] and [7], respectively. They assign each net to a layer, and realize nets in respective layer.

All of them divide the package into several sectors, and nets are realized within each sector. Basically, each sector consists of bonding fingers on the same boundary of the package and solder balls, and a net in each sector consists of a bonding finger and a solder ball. Namely, their approach cannot be applied if it is impossible to divide the package into such sectors. So, we propose an approach for the region consisting of solder balls and bonding fingers on two boundaries of the package as shown in Fig. 2.

Section II introduces routing model, and gives some definitions for analysis. Connection requirements are the set of nets and are called a netlist. A parallel netlist, in which bonding fingers are on two parallel boundaries of the package, is shown in Fig. 2(a). In section III, we give

0-7803-9451-8/06/$20.00 ©2006 IEEE.

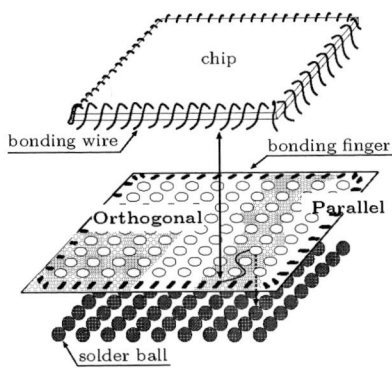

Fig. 1. A Ball Grid Array package

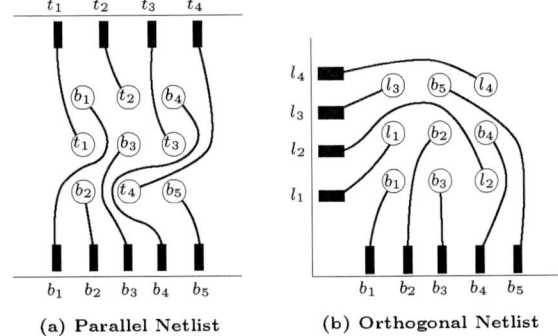

(a) Parallel Netlist   (b) Orthogonal Netlist

Fig. 2. Monotonic Netlists Decision Problem

the necessary and sufficient condition that all nets in the parallel netlist can be realized by monotonic routes, and propose a monotonic routing method based on this condition. An orthogonal netlist, in which bonding fingers are on two orthogonal boundaries of the package, is shown in Fig. 2(b). In section IV, we give a necessary condition and a sufficient condition that all nets in the orthogonal netlist can be realized by monotonic routes, and propose a monotonic routing method based on the proposed necessary condition. There may exist more than one monotonic routing pattern that corresponds to a parallel netlist and an orthogonal netlist, respectively. How to select one among them that meets the design rule, if it exists, is in our future works. Since it is not guaranteed that our routing method for orthogonal netlists completes routing, we implement our method for orthogonal netlists with C++ language, and applied it to orthogonal netlists in section V. Section VI concludes this paper.

II. PRELIMINARY

A. Definitions

In this paper, we assume that the BGA package has connection requirements between bounding fingers placed on boundaries of the package and solder balls placed in a grid array pattern. A solder ball, which we will refer to as a ball, is an I/O pin of the package and is connected to the PCB. A bonding finger, which we will refer to as a finger, is connected to the chip by a bonding wire. In this paper, we assume that all nets are two-terminal nets connecting a finger to a ball. A netlist is the set of such nets and is represented by $\mathbf{N}$. We refer to a finger placed on a bottom boundary of the package as a bottom finger, and refer to a net which consists of a bottom finger and a ball as a bottom net. Similarly, a top net and a left net are defined. Bottom nets and top nets are labeled according to the order of fingers from the left to the right as $b_1, b_2, b_3, \ldots$ and $t_1, t_2, t_3, \ldots$, respectively. Left nets are labeled according to the order of fingers from the bottom to the top as $l_1, l_2, l_3, \ldots$. Let $\mathbf{B}, \mathbf{T}$ and $\mathbf{L}$ be the sets of bottom, top and left nets, respectively.

We define the relation between nets according to their ball positions as shown in Fig. 3. Let (x_a, y_a) and (x_b, y_b) be the coordinates of the balls of nets a and b, respectively. The relation between a and b is defined as follows.

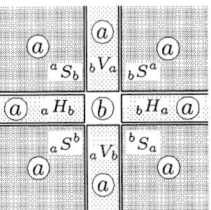

Fig. 3. Relationship between a and b

- If $x_a < x_b, y_a = y_b$ then a is said to be to the left of b and the relation is represented by $_aH_b$.

- If $x_a = x_b, y_a < y_b$ then a is said to be below b and the relation is represented by $_aV_b$.

- If $x_a < x_b, y_a < y_b$ then a is said to be to the lower-left of b and the relation is represented by $_aS^b$.

- If $x_a < x_b, y_a > y_b$ then a is said to be to the upper-left of b and the relation is represented by ^aS_b.

- $_bH_a, _bV_a, _bS^a$ and ^bS_a are defined symmetrically.

B. Order Graphs

We use some order graphs where a vertex v corresponds to a net $v \in \mathbf{N}$. The edge from a vertex u to a vertex v is represented by the ordered pair (u, v). In this paper, every order graph has edges corresponding to the order of fingers in each boundary. E_f^b, E_f^t and E_f^l are the sets of edges corresponding to the order in bottom, top and left boundaries, respectively. Formally, they are given as follows:
$$E_f^b = \{(b_i, b_j) \mid b_i, b_j \in \mathbf{B}, i < j\},$$
$$E_f^t = \{(t_i, t_j) \mid t_i, t_j \in \mathbf{T}, i < j\},$$
$$E_f^l = \{(l_i, l_j) \mid l_i, l_j \in \mathbf{L}, i < j\}.$$

C. Monotonic Routes

A boundary, where the finger of a net is placed, is called the finger boundary of the net. For example the finger

6C-5

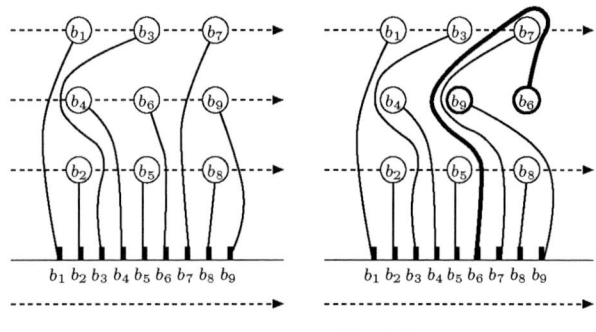

(a) All routes are monotonic   (b) $R(b_6)$ is non-monotonic

Fig. 4. Monotonic and non-monotonic routes

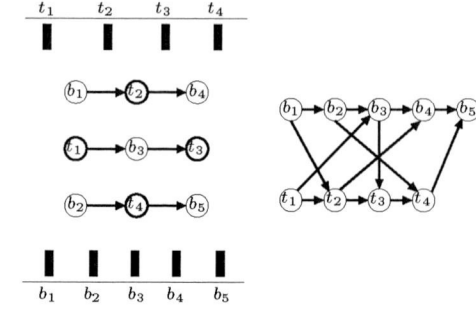

Fig. 5. MPN Decision Problem

boundary of b_1 in Fig. 2(a) is the bottom boundary. In this paper, a monotonic route and a non-monotonic route are defined as follows:

Definition 1 *If the route from a finger to a ball intersects any straight lines running parallel with the finger boundary at most once, then the route is said to be monotonic. Otherwise the route is said to be non-monotonic.*

Let $R(v)$ be the route of a net $v \in \mathbf{N}$. All routes are monotonic in Fig. 4(a), but $R(b_6)$ is non-monotonic in Fig. 4(b).

If all nets in a netlist can be realized by monotonic routes without intersecting each other, the netlist is said to be monotonic. A netlist is said to be single if the fingers of nets in the netlist are placed on the same boundary.

Consider that nets consist of bottom fingers and balls as shown in Fig. 4. A single netlist is monotonic if and only if nets on each row are in increasing order. Since a netlist in Fig. 4(a) satisfies this condition, it is monotonic. On the other hand, either $R(b_6)$ or $R(b_9)$ is non-monotonic in Fig. 4(b) since b_6 and b_9 are in decreasing order. A monotonic routing pattern for a monotonic single netlist, in which all routes are monotonic, is unique. Similar observations are found in [3, 4, 5].

Whether a single netlist is monotonic is decided by the order graph G_S. $E(G_S)$ consists of E_f^b and edges that correspond to the order of nets on each row. For example in Fig. 4(a), G_S has edges $(b_2, b_5), (b_2, b_8)$ and (b_5, b_8) corresponding to the bottom row. Similarly, G_S has edges for other rows. Clearly, G_S is cyclic if and only if there exist nets on a row which are in decreasing order, such as b_6 and b_9 in Fig. 4 (b).

III. PARALLEL NETLISTS

A parallel netlist is a netlist in which fingers are placed on the two parallel boundaries of the package. In this section, we analyze parallel netlists.

A. Monotonic Parallel Netlists

The Monotonic Parallel Netlist (MPN) Decision Problem is defined as follows:

Definition 2 *MPN Decision Problem*
Input:
   A parallel netlist.
Question:
   Is it possible to realize all connection requirements by monotonic routes?

An example of MPN Decision Problem is given in Fig. 2(a). In this case, $\mathbf{N} = \mathbf{B} \cup \mathbf{T}$. The necessary and sufficient condition for being monotonic is that nets on each row are in increasing order without distinguishing bottom and top nets. This condition is represented by the order graph G_P. $E(G_P)$ consists of E_f^b, E_f^t and edges corresponding to the order of nets on each row. In Fig. 5, G_P has edges $(b_2, t_4), (b_2, b_5)$ and (t_4, b_5) corresponding to the bottom row. Similarly, G_P has edges for other rows. In Fig. 5, the transitive edges like (b_2, b_5) are omitted. A parallel netlist is monotonic if and only if G_P is acyclic.

Theorem 1 *A parallel netlist is monotonic if and only if the order graph G_P is acyclic, where $E(G_P) = E_f^b \cup E_f^t \cup E_p$ and $E_p = \{(x, y) \mid x, y \in \mathbf{N}, {}_x H_y\}$.*

Proof. If the order graph G_P is acyclic, then an order can be obtained by G_P. A monotonic routing pattern can be realized according to the order as shown in section III.B. Conversely, consider that G_P has a cycle C. If C consists of only bottom nets, then non-monotonic routes are needed since it means that bottom nets are in decreasing order on a row. The same discussion is possible for top nets. So we assume that C consists of bottom nets and top nets. Without loss of generality, we assume that C has (b_j, t_p) and (t_q, b_i), where $b_i, b_j \in \mathbf{B}$ $(i < j)$ and $t_p, t_q \in \mathbf{T}$ $(p < q)$. Since ${}_{b_j} H_{t_p}$ and ${}_{t_q} H_{b_i}$, non-monotonic routes are needed by at least one of them as shown in Fig. 6. $\square$

B. A Parallel Routing Method

A partial order is defined by G_P, and some orders are obtained by the partial order. This order corresponds to an order such that the sources in G_P are removed one by one. For example, the following order

$$t_1 \rightarrow b_1 \rightarrow b_2 \rightarrow t_2 \rightarrow b_3 \rightarrow b_4 \rightarrow t_3 \rightarrow t_4 \rightarrow b_5$$

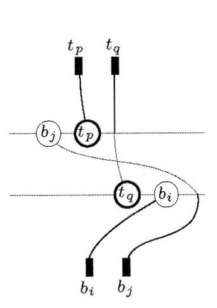

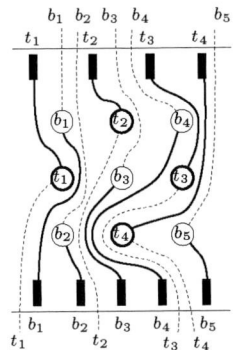

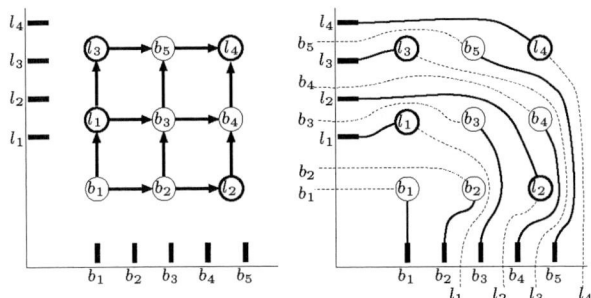

Fig. 8. Examples of MON Decision Problem and it's monotonic routing pattern

Fig. 6. $_{b_j}H_{t_p} \wedge {}_{t_q}H_{b_i}$ Fig. 7. An example of monotonic routing pattern for MPN

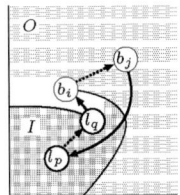

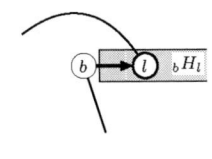

Fig. 9. A directed graph G_R Fig. 10. A constraint between two nets

is obtained for Fig. 5. According to the obtained order, we put virtual fingers on bottom boundary of the package for top nets and put virtual fingers on top boundary for bottom nets. All nets can be realized because we can connect a finger to its virtual finger via its ball by monotonic route one by one from the left. An example of solution are given in Fig. 7.

For a monotonic parallel netlist, there are several monotonic routing patterns since the order obtained by G_P is not unique in general. The selection of the order that reduces the density is in our future work.

IV. ORTHOGONAL NETLISTS

An orthogonal netlist is a netlist in which fingers are placed on the two orthogonal boundaries of the package. In this section, we analyze orthogonal netlists.

A. Monotonic Orthogonal Netlists

The Monotonic Orthogonal Netlist (MON) Decision Problem is defined as follows:

Definition 3 *MON Decision Problem*
Input:
   An orthogonal netlist.
Question:
   Is it possible to realize all connection requirements by monotonic routes?

An example of MON Decision Problem is given in Fig. 2(b). In this case, $\mathbf{N} = \mathbf{B} \cup \mathbf{L}$.

A.1   A Sufficient Condition

An orthogonal netlist is monotonic if the order graph G_s, which has edges corresponding to the order of nets on each row and column without distinguishing bottom and left nets, is acyclic. According to the order given by G_s, we put virtual fingers. Nets are realized by connecting each finger to its virtual finger via its ball from the lower left. Examples of MON Decision Problem and it's monotonic routing pattern are given in Fig. 8.

Theorem 2 *An orthogonal netlist is monotonic if the order graph G_s is acyclic, where $E(G_s) = E_f^b \cup E_f^l \cup E_s$ and $E_s = \{(x, y) \mid x, y \in \mathbf{N}, {}_xH_y \vee {}_xV_y\}$.*

If b_2 in the second column in Fig. 8 is swapped for b_3, G_s becomes cyclic. But, the netlist is monotonic since a monotonic routing pattern for the netlist is given in Fig. 2(b). So, this condition is not a necessary condition.

A.2   A Necessary Condition

A route can be regarded as the set of points. Let b be a bottom net and v be a net. If there exists a point (x_b, y_b) on $R(b)$ and a point (x_v, y_v) on $R(v)$ such that $x_b < x_v$ and $y_b = y_v$, then $R(v)$ is said to be to the right of $R(b)$. In other words, $R(v)$ is said to be to the right of $R(b)$ if there exists a point on $R(v)$ which is to the right of $R(b)$. Similarly, $R(v)$ for net v is said to be above $R(l)$ for left nets if there exists a point on $R(v)$ which is above $R(l)$.

Theorem 3 *Let G_R be the directed graph constructed for a routing pattern, where the vertices correspond to the nets, and edge set is defined as follows:*

- *An edge (b, v) $(b \in \mathbf{B}, v \in \mathbf{N})$ exists if and only if $R(v)$ is to the right of $R(b)$.*

- *An edge (l, v) $(l \in \mathbf{L}, v \in \mathbf{N})$ exists if and only if $R(v)$ is above $R(l)$.*

If the routing pattern is monotonic, then G_R is acyclic.

Proof. Consider that G_R is cyclic. Let C be a cycle in G_R. The cycle cannot consist of only bottom nets or only left nets since there is no non-monotonic route. So, there is an edge from a bottom net to a left net and an

6C-5   2006 Asia and South Pacific Design Automation Conference

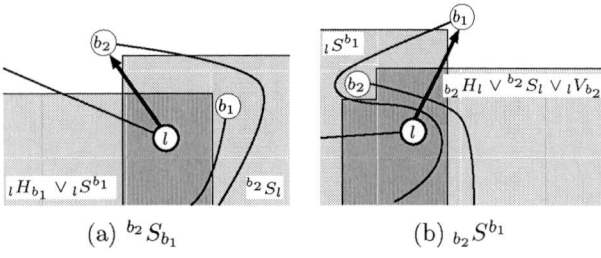

(a) $^{b_2}S_{b_1}$ (b) $_{b_2}S^{b_1}$

Fig. 11. Constraints between three nets

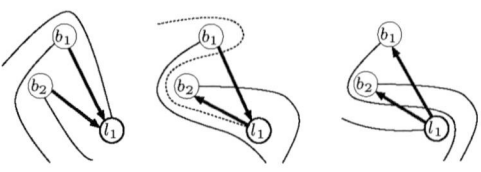

Fig. 12. An example of alternative constraints $(l_1, b_1) \oplus (b_2, l_1)$

edge from a left net to a bottom net. Without loss of generality, we assume that C includes (b_j, l_p) and (l_q, b_i), and that $R(b_i)$ is above $R(l_q)$, where $b_i, b_j \in \mathbf{B}$ ($i \leq j$) and $l_p, l_q \in \mathbf{L}$ ($p \leq q$). See Fig. 9. Since all routes are monotonic, the ball of b_j and $R(b_j)$ need to exist in region O shown in Fig. 9. Similarly, the ball of l_p and $R(l_p)$ need to exist in region I. Therefore, $R(l_p)$ is not to the right of $R(b_j)$, since $x_i < x_o$ if $y_i = y_o$ where (x_i, y_i) and (x_o, y_o) are points in region I and O, respectively. However, G_R has the edge (b_j, l_p). It contradicts definition of $E(G_R)$. So, G_R is acyclic if all routes are monotonic. □

G_R is not defined when a monotonic routing pattern is not given. However, depending on the relationship between bottom net b and left net l, there are cases such that it is decided that $R(l)$ is to the right of $R(b)$ or $R(b)$ is above $R(l)$ in any monotonic routing pattern. In such cases, (b, l) or (l, b) exist in G_R for any monotonic routing pattern. A graph where only such edges exist is the graph obtained from G_R by removing some of edges. Therefore, we consider such order graph G_n. Clearly, an orthogonal netlist is not monotonic if G_n is cyclic.

For example, if $R(b)$ and $R(l)$ are monotonic and $_bH_l$ as shown in Fig. 10, then an edge (b, l) is in G_n since $R(l)$ is always to the right of $R(b)$. Similarly, $R(b_2)$ is always above $R(l)$ if three balls of nets b_1, b_2 and l are placed as shown in Fig. 11(a). In Fig. 11(b), $R(b_1)$ and $R(b_2)$ are always above $R(l_1)$.

A necessary condition is given focusing on two or three nets. G_n can be constructed by necessary conditions for being monotonic.

Theorem 4 *An orthogonal netlist is not monotonic if the order graph G_n is cyclic, where $E(G_n) = E_f^b \cup E_f^l \cup E_h \cup E_v \cup E_1^b \cup E_1^l \cup E_2^b \cup E_2^l$,*
$E_h = \{(b, x) \mid {}_bH_x\}$,
$E_v = \{(l, x) \mid {}_lV_x\}$,
$E_1^b = \{(l, b_j) \mid {}^{b_j}S_{b_i} \wedge ({}_lH_{b_i} \vee {}_lS^{b_i}) \wedge {}^{b_j}S_l\}$,
$E_1^l = \{(b, l_j) \mid {}^{l_i}S_{l_j} \wedge ({}_bV_{l_i} \vee {}_bS^{l_i}) \wedge {}^bS_{l_j}\}$,
$E_2^b = \{(l, b_i) \mid {}_{b_j}S^{b_i} \wedge {}_lS^{b_i} \wedge ({}_{b_j}H_l \vee {}^{b_j}S_l \vee {}_lV_{b_j})\}$,
$E_2^l = \{(b, l_i) \mid {}_{l_j}S^{l_i} \wedge {}_bS^{l_i} \wedge ({}_bH_{l_j} \vee {}^bS_{l_j} \vee {}_{l_j}V_b)\}$,
where $x \in \mathbf{N}$, $b, b_i, b_j \in \mathbf{B}$, $l, l_i, l_j \in \mathbf{L}$ and $i < j$.

In addition, there are alternative constraints. When three balls of nets b_1, b_2 and l_1 are placed as shown in Fig. 12, $R(l_1)$ is non-monotonic if $R(l_1)$ is to the right of $R(b_1)$ and below $R(b_2)$. So $R(l_1)$ is either below or to

the right of $R(b_1)$ and $R(b_2)$. Therefore, either (l_1, b_1) or (b_2, l_1) should exist in G_n. These constraint is represented by $(l_1, b_1) \oplus (b_2, l_1)$.

There exists an alternative constraint $(l, b_i) \oplus (b_j, l)$ if $({}_{b_j}V_{b_i} \vee {}_{b_j}S^{b_i}) \wedge {}^{b_i}S_l \wedge {}^{b_j}S_l$, where $b_i, b_j \in \mathbf{B}$ ($i < j$) and $l \in \mathbf{L}$. Similarly, alternative constraints for two left nets and a bottom net are defined.

Assume that there is an alternative constraint $(l, b_i) \oplus (b_j, l)$. If G_n becomes cyclic when (l, b_i) is added in G_n, (b_j, l) should be selected. But if G_n does not become cyclic for either edge, then it is not easy to decide which is to be selected. When there are some alternative constraints, an orthogonal netlist is not monotonic if G_n are cyclic for all combinations of alternative constrains. This constraints should be analyzed thoroughly in our future work since the number of combinations is exponential for the number of alternative constraints.

B. An Orthogonal Routing Method

An initial order graph is constructed by necessary conditions without alternative constraints. If decision is possible for an alternative constraint, the corresponding edge is added. The order of nets is determined by G_n by removing the source one by one. The order graph is updated if removal of the source forces an alternative constraint and the decision is possible for it. In this method, the updated order graph might become cyclic depending on the selection of source.

According to the obtained order, fingers are connected to balls by monotonic routes one by one from the lower left. Formally, routing of a bottom net is defined as follows: A ball is said to be connected if its route is completed. Otherwise, a ball is said to be unconnected. Let b be a bottom net. $R(b)$ passes as the left as possible on condition that $R(b)$ passes to the right of the unconnected left net balls in the lower-left region of b and connected balls. For example, consider $R(b_3)$ in Fig. 13. $R(l_3)$ and $R(l_4)$ become non-monotonic if $R(b_3)$ passes to the left of them. Therefore, $R(b_3)$ needs to avoid balls of nets l_3 and l_4 as shown in Fig. 13(b). Similarly, routes of left nets can be decided.

V. Experiments and Results

We implemented our method for orthogonal netlists with C++ language and applied it to monotonic orthogonal netlists since it is not guaranteed that our method completes routing. Monotonic orthogonal netlists are

646

2006 Asia and South Pacific Design Automation Conference   6C-5

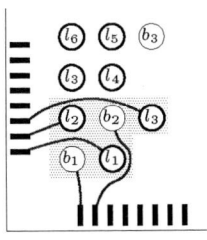 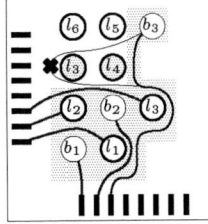

(a) Before b_3 is realized.   (b) After b_3 is realized.

Fig. 13. An example of making routes

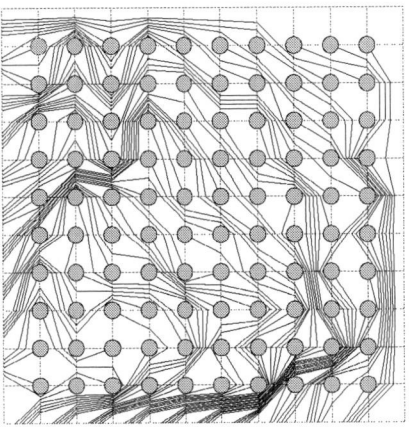

Fig. 15. An example of output (100 nets)

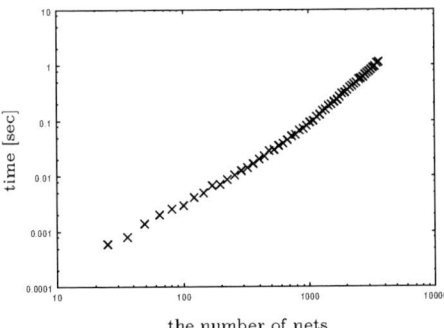

Fig. 14. The relationship between the number of nets and time for monotonic orthogonal netlist

over, the method taking non-monotonic routes into account should be proposed. In the method, the monotonic routes obtained by our proposed method is used as an initial solution and is improved iteratively to satisfy the design rule.

generated by relaxing the sufficient condition in section IV.A.1.

In this experiment, we used the order graph G, where $E(G) = E_f^b \cup E_f^l \cup E_h \cup E_v \cup E_1^b \cup E_1^l$. We applied it to problems of 56 size from 5×5 to 60×60. 100 patterns were generated in each size. Two instances in $44 \times 44, 45 \times 45$ could not be completed, since G became cyclic due to the alternative edges added in routing process. The others were completed, and monotonic routing patterns were generated. The graph between the number of nets and average execution time are shown in Fig. 14 , and an example of output is shown in Fig. 15. The algorithm generates a monotonic routing pattern within 1 second even for 3000 nets problem. The practical algorithm will be obtained if the density is taken into account in order selecting.

VI. Conclusion

We gave the necessary and sufficient condition for parallel netlist being monotonic, and proposed a routing method for monotonic parallel netlists based on this condition. Moreover we gave a necessary condition and a sufficient condition for orthogonal netlists being monotonic, and proposed a routing method for monotonic orthogonal netlists based on the necessary condition.

As our future work, we need to investigate alternative constraints and whether there are constraints between four or more nets. Routing methods that take routing density into consideration should be proposed. More-

References

[1] E. S. Kuh, T. Kashiwabara, and T. Fujisawa, "On Optimum Single-Row Routing," *IEEE Transactions on Circuits and Systems*, vol. CAS-26, no. 6, pp. 361–368, 1979.

[2] S. Tsukiyama and E. S. Kuh, "Double-Row Planar Routing and Permutation Layout," *networks*, vol. 12, no. 3, pp. 287–316, 1982.

[3] M.-F. Yu and W. W.-M. Dai, "Single-Layer Fanout Routing and Routability Analysis for Ball Grid Arrays," in *Proceedings of International Conference Computer-Aided Design*, pp. 581–586, 1995.

[4] S. Shibata, K. Ukai, N. Togawa, M. Sato, and T. Ohtsuki, "A BGA Package Routing Algorithm on Sketch Layout System," *The journal of Japan Institute for Interconnecting and Packaging Electronic Circuits*, vol. 12, no. 4, pp. 241–246, 1997. (In Japanese).

[5] Y. Kubo and A. Takahashi, "A Global Routing Method for 2-Layer Ball Grid Array Packages," in *Proceedings of ACM International Symposium on Physical Design*, pp. 36–43, April 3-6 2005.

[6] C.-C. Tsai, C.-M. Wang, and S.-J. Chen, "NEWS: A Net-Even-Wiring System for the Routing on a Multilayer PGA Package," *IEEE Transactions on Computer-Aided Design of Integrated Circuits and Systems*, vol. 17, no. 2, pp. 182–189, 1998.

[7] S.-S. Chen, J.-J. Chen, C.-C. Tsai, and S.-J. Chen, "An Even Wiring Approach to the Ball Grid Array Package Routing," in *Proceedings of International Conference on Computer Design*, pp. 303–306, 1999.

647

A Routability Constrained Scan Chain Ordering Technique for Test Power Reduction[*]

X.-L. Huang and J.-L. Huang

Graduate Institute of Electronics Engineering
Department of Electrical Engineering
National Taiwan University
Taipei 106, Taiwan

Abstract— For scan-based testing, the high test power consumption may cause test power management problems, and the extra scan chain connections may cause routability degradation during the physical design stage. In this paper, a scan chain ordering technique for test power reduction under user-specified routability constraints is presented. The proposed technique allows the user to explicitly set the routing constraints and the achievable power reduction is rather insensitive to the routing constraints. The proposed method is applied to six industrial designs. The achievable power reduction is in the range of 37–48% without violating any user-specified routing constraint.

I. INTRODUCTION

The growing IC capacity and complexity have caused several manufacturing testing challenges one of which is the elevated power consumption during testing. Without careful test power management, the excess heat dissipation during testing may immediately damage the circuit under test (CUT) or cause reliability degradation problems.

It has been shown in [13] that the power in the test mode can be as much as twice that in the functional mode. The two major causes of the high test power consumption are:

Test pattern characteristics: ATPG and LFSR generated patterns are usually less correlated than functional patterns [12] and thus cause higher logic switching activities.

DfT circuitry operation: The scan flip-flops toggle at a higher rate during the scan chain shifting operations than in the normal mode. In [7], it is shown that, using traditional MUX-based scan flip-flops, the flip-flop short-circuit power alone already accounts for about 25% of the test power. As the IC capacity keeps growing, the situation will get worse due to the growing number of scan flip-flops.

Several approaches have been proposed to reduce test power consumption, and a detailed review is available in [9]. In general, the reported methods can be divided into the following categories: (1) power-aware test pattern generation (for external ATE or BIST), (2) test pattern and/or scan chain ordering, (3) primary input control to suppress logic transitions, and (4) scan chain and/or clock scheme modification to suppress logic transitions.

In this paper, we are interested in the scan chain ordering technique because it (1) has no negative impact on the test time and fault coverage, (2) impacts less on the design flow, and (3) can be easily combined with other power reduction techniques. In the past, when area and performance are the major design concerns, scan chain ordering techniques were proposed to reduce the total or average scan chain length [1, 6, 2, 10] so that routability and timing constraints are satisfied. However, as IC complexity multiplies, test power management has become the focus of recently reported scan chain ordering techniques.

In [8], a random ordering and a simulated annealing scan chain ordering algorithms are proposed for test power reduction. However, the proposed heuristics are not capable of handling large designs. Another simple yet efficient scan cell ordering method for test power reduction is reported in [5]. This technique first determines the scan cell chaining order to minimize scan cell transitions, and then identifies the input and output scan cells.

Routability-constrained test power reduction technique is reported in [3]. Based on the power reduction heuristic in [5], [3] divides the scan flip-flops into clusters. (An improved clustering method is proposed in [4].) By clustering, the inter-cell connections are limited to be within the clusters except for the inter-cluster connections that connect the scan chain segments of adjacent clusters into a single scan chain. One limitation, however, is the loose correlation between the clustering scheme and the commonly used routing constraints.

The main contribution of this paper is a scan chain ordering technique that reduces the scan chain test power under the user specified routing constraints. Compared to [3, 4], the advantages of our technique are as follows. First, the user can explicitly specifies the routing constraints, including the maximum distance between successive scan cells, and the total scan chain length. Secondly, the negative impact of routing constraints on power reduction is much less significant. Simulation results on six industrial designs are shown. Our technique achieves 37–48% test power reduction without violating any routing constraints. Furthermore, the power reduction performance is rather insensitive to the routing constraints.

The remaining of this paper is organized as follows. In Section II, we will briefly review the background and related researches. Details of the proposed technique is depicted in Section III. Experimental results on six industrial designs are presented in Section IV. Finally, we conclude this paper in Sec-

---

[*]This work was partially supported by the National Science Council of Taiwan, R.O.C., under Grant No. NSC94-2220-E-002-006 and NSC94-2220-E-002-011.

tion V.

II. PRELIMINARIES

The goal of scan chain ordering is to find a scan cell chaining order such that the power dissipation during the scan chain shifting operations is minimized. Since only the chaining order is modified, scan chain ordering techniques cause no negative impact on the fault coverage and test time, and can be integrated into current design flow more easily. One limitation of scan chain ordering techniques is that the generated scan chain order targets a fixed set of test patterns (and output responses). However, as IP reuse becomes common practice, this problem will diminish. Under the core-based design methodology, the IP users' responsibility is to integrate the individual IP's test plan, and it is the IP designer that determines which test vectors, whether from external ATE or on-chip LFSR, to apply to achieve the specified fault coverage.

In the following, we will briefly review the background and related research on scan chain ordering.

A. Power dissipation definitions

Among the several definitions of test power consumption, e.g., total power (energy), average power, instantaneous power, peak power, the proposed technique concerns the total power and the peak power. Total power consumption is the sum of power during test application, and peak power is the highest value of power at any given instant.

B. Power dissipation of scan-based testing

Power dissipation during scan shifting can be divided into two parts: the scan cell switching activities and the induced logic switching activities. In practice, exact evaluation of the total switching activities during shift operations is time consuming. In [11], it is shown that the number of scan chain transitions and the induced logic element transitions are fairly closely correlated. Thus, the number scan chain switching activities is a good indication of the overall power consumption during scan chain shift operations.

C. Weighted scan chain transition [11]

For convenience, we will first list the notations used throughout the rest of the paper.

$c_1, c_2, \ldots, c_f$: The f scan cells in the circuit under test.

$O = (o_1, o_2, \ldots, o_f)$: A scan chain order that contains the f scan cells with the scan-in port at the first position. o_i's must be such that $o_i \in \{1, 2, \ldots, f\}$ and $o_i \neq o_j$ if $i \neq j$. For example, $O = (2, 5, 3, \cdots, 7)$ corresponds to the following scan chain configuration:

scan in $\rightarrow c_2 \rightarrow c_5 \rightarrow c_3 \cdots \rightarrow c_7 \rightarrow$ scan out

$V = (v_1, v_2, \ldots, v_f)$: An f-bit test pattern, where bit v_i is to be shifted to cell c_i for test application. For a given scan chain order $O = (o_1, o_2, \ldots, o_f)$, V is scanned in in the order $(v_{o_f}, v_{o_{f-1}}, \cdots, v_{o_2}, v_{o_1})$—$v_{o_f}$ first.

$R = (r_1, r_2, \ldots, r_f)$: An f-bit test response, where bit r_i is the test response captured at cell c_i. R is scanned out in the order $(r_{o_f}, r_{o_{f-1}}, \cdots, r_{o_2}, r_{o_1})$—$r_{o_f}$ first.

The weighted transition of a test pattern V, denoted by $WT(V)$ is defined as

$$WT(V) = \sum_{i=1}^{f-1} i \cdot \left(v_{o_i} \oplus v_{o_{i+1}} \right) \qquad (1)$$

In this equation, the $\oplus$ operator detects if a transition exists between two successive scan-in values v_{o_i} and $v_{o_{i+1}}$. (We associate the two output values, *true* and *false*, of the $\oplus$ operator with the two numerical values, 1 and 0, respectively.) It can be easily shown that a transition between v_{o_i} and $v_{o_{i+1}}$, if exists, will cause a total of i scan cell state transitions. i is thus referred to as the weight of the transition between v_{o_i} and $v_{o_{i+1}}$.

For a set of m test vectors, $V^1, V^2, \ldots, V^m$, the number of total weighted transitions is

$$WT\left(\{V^1, V^2, \ldots, V^m\}\right) = \sum_{j=1}^{m} \sum_{i=1}^{f-1} i \cdot \left(v_{o_i}^j \oplus v_{o_{i+1}}^j \right) \quad (2)$$

where $v_{o_i}^j$ is the o_i-th bit of the vector V^j. Similarly, the number of weighted transitions associated with a set of m output responses, $R^1, R^2, \ldots, R^m$, where $R^i = \left(r_1^i, r_2^i, \ldots, r_f^i \right)$, is

$$WT\left(\{R^1, R^2, \ldots, R^m\}\right) = \sum_{j=1}^{m} \sum_{i=1}^{f-1} (f-i) \cdot \left(r_{o_i}^j \oplus r_{o_{i+1}}^j \right)$$

(3)

The only difference between Eq. 2 and Eq. 3 is the transition weight assignment, which reflects the fact that one is scanned into and the other is scanned out of the chain.

Consider the example in Fig. 1(a) in which $O = (1, 2, 3, 4)$, $V = (1011)$, and $R = (0101)$. For the test vector V, there are two transitions whose weights are 1 and 2, respectively. Thus, $WT(V) = 1 + 2 = 3$. Similarly, for the output response R, there are three transitions and $WT(R) = 3 + 2 + 1 = 6$. In Fig. 1(b), the scan chain order is modified to $O = (2, 4, 3, 1)$, and the weighted transitions are reduced to $WT(V) = 1$ and $WT(R) = 2$, which corresponds to a 66% reduction.

D. Past related works

In [3], routing constrained scan chain ordering for power reduction is investigated. The reported technique consists of three steps: (1) dividing scan cells into clusters according to their locations, (2) preforming scan chain ordering within each cluster to reduce test power, and (3) connecting adjacent clusters according to a pre-determined order. Because routing constraints are enforced by clustering, this approach has the following limitations:

Degraded power reduction: Clustering substantially sacrifices the achievable power reduction because most of the scan flip-flops are connected based on the closest neighbor criteria and only a few of them are connected according to test power optimization. In [4], the clustering method is modified for more evenly distributed scan flip-flops per cluster. Although the

7A-1

2006 Asia and South Pacific Design Automation Conference

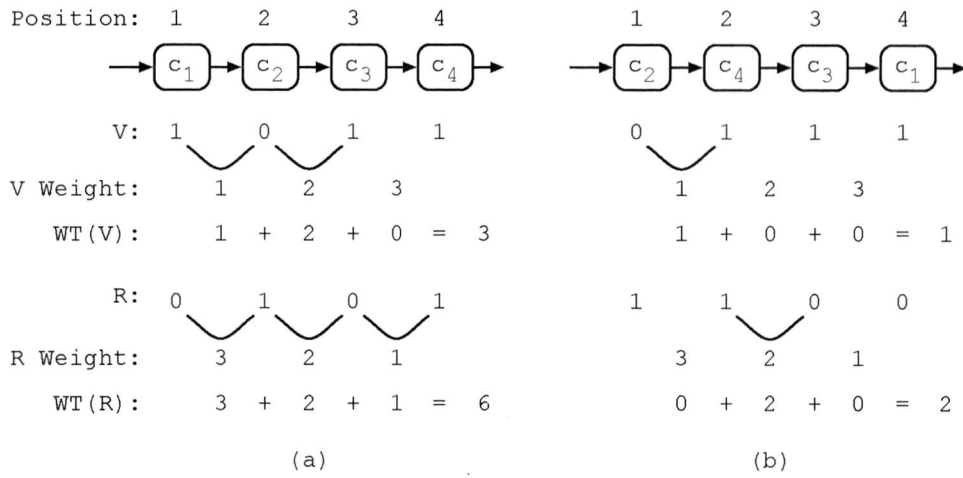

Fig. 1. Weighted transition.

achievable power reduction is improved by 3%, the above mentioned limitation still exists.

Difficulties in determining the clustering schemes: It is nontrivial to determine how the clusters should be formed to satisfy the desired routing constraints, e.g., the maximum distance between successive scan cells and the total scan chain length. The designer may have to try several different clustering schemes before the routing constraints are satisfied.

III. THE PROPOSED TECHNIQUE

In this paper, we propose a novel scan cell ordering technique that minimizes the shift operation power dissipation while satisfying the given routing constraints. The proposed algorithm is shown in Algorithm 1. Its inputs consist of

Test information: The sets of test vectors and responses.

FF information: The position (x_i, y_i) and power consumption factor p_i of each scan cell c_i. The reason to consider the individual flip-flop power consumption is to reflect the fact that the power consumption of each cell is affected by its size and load.

Routing constraints: The routing constraints include L_{max}, the maximum allowable scan chain length, and l_{max}, the maximum allowable Manhattan distance between two successive scan cells.

The output of the algorithm is the scan chain order O.

A. The algorithm flow

At the beginning of the algorithm, the routability bias factor β is set to zero and the initial order O is empty (lines 1–2). Setting β to zero causes the routing constraints to be ignored. Then, the cost graph G is constructed (line 3). In G, there is a vertex corresponding to each scan cell, and there is an undirected edge between a pair of scan cells if the Manhattan distance between the two cells is less than l_{max}. Each edge is associated with the power consumption and routability information of the cells it is connected to. In line 4, the first scan cell, i.e., the scan-in port, is selected. Choice of this first cell is

application dependent. It may be selected randomly if no constraint is given, or be such that the resulting weighted transition is further minimized (as in [5, 3, 4]). Here, we choose the one closest to the assigned scan in pin location as the first scan cell. The selected SI cell then becomes the current flip-flop, CF, and is appended to O (line 5).

From line 6 to 20 is the main ordering algorithm which terminates when all the scan cells are connected without violating any routing constraint. Each time, the next scan cell NF is selected using a greedy search heuristic (line 7). Note that the greedy heuristic may not return any next cell (line 8) if CF is not connected to any un-ordered scan cell due to the l_{max} constraint. For such cases, the algorithm will abandon the current results, adjust the routability bias factor β, empty O, and then start a new iteration (lines 9–11). After appending NF to O and making it CF, the L_{max} constraint is examined (line 15). In case of L_{max} violation, the same procedure as l_{max} violation is taken (lines 16–18).

B. Constructing the cost graph

As shown in Algorithm 1, our scan chain ordering heuristic relies on the cost graph to greedily add scan cells to the scan chain. The three steps to construct the cost graph G are:

1. For each scan cell c_i, add a vertex n_i to G.

2. For each pair of scan cells (c_i, c_j), $i \neq j$, add an edge e_{ij} between (n_i, n_j) if the Manhattan distance D_{ij} between c_i and c_j, defined as

$$D_{ij} = |x_i - x_j| + |y_i - y_j|$$

is less than l_{max}.

3. Associate with each edge e_{ij} the transition frequency T_{ij} and Manhattan distance D_{ij} between c_i and c_j. To obtain the transition frequency T_{ij}, a vector B_i is first derived. B_i consists of the test vector bits to appear and the response bits to be captured in c_i, i.e., $B_i =$

650

Test Information :
  Test vectors $V^1, V^2, \ldots, V^m$
  Test responses $R^1, R^2, \ldots, R^m$
FF Information :
  FF locations (x_i, y_i)
  FF power consumption factor p_i
Routing Constraints:
  Maximum successive scan cell distance l_{max}
  Maximum total scan chain length L_{max}

Output :
  Scan chain order O

```
1  β ← 0;
2  O ← φ;
3  G ← constructCostGraph();
4  CF ← identifySI();
5  append(CF, O);
6  while size(O) < f do
7      NF ← greedyNext(CF,O,G);
8      if NF == null then
9          β ← adjustBias();
10         O ← φ;
11         goto 4;
12     end
13     append(NF, O);
14     CF ← NF;
15     if chainLength(O) > L_max then
16         β ← adjustBias();
17         O ← φ;
18         goto 4;
19     end
20 end
21 return O;
```

Algorithm 1: The proposed algorithm

$\left(v_i^1, r_i^1, v_i^2, r_i^2, \ldots, v_i^m, r_i^m\right)$ where m is the number of test vectors. T_{ij} is then defined as

$$T_{ij} = \frac{H\left(B_i, B_j\right)}{2 \cdot m}$$

where H denotes Hamming distance.

C. Finding next scan cell

In our algorithm, the `greedyNext()` function determines the next scan cell to be appended to the scan chain. First, the scan cells that are adjacent to CF and not ordered are identified and form the candidate set. Let c_i be the current flip-flop CF. For each cell c_j in the candidate set, the cost of appending c_j to the scan chain is defined as

$$cost(i, j) = \alpha \cdot T_{ij} \cdot p_j / p_{max} + \beta \cdot D_{ij} / l_{max} \qquad (4)$$

where p_{max} is the maximum power consumption factor, α is set to 100, and β is automatically adjusted by the algorithm. Once all the costs are available, the lowest cost scan cell is selected as the next cell NF.

D. Bias adjustment

In the cost function (Eq. 4), one can adjust the two scaling factors α and β to control the bias of the scan chain ordering towards power reduction or routability. Since the goal is to optimize power consumption, in our algorithm, α is fixed at 100. The initial value of β is set to zero at the beginning and incremented by one each time any of the routing constraints is violated and thus the `adjustBias()` function is called. The underlying idea of this strategy is to ignore the routing constraints unless necessary so that we can arrive at better power reduction.

E. Extension to multiple scan chains

The proposed algorithm can also be applied to multiple scan chain designs. First, we can group the scan cells that belong to the same clock domain and/or the same sub-circuit into clusters. Then, the proposed technique is applied to each cluster for test power reduction.

IV. EXPERIMENTAL RESULTS

To validate our idea, we use a set of six industrial designs to perform various experiments. Each of the test cases comes with an initial scan chain ordering. Statistics of the designs are shown in Table I. The number of scan cells of each design is listed in column two. It ranges from 596 to 53,946. In columns three and four, the x and y spans correspond to the width and height of the smallest rectangle that encloses all the flip flops. In column five, the maximum allowable successive scan cell distance l_{max} is listed. The other routing constraint, maximum allowable scan chain length L_{max}, is listed in column six. Not shown in the table, the position and power consumption factor of each flip-flop are also provided.

TABLE I
CIRCUIT STATISTICS

design	# cell	x span	y span	l_{max}	L_{max}
1	596	643	1,252	648	147,124
2	596	990	1,188	748	130,269
3	8,755	2,810	2,338	1,146	1,745,662
4	6,389	3,781	1,261	1,536	4,158,421
5	5,994	1,670	1,774	1,016	1,571,524
6	53,946	5,670	5,774	2,693	22,638,289

The proposed technique is applied to the six designs and the experimental results are shown in Table II. In the total power column, the optimized test power and the improvement compared to the original ordering are shown. The power reduction is in the range of 37–48%. In the peak power column, the peak power of the optimized scan chain and the improvement are shown. Peak power reduction is in the range of 10–22%. In the l column, the maximum successive scan cell distances of the optimized scan chains are shown. Not only the constraint is satisfied, but also the maximum lengths are substantially reduced. In the L column, the scan chain lengths of the optimized scan chains are listed. Again, our algorithm actually

TABLE II
EXPERIMENTAL RESULTS

design	Total power		Peak power		l		L		β	CPU (sec)
	optimized	gain (%)	optimized	gain (%)	optimized	gain (%)	optimized	gain (%)		
1	16.64e06	37.99	597	11.94	643	60.98	95,341	65.77	11	2
2	15.02e06	39.57	567	10.88	740	60.99	129,553	54.34	9	2
3	3.04e09	43.69	7,201	16.37	1,060	72.06	1,398,054	89.17	12	368
4	1.59e09	44.69	5,308	15.20	1,536	67.63	1,553,632	85.05	11	229
5	1.43e09	44.08	5,041	15.14	969	65.31	967,173	75.38	12	280
6	71.33e09	48.19	40,455	22.09	2,456	77.20	12,190,595	94.21	23	16,005

significantly reduces the total length. The last two columns are the final β values and the CPU times, respectively.

To investigate the effect of routing constraints on the power reduction performance, we perform further experiments on the second design. In the first experiment, we vary l_{max} from 2,000 to 400 with L_{max} fixed at 240,000 and the results are shown in Table III. In columns two and three, the total and peak power reduction percentages are shown. No apparent degradation is observed until l_{max} is reduced to 400, a rather stringent constraint. On the other hand, columns four and five show that both l and L are reduced as l_{max} decreases.

TABLE III
IMPACT OF l_{max} CONSTRAINT ON POWER REDUCTION PERFORMANCE

l_{max}	Gain		Gain		β
	Total(%)	Peak (%)	l (%)	L (%)	
2,000	41.85	13.93	6.59	16.42	7
1,600	41.69	13.67	16.34	23.60	6
1,200	41.44	10.06	37.06	30.77	6
800	41.35	12.62	57.83	21.12	2
400	34.18	6.58	79.70	77.66	24

In the second experiment, l_{max} is fixed at 800 while L_{max} is gradually decreased from 240,000 to 80,000. The results are shown in Table IV. The same trend as Table III is observed.

TABLE IV
IMPACT OF L_{max} CONSTRAINT ON POWER REDUCTION PERFORMANCE

L_{max}	Gain		Gain		β
	Total(%)	Peak (%)	l (%)	L (%)	
240,000	41.35	12.62	57.83	21.18	2
200,000	40.91	7.29	58.14	34.32	4
160,000	39.56	10.70	58.14	54.07	9
120,000	39.00	10.55	58.14	57.82	11
80,000	36.22	6.86	60.57	71.94	28

V. CONCLUSION

In this paper, a novel routability constrained scan chain ordering technique for test power reduction is proposed. Simulation results on six industrial designs show significant power reduction. Furthermore, the algorithm is rather insensitive to routing constraints. Our future work will be to get more accurate power consumption information using commercial tools.

REFERENCES

[1] S. Barbagallo, M. L. Bodoni, D. Medina, F. Corno, P. Prinetto, and M. S. Reorda. Scan insertion criteria for low design impact. In *VLSI Test Symposium*, pages 26–31, 1996.

[2] S. Barbagallo, G. Borgonovo, D. Grassi, D. Medina, F. Corno, P. Prinetto, and M. S. Reorda. Scan chain partitioning and reordering based on layout information: an industrial experience. In *Design, Automation and Test in Europe*, 1998.

[3] Y. Bonhomme, P. Girard, L. Guiller, C. Landrault, and S. Pravossoudovitch. Efficient scan chain design for power minimization during scan testing under routing constraint. In *International Test Conference*, pages 488–493, 2003.

[4] Y. Bonhomme, P. Girard, L. Guiller, C. Landrault, S. Pravossoudovitch, and A. Virazel. Design of routing-constrained low power scan chains. In *Design, Automation and Test in Europe*, pages 16–20, 2004.

[5] Y. Bonhomme, P. Girard, C. Landrault, and S. Pravossoudovitch. Power driven chaining of flip-flops in scan architectures. In *International Test Conference*, pages 796–803, 2002.

[6] C. S. Chen, K. H. Lin, and T. T. Hwang. Layout driven selection and chaining of partial scan flip-flops. In *Design Automation Conference*, pages 262–267, 1996.

[7] M.-H. Chiu and J. C.-M. Li. Jump Scan: A DFT technique for low power testing. In *VLSI Test Symposium*, pages 277–282, 2005.

[8] V. Dabholkar, S. Chakravarty, I. Pomeranz, and S. Reddy. Techniques for reducing power dissipation during test application in full scan circuits. *IEEE Transactions on CAD*, 17(12):1325–1333, December 1998.

[9] P. Girard. Survey of low-power testing of VLSI circuits. *IEEE Design & Test of Computers*, 19(3):82–92, May–June 2002.

[10] M. Hirech, J. Beausang, and X. GU. A new approach to scan chain reordering using physical design information. In *International Test Conference*, pages 348–355, 1998.

[11] R. Sankaralingam, R. R. Oruganti, and N. A. Touba. Static compaction techniques to control scan vector power dissipation. In *VLSI Test Symposium*, pages 35–40, 2000.

[12] S. Wang and S. K. Gupta. DS-LFSR: A new BIST TPG for low heat dissipation. In *International Test Conference*, pages 848–857, 1997.

[13] Y. Zorian. A distributed BIST control scheme for complex VLSI design. In *VLSI Test Symposium*, pages 4–9, 1993.

FCSCAN: An Efficient Multiscan-based Test Compression Technique for Test Cost Reduction

Youhua Shi, Nozomu Togawa, Shinji Kimura*, Masao Yanagisawa, and Tatsuo Ohtsuki

Dept. of Computer Science
Waseda University, Japan

*Grad. School of IPS
Waseda University, Japan

e-mail: shi@yanagi.comm.waseda.ac.jp

Abstract— This paper proposes a new multiscan-based test input data compression technique by employing a Fan-out Compression Scan Architecture (FCSCAN) for test cost reduction. The basic idea of FCSCAN is to target the minority specified 1 or 0 bits (either 1 or 0) in scan slices for compression. Due to the low specified bit density in test cube set, FCSCAN can significantly reduce input test data volume and the number of required test channels so as to reduce test cost. The FCSCAN technique is easy to be implemented with small hardware overhead and does not need any special ATPG for test generation. In addition, based on the theoretical compression efficiency analysis, improved procedures are also proposed for the FCSCAN to achieve further compression. Experimental results on both benchmark circuits and one real industrial design indicate that drastic reduction in test cost can be indeed achieved.

I. INTRODUCTION

The central issue in manufacturing test has always been how to apply sufficient test data to a design to ensure that the highest quality is reached, while at the same time to minimize test cost. Nowadays, almost all of the current design-for-test (DFT) techniques start with a baseline of scan technology. However, the chip complexity continuous increasing, which results in excessive test data volume even for single-stuck-at faults with single-detection [1]. In conventional external testing, this huge amount of test data must be stored on the external automatic test equipment (ATE) and be transferred to and from the circuit-under-test (CUT) through the limited test channels. This poses a serious problem on manufacturing test because, as test data volume increases, it takes more tester buffer space to hold the complete test set, and longer to deliver the test set through limited test channels, both leading to higher test cost. Therefore, to reduce tester storage and tester channel bandwidth requirements for million-gates designs is recognized as an extremely important problem and, consequently, is receiving a lot of attention in the past five years.

In the literature numerous papers have been published on test cost reduction using test data compression, which involves employing on-chip DFT structures to reduce the amount of data stored on the ATE and thereby shorten the time it takes to load or observe the scan chains. An important class of test data compression (TDC) techniques involves using an on-chip linear decompressor to decompress test vectors. This includes techniques based on linear feedback shift register (LFSR) reseeding [2, 3] and combinational linear expansion circuits [4]. There are also some commercial tools based on LFSR reseeding combined with on-chip decompression developed recently including Mentor Graphics' TestKompress [5], SmartBIST from IBM/Cadence [6] and Synopsys' DBIST [7]. Unfortunately, although these techniques can achieve high compression ratios, for an efficient implementation, most of them need to be combined with fault simulation or interleaved with automatic test pattern generation (ATPG). Another group of compression techniques uses lossless source coding for test data reduction, such as selective Huffman coding [8], run-length coding [9], and dictionary coding [10]. However, when applied to multiscan-based designs, the effectiveness of these methods is limited either by the additional hardware overhead or or the increased test time.

In addition, there are some methods focused on reducing external test channels to achieve great compression for designs with multiple scan chains. A technique using a single input supporting multiple scan chains was proposed in [11], but its application is limited to test multiple independent full scan circuits in parallel. The Illinois scan architecture [12] overcomes this limitation by using two modes of scan operation, parallel scan and serial scan. CircularScan [13] configures the scan chains in a circular form, enabling the generation of the next pattern from the captured response. CircularScan efficiently overcomes the tester channel bandwidth limitation, however it introduces a new problem on the test diagnosis process due to the undeterministic property of the test response. The approaches proposed in [14] and [15] explored the logic dependencies of the internal scan chains to construct a simple logic gates based decompression network, so that a great number of scan chains could be driven by a limited number of external scan channels and test cost is reduced.

In this paper we propose a new DFT technique – Fan-out Compression Scan Architecture (FCSCAN), to drastically reduce test cost for multiscan-based designs. The basic idea of the proposed FCSCAN technique is to only encode the minority specified 1 or 0 bits (either 1 or 0) (to be referred as *coded bits* in this paper) in scan slices for compression. While retaining the original number of scan chains, it drastically reduces input test data volume as well as the number of required test channels for precomputed test sets of multiscan-based designs so as to reduce test cost. To be mentioned that we assume

7A-2

2006 Asia and South Pacific Design Automation Conference

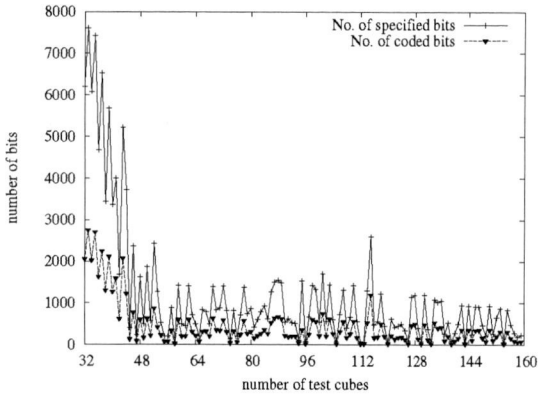

Fig. 1. Test cube profile for one industrial circuit - ASIC 1

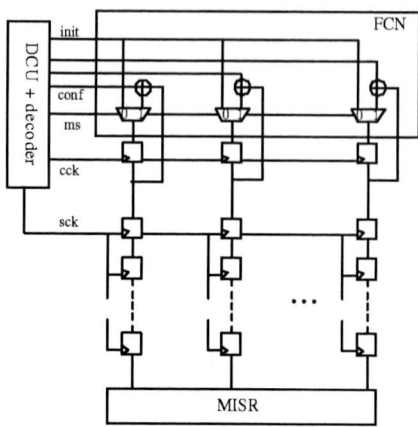

Fig. 2. The proposed fan-out compression scan (FCSCAN) architecture

the test response could be compacted using a Multiple Input Signature Register (MISR) or some other techniques [16], and we only consider the input test data reduction in this work. The proposed FCSCAN technique is very easy to implement with small hardware overhead and it does not need any special ATPG for test generation. In addition, starting with the analysis on the compression efficiency of FCSCAN, improved procedures by exploring the linear dependencies of the internal scan chains are also proposed for further compression. Unlike previous works, the computation complexity is greatly reduced with better compression. Experimental results indicate that drastic reduction in test cost can be indeed achieved using the proposed FCSCAN technique.

The rest of this paper is organized as follows: Section 2 presents the FCSCAN technique. Section 3 makes an analysis on the compression efficiency of the FCSCAN scheme and then describes the procedures for improvement. Finally experimental results and conclusions are given in Section 4 and 5, respectively.

II. FCSCAN TECHNIQUE

In practice, in an ATPG-based test set even with the application of state-of-the-art dynamic and static test pattern compaction techniques, only 1%-10% of the total test data is specified with logic values for fault detection, and the other larger part of bits are don't cares. One example is shown in Figure 1, where ASIC 1 is a real industrial circuit used in our work. Due to the large fraction of don't-cares in test set, most of the test input data compression techniques as referred above focus on the don't-cares and utilize them for test data compression.

In our work, in addition to low specified bit densities, we have further studied the care bits information and found some other useful scan test properties. Specifically, in each scan slice, the number of the minority specified 1 or 0 bits (either 1 or 0) is definitely to be less than half of the total number of specified bits. In this paper we call the specified bits with the minority specified density in the current slice as **coded bits** and the coded bit density in the current slice, P_{ci}, is equal to $min(P_i(0), P_i(1))$. Figure 1 also illustrates this observation for ASIC 1, where there are 128 scan chains and the specified bit density is 12%. It can be seen that the total number of

coded bits is much less than that of specified bits. With this important observation greater compression can be expected by loading only the coded bits information other than loading all the specified bits. Details will be illustrated as follows.

Figure 2 shows the proposed FCSCAN architecture, which enables the scan chains to be assigned proper data through one of the two loading mode: broadcast mode and configuration mode.

- Broadcast mode – all the internal scan chains are loaded the same value, which is similar to the parallel scan mode in Illinois Scan Architecture [12].

- Configuration mode – in this mode according to the values supplied from the test channels, an individual scan chain is selected to be inverted on the pre-loaded values in Broadcast mode.

The proposed FCSCAN architecture mainly consists of a decompression control unit (DCU), a flip configuration network (FCN) and a $\log_2 N$-to-N decoder. The DCU is a finite-state machine, which is responsible for controlling the decompression process, generating the mode selection signal *ms* and controlling the scan clock *sck* for the CUT. The flip configuration network is a combinational block composed of one XOR gate and one MUX for each internal scan chain. The multiplexers are controlled by the *ms* signal provided from DCU. When *ms* is 0, the FCSCAN is operated in broadcast mode, otherwise it runs in configuration mode. The ith bit of the $\log_2 N$-to-N decoder is XORed with the feedback of the first scan cell in the ith scan chain that allows for the pre-existing values to be kept or inverted in the scan cells of the associated chain.

Based on the FCSCAN architecture, a compressed form of a scan slice contains the following:
(1) an initial vector: to indicate the initial value for the scan cells (1 bit) and the number of coded bits in the current slice (M-1 bits). Since the number of internal scan chains is N_{sc} and the number of coded bits in any scan slice would be less than $0.5 * N_{sc}$, when $M = \lceil \log_2 N_{sc} \rceil$, it can guarantee a regular test application.
(2) configuration vectors: to indicate the positions of the coded bits needed by the decompressor to specify an inversion on any bits in the current slice.

654

TABLE I
FCSCAN COMPRESSION EXAMPLE

	c_1	c_2	c_3	c_4	c_5	c_6	c_7	c_8	c_9	c_{10}
s1:	0	0	0	1	0	0	X	X	X	X
s2:	1	0	X	0	0	1	1	X	X	X
s3:	1	X̄	0	1	X̄	X	X	X	X	X
s4:	0	0	1	1	1	X	1	X	0	X
s5:	X̄	0	X	1	X	X	1	1	X̄	1
s6:	0	X̄	0	X	X	0	X	X	0	X
s7:	X	1	1	X	X	X	X	X	X	X
s8:	1	X	X	0	X	1	1	X	1	X

(a)

	init vector	conf. vector	final vec
s1	0001	0100	0001000000
s2	1011	0010	
		0100	
		0101	1010011111
s3	1001	0011	1101111111
s4	1011	0001	
		0010	
		1001	0011111101
s5	1001	0010	1011111111
s6	0000	- - - -	0000000000
s7	1000	- - - -	1111111111
s8	1001	0100	1110111111

(b)

Example I - Table I (a) shows an example of test data for multiple scan chains. There are 10 scan chains ($c_1, c_2, \ldots, c_{10}$) and 8 scan slices ($s_1, s_2, \ldots, s_8$). In the proposed method, instead of using 10 external test channels, only 4 are used. Look at the second slice (s_2), in the first stage, we load an initial vector $i_2 = 1011$ to the decompressor. In this paper we assume the leftmost bit of the initial vector is the initial value for the current slice, and the other three bits are used to indicate the number of coded bits, which is 011 (i.e. there are three bits to be flipped). Thus by clocking *cck* as shown in Figure 2, the scan cells are filled with the same logic values as the initial data (i.e. the current content in the scan cells is $v = 1111111111$). In the second stage, we load the configuration vector $vc_1 = 0010$ into the decompressor, and clock *cck* again. Thus the *conf* signal is 0100000000, and an XOR operation is performed on the *conf* signal with the previously loaded scan slice v. Then the second bit position of the v is flipped while at the same time the counter is decremented. The configuration process is repeated until the counter reaches zero. Then in the next cycle, shift the current slice into the next scan cells by enabling the clock *sck*; while at the same time comes the next initial block $i_3 = 1001$. The compressed test data is shown in Table I (b). In our work the decompression logic is implemented in VHDL and synthesized using Synopsys' design compiler to access the hardware overhead of the decompressor. The synthesized circuit (DCU) for the above example only contains two flip flops and 19 combinational gates, which is very small.

III. OPTIMIZATION FOR FURTHER COMPRESSION

In this section we first make an analysis on the compression efficiency of the proposed FCSCAN technique, from which we derive that the compression efficiency is affected by two parameters, such as the number of test channels and the total number of coded bits. Consequently, an optimized solution is proposed to further the compression through reducing the two parameters.

A. Compression Analysis

In our work, we use the compression ratio γ to analyze the compression efficiency, which is defined as the ratio of the compressed data volume ($|T_E|$) to the original uncompressed data size ($|T_D|$) (i.e. $\gamma = |T_E|/|T_D|$), where γ must always be less than unity for the compression method to be effective.

Equation 1 computes the number of bits in the compressed test set ($|T_E|$) using the proposed FCSCAN technique.

$$|T_E| = \sum_{i=1}^{N_v} \sum_{j=1}^{N_{sl}} M * (1 + n_{i,j}) \qquad (1)$$

Where M is the number of external scan channels, N_v is the total number of test patterns, N_{sl} represents the maximum scan length, and $n_{i,j}$ is equal to the number of coded bits in slice $s_{i,j}$. The compression ratio γ therefore can be calculated as following.

$$
\begin{aligned}
\gamma &= \sum_{i=1}^{N_v} \sum_{j=1}^{N_{sl}} \frac{M * (1 + n_{i,j})}{N_v * N_{sl} * N_{sc}} \\
&= \frac{M}{N_{sc}} + M * P_c
\end{aligned}
\qquad (2)
$$

Where the product of $N_v * N_{sl} * N_{sc}$ amounts to the uncompressed test data volume $|T_D|$, while P_c is the coded bit density that is equal to the total number of coded bits divided by $|T_D|$.

Look at the example shown in Table I again, where $N_{sc} = 10$, $M = 4$ and for the 8 test cubes the number of coded bits is (1, 3, 1, 3, 1, 0, 0, 1) respectively. Thus the coded bit density is $10/80 = 12\%$, the total number of test data after compression is 72 bits and for this example we can achieve 10% savings in the test data volume. Since this test data example is very small and the filling rate ($P_s = 48\%$) is relatively high when compared with typical industry designs ($P_s \approx 1\% - 10\%$), greater compression could be expected for the larger circuits. For example, if we have a CUT with 1000 scan chains and the average filling rate is 10%, using the proposed FCSCAN scheme only 10 external test channels are required for a regular test application. Furthermore, in the worst case (i.e. $P_c \leq 0.5 * P_s = 5\%$) our method can still achieve 49% reduction in test input data volume no matter whether the specified bit distribution is uniform or non-uniform.

As can be seen from Equation 2, the compression efficiency of FCSCAN depends on two parameters, such as the number of test channels (M) and the total number of coded bits ($N_c = \sum_{i=1}^{N_v} \sum_{j=1}^{N_{sl}} n_{i,j}$). If the required test channels M and/or the total number of coded bits decreases, then the compression level will increase provided further reduction in test cost. Reducing M can help to reduce the number of bits for both initial vectors and configuration vectors, while reducing total number of coded bits could be used to decrease the number of configuration vectors. An improved solutions based on the FCSCAN scheme is presented in the following subsections, while keep both of these parameters intact.

7A-2

1. For a precomputed test set, descendingly sort the scan chains according to the number of specified bits $S = (c'_1, c'_2, \ldots, c'_n)$ and $SC = \emptyset$;
2. Loop until no entry left in S
   i) select the first entry c_k in S;
   ii) for each c_i in SC
   iii) if $(d(c_k, k_i) = 0$ or $d(c_k, k_i) = N_{sc})$
       // d: hamming distance
       update k_i and delete c_k from S;
   iv) else
       copy c_k to SC and delete it from S;
3. According to the logic dependencies between scan chain inputs, build the fan-out network only with inverters.

Fig. 3. Scan chain clustering algorithm

B. Minimizing Required Test Channels

The first step is to minimize the required test channels (M). This is achieved by exploiting the compatibilities among the internal scan chains for a given test set to construct a simple fan-out structure between the decompressor output and the scan chains.

In the literature, there are a number of published papers based on constructing such a simple logic fan-out network structure through compatibility analysis for test cost reduction. Our method generalizes the design techniques introduced in [17, 14] by constructing a single-level fan-out structure with inverters. This structure is chosen because it is very low cost, but still capable of achieving significant compression for the proposed FCSCAN technique. Figure 4(a) shows an example of such a network through exploiting the compatibility and inverse compatibility of the internal scan chains.

To build such a fan-out structure needs to divide the internal scan chains into scan clusters, which can be viewed as a clique-partitioning problem such as illustrated in [15, 14]. However due to the $NP-$ completeness of the clique covering problem, finding an optimal solution might require exponential time. Therefore a simple heuristic summarized in Figure 3 has been developed.

The proposed scan chain clustering (SCC) algorithm has two steps. In the first step, it descendly sorts the scan input of each internal scan chain into S according to the corresponding number of specified bits. The computation complexity of this step is $O(n^2)$, where n is the number of scan chains. In the second step, initially the scan cluster set SC is empty, and the first scan chain input c'_1 in S is moved from S to SC as the first entry (k_1). Then the remaining scan chain inputs (c'_i) are analyzed one by one according to their corresponding hamming distance with the entries in SC. If the hamming distance between c'_i and k_k is 0 or N_{sc}, i.e. c'_i and k_k are directly/inversely compatible, then remove c'_i from SC and properly fill the don't-cares in k_k; otherwise copy c'_i to SC and delete it from S. This process is repeated until the set S is empty. In addition, we also set some constraints in our algorithm because our goal is to divide the N_{sc} internal scan chains into m scan clusters, where $\lceil \log_2 m \rceil < \lceil \log_2 Nsc \rceil$, so as to reduce M. Thus when the number of entries in SC is larger than $2^{\lfloor \log_2 N_{sc} \rfloor}$, the process is stopped. To do this is very useful, because it can not only help to reduce the useless computation but also preserve the compression space. The computation complexity

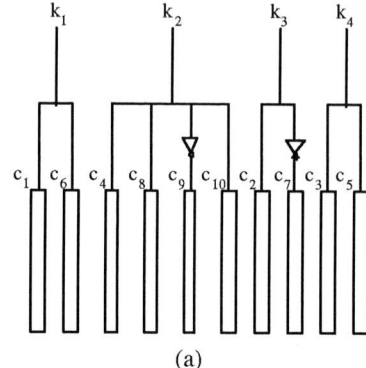

(a)

	init vec	conf. vec	scan in k_i 1 2 3 4	final vec c_i 1 2 3 4 5 6 7 8 9 10
s1	0 0 1	0 1 0	0 1 0 0	0 0 0 1 0 0 1 1 0 1
s2	0 0 1	0 0 1	1 0 0 0	1 0 0 0 0 1 1 1 0 1 0
s3	1 0 1	1 0 0	1 1 1 0	1 1 0 1 0 1 0 1 0 1
s4	0 1 0	0 1 0		
		1 0 0	0 1 0 1	0 0 1 1 1 0 1 1 0 1
s5	0 0 1	0 1 0	0 1 0 0	0 0 0 1 0 0 1 1 0 1
s6	0 0 1	0 1 0	0 1 0 0	0 0 0 1 0 0 1 1 0 1
s7	1 0 0	- - -	1 1 1 1	1 1 1 1 1 1 0 1 0 1
s8	0 0 1	0 0 1	1 0 0 0	1 0 0 0 0 1 1 0 1 0

(b)

Fig. 4. Compression example using scan chain clustering (a)single-level fan-out structure with inverters and (b) the compressed data with final test set

of this step is $O(n \log_2 n)$, where $n = N_{sc}$, therefore the total computation complexity of the proposed SCC algorithm is: $O(n^2) + O(n \log_2 n) \approx O(n^2)$. While for the previous approaches based on scan chains clustering such as [14, 15], the computation complexity is $O(n^4)$ and $O(n^3)$ respecively.

Example II - We use the same test data as in Table I. Initially, N_{sc} is 10, so we set the constraint to the number of scan clusters, when $m > 7$ the procedure is terminated. Then the SCC algorithm is applied, which divides the 10 scan chains into 4 scan clusters $k_1 = \{c_1, c_6\}$, $k_2 = \{c_4, c_8, c_9, c_{10}\}$, $k_3 = \{c_2, c_7\}$ and $k_4 = \{c_3, c_5\}$, as shown in Figure 4(a). Three scan inputs can fan out to 10 scan chains, thus M is reduced from 4 to 3. Figure 4(b) shows the corresponding compressed data with the final fully specified test vectors. In this example, the total number of test data after compression is 48 bits, and we can achieve 40% savings in the test data volume. When compared with that in Example I, 24 bits are reduced. In addition, the number of coded bit is reduced from 10 to 8. This is an additional benefit of the scan chain clustering algorithm.

C. Reducing Coded Bits

While dividing scan chains into directly/inversely compatible scan clusters provides a lot of benefits for test cost reduction, we can achieve additional compression by reducing the number of coded bits for the FCSCAN scheme. Since the basic idea of the FCSCAN is to encode the minority specified 1 or 0 bits (either 1 or 0) in scan slices for compression, we can alter and reshape the space of the scan inputs (i.e. the output space of the decompressor) by inserting some appropriate inverters.

Figure 5 shows the procedure how to reshape the scan inputs to reduce the total number of coded bits. Our heuristic seeks to

656

1. Counting the number of coded bits for each scan inputs (k_i);
2. Loop until for each scan input, the number of coded bits is less than half of the number of specified bits ($k_i < 0.5 * s_i$)
   i) select k_j with the maximum coded bits;
   ii) invert k_j;
   iii) add an inverter to the fan-out structure;
   iv) re-counting k_i for each scan inputs;
3. Optimization for reducing the number of inverters.

Fig. 5. Heuristic for coded bits reduction

$k_1k_2k_3k_4$	$k_1k_2k_3k_4$	$k_1k_2'k_3k_4$
0100	0100	0000
1000	1000	1100
11X0	1110	1000
0101	0101	0001
X10X	0100	0000
01X0	0100	0000
XX11	1111	1111
100X	1000	1101
(a)	(b)	(c)

(d)

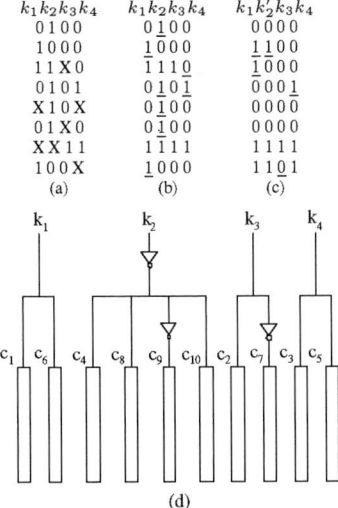

Fig. 6. Example for reducing coded bits

reduce the number of coded bits in the scan clusters extracted from Section 3.2 by adding appropriate inverters between the decompressor and the fan-out structure.

Example III - Figure 6 illustrates the example how to reduce coded bits for the previous test data after scan clustering. Figure 6 (a) shows the test data after scan chain clustering, in which there are some don'-cares left. After carefully filling the don't-cares, the number of coded bits is 7 as shown in Figure 6 (b). Since k_2 has 4 coded bits, which is larger than half of its specified bits (i.e. 7 bits). Then k_2 is inverted, and now the total number of coded bits for the whole test set is reduced to 5 when carefully fill the left don't-cares as shown in Figure 6 (c). When compared with Example II, three coded bits are reduced with only one additional inverter.

IV. EXPERIMENTAL RESULTS

In this section, we present experimental results for the five largest ISCAS'89 benchmark circuits and one real industry design (ASIC 1) to validate the effectiveness of the proposed FCSCAN technique. For each circuit, a commercial ATPG tool is used to generate test cubes with dynamic and static compaction providing 100% coverage of detectable faults. In our work, We first apply the scan clustering technique to minimize the number of required channels based on the test cube set, and then use the method proposed in Section 3.3 to reduce the number of coded bits and construct the fan-out structure using inverters. Finally the basic FCSCAN technique is applied.

Table II presents the results on the compression efficiency of the proposed FCSCAN technique for the six circuits with varying number of scan chains. In the table, the name of circuits, the internal scan cells, and the number of test vectors are shown in the first three columns, respectively. Because we focus on test cost reduction for large designs with multiple scan chains, thus we set the number of scan chains to be 50, 100 and 200 as representatives for the benchmark circuits and 100,200 and 400 for ASIC 1. The size of uncompressed test data ($|T_D|$) is shown followed by the average specified bit density (P_s) and the varying number of scan chains (N_{sc}). The compressed results for the basic FCSCAN scheme and the improved technique proposed in Section 3 are also listed, where $|T_E|$ and M are the compressed data size and the number of external test channels, respectively. In the improved scheme column, *cluster* indicates the number of scan clusters after applying the heuristic of Figure 3, and the number of added inverters is also summarized.

Because we assume that the scan chains already exist in the CUT and scan chain reordering is prohibited, the data shown for FCSCAN do not use the scan chain reordering methods. From the table, we can observe that as the number of scan chains increases, FCSCAN leads to greatly reduced input test data volume with small hardware overhead. It is notable that for the four largest benchmark circuits (s15850, s35932, s38417 and s38584), the volume of compressed test data is less than the number of specified bits in the original test set.

Table 3 shows a comparison of the results for the proposed method with CircularScan [13], 9C coding [19], Dictionary Coding with Correction (DCC) [20] and Frugal Linear network (FLN) [14], which are the representatives of the recently introduced test data compression schemes. The result of the Mintest ATPG-compacted test sets [18] is also listed for the sake of comparison. The result listed for the proposed method is the minimum size shown in boldface in Table II. Since different ATPG tools may be used, Table 3 shows both the number of vectors and the total number of compressed bits for each case. As can be seen from the table, the results obtained using the proposed FCSCAN scheme are better than those of CircularScan [13] and DCC [20] for all the circuits. When compared with 9C coding [19] and FLN [19], our method achieves higher compression for all but one of the circuits (s35932). We should note that when compared with industrial designs, the used benchmark circuits are very smaller in size and the specified bit density is relatively higher, which will limit the compression efficiency of the proposed FCSCAN technique.

Finally, it must be mentioned that although the proposed FCSCAN is very easy to implement, it really leads to drastic reduction in test cost, reducing both test data volume and the test channel requirement, which is especially effective for the designs with a great number of scan chains.

V. CONCLUSIONS

We have presented a FCSCAN technique that can reduce both input test data volume and external test channel requirement with small hardware overhead for large designs. Improved procedures have also been proposed to help the FCSCAN to further the compression. Experimental results for the

TABLE II

COMPRESSION RESULTS OF FCSCAN

Circuits	No. of FFs	No. of vectors	Size of original bits	P_s (%)	N_{sc}	FCSCAN		Improved FCSCAN				
						M	T_E	cluster	M	T_E	P_c	No. of inverters
s13207	700	251	175,700	4.36%	50	6	30,564	50	6	29,520	0.8%	14
					100	7	25,830	61	6	18,132	0.72%	21
					200	8	26,304	30	5	10,290	0.6%	46
s15850	611	148	90.428	12.40%	50	6	24,024	49	6	20,766	1.7%	11
					100	7	26,873	28	5	12,415	1.6%	39
					200	8	30,056	14	4	7,072	1.3%	48
s35932	1763	35	61,635	14.10%	50	6	22,722	50	6	17,916	2.8%	10
					100	7	23,828	58	6	13,794	2.7%	20
					200	8	27,176	27	5	8,045	2.1%	35
s38417	1664	183	304,512	13.40%	50	6	88,488	31	5	58,515	1.8%	20
					100	7	92,120	60	6	53,382	1.9%	22
					200	8	10,5752	22	5	29,550	1.4%	29
s38584	1464	288	421,632	6.01%	50	6	84,726	29	5	53,740	0.5%	17
					100	7	80,416	22	5	32,190	0.5%	24
					200	8	85,896	13	4	21,020	0.7%	47
ASIC 1	8017	246	1,972,182	12%	100	7	733,110	60	6	510,048	3.3%	21
					200	8	806,448	59	6	391,842	2.8%	35
					400	9	951,723	27	5	292,075	2.7%	58

TABLE III

COMPARISON WITH PREVIOUS WORKS

Circuits	Mintest [18]		Circular Scan [13]		9C [19]		DCC [20]		FLN [14]		Proposed	
	vectors	T_E	vectors	T_E	vectors	T_E	vectors	T_E	vectors	T_E	vectors	T_E
s13207	233	163,100	299	62,415	236	28,893	236	31,772	317	13,948	251	10,290
s15850	94	57,434	186	62,408	126	25,143	126	27,721	309	13,596	148	7,072
s35932	11	19,393	34	46,593	16	3,029	N/A	N/A	38	836	35	8,045
s38417	68	113,152	270	250,016	99	59,024	99	84,896	678	63,732	183	29,550
s38584	110	161,040	251	162,909	136	74,863	136	65,396	477	25,758	288	21,020

larger benchmark circuits and a real industrial ASIC demonstrate that FCSCAN provides significant improvement over other recent works. One notable result is that the test data volume could be reduced to less than or close to the number of specified bits in the original test set.

REFERENCES

[1] H. Furukawa F. Hsu S. Lin S. Tsai K. Abdel-hafez L. Wang, X. Wen and S. Wu. VirtualScan: A New Compressed Scan Technology for Test Cost Reduction. In *Proceedings IEEE International Test Conference (ITC)*, pages 916–925, October 2004.

[2] J. Rajski S. Hellebrand and el. Built-in test for circuits with scan based on reseeding of multiple-polynomial linear feedback shift registers. *IEEE Transactions on Computers*, 44(22):223–233, February 1995.

[3] B. Pouya A. Jas and N. A. Touba. Virtual scan chains: a means for reducing scan length in cores. In *Proceedings IEEE VLSI Test Symposium (VTS)*, pages 73–78, May 2000.

[4] I. Bayraktaroglu and A. Orailoglu. Test volume and application time reduction through scan chain concealment. In *Proceedings ACM/IEEE Design Automation Conference (DAC)*, pages 151–155, June 2001.

[5] M. Kassab J. Rajski, J. Tyszer and N. Mukherjee. Embedded Deterministic Test. *IEEE Transactions on Computer-Aided Design of Integrated Circuits and Systems*, 23(5):776–792, May 2004.

[6] B. Koenemann et al. A smartBIST variant with guaranteed encoding. In *Proceedings IEEE Asian Test Symposium (ATS)*, pages 325–330, November 2001.

[7] S.Patel P.Wohl, J.Waicukauski and M.Amin. Efficient Compression and Application of Deterministic Patterns in a Logic BIST architecture. In *Proceedings ACM/IEEE Design Automation Conference (DAC)*, pages 566–569, June 2003.

[8] M.-E. Ng A. Jas, J. Ghosh-Dastidar and N. A. Touba. An Efficient Test Vector Compression Scheme Using Selective Huffman Coding. *IEEE Transactions on Computer-Aided Design*, 22(6):797–806, November 2003.

[9] A. Chandra and K. Chakrabarty. Frequency-directed run-length (FDR) codes with application to system-on-a-chip test data compression. In *Proceedings IEEE VLSI Test Symposium (VTS)*, pages 42–47, May 2001.

[10] K. Chakrabarty L. Li and N. Touba. Test data compression using dictionaries with selective entries and fixed-length indices. *ACM Transactions on Design Automation of Electronic Systems*, 8(4):470–490, October 2003.

[11] J.-J. Chen K.-J. Lee and C.-H. Huang. Using a single input to support multiple scan chains. In *Proceedings International Conference on Computer-Aided Design (ICCAD)*, pages 74–78, November 1998.

[12] K. Butler F. Hsu and J. Patel. A case study on the implementation of the Illinois scan architecture. In *Proceedings IEEE International Test Conference (ITC)*, pages 538–547, October 2001.

[13] B. Arslan and A. Orailoglu. CircularScan A Scan Architecture for Test Cost Reduction. In *Proceedings Design, Automation, and Test in Europe (DATE)*, pages 1290–1295, March 2004.

[14] A. Orailoglu W. Rao and G. Su. Frugal Linear Network-Based Test Decompression for Drastic Test Cost Reduction. In *Proceedings International Conference on Computer-Aided Design (ICCAD)*, pages 721–725, November 2004.

[15] S. Kajihara L. Li, K. Chakrabarty and S. Swaminathan. Efficient Space/Time Compression to Reduce Test Data Volume and Testing Time for IP Cores. In *Proceedings IEEE VLSI Design (VLSID)*, pages 53–58, January 2005.

[16] S. Mitra and K. S. Kim. X-Compact: An efficient response compaction technique for test cost reduction. In *Proceedings IEEE International Test Conference (ITC)*, pages 311–320, October 2002.

[17] C. Chen and S. Gupta. A methodology to design efficient BIST test pattern generators. In *Proceedings IEEE International Test Conference (ITC)*, pages 814–823, October 1995.

[18] I. Hamzaoglu and J. H. Patel. Test set compaction algorithms for combinational circuits. In *Proceedings IEEE International Test Conference (ITC)*, pages 283–289, October 1998.

[19] M. Nourani M. Tehranipour and K. Chakrabarty. Nine-coded Compression Technique with Application to Reduced Pin-count Testing and Flexible On-chip Decompression. In *Proceedings Design, Automation, and Test in Europe (DATE)*, pages 1284–1289, March 2004.

[20] C. S. Tautermann A. Wurtenberger and S. Hellebrand. Data Compression for Multiple Scan Chains Using Dictionaries with Corrections. In *Proceedings IEEE International Test Conference (ITC)*, pages 926–935, October 2004.

2006 Asia and South Pacific Design Automation Conference

7A-3

Compaction of Pass/Fail-based Diagnostic Test Vectors for Combinational and Sequential Circuits *

Yoshinobu Higami, Kewal K. Saluja*, Hiroshi Takahashi, Shin-ya Kobayashi and Yuzo Takamatsu

Department of Computer Science, Ehime University
*Department of Electrical and Computer Engineering, University of Wisconsin-Madison

Abstract— Substantial attention is being paid to the fault diagnosis problem in recent test literature. Yet, the compaction of test vectors for fault diagnosis is little explored. The compaction of diagnostic test vectors must take care of all fault pairs that need to be distinguished by a given test vector set. Clearly, the number of fault pairs is much larger than the number of faults thus making this problem very difficult and challenging. The key contributions of this paper are: 1) to use techniques for reducing the size of fault pairs to be considered at a time, 2) to use novel variants of the fault distinguishing table method for combinational circuits and reverse order restoration method for sequential circuits, and 3) to introduce heuristics to manage the space complexity of considering all fault pairs for large circuits. Finally, the experimental results for ISCAS benchmark circuits are presented to demonstrate the effectiveness of the proposed methods.

I. INTRODUCTION

Increase in testing cost is one of the most significant problems in testing LSI circuits. This is primarily due to the fact that testing large scale circuits requires a large number of test vectors which in turn increase the test application time, tester memory space, and hence the total test cost. Numerous research papers have proposed various compaction techniques to reduce the number of test vectors and the volume of test data [4]. In particular, [8] studied the effect of detection-oriented test compaction on fault diagnosis experimentally and showed the loss of diagnosis capability of such methods. Little or no attention has been paid to reduce the number of diagnostic vectors. We believe that with respect to fault diagnosis, reducing the number of test vectors is also important. Reduction of the number of diagnostic test vectors will shorten real execution time for fault diagnosis, reduce tester memory space, and thus reduce the cost of design debug. Further, it will also allow a reduction in the size of fault dictionary, which is an important method for fault diagnosis.

Below, we briefly explain the relation between the compaction for fault detection and fault diagnosis. This is included here for the sake of completeness of this paper and can be

*The work was supported in part by JSPS under the Grant-in-Aid for Scientific Research

found in [5]. While compacting detection test vectors, it is important to find redundant vectors efficiently. An approach to find redundant vectors is to use a fault detecting table which contains information about detection of all faults by every test vector. If the fault detecting table is given, compaction of detection test vectors can be formulated as a set-cover problem.

Compaction of diagnostic test vectors can also be similarly formulated by using a fault distinguishing table, which contains information about distinguishability of all fault pairs by every test vector. Once the fault distinguishing table is obtained, the minimum number of diagnostic test vectors can be obtained by solving the set-cover problem as before. However, it is important to note that number of fault pairs is a square function of the number of faults in the circuit and hence it may be rather difficult, if not impossible, to store a complete fault distinguishing table to arrive at an optimal solution. How to reduce the size of fault pairs that need to be stored in memory is one of the very important components for compaction of diagnostic test vectors

In this paper, we propose algorithms to compact diagnostic test vectors for combinational and sequential circuits. The proposed algorithms introduce some heuristics for reducing the size of the fault distinguishing table to be considered at any time. The fault distinguishing table is constructed by considering a chosen subset of fault pairs. A nearly minimum set of test vectors that can distinguish the fault pairs is selected by solving the set-cover problem. After that, fault simulation is performed to find the fault pairs that are distinguished by the selected test set. If not all the fault pairs are distinguished, then further construction of the fault distinguishing table and selection of test vectors is performed. In the present work, we only consider the pass/fail information, that is, we do not take into account the locations of primary outputs where a fault effect is propagated. Thus our method is well suited for BIST-based fault diagnosis [2, 9]. None the less we must add that our method is general and is also applicable for diagnosis methods that contain the information on locations of primary outputs as well as extra observation points [7] to which a fault effect is propagated. Note also that this work is different from the method that deal with generation of diagnostic test sets as in [1].

Compaction of test vectors for sequential circuits is even more difficult than that for combinational circuits, just as di-

659

agnostic fault simulation and test generation for sequential circuits are more difficult than for combinational circuits. The proposed method uses a variant of reverse order restoration (ROR) technique proposed in [3] and found to be very efficient for detection-oriented test compaction for sequential circuits. When diagnostic test sequences are compacted using ROR, the list of fault pairs distinguished by the original test sequence is required. However, it is difficult to store information about all fault pairs since the number of fault pairs is much larger than the number of faults. Therefore, we also propose a method to reduce the number of target fault pairs for ROR.

The rest of the paper is organized as follows. In Section 2 we give the necessary definitions. In Section 3 a compaction algorithm for combinational circuits is explained and the experimental results for implementing our algorithm are given. In Section 4, compaction algorithms for sequential circuits are explained, and experimental results for sequential circuits are given. Finally in Section 5, conclusions are described.

II. PRELIMINARIES

The problem we consider in this paper is to find a subset of test vectors or test subsequences from a given test set or test sequence such that the new set or sequence is as small as possible while providing the same level or better diagnosis as the original test set or test sequence. We need the following definitions for developing algorithms.

[**Definition 1**]: Two faults f_1 and f_2 are said to be distinguished by a test vector v if there exists at least one test vector v such that v detects f_1 but not f_2 or that v detects f_2 but not f_1.

[**Definition 2**]: A **fault detecting table (FDETT)** for a test set T contains information about faults detected by each test vector. Similarly, a **fault distinguishing table (FDIST)** for T contains information about fault pairs distinguished by each test vector.

Note that we do not take into account the faults that are not detected by the original test set or the original test sequence. This is because a circuit with such an undetected fault can not be identified to be a faulty circuit, and thus such a fault can not be the target of fault diagnosis.

III. COMPACTION FOR COMBINATIONAL CIRCUITS

A. Basic idea

If we can construct a complete FDIST, that is, if the information about all the pairs of detected faults is available, the minimum test vectors can be selected by solving the set-cover problem. However, since the number of fault pairs is generally very large, it is difficult to store a complete FDETT. Therefore we must work with a partial FDIST. The partial FDIST includes information only about a subset of all possible candidate fault pairs. Using such a partial FDIST, a small number

of test vectors are selected, and then fault simulation is performed with the selected test vectors. If the selected test vectors do not achieve the original diagnostic resolution, another subset of fault pairs is selected and a partial FDIST is obtained to select additional test vectors. Such process is repeated until all distinguishable fault pairs are distinguished. The following theorem provides conditions for an initial choice of test vectors.

[**Theorem 1**]: Let V_0 be a set of test vectors, and Fd_1 be the set of those faults each of which is detected by only one test vector in V_0. Further, let FP be a set of fault pairs constructed from faults in Fd_1 and distinguished by $^\forall v \in V_0$. Now collect the test vectors that detect $^\forall f \in Fd_1$, and the set of the test vectors is denoted by V_1. If any two test vectors are removed from V_1 and the resulting vector set is V_1', then at least one fault pair in FP can not be distinguished by $^\forall v \in V_1'$.

(Proof:) Let v_1 and v_2 be test vectors in V_1, and f_1 and f_2 be faults detected only by v_1 and v_2, respectively. If v_1 and v_2 are removed from V_1, then fault pair $< f_1, f_2 >$ is not distinguished by the remaining vectors in V_1. Since V_1 contains only test vectors that detect $f \in Fd_1$, the above statements are true no matter which pair of test vectors are selected as v_1 and v_2.

B. Proposed algorithm

We now state an algorithm for diagnostic compaction for combinational circuits, called DCOMP-C. This algorithm makes use of the above result in selecting the initial vector set, thus keeping the FDIST to a manageable size. The steps of the algorithm are explained following the algorithm proper and an example is also given.

Algorithm: DCOMP-C
/* V_0: a given test set */
1) Set $V_s = \phi$.
2) Perform fault simulation with V_0 and obtain Fd_1, which is a set of faults detected by only one test vector in V_0.
3) Collect test vectors that detect faults in Fd_1, add the test vectors to V_s, and remove the test vectors from V_0.
4) Perform fault simulation with V_s and select n_p fault pairs among the fault pairs that are not distinguished by any $v \in V_s$. Let P be a set of the selected fault pairs.
5) Perform fault simulation with V_0 and P in order to construct an FDIST.
6) If no fault pairs in P are distinguished by any $v \in V_0$, then go to 8).
7) Select test vectors among V_0 that distinguish fault pair $p \in P$ by solving the set-cover problem, add the selected test vectors to V_s, remove them from V_0, and go to 4).
8) Obtain all the fault pairs that are not distinguished by $v \in V_s$. Let P be a set of selected fault pairs.
9) Make an FDIST with V_0 and P.
10) Select test vectors among V_0 that distinguish fault pair $p \in P$ and add them to V_s. (V_s is a resultant compacted test

set.)

In steps 2) and 3), test vectors that detect the faults detected once are collected. Theorem 1 implies that all such test vectors are necessary even in the compacted test set for diagnosis, precisely speaking, all except for one test vector are necessary. In steps 4) and 5), n_p fault pairs are selected among the fault pairs that are not distinguished by the currently obtained test set, and an FDIST is constructed, where n_p is a predetermined number. After the FDIST is constructed, approximately minimum number of test vectors can be obtained by solving the set-cover problem so that they can distinguish the selected fault pairs in step 7). This process is repeated until no fault pair among the selected fault pairs is distinguished by the remaining vectors in V_0, in step 6). Usually, set P includes faults distinguished and not distinguished by V_0. After selecting sufficiently many test vectors, P includes no fault pairs distinguished by V_0. In such a case, the process goes to step 8), where fault pairs undistinguished by V_s are all collected. By selecting test vectors that can distinguish such fault pairs except for indistinguishable ones, the obtained test set V_s can distinguish all the fault pairs distinguished by the initial V_0.

Example1: Consider a test set $V_0 = \{v_1, v_2, v_3, v_4, v_5\}$ and faults $f_1, f_2, f_3, f_4, f_5, f_6$ and f_7 that are detected by $v \in V_0$. Table I shows an FDETT for this example. (Note that DCOMP-C does not use such a table explicitly.) Consider the case where DCOMP-C is applied to V_0. In step 2), faults that are detected only by one test vector are collected, and $Fd_1 = \{f_1\}$ is obtained. In step 3), test vectors that detect the fault in Fd_1 are collected, and $V_s = \{v_1\}$ is obtained. In step 4), fault simulation is performed with v_1, and n_p fault pairs are selected among the fault pairs that are not distinguished by v_1. Now suppose that n_p is set to 2 and $P = \{< f_2, f_3 >, < f_4, f_5 >\}$ is obtained. In step 5), fault simulation is performed with $V_0 = \{v_2, v_3, v_4, v_5\}$, and an FDIST is constructed. In step 7), the set-cover problem is solved. In this case, suppose that v_2 and v_3 are selected. The process goes to step 4) again. Next suppose that $P = \{< f_3, f_4 >, < f_5, f_6 >\}$ is obtained in step 4). In step 6), since no fault pair in P is distinguished by V_0, the process goes to step 8). Fault pairs that are not distinguished by current V_s are all collected, and $P = \{< f_3, f_7 >, < f_4, f_7 >\}$ is obtained. In step 9), the FDIST is constructed, and in step 10), v_5 is selected. Finally the compacted test set $V_s = \{v_1, v_2, v_3, v_5\}$ is obtained.

C. Experimental results

We implemented DCOMP-C in C programming language and ran it on a Pentium IV 2.6GHz platform targeting IS-CAS'85 and ISCAS'89 (scan version) benchmark circuits. In these experiments, 1024 random vectors were used as a given test set. Table II shows results by DCOMP-C, where name of the circuit (circuit), fault coverage (cov), the number of fault pairs (pair), the number of undistinguished fault pairs (undis), the number of compacted test vectors (vect), the percentage

TABLE I
EXAMPLE OF AN FDETT

	v_1	v_2	v_3	v_4	v_5
f_1	d				
f_2		d	d	d	
f_3			d	d	
f_4			d	d	
f_5				d	d
f_6				d	d
f_7			d	d	d

d: detected

of removed test vectors for the number of original test vectors (%) and CPU time in seconds (cpu) are shown. The predetermined number n_p was set to 100,000 and 1,000. It is evident from these results that the compaction algorithm provides substantial test compaction without loss of diagnostic resolution. In comparison of results between $n_p = 100,000$ and $n_p = 1,000$, fewer test vectors and shorter CPU time are achieved with $n_p = 100,000$ than $n_p = 1,000$. For the circuit s38584, the test vectors were compacted to 509 test vectors, while handling over half a billion fault pairs. In general, as the table shows, the proposed algorithm can deal with circuits having several hundred million fault pairs in an efficient manner.

IV. COMPACTION FOR SEQUENTIAL CIRCUITS

A. Reverse order restoration

Test compaction for sequential circuits is substantially more difficult than that for combinational circuits. In case of sequential circuits when a test vector is deleted from a test sequence, faults detected by the original test sequence may become undetected by the compacted test sequence as well as new faults may be detected by the compacted sequence. Therefore in order to keep fault coverage and diagnostic resolution as high as the original one, fault simulation must be performed or state transition must be checked. Reverse order restoration (ROR) has been proposed [3] as an efficient detection oriented test compaction method. ROR first removes all the test vectors except for initialization vectors. Next it restores test vectors so that a subset of all faults are detected. The restoration is repeated until all the originally detected faults are detected. As this method forms the basis of our approach, we briefly explain how ROR works through the following example.

Example 2: Suppose that a test sequence T_0 consisting of seven vectors $v_1, v_2, ..., v_7$ is given. Sequence T_0 detects 3 faults f_1, f_2 and f_3, and f_1, f_2 and f_3 are detected by v_3, v_4 and v_7, respectively. Table III shows an FDETT for T_0. ROR first removes all the test vectors except for the initialization vectors. Now suppose that v_1 and v_2 are restored. Compacted test sequence T_c initially consists of v_1 v_2. Next, faults that are detected at the latest time are collected. In this example this is fault f_3 which is detected by the test vector v_7. Therefore, v_7 is restored. After that, fault simulation is performed to

7A-3   2006 Asia and South Pacific Design Automation Conference

TABLE II
RESULTS BY DCOMP-C

circuit	cov	pair	undis	$n_p = 100,000$			$n_p = 1,000$		
				vect	%	cpu(s)	vect	%	cpu(s)
c432	97.52	$1.30 * 10^5$	93	68	93.4	0.1	71	93.1	0.1
c880	97.52	$4.45 * 10^5$	104	63	93.8	0.2	70	93.2	0.3
c1355	98.57	$1.25 * 10^6$	878	88	91.4	0.8	89	91.3	1.3
c1908	94.12	$1.84 * 10^6$	1208	139	86.4	2.1	144	85.9	2.8
c2670	84.40	$3.08 * 10^6$	1838	79	92.3	2.8	79	92.3	3.9
c3540	94.49	$5.95 * 10^6$	1585	205	80.0	10.6	207	79.8	16.6
c5315	98.83	$1.57 * 10^7$	1579	188	81.6	15.4	199	80.6	25.1
c6288	99.56	$2.97 * 10^7$	4491	37	96.4	1659	38	96.3	3560
c7552	91.97	$2.76 * 10^7$	4438	198	80.7	33.8	208	79.7	110.5
s5378	94.55	$9.47 * 10^6$	1802	231	77.4	12.2	245	76.1	21.0
s9234	73.86	$1.31 * 10^7$	6798	246	76.0	36.1	261	74.5	100.6
s15850	85.05	$4.97 * 10^7$	9368	261	74.5	141.8	279	72.8	389.4
s35932	89.81	$6.16 * 10^8$	16655	91	91.1	962.2	96	90.6	2023
s38417	86.58	$3.64 * 10^8$	12480	436	57.4	1848	453	55.8	7021
s38584	90.60	$5.41 * 10^8$	31607	509	50.3	2069	537	47.6	10030

check if $T_c = v_1 \, v_2 \, v_7$ detects f_3. If it does, fault simulation is performed again in order to drop faults detected by T_c from the fault list. Otherwise, test vector v_6 is restored, and it is checked whether $T_c = v_1 \, v_2 \, v_6 \, v_7$ detects f_3. After f_3 is detected by T_c, test vectors are restored while targeting fault f_2. These steps are repeated until all the target faults are detected.

TABLE III
EXAMPLE OF AN FDETT

	v_1	v_2	v_3	v_4	v_5	v_6	v_7
f_1			d				
f_2				d			
f_3							d

d: detected

B. Basic algorithm

We now present a diagnostic compaction algorithm for sequential circuits using ROR, called DCOMP-S. Similar to the detection-oriented ROR, DCOMP-S first removes all the test vectors, except for initialization vectors. After that, it restores test vectors such that a targeted subset of fault pairs are distinguished. The restoration is repeated until all fault pairs distinguished by the original test sequence are distinguished. The steps of the algorithm are explained following the algorithm and an example is given to further clarify the steps.

Algorithm: DCOMP-S
/* T_0: a given test sequence */
/* v_i: i-th test vector in T_0 */
/* F_0: a set of faults detected by T_0 */
/* FP_0: a set of fault pairs that are constructed from $f \in F_0$ and distinguished by T_0 */
1) Set $T_c = \phi$ and $FP_{trg} = \phi$.
2) Restore test vectors from v_1 to v_{init} as initialization vectors.

3) Drop fault pairs distinguished by T_c from FP_0.
4) Find the latest time among the times when fault pairs in FP_0 are distinguished. Let the time to be t.
5) Add fault pairs that are distinguished by v_t to FP_{trg}.
6) Restore v_t to T_c.
7) Check whether fault pairs in FP_{trg} are distinguished by T_c.
8) If at least one fault pair is not distinguished, then $t = t - 1$ and go to 5).
9) Drop fault pairs distinguished by T_c from FP_0, and set $FP_{trg} = \phi$.
10) If $FP_0 \neq \phi$, then go to 5).

First, the algorithm removes all the test vectors and restores $init$ test vectors as initialization vectors, where $init$ is a predetermined number and it is usually a small number. At this time, compacted test sequence $T_c = v_1 \, v_2 \ldots v_{init}$. Fault pairs that are distinguished by T_c are dropped from FP_0 in step 3). Next the algorithm investigates the time when each fault pair is distinguished in T_0, and finds the latest time t among them in step 4). Note that when a fault pair is distinguished at more than one time, only the latest time is used and all other times are ignored. Fault pairs that are distinguished by v_t in T_0 are collected in step 5), and they are targeted for vector restoration. In step 6), test vector v_t is restored to T_c. If FP_{trg} includes only fault pairs distinguished by v_t, then v_t is concatenated at the end of T_c. Otherwise, v_t is inserted at the one time earlier position than v_{t+1} in T_c. If the current T_c does not distinguish at least one fault pair in FP_{trg}, time t is decreased in step 8) and v_t is further restored in step 6). If T_c distinguishes all the pairs in FP_{trg}, other fault pairs that are distinguished by T_c are dropped from FP_0 in step 9).

Example 3: Suppose that a test sequence T_0 consisting of 7 vectors is given, and that T_0 distinguishes 5 fault pairs fp_1, fp_2, fp_3, fp_4 and fp_5. Table IV shows an FDIST of T_0.

TABLE IV
AN FDIST OF T_0

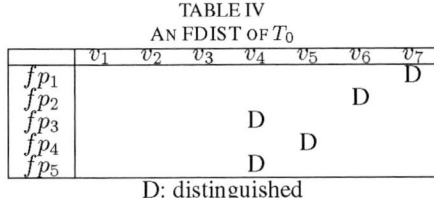

D: distinguished

Now suppose that v_1 and v_2 are first restored as initialization vectors. Fault simulation is performed to check if each fault pair is distinguished by $T_c = v_1\ v_2$. In this case no fault pair is distinguished. Next the time when each fault pair is distinguished is investigated, and it is found that the latest one t is set to 7 in step 4). In step 5) fault pairs distinguished by v_7 are collected, and $FP_{trg} = \{fp_1\}$ is obtained. Vector v_7 is restored in step 6), and it is checked whether fp_1 is distinguished by $T_c = v_1\ v_2\ v_7$. If it is not distinguished, then t is decreased and the process goes to step 5). In step 5), fault pair fp_2 is added to FP_{trg}, and in step 6), v_6 is restored. It is inserted between v_2 and v_7, thus $T_c = v_1\ v_2\ v_6\ v_7$ is obtained. In step 7), it is checked whether fp_1 and fp_2 are both distinguished by T_c. If both of them are distinguished, then it is checked whether other fault pairs fp_3, fp_4 and fp_5 are distinguished by T_c. In this case suppose that fp_4 and fp_5 are distinguished. Next the process goes to step 4), and $t = 4$ is obtained, because the remaining fault pair fp_3 is distinguished by v_4 in T_0. In step 6), v_4 is restored, and it is concatenated at the end of T_c. It is checked whether fp_3 is distinguished by T_c in step 7). If it is distinguished, then the process is terminated and finally $T_c = v_1\ v_2\ v_6\ v_7\ v_4$ is obtained.

C. Results by DCOMP-S

We implemented DCOMP-S algorithm using C programming language, and experimented with ISCAS'89 benchmark circuits on a Pentium IV 2.6GHz platform. We used test sequences generated by HITEC[6] as initially given test sequences. Table V shows the results by DCOMP-S. In the table, circuit name (circuit), the original test length (len), compacted test length (comp), the percentage of removed test vectors for the original test length (%), the total number of target fault pairs (pair), the number of fault pairs undistinguished by the original test sequences (undis) and CPU time in seconds (cpu) are given. Variable $init$ was set to 2% of the original test length. It is found that for every circuit, more than 10% test vectors could be removed from the original test sequences. In the best case, for s526, about 72% test vectors could be removed. These are all relatively small circuits. DCOMP-C can not be used for large circuits as the size of the FDIST will be too large to store. This aspect is dealt in the next subsection.

TABLE V
RESULTS BY DCOMP-S

circuit	len	comp	%	pair	undis	cpu(s)
s344	127	82	35.4	50721	367	0.3
s349	134	85	36.6	52650	402	0.3
s382	2074	846	59.2	48516	1806	16.4
s386	286	164	42.7	49141	106	0.7
s400	2214	845	61.8	61075	2343	24.9
s444	2240	956	57.3	75466	2994	21.3
s526	2258	631	72.1	64980	6617	19.4
s641	209	123	41.1	80601	129	0.7
s713	173	113	34.7	113050	383	0.8
s820	1115	745	33.2	330078	284	33.9
s832	1137	761	33.1	333336	270	42.6
s1196	435	295	32.2	766941	167	18.2
s1238	475	316	33.5	822403	191	22.9
s1423	150	134	10.7	261003	3823	4.9
s1488	1170	787	32.7	1041846	355	75.7
s1494	1245	772	38.0	1054878	411	106.2

D. Algorithm applicable for large circuits

DCOMP-S algorithm needs the information of all the fault pairs that are distinguished by the original test sequence, and hence can compact test sequences only for small circuits. For large circuits, FP_{trg}, a set of target fault pairs, can not be stored. We now propose another algorithm applicable for large circuits, called DCOMP-LS. DCOMP-LS has similar steps as DCOMP-S. It first constructs a compacted test sequence T_c by selecting S test vectors v_1 to v_S among the original test sequence T_0, where S is predetermined. Fault simulation is performed in order to collect fault pairs that are distinguished by T_0 but not distinguished by the initial T_c. The set of the collected fault pairs is referred to by FP_{trg}. By making S large, FP_{trg} can be small enough to store in a computer memory. Although DCOMP-S requires the list of all the target fault pairs, denoted by FP_0, DCOMP-LS requires only a subset of fault pairs FP_{trg}, which can be made much smaller than FP_0. Only the fault pairs in FP_{trg} are targeted during vector restoration. Next test vectors are restored from T_0 such that fault pairs in FP_{trg} are all distinguished. DCOMP-LS algorithm is described below.

Algorithm: DCOMP-LS
/* T_0: a given test sequence */
1) Set $T_c = \phi$.
2) Restore test vectors from v_1 to v_S in T_c.
3) Perform fault simulation, and collect fault pairs distinguished by T_0 but not distinguished by T_c. Let FP_{trg} be a set of the collected fault pairs.
4) Same as step 4) to 10) in DCOMP-S algorithm.

E. Results by DCOMP-LS

DCOMP-LS algorithm was implemented and experiments were performed on ISCAS'89 benchmark circuits. First we compare the results by DCOMP-LS with those by DCOMP-

7A-3

S. In this experiment, test sequences generated by HITEC[6] were also used as original test sequence, and variable S in DCOMP-LS was set to 50% of the original test length. Table VI shows the results, where column "S" and "LS" show the results by DCOMP-S and DCOMP-LS, respectively. The results by DCOMP-S are the same as in Table V. Columns "compacted vectors" show the number of compacted test vectors, column "%" shows the percentage of the number of deleted test vectors for the original test length, and columns "cpu(s)" show CPU times in seconds. Although DCOMP-LS could not achieve as short test sequences as DCOMP-S, it still deleted about 3% to 35% test vectors. CPU times by DCOMP-LS were shorter than DCOMP-S for most of circuits. For s382, s400 and s444, DCOMP-LS restored much more test vectors than DCOMP-S, and this increased the fault simulation time, and thus increased CPU time.

TABLE VI
COMPARISON BETWEEN DCOMP-S AND DCOMP-LS

circuit	compacted vectors				cpu(s)	
	S	%	LS	%	S	LS
s344	82	35.4	98	22.8	0.3	0.1
s349	85	36.6	111	17.2	0.3	0.1
s382	846	59.2	1580	23.8	16.4	18.5
s386	164	42.7	233	18.5	0.7	0.6
s400	845	61.8	1865	15.8	24.9	42.4
s444	956	57.3	1648	26.4	21.3	34.4
s526	631	72.1	1468	35.0	19.4	19.6
s641	123	41.1	177	15.3	0.7	0.4
s713	113	34.7	154	11.0	0.8	0.4
s820	745	33.2	940	15.7	33.9	19.1
s832	761	33.1	998	12.2	42.6	31.5
s1196	295	32.2	366	15.9	18.2	12.5
s1238	316	33.5	398	16.2	22.9	17.7
s1423	134	10.7	145	3.3	4.9	2.6
s1488	787	32.7	992	15.2	75.7	35.2
s1494	772	38.0	1074	13.7	106.2	47.1

TABLE VII
RESULTS BY DCOMP-LS FOR LARGE CIRCUITS

circuit	len	comp	%	pair	undis	cpu(s)
s5378	912	839	8.0	$5.064 * 10^6$	$5.912 * 10^3$	20.0
s35932	496	496	0.0	$6.090 * 10^8$	$6.641 * 10^6$	3409
s35932	546	536	1.8	$6.098 * 10^8$	$6.641 * 10^6$	2502
s35932	1024	921	10.1	$3.753 * 10^8$	$1.121 * 10^8$	3522

Table VII shows the results for s5378 and s35932. In the second and the third row, the results for HITEC test sequences are shown, in the fourth row, the results for HITEC sequence plus 50 random test vectors are shown, and in the fifth row, the results for 5 initialization vectors plus 1019 random test vectors are shown. Variable S in DCOMP-LS was set to 90% of the original test length. Each column in Table VII has the same meaning as that in Table V. For s5378 with the HITEC sequence and s35932 with initialization vectors plus

random vectors, 8.0% and 10.1% test vectors could be deleted by DCOMP-LS, respectively, but for s35932 with only HITEC sequence, no compaction was achieved. It should be noted, however, that the proposed algorithm can deal with large circuits that have more than 600 million fault pairs.

V. CONCLUSION

In this paper, we proposed compaction algorithms of diagnostic test vectors for combinational and sequential circuits. The algorithm for combinational circuits uses an FDIST table and finds a subset of test vectors that achieve the same diagnostic resolution as that for the original test set. Since for large circuits a complete FDIST can not be stored, the algorithm uses a partial FDIST and repeats selection of test vectors.

The algorithm for sequential circuits uses a ROR technique. Since the original method needs a complete list of target fault pairs, it is difficult to apply it to large sequential circuits. Therefore, we developed heuristics for use in the improved algorithm for reducing the target fault pairs, and it is applicable to large circuits.

REFERENCES

[1] D. H. Baik, Y. C. Kim, K. K. Saluja, and V. D. Agrawal, "Exclusive test and its applications to fault diagnosis," *Proc. Int. Conf. on VLSI Design*, pp. 143–148, January 2003.

[2] J. G.-Dastidar, d. Das and N. A. Touba, "Fault Diagnosis in Scanbased BIST using both Time and Space Information," in *Proc. Int. Test Conf.*, pp. 95–102, 1999.

[3] R. Guo, I. Pomeranz, and S. M. Reddy, "On speed-up vector restoration based static compaction of test sequences for sequential circuits," in *Proc. Asian Test Sympo.*, pp. 467–471, Dec. 1998.

[4] Y. Higami, S. Kajihara, H. Ichihara and Y. Takamatsu, "Test Cost Reduction for Large Circuits: Reduction of Test Data Volume and Test Application Time," in *Wiley Interscience Journal of Systems and Computers in Japan*, vol. 36, No. 6, pp. 69-83, 2005.

[5] W. H. Kautz, "Fault Testing and Diagnosis in Combinational Digital Circuits," in *IEEE Trans. on Computers*, vol. EC-15, pp 352-366, April 1968.

[6] T. M. Niermann and J. H.Patel, "HITEC: A test generation package for sequential circuits," in *Proc. European Conf. on Design Automation*, pp. 214–218, Feb. 1991.

[7] I. Pomeranz, S. Venkataraman, and S. M. Reddy, "Z-DFD: Design-for-diagnosability based on the concept of Z-detection," in *Proc. Int. Test Conf.*, pp. 489–497, 2004.

[8] Y. Shao, R. Guo, I. Pomeranz, and S. M. Reddy, "The effects of test compaction on fault diagnosis," in *Proc. Int. Test Conf.*, pp. 1083–1089, 1999.

[9] H. Takahashi, Y. Yamamoto, Y. Higami, and Y. Takamatsu, "Enhancing BIST Based Single/Multiple Stuck-at Fault Diagnosis by Ambiguous Test Set," in *Proc. Asian Test Symp.*, pp. 216–212, 2004.

2006 Asia and South Pacific Design Automation Conference

Low-Overhead Design of Soft-Error-Tolerant Scan Flip-Flops with Enhanced-Scan Capability*

Ashish Goel, Swarup Bhunia[1], Hamid Mahmoodi[2], and Kaushik Roy

Dept of ECE, Purdue University, West Lafayette, IN-47907, USA, email: <ashishg, kaushik>@ecn.purdue.edu
[1]Dept of EECS, Case Western Reserve University, Cleveland, OH-44106, USA, email: Swarup.Bhunia@case.edu
[2]School of Engineering, San Francisco State University, San Francisco, CA-94132, USA, email: mahmoodi@sfsu.edu

Abstract – With technology scaling, soft error resilience is becoming a major concern in circuit design. This paper presents a class of low-overhead flip-flops suitable for soft error detection and correction. The proposed design reuses logic elements typically available in a standard-cell implementation of a flip-flop to reduce hardware overhead. We demonstrate that the proposed flip-flops are also suitable for enhanced scan based delay fault testing, which allows arbitrary two-pattern test application for the best combinational path testability. The proposed flip-flops show an average power reduction of 16% and area improvement of 17% compared to the best alternative techniques with no additional delay overhead.

I. INTRODUCTION

Device dimensions are scaled down aggressively every technology generation. With scaling, the area per bit scales down and is about $1\mu m^2$ in a 90nm technology SRAM cell [1]. This results in reduced average node capacitance. To maintain the electrostatics of the device and prevent breakdown caused by high electric fields, the supply voltage has also been scaled down with transistor dimensions. The net effect of reduction in node capacitance and the supply voltage is that the amount of charge stored at a particular node is going down every generation. This results in increased susceptibility of latches and flip-flops to radiation induced soft errors [1].

Soft errors are caused by alpha particles or neutrons emitted by packaging materials, and cosmic rays from deep space [1]. The circuit however, is not permanently damaged by these radiations. The energetic particles create minority carriers that are collected by parasitic source and drain diodes. This collected charge causes a transient voltage pulse and is often large enough to alter the logic state of a node. If the voltage fluctuation is smaller than noise margin, then the circuit will continue to perform properly. With the reduction in node capacitance, the *same-energy particle* can create a larger voltage fluctuation at a node. Hence, the susceptibility of a circuit to soft errors increases every technology generation.

Soft error affects memory, sequential elements as well as logic. Conventionally, ECC (Error Correcting Code) is used to detect and correct any soft errors in SRAM memories [2]. However, latches and flip-flops are more difficult to protect using parity and error correcting codes [3] [4]. Several hardening techniques have been proposed to increase the soft error resilience of flip-flops and latches [3] [5]. These techniques rely on increasing the node capacitance and using redundancy to improve the tolerance to soft errors. However, these techniques come at the cost of large delay and area overhead.

In today's design, all flip-flops and latches are expected to be in a scan path so that efficient testing (allowing easy access to the internal nodes of a logic block) is possible. Hence, it is important to consider integration of any soft error tolerant flip-flop or latch into the scan path. Enhanced scan is a delay fault test method, which allows easy application of state transition and enables deterministic choice of any launching pattern in the scan flip-flops for the best possible combinational path testability [6] [7]. A recently proposed scan design [8], referred as HSSG (Hold Scan using Scan Gadget), uses a *scan gadget element* along with the system latch to implement enhanced scan based delay testing.

As mentioned above, it is becoming necessary to have soft error resilience as well as delay fault testing ability in a system flip-flop. In [9], authors have proposed a technique where they use the existing on-chip scan design-for-testability resources for soft error protection during normal operation (referred to as ISR technique in this paper).

In this paper, we propose two novel flip-flop designs that allow:
- Soft error detection/correction along with enhanced scan delay testing with lower area and power overhead compared to the existing approach (ISR).
- Arbitrary two-pattern test application (i.e. enhanced scan based delay testing) with lower design overhead compared to HSSG.

A typical standard-cell flip-flop has buffers to drive the output stage [12]. These drivers are sized up to drive a high number of fanouts. We make use of these drivers and convert them into a latch. This modification helps us in reducing the number of

---

* This work is sponsored in part by Marco Gigascale Systems Research Center(GSRC) and Semiconductor Research Corp.

0-7803-9451-8/06/$20.00 ©2006 IEEE.

latches from four (as required in HSSG and ISR) to three for implementing the scan latch and system flip-flop. The proposed flip-flops are suitable for use in muxed-scan designs. Experiments performed on ISCAS89 benchmarks show better results over the competing techniques in terms of area and power overhead.

The rest of the paper is organized as follows: Section II shows the proposed flip-flop design for soft error detection and correction. Section III illustrates the design to be used for two-pattern delay testing. Section IV presents the simulation results in terms of area, delay and power for a set of benchmark circuits. Section V concludes the paper.

II. SOFT ERROR DETECTION/CORRECTION

Soft error can result in a change in the state of a particular node in a circuit. If soft error occurs at a node of a latch, it can lead to a wrong value being stored in the latch. In a master-slave flip-flop, soft error can occur either in the master stage or in the slave stage. If soft error occurs in any one of them, it will corrupt the data stored in the flip-flop. Both master and slave stages are susceptible to soft errors during the data storing state of the clock. Master stage stores the data when the clock is '1' and slave stage stores the data when the clock is '0'.

One of the basic methodologies for soft error detection/correction is to have a redundant copy of the stored data. The redundant copy is used to determine whether a soft error has occurred or not. This is based on the assumption that only one of the copies of data is affected by soft error. Self correcting flip-flops have been designed using hardening techniques [3]. These schemes use data redundancy and feedback topology to correct the value stored in the event of a soft error. However, the hardware used for redundancy is not used for any other functions and add to the area, power and delay overhead.

Delay test schemes like HSSG have redundant scan resources that are unused during normal mode of operation. These scan resources add to the area of the chip as well as to the leakage power in normal mode of operation. In [8], the authors proposed a method, where they use these scan resources as a shadow of the system flip-flop However, the design uses four latches (two for system flip-flop and two for scan flip-flop) to store the redundant copy of the data. Therefore, it has considerable area overhead.

II. a. Soft Error Detection

Pipelines having flush capability can work fine with soft error detection only. Once a soft error is detected in a particular stage of the pipeline, the pipeline can be

stalled and started from a particular check point. In such kind of a design, soft error correction is not required. Fig. 1 shows the proposed flip-flop design having enhanced scan and soft error detection capability. To reduce the overhead involved, we convert the inverters used to drive the outputs (I_1 and I_2) into a latch. This is done to store a copy of the data stored in the flip-flop. We also combine the master stages of the scan flip-flop and the system flip-flop into one master stage. Using this strategy we need just three latches to store the two copies of the data. The two drivers I_1 & I_2 are converted into a latch using transmission gates T_2, T_3, T_4 and T_5. The multiplexer at the input of the flip-flop is controlled by the Test Control (TC) signal. The multiplexer selects the input SI (which is connected to output SO of the previous scan flip-flop) in the test mode (TC = '0') to enable the loading of the scan-chain. During normal mode of operation (TC = '1'), the multiplexer selects the input D coming from the previous stage. The transmission gate (T_6) is added in the circuit to separate latch L_2 from L_3. The transmission gates T_3 and T_4 are controlled by a signal HOLD which is generated using Test Control signal (TC) and clock. During test mode (TC = '0'), HOLD becomes '0, turning the transmission gate T_3 OFF and T_4 ON, thereby closing latch L_3. As T_3 is OFF in test mode, L_3 gets disconnected from L_1 and L_2. Latch L_3 thus acts as a hold latch during test mode. The first test vector is scanned in serially in all the flip-flops. When scan-shifting is completed for the first pattern, it is applied to the combinational circuit by making TC = '1'. After the combinational circuit stabilizes, the second pattern is scanned-in and the first pattern is held in the hold latch L_3. Next, the transition is launched by making TC

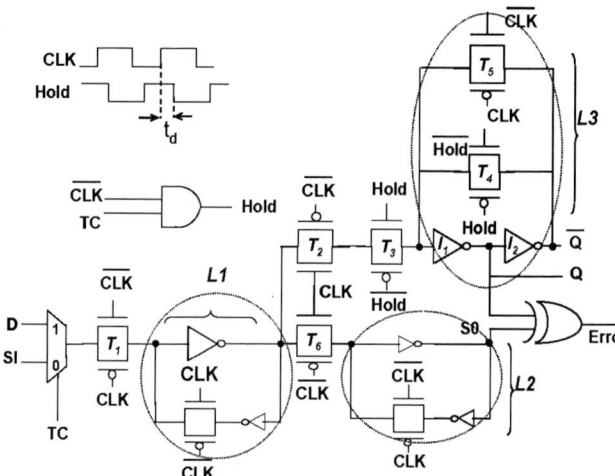

Fig. 1. Flip Flop design with Enhanced Scan Capability and Soft Error Detection (ESFF-SED)

= '1' and the results are latched after one rated clock period. During normal mode of operation (TC = '1'), HOLD is a delayed and inverted version of the clock signal. An XOR gate is used to detect the soft error by comparing the voltages at nodes Q and SO. To keep the area overhead minimum, we remove the inverter in a normal XOR gate design and use $\overline{Q}$ which is already available in the latch L_3.

We need to detect the occurrence of soft error in master and slave stage. To do this, we need to sample the data from master stage and store it in latch L_3. This sampling has to be done at the rising edge of the clock and we need to disconnect L_3 from L_1 and L_2 for the rest of the clock period. This has to be done to prevent a soft error in any of these latches from affecting the data stored in L_3. This functionality can be achieved by generating a pulse at the rising clock edge. This pulse can control the transmission gate at the input of latch L_3. However, pulse generation is usually not desired in a circuit because of the large power overhead involved. To achieve the required functionality we use the concept of implicit pulsing [10]. An implicit pulse can be generated using two transmission gates connected in series. The controlling signals of the gates are designed in such a way that both the gates are ON together only for a short period of time t_d. Thus, there is a direct path from the input to the output only during the time interval t_d. Using this method we can achieve the functionality of a pulse of duration t_d without generating the pulse signal explicitly.

In our circuit, implicit pulse is generated by the combination of transmission gates T_2 and T_3. T_2 turns ON at the rising edge of the clock and starts propagating. As HOLD is just a delayed and inverted

version of the clock, T_3 is ON at the rising edge of the clock and it turns OFF after a small time delay. This time delay is determined by the delay of the AND gate used in generating the HOLD signal. During this time period after the rising edge of the clock, both T_2 and T_3 are ON and there exists a path between L_1 and L_3. Transmission gates T_4 and T_5 are OFF during this period of time, thus opening latch L_3. At the rising edge of the clock the value stored in the master stage is sampled by L_3. After T_3 turns OFF, L_3 is disconnected from L_1. During rest of the clock cycle, either T_3 or T_2 is OFF while T_4 or T_5 is ON keeping the cross coupled inverter loop (I_1 and I_2) closed. When clk='0', T_5 is ON and T_4 is OFF and vice versa when clk = '1'. We compare our design to the ISR design with the C-element removed and an XOR gate added for detecting soft error (referred to as ISR-WC).

II. b. Soft Error Correction

The method described in [9] (ISR) inserts a C-element after the flip-flops. The C-element compares the output of the shadow latch and the system flip-flop. If the values are same it just functions as an inverter. If the two values are different it blocks the propagation of the wrong value. A keeper is used after the C-element to store the value. This design has a very good soft error tolerance as it does not allow the wrong value to propagate. However, the shadow latch has to be upsized to match the timing of the system flip-flop otherwise the delay of the flip-flop will be the delay of the slower flip-flop. This leads to a large area and power overhead.

We use the C-element and include it in our design as shown in Fig. 2. Inverters I_1 and I_2 are no longer acting as drivers because they are no longer driving the output load. Thus they are sized down to reduce the overhead. Table 1 shows the comparison between the proposed (ESFF-SEC) design and the ISR design. The proposed flip-flop has a 20% improvement in power for the same C-to-Q delay. This is due to reduction in the number of latches from four to three. The setup time also shows an improvement of 40%. This is due to upsizing of the shadow latch in ISR flip-flop. Upsizing increases the load on the data driver which results in an increased setup time.

As described in the last sub-section, latch L_3 is

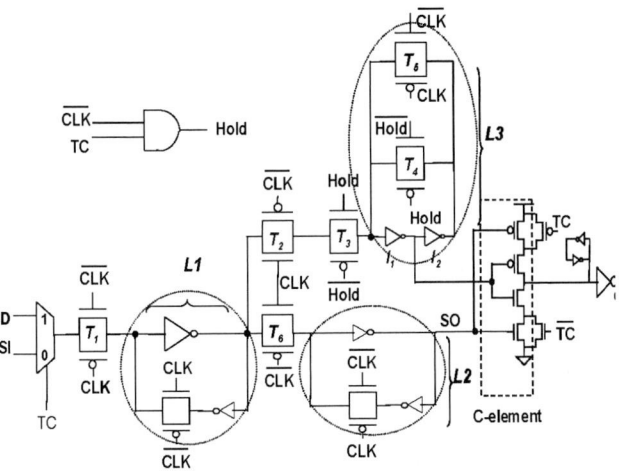

Fig. 2. Flip Flop design with Enhanced Scan Capability and Soft Error Correction (ESFF-SEC)

Table I. Normalized Power, Delay and Area comparison for flip-flops used for soft error correction

	ESFF-SEC	ISR FF
Power	1	1.26
C-to-Q delay	1	1
Setup time	1	1.7
Area	1	1.61

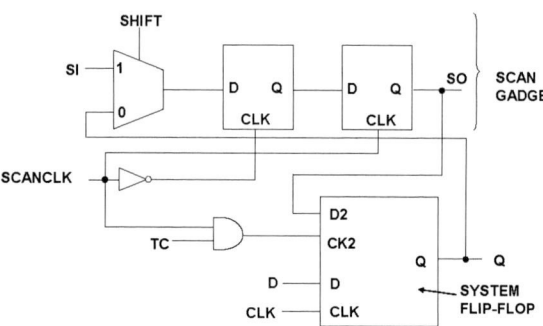

Fig. 3. Scan Gadget Scheme (HSSG)

open to the master stage during a small period of time fter the rising edge of clock. During this time, if soft error occurs in the master stage, then wrong value is written into L_2. In such a case, both the copies of data (L_1 and L_3) are wrong and the flip-flop is not able to detect the soft error. However, this time period t_d is very small as compared to the clock period. Thus the overall soft error resilience does not change much and is comparable to the ISR design.

III. ENHANCED SCAN APPROACH TO DELAY TESTING

The previous flip-flop design can provide delay fault testing capability along with soft error resilience in terms of detection and correction. However, adding the soft error resilience to all the flip-flops in the system to achieve 100 percent soft error protection is not really needed for most ground-level applications [9]. If soft error resilience is not a requirement, then the design proposed in Fig.1 and Fig. 2 can be simplified to include only delay fault testing functionality.

Enhanced scan based delay fault testing requires application of a transition at the state inputs of a combinational block by holding its output state in response to the initial pattern before applying the second pattern. HSSG uses a scan gadget along with the system latch as shown in Fig. 3 [8]. The scan chain is implemented by scan gadget element, which

provides the basic test functions (shift, load and capture). This design has an overhead of using two extra latches, adding to the leakage power and area overhead. Moreover, in this scheme the system flip-flop has a more complicated design, with two clocks and two data inputs, one for system operation and the other for loading to/from the scan chain. The extra circuitry adds to the flip-flop power in the normal mode of operation.

Fig. 4 shows the proposed master-slave flip-flop design with enhanced scan capability. The two inverters I_1 & I_2 are converted into a latch using transmission gates T_3 and T_4. Latch L_3 acts as a hold latch during test mode and drives the combinational circuit. Transmission gates T_3 and T_4 combined together, acts as the controlling circuit to enable operation of the circuit during test mode and normal mode. The operation of the flip-flop is as follows: In normal mode of operation (TC = '1'), transmission gate T_4 is OFF and T_3 is ON. The inverters I1 and I2 act as drivers and propagate the value stored in the slave latch L_2. During the test mode of operation (TC = '0'), T_3 is OFF and T_4 is ON. This converts the two inverters (I_1 and I_2) into a cross coupled loop to form the hold latch L_3. As T_3 is open, L_3 gets disconnected from latch L_1 and L_2.

The ESFF design does not require any extra timing control signals as required in a conventional enhanced scan test method. It only uses the test control signal (TC) and its complement ($\overline{TC}$), which are used in conventional scan-based testing. The proposed design does not involve the overhead due to extra latches and the complex system flip-flop used in HSSG. It is worth noting that the proposed design also maintains the power-saving advantage of enhanced scan in the test mode, since it prevents redundant switching in the combinational block by isolating it from the activity in scan register.

IV. SIMULATION RESULTS AND COMPARISON

To estimate the effectiveness of the proposed designs, we simulated a set of ISCAS89 benchmark circuits and obtained area, power, and performance overhead in case of proposed design and competing techniques. The simulations were performed using the 70nm BPTM models [11] to observe the effects in a sub-100nm scaled technology. The gate-level netlists were first technology-mapped to *LEDA* 0.25θm standard cell library using Synopsys design compiler by setting the mapping effort to medium. The library contains complex gate types such as "aoi" (and-or-invert) and "mux", and hence, the total number of logic gates is reduced from that in original benchmark. The benchmark circuits are then translated to *Hspice* netlists and scaled to 70nm. We assumed full-scan

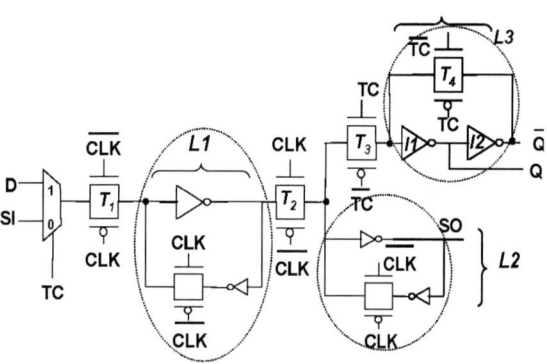

Fig. 4. Flip Flop design with Enhanced Scan Capability (ESFF)

2006 Asia and South Pacific Design Automation Conference

Table II. Comparison of Power, Delay and Area for ISR-WC and ESFF-SED (Normalized to scale of 100)

ISCAS89 Ckt	# Flip-Flops	Power			Delay			Area		
		ISR-WC	ESFF-SED	%imp. over ISR-WC	ISR-WC	ESFF-SED	%imp. over ISR-WC	ISR-WC	ESFF-SED	%imp. over ISR-WC
s298	14	0.837	0.655	21.74	23.648	23.600	0.20	0.878	0.723	17.69
s344	15	0.914	0.719	21.32	27.924	27.88	0.169	0.9541	0.787	17.45
s641	19	1.054	0.808	23.41	51.139	51.091	0.092	1.247	1.036	16.91
s838	32	1.773	1.358	23.44	52.813	52.766	0.089	2.087	1.732	17.01
s1196	18	1.563	1.3297	14.96	40.077	40.030	0.118	1.859	1.659	10.74
s1423	74	4.604	3.642	20.88	100.00	99.953	0.047	4.648	3.827	17.67
s5378	179	10.514	8.188	22.12	32.312	32.265	0.146	10.913	8.928	18.20
s9234	211	12.304	9.562	22.28	45.044	44.997	0.105	13.315	10.975	17.58
s13207	638	36.182	27.891	22.91	56.259	56.212	0.084	26.522	19.443	26.69
s15850	534	30.740	23.80	22.57	67.122	67.075	0.070	25.174	19.249	23.53
s35932	1728	100.00	77.55	22.45	22.809	22.762	0.207	100.00	80.827	19.17

implementation of the benchmarks. Power is measured in *NanoSim* by applying 100 random vectors to the inputs and delay is measured by *Hspice* simulation of the critical path of a circuit. Since the layout rules for the 70nm node are not available, the measure used for area is the total transistor active area ($W\psi * L\psi$ for a transistor).

Table II shows comparison of power, delay and area for our proposed soft error detection design (ESFF-SED) (Fig. 1) with ISR-WC. All the power, delay and area values have been normalized with respect to the maximum value among all the benchmarks respectively (where maximum value is given the value of 100). In the ESFF-SED design, on an average we observe a 21% improvement in power and 17% improvement in area over ISR-WC with no additional delay overhead. However, it is interesting to note that improvement in area over ISR-WC drops from 26.69% in benchmark s13207 down to 10.74% in benchmark s1196. This is due to the fact that sequential elements contribute to just 22% of the total area in benchmark s1196, whereas they contribute to about 70% of the total area in s13207. Similarly power improvement also drops down to 14.96% in benchmark s1196 from 22.91% in benchmark s13207. The delay is comparable in both the designs as the path for C-to-Q delay is same for both the flip-flops. This is because both the designs have the same number of gates in the C-to-Q delay path.

Table III. Comparison of Power, Delay and Area for ISR and ESFF-SEC (Normalized to scale of 100)

ISCAS89 Ckt	# Flip-Flops	Power			Delay			Area		
		ISR	ESFF-SEC	%imp. over ISR	ISR	ESFF-SEC	%imp. over ISR	ISR	ESFF-SEC	%imp. over ISR
s298	14	0.836	0.701	16.21	25.625	25.617	0.031	0.869	0.704	18.99
s344	15	0.913	0.768	15.90	29.790	29.782	0.026	0.943	0.766	18.76
s641	19	1.055	0.871	17.43	52.404	52.396	0.015	1.228	1.004	18.25
s838	32	1.775	1.465	17.45	54.034	54.027	0.015	2.058	1.680	18.34
s1196	18	1.553	1.378	11.23	41.629	41.621	0.019	1.759	1.547	12.07
s1423	74	4.597	3.880	15.59	100.00	99.993	0.007	4.603	3.731	18.97
s5378	179	10.511	8.778	16.49	34.065	34.057	0.023	10.846	8.734	19.47
s9234	211	12.302	10.259	16.73	46.468	46.460	0.032	13.181	10.691	18.89
s13207	638	36.197	30.019	16.61	57.391	57.384	0.017	27.789	20.261	27.09
s15850	534	30.743	25.572	16.82	67.974	67.966	0.012	25.872	19.571	24.35
s35932	1728	100.00	83.266	16.73	24.808	24.800	0.014	100.00	79.610	20.39

Table IV. Comparison of Power, Delay and Area for HSSG and ESFF (Normalized to scale of 100)

ISCAS89 Ckt	# Flip-Flops	Power			Delay			Area		
		HSSG	ESFF	%imp. over HSSG	HSSG	ESFF	%imp. over HSSG	HSSG	ESFF	%imp. over HSSG
s298	14	0.845	0.638	24.4	23.128	23.093	0.15	0.831	0.719	13.5
s344	15	0.928	0.707	23.8	27.433	27.398	0.13	0.904	0.784	13.3
s641	19	1.040	0.759	26.9	50.806	50.771	0.07	1.184	1.032	12.9
s838	32	1.748	1.276	27.0	52.491	52.457	0.07	1.981	1.724	13.0
s1196	18	1.728	1.462	15.4	39.669	39.634	0.09	1.816	1.671	8.0
s1423	74	4.705	3.612	23.2	100.00	99.970	0.03	4.400	3.807	13.5
s5378	179	10.563	7.919	25.0	31.851	31.816	0.11	10.306	8.870	13.9
s9234	211	12.333	9.217	25.3	44.670	44.635	0.07	12.610	10.912	13.4
s13207	638	35.948	26.525	26.2	55.961	55.926	0.06	24.275	19.154	21.1
s15850	534	30.688	22.801	25.7	66.898	66.863	0.05	22.985	18.700	18.6
s35932	1728	100.00	74.500	25.5	22.284	22.249	0.15	100.00	80.200	19.8

Table III compares the power, delay and area for the proposed soft error correction design (ESFF-SEC) (Fig. 2) with ISR. On an average, we get a 16% improvement in power and 18% improvement in area. The delay is again comparable in both the designs. The power savings goes down from 21% in ESFF-SED to 16% in ESFF-SEC. This can be attributed to the fact that with the addition of the C-element to both the designs the percentage savings in a single flip-flop in ESFF-SEC over ISR goes down. As a result, the overall savings also decrease.

Table IV compares the power, delay and area for the ESFF design and HSSG design. On an average we get a 24% improvement in power and 14% improvement in area at no additional delay overhead. The results indicate the effectiveness of the proposed flip-flops in reducing the overhead involved with addition of soft error tolerance and enhanced scan capability in real circuits.

V. CONCLUSION

In this paper, we have proposed novel flip-flop designs, which are soft error resilient and at the same time have enhanced scan based delay fault testing capability. A simplified version of the flip-flop with enhanced scan delay fault testing capability alone is also presented, which can be used in application where soft error resilience is not required. The proposed designs achieve low overhead by utilizing existing hardware resources in a typical flip-flop to realize soft error detection/correction and enhanced-scan-like delay fault testing. Compared to the existing techniques, the proposed designs have considerably less power and area overhead with no additional delay penalty.

VI. REFERENCES

[1] T. Karnik, P. Hazucha, J. Patel, "Characterization of Soft Errors Caused by Single Event Upsets in CMOS Processes," *IEEE Transactions on Dependable and Secure Computing*, Vol. 1, No. 2, April-June 2004.

[2] P. Shivakumar, M. Kistler, S.W. Keckler, D. Burger, L. Alvisi, "Modeling the effect of technology trends on the soft error rate of combinational logic," *International Conference on Dependable Systems and Networks*, 2002, pp. 389-398.

[3] P. Hazucha et al, "Measurements and Analysis of SER-Tolerant Latch in a 90-nm Dual-V_T CMOS Process", *IEEE Journal of Solid-State Circuits*, Vol. 39, No. 9, September 2004.

[4] R. Ramanarayanan, V. Degalahal, N. Vijaykrishnan, M.J. Irwin, D. Duarte, "Analysis of Soft Error Rate in Flip-Flops and Scannable Latches," *IEEE SOC Conference*, Sept. 2003, pp. 231-234.

[5] Q. Zhaou, K. Mohanram, "Cost-Effective Radiation Hardening Technique for Combinational Logic," *ICCAD, 2004*, pp. 100-106.

[6] W. Mao et al., "Reducing correlation to improve coverage of delay faults in scan-path design," *IEEE Transactions on CAD*, Vol. 13, No. 5, May 1994 pp. 638-646.

[7] M. L. Bushnell and V. D. Agrawal, *Essentials of Electronic Testing for Digital, Memory, and Mixed-Signal VLSI Circuits*, Kluwer Academic Publishers, 2000.

[8] R. Kuppuswamy et al., "Full Hold-Scan Systems in Microprocessors: Cost/Benefit Analysis," *Intel Technology Journal*, Vol.8, Issue 1, Feb. 2004.

[9] S. Mitra, N. Seifert, M. Zhang, Q. Shi, K. Kim. "Robust System Design with Built-In Soft-Error Resilience," *Computer*, vol. 38, No. 2, Feb. 2005, pp. 43-52.

[10] J. Tschanz et al, "Comparative Delay and Energy of Single Edge-Trigerred & Dual Edge-Triggered Pulsed Flip-Flops for High-Performance Microprocessors," *ISLPED*, 2001, pp. 147-152

[11] University of California, Predictive Technology Model, *http://www.device.eecs.berkeley.edu/~ptm*, 2001.

[12] Artisan Standard Cell Library for 0.13-micron TSMC process, *http://www.artisan.com/products/standard_cell.html*

A Memory Grouping Method for Sharing Memory BIST Logic

Masahide Miyazaki, Tomokazu Yoneda, and Hideo Fujiwara

Graduate School of Information Science, Nara Institute of Science and Technology (NAIST),

8916-5 Takayama, Ikoma, Nara 630-0101, Japan

Email: {masah-mi, Yoneda, fujiwara}@is.naist.jp

Abstract - With the increasing demand for SoCs to include rich functionality, SoCs are being designed with hundreds of small memories with different sizes and frequencies. If memory BIST logics were individually added to these various memories, the area overhead would be very high. To reduce the overhead, memory BIST logic must therefore be shared. This paper proposes a memory-grouping method for memory BIST logic sharing. A memory-grouping problem is formulated and an algorithm to solve the problem is proposed. Experimental results showed that the proposed method reduced the area of the memory BIST wrapper by up to 40.55%. The results also showed that the ability to select from two types of connection methods produced a greater reduction in area than using a single connection method.

I. Introduction

With the increasing number of functions being included in SoCs, many memories with different sizes and frequencies are being used. The latest SoCs contain hundreds of memories. Testing all the memories in these SoCs sequentially would take a long time. Therefore, a memory BIST design that allows two or more memories to be tested simultaneously is needed. However, due to power-consumption constraints, not all memories can be activated at the same time. To solve this problem, a scheduling technique for minimizing the test application time under power-consumption constraints is needed. Adding individual circuits for memory BISTs to lots of small memories would result in huge area overheads. To reduce these overheads, memory BIST logic must be able to be shared.

A BIST architecture, based on a single micro-programmable BIST processor and a set of memory wrappers, was proposed to simplify the testing of systems containing many distributed SRAMs of different sizes [1]. To reduce the BIST area overhead, it was proposed to share a single wrapper between a cluster of SRAMs (same type, width, and addressing space). However, in some cases, memories that have different widths or addressing spaces can be connected and share BIST logic. There can also be two or more connection methods. To achieve a satisfactory solution, the memory-connection type should be considered along with decisions on memory groups.

In this paper, we propose two types of memory-connection methods for BIST wrapper sharing. A memory-grouping problem for test circuit minimization under constraints of power consumption and test application time is also formulated together with an algorithm that solves the problem. In addition, the effectiveness of this technique is demonstrated experimentally. This paper is organized as follows. In section II, our method for memory BIST logic sharing is described. In section III, the memory-grouping problem and an algorithm to solve the problem are presented. The experimental results are shown in section IV.

II. Memory BIST Logic sharing

In this section, we describe our method of BIST logic sharing for single port and word access memory. Figure 1 shows an example of a memory BIST wrapper. The data generator generates input test sequences. The address generator generates read and write addresses and the response analyzer captures test output responses and detects faults. The by-pass FFs are not used to test memory, but are used to care the memory interface signal during a scan test. The area of the address generator, data generator, and response analyzer are almost proportional to the bit width of the address, input data, and output data, respectively. However, some of these logics can be shared by different memories wherever the number of words or the data bit width are the same; hence, the area of test circuits can be reduced. In this paper, we treat the following two memory connection methods for memory BIST logic sharing: parallel connection and serial connection. Parallel connection can be used to connect memories that have the same number of words. Figure 2 shows an example of parallel connection.

In this example, three data and address generators are reduced to one by distributing the same test data and address signals from a couple of data and address generators to (1) - (4), enabling four memories to be tested simultaneously.

0-7803-9451-8/06/$20.00 ©2006 IEEE.

7A-5

2006 Asia and South Pacific Design Automation Conference

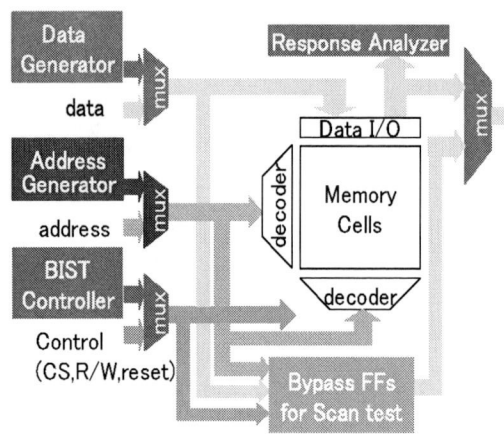

Fig.1 Memory BIST Wrapper

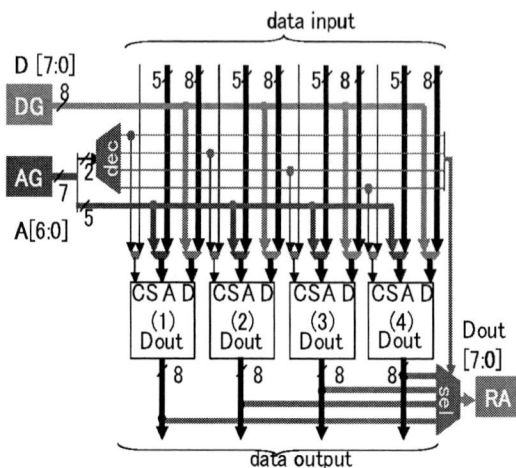

Fig.3 Serial connection of memories

III. Memory-Grouping Problem and Algorithm

A. Formulation of Memory-Grouping Problem

In this subsection, we present a memory-grouping problem. We assume that the following information for each memory mi is given:

- b_i: data bit width of m_i
- w_i: word depth of m_i
- p_i: maximum power consumption of testing m_i
- f_i: operating frequency of m_i
- x_i: X coordinate of m_i, y_i: Y coordinate of m_i

We define two types of compatibility, namely p-compatibility and s-compatibility, as follows:

Given a set of memories $V=\{m_1, m_2, ...m_n\}$, a pair of memories $m_i, m_j \in V$ is **p-compatible** if they satisfy the following conditions:

$$w_i = w_j \tag{1}$$

$$f_i = f_j \tag{2}$$

$$\sqrt{(x_i - x_j)^2 + (y_i - y_j)^2} < D \tag{3}$$

D is a constraint value that the designer decides according to the design condition.

P-compatibility is represented by a graph $G_p = (V, E_p)$, where V is a set of a memory and the edge between a pair of vertices $(m_i, m_j) \in E_p$ exists if m_i and m_j are **p-compatible**. If a set of memories can be connected in parallel, the graph induced on G_p by the memories has to be a clique.

In the same way, a pair of memories $m_i, m_j \in V$ is **s-compatible** if they satisfy the following conditions:

$$b_i = b_j \tag{4}$$

$$f_i = f_j \tag{5}$$

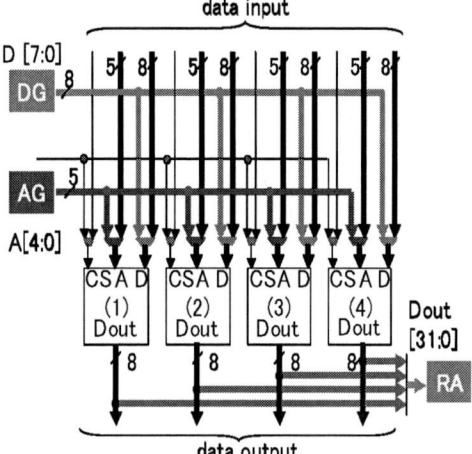

Fig.2 Parallel connection of memories

Serial connection allows memories with the same bit width to be connected. Figure 3 shows an example of four serially connected 8x32 word memories. In this example, the four memories are tested as an 8x128 word memory. The address generator generates an additional 2bit signal, and the signal is used to select the memories from (1) - (4), enabling the four memories to be tested serially. If all the memories have individual BIST logic, a 32-bit data generator and response analyzer are required, but in this example, all the memories can be tested using a shared 8bit generator and 8bit response analyzer.

Serial connection reduces the area more than parallel connection and also uses less power than parallel connection. However, the time required for serial connection testing is longer than that for parallel connection testing. To achieve the minimum area and a reasonable test application time under power consumption constraints, the type of memory connection should be considered during decisions on memory grouping. The layout design must also take into account distance constraints in relation to these connections.

672

$$\sqrt{(x_i - x_j)^2 + (y_i - y_j)^2} < D \qquad (6)$$

S-compatibility is represented by a graph $G_s = (V, E_s)$, where V is a set of memories and the edge between a pair of vertices $(m_i, m_j) \in E_s$ exists if m_i and m_j are *s-compatible*. If a set of memories can be connected serially, the graph induced on G_s by the memories has to be a clique.

To design memory BIST wrappers using these techniques for memory BIST logic sharing, we have to find a partition of V such that the memories that share the wrapper are included in the same block. Moreover, the partition $\pi = \{B_1, B_2, \dots B_k\}$ has to satisfy the following conditions:

G_{ip} is the graph induced on G_p by block B_i.

G_{is} is the graph induced on G_s by block B_i.

G_{ip} or G_{is} is a clique.

When only the graph G_{ip} (G_{is}) is a clique, the memories included in B_i are connected in parallel (serially). If G_{ip} and G_{is} are both clique, we have to select the type of connection.

For a partition π, we can calculate the area of the BIST wrapper, test application time, and power consumption of each block. The area and test application time depend on the test-pattern algorithm. In this work, these were calculated according to a published design [4] using an 8N algorithm as follows.

If the connection type of block $B_i = \{m_1, m_2, \dots m_k\}$ is a parallel connection,

$$\text{Area } S_{Bi} = 0.75\left(\log_2\left(w_{Bi}\right)\right)^2 + 2k \log_2\left(w_{Bi}\right) + 18\sum_{l=1}^{k} b_l$$
$$+ 25\log_2\left(w_{Bi}\right) + 3\max_{l}(b_i) + 66 \qquad (7)$$

$$\text{Power consumption } P_{Bi} = \sum_{l=1}^{k} p_l \qquad (8)$$

$$\text{Test application time } T_{Bi} = 8 \times w_{Bi} / f_{Bi} \qquad (9)$$

$$(f_{Bi} = f_1 = f_2 = \dots = f_k)$$

If the connection type of block $B_j = \{m_1, m_2, \dots m_k\}$ is a serial connection,

$$\text{Area } S_{Bj} = 0.75\left(\log_2\left(\sum_{l=1}^{k} w_l\right)\right)^2 + 2k \log_2\left(\sum_{l=1}^{k} w_i\right)$$
$$+ 25\log_2\left(\sum_{l=1}^{k} w_i\right) + k\left(\log_2 k\right) + 9b_{Bi}k + 14b_{Bi} + 8k + 61 \qquad (10)$$

$$(b_{Bi} = b_1 = b_2 = \dots = b_k)$$

$$\text{Power consumption } P_{Bj} = \max_{l}(p_l) \qquad (11)$$

$$\text{Test application time } T_{Bj} = 8 \times \left(\sum_{l=1}^{k} w_l\right) / f_{Bj}$$

$$\times (\text{number of background patterns}) \qquad (12)$$

The expressions for area calculation (7) and (10) do not consider the influence of timing conditions, but feedback is available from previous designs.

Parallel-connected memories are tested concurrently, and the power consumption is the sum of the power consumption of each memory. In contrast, serial-connected memories are activated one by one. Therefore, the power consumption is the maximum power consumption of the connected memories.

When a partition π is found, the area, power consumption and test application time of each block are calculated using the above expression.

The total area of the memory BIST wrappers S_{total} is calculated as the sum of S_{Bi}.

$$S_{total} = \sum_{i=1}^{k} S_{Bi} \qquad (13)$$

To control each memory BIST wrapper, at least one BIST controller must be used. In this study, the number of memory BIST wrappers was reduced by using the proposed connections. There was therefore no increase in the number of controllers. In addition, our target design includes a lot of memories so that the area of the memory BIST wrappers is predominant. Therefore the area of the BIST controllers is disregarded.

To calculate the total test application time of a memory BIST under a power-consumption constraint, we used a rectangle packing algorithm that has been described elsewhere [5]. The algorithm optimizes the test schedule of each core so that the total test application time of an SoC is minimized under maximum power constraints. The inputs of the scheduling algorithm are the maximum allowed power consumption, the test application time, and the power consumption of each core. In this study, we considered a block to be a core. Therefore, we input $\{P_{Bj}\}$ $\{T_{Bj}\}$ as the information for each core. In addition, we assumed the bit width of the inter-connect between each wrapper and control logic remained unchanged. We therefore disregarded the maximum TAM width.

To reduce the total area of memory BIST wrappers by memory BIST logic sharing, we formulated the following memory-grouping problem.

Inputs:

    a) A set of memories S and
       Information for each memory:
       $M = M_i (b_i, w_i, p_i, f_i, x_i y_i)$
       where, b_i, w_i, p_i, f_i, x_i and y_i are as follows:
       b_i: data bit width of m_i
       w_i: word depth of m_i
       p_i: maximum power consumption of testing m_i
       f_i: frequency of m_i
       x_i: X coordinate of m_i

y_i: Y coordinate of m_i

Outputs:

  a) A partition π of a given set of memories S for which all the blocks satisfy the following conditions:

    G_{ip} is the graph induced on G_p by block B_i.

    G_{is} is the graph induced on G_s by block B_i.

    G_{ip} or G_{is} is a clique.

  b) Type of connection of each block

  c) Test schedule of each memory

Constraints:

  a) Maximum distance of memory connection: D

  b) Maximum available peak power of the SoC: P

  c) Maximum test application time of memory: T

Objective:

To minimize S_{total}.

To solve this problem, an algorithm is proposed below.

B. Memory Grouping Algorithm

Fig.4 shows the pseudo code of the Memory Grouping Algorithm. Our proposed algorithm repeats division from 0-partition that only one block includes all memory to obtaining a target partition. As the algorithm divides the block, S_{total} increases. The min-cut method [2][3] is used to leave the possibility of the area reduction as much as possible. Moreover, it uses the following strategies to decide the compatibility of each block of the partition. Serial connection can reduce the area than parallel connection, and the power consumption is smaller than that of parallel connection. Therefore, it is possible that giving priority to serial connection reduces S_{total}. Based on this prospect, proposed algorithm searches for the partition that minimizes S_{total} only using s-compatibility in the first search.

First, the algorithm initializes variables. The minimum value of S_{total} is stored into S_{min}, and, in the first step, S_{min} is set to the total area of memory BIST wrapper without sharing. The partition of a set of memory S is stored into π, and the initial partition is set to 0-partition of S. (line 1-2).

Next, the algorithm creates two compatibility graphs (line 3), and select s-compatibility graph as the graph G that is used to find partition (line 4).

In order to check the compatibility of each block, the algorithm construct a set of graph C_{all} (line 6). Each graph G_i that is the member of C_{all} is induced on G by block B_i that is the member of π.

Then, for all B_i that include two or more memories, execute the following operations (line 7-21).

The minimum cut edge is calculated and delete them from G_i. By this operation, the vertex set B_i is divided into two blocks, leaving much possibility of the area reduction. If all the graph of new graph set C_{all} are clique, calculate S_{total} and test schedule of the new partition π_{tmp}. If $S_{min} > S_{total}$ and the test scheduling succeeded, π_{tmp} is stored into π_{best} as the best partition, and S_{total} is stored into S_{min} (line 8-17). If there is a graph G_i that is not a clique, or the test scheduling failed, π_{tmp} is stored into π_{next} (line 18-20).

If there is no partition that should be tried, the first search is end (line22-24). Then the algorithm stores p-compatibility graph into G, and collects the blocks that have only one memory into one block (line25-29). Then, the algorithm searches for the partition that S_{total} is minimized using p-compatibility (line5-24). In the second search, it doesn't touch the blocks in which two or more memories are included after first search. Their connection type is fixed to serial connection. The connection type of the rest is determined to be parallel connection.

This algorithm performs n(n-1) times division and scheduling in the worst case. The complexity of the scheduling algorithm and min-cut algorithm are $O(VlogV)$ and $O(V^2logV)$, respectively. Therefore the complexity of this algorithm is $O(V^3logV)$.

In this paper, we described our method for a single port and the word access memory. However, this method is applicable to other memories if the compatibility is defined about the memory type and the connecting method, and the area, power consumption, test application time can be shown by expression.

IV. Experimental Results

We carried out experiments to evaluate the proposed method. The proposed algorithm was implemented in C and the experiments were conducted on a 600-MHz Windows PC. Table 1 shows the information in each memory used in the experiment. The 2-4th columns denote the data bit width, word depth, and operating frequencies, respectively. The 5th column shows the power consumption. In this experiment, the power consumption of each memory was a relative value in which memory No. 1 was assumed to be 100 under the following assumption:

  a) The area is proportional to (number of words × number of bits).

  b) The power consumption is proportional to the area.

  c) The power consumption is proportional to the frequency.

Table1. Information on Memories

No.	# data bit width	#Words	Frequency (MHz)	Power *1	Location X	Location Y
1	16	128	133	100	10	
2	16	128	133	100	20	
3	16	128	266	200	30	
4	16	128	266	200	40	10,
5	16	256	133	200	50	20,
6	16	256	133	200	60	30,
7	16	256	133	200	70	40,
8	16	256	133	200	80	50
9	32	512	133	400	90	
10	32	512	133	400	100	

*1 Relative values in which memory No.1 is assumed to be 100

	*Procedure Memory_Grouping (**M**, P, T, D){*
1	S_{min}= the total area of memory BIST wrapper without sharing; maxedgenum=0; edgenum=0;
2	π ={B}, B={m_1, m_2, ... m_n}; $\pi_{tmp} = \phi$; $\pi_{next} = \phi$; $\pi_{best} = \phi$; $\pi_{s-compatible} = \phi$;
3	G_s = s-compatibility_graph of B; G_p = p-compatibility_graph of B;
4	$G = G_s$;
5	**loop:**
6	Construct a set of graph C_{all}={ G_i| G_i is induced graph on G by $B_i \in \pi$ }
7	for({$B_i \in (\pi - \pi_{s-compatible})$|which includes two or more memories}){
8	delete min-cut edge from G_i, make a set of graph C_{min}={G_{i1}, G_{i2}};
9	C_{all}= (C_{all}- G_i)$\cup C_{min}$;
10	Set edgenum=$\sum_j$ (the number of edges of $G_j \in C_{all}$);
11	Set a partition π_{tmp} ={B_j| vertex set of $G_j \in C_{all}$ };if all G_j are clique,calculate S_{total} of π_{tmp}
12	if(($\forall G_j \in C_{all}$, G_j is clique) $\wedge$ (S_{min}>S_{total} of π_{tmp})){
13	calculate T_{total}=Schedule(P, {P_{Bj}}, {T_{Bj}});
14	if((Schedule succeeded) $\wedge$ ($T_{total} \leq T$)){
15	$S_{min} = S_{total}$; $\pi_{best} = \pi_{tmp}$;
16	}
17	}
18	if(edgenum > maxedgenum $\wedge$ ((Schedule failed, or $T_{total} \leq T$) $\vee$ ($\exists G_j \in C_{all}$, G_j is not a clique))){
19	$\pi_{next} = \pi_{tmp}$; maxedgenum=edgenum;
20	}
21	}
22	if($\pi_{next} \neq \phi$){
23	$\pi = \pi_{next}$; $\pi_{next} = \phi$; go to **loop**;
24	}
25	else if(G=G_s){
26	$G = G_p$;
27	$\pi_{s-compatible}$ ={ $B_j \in \pi_{best}$	which includes two or more memories };
28	B_s=$\bigcup_k B_k$ ($B_k \in (\pi_{best} - \pi_{s-compatible})$);
29	π ={B_s}$\cup \pi_{s-compatible}$; go to **loop**;
30	}else{end;}

Fig.4 Memory Grouping Algorithm

The 6th and 7th columns show location. In this experiment, the number of memories was varied between 3 and 50, and the program was executed respectively. When the number of memories was N<11, we used No. 1 to N, and for the rest, we extended the same set of No. 1-10, with the Y coordinate changing between 20 to 50.

In an actual test, several background patterns (e.g. marching, checker, checker-bar) are used, but in this experiment, the test application time was calculated by assuming the number of background patterns=1. In addition, the following constraint values were used:

Maximum distance of memory connection: D=40

Maximum available peak power of the SoC:

P=5000

Maximum test application time of memory:

T=300 μs

Experiments were carried out for the following five cases: (1) Not shared (all the memories had individual BIST wrappers); (2) parallel connection (memory BIST logic was shared using only parallel connection as described in the proposed technique); (3) serial connection (memory BIST logic was shared using only serial connection as described in the proposed technique); (4) parallel and serial connection (memory BIST logic was shared using both parallel and serial connection as described in the proposed technique); and (5) exhaustive search (memory BIST logic was shared using only parallel connection after an exhaustive search). Table 2 shows the experimental results. The first column shows the number of memories and the second column shows the total area of memory BIST wrappers without sharing. Columns 3-5 shows the total area of memory BIST wrappers using the proposed techniques. The third column shows the results of using only parallel connection, while the fourth column shows the results of using only serial connection. The fifth column shows the results of using both parallel and serial connection and the sixth column shows the minimum solution obtained using an exhaustive search.

We were only able to complete an exhaustive search when the number of memories was less than 7. In these cases, the results of the exhaustive search showed that the memory BIST logic sharing technique reduced the area of the BIST wrappers by between 21.59 and 47.83% as minimum solutions. However, the technique achieved only 64.45% of the minimum solution in these cases, so there is room for improving the quality of the solution.

The average reduction ratio for parallel connection, serial connection, and parallel and serial connection were 21.08%, 37.25%, and 40.55%, respectively. In all cases, parallel and serial connection achieved the best solution. This result demonstrates that selection from two types of connection methods reduces the area more than using a single connection method.

Finally, Figure 5 shows the execution time of the implemented memory-grouping program. In all cases, the program was executed within 10 seconds using the proposed algorithm. The technique thus obtained good results within a very short CPU time so it is suitable for practical application.

Table2. Area of Memory-BIST Logic

#mem	not shared	Proposed algorithm			exhaustive
		P only	S only	S&P	
3	2289	1967	1660	1660	1660
4	2913	2591	2284	2284	2284
5	3537	2893	2279	2279	2279
6	4203	3559	3044	2722	2415
7	4869	3863	3690	3368	2540
8	5535	4529	3719	3397	
9	6201	4793	3828	3506	
10	7242	5427	3122	3122	
11	8283	6021	6447	5678	
12	8907	7539	4703	4703	N/A
13	9531	7394	5411	5089	
14	10155	7696	5406	5406	
15	10779	7998	5401	5401	
20	14484	10854	6769	6769	
30	21726	16281	10455	10455	
40	28968	22070	23784	19662	
50	36210	27497	22964	21551	

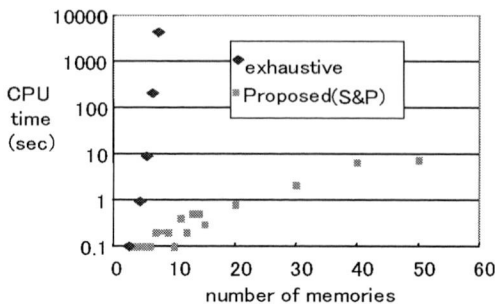

Fig.5 CPU Time of Memory Grouping program

V. Summary and Conclusions

A memory grouping problem was formulated and an algorithm to solve the problem was proposed. Experimental results showed that the proposed method reduced the area of memory BIST wrappers by up to 40.55%. It was also shown that the ability to select from two types of connection methods reduced the area more than using a single connection method.

In future work we will investigate improving the quality of the solution and minimizing the test application time.

Acknowledgements

Authors would like to thank Prof. Michiko Inoue and Prof. Satoshi Ohtake and members of Computer Design and Test Lab. (Nara Institute of Science and Technology) for their valuable discussions.

References

[1] A.Benso, S.Di Carlo, G. Di Natale and P. Prinetto, "A Programmable BIST Architecture for Clusters of Multiple-Port SRAMs," in *Proc. International Test Conf.*, pp557-566, October 2000.

[2] H. Nagamochi and T. Ibaraki, "A linear-time algorithm for finding a sparse k-connected spanning subgraph of a k-connected graph," Algorithmica, vol. 7, 1992, pp. 583--596.

[3] H. Nagamochi and T. Ibaraki, "Computing the edge-connectivity of multigraphs and capacitated graphs, " SIAM J. Discrete Mathematics, vol. 5, 1992, pp. 54--66.

[4] Charles E. Stroud, A Designer's Guide to Built-In Self-Test, Kluwer Academic Publishers, The Netherlands, 2002.

[5] V. Iyengar, K. Chakrabarty and E. J. Marinissen, "On using rectangle packaging for SOC wrapper/TAM co-optimization," in *Proc. VLSI Test Symposium*, pp. 253-258, May 2002.

[6] Y. Huang, N. Mukherjee, S. Reddy, C. Tsai, W. Cheng, O. Samman, P. Reuter, and Y. Zaidan, "Optimal Core Wrapper Width Selection and SOC Test Scheduling Based On 3-Dimensional Bin Packing Algorithm," in *Proc. International Test Conf.*, pp. 74-82, October 2002.

Equivalent circuit modeling of guard ring structures for evaluation of substrate crosstalk isolation

Daisuke Kosaka   Makoto Nagata

Department of Computer and Systems Engineering, Kobe University
1-1 Rokkodai-cho, Nada-ku, Kobe 657-8501, Japan
e-mail: {kosaka, nagata}@cs26.scitec.kobe-u.ac.jp

Abstract— A substrate-coupling equivalent circuit can be derived for an arbitrary guard ring test structure by way of F-matrix computation. The derived netlist represents a unified impedance network among multiple sites on a chip surface and allows circuit simulation for evaluation of isolation effects provided by guard rings. Geometry dependency of guard ring effects attributes to layout patterns of a test structure, including such as area of a guard ring as well as location distance from the circuit to be isolated by the guard ring. In addition, structural dependency arises from vertical impurity concentrations such as p^+, n^+, and deep n-well, which are generally available in a deep-submicron CMOS technology. The proposed simulation based prototyping technique of guard ring structures can include all these dependences and thus can be strongly helpful to establish isolation strategy against substrate coupling in a given technology, in an early stage of SoC developments.

I. INTRODUCTION

Systems-on-a-chip (SoC) integrated circuits for mobile electronics often demands reconciliation of rich functionality and low cost, or even that of high performance and low power, where trends require to establish a successful design solution to incorporate CMOS RF front-end, baseband mixed-signal signal processing units, as well as application processors in a single die. Accurate prediction and reduction of substrate crosstalk have been technology challenges of quite importance in such CMOS mixed-signal/RF SoC designs [1]. Here, the substrate crosstalk is a phenomenon where noises injected by digital circuits into a common substrate propagate toward embedded analog circuits and leak to analog signal paths, which finally interfere with analog operation and degrade analog circuit performance.

Realization of chip-level substrate coupling analysis has been successfully achieved in several ways, where parasitic coupling in a silicon substrate is modeled as a lumped resistance [2][3], calculated from resistive network of meshed substrate media [4][5][6] or from integral of Green's func-

tion [7]. It should be noted that the models extracted by these techniques are represented in the form of a circuit description, for the purpose of simulating a circuit response including the substrate coupling. On the other hand, substrate-coupling reduction techniques have also been widely discussed, which include low-noise modification applied to CMOS digital logic cells [8] or layout/device level approaches [9][10]. Obviously, chip-level simulation is helpful to minimize substrate-coupling in a design, however, it is necessary to establish the way to model each substrate-coupling reduction techniques at layout and/or at circuit levels.

As is well known, the placement of guard bands in between circuits to be isolated among each other and/or guard rings surrounding substrate-noise sensitive devices must be a baseline measure to achieve isolation against substrate coupling, since this approach does not necessitate changes of a circuit design. Therefore, how large isolation can be achieved by a guard band/guard ring in a given technology should be evaluated at the initial stage of any SoC developments. This paper discusses a methodology to derive an equivalent circuit expression of a guard ring based on substrate coupling analysis using F-matrix computation in Section 2, and demonstrates evaluation of various guard ring structures in Section 3. A brief conclusion will be given in Section 4.

II. SUBSTRATE-COUPLING EQUIVALENT CIRCUIT

A. substrate-coupling in a CMOS technology

Representative substrate coupling paths parasitic to CMOS devices in a standard p-type bulk silicon technology include resistive connection of P-P-P and capacitive-resistive combined connections of N-P-P and N-N-P, as shown in Figure 1. Here, P-P-P couples arrays of substrate contacts at different locations, while N-P-P and N-N-P correspond to a capacitive coupling at source to bulk junction of N-channel MOSFET and well to bulk junction of P-channel MOSFET, respectively.

0-7803-9451-8/06/$20.00 ©2006 IEEE.

7B-1

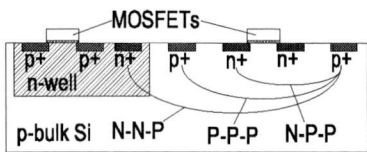

Fig. 1. Substrate coupling paths in CMOS technology.

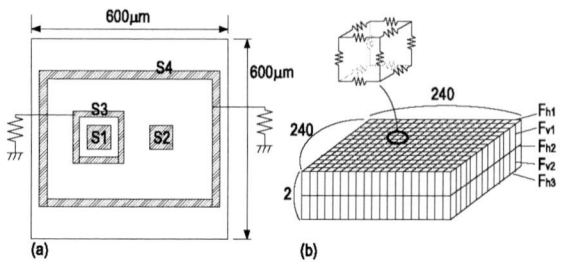

Fig. 2. (a) Test structure of substrate coupling, (b) meshing for extracting equivalent circuit.

B. F-matrix computation of substrate-coupling test structure

Figure 2(a) shows a test structure for evaluating substrate coupling, where two signal ports of S1 and S2, inner guard ring of S3 surrounding S1 port, and outer guard ring of S4 locate within the area of 600 μm $\times$ 600 μm. Here, isolation of S2 from S1 is measured as AC scattering parameter of S21. The effect of guard ring in reducing substrate coupling can be evaluated as the difference of S21 when S3 is connected to AC ground from that when S3 is floated. The outer guard ring (S4) is fixedly connected to the system ground in order to supply DC bias voltage of 0 V to the body of test structure formed on a p-type substrate. Here, a termination resistance of 50 Ω is inserted between S3 and AC ground as well as S4 and system ground, since isolation provided by guard-rings has to be effectively evaluated as is in an assembled chip.

A silicon substrate can be regarded as an equivalent resistive mesh, as long as the frequency of interest is within a few GHz. Therefore, we have applied 240 $\times$ 240 isotropic meshing to the test structure as shown in Figure 2(b), namely, mesh nodes are placed in every 2.5-μm distance in both x and y directions. In z direction, three layers with the same horizontal meshes are stacked with identical vertical separation. The resistivity of an individual mesh branch is determined from the impurity concentration at its position in a bulk.

In order to generate a compact equivalent circuit of the mesh stacks, mathematical preprocessing for network reduction such as [11] is necessary. We have applied a network reduction methodology using fundamental matrix (F-matrix) computation [12], where the test structure is dealt with as a resistive network in alternate piles of horizontal and vertical layers, as shown in Figure 2(b). Here, a multi-terminal F-matrix relates voltage V_2 and current I_2 of n output terminals to those of n input terminals V_1 and I_2 as follows:

$$\begin{bmatrix} \mathbf{V_1} \\ \mathbf{I_1} \end{bmatrix} = \mathbf{F} \begin{bmatrix} \mathbf{V_2} \\ \mathbf{I_2} \end{bmatrix}. \qquad (1)$$

Cascading F-matrices gives a synthesized system F-matrix where all of the intermediate nodes are included to a single F-matrix, which is the most attractive feature of the F-matrix computation, as shown in the following equation.

$$\begin{bmatrix} \mathbf{V_{top}} \\ \mathbf{I_{top}} \end{bmatrix} = \mathbf{Fh_1 Fv_1 Fh_2 Fv_2 Fh_3} \begin{bmatrix} \mathbf{V_{btm}} \\ \mathbf{I_{btm}} \end{bmatrix}, \qquad (2)$$

where (V_{top}, I_{top}) stand for the current and voltage of n nodes on the chip surface, (V_{btm}, I_{btm}) those on the chip bottom, $Fh_{1,2,3}$ and $Fv_{1,2}$ horizontal and vertical F-matrices within a substrate mesh, respectively. From the Kirchhof's laws, we can find $I_1 = I_2$ in determining vertical and $V_1 = V_2$ in determining horizontal F-matrices, respectively, and obtain general forms of F-matrices as follows:

$$\mathbf{Fv} = \begin{bmatrix} \mathbf{E} & \mathbf{A} \\ \mathbf{0} & \mathbf{E} \end{bmatrix}, \mathbf{Fh} = \begin{bmatrix} \mathbf{E} & \mathbf{0} \\ \mathbf{B} & \mathbf{E} \end{bmatrix}. \qquad (3)$$

Here, A, B are sub-matrices representing vertical resistive elements determining vertical voltage differences and horizontal resistive elements determining horizontal currents induced from voltage differences from four neighboring nodes, respectively, and E an identity matrix. Although the synthesized system F-matrix relates (V, I) of all the surface nodes to those of the bottom nodes, we often represent a few of the nodes as explicit observation nodes and leave other nodes floated. Further network reduction can be performed in converting the F-matrix to a Y-matrix of the observation nodes under a condition where $I_{top-floated} = 0$, $I_{btm-floated} = 0$ for the other floated nodes. The final form of a substrate model is a SPICE compatible sub-circuit netlist, where the ports equivalent to the observation nodes are fully connected each other with resistors constituting the Y-matrix. Finally, three-dimensional test structure is well reduced to a two-dimensional equivalent circuit.

Figure 3 shows simulated frequency responses of S21 for P-P-P and N-P-P test structures. S1 port is covered with p$^+$ in P-P-P while with n$^+$ in N-P-P, on the other hand, S2 port and S3 guard ring surrounding S1 port are covered with p$^+$ in both structures. A lumped capacitor corresponding to junction capacitance of n$^+$ to p-type substrate in S1 port is estimated from cross-sectional

678

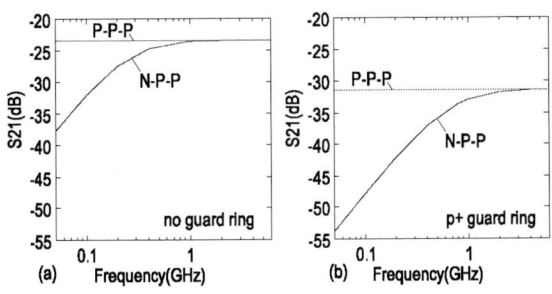

Fig. 3. Simulated S21 versus frequency dependence. Resistive (P-P-P) and capacitive (N-P-P) substrate coupling. (a)p+ guard ring (S3) is floated, (b) p+ guard ring (S3) is AC grounded.

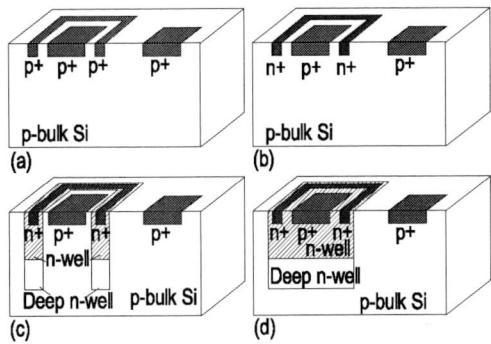

Fig. 4. Guard ring (GR) structures. (a) p^+ GR; (b) n^+ GR, (c) deep n-well GR, and (d) deep n-well pocket for comparison.

impurity profile and connected to a single observation node in S1 port of the computed N-P-P equivalent circuit. Here, the N-P-P test structure exhibits much higher isolation of S2 from S1 in low frequency range compared with the P-P-P counterpart, however, substrate coupling increases with 20 dB/decade against frequency due to capacitive coupling and finally dominated by frequency independent resistive coupling same as P-P-P for frequency higher than 2 GHz. It is also shown that the p^+-guard ring at S3 can effectively isolate S2 from S1 for substrate couplings in both structures, where the inclusion of guard ring structures in the test structure equivalent circuit will be detailed in the next section.

III. EVALUATION OF GUARD RING STRUCTURE

A. Equivalent circuit modeling of guard ring structures

We have developed an enhanced F-matrix computation flow that can include various guard ring structures in an equivalent circuit of the test structure shown in Figure 4, where S1 port is covered with p^+ and surrounded by a guard ring or included in a deep n-well pocket, while S2 port is covered with p^+ and connected to a common p-type substrate. Here, a guard ring can be resistive such as p^+ or capacitive such as n^+ and deep n-well (DNW). While the former absorbs and drains out the currents flowing through a substrate, the latter inserts a high impedance cut on the current flow and forces to detour, both result in the increase of isolation between ports. Since these guard ring structures introduce impedance components that are sharply localized in space and also accompany exponential difference in magnitude from a bulk resistivity, F-matrix computation becomes almost impracticable because of very dense meshing in F-matrices and explosion of computation time as well as memory usage. In order to solve this issue and alleviate computation requirements, following two modifications are made to the basic F-matrix computation flow described in Section 2.

[Short or cut of observation points]

Since most of the area in the test structure of Figure 2(a) has identical vertical impurity profile, F-matrix computation accordingly to (2) is performed under the assumption of uniform impurity. Then, an equivalent circuit is derived from F-to-Y matrix conversion with observation points assigned along the periphery as well as within the surface of S1, S2, S3, and S4 areas as shown in Figure 5(a).

Here, we can assume that the observation points in each of S1, S2, S3, and S4 areas show identical node voltages when each of the area is entirely covered by a highly conductive sheet, roughly $10^3 \times$ higher conductivity than the substrate, formed by selectively implanted high-density impurities and metal wirings. In this case, all the observation nodes within each of such areas are shorted together and provides a single representative port in a modified equivalent circuit as shown in Figure 5(b).

On the other hand, we can expect infinite isolation at DC from each of the S1, S2, S3, and S4 areas to the bulk when each of the area is entirely covered by junction capacitance against the bulk. In this case, the observation nodes within each of the areas are cut out and thus corresponding nodes are eliminated from a modified equivalent circuit, however, the observation nodes on the continuous periphery facing to the bulk are united to another single port as shown in Figure 5(c).

Note that mesh nodes other than the observation nodes are all included in the F-matrix computation to represent resistive networks in a bulk.

[Three level stacks of F-matrix cascade sub-models]

Impurity profile in a bulk-silicon CMOS technology shows strongly localized high concentration within a few μm depth from the surface and mostly constant low-level concentration in the rest of a bulk, typically with more than 500-μm thickness depending on assembly. In order

7B-1

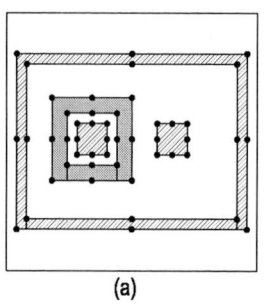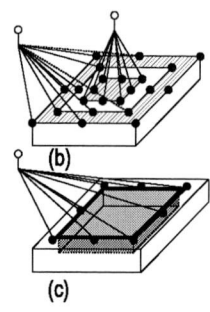

Fig. 5. (a) Assignments of observation points and providing (b) shorts or (c) cuts among observation points.

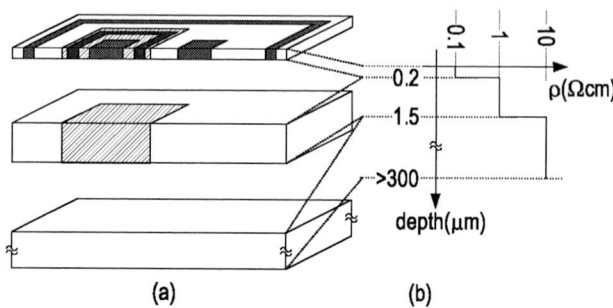

Fig. 6. (a) Stack of three F-matrix cascade sub-models corresponding to chip surface, well, and deep bulk, and (b) assumed impurity profile of guard ring test structure.

to include the surface impurity concentration properly in a full-depth F-matrix computation, we have divided the test structure vertically into three-level sub-models of chip surface, well, and deep bulk, as shown in Figure 6(a), where the chip surface sub-model has a depth identical to channel stop implant or p^+/n^+ active diffusions, the well sub-model has the depth corresponding to p-well or n-well, and the deep bulk sub-model covers the rest. Here, the areas of S1, S2, S3, and S4 defined in the layout of the surface sub-model are identically copied to the well and deep bulk sub-models for modeling purpose. Assumed vertical impurity is also shown in Figure 6(b).

A three-dimensional F-matrix cascade with the assumed uniform impurity concentration is built in each of the sub-models and then reduced to a two-dimensional equivalent circuit through F-matrix computation accordingly to (2) followed by F-to-Y matrix conversion with the short or cut of observation points in each of the S1, S2, S3, and S4 areas, as mentioned. The resultant sub-model equivalent circuit is described as a sub-circuit netlist with explicit ports relating to S1, S2, S3, and S4 areas. Finally, the entire test structure is modeled as a single equivalent circuit by stacking the three sub-models, namely, by connecting the explicit ports of the three sub-circuits, with intermediate lumped passive elements.

B. Evaluation of various guard ring structures

We have tailored the enhanced F-matrix computation flow so as to fit various guard ring structures and demonstrated that the three-level stacks of F-matrix cascade sub-models could successfully capture isolation characteristics. Here, all the p^+ guard ring structures are assumed to have p-well sub-models with uniform resistivity of 1 Ωcm and deep bulk sub-models with uniform bulk resistivity of 10 Ωcm as shown in Figure 6(b).

Figure 7 shows modeling of p^+ guard-ring test structure. The surface sub-model includes p^+ thus highly conductive areas of S1, S2, S3, and S4, where each of the areas can be united to a single port by shorting obser-

vation points, as was shown in Figure 5(b). The p-well and deep bulk sub-models also include S1, S2, S3, and S4 areas at the same position as the surface sub-model, and the areas each is similarly united to a corresponding single port in each sub-models. Finally, the three equivalent circuits are connected without intermediate elements (Figure 7(b)).

Figure 8 shows modeling of n^+ guard-ring test structure. The observation points within n^+ areas of S3 in the surface sub-model are cut out, as was shown in Figure 5(b). However, both inner and outer continuous peripheries of S3 area are united to each single port named $S3_{inner}$ and $S3_{outer}$, respectively. On the other hand, $S3_{top}$ is virtually re-defined as a single unified port corresponding to the eliminated n^+ area. In the well sub-model, S1, S2, S3, and S4 areas identically defined as the surface sub-model are filled with p-type dopant and thus each is united to a corresponding single port, where the S3 port is named $S3_{bottom}$. These ports relating to S3 area: $S3_{inner}$, $S3_{outer}$, and $S3_{bottom}$, are connected to $S3_{top}$ through lumped capacitors parasitic to peripheral junction C_{per} and bottom junction C_{btm}, in order to express the placement of n^+ guard ring in a final netlist (Figure 8(b)).

Figure 9 shows modeling of deep n-well guard-ring test structure. Although the modeling steps are same as in the n^+ guard-ring test structure, both of the surface and well sub-models are applied with the elimination of the observation points within S3 area and the creation of $S3_{inner}$ and $S3_{outer}$ in the peripheries of S3 area, since S3 area is deeply isolated from the other areas by the deep n-well guard ring. The re-definition of $S3_{top}$ is also performed in both sub-models, however, $S3_{bottom}$ is defined for S3 areas in the deep bulk sub-model. Again, the ports relating to S3 area are connected to $S3_{top}$ through lumped capacitors, in order to express the placement of deep n-well guard ring in a final netlist (Figure 9(b)).

Figure 10 shows modeling of deep n-well pocket test structure. In this case, the observation points within n^+

680

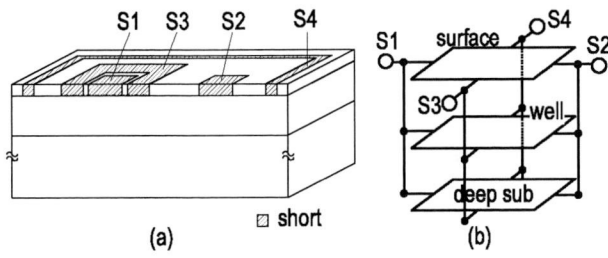

Fig. 7. (a) p$^+$ guard-ring test structure and (b) equivalent circuit.

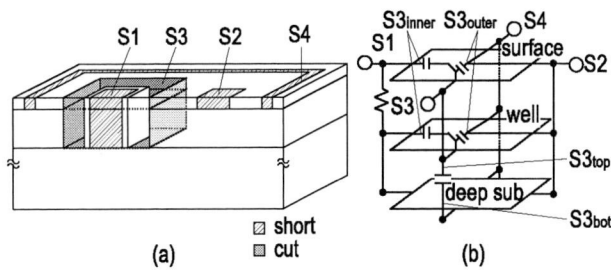

Fig. 9. (a) Deep n-well guard-ring test structure and (b) equivalent circuit.

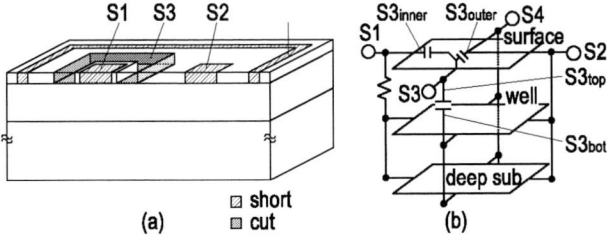

Fig. 8. (a) n$^+$ guard-ring test structure and (b) equivalent circuit.

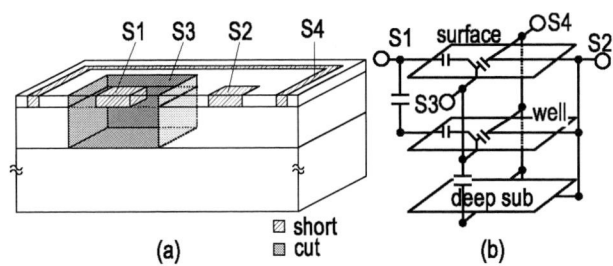

Fig. 10. (a) Deep n-well pocket test structure and (b) equivalent circuit.

areas of S3 in the surface sub-model are cut out, and moreover, those within the areas covered by a single deep n-well pocket in the well sub-model are also eliminated. On the other hand, the single port of S3$_{outer}$ is defined by shorting observation points along the periphery of the deep n-well pocket. The bottom plate is defined in the deep bulk sub-model as a single port of S3$_{bottom}$ covering the same footprint of the single deep n-well pocket.

Then, the re-defined S3 port of the single deep n-well is connected to S3$_{outer}$ and S3$_{bottom}$ and also to the S1 port defined for S1 through lumped parasitic capacitors, in order to express the placement of deep n-well pocket in a final netlist (Figure 10(b)).

Figure 11 shows simulated and measured results of isolation effectiveness of p$^+$ guard ring with geometry dependency, where the models with different distance between S2 and S1 ports and those with various areas of S1 ports (and S3 guard rings as well) are evaluated. The measurement results of these structures shown in Figure 11 are reported in [13]. The results of simulation and measurement are well consistent.

Figure 12 compares isolation effectiveness achieved by various guard ring structures of p$^+$, n$^+$, and deep n-well, as well as by deep n-well pocket. It is as expected that p$^+$ guard ring provides moderate isolation within the entire frequency range. The most effective isolation for frequency beyond 1 GHz is achieved by deep n-well guard ring.

Figure 13 shows geometry dependence of isolation achieved by deep n-well guard ring and by deep n-well pocket, evaluated at 100 MHz for representing low frequency range as well as at 2 GHz for radio frequency (RF) range. The superiority of the deep n-well guard ring in RF isolation is quite obvious, however, the effectiveness degrades due to capacitive couplings as S1 as well as S3 area enlarge. Here, the relatively small isolation in low frequency range can be supplemented by the combinational use with p$^+$ guard ring as is implied from Figure 11.

The derivation of test-structure equivalent circuits by the enhanced F-matrix computation flow took 30 minutes in average over various guard band structures, on a work station incorporating dual Ultra SPARC III running at 750 MHz and 2 GByte memory. In addition, a few minutes was required for S21 simulation by commercial SPICE circuit simulator.

IV. CONCLUSION

The proposed F-matrix computation flow realizes equivalent circuit modeling of various guard ring structures with geometry dependency as well as structural differences, by incorporating two novel techniques: three-level stacks of F-matrix cascade and short-or-cut of observation points. The isolation difference among p$^+$, n$^+$, and deep n-well guard rings is successfully evaluated along with distance as well as area dependencies. The combinational

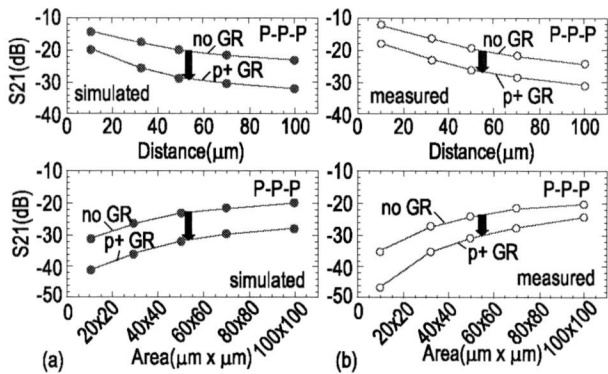

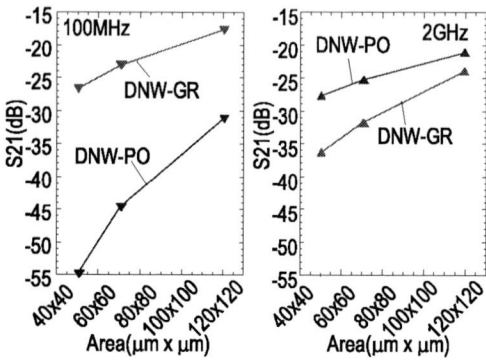

Fig. 11. (a) Simulated and (b) measured geometry dependency of GR isolation effects. The measurement results have been reported in [13].

Fig. 13. Simulated geometry dependency of deep N-well guard ring (DNW-GR) and deep N-well pocket (DNW-PO).

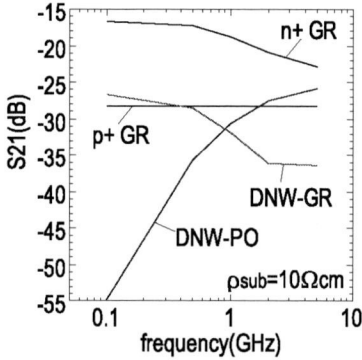

Fig. 12. Simulated S21 versus frequency dependence, comparing p+-guard ring (GR), n+-guard ring (GR), deep N-well guard ring (DNW-GR) and deep N-well pocket (DNW-PO). S1 and S2 has area of 50 μm × 50 μm.

use of p+ and deep n-well guard rings are suggested by those simulations in terms of RF isolation. The proposed technique enables simulation based prototyping of guard ring structures, which can be strongly helpful to establish isolation strategy against substrate coupling in a given technology, in an early stage of SoC developments.

REFERENCES

[1] N. Verghese and D. Allstot, "Computer-Aided Design Considerations for Mixed-Signal Coupling in RF Integrated Circuits," *IEEE J. Solid-State Circuits*, pp. 314-323, Mar. 1998.

[2] D. K. Su, M. J. Loinaz, S. Masui, and B. A. Wooley, "Experimental results and modeling techniques for substrate noise in mixed-signal integrated circuits," *IEEE J. Solid-State Circuits*, Vol. 28, pp. 420–430, Apr. 1993.

[3] A. Samavedam, A. Sadate, K. Mayaram, T. S. Fiez, "A Scalable Substrate Noise Coupling Model for Design of Mixed-

Signal IC's," *IEEE J. Solid-State Circuits*, pp. 895-904, June 2000.

[4] N. K. Verghese, T. J. Schmerbeck, and D. J. Allstot. "*Simulation Techniques and Solutions for Mixed-Signal Coupling in Integrated Circuits*," Boston, MA: Kluwer Academic Publishers, 1995.

[5] I. L. Wemple and A. T. Yang. "Integrated circuit substrate coupling models based on voronoi tessellation," *IEEE Trans. Computer-Aided Design of Integrated Circuits and Systems*, Vol. 14, pp. 1459–1469, Dec. 1995.

[6] M. Nagata, Y. Murasaka, Y. Nishimori, T. Morie, and A. Iwata "Substrate Noise Analysis with Compact Digital Noise Injection and Substrate Models" in *Proc. ASP-DAC*, pp. 71–76, Jan 2002.

[7] E. Charbon, R. Gharpurey, R. G. Meyer, and A. Sangiovanni-Vincentelli, "Substrate optimization based on semi-analytical techniques," *IEEE Trans. Computer-Aided Design of Integrated Circuits and Systems*, Vol. 18, pp. 172–190, Feb. 1999.

[8] M. Nagata, J. Nagai, K. Hijikata, T. Morie, and A. Iwata, "Physical Design Guides for Substrate Noise Reduction in C-MOS Digital Circuits," *IEEE J. Solid-State Circuits*, pp. 539–549, Mar. 2001.

[9] T. Blalack, Y. Leclercq, and C. P. Yue, "On-chip RF Isolation Techniques," in *Bipolar/BiCMOS Circuits and Tech. Mtg.*, pp. 205-211, Oct. 2002.

[10] M. Pfost, P. Brenner, T. Huttner, A. Romanyuk, "An Experimental Study on Substrate Coupling in Bipolar/BiCMOS Technologies," *IEEE J. Solid-State Circuits*, pp. 1755-1763, Oct. 2004.

[11] T. A. Johnson, R. W. Knepper, V. Marcello, and W. Wang, "Chip substrate resistance modeling technique for integrated circuit design," *IEEE Trans. Computer-Aided Design of Integrated Circuits and Systems*, Vol. CAD-3, pp. 126–134, Apr. 1984.

[12] Y. Murasaka, M. Nagata, T. Ohmoto, T. Morie, and A. Iwata, "Chip-Level Substrate Noise Analysis with Network Reduction by Fundamental Matrix Computation," in *IEEE Int. Symp. Quality Electronic Design*, pp. 482–487, Mar. 2001.

[13] D. Kosaka, M. Nagata, Y. Hiraoka, I. Imanishi, M. Maeda, Y. Murasaka, and A. Iwata, "Isolation Strategy against Substrate Coupling in CMOS Mixed-Signal/RF Circuits," in *Symp. VLSI Circuits*, pp. 276–279, Jun. 2005.

A New Boundary Element Method for Accurate Modeling of Lossy Substrates with Arbitrary Doping Profiles[*]

Xiren Wang, Wenjian Yu, and Zeyi Wang

Eda Lab, Dept. Computer Science & Technology,
Tsinghua Univ., Beijing, 100084, China

Abstract - It is important to model substrate couplings for SoC/mixed-signal circuit designs. After introducing the continuation equation of full current in lossy substrates, we present a new direct boundary element method (DBEM), which can handle the substrates with arbitrary doping profiles. Three techniques can speed up the DBEM remarkably, which include reusing coefficient matrices for multiple-frequency calculation, condensing the linear system, and sparsifying coefficient matrix. Numerical experiments illustrate that DBEM has high accuracy and high efficiency, and is versatile for arbitrary doping profiles.

I Introduction

There are currently increasing demands for high-integration circuits [1]. High-speed digital components and highly-sensitive analog components are often built on a common substrate. Although the high integration has some advantages, such as low power dissipation [2], there is a problem that the current noises injected by digital components travel the shared substrate and impact sensitive analog components severely [3]. Besides, substrate losses also impact circuit performances considerably. For instance, the quality factor of such passive devices as inductors is important for circuit performances, especially in wireless communication applications. But it is limited by substrate losses, especially at high frequencies. Thus acknowledge of lossy substrates is very important even critical for designs [4].

A resistance network model of a substrate is often efficient at frequency of up to several gigahertz. Some numerical methods are presented for the resistive simulation, including finite element method (FEM), finite difference method (FDM) [5], and methods based on Green's function (sometimes also called boundary element method) [2, 4, 6, 7]. FEM and FDM are too slow, since they discretize entire volume of a substrate. However, they have the advantage of versatility, i.e., they can handle substrates with arbitrary doping profiles rather than only layered substrates.

The Green's function based methods [2, 6, 7] are generally faster than FDM and FEM, because they only discretize contact surfaces. A suitable Green's function that satisfies boundary conditions needs to be found. For multilayered substrates, the function consists of multiple infinite series,

and converges slowly. In [4], a 'numerically stable' method was proposed to calculate the Green's function with the acceleration of discrete cosine transform (DCT) [7]. However, it is not actually stable, with further remedy presented in [8]. In [2], there is an excellent idea of eigende-composition, which results in a speedup of a dozen over the above DCT-Green's function method. The coefficient matrix is dense for these methods, however. Storing and solving the linear system are memory- and time-consuming.

On the other hand, the resistive model becomes invalid at high frequencies, and a comprehensive frequency-dependent impedance model is desired. The method in [4] is able to give such impedances. However, it also needs to find an expensive Green's function, whose derivation is based on the layered structure of substrates.

Note that substrates are not always stratified. In fact, there are lots of realistic substrates with layout-dependent doping profiles, such as oxide wells, trenches, sinkers, buried diffusions and etc [9]. There are also special structures like Faraday shields and junction shields [10] for noise reduction. To simulate such substrates, the methods based on Green's function meet much difficulty.

In this paper, we bring forward a direct boundary element method (DBEM) [3, 11, 12] for lossy substrate modeling. Conventionally it is applied in electrostatic DC capacitance extraction [12] or resistance extraction [11]. We will modify it to model lossy substrates at any frequency with the help of the concept of complex permittivity [13].

In DBEM, only substrate boundary is discretized, then the variables are fewer than in FDM/FEM. Only the free-space Green's function is used, which brings two advantages. One is that no derivation of the function is needed, and its computation is also straightforward. The other is that the function is independent of structures, so DBEM can handle substrates with arbitrary doping profiles. Besides, three kinds of accelerating techniques are presented, which are re-using coefficient matrices for multiple-frequency calculation, condensing the linear system before solving, and sparsifying the coefficient matrix. They can enhance the efficiency of DBEM considerably, but preserve the accuracy exactly.

The rest is organized as follows. In Section II, DBEM principle for impedance modeling is presented. In Section III, the accelerations for DBEM are described separately. Numerical experiments follows, so as to demonstrate the accuracy and efficiency of the method. A special case is a non-stratified substrate with lateral resistivity variation, and it can illustrate versatility of the method. The conclusions are given at last.

---

[*] This work is supported by the China National Science Foundation under Grant 60401010. It is also partly supported by National High Technology Research and Development Program of China (No. 2004AA1Z1050).

II. Frequency-Dependent Extraction of Substrate Coupling with DBEM

Fig. 1 shows an example of lossy substrate, which is constituted by layers of mediums M_i (with finite resistivity) and contacts C_j. There is usually a grounded plane on the bottom. In many cases, contacts are assumed to be on the top. Here, however, they can be placed anywhere if necessary.

For lossy substrates (such as Silicon), the coupling between contacts can be modeled as frequency-dependent impedances. To get the impedances needs to pre-set voltages of contacts, and then to calculate the full current flowing through contacts. The reciprocal of the current is related to the desired impedance.

Assume electric potential of contacts to be in $u = \overline{u}e^{j\omega t}$ form, where $\overline{u}$ is the maximum voltage, ω is the angular frequency, and t is time. Electric field within the substrate is also in $E = \overline{E}e^{j\omega t + \theta}$ form. Start from the Maxwell equation

$$\nabla \times H = J + \frac{\partial D}{\partial t}, \tag{1}$$

where H is the magnetic filed intensity, J is the conductance (ohmic) current density, D is the electric displacement. For mediums with constant conductivity σ and permittivity ε,

$$(\sigma + j\omega\varepsilon)\nabla \cdot E = 0. \tag{2}$$

Because $E = -\nabla u$ [1], the following Laplace equation holds:

$$\nabla^2 u = 0, \qquad \text{in medium } M_i \tag{3}$$

with the mixed boundary conditions of

$u = \overline{u}$, on contact surfaces (Dirichlet boundary) (4a)
$E_n = 0$, on natural boundary (Neumann boundary) (4b)

where $\overline{u}$ is pre-set voltage of contacts, usually 1V or 0V. E_n is normal electric field intensity. Besides, the potential and full current is continuous on the interface of adjacent mediums a and b:

$$u_a = u_b, \tag{5a}$$
$$(\sigma_a + j\omega\varepsilon_a)E_{n,a} = (\sigma_b + j\omega\varepsilon_b)E_{n,b}. \tag{5b}$$

where σ_a and σ_b (ε_a and ε_b) are the conductivity (permittivity) of regions a and b, respectively.

---

[1]Generally speaking, $E = -\nabla u - \dfrac{\partial A}{\partial t}$, where A is the magnetic vector potential. Here we adopt the Coulomb Gauge [15], which defines $\nabla \cdot A = 0$. Thus $E = -\nabla u$.

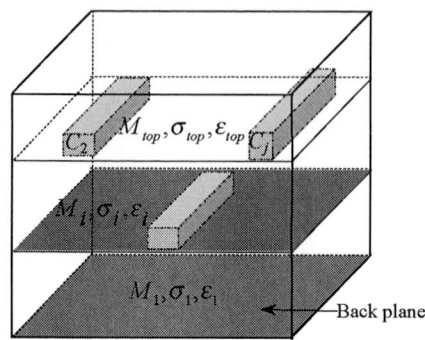

Fig. 1 An example substrate with a back plane. The non-zero conductivity of medium M_i is σ_i, and the permittivity is ε_i.

If E_n is known, we can get the current between contacts m and k. If the pre-set voltage of contacts m is 1V, and contact k is of 0V, the impedance between m and k will be

$$Z_{mk} = \frac{1}{\int_{\Gamma_k} (\sigma + j\omega\varepsilon)E_n d\Gamma}. \tag{6}$$

where Γ_k is the surface of contact k. σ and ε are the conductivity (reciprocal of resistivity) and permittivity of the medium surrounding contact k, respectively.

Note that only E_n is needed in (6). This can be made use of later.

Now we start solving E_n from (3). Utilizing the Green's identities, and selecting free-space Green's function as weighting function, we transform (3) into a boundary integral equation (BIE), which is defined on the boundary of medium M_i. Partition the boundary into N_i elements Γ_{ij}. Then we get a discretized BIE [11]:

$$c_s u_s + \sum_{j=1}^{N_i} \int_{\Gamma_{ij}} q^*_{(s)} u d\Gamma = \sum_{j=1}^{N_i} \int_{\Gamma_{ij}} u^*_{(s)} E_n d\Gamma, \text{ for medium } M_i \tag{7}$$

where collocation point s is on the boundary of M_i. c_s is a constant. $u^*_{(s)}$ is the fundamental solution of Laplace equation which is related with point s, and $q^*_{(s)}$ is its normal derivative. Here, u^* is equal to the free-space Green's function.

List the equations (7) for each of the mediums in the substrate. Combine these equations with the interface conditions (5). Substitute the conditions (4), and we get an overall system

$$Ax = b, \tag{8}$$

where A is composed of complex entries. Vector x is constituted by u and E_n unknowns of all mediums. b is created after moving all the known values related to pre-set voltage $\overline{u}$ to the right side. Solving the system, we can directly get E_n, and in turn the impedance in (6). When the impedances between many pairs of contacts are desired, the system (8) will become

684

$$AX = B, \tag{9}$$

where B is also a matrix.

III. Effective Techniques for DBEM

Although DBEM has much fewer variables in some sense, it seems still too slow for substrates with many contacts. Accelerating techniques are preferred.

A. Reuse Matrix

To evaluate a lossy substrate, we usually need to calculate coupling impedances at many frequencies. Usually we need to build the system (9) for many times. However, most entries of matrix A and vector B can be reused so as to save the computational time.

For an interface element, there is a pair of variables $E_{n,a}$ and $E_{n,b}$ related to it. In medium a, the coefficient of $E_{n,a}$, denoted by I_a, can be determined by the integral in (7). In medium b, I_b, the coefficient of $E_{n,b}$ can be similarly obtained. $E_{n,a}$ and $E_{n,b}$ have the relationship of $E_{n,b} = \frac{\sigma_a + j\omega\varepsilon_a}{\sigma_\square + j\omega\varepsilon_\square} E_{n,a}$. Actually, to overcome the singularity of the matrix, we keep one of them in the system, say $E_{n,a}$. Then the $E_{n,a}$ coefficient in medium b becomes $I_b \times \frac{\sigma_a + j\omega\varepsilon_a}{\sigma_b + j\omega\varepsilon_b}$, which depends on ω.

In a word, only those matrix entries related to the interface variables like $E_{n,a}$ are frequency-dependent. The other entries remain the same for any frequency. Besides, B in (9) is also independent of frequency, because it is merely related to the pre-set voltage $\overline{u}$ and the integral similar to I_a.

When impedances at multiple frequencies are needed, we need to build the system (9) for each frequency as usual. However, a better choice is to compute the aforementioned frequency-independent entries only once, and reuse them for any frequency. The other entries are re-calculated at each frequency. In this way some CPU time can be saved. For the first test case in Section IV, this reusing technique reduces the total running time for 20 frequencies from 144.91 to 124.70 seconds, or by 14%.

B. Condense the Linear Equation System

Solving the complex system (9) consumes a great deal of CPU time. Direct solutions have too high computational complexity. As we will see in subsection C, the coefficient matrix is sparse for multi-medium problems, so we choose a GMRES solver [14]. Moreover, reordering unknowns and corresponding collocation points in the same sequence can make the matrix diagonally dominant [3]. Reorder unknowns as in Fig.2 can arrange non-zero blocks as close to

the diagonal as possible [3, 12]. These can shorten solution time to some extent. But solving a large linear system takes still too much time.

However, if we discard inessential variables at first, and the consequent solving the condensed system will be much easier. Remind that only E_n variables are needed in (6), so we can discard some u variables. Rewrite a system $Ax = b$ in terms of sub-matrices (for (9), we can do it similarly):

$$\begin{bmatrix} A_{00} & A_{0T} \\ A_{T0} & A_{TT} \end{bmatrix} \cdot \begin{bmatrix} \overline{x} \\ u_T \end{bmatrix} = \begin{bmatrix} b_0 \\ b_T \end{bmatrix}, \tag{10}$$

where $\overline{x}$ and u_T are two subsets of x. Remove the inessential u_T equivalently:

$$\overline{A}\,\overline{x} = \overline{b}, \tag{11}$$

where

$$\overline{A} = A_{00} - A_{0T} A_{TT}^{-1} A_{T0}, \tag{12}$$

and

$$\overline{b} = b_0 - A_{0T} A_{TT}^{-1} b_T. \tag{13}$$

The condensing procedure (12) (13) is generally expensive in computation, because it involves the inversion of A_{TT} and matrix-matrix multiplications.

We start to reduce the computational complexity of (12) (13). For an example three-layer substrate, reordering the non-interface variable v_{33} of medium 3:

$$v_O \rightarrow E_{nC} \rightarrow u_T, \tag{14}$$

where E_{nC} represents the unknown E_n on contact surfaces, u_T denotes u unknowns on the top surface of medium 3, and v_O denotes the other unknowns in v_{33}.

With the ordering (14) and Fig.2 applied, the DBEM matrix A for a substrate is shown in Fig.3a, where zero entries are in white. The condensed matrix $\overline{A}$ is in Fig.3b. A_{TT} is diagonal; A_{0T} and b_T are sparse. These properties can be analytically proven. Refer to [3] for more details. Thus calculating $\overline{A}$ and $\overline{b}$ through (12) (13) is easier.

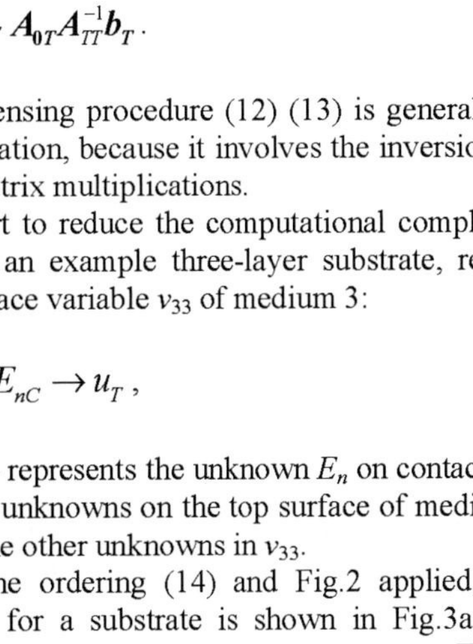

Fig. 2 Reorder unknowns. M is the number of mediums. v_{ii} includes u on Dirichlet boundary and q on Neumann boundary of medium i. u_{ij} and q_{ji} ($i < j$) are on the interface of mediums i and j.

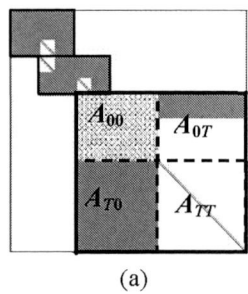 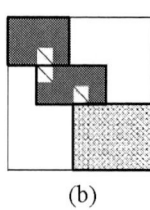
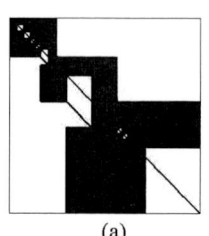 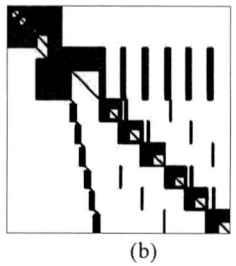

(a)   (b)

Fig. 3. Distribution of nonzero entries in A: (a) original; (b) after using the condensing technique. Zero entries are in white.

(a)   (b)

Fig. 4. (a) Original matrix, with 248,700 nonzero entries. (b) Matrix after the QMM idea is used, with 142,716 nonzero entries.

In the first experiment in Section IV, it takes 4.1 seconds to solve (8). If the condensing technique is applied, the solution takes merely 1.5 seconds, or 60% is saved. Unfortunately, the condensing procedure (12) (13) itself takes 5.0 seconds. But when the sparsifying technique in the next subsection is applied, the procedure needs only 0.58 seconds.

On the other hand, there is only one right-hand side here. If there are many sides as (9) indicates, the condensing technique will bring higher efficiency [3, 11], since solution time for each side is considerably reduced.

C. Sparsify Matrix

From Fig.3, we learn that coefficient matrix A is blocked sparse. The sparsity can be analytically deduced from equation (7), where collocation points and variables related to each other are within the same medium. In other words, collocation points and variables in different mediums have no direct numerical relationship. Therefore, A is blocked sparse for multi-medium problems.

If a physical medium is quasi-cut into fictitious components, the matrix sparsity can be enlarged. This is the basic idea of a quasi-multiple medium (QMM) technique in [11, 12]. For an example substrate, if we cut the third medium into 6 smaller components, the original matrix Fig.4a turns into a sparser one Fig.4b. The new matrix is a bit larger, because the quasi-cutting will bring additional variables on interface of adjacent fictitious components. What's most important is the total number of non-zero entries is remarkably reduced. Since most time is spent in matrix-vector multiplication in a GMRES solver, the enhanced sparsity is certain to speed up the solution. Similarly, the technique can also speed up the condensing procedure (12) (13).

For instance, in the first experiment in Section IV, solution time of (9) is reduced from 4.1 seconds to 2.0 seconds, and the speed up is 2.

As we have seen, each of the three techniques above can enhance the efficiency of DBEM without any accuracy loss. What's more, if they are combined together, the enhancement will be more obvious, as shown in the next section.

IV. Numerical results

The first experiment is a substrate involving a top contact and a back plane, as shown in Fig.5a. And the second case is shown in Fig.5b, involving a back plane and two top contacts separated by s μm. The test cases will be configured with two types of substrate processes [4], as shown in Fig.6. In order to be compatible with literature data, the contacts are assumed to be of zero heights.

For the first test case, the impedances obtained with the presented DBEM are depicted in Fig.7 (for the LR process) and Fig.8 (for the HR process), respectively. The data in [4] are also depicted. The figures indicate that with any process, the magnitude of impedance decreases as the frequency increases. This is because the substrate current injection is capacitive in nature [4]. With the low-resistivity (LR) process, the impedance is smaller in magnitude than with the high-resistivity (HR) process. This is because the bulk is with much smaller resistivity in the LR process.

Both the magnitude and the phase of the impedances obtained with DBEM are very close to literature data. The discrepancy is within 1.0%, which also includes the possible error of experiment setups (we did not find the exact setups in [4]). In a word, DBEM has a high accuracy.

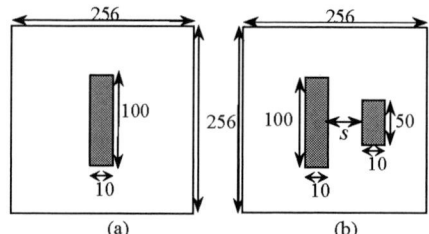

(a)   (b)

Fig. 5. (a) A single contact at center. (b) Two contacts located symmetrically along the center with distance of s μm. There is a back plane in the two substrates.

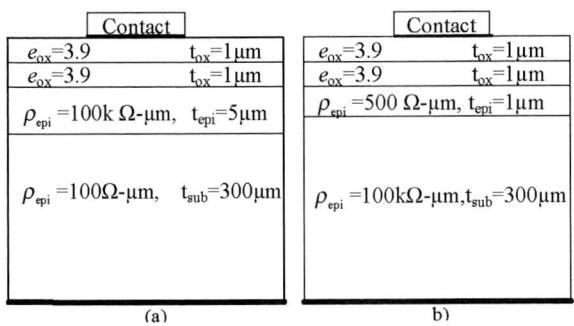

(a)   b)

Fig. 6. Side views of two doping profiles [4]. (a) Low-resistivity (LR) process; b) high-resistivity (HR) process. Bottom bold lines denote back planes.

Take a look at the efficiency. Since running parameters of the method in [4] can not be obtained, we do not compare DBEM with it directly. But the superiority of DBEM over Green's function methods for substrate impedance modeling can be roughly deduced according to [3] and [11], where DBEM is up to hundreds of times faster than the excellent eigendecompostion method in [2] for resistance extraction.

The total running time to get the impedances at 20 frequencies is listed in Table 1. All DBEM programs run on a Sun Fire 880 workstation, with CPU frequency of 750 MHz. The table tells us that each of the techniques in section III can bring enhancement of efficiency. Furthermore, their combination performs much better, with a speed-up of 2.7 over pure DBEM.

For the two-contact substrate Fig.5b, the contact-contact impedance depends on the spatial separation. The impedance of this substrate with the LR process is depicted in Fig. 9. We learn that the impedance is very sensitive to the distance s. As s extends from 2μm to 10μm, the impedance increases by three times in magnitude. When s turns from 10μm to 50μm, the impedance rises by ten times. The same tendency is found in [4].

Note that the both doping profiles in Fig. 6 are stratified. However, these methods based on Green's functions have

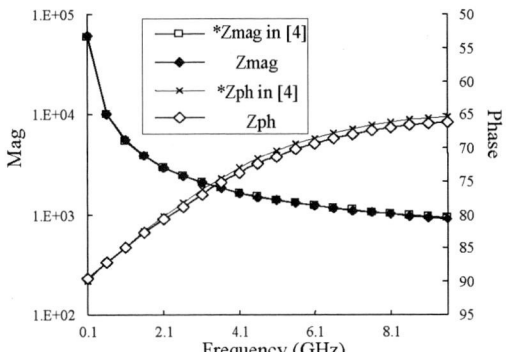

Fig. 8. Magnitude and phase of the contact-ground impedance of the single-contact substrate with the HR process.

much difficulty in handling substrates with non-stratified doping profiles. In reality, there are many such substrates. For example, there are many lateral variation components in the substrates, such as oxide trenches and wells [9]. In order to illustrate the versatility of DBEM, examine the simple example shown in Fig. 10, where the top central block in gray has a distinct resistivity from its lateral neighbor. We call the block "LVB" (lateral variation block).

TABLE I.
Running time for substrate Fig.5a with LR process

Pure DBEM	**144.91**
DBEM + Reusing (Sect III.A)	124.70
DBEM + Condensing (Sect III.B)	74.06
DBEM + Sparsifying (Sect III.C)	70.54
DBEM+Reusing+Condensing+Sparsifying	**52.90**

Since the conductivity of epi layer (2000 s/m) is large, and at frequency of up to 10 GHz, the ohmic current is dominant over the displacement current in the layer. Thus changing the dimension of the higher-resistivity LVB block will have limited impact on the contact-contact impedance, as shown in Fig. 11.

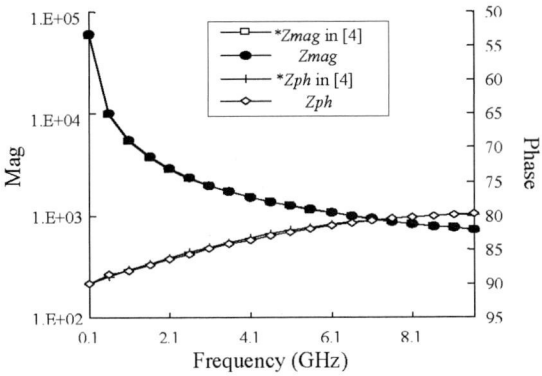

Fig. 7. Magnitude and phase of the contact-ground impedance of the single-contact substrate with the LR process.

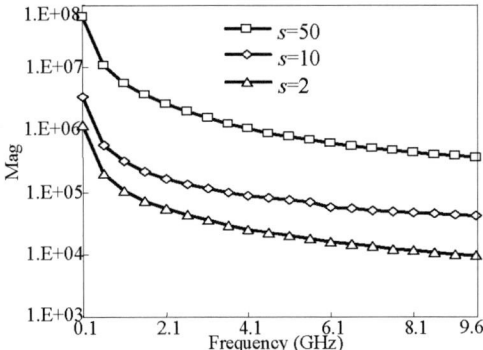

Fig. 9. Magnitude of the contact-contact impedance of the two-contact substrate Fig. 5b, with the LR process. s is the distance between the contacts.

7B-2

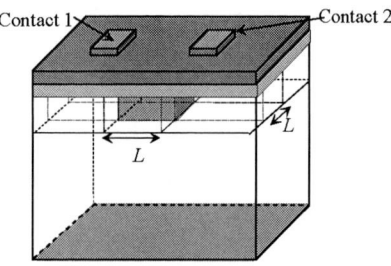

Fig. 10. Substrate with lateral resistivity variation in the epi layer of the HR process. The width and length of the substrate are 200 μm. The resistivity of the epi layer and the gray block are 5e-4 and 0.5 Ωm, repectively. Note that each contact here is modeled as a block with 3-D shape, as well as the LVB block.

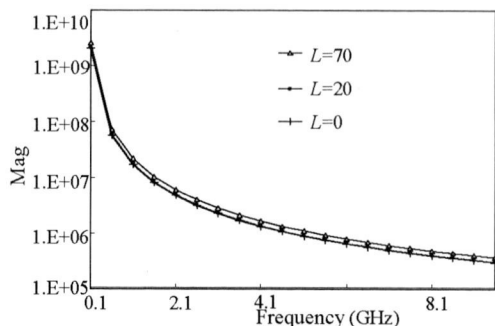

Fig. 11. Magnitude of the contact-contact impedance. Since the epi layer of HR process is very small, the central block is not possible to be large, and thus the curves are close to each other.

Fig.11 illustrates that when the LVB block is small (L=0, or L=20), the impedance is relatively small. As the block grows large (L=70), the impedance increases, since it obstructs the ohmic current flow considerably. Besides, for a specified frequency, the difference between the impedances under different L is small. A reason is that the obstructing LVB block can not be very large due to spatial limitation.

The results tell us that DBEM is able to simulate such complex substrates without any difficulty.

Note that the accuracy of DBEM can be further improved by partitioning the substrate boundary more densely. On the other hand, the efficiency of it may also be enhanced by partitioning more coarsely. This allows a tradeoff of them, which makes the method adaptive in various circumstances.

V. Conclusions

In this paper, we present a new direct boundary element method (DBEM) to extract the impedances of lossy substrates at any frequency. Based on Maxwell equations, we deduce the Laplace equation that holds in the substrate and its boundary conditions. Utilizing the complex current continuity conditions on the interfaces of lossy mediums, the new DBEM becomes suitable for impedance modeling. Three accelerating techniques are proposed without any sacrifice of accuracy. Reusing matrix entries can save lots of time for building matrices for impedance extraction at multiple frequencies. Condensing the linear system and sparsifying the coefficient matrix can speed up system solution remarkably. Numerical experiments illustrate that DBEM can be of high accuracy and high efficiency. DBEM is able to simulate substrates with arbitrarily processes, and numerical experiment results are reasonable.

References

[1] R. Gharpurey and E. Charbon, "Substrate coupling: Modeling, simulation and design perspectives," *Proc. 5th International Symposium on Quality Electronic Design*, pp. 283-290, 2004.
[2] J.P. Costa, M. Chou, and L.M. Silveira, "Efficient techniques for accurate modeling and simulation of substrate

coupling in mixed-signal IC's," *IEEE Trans. Comput. Aided Design*, Vol. 18, pp. 597-607, 1999.
[3] X. Wang, W. Yu, and Z. Wang, "Substrate Resistance Extraction with Direct Boundary Element Method," *Asia and South Pacific Design Automation Conference*, pp.208-211, 2005.
[4] A.M. Niknejad, R. Gharpurey, and R.G. Meyer, "Numerically stable Green function for modeling and analysis of substrate coupling in integrated circuits," *IEEE Trans. Comput. Aided Design*, Vol. 17, pp. 305-315, 1998.
[5] N.K. Verghese and D.J. Allstot, "Fast simulation of substrate coupling effects in mixed-mode ICs," *IEEE Custom Integrated Circuits Conference*, pp. 18.3.1–18.3.4, 1993.
[6] T. Smedes, N.P. van der Meijs, and A.J. van Genderen, "Extraction of circuit models for substrate cross-talks," *IEEE/ACM International Conference on Computer-Aided Design*, pp. 199-206, 1995.
[7] R. Ghapurey and R.G. Meyer, "Analysis and simulation of substrate coupling in integrated circuits," *International Journal of Circuit Theory and Application*, Vol. 23, pp. 381–394, 1995.
[8] C. Xu, T. Fiez and K. Mayaram, "On the numerical stability of Green's function for substrate coupling in integrated circuits," *IEEE Trans. Computer-Aided Design*, Vol. 24, pp. 653–658, 2005.
[9] S. Donnay and G. Gielen, *Substrate Noise Coupling in Mixed-Signal ASICs*. Kluwer Academic Publishers, 2003.
[10] S. Ardalan and M. Sachdev, "An overview of substrate noise reduction technieques," *Proc. 5th International Symposium on Quality Electronic Design*, pp. 291-296, 2004.
[11] X. Wang, W. Yu, Z. Wang, and X. Hong, "An improved direct boundary element method for substrate coupling resistance extraction," *Proceedings of the ACM Great Lakes Symposium on VLSI*, pp. 84-87, 2005.
[12] W. Yu, Z. Wang, and J. Gu, "Fast capacitance extraction of actual 3-D VLSI interconnects using quasi-multiple medium accelerated BEM," *IEEE Trans. Microwave Theory Tech.*, Vol. 51, pp. 109-199, 2003.
[13] S. Ramo, J.R. Whinnery, and T. van Duzer, *Fields and Waves in Communication Electronics,* 2nd ed. New York: Wiley, pp. 279-283, pp. 572-676, 1984.
[14] Y. Saad and M.H. Schultz, "GMRES: A generalized minimal residual algorithm for solving nonsymmetric linear systems," *SIAM J. Sci. Statis. Comput.*, Vol. 7, pp. 856-869, 1986.
[15] S. Ramo, J.R. Whinnery, and T. van Duzer, *Fields and Waves in Communication Electronics,* 3rd ed. New York: Wiley, pp. 283-287, pp. 572-676, 1994.

688

Parasitics Extraction Involving 3-D Conductors Based on Multi-layered Green's Function

Zuochang Ye, Zhiping Yu

Institute of Microelectronics, Tsinghua University, Beijing, 100084, China

yzc02@mails.tsinghua.edu.cn

Abstract— An efficient algorithm for three-dimensional (3-D) capacitance extraction on multi-layered and lossy substrate is presented. The new algorithm represents a major improvement over the quasi-3D approach used in Green's function-based solvers by taking into consideration of the side-wall effects of the conductors. The accuracy and efficiency of the new algorithm is tested by examples.

I. INTRODUCTION

It has become more critical that 3-D effects should be considered accurately in modeling capacitances on RF CMOS chips because of shrunk feature size and increasing operation frequency. Boundary element method (BEM) is an efficient method for this problem, since only surfaces and interfaces between regions need to be discretized, which greatly reduced the problem compared to volume-discretized methods such as finite difference method (FDM) and finite element method (FEM), e.g., Raphael [1] and HFSS [2].

Great efforts have been made in the BEM-based 3-D capacitance extraction, such as fast multi-pole (FMM) [3], singular value decomposition (SVD) [4], hierarchical [5], and quasi-multiple medium (QMM) [6] methods . All these methods, however, mainly focus on the fast solving of the linear system. In this paper, we study the computation of potential coefficient matrix $\mathbf{P}$ (to be defined later) for multi-layered and lossy substrate based on Green's function [7, 8]. In this method, the Green's function is calculated and stored based on the process technology only, thus the dielectric interfaces don't need to be discretized. As a result, the size of the linear system is greatly reduced.

In [7], the chip structure is assumed to be confined by rectangle electric and magnetic walls. Considering the Neumann boundary condition at the magnetic walls, the Green's function can be expanded to cosine series, such that fast Fourier transform (FFT) can be applied to accelerate the calculation of cosine series summation. By introducing the concept of complex permittivity [9], this method can handle lossy substrate easily with little extra efforts. The original algorithm was improved in [8] and implemented in software ASITIC [10].

In ASITIC, however, since only analytical integration over horizontal panels is derived, the conductors are restricted to two dimensional sheets. To extend the capacitance extraction

to 3-D domain, the Green's function should be integrated analytically in z-direction. However, this meets some difficulty. Because in [8], to preserve the numerical stability, the Green's function in z-direction is obtained through a complicated recursive procedure. The resulted expression is a continual multiplication, which cannot be integrated analytically. Although measures are taken in [11] by considering the bottom and top plates of conductors, the sidewall capacitance still cannot be handled. In [12], the conductors are cut horizontally into slices in order to include 3-D effects, but this amounts to volume discretization, leading to prohibitive extra cost.

In this paper, we use a new formula for recursive computation of the Green's function, such that it can be integrated stably in z-direction. The formula for the computation of the potential coefficients involving sidewall integration is derived to accomplish the true 3-D parasitics extraction.

II. GREEN'S FUNCTION APPROACH TO CAPACITANCE EXTRACTION WITH LOSSY SUBSTRATE

The substrate is characterized in [7] as a multi-layered structure as shown in Fig. 1. Each layer has a thickness and a uniform permittivity and conductivity. Conductors are embedded in the layers. The objective of the parasitics extraction is to compute the $m \times m$ capacitance matrix $\mathbf{C}$ for an m-conductor geometry.

To determine the j-th column of $\mathbf{C}$, we need only to solve for the surface charges on each conductor produced by raising conductor j to unit potential while grounding rest of the conductors. Then C_{ij} is numerically equal to the charge q_i on conductor i. This procedure is repeated m times to compute all columns of $\mathbf{C}$. The charge q_i is obtained by solving the linear system

$$\mathbf{Pq} = \mathbf{v}, \tag{1}$$

where $\mathbf{v}$ is the given potential vector and $\mathbf{P}$ is the so-called potential coefficient matrix. Each entry of $\mathbf{P}$, i.e. p_{ij}, is computed by convolving the charge distribution with the Green's function as

$$p_{ij} = \frac{1}{A_i A_j} \int_{\mathbf{r}' \in A_i} \int_{\mathbf{r} \in A_j} G(\mathbf{r}', \mathbf{r}) d\mathbf{r} d\mathbf{r}', \tag{2}$$

where A_1 and A_2 are the areas of the two panels. The Green's function $G(\mathbf{r}', \mathbf{r})$ can be computed by solving the Laplace equation

$$\nabla G(\mathbf{r}', \mathbf{r}) = \frac{-\delta_m(\mathbf{r} - \mathbf{r}')}{\dot{\varepsilon}_k}, \qquad (3)$$

where

$$\dot{\varepsilon} = \varepsilon + \frac{\sigma}{j\omega} \qquad (4)$$

is the complex medium permittivity [9]. By the use of $\dot{\varepsilon}$ instead of ε, both ohmic and displacement currents are accounted for, thus the frequency-dependent effect for conductive substrate is included.

The Greens function for the boundary condition shown in Fig. 1 can be solved as a double infinite summation

$$G(x', y', z', x, y, z) = \sum_{m=0}^{\infty} \sum_{n=0}^{\infty} f_{mn}(z', z) \times$$
$$\cos(\delta_m x') \cos(\xi_n y') \cos(\delta_m x) \cos(\xi_n y), \qquad (5)$$

where $\delta_m = m\pi/a$, $\xi_n = n\pi/b$, a, b are the substrate lateral dimensions. f_{mn} is obtained by solving the Laplace equations in the z-direction along with the boundary conditions in this dimension. In [7], f_{mn} is composed of hyperbolic functions, while in this paper, it is expressed in an exponential manner as

$$f_{mn}(z', z) = \frac{C_{mn}}{ab\dot{\varepsilon}_s\gamma} \times$$
$$\frac{\left(\alpha_s^{u,l}e^{\gamma z'} + \beta_s^{u,l}e^{-\gamma z'}\right)\left(\alpha_f^{l,u}e^{\gamma z} + \beta_f^{l,u}e^{-\gamma z}\right)}{\alpha_s^u\beta_s^l - \alpha_s^l\beta_s^u}$$
$$= (\alpha_s e^{\gamma z'} + \beta_s e^{-\gamma z'})(\alpha_f e^{\gamma z'} + \beta_f e^{-\gamma z'}), \qquad (6)$$

where $C_{mn} = 4$ for $(m, n > 0)$, $C_{mn} = 2$ for $(m = 0$ or $n = 0$ but not both being zero), $C_{00} = 1$ and $\gamma = \sqrt{(m\pi/a)^2 + (n\pi/b)^2}$. The subscript 's' denotes the source layer, and the subscript 'f' denotes the field layer. The coefficients $\alpha_s^{u,l}$, $\alpha_f^{u,l}$, $\beta_s^{u,l}$, $\beta_f^{u,l}$ can be derived with the help of recursive formula, similar to the one in [7].

The major advantage of using the exponential function to express f_{mn} is that it makes it feasible to integrate f_{mn} in z-direction analytically and stably, which cannot be done using the expression in [8]. In the next section, the computation involving integration in z-direction will be presented in detail.

III. DERIVATION OF THE PANEL INTEGRATIONS FOR 3-D EXTRACTION

Suppose the coordinates for two conductors are as shown in Fig. 1. Each of the conductors' faces is treated as an independent panel. To find p_{ij}, (5) is integrated over the surfaces of panel i and panel j. If one substitutes (5) in (2), and interchanges the order of the operations, the integral can be computed analytically, leaving a 2-D infinite summation.

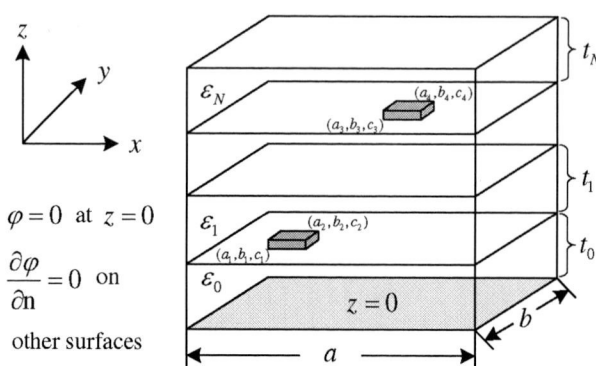

Fig. 1. Geometry and boundary condition of multi-layer substrate.

To compute the 3-D capacitance, the panels of a cuboid conductor is grouped into three classes, horizontal (in x-y plane), vertical I (in x-z plane) and vertical II (in y-z plane). The positional combination of panel pairs can be classified as four types as shown in Fig. 2. In type A, the two panels are both horizontal ones, i.e., on the bottom or the top surfaces of the conductors. In type B, one of the panels is horizontal, and the other is vertical. In types C and D, both panels are vertical. They are distinguished by whether the panels are parallel or perpendicular to each other. In the following subsections, each type of panel pairs is discussed in turn.

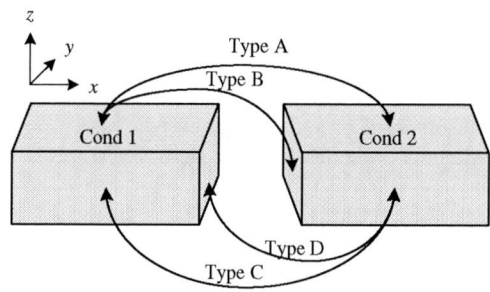

Fig. 2. Positional relations of two panels

A. Horizontal-Horizontal (Type A)

To compute the potential coefficient of two panels in x-y plane, the Green's function is integrated over them. Excluding the $m = 0$ and $n = 0$ case, it can be written as [8]

$$p_{ij} = \sum_{m=1}^{S-1} \sum_{n=1}^{S-1} \frac{f_{mn}(z, z')}{A_1 A_2} \int_{a_1}^{a_2} \int_{b_1}^{b_2} \int_{a_3}^{a_4} \int_{b_3}^{b_4} \times$$
$$\cos(\delta_m x')\cos(\xi_n y')\cos(\delta_m x)\cos(\xi_n y)dydxdy'dx', \qquad (7)$$

where S is the upper bound of the truncated infinite series. The integration in (7) can be performed easily, resulting in sine functions, and it can be cast into a sum of 64 terms in the form

$$p_{ij} = \sum_{m=1}^{S-1} \sum_{n=1}^{S-1} \frac{f_{mn}}{A_1 A_2 \delta_m^2 \xi_n^2} \cos[\delta_m(a_{1,2} \pm a_{3,4})] \cos[\xi_n(b_{1,2} \pm b_{3,4})].$$
(8)

Thus, for a specific technology, the matrix of DCT (discrete cosine transform), i.e.

$$K_{pq} = \sum_{m=1}^{S-1} \sum_{n=1}^{S-1} \frac{f_{mn}}{\delta_m^2 \xi_n^2} \cos(m\pi\frac{p}{S}) \cos(n\pi\frac{q}{S}),$$
(9)

can be computed beforehand and stored in a database for $1 \leq p, q \leq S$. When the layout information, i.e. the coordinates a_1-a_4, b_1-b_4, is given, the potential coefficients can be obtained by simply access in matrix $\mathbf{K}$ and summating the 64 terms.

The computation of (9) need to be computed only once for a given technology, and it can be very efficient by use of the fast Fourier transform. For more detail refer to [7].

B. Horizontal-Vertical (Type B)

For the case of type B, since a vertical panel is involved, the computation of potential coefficients consists of three integrations in x- and y- directions and one integration in z-direction. With the presence of (6), it is very easy to perform the analytical integration of $f_{mn}(z', z)$ in z-direction:

$$f_{mn}^{(1)} = \int_{c_1}^{c_2} f_{mn}(z', z) dz$$
(10)

The remaining three integrations in x-y plane are the same as in type A. Suppose the vertical panel is field panel, and it is in x-z plane with y-coordinate being b_f, the integration of potential coefficient can be written as

$$p_{ij} = \sum_{m=1}^{S-1} \sum_{n=1}^{S-1} \frac{\cos(\xi_n b_f)}{A_1 A_2} \int_{a_1}^{a_2} \int_{b_1}^{b_2} \int_{a_3}^{a_4} \int_{c_3}^{c_4} \times$$
$$\cos(\delta_m x')\cos(\xi_n y')\cos(\delta_m x) f_{mn}(z, z') dz' dx dy' dx'$$
$$= \sum_{m=1}^{S-1} \sum_{n=1}^{S-1} \frac{f_{mn}^{(1)}}{A_1 A_2 \delta_m^2 \xi_n} [\sin(\xi_n b_2) - \sin(\xi_n b_1)] \cos(\xi_n b_f) \times$$
$$[\sin(\delta_m a_2) - \sin(\delta_m a_1)] [\sin(\delta_m a_4) - \sin(\delta_m a_3)].$$
(11)

It can be cast into a sum of 32 terms just like in the case of type A, and it can be recast into the form of discrete cosine-sine transform (DCST)

$$K_{pq} = \sum_{m=1}^{S-1} \sum_{n=1}^{S-1} \frac{f_{mn}^{(1)}}{\delta_m^2 \xi_n} \cos(m\pi\frac{p}{S}) \sin(n\pi\frac{q}{S}),$$
(12)

which is very similar to (9), and it can also be computed rapidly by the use of the fast Fourier transform.

C. Vertical-Vertical (Types C, D)

For types C and D, the two panels are both vertical. The computation of p_{ij} consists of two integrations in x-y plane and two integrations in z-direction. Since the coefficients are discontinuous at the source point, depending on whether the two panels are in the same layer or not, the analytical integration in z-direction

$$f_{mn}^{(2)} = \int_{c_1}^{c_2} \int_{c_3}^{c_4} f_{mn}(z', z) dz dz'$$
(13)

should be derived individually, and it can be done with the help of mathematical tools, such as Mathematica [13].

The remaining derivation of p_{ij} includes two integrations in x-y plane, and the resulted matrix $\mathbf{K}$ are DCT (for type C) and DST (discrete sine transform, for type D), respectively. Due to length limit, the detail will not be presented here.

IV. NUMERICAL RESULTS

The algorithm described above is implemented in a solver named SCAPE (Substrate Coupling Analyzer for Passive Elements). In this section, several examples will be shown to verify SCAPE and to compare the accuracy and efficiency with other methods. The simulations are done on a Sun Blade 2000 with Ultra SPARC III Cu processors at 900 MHz and 2 GB memory.

The test cases are $k \times k$ ($k = 2 \sim 5$) crossing conductors in five dielectrics as shown in Fig. 3. The size of each conductor is $(2k + 1) \times 1 \times 1$ (unit in micron). The spacing between neighboring conductors in the same layer is 1μm. The relative permittivity of every dielectric layer is 3.9. The thickness is 28 for the top layer and 1 for the rest layers. Each of the structures are surrounded by a $32 \times 32 \times 32$ box. The bottom of the box is a perfect ground plane, and the other five surfaces of the box are Neumann boundaries.

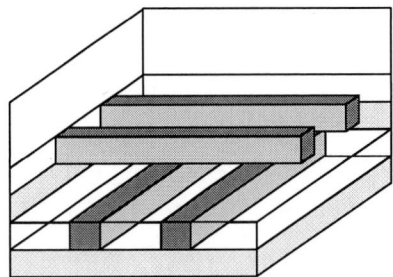

Fig. 3. 2×2 crossover in dielectrics

The above crossover problems are computed by FastCap with very fine meshing, denoted by FastCap I, FastCap with relative coarse meshing, denoted by FastCap II, ASITIC, and SCAPE. Both the expansion order of FastCap I and FastCap II is 2. It is the most accurate version of FastCap, within a reasonable amount of time. In the input of FastCap, 0 is assigned to the permittivity of outer space to handle the boundary condition, and each interface between the dielectric layers is specified to make the comparisons equitable.

In ASITIC, since it can not handle conductors with finite thickness, to model thickness effect to the best of its abilities, the upper and lower surfaces of conductors are treated independently as conductors of zero thickness and extracted as such [11]. After extraction, the resulted capacitance matrix is reduced by combining together the top and the bottom plates of the conductors. This is electrically equivalent to shorting the top and the bottom plates together.

Using the capacitance matrix $\mathbf{C}$ computed by FastCap I, i.e. FastCap with very fine meshing, as the standard, the error of the capacitance matrix $\mathbf{C}'$ computed by another program is estimated in the two-norm: $||\mathbf{C}' - \mathbf{C}||/||\mathbf{C}||$. The error of FastCap II is set to around 3%, controlled by adjusting the mesh size.

Table I compares SCAPE with ASITIC and FastCap. The following is a summary of the comparison.

1. Using FastCap I's result as criterion, SCAPE is more accurate than FastCap II, which is within 2.5% and around 3%, respectively. ASITIC does not account for the sidewall effect, so its error is unacceptably large ($> 20\%$).

2. Since SCAPE does not need to discretize the ground plane, the boundary, and the dielectric interfaces, it uses much fewer panels than FastCap I and II, thus it is much faster than them. The speedup compared to FastCap I is 169 to 42, and 36 to 6 to FastCap II.

3. SCAPE uses 1/14 to 1/6 of the memory used by FastCap I, and about 1/3 of the memory used by FastCap II. The memory required by SCAPE grows slower than FastCap does with the problem size grows. This is because in SCAPE most of the memories are used to store the Green's function, which is fixed when the problem size grows. So it is better for large scale problems.

It is worth noting that, in the above simulations with FastCap, the substrate is only 32 μm$\times$32 μm, which is much smaller than reality. For an actual technology, the problem size for FastCap will increase drastically, while for the new algorithm it is fixed, since the dielectric interface charge is accounted for during the computation of Green's function.

V. CONCLUSION

The multi-layered Green's function-based algorithm proposed in [7] for substrate coupling analysis is extended to 3-D domain. To accomplish this, a new recursive formula for computing the Green's function is derived and it is integrated analytically over the sidewall surfaces. Test examples show that it brings prominent improvement in accuracy, compared to ASITIC.

TABLE I
COMPARISON FOR $k \times k$ BUS PROBLEMS.

Test problem				
	2×2	3×3	4×4	5×5
FastCap I (with fine mesh)				
CPU Time (s)	54	120	218	349
Memory (MB)	111	131	160	197
Panel #	7812	8724	9948	11492
FastCap II (with coarse mesh)				
CPU Time (s)	11.6	19.1	31.1	46.6
Memory (MB)	56	64	75	89
Panel #	3628	4080	4684	5448
Error (%)	2.89	2.84	2.85	2.76
ASITIC				
Panel #	50	84	144	220
Error (%)	23.7	24.8	25.8	25.4
SCAPE				
CPU time (s)	0.32	1.14	3.41	8.39
Memory (MB)	19	20	21	25
Panel #	146	276	464	700
Error (%)	0.99	0.91	1.60	2.38

VI. ACKNOWLEDGMENTS

This work is supported by a research grant (2004AA1Z11050) from the National 863 Plan sponsored by the Ministry of Science and Technology in China. The collaboration with Cadence in San Jose, U.S. is greatly appreciated.

REFERENCES

[1] *Raphael User's Manual*, Synopsys, 2003.

[2] *HFSS User Manual*, Ansoft, 2002.

[3] K. Nabors and J. White, "FastCap: a multipole accelerated 3-d capacitance extraction program," *IEEE Trans. Computer-Aided Design*, vol. 10, no. 11, pp. 1447–1459, Nov. 1991.

[4] S. Kapur and D. E. Long, "IES3 : A fast integral equation solver for efficient 3-dimensional extraction," in *Proceedings of International Conference on Computer Aided Design*, San Jose, CA , USA, Nov. 1997, pp. 448–455.

[5] W. Shi, J. Liu, N. Kakani, and T. Yu, "A fast hierarchical algorithm for three-dimensional capacitance extraction," *IEEE Trans. Computer-Aided Design*, vol. 21, no. 3, pp. 330–336, Mar. 2002.

[6] W. Yu and Z. Wang, "Enhanced QMM-BEM solver for three-dimensional multiple-dielectric capacitance extraction within the finite domain," *IEEE Trans. Microwave Theory Tech.*, vol. 52, no. 2, pp. 560–566, Feb. 2004.

[7] R. Gharpurey and R. G. Meyer, "Modeling and analysis of substrate coupling in integrated circuits," *IEEE J. Solid-State Circuits*, vol. 31, no. 3, pp. 344–353, Mar. 1996.

[8] A. M. Niknejad, R. Gharpurey, and R. G. Meyer, "Numerically stable Green function for modeling and analysis of substrate coupling in integrated circuits," *IEEE Trans. Computer-Aided Design*, vol. 17, no. 4, pp. 305–315, Apr. 1998.

[9] S. Ramo, J. R. Whinnery, and T. V. Duzer, *Fields and Waves in Communication Electronics, 3rd ed.* New York: Wiley, 1994.

[10] A. M. Niknejad and R. G. Meyer, "Analysis, design, and optimization of spiral inductors and transformers for Si

RF IC's," *IEEE J. Solid-State Circuits*, vol. 33, no. 10, pp. 1470–1481, Oct. 1998.

[11] R. Gharpurey and S. Hosur, "Transform domain techniques for efficient extraction of substrate," in *Proceedings of International Conference on Computer Aided Design*, San Jose, California, 1997, pp. 461–467.

[12] R. Gharpurey, "Modeling and Analysis of Substrate Coupling in Integrated Circuits," Ph.D. dissertation, Collage of Engineering, University of California at Berkeley, 1999.

[13] Mathematica User's Guide, Wolfram Research, 2003.

Signal-Path Driven Partition and Placement for Analog Circuit

Di Long, Xianlong Hong, Sheqin Dong

EDA Lab, Department of Computer Science and Technology, Tsinghua University, Beijing 100084, China

longd02@mails.tsinghua.edu.cn; hxl-dcs@mail.tsinghua.edu.cn; dongsq@mail.tsinghua.edu.cn

Abstract-This paper advances a new methodology based on signal-path information to resolve the problem of device-level placement for analog layout. This methodology is mainly based on three observations: thinking of hierarchical design for analog, structural feature of circuit based on signal-path, requirements of matching/symmetry constraint and the reduction of parasitics. The thinking of hierarchical design makes the whole analog circuit divided into core-circuit and bias-circuit. So, the algorithm is designed as two independent steps: core-circuit is placed firstly, and then bias-circuit. The structural feature of circuit based on signal-path and the requirement of matching/symmetry constraint decide the placement pattern of core-circuit. The reduction of parasitics requires the algorithm to select the optimal variants to realize the placement. Experimental results demonstrate that this algorithm can generate the compact layout with high performance and it is universal and effective.[1]

Keywords: signal-path, analog placement, layout automation, circuit partition, symmetry constrain, device merging

I Introduction

Nowadays, SOC integrates all of the circuits on one chip, including digital and analog parts. The physical design of digital circuits is automated to a large extent but the layout of analog circuits is still a manual, time-consuming and error-prone, which makes custom analog layout be a bottleneck in the mixed-signal design flow. The fast changes of demands in ASIC market also require analog layout automation tool to accelerate the whole design process. Thus, the time from product demands to market can be greatly shortened. Analog placement is a vital step in the design flow from circuit schematics to layout. Analog placement automation tool must not only provide a good rectangle packing functionality, but also satisfy analog specifications. Mismatch and parasitics induced by the layout are the most important factors to affect the performance of analog circuit.

In the past two decades, researches have done many researches about analog placement, but no successful commercial products have existed. Generally speaking, the existing methodologies for analog placement automation can be classified into the following several categories.

- The constructive placement techniques, which adopts the increase placement thinking that one module is selected at a time and positioned in the best available location, which is calculated by an evaluation function or directed by an expert system. A placement tool based on expert knowledge is developed in [1]. These methods are very fast, and scales

well with the problem size, but the final placement can be poor because there are no effective methods to decide the order of modules and the view of global optimization and the costs of supporting expert system are expensive.

- A placement technique iteratively combining min-cut partitioning and force-directed placement (FDP) has been employed in an interactive environment for full-custom designs [2]. This method can fast obtain a feasible placement solution satisfying geometrical constraints but not ensure that the placement is a good placement.

- Nowadays, the simulated annealing [3] and genetic algorithms [4], which are global and stochastic optimization algorithms, are the most effective engines to solve the analog placement problem. There is no mathematical limit for the solution space of the problem and the cost function. Analog constraints can be easily implemented by adding the extra punishment items in the cost function. ILAC [5], KOAN/ANAGRAM II [6], PUPPY-A [7] and LAYLA [8] all adopt these optimization algorithms. But these methods all adopt absolute coordinate to represent placement and explore an extremely large search space including feasible and unfeasible placement solutions.

- Analog placement method based on topological representations, which still adopts the optimization engines of *SA* or *GA*. Topological representations are used to describe the placement, such as BSG [9], SP [10], CBL [11]. Solutions in search space expressed by topological representations are all feasible, so this solution space is much smaller than that represented by absolute coordinate. In [12], the analog constraint of matching/symmetry is realized by the topological representation, which further reduces the solution space. But, from the analysis of solution space in [9], [10], [11] and [12], we can find these solution spaces are still very big and [12] only considers the symmetry constraint.

This paper advances a novel methodology of signal-path driven partition and placement for analog circuit, which can automatic generate a compact placement with high performance from a circuit schematics based on signal-path information. The methodology is mainly based on three observations: thinking of hierarchical design for analog circuit, structural feature of circuit based on signal-path, requirements of matching/symmetry constraint and the reduction of parasitics. The above heuristic information about circuit make solution space greatly decreased.

This paper is organized as follows. Section II will give some necessary definitions and the algorithm of signal-path driven circuit partition for analog circuit. Section III will analyze foundation why this methodology can direct placement of core-circuit and then present the algorithm of core-circuit placement in details. Section IV will present the

---

[1] This work is supported by "National Nature Science Foundation of China: 90307005" and "NSFC: 60473126" and "Hi-Tech Research & Development (863) Program of China 2004AA1Z1050".

0-7803-9451-8/06/$20.00 ©2006 IEEE.

algorithm of bias-circuit placement. In the last section, we will illustrate some layouts generated by the algorithm according to the industrial circuit schematics.

II. Signal-Path Driven Circuit Partition for Analog

Generally speaking, the design of analog circuit is hierarchical. Firstly, according to the whole specifications of analog circuit, designers divide it into several sub-functional modules, and then design sub-circuit schematics for each sub-functional module. In each sub-circuit, designers separately design core-circuit and bias-circuit. The layout design is also hierarchical. The methodology of hierarchical design and layout is illustrated in Fig.1.

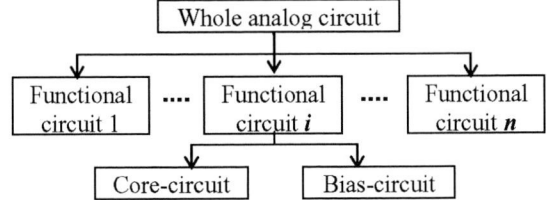

Fig.1 hierarchical design and layout for analog circuit

The hierarchical methodology of manual design and layout for analog circuit indicates that automation algorithm of analog placement should follow the methodology. Firstly each sub-circuit are placed separately and then combined together. For each sub-circuit, firstly core-circuit are placed and then bias-circuit are placed with core-circuit, which are regarded as a black box. Before introducing the methodology of signal-path driven partition and placement for analog circuit, it is necessary to give some definitions.

Definition 1: core-circuit, which is the main circuit in each unit-functional circuit, is responsible for transporting and processing analog signal. There are very strict constraints of matching, symmetry and parasitics minimization on it.

Definition 2: bias-circuit is responsible for providing bias voltages and bias currents for some MOS transistors in *core-circuit*.

Definition 3: power-earth transistor chain is a chain composed of transistors satisfying the following conditions:

● The source/drain of each transistor in the chain must be connected to the power net or the earth net or the source/drain of another transistor in the same chain.

● There is one and only one transistor in the chain, source/drain of which is connected to the power net and there is one and only one transistor in the chain, source/drain of which is connected to the earth net.

In analog circuit, there often exist several *power-earth transistor chains*, each of which is named *pec* and denoted by $\langle Mi_1, Mi_2, ..., Mi_n \rangle$. There is an example for *pecs* in Fig.2, where different color lines stand for different *pecs*. In Fig.2 there are total six *pecs*, which are $pec_1 = \langle M5, M1, M3 \rangle$, $pec_2 = \langle M5, M2, M4 \rangle$, $pec_3 = \langle M8, M6, M10, M3 \rangle$, $pec_4 = \langle M9, M7, M11, M4 \rangle$. $pec_5 = \langle M12, M14 \rangle$ and $pec_6 = \langle M13, M15 \rangle$. It is also possible that the same transistor belongs to different *pecs*. In Fig. 2, the

transistor **M5** belongs to pec_1 and pec_2 at the same time and the transistor **M3** belongs to pec_1 and pec_3 at the same time.

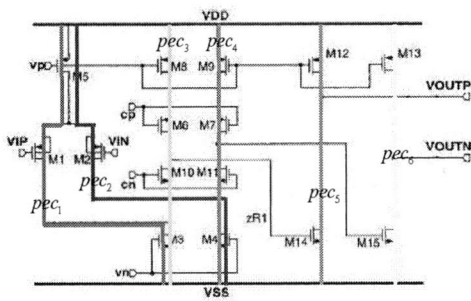

Fig.2 Schematics of full-differential Miller-compensated two-stage amplifier and the diagram of all *pecs*

Definition 4: signal-path is a special type of *pec*, the generation algorithm of which is listed as follows:

● Define the set of signal input nets as I and the set of all *pecs* is Γ_{pec} and transistor $Mj = \{g_j, s_j, d_j\}$ is regarded as a set of its gate net, source net and drain net.

● Define the set of initial input transistors as $\Phi = \{Mj \mid Mj \cap I \neq \phi\}$, so the initial set of signal-paths can be defined as $\Psi_0 = \{pec_i \mid pec_i \cap \Phi \neq \phi\}$;

● This step is a loop procedure: $\Psi_{n+1} = \Psi_n \bigcup \Psi_{sd_n} \bigcup \Psi_{g_n}$, where

$\Psi_{sd_n} = \left\{ pec_i \mid \left(pec_i \in \Gamma_{pec} - \Psi_n \right) \wedge \left(\exists pec \in \Psi_n \right) pec_i \cap pec \neq \phi \right\}$ and

$$\Psi_{g_n} = \left\{ pec_i \left| \begin{array}{l} \left(pec_i \in \Gamma_{pec} - \Psi_n \right) \wedge \left(\exists pec \in \Psi_n \right) \\ \left(\exists Mj \in pec_i \right) \left(\exists Mh \in pec \right) s_h = g_j \vee d_h = g_j \end{array} \right. \right\}.$$

In fact, the current of pec_i in the set Ψ_{sd} is controlled by the source/drain current of pec, which is the direct front-stage of pec_i and the current of pec_i in the set Ψ_g is controlled by the voltages of the source/drain nets of pec, which is also the direct front-stage of pec_i.

● The stop condition of the above loop procedure is $\Psi_{sd_n} = \phi \wedge \Psi_{g_n} = \phi$.

We still take the schematics in Fig. 2 as an example to demonstrate the *signal-paths* generation algorithm.

1. $\Gamma_{pec} = \{pec_1, pec_2, ..., pec_6\}$ and $I = \{VIP, VIN\}$, so $\Phi = \{M1, M2\}$ and $\Psi_0 = \{pec_1, pec_2\}$;

2. when $n = 0$, $\Psi_{sd_0} = \{pec_3, pec_4\}$ and $\Psi_{g_0} = \phi$. So $\Psi_1 = \Psi_0 \bigcup \Psi_{sd_0} \bigcup \Psi_{g_0} = \{pec_1, pec_2, pec_3, pec_4\}$;

3. when $n = 1$, $\Psi_{sd_1} = \phi$ and $\Psi_{g_1} = \{pec_5, pec_6\}$. So $\Psi_2 = \Psi_1 \bigcup \Psi_{sd_1} \bigcup \Psi_{g_1} = \{pec_1, pec_2, pec_3, pec_4, pec_5, pec_6\}$.

4. when $n = 2$, $\Psi_{sd_2} = \phi$ and $\Psi_{g_2} = \phi$, stop.

Now, we can find $\Gamma_{pec} = \Psi_2$, which explains that the schematics in Fig. 2 only include *core-circuit* because the *bias-circuit* providing bias voltage for nets of **vp, cp, cn, vn** is not drawn in the schematics. Generally speaking, when the generation algorithm stops the set Ψ_n is *core-circuit* and the set $\Gamma_{pec} - \Psi_n$ is *bias-circuit*.

III. Algorithm of Core-Circuit Placement

In the two phases of *core-circuit* placement and *bias-circuit* placement, we must orient all the MOS transistors in the same direction, because MOS transistors that do not lie parallel to one another become vulnerable to stress and tilt induced mobility variations, which can cause several percent variations in their transconductance. In this method, the gate of each MOS transistor is kept vertical.

Before placing the devices on the die, we must know what placement constraints are required on these devices. These constraints can be appointed by customs or distinguished by circuit analysis algorithm. Matching and symmetry are the most important constraints for placement, which can be extracted by the algorithm in [13].

Through the generation of the set Ψ_n, we obtain circuit structure information that the *core-circuit* is composed of *signal-paths* and these *signal-paths* have obvious sequence feature from input to output. So, the following placement pattern can be used to realize the *core-circuit* placement:

- Firstly, realize the inner placement of each *signal-path* in the set Ψ_n. Pay attention to the feature that the MOS transistors in the same *signal-path* all have the sequential drain/source connection relationship and the signal is transported in the form of drain/source current, so minimization of connection capacitor and resistor requires these MOS transistors are placed close to each other. Based on the above consideration, Layout all the MOS transistors in each *signal-path* from left to right according to the connection sequence from the power net to earth net. To obtain a compact layout, the height of these MOS transistor layouts must approximate to each other.

- Realize the whole placement of the *core-circuit* composed of *signal-paths*. There are following three different placement patters according to different situations of symmetry constraints: (here, assume $\Psi_n = \{pec_1, ..., pec_j\}$, and according to generation algorithm of *signal-paths*, the sets of $\Psi_0, \Psi_{sd_0}, \Psi_{g_0}, ..., \Psi_{sd_{n-1}}, \Psi_{g_{n-1}}$ are arranged according to generation sequence)

1. If the *core-circuit* is full-symmetrical structure, the integer j must be even number. We assume pec_k is symmetrical with pec_{j-k+1}. So, the set Ψ_n is divided into two sets: $\Psi S_n = \{pec_1, ..., pec_{j/2}\}$ and $\Psi M_n = \{pec_{1+j/2}, ..., pec_j\}$. Place all the *signal-paths* in set ΨS_n from bottom to top according to

the sequence of elements in the sets $\Psi_0 - \Psi M_n, \Psi_{sd_0} - \Psi M_n, \Psi_{g_0} - \Psi M_n, ..., \Psi_{sd_{n-1}} - \Psi M_n, \Psi_{g_{n-1}} - \Psi M_n$. At last, mirror the placement of ΨS_n with the horizontal symmetry axis. Fig. 6 is an example for this case.

2. If the *core-circuit* is not full-symmetrical structure, the placement of the symmetrical part is realized according to method in item 1 and the placement of the non-symmetrical part is realized according to the placement method for ΨS_n in item 1. Finally, place the non-symmetrical part above the symmetrical part. Fig. 7 is an example for this case.

3. If part but not all of transistors in pec_i have symmetry constraints with others in pec_j, the pec_i and pec_j are grouped together and the transistors without symmetrical counterparts are regarded as self-symmetry. Fig. 7 is an example for this case.

- If $Mi_n \in pec_h$ and $Mi_m \in pec_k$ need to be placed according to common centroid style [15], pec_h and pec_k, must be placed in the most front of the generation sequence. pec_j, in which there are the output nets, must be placed in the most back of the generation sequence.

To make the outline of *core-circuit* layout approximate to a rectangle, the width of all *signal-paths* must approximate to each other.

The thinking of the above placement pattern is mainly based on the following two observations. Firstly, the metals used to connect the inner nets in *signal-path* are to transport the currents, so these metals are required very short. What's more, it is possible that the adjacent transistors can be merged together. Secondly, the metals used to connect the nets among *signal-paths* are to transport the voltage, so these metals are not required very short.

After the pattern of *core-circuit* placement is determined, we will introduce the algorithm to realize the equal height of MOS transistors in the same *signal-path* and the equal width of all the *signal-paths*. Before introducing the algorithm, we must firstly introduce the fundamental element to realize the algorithm. That is the variants of MOS transistors.

The transistor is often divided into several sections called fingers. Different partition methods will generate different variants with different aspect ratio, finger number and parasitic capacitor. Any $Variant_m$ and $Variant_n$ of the same MOS transistor must hold the equation $W_m \times F_m = W_n \times F_n$, where W and F stand for channel width and the finger number respectively. Different layouts for the same transistor are illustrated in Fig.3.

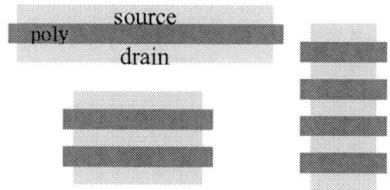

Fig. 3 layouts of varieties of the same MOS transistor

In COMS circuit, the parasitic capacitor of MOS transistor is mainly composed of source/drain bulk capacitance, which can be calculated by formula (1), where C_{jSBt} refers to the total source/drain bulk capacitance, A and P to the source/drain area and perimeter, C_j and C_{jsw} to the bottom and sidewall junction capacitances in absence of any junction voltage and ϕ_j to the built-in junction potential. m_j and m_{jsw} depend on the doping profile of the junction.

$$C_{jSBt} = \frac{AC_j}{\left(1 - \frac{V_{BS}}{\phi_j}\right)^{m_j}} + \frac{PC_{jsw}}{\left(1 - \frac{V_{BS}}{\phi_j}\right)^{m_{jsw}}} \tag{1}$$

For the given MOS transistor, channel width W and length L are decided and the only variable is finger number F. For simple we assume $\alpha = \left(1 - \frac{V_{BS}}{\phi_j}\right)^{m_j}$ and $\beta = \left(1 - \frac{V_{BS}}{\phi_j}\right)^{m_{jsw}}$, so C_{jSBt} can be expressed as the function of F in formula (2), which can be simplified as formula (3). It is obvious that the formula (3) exits the minimum. The meaning of δ_{gap} is illustrated in Fig. 5.

$$C_{jSBt}(F) = \alpha \frac{W}{F} \times \left[FL + (F+1)\delta_{gap} \right] + 2\beta \left[\frac{W}{F} + FL + (F+1)\delta_{gap} \right] \tag{2}$$

$$C_{jSBt}(F) = \left(\alpha WL + \alpha W\delta_{gap} + 2\beta\delta_{gap} \right) + 2\beta \left(L + W\delta_{gap} \right)F + W\left(\alpha\delta_{gap} + 2\beta \right)\frac{1}{F} \tag{3}$$

For a given group of parameter values, we will obtain a graph of function $C_{jSBt}(F)$ in Fig.4, where the horizontal coordinate represents the finger number and the vertical coordinate represents the source/drain bulk capacitance.

The algorithm of placement for *core-circuit* utilizes the feature that variants of the same MOS transistor have different height and width to realize the requirements of equal height of MOS transistors in *signal-path* and equal width of *signal-paths* in *core-circuit*.

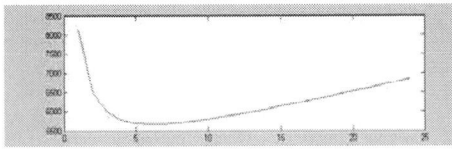

Fig.4 graph of function $C_{jSBt}(n)$

Though the pattern of *core-circuit* placement is determined there are still many different placements because the same MOS transistor has many variants. The *core-circuit* placement algorithm is to realize the following objectives:
- minimizing the differences of the height of all MOS transistor layouts in the same *signal-path*;
- minimizing the differences of the width of all *signal-paths* in the core-circuit;
- minimizing the total capacitance parasitics of all the MOS transistors;
- maximizing the area utility.

So, in fact the *core-circuit* placement problem is an optimization problem. We use the simulated annealing

algorithm as optimization engine. The most important things to design simulated annealing algorithm are the definition of cost function and the strategy of new solution generation.

For the facility of defining the cost function, we give the following rules: according to placement pattern, we number the MOS transistors in the set pec_i belonging to the set Ψ_n form left to right as $M_{i,1}, M_{i,2}, ..., M_{i,h(i)}$ and number the *signal-paths* from top to bottom as $pec_1, pec_2, ..., pec_l$.

The cost function is defined in formula (4), where H_{max_diff} denotes maximum in all values of the maximum height differences between any two MOS transistors in the same *signal-path*. W_{max_diff} denotes maximum width difference between any two *signal-paths* in *core-circuit*. P_{cap} denotes the sum of source/drain parasitical capacitance of all MOS transistors. U_{area} denotes the utility of layout area. Customs can tune the coefficients of $\alpha, \beta, \gamma, \delta$ to obtain different *core-circuit* layout as expected.

$$CP_{core} = \alpha H_{max\_diff} + \beta W_{max\_diff} + \gamma P_{cap} + \delta(1 - U_{area}) \tag{4}$$

$$H_{max\_diff} = \max_{pec_k \in \Psi_n} \left[\max_{M_{k,i}, M_{k,j} \in pec_k} \left| height(M_{k,i}) - height(M_{k,j}) \right| \right] \tag{5}$$

$$W_{max\_diff} = \max_{pec_i, pec_j \in \Psi_n} \left| width(pec_i) - width(pec_j) \right| \tag{6}$$

$$P_{cap} = \sum_{i=1}^{n} \sum_{j=1}^{h(i)} C_{jBSt}(F_{i,j}) \tag{7}$$

$$U_{area} = \frac{\sum_{i-1}^{l} \sum_{j-1}^{h(i)} height(M_{i,j}) \times width(M_{i,j})}{\max_{pec_i \in \Psi_n} \left[width(pec_i) \right] \times \left[\sum_{i-1}^{l} \max_{M_{i,j} \in pec_i} \left[height(M_{i,j}) \right] + \sum_{i-1}^{l-1} RSS(pec_i, pec_{i+1}) \right]} \tag{8}$$

$$height(M_{i,j}) = W_{i,j} / F_{i,j} + 2\delta_{head} \tag{9}$$

$$width(pec_i) = \left[\sum_{j-1}^{h(i)} F_{i,j} L_{i,j} + (F_{i,j}+1)\delta_{gap} \right] + \sum_{j-1}^{h(i)-1} RSM(M_{i,j}, M_{i,j+1}) \tag{10}$$

In the above formulas, δ_{gap} and δ_{head} are the constant parameters decided by design rule. Function $RSM(M_{i,j}, M_{i,j+1})$ denotes the spacing between two MOS transistor layouts and function $RSS(pec_i, pec_{i+1})$ denotes the spacing between two *signal-paths*. These two functions can be assigned two different constants or decided by some interconnection area estimation model. For clarity, the diagram of these parameters is illustrated in Fig. 5. If the transistor $M_{i,j}$ and the transistor $M_{i,j+1}$ can be merged in the horizontal direction, the value of the function $RSM(M_{i,j}, M_{i,j+1})$ is zero.

Firstly, the function $height(\ )$ and $width(\ )$ in formulas (5), (6), (8) are substituted with formula (9) and (10) respectively. Secondly, the function $C_{jBSt}(\ )$ in formula (7) is substituted with formula (3). Lastly, substitute all the variable items in formula (4) with formulas (5), (6), (7), (8). Thus, we can find that the formula (4) is the function of finger numbers of all MOS transistors. So, the generation of new solution is to change finger number of some MOS transistor. Firstly, the algorithm randomly selects a MOS transistor assumed to be $M_{i,j}$ and then randomly selects an integer from the range $\left[1, \lceil W_{i,j}/L_{i,j} \rceil\right]$ as finger number of $M_{i,j}$.

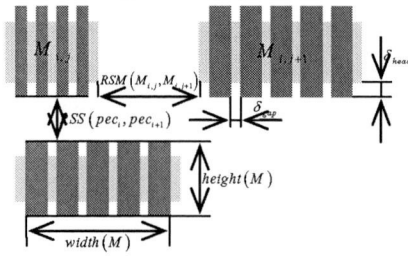

Fig. 5 diagram of parameters in formula (5)

Fig.6 is the *core-circuit* placement of schematics in Fig.2, which is generated by the above *core-circuit* placement pattern and optimization algorithm.

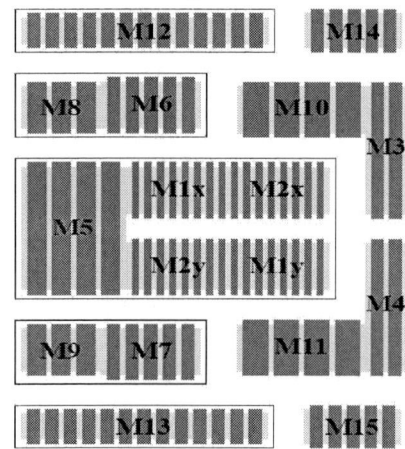

Fig. 6 core-circuit placement of schematics in Fig.2

IV. Placement of Bias-Circuit

After completing the placement of *core-circuit*, we will realize the placement of whole circuit including *core-circuit* and *bias-circuit*, which is called the second phase placement. In this phase of placement, the *core-circuit* layout is regarded as a black box. There is no strict feature of circuit structure in *bias-circuit*, so the placement of *bias-circuit* has no determined pattern and the stochastic placement algorithm is adopted to realize the placement of *bias-circuit*. Of cause, in the phase of *bias-circuit* placement, there are the following objectives to be optimized:
- minimizing the total parasitic capacitance of all the MOS transistors belonging to the *bias-circuit*;
- minimizing the total length of routing metals;
- maximizing the area utility.

$$CP_{bias} = \omega P_{cap} + \theta L_{routing} + \lambda\left(1 - U_{area}\right) \quad (11)$$

It is possible there is matching/symmetry constraint forced on some MOS transistors in *bias-circuit*, so the topological representation based on SP [10] is adopted to describe the placement of *bias-circuit*, which is easy to realize symmetry constraint during the process of stochastic placement [12]. The MOS transistors in *bias-circuit* are all oriented in the same direction with those in *core-circuit*. The selection of variants and device merging is also considered in this phase of placement. The selection of variants is realized when new solution is generated and device merging is settled by the algorithm in [14], which is also based on the topological representation of SP. The only difference is only horizontal merging is considered in *bias-circuit* placement because the MOS transistors are oriented in one direction.

V. Experiment and Conclusion

In Fig. 7 (a), $\Gamma_{pec} = \{pec_1, pec_2, ..., pec_6\}$ and $I = \{inn, inp\}$, through the *signal-paths* generation algorithm, we can obtain $\Psi_n = \{pec_3, pec_4, pec_2, pec_5, pec_6\}$. The elements in the set Ψ_n are arranged by the generation sequence. Elements in the set $\Gamma_{pec} - \Psi_n = \{pec_1\}$ are bias-circuit. The algorithm in [12] obtains pec_3 and pec_4 are full-symmetry pair; pec_2 and pec_5 are partial-symmetry pair; pec_6 is non-symmetry *signal-path*.

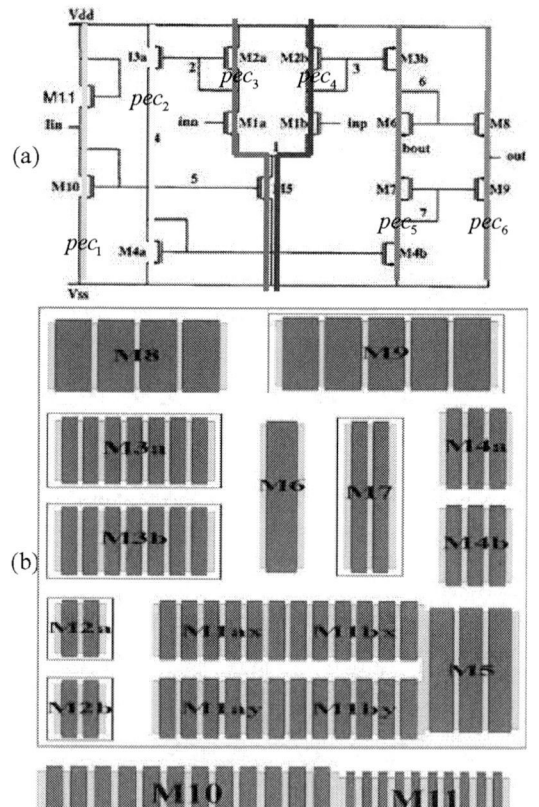

Fig.7 test case-1 (a) Schematics of a high-speed CMOS comparator (b) Placement of the schematics

In Fig. 8, for clarity core-circuit and bias-circuit are drawn separately. When (a) and (b) are regarded as an integer, we obtain $\Gamma_{pec} = \{pec_1, pec_2, ..., pec_{12}\}$ and $I = \{I_{in+}, I_{in-}\}$. Through the *signal-paths* generation algorithm, we obtain $\Psi_n = \{pec_4, pec_5, pec_6, pec_7, pec_1, pec_2, pec_3, pec_8, pec_9, pec_{10}\}$. The elements in the set Ψ_n are arranged by the generation sequence. After adjusting the elements sequence, we obtain sequence $pec_5, pec_6, pec_4, pec_7, pec_2, pec_3, pec_8, pec_9, pec_{10}, pec_1$. The algorithm in [12] can obtain the full-symmetry group of pairs $(pec_5, pec_6), (pec_4, pec_7), (pec_1, pec_{10}), (pec_8, pec_2), (pec_3, pec_9)$. Elements in set $\Gamma_{pec} - \Psi_n = \{pec_{11}, pec_{12}\}$ are bias-circuit.

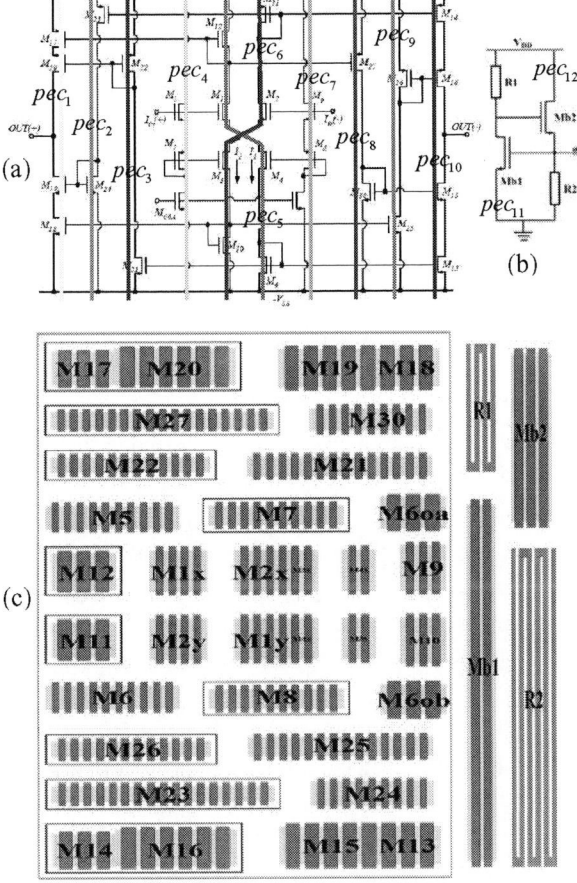

Fig. 8 test case-2 (a) core-circuit of class AB open-loop opamp (b) bias-circuit providing bias-voltage for **M6oa** and **M6ob** (c) the whole layout of the schematics

In this paper, a new methodology of signal-path driven partition and placement for analog circuit is proposed, which sufficient utilizes the thinking of hierarchical design, structural feature of analog circuit based on signal-path and variants of MOS transistors. Experimental results demonstrate this algorithm can generate compact layout with high performance and it is universal and effective.

References

[1] M. Kayal, S. Piguet, M. Declerq, and B. Hochet, "SALIM: A layout generation tool for analog ICs," in Proc. IEEE Custom Integrated Circuits Conf., 1988, pp. 7.5.1–7.5.4.

[2] E. Malavasi, J. L. Ganley, and E. Charbon, "Quick placement with geometric constraints," in Proc. IEEE Custom Integrated Circuits Conf., 1997, pp. 561–564.

[3] S. Kirkpatrick, C. D. Gelatt, and M. P. Vecchi, "Optimization by simulated annealing," Science, vol. 220, no. 4598, pp. 671–680, May 1983.

[4] J. Cohoon and W. Paris, "Genetic placement," IEEE Trans. Computer-Aided Design, vol. CAD–6, pp. 956–964, Nov. 1987.

[5] J. Rijmenants, J. B. Litsios, T. R. Schwarz, and M. Degrauwe, "ILAC: An automated layout tool for analog CMOS circuits," IEEE J. Solid- State Circuits, vol. SC-24, no. 2, pp. 417–425, Apr. 1989.

[6] J. Cohn, D. Garrod, R. Rutenbar, and L. Carley, "KOAN/ANAGRAMII: New tools for device-level analog layout," IEEE J. Solid-State Circuits, vol. 26, pp. 330–342, Mar. 1991.

[7] E. Malavasi, E. Charbon, E. Felt, and A. Sangiovanni-Vincentelli, "Automation of IC layout with analog constraints," IEEE Trans. Computer-Aided Design, vol. 15, pp. 923–942, Aug. 1996.

[8] K. Lampaert, G. Gielen, and W. Sansen, "A performance-driven placement tool for analog integrated circuits," IEEE J. Solid-State Circuits, vol. 30, pp. 773–780, July 1995.

[9] S. Nakatake, K. Fujiyoshi, H. Murata, and Y. Kajitani, "Module packing based on the BSG-structure and IC layout applications," IEEE Trans. Computer-Aided Design, vol. 17, pp. 519–530, June 1998.

[10] H. Murata, K. Fujiyoshi, S. Nakatake, and Y. Kajitani, "VLSI module placement based on rectangle-packing by the sequence-pair," IEEE Trans. Computer-Aided Design, vol. 15, pp. 1518–1524, Dec. 1996.

[11] Xianlong Hong, Sheqin Dong, et al, "Corner Block List Representation and Its Application to Floorplan Optimization", IEEE Transactions on Circuits and Systems, II, 51(5):228-233.

[12] Florin Balasa, Koen Lampaert, "Module Placement for Analog Layout Using the Sequence-Pair Representation", in Proc. DAC, pp. 274-279, 1999.

[13] Su Yi, Sheqin Dong, Qingsheng Hao, Xiangqing He, Xianlong Hong, "Automated Analog Circuits Symmetrical Layout Constraint Extraction by Partition", in Proc. ASICON, pp. 166-169, Oct. 2003.

[14] Liu Rui, Dong Sheqin, Hong Xianlong, Long Di, Gu Jun, "Algorithms for analog VLSI 2D stack generation and block merging", Proc. ISCAS, vol. IV, pp. 716-719, 2003.

[15] Di Long, Xianlong Hong, Sheqin Dong, "Optimal Two-Dimension Common Centroid Layout Automation for MOS Transistor Unit-Circuit", Proc. ISCAS, pp 2999-3002, 2005.

An Approach to Topology Synthesis of Analog Circuits Using Hierarchical Blocks and Symbolic Analysis *

Xiaoying Wang   Lars Hedrich

Department of Computer Science, University of Frankfurt

wang@em.informatik.uni-frankfurt.de

hedrich@em.informatik.uni-frankfurt.de

Abstract— This paper presents a method of design automation for analog circuits, focusing on topology generation and quick performance evaluation. First we describe mechanisms to generate circuit topologies with hierarchical blocks. Those blocks are specialized by adding terminal information. The connection between blocks is in compliance with a set of synthesis rules, which are extracted from typical schematics in the literature. Symbolic analysis has been used to select an appropriate topology quickly and to help the designer gain a better understanding of a circuit's behavior. Finally, experimental results show the creativity and efficiency of our method.

I. INTRODUCTION

Chips based on future nanotechnologies make it possible that more and more complex systems can be integrated on one single chip, which includes digital, analog and mixed-signal sections. Although most functions of such systems are realized in a digital/mixed-signal section, the analog circuits cannot be neglected, because of fundamental necessity (e.g. interface with the outside world).

EDA tools for analog design compared to those for digital design are not yet fully mature and still in the process of exploration and development, because analog design is mostly heuristic and knowledge-intensive. More and more circuits with new topology are required to fit the growing complexity of systems. Hence, analog circuit synthesis is one critical step, which includes topology generation/selection and circuit sizing [1]. Up to now there are several approaches providing ideas and solutions for automated topology synthesis of analog circuits in different ways: OASYS [2], BLADES [3] are based on a library with a set of well-defined topologies, from which a proper topology will be selected. Some sized topology generation approaches e.g. [4] and [5] use genetic algorithms to find a lot of variety of circuits topologies. Another approach from Klumperink [6] explores all elementary circuits with one or two Voltage Controlled Current Sources (VCCS).

In this paper we present a new method in topology synthesis that can be roughly divided into two lines of work, generation and selection of circuit topology:

- Hierarchical design methodology for generation of circuit topology
  - A set of well defined blocks, such as blocks for current mirrors or for differential pair, with specialized signal information of terminals.
  - A set of extracted rules for combination between blocks in consideration of the signal information of terminals.

- Symbolic analysis methodology for selection of circuit topology
  - The circuit topologies performances are analyzed with linear symbolic analysis methods.
  - The symbolic formulas enable fast but reliable estimation of circuit performances taken matching, and the dependencies of linear device parameters in terms of size and biasing into account.

The paper is organized as follows. Section II describes the hierarchical topology generation system. Section III focuses on the symbolic analysis. Section IV presents experimental results, and finally, in section V conclusions are drawn.

II. HIERARCHICAL TOPOLOGY SYNTHESIS

It is known that analog circuit topologies can be represented in many small subcircuits, e.g. differential pair, current mirror, cascode stages, etc. [2][7]. The designer can catch the functions of circuits more easily with help from identification of small subcircuits. The basic idea of our method is: these subcircuits are just like "building blocks" of analog circuits and our aim is how to get analog circuits using these building blocks. To do this, we have defined new blocks with terminal information for building blocks and a set of synthesis rules,

---

*This work is supported by BMBF/edacentrum project SAMS under No. 01M3070D.

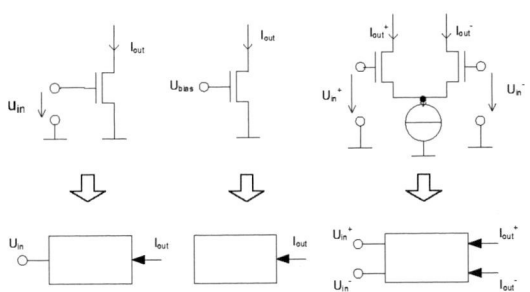

Fig. 1. Some examples of subcircuits and blocks: NMOS common-source stage, NMOS current source and NMOS differential pair

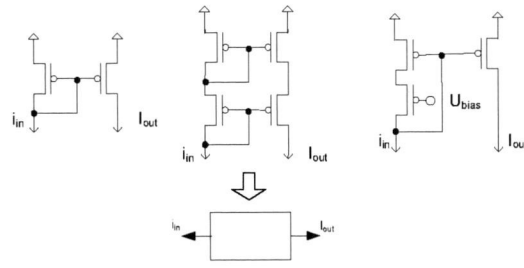

Fig. 2. Current mirrors and their block

and then we can create plenty of block-chains (the connection of blocks in one dimension) and block-nets (the connection of blocks in two dimensions) with respect to the defined rules. These block-chains/-nets can represent the topology of circuits.

A. Features of block

The following features of terminals in each block should be defined for the later topology synthesis:

- *Type of signal*, which tells us if a voltage (U) or current (I) signal is applied to the terminal. Direction of a current signal should be given, which is determined by the bias current direction. Current signal may flow into a block or from a block.

- *Type of terminal*: which is divided into two parts - input or output terminal. The input terminal lies on the left side of the block, and the output on the right side.

- *Impedance*, which is relevant for our method. The size of impedance can be divided into two sections - low or high. For low impedance the type of signal is in lower case, and for high impedance in upper case, e.g. u means a voltage signal with low impedance, and I a current signal with high impedance.

Fig. 1 shows some examples of blocks. These blocks contain not only the transistors but also terminals of power supply (V_{dd}), bias voltage or bias current. Hence, we do not need to worry about how to connect these no signal terminals in topology synthesis. One block can be represented by different transistor level netlist (see Fig. 2). For example, basic, cascode and low power current mirror have the same type of block. After classification of basic subcircuits [8] we have got 24 types of subcircuits and 14 types of block.

B. Rules of Topology Synthesis

In this approach we have defined a set of rules conforming with the standard circuits in literature [8][9] . They describe

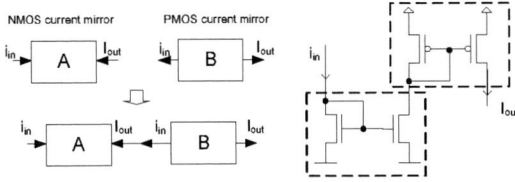

Fig. 3. Example of current rule

the possible cases in which blocks can be connected. Roughly we can divide these rules into 3 categories: *general rules*, *current source rule* and *split & combination rules*.

General Rules make a reasonable connection between two blocks und create various block-chains in one dimension.

- *signal type rule*: block B can be connected to block A, if the type of input signal of block B is the same as the type of output signal of block A.
  It is obvious that a current signal from one block can flow forward into another block, only when the input signal of the second block is also current. The same is valid for voltage signals.

- *current rule*: block B can be connected to block A, if the directions of current signal of both blocks are matched and block B has low input impedance.
  As is known, an NMOS current mirror can connect to a PMOS but not to other NMOS current mirror; a PMOS diode-connected load can connect to an NMOS but not to other PMOS common-source stage. Such cases can be explained by this rule. The current of an NMOS current mirror flows into the block, and that of a PMOS in contrast from the block. At a node where an NMOS and a PMOS current mirror are connected as in Fig. 3, the directions are matched, in other words there are one current entering and one current leaving that node. Block B should also have low input impedance for proper transfer of the current signal (without bias current, i.e. small-signal).

- *Voltage rule*: block B can be connected to block A, if the output signal of block A and the input signal of block B

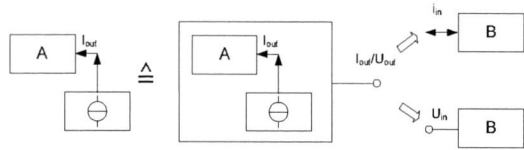

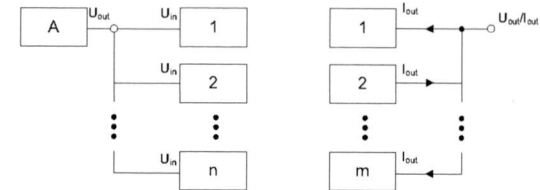

Fig. 4. Current source rule

Fig. 6. Split & combination rules

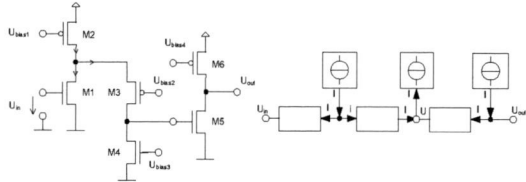

Fig. 5. Example for current source rule: folded cascode

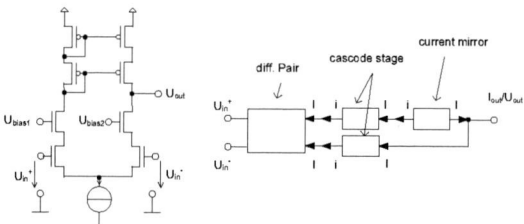

Fig. 7. cascode op amp and its block-net

are voltages and block B has high input impedance.
A high input impedance of a block ensures that a voltage signal will not be damped by a heavy load.

Current Source Rule can create block-chains in quasi-one dimension: The current output terminal with high impedance of a block can be connected with a current direction matched current source. Additionally, the connection node can be treated not only as a current terminal but also as a voltage terminal.

This dual feature of an output terminal is shown in Fig. 4. Treating the terminal as current terminal, we can connect a block with current input to it. The current direction of this block is irrelevant, because there are at least one current entering and one current leaving the node where blocks are connected. If it is treated as a voltage terminal, we can connect a block with voltage input to it. An example is given in Fig. 5. The structure is called folded cascode. M_1 (NMOS common-source stage) connected with M_2 (PMOS current source) is able to connect to block M_3 (NMOS cascode stage), because the current direction of M_1 and M_2 are matched. Let us see how important to have a current source in this circuit. According to the current rule in general rules a direct connection between M_1 and M_3 without M_2 is not allowed, because there are only two current leaving but none current entering the node where the blocks are connected. In order to bias M_1 and M_3, a current source must be added. Therefore this rule expands the design space. Another feature of this rule can also be explained using the example in Fig. 5. The output terminal of M_3 connected with M_4 (NMOS current source) can be treated as a voltage terminal, so that this folded cascode works as a single-stage voltage amplifier. At the right of Fig. 5 is the corresponding block-chain of folded cascode. Blocks in such chains do not really keep in line, so that we call them block-

chains in quasi-one dimension.

Split & Combination Rules expand the connection between blocks in two dimensions and we get a plenty of block-nets, which represent analog circuits.

- *Voltage split rule*: a voltage output signal of a block can be split to feed several blocks.

- *Current combination rule*: output current signals of blocks can be combined to an input of a block. If there is no bias current entering or leaving that node where all blocks are connected, one current source should be added.

These rules are described in Fig. 6. In order to control the complexity of synthesized circuits a reasonable number of blocks, which the voltage signal spilt to or whose current signals can be combined, is smaller then 3.

A common application area of these rules is to determine, how to connect the blocks behind the differential pair. As we see in Fig. 1, there are two current outputs for differential pair block. They can connect to other blocks separately. It seems like that differential pair block is followed by two block-chains. If both block-chains have current outputs with high impedance, both block-chains can be connected together. If both are voltage outputs, they build a differential signal, which can be connected to another differential pair. An example for the first stage of a cascode operational amplifier (schematic and its block-net), also called unbuffered op amp or OTA [9] , is given in Fig. 7.

For a further synthesis with differential pairs there are some other rules to define: one of the differential pair outputs with a bias current flowing out of the block can directly be connected to GND (ground), and one with a bias current flowing into block can be connected to V_{dd} (power supply); one of differential pair inputs can directly be connected to GND.

C. Topology Generation Step

In this section, we will describe how to create circuit topologies by using predefined hierarchical blocks and specification.

- *Block specification*: If we treat block-chain or -net as "black box", then all what we know about this black box are only input and output terminals but not its structure or function. We need the following input information for the op amp in Fig. 7: differential voltage signal at input terminals with high impedance and voltage signal at output terminal with high impedance. The structure of block-chain/-net can not be influenced by user, because they are generated according to synthesis rules. Therefore the input for the topology generator are just the attributes of terminals. This input information can be applied to create a large variety of topologies, but it is also a challenge to select an appropriate circuit from them. This task will be done in the topology selection phase using symbolic analysis. Additional input information is the maximum number of blocks in block-chain or -net, so that the size of a topology and the number of synthesized topologies can be limited. The block size of the op amp shown in Fig. 7 is for example 4.

- *Algorithm*: A structure chart of chaining up blocks is shown in Fig. 8. The algorithm writes one or several corresponding circuits in VHDL-AMS or Spice netlist from a block-chain/-net, since a block can represent one or more analog subcircuits. If we have got a block-net for Fig. 7 from the generator, another op amp with basic current mirror can be also a variant.

III. PERFORMANCE ESTIMATION BY SYMBOLIC ANALYSIS

The topology generation steps generate all possible topologies which meet the rules defined in previous section. The rules restrict the possible structure space implicitly or explicitly to more or less usable circuits. However, after this step a very high number of circuit are good candidates for the design goal. This number has to be reduced to the real feasible circuits. Common topology synthesis approaches use Spice simulations in the loop [5]. Furthermore a full, time consuming sizing of circuit parameters is needed to be able to evaluate the performance of each circuit.

We propose a stepwise method based on symbolic analysis to come up with these time consuming task and split it into several hierarchical steps to detect at a very early stage non sufficient circuit topologies:

a) Calculate symbolic formulas for linear performances in terms of linear device parameters.

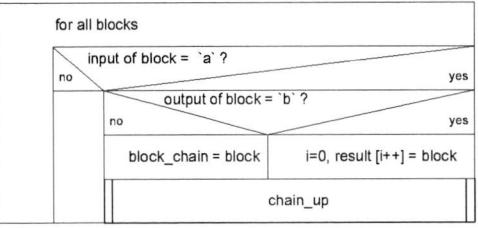

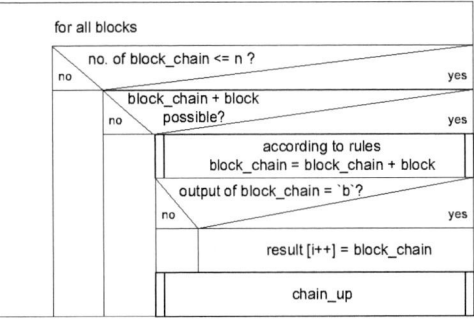

Fig. 8. structure chart of Algorithm

b) Compute for each individual performance formulas in terms of the independent sizing parameters width, length and bias current of (pairs of) transistors.

c) Check the performances against the specification.

d) Do detailed sizing including biasing for the remaining circuits. This step is not included in this paper and will be future work.

Step a), b) and c) can be executed together for each performance to quickly get false results and prevent the time consuming calculation of other performances. In the following, each step is explained for the example of an operational amplifier.

A. Specification

Input for the analysis are n_{Spec} specifications in terms of performances using inequalities:

$$S = \begin{bmatrix} Gain & > & 60\,dB \\ GBW & > & 1 \cdot 10^9\,Hz \\ PSRR & > & 40\,dB \\ R_{out} & > & 100\,k\Omega \end{bmatrix} \qquad (1)$$

On the other hand we have basic device and biasing parameters in predefined ranges according to process technology, which are used for sizing.

$$P = [W_1, L_1, I_{Bias1}, ..., W_n, L_n, I_{Biasn}, ...]$$
$$\text{with} \quad \forall i \quad W_i \in [W_{min}..W_{max}], L_i \in [L_{min}..L_{max}],$$
$$I_{Biasi} \in [I_{Biasmin}..I_{Biasmax}] \tag{2}$$

where W_i, L_i, I_{Biasi} are the width, length and bias current of the transistors of the i-th block and n is the block count. Hence, matching is automatically ensured for each block or transistor pair.

B. Linear Symbolic Analysis

Linear symbolic analysis is a technique for calculating transfer functions in frequency domain in symbolic form [10] [11]. The basis for the analysis is a netlist with transistors being replaced by their linear model using small signal parameters like transconductance g_{mi}, output resistance r_{DSi} and capacitances. Additionally, a test bench is provided to correctly model the environment of the circuit. Using a modified nodal approach a system of differential algebraic equations is setup. We use a technique described in [12] to simplify this system and finally solve it for the output variables. The simplification process needs assumptions about numerical values of the small signal parameters which can easily estimated from technology parameters. We use center values of ranges in equation (2) calculate the correponding small signal values using a nonlinear device model. The result is a transfer function in terms of small signal parameters:

$$H_l(s) = \frac{Y(s)}{X(s)} =$$
$$f(g_{m1}, ... g_{mn}, r_{DS1}, ... r_{DSn}, c_{GS1}, ... c_{GSn}, ...) \tag{3}$$

where $Y(s)$ is the output and $X(s)$ the input of a performance function, for example U_{output} and I_{output} respectively for calculating the output resistance. A symbolic formula for the corner frequency f_c could be derived in a similar way.

C. Performance evaluation

In order to correctly check the performances with respect to matching, same bias currents and dependence between g_m, r_{DS}, W, L the linear transfer function is modified by inserting basic formulas for the small signal parameters in terms of basic device parameters derived from a nonlinear device model.

$$H_d(s) = \frac{Y(s)}{X(s)} =$$
$$f(W_1, ... W_n, L_1, ... L_n, I_{bias1}, ... I_{Biasn}, ...) \tag{4}$$

Finally the transfer functions is checked in ranges for fulfilling the specification. That means evaluating for each specification the performance equation (4) for all corners of a lower dimensional subset of the parameter space P and checking if a feasible solution exists.

These method is quick because in general we have restricted number of parameters in the transfer function, which reduce the dimension of the parameter space. In contrast, a full Spice-based performance evaluation needs to check all corners of all parameters resulting in a basic sizing step (at least the bias voltages should be sized), DC-analysis and AC-analysis step for

each corner, which seems to be prohibitive in runtime. In our case the runtimes for symbolic analysis is much larger than the following corner analysis. A small example of the symbolic analysis will be shown in Section IV.

IV. SYNTHESIS RESULTS

The proposed method was applied to generate and select different circuits types, such as unbuffered op amp and differential-input current amplifier. Using PC with 3GHz prozessor we need $1-2$ minutes to create all circuits and about 2 seconds for analysis one circiuts.

A. Unbuffered op amp

A general application area of analog synthesis is the design of operational amplifiers. From these essential elementary circuits several useful circuits, such as comparator, differential ring oscillator and filters can be derived. The op amps with a differential input and a high output impedance are classified as unbuffered op amps. They are always used as the first stages of op amps. In this section, we synthesized thousand of op amps and selected several appropriate op amps.

The input information for hierarchical method is listed as follows:

- *Input terminals*: A differential voltage signal with high impedance

- *Output terminal*: A voltage signal with high impedance

- *No. of blocks*: 4

- *Rules*: Without voltage split rule, in order to reduce the complexity of the circuits

The following values of technology parameters were chosen for symbolic analysis: $W, L = [1\,10^{-7}..2\,10^{-5}]$, $I_{Bias} = [0.01\,10^{-6}..0.2\,10^{-6}]$, $K_p = 400\,10^{-6}$, $V_E = 5\,10^5$, $c_{gd} = 1pF$, $c_{gs} = 1pF$. The electrical specification for topology selection is described as follows: *DC gain* is more than $80\,dB$, *Corner frequency* is more then $1\,MHz$ and *Output impedance* is more then $400\,k\Omega$.

In topology generation step we have got total 448 block-chains/-nets, which can represent 2732 unsized circuits, because some blocks respond to more then one subcircuits (e.g. basic and cascode current mirror). In the next selection step all these circuits were analyzed and evaluated. Finally 50 circuits were chosen. Some of them have familiar topologies because of complimentary characteristics, i.e. PMOS subcircuits instead of NMOS subcircuits or vice versa. The reason why we didn't choose only one circuit but several circuits is, that the designer can get various supply of circuit topologies for the later sizing step. Here one circuit of them is shown in Fig. 9. And its performance was calculated: DC gain $81.3\,dB$, corner frequency $1.01\,MHz$, output impedance $5\,M\Omega$.

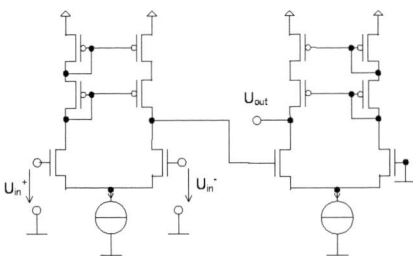

Fig. 9. Op amp generated by synthesis method

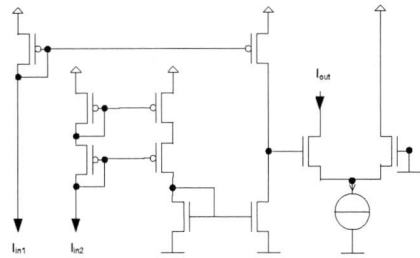

Fig. 10. Differential-input current amplifier generated by synthesis method

B. Differential-input current amplifier

A current amplifier is an amplifier with a defined relationship between the input and output currents, which offers wide signal dynamic range and will find applications in low-voltage and in switched-current circuits. We used the following input information for topology generation:

- *Input terminals*: A differential current signal with low impedance and the bias current flows from the circuits

- *Output terminal*: A current signal with high impedance and the bias current flows into the circuits

- *No. of blocks*: 4

- *Rules*: Without voltage split rule, in order to reduce the complexity of the circuits

The following values of small-signal CMOS model parameters were chosen for symbolic analysis: $g_m = 0.0021S$, $r_{ds} = 100k\Omega$, $c_{gd} = 1pF$, $c_{gs} = 1pF$. In this example we do not use range values for device parameters resulting in only nominal point analysis.

The electrical specification for topology selection is described as follows: *DC gain* should be as high as possible, *Corner frequency* is more then $2MHz$, *Input impedance* is less then $1k\Omega$ and *Output impedance* is more then $100k\Omega$.

There were total 1157 generated block-chains/nets according to 4718 circuits. In Fig. 10 we show one example circuit of 37 selected circuits.

V. CONCLUSION

This paper has introduced a new concept of analog circuit synthesis using hierarchical blocks for topology generation and using symbolic analysis for topology selection. This hierarchical method enables the designer to completely search the topology space with restrictions given by rules. The rules extracts the essential standard from human circuit design. Symbolic analysis helps designer choose some appropriate circuits from a large number of generated topologies. A further step towards an completely automated method would be a subsequent automated sizing process, which will be part of our future work.

We have successfully demonstrated two applications, op amp and current amplifier.

REFERENCES

[1] Georges G. E. Gielen and Rob A. Rutenbar, "Computer-Aided design of analog and mixed-Signal integrated circuits" *Proceedings of the IEEE*, Vol. 88, No. 12, Dec. 2000, pp. 1825-1852

[2] R. Harjani, R. Rutenbar and L. R. Carley, "OASYS: A framework for analog circuit synthesis" *IEEE Trans. Computer-Aided Design*, vol. 8, pp. 1247-1265, Dec.1989

[3] F. El-Turky and E. Perry, "BLADES: An artificial intelligence approach to analog circuit design" *IEEE Trans. Computer-Aided Design*, vol. 8, pp. 680-691, June 1989

[4] W. Kruiskamp and D. Leenaerts, "DARWIN: CMOS opamp synthesis by means of a genetic algorthm" in *Proc. ACM/IEEE Design Automation Conf. (DAC)*, 1995, pp. 550-553

[5] T. R. Dastidar, P. P. Chakrabarti and R. Ray, "A synthesis system for analog circuits based on evolutionary search and topological reuse" *IEEE Trans. Evol. Comput.*, vol. 9, no. 2, Apr. 2005, pp. 211-224

[6] E.A.M. Klumperink, F. Bruccoleri, B. Nauta, "Finding all Elementary Circuits Exploiting Transconductance", *Circuits and Systems, 2001. IS-CAS 2001. The 2001 IEEE International Symposium on*, vol. 1, 6-9 May 2001 pp. 667 - 670

[7] H. Graeb, S. Zizala, J. Eckmueller and K. Antreich, "The sizing rules method for analog integrated circuit design" *ICCAD*, 2001

[8] Behzad Razavi, *Design of Analog CMOS Integrated Circuits* [Boston, Mass.]: McGraw-Hill, 2000

[9] P.E. Allen and D.R. Holberg, *CMOS Analog Circuit Design* London, U.K.: Oxford Univ. Press, 2002

[10] F.V. Fernandez, A. Rodriguez-Vazquez, J.L. Huertas and G. Gielen: "Symbolic Analysis Techniques - Applications to Analog Design", Automation. *IEEE Press*, 1998

[11] G. Gielen and W. Sansen: "Symbolic Analysis for Automated Design of Analog Integrated Circuits" *Kluwer Academic Publishers, Boston/ Dordrecht/London* , 1991

[12] Popp, R. and Barke, E.: "Symbolic Analysis of Nonlinear Analog Circuits by Simplification of Nested Expressions" *SMACD 00: Proc. of the InternationalWorkshop on Symbolic Methods and Applications in Circuit Design*, 2000

7C-1

Statistical Corner Conditions of Interconnect Delay (Corner LPE Specifications)

Kenta Yamada and Noriaki Oda

NEC Electronics Corporation
1753, Shimonumabe, Nakahara-Ku, Kawasaki, Kanagawa, 211-8668, Japan
TEL: +81-44-435-1255 FAX: +81-44-435-1878 E-mail: kenta.yamada@necel.com

Abstract - **Timing closure in LSI design becomes more and more difficult. But the conventional interconnect RC extraction method have over-margins caused by its corner conditions settings. In this paper, statistical corner conditions using the independence of variations between process parameters and between interconnect layers are proposed. As a result, the fast-to-slow guardband decreases by half in average, compared to the conventional method. The proposed method is ready for implementation to LPE tools.**

I. Introduction

Timing closure becomes difficult in LSI design as device scaling proceeds[1]. The demands for reducing over-margin caused by corner condition settings become strong.

An ordinary interconnect RC extraction flow is shown in Fig.1. Firstly, the RC library for various interconnect patterns is generated from the cross-section, once for one technology. On a RC extraction run for a design, the netlist with parasitic RCs is extracted from the layout data by using this RC library. This is generally called LPE (Layout Parameters Extraction). The netlist with parasitic RCs is used for timing verifications in subsequent design steps.

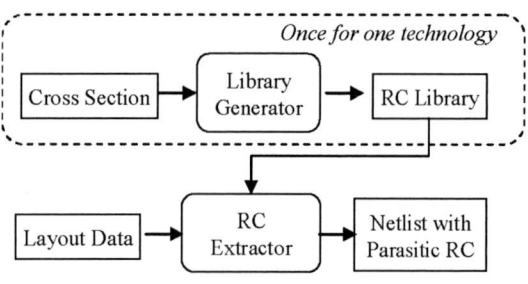

Fig. 1.   Ordinary LPE Flow

Recently, the layout pattern dependence of interconnect shape, such as the fluctuation of width and thickness caused by optical proximity effects, an etching process or CMP process, can be taken into account at LPE[2]. This means RC extraction becomes accurate in the center conditions.

In contrast, no effective method has been proposed that determines statistical and accurate corner conditions, because there is not certain knowledge on how to set the variations of the various process parameters statistically to maximize or minimize the propagation delay of signals (τpd). The details are explained in the next section.

II. Problems of Conventional Corner Conditions

Fig.2 shows the LPE flow for conventional corner conditions[3],[4]. Two to six cross-sections in which the variations of the process parameters are set to maximize or to minimize τpd are input to the library generator and the RC libraries corresponding to each cross-section are generated. Here, each process parameter is set to be the maximum or minimum value in the variation range. Then the netlists with parasitic RCs corresponding to each library are extracted.

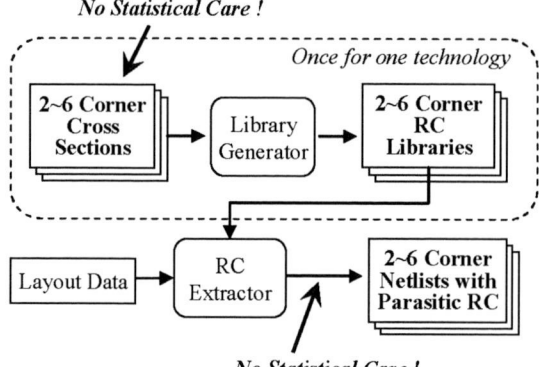

Fig. 2.   LPE flow for Conventional Corner Conditions

Surely the timing verification would be accomplished by using the all extracted netlists. But this method is too pessimistic from a statistical viewpoint. As the variations of each process parameter come from different processes, these are independent from each other. Therefore, the corner conditions in which each process parameter is set to be the maximum or minimum value in the variation range are statistically unlikely to happen and the guardband spanning from one corner to another, i.e., fast to slow corners becomes needlessly wide, and causes over-margins.

The same can be applied to the relation between interconnect layers. As different interconnect layers come from different processes, their variations are independent from each other. Therefore, the conventional corner conditions that do not take the statistical independence into considerations is too pessimistic from a statistical viewpoint, and cause over-margins.

These over-margins make timing closure difficult, or enlarge chip sizes, or increase power consumption or waste the performance of the present process technologies. As timing closure, in particular, becomes more and more difficult in recent designs, this situation is fatal.

III. Method to solve the problems

0-7803-9451-8/06/$20.00 ©2006 IEEE.

In this work, the reduction of over-margins caused by the conventional corner conditions is examined from statistical concepts. Here are two concepts:

The first one is the reduction of over-margins by using the independence of the variations of process parameters in a single interconnect layer. It is examined whether the statistical input cross-sections (corner conditions) can be set or not. And a simple method of finding the corner conditions is also examined.

The second one is the reduction of over-margins by using the independence between the variations of interconnect layers. The input cross-sections to be fed in the RC library generator as the corner conditions are constructed from a combination of the corner conditions of involved layers. The resultant output netlists with parasitic RCs would have over-margins. To overcome this, the concept of statistical independence will be applied to the nodes shared with different interconnect layers; the extracted resistances or capacitances will be adjusted appropriately.

IV. Statistical Corner Conditions in Single Layer

A. Handling of Process Variations

In an interconnect process, the width (W), the thickness (T), the thickness of insulators (D_1, D_2) and the dielectric constant (ε) are process parameters, as shown in Fig.3. These parameters vary independently because of independent origins, i.e. different process steps.

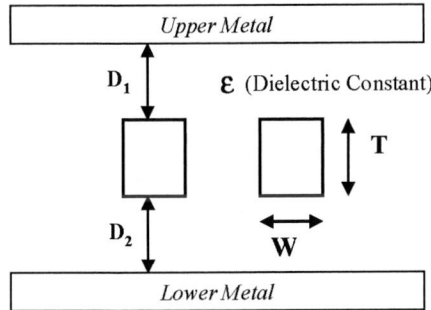

Fig. 3. Process Parameters in Interconnect Cross-Section

Because it is too complicated and impractical to treat all these parameters statistically, target to treat statistically should be limited. W and T mainly influence resistance and capacitance, and the other parameters are secondary. The variations of D1 or D2 influence the capacitance of interconnects without neighboring interconnects, but signal lines usually have neighboring interconnects. Therefore the ratio of the capacitance to under or upper interconnects in the total capacitance is small. As a result, the influence of the variations of D1 or D2 to total capacitance is small. In addition, these secondary parameters can clearly be set to either end of the full variation range to maximize or minimize τpd. Therefore, W and T are treated statistically and the secondary parameters are set to be either end of the full variation range (not statistically). There is another reason. If the secondary parameters also treated statistically, the corner conditions cannot be fixed to limited number. It can be easily understood to extend the statistical examinations on W and T described in this paper to these secondary parameters.

B. Variations of R & C due to the change of W & T

The relation between W variations and T variations is shown in Fig.4. The x-axis is $\Delta W/\sigma(W)$, and the y-axis is $\Delta T/\sigma(T)$. The variation ranges are set to +/-3σ here (as an example), but these can be set to any value at a designer's discretion.

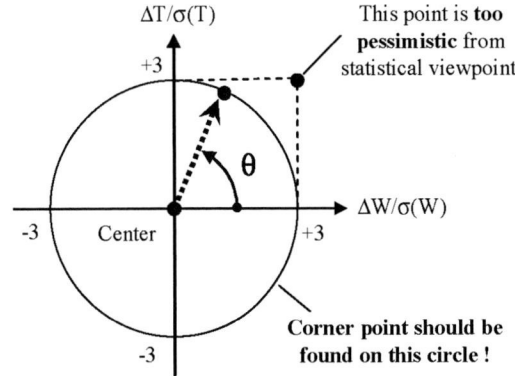

Fig. 4. Variations of W & T

It is pessimistic to set both W and T to the maximum or the minimum and it is sufficient to find the corner points on the circle $((\Delta W/\sigma(W))^2 + (\Delta T/\sigma(T))^2)^{1/2} = 3$ from a statistical viewpoint. It matters whether the point that maximizes or minimizes τpd can be fixed on this circle or not.

Firstly, the resistance and capacitance variations simulated by TCAD for various θ along the circle are shown in Fig.5 and Fig.6. The target is a local interconnect of a 90nm CMOS process. The interconnect pitch was varied when the upper and lower interconnects exist (Case1) or not (Case2). The x-axis is θ, and the y-axis is the ratio to the center condition.

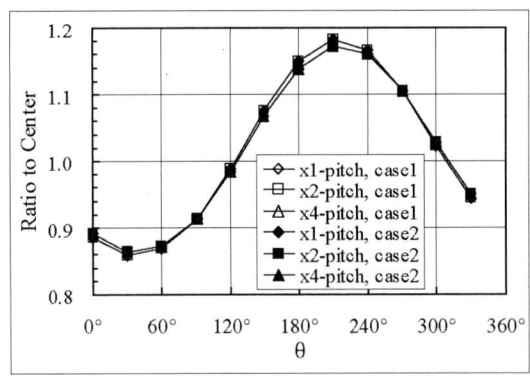

Fig. 5. Resistance Variation

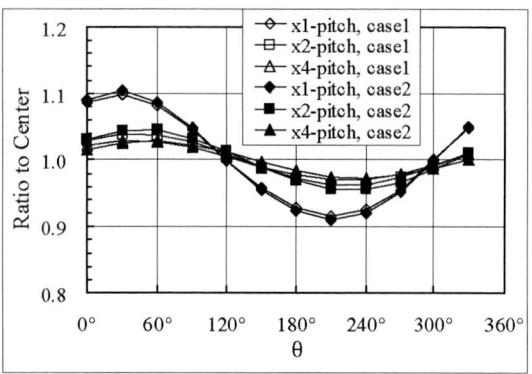

Fig. 6. Capacitance Variation

It can be seen that the resistances are minimized and the capacitances are maximized at $\theta=30°$, and the resistances are maximized and the capacitances are minimized at $\theta=210°$. These results are reasonable because resistances are maximized when the sectional area of the interconnect is minimized and capacitances are minimized in the same case. Note that θ maximizing or minimizing capacitance doesn't change for all interconnect pitches or cases.

C. τpd Variations due to the change of θ

The variations of τpd with respect to θ are analyzed. Various inverter chains were simulated and τpd per one stage was measured. The inverter chains were loaded with interconnect of various patterns and lengths. The inverter size was also changed.

Fig.7 shows the inverter size dependence of the τpd variations. The interconnect length is fixed at 2mm and the interconnect pattern is fixed to the minimum pitch of Case2. τpd is maximized at $\theta=30°$ and minimized at $\theta=210°$ when the inverter is smaller. Oppositely, τpd is minimized at $\theta=30°$ and maximized at $\theta=210°$ when the inverter is larger. It is remarkable that τpd is always minimized or maximized at $\theta=30°$ or $\theta=210°$ in any case. In this case, it can be observed that the variations of W and T do not change the delay with X4-size inverter.

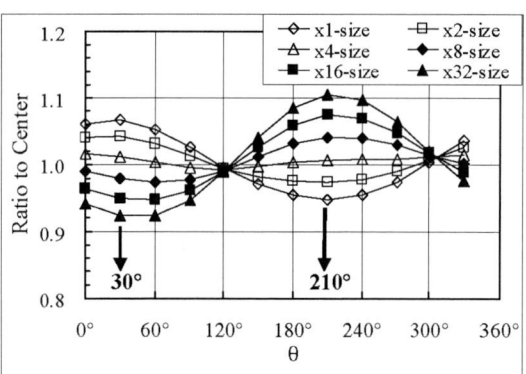

Fig. 7. τpd Variation (Inverter Size Dependence)

Fig.8 shows the pattern dependence of τpd variations. The inverter size is "x2" and the interconnect length is fixed at 2mm. τpd is maximized at $\theta=30°$ and minimized at $\theta=210°$ when the interconnect pitch is smaller. τpd is minimized at $\theta=30°$ and maximized at $\theta=210°$ when the interconnect pitch is larger. Whether the upper and lower interconnects exist or not does not change the essence of the results. Also in this case, τpd always minimized or maximized at $\theta=30°$ or $\theta=210°$.

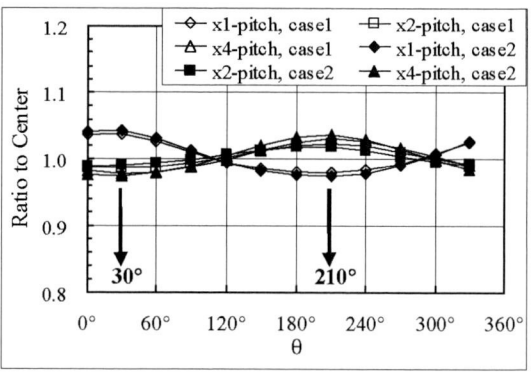

Fig. 8. τpd Variation (Interconnect Pattern Dependence)

Fig.9 shows the interconnect length dependence of τpd variations. The inverter size is "x1" and the interconnect pattern is fixed to minimum pitch in case2. τpd is maximized at $\theta=30°$ and minimized at $\theta=210°$ in all cases.

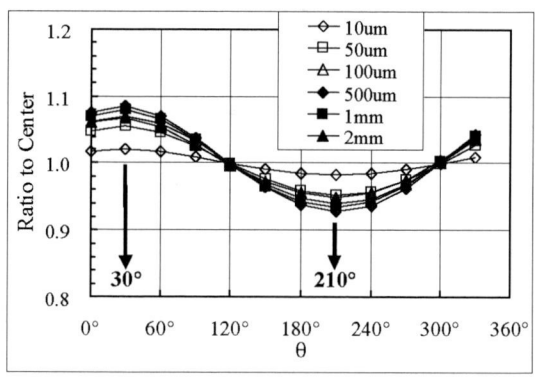

Fig. 9. τpd Variation (Interconnect Length Dependence)

D. Summary of τpd Variations

From the results above, it is understood that τpd is always minimized or maximized at $\theta=30°$ or $\theta=210°$. This means that the corner conditions of this local interconnect layer are limited to two cases only ($\theta_1=30°$ and $\theta_2=210°$).

This can be explained as follows. Resistance and capacitance govern τpd. Each of them varies alternately in the opposite phase with respect to θ. As either of them mainly governs τpd in any case, θ maximizing or minimizing the resistance or capacitance ultimately maximizes or minimizes τpd.

For a semi-global interconnect layer of a 90nm CMOS process, either of θ=60° or θ=240° is found to be the corner condition from a similar analysis. The reason why the θ that achieve corner conditions is different from the local interconnect layer is the difference of the aspect ratio and the difference of the ratio of W or T variation ranges to W or T themselves.

E. Determination method of Corner Conditions (θ_1 and θ_2)

The above examinations are detailed to show the validity of the method. But it can be understood from these examinations that only the resistance (or capacitance) simulations with varying θ are necessary to find the θ maximizing or minimizing resistance (or capacitance), and these are the corner conditions (θ_1 and θ_2). Therefore, the corner conditions can be automatically found by improved LPE tools.

F. Statistical Corner Conditions with All Parameters

As already mentioned, the secondary parameters (except for W and T) are set to the either end of the full variation range to maximize or minimize τpd. The thickness of insulators among layers(D_1, D_2) are set to the maximum in the fast corner conditions, and the minimum in the slow corner conditions. On the other hand, the dielectric constant(ε) is set to the minimum in the fast corner conditions, and the maximum in the slow corner conditions. In addition, the resistance of via (R_{via}) is set by the similar way. After all, as these parameters should be set to both ends to each corner of W and T (θ_1 and θ_2), it ends up with $2 \cdot 2 = 4$ corner conditions (Fig.10 and TABLE I).

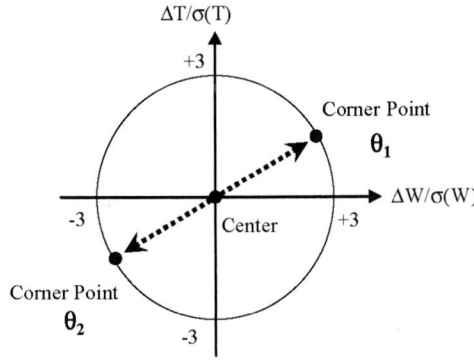

Fig. 10. Statistical Corner Conditions (θ_1 and θ_2)

TABLE I
Statistical Corner Conditions with All Parameters

Corner Conditions	W	T	D_1, D_2	ε	R_{via}
RCmax	$3\cos(\theta_2)\,\sigma$	$3\sin(\theta_2)\,\sigma$	-3σ	$+3\sigma$	$+3\sigma$
Cmax	$3\cos(\theta_1)\,\sigma$	$3\sin(\theta_1)\,\sigma$	-3σ	$+3\sigma$	$+3\sigma$
RCmin	$3\cos(\theta_1)\,\sigma$	$3\sin(\theta_1)\,\sigma$	$+3\sigma$	-3σ	-3σ
Cmin	$3\cos(\theta_2)\,\sigma$	$3\sin(\theta_2)\,\sigma$	$+3\sigma$	-3σ	-3σ

The names of the corner conditions RCmax or RCmin signify that these conditions maximize or minimize R·C. The names of the corner conditions Cmax or Cmin signify that these conditions maximize or minimize C. In general, RCmax and RCmin become the corner conditions for long interconnects, and Cmax and Cmin become the corner conditions for short interconnects.

G. Implementation in LPE

On implementation in LPE, each corner condition of all layers is combined to one cross-section. Therefore, there are four cross-sections of corner conditions. Then these four cross-sections and the center cross section are input to the RC library generator, and the RC parameters in the center condition and the ratio of RC parameters in each corner condition to those at the center condition is output to the RC library as coefficients. On a RC extraction for a design, the RC parameters are extracted at the center condition and those at the corner conditions are calculated by using the coefficients. Fig.11 shows the flow of the RC library generation along with the proposed method and TABLE II is the schematic view of the RC library.

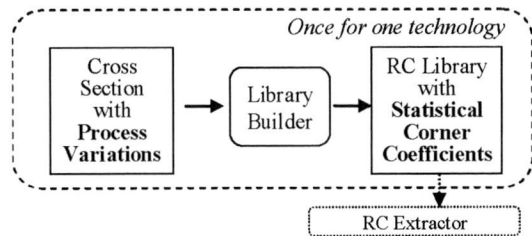

Fig. 11.   Library Generator with Statistical Corner Conditions

TABLE II
Schematic View of RC Library with Statistical Corner Coefficients

Patterns	Center Value		Corner Coefficients							
			RCmax		Cmax		RCmin		Cmin	
	R	C	β_R	β_C	β_R	β_C	β_R	β_C	β_R	β_C
1										
2	Center : R, C									
3	Corner : $\beta_R \cdot$ R, $\beta_C \cdot$ C									
...										

V. Statistical Corner Correction between Layers

The problem is that the variations of different interconnect layers are independent from each other because they come from different process steps. This means that the RC extraction from the cross-sections of corner conditions constructed by combining corner conditions of each layer would take into account situations that are statistically unlikely to happen, and would cause over-margins. The corner conditions can be brought even closer to the center condition by introducing statistical ideas to this problem.

Fig.12 shows the relation between the variations of two layers (M1 and M2) when an interconnect node is composed of these two layers. Fig.13 shows an extracted netlist with parasitic RCs involved with this node.

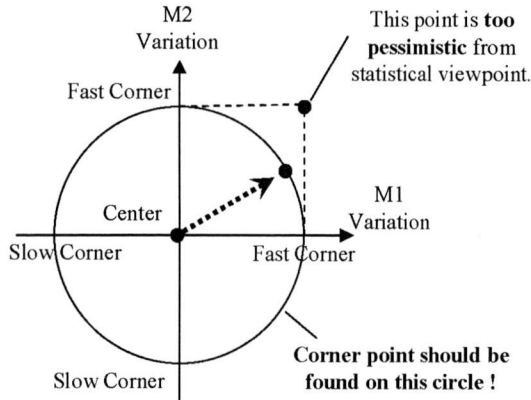

Fig. 12. Variation between Interconnect Layers

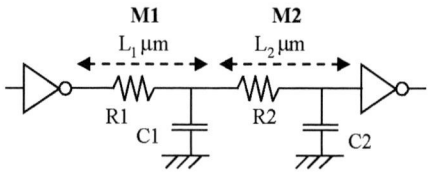

Fig. 13. Node composed of M1 and M2

A. Calculation of Statistical Corner Correction

The extracted R and C in one corner condition by the method proposed above (without statistical correction yet) can be expressed as " coefficient(β) · center value(R or C) ", shown in TABLE II. Note that only one R or C for one layer is described in Fig.13 for simplification but there can be many R or C (for each interconnect segment or pattern) actually, and the coefficients are also different.

Here, all these coefficients (β) are changed to new coefficients (β') which is closer to 1 (= center condition) than β itself, by using γ shown in TABLE II, as follows;

$$\beta' = 1 + (\beta - 1) \cdot \gamma \quad (0 < \gamma \leq 1).$$

TABLE III
Statistical Correction of Corner Coefficients

Center	Corner Condition without Statistical Correction	Corner Condition with Statistical Correction
R1	$\beta_R1 \cdot R1$	$\beta_R1' \cdot R1$
C1	$\beta_C1 \cdot C1$	$\beta_C1' \cdot C1$
R2	$\beta_R2 \cdot R2$	$\beta_R2' \cdot R2$
C2	$\beta_C2 \cdot C2$	$\beta_C2' \cdot C2$

$$\beta*' = 1 + (\beta* - 1) \cdot \gamma$$

B. Determination Method of Correction Coefficient

When the total length of M1 is L_1 and that of M2 is L_2 at this node as shown in Fig.13, γ is calculated as follows;

$$\gamma = (L_1^2 + L_2^2)^{1/2} / (L_1 + L_2).$$

The reason of this expression is explained as follows. This explanation is for capacitance but the same explanation can be applied to resistance. Two assumptions are set here.

Assumption 1) Capacitance for each unit length is constant for all patterns and layers. This capacitance value is temporarily called C_0.

Assumption 2) Coefficient of the corner condition (β) is constant for all patterns and layers. This coefficient is temporarily called β_0.

Under these assumptions, the capacitance in the corner condition without correction is expressed by center value C_{total} as follows;

$$\beta_0 \cdot C_{total}$$

And the capacitance variations of each layer are expressed as follows;

M1: $\Delta C_1 = (\beta_0 - 1) \cdot C_0 \cdot L_1$
M2: $\Delta C_2 = (\beta_0 - 1) \cdot C_0 \cdot L_2$

From a statistical viewpoint, using the independence of the variations between layers, the total capacitance variation is calculated as follows;

$$\Delta C_{total} = C_{total} \cdot ((\Delta C_1/C_{total})^2 + (\Delta C_2/C_{total})^2)^{1/2}$$
$$= C_{total} \cdot (\beta_0 - 1) \cdot \gamma.$$

Therefore, the capacitance in the corner condition becomes;

$$C_{total} \cdot (1 + (\beta_0 - 1) \cdot \gamma) = \beta_0' \cdot C_{total}$$

It should be noted that an estimation error will be likely to grow when an actual interconnect does not satisfy these two assumptions. But these assumptions can be easily met for an interconnect node composed of long interconnect segments because they usually consist of various patterns or layers. A node that needs accuracy is usually a long one because it causes a large interconnect delay.

The similar calculation can be applied to nodes composed of more than two layers. γ is calculated as follows;

$$\gamma = (L_1^2 + L_2^2 + L_3^2 + ...)^{1/2} / (L_1 + L_2 + L_3 + ...)$$

For this corner correction, information of all nodes with interconnect length and layers is necessary. But LPE tools may need not to make this data because this can be output

710

from the "Routing and Placing" phase of design.

C. Care for Branched Nodes

To extract resistances of a branched node, the calculation of γ for resistance becomes as follows. Resistances that contribute to a signal are limited to those on the path of the signal. Then, resistances should be limited to the path of the signal on γ calculation. As there would be plural γ for resistance of the root of nodes, maximum of these γ should be adopted.

On the other hand, all capacitance of a node contribute to a signal passing this node. Then, the calculation already shown doesn't need to be changed.

D. Care for Coupling Capacitance

On coupling mode extraction, γ of a capacitance would be extracted from the both nodes to which this capacitance is connected, and these γ are generally different from each other. In these case, maximum of these two γ should be adopted.

Finally, the LPE flow becomes Fig. 14, following Fig. 12.

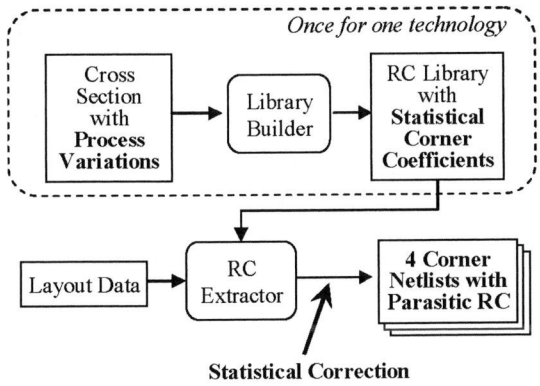

Fig. 14.   LPE flow with Statistical Corner Conditions

VI. Effects

By using the statistical corner conditions in a single layer, the guardband width from the fast corner to the slow corner decreases to a factor of $1/(2^{1/2}) \approx 0.7$, compared to that by conventional corner conditions. This factor of "0.7" is the ratio of the diameter of the circle and the diagonal of the square enclosing this circle in Fig.4. In other words, the corner conditions are brought closer to the center condition.

Moreover, the corner conditions are brought still closer to the center condition by using the statistical corner correction between layers. For example, suppose an interconnect node be composed of nine segments, all of which are different metal layers, but have the same length. In this case, the guardband width decreases to a factor of $(9^{1/2})/9 \approx 0.33$ compared to that without the statistical corner correction. On the other hand, there is no effect to a node composed of a single layer. In average, the guardband width is 0.7 times of that without the corner correction.

With those two types of statistical aspects combined, the guardband width reduces to the half of the conventional one. Note both intralayer and interlayer reduction factors are 0.7, leading to an overall reduction factor of $0.7 \cdot 0.7 \approx 0.5$. In other words, if the capacitance or resistance variation range is +/-20% in the conventional corner conditions, it shrinks to +/-10% in the proposed statistical corner conditions. With narrower guardband, timing closure becomes easier. Narrower guardband allows designers the reduction of chip size and power consumption, and the full advantage of the present process technology in its performance.

VII. Conclusions

Statistical ideas are introduced to the corner conditions for interconnect RC extraction. These ideas are utilized for trimming the over-margin caused by the conventional corner conditions. The basic idea is to exclude situations that are statistically unlikely to happen. By using two ideas, which are the statistical corner conditions in a single layer and the statistical corner correction between layers, the guardband width from the fast corner to the slow corner decreases by half, compared to that by the conventional corner conditions, in average. In other words, if the capacitance or resistance variation range is +/-20% in the conventional corner conditions, it shrinks to +/-10% in the proposed statistical corner conditions.

Acknowledgements

The author would like to thank T. Saito, H. Kuge, T. Akimoto, K. Horiuchi, S. Yoshioka, H. Futami, T. Iizuka and Y. Asai for their supports and useful discussions.

References

[1] S. Takahashi, Tech. Dig. IEDM, 1998, p.833-836
[2] K. Yamada, VLSI Tech. Dig., 2003, p.111-112
[3] S. Inoue, Japanese patent pending, No. P2001-265826A
[4] A. Kuroda, Workshop on Circuits and Systems in Karuizawa, 2005, p.25-30

7C-2

2006 Asia and South Pacific Design Automation Conference

Speed Binning Aware Design Methodology to Improve Profit under Parameter Variations*

Animesh Datta, Swarup Bhunia[1], Jung Hwan Choi, Saibal Mukhopadhyay, and Kaushik Roy

School of ECE, Purdue University, IN 47907, USA, <adatta, choi56, sm, kaushik> @ecn.purdue.edu

Dept EECS, Case Western Reserve University, Cleveland, OH 44106, USA, Swarup.Bhunia@case.edu

Abstract—*Designing high-performance systems with high yield under parameter variations has raised serious design challenges in nanometer technologies. In this paper, we propose a profit-aware yield model, based on which we present a statistical design methodology to improve profit of a design considering frequency binning and product price profile. A low-complexity sensitivity-based gate sizing algorithm is developed to improve the profitability of design over an initial yield-optimized design. We also propose an algorithm to determine optimal bin boundaries for maximizing profit with frequency binning. Finally, we present an integrated design methodology for simultaneous sizing and bin placement to enhance profit under an area constraint. Experiments on a set of ISCAS85 benchmarks show up to 26% (36%) improvement in profit for fixed bin (for simultaneous sizing and bin placement) with three frequency bins considering both leakage and delay bounds compared to a design optimized for 90% yield at iso-area.*

1. INTRODUCTION

Aggressive technology scaling has led to large uncertainties in device and interconnects characteristics for nano-scaled circuits [1]. Increasing variations (both inter-die and intra-die) in device parameters (channel length, gate width, oxide thickness, device threshold voltage etc.) produce large spread in the speed and leakage power consumption of integrated circuits (ICs) [1, 3]. Fig. 1 shows the distribution of operating frequency and leakage current over a large number of high-end micro-processor chips [3]. From the figure, it can be observed that a mature silicon technology like 130nm suffers from about 30% variation in maximum allowable frequency of operation and about 5X variation in leakage power. For newer technologies, the variations are reported to be much higher with about 20X variation in leakage power for 90nm technology [3]. Consequently, parametric yield of a circuit (probability to meet the desired performance or power specification) is expected to suffer considerably, unless an overly pessimistic worst-case design approach is followed. Therefore, design of high-performance circuits maintaining or enhancing yield under parameter variations has emerged as a major challenge in nano-scaled technologies [3].

In recent years, statistical analysis of timing and power has been extensively explored [2, 5, 8]. Several parametric yield models have been proposed to consider impact of different sources of variations on circuit delay and power [5, 6]. At the same time, multiple efforts have been made to develop statistical design methodology that either ensures or enhances parametric yield (e.g. with respect to delay or power) under specific design constraint (e.g. on area or power) [6, 10, 13].

Profitability of a design is conventionally equated with yield [1, 3]. However, large spread in the frequency distribution due to increasing uncertainties has led to the concept of speed-binning to improve the design profit [4]. Presently, speed-binning is widely used during manufacturing test to qualitatively sort the working (i.e. free from manufacturing defects) ICs based on their highest permissible frequency of operation. During the speed-binning process, functional or structural tests are run at multiple

---

* The work is sponsored in part by Marco Gigascale Systems Research Center (GSRC) and Semiconductor Research Corp.

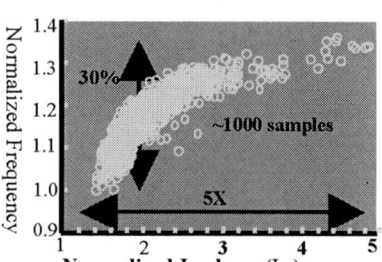

Figure 1: Leakage and frequency variations in 130nm technology (source: Intel)

frequencies and parts are binned according to the highest speed test they pass. Working ICs are then priced based on their respective frequency bins [4].

Since high-frequency ICs correspond to higher price points, maintaining yield at a target circuit delay (i.e. frequency) under statistical delay distribution does not ensure high profit. Here, profit refers to the cumulative sum of price-weighted yield of the ICs across all frequency bins. Hence, there is a need to develop design methodology that can either maintain or improve the design profit (instead of yield at a target delay) under statistical delay variations. In [10] authors have used a linear profit function to capture the performance levels of different ICs. However, they did not consider the impact of design choices or delay distribution change on the profit function.

To the best of our knowledge, there is no design technique addressing optimization of design profit (instead of parametric yield at a fixed target delay), considering frequency bins and product price profile. In particular, this paper makes the following contributions:

- A weighted yield model to represent profit that considers price of ICs running at different frequencies and satisfying specific power dissipation requirement.
- A statistical design methodology to improve profitability of a design using a sensitivity-based low-complexity gate sizing under both delay and power bounds.
- An algorithm to determine optimal bin boundaries for a design to maximize profit for a given price profile and design specification.
- An integrated design flow for simultaneous gate sizing and optimal bin boundary placement to improve design profit for a given price profile with a constraint on area.

2. BACKGROUND AND MOTIVATION

In this section, we describe a parametric yield model of a design based on a target delay and leakage power constraint.

2.1 Modeling Yield with respect to a Target Delay

Under parameter variations, circuit delay is the maximum of all path delays in the circuit [1]. The overall circuit delay, T_{ckt} thus follows a distribution and can be modeled as a random variable with mean μ and standard deviation (STD) σ (i.e. $T_{ckt} \sim N(\mu, \sigma)$) [1]. Hence, for a given target delay, the circuit will have a certain

---

0-7803-9451-8/06/$20.00 ©2006 IEEE.

712

probability to meet it depending on its delay distribution parameters. Conventionally, yield of a design is defined as its probability to meet the target delay (T_D) [6]:

$$Y = P_D = \Pr\{T_{ckt}(\mu,\sigma) \le T_D\} \qquad (1)$$

However, in nano-scaled circuits, power consumption of the system also varies from chip to chip along with performance due to variations in transistor threshold voltage, channel length, gate width and gate oxide thickness, resulting in parametric yield loss. The yield loss occurs due to the fact that devices with lower V_t (and/or lower channel length) suffer from an exponential increase in sub-threshold leakage. Therefore, there exists a strong correlation among the maximum operating frequency and leakage power of a system. In [5], authors show that when both power and performance constraints are considered, maximum yield loss occurs in the highest frequency bin, while negligible yield loss occurs in other frequency bins. Hence, we can effectively use a minimum delay $T_{leakage}$ (or maximum operating frequency) value as a bound on the leakage power dissipation of the design. Mathematically, when $T_{leakage}$ bound is used together with T_D bound in (1), effective yield of a design can be given by:

$$Y_{effective} = \Pr\{T_D \ge T_{ckt}(\mu,\sigma) \ge T_{leakage}\} \qquad (2)$$

2.2 Motivation

Fig. 2 shows two possible circuit delay distributions (Gaussian) with the corresponding three frequency bin boundaries for a small test circuit. In this figure, Yield$_{optimized}$ distribution corresponds to a design obtained by optimizing for certain yield constraint at a target delay, T_D. The other distribution (Profit$_{optimized}$) corresponds to a design where the distribution is changed (by properly sizing the netlist) to improve profit with respect to an exponential price profile (i.e. price point of ICs have exponential dependence on their operating frequency). Fig 2(a) shows that although both distributions meet the yield constraint at T_D, the process of improving profit deterministically changes the distribution (μ increases, while σ decreases) in a way that increases the profit with respect to the given price-profile. In this case, yield loss suffered by profit-optimized design in the highest frequency bin, is easily amortized by significant gain in yields at the other two lower frequency bins. In Fig. 2(b), interestingly, yield degrades due to increase in σ though profit increases due to adequate increase in the number of highest frequency ICs (that more than offsets yield loss at the two lower frequency bins).

Thus, instead of modeling and optimizing yield at a target delay, it is important to consider modeling of profit and to explore design space to optimize profit for high performance systems, which employ frequency binning.

3. PROFIT-AWARE YIELD MODEL

We propose a new profit-aware yield model that represents product profitability considering both frequency binning and

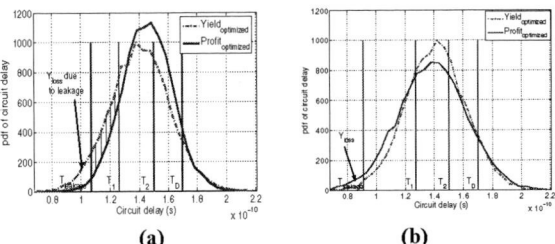

(a)   (b)

Figure 2: Circuit delay distributions for yield- and profit-optimized design with a) iso-yield at the target delay, and b) profit-optimized design of smaller yield

product price profile.

3.1 Yield Model Considering Frequency binning and Price Profile

Fig. 3(a) shows a normalized price vs. frequency specification of two recent high-end processors [14]. We observe that for both products, price of the highest frequency part is about three times higher than that of the lowest frequency parts. Hence, to capture the quality of different chips in the design objective function, we model design profit as a price-weighted cumulative sum of yields at different frequency bins. Thus, considering N frequency bins, the profit-aware design yield can be expressed as:

$$Y_P = \sum_{i=1}^{N} C(T_i) Y_{bin_i}; \text{ where } T_N = T_D = \text{Target design delay}$$

$$Y_P = \sum_{i=1}^{N} w_i Y_i; \text{ where } w_i = C(T_i) \qquad (3)$$

where, weighing parameter $w_i = C(T_i)$ is the price of a chip in the ith frequency bin. It can be noted that the weighted yield Y_P directly represents the profitability of the design. Fig. 3(b) shows delay distribution vs. exponential product price profile with three frequency bins (i.e. $N = 3$) for an ISCAS85 benchmark circuit (c499) realized in 70nm BPTM technology [9]. The delay distribution is computed considering both systematic and random variations in threshold voltage. Since all the ICs in a particular frequency bin are sold at the same price, product price profile becomes a stair-case function of delay (or frequency) irrespective of the nature of the price function (Fig. 3(b)). As different circuits have different delay distribution parameters (μ, σ), for a given circuit, we choose price weights (w_i) in such a way that the ratio of the prices at the highest and lowest frequencies is constant for all circuits. Mathematically this can be represented as:

$$\frac{w_{max}}{w_{min}} = \text{Constant} = R_{price\_profile}$$

$$\text{where } w_{min} = C(f_{min}), f_{min} = \frac{1}{T_D}; \quad w_{max} = C(f_{max}), f_{max} = \frac{1}{T_{leakage}} \qquad (4)$$

Four delay specifications ($T_{leakage}$, T_1, T_2, T_D) are used to consider three frequency bins (Fig. 3(b)). Yield of a frequency bin is defined as the fraction of the chips that lies within a specified

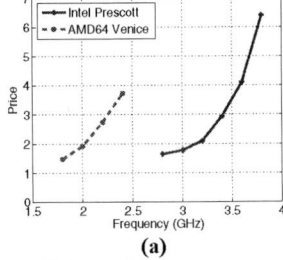 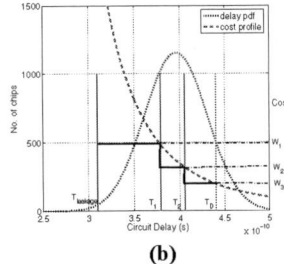 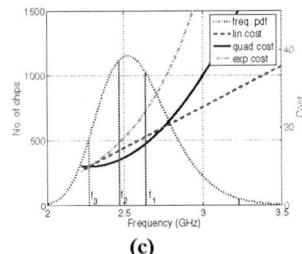

(a)   (b)   (c)

Figure 3: (a) Price and frequency comparisons of two recent high-end processors; (b) Exponential price profile vs. delay distribution for benchmark c499 and (c) Frequency distribution vs. different cost profile

713

delay (frequency) range. Assuming a Gaussian circuit delay distribution with mean (μ) and standard deviation (σ), yields of different bins can be expressed as:

$$Y_1 = Y(T_1) - Y_{leakage} = \Phi\left(\frac{T_1 - \mu}{\sigma}\right) - \Phi\left(\frac{T_{leakage} - \mu}{\sigma}\right)$$

$$Y_2 = \Phi\left(\frac{T_2 - \mu}{\sigma}\right) - \Phi\left(\frac{T_1 - \mu}{\sigma}\right); \quad Y_3 = \Phi\left(\frac{T_D - \mu}{\sigma}\right) - \Phi\left(\frac{T_2 - \mu}{\sigma}\right) \quad (5)$$

where, Φ is the Cumulative Distribution Function of circuit delay. In (3) and (4), price function 'C' can represent any price profile depending on the specific product. In our experiments, we have considered a variety of price profiles with linear, quadratic, exponential dependence on operating frequency. Any other price profiles and even discrete bin prices can also be modeled in the similar manner to compute the profit. Typical example of an exponential price profile vs. delay distribution for an ISCAS85 benchmark c499 is shown in Fig. 3(b). Fig. 3(c) shows different price profiles vs. operating frequency for this circuit.

The profit-aware yield model (3) can help us to design high-performance circuits under variations such that design profit is maximized instead of the yield at a specific target delay. Using (3), profit optimization problem with respect to a given price profile and the number of frequency bins N can be formulated as:

$$\text{Maximize } Y_R = \sum_{i=1}^{N} C(\frac{1}{T_i}) Y_i, \text{ where } Y_i \text{ and } T_i \text{ are defined in (3)}$$

$$\text{Subject to}: A = \sum_{i=1}^{n} x_i = const, \text{ where } A = \text{total area of circuit}; \quad (6)$$

$$x_i = \text{size for the } i^{th} \text{ gate}; n = \text{total number of gates}$$

3.2 Statistical Delay Model

To compute the delay distribution of a circuit based on the information of parameter variations, we have employed the statistical static timing analysis (SSTA) algorithm proposed in [8], where delay distribution of a circuit is calculated using Levelized Covariance Propagation (LCP). It was shown in [8] that, using this technique, the effect of both inter-die and intra-die variations can be taken into account. The simulation results on several ISCAS85 benchmarks show average error of 0.21% and 1.07% compared to the Monte-Carlo analysis for mean and standard deviation of delay, respectively. Gate delays are modeled with 70nm BPTM [9] parameters using analytical expression as in Sakurai et al. [11]. For simplicity, we ignore interconnect delays and assume a constant capacitive load for each net. However, our algorithm can be easily extended to incorporate interconnect delays using conventional π-type RC model as used in [7].

4. PROFIT-AWARE DESIGN OPTIMIZATION

Using our profit-aware yield model presented in section 3.1, we propose a statistical design flow for profit optimization under an area constraint.

4.1 Yield Optimization using Gate Sizing

Gate sizing is conventionally used in different circuit synthesis tools for area/power optimizations while meeting the desired timing constraint, or for minimizing the maximum delay under constraint on area/power [6, 7, 10, 13]. Mathematically, gate sizing problem for achieving mean delay A_0 with minimum active area can be formulated as [7]:

$$\text{Minimize } \sum_{i=1}^{n} x_i \quad \text{where } L_i \le x_i \le U_i, \text{ for } i = 1, \ldots, n$$

$$\text{Subject to}: \sum_{i \in p} \text{mean}(D_i) \le A_0, \quad \forall p \in \text{set of Paths}, \quad (7)$$

$$\text{Yield } Y \ge Y_T; \text{ where } Y = \Phi(\frac{T_D - \mu}{\sigma})$$

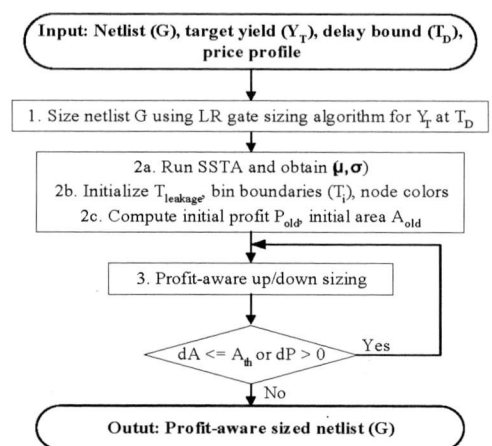

Figure 4: Profit-aware statistical gate sizing algorithm

where, U_i, L_i are the bounds of maximum and minimum gate size, respectively, the value of A_0 depends upon the target yield (Y_T), target delay T_D, and (μ, σ) of the delay distribution. In [7], a solution for convex gate-level sizing problem is proposed to minimize maximum delay under an area constraint. Starting from the minimum-sized netlist, we iteratively use the Lagrangian Relaxation (LR) based sizing algorithm [7] in conjunction with the statistical timing analysis to achieve the yield target with minimum area. In the mth iteration mean target delay is set to μ_m and at the next iteration we update the target delay by small steps to μ_{m+1} as:

$$\Phi(\frac{T_D - \mu_{m+1}}{\sigma_{m+1}}) \ge Y_T \Rightarrow \mu_{m+1} \le T_D - k\sigma_m\Phi^{-1}(Y_T); \text{ where } \sigma_{m+1} = k\sigma_m \quad (8)$$

where, σ_{m+1} is the STD of the circuit delay after sizing in the next iteration, σ_m is STD of circuit delay with current sizes and $k < 1$ is a constant. The solution in [7] provides a globally optimal solution for the problem of area minimization under a static delay constraint. We have observed that with up/down sizing of a logic gate, the mean and standard deviation of the circuit delay shift in the same direction. Based on this observation, we obtain a yield optimized initial design with the minimum area.

4.2 Profit-Aware Gate Sizing

In this section, we propose a sensitivity-based profit-aware gate sizing methodology for the optimization problem proposed in (6). Fig. 4 shows principal steps of the proposed profit-aware sizing methodology. Step 1 has been detailed in section 4.1. Once the design is optimized for a target yield with minimum area, we perform SSTA to determine the delay distribution parameters (step 2a). We define leakage bound based on delay distribution parameters (μ, σ) of the design as:

$$T_{leakage} = f(\mu, \sigma) = \mu - (l * \sigma); \text{ where, } l = \text{constant} \quad (9)$$

Since the delay parameters (μ, σ) vary with sizing, the leakage bound also changes with sizing. We define the fixed bin boundaries based on these delay parameters so that $Y_{effective}$ (effective yield between the highest and lowest permissible frequencies) is equally distributed among N frequency bins as:

$$Y_T = \Phi\left(\frac{T_D - \mu}{\sigma}\right) \Rightarrow T_D = \Phi^{-1}(Y_T, \mu, \sigma); \quad Y_{bin} = \frac{Y_T - Y_{leakage}}{N}$$

$$\Rightarrow T_i = \Phi^{-1}((Y_{leakage} + i * Y_{bin}), \mu, \sigma), \quad \forall i = 1, \ldots, N \quad (10)$$

Then we compute initial design profit P_{old} (using (3)), initial design area A_{old} (computed as active area). In step 3, we compute sensitivity of logic gates with respect to profit (dP/dx) for up and down sizing. Next, starting from the most sensitive gate, we perform up/down sizing of logic gates to improve the overall profit of the circuits. This process is performed iteratively until there is no improvement in profit ($dP \le 0$) or the area constraint cannot be satisfied ($dA > A_{th}$) for further improvement in profit.

Fig. 5 shows pseudo-code for our sensitivity-based up/down sizing routine. We apply sizing step (satisfying the bounds on maximum and minimum gate size) to these nodes and perform SSTA to re-compute the delay distribution. A node with higher sensitivity is sized before the node with lower sensitivity. The runtime complexity of the routine depends on the number of SSTA calls in sensitivity analysis during an up/down sizing iteration. We have employed three optimization techniques to improve the runtime of the proposed gate sizing method.

1. Considering the fact that gates lying in the critical path are most sensitive in terms of delay variation and hence, change in profit, we select a set of gates $\{S_C\}$ on the critical paths. After the SSTA run, we choose a fixed number of critical paths (10,000 in our case) and compute their delay failure probabilities (with respect to T_D) from the path delay parameters (mean and STD). Next, we compute the worst-case delay failure probability for each gate based on the paths crossing it. Gates with high delay failure probability are chosen in an iteration to improve profit (line 1, Fig. 5).

2. We determine profit sensitivity (S_P) of all gates in $\{S_C\}$ by changing one gate size at a time, with a small step (dx) and computing the corresponding change in profit (i.e. $S_P = dP/dx$). The sensitivity analysis is performed for both sizing directions (i.e. up and down). If a logic gate has unacceptable profit sensitivities (i.e. profit drops with upsizing or degrades too much with down sizing) in an iteration, we remove it from $\{S_C\}$ in the subsequent calls of up/down sizing (line 3, Fig. 5).

3. Multiple up/down sizing steps are performed after each sensitivity analysis by selecting successive gates not lying in the fan-in and fan-out logic cone of the previously sized gates. In each iteration, we color the fan-in and fan-out cone of a sized logic gate in the graph (G) (line 8, Fig. 5). The colored nodes are not considered for sizing in that iteration. When no suitable uncolored gate exists in $\{S_C\}$, the iteration terminates. We then mark all gates uncolored and perform a SSTA to update the increment of profit (dP) and area (dA) values of the given circuit (line 10-15, Fig. 5).

up-downSizing ( )
Input: Netlist (G)
Output: Sized netlist (G) after one sizing iteration
Direction: down = 0; up = 1
1. Select gates $\{S_C\}$ from critical set of paths with high failure probability;
2. Compute profit sensitivity of all gates $\{S_i\}$ for up/down sizing;
3. Remove the gates with unacceptable profit sensitivity form $\{S_i\}$;
4. Sort gates of set $\{S_i\}$ in the descending order of profit sensitivity;
5. while (not all gates are colored)
6.   Choose the most sensitive uncolored gate;
7.   Size the gate by dx in proper direction;
8.   Color its fan-in and fan-out logic cone;
9. end //while
10. Reset color of all gates;
11. Perform SSTA (G) to obtain μ', σ';
12. $P_{new} = P(T_0, T_1, ..., T_N, \mu', \sigma')$;
13. $dP = P_{new} - P_{old}$;
14. $P_{old} = P_{new}$;
15. $dA = Area(G) - A_{old}$;

Figure 5: Pseudo code for up/down sizing

Table I: Profit-aware design results compared to 90% yield-optimized design (N = 3)

Circuit	Target Delay T_D (ps)	Profit improvement (%)			Runtime (sec)
		R_{Lin}	R_{Quad}	R_{Exp}	
c432	520	9.53	9.20	12.80	1.12
c499	440	7.75	8.58	9.66	15.5
c880	400	16.28	15.16	20.89	11.88
c1908	520	7.02	6.50	9.07	51.69
c2670	425	8.11	7.15	10.32	9.91
c3540	640	18.78	18.04	26.18	56.71
c6288	1725	12.98	12.04	17.89	55.74
c74181	200	7.27	6.49	9.28	0.24
c74L85	150	13.83	13.29	19.06	0.04
c74283	170	2.77	2.57	3.53	0.08
Avg.		**10.43**	**9.92**	**14.15**	**18.45**

These three techniques reduce the number of gates used to compute profit sensitivity in an iteration with negligible degradation ($< 1\%$) in the over-all profit improvement compared to a case where sensitivity analysis is performed for all gates. It is important to note that, we perform downsizing of least profit sensitive (by assigning higher weight to the downsizing sensitivity) gates to recover from the area overhead incurred during upsizing phase and to satisfy area overhead constraint A_{th}. Note that the proposed profit-aware sizing methodology (up/down sizing as described in Fig. 5) can be used to improve the profit independent of initial sizing of the design. For example, profit optimization can be performed starting from a minimum-sized design instead of a yield optimized design (Section 4.1).

4.3 Experimental Results
We have applied the proposed profit-aware statistical design on several ISCAS85 benchmarks for different number of frequency bins. We have considered design profit improvement for three different price profiles with different price ratios as defined in (4) (e.g. $R_{lin} = 3$, $R_{quad} = 5$, $R_{expo} = 10$). The final product profit (Y_P) is then computed using (3). The area overhead threshold (A_{th}) is taken to be a very small value (0.3%) to observe the scope of profit optimization at iso-area over a yield-optimized design. In Table I columns 3 to 5 present profit improvement for different price profiles as a percentage of profit for a 90% yield-optimized design with $T_{leakage} = \mu - 2.5 * \sigma$.

Using the proposed method, we obtain up to 26% profit improvement (for c3540). On average, we observe profit improvements of 10.4% for the linear, 9.9% for the quadratic, and 14.2% for the exponential price profile at iso-area (Table I). Smaller slope of quadratic price function than the linear one (Fig 3(c)) for lower frequency bins is responsible for smaller profit improvement under quadratic price profile over linear one, considering equal yield bin boundaries. However, profit improvement for a particular price profile varies widely across the benchmarks and it largely depends on the design specifications as well as on the circuit topology (Table I).

Profit improvement at the iso-area comes from the proper distribution of the right amount of yield in the right frequency bin (depending on the price profile). Average runtimes of a sizing iteration considering linear price profile are reported in column 6. Similar runtimes are observed for the other price profiles since they are dominated by the number of calls to the SSTA routine. The number of sizing iterations required by the proposed sizing scheme varies from 3 to 21.

Fig. 6(a) plots the delay distributions for c2670 circuit after initial yield optimized design and profit-aware sizing along with its exponential price profile. Note that considerable profit

7C-2

2006 Asia and South Pacific Design Automation Conference

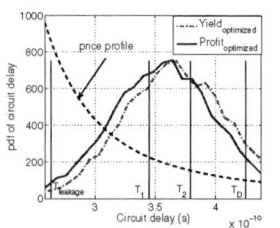
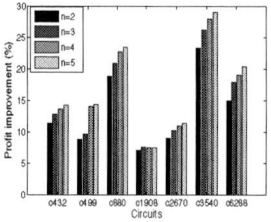

(a)   **(b)**

Figure 6: (a) Delay distributions change for c2670 with an exponential price profile. (b) Profit improvement of different number of bins for fixed bin boundaries

improvement (10.3%) is achieved for increased high frequency bin yield (due to reduction in mean). Figure 6(b) shows profit improvements of the various benchmarks as the number of frequency bins is varied under an exponential price profile. Note that we have not considered the smaller benchmarks (i.e. c74 series) in this plot because they have small delay spread (Table I). The trend of increasing profit improvement with number of bins can be attributed to the fact that with fine-grained frequency binning, the high frequency bin prices (and the average bin price) increase considerably under a given price profile.

We have obtained three sets of average profit improvement results for three different target yields (Y_T), while keeping other design specifications ($T_{leakage}$, T_D, N) unchanged (Fig. 7(a)). We observe that profit improvement decreases with the increase in Y_T (Fig. 7(a)). This is due to the fact that with low initial yield, design profit is usually low for the specific price profile, thus has good scope of improvement. Fig. 7(b) shows the profit improvement trend with different leakage bounds. We observe that when we change the leakage bound ($T_{leakage}$) from $\mu - 2*\sigma$ to $\mu - 3*\sigma$ for $N = 3$, average profit improvement increases for all the cost functions. This happens because scope of change in high frequency bin yields and thus scope of profit improvement for a given price profile increase with the smaller leakage bounds.

4.4 Complexity of Profit-Aware Sizing Algorithm

In our proposed sensitivity-based gate sizing approach, we run SSTA multiple times during sensitivity computation of different gates. The complexity of the SSTA is $O(m)$ [7], where m is the maximum number of gates in any level in the levelized netlist of the given circuit. It is shown that even for a very large circuit this number grows very slowly [7]. The complexity of a LR sizing run is $O(n^a)$, where n is the number of gates in the circuit and $a \approx 1.7$ [8]. Moreover, since after the 1st iteration of up/down sizing, the subsequent sizing steps start with a reduced set of gates, the effective runtime of the subsequent up/down sizing iteration reduces. Hence, the over-all complexity of the design flow is dominated by the runtime complexity of the LR-based sizing

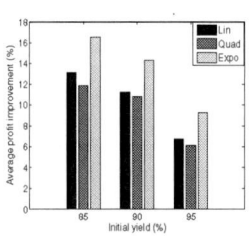
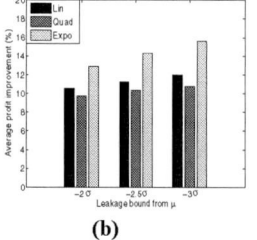

(a)   **(b)**

Figure 7: Average profit improvement for different (a) initial yield targets and (b) leakage bounds for (N = 3)

(section 4.1) and SSTA routine. Let us assume that the LR-sizing routine is called r times and the number of iterations for profit improvement is s. Hence, the total runtime complexity can be given by $O(r*n^a + r*m + s*m*l)$ for a design with average number of sensitive gates in set $\{S_C\}$ equal to l.

The complete profit-aware design flow takes 0.12 second for the smallest benchmark (c741L85, 33 gates) and 15 minutes for the largest benchmark (c6288, 2503 gates). All the simulations have been run on a Linux server with 3.06 *GHz* Pentium Xeon processor and 2*GB* RAM. Hence, this sensitivity-based profit-aware gate sizing can be efficiently applied to improve design profit of large scale industrial circuits containing large number of gates. Note that we can apply incremental timing analysis (realized by incremental timing refinement considering only the gates with modified size) [2] for a large number of SSTA runs to further improve the runtime.

5. SIMULTANEOUS GATE SIZING AND OPTIMAL BIN PLACEMENT

In this section we present key observations on how the choice of frequency bin boundaries affects design profitability under process parameter variations. Next, we propose a design method, which maximizes design profit by simultaneous gate sizing and optimal bin-boundary placement.

5.1 Optimal bin placement

In case the bin boundaries are not available or designers are allowed to change it, they can be chosen appropriately such that the profit metric is optimized for a given price profile. Assuming fixed number of frequency bins (say N), given a delay distribution $D \sim N(\mu, \sigma)$ and price profile C, the problem of finding optimal bin boundaries can be expressed as:

$$\text{Maximize } Y = \sum_{i=1}^{N} C(T_i)Y(T_i),$$
$$\text{Subject to}: T_{leakage} \leq T_i \leq T_D, \text{ for } i = 1 \text{ to } N \quad (11)$$

In order to solve the problem of optimal bin boundary determination (11), we start with equal yield bin boundaries as defined in (10). We employ a greedy approach and at a time we search for one bin boundary that optimizes the design profit keeping other boundaries fixed. First, we search the optimal boundary for the highest allowable frequency bin. We repeatedly perform such optimization for all other bins in descending order of bin frequencies. At the end, we obtain modified bin boundaries

findOptBinBoundary ( )
Input: Number of bins (N), **delay params:** μ, σ
Output: Optimal bin boundaries ($T_0, \ldots, T_N = T_D$)

1. Find equal yield bin boundaries ($T_0, T_1 \ldots, T_N$) for N bins
2. $P_{old} = P(T_0, T_1 \ldots, T_N, \mu, \sigma)$
3. while (no change in T_i)
4.   for each T_i (0 < i < N)
5. $P_{new+} = P(\ldots, T_i + dT_i, \ldots, T_N, \mu, \sigma)$;
6. $dP+ = P_{new+} - P_{old}$;
7. $P_{new-} = P(\ldots, T_i - dT_i, \ldots, T_N, \mu, \sigma)$;
8. $dP- = P_{new-} - P_{old}$;
9.   if (dP+ > 0)
10. $T_i = T_i + dT$; $P_{old} = P_{new+}$;
11.   else if (dP- > 0)
12. $T_i = T_i - dT$; $P_{old} = P_{new-}$;
13.   end //if
14.   end //for
15. end //while

Figure 8: Pseudo-code to find optimal bin boundaries

716

for all the bins that locally optimize the profit. The pseudo-code for optimal bin boundary determination is given in Fig. 8. Results obtained from this algorithm match very closely with optimal bin placement results obtained from an exhaustive search using an implementation in MATLAB.

It is important to note that this technique does not have any design overhead. It only requires an extra design step to be incorporated after the final up/down sizing routine (Fig. 4).

5.2 Simultaneous Sizing and Optimal Bin Placement

Finally, we integrate optimal bin placement procedure (section 5.1) and our profit-aware design flow (section 4.2) to develop an integrated design methodology that simultaneously perform gate sizing as well as optimal bin placement. Basic steps of the design flow are similar to that shown in Fig. 4, except that now we use optimal bin placement routine for profit computation during sensitivity analysis in each up/down sizing routine (step 2, Fig. 5). This means that after obtaining μ and σ corresponding to sizing of a logic gate we perform optimal bin placement to compute the profit sensitivity of the gate. This method when employed to different ISCAS85 benchmarks shows up to 36% profit improvement with three frequency bins (Table II), considering a leakage bound of $\mu - 2.5*\sigma$ for an 90% yield-optimized design at equal area. For all price profiles, additional improvements in profit with simultaneous sizing and bin placement over fixed bin boundaries are also shown in Table II under the columns labeled **Imp**. We observe that using integrated approach up to 10.1% (c3540) more improvement in profit over fixed bin boundaries can be achieved for exponential price profile. Note that profit improvements do not change much with optimal bin-placement for a linear price profile, since linear price profile has the smallest bin price ratio (i.e. $R_{Lin} < R_{Quad} < R_{Expo}$). However, the integrated approach shows significant average profit improvement compared to fixed bin boundary results for both quadratic (from 10% to 16.7%) and exponential price profiles (from 14.1% to 18.8%) for a set of ISCAS85 benchmarks (Table I, and II).

Fig. 9(a) shows effectiveness of two proposed profit-aware sizing methodologies (i.e. fixed bin and simultaneous sizing and optimal bin placement) for different price functions with $N = 3$ and $T_{leakage} = \mu - 2.5*\sigma$. Fig. 9(b) shows a consistent increasing trend in average profit improvements for both the methods under an exponential price profile as the number of bins is increased. Note that the runtime of the simultaneous sizing and optimal bin placement algorithm is almost indistinguishable from the runtime of profit optimization with fixed bin boundaries (Table I). This is because optimal bin placement routine has negligible impact on overall runtime, which is dominated by the number of SSTA runs. It is worth noting that although our delay models follow Gaussian (normal) distribution, the proposed methods can be easily

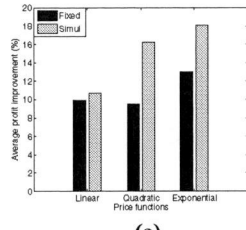 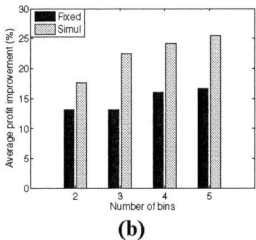

(a)   (b)

Figure 9: Average profit improvements in different methods (a) for different price functions by (N=3), and (b) for different N (exponential price profile)

extended to a recently-proposed non-normal delay distribution [12], which is reported to have higher accuracy in representing delay variations. With non-normal distribution, the steps on making sizing decision using profit sensitivity do not change. However, computation of effective yield in each frequency bin needs to be modified based on the nature of the distribution. Determination of optimal bin boundaries using the greedy approach section 5.1) also remains valid for non-normal delay distributions.

6. CONCLUSIONS

We have proposed a profit-aware yield model and a statistical design methodology to optimize design profit for a given price profile under an area constraint. The methodology can be applied to any price profile and any delay distribution models (normal/non-normal). We have demonstrated that optimal bin-boundary determination can be used to increase the design profit. Experimental results on ISCAS85 benchmarks show that the proposed profit-aware design flow that incorporates information on price profile and frequency binning during the design phase can be very effective to improve design profit under parameter variations.

REFERENCES

[1] K. A. Bowman et al., "Impact of Die-to-Die and Within-Die Parameter Fluctuations on the Maximum Clock Frequency Distribution for Gigascale Integration", *JSSC*, 2002, pp. 183-190.

[2] L.-C. Chen et al., "A New Framework for Static Timing Analysis, Incremental Timing Refinement, and Timing Simulation", *ATS*, 2000, pp. 102-107.

[3] S. Borkar et al., "Parameter Variations and Impact on Circuits and Micro-architecture", *DAC*, 2003, pp. 338-342.

[4] B. Cory et al., "Speed binning with path delay test in 150-nm technology", *IEEE Design and Test of Computers*, 2003, pp. 41-45.

[5] R. R. Rao et al., "Parametric Yield Estimation Considering Leakage variability", *DAC*, 2004, pp. 442-447.

[6] S. Choi et al., "Novel Sizing Algorithm for Yield Improvement under Process Variation in Nanometer Technology", *DAC*, 2004, pp. 454-459.

[7] C. P. Chen et al., "Fast and Exact Simultaneous Gate and Wire Sizing by Lagrangian Relaxation," *IEEE TCAD*, 1999, pp. 1014-1025.

[8] K. Kang et al., "Statistical Timing Analysis using Levelized Covariance Propagation", *DATE*, 2005, pp. 764-769.

[9] "Technology Models", *http://www-device.eecs.berkeley.edu/~ptm*.

[10] A. Agarwal et al., "Circuit Optimization using Statistical Timing Analysis", *DAC*, 2005, pp. 321-324.

[11] T. Sakurai et al., "Delay Analysis of Series-connected MOSFET Circuits", *IEEE JSSC*, vol. 26, no. 2, 1991, pp. 122-131.

[12] X. Li et al., "Asymptotic Probability Extraction for Non-Normal Distributions of Circuit Performance", *ICCAD*, 2004, pp. 2-9.

[13] X. Bai et al., "Uncertainty-Aware Circuit Optimization", *DAC*, 2002, pp. 58-63.

[14] "Processor Retail prices", *http://www.newegg.com*.

Table II: Simultaneous profit-aware sizing and optimal bin placement compared to 90% yield-optimized design (N = 3)

Circuit	Profit improvement (%)					
	R_{Lin}	Imp.	R_{Quad}	Imp.	R_{Exp}	Imp.
c432	9.98	0.33	15.58	6.38	17.98	5.18
c499	8.01	0.26	14.58	6.00	13.13	3.47
c880	17.01	0.73	22.63	7.47	27.17	6.28
c1908	7.13	0.11	8.32	1.82	9.52	0.45
c2670	8.59	0.48	13.17	6.02	14.68	4.36
c3540	20.97	2.19	30.00	11.96	36.31	10.13
c6288	14.02	1.04	20.69	8.65	26.07	8.18
c74181	7.66	0.39	11.93	5.44	13.30	4.02
c74L85	14.85	1.02	22.48	9.19	27.60	8.54
c74283	2.93	0.16	6.59	4.02	6.06	2.53
Avg.	**11.11**	**0.67**	**16.70**	**6.70**	**18.78**	**5.31**

Yield-Area Optimizations of Digital Circuits Using Non-dominated Sorting Genetic Algorithm（YOGA）

Vineet Agarwal
Electrical and Computer Engineering
The University of Arizona,Tucson. AZ
vagarwal@ece.arizona.edu

Janet Wang
Electrical and Computer Engineering
The University of Arizona,Tucson. AZ
wml@ece.arizona.edu

ABSTRACT

With shrinking technology, the timing variation of a digital circuit is becoming the most important factor while designing a functionally reliable circuit. Gate sizing has emerged as one of the efficient way to subside the yield deterioration due to manufacturing variations. In the past single-objective optimization techniques have been used to optimize the timing variation whereas on the other hand multi-objective optimization techniques can provide a more promising approach to design the circuit. We propose a new algorithm called YOGA, based on multi-objective optimization technique called Non-dominated Sorting Genetic Algorithm (NSGA). YOGA optimizes a circuit in multi domains and provides the user with *Pareto-optimal set* of solutions which are distributed all over the optimal design spectrum, giving users the flexibility to choose the best fitting solution for their requirements. YOGA overcomes the disadvantages of traditional optimization techniques, while even providing solutions in very stringent bounds.

1. INTRODUCTION

With recent shrinking of transistor feature size, uncertainty in circuit response has become the most menacing issue for the circuit designers. Data statistics indicate that delay variation of a simple gate can go as high as 10%-15% of the expected value. If no compensating techniques are employed to encounter these variations, the yield of nanometer digital circuits can decrease drastically, thereby increasing the manufacturing cost by multiple folds. In recent past the technique of scaling the transistor W/L ratio, also known as *gate sizing*, has emerged as one of the convincing technique for combating these uncertainty due fickle fabrication processes. In this technique, various algorithms and heuristics are used to scale transistor sizes, so as to decrease the timing variations of the circuit and increase the yield. Thus gate sizing presents a promising approach to counter the diabolic affects of manufacturing process uncertainty.

A survey of previous literature reveals that gate sizing techniques have been in lime light for the past decade. The issue of calculating the timing yield and optimizing it has been addressed numerous number of times. Most generally gate sizing problem has been viewed as minimizing a tim-

ing variability while having constraints on area or delay [1]. Yield optimizations has also carried out using optimization tools [2] under similar constraints, through modular neural networks [3], geometric programming [4] and also in presynthesis steps [5]. Yield issues for sequential circuits has also been addressed in [6]. Song in [7] proposed an approach in which gate sizing is done on statistical critical paths and offers a trade off between area and variation deduction. But most of these work concentrated on trade-off between area overhead and timing variance reduction. Recently Orshansky *el al* proposed an stochastic algorithm for power minimization under variability with a guarantee of power and timing yield [8]. Thus power now played an vital role in designing the circuit, while keeping an constraint on timing yield. Thus the traditional framework for optimization can be summarized as:

$$
\begin{aligned}
\min \quad & f(x) \\
Subject\ to \quad & g(x) \leq g_{\max} \\
& h(x) \leq h_{\max}
\end{aligned}
\tag{1}
$$

Doing optimization through techniques just mentioned in Equation (1), which has a single objective function, have various disadvantages. First and foremost, the final solution of single objective optimization process will depend on the constraints placed on other parameter (g_{max}, h_{max}), which are user defined. Thus if these values are not properly chosen then the resulting solution may not be the most appropriate one. For example if we define penalty as the increase in $g(x)$ and $h(x)$, and if the bounds are set very loose, the optimization may minimize $f(x)$ while using all the penalty it is allowed to incur. A second solution may be present with similar objective function value but with considerable less penalty. In those cases the traditional techniques may not deliver the second solutions and thus may not spot the most 'attractive' solution. Secondly, all the optimization technique have been only single-objective function. Thus if more than one objective have to be optimized, the traditional techniques, will optimize the first objective function first and then redo the optimization on this solution with second objective as primary aim. Thus it has to be done in a 'sequential' sense. Multi-objective optimization pose a better approach in which multi objectives are simultaneously optimized. Thus using multi-objective optimization both $f(x)$ and penalties on $g(x)$, $h(x)$ can be optimized (minimized) simultaneously. Thirdly, if the bounds on $g(x)$ and $h(x)$ are set too tight, then there may not exist any feasible solution which satisfies all the constraints. In such cases, the traditional techniques will flounder ending up in giving no solution. Thus optimization techniques are required which can minimize multiple objectives, which are user independent and which don't crumple under too strin-

Permission to make digital or hard copies of all or part of this work for personal or classroom use is granted without fee provided that copies are not made or distributed for profit or commercial advantage and that copies bear this notice and the full citation on the first page. To copy otherwise, to republish, to post on servers or to redistribute to lists, requires prior specific permission and/or a fee.
Copyright 200X ACM X-XXXXX-XX-X/XX/XX ...$5.00.

0-7803-9451-8/06/$20.00 ©2006 IEEE.

gent constraints.

Thus we propose our technique, Yield-area Optimization using Genetic Algorithm, YOGA, which encounters the above stated disadvantages. YOGA attempts to minimize all criterion i.e, area, mean delay and delay variance simultaneously. YOGA at the simulation end provides the user with multiple solution spread over the optimum design spectrum. All the solutions in the final set are called Pareto-optimal solutions [9, 10], in which none of the solutions are better than any other in the set, thus any one solution can be chosen without hesitation of optimality. Thereby it gives the designers the flexibility to choose the most appropriate solution according to their needs. YOGA's application can be easily extended to simultaneously optimize power consumption of a circuit. Just by including an extra objective of minimizing power stochastically [8] the designers can get similar trade-off information between power and timing yield and find the best fitting solution or otherwise get trade-off information of a larger picture including area, power and timing yield all together.

Our rest of the paper is organized as follows. Section 2 provides a preliminary introduction to technique of gate sizing. In Section 3, we describe the backbone algorithm of *Non-dominated Sorting Genetic Algorithm*. Section 4 provides a demonstration of the technique while YOGA implementation for gate sizing is discussed in Section 5. The experimental results are listed in Section 6 and Section 7 concludes our paper.

2. BASIC GATE SIZING TECHNIQUE

The traditional framework of gate sizing technique can be summarized as:

$$
\begin{array}{ll}
find & \mathbf{s} \\
\min & \sigma^2(d_\mathcal{O}) \\
Subject\ to & \mu(d_\mathcal{O}) \leq \mu_{\max} \\
& \pi(\mathcal{G}) \leq \pi_{\max}
\end{array} \tag{2}
$$

where

1. $\mu(x)$ is the mean and $\sigma^2(x)$ is the variance of the random variable x.
2. $\mathbf{s} = \{s_1, s_2, \ldots, s_n\}$, where s_i is the size of gate g_i, $\mathcal{G}$ is the set of all gates and $g_i \in \mathcal{G}$. Size s for any gate denotes, the ratio of area of the gate to that of a minimum sized inverter.
3. $\mathcal{O}$ is the set of all the outputs of the circuit.
4. d_i is the delay of the ith output of the circuit and $i \in \mathcal{O}$
5. $\pi(\mathcal{G})$ denotes the area of circuit containing gates $\mathcal{G}$
6. $\pi_{\max}$ and $\mu_{\max}$ are the maximum allowable Area and Mean output delay of the circuit.

Thus a regular gate sizing algorithm will solve the non-linear optimization problem stated in Equation 2 and provide the end user with a single $\mathbf{s}$. The constraints place an upper bound on the *penalty* entailed while trying to optimize the objective function, where *penalty* can be in form of area or delay overhead.

3. NON-DOMINATED SORTING GENETIC ALGORITHM

In this section we provide an overview of Non-dominated Sorting Genetic algorithm (NSGA) which will be used as the base for our gate sizing algorithm. Extensive study has been done in multi-objective programming using genetic algorithms in the past [9, 10, 11, 12]. One of the most influential work has been done by Deb in [13], in which he has coined the primary prototype of NSGA.

Algorithm 1 NSGA Algorithm

1: $pop \leftarrow$ GenerateInitialPopulation
2: **if** $generation \leq \max generation$ **then**
3:    $rank \leftarrow$ NonDominatedRanking(pop)
4:    $fitness \leftarrow$ FitnessAssignment$(pop, rank)$
5:    **for** $i = 1$ to N step 2 **do**
6:       $parent_1 \leftarrow$ Selection$(pop, fitness)$
7:       $parent_2 \leftarrow$ Selection$(pop, fitness)$
8:       $(child_1, child_2) \leftarrow$ Crossover$(parent_1, parent_2)$
9:       $newpop_i \leftarrow$ Mutation$(child_1)$
10:      $newpop_{i+1} \leftarrow$ Mutation$(child_2)$
11:    **end for**
12:    $pop \leftarrow newpop$
13:    $generation \leftarrow generation + 1$
14: **end if**
15: $final\_rank \leftarrow$ NonDominatedRanking(pop)
16: NonDominatedSolutions $\leftarrow pop_i, \forall i \in final\_rank_i = 1$
17: **return** NonDominatedSolutions

3.1 Pareto-optimal Solution

A solution is called Pareto-optimal solution, if there exists no other solution for which at least one of its criterion has a better value while values of remaining criteria are the same or better. In other words, one can not improve any criterion without deteriorating a value of at least one other criterion. Thus a design vector x^* is Pareto-optimum if and only if, for an minimization problem and for any x and i,

$$
\begin{array}{l}
f_j(x) \leq f_j(x^*) \quad \forall j = 1, \ldots, m, j \neq i \\
\Rightarrow f_i(x) \geq f_i(x^*)
\end{array}
$$

where $f_i(x)$ is the value of ith objective function.

NSGA gives rise to a set of such *Pareto-optimal* solutions. In a multi-optimization problem, the designer may be interested in more than one of such solutions because of the equal *quality* of the solutions. The concept of NSGA is discussed in next section.

3.2 NSGA Algorithm

The algorithmic flow of NSGA is summarized in Algorithm 1. A Model multi-objective optimization problem can be summarized as.

$$
\begin{array}{lll}
\min f_i(\mathbf{x}) & \forall i \in \{1, M\} \\
Subject\ to: & x_j^L \leq \mathbf{x}_j \leq x_j^U & \forall j \in \{1, P\}
\end{array} \tag{3}
$$

where f_i is the ith objective function for $i = 1, \ldots, M$ and $\mathbf{x} = \{x_1, \ldots, x_P\}$ where x_j is the jth design variable and $x_j \in [x_j^L, x_j^U]$. Henceforth all optimization will pertain to a minimization problem, unless otherwise stated. The various steps in Algorithm 1 are explained below.

3.2.1 GenerateInitialPopulation *Operator*

This process generates *pop*, an array of size $N \times P$, which contains N (population size) members, each of which is a vector for length P. This population is created by generating uniformly distributed random numbers bounded by $[x_j^L, x_j^U]$ for each x_j.

3.2.2 NonDominatedRanking *Operator*

DEFINITION 1. *Say operator '$\lhd$' denotes worse and '$\rhd$' denotes better solution i.e, for a minimization problem $a \lhd b$ means $a > b$ while for a maximization problem $a \lhd b$ means $a < b$ and vice-versa for '$\rhd$' operator. Thus a solution $\mathbf{x}^a$ is said to dominate $\mathbf{x}^b$ if both of following condition is true:*

7C-3

Algorithm 2 Non-Dominated Ranking Algorithm

$\chi \leftarrow 1$ {comment: *pop* is the current population and $\{x^a, x^b\} \in pop$}
repeat
  $N \leftarrow length(pop)$
  for $a = 1$ to N **do**
    if $\exists\, b$ such that $x^b \rhd x^a$ **then**
      $push(x^a, dominated)$
    end if
  end for
  for $a = 1$ to N **do**
    if $x^a \notin dominated$ **then**
      $rank_{x^a} \leftarrow \chi$
      $remove(x^a, pop)$
    end if
  end for
  $\chi \leftarrow \chi + 1$
  clear dominated
until $pop = \{\emptyset\}$
return NonDominatedRanking

1. $\mathbf{x}^a$ is no worse than $\mathbf{x}^b$ for any of the objective functions. In other words, $f_i(\mathbf{x}^a) \not\lessdot f_i(\mathbf{x}^b)$ for all $j = 1, \ldots, M$.

2. The solution $\mathbf{x}^a$ is strictly better than $\mathbf{x}^b$ in at least one objective. In other words, $\exists j$ for which $f_j(\mathbf{x}^a) \rhd f_j(\mathbf{x}^b)$.

Now for a set of N solution vectors, the non-dominated ranking can be done according to Algorithm 2.

3.2.3 FitnessAssignment *Operator*

After the population of size N has been ranked, fitness values are assigned each of the solution. At first all the solutions in a particular front (having same rank) are assigned same fitness value (f_k) and then those fitness value is *shared* with other solutions in the same front. The sharing procedure for a solution x^a in the kth front is performed as follows:

i. Compute Euclidean distance measure with another solution x^b in the kth front as:

$$d_{ab} = \sqrt{\sum_{p=1}^{P} \left(\frac{x_p^a - x_p^b}{x_p^U - x_p^L} \right)^2}$$

ii. The Sharing function value is then computed as [12]:

$$S(d_{ab}) = \begin{cases} 1 - \left(\frac{d_{ab}}{\xi} \right)^2, & \text{if } d_{ab} \leq \xi \\ 0 & \text{otherwise} \end{cases}$$

where $\xi \approx \frac{0.5}{\sqrt[P]{10}}$

iii. The niche count is then computed as:

$$n_a = \sum_{b=1}^{p_k} S(d_{ab})$$

where p_k is the number of solutions in the kth front.

iv. Then the new fitness value is computed as:

$$f_i' = \frac{f_k}{n_a}$$

This procedure is repeated for all the fronts with $f_{k+1} = \min(f_k) - \kappa$ where κ is a small positive value.

3.2.4 Selection *Operator*

After fitness has been assigned to all the solutions of the current population, one solution is randomly picked up from the pool. The probability of selection of a solution x^a with fitness value f_a is given by [12]:

$$P(x^a) = \frac{f_a}{\sum_{j=1}^{N} f_j}$$

3.2.5 Crossover *Operator*

Crossover is the fundamental mechanism of genetic rearrangement for both real organisms and genetic algorithms. After two parents are selected by the Selection operator, these two parents are crossed over with a probability of 0.9 to give rise to two new children. The Crossover operation can be summarized as:

i. Calculate β_q as:

$$\beta_q = \begin{cases} (u\alpha)^{\frac{1}{\eta_c+1}}, & \text{if } u \leq \frac{1}{\alpha} \\ \left(\frac{1}{2-u\alpha} \right)^{\frac{1}{\eta_c+1}} & \text{otherwise} \end{cases}$$

where $\eta_c = 30$, u is a random number between 0 and 1 and $\alpha = 2 - \beta^{-(\eta_c+1)}$ where β is calculated as follows:

$$\beta = 1 + \frac{2}{x_i^b - x_i^a} \min[(x_i^a - x_i^L), (x_i^U - x_i^b)]$$

ii. The children are then calculated as follows:

$$y_i^a = 0.5[(x_i^a + x_i^b) - \beta_q \left| x_i^b - x_i^a \right|]$$
$$y_i^b = 0.5[(x_i^a + x_i^b) + \beta_q \left| x_i^b - x_i^a \right|]$$

3.2.6 Mutation *Operator*

Mutations modify a small fraction of the solution variables: roughly one in every 10,000. Mutation alone does not generally advance the search for a solution, but it does provide insurance against the development of a uniform population incapable of further evolution. After the two children are generated by the Crossover operator, they are mutated by using the polynomial probability distribution [13]. The following steps are used on each y_i for $i \in \{1, P\}$ with a probability p_m:

i. Parameter δ_q is calculated as:

$$\delta_q = \begin{cases} [2u + (1-2u)(1-\delta)^{\eta_m+1}]^{\frac{1}{\eta_m+1}} - 1 \\ \qquad \text{if } u \leq 0.5 \\ 1 - [2(1-u) + 2(u-0.5)(1-\delta)^{\eta_m+1}]^{\frac{1}{\eta_m+1}} \\ \qquad \text{otherswise} \end{cases}$$

where u is a random number between 0 and 1, $\delta = \min[(y_i^a - y_i^L), (y_i^U - y_i^a)]$, $\eta_m = 100 +$ iteration number.

ii. The mutated child portion is calculated as follows:

$$z_i^a = y_i^a + \delta_q(y_i^U - y_i^L)$$

The mutation probability p_m is linearly varied from $\frac{1}{P}$ till 1.0.

For minimization problems in NSGA, the objective function of the form $f_j(\overrightarrow{x}) \leq t_j$ is modified to $\min\langle f_j(\overrightarrow{x}) - t_j \rangle$ where the operator $\langle x \rangle$ returns x if x is positive or 0 otherwise. A much detailed explanation for these operators can be found in [10]

4. NSGA DEMONSTRATION

We shall now demonstrate the application of NSGA on a sample non-linear multi objective optimization problem. The sample problem is formulated as:

$$goal \quad f_1 = 10x_1 \leq 2 \equiv \min \langle f_1 - 2 \rangle$$
$$goal \quad f_2 = \frac{10+(x_2-5)^2}{10x_1} \leq 2 \equiv \min \langle f_2 - 2 \rangle$$
$$Subject\ to \quad \mathbb{F} \equiv \quad (0.1 \leq x_1 \leq 1, 0 \leq x_2 \leq 10)$$

$$(4)$$

Figure 1 shows a values of f_1 and f_2 for randomly chosen values of x_1 and x_2. It is noticed that there exists no feasible solution to this problem, since no values of (x_1, x_2) can make both the objective function meet their goal. In this situation

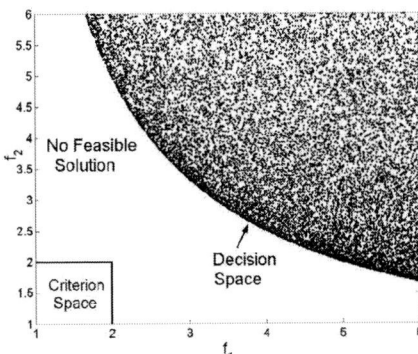

Figure 1: Non-overlapping Criterion Space and Decision space of the problem stated according to Equation 4

the traditional optimization technique will cease to give any fruitful results, and on the other hand the solutions provided by NSGA are plotted in Figure 2. Thus NSGA provides solutions which are concentrated in a certain neighborhood and tries to bring both the objective functions near their goals and also the tries to keep the penalty of the solutions to a minimum. Thus instead of providing no-feasible-solution, it provides the solutions from the decision space which are closest to the required non-feasible objective function point. Thus the advantages of NSGA over the traditional techniques

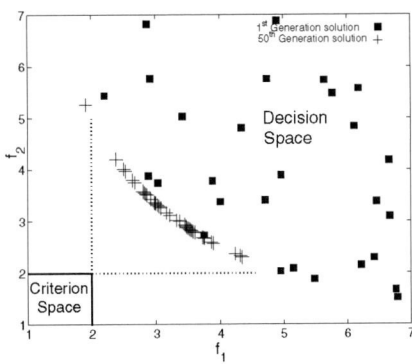

Figure 2: NSGA solution to the Equation 4

are evident from this example. YOGA can thus use NSGA in order to do circuit optimization in the similar way, which is explained in further details in next section.

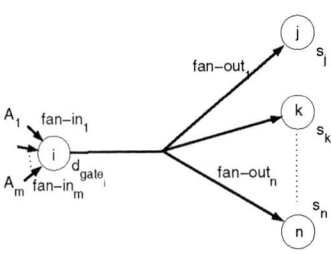

Figure 3: Depth based Nodal delay model for STA

5. YOGA IMPLEMENTATION

NSGA in [13] concentrated on convergence of such Pareto-optimal set, while YOGA maintains wide diversity in the solution set, so as to provide the designer with ample different solutions distributed over design spectrum. Thus for using NSGA in YOGA, the problem is formulated as in Equation 5.

$$find \quad \mathbf{s}$$
$$Minimize \quad \max(\overrightarrow{\sigma}_{\mathcal{O}}/\overrightarrow{\mu}_{\mathcal{O}})$$
$$Minimize \quad \pi(\mathcal{G})$$
$$Subject\ to \quad s_j^L \leq s_j \leq s_j^U, \quad \forall j \in [1, N], g_j \in \mathcal{G}$$

$$(5)$$

where $\mathcal{O}$ is the set of all the outputs of the circuit, $\mathcal{G}$ is the set of all the N gates in the circuit, $\overrightarrow{\mu}_{\mathcal{O}}$ is a vector of mean delay values of all the outputs in $\mathcal{O}$, $\overrightarrow{\sigma}_{\mathcal{O}}$ is a vector of delay deviation of all the outputs in $\mathcal{O}$ and g_i is the ith gate with a sizing factor of s_i.

Thus the solution vector of YOGA will be set of vectors each of length N, where each element of each solution will correspond to the sizing factor of the respective gates.

5.1 Gate Delay model

In the past, for the purpose of Statistical Timing Analysis (STA) of digital circuit, the delay was modelled as a random variable to incorporate the manufacturing uncertainty. Song in [7] modelled the variation in delay as sum of a factor directly proportional to delay and another random factor whereas Raj in [2] assumed variation to be 15% of the expected value. In our work we model the delay as:

$$d_i = \max(A_{i1}, \ldots, A_{in}) + d_{gate_i} + d_{load_i} \quad (6)$$

where d_i is the delay at wire i, $(A_{i1}, \ldots, A_{in})$ are the arrival times of fan-ins of node i, d_{gate_i} is the intrinsic delay of $gate_i$ and d_{load_i} is the delay incurred due to fan-out loads of node i. The situation is depicted in Figure 3. Here the scaling factors $(s_1, \ldots, s_N)$ are modelled as random variables. Intuitively it can be comprehended that a smaller the sizing factor, larger is the uncertainty in manufacturing. Thus the factor (σ/μ) decreases with increasing size (where σ_s and μ_s are the variance and mean of sizing factor respectively). With this kind of delay model and delay characterization (Section 5.2.1), delay at node i can be computed fairly accurately by considering effects of arrival time, intrinsic delay, the loading effect.

5.2 Statistical Timing Analysis (STA)

Block based and event propagation techniques [14, 15] are efficient ways for doing the STA for the circuit. Our methodology is based on techniques mentioned in [15]. At first arrival times are assigned to all the primary inputs of the circuits depending on the fan-out load for each of the inputs. Then the whole circuit is divided into different depths.

The gates which receive their inputs from the primary inputs of the circuit are assigned a depth 1. If $\mathcal{F}_i$ is the set of fan-ins of the gate g_i, and $\mathcal{D}_i$ is the depths of members of $\mathcal{F}_i$, then the depth of a gate g_i, is given by $\max(\mathcal{D}_i) + 1$. A gate is not assigned its depth unless and until all the elements in $\mathcal{F}_i$ have been assigned their corresponding depths. After the depth assignment is completed, the mean and variance values of delay are computed for all the gates having depth 1. Then the delay values are propagated from depth 1 to depth 2 and so on from depth k to depth $k+1$. The delay at the output of any gate g_i is given by Equation 6. The $\max(A_{i1}, \ldots, A_{in})$ operation is performed with the use of formulation for $z = \max(x, y)$ [16] according to Equation 7.

$$
\begin{aligned}
a^2 &= \sigma_x^2 + \sigma_y^2 - 2\sigma_x\sigma_y\rho \\
\alpha &= (\mu_x - \mu_y)/a \\
\nu_1 &= \mu_x\phi(\alpha) + \mu_y\phi(-\alpha) + a\ \varphi(\alpha) \\
\nu_2 &= (\mu_x^2 + \sigma_x^2)\phi(\alpha) + (\mu_y^2 + \sigma_y^2)\phi(-\alpha) + (\mu_x + \mu_y)\ a\ \varphi(\alpha)
\end{aligned}
$$
$$(7)$$

where $x \sim N(\mu_x, \sigma_x)$ and $y \sim N(\mu_y, \sigma_y)$ and $\rho = r(x, y)$ where $r()$ is the correlation operator. Thus we can get $\mu_z = \nu_1$ and $\sigma_z = \sqrt{\nu_2 - \nu_1^2}$.

Thus the delay values are propagated from depth 1 till the last front which are the primary outputs of the circuit. After the mean and variance values are computed at the primary output, the factor $\max(\sigma_1/\mu_1, \ldots, \sigma_L/\mu_L)$, where L is the number of outputs of the circuit, is computed and returned to YOGA as the value of one of it objective function. Though this is not a very accurate technique of doing STA, but since our main concentration in this work is the effectiveness of YOGA, we can compromise with a less complicated model for doing STA.

The area of the whole circuit is simply calculated as :

$$
Area(\mathcal{G}) = \sum_{j=1}^{N} \alpha_j s_j
$$

where N is the number of gates in the circuit. s_j is the size of gate i which has a weighting factor of α_j. The value of α_j for any particular gate depends on the number of its input and the type of the gate. For example a 2-input XOR gate has an α value of 24 units while a 3-input NAND gate has α value of 15 units. The weighting factor α depicts the size of a minimum sized gate of that type.

5.2.1 Delay Characterization

For shorter run time while doing the STA of any circuit, delay characterization is done before hand. In this technique, the effect on delay at node i due to variations in s_i of gate g_i and size of fan-outs can be captured simultaneously. Initially a sample circuit is constructed consisting a chain of 2 inverters which have size as s_D, s_R respectively. Then the sample circuit is simulated using SPICE for some set of random values pairs of s_D and s_R and the delay of 1st inverter is computed for each value pair. From Elmore delay analysis we conclude that the delay is of the form:

$$
d = \kappa_1 + \kappa_2\left(\frac{s_R}{s_D}\right) + \kappa_3\left(\frac{s_R}{s_D}\right)^2. \tag{8}
$$

Any curve fitting technique, like Response Surface Method (RSM) or Singular Value decomposition (SVD) can be used to estimate the three coefficients of Equation 8. This delay model can be used for STA. If 50 different values are used for characterization, then the error in estimated delay is around 0.0001% of the actual value, which is fairly acceptable.

6. EXPERIMENTAL RESULTS

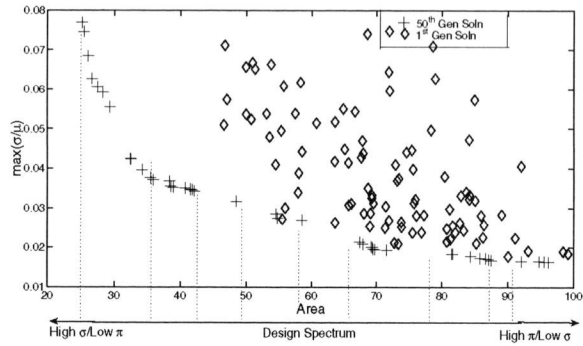

Figure 4: Results of YOGA on ISCAS C17 Benchmark

This section presents the results of application of YOGA algorithm on the variety of benchmark circuits of ISCAS family [17] and few circuits for which the types of gate in it were chosen arbitrarily. The STA was done using Berkeley PTM [18] $70nm$ model cards for NMOS and PMOS transistors. The results of YOGA on c17 circuit is shown in Figure 4. It can be inferred from the figure that given a random set of gate sizes, YOGA delivers a variety of solutions at the designers disposal. All the solutions marked with '+' in figure 4 consist the *Pareto-optimal set* in which no one solution is better than any other. Few of the Pareto-optimal solutions for other benchmarks circuits are listed in Table 1.

If traditional single objective function techniques were used to solve this optimization problem, in a way to minimize $\max(\sigma/\mu)$ and to keep total area under 120 units, then the solution provided will be a single set of gate sizes which will lie on the right extreme section of design spectrum of Figure 4. But it can be noticed from the figure, that other solutions exist which can minimize objective function to the same amount but can still account for lesser area of the circuit. Single objective function techniques fail to see these solutions and thus may not provide solutions which are optimum in multi-sense. On the other hand, YOGA gives a wide range of solutions and designers can pick any of the pareto-optimal solution according to their need. For example low area solutions can be selected from left section of design spectrum, while least variation solutions can be picked from right section. The pareto-optimal solutions provide the users with visible trade-offs for choosing their solutions. If they don't wish to loose much on both end of area and delay variability, then they can choose solutions from middle sections of spectrum. This flexibility is be provided by traditional techniques.

For the case of c17 benchmark the designer has varied solutions to pick any one of them according to their requirements. For example if the area is of more concern then he can choose solutions such as (area = 25.1485, $\max(\sigma/\mu)$ = 0.0769) or (area = 32.4708, $\max(\sigma/\mu)$ = 0.0425), where he gets low area solution but has to compensate by getting a high variance circuit along with. But on the other hand of timing variance is of more importance then he can choose solution from the other end of the spectrum, such as (area = 71.5729, $\max(\sigma/\mu)$ = 0.0193) or (area = 96.1451, $\max(\sigma/\mu)$ = 0.0163) where he has to compromise on area to get a low variance circuit. Whereas if he needs to keep both in objective in control then he can choose the solution which are midway in the design spectrum such as (area=54.8, max(σ/μ) = 0.0275).

Similarly, solutions from different portion of design spectrum for any given circuit can be provided to suit the designers needs and requirements. The only input which is required is circuit topology and the minimum and maximum allowable gate sizes.

7. CONCLUSION

In this paper, we have shown the advantages of Multi-objective optimization techniques over the traditional single objective techniques. A new algorithm YOGA was proposed to use multi-objective optimization technique for gate sizing domain of digital circuits and flexibility obtained by it was demonstrated for various benchmark circuits.

8. REFERENCES

[1] S. H. Choi, B. C. Paul, and K. Roy, "Novel sizing algorithm for yield improvement under process variation in nanometer technology," in *DAC*, 2004.

[2] S. Raj, S. Vrudhula, and J. Wang, "A methodology to improve timing yield in the presence of process variations," in *DAC*, 2004, pp. 448–453.

[3] A.A.Ilumoka, "Optimal transistor sizing for cmos vlsi circuits using modular artificial neural networks," in *Twenty-Ninth Southeastern Symposium on System Theory*, 1997.

[4] J. Singh, V. Nookala, Z.-Q. Luo, and S. Sapatnekar, "Robust gate sizing by geometric programming," in *DAC*, 2005.

[5] A. Nardi and A. L. Sangiovanni-Vincentelli, "Synthesis for manufacturability- a sanity check," in *Design Automation and Test in Europe*, 2004.

[6] M. Pan, C. C. N. Chu, and H. Zhou, "Timing yield estimation using statistical static timing analysis," in *Intl Symposium on Circuits and Systems*, 2005.

[7] O. Neiroukh and X. Song, "Improving the process-variation tolerance of digital circuits using gate sizing and statistical techniques," in *Design Automation and Test in Europe*, 2005.

[8] M. Mani, A. Devgan, and M. Orshansky, "An efficient algorithm for statistical minimization of total power under timing yield constraints," in *DAC*, 2005.

[9] C. M. Foncesa and P. J. Flemming, "Genetic algorithms for multi-objective optimization: Formulaton, discussion and generalization," in *5th Intl Conf on Genetic Algorithms*, 1993, pp. 416–423.

[10] N. Srinivas and K. Deb, "Multi-objective function optimization using non-dominated sorting genetic algorithms," in *Evolutionary Computation*, 1994.

[11] A. G. Cunha, P. Oliveira, and J. A. Covas, "Use of genetic algorithms in multicriteria optimization to solve industrial problems," in *Proceedings of the Seventh International Conference on Genetic Algorithms*, 1997, pp. 682–688.

[12] D. E. Goldberg, *Genetic algorithms for search, optimization, and machine learning.* MA: Addison-Wesley., 1989.

[13] K. Deb, "Non-linear goal programming using multi-objective genetic algorithms," *Evolutionary Computation Journal*, 1994.

[14] A. Devgan and C. Kashyap, "Block-based static timing analysis with uncertainty," in *ICCAD*, 2003.

[15] J. Liou, K. Cheng, S. Kundu, and A. Krstic, "Fast statistical timing analysis by probabilistic event propagation," in *DAC*, 2001, pp. 661–666.

[16] C. Clark, "The greatest of a finite set of random variables," *Operations Research*, vol. 9, no. 2, 1961.

[17] M. Hansen, H. Yalcin, and J. P. Hayes, "Unveiling the iscas-85 benchmarks: A case study in reverse engineering," *IEEE Design and Test*, 1999.

[18] "Berkeley ptm." [Online]. Available: http://www-device.eecs.berkeley.edu/~ptm/

Table 1: Few solutions of *Pareto-optimal set* obtained by YOGA for ISCAS benchmark circuits

Circuit Name	No. of gates	Area $= \sum\limits_{j=1}^{N} \alpha_j s_j$	max (σ/μ)	Run time(s)
c17	6	25.1485	0.0769	4.42
		32.4708	0.0425	
		54.7947	0.0275	
		71.5729	0.0193	
		96.1451	0.0163	
c432	160	769.5831	0.0538	43.43
		1054.2467	0.0317	
		2131.5073	0.0169	
		3028.1891	0.0142	
		3798.7005	0.0121	
c499	202	1418.9510	0.0231	51.37
		2025.2311	0.0189	
		3869.6068	0.0145	
		4932.4860	0.0135	
		6420.8214	0.0126	
c880	383	1818.4447	0.0618	94.25
		3102.6154	0.0432	
		4756.7919	0.0281	
		6281.9218	0.0228	
		7554.7382	0.0205	
c1355	546	2681.1763	0.0117	137.18
		3817.4057	0.0085	
		3965.4355	0.0068	
		8815.9168	0.0057	
		10672.4103	0.0042	
c1908	880	3218.8936	0.0247	227.20
		5476.0578	0.0202	
		8981.3529	0.0163	
		12692.8132	0.0148	
		15834.0099	0.0131	
c270	1193	6973.1389	0.0687	341.68
		9281.2028	0.0534	
		16315.1987	0.0323	
		22180.6038	0.0241	
		24872.2772	0.0213	
c3540	1669	7281.1988	0.0357	470.75
		14672.0153	0.0248	
		21915.7468	0.0162	
		29180.4415	0.0151	
		34971.9318	0.0138	
circuit$_1$	20	172.4660	0.0631	7.67
		275.4186	0.0421	
		438.8642	0.0284	
		569.5252	0.0231	
		661.2028	0.0208	
circuit$_2$	30	252.6721	0.0648	10.20
		628.8381	0.0408	
		912.0196	0.0256	
		1181.6013	0.0237	
		1263.3795	0.0228	

A Probabilistic Analysis of Pipelined Global Interconnect Under Process Variations

Navneeth Kankani   Vineet Agarwal   Janet Wang

Electrical and Computer Engineering Department

The University of Arizona, Tucson, AZ

{kankani, vagarwal, wml}@ece.arizona.edu

Abstract—**The main thesis of this paper is to perform a reliability based performance analysis for a shared latch inserted global interconnect under uncertainty. We first put forward a novel delay metric named DMA for estimation of interconnect delay probability density function considering process variations. Without considerable loss in accuracy, DMA can achieve high computational efficiency even in a large space of random variables. We then propose a comprehensive probabilistic methodology for sampling transfers, on a shared latch inserted global interconnect, that highly improves the reliability of the interconnect. Improvements up to 125% are observed in the reliability when compared to deterministic sampling approach. It is also shown that dual phase clocking scheme for pipelined global interconnect is able to meet more stringent timing constraints due to its lower latency.**

I. INTRODUCTION

In multi-gigahertz system-on-chip designs, global interconnect wire proves to be a major bottleneck to the continual increase in clock frequency as it requires several clock periods of time to propagate a signal from source to sink. Extensive amount of work has been done in recent past to alleviate the global interconnect from the clock frequency constraints. And, it is an accepted fact that pipelining of global interconnects offers a promising solution to this problem [1], [2].

Although pipelining increases the throughput of interconnect, it comes with practical difficulties like the cycle level behavior changes in the RTL level which requires a lot of manual rework in design. Scheffer in [3] listed the challenges faced by the designer while pipelining an interconnect. In [4] Cong et al addressed the problem of automatic interconnect pipelining at RTL level by proposing a RDR-pipe(Regular Distributed Register) approach to efficiently support the multi-cycle on chip communication with interconnect pipelining.

For instance, consider a scheme where a single pipelined interconnect is shared between 3 computational units that need to transfer the data to another computational block. In a real-time system operated at a clock period of $285ps$, let the arrival time set of these 3 transfers to be (170, 350, 690)ps; each having a corresponding deadline. Now, if these transfers are scheduled to be sent on the interconnect such that at most one transfer is issued per clock cycle, then they can share the same interconnect, provided their respective deadlines are met.

However, it should be noted that the uncertainties in the arrival times and interconnect delay due to manufacturing variations causes considerable bit error rate (BER) on the global interconnect. For the above example, we observe that deterministic sampling of transfers at the μ or $(\mu + 3\sigma)$ values, can give an error rate as high as 59%, which is highly undesirable. However, there may exist other set of sampling

times which can yield much higher reliability (as high as 99%). Hence, there is a need of a formal methodology that can find such sampling sets to address the problem of increasing error rates on shared global interconnects.

To this purpose, we propose a comprehensive probabilistic methodology for sampling transfers, to increase the reliability of shared latch inserted global interconnect under process variations. The proposed technique meets the timing constraints of the global interconnect with a much higher reliability compared to the deterministic sampling approach. Note that the overall reliability of each transfer depends upon its sampling reliability and transmission reliability. The latter depends upon the BER encountered by transfer on the pipelined interconnect. In order to compute the BER , we perform the statistical timing analysis on the interconnect. While doing so, we support a simple dual phase clocking scheme for pipelined global interconnect, as it proves to be robust in noisy environment. We also propose a novel delay metric based on ANOVA (DMA) for estimating the probability density function of the interconnect delay. It is an efficient and accurate metric when compared to other variational delay metrics and provides the designer with an explicit polynomial approximate expansion for the delay response.

To begin, we explain the new delay metric for interconnect and providing the results to prove its efficiency compared to other delay metrics in Section II. A probabilistic methodology for sampling transfers, to reliably transmit the data on a shared global interconnect is presented in Section III. The results of our technique are listed in Section IV and Section V concludes our paper.

II. DELAY METRIC BASED ON ANOVA (DMA)

DMA serves as an efficient metric for finding the pdf of the global interconnect delay considering process variations. Given the uncertainty in parameters and degree p of the required model, the prototype DMA returns the pth degree polynomial for the delay of the global interconnect along with its mean and variance.

In our formulation, we first approximate the delay response of a single RC model of an interconnect as a function of uncertainty in geometric parameters such as wire width(w), spacing(s), thickness(t) and inter-layer dielectric thickness(h). Our aim is to find a computationally efficient approximation $\hat{y}$ of actual output ($y = f(w, s, t, h)$), with a very small error margin.

$$\hat{y} = \hat{f}(w, s, t, h) \qquad (1)$$

where $\hat{f}$ is a finite order polynomial function that approximates the behavior of the model and w, s, t, h are random variables that can be expressed in terms of zero mean and unit variance vector $\overline{\xi} = (\xi_1, \xi_2, \xi_3, \xi_4)$ such that

$$
\begin{aligned}
w &= \mu_w + \sigma_w \xi_1 & s &= \mu_s + \sigma_s \xi_2 \\
t &= \mu_t + \sigma_t \xi_3 & h &= \mu_h + \sigma_h \xi_4
\end{aligned}
$$

In this work, we assume the pdf of geometric variations to be gaussian. Thus, without the loss of generality, this delay metric can be used for any random distribution, and hence we can approximate the output using a set of hermite polynomials [5]. Thus equation 1 can be written as

$$
\hat{y} = \sum_{i=0}^{N} \alpha_i g_i \left(\overline{\xi} \right) \tag{2}
$$

where $g(\overline{\xi})$ is a set of hermite polynomials, $\overline{\alpha}$ represents a coefficient vector, N is the order of approximation. To compute $\overline{\alpha}$ we use probabilistic collocation method (PCM) proposed in [6].

The polynomial constructed in (2) is expressed in terms of the input parameters and their interactions, not all of them may be significant in the approximation of the response. In fact, considering a large space of random parameters there may be certain parameters that have negligible effect on the response. Therefore we advocate that by detecting and eliminating such parameters from our design, we can reduce the computational complexity involved in evaluating the response without significant loss of accuracy.

We apply a technique named ANOVA (Analysis of Variance)[7] on this RC model to quantify the importance of variables on the variability of the response. As its name suggests, ANOVA analyzes the variances to test for significant differences between means by partitioning the total variability into component parts. The proportion of variance due to each input (or its correlation) towards the total variance can be used as a statistical significance parameter (F) of that particular input.

A. Overview of ANOVA

The statistical significance parameter (F) can be computed by applying the underlying notion of ANOVA. We explain the notation and implementation details by considering a simple model with single uncertain input ξ and response $y = f(\xi)$. As the input is random, let us assume that we consider 'm' different values for the input variable at which we will observe the response. Then the observed response 'y_{ij}' represents the jth observation taken under i^{th} instance of ξ. The observation can be described by the linear statistical model as

$$
\begin{aligned}
y_{ij} = \mu + \tau_i + \varepsilon_{ij} \quad & i = 1, 2, \ldots, m \\
& j = 1, 2, \ldots, n
\end{aligned} \tag{3}
$$

μ is the overall mean, τ_i is called i^{th} instance effect and is unique to that instance, and ε_{ij} is the error component. The basic idea in ANOVA is the comparison of the variance in the response due to intra-instance and inter-instance variability. The null hypothesis defined as inter-instance variability (σ_t^2) is zero, implies that the response means are same for different instances.

$$
H_0 : \sigma_t^2 = 0 \tag{4}
$$
$$
H_1 : \sigma_t^2 > 0 \tag{5}
$$

where σ_t is variance of τ_i. $\sigma_t > 0$ implies that the variability exists between 'm' instances.

In order to calculate variance, we define between-instance sum of squares(SSF) and within-instance sum of squares (SSE) such that:

$$
\text{SSF} = n \sum_{i=1}^{m} (\overline{y}_{ei} - \overline{y}_t)^2 \tag{6}
$$

$$
\text{SSE} = \sum_{i=1}^{m} \sum_{j=1}^{n} (y_{ij} - \overline{y}_{ei})^2 \tag{7}
$$

$$
\overline{y}_{ei} = \left(\sum_{j=1}^{n} y_{ij} \right) / n \tag{8}
$$

$$
\overline{y}_t = \left(\sum_{i=1}^{m} \sum_{j=1}^{n} y_{ij} \right) / N \tag{9}
$$

Here $N = mn$ is the total number of observations. $\overline{y}_{ei}$ represents the average of the observations taken under i^{th} instance whereas $\overline{y}_t$ represents the average of observations taken under all instances. The significance of SSF is that it explains the variability in response due to difference in mean of instances whereas SSE is referred to as error variance. The variance or the mean square for between and within instance is defined as:

$$
\text{MSF} = \text{SSF}/(m-1) \tag{10}
$$
$$
\text{MSE} = \text{SSE}/(N-m) \tag{11}
$$

To test the null hypothesis H_0, a F-ratio is defined as F = MSF/MSE. If the null hypothesis is true, then both MSF and MSE estimate the same quantity and thus F-ratio must be 1. Assuming the observations are normally distributed, it can be shown that SSF/σ^2 and SSE/σ^2 are independently distributed chi-square random variables [8]. Thus, if null-hypothesis is true then the F-ratio must also be chi-square distributed with $(m-1, N-m)$ number of samples. We can find this F-ratio (F_0) using a look up table [7] for a given significance level (α). The probability P of obtaining the computed F-ratio (MSF/MSE) greater than F_0 is used as a significance parameter such that the null hypothesis is rejected if P is lower than α (set here as 0.05). Another important quantity that is used in determining the proportion of variability in response explained by the model is defined as R^2 = SSF/SST.

B. Reduction using ANOVA

We apply ANOVA technique on the polynomial generated by PCM in equation (2) to find the insignificant terms in the model. The analytical expression for the delay of single RC segment, generated by PCM is given as:

$$
\begin{aligned}
delay = \ & 19.65 - 2.28\xi_1 - 0.9\xi_2 - 1.82\xi_3 - 0.32\xi_4 \\
& + 0.28(\xi_1^2 - 1) + 0.1(\xi_2^2 - 1) + 0.12(\xi_3^2 - 1) \\
& + 0.05(\xi_4^2 - 1) + 0.17(\xi_1\xi_2) + 0.03(\xi_1\xi_4) \\
& + 0.2(\xi_2\xi_3) - 0.17(\xi_2\xi_4) + 0.17(\xi_3\xi_4) \ ps
\end{aligned} \tag{12}
$$

TABLE I
COMPARISON OF MONTE CARLO(MC), PCM AND DMA RESULTS

Number of segments	Mean delay (ps)			Delay Variation (ps)			Number of Spice Runs			Error%	
	MC	PCM	DMA	MC	PCM	DMA	MC	PCM	DMA	μ	σ
2	40.06	39.92	40.11	3.59	3.45	3.63	10000	45	38	0.12	1.11
4	163.67	163.94	164.04	9.24	9.37	9.42	15000	153	86	0.22	1.94
8	496.36	496.66	494.62	26.59	26.36	27.16	20000	561	261	0.35	2.15
16	1616.29	1617.15	1625.82	43.45	42.81	44.86	25000	2145	912	0.59	3.26

The mean and standard deviation of the delay are $19.62ps$ and $3.15ps$ respectively. At first, we apply a primary level of screening to determine the individual effect of each ξ_i on the delay. Figure 1 shows the individual significance of w, s, t, h in the delay of a RC segment which is used in primary level of screening. We then compute the delay gradient of individual effects of ξ_i. The set of ξ_i, for which the delay gradient is below a certain threshold, is used in secondary level of screening. For example, it is noted that h has negligible effect on delay compared to other parameters. And thus, h will be passed on to secondary level of screening.

We use ANOVA on the Equation (12) in the secondary level of screening to remove insignificant terms. Removing the insignificant terms(Ω), we generate a reduced analytical equation such that its R^2 value is at least 98.5%. In this case ANOVA gives us that the terms $\{\xi_4, \xi_2^2, \xi_4^2, \xi_1\xi_2, \xi_1\xi_3, \xi_1\xi_4, \xi_2\xi_4, \xi_3\xi_4\}$ are insignificant and the ANOVA table for the corresponding reduced model is given in Table II. The reduced analytical equation is of the form

$$delay = 19.65 - 2.28\xi_1 - 0.9\xi_2 - 1.82\xi_3 + 0.28(\xi_1^2 - 1) \\ +0.12(\xi_3^2 - 1) + 0.2(\xi_2\xi_3) \; ps$$

(13)

which has a mean and standard deviation of $19.64ps$ and $3.13ps$ respectively, there by giving a mere error of 0.02% in mean value and 1.2% in standard deviation. The definitions of the terms in Table II can be found in Section II-A.

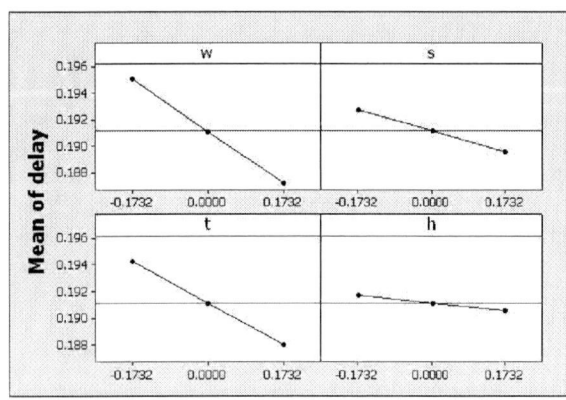

Fig. 1. Significance of geometric parameters on delay

C. DMA Implementation

To begin our DMA analysis, we divide the global interconnect into smaller identical modules (Figure 2) where each smaller module has n random variables. To find the delay equation for interconnect, we use PCM technique in which, the

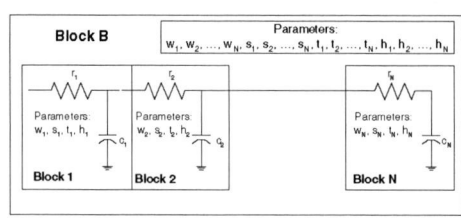

Fig. 2. RC tree Model for Global Interconnect

TABLE II
ANALYSIS OF VARIANCE FOR DELAY

Source	Υ	Sum of Squares	Mean Squares	F value	P
Model	7	454×10^{-6}	76×10^{-6}	8059.65	0.0
Residual Error	8	12×10^{-9}	26×10^{-10}	5.28	
Total	15			$R^2 = 99.73\%$	

$\Upsilon \rightarrow$ Degree of Freedom

collocation points are computed for these n random variables, for a given degree p.

An important step in DMA is analysis of this smaller blocks using ANOVA to determine the parameters that have insignificant effect on the variability of the response. This is performed by the technique outlined in Section II-B. As all the segments of the macro-model (Block B) are identical we can extrapolate the information about insignificant parameters of the other smaller blocks based on the ANOVA results of the first block. A set of all insignificant parameters $\Gamma = \{\Omega_1 \cup \Omega_2 \ldots \cup \Omega_n\}$ is then used in evaluation of response of larger block. It should be noted that DMA preserves significant correlations among the design parameters. Since we know from [9] that the computational complexity is directly proportional to the number of significant terms in the model, the information in Γ can be used to decrease the complexity. Hence, we perform the model runs on only those input sampling points that correspond to the significant parameters in Block B. In this way, by hierarchically removing the insignificant terms and thus reducing the model runs, the computational complexity of DMA is decreased. A comparison of delay values of Monte Carlo, PCM and DMA is done in Table I.

III. RELIABILITY AWARE GLOBAL INTERCONNECT SHARING

The latch inserted global interconnect can be shared between two computational units in order to achieve higher levels of resource utilization. As shown in Figure III, the three transfers from different computational units in Block A can be transferred to Block B using a shared pipelined interconnect.

Each block has a sampler (multiplexer) that sends the transfers on the interconnect according to a priori sampling times such that at most one transfer is issued per clock cycle on the interconnect.

A feasible schedule would be the one in which each transfer starts after its arrival time and completes before the specified deadline. While constructing such a schedule of transfers on the interconnect, it is also necessary to keep the BER below a minimum allowable value, to ensure correct signal transmission in the presence of uncertainty in the arrival times and global interconnect delay. Hence, in this section we formalize a probabilistic methodology to reliably transmit the data on the interconnect with the consideration of process variations. Before discussing our approach, the basic notation

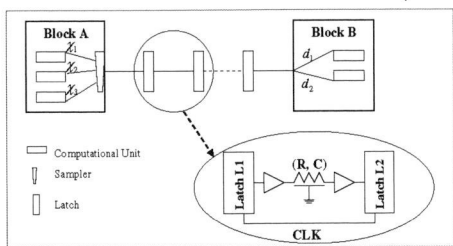

Fig. 3.   Global Interconnect Sharing Stage

and terminology is presented.

- $\overline{\chi}$: Arrival Times$\rightarrow N(\overline{\mu}_A, \overline{\sigma}_A)$
- $\overline{\mu}_A$: Mean
- $\overline{\sigma}_A$: Standard Deviation
- $\overline{\lambda}$: Deadline Time Vector
- clk : Clock period of Global Interconnect
- n : Number of Pipelined stages
- $\overline{\zeta}$: Sampling Times
- $\overline{D}$: Delivery Times
- σ_p : Standard Deviation of Sampler
- $\overline{\omega}$: Bit Error Rate
- $\overline{\psi}_s$: Sampling Confidence
- $\overline{\psi}_b$: Transmission Confidence = $(1 - \overline{\omega})$
- $\overline{\psi}_t$: Total Confidence Level = $\overline{\psi}_s * \overline{\psi}_b$
- Ψ_f : Root Mean Square $(\overline{\psi}_t)$

We assume that each arrival time vector is associated with a given deadline time vector where deadline time is defined as the latest time by which the data must be received by the other end of the interconnect. The sampling time vector is the time when the data is sent over interconnect by the sampler. The sum of sampling time and the interconnect latency is termed as delivery time.

Definition 1: (Sampling Confidence = ψ_s): The Sampling Confidence level ψ_s of x is defined as

$$\psi = 100 * \int_{-\infty}^{\pi_\psi} f(x)dx \qquad (14)$$

where f(x) is the pdf of x and π_ψ (or $\pi_\psi(x)$) is ψ^{th} percentile of x, meaning that a designer can be $\psi\%$ assured that the random parameter x will be less than π_ψ (or $\pi_\psi(x)$).
Transmission confidence $(\overline{\psi}_b)$ gives us the reliability of transmitting data over pipelined interconnect considering the effect

Algorithm 1 probabilistic Scheduling $(clk, \overline{\chi}, \overline{\lambda}, \omega_{min}, \sigma_P)$

$\overline{\tau} \leftarrow \text{MOD}(\overline{\mu}_A, clk) - 0.5 * clk \qquad \overline{\gamma} \leftarrow N(\overline{\tau}, \overline{\sigma}_A)$
$p \leftarrow \text{LUT}(clk, \sigma_p) \qquad\qquad\quad \alpha \leftarrow \text{QUOTIENT}(\overline{\mu}_A, clk)$
$\{\overline{\alpha}_{new}, \overline{\delta}\} \leftarrow \text{COMPUTESLACK}(\omega_{min}, \alpha)$
if $\exists i$ such that $\delta_i < 0$ **then**
$\quad \overline{\beta} \leftarrow \text{FINDSAMPLINGCYCLE}(\alpha_{new}, \overline{\delta})$
$\quad$ **if** $\exists (i, j, i \neq j)$ such that $\beta_i = \beta_j$ **then**
$\qquad$ **for** $i = 1$ to N **do**
$\qquad\quad$ **if** $\alpha_i = \beta_i$ **then**
$\qquad\qquad$ **if** $\pi_{99}(\gamma_i) \leq p$ **then**
$\qquad\qquad\quad k \leftarrow 3$
$\qquad\qquad$ **else if** $\pi_{50}(\gamma_i) \leq p$ **then**
$\qquad\qquad\quad k \leftarrow \text{SOLVE}(\tau_i + k\sigma_i = p)$
$\qquad\qquad$ **else**
$\qquad\qquad\quad k \leftarrow \text{SOLVEFORMAXIMUM}(\psi_{t_i}(k))$
$\qquad\qquad$ **end if**
$\qquad\qquad \zeta_i \leftarrow \mu_i + k\sigma_i$
$\qquad\quad$ **else**
$\qquad\qquad \zeta_i \leftarrow ((\beta_i - 0.5) * clk + p)$
$\qquad\quad$ **end if**
$\qquad$ **end for**
$\qquad \overline{\omega} \leftarrow \text{LUT}(clk, \overline{\zeta})$
$\qquad \Psi_f \leftarrow \text{COMPUTECONFIDENCE}(\overline{\zeta}, \overline{\omega})$
$\quad$ **else**
$\qquad$ **print** `Abort: Deadline Constraint Violation'
$\quad$ **end if**
else
$\quad$ **print** `Error: No feasible schedule possible'
end if
return $(\Psi_f, \overline{\zeta})$

of process variations. It is computed using BER encountered while transferring data over interconnect. We calculate the root mean square confidence (Ψ_f) and choose it as an optimization criteria because it maximizes both sampling confidence and transmission confidence (which in-turn means BER (ω) is minimal).

A. Problem Statement

Given a set of n arrival times $\overline{\chi} = \{\chi_1, \chi_2, \ldots, \chi_n\}$, their corresponding deadlines $\overline{\lambda} = \{\lambda_1, \lambda_2, \ldots, \lambda_n\}$, and a global interconnect clock period clk, the sampling time set $\overline{\zeta} = \{\zeta_1, \zeta_2, \ldots, \zeta_n\}$ is found such that the Ψ_f is maximized. The problem can thus be formulated as formulated as:

$$\max \quad \Psi_f$$
$$\text{subject to:} \quad \omega_i \leq \omega_{min} \quad \forall i \in n$$
$$d_i \leq \lambda_i \quad \forall i \in n$$
$$\psi_{s_i} \geq 50\% \quad \forall i \in n$$

Our goal is to maximize the total confidence Ψ_f so that the timing constraints are met with some guaranteed probability in the presence of process variations. In order to maximize Ψ_f, we have to optimize sampling confidence ψ_s.

It is evident from the third constraint $(\psi_{s_i} \geq 50\%)$ that we want the sampling confidence of each arrival time to be higher than 50%. For simplicity, we hereafter refer to each arrival time as an event to be scheduled on the interconnect. Note

727

that the overall reliability of each transfer depends upon its sampling reliability and transmission reliability. The sampling reliability is evaluated using the confidence level at which the sampler selects the arrival times. The latter depends upon the BER encountered by transfer on the pipelined interconnect. In order to compute the BER associated with each event, we need to perform the statistical timing analysis on the interconnect which is found using the following section.

B. Bit Error Rate Computation

The statistical timing analysis is performed on the latch inserted global interconnect for computation of BER as formulated in [10]. The notations which are used in section are also kept same. For a given latch of stage i, the opaque region (R_{opaque}) is defined as the time at which clock to the latch is low. And the region of high-clock where data can be sampled correctly by latch is termed as transparent Region (R_{tran}). All the other region of clock apart from R_{opaq} and R_{tran} is termed as faulty region (R_{faulty}). Based on the region where the propagation delay of previous stage lies, the propagation delay of stage i is written as:

$$p_i = \begin{cases} \tau_{wire} + \tau_{data} - T_{clk} & p_i \in R_{opaque} \\ p_{i-1} + \tau_{wire} + \tau_{prop} - T_{clk} & p_i \in R_{tran} \end{cases} \quad (15)$$

Using, the pdf of propagation delays the total probability of correct transmission and thereby the BER on the pipelined interconnect can be found using the following equation:

$$BER = 1 - \underbrace{q_1 q_2 \ldots q_N}_{N} = 1 - \prod_{i=1}^{N} q_i$$

where q_i is the probability that stage i of the pipeline will transmit the data correctly. The details for the notation and analysis can be found in [10].

With the use of single phase clocking in latches, the flexibility in timing provided by latch based methodology is not fully utilized. A dual-phase clock system can be used instead of single phase clocking. The advantages of using such a scheme are reduced latency, clock skew tolerance and higher performance. Nevertheless, these advantages come along with area overhead for generating two different clocks.

C. Estimation of Interconnect Reliability

Once the BER is computed, we build a Look-up table (LUT) for a given clock frequency of the pipelined global interconnect and the delay variation of the sampler (σ_s). The inputs to our algorithm are the pdf of the events, their deadlines,

and the maximum clock frequency at which the pipelined interconnect can be operated for a given BER (ω_{min}). At this clock frequency, and for a particular sampler delay variation (σ_s), we find the value p using LUT (figure 4, where p is the mean arrival time beyond which BER starts increasing and approaches one. Thus, it is important to select an optimum value of p for obtaining minimum BER.

The throughput of pipelined interconnect is one event per clock cycle for a single phase clocking scheme. Hence, we must find a sampling clock number for each event such that there is at least one clock period difference between any two events in the set to guarantee no overlap in pipelined stages. We define $\alpha = \text{quotient}(\overline{\mu}_A, clk)$ as the earliest sampling clock number for an event. If $\omega_i > \omega_{min}$ for this α_i, we assign $\alpha_{new,i} = \alpha_i + 1$. Based on the deadline ($\overline{\lambda}$) and number of pipelined stages, we compute slack $\overline{\delta}$ which is defined as:

Definition 2: (Slack $= \overline{\delta}$): It is the difference between the deadline time and the delivery time, that is, the absolute time at which the event reaches the other end of interconnect.

$$\overline{\delta} = \left(\overline{\lambda} - \overline{\alpha}_{new} - n * clk\right) \quad (16)$$

The foremost requirement for any event to be eligible for scheduling is that the $\overline{\delta}$ must be non-negative because it otherwise violates the deadline constraint. We then compute the actual sampling clock number β using available slack for each event, such that there is no overlap in clock number among any two events. If there is an overlap of actual sampling clock numbers, we conclude that there cannot be a feasible schedule that meets the constraint of minimum BER (ω_{min}). Now, for this sampling clock number (β) we want to find an optimum sampling time (ψ_s) for each event in order to maximize Ψ_f. It is assumed that the sampler that is used can select these ψ_s from the continuous arrival time pdf.

If α_i remains unmodified ($\alpha_i = \beta_i$), then we calculate the sampling time $\overline{\zeta} = \overline{\mu}_A + k\sigma_A$, where k is found on the basis of 50th and 99th percentile values of pdf of event set and p. The value of k is computed for the three cases as shown in the algorithm. For instance, when 50th% $\geq p$, we solve for k using LUT for finding transmission confidence (ψ_b) and sampling confidence simultaneously, until ψ_t is maximized. And in case α is modified (provided there is slack available), $\overline{\zeta}$ is found using sampling clock number (β) and mean arrival time (p). Once the sampling time ($\overline{\zeta}$) is computed, we check for the BER (ω) at these sampling times. The total root mean square confidence is finally computed using sampling times and BER.

IV. RESULTS

We perform our analysis on global interconnect based on the $0.18\mu m$ technology parameters given by Berkeley PTM model [11]. There are 8 pipelined stages and the length of wire in each stage is taken to be $1.4\mu m$. The mean and variance of delays are computed using DMA and are listed in Table IV.

Next, we use Section III-B to find the maximum operable clock clk so that $\omega_i < \omega_{min}$. We proceed in our analysis using this clock period and assuming that $\omega_{min} = 2\%$. For comparison of deterministic sampling and our approach, we consider three cases of 4 randomly generated events that needs to be transferred between two blocks. The 3 cases correspond to $\overline{\delta} > 0$, $\overline{\delta} = 0$ and $\overline{\delta} < 0$. Without the loss

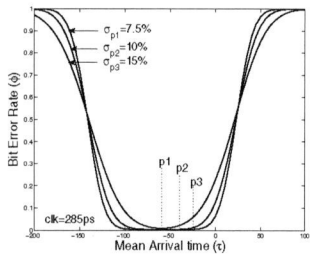

Fig. 4. Graph showing maximum allowable arrival time

TABLE III

COMPARISON OF DETERMINISTIC AND PROBABILISTIC SCHEDULING

| | μ_A | λ | δ | | ζ | | Ψ_f | | | | | | Gain % | | | |
| | | | | | | | μ Sampling | | $\mu+3\sigma$ Sampling | | probabilistic Sampling | | μ Sampling | | $\mu+3\sigma$ Sampling | |
	ps	ps	SP	DP	SP	DP	SP	DP	SP	DP	SP	DP	SP	DP	SP	DP
Case 1																
E1	70	2850	1	1	385	650	44.64	43.13	69.99	70.62	99.47	99.79	122.77	131.37	42.09	41.29
E2	395	3700	3	1	1240	1110										
E3	690	3450	1	2	955	1570										
E4	1020	4100	2	2	1810	2030										
Case 2																
E1	70	2600	0	0	90	90	49.47	40.65	0	81.54	73.52	99.74	48.62	145.36	-	22.30
E2	395	2900	0	0	405	650										
E3	690	3150	0	1	695	1110										
Case 3																
E1	70	2600	0	0	-	90	×	43.13	×	0	×	89.38	-	107.23	-	-
E2	395	3450	2	2	-	1570										
E3	690	2850	-1	0	-	710										
E4	1020	3100	-2	0	-	1165										

SP → Single Phase clocking   DP → Dual Phase clocking   × → Deadline Constraint Violation

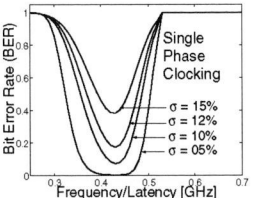

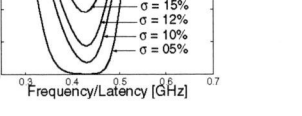

Fig. 5. Effects of σ_{clk} on BER for single phase scheme

Fig. 6. Effects of σ_{clk} on BER for dual phase scheme

of generality, the events are selected such that they cover all possible regions (opaque, transparent, faulty) of sampling within a clock period. The results are tabulated in Table III. We consider two types of deterministic samplings which are 50% and 99% sampling. 50% and 99% samplings corresponds to sampling at μ and $\mu+3\sigma$ values respectively. For deterministic and probabilistic sampling, we show the results when either single phase or dual phase clocking is used. It should be noted in Case 3 that for the same set of arrival times and deadlines, the single phase clocking succumbs because the deadline constraint does not meet. However, its dual phase clocking counterpart gives a feasible schedule with a much higher confidence than mean sampling value. This is possible because the latency of shared global interconnect in dual phase clocking scheme is lower than that of single phase clocking. Thus that more stringent timing constraints can be met with dual phase clocking yielding a low error rate. Furthermore a comparison of dual and single phase clocking, in Figure 5 and 6, for various values of σ_{clk} proves that dual phase is more robust in noisy environments.

TABLE IV

DELAY VALUES USED FOR 0.18μM TECHNOLOGY

| | DMA Delay | |
	$\mu(ps)$	$\sigma(ps)$
τ_{wire}	109.12	9.13
τ_{data}	119.60	13.19
τ_{prop}	115.42	13.62
τ_{setup}	119	13
T_{clk}	μ_c	22

V. CONCLUSIONS

The major contribution of this work is two-fold. First, we provided an efficient and accurate delay metric DMA for estimating the delay pdf of an interconnect under process variations. This delay metric uses ANOVA technique to reduce its computational complexity. Then, a probabilistic methodology for sampling events to reliably transmit the data on the shared latch inserted global interconnect was presented. Using our approach, the error rates are dramatically reduced and significant improvements are observed in total confidence compared to 50th or 99th percentile deterministic sampling approach.

REFERENCES

[1] R. McInerney, K. Leeper, T. Hill, H. Chan, B. Basaran, and L. McQuiddy, "Methodology for repeater insertion management in the rtl layout, floorplan and fullchip timing databases of the itanium microprocessor," in *Proc. of 2000 International Symposium on Physical Design*, 2000.

[2] "Semiconductor industry association," *International Technology Roadmap for Semiconductors*, 2004.

[3] L.Scheffer, "Methodologies and tools for pipelined on-chip interconnect," in *Proc. of International Conference on Computer Design*, 2002, pp. 152–157.

[4] J. Cong, Y. Fan, X. Yang, and Z. Zhang, "Architecture and synthesis for multi-cycle communication," in *ISPD '03: Proceedings of the 2003 international symposium on Physical design*, 2003.

[5] J. M. Wang, P. Ghanta, and S. Vrudhula, "Stochastic analysis of interconnect performance in the presence of process variations," in *ICCAD*, 2004, p. 880.

[6] M. Webster, M. A. Tatang, and G. J. McRae, "Application of the probabilistic collocation method for an uncertainty analysis of a simple ocean modeltesting multivariate uniformity and its applications," *MIT Joint Program on the Science and Policy of Global Change*, 1996.

[7] D. C. Montgomery, *Design and Analysis of Experiments*. John Wiley and Sons, 1997.

[8] Cochran, G. William, and G. M.Cox, *Experimental Designs*, 2nd ed. John Wiley Sons, Inc., 1957.

[9] Y. S. Kumar, J. Li, C. Talarico, and J. Wang, "A probabilistic collocation method based statistical gate delay model considering process variations and multiple input switching." in *DATE*, 2005.

[10] L. Zhang, Y. Hu, and C. C. Chen, "Statistical timing analysis in sequential circuit for on-chip global interconnect pipelining," in *Proc. Design Automation Conf.*, 2004, pp. 904–907.

[11] "Berkeley ptm-interconnect." [Online]. Available: www-device.eecs.berkeley.edu/~ptm/interconnect.html

Yield-Preferred Via Insertion Based on Novel Geotopological Technology

Fangyi Luo

Department of Computer Engineering
University of California at Santa Cruz
Santa Cruz, CA 95064
luofy@soe.ucsc.edu
Tel:408-844-8558
Fax:408-844-8566

Yongbo Jia

Nannor Technologies, Inc.
4699 Old Ironsides Dr., STE 270
Santa Clara, CA 95054
jiayb@nannor.com

Wayne Wei-Ming Dai

Department of Computer Engineering
University of California at Santa Cruz
Santa Cruz, CA 95064
dai@soe.ucsc.edu

Yield-preferred via insertion is an effective method to reduce the yield loss caused by via failures. The existing methods to apply the redundant-cut vias in metal layers are not efficient nor adequate. In this paper, we present an effective and efficient yield-preferred via insertion method based on a novel geotoplogical layout platform, GEOTOP. Our method chooses the most yield-favored via candidate and insert it into the layout without causing any design rule violations. Experiments with real industry designs show that our method can achieve very high rate of yield-preferred via without increasing the design die size within acceptable running time.

I. INTRODUCTION

As the feature size of integrated circuit continues exponentially scaling down, many new problems appear in recent process generations [1]. These new problems make it a very hard task for both designers and foundries to maintain high yield as the design size grows up. Among them, one of the most important is the yield loss due to via open failures.

Via is the component in VLSI designs to connect metal wires on different metal layers. Via open failure may occur, causing unwanted loss in yield and performance. A complete via open failure will lead to a complete broken net, which will fail the entire design. A partial via open failure will increase the resistance of the interconnect on the signal net and bring undesired delay and damage the performance. As the VLSI feature size continues shrinking to deep sub-micron regime, vias become more and more sensitive to various variations, such as cut misalignment, electron migration and thermal stress induced voiding [2] [3]. In this background, yield loss due to via open failure becomes more and more important and requires better control.

Several types of yield-preferred vias were proposed to reduce the yield loss caused by via failures [2][4]. One of the most talked about type is the redundant-cut via. A redundant-cut via is a via with 1 or more redundant cuts which are not required in functionality but will greatly reduce the chance of the via open failure. Another one is the fat via, which is a via with enlarged metal coverage. Some major foundries [5] are already recommending the massive usage of yield-preferred vias in their 130nm and 90nm processes to their customers for better yield. And it is likely that a certain rate of yield-preferred vias will be a required rule in the more advanced processes.

How to use yield-preferred vias in designs becomes a new challenge. Several EDA tool vendors claim that their latest products, such as [6][7], can apply redundant-cut vias in the layouts, while the fat vias are seldom spoken of. Even only dealing with redundant-cut vias, they can not do it well. These tools can be classified into two categories: detail router based tools and GDSII based tools. It seems like a good choice to apply yield-preferred vias at the detail routing stage. However, the yield-preferred vias need more die area and make routing problem more complicated. The routers' built-in grid-base nature has limited their ability to exploit the routing resource thoroughly. On the other hand, it is very hard for the routers to make the online decisions, such as where to use the yield-preferred vias, which via is the most yield preferred, since the whole layout routing has not been finished yet. Therefore, the routers will encounter either a low insertion rate or a loss of routable nets when trying to insert yield- preferred vias. Xu's [4] effort with maze router has achieved a redundant-cut via rate around 65% with a non-negligible loss (10%) of routable nets, which is not acceptable for industry applications. More die size area will be required as the cost for a better redundant-cut via rate and less loss of routable nets. After all the major task of the router is to determine the paths in a limited area for as many nets as possible. Cadence Nanoroute, for example, belongs to this category. On the other hand, the GDSII based tools work on GDSII files. They look around polygons and replace single-cut vias with redundant-cut vias. However, due to the huge size of GDSII data, they are very slow. Furthermore, since they are working at a post tapeout stage, their ability is quite limited so that they can only achieve a relatively lower rate. Even worse, it is almost impossible to feed their result back to earlier design stages for verification and further optimization. Sagantec SiFix, for example, belongs to this category.

In this paper, we step further and propose a new postlayout yield-preferred via insertion method based on a novel GEOTOPological layout encoding platform, GEOTOP. Our method selects the most yield favored via according to the context and insert it into the layout. GEOTOP can achieve a very high yield-preferred via rate without sacrificing the routable nets in the given routed design. This method also provides the option for designers to keep selected nets untouched, which is desired by designers to minimize the disturbance on the design due to the insertion of the yield-preferred vias.

0-7803-9451-8/06/$20.00 ©2006 IEEE.

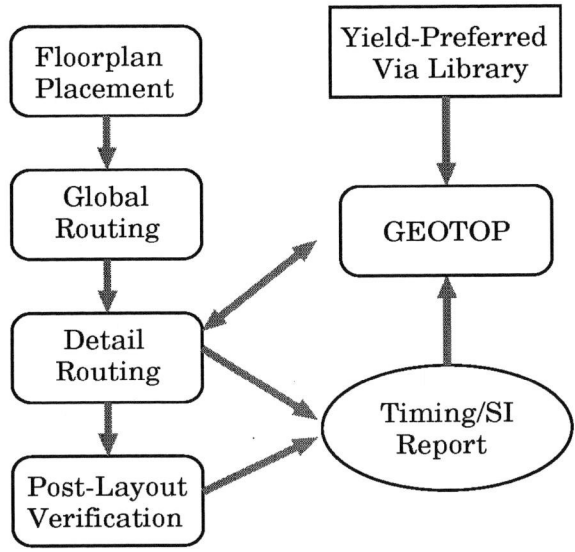

Fig. 1. Geotopological layout optimization platform in the major design flow

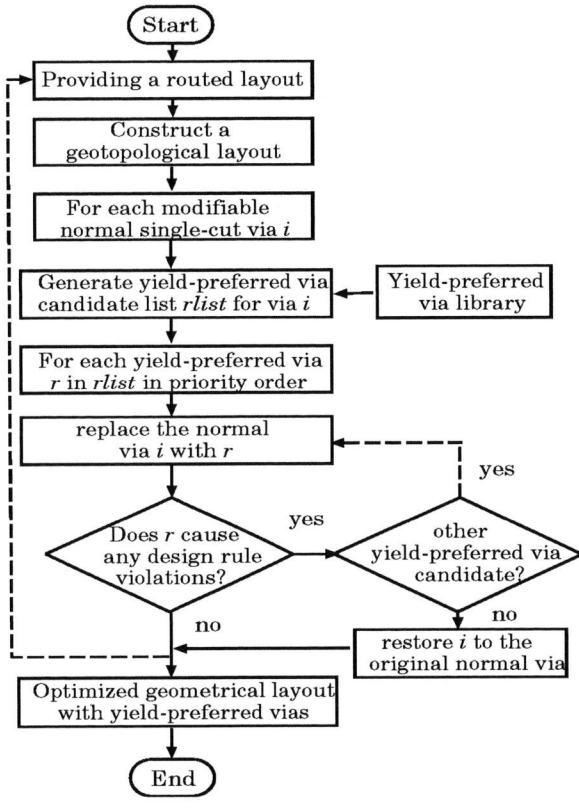

Fig. 2. Yield-preferred via insertion flow

II. METHOD OVERVIEW

Our motivation is very straightforward. Since the major task for the routers is to solve the problem of routability, just let them focus on producing a routed layout with the limited routing resource. After a valid layout is generated, we insert the most yield-favored vias wherever applicable in the layout.

We use industry standard exchange files for input and output. Different from the GDSII based tools which work in the post tapeout stage, GEOTOP works in the post routing stage. Our results can be fed back to the routers for more operations. The designer can iterate between the router and GEOTOP until a satisfactory result is achieved. Figure 1 shows the position of GEOTOP in the major design flow.

Figure 2 illustrates our method to insert yield-preferred vias. Given a routed metal layout with geometrical routing paths, we transform it into a geotopological layout in which the geometrical paths of unmodifiable nets are recorded and the paths of modifiable nets are extracted to topological equivalent paths. A yield-preferred via candidate list is created with a priority order according to the yield preference. For each modifiable normal single-cut via, the candidate list is scanned to get the most yield-favored via to insert into the layout without introducing any design rule violations. Finally, a fast geometrical engine produces the optimized geometrical wiring layout according to the geotopological layout with the yield-preferred vias. The following section will give more details.

III. METHOD DETAILS

This section will give more details on our geotopological platform, GEOTOP, yield-preferred vias selection/insertion and the geometrical engine which produces the optimized layout.

A. Geotopological Technology

Topological layout representation technologies were developed in the last decade for various purposes [8][9][10]. TEG [11] is the first post-layout optimization system based on the topological layout representation. The topological layout only captures the relative positions and connections of the layout elements. This built-in nature makes this technology very flexible in modifying routed layouts. However, since there is no geometrical information of the wires kept in the topological layout, it may suffer from the inconsistency with the original geometry layout. Furthermore, it may change wiring paths which the designers want to keep untouched, such as timing critical nets, etc. Based on the concerns above, we developed our new geotopological technology.

Our new geotopological technology is designed to keep the power of topological technology for post-layout optimization and, at the same time, avoid the shortcomings by keeping necessary geometrical information from the original geometrical layout.

The necessary geometrical information may be included in a "hard" net list which comes along with the routed design. The designers can put the nets they want to keep untouched in the list, for example:

- Timing critical nets. Any change in the wiring path or the wire length of these timing critical may result in the timing failure of the design. These nets should be kept unchanged absolutely.

- The nets specified by customers. Some patterns in the layout may be designed according to some certain requirements, such as antenna rule. The designers will, of course, hope to keep these features.

- Sometimes, the designers only want a local optimization in a specified area or an optimization among several specified nets. Extracting the whole layout to topological layout and optimization will be highly costly and risky. The uncertainty introduced may demand for extra ECO loops and longer design-to-market time. In this case, all the nets which are NOT specified for optimization will be "hard" nets and kept untouched.

Given a routed layout, TEG uses the vias and the corners of the boundary as the vertices of the triangles to encode the whole layout. Each wiring path is represented by the sequence of the triangle edges it goes across and the relative position with other wiring paths on each triangle edge it goes across. This encoding only captures the topological information of the layout. The geometrical information which is necessary and critical may be totally lost.

In our geotopological method, GEOTOP first goes through the wiring paths of the nets in the "hard" net list when importing a routed design. In the geotopological layout, every segment of the "hard" wiring path is exactly the same as in the original geometrical layout. All the wiring paths of the nets not in the "hard" net list are represented topologically. In this way, all unmodifiable nets defined in the list are represented by their respective geometrical wiring paths while all other nets are simultaneously represented by their respective topological wiring paths which are extracted from the initial geometrical layout. Given a routed design, every metal layer is processed and a corresponding geotopological layout is generated. Figure 3 gives a pair of the original geometrical layout and the corresponding geotopological layout. In the example, the solid geometrical wiring paths stand for the wiring paths of the nets in the "hard" net list.

In the following section, we will describe how we insert the most yield favored vias into the geotopological layout.

B. Yield-preferred Vias Insertion

Based on the geotopological encoded layouts, yield-preferred vias are inserted according to applicable design rules, leaving the unmodifiable nets intact.

Referring back to Figure 2, a yield-preferred via candidate priority list is generated for each normal single-cut via which is modifiable. A yield-preferred via is a via with a certain geometrical shape configuration. Figure 4 gives the examples for a fat single via, a normal redundant-cut via and a fat redundant-cut via. Different geometrical shape configuration will result in different yield influence on the layout. In the fat vias, the enlarged metal covering over the cut will reduce the yield loss caused by misalignment. The redundant-cut vias will greatly reduce the chance of net breaking fault caused by via open failure. From the manufacturability viewpoint, the fat redundant-cut via has the least possibility of via break failure and the best yield improvement, followed by the normal redundant-cut via, and then the fat single via. On the other hand, for a specified normal single-cut via, different insertion direction of the

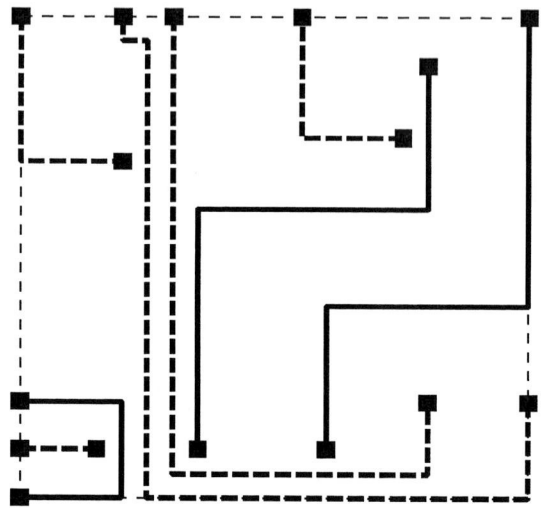

(A) Original geometrical layout

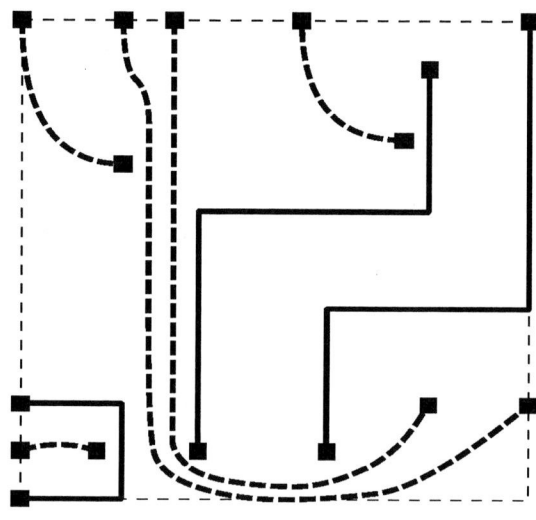

(B) Geotopological layout

Fig. 3. (A) Original geometrical layout (B) Equivalent geotopological layout

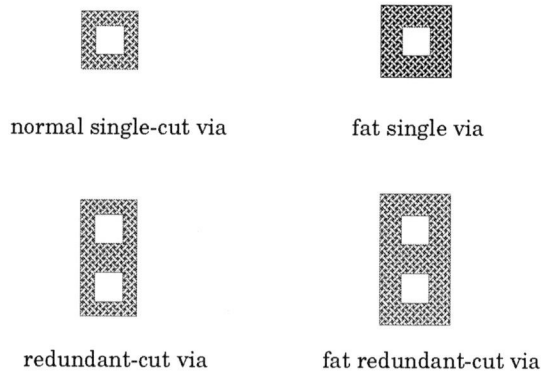

Fig. 4. Yield-preferred vias: fat single via, normal redundant-cut via and fat redundant-cut via

redundant-cut via also has different disturbance on the layout [12]. Figure 5 illustrates a group of possible redundant-cut vias insertion direction for replacing a normal single-cut via which connects to two metal layers, M1 and M2. M1 is the lower metal layer in the vertical preferred routing direction. M2 is the upper metal layer in the horizontal preferred routing direction. Via A, B, C are in the vertical direction. Since M1 is in vertical direction and M2 is in horizontal direction, these three placements cause more changes in M2 layer than in M1 layer. For the same reason, Via D, E, F will cause more changes in M1 layer then in M2 layer. Since M1 is the lower layer, A, B, C have higher priority than D, E, F. Among Via A, B and C, A brings the least disturbance to M1, C brings the most. Therefore, A is the most yield-preferred, followed by B, and then C. For the same reason, we have the priority order of D, E, F. Every redundant-cut via has a fat version with a higher priority. Therefore, we have 6 fat redundant-cut vias with the highest priority, and 6 normal redundant-cut vias, plus 1 fat single-cut via with the lowest priority, 13 yield-preferred vias in total in the candidate list.

Please note that our configuration of the yield-preferred via candidate priority list may not be the optimal one. Here in this paper, what we want to present is the concept of the yield-preferred via candidate priority list and the method to insert as many yield-preferred vias into the layout as possible. The designers are strongly recommended to consult with the foundries they are working with to figure out the optimal yield-preferred via candidate priority list.

Referring back to Figure 2, after the yield-preferred via candidate list is generated for a specified modifiable normal single via, the normal single via is replaced with the yield-preferred via from the candidate list with the highest priority. Replacing a normal single via with a redundant via would cause changes on both metal layers. A incremental geotopological design rule checker then checks the affected area for design rule violations in the 2 relevant geotopological layouts. If the replacement does introduce new design rule violations, another attempt is made to replace this normal single via with the next yield-preferred via in the candidate list. If not, the current yield-preferred via replaces this normal single via and the process goes back to replace the next normal single modifiable via. In the situation that every yield-preferred via in candidate list would cause design rule violations, the original via will not be changed.

C. Optimized Geometrical Layout

After all modifiable normal single-cut vias have been processed and replaced with yield-preferred vias where applicable, the geometrical engine produces the optimized geometrical layouts according to the geotopological layouts with the yield-preferred vias.

This geometrical optimization engine is derived from a previous studied topology-to-geometry transformation engine [13]. In geotopological layout, the "hard" wire paths are kept as their original geometrical paths. When producing the optimized geometrical layout, the geometrical engine sweeps through the geotopological layout and optimizes the geometrical wiring path for each wiring segment one by one. The opti-

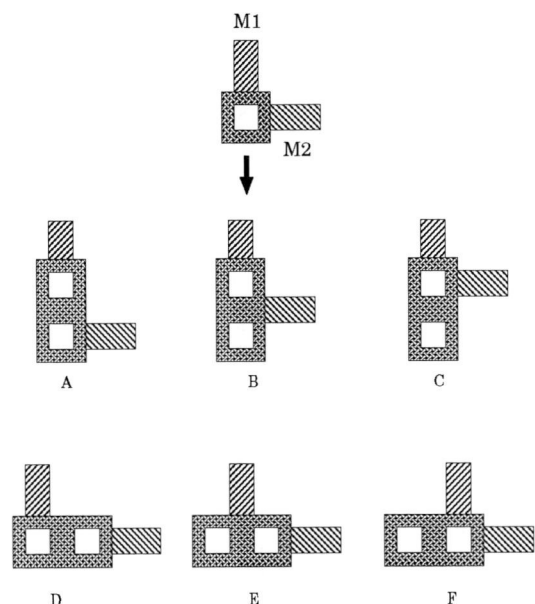

Fig. 5. Redundant-cut vias in different directions

mized paths of the "hard" wiring paths are directly copied from the geotopological representation, while those of the other wiring paths are adjusted according to the nearby geometrical shapes. Another improvement is a net-based spacing rule lookup table. The spacing rule lookup table enables the engine to optimize the valid geometrical layout following a complex rule set. The spacing rule lookup table is developed at the time the original layout is read, recording the special spacing rules other than the minimum spacing requirement. When producing the optimized geometrical layout, the lookup table is referred for the spacing requirement between the specified nets.

After an optimized geometrical layout has been produced for each metal layer, the engine exports an exchange file in industry standard which has all the geometrical information in the design, including all the yield-preferred vias. The designers can feed this exchange file back to routers for more operations or as the input for the next design stage.

IV. Experiment Results

All the experiments are running on a Linux machine with 2 P4 2.8G CPUs and 2G memory. We choose our test cases from industry with different sizes. Table I shows the list of a group of our test cases. Column 1 of Table I shows the number of routing layers of each design. Column 2 and Column 3 give the number of the nets and the number of the gates in each design, giving a basic concept of size of the designs. Column 6 gives the overall yield-preferred via insertion rate of the design.

In this group of experiments, we input each routed design, insert yield-preferred vias into the routing layers, then produce the optimized wiring paths in the routing layer and output the design in the same format. To make a comparison with other methods, the yield-preferred vias are chosen from the 6 normal redundant-cut vias. The usage of the 6 fat redundant-cut vias will slightly lower the insertion rate, since they are using

TABLE I
YIELD-PREFERRED VIA INSERTION RATE

Design	# of routing layers	# of nets	# of gates	# of total vias	# of yield-preferred vias	yield-preferred via rate
case1	2	5k	22k	25.1k	24.2k	96.4%
case2	3	7.8k	35k	45k	40.2k	89.3%
case3	4	15k	44k	74.7k	68.0k	91.4%
case4	4	17k	58k	87.3k	80.8k	92.6%
case5	3	24k	83k	149.1k	143.1k	95.95%
case6	3	37.8k	285k	280.1k	251.5k	89.8%
case7	7	64.7k	256K	890.8k	813.2k	91.3%
case8	3	115k	1020k	1146k	1028k	89.7%

more area. When importing each design, the PrimeTime report is also imported to identify the timing critical nets and freeze them. These timing critical nets are treated as "hard" nets, their geometrical wiring paths are kept unchanged during the whole optimization. We use commercial DRC tools from major EDA tool vendors to check through the optimized layouts GEOTOP produced to ensure they have no design rule violations.

Our experiments show a very high yield-preferred via insertion rate. Compared with Xu's experiments [4], our experiments use real industry designs which are not randomly generated. We get a much higher insertion rate. Even in the design with the highest metal density, our method still achieves a rate over 89%. Please keep in mind that we didn't increase the die size area nor sacrifice even a single routable net when achieving these impressive insertion rates. Our method is also very fast. For case 8, a 1 million gate design, the total turn-around time is around 1 hour and a half.

Figure 6 and Figure 7 give a pair of example of the original layout and the layout after yield-preferred via insertion in test case 6. Two metal layers are displayed in the figures.

We have another experiment with test case 9. It is also a real industry design with 148k nets and 1.2M gates in 90nm technology. The total number of vias is about 1.2M. The original routing is generated by some major commercial detail router with redundant-cut via option. The original redundant-cut via rate is about 58%. First, we directly insert normal redundant-cut vias into the original layout. We get 92k more redundant-cut vias, the rate rises to 66%. We make another attempt by restoring the redundant-cut vias in the original layout back to normal single vias and then inserting the redundant-cut vias into the layout through GEOTOP. The rate jumps to 84%, which is about 310K MORE redundant-cut vias than the commercial router without increasing die size or loss of routable nets. According to Poisson yield model [14], 2,000,000 more vias in a design with an average rate of 5 break failures per 1 billion cuts, which is a very realistic data, will bring at least 1% yield improvement. If scaled up to multi-million gate design, our result will bring significant yield enhancement.

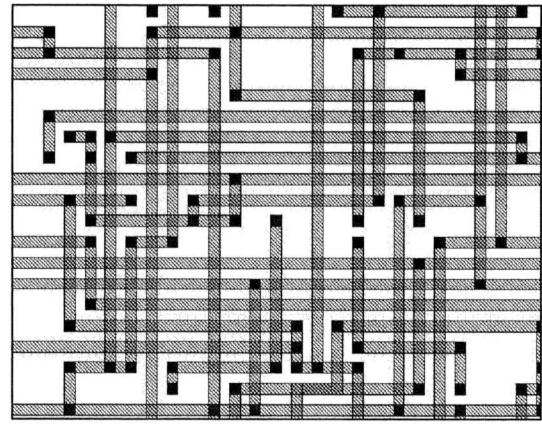

Fig. 6. Original layout without any yield-preferred vias

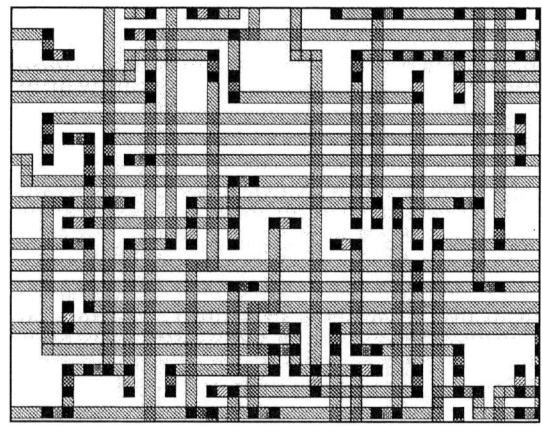

Fig. 7. Optimized layout with yield-preferred vias

V. CONCLUSION

In this paper, we present a yield-preferred via insertion method based a novel geotopological platform, GEOTOP. The objective layout is first extracted to a geotopological layout, then the most suitable yield-preferred vias are inserted into the geotopological layout where applicable, finally the optimized geometrical layout is produced. This method guarantees the highest insertion rate and keeps the specified wire paths untouched. Experiments with real industry designs show that over 89% insertion rate can be achieved even in the densest design without introducing any design rule violations, which will bring profitable yield improvement.

ACKNOWLEDGEMENTS

This work was sponsored by Nannor Technologies, Inc.

REFERENCES

[1] Andrew B. Kahng and Y. C. Pati, "Subwavelength lithography and its potential impact on design and EDA," *IEEE/ACM Design Automation Conf.*, pp. 799-804, 1999.

[2] T.C. Huang and C.H. Yao and W.K. Wan and Chin C. Hsia and M.S. Liang, "Numerical modeling and characterization of the stress migration behavior upon various 90 nanometer Cu/Low k interconnects," *IEEE International Interconnect Technology Conference*, pp. 207-209, 2003.

[3] Kelvin Y.Y, Doong and Robin C.J. Wang and S.C. Lin and L.J. Huang and S.Y. Lee and C.C. Chiu and David Su and Kenneth Wu and K.L. Young and Y.K. Peng, "Stress-induced voiding and its geometry dependency characterization," *IEEE International Interconnect Technology Conference*, pp. 156-160, 2003.

[4] Gang Xu, Li-Da Huang. David Z. Pan and Martin D.F. Wong, "Redundant-via enhanced maze routing for yield improvement," *Proc. Asia South Pacific Design Auto. Conf.*, 2005.

[5] *TSMC Symposium*, 2004.

[6] NanoRoute, *Cadence Design Systems.*

[7] SiFix, *Sagantec.*

[8] Wayne Wei-Ming Dai, David Staepelaere, Jeffrey Jue and Tal Dayan, "Cost-driven layout for thin film MCMs," *Proc. IEEE MCM Conf.*, pp. 174-178, 1993.

[9] F. Miller Maley, *Single-layer wire routing and compaction*, MIT Press, Cambridge, MA, 1990.

[10] Man-Fai Yu, Joel Darnauer, Wayne Wei-Ming Dai "Interchangeable pin routing with application to package layout," *Proc. Intl. Conf. Computer-aided Design*, Santa Clara, CA, Nov., 1996.

[11] S. Zhang and W. Dai, "TEG: a new post-Layout optimization method," *IEEE Transactions on Computer-Aided Design of Integrated Circuits and Systems*, Vol. 22, Num. 2, pp. 1-12, 2003.

[12] Hardy Kwok-Shing Leung, "Advanced routing in changing technology landscape," *International Symposium on Physical Design*, pp. 118-121, 2003.

[13] Fangyi Luo, "Layout Wiring Generation," *Master's Thesis*, University of California, Santa Cruz, 2004.

[14] R. M. Warner, Jr., "Applying a composite model to the IC yield problem," *IEEE J. Solid-State Circuits*, Vol. SC-9, pp. 86-95, 1974.

Introduction to H.264 Advanced Video Coding

Jian-Wen Chen   Chao-Yang Kao   Youn-Long Lin

Department of Computer Science
National Tsing Hua University
Hsin-Chu, TAIWAN 300
Tel : +886-3-573-1072
Fax : +886-3-572-3694
e-mail : ylin@cs.nthu.edu.tw.

Abstract - We give a tutorial on video coding principles and standards with emphasis on the latest technology called H.264 or MPEG-4 Part 10. We describe a basic method called block-based hybrid coding employed by most video coding standards. We use graphical illustration to show the functionality. This paper is suitable for those who are interested in implementing video codec in embedded software, pure hardwired, or a combination of both.

I. Introduction

Digitized video has played an important role in many consumer electronics applications including VCD, DVD, video phone, portable media player, video conferencing, video recording, e-learning etc.   In order to provide solutions of high quality (high frame resolution, high frame rate, and low distortion) or low cost (low bit rate for storage or transmission) or both, video compression is indispensable. Advancement in semiconductor technology makes possible efficient implementation of effective but computationally complicated compression methods.

Because there are a wide range of target applications from low-end to high-end under various constraints such as power consumption and area cost, an application-specific implementation may be pure software, pure hardwired, or something in between. In order to do an optimal implementation, it is essential to fully understand the principles behind and algorithms employed in video coding. Starting with MPEG-1[1] and H.261 [2], video coding techniques/standards have gone through several generations.

The latest standard is called H.264 (also called MPEG-4 AVC, Advanced Video Coding defined in MPEG-4 Part 10) [3][4][5]. Compared with previous standards, H.264 achieves up to 50% improvement in bit-rate efficiency. It has been adopted by many application standards such as HD DVD [6], DVB-H [7], HD-DTV [8], etc. Therefore, its implementation is a very popular research topic to date. In this tutorial paper, we introduce the essential features of H.264.

The rest of this paper is organized as following. In Section II, we give an outline of the block-based hybrid video coding method. In Section III, we describe in more detail each basic coding function. Finally, in Section IV, we draw some conclusions.

II. Block-Based Hybrid Video Coding

A digitized video signal consists of a periodical sequence of images called frame. Each frame consists of a two dimensional array of pixels. Each pixel consists of three color components, R, G and B. Usually, pixel data is converted from RGB to another color space called YUV in which U and V components can be sub-sampled. A block-based coding approach divides a frame into macroblocks each consisting of say 16x16 pixels. In a 4:2:0 format, each MB consists of 16x16 = 256 Y components and 8x8 = 64 U and 64 V components. Each of three components of an MB is processed separately.

Fig. 1 shows a pseudo-code description of how to compress a frame MB by MB. To compress an MB, we use a hybrid of three techniques: prediction, transformation & quantization, and entropy coding. The procedure works on a frame of video. For video sequence level, we need a top level handler, which is not covered in this paper. In the pseudo code, f_t denotes the current frame to be compressed and mode could be I, P, or B.

```
procedure encode_a_frame (ft, mode)

    for I = 1, N        //** N: #rows of MBs per frame
        for J = 1, M    //** M: #columns of MBs per frame
        Curr_MB = MB(ft, I, J);
        case (mode)
            I: Pred_MB = Intra_Pred (f't, I, J);
            P: Pred_MB = ME (f't-1, I, J);
            B: Pred_MB = ME (f't-1, f't+1, I, J);

        Res_MB = Curr_MB – Pred_MB;
        Res_Coef = Quant(Transform(Res_MB));
        Output(Entropy_code(Res_Coef));

        Reconst_res = ITransform(IQuant(Res_Coef)) ;
        Reconst_MB = Reconst_res + Pred_MB;
        Insert(Reconst_MB, f't) ;

end encode_a_frame;
```

Fig. 1. Pseudo Code for Block-Based Hybrid Coding
a Video Frame

Prediction tries to find a reference MB that is similar to the current MB under processing so that, instead of the whole current MB, only their (hopefully small) difference needs to be coded. Depending on where the reference MB comes from, prediction is classified into inter-frame prediction and intra-frame prediction. In an inter-predict (P or B) mode, the reference MB is somewhere in a frame before or after the current frame, where the current MB resides. It could also be some weighted function of MBs

0-7803-9451-8/06/$20.00 ©2006 IEEE.

from multiple frames. In an intra-predict (I) mode, the reference MB is usually calculated with mathematical functions of neighboring pixels of the current MB.

The difference between the current MB and its prediction is called residual error data (residual). It is transformed from spatial domain to frequency domain by means of discrete cosine transform. Because human visual system is more sensitive to low frequency image and less sensitive to high frequency ones, quantization is applied such that more low frequency information is retained while more high frequency information discarded.

The third and final type of compression is entropy coding. A variable-length coding gives shorter codes to more probable symbols and longer codes to less probable ones such that the total bit count is minimized. After this phase, the output bit stream is ready for transmission or storage.

There is also a decoding path in the encoder. Because in the decoder side only the reconstructed frame instead of the original frame is available, we have to use a reconstructed frame as the reference for prediction. Therefore, in the bottom part of Fig. 1, we obtain the restored residual data by performing inverse quantization and then inverse transformation. Adding the restored residual to the predicted MB, we get the reconstructed MB that is then inserted to the reconstructed frame f'$_t$. Now, the reconstructed frame can be referred to by either the current I-type compression or future P-type or B-type prediction.

In the next section, we explain in more detail each of the video coding functions invoked in the pseudo code.

III. Basic Video Coding Functions

A. Prediction

Prediction exploits the spatial or the temporal redundancy of a video sequence so that only the difference between actual and predict instead of the whole image data need to be encoded. There are two types of prediction: intra prediction for I-type frame and inter prediction for P-type (Predictive) and B-type (Bidirectional Predictive) frame.

Intra Prediction -- There exists high similarity among neighboring blocks in a video frame. Consequently, a block can be predicted from its neighboring pixels of already coded and reconstructed blocks. The prediction is carried out by means of a set of mathematical functions.

In H.264/AVC, an I-type 16x16 4:2:0 MB has its luminance component (one 16x16) and chrominance components (two 8x8 blocks) separately predicted. There are many ways to predict a macroblock as illustrated in Fig. 2. The luminance component may be intra-predicted as one single INTRA16x16 block or 16 INTRA4x4 blocks. When using the INTRA4x4 case, each 4x4 block utilizes one of nine prediction modes (one DC prediction mode and eight directional prediction modes). When using the INTRA16x16 case, which is well suited for smooth image area, a uniform

prediction is performed for the whole luminance component of a macroblock. Four prediction modes are defined. Each chrominance component is predicted as a single 8x8 block using one of four modes.

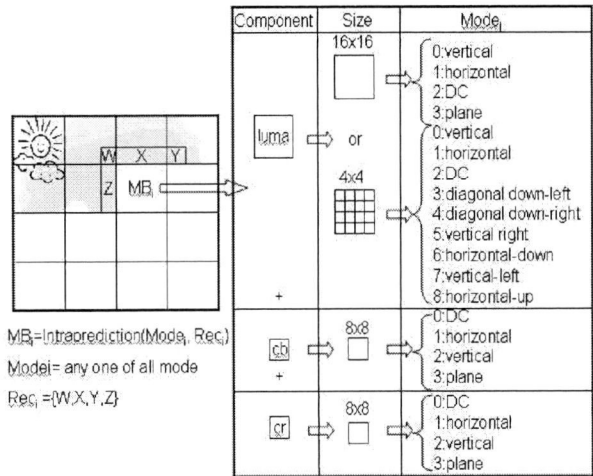

Fig. 2. Overview of H.264 intra prediction modes

Inter Prediction (Motion Estimation) -- High quality video sequences usually have high frame rate at 30 or 60 frames per second (fps). Therefore, two successive frames in a video sequence are very likely to be similar. The goal of inter prediction is to utilize this temporal redundancy to reduce data need to be encoded. In Fig. 3, for example, when encoding frame t, we only need to encode the difference between frame t-1 and frame t (i.e., the airplane) instead of the whole frame t. This is called motion estimated inter-frame prediction.

Fig. 3. Successive video frames

In most video coding standards, the block-based motion estimation (BME) [9] is used to estimate for movement of a rectangular block from the current frame. For each M x N-pixel current block in the current frame, BME compares it with some or all of possible M x N candidate blocks in the search area in the reference frame for the best match, as shown in Fig. 4. The reference frame may be a previous frame or a next frame in P-type coding, or both in B-type coding. A popular matching criterion is to measure the residual calculated by subtracting the current block from the candidate block, so that the candidate block that minimizes the residual is chosen as the best match. The cost function is called sum of absolute difference (SAD), which is the sum of pixel by pixel absolute difference between predicted and actual image.

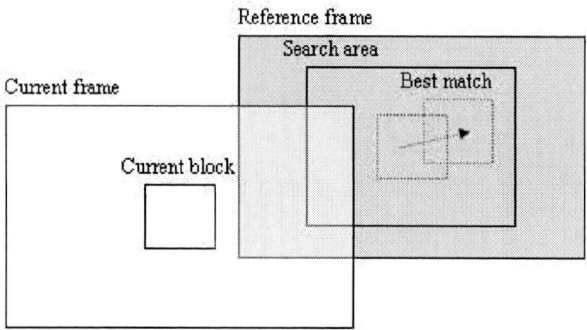

Fig. 4. Block-based motion estimation

There are three new features of motion estimation in H.264: variable block-size, multiple reference frames and quarter-pixel accuracy.

Variable block-size – Block size determines tradeoff between the residual error and the number of motion vectors transmitted. In the previous video coding standards, the block-size of motion estimation is fixed, such as 8 x 8 (MPEG-1, MPEG-2) or 16 x 16 (MPEG-4). Fixed block-size motion estimation (FBSME) spends the same efforts when estimating the motion of moving objects and background (no motion). This method causes low coding efficiency. In H.264, each macroblock (16 x16 pixels) may be split into sub-macroblocks in four ways: one 16 x 16 sub-macroblock, two 16 x 8 sub-macroblocks, two 8 x 16 sub-macroblocks, or four 8 x 8 sub-macroblocks. If the 8 x 8 mode is chosen, each of the four 8 x 8 sub-macroblocks may be split further in four ways: one 8 x 8 partition, two 8 x 4 partitions, two 4 x 8 partitions or four 4 x 4 partitions. Therefore, variable block-size motion estimation (VBSME) uses smaller block size for moving objects and larger block size for background, as shown in Fig. 5, to increase the video quality and the coding efficiency.

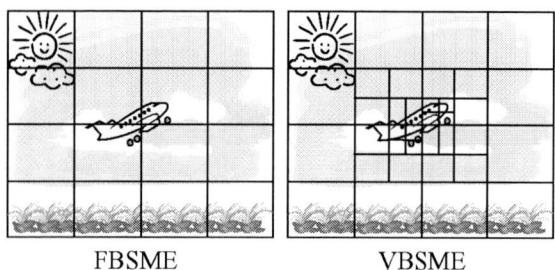

FBSME   VBSME

Fig. 5. Comparisons between FBSME and VBSME

Multiple reference frames -- In previous video coding standards, there is only one reference frame for motion estimation. In H.264, the number of reference frames increases to 5, as shown in Fig. 6, for P frame and to 10 (5 previous frames and 5 next frames) for B frame [10]. More

reference frames result in smaller residual data and, therefore, lower bit rate. Nevertheless, it requires more computation and more memory traffic.

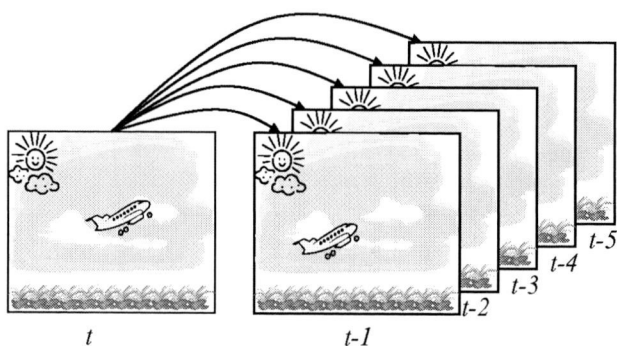

Fig. 6. Multiple reference frames for motion estimation

Quarter-pixel accuracy -- In previous video coding standards, motion vector accuracy is half-pixel at most. In H.264, motion vector accuracy is down to quarter-pixel and results in smaller residual data.

B. Compensation

Corresponding to prediction, there is also two kinds of compensation, intra compensation for I-type frame and inter compensation for P-type and B-type frame.

Intra Compensation – According to the encoding process, intra compensation regenerates the current block pixels by one of 13 modes (9 for Intra4x4 and 4 for Intra16x16) for luminance component and one of 4 modes for chrominance components.

Inter Compensation (Motion Compensation) -- Inter compensation is used in a decoding path to generate the inter-frame motion predicted (estimated) pixels by using motion vectors, reference index and reference pixel from inter prediction, as shown in Fig. 7. In H.264, inter compensation [11] also allows variable block-size, multiple reference frames and quarter-pixel accurate motion vector. Its luminance interpolation uses a 6-tap filter for half-pixel and a 2-tap filter for quarter pixel while the chrominance one uses neighboring four integer pixels to predict pixels up to accuracy of 1/8 pixel. It can refer to forward frames for P macroblocks and both forward and backward frames for B macroblocks. It allows arbitrary weighting factors for bidirectional weighted prediction.

C. Transformation and Quantization

The difference between the actual and predicted data is called residual error data. Discrete Cosine Transform (DCT) is a popular block-based transform for image and video compression. It transforms the residual data from time domain representation to frequency domain representation. Because most image and video are low frequency data,

DCT can centralize the coding information.

The main functionality of quantization is to scale down the transformed coefficients and to reduce the coding information. Because human visual system is less sensitive to high frequency image component, some video and image compression standards may use higher scaling-value (quantization parameter) for high frequency data.

The H.264 standard employs a 4x4 integer DCT [12]. Fig. 8 illustrates transformation and quantization in H.264 with an example.

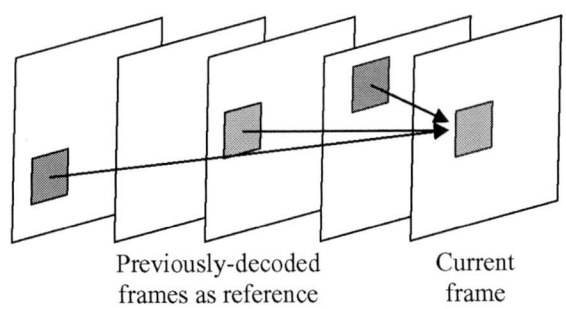

Previously-decoded
frames as reference

Current
frame

Fig. 7. Inter Compensation

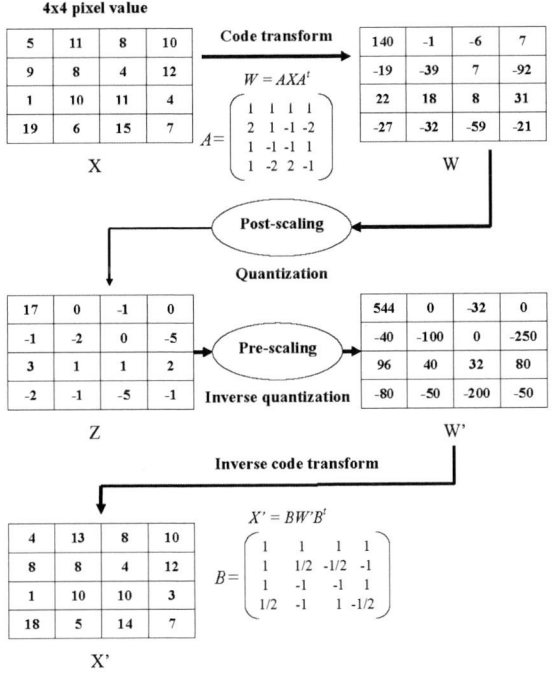

Fig. 8. Illustrating Transformation and Quantization

In Fig. 8, X is a 4x4 block of residual data. After integer DCT, we get W, a 4x4 coefficient matrix. Its upper left portion represents lower frequency components of X while its lower right portion gives higher frequency

components. Z is the quantized version of W. We can see that the amount of data is much smaller than that of X, the original residual data. Z is the information to be entropy-coded and passed to the decoder part. W' is the scale-up (inversely quantized) version of Z. After applying inverse integer DCT (IDCT) on W', we get X', which is the decoded residual. Note that X' is not exactly identical to X. That is, this process is lossy due to the irreversibility of quantization.

D. In-loop filter

One of the disadvantages of block-based video coding is that discontinuity is likely to appear at the block edge. In order to reduce this effect, the H.264 standard employs the deblocking filter [13] to eliminate blocking artifact and thus generate a smooth picture.

In the encoder side, deblocking filter can reduce the difference between the reconstructed block and the original block. According to some experiments, it can not only improve PSNR, but also achieve up to 9% bit-rate saving. Fig. 9 depicts the input and output of deblocking filter.

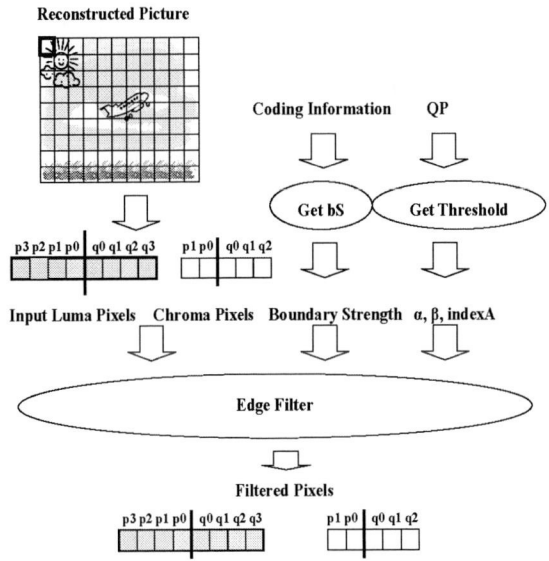

Fig. 9. Deblocking Filter Illustration

The deblocking filter works on one 16x16 MB at a time. It filters every boundary defined by 4x4 blocks within the MB. The deblocking filter consists of a horizontal filtering across all vertical edges and a vertical filtering across all horizontal edges. Therefore, for the luma component, it goes through 4 vertical boundaries and 4 horizontal boundaries with each boundary requiring 16 filtering operations. For both chroma components, it goes through 2 vertical boundaries and 2 horizontal boundaries with each boundary consisting of 8 filtering operations. As depicted in Fig. 9, inputs to a filtering operation includes eight luma pixels (p3, p2, p1, p0, q0, q1, q2, q3) or five chroma pixels (p0, q0, q1, q2, q3), boundary strength, and threshold variables. At most six luma pixels (p2, p1, p0, q0, q1, q2) or two chroma pixels (q0, q1) will be modified by the filter. After the whole reconstructed frame is filtered, it is ready for display as well as being a reference picture.

The boundary strength (bS) is used to set the filter strength. As the boundary strength increasing, it eliminate more blocking artifact. The threshold variables are used to distinguish the true edge from the false edge.

E. Entropy Coding

The entropy encoder is responsible of converting the syntax elements (quantized coefficients and other information such as motion vectors, prediction modes, etc) to bit stream and then the entropy decoder can recover syntax elements from bit stream. Its function is similar to that of WinZip, which is commonly used for compressing files in the Windows Operating System.

There are two popular entropy coding methods, variable length coding and arithmetic coding. The former encodes symbol by looking up a Huffman table. Therefore, it must represent a symbol with one or more integer number of bits. On the other hand, arithmetic coding encodes a symbol by its appearance probability. So, it can represent a symbol with fractional number of bits and, thus, achieve higher compression efficiency than variable length coding does.

The H.264 standard defines two entropy coding methods: context adaptive variable length coding (CAVLC) and context based adaptive arithmetic coding (CABAC) [14]. For baseline profile, only CAVLC is employed. For main profile, both CAVLC and CABAC must be supported. According to our experiments, CABAC can achieve up to 7% bit rate saving at the expense of more computation complexity in comparison with CAVLC. Fig. 10 shows the coding flow of CABAC.

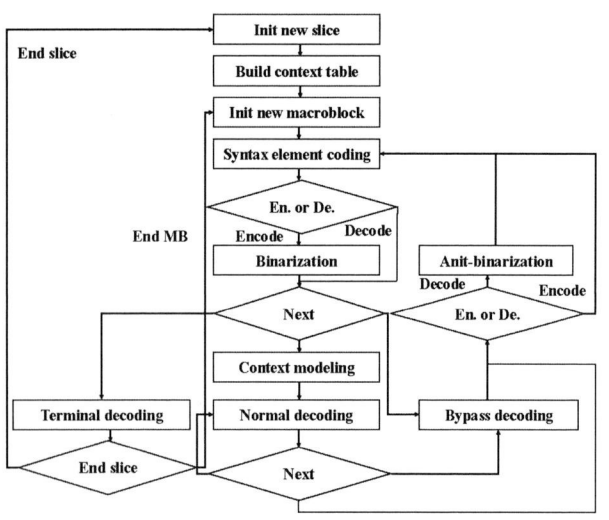

Fig. 9. The CABAC decoding flow

When the CABAC circuit processes a new slice, it first builds the context table before processing the first macroblock of the current slice. The basic information unit is called syntax element. For encoding, it will binarize these syntax elements before calculating the context value and then goes on to arithmetic coding. CABAC defines three arithmetic coding methods: normal decoding, bypass decoding and terminal decoding. After arithmetic coding, it proceeds to decode the next syntax element. For CABAC decoding, we have to convert the decoding result back to real syntax element value.

Most syntax elements go through the normal decoding process as shown in Fig. 11. Before decoding we get the context value through context modeling. Then, the decoder can look up the context table and get MPS value and pState. With these variable value, it goes to arithmetic coding. After arithmetic coding, it will update the context table by looking up TransIdxLPS table or TransIdxMPS table depending on whether the decoding result is equivalent to the MPS value. After entropy encoding, the bit stream is ready for output to a storage media or transmission medium.

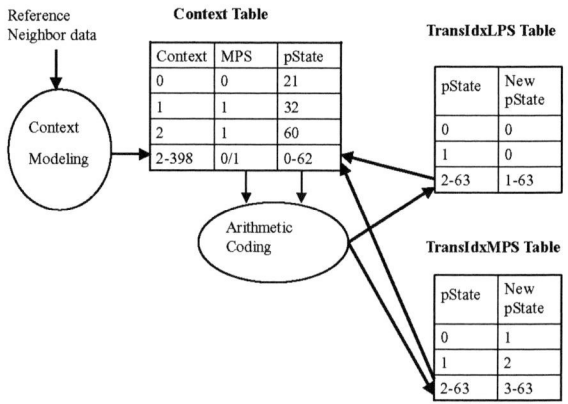

Fig. 11. Normal Decoding Process in CABAC

IV. Summary and Conclusions

We have given a brief introduction to video coding with emphasis on H.264, the latest international video coding standards. Starting from MPEG-1 and H.261, most video coding standards follow the block-based hybrid coding approach. Great impact has been achieved with early generations of standards such as MPEG-1 for VCD and MPEG-2 for DVD and Digital TV.

Requirement for high quality applications drives continue development of the next generation standards. As shown in Table 1, compared with MPEG-4, H.264 calls for more sophisticated implementation of every part of the coding process. Fortunately, advancement in semiconductor manufacturing technology has made their low cost implementation possible.

Because H.264 was defined targeted towards a wide range of applications from low bit-rate low-resolution such mobile video conferencing to high-rate high-definition such as Ultra HDTV, there will not be a single implementation method that fits all. One can implement a baseline version for low end application with software running on an embedded microprocessor or a DSP core with possible video-specific instruction extension. For very high end application, hardwired acceleration of critical functions such as motion estimation/compensation and deblocking filter, or even the whole system, might be necessary. There is yet another approach called application-specific instruction set processor (ASIP) that defines custom instructions based on the function of video coding.

No matter which method is employed, there is yet another tradeoff between implementation completeness and coding efficiency. Quite often we see an implementation that trade quality for simplification of implementation. For example, a deblocking filter may treat all variable-size blocks as composition of 4x4 blocks or a simple bilinear interpolation is substituted for the standard 6-tap filter during sub-pixel motion estimation. This is not encouraged because it will lose the original spirit of H.264. That is, achieving big gain by accumulating small gain in every part of the coding process.

Acknowledgements

This work is supported in part by the National Science Council of Taiwan under grants no. NSC94-2215-E-007-029, NSC94-2220-E-007-007, NSC94-2220-E-007-009 and NSC94-2220-E-007-017, the Ministry of Economic Affairs of Taiwan under grant no. 94-EC-17-A-01-S1-038, and Taiwan Semiconductor Manufacturing Company under grant no. NTHU-0416. The authors would like to thank their colleagues, C. R. Chang, S. Y. Shih, H. C. Tzeng, C. L. Chiu, and Y. H. Chang, for collaborating in the NTHU video decoder project.

References

[1] MPEG (Moving Pictures Expert Group), Final Text for ISO/IEC11172, "Information technology − Coding of moving pictures and associated audio for digital storage media at up to about 1.5 Mbit/s" , ISO/IEC, 1993.

[2] CCITT Recommendation H.261, International Telecommunication Union, "Video Codec for Audiovisual Services at px64 kbit/s", 1993

[3] "Draft ITU-T recommendation and final draft international standard of joint video specification (ITU-T Rec. H.264 | ISO/IEC 14496-10 AVC), "JVT G050, 2003.

[4] T. Wiegand, G. J. Sullivan, G. Bjontegaard, and A. Luthra, "Overview of the H.264/AVC video coding standard", *IEEE Transactions on Circuit and System*

for Video Technology, pp. 560-576, Jul. 2003.

[5] Iain E. G. Richardson, *H.264 and MPEG-4 Video Compression: Video Coding for Next-generation Multimedia,* John Wiley & Sons, 2003.

[6] Kobota, T.; "HD DVD-overview of next generation optical disc format", IT to HD: Visions of Broadcasting in the 21st Century, The IEE 2-Day Seminar on (Ref. No. 2004/10760), pp. 213-224, 2004

[7] European Telecommunications Standards Institute (ETSI) ETS 300 744v1.1.2 (1997-08) : Digital Video Broadcasting (DVB) ; Framing Structure, Channel Coding and Modulation for Digital Terrestrial Television,1997

[8] Code of Federal Regulations. Title 47, Telecommunications, Parts 70-79 Revised as of Oct.1, 2003

[9] M. Flierl, T. Wiegand, and B. Girod, "A Locally Optimal Design Algorithm for Block-Based Multi-Hypothesis Motion-Compensated Prediction", *in Data Compression Conference,* pp: 239-248, 1998.

[10] T. Wiegand and B. Girod, "Multi-frame Motion-Compensated Prediction for Video Transmission", Kluwer Academic Publishers, Sept. 2001.

[11] T. Wedi, "Motion Compensation in H.264/AVC", *in IEEE Transactions on Circuits and Systems for Video Technology.*

[12] H. Malvar, A. Hallapuro, M. Karczewicz, and L. Kerofsky, "Low-Complexity Transform and Quantization in H.264/AVC", *in IEEE Transactions on Circuits and Systems for Video Technology,* pp: 598-603, Jul. 2003.

[13] D. Marpe, H. Schwarz, T. Wiegand, "Context-based adaptive binary arithmetic coding in the H.264/AVC video compression standard", *IEEE Transactions on Circuits and Systems for Video Technology,* pp: 620-636, Jul. 2003.

[14] P. List, A. Joch, J. Lainema, G. Bjntegaard, and M. Karczewicz, "Adaptive deblocking filter," *IEEE Transactions on Circuits and Systems for Video Technology,* vol. 13, pp. 614-619, 2003.

Table 1. Comparison between MPEG-4 and H.264

Standard	MPEG-4	H.264
Block size	16*16 or 8*8	16*16 to 4*4
Transform	8*8 DCT	4*4 integer DCT
Entropy coding	VLC	VLC,CAVLC, CABAC
Ref frame	1 frame	Multiple (5) frames
Picture type	I, P, B	I, P, B, SI, SP
Coding efficiency	1	2
Decoder complexity	1	2.6
Target applications	Mobile devices	DTV, HD-DVD, Mobile devices

7D-2

Algorithms and DSP Implementation of H.264/AVC

Hung-Chih Lin, Yu-Jen Wang, Kai-Ting Cheng, Shang-Yu
Yeh, Wei-Nien Chen, Chia-Yang Tsai, Tian-Sheuan Chang, Hsueh-Ming Hang

Dept. Electronics Engineering, and Institute of Electronics, Hsinchu 300, Taiwan

e-mail: hclin.ee93g@nctu.edu.tw, cosbe@twins.ee.nctu.edu.tw, kt34.ece90@nctu.edu.tw, chucky1984820.ee91@nctu.edu.tw,
tpht78@hotmail.com, cytsai.ee90g@nctu.edu.tw, tschang@twins.ee.nctu.edu.tw, hmhang@mail.nctu.edu.tw

Abstract - This survey paper intends to provide a comprehensive coverage of the techniques that are pertinent to the processor-based implementation of H.264/AVC video codec, particularly on DSP. Most of this paper is devoted to the computationally efficient algorithms, or the *fast algorithms*. Fast algorithms for motion estimation, intra-prediction and mode decision are described to reduce the computational complexity. In addition, in order to port the H.264/AVC codec to DSP, we also outline the basic principles of DSP code optimization.

I. Introduction

ITU H.264 Advance Video Coding (AVC), also known as the MPEG-4 part 10 [1], offers the highest coding efficiency among all the existing video compression standards for, particularly, very low rate video transmission. However, it also has the highest computational complexity. Therefore, reducing its implementation complexity becomes a very challenging subject.

Numerous studies on reducing H.264/AVC codec implementation complexity have been published in the past 3 years since this standard was finalized in late 2002. The purpose of this survey is to give a comprehensive treatment of the techniques that are pertinent to the processor-based implementation of H.264 codec. Although H.264 is a very new standard, its literature is abundant. Limited by space and our knowledge, we will describe the approaches that, based on our experiences, have good potential in constructing a DSP-based codec. In general, the encoder part can be speedup by various fast algorithms to save the computation, while both encoder and decoder can be accelerated by the processor-dedicated parallel processing instructions.

The rest of the paper is organized as follows. In Section II, we first brief review the H.264 video standard and its computational profile. Section III contains a short discussion on the general principles of accelerating an algorithm implemented on a processor. Then, we present the fast algorithms for the intra prediction, motion estimation, mode decision and other parts in Section II to Section VI, respectively. Then we show the code speed-up tips for DSP in Section VII. Finally, a few conclusion remarks are made in Section VIII.

II. Overview of H.264 Video Coding

A. Overview

H.264 consists of a number of tools. Its basic structure is the so-called motion-compensated transform coder. Compared to the prior video coding standards, many important and new techniques are employed in H.264 and they together bring significant improvement on coding performance. Some of these techniques are highlighted here [2]. We may want to add that the concepts of some of these tools have existed for some time but they are nicely tuned and integrated together to form a good compression scheme in H.264.

1) Variable block-size motion compensation with multiple references

The basic unit in H.264 motion estimation is the 16x16 macroblock. It can be further split into a tree structure, with a minimum motion compensation block size as small as 4x4. Also, up to five reference frames may be used for motion compensation.

2) Directional spatial intra coding

To reduce the correlation inside a block, H.264 adopts the intra-prediction technique, which estimates the current block pixel values based on the known pixels of its neighbor blocks. The prediction results implicitly follow the edge direction, and often bring significant improvements.

3) In-loop deblocking filter

Block-based video coding produces artifacts known as blocking artifacts at low bit rates. This in-loop deblocking filter adjusts its filter strength adaptively according to the image local characteristics, and thus it provides better quality pictures at the decode end.

4) Context adaptive entropy coding

Two entropy coding methods, Context-based Adaptive Binary Arithmetic Coding (CABAC) and Context-based Adaptive Variable Length Coding (CAVLC), are provided in H.264. Both methods use context-base adaptivity to improve the entropy coding performance and the results show this approach is quite successful.

A simplified encoding flow of H.264 is shown in Fig. 1. A video frame is first partitioned into a number of 16x16 macroblocks. Then, each macroblock goes through the intra-prediction or the inter-prediction unit. The intra prediction unit uses the neighboring block data to predict the current block. The inter-prediction uses reference frames to predict the current frame. Each predictor has a number of modes. A good design should pick up the best mode with the lowest rate and distortion. The prediction residuals are then transformed, quantized and further entropy-coded into the output bitstream. In order to continue operating on the next incoming frame, the quantized current frame is reconstructed and stored. The decoder data flow is the reverse of the encoder flow.

B. Computational profile

The H.264 encoder reference software provided by the ITU/MPEG standard committee is known for its high computational complexity. A typical computational profile of the

---

0-7803-9451-8/06/$20.00 ©2006 IEEE.

H.264 encoder (ITU/MPEG reference software) running on Intel PC, is shown in Fig. 2. It shows that the tools of (a) motion estimation, (b) entropy coding, (c) transform and quantization, (d) interpolation, and (e) mode decision and intra-prediction are the most time-consuming modules. Although the other processors would have somewhat different architectures from the Intel processor, by and large, the trend is pretty much the same. As for the decoder, the tools of (a) motion compensation (including interpolation), (b) entropy decoding, and (c) intra-prediction have the CPU load.

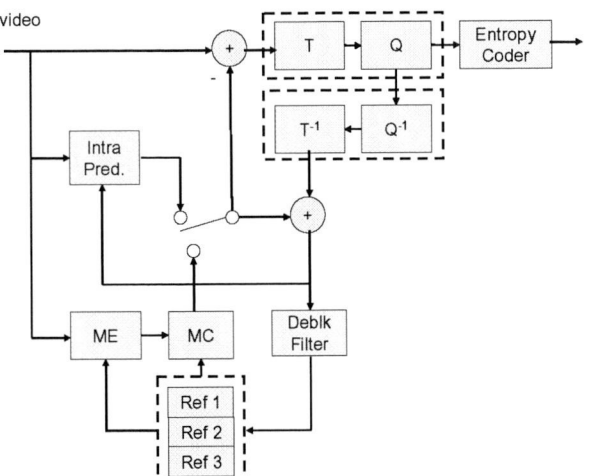

Fig. 1. Block diagram of H.264 encoder

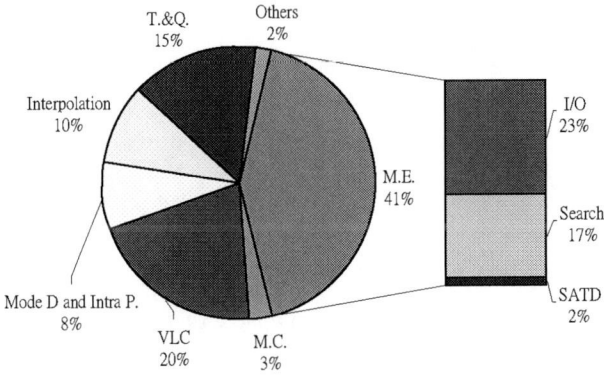

Fig. 2 Computational profile of H.264 video encoding.

III. Acceleration Methodology

The focus of this paper is efficient implementations of H.264 on a DSP system. Limited by the computing power and memory size of DSP, we need to modify the original software to reduce its computational complexity and to match the DSP computing architecture.

Essentially there are two types of calculation acceleration steps. The first type is programming techniques that reduce the redundancy in the execution codes and alter the program to match the DSP structure, for example, loop unrolling. The second type is to replace certain complicated modules by

their approximation counter parts. That is, in contrast to the first type of speed-up process that does not change the output values, the second type speed-up process changes the output values. Our target is to find the fast algorithm modules that approximate the original modules well and, therefore, little performance degradation is encountered.

Typically, we first analyze the current program complexity by profiling its execution as we did in the last section. After identifying the most computational intensive modules, we look for proper acceleration steps. In the case of H.264, the decoder is rigidly specified by the standard and thus generally only the first type acceleration steps can be used so that the output values are precisely reserved after acceleration. On the other hand, the encoder is not completely specified by the standard and thus there is quite a lot of flexibility in the encoder. We thus look for good fast algorithms that replace the original modules without sacrificing the compression efficiency.

IV. Fast Algorithms for Intra Prediction

A. Overview

Intra-prediction uses the high correlation property of neighboring samples in spatial domain to predict the current encoded samples. For the luma samples, each prediction block may be formed for each 4x4 block (denoted as I4MB) or for an entire MB (denoted as I16MB). When utilizing Intra_4x4 prediction, each 4x4 block chooses one of the nine prediction modes, which include one DC mode plus eight directional prediction modes, as shown in Fig. 3 (left), as the best one. In the luma component of an MB, the Intra_16x16 prediction is typically chosen for smooth image areas, and thus, only four prediction modes are specified as shown in Fig.3 (right) except for the DC mode. The chroma samples of an MB are predicted using a similar prediction pattern, Intra_8x8, which is similar to the luma Intra_16x16 prediction.

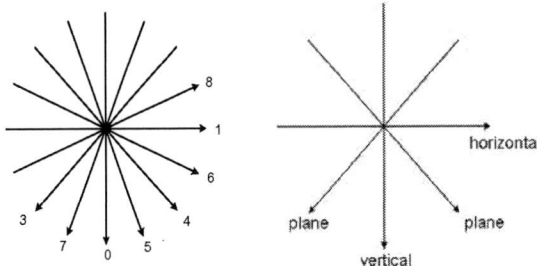

Fig. 3 Intra prediction modes for Intra_4x4 (left) and Intra_16x16 (right).

B. Fast algorithms

The fast algorithms of intra prediction can be classified into several types. The first approach is "early termination", which ends the search operation when the calculated distortion is smaller than a pre-chosen threshold. The selection of a proper measure for deciding termination is critical to the performance. It may be derived based on the macroblock smoothness [3][4] or the most probable mode [5]. The early termination based on the macroblock smoothness calculates a smoothness measure of a macroblock to determine the block type. For example, the large block type such as Intra_16x16 is chosen often for the flat image areas [3][4]. "Smooth" means that all the pixel values in a MB are similar; that is, their variance is small. The variance computation shall be simple to save computation. Therefore, the Mean-Absolute-Difference (MAD) operation [3] or the AC/DC ratio [4] is often used. If the variable is smaller than a pre-selected threshold value, the Intra_16x16 mode is chosen and thus the costly Intra_4x4 can be skipped.

Another kind of early termination proposal examines the most probable mode first. For example, in searching for the best Intra_4x4 mode, if its residual is smaller than a threshold, then the other eight Intra_4x4 modes are skipped (not chosen). Otherwise, all nine modes have to be tested. Then, we set another threshold to decide whether to keep on checking the Intra_16x16 prediction or not. It was reported that in one case, this method together with the 2:1 down-sampling and rate-distortion optimization (RDO) can reduce 68.8% of total computation time with only 1.35% of bit rate increase comparing to the reference software [5]. The major issue in this type of algorithms is how to determine the threshold. The threshold value can be adjusted according to the quantization parameters for instance. To construct a more efficient scheme, we propose a mixed fast intra prediction algorithm. It first examines both the most probable mode and the DC mode to determine if it meets the early termination criterion. The threshold value is decided by the average of SATD (sum of absolute transformed difference) of all the previous Intra_4x4 blocks in this frame. Once the 16 Intra_4x4 blocks are done, their total cost will be used as the threshold for deciding Intra_16x16 mode. These threshold values seem to be able to match the video local characteristics and provide good results. Even when RDO is turned off, we can achieve around 30% computational savings for the intra prediction module.

The second approach uses the edge analysis to quickly identify the edge direction since the intra prediction is basically a directional prediction [6][7]. Often the Sobel operators or the first order derivative are used as the edge analysis tool to find the most probable edge, which will be used as one of the final edge candidates. The final mode candidate list includes the one selected by the edge detector together with the other highly probable modes. In the case Intra_4x4, this would mean two modes of the neighboring blocks and the DC mode; and in the Intra_16x16 and Intra_8x8 cases, only the DC mode is considered highly probable. Therefore, only four candidate modes (for Intra_4x4) or two candidate modes (other types) are needed to be examined. The result shows that 60% of intra_only computation time reduction is observed with RDO and the bit rate increase is around 2~3% [6]. The bit rate increase may be owing to the irregular edges within a block. On the other side, the extra computation needed for edge analysis can be a computation burden and reduce the overall saving significantly.

The third approach uses the so-called three step approach

[8]. It first tests the horizontal and vertical directions, it then tests the neighboring 22.5 degree modes close to the better one from the previous step, and finally the best mode up-to-now is checked against the DC mode for the final winner. This approach has the advantage of a fixed number of modes are examined for all cases. However, computation time reduction is around 33% with about 1% bit rate increase.

The last approach makes use of the correlation in the temporal domain [9] since the best prediction mode in the current macroblock is likely similar to that in the reference macroblock in the previously coded frame(s). Thus, the primary intra prediction mode is selected from the mode of the most overlapped block in motion estimation. The computational overhead is nearly zero since all information is obtained during the inter-prediction operation. It is reported that the coding performance is nearly unchanged while the computational savings is about 50% assuming the intra-frame period is 10 [9].

In summarizing various fast intra-prediction algorithms, although we cite the experimental results from the proposed documents, a fair comparison among all methods is difficult because their simulation environments are quite different. One important element affecting computation is the option of RDO in the reference software. This is particularly true for the early termination method with thresholds.

The algorithms described in the above can be combined together to achieve further speed-up. For example, the first step could be the decision on Intra_4x4 or Intra_16x16. The second step could be the early termination for the chosen intra type. Finally, the rest of mode tests could be a fast algorithm to select one from the nine or four candidate modes.

V. Fast Algorithms for Motion Estimation

A. Overview

Block matching based motion estimation and compensation is a fundamental process in the current international video compression standards. It can efficiently remove interframe redundancy. A direct implementation is the full search algorithm that examines exhaustively every candidate motion vector in the search window to find the globally best matched block in the reference frame. However, its computationally intensive nature prevents it from practical implementation on a processor for real-time applications. The computation burden is increased drastically for the H.264 encoder because there are a number of combinations of partitioning a macroblock into sub-block(s) ranging from 4x4 to 16x16. Potentially each sub-block can have its own motion vector. This feature significant increases the computational complexity in motion estimation. Thus, many fast motion estimation algorithms have been proposed to alleviate the computational load.

Most of the fast algorithms are based on the well-known a priori knowledge, "the motion field of a real world image sequence is usually gentle, smooth and varies slowly". Fast motion estimation algorithms can be categorized into roughly three families as described below.

B. Fast algorithms for motion estimation

1) Searching over a subset of possible candidate points with certain search patterns

Based on the assumption of convexity of the unimodal error surface, i.e., block matching distortion increases monotonically away from the global minimum point, many gradient-based search methods with carefully designed search patterns have been developed to limit search points to a small subset of all possible candidates. This category includes the well-known three-step search (3SS) [10], the new three-step search (N3SS) [11], the cross search (CS) [12], the one-dimensional gradient descent search (1DGDS) [13], the block-based gradient descent search (BBGDS) [14], the four-step search (4SS) [15], the diamond search (DS) [16], the cross-diamond search (CDS)[17] and the hexagon-based search (HEXBS) [18]. Although this category of algorithms may be trapped into a local minimum point and hence the efficiency of the motion compensation may drop, they can considerably reduce the number of block matching computations.

2) Prediction of the motion vector based on correlations among motion vectors

Motion in most natural image sequences involves a few blocks and lasts for a few frames. Therefore, spatially or temporally adjacent blocks often have similar motion vectors. Taking the advantage of the correlation among neighboring motion vectors, the search window can be constrained to a small clique surrounding the "predicted vector", a candidate position predicated based on the known neighboring motion vectors. Many prediction algorithms have been developed with different complexities. The prediction search algorithm (PSA) [19] simply predicts the current block motion vector as the mean value of its neighboring blocks' motion vectors. Fuzzy search [20] applies fuzzy logic to predict the motion vector. In [21], motion vectors are predicted by integral projections. In [22], a spatial-temporal AR model of motion vectors is constructed and an adaptive Kalman filter is employed. The multi-resolution search [23] down-samples a picture to obtain raw motion vectors at different resolution levels, then it estimates finer motion vectors from the coarser ones. The multiresolution-spatiotemporal (MRST) scheme [23] modifies the normal raster scan order so that some blocks can reference more motion information by increasing their neighboring blocks along more directions. It then combines a multiresolution scheme and spatiotemporal correlation to predict motion vectors. For burst motions and blocks at the top-left corner, which has little correlation information, the performance of this category of algorithms may deteriorate because the refinement of prediction is restricted to a small search region. Moreover, the prediction overhead may reduce the speed gain.

3) Low complexity block matching criteria

The majority of the computations in motion estimation originate from computations of block matching distortion. In general, block matching metrics, such as the mean absolute difference (MAD) and the mean square error (MSE), involve pixel-wise operations, which are highly computationally intensive. Some methods try to simplify distortion computation by substituting the distortion defined on a subset of pixels for the whole block distortion. For instance, the MAD of 128 pixels is used as the matching distortion for a 16x16 macroblock in [23]; the computations can be reduced by one half with little performance loss. However, this method is not suitable for small blocks such as 4x4 blocks. Partial distortion elimination (PDE) in [24] compares every line's distortion in a block to avoid computing the distortion of the entire block. In [25], hypothesis testing is used to estimate the MAD from the partial mean absolute difference (PMAD), and the estimated MAD value is used to judge the matching result.

When fast algorithms in the above three categories are put together, the motion estimation accuracy may degrade. Additional calculations such as the initial motion vector prediction could lead to a considerable amount of computational overhead.

An approach proposed without quality degradation is the successive elimination algorithm (SEA) suggested by Li and Salari [26], which pre-excludes some impossible candidate points before completing the matching distortion calculation. SEA is a fast full search algorithm having a performance identical to FS while it speeds up the search process approximately by 10 times for 16x16 macroblock based motion estimation. Some further improvements have been made in subsequent research [24][27]-[30].

4) Fast fractional motion estimation

In the H.264 video coding scheme [1], the inter prediction (motion vectors) precision has been increased to quarter pixel. Typically, people perform the integer pixel motion estimation (IME) first. Then, the sub-pixel motion estimation or fractional motion estimation (FME) is applied to achieve refinement. As compared to the integer-value search, FME has a somewhat different statistical character. This may due to the facts that the search window of FME refinement is much smaller than that of IME and that the referenced sub-pixels are interpolated from the integer-coordinate pixels. Consequently, the error surface of FME is much closer to a uni-modal one, which favors fast algorithms.

Therefore, traditional fast algorithms in IME can also be used and can be more effective. The scheme adopted by the H.264 reference software is a three-step-like fast algorithm. It first checks the nine candidates surrounding the best match of IME, and then checks further the nine candidates surrounding the best match from the previous step. However, to take even more advantage of the uni-modal surface property and the highly centralized distribution of sub-pixel motion vectors, several fast FME algorithms with additional features are proposed. In [31], a gradient based search algorithm is brought up. The search direction is determined first and looks for the best motion vector along that direction. In [32], an adaptive search-pattern algorithm is proposed. The search-pattern is determined by outcome of the previous step and it biased towards the search center. This method saves half of the computations when compared to the reference software.

5) Some recent approaches

The recent trend to further reduce the motion estimation calculations is to combine the techniques mentioned before. The idea is each technique, a fast algorithm, is placed its most suitable target area. Thus, how to find a specific combination that achieves the optimal solution for a specific application becomes the most important issue. In [33], a fast algorithm with better coding efficiency on residuals is proposed, which leads to a lower bit rate compared to the full search algorithm. The method proposed in [34] produces larger residuals (due to fewer search points) but less motion information. Overall, it has a better encoding efficiency and a rather fast coding speed. This type of solutions seems to the target now researchers are aiming at.

VI. Fast Mode Decision Algorithms

A. Overview

The mode decision algorithm determines the best mode of the macroblock from various combinations of inter-prediction and intra-prediction. It can be coded with seven different block sizes for motion-compensation in the inter mode, and various spatial directional prediction modes in the intra mode. To achieve the highest coding efficient as close as possible, the reference software calculates the rate distortion costs of all possible modes and the it chooses the best one that has the minimum cost. This is a very time-consuming process. To reduce the computation load, a fast mode decision algorithm is necessary, which can do a quick screening to drop most poor modes and then it examines the reminders and identifies the (nearly) best one.

B. Fast mode decision algorithm

The fast mode decision algorithm can be divided into two types. The first type uses an early termination threshold to terminate the lengthy mode decision process. The early termination step can be placed between the intra and inter prediction processes [35][36] or inside the inter prediction process [37].

The scheme proposed in [35][36] uses the fact that intra mode needs more bits for coding and thus has a lower priority than the inter mode. Thus, if the best inter mode cost is smaller than a threshold, the intra prediction mode is skipped. The threshold can be the average of rate distortion cost of a number of previously coded intra blocks [35] or a ratio between the average boundary error (ABE) and average rate (AR) [36], where AR is the average bits for encoding the motion-compensated residuals and ABE is the average pixel error between the pixels at boundary of the current and its adjacent blocks in the best inter mode. The simulation results show that it can achieve about 20% reduction of com-

putational time with a slight bitrate increase.

In [37], it observes the fact that the 16x16 block usually is the best block size for large areas of background with still or uniform motion since it has less motion vector overhead. Thus, it first checks the cost of 16x16 block size. If it is smaller than a threshold, say, an average value of previous 16x16 blocks, the inter prediction process is terminated. Otherwise, a similar procedure is applied to the 8x8 block size.

The second type of the mode decision algorithms is to reduce the number of candidate modes. Intuitively, if the cost of a larger block-size mode is higher than the cost of the current block-size mode, the even larger block-size modes can be excluded. Similarly, if the cost of a smaller block-size is higher than that of the current block-size mode, the even smaller block-size modes can be excluded. Following this argument, we give different priority to each mode. If the mode with higher priority can provide sufficient image quality, we can skip the other lower priority modes. A specific case is the SKIP mode. The SKIP mode refers to the 16x16 mode of which no motion and residual information is coded. Thus, no motion search is required and it has the lowest complexity. Therefore, many algorithms assign the highest priority to the SKIP mode and thus a large percentage of macroblocks would get the SKIP mode based on spatial-temporal neighborhood information [38]-[40]. This approach can save a significant proportion of the encoding time with a slightly bit rate increase and quality drop.

In summary, the fast mode decision algorithms can be combined with the other fast intra and inter prediction algorithms to achieve further speedup. In all these algorithms, the SKIP mode first approach can save significant computational time. How to determine proper threshold values in a simple and automatic way is one critical issue for research and many proposals have been suggested.

VII. DSP Optimization for Video Codec

A. Overview

DSP processors made by different manufacturers vary in their functionality and capability. For real-time video codec implementation, the DSP processor shall have the parallel processing units and a wide data bus bandwidth to support the huge computational and memory access requirements. Almost all the high-end DSPs offered by several well-known vendors can meet these requirements. To make the following discussions more concrete, the popular TI's DSP is chosen as an example in this paper.

Tuning the video codec software for DSP implementation involves several steps. Traditional development flows in the DSP industry includes the following. Construct a C model for validating purpose. The C model is first run on a host PC or a UNIX workstation. Then, port the C codes to the DSP assembly language. Years ago, this is done manually and thus is a painstaking task. As the modern DSP compilers become more mature, they can do part of the laborious work of instruction selection, parallelizing, pipelining, and register allocation. However, we still often find that the compilers are making mistakes from time to time. In addition, in order to make the final code more compact in size and faster in speed, the C codes have to be tuned to match the DSP architecture. Fig. 4 shows the typical three-step DSP code development flow [41]. With the help of DSP compiler and optimization tools provided by the venders, the programmer can now focus on high level algorithm development first, and then further fine-tune the DSP codes only when necessary.

For porting to DSP, the data type shall be first considered since the definition of data such as integer can be different for different processors. For example, on TI C6000, the long integer means 40 bits. Since the H.264 codec deals with 8-bit pixels, the programmer can use the short data type for fixed-point multiplication, which takes only one cycle. Further optimizations shall make maximal use of all the hardware resources in the critical loops.

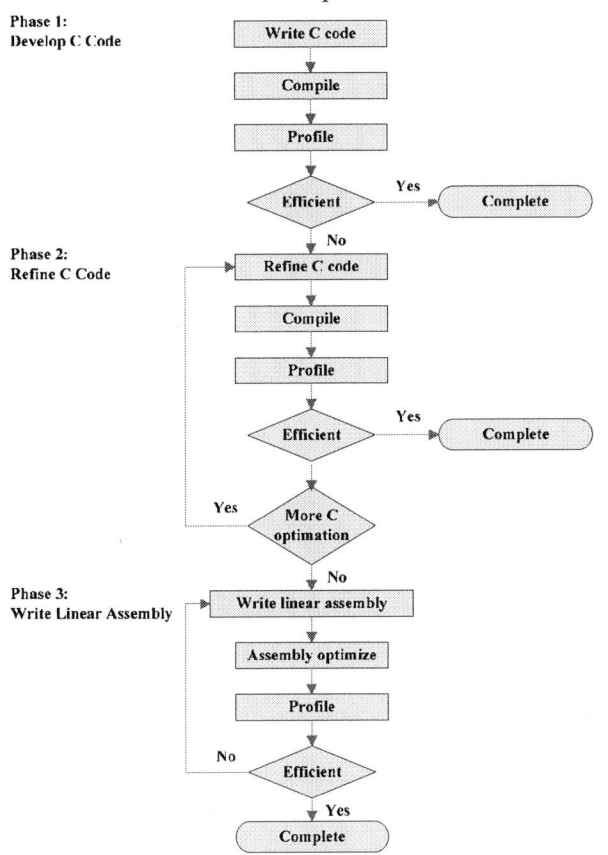

Fig. 4 DSP code development flow

B. Optimization for DSP architecture

To maximize C/C++ performance, the following optimization methods can be used [41].

1) C/C++ language level optimization

Since the DSP processor has less hardware resources than the PC environment, unnecessary operations has a strong impact on the processing speed. To achieve the best software performance, we use programming tricks to speed up the software at C/C++ language level. For example, we can use look-up table to reduce the arithmetic operations in following decoding steps: inverse transform, de-quantization, and entropy decoding [42].Also, to improve the loop operation, loop unrolling and software pipelining are exploited.

Loop unrolling eliminates or reduces loop management overhead by "unrolling" the loop. Unrolling loops involves replacing iterations of the loop by creating additional copies of the loop itself. Often it leads to faster but larger codes. However, the trade-off between code size and execution performance should be carefully balanced. Unrolling only speeds up code to a certain point, i.e., the law of diminishing returns prevails.

Software pipelining is used to schedule instructions from a loop so that multiple iterations of the loop execute in parallel. In C6000 compiler, the programmer can use "-o2" and "-o3" compiler options. The compiler then attempts to arrange software pipelines for the codes with the information that it gathers from the program.

2) Intrinsic operator

The C6000 compiler provides intrinsics, special functions that map certain high-level operations directly to the inline C6x instructions to speed up the C codes. All instructions that are not easily expressed in C codes are supported as intrinsics [41]. For example, we can use the intrinsic operator "_abs" to calculate the saturated absolute value.

3) Wider memory access for smaller data width

In order to maximize data throughput, it is often desirable to use a single load or store instruction to access multiple data values consecutively located in the memory. For example, C6x have instructions with associated intrinsics, such as "_add2()", "_mpyhl()", "_mpylh()", etc, that operate on the 16-bit data stored in the high and low parts of a 32-bit register. When operating on a stream of 16-bit data, we can use word accesses to read two 16-bit values at a time, and then use another C6x intrinsic to operate on the data. In the ideal case, we like to get all the units simultaneously operating on all individual instructions. This parallelism is still hard to achieve by the compiler and may still need hand–code in some cases.

4) Memory management

To maximize the processing speed, we prefer using internal memory to store instructions and data. However, the internal memory is generally quite small on the DSP processor. For example, on TMS320DM642 processor, there are only 16 Kbytes program memory and 16 Kbytes data memory in L1 cache, and 256 Kbytes unified memory in L2 cache. Therefore, memory management becomes very important. In managing the program memory, we need to delete unused codes and re-write certain functions to decrease the program code size. Next, we can use the compiler options to optimize the execution speed. In managing the data memory, we put all dynamically allocated memory sections into the external SDRAM and put the frequently used data in the internal data memory. This highly efficient memory management can effectively reduce memory stalls.

VIII. Conclusions

H.264/AVC is an efficient video compression scheme but this codec, particularly the encoder, has a very high computational complexity. After a short introduction to the H.264 standard, this paper summarizes a number of existing fast algorithms that can potentially be implemented on a DSP-like processor. These algorithms are mainly aiming at accelerating the computational speed of motion estimation, intra prediction and mode decision modules. In addition, we also describe the typical tricks used to speed up the C codes running on DSP.

Acknowledgements

This work was partially sponsored by ZyXEL Communications Corp., Taiwan, R.O.C., under Grant NCTU-94C031.

References

[1] ITU-T Rec.H.264, ISO/IEC 14496-10 "Advanced video coding", Final Draft International Standard, JVT-G050r1, Geneva, Switzerland, May 2003.

[2] T. Wiegand, G. J. Sullivan, G. Bjontegaad, and A. Luthra, "Overview of the H.264/AVC video coding standard", *IEEE Trans. Circuits Syst. Video Technol.*, vol. 13, pp. 560-575, July 2003.

[3] C.-L. Yang, L.-M. Po, and W.-H. Lam, "A fast H.264 intra prediction algorithm using macroblock properties," in *Proc. ICIP*, vol. 1, pp. 461 – 464, Oct. 2004

[4] Y.-K. Lin and T.-S. Chang, "Fast block type decision algorithm for intra prediction in H.264 FRext," in *Proc. ICIP*, Oct. 2005

[5] B. Meng, O.C. Au, C.-W. Wong, and H.-K. Lam, "Efficient intra-prediction algorithm in H.264," in *Proc. ICIP*, vol. 3, pp. 837-840, Sept. 2003.

[6] F. Pan, X. Lin, S. Rahardja, K. P. Lim, Z. G. Li, G. N. Feng, D. J. Wu, and S. Wu, "Fast mode decision algorithm for JVT intra prediction," JVT-G013, 7th JVT Meeting, Pattaya, Thailand, March 2003.

[7] Y.-D. Zhang, F. Dai, and S.-X. Lin, "Fast 4x4 intra-prediction mode selection for H.264," in *Proc. ICME*, vol. 2, pp.1151 – 1154, June 2004.

[8] C. C. Chen, T. S. Chang, "Fast three step intra prediction algorithm for 4x4 blocks in H.264," in *Proc. ISCAS*, 2005.

[9] M.-C. Hwang, J.-K. Cho, J.-H. Kim, and S.-J. Ko, "A fast intra prediction mode decision algorithm based on temporal correlation for H.264," in *Proc. of 2005 Int'l Tech. Conf. on Circuits Systems, Computers and Communications,* vol. 4, pp. 1573-1574, Jeiu, July 2005.

[10] J. Jain and A. Jain, "Displacement measurement and its application in interframe image coding," *IEEE Trans. Commun.*, Vol.29, (12), pp. 1799–1808, 1981.

[11] R. Li, B. Zeng, and M.L. Liou, "A new three-step search algorithm for block motion estimation," *IEEE Trans.*

Circuits Syst. Video Technol., vol. 4, (4), pp. 438–443, 1994

[12] M. Ghanbari, "The cross-search algorithm for motion estimation," *IEEE Trans. Commun.*, 38, (7), pp. 950–953, 1990.

[13] O.T.-C. Chen, "Motion estimation using a one-dimensional gradient descent search," *IEEE Trans. Circuits Syst. Video Technol.*, 10, (4), pp. 608–616, 2000

[14] L.-K. Liu and E. Feig, "A block-based gradient descent search algorithm for block motion estimation in video coding," *IEEE Trans. Circuits Syst. Video Technol.*, 6, (4), pp. 419–422, 1996.

[15] L.-M. Po and W.-C. Ma, "A novel four-step search algorithm for fast block motion estimation," *IEEE Trans. Circuits Syst. Video Technol.*, 6, (2), pp. 313–317, 1996.

[16] S. Zhu and K.-K. Ma, "A new diamond search algorithm for fast block matching motion estimation," *IEEE Trans. Image Process.*, 9, (2), pp. 287–290, 2000

[17] C.-H. Cheung and L.-M. Po, "A novel cross-diamond search algorithm for fast block motion estimation," *IEEE Trans. Circuits Syst. Video Technol.*, 12, (12), pp. 1168–1177, 2002

[18] C. Zhu, X. Lin, and L.-P. Chau, "Hexagon-based search pattern for fast block motion estimation," *IEEE Trans. Circuits Syst. Video Technol.*, 12, (5), pp. 349–355, 2002

[19] L. Luo, C. Zou, X. Gao, and Z. He, "A new prediction search algorithm for block motion estimation in video coding," *IEEE Trans. Consumer Electron.*, 43, (1), pp. 56–61, 1997.

[20] Y.-T. Roan and P.-Y. Chen, "A fuzzy search algorithm for the estimation of motion vectors," *IEEE Trans. Broadcast.*, 46, (2), pp. 121–127, 2000

[21] J.H. Lee and J.B. Ra, "Block motion estimation based on selective integral projections," *Int. Conf. on Image Processing*, vol. 1, pp. 689–692, Sept. 2002.

[22] C.-M. Kuo, C-P. Chao, and C-H Hsieh, "A new motion estimation algorithm for video coding using adaptive Kalman filter," *Real-Time Imaging*, 8, pp. 387–398, 2002

[23] J. Chalidabhongse and C.-C.J. Kuo, "Fast motion vector estimation using multiresolution-spatio-temporal correlations," *IEEE Trans. Circuits Syst. Video Technol.*, 7, (3), pp. 477–488, 1997

[24] H.-S. Wang and R.M. Mersereau, "Fast algorithms for the estimation of motion vectors," *IEEE Trans. Image Process.*, 8, (3), pp. 435–438, 1999

[25] K. Lengwehasatit and A. Ortega, "Probabilistic partial-distance fast matching algorithms for motion estimation," *IEEE Trans. Circuits Syst. Video Technol.*, 11, (2), pp. 139–152, 2001

[26] W. Li and E. Salari, "Successive elimination algorithm for motion estimation," *IEEE Trans. Image Process.*, 4, (1), pp. 105–107, 1995

[27] S.-M. Jung, S.-C. Shin, H. Baik, and M.-S. Park, "New fast successive elimination algorithm," *Proc. 43rd IEEE Midwest Symp. on Circuits and Systems*, vol. 2, pp. 616–619, Aug. 2000

[28] X.Q. Gao, C.J. Duanmu, and C.R. Zou, "A multilevel successive elimination algorithm for block matching motion estimation," *IEEE Trans. Image Process.*, 9, (3), pp. 501–504, 2000

[29] S.-M. Jung, S.-C. Shin, H. Baik, and M.-S. Park, "Efficient multilevel successive elimination algorithms for block matching motion estimation," *IEE Proc., Vis., Image Signal Process.*, 149, (2), pp. 73–84, 2002

[30] M. Yang, H. Cui, and K. Tang, "Efficient tree structured motion estimation using successive elimination," *IEE Proc. Vis., Image Signal Process.*, Vol. 151, No. 5, Oct. 2004

[31] H.-M. Wong, O. C Au, and A. Chang, "Fast sub-pixel inter-prediction – based on the texture direction analysis," Proc. *IEEE International Symposium, Circuits and Systems*, Oct. 2005.

[32] C.-C. Cheng, Y.-J. Wang, and T.-S. Chang, "A fast fractional pel motion estimation algorithm for H.264/AVC," in *Proc. VLSI/CAD Conf.*, 2005.

[33] Z. Chen, P. Zhou, and Y. He, "Fast motion estimation for JVT", JVT G-016, 2003

[34] X. Yi, J. Zhang, N. Ling, and W. Shang, "Improved and simplified fast motion estimation for JM," JVT P-021, Oct. 2005

[35] K.-H. Han and Y.-L. Lee, "Fast macroblock mode determination to reduce H.264 complexity," IEICE Trans. Fundamentals, Vol.E88–A, 3, pp.800-804, March 2005

[36] J. Lee and Y. Jean, "Fast mode decision for H.264", LG Electronics Inc, Digital media research laboratory

[37] Z. Zhou and M.-T. Sun, "Fast macroblock inter mode decision and motion estimation for H.264/MPEG-4 AVC," *IEEE International Conference on Image Processing*, Oct. 2004.

[38] P. Yin, A. M. Towropes, and J. Boyce, "Fast mode decision and motion estimation for JVT/H.264," *Pro. ICIP*, pp.853-856, 2003

[39] A. C. Yu and G.R. Martin, "Advanced block size selection algorithm for inter frame coding in H.264/MPEG-4 AVC," *Proc. ICIP*, pp. 95-98, 2004

[40] C. Grecos and M.Y. Yang "Fast inter mode prediction for P Slices in the H264 video coding standard," *IEEE Trans on Broadcasting*, Vol. 51, 2, pp.256-263, June 2005.

[41] Texas Instruments, *TMS320C6000 Programmer Guide*, 2001.

[42] S.-W. Wang, Y.-T. Yang, C.-Y. Li, Y.-S. Tung, and J.-L. Wu, "An optimization of H.264/AVC baseline decoder on low-cost TriMedia DSP processor", *Proc. of 49th SPIE Annual Meeting*, 2004.

Hardware Architecture Design of an H.264/AVC Video Codec

Tung-Chien Chen, Chung-Jr Lian, and Liang-Gee Chen

DSP/IC Design Lab, Graduate Institute of Electronics Engineering,
Department of Electrical Engineering, National Taiwan University, Taipei, Taiwan
(e-mail: {djchen, cjlian, lgchen} @video.ee.ntu.edu.tw)

Abstract—H.264/AVC is the latest video coding standard. It significantly outperforms the previous video coding standards, but the extraordinary huge computation complexity and memory access requirement make the hardwired codec solution a tough job. This paper describes the design methodology for H.264/AVC video codec. The system architecture and scheduling will be addressed. The design consideration and optimization for its significant modules including bandwidth optimized motion compensation engine, reconfigurable intra predictor generator, low bandwidth parallel integer motion estimation will be mentioned. Due to the complex, sequential, and highly data-depended characteristics of all essential algorithms in H.264/AVC, not only the pipeline structure but also efficient memory hierarchy is required. The design case with a hybrid task pipelining scheme, a balanced schedule with block-level, MB-level, and frame-level pipelining, will be presented. By combining with many bandwidth reduction techniques and data reused schemes, very efficient architecture and implementation for plate-form based system is proved by the prototype chips.

I. INTRODUCTION

H.264/AVC is the new video coding standard. It can save 25%-45% and 50%-70% of bitrates when compared with MPEG-4 Advanced Simple Profile (ASP) [1] and MPEG-2 [2], respectively [3]. Although motion compensated transform coding is still adopted, many new features are used to achieve much better compression performance and subjective quality, such as quarter-pixel Motion Estimation (ME) with Multiple Reference Frames (MRF) and Variable Block Sizes (VBS), intra prediction, Context-based Adaptive Variable Length Coding (CAVLC), and in-loop deblocking filter [4][5][6]. The rate-distortion optimized mode decision [3] is also included in reference software [7] to improve rate-distortion efficiency.

There are many potential applications of H.264/AVC. Ongoing applications range from High Definition Digital Video Disc (HD-DVD) or BluRay for living room entertainment with large screens to Digital Video Broadcasting for Handheld terminals (DVB-H) with small screens. However, the H.264/AVC coding performance comes at the price of computation complexity. According to the instruction profiling with HDTV1024P (2048×1024, 30fps) specification, H.264/AVC decoding process requires 83 Giga-Instructions Per Second (GIPS) computation and 70 Giga-Bytes Per Second (GBPS) memory access. As for H.264/AVC encoder, up to 3600 GIPS and 5570 GBPS are required for HDTV720P (1280×720, 30fps) specification. For real-time applications, accelerating by the dedicated hardware is a must.

However, it is difficult to design an system architecture for the H.264/AVC codec. The design for the significant modules are also very challenging. Besides high computation complexity and memory access, the coding path is very long, which includes prediction, reconstruction, and entropy coding. The reference software adopts many sequential processing of each block in the MacroBlock (MB), which restricts the parallel processing. The block-level reconstruction loop caused by intra prediction in-

duces the bubble cycles and decreases the hardware utilization and throughput. The coding tools involve with many data dependencies to enhance the coding performance, but the considerable storage space is the penalty. There are functionalities that have multiplex modes, and the re-configurable engine is essential to achieve resource sharing.

To address these difficulties, the hardware design methodology is described for H.264/AVC video coding system in this paper. There are three critical issues to be addressed. First, for H.264/AVC encoder, the traditional two-stage MB pipelines cannot be efficiently applied because of the long critical path and feedback loop. According to our analysis, five major functions are extracted and mapped into four stage MB pipelining structure with suitable task scheduling. Second, a hybrid task pipelining scheme is presented with a balanced schedule with block-level, MB-level, and frame-level pipelining to greatly reduce the internal memory size and bandwidth. Third, the design consideration and optimization for the significant modules, including bandwidth optimized Motion Compensation (MC) engine, reconfigurable intra predictor generator, low bandwidth parallel IME are involved. The design cases show that efficient implementation for H.264/AVC video coding system is achievable by combining these efficient architectures.

The rest of this paper is organized as follows. In Section II, the profiling and the design considerations are described. Then, the architecture optimization of H.264/AVC encoding system will be addressed in Section III. Those of H.264/AVC decoding parts are mentioned in Section IV. These architectures are proved by the prototype chips, which will be described in Section V. Finally, we will make a conclusion in Section VI.

II. ANALYSIS AND DESIGN SPACE EXPLORATION

A. Profiling

We use instruction profile to show the high computation complexity and memory access of H.264/AVC, and find the critical parts for hardware implementation. The iprof [8], a software analyzer at the instruction level, is used to profile an H.264/AVC encoder at a processor-based platform (SunBlade 2000 workstation, 1.015GHz Ultra Sparc II CPU, 8GB RAM, Solaris 8). To focus on the target specification, a software C model is developed by extracting all baseline profile compression tools from reference software [7]. Our focused design case is mainly targeted for SDTV (720×480, 30fps)/HDTV720P videos with four/one reference frame. The computational complexity and memory access for SDTV/HDTV720P are 2470/3600 GIPS and 3800/5570 GBPS. As for decoder with HDTV-1024P video formats. 83 GIPS and 70 GBPS of computation and memory access are required. The huge computational loads are far beyond the

2006 Asia and South Pacific Design Automation Conference

7D-3

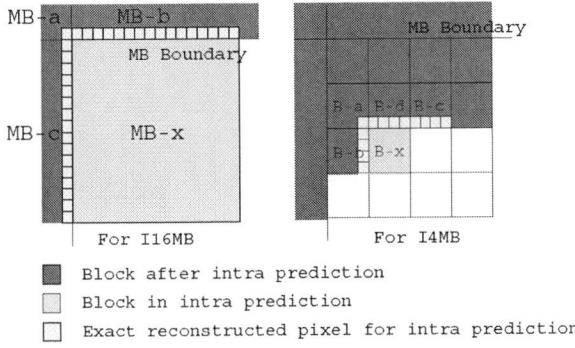

Fig. 1. (a) MB-level reconstruction loop; (b) block-level reconstruction loop. The intra prediction requires the reconstructed pixels of the left and top neighboring blocks, which induce the MB-level and block-level reconstruction loops.

capability of today's general purpose processors (GPPs). The dedicated hardware is essential for real-time applications.

B. Design Space Exploration

The considerations of hardware design are analyzed with the H.264/AVC compression algorithms. The major challenges are described as follows.

- **Computation Complexity and Bandwidth Requirement**: According to the profiling, H.264/AVC requires much more computation complexity than previous coding standards. this will greatly increase the hardware cost especially for the HDTV applications. The bandwidth requirement of H.264/AVC encoding system is also much higher than previous coding standard. The MRF-ME contributes the most traffic for loading reference pixels. Neighboring reconstructed pixels are required by intra prediction, and are also required by deblocking filter. Besides, Lagrangian mode decision and context-adaptive entropy coding have data dependencies between neighboring MB, and transmitting related information contribute considerable bandwidth as well. An efficient memory hierarchy combined with data sharing and Data Reuse (DR) scheme must be designed to reduce the system bandwidth.

- **Sequential Flow**: The H.264/AVC reference software adopts many sequential process to enhance the compression performance. It is hard to efficiently map the sequential algorithm to parallel hardware. For system architecture, we partition the sequential encoding process (prediction, reconstruction, and then entropy encoding) into several tasks and process them in MB-based pipelining structure, which improves the hardware utilization and the throughput. For module architecture, this problem is critical for ME since ME is the most computationally intensive part and requires the most degrees of parallelism. The inter Lagrangian mode decision takes MV costs into consideration. The MV of each block is generally medium predicted by left, top, and top-right neighboring blocks. The cost function can be computed only after prediction modes of neighboring blocks are determined, which also causes inevitable sequential processing. The modified hardware-oriented algorithms can be designed to enable parallel processing.

- **Coding Loops**: In traditional video coding standard, there

is a frame-level reconstruction loop generating the reference frames for ME and MC. In H.264/AVC, the intra prediction requires the reconstructed pixels of the left and top neighboring blocks, which induce the MB-level and block-level reconstruction loops. For the MB-level reconstruction loop as shown in Fig. 1 (a), the reconstructed pixels of MB-a, MB-b, and MB-c are used to predict the pixels in MB-x for I16MB. Not until MB-a, MB-b, and MB-c are reconstructed can MB-x be predicted. Similarly, as Fig. 1 (b) shows, in order to support I4MB, not until 4×4-intra mode of B-a, B-b, B-c, and B-d are decided and reconstructed can B-x be processed. The reconstructed latency is harmful to hardware utilization and throughput if the intra prediction and reconstruction are not jointly considered and scheduled.

- **Data Dependency**: The new coding tools improve the compression performance with many data dependencies. The frame-level data dependencies contribute the considerable system bandwidth. The dependencies between neighboring MBs constrain the solution space of MB pipelining, and those between neighboring blocks limit the possibility of parallel processing.

- **Abundant Modes**: There are many algorithms of H.264/AVC that have multiplex modes. For example, there are 17 different modes for intra prediction while 259 kinds of partitions for inter prediction. Six kinds of 2-D transform, 4×4/2×2 DCT/IDCT/Hadamard transform, are involved in reconstruction loops. The reconfigurable processing engine, reusable prediction core, and appropriate pipeline system design are important to efficiently support all these functions.

C. Related Works

The conventional two-stage MB pipelining architecture [9][10] is widely adopted in prior video encoding hardware designs. Two MBs are processed simultaneously by prediction engine (ME only) and Block Engine (BE, including MC, reconstruction loop, and entropy coding) in pipeline manner. Several problems will be encountered if the two-stage MB pipelining is directly applied to H.264/AVC encoding. The prediction stage includes IME, FME, and intra prediction in H.264/AVC. The sequential prediction flow will make high operation frequency and low hardware utilization. Besides, because of MB-level and block-level reconstruction loops, it is impossible to completely separate the prediction and BE stages. The large bandwidth must also be reduced by efficient memory hierarchy and data reuse scheme.

Furthermore, because of the new funcationalities of H.264/AVC, the advanced module architectures are demanded for H.264/AVC encoder. IME is the most computationally intensive part in the encoder. Several IME architectures are proposed for VBSME [11] [12] [13][14] with different specifications. The main challenge for FME is to achieve parallel processing under the constraints of sequential Lagrangian mode decision. Besides, the functionalities of the VBS, MRF, 6-tap FIR, and Hadamard transform are involved. In [15], FME procedure is analyzed and decomposed into several loops. The fully pipelined architecture is designed with unfolding techniques and efficient scheduling. In [16], the forward/inverse multi-transform are designed to support 4×4 DCT, IDCT, and Hadamard transforms. The first deblocking filter propose the ad-

751

7D-3   2006 Asia and South Pacific Design Automation Conference

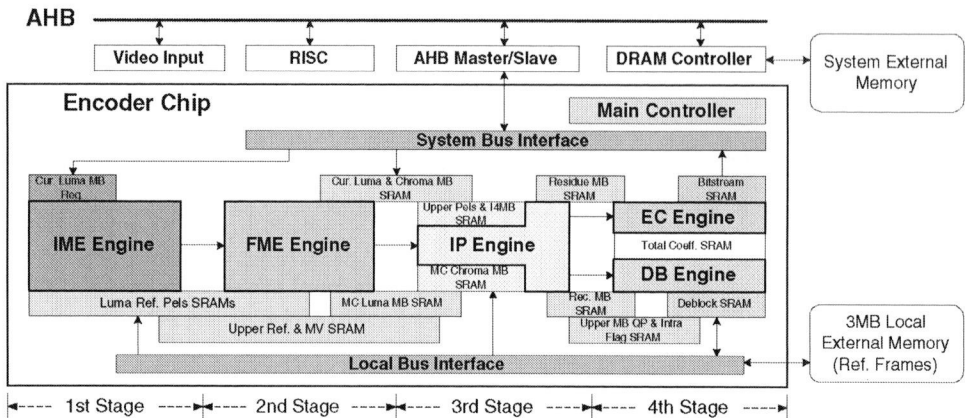

Fig. 2. Block diagram of the H.264/AVC encoding system. Five major tasks, including IME, FME, IP, EC, and DB, are partitioned from the sequential encoding procedure and processed MB by MB in pipeline structure.

vanced filtering schedule to save 50% on-chip bandwidth [17]. A column addressing technique used to favor the direction of vertical filtering is then proposed in [18] to increase the throughput. In [19], the architecture is further improved by interleavingly filtering the the horizontal and vertical edges. As for entropy encoder, the single buffer CAVLC architecture [20] is designed for SDTV specification. The dual-buffer architecture with block pipelining is proposed in [21] to double the hardware utilization and throughput.

III. H.264/AVC ENCODING SYSTEM

This section describes a new MB pipelining scheme for H.264/AVC encoder. The traditional two-stage MB pipelining [9][10], prediction (ME only) and BE, cannot be efficiently applied to H.264/AVC. In this encoding system, five major functions are extracted and mapped into four MB pipelining stages with suitable task scheduling [22]. Furthermore, the design consideration and optimization for the significant modules is described to enable the whole system. The efficient implementation for H.264/AVC encoding system is achieved by combining these techniques.

A. Four-Stage Macroblock Pipelining

The system architecture is shown in Fig. 2. Five major tasks, including IME, FME, Intra Prediction with reconstruction loop (IP), Entropy Coding (EC), and in-loop DeBlocking filter (DB), are partitioned from the sequential encoding procedure and processed MB by MB in pipeline structure.

Several issues are described as follows. The prediction, which is ME only in previous standards, includes IME, FME, and intra prediction in H.264/AVC. Because of the algorithms diversities and different computation complexity, it is difficult to implement IME, FME, and intra prediction by the same hardware. Putting IME, FME, and intra prediction in the same MB pipelining stage leads to very low utilization. Even if the resource sharing is achieved, the operating frequency becomes too high due to the sequentially processing. Therefore, FME is firstly pipelined MB by MB after IME to double the throughout. As for intra prediction, because of MB-level and block-level recon-

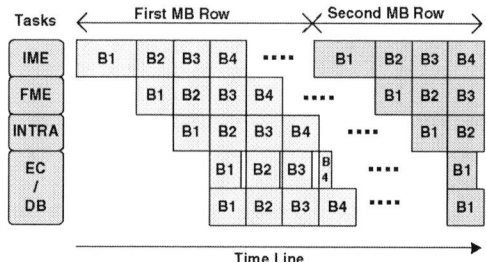

Fig. 3. The MB schedule of four stage MB pipelining. The horizontal arrow denotes the time line. One horizontal column indicates the MBs with different tasks that are processed in parallel.

struction loop, it cannot be separated with the reconstruction engine. Besides, the reconstruction process should be separated from ME and pipelined MB by MB to achieve highest hardware utilization. Therefore, engines of intra prediction together with forward/inverse transform/quantization should be located in the IP stage. In this way, the MB-level and block-level reconstruction loops can also be isolated in this pipelining stage. The EC encodes the MB header and residues after mode decision and prediction. The DB generates the standard-compliant reference frame after reconstruction. Since the EC/DB can be processed in parallel, they are placed at the 4th stage. The reference frame will be stored in external memory for the ME of the next frame, which constructs the frame-level reconstruction loop. Please note that, the luma MC is placed in FME stage for reuse of *Luma Ref. Pels SRAMs* and interpolation circuits. The compensated MB is transmitted to IP stage for generation of residues after intra/inter mode decision. On the other hand, chroma MC is implemented in IP stage since it is not required before intra/inter mode decision.

MBs within one frame are coded in raster order with schedule in Fig. 3. The horizontal arrow denotes the time line. One horizontal column indicates the MBs with different tasks that are processed in parallel. As for reduction in system bandwidth, many on-chip memories are used for three purposes. First, in

752

order to find the best matched candidate, a huge amount of reference data are required for both IME and FME. Since pixels in neighboring candidate blocks are considerably overlapped, and so are the SWs of neighboring current MBs, the bandwidth of system bus can be greatly reduced by designing local buffers to store reusable data. Second, instead of transmitted by system bus, the raw data such as luma motion compensated MB, transformed and quantized residues, and reconstructed MB are shifted forward via shared memories. Third, because of data dependency, a MB is processed according to the upper and left MBs. The local memories are used to store the related data during the encoding process. For software implementation, the external bandwidth requirement is up to 5570 TBPS. As for hardware solution with local search window buffer embedded, the external bandwidth requirement is reduced to 700 MBPS. After all three techniques are applied, the final external bandwidth requirement is about 280 Mbytes/sec.

B. Low-Bandwidth Parallel Integer Motion Estimation

IME requires the most computational complexity and memory bandwidth in H.264/AVC. Besides, it is a kind of sequential flow due to the Lagrangian mode decision flow. However, a large degree of parallelism is required for the SDTV/HDTV specifications. In the following, techniques in algorithmic and architectural levels are used to enable parallel processing and to reduce the required hardware resources.

B.1 Hardware-Oriented Algorithm

The Motion Vector (MV) of each block is generally predicted by the medium values of MVs from the left, top, and top right neighboring blocks. The rate term of the Lagrangian cost function can be computed only after MVs of neighboring blocks are determined, which causes inevitable sequential processing. The blocks and subblocks in a MB cannot be processed in parallel. Moreover, when a MB is processed at the IME stage, its previous MB is still in the FME stage. The MB mode and the best MVs of the left MB cannot be obtained in the four-stage MB pipelining architecture. To solve these problems, the exact MVPs are replaced by modified MVPs, which is the medium of MVs from top-left, top, and top-right MBs. In addition, the modified MVP is applied for all of the 41 blocks in MB, as shown in Fig. 4. For example, the exact MV cost of the C22 4×4-block is the medium of the C12, C13, and C21 MVs. The MVPs of all 41 blocks are changed to the medium of MV0, MV1, and MV2 in order to facilitate the parallel processing and MB pipelining.

As for searching algorithm, FS is adopted to guarantee the highest compressing performance. The regular searching pattern is suitable for parallel processing. Besides, FS can effectively support VBS by reusing each 4×4-block SADs for larger blocks. Pixel truncation [23] of five-bit precision and subsampling [24] of half pixel rate are applied to reduce the hardware cost. Moreover, adaptive search range adjustment [25] is also applied to save computations.

B.2 Architectures Design of IME

In IME, in order to find the best matched candidate, a Search Window (SW) within one reference frame has to be searched.

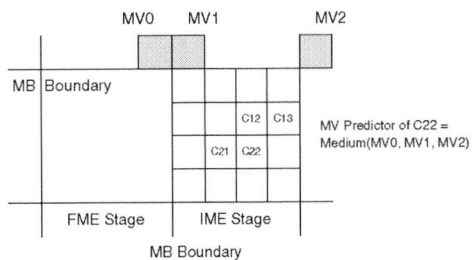

Fig. 4. Modified MVPs. In order to facilitate the parallel processing and MB pipelining, the MVPs of all 41 blocks are changed to the medium of MV0, MV1, and MV2.

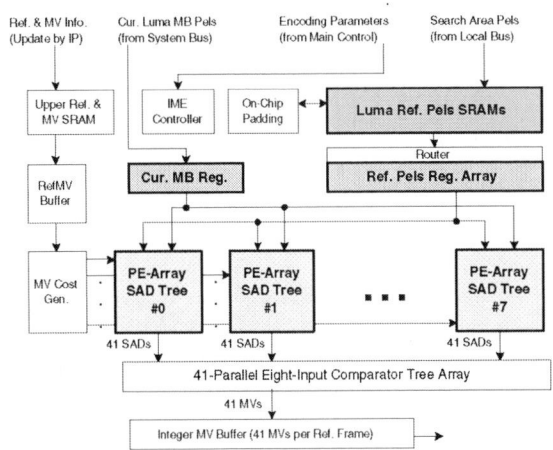

Fig. 5. Block diagram of the low bandwidth parallel IME engine. It mainly comprises eight *PE-Array 2-D SAD Tree*, and eight horizontally adjacent candidates are processed in parallel.

Figure 5 shows the low bandwidth parallel IME architecture, which mainly comprises eight *PE-Array 2-D SAD Tree*. The Current MB (CMB) is stored in *Cur. MB Reg.* The reference pixels are read from external memory and stored in *Luma Ref. Pels SRAMs*. Each PE array and its corresponding 2-D SAD tree compute the 41 SADs of VBS for one searching candidate each cycle. Therefore, eight horizontally adjacent candidates are processed in parallel.

Because SWs of neighboring current MBs are considerably overlapped, and so are the pixels of neighboring candidate blocks, a three-level memory hierarchy, including external memory, *Luma Ref. Pels SRAMs*, and *Ref. Pels Reg. Array*, is used to reduce bandwidth requirement by data reuse (DR). Three kinds of DR are implemented—MB-level DR, inter-candidate DR, and intra-candidate DR. The *Luma Ref. Pels SRAMs* are firstly embedded to achieve MB-level DR. when ME process is changed from one CMB to another CMB, there are overlapped area between neighboring SWs. Therefore, only a part of SW must be loaded from system memory, and the system bandwidth can be reduced [26].

The *Ref. Pels Reg. Array* acts as the cache between *PE-Array 2-D SAD Tree* and *Luma Ref. Pels SRAMs*. It is designed to achieve inter-candidate DR. A horizontal row of reference pixels, which is read from SRAMs, is stored and shifted

7D-3   2006 Asia and South Pacific Design Automation Conference

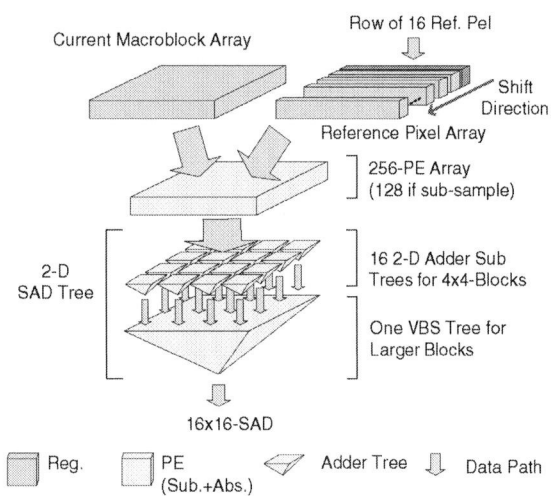

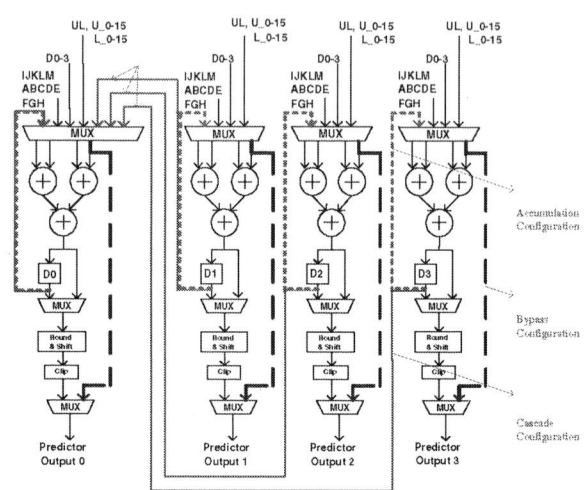

Fig. 6. *PE-array 2-D SAD Tree* architecture. The costs of sixteen 4×4-blocks are separately summed up by sixteen *2-D Sub-trees*, and then reused by one *VBS Tree* for larger blocks.

Fig. 7. Four-parallel reconfigurable intra predictor generator. Four different configurations are designed to support all intra prediction modes in H.264/AVC.

downward in *Ref. Pels Reg. Array*. When one candidate is processed, 256 reference pixels are required. When eight horizontally adjacent candidates are processed in parallel, not (256×8) but (256+16×7) reference pixels are required. Besides, when the ME process is changed to the next eight candidates, most data can be reused in *Ref. Pels Array*. The parallel architecture achieves inter-candidate DR in both horizontal and vertical directions and reduce the on-chip SRAM bandwidth.

Fig. 6 shows the architecture of *PE-array 2-D SAD Tree*. The costs of sixteen 4×4-blocks are separately summed up by sixteen *2-D Sub-trees*, and then reused by one *VBS Tree* for larger blocks. This is so-called intra-candidate DR. All 41 SADs for one candidate are simultaneously generated and compared with the 41 best costs. No intermediate data are buffered. Therefore, this architecture can support VBS without any partial SAD registers.

C. Reconfigurable Intra Predictor Generator

The intra prediction supports the most various prediction modes, which includes four I16MB modes, night I4MB modes, and four Chorma modes. For the RISC-based solution, the required operation frequency will become too high. For the dedicated hardware, 17 kinds of PEs for the 17 modes make the hardware cost high. Therefore, the reconfigurable circuit and resource sharing for all intra prediction modes are the efficient solutions.

The hardware architecture of the four-parallel reconfigurable intra predictor generator is shown in Fig. 7. Capital letters (A, B, C, ...) are the neighboring 4×4-block pixels. UL, L0-L15, and U0-U15 denote the bottom right pixel from the upper left MB, the 16 pixels of the right most column from the left MB, and the 16 pixels of the bottom row from the upper MB, respectively. Four different configurations are designed to support all intra prediction modes in H.264/AVC. Firstly, the I4MB/I16MB horizontal/vertical modes use the bypass data path to select the predictors extended from the block boundaries.

Secondly, multiple PEs are cascaded to sum up the DC value for I4MB/I16MB/chroma DC mode. Thirdly, the normal configuration is used for I4MB directional modes 3–8. The four PEs select the corresponding pixels multiple times according to the weighted factors, and process independently. Finally, the recursive configuration is designed for I16MB plan prediction. The predictors are generated by adding the gradient values to the result of previous cycles.

IV. H.264/AVC DECODING SYSTEM

In this section, a methodology to determine a suitable pipelining structure of a H.264/AVC decoder is presented. The target resolution is HDTV1024P 30fps videos. The design goals of this work are low area cost and low system bandwidth. In the following sections, the scheduling as well as the key modules of the decoding system will be elaborated.

A. Hybrid Task Pipelining

The overall system architecture is shown in Fig. 8 [27]. The sequence parameter set, picture parameter set, and slice headers are parsed by system processor. The MB-level information including MB headers and transformed/quantized residues are decoded by the *Parser Engine*. The predicted pixels are generated by the *Inter Pred. Engine* or *Intra Pred. Engine* according to the MB mode. The residues are recovered by the *IQ/IT Engine*. The MB is reconstructed by *Sum and Clipping Engine*. Finally, *Deblocking Engine* filters MB pixels and outputs them to the external memory. The buffers between the processing engines are required to separate pipelining stages.

Previous designs of video decoders are usually based on MB pipelining scheme [28]. Our system architecture is based on a hybrid task pipelining scheme including 4×4-block-level pipelining, MB-level pipelining, and frame-level pipelining. The reasons are stated as following. In H.264/AVC, 4×4-block is the smallest element of the prediction block mode. The transforms and entropy coding are also based on 4×4-blocks.

754

2006 Asia and South Pacific Design Automation Conference

7D-3

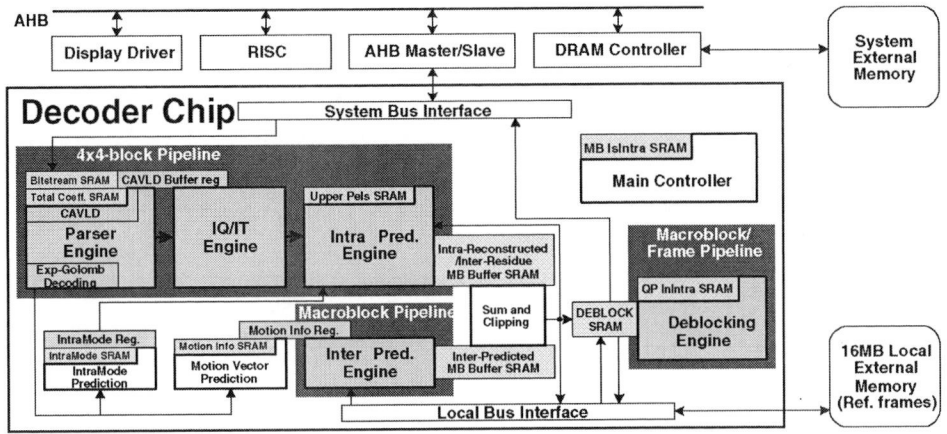

Fig. 8. Hybrid task pipelining system architecture.

Therefore, a 4×4-block pipelining scheme can be designed for CAVLD, inverse quantization/inverse transformation, and intra prediction with the benefit of less coding latency. It requires about 1/24 of buffer size compared to the traditional MB pipelining architecture.

Inter prediction produces the predicted MB pixels from previously decoded reference frames. As with intra prediction, the basic processing element of inter prediction is also a 4×4-block. Due to the six-tap FIR filter for interpolation, 9×9 integer reference pixels are required for a current 4×4-block. If the blocks of prediction mode are larger than 4×4, overlapped reference frame pixels of these 4×4-blocks can be reused to reduce the system bandwidth. The inherent order of 4×4-blocks in the bitstream is the double-z-scan order. Reference frame DR will be less efficient if *Inter Pred. Engine* adopts the 4×4-block pipelining scheme and follows the double-z-scan order. Therefore, *Inter Pred. Engine* should be scheduled to MB-level pipelining with a customized scan order to exploit the reference frame DR. All reference pixels necessary to predict a MB are read from memory at once to reduce memory bandwidth.

Deblocking Engine is another special case that does not suit to the double-z scan order. *Deblocking Engine* filters the edges of each 4×4 block vertically then horizontally. Besides, one 4×4-block cannot be completely filtered until its neighboring blocks are reconstructed. This data dependency makes it impractical to fit the deblocking operation into a 4×4-block pipelining, since the buffer cannot be efficiently reduced and serious control overhead is required. If the decoder has to support FMO and ASO, where the MBs of a frame may not be coded in raster-scan order, the DB unit has to be scheduled to frame level pipelining because the filtering order of one frame can not either be violated in MB boundaries. Otherwise, the MB pipelining schedule is adopted.

B. Low-bandwidth Motion Compensation Engine

According to the analysis in decoder system, MC should be scheduled to MB-level pipelining with a customized scan order to exploit the reference frame DR. The 4×4-based MC is firstly adopted. All VBS are decomposed into several 4×4 element

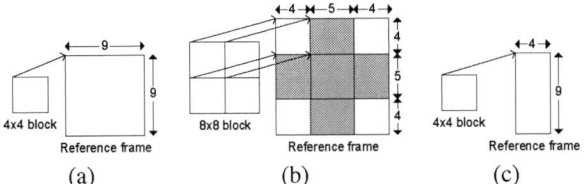

Fig. 9. (a) General case interpolation window; (b) Four interpolation windows for an 8×8 block (shaded region means reusable); (c) Interpolation window when MV pointing to horizontal integer pixels.

blocks, and processed by the MC engine with full hardware utilization. The straightforward memory access scheme processes every decomposed 4×4 element blocks independently, and always loads 9×9 pixels for interpolation as shown in Fig. 9(a). The bandwidth requirement of 4×4-based MC can be reduced by two bandwidth reduction techniques [29].

The first technique is *Interpolation Window Reuse* (IWR). As shown in Fig. 9(b), there are overlapped regions between interpolation windows for neighboring 4×4 element blocks when the block mode is larger than 4×4. The shaded regions can be reused. The second scheme is *Interpolation Window Classification* (IWC). The interpolation window is not always $(X+5)\times(Y+5)$ for a $X\times Y$ block. As shown in Fig. 9(c), a 4×4 block with integer MV in horizontal direction does not require horizontal filtering. A 4×9 interpolation window is read. In brief, the IWR and IWC scheme aim to precisely control the MC hardware to load a smaller and exact interpolation window. These techniques can provide about 60–80% bandwidth reduction for the 4×4-based MC.

Figure 10 shows the MC architecture. The efficient vertical scheduling in [15] is applied with *Down Shift Register Array* to support vertical IWR. Besides, an *Horizontal Reuse Memory* is designed for horizontal IWR. The IWC is implemented by *Control FSM* and *Address Generator*. The *Shift & Combine* circuit packs the required integer pixels inputted from external frame memory and *Horizontal Reuse Memory*. The *2-D IP Unit* performs the interpolation, and the compensated MB is buffered in the MC memory.

755

7D-3

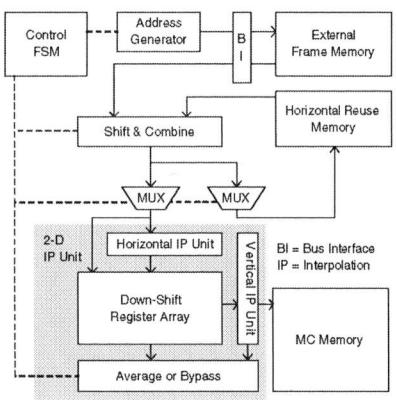

Fig. 10. Block diagram of MC hardware.

TABLE I

HARDWARE RESOURCE OF H.264/AVC ENCODER

Functional Block	Gate Counts	Memory (KB)
IME Module	305211	13.71
FME Module	401885	13.82
IP Module	121012	5.01
EC Module	29332	1.27
DB Module	20152	0.91
Others	45176	0.00
Total	922768	34.72

V. IMPLEMENTATION RESULTS

A. Implementation results of H.264/AVC SDTV/HDTV720P Encoder

The specification of this H.264/AVC encoder is baseline profile with level up to 3.1. The maximum computational capability is to real-time encode SDTV 30fps video with four reference frames or HDTV720P 30fps video with one reference frame. Table I shows the logic gate count profile synthesized at 120 MHz. The total logic gate count is about 922.8K. The prediction engine, including IME, FME, and IP stages, dominates 90% of logic area. As for on-chip SRAM requirement, 34.88 KB are required. The chip is fabricated with UMC 0.18 μm 1P6M CMOS process. Figure 11 shows the die micrograph. The core size is 7.68×4.13 mm^2. The power consumption is 581 mW for D1 videos and 785 mW for HDTV720p videos at 1.8 V supply voltage with 81/108 MHz operating frequency. The

TABLE II

SPECIFICATION OF THE DEVELOPED H.264/AVC BASELINE PROFILE ENCODER CHIP

Technology	UMC 0.18 μm 1P6M CMOS
Pad/Core Voltage	3.3/1.8 V
Core Area	7.68×4.13 mm^2
Logic Gates	922.8 K (2-input NAND gate)
SRAM	34.72 KB
Operating Frequency	81/108 for D1/HDTV720P
Power Consumption	581/785 mW for D1/HDTV720P
Encoding Features	All Baseline Profile Compression Tools
Max. # of Ref. Frames	4/1 for D1/HDTV720P
Max. SR (Ref. 0)	H[-64,+63] V[-32,+31]
Max. SR (Ref. 1-3)	H[-32,+31] V[-16,+15]

TABLE III

GATE COUNT PROFILE OF THE H.264/AVC DECODER

Functional Block	Gate Counts	Memory (KB)
Main Control	22695	0.48
Parser Engine	21121	1.02
Inter Pred. Engine	69695	2.43
Intra Pred. Engine	28707	4.93
IQ/IT Engine	19792	0.00
Deblocking Engine	35437	1.12
Others	19980	0.00
Total	217428	9.98

TABLE IV

CHIP FEATURES OF H.264/AVC DECODER

Technology	TSMC 0.18 μm 1P6M CMOS
Pad/Core Voltage	3.3/1.8 V
Core Area	2.19×2.19 mm^2
Logic Gates	21.743 K (2-input NAND gate)
SRAM	9.98 KByte
Profile	Baseline
Operating Frequency	120/1.5 MHz for HDTV1024P 30fps/QCIF 15fps
Power Consumption	186.4/1.18 mW for HDTV1024P 30fps/QCIF 15fps

detailed chip features are shown in Table II.

B. Implementation results of H.264/AVC HDTV1024P Decoder

The specification of this H.264/AVC decoder is baseline profile with level 4.1. It can real-time decode HDTV1024P 30fps video with operational frequency of 120 MHz. Table III shows the hardware resource requirement. The total logic gate count is 217K, and the on-chip SRAM requirement is 10 KB. The chip is fabricated with TSMC 0.18μm 1P6M CMOS technology. The core size is 2.19×2.19 mm^2. The power consumption is 186.4 mW at 1.8V and 120MHz for decoding HDTV1024P 30fps videos, and is 1.18 mW at 1.8V and 1.5MHx for decoding QCIF (176×144) 15fps videos. The detailed chip features are shown in Table IV.

VI. CONCLUSION

In this paper, state-of-the-art hardware architectures for H.264/AVC video coding core have been presented. First, five major functional blocks are mapped into four-stage MB pipelining structure to highly increase the processing capability and hardware utilization. Besides, a hybrid task pipelining scheme, a balanced schedule with block-level, MB-level, and frame-level pipelining, is used for decoder to greatly reduce the internal memory size. Combined with many bandwidth reduction techniques and DR schemes, these two system architectures are all characterized by high throughput and low system bandwidth requirement. Moreover, efficient modules are designed to support the new H.264/AVC functionalities. A parallel IME architecture comprising eight PE arrays and adder trees is applied to dramatically reduce the memory bandwidth by three-level memory hierarchy and DR scheme. A reconfigurable intra predictor generator was designed to achieve resource sharing for all intra prediction modes. A bandwidth optimized MC engine highly exploits DR between interpolation windows of neighboring blocks. The hardwired coding system can efficiently support HDTV videos with real-time constraint, which is proved by the

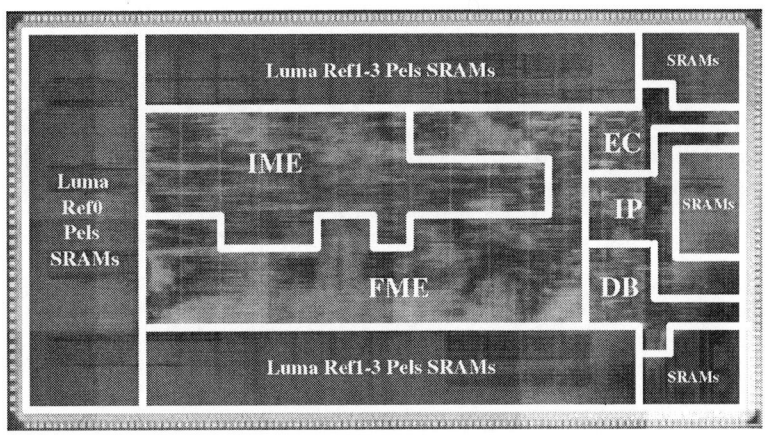

Fig. 11. Die micrograph of the H.264/AVC encoder [30].

prototype chips.

REFERENCES

[1] *Information Technology - Coding of Audio-Visual Objects - Part 2: Visual*, ISO/IEC 14496-2, 1999.

[2] *Information Technology - Generic Coding of Moving Pictures and Associated Audio Information: Video*, ISO/IEC 13818-2 and ITU-T Rec. H.262, 1996.

[3] T. Wiegand, H. Schwarz, A. Joch, F. Kossentini, and G. J. Sullivan, "Rate-constrained coder control and comparison of video coding standards," *IEEE Transactions on Circuits and Systems for Video Technology*, vol. 13, no. 7, pp. 688–703, July 2003.

[4] T. Wiegand, G. J. Sullivan, G. Bjøntegaard, and A. Luthra, "Overview of the H.264/AVC video coding standard," *IEEE Transactions on Circuits and Systems for Video Technology*, vol. 13, no. 7, pp. 560–576, July 2003.

[5] J. Ostermann, J. Bormans, P. List, D. Marpe, M. Narroschke, F. Pereira, T. Stockhammer, and T. Wedi, "Video coding with H.264/AVC: tools, performance, and complexity," *IEEE Magazine on Circuits and Systems Magazine*, vol. 4, pp. 7–28, 2004.

[6] Atul Puri, Xuemin Chen, and Ajay Luthra, "Video coding using the H.264/MPEG-4 AVC compression standard," *Signal Processing: Image Communication*, vol. 19, pp. 793–849, Oct. 2004.

[7] *Joint Video Team Reference Software JM7.3*, http://bs.hhi.de/ suehring/tml/download/, Aug. 2003.

[8] ftp://ftp.lis.e-technik.tu muenchen.de/pub/iprof/, "Iprof ftp server," .

[9] M. Takahashi and et.al., "A 60-MHz 240-mW MPEG-4 videophone LSI with 16-Mb embedded DRAM," *IEEE Journal of Solid-State Circuits*, vol. 35, pp. 1713–1721, Nov. 2000.

[10] H. Nakayama and et.al., "A MPEG-4 video LSI with an error-resilient codec core based on a fast motion estimation algorithm," in *Proceedings of IEEE International Solid-State Circuits Conference (ISSCC'02)*, Feb. 2005, vol. 2, pp. 296–512.

[11] Y.-W. Huang, T.-C. Wang, B.-Y. Hsieh, and L.-G. Chen, "Hardware architecture design for variable block size motion estimation in MPEG-4 AVC/JVT/ITU-T H.264," in *Proceedings of IEEE International Symposium on Circuits and Systems (ISCAS'03)*, 2003, pp. 796–799.

[12] J.-H. Lee and N.-S. Lee, "Variable block size motion estimation algorithm and its hardware architecture for H.264," in *Proceedings of IEEE International Symposium on Circuits and Systems (ISCAS'04)*, May 2004, vol. 3, pp. 740–743.

[13] Swee Yeow Yap and J. V. McCanny, "A VLSI architecture for variable block size video motion estimation," *IEEE Transactions on Circuit and System II*, vol. 51, pp. 384–389, 2004.

[14] Minho Kim, Ingu Hwang, and Soo-Ik Chae, "A fast vlsi architecture for full-search variable block size motion estimation in MPEG-4 AVC/H.264," in *Proc. of 2005 Asia and South Pacific Design Automation Conference*, Jan. 2005, vol. 1, pp. 631–634.

[15] T.-C Chen, Y.-W. Huang, and L.-G. Chen, "Fully utilized and reusable architecture for fractional motion estimation of H.264/AVC," in *Proceedings of IEEE International Conference on Acoustics, Speech, and Signal Processing (ICASSP'04)*, 2004, pp. V9–V12.

[16] T.-C. Wang, Y.-W. Huang H.-C. Fang, and L.-G. Chen, "Parallel 4x4 2D transform and inverse transform architecture for MPEG-4 AVC/H.264," in

Proceedings of IEEE International Symposium on Circuits and Systems (ISCAS'03), 2003, pp. 800–803.

[17] Y.-W. Huang, T.-C. Wang, B.-Y. Hsieh, T.-C. Wang, T.-H. Chang, and L.-G. Chen, "Architecture design for deblocking filter in H.264/JVT/AVC," in *Proceedings of IEEE International Conference on Multimedia and Expo (ICME'03)*, 2003, pp. 1693–1696.

[18] S.-Y Shih, C.-R. Chang, and Y.-L. Lin, "An AMBA-compliant deblocking filter IP for H.264/AVC," in *submission to IEEE International Symposium on Circuits and Systems (ISCAS'05)*, 2004.

[19] Tsu-Ming Liu, Wen-Ping Lee, and Chen-Yi Lee, "An area-efficient and high-throughput de-blocking filter for multi-standard video applications," in *Proceedings of IEEE International Conference on Image Processing (ICIP'05)*, 2005, pp. III–1044–1047.

[20] Y.-W. Huang, B.-Y. Hsieh, T.-C. Chen, , and L.-G. Chen, "Analysis, fast algorithm, and vlsi architecture design for H.264/AVC intra frame coder," *IEEE Transactions on Circuits and Systems for Video Technology*, 2004.

[21] T.-C. Chen, Y.-W. Huang, C.-Y. Tsai, , and L.-G. Chen, "Dual-block-pipelined VLSI architecture of entropy coding for H.264/AVC baseline profile," in *Proceedings of IEEE International Symposium on VLSI Design, Automation and Test (VLSI-TSA-DAT'05)*, 2005, pp. 271–274.

[22] T.-C Chen, Y.-W. Huang, and L.-G. Chen, "Analysis and design of macroblock pipelining for H.264/AVC VLSI architecture," in *Proceedings of 2004 International Symposium on Circuits and Systems (ISCAS'04)*, 2004, pp. II273–II276.

[23] Z.-L. He, C.-Y. Tsui, K.-K. Chan, and M.-L. Liou, "Low-power VLSI design for motion estimation using adaptive pixel truncation," *IEEE Transactions on Circuits and Systems for Video Technology*, vol. 10, no. 5, pp. 669–678, Aug. 2000.

[24] B. Liu and A. Zaccarin, "New fast algorithms for the estimation of block motion vectors," *IEEE Transactions on Circuits and Systems for Video Technology*, vol. 3, no. 2, pp. 148–157, Apr. 1993.

[25] S. Saponara and L. Fanucci, "Data-adaptive motion estimation algorithm and VLSI architecture design for low-power video systems," *Proc. IEE on Computers and Digital Techniques*, vol. 151, pp. 51–59, 2004.

[26] J.-C. Tuan, T.-S. Chang, and C.-W. Jen, "On the data reuse and memory bandwidth analysis for full-search block-matching VLSI architecture," *IEEE Transactions on CSVT*, vol. 12, pp. 61–72, Jan. 2002.

[27] T.-W. Chen, Y.-W. Huang, T.-C. Chen, Y.-H. Chen, C.-Y. Tsai, and L.-G. Chen, "Architecture design of H.264/AVC decoder with hybrid task pipelining for high definition videos," in *Proceedings of 2005 International Symposium on Circuits and Systems (ISCAS'05)*, 2005, pp. 2931–2934.

[28] H.-Y. Kang, K.-A. Jeong, J.-Y. Bae, Y.-S. Lee, and S.-H. Lee, "MPEG4 AVC/H.264 decoder with scalable bus architecture and dual memory controller," in *Proc. of Int. Symposium on Circuits and Systems (ISCAS'04)*, 2004.

[29] C.-Y. Tsai, T.-C. Chen, T.-W. Chen, , and L.-G. Chen, "Bandwidth optimized motion compensation hardware design for H.264/AVC HDTV decoder," in *Proceedings of 2005 International Midwest Symposium on Circuit and Systems (MWSCAS'05)*, 2005.

[30] Y.-W. Huang and etc., "A 1.3TOPS H.264/AVC Single-Chip Encoder for HDTV Applications," in *Proceedings of IEEE International Solid-State Circuits Conference (ISSCC'05)*, 2005, pp. 128–130.

ASIP Approach for Implementation of H.264/AVC

Sung Dae Kim, Jeong Hoo Lee, Chung Jin Hyun and Myung Hoon Sunwoo

School of Electrical and Computer Engineering, Ajou University
San 5, Wonchun-Dong, Yeungtong-Gu
Suwon, 443-749 Korea
Tel : 82-31-219-2390
Fax : 82-31-212-9531
e-mail : sunwoo@ajou.ac.kr

Abstract - This paper presents an Application-Specific Instruction Set Processor (ASIP) approach for implementation of H.264/AVC. The proposed ASIP has special instructions for intra prediction, deblocking filter, integer transform, etc. In addition, the proposed ASIP has hardware accelerators for inter prediction and entropy coding. Performance comparisons show a significant improvement compared with existing DSPs. The proposed hardware accelerators have small size and can support real-time video processing. Moreover, the proposed ASIP can handle various multimedia standards. The results indicate that the ASIP approach is one of promising solutions for H.264/AVC.

I. INTRODUCTION

With the rapid progress of semiconductor technology, the market of ASIP is dramatically growing. Once algorithms have been fixed, custom Application-Specific Integrated Circuit (ASIC) chips have been implemented to reduce the cost, size, and power consumption of systems. However, custom ASIC solutions have been found inadequate to upgrade standards since they should be redesigned. With the rapid increase in clock speed it has become feasible to keep the functionality entirely in a programmable DSP, greatly improving time-to-market and allowing faster changes and upgrades. However, a programmable DSP should solve the disadvantages, such as cost, size, and power consumption. ASIP can compromise advantages of custom ASIC chips and general DSP chips [1]-[5]. In other words, ASIP chips adopt high performance and low power of ASIC chips and flexibility of DSP chips.

Multimedia signal processing technology has been developed with the progress of semiconductor technology. Technology related to multimedia signal processing has been standardized as MPEG-2, MPEG-4, H.261, H.263, etc. The Joint Video Team (JVT) announced H.264/AVC in Dec. 2003 [6]. The new video coding standard H.264/AVC can provide twice as much as higher compression efficiency than MPEG-4. However, it has about 2 times more hardware complexity for a decoder, and about 10 times more hardware complexity for an encoder than the MPEG-4 visual simple profile codec [7]. Because of the hardware complexity, the H.264/AVC codec is usually implemented with ASIC chips

or multiple processors, such as ARM and multiple programmable DSPs.

In mobile communication, the implementation of H.264/AVC needs high performance, low power consumption and low cost. It also requires the flexible system which can upgrade without replacing the system. The ASIP approach can be suitable for these requirements. This paper proposes the ASIP implementation of H.264/AVC. Computation-intensive parts in H.264/AVC have been implemented using hardwired accelerators and other parts have been implemented using a programmable DSP.

This paper is organized as follows. Section II analyzes H.264/AVC and describes existing DSP instructions to implement multimedia standards. Section III proposes novel instructions and their hardware architectures and introduces hardware accelerators, and Section IV explains performance comparisons. Finally, Section V contains concluding remarks.

II. ANALYZING H.264/AVC AND EXISTING DSP INSTRUCTIONS FOR VIDEO SIGNAL PROCESSING

A. Analyzing H.264/AVC

H.264/AVC has adopted new features to improve code efficiency. The new features are as follows. H.264/AVC uses several reference frames, variable block size, and a quarter picture element in Motion Estimation (ME)/Motion Compensation (MC). These features enable the encoder to search for the best match for the current frame. However, the memory access and hardware complexity are significantly increased. The past standards should be transmitted the first frame without compression. On the other hand, the H.264/AVC encoder adopts intra prediction, which eliminates the redundancy of intra frame.

The block based structure causes blocking artifacts. H.264/AVC adopts the in-loop deblocking filter to eliminate blocking artifacts. The Exponential Golomb Coding (EGC) and Context Adaptive Variable Length Coding (CAVLC) are also newly adopted features of the H.264/AVC baseline

0-7803-9451-8/06/$20.00 ©2006 IEEE.

profile. EGC is variable length codes with a regular construction [8]. CAVLC is the method used to encode residual data, 4 x 4 blocks of transform coefficients [9]-[11].

Fig. 1 shows the operation complexity of the H.264/AVC baseline profile [12]. As shown in Fig. 1, MC takes 53% and VLC takes 18.20% of the operation complexity. Since these features require heavy computational loads, the hardware accelerator of each feature is required to implement the H.264/AVC system. Therefore, the proposed ASIP employs the hardware accelerators for these tasks, which can efficiently perform real-time video processing.

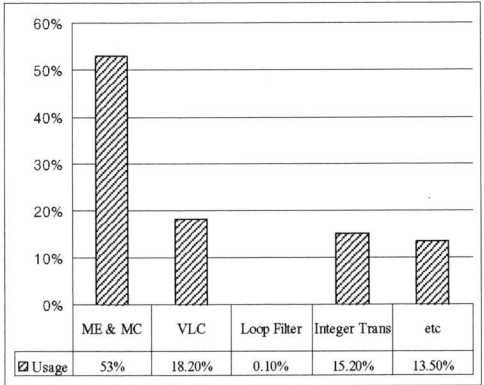

Fig. 1. Complexity analysis of the H.264/AVC baseline profile

B. Existing DSP instructions for video signal processing

Existing DSPs support various instructions to execute packed operations between two registers. These operations are used for various video signal processing, such as DCT, IDCT, ME/MC, etc. TMS320c6x of Texas Instruments supports special instructions for multimedia signal processing, such as SUBABS4, AVGx, etc. [13]. The SUBABS4 instruction calculates absolute differences of four pairs of the packed data. The AVG4 instruction calculates averages of the packed data in two registers. After additions of four packed data, four results are shifted a bit to the left for division, and 0.5 is added to each result for rounding. The TMS320c6x series also support the DOTPU4 instruction which calculates the dot product between four sets of packed 8 bit values. Fig. 2 shows the operation flow of the DOTPU4 instruction. The values in both src1 and src2 are treated as the unsigned 8 bit packed data. The 32 bit unsigned result is written into dst. Four clock cycles are required to execute this instruction.

DCT has a regular computation flow, while ME/MC and entropy coding have control based computations. TMS320c55x has a coprocessor for DCT computations, and it requires 2.8 MIPS for DCT computations to achieve the processing speed of 30 fps for the QCIF format. TMS320c6x having eight function units requires 1.1 MIPS to implement DCT of 30 QCIF fps video data using DSP instructions [14].

In entropy coding, the code word table is referred according to the number of successive zeros in the input bit stream. Moreover, packed compare operations are required. To execute these operations, TMS320c64x supports the LMBD and CMPEQ/GT/LT instructions, and the Blackfin DSP of Analog Device supports the ONES instruction [14] [15]. The LMBD instruction counts the number of zeros in a register. The CMPEQ/GT/LT instructions compare pairs of 8 bits or 16 bit packed data.

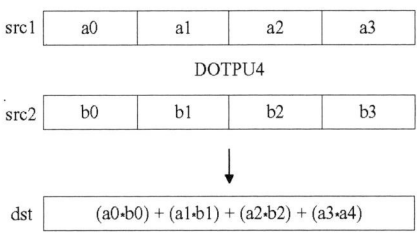

Fig. 2. DOTPU4 instruction in TMS320c64x

III. NOVEL INSTRUCTIONS AND HARDWARE ACCELERATORS

This section presents an overall architecture, new instructions and hardware accelerators for the H.264/AVC codec.

A. Overall architecture of the proposed ASIP

Fig. 3 shows the overall architecture of the proposed ASIP. The proposed ASIP consists of two parts, a programmable DSP part and a hardware accelerator part. The DSP part has a program control unit (PCU), a data processing unit (DPU), and an address unit (AU). The hardware accelerator part has an Inter Prediction Accelerator (IPA) and an Entropy Coding Accelerator (ECA). IPA consists of an ME accelerator and an MC accelerator. ECA has a CAVLC accelerator and an EGC accelerator. The hardware accelerators can operate in parallel with the DSP units.

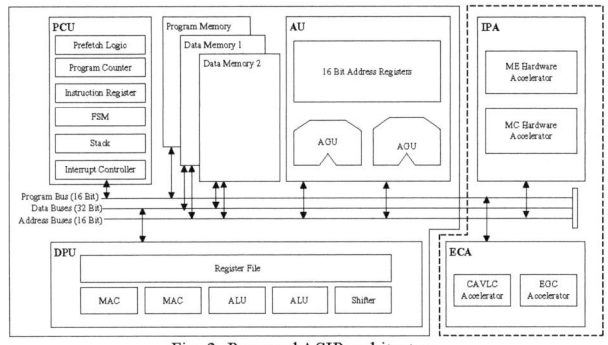

Fig. 3. Proposed ASIP architecture

PCU consists of a prefetch logic, a program counter, an instruction register, an FSM (Finite State Machine), a stack, and an interrupt controller. DPU consists of two Multiply

759

and Accumulate (MAC) units for two 16-bit by 16-bit multiplications and accumulations, two Arithmetic Logic units (ALU), a barrel shifter and a register file. AU has two address generation units (AGU) for load and store. Each of the internal word lengths is 32 bit. The instruction pipeline consists of six stages, that is, pre-fetch, fetch, decode, execute1, execute2, and execute3. The proposed ASIP has 35 arithmetic instructions, 11 logical and shift instructions, 6 program control instructions, 4 move instructions and 16 special instructions including instructions for H.264/AVC, which will be discussed next.

B. Proposed instructions for in-loop deblocking filter and intra prediction

The in-loop deblocking filter is used to eliminate blocking artifacts as mentioned in Section II. Fig. 4 shows 8 pixels of neighboring 4 x 4 blocks. The 8 pixel values are decided according to the boundary strength (bS), which represents the difference of two neighboring blocks, using p0 ~ p3 and q0 ~ q3. The equations calculating pixel values are defined in [6]. The equations can be classified into five categories as follows.

$$p2 + p1 + p0 \tag{1}$$
$$p2 + 2 \times p1 + 2 \times p0 \tag{2}$$
$$2 \times p3 + 3 \times p2 + p1 + p0 \tag{3}$$
$$2 \times p1 + p0 \tag{4}$$
$$(p0 + q0 + 1) >> 1 \tag{5}$$

p0 ~ p3 are the packed data in a register, and q0 ~ q3 are also the packed data in another register. Then, equation (1) shows additions of three packed data in one register. Equation (2) represents one bit shift left operations of two data followed by additions of three packed data in the same register. Equation (3) shows one bit shift left operation of data and a multiplication operation of data followed by additions of four packed data. Equation (4) shows one bit shift left operation of the packed data followed by an addition of two packed data. Equation (5) shows an addition of the most significant byte (MSB) of one register and the least significant byte (LSB) of the other register followed by one bit shift operation.

Even though these computations are packed operations, these operations do not occur between two registers as shown in Fig. 4, but they occur between the packed data within the same register.

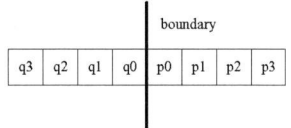

Fig. 4. Block boundary

As mentioned in Section II, the intra prediction eliminates the redundancy of intra frame and inter frame, which has few redundancies between two frames. Fig. 5 shows an identification of samples for 4 x 4 intra prediction. a ~ p in Fig. 5 are predicted using A ~ Q according to the equations defined in [6] and some of equations are represented in equation (6), where A, B, and C represent pixel values, and a pixel value is represented using 8 bits. For a 32 bit architecture, A, B, C and D are stored in one register since a ~ p and A ~ Q in Fig. 5 are 8 bit values.

$$(A + 2 \times B + C + 2) >> 2$$
$$(A + B + 1) >> 1 \tag{6}$$
$$(A + 2 \times B + 1) >> 1$$

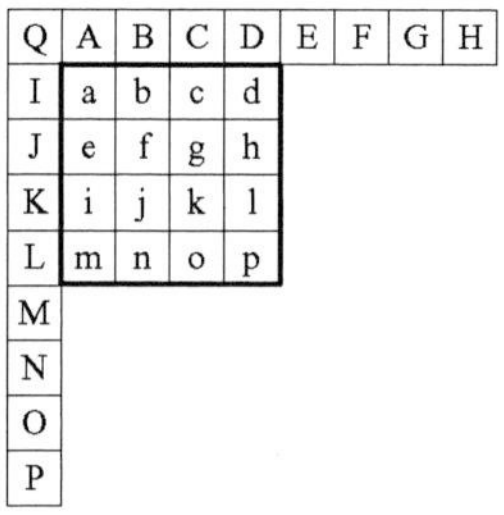

Fig. 5. Identification of samples for 4 x 4 intra prediction

As described in Section II, existing DSPs support only packed operations between two registers. A large number of instruction cycles is required to implement the in-loop deblocking filter and intra prediction with the existing packed instructions that execute packed operations between two registers. Hence, H.264/AVC may require a new instruction to execute packed operations within a register.

Fig. 6 shows the proposed three horizontal addition (hadd) instructions. Three hadd instructions are as follows. The proposed instruction in Fig. 6(a) packs a 32 bit register into four 8 bit data, adds four packed data, and then saturates the result to 8 bit data. Fig. 6(b) is similar with Fig. 6(a). However, the packed data, which is selected by a mask, is one bit shifted to the left. In Fig. 6(c), mask1 selects the data to be added, and mask2 selects the data to be shifted. Intra prediction and equations (1), (2), (4), (5) of the in-loop deblocking filter can be implemented using the proposed instructions. Equation (3) can be implemented using the packed multiplication instruction, such as the DOTPU4 instruction of TMS320c6x or the PMUL instruction of the proposed ASIP. Equations (1), (2), (4), (5) can also be implemented using the existing packed multiplication instructions. However, the DOTPU4 instruction in TMS320c6x requires four clock cycles since a multiplication should be executed.

$$dst = hadd(src)$$
$$dst = hadd(src:mask)$$
$$dst = hadd(src:mask1.mask2)$$

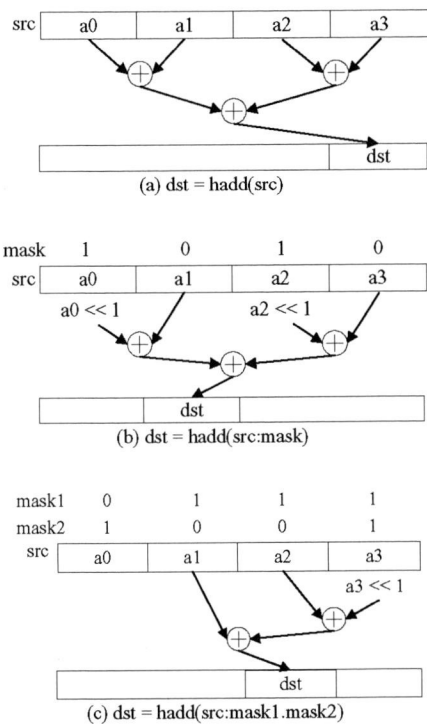

(a) dst = hadd(src)

(b) dst = hadd(src:mask)

(c) dst = hadd(src:mask1.mask2)

Fig. 6. Proposed instructions for packed additions within one register

C. Proposed instructions for integer transform

The 4 x 4 integer transform can be operated using the forward transform as shown in Fig. 7(a). The forward transform is executed with four rows of four packed data. Then, the forward transform is performed again with four columns of four packed data to get the results of the 4 x 4 integer transform. Fig. 7(b) represents an inverse transform. Similarly, the 4 x 4 inverse integer transform can be executed using the operations in Fig. 7(b).

This paper proposes novel instructions to efficiently execute the forward/inverse 4 x 4 integer transform as follows.

$$dst = fTRAN (src)$$
$$dst = iTRAN (src).$$

Each instruction performs the operations of Fig. 7(a) and (b). These instructions read a 32 bit general register in one register file, which consists of four 32 bit registers, and execute the operation flow in Fig. 7. Then, the results are written in another register file consisting of four 32 bit registers. These instructions can be implemented using the adders and eight additional 2 x 1 multiplexers.

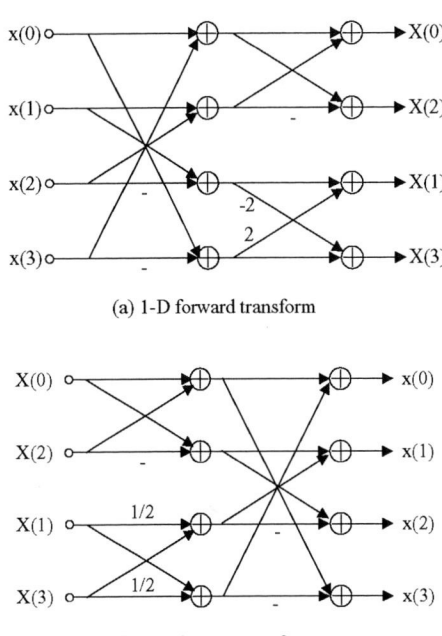

(a) 1-D forward transform

(b) 1-D inverse transform

Fig. 7. Operation flow of 4 x 4 integer transform

D. Hardware architecture for proposed instructions

Fig. 8 shows the ALU for the proposed instructions. Switching Logic 1, 2, and 3 which only consist of eight 2 x 1 multiplexers and two 1 x 2 de-multiplexers, are only the additional hardware for the proposed instructions. One ALU can perform one horizontal addition instruction and two ALUs can execute fTRAN and iTRAN instructions.

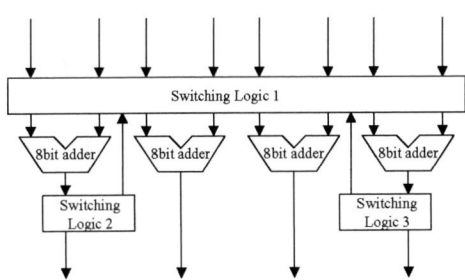

Fig. 8. ALU for the proposed instructions

E. Proposed hardware accelerator for inter prediction

As described in Section II, ME/MC should frequently access memory. From a performance point of view and a low power point of view, it is a serious problem. Thus, the sliding window method is used to alleviate this problem [16]. Fig. 9 illustrates the proposed ME operation flow.

761

7D-4

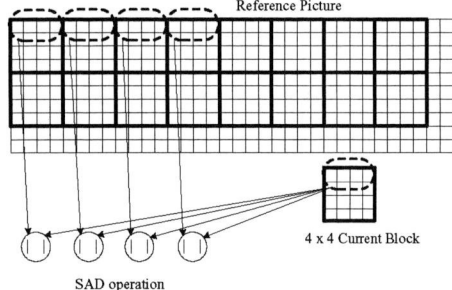

(a) ME operation in the first cycle

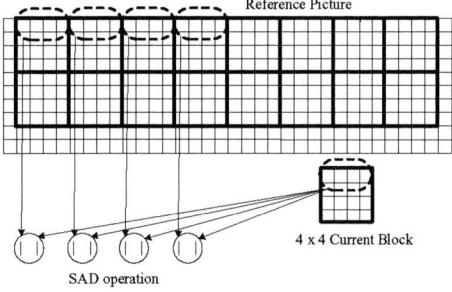

(b) ME operation in the second cycle

Fig. 9. Operation flow of the proposed motion estimation

The proposed ME architecture supports the [+16, -15] search window. In the [+16, -15] search window, 32 4 x 4 blocks exist in a row. In the first cycle, four SADs are simultaneously calculated as shown in Fig. 9(a). Next, the search window shifts left and each operation unit repeats the SAD calculation as shown in Fig. 9(b). SADs of upper four pixels of every block in a row can be obtained after 8 cycles and 32 SADs are stored in buffers. SADs of the second upper are calculated in the same way, and 32 SADs are accumulated with the 32 SADs in buffers, respectively. Then, after 32 cycles, 32 SADs of 4 x 4 blocks can be obtained.

Fig. 10 shows the ME computation flows of existing architectures and the proposed architecture. In Fig. 10(a), the pixel values in the dotted block should be fetched again to calculate SAD of the dotted block after SADs of two adjacent blocks (block 1 and block 2) are obtained. However, if a 4 x 4 block is shifted pixel by pixel as shown in Fig. 10(b), the data in the dotted block in Fig. 10(a) can be reused.

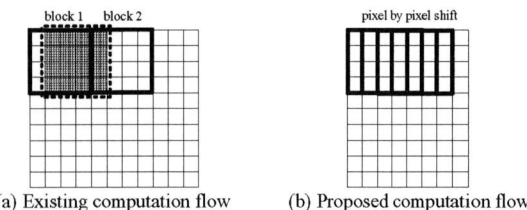

(a) Existing computation flow   (b) Proposed computation flow

Fig. 10. ME computation flow

F. Proposed hardware accelerator for entropy coding

The encoder of CAVLC finds the value of the codeword and the length of the codeword in memory according to the data. Therefore, the efficient memory address generator is needed. The decoder of CAVLC is usually implemented with a lookup table. In the decoding process, the level of the nonzero coefficient decoding is an iterative method, which can be implemented without a lookup table.

A generic decoding process for the level of the nonzero coefficient is as follows. First, the decoder obtains the number of successive ones in the input bit stream. Next, the decoder calculates the current symbol length and decodes the current symbol. Finally, the decoder updates the table information used for next symbol decoding. The decoder cannot decode the next symbol until the table information is decided. The generic level of the nonzero coefficient decoding process is shown in Fig. 11(a). Table updating is decided whether the current symbol value is more than the threshold value. Since each table's threshold value has a regular form, we can update the table before current symbol decoding. Hence, the Level Decode stage in the current symbol and the First 1 detect stage in the next symbol can be executed in parallel. Fig. 11(b) shows the proposed level of the nonzero coefficient decoding process. As you can see, we can reduce the computation cycles for level decoding.

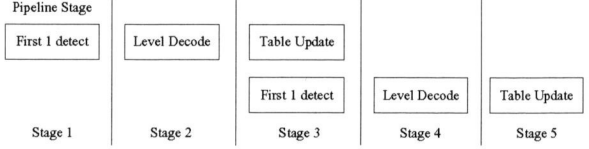

(a) Generic level of the nonzero coefficient decoding flow

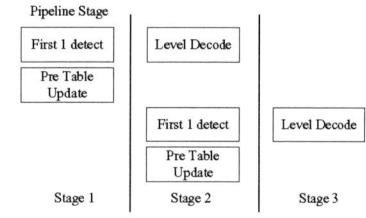

(b) Proposed level of the nonzero coefficient decoding flow

Fig. 11. Comparison of flows for the level of the nonzero coefficient decoding

If a ROM based look up table is used to implement the runs of zero decoder, the area cost of the table will be expensive. Hence, the ZTEBA (Zero-left Table Elimination by Arithmetic) method was introduced by Chang [10]. In our design, we improve the ZTEBA method. We eliminate the SUB unit and the Saturation unit used in [10] by adding some multiplexers and wires. Fig. 12 shows the block diagram for the proposed runs of zero decoder.

762

2006 Asia and South Pacific Design Automation Conference

7D-4

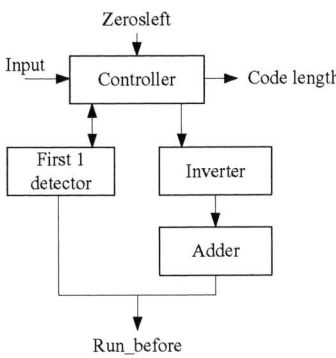

Fig. 12. Proposed runs of zero decoder block diagram

IV. PERFORMANCE COMPARISONS

H.264/AVC can be implemented using ASIP and hardware accelerators. Several core blocks for generating an intra predictor and an in-loop deblocking filter are coded using the proposed application specific instructions and the same blocks are also coded using the existing instructions of TMS320c64x. The proposed architecture can reduce the number of clock cycles for several core blocks for generating an intra predictor about 40% than TMS320c6x. Moreover, the total number of clock cycles to execute the in-loop deblocking filter can be reduced about 20 ~ 25% than TMS320c6x. TMS320C64x supports the DOTPU4 instruction that executes packed multiplications of two registers and adds four results in four cycles. The computation cycles of TMS320c64x can be reduced, since it supports the DOTPU4 instruction. Hence, other DSPs, which do not support the special instruction, require more instructions.

The fTRAN and iTRAN instructions can be executed in one cycle. Hence, 12 clock cycles are required to execute the 4 x 4 integer transform using the proposed instructions and about 1,140,480 instructions for 30 frames ((24 cycles x 16 blocks) x 99 macro blocks x 30 frame) are required for QCIF images, since a QCIF image has 99 16 x 16 macro blocks. Table I shows the number of the required instructions for 30 frames on existing DSPs [14] [17] and the proposed ASIP. The proposed ASIP having the special instructions can be more efficient than the implementation using instructions of TMS320c55x (SW) and using a coprocessor of TMS320c55x (HW) for the integer transform. TMS320c64x has a VLIW architecture and has eight function units while the proposed ASIP architecture requires only two 32 bit adders.

Table I. Performance comparisons of the 4 x 4 integer transform

	TMS320c55x (SW) [17]	TMS320c55x (HW) [17]	TMS320c64x [14]	Proposed ASIP
MIPS	12.8	2.8	1.1	1.1

The proposed hardware accelerators have been modeled by Verilog HDL and synthesized using the Samsung SEC 0.18 μm standard cell library by the Synopsys Design Compiler. The proposed ME accelerator can significantly reduce the gate count and the required computation cycles compared with the Samsung architecture [18] and can support much larger search range than the Amphion architecture [19]. Table II shows the comparisons among Samsung, Amphion and our architecture. The Samsung architecture has 64 Processing Elements (PEs) and can support much larger search range than the other architectures. However, it requires much larger computation cycles than the other architectures. The total gate count and the required computation cycles of Amphion are comparable with our architecture. However, the search range of Amphion is much smaller than our architecture.

The proposed hardware accelerator for CAVLC takes average 368 clock cycles for a macro block. In order to achieve the real-time processing requirement for H.264/AVC decoding with HD1080i format, the proposed design should run over 90 MHz. The proposed design can support real-time processing since the maximum operating frequency of the proposed design is about 130 MHz.

Table II. Performance comparisons of the hardware ME architectures

	Clock cycles / frame	Search range	Supported block size	Gate counts
Samsung [18]	12,103,740	H :[-64, +63] V :[-32, +31]	Variable block support	64 PEs
Amphion [19]	406,077	[-8, +7]	Variable block support	61K
Proposed architecture	**456,192**	**[-16, +15]**	**Variable block support**	**76K**

Fig. 13 shows computation times of DSP, ASIC, and ASIP implementations according to the profiling results. Fig. 13(a) shows the computation times of the DSP implementation. If we assume that a single core is used to implement the H.264/AVC algorithm, DSP serially executes all of the algorithm blocks.

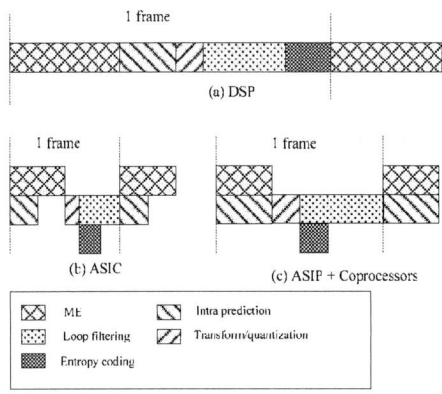

Fig. 13. Computation times of various implementations

763

Fig. 13(b) shows the computation times of the ASIC implementation. Each block is executed using the dedicated hardware. However, all of the blocks cannot be executed in parallel, since the ME block needs the reconstructed results of the previous frame and the transform block uses ME results. Fig. 13(c) shows the ASIP having accelerators implementation. ME/MC and entropy coding are implemented using the accelerators. The ASIP having accelerators implementation requires more computation times than the ASIC implementation. However, it requires much less computation times than DSP and can support various profiles and standards.

V. Conclusions

This paper has presented efficient instructions to implement the in-loop deblocking filter, intra prediction and integer transform of H.264/AVC. This paper has proposed the hardware accelerators for ME/MC and entropy coding. Three hadd instructions can execute various packed additions within a register. Performance comparisons have shown that the number of clock cycles can be reduced about 20 ~ 25% compared with the existing DSP for the in-loop deblocking filter. The fTRAN and iTRAN instructions can perform the 4 x 4 integer transform in 12 clock cycles. The integer transform can be implemented using much smaller hardware size compared with existing DSPs. The proposed hardware accelerators can efficiently perform ME/MC and entropy coding of H.264/AVC and require minimal hardware. Since the proposed ASIP having the hardware accelerators can concurrently operate, it can handle real-time video processing and can support various multimedia algorithms.

Acknowledgements

This work was supported in part by the NRL (National Research laboratory) program of MOST (Ministry of Science & Technology), in part by the ITSOC program of MIC (Ministry of Information and Communication), and in part by IDEC.

References

[1] Jae S. Lee, Young S. Jeon, and Myung H. Sunwoo, "Design of new DSP instructions and their hardware architecture for high-speed FFT," in *Proc. IEEE Workshop on Signal Processing Syst.,* Sept. 2001, pp. 80-90,.

[2] J. Glossner, J. Moreno, M. Moudgill, J. Derby, E. Hokenek, D. Meltzer, U. Shavadron, and M. Ware, "Trends in compilable DSP architecture," in *Proc. IEEE Workshop on Signal Processing Syst.,* 2000, pp. 181-199.

[3] Jeong H. Lee, Jong H. Moon, Kyung L. Heo, Myung H. Sunwoo, Sung K. Oh, and In H. Kim, "Implementation of Application Specific DSP for OFDM Systems," in *Proc, IEEE IEEE Int. Symp. Circuit Syst.,* May 2004.

[4] Suk Hyun Yoon, Jong Ha Moon, and Myung Hoon Sunwoo, "Efficient DSP Architecture for High-Quality Audio Algorithms," in *Proc. IEEE Int. Symp. Circuits Syst.,* May 2005

[5] Kim, S.D., Lee, J.H, Yang, J.M., Sunwoo, M.H., and Oh, S.K., "Novel Instructions and Their Hardware Architecture for Video Signal Processing," in *Proc. IEEE Int. Symp. Circuits Syst.,* May 2005

[6] Draft ITU-T Recommendation and Final Draft International Standard of Joint Video Specification (ITU-T Rec. H.264/ISO/IEC 14496-10 (E) AVC). July, 2004.

[7] J. Ostermann, T. Wedi, et al.,"Video coding with H.264/AVC: tools, performance, and complexity," *IEEE Circuits and Systems Magazine,* vol. 4, pp. 7-28, 2004.

[8] Wu Di, Gao Wen, Hu Mingzeng and Ji Zhenzhou, "An Exp-Golomb encoder and decoder architecture for JVT/AVS," in *Proc. 5th International Conference on ASIC,* 21-24 Oct. 2003 vol. 2, pp. 910-913.

[9] Gisle Bjontcgaard and Karl Lillcvold, "Context-adaptive VLC (CAVLC) coding of coefficients," Doc. JVT-028, JVT of ISO/IEC MPEG & ITU-T VCEG 3rd Meeting, Virginia, USA, May. 2002.

[10] Hsiu-Cheng Chang, Chien-Chang Lin, and Jiun-In Guo, "A Novel Low-Cost High-Performance VLSI Architecture for MPEG-4 AVC/H.264 CAVLC Decoding," in *Proc. IEEE Int. Symp. Circuits Syst.,* May 2005.

[11] Yeong-Kang Lai, Chih-Chung Chou, and Yu-Chieh Chung, "A simple and cost effective video encoder with memory-reducing CAVLC," in *Proc. IEEE Int. Symp. Circuits Syst.,* May 2005.

[12] Woong IL Choi, Byeungwoo Jeon and Jechang Jeong, "Fast motion estimation with modified diamond search for variable motion block sizes," in *Proc. International Conference on Image Processing,* Sept. 2003, vol. 3, pp. 14-17.

[13] *TMS320C6000 CPU and Instruction Set Reference Guide,* Texas Instruments Inc., Dallas, TX, 2000.

[14] *TMS320C64x Image/Video Processing Library,* Texas Instruments Inc., Dallas, TX, 2003.

[15] *BlackfinTM DSP Instruction Set Reference,* Analog Device Inc., Norwood, Mass. 2002.

[16] Thomas Wiegand, Xiaozheng Zhang, and Bernd Girod, "Long-Term Memory Motion-Compensated Prediction," *Trans. Circuit Syst. Video Technol.,* vol. 9, no. 1, pp. 70-84, Feb. 1999.

[17] *TMS320C55x Hardware Extensions for Image/Video Applications Programmer's Reference,* Texas Instruments Inc., Dallas, TX, 2002

[18] Jae H. Lee and Nam S. Lee, "Variable Block Size Motion Estimation Algorithm and Its Hardware Architecture for H.264/AVC," in *Proc. IEEE Int. Symp. Circuits Syst.,* May 2004.

[19] Swee Yeow Yap and john V. McCanny, "A VLSI Architecture for Variable Block Size Video Motion Estimation," *Trans. Circuit Syst. Video Technol.,* vol. 51, no. 7, July 2004.

2006 Asia and South Pacific Design Automation Conference

8A-1

Fast Substrate Noise-Aware Floorplanning with Preference Directed Graph for Mixed-Signal SOCs

Minsik Cho, Hongjoong Shin and David Z. Pan

Dept. of ECE, The University of Texas at Austin, Austin, TX 78712

thyeros@cerc.utexas.edu, unishin@cerc.utexas.edu, dpan@ece.utexas.edu

Abstract—**In this paper, we introduce a novel substrate noise estimation technique during early floorplanning, based on the concept of Block Preference Directed Graph (BPDG) and the classic Sequence Pair (SP) floorplan representation. Given a set of analog and digital blocks, the BPDG is constructed based on their inherent noise characteristics to capture their preferred relative orders for substrate noise minimization. For each sequence pair generated during floorplanning evaluation, we can measure its violation against BPDG very efficiently. We observe that by simply counting the number of violations obtained in this manner, it correlates remarkably well with accurate but computation-intensive substrate noise modeling. Thus, our BPDG-based model has high fidelity to guide the substrate noise-aware floorplanning and layout optimization, which become a growing concern for mixed-signal/RF system on chips (SOC). Our experimental results show that the proposed approach is over 60x faster than conventional floorplanning with even very compact substrate noise models. We also obtain less area and total substrate noise than the conventional approach.**

I. INTRODUCTION

Continuing demand for data and telecommunication application is driving tighter integration of many heterogeneous functions into a single system-on-chip. These components can be pre-designed IP cores of different natures such as sensitive front-end RF circuits, high-precision analog/mixed-signal circuits, and high-performance digital circuits. Therefore, the interference between these heterogeneous components has to be considered during layout planning and optimization [1]. A key interference is the substrate noise caused by large amount of switching activities in high speed digital cores, to the analog/RF components. It may degrade the reliability and performance of these sensitive analog/mixed-signal/RF IPs [2]. The problem is becoming a growing concern due to higher clock frequency, more accurate analog precision, deeper technology scaling, and the integration of front-end RF with digital blocks [3]–[5]. Many effects that corrupt RF signal such as DC offset, oscillator pulling and pushing, local oscillator leakage can be traced to the substrate-coupled noise [2].

Therefore, fast yet accurate evaluation and optimization of substrate noise in physical design has become a crucial part of mixed-signal SOC designs, in order to avoid expensive over-design and excessive design iterations. A key step of such layout optimization is the floorplanning stage [6]. Although

This work is supported in part by SRC under contract 2005-TJ-1321, IBM Faculty Award, and equipment donations from Intel.

many works have been done in modeling and simulation of substrate noise [2], [3], [5], [12]–[15], there is not much in the literature on substrate noise optimization in early floorplanning stage. Lin et al. [7] proposed an optimization technique that incorporates substrate noise minimization into optimization loops. This technique, however, requires detailed and expensive circuit simulations to estimate the coupled substrate noise. Blakiewicz et al. [8] proposed a floorplanning algorithm with substrate noise as a cost function. A more scalable substrate model with frequency-dependent sensitivity function of analog and digital blocks is used, but still it requires significant computational overhead to evaluate the substrate noise cost function during floorplanning.

In this paper, we propose a novel concept of block preference directed graph (BPDG) to overcome the modeling bottleneck for substrate noise-aware floorplanning. Our BPDG-based model has high fidelity compared with accurate but much more expensive substrate noise modeling. Thus, it is suitable to guide MS-SOC floorplanning. The major contributions of this paper include the following.

- We introduce the novel concept of block preference directed graph (BPDG) to represent the preferred relative block locations in floorplanning. In BPDG, all the preferences are decided to minimize the substrate noise, and each preference is specified as a directed edge.
- We propose a fast substrate noise estimation algorithm by combining BPDG and sequence pair. We simply count how many preferences in BPDG are not held in a sequence pair with simple bitwise-OR operation.
- We show that our approach has surprisingly high fidelity to the substrate noise calculated by a most recent, accurate substrate noise model [5].
- We propose a fast substrate noise-aware floorplanning algorithm with BPDG and sequence pair. Our experimental results show the proposed approach is significantly (at least 60x) faster than a conventional simulation-based, substrate noise-aware floorplanning.

The rest of the paper is organized as follows. In Section II, preliminaries are described. In Section III, the concept of block preference directed graph is introduced. The fast substrate noise estimation algorithm is proposed in Section IV, and the overall floorplanning flow is described in Section V. Experimental results are discussed in Section VI. Section VII concludes this paper with future work.

0-7803-9451-8/06/$20.00 ©2006 IEEE.

765

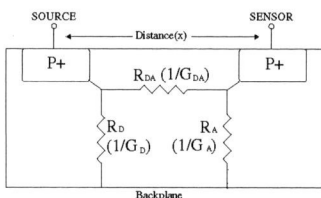

Fig. 1. Macromodel for the substrate

II. PRELIMINARIES

A. Sequence Pair and Block Alignment

A sequence pair [9] is a pair of sequences of n elements representing a list of n blocks. Two sequences specify the geometric relations between each pair of blocks. For example, (..A..B.., ..A..B..) means that a block A is to the left of a block B, and (..B..A.., ..A..B..) implies that A is below B. A sequence pair can be translated into a floorplan by horizontal and vertical constraint graph [9]. Conditions for block alignments in sequence pair are studied in [10]. H/V alignment constraints and abutting constraint between blocks are introduced and applied for performance-aware floorplanning.

B. Substrate Noise Model

Several techniques have been proposed to model and analyze substrate noise accurately in integrated circuit level [11]–[13]. However, these techniques require the detailed implementation information in transistors and time-intensive transistor-level simulation. In this paper, we use compact substrate coupling model [5] to evaluate an instance of floorplan from conventional approach, and to verify the final floorplan. The model in [5] is known to be highly scalable with dimensions and separations.

A two-port lumped resistor network, modeling substrate is illustrated in Fig. 1. The resistance R_{DA}, models the coupling between two blocks, and R_A and R_D model the coupling from the blocks to the backplane. The resistances, R_{DA}, R_A and R_D can be derived from the scalable macromodel, which is based on Z-parameters.

$$Z = \begin{bmatrix} Z_{11} & Z_{12} \\ Z_{21} & Z_{22} \end{bmatrix} = \frac{1}{\triangle} \begin{bmatrix} G_D + G_{DA} & G_{DA} \\ G_{DA} & G_A + G_{DA} \end{bmatrix} \tag{1}$$

where $\triangle = G_A G_{DA} + G_D G_{DA} + G_A G_D$ and any Z_{ij} can be calculated with equations in [5], [14], [15].

The coupling gain (propagation gain) of the substrate can be calculated from the value of resistors in the two-port lumped network shown in Fig. 1. The coupling gain of i-th digital block to j-th analog block, $CG_{i,j}$ is given by:

$$CG_{i,j} = \frac{R_A}{R_A + R_{DA}} = \frac{G_{DA}}{G_{DA} + G_A} = \frac{Z_{12}}{Z_{22}} \tag{2}$$

Although $CG_{i,j}$ exhibits frequency-dependent characteristics, it is constant under a few gigahertz [4]. In this paper, we assume that the bands of interest are within this limit. The quantity of the substrate noise can be estimated using frequency-dependent characteristics of noise source and sensor block, and a simple analytical formula based on $CG_{i,j}$. The

substrate noise of j-th analog block from switching of i-th digital block, $N_{i,j}$ can be approximated by [8]:

$$N_{i,j} = (CG_{i,j}) \cdot \sqrt{\int_0^\infty (S_i(f) \cdot H_j(f))^2 df} \tag{3}$$

where $S_i(f)$ and $H_j(f)$ are Power Spectral Density (PSD) of noise source and transfer function of noise sensor respectively. Also, the total noise from all digital blocks is:

$$N_{total} = \sum_i \sum_j N_{i,j} \tag{4}$$

As shown in Eqn. (3), $CG_{i,j}$ is scaled by average power of noise with regard to the frequency. The frequency-dependent noise generated by a digital block, $S_i(f)$ is shaped by the transfer function of the noise sensor. The integration of the shaped power of noise represents the quantity of noise injected into analog block, when $CG_{i,j}$ is equal to 1. We use a piecewise-linear approximation of PSD to estimate $S_i(f)$, and Power/Ground bounce limits to determine its parameters.

III. BPDG: BLOCK PREFERENCE DIRECTED GRAPH

The substrate noise model in Section II-B is one of the most compact models with high scalability and accuracy. However, it is still computationally expensive to perform a substrate noise estimation even with such an efficient model during simulated annealing-based floorplanning, because every noise estimation after a movement requires the accurate location of every block (substrate noise is exponentially sensitive to geometric distance [5], [14], [15]), whereas area and wire-length can be calculated approximately. Furthermore, computing noise itself with Eqn. (2, 3) is not computationally trivial.

For fast substrate noise estimation, a new concept of *block preference directed graph*, BPDG is introduced and described in this section. BPDG represents preferred relative locations of blocks to guide substrate noise-aware floorplanning. BPDG construction consists of three steps.

1) A table of substrate noises ($N_{i,j}$) between all analog and digital blocks is constructed.
2) Analog block orderings and digital block orderings are created separately with the substrate noise table.
3) BPDG is constructed by finding common orders from block orderings.

The following subsections illustrate each step with detailed examples in Table I and Fig. 2, 3, 4.

A. Substrate Noise Table Construction

Since substrate noise is heavily related to the distance between blocks, we assume that the nominal distance is fixed to normalize the effect of distance. With such fixed distance, the substrate noise between a digital block and an analog block purely depends on frequency coupling and geometric properties like area and perimeter [5], [14], [15]. Under such condition, for each digital block D_i and analog block A_j, a substrate noise on A_j due to D_i, $N_{i,j}$ can be computed from Eqn. (3). Table I shows an example of substrate noise table of between digital blocks ($D_1, D_2, D_3, D_4, D_5, D_6$) and analog blocks ($A_1, A_2, A_3$).

TABLE I
Substrate Noise Table

	D_1	D_2	D_3	D_4	D_5	D_6
A_1	5	2	6	3	10	1
A_2	2	1	3	10	8	5
A_3	3	8	7	11	9	12

B. Analog Block Ordering

Based on the substrate noise table, analog blocks can be sorted for each digital block by the *descending* order of substrate noise. Consider the example in Table I. Analog block A_1, A_3 and A_2 can be ordered by the substrate noise from D_1, as $N_{1,1} = 5 > N_{1,3} = 3 > N_{1,2} = 2$. The other five orderings can be obtained in the same manner, as shown in Fig. 2. Basically, this ordering pushes more noise-sensitive analog blocks to the head, and less sensitive ones to the tail of block orderings.

$$D_1 : A_1 \leftarrow A_3 \leftarrow A_2 \quad , \quad D_2 : A_3 \leftarrow A_1 \leftarrow A_2$$
$$D_3 : A_3 \leftarrow A_1 \leftarrow A_2 \quad , \quad D_4 : A_3 \leftarrow A_2 \leftarrow A_1$$
$$D_5 : A_1 \leftarrow A_3 \leftarrow A_2 \quad , \quad D_6 : A_3 \leftarrow A_2 \leftarrow A_1$$

Fig. 2. Analog block orderings

C. Digital Block Ordering

In similar way, digital blocks can be sorted for each analog block by the *ascending* order of substrate noise. Again considering the example in Table I, digital block D_6, D_2, D_4, D_1, D_3 and D_5 can be ordered such that the substrate noise on A_1 is increasing. All digital block orderings are shown in Fig. 3. This pushes less aggressive blocks to the head and more aggressive blocks to the tail of block orderings.

$$A_1 \ : \ D_6 \leftarrow D_2 \leftarrow D_4 \leftarrow D_1 \leftarrow D_3 \leftarrow D_5$$
$$A_2 \ : \ D_2 \leftarrow D_1 \leftarrow D_3 \leftarrow D_6 \leftarrow D_5 \leftarrow D_4$$
$$A_3 \ : \ D_1 \leftarrow D_3 \leftarrow D_2 \leftarrow D_5 \leftarrow D_4 \leftarrow D_6$$

Fig. 3. Digital block orderings

D. BPDG Construction

The two key ideas behind BPDG construction are: first, to find common block order patterns in order to minimize the substrate noise; second, to make less aggressive digital blocks and less sensitive analog blocks interfaced. An analog BPDG and a digital BPDG are constructed with analog and digital block orderings by Algorithm 1. The reason to create a virtual vertex in Algorithm 1 is to force analog blocks isolated from digital blocks, which is common in real mixed-signal design.

Consider the final BPDG in Fig. 4 as an example. Since A_3 is before A_2 for all analog block orderings in Fig. 2, vertices A_3 and A_2 are inserted into G_a (Analog BPDG), and connected with a directed edge. Again, vertices D_1 and D_3

Algorithm 1 BPDG Construction

Input: Analog, Digital block orderings O_a and O_d
1: Analog BPDG $G_a \leftarrow \phi$, Digital BPDG $G_d \leftarrow \phi$
2: **for** each analog block A_i, A_j, i $\neq$j **do**
3:     **if** A_i is before A_j in all O_a **then**
4:        Add a directed edge from A_j to A_i to G_a
5:     **end if**
6: **end for**
7: **for** each digital block D_i, D_j, i $\neq$j **do**
8:     **if** D_i is before D_j in all O_d **then**
9:        Add a directed edge from D_j to D_i to G_d
10:     **end if**
11: **end for**
12: Add a virtual vertex D_0 for G_a to G_d
13: Add directed edges from all root vertices to D_0
Output: G_d

---

are inserted into G_d (Digital BPDG) with a directed edge from D_3 to D_1, as D_1 is before D_3 for all digital block orderings in Fig. 3. Note that A_1 does not have any edge, as there is no common order regarding A_1 in Fig. 2, and D_6 only has an edge to D_0 for analog-digital separation. Lastly, G_a and G_d are merged via virtual vertex D_0.

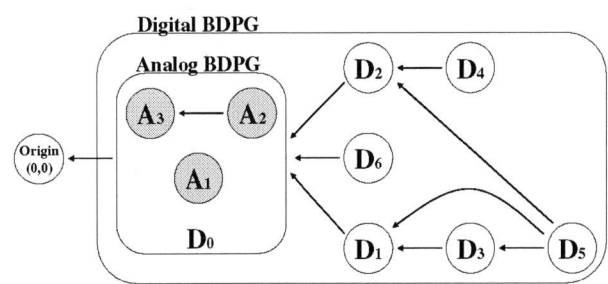

Fig. 4. Block preference directed graph (BPDG)

IV. Substrate Noise Estimation with BPDG

The BPDG in Section III can be used to estimate substrate noise quickly by comparing it against a sequence pair which is one of the most popular floorplan representations. In this section, a theorem which returns the number of violations against preferences in a BPDG from a sequence pair is presented, and its high fidelity to substrate noise is shown. The number of violations is highly correlated to substrate noise quantity; intuitively, more violations indicate more noise, because each directed edge from a block B_a to a block B_b in the BPDG means that B_b is preferred to be closer to the origin (left-bottom corner of floorplan) than B_a to reduce substrate noise.

A. Sequence Pair with BPDG

In [10], the concept of *strictly ahead* is defined for block alignment in a floorplanning with sequence pair. When there is no block between B_a and B_b in a floorplan, B_a is strictly ahead of B_b. Fig. 5(a) shows a floorplan with several blocks. In this example, B_a is strictly ahead of B_1, B_2, B_3 and B_4. In fact, *strictly ahead* is a necessary condition for two blocks to be abutted (only B_1 and B_3 are abutted to B_a). In the following, we introduce several definitions based on *strictly ahead* for easier explanation of this section.

DEFINITION α. Given a block B_a and a sequence pair (P, N), all the blocks which are both strictly ahead of B_a and below B_a form a **strictly below set** of B_a.

DEFINITION β. Given a block B_a and a sequence pair (P, N), all the blocks which are both strictly ahead of B_a and to the left of B_a form a **strictly left set** of B_a.

DEFINITION γ. Given a block B_a and a sequence pair (P, N), any block in a strictly below/left set of B_a and abutting to B_a is a **reference block**.

In Fig. 5(a), B_2, B_3 and B_4 are strictly below set of B_a, because they are strictly ahead of B_a as well as below B_a. Also, B_3 is a reference block of B_a. One intuitive property of the reference block is stated in Lemma α referring to [10].

Lemma α: *If a block B_a has a strictly below/left set S, there must exist some reference block B_x in S under a completely packed floorplan.*

Based on Lemma α, the relative locations of two blocks can be determined. Consider a specific floorplan in Fig. 5(b) where B_a is to the left of B_b, and B_x is a reference block of B_a. It can be easily proved that if a block such as B_x exists below B_b, it is guaranteed that B_a has a shorter distance to the origin (0,0) than B_b. This key idea to compare the relative location of two blocks **conservatively** with a sequence pair is presented as Theorem 1 by extending Theorem 1 and 3 in [10]. Note that the conditions 1) and 2) of Theorem 1 are corresponding to Fig. 5(b) and (c) respectively.

Theorem 1: *Let S_b be a strictly below set of B_a and S_l a strictly left set of B_a. A block B_a is guaranteed to have shorter distance to the left bottom corner than a block B_b under a completely packed floorplan, if either of following conditions is satisfied.*

1) *for any block B_s in S_b, a sequence pair (P, N) is $(..B_a X_1 B_b X_2 B_s.., ..B_s Y_1 B_a..B_b..)$.*
2) *for any block B_s in S_l, a sequence pair (P, N) is $(..B_s X_3 B_b X_4 B_a.., ..B_s Y_2 B_a..B_b..)$.*

Thus, when a sequence pair (P, N) and a BPDG G are given, the preferred relative block location (an edge) in G can be examined with Theorem 1 to see if such preference is held in (P, N). Theorem 1 can be further simplified into Theorem 2 with the longest common string (LCS) search by narrowing down the size of subsequences to scan.

Theorem 2: *A block B_a is guaranteed to have shorter distance to the left-bottom corner than a block B_b under a completely packed floorplan, if either of following conditions is satisfied.*

1) *there is no block B_s satisfying $LCS(X_1, Y_1)=\phi$ in a sequence pair $(P, N)=(..B_a X_1 B_s..B_b.., ..B_s Y_1 B_a..B_b..)$.*
2) *there is no block B_s satisfying $LCS(X_2, Y_2)=\phi$ in a sequence pair $(P, N)=(..B_b..B_s X_2 B_a.., ..B_s Y_2 B_a..B_b..)$.*

The following sequence pairs show examples with the BPDG in Fig. 4. Note that the blocks one need to pay attention to are marked with *, and we highlight one violation, even though there can be more.

- $(D_0 D_6 D_1 D_2^* D_3 D_4^* D_5, D_4^* D_0 D_6 D_1 D_2^* D_3 D_5)$
  This case has $D_2 \leftarrow D_4$ violation, because D_2 is after D_4 in the second sequence which does not match either one of required sequence pair patterns in Theorem 1.
- $(D_4 D_5 D_1^* D_2 D_6 D_0^* D_3^*, D_0^* D_5 D_4 D_1^* D_2 D_6 D_3^*)$
  This case has $D_1 \leftarrow D_3$ violation. D_0 is below D_1 and $LCS(D_2 D_6, D_5 D_4) = \phi$ But, D_0 is before D_3 in the first sequence which violates the required sequence pair pattern in condition 1) of Theorem 1.

The significance of Theorem 1 and 2 is that a geometric distance from the origin to any two blocks in a sequence pair can be compared conservatively *without other geometric information*. Thus, whether an edge (preference) in a BPDG is held in a sequence pair can be checked extremely efficiently. Note that in a real implementation, bitwise-OR can be used instead of LCS computation, since we are only interested in whether there is a common sequence.

B. Fidelity of BPDG

In order to measure the fidelity of BPDG-based model for substrate noise estimation, *ami33* from MCNC benchmarks [16] was simulated with carefully generated noise characteristics. Fig. 6 shows the normalized substrate noise on all analog blocks by the number of violations counted by Theorem 2 with different total number of violations. It shows that normalized substrate noise increases near linearly as the

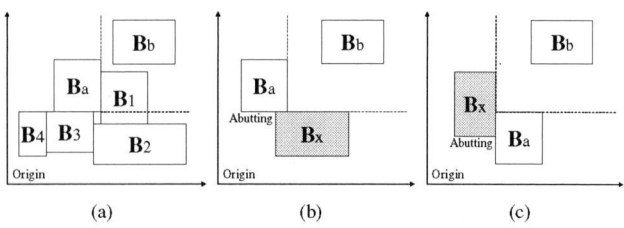

(a)   (b)   (c)

Fig. 5.   Floorplan examples

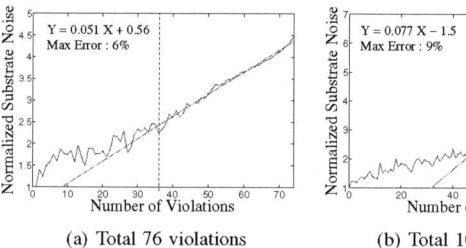

(a) Total 76 violations   (b) Total 100 violations

Fig. 6.   Number of violations vs. Substrate noise

TABLE II

EXPERIMENTAL RESULTS

Name	Algorithm Description	Cost[a] Function	Input (node)	Area (mm^2)	White Space(%)	Normalized Noise	CPU (sec)	Overhead (%)	
								CPU	Area
parq	Pure Parquet with Sequence Pair	$\frac{A}{A_r}$	ami33	1.19	3.2	821.1	0.8	0.0	0.0
			ami49	36.7	3.6	1629.9	2.6	0.0	0.0
			n75	42.04	4.0	3559.9	8.6	0.0	0.0
			n100	18.86	5.1	4697.5	24.6	0.0	0.0
bpdg[b]	BPDG with Sequence Pair	$0.6\frac{A}{A_r} + 0.4\frac{NV}{NV_r}$	ami33	1.24	6.9	121.2	0.9	15.5	3.8
			ami49	37.9	7.1	72.2	2.7	6.6	3.4
			n75	43.12	6.6	173.1	9.2	7.1	2.6
			n100	19.22	6.9	202.5	26.4	7.1	1.9
modl	Substrate Noise Model with Sequence Pair	$0.6\frac{A}{A_r} + 0.4\frac{SN}{SN_r}$	ami33	1.23	6.1	143.9	73.0	8782.5	2.8
			ami49	38.4	8.4	90.8	158.3	6103.3	4.6
			n75	44.08	9.1	322.4	666.9	7692.5	4.9
			n100	19.94	11.1	696.1	1956.3	7844.1	5.8

[a] A, NV and SN denote total area, the number of violations and total substrate noise on analog blocks respectively. A_r, NV_r and SN_r are the reference values of A, NV and SN respectively.

[b] for **bpdg**, each side of the virtual analog block is inflated by 0.6% as a whitespace(guard ring) insertion.

number of violations increases. Notice that the range over 50% of maximum violations shows high fidelity with less than 6% error in Fig. 6(a), and 9% in Fig. 6(b). Since the typical number of violations during simulated annealing falls in this high fidelity range, the number of violations in sequence pair is a good indicator of substrate noise. Thus, by comparing BPDG of Section III against sequence pair, substrate noise can be estimated very fast with high fidelity.

- Our approach estimates substrate noise without accurate geometric information such as x and y coordinates. As a result, all the efforts to compute the accurate locations of all blocks can be saved.
- Our approach needs only integer and bitwise-OR operations whereas model-based noise estimation requires floating point operations and transcendental functions like $\exp(x)$.

V. FAST SUBSTRATE NOISE-AWARE FLOORPLANNING

Our floorplanning algorithm takes advantage of the BPDG-based fast substrate noise estimation, efficiently examining discrepancy between BDPG in Section III and sequence pair in Section IV. Also, using block inflation, we insert whitespace around analog blocks as a guard ring. Note that our white space allocation is done as a preprocessing to minimize area overhead. The overall algorithm is described in Algorithm 2.

VI. EXPERIMENTAL RESULTS

We implemented the proposed algorithm in C++ by modifying Parquet [17] which is a simulated annealing-based floorplanning package available in [18]. In order to compare our approach (**bpdg** in Table II) with other approaches, we also implemented the conventional model-based, substrate noise-aware floorplanning algorithm (**modl** in Table II). After every movement inside the simulated annealing loop, to estimate current floorplan instance's substrate noise on the analog blocks, the number of violations by Theorem 2 was *counted* for **bpdg**, whereas substrate noise was *computed* for **modl** based on the substrate noise model, i.e., Eqn. (4) in Section II-B. The

---

Algorithm 2 Fast Substrate Noise-Aware Floorplanning

Input: Analog BPDG Ag, Digital BPDG Dg
1: Do floorplanning with analog blocks with Ag
2: Inflate the analog block floorplan
3: Make the analog floorplan as a virtual block B_v
4: Do floorplanning with digital blocks and B_v with Dg
Output: Final floorplan

---

cost functions we used for each algorithm are summarized in Table II. Note that we disabled wirelength optimization, since real implementation of mixed-signal SOCs has sparse interconnection between analog blocks and digital blocks, which is not well reflected in MCNC benchmarks. However, our approach can be readily extended to include wirelength optimization, maintaining high computational efficiency.

All algorithms were tested on a Pentium4 Linux machines with MCNC [16] benchmarks (ami33, ami49) and two randomly generated larger benchmarks (n75 with 75 blocks, n100 with 100 blocks). About 30% of the blocks in each benchmark were chosen as analog blocks, and noise characteristics of all the blocks were carefully generated. All process dependent parameters were the same as in [5], [14] and [15].

Table II shows experimental results for all benchmarks with three algorithms. Each number in the table is generated by taking the average of numbers obtained over 250 floorplans. The simulated annealing of each floorplanning is scheduled by Parquet, and stopped after the same number of movements for each benchmark. The final noise quantities for all algorithms were computed based on Eqn. (4) for fair comparison.

The last two columns show the overhead of each algorithm in terms of cpu time and area with respect to **parq**. From the table, **parq** shows the best area and cpu time (thus, 0% overhead), but the worst noise for all benchmarks as expected. The cpu time of **bpdg** is significantly smaller than that of **modl** for all benchmarks; **bpdg** is approximately 60-80 times faster than **modl**. The area overhead of **bpdg** is slightly smaller for the three larger benchmarks as well than

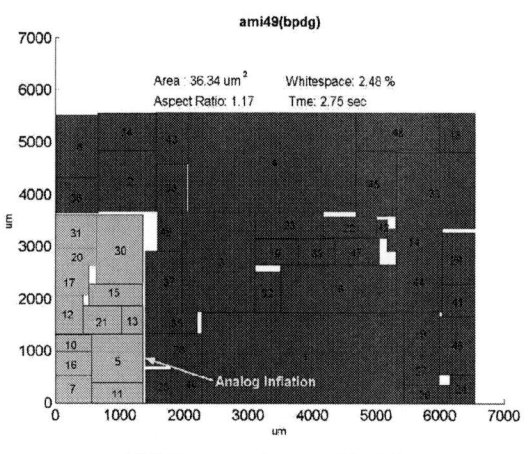 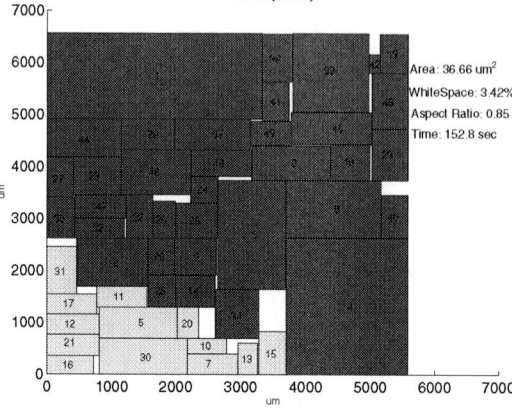

(a) With proposed approach(**bpdg**)   (b) With model-based simulation approach(**modl**)

Fig. 7. Result of packing ami49

modl. Lastly **bpdg** shows less total substrate noise than **modl**. The reason why the proposed algorithm overall shows both smaller area and less substrate noise is that whitespace is more efficiently utilized. By making an analog floorplan inflated as a preprocessing step as in Section V, the substrate noise becomes less in the beginning of annealing, and this allows the simulated annealing engine to optimize the area further without increasing substrate noise. An analogy of this kind of effect can be found in congestion-aware placement [19].

VII. CONCLUSION

In order to cope with significant substrate noise impact on analog circuits from digital circuits, we propose substrate noise-aware floorplanning with fast substrate noise estimation powered by block preference directed graph (BPDG) and sequence pair. Compared with Parquet [17], the proposed approach has on average only 9% cpu time overhead, whereas naive model-based simulation approach shows over 6000% overhead. Also, the proposed approach shows smaller area overhead due to the efficient utilization of whitespace.

Since BPDG is a general concept for fast cost evaluation, it will be extend to deal with temperature or performance estimation in the future.

VIII. ACKNOWLEDGEMENT

The authors would like to thank Prof. Karti Mayaram from Oregon State Univ. and Prof. Ranjit Gharpurey from the Univ. of Texas at Austin for helpful discussions.

REFERENCES

[1] A. Nardi, H. Zeng, J. L. Garrett, L. Daniel, and A. L. S-Vincentelli, "A Methodology for the computation of an upper bound on noise current spectrum of CMOS swichting activity," in *Proc. Int. Conf. on Computer Aided Design*, 2003, pp. 778–785.

[2] A. Koukab, K. Banerjee, and M. Declercq, "Modeling Techniques and Verification Methdologies for Substrate Coupling Effects in Mixed-signal System-on-Chip designs," *IEEE Trans. on Computer-Aided Design of Integrated Circuits and Systems*, vol. 23, no. 6, Jun 2004.

[3] M. V. Heijingen, M. Badarouglu, S. Donnay, G. G. E. Gielen, and H. J. D. Man, "Substrate Noise Generation in Complex Digital systems: efficient modeling and simulation methodology and experiemenal verification," *IEEE J. Solid-State Circuits*, vol. 37, Aug 2002.

[4] H. Lan, Z. Yu, and R. W. Dutton, "A CAD-oriented Modeling Approach of frequency-dependent behavior of Substrate Noise Coupling for Mixed-Signal IC Design," in *Proc. Int. Symp. on Quality Electronic Design*, Mar 2003, pp. 195–200.

[5] B. Owens, S. Adluri, P. Birrer, R. Shreeve, S. K. Arunachalam, and K. Mayaram, "Simulation and Measurement of Supply and Substrate Noise in Mixed-Signal ICs," *IEEE J. Solid-State Circuits*, vol. 40, no. 2, Feb 2005.

[6] T. Blalack, Y. Leclercq, and C. P. Yue, "On-chip RF isolation techniques," in *Proc. IEEE BCTM.*, 2002, pp. 205–211.

[7] C. Lin and D. Leenaerts, "A New Efficient Method Substrate-Aware Device-Level Placement," in *Proc. Asia and South Pacific Design Automation Conf.*, 2000, pp. 533–536.

[8] G. Blakiewicz, M. Jeske, M. Chrzanowska-Jeske, and J. S. Zhang, "Substrate Noise Modeling in Early Floorplanning of Mixed-Signal SOCs," in *Proc. Asia and South Pacific Design Automation Conf.*, Jan 2005, pp. 819–823.

[9] H. Murata, K. Fujiyoshi, S. Nakatake, and Y. Kajitani, "VLSI Module Placement Based on Rectangle-Packing by the Sequence-Pair," vol. 15, Dec 1996.

[10] X. Tang and D. Wong, "Floorplanning with Alignment and Performance Constraints," in *Proc. Design Automation Conf.*, Jun 2002.

[11] W. K. Chu, N. Verghese, K. S. H. Cho, H. Tsujikawa, S. Hirano, S. Doushoh, M. Nagata, A. Iwata, and T. Ohmoto, "A Substrate Noise Analysis Methodology for Large-Scale Mixed-Signal ICs," in *Proc. IEEE Custom Integrated Circuits Conf.*, 2003.

[12] N. K. Verghese and J. J. Allstot, "Computer-aided design considerations in Mixed-signal coupling in RF integration circuits," *IEEE J. Solid-State Circuits*, vol. 33, Mar 1998.

[13] J. P. Costa, M. Chou, and L. M. Silveria, "Efficient techniques for accurate modeling and simulation of substrate coupling in Mixed-signal ICs," *IEEE Trans. on Computer-Aided Design of Integrated Circuits and Systems*, vol. 18, no. 5, pp. 597–607, May 1999.

[14] D. Ozis, T. Fiez, and K. Mayaram, "An Efficient Modeling Approach for Substrate Noise Coupling Analysis," in *Proc. IEEE Int. Symp. on Circuits and Systems*, 2002.

[15] ——, "Comprehensive geometry-dependent macromodel for substrate noise coupling in heavily doped cmos processes," in *Proc. IEEE Custom Integrated Circuits Conf.*, 2002.

[16] http://www.cse.ucsc.edu/research/surf/GSRC/MCNC.

[17] S. N. Adya and I. L. Markov, "Fixed-outline Floorplanning : Enabling Hierarchical Design," *IEEE Trans. on Very Large Scale Integration (VLSI) Systems*, vol. 11, pp. 120–1135, Dec 2003.

[18] http://vlsicad.eecs.umich.edu/BK/parquet.

[19] U. Brenner and A. Rohe, "An effective congestion driven placemnet framework," in *Proc. Int. Symp. on Physical Design*, 2002, pp. 6–11.

A Fixed-die Floorplanning Algorithm Using an Analytical Approach

Yong Zhan, Yan Feng, and Sachin S. Sapatnekar

Department of Electrical and Computer Engineering

University of Minnesota

Minneapolis, MN 55455, USA

Abstract—**Fixed-die floorplanning is an important problem in the modern physical design process. An effective floorplanning algorithm is crucial to improving both the quality and the time-to-market of the design. In this paper, we present an analytical floorplanning algorithm that can be used to efficiently pack soft modules into a fixed die. The locations and sizing of the modules are simultaneously optimized so that a minimum total wire length is achieved. Experiments on the MCNC and GSRC benchmarks show that our algorithm can achieve above a 90% success rate with a 10% white space constraint in the fixed die, and the efficiency is much higher than that of the simulated annealing based algorithms for benchmarks containing a large number of modules.**

I. INTRODUCTION

Floorplanning is a crucial step in early stages of the physical design process. A high quality floorplan with small wire lengths and low white space will have a positive impact on both the performance and the yield of the final manufactured ICs. The high complexity of modern VLSI systems has made hierarchical design the preferred methodology even for the floorplanning stage. Hence, as pointed out in [1], the problem of fixed-die floorplanning, in which the outline of the floorplan is pre-determined, has become more relevant than outline-free floorplanning, because at a lower level of the design hierarchy, the floorplan of a sub-system must be confined to the outline set by the higher level of hierarchy that is immediately above it.

During the past few years, several works have been performed in the direction of solving the fixed-die floorplanning problem. In [2], a simulated annealing based algorithm was presented, and slack-based moves were introduced to facilitate the reduction of the floorplan span in a given direction, and in [3], an evolutionary search approach was used to handle the fixed-die floorplanning problem, based on normalized Polish expressions. In [4], Chen and Chang proposed a novel cooling scheme for the simulated annealing process such that the runtime of the algorithm was significantly reduced, while at the same time, the quality of the resulting floorplan was improved. In [5], partitioning was effectively combined with the simulated annealing algorithm to make the later much more scalable with respect to problem size.

Another direction in the research of floorplanning problems is to study the representations of the geometric relationships among modules so that the algorithms such as the simulated annealing can be implemented more effectively. Examples of some of the recent works concerning the floorplan representations include the sequence pair method [6], BSG [7], O-tree [8], CBL [9], B*-tree [10], and TCG [11].

The fixed-die floorplanning problem can be considered as one of packing rectangular-shaped modules into a fixed outline. Two types of modules can be involved in a floorplanning problem, i.e., hard modules, whose shape cannot change during the floorplanning process, and soft modules, whose area remains the same but whose aspect ratio can vary. Over the years, the annealing-based algorithms have made remarkable progress in the floorplanning of hard modules. Nowadays, a state-of-the-art annealing-based floorplanning algorithm can achieve a good floorplan containing hundreds of hard modules within minutes. For soft modules, however, the results from the annealing-based algorithms are not as satisfactory. This either shows up as a long runtime to

execute the algorithm or a low success rate[1]. One of the reasons for the low performance of annealing algorithms on the problem of floorplanning with soft modules is that the sizing of the modules adds more dimensions to the optimization problem, and hence increases the difficulty of the annealing process.

The floorplanning of soft modules remains an important problem because at the floorplanning stage, the detailed layout of modules has not usually been obtained yet. Hence, the rectangular-shaped modules still have certain flexibility in changing their aspect ratios. Several previous works have tackled the problem of effective floorplanning and sizing of soft modules [12] [13] [14]. However, these works were not performed in the fixed-die context. In [15], a highly efficient bipartitioning-based algorithm was proposed that can effectively deal with the soft modules in the fixed-die floorplanning problem. Nevertheless, the success rate of this algorithm is sensitive to the input benchmark and the constraints such as the maximum allowed aspect ratios of the modules. In this paper, we present a fixed-die floorplanning algorithm based on an analytical approach that can be used to efficiently pack soft modules into the fixed die while minimizing the total wire length. The success rate of the algorithm is benchmark and constraint-insensitive. The floorplanning problem is formulated into a constrained optimization problem, and the location and sizing for each of the modules are obtained simultaneously. The optimization algorithm is divided into two stages, i.e., rough floorplanning, followed by overlap reduction and final legalization. In the first stage, we have adopted a method similar to that used in [16] and [17] for the placement problem of standard cells to spread the modules relatively uniformly across the die. The difference between the floorplanning problem of soft modules and the placement problem of standard cells is that the modules in the floorplanning stage can have significant difference in both the widths and heights, and the aspect ratios of the modules can vary during the optimization process. Hence, besides the center coordinates of the modules, the widths of the modules also enter the floorplanning problem as optimization variables to take care of the module-sizing issue. In the second stage, we first use an optimization-based approach to effectively reduce the overlaps between modules in the rough floorplan. Then we send the improved floorplan, which already has little or no overlap between modules, to the $pl2sp()$ routine in Parquet-4 [2], whose function is to shift some of the modules so that a overlap-free floorplan is obtained. No further sizing or switching the order between modules is performed in $pl2sp()$. The operations that take place in the second stage of the algorithm are in contrast to the simple greedy algorithm used in [17] for the legalization of the placement of standard cells immediately after the rough placement, which usually fails to generate a legal floorplan in our situation because of the significant difference in both the widths and heights of the modules. Experimental results on the MCNC and GSRC benchmarks show that our method can achieve above a 90% success rate with a 10% white space constraint in the fixed die, while the runtime is almost linear with respect to the number of modules in the design.

The rest of the paper will be organized as follows. Section II formulates the fixed-die floorplanning problem and presents the two-stage optimization algorithm using an analytical approach. Section III shows the experimental results in terms of the runtime, success rate, and the total half perimeter wire length (HPWL) in the final floorplan, and the conclusions are provided in section IV.

---

[1]The success rate of a set of annealing runs is defined as the ratio of the number of runs that result in a floorplan that can fit into the fixed die, to the total number of annealing runs.

This work was supported in part by DARPA under grant N66001-04-1-8909, SRC under contract 2003-TJ-1092, and NSF under award CCR-0205227.

II. PROBLEM FORMULATION AND THE ANALYTICAL FLOORPLANNING ALGORITHM

A. Problem formulation

As stated previously, our goal is to develop an efficient algorithm to pack rectangular-shaped soft modules into a fixed die while minimizing the total wire length. The input to the algorithm includes the width and height of the die, the area of each module, and the maximum allowed aspect ratios of the modules. The output of the algorithm is the final floorplan within the fixed die which includes the location and sizing for each of the modules. Hence, our fixed-die floorplanning problem for soft modules is formulated as

Given the dimensions of the fixed die, L_x and L_y, and assuming that the area and the maximum allowed aspect ratio of the ith module are A_i and R_i, respectively, find the location and sizing for each of the modules such that the total wire length is minimized and the final floorplan is within the fixed die.

Our algorithm achieves the objective in two stages, i.e., rough floorplanning, followed by overlap reduction and final legalization. The objective of the first stage is to find the approximate location and sizing for each of the modules such that the total wire length is small, while the modules are spread relatively uniformly across the fixed die. Some overlap is allowed in this stage. The objective of the second stage is to fine-tune the location and sizing for each of the modules such that a floorplan without any overlap between modules is obtained.

B. Rough floorplanning

In the rough floorplanning stage, we try to minimize the total wire length, WL, and spread the modules as uniformly as possible across the fixed die. The spreading of modules is characterized by a penalty term P_D, which describes the non-uniformity of the module distribution.

Let x_i, y_i, and w_i be the center coordinates and the width of module i, respectively. Then the optimization problem that we solve in the rough floorplanning stage is

$$
\begin{aligned}
\text{minimize} \quad & WL + \alpha \cdot P_D & (1)\\
\text{such that} \quad & 0 \le x_i \le L_x\\
& 0 \le y_i \le L_y\\
& \sqrt{\frac{A_i}{R_i}} \le w_i \le \sqrt{A_i R_i}\\
& i = 1, 2, \ldots, n
\end{aligned}
$$

where n is the total number of modules, and the weighting factor α is used to determine the relative significance of the penalty term P_D with respect to the wire length WL. The first two constraints state that the center of each module should not exceed the boundary of the fixed-die while the last constraint ensures that the real aspect ratio of each module is within its maximum allowed range. Note that these constraints do not exclude the possibility that some of the modules may lie partly outside the fixed-die. This issue is taken care of by the P_D term in our implementation and will be discussed further when we present the calculation of P_D.

The constrained optimization problem (1) can be conveniently transformed into an unconstrained optimization problem and solved using the Sequential Unconstrained Minimization Technique (SUMT) [18]. The unconstrained optimization problem has the form

$$
\text{minimize} \quad WL + \alpha \cdot P_D + \beta \cdot B \qquad (2)
$$

where B is the barrier term created to take care of the removed constraints in (1) and β is the weighting factor associated with B.

One form of the barrier term B is given by

$$
\begin{aligned}
B = {} & \sum_{i=1}^{n} \frac{1}{x_i} + \sum_{i=1}^{n} \frac{1}{L_x - x_i} & (3)\\
& + \sum_{i=1}^{n} \frac{1}{y_i} + \sum_{i=1}^{n} \frac{1}{L_y - y_i}\\
& + \sum_{i=1}^{n} \frac{1}{w_i - \sqrt{\frac{A_i}{R_i}}} + \sum_{i=1}^{n} \frac{1}{\sqrt{A_i R_i} - w_i}
\end{aligned}
$$

The key property of the objective function in (2) and the barrier function B in (3) is that if we start from a point within the feasible region determined by the constraints in (1) and use any kind of iterative searching algorithm, e.g., the conjugate gradient method, to solve the unconstrained optimization problem, the points being searched will always remain feasible. This is because B approaches ∞ as the boundary of the feasible region is approached, and hence, the search process will never go outside the feasible region. A convenient starting point is obtained by choosing x_i, y_i, and w_i randomly within their corresponding bounds.

The total wire length WL is calculated using a quadratic formula. Since the quadratic formula only works with two-pin nets, we first use the *clique* model to decompose all of the multi-pin nets into two-pin nets. Assume modules i_1, i_2, $\ldots$, i_j are connected by multi-pin net i with weight k_i. The clique model removes the multi-pin net and reconnects every pair of modules in i_1, i_2, $\ldots$, i_j with a two-pin net with a weight of $\frac{2k_i}{j(j-1)}$. After all of the multi-pin nets have been replaced by the corresponding two-pin nets, the total quadratic wire length WL can be calculated using

$$
WL = \sum_{i,j} k_{ij} [(x_i - x_j)^2 + (y_i - y_j)^2] \qquad (4)
$$

where the indices i and j run through all of the modules connected by two-pin nets, and the k_{ij}'s are the net weights obtained from the clique model.

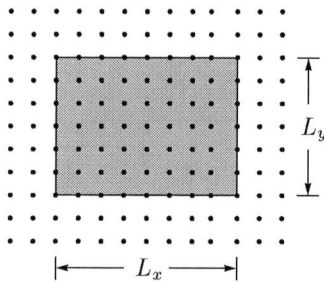

Fig. 1: Grid for calculating the penalty term P_D. The shaded region represents the fixed die. The points inside the die constitute the set C_{in}, and the points outside the die constitute the set C_{out}.

As stated previously, the penalty term P_D is used to characterize the non-uniformity in the distribution of modules across the fixed die. We have adopted a method similar to that presented in [16] and [17] for the placement problem to calculate P_D. Specifically, we first define the concept of area density, D_R, within a region R, which is given by

$$
D_R = \frac{S_A}{S_R} \qquad (5)
$$

where S_R is the area of region R, and S_A is the sum of the overlap areas of all of the modules with region R. Note that as $S_R \to 0$, we obtain the area density of a point. Under this definition, if the modules are uniformly distributed within the fixed die, and a legal floorplan is obtained, the area density inside the die should be $\bar{D} = (\sum_{i=1}^{n} A_i)/(L_x L_y)$ while the area density outside the die should be 0.

During the optimization process, to keep track of the non-uniformity of the area density, we superimpose an array of monitoring points over the die area and its surrounding region as shown in Fig. 1. The penalty term P_D is then calculated as the sum of squares of the variations of the densities from their ideal values, over all of the monitoring points. Specifically, let C_{in} and C_{out} represent the sets of monitoring points inside and outside the die area, respectively, and let D_i be the actual area density at monitoring point i, then the penalty term P_D can be calculated by

$$P_D = \sum_{i \in C_{in}} (D_i - \bar{D})^2 + \sum_{i \in C_{out}} D_i^2 \tag{6}$$

The second term in the expression of P_D penalizes the modules that go partly outside the fixed die during the optimization process.

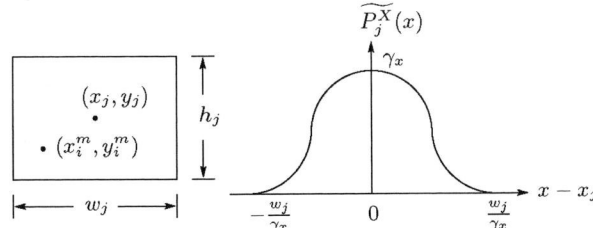

Fig. 2: Calculation of the area density function (a) coordinates of the module and the monitoring point (b) bell-shaped function used to approximate the contribution to the area density from a module.

In equation (6), the actual area density D_i at monitoring point i is calculated by summing up the contributions from all of the modules in the floorplan, i.e.,

$$D_i = \sum_{j=1}^{n} P_j(x_i^m, y_i^m) \tag{7}$$

where $P_j(x, y)$ is the contribution to the area density distribution from module j and (x_i^m, y_i^m) is the location of monitoring point i as shown in Fig. 2(a). Ideally, $P_j(x, y)$ is 1 if module j overlaps with point (x, y), and 0 otherwise, i.e.,

$$P_j(x, y) = \begin{cases} 1 & \text{if } |x - x_j| \leq \frac{w_j}{2} \text{ and } |y - y_j| \leq \frac{h_j}{2} \\ 0 & \text{otherwise} \end{cases} \tag{8}$$

where $h_j = \frac{A_j}{w_j}$ is the height of module j. However, the D_i obtained this way is not a smooth function of the optimization variables x_j and y_j, which may cause practical difficulties in applying gradient-based optimization algorithms. To resolve this difficulty, we can use a bell-shaped smooth function $\widetilde{P_j}(x, y)$ to approximate the effect of module j on the area density distribution [16] [17]. The function $\widetilde{P_j}(x, y)$ can be decomposed into

$$\widetilde{P_j}(x, y) = \widetilde{P_j^X}(x) \times \widetilde{P_j^Y}(y) \tag{9}$$

where

$$\widetilde{P_j^X}(x) = \begin{cases} \gamma_x \times \left(1 - \frac{2(x - x_j)^2}{(\frac{w_j}{\gamma_x})^2}\right) & \text{if } 0 \leq |x - x_j| \leq \frac{w_j}{2\gamma_x} \\ \gamma_x \times \frac{2(|x - x_j| - \frac{w_j}{\gamma_x})}{(\frac{w_j}{\gamma_x})^2} & \text{if } \frac{w_j}{2\gamma_x} < |x - x_j| \leq \frac{w_j}{\gamma_x} \end{cases} \tag{10}$$

and

$$\widetilde{P_j^Y}(y) = \begin{cases} \gamma_y \times \left(1 - \frac{2(y - y_j)^2}{(\frac{h_j}{\gamma_y})^2}\right) & \text{if } 0 \leq |y - y_j| \leq \frac{h_j}{2\gamma_y} \\ \gamma_y \times \frac{2(|y - y_j| - \frac{h_j}{\gamma_y})}{(\frac{h_j}{\gamma_y})^2} & \text{if } \frac{h_j}{2\gamma_y} < |y - y_j| \leq \frac{h_j}{\gamma_y} \end{cases} \tag{11}$$

are the shape factors along the x and y directions, respectively, and γ_x and γ_y are two parameters used to control the spreading of the bell-shaped function $\widetilde{P_j}(x, y)$. Smaller values of γ_x and γ_y will cause the

bell shaped function to become wider, which can make the spreading of highly clustered modules faster. However, if γ_x and γ_y are too small, then the approximation to the real area density function $P_j(x, y)$ will be very inaccurate. Hence, tradeoff values are chosen for these two parameters in our implementation of the algorithm. The functions $\widetilde{P_j^X}(x)$ and $\widetilde{P_j^Y}(y)$, as shown in equation (10), (11), and Fig. 2(b), have the attractive property of being normalized. Specifically, the integral of $\widetilde{P_j^X}(x)$ with respect to x over the entire real axis $\Re$ gives w_j while the integral of $\widetilde{P_j^Y}(y)$ with respect to y over $\Re$ gives $h_j = \frac{A_j}{w_j}$. Hence, the total contribution of module j to the overall area density distribution is still A_j although the smooth bell-shaped function $\widetilde{P_j}(x, y)$ is used to approximate the original rectangular-shaped density function $P_j(x, y)$.

The optimization problem (2) is solved using the conjugate gradient method [19]. There are numerous descriptions of this method in the literature and interested readers are referred to [19] for details. We emphasize here that in our implementation, the conjugate gradient algorithm is executed multiple times. Each run is terminated if the improvement between consecutive iterations becomes insignificant or a pre-determined maximum number of iterations is reached. The final solution of each run is used as the starting point of the new run. The original value of the parameter α is chosen to be small such that a higher weight is assigned to the wire length term WL to ensure the quality of the solution. Then, in successive runs, the value of α is increased, while the value of β is decreased. This makes the module distribution more uniform, and at the same time, allows the exploration of the optimum to go closer to the boundary of the feasible region.

C. Overlap reduction and final legalization

After the rough floorplanning stage, the modules are relatively uniformly distributed across the fixed die. However, some overlaps between modules are often present after this stage. Although increasing the density of the monitoring-point grid in the rough floorplanning stage is beneficial to reducing the overlap, it will invariably increase the computational cost. In addition, the overlap can rarely be removed completely in the rough floorplanning stage even if a rather dense grid of monitoring points is used. Hence, instead of using a very dense grid in rough floorplanning and sacrificing the runtime, we add an explicit overlap reduction and final legalization stage to the algorithm such that a legal floorplan within the given fixed die is obtained.

Due to the fact that the modules in the floorplanning stage can have significant difference in both the widths and heights and the allowed white space with respect to the die area is relatively small, the simple greedy algorithm used in [17] for legalizing the placement of standard cells will usually not be able to result in a legal floorplan within the fixed die. In our work, we choose to first reduce the overlap between modules through solving the following optimization problem

$$\text{minimize} \quad WL + \eta \cdot P_O + \beta \cdot B \tag{12}$$

where the definitions of WL and B are the same as those shown in section II B. The term P_O in (12) is used to represent the penalty due to the overlaps between modules and the overlap between a module and the outside of the fixed die. Hence, P_O can be written as

$$P_O = \sum_{i=1}^{n-1} \sum_{j=i+1}^{n} S_{ij}^M + \sum_{i=1}^{n} (A_i - S_i^D) \tag{13}$$

where S_{ij}^M is the overlap area between modules i and j, S_i^D is the overlap area between module i and the fixed die, and $A_i - S_i^D$ is the overlap area between module i and the *outside* of the fixed die. Note that both S_{ij}^M and S_i^D can be considered as the overlap area between two rectangles. Hence, we discuss below in detail how to quickly find this overlap area given the center coordinates and dimensions of the two rectangles, and how to use smooth functions to approximate the overlap area.

In Fig. 3, we show two overlapping rectangles, located at (x_i, y_i) and (x_j, y_j), and with dimensions $w_i \times h_i$ and $w_j \times h_j$, respectively. The overlap area can be calculated as

$$S = \Delta x \cdot \Delta y \tag{14}$$

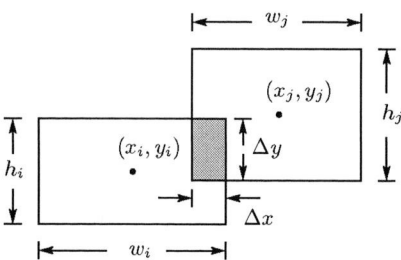

Fig. 3: Overlap between two rectangles.

where

$$\Delta x = (\min\{x_i + \frac{1}{2}w_i, x_j + \frac{1}{2}w_j\} - \max\{x_i - \frac{1}{2}w_i, x_j - \frac{1}{2}w_j\})$$
$$\times U(\min\{x_i + \frac{1}{2}w_i, x_j + \frac{1}{2}w_j\} - \max\{x_i - \frac{1}{2}w_i, x_j - \frac{1}{2}w_j\})$$
(15)

and

$$\Delta y = (\min\{y_i + \frac{1}{2}h_i, y_j + \frac{1}{2}h_j\} - \max\{y_i - \frac{1}{2}h_i, y_j - \frac{1}{2}h_j\})$$
$$\times U(\min\{y_i + \frac{1}{2}h_i, y_j + \frac{1}{2}h_j\} - \max\{y_i - \frac{1}{2}h_i, y_j - \frac{1}{2}h_j\})$$
(16)

Here, $U(x)$ is the unit step function, i.e.,

$$U(x) = \begin{cases} 1 & \text{if } x \geq 0 \\ 0 & \text{if } x < 0 \end{cases}$$
(17)

Equation (14)-(17) can be used to calculate the overlap area between two rectangles exactly. However, none of the functions $\min\{\}$, $\max\{\}$, or $U(x)$ is smooth, and this will again create problems for the gradient-based optimization algorithm. These non-smooth functions are replaced by the following smooth approximating functions in our implementation of the floorplanner.

$$\min\{x, y\} \approx \frac{x \cdot e^{k(y-x)} + y \cdot e^{k(x-y)}}{e^{k(x-y)} + e^{k(y-x)}}$$
(18)

$$\max\{x, y\} \approx \frac{x \cdot e^{k(x-y)} + y \cdot e^{k(y-x)}}{e^{k(x-y)} + e^{k(y-x)}}$$
(19)

$$U(x) \approx \frac{1}{2}(1 + \tanh(k'x))$$
(20)

where k and k' are parameters used to control the accuracy of the approximating functions. Larger values of k and k' will increase the accuracy of the approximation, but at the same time reduce the smoothness of the approximating functions, which may increase the difficulty of the optimization process. These two parameters are tuned in our implementation to achieve the best balance between the runtime of the algorithm and the quality of the resulting floorplans. The optimization problem (12) is again solved using the conjugate gradient method.

After solving the optimization problem (12), the resultant floorplan will contain little, if any, overlap between modules. We then send this improved floorplan to the $pl2sp()$ function in Parquet-4 such that a sequence pair corresponding to a floorplan without any overlap between modules is obtained. The core functionality of the $pl2sp()$ routine is to legalize a floorplan without performing module sizing or switching the order between modules. The reason why we cannot call $pl2sp()$ directly after the rough floorplanning stage is that the overlap between modules is not small enough at that time, and the simple legalization procedure used in $pl2sp()$ will generally not be able to fit the floorplan into the fixed-die.

D. Time complexity analysis

The runtimes for solving the unconstrained optimization problems (2) and (12) are determined primarily by the time required to calculate the objective functions and their gradients. Let n, N, N_2, and M be the total number of modules, the total number of nets in the original netlist, the total number of decomposed two-pin nets, and the total number of monitoring points in the grid superimposed on the die area and its surrounding region, respectively. The time complexity of calculating the barrier term B and its gradient is $O(n)$, and the time complexity of calculating the wire length term WL and its gradient is $O(N_2)$. However, under the reasonable assumption that the fanout of each net in the original netlist is upper-bounded by a constant, we obtain $O(N_2) = O(N)$.

From (6) and (7), it may seem that the time complexity of calculating P_D is $O(nM)$. However, a better analysis will show that the actual complexity is only $O(M)$. This is because to calculate P_D, we use a two dimensional array to store the D_i's corresponding to the monitoring points. Initially, all of the D_i's are set to 0. Then we go through the n modules, and for each module, we *only* update the D_i's corresponding to the monitoring points covered by this module. Since the monitoring points form a regular grid, it takes constant time to find the range of the array indices corresponding to the points covered by each module, and each updating of a D_i also takes constant time. Hence, if M' represents the total count of the monitoring points covered by all of the modules (note that if a point is covered by multiple modules, it should be counted multiple times.), then the total cost of calculating all of the D_i's is $O(M')$. Finally, we sum up all of the D_i's to obtain P_D, which takes $O(M)$ time. Since $M' = O(M)$, the total cost of calculating P_D becomes $O(M') + O(M) = O(M)$. Similarly, we can show that the time complexity of calculating the gradient of P_D is also $O(M)$.

To calculate the P_O term in (12) using expression (13), the apparent time complexity is $O(n^2)$, because of the double summation involved. However, this cost can be reduced significantly by observing that two modules i and j that are separated far apart from each other after the rough floorplanning stage will have no interaction in the overlap removal stage that follows. Hence, the S_{ij}^M term corresponding to these two modules can be dropped from the double summation in (13). Our strategy of calculating the P_O term efficiently is to associate with each module an interaction range box as shown in Fig. 4, and the S_{ij}^M term enters the double summation in (13) only if the interaction range boxes associated with modules i and j overlap with each other after the rough floorplanning stage. There is no unique way of determining the size and shape of each of the n interaction range boxes. A good heuristic is to associate with each module i a square-shaped interaction range box with the same center coordinate as the module itself and a side length of $2\sqrt{A_i R_i}$. An interaction list is established for each module after the rough floorplanning stage to store the indices of the modules that have interactions with it. The time it takes to build all of the lists is $O(n \times \log(n) + K)$ using the interval tree and range tree [20], where K is the total number of the pairs of modules that have interactions with each other. However, since the lists only need to be built once, and then they can be used many times in solving the optimization problem (12), the amortized cost of this step of the algorithm can be practically ignored, considering the problem sizes encountered in the floorplanning stage of the design, i.e., a few hundred to a few thousand modules. After the lists are established, each calculation of P_O and its gradient is reduced from $O(n^2)$ to approximately $O(n)$ time.

From the above analysis, we see that the calculation of the objective function and its gradient in the optimization problem (2) has a time complexity of $O(N + n + M)$, and the corresponding cost for the optimization problem (12) is $O(N + n)$ after the interaction lists are established. The building up of the interaction lists takes $O(n \times \log(n) + K)$ time but has an extremely small amortized cost that can be ignored in practice. We emphasize here that because the operations involved in the calculations of WL, B, and their gradients are all very simple, the actual costs of solving the problems (2) and (12) are dominated by the calculations of P_D, P_O, and their gradients.

Finally, the $pl2sp()$ function from Parquet-4 that we use to obtain the final overlap-free floorplan has a time complexity of $O(n^3)$, where n is the total number of modules. However, in practice, we find that the runtime associated with this function call is negligibly small compared with that of solving the optimization problems (2) and (12). This is because all of the operations involved in the $pl2sp()$ function are very

Die Aspect Ratio	1:1		2:1		3:1		4:1	
	Analytical	Parquet-4	Analytical	Parquet-4	Analytical	Parquet-4	Analytical	Parquet-4
ami33	10/10	4/10	9/10	2/10	10/10	1/10	10/10	0/10
ami49	10/10	5/10	10/10	2/10	10/10	1/10	9/10	0/10
n100	10/10	2/10	10/10	1/10	10/10	0/10	10/10	0/10
n200	9/10	0/10	10/10	0/10	9/10	0/10	9/10	0/10
n300	10/10	0/10	10/10	0/10	10/10	0/10	9/10	0/10

(a)

Die Aspect Ratio	1:1		2:1		3:1		4:1	
	Analytical	Parquet-4	Analytical	Parquet-4	Analytical	Parquet-4	Analytical	Parquet-4
ami33	10/10	5/10	10/10	6/10	10/10	2/10	10/10	2/10
ami49	10/10	8/10	10/10	6/10	10/10	4/10	9/10	1/10
n100	10/10	4/10	10/10	2/10	10/10	2/10	10/10	1/10
n200	10/10	7/10	10/10	3/10	10/10	1/10	9/10	2/10
n300	10/10	1/10	10/10	2/10	10/10	1/10	10/10	1/10

(b)

TABLE I: Success rate of the floorplanning algorithms under (a) a 10% white space constraint and (b) a 15% white space constraint.

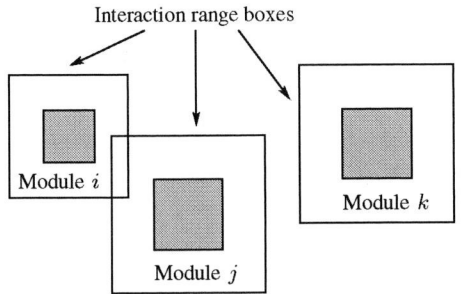

Fig. 4: Interaction range boxes of modules. The shaded areas represent modules. S_{ij}^M is included in the double summation in (13) because the corresponding interaction range boxes overlap with each other. S_{ik}^M and S_{jk}^M are excluded from the double summation in (13) because the corresponding interaction range boxes do not overlap.

simple, and n is generally much smaller than N and M, which are the total numbers of nets and monitoring points, respectively.

III. EXPERIMENTAL RESULTS

Our algorithm is implemented in C++, and the experiments are performed on a desktop with a 3.2GHz Intel(R) Pentium(R)-4 CPU running the Red Hat Linux 8.0 operating system.

As stated previously, this work has focused on the problem of floorplanning with soft modules. The maximum allowed aspect ratio of each module is assumed to be 3 in our experiments. We have tested our algorithm on the two largest MCNC benchmarks ami33 and ami49 and the three largest GSRC benchmarks n100, n200 and n300, and the aspect ratio of the fixed die ranged from 1:1 to 4:1. The experimental results have been compared with those obtained from Parquet-4, which we see as a representative of the current state-of-the-art fixed-die floorplanners. Note that Parquet-4 is specifically tuned for the efficient floorplanning of hard modules, although it can also deal with soft modules [21]. The reason why we choose Parquet-4 is that, to the best of our knowledge, Parquet is the only free floorplanning package online that can handle both fixed die and soft modules, and Parquet-4 is the newest release in the Parquet series.

Tables I (a) and (b) show the comparison results of the success rate between our algorithm and Parquet-4 for the 10% and 15% white space constraints, respectively. To obtain each data value in the two tables, the corresponding floorplanning program was executed 10 times, the number of the resulting floorplans that could fit into the fixed die was counted, and the success rate was calculated. We see clearly from the tables that our analytical floorplanning algorithm can achieve above a 90% success rate for both the 10% and 15% white space constraints, and the success rate improves as more white space is allowed in the

floorplan[2]. As a comparison, Parquet-4 achieves a maximum of 80% success rate and it fails most of the tests for the GSRC benchmarks when the white space constraint is set to 10%. In Fig. 5, we show some examples of the final floorplans of the n100 benchmark obtained by our algorithm under a 10% white space constraint. The outer rectangles represent the outlines of the fixed dies. Each floorplan is shifted towards the lower-left corner of its corresponding fixed die for the purpose of easy comparison. Due to the space limit, the floorplan examples of other benchmarks are omitted here.

In Fig. 6, we show the average runtimes of both our algorithm and Parquet-4 with respect to the number of modules in the floorplan. Parquet-4 performs better for the small MCNC benchmarks but its runtime increases rapidly as the number of modules increases. On the contrary, the runtime of our algorithm increases at a much slower rate and it beats Parquet-4 by about 2X for n200 and 4X for n300. Note that Fig. 6 only compares the runtime of a single execution of the algorithms. The evaluation of the real efficiency of each algorithm should also be based on the success rate because any time spent on the unsuccessful runs is wasted. Considering the much higher success rate of the analytical-based approach, we can see that our algorithm can achieve an order of magnitude effective improvement in runtime, as compared with Parquet-4, for the large GSRC benchmarks.

In Table II, we compare the average total wire length obtained from our analytical approach and that from Parquet-4 for the successful runs. Because Parquet-4 fails to find a legal floorplan for n200 and n300 when the white space constraint is set to 10%, we only compare the two algorithms for the case of 15% white space constraint. We can see from the table that our algorithm achieves better wire length than Parquet-4 and the average improvement is about 12%.

IV. CONCLUSIONS

In this paper, we presented a soft-module floorplanning algorithm based on an analytical approach. The algorithm is divided into two stages, i.e., rough floorplanning , followed by overlap reduction and final legalization. In the rough floorplanning stage, an optimization problem is solved where the objective function is a linear combination of the total wire length and the area distribution density of modules. In the overlap reduction and final legalization stage, we first solve an optimization problem to minimize the linear combination of the total wire length and the overlap area, then we call the $pl2sp()$ function in Parquet-4 to obtain a sequence pair corresponding to a floorplan without any overlap between modules. Experimental results on the MCNC and GSRC benchmarks show that our algorithm can achieve

---

[2]For our algorithm, the initial values of the optimization variables, i.e., x_i, y_i, and w_i are chosen randomly. Hence, due to the inherent non-convexity of the problem, the obtained final floorplans also differ from run to run of the algorithm. To take this effect into consideration, we use the success rate, the average runtime, and the average total wire length to characterize the performance of our algorithm.

8A-2

Die Aspect Ratio	1:1			2:1			3:1			4:1		
	Analytical	Parquet-4	Improve	Analytical	Parquet-4	Improve	Analytical	Parquet-4	Improve	Analytical	Parquet-4	Improve
ami33	74072	82149	9.8%	75168	79131	5.0%	75180	91721	18.0%	79529	101274	21.5%
ami49	799239	928597	13.9%	829888	942117	11.9%	880387	1092771	19.4%	939049	1003220	6.4%
n100	291628	342103	14.8%	290158	351542	17.5%	298894	351338	14.9%	313060	392118	20.2%
n200	572145	630014	9.2%	565927	645219	12.3%	583282	639803	8.8%	608074	685057	11.2%
n300	702822	770354	8.8%	722527	780406	7.4%	793771	838600	5.3%	858346	872501	1.6%

TABLE II: Average total wire length (HPWL) of the floorplans obtained using the analytical approach and Parquet-4 under a 15% white space constraint.

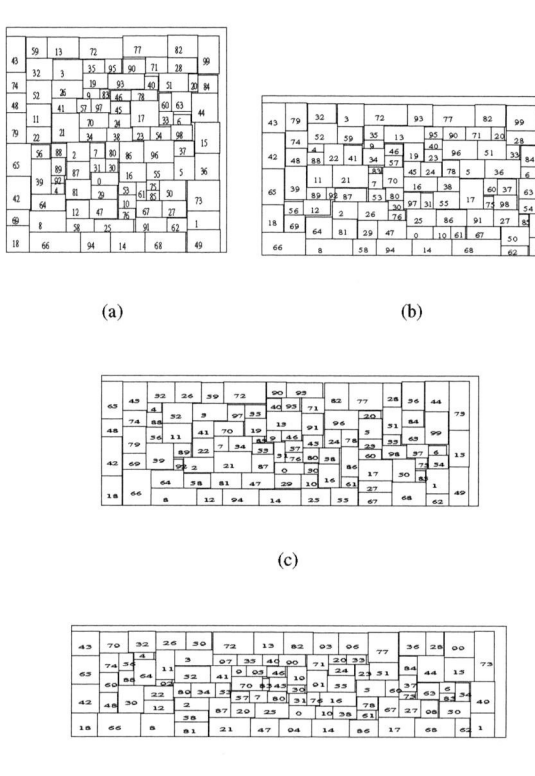

(a)　　　　　　　(b)

(c)

(d)

Fig. 5: The floorplans of the n100 benchmark under a 10% white space constraint with the aspect ratio of the fixed die set to (a) 1:1, (b) 2:1, (c) 3:1, and (d) 4:1. Each floorplan is shifted towards the lower-left corner of its corresponding fixed die for the purpose of easy comparison.

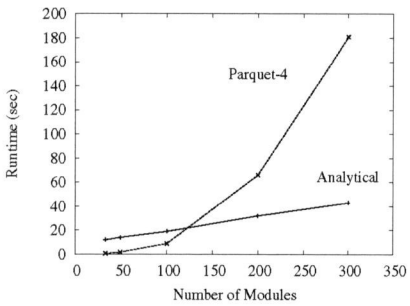

Fig. 6: Comparison of the runtimes of our algorithm and Parquet-4.

above a 90% success rate for a white space constraint of as low as 10%, while the average improvement in wire length is about 12% compared with Parquet-4. In addition, our algorithm can achieve an order of magnitude effective speedup compared with Parquet-4 for the large GSRC benchmarks.

REFERENCES

[1] A. B. Kahng, "Classical Floorplanning Harmful?" *Proceedings of the ACM International Symposium on Physical Design*, pp. 207-213, Apr. 2000.

[2] S. N. Adya and I. L. Markov, "Fixed-outline Floorplanning: Enabling Hierarchical Design," *IEEE Transactions on VLSI Systems*, vol. 11, no. 6, pp. 1120-1135, Dec. 2003.

[3] C. T. Lin, D. S. Chen, and Y. W. Wang, "Robust Fixed-outline Floorplanning through Evolutionary Search," *Proceedings of the IEEE/ACM Asia and South Pacific Design Automation Conference*, pp. 42-44, Jan. 2004.

[4] T. C. Chen and Y. W. Chang, "Modern Floorplanning Based on Fast Simulated Annealing," *Proceedings of the ACM International Symposium on Physical Design*, pp. 104-112, Apr. 2005.

[5] S. N. Adya, S. Chaturvedi, J. A. Roy, D. A. Papa, and I. L. Markov, "Unification of Partitioning, Placement, and Floorplanning," *Proceedings of the IEEE/ACM International Conference on Computer-Aided Design*, pp. 550-557, Nov. 2004.

[6] H. Murata, K. Fujiyoshi, S. Nakatake, and Y. Kajitani, "Rectangle-Packing-Based Module Placement," *Proceedings of the IEEE/ACM International Conference on Computer-Aided Design*, pp. 472-479, Nov. 1995.

[7] S. Nakatake, K. Fujiyoshi, H. Mirata, and Y. Kajitani, "Module Packing Based on the BSG-Structure and IC Layout Applications," *IEEE Transactions on Computer-Aided Design of Integrated Circuits and Systems*, vol. 17, no. 6, pp. 519-530, Jun. 1998.

[8] P Guo, C. Cheng, and T. Yoshimura, "An O-Tree Representation of Non-Slicing Floorplan and Its Applications," *Proceedings of the IEEE/ACM Design Automation Conference*, pp. 268-273, Jun. 1999.

[9] X. Hong, S. Dong, G. Huang, Y. Ma, Y. Cai, C. Cheng, and J. Gu, "A Non-Slicing Floorplanning Algorithm Using Corner Block List Topological Representation," *Proceedings of the IEEE Asia-Pacific Conference on Circuits and Systems*, pp. 833-836, Dec. 2000.

[10] Y. C. Chang, Y. W. Chang, G. M. Wu, and S. W. Wu, "B*-Tree: A New Representation for Non-Slicing Floorplans," *Proceedings of the IEEE/ACM Design Automation Conference*, pp. 458-463, Jun. 2000.

[11] J. M. Lin and Y. W. Chang, "TCG: A Transitive Closure Graph-Based Representation for Non-Slicing Floorplans," *Proceedings of the IEEE/ACM Design Automation Conference*, pp. 764-760, Jun. 2001.

[12] A. Ranjan, K. Bazargan, S. Ogrenci, and M. Sarrafzadeh, "Fast Floorplanning for Effective Prediction and Construction," *IEEE Transactions on Very Large Scale Integration (VLSI)*, vol. 9, no. 2, pp. 341-351, Apr. 2001.

[13] T. C. Wang and D. F. Wong, "Optimal Floorplan Area Optimization," *IEEE Transactions on Computer-Aided Design of Integrated Circuits and Systems*, vol. 11, no. 8, pp. 992-1001, Aug. 1992.

[14] F. Y. Young, C. C. N. Chu, W. S. Luk, and Y. C. Wong, "Handling Soft Modules in General Non-Slicing Floorplan using Lagrangian Relaxation," *IEEE Transactions on Computer-Aided Design of Integrated Circuits and Systems*, vol. 20, no. 5, pp. 687-692, May 2001.

[15] J. Cong, M. Romesis, and J. R. Shinnerl, "Fast Floorplanning by Look-Ahead Enabled Recursive Bipartitioning," *Proceedings of the IEEE/ACM Asia and South Pacific Design Automation Conference*, pp. 1119-1122, Jan. 2005.

[16] W. Naylor *et al.*, "Non-Linear Optimization System and Method for Wire Length and Delay Optimization for an Automatic Electric Circuit Placer," *US Patent 6301693*, Oct. 2001.

[17] A. B. Kahng and Q. Wang, "Implementation and Extensibility of an Analytical Placer," *Proceedings of the ACM International Symposium on Physical Design*, pp. 18-25, Apr. 2004.

[18] F. S. Hillier and G. J. Lieberman, "Introduction to Operations Research," McGraw-Hill, New York, NY, 1995.

[19] W. H. Press *et al.*, "Numerical Recipes in C++," Cambridge University Press, Cambridge, UK, 2002.

[20] M. de Berg, M. van Kreveld, M. Overmars, and O. Schwarzkopf, "Computational Geometry," Springer-Verlag, Berlin, Germany, 2000.

[21] I. Markov, Private communication, 2005.

2006 Asia and South Pacific Design Automation Conference

8A-3

A Multi-Technology-Process Reticle Floorplanner and Wafer Dicing Planner for Multi-Project Wafers*

Chien-Chang Chen and Wai-Kei Mak
Department of Computer Science
National Tsing Hua University
Taiwan 300 R.O.C.

Abstract—As the VLSI manufacturing technology advances into the deep sub-micron(DSM) era, the mask cost can reach one or two million dollars. Multiple project wafers (MPW) which put different dies onto the same set of masks is a good cost-sharing approach. Every design needs to be produced by its desired technology process, such as 1 poly with 4 metal layers (1P4M), or 1 poly with 5 metal layers (1P5M). Dies with different desired manufacturing processes cannot be produced from the same wafer, but they can be put onto the same set of masks in order to reduce the total cost of the used masks and wafers. In this paper, we propose a novel integer linear programming (ILP)-based floorplanner for shuttle runs consisting of projects requiring different desired processes. Two simulated annealing-based side-to-side wafer dicing planners are also presented. Experimental results show that our approach achieves 28% wafer reduction on average compared to a previous simulated annealing-based reticle floorplanner.

I. INTRODUCTION

As the VLSI manufacturing technology advances into the deep sub-micron era, mask cost increases at an accelerating rate. The mask cost is around $700k dollars for 130nm and $1 million dollars for 90nm. Reticle enhancement technologies(RET) such as optical proximity correction (OPC) and phase shifting mask (PSM) cause the complexity and the cost of mask to grow dramatically [1, 2, 3]. Multiple project wafers(MPW), or shuttle run, allows customers to share the expensive cost of a common mask tooling set for an engineering-run and obtain their samples quickly for fast prototyping and low volume designs. MPW vendors like Taiwanese Semiconductor Manufacturing Company (TSMC) and IBM provide shuttle services to their customers.

MPW involves two key problems : (1) shuttle mask (reticle) floorplanning and (2) wafer dicing planning. Unlike the traditional floorplanning problem which is to pack the blocks as closely as possible in order to minimize the total area and minimize the wirelength between the blocks, the objective of shuttle mask floorplanning is to reduce the total cost of the used masks and wafers. A

---

*This work was supported in part by NSC 94-2220-E-007-014 and MOEA 94-EC-17-A-01-S1-038.

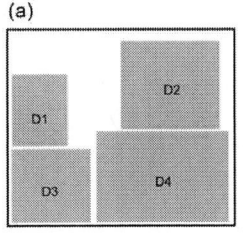 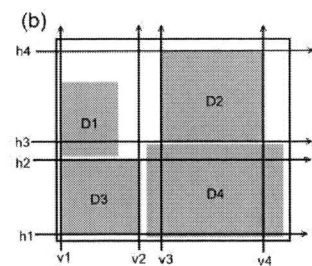

Fig. 1. (a) A reticle floorplan consisting of four dies. (b) $\{h1, h2, h3, h4\}$ is a row dicing plan and $\{v1, v2, v3, v4\}$ is a column dicing plan. Only D2 and D3 are diced out.

better shuttle mask floorplan provides better wafer yield, in other words, more dies can be produced by less wafers. A simple reticle floorplan example is shown in figure 1(a). The fabricated wafers must be cut to obtain the bare dies. Usually the dies on a wafer are cut out by a cutting saw that traverses the whole wafer horizontally and vertically. This is called side-to-side wafer dicing. A set of horizontal (vertical) cut lines, called a row (column) dicing plan must be assigned for each row (column) of reticles on the wafer as shown in figure 2. A die can be diced out successfully only when the cut lines are along the margins of the die in the reticle floorplan and no cut lines cut across the die. Only a fraction of dies can be diced out according to a dicing plan because cutting out one die may destroy another die. For example, in figure 1(b), only two dies are diced out. Packing the different dies on a reticle and choosing the right wafer dicing plans are two interesting and challenging problems.

Recently, a number of papers considered the reticle floorplanning problem and the wafer dicing problem [4]-[10]. Many works (eg. [4, 5, 6, 7, 9]) only considered how to minimize the reticle area and/or maximize the minimum number of successfully extracted copies of the same die type in a wafer without considering the actual demand of each type. Some works are based on the simple but inaccurate assumption that a wafer is rectangular when performing optimization [7, 9]. The use of highly expensive wafer dicing equipment other than a side-to-side wafer dicing equipment was considered in [6]. A simulated annealing-based floorplanner considering a weighted sum of various objectives was proposed in [8]. Wu et al. [10]

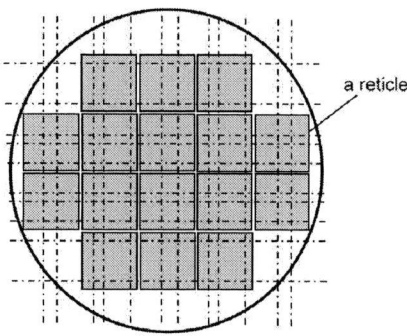

Fig. 2. A side-to-side wafer dicing plan.

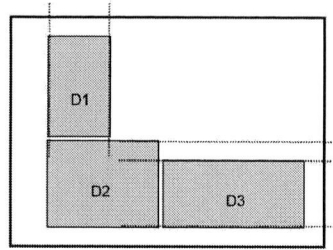

Fig. 3. D1 and D2 are in vertical conflict. D2 and D3 are in horizontal conflict. D1 and D3 are conflict-free.

dies simultaneously. Figure 3 illustrates these concepts. If many dies are in vertical/horizontal conflicts, the yield will be seriously affected.

In this paper, we consider the fact that the projects in a shuttle run may require different technology processes. We note that the dies of different technology processes cannot be extracted from the same wafer. Therefore, we have the following key observation.

Observation 1 : *Even if two dies with different desired processes are conflict-free in the reticle floorplan, they still cannot be produced at the same time.*

An example is shown in figure 4, where Die A must be produced by 1P4M wafers and Die B by 1P5M wafers. The two dies can be diced out at the same time from the reticle floorplan. But if a wafer is targeted to fabricate 1P4M dies, Die B extracted from this wafer will malfunction and must be discarded. On the other hand, if a wafer is targeted to fabricate 1P5M dies, Die A extracted from this wafer will malfunction and must be discarded.

considered the circular shape of the wafers and the production volume requirement of each die type when performing optimization, but their approach required a long run time.

In practice, in order to maximize the MPW utilization and reduce the total cost of the used masks and wafers, projects requiring different number of metal layers can be put on the same shuttle. If a wafer is used to fabricate the projects with 4 metal layers, the wafer is taken off before processing the 5th metal layers, so on and so forth. Non-1P4M projects that are diced out from 1P4M wafer will malfunction and must be discarded. In this paper, we propose a novel integer linear programming-based floorplanner for shuttle runs consisting of projects requiring different desired technology processes.

Our goal is to minimize the number of wafers needed to satisfy the demands of all die types. Our floorplanner incorporates die replication on a reticle to reduce the total number of wafers needed to meet the different demands of different die types. Moreover, we propose two simulated annealing-based wafer dicing planners to minimize the number of required wafers.

The rest of this paper is organized as follows. We formulate and analyze the problem in section II. A novel integer linear programming-based floorplanner is given in section III. A simulated annealing-based wafer dicing planner is presented in section IV. Experimental results are reported in chapter V.

II. PROBLEM ANALYSIS

The problem considered in this paper is as follows. Given (1) a set of N projects with their desired technology processes and demands, (2) the maximum dimensions of a reticle, (3) the wafer size, we want to find a reticle floorplan and a set of side-to-side wafer dicing plans in order to satisfy the demand of each die type while minimizing the required wafers.

We say that two dies on the reticle are in *vertical (horizontal) conflict* if no set of vertical (horizontal) cut lines can dice the two dies simultaneously. On the contrary, two dies are *conflict-free* if there exists a set of vertical cut lines and a set of horizontal cut lines to extract both

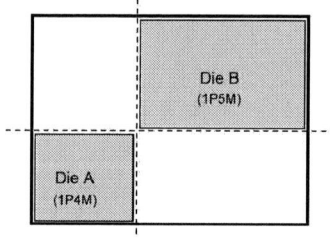

Fig. 4. Die A must be produced by technology process 1P4M, but Die B needs 1P5M.

In order to satisfy the different demands of the dies and reduce the total number of wafers needed, we have another observation as follows.

Observation 2 : *We can put multiple instances of the same die on a reticle during floorplanning depending on the demand of the die.*

III. INTEGER LINEAR PROGRAMMING-BASED FLOORPLANNER

As explained in section II, if the dies are conflict-free on the reticle, they can be diced out at the same time.

778

However, we cannot put all the dies in conflict-free positions on the reticle due to the limitation on the size of the reticle. The above observation motivates us to place the dies of the same technology process in conflict-free positions on the reticle so that they may be extracted from the same wafer at the same time. In other words, we must minimize the horizontal conflict or vertical conflict situations for the dies of the same technology process but we are not concerned about the horizontal/vertical conflicts for the dies of different technology processes.

We assume that a grid structure with p rows and q columns is imposed on the reticle as shown in figure 5. We assume that there is at most one die allocated to each grid cell and the die is aligned with left-bottom corner in the grid cell. The width of column j in the grid structure is determined by the width of the widest die in column j. Similarly, the height of row i in the grid structure is determined by the height of the tallest die in row i.

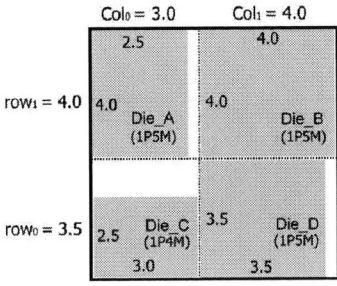

Fig. 5. A grid structure with two rows and two columns imposed on a reticle.

Variables used in our ILP formulation are:

- x_{ijk} denotes whether die k is allocated to row i and column j in the grid or not. $x_{ijk} = 1$ if die k is allocated to row i and column j, and $x_{ijk} = 0$ otherwise.

- r_k denotes whether die k is rotated or not. $r_k = 1$ if die k is rotated, and $r_k = 0$ otherwise.

- row_i denotes the height of row i in the grid. For the example in figure 5, $row_0 = 3.5$ and $row_1 = 4.0$.

- col_j denotes the width of column j in the grid. For the example in figure 5, $col_0 = 3.0$ and $col_1 = 4.0$.

- rc_i denotes the maximum number of the same technology process dies in row i. For the example in figure 5, $rc_0 = 1$ and $rc_1 = 2$.

- cc_j denotes the maximum number of same technology process dies in column j. For the example in figure 5, $cc_0 = 1$ and $cc_1 = 2$.

Constants for the ILP formulation are as follows,

- W_k, H_k, D_k denote the width, height and demand of die k, respectively, where $1 \leq k \leq N$.

- $Total\_Dmd$ denotes the total demand of the dies, i.e., $Total\_Dmd = \sum_{k=1}^{N} D_k$.

- T_m denotes the set of dies which must be produced by the same technology process 1PmM. For example in figure 5, $T_4 = \{Die\_C\}$ and $T_5 = \{Die\_A,\ Die\_B,\ Die\_D\}$

- Rw, Rh denote the given maximum width and height of the reticle.

We formulate the shuttle mask floorplan problem as an integer linear program as follows,

$$min\ (\sum_{i=1}^{p} rc_i + \sum_{j=1}^{q} cc_j) - \sum_{k=1}^{N} (\frac{D_k}{Total\_Dmd} \sum_{i=1}^{p} \sum_{j=1}^{q} x_{ijk})$$

s.t.

$$\sum_{i=1}^{p} \sum_{j=1}^{q} x_{ijk} \geq 1 \qquad \forall k \quad (1)$$

$$\sum_{k=1}^{N} x_{ijk} = 1 \qquad \forall i,j \quad (2)$$

$$H_k(x_{ijk} - r_k) + W_k(x_{ijk} + r_k - 1) \leq row_i \qquad \forall i,j,k \quad (3)$$

$$W_k(x_{ijk} - r_k) + H_k(x_{ijk} + r_k - 1) \leq col_j \qquad \forall i,j,k \quad (4)$$

$$row_i \geq 0 \qquad \forall i \quad (5)$$

$$col_j \geq 0 \qquad \forall j \quad (6)$$

$$\sum_{i=1}^{p} row_i \leq Rh \qquad (7)$$

$$\sum_{j=1}^{q} col_j \leq Rw \qquad (8)$$

$$\sum_{k \in T_m} \sum_{j=1}^{q} x_{ijk} \leq rc_i \qquad \forall i,m \quad (9)$$

$$\sum_{k \in T_m} \sum_{i=1}^{p} x_{ijk} \leq cc_j \qquad \forall j,m \quad (10)$$

$$x_{ijk} \in \{0,1\} \qquad \forall i,j,k \quad (11)$$

$$r_k \in \{0,1\} \qquad \forall k \quad (12)$$

(1) guarantees that each die type must be allocated at least one grid cell, in other words, there is at least one instance for each die type in the reticle. Moreover, multiple instances of the same die type can be assigned to multiple grid cells. (2) ensures that there is at most one die allocated to a grid cell. The variables x_{ijk} and r_k in the LHS of (3) and (4) have four possible combinations:

$$\begin{cases} case1: & x_{ijk} = 1,\ r_k = 1; \\ case2: & x_{ijk} = 1,\ r_k = 0; \\ case3: & x_{ijk} = 0,\ r_k = 1; \\ case4: & x_{ijk} = 0,\ r_k = 0. \end{cases}$$

If x_{ijk} is equal to 1, it means that die k is allocated to row i, column j in the grid. The LHS of (3) is equal to the width of die k if $r_k = 1$, otherwise it is equal to the height

779

of die k. However, if $x_{ijk}=0$, no matter $r_k=0$ or 1, the LHS of (3) and (4) is negative. So, when $x_{ijk}=0$, constraints (3) and (4) will become non-binding. Row_i and col_j must be larger than or equal to zero by (5) and (6). If $row_i = 0$ ($col_j = 0$), it means that no die is allocated to row i (column j). (7) and (8) guarantee that the sum of all row heights and the sum of all column widths are not greater than the given maximum reticle dimensions. (9) and (10) are used to calculate the maximum number of dies of the same process in each row and column.

The objective function is set up according to two key factors, the maximum number of dies of the same process in each row(column) and the demand of each die type. By minimizing the maximum number of dies of the same process in each row(column), we can maximize the number of conflict-free dies of the same process. If more dies of the same technology process are arranged in conflict-free positions, more dies can be extracted at the same time. Secondly, we can put more than one instance of a die type on the reticle according to its demand. We set the weight for die type k as $\frac{D_k}{Total\_Dmd}$. The higher the demand of die type k, the larger is its weight. The objective function is set up to encourage replication of die types with higher weights.

After solving the integer linear programming problem, we get a reticle floorplan based on a grid graph. There may be much wasted reticle area because of the differing heights(widths) of the dies arranged on the reticle as shown in figure 6(a). The bottom-left grid cell of figure 6(a) contains a small die, Die_B, and resulted in much wasted area. In order to maximize the area utility of the reticle, we do the following refinement. We duplicate a die as many times as possible within its allocated grid cell as in figure 6(b).

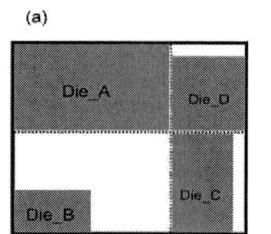 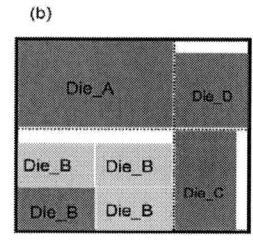

Fig. 6. An example of die replication after floorplanning.(a) The bottom-left grid cell has much wasted area. (b) If we duplicate Die_B in its allocated grid cell, the area utility of the reticle is increased.

IV. SIMULATED ANNEALING-BASED WAFER DICING

In this section, we propose two simulated annealing-based dicing planners. A set of horizontal (vertical) cut lines, called a row (column) dicing plan must be assigned for each row (column) of reticles on the wafer as shown in figure 2. We note that different technology processes must use different wafers. Instead of using a single wafer

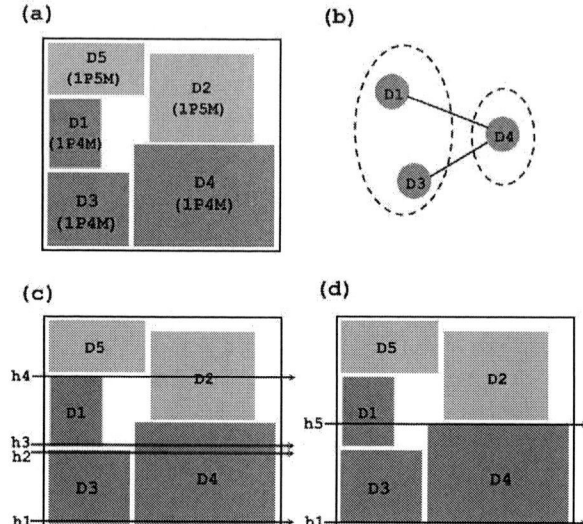

Fig. 7. Finding candidate row dicing plans. (a) A reticle floorplan. (b) A horizontal conflict graph of technology process 1P4M corresponding to the floorplan in (a). {D1, D3} is a maximal independent set and {D4} is another maximal independent set. (c) D1 and D3 are simultaneously dicable by row dicing plan {h1, h2, h3, h4}. (d) D4 is dicable by row dicing plan {h1, h5}.

dicing plan for all wafers, we choose to optimize the dicing plans of different technology process wafers independently. For example, when we compute the wafer dicing plan for 1P4M wafer, our objective is to be able to use the minimum number of 1P4M wafers to satisfy the demands of all 1P4M projects.

We find all the row (column) dicing plans for each technology process m by computing all the maximal independent sets in a horizontal (vertical) conflict graph. Figure 7 shows an example for finding row dicing plans of 1P4M wafer.

Each die of technology process m on a reticle is represented by a node in the horizontal (vertical) conflict graph, there is an edge between two nodes if the corresponding dies are in horizontal (vertical) conflict in the reticle floorplan. An independent set in a conflict graph is a set of nodes without any edge between them. A maximal independent set is an independent set such that no node can be added to the set without breaking the independence property. The dies corresponding to the nodes in the same maximal independent set can be diced out at the same time. We construct an initial wafer dicing plan by assigning a row (column) dicing plan to each row (column) of reticles on the wafer randomly. Then we apply simulated annealing to search for a good wafer dicing plan with the following moves:

- Exchange the row (column) dicing plans of two rows (columns) where the number of printed reticle images of the rows (columns) are not equal.

- Change the row (column) dicing plan of a row (column).

We propose two slightly different wafer dicing planners as follows.

- **D1**: Use the same wafer dicing plan for the wafers of the same technology process. And we use simulated annealing to search for a single wafer dicing plan for each technology process m. For each technology process m, the objective is to search for a wafer dicing plan such that $\max_{k \in T_m} \frac{D_k}{d_k}$ is minimized (i.e., the number of technology process m wafers needed is minimized), where d_k denotes the number of die k obtained from each wafer.

- **D2**: Use different wafer dicing plan for each wafer (even for wafers of the same technology process). For example, for the first wafer of technology process m, we use simulated annealing to compute a wafer dicing plan to minimize $\sum_{k \in T_m} D'_k$ where $D'_k = D_k - d_k$ if $d_k \leq D_k$, and $D'_k = 0$ otherwise. If $\sum_{k \in T_m} D'_k = 0$, then the demands of all dies k of technology process m have been met. Otherwise, there is still some die k of technology process m whose demand has not been met and we need another wafer of technology process m. In this case, we repeat the step above after updating the demand D_k of die k to D'_k. This can be repeated until the demands of all dies k of technology process m are met.

V. Experimental Results

We compared our ILP-based floorplanner with the simulated annealing-based floorplanner in [7]. The characteristics of ten benchmarks are shown in Table I. The smaller benchmarks are provided by [11] and we combined the projects in the smaller benchmarks randomly to generate the larger benchmarks. The demands of the dies in each benchmark vary from 100 to 200. We employed lp_solve 5.1[12] to solve the integer linear programs and implemented the SA-based floorplanner of [7] in C++. All experiments were run on an AMD Opteron processor with 4GB memories. While performing ILP-based floorplanning, we set up the grid size with the smallest p such that p=q and $p * q \geq N$.

In the first experiment, we wanted to compare the performance of our floorplanner and the floorplanner in [7] if all dies could use the same technology process. So, we computed the number of wafers needed using the simulated annealing-based method in [7], and we also computed the number of wafers required using our floorplanner. The results are reported in Table II. For all benchmarks, the minimum required wafers based on our ILP-based_floorplanner is less than the floorplanner in [7]. The maximum wafer reduction is 52% and the average reduction is 30%. Moreover, the run-time of our ILP-based_floorplanner is also faster than the previous method.

In the second experiment, we assumed that different technology processes must be used for the dies in each benchmark as shown in Table I. The minimum required wafers for each benchmark is reported in Table III. Our

ILP-based_floorplanner is again better than the simulated annealing-based floorplanner and resulted in 50% maximum wafer reduction and 28% reduction on average. The run-time of our floorplanner is again faster.

Finally, we tried two different mechanism of wafer dicing. We used the same floorplanner (ILP-based_floorplanner) with the two dicing planners D1 and D2. The minimum required wafers for each benchmark is given in Table IV. The results of using a different wafer dicing for each wafer is clearly better than the results of using the same wafer dicing for the same process wafers.

VI. Conclusion

In this paper, we proposed an integer linear programming (ILP)-based floorplanner for shuttle runs consisting of projects requiring different desired processes. Our floorplanner incorporate die replication on a reticle during floorplanning to reduce the total number of wafers needed to meet the different demands of different die types. Two simulated annealing-based wafer dicing planners were also presented. Experimental results show that our approach achieves 28% wafer reduction on average compared to a previous SA-based floorplanner and our method is computationally efficient.

References

[1] W. Gorbman, R. Boone, C. Phibin, and B. Jarvis, "Reticle Enhancement Technology Trends: Resource and Manufacturability Implication for the Implementation of Physical Design" in *Proc. of ACM/IEEE on ISPD*, pp. 45-51, 2001

[2] John, and B. Lin, "Mask Cost and Cycle Time Reduction" http://www.sematech.org/resources/litho/meetings/mask/20011001/E_TSMC.PDF

[3] C. Yang, "Challenges of Mask Cost & Cycle Time" http://www.sematech.org/resources/litho/meetings/mask/20011001/K_Mask_cost_Intel.pdf

[4] S. Chen and E. C. Lynn, "Efficient Placement of Chips on a Shuttle Mask", in *Proc. of SPIE*, Vol. 5130, 2003, pp.681-688.

[5] M. Andersson, C. Levcopoulos, and J. Gudmundsson, "Chips on Wafers" in *Proc. Workshop on Algorithms and Data Structures*, 2003.

[6] G. Xu, R. Tian, D. F. Wang, and A. Reich "Shuttle Mask Floorplanning" in *Proc. of SPIE*, Vol. 5256, 2003, pp.185-194.

[7] A. B. Kahng, I. Mandoiu, Q. Wang, X. Xu, and A. Zelikovsky, "Multi-Project Reticle Floorplanning and Wafer Dicing" in *Proc. of ACM/IEEE on ISPD*, 2004, pp. 70-77.

[8] G. Xu, R. Tian, D. Z. Pan, and D. F. Wang "A Multiple-objective Floorplanner for Shuttle Mask Optimization" in *Proc. of SPIE*, Vol. 5567 , 2004, pp.185-144.

[9] A. B. Kahng, and S. Reda, "Reticle Floorplanning With Guaranteed Yield for Multi-Project Wafers" in *Proc. of ACM/IEEE on ICCD*, 2004, pp. 106-110.

[10] M. C. Wu and R. B. Lin, " Reticle Floorplanning and Wafer Dicing for Multiple Project Wafers" in *Proc. of ISQED*, 2005.

[11] Global UniChip Corp. http://www.globalunichip.com/

[12] M. Berkelaar, K. Eikland, P. Notebaert, *lp_solve*, available from http://groups.yahoo.com.tw/group/lp_solve

TABLE I
THE CHARACTERISTICS OF EACH BENCHMARK.

Benchmark	M1	M2	M3	M4	M5	M6	M7	M8	M9	M10
No. of die types	10	10	14	15	15	16	18	18	20	20
No. of Technology processes	4	5	4	4	4	4	5	3	5	4

TABLE II
THE EXPERIMENTAL RESULTS OF EACH BENCHMARK IF ALL DIES USED THE SAME TECHNOLOGY PROCESS. THE REQUIRED WAFERS MEANS THE NUMBER OF 200MM WAFERS REQUIRED TO SATISFY THE DEMAND OF EACH DIE IN EACH BENCHMARK.

Benchmark	**SA**-based_floorplanner+D1		**ILP**-based_floorplanner+D1		Reduction of
	Required wafers	run-time(sec)	Required wafers	run-time(sec)	Required wafers (%)
M1	9	325	7	18	22
M2	14	132	11	2	21
M3	16	122	14	11	12
M4	17	426	8	24	52
M5	26	281	14	26	46
M6	16	409	12	13	25
M7	17	646	10	217	41
M8	17	941	12	104	29
M9	17	1229	16	105	5
M10	25	1136	14	656	44

TABLE III
THE EXPERIMENTAL RESULTS OF EACH BENCHMARK FOR THE NEW MPW PROBLEM WHICH CONSIDER THE DESIRED TECHNOLOGY PROCESS OF EACH DIE.

Benchmark	**SA**-based_floorplanner+D1		**ILP**-based_floorplanner+D1		Reduction of
	Required wafers	run-time(sec)	Required wafers	run-time(sec)	Required wafers (%)
M1	11	229	8	25	27
M2	14	345	14	2	0
M3	24	359	14	24	41
M4	24	401	12	107	50
M5	28	461	14	26	50
M6	32	248	21	243	37
M7	18	853	15	717	16
M8	18	988	13	693	27
M9	18	944	17	95	5
M10	26	1348	18	703	30

TABLE IV
THE EXPERIMENTAL RESULTS OF EACH BENCHMARK WITH THE TWO WAFER DICING PLANNERS

Benchmark	ILP-based_floorplanner+**D1**		ILP-based_floorplanner+**D2**		Reduction of
	Required wafers	run-time(sec)	Required wafers	run-time(sec)	Required wafers (%)
M1	8	25	7	393	12
M2	14	2	8	65	42
M3	14	24	10	2	28
M4	12	107	9	587	25
M5	14	26	11	6	21
M6	21	243	19	14	10
M7	15	717	10	395	33
M8	13	693	13	2814	0
M9	17	95	15	487	11
M10	18	703	16	723	11

Design Space Exploration for Minimizing Multi-Project Wafer Production Cost

Rung-Bin Lin, Meng-Chiou Wu, Wei-Chiu Tseng, Ming-Hsine Kuo, Tsai-Ying Lin and Shr-Cheng Tsai

Computer Science and Engineering
Yuan Ze University
Chung-Li, 320 Taiwan
csrlin@cs.yzu.edu.tw, {mcwu, cyber, aman, tsaiyin, hamdo}@vlsi.cse.yzu.edu.tw

Abstract - Chip floorplan in a reticle for Multi-Project Wafer (MPW) plays a key role in deciding chip fabrication cost. In this paper[1], we propose a methodology to explore reticle flooplan design space to minimize MPW production cost, facilitated by a new cost model and an efficient reticle floorplanning method. It is shown that a good floorplan saves 47% and 42% production cost with respect to a poor floorplan for small and medium volume production, respectively.

I. Introduction

Multi-project wafer (MPW) has long been used for low-volume IC production and for fabrication of educational chips [1] so that the sky-rocketing mask cost [2] can be shared among the chips. The designs participating MPW production are first placed into a reticle. This task is called reticle floorplanning [3,4]. The reticle is then repetitively exposed using lithographic equipments to form design patterns on a wafer during fabrication. Once a wafer is fabricated, dice must be cut from the wafer. This task is called wafer dicing. Much work on reticle floorplanning and wafer dicing has been carried out recently to optimize a metric called dicing yield [3]. Dicing yield is defined as a ratio of good bare dice to the required production volume, obtained by dicing a given number of wafers. The larger the dicing yield is, the fewer wafers are needed. Although it is a good metric to evaluate the quality of a reticle floorplanning and wafer dicing method, it does not always correspond well to the mask and wafer production cost, especially for larger volume production. The reason for this is that a reticle floorplan with a larger dicing yield may also use more reticle area which incurs a higher mask cost. Yet another reason is that a reticle floorplan with larger dicing yield may be mainly due to the use of a smaller reticle which would incur more exposure cost during wafer production. As a consequence, it may not be wise simply to develop a reticle floorplanning method that only maximizes dicing yield. Actual mask and wafer production cost should be used to qualify a reticle floorplan.

In this paper, we propose a methodology to explore reticle floorplan design space to minimize MPW fabrication cost. We present a revised formula to compute the MPW fabrication cost assumed by each individual project. We also develop an efficient reticle floorplanning method based on simulated annealing (SA) to facilitate design space exploration. We find that a good floorplan has 47% and 42% saving in production cost with respect to a poor floorplan for small volume production and medium volume production, respectively. All this can be done with a 5-hour overnight run on a 64-bit PC. Our study also finds that reticle area generally corresponds well to production cost especially for small volume production, but a design space exploration is strongly recommended for achieving minimal-cost production.

The rest of this paper is organized as follows. Section II reviews wafer dicing and reticle floorplanning problems. Section III depicts an MPW cost model. Section IV presents a compatibility and area driven floorplanner. Section V proposes a design space exploration methodology. Section VI carries out experimental studies with two industry test cases. The last section draws some conclusions.

II. MPW Dicing and Reticle Floorplanning

Fig. 1 shows an example of an MPW where 10 chips are placed on a reticle [5]. Forty replications of the same reticle are on the wafer, i.e., 40 layouts on the wafer per chip. However, with side-to-side dicing constraints (a dicing line starts from one side of a wafer and must stop at the other side of the wafer), we can not obtain the above number of bare dice because chips 6 and 7 will be destroyed and chip 1 will be discarded when dicing lines h2, h3, v1, and v2 are used to obtain chip 8. A good bare die considered in this paper is a die with four dicing lines located on its four borders, respectively, and without any other dicing lines across it. Two chips are said to be *compatible* if they can be good bare dice at the same time. The dicing lines used to obtain some good bare dice in a wafer (reticle) form a wafer (reticle) dicing plan. A wafer dicing problem can be formulated as follows:

Given a reticle floorplan of N chips and the required production volume V_p for chip $p = 1..N$, find the dicing plans for a minimal number of wafers such that the dicing outcomes can attain the production volumes of all chips.

A dicing yield z_k is defined as that at least $z_k V_p$ good dice for each chip p must be obtained from dicing k wafers. The number of wafers for achieving the production volume is

$$Q = k \lceil 1/z_k \rceil. \tag{1}$$

The advantage of this dicing approach is that we need only decide the dicing plans for a few wafers and repeatedly apply these dicing plans to the rest of wafers. However, (1) gives only an upper bound on the number of needed wafers. A tighter lower bound is as follows.

---

[1] This work is supported in part by the National Science Council, Taiwan, under Grant NSC 94-2215-E-155-005

$$Q = \lceil k/z_k \rceil \qquad (2)$$

This lower bound can be employed by the dicing method such as *HVMIS-SA-Z* to obtain a better solution [6]. Here, we will use *HVMIS-SA-Z* to perform wafer dicing.

A reticle floorplanning problem is formulated as follows:

Given a set of N chips and their required production volumes, decide the coordinates of the chips such that the number of wafers used to attain the required production volumes is minimized on the condition that no chips overlap and all chips are inside the reticle whose dimensions are not larger than maximally allowable values.

The floorplanning objective function is very difficult to evaluate exactly if wafer dicing is not performed. In [5], the authors use compatibility to account for dicing yield during floorplanning. A higher compatibility score for a floorplan implies that more dice in the same reticle can be good at the same time. However, this does not necessarily mean that the floorplan would render a higher dicing yield. The number of reticles replicated in a wafer also plays an important role in getting higher dicing yield. Reticle size and compatibility are two competing factors that make reticle floorplanning a hard problem. Thus, in this paper we propose to explore the reticle floorplan design space to find an answer to this problem.

III. MPW Fabrication Cost

Mask cost is mainly incurred by data preparation, mask write, mask inspection, etc [7]. Mask yield highly depends on the number of (very) critical layers used in a chip and the chip area. The materials used for mask tooling also incur a considerable portion of mask cost, especially for advanced technology nodes. The main contributors to wafer costs include exposure, hot process, etch, sputter, polish, etc [8]. Among them, exposure cost highly depends on the type of layers employed in a chip. An exposure on a very critical layer may cost as much as five times than an exposure on a non-critical layer [7]. Exposure cost also depends on the number of reticles on a wafer, i.e., depends on the wafer field size (equivalent to 1X reticle size). This part of cost is called *field-size dependent cost* in [9]. The part of wafer cost other than exposure cost is called *field-size independent wafer cost*.

There are not too many MPW cost model found in the open literature [1,10]. The formulas given in MOSIS web site for calculating the cost based on chip area is in fact a pricing model which often includes a targeted profit margin. Based

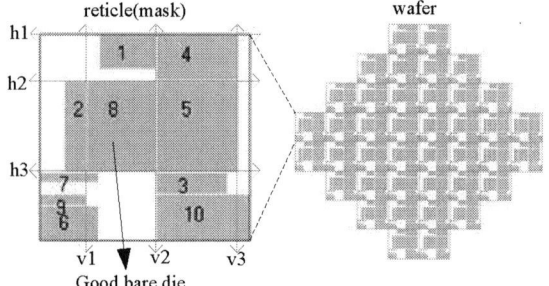

Fig. 1. A multi-project wafer.

on our study, the price given in MOSIS web site is generally a piece-wise linear function of chip area for low and equal volume production. Several pricing models have been proposed in [11] to entice customers purchasing more reticle area with lower cost. The first MPW fabrication cost model is presented in [9]. This cost model includes two parts. The first part calculates the total MPW fabrication cost for a reticle floorplan with reticle area A, which is

$$T_{mpw}(A) = C_m(A) + Q(A)C_e(A) + Q(A)C_w \qquad (3)$$

Where $C_m(A)$, $C_e(A)$ and C_w are the mask cost, exposure cost per wafer, and field-size independent wafer cost, respectively; $Q(A)$ is the number of wafers needed to satisfy volume requirements.

The second part of the cost model [11] gives a formula to compute the cost assumed by each individual project. This cost model is generally correct except that it assumes the wafer exposure cost shared by a chip is inversely proportional to its area. This assumption causes an unfair share of cost, i.e., a smaller chip may pay more than a larger chip does for equal production volume. Here we present a revised formula to correct this deficiency. Given N chips whose areas and required production volumes are respectively A_p and V_p for $p = 1..N$, the cost assumed by chip p is

$$C_{mpw}(p) = C_m(A)A_p \Big/ \sum_{i=1..N} A_i +$$
$$C_e(A)Q(A)V_p \Big/ \sum_{i=1..N} V_i + Q(A)C_w A_p V_p \Big/ \sum_{i=1..N} A_i V_i \qquad (4)$$

The first term is the share of mask cost, the second term is the share of exposure cost, and the third term is the share of field-size independent wafer cost. Our reasoning is that no matter how large a chip is, an exposure of a reticle will yield a copy of the chip. Therefore, exposure cost shared by a chip should be independent of chip area and is proportional to its production volume.

We observe that a reasonable cost model should possess the following two properties:

- A smaller chip should pay less than a larger chip if they have the same production volume.

- A chip with larger production volume should pay more than a chip with smaller production volume if they have the same area.

We need to check whether (4) possesses these two properties. We first obtain a reticle floorplan for the test case I6 (see TABLE II) using the SA approach presented in the next section. The dicing algorithm HVMIS-SA-Z [6] is used to obtain the number of wafers needed. All data used to compute the cost can be found in the experimental section. To check the first property, we arbitrarily set the required volume to 5500 dice per chip. As one can see from Fig. 2, the MPW fabrication cost is almost a linear function of chip area if the chips have same production volume. Thus, (4) possesses the first property. To check the second property, we scale up the volume requirement given in TABLE II by 50 times to make volume dependent cost more obvious. As one can see from Fig. 3, the chip with dimension (4.5,5.0) and the chip with dimension (6.5,3.5) have almost the same area, but

784

the later pays considerable more money for its fabrication owing to its larger volume. Thus, (4) also possesses the second property.

From the cost point of view, we want mask cost as small as possible for small volume production, but for large volume production we want to balance the three cost factors in (3). Given a set of chips and their production volumes, what should their minimal-cost floorplan look like? This is a difficult question. As it was said previously, answering this question requires a design space exploration. To perform such a task, we need an efficient reticle floorplanner and dicing method. However, the reticle floorplanners found in the literature are either too restricted or too time consuming to prohibit design space exploration. Thus, we will present an efficient floorplanner in the next section.

IV. Compatibility and Area Driven Floorplanner

Our reticle floorplanner can adjust a coefficient to favor either reticle area minimization or compatibility maximization. It uses Simulated Annealing (SA) to place chips in a reticle which is divided into a matrix of grids as shown in Fig. 4, where W_{max} and H_{max} are maximum reticle width and height. Grids are sized to a number such that the boundary of a chip will align the grid lines when the chip is placed in the reticle. The objective function is

$$Max \ (1-\delta_1-\delta_2)\sum_{p=1}^{N-1}\left(\sum_{q=p+1}^{N}E_{pq}(V_p+V_q)\right)-\delta_1\beta WH-\delta_2\beta R \quad (5)$$

Where $\delta_1+\delta_2 \leq 1$. The first term accounts for compatibility. It tries to maximize the compatibility among chips whose production volumes are large. The second term accounts for

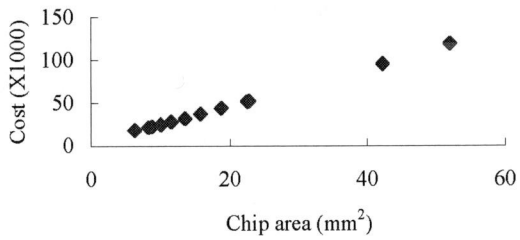

Fig. 2. Cost shared by chips with same production volume.

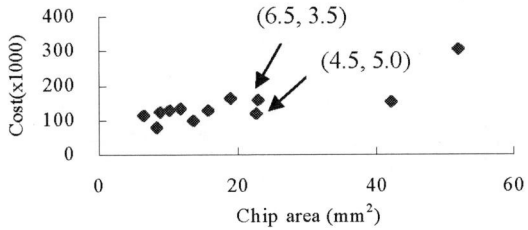

Fig. 3. Cost shared by chips with different production volumes.

reticle area which should be minimized. The third term accounts for chip overlap which should be eliminated to obtain a feasible solution. E_{pq} is 1 if chips p and q are compatible in a reticle. Otherwise, E_{pq} is 0. $\beta=(N-1)\sum_{p=1}^{N}V_p/(W_{max}H_{max})$ is a normalizing factor; W and H are reticle width and height, respectively; R is the total overlap area of chips. Our objective function is similar to the one used in the ILP model presented in [5]. The notable difference is that our objective function has a penalty on chip overlap. Allowing chip overlap during floorplanning would render SA more ropes to find better solutions. Four neighborhood structures are employed.

● Move a chip to a new location.
● Rotate a chip.
● Move a chip and then rotate it.
● Move a chip and align it with another chip.

To move a chip, we randomly select a chip and then a legal location for that chip. Any move or rotation is prohibited if its change would place a chip outside a predefined maximum reticle area. The third neighborhood structure becomes the first one if rotation of a chip at its new location is prohibited. Basically, we don't need the third neighborhood structure if we have the first two. Since we penalize chip overlap heavily to maintain a solution as feasible as possible, it is sometimes hard to place a chip at a location that first causes overlapping which is removed subsequently by a rotation. For example, as shown in Fig. 4, an attempt to move chip F to (0,3) might not be successful because of overlapping with chips C and E so that F has no chance being rotated at (0,3). However, the third neighborhood structure works just fine for this. A typical cooling schedule with cooling coefficient 0.95 is used. Simulated annealing is terminated when the best solution is not improved for a number of consecutive inner loops. The coefficient δ_2 for chip overlap is set to a very large number to avoid obtaining an infeasible solution. δ_1 can be adjusted to bias the optimization toward reticle minimization or compatibility maximization. This allows us to perform an exploration on reticle floorplan solution space. Our experiments show that the best floorplans obtained by the SA floorplanner is as good as or even better than those obtained by the approaches presented in [5].

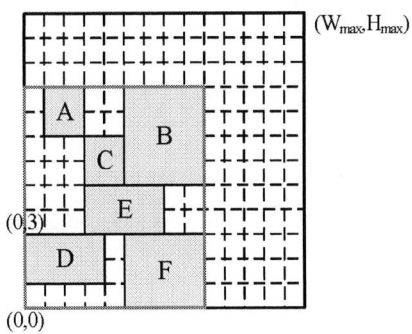

Fig. 4. Discretized reticle for floorplanning.

V. Space Exploration Methodology

Here we propose a methodology to obtain a floorplan that would minimize the total MPW fabrication cost. The cost model proposed in Section III is used for cost calculation. This methodology consists of three steps:

- Use the SA floorplanner presented in Section IV to perform design space exploration by varying the coefficient δ_1 from 0.000001 to 0.000009 with δ_2 =0.99999.
- Employ the *HVMIS-SA-Z* dicing method [6] to perform wafer dicing for each floorplan.
- Use the cost model to calculate the total fabrication cost based on the reticle size and the number of wafers used. The reticle floorplan incurring least total MPW fabrication cost is selected as the final solution. The cost assumed by each project is computed using (4) accordingly.

An integrated tool that combines SA floorplanner, *HVMIS-SA-Z* dicing method, and MPW cost calculator has been implemented to facilitate design space exploration. Note that other robust and efficient floorplanners can also be used in our methodology as well.

VI. Experimental Results

In this section we perform a case study on the design space exploration using the methodology proposed in Section V. All experiments are done on a 1.8GHz 64-bit PC. Our study is made on 300 mm wafers. Although there exists much work about mask cost of ownership analysis, for simplicity we will use the mask cost data from [7]. These data given in TABLE I are calculated for 90nm technology node assuming that a chip has 8 very critical layers, 8 critical layers, and 12 non-critical layers. They are originally estimated for a reticle containing a number of dice of the same design with 8*8 mm² wafer field size (equivalent to 1X reticle size). Note that the mask cost of MPW is somewhat underestimated due to underestimating its data preparation time and inspection time. We also use the data about cost per exposure in [7] to compute the exposure cost per wafer. The cost per exposure is $2.5 for very critical layer, $1.5 for critical layer, and $0.5 for non-critical layer. For the time being, we assume that wafer field-size independent wafer cost [12] is C_w=$2500. Two test cases in TABLE II obtained from the industry [13] are used in our experiments. These two test cases have drastically different characteristics. All chips in I5 have different dimensions with total chip area much smaller than the maximum reticle size. Many chips in I6 have same widths and heights with total chip area more than half the maximum reticle size.

We use the method proposed in Section V to find a floorplan with minimum MPW fabrication cost. Figs. 5 and 6 show total MPW fabrication costs for different reticle floorplans for I5 with 1X and 50X volumes, respectively. Figs. 7 and 8 serve a similar purpose for I6. Each bar consists of mask cost, exposure cost, and field-size independent wafer cost. As shown in these figures, mask cost dominates the total wafer fabrication cost for low (1X) production volume. With larger (50X) production volume, exposure cost takes a lion share of the total cost so that the number of exposures per wafer and the number of wafers needed to satisfy production

volumes are key to the total fabrication cost. These data show that a good floorplan for I5 (I6) has a 47% (30%) saving in production cost with respect to a poor floorplan for small volume production and 41% (42%) saving for medium volume production. The time takes to perform such a design space exploration for a case is about 5 hours. Compared to the amount of cost saving, this overnight run is indeed a great pay-off. Once we have a reticle floorplan that results in minimum MPW fabrication cost, we can easily compute the cost assumed by each chip using (4).

TABLE I
Mask set cost for different wafer field sizes

Wafer field size	25*25 625 mm²	16*24 384 mm²	16*16 256 mm²	8*16 128 mm²	8*8 64 mm²
Mask cost	1,240,000	728,000	532,000	352,000	296,000

TABLE II
Test cases

(w, h | 1X required volume) W_{max}=20 mm, H_{max}=20 mm
I5 (2.5,6.25 | 100]), (1.8,5.5 | 200), (2,1.25 | 300), (2.2,1.75 | 200), (1.7,2.25 | 200), (1.5,1.55 | 200), (2.3,3.75 | 200), (1,3.25 | 200), (1.3,4.25 | 80), (2.7, 1.1 | 60)
I6 (6.5, 6.5 | 60), (4.5, 5.0 | 100), (5.5, 1.5 | 120), (4.5, 3.0 | 120), (6.5, 3.5 | 160), (4.5, 3.5 | 160), (6.5, 8.0 | 200), (3.3, 3.5 | 200), (2.5, 3.5 | 200), (3.5, 2.5 | 200), (7.5, 2.5 | 200), (4.0, 2.5 | 200), (2.5, 2.5 | 200)

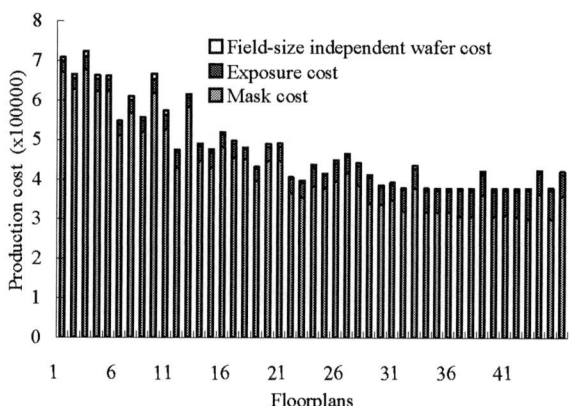

Fig. 5. MPW fabrication cost for I5 with 1X volume.

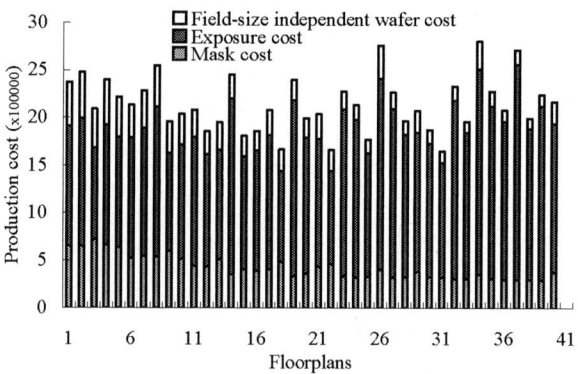

Fig. 6. MPW fabrication cost for I5 with 50X volume.

2006 Asia and South Pacific Design Automation Conference

8A-4

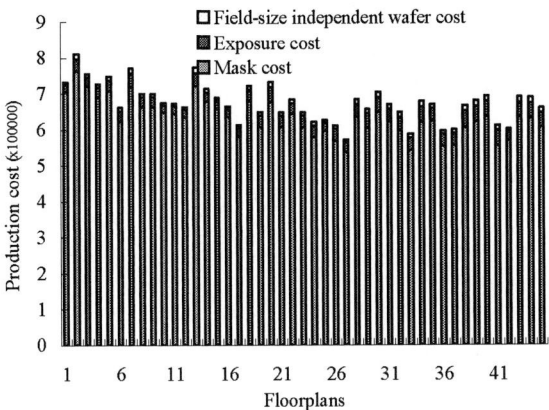

Fig. 7. MPW fabrication cost for I6 with 1X volume.

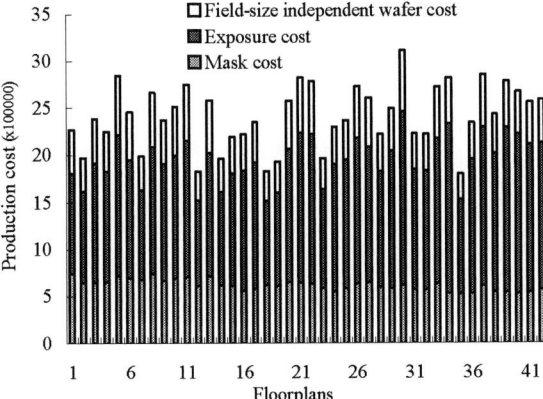

Fig. 8. MPW fabrication cost for I6 with 50X volume.

Figs. 9 and 10 show the relations among reticle area, total fabrication cost, and dicing yield for I5 with 1X and 50X volume requirement, respectively. Figs. 11 and 12 serve a similar purpose for I6. Dicing yield is obtained from dicing three wafers at the same time. Normally, a smaller reticle corresponds to a higher dicing yield for I5, but this phenomenon is not that obvious for I6. Given the same reticle area, the dicing yields of different floorplans for I5 do not spread as widely as that for I6. This is due to the fact that the chips in I5 have different dimensions so that different floorplans within a given reticle area will not have a large difference in compatibility and thus in dicing yield. However, this is not true for I6 where the alignment of chip placement would make a bigger difference in dicing yield. For I5 with 1X production volume, the larger the yield, the smaller the cost. For I6 this is generally true, but a larger spread in production cost for the floorplans with the same yield is observed. Since mask cost dominates the total cost for low-volume production, it is the smaller reticle area rather than the higher yield that decides the total cost. This can be clearly observed from the data given in Figs. 9 and 11. Therefore, for low-volume production, the smaller the reticle area, the lower the total MPW fabrication cost. However, this does not always happen for 50X production volume. We have

observed that a higher dicing yield of many floorplans is mainly derived from more exposures per wafer. If we measure the yield per reticle exposure, the floorplans with a higher dicing yield are commonly seen to score low for this metric than those floorplans with a smaller dicing yield. This is especially true for I5. We have observed an instance for I5 that a floorplan with 1/3 of the reticle area, twice the dicing yield, and more than three times the number of exposures per wafer of another floorplan incurs 20% more cost. This gives the reason why the points in the center picture of Fig. 10 are spread much widely than those in Fig. 12. The degree of spreading highly depends on the underlying problem instance. Based on the above observations, it is clear that neither reticle area nor dicing yield can solely decide the total production cost, especially for large-volume production. Thus, performing a reticle floorplan design space exploration is required for finding a minimal-cost reticle floorplan.

VII. Conclusions

This paper has presented a methodology to explore MPW reticle floorplan design space facilitated by a compatibility and area driven floorplaner. A new formula is introduced to compute the MPW fabrication cost assumed by each chip. It is shown that a good floorplan saves 47% production cost with respect to a poor floorplan for small volume production and saves 42% cost for medium volume production. Although reticle area generally corresponds well to production cost especially for small volume production, a design space exploration is strongly recommended for achieving minimal-cost production irregardless of volume requirements.

References

[1] C. A. Pina, "MOSIS: IC prototyping and low volume production service," Proc. of Intl. Conf. on Microelectronic Systems Education, 2001.
[2] M. LaPedus. Is IC industry heading to the $10 million photomask?. Semiconductor Business News, Oct. 7, 2002.
[3] A. B. Kahng, I. Mandoiu, Q. Wang, X. Xu, and A. Z. Zelikovsky, "Multi-project reticle floorplanning and wafer dicing," Proc. of ISPD, pp.70-77, 2004
[4] G. Xu, R. Tian, D. Z. Pan, D.F. Wong, "A multi-objective floorplanner for shuttle mask," Proc. of SPIE, vol 5567, pp. 340-350, 2004.
[5] M. C. Wu and R. B. Lin, "Reticle floorplanning and wafer Dicing for Multiple Project Wafers," ISQED, pp. 610-615, 2005.
[6] M. C. Wu and R. B. Lin, "A comparative study on dicing of multiple project wafers," IEEE Computer Society Annual Symposium on VLSI, pp. 314-315, 2005
[7] D. Pramanik, H. Kamberian, C. Progler, M. Sanie, and D. Pinto, "Cost effective strategies for ASIC masks," Proc. of SPIE, vol. 5043, pp. 142-152, 2003.
[8] S. Miraglia, C. Blouin, G. Boldman, S. Judd, T. Richardson, and D. Yao, "ABC modeling: advanced features," Advanced Semiconductor Manufacturing Conf., pp. 336-339, 2002.
[9] M. C. Wu and R. B. Lin, "Multiple project wafers for medium-volume IC production," ISCAS, pp. 4725-4728, 2005.

[10] J. Bonn, S. Sisler, and P. Tivnan, "Balancing mask and lithography cost," Advanced Semiconductor Manufacturing Conf., pp. 25-27, 2001.

[11] T. Y. Yang, L. I. Tong, B. J.C. Yuan, "An innovative model of multi-project wafer service in the foundry industry," International Journal of Technology Management, Vol. 30, No.1/2 pp. 172 – 187, 2005.

[12] "Optical lithography cost of ownership(COO) – Final Report for LITG501," International SEMATECH.

[13] Global UniChip, http://www.globalunichip.com

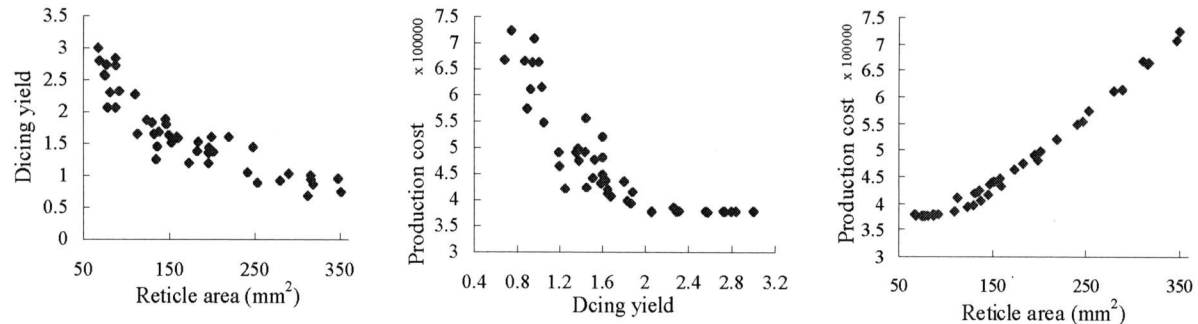

Fig. 9. Area, yield, and cost for I5 with 1X volume.

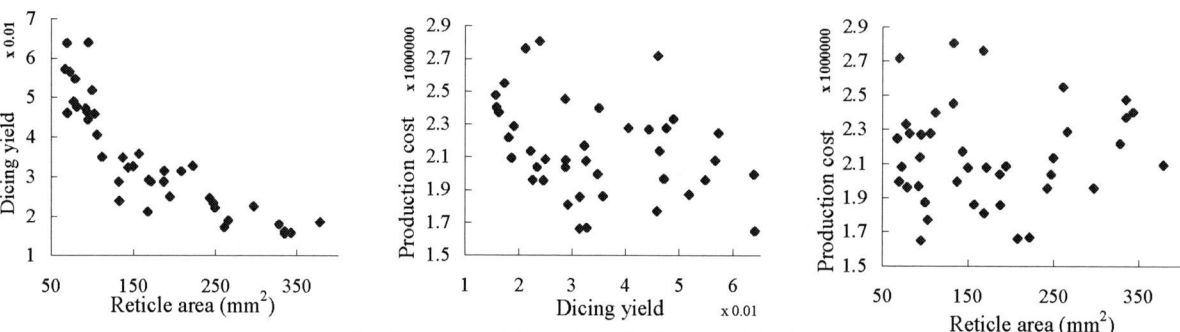

Fig. 10. Area, yield, and cost for I5 with 50X volume.

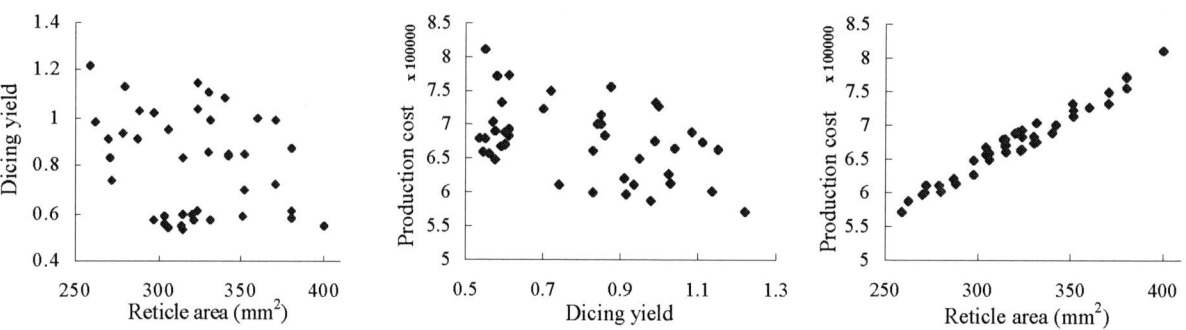

Fig. 11. Area, yield, and cost for I6 with 1X volume.

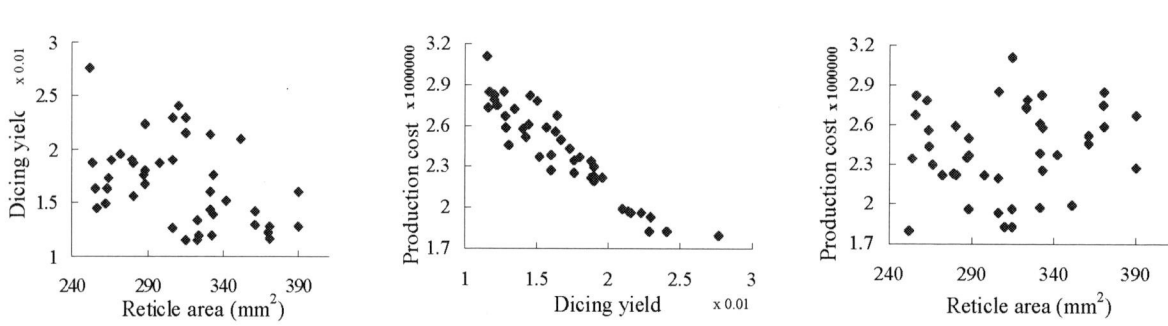

Fig. 12. Area, yield, and cost for I6 with 50X volume.

SAT-Based Optimal Hypergraph Partitioning with Replication

Michael G. Wrighton[1]
Tabula, Inc.
3250 Olcott St.
Santa Clara, CA 95054
mwrighton@tabula.com

André M. DeHon
California Institute of Technology
Computer Science, 256-80
Pasadena, CA 91125
andre@cs.caltech.edu

ABSTRACT

We propose a methodology for optimal k-way partitioning with replication of directed hypergraphs via Boolean satisfiability. We begin by leveraging the power of existing and emerging SAT solvers to attack traditional logic bipartitioning and show good scaling behavior. We continue to present the first optimal partitioning results that admit generation and assignment of replicated nodes concurrently. Our framework is general enough that we also give the first published optimal results for partitioning with respect to the *maximum subdomain degree* metric and the *sum of external degrees* metric.

We show that for the bipartitioning case we can feasibly solve problems of up to 150 nodes with simultaneous replication in hundreds of seconds. For other partitioning metrics, we are able to solve problems up to 40 nodes in hundreds of seconds.

1 INTRODUCTION

Balanced, k-way hypergraph partitioning is a fundamental problem in the design of integrated circuits. The precise details of the partitioning problems vary by application [1], but all known useful formulations of balanced partitioning result in NP-hard optimization problems.

Although effective heuristics exist to solve many partitioning problems, few provably optimal solution techniques have been explored. Where optimal techniques have been applied, significant quality per given runtime improvements have been observed over heuristics for small problem sizes [2]. Rather than focus on time-quality tradeoffs, we decided to explore the limits of solving partitioning optimally.

A key definition in computer science is that any NP-complete problem can be transformed into any other NP-complete problem given only a polynomial amount of time and space [3]. Seminal work nearly two decades ago [4] suggested that partitioning can be efficiently remapped to Boolean satisfiability. Recent advances in SAT solvers allow enormous SAT instance to be solved (or proven UNSAT) within seconds [5]. These results have been obtained across a range of benchmarks represented in a canonical "conjunctive normal form" (CNF). Annual competitions have yielded rapid SAT solver progress in recent years.

A key limitation of the early SAT-mapped partitioning work is that the published results handled a very traditional and specific partitioning formulation. "Real-world" partitioning problems may have complex formulations [6]. The author of the early work, however, recognized a broader potential of the SAT-mapped approach [4]:

"An attractive feature of this approach is that the entire space of feasible solutions can be represented in a compact way, facilitating the search for optimal solutions under complex cost functions and associated constraints."

We developed and benchmarked a framework for SAT-mapping more complex cost functions and constraints than considered in prior work on optimal partitioning. This framework allows us to construct optimal k-way partitions. Further, we consider partitioning balance constraints that allow nodes to appear in multiple partitions (i.e. replication). To illustrate cost function flexibility, this work considers three well-established metrics for partitioning quality: The traditional *total cut hyperedges* metric penalizes every edge that is not fully contained within a single partition. The *sum of external degrees* function penalizes every entry or exit of a wire from a partition. In multi-way partitioning problems, this may be appropriate as solvers targeting this cost function prefer solutions where edges interact with small numbers of partitions over solutions where the same number of edges are split over all of the partitions. Finally, the *maximum subdomain degree* metric limits the maximum IO into any given partition as opposed to the average IO over all of the partitions. Though runtime remains an issue for the more sophisticated metrics, we show techniques that are sufficiently general to consider cost functions that are tightly defined by their application domain.

With this work, we expand the literature on optimal partitioning with the following innovations:

- SAT formulations for three distinct partitioning metrics

- An enhancement for optimal k-way partitioning (as opposed to simple bipartitioning)

- Concurrent replication and partitioning of nodes as a natural part of problem formulation

We show that when considering the traditional bipartitioning problem, our technique scales to more difficult problems than a branch-and-bound implementation. Within that formulation, we can consider integrated replication set generation for not more than an order of magnitude runtime penalty. Finally, for some small netlists, we show the first published optimal results for two sophisticated cost metrics, with and without replicated nodes.

2 PRIOR WORK

We are aware of a single published work on hypergraph partitioning via Boolean SAT [4]. Devadas considers the traditional formulation of hypergraph bipartitioning. He shows that within a reasonable amount of time (14 minutes in the longest case), the hardware and SAT solvers of 1989 could optimally bipartition a benchmark netlist of 32 nodes under the *total cut hyperedges* metric.

Instead of SAT, the recent work to date on optimal netlist partitioning has focused on branch-and-bound techniques — an approach to which the cut hyperedges metric avails itself due to several clever techniques available in that formulation [2]. These optimizations, by their nature, are limited to the traditional bipartitioning formulation.

Previous implementations of k-way and broader cost metric partitioning have been limited to heuristic techniques. For example, Karypsis's group at the University of Minnesota has developed partitioners that generate k-way partitionings to minimize the *sum of external degrees*, *total cut hyperedges*, and *maximum subdomain degree* metrics [7, 8].

The best-known techniques for allowing replication in order to improve partitioning quality are based on network flows [9, 10]. These techniques offer exact solutions to several unbalanced formulations of the bipartitioning problem. At the cost of optimality, they may be used as kernels in heuristics for k-way, balanced hypergraph partitioning and replication.

---

[1] This work was done while the first author was at Caltech.

The partitioning problem is fundamentally no different from any other finite-domain constraint satisfaction problem. This fact led us to apply unsuccessfully a Prolog-based constraint solver [11] to the task. We are not the first to note that constraint solvers which remap their problems to CNF can be significantly faster than solvers that operate on a more direct problem formulation [12].

There are numerous solvers available for SAT instances, a complete exploration of which is beyond the scope of this paper. Readers may refer to annual SAT Solver Competitions at ICSAT for the current state-of-the-art in solvers. We found that `siege_v4` [13] (which does not participate in the competition due to a 'black-box' restriction) generally provided the best performance for our SAT instances.

3 SOLUTION OUTLINE

A well-known technique for solving an NP-hard optimization problem is to transform it into a series of NP-complete decision problems. Generically, we represent this as:

```
while (upperBound != lowerBound) {
 thisTry = (upperBound + lowerBound) / 2;
 if (existsSolutionLessThan(thisTry)) {
   upperBound = thisTry;
 }
 else
   lowerBound = thisTry;
}
```

The kernel of our approach is that we build a SAT problem instance which is satisfiable if, and only if, the circuit netlist can be partitioned with a goodness metric less than or equal to some target (an NP-complete decision problem). In order to be a valid solution which meets a particular cost metric we will assert that:

$$SAT = AllNodesRepresented$$
$$\wedge PartitionsBalanced$$
$$\wedge MetricMet$$

Then after obtaining a SAT/UNSAT result, we have a new upper or lower bound on the optimal solution to use in a binary search for the minimum cost partitioning. Once the bounds are tight, the satisfying assignment to the SAT instance associated with the best goodness metric implies a solution to the NP-hard optimization problem. If we wish to relax the optimality constraint, we can treat problem instances where the solver times out as UNSAT. Otherwise, we must report the optimization problem as unsolved. Our experience has shown that SAT instances asserting a cost metric some distance from the minimum can be solved or proven UNSAT much more quickly than those very near the minimum. We found that this effect was more significant than the count of CNF clauses and variables.

The SAT instances we generate have kN binary inputs, representing which nodes appear in each of the partitions. For 3-way partitioning, the graph in Figure 1 implies these inputs to a satisfiability instance:

$$A_0, A_1, A_2, B_0, B_1, B_2, C_0, C_1, C_2, D_0, D_1, D_2, E_0, E_1, E_2$$

A '1' in the satisfying input assignment indicates that a node appears in a particular partition. This contrasts with Devadas' approach [4], which encodes the partition assignment with a single bit per node.

Having outlined the approach, we have several natural questions.

- How do we assert a valid partitioning?
- How are the various metrics specified in the SAT instance?
- How does the SAT approach scale against other solutions?
- What size problems are feasible to solve for the various cost metrics?

The following sections answer these questions.

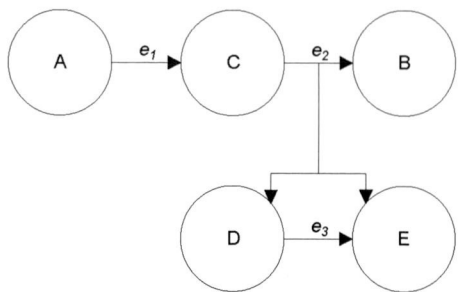

Figure 1. An Example Hypergraph

4 THE DETAILS

As stated in the previous section, we construct a SAT instance (in the form of a Boolean logic circuit). We show how each higher-level asserted variable in our circuit is constructed.

4.1 Cardinality Constraints

At several points, we will need to make an assertion about the number of '1's set over some number of variables. Our notation in this work represents '1's counters with the Σ symbol, both in equations and diagrams. In text, we refer to the '1' counter as a totalizer. We expand on our technique for asserting cardinality constraints in Section 5.1.

4.2 All Nodes Represented

In order to be a valid solution we assert that each node appears in at least one partition. Continuing with the example of a three-way partitioning of the netlist from Figure 1:

$$AllNodesRepresented = \begin{matrix} (A_0 \vee A_1 \vee A_2) & \wedge \\ (B_0 \vee B_1 \vee B_2) & \wedge \\ (C_0 \vee C_1 \vee C_2) & \wedge \\ (D_0 \vee D_1 \vee D_2) & \wedge \\ (E_0 \vee E_1 \vee E_2) & \end{matrix}$$

Or more generally, for a k-way partitioning:

$$AllNodesRepresented = \bigwedge_{A \in Nodes} \left(\bigvee_{0 \le i < k} (A_i) \right)$$

If the problem formulation we are considering does not admit replication, then we can assert that no node appears in two (or more) partitions:

$$NoReplicants = \bigwedge_{A \in Nodes} \bigwedge_{\substack{0 \le i < k \\ i+1 \le j < k}} \overline{A_i \wedge A_j}$$

4.3 Partitions Balanced

We make an assertion on the cardinality of each of the partitions by means of the totalizer described above. If the partitions are balanced then no partition has more than some fraction of the total nodes in the graph.

$$PartitionsBalanced = \bigwedge_{0 \le i < k} \left(\left(\sum_{A \in Nodes} A_i \right) \le MaxSize \right)$$

If replicated nodes are allowed, we may also construct a totalizer for the 'excess' nodes in the design and limit them as desired.

4.4 Metric Met

The subtleties of SAT-mapped partitioning occur when we assert that a metric is less than or equal to some given value. We begin this work by examining a single cost metric. We discuss additional metrics in Section 7.

4.4.1 Total Cut Hyperedges

The *total cut hyperedges* metric is the easiest to describe. We define an edge (e) as a source (e.Source) with a set of sinks (e.Sinks) and assert that it is cut if any of its sink nodes appear in a partition without the source node.

$$Cut(e) = \bigvee_{0 \leq k} \bigvee_{s \in e.Sinks} \left(\overline{e.Source_i} \wedge s_i\right)$$

We employ a totalizer to sum over all the potentially cut edges, and assert that their cardinality is less than our current target.

$$MetricMet = \sum_{e \in Edges} Cut(e) \leq MaxMetric$$

Figure 2 shows how we would build the MetricMet function using the *total cut hyperedges* metric for a three-way partitioning of the sample hypergraph shown in Figure 1.

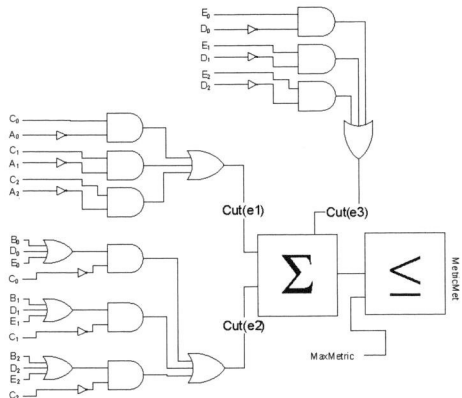

Figure 2. Cut Hyperedges Metric

5 OPTIMIZING IMPLEMENTATION

Having identified problem formulations that guarantee optimal results given unlimited SAT runtime, we move on to consider a few techniques that offer the potential of reducing the total amount of time that the SAT solver will require to solve the generated instances. Intuitively, the key to these techniques is the notion that we can improve the tractability of SAT by assuring that partial assignments of variables allow maximum implications to assign other variables in the instance. We show, on a few sample netlists, how two techniques improve our runtimes.

5.1 Cardinality Constraints

One way to assert cardinality constraints in our SAT instances is via a tree of adders. This is the technique employed by Devadas [4]. A desirable property of this approach is that it requires few clauses or variables in the generated SAT instance. However, SAT solvers operate on a partial assignment of variables. Figure 3 shows that if we limit the '1's cardinality over a set of four variables to one, even after three of the input variables are set to '1', the solver will not be able to prune the search space.

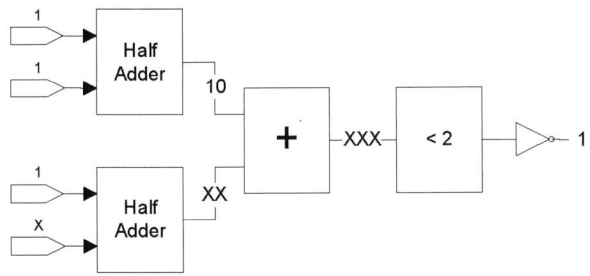

Figure 3. Cardinality constraint expressed in binary arithmetic

We improve on Devadas' cardinality constraints by sacrificing the brevity of the binary approach for the strategy of Bailleux and Boufkhad [12, 14]. They construct binary trees of *totalizers*, each

operating on pre-unate[2] input. The pre-unate representation allows the solver to more quickly discover conflicts and prune the search space with fewer input variables assigned.[3] We then assert SAT on the appropriate inverted output from the top-level totalizer. Figure 4 shows how a SAT solver would interpret the cardinality constraint from before with the same partial assignment of inputs. In fact, as soon as any two input pins are assigned '1', the totalizer tree will produce a contradiction. The complexity of the CNF encoding is $O(N^2)$ clauses and $O(N \log (N))$ variables if we are constraining N variables to a sum of N-1 (the most pathological case).

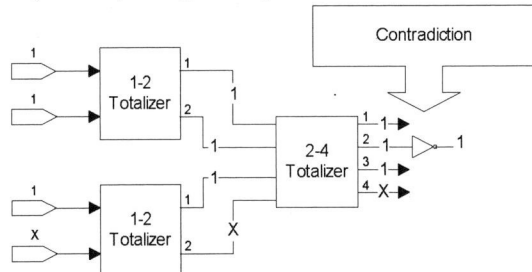

Figure 4. Cardinality constraint expressed with pre-unate totalizers

5.2 Symmetry Breaking

In a k-way partitioning, we have *k!* potential orderings (i.e. symmetries) of the partition sets. When we generate the satisfiability instance, we can break one degree of symmetry by preassigning a single node to an arbitrary partition. We considered several potential heuristics to select the breaking node: The node connected to the most other nodes; the node connected to the least other nodes; and a random node. Table 2 shows that *not* breaking the symmetry reliably causes the longest runtimes over our selected netlists. When we do break the symmetry, choosing a random node seems to provide as good a result as any of the more crafted selection heuristics.

For k > 2 partitioning, preassigning a single node for symmetry breaking leaves obvious symmetries unaddressed. Adding another hard symmetry breaking preassignment (assigning another node to another arbitrary partition) would destroy our optimality guarantee. However, we can safely preassign a second node to *either* the same partition as the first *or* another arbitrary partition. We generalize the technique by inserting a "weak backbone" as a symmetry constraint. We construct a weak backbone for the netlist as follows:

```
weakBackbone(k, netlist) {
  P = heuristicPartition(k, netlist)
  for i ← 0 to k - 1
  do
    Node = randomSelection(P[i])
    assertInOneOf(Node, 0..i)
```

The power of this approach is that it allows us to imply an ordering of the partitions, but without removing the optimality guarantee from the result. Table 2 shows that the weak-backbone approach provides better scalability than simply constraining a single node as the number of partitions generated increases.

---

[2] "Pre-unate" represents k in N bits by setting the first k bits to '1' and the trailing N - k bits to '0'. For example, if N = 5, three is expressed as '00111'.

[3] More formally, unit propagation on any subset of assigned variables restores generalized arc-consistency.

Table 1. Effect of Cardinality Constraints

| Netlist | Size | k | SAT Runtime (ms) | | Speedup |
			Binary	Bailleux & Boufkhad	
ex4	55	2	16360	2762	5.9
		3	20130	4168	4.8
		4	42877	10960	3.9
		5	93999	15332	6.1
		6	174480	22921	7.6
misex2	97	2	16098	1246	12.9
		3	59514	14344	4.1
		4	105858	14678	7.2
		5	160268	16358	9.8
		6	524047	62035	8.4
5xp1	100	2	73936	8631	8.6
		3	207867	40410	5.1
		4	716010	102163	7.0
		5	Timeout	243489	-
		6	Timeout	652259	-
f51m	114	2	7416	2010.25	3.7
		3	25970	3049	8.5
		4	27988	2694	10.4
		5	135951	9881	13.8
		6	533034	34044	15.7
kirkman	151	2	400696	22714	17.6
		3	742069	75493	9.8
		4	1442291	160343	9.0
		5	Timeout	Timeout	-
		6	Timeout	Timeout	-

6 BENCHMARKING RESULTS

In order to show the potential for practical usefulness of our technique, we quantify the performance of our SAT mappings. In this section, we examine our benchmark results for traditional bipartitioning with cut hyperedges as the metric – we perform a side-by-side comparison against an optimized and widely-used branch-and-bound partitioner. We continue to show that adding replication to this formulation does not cause runtime to increase too much.

6.1 Methodology

Our implementation of the approach generates ISCAS89 format files, which represent the satisfiability instances. We convert these to CNF via a Perl script [15]. As benchmark hypergraphs, we employed 4-LUT FPGA mappings (generated via Flowmap [16]) of the small IWLS93 benchmark circuits. These netlists are appropriate to explore this methodology as they consist of nodes of equal size – standard-cell mapped circuits would complicate our construction of balance constraints. We ran our flow on dual-processor 2.8 GHz Intel Pentium 4 machines with 512 KB L2 cache and 4 gigabytes of RAM. Whenever we report a CPU time, it is the total time spent in the SAT solver.

6.2 SAT Solver Choice

The experience of the SAT community is that it is unusual for a given solver to be ideal across a range of problem types. Therefore, we conducted an evaluation of several SAT solvers [13, 17, 18]. We determined that a solver would receive a 'pass' for a partitioning-derived satisfiability instance if it could solve (SAT/UNSAT) the problem within a reasonably long amount of time (which we arbitrarily set at two hours of CPU time). At this stage, we deemed the higher priority to find a solver that would solve many problems than to reduce the average runtime of solved instances. Over many SAT instances, we found that siege_v4 consistently solved more instances than the other solvers.

If a solver superior to siege_v4 appears, it is a simple matter to adjust our flow to leverage the new tool because the SAT community

has widely standardized upon the CNF representation of problem instances.

Table 2. Effect of various symmetry-breaking choices.

| Netlist | Size | k | SAT Runtime (ms) | | | | |
			None	Least Conn.	Most Conn.	Rand.	Weak Back.
ex4	55	2	2782	1448	1202	1622	1678
		3	4226	2687	2238	3311	1851
		4	11480	7361	3440	7226	3079
		5	15652	9491	6238	11006	6422
		6	23563	15420	9169	16311	8840
misex2	97	2	1237	1343	998	928	1092
		3	15082	9293	4291	5481	4443
		4	14740	10695	7269	7095	4814
		5	16671	14101	9831	8530	5463
		6	64989	56707	45003	38717	23974
5xp1	100	2	8692	8477	5663	5347	7317
		3	42566	40228	22779	18809	27546
		4	103773	91713	70658	72145	41012
		5	248916	240253	172330	192982	113177
		6	665235	846848	651781	441434	490670
f51m	114	2	2022	1729	1021	2540	1503
		3	3091	2358	1939	2259	2060
		4	2713	2941	2623	2004	2059
		5	10000	13515	12639	12848	10118
		6	34828	32823	31933	25315	21705
kirkman	151	2	23141	18457	18788	16561	17206
		3	77397	66562	71601	52921	47441
		4	164108	136261	172171	97556	60365
		5	Timeout	574638	687024	876982	501577
		6	Timeout	Timeout	Timeout	Timeout	Timeout

6.3 Scalability Against Branch and Bound

We used our approach and the branch-and-bound solver from Capo [2] to optimally bipartition our benchmark circuits (which have from 10 to 255 nodes). We allowed up to a 10% unbalanced partitioning. We validated that the partitioning metrics generated between the two solvers were identical. Figure 5 shows that while the branch-and-bound approach is superior for many netlists, as the complexity increases, our approach dominates. We present plots sorted by both SAT runtime and branch-and-bound runtime because merely considering node count is not a strong enough predictor of problem complexity to give a clear visualization.

We observe that as branch-and-bound's runtime increases, our SAT-mapped formulation offers much better scaling properties. Branch-and bound times out on many of the benchmarks that SAT completes (even though we allowed the branch-and-bound implementation 10x longer runtime). There are no examples in the benchmark set however where SAT times out and branch-and-bound does not.

6.4 Bipartitioning with Integrated Replication

Typically, adding replication to the logic bipartitioning SAT formulation does not increase the runtime by more than an order of magnitude. We considered the case of allowing two partitions, each 60% of the total size of the netlist. Table 3 shows that, for large benchmarks of fewer than 152 nodes (the largest netlists we could reliably partition without timing out), we can obtain improvements in cutsize with a modest overhead in compute time over the non-replicated case. In many cases, the freedom to replicate nodes allows us to find an optimal solution even more quickly.

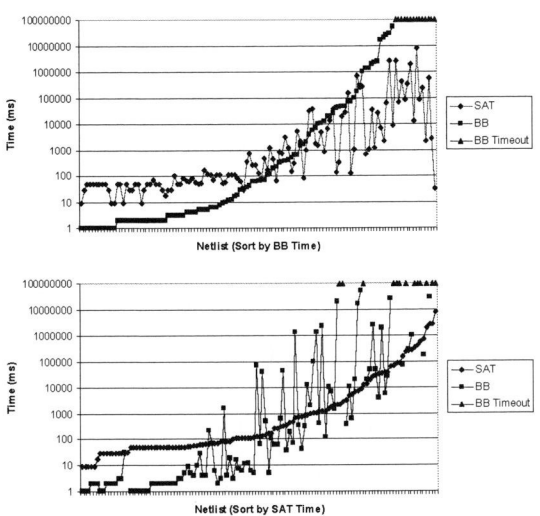

Figure 5. SAT vs. Branch-and-Bound Scaling

7 BEYOND CUT HYPEREDGES

In principle, a key feature of the SAT formulation is that we need not be limited to simple formulations of the partitioning problem. It is not difficult to construct satisfiability instances that represent more sophisticated cost metrics than total cut hyperedges. We conducted experiments on two such metrics (which are formally described in the appendix) and show results for three benchmarks of forty nodes. We report the first results from an optimal algorithm to solve partitioning for these metrics (and allow replicated nodes in the formulation) – unfortunately, our results to date indicate that the SAT formulation appears to be a very slow method of attacking these problems. Our results consider dividing the nodes into partitions of maximum size = (1.2 / k × total node area). In the replicated case we allow nodes to appear in multiple partitions while keeping the maximum size fixed. We report a timeout when the SAT solver times out at 1200 s.

Table 3. Optimal Bipartitioning with Simultaneous Replication

Netlist	Size	No Replication		Replication		Slowdown	% Cutsize
		Cut	ms	Cut	ms		Impr.
c8	131	8	1413	8	2228	1.58	0
sao2	133	15	188887	10	7401	0.04	33
s641	135	13	55061	10	16559	0.30	23
s713	137	13	56494	10	12840	0.23	23
mm9b	141	17	344367	15	3348853	9.72	12
C1355	147	16	32097	16	117767	3.67	0
C499	147	16	28155	16	292111	10.38	0
cse	148	18	1522416	11	221276	0.15	39
cht	151	5	170	5	145	0.85	0
kirkman	151	12	11317	9	15006	1.33	25
Avg.						2.82	15.5

7.1 Sum of External Degrees

When we consider the *sum of external degrees cost* ("SOED") function, we must consider every hyperedge as a potential external degree of one or more partitions. If a net is cut, it will appear as an output degree on exactly one partition and an input degree on at least one partition. A totalizer tree sums over all the potential partition pins (all the hyperedges in the hypergraph for every partition for inputs, and all the hyperedges again for outputs).

$$MetricMet = \sum_{\substack{e \in Edges \\ 0 \le i < k}} \left(Cut(e) \wedge \overline{e.Source_i} \wedge \left(\bigvee_{s \in e.Sinks} (s_i) \right) \right) + \sum_{e \in Edges} Cut(e) \le MaxMetric$$

Table 4 shows that our SAT formulation is not yet efficient enough to reliably optimize this cost function, even for small, forty node netlists. We present the results to show our current progress on optimizing this metric.

Figure 6 shows how we assert the sum of external degrees function for a three-way partitioning of the example circuit in Figure 1.

Table 4. Sum of External Degrees Optimization

Netlist	k	No Replication		Replication	
		SOED	ms	SOED	ms
misex1	2	25	195	21	79
	3	33	6538	27	1434
	4	35	5824	34	220448
	5	42	895024		Timeout
	6		Timeout		Timeout
bbara	2	22	428	18	423
	3	31	45958	28	640239
	4	38	2515958		Timeout
	5		Timeout		Timeout
	6		Timeout		Timeout
ex7	2	22	897	18	1647
	3	32	119169		Timeout
	4		Timeout		Timeout
	5		Timeout		Timeout
	6		Timeout		Timeout

7.2 Maximum Subdomain Degree

Minimizing the *maximum subdomain degree* ("MSD") requires the most intricate SAT formulation. At first, it appears sufficient to modify slightly the SOED formulation. However, the replicated nodes create an additional complexity. We must only charge one partition for the output from a replicated node.

We employ totalizers at several points in the assertion of the MSD value. First, we employ totalizers for each partition to sum the number of input pins into each partition – this is similar to the SOED metric. Then we totalize the outputs.

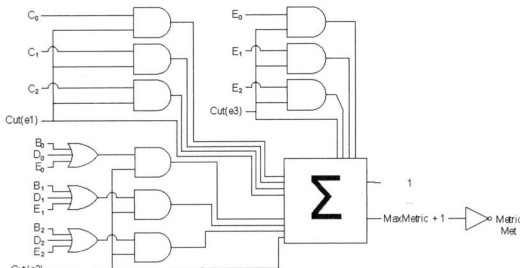

Figure 6. Sum of External Degrees Metric

We assert that if an edge is cut, its output node must count as an output in exactly one of the partitions where it appears. We employ separate totalizer trees to sum the inputs and outputs over each of the partitions (subdomains). We assert that for every partition, $Sum_{MaxMetric+1}$ is false. If we wish, this technique is easily extended to to solve the FPGA clustering problem (where each cluster would typically have a particular number fixed inputs and fixed outputs as opposed to general IO pins [6]). Figure 7 shows the satisfiability instance we construct for this metric on the example hypergraph in Figure 1.

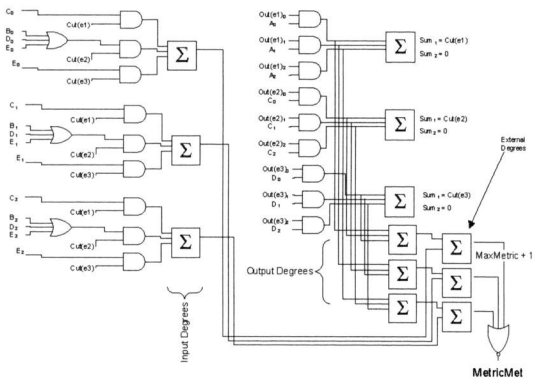

Figure 7. Maximum Subdomain Degree Metric

Table 5 shows that our formulation of the MSD problem scales better than the SOED formulation. Further, we see at least anecdotally, that allowing replication improves results for this metric.

8 FUTURE DIRECTIONS

We believe that the flexibility introduced by representing partitioning as a SAT problem creates new opportunities for optimal solutions with diverse notions of partitioning quality. Future work will broadly consider how we can improve the scaling properties.

8.1 Higher K Partitioning

So far, we have found that as K is increased, the time to compute an optimal partition set goes up quickly. We believe that this could be somewhat alleviated if more of symmetries in partition ordering could be broken. If we were to optimize a partitioning metric where the partitions naturally form a spatial order, we could break this symmetry.

8.2 Hybrid Cost Functions

We have seen that two cost functions other than cut-hyperedges are very difficult to solve optimally with SAT. The cut-hyperedges solutions may be very helpful in identifying the gross locality in the hypergraphs. We believe that by creating a hybrid cost function, we can significantly speedup the search for solutions to more complex partitioning problems. For example, at the cost of our optimality guarantee, we might seed an MSD or SOED problem instance with some hard preassignments from a cut hyperedges solution.

9 CONCLUSION

We have shown several new problem formulations and speedup techniques for optimally solving hypergraph partitioning by remapping the problem to Boolean satisfiability. Combining these techniques with a leading edge SAT solver allows us to perform optimal multiway partitioning with integrated optimal replication for small benchmark netlists. For the traditional logic bipartitioning cost function, our method scales better than the optimal branch-and-bound algorithm in an academic placer. Under this model, we can add integrated generation of replicated nodes to the problem formulation for relatively small additional runtime.

We further show how our framework allows us to produce the first published optimal results of k-way logic partitioning under two more sophisticated cost metrics. We generate these with and without replicated nodes. Our runtime results to date in this effort suggest that Devadas' claim of SAT's applicability to broader cost functions may have been premature. Future work will consider how we can improve the scaling properties of SAT-based solutions to these problems.

Appendix

In this work, we consider several well-known partitioning cost metrics. For the reader's convenience, we formalize our definitions in this appendix. We adapt the descriptions given in [8] to describe our hypergraph model and partitioning metrics along with their evaluation. We make appropriate modifications to expand the model to include a notion of node replication appropriate for VLSI layout applications.

Table 5. Maximum Subdomain Degree Optimization

Netlist	k	No Replication		Replication	
		MSD	**ms**	**MSD**	**ms**
misex1	2	14	470	11	120
	3	12	1490	10	2667
	4	11	6734	9	13478
	5	10	14052		Timeout
	6	9	8616	8	1780920
bbara	2	12	525	9	927
	3	11	3979	10	18182
	4	11	13421	9	138265
	5	10	23967	9	107820
	6	10	45501		Timeout
ex7	2	12	3496	9	4586
	3	12	8490	9	99578
	4	11	25455		Timeout
	5	11	122051		Timeout
	6	10	218037		Timeout

A.1 Netlist and Partitioning Definitions

A *netlist hypergraph* $G = (V, E)$ consists of a set of vertices V ("logic elements") and directed hyperedges E ("wires"). Every hyperedge $e = (d, R)$ has exactly one source vertex (d) and a set of at least one sink vertex R. This model is not appropriate for all applications – for example, circuit models which include tristate drivers would not include the limitation of one source node per hyperedge.

A decomposition of V into k subsets $V_1, V_2, ..., V_k$, such that $\cap_i V_i = V$ is called a k-way partitioning of V. We refer to each of these subsets as a partition or subdomain. The partitions need not be disjoint if replication is permitted. A k-way partitioning of V satisfies a balance constraint specified by $[l, u]$ if, for each partition $l \leq |V_i| \leq u$.

A.2 Total Cut Hyperedges

The total cut hyperedges metric counts the number of hyperedges that are cut between the partitions. A hyperedge is cut if it has at least one sink vertex $r \in R$ in a partition where the driving vertex (d) is not assigned. Formally, a hyperedge $e = (d, R)$ is cut if there exists V_i s.t. $R \cap V_i \neq \varnothing$ and $d \notin V_i$.

For the example partitioning in Figure 8, the cut hyperedges metric is 4, because edges sourced by vertices A, B, E, and F are shared between multiple partitions.

A.3 Sum of External Degrees

The *sum of external degrees* cost function (SOED) counts the total 'pins' required over the partitions. Partitions have two types of external degrees: input and output. The input set I_i of a partition V_i is the set of hyperedges which form its input. V_i includes at least one sink vertex from i and does not include the source vertex from i. We compute the set of output degrees overall by assigning an output degree to every input degree $\cap_i I_i = O$. Note that we do not assign the output edges to particular partitions – this is significant as the source node might appear in multiple partitions. The SOED cost function is SOED $= |O| + \sum_i |I_i|$.

For the example in Figure 8: V_1 has zero inputs, V_2 has inputs from B, E, and F. V_3 has inputs from A and B. We then have that A, B, E, and F are outputs. So the SOED metric is $0 + 3 + 2 + 4 = 9$.

A.4 Maximum Subdomain Degree

The most intricate metric is the *maximum subdomain degree* (MSD) criteria. We expand from the computation of the SOED metric. We separate the output hyperedge set O into disjoint subsets $O_1,...,O_k$ s. t. $\cap_i O_i = O$ and MSD $= \text{Max}_i (|O_i| + |I_i|)$ is minimized. In the non-replicated case, the computation of the MSD metric for a given partitioning is trivial (the output hyperedges are constrained to assignment in the partitions where their source nodes are assigned). If we allow node replication, then assigning the output hyperedges to minimize the MSD for a given partitioning is more involved. Each output hyperedge o with source vertex d may be feasibly assigned to a partition $Q \in \{V_1, ..., V_k\}$ s.t. $d \in Q$.

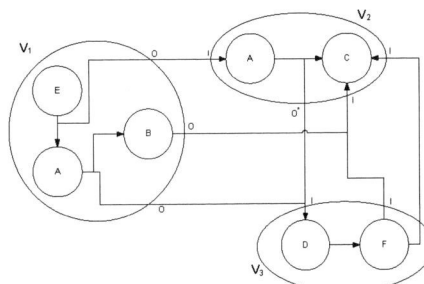

Figure 8. A Partitioned Hypergraph

For the example in Figure 8 we begin with the input assignments from SOED: V_1 has zero inputs, V_2 has inputs from B, E, and F. V_3 has inputs from A and B. We must assign the output E to V_1 and the output F to V_3. Before considering output A, V_1 has degree 2, V_2 degree 3 and V_3 degree 3. By choosing to assign the output A to V_1 instead of V_2, the MSD metric is 3.

In debugging our work and examining non-SAT generated partitions, we found that evaluating MSD with of replicated nodes is non-trivial.

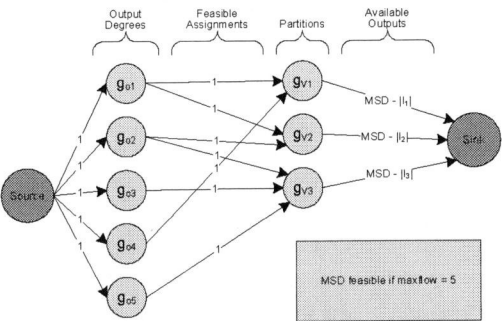

Figure 9. Max-flow Output Assignment Formulation

Our technique for evaluated the MSD metric without SAT is as follows: We perform a binary search on potential MSD values to determine the minimum MSD. We formulate the decision problem as max-flow problem as follows:

For a particular value of MSD, we allow each partition V_i to have output degree of $o_i = \text{MSD} - |I_i|$. We construct a directed flow graph that consists of a source with unit capacity flows leading to nodes representing each output hyperedge to assign. We create nodes g_{vi} for each of the potential partitions to assign outputs. We create unit flow edges from the nodes representing outputs to each of their feasible partitions. We connect every node g_{Vi} representing partitions to a sink via edges of capacity o_i. We compute the maxflow. If the max flow is equal to the total output degree, then the links with positive flow imply an assignment of output hyperedges to partitions and a demonstration that the potential MSD is feasible. If the maxflow is less than the total output degree, the specified MSD is infeasible. Figure 9 shows an example of flow graph constructed for assigning five output edges (two of which are driven by replicated nodes) to three partitions.

Acknowledgements

This research was funded in part by the NSF CAREER program under Grant CCR-0133102. We are grateful for the Capo source code, provided by Igor Markov. Discussions with Michael deLorimier were key in understanding how to evaluate the MSD metric.

REFERENCES

[1] F. M. Johannes, "Partitioning of VLSI Circuits and Systems," presented at ACM/IEEE Design Automation Conference, 1996.

[2] A. E. Caldwell, A. B. Kahng, and I. L. Markov, "Optimal Partitioners and End-case Placers for Standard-cell Layout," *IEEE Transactions on Computer-Aided Design*, vol. 19, pp. 1304-1313, 2000.

[3] M. R. Garey and D. S. Johnson, *Computers and Intractability: A Guide to the Theory of NP-Completeness*: W. H. Freeman, 1979.

[4] S. Devadas, "Optimal Layout via Boolean Satisfiability," presented at International Conference on Computer-Aided Design, 1989.

[5] L. Zhang and S. Malik, "The Quest for Efficient Boolean Satisfiability Solvers," presented at International Conference on Computer Aided Verification, Copenhagen, 2002.

[6] V. Betz and J. Rose, "Cluster-Based Logic Blocks for FPGAs: Area-Efficiency vs. Input Sharing and Size," presented at IEEE Custom Integrated Circuits Conference, 1997.

[7] G. Karypis and V. Kumar, "Multilevel K-way Hypergraph Partitioning," presented at Design Automation Conference, 1999.

[8] N. Selvakkumaran and G. Karypis, "Multi-Objective Hypergraph Partitioning Algorithms for Cut and Maximum Subdomain Degree Minimization," presented at International Conference on Computer-Aided Design, 2003.

[9] W.-K. Mak and D. F. Wong, "Minimum Replication Min-Cut Partitioning," *IEEE Transactions on Computed-Aided Design for Integrated Circuits and Systems*, vol. 16, pp. 1221-1227, 1997.

[10] L. J. Hwang and A. E. Gamal, "Min-Cut Replication in Partitioned Networks," *IEEE Transactions on Computer-Aided Design of Integrated Circuits and Systems*, vol. 14, pp. 96-106, 1995.

[11] M. Wallace and A. Veron, "Two Problems - Two Solutions: One System - Eclipse," presented at IEE Colloquium on Advanced Software Technologies for Scheduling, 1993.

[12] O. Bailleux and Y. Boufkhad, "Full CNF-Encoding: The Counting Constraints Case," presented at International Conference on Theory and Applications of Satisfiability Testing, 2004.

[13] L. Ryan, *Efficient Algorithms for Clause-Learning SAT Solvers*. Masters thesis: Simon Fraser University, 2004.

[14] O. Bailleux and Y. Boufkhad, "Efficient CNF Encodings of Boolean Cardinality Constraints," presented at International Conference on the Principles and Practice of Constraint Programming, 2003.

[15] J. M. Silva, Personal Website. http://sat.inesc-id.pt/~jpms/scripts/.

[16] J. Cong and Y. Ding, "FlowMap: An Optimal Technology Mapping Algorithm for Delay Optimization in Lookup-Table Based FPGA Designs," *IEEE Transactions on Computer-Aided Design*, vol. 13, pp. 1-12, 1994.

[17] A. Nadel, *Backtrack Search Algorithms for Propositional Logic Satisfiability*. Masters thesis: Tel-Aviv University, 2002.

[18] N. Eén and N. Sörensson, "An Extensible SAT-solver," presented at International Conference on Theory and Applications of Satisfiability Testing, 2003.

Finding Optimal L1 Cache Configuration for Embedded Systems

Andhi Janapsatya[†], Aleksandar Ignjatović[††], Sri Parameswaran[††]

†School of Computer Science and Engineering, The University of New South Wales
Sydney, NSW 2052, Australia
‡NICTA, The University of New South Wales
Sydney, NSW 2052, Australia
{andhij,sridevan,ignjat}@cse.unsw.edu.au

ABSTRACT

Modern embedded system execute a single application or a class of applications repeatedly. A new emerging methodology of designing embedded system utilizes configurable processors where the cache size, associativity, and line size can be chosen by the designer. In this paper, a method is given to rapidly find the L1 cache miss rate of an application. An energy model and an execution time model are developed to find the best cache configuration for the given embedded application. Using benchmarks from Mediabench, we find that our method is on average 45 times faster to explore the design space, compared to Dinero IV while still having 100% accuracy.

1. Introduction

Today, cache memory is an integral component of mid to high end processor based embedded systems. The inclusion of cache significantly improves system performance and reduces energy consumption. Current processor design methodologies rely on reserving large enough chip area for caches while conforming with area, performance, and energy cost constraints. Recent application specific processor design platforms (such as the Tensilica's Xtensa platform [1]) allows a cache to be customized for the processor. This allows a design which can meet tighter energy consumption, performance, and cost constraints.

In existing low power processors, cache memory is known to consume a large portion of the on-chip energy. For example, in [2] Montanaro et al. report that cache consumes up to 43% of the total energy of a processor. In embedded systems where a single application or a class of applications are repeatedly executed on a processor, the memory hierarchy could be customized such that an optimal configuration is achieved. The right choice of cache configuration for a given application could have a significant impact on overall performance and energy consumption.

Choosing the correct cache configuration for an embedded system is crucial in reducing energy consumption and improving performance. To find the correct configuration the hit and miss rates must be evaluated, and the resulting energy consumption and execution times must be accurately estimated. Estimating the hit and miss rates (for a particular application with sample input data) is fairly easy using tools such as Dinero IV [7], but enormously time consuming to do so for various cache sizes, associativities and line sizes. The resulting energy consumption and execution times are difficult to examine due to the non uniform nature of memory, where the first memory access takes far greater time compared to subsequent accesses (which are sequential to the first access). Energy and access times are further complicated by differing cache configurations consuming energy at different rates and taking differing amounts of time to access.

Our research results demonstrate that low miss rates do not necessarily mean a faster execution time. Figure 1 shows the effect of different cache configurations have on the number of total cache misses and the total execution time for the G721 encode application. The graph in Figure 1 shows that higher total cache miss rates can possibly provide the fastest execution time. This is due to large or more

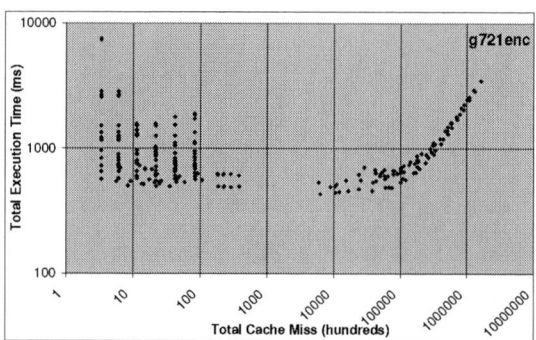

Figure 1: Total Cache miss vs. Total Execution Time

complex (higher associativity) caches having significantly longer access times.

Existing methodologies for cache miss rate estimation, use heuristics to search through the cache parameter design space [3, 4, 5, 6]. Other existing cache miss rates estimation tools, such as Dinero IV [7] can accurately determine the cache miss rate for a single cache configuration. To use Dinero IV to estimate cache miss rate for a number of cache configurations means that a large program trace needs to be repeatedly read and evaluated which is time consuming.

In this paper, we present a methodology to rapidly and accurately explore the cache design space by: estimating cache miss rates for many different cache configurations simultaneously; and investigate the effect of different cache configurations on the energy and performance of a system. The method performs simultaneous evaluation of multiple cache configurations by reading a program trace just once. Simultaneous evaluation can be rapidly performed by taking advantage of the high correlation between cache behavior of different cache configurations. The idea of utilizing correlation in cache behavior comes from the following observations: one, given two caches with the same associativity using the Least Recently Used (LRU) replacement policy, whenever a cache hit occurs, all caches that have larger set sizes will also guarantee a cache hit; two, a hit on a set-associative cache means a cache hit is guarantee on all caches with larger associativity (Explained in greater detail in Section 3).

These observations were also described [8] and [9]. Gecsei et al. in [8] introduce the "stack algorithm" for finding the exact frequency of accesses to each level in a memory hierarchy system. In [9], Hill and Smith analyze the effect of different cache associativity on cache miss rates. The benefit of our method is a quick exploration of different cache parameters from a single read of a large program trace. This will then allow an embedded system to be tested with multiple input sets and with multiple architectures, where it may not be possible to store the large program traces.

The rest of this paper is structured as follow, Section 2 presents existing cache exploration methodologies; Section 3 presents our cache parameters exploration space methodology; Section 4 describes the performance and energy model for the system architecture considered in this work; Section 5 describes the experimental setup and discusses the results; and Section 6 concludes the paper.

2. Related Work

Cache simulation is required for tuning cache parameters to ensure maximal performance and minimal energy consumption. In the past, various methodologies have been researched for cache simulation. These cache simulation methodologies can be divided into two classes: one, estimation techniques, and the other, exact simulation.

Cache estimation techniques use heuristics to predict the cache misses for multiple cache configurations. Pieper et al. in [3] developed a metric to represent cache behavior independently of the cache structure. Their metric based result is within 20% accuracy of a uniprocessor trace-based simulation and can be applied for estimating multiprocessor architectures.

In [4], Fornaciari proposed a heuristic method for configuration of cache architecture without exhaustive analysis of the space of parameters. Their analysis looked at the sensitivity of individual cache parameters on the energy delay product. Maximum error is less than 10%.

Ghosh et al. described a method to generate and solve a cache miss equations (CME) to represent the cache behavior [5]. In [6], Vera et al. proposed a fast and accurate method to solve the cache miss equation (CME).

For exact cache simulation techniques, there exists a tool called Dinero IV [7]. Dinero IV is a single processor cache simulation tool developed by Jan Edler and Mark Hill. Its purpose is to estimate the number of cache misses given a cache configuration; its features include simulating separate or combined instruction and data caches, and simulating multiple levels of cache.

For simulating multiple cache configurations, exact cache simulation techniques rely on exploiting the inclusion property of caches. Inclusion means that given two cache configuration, $Cache\ C_2 \subset Cache\ C_1$ if all the content of $Cache\ C_2$ is a subset of the content of $Cache\ C_1$.

In 1970, Gecsei et al. [8] introduced the 'Stack' algorithm for performing simulation of multiple levels of storage systems. In 1989, Hill in [9] investigated the effects of associativity of caches. They briefly described the methodology of forest simulation for quick simulation of alternate direct-mapped caches. They also introduced the all-associative methodology for simulating alternate direct-mapped, set-associative, and fully-associative caches based on the 'Stack' algorithm. The space complexity of the all-associativity simulation is $O(N_{unique})$, where N_{unique} is the number of unique blocks referenced in an address trace. In their experiment, they showed that to simulate alternate direct-mapped caches, the forest simulation method is faster than the all-associative simulation.

Sugumar et al., introduced a cache simulation methodology by using a binomial trees [10] to improve the method described in [9]. The time complexity of their algorithm for searching procedure is $O((log_2(N)+1) \times A)$ and the time complexity for maintaining the binomial tree is $O((log_2(N)+1) \times A)$, where N is the size of the cache and A is the associativity of the cache. Li in [11] then extends the work in [10] by introducing a method to compress the program trace for reducing the cache simulation time.

Other existing exact cache simulation methodologies uses parallel processing units and/or multiprocessor systems. Nicol in [12] presented a parallel methodology to simulate cache using SIMD and MIMD hardware units.

Heidelberger in [13] presented a method of to analyze a cache trace using a parallel processor system. They split long traces into several shorter traces, and the shorter traces are then executed on parallel independent processors. The sum of the individual results are not accurate, but by executing a re-simulation phase, it is possible to accurately count the exact number of cache misses.

Our simulation methodology was created as a forest simulation data structure. Our methodology extends the idea of forest simulation described in [9]. The space complexity of our methodology is fixed depending on the number of cache configurations to be evaluated. The required space of our cache simulation method is larger than the

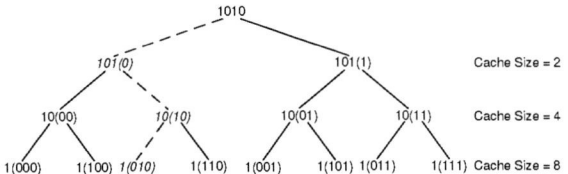

Figure 2: The cache tree data structure.

space needed for the all-associativity simulation described in [9] and the binomial tree simulation described in [10]. The time complexity for searching our data structure is $O((log_2(N)+1) \times A)$, and the time complexity for updating the data structure is $O(log_2(N)+1)$. This is faster compared to the method described in [10].

Other research looked at the use of heuristics for predicting cache behavior. Ghosh et al, in [14] presented an algorithm for simulating cache parameters and finding cache configurations that guaranteed cache miss rates lower than a desired cache miss rate. Their space complexity is in the order of the size of the trace file.

2.1 Our Contribution

- For the first time a methodology is proposed which allows an L1 cache to be chosen for an Application Specific Instruction Set Processor (ASIP) based upon the energy consumption and execution time.

- We also propose a modified forest algorithm based on a simplified data structure, for fast and accurate simulation of the cache. The time taken by the algorithm for both simulation and updating of the data structure is considerably quicker than previous methods. In addition, this method allows parallelization.

3. Cache Parameters Exploration Methodology

A cache configuration is dependent on the cache parameters: cache size N, cache associativity A, and cache line size L. The cache size refers to the total number of bits that can be stored in the cache. The cache associativity refers to the number of ways a data can be stored within the same address of the cache. For a direct-mapped cache ($A = 1$), each datum has a single location where it can be stored within the cache. The total cache size divided by associativity of the cache is called the cache set size, $M = N/A$. In our simulations we will consider cache configurations with the cache set in the range from $2^{m_{min}}$ to $2^{m_{max}}$, where m_{min} and m_{max} refer to the number of address bits needed to address $2^{m_{min}}$ and $2^{m_{max}}$ locations.

We perform design space exploration on the cache parameters by accurately and efficiently simulating the number of cache misses that would occur for a given collection of cache configurations. We optimize the run time of our cache simulation by replacing multiple readings of large program traces with a single reading and simulating multiple cache configurations simultaneously. This is possible due to the following observations.

First, assume that two cache configurations have the same associativity and the same cache line size but that one cache is twice the size of the other cache. In this case if a cache hit occurs on the smaller size cache, a cache hit will also occur on the larger cache size.

This is illustrated in Figure 2. For a memory address request of '1010', cache location pointed to by the address '1010' can be found on the cache locations shown with the dotted line branches in Figure 2. The numbers inside the parentheses shown in Figure 2 indicate the cache address for that location. From the figure, it can be seen that the entries within $cache\_size = 2$ are a subset of the entries within $cache\_size = 4$, and the entries within $cache\_size = 4$ are a subset of the entries within $cache\_size = 8$.

The second observation is that if a cache miss occurs for a cache of size N, associativity A, and of cache set size M, then for all other cache configurations with the same cache set size M and associativity larger than A, a cache hit will also occur. Hence, evaluation for all values larger than A is not required to determine the number of cache

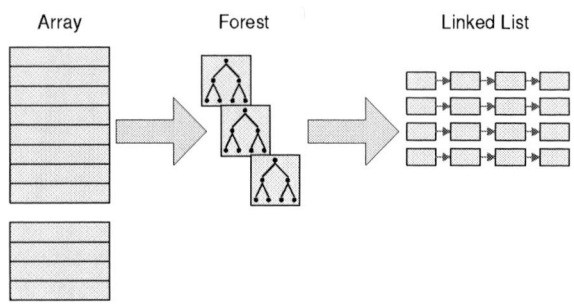

Figure 3: The cache simulation data structure.

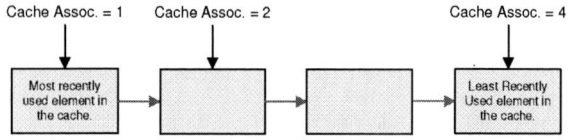

Figure 4: The linked list data structure.

misses for such configurations. This observation is also known as 'inclusion' [8].

3.1 Cache Simulation Methodology

Based on the two observations described above, we designed and implemented the following procedure for simultaneous simulation of multiple cache configurations using a single read of the program trace.

We created an array of cache miss counters to record the cumulative cache misses that occur with different cache configurations. The size of this array is dependent on the number of cache configurations to be simulated and the array is indexed using cache configuration parameters. To simulate the cache, we created a collection of forest data structures as follows.

Each forest corresponds to a collection of cache configurations that have the same cache line size. Each tree in such a forest is used as a convenient data structure to store pointers to locations in cache sets for caches of different sizes. An array of size equal to minimum cache set size $2^{m_{min}}$ is created to store addresses of such trees; note that m_{min} bits are required to store address $2^{m_{min}}$ such locations of the array. The kth level of the tree corresponds to the cache configuration with the cache set size $2^{k+m_{min}}$. Each node in the tree corresponds to a cache location and points to a linked list. Elements in this list correspond to cache ways, and will be used to store the tag address, the valid bit, and the pointer to the next element. The total number of elements in each linked list corresponds to the largest associativity of the family of cache configurations considered. The head element of the linked list corresponds to the most recently used data in the cache; the tail corresponds to the least recently used data. The linked list is illustrated in Figure 4, and the whole data structure is illustrated in Figure 3. Searches of each linked list on different nodes of the tree are independent of each other; this property allows parallelization to optimize the simulation procedure.

To simulate different cache line sizes, we replicate the forest for each cache line size parameter that need to be simulated. For the sake of explanation, we first assume that the cache line size is fixed to one byte; as this assumption does not change the nature of the algorithm.

The cache simulation methodology is shown in Figure 5 and illustrated in Figure 6. The methodology takes its input from the program trace. For each address x read from the program trace, we use the least significant m_{min} bits of the address x to locate in the array the pointer to the appropriate tree. For each cache set size of the form $2^{k+m_{min}}$, thus ranging in powers of two from $2^{m_{min}}$ to $2^{m_{max}}$ we take the next k bits of the memory address and use these bits to find the node in the tree that corresponds to the cache set location in the cache of this size (i.e., $2^{k+m_{min}}$). This is done by traversing down the

---

For each address x from the trace {
    Use the least significant m_{min} bits of the address x to locate in the array, the pointer to the corresponding tree, and go to the root of the tree.
    For $k = 0$ to $(m_{max} - m_{min})$ { // k corresponds to the level of the tree;
                           // the total cache set size is $2^{k+m_{min}}$
        Go to the head of the linked list pointed by this node of the tree
        Search linked list to find the tag entry that
        is equal to the tag of the address x.
        If a cache hit occurs in an element s of the linked list {
            Increment cache miss counters for all cache configuration with
            cache set size equal to $2^{k+m_{min}}$ and lower associativity than s.
            Move element s to be the head element of the linked list.
        } **else** {
            Increment cache miss counter for all cache
            configuration with cache set equal to $2^{k+m_{min}}$
            Replace entry in the tail element of the
            linked list with the current tag address.
            Move this tail element to become the head
            element of the linked list.
        }
        Increment the value of k.
        Step down the tree according to the value of the bit $k + m_{min}$
        taking the left child if this bit is 0 else take the right child.
    }
    Scan the next address $x + 1$ from the trace.
}

Figure 5: Cache Simulation Procedure

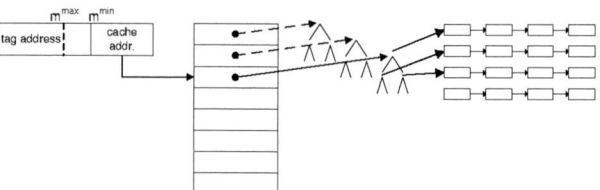

Figure 6: Cache Simulation Procedure.

tree depending on the m_{min} to $m_{min} + k$ bits of the address to choose whether to traverse the left branch or the right branch. The node has a pointer to the head element in the linked list that corresponds to the multiple way associativity of this cache set location. The remaining bits of the address x form the tag address that is to be searched for in the linked list. If the tag is found at position s, this indicates that a cache hit would occur in all s-way or higher-way associative caches. If this happens, the sth element of the list is moved to the head of the list. In this case, all cache miss counters for caches with the same cache set size and associativity less than s are incremented. If the tag is not found, the tag in the tail element is replaced by the new tag, and this tail element is moved to the head of the list. In such a case, cache miss counters for caches with the same cache set size and any values of associativity are incremented.

This is illustrated for a 4-way set-associative cache in Figure 4. If the tag comparison results in a hit with the second element in the linked list, then this indicates that a cache hit would occur in the 4-way set-associative cache and the 2-way set-associative cache. A cache miss would occur in the direct mapped cache, hence it is not necessary to continue the search until the tail element. The second element is then moved to be the head element to conform with the Least Recently Used (LRU) replacement policy.

The procedure continues by traversing down the trees to find the linked list corresponding to the larger cache set size. The procedure for tag comparison and updating the associativity replacement policy is then repeated. An example is shown in Figure 2, once the cache estimation for the level $cache\_size = 2$ is completed, the algorithm will look at the least significant bit of the current tag part of the address which will make up the most significant bit of the cache address for the $cache\_size = 4$ tree level to determine whether to traverse the left or right branch. In Figure 2, the dotted line indicates the traversing path given the address '1010'.

The algorithm then continues by using the replicated forest data

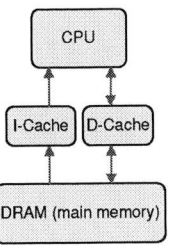

Figure 7: System Architecture

Application	Trace size	Total Execution Time (sec)		ratio
		Dinero IV	Our methodology	
cjpeg	15531057	1592	41	38.83
djpeg	4616784	363	20	18.15
pegwitenc	33070433	2831	136	20.82
pegwitdec	18903066	1642	70	23.46
epic	52801084	4017	30	133.90
unepic	6746679	512	6	85.33
g721enc	3.15E+08	(7.5 hr) 26990	(26.5 min) 1595	16.92
g721dec	3.03E+08	(7 hr) 25132	(25 min) 1503	16.72
mpeg2enc	1.13E+09	(25.2 hr) 90431	(53 min) 3186	28.38
mpeg2dec	35398584	3377	49	68.92
			Average	45.14

Table 1: Total execution time comparison of our methodology with Dinero IV.

structures for simulating the different cache line sizes. Different parts of the current address are used for locating the appropriate tree data structure.

3.2 Efficiency of The Methodology

With the implementation described above, we can limit the search space for all the different cache configurations and reduce the time taken to estimate cache misses by increasing the space requirements.

Each linked list entry is used to store the tag address (32 bits), a valid bit (1 bit), and a pointer to the next element in the linked list (32 bits). In total, each linked list entry needs to keep 65 bits of data. Each node in the tree needs to store pointers to the head element of the linked list (32 bits), a pointer to the left branch (32 bits), and a pointer to the right branch (32 bits); giving a total of 96 bits per node.

For each cache way, a linked list entry needs to be kept, this gives a total size of $list_{size} = A \times 65 bits$. The number of nodes created is dependent on the number of cache sets, cache line size, and cache size range. The number of nodes is calculated by $node_{size} = (2^{m_{max}} - 2^{m_{min}}) \times (log_2(L) + 1) \times 96 \ bits$.

Space complexity is calculated by summing all the space needed for each node and each linked list entry. In terms of space, our data structure is optimal for storing all the necessary parameters with exception of the redundancy of the content of the linked lists. However, the space requirement is fully manageable by standard desktop computers. For the work described in this paper, we simulated cache sizes ranging from 512 bytes up to 2M bytes, cache associativity of 1 up to 32, and cache line size of 8 bytes up to 256 bytes. In total, we have simulated 268 cache configurations requiring 9.3 megabytes. This reasonable redundancy simplifies the maintenance of the data structure and the associated algorithms.

4. System Energy and Performance Model

To facilitate the design space exploration steps, we created crude performance and energy models for the system. The model of the embedded system architecture consisted of a processor with an instruction cache, a data cache, and embedded DRAM as main memory. The data cache uses a write-through strategy. The system architecture is illustrated in Figure. 7.

The equation for calculating the system's total execution time is given by:

$$
\begin{aligned}
Exec_{time} = & Icache_{access} \times Icache_{access\_time} + \\
& Icache_{miss} \times DRAM_{access\_time} + \\
& Icache_{miss} \times Icache_{linesize} \times \frac{1}{DRAM_{bandwidth}} + \\
& Dcache_{access} \times Dcache_{access\_time} + \\
& Dcache_{miss} \times DRAM_{access\_time} + \\
& Dcache_{miss} \times Dcache_{linesize} \times \frac{1}{DRAM_{bandwidth}}
\end{aligned} \tag{1}
$$

where,

- $Icache_{access}$ and $Dcache_{access}$ is the total number of memory accesses to the instruction and data cache, respectively.

- $Icache_{access\_time}$ and $Dcache_{access\_time}$ is the access time of the instruction and data cache, respectively.

- $Icache_{miss}$ and $Dcache_{miss}$ is the total number of cache misses for the instruction and data cache, respectively.

- $Icache_{linesize}$ and $Dcache_{linesize}$ is the cache line size of the instruction and data cache, respectively.

- $DRAM_{access\_time}$ is the DRAM latency time.

- $DRAM_{bandwidth}$ is the bandwidth of the DRAM.

There exists six components in the system's execution time shown in equation 1. The first and fourth terms $Icache_{access} \times Icache_{access\_time}$ and $Dcache_{access} \times Dcache_{access\_time}$ are for calculating the amount of time taken for the processor to access the instruction cache or the data cache. The second and fifth terms $Icache_{miss} \times DRAM_{access\_time}$ and $Dcache_{miss} \times DRAM_{access\_time}$ calculate the amount of time required for the DRAM to respond to each cache miss. The third and sixth terms $Icache_{miss} \times Icache_{linesize} \times \frac{1}{DRAM_{bandwidth}}$ and $Dcache_{miss} \times Dcache_{linesize} \times \frac{1}{DRAM_{bandwidth}}$ calculates the amount of time taken to fill a cache line on each cache miss.

In our execution time model, we assume that all data cache misses will cause a pipeline stall, and we ignore the bus communication time cost. As the bus communication time is expected to be similar to other systems, ignoring this will not adversely affect the final results.

Energy equation of the system is given by the following equation:

$$
\begin{aligned}
Energy_{total} = & Exec_{time} \times CPU_{power} + \\
& Icache_{access} \times Icache_{access\_energy} + \\
& Dcache_{access} \times Dcache_{access\_energy} + \\
& Icache_{miss} \times Icache_{access\_energy} \times Icache_{linesize} + \\
& Dcache_{miss} \times Dcache_{access\_energy} \times Dcache_{linesize} + \\
& Icache_{miss} \times DRAM_{access\_power} \times \\
& (DRAM_{access\_time} + Icache_{linesize} \times \frac{1}{DRAM_{bandwidth}}) + \\
& Dcache_{miss} \times DRAM_{access\_power} \times \\
& (DRAM_{access\_time} + Dcache_{linesize} \times \frac{1}{DRAM_{bandwidth}})
\end{aligned} \tag{2}
$$

where,

- CPU_{power} is the total processor power excluding the instruction and data cache power.

- $Icache_{access\_energy}$ and $Dcache_{access\_energy}$ is the instruction cache and data cache access energy, respectively.

- $Dcache_{access\_power}$ is the active power consumed by the DRAM.

There exist seven components in the energy equation 2. The first term $Exec_{time} \times CPU_{power}$ calculates the processor energy given that execution time takes $Exec_{time}$ amount of time. The second and third

Processor Energy	168mW @ 100MHz
Embedded DRAM energy	@ 100MHZ 19.5mW
Latency	19.5 ns
Bandwidth	50MB/sec

Table 2: System Specification

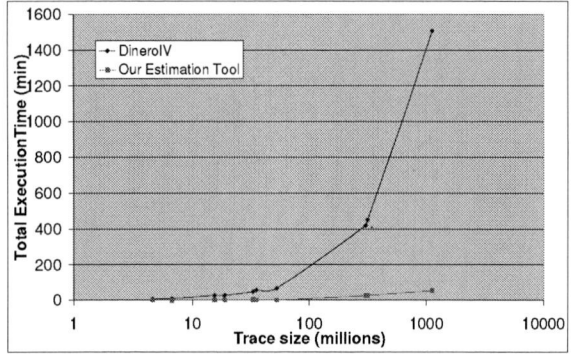

Figure 8: Total execution time compared to increasing program trace size

terms, $Icache_{access} \times Icache_{access\_energy}$ and $Dcache_{access} \times Dcache_{access\_energy}$ calculate the amount of energy consumed by the instruction and data cache, respectively. The fourth and fifth terms, $Icache_{miss} \times Icache_{access\_energy} \times Icache_{linesize}$ and $Dcache_{miss} \times Dcache_{access\_energy} \times Dcache_{linesize}$ calculate the energy cost of writing to cache for each cache miss. The sixth and seventh terms, $Icache_{miss} \times DRAM_{access\_energy} \times (DRAM_{access\_time} + Icache_{linesize} \times DRAM_{bandwidth})$ and $Dcache_{miss} \times DRAM_{access\_energy} \times (DRAM_{access\_time} + Dcache_{linesize} \times DRAM_{bandwidth})$ calculate the energy cost of the DRAM to service all the cache misses. The fourth, fifth, sixth, and seventh terms vary depending on the cache line size, as larger line size means more data need to be read from the main memory and written into the respective caches.

Units for time variables in the equations are in seconds, bandwidth is in bytes/sec., cache line size is in bytes, power variable is in Watts, and energy unit is in Joules.

5. Experimental Procedure and Results

We compiled and simulated programs from Mediabench [15] with SimpleScalar/PISA 3.0d [16]. Program traces were generated by SimpleScalar and fed into both Dinero IV[7] and our estimation tool. Our results were completely consistent with the ones produced by Dinero IV. Our estimation tool is written in C and compiled with GNU/GCC version 3.4.3 build 20050227 with -O1 optimization. Simulations were performed on a dual Opteron64 2GHz machine with 2GBytes of memory. We simulated 268 different cache configurations, with cache sizes ranging from 512 Bytes up to 2M bytes, cache associativity ranging from 1 up to 32, and cache block sizes ranging from 8 bytes per cache line up to 256 bytes per cache line.

Table 1 shows the execution time comparison of executing Dinero IV multiple times with different cache configurations and execution of our estimation tool once. Column 1 in Table 1 shows the application name, column 2 shows the trace size of the benchmark, column 3 shows the total time taken for executing Dinero IV multiple times, column 4 shows the execution time of our estimation tool, and column 5 shows the ratio of time savings of our tool when compared to executing Dinero IV multiple times. The trace size shown in Column 2 in Table 1 only shows the size of the instruction memory access trace. From column 5 in Table 1, it can be observed that, on average, our tool is approximately 45 times faster than Dinero IV. Plotting the total execution time versus the size of the trace in Figure 8 shows that as the trace size grows exponentially, our methodology shows a linear increase in total execution time required while Dinero IV shows an exponential increase in the time required.

5.1 Result Analysis

To analyze the effect of cache miss rates on system's performance and energy consumption, we utilized cache models from CACTI [17] for the cache access time and cache access energy. Processor energy is taken from [2]. The main memory model is taken from the embedded DRAM described in [18]. The processor and memory specification is described in Table 2. System total execution time and its energy consumption is calculated using equation 1 and equation 2, respectively.

In Figure 9, We plot the number of cache misses versus the total energy consumption for different cache configurations. The plot for cache misses versus execution time for g721enc application was shown in Figure 1. The three plots in Figure 9 show the same plot with different coloring to pick out the effect of differing cache line sizes, cache associativity, and cache sizes on the energy consumption. Figure 9(b) highlights the effect of different cache line size on the total energy consumption. Figure 9(a) highlights the effect of changing associativity on the total energy consumption. Figure 9(c) highlights the effect of varying cache sizes on the total energy consumption. Due to space constraints, we are unable to show the cache miss versus total execution time graphs with different coloring to highlight the effect of cache associativity, cache line size, and cache size on the total execution time.

5.2 Model Validation

We validate the energy model of the processor by comparing with output results of Wattch [19]. The mediabench applications were executed in Wattch and the total energy results is plotted against the total cache miss number. Figure 10 shows the total energy versus total cache miss number for g721enc application with energy figures obtained from Wattch output.

Comparing the energy versus cache miss number graphs in Figure 10 and in Figure 9(c) show that energy results from Wattch and from equation 2 display similar patterns. As cache size gets larger, the cache miss number decreases and the energy consumption decreases; but when the cache size reaches a certain size, the energy consumption starts to increase due to compulsory misses.

It should be noted that simulation time for the 268 cache configurations with Wattch took 2.5 days. The energy values obtained from Wattch simulation has a unit of *Watts.cycle* and the energy values should not be compared directly against the energy values obtained from Equation 2. The energy graphs obtained using the Equation 2 and from Wattch is not an exact copy of each other. This error is due to several reasons, such as, the different processor parameters of the two processors and the inaccuracy of the simplescalar model (Wattch is built on top of Simplescalar) for reporting cache misses. In addition, it is also known that processor energy calculation using processor model derived from simplescalar is inaccurate; for example, simplescalar modeled the issue queue, reorder buffer, and the physical register file as a unified structure called Register Update Unit (RUU), unlike in real implementations where the number of entries and the number of ports in all these components are quite disparate [20].

We also performed simulation with Wattch for the remaining Mediabench benchmarks and obtained similar energy graph results in comparison to energy graph obtain from using Equation 2. Due to space constraints, we are unable to include the energy graphs obtained with Wattch for all other benchmarks.

5.3 Design Space Exploration

Looking at the Pareto optimal points in the three plots (Figure 9), it can be concluded that for g721enc application a 16K direct-mapped cache with line size of 16 bytes would be the best cache configuration in terms of lowest energy consumption.

For design space exploration purposes, the best cache configuration based on performance or energy consumption can then be chosen from the performance plots and the energy plots. Best choices of cache configurations for the Mediabench benchmark is shown in Table 3.

For g721enc application, it can be seen in Table 3 that for best performance a 16K bytes direct-mapped cache with line size of 256

	Cache configuration corresponding to					
	Best Performance			Lowest energy consumption		
Application	Cache size	Cache Assoc.	Cache line size	Cache size	Cache Assoc.	Cache line size
cjpeg	16384	1	256	16384	1	16
djpeg	8192	1	128	8192	1	16
pegwitenc	16384	1	128	16384	1	16
pegwitdec	8192	1	128	8192	1	8
epic	8192	1	128	8192	1	32
unepic	8192	1	128	4096	1	32
g721enc	16384	1	256	16384	1	32
g721dec	16384	1	16	8192	1	64
mpeg2dec	4096	1	64	4096	1	16

Table 3: Best cache configuration choice in terms of performance or energy consumption.

bytes should be used. This indicates that the lowest energy consuming cache configurations does not translate to the fastest execution time.

6. Conclusions

In this paper, we presented a cache selection method for configurable processors. This method uses a cache simulation procedure to perform fast and accurate simulation of multiple cache configurations simultaneously using a single reading of a program trace. Our method is 45 times faster compared to existing methods of cache simulation. Fast and accurate cache miss calculations allow rapid design space exploration of optimal cache parameters for desired performance and/or energy consumption.

7. References

[1] Xtensa Processor, *http://www.tensilica.com.*
[2] J. Montanaro et al., "A 160MHz, 32b, 0.5W CMOS RISC microprocessor," *JSSC,* vol.31(11), pp. 1703-1712, 1996.
[3] J. J. Pieper et.al., "High level Cache Simulation for Heterogeneous Multiprocessors," *DAC,* 2004.
[4] W. Fornaciari et.al., "A Design Framework to Efficiently Explore Energy-Delay Tradeoffs," *CODES,* 2001.
[5] S. Ghosh et.al., "Cache Miss Equations: A Compiler Framework for Analyzing and Tuning Memory Behavior," *TOPLAS,* 1999.
[6] X. Vera et.al., "A Fast and Accurate Framework to Analyze and Optimize Cache Memory Behavior," *TOPLAS,* 2004.
[7] J. Edler and M. D. Hill, "Dinero IV Trace-Driven Uniprocessor Cache Simulator," *http://www.cs.wisc.edu/~markhill/DineroIV/.*
[8] R. L. Mattson et.al., "Evaluation Techniques for Storage Hierarchies," *IBM System Journal,* vol.9, no.2, pp. 78-117, 1970.
[9] M. D. Hill and A. J. Smith, "Evaluating Associativity in CPU Caches," *IEEE Transactions on Computer,* 1989.
[10] R. A. Sugumar et.al., "Set-Associative Cache Simulation Using generalized Binomial Trees," *ACM Transactions on computer Systems,* vol. 13, No. 1, 1995.
[11] X. Li et.al., "Design Space Exploration of Caches Using Compressed Traces," *ICS,* 2004.
[12] D. Nicol and E. Carr, "Empirical Study of Parallel Trace-Driven LRU Cache Simulator, *ACM SIGSIM Simulation Digest, Proceedings of the ninth workshop on Parallel and distributed simulation,* 1995.
[13] P. Heidelberger and H. S. Stone, "Parallel Trace-Driven Cache Simulation by Time Partitioning," *Winter Simulation Conference,* 1990.
[14] A. Ghosh and T. Givargis, "Analytical Design Space Exploration of Caches for Embedded Systems," *DATE,* 2003.
[15] C. Lee et.al., "MediaBench: A Tool for Evaluating Multimedia and Communications Systems," *IEEE MICRO 30,* 1997.
[16] D. C. Burger and T. M. Austin, "The SimpleScalar tool-set, Version 2.0," *Technical Report 1342,* Department of Computer Science, UW, June, 1997.
[17] P. Shivakumar and N. P. Jouppi, "Cacti 3.0: An Integrated Cache Timing, Power, and Area Model," *Technical Report 2001/2,* Compaq Computer Corporation, August, 2001. 2001.
[18] K. Hardee et.al., "A 0.6V 205MHz 19.5ns tRC 16Mb Embedded DRAM," *ISSCC,* 2004.
[19] D. Brooks et.al., "Watch: A Framework for Architectural-Level Power Analysis and Optimizations," *ISCA,* 2000.
[20] D. Ponomarev, G. Kucuk, and K. Ghose, "AccuPower: An Accurate Power Estimation Tools for Superscalar Microprocessors," *DATE,* 2002.

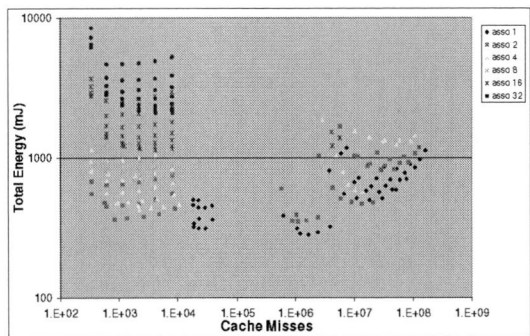

(a) Energy comparison for different cache associativity

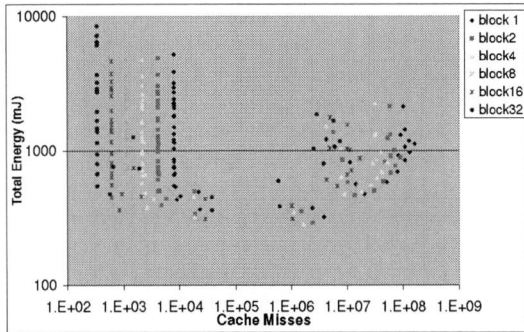

(b) Energy comparison for different cache line size

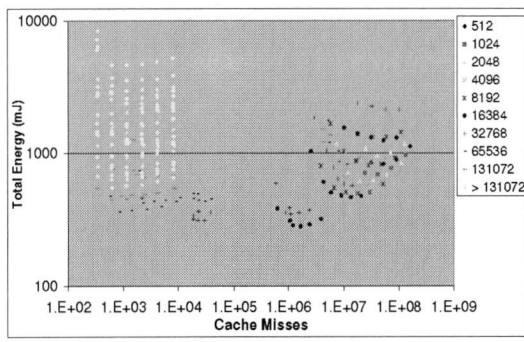

(c) Energy comparison for different cache size

Figure 9: Total energy consumption compared to cache miss number

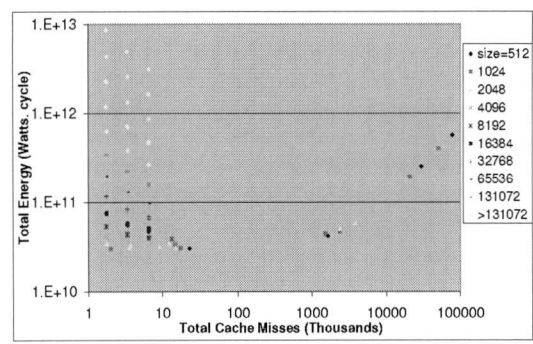

Figure 10: Total energy consumption of g721enc application measured using Wattch.

Memory Size Computation for Multimedia Processing Applications[*]

Hongwei Zhu   Ilie I. Luican   Florin Balasa

Dept. of Computer Science, University of Illinois at Chicago, Chicago, IL 60607, U.S.A.

Abstract – In real-time multimedia processing systems a very large part of the power consumption is due to the data storage and data transfer. Moreover, the area cost is often largely dominated by the memory modules. The computation of the memory size is an important step in the process of designing an optimized (for area and/or power) memory architecture for multimedia processing systems. This paper presents a novel non-scalar approach for computing *exactly* the memory size in real-time multimedia algorithms. This methodology uses both algebraic techniques specific to the data-flow analysis used in modern compilers, and also recent advances in the theory of integral polyhedra. In contrast with all the previous works which are only *estimation* methods, this approach performs *exact* memory computations even for applications with a large number of scalar signals.

1   Introduction

In real-time multimedia processing systems – including video and image processing, medical imaging, artificial vision, real-time 3D rendering, advanced audio and speech coding – a very large part of the power consumption is due to the data storage and data transfer. A typical system architecture includes custom hardware (application-specific accelerator datapaths and logic), programmable hardware (DSP core and controller), and a distributed memory organization which is usually expensive in terms of power and area cost. Data transfer and memory access operations typically consume more power than a datapath operation. For instance, fetching an operand from an off-chip memory for an addition consumes 33 times more power than the actual computation; even a transfer from an on-chip memory consumes about 4 to 10 times more power than the addition itself [4]. Moreover, the area cost is often largely dominated by memories. Hence, the optimization of the memory architecture is a crucial step in the design methodology for this type of applications. In deriving an optimized memory architecture, memory size estimation/computation is an important step in the data transfer and storage exploration stage. This problem has been tackled in the past both in register transfer-level (RTL) programs at scalar level [8, 11, 14] and in behavioral specifications at non-scalar level [2, 7, 16, 17]. Good overviews of these techniques can be found in [4, 10].

This paper presents a non-scalar method for computing *exactly* the memory size in real-time multimedia algorithms (assuming

the code is procedural). This approach uses both algebraic techniques specific to the data-flow analysis used in modern compilers [9], and recent advances in the theory of integral (n-dimensional) polyhedra. In contrast with previous works which utilize only *approximate* methods due to the size of the problems (in terms of number of scalars and number of array references), *this approach obtains exact determinations even for applications significantly large*. Since the mathematical model is very general, this novel approach is able to handle the entire class of "affine" specifications (see Section 2), therefore practically the entire class of real-time multimedia applications.

The paper is organized as follows. Section 2 explains the problem of memory size computation. The core of the paper – Sections 3 and 4 – presents the technical aspects of this novel approach. Section 5 will briefly discuss implementation aspects and present several experimental results. Section 6 will summarize the main conclusions of this work.

2   The memory size computation problem

The (real-time) multimedia processing algorithms are typically specified in a high-level programming language, where the code is organized in sequences of loop nests having as boundaries linear functions of the outer loop iterators, conditional instructions where the conditions may be both data-dependent or data-independent (relational and/or logical operators of linear functions of loop iterators), and multidimensional signals which array references have (complex) linear indices. This class of specifications is often referred to as *affine*. Sometimes, in image and video processing, there may be also indices containing modulo operators, but these situations can be brought into the affine specification class [4].

Real-time multimedia algorithms describe the processing of streams of data samples. The source code of these algorithms can be imagined as surrounded by an implicit loop having the *time* as iterator. Consequently, each signal in the algorithm has an *implicit* extra dimension corresponding to the *time* axis. These algorithms often contain *delayed* signals, i.e., signals produced (or inputs) in previous data-sample processings, which are consumed during the current sample processing. The delay operator "@" indicates such delayed signals, the following argument signifying the number of previous samples. The delayed signals must be kept "alive" during several *time* iterations, i.e., they must be stored in the background memory during one or several data-sample processings.

An illustrative example, derived from a motion detection algorithm [4], is given below:

$optDlt[0] = 0;$

---

[*]This research was sponsored by the U.S. National Science Foundation (DAP 0133318).

0-7803-9451-8/06/$20.00 ©2006 IEEE.

2006 Asia and South Pacific Design Automation Conference

8B-2

```
for (i = 8; i ≤ 120; i + +)
  for (j = 8; j ≤ 120; j + +)
  { Delta[i][j][0] = 0;
    for (k = i − 8; k ≤ i + 8; k + +)
      for (l = j − 8; l ≤ j + 8; l + +)
        Delta[i][j][17(k − i) + l − j + 145] = A[i][j] − A[k][l]@1
                        +Delta[i][j][17(k − i) + l − j + 144];
      optDlt[113i+j−911] = Delta[i][j][289]+optDlt[113i+j−912];
  }
opt = optDlt[12769];
```

The problem is to determine the *minimum* amount of memory locations necessary to store the signals of a given multimedia algorithm during its execution, or, equivalently, the *maximum* storage occupancy assuming any scalar signal must be stored only during its lifetime. The total number of scalars in the algorithm above is 3,749,063. But due to the fact that scalars having disjoint lifetimes can share the same memory location, the amount of storage can be much smaller than the total number of scalar signals. Actually, only 33,284 memory locations are necessary for this example.

It must be emphasized that image and video processing applications contain deeper loop nests with iterators having typically large ranges, resulting in extremely large numbers of scalar signals. Enumerative techniques or RTL approaches based on the left edge algorithm [8], although appealing by means of simplicity, are too computationally expensive in such cases, often prohibitive to use. For multimedia algorithmic specifications, the algebraic techniques are the only hope.

All the past works, for instance [16, 2, 17, 7], achieved only a *memory size estimation*, rather than an *exact size computation*. The algorithm presented in this paper is the first one – to the best of our knowledge – able to compute *exactly* the storage requirements for multimedia applications, even when the number of scalar signals is very large. The basic reasons of its efficiency are: (a) the use of a relatively recent mathematics advance – the polynomial-time decomposition of an n-dimensional polyhedron into an algebraic sum of unimodular cones [3], (b) the efficient decomposition of the array references of the multidimensional signals in disjoint *linearly bounded lattives* (LBL's) [15], and (c) an efficient mechanism of pruning the code of the algorithmic specification.

Note that the problem of organizing the signals in a distributed (hierarchical) memory architecture, often referred to as the problem of *memory allocation* is beyond the scope of this paper.

3 Computation of array reference size

Definitions A *polyhedron* is a set of points $P \subset \Re^n$ satisfying a finite set of linear inequalities: $P = \{ \mathbf{x} \in \Re^n \mid \mathbf{A} \cdot \mathbf{x} \geq \mathbf{b} \}$, where $\mathbf{A} \in \Re^{m \times n}$ and $\mathbf{b} \in \Re^m$. If P is a bounded set, then P is called a *polytope*. If $\mathbf{x} \in \mathbf{Z}^n$, then P is called an *integral* polyhedron/polytope. The set $\{ \mathbf{y} \in \Re^m \mid \mathbf{y} = \mathbf{A}\mathbf{x}, \mathbf{x} \in \mathbf{Z}^n \}$ is called the *lattice* generated by the columns of matrix $\mathbf{A}$.

Each *array reference* $M[x_1(i_1, \ldots, i_n)] \cdots [x_m(i_1, \ldots, i_n)]$ of an m-dimensional signal M, in the scope of a nest of n loops having the iterators $i_1, \ldots, i_n$, is characterized by an *iterator*

space and an *index space*. The iterator space signifies the set of all iterator vectors $\mathbf{i} = (i_1, \ldots, i_n) \in \mathbf{Z}^n$ in the scope of the array reference. The index space is the set of all index vectors $\mathbf{x} = (x_1, \ldots, x_m) \in \mathbf{Z}^m$ of the array reference. When the indices of an array reference are linear mappings with integer coefficients of the loop iterators, the index space consists of one or several *linearly bounded lattices* (LBL) [15] – the image of an affine vector function over the iterator polytope $\mathbf{A} \cdot \mathbf{i} \geq \mathbf{b}$:

$$\{ \mathbf{x} = \mathbf{T} \cdot \mathbf{i} + \mathbf{u} \in \mathbf{Z}^m \mid \mathbf{A} \cdot \mathbf{i} \geq \mathbf{b}, \mathbf{i} \in \mathbf{Z}^n \} \qquad (1)$$

where $\mathbf{x} \in \mathbf{Z}^m$ is the index vector of the m-dimensional signal and $\mathbf{i} \in \mathbf{Z}^n$ is an n-dimensional iterator vector (see the example below).

In order to address the computation of the memory size necessary for the execution of a multidimensional signal processing algorithm, a simpler problem must be addressed first: the computation of the number of distinct scalars in an array reference, that is, how many locations are needed to store one array reference.

If the rank of matrix $\mathbf{T}$ is equal to its number of columns, then the vector function $\mathbf{x} = \mathbf{T} \cdot \mathbf{i} + \mathbf{u}$ between the iterator and index spaces is proven to be a one-to-one mapping [2], and the computation of the number of distinct signal indices (i.e., the amount of memory necessary to store the scalars covered by the array reference) is hence reduced to counting the number of iterator vectors or, equivalently, the number of lattice points (i.e., points having integer coordinates) in the iterator polytope $\mathbf{A} \cdot \mathbf{i} \geq \mathbf{b}$ in (1). In such a situation, a computation technique based on Barvinok's decomposition of a simplicial cone into unimodular cones [3] is used.

An example is given below, illustrating both the concepts and the technique. Note that due to space limitation, several details of the computation had to be skipped, along with part of the theoretical justifications. However, this example succeeds to illustrate well enough the main steps of the computation flow. Moreover, it will show clearly why this approach can handle algorithmic specifications typical to multimedia applications.

Example 1: `for (i = 0; i ≤ 4; i + +)`
  `for (j = 0; j ≤ 2i && j ≤ −i + 6; j + +)`
    `··· A[2i + 3j][5i + j] ···`

How many memory locations are necessary to store the array reference $A[2i + 3j][5i + j]$? The linearly bounded lattice (LBL) corresponding to this array reference is $\{\mathbf{x} = \mathbf{Ti} + \mathbf{u} \mid \mathbf{Ai} \geq \mathbf{b}\} =$

$$\left\{ \begin{bmatrix} x \\ y \end{bmatrix} = \begin{bmatrix} 2 & 3 \\ 5 & 1 \end{bmatrix} \begin{bmatrix} i \\ j \end{bmatrix} \; \middle| \; \begin{bmatrix} 1 & 0 \\ -1 & 0 \\ 0 & 1 \\ 2 & -1 \\ -1 & -1 \end{bmatrix} \begin{bmatrix} i \\ j \end{bmatrix} \geq \begin{bmatrix} 0 \\ -4 \\ 0 \\ 0 \\ -6 \end{bmatrix} \right\}$$

($\mathbf{u} = \mathbf{0}$ here) and the problem is equivalent to computing the size of this set. Since post-multiplying $\mathbf{T}$ with the unimodular matrix[1] $\mathbf{S} = \begin{bmatrix} -1 & 3 \\ 1 & -2 \end{bmatrix}$ results in $\mathbf{T} \cdot \mathbf{S} = \begin{bmatrix} 1 & 0 \\ -4 & 13 \end{bmatrix}$, the rank of matrix $\mathbf{T}$ is 2 – equal to its number of columns; therefore, the vector function $\mathbf{x} = \mathbf{Ti} + \mathbf{u}$ is a one-to-one mapping. The computation of the number of scalars covered by $A[2i + 3j][5i + j]$ is equivalent to counting the number of lattice points in the iterator

---

[1] A square matrix with integer elements having the determinant equal to ± 1.

803

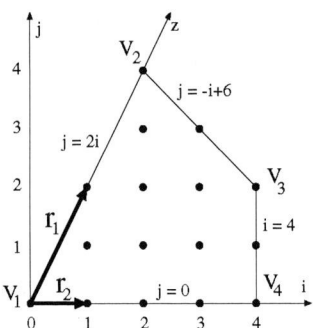

Figure 1: Convex polytope representing the iterator space in *Example 1*

polytope $\mathbf{Ai} \geq \mathbf{b}$ shown in Fig. 1. This latter operation is done as explained below. But, first, a few definitions are necessary.

Definitions Let $r_1, \ldots, r_d \in \mathbf{Z}^d$ be linearly independent integer vectors. The (rational polyhedral) *cone* generated by the *rays* $r_1, \ldots, r_d$ is the set $C(r_1, \ldots, r_d) = \{\sum_1^d \alpha_i r_i, \ \alpha_i \geq 0\}$. For instance, the set of points inside the angle iV_1z is a 2-dimensional cone generated by the rays $r_1 = [1\ 2]^T$ and $r_2 = [1\ 0]^T$ (see Fig. 1). To each vertex of a polyhedron corresponds a *supporting* cone. The supporting cone of the vertex V_1 (Fig. 1), denoted $C(V_1)$, is the one generated by the rays r_1 and r_2. A cone is called *unimodular* if the matrix of the rays $[r_1 \cdots r_d]$ is unimodular (i.e., its determinant is ± 1).

Step 1 Find the vertices of the iterator polytope $\mathbf{Ai} \geq \mathbf{b}$ and their supporting cones.

Given the inequalities $\{0 \leq i \leq 4, \ 0 \leq j \leq 2i, \ j \leq -i + 6\}$ defining the iterator space, the vertices and the rays are computed using the *reverse search* algorithm [1]. The supporting cones corresponding to the vertices $V_1, \ldots, V_4$ of the iterator polytope, as well as their generating rays shown below as column vectors, are:

$$C(V_1) = \left\{ \begin{pmatrix} 1 \\ 2 \end{pmatrix}, \begin{pmatrix} 1 \\ 0 \end{pmatrix} \right\}, \ C(V_2) = \left\{ \begin{pmatrix} -1 \\ -2 \end{pmatrix}, \begin{pmatrix} 1 \\ -1 \end{pmatrix} \right\},$$

$$C(V_3) = \left\{ \begin{pmatrix} -1 \\ 1 \end{pmatrix}, \begin{pmatrix} 0 \\ -1 \end{pmatrix} \right\}, \ C(V_4) = \left\{ \begin{pmatrix} -1 \\ 0 \end{pmatrix}, \begin{pmatrix} 0 \\ 1 \end{pmatrix} \right\}$$

Step 2 Apply Barvinok's algorithm [3] to decompose the supporting cones into unimodular cones.

The first two cones in our example are not unimodular. Their decomposition is given below, without any additional explanation due to lack of space:

$$C(V_1) = \oplus \left\{ \begin{pmatrix} 0 \\ 1 \end{pmatrix}, \begin{pmatrix} 1 \\ 0 \end{pmatrix} \right\} \oplus \left\{ \begin{pmatrix} 1 \\ 2 \end{pmatrix}, \begin{pmatrix} 0 \\ -1 \end{pmatrix} \right\} \quad (2)$$

$$C(V_2) = \oplus \left\{ \begin{pmatrix} 0 \\ -1 \end{pmatrix}, \begin{pmatrix} -1 \\ -2 \end{pmatrix} \right\} \ominus \left\{ \begin{pmatrix} -1 \\ 1 \end{pmatrix}, \begin{pmatrix} 0 \\ -1 \end{pmatrix} \right\}$$

Step 3 Find out the generating function of each supporting cone.

The above decomposition is performed since any unimodular cone $C(V)$ has associated a generating function [3] of the form

$$F(V) = \frac{z^V}{\Pi_i(1 - z^{r_i})}$$

where $z^V = x^a y^b$ (since a vertex $V(a, b)$ has two coordinates a, b in this case), and the product is over all the generating rays r_i. For instance, if the vertex of the cone is $V = V_1(0,0)$ then $z^V = x^0 y^0 = 1$; if the ray $r_i = [1\ 2]^T$ then $z^{r_i} = x^1 y^2$. The generating function of any cone is obtained making the summation of the functions of all the unimodular component cones. Therefore, from (2), the generating function of the cone $C(V_1)$ is

$$F(V_1) = \frac{1}{(1 - y)(1 - x)} + \frac{1}{(1 - xy^2)(1 - y^{-1})}$$

With similar computations, the generating functions for the other supporting cones are:

$$F(V_2) = \frac{x^2 y^4}{(1 - y^{-1})(1 - x^{-1}y^{-2})} - \frac{x^2 y^4}{(1 - x^{-1}y)(1 - y^{-1})}$$

$$F(V_3) = \frac{x^4 y^2}{(1 - x^{-1}y)(1 - y^{-1})} \ , \ F(V_4) = \frac{x^4}{(1 - x^{-1})(1 - y)}$$

The sum of these rational functions yields the generating function F of the whole quadrilateral in Fig. 1.

Step 4 Compute the number of lattice points from the generating function $F = \sum_i F(V_i)$ of the whole polytope.

In order to obtain a one-variable generating function F, we make the substitution $z \longrightarrow t^\lambda$, where λ is an integer vector chosen such that no dot product of λ with any generating ray is zero [6]. In this example, we choose $\lambda = [1\ -1]^T$ (there is an algorithm for this as well). With the substitution $x = t$, $y = t^{-1}$, the generating function of the iterator polytope becomes:

$$F = \frac{1}{(1 - t^{-1})(1 - t)} + \frac{1}{(1 - t^{-1})(1 - t)} + \frac{t^{-2}}{(1 - t)^2}$$
$$- \frac{t^{-2}}{(1 - t^{-2})(1 - t)} + \frac{t^2}{(1 - t^{-2})(1 - t)} + \frac{t^4}{(1 - t^{-1})^2}$$

After eliminating the negative exponents in the denominators, and after factorizing t^{-2}, we substitute $t = s + 1$, obtaining rational terms of the form $\frac{P(s)}{s^d Q(s)}$, where $P(s)$ and $Q(s)$ are polynomials, and $d = 2$ is the dimension of the iterator space:

$$F = \frac{-(s+1)^3}{s^2} + \frac{-(s+1)^3}{s^2} + \frac{1}{s^2} - \frac{(s+1)^2}{s^2(s+2)} + \frac{-(s+1)^6}{s^2(s+2)} + \frac{(s+1)^8}{s^2}$$

If $P(s) = a_0 + a_1 s + a_2 s^2 + \ldots$ and $Q(s) = b_0 + b_1 s + b_2 s^2 + \ldots$, the coefficients of the quotient $P(s)/Q(s) = c_0 + c_1 s + c_2 s^2 + \ldots$ can be obtained recursively as follows [6]:

$$c_0 = \frac{a_0}{b_0} \text{ and } c_k = \frac{1}{b_0}(a_k - b_1 c_{k-1} - b_2 c_{k-2} - \ldots - b_k c_0) \text{ for } k \geq 1$$

The algebraic sum of the coefficients c_2 (since here the space dimension is 2) after the polynomial divisions in all the terms of F is the number of lattice points [3]. In this example, the 6 coefficients c_2 (one for each term of F) are $\{-3, -3, 0, \frac{1}{8}, -\frac{49}{8}, 28\}$. Their sum yields 16, which is indeed the number of lattice points inside (or on the border of) the iterator polytope in Fig. 1, and it is also the number of memory locations to store the array reference $A[2i + 3j][5i + j]$ since the vector function $\mathbf{x} = \mathbf{Ti} + \mathbf{u}$ from the iterator to the index space is a one-to-one mapping.

Assume now that the range of the first iterator in *Example 1* is 0 to 400 (rather than 0 to 4) and in the second loop the condition $j \leq -i + 6$ is replaced by $j \leq -i + 600$. The iterator polytope is a quadrilateral similar with the one in Fig. 1, but much larger, the similarity ratio being 100. The computation effort necessary to find the number of memory locations for the array reference $A[2i + 3j][5i + j]$ is not affected by the very significant increase in size of the iterator space. Indeed, the 4 supporting cones are generated by the same rays, the decompositions are the same, the generating functions are almost the same. The only difference appears at the numerators of $F(V_2)$, $F(V_3)$, and $F(V_4)$ due to the modifications of the coordinates of these vertices. For instance, the numerator of $F(V_3)$ becomes $x^{400}y^{200}$ since the new coordinates of V_3 are (400,200). The storage requirement for this case is 100,501 locations. Note that the number of lattice points does not scale up with the square of the similarity ratio like, for instance, the area of the quadrilateral.

Moreover, the technique sketched above, although illustrated for a 2-dimensional signal in the scope of an iterator space of dimension 2, works for arbitrary numbers of dimensions of both the index and iterator spaces. Therefore, it is well-suited to address the size of array references typical to multimedia applications. □

The example above illustrated the case when there is a one-to-one mapping between the iterator and index spaces. But this is not always true. When the rank r of matrix $\mathbf{T}$ is smaller than n, the number of columns of $\mathbf{T}$, the memory occupied by the array reference is upper bounded by the number of lattice points in the r-dimensional polytope $pr_r(\mathbf{Ai} \geq \mathbf{b})$ – the real projection of $\mathbf{Ai} \geq \mathbf{b}$ on $\mathcal{R}^r$ along the first r coordinates. $pr_r(\mathbf{Ai} \geq \mathbf{b})$ can be easily computed by eliminating the last $n - r$ iterators in $\mathbf{Ai} \geq \mathbf{b}$ with the Fourier-Motzkin technique [5]. It must be noticed that not necessarily all the lattice points in $pr_r(\mathbf{Ai} \geq \mathbf{b})$ represent projections of lattice points from $\mathbf{Ai} \geq \mathbf{b}$ [12]. These invalid projections are detected by replacing the r coordinates of the projection point under question in the polytope $\mathbf{Ai} \geq \mathbf{b}$ and checking if the resulting $(n - r)$-dimensional integral polytope is empty.

Example 2: for $(i = 0; i \leq 4; i + +)$
$\qquad for (j = 0; j \leq 2i \,\&\&\, j \leq -i + 6; j + +) \cdots A[3i + j]$

Since post-multiplying $\mathbf{T} = \begin{bmatrix} 3 & 1 \end{bmatrix}$ with the unimodular matrix $\mathbf{S} = \begin{bmatrix} 0 & 1 \\ 1 & -3 \end{bmatrix}$ results in $\mathbf{T} \cdot \mathbf{S} = \begin{bmatrix} 1 & 0 \end{bmatrix}$, the rank of matrix $\mathbf{T}$ is $r = 1$ – less than the number of columns $n = 2$ of $\mathbf{T}$; in this case, the vector function $\mathbf{x} = \mathbf{Ti} + \mathbf{u}$ may not be a one-to-one mapping. Indeed, the iterator vectors $[i\ j]^T = [2\ 3]^T$ and $[3\ 0]^T$ from the iterator space in Fig. 1 are mapped to the same index $3i + j = 9$. The transformation $\mathbf{S}$ modifies the iterator space into $\{\mathbf{AS} \cdot \mathbf{j} \geq \mathbf{b}\} = \{\ 3l \leq k \leq 5l,\ k \leq 2l + 6,\ l \leq 4\ \}$ ($\mathbf{j} \stackrel{not}{=} [k\ l]^T = \mathbf{S}^{-1}[i\ j]^T$ is the new iterator vector after the transformation $\mathbf{S}$) which real projection is $\{\ 0 \leq k \leq 14\ \}$, obtained eliminating l in the inequalities above. But not all these 15 points are valid projections. $k = 1, 2$ result to be invalid: replacing these values in the modified iterator space, no integer solution for l can be found. Therefore, storing $A[3i + j]$ requires 15-2=13 locations.

The algorithm described above is implemented in our memory computation tool (see Section 5).

4 Memory size computation algorithm based on data-dependence analysis

The main steps of the memory size computation algorithm will be discussed below.

Step 1 Extract the array references from the given algorithmic specification of the multimedia application and decompose the array references for every indexed signal into *disjoint* linearly bounded lattices.

The analytical partitioning of the array references of every signal into disjoint LBL's can be performed by a recursive intersection, starting from the array references in the code. Let
$\{\mathbf{x} = \mathbf{T}_1\mathbf{i}_1 + \mathbf{u}_1 \mid \mathbf{A}_1\mathbf{i}_1 \geq \mathbf{b}_1\}$, $\{\mathbf{x} = \mathbf{T}_2\mathbf{i}_2 + \mathbf{u}_2 \mid \mathbf{A}_2\mathbf{i}_2 \geq \mathbf{b}_2\}$
be two LBL's derived from the same indexed signal, where $\mathbf{T}_1$ and $\mathbf{T}_2$ have obviously the same number of rows – the signal dimension. Intersecting the two linearly bounded lattices means, first of all, solving a linear Diophantine system[2] $\mathbf{T}_1\mathbf{i}_1 - \mathbf{T}_2\mathbf{i}_2 = \mathbf{u}_2 - \mathbf{u}_1$ having the elements of $\mathbf{i}_1$ and $\mathbf{i}_2$ as unknowns. If the system has no solution, the intersection is empty. Otherwise, let

$$\begin{bmatrix} \mathbf{i}_1 \\ \mathbf{i}_2 \end{bmatrix} = \begin{bmatrix} \mathbf{V}_1 \\ \mathbf{V}_2 \end{bmatrix} \mathbf{i} + \begin{bmatrix} \mathbf{v}_1 \\ \mathbf{v}_2 \end{bmatrix}$$

be the solution of the Diophantine system. If the set of coalesced constraints of the two LBL's (denoted Lbl_1 and Lbl_2)

$$\mathbf{A}_1\mathbf{V}_1 \cdot \mathbf{i} \geq \mathbf{b}_1 - \mathbf{A}_1\mathbf{v}_1 \qquad (3)$$
$$\mathbf{A}_2\mathbf{V}_2 \cdot \mathbf{i} \geq \mathbf{b}_2 - \mathbf{A}_2\mathbf{v}_2$$

has integer solutions, then the intersection is a new LBL:

$Lbl_1 \cap Lbl_2 = \{\mathbf{x} = \mathbf{T}_1\mathbf{V}_1 \cdot \mathbf{i} + \mathbf{T}_1\mathbf{v}_1 + \mathbf{u}_1\mid$ s.t. constraints (3)$\}$

However, the real difficulty is the decomposition of the differences $Lbl_1 - (Lbl_1 \cap Lbl_2)$ and $Lbl_2 - (Lbl_1 \cap Lbl_2)$, and the reason is that the difference of two LBL's is not necessarily an LBL. Due to the present space limitation, the full LBL decomposition algorithm will be published elsewhere.

Example: The disjoint LBL's of signal *Delta* from the illustrative example in Section 2 are (in non-matrix format):
$Delta_1 = \{\ x = i,\ y = j,\ z = 0 \mid 120 \geq i,\ j \geq 8\ \}$
$Delta_2 = \{\ x = i,\ y = j,\ z = 289 \mid 120 \geq i,\ j \geq 8\ \}$
$Delta_3 = \{\ x = i,\ y = j,\ z = 17(k-i)+l-j+145 \mid 120 \geq i,\ j \geq 8$
$\qquad\qquad 8 \geq k - i,\ l - j \geq -8\ $ and $\ 143 \geq 17(k - i) + l - j\ \}$

Figure 2 shows a polyhedral dependence graph built from the illustrative example in Section 2, where the nodes are the disjoint LBL's determined at this step and the arcs are the dependence relations between them derived from the code. The nodes are labeled with the number of scalar signals they cover and the arcs are labeled with the number of dependencies (both computed using the algorithm from Section 3).

---

[2]Finding the integer solutions of the system. Solving a linear Diophantine system was proven to be of polynomial complexity, all the known methods being based on bringing the system matrix to the Hermite Normal Form [13].

8B-2

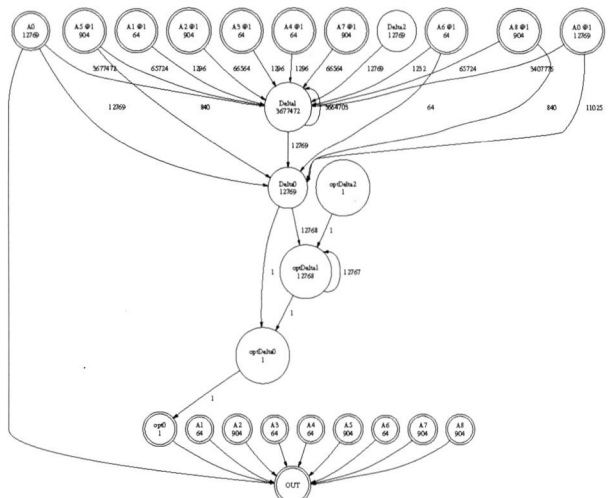

Figure 2: Polyhedral dependence graph having as nodes the disjoint linearly bounded lattices from the example in Section 2

Step 2 Determine the memory size at the boundaries between the blocks of code.

The algorithmic specification is a sequence of nested loops, referred also as *blocks*. After the decomposition of the array references, for each disjoint LBL it is determined the block where the LBL is created (i.e., *produced*), and the block where it is used as an operand for the last time (i.e., *consumed*). Based on this information, the memory size between the blocks can be determined *exactly*, since the storage requirement of each disjoint LBL can be computed *exactly* – using the algorithm explained in Section 3.

Step 3 Pruning the algorithmic specification.

If in a block signals are produced but no signal is last time consumed, that block is irrelevant in memory size point of view and can be skipped from further analysis since the memory will only increase to the amount at the end of the block – which is already known from *Step 2*. Similarly, if the amount of storage required by the newly created LBL's, together with the amount of memory at the beginning of the block, is not larger than the maximum storage at the boundary level, then that block can be pruned as well. This pruning speeds up the tool, concentrating the analysis on those portions of code where the memory increase is likely to happen.

Step 4 For each of the remaining blocks of code, compute the maximum memory size inside the block. This operation is based on the computation of min/max iterator vectors relative to the lexicographic order.

Definition Let $\mathbf{i} = [i_i, \ldots, i_n]^T$ and $\mathbf{j} = [j_i, \ldots, j_n]^T$ be two iterator vectors in the scope of n nested loops, which may be assumed "normalized" (i.e., all the iterators are increasing with the step 1). Iterator vector $\mathbf{j}$ is larger lexicographically than $\mathbf{i}$ (written $\mathbf{j} \succ \mathbf{i}$) if $(j_1 > i_1)$, or $(j_1 = i_1$ and $j_2 > i_2)$, or ... $(j_1 = i_1, \ldots, j_{n-1} = i_{n-1}$, and $j_n > i_n)$. The min/max iterator vector from a set of such vectors is the smallest/largest vector in the set relative to the lexicographic order.

Example 3: $for\ (i = 0;\ i \leq 3;\ i++)$
$\qquad for\ (j = 0;\ j \leq 3;\ j++)$
$\qquad\qquad for\ (k = 0;\ k \leq 3;\ k++)\ \ \cdots A[i + j + k] \cdots$

The max iterator vector addressing the scalar, say, $A[5]$ is $[i\ j\ k]^T_{max} = [3\ 2\ 0]^T$, while the min iterator vector is $[0\ 2\ 3]^T$.

Our algorithm finds the LBL's produced and consumed in the current block, computing the min and, respectively, max iterator vectors for the scalar signals covered by these LBL's since these iterator vectors correspond to the increase and, respectively, decrease of the memory. Knowing the number of flops (i.e., elementary iterations) in the loop nest (by counting the lattice points in the iterator spaces with the algorithm in Section 3), one can then determine *exactly* the memory variation and, in particular, the maximum storage amount in each of the blocks. Actually, part of the LBL's produced or consumed in the block can be conveniently skipped if their effect on the memory variation can be taken into account without generating the scalars they cover. For instance, in the illustrative example from Section 2, each iterator vector $[i\ j\ k\ l]^T$ corresponds to a unique produced scalar $Delta[i][j][17(k-i)+l-j+145]$ and a unique consumed scalar $Delta[i][j][17(k-i)+l-j+144]$. The effect of the two array references on the memory variation is +1-1=0 in each iteration and, therefore, these operands can be skipped from further analysis, pruning that increases significantly the computation speed.

5 Experimental results

A memory size computation tool (named $K2$ after the famous peak which climbing adversity intends to suggest the difficulty of its implementation) has been implemented in C++, incorporating the ideas and algorithms described in this paper. For the syntax of the algorithmic specifications, we adopted a subset of the C language (see, e.g., the illustrative example in Section 2). This is not a restrictive feature of the theoretical model since any modification in the specification language would affect only the front-end of the tool. In addition to the computation of the minimum memory size requirements and different statistical data on the memory usage by the multidimensional signals in the multimedia specification, the tool can optionally generate dependence graphs (like the one in Fig. 2) at different granularity levels, which provide information about the relations between different groups of signals, and also the trace of the memory occupancy during the execution of the input specification. Such a memory trace is shown in Fig. 3.

Table 1 summarizes the results of our experiments, carried out on a Sun Blade 100 workstation. The benchmarks used are: (1) Durbin's algorithm which solves a Toeplitz system with N unknowns [16], (2) a real-time regularity detection algorithm used in robot vision, (3) a 2D Gaussian blur filter from a medical image processing application which extracts contours from tomograph images in order to detect brain tumors, (4) a motion detection algorithm used in the transmission of real-time video signals on data networks [4], and (5) the kernel of a voice coding application – essential component of a mobile radio terminal.

This tool can process large specifications in terms of number

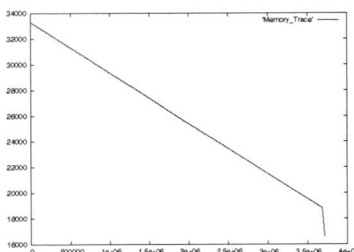 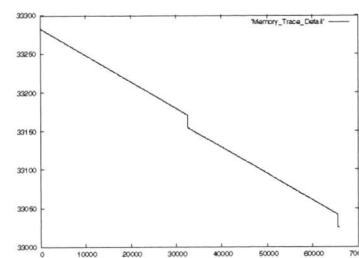 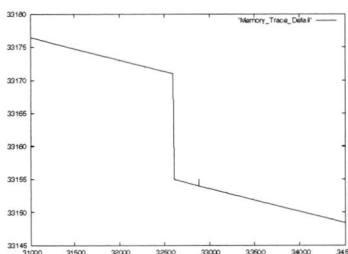

Figure 3: Memory trace for the illustrative example in Section 2. The abscissae are the numbers of datapath instructions in the code, the ordinates are memory locations. The first graphic represents the entire trace. The second graphic is a detailed trace in the interval [0 : 65767], which corresponds to the first two iterations of the outer loop ($i = 8$ and 9). The third graphic is a detailed trace in the zone covering the end of the first outer-loop iteration and the start of the second one. The global maximum is at the point (x=2, y=33284).

Application	Parameters	Memory	CPU
Durbin alg.	N=100	249	< 1s
Regularity detection	MaxGrid=5, L=64	960	< 1s
2D Gauss. blur filter	M=N=50	14451	2s
Motion detection	M=N=32, m=n=4	2740	16s
Vocoder kernel	–	11890	7s

Table 1: Experimental results (col. 3 shows the memory locations)

of loop nests, lines of code, number of array references. For instance, the voice coding application contains 236 array references organized in 40 loop nests. In one of our experiments, the illustrative example (Section 2) was unrolled one loop level, resulting in a code with 113 loop nests 3 level deep, and a total of 906 array references, many having complex indices. The tool processed this example in about 8 minutes.

A comparative evaluation with previous works performing memory size *estimation* is hard to do in the absence of their benchmark algorithms. For instance, the test of the motion detection kernel is referred also in [17]. If the authors used like us the algorithm given in [4], then their result (of 1372 memory locations) is a poor estimation since the correct *exact* result for the same parameters is 2740 (see Table 1).

6 Conclusions

This paper has presented a non-scalar approach for computing the memory size in real-time multimedia algorithms, where the storage of large multidimensional signals causes a significant cost in terms of both area and power consumption. This method uses modern elements in the theory of polyhedra and algebraic techniques specific to the data-flow analysis used nowadays in compilers. Different from past works which were only performing a *memory size estimation*, our approach does *exact* computations and it is the first one to do so.

References

[1] D. Avis, "lrs: A revised implementation of the reverse search vertex enumeration algorithm," in *Polytopes – Combinatorics and Compu-*

tation, G. Kalai, G. Ziegler, Birkhauser-Verlag, pp. 177-198, 2000.

[2] F. Balasa, F. Catthoor, H. De Man, "Background memory area estimation for multi-dimensional signal processing systems," *IEEE Trans. VLSI Syst.*, vol. 3, no. 2, pp. 157-172, June 1995.

[3] A.I. Barvinok, "A polynomial time algorithm for counting integral points in polyhedra when the dimension is fixed," *Mathematics of Operations Research*, vol. 19, no. 4, pp. 769-779, Nov. 1994.

[4] F. Catthoor, S. Wuytack, E. De Greef, F. Balasa, L. Nachtergaele, A. Vandecappelle, *Custom Memory Management Methodology: Exploration of Memory Organization for Embedded Multimedia System Design*, Kluwer Academic Publishers, Boston, 1998.

[5] G.B. Dantzig, B.C. Eaves, "Fourier-Motzkin elimination and its dual," *J. Combinatorial Theory (A)*, vol. 14, pp. 288-297, 1973.

[6] J.A. De Loera, R. Hemmecke, J. Tauzer, R. Yoshida, "Effective lattice point counting in rational convex polytopes," *http://www.math.ucdavis.edu/~latte/pdf/lattE.pdf*, 2003.

[7] P.G. Kjeldsberg, F. Catthoor, E.J. Aas, "Data dependency size estimation for use in memory optimization," *IEEE Trans. CAD of IC's and Syst.*, vol. 22, no. 7, pp. 908-921, July 2003.

[8] F.J. Kurdahi, A.C. Parker, "REAL: A program for register allocation," *Proc. 24th Design Automation Conf.*, pp. 210-215, 1987.

[9] S.S. Muchnick, *Advanced Compiler Design and Implementation*, Morgan Kaufmann, San Francisco, 1997.

[10] P.R. Panda, F. Catthoor, N. Dutt, K. Dankaert, E. Brockmeyer, C. Kulkarni, P.G. Kjeldsberg, "Data and memory optimization techniques for embedded systems," *ACM Trans. Design Automation of Electronic Syst.*, vol. 6, no. 2, pp. 149-206, April 2001.

[11] K.K. Parhi, "Calculation of minimum number of registers in arbitrary life time chart," *IEEE Trans. Circ. & Syst. - II: Analog and Digital Signal Processing*, vol. 41, no. 6, pp. 434-436, 1994.

[12] W. Pugh, "A practical algorithm for exact array dependence analysis," *Comm. of the ACM*, vol. 35, no. 8, pp. 102-114, Aug. 1992.

[13] A. Schrijver, *Theory of Linear and Integer Programming*, John Wiley, New York, 1986.

[14] L. Stok, J. Jess, "Foreground memory management in data path synthesis", *Int. J. Circuit Theory and Appl.*, vol. 20, pp. 235-255, 1992.

[15] L. Thiele, "Compiler techniques for massive parallel architectures," in *State-of-the-art in Computer Science*, P. Dewilde (ed.), Kluwer Acad. Publ., 1992.

[16] I. Verbauwhede, C. Scheers, J.M. Rabaey, "Memory estimation for high level synthesis," *Proc. 31st ACM/IEEE Design Automation Conf.*, pp. 143-148, June 1994.

[17] Y. Zhao, S. Malik, "Exact memory size estimation for array computations," *IEEE Trans. VLSI Syst.*, vol. 8, no. 5, pp. 517-521, 2000.

Maximizing Data Reuse for Minimizing Memory Space Requirements and Execution Cycles *

M. Kandemir, G. Chen, and F. Li
Computer Science and Engineering Department
Pennsylvania State University
e-mail: {kandemir,gchen,feli}@cse.psu.edu

Abstract— Embedded systems in the form of vehicles and mobile devices such as wireless phones, automatic banking machines and new multi-modal devices operate under tight memory and power constraints. Therefore, their performance demands must be balanced very well against their memory space requirements and power consumption. Automatic tools that can optimize for memory space utilization and performance are expected to be increasingly important in the future as increasingly larger portions of embedded designs are being implemented in software. In this paper, we describe a novel optimization framework that can be used in two different ways: (i) deciding a suitable on-chip memory capacity for a given code, and (ii) restructuring the application code to make better use of the available on-chip memory space. While prior proposals have addressed these two questions, the solutions proposed in this paper are very aggressive in extracting and exploiting all data reuse in the application code, restricted only by inherent data dependences.

I. INTRODUCTION

Applications running on embedded/mobile systems face very different constraints than their counterparts executing on high-end systems. For example, power consumption, and memory size and form factor limitations can severely limit the datasets that can be handled. While sophisticated packaging techniques help squeeze large datasets in small memories, such techniques are costly and not scalable. Also, one does not have the option of arbitrarily increasing memory size due to power and form factor considerations. Therefore, techniques that reduce memory space requirements of embedded applications and those that can help designers to maximize the utilization of the available memory space are very important. In addition, the fact that the embedded/mobile applications keep increasing in both size and complexity makes the overall memory optimization problem a very challenging one.

Regarding the memory management of embedded devices, there are two critical issues: (i) designing the most suitable on-chip memory configuration and (ii) restructuring the application code to make better use of the available on-chip memory space. In particular, exploiting small, fast, and power-efficient on-chip memories is critical from both performance and power angles. While it is conceivable that a knowledgeable application programmer can restructure her application code for the best memory behavior; in general this is a very hard task and can benefit a lot from automation. Similarly, an automatic tool

that helps the designer decide on the on-chip memory configuration (in particular, memory capacity) can be very important. Therefore, recent studies have addressed memory architecture design and automated software optimization for data reuse [4–6, 8, 13, 15, 18]. This paper mainly addresses two important questions:

- What is the minimum on-chip memory capacity (measured in terms of fixed size blocks) that minimizes the frequency and volume of data transfers between off-chip and on-chip memory in a two-level memory hierarchy? This is a particularly important problem when one wants to design an application-specific on-chip memory for a given data-intensive embedded application.

- How can we restructure an application code to make better use of the available memory hierarchy? This is a relevant problem in cases where one wants to adapt the behavior of a given application to an existing memory hierarchy that has both on-chip and off-chip components.

While prior proposals have addressed these two questions, the solutions proposed in this paper are unique because they are very aggressive in extracting and exploiting data reuse, to the extent allowed by intrinsic data dependences. This helps improve the quality of the solution for both the problems posed above. That is, by maximizing data reuse we can make better use of the available memory space and can also reduce the total required storage space. Our approach is based on dividing large data structures into logical blocks and clustering computations that access the same data block using a scheduler. While our current implementation uses a software-managed on-chip memory (e.g., a scratch-pad memory [15]), we believe that the proposed techniques can also be adapted to work with hardware-managed cache memories.

The rest of this paper is organized as follows. The next section presents the details of our approach to determining the minimum on-chip storage capacity for reducing the number of data transfers between off-chip and on-chip memories. Section III discusses our solution to code restructuring for an existing on-chip memory space. Section IV concludes the paper by summarizing our major observations.

II. DETERMINING MINIMUM ON-CHIP MEMORY CAPACITY

In this section, we present the details of our approach to determining the minimum on-chip memory capacity. While the other aspects of the on-chip memory are also important, in this section, we focus our attention on the memory capacity problem.

---

*This work is supported in part by NSF Career Award #0093082 and by a grant from GSRC.

```
do i = 1, N
    do j = 1, N
        ... = U[i][j]
    end do
    do k = 1, N
        do l = 1, N
            ... = U[k][l]
        end do
    end do
(a)
```

```
do t1 = 1, N
    do t2 = 1, N
        ... = U[t1][t2]
        ... = U[t1][t2]
    end do
(b)
```

Fig. 1. An example fragment (a) and its transformed version (b).

A. Data-Centric View for Reuse

Our approach focuses on array-based data-intensive embedded applications and operates with the concept of a *data block,* which represents both the minimum amount of data transferred from the off-chip memory in a single transfer and the minimum amount of on-chip storage. We use the term *on-chip block* to refer to the unit (i.e., the building block) for the on-chip storage, which is the same size as a data block. Then, our problem can be expressed as one of determining the optimal number of on-chip blocks for a given embedded application. At the high-level, the goal behind our approach is to determine the minimum on-chip memory size (capacity) in terms of the number of on-chip blocks so that increasing that number will not bring any additional performance benefits. To illustrate the idea let us consider the code fragment in Figure 1(a), where two separate loop nests iterate over a two-dimensional array (U). Assuming that there are no data dependences between the iterations of these loop nests, the entire code fragment can be restructured as shown in Figure 1(b). In this case, the references to the array are brought together; similar to the way a loop fusion algorithm [9] would perform. Now, to execute this fragment, only one on-chip memory block would be sufficient. For example, if we have an on-chip block that can hold K elements, we bring the first K elements of the array to the on-chip block and process them. Since these K elements are not needed for subsequent computation, once we are done with them, we can remove them from the on-chip storage and bring the next group of K elements, and so on. With the transformed code in Figure 1(b), having another on-chip block would not help futher improve performance at all. In contrast, in the original code fragment in Figure 1(a), the successive accesses to a given data block are far apart from each other; in fact, they occur in different nests. Therefore, if one wants to exploit this reuse, a total of N^2/K on-chip blocks are needed (so that each data block can be kept on-chip until its reuse takes place). As a more complex scenario, consider the code fragment in Figure 2(a). While the two nests shown in this example share a lot of data elements (i.e., they exhibit high data reuse), the code fragment needs many on-chip blocks to exploit this data reuse. In comparison, the transformed code depicted in Figure 2(b) – obtained automatically through our approach to be explained in this paper – needs only a single on-chip data block (though, this fact is not very clear from this transformed code description!). Note that, the transformation performed to obtain Figure 2(b) from Figure 2(a) is not a simple application of loop fusion.

In this section of the paper, we want to determine the *minimum* number of on-chip blocks such that increasing this number does not bring any additional performance benefits. Consequently, our approach comes up with the minimum on-chip memory capacity with the high performance, and is thus suit-

```
do i = 1, N
    do j = 1, N
        ... = U[i+j][j+3]
    end do
    do k = 2, N
        do l = 1, N
            ... = U[k-1][k+l+2]
        end do
    end do
(a)
```

```
do G = 1, N
    do F = 1, N, K
        do t1 = max(G-N,G-K-F+4,1),
                min(G-1,N,G-F+3)
            ... = U[G][G-t1+3]
        end do
        if(G ≤ N-1 && G ≥ 0) {
            do t1 = max(-G+F-3,1), min(N,-G+K+F-4)
                ... = U[G][G+t1+3]
            end do
        }
    end do
end do
(b)
```

Fig. 2. An example fragment (a) and its transformed version (b).

able from the viewpoint of energy consumption (since a larger on-chip storage would only increase the power consumption without reducing the execution time any further). In doing so, we also modify the application code (re-schedule its loop iterations) as well. However, there exist several issues that make this problem very challenging:

● In general, there can be both intra-loop and inter-loop data dependences that prevent us from reusing a data block brought from the off-chip memory completely before being replaced (by another block). Therefore, one needs a suitable representation of data dependences across different loop nests. Most of the prior research focuses only on intra-loop dependences.

● Since there can be multiple arrays accessed by the application, this can increase the number of on-chip blocks required. However, one can potentially reduce this number by considering the lifetimes of the data blocks of different arrays.

● Array access patterns can in general be very complex. Consequently, the loop transformations required can be much more complex than the simple fusion-like transformation illustrated in Figure 1(b). In fact, the loop transformation used for obtaining the transformed code in Figure 2(b) from the one in Figure 2(a) is not easy to derive using the conventional (linear algebra based) loop transformation theory [14].

In the rest of this section, we present a mathematical model, within which the potential problems posed above can be addressed.

B. Data Block Graph

We assume all array indices and loop bounds are affine functions of enclosing loop indices. Let us assume, without loss of generality, that the program to be optimized has v loop nests, and $I_1, I_2, I_3, \cdots, I_v$ denote iteration spaces of these loop nests. Each iteration space is a set that contains the iteration points executed by a loop nest. For example, a loop nest with two loops each iterating N times has a total of N^2 iteration points. We say that these iteration spaces collectively define the *computation domain* of the application; that is:

$$I_D = \bigcup_{1 \leq i \leq v} I_I,$$

where I_D is the computation domain. We use $I_{U,i,j}$ to represent the set of iterations from loop nest i that access data block j of array U. In formal terms, an iteration (point) l belongs to $I_{U,i,j}$ if and only if the following holds:

$$\exists R \text{ and } \exists d \in \text{ data block } j \text{ of } U \text{ such that } R(l) = d,$$

where R is a reference in loop nest i to array U. Note that, we have:

$$\bigcup_i \bigcup_U \bigcup_j I_{U,i,j} = I_D.$$

That is, all $I_{U,i,j}$s when combined together cover the entire computation domain. A data dependence is said to exists between $I_{U,i,j}$ and $I_{U,m,n}$ if an iteration that belongs to $I_{U,m,n}$ depends on (the result generated by) an iteration of $I_{U,i,j}$. A data dependence imposes an execution order for the loop iterations in $I_{U,i,j}$ and $I_{U,m,n}$.

Conceptually, we can use a graph, called *data block graph* (*DBG*), to represent $I_{U,i,j}$ nodes and the data dependence relationships between them. Specifically, each node of this graph corresponds to a $I_{U,i,j}$, and a directed edge from $I_{U,i,j}$ to $I_{U,m,n}$ indicates a data dependence between them (which we extract using well-known techniques [2, 9]). Note that an execution of the computation domain means visiting each node of the corresponding DBG. Any *legal* execution is a traversal of this graph that respects all data dependences between them. That is, if there is a dependence from $I_{U,i,j}$ to $I_{U,m,n}$, the latter can be executed only after the former is finished. The subsections below discuss this scheduling problem in detail.

C. Scheduling Problem

In order to generate code for the application being optimized, the entire computation domain must be covered. As mentioned earlier, this can be achieved by visiting each node of the DBG and by observing data dependences during this traversal. While there may be many different traversal orders that can generate legal (dependence-observing) results, we are interested in ones that enhance *data block reuse;* that is, when a data block is accessed, we want to reuse its contents as much as possible before accessing the next data block. While conventional loop transformations attempt to achieve this by transforming each loop nest individually (and independently of the others) as in the case of linear loop transformations [14], or considering only neighboring nests as in the case of loop fusion [9, 14], in this paper, we go beyond these techniques, and consider the entire computation domain (i.e., the entire application), denoted by I_D, the entire computation domain. Our goal is to extract and exploit much more reuse than what is possible by well-known locality-enhancing techniques such as loop tiling, linear loop transformations, and loop fusion.

It is to be noted that the two nodes in a DBG, $I_{U,i,j}$ and $I_{U,i',j}$ (where $i \neq i'$), exhibit data reuse between them. More specifically, both these nodes access the same block j of array U. Consequently, if one wants to exploit this reuse, these two nodes should be scheduled one after another. However, as mentioned earlier, data dependences should also be accounted for. To simplify the problem, we first focus on the special case (Section D), where we have only one array in the code, and each loop nests operates on this array using an arbitrary number of references. We subsequently discuss how this scheme is extended to the more general case, where we have multiple arrays, each can be accessed by any loop nest using an arbitrary number of references (Section E).

D. Solution for Single Array Case

While it is possible to formulate the scheduling problem using known techniques such as integer linear programming

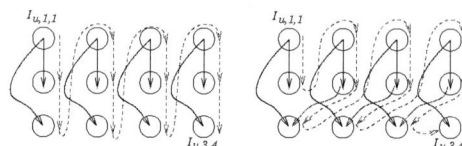

Fig. 3. Two different data block graphs (DBGs) and potential schedulings. The solid arrows denote data dependences, while the dashed arrows indicate scheduling (execution) order. For clarity, only two nodes are explicitly labeled in each DBG.

(ILP) or genetic algorithms (GA), these approaches would be very expensive unless one focuses only on very small-sized graphs/ problems. Therefore, our goal is to come up with a heuristic solution that generates good results most of the time. As mentioned earlier, since DBG nodes $I_{U,i,j}$ and $I_{U,i',j}$ exhibit locality (as they access the same data block), the final schedule should place such nodes into consecutive slots as much as possible. In other words, the access to $I_{U,i,j}$ should be followed by an access to $I_{U,i',j}$. We propose a heuristic scheduling algorithm based on *list scheduling,* a scheduling paradigm used in the past by optimizing compilers [14] and high-level synthesis tools [7].

Our approach is a greedy heuristic that selects one node from the DBG at a time and schedules it. In selecting the next node to schedule, the proposed algorithm observes the following two rules: (1) all the DBG nodes on which this node depends must be already scheduled (data dependence constraint), and (2) this node should access the same data block as the previous node if possible (data reuse constraint). Note that, while breaking the first rule would lead to incorrect execution, breaking the second rule would reduce data reuse. If, at any step during scheduling, the algorithm could not satisfy the second rule, that means we are moving to another data block. This may happen due to two reasons. First, it is possible that all the nodes that access the current data block have already been scheduled. This is the preferable case as this means we can now use the same on-chip block for another data block, i.e., we do not need a new on-chip block for the new data block that needs to be accessed. The second potential reason is that, although we still have some unscheduled nodes that access the current data block, we cannot schedule any of them as the next node due to data dependences. As opposed to the previous one, in this case, we need another on-chip block if we do not want to incur any performance degradation.[1] Each time this second case occurs, we increment the number of on-chip blocks by 1. The algorithm terminates when all the DBG nodes have been scheduled. When this happens, the current number of on-chip blocks gives us the total minimum on-chip memory capacity required. However, during scheduling, whenever the last node that accesses a data block is scheduled, the on-chip block reserved for it is deallocated (recycled). Notice that, our approach takes care of data dependences both across the different loop nests and across the different iterations of the same loop nest.

We consider the following code fragment to illustrate how our approach schedules the nodes of a DBG. The DBG cor-

---

[1]The alternative would be displacing the current data block from the on-chip block, and placing the new data block into it. However, this incurs performance penalty when the current block needs to be reused in the future. Recall that our objective in this section is to determine the minimum number for on-chip blocks that generate the best performance.

responding to this code fragment is given in Figure 3(a), assuming (for illustration purposes) that each array occupies four data blocks.

```
do i = 1, N
    U[i] = ...
end do
do j = 1, N
    ... = U[j]
end do
do k = 1, N
    ... = U[k]
end do
```

The solid arrows in Figure 3(a) indicate the data dependences between the nodes. A possible schedule determined by our approach is also shown in the figure using dashed arrows. This schedule finishes one data block before moving to the next one. Consequently, it schedules the entire graph using only a single on-chip block. This on-chip block holds each data block in turn, and when a data block is resident in it, all the 3 nodes (coming from the different nests) are scheduled one after another. In other words, in this particular case, we achieve perfect data block reuse, i.e., even if we have more on-chip blocks, we could not achieve a better performance. To illustrate the impact of data dependences in preventing the scheduling from exploiting the maximum data locality, let us now consider the DBG in Figure 3(b). This DBG is similar to the one in (a), the difference being the additional dependence arcs entering into the nodes that belong to the third nest. These dependences prevent us from scheduling the third node (that accesses the same data block) right after the first two nodes. Consequently, one possible schedule is the one shown in the figure (using dashed lines/curves). To demonstrate our approach, let us trace the initial portion of the schedule shown in Figure 3(b). We use OB1, OB2, ..., etc to denote the on-chip blocks used. We start with $I_{U,1,1}$ and assign it (actually the data accessed by it) to OB1. Next, we move to $I_{U,2,1}$ and still operate on OB1. However, after this, we cannot execute $I_{U,3,1}$ from OB1 at this moment due to the data dependence. Instead, we access $I_{U,1,2}$ and assign it to OB2. We subsequently proceed with $I_{U,2,2}$ and continue to use OB2. Once we are done with $I_{U,2,2}$, we can execute $I_{U,3,1}$ using OB1. Now, since all the nodes that access the first data block of the array have been processed, we can discard the contents of OB1 (that is, it can be overwritten/recycled). So, we now use OB1 to load the third data block and schedule $I_{U,1,3}$ and $I_{U,2,3}$, at which point $I_{U,3,2}$ can get scheduled, and so on. To sum up, in this example, with only two on-chip blocks we can schedule the entire DBG and the total number of off-chip memory accesses (on-chip block loads/updates) is 4. Increasing the number of on-chip blocks further would not bring any performance benefits. For example, even if we have 4 on-chip blocks, we would need 4 off-chip memory accesses. We discuss in Section III how such graphs can be scheduled with a given (fixed) number of on-chip blocks.

E. Solution for Multiple Array Case

In this paper we explore two different methods for handling the multiple array case. The first method, called *single array centric*, is based on two observations: (i) although an embedded application can access multiple arrays, there is typically one or two arrays whose accesses constitute a large fraction of overall memory accesses. (ii) Given a schedule in the data block graph, it is possible to estimate the number of on-chip blocks required to minimize the number of off-chip memory accesses (on-chip block updates). This is possible as the compiler can determine the data access patterns and on-chip block updates easily for array-based embedded applications. Based on these two observations, the single array centric method considers each array in the application in turn. When considering an array, it builds the DBG for this array (omitting the other arrays in the application), and performs the scheduling explained above for the single array case. After the scheduling, we compute the number of on-chip blocks required to minimize the number of on-chip block updates. The important point here is that, in determining this number (denoted L_u for array U), the accesses to other arrays are also accounted for. It is to be noted that this process restructures the entire computation around a single array. It is repeated for each array, and array V that generates the minimum L_v among all alternatives is selected to be the one, around which the application code (the computation domain) is restructured and the output code is generated. The main advantage of this method is its simplicity since it makes use of the scheme explained in the single array case as a subcomponent. Its main drawback is that it fails to capture the coupling between references to different arrays, and this can lead to inefficiencies for applications with more than a single dominant array.

Our second method, called *global*, tries to capture the interaction between different arrays by using a separate DBG for each array. Specifically, this method operates with a *combined DBG* (or CDBG for short), which is constructed as follows. First, we build a DBG for each array. Then, we add some extra edges to connect these DBGs. These extra edges, referred to as *locality edges*, indicate how the data blocks of different arrays are accessed together (by the same computation). To explain the idea, we first consider the simple code fragment shown below:

```
do i = 1, N
    U[i] = V[N-i]
end do
```

Let $I_{U,i,1}$ denote the DBG node that contains the iterations that access the first data block for array U, i.e., the block that contains the elements $U[1], U[2], ..., U[K]$, assuming that the an on-chip block can hold K elements. Note that, when these elements of array U are accessed by $I_{U,i,1}$, the elements $V[N-1], V[N-2], ..., V[N-K+1]$ are accessed from array V. Further, let us assume that there is a node $I_{V,i,f}$ in the DBG for array V that accesses these array elements. Consequently, in our CDBG, we put an (undirected) locality edge between nodes $I_{U,i,1}$ and $I_{V,i,f}$. In a sense, this edge should be visited whenever $I_{U,i,1}$ (or $I_{V,i,f}$) is accessed. After building CDBG, the goal of the global method is to come up with a scheduling such that the current contents of the on-chip blocks are maintained (i.e., reused) as much as possible. It should be noted that, in a sense, the locality edges impose constraints (for scheduling) similar to those imposed by dependence edges.

To illustrate how the global method works, let us consider the code fragment shown below, assuming for clarity that both arrays have two data blocks:

```
do i = 1, N
    U[i] = V[N+1-i] + ...
end do
do i = 1, N
```

811

8B-3

2006 Asia and South Pacific Design Automation Conference

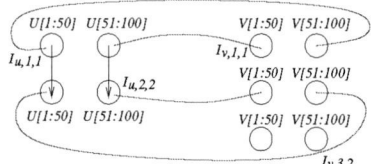

Fig. 4. A combined DBG (CDBG). The array region identifier attached to a node indicates the array elements accessed by that node. For clarity, only four nodes are labeled.

```
    ... = V[i] + ...
end do
do i = 1, N
    ... = V[N+1-i] + ...
end do
```

The CDBG for this fragment is illustrated in Figure 4. Attached to each node are the elements of the data block accessed. The locality edges are between the DBGs of the two arrays. Now, let us trace the scheduling using this CDBG. We start with $I_{U,1,1}$ and assign it to OB1. As a result, the locality edge forces us to assign $I_{V,1,2}$ to OB2. We next move to $I_{U,2,1}$ which reuses OB1. We can now schedule $I_{V,2,2}$ and $I_{V,3,2}$ as well. Since, at this point, all accesses to the first block of U and the second block of V are completed, OB1 and OB2 can be deallocated. In the remaining part of the schedule, we assign OB1 to the second block of U and OB2 to the first block of V. Thus, we can schedule the entire CDBG with only two on-chip blocks.

F. Code Generation

Once the DBG (or CDBG) has been scheduled (i.e., an order in which its nodes are to be traversed has been determined) as explained above, the next task is *code generation*. To do this, we propose to use a polyhedral tool such as the Omega Library [11]. The Omega library manipulates integer tuple relations and sets, which are described using Presburger formulas, a class of logical formulas which can be built from affine constraints over integer variables, the logical connectives ($\vee$, $\wedge$, and $\neg$), and the existential and universal quantifiers ($\exists$ and $\forall$). In our context, the Omega library is used for generating a loop that iterates over the elements (iteration points) that access a data block; that is, the iterations that constitute a node in the DBG. For example, let us consider an array reference $U[i + j - 1][j + k + 3]$ that appears within a nest with three loops (i, j, and k from outermost to innermost). Assume further that LB_i and UB_i denotes the lower and upper bounds, respectively, for loop index i, and similar lower/upper bounds exist for the remaining two loops as well. The iteration space of this loop nest can thus be described as:

$$IS = \{[i,j,k] : (LB_i \leq i \leq UB_i) \wedge (LB_j \leq j \leq UB_j) \wedge (LB_k \leq k \leq UB_k)\}.$$

Let us focus on a particular data block that contains array elements $U[G][F], U[G][F+1], U[G][F+2], ... , U[G][F+K-1]$. Within the Omega framework, this block can be defined as:

$$DB = \{[a,b] : (a = G) \wedge (F \leq b \leq F+K-1) \wedge ([a:b] \in DS)\}.$$

Here, DS represents the data (array) space, i.e., the elements of the data block should be within the array bounds. Then,

the iterations in the DBG node that access this block can be expressed as the following Presburger formulation:

$$ND = \{[i,j,k] : \exists a \exists b \text{ such that } (a = i + j - 1) \wedge (b = j + k + 3) \wedge ([i,j,k] \in IS) \wedge ([a,b] \in DB)\}.$$

The last condition here forces the accessed element to be within the data block, and the condition before the last one makes sure that any iteration included in ND is legal, i.e., within the loop boundaries. Since an array has typically multiple data blocks, to iterate over all of them in sequence, we can construct the following code (assuming that K – the number of elements in an on-chip block – divides N evenly and omitting the condition on array bounds). In this loop nest, which has been generated with the help of the "codegen" utility in the Omega library, the first two loops iterate over the blocks of the array, whereas the inner two loops visit the iterations that access a given data block (indexed by the upper two loops).

```
do G = 1, N
  do F = 1, N, K
    if (K ≥ 1 && UB3 ≥ LB3) {
      do t1 = max(G-F-K+LB3+5,LB1,G-UB2+1),
              min(G+UB3-F+4,UB1,G-LB2+1)
        do t3 = max(LB3,-G+t1+F-4),
                min(UB3,-G+t1+F+K-5)
          ...U[G,G-t1+t3+4]...
        end do
      end do
    }
  end do
end do
```

III. SCHEDULING UNDER CAPACITY CONSTRAINTS

In Section II, our main focus was on determining the minimum number of on-chip blocks such that the number of on-chip block updates is minimized (i.e., the memory capacity problem). This is an important problem if one targets at determining the minimum capacity of the on-chip storage for extracting the best performance from a given embedded application. Another important question that needs to be answered, though, is how can we schedule a DBG with a fixed number of on-chip blocks (i.e., fixed storage capacity)? Obviously, this problem is more relevant in cases where the on-chip memory capacity is fixed. Our objective in this case is to minimize the number of on-chip block loads/updates as much as possible.

We attack this (scheduling) problem by adopting a greedy strategy, which works as follows. Let us assume that we have r on-chip blocks of equal sizes, and that we have only one array. In the first step of our approach, we determine a set of *chains* in the DBG such that each chain consists of nodes that access the same data block. Let us assume that the number of such chains is t. If $r \geq t$, that is, the number of on-chip blocks is larger than the number of chains, we assign a private on-chip block to each chain. Note that, with such an assignment, the contents of an on-chip block are updated only for the initial load. Therefore, the total cost of such an assignment is t. Consider for example, the DBG depicted in Figure 5(a). We have five six chains in this DBG, i.e., $t = 6$. Assuming that we have six on-chip blocks (i.e., $r = 6$), we assign a private on-chip block per chain. Consequently, the DBG can be easily scheduled with

812

only 6 (on-chip block) loads; i.e., each OB is reserved for a data block. The difficult scenario occurs when $r < t$. In this case, we still assign one on-chip block per chain. However, an on-chip block is *reassigned* to another chain (temporarily or permanently) during scheduling. More specifically, when $r < t$, we need to consider two cases:

• *When the first node of a chain is scheduled, all the remaining nodes can also be scheduled.* In other words, there are either no any inter-chain data dependences, or the existing inter-chain dependences are such that all the chains can be scheduled one after another. Consider the DBG in Figure 5(b), which is the same in Figure 5(a). If we only have 3 on-chip blocks ($r = 3$), the different chains need to share on-chip blocks. In this particular example, however, all the chains are independent of each other. Consequently, we can schedule the chains one after another using only 1 on-chip block. In this case, the total number of loads is 6. Even if we try to use all three on-chip blocks, the total number of loads would not change.

• *Due to inter-chain dependences, all the nodes in a given chain cannot be scheduled one after another.* In this case, we use a greedy heuristic. More specifically, we start with a chain and try to schedule all the nodes (in that chain) one after another until it is not possible to proceed further due to dependences. When this occurs, we select another node among the schedulable ones, and schedule it using another on-chip block if an unoccupied one is available. If not, we select an on-chip block among the ones that are already occupied (by another chain), and use it for the currently required block. While selecting this victim on-chip block can be done using different algorithms, in this paper we use a simple LRU-based one. That is, we reassign the on-chip block that has not been touched for the longest duration of time. This process is repeated until all the nodes in the DBG are scheduled. Figure 5(c) illustrates such a scenario with $r = 2$. The total number of on-chip block loads is 6. In contrast, assuming that r is 1, we would need a total of 12 on-chip block loads/updates.

This approach can also be extended to the multiple array case. While it is conceivable that both the methods described in Section E can be adapted to work for this case as well, our current implementation uses only the first method, i.e., the one tries each array in turn, and selects the best one to restructure the entire computation (single array centric). Implementing the global method and comparing it with the single array centric method are in our future agenda as such an implementation is more complex.

It should be noted that, the approach presented in this section is very different from the one discussed in Section II. This is because in the on-chip memory capacity problem our objective is to determine the *minimum* number of on-chip blocks. Therefore, during the scheduling process, we *increment* the number of on-chip blocks when needed. In contrast, in the scheduling problem addressed in this section, our objective is to schedule the DBG with a *fixed* number of on-chip blocks (i.e., under fixed capacity). Therefore, when we are short of on-chip blocks, we need to vacate one that is currently in use to open space for the new incoming block.

IV. CONCLUSION

We witness an unprecedented proliferation of embedded/mobile applications. Most of the embedded environ-

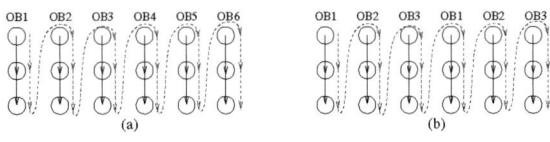

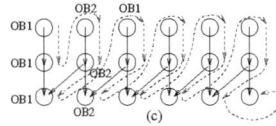

Fig. 5. Three different schedulings. The solid arrows denote data dependences, while the dashed arrows indicate scheduling (execution) order. OBi denotes the ith on-chip block. In (c), the nodes in the initial portion of the schedule are marked with the associated on-chip block.

ments that execute these applications have severe power, performance, and memory space constraints that need to be accounted for. This paper presents an optimization framework that maximizes data reuse to the greatest extent possible, by restructuring the application code based on data access patterns. The proposed infrastructure can be used for (i) deciding the capacity for on-chip storage and (ii) effective use of the available on-chip memory space.

REFERENCES

[1] T. M. Austin. The SimpleScalar/ARM Toolset. http:// www.eecs.umich.edu/~taustin/simplescalar

[2] S. P. Amarasinghe et al. The SUIF Compiler for Scalable Parallel Machines. In Proc. the 7th SIAM Conf. on Parallel Proc. for Sci. Comp., 1995.

[3] F. Balasa et al. Background memory area estimation for multidimensional signal processing systems. IEEE Transactions on VLSI Systems, 3(2):157-172, 1995.

[4] L. Benini et al. Increasing Energy Efficiency of Embedded Systems by Application-Specific Memory Hierarchy Generation. IEEE Design & Test, 2000.

[5] E. Brockmeyer et al. Layer Assignment Techniques for Low Energy in Multi-Layered Memory Organizations. In Proc. Design Automation and Test in Europe, Messe Munich, Germany, 2003.

[6] E. Brockmeyer et al. Low Power Memory Storage and Transfer Organization for the MPEG-4 Full Pel Motion Estimation on a Multimedia Processor. IEEE Trans. on Multimedia, 1(2): 202–216, 1999.

[7] G. De Micheli. High-level synthesis of digital circuits. Advances in Computers, 37: 207–283, 1993.

[8] E. G. Hallnor and S. K. Reinhardt. A fully-associative software-managed cache design. In Proc. International Conference on Computer Architecture, pp. 107–116, Vancouver, British Columbia, Canada, 2000.

[9] K. Kennedy and K. McKinley. Optimizing for Parallelism and Data Locality. In Proc. the 1992 ACM International Conference on Supercomputing.

[10] I. Kodukula et al. Data-Centric Multi-Level Blocking. In Proc. ACM PLDI, 1997.

[11] W. Kelly and W. Pugh. Finding Legal Reordering Transformations Using Mappings. In Proc. LCPC, pp. 107–124, 1994.

[12] S.-T. Leung and J. Zahorjan. Optimizing Data Locality by Array Restructuring. Technical Report TR 95–09–01, Dept. of Computer Science and Engineering, University of Washington, September 1995.

[13] M. Miranda et al. Systematic Speed-Power Memory Data Layout Exploration for Cache Controlled Embedded multimedia applications. In Proc. ISSS'01, Montreal, 2001.

[14] S. S. Muchnick. Advanced Compiler Design and Implementation. Morgan-Kaufmann Publishers, San Fransisco, CA, 1997.

[15] P. R. Panda et al. Efficient Utilization of Scratch-Pad Memory in Embedded Processor Applications. In Proc. the European conference on Design and Test, 1997.

[16] G. Rivera and C.-W. Tseng. Data Transformations for Eliminating Conflict Misses. In Proc. ACM SIGPLAN Conference on Programming Language Design and Implementation, Montreal, Canada, June 1998.

[17] X. Vera et al. A Fast and Accurate Framework to Analyze and Optimize Cache Memory Behavior. ACM Transactions on Programming Languages and Systems, March 2004.

[18] L. Wang et al. Optimizing On-Chip Memory Usage through Loop Restructuring for Embedded Processors. In Proc. 9th International Conference on Compiler Construction, March 30–31 2000, pp. 141–156, Berlin, Germany.

Compiler-Guided Data Compression for Reducing Memory Consumption of Embedded Applications *

O. Ozturk, G. Chen, and M. Kandemir

Computer Science and Engineering Department
Pennsylvania State University
e-mail: {ozturk,gchen,kandemir}@cse.psu.edu

I. Kolcu

Computation Department
University of Manchester
e-mail: ikolcu@umist.ac.uk

Abstract— Memory system presents one of the critical challenges on embedded system design and optimization. This is mainly due to ever-increasing code complexity of embedded applications and exponential increase witnessed in the amount of data they manipulate. As a result, reducing memory space occupancy of embedded applications is very important and will be even more important in the next decade. Motivated by this observation, this paper presents and evaluates a compiler-driven approach to data compression for reducing memory space occupancy. Our goal in this paper is to study how automated compiler support can help in deciding the set of data elements to compress/decompress and the points during execution at which these compressions/decompressions should be performed. The proposed compiler support achieves this by analyzing the source code of the application to be optimized and identifying the order in which the different data blocks are accessed. Based on this analysis, the compiler then automatically inserts compression/decompression calls in the application code. The compression calls target the data blocks that are not expected to be used in the near future, whereas the decompression calls target those data blocks with expected reuse but currently in compressed form.

I. INTRODUCTION

Most embedded systems have very tight constraints on memory space, power consumption, and performance. In particular, memory constraints are getting increasingly important as both code complexity of embedded applications and amount of data they process in a typical execution are increasing. Prior research on memory systems proposed and evaluated several techniques which can potentially improve the memory performance of embedded software. Power and memory space efficiency, on the other hand, took relatively less attention so far.

One of the techniques that can be used to reduce memory space consumption (occupancy) of embedded applications is data compression. The goal of data compression is to represent an information source (e.g., a data file, a speech signal, an image, or a video signal) as accurately as possible using the fewest number of bits. Previous research considered efficient hardware and software based data compression techniques and applied compression to different domains. While data compression can be an effective way of reducing memory space consumption of embedded applications, it needs to be invoked with care since performance and power costs of decompression (when we need to access data stored currently in

the compressed form) can be overwhelming. Therefore, compression/decompression decisions must be made based on a careful analysis of data access pattern of the application.

Our goal in this paper is to study how automated compiler support can help in deciding the set of data elements to compress/decompress and the points during execution at which these compressions/decompressions should be performed. The proposed compiler support achieves this by analyzing the source code of the application to be optimized and identifying the order in which different data blocks are accessed. Based on this analysis and data reuse information, the compiler then automatically inserts compression and decompression calls in the application code. The compression calls target the data blocks that are not expected to be used in the near future, whereas the decompression calls target those data blocks with expected reuse but currently in compressed form. We discuss our compiler algorithm in detail and explain how it operates in practice. In providing compiler support for data compression, we want to achieve two objectives. First, we want to enable memory space savings without the involvement of the application programmer; that is, the programmer gets the benefits of reducing the memory space consumption without any effort on her part. Second, we want to minimize the potential impact of compression on performance. To do this, our automated approach makes use of data reuse analysis. It needs to be emphasized that in this paper we do not propose a new data compression technique. Instead, we demonstrate how an optimizing compiler for embedded systems can schedule data compressions and decompressions intelligently to minimize memory space occupancy at runtime while also minimizing the associated performance overheads.

The remainder of this paper is structured as follows. In the next section, we give a discussion of the related work on compression. In Section III, we define concept of memory space consumption (occupancy). In Section IV, we present our compiler algorithm. Finally, we conclude the paper in Section V with a summary.

II. RELATED WORK

Compression techniques have been used for both program code and application data to reduce the memory footprint and the energy consumption. RISC processors have been the main focus for code compression techniques. To reduce the size of a program's code segment, [9] uses pattern-matching techniques that identify and coalesce together repeated instruction sequences. A RISC system that can directly execute com-

---

*This work is supported in part by NSF Career Award #0093082 and by a grant from GSRC.

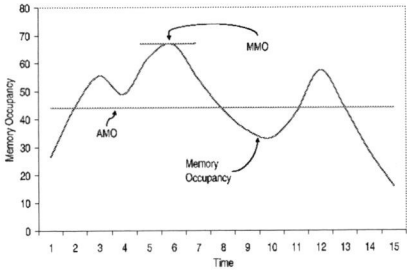

Fig. 1. An example memory space consumption curve and MMO and AMO metrics.

pressed programs is presented in [17]. VLIW (Very Long Instruction Word) processors are now being considered in this area as well. Ros and Sutton [15] present code compression algorithms applied to instruction words. Profile-driven methods that selectively compress code segments in VLIW architectures have also been proposed [18, 10]. Code compression schemes for VLIWs that use Variable-to-fixed (V2F) coding was investigated by [16, 19]. Variable-sized-block method for VLIW code compression has been introduced by Lin et al [12]. Data compression has been used to reduce storage requirements, bus bandwidth, and energy consumption in the past [1, 5, 6]. In [5], a hardware-assisted data compression for on-the-fly data compression and decompression has been proposed. In their work, compression and decompression takes place on the cache-to-memory path. Lee et al [11] use compression in an effort to explore the potential for an on-chip cache compression to reduce cache miss ratio and miss penalty. Abali et al [1] investigate the performance impact of hardware compression. Compression algorithms that would be suitable to use in a compressed cache have been presented in [2]. Apart from memory subsystems, data compression has also been used to reduce the communication volume. For example, data compression has been proposed to reduce the communication latency and energy consumption in sensor networks [7]. Our work is different from all these prior efforts since we give the task of management of the compressed data blocks to the compiler. In deciding the data blocks to compress and decompress, our compiler approach makes use of the data reuse information extracted from the array accesses in the application source code.

III. MEMORY SPACE OCCUPANCY

Memory space occupancy indicates the memory space occupied by an application data at each point during the course of execution. There are two important metrics associated with memory space occupancy. The first one is the *maximum memory occupancy* (MMO), which gives the maximum memory space occupied by data during the course of execution when considering all execution cycles. The second metric is the *average memory occupancy* (AMO), which gives the memory occupancy when averaged over all execution cycles. Figure 1 illustrates these two metrics for an example case. Note that, the drops in the memory occupancy curve indicate either some application-level dead memory block recycling or system-level garbage collection. Both these met-

rics, MMO and AMO, can be important targets for optimizations. MMO is critical since it captures the amount of memory that needs to be allocated for the application if the application is to run successfully without an out-of-memory exception. The AMO metric on the other hand can be important in a multi-programming based embedded environment where multiple applications compete for the same memory space. The goal behind our compiler-directed approach is to reduce both MMO and AMO for array/loop-intensive embedded applications. Note that, array/loop-intensive applications are frequently used in embedded image/video processing.

IV. COMPILER ALGORITHM

Employing data compression in managing the memory space of the system requires a careful analysis of the data access pattern of the application being considered. This is because using data compression in an untimely manner can cause significant performance and power penalties. For example, compressing data blocks with short reuse distances can increase the number of decompressions dramatically. Also, decompressing data blocks with long reuse distances prematurely can increase memory space consumption unnecessarily. Therefore, one needs to be very careful in deciding both the set of data blocks to compress/decompress and the points in execution to compress/decompress them. Clearly, this is an area that can benefit a lot from automation. Our goal is to reduce MMO and AMO as much as possible, with as little performance penalty as possible.

A. Data Tiling and Memory Compression

Our scheme compresses only the arrays that can benefit from data compression (this can be determined either through profiling or via programmer annotations). These arrays are referred to as the "compressible" arrays in this paper. We do not compress scalar variables or incompressible arrays (i.e., the arrays that cannot benefit from data compression). Figure 2 shows the organization of the memory space for supporting our compiler-directed data compression approach. We divide the memory space into three parts: *compressed area, decompression buffer,* and *static data area.* The static data area contains scalar variables, incompressible arrays, and the directories for compressible arrays. The data entities in the static area are statically allocated at compilation time. The compressed area and the decompression buffer, however, are dynamically managed at runtime based on the compiler-determined schedule for compressions and decompressions.

We divide each compressible array into equal-sized *tiles* (*blocks*). An element of a tiled array X can be indexed using the following expression: $X[\vec{I}][\vec{J}]$, where $\vec{I}$ is the tile subscript vector, which indexes a tile of array X; and $\vec{J}$ is the intra-tile subscript vector, which indexes an element within a given tile. For example, Figure 3 shows an array X that has been divided into nine (3×3) tiles, and each tile contains sixteen (4×4) elements. $X[2,3]$ refers to the tile at the second row, third column; and $X[2,3][3,2]$ refers to the data element at the third row, second column of tile $X[2,3]$.

Figure 2 shows how we store tiled arrays in the memory. For each compressible array X, our compiler creates a directory, each entry of which corresponds to a tile of array X, and can be indexed using a tile subscript vector. Each entry in the directory of array X, denoted as $X\llbracket \vec{I} \rrbracket$ ($\vec{I}$ is a tile subscript vector), contains a pointer to the memory location where the corresponding tile is stored. As mentioned above, the directory of each array is stored in the static data area of the memory space.

An array tile can either be compressed or uncompressed. Uncompressed tiles are stored in the decompression buffer, and compressed tiles are stored in the compressed area. The decompression buffer is divided into equal-sized blocks, and the size of a block is equal to that of a tile. We use a *free table* to keep track of the free blocks in the decompression buffer. When we need to decompress a tile, we first need to allocate a free block in the decompression buffer.

Compressed tiles are stored in the compressed area. The memory in this area is divided into equal-sized *slices*. The size of a slice is smaller than that of a block in the decompression buffer. In our implementation, a slice is equal to a quarter of a block. Although the size of tiles is a constant, the compression ratio depends on the specific tile. Therefore, the number of slices required to store a compressed tile may vary from one tile to another. In Figure 2, we can observe that the slices of the same tile form a link table. Like in the case of the decompression buffer, the compressed area also has a free table keeping all free slices.

Figure 4 shows the architecture of our system. When the program starts its execution, all tiles are in the compressed format and are stored in the compressed area. A compressed tile is decompressed and stored in the the decompression buffer by a *decompressor* before it is accessed by the program. If this tile belongs to an array that is not written (updated) by the current loop nest, the compressed version of this tile remains in the compressed area. On the other hand, if this tile belongs to an array that might be written by the current loop nest, we discard the compressed version of this tile, and return the slices occupied by this tile to the free table of the compressed area. When we need to decompress a new tile but there is no free space in the decompression buffer, we select a set of old tiles in the decompression buffer and discard them to make space for the new tile. If a victim tile (the tile to be evicted) belongs to an array that might be written by the current loop nest, we must decompress and store its compressed version in the compressed area before we evict its uncompressed version. On the other hand, if this tile belongs to an array that is not written by the current loop nest, we can discard the uncompressed version of this tile without recompressing it. The important point to emphasize here is that our approach is not tied to any specific compression/decompression algorithm, and the compressor and decompressor can be implemented either in software or hardware. In our current implementation, however, we use only software compression/decompression.

It should be emphasized that data tiling is required by our implementation of memory compression, not by the logic of the application. Therefore, we do not require the programmer to be aware of data tiling. Our compiler automatically (in a user-transparent manner) tiles every array that needs to be compressed. Data tiling requires two mapping functions p and q that map an original array subscript vector to a tile subscript vector and an intra-tile subscript vector, respectively. That is, we map $X[\vec{I}]$ into $X\llbracket p(\vec{I}) \rrbracket [q(\vec{I})]$. In this paper, given a tile

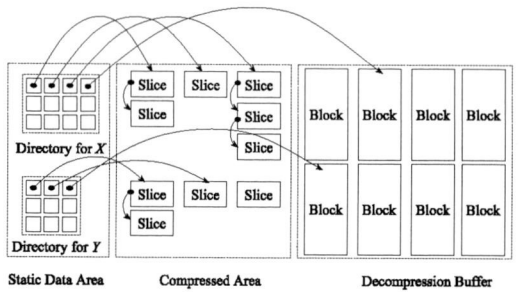

Fig. 2. Memory organization.

size T, we use the following mapping functions:

$$p((i_1, i_2, ..., i_n)) = (\lfloor i_1/N_1 \rfloor, \lfloor i_2/N_2 \rfloor, ..., \lfloor i_n/N_n \rfloor);$$
$$q((i_1, i_2, ..., i_n)) = (i_1 \mod N_1, i_2 \mod N_2, ..., i_n \mod N_n);$$

where $N_1, N_2, ..., N_n$ are magnitudes (extents) of each dimension subscript vector such that $N_1 N_2 ... N_n = T$. When an array is tiled, our compiler also *rewrites* the program statements that access this array accordingly.

B. Loop Tiling

While data tiling transforms memory layout of each compressible array, loop tiling (iteration space blocking) transforms the order in which the array elements are accessed within a loop nest. If used appropriately, loop tiling can significantly reduce the number of decompressions invoked during the execution of a loop nest. Figure 5 gives such an example. Figure 5(a) is the original code of a loop nest, which accesses a 600×600 array X. We apply data tiling to array X such that the size of each tile is 100×100. Figure 5(b) shows the code after data tiling. For illustration purposes, let us assume that the decompression buffer can contain up to three tiles, and that we use an LRU based policy to select the victim tiles in the decompression buffer. We can compute that, during the execution of this loop nest, we need to invoke the decompressor 100 times for each tile. Hence, the decompressor is invoked $100 \times 36 = 3600$ times. By applying loop tiling to the loop nest shown in Figure 5(b), we obtain the tiled loop nest in Figure 5(c). In this tiled code, loop iterators i and j are the *inter-tile iterators* and the loop nest formed by them is referred to as the *inter-tile loop nest*. Similarly, the iterators ii and jj are the *intra-tile iterators* and the loop nest formed by them is referred to as the *intra-tile loop nest*. During the execution of this loop nest, the decompressor is invoked only 36 times, once for each i, j combination. Loop tiling has been widely studied in the literature (e.g., [20]). Due to the space limitation we have, we do not discuss the details of loop tiling. In the rest of this paper, we assume that the loop nests in the application program have been appropriately tiled according to the layout of the arrays imposed by data tiling. It is to be mentioned however that our compiler uses loop tiling for a different purpose than most of the commercial and academic compilers.

C. Compression-Based Space Management

Our compiler inserts buffer management code at each loop nest that uses the decompression buffer. For ease of discussion,

2006 Asia and South Pacific Design Automation Conference

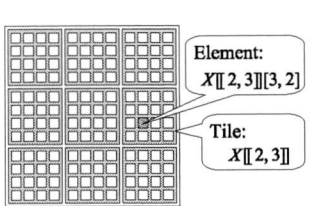

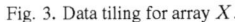

Fig. 3. Data tiling for array X.

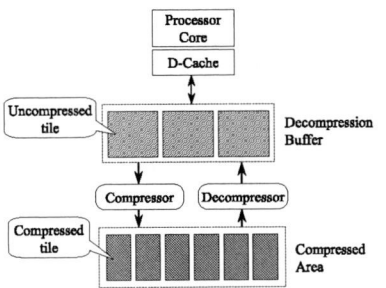

Fig. 4. Architecture supporting memory compression.

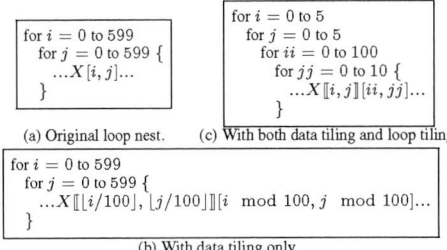

(a) Original loop nest. (c) With both data tiling and loop tiling.

(b) With data tiling only.

Fig. 5. Code transformation for data tiling and loop tiling.

we use the following abstract form to represent a tiled loop nest:

$$
\begin{aligned}
&\text{for } \vec{I} = \vec{L} \text{ to } \vec{U} \ \{ \\
&\quad T_1(R_1(\vec{I}), W_1(\vec{I})); \\
&\quad T_2(R_2(\vec{I}), W_2(\vec{I})); \\
&\quad \ldots\ldots \\
&\quad T_n(R_n(\vec{I}), W_n(\vec{I})); \\
&\}
\end{aligned}
$$

where $\vec{I}$ is the iteration vector, $\vec{L}$ and $\vec{U}$ are the lower and upper bound vectors for the loop nest. T_i ($i = 1..n$) represents an intra-tile loop nest. Since we focus on the access pattern of each array at a tile level, we treat each intra-tile loop nests as an atomic operation. $R_i(\vec{I})$ is the set of tiles that might be read, and $W_i(\vec{I})$ is the set of tiles that might be written in the intra-tile loop nest T_i at the inter-tile iteration $\vec{I}$. $R_i(\vec{I})$ and $W_i(\vec{I})$ can be computed as follows:

$$
R_i(\vec{I}) = \{X_k[\![f_j^{(i)}(\vec{I})]\!] \mid \text{``} \ldots = \ldots X_k[\![f_j^{(i)}(\vec{I})]\!][\ldots]\ldots\text{'' appears in } T_i\},
$$

$$
W_i(\vec{I}) = \{X_k[\![f_j^{(i)}(\vec{I})]\!] \mid \text{``} X_k[\![f_j^{(i)}(\vec{I})]\!][\ldots] = \ldots\text{'' appears in } T_i\},
$$

where X_k is an array accessed by T_i, and $f_j^{(i)}(\vec{I})$ is a mapping function that maps inter-tile iteration vector $\vec{I}$ into a tile of array X_i. Note that, we must be conservative in computing $R_i(\vec{I})$ and $W_i(\vec{I})$.

For intra-tile loop nest T_i, let us assume:

$$
W_i(\vec{I}) = \{X_{k_1}[\![f_1^{(i)}(\vec{I})]\!], X_{k_2}[\![f_2^{(i)}(\vec{I})]\!], X_{k_j}[\![f_j^{(i)}(\vec{I})]\!]\}
$$

$$
R_i(\vec{I}) - W_i(\vec{I}) = \{X_{k_{j+1}}[\![f_{j+1}^{(i)}(\vec{I})]\!], X_{k_{j+2}}[\![f_{j+2}^{(i)}(\vec{I})]\!], X_{k_m}[\![f_m^{(i)}(\vec{I})]\!]\}.
$$

Figure 6 shows a transformed loop nest augmented with the decompression buffer management code. In the transformed code, we use counter c to count the number of intra-tile loop nests that have been executed up to a specific point. B is the set of tiles that are currently in the decompression buffer. Before entering intra-tile loop nest T_i, we need to decompress all the tiles in the set $R_i(\vec{I}) \cup W_i(\vec{I}) - B$. That is, all the the tiles that will be used by T_i must be available in the uncompressed format before we start executing T_i. When decompressing a tile t, we may need to evict a tile from the decompression buffer if there is no free block in the decompression buffer (indicated by $|B| = D$). Each tile t in the decompression buffer is associated with an integer $t.r$, indicating when this tile will be reused in the future. Specifically, $t_1.r < t_2.r$ indicates that tile t_1 will be reused earlier than t_2. When evicting a tile, we select the one that will be reused in the furthest future. Each tile t in the decompression buffer also has a flag $t.w$ indicating whether this block has been written

since its last decompression. If this victim tile has been written, we need to recompress this tile. Before entering T_i, we also update the *next reuse time* ($t.r$) for each tile (t) used by T_i. The next reuse time of tile t is computed using $t.r = c + d_i(t)$, where c is number of intra-tile loop nests that have been executed, and $d_i(t)$ is the *reuse distance* of tile t at intra-tile loop nest T_i. The reuse distance of tile is the number of intra-tile loop nests executed between the current and the next accesses to this tile.

We use a compiler-based approach to compute $d_i(t)$–the reuse distance of tile t at intra-tile loop nest T_i. In the following discussion, we explain how $d_i(t)$ can be computed at compilation time. The tiles in $R_i(\vec{I}) \cup W_i(\vec{I})$, the set of tiles used by the ith intra-tile loop nest at inter-tile loop iteration $\vec{I}$, can be divided into three types: (1) the tiles that will be reused by another intra-tile loop nest at the same inter-tile iteration $\vec{I}$, (2) the tiles that will be reused by some intra-tile loop nest at another inter-tile iteration $\vec{I}'$ ($\vec{I} \prec \vec{I}'$), and (3) the tiles that will never be reused. The set of tiles belonging to the first type, i.e., the tiles that will be reused at the same inter-tile loop iteration, can be computed as follows:

$$
U_i(\vec{I}) = (R_i(\vec{I}) \cup W_i(\vec{I})) \cap \bigcup_{j=i+1}^{n}(R_j(\vec{I}) \cup W_j(\vec{I})),
$$

where n is the number of intra-tile loop nests in the body of the inter-tile loop nest. For each tile $t \in U_i(\vec{I})$, the reuse distance can be computed as:

$$
d_i(t) = j - i,
$$

where j is the minimum integer greater than i such that $t \in R_j(\vec{I}) \cup W_j(\vec{I})$.

For the tiles in the set $V_i(\vec{I}) = (R_i(\vec{I}) \cup W_i(\vec{I})) - U_i(\vec{I})$, i.e., the tiles of types (2) and (3), our compiler computes their reuse distances by executing the loop nest below (this loop nest is generated by the compiler using Omega library [14],[1] and it is executed only once at the compilation time):

```
V = V_i(I_0);   c = 0;
for I = I_0 + (0, 0, ..., 0, 1)^T to I_N {
    c = c + n;
    for i = 1 to n {
        for each t ∈ V ∩ (R_j(I) ∪ W_j(I)) { d_j(t) = c + j - 1; }
        V = V - (R_j(I) ∪ W_j(I));
    }
}
for each t ∈ V { d_i(t) = ∞; }
```

[1]The Omega Library is a tool that provides functions for manipulating sets and relations that are defined using Presburger formulas. Presburger formulas are logical formulas that are built using affine expressions and universal/existential quantifiers.

817

```
for I = L to U {
    … …
        T_i(R_i(I), W_i(I));
    … …
}
```

(a) Original loop nest.

```
B — the set of tiles in the buffer;
D — the size of buffer;
t — the tile to be loaded;
⇒ procedure load(t) {
⇒    if(t ∉ B) {
⇒        if(|B| = D) {
⇒            for each v ∈ B such that v.r < c
⇒                v.r = ∞; // the next use time of v has been mispredicted
⇒            select v ∈ B such that
⇒                the directory for v is not marked and v.r is maximized;
⇒            if(v.w = 1) compress(v);
⇒            evict(v); B = B − {v};
⇒        }
⇒        decompress(t); B = B + {t};
⇒    }
⇒    t.r = c + d_i(t);
⇒ }

for I = L to U {
    … …
⇒    c = c + 1;
⇒    // mark the tiles in W_i(I) ∪ R_i(I),
⇒    // preventing these tiles from being evicted.
⇒    mark_directory_entry(X_{k_1} [[f_1^{(i)}(I)]]);
⇒    mark_directory_entry(X_{k_2} [[f_2^{(i)}(I)]]);
⇒    … …
⇒    mark_directory_entry(X_{k_m} [[f_m^{(i)}(I)]]);

⇒    // load tiles in W_i(I)
⇒    load(X_{k_1} [[f_1^{(i)}(I)]]); X_{k_1} [[f_1^{(i)}(I)]].w = 1;
⇒    load(X_{k_2} [[f_2^{(i)}(I)]]); X_{k_2} [[f_2^{(i)}(I)]].w = 1;
⇒    … …
⇒    load(X_{k_j} [[f_j^{(i)}(I)]]); X_{k_j} [[f_j^{(i)}(I)]].w = 1;

⇒    // load tiles in R_i(I)
⇒    load(X_{k_{j+1}} [[f_{j+1}^{(i)}(I)]]);
⇒    load(X_{k_{j+2}} [[f_{j+2}^{(i)}(I)]]);
⇒    … …
⇒    load(X_{k_m} [[f_m^{(i)}(I)]]);

⇒    // unmark the tiles in W_i(I) ∪ R_i(I),
⇒    // allowing these tiles to be evicted.
⇒    unmark_directory_entry(X_{k_1} [[f_1^{(i)}(I)]]);
⇒    unmark_directory_entry(X_{k_2} [[f_2^{(i)}(I)]]);
⇒    … …
⇒    unmark_directory_entry(X_{k_m} [[f_m^{(i)}(I)]]);
        T_i(R_i(I), W_i(I));
    … …
}
```

(b) The transformed loop nest augmented with the decompression buffer management code. The lines marked with "⇒" are inserted by our compiler.

Fig. 6. Code transformation employed by our compiler.

In the above compiler-generated code, $\vec{I}_0$ and $\vec{I}_N$ are two vectors such that $|\vec{I}_N - \vec{I}_0| = N$ and $\vec{L} \preceq \vec{I}_0 \prec \vec{I}_n \preceq \vec{U}$, where $|\vec{I}_N - \vec{I}_0|$ denotes the number of loop iterations between $\vec{I}_0$ and $\vec{I}_N$, and $\vec{L}$ and $\vec{U}$ are, respectively, the lower and upper bound vectors for the target inter-tile loop nest for which we compute the reuse distances. $\vec{I}_0$ can be any vector between $\vec{L}$ and $\vec{U}$. Note also that integer N is a threshold, and n is the number of intra-tile loop nests in the body of the inter-tile loop nest. The reuse distances larger than nN are treated as infinity (∞). Note that an inaccuracy in computing the reuse distance may lead to performance penalties when the program is executed; however, it does not cause any error in the program execution

(i.e., it is not a correctness issue), since we are conservative in computing the set of of tiles that are used in each intra-tile loop nest.

D. Exploiting Extra Resources

While the compiler approach presented above schedules compressions and decompressions such that memory space consumption is reduced without excessively increasing the original execution time, we can still incur performance penalties. This is because the decompression activities can occur on the critical path of execution and this in a sense symbolizes the tradeoff between memory savings and performance overheads due to data compression. In this section, we show how we can make use of extra resources available in our approach.

In a multiprocessor environment, we can reduce the performance overheads incurred by data compression by overlapping the execution of compression and decompression procedures with that of the computing loop nest. Specifically, for each inter-tile loop nest, our compiler generates two threads: the *computing thread* and the *buffer management thread*. In a multiprocessor based environment, these two threads can be executed in parallel. Figure 7 shows the code our compiler generates. Figure 7(b) gives the code for the buffer management thread. For each intra-tile iteration T_i at each inter-tile iteration $\vec{I}$, the buffer management thread decompresses each tile t in the set $R_i(\vec{I}) \cup W_i(\vec{I})$ if t is not in the buffer. The management thread increases $t.c$, the *reference counter* associated with tile t, by one for each $t \in R_i(\vec{I}) \cup W_i(\vec{I})$. The reference counter associated with each tile $t \in R_i(\vec{I}) \cup W_i(\vec{I})$ will be decreased by the computing thread after the execution of intra-tile T_i. A non-zero value in the reference counter $t.c$ indicates that tile t is being used or will be used by the computing thread, and consequently, t cannot be evicted from the buffer. On the other hand, when the buffer management thread needs to evict a tile from the buffer to make space for a new tile, it can evict any tile whose reference counter is zero (nevertheless, for better performance, as discussed in Section C, we also require $v.r$ be maximized). After increasing the reference counter for each tile in $R_i(\vec{I}) \cup W_i(\vec{I})$, the management thread performs a V operation on a counting semaphore named "Iteration". The value of this semaphore (Iteration.v) indicates the number of intra-tile loop nests that the computing thread can execute without being blocked. If the value of this semaphore is zero, the computing thread cannot continue with its execution due to the fact that some tiles required by the computing thread are not yet ready in the buffer. After the V operation on semaphore "Iteration", the management thread starts to decompress the tiles that will be used by the next intra-tile loop nest, without further synchronization with the computing thread.

Figure 7(c) gives the code (with necessary instructions inserted by our compiler) for the computing thread. Before executing each intra-tile loop nest T_i, the computing thread performs a P operation on the semaphore "Iteration". This P operation blocks the computing thread if some tiles that will be used by T_i are not ready in the buffer. In this case, the computing thread has to wait until the management thread decompresses all the required data tiles. After executing intra-tile loop nest T_i, the computing thread decreases the reference counter for each tile used by T_i. As discussed above, if the

```
// definition of semaphore operations
Struct Semaphore {
    int v;
    ThreadQueue q;
}

P(s) {
    s.v = s.v - 1;
    if(s.v < 0)
        block in s.q;
}

V(s) {
    s.v = s.v + 1;
    if(s.v ≤ 0)
        unblock one thread in s.q;
}


// semaphores for synchronizing the
// computing and management threads
Iteration – counting semaphore,
        Iteration.v is initialized to 0;
Counters – binary semaphore,
        Counters.v is initialized to 1;
```
(a) Semaphores and their operations.

```
for I = L to U {
    for i = 1 to n {
        c = c + 1;
        for each t ∈ R_i(I) ∪ W_i(I) {
            if(t ∉ B) {
                if(|B| = D) {
                    repeat
                        S = {v | v ∈ B ∧ v.c = 0}
                    until S ≠ φ;
                    select v ∈ S such that
                        v.r is maximized
                    if(v.w = 1) compress(v);
                    evict(v); B = B - {v};
                }
                decompress(t);
                t.c = 0; B = B ∪ {t};
            }
            t.r = c + d_i(t);
            if(t ∈ W_i(I)) t.w = 1;
            P(Counters); t.c = t.c + 1;
            V(Counters);
        }
    V(Iteration);
    }
}
```
(b) The buffer management thread.

```
for I = L to U {
    P(Iteration);
    T_1(R_1(I), W_1(I));
    P(Counters);
    for each t ∈ R_1(I) ∪ W_1(I)
        t.c = t.c - 1;
    V(Counters);
    ... ...

    P(Iteration);
    T_i(R_i(I), W_i(I));
    P(Counters);
    for each t ∈ R_i(I) ∪ W_i(I)
        t.c = t.c - 1;
    V(Counters);
    ... ...
    ... ...
    P(Iteration);
    T_n(R_n(I), W_n(I));
    P(Counters);
    for each t ∈ R_n(I) ∪ W_n(I)
        t.c = t.c - 1;
    V(Counters);
}
```
(c) The computing thread.

Fig. 7. The buffer management thread and the computing thread generated by our compiler.

value of the reference counter of tile t is reduced to zero, we allow the buffer management thread to reuse the memory space occupied by t.

The management thread is blocked when it needs to decompress a new tile but the value of the reference counter for each tile in the buffer is greater than 0 (that is, none of the tiles have been used yet). In this case, the management thread has to wait for the computing thread to release some tiles by reducing their reference counters. If the computing thread is also blocked at the P operation on semaphore "Iteration", the system is deadlocked. Fortunately, this deadlock cannot happen as long as the following condition is satisfied:

$$D \geq \max_{\forall i, \vec{I}} |R_i(\vec{I}) \cup W_i(\vec{I})|, \tag{1}$$

where D is size of the buffer (in terms of the number of tiles). Note that, if this condition is not satisfied, the single-threaded approach discussed in Section C cannot work properly, either. It is important to note that, in our approach, the condition expressed by 1 is always satisfied.

V. CONCLUDING REMARKS

This paper presents a compiler-directed approach that inserts compression and decompression calls in the application code to reduce maximum and average memory space consumption. In this approach, the compiler analyzes a given application code and extracts data reuse information at the data block level. It then uses this information in deciding the set of data blocks to be compressed/decompressed as well as the points at which these actions need to be invoked. Our preliminary results are encouraging and motivate further research on compiler-directed data compression.

REFERENCES

[1] B. Abali, M. Banikazemi, X. Shen, H. Franke, D. E. Poff, and T. B. Smith. Hardware Compressed Main Memory: Operating System Support and Performance Evaluation. *IEEE Trans. on Computers*, Nov 2001.

[2] E. Ahn, S. Yoo, S. Mo, and S. Kang. Effective algorithms for cache-level compression. In Proc. The 11th Great Lakes symposium on VLSI, 2001.

[3] S. P. Amarasinghe, J. M. Anderson, M. S. Lam, and C. W. Tseng. The SUIF compiler for scalable parallel machines. In Proc. *SIAM Conference on Parallel Processing for Scientific Computing*, February, 1995.

[4] T. M. Austin and D. Burger. The SimpleScalar architectural research tool set. http://www.cs.wisc.edu/~mscalar/simplescalar.html

[5] L. Benini, D. Bruni, A. Macii, and E. Macii Hardware-Assisted Data Compression for Energy Minimization in Systems with Embedded Processors. In Proc. *Conf. on Design, Automation and Test in Europe*, 2002.

[6] C. D. Benveniste, P. A. Franaszek, and J. T. Robinson. Cache-Memory Interfaces in Compressed Memory Systems. In Proc. *Workshop on Solving the Memory Wall Problem*, June 2000.

[7] M. Chen and M. L. Fowler. The importance of data compression for energy efficiency in sensor networks. In Proc. *2003 Conference on Information Sciences and Systems*, The Johns Hopkins Univ., March 2003.

[8] S. Coleman and K. S. McKinley. Tile size selection using cache organization and data layout. In Proc. *the SIGPLAN Conference on Programming Language Design and Implementation*, La Jolla, CA, June 1995.

[9] K. D. Cooper, and N. McIntosh. Enhanced code compression for embedded RISC processors. In Proc. *Conference on Programming Language Design and Implementation*, 1999.

[10] S. Debray, and W. Evans. Profile-Guided Code Compression. In Proc. *Conf. on Programming Language Design and Implementation*, 2002.

[11] J. S. Lee, W. K. Hong, and S. D. Kim. Design and Evaluation of a Selective Compressed Memory System. In Proc. *The 1999 IEEE International Conference on Computer Design*, 1999.

[12] C. H. Lin, W. Wolf, and Y. Xie. LZW-based code compression for embedded systems. In Proc. *Conf. on Design, Automation and Test in Europe*, 2004.

[13] LZO Algorithm. http://gnuwin32.sourceforge.net/packages/lzo.htm

[14] The Omega Project. http://www.cs.umd.edu/projects/omega/

[15] M. Ros, and P. Sutton Code Compression Based on Operand-Factorization for VLIW Processors. In Proc. *The Conference on Data Compression*, 2004.

[16] B. P. Tunstall. Synthesis of Noiseless Compression Codes. *PhD Thesis, Georgia Institute of Technology*, Sept. 1967.

[17] A. Wolfe, and A. Chanin. Executing compressed programs on an embedded RISC architecture. In Proc. *The 25th International Symposium on Microarchitecture*, 1992.

[18] Y. Xie, W. Wolfe, and H. Lekatsas. Profile-Driven Selective Code Compression. In Proc. *DATE*, 2003.

[19] Y. Xie, W. Wolfe, and H. Lekatsas. Code compression for VLIW using variable-to-fixed coding. In Proc. *The 15th International Symposium on System Synthesis*, 2002.

[20] M. Wolfe. *High Performance Compilers for Parallel Computing*. Addison-Wesley Publishing Company, CA, 1996.

Analysis of Scratch-Pad and Data-Cache Performance Using Statistical Methods

Javed Absar[*][†] and Francky Catthoor[*]

[*]IMEC vzw., Katholieke Universiteit Leuven, Belgium.
[†] STMicroelectronics Asia Pacific, Singapore.
{javed.absar, francky.catthoor}@imec.be

Abstract— An effectively designed and efficiently used memory hierarchy, composed of scratch-pads or cache, is seen today as the key to obtaining energy and performance gains in data-dominated embedded applications. However, an unsolved problem is – how to make the right choice between the scratch-pad and the data-cache for different class of applications? Recent studies show that applications with regular and manifest data access patterns (e.g. matrix multiplication) perform better on the scratch-pad compared to the cache. In the case of dynamic applications with irregular and non-manifest access patterns, it is however commonly and intuitively believed that the cache would perform better. In this paper, we show by theoretical analysis and empirical results that this *intuition* can sometimes be misleading. When access-probabilities remain fixed, we prove that the scratch-pad, with an optimal mapping, will always outperform the cache. We also demonstrate how to map dynamic applications efficiently to scratch-pad or cache and additionally, how to accurately predict the performance.

I. INTRODUCTION

Multimedia and network applications are well-known to be highly data dominated [5]. To reduce memory-access related energy and performance costs in such applications, the memory hierarchy must be effectively designed (with respect to capacity and number of levels) and efficiently used [18][16] (i.e. data mapping to carefully exploit the hierarchy). Most multimedia and network platforms today have at least one, two or more levels of cache. Lately, software-controlled cache, also known as scratch-pad memory (SPM), have also been viewed as an alternative to the data and instruction cache [9][12][23].

SPM is software controlled. Therefore, the compiler or the application developer must make, and instrument, all the decisions about which data should reside on SPM at any time. This situation is quite different from the cache, where the hardware is in charge of exploiting the locality using a circuit that implements the *least recently used* algorithm. The hardware simplicity of SPM enables it to provide a low power-consumption and access-time number, on a per access basis, compared to cache [4]. Applications in which the data access pattern is regular and manifest (i.e. known at compile time) can be mapped easily and efficiently to SPM [1][9][11]. Such applications include matrix multiplication, filtering and large portions of audio and video compression algorithms. Since SPM is cheaper than cache, in such cases the better choice is, indeed, the SPM. If, however, the applications exhibits more dynamism – objects being accessed in a data-dependent and seemingly random fashion – mapping to SPM requires more

ingenuity. Applications involving trees, heaps, tries, graphs and linked lists often exhibit this kind of dynamism. One may reckon that the cache would do a better job in such cases and, therefore, decide to let such objects be handled by the cache. This may or may not hold true. We, therefore, require more sound techniques for deciding between SPM and cache.

In this paper, we study the performance of dynamic applications, when mapped to SPM or cache, using models of *access probability*. This allows us to analyze and predict whether the cache would outperform the SPM, and by how much. We back our theoretical conclusions with empirical results. To the best of our knowledge, this is the first work that analytically compares the performance of SPM with cache. Previous studies [7][12] have focused on finding good mapping techniques for SPM. As such, the comparison of SPM performance with cache was limited to *simulation results*. Unfortunately, that does not provide broader insight.

The remainder of this paper is organized as follows. Section 2 gives a motivating example. Section 3 reviews related work. Section 4 gives a brief description of the probability model that we employ. Section 5 compares SPM and cache performance with this model. And last, section 6 presents empirical results.

II. MOTIVATING EXAMPLE

The function search in the program section below, does a spell-check by performing a search of the given word, of n letters, against its internal dictionary. The dictionary is implemented as a trie (tree with variable number of child-nodes) which enables fast searches.

```
typedef struct node{ //trie data-structure
  struct node *next, *down; char letter;
}Node;

bool search(Node *nptr, char *word, int n){
  for( i = 0 ; i < n ; i++ ){
    while(nptr && (nptr->letter < word[i])){
      nptr = nptr->next;
    }
    if(nptr){
      if(nptr->letter == word[i])
        nptr = nptr->down;
      else  return false; //word not found
    }else  return false; //word not found
  }
  return true;        //word found
}
```

0-7803-9451-8/06/$20.00 ©2006 IEEE.

This application can be mapped to the SPM or the cache. For the case of data-cache, since the tracking and migration of data to and from the cache is handled by hardware, no changes are required on the code. For the case of SPM, however, additional instructions are necessary to place some of the nodes on SPM. The rest of the nodes remain in external memory from where they are directly accessed. Access to the nodes mapped to SPM will be quick and energy efficient. If the nodes are mapped intelligently such that most of the search is to nodes on the SPM, then overall good performance and energy efficiency can be expected. The question is: Can the SPM really compete with the cache in such a dynamic and data-dependent application. We will address such questions both analytically and with empirical results.

As a side remark, note that pointer based data-structures (such as trie) allow placement (of some of the nodes) on SPM in a seamless way without requiring checks with each access. On the other hand, if we place parts of an array on SPM and the rest on the external memory, then each access to the array (using an index expression) will require first a check to see where that segment resides.

III. RELATED WORK

Banakar et al. [4] did a detailed study of the energy and area advantage of the SPM over the cache. On a per access basis, they found that a 2KB, 2-way cache consumed 4.57 nJ., while a 2KB SPM consumed only 1.53 nJ. Initial work by Panda et al. [17] on utilizing the SPM focused on scalars and highly used, small-sized arrays. Their approach was extended and improved upon by other researchers [3][21][23] who applied knapsack formulation and ILP solvers to find the best set of data objects (globals, stack variables and arrays) and program routines that would still fit into the SPM and yet save the highest amount of energy. For arrays larger than the SPM size, these solutions do not work well. However, for arrays accessed in a regular pattern – inside nested-loops by index expressions which are linear functions of the loop iterators – several additional SPM mapping techniques using data-space and iteration space tiling have been proposed [9][11][12]. They work well irrespective of the array size, and always outperform the cache.

Dominguez et al.[7] explain a technique for mapping dynamic data structures, e.g. linked lists and trees, to SPM. The size allocated in the SPM to each set, e.g. nodes of a particular tree, is made proportional to the overall number of access to that set. They do not compare between SPM and cache but are only concerned with finding the best allocation of the SPM for the different dynamic objects, each vying for space on the SPM.

For the cache, itself, there have been numerous studies to estimate its performance for regular and non-regular applications. Several studies have tried to quantify the cache performance by summarizing or analyzing actual memory access trace [2][20]. From analytical comparison perspective, however, trace analysis is not fruitful. On the other hand, the Independent Reference Model (IRM) of Rao [19][13] is more suited to our purpose of analyzing cache behavior for comparison with SPM. This model was recently extended [8][15] to algorithmic analysis. Rao's equations assume a given (specified) data-layout. However, the results in our paper allow conclusions to be drawn across all possible layouts.

IV. MODEL DESCRIPTION

Computations such as matrix multiplication generate *memory reference strings*, i.e. sequence of data memory addresses, that are regular and input-independent, and can be determined without even running the program. This enables good SPM mapping, and as such the SPM is able to outperform the cache for all these types of computations [12].

On the other hand, dynamic applications generate reference strings that are irregular and input-dependent. Therefore, predicting SPM and cache performance (hit-rate) with the techniques used for the regular case turns out to be unwieldy and complicated. In this paper we use statistical methods [22] to compare the SPM performance with the cache. We start by characterizing the reference string using the Independent Reference Model (IRM) [15][19].

Consider a set of objects $O_1, O_2, \ldots, O_n$ and a set of corresponding *access probabilities* $p_1, p_2, \ldots, p_n$. The reference string can be denoted as $r_1, r_2, \ldots, r_t, \ldots, r_N$, where r_t is the object referenced at time t. Under IRM:

$$Pr[r_t = O_i] = p_i, \, 1 \leq i \leq n, \, t > 0$$

That is, the probability of r_t being O_i is p_i. Though this model does not take into consideration the correlation between accesses, studies [10] show that the behavior modeled assuming *independent reference* gives results that are very close to those obtained using models that do indeed incorporate such correlation. We will also show this with empirical results which reconfirm that IRM is indeed able to accurately predict SPM and cache performance for data-structures such as trees where the access-pattern is clearly data-dependent and correlated.

Now, we will compute the cache hit-rate for the set of access probabilities given above. Suppose we have a cache with just one block and so it contains only the last object accessed. It can been shown [13] that the states of this cache forms a homogeneous Markov chain, where each state S_i is defined as having the object O_i inside the cache block. The equilibrium probability of state S_i equals p_i [19] and hence, the probability that object O_i is inside the cache, at anytime, equals p_i. Therefore, if object O_i is accessed, then the probability that the access results in a hit (object found in cache) equals p_i. Averaging the hit-rate across all possible accesses, we get the expected hit-rate η for a cache of size one as:

$$\eta = \sum_{i=1}^{n} p_i^2 \qquad (1)$$

Embedded systems usually contain caches with low associativity to reduce the energy and area. The results that we derive are, therefore, in the context of the direct-mapped cache (DM-Cache). Caches with low associativity perform similar to the

8B-5

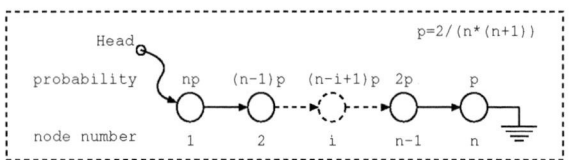

Fig. 1. Access-probabilities in a linked list, when each node is equally likely to be the target of the next search. A search starts at head and stops when target is found.

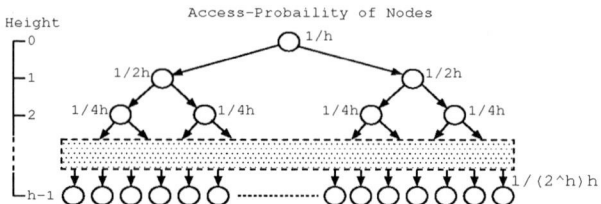

Fig. 2. Access-probabilities of nodes in a binary tree.

direct-mapped [19]. We will later present empirical results that confirm this as well.

The expected hit-rate for a DM-Cache of more than one block can be computed by placing the objects into disjoint groups. Assume that the DM-Cache contains m cache blocks, and each block can hold only one object. Let G_i denote the set of objects which map to cache block i. Out of the n objects $O_1, O_2, \ldots, O_n$, assume (for simplicity) that $k = n/m$ objects map to each cache block. Objects in G_i are denoted as $O_1(i), O_2(i), \ldots, O_k(i)$, and the corresponding probabilities as $p_1(i), p_2(i), \ldots, p_k(i)$, respectively. Let $D_i = p_1(i) + p_2(i) + \ldots + p_k(i)$. The following result, from Rao [19], gives the expected hit-rate of a DM-Cache:

$$\eta_{\text{DM}} = \sum_{i=1}^{m} \frac{1}{D_i} \sum_{j=1}^{k} p_j^2(i) \tag{2}$$

Essentially, a DM-Cache behaves as m disjoint, fully-associative caches, each of size one. So, Eq. 1 can be applied to each cache block, but with conditional probabilities $p_j(i)/D_i$. The overall hit-rate equals the weighted sum of hit-rates of each individual block. The weight for a block equals the probability that the next access would be to that block. For block i, this equals D_i.

Let us now see how to compute access probabilities. A linked list is shown in Fig. 1. Each node contains a key and some data. The search for a node, with a certain key, starts at the head and continues till that node is found. Each key is equally likely to be the target of the next search. Now, if the nth node has access probability p, then the $(n-1)^{th}$ node has access probability $2p$. Reason: the $(n-1)^{th}$ node is referenced when the nth node is the target of a search, and it is also referenced when it is, itself, the target of a search. Probabilities for all the nodes is shown in Fig. 1. It is also possible to assign probabilities in an application-specific way or based on profiling as we will show later.

Next, consider a binary search-tree. A search starts at the root and proceeds downward to the target leaf-node. From any parent, the search-path has an equal chance of moving to the left child-node or to the right child-node. Therefore, if the access probability of a parent is p, each child-node has access probability $p/2$. Fig. 2 shows the access probabilities of nodes in a binary tree.

V. ANALYTICAL STUDY

In this section, we analyze SPM and DM-Cache performance using the model discussed before.

To recapitulate, we have n objects $O_1, O_2, \ldots, O_n$ with access probabilities $p_1, p_2, \ldots, p_n$, respectively. Without loss of generality, let $p_1 \geq p_2 \geq \ldots \geq p_n$. To maximize the SPM hit-rate, objects $O_1, O_2, \ldots, O_m$, where m is SPM size, must be placed on SPM. The rest of the objects remain in the memory, from where they are accessed directly. Hence, each access to O_i, where $i > m$, constitutes a miss. The expected hit-rate of this optimal SPM mapping is:

$$\eta_{\text{SPM}} = \sum_{i=1}^{m} p_i \tag{3}$$

Let us now compare this with the hit-rate of a DM-Cache, also of size m, using Eq. 2. As before, objects $O_1(i), O_2(i), \ldots, O_k(i)$, with access probabilities $p_1(i), p_2(i), \ldots, p_k(i)$, respectively, map to cache block i. Again, without loss of generality, let $p_1(i) \geq p_2(i) \geq \ldots \geq p_k(i)$. Although we assume that exactly $k = n/m$ objects map to each block, it is not a limitation of the proof, but is done so as to simplify the notation.

Since $D_i = p_1(i) + p_2(i) + \ldots + p_k(i)$ in Eq. 2, we have $p_1(i) = D_i - \sum_{j=2}^{k} p_j(i)$ Now, Eq. 2 can be rewritten as:

$$
\begin{aligned}
\eta_{\text{DM}} &= \sum_{i=1}^{m} \frac{1}{D_i} \left[p_1^2(i) + \sum_{j=2}^{k} p_j^2(i) \right] \\
&= \sum_{i=1}^{m} \frac{1}{D_i} \left[p_1(i) \left(D_i - \sum_{j=2}^{k} p_j(i) \right) + \sum_{j=2}^{k} p_j^2(i) \right] \\
&= \sum_{i=1}^{m} p_1(i) - \sum_{i=1}^{m} \frac{1}{D_i} \sum_{j=2}^{k} p_1(i) p_j(i) - p_j^2(i) \\
&= \sum_{i=1}^{m} p_1(i) - \sum_{i=1}^{m} \frac{1}{D_i} \sum_{j=2}^{k} p_j(i) \big(p_1(i) - p_j(i) \big) \tag{4}
\end{aligned}
$$

Since $p_1(i) \geq p_j(i)$, $1 < j \leq k$, in Eq. 4 each expression $p_j(i)\big(p_1(i) - p_j(i)\big)$ is always positive. Hence:

$$\eta_{\text{DM}} \leq \sum_{i=1}^{m} p_1(i) \tag{5}$$

The expression $\sum_{i=1}^{m} p_1(i)$ in Eq. 5 attains its maximum value when the objects $O_1(1), O_1(2), \ldots, O_1(m)$, with probabilities $p_1(1), p_1(2), \ldots, p_1(m)$, respectively, are any permutation of the objects in the set $\{O_1, O_2, \ldots, O_m\}$. Note that $O_1, O_2, \ldots, O_m$ are the objects with the highest access probabilities among all the n objects. Therefore, the highest

822

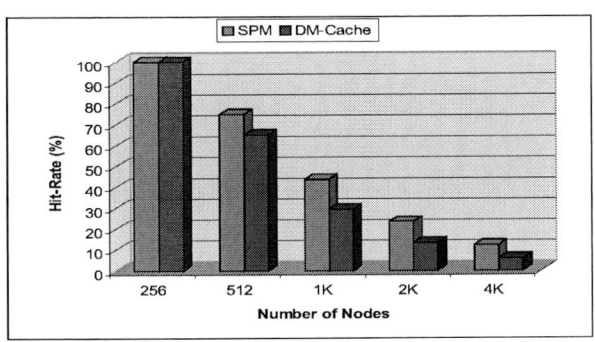

Fig. 3. Hit-rates for key search on the linked list. Cache performance worsens, compared to SPM, with increasing problem size.

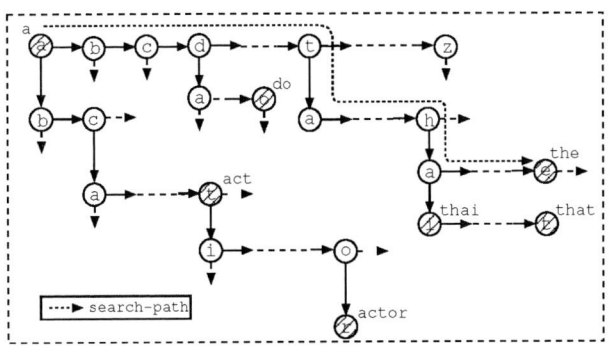

Fig. 4. Spell-checker implemented with trie data-structure. Search-path to *the* is delineated with dotted line.

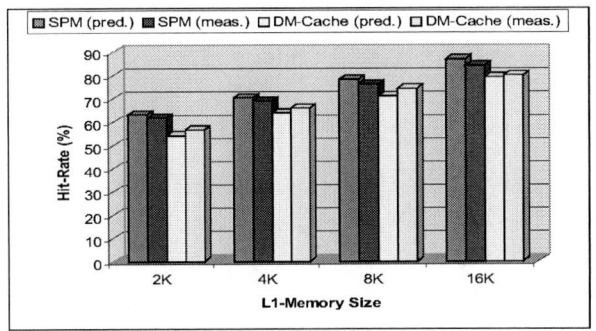

Fig. 5. Spell-Checker: Predicted (pred.) and measured (meas.) hit-rates. Predictions are within 2.6% accuracy.

value attained by η_{DM} is $\sum_{i=1}^{m} p_i$. Comparing this with Eq. 3, we conclude that $\eta_{\text{DM}} \leq \eta_{\text{SPM}}$.

Therefore, under the given assumptions, we see that the SPM with an optimal mapping can always outperform the DM-Cache. The cache can have any data-layout, whatsoever, and yet the SPM, with optimal mapping, will still perform better. In the next section, we will validate this conclusion with experiments. We will also show that our conclusion holds even for set associative caches.

In this paper, we have looked at dynamic data structures whose topology does not change or changes gradually over time. For example, our result holds well for a tree on which the basic operation is traversal. The tree's topology may change only very slowly over time through re-balancing, deletion and insertions. Another class of problems is when the topology changes very fast. In that case, the comparison between SPM and cache has to take into consideration the exact replacement policy of SPM. We are currently studying this class of applications but do not discuss the solution any further in this paper.

VI. EMPIRICAL VALIDATION

Let us now verify the theory with experiments. We will look at three applications and see how they fare on SPM and cache.

A. Linked List

Fig. 3 plots the measured hit-rates for searches conducted on the linked list of Fig. 1. In the experiment, the L1-memory size is 4KB, with cache block-size 16B. Each node is 16 bytes. In the case of SPM, the first m (256) nodes from the head of the list were placed on SPM. For the DM-Cache case, the first m nodes were placed in different cache blocks. From Fig. 3 we see that SPM does indeed outperform the direct-mapped cache. The cache performance, with respect to SPM, worsens for increasing problem size because of increasing conflicts.

B. Spell-Checker

The *spell-checker* [14] checks and reports whether the given word exists in its dictionary. To enable fast searches, the dictionary is built as a *trie* (tree with variable number of child-nodes). A trie is shown in Fig. 4. The search-path to the word *the* is delineated with a dotted line in the figure.

At first, one might be tempted to map the spell-checker onto the cache because it involves data-dependent traversal of a pointer-based dynamic data structure. However, we will see that with a smart mapping the spell-checker actually does better on SPM. In our experiment, the trie contains over five thousand commonly used words. By performing a *mock* spell-check on a *training-essay*, the access probability of each trie-node is estimated. The access probabilities are then used to predict the hit-rates using Eq. 2 and 3. For SPM, it assumes that the nodes with highest access probabilities are mapped to SPM, while for the DM-Cache case it assumes that they map to different cache-blocks. The actual hit-rates are measured by running the spell-checker over another document of more than hundred thousand words.

Fig. 5 shows the predicted and measured hit-rates. The first observation is that the predictions, both for SPM and DM-Cache, is close to the measured values. Therefore, IRM is indeed able to model SPM and cache behavior for data-dependent traversals to a high degree of accuracy. The second observation is that, as proven in previous section, the SPM with optimal mapping does indeed outperform the DM-Cache – albeit by a small margin. However, since SPM is more energy efficient than cache, SPM is better suited for this application.

Let us now compare the previous SPM and DM-Cache mappings with other mapping techniques. Typically, the trie is *grown* by inserting new words into it. In the first version

8B-5   2006 Asia and South Pacific Design Automation Conference

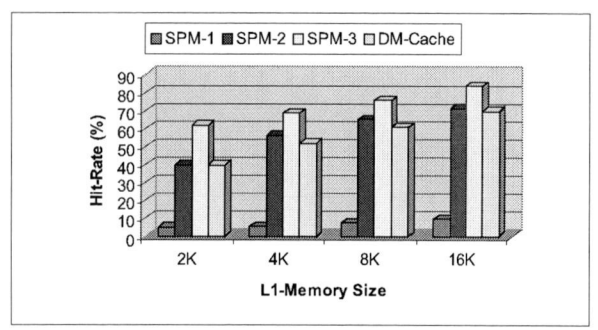

Fig. 6. Spell-Checker: Hit-rate comparison of SPM with DM-Cache. Using access-probabilities (SPM-3) gives better results over conventional mapping schemes (SPM-1 & 2).

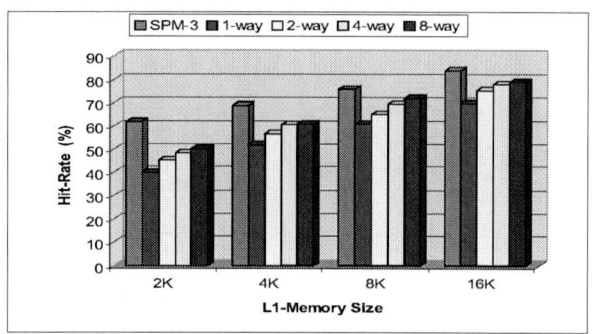

Fig. 7. Spell-Checker: SPM performance compared to 1, 2, 4 and 8 way set-associative caches. Increasing associativity does not improve performance significantly.

SPM-1, the words are inserted into the trie in lexicographical order and the nodes are allocated space in the SPM on a first come first serve basis, till the SPM gets full. In the SPM-2 version, the trie is built by inserting the most commonly used words (e.g. *the*, *as*, and *and*) first, and then inserting the remaining words. Therefore, words which are searched most often have their paths *almost* entirely on the SPM. We say *almost* because words inserted later-on could potentially add new nodes in the paths to the commonly used words. Fig. 6 shows the hit-rates for SPM-1 and SPM-2, and compares them with SPM-3 and DM-Cache. The SPM-3 version puts the nodes with highest access probabilities on SPM. Therefore, SPM-3 is identical to SPM (meas.) in Fig. 5 but is shown again for convenience. DM-Cache in Fig. 6 shows the hit-rate when no customized mapping of nodes is done. This is different from the experiment conducted for DM-Cache (meas.) in Fig. 5 where nodes with highest access probabilities were mapped to different cache-blocks. Therefore, as expected, the hit-rate values for DM-Cache in Fig. 6 are less than those of DM-Cache in Fig. 5. From Fig. 6, we therefore conclude that mapping using access probabilities can be superior compared to conventional mapping techniques.

Next, we study the impact of increasing associativity on the performance of the cache. In particular, we would like to see if set-associative caches can outperform the SPM.

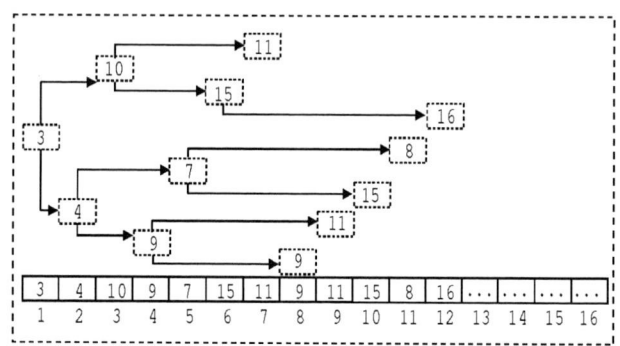

Fig. 8. A binary-heap implementation of the priority-queue for the minimum spanning tree algorithm.

Fig. 7 compares direct mapped (1-way) with 2, 4 and 8-way set-associative cache. The SPM hit-rate (columns for SPM-3) are also shown for comparison. We see that increasing the associativity does not tilt the performance toward cache. Moreover, set-associative caches are power hungry and hence any performance gains comes at high energy penalty.

C. MST - Prim's Algorithm

Prim's algorithm finds a minimum spanning tree (MST) for the given connected graph [6]. The algorithm starts with a single node and adds one edge at a time, till all the nodes have been connected. At any time, during the building of the tree, there is a set of nodes T already in the tree, and another set of nodes T' currently not in the tree. Each node in T' is assigned a weight that is equal to the cheapest edge connecting it to some node in T. Organizing the set of weights in T' such that the cheapest edge can be found quickly is done using a binary-heap. A binary-heap, such as in Fig. 8, is a complete binary-tree embedded into an array. Each node in the heap has a weight which is less than or equal to that of its two children. Therefore, the node h with the least weight is at the top (root). When h is removed, the last heap-node is moved to the top and then it *percolates* down to its new appropriate place.

In estimating the hit-rates (using Eq. 2 and 3), we assume that a node percolating downward, ultimately reaches the bottom. This simplification has a certain degree of associated error. Secondly, note that the tree shrinks in size as nodes are removed from it. Therefore, the estimated hit-rate is the mean for the entire range. Fig. 9 shows the predicted and measured hit-rates on a complete graph of over thousand nodes. We observe that the predicted values are slightly below the measured values. The reason is that since some nodes do not percolate all the way down, the access probabilities are actually higher for nodes near the top of the heap as compared to what was assumed in our calculations.

Fig. 10 gives the hit-rates for different set associative caches. We see that increasing the associative does not have a big impact on the cache performance and the SPM still outperforms the cache.

824

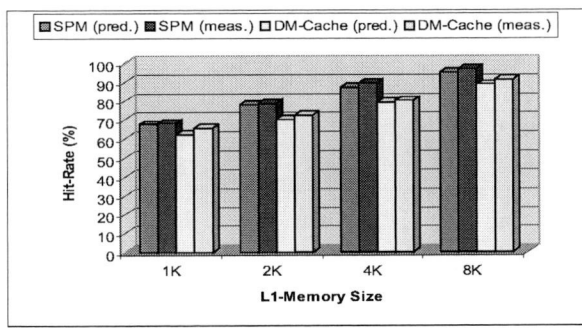

Fig. 9. Minimum Spanning Tree: Predicted (pred.) and measured (meas.) hit-rates of SPM and DM-Cache.

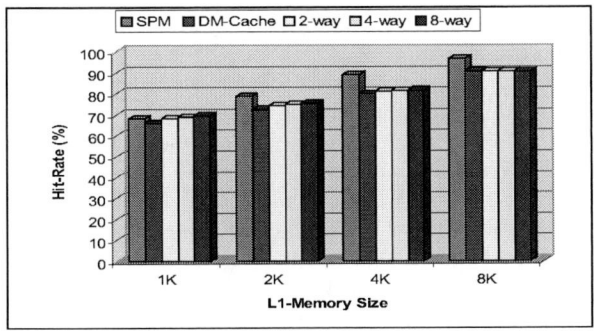

Fig. 10. Minimum Spanning Tree: SPM performance compared to 1, 2, 4 and 8 way set-associative caches. Increasing associativity does not improve performance significantly.

VII. CONCLUSION AND FUTURE WORK

Data-dependent access to data structures such as tries, trees, heaps and linked lists can be modeled to a reasonable level of accuracy in Independent Reference Model (IRM). We use IRM to prove that scratch-pad memories, with an optimal mapping based on access probabilities, can outperform the direct-mapped cache, irrespective of the layout influencing the cache behavior. This analytical result is then verified with experiments. Increasing the associativity in the cache is shown not to improve the cache performance in any significant way. We, therefore, see our main contribution as demonstrating, theoretically and empirically, that scratch-pad memories can be effectively used more than just for regular applications.

This paper does not address the issue of SPM and cache behavior when the topology of the data structure changes rapidly with time – resulting from insertions, deletions and restructuring of the nodes. In that case, scratch-pad and cache performance evaluation has to take into consideration the replacement policy of the scratch-pad, and the conflicts in the cache from different placement (layout) of objects in memory. Scratch-pad replacement policy can be application dependent or be independent (using generic allocators).

REFERENCES

[1] Mohammed Javed Absar and Francky Catthoor. Compiler-based approach for exploiting scratch-pad in presence of irregular array access.

In *Design Automation and Test in Europe (DATE)*, pages 1162–1167, March 2005.

[2] A. Agarwal, J. Hennessy, and M. Horowitz. An analytical cache model. *ACM Transactions on Computer Systems*, 7(2):184–215, 1989.

[3] O. Avissar and R. Barua. An optimal memory allocation scheme for scratch-pad based embedded systems. *ACM Transactions on Embedded Computing Systems*, pages 6–26, November 2002.

[4] Rajeshwari Banakar, Stefan Steinke, Bo-Sik Lee, M. Balakrishnan, and Peter Marwedel. Scratchpad memory: design alternative for cache on-chip memory in embedded systems. In *CODES '02: Proceedings of the tenth international symposium on Hardware/software codesign*, pages 73–78, New York, NY, USA, 2002. ACM Press.

[5] F. Catthoor, F. Balasa, E. D. Greef, and L. Nachtergaele. *Custom Memory Management Methodology: Exploration of Memory Organization for Embedded Multimedia System Design*. Kluwer Academic Publisher, 1998.

[6] T. Cormen, C. E. Leicerson, and R. Rivest. *Introduction to Algorithms*. Prentice Hall, 1998.

[7] A. Dominguez, S. Udayakumaran, and R. Barua. Heap data allocation to scratch-pad memory in embedded systems. *Journal of Embedded Computing, Cambridge Publishing*, 2005.

[8] J. D. Fix. Cache performance analysis of algorithms. *PhD Dissertation, University of Washington*, pages 22–36, 2002.

[9] I. Issenin, E. Brockmeyer, M. Miranda, and N. Dutt. Data reuse analysis technique for software-controlled memory hierarchies. In *Design Automation and Test in Europe (DATE)*, pages 202–207, March 2004.

[10] P. R. Jelenkovic and A. Radovanovic. Least-recently-used caching with dependent requests. *Theoretical Computer Science*, 326(1-3):293–327, 2004.

[11] M. T. Kandemir, I. Kadayif, and U. Sezer. Exploiting scratch-pad memory using preseburger formulas. *International Symposium on System Synthesis (ISSS)*, 7(12), 2001.

[12] M. T. Kandemir and J. Ramanujan. A compiler-based approach for dynamically managing scratch-pad memories in embedded systems. *IEEE Transaction on Computer Aided Design of Integrated Circuits and Systems*, 23(2):243–259, March 2004.

[13] W. F. King. Analysis of paging algorithm. *Proceedings of IFIP Congress*, pages 485–490, August 1971.

[14] Karen Kukich. Technique for automatically correcting words in text. *ACM Comput. Surv.*, 24(4):377–439, 1992.

[15] Richard E. Ladner, James D. Fix, and Anthony LaMarca. Cache performance analysis of traversals and random accesses. In *SODA '99: Proceedings of the tenth annual ACM-SIAM symposium on Discrete algorithms*, pages 613–622, Philadelphia, PA, USA, 1999. Society for Industrial and Applied Mathematics.

[16] Peter Marwedel. *Embedded System Design*. Kluwer Academic Publishers (Springer), Norwell, MA, USA, 2003.

[17] Preeti Ranjan Panda, Nikil D. Dutt, and Alexandru Nicolau. Efficient utilization of scratch-pad memory in embedded processor applications. In *EDTC '97: Proceedings of the 1997 European conference on Design and Test*, page 7, Washington, DC, USA, 1997. IEEE Computer Society.

[18] Preeti Ranjan Panda, Alexandru Nicolau, and Nikil Dutt. *Memory Issues in Embedded Systems-on-Chip: Optimizations and Exploration*. Kluwer Academic Publishers, Norwell, MA, USA, 1998.

[19] Gururaj S. Rao. Performance analysis of cache memories. *J. ACM*, 25(3):378–395, 1978.

[20] Jaswinder Pal Singh, Harold S. Stone, and Dominique F. Thibaut. A model of workloads and its use in miss-rate prediction for fully associative caches. *IEEE Transactions on Computers*, 41(7):811–825, 1992.

[21] S. Steinke, L. Wehmeyer, B. Lee, and P. Marwedel. Assigning program and data objects to scratchpad for energy reduction. *Design Automation and Test in Europe (DATE)*, pages 409–414, March 2002.

[22] Kishore S Trivedi. *Probability and Statistics with Reliability, Queuing and Computer Science Applications*. John Wiley and Sons, New York, USA, 2002.

[23] Manish Verma, Lars Wehmeyer, and Peter Marwedel. Dynamic overlay of scratchpad memory for energy minimization. In *CODES+ISSS '04: Proceedings of the 2nd IEEE/ACM/IFIP international conference on Hardware/software codesign and system synthesis*, pages 104–109, New York, NY, USA, 2004. ACM Press.

Efficient Early Stage Resonance Estimation Techniques for C4 Package[*]

Jin Shi[1], Yici Cai[1], Shelton X-D Tan[2]，Xianlong Hong[1]

[1] Department of Computer Science and Technology,
Tsinghua University, Beijing, P.R.China, 100084
Tel: +86-10-62785564
e-mail: shi-j03@mails.tsinghua.edu.cn
caiyc@mail.tsinghua.edu.cn
hxl-dcs@mail.tsinghua.edu.cn

[2] Department of Electrical Engineering,
University of California at Riverside,
CA 92521, USA
e-mail: stan@ee.ucr.edu

Abstract - In this paper, we study the relationship between C4 package resonance effects and logical switching timing correlations, which has not been thoroughly investigated in the past. We show that improper logic designs with some special timing correlations can lead to adverse large voltage drops, which are due to resonance effects in the widely used C4 package. We first present the numerical analysis results on industry C4 package circuits to demonstrate resonance phenomenon. Then we propose a simple algorithm to compute the worst-case logical timing correlations among cells leading to resonance. Finally, we develop an efficient technique in early logic design stage to estimate the resonance risk. Experiment results demonstrate the effectiveness of the proposed method for the accurate prediction of the resonance effect in C4 package.

I. Introduction

As the power consumption of modern high-performance VLSI chips is increasing rapidly, reliable on-chip power delivery and associated verification methods have become major design challenges. On top of this, wire-bond based package is becoming more difficult to accommodate the large quantities of I/O pins and associated power dissipation, which will introduce obvious package resistance and inductance. Flip-chip packages such as C4 (Controlled Collapse Chip Connection) are becoming widely used in high-performance power-hungry designs because of their superior electrical performance [1][2]. However, C4 package may suffer large voltage drops due to package resonance caused by improperly designed logical switching timing correlations. Traditionally, the design and optimization of packages are independent of the design and optimization of power delivery networks as well as the logics (their timing). This separation in design flow may cause potential problems such as resource-induced drops due to the strong interplay among package, power delivery networks and logical timing. To the best knowledge of the authors, only a few research

works are reported to consider package and power/ground network co-design [3][4], seldom research is reported on logic-package co-design. The relationship between the package electrical characteristic and logic timing was meagerly investigated in the past.

In this paper, we first perform some numerical analysis on real industry C4 package circuits in frequency domain. We show that that if the dynamic current flowing through a C4 bump happens to contain special frequency harmonics, the voltage drop on the package will become significant due to adverse resonance effects. Further more, if resonance happens in a local area, based on the 'locality' property of C4 package [5], the power supply in this area will suffer large drops.

Based on our resonance analysis on the C4 package, we further investigate the conditions for generating the required resonance-induced harmonics in the dynamic currents. Our study shows that certain logic switching timing correlations among different gates or cells can lead to such resonance-induced harmonics. The significance of such study is that we need to estimate the resonance risk in the early design stage (logic design stages) to avoid the resonance risk. To the line, we develop an efficient technique to estimate resonance in the early logical design stage. Our experimental results validate the proposed method. This paper is organized as follows: Section II introduces an industrial C4 package model and its frequency response. Section III analyzes the 'locality' property of C4 and provides a power supply failure example caused by the resonance effects. Section IV discusses the conditions for the resonance in package. Section V presents a fast estimation method. Experiment results are introduced in Section VI while Section VII summarizes the paper and presents forthcoming studies.

II. Frequency Response of C4 Package Model

In this paper, we use a C4 package model provided by

---

[*] This work is supported by National Natural Science Foundation of China (NSFC) 60476014, National Hi-tech R&D Program of China 2005AA1Z1230, and Foundation of Intel Corporation

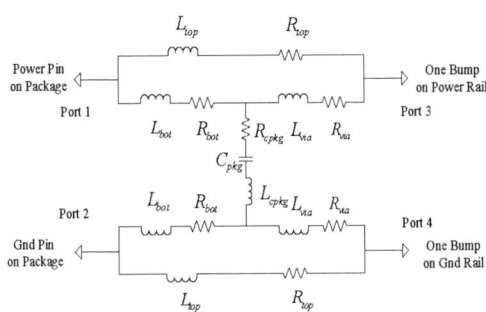

Fig. 1. Two metal layer C4 package model

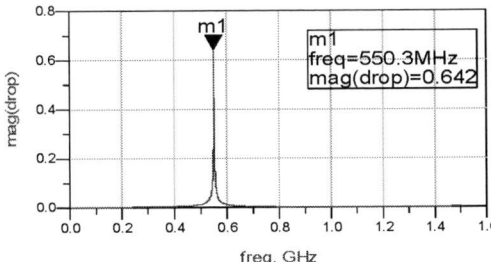

Fig. 2. Frequency response of package model

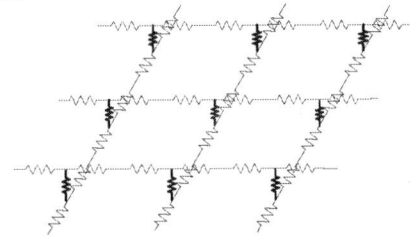

Fig. 3. Transient analysis of resonance

our industry partner to discuss resonance problem. Figure 1 demonstrates a lump model for a two metal layer C4 package. In this model, each bump on the die connects to a package pin through two metal layers and one via. In Figure 1, L_{top}, R_{top}, L_{bot} and R_{bot} represent the inductance and resistance of metal trace in the package while L_{via} and R_{via} represent the inductance and the resistance of the via in the package. There also exist package decoupling capacitors, which are represented by C_{pkg}, R_{cpkg}, L_{cpkg}. Here we treat our package model as a four port RLC network, because port1 and port2 will connect to voltage regulation model (VRM), which has very low internal resistance. We can ground port 1 and 2 and then obtain the package frequency response from port 3 and 4. The response is shown in Figure 2. From Figure 2, we can see that the frequency response has a single-peak around 550 Mhz, which indicates that if the current flowing from port 3 to port 4 (bump current) contains large amplitude harmonics around this frequency, even only 100mA in amplitude, the voltage

drop on the package can be as large as 0.6 V. Figure 3 shows a transient analysis result when the bump current is similar to a sine wave having resonance frequency. In this situation, the voltage drop is obvious.

III. Locality and Power Supply Failure

In Section II, we observed that if the bump current contains the significant harmonics which are close to the resonance frequency of the package model, the voltage drop on the package would become significant. However, what will happen if package sees such current in a local area? We use a 7-layer power/ground grid with vias for a test. The circuit model used is shown in Figure 4. The test analyzes a 300um x 300um local area with 1.5V voltage supply through a C4 package with dissipation of 20-watt power. The result is that we have local resonance in C4 package as shown in Figure 5. We observe that when drop on the package is large in area A on M7, the voltage supply in area B right below A on M2 is large as well. We find that circular area B is larger than circular area A with the ratio of radius B to A equaling or less than 1.5.

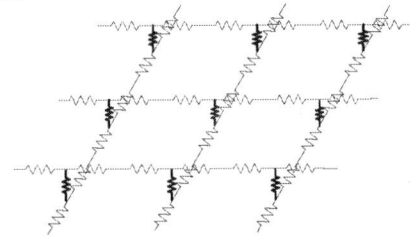

Fig. 4. Resistance model of 7 layer grid with vias

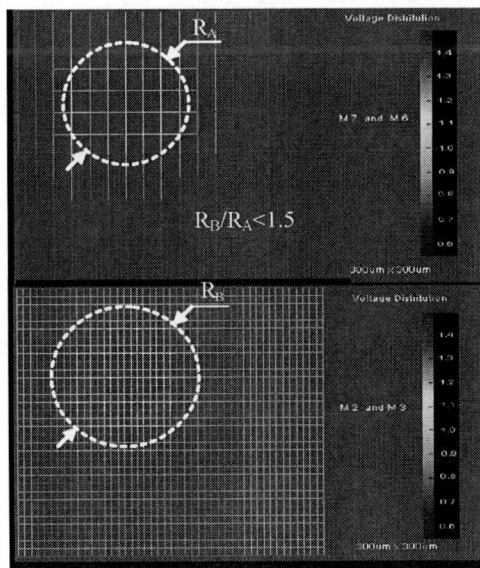

Fig. 5. Resonance in local area

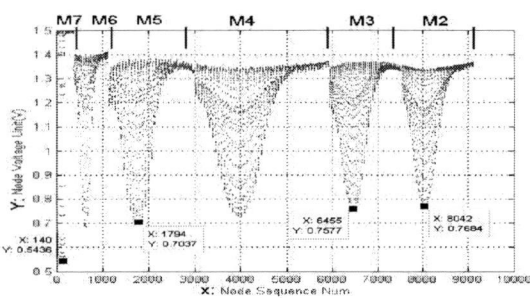

Fig. 6. Voltage Distribution on Different Layer

The reason for this phenomenon is due to locality [5]. Usually, C4 package can provide a relative clean voltage supply because of the abundance of the power/ground around any small blocks, each of which can be supplied sufficiently and separately. However, if one area happens to suffer large droop, it is less likely to get power supply from an adjacent area. This explains why the package may suffer a power failure in local area. Further, if we draw the voltage distribution on different metal lagers in Figure 6, we can see that the distribution of lowest voltage points is almost the same in each layer.

Another interesting point in Figure 6 is that as the layer goes down from M7 to M2, the minimal voltage value in a certain layer increases. However, this improvement is not obvious above M6, which indicates that the 'locality' highly depends on the density of vias in each layer. In M7 and M6, because vias are sparse, voltage potential between two adjacent points can be relatively large. Therefore, points, which suffer low voltage, are more likely to get compensation from their neighbors. However, as vias become dense in M5-M2, potential between any two local points becomes smaller, so current almost flows from a vertically through vias, which suggests that the current flow direction is the main reason of the 'locality'.

IV. Resonance Caused by Logical Correlation

In Section II, we showed that current with special harmonic components on the bump could cause the package resonance. But we don't analyze the conditions for generating such currents. In this section, we show such conditions can be linked to certain logical correlations among different gates or cells.

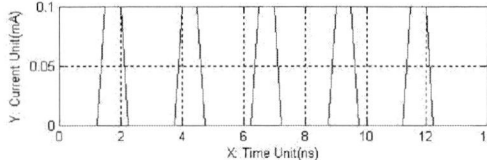

Fig. 7. Basic TW current waveform

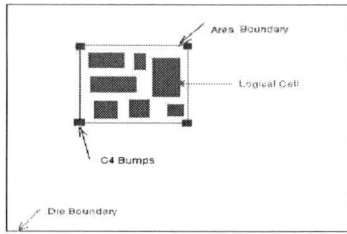

Fig. 8. Geometry relationships of discussing elements

Usually, the working frequency of logical cells is magnitudes higher than the resonance frequency of the package, in turn high frequency current typically will not cause resonance problem. This is the case especially when sufficient decoupling capacitors are placed in the package. On the other hand, the logical timing correlation among cells may generate much lower frequency components for potential resonance. Let's examine how the timing correlation can cause resonance.

First, we assume that the current generated by gates has trapezoidal waveform (TW) and the duty ratio is 50%, as shown in figure 7. This situation is common in digital units, especially in synchronous logical units. Secondly, we define a local area and have some logical cells in the area shown in Figure 8. Also we assume each cell contains many gates, and gates in the same cell share the same current waveform. Thirdly, we assume all gates are controlled by a synchronous clock while gates in different cells can have phase deviation either 0 degree or 180 degree. Lastly the clock frequency f_{clk} should be higher than the package resonance frequency f_{res} or at least, equal to it.

Then we can obtain an algorithm in Figure 9 which can give out a worst logical correlation for resonance problem. The main idea of this algorithm is that we arrange the on/off timing of the cells so that the resulting currents will be similar to a sinusoidal wave including the resonance frequency.

Specifically, we first divide cells in two classes, one has 0 degree phase deviation and the other one has 180 degree phase deviation. Then we can set certain number of cells active according to the estimation value, which is obtained from the minimal consumption current of all the cells at a certain time step. If we have Eq.(1) satisfied, we can have sufficient cells to undertake allocation, which means that after adding the currents of each cell together, a sinusoidal-like waveform current can be generated, as shown in Figure 10.

$$k \leq \left\lceil \frac{A}{\min\{IC_i\}} \right\rceil \leq \frac{N}{2} \qquad (1)$$

The notations in the algorithm are explained below:

T_{res} : the reciprocal of package resonance frequency f_{res}

T_{clk} : the clock cycle time ; $\sigma = \left\lceil \dfrac{T_{res}}{2T_{clk}} \right\rceil$;

N : the number of cells

A : the tolerance of harmonic amplitude around f_{res}

IC_i : the current when all gates in cell i is turned on by clock

k : sample ratio factor, an integer bigger than 1 and less than σ

Here, variable k is used to control the similarity between an ideal sinusoidal wave and the actual current waveform we can obtain. Because the total current only has a positive part, it contains dominant harmonics around its secondary harmonic frequency that is equal to the package's resonance frequency. Therefore, according to 'locality' property, resonance will occur on package in this area. As an example, Figure 11 illustrates a simple timing correlation among 4 cells. The rising edge of cell A causes one active cycle of cell B and C, then cell D. We can observe that the total current on bump is very similar to a sine wave.

In theory, the timing correlation generated by this algorithm is the worst one because the amplitude of harmonics around the resonance frequency can be set as large as possible. In practical, situations are more complicated than the worst case discussed in our algorithm. The reason is that Eq.(1) may not be satisfied, also decoupling effect should be considered, which can alter the resonance frequency. However, we believe that the actual risk exists in logic design and we should develop methods to estimate the resonance risk.

V. Fast Resonance Estimation in Early Logic Design

Complex logical subsystems, such as 'clock gating' controller [6], bus controller, memory controller and DMA controller, usually contain many function cells; therefore, they are more likely to have resonance problems. However, in today's design flows, designers typically ignore the resonance risk due to lack of efficient estimation tools for the resonance verification. In this section, we present a novel technique to help logic designers estimate the resonance risk. Before introducing our algorithm, let's define some parameters:

R: logical correlations described in regular expression

N: cells in a local area

D: number of cycles of a time window

k: a sample ratio factor, which is the same as defined in section IV

Algorithm Name: resonance trigger

Input: time window, A, k, N, IC_i

Output on/off matrix T

1) divide all the cells into two group according to their phase derivation

2) map the time variable t within the time window we concerned into a standard area ranging from 0 to π

3) use sample ratio factor k to get a vector containing k elements each of which is equal to x_j

$$x_j = \frac{j}{\pi} \quad 0 \le j \le k-1$$

4) estimate the maximal number of cells which should be activated at time step j according to formula below

$$f_j = \left\lceil \frac{A \sin(x_j)}{\min\{IC_i\}} \right\rceil$$

5) generate a zero matrix T which contains N rows and k columns

6) for each time step j, activate half of f_j cells in each cell group, which means to set T(i,j)=1 in T matrix

7) output matrix T

Fig. 9. Resonance trigger algorithm

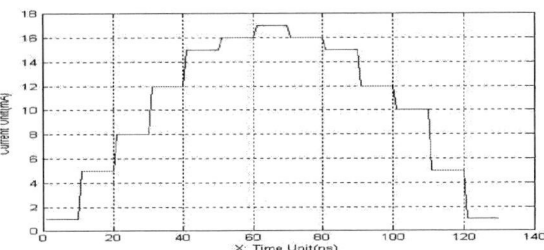

Fig. 10. Total current of all cells

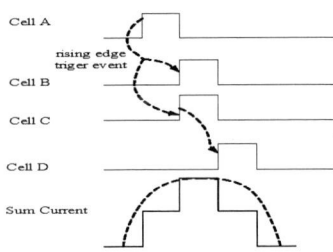

Fig.11. Logical correlation to cause resonance

T: an $N{\times}k$ zero matrix

$H(f,a)$: package frequency response function, where input parameter f represents harmonic frequency while a represents the harmonic amplitude; the out put of this function is the voltage droop on package under f and a.

$\{\alpha_1 \cdots \alpha_N\}$: a vector to computing current on vias

$\{\beta_1 \cdots \beta_N\}$: a vector to compensate blank area effect

f_{res} : the resonance frequency

f_{clk} : the clock frequency

Our algorithm accepts a given logics as the inputs but requires a regular expression description for the timing correlation which is defined in Eq.(2). It can represent any logic timing correlation among different cells. For instance, Eq. (3) below reveals that cell 13 will be turned on 5 clock cycles after cell 12's raising edge and 6 clock cycles after cell 11's raising edge and it will be always turned on at cycle 5 and 6.

$$C[0-9]* \{[0-9]* +\}\{C[0-9]*+[0-9]*\} \qquad (2)$$
$$C13\ 5+6+\ C12+5\ C11+6 \qquad (3)$$

Then we can have the estimating algorithm named *FFT* (Fast Fourier Transformation) *estimator* in Figure 12. This algorithm is mainly consisted of three parts. First, it attempts to construct the actual current waveform of all cells according to logic timing constraints in a given time window. Second, it computes the frequency spectrum of each cell's current waveform and compensates decoupling effect and blank area effect via FFT method. Finally, all the modified spectrums are added together, and the estimation of package drop is made according to package's frequency response.

Usually, decoupling capacitors are placed in each cell, and they will make the current waveform of supply nodes different from the waveform of the current sources connected to it at a certain time point. Here we assume each current source in the cell shares the same on/off waveform. This is because all gates in a cell are usually controlled by the same clock signal. Then we can obtain our decoupling model shown in Figure 13(1). Remember that because of the 'locality', current flows almost directly from via to via without diverging. Therefore, Figure 13(1) can be simplified to Figure 13(2). Now we can estimate the equivalent resistance of the power grid and ground grid to get Figure 13(3) (according to 'locality', the sum of via resistance from M7 to M3 can be a roughly estimated as R_P and R_G). Finally, a linear transform can be performed to acquire the coefficient α in order to get the current flowing through via resistor R_v.

Another effect, which should be compensated, is blank area effect. A blank area is the area that contains decoupling capacitors but is not within the cell boundary. In our algorithm, we only consider decoupling capacitors in each cell, and add all current together to do estimation. However, blank area exists in place where decoupling capacitors also are placed. Therefore, when actual current is flowing through these blank areas additional current is supplied by these decoupling capacitors. This is why we use a coefficient β to compensate this effect in our algorithm. The simplest way to estimate β is to use the area ratio of blank area to sum up all cell areas when decoupling capacitors are evenly distributed.

One advantage of this algorithm is that in early design stage, logic correlations, total decoupling value as well as via resistance are easy to obtain from both product specification and experiences of the existing products. Also, the compensation parameters in the algorithm are relatively easy to compute for C4 package model. Further, because the number of cells in a local area is usually small and the timing window is less likely to be very wide, the computation cost of this algorithm is very low.

VI. Experimental Results

We implement our algorithm using C++ language and test it under a 1Ghz Linux workstation with 512MB memory. Table I gives the performance of our algorithm when estimating package drop considering a 7-layer P/G grid with decoupling capacitors in a local area. Here we pass the timing information constructed by our algorithm to Hspice to produce a comparison. From the table, we can see that our algorithm can give a valid estimation of the droop on package when resonance is likely to happen while it is less accurate when current contains harmonics far away from resonance frequency. However, comparing with [7]-[10], which tries to get more accurate solution in dynamic simulation, accurate is not a serious problem

Algorithm Name: FFT estimator

Input: R, N, D, T, k, $H(f,a)$ $\{\alpha_1 \cdots \alpha_N\}$ $\{\beta_1 \cdots \beta_N\}$

Output: estimation of droop on the package

1) partition all regular expression to two-tuples (Ci, Cj+m) or (Ci, t)
2) do logical assignment until no changes in T happen, that is if a check process find that the column j of matrix T exist nonzero elements, it begin to check whether tuple(Ci, Cj+m) exists, if it is true, then it set element T(i,j+m) to 1. Also, if tuple (Ci, t) exists, it set T(i,j) to 1 directly.
3) Transform each row of matrix T to a time domain vector. This process replace any nonzero element of T, i.e., T(i,j) with a nonzero vector of length k, which is got by sample the trapezoid waveform of cell i using k time points while replace any zero element of T with a zero vector of length k.
4) After 3) we can get a new Nxp matrix T1, here p is the total number of sample points. Do Fast Fourier Transform (FFT) on T1 to get spectrum matrix Z.
5) After 4), $|Z(i, j)|$ represents the harmonic amplitude of cell i at frequency $j \cdot f_s$, where $f_s = \dfrac{f_{clk}}{D}$
6) for each element of Z, i.e, Z(m, n), multiply factor $\alpha_m (2\pi n \cdot f_s)$ to compensate decoupling effect
7) for each row in matrix Z, i.e. Z(m,n), multiply factor β_m to compensate blank area effect.
8) sum all the row vector of Z to get a 1xp vector S, then divide S by the number of bumps in the area
9) compute (4)
$$\max\{H(j \cdot f_s, |S(j)|\} \quad j \cdot f_s \in \left[\frac{4f_{res}}{5}, \frac{5f_{res}}{4}\right] \qquad (4)$$
10) output (4) as result

Fig. 12. FFT estimator algorithm

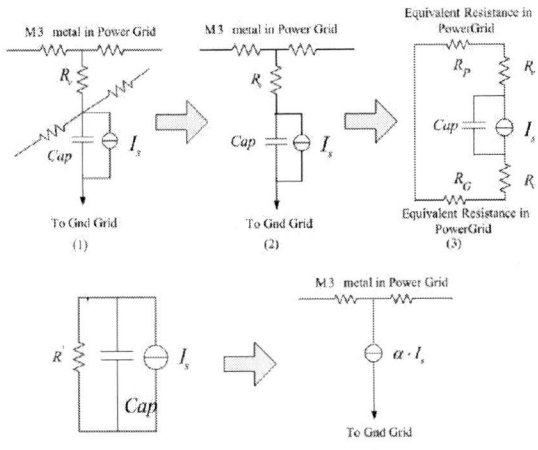

Fig. 13. Estimation of decoupling effect

because we are concerned with the resonance risk. Therefore, it remains an algorithm in early estimation with low computation complexity.

VII. Summary and Future Work

In this paper, we analyzed the resonance effect and the timing correlations of logic cells in the C4 package. We proved that the resonance effects do exist in the local area of C4 package. We then analyzed the conditions for generating the resonance effects from logic cell's timing correlation perspective. We consequently proposed an efficient algorithm to perform the early estimation of resonance effects. The experiment results show that the proposed algorithm can predict the resonance risk very well while the run time is reasonable.

The resonance problem we studied is quite different from the usual P/G droop problems; it is impossible to overcome by simply allocating more metal resources. To remove or reduce the resonance effects, further investigations are needed. For instance, one method is to optimize the placement of decoupling capacitors to minimize the harmonics amplitude around resonance frequency at bump nodes.

Also the simple resonance trigger algorithm only can give an ideal correlation among cells to cause resonance.

Anticipated research can include further investigations on how to give more actual constraints to help logical designers.

Acknowledgements

The author would like to thank Dr. Eli. Chiprout from Intel Strategy CAD lab for his great help to this research work. He gave a lot of insightful suggestions and we've learned a lot from weekly discussions.

References

[1]. "Performance characteristics of IC packages", Intel Corp., 2000 http://www.intel.com/design/packtech/ch_04.pdf

[2]. D. Tönnies, "A review and trends in flip-chip technology", Chip scale review, April 2004.

[3]. A. Dubey, "P/G pad placement optimization: problem formulation for best IR droop", ISQED 2005 Proceeding, pp 340-345

[4]. N. Srivastava, X. Qi, K. Banerjee, "Impact of on-chip inductance on power distribution network design for nanometer scale integrated circuits", ISQED 2005 Proceeding, pp 346-351

[5]. E. Chiprout, "Fast flip-chip power grid analysis via locality and grid shells", ICCAD 2004 Proceeding, pp 485-488

[6]. H. Li, S. Bhunia, Y. Chen, T.N. Vijaykumar, K. Roy, "Deterministic clock gating for microprocessor power reduction," Proceedings of the The Ninth International Symposium on High-Performance Computer Architecture (HPCA'03), pp 113.

[7]. T. Chen and C. C. Chen: "Efficient large-scale power grid analysis based on preconditioned Krylov-subspace iterative methods", DAC2001 Proceedings, pp. 559-562

[8]. W. Guo, S. X. D. Tan: "Circuit level alternation-direction-implicit approach to transient analysis of power distribution networks", International Conference on ASIC Proceedings, 2003, Beijing, pp. 246-249

[9]. H. Qian, S. R. Nassif, S. S. Sapatnekar: "Random walks in a supply network", DAC2003 Proceedings, pp 93-98

[10]. J. N. Kozhaya, S. R. Nassif and F. N. Najm: "A multigrid-like technique for power grid analysis", IEEE Trans. Computer-Aided Design, vol.21, no.10, Oct. 2002, pp 1148-1160

TABLE I Performance of FFT Estimator Algorithm (Obtained in a 1Ghz Linux Workstation with 512MB Memory)

Cell Num	Area size um x um	Area Decap pF/um^2	Time Window	Total Power	Worst Droop on Package				Relative Error of Estimation
					Hspice	Run Time	FFT estimator	Run Time	
resonance is less likely to happen, contains harmonics away from resonance frequency									
20	100x100	400	30 ns	10W	8 mv	11 s	5 mv	<2 s	54%
45	300x300	400	30 ns	20W	13 mv	26 min	7 mv	<2 s	46%
resonance is likely to happen, contains harmonics near resonance frequency									
20	100x100	100	30 ns	10W	0.43 v	11 s	0.38 v	<2 s	11.6%
45	300x300	100	30 ns	20W	0.46 v	26 min	0.49 v	<2 s	6.5%

Parallel-Distributed Time-Domain Circuit Simulation of Power Distribution Networks with Frequency-Dependent Parameters

Takayuki WATANABE

School of Administration and
Informatics,
University of Shizuoka,
52-1, Yada, Suruga-ku, Shizuoka,
422-8526,Japan.
Tel/Fax: +81-54-264-5433
e-mail: watanat@u-shizuoka-ken.ac.jp

Yuichi TANJI

Dept. of Reliability-based Information
Systems Engineering, Faculty of
Engineering, Kagawa University,
2217-20, Hayashi, Takamatsu,
761-0396, Japan.
Tel/Fax: +81-87-864-2235/2262
e-mail: tanji@eng.kagawa-u.ac.jp

Hidemasa KUBOTA   Hideki ASAI

Department of Systems Engineering,
Faculty of Engineering,
Shizuoka University
3-5-1, Johoku, Hamamatsu,
432-8561, Japan
Tel/Fax: +81-53-478-1237/1269
e-mail: hideasai@sys.eng.shizuoka.ac.jp

Abstract - **In this paper, we focus on the verification of the PCB/Package power integrity, which becomes very important for the design of state-of-art high speed digital circuits. The simulation of power distribution networks (PDNs) of the PCB/Package, which can be modeled as a large number of RLC lumped components, is a time-consuming task for using the conventional circuit simulator, such as SPICE. For this problem, we propose a parallel-distributed time-domain circuit simulation algorithm based on LIM. Furthermore, an effective modeling of frequency-dependencies of the PDNs, such as skin effects and dielectric losses, to solve by LIM is proposed.**

I. Introduction

In the design of high-speed digital circuits, it is important to model the power distribution networks (PDNs) of the PCB/Package in order to estimate and analyze unwanted noises, such as ground bounce, delta-I noise, and simultaneous switching noise (SSN). Usually, the PDNs of the PCB/Package are designed using multilayered power/ground plane pairs. Due to the transient switching current of the CMOS transistors, voltage fluctuations are generated by the parasitic inductances/capacitances of the planes, and the power plane resonance causes a performance degradation of the system. Then, taking measures in the early stages of design enables improvement of quality and reduction of cost [1].

Detailed analysis of the PDN using full-wave electromagnetic simulators provides accurate results. However it takes enormous CPU time and huge memory capacity. In the case of on-chip power distribution grids, the PDN can be modeled as power and ground (P/G) lines. On the other hand, in the case of the PCB/Package, the PDN can be modeled as two-dimensional P/G planes in many cases. As is well known, each P/G plane pair can be discretized spatially into (M−1) × (N−1) unit cells as shown in Fig. 1 [2]. Each RLGC parameter of the equivalent circuit of the unit cell is derived by dimensions and medium coefficients. Therefore, instead of full-wave simulators, if the power plane is modeled as a large number of lumped RLC components, the conventional circuit simulator, such as SPICE, is available. However, it is still difficult to analyze them using SPICE, because of the large scale of the PDN circuit.

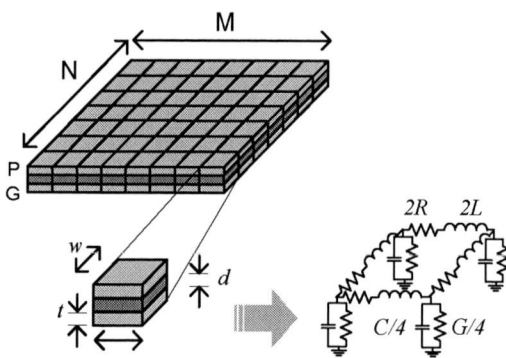

Fig. 1. Unit cell and equivalent circuit of PDNs.

For these problems, Latency Insertion Method (LIM), which is an algorithm for the time-domain simulation of large networks, is effective [3][4]. LIM is the derivative method in a class of the algorithms such as "leapfrog" finite-difference time-domain (FDTD) method [5][6]. Hence, LIM is sometimes referred to as the "circuit-based FDTD" method. That is to say, the node voltage vector and the branch-current vector are computed alternately. It can analyze large RLC networks very efficiently with much lower calculation cost of solving simultaneous equations in contrast to the implicit numerical integration used in SPICE-like simulator.

Because LIM is based on the basic leapfrog time stepping scheme, several parallel computation techniques can be applied to it [7]. Using these techniques, the whole circuit to be analyzed is divided into several subcircuits, and each subcircuit is simulated by each PE (Processing Element). In this paper, a parallel-distributed LIM algorithm is proposed to achieve a faster time-domain simulation of larger scale PDNs.

Also, actual PDNs have some frequency-dependent properties, such as skin effects and dielectric losses. In [8], high-speed interconnects having these frequency-dependent parameters were represented as the lumped segmentation model. Each segment can be obtained from the Debye rational function which approximates the frequency-dependent parameters. This segmentation model was solved by an LIM-like leapfrog time stepping scheme. However, this scheme has a limitation of the circuit structure

0-7803-9451-8/06/$20.00 ©2006 IEEE.

to be analyzed. That is to say, every branch must have an inductor and every node must be connected with the grounded capacitor [9]. Because the model of [8] does not meet these conditions, every first-order Debye model of each segment had to be solved via matrix inversion in order to apply LIM to solve the whole interconnect model. In this paper, we propose an effective modeling of frequency-dependent parameters for LIM simulation. In our modeling, the circuit representation of the first-order Debye function is modified to a suitable form for LIM. Finally, some PDN examples are simulated and the efficiency of our method is verified.

II. Latency Insertion Method

In this section, we briefly introduce LIM [3][4]. The circuit to be analyzed by LIM requires that each branch has an inductor and each node has a grounded capacitor as shown in Fig. 2. For the branch which consists of a series of an inductor, a resistor and a voltage source as shown in Fig. 3(a), the KVL equation is obtained by

$$V_i^{n+1/2} - V_j^{n+1/2} = L_{ij}\left(\frac{I_{ij}^{n+1} - I_{ij}^n}{\Delta t}\right) + R_{ij}I_{ij}^n - E_{ij}^{n+1/2}$$

(1)

Then, the branch current is updated, as in

$$I_{ij}^{n+1} = I_{ij}^n + \frac{\Delta t}{L_{ij}}\left(V_i^{n+1/2} - V_j^{n+1/2} - R_{ij}I_{ij}^n + E_{ij}^{n+1/2}\right),$$

(2)

Each node has a parallel combination of a capacitor, a conductance and a current source to the ground as shown in Fig. 3(b). Then, KCL leads to

$$C_i\left(\frac{V_i^{n+1/2} - V_i^{n-1/2}}{\Delta t}\right) + G_i V_i^{n+1/2} - H_i^n = -\sum_{k=1}^{M_i} I_{ik}^n,$$

(3)

where M_i is the number of branches connected to the node. Then, the node voltage is updated as

$$V_i^{n+1/2} = \frac{\frac{C_i}{\Delta t}V_i^{n-1/2} + H_i^n - \sum_{k=1}^{M_i} I_{ik}^n}{\frac{C_i}{\Delta t} + G_i}.$$

(4)

Time-domain simulation is done by the alternate "leapfrog" updates of branch-currents and node-voltages according to (2) and (4). To simulate stably, the time step Δt is determined based on the minimum values of inductance and capacitance in the circuit as follows,

$$\Delta t \le \sqrt{LC}.$$

(5)

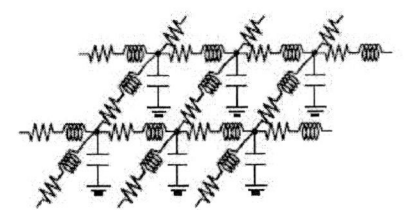

Fig. 2. Circuit structure suitable to LIM.

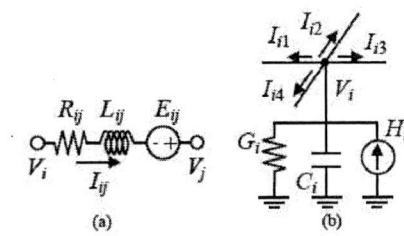

Fig. 3. Branch and node in LIM. (a): branch. (b): node.

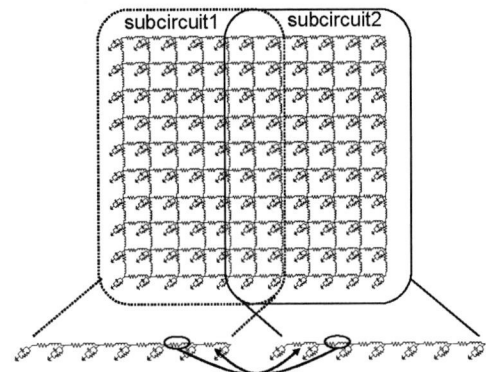

the current data transfer of the boundary branches

Fig. 4. Partitioning the PDN circuit.

III. Parallel-Distributed LIM Algorithm

Because LIM is based on the basic leapfrog time stepping scheme, it is faster than the conventional circuit simulator such as SPICE which requires solving the large sparse system of equations. However, there is still a limit to the size of the circuit which can be solved by one PE (Processing Element) even in LIM. To address this problem, the parallel-distributed LIM algorithm is proposed in this section. Actually, we have already developed the parallel-distributed full-wave FDTD simulator [7]. Because of similarities between LIM and full-wave FDTD, several parallel computation techniques of full-wave FDTD could be applied to LIM.

First of all, our algorithm divides the whole circuit into some subcircuits, and each subcircuit is assigned to different computer resource and analyzed by each PE on a PC-cluster. In the case of a planer power/ground structure, it is easy to divide the circuit into subcircuits as illustrated in Fig. 4.

Next, for calculating voltage of a node located at the boundary part of each subcircuit, the past values of the

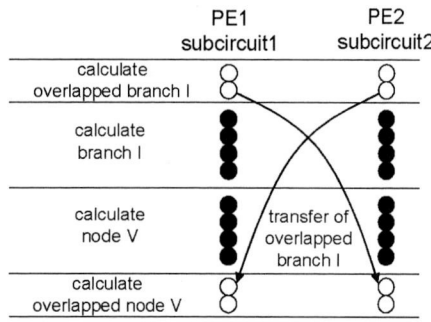

Fig. 5. Parallel-distributed LIM algorithm

branch current on the neighboring subcircuit, namely, on the adjoining PC, is required in each time step. Therefore, each subcircuit is chosen so that the boundary nodes overlap as illustrated in Fig. 4. While the communication speed between computers is usually slower as compared with the speed of the memory transfer in a computer, the time required for a current data transfer of the boundary branches can be ignored because the number of boundary nodes is relatively small, and the data transfer of the boundary branches and the calculation of node voltages and branch currents except the boundary can execute simultaneously. Finally, our algorithm is summarized in Fig. 5.

IV. Frequency-Dependent Parameters

A. first-order Debye model

In this section, we discuss the modeling of frequency-dependent parameters of PDNs. Actual PDNs have some frequency-dependent properties, such as skin effects and dielectric losses. These properties can not be simulated by the frequency-independent unit cell as illustrated in Fig. 6(a). Instead of this model, in order to simulate frequency-dependent effects, we can use the frequency-dependent transmission line model, such as the W-element of the Synopsys's Star-Hspice as illustrated in Fig. 6(b) [2]. In this case, the distributed series impedance and shunt admittance of the transmission line are defined as

$$Z(\omega) = R_{dc} + j\omega L_{ext} + R_{ac}\sqrt{\omega}(1+j), \quad (6)$$
$$Y(\omega) = G_{dc} + \omega G_d + j\omega C, \quad (7)$$

where

$$R_{dc} = \frac{2}{\sigma_c t}, \quad R_{ac} = \sqrt{\frac{2\mu_0}{\sigma_c}},$$
$$L_{ext} = \mu_0 d, \quad (8)$$
$$G_d = \varepsilon_0 \varepsilon_r \frac{w^2}{d} \tan(\delta),$$

where σ_c is the conductivity of the conductor, μ is the permeability, ε is the electrical permittivity, $\tan(\delta)$ is a loss tangent of the material. Also w, t and d are defined in Fig. 1. As a result, frequency-dependent RLGC parameters of

PDNs are analytically obtained as following forms:

$$R = R_{dc} + R_{ac}\sqrt{\omega},$$
$$L = L_{ext} + \frac{R_{ac}}{\sqrt{\omega}},$$
$$G = G_{dc} + \omega G_d, \quad (9)$$
$$C = \varepsilon_0 \varepsilon_r \frac{w^2}{d}.$$

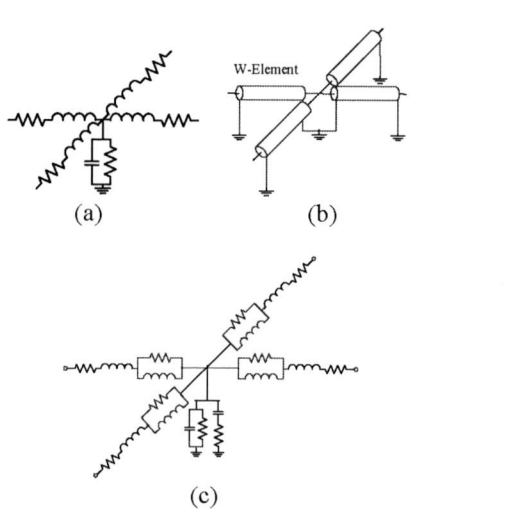

Fig. 6. Several types of unit cell model.
(a): frequency-independent model. (b): W-element model.
(c): first-order Debye model.

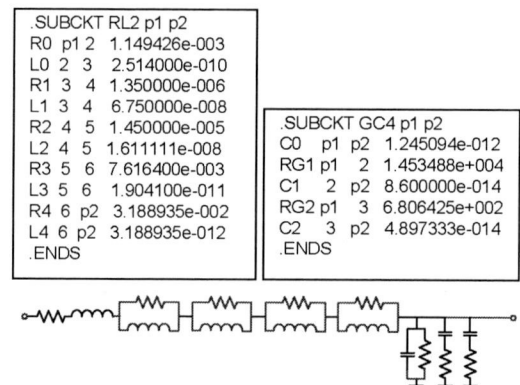

Fig. 7. a concrete example of the Debye model.

Usually, it takes long CPU time and large memory capacity to solve the PDN circuit consisting of the W-element. Therefore, the first-order Debye rational function is frequently-used to approximate the distributed series impedance (6) and shunt admittance (7):

$$Z(s) = R_0 + sL_0 + \frac{R_1 L_1}{R_1 + sL_1} + \frac{R_2 L_2}{R_2 + sL_2} + \frac{R_3 L_3}{R_3 + sL_3} + \cdots,$$
$$(10)$$

$$Y(s) = G_0 + sC_0 + \frac{G_1 C_1}{G_1 + sC_1} + \frac{G_2 C_2}{G_2 + sC_2} + \frac{G_3 C_3}{G_3 + sC_3} + \cdots . \tag{11}$$

From Eq.(10) and (11), the frequency-dependent unit cell model can be derived as illustrated in Fig. 5(c). In Fig. 6(c), the number of RL parallel networks at the portion of the distributed series impedance is determined from the number of poles of Eq. (10). the number of GC series networks is also determined from the number of poles of Eq. (11).

The unknown parameters of the Debye functions, such as R_i, L_i, G_i, and C_i ($i = 1 \dots N$), have to be chosen by fitting Eq.(10) and (11) to sampled-data calculated by Eq. (6) and (7). This procedure can be performed by several fitting and optimizing routines, such as the fitting functions in the MathWorks MATLAB and vector-fitting algorithm [10], et al. For example, we obtained the first-order Debye model as shown in Fig. 7. In this example, the dimensions and medium coefficients of the unit cell are w=2.5mm, d=0.2mm, t=0.03mm, σ_c =5.8×107, ε_r=4.5 and tan(δ)=0.02. Fig. 8 indicates that there is a good correlation between the Debye model and analytical data.

B. LIM simulation

In [8], the first-order Debye model of high-speed interconnects was solved by a LIM-like leapfrog time stepping scheme. However, as has been mentioned, LIM scheme has a limitation of the circuit structure to be analyzed. Every branch must have an inductor and every node must be connected with the grounded capacitor. The Debye model as illustrated in Fig. 7 does not meet these conditions. Therefore, in order to obtain I_{ij} from V_i and V_j, every first-order Debye model of the distributed series impedance had to be solved via matrix inversion in [8]. Also the shunt admittance has to be solved in a similar manner. This procedure is an inefficient even if LIM is used.

In this paper, we propose an effective modeling of frequency-dependent parameters for LIM simulation. In our method the RL parallel network is transformed into RL series network. From Eq. (10), the KVL is given by:

$$V_i - V_j = \left(R_0 + sL_0 + \sum_{m=1}^{4} \frac{R_m L_m}{R_m + sL_m} \right) I_{ij} . \tag{12}$$

Next, Eq. (12) can be rewritten as

$$(V_i - V_j) \sum_{m=1}^{5} \frac{k_m}{s - p_m} = I_{ij} , \tag{13}$$

where,

$$(V_i - V_j) \frac{k_m}{s - p_m} = I_{ij,m} \quad (m = 1, \cdots, 5) . \tag{14}$$

From Eq.(14), each parameter of the RL series network can be obtained from:

$$V_i - V_j = \frac{s - p_m}{k_m} I_{ij,m} = (sL_{ij,m} + R_{ij,m}) I_{ij,m} . \tag{15}$$

In our method, the GC series network is also transformed into GC parallel network in a similar manner. Finally, the procedure of our transformations is summarized in Fig. 9. Through this transformation, we do not have to perform any matrix inversions. For instance, we can obtain I_{ij_m} from V_i and V_j using only Eq.(15) even in time-domain.

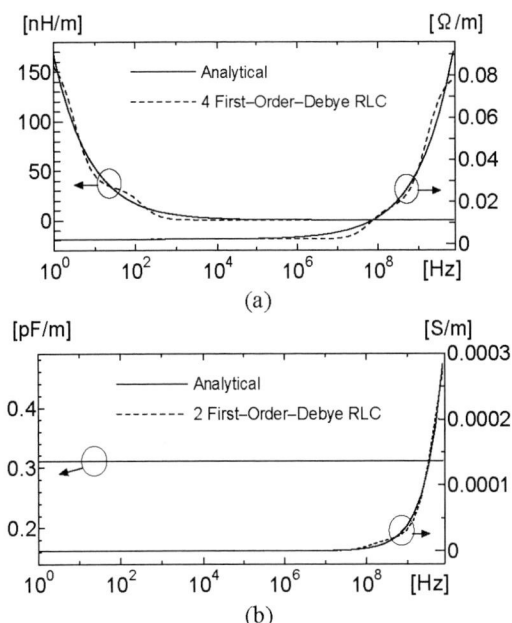

(a)

(b)

Fig. 8. Characteristics of Frequency-dependent parameters. (a) R(orms/m) and L(H/m). (b) G(S/m) and C(F/m).

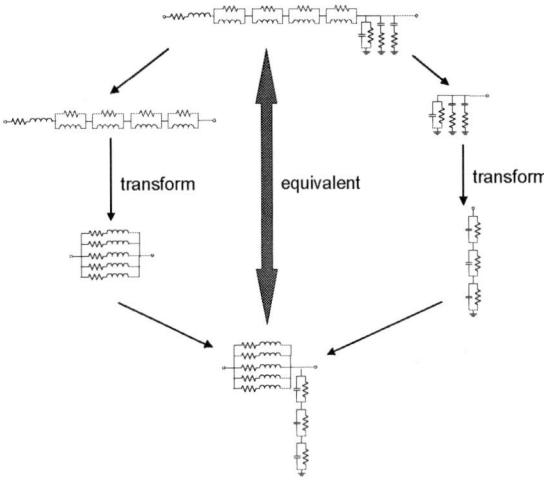

Fig. 9. Transformation of the first-order Debye model.

V. Numerical Results

First of all, in order to verify the efficiency of our parallel-distributed LIM algorithm, we simulated a transient response of a power/ground plane, as illustrated in Fig. 10, using 1PE and 2PE. The source point was excited with a Gaussian pulse. All simulations were performed on Intel Pentium M 1.7GHz personal computer. From Fig. 11, the voltage fluctuations at the observation point (60cm × 60cm) show the good agreement between results using 1PE and 2PE. Furthermore, Table 1 indicates that the CPU time of the 2PE simulation is almost half of the 1PE's CPU time.

Next, we simulated a power/ground plane which has frequency-dependent properties as illustrated in Fig. 12. The source point P01 was excited with a triangular waveform. All simulations were performed on SUN Blade Workstation. To verify our modeling validity and efficiency, we compared the transient responses simulated using the Star-Hspice and our LIM simulator. From Fig. 13, our first-order Debye model can be correctly solved by the LIM simulator. From Table 2, our LIM simulation is fastest among all simulations.

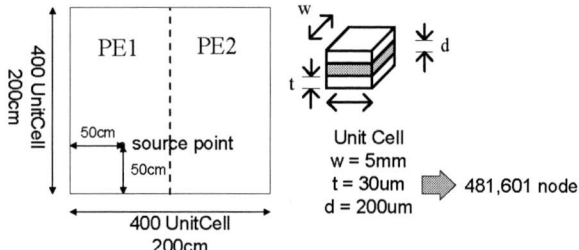

Fig. 10. First example of power/ground plane.

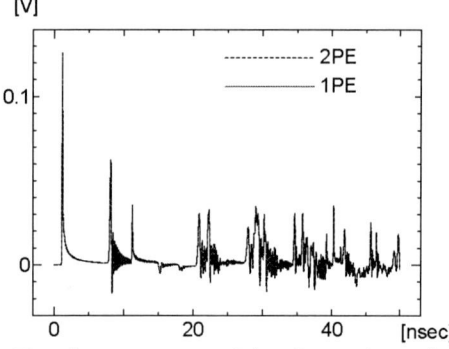

Fig. 11. Transient responses of the observation point (60cm × 60cm) in the first example.

Table 1: The CPU time comparisons of the first example.

Number of PE	CPU Time (sec)
1	110.3
2	57.5

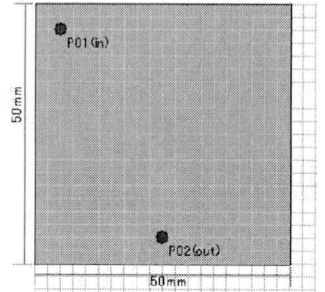

Fig. 12. Second example of power/ground plane.

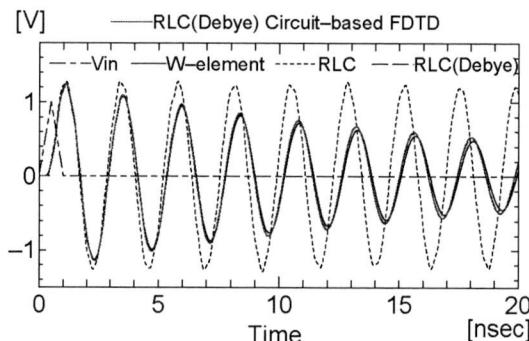

Fig. 13. Transient responses of the observation point P02 in the second example.

Table 2: The CPU time comparisons of the second example.

Simulator	Star-Hspice			LIM
Model	RLC	Debey	W-element	Debey
Problem Size	1282 nodes	5524 nodes	442 nodes	400 unit cells
CPU Time (sec)	3.3	19.7	365.6	0.47

VI. Conclusions

In this paper, we proposed the parallel-distributed LIM algorithm and an effective modeling of frequency-dependencies of the PDNs to solve by LIM. From the numerical results, it is obvious that the parallel computation is very efficient for the LIM algorithm. Also our transformed Debye model is quite effective to the LIM simulation.

In the future work, we have to consider the optimal partitioning of any irregular shaped power/ground plane for parallel computing in case of using many PEs.

Acknowledgements

This work was supported, in part, by Semiconductor Technology Academic Research Center (STARC), Japan. The authors would like to thank to Dr. K. Mashiko,

STARC, Mr. G. Yokomizo, Renesas, Mr. M. Utsuki, Sony, Mr. S. Hatasa, ROHM, for their useful suggestions and comments.

References

[1]. T. Watanabe and K. Srinvasan, H. Asai, M. Swaminathan, "Modeling of Power Distribution Networks with Retardation Using the Transmission Matrix Method," *Proc. IEEE Topical Meeting on Electrical Performance of Electronic Packaging (EPEP)*, pp.233-236, Oct. 2004.

[2]. L.D. Smith, R. Anderson, and T. Roy, "Power Plane SPICE Models and Simulated Performance for Materials and Geometries, " *IEEE Trans. Advanced Packaging*, vol. 24, no. 3, pp.277-287, Aug. 2001.

[3]. J. E. Shutt-Ainé, "Latency Insertion Method (LIM) for the Fast Transient Simulation of Large Networks," *IEEE Trans. Circuits and Systems-I*, vol. 48, no.1 pp. 81-89, Jan. 2001.

[4]. Z. Deng and J. E. Schutt-Ainé, "Stability analysis of latency insertion method (LIM)," *Proc. IEEE 13th Topical Meeting on Electrical Performance of Electronic Packaging*, pp. 167-170, 2004.

[5]. K. S. Yee, "Numerical Solution of Initial Boundary Value Problems Involving Maxwell's Equations in Isotropic Media," *IEEE Trans. Antennas Propagat.*, 14, 4, pp.302-207, 1966.

[6]. T. Watanabe and H. Asai, "Synthesis of time-domain models for interconnects having 3-D structure based on FDTD method", *IEEE Trans. Circuits and Syst. –II*, vol. 47, no. 4, pp. 302-305, April 2000.

[7]. T. Watanabe, H. Asai, T. Sasaki, and K. Araki, "Parallel-Distributed FDTD-Based Full-Wave Simulator for Large-Scale Printed Wiring Boards," *Proc. IEEE Topical Meeting on Electrical Performance of Electronic Packaging (EPEP)*, pp.287-290, Oct. 2002.

[8]. Alberto Scarlatti and Christopher L. Holloway,"An Equivalent Transmission-Line Model Containing Dispersion for High-Speed Digital Lines—With an FDTD Implementation", *IEEE Trans. Electromagnetic Comp.*, vol. 43, no. 4, Nov. 2001

[9]. H. Kubota, Y. Tanji, T. Watanabe, and H. Asai , "Generalized Method of the Time-Domain Circuit Simulation based on LIM with MNA Formulation", *Proc. The IEEE Custom Integrated Circuits Conference (CICC)*, Sept. 2005, in press.

[10].B. Gustavsen and A. Semlyen, "Rational approximation of frequency domain responses by vector fitting," *IEEE Trans. Power Delivery*, vol. 14, no. 3, pp. 1052-1061, July 1999.

8C-3

Power Distribution Techniques for Dual VDD Circuits

Sarvesh H. Kulkarni
EECS Department, University of Michigan
Ann Arbor, MI 48109, USA
shkulkar@eecs.umich.edu

Dennis Sylvester
EECS Department, University of Michigan
Ann Arbor, MI 48109, USA
dennis@eecs.umich.edu

ABSTRACT

Extensive research has proposed the use of multiple on-die power supplies (VDD) for reducing power consumption in CMOS circuits. We present a detailed study and design techniques for power delivery systems in dual VDD CMOS circuits. We first show that the total current to be delivered by the voltage supplies is significantly reduced (by 27%–46%) in dual VDD circuits. This current reduction prompts various design strategies that can be employed to design the power delivery system. We describe issues that arise at the system, board and package levels and propose a high-level model for the same. We then provide a new placement driven approach for designing on-die dual VDD power grids. Compared to already existing methods, the dual VDD grids generated by our approach reduce the worst case and average voltage drop by up to 12.3% and 6.8% respectively with no area overhead and sometimes improving wire congestion. We also show that dual VDD circuits can afford lower on-die decoupling capacitance budgets.

1. INTRODUCTION

Reducing power consumption at high speed is a critical goal for VLSI designers today. The dynamic and static power consumption in CMOS circuits have a quadratic and roughly cubic dependence on the power supply voltage (VDD) respectively [1]. There is extensive work in literature that exploits this concept for reducing power by using dual (or in general, multiple) power supplies in the design. Multiple VDD design applies higher voltages to gates on critical paths and lower voltages to gates on paths with slack. In this way the power consumption reduces while timing is met. Most earlier work in this area focuses on the power supply assignment problem. In particular, [2]-[8] provide algorithms that select gates to be assigned to the available supplies. Ref. [8] and recently [9] and [10], detailed designs based on multiple power supplies.

Ref. [12] shows that using two different supplies provides near optimal power savings and adding a third supply yields little additional power reduction while further worsening the power delivery challenges. We therefore focus on dual VDD designs in this work. Also, voltage assignment can be performed at a fine-grained level (gate-level) or at the module-level. The module-level assignment somewhat simplifies the physical design and power delivery problems; however, its power savings are curtailed due to less freedom in low VDD assignment. Hence, we focus on fine-grained dual VDD assignment and also assume a standard cell row-based layout style that is commonly used in ASICs. In the remainder of this paper we refer to the lower supply in a dual VDD design as VDDL and to the higher supply as VDDH. The power supply of a reference single VDD design will simply be referred to as VDD.

Issues such as level conversion and physical design for dual VDD also arise when using this technique. Ref. [11] first proposed techniques to enable the physical design of standard cell based dual VDD circuits. The focus was on developing a new placement tool & proposing a new cell layout style that facilitated dual supply usage.

Although some dual VDD issues have received greater attention, the important issue of power delivery discussed in our paper has received little attention. Related work in this area includes approaches presented in [8], [11], [13]-[15]. Since two power supplies need to be now supplied across the die, [8] and [13] proposed physical design approaches to partition cells into regions

(rows/blocks) that contain either only VDDL cells or only VDDH cells. This allows existing standard cell libraries to be used but needs a modified standard cell placer that leads to high wirelength and core area overheads. Ref. [11] and [14] thus proposed the use of a modified standard cell layout where an extra rail is added for the added supply voltage. The existing placer can then be freely used with its full optimization power. In all approaches mentioned so far, VDDL and VDDH cells share a common ground (GND). Recently, [15] proposed the use of a separate ground GNDL (for VDDL cells) and GNDH (for VDDH cells) respectively.

Before introducing our approach for designing dual VDD power grids, we first make the observation that the power supply current required for the operation of dual VDD circuits is greatly reduced compared to a single VDD circuit with the same timing. This is so for two reasons: (i) gates assigned to VDDL need to deliver less current to charge up their load capacitances to logic state 1, and (ii) current demand on the VDDH power supply reduces since only a subset of the initial gates now draw current from it. In [6] the authors suggest that as high as 60% of total gates are assigned to the lower supply for stringent timing constraints, strengthening the claim that current drawn from the VDDH power supply should fall substantially. We use this observation and show that this quality can be harnessed to design robust power distribution systems for dual VDD circuits. Power grid design approaches should take into account the actual placement of the VDDH and VDDL cells while sizing the grid wires, i.e., if a region of the die contains more VDDH cells, more wiring resources should be dedicated to the VDDH grid, while recovering resources from the VDDL grid. The approaches in literature failed to take the placement into account. Using such ideas we show that dual VDD grids can be design to be as robust as their single VDD counterparts for no area or wire congestion penalties. Interestingly, we also show that dual VDD circuits can afford reduced decoupling capacitance due to their reduced current demand. This reduction in decoupling capacitance will improve leakage, die area and yield.

To summarize, the main contributions of this paper are:
1. We present the first detailed study of power distribution for dual VDD circuits. We explore solutions for package/board level issues as well as issues for on-die power grids.
2. We present a new placement driven power grid design methodology, *D-Place*, which improves power grid integrity.
3. We demonstrate that dual VDD power grids can be designed to be as robust as their single VDD counterparts. In fact, we show that dual VDD designs can afford lower decoupling capacitance budgets.

Our paper is organized as follows. In Section 2, we describe our simulation setup and general framework. In Section 3, we demonstrate that dual VDD circuits have significantly lower supply current demands. In Section 4, we describe a study of the system board and package level issues when working with dual supplies. Section 5 presents our work for on-die power distribution grids. In Section 6 we show that dual VDD circuits can often afford lower decoupling capacitance budgets. Section 7 concludes the paper.

2. SIMULATION SETUP AND FRAMEWORK

Our work is based upon a 6 metal layer 0.13μm CMOS technology. The nominal voltage for this technology is 1.2V and two threshold voltages (VTH) are available; 0.2V/0.1V and -0.2V/-0.1V

0-7803-9451-8/06/$20.00 ©2006 IEEE.

Table I – Dual VDD power savings and VDDL assignments
(the reported % are with respect to the original single VDD design)

	VDDL = 0.8V		VDDL = 0.6V	
	% Savings	%VDDL	% Savings	%VDDL
c880	28	65	31	55
c2670	32	65	37	56
c5315	35	58	37	49
c7552	44	91	49	71

for NMOS and PMOS respectively. Gates in the original single VDD design are sized based on an algorithm similar to TILOS [16]. We developed a static timing analysis engine and use look-up table based power/delay standard cell libraries for timing analysis and power estimation (Synopsys Library Compiler format).

For obtaining the dual VDD design from the single VDD design, we adopted the method outlined by the authors in [6] because of its simplicity in implementation; the concepts outlined in this paper are applicable to all other dual VDD assignment algorithms (such as [2]-[5], [7] or [8]). We extended the work in [6] by adding sensitivity based dual VTH assignment since dual VTH assignment is widely used in practice for further optimizing power. The details of the power optimization flow itself have been omitted due to lack of space. Literature suggests that a VDDL value that is about 50–70% of VDDH is ideal for minimizing power [17], [18]. So, for this work, we tested our algorithms for VDDL = 0.6V and 0.8V. Table I summarizes the power savings obtained for several ISCAS85 benchmark circuits [19]. The column marked "%VDDL" indicates the fraction of gates from the original single VDD design that are mapped to VDDL. The single and dual VDD designs meet the same stringent timing constraint. These results confirm that significant power savings are possible using dual VDD design and that a significant number of gates get assigned to the lower supply.

3. DUAL VDD SUPPLY CURRENT DEMAND

This section demonstrates that dual VDD circuits have significantly reduced power supply current demands. Since we use a dual VTH dual VDD process, each gate in the design can be one of four combinations, namely VDDH-low VTH, VDDH-high VTH, VDDL-low VTH and VDDL-high VTH. As a cell moves from VDDH to VDDL, it has a significantly reduced current demand. A subtle point however, is that short-circuit current that flows during switching events is also significantly reduced by such a VDD change; this also holds when a gate moves from low to high VTH. Table II summarizes these reductions in current demands for a few gates in the library. The average over all 160 cells in the library is also reported. This data was obtained through transient simulations in SPICE and thus accurately includes the changes in short-circuit current and currents charging the load.

We next extend this concept from the gate level to the circuit level. The circuit-level current demands follow the gate-level numbers presented in the previous table. Using SPICE simulations over 1000 randomly selected input vectors, Table III reports the current load on each of the power supplies (VDDL and VDDH). These results confirm that the current to be supplied by the power supplies is significantly reduced in dual VDD designs.

Table II – Normalized gate-level supply current reduction

	Single VDD		Dual VDD: VDDL=0.8V		Dual VDD: VDDL=0.6V	
	Low VTH	High VTH	Low VTH	High VTH	Low VTH	High VTH
INVX10	1.00	0.90	0.57	0.49	0.36	0.27
NAND2X2	1.00	0.85	0.54	0.45	0.34	0.23
NAND3X6	1.00	0.88	0.55	0.47	0.35	0.24
NOR2X1	1.00	0.86	0.52	0.39	0.30	0.19
NOR3X4	1.00	0.85	0.50	0.37	0.29	0.18
AVERAGE	1.00	0.88	0.54	0.44	0.33	0.23

Table III – Circuit-level current (mA) drawn from the power supplies

	Single VDD	Dual VDD: VDDL=0.8V		Dual VDD: VDDL=0.6V	
	VDD	VDDH	VDDL	VDDH	VDDL
c880	9.7	5.6	2.2	5.9	1.3
c2670	23.6	11.9	6.5	10.1	3.0
c5315	36.7	20.9	7.2	20.9	3.6
c7552	47.9	13.9	19.4	20.4	8.5
AVERAGE %	100.0	48.5	27.7	50.7	13.5

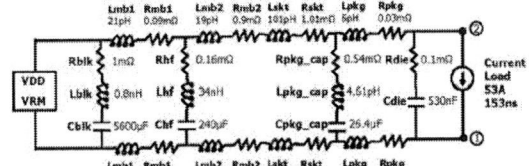

Fig. 1. Single VDD power delivery model [20].

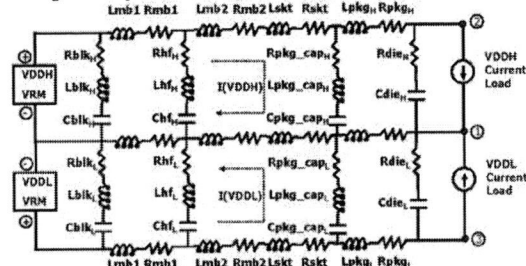

Fig. 2. Dual VDD power delivery model.

4. SYSTEM BOARD/PACKAGE DESIGN

4.1 Single VDD board

Fig. 1 shows a high level power delivery model for the system board & package of a typical integrated circuit (IC) [20]. The voltage regulator module (VRM) is situated on the motherboard and is the primary voltage source. Capacitors Cblk/Chf, Cpkg_cap and Cdie are motherboard, package and on-die decoupling capacitors respectively. The resistances and inductances in series with the decoupling capacitors model the effective parasitic series resistance and inductance. Lmb1-Rmb1/Lmb2-Rmb2 and Lskt-Rskt represent the inductance and resistance of signal tracks on the motherboard and cables connecting the motherboard to the package respectively. Lpkg and Rpkg represent the inductance & resistance of the package/C4s. The values for the parameters used in this figure are from [20].

4.2 Dual VDD board

The single VDD model in Fig. 1 is extended for the dual VDD case as shown in Fig. 2. Two separate VRMs are now needed (for VDDH and VDDL). We see that this model provides two on-chip voltages (VDDH and VDDL) at nodes 3 and 2 respectively. The ground can be shared between the two voltages (node 1). In this model, the current loop shown in the upper half of the figure corresponds to that seen by the VDDH VRM and the lower one corresponds to the VDDL VRM. The current loads for each of the power supplies are also shown. As demonstrated in Section 3, each of these individual current sources is a fraction of the original current source shown in Fig. 1. On average across the benchmarks we studied, we found that the VDDH and VDDL current loads are about 49% and 28% of the original current load (VDDL = 0.8V). Since the ground path is shared, current flowing through the ground path is about 77% of that in the original design. These numbers reduce further when VDDL = 0.6V. The parameters corresponding to the motherboard will not change since similar track/planes will be still be followed by each of the currents. The socket parameter (Rskt and Lskt) are also kept unchanged since they model the cables connecting the motherboard to the package. The package inductance & resistance (Lpkg and Rpkg), depend on the allocation of the C4s (assumed equal in number as the single VDD case). The C4s should be split equally between VDDL and VDDH, since although fewer gates remain assigned to VDDH (Table I), the current demand on VDDH still remains much more than on VDDL (Table III). Keeping the same number of C4s for the ground, these package parameters for VDDL and VDDH will be in *inverse* proportion to the fraction (=0.5) of C4s assigned to each supply.

Assuming that the dual VDD board/package allows the same real

Table IV. Power delivery model simulations.

			VDD or VDDH & VDDL		GND	
			PK	QS	PK	QS
Single VDD	VDD	mV	92.7	65.0	92.7	65.0
		%	7.7	5.4	7.7	5.4
Dual VDD	VDDH	mV	63.0	34.0	68.9	40.7
		%	5.3	2.8	5.7	3.4
VDDL = 0.6V	VDDL	mV	18.0	9.0	68.9	40.7
		%	3.0	1.5	11.5	6.8
Dual VDD	VDDH	mV	63.0	32.0	77.8	46.0
		%	5.3	2.7	6.5	3.8
VDDL = 0.8V	VDDL	mV	37.0	18.0	77.8	46.0
		%	4.6	2.3	9.7	5.7

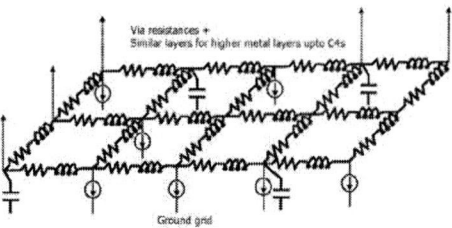

Fig. 3. PEEC model of power grid.

estate area for decoupling capacitors (as the single VDD board/package), we propose splitting the capacitances in the ratio of the current loads. The reduced current demand in dual VDD circuits proves useful here since it becomes feasible to use lesser decoupling capacitors for each of VDDL and VDDH as compared to the single VDD case. Thus, for Cblk (and similarly for others), we have,

$$Cblk = Cblk_H + Cblk_L, \; and \qquad (1)$$
$$Cblk_H \, / \, Cblk_L = I(VDDH) \, / \, I(VDDL)$$

where, $Cblk_H/Cblk_L$ are the decoupling capacitances attached to VDDH/VDDL, and $I(VDDH)/I(VDDL)$ are the current demands on VDDH/VDDL respectively.

4.3 Results

Table IV reports the peak (PK) droop/bounce voltages at nodes 1, 2 and 3. The quiescent values (QS) of the voltages at these nodes are also reported and correspond to the resistive drop from the VRM up to the C4s. Absolute values in mV as well as percentages of total nominal swing (0.6V/0.8V for VDDL, and 1.2V for VDDH) are reported. The percentages reported here carry more relevance since they are essentially the fraction of the total nominal swing of the voltage supply that the on-die circuitry actually receives from the board. This percentage directly translates into delay degradation in the standard cells and is hence a good metric to follow. From Table IV, the absolute values of the droop/bounce for the dual VDD case are always better than the single VDD case. Based on a percentage metric as well, the dual VDD case does better than the single VDD case, except for the ground bounce afflicting the VDDL cells (boxes shaded gray). This subtle difference arises since, although the actual ground bounce in mV has reduced, it remains a large percentage when normalizing to VDDL (0.6V or 0.8V). The reason for this behavior lies solely in the fact that the ground path is shared by both the supplies. This points to the fact that the dual VDD board may require special care for the ground path (such as a reduced resistivity return path for ground) as compared to the single VDD board.

We also used the HSPICE circuit optimizer to optimize this circuit for minimum PK/QS values. We found that the results using the simple intuitive approach outlined above gave us results that match the optimal solution. However, from this study we also found that PK/QS are fairly insensitive to the exact ratio in which parameters such capacitors or C4s are split among VDDL and VDDH. This is a favorable finding, since it gives the board/package designer considerable flexibility in allocating the resources to each of the two power supplies. This can also help in designing for desired values of resonance frequencies between the package inductance and on-die decoupling capacitance.

5. DUAL VDD POWER GRID DESIGN

5.1 Framework

Our process technology has 6 metal (Cu) layers and has flip-chip package technology. The standard cells are placed in rows on the bottommost metal layer and are represented as current sources. We assume a partial electrical equivalent circuit (PEEC) model for the grid [21]. In this model, each line of each metal layer is fractured into smaller segments, and each segment is then modeled using a resistor, self inductance, mutual inductance to other segments and ground and coupling capacitances. We used the following methods in calculating the parameters for each segment: (a) Resistance: is calculated simply once the length, width and sheet resistance are known. (b) Capacitance: is calculated using Wong's model [22]. Our estimation of decoupling capacitance to be added is described below. (c) Inductance: Partial self and mutual inductance depend on the geometry of the wire segments and are calculated using the method described in [23].

For simulating the grid we follow the fast and accurate R/L/C simulation method presented in [24]. The PEEC models of the grids we worked on typically have about 600,000 R/L/C elements. On a 3GHz Pentium4 2GB RAM computer, our implementation takes about 160s to generate the PEEC model and about 40s for simulation (single as well as dual VDD grids). While the PEEC model can become computationally expensive for large die sizes, we have used it for its suitability in modeling on-die power grids - the ideas presented by us can also be applied to simpler R/L/C models such as those obtained through FastHenry for the price of accuracy.

Fig. 3 shows an example of the model for the bottommost layer of the grid. The current sources shown represent the gates of the design. The vias connecting the various metal layers are modeled as resistors. The resistances and the inductances of the wire segments and some decoupling capacitors are also shown in the figure.

5.2 Single VDD grid design

Single VDD grids were held as the reference for comparison with the dual VDD grids. The single VDD grid is assumed to be regular in structure. The VDD and ground lines alternate each other and have increasing thicknesses as we move higher up in the metal layer hierarchy. The C4 diameter and pitch was assumed to be 30μm and 150μm respectively. The C4s are placed on 30μm wide straps on the topmost metal layer. Alternate straps for VDD and GND arrange all the C4s in a checker-board fashion such that the pitch of 150μm is met. We simulate grids that are ~0.5mm² in area and allow for 24 C4 locations (12 for VDD and 12 for GND).

Decoupling capacitors help mitigate voltage drop in the grid. We follow a simple method described in [25] to estimate the decoupling capacitance needed. We first fix a tolerance level for the permitted voltage drop on the grid ($V_{noise-limit}$). The decoupling capacitance is then estimated using Eq. 2. While this estimate appears simple, it is commonly used in practice [25]. Approaches such as [26] can be used alternatively.

$$C_{decap} = \frac{\int_0^{\tau} I(t)dt}{V_{noise-lim}} \qquad (2)$$

where, I(t) represents the current sunk by the switching events and τ is the switching period.

5.3 Dual VDD grid design

We first point out that we have constrained the areas of all grids studied to be the same. Since dual VDD grids need to route an additional voltage, this may imply worsened wire congestion. Thus under constant area, we finally compare the quality of the grids with respect to the voltage droop/bounce as well as wire congestion. We also assume the same number of C4s as in the single VDD case & distribute them equally between VDDH/VDDL (for the same reason

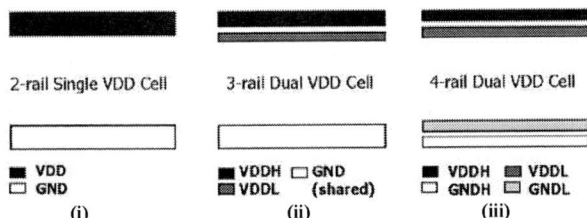

Fig. 4. Different standard cell layouts.

as described in Section 4). Since, the VDDL and VDDH cells are evenly dispersed across the die, the VDDL and VDDH C4s are also evenly dispersed. Our C4 assignment scheme is in keeping with [27] where a strong spatial locality effect is shown. The values of the current sources connected to model the gates are obtained using the analysis described in Section 3.

As mentioned briefly in Section 1, [8] and [13] proposed partitioning cells into blocks or rows, such that each block or row contained only one kind (VDDL or VDDH) of cells. This allows existing standard cell libraries to be used. However, [11] and [14] showed that this constrains the placement tool and can lead to an increase in wire length as well as core area (by up to 23% and 15% respectively) and higher post-route power dissipation. To overcome this problem, modified standard cells shown in Fig. 4 can be used. We now describe three methods (referred to as *D-Vanilla*, *DSDG* and *D-Place*) for designing dual VDD power distribution grids. We compare our approach *D-Place* against *D-Vanilla* and *DSDG*, since all these techniques allow the use of the same placement tool as used for the single VDD design (thus enabling a fair comparison).

5.3.1 D-Vanilla:

Authors in [11] proposed the 3-rail standard cell layout shown in Fig. 4(ii). (Fig. 4(i) shows a conventional 2-rail single VDD cell.) The new cell library now has two copies of each cell from the old library: one copy is powered from VDDL and the other is powered from VDDH (when using a dual VTH process there will be two more copies of each cell for low VTH and high VTH). The existing placement tool can now be used with its full optimization power since there are no constraints about where to place each kind of cell. The authors in [11] observed that since the current requirement of dual VDD circuits is reduced, the VDDL rail can be reduced in width compared to the VDDH rail, mitigating wire congestion. Thus, the rails widths for GND and VDDH are kept the same as the single VDD cell, and the rail width of the VDDL rail is scaled down by the ratio of current demand of the design when powered by VDDL to the current demand when powered by VDDH. This ratio is quite design invariant and is about 0.32 for VDDL=0.6V and 0.54 for VDDL=0.8V. Grids designed using this work look like Fig. 5(ii).

5.3.2 DSDG (Dual Supply Dual Ground) [15]:

D-Vanilla discussed above shares ground between VDDL and VDDH cells. DSDG proposes the use of separated grounds (GNDL and GNDH) for VDDL and VDDH. Every alternate rail (from the single VDD floorplan) is now assigned to VDDL/GNDL and VDDH/GNDH. This work however did not discuss the standard cell layout that should be used. We hence propose the 4-rail standard cell layout shown in Fig. 4(iii). Each of the rails in this layout (VDDH, VDDL, GNDH and GNDL) is now half as wide as the rails in the cell shown in Fig. 4(i). Not shrinking each of the rails this way would lead to very high wire congestion for fixed area (or about 23% higher area for the same wire congestion in our studies). Grids designed using this work look like Fig. 5(iii). While the authors used simplified FastHenry based models, our analysis is more accurate since we use the PEEC model. Also, effects such as the sharing of C4s among VDDL and VDDH are considered in our implementation.

5.3.3 D-Place:

Before moving on to describing our placement driven approach, we list important points that the earlier approaches did not consider:

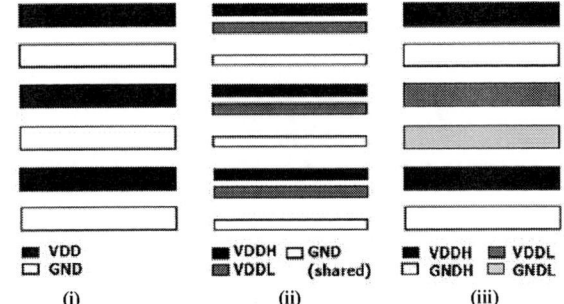

Fig. 5. Texture of grids designed using cell layouts in Fig. 4.

- Firstly, it is important to control voltage droop/bounce on the VDDL grid to a value that is lower than that for the VDDH grid (e.g., to ensure that <10% of rail-to-rail swing is lost due to grid losses, we need to limit droop on the VDDH (1.2V) grid to 120mV; however, for the VDDL (say 0.6V) grid, we need to limit it to only 60mV).
- Secondly, the method followed in sizing wires of each grid should take into account how much current needs to be delivered.
- Thirdly, the method followed in sizing wires should also consider the placement of the two kinds of cells.
- Fourthly, when designing the power distribution system, effects arising at the system board and package level should also be accounted for. While our work addressed these issues in Section 4, earlier work such as [15] failed to do so. This becomes especially important since [15] requires two separate grounds (GNDL and GNDH) to be supplied from outside the chip which can greatly complicate the design at the system board and package level.

The details of D-Place are as follows. We use 3-rail standard cells shown in Fig. 4(ii). Conventional placement tools can thus be used for the layout of the dual VDD design and the placement of the dual VDD cells is very close to the placement of the cells in the original single VDD design. The ground is common for VDDL and VDDH. Let α and β respectively be the ratios of the current demands on the VDDH and VDDL grids to the current demand of the single VDD grid, i.e., $\alpha = I(VDDH)/I(VDD)$ and $\beta = I(VDDL)/I(VDD)$. The grid droop/bounce is a result of two mechanisms: a resistive IR drop and an inductive LdI/Dt drop, where the IR drop usually dominates the on-chip inductive effect. Let 'W'μm be the wire width that was assigned to some wire in the VDD/GND grid in the original single VDD design. We now assign the widths to corresponding VDDH, VDDL and GND wires in the dual VDD grids as follows:

$$W_{VDDH} = \alpha W$$

$$W_{VDDL} = \beta \frac{VDDH}{VDDL} W \qquad (3)$$

$$W_{GND} = (\alpha + \beta) \frac{VDDH}{VDDL} W$$

The reasoning behind this sizing becomes clear when we recall our earlier comment that droop/bounce on VDDL grids needs to be very well controlled. Assuming that the original single VDD grid met a certain percentage droop/bounce budget, the VDDH grid will meet the same budget with wire widths scaled down by α. This is so since although the grid resistance goes up by $1/\alpha$, the current demand reduces by α. The IR product thus remains same. Also, the LdI/dt effect is controlled by this wire sizing since dI/dt goes down by α and inductance exhibits a sub-linear dependence on wire width.[1] Wire sizing for the VDDL grid is done similarly with the addition of the scaling factor *VDDH/VDDL*. This scaling factor accounts for the fact that the VDDL grid requires tighter absolute droop/bounce

---

[1] Although on-chip L is typically dominated by IR, we have included it in our analysis for scenarios where this might not hold. LdI/dt effect arising from the package is significant & was addressed by Section 4.

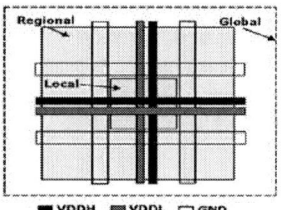

Fig. 6. Local and regional areas.

control (i.e. more wire width) in order to meet the same relative voltage drop budget. The tighter design requirement of the VDDL grid must be imposed on the GND grid too (Eq. 3), since the ground path is shared by VDDL and VDDH cells, and VDDL and VDDH cells are evenly interspersed on the die.

α and β discussed up to this point are chip level (global) current ratios and do not include placement information. Indeed, the current demands across various regions of the chip can differ substantially. In order to include the placement information while sizing each wire segment, we thus introduce local and regional variations of these α and β. We first divide the die into several small "local" areas. The exact size of this local area can be freely chosen; we took it to be the area bound by adjacent ground (or equivalently power) lines on consecutive metal layers. The "regional" area around each "local" area is then defined as the area bounded by its neighboring "local" areas (Fig. 6). Now we compute α and β for all local and regional areas. Finally, each wire segment inside each local region is sized using Eq. 3, where the α and β are replaced by "effective" α and β. The effective α (similarly β) is defined as follows:

$$\alpha_{effective} = \frac{\alpha_{local} + \alpha_{regional}\dfrac{Area_{local}}{Area_{regional}} + \alpha_{global}\dfrac{Area_{local}}{Area_{global}}}{1 + \dfrac{Area_{local}}{Area_{regional}} + \dfrac{Area_{local}}{Area_{global}}} \quad (4)$$

The ratio of the areas used in Eq. 4 act as scaling factors in order to ensure that the local α are weighted more while allowing for neighboring regions to be taken into account when sizing wires. This heuristic approach effectively guides the sizing of the wires by thickening wires in areas of higher current demand and shrinking them down in areas of lower current demand for each grid (VDDH/VDDL/GND). Fig. 7 shows the flowchart for D-Place.

Although the newly proposed 3-rail/4-rail standard cells require library modifications, this can be accomplished using existing design automation tools such as Cadence Abstract Generator. We emphasize that as wires are sized in proportion to the current (maintaining current density), they will not violate electromigration constraints.

5.4 Results

5.4.1 Voltage drop across grids

Table V presents results for the grids when VDDL = 0.6V and 0.8V. The MAX/AVG rows correspond to maximum/average voltage drops across all nodes in the design. The percentages reported are the sum of the power droop and the ground bounce, representing the potential difference available to the cell. Also, the percentages for each cell are taken with respect to its nominal rail to rail swing (1.2V for VDDH gates and 0.6V or 0.8V for VDDL gates).

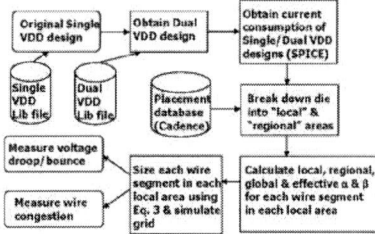

Fig. 7. Flowchart for D-Place.

Table V. Power grid % voltage drop comparisons.

(A) VDDL = 0.6V

		Single VDD	DSDG	D-Vanilla	D-Place
c880	MAX	16.9%	30.9%	16.4%	18.6%
	AVG	9.5%	14.7%	9.6%	9.5%
c2670	MAX	25.6%	35.5%	32.2%	25.5%
	AVG	15.9%	19.8%	15.2%	14.5%
c5315	MAX	29.6%	38.2%	37.4%	32.0%
	AVG	21.6%	23.4%	20.2%	19.8%
c7552	MAX	26.8%	34.2%	34.5%	29.4%
	AVG	22.2%	21.0%	21.1%	18.7%

(B) VDDL = 0.8V

		Single VDD	DSDG	D-Vanilla	D-Place
c880	MAX	16.9%	30.3%	16.3%	19.5%
	AVG	9.5%	15.9%	9.7%	9.8%
c2670	MAX	25.6%	36.1%	27.6%	27.0%
	AVG	15.9%	22.1%	15.8%	15.3%
c5315	MAX	29.6%	38.1%	33.0%	31.8%
	AVG	21.6%	25.4%	20.1%	20.3%
c7552	MAX	26.8%	31.4%	31.6%	28.7%
	AVG	22.2%	24.9%	22.3%	20.1%

From these tables, it can be seen that dual VDD grids can be designed to be as robust as their single VDD counterparts in terms of average voltage drop with some cases showing better results in the dual VDD design. D-Place has slightly inferior (<2.6%) results compared to single VDD in terms of the MAX values (in terms of absolute values in mV this corresponds to <15mV and can be easily compensated by techniques such as locally widening wires if desired by the designer). Also, since the MAX values are singularities, the AVG values as discussed above better depict the general trend. We found that, although D-Place has poorer MAX values in rare cases, the voltage at a majority of other locations on the grid are in fact better for D-Place than for single VDD. This fact is borne out by the fact that although the MAX values for D-Place can be poorer, the AVG values are in fact better (e.g., c7552 in Table V.A). Voltage drop contours shown in Fig. 8 (please note the different scales) show that the dual VDD grid is better off across most of the die. This is also evident in the gate count histogram in Fig. 9.

With respect to the AVG values, D-Place outperforms DSDG and D-Vanilla by up to 6.8% and 2.4% respectively. Looking at the MAX values, the results obtained using D-Place are generally comparable to the single VDD case. On the other hand, D-Vanilla and in particular DSDG frequently have poor performance as compared to single VDD. D-Place outperforms DSDG and D-Vanilla by up to 12.3% and 6.7% with respect to MAX values.

Finally, comparing the VDDL = 0.8V and VDDL = 0.6V cases, we see that the 0.6V case behaves better than 0.8V. This is a favorable finding as the 0.6V VDDL also has lower power (Table I).

We attribute the poor performance of D-Vanilla and DSDG to their failure in considering the important points listed in Section 5.3.

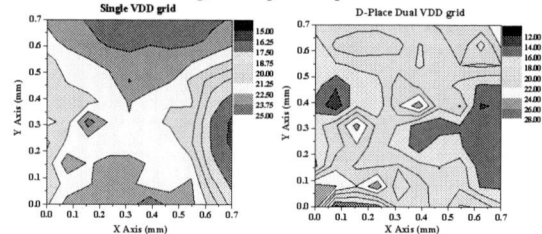

Fig. 8. Voltage drop (%) contours across die for c7552 [VDDL = 0.6V].

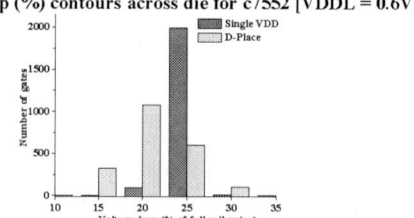

Fig. 9. Statistics of gate voltage drop for c7552 [VDDL = 0.6V].

Table VI. Additional power grid metrics.
(A) Voltage variation metric.

	Single VDD	DSDG		D-Vanilla		D-Place	
		0.6V	0.8V	0.6V	0.8V	0.6V	0.8V
c880	10.4%	24.5%	21.1%	11.2%	11.0%	13.8%	13.5%
c2670	14.9%	26.6%	25.2%	26.3%	22.4%	18.7%	19.7%
c5315	13.7%	28.2%	23.8%	28.4%	22.6%	21.9%	20.2%
c7552	10.8%	19.9%	16.3%	24.5%	23.9%	19.1%	18.3%

(B) Wire congestion metric.

	Single VDD	DSDG		D-Vanilla		D-Place	
		0.6V	0.8V	0.6V	0.8V	0.6V	0.8V
c880	0.17	0.17	0.17	0.19	0.20	0.17	0.16
c2670	0.17	0.17	0.17	0.19	0.20	0.16	0.16
c5315	0.17	0.17	0.17	0.19	0.20	0.18	0.16
c7552	0.17	0.17	0.17	0.19	0.20	0.15	0.15

In addition, referring to Fig. 5, we can observe that DSDG (in contrast to D-Place and D-Vanilla) results in longer current return paths thus leading to more severe LdI/dt effects.

5.4.2 Additional comparison metrics

Other metrics usually followed when studying power grid performance consider wire congestion and the variation of the voltage across the die. The variation metric is defined as the difference between the maximum and minimum voltage droop/bounce at a given time and is important when performing static timing analysis. This metric should ideally be small. The wire congestion metric amounts to comparing the fraction of routing tracks used by power grid. Again, this metric should ideally be small since the remaining signal wires will have more space for routing. We have ensured that technology imposed rules on minimum wire width and spacing are obeyed by all grids studied. Table VI.A and VI.B compare the various grids with respect to the voltage variation metric and the wire congestion metric for VDDL = 0.6V and 0.8V. From Table VI.A, among the dual VDD grids, D-Place grids have minimum variation. The voltage variation due to D-Place is somewhat inferior compared to single VDD. From Table VI.B, the wire congestion in the D-Place grids is seen to be superior. This is due to the fact that D-Place adaptively shrinks and widens the grid wires depending on the current demand. DSDG on the other hand is invariant to the value of the VDDL (and the design itself) and hence has uniformly worse congestion. Since D-Vanilla only shrinks down the VDDL rail, it has more congestion.

6. DECOUPLING CAPACITANCE BUDGET

Up to this point we have calculated the decoupling capacitance for the single VDD grid once and held it fixed across all the dual VDD grids. We now relax this constraint to examine how the reduced current demand in dual VDD circuits can be used to reduce decoupling capacitance. Recalling Eq. 2, due to the lower switching current for VDDL gates the required decoupling capacitance corresponding to VDDL gates is also lower. Care must be taken when dealing with the denominator of Eq. 2 however. $V_{noise-lim}$ is an absolute voltage value, and if a constraint of 10% of nominal is considered, the value of $V_{noise-lim}$ differs between VDDH and VDDL gates. We employed this technique and scaled the decoupling capacitances of c2670 accordingly. Table VII summarizes the results for this case including the MAX/AVG voltage droop/bounce of the resultant power grids. Numbers reported in brackets are the values from Table V.B (i.e., with the original decoupling capacitance).

From this table and Table V.B, we see that decoupling capacitance in the dual VDD grid can be reduced from 2.36nF to 1.93nF (18%) while resulting in only 0.6% and 1.6% increase in MAX and AVG

Table VII. Decoupling capacitance (Decap) reduction [VDDL = 0.8V].

		Scaled Decap Dual VDD
Decoupling Capacitance	Decap (VDDH)	1.02nF (1.06nF)
	Decap (VDDL)	0.91nF (1.30nF)
	Total Decap	1.93nF (2.36nF)
Grid integrity metrics	MAX	27.6% (27.0%)
	AVG	16.9% (15.3%)

voltage droop/bounce respectively. This reduction will improve leakage, area and yield (arising from oxide defects).

7. CONCLUSIONS

In conclusion, we presented the first detailed study of power delivery issues in dual VDD design. We first showed that dual VDD circuits lead to highly reduced current demands on the power supplies mitigating the power delivery problem. We began with board/package level issues and moved on to describe a placement driven method for designing on-die power grids. We demonstrated that the dual VDD power delivery scenario is no worse than for single VDD circuits. We also showed that dual VDD circuits can afford reduced decoupling capacitance budgets.

We have presented a practical approach for dual VDD grid design that provides superior results as compared to prior approaches. Future work could include the application of more rigorous single VDD power grid optimization approaches such as [28] to dual VDD designs, further enabling this powerful power optimization technique.

8. ACKNOWLEDGMENTS

The authors would like to thank Y. Kim and S. Pant for their valuable inputs and assistance with the PEEC model extractor.

9. REFERENCES

[1] R. Krishnamurthy, et al., "High-performance and low-power challenges for sub-70nm microprocessor circuits," Proc. CICC, pp. 125-128, 2002.
[2] K. Usami and M. Horowitz, "Clustered voltage scaling technique for low-power design," Proc. ISLPED, pp. 3-8, 1995.
[3] C. Chen, A. Srivastava, and M. Sarrafzadeh, "On gate level power optimization using dual-supply voltages," IEEE TVLSI, vol. 9, pp. 616-629, 2001.
[4] C. Yeh, et al., "Gate-level design exploiting dual supply voltages for power-driven applications," Proc. DAC, pp. 68-71, 1999.
[5] D. Nguyen, et al., "Minimization of dynamic & static power through joint assignment of threshold voltages and sizing optimization," Proc. ISLPED, pp. 158-163, 2003.
[6] S. H. Kulkarni, et al., "A new algorithm for improved VDD assignment in low power dual VDD systems," Proc. ISLPED, pp. 200-205, 2004.
[7] W. Hung, et al., "Total power optimization through simultaneously multiple-VDD multiple-VTH assignment and device sizing," Proc. ISLPED, pp. 144-149, 2004.
[8] K. Usami, et al., "Automated low-power technique exploiting multiple supply voltages applied to a media processor," IEEE JSSC, pp. 463-472, 1998.
[9] S. Mathew, et al., "A 4GHz 300mW 64b integer execution ALU with dual supply voltages in 90nm CMOS," Proc. ISSCC, pp. 162-519, 2004.
[10] K. Zhang, et al., "A 3GHz 79Mb SRAM in 65nm CMOS technology with integrated column based dynamic power supply," Proc. ISSCC, 2005.
[11] C. Yeh, et al, "Layout techniques supporting the use of dual supply voltages for cell-based designs," Proc. DAC, pp. 62-67, 1999.
[12] M. Hamada, Y. Ootaguro, and T. Kuroda, "Utilizing surplus timing for power reduction," Proc. CICC, pp. 89-92, 2001.
[13] M. Igarashi, et al., "A low-power design method using multiple supply voltages," Proc. ISLPED, pp. 36-41, 1997.
[14] J.-S. Wang, et al, "Design of standard cells used in low-power ASIC's exploiting the multiple-supply-voltage scheme," Proc. IEEE ASIC Conf., pp. 119-123, 1998.
[15] M. Popovich, et al, "On-chip power distribution grids with multiple supply voltages for high performance integrated circuits," Proc. GLVLSI, pp. 2-7, 2005.
[16] J. Fishburn and A. Dunlop, "TILOS: a posynomial programming approach to transistor sizing," Proc. ICCAD, pp. 326-328, 1985.
[17] M. Takahashi, et al., "A 60-mW MPEG4 video codec using clustered voltage scaling with variable supply-voltage scheme," IEEE JSSC, pp. 1772-1780, 1998.
[18] T. Kuroda and M. Hamada, "Low-power CMOS digital design with dual embedded adaptive power supplies," IEEE JSSC, pp. 652-655, 2000.
[19] F. Brglez and H. Fujiwara, "A neural netlist of 10 combinational benchmark circuits and a target translator in Fortran," Proc. ISCAS, pp. 695-698, 1985.
[20] Intel, "Intel Pentium 4 processor in the 423 pin/Intel 850 Chipset Platform," 2002.
[21] A. E. Ruehli, "Inductance calculations in a complex integrated circuit environment," IBM J. R&D, pp. 470-481, 1972.
[22] S. C. Wong, G. Y. Lee, and D. J. Ma, "Modeling of interconnect capacitance, delay and crosstalk in VLSI," IEEE Trans. Semiconductor Manufacturing, pp. 108-111, 2000.
[23] C. Hoer and C. Love, "Exact inductance equations for rectangular conductors with applications to more complicated geometries," J. Res. Nat. Bureau Standards, 1965.
[24] T.-H. Chen, et al, "INDUCTWISE: inductance-wise interconnect simulator and extractor," Proc. ICCAD, pp. 215-220, 2002.
[25] S. Zhao, K. Roy, and C-K Koh, "Decoupling capacitance allocation and its application to power-supply noise-aware floorplanning," IEEE TCAD, pp. 81-92, 2002.
[26] H. Su, S. S. Sapatnekar, and S. R. Nassif, "Optimal decoupling capacitor sizing and placement for standard-cell layout designs," IEEE TCAD, pp. 428-436, 2003.
[27] E. Chiprout, "Fast flip-chip power grid analysis via locality and grid shells," Proc. ICCAD, pp. 485-488, 2004.
[28] J. Singh and S. Sapatnekar, "Congestion-aware topology optimization of structured power/ground networks," IEEE TCAD, pp. 683-695, 2005.

Calculating Frequency-Dependent Inductance of VLSI Interconnect by Complete Multiple Reciprocity Boundary Element Method*

Changhao Yan

Wenjian Yu

Zeyi Wang

Dept. of Comp. Sci. & Tech.
Tsinghua University
Beijing, 100084
Tel: 8610-6279-5428
Fax: 8610-6278-1489
yanch02@mails.tsinghua.edu.cn

Dept. of Comp. Sci. & Tech.
Tsinghua University
Beijing, 100084
Tel: 8610-6279-5428
Fax: 8610-6278-1489
yu-wj@tsinghua.edu.cn

Dept. of Comp. Sci. & Tech.
Tsinghua University
Beijing, 100084
Tel: 8610-6279-5428
Fax: 8610-6278-1489
wangzy@mail.tsinghua.edu.cn

Abstract— A complete multiple reciprocity method (CMRM), usually for the eigenvalue analysis of Helmholtz equation, is introduced to the BEM for frequency-dependent inductance extraction. Several approaches are proposed to resolve the problem of "ill-conditioned" series encountered when applying the CMRM practically. Using the BEM combined with CMRM, the major operations of calculating the numerical integrals for a frequency point become reusable, so that inductance extraction for a frequency range is greatly accelerated. Numerical results verify the accuracy and efficiency of the proposed method.

I. INTRODUCTION

The parasitic inductance of VLSI interconnect has become very important as the operating frequency of circuit keeps increasing [1]. Since the inductance is frequency-dependent, inductance extraction for a range of frequency points is usually needed for accurate simulation and verification. Besides, with the inductances and resistances for a frequency range, the quality factor of an inductor component in RF circuit can be calculated [2]. Therefore, inductance extraction for a frequency range, not only for a single frequency point, becomes another important research topic.

There are two categories of methods for calculating frequency-dependent inductance. The first category employs the interpolation method [3, 4]. For example, paper [4] uses a coupled circuit method to calculate the low-frequency inductance and the inversion of capacitance matrix for high-frequency inductance, and then forms the full-frequency inductance curve with an interpolation formula. Although this category of methods has high computational speed, their reliability and accuracy is very limited because only several interpolation points for very low frequency and very high frequency are computed accurately.

The other category is straightforward. To obtain the inductance for a range of frequency, a method to extract the inductance must run repeatedly for all sampling points throughout the frequency range. The methods suitable for inductance extraction at arbitrary frequency include FastHenry, a volume integral method [5], FastImp, a boundary element method (BEM) [6], etc. These methods have high accuracy for inductance extraction at any frequency point, but computational speed is relatively slow. The BEM for inductance extraction proposed recently employs much fewer unknowns and is suitable for a wideband simulation [6]. This method has been implemented as a software prototype named FastImp, available via the Internet [8]. Compared with FastHenry, the FastImp is faster. However, using the FastImp, the time of inductance extraction for a frequency range is still very long, which approximates to the number of frequency points times the computational time for one frequency point. Speeding up the inductance extraction based on BEM for a range of frequency is the main target of this paper.

In the inductance extraction based on BEM, the time is mainly spent on calculating all kinds of numerical integrations and solving the linear equation system $Ax = b$. The CPU time of solving $Ax = b$ can be reduced greatly, if an efficient iterative method, such as GMRES, is applied. So, the time of numerical integrations occupies the major part of the whole. For example, computing the example *wire.inp* in one frequency point, FastImp needs $4.62s$ in total, where solving $Ax = b$ costs only $0.58s$ and the frequency-dependent numerical integration (with pFFT acceleration) costs $3.95s$ [8]. These numerical integrals can be divided into two types. One type is frequency-independent, where the integral kernel is $1/r$ or its derivative, coming from the fundamental solution of the Laplace equation in dielectrics. Their results can be reused among all sampling frequencies. The other type is frequency-dependent, where the integral kernel is e^{-jkr}/r or its derivative, coming from the fundamental solution of the Helmholtz equation in conductors, where k is a complex number and means the wave number in physics. The latter type must be recomputed in every

---

*This work is supported by the China National Science Foundations under Grant 90407004 and 60401010. And, it is partly supported by National High Technology Research and Development Program of China (No. 2004AA1Z1050).

0-7803-9451-8/06/$20.00 ©2006 IEEE.

frequency point. Furthermore, the frequency-dependent integrations are much more expensive than the frequency-independent integrations in CPU time, since computing the e^{-jkr} is very arduous. For example, on a PC with an Intel Pentium4 1.8G CPU, the time ratio of integrating e^{-jkr}/r to $1/r$ is about 7:1. So how to shorten the computing time of the second type of integration becomes a key problem for inductance extraction based on BEM, especially for many frequency points.

On the other hand, a multiple reciprocity method (MRM) was proposed in [7] for boundary value problems of the Helmholtz equation. Then, it is applied to the eigenvalue analysis of Helmholtz equation successfully [9], which reduces the computational expense greatly. In MRM, the integral $\int e^{-jkr}/r$ can be transformed into a series $\sum_{j=0}^{N} g_j(k) \int f_j(r)$, where $\int f_j(r)$ is a frequency-independent series and $g_j(k)$ is a frequency-dependent series. Based on this transformation, one can compute and save the numerical integrals of $\int f_j(r)$ at first, and then for a different sampling frequency, i.e., a different k, $\int f_j(r)$ need not to be computed again. Instead, only the coefficients of $g_j(k)$ are calculated for each frequency point, and the integral $\int f_j(r)$ need to be computed only once. Therefore, the computational time of the frequency-dependent integral $\int e^{-jkr}/r$ can be greatly reduced for the other frequency points except the first one.

In this paper, the MRM is applied to the interior boundary value problem (BVP) of Helmholtz equation in 3-D inductance extraction. To avoid the spurious eigenvalue problem caused by the MRM formulation [10], a complete MRM (CMRM) formulation proposed by [11] is actually adopted in our method. Although it was proved that the series in MRM or CMRM are convergent, a problem of "ill-conditioned" series is still encountered when we apply the MRM formulation to 3-D inductance extraction. Because the k (frequency) and r (distance) are both near to 1 in the eigenvalue analysis with MRM [10, 12, 13], this problem did not emerge in literatures. But in the inductance extraction of VLSI interconnect, the r usually varies from $1mm$ to $90nm$ and the frequency f varies from 1 Hz to 10 GHz. Therefore, several approaches are proposed in this paper to handle the problem of "ill-conditioned" series. Several interconnect structures are calculated with the BEM combined with CMRM. Numerical results show that our method has high accuracy as that of FastImp, and the CMRM exhibits large speedup for inductance extraction with many frequency points.

II. MULTIPLE RECIPROCITY BOUNDARY ELEMENT METHOD

In this section, we will briefly introduce the MRM [9] and CMRM for the BEM used for inductance extraction.

A. Conventional MRM Formulation

In a 3-D closed domain Ω surrounded by the boundary Γ, potential u satisfies the Helmholtz equation:
$$\nabla^2 u + k^2 u = 0, \qquad (1)$$

where k is the wave number. In the inductance extraction problem with MQS (magnetoquasistatics) assumption, $k = \sqrt{-j\omega\mu\sigma}$, where j, ω, μ and σ are the imaginary unit, angular frequency, permeability and conductivity respectively. μ and σ are constant.

From (1), it is easy to obtain the boundary integral equation (BIE) as follows:
$$cu + \int_{\Gamma} (uq_1^* - qu_1^*)d\Gamma = 0, \qquad (2)$$

where $u_1^* = \frac{e^{-jkr}}{4\pi r}$, $q_1^* = \frac{\partial u_1^*}{\partial n}$, c is constant depending on the position of the source point and where the equation is considered, r is the distance between the source and filed point and $\overrightarrow{n}$ is the outward normal on the boundary. The Eq. (2) is called complex-value formulation for the Helmholtz equation in this paper. The FastImp belongs to a direct BEM because it utilizes this formula [15].

Moving the second term on the left hand side (l.h.s.) of (1) to the right, one gets a Poisson equation as:
$$\nabla^2 u = -k^2 u. \qquad (3)$$
Similarly, a BIE can be obtained:
$$cu + \int_{\Gamma} (uq_0^* - qu_0^*)d\Gamma = k^2 \int_{\Omega} uu_0^* d\Omega, \qquad (4)$$
where $u_0^* = \frac{1}{4\pi r}$, $q_0^* = \frac{\partial u_0^*}{\partial n}$.

Although in (4), the Laplace fundamental solution u_0^*, q_0^* is frequency-independent, for the volume integral term on the r.h.s. of (4), directly applying it leads to volume discretization needed, while the absence of the volume discretization is thought as the key advantage of the BEM compared with other volume integral methods.

The MRM is known as a powerful conversion scheme from the domain integral to the boundary [7, 9]. Adopting the higher order fundamental solution of the Laplace equation:
$$\nabla^2 u_{j+1}^* = u_j^* = \frac{1}{4\pi r}\frac{r^{2j}}{(2j)!}, \quad q_j^* = \frac{\partial u_j^*}{\partial n}, \quad j = 0, 1, 2, ...,$$
the domain integral of (4) can be transformed into:
$$\int_{\Omega} u\nabla^2 u_1^* d\Omega = \int_{\Gamma} (uq_1^* - qu_1^*)d\Gamma + \int_{\Omega} u_1^* \Delta u d\Omega$$
$$= \int_{\Gamma} (uq_1^* - qu_1^*)d\Gamma - k^2 \int_{\Omega} u_1^* u d\Omega, \qquad (5)$$
where the first transformation uses the second Green theorem and the second transformation uses (3). Substituting (5) into the r.h.s. of (4), and arranging it, one obtains:
$$cu + \int_{\Gamma} (uq_0^* - qu_0^*)d\Gamma +$$
$$-k^2 \int_{\Gamma} (uq_1^* - qu_1^*)d\Gamma = (-k^2)^2 \int_{\Omega} u_1^* u d\Omega.$$
Repeating similar computation leads to
$$cu + \sum_{j=0}^{N} (-k^2)^j \left[\int_{\Gamma} (uq_j^* - qu_j^*)d\Gamma \right]$$
$$= (-1)^N (k^2)^{N+1} \int_{\Omega} u_N^* u d\Omega \approx 0. \qquad (6)$$
For sufficient large N, if r and k are bounded, the domain integral on the r.h.s. of (6) becomes negligible [7]. Eq. (6) is the basic formula in the MRM.

Note that (6) does not satisfy the Sommerfeld radiation condition at infinity and cannot be employed for an unbounded domain such as: exterior BVP, But this is not a problem in inductance extraction for that all the domains in consideration are within bounded conductors.

B. Complete MRM Formulation

The MRM formulation mentioned above is called conventional MRM formulation, where both u_j^*, q_j^* are real. If the conventional MRM formulation is used for eigenvalue problem, *spurious* eigenvalue will occur [10]. Ref. [11] proposed a complete MRM formulation as follows:

$$u_j^* = \frac{1}{4\pi}\left[\frac{1}{r}\frac{r^{2j}}{(2j)!} - ik\frac{r^{2j}}{(2j+1)!}\right], \quad q_j^* = \frac{\partial u_j^*}{\partial n}, \quad j = 0, 1, 2, \ldots,$$

$$u^* = \sum_{j=0}^{N}(-k^2)^j \int_\Gamma u_j^* d\Gamma, \quad q^* = \sum_{j=0}^{N}(-k^2)^j \int_\Gamma q_j^* d\Gamma,$$

(7)

by which this *spurious* eigenvalue problem can be solved perfectly. Moreover, they pointed out that the final series form of the kernel in (complete) MRM simply converges to corresponding kernels in the complex-value formulation [11].

To avoid the possible *spurious* eigenvalue problem caused by the MRM formulation, the CMRM formulation is adopted in our method for inductance extraction. When applying the CMRM to the direct BEM in [15], we only need to replace the u_1^* and q_1^* in (2) with the u^* and q^* in (7), respectively. It should also be pointed out that the CMRM formulation actually expends more computational resources than the conventional MRM, but its good numerical stability makes us choose it.

III. Difficulties and Solutions of Inductance Extraction using CMRM

The numerical difficulties produced by directly applying CMRM formulation will be discussed in the first subsection. Then a set of methods to overcome these difficulties are proposed in next three subsections. In the last subsection a recursion formulation for accelerating computation is proposed.

A. Ill-conditioned Series

In numerical methods textbooks, the series like:

$$e^{-x} = \sum_{j=0}^{n}\frac{(-x)^n}{n!}$$

(8)

is called *ill-conditioned* series [14]. In such a kind of series, *large terms* which are many orders of magnitude larger than the sum of the series cancel out with each other. So very small errors in these *large terms* produce large errors in the final result.

Reexamining the CMRM formulation, one will find the series u_j^*, q_j^* and $(-k^2)^j$ are all ill-conditioned. For a slightly large k or r, numerical computations are impossible. Eigenvalue analysis doesn't encounter this difficulty because it guarantees both the k and r are near to 1 [10]. But in VLSI circuits, this cannot be satisfied. The numerical difficulties are in three aspects. First, The very small absolute value of r. In VLSI circuits, the unit of distances is micron or nanometer. For avoiding the fussy unit conversion, the International System of Units is usually adopted. Then r will be a very small number. Substituting it into ill-conditioned series directly will cause the *underflow*. Second, The wide variation range of the r. In VLSI interconnects, the ratio of the dimension in the length to the one in the section is up to several decades. This problem cannot be solved by simple unit conversion.

Finally, The wide variation range of the k. The frequencies from $1hz$ to $10Ghz$ are taken into consideration. If conductors are coppers, the k varies from 0 to 10^6.

If x is a large number, computing the series (8) directly is not feasible, but its analysis result is known and approaches zero. In physics, the series is called *exponential decay*. So it is possible to compute it by other approximate method or even to ignore it. If x nears to 1, however, the series converges quickly. This feature is valuable.

B. Normalization of the Distance r

Without loss of generality, let's consider the real part of the u^*. Introducing a average distance r_{avg}, one obtains:

$$\begin{aligned} u_{real}^* &= \frac{1}{4\pi}\sum_{j=0}^{N}(-k^2)^j \int \frac{1}{r}\frac{r^{2j}}{(2j)!}d\Gamma \\ &= \frac{1}{4\pi}\sum_{j=0}^{N}(-1)^j(kr_{avg})^{2j}\int\frac{1}{r}\frac{r^{2j}}{(2j)!}d\Gamma, \end{aligned}$$

(9)

where: $r_{rel} = \frac{r}{r_{avg}}$.

From the discussion above, the series $\sum_{j=0}^{N}\int\frac{1}{r}\frac{r_{rel}^{2j}}{(2j)!}d\Gamma$ can converge very quickly if choosing an appropriate r_{avg} to keeping r_{rel} near to 1. This transformation has two effects. The one is solving the first problem, i.e., the very small absolute distance. The other is partially solving the second and third problem, combining the variation of the k or r respectively into the variation of the kr_{avg} together.

Note that a different r_{avg} is needed between any two panels. That means r_{avg} is a matrix indeed, but it is frequency-independent. In the program of this paper, the formula of computing r_{avg} is the average distance among the distances between any corner of the integral panel and the source point.

C. Localization of the Near Field

If kr_{avg} is a large number, a *window* with size W is given. When $kr_{avg} < W$, u_{real}^* can be computed by series formula (9) directly. Then two questions should be answered, that is how large of the window is big enough and how many terms of the series are needed for a given windows size W.

Let us consider the meaning of kr_{avg} in physics. Under MQS assumption, one obtains $k = (1 - j)k_{real}$. For simplicity, only considering $k_{real}r_{avg}$, based on the fact that:

$$k_{real} = \sqrt{\frac{2}{\omega\mu\sigma}} = \frac{1}{\delta},$$

where δ is the skin depth in the angle frequency ω, we get:

$$k_{real}r_{avg} = \frac{r_{avg}}{\delta}.$$

(10)

In physics, the Eq. (10) indicates that the $k_{real}r_{avg}$ is the ratio of the average distance between the panel and the source point to the skin depth in current frequency. It is nondimensional. In the electromagnetic wave theory, the k_{real} is termed *attenuation constant*. If the filed point is five times skin depth away from the source, the amplitude attenuation is under one percent [2](pp.253). So for the engineering application, letting $W = 5 \sim 6$ is enough.

Next, considering that the real part of u^* simply converges to $cos(kr)$, from the convergence curve of $cos\theta$, one obtains the relationship between the length N of the series and the windows size W in Table I.

From the Table I, if $W = 2\pi$ (a wave length), only letting $N = 11$, four correct decimals can be obtained. If W is smaller, for the same accuracy, the N is smaller too.

D. Approximate Calculation of the Far Field

If $kr_{avg} > W$, from the discussion above, the effect of the source is feeble, so it is possible to sum this effect in a coarse way.

Considering the fact that the complete MRM formulation simple converges to the kernel in the complex-value formulation, one can directly compute u^* by the integral of the kernel of the Helmholtz equation as follow:

$$u^* = \frac{1}{4\pi} \int_\Gamma \frac{e^{-jkr}}{r} d\Gamma \approx \frac{e^{-jkr_{avg}}}{4\pi} \int_\Gamma \frac{1}{r} d\Gamma = e^{-jkr_{avg}} \cdot u_0.$$
(11)

The reason of the $\approx$ in (11) is that now that the source point is far away from the integral panel, it is meaningless to exactly sum the integration on every r which is the distance between a guass point in the integral panel and the source point, so a *average* distance between the integral panel and the source point is enough.

Notice that this approximate formulation also includes the arduous $e^{-jkr_{avg}}$, but it is computed only on a distance r_{avg}, not like the direct integral formula where computation of e^{-jkr} is needed on every gauss point. More convenient, the u_0 is right the first term of the series u_j^*, so the additional computation is unnecessary.

E. Fast Computation of the Series

Direct computing the power series as the formulation proposed above will waste lots of CPU time. Fortunately, it is easy to transform the direct form of the series into recursion formula from which amounts of time can be saved. Taking the real part of u_j^* as example, one obtain the recursion form as follow:

$$u_{j+1,real}^* = \frac{1}{r} \frac{r_{rel}^{2j}}{(2j)!} = \frac{r_{rel}^2}{(2j+2)(2j+1)} u_{j,real}^*,$$

where $u_{0,real}^* = \frac{1}{r}$.

Moreover, in order to accelerate the computation, the coefficient $\frac{1}{(2j+2)(2j+1)}$ can be computed and saved in a table before. By now, on a gauss point, one term of the series added needs only two extra multiply operators.

Note that when the source point locates in the integral panel, $u_{0,real}^*$ is weak singular integral and q_0^* is strong singular integral. Both of them need special treatments.

IV. NUMERICAL RESULTS

The proposed method based on CMRM is implemented in C++ language, and the original BEM for inductance extraction is as same as that proposed in [15]. In the following experiments, our method is compared with the original direct BEM. The former is denoted by CMRM

and the latter by ODBEM. In the FastImp, which is also based on the direct BEM of [15], a precorrected-FFT (pFFT) acceleration algorithm is employed [6]. However, pFFT also handles the boundary integrals, and conflicts with the CMRM. Therefore, in the ODBEM, only a direct solver of linear equation system is implemented, which does not affect the performance evaluation of the CMRM for extraction with many frequency points. Our CMRM and ODBEM are used to extraction frequency-dependent inductance for three interconnect structures. In the computations, conductor surface is discretized into constant rectangular elements, with collocation point located in the center. To guarantee the accuracy of near singular integrals, a 2D Guass-Legendre integral scheme with 20×20 integral points is used.

A. A Single Rectangle Wire

A straight conductor wire is $8\mu m$ long, $1\mu m$ wide and $1\mu m$ thick. The surface of this wire is discretized into $160(4 \times 4 \times 8)$ panels, as shown in Fig. 1. The number of unknowns is 1122.

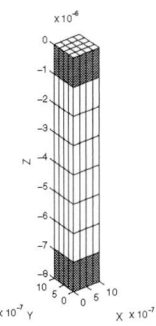

Fig. 1. A Single Rectangle Wire

The red panels surrounding the *contacts* are for summing the current in the conductor with the high frequency method and the $F_h = 15Ghz$ [15]. Calculating the inductance from 1.5Mhz to 40Ghz with the ODBEM, one obtains the inductance curve in Fig. 2. Fig. 3 is the result from [15](pp.110), where the frequency range is from 1Hz to 10^{20}hz. Note that the highest frequency compared is only up to 40Ghz.

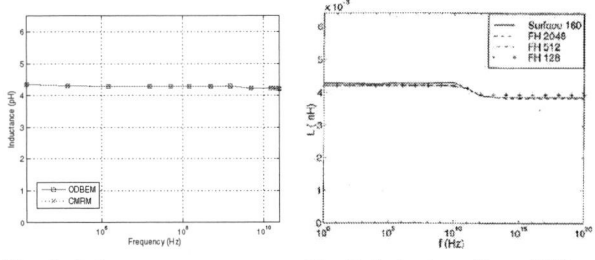

Fig. 2. Inductance   Fig. 3. Inductance(from [15])

In CMRM, the window size $W = 2\pi$ and the length of the series $N = 12$. From Fig. 2 both results from the CMRM and the ODBEM coincide with each other very well on the whole frequency range, and their results coincide with the data from [15], i.e. Fig. 3.

TABLE I
Relation between W and N

N	1	2	3	4	5	6	7	8	9	10	11	12	exact
W=$\pi/2$	1.0000	-0.2337	0.0200	-0.0009	0.0000								0.0000
W=π	1.0000	-3.9348	0.1239	-1.2114	-0.9760	-1.0018	-0.9999	-1.0000					-1.0000
W=$3\pi/2$	1.0000	-10.1033	10.4439	-4.7656	1.2657	-0.2224	0.0279	-0.0026	0.0002	-0.0000			-0.0000
W=2π	1.0000	-18.7392	46.2002	-39.2566	20.9880	-5.4382	2.4653	0.7509	1.0329	0.9965	1.0003	0.99998	1.0000

Note that when $f = 15Ghz$, the skin depth $\delta = 0.54\mu m$ and $r_{avg} < 3.3\mu m$. That means the approximate formulation has been employed when the distance between two panels is larger than $3.3\mu m$. At this frequency, the results from the ODBEM and the CMRM still have the accuracy of three correct decimals.

In order to compare the speed between the ODBEM and the CMRM clearly, we briefly summarize the whole steps in the program listed as follows.

1. calculate the integral of the Laplace kernel related
2. calculate the integral of the Helmholtz kernel related (repeated in ODBEM, partly repeated in CMRM)
3. generate the linear system (repeated)
4. solve the linear system (repeated)
5. postporcess. Obtain impedance matrix (repeated)

The main differences lie in step 2. The ODBEM directly compute numerical integrals. The CMRM divides this step into three steps:

1. compute r_{avg} matrix
2. generate and save the series
3. sum the series for a given k (repeated)

The time result is listed in Table II. The unit is second. This program is run on a desktop PC with a Pentium4 1.8G CPU and 256M RAM. The symbol asterisk indicates that the step should be recomputed at different frequency points.

TABLE II
CPU Time of Single Wire Example

	Lap. Int.	Helm Int			Gen. Ax=b	Sol. Ax=b	Post proc.
ODBEM		11.141*			0.125*	2.766*	0.000*
CMRM	1.766	R_{avg}	Serial	Combine			
		0.000	9.234	0.156*			

From the Table II, we can conclude:

1. The time ratio of the Helmholtz integral to the Laplace integral is about 7:1(11.141:1.766). For only one conductor in the example, this ratio is also the CPU time ratio of computing the kernel e^{-jkr}/r to $1/r$. Obviously, the Helmholtz kernel is much more expensive than the Laplace kernel on the CPU time.
2. From the second frequency point, the time of the integral is near to zero (0.156s) by the ODBEM, and the rest is only the time of solving $Ax = b$.
3. Even for the first frequency point, CMRM (9.390s) is still a little faster than ODBEM (11.141s).

The same conclusion can be obtained from the Fig. 4, where the horizontal coordinate axis stands for the number of the sampling frequency points and the vertical coordinate axis stands for the total time. The inductances on twelve sampling frequency points are computed. The advantage of saving time by CMRM is very apparent.

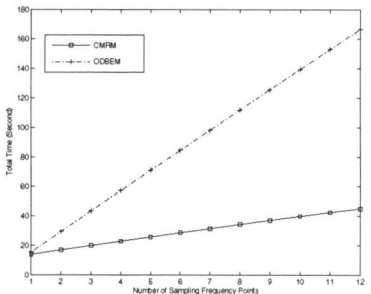

Fig. 4. CPU Time Comparison

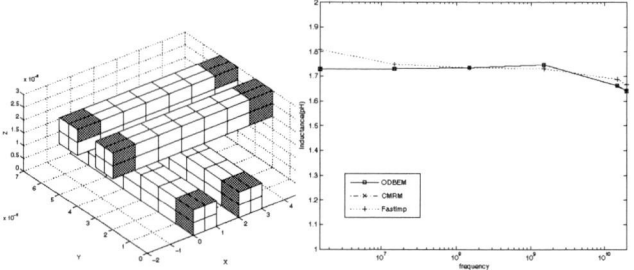

Fig. 5. A 2x2 Bus

Fig. 6. Mutual Inductance of a Parallel Wire in 2x2 Bus

B. A 2x2 Bus

The dimension and discretization of the bus2x2 example is shown in Fig. 5.

The numbers of panels and unknowns are 256 and 1800 respectively. Because there are four conductors, the result of the inductance should be a 4×4 matrix.

Table III gives this results at $f = 15Ghz$. The unit is pH. For comparison, the result of the Fastimp is introduced. Fig. 6 shows the mutual inductance of parallel conductors i.e. the location $(1,2)$ in inductance matrix from $1.5Mhz$ to $40Ghz$. All of them coincide with each other very well. The result of CPU time Comparison is not listed here because it is similar to the first example.

C. A 1x1 Bus

This example describes the relation between accuracy and window size W. The dimension and discretization of the bus1x1 example is shown in Fig. 7

The number of panels and unknowns are 320 and 2240 respectively. At $f = 15Ghz$, applying different window size W and length of series N, the result of the self-inductance in a conductor are listed in Table IV.

Take the result from the ODBEM as the exact result. From the Table IV, one concludes that at high frequency, even if $N = 5$, the resistance has a relative error about 10% and the inductance has a relative error about 3%. It

TABLE III
INDUCTANCE MATRIX OF A 2X2 BUS WHEN $f = 15Ghz$

CMRM				ODBEM				FastImp			
3.41942	1.66187	7.16e-4	7.32e-4	3.41989	1.66161	7.16e-4	7.32e-4	3.46971	1.68804	7.81e-4	7.10e-4
1.66187	3.41942	7.16e-4	7.32e-4	1.66161	3.41989	7.16e-4	7.32e-4	1.68802	3.4697	7.73e-4	7.08e-4
7.32e-4	7.16e-4	3.41942	1.66187	7.32e-4	7.16e-4	3.41989	1.66161	8.02e-4	6.59e-4	3.47036	1.68878
7.32e-4	7.16e-4	1.66187	3.41942	7.32e-4	7.16e-4	1.66161	3.41989	8.24e-4	6.76e-4	1.68877	3.47036

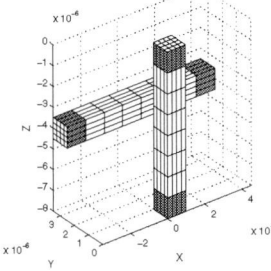

Fig. 7. A 1x1 Bus

TABLE IV
RELATION BETWEEN ACCURACY
AND (W, N)

(N, W)	$(R(\Omega), L(pH))$
$(12, 2\pi)$	$(0.143694, 4.21306)$
$(9, 2\pi)$	$(0.147152, 4.18193)$
$(7, \pi)$	$(0.157814, 4.31280)$
$(6, \pi)$	$(0.157947, 4.31319)$
$(5, \pi)$	$(0.158561, 4.30941)$
ODBEM	$(0.143738, 4.21392)$

is meaningful for saving double size of spaces with acceptable loss of the accuracy.

V. CONCLUSIONS

In the inductance extraction with BEM, much of the CPU time is spent on numerical integrals. A CMRM, usually for eigenvalue analysis of Helmholtz equation, is introduced to the BEM based inductance extraction. To solve the problem of ill-conditioned series caused by large variation of parameter k and r in inductance extraction, several approaches are proposed for CMRM, including the normalization of the distance r, the localization of the near field and the approximate computation for the far field. With the CMRM and proposed efficient techniques, the computation of frequency-dependent integrals in BEM becomes reusable. Numerical experiments show that the proposed method greatly speeds up the inductance extraction for a frequency range, while preserving high accuracy.

ACKNOWLEDGEMENTS

The authors would like to thank Prof. J. T. Chen, Dr. W. Yeih of Taiwan Ocean University for useful discussion.

REFERENCES

[1] A. B. Kahng, S. Muddu, "Analytical delay model for RLC interconnects," *IEEE Trans. on CAD*, vol. 16, pp. 1507-1514, 1997.

[2] Bhag Singh Guru, Huseyin R. Hiziroglu, *Electromagnetic Field Theory Fundamentals*, pp. 365–367. PWS Publishing Company, 1998.

[3] M. J. Tsuk, J. A. Kong, "A hybid method for the calculation of the resistance and inductance of transmission lines with arbitrary cross sections," *IEEE Trans. on MTT.*, vol. 39, pp. 1338–1347, 1991.

[4] S. Fang, X. Tang, Z. Wang and X. Hong, "A simplified hybrid method for calculating the frequency-dependent inductances of transimission lines with rectangular cross section," *Proc. of ASP-DAC*, pp. 453–456, 2000, Yokohama, Japan.

[5] M. Kamon, M. J. Tsuk, J. White, "Fasthenry: a multipole-accelerated 3-d inductance extraction program," *IEEE Trans, on MTT*, vol. 42, no. 9, pp. 1750–1758, 1994.

[6] Z. Zhu, B. Song, J. White, "Algorithms in fastimp: a fast and wideband impedance extraction program for complicated 3-d geometries," *DAC*, pp. 712–717, 2003.

[7] A. J. Nowak and C. A. Brebbia, "Solving helmholtz equation by boundary elements using the multiple reciprocity method," *Computers and Experiments in Fluid Flow*, pp. 265–270, 1989.

[8] Z. Zhu, B. Song, J. White, *FastImp user's guid*, http://relweb.mit.edu/vlsi/codes.htm.

[9] N. Kamiya and E. Andoh, "Helmholtz eigenvalue analysis by boundary element mehtod," *Journal of Sound and Vibration*, vol. 160, no. 2, pp. 279–287, 1993.

[10] N. Kamiya, E. Andoh and K. Nogae, "Three-dimensional eigenvalue analysis of the helmholtz equation by multiple reciprocity boundary element method," *Advances in Engineering Software*, vol. 16, pp. 203–207, 1993.

[11] W. Yeih, J. T. Chen, K. H. Chen and F. C. Wong, "A study on the multiple reciprocity method and complex-valued formulation for the helmholtz equation," *Advances in Engineering Software*, vol. 29, no. 1, pp. 1–6, 1998.

[12] N. Kamiya, E. Andoh, K. Nogae, "A new complex-valued formulation and eigenvalue analysis of the helmholtz equation by boundary element method," *Advances in Engineering Software*, vol. 26, pp. 219–227, 1996.

[13] J.T. Chen and F.C. Wong, "Analytical derivations for one-dimensional eigenproblems using dual boundary element method and multiple reciprocity method," *Engineering Analysis with Boundary Elements*, vol. 20, pp. 25–33, 1997.

[14] Germund Dahlquist, Ake Bjorck, Translated by Ned Anderson, *Numerical Methods*, pp. 75–78. Englewood Cliffs, N.J.: Prentice-Hall, Inc., 1974.

[15] J. F. Wang, *A new surface integral formulation of EMQS impedance extraction for 3-D structures*. PhD thesis, MIT, 1999.

Controlling Inductive Cross-Talk and Power in Off-chip Buses using CODECs

Brock J. LaMeres

Design Validation Division
Agilent Technologies Inc.
Colorado Springs, CO 80907
brock_lameres@agilent.com

Kanupriya Gulati

Electrical Engineering Department
Texas A&M University
College Station, TX 77843
kanu.gulati@gmail.com

Sunil P. Khatri

Electrical Engineering Department
Texas A&M University
College Station, TX 77843
sunil@ee.tamu.edu

Abstract— The parasitic inductances within IC packaging cause supply bounce as well as glitches on the signal pins, significantly limiting the frequency of high-speed inter-chip communication. Also, off-chip communication contributes a large fraction of the total system power. Until recently, the parasitic inductance problem was addressed by aggressive package design, which is expensive. In this work we present a technique to encode the off-chip data transmission to i) limit bounce on the supplies ii) reduce glitching caused by inductive signal coupling from neighboring signals iii) limit the edge degradation of signals due to mutually inducted voltages from neighboring switching signals and iv) control the total power consumption of the I/O logic. All these factors are modeled in a unified mathematical framework. Our experimental results show that the proposed encoding based techniques result in reduced supply bounce and signal glitching due to inductive cross-talk, closely matching the theoretical predictions. Also, we show that the bus size overhead is reasonable even after stringent power reduction constraints are imposed. We demonstrate that the overall bandwidth of a bus actually increases by 100% over an unencoded bus, using our technique with inductive constraints *only* (even after accounting for the encoding overhead). When the power constraints were added (to limit the power to 20% of worst case switching power) in addition to the inductive constraints, the bandwidth was again 100% improved over the unencoded bus. The asymptotic bus size overhead depends on how stringent the user-specified power and inductive cross-talk parameters are. We have validated our approach by *simulating it in an ASIC setting as well as prototyping and testing it in an FPGA environment.*

1 Introduction

Advances in VLSI fabrication technologies have led to a dramatic increase in the on-chip performance of integrated circuits. The increase in IC performance is predicted by the International Technology Roadmap for Semiconductors (ITRS) [1] to continue doubling every 18 months, following Moore's Law, for at least the next several years [2]. However, package performance is predicted by the ITRS to only double over the next decade. This imbalance in performance expectations between the IC and the package is a major concern for system designers. The main limitation of the package performance is the parasitic inductance present in the level 1 (from IC die to package) and level 2 (from package to board) interconnects [3, 4, 5]. The inductance factors that affect signal speed and integrity are as follows:

- Supply bounce. Typically supply (V_{SS} and V_{DD}) pins are interspersed at regular intervals between signal pins. Every n^{th} pin is a V_{SS} or V_{DD}. The supply bounce is proportional to the number of pins switching low or high. Ground bounce is expressed as:

$$V_{bnc} = L \sum_i (\frac{di}{dt}) \qquad (1)$$

Where L is the self-inductance of the V_{SS} pin, and $\sum_i(\frac{di}{dt})$ is evaluated over the number of signal pins switching low. Since the placement of power and signal pins is regular, we can compute this quantity as half the number of signal pins switching low to the immediate right of the V_{SS} pin plus half the number of signal pins switching low to the immediate left of the V_{SS} pin. Since each signal always has a V_{SS} pin to the

left and to the right, we assume that if it switches high, then half the switching current is supplied by the V_{SS} pin to its left, and the other half by the V_{SS} pin to its right.

In a similar manner, a supply voltage droop is encountered on V_{DD} pins as well.

- Glitching. If a signal pin j is static, then a glitch may be induced in its voltage due to neighboring pins which switch. This is governed by the expression

$$V_{glitch}^j = \sum_k \pm (M_{jk} \frac{di_k}{dt}) \qquad (2)$$

where i_k is the current in the kth pin, and M_{jk} is the mutual inductance between the jth pin being considered and the kth pin. The sign of the coupled voltage is positive or negative depending on whether the kth neighboring pin undergoes a rising or falling transition.

- Switching speed. When a signal is switching, its transition can be sped up if the coupled voltage induced by its neighbors' mutual inductance aids the transition. We would like that a signal is not slowed down (i.e. either sped up, or unhindered) in its transitions due to this effect. We desire that when a signal j is rising (falling), the coupled voltage on this signal (Equation 2) due to its neighbors' transitions is zero or positive (negative). In this way, the transitions of signals are not slowed down due to inductive cross-talk.

The traditional approach to reducing the parasitic inductance within the package has been through aggressive package design. We are currently seeing success in the application of chip-scale and flip-chip (solder bump) technologies in level 1 interconnect for high-end applications. While such technologies decrease the above mentioned inductive effects, they are still prohibitively expensive for the majority of ICs. Further, they do not completely eliminate the inductive problems. Level 2 interconnect has been improved by moving toward surface mount and grid array style packaging. While these technologies are becoming affordable due to process improvements, they do not completely eliminate the inductance problem either. While aggressive package design assists in the problem, it is a slow and expensive process to develop new packages.

Another pressing design issue in modern VLSI design is power [1]. The high power consumption of devices has been a significant stumbling block for designers. Approaches which reduce the power consumption of the I/O structures could therefore contribute significantly to the goal of reducing chip-level and system-level power consumption. Typical off-chip output drivers are rated to drive a typical capacitive load of 5pF. Assuming a supply voltage of 1.5V and a switching frequency of 2Gb/s, each output driver requires consumes 22.5mW of power.

In this paper, we present a technique to avoid the inductive cross-talk in the interconnect, and also bound the I/O power of the IC, by encoding the data being transmitted off-chip. The receiving IC decodes this encoded data to recover the original un-encoded information. The implementation of the interface between the two ICs is unaltered, other than the need to utilize additional bits for the encoding. We construct a set of equations which reflect the constraints that any legal vector sequence must satisfy to avoid supply bounce, signal glitching, and signal edge speed degradation. We also construct equations which reflect the condition that the maximum power consumption of the I/O structure is bounded by some

0-7803-9451-8/06/$20.00 ©2006 IEEE.

user-specified quantity. The degree of supply bounce, glitching and edge speed degradation that can be tolerated are expressed by means of user-specified parameters as well. From this set of constraint equations, we construct a set of legal vector sequences for the bus. We use this set to find the largest effective size of the bus that can be achieved by encoding, for a given physical size of the bus. A Reduced Ordered Binary Decision Diagram [6] (ROBDD) based algorithm is used for this purpose.

We note that the proposed approach is applicable to arbitrary-sized buses. In practice, when a wide off-chip bus is implemented on a VLSI IC, it is decomposed into smaller bus segments as described in Section 4. Typically the size of these segments does not exceed 7 or 8 bits. The analysis which is described in the sequel is performed on bus segments of size up to 14 bits.

We show that the inter-chip bus throughput is increased as much as 100% compared to an unencoded bus, by using our inductive encoding techniques alone. By adding power constraints (limiting the power to 20% of the maximum switching power) the bus throughput was still 100% improved over the unencoded bus, with significantly lowered inductive effects as well. The asymptotic bus overhead varies depending on how aggressive the user-specified inductive cross-talk and power constraints are.

The rest of this paper is organized as follows. Section 2 provides the definitions used in the rest of this paper. Section 3 describes previous work on this topic. Section 4 presents our encoding scheme to reduce inductive cross-talk. Experimental results are presented in Section 5, and conclusions are drawn in Section 6.

2 Preliminaries and Terminology

Consider k bus segments with n bus bits each, with the jth segment consisting of signals $b_0^j, b_1^j, b_2^j \cdots b_{n-1}^j$. Let the vector sequence on segment j be denoted as v^j.

For example, if we had a V_{SS} and V_{DD} pin repeating after every 4 signal pins, the segments would consist of $n = 6$ pins. If the bus consisted of 20 signal pins, then we would implement it using 5 such segments.

- **Definition 1 :** *A Vector Sequence v^j is an assignment of values to the signals b_i^j as follows:*

  $b_i^j = v_i^j$, (where $0 \leq i \leq n-1$ and $v_i^j \in \{0, 1, -1\}$).

  Note that $v_i^j = 1(-1)$ indicates that the ith signal of the jth bus segment is rising (falling), while $v_i^j = 0$ indicates that it is either statically low or high.

- **Definition 2 :** *A Legal Vector Sequence (modulo inductive cross-talk) v is an assignment to the signals b_i^j such that:*

  - If b_i^j is a supply pin, the total bounce on this pin is bounded by P_{bnc} volts, where P_{bnc} is a user-specified constant.

  - if b_i^j is a signal pin which is static during the vector sequence, the glitch on this pin has a magnitude bounded by P_0 volts, where P_0 is a user-specified constant.

  - if b_i^j is a signal pin which is switching during the vector sequence, the switching speed of this pin is not degraded due to the effect of inductive cross-talk. Note that we can make this restriction stricter – by specifying that b_i^j's transition is in fact *sped up* due to inductive cross-talk.

The power consumed when a capacitance C is charged at frequency f over a voltage range V is $P = C \cdot V^2 \cdot f$. We assume that our I/O drivers are rated to drive a 5pF load at a frequency of 2Gb/s, and a power supply voltage of 1.5V. This results in a power consumption of 22.5mW per output driver.

3 Previous Work

There has been much work into the reduction of parasitic inductance through package advancement [7, 5]. Since the performance limitation is caused by the parasitic inductance in the level 1 and level 2 interconnects of the IC package, many packaging technologies have been developed. Table 1 shows the parasitic inductance values for three industry standard packages (a Quad Flat Pack (QFP) with wirebonding, a Ball Grid Array (BGA) with wirebonding, and a flip-chip BGA package). The last approach is an example of solder bump technology. In this table, L_{self} is the self-inductance of a pin, and the columns to its right are the mutual inductive coupling coefficients of successive neighbors of this pin. We observe

that while solder bump approaches reduce the parasitic inductances, *their significant cost makes them cost-effective only for the highest performance designs.*

Bus encoding algorithms have been developed to overcome the capacitive cross-talk for on-chip buses [8, 9, 10]. However, the problem of on-chip capacitive cross-talk minimization for buses is very different from that of off-chip inductive cross-talk minimization. Although our approach also constructs (inductive) cross-talk resistant CODECs algorithmically, in contrast to [8, 9], we utilize memory-based CODEC solutions.

There has been some recent interest in bus encoding to reduce power in buses [11, 12, 13]. These approaches target on-chip buses, in contrast to our work, and as a result they do not consider inductive effects in the problem formulation.

IO signals are often intentionally skewed to avoid inductive cross-talk effects. However, with increasing process variations [1] in recent technologies, these approaches may incur worst-case inductive cross-talk effects. Further, our approach is able to aid signal transitions by *exploiting* inductive cross-talk effects, something that skewing based techniques are unable to do.

In [14], an approach to encode and decode bus data to avoid inductive cross-talk was presented. In contrast to this approach, our work reduces bus power as well, all under a unified mathematical framework. Further, we employ an *implicit, ROBDD [6]* based formulation to compute the legal vectors on the bus, as opposed to the explicit approach of [14]. Finally, we have simulated our approach in an ASIC setting, and prototyped/tested it in an FPGA framework. The work of [14] did not provide such implementation results.

Techniques have been presented to minimize the inductive problems due to packaging. Pipeline damping was presented in [15]. In this approach, the authors attempt to minimize peak current levels by using a multi-valued output driver. While this approach improves performance by reducing the inductive ringing, it requires complex circuitry to implement the multi-valued output driver.

CODECs have also been presented [16] that limit the total number of simultaneously switching signals with the same transition direction. This has the effect of reducing the power supply bounce by limiting the total amount of current flowing through the power supply pins at any given time. This technique reported performance improvements but only considered the supply bounce and not the signal-to-signal cross-talk. Our work improves upon previous techniques by additionally considering signal rise-time degradation and glitching due to inductive cross-talk. Our approach is the first to include all the inductive and power effects, and model them in a common mathematical framework.

4 Our Approach

Consider a bus consisting of k identical segments, each of width n. For any segment j, let $j-1$ represent the segment to the immediate left of j, and let $j+1$ represent the segment to its immediate right. Let us also denote the values of the n bits of segment j as v_i^j ($0 \leq i \leq n-1$). Figure 1 shows an example of a bus configuration with $k = 3$ and $n = 5$. The signal-to-power ratio for this bus configuration is $\alpha = \frac{\# \text{ of pins in each segment}}{\# \text{ of supply pins in each segment}} = \frac{5}{2}$.

In general, when assigning package pins for an off-chip bus, V_{DD} and V_{SS} pins are interspersed among the signal pins in a regular fashion. The overall bus arrangement consists of a repetitive pattern of segments, each with their V_{DD} and V_{SS} pins in the same relative position within the segment (as shown in Figure 1).

In our approach, we write equations to encode the power and inductive cross-talk constraints for all bits of the jth bus segment. The constraints are different for signal, V_{DD}, and V_{SS} pins. Depending on the number of neighboring pins whose mutual inductance effects we want to model, the constraint equations will include pins belonging to neighboring segments as well. Since the segments are arranged in a repetitive manner, the en-

Package	L_{self}	K_1	K_2	K_3	K_4	K_5
QFP-wb	4.550nH	0.744	0.477	0.352	0.283	0.263
BGA-wb	3.766nH	0.537	0.169	0.123	0.097	0.078
BGA-fc	1.244nH	0.630	0.287	0.230	0.200	0.175

Table 1: Self and Mutual Inductance Values for Modern Packages

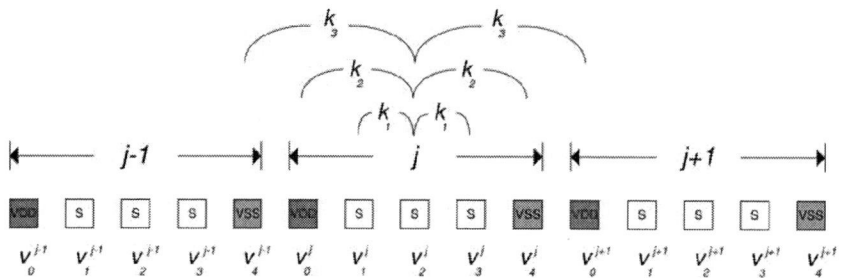

Figure 1: Example Bus Configuration

coding obtained for any segment will be valid for all k segments within the bus.

Having written these constraints, we then determine the vector sequences which satisfy these constraints. The valid sequences are used to construct a ROBDD [6] which encodes legal transitions between bus vectors. From this ROBDD, we construct a memory-based CODEC which is used during the bus data transfer.

4.1 Signal Pin Constraints

Consider the coupled voltage on a pin i (in bus segment j), due to a transition on its neighbor p (which is q pins away from i, and called the qth neighbor of i). This voltage is expressed as $v_i = \pm M_{ip} \frac{di_p}{dt}$. The sign of the coupled voltage depends on the direction of the transition on the qth neighbor p. Since output drivers in a bus all have the same drive strength (i.e. $\frac{di_p}{dt} = \frac{di_q}{dt}$ for any pair of bus signal pins p and q), let $k_q = |M_{ip} \frac{di_p}{dt}|$. As a result, we can write $v_i = k_q \cdot v_{i \mid p}^j$, where $v_{i \mid p}^j \in \{-1, 0, 1\}$ as per Definition 1. Also, the arithmetic in the subscript of v_{i+p}^j is performed modulo n. For example, if $n = 5$, $j = 4$, and $i = 0$, then v_{i-3}^j is the same as v_2^3 (i.e. the second bit of the adjacent bus segment to the left). Using this notation allows us to write the inductive cross-talk constraints very compactly.

We can write the mutual inductive coupling of any signal pin to its immediate neighbor signal pin as k_1. Further, let the mutual inductive coupling of a signal pin to its neighbor's neighbor be expressed as k_2 (likewise k_3, k_4, etc.). We assume that $k_x = 0$ for $x > p$. In other words, if $p = 3$, then we ignore the inductive cross-talk due to the 4th neighbor and beyond, by setting $k_4 = k_5 = ... = k_n = 0$. As a consequence, we include the mutual inductive contributions of three neighboring pins on either side of the pin under consideration. The k_i labels in Figure 1 illustrate the mutual inductive signal coupling for $p = 3$. Note that each signal pin within the bus will experience coupling from pins on either side. This symmetry allows for encoding to reduce or cancel out the net mutual inductive effect experienced on a victim signal. For this work, any K_j value less than 0.15 is ignored, and the corresponding k_j values are set to 0.

The polarity of the mutual inductive coupling on the victim signal will depend on whether the neighboring signals are rising ($v_i^j = 1$) or falling ($v_i^j = -1$). Constraints for the victim signal are written for all three possible transitions, those being rising ($v_i^j = 1$), falling ($v_i^j = -1$), or static ($v_i^j = 0$). Using the notation described above, a constraint equation can be written for each victim signal, to limit the mutual inductive coupling effect. The inductive cross-talk requirements for a signal pin i in segment j are expressed below.

We must also guarantee that the total switching power of each segment j is less than the user-specified upper bound p_{max}. Given that the power consumption per pin is p_{pin}[1], we know that for any segment j:

$$\frac{p_{pin}}{2} \cdot (\text{\# of } v_i^j \text{ pins that are } -1 \text{ or } 1) \leq p_{max}$$

or, alternately,

$$(\text{\# of } v_i^j \text{ pins that are } -1 \text{ or } 1) \leq P_{power}$$

Where $P_{power} = \frac{2 \cdot p_{max}}{p_{pin}}$, a user-supplied parameter.

- If signal i rises in segment j, then the cumulative inductive cross-talk on this signal should not deter (or should aid) its transition by inducing a mutually coupled voltage which is greater than or equal to a user-specified quantity P_1:

$$v_i^j = 1 \Rightarrow$$
$$k_1 \cdot (v_{i-1}^j + v_{i+1}^j) + k_2 \cdot (v_{i-2}^j + v_{i+2}^j) + ... + k_p \cdot (v_{i-p}^j + v_{i+p}^j) \geq P_1$$

Note that P_1 has units of voltage and represents the minimum amount of inductive signal coupling allowed for the pin i in segment j. If $P_1 = 0$ and the inequality in the above expression is changed to an equality, then all the mutual inductive cross-talk is canceled out (i.e. $v_{i-1}^j = -v_{i+1}^j$, etc.). If we wish to speed up the transition of pin i in segment j, then we simply set $P_1 > 0$. This would force the mutually induced voltage on pin i of segment j to speed up its rising transition.

Also note that by definition v_i^j for any supply pin is 0. This eliminates any mutual induced voltage on a victim signal pin i, due to V_{SS} and V_{DD} pins, as required. Likewise, any signal pin which remains static will also have $v_i^j = 0$ and hence will not cause in any mutually induced voltage on any neighboring victim pins.

- If signal i falls in segment j, then the cumulative inductive cross-talk on this signal should not deter (or should aid) its transition by inducing a mutually coupled inductive voltage which is less than or equal to a user-specified quantity P_{-1}:

$$v_i^j = -1 \Rightarrow$$
$$k_1 \cdot (v_{i-1}^j + v_{i+1}^j) + k_2 \cdot (v_{i-2}^j + v_{i+2}^j) + ... + k_p \cdot (v_{i-p}^j + v_{i+p}^j) \leq P_{-1}$$

Again, P_{-1} has units of voltage, and $P_{-1} \leq 0$. Note that for symmetric rise and fall times we set $|P_1| = |P_{-1}|$. However, $|P_1|$ and $|P_{-1}|$ can be set to different values, to aid in only a rising or falling transition. In this way, the designer could compensate for differences in the rise and fall times of off-chip drivers.

- If signal i is static in segment j, then the cumulative inductive cross-talk on this signal should not result in a glitch greater than P_0.

$$v_i^j = 0 \Rightarrow$$
$$-P_0 \leq k_1 \cdot (v_{i-1}^j + v_{i+1}^j) + k_2 \cdot (v_{i-2}^j + v_{i+2}^j) + ... + k_p \cdot (v_{i-p}^j + v_{i+p}^j) \leq P_0$$

Again, P_0 has units of voltage, just like P_1 and P_{-1}.

- For all signal pins in segment j, ensuring that the power is bounded by p_{max} means that

$$(\text{\# of } v_i^j \text{ pins that are } -1 \text{ or } 1) \leq P_{power}$$

Where $P_{power} = \frac{2 \cdot p_{max}}{p_{pin}}$, as derived earlier. The factor of 2 arises due to the fact that only one bus transition happens per clock cycle. For example, if $n = 7$ (i.e. there are 5 signal pins per segment) and $p_{max} = 20\%$ of the maximum (i.e. $p_{max} = 0.2 \cdot p_{pin} \cdot 5$) then $P_{power} = 2$. In the sequel we refer to the power constraint as a percentage of the maximum possible value for

---

[1] For an output pin, p_{pin} is $C \cdot VDD^2 \cdot f$, where C is the trace capacitance (typically 5pF), VDD is the supply voltage (assumed to be 1.5V), and f is the switching frequency (assumed to be 2Gb/s).

4.2 Power Pin Constraints

If a pin i in segment j is a V_{SS} (V_{DD}) pin, we require that the bounce due to its self inductance be limited by P_{bnc}, the absolute bounce (droop) voltage that can be tolerated. P_{bnc} is a user-specified quantity.

Let $z = |L\frac{di}{dt}|$ in Equation 1. Note that since all output drivers of the bus are identically sized, $\frac{di}{dt}$ is identical for all drivers. Using this notation, we can write the constraint equation for V_{DD} and V_{SS} pins as follows:

- If signal i is V_{DD} in segment j, then the cumulative supply bounce should be less than P_{bnc}.

  $v_i^j = V_{DD} \Rightarrow \frac{z}{2} \cdot (\text{\# of } v_i^j \text{ and } v_i^{j-1} \text{ pins that are } 1) < P_{bnc}$

  Note that this assumes that any V_{DD} pin supplies switching current for half the signal pins in its segment j, and half the signal pins in the segment to its left. Since each signal always has a V_{DD} pin to the left and to the right, we assume that if it switches high, then half the switching current is supplied by the V_{DD} pin to its left, and the other half by the V_{DD} pin to its right. This explains the presence of the $\frac{z}{2}$ term in the constraint equation above.

- If signal i is V_{SS} in segment j, then the cumulative ground bounce should be less than P_{bnc}.

  $v_i^j = V_{SS} \Rightarrow \frac{z}{2} \cdot (\text{\# of } v_i^j \text{ and } v_i^{j-1} \text{ pins that are } -1) \leq P_{bnc}$

  It should be noted that the constraints for supply pins are solved to find the maximum number of signals that are allowed to transition in the same direction at once.

Once the configuration of V_{DD}, V_{SS} and signal pins is known for the bus, the above constraints can be greatly simplified. For example, in Figure 1, setting $v_0^{j-1} = v_4^{j-1} = v_0^j = v_4^j = v_0^{j+1} = v_4^{j+1} = 0$ would encode the supply constraints. In this manner, a single mathematical framework encodes all the required inductive cross-talk constraints, which are i) that switching signals should not have their slew-rates degraded, ii) that the glitch magnitude on static signal pins should be limited, iii) the bounce on V_{DD} and V_{SS} pins should be bounded and iv) the power in the bus segment is bounded.

4.3 Constructing Legal Vector Sequences

Consider a particular bus configuration (n, k, and α) and user-specified inductive cross-talk constraints (P_1, P_{-1}, P_0 and P_{bnc}) and power constraint P_{power}. For each signal pin i within the segment j, three constraints equations are written (for $v_i^j = 1, -1, and\ 0$, per Section 4.1). For each power supply pin, one constraint expression is written, per Section 4.2. For each bus segment, we write one power constraint equation as described in Section 4.1. This results in a total of $3n - 3$ constraint equations for an $n - bit$ bus segment. These equations may refer to v_i^j values from neighboring bus segments as well.

Each possible vector sequence is evaluated for legality by testing if it satisfies each of the $3n - 3$ constraint equations. The total number of signal pins that need to be considered depends on p. Since the v_i^j values for V_{DD} and V_{SS} pins are always zero, the number of evaluations is significantly reduced. Since there are three possible signal transitions ($v_i^j = 1, -1, and\ 0$) per signal bit, the total number of vector sequences that need to be tested for legality is $3^{(n+2 \cdot p-6)}$. *Note that the values of n and p for realistic buses is small, so these tests (which need to be done exactly once for a design) can be performed easily.* In our experiments, $n = 7$ and $p = 2$, which is reasonable for real-life buses.

After testing the vector sequences for legality modulo inductive cross-talk and power, we create a set of legal vector sequences for the segment j. The size of this subset depends on how aggressively the parameters P_1, P_{-1}, P_0, P_{bnc} and P_{power} are selected. The final list of legal vector sequences refers to $n + 2p - 6$ signal pins ($n - 2$ pins within the segment being considered, and $2p - 4$ pins on either side of the segment under consideration).

4.4 Constructing the CODEC

From the set of legal vector sequences, we next create a ROBDD [6] G, to encode legal bus transitions. We then find the effective size m of the bus that can be encoded using the transitions in G, using a ROBDD based algorithm. Note that the ROBDD G has $2n$ variables. The first n variables refer to the *from* vertices and the next n variables refer to the *to* vertices

of the vector transition. There is a legal edge between vertices v_1 and v_2 iff $G(v_1, v_2) = 1$.

Note that for a vector sequence v^j, we can construct minterms in G to encode transitions between vectors w_{from}^j and w_{to}^j. The end-points of this edge (w_{from}^j and w_{to}^j) can be constructed given v^j, as follows:

$w_{from.i}^j = 1$ if $v_i^j = -1$ (i.e. the signal is falling) or if $v_i^j = 0$ (i.e. the signal is static).

$w_{from.i}^j = 0$ if $v_i^j = 1$ (i.e. the signal is rising) or if $v_i^j = 0$ (i.e. the signal is static).

Similarly, we can write

$w_{to.i}^j = 0$ if $v_i^j = -1$ or if $v_i^j = 0$.

$w_{to.i}^j = 1$ if $v_i^j = 1$ or if $v_i^j = 0$.

$G(w_{from}^j, w_{to}^j) = 1$ indicates the legality (from an inductive cross-talk and power viewpoint) of the transition from vector w_{from}^j to w_{to}^j. Therefore, given a set of vector sequences $\{v^j\}$ which are *legal* from a inductive cross-talk and power standpoint, we can construct a ROBDD G whose minterms ($w_{from} : w_{to}$) are vectors in B^{2n}, such that they indicate a legal transition (from an inductive cross-talk and power viewpoint) between the source (w_{from}) and sink (w_{to}) vertices. Note that the ":" symbol above refers to the concatenation operator.

If an m-bit bus can be encoded using the legal transitions in G, then there must exist a set of vertices $V_c \subseteq B^n$ such that

- Each $v_s \in V_c$ has at least 2^m outgoing edges $e(v_s, v_d)$ (including the self edge), such that the destination vertex $v_d \in V_c$.
- The cardinality of V_c is at least 2^m.

The resulting encoder is memory based. Note that the physical size of the bus n is obviously greater than or equal to m.

Given G, we find m using Algorithm 1. The input to the algorithm is m and G. We first find the out-degrees (self-edges are counted) of each $v_s \in B^n$. This is done by logically ANDing the ROBDD of the vertex v_s with G. We find the cardinality of the resulting ROBDD – it represents the out-degree of v_s. If the number of out-edges of any v_s is greater than 2^m, we add v_s (and its out-degree) into a hash table V.

For each $v_s \in V$, we next check if each of its destination nodes v_d are in V. If $v_d \notin V$, we decrement the out-degree of v_s by 1. If the out-degree of v_s becomes less than 2^m, we remove v_s from V.

These operations are performed until convergence. If at this point, the number of surviving vertices in V is 2^m or more, then an m-bit memoryless CODEC can be constructed from G.

We initially call the algorithm with $m = n - 1$ (where n is the physical bus size). If an m bit bus cannot be encoded using G, then we decrement m. We repeat this until we find a value of m such that the m-bit bus can be encoded by G.

Algorithm 1 Testing if G can encode an m-bit bus

```
test_encoder(m, G)
    find out-degree(v_s) of each node v_s, insert (v_s.out-degree(v_s)) in V if out-degree(v_s) ≥ 2^m
    degrees_changed = 1
    while degrees_changed do
        degrees_changed = 0
        for each v_s ∈ V do
            for each v_d S.T. G(v_s, v_d) = 1 do
                if v_d ∉ V then
                    decrement out-degree(v_s) in V
                    degrees_changed = 1
                end if
                If out-degree(v_s) < 2^m then
                    V ← V \ v_s
                    break
                end if
            end for
        end for
    end while
    if |V| ≥ 2^m then
        print(m bit bus may be encoded using G)
    else
        print(m bit bus cannot be encoded using G)
    end if
```

Note that this entire analysis needs to be performed for a representative bus segment. In other words, even if the bus is very wide, the analysis is performed for a single segment (which is typically very small). The experimental results we report next consider a typical bus segment (n = 7, k = 3).

This segment could be part of a much larger bus, and the analysis would be valid for all segments of the bus.

5 Experimental Results

To validate the technique presented, we encoded an example bus segment to avoid inductive cross-talk and limit power consumption. The bus segment configuration is shown in Figure 1, except that our experimental bus segment had 5 signal bits (i.e. $n = 5 + 2$). We used electrical parameters from a standard BGA-wb package in our simulations. This bus segment was encoded using P_0, P_1, P_{-1} and P_{bnc} set to 12.5% of V_{DD}. We compared three configurations – i) an unencoded bus segment ii) a bus segment encoded only for inductive constraints and iii) a bus segment encoded for inductive as well as power constraints.

The first step consists of writing the constraint equations for every pin in the bus segment. Example constraints for $n = 5$ (i.e. 3 signal pins), $k = 3$, and $\alpha = 5/2$ are provided below. From the inductive coupling values in Table 1, we set $p = 2$ to ignore inductive coupling with a magnitude less than 0.15. For $p > 2$, the mutual inductive coupling drops off rapidly, justifying our choice. This exercise yields 12 constraint equations, shown below. Note that these constraints have been simplified by removing terms with $v_i^j = 0$.

1) $v_0^j = V_{DD} \Rightarrow \frac{k}{2} \cdot (\# \ of \ v_i^j \ (or \ v_i^{j-1}) \ pins \ that \ are \ 1) \leq P_{bnc}$

2) $v_1^j = 1 \Rightarrow k_1 \cdot (v_2^j) + k_2 \cdot (v_3^j) \geq P_1$

3) $v_1^j = -1 \Rightarrow k_1 \cdot (v_2^j) + k_2 \cdot (v_3^j) < P_{-1}$

4) $v_1^j = 0 \Rightarrow \quad P_0 \leq k_1 \cdot (v_2^j) + k_2 \cdot (v_3^j) \leq P_0$

5) $v_2^j = 1 \rightarrow k_1 \cdot (v_1^j) + k_1 \cdot (v_3^j) \geq P_1$

6) $v_2^j = -1 \Rightarrow k_1 \cdot (v_1^j) + k_1 \cdot (v_3^j) < P_{-1}$

7) $v_2^j = 0 \Rightarrow -P_0 \leq k_1 \cdot (v_1^j) + k_1 \cdot (v_3^j) \leq P_0$

8) $v_3^j = 1 \Rightarrow k_2 \cdot (v_1^j) + k_1 \cdot (v_2^j) \geq P_1$

9) $v_3^j = -1 \Rightarrow k_2 \cdot (v_1^j) - k_1 \cdot (v_2^j) \leq P_{-1}$

10) $v_3^j = 0 \Rightarrow -P_0 \leq k_2 \cdot (v_1^j) + k_1 \cdot (v_2^j) \leq P_0$

11) $v_4^j = V_{SS} \Rightarrow \frac{k}{2} \cdot (\# \ of \ v_i^j \ (or \ v_i^{j-1}) \ pins \ that \ are \ -1) \leq P_{bnc}$

12) $(\# \ of \ v_i^j \ pins \ that \ are \ -1 \ or \ 1) \leq P_{power}$

Note that the k_i values depend on the magnitude of $\frac{di}{dt}$. This means that as $\frac{di}{dt}$ is changed, the k_i parameters will also change. However, the absolute voltage that the P_x parameters represent (i.e., 12.5% of V_{DD}) will remain fixed.

We next find the set of legal vector sequences. We note that the supply bounce and power constraints were violated most frequently. Using the remaining (legal) vector sequences, we construct the ROBDD G as described in Section 4.4. We then find the effective bus width m which can be encoded using the legal transitions in G, as described in Algorithm 1.

We found the value of the effective bus size m as a function of the physical bus size $n - 2$ (since 2 pins are VDD and VSS). The results are shown in Table 2, where we list the effective bus size as a function of $n - 2$. Note that the second column of this table indicates the effective bus size assuming no power constraints are specified. The third, fourth and fifth columns were generated assuming a power constraint of 33%, 20% and 18% of the maximum bus power. These columns include inductive constraints just as in column 2. Note that the effective bus width reported in Table 2 refer to the effective number of *signal* pins in the corresponding bus segment. Also, note that when the value of n and the power constraint are both small, it is impossible to find m, since it is possible that all transitions on the bus segment become illegal.

Table 2 indicates that for bus segments with 7 or more signal pins, the effective bus sizes are comparable when we utilize encoding with or without power constraints. This suggests that if we were to use bus segments with 7 or more signal pins, we can curtail inductive effects and also limit power with a small bus size penalty. In fact the bus throughput increases significantly with encoding, as we will discuss shortly.

5.1 Case 1: Fixed $\frac{di}{dt}$

The first bus segment considered has a fixed $\frac{di}{dt} = 33 \frac{MA}{s}$. This corresponds to a data rate of 550 Mb/s in a 50 Ω system using the rule of thumb that $datarate = \frac{1}{3 \cdot risetime}$.

SPICE simulations were conducted to quantify the increased performance of the encoded bus segment. We utilized a TSMC 0.13μm process for this purpose. We compared the *original* unencoded bus segment with a *non-aggressive* encoded segment (which represents the case when only inductive constraints were used) and an *aggressive* encoded segment (which represents the case when both inductive and power (limiting the power to 20% of the maximum) constraints were applied).

The simulation results confirm a reduction in the inductive cross-talk on the bus segment, while power is restricted within its specified bound (20% of maximum). SPICE plots are not shown due to lack of space. We observed that the ground bounce magnitude and the glitch magnitude for both versions of the encoded bus are exactly at or below the limit specified (12.5% of V_{DD}), indicating that the *experimental results track closely with the theory*. The aggressive constraints further reduced the glitching and supply bounce magnitudes. We also found that the edge degradation constraints do not play a major part in determining the final solution. This was because satisfying the remaining constraints (particularly the power constraint and the supply bounce constraint) typically ensured that the edge degradation was severely limited.

5.2 Case 2: Varying $\frac{di}{dt}$

Using the same analysis technique described in Case 1, we can sweep $\frac{di}{dt}$ to find the data rate at which the bus reaches the power and inductive cross-talk limits. For this example, we use the same bus configuration as in the previous section. The *original*, *non-aggressive* and *aggressive* conditions are also as described earlier.

The $\frac{di}{dt}$ for the *original* bus and the *non-aggressive* and *aggressive* encoded bus is increased until the coupling limits are reached. The maximum di/dt values are 8.0 MA/s (original), 19.9 MA/s (non-aggressive) and 37.0 MA/s (aggressive). The 5-bit bus without encoding operates at 133 Mb/s (for a total throughput of 665 Mb/s), while our non-aggressive encoded 4-bit (effective) bus operates at 333 Mb/s (for a total throughput of 1332 Mb/s). The aggressive encoded 2-bit (effective) bus operates at 666 Mb/s, for a total throughput of 1332 Mb/s. Hence, encoding the bus increases the total throughput by 100% using the same physical size of the bus, and considering the encoder overhead. The power reduction methodology helps further, allowing us to reduce power to 20% of the worst case, while retaining the throughput of the non-aggressive case.

5.2.1 TSMC 0.13um ASIC Process

The CODECs were implemented using the TSMC 0.13um CMOS IC process to understand their impact on delay and area of the IC. Bus sizes of 2, 4, 6, and 8 were used. For each of these sizes, both the aggressive and the non-aggressive CODECs were synthesized, placed and routed.

$n - 2$	no power constraint	33%	20%	18%
3	2	0	0	0
4	3	2	0	0
5	4	2	2	0
6	5	2	2	2
7	5	4	3	3
8	6	5	3	3
9	7	5	3	3
10	7	6	5	3
11	8	7	6	3
12	8	7	6	6

Table 2: Effective Bus Width for Different Power Constraint Values

	Bus Size (m)	Style	
	-	aggressive	non-aggressive
Delay (ns)	2	0.170	N/A
	4	0.670	0.503
	6	1.150	0.955
	8	1.310	0.983
Area (um^2)	2	22	N/A
	4	152	114
	6	614	509
	8	1,181	886

Table 3: Encoder in a TSMC 0.13um Process

Table 3 lists the delay and area impact of the CODECs implemented in a TSMC 0.13um process. The delays in this table represent the delay of the encoder. *Note that encoding and decoding delays are unimportant for heavily pipelined systems, where these delays can be hidden.* In case of heavily pipelined systems, the maximum data-rate is significantly improved by using our encoding based schemes.

This table illustrates the negligible impact of our approach on a modern VLSI design.

5.2.2 Xilinx 0.35um FPGA Experiment

The CODECs were also *synthesized, mapped, prototyped and tested* for a *Xilinx VirtexIIPro*, Field Programmable Gate Array (FPGA) which used a 0.35um CMOS process.

CODECs for bus sizes of 2, 4, 6, and 8 were implemented using both the aggressive and non-aggressive constraints. Table 4 lists the delay and area impact of our CODECs when implemented in the FPGA environment. In all cases, the CODEC designs occupied less than 1% of the FPGA to be implemented. The CODECs were implemented using standard Function Generators (FGs) within the FPGA which resulted in minimal propagation delay through the circuit. As noted earlier, in the case of *pipelined data transfers, the actual delay of the encoding and decoding process can be hidden.*

The outputs of the FPGA were monitored using the 16950A Logic Analyzer from *Agilent Technologies Inc*. The logic analysis measurements verified that the CODECs could be taken from the conception stage to final implementation using standard IC design practices. Logic analyzer measurement results and a photograph of the FPGA test setup are not shown for lack of space.

6 Conclusions

Inductive cross-talk within IC packages is an important factor limiting off-chip I/O throughput. Addressing this issue with aggressive package design is slow and often too expensive for a majority of applications. Another important design issue in modern VLSI design is power. Approaches which limit the power consumption of the I/O structures are therefore important to achieve the goal of reduced chip-level and system-level power consumption.

	Bus Size (m)	Style
	-	aggressive & non-aggressive
Delay (ns)	2	0.351
	4	1.020
	6	1.450
	8	1.610
FPGA Usage	2	< 1%
	4	< 1%
	6	< 1%
	8	< 1%
FPGA Implementation	2	3x, 2-Input FG's
	4	6x, 4-Input FG's
	6	9x, 6-Input FG's
	8	12x, 8-Input FG's

Table 4: Bus Expansion Encoder Implementation Results using a 0.35um, CMOS FPGA Process

In this work, we presented a technique to encode off-chip bus data to avoid inductive cross-talk effects as well as to limit the power consumption of the I/O. Our technique involves writing constraint equations which express the user-specified bounds on the amount of edge speed degradation, glitch magnitude, supply bounce and power consumption that can be tolerated. We express all these constraints in a common mathematical framework. We construct a set of legal vector sequences with respect to inductive cross-talk and power, and use these to develop a CODEC for inductive cross-talk avoidance. The CODEC is constructed using a ROBDD based computation.

Experimental results track very closely with the theory, and demonstrate an improvement of 100% in the bus throughput for an example 5-bit bus when only inductive constraints are applied. When power constraints (limiting the power of the bus to 20% of the worst case) are applied, the bus throughput is still 100% improved over the unencoded bus. The reduced switching results in improved glitching and supply bounce performance as well. We have validated our approach by *simulating it in an ASIC setting as well as prototyping and testing it in an FPGA environment.*

References

[1] "The International Technology Roadmap for Semiconductors." http://public.itrs.net, 2003.

[2] R. Tummalo, *Fundamentals of Microsystem Packaging*. McGraw-Hill, 2001.

[3] M. Miura, N. Hirano, Y. Hiruta, and T. Sudo, "Electrical characterization and modeling of simultaneous switching noise for leadframe packages," in *Proceedings of 45th Electronic Components and Technology Conference*, pp. 857–864, May 1995.

[4] B. Young, "Return path inductance in measurements of package inductance matrixes," in *IEEE Transactions on Components, Packaging, and Manufacturing Technology*, vol. 20, Feb 1997.

[5] N. Hirano, M. Miura, Y. Hiruta, and T. Sudo, "Characterization and reduction of simultaneous switching noise for a multilayer package,"

[6] R. E. Bryant, "Graph based algorithms for Boolean function representation," *IEEE Transactions on Computers*, vol. C-35, pp. 677–690, August 1986.

[7] M. Lopez, J. Prince, and A. Cangellaris, "Influence of a floating plane on effective ground plane inductance in multilayer and coplanar packages," in *IEEE Transactions on Advanced Packaging*, vol. 22, pp. 182–188, May 1999.

[8] C. Duan, A. Tirumala, and S. Khatri, "Analysis and avoidance of cross-talk in on-chip buses," *IEEE Symposium on High-Performance Interconnects (HOT Interconnects)*, pp. 133–138, Aug 2001.

[9] C. Duan and S. Khatri, "Exploiting crosstalk to speed up on-chip buses," *Design Automation and Test in Europe Conference*, Feb 2004.

[10] B. Victor and K. Keutzer, "Bus encoding to prevent crosstalk delay," in *Proceedings, IEEE/ACM International Conference on Computer Aided Design*, (San Jose, CA), pp. 57–63, Nov 2001.

[11] M. Stan and W. Burleson, "Bus-invert coding for low-power I/O," *IEEE Transactions on Very Large Scale Integration (VLSI) Systems*, vol. 3, pp. 49–58, Mar 1995.

[12] P. Sotiriadis and A. Chandrakasan, "Reducing bus delay in submicron technology using coding," in *Proceedings of the Asia and South Pacific Design Automation Conference*, pp. 109–114, Jan-Feb 2001.

[13] T. Lv, J. Henkel, H. Lekatsas, and W. Wolf, "A dictionary-based en/decoding scheme for low-power data buses," *IEEE Transactions on Very Large Scale Integration (VLSI) Systems*, vol. 11, pp. 943–951, Oct 2003.

[14] B. LaMeres and S. Khatri, "Encoding-based minimization of inductive cross-talk for off-chip data transmission," in *Proceedings, Design Automation and Test in Europe (DATE) Conference*, (Munich, Germany), Mar 2005.

[15] M. Powell and T. Vijaykumar, "Pipeline damping: a microarchitectural technique to reduce inductive noise in supply voltage," in *Proceedings of 30th International Symposium on Computer Architecture*, pp. 72–83, June 2003.

[16] C. Chen and B. Curran, "Switching codes for delta-i noise reduction," in *IEEE Transactions of the 43rd IEEE Midwest Symposium on Circuits and Systems*, vol. 45, pp. 1017 – 1021, Sept 1996.

[17] E. Mejia-Motta, F. Sandoval-Ibarra, and J. Santana, "Design of cmos buffers using the settling time of the ground bounce voltage as a key parameter," in *Proceedings of 43rd IEEE Midwest Symposium on Circuits and Systems*, vol. 2, pp. 718–721, Aug 2000.

8D-1

2006 Asia and South Pacific Design Automation Conference

A new test and characterization scheme for
10+ GHz Low Jitter Wide Band PLL

Kazuhiko Miki, David Boerstler[1], Eskinder Hailu[1], Jieming Qi[1],
Sarah Pettengill[1], Yuichi Goto

Toshiba Corporation, 580-1, Horikawa-Cho, Saiwai-Ku, Kawasaki, 212-8520, Japan
[1]IBM Microelectronics, 11501 Burnet Rd, Austin TX 78758, USA
Email: kazuhiko.miki@toshiba.co.jp

Abstract - This paper presents a new test and characterization scheme for 10+ GHz low jitter wide band PLL in 90 nm partially depleted (PD) Silicon-On-Insulator (SOI) CMOS technology. We measure the frequency range of VCOs without adding any devices for test between charge-pump (CP) and voltage- controlled oscillator (VCO). That test scheme gives us the intermediate frequency of VCO as well as the maximum and the minimum frequency. This paper also describes circuitry to observe the duty cycle of 4.2GHz clock directly on a wafer probe station, including a method to verify the measured duty cycle.

I. Introduction

A high performance microprocessor with multigigahertz operating frequencies becomes more ubiquitous than before since even consumer appliances like game consoles or low-end servers recently require it in addition to personal computers and high-end servers. In order to generate a very high frequency system clock, the microprocessor must implement phase locked loops (PLL's), the performance of which reach 10+ GHz operating frequency in some cases [1].

As a frequency of PLL becomes higher, the PLL should have more superior characteristics like a lower jitter, a higher and wider frequency range of VCO, and a well-controlled duty cycle of an output clock. It, however, is very hard to test or measure those characteristics, especially more difficult on manufacturing because a tester has a restricted specification mainly.

Some good schemes to characterize a high performance PLL are already proposed, and they try to add some devices or circuits in order to make the PLL open-loop in measurement. Those appendages, however, may negatively affect the characteristics of PLL themselves; for instance, induced noise from additional devices on an input node of voltage-controlled oscillator (VCO) gets a jitter of PLL high, or an adding multiplex circuit at an output of PLL makes the duty cycle of that output vary. We have a lot of interests in the scheme to characterize a high performance PLL without giving it negative influence, and that scheme should be able to use on manufacturing in order to screen the PLL. A frequency range of VCO should be focused on since it is one of the most important parameters to determine the characteristics of PLL.

It is generally desirable for a duty cycle of a system clock bring it close to 50% as much as possible. A high performance microprocessor, however, has a situation demanding the duty cycle that is not 50% precisely since it uses both the edges of a system clock though it is not a DDR (Double Data Rate) system. One of the greatest concerns about the duty cycle of the high frequency system clock is to know what duty cycle is optimum for the microprocessor or the circuit. We propose circuitry to observe the actual duty cycle of the high frequency clock directly.

The proposed scheme uses unique circuits and I/Os for observing divided clocks and a duty cycle of the clock has easily varied just by going through a simple circuit like an inverter. So we try to verify the duty cycle that is measured by that scheme.

II. How to characterize PLL

A. A frequency range of VCO

A simplified block diagram of the PLL architecture with VCO test mode is shown in Fig. 1, the power of which is supplied by analog vdd (VDDA). Outputs of a phase-frequency detector (PFD), "UP1B" and "DN1", are directly fixed to VDDA or analog gnd (GNDA) by control signals, "UP_ctrl", "DN_ctrl", "VCO_test" and "VCO_ini" instead of implementing additional devices between charge-pump (CP) and VCO in order to apply some voltages to "CTRL" node.

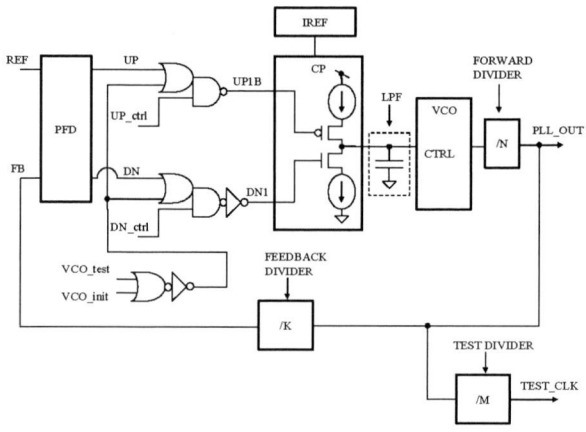

Fig. 1. Simplified PLL architecture with VCO test mode.

In a normal operation mode, both UP_ctrl and DN_ctrl are set to high level ("H"), and both VCO_test and VCO_ini are set to low level ("L"). When VCO_ini is asserted and UP_ctrl is set to "L" in the normal operation mode, CP is in state of discharge then the frequency of VCO becomes the minimum eventually. This function is used during power on reset (POR) sequence in order to initialize the PLL.

0-7803-9451-8/06/$20.00 ©2006 IEEE.

856

As VCO_test is asserted, the PLL gets into VCO test mode. By setting UP_ctrl to "L" and DN_ctrl to "H", the frequency of VCO becomes the minimum like PLL initialization mode mentioned above. "PLL_OUT" goes to not only a tree of clock distribution but also another divider for PLL characterization, and the divided clock is observed out of a chip. By setting UP_ctrl to "H" and DN_ctrl to "L", CP is in state of charge then the maximum frequency of VCO is measured. When both UP_ctrl and DN_ctrl are set to "H", CP makes crowbar current and outputs intermediate voltage at "CTRL" node. It means that the intermediate frequency of VCO is observed.

B. Duty cycle of PLL output

Since PLL power is supplied by VDDA and the logic circuits that work with the PLL output are powered by digital vdd (VDD), PLL-to-logic level shifters are necessary between the PLL and the logic circuits. The level shifter (LS), however, may make duty cycle variation, so we have to make great attention to design it and try to measure the duty cycle of the LS output.

Fig. 2 and Fig. 3 show a simplified block diagram of a duty cycle measurement apparatus and a logic waveform of it respectively. There are two identical counters with clear mode in that apparatus, one operates with "CLK" and the other does with "/CLK". Outputs of those counters, "N18" and "N28", have a same period (16 times of CLK period), and the phase difference between N18 and N28 is the same time period as the pulse width of CLK. In other words, a pulse width of a very high frequency clock is taken out as a phase difference between two lower frequency clocks that can be more easily observed. It, however, is not easy to measure a phase difference between two clocks precisely on a basic digital oscilloscope. So "OUT160" and "OUT155" are prepared to measure the pulse width of CLK actually. OUT160 has the same pulse width as N18, while OUT155 has a short pulse width than N28 for the pulse width of CLK. It is easier to measure a pulse width of a clock on a basic digital oscilloscope and a difference of pulse width between OUT160 and OUT155 is the same time period as the pulse width of CLK. By using this scheme, a very short time period like 125ps (equivalent pulse width of 4GHz clock) can be observed on even a wafer probe station, the pass bandwidth of which is lower than 1GHz generally.

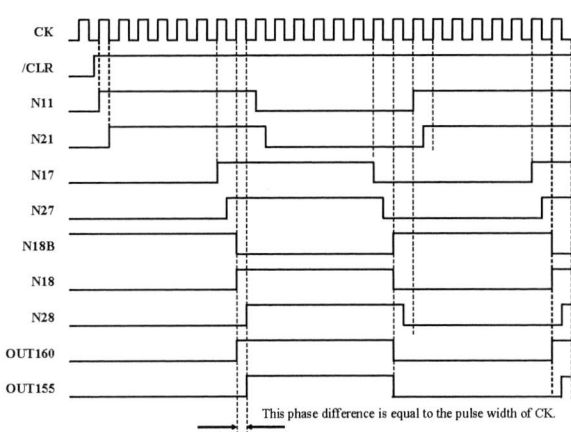

Fig. 3. Logic waveforms of duty cycle measurement apparatus.

C. Verification method of measured duty cycle

The duty cycle measurement apparatus described in the preceding paragraph is implemented in order to observe the duty cycle variation that is made by LS. We try to take out the duty cycle of the clock correct as possible; for instance making the parasitic resistance and capacitance on the clock path of OUT160 and one of OUT155 the same, maintaining the identity of two counters, and so on. The results, however, may not be correct because of an unexpected factor. So another circuitry to verify a measured duty cycle of the clock is also implemented.

Simplified duty cycle verification circuitry is shown in Fig. 4, which contains 4+ GHz PLL, three candidates of LS and the duty cycle measurement apparatus, too. First candidate of LS is a simple inverter, the output of which is "CK1". Second one is an AC-coupled type, the output of which is "CK2", and last one is an inverter with duty cycle adjuster (DCA), which can change a duty cycle of the output by applying analog voltage (Vbias0). The output of last one is "CK3". One of three clocks is selected at MUX1 that can also invert the input clock.

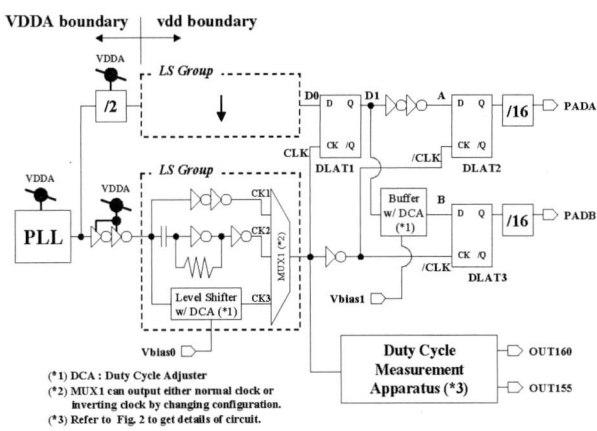

Fig. 4. Simplified duty cycle measurement and verification circuitry.

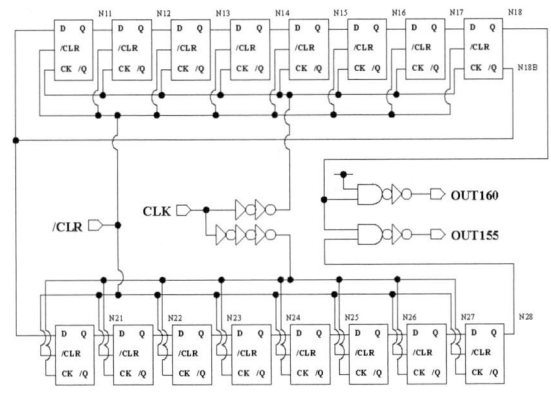

Fig. 2. Simplified duty cycle measurement apparatus.

857

"D0" is a divided clock by two and a data signal for D-latch (DLAT1). "D1" is a latched clock by CLK that is an output clock of MUX1. "A" is just a buffered clock and a data signal for another D-latch (DLAT2), while "B" is a buffered clock changing duty cycle of "D1" by applying another analog voltage (Vbias1) and a data signal for DLAT3. As Vbias1 is lower, the pulse width of B is narrower. When Vbias1 is swept, latching the data (B) begins to fail at certain voltage, which is defined as V_{fail}. Since B is latched by "/CLK", V_{fail} is related to the duty cycle of CLK.

Fig. 5 shows logic waveforms of duty cycle verification circuitry in Fig. 4. CLK' describes a clock that has a smaller duty cycle than 50%. When /CLK', which is the inverted clock of CLK', is input to "CK" of DLAT3, V_{fail} (=VH) is higher than the voltage (V_{fail}=V50) which latching the data (B) begins to fail at in the case of 50% duty cycle. In contrast CLK" describes a clock that has bigger duty cycle than 50% and V_{fail} (=VL) is lower than V50. If the duty cycle of CLK is around 50%, VH is very close to VL at least. Expressed in another way, as the voltage difference between VH and VL is big, the duty cycle of CLK is off 50%.

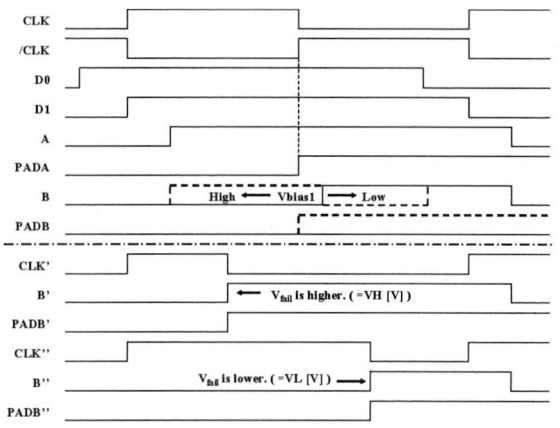

Fig. 5. Logic waveforms of duty cycle verification circuit.

III. Experimental Results

Both a tracking range of 10+ GHz low jitter wide band PLL and a frequency range of VCO implemented in that PLL were characterized on the actual products using the customized module test station with custom pattern generator hardware. In contrast, the duty cycle of the PLL output was tested on unpackaged PLL testsite die using a Cascade Microtech Summit probe station, 25 pad GGB Industries multi-contact wedge probes with integral bypass capacitors and coax cables, and assorted power supplies. Agilent 8133A pulse generator and 54855A sampling scope, and custom pattern generator hardware were used in all measurements.

A. Compare a frequency range of VCO with a tracking range of PLL

Fig. 6 provides the tracking range of 10+ GHz low jitter wide band PLL and the frequency range of the VCO versus the power supply voltage of the PLL (VDDA). The range

between "Tracking_Low" and "Tracking_High" is the tracking range of the PLL, and the range between "VCO_Min" and "VCO_Max" is the frequency range of the VCO that was measured by using VCO test mode. Here are the dividing ratios in the measurement; the dividing ratio (=N) of FORWARD DIVIDER in Fig.1 is 2, the dividing ratio (= K) of FEEDBACK DIVIDER is 8, and the dividing ratio (=M) of TEST DIVIDER is 8. A clock more than 1GHz was input as a reference clock in order to measure the maximum tracking frequency in this condition.

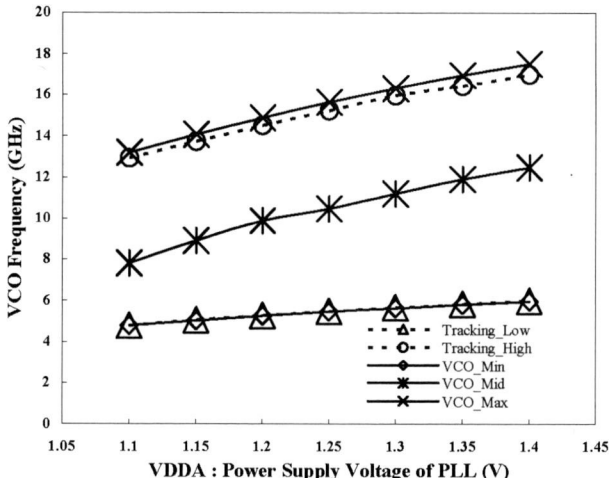

Fig. 6. Tracking range of PLL and frequency range of VCO versus power supply voltage of PLL.

A tracking range is one of most important parameters to know an actual performance and characteristics of PLL or to screen PLL. It, however, takes a long time to measure it generally since it is necessary to sweep a reference clock ("REF" in Fig. 1) and to change a power supply voltage of PLL every measurement. It, furthermore, is difficult to input a very high frequency clock more than 500MHz into REF on manufacturing because of the restriction that a basic tester has a lower bandwidth of the input.

Measuring the frequency range of the VCO by using VCO test mode on manufacturing is much easier than measuring the tracking range since VCO test mode does not require any clocks as an input. According to Fig. 6, the tracking range of the PLL is almost the same as the frequency range of the VCO, and the ratio between VCO maximum and minimum frequency was very close to the design value, 2.7 to 2.9. It means that the VCO test mode can be used in order to screen the PLL on manufacturing. This is very useful for reducing test time.

B. Actual duty cycle of a 4.2GHz clock

Fig. 7 shows an oscilloscope view in which the waveforms of the low frequency clocks, which is OUT160 and OUT155 in Fig. 4, were observed in order to measure the pulse width of the very high frequency clock, which is CLK in Fig. 4. The difference between rising edges of two clocks (265MHz) is equal to pulse width of a 4.2GHz clock, the actual time period is 104.8ps and the duty cycle is 44%.

2006 Asia and South Pacific Design Automation Conference

8D-1

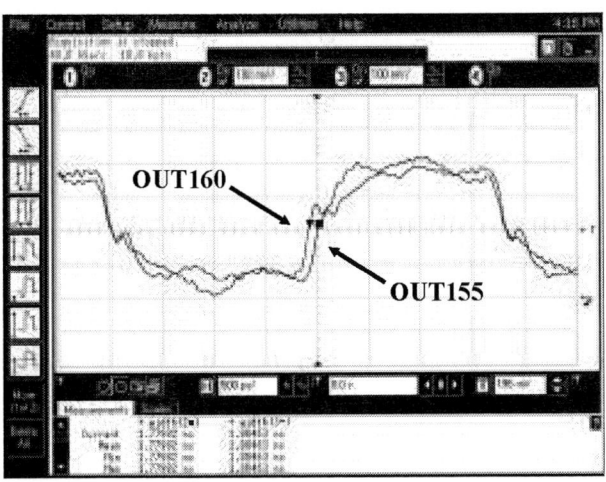

Fig. 7. An oscilloscope view of the pulse width measured for 4.2 GHz clock. The difference between rising edges of two clocks is equal to the pulse width of 4.2GHz clock.

Fig. 8 provides the actual duty cycle of the 4.2GHz clocks, which are CK1, CK2 and CK3 in Fig. 4, versus a voltage difference between VDDA and VDD. We try to describe the duty cycle variation that is made at LS in Fig. 4 in relation to the voltage difference in addition to the actual duty cycle in that chart. The duty cycle of CK2 must be the same as the output of the PLL in Fig. 4 because AC-coupled LS transfers a duty cycle of the input to the output precisely. It means that the output of the PLL in this testsite die seems to have roughly 10% offset of duty cycle originally since the output of VCO is divided by two at FORWARD DIVIDER and the duty cycle of the PLL output must be close to 50%. In addition, LS with DCA can correct that offset and the duty cycle variation at AC-coupled LS has no dependency on the difference between two power supply voltages. This shows a right result according to AC-coupled LS characteristics.

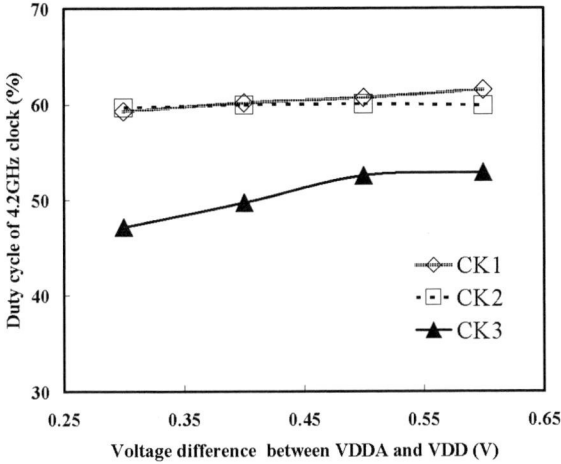

Fig. 8. The duty cycle of 4.2GHz clocks versus a difference between two power supply voltages. CK1, CK2 and CK3 are outputs from different type of level shifters.

We strongly believe that all of the relative information is right in Fig. 8, and it is necessary to verify the actual value of duty cycle as the next step.

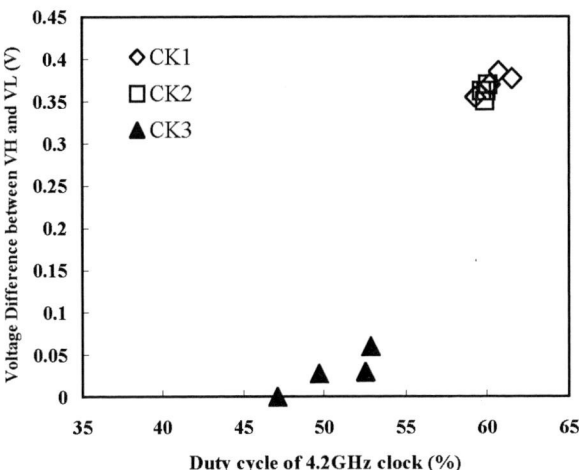

Fig. 9. The duty cycle of 4.2GHz clock versus a voltage difference between VH and VL that are applied biases for DCA. VH is V_{fail} in the case of using a normal clock, and VL is V_{fail} in the case of using the clock inverting the normal clock.

Fig. 9 provides the duty cycle of 4.2GHz clock versus a difference between two voltages of V_{fail}. At this point, VH is the measured voltage in the case of using certain CLK (=Clk_A) and VL is the measured voltage in the case of using the clock that is made by inverting Clk_A. If the duty cycle of CLK is around 50%, the voltage difference between VH and VL is close to 0V. It means that the duty cycle of CK3 must be around 50%, and it agrees with the duty cycle in Fig. 8. On the other hand, the voltage difference between VH and VL in the case of CK1 or CK2 is higher than 0.35V. A big voltage difference means that the duty cycle of the clock is off 50%, and in fact the duty cycle of CK1 or CK2 is around 60% in Fig. 8. As a result, the duty cycle that is measured by the new scheme in Fig. 2 is verified by another new scheme in Fig. 4.

IV. Summary and Conclusions

We have reported a new test and characterization scheme for a high performance PLL. By using VCO test mode, a frequency range of the VCO can be easily measured without jitter penalty on even manufacturing test and a tracking range of the PLL will be able to be predicted from that measurement result. The duty cycle measurement apparatus is suitable for characterizing a high performance PLL or a system clock of a microprocessor with multigigahertz operating frequencies under a general test environment, which needs lower setup cost. We also verified the accuracy of measured duty cycle at 4+ GHz.

References

[1] D.Boerstler, K.Miki, E.Hailu, H.Kihara, E.Lukes, S.Pettengill, J.Qi, J.Strom and M.Yoshida: "A 10+ GHz Low Jitter Wide Band PLL in 90 nm PD SOI CMOS Technology," Symposium on VLSI Circuits Digest of Technical Papers, pp.228-231, 2004

859

An SPU Reference Model for Simulation, Random Test Generation and Verification

Yukio Watanabe

Toshiba Corporation
Semiconductor Company
580-1 Horikawa-Cho, Saiwai-Ku,
Kawasaki 212-8520, Japan
yukio.watanabe@toshiba.co.jp

Balazs Sallay, Brad Michael,
Daniel Brokenshire, Gavin Meil,
Hazim Shafi

IBM
11501 Burnet Rd,
Austin, TX 78758, U.S.A.
{balazs, bradmich, brokensh, meil, hshafi}
@us.ibm.com

Daisuke Hiraoka

Sony Computer Entertainment Inc.
2-6-1 Minami-Aoyama, Minato-ku,
Tokyo 107-0062, Japan
hiraoka@rd.scei.sony.co.jp

Abstract – An instruction set level reference model was developed for the development of synergistic processing unit (SPU) , which is one of the key components of the cell processor [1][2]. This reference model was used for the simulators to define the instruction set architecture (ISA), for the random test case generator, for the reference in the verification environment and for the software development. Using the same reference model for multiple purposes made it easier to keep up with the architecture changes at the early stage of the microprocessor development. Also including the reference model in the simulation environment increased the robustness for the random test executions and made it possible to find bugs that are usually difficult to catch.

I Introduction

The Synergistic Processing Unit (SPU) is the first implementation of a new processor architecture designed to accelerate media and streaming workloads. The SPU instruction set architecture (ISA) was defined considering the physical implementation such as area, timing and power efficiency as well as the efficiency to run media and streaming applications. To achieve the target performance, ISA definition was frequently changed as the logic and physical design advanced.

To define the ISA, it was very important to write workload codes using the candidate ISA and evaluate its performance. The first reference model was implemented for this purpose as a simulator along with the assembler program. As the project advanced, the ISA definition changes were implemented in the reference model first, and then the updated reference model was used by various applications.

Since ISA changes continued even after verification started, various teams such as design team, performance analysis team and verification team were affected by these changes. However, since the SPU reference model could be included and used in those applications, only the reference model had to be modified and it was considered as the 'golden' ISA definition. Using the same reference model in various applications could reduce the mistakes when the applications needed to be updated because of the change, and it also became easier to keep up with the ISA changes.

In this paper, the basic structure of the reference model is described and followed by the explanation of applications that use the reference model. Among those applications, the verification environment is described in detail in a separate chapter because the reference model played a very important role to build up an effective verification environment.

II Instruction Set Reference Model

The SPU reference model is a set of C programs that perform SPU instruction execution. The reference model is comprised of following components.

- All the architected memory/register resources such as a register file, a local storage and status registers.
- Instruction decoding.
- Instruction execution and updating architected memory/register resources as the result of the instruction execution.

All the architected memory/register resources are defined in a C struct as 'struct APU_t' and this struct is usually memory allocated in each application program and used as the argument to the reference model. However, the random test case generator uses its own memory/register resource structure for some reasons, which will be described later.

The major functions defined in the reference model are following two functions.

void INTERPRETER_init(void);
void INTERPRETER_exec(APU_t *apu, unsigned int inst);

The first function is used for initialization of the reference model. The second one is used for instruction execution. The first argument of the second function is a pointer to the struct of architected memory/register resources APU_t, which represents the entity of the SPU. The second argument is the instruction encoded in a 32bit integer. When this function is invoked, the instruction of the second argument is decoded and a proper function that corresponds to the instruction is called in the reference model. The following C program is an example of the function that is

0-7803-9451-8/06/$20.00 ©2006 IEEE.

called when 'a' instruction is decoded.

```
/* add word: a rt, ra, rb */
void INTERPRETER_inst_a(APU_t *apu,
                        unsigned int inst)
{
  RTW(0)  = RAW(0)  + RBW(0);
  RTW(1)  = RAW(1)  + RBW(1);
  RTW(2)  = RAW(2)  + RBW(2);
  RTW(3)  = RAW(3)  + RBW(3);
}
```

RTW(n), RAW(n) and RBW(n) are macros which specify the register files whose numbers are determined by the portions of the instruction 'unsigned int inst'. These macros specify the memory elements in the SPU memory/register struct 'APU_t *apu'.

As for the floating instruction implementations, an SPU specific floating-point model was written and used in the functions that represent the floating instructions. This is because the SPU introduced a new floating-point architecture that differed in some ways from the commonly used IEEE 754 floating-point standard, in order to enhance performance for media-oriented applications. One method for calculating floating-point results in a C-language reference model is to use C's "float" and "double" data types. However, operations using such data types will be calculated using the floating-point hardware native to the machine executing the reference model, which may not match the SPU's architected behavior in certain cases.

To avoid any dependency on the underlying machine's floating-point hardware, the floating-point model within the SPU reference model was written using only integer data types, using separate integer variables to store the sign, exponent, and significand parts that make up a floating-point value. A 64-bit integer type is used to store the significand, to support the precision required by single- and double-precision floating-point instructions. The floating-point reference model includes functions to normalize, align, add, multiply, multiply-and-add, and round, all using integer arithmetic; as well as functions to "pack" the final results into the 32- or 64-bit representation called for by the architecture. The reference model detects and handles all exception cases such as overflow, underflow, divide by zero and loss of precision.

Since there can be multiple SPUs in one system, the SPU memory struct might be allocated multiple times and each one is used to represent the resources of each SPU. As shown in Figure 1, when the reference model is used in an application, the application program allocates SPU architected memory/register resource struct as many as the number of the SPUs that the application program needs to handle. The application program can directly access the resources in each SPU, and can execute the instruction by passing one of the pointer of the SPU memory/register struct in INTERPRETER_exec() function.

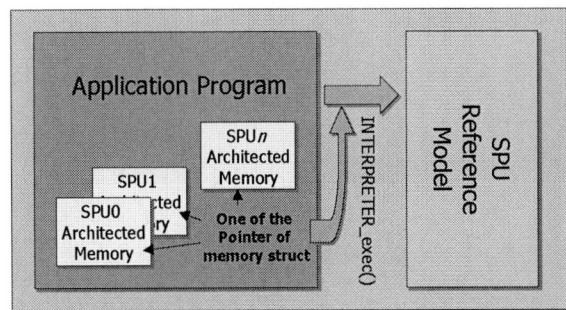

Fig. 1. Basic usage of the SPU reference model in an application

The definition of each instruction except for its function is written in a common definition file as a C macro. Following is an example of the definition of 'a' instruction.

```
APUOP(M_A,  RR, 0x0c0, "a", ASM_RR, 00112, FX2)
                    /* Add% RT<-RA+RB */
```

APUOP is the name of the macro that defines each instruction. The first argument of the macro is an identifier of the instruction. The second argument is the instruction format. This is followed by the op-code, the mnemonic, the assembler format, the register file usage and the kind of the pipeline used to execute the instruction.

The reference model itself does not use all the information in the macro. However, other applications such as the assembler, the pipeline simulator, the logic RTL and the verification environment refer the same common definition macro as well as the reference model, and each item in the macro is used by at least one application.

For example, the 6th argument in the macro indicates which register is the source register and which register is the target register. Each SPU instruction can take up to 4 registers as the arguments of the instruction depending on the instruction format. The four registers are represented as RA, RB, RC and RT. There are five digits in the 6th argument. From the most significant digit, the first one is always 0. The second digit corresponds to RC, and then the following digits correspond to RB, RA and RT, respectively. If the digit is 0, the register is not used by the instruction. If the digit is 1, that means the register is used as a source. If the digit is 2, the register is used as a target, and if the digit is 3, the register is used both as a source and a target. In case of 'a' instruction, two values of register RA and register RB are the source registers and their values are added and the result is put into the target register RT. So '00112' is put into the 6th argument of the macro. The RTL for register dependency checking logic uses this information. A script converts the macro description into the RTL description.

8D-2

III Usages of the Reference Model

The SPU reference model is used for various applications. In this section, some of them are introduced.

A. Instruction Simulator

Since the SPU is a new architecture, it was very important to implement and use the instruction simulator to evaluate the new ISA. Actually, the initial reference model was developed for this purpose. Figure 2(a) shows the configuration of the instruction simulator. In this figure, assembler is also included. The assembler program uses the common definition file described in section II as well as the reference model, and only changing the common definition file is required for the instruction changes. In the instruction simulator, a COFF format file generated by the assembler is read and the information (memory contents and the program counter value) are written into the SPU architected memory/register struct. The simulator gets commands via command prompt or command files. Commands for the instruction simulator includes

- Memory/Register dump
- Breakpoint setting
- Running program
- Step execution
- Show statistics such as number of instructions executed

The instruction simulator was mainly used to study the completeness of the instructions as a set and the correctness of the definition of each instruction. For example, at the beginning of the ISA study, the SPU ISA had byte/half word/word/double word-wise load/store instructions, but to realize a high frequency microprocessor, those instructions were divided into multiple instructions --- load quad word, generate control and shuffle byte instructions. Writing a code and running simulations with combining those instructions proved the completeness of the instruction set. Another example is that SPU has a series of right shift/rotate instructions that are represented as 'rotate mask' and 'rotate mask algebraic' instructions. The definitions are complicated and simulating those instructions found the bugs in the instruction definitions by running workload programs.

Extended Programming Features

In order to validate the SPU Instruction Set Architecture and implementation, real application workloads were developed. Since the targeted, media rich applications are visual in nature, verification of correct and efficient program execution required the development of several extended programming features. These features, which include check-pointing, file I/O, and streaming I/O, where added as special simulator extensions to the architected instruction set.

- Check-pointing

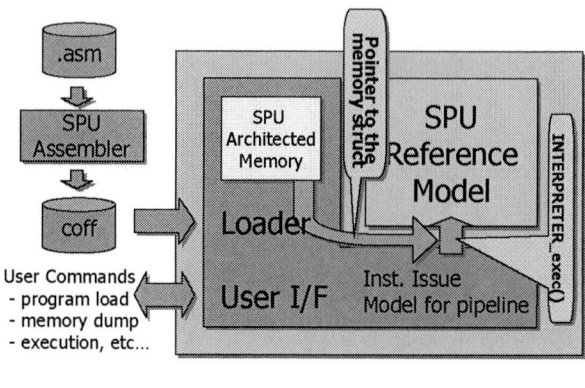

(a)

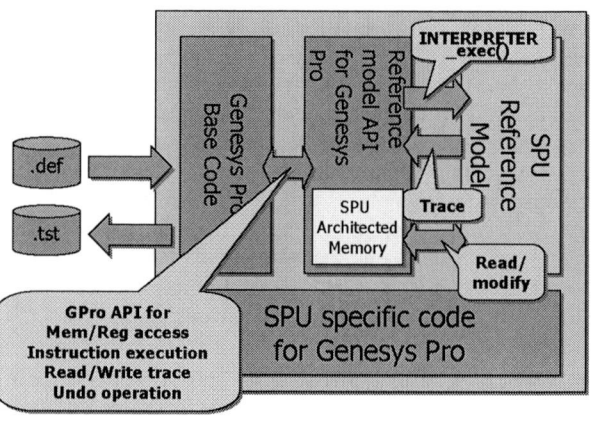

(b)

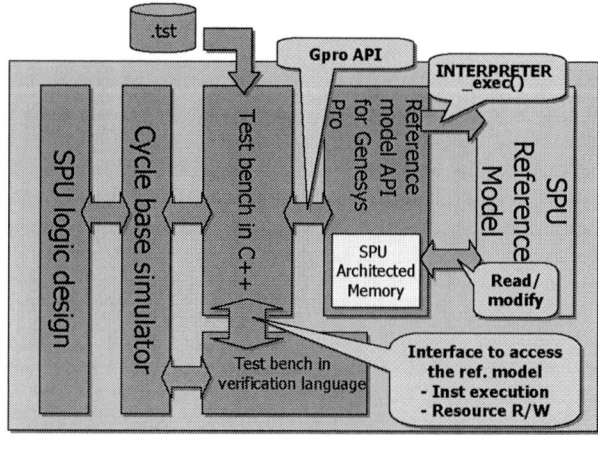

(c)

Fig. 2. Applications using the SPU reference model. (a)Instruction/Cycle-base simulator. (b) Random test case generator, (c) SPU verification environment.

862

For the purposes of computing cycle times for user specified code sections, software runtime check-pointing was provided by adding extended simulator operations for AND instructions in which all register fields are equivalent. For example:

- and r0, r0, r0 – clear the current instruction/cycle counts.
- and r30, r30, r30 – start counting instructions and cycles.
- and r31, r31, r31 – stop counting instructions and cycles.
- and r#, r#, r# (where # is a number 1-29) – output check-point # including the current instruction and cycles counts.

These instructions can easily be added to C language programs using inline assembly.

● File I/O

To support rudimentary debugging of SPE programs, the simulator was enhanced to support a full-function, file print (printf) subroutine. Since the SPU has architected Special Purpose Registers (SPRs) but its implementation contains no SPRs, the mtspr instruction simulation was extended such that the unused RB opcode field is used as an indicator of the extended function. The simulator's printf function is serviced by fetching parameters in accordance with the SPU's ABI (Application Binary Interface) standard.

● Streaming I/O

To support the display of graphical workloads, the simulator was also extended to enable external socket communications. The mtspr instruction was again used to make Unix sockets requests so that graphical data can be streamed to an external display program. The streams (connection sockets) are identified by a filename and its stream direction (outgoing or incoming). The supported socket requests include opening, closing and selecting a socket connection, as well as sending and receiving data on the current (outgoing and incoming, respectively) socket connection.

B. Pipeline Simulator

The pipeline simulator can count the actual cycle number when a SPU program is executed. This is used to evaluate the performance of the SPU micro architecture and to optimize the SPU application programs. The basic structure of the pipeline simulator is almost same as the instruction simulator case in Figure 2(a) except that it has an instruction fetch and issue model. This model takes into account the instruction fetch flow including local store access, instruction line buffers and other pipeline buffers, branch hint instructions and instruction issue controls considering the register dependencies and structural hazards and so on. There is a common pipeline definition file that describes the depth of each pipeline and the resources that each pipeline uses, which made it easier to evaluate various pipeline configurations.

The same user interface as the instruction simulator is used in the pipeline simulator as well. In the actual simulator program, both the instruction simulator and pipeline simulator are consolidated in one program and the simulation mode (instruction or pipeline) can be switched by a simulator command.

By modifying the common pipeline definition file, the impact of the pipeline depth change could be easily estimated for each workload program. This played an important role when the pipeline depth of the floating-point instruction was determined. Also the pipeline simulator was used to optimize workload programs. This includes the evaluation of the SPU C/C++ compiler optimization. By utilizing the pipeline simulator, one workload program that calculates the product of two large matrices was fully optimized and could achieve almost 0.5 cycles per instruction (CPI), which is the best case for the dual issue microprocessor.

Since the reference model covered the entire instruction execution portion, by using the reference model, the pipeline simulator programmer did not have to work for the instruction changes at all and could concentrate on the instruction fetch/issue logic changes.

C. Random Test Case Generator

In the SPU verification, random test cases are used to stimulate the logic function. To guarantee the quality of the verification, function coverage is used as the metric. It is important to generate high quality random test cases to hit all the function coverage items. To generate directed random test cases, Genesys Pro (GPro)[3], a tool developed by IBM, was applied to the SPU. GPro has a base core code that uses a specific application program interface (API) to communicate with various kinds of reference models. So Gpro becomes able to communicate with the SPU reference model by preparing the specific API functions that GPro uses for the SPU reference model. The API functions include

- Accessing the architected memory resources.
- Executing an instruction at the program counter and get the information what memory/register resources were read to execute the instruction and what memory/register resources were updated by executing the instruction.
- Setting undo points and undoing to handle mispredicted branches and recovery from test case generation failure.

To record the resource access traces and manage the undoing automatically using the reference model, all the SPU memory and register resources were defined in the random test case generator without using the struct APU_t which was described in section II. A C++ class was defined to access those memory/register resources and the resource access trace generation and undoing management. To utilize the reference model, the macros used in each instruction implementation such as RTW(n), RAW(n) and RBW(n) described in section II in the 'a' instruction example were redefined to use the class to access the appropriate memory/register resources. By using the operator overloading of the class, read trace information is added automatically when the resource was evaluated in the right hand side of the '=' operator, and the write trace and undo information are added automatically when the resource was placed in the left hand side of the '=' operator and updated with the value of right hand side. Thus, by redefining the macros in the reference model to use the C++ class, it became possible to use the same reference model and add the resource access trace and undoing functions.

Figure 2(b) shows the structure of the random test case generator. A definition file (.def file) is given to the random test case generator and a test case file (.tst file) is randomly generated with constraints written in the definition file. To generate the test case, GPro base code communicates with the SPU reference model. In this case, the API portion has the entity of the SPU memory/register resources. When GPro accesses the memory resources using the API, the content of the resources in the API is directly accessed. When GPro calls a function to execute an instruction specified by the program counter, the API portion picks up the instruction at the program counter in the SPU memory resources and invokes the function INTERPRETER_exec() with passing the instruction. The memory/register resources in the API are updated properly and the resource access information is generated and passed to the GPro base code and the information is used for the test case generation.

D. Reference in the Verification Environment

Figure 2(c) shows the SPU verification environment structure. There are two separate test benches written in C++ language and other verification language. The reference model is used to obtain the expected results for the checkers in the test bench written in the verification language. The C++ test bench is used to provide the API to the other test bench to communicate with the reference model to get and put the values of memory/register resources and get the expected values by executing instructions in it. In this application, the same API used by GPro, the random test case generator is reused as the interface to the reference model.

A test case file (.tst file) generated by GPro is given to the C++ test bench and the initial values in the test case are written into both SPU reference model memory/register resources and the logic design under the RTL simulator. When an instruction is executed in the SPU logic, it is observed by the monitor in the test bench written in the verification language and the test bench executes one instruction in the reference model and gets the expected

results through the API.

There are also drivers in the test bench which stimulate the SPU external interface buses. Monitors in the test bench observe the external interface buses. When there is a stimulus on the external interface buses which is considered to update the SPU memory/register resources, the resources in the reference model is updated with the value on the buses at a proper timing using the API. An example is a DMA write access. When a DMA write access happens, the content on the data bus is written into the SPU local storage with the address specified by the address bus. If a memory load instructions refers the address updated by the DMA transaction later, actual logic will use the updated value and the load instruction executed in the reference model also uses the same value. Thus, even if the driver issues random external bus stimulus, the reference model can generate accurate expected values.

E. Cell System Simulator

The SPU reference model was integrated with the IBM full system simulator known as Mambo [4] at an early stage of the development program. Mambo is an execution-driven full system simulator that allows multiple system configurations to be simulated. For example, Mambo was adopted as the software bring-up environment during the development of the CELL architecture. It has been used for operating system development, programming model investigations, compiler development, porting of important applications and libraries, and performance tuning. During the early stages of the Cell project, Mambo was intended to be functionally accurate and as such did not include cycle-accurate models of the full system. As the development project matured, so did the Mambo model, which was enhanced to provide more cycle-accurate feedback about the interactions between the SPUs, system memory, and the PowerPC core. The availability of such a model was crucial for studying the performance of the architecture at an early stage and to allow the performance tuning of applications as the project progressed.

IV. Reference model usage in the verification environment

As briefly described in section III-D, the SPU simulation environment incorporates the reference model, which made it possible to generate the correct expected values on the fly even with the asynchronous random external stimuli. This feature is very important for the SPU verification environment because lots of corner cases can be covered by the combination of random instruction sequences and random external transactions. The important thing here is that the instruction sequences and external transactions can be generated completely independent. The reference model can keep the same state as the actual logic with the help of the test bench and can continue the verification generating the correct expect values.

However, since the reference model is not a pipeline model, there are lots of things to be solved to integrate the reference model and make it work properly in the simulation environment. In this section, some of the items that need consideration are described.

A. Timing to execute the instruction

To integrate the reference model in the simulation environment, what had to be considered first was the timing when to execute the instruction in the reference model. Since the SPU is an in-order issue in-order completion processor, an instruction can be executed in the reference model to generate expected values whenever after an instruction is committed in the actual logic unless there are external transactions. However, since there are interactions between the instruction execution and external transactions they have to be handled properly to generate the correct expected values.

There are two kinds of external transactions. One is a DMA transaction that reads or writes local store memory. The other is a channel access transaction which reads or writes channel registers. The verification test bench monitors the external transactions and checks if the transactions were properly executed. At the same time, the test bench updates the reference model memory/register resources at a proper timing so that the instruction execution can use the proper memory/register values. As shown in figure 3, the actual logic accesses the local store memory array at the 'p' stage of the pipeline which is 2 cycles after the load/store instruction committed at 'n' stage. In case of channel instructions, channel registers are accessed at 'q' stage, which is 3 cycles after the channel instruction committed. To cope with the pipeline stage differences to access the resources with the reference model which doesn't have the concept of pipeline, a decision was made to make all the updates of the memory/register resources including external transactions and instruction executions by the test bench happen at 'q' stage. This made it possible to execute instructions in the reference model with asynchronous external transaction occurrences. However, since the timing to update the local memory array is different between the actual logic and the reference model, the checkers have to take care of this. For example a checker which checks if a store instruction updated the memory array properly has to get the actual logic value at 'p' stage, then after the store instruction is executed in 'q' stage in the reference model, the expected value is obtained from the reference model and it is compared against the actual value obtained at 'p' stage.

B. Resource confliction case

Channel instructions and external channel transactions can access the same channel registers exactly at the same timing.

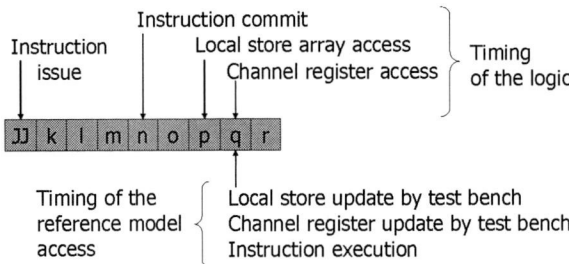

Fig. 3. Pipeline diagram of the actual logic pipeline and the timing to access the reference model.

Since what will be written into or what will be read out from a particular channel register in such a conflict case depends on the hardware implementation and there is no regularity, all the rules are described in the test bench and when a resource conflict happens, the test bench arbitrates this to generate the correct expect values. For example, there are cases that both of an instruction execution and an external transaction try to write into the same channel resource. The instruction is always executed in the reference model at 'q' stage and the channel register in the reference model is updated and then the expected value for the channel instruction result is obtained from the reference model. When the channel instruction has a higher write priority than the external transaction, the test bench does not update the channel register in the reference model with the value supposed to be written by the external transaction, but still obtain the expected value for the external transaction from the reference model. On the other hand, when the external transaction has a higher write priority, the test bench updates the channel register in the reference model at the same 'q' stage but after the instruction execution with the value that the external transaction is supposed to write. Then the test bench obtains the expected value for the external transaction from the reference model, and also the test bench overwrites the expected value for the channel instruction that was obtained right after the instruction execution in the reference model. Thus, with the help of the test bench, reference model can keep the same channel register values as the actual logic and continue to provide the correct expected results.

C. Self modifying code case

In the actual logic, instructions are fetched from the local store memory and travel through the pipeline stages. What will happen if a 'store' instruction writes into an address of the instructions following the store instruction? Since the instructions had been fetched a while ago, the next instructions would be the instructions before modification. On the other hand, if the same store instruction was executed in the reference model, the next instruction executed in the reference model would be a modified instruction because there is no concept of pipeline in the reference model. Therefore, if a self-modifying code is executed, the reference model cannot continue executing the same sequence as the actual logic. To handle this case, test bench keeps information which memory addresses were written by either store instruction or DMA write transactions. Before the instruction is executed in the reference model, the test bench compares the instruction to be executed with the instruction committed by the logic. If they are different, it is usually reported as a test failure, but if the instruction address was modified recently enough to make a difference, the content of the memory of the instruction address in the reference model is saved and replaced by the instruction value that was committed in the actual logic, and the instruction is executed in the reference model. Then unless the executed instruction is not a store instruction that modifies itself, the saved instruction is restored into the reference model. Thus, with the help of the test bench, the reference model can keep up with the actual logic state even with the self modifying code case which instruction set reference model cannot handle properly only by itself.

D. ECC error handling

The SPU has an ECC facility that can correct 1-bit error and detect 2-bit errors in 128-bit data in the local store memory. Since the reference model doesn't have the ECC information in it, the test bench has to support the reference model to handle the ECC. Basically, the ECC errors are injected at the beginning of the simulation or injected by the DMA transactions. The information which address has what kind of ECC errors is kept in the test bench. When a correctable error happens in the logic, the test bench doesn't have to do anything against the reference model because the correctable error should be corrected in the actual logic and the corrected data, which should be the same value in the reference model, will be used. When the instruction that has an uncorrectable ECC error is executed in the actual logic, the test bench can detect it because the test bench has the information of the addresses that have ECC errors. In this case, the test bench saves the original instruction and overwrites the address of the instruction in the reference model with the value modified by the uncorrectable ECC error. Then, as in the case of self-modifying code, the modified instruction is executed in the reference model and the original instruction is restored unless the modified instruction is the store instruction that modifies the instruction itself.

E. Others

There are several other cases such as asynchronous interrupt handling, error handling, undefined instruction handling and so on that cannot be handled only by the instruction level reference model. However, all those cases were implemented in the SPU verification environment with the help of the test bench. So the SPU verification environment has a good robustness that any kinds of combinations of instruction sequence and external transaction can be treated properly and the model generates the correct expected values.

If the reference model were implemented as the complete pipeline model, the reference model itself could cover all the cases. But it is usually very difficult to implement the complete pipeline model, and it also slows down the simulation speed. Using the instruction level reference model realized the verification environment with a good quality and performance.

V. Summary and Conclusions

In this paper, the structure and usage of the SPU instruction level reference model was described. The reference model played important roles for the development of the SPU, a novel high performance processor. The reference model was used for the purpose of defining the ISA, analyzing performance, verification environment development and software development. The same reference model could be used for all these purposes, which reduced the burden to keep up with the ISA changes for each application developer and significantly reduced the likelihood of mistakes in the implementation. As for the verification environment, with the help of the test bench, the instruction level reference model could cover pipeline related items and it realized an ability to continue the simulation and generate correct expected values with asynchronous external events. As a result, though the SPU is a very novel processor, only one bug was found in first silicon, which was a mistake of the specification of the asynchronous interrupt. And a demo program that utilizes the SPU was successfully run on first silicon with the expected good performance.

Acknowledgements

The authors would like to thank Seiichiro Saito, Hiroko Fujii and Nobuhiro Kondo of Toshiba Corporation for the development of the prototype of the SPU reference model, and Kanna Shimizu and Peter Hofstee of IBM for useful suggestions.

References

[1] Pham D. et al, "The Design and Implementation of a First-eneration CELL Processor," 2005 IEEE International Solid-State Circuits Conference Digest of Technical Papers, pp. 184-185, Feb. 2005.

[2] Flachs B. et al, "A streaming Processing Unit for a CELL Processor," 2005 IEEE International Solid-State Circuits Conference Digest of Technical Papers, pp. 134-135, Feb. 2005

[3] Allon A. et al, "Genesys-Pro: Innovations in Test Program Generation for Functional Processor Verification," IEEE Design & Test of Computers 21(2): 84-93 (2004)

[4] Patrick B. et al, "Mambo -- A Full System Simulator for the PowerPC Architecture," ACM SIGMETRICS Performance Evaluation Review, 31(4): 8-12, Mar. 2004.

A Cycle Accurate Power Estimation Tool

Rajat Chaudhry

STI Design Center
IBM Corporation
Austin, TX 78758
Tel : 1 512-838-8794
Fax : 1 512-838-2132
rajat@us.ibm.com

Daniel Stasiak

STI Design Center
IBM Corporation
Austin, TX 78758
Tel : 1 512-838-1952
Fax : 1 512-838-2132
stasiak@us.ibm.com

Stephen Posluszny

STI Design Center
IBM Corporation
Austin, TX 78758
Tel : 1 512-838-6508
Fax : 1 512-838-2131
Stephen.Posluszny@us.ibm.com

Sang Dhong

STI Design Center
IBM Corporation
Austin, TX 78758
Tel : 1 512-838-0106
Fax : 1 512-838-1272
dhong@us.ibm.com

Abstract - Power consumption is one of the major challenges in VLSI Design. Power constrained designs need tools to accurately predict the power consumption and provide feedback to designers on the efficiency of the power management logic. In this paper we present the methodology behind a cycle accurate power estimation tool. This tool was used to estimate the power of a first generation CELL Processor. The tool extracts switching and clock activity from RTL simulations and applies them to transistor level macro power models to calculate the power for every cycle of the simulation trace.

1. Introduction

Power Consumption is one of biggest challenges in VLSI design. In the past, power has mainly been a concern for chips used in battery-powered devices. But due to the continuous increase in frequency and increase in leakage current due to scaling, it is becoming a constraint even in wall socket powered devices. Packaging and thermal cooling costs are the biggest drivers for reducing power in such chips, especially chips that are manufactured in large quantities for price sensitive products.

Due to the tight power constraints, chip designers require power estimation tools that can provide them with accurate power estimates and timely feedback on the efficacy of their power management design. As the margin of error in power budgets reduces, the need for dynamic power estimation increases. Not only is dynamic power estimation required to verify power management logic, but also because static estimates are too pessimistic. It is important to concentrate on realistic high power test cases. Pathologically high power workloads can be handled by hardware based thermal management solutions, by reducing the frequency or shutting down the system.

2. Previous Work

Power estimation has been done at different level of abstraction. The Most accurate estimates are done by running a SPICE like simulator on a transistor level netlist.

But this is feasible only for a small circuit. Many authors have worked on building accurate macro power models [1]. Macro power models are not very useful in estimating total chip power unless we have accurate information on the switching and clocking activity of the macros. A lot of work has been also been done at measuring power at the architectural level [2]. Although architectural level solutions are very fast, they are not very accurate and cannot give feedback on clock gating at the RTL level. In ASICs, gate level power estimation methodology has been employed. This does not work well for chips with a lot of custom blocks.

In the first generation CELL Processor (the first generation CELL processor will be referred to as the Processor for the rest of this paper), designers have employed a lot of fine-grained clock gating. We needed a tool that could give accurate power estimates as well as provide information on the power tradeoff of adding clock gating logic. The processor consists of custom blocks and synthesized logic [4]. In our cycle accurate power estimation tool (which will be referred to as CAPET in this paper) we combine the benefit of accurate transistor level macro power models coupled with switching and clocking activity from RTL level simulations. This is essential, because gate level techniques cannot accurately model power for custom designed blocks. We also model the power consumed due to the switching of signal interconnect capacitance.

3. CAPET Approach

The biggest components of the power dissipation of a circuit consists of 1) the leakage current, which depends on the state of the transistors and 2) the switching power which consists of the power dissipated by the switching of node capacitances due to circuit activity. CAPET estimates switching power, so we will limit our discussion to switching power.

The switching power of a circuit can be calculated by the function

$$P = \frac{1}{2}\,CV^2 f. \tag{1}$$

Where C is the total node Capacitance switched, V is the power supply voltage and f is the frequency of switching. For a circuit design, V and f are generally fixed.

To reduce the power consumption, designers have to work on reducing the switched capacitance. In a power efficient processor like our Processor, designers employ intensive amount of fine-grained clock gating to reduce power. This reduces the switched capacitance by limiting the amount of switching in the latches and unnecessary switching of combinational logic. Therefore the key determinants of power estimation are monitoring the switching and clocking activity in a design.

The CAPET methodology consists of monitoring the switching and clock activity of each macro in an RTL simulation and then applying that information to a transistor level macro model to estimate the power. This is done for each clock cycle to get a cycle-by-cycle power estimate. The power consumed by switching interconnect capacitance is also estimated by monitoring the switching of each global net. The gate capacitance of buffers on the net is included as part of interconnect capacitance.

The minimum requirements for running CAPET are a functional VHDL model and a chip floorplan. In the early phase of design, macro power models can be roughly estimated using macro area and net capacitance is calculated by Steiner estimates. This way CAPET can be run early in the design for designer feedback and the estimates get more refined as the design progresses as shown in figure 1.

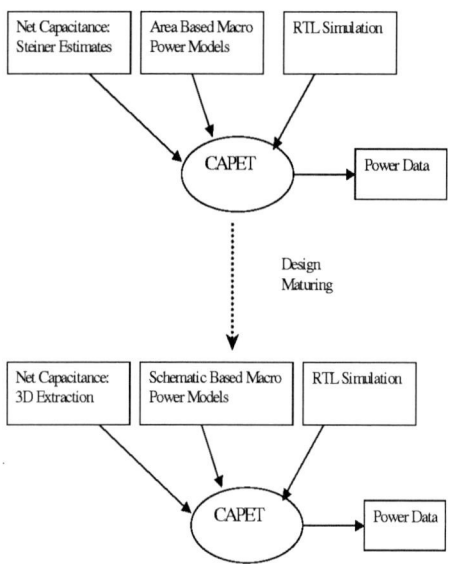

Figure 1: CAPET flow at different stages of design

3.1 Macro Power Models

The node switching in a circuit block in a given cycle is proportional to the percent of inputs switching and the amount of clock activity in the circuit. Switching factor of a circuit is defined as the percent of inputs changing state between two consecutive clock cycles. Clock activity is defined as the percent of capacitive load being driven in a given cycle with respect to the total clock load in the circuit. We build our macro power models at the transistor level using IBM's CPAM tool [3]. CPAM runs random vectors with different Switching Factors on the schematic under two conditions 1) with all local clock buffers turned off and 2) all local clock buffers turned on. The power model assumes that power is linear with switching factor and clock activity. CPAM provides power information at simulated switching factors

If for a given clock cycle, SF is the Switching factor of a circuit block and CLK is the clock activity of the circuit then the Power consumed by the circuit in a given cycle C is

$$P(C) = P_{clk0}(SF) + (P_{clk100}(SF) - P_{clk0}(SF)) * CLK.$$

Where

$P_{clk0}(SF)$ is the power at input switching factor SF when clock activity is 0%

$P_{clk100}(SF)$ is the power at input switching factor SF when clock activity is 100%

The 2 curves P_{clk0} and P_{clk100} are shown in figure 2.

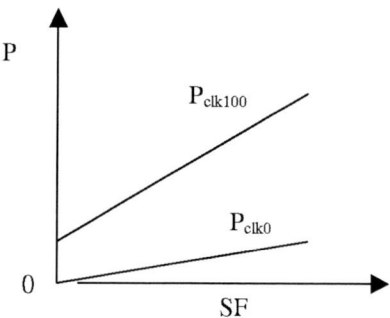

Figure 2: Power dependence on Switching Factor at 100% clock activity and 0% Clock Activity

Currently we simulate macros only at a SF of 0% and 50%; therefore our current models are purely linear with respect to SF. The accuracy of the power model can be improved by simulating at more number of Switching factors. In the early part of the design phase the 2 power curves P_{clk0} and P_{clk100} are estimated using the area of the macro by using a power density derived by scaling from previous technology or from similar macros whose schematics are complete.

3.2 Global Signal Net Power

The macro power models characterize the power switching power inside macros. To estimate the power due to switching of global interconnect capacitance, we calculate the interconnect capacitance using Steiner estimates in the early part of the design phase and use a 3D extraction tool in the later part of the design. The power of switching global nets can be calculated using equation (1). The power of the buffers inserted on the global nets is estimated by adding the gate capacitance of the buffers to the net capacitance. This approach does neglect the shoot through current from VDD to GND while the buffer is switching. Our experiments show that this shoot through current is a negligible part of switching power.

3.3 Measuring Switching Factors and Clock Activity

To calculate the total power, Input switching factors and clock activity for each block instance are monitored in an RTL level simulation of a given workload. For every macro instance the input Switching Factor is calculated by observing the percent of inputs that have changed state from the previous cycle. The Clock Activity is measured by observing the number of clock buffers that are turned on in a given cycle.

Since power is dependent on both Switching factor and clock activity, it needs to be calculated for every cycle of the simulation. The switching factor and clock activity cannot be averaged over the entire simulation.

In a given cycle the total power is calculated as

$$TotalPower(C) = \Sigma BlockPower(SF, CLK) + \tfrac{1}{2}\, C_{net}(C)V^2 f$$

Where C_{net} (C) is the amount of global net capacitance switched in the given cycle.

The measure of clock activity is dependent on the type of macro. Since each local clock buffer drives different amount of load, they cannot be treated as equal. For custom macros the designers provide a table, which puts relative weights on each local clock buffer. In case of synthesized blocks the relative weights are calculated using the number of latch bits that are driven by each local clock buffer.

4. CAPET Usage

CAPET methodology was used to estimate the power consumption of the first generation CELL processor and for refining and verifying the power management logic of the chip.

4.1 RTL Workloads

Each core or functional unit on the chip was required to run at least 3 different type of workloads, 1) idle, 2) typical and

3) high power. The idle workload is very useful in making sure that when the core is supposed to be at the lowest power state, it shuts off as many clock buffers as possible. Analyzing the results of the idle test case is very useful in catching the most obvious errors. After running CAPET, designers look at the macros that have the highest power and highest clock activity and try to reduce them. Some of the cores also used the tool to verify the power savings by issuing instructions at a slower rate.

4.2 Power Grid and Thermal Analysis

We used the results of the high power test case as a stimulus for Power Grid integrity analysis and thermal analysis. Each core is assigned the average power for the high power workload for IR drop analysis. For *di/dt* analysis on the package the cycle-by-cycle power for the high power workload was used for each core. The ability to have a cycle-by-cycle power for each macro for a realistic workload makes CAPET very useful for package analysis. For thermal analysis the average power for the high power workloads are used to calculate the temperature map for the chip.

4.3 Full Chip Estimates

Since RTL simulation is not fast enough to run a complete program at the chip level, the chip level estimates were calculated by getting utilization information on the different cores from architectural simulation. Based on realistic high power programs, utilization rates for the 3 different work loads, idle, typical and high power were assigned to each core and this was aggregated up to the chip level.

5. Results and Conclusion

In this section we present the results of CAPET on one of the cores of the Processor. Figure 4 shows the power waveforms for the stop, typical and high power test cases. Figure 3 shows the macro power and net power of typical power workload. Figure 5 shows the number of clock buffers active for the 3 different workloads. The runtime for CAPET is similar to the runtime for an RTL simulation. The runtime for 4000 clock cycles on a core with 20.9 million transistors is approximately 30 minutes.

We have correlated the results of CAPET with hardware and the estimates are within plus or minus 10%. Multiple chips were carefully run under the exact same conditions as simulation to correlate the results.

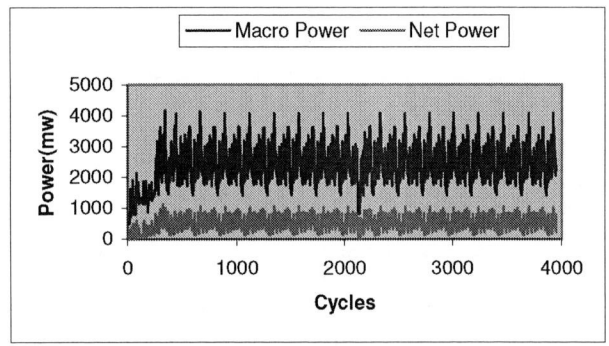

Figure 3: Macro Power and Net Power for typical power workload

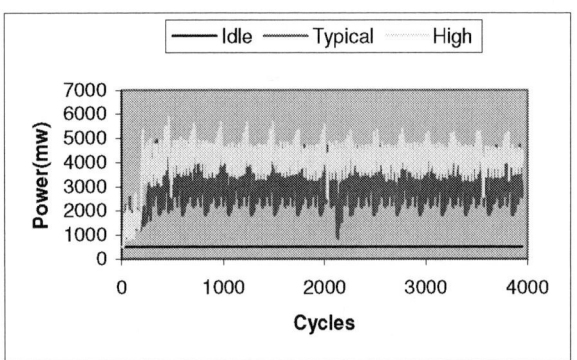

Figure 4: Total Power for Idle, Typical and High power workloads

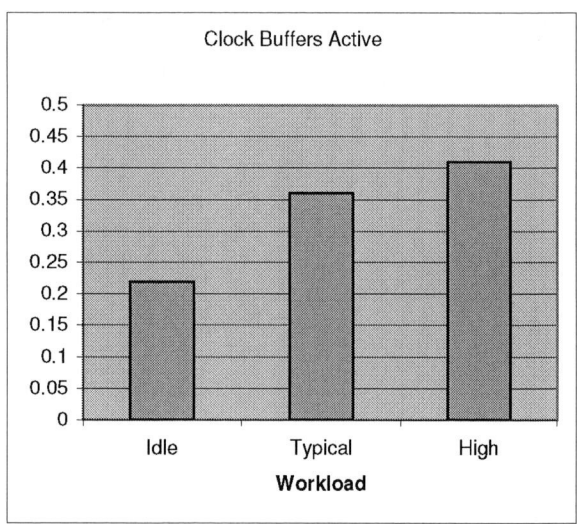

Figure 5: Percent of local clock buffers active for idle, typical and high power workload

We have presented a power estimation methodology that helps in providing accurate power estimates and can provide designers with feedback on the clock gating logic. The hardware measurements show that our methodology is accurate. We are working on improving our macro models by measuring power of macros at multiple switching factors and making power a piece wise linear function with respect to switching factor.

6. REFERENCES

[1] Subodh Gupta and Farid N Najim, *Power Modeling for High Level Power Estimation*, IEEE transaction on VLSI, vol 8. No 1, Feb 2000.

[2] Brooks et al., *Power-aware microarchitecture: design and modeling challenges for next-generation microprocessors,* Micro, IEEE, Volume: 20, Issue: 6, Nov.-Dec. 2000 Pages: 26 - 44

[3] Neely, J.S et al., ***CPAM: a common power analysis methodology for high-performance VLSI design,*** Electrical Performance of Electronic Packaging, 2000, IEEE Conference on. , 23-25 Oct. 2000

[4] E Behnen et al, *Ontology of The First Generation CELL Microprocessor*, Submitted to DAC 2005.

2006 Asia and South Pacific Design Automation Conference

Key Features of the Design Methodology Enabling a Multi-Core SoC Implementation of a First-Generation CELL Processor

Dac Pham, Hans-Werner Anderson, Erwin Behnen, Mark Bolliger, Sanjay Gupta, Peter Hofstee, Paul Harvey, Charles Johns,
Jim Kahle, Atsushi Kameyama[1], John Keaty, Bob Le, Sang Lee, Tuyen Nguyen, John Petrovick, Mydung Pham, Juergen Pille,
Stephen Posluszny, Mack Riley, Joseph Verock, James Warnock, Steve Weitzel, Dieter Wendel

IBM Systems and Technology Group, Austin, TX
[1]Toshiba America Electronic Components, Austin, TX

Abstract-- **This paper reviews the design challenges that current and future processors must face, with stringent power limits and high frequency targets, and the design methods required to overcome the above challenges and address the continuing Giga-scale system integration trend. This paper then describes the details behind the design methodology that was used to successfully implement a first-generation CELL processor - a multi-core SoC. Key features of this methodology are broad optimization with fast rule-based analysis engines using macro-level abstraction for constraints propagation up/down the design hierarchy, coupled with accurate transistor level simulation for detailed analysis. The methodology fostered the modular design concept that is inherent to the CELL architecture, enabling a high frequency design by maximizing custom circuit content through re-use, and balanced power, frequency, and die size targets through global convergence capabilities. The design has roughly 241 million transistors implemented in 90nm SOI technology with 8 levels of copper interconnects and one local interconnect layer. The chip has been tested at various temperatures, voltages, and frequencies. Correct operation has been observed in the lab on first pass silicon at frequencies well over 4GHz.**

Index Terms—CELL Processor, multi-core, SOC, SOI, modularity, re-use, 64-bit Power Architecture, multi-threading, synergistic processor, flexible IO, Linux, multi-operating system, virtualization technology, real-time system, hardware content protection, correct-by-construction, thermal management, power management, clock distribution, high-performance latch, local clock buffer, design hierarchy, design environment, design dependency solution, linear sensor, digital thermal sensor.

I. INTRODUCTION

The architectural vision of "bringing supercomputer power to everyday life" is the driving force behind the CELL processor design, setting a new performance standard by exploiting parallelism while achieving high frequency [1]. CELL is designed for natural human interactions: photo realistic, predictable real time response, and virtualized resource for concurrent activities. CELL supports multiple operating systems including Linux, and is designed for flexibility with a wide variety of application domains. Other attributes include hardware content protection, and extensive single-precision floating-point capability. By extending the Power Architecture with Synergistic Processor Elements (SPE) having coherent DMA access to system storage and with multi-operating system resource management, CELL supports concurrent real time and conventional computing.

With a dual-threaded Power Processor Element (PPE) and eight SPEs this implementation is capable of 10 simultaneous threads and over 128 outstanding memory requests.

The First-Generation CELL processor consists of the PPE and its L2 cache, eight SPEs [2] each with its own local memory (LS) [3], a high bandwidth internal Element Interconnect Bus (EIB) [4], two configurable non-coherent I/O interfaces, a Memory Interface Controller (MIC), and a Pervasive unit that supports extensive test, monitoring, and debug functions. The high level chip diagram is shown in figure 1 below.

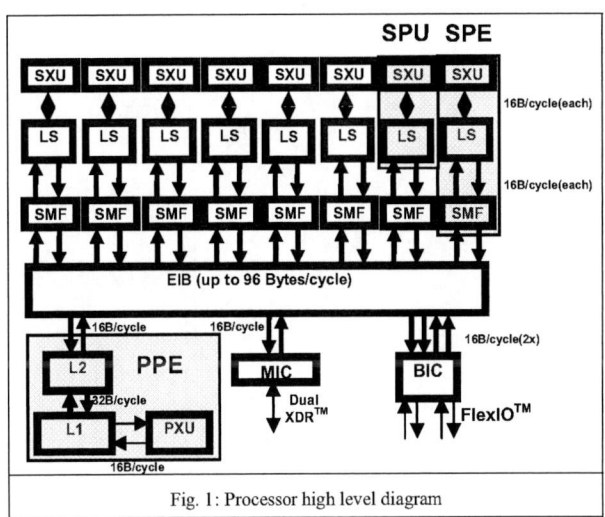

Fig. 1: Processor high level diagram

II. THE DESIGN CHALLENGES FOR GIGA-SCALE INTEGRATION

II.1. *Power & Frequency Walls*

Over the last decade, technology scaling has resulted in leakage power increases of over 1000X (fig. 2). With gate dielectrics and other device features fast approaching fundamental limits, a continuation of historical trends would see passive power surpassing active power within the next few years. Furthermore, the technique of increasing frequency by deepening the pipeline has reached a point of diminishing performance returns if power is taken into consideration. In the face of this power/performance wall, increased design efficiency becomes essential. These

0-7803-9451-8/06/$20.00 ©2006 IEEE.

871

factors drove the decision to support a wider processor issue width (e.g. multi-threading) and to increase the number of architected registers.

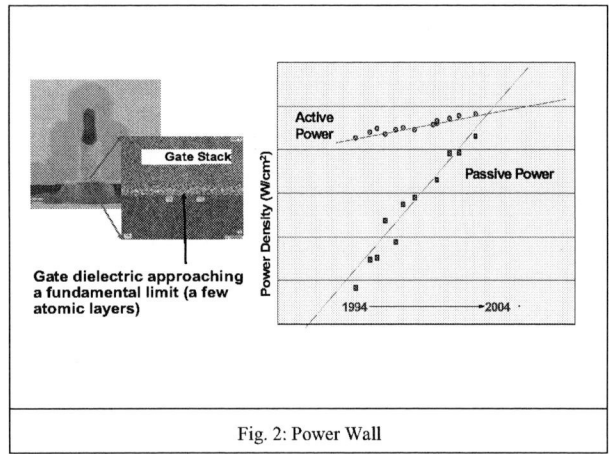

Fig. 2: Power Wall

II.2. System Trends and Giga-Scale Integration

Increased system integration is driving processors to take on many of the functions typically associated with the system: off load and acceleration, and integration of bridge chips as shown in figure 3.

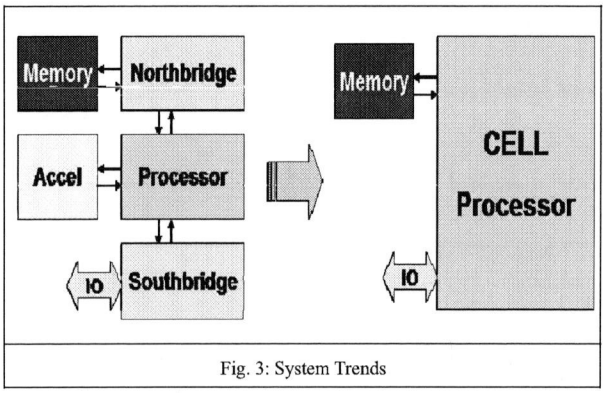

Fig. 3: System Trends

III. DESIGN IMPLEMENTATION TO ADDRESS POWER AND FREQUENCY WALLS

III.1. Components and Libraries Design

Given a short cycle time target, a significant amount of the chip power is consumed by latches, flip-flops, and other clocked elements. However, the delay overhead imposed by standard flip-flops is considerable. Therefore, a rich set of latches and flip-flops were developed to allow for both power and delay optimizations. The basic local clock splitter components are shown in figure 4. In addition to test controls, the base block accepts a local clock gating signal, with a small setup time relative to the falling global clock (cycle boundary). Input setup and hold times are specified against the falling clock edge, as a result of the built-in latching action of the base block. Local clocks, to drive

typical master-slave flip-flops, are derived from the common output point of the base block.

For timing critical paths, a high-performance latch (HPL) [5, 6] was designed which combines a wide mux (up to 10-way), relying on a dynamic NOR gate, with a set-reset latch (fig. 5). The dynamic NOR starts evaluating with the launch clock, and the input data hold time is limited when all sel_b inputs are forced high after a fixed delay.

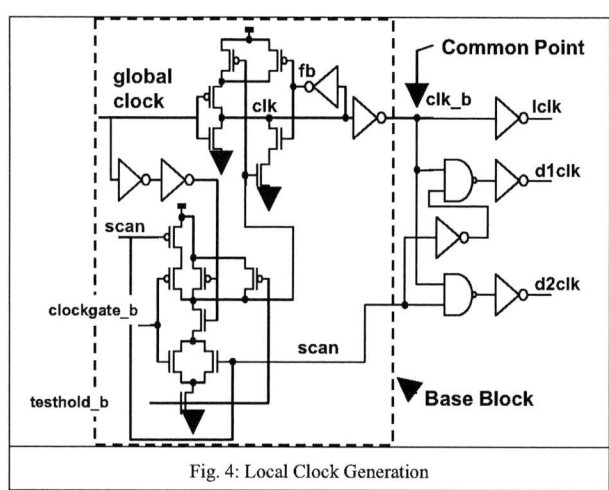

Fig. 4: Local Clock Generation

Dynamic circuits were used in several critical macros, in the arrays, and in PLAs. All dynamic macros were latch-bounded (macro-to-macro signals are static). Signals feeding dynamic logic were usually launched from the master portion of a flip-flop, and ANDed with the slave clock (lclk) to provide a signal which resets to 0 every cycle when lclk is low. Dynamic logic was always followed by a set-reset latch similar to that used for the HPL shown earlier.

In addition, various rules were adopted to ensure a "correct-by-construction" design methodology. All circuits used a common set of clocking components to ensure uniformity across the design, with no rotation of the components allowed. An extensive set of electrical and physical checks and audits were put in place. Finally, a customized series of yield-related checking rules was employed to ensure manufacturability of the chip.

The 90nm PD SOI technology offers three oxide thicknesses (thin oxide, thick oxide for high voltage device, and decoupling capacitor) and four different V_T settings for the thin oxide devices. Since power was such a critical design issue, static circuit implementations were favored for the majority of the design. A variety of static circuit families were used in full custom designs, with tuners and device width optimizers used for power-performance tuning. Higher threshold voltage devices were inserted wherever possible to cut down on leakage current (no low V_T devices were used), and the threshold voltage for the array devices was adjusted independently from that of the logic devices. Approximately 40% of the logic was implemented as synthesized random logic macros (RLMs), with the rest being full custom design.

The local clocking described in figure 4 has several important features. Overall clock latency and absolute clock uncertainty is minimized by this scheme since there are only three gate delays between the global clock input and the data launch clock (lclk). Also, the common point for both launch and capture clocks are at the output of the base block, minimizing the relative uncertainty between launch and capture clocks. When clocks are in the gated state, lclk is held inactive, and the capture clock is held high. The system state is therefore stored in the slave latch.

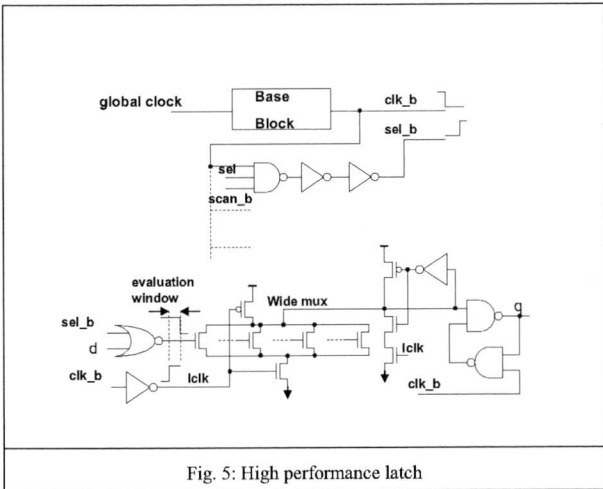

Fig. 5: High performance latch

For power reduction, the standard flip-flop can be run in pulsed-mode, with a clock configuration shown in figure 6. In this case the slave clock is pulsed in normal operation, and master clock is held high. There is also a "chicken switch" which allows running in normal master-slave clocked mode if race problems are seen in the hardware. A non-scannable pulsed latch was also supported, minimizing area, power, and latency in situations where a longer hold time could be tolerated.

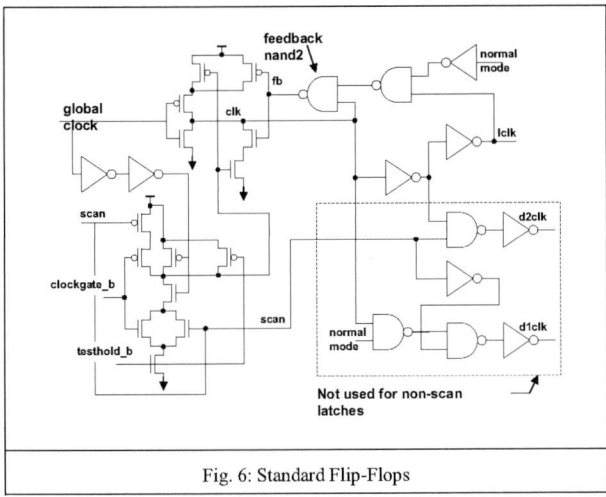

Fig. 6: Standard Flip-Flops

With the widespread use of pulsed latches, and the controlled use of clock delay elements, it was very important to have a robust methodology to check for race conditions.

The timing methodology required a design margin to be applied which scaled with the total path delay of the racing paths (in addition to a certain fixed margin), as measured from the common point of divergence. This ensured that race conditions with larger uncertainties were designed with correspondingly larger margins.

III.2. Clock Distribution

The chip contains three distinct clock distribution systems, each sourced by an independent PLL, to support processor, bus interface, and memory interface requirements. A main high frequency clock grid covers over 85% of the chip, delivering the clock signal to the processors and miscellaneous circuits. Second and third clock grids, each operating at fractions of the main clock signal are interleaved with the main clock grid structure, creating multiple clock frequency islands within the chip. All clock grids were constructed on the lowest impedance final two layers of metal, and were supported by a matrix of over 850 individually tuned buffers. This enabled control of the clock arrival times and skews, especially on the main clock grid, which supports regions of widely varying clock load densities. As shown in figure 7, final worst-case clock skew across chip was less than 12ps. High frequency clock distribution optimization and verification needed models which included frequency sensitive inductance and resistance phenomena [7]. These models were built from data extracted from combined clock and chip power distributions, two dimensional cross sections, and capacitance models extracted from three dimensional sections. Reduced clock grid power dissipation was achieved through optimization of buffer drive strengths, grid wire periodicity, clock wire to return path spacing, and clock twig wire widths. Together, these techniques lowered clock distribution power dissipation by more than 20% compare to previous design [8].

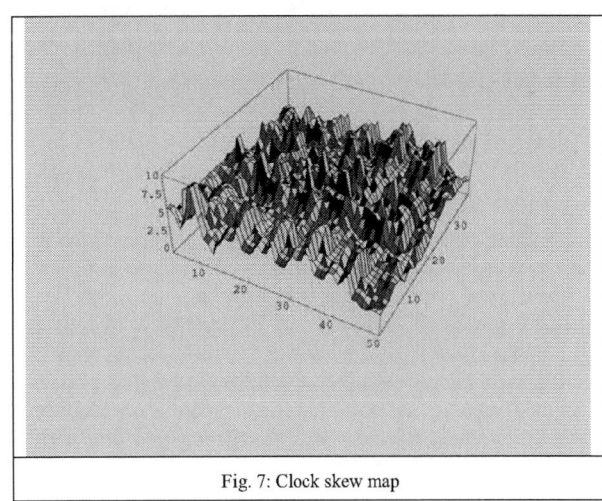

Fig. 7: Clock skew map

III.3. Thermal and Power management

This SOC presented new challenges in chip thermal design. The higher heat flux from smaller hot spots hindered spreading of the heat across the silicon substrate [9]. Extensive thermal analysis carried out early in the design cycle ensured that the maximum junction temperature, as well as the average temperature of the die, would end up within design specifications. Various workloads were simulated for each component and power maps were constructed. From these maps, a matrix of small power sources was created, for use with package and heat sink models. Thermal models were then created and used to simulate both steady state and transient thermal behavior. These data were analyzed to improve the design and floor plan of the chip, and also provided feedback for improved thermal sensor design (fig. 8).

Due to local heating caused by individual processing units, sophisticated local thermal sensing strategies and thermal control mechanisms were used to allow an aggressive low cost thermal design. The processor contains a linear sensor and 10 local digital thermal sensors. The linear sensor is essentially a diode connected to two external I/Os, used to measure the die's global temperature and to adjust the system cooling. The digital thermal sensors provide for early warning of any temperature increase and for thermal protection.

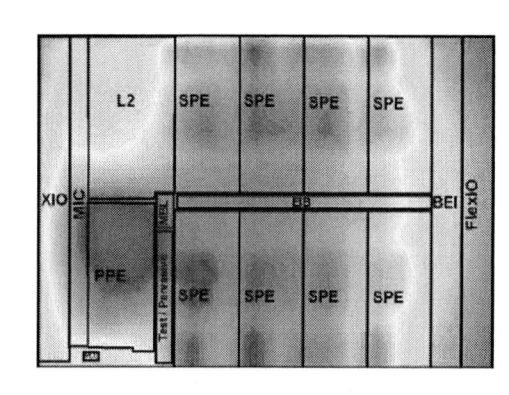

Fig. 8: Chip thermal map

III.4. Pervasive Design

The pervasive logic comprises all the function necessary for initialization, clock control, test, performance monitoring, and error checking and reporting. For a complex multi-core processor, the design of the pervasive logic is a key emphasis early in the design cycle. The pervasive function is implemented as a centralized controller and in distributed units across the chip. A Performance Monitor (PFM) is provided to assist with the debug and tuning of software applications, and an on board logic analyzer (LA) assists with hardware debug. Both the PFM and LA are capable of capturing information at speed from all units across the chip. The PFM and LA also provide the capability to view single or multiple units. Extensive debug and control capabilities are provided that can be accessed via an IEEE 1149.1 interface. For manufacturing support, the

pervasive unit provides array and logic built in self test (BIST) engines. The ability to scan at speed is provided to assist with detection of AC related faults. Electronic fuses are used extensively for array repair and selected chip personalization.

III.5. Physical Design

Figure 9 shows the die photo with roughly 234M transistors from 17 physical entities, 580K repeaters and 1.4M nets implemented in 90nm SOI technology with 8 levels of copper interconnects and one local interconnect layer.

Fig. 9: Die Photo

At the center of the chip is the EIB, composed of four 128-bit data rings plus a 64-bit tag operated at half the processor clock rate. The wires were arranged in groups of four, interleaved with GND and VDD shields twisted at the center to reduce coupling noise on the two unshielded wires. To ensure signal integrity, over 50% of global nets were engineered with 32K repeaters. The SOC uses 2965 C4s with four regions of different row column pitches attached to a low cost organic package. This structure supports 15 separate power domains on the chip, many of which overlap physically on the die. The processor element design, power and clock grids, global routing, and chip assembly support a modular design in a building block like construction.

IV. KEY FEATURES OF DESIGN METHODOLOGY

IV.1. Hierarchical Design and Rule-Based Optimization Methodology

IV.1.1. Design Environment and Database Structure

There were many challenges in meeting the defined objectives for setting up the design environment and database system for the first-generation CELL processor project. First, the methodology had to support concurrent design execution of each *partition* (major core of the design such as PPE or SPE); meaning design work had to be done simultaneously and independently by different teams located in different geographical areas. The existing inherently hierarchical nature of the design was carefully considered when defining the physical partitions in order to minimize the impact of creating discrete physical partitions. Strict *naming convention* schemata were applied to the entire *design hierarchy* to facilitate parallelization and to prevent collisions.

Second, the *database structure* had to support both the hierarchical objective and also multiple design disciplines, namely logic design and verification, physical design and verification, integration, and timing, etc., to allow for efficient schedule interlocking. An AFS network file system was used to allow transparent access to design data by all team members across multiple geographical locations. Additionally, the database structure had to support common design libraries and many "shared" macros used in multiple units or partitions. Any dependency conflicts caused by usage of different levels of these libraries and macros across the design hierarchy were resolvable by *design dependency solution* algorithms, supported by the design environment.

Third, the *design environment* had to fully support the custom processor design methodology by providing tools, processes, and a workspace for every designer. The design environment fills the vital linkage between designers and the supporting database and must do so in a simple and effective way. Design environment initialization was simply done with a single command to set up all required tools and environment variables necessary for design work and the database interface.

IV.1.2. Front End Design and Verification Methodology

The front end logic design is captured in VHDL with all the verification done at the behavioral level. The chip verification uses Top down Specification / Bottom up Implementation strategy. For custom circuits, the schematic netlists and behavioral VHDL are verified for correctness with equivalency checking tools.

The design is divided into partitions, islands, units, and modules. All the verification environments and test coverage needed to create a high quality chip is planned during *High Level Design* phase. The verification process is hierarchical with all the environments and checkers created at lower level being used in the higher level environments. For performance and throughput purposes, there are options to turn off some checkers during run time. The test plan is based on the coverage plan to guarantee 100% coverage with written tests. The coverage is also hierarchical i.e. lower environments designate what portions of the lower level coverage needs to be hot at higher levels. Extra coverage and checkers are also added at higher levels for corner cases.

For a complex and large design such as CELL, a cycle based simulator is used for all the simulation. Both C++ based and Specman languages are support in the verification flow. Apart from the specialized test case generators used in processor core verification; Specman, C++ and Perl test cases are used for the rest of the design. Formal verification is also done at module level at various parts of the chip. Special tools were employed for

Asynchronous clock boundary verification since the simulator used is cycle based.

In addition to functional verification, pervasive design is also verified at various levels. This includes Scan verification, POR verification, Test mode verification, RAS verification, Trace and Debug Bus verification, etc. Hardware based accelerators are also used for software workloads, Boot code, and OS boot verification.

The Grid computing usage for processor design is demonstrated in this project: over 1.5 trillion simulation cycles or about 2 million hours of simulation was completed over multiple Sim farms spanning throughout IBM US & Germany. This is one of the key attribute for over 98% of total logic bugs found, the processor core VHDL model booted Linux, and Chip Bring up exercisers ran in simulation prior to design tape out.

IV.1.3. Physical Synthesis

The increased volume of synthesized logic on the CELL processor requires maximizing the productivity of the random logic macro (RLM) designers. This is accomplished by accelerating timing closure and automating the build process.

The design of RLMs used physical synthesis to accelerate timing closure. Physical footprints were imported into the synthesis tool to allow accurate timing estimations during netlist creation and placement. The placed, optimized netlists were then fed into the physical build process. Early estimated abstracts allowed for synthesis and sizing before final contracts were available from the unit integrator.

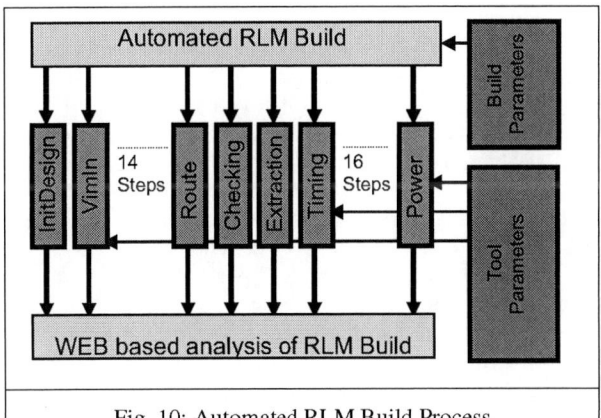

Fig. 10: Automated RLM Build Process

The RLM physical build process was streamlined and automated as shown in figure 10. This was accomplished by creating a supervisor program to "drive" over 30 individual design steps from netlist import through final checking. The supervisor script used customizable templates to control the individual tool interfaces, allowing designers an automated solution with the flexibility of a manual build flow. Job management was further improved by the *Report Generation's Tool* (XRG), which generated web based reports that allowed designers to easily identify failed job steps and quickly access log files.

To ensure high-availability of the tool set, automated daily regression tests were performed that exercised the build process and evaluated the results. This helped identify problems before they were encountered by designers.

Custom methodology checks were implemented to ensure that RLMs met design specifications before being delivered to unit integrators.

IV.1.4. Static Timing Methodology

To simplify timing closure and reduce runtime, all latches were modeled for the late mode timing run in the nominal process corner as non-transparent to remove timing loops. Custom designers; however, were still able to use cycle stealing techniques with an internally developed algorithm, which allowed the designer to specify the effective cycle boundary point within a given window of transparency. A timing adjust could be applied for all latches connected to a given LCB, allowing for an improved setup time, but delaying the launch of the data out of the L2 latch by a corresponding amount beyond that which the actual non-transparent latch modeling would require. To lower the cost for high-volume production, all Local Clock Buffers (LCB) and latches are designed to support at-speed scan to reduce manufacturing test time. For power management, each LCB included global and local clock gating signals. These signals have to work correctly on a cycle-by-cycle basis to allow switching from scan to functional mode in one clock cycle. At-speed scan operation allows us to time both functional and scan paths in a single timing run without the need to apply different phases to distinguish scan signals from regular ones.

For the early mode timing run in the fast corner, we wanted to ensure enough margins to cover a wide process range window needed for a high-volume product. To achieve this, we used the Linear Combination of Delays (LCD) feature of our Gate-level Static Timer. This feature allows combining different process corners [10]. Usually, the coefficients for the three corners, best, nominal, and worst, add up to be 1, e.g. 15% best, 70% nominal, and 15% worst case. We used a coefficient of 1.27 plus a fixed amount of offset for the worst case calculated timing delays. This allows an increased hold time margin by slowing down the clock propagation.

IV.1.5. Chip Integration & Physical Verification Methodology

The chip integration methodology was created to support parallel, concurrent design at high clock frequencies. Multiple levels of hierarchies were used to manage the design problem and enable concurrency. The high level design process consisted of top-down constraint setting which lead to the division of the design into functional islands and units. The constraints became a design budget for each floorplannable object. Those budgets dictated the size, aspect ratio, rectilinear outline, pin locations, and routing layers used for each object. The implementation process fulfilled the constraints passed down the hierarchy.

The integration methodology was tightly woven with timing throughout the design process. Very early in the Floorplanning process, timing shells represented each object in the hierarchy. These shells enable early timing feedback to drive partitioning, pipelining, and buffering decisions from the outset. As the data evolved through the design process, shell timing rules and Steiner estimates became schematic based timing rules with 2-D extracted parasitic and finally fully extracted timing rules and 3-D extracted routing parasitic. Buffering of signals is performed by an internally developed algorithm. Unit floorplans are filled with 4, 8, 16, or 32 bit buffer packs with all bits initially unused. A process based on Dijkstra's Algorithm finds the shortest path from source to sinks across available buffer packs. Routing was performed using a gridded router and 13 distinct non-default routing rules. Timing estimations that used particular non-default rules carried directly into the routing process, insuring that actual routes would mirror the estimation.

Later in the design cycle, each partition would analyze and correct coupled noise events predicted on closely routed nets. Noisy nets were fixed either through rerouting or by buffering. Electro-migration and missing via analysis on the power bus was also performed to insure that the power distribution met design requirements.

Physical verification of all Floorplan blocks consisted mainly of LVS, DRC, methodology checks, and formal Netlist verification. All physical verification is done with cover cells that represent fixed obstructions pushed down from the parent or the routing contract. Checking with these views insures that the object will not create a conflict when stitched into each level of hierarchy. Special methodology checks enforce specific design requirements beyond traditional design rules. This would include checks for pin accessibility, design shapes properly within the boundary, and power pins on proper pitch, among others. The formal verification process insured that the final, buffered Netlist was Boolean equivalent to the original vhdl description.

IV.2. Transistor Level Analysis

IV.2.1. Circuit & Array Methodology for an 11FO4 Design

For an 11FO4 design within an air-cooled power envelop, special emphasis was placed on power distribution, power consumption, clock distribution, signal distribution, variation due to hot spots, and inductance effects. Furthermore the chip team also had to plan for multiple clock domains, cross chip variations in delay, leakage, intra-chip interconnections, and array bit cell stability early in the design cycle. Strict design guidelines in layout and circuit topology were enforced to minimize design variations.

A major focus of the circuit methodology is on array design since memory arrays occupy an increasingly larger share of chip area and it is where more aggressive design techniques are used to ensure performance. There are three major challenges for array design at low voltage levels: stable cell operation (for functionality), leakage current reduction (for low power), and management of speed variations (for yield).

A critical part of the circuit/array methodology is a detailed statistical analysis of cell stability, leakage, and yield in the early design phase. This analysis will determine the optimal cell size for a given technology to achieve stability, power, and yield goals while reducing chip area. The analysis also helps guide the design team and the manufacturing team to decide on the device menu for the technology. A sample result of the statistical analysis is shown in Figure 11 below. This figure plots the failures of the cells at various voltage levels for the peripheral logic circuit (Vdd1) and the core cells (Vdd2). Note that as the design enters the sub-1V operating voltage range, it may be necessary to have a separate supply for the array core cells. This design decision will have significant impact across the whole methodology. Chip planning/integration, packaging, libraries, and tools will have to be adapted to support multiple supply domains.

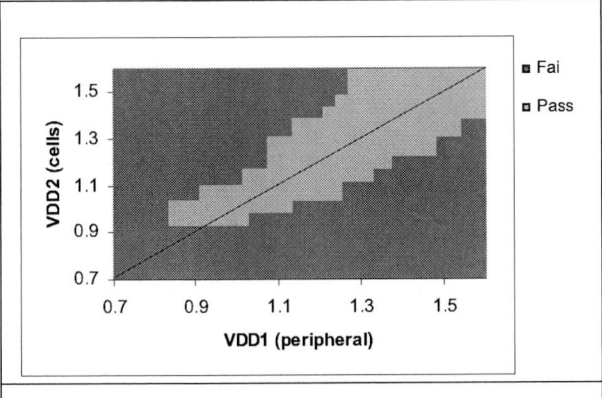

Fig. 11: Statistical analysis of SRAM cells – Vdd1 versus Vdd2 data

Transistor level analysis also plays an important role in the array verification methodology. For arrays, the high level design begins will the RTL and the implementation begins with the schematic design. There will be long lead time before layout is completed. So it is very important to have a methodology to provide accurate parasitic and interconnect models at the schematic level. The design methodology allows for a structured early floor-planning with accurate wire load models or Steiner based routing approximation to provide sufficient accuracy for schematic transistor level analysis. Logic extraction from array schematic is performed to build the test model, except for the array core which is synthesized from the high level RTL to reduce the model size. Then symbolic switch level simulation is run on the schematic and verified against the RTL as well as the test model. ATPG and marching patterns are also run on the array schematic using fast circuit simulator to verify the test patterns against the schematic.

IV.2.2. Transistor Level Timing

Static transistor level timing (TLT) was an integral part of the design methodology, with all custom macros, arrays, and even standard cell-based RLMs running through the tool, thereby providing comprehensive and consistent timing analysis and models. To meet aggressive frequency goals while satisfying area and power constraints, designers need to be able to quickly determine critical paths and delays in a circuit. Static timing at the transistor level using TLT helped achieve these goals [11]. The TLT team for the CELL project has improved the existing transistor-level timing methodology [12] in these four areas: improved timing margin calculation [13], local latch transparency modeling [14], pulse waveform timing in TLT templates [15], and improved method of timing model abstraction for simultaneously switching signals [16].

TLT is a transistor-level static timing tool that extends the capabilities of a Gate Level Static Timer to the transistor level. These extensions include a state analysis engine that is used to understand the timing behavior of groups of transistors and build timing models for them, and a fast circuit simulator [17] that is used for calculating propagation delays/waveforms through these transistor groups.

TLT uses piecewise-linear waveforms (rather than ramps) for timing accuracy and a modified version of AWE/RICE [18, 19] to propagate these waveforms through RC interconnect. TLT runs on flattened netlists from either schematics or extracted physical data. In addition to generating transistor level timing reports, it compiles a timing model or rule that is used for static timing at higher levels of the design hierarchy.

IV.2.3. Modularity and Integration of Black Box IP

The architectural modularity of the CELL processor also projects into the physical domain, where all 8 SPEs are instantiations of a single SPE design partition. To make this work correctly, the interaction between the local SPE layout and the global physical design structures had to be identical at all 8 locations where each SPE is instantiated. The C4 footprint, power busses, clock sector buffers, pervasive elements, and EIB components all had to be designed upfront to fit into this scheme in a modular way. The other extreme was taken with the integration of the high-speed IO and Memory interfaces on the left and right side of CELL [20]. These partitions were designed by a 3rd party vendor as "black box" IP, and used all layout resources from the silicon up to the C4 pins over their area. The only interaction with the core/chip happened at the boundary where predetermined power and signal pins were provided to cross the interface.

V. CONCLUSION

In conclusion, special circuit techniques, rules for modularity and reuse, customized clocking structures, and unique power and thermal management concepts were applied to optimize the design [21]. Correct operation has been observed in the lab on first pass silicon at frequencies well over 4GHz as shown in figure 12.

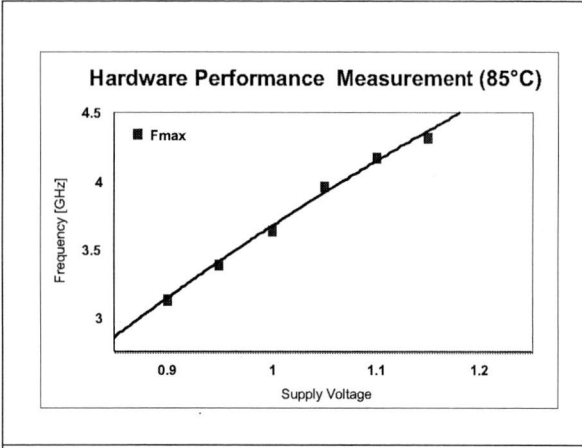

Fig. 12: First pass hardware in the Lab

VI. ACKNOWLEDGMENTS

The authors gratefully acknowledge the many contributions from the entire Sony-Toshiba-IBM team who worked tirelessly side-by-side on the design of this processor.

VII. REFERENCES

[1] D. Pham et al, "The Design and Implementation of a First-Generation CELL Processor", ISSCC 2005 Digest of Technical Papers, Feb. 2005, pp. 184-185.

[2] B. Flachs et al, "The Microarchitecture of the Streaming Processor for a CELL Processor", ISSCC 2005 Digest of Technical Papers, Feb. 2005, pp. 134-135.

[3] T. Asano et al, "A 4.8GHz Fully Pipelined Embedded SRAM in the Streaming Processor of a CELL Processor", ISSCC 2005 Digest of Technical Papers, Feb. 2005, pp. 486-487.

[4] S. Clark et al, "IBM CELL Interconnect Unit, Bus and Memory Controller", Hot Chip'05, Aug. 2005, Paper #1.2

[5] F. Klass, C. Amir, A. Das, K. Aingaran, C. Truong, R. Wang, A. Mehta, R. Heald, G. Yee, "A New Family of Semi-dynamic and Dynamic Flip-Flops with Embedded Logic for High-Performance Processors", IEEE J. Solid State Circuits, vol. 34, pp. 712-716 (1999).

[6] L. Sigal, J.D. Warnock, B.W. Curran, Y.H. Chan, P.J. Camporese, M.D. Mayo, W.V. Huott, D.R. Knebel, C.T. Chuang, J.P. Eckhardt, and P.T. Wu, "Circuit Design Techniques for the High-Performance CMOS IBM S/390 Parallel Enterprise Server G4 Microprocessor", IBM J. Res. & Dev. Vol 41 pp. 489-503 (1997).

[7] P. J. Restle, et al, "A Clock Distribution Method for Microprocessors", IEEE J. Solid-State Circuits, vol. 36, pp 792-799, May 2001

[8] P. J. Restle, et al, "The Clock Distribution of the Power4 Microprocessor", IEEE International Solid-State Circuits Conference 2002 Digest of Technical Papers, vol. 45, pp 144-145

[9] K. Yazawa and M. Ishizuka, "Thermal Modeling with Transfer Function for the Transient Chip-On-Substrate Problem", Thermal Science and Engineering, vol. 13, No. 1, Heat Transfer Society of Japan, 2005, pp. 37–40

[10] Posluszny, S. et al. "Timing Closure by Design," Proceedings for the 37th Conference on Design Automation, vol.37, pp.712-717, June 2000.

[11] Rao, V., J. Soreff, T. Brodnax, and R. Mains, "EinsTLT: Transistor Level Timing with EinsTimer," Proc. Of Int. Workshop on Timing Issues (TAU), 1999.

[12] Lee, Sang Y., J. Warnock, E. Behnen, J. Soreff, V. Rao, and S. Posluszny, "Improved Transistor-Level Timing Methodology for a CELL Microprocessor," ASPDAC 2006 (submitted for publication)

[13] Warnock, J.D., Erwin Behnen, Sang Y. Lee, and Jeffrey Soreff, "Improved Method for Timing Margin Calculation," IBM Invention Publish, Feb. 2004.

[14] Behnen, E., Jeffrey Soreff, James D. Warnock, and Dieter Wendel, "Method to Apply Latch Transparency Locally While Avoiding It Globally During Timing," Filed with U.S. Patent Office, May 2004.

[15] Soreff, J., Vasant Rao, James D. Warnock, Sang Y. Lee, and David Winston, "Pulse waveform timing in EinsTLT templates," Filed with U.S. Patent Office, May 2004.

[16] Warnock, J.D. and Jeffrey Soreff, "Improved Method of Timing Model Abstraction for Circuits Potentially Simultaneously Switching Internal Signals," Filed with U.S. Patent Office, May 2004.

[17] Devgan, A. and R.A.Rohrer, "Adaptively controlled explicit simulation," IEEE Trans. Computer-Aided Design, vol. 13, pp.746-762, June 1994.

[18] Pillage, L.T. and R.A. Rohrer, "Asymptotic waveform evaluation for timing analysis," IEEE Trans. Computer-Aided Design, vol. 9, No. 4, pp. 352-366, April 1990.

[19] Ratzlaff, C.L, N. Gopal, and L.T. Pillage, "RICE: Rapid interconnect circuit evaluator," IEEE Trans. Computer-Aided Design, vol. 13, No. 6, pp. 763-776, June 1994.

[20] K. Chang et al, "Clocking and Circuit Design for a Parallel I/O on a First-Generation CELL Processor", ISSCC'05 Paper #28.9

[21] Pham, D. et al. "Overview of the Architecture, Circuit Design, and Physical Implementation of a First-Generation CELL Processor," JSSCC, October. 2005 Special issue (submitted for publication).

2006 Asia and South Pacific Design Automation Conference

TAPHS: Thermal-Aware Unified Physical-Level and High-Level Synthesis

Zhenyu (Peter) Gu†   Yonghong Yang‡   Jia Wang†   Robert P. Dick†   Li Shang‡

<table>
<tr>
<td align="center">†EECS Department
Northwestern University
Evanston, IL 60208, U.S.A.
{zgu646, jwa112, dickrp}@ece.northwestern.edu</td>
<td align="center">‡ECE Department
Queen's University
Kingston, ON K7L 3N6, Canada
4yy6@qlink.queensu.ca, li.shang@queensu.ca</td>
</tr>
</table>

Abstract— Thermal effects are becoming increasingly important during integrated circuit design. Thermal characteristics influence reliability, power consumption, cooling costs, and performance. It is necessary to consider thermal effects during all levels of the design process, from the architectural level to the physical level. However, design-time temperature prediction requires access to block placement, wire models, power profile, and a chip-package thermal model. Thermal-aware design and synthesis necessarily couple architectural-level design decisions (e.g., scheduling) with physical design (e.g., floorplanning) and modeling (e.g., wire and thermal modeling).

This article proposes an efficient and accurate thermal-aware floorplanning high-level synthesis system that makes use of integrated high-level and physical-level thermal optimization techniques. Voltage islands are automatically generated via novel slack distribution and voltage partitioning algorithms in order to reduce the design's power consumption and peak temperature. A new thermal-aware floorplanning technique is proposed to balance chip thermal profile, thereby further reducing peak temperature. The proposed system was used to synthesize a number of benchmarks, yielding numerous designs that trade off peak temperature, integrated circuit area, and power consumption. The proposed techniques reduces peak temperature by 12.5 °C on average. When used to minimize peak temperature with a fixed area, peak temperature reductions are common. Under a constraint on peak temperature, integrated circuit area is reduced by 9.9% on average.

1. Introduction

Increasing performance requirements and system integration are dramatically increasing integrated circuit (IC) power density and, therefore, cooling costs. Energy consumption and thermal issues are now central to the design of ICs, including both high-end instruction processors in general-purpose computers and application-specific integrated circuits (ASICs) in low-cost portable electronic consumer devices. Peak local temperature influences the reliability, packaging costs, cooling costs, bulk, and performance of IC. Many of these considerations are particularly important for portable devices.

Increasing IC power consumption raises average and peak temperatures. Temperature variations and hot spots account for over 50% of electronic failures [1], most of which are due to electromigration, hot carrier effects, thermal stress, and oxide thermal breakdown. Power and thermal variation can also lead to significant timing uncertainty, requiring more conservative timing margins, thereby reducing performance. Designers must frequently trade off other design metrics, such as performance, area, and cooling costs, to meet tight thermal constraints. The interaction of power and thermal constraints with other design metrics further increases system complexity. As projected by the International Technology Roadmap for Semiconductors (ITRS) [2], further process scaling will be bounded by power consumption and heat dissipation below 65 nm: it is critical to address energy and thermal issues during IC design to meet the urgent needs of the semiconductor industry and enable future technology scaling.

Thermal problems cannot be well solved at any single level of the design process. Thermal characterization requires detailed physical information, including an IC floorplan and power profile as well as interconnect and chip-package thermal models. Thermal optimization requires a unified high-level and physical-level design flow. At the architectural level, reducing supply voltage can reduce IC power consumption, hence the temperature, while at the physical level, peak temperature can further be reduced by modifying the IC floorplan to balance the thermal profile. Furthermore, the evaluation and optimization of the tradeoff between IC temperature and other design metrics, such as performance, area, and cooling cost, requires a comprehensive architectural-level and physical-level infrastructure.

Incremental synthesis is a promising design technique that may be used to unify high-level synthesis and physical design. It improves the quality of results by maintaining important physical-level properties across consecutive physical design changes, many of which are triggered by architectural changes [3–5]. Moreover, it dramatically improves synthesis time by reusing and building upon high-quality, previous physical design solutions that required a huge amount of time and effort to produce.

This paper presents a thermal-aware, floorplanning, incremental high-level synthesis system called TAPHS. The proposed incremental synthesis techniques rapidly determine the impacts of architectural changes on floorplan-dependent characteristics and concurrently optimize IC thermal profile, area, and energy consumption under performance constraints.

2. Related work

In this section, we survey related work in the two main research areas in which TAPHS is rooted: (1) high-level and physical-level co-synthesis and (2) thermal-aware analysis and design.

As a result of technology scaling, it is becoming increasingly important to consider the physical design impacts of high-level design decisions. This requires floorplanning and interconnect estimation at the highest levels of design. A few researchers have previously considered incremental floorplanning [6] and the impact of incorporating loosely-coupled constructive floorplanners within high-level synthesis [7], [8]. Other researchers subsequently used incremental floorplanning and synthesis [4] to tightly couple high-level and physical synthesis [3].

Recent studies on thermal issue focus on analysis and optimization. A number of thermal analysis approaches that try to efficiently model chip-package designs have been proposed [9–13]. Thermal and thermal-reliability issues are becoming increasingly important for IC interconnection networks due to their influence on electromigration and stress migration voiding. Recent studies [14], [15] have proposed numerical and analytical modeling techniques to characterize the thermal profile of on-chip interconnect layers. Thermal issues have also been considered during chip cell-level placement [16], [17], three-dimensional IC floorplanning [18], and high-level synthesis resource sharing [19].

This work is supported in part by the NSF under awards CNS-0347941 and CCR-0238484, and in part by an NSERC Discovery Grant #388694-01.

0-7803-9451-8/06/$20.00 ©2006 IEEE.

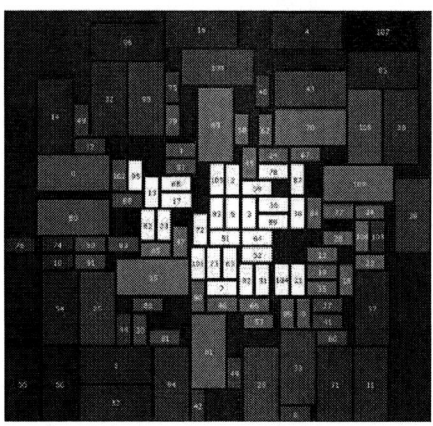

Figure 1. Post-synthesis thermal profile without voltage islands.

3. Motivating example

In this section, we use an example IC design to demonstrate the challenges of thermal optimization in high-level synthesis. Figure 1 shows an IC floorplan produced by an integrated high-level synthesis and floorplanning algorithm. In this figure, the numbered rectangles are functional units, e.g., adders, multipliers, dividers, or registers. Using thermal analysis, as described in Section 8, the IC thermal profile is determined. The temperature of each functional unit is indicated by its brightness: brighter functional units are hotter. 85 °C is a typical thermal emergency threshold to ensure reliable operation. In this example, functional units temperatures higher than 85 °C are white. 29 of the functional units are operating at dangerously high temperatures: this chip is likely to suffer from failure caused by thermal-related reliability problems, e.g., electronmigration. Note that producing the detailed chip thermal profile in Figure 1 requires detailed physical information, i.e., a floorplan, a power profile, and a chip-package thermal model. Therefore, stand-alone high-level synthesis algorithms have no means of detecting, let alone correcting, thermal crises.

High-level synthesis provides numerous thermal optimization opportunities. Reducing supply voltage reduces power consumption, hence temperature, but may also impair performance. Recent work on voltage islands has proposed operating different regions of an IC at different voltages. Figure 2 illustrates the floorplan of an IC using voltage islands. In this design, functional units are assigned to contiguous voltage islands with different supply voltages. The brightnesses of the thick functional unit boundaries indicate their voltages. In this example, three voltage islands are used. As in Figure 1, functional units violating the 85 °C temperature constraint are white.

A comparison of Figures 1 and 2 indicates that voltage islands can dramatically improve thermal conditions. The number of functional units with temperatures above the thermal constraint decreased from 29 to 19. However, as shown in Figure 2, localized hot spots still exist. The remaining hot spots are primarily the result of local concentrations in power density. Although changing the assignment of operations to functional units may improve local temperatures in some circumstances, in practice we found that the vast majority of thermal problems in designs already using voltage islands could only be resolved with thermal-aware physical design techniques, especially for designs with tight deadlines.

Our study suggests a rich set of high-level and physical-level thermal optimization techniques, including multiple operating voltages, appropriate scheduling, and thermal-aware floorplanning. Many of the techniques to optimize IC thermal properties also impact other design metrics such as area and power consumption. We have considered the side effects of a number of techniques, proposing those that allow improvements to thermal properties while maintaining good area, performance, and power consumption.

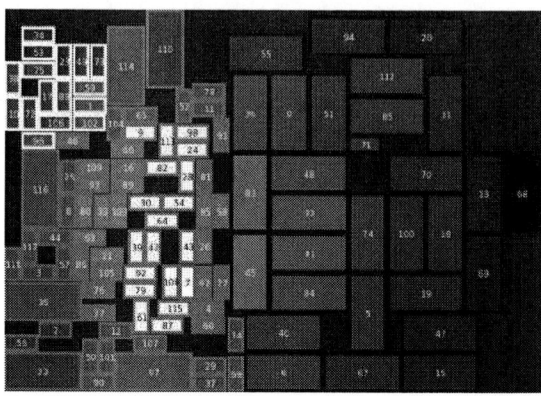

Figure 2. Post-synthesis thermal profile with voltage islands.

Using voltage islands has a significant impact on chip area and performance as well as increasing the complexity of floorplanning. Voltage islands require the addition of voltage converters and delivery circuits, as well as on-chip level shifters to support communication among functional units in different voltage islands. Moreover, reduced supply voltage requires a longer clock period to compensate for reduced switching speeds. In order to use voltage islands, a synthesis algorithm must wisely choose the island for each functional unit, appropriately allocate timing slack to allow scheduling, and generate floorplans in which functional units in the same voltage island are contiguous. This tightly couples the architectural and physical levels of design.

Thermal-aware floorplanning is challenging because it must trade off multiple conflicting objectives: peak temperature, IC area, and power consumption. Adjusting functional unit positions to balance power density, thereby reducing peak IC temperature, may increase chip area. Moreover, if the high-power functional units are also high-activity functional units, as would be expected as a result of resource sharing, they are also likely to frequently communicate with other functional units. In general, high-connectivity functional units hosting operations on critical timing paths ought to be placed near each other to minimize interconnect delay and power consumption. However, doing so can have the side effect of increasing peak IC temperature.

Facing these design challenges, a high-quality thermal-aware synthesis system must incorporate thermal optimization techniques into a unified high-level and physical level design flow, as well as striking wise tradeoffs among conflicting design goals.

4. Overview of TAPHS

In this section, we give an overview of TAPHS: our incremental thermal-aware physical and high-level synthesis system. TAPHS considers the thermal impact of both logic and interconnect power dissipation. It automatically generates voltage islands and schedules operations to reduce IC power consumption and peak temperature. In addition, it does thermal-aware floorplan optimization.

Figure 3 illustrates the main algorithms used in TAPHS. First, the control and data flow graph is simulated with typical input traces in order to profile each operation and data transfer edge. The profile information, an RTL design library, floorplanner, and thermal model are used to evaluate the IC temperature profile, power, area, and performance. Slack distribution, voltage clustering, and voltage island aware floorplanning are used to generate voltage islands for use in the initial solution: a fully parallel implementation. There are two loops within the high-level synthesis algorithm. In the outer loop, the clock period of the design is iteratively changed from the minimum to maximum potentially feasible values. Incremental rescheduling, resource sharing, resource splitting (i.e., the opposite of resource sharing), and slack distribution are used to generate valid solutions. In the inner loop, back-tracking iterative improvement is used to op-

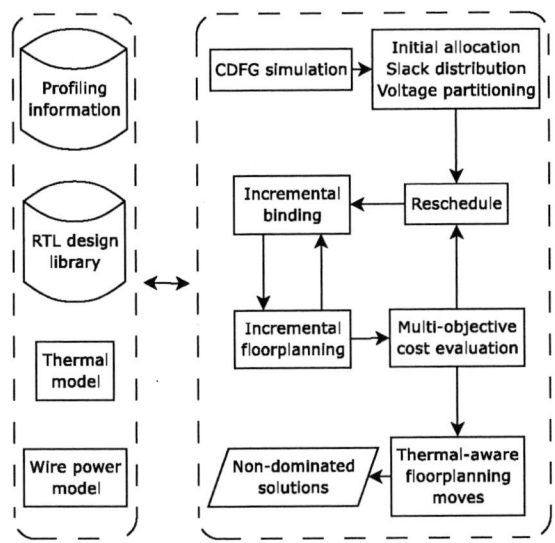

Figure 3. Incremental high-level synthesis algorithm

timize the RTL architecture, considering multiple objectives, e.g., peak temperature, area, and power consumption. A *dominated* solution is inferior to some other previously encountered solution in all costs. Non-dominated solutions are preserved in a solution cache, from which the designer may choose based upon the desired trade-offs among costs.

A high-quality thermal-aware incremental floorplan was developed and incorporated into TAPHS. Each time the high-level synthesis algorithm needs thermal and physical information to guide its moves, it extracts that information from the current, incrementally generated, floorplan. In addition, costs derived from the floorplan are used to guide high-level synthesis moves. By using incremental floorplanning, closer interaction between high-level synthesis and physical design is possible, i.e., the high-level synthesis algorithm may determine the impact of potential changes to binding upon physical attributes such as IC thermal profile, area, and interconnect energy consumption.

5. Slack distribution

To allow voltage scaling, it is necessary to appropriately distribute scheduling slack among operations. *Slack* is the difference between latest and earliest start time. Determining whether it is possible, and desirable, to assign an operation to a lower-voltage functional unit is not possible based on as soon as possible (ASAP) operation start times. TAPHS redistributes slack among operations in order to support more energy-optimal assignment of functional units to voltage islands.

Assume that control and data flow graphs have been partitioned into same-slack paths, as described later in this subsection. Given a single path composed of sequential operations, the slack distribution problem is equivalent to deciding the execution time of each operation such that energy consumption is minimized under a hard constraint on path execution time. We shall use the following variables and constants: D is the bound on path execution time; p is the set of all operations on the path; d_i is the delay of an operation's functional unit; v_i is the voltage of an operation's functional unit; V_t is the threshold voltage constant; K is an execution time constant; E is the total path energy consumption; e_i is the energy required for an operation; C_i is the switched capacitance constant of an operation's functional unit; and α is the alpha power law constant [20].

$$d_i = \frac{Kv_i}{(v_i - V_t)^\alpha} \quad \text{subject to the constraint } D \ge \sum_{i \in p} d_i \quad (1)$$

However, V_t is small and a very low value of v will generally imply an unacceptable path delay that will be prevented by the constraint on line 1. Therfore, we may assume V_t is small, thus

$$d_i \simeq \frac{Kv_i}{v_i^\alpha} \quad \text{and} \quad v_i = \left(\frac{d_i}{K}\right)^{\frac{1}{1-\alpha}} \quad (2)$$

$$e_i = C_i v_i^2 = C_i \left(\frac{d_i}{K}\right)^{\frac{2}{1-\alpha}} \quad \text{and} \quad E = \sum_{i \in p} C_i \left(\frac{d_i}{K_i}\right)^{\frac{2}{1-\alpha}} \quad (3)$$

$$\min_{\substack{\forall i \in p \\ v_i}} \quad \sum_{i \in p} C_i \left(\frac{d_i}{K_i}\right)^{\frac{2}{1-\alpha}} \quad (4)$$

Note that a decrease in v_i implies a decrease in e_i, which implies an increase in d_i. Therefore, for minimal E, $D = \sum_{i \in p} d_i$. Consider the delay and energy trade-off for an arbitrary pair of operations:

$$d_{12} = d_1 + d_2 \quad \text{and} \quad e_{12} = e_1 + e_2 \quad (5)$$

$$e_{12} = \frac{C_1}{K_1^{\frac{2}{1-\alpha}}} (d_1)^{\frac{2}{1-\alpha}} + \frac{C_2}{K_2^{\frac{2}{1-\alpha}}} (d_{12} - d_1)^{\frac{2}{1-\alpha}} \quad (6)$$

Take the derivative of e_{12} with respect to d_1, set to zero, and solve to find d_2/d_1 for minimal E.

$$\frac{d_2}{d_1} = \left(\frac{\frac{C_1}{K_1^{\frac{2}{1-\alpha}}}}{\frac{C_2}{K_2^{\frac{2}{1-\alpha}}}}\right)^{\frac{1-\alpha}{1+\alpha}} \quad (7)$$

This optimal delay ratio for two operations may be used to compute the optimal delay ratio for an arbitrary pair of operations. These ratios can be scaled by a dynamically computed value, N, to ensure that the constraint on line 4 is honored.

$$N = \sum_{i \in p} \frac{d_i}{d_1} \quad (8)$$

$$\forall_{i \in p} \ d_i = \frac{D}{N} \left(\frac{\frac{C_1}{K_1^{\frac{2}{1-\alpha}}}}{\frac{C_i}{K_i^{\frac{2}{1-\alpha}}}}\right)^{\frac{1-\alpha}{1+\alpha}} \quad \text{or} \quad \frac{D}{N} \sqrt[3]{\frac{C_i K_i^2}{C_1 K_1^2}} \quad \text{for } \alpha = 2 \quad (9)$$

Equations 6 and 9 yield the optimal time, d_i, to dedicate to each operation. By granting slack to each operation in the path such that its time is proportional to its time share, we allow the voltage island generation algorithm the opportunity to assign functional units to voltage islands such that energy consumption may be minimized under a hard constraint on path execution time (please see Section 6).

Thus far, we have discussed individual operation paths. However, it is necessary for TAPHS to determine slack distributions along numerous paths in arbitrary directed acyclic graphs of operations. Assigning time shares eventually has the effect of (temporarily) fixing operation start times. These start times may influence the earliest start times and latest finish times of operations on other paths; in order to avoid deadline violations, slack distribution is conducted on operation paths in order of increasing path slack. In order to generate paths, a modified depth-first search is conducted on a graph in which each vertex is an operation labeled with its slack and each edge is a data dependency. Vertex children are visited in increasing order of slack, thereby guaranteeing that vertices on multiple paths will be included in minimal-slack paths.

As shown in Algorithm 1, starting from the minimal-slack path, TAPHS incrementally assigns extra clock cycles to operations. At each step, it locates the operation, j, for which the current allocated time, t_j, differs most from d_j (Step 7) and assigns it an additional

9A-1

Algorithm 1 Slack distribution procedure

1: Compute all operation slacks
2: Group operations into same-slack paths, P
3: Sort paths P in order of increasing slack
4: **for all** $p \in P$ **do**
5:     **while** slack remains on p **do**
6:         $\forall_{i \in p}\, t_i$ is the time assigned to operation i
7:         Operation $i = \min_j^{\arg} \sqrt[3]{\dfrac{C_j K_j^2}{C_1 K_1^2}} \cdot \dfrac{D}{N} - t_j$ by Equation 9
8:         Assign one additional clock cycle to operation i
9:     **end while**
10:    Recompute all operation slacks
11: **end for**

clock cycle (Step 8). It is guaranteed that this will not result in deadline violations on other paths because slack distribution is carried out on paths in order of increasing slack. Therefore, slack distribution on a given path is prevented from delaying any node so much that slack becomes negative on other paths on which the node lies. After slack sharing is done for a given path, the slacks of all nodes are recomputed and slack distribution proceeds for the next path.

6. Voltage partitioning

TAPHS uses on-chip voltage islands to optimize IC thermal profiles and energy consumption. On-chip voltage islands are generated in two stages. *Voltage partitioning* classifies functional units into different voltage levels to maximize overall power and energy savings hence potential IC temperature reduction. *Voltage island generation* is then conducted via incremental floorplanning to produce and optimize voltage islands.

In this section, we focus on voltage partitioning under two design constraints. First, reducing supply voltage increases circuit propagation delay. Hence, the minimal supply voltage of a functional unit is constrained by its available time slack. Second, increasing the number of on-chip voltage levels introduces significant overhead to off-chip and on-chip power supply and delivery circuits. Therefore, only a limited, design-dependent, number of voltage levels should be generated.

In this work, we propose an efficient voltage partitioning algorithm. It conducts optimal voltage allocation and assignment to maximize overall power savings and strike judicious trade-offs among different design metrics.

Motivating example

We next present an example to illustrate the voltage partitioning problem. Consider a circuit design with five functional units as shown in Fig. 4. For each functional unit, FU_i, the minimal allowed supply voltage, $V_{FU_i}^{min}$, is uniquely determined by the ratio of its time slack to its propagation delay under the initial (maximum) supply voltage. In a voltage partition Ψ_i^S with S clusters, to satisfy the deadline constraints of functional units, for each cluster, $\psi_j = \{FU_{j1}, \dots, FU_{jn}\}$, $\psi_j \in \Psi_i^S$, its supply voltage, V_{ψ_j}, is greater than or equal to $max\{V_{FU_{jt}}^{min}\}_{t=1,\dots,n}$, i.e., the minimal supply voltage of the functional unit with the lowest slack-delay ratio inside this cluster. Consider the voltage partitioning shown in Fig. 4(a). This partitioning contains two voltage clusters, $\psi_1 = \{FU_1, FU_2\}$ and $\psi_2 = \{FU_3, FU_4, FU_5\}$. The supply voltage of ψ_1, V_{ψ_1} is 1 V, which is the minimal allowed supply voltage of FU_2. The supply voltage of ψ_2, V_{ψ_2} is 2 V, which is the minimal allowed supply voltage of FU_5.

Fig. 4(c) shows the energy consumptions of different voltage partitions, which are derived using a linear scan along the functional unit list. This list is sorted in order of increasing slack-delay ratios (or minimal allowed supply voltages) of functional units. This figure shows that, using linear scan, the energy curve is not monotonic, implying that an algorithm with $O(N)$ time complexity is necessary to find a single voltage partitioning cut with minimal energy consumption.

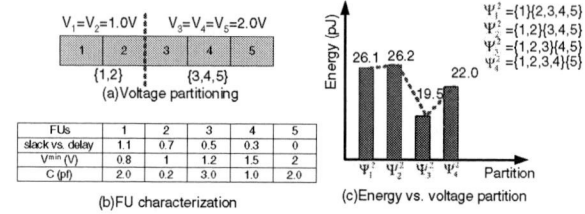

Figure 4. Voltage partitioning example.

We now define the optimal voltage partitioning problem.

Problem Definition *Given N functional units, $\{FU_1, \dots, FU_n\}$, and an input M, find an optimal voltage partition, Ψ_{opt}^M, containing M voltage clusters, $\{\psi_{opt\,j}\}_{j=1,\dots,M}$, such that its energy consumption, $E(\Psi_{opt}^M) \leq E(\Psi_i^M), \forall \Psi_i^M$, in which $E(\Psi_{opt}^M) = \sum_{l=1}^{N} C_l \times V_{\psi_{opt\,j}}^2$. C_l is the capacitance of $FU_l, FU_l \in \psi_{opt\,j}, j = 1, \dots, M$.*

For each functional unit, FU_i, to satisfy its deadline constraint, its minimal allowed supply voltage, $V_{FU_i}^{min}$, is uniquely determined by the ratio of its slack time to its propagation delay under the initial (maximum) supply voltage. Then, for each cluster, $\psi_j = \{FU_{j1}, \dots, FU_{jn}\}$, $\psi_j \in \Psi_i^M$, its supply voltage, $V_{\psi_j} \geq max\{V_{FU_{jt}}^{min}\}_{t=1,\dots,n}$, i.e., the minimal supply voltage of the functional unit with the lowest slack to delay ratio inside this cluster.

An optimal voltage partitioning is derived using the following approach. Functional units are first sorted by their slack to propagation delay ratios. Then, linear scans along the sorted functional unit list determine the optimal partitioning. Note that the energy saving curve is not monotonic, implying that an algorithm with $O(N)$ time complexity is required to find an energy-optimal voltage partitioning. For M partitions, the time complexity of this algorithm is $O(N^M)$.

An optimal voltage partitioning algorithm of $O(N^2)$ complexity

We introduce an optimal voltage partitioning algorithm of $O(N^2)$ time complexity. Its pseudo-code is shown in Algorithm 2, which is described in a recursive form. *Partition()* has five input/output parameters. $*FU\_list$ points to the sorted functional unit list. *Start* and *End* designate the sub-list: the portion of the original list that needs to be partitioned. Initially, $Start = 0$ and $End = N$ denote the voltage partitioning targets on the whole sorted list. M defines the targeted number of partition cuts. *OptTable* stores intermediate optimal partitions of sub-lists.

Partition() is invoked recursively when $M > 1$ (line 1-4). For each sub-partitioning (M cuts) on a sub-list (from *Start* to *End*), the optimal solution is derived using a linear scan to examine the Mth cut from *Start* to *End*, which is combined with the optimal solution of the sub-partitioning ($M-1$ cuts) on its sub-list (from i to *End*). When $M = 1$, the algorithm uses a linear scan to find the optimal cut in the targeted sub-list (line 7).

Lemma 1 *In an optimal partition Ψ_{opt}^M with M voltage clusters, $\{\psi_{opt,1}, \dots, \psi_{opt,M}\}$, and $V_{\psi_{opt,1}} \geq V_{\psi_{opt,2}} \geq \dots \geq V_{\psi_{opt,M}}$, then $V_{\psi_{opt,t}} \leq V_{FU_j}^{min} \leq V_{\psi_{opt,t-1}}, \forall FU_j \in \psi_{opt,i}$.*

Lemma 1 implies that the optimal partitioning can be found by partitioning the sorted functional unit list. This lemma guarantees the optimality of the algorithm: it uses a linear scan to explore all the possible partitioning combinations of the sorted list, including the optimal solution. Using linear scan to find the optimal M partitions on a sorted list with N functional units, the computational complexity is $O(N^M)$. To improve computation efficiency, we use a data structure, called *OptTable*, to store optimal sub-partitions. The time complexity of partitioning M results from a linear scan of the Mth cut multiplied by the time complexity of finding the optimal $M-1$ partitions, which only requires a linear search in *OptTable* table (line 5) with complexity $O(N)$. In total, there are M recursive

Algorithm 2 $Partition(*FU\_list, Start, End, M, *OptTable)$

1: **if** $M > 1$ **then**
2:    $C \leftarrow 0$
3:    **for** $(i \leftarrow Start; i \leq End; i++)$ **do**
4:      $Partition(*FU\_list, i, End, M--, *OptTable)$
5:      $E_{M^{th}=i}^{M} \leftarrow C \times (V_{FU_{i-1}}^{min})^2 + OptTable[M-1][i]$
6:      $C += C_{FU_i}$
7:    **end for**
8:    $E_{opt}^{M}(Start, End) \leftarrow min\{E_{M^{th}=i}^{M}\}_{i \leftarrow Start, \ldots, End}$
9:    $cut_{opt}^{M}(Start, End) \leftarrow i$ if $E_{M^{th} \leftarrow i}^{M} = E_{opt}^{M}(Start, End)$
10:    $OptTable[M][Start] \leftarrow pair(E_{opt}^{M}(Start, End), cut_{opt}^{M}(i, End))$
11: **else**
12:    $Linear\_Scan(*FU\_list, End, \&E_{opt}^{1}(Start, End),$
      $\&cut_{opt}^{1}(Start, End))$
13:    $OptTable[1][Start] \leftarrow pair(E_{opt}^{1}(Start, End),$
      $cut_{opt}^{1}(Start, End))$
14: **end if**

---

layers. Since M is much smaller than N, the overall time complexity of this optimal voltage partitioning algorithm is $O(N^2)$.

7. Thermal-aware floorplanning

In order to support thermal-aware, incremental, unified high-level and physical-level optimization, it was necessary to incorporate a high-quality, incremental floorplanner within TAPHS. New algorithms were developed and incorporated into this floorplanner to directly support physical-level thermal optimization and indirectly support architectural-level thermal optimization.

The floorplanner within TAPHS is based on the Adjacent Constraint Graph (ACG) representation [21]. An ACG is a constraint graph with exactly one geometric relationship between every pair of modules. ACGs have invariant structural properties that allow the number of edges in the graph to be bounded. Operations on ACGs have straightforward meanings in physical space and change graph topology locally; they require few, if any, global changes. The operations of removing and splitting modules are designed to reflect high-level operation to functional unit binding decisions. To obtain the physical position of each module, packing based on longest path computation is employed. Simulated annealing is used to obtain an initial floorplan. A weighted sum of the area and the interconnect power consumption is calculated for use as the floorplanner cost function, i.e.,

$$A + w \sum_{e \in E} C_e D_e \qquad (10)$$

where A is the area, w is the power consumption weight, E is the set of all wires, e is an interconnect wire, C_e is the unit-length switched capacitance for the data transfer along e, and D_e is the length of e, which is calculated as Manhattan distance between the two modules connected by the wire. Using this cost function, the floorplanner optimizes the interconnect power consumption, interconnect delay, and area. The resulting floorplan will be improved during the subsequent incremental floorplanning high-level synthesis moves. Therefore, the number of simulated annealing iterations is bounded to reduce synthesis time.

After each high-level synthesis move, the previous floorplan is modified by removing or splitting a module. The modules and switched capacitances are updated based upon the impact of these merges and splits. The floorplan is then re-optimized with a greedy iterative improvement algorithm using the same cost function as the simulated annealing algorithm. There are two reasons to use a greedy algorithm during this stage of synthesis: (1) re-optimization requires fewer global changes and less hill climbing and (2) perturbations resulting from high temperatures may disrupt high-quality floorplans.

After determining the best binding across all the possible clock frequencies, another simulated annealing floorplanning run is used for that binding. This final floorplanning stage occurs only once for every synthesis run. Therefore, it is acceptable to use a slower, but higher-quality, annealing schedule than those in the inner loop of high-level synthesis, thereby improving IC area and interconnect power consumption.

During the annealing schedule, we use a constant cooling factor, r, i.e.,

$$T^+ = r \times T \qquad (11)$$

where T is the current temperature and $T+$ is the temperature during the next iteration. The number of the perturbations for the initial floorplanning run, the floorplanning for each clock frequency, and the final floorplanning are related as follows: $1 : 2 : 20$. The number of perturbations per round for the greedy iterative improvement algorithm is the same as that for final floorplanning run.

7.1. Voltage island implementation in floorplanning

As described in Section 6, voltage island generation was introduced into the high-level synthesis system in order to improve thermal profiles and reduce energy consumption. Therefore, the floorplanner must attempt to keep functional units assigned to the same voltage level contiguous in order to minimize the need for level converters and simplify power distribution. The floorplanner must still honor the elements in the original cost function shown in Equation 10. Pair-wise weighted edges were added between all pairs of functional units operating at the same voltage, yielding the following updated cost function:

$$n\sqrt{A} + 2n \sum_{v \in V} L_v + \sum_{e \in E} C_e D_e \qquad (12)$$

where A is the area, n is the number of functional units, V is the set of all functional unit pairs sharing the same voltage, v is a pair of functional units sharing the same voltage, L_v is the separation between a pair of functional units sharing the same voltage, E is the set of all interconnects, e is an interconnect line, C_e is the unit-length switched capacitance for the data transfer along e (zero in the case of no communication), and D_e is the length of e. This approach generates contiguous voltage islands, as well as optimizing area and interconnect power consumption.

Figure 2, described in Section 3, shows an example of the results produced by this floorplanning algorithm. TAPHS rapidly generated this result using only pair-wise edges for functional unit clustering, i.e., hierarchical floorplanning was not required. Note that functional units operating at the same voltage are contiguous. In some cases, keeping functional units within voltage islands contiguous and minimizing wire length results in a slight area penalty. This is to be expected, regardless of the quality of a floorplanner, because it is rare for a minimal-area solution to maintain contiguous voltage levels and minimal interconnect power consumption. During incremental improvement, operation merging (functional unit resource sharing) combines functional units with other compatible functional units, always merging from the lower-voltage functional unit to the higher-voltage functional unit in order to honor performance constraints (please see Section 4).

7.2. Thermal-aware swap operation

As explained in Section 3, hot spots may occur because a number of functional units with high power densities are physically close to each other. Such concentrations are natural. It is common for high-activity, high-power functional units to frequently communicate with other high-activity functional units. This causes the floorplanner to position the functional units near each other in order to reduce interconnect power consumption. However, in some cases, the objectives of minimizing average power consumption and minimizing peak temperature conflict with each other.

9A-1

We propose a thermal-aware swap operation that exchanges hot, generally high power density, functional units with cool, generally low power density, functional units within the same voltage island. This heuristic sorts compatible functional units in a voltage island in order of increasing temperature. The positions of the highest and lowest temperature functional units are the exchanged, after which the exchanged functional unit positions are locked and the operation is repeated. The thermal-aware functional unit swapping heuristic halts after some proportion of the functional units have been moved. In practice, a proportion of 1/3 allowed a significant reduction in peak temperature for most examples.

8. Experimental results

In this section, we present experimental results for the TAPHS thermal-aware high-level synthesis system, including the thermal optimization techniques described in Sections 5, 6, and 7. The circuits described in this section were synthesized using a register transfer level (RTL) design library based on the TSMC $0.18\,\mu m$ process. The experiments were conducted on AMD Athlon-based Linux workstations with 512 MB–1 GB of random access memory. All IC synthesis runs required less than 1,195 s of CPU time.

8.1. Thermal model

As mentioned in Section 4, thermal modeling and analysis are used in the inner loop of the optimization flow to provide direct guidance for thermal optimization. Therefore, in order to determine the thermal profile of our system, we integrated our original work, a compact chip-package thermal model [13], into TAPHS. The thermal model has been validated against FEMLAB [22], an accurate but slow commercial finite-element based simulator, with less than 2.5% estimation error on the Kelvin scale. In the following experiments, each chip design is attached to a copper heat sink using forced air-cooling. We model two thermally conductive paths: heat dissipates from the silicon die through the cooling package to the ambient environment and through the package to the printed circuit board. We use an ambient temperature of 45 °C and a silicon thickness of $200\,\mu m$. In high-end microprocessor systems more than 80% of heat is dissipated through the first conductive path. In portable consumer electronic devices, due to the tight cooling budget and limited cooling space, the impact of the secondary conductive path becomes significant.

8.2. Benchmarks

We used TAPHS to synthesize 13 synthesis benchmarks. *Chemical* and *IIR77* are infinite impulse response (IIR) filters used in industry. *DCT_IJPEG* is the Independent JPEG Group's implementation of discrete cosine transform (DCT). *DCT_Wang* and *DCT_Lee* are DCT algorithms named after their inventors. All DCT algorithms work on 8×8 pixel of arrays. *Elliptic*, an elliptic wave filter, comes from the NCSU CBL high-level synthesis benchmark suite [23]. *Jacobi* is the Jacobi iterative algorithm for solving a fourth order linear system. *WDF* is a finite impulse response (FIR) wave digital filter. The largest benchmark, Jacobi, has 24 multiplications, 8 divisions, 8 additions, and 16 subtractions. In addition, we generated two CD-FGs using a pseudo-random graph generator [24]. Random100 has 20 additions, 15 subtractions, and 19 multiplications. Random200 has 39 additions, 44 subtractions, and 36 multiplications. The same sample periods (deadlines) were used for the benchmarks when evaluating each synthesis technique.

8.3. Multiobjective results

Table 1 shows the results of doing full multiobjective optimization of peak temperature, area, and energy consumption. In total, we compared 13 benchmarks. For each benchmark, the table shows non-dominated solutions produced by TAPHS. Due to space con-

Table 1. Comparison of non-dominated (multiobjective) results

Example	No voltage islands			Voltage islands			Thermal FP	
	Peak T (°C)	Area (%)	Power (W)	Peak T (°C)	Area (%)	Power (W)	Peak T (°C)	Power (W)
chemical	123.4	116.6	2.18	98.0	142.4	1.60	93.6	1.48
	123.6	112.0	2.18	100.4	121.7	1.62	96.2	1.51
	123.7	109.3	2.18	103.3	112.7	1.59	99.7	1.50
	128.6	112.9	2.24	110.3	95.0	1.62	105.3	1.53
dct_dif	79.0	87.9	0.85	67.3	92.5	0.60	65.6	0.55
	79.7	78.6	0.83	67.6	81.5	0.58	66.1	0.54
	80.3	83.7	0.85	69.8	83.4	0.61	67.4	0.57
	80.1	81.4	0.84	69.3	74.9	0.57	67.6	0.53
	81.7	80.7	0.86	69.9	80.0	0.60	68.4	0.56
	82.9	76.0	0.87	71.3	78.8	0.63	68.5	0.57
	84.5	68.8	0.87	71.4	75.8	0.62	68.7	0.57
dct_ijpeg	126.0	118.2	2.44	113.6	117.6	1.99	106.9	1.79
	129.4	107.2	2.39	115.8	114.9	2.03	107.4	1.81
	129.5	104.5	2.41	118.6	99.9	2.00	110.6	1.80
	130.6	104.7	2.40	118.9	102.0	2.03	111.2	1.82
	140.9	93.6	2.45	122.8	92.0	2.04	113.3	1.84
dct_lee	71.5	98.9	0.79	63.7	106.4	0.59	62.3	0.54
	71.8	95.6	0.79	65.5	119.2	0.61	62.4	0.55
	71.9	99.9	0.79	64.6	106.3	0.59	63.1	0.54
	75.0	87.8	0.80	65.2	100.4	0.60	63.6	0.55
	73.8	91.9	0.79	65.8	107.9	0.60	64.0	0.54
	73.7	91.7	0.79	66.8	106.4	0.62	65.0	0.57
	73.9	102.0	0.82	68.1	112.3	0.64	65.8	0.57
	73.6	102.0	0.81	68.7	101.2	0.64	66.3	0.58
dct_wang	70.7	101.3	0.70	59.8	109.8	0.42	57.6	0.39
	68.2	97.5	0.68	59.1	116.0	0.43	57.9	0.40
	68.5	108.1	0.68	60.1	108.0	0.42	58.1	0.39
	70.4	89.1	0.70	59.8	102.8	0.44	58.3	0.41
	71.3	100.5	0.69	61.1	100.1	0.45	59.3	0.42
	70.3	101.0	0.70	61.2	113.0	0.45	59.6	0.42
	72.0	85.1	0.72	63.1	109.8	0.48	60.7	0.43
	72.4	77.4	0.70	65.2	91.8	0.47	61.4	0.42
	72.0	88.9	0.72	66.3	90.8	0.48	63.6	0.43
	70.8	86.6	0.70	66.7	78.2	0.47	64.5	0.43
elliptic	136.8	105.5	2.55	111.6	122.6	2.04	108.0	1.93
iir77	94.5	105.0	1.57	73.7	119.7	0.94	72.0	0.89
	97.7	93.1	1.56	74.6	115.7	0.94	72.9	0.90
	99.0	93.1	1.57	76.5	94.9	0.96	75.2	0.92
jacobi	54.2	64.4	0.25	51.8	81.5	0.20	51.3	0.19
	53.9	65.5	0.25	52.1	77.7	0.20	51.5	0.19
	53.8	63.2	0.24	52.9	69.2	0.21	52.1	0.20
	54.9	59.4	0.25	52.5	69.4	0.21	52.2	0.20
	54.2	60.0	0.24	53.1	65.2	0.21	52.5	0.20
	54.4	66.0	0.25	53.1	61.7	0.21	52.6	0.20
	54.8	58.7	0.25	53.3	62.4	0.22	52.7	0.21
	54.6	58.0	0.25	53.6	64.5	0.22	53.3	0.21
	55.0	57.4	0.25	53.6	61.4	0.22	53.4	0.21
	55.5	52.9	0.25	54.5	61.5	0.22	53.5	0.20
pr1	98.0	104.0	1.49	82.5	106.1	1.10	80.3	1.02
	97.4	103.1	1.52	84.8	92.6	1.10	81.9	1.02
pr2	95.4	103.8	1.67	87.9	110.2	1.44	83.5	1.32
	97.3	89.2	1.68	87.5	100.6	1.45	83.6	1.33
	95.8	98.4	1.67	88.0	105.4	1.45	84.2	1.32
	99.3	91.2	1.68	88.4	98.4	1.44	84.6	1.32
	98.7	90.5	1.68	90.3	102.3	1.47	85.5	1.34
	99.9	79.8	1.71	91.7	83.0	1.47	86.6	1.34
	98.6	84.5	1.70	92.2	88.7	1.47	86.7	1.34
random100	71.6	100.0	0.85	66.0	98.8	0.63	63.6	0.57
	72.1	99.2	0.85	65.7	99.6	0.62	64.2	0.58
	72.7	99.7	0.86	67.6	85.1	0.67	64.6	0.62
	73.2	85.4	0.86	67.2	87.3	0.64	64.6	0.60
	74.1	91.5	0.86	66.5	92.3	0.63	64.7	0.58
	73.8	89.5	0.88	66.3	98.7	0.63	64.7	0.58
	73.7	94.5	0.88	68.6	87.2	0.66	65.3	0.61
	73.7	87.8	0.86	68.4	82.2	0.64	65.3	0.59
	75.0	89.1	0.88	69.4	86.5	0.68	65.3	0.62
	74.4	86.3	0.87	68.1	79.3	0.64	65.5	0.60
	73.9	87.0	0.85	68.9	80.4	0.66	66.1	0.61
	74.2	85.5	0.87	74.2	76.8	0.72	67.2	0.66
	76.5	84.1	0.87	73.5	62.9	0.73	69.8	0.66
	79.1	68.3	0.88	73.1	71.2	0.72	70.1	0.66
random200	90.8	90.2	1.77	81.4	112.0	1.37	76.2	1.20
	91.1	93.0	1.77	83.2	90.2	1.37	78.6	1.20
wdf	75.6	108.0	0.75	68.0	104.5	0.59	65.4	0.55
	74.8	96.9	0.73	67.8	101.8	0.59	67.0	0.55

884

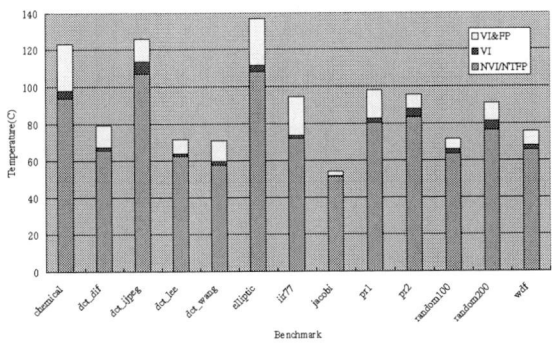

Figure 5. Peak temperature comparison

straints, we sorted the solutions for each problem in order of increasing peak temperature and uniformly eliminated all but seven solutions.

For each solution, the left column indicates the name of the benchmark. The next three columns show the peak temperatures, areas, and power consumptions of solutions produced without using voltage islands or thermal-aware floorplanning moves. Area is reported as a percentage of the area of the an initial solution without resource sharing or voltage islands. The floorplanner typically has an area efficiency ranging from 75%–90% for these benchmarks. From these columns, it should be clear that it is possible to trade off peak temperature for area as long as a thermal model is available during multiobjective synthesis. However, improving both objectives requires architectural-level and physical-level thermal optimization techniques.

The next three columns show the results produced using voltage islands, but without using thermal-aware floorplanning moves. From these columns, it is clear that voltage islands yield significant improvements in peak temperature, area, and power consumption. For example, the peak temperatures of the lowest peak temperature solutions to each problem were reduced by an average of 12.5 °C.

The next two columns show the results produced using both voltage islands and thermal-aware floorplaning moves. Note that the areas of these solution are the same as those without thermal-aware floorplanning moves. Combined with voltage islands, this technique allowed an average of 3.1 °C reduction in peak temperature.

Figure 5 shows only the lowest peak temperature for each benchmark after synthesis with voltage islands and thermal-aware floorplanning moves, with voltage islands but without thermal-aware floorplanning moves, and without voltage islands. As this figure indicates that both voltage islands and thermal-aware floorplanning moves can substantially reduce IC peak temperature, and that the relative contribution of each technique depends on the benchmark. In general, the best results were produced when these techniques were used together.

In addition, given the same area, TAPHS achieves lower peak temperatures for most benchmarks. For example, the peak temperature of *pr2* was reduced from 95.8 °C to 88.4 °C with the same area. Similar reduction were possible for *dct_dif*, *dct_ijpeg*, and *dct_lee*. In addition to reducing peak temperature, the proposed techniques can also be used to reduce area given a fixed peak temperature. When constraining temperature to the lowest temperature solution found without thermal optimization techniques, using voltage islands and thermal-aware floorplanning reduced area by, on average, 9.9%.

9. Conclusions

In this paper, we have described TAPHS, a thermal-aware high-level synthesis system that uses a tightly integrated thermal model and incremental floorplanner to optimize ICs peak temperatures, areas, and power consumptions, while meeting performance constraints. In order to optimize peak temperature, it was necessary to tightly integrate floorplanning, wire modeling, power profile gen-

eration, and chip-package thermal analysis with high-level synthesis. Experimental results indicate that TAPHS is able to trade off peak temperature, IC area, and power consumption. The proposed techniques allowed a reduction in peak temperature of 12.5 °C, on average. Peak temperature was also reduced under a fixed area constraint. Moreover, we have found that thermal optimization can allow significant improvements in IC area under temperature constraints. We conclude that it is important to incorporate thermal optimization in high-level synthesis to support continued increases in device and power density.

References

[1] L.-T. Yeh and R. C. Chu, *Thermal Management of Microelectronic Equipment: Heat Transfer Theory, Analysis Methods, and Design Practices.* New York, NY: ASME Press, 2002.

[2] International Technology Roadmap for Semiconductors, http://public.itrs.net.

[3] Z. P. Gu, *et al.*, "Incremental exploration of the combined physical and behavioral design space," in *Proc. Design Automation Conf.*, June 2005, pp. 208–213.

[4] O. Coudert, *et al.*, "Incremental CAD," in *Proc. Int. Conf. Computer-Aided Design*, Nov. 2000, pp. 236–244.

[5] J. Cong and M. Sarrafzadeh, "Incremental physical design," in *Proc. Int. Symp. Physical Design*, Apr. 2000.

[6] W. Choi and K. Bazargan, "Incremental placement for timing optimization," in *Proc. Int. Conf. Computer-Aided Design*, Nov. 2003.

[7] L. Zhong and N. K. Jha, "Interconnect-aware high-level synthesis for low power," in *Proc. Int. Conf. Computer-Aided Design*, Nov. 2002, pp. 110–117.

[8] A. Stammermann, *et al.*, "Binding, allocation and floorplanning in low power high-level synthesis," in *Proc. Int. Conf. Computer-Aided Design*, Nov. 2003.

[9] Y. Cheng, *et al.*, "ILLIADS-T: An electrothermal timing simulator for temperature-sensitive reliability diagnosis of CMOS VLSI chips," *IEEE Trans. Computer-Aided Design of Integrated Circuits and Systems*, vol. 17, no. 8, pp. 668–681, Aug. 1998.

[10] P. Li, *et al.*, "Efficient full-chip thermal modeling and analysis," in *Proc. Int. Conf. Computer-Aided Design*, Nov. 2004, pp. 319–326.

[11] H.-S. Wang, *et al.*, "Orion: A power-performance simulator for interconnection networks," in *Proc. Int. Symp. Microarchitecture*, Dec. 2002, pp. 294–305.

[12] K. Skadron, *et al.*, "Temperature-aware microarchitecutre," in *Proc. Int. Symp. Computer Architecture*, June 2003, pp. 2–13.

[13] L. Shang, *et al.*, "Thermal modeling, characterization and management of on-chip networks," in *Proc. Int. Symp. Microarchitecture*, Dec. 2004, pp. 67–80.

[14] T.-Y. Chiang, K. Banerjee, and K. C. Saraswat, "Analytical thermal model for multilevel VLSI interconnects incorporating via effect," *IEEE Electron Device Letters*, vol. 23, no. 1, pp. 31–33, Jan. 2002.

[15] Z. Lu, *et al.*, "Interconnect lifetime prediction under dynamic stress for reliability-aware design," in *Proc. Int. Conf. Computer-Aided Design*, Nov. 2004, pp. 327–334.

[16] C. Tsai and S. Kang, "Cell-level placement for improving substrate thermal distribution," *IEEE Trans. Computer-Aided Design of Integrated Circuits and Systems*, vol. 19, no. 2, pp. 253–266, Feb. 2000.

[17] B. Goplen and S. Sapatnekar, "Efficient thermal placement of standard cells in 3D ICs using a force directed approach," in *Proc. Int. Conf. Computer-Aided Design*, Nov. 2003, pp. 86–89.

[18] J. Cong, J. Wei, and Y. Zhang, "A thermal-driven floorplanning algorithm for 3D ICs," in *Proc. Int. Conf. Computer-Aided Design*, Nov. 2004, pp. 306–313.

[19] R. Mukherjee, S. O. Memik, and G. Memik, "Temperature-aware resource allocation and binding in high-level synthesis," in *Proc. Design Automation Conf.*, June 2005.

[20] K. A. Bowman, *et al.*, "A physical alpha-power law MOSFET model," *J. Solid-State Circuits*, vol. 34, pp. 1410–1414, Oct. 1999.

[21] H. Zhou and J. Wang, "ACG–Adjacent constraint graph for general floorplans," in *Proc. Int. Conf. Computer Design*, Oct. 2004.

[22] "COMSOL multiphysics ," http://www.comsol.com/products/multiphysics.

[23] "NCSU CBL High-Level Synthesis Benchmark Suite," www.cbl.ncsu.edu/benchmarks.

[24] R. P. Dick, D. L. Rhodes, and W. Wolf, "TGFF: Task graphs for free," in *Proc. Int. Wkshp. Hardware/Software Co-Design*, Mar. 1998, pp. 97–101.

An Automated, Efficient and Static Bit-width Optimization Methodology Towards Maximum Bit-width-to-Error Tradeoff With Affine Arithmetic Model

Yu Pu   and   Yajun Ha

Department of Electrical and Computer Engineering
National University of Singapore, Singapore, 119260
{Yu_Pu, elehy}@nus.edu.sg

Abstract – Ideally, bit-width analysis methods should be able to find the most appropriate bit-widths to achieve the optimum bit-width-to-error tradeoff for variables and constants in high level DSP algorithms when they are implemented into hardware. The tradeoff enables the fixed-point hardware implementation to be area efficient but still within the allowed error tolerance. Unfortunately, almost all the existing static bit-width analysis methods are *Interval Arithmetic* (IA) based that may overestimate bit-widths and enable fairly pessimistic bit-width-to-error tradeoff. We have developed an automated and efficient bit-width optimization methodology that is *Affine Arithmetic* (AA) based. Experiments have proven that, compared to the previous static analysis methods, our methodology not only dramatically reduces the fractional bit-width by more than 35% but also slightly reduces the integer bit-width. In addition, our probabilistic error analysis method further enlarges the bit-width-to-error tradeoff.

I. Introduction

Most of the computational-intensive applications are initially developed with high-level standard programming languages where the variables and constants are possible to be described in high precision data types. While being implemented into hardware subjected to fixed-point restriction, these variables and constants prototyped in high precision must be re-defined to the fixed-point. The translation process from full to limited precision must be optimized to tradeoff precision for silicon area, latency, power consumption and etc. Unfortunately, choosing the most appropriate bit-width is a non-trivial task. Moreover, it is too burdensome for designers to translate them manually.

Many researchers have focused their interests in developing tools that may help decide the bit-widths automatically. One category of techniques is based on detailed simulations. Kum et. al. [7] have used Monte Carlo-style statistical simulation to optimize word-length iteratively. This technique can get the optimum results but is quite inefficient since there is often a huge searching space to simulate with a full coverage of input vectors. Chang et al. [2] have developed a design-time tool, *Precis*, which provides designers with area-precision information by doing slack analysis and aids the designer in focusing their manual precision optimization. They have also presented a method that broadens their work in [2] to explore area-to-error tradeoffs by precision steering [3]. However their technique is semi-automated. The other category of techniques is based

on static analysis. Stephenson et. al. [1] have formulated the bit-width analysis as a value range propagation problem and introduced a compiler *Bitwise* that extracts the value of each variable by seamlessly performing bi-directional propagation, but their compiler is limited to integers and pointers. Nayak et. al. [4] have used a precision analysis based on the method presented in [1] to infer the minimum number of bits for the integer part and a combined precision and error analysis to infer the fractional bit-width. However, their algorithm is restricted by the assumption that the number of fractional bits is constant for all variables.

We observe that almost all the existing static bit-width optimization methods are *Interval Arithmetic*(IA) based, which may lead to the pessimism because the overestimation accumulates exponentially along the computation path. Fang et. al. [5][6] have introduced the *Affine Arithmetic* model into the verification of the finite-precision effects in DSP applications. Inspired by their work, we have extended the AA model to the bit-width analysis. A closely related but independent research has been simultaneously conducted by Lee et. al. [8], but their method uses the adaptive simulated annealing algorithm to find the optimum number of fraction bits, which is quite time consuming. Compared to their methodology, ours has the following features:

- We use the hard error analysis to insure the output error not to exceed the designer-specified error tolerance.
- In most of the DSP designs, the designer allows certain degree of error rate, which motivates us to explore a larger bit-width-to-error tradeoff than using hard error analysis. Our probabilistic error analysis almost insures that the probability for the output error to lie in the specified error tolerance is higher than the designer specified parameter λ.
- Decision time is greatly reduced since our approach iterates only once.
- Our approach performs on the algorithm level, followed by high-level synthesis which is based on the bit-width optimization results. Therefore, our approach can be easily incorporated into the existing synthesis tools.

The remainder of the paper is organized as follows: Section II briefly gives the background of IA and AA models. Section III goes through our bit-width analysis methodology in detail. Section IV gives and analyzes the experimental results. Section V draws the conclusion of the work.

II. Background

Interval arithmetic has been widely used in the range propagation analysis. However, IA model could lead to overestimations and such overestimations are exponentially accumulated along a computation path. As a consequence, the final intervals may be too large to be useful.

Affine Arithmetic (AA) model can overcome the overestimation explosion problem. In AA model, the uncertain variable x is expressed as:

$$\hat{x} = x_0 + x_1\varepsilon_1 + x_2\varepsilon_2 + \cdots + x_n\varepsilon_n, \quad -1 \le \varepsilon_i \le 1$$

where x_0 is the *central value* of the affine form of $\hat{x}$, ε_i is an independent *noise symbol* multiplied by the corresponding *coefficient* x_i. Along the computation path, one symbol of ε_i may contribute to be uncertainties of two or more variables. When these variables are combined, the uncertainties may be cancelled out so as to make the results of range propagation analysis tighter than that of IA model.

In order to present the mathematical reasoning behind our methodology, we briefly introduce the AA based operation models. Assume that there are variables x, y and constant c, we have $x = x_0 + \sum_{i=1}^{n} x_i\varepsilon_i + E_x$, $y = y_0 + \sum_{i=1}^{n} y_i\varepsilon_i + E_y$, and $c = c_0 + E_c$, where E_x, E_y and E_c are the error terms generated from quantization. Each of the error terms is expressed in the affine form as $E = |\Delta|\varepsilon_e$, where $|\Delta|$ is the upper bound for the quantization error. If we denote f as the bit-width for the fractional part, we have $|\Delta| = 2^{-f}$ in the case of truncation and $|\Delta| = 2^{-(f+1)}$ in the case of rounding. The noise symbol ε_e indicates the uncertainty of the quantization error.

The AA operation models for the addition, subtraction and multiplication are shown below. With the AA form operation models, the results are also in affine forms. The error that occurs after each operation is linear with respect to errors of its two operands and it shifts with a new quantization error σ.

Addition and Subtraction:

$$x \pm c = (x_0 \pm c_0) + \sum_{i=1}^{n} x_i\varepsilon_i + \Phi,$$

Where $\Phi = E_x \pm E_c + \sigma$

$$x \pm y = (x_0 \pm y_0) + \sum_{i=1}^{n} (x_i \pm y_i)\varepsilon_i + \Phi,$$

Where $\Phi = E_x \pm E_y + \sigma$

Multiplication:

$$x \cdot c = c_0 x_0 + c_0 \sum_{i=1}^{n} x_i\varepsilon_i + \Phi,$$

Since we know that $E_x \in [-1,1]$ and $E_c \in [-1,1]$, so $E_x E_c \le E_x^2 + E_c^2 / 2 \le (|E_c| + |E_x|)/2$ □then we have

$$\Phi \le E_x \cdot c_0 + E_c(x_0 + \sum_{i=1}^{n} x_i\varepsilon_i) + (|E_c| + |E_x|)/2 + \sigma$$

Similarly,

$$x \cdot y = x_0 y_0 + \sum_{i=1}^{n} (y_0 x_i + x_0 y_i)\varepsilon_i + \sum_{i=1}^{n} |x_i| \sum_{i=1}^{n} |y_i|\varepsilon_k + \Phi$$

$$\Phi \le E_y(x_0 + \left|\sum_{i=1}^{n} x_i\right|) + E_x(y_0 + \left|\sum_{i=1}^{n} y_i\right|) + (|E_x| + |E_y|)/2 + \sigma$$

A quantity in the affine form can be bounded by

$$x = x_0 + \sum_{i=1}^{n} x_i\varepsilon_i \le x_0 + \sum_{i=1}^{n} |x_i| = |x| \quad (1)$$

where the term $\sum_{i=1}^{n} |x_i|$ is called the *total deviation* of x.

III. Bit-width Analysis Methodology

Figure 1 gives a brief overview of our methodology. First, we accept the designer-specified constraints via a user-interface, and convert the input range into the AA format. Second, we run the program "symbolically" to get the output results in AA form. Third, we use the range analysis to determine the range of each variable (or constant) and the corresponding integer bit-width. The subsequent error analysis will return the optimized fractional bit-width.

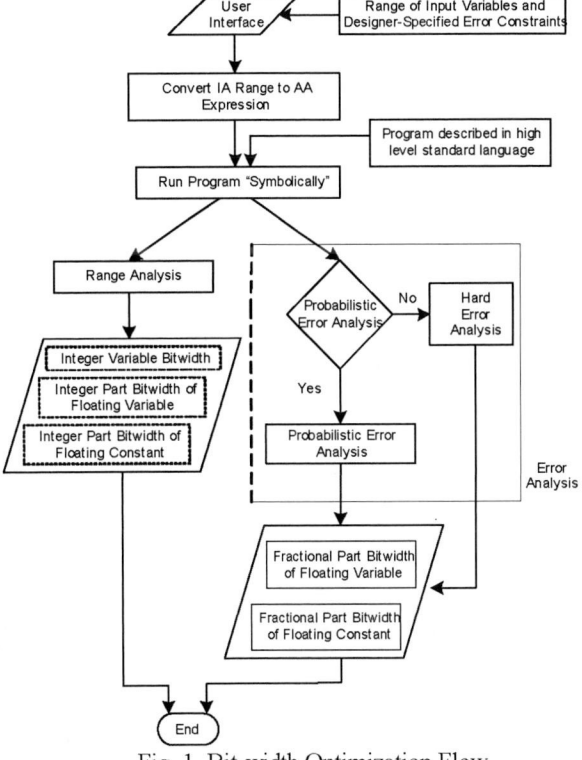

Fig. 1. Bit-width Optimization Flow

Information Provided by Designer:

To limit the complexity, we implemented our method using the multiple-input to single-output computational model. The designer provides the following 4 types of information using the interface shown in Figure 2: (1) the value range of all the input variables; (2) the output variable; (3)the absolute maximum error tolerance E_{spec} at the output; (4)optionally the probability parameter λ.

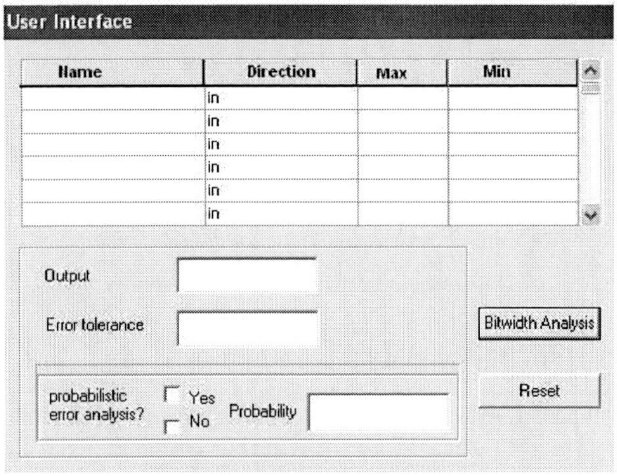

Fig. 2. The User Interface

Convert IA Range to AA Range:

Since the ranges of input variables given by designers are in the form of interval range, for example, an input variable x is ranged in $[a,b]$. We first convert them to the AA form, so x is expressed as $x = (a+b)/2 + (a+b)\varepsilon_x/2$, $(-1 \le \varepsilon_x \le 1)$, and ε_x is the noise symbol introduced for representing the uncertainty of x.

Run Program "Symbolically":

We "symbolically" traverse the whole program from the entry to the exit in *forward* direction. Assume that there are totally N floating-point variables and constants whose upper quantization error bounds are expressed as $|\Delta_1|, |\Delta_2|, |\Delta_3| \cdots |\Delta_N|$, respectively. When they are transformed to be fixed-point, the actual quantization errors incurred at each of them are expressed in affine forms as $|\Delta_1|\varepsilon_1, |\Delta_2|\varepsilon_2, |\Delta_3|\varepsilon_3 \cdots |\Delta_N|\varepsilon_N$. With the operation models outlined in the background part, we get the results in symbolical affine form for all variables and floating-point constants.

There are some rules of running the program:
- For loop construct whose loop count has been prescribed, the variables in the body of the loop are calculated by thoroughly traversing the loop.
- For a variable that is condition dependent, the variable's range is refined based on the outcome of the conditional branch. For example, in Figure 3 the variable x is condition dependent, so at the output of the condition branch($x \le 0$?), the range of x is refined with newly introduced affine form expressions.

- We do not do *backward* propagation because the affine form itself is not an efficient form for performing backward propagation.

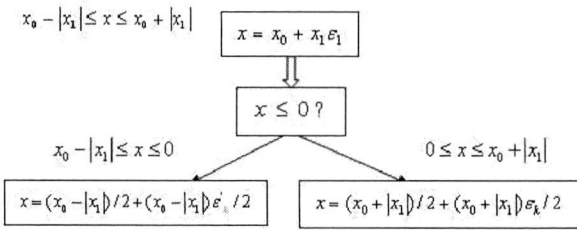

Fig. 3. Range Refinement of Condition Dependent Variable

Range Analysis:

After "symbolically" running the program, we get all the variables and floating-point constants in the affine forms. By Equation (1), their ranges are bounded. Hence, the bit-width of the integer variables, the bit-width for the integer part of the floating-point variables and the floating-point constants are determined. Assume that the quantity x (either variable or constant) is translated to be fixed-point with m bits to be its integer part bit-width, we obtain the minimum m by the following expression:

$$2^{m-1} - 1 \ge |x| \tag{2}$$

Error Analysis:

After "symbolically" running the program, we also get the error of the output variable. This output error is in the affine form as

$$\Delta_{output} = A_1|\Delta_1|\varepsilon_1 + A_2|\Delta_2|\varepsilon_2 + A_3|\Delta_3|\varepsilon_3 + \cdots + A_N|\Delta_N|\varepsilon_N \tag{3}$$

$$= |A_1||\Delta_1|\varepsilon_1' + |A_2||\Delta_2|\varepsilon_2' + |A_3||\Delta_3|\varepsilon_3' + \cdots + |A_N||\Delta_N|\varepsilon_N' \tag{4}$$

$$\le |A_1||\Delta_1| + |A_2||\Delta_2| + |A_3||\Delta_3| + \cdots + |A_N||\Delta_N| \tag{5}$$

Each independent term $A_i|\Delta_i|\varepsilon_i$ in Equation (3) tells the contribution that a floating point variable or constant's quantization error makes to the overall error at the output, where A_i is an affine form expression. Since the designer-specified output error tolerance E_{spec} is the upper bound of $|\Delta_{output}|$, we have:

$$|\Delta_{output}| \le E_{spec} \tag{6}$$

And we assign non-negative weights to each term in Equation (5):

$$\sum_{i=1}^{i=N} W_i |A_i||\Delta_i| = |\Delta_{output}| \tag{7}$$

$$\sum_{i=1}^{N} W_i = 1 \tag{8}$$

Through Equations (5) (6) and (7), once W_i is known, we can infer $|\Delta_i|$. At this moment, we simply assign each term with an equal weight, so $|A_i||\Delta_i| \le E_{spec}/N$ (9)

Now we have $|\Delta_i| \le Espec/(N \cdot |A_i|)$, and each variable's fractional bit-width $f_1, f_2, ... f_N$ can be decided:

If the quantization is real rounding based,

$$2^{-(f_i+1)} \leq |\Delta_i| \Leftrightarrow f_i \geq -\log_2|\Delta_i| - 1 (f \geq 0) \qquad (10)$$

if the quantization is truncation based,

$$2^{-f_i} \leq |\Delta_i| \Leftrightarrow f_i \geq -\log_2|\Delta_i| (f \geq 0) \qquad (11)$$

For the experimental results presented in this paper, we employ real rounding.

To further explore the bit-width-to-error tradeoffs, after $f_1, f_2 ... f_N$ are obtained, we use the following "greedy algorithm" to optimize the bit-width:

---

1. Sort $f_1, f_2, ... f_N$ and map them to a new array $X[N]$ according to their decreasing order.

2. for (j=0;j++j<N)
   {
       for (i=0;i++;i<N)
       {
   $$E_{residue} = Espec - \sum_{i=1}^{N} 2^{-(X[i]+1)} ;$$
       }
   $$2^{-(X[j]+1)} = 2^{-(X[j]+1)} + (E_{residue} / |A_K|);$$
       // assume that f_k is mapped to $X[j]$
       Recalculate $X[j]$ → Recalculate f_k ;
   }

We call the aforementioned fractional bit-width analysis method as the "hard" error analysis because it fully insures the output error not to exceed the designer-specified error tolerance.

However, we note that in Equation (4) when N gets large, it is quite unlikely for all the ε_i' to simultaneously take extreme values and push the output error to its upper bound. [6] has introduced a probabilistic model to estimate the error of a affine form variable. Our probabilistic error analysis method is partially based on their idea.

Since we have $|A_i||\Delta_i| = |A_i||\Delta_i| = \cdots = |A_i||\Delta_i| = k$, so Equation (4) depicts a sum of many statistically independent and identically distributed terms. By the *central limit theorem*, the Equation (4) approaches a Gaussian CDF as the parameter N increases. We have

$$\frac{\Delta_{output}}{\sqrt{N}\sqrt{Variance}} \to N(0,1) \qquad (12)$$

Where $N(0,1)$ is a standard normal distribution, and the

$$Variance = k^2/3 = (|A_i||\Delta_i|)^2/3 \qquad (13)$$

We have the designer-specified parameter λ , which bounds the probability that the output error lies within the designer-specified error tolerance. For example, if the designer sets $\lambda = 0.999$, then the chance for the output error to exceed E_{spec} is only once in 10^3 times of simulation.

$$prob(|\Delta_{output}| \leq E_{spec}) \geq \lambda \qquad (14)$$

The probabilistic error analysis is quite straightforward:

Step 1:

Look up the Gaussian approximation statistical table according to λ . Since $|\Delta_{output}|$ and N are known, we can get the variance and k by Equation (12).

Step 2:

Determine $|\Delta_i|$ by Equation (13), so f_i can be determined.

IV. Experimental Results and Analysis

We use an example to prove the strength of our methodology. The signal flow of a butterfly part in IDCT is shown in Figure 4 and such kinds of computations are common in DSP. To explain explicitly, the data flow graph is shown in Figure 5. We set $Const_1 = Const_2 = \sqrt{2}/2$. Since C/C++ is the most popular general purpose language [13-15], the algorithm is initially described in C++ with all the variables and constants set to be in 64-bit long double precision.

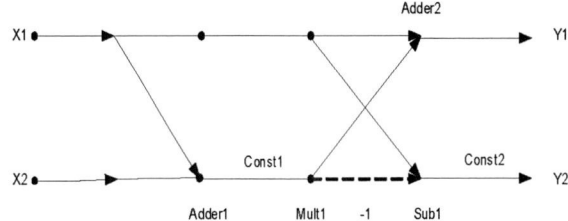

Fig. 4. Signal Flow Graph

We apply our bit-width analysis methodology to this example. We first provide the value range of all the input variables X_1 and X_2 : $X_1 \in [-128,127]$ and $X_2 \in [-128,127]$, then specify the maximum error tolerance at Y_2 to be 1.0. In probabilistic error analysis, the probability λ is 0.999. We compared our bit-width optimization results with the results generated from using the method presented in [4] which is considered so far as the best fully automated IA based static bit-width analysis approach to optimize the word-length for the algorithms described in MATLAB.

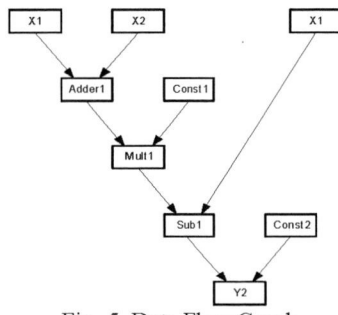

Fig. 5. Data Flow Graph

Bit-width optimization results and Analysis:

TABLE I
Bit-Width Analysis Results

Node		X_1	X_2	$Adder_1$	$Const_1$	$Mult_1$	Sub_1	$Const_2$	Y_2
Integer	IA	8	8	9	1	9	10	1	9
	AA	8	8	9	1	9	8	1	8
Fractional	IA	8	8	8	8	8	8	8	8
	Hard	0	3	3	9	3	3	8	3
	Prob.	0	2	2	10	2	2	9	2

Table I lists the bit-width analysis results. For the integer part, we see that at nodes Sub_1 and Y_2, the IA based range analysis cannot consider the range cancellation among variables such that the pessimistic results are occurred. With our AA based method, the integer bit-width can be optimized. Considering the efficiency reflected by this simple example, it is predicted that in a large scale computation where many variables may have correlations, our method can save considerable integer bit-width. For the fractional part, we see that IA based bit-width analysis returns a large overestimation in terms of the bit-width. At most of the nodes, our method results in much less bit-width than the IA method. However, we notice that at the nodes $Const_1$ and $Const_2$ our method results in a little larger or equal bit-width compared to the IA based method. Note that in the current stage we simply assign the same weight to each node, but actually we can adjust the weights. For example, we assign larger weights to the nodes who need more bit-width, such as $Const_1$ and $Const_2$, and the bit-width for these nodes can be reduced while the bit-width of other nodes only suffer slight increase.

With the similar approach used for the first example, we have experimented on a 4th order polynomial and a FIR low-pass filter to test our methodology. The polynomial equation in our benchmark is $y = x^4 + x^3 + x^2 + x + 1$, where x is initialed as a floating point variable and $x \in [-16, 15]$. The FIR filter is implemented according to the architecture shown in Figure 6, where we set all X_n to be in the floating format, $X_n \in [-64, 63]$ and the floating constants a0 to a4 to be in *long* format obtained from the command *firpm()* in MATLAB. For each example, we specify the error tolerance to be 1.0.

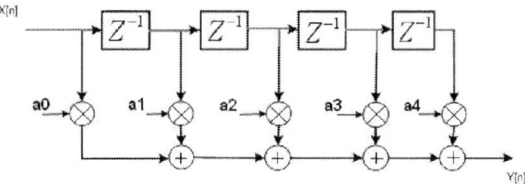

Fig. 6. Architecture of the FIR filter

The average bit-widths for the fractional part of these benchmarks are listed in Figure 7. They are normalized with respect to the IA based fractional bit-width. We see that with our hard error analysis, we can achieve a more than 35% resource saving in terms of the bit-width for the fractional

part compared to the IA based fractional bit-width analysis. Therefore, our method has achieved a much larger bit-width-to-error tradeoff than previous IA based method. In addition, with our probabilistic error analysis, the bit-width can be even reduced by 50%. As we gradually release the probabilistic restriction, the tradeoff keeps on going upwardly.

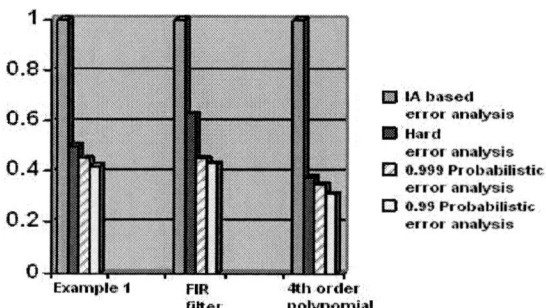

Fig. 7. Average Bit-width for Fractional Part

Verification:

To verify the correctness of our method, for the hard error analysis, we check whether the maximum output error lies within the specified error tolerance after implementing the algorithms into hardware that is subjected to fixed-point restriction. For the probabilistic error analysis, we check that whether the probability for the output error to lie in the specified error tolerance can be higher than the probability specified by the designer. We execute the program written in C++ with 2×10^3 random input data sets that are uniformly distributed within their ranges. Then we modify the C++ programs linking with the compiled SystemC library, set the quantization option to be real rounding and set each variable to be fixed-point data type with the bit-width calculated by our method. We use the same input data sets to get the results. We compare the results to obtain the simulated maximum output error and the simulated probability that the error lies in the specified error. The experimental results for all the examples are shown in Table II.

TABLE II
Verification Results

Test-bench	Bit-width analysis methods	Simulated output error	Specified probability λ	simulated probability λ'
Example1	IA based	0.02	N/A	1
	AA based	0.22	N/A	1
		0.47	0.99	1
		0.33	0.999	1
FIR filter	IA based	0.09	N/A	1
	AA based	0.15	N/A	1
		0.58	0.99	1
		0.58	0.999	1
4th order Polynomial	IA based	0.88	N/A	1
	AA based	0.93	N/A	1
		1.10	0.99	0.999
		0.95	0.999	1

In Table II, the 3rd column gives the simulated maximum

output error. The 4th column gives the parameter λ, which is the designer-specified probability for the output error to lie in the specified error tolerance. Note that when the parameter is not available (N/A), we use the hard error analysis. The 5th column gives the simulated probability λ'. Table II shows that, while our hard error analysis fully insures that the output error will not exceed the designer specified error, the probabilistic error analysis can almost guarantee that the probability for the output error to lie in the specified error will be higher than the specified probability. However, since we have used Gaussian approximation during the analysis, theoretically it is difficult to fully guarantee the error probability to be bounded by a designer, even in our experiments it works quite well. Therefore, we suggest that the specified probability should be flexibly restricted.

V. Summary and Conclusions

We have presented an automated and efficient static bit-width optimization methodology which is based on the affine arithmetic model. While our AA based range analysis can slightly reduce the integer part bit-width, the AA based hard error analysis can dramatically reduce the fractional bit-width. In addition, we have proposed the probabilistic error analysis method which can further shorten the bit-width. Our experimental results have proven that our approach can explore a larger bit-width-to-error tradeoff.

However, we also note the limitations of our methodology. First, we bear the same drawback as mentioned in [6] that we assume that all the input variables are independent while this is not always the case. The future work can model the correlations explicitly from start. Second, our method lies at the algorithm level which does not consider the hardware sharing and hardware cost function. Our ongoing research will further explore the solutions for these problems.

References

[1] M. Stephenson, J. Babb, and S. Amarasinghe, "Bitwidth analysis with application to silicon compilation," *Proceedings of the SIGPLAN conference on Programming Language Design and Implementation*, June 2000.

[2] Mark L. Chang and S. Hauck, "Precis: A design-time precision analysis Tool," *IEEE Symposium on Field-Programmable Custom Computing Machines*, pp.229-238, 2002.

[3] M. L. Chang and S. Hauck, "Variable Precision Analysis for FPGA Synthesis," Nasa Earth Science Technology Conference 2003.

[4] A. Nayak, M. Haldar, A. Choudhary and P. Banerjee, "Precision and error analysis of MATLAB applications during automated hardware synthesis for FPGAs," *Design Automation & Test*, March 2001.

[5] C. F. Fang and R. A. Rutenbar and M.Puschel and T. Chen, "Towards efficient static analysis of finite precision effects in DSP applications via affine arithmetic modeling," *Design Automation Conference*, 2003.

[6] C. F. Fang, R. A. Rutenbar, M. Puschel and T. Chen, "Fast, accurate static analysis for fixed-point finite-precision effects in DSP designs," *Proceedings of the International Conference on Computer Aided Design* (ICCAD'03).

[7] K. Kum and W. Sung, "Combined word-length optimization and high-level synthesis of digital signal processing systems," *IEEE Trans. on Computer-Aided Design*, Aug, 2001.

[8] D-U. Lee, A. A. Gaffar, O. Mencer, W. Luk, "MiniBit: bitwidth optimization via affine arithmetic," *Proceedings of the 42nd annual conference on Design automation*.

[9] S. Kim, K.-I. Kum, and W. Sung, "Fixed-point optimization utility for C and C++ based digital signal processing programs," *Workshop on VLSI and Signal Processing*, Osaka, 1995.

[10] N. Shirazi, A. Walters and P. Athanas, "Quantitative analysis of floating point arithmetic on FPGA based custom computing machines," *IEEE Symposium on FPGAs for Custom Computing Machines*, April, 1995.

[11] J. Stolfi and L.H. de Figueiredo, "An introduction to affine arithmetic," *TEMA Tend. Mat. Apl. Comput.*, 4, Vol.3, pp. 297-312, 2003.

[12] http://www.systemc.org

[13] D. Galloway, "The Transmogrifier C Hardware Description Language and Compiler for FPGAs," *FCCM'95*

[14] G. Doncev, M. Leeser and S. Tarafdar, "High-Level Synthesis for Designing Custom Hardware," *Proc. Field-Programmable custom Computing Machines*, April 1998.

[15] B. L. Hutchings and B. E. Nelson, "Using General - Purpose Programming Languages for FPGA Design," *Proc.37th Design Automation Conference*, June 2000.

[16] Mark Stephenson, "Bitwise: Optimization Bitwidths Using Data-Range Propagation," Master's thesis. Massachusetts Institute of Technology, May 2000.

[17] K. Bondalapati and V. K. Prasanna, "Dynamic precision management for loop computations on reconfigurable architectures," *IEEE Symposium on Field-Programmable Custom Computing Machines*, April 1999.

[18] W. Sung and K. I. Kum. "Simulation-based word-length optimization method for fixed-point digital signal processing systems," *IEEE Transactions on Signal Processing*, vol. 43, no.12, pp.3087-3090, December 1995.

[19] J. Patterson, "Accurate Static Branch Prediction by Value Range Propagation," *Proc.of the ACM SIGPLAN Conference on Programming Language Design and Implementation*, pp. 67 – 78, June 1995.

[20] R. Cmar, L. Rijnders, P. Schaumont, S. Vernalde, and I. Bolsens, "A methodology and design environment for DSP asic fixed-point refinement," *Design, Automation and Test in Europe Conf.*, 1999.

Abridged Addressing: A Low Power Memory Addressing Strategy

Preeti Ranjan Panda

Department of Computer Science and Engineering
Indian Institute of Technology Delhi, Hauz Khas, New Delhi 110016, INDIA

Abstract— The memory subsystem is known to comprise a significant fraction of the power dissipation in embedded systems. The memory addressing strategy, which determines the sequence of addresses appearing on the memory address bus as well as the switching activity in the addressing logic, has a major impact on the memory subsystem power dissipation. We present a novel addressing strategy, *Abridged Addressing*, that helps reduce system power dissipation by substantially reducing both the address bus switching as well the addressing logic power. The strategy, which relies on minimizing register accesses in the addressing logic, helps overcome some of the limitations of existing approaches: the address bus switching is low; there is very little area, performance, and power overhead; and the addressing hardware is simpler, making the technique suitable for both on-chip and off-chip memory, as well as single-port and multi-port memories.

I. INTRODUCTION

Memory accesses constitute an important target for power optimizations since memory accounts for an increasing fraction of the total area and power dissipation of embedded systems. Since dynamic power dissipation in CMOS circuits is a function of the total switching capacitance, particular attention has been paid to the memory address and data buses because these are wide, and are typically long with high capacitance. Lower switching on these buses leads to lower overall power dissipation. Although it may not be possible to optimize well for the data bus because the relevant data values are typically not known in advance, the switching on the address is often known statically and is controllable by the synthesis process. In typical design descriptions, the access patterns of the arrays – the loop induction variables and the array index expressions are known and usually exhibit some regularity. This regularity results in a strong correlation in the sequence of addresses placed on the address bus, which can be exploited by a power optimizing synthesis tool.

The idea of encoding the address bus with the objective of minimizing switching has been extensively studied [1]. In Bus-invert coding [2], an additional bit is appended to the bus to indicate that the bus is inverted – this is invoked if too many bits (more than half) are switching on the bus. In [3], the instruction address bus was encoded using Gray code. The T0 encoding scheme [4], where an extra bit encodes the information that the following address is the successor of the previous address, results in zero transitions in the best case – i.e., where addresses are strictly sequential.

Recognizing that programs tend to spend a lot of time accessing arrays inside loops, [5] proposed an address encoding technique called Working Zone Encoding (WZE). The technique involves the construction of encoding and decoding circuits that keep track of currently active memory zones. During memory accesses, the zone is selected and an encoded offset is transmitted on the address bus, which is used by the decoder to recover the actual address. More general correlations on memory data buses have been studied, which rely on having static access to data streams so that frequently occurring patterns can be identified and assigned appropriate codes [6, 7]. In [8], the authors present an encoding/decoding strategy consisting of adaptive self-organizing lists to handle complex interactions arising out of multiplexed address buses. In [9], the authors first profile an application to determine which regions of memory are most accessed. To reduce power dissipation, these regions are assigned to smaller memory modules. In [10], strategies including array interleaving were described for arranging data in memory so that switching on the address bus is minimized. However, this strategy does not generalize well because interleaving requirements in different loops might conflict. In this paper, we assume that the start addresses of arrays have been determined, and the structure of the individual arrays is not modified; interleaving could be combined with our proposed strategy.

For data memory, the encoding strategies indicated above work well for sequential addresses, but not for correlated addresses that are not necessarily sequential, although WZE and [8] address this deficiency to some extent. In [11], the authors present a customized data memory interface (CDMI) that uses static information about the loop strides and the order of memory accesses to generate addresses within the data memory subsystem, thereby minimizing communication across the processor-memory address bus. Although proposed in a processor context, a similar strategy can also be applied to synthesized hardware. The technique requires a relatively power-expensive decoding circuit in the memory subsystem, which is justified in case of off-chip power-memory bus, because power savings on the address bus dwarfs the decoding overhead. However, in case of on-chip memory, the address bus capacitances are relatively smaller, and the decoder hardware's power overhead becomes significant. We present a novel strategy, *Abridged Addressing*, where we optimize register file accesses and simplify the addressing logic by harnessing memory address correlations. This not only reduces address bus energy, but also significantly reduces power dissipation in the addressing logic, thereby making the addressing mechanism equally applicable to both the on-chip as well as off-chip memory interfaces.

II. MEMORY ADDRESSING

When designing optimizations targeting the data memory interface, particular attention is paid to loops in the specification program because most of the execution time is known to be spent inside loops, irrespective of whether the target is software or hardware; consequently, most memory accesses occur during loop executions. Since typical loops in a large variety of application domains (such as DSP, graphics, vision, image processing, etc.) have regular loop strides and array index expressions that are affine (linear) in the loop induction variables, it is possible to analyze the array access patterns statically and synthesize power-efficient address generation, encoding, and decoding circuits that minimize the overall power consumption during memory accesses. Consider a loop of the form:

```
for (i = 0; i < n; i = i + 2)
    x = a[i] + b[i+1];
    ...
```

Assume that the implementation target uses a dual-port memory because of performance constraints. The addressing logic now drives two memory address buses. A straightforward implementation of the addressing logic section of the datapath is shown in Figure 1. This architecture suffers from some serious disadvantages when the memory is physically far apart from the datapath. For example, when the memory is off-chip, the power dissipation in the address buses is large due to the transitions in the off-chip wires with large capacitances. Several optimization strategies target this interface typically with encoders on the datapath side, and decoders on the memory side to minimize the transitions in the address bus. A simplified version of an example efficient implementation, based on the CDMI proposal in [11], is shown in Figure 2. Because of the regular memory access pattern in the loop, there is no need to send the addresses explicitly on the address bus in the steady state. Instead, the initial addresses are sent to registers in the decoders only once, and the registers are updated after every access so that subsequent memory addresses are generated in the decoder itself. A small counter-based FSM (which is initialized before each loop begins) keeps track of the register updates. The technique is very effective in minimizing address bus traffic on the datapath-memory interface because the datapath updates the address only once for every loop.

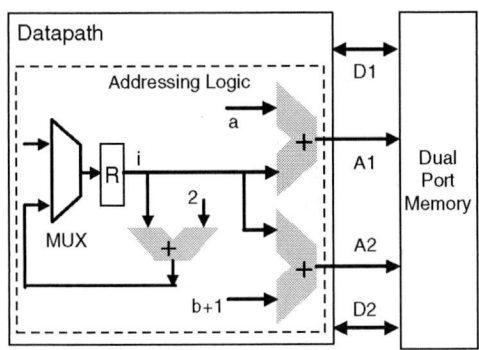

Fig. 1. The datapath/memory interface

The above strategy yields significant power savings when the datapath-memory interface consists of high capacitance wires, especially for off-chip memories where the switching capacitances are about three orders of magnitude higher than typical on-chip wire capacitances. However, if the memory is on-chip, then the power overheads of the decoding circuitry may overwhelm the power savings due to reduced address bus switching; the overall power savings depends on the physical placement of the datapath and memory blocks. Since the addressing decisions are made very early in the design phase, when the physical information is not known, it is important to design the addressing logic in such a way that power overheads are minimized.

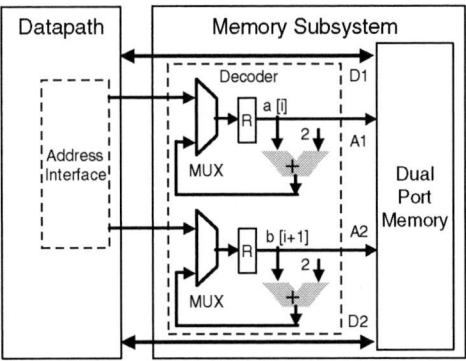

Fig. 2. Minimizing address bus transitions

Figure 3 shows an alternate address generation mechanism, *Abridged Addressing*, where the address for $a[i]$ is generated using a register as before, but the address for $b[i + 1]$ is generated from that of $a[i]$ by just adding $(B - A + 1)$ where A and B are the start addresses of a and b respectively. For simplicity of discussion, we have used only word addresses in this paper. Since $(B - A + 1)$ is a constant known during synthesis, the addressing logic becomes simpler. In the previous approach (Figure 2), we need to keep a register in the decoder for every memory access, and if there are multiple ports, we may organize them into one register file for each port. However, reading from and writing to these register files in each cycle is power-inefficient. Accesses to the decoder's register file results in extra power consumption during read/write of the individual registers as well as during the register file address decoding. The alternative we propose in Figure 3 maintains only one address register for a loop body, and derives the remaining addresses from this register using simple arithmetic operations. Our experiments show that replacing the expensive register file access operations by simple additions cause a 23% reduction in power dissipation for the above example.

III. ADDRESS DECODER CIRCUITS

The CDMI-based approach, although proposed in the processor context in [11], can be extended in a straightforward way to synthesized hardware. Consider a general n-level nested loop accessing an r-dimensional array $A[N_1][N_2]...[N_r]$ as follows:

$$\textbf{for } (i_1 = l_1; i_1 < h_1; i_1 = i_1 + s_1)$$

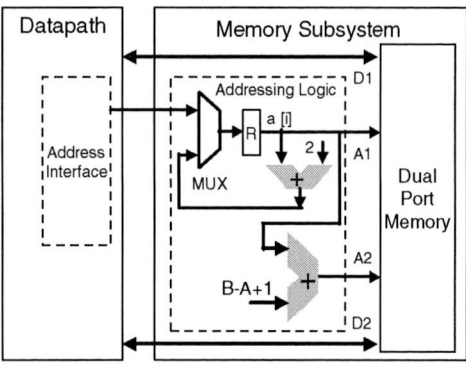

Fig. 3. Abridged Addressing: minimizing address bus transitions and addressing logic

$$\textbf{for } (i_2 = l_2; i_2 < h_2; i_2 = i_2 + s_2)$$

$$\ldots$$

$$\textbf{for } (i_n = l_n; i_n < h_n; i_n = i_n + s_n)$$
$$\text{READ } A[c_{11}i_1 + c_{12}i_2 + \ldots + c_{1n}i_n + k_1]$$
$$[c_{21}i_1 + c_{22}i_2 + \ldots + c_{2n}i_n + k_2]$$
$$\ldots$$
$$[c_{r1}i_1 + c_{r2}i_2 + \ldots + c_{rn}i_n + k_r]$$

The difference in address locations between two successive iterations of the innermost loop is:

$$A_d = (N_2 N_3 \ldots N_r)c_{1n} + (N_3 \ldots N_r)c_{2n} + \ldots + N_r c_{r-1n} + c_{rn}$$

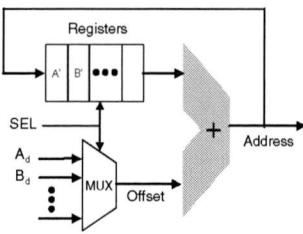

Fig. 4. Address decoder for linear array indices

For linear indices, all the c_{ij}'s and N_i's are constants, so A_d can be computed at compile time and set as an input to the MUX shown in the generalized addressing circuit shown in Figure 4. The MUX inputs A_d, B_d, etc., represent the address offsets to the respective arrays in successive iterations of the i_n-loop. The values in the registers, A', B', etc., represent the addresses for arrays A, B,...respectively from the previous iteration. The address for the current access is determined by the SEL signal from a small counter-based FSM selecting both the appropriate register and the offset input to the adder. The register file is initialized before the loop execution begins from the address bus in the datapath interface, but in the steady state, the bus from the datapath never changes and addresses to the memory are generated within the decoder itself. This is the result of an application of a combination of the strength reduction and induction variable elimination compiler optimizations.

IV. ABRIDGED ADDRESSING LOGIC STRUCTURE

The structure of the typical addressing logic in Abridged Addressing is shown in Figure 5. $A[i]$, $B[3i]$, $C[11i]$, and $D[11i + 1]$ represent the address computations required; there is one address computation output corresponding to every array access in the loop body. There will be a final level of multiplexing between these independent computations depending on the number of ports in the memory and on the assignment of individual addresses to ports. As mentioned earlier, we store only one value in a register and derive all the address outputs from it. In this case, we choose to store i, the loop induction variable, which is incremented by the loop stride every iteration. A network of adders is required to generate all the memory addresses, as shown. If the array indices are linear, the multiplicative coefficients are constants and there is no need to instantiate expensive multipliers. Also, the shift circuits shown in the figure do not correspond to any actual hardware; shifting is realized though concatenation of zeros. Such a structure results in considerably lower power dissipation than the CDMI-based strategy, which would consist of a register file with four registers – each storing the updated memory address; the CDMI decoder performs reads and writes to the register file during every memory access. In Abridged Addressing, the i register is updated only once per loop iteration. Even though there may be more adders in Abridged Addressing, each adder is activated only once per loop iteration (and not on every memory access), which may result in only slightly worse power dissipation in adders than CDMI, where a single adder is activated on every access. In the example of Figure 5 there are six adders activated once whereas in CDMI, we would have one adder activated four times. However, there is a very considerable power saving due to avoiding the expensive register file accesses.

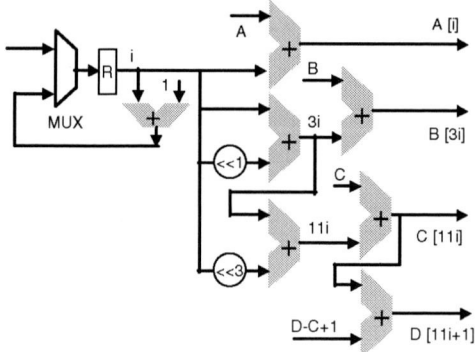

Fig. 5. Addressing logic structure

Although we chose the loop induction variable i for storage in Figure 5, there are other cases when this choice does not necessarily result in the most efficient design. A simple example is shown in Figure 6. To generate the $A[11i]$ and $D[11i]$ addresses in every loop iteration, we can just store $A[11i]$ in the register and simply derive $D[11i]$ from it by adding the constant $D - A$. If the loop stride was '1', then we add '11' to the current $A[11i]$ address to generate the value in the next iteration.

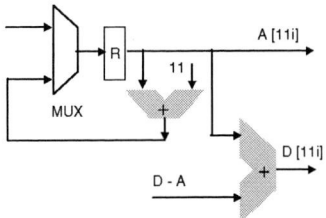

Fig. 6. Addressing logic for a different example

V. EFFICIENT ADDRESSING CIRCUITS

We formulate and solve the problem of determining an efficient addressing logic structure from a given set of array references, each translating to a memory address output. The overall strategy is similar in principle to that used in designing multiplier-less filters with minimum resources (e.g., [12]), but differs in several important steps.

We construct an adder network graph $G(V, E)$, with a set of nodes V and edges E. Each node corresponds to an array access, and there is an edge $i \rightarrow j$ with edge weight $w(i, j)$ if it is possible to derive the address corresponding to j from the address corresponding to i using $w(i, j)$ adders. An example adder network graph is shown in Figure 7(a). The edges $A[k] \rightarrow B[k]$, $B[k] \rightarrow A[k]$, $B[k] \rightarrow B[k + 1]$, etc., have weight 1 because it is possible to derive the target address from the source address using just one adder. The edge $A[k] \rightarrow B[k + 1]$, has weight 1 because we also need only one adder to generate $B[k + 1]$ from $A[k]$ (add the constant 'B − A + 1' to $A[k]$). The edge $A[k] \rightarrow B[3k]$, has weight 3 because we need three additions – first subtract A to get k, then add $k + 2k$ to get $3k$, finally add B to get $B[3k]$. Edges such as $B[3k] \rightarrow B[k]$, have a higher weight because the derivation requires more expensive circuitry – in the worst case, an appropriate divider. The multiplication operation corresponds to an appropriate number of adders, since the multiplicative coefficients are all constants. The edge weights in the graph are a measure of how many adders would be activated to get from one memory address to another, and reflect the relative power dissipation of the circuit compared to a simple adder. The same metric can be used for modeling the area of the addressing logic, and with small modifications, the delay also.

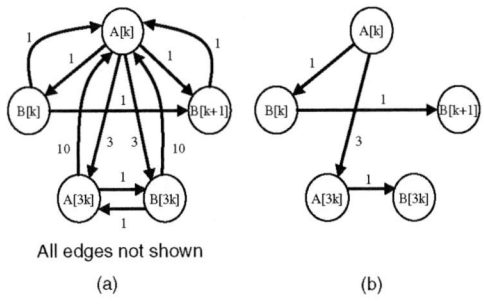

All edges not shown

(a)   (b)

Fig. 7. (a) Adder network graph (b) Directed minimum spanning tree

From the adder network graph, we can derive the simplest

adder network covering all the nodes by finding the *minimum spanning tree* in the (directed) graph. The minimum spanning tree is a subset of the edges of the graph that connects all the nodes and minimizes the sum of the weights of the edges selected. The standard minimum spanning tree algorithms (which operate on undirected graphs) cannot be used for this purpose because our adder network graph is directed. This is because the circuit to generate i from j may be more complex than the reverse. For example, $3i$ can be generated from i by just one addition, but generating i from $3i$ requires more complex hardware. In other words, $w(i, j) \neq w(j, i)$ in general.

To obtain the minimum spanning tree of a directed graph, we can use Chu-Liu's algorithm [13], an efficient polynomial time algorithm for computing the directed minimum spanning tree (DMST). The algorithm works by first selecting, in an outer loop, different nodes of the graph as the possible root, deleting all incoming edges to it, and then choosing the lowest cost incoming edge to all the other nodes. If no cycle is formed, we have the DMST. If there is a cycle, it is collapsed into a super-node x and the incoming edge weight to each node y in the cycle from nodes z outside the cycle is recomputed according to the equation:

$$w(z, x) = w(z, y) - (w(\mathrm{pred}(y), y) - \min_i(w(\mathrm{pred}(i), i))$$

where $\mathrm{pred}(j)$ is the predecessor node of j in the cycle. For each super-node, we select the incoming edge with smallest updated weight, which replaces the original incoming edge. We repeat the procedure for the collapsed graph. The root of the resulting tree corresponds to the element we wish to store in the register of the addressing circuit. The DMST for the adder network graph of Figure 7(a) is shown in Figure 7(b).

There may be cases when the addition of an auxiliary node to the adder network graph results in a more efficient addressing logic structure (i.e., a lower cost DMST). In the example graph shown in Figure 8(a), the cost of deriving each node from the other is high. In such cases, it is beneficial to augment the graph with the node k corresponding to the loop induction variable, add the edges from this node to the others, and compute the DMST, which may result in lower overall cost. The addition of the new node results in the graph of Figure 8(b), and the resulting DMST is shown in Figure 8(c), which has lower overall cost than the DMST of the original graph. Note that the addition of the loop variable k node automatically results in the implicit (zero cost) addition of nodes $2k$, $4k$, $8k$, etc., since no extra hardware is required to generate them. The weights of edges from k to the other nodes takes this into account. There is no need to explicitly add these other nodes.

The overall strategy in Abridged Addressing is to find the DMST for both the original graph G and the augmented graph G' (with the loop variable node), and choose the solution with the lower overall cost. The addressing logic structure is inferred directly from the DMST.

A. Analysis

The key idea in Abridged Addressing is to minimize the storage of redundant information in the addressing logic. For example, the continuous updation of two address registers $a[i$

9A-3

2006 Asia and South Pacific Design Automation Conference

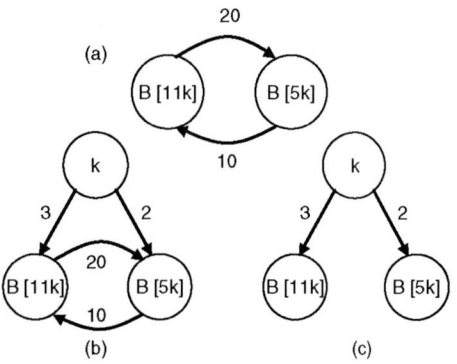

Fig. 8. (a) Example (b) Augmented graph (c) DMST of augmented graph

and $b[i]$ within a loop is unnecessary because the same pattern is repeated. Minimizing the storage helps reduce power-expensive register file access operations. It is important to note that the addressing logic is not an overhead – the computation would performed in the datapath in any case. We have merely transferred the address computation to the memory subsystem. The Abridged Addressing strategy is independent of the number of ports in the memory. There is only one address bus in the interface between the datapath and the memory subsystem. There is a final level of multiplexing of the computed addresses depending on the number of ports and the port assignment.

In the addressing logic, we have relied on the start addresses of the arrays and loop strides being constants known at synthesis time. If this is not the case, then the inputs to the adders come from registers which are initialized by the datapath through the single address bus. Note that although this increases the number of registers in the addressing logic, it does not affect the power dissipation adversely because these registers are not updated during the loop iterations. When there are multiple loops, the analysis of each loop is performed independently and sharing decisions of common parts of the addressing are taken later. For example, in Figure 5, if the strides of the two loops are different (say 1 and 2), the input to the adder that updates the register comes from a multiplexer whose inputs are 1 and 2, and whose select signal comes from the small FSM in the decoder that also controls the final address output. The power overhead of this FSM is small in comparison the typical power dissipation in adders. Conditionals in loop bodies do not require any additional support. The addresses of all memory accesses in the loop are computed, and the FSM selects the final output depending on the current state of the datapath subsystem, which is presented to the memory interface through the single address bus (as in [11]). When we have array accesses that are not part of loop bodies, then the address is sent over the address bus and presented as it is to the memory. In this case, there might be a small theoretical performance overhead because Abridged Addressing presents only one address bus to the datapath instead of two for a dual-port memory, but such cases occur a relatively small number of times – most memory accesses occur within loop bodies.

On investigating the possible performance overhead due to the addressing structure, we found that: (1) the cascaded structure leads to a very minimal increase in delay over a single adder. Note that the critical path delay of two cascaded adders

is only marginally more than that of a single adder – e.g., the delay of two cascaded 32-bit ripple carry adders is that of one 33-bit adder; (2) more importantly, the critical path in a loop body never passes through the addressing structure, since the addresses are generated as soon as the loop index is updated, independent of when they are needed. In our experiments, we found that the addresses were generated in the first one or two cycles of a loop, whereas the remaining computation (which was on the critical path) took much longer.

One limitation of abridged addressing is that, if different loop nests operate on the same array in very different ways, then it might lead to different addressing circuits. Usually, this can be accommodated by first generating the addressing circuit for the dominant loop nest, and then introducing minor MUX additions to handle the others. A separate addressing logic structure for each loop nest leads to area overhead, but is nevertheless feasible because, as mentioned before, the addressing logic is seldom on the critical path. One limitation that still remains is that, we currently consider all memory accesses in the inner loop, ignoring the presence of conditionals, which leads to more computation than strictly necessary. Array accesses occurring outside the innermost loop need not be covered by the abridged addressing structure as these are relatively infrequent, and we can compute them in the datapath itself. Multidimensional arrays, are treated as single dimensional arrays, where the array index expression involving all but the innermost loop index are hoisted out of the loop (and computed in the datapath), and abridged addressing applies only to the expression involving the inner loop index.

VI. EXPERIMENTS

We verified the addressing optimization strategy using several examples from literature, a 0.18μ ASIC library with a dual-port RAM as the data storage element, and the Synopsys synthesis tools. Behavioural synthesis with a 100 MHz clock was first performed to determine the schedule of memory accesses and the memory port assignment. Following this, the addressing logic was generated and synthesized. The synthesized designs were simulated, and the resulting switching activity file was fed into a power simulation framework (Synopsys Prime Power) to generate the total power dissipation for the application. *SOR* (Succesive Over Relaxation) and *Compress* are popular examples from numerical/scientific computing. *MM* and *Dprod* are the matrix multiplication and dot product functions. *Laplace* and *Lowpass* (accentuating low frequencies in an image) are frequently used in the image processing domain. The examples are data-intensive and have arrays indexed by linear expressions accessed in the inner loop bodies.

In our first experiment, we compared the relative power dissipation of the synthesized design examples. Here, the memory is on-chip, and the address bus capacitances, while still significant, are not orders of magnitude larger than typical nets. The results are summarized in Figure 9. For each example, we compare the power dissipation for three different implementation strategies: WZE, the CDMI method, and Abridged Addressing. WZE is used because it is tuned for data memory accesses and we found that it performs better than other sim-

896

pler techniques such as Gray code, T0, etc. for data memory.

We notice that the CDMI, in spite of eliminating address bus switching activity during the inner loop array accesses to a large extent, gives better results than WZE in only 3 out of the 6 examples. This is primarily because the decoder overhead is significant compared to the address bus power saved. Abridged Addressing results in average power savings of 40% over WZE and 44% over CDMI. The difference is significant in examples such as SOR with a large number of array accesses in the inner loop because in such cases, CDMI maintains a register file with many registers, thereby incurring a larger overhead due to the register accesses. We found that the addressing logic occupied, on an average, almost 65% less area than the CDMI decoder. This shows that, although there may be some extra adders in the addressing logic, the area saved due to reduced registers is more significant. The details are omitted due to lack of space. There was no performance overhead because of the reasons described in Section A.

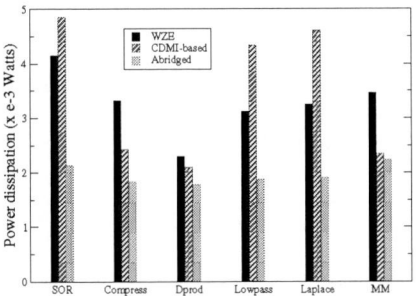

Fig. 9. Summary of experimental results

In our second experiment, we observed the variation of the power dissipation due to changes in the address bus capacitance for one of the examples (SOR). In the power simulation, the bus capacitances were set to different values to simulate longer wires. As the wire capacitance increases, we expect that the decoder overhead in techniques such as CDMI will be a relatively smaller fraction of the address bus power. This variation is observed in the comparison of WZE and CDMI curves in Figure 10. Abridged Addressing performs well in comparison, because on one hand it uses the same strategy as CDMI to reduce the address bus power, and on the other, it uses a simpler addressing logic that results in lower overall power. Lower capacitance values in Figure 10 correspond to on-chip memory, and higher capacitances correspond to either off-chip memory, or long on-chip address buses. Abridged Addressing performs well throughout the range because the technique results in just a logical transfer of the addressing computation from the datapath into the memory subsystem; there is very little extra encoding and decoding involved. This makes the technique suitable for both on-chip and off-chip memory.

VII. CONCLUSION

We presented Abridged Addressing, a strategy for generating efficient memory addressing circuits that minimize power in both the address buses and the addressing logic. Instead

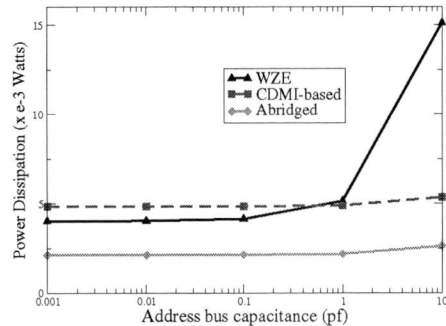

Fig. 10. Variation of power dissipation with address bus capacitance

of updating address registers during each array access, we maintain only one address register throughout the loop execution and generate the other addresses through simple arithmetic computations off this register, which minimizes power-expensive register file accesses. Address bus switching is minimized by generating the addresses in the memory subsystem itself. Experimental results indicate that the addressing technique is suitable for both the off-chip as well as on-chip memory because there is very little decoding overhead. Future work in this direction includes handling multiple memory modules in a design, and analyzing the impact of static power.

REFERENCES

[1] L. Benini and G. De Micheli, "System level power optimization: Techniques and tools," *ACM TODAES*, Apr. 2000.

[2] M. R. Stan and W. P. Burleson, "Bus-invert coding for low power I/O," *IEEE TVLSI*, Mar. 1995.

[3] C.-L. Su and A. M. Despain, "Cache design trade-offs for power and performance optimization: a case study," in *ISLPD*, 1995.

[4] L. Benini and G. de Micheli, *Dynamic power management: Design Techniques and CAD Tools*, Kluwer Acad. Publ., 1998.

[5] E. Musoll et al., "Working-zone encoding for reducing the energy in microprocessor address buses," *IEEE TVLSI*, Dec. 1998.

[6] S. Ramprasad et al., "A coding framework for low power address and data buses," *IEEE TVLSI*, July 1999.

[7] L. Benini et al., "Power optimization of core-based systems by address bus encoding," *IEEE TVLSI*, Dec. 1998.

[8] M. Mamidipaka, D. Hirschberg, and N. Dutt, "Low power address encoding using self-organizing lists," in *ISLPED*, Aug. 2001.

[9] L. Benini, A. Macii, and M. Poncino, "A recursive algorithm for low-power memory partitioning," in *ISLPED*, Aug. 2000.

[10] P. R. Panda and N. D. Dutt, "Low-power memory mapping through reducing address bus activity," *IEEE TVLSI*, Sept. 1999.

[11] P. Petrov and A. Orailoglu, "Low-power data memory communication for application-specifi c embedded processors," in *ISSS*, 2002.

[12] K. Muhammad and K. Roy, "A novel design methodology for high performance and low power digital fi lters," in *ICCAD*, Nov. 1999.

[13] E. Lawler, *Combinatorial optimization: networks and matroids*, Saunders College Publishing, Cambridge, MA, 1976.

Using Speculative Computation and Parallelizing techniques to improve Scheduling of Control based Designs

Roberto Cordone[‡]

Fabrizio Ferrandi[#]   Gianluca Palermo[#]
Marco D. Santambrogio[#]   Donatella Sciuto[#]

[‡]Università Statale di Milano - DTI
via Bramante, 65
26013, Crema, ITALY
Tel: +39-0373-898-054
Fax: +39-0373-898-010
e-mail: cordone@dti.unimi.it

[#]Politecnico di Milano - DEI
pza Leonardo da Vinci, 32
20133, MILANO, ITALY
Tel: +39-02-2399-3479
Fax: +39-02-2399-3411
e-mail: {ferrandi—gpalermo—santambr—sciuto}@elet.polimi.it

November 18, 2005

ABSTRACT

Recent research results have seen the application of parallelizing techniques to high-level synthesis. In particular, the effect of speculative code transformations on mixed control-data flow designs has demonstrated effective results on schedule lengths. In this paper we first analyze the use of the control and data dependence graph as an intermediate representation that provides the possibility of extracting the maximum parallelism. Then we analyze the scheduling problem by formulating an approach based on Integer Linear Programming (ILP) to minimize the number of control steps given the amount of resources. We improve the already proposed ILP scheduling approaches by introducing a new conditional resource sharing constraint which is then extended to the case of speculative computation. The ILP formulation has been solved by using a Branch and Cut framework which provides better results than standard branch and bound techniques.

I. INTRODUCTION

Today high-level synthesis systems need to deal with designs much more complex than a few years ago. Synthesis results were improved by applying standard optimizations such as re-timing and algebraic transformations, as shown in [21], while nowadays speculative code motion techniques demonstrate their effectiveness on scheduling lengths.

Those *new* techniques are well known in the software compilers area [7, 5], and their application to the high level synthesis problem has been proposed by Santos et al. [4] and Rim et al. [23]. Several works such as the Waveschedule approach [12] have introduced the speculative execution as an efficient method to achieve the goal of the minimization of the expected number of cycles. With the exception of [22], no exact methods provide support to speculation of control-dependent specifications. More recently, Gupta et al.[10] have defined a methodology based on code motion techniques developed for parallelizing compilers for high-level synthesis of C-based specifications.

Starting from previous works on parallelizing compilers [8, 9], this paper presents a methodology that extracts from the control and data flow graph of a sequential program a data structure that exposes the parallelism inherent in the specification. The analysis of this data structure commonly used by parallelizing compilers, represents the starting point for the definition of a scheduling technique. In this paper we do not address the scheduling of specifications with loop control structures. Therefore, the proposed algorithm works on specifications obtained by removing all feedback control edges.

We analyze the problem by formulating an approach based on Integer Linear Programming (ILP) to minimize the number of control steps given the amount of resources. We improve the already proposed ILP scheduling approaches by introducing a new conditional resource sharing constraint which is then extended to the case of speculative computation.

Section II presents a compared analysis of the proposed data structure with respect to the state of the art for speculative code transformations on mixed control-data flow designs. Section III describes the integer linear programming (ILP) model that computes a scheduling taking into account the speculative computation issue. The ILP has been solved by a Branch and Cut framework [15]. Section IV presents the results obtained by the proposed approach, while section V concludes by giving an overview of the future directions of the proposed approach.

II. INTERMEDIATE REPRESENTATION

Language based specifications are usually translated into intermediate representations to efficiently manage and analyze the design specification. Several types of intermediate representations have been proposed in literature, each one targeting different types of applications: data flow graph, control flow graph, hierarchical task graph (HTG) [8].

HTGs have been defined as intermediate parallel program representations that encapsulate minimal data and control dependences, and can be used to extract and exploit functional and task-level parallelism. In particular, the hierarchical task graph, as defined in [8], is a directed graph HTG whose vertices can be: simple, representing a task with no subtasks, compound, representing a task that consists of other tasks in an HTG (e.g., higher level structures such as subroutines or loops), loop, representing a task that is a loop whose iteration body is an HTG.

The hierarchical task graph can be extracted from the control flow graph of a sequential program, by identifying the edges through data and control dependences analysis [8, 9].

Let us first consider *control dependences*. A node B is control dependent on A if A can control whether or not B will be executed. Ferrante in [6] defines how control dependences can be identified:

A node *B* is control dependent on *A* if, and only if, *A* is not post-dominated by *B* in the CFG, and there exists a directed path from *A* to *B* in the CFG such that every node other than *A* on the path is postdominated by *B*.

Postdominance [14] is the relation defined as follows: in a directed graph with a distinguished node *STOP*, a node *V* is *postdominated* by another node *W* if, and only if, every directed path from *V* to *STOP* contains *W*.

On the other hand, we can define *data dependence* edges in this way: a node B is data dependent from node A when a data transfer from A towards B exists. They can be further subdivided into three main types: flow dependences (*RAW* dependences), anti dependences (*WAR* dependences) and output dependences (*WAW* dependences).

In [8, 9] control and data analysis are used to define the notion of *precedence* giving some conditions on when a node precedes another, with the aim of maximizing the overall parallelism. Moreover, the precedence notion can be used during the scheduling of the operations since it only considers the actual constraints on the execution order of the operations whatever is the considered granularity (i.e, instruction, functional or task level parallelism).

Gupta et al. [10] reconsider these works on parallelism extraction by using HTG as intermediate representation for high-level synthesis. In particular, they exploit the structural nature of the HTG to perform the scheduling of the operations with code motion and speculation. They consider a particular level of granularity, basic-block, and they slightly modify the definition of the compound node by adding to that node the following control structures: if-then-else, switch-case and sequence of HTGs. With these modifications they loose the power of the control dependence graph, the edges between the nodes are the same of the control flow graph (CFG), but they gain the ability to perform code motion transformations that improve the synthesis results in control intensive designs.

To better understand the differences between control flow graph and its corresponding control dependence graph (CDG), let us consider the example reported in Figure 1, assuming that no data dependences are present. Note that, the CFG imposes more constraints on the order of nodes than the CDG. In fact, without data dependences, the edges of the CFG A-E, B-E, C-D and D-E do not express true precedences between the operations of the specification.

[10] defines several transformations which can be classified into the following four types: *Across Hierarchical Blocks*, *Speculation*, *Reverse Speculation* and *Conditional Speculation*. *Across Hierarchical Blocks*: movement of operations across entire hierarchical blocks, *Speculation*: unconditional

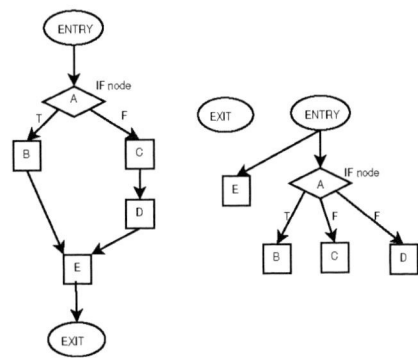

Fig. 1. Comparing control flow graph (a) and control dependence graph (b).

execution of operations that were originally supposed to have executed conditionally, *Reverse Speculation*: where operations before conditionals are moved into subsequent conditional blocks and executed conditionally, *Conditional Speculation*: in which an operation is moved up and duplicated into preceding conditional branches and executed conditionally.

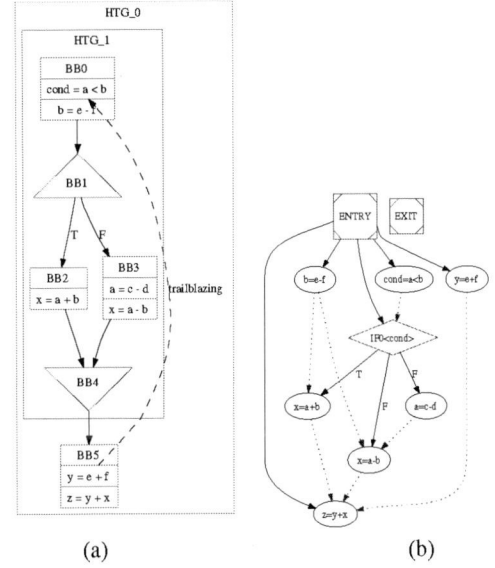

Fig. 2. (a) Code transformation of type 1, starting from the HTG of Figure 8 of [10]. (b) Corresponding control and data dependences graph.

The first transformation considered (i.e., code transformation of type 1) moves blocks following the Trailblazing code motion technique [18] across the nodes of the HTG. Consider for example the operation $y = e + f$ of basic block BB5 reported in Figure 2(a). This operation does not have any data dependence with any node of the if-HTG node, therefore exploiting the hierarchy of the HTG it can be easily moved from BB5 to BB0.

Let us now analyze the same example but considering a different data structure. Starting from the control flow graph we build the CDG and from the flow dependences of the specification, we build the data dependence graph (DDG). The graph built by joining the CDG and the DDG for the example of Figure 2(a) is reported in Figure 2(b). On this graph, the identification of the control step of the operation $y = e + f$ does not require any move across the hierarchy. The graph imposes only two precedence constraints: one specifying that the node must be executed after the $ENTRY$ node and the other requiring that the node must be executed before the $z = y + x$ operation. As shown by [10] anti and output data dependences are required to correctly build the data-path after the scheduling step. [10] performs dynamic variable renaming during the scheduling. We perform this step after the scheduling has been performed. Therefore, if the operation $b = e - f$ has been scheduled before the operation $cond = a < b$, the anti dependence edge will require a renaming of variable b.

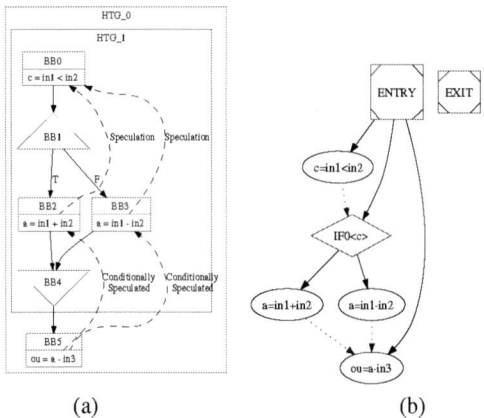

Fig. 3. (a) Code transformation of type 2 and 4, starting from the HTG of Figure 6 of [10]. (b) Corresponding control and data dependences graph.

Speculation (code transformation of type 2) is shown through the example of Figure 3. The HTG based speculation solves the problem of executing an operation before the branch condition by performing a transformation and then the scheduling of the graph. In our approach, we remove the control edges from the graph and then we change the scheduling algorithm with respect to [3]. In particular, if an operation is scheduled before the branch condition no sharing of resources is allowed, while a resource sharing can be exploited if the operation is scheduled after the branch condition. Next section details the ILP-formulation of the resource sharing and of the speculation constraint. Figure 3(a) shows also how the transformation of type 4 can be performed on the HTG. The graph corresponding to the combined CDG and DDG of Figure 3(b) imposes only relative constraints on the operation $ou = a - in3$. Therefore, if speculation moves the operations $a = in1 + in2$ and $a = in1 - in2$ one step earlier, the same can be done on operation $ou = a - in3$. Similar conclusions can be drawn also for reverse speculation. Unfortunately, there are some cases in which parallel execution of an operation does not give the same result as performing conditional or reverse speculation.

In general, transformations of type 3 and 4 may require CFG transformations not easily manageable by an ILP-formulation or by a standard scheduling algorithm. Therefore the methodology proposed in this paper performs a scheduling on the CDG+DDG as extracted from the sequential specification per-

forming code transformations of type 1 and 2. In any case the proposed methodology can take advantage of transformations of type 3 and 4 by coupling the proposed scheduling approach with a transformation toolbox as the one proposed in [10].

III. SCHEDULING MODEL

This section presents an integer linear programming model for the scheduling problem. Gebotys *et al.* in [3] analyzed the classical ILP model of the scheduling problem, which consists of assignment, precedence and capacity constraints. They also developed a family of additional constraints that, though redundant for the ILP formulation, remove a large subset of fractional solutions when the integrality constraints on the decision variables are relaxed. As a consequence, the continuous relaxation provides a much tighter bound and a nearly integer, if not even integer, solution. This information can be exploited by an ILP solver to compute an optimal solution in shorter time. Next section presents a new way to express the capacity constraints in order to better support the scheduling of control intensive designs. In fact, the new formulation also allows the transformation of type 2.

A code is modeled as a directed acyclic graph, whose nodes represent single operations (or blocks of operations), while the arcs represent precedence constraints due to data or control dependences between nodes. A *branching block* is defined as a condition statement and a number of alternative paths P, one of which is performed, according to the outcome of the condition statement. Each path is a set of code operations and, possibly, branching blocks.

Since only one of the alternative paths actually needs to be followed (based on the outcome of the condition statement), operations belonging to different alternative paths can be assigned to the same functional unit in a given control step. Notice that the operations included in a branching block could also be performed before the condition statement, if no precedence constraint forbids it. In that case, however, no pair of operations can share the same functional unit in a control step.

Let I denote the set of all functional unit types available, K the set of operations and J the set of control steps available (from 0 to the length of a heuristic schedule). L_{ik} is the number of control steps that operation k, when mapped on a functional unit of type i, takes to return ready to accept successive data inputs after a previous execution. C_{ik} is the number of control steps that operation k, when mapped on a functional unit of type i, needs to be executed. Note that $L_{ik} \le C_{ik}$. N_i is the number of functional units of type i available.

The Boolean variables x_{ijk} model the assignment of code operations to control steps and functional units: when $x_{ijk} = 1$ operation k starts executing at control step j and it is assigned to a functional unit of type i; otherwise $x_{ijk} = 0$. All variables x_{ijk} concerning functional units or control steps incompatible with operation i are undefined.

The integer variables z_{ijB} provide the number of resources of type i occupied in the control step j by branching block B. The integer variable w is the makespan, that is the last control step in which a code operation is performed.

The aim of the scheduling problem is to minimize the

makespan:

$$\min(w)$$

subject to

$$w \geq \sum_{ij}(j+C_{k,i}-1)x_{i,j,k} \qquad k \in K$$

Each operation is assigned to a specific control step and functional unit type:

$$\sum_{ij}x_{ijk}=1 \qquad k \in K$$

For each precedence relation $k \prec k'$, operation k cannot be scheduled after operation k'.

$$\sum_{i}\left(\sum_{j\leq j_c}x_{ijk'}+\sum_{j\geq j_c-C_{ik}+1}x_{ijk}\right)\leq 1$$

where

$$k \prec k', j_c \in \Gamma_{kk'}$$

The first sum states whether operation k' starts before step j_c, while the second sum states whether operation k ends after it. These two events are mutually exclusive. The condition needs to be checked only in the time interval $\Gamma_{kk'}$ in which both events are feasible:

$$\Gamma_{kk'}=[L_{kk'};R_{kk'}]$$

where

$$L_{kk'}=\max\left(\text{asap}(k'),\text{asap}(k)+\min_{i}C_{ik}-1\right)$$

$$R_{kk'}=\min\left(\text{alap}(k'),\text{alap}(k)+\max_{i}C_{ik}-1\right)$$

As in [3], the elementary precedence constraints are combined through a node packing approach, in order to restrict the search space.

The maximum number of functional units used by the whole specification is known:

$$z_{ijB_0}\leq N_i \qquad i \in I, j \in J$$

For each branching block B, each alternative path P of B employs at most z_{ijB} functional units of type i in control step j.

$$\sum_{k\in P}\sum_{j-L_{ki}+1\leq j'\leq j}x_{ij'k}+\sum_{B'\in P}z_{ijB'}\leq z_{ijB} \qquad i \in I, j \in J, P \in B, B \in \mathcal{B}$$

where $\mathcal{B}$ is the set of all branching blocks, and the sum over j' takes care of the fact that operation k could occupy functional unit i in control step j, even if it starts before, due to its latency.

Previous ILP approaches ([3]) support conditional branches descriptions generating a similar capacity constraint for each set of mutually exclusive code operations or code operations from each possible path generated by conditional branches. Since the proposed constraint is local to the branching block B, the number of constraints required by the model is not exponential but linear in the number of conditional branches.

The previous capacity constraints operate separately on each alternative path in block B. However, if the operations considered are performed before the condition statement which defines block B, they cannot share the same resources. Therefore, in that case the left-hand-side term must be summed over all paths:

$$\sum_{P\in B}\left(\sum_{k\in P}\sum_{j'=j}^{j-L_{ki}+1}x_{ij'k}+\sum_{B'\in P}z_{ijB'}\right)\leq z_{ijB}+M\left(1-\sum_{i'\in I_B}\sum_{j'\geq j-C_{k_Bi'}+1}x_{ij'k_B}\right)$$

$$i \in I, j \in J, B \in \mathcal{B}\setminus\{B_0\}$$

where k_B is the condition statement associated with block B and I_B the subset of unit types which can perform k_B. The sum on the right hand side states whether the condition statement terminates after control step j or not. In the second case, M is a constant large enough to make the constraint redundant: a sufficient value is $M = N_i$.

IV. EXPERIMENTAL RESULTS

The approach presented in this paper has been implemented in a high-level synthesis framework called *PandA*. The framework takes as input C-code and generates optimized scheduled code. *PandA* uses as front-end a customized interface to the GNU *GCC* compiler [1]. Starting from version 3.4, *GCC* provides the possibility of dumping on file the syntax tree structure representing the compiled source code. The combined CDG+DDG data structure is built starting from this syntax tree structure. The use of *GCC* allows the introduction of several compiler optimization techniques into a high-level synthesis framework, such as loop unrolling, constant propagation, dead code elimination, common subexpression elimination, etc. Moreover, the GCC front-end provides an internal representation (i.e., GIMPLE [16]) which is a language-independent tree representation, thus allowing future partial support of languages such as C++ and Java. The ILP formulation has been solved by using a Branch and Cut framework. Branch and cut is a refinement of the standard linear programming based branch and bound approach[17]. The branch and cut approach, with respect to the branch and bound technique, looks for linear inequalities which are violated by the current fractional optimal solution but are respected by all feasible integer solutions of the problem. By adding these inequalities (named cuts or valid inequalities) the continuous relaxation achieves a tight bound and a less fractional solution. There are several standard techniques to generate valid inequalities, both for general ILPs and for specific families of problems. The node packing approach of Gebotys is one of the latter. The open source package *COIN-OR* [15] provides a set of tools among which an ILP solver with the capability of generating the most important families of valid inequalities. Those which are proved effective to solve the scheduling problem are the *Probing*, *Gomory* and *Clique* inequalities.

The validation of the presented approach has been performed using the results obtained by the *Spark* [10] framework as a comparison. All the computational times reported in the result tables refer to a 1.7GHz Pentium IV Linux Workstation.

To produce the experimental results, we have chosen a set of 6 well known standard benchmarks for the problem of high level synthesis. A first subset is composed of small benchmarks derived from [19] (*sehwa*), [20] (*maha*) and [11] (*kim*), while the second one is composed of a set of multimedia applications extracted from the Mediabench suite [13] (*adpcm encode, adpcm decode, motion vector* for Mpeg2) taken from [2]. Table I reports the number of operations and branching blocks for the above mentioned set of benchmarks (upper part of the table) as well as a further set used in a following analysis on ILP complexity (lower part of the table).

The experimental results compare three different scheduling techniques: *SPARK, LIST* and *ILP. SPARK*: shows the results obtained by the *Spark* framework enabling all its features, e.g. code motion, speculation; *LIST*: represents the results obtained by applying the *List-Based scheduling* techniques with the presented intermediate representation[1]; *ILP*: represents the results obtained by applying the *ILP* formulation presented in section III using our intermediate representation;

The results for each technique and each target benchmark are shown in terms of the computational time needed to generate the schedule and the number of control steps. The computational times are average values obtained by applying the scheduling processes for ten times.

The table II shows the obtained results with the following different configurations:

- *ARCH-1 - 1-Add, 1-Sub, 1-Mul, 1-Cmp, 1-Sh, 2-[]*

- *ARCH-2 - 1-Add, 1-Sub, 1-Mul, 2-Cmp, 1-Sh, 2-[]*

- *ARCH-3 - 2-Add, 2-Sub, 1-Mul, 2-Cmp, 1-Sh, 2-[]*

where $X - <res>$ means that X resources for $<res>$ are allocated. *Add, Sub, Mul, Cmp, Sh* and *[]* stand for adder, subtractor, multiplier, comparator, shifter and array address decoder, respectively.

Three scheduling techniques are applied using three different architectures. With the first architecture (*ARCH-1*) the number of control steps obtained using the *Spark* framework is always greater or equal to the value obtained by the two methods that use the intermediate representation presented in the paper. Moreoever, our approach increases its effectiveness on applications of increasing complexity. The ILP approach improved the results obtained by LIST, in terms of control steps for *Motion Vector* (11 w.r.t. 12) and *Sehwa* (8 w.r.t. 9).

Similar results are shown with the other two configurations *ARCH-2* and *ARCH-3*, where the advantage obtained by the proposed scheduling over SPARK is up to 27% for *ARCH-2* and up to 33% for *ARCH-3*, both for the *AdpcmEncode* benchmark.

Experimental results that take into account the time needed for the scheduling show that this value is comparable for SPARK and LIST. On the other side, the ILP approach requires a time which is one order of magnitude greater with respect to SPARK and LIST, excluding the *Motion Vector* benchmark where for the architectures *ARCH-1* and *ARCH-2* it is up to two order of magnitude larger. As we expected, by increasing the number of functional units available for the scheduling, the problem becomes easier to solve and the time needed for the scheduling decreases.

---

[1]List-based scheduling can be directly adapted to the combined CDG+DDG data structure

To better understand the power of the branch and cut approach with respect to the branch and bound technique we performed further experiments on a larger set of benchmarks. In particular, we have enriched the set with some well known data-intensive high level synthesis benchmarks.

Table III compares the results obtained by the LIST approach with a branch and bound (*B&B*) and branch and cut (*B&C*) applied to the ILP formulation described in section III, reporting the number of control steps obtained and the computational time required. Note that, the LIST approach is heuristic while the other two also proved the optimality of the solution if the required time is less than the time limit of 1000 seconds.

Table III clearly shows that branch and cut is more scalable than the standard branch and bound, solving several problems in a reasonable computation time with the exception of two cases. In fact, on small benchmarks the overhead due to the generation of valid inequalities makes branch and cut slower. On the other hand, the branch and bound is not able to solve 11 of the 26 scheduling problems while branch and cut solves all apart 2 benchmarks.

Benchmark	#Operations	Branching Blocks
Kim	33	3
Sehwa	29	6
Maha	29	6
MotionVector	100	11
AdpcmDecode	87	11
AdpcmEncode	108	15
Chemical	38	1
Dct_wang	57	1
Ewf	39	1
Paulin	15	1
Pr1	51	1
Tseng	13	1
Wdf	44	1

TABLE I

OPERATIONS AND BRANCHING BLOCKS.

V. CONCLUDING REMARKS

The complexity of design for modern applications has extremely grown in recent years. This means that the standard techniques for high level synthesis can be considered obsolete for a certain number of new designs. To cope with this problem, recent research results have demonstrated, for example, the effectiveness of speculative code transformations on mixed control-data flow design to reduce the length of the resulting schedules. Our work proposes an approach based on a new data structure, the control and data dependence graph, that allows a better exploitation of parallelism present in the original specification. Moreover, this work introduces an Integer Linear Programming formulation of the scheduling problem, which minimizes the number of control steps given the amount of resources. The capacity constraint introduced has been verified to be well suited to manage resource sharing constraints in case of speculative computations. The experimental results show the validity of the proposed methodology. In general, the quality of the solution provided by the heuristic approach is nearly optimal showing that the data structure based on the

ARCH-1

	SPARK		LIST		ILP	
	Control Steps	Time [sec]	Control Steps	Time[Sec]	Control Steps	Time[sec]
Kim	10	0.030	10	0.013	10	0.169
Sehwa	8	0.046	9	0.015	8	0.332
Maha	9	0.049	9	0.013	9	0.125
MotionVector	15	0.309	12	0.173	11	28.2
AdpcmDecode	18	0.110	13	0.212	13	3.357
AdpcmEncode	18	0.144	14	0.210	14	2.011

ARCH-2

	SPARK		LIST		ILP	
	Control Steps	Time [sec]	Control Steps	Time[Sec]	Control Steps	Time[sec]
Kim	10	0.054	10	0.012	9	0.163
Sehwa	7	0.045	7	0.014	7	0.132
Maha	9	0.054	9	0.012	9	0.129
MotionVector	13	0.274	12	0.177	11	26.2
AdpcmDecode	15	0.122	11	0.190	11	0.807
AdpcmEncode	18	0.145	13	0.519	13	1.237

ARCH-3

	SPARK		LIST		ILP	
	Control Steps	Time [sec]	Control Steps	Time[Sec]	Control Steps	Time[sec]
Kim	8	0.036	8	0.010	8	0.074
Sehwa	6	0.049	6	0.013	6	0.085
Maha	9	0.054	9	0.011	9	0.123
MotionVector	10	0.323	10	0.162	9	0.968
AdpcmDecode	15	0.114	10	0.164	10	0.474
AdpcmEncode	18	0.148	13	0.367	13	0.843

TABLE II

EXPERIMENTAL RESULTS OBTAINED USING DIFFERENT ARCHITECTURES.

ARCH-1

	LIST		ILP-B&B		ILP-B&C	
	Control Steps	Time [sec]	Control Steps	Time[Sec]	Control Steps	Time[sec]
Kim	10	0.013	10	0.099	10	0.169
Sehwa	9	0.015	8	0.201	8	0.332
Maha	9	0.013	9	0.080	9	0.125
MotionVector	12	0.173	12	>1000	11	28.2
AdpcmDecode	13	0.212	13	>1000	13	3.357
AdpcmEncode	14	0.210	14	2.583	14	2.011
Chemical	19	0.021	19	>1000	19	56.4
Dct_wang	25	0.031	25	>1000	25	>1000
Ewf	21	0.021	21	>1000	21	>1000
Paulin	8	0.005	8	0.044	8	0.137
Prl	19	0.022	19	>1000	19	48.56
Tseng	6	0.006	6	0.021	6	0.057
Wdf	28	0.020	28	>1000	28	84.21

ARCH-3

	LIST		ILP-B&B		ILP-B&C	
	Control Steps	Time [sec]	Control Steps	Time[Sec]	Control Steps	Time[sec]
Kim	8	0.010	8	0.056	8	0.074
Sehwa	6	0.013	6	0.059	6	0.085
Maha	9	0.011	9	0.078	9	0.123
MotionVector	10	0.162	9	5.507	9	0.968
AdpcmDecode	10	0.164	10	0.317	10	0.474
AdpcmEncode	13	0.367	13	0.783	13	0.843
Chemical	19	0.020	19	>1000	19	49.03
Dct_wang	24	0.031	24	>1000	23	342.7
Ewf	19	0.020	19	>1000	19	182.7
Paulin	8	0.005	8	0.047	8	0.129
Prl	18	0.022	18	>1000	18	36.717
Tseng	5	0.005	5	0.014	5	0.033
Wdf	17	0.016	17	0.145	17	0.620

TABLE III

COMPARISON OF RESULTS OBTAINED WITH LIST BASED SCHEDULING, STANDARD BRANCH AND BOUND AND BRANCH AND CUT.

combined CDG+DDG is better than other intermediate representations previously proposed in literature. Therefore, future work will consider the introduction of speculative computation into the List based algorithm. The ILP approach has shown reasonable execution time and can be fruitfully used to optimize kernel functions, where a larger computation time can be afforded.

VI. ACKNOWLEDGMENTS

This publication has been part funded by the European Commission's Sixth Framework Programme.

REFERENCES

[1] GCC - GNU Compiler Collection. http://gcc.gnu.org.

[2] Spark synthesis benchmarks ftp site. ftp://ftp.ics.uci.edu/pub/spark/benchmarks.

[3] M. I. H. Catherine H. Gebotys. Global optimization approach for architectural synthesis. *IEEE Transactions on Computer-Aided Design of Integrated Circuits and Systems*, 12(9):1266–1278, September 1993.

[4] L. dos Santos and J. Jess. A reordering technique for efficient code motion. In *Design Automation Conference*, 1999.

[5] K. Ebcioglu and A. Nicolau. A global resource-constrained parallelization technique. In *3rd International Conference on Supercomputing*, 1989.

[6] J. Ferrante, K. J. Ottenstein, and J. D. Warren. The program dependence graph and its use in optimization. *ACM Trans. on Programming Language and Systems*, 9(3):319–349, 1987.

[7] J. Fisher. Trace scheduling: A technique for global microcode compaction. *IEEE Transactions on Computers*, July 1981.

[8] M. Girkar and C. Polychronopoulos. Automatic extraction of functional parallelism from ordinary programs. *IEEE Trans. on Parallel and Distributed Systems*, 3(2):166–178, March 1992.

[9] M. Girkar and C. Polychronopoulos. Extracting task-level parallelism. *ACM Trans. on Programming Language and Systems*, 17(4):600–634, July 1995.

[10] S. Gupta, N. Savoiu, N. Dutt, R. Gupta, and A. Nicolau. Using global code motions to improve the quality of results for high-level synthesis. *IEEE Transactions on CAD*, 23(2), February 2003.

[11] T. Kim, N. Yonezawa, J. Liu, and C. Liu. A scheduling algorithm for conditional resource sharing - a hierarchical reduction approach. *IEEE Transactions on CAD*, April 1994.

[12] G. Lakshminarayana, A. Raghunathan, and N. Jha. Wavesched: a novel scheduling technique for control-flow intensive designs. *IEEE Transactions on CAD*, May 1999.

[13] C. Lee, M. Potkonjak, and W. H. Mangione-Smith. Mediabench: A tool for evaluating and synthesizing multimedia and communicatons systems. In *International Symposium on Microarchitecture*, 1997.

[14] T. Lengauer and R. E. Tarjan. A fast algorithm for finding dominators in a flowgraph. *ACM Trans. on Programming Language and Systems*, 1(1):121–141, 1979.

[15] R. Lougee-Heimer, F. Barahona, B. Dietrich, J. P. Fasano, J. Forrest, R. Harder, L. Ladanyi, T. Pfender, T. Ralphs, M. Saltzman, and K. Schienberger. The coin-or initiative: Open-source software accelerates operations research progress. *ORMS Today*, 28(5):20–22, October 2001 2001.

[16] J. Merril. Generic and gimple: A new tree representation for entire functions. In *Proceedings of GCC Developers Summit*, pages 171 – 180, 2003.

[17] G. Nemhauser and L. Wolsey. *Integer and Combinatorial Optimization*. Wiley-Interscience Series in Discrete Mathematics and Optimization. Wiley, 1988.

[18] A. Nicolau and S. Novack. Trailblazing: A hierarchical approach to percolation scheduling. In *International Conference on Parallel Processing*, 1993.

[19] N. Park and A. Parker. Sehwa: A software package for synthesis of pipelines from behavioral specifications. *IEEE Transactions on Computer-Aided Design*, March 1988.

[20] A. Parker, J. Pizarro, and M. Mlinar. MAHA: A program for datapath synthesis. In *Design Automation Conference*, 1986.

[21] M. Potkonjak and J. Rabaey. Optimizing resource utlization using tranformations. *IEEE Trans. on CAD*, March 1994.

[22] I. Radivojevic and F. Brewer. A new symbolic technique for control-dependent scheduling. *IEEE Transactions on CAD*, January 1996.

[23] M. Rim, Y. Fann, and R. Jain. Global scheduling with code-motions for high-level synthesis applications. *IEEE Transactions on VLSI Systems*, September 1995.

Worst Case Execution Time Analysis for Synthesized Hardware

Jun-hee Yoo
ihavnoid@poppy.snu.ac.kr
Seoul National University,
Seoul, Republic of Korea

Xingguang Feng
fengxg@poppy.snu.ac.kr
Seoul National University,
Seoul, Republic of Korea

Kiyoung Choi
kchoi@snu.ac.kr
Seoul National University,
Seoul, Republic of Korea

Eui-young Chung
euiyoung.chung@samsung.com
Samsung Electronics,
Yongin, Republic of Korea

Kyu-Myung Choi
kmchoi@samsung.com
Samsung Electronics,
Yongin, Republic of Korea

Abstract - We propose a hardware performance estimation flow for fast design space exploration, based on worst-case execution time analysis algorithms for software analysis. Test cases on some real-world applications show that our flow provides a tight upper bound of the execution time, and many useful hints to the designer.

I. Introduction

As the Moore's law continues to apply to the embedded system industry, systems become more complex every year. Unfortunately, designer productivity doesn't continue to grow as fast as system complexity, making design cost grow rapidly every year. Especially, failing to meet the given constraints (cost, performance, power consumption, etc) in late stages of design and repeating the whole design cycle can be catastrophic. Therefore, accurately estimating the final design in early design stages becomes more and more important.

Although many recent designs try to rely more on software, many modern embedded systems use hardwired logic for performance-critical portions of the given algorithm, which is mainly because hardwired logic provides performance higher than software. However, the performance is achieved at higher manufacturing cost. To find an optimal design in early design stages, it is crucial to estimate accurately the performance and cost of the hardware implementation of a given algorithm.

Since it is difficult and time-consuming to estimate the final design manually considering the exceedingly large design size of modern systems, there has to be some automated method. This problem has been a research issue for more than 10 years and there exist many hardware estimation and analysis tools, along with tools that generate hardware implementations from behavioral models. However, most of the previous approaches evaluate the hardware's performance based on simulation.

Simulation-based estimation flows have many limitations. As the design size grows, simulation speed gets slower. Moreover, the number of test cases needs to increase exponentially as the system size grows, since there can be many corner cases. Even with all those efforts, there's no way to guarantee that every corner case has been tested, so there always exists some possibility of missing tests for worst case performance.

In this paper, we present a hardware estimation flow based on static performance analysis techniques. The proposed estimation flow translates a given C function into CDFG, synthesizes a hardware structure from the CDFG, and statically analyzes the generated hardware to estimate the performance. Although the proposed estimation flow analyzes the hardware implementation statically to obtain worst case performance, it also does simulation-based estimation for average case performance.

II. Related Work

A. Behavioral Synthesis and Estimation

Behavioral synthesis (commonly known as high-level synthesis or architectural synthesis)[1] is a core part of our estimation flow. It has been a research topic for more than a decade and there are some real-world products that perform behavior-level synthesis on C-based input description. Catapult C-synthesis[2] from Mentor Graphics, an example of such product, generates synthesizable HDL code out of some restricted form of C/C++ code. The user can explore the hardware design space by interactively specifying how to implement some portions of codes, such as by setting a loop to be always unrolled, or by modifying the resource constraints. Additionally, the user can select what kind of external interface the hardware will have.

Forte's Cynthesizer[3] is another such synthesis flow which starts from SystemC behavioral description. By using Cynthesizer, users can generate many RTL descriptions of an algorithm with different constraints, and choose the appropriate one for the whole design.

However, these commercial tools focus on fast design implementation by automating RTL coding, neglecting tight worst-case execution time (WCET) analysis techniques. The user can figure out how many cycles it takes by simulation, or by figuring out how many times a loop might iterate in the worst case, and multiply it by the worst-case execution cycles of the loop body.

B. Static Estimation

There are many approaches to static software analysis. Our work is inspired from those software estimation flows.

The Cinderella system[4] is a static approach to estimating the performance of real-time software. The goal

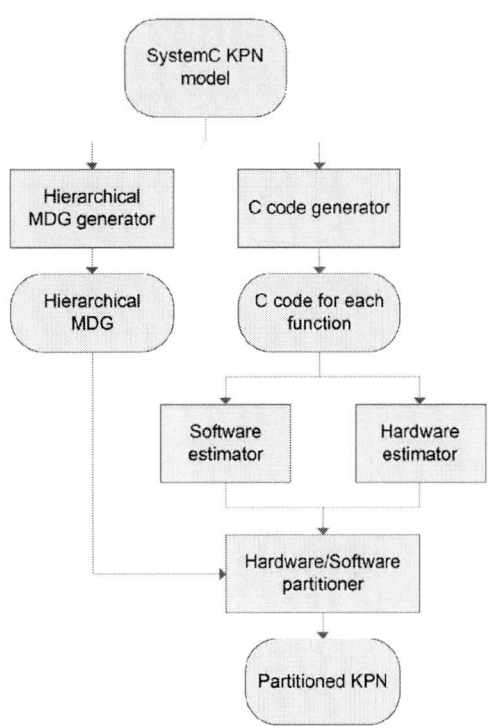

Figure 1. The KPN-based SoC design space exploration flow.

of this approach is to estimate the worst-case execution time (WCET) of a program. Based on basic block analysis, the authors generate a series of linear constraints about the execution counts of each basic block. Thus, the problem of finding the WCET of a program is reduced to an integer linear programming (ILP) problem. Together with the delay of each basic block, the WCET is obtained by solving an ILP with the objective of maximizing the total delay. Our static hardware estimation flow is based on the idea from this work.

SymTA/S[5] is another work based on static analysis. It uses many approaches from real-time analysis theory to the system level. The task execution times are obtained by using the approaches from Cinderella. Furthermore, it extends the Cinderella's work to system level WCET analysis. Users can explore the design space of the given system specification by trying different schedules and different implementations.

There have been many researches on WCET analysis for various microprocessor architectures, such as microprocessors with cache, or microprocessors with branch prediction[6][7]. However, the authors are not aware of any prior research on applying WCET analysis on synthesized hardware.

III. Application of the Estimation Flow

The hardware estimation flow described in this paper is based on the hardware estimator of our prior work[8]. Figure 1 illustrates our SoC design space exploration flow. Our prior work is an interactive SoC design space exploration tool, which uses a KPN-modeled SystemC description as its starting point. The steps of our SoC design space exploration tool is as follows:
1) The input KPN model is translated to an HMDG

(hierarchical module dependency graph) consisting of many MDGs. (module dependency graphs) and the behavior of each node of the MDGs is extracted into C code.
2) The C code is sent to the hardware estimator and software estimator, which estimates the execution time and implementation cost of each function.
3) Based on the estimation information, the hardware/software partitioner decides whether an MDG node should be implemented in hardware or software. It may report mixed implementation of an MDG, if some of its sub-MDGs are implemented in hardware and the rest is implemented in software.
4) Since there can be many different implementations of an algorithm, the optimal system implementation may not be obtained unless we examine various designs with various constraints. In our case, the partitioner runs the hardware estimator multiple times on the same input C function, each with different cost constraints. However, since just estimating the hardware design for all possible constraints for each functional block of the system takes infeasible amount of time, our partitioner uses a heuristic algorithm to decide which subset of constraints will be tried for a given function. However, it still takes a huge amount of time with a naive approach such as simulation, and thus requires a fast hardware estimation method.

IV. Hardware Estimation Flow

A. Hardware Model

We assume that the generated hardware will be connected using a bus, and communicate using a DMA controller. We also assume that a hardware block will have its own private memory space, which can be accessed by other processors via the bus while the hardware block is idle.

We assume the system operates as follows:
1) The caller (microprocessor or another hardware block) checks to see if the hardware block is available, and acquires the control over the generated hardware. This control can be implemented using mutex.
2) The caller uses the DMA controller to transmit all data required for the hardware to complete its execution to the hardware's private memory space.
3) The caller writes a 'start' command on the hardware's 'command' register, which starts the execution of the hardware.
4) After the execution completes, the caller uses a DMA controller to fetch the computed result from the hardware's private memory.

Figure 2 shows the hardware model we assumed. We used this model because of the following reasons:
1) The hardware estimation flow is expected to estimate hardware blocks that will run as a microprocessor accelerator, thus, assuming a DMA using general buses for communication will be appropriate.

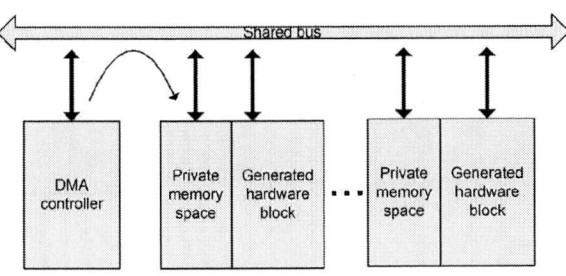

Figure 2. Hardware model.

2) We assumed a private memory block because it's difficult to predict how long a bus access will take. If we assumed shared memory space, many kinds of hard-to-predict latencies such as arbitration has to be added, and it will be difficult to predict the worst-case execution time correctly.

Although the model limits the generality of the synthesized hardware, we believe that the proposed approach can be applied to a more general model with proper adjustment.

B. Overview of the analysis flow

Figure 3 illustrates our hardware estimation flow. The input of the current estimation flow is a C function with some restrictions on the constructs, including pointer access and some control flow statements such as *goto* or *break*. First, the input code is translated into an in-house control-flow data graph (CDFG) implementation[9]. Next, many target-independent optimizations, such as common subexpression elimination and constant propagation, are applied to the CDFG. The optimized CDFG is then scheduled, and appropriate hardware resources are allocated to meet the hardware constraints given by the user or some other tool. This generates a CDFG that's ready to be synthesized into hardware. These steps are a typical behavioral synthesis flow.

However, no hardware is generated after this step. The 'synthesized' CDFG, which contains cycle-accurate scheduling information, can be either analyzed statically or simulated. The analysis (or simulation) result provides a feedback such that the user can apply different constraints to the CDFG and obtain different results.

Although this paper focuses on static analysis of worst-case execution, simulation-based analysis is still useful for analyzing the average-case execution cycles. The CDFG simulator does cycle-accurate simulation for obtaining the performance for some test case. Test cases are generated by extracting the input data applied to the function. The testbench generator library gets linked to the original C code, to log the C code's input data to be used later as the input test vector for simulation.

The following two sections describe the two blocks - the constraint extractor and the CDFG extractor - which are tightly related to static analysis.

C. Constraint Extractor

The constraint extractor analyzes the C code's structure,

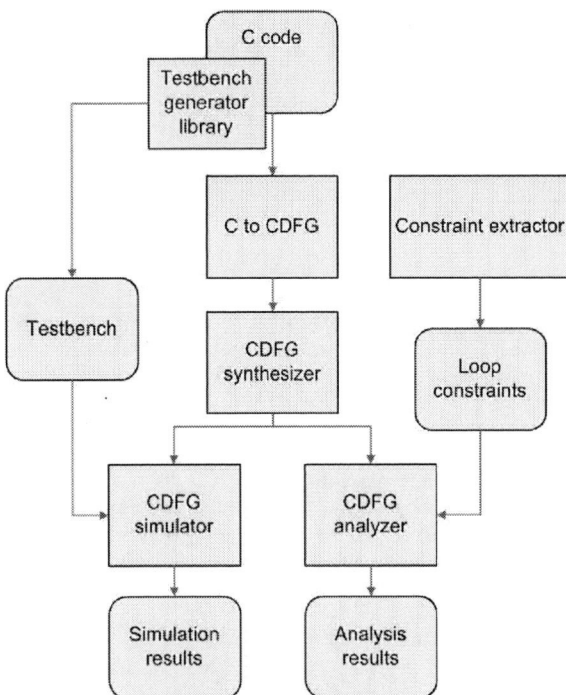

Figure 3. Hardware estimation flow.

```
int some_function(int a, int data[1200]) {
    /*##constraints
    loop1 < 1200;
    b1(true) < b2(true);
    */
    int i;
    for(i=0; i<a; i++) { //##label:loop1
        if(data[i] == TYPE_A) { //##label:b1
            /* .... some code .... */
        }
        //##label:b2
        else if(data[i] == TYPE_B) {
            /* .... some code .... */
        }
    }
    /* .... some code .... */
}
```

Figure 4. An example code with user constraints inserted.

and generates constraints that are always fulfilled regardless of the input condition. Currently, our constraint extractor looks for trivial loops with fixed number of iterations by using the SUIF compiler infrastructure's code analyzer.[10] Although our constraint extractor looks for trivial constraints only, this can be further improved in our later versions of the flow. Since constraint analysis has been an active research field, we believe that we can integrate most of these approaches to our flow.[11]

Even with the most sophisticated analysis algorithms, there can be many constraints that are difficult to determine automatically, or that are input-dependent. In this case, the user may add constraints by annotating C code. The constraint extractor extracts all constraints including the user-specified constraints, and sends them to the CDFG analyzer.

Constraints can be specified as a number of equations, using variables as the execution count of the corresponding control flow. Figure 4 shows an example function with the user-given constraints added in. The user can add simple constraints such as the maximum number of iterations a loop will run (line 3). Moreover, the user can add complicated dependencies between many execution counts, such as the relationship between two branches' execution paths (line 4).

D. Static CDFG Analyzer

The static CDFG analyzer generates a large list of integer linear programming (ILP) constraints from the synthesized CDFG. The constraints are then solved using an ILP solver. For our flow, we have used GLPK[12], an open-source ILP solver. Since we are interested in the worst case performance, the ILP solver is invoked to find the worst case.

The method of generating the ILP is adopted from software estimation flows[13]. Although the approach has been originally developed for static software analysis, it is also possible to apply this technique to CDFG analysis, since CDFGs also represent an algorithm's operations and control flow.

The total execution time can be modeled as:

$$\sum_i^N c_i x_i$$

where N is the total number of basic blocks, c_i is the execution time of the basic block, and x_i is the number of times that the basic block is executed. This is the object value that must be maximized when solving the ILP. Since c_i is the number of cycles a basic block takes to execute, it is determined when the hardware is synthesized.

Constraints by the control path is generated by the fact that for each basic block, the number of times the control enters the basic block equals to the number of times the control exits the basic block. However, unlike the typical 'flat' basic block graph structure, our synthesis flow represents the control flow by using hierarchical nodes which contains many basic blocks. This approach can be found from some other high-level synthesis flows[14].

Our control flow representation contains two kinds of hierarchical nodes - loop nodes which represents *do-while* loops, and condition nodes which represents *if-then-else* statements. A loop node contains one basic block which represents the body and the condition expression of the loop, and a condition node contains three basic blocks where each of them represents the condition block, true block and the false block. Each basic block can contain other hierarchical nodes, and the whole function is represented as a single basic block.

We modified the control path modeling as the following: for all condition nodes C_i, if we let x_i be the execution count of that node, and $x(true)_i$ and $x(false)_i$ be the execution count of the 'true' basic block and the 'false' basic block of that condition node, a condition node can be modeled as:

$$x_i = x(true)_i + x(false)_i$$

For loops, since we are modeling *do-while* loops, we generate the constraint that the loop body should execute more or same times than the loop node itself. This can be

modeled as:

$$x_i \leq x(body)_i$$

Additionally, all nodes within a basic block has the same execution count. Therefore, the constraint for that should also be added.

For example, the constraints by the control path of the example on figure 4 is generated as:

$$body = 1$$
$$loop1 = body$$
$$loop1 \leq loop1(body)$$
$$b1 = loop1(body)$$
$$b1 = b1(true) + b1(false)$$
$$b2 = b1(false)$$
$$b2 = b2(true) + b2(false)$$

where $body$ is the execution count of the function itself, and the other terms equal to the execution count of the corresponding control block annotated on the code.

The final ILP formulation is generated by adding user-specified constraints and automatically extracted constraints. For the example on figure 4, the following user-specified constraints can be added:

$$loop1(body) < 1200$$
$$b1(true) < b2(true)$$

In order to make the static estimator give accurate results, the execution counts of all loops in the CDFG have to be known prior to performing static estimation. The loop information generated by the constraint analyzer and constraint extractor is used in this stage. However, in some cases, the user may fail to specify all required loop information, and that will result in failing to find tight upper bound of execution time. In that case, users can incrementally improve the analysis results by adding more constraints on performance-critical code first. Additionally, the user can compare the static analysis results with the simulation results to see if there are unacceptably loose estimation results.

Currently, the static analyzer can only analyze the execution time. Other performance metrics such as energy consumption per execution are obtained via simulation.

V. Test Case and Experimental Results

We have done experiments using two real-world multimedia applications: the h.263 video encoder, and the Karplus-Strong algorithm.

A. H.263 Encoder

We have experimented with the h.263 encoder, a widely adopted video compression application. We analyzed the two most heavily used functions: SAD_Macroblock and Quantize.

Figure 5 and 6 show the area-performance tradeoff of SAD_Macroblock function and Quantize function, respectively. We have used different number of ALUs as the area constraint of the implementation.

Figure 7 and 8 show the simulation results of the hardware implementation with the maximum number of

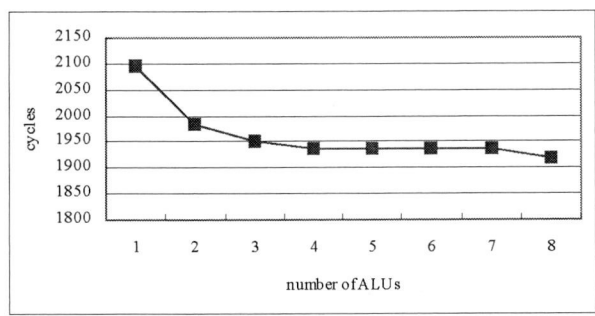

Figure 5. Area-performance tradeoff of SAD Macroblock.

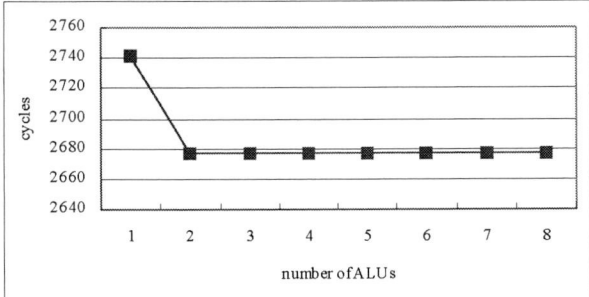

Figure 6. Area-performance tradeoff of Quantize.

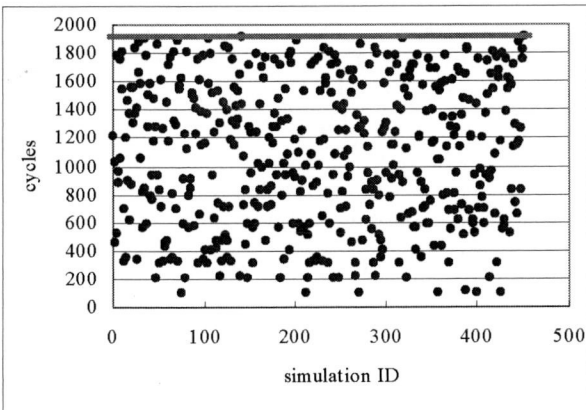

Figure 7. Simulation result of SAD Macroblock.

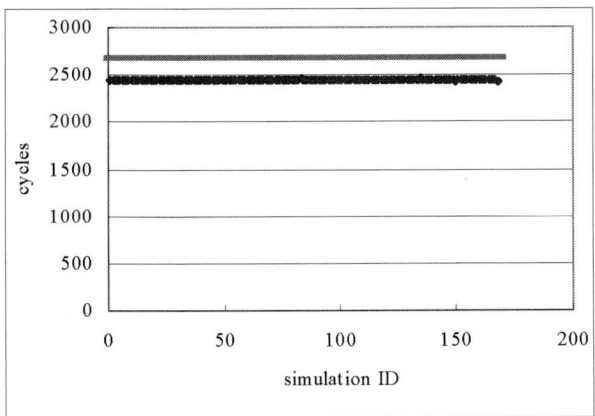

Figure 8. Simulation result of Quantize.

```
#define NUM_SAMPLES 1024
#define COMB_FILTER(cn,cn1,v0,vn,vn1) ¥
    ((((v0)-MID)*NSF + ((vn)-MID)*(cn)
    +((vn1)-MID)*(cn1)
    /256) + MID)

void karplus_strong(int cn, int cn1,
            unsigned int n, short block[NUM_SAMPLES],
            short blockprev[NUM_SAMPLES]) {
    int i;
    for (i = 0; i < n; i++){
        block[i] =
            COMB_FILTER(cn, cn1, MID,
                blockprev[NUM_SAMPLES + i - n],
                blockprev[NUM_SAMPLES + i - n - 1] );
    }
    block[n] =
        COMB_FILTER(cn, cn1, MID, block[0],
            blockprev[(NUM_SAMPLES - 1)] );
    for (i = n + 1; i < NUM_SAMPLES; i++)    {
        block[i] =
            COMB_FILTER(cn, cn1, MID, block[i - n],
                block[i - n - 1]);
    }
}
```

Figure 9. The Karplus-Strong code for this experiment.

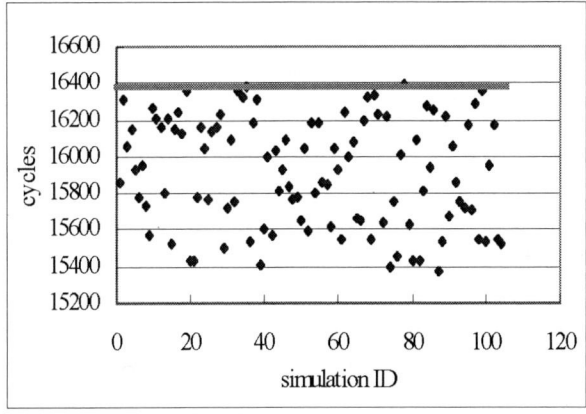

Figure 10. Simulation result of Karplus-Strong.

functional units. The X axis represents the simulation run ID number, and the Y axis represents the number of cycles. We have done 451 simulations for SAD_Macroblock and 168 for Quantize, using different real-world testbenches. From the figures, we can make sure that the execution time obtained by simulation is always within the upper bound obtained by static analysis.

For SAD_Macroblock, the worst case execution time obtained by the simulation exactly matches the upper bound. However, for Quantize, the worst case execution time by the simulation is 2,458 cycles, whereas the static estimation result gives 2,677 cycles. Through a careful analysis, we have found that our simulation-based estimation has missed running the h.263 encoder with different modes, which takes exactly the same cycles as the static analysis reports.

Even though ILP is known to be an NP-complete problem, the execution of the ILP solver is done almost instantly on an average workstation. (Intel Xeon 2.4GHz). This is because the ILP problem is reasonably small - the set of equations from Quantize has 49 variables and 55 equations and that from SAD_Macroblock has 54 variables and 39 equations.

B. Karplus-Strong

Karplus-strong[15] is a method of generating synthesized sound waveforms, which is based on physical modeling of a hammered or plucked string, or some type of percussion. We have used our implementation of Karplus-strong, which does double-buffering on the output buffer. Figure 9 shows the C code of our implementation.

By observing the C code, it's easy to find the constraint that the two *for* loops have dependencies on loop iteration numbers - the sum of the number of iterations on the two loops are 1,023. These kind of constraints can be added by formulating it as an ILP. In this case, the estimated worst case was 16,385 cycles.

However, when using the simple method - multiplying the number of worst-case iterations to the cycles that takes to execute the loop body, we have to use the worst case for both loops, and the two loops are assumed to iterate 1,023 times. In this case, the estimated worst case is 31,731 cycles.

Figure 10 shows the execution cycles of the simulation. The X axis represents the simulation run ID number, and the Y axis represents the number of cycles. The worst case execution cycle of the simulation equals to our analysis result. This shows that the analysis method of our flow gives a tight upper bound.

VI. Conclusion and Future Work

In this paper, we present a hardware estimation flow that can be used for design space exploration. The test cases and experiments show that our flow can help the designer to understand many possible issues that can happen in the hardware implementation.

We summarize our contribution as:
1) Presenting a hardware estimation flow based on static analysis of the execution pattern, and
2) presenting a method of adding complex execution path constraints to a C function, and using them for worst-case execution analysis.

However, our estimation flow has some more points to improve. Our estimation flow made many on the restrictions made in the input C code. These limitations have added much labor on modifying the reference code to work with the estimation flow. However, we expect to remove most of these limitations in our future work.

Some additional features that might help the users of this flow can be static energy consumption analysis. By predicting a reasonable upper bound of energy consumption, it would be possible to make a reasonable power budget.

Additionally, the hardware model that we used assumes a local buffer memory, so that there would be no unpredictable memory accesses delays. We believe this can be improved by adding information about the external bus into the static analyzer, and generate the appropriate ILP formula based on those information.

References

[1] G. De Micheli, *Synthesis and Optimization of Digital Circuits*, McGraw-Hill, 1994

[2] Catapult C synthesis, Mentor Graphics, http://www.mentor.com/products/c-based_design/

[3] Forte Cynthesizer, http://www.forteds.com/products/cynthesizer.asp

[4] Cinderella 2.0 http://www.princeton.edu/~yauli/cinderella-2.0/

[5] SymTA/s, http://www.symta.org/

[6] S. Kim, S. Min, R. Ha, "Efficient worst case timing analysis of data caching", *In Proceedings of the IEEE Real-Time Technology and Applications Symposium,* Pages 230 - 240, June 1996

[7] A. Colin, I. Puaut, "Worst case execution time analysis for a processor with branch prediction", *Real-Time Systems.* Vol. 18, no. 2, Pages 249-274. Kluwer Academic Publishers, 2000,

[8] Y. Ahn et al, "An Interactive Environment for SoC Design Starting from KPN in SystemC", Global Signal Processing Expo., Oct. 2004

[9] Control Data Flow Graph Toolset, http://poppy.snu.ac.kr/CDFG/

[10] The SUIF 1.x compiler system, http://suif.stanford.edu/suif/suif1/index.html

[11] C. Healy and D. Whalley, "Tighter Timing Predictions by Automatic Detection and Exploitation of Value-Dependent Constraints", *In Proceedings of Fifth IEEE Real-Time Technology and Applications Symposium,* pages 79-88, 1999

[12] GNU Linear Programming Kit, http://www.gnu.org/software/glpk/glpk.html

[13] Y. Li, S. Malik, and A. Wolfe, "Efficient microarchitecture modeling and path analysis for real-time software," *In Proceedings of the 16th IEEE Real-Time Systems Symposium,* pages 298-307, 1995

[14] S. Gupta, N. Dutt, R. Gupta, A. Nicolau, "SPARK: a high-level synthesis framework for applying parallelizing computer transformations", *In Proceedings of the 16th International Conference on VLSI Design,* pages 461-466, 2003

[15] Karplus-strong string synthesis, from Wikipedia http://en.wikipedia.org/wiki/Karplus-Strong_string_synthesis

Workload Prediction and Dynamic Voltage Scaling for MPEG Decoding

Ying Tan   Parth Malani   Qinru Qiu   Qing Wu

Department of Electrical and Computer Engineering
State University of New York at Binghamton
Binghamton, NY 13902-6000
E-mail: {ying, parth, qqiu, qwu}@binghamton.edu

Abstract – In this paper we present three efficient DVS techniques for an MPEG decoder. Their energy reduction is comparable to that of the optimal solution. A workload prediction model is also developed based on the block level statistics of each MPEG frame. Compared with previous works, the new model exhibits a remarkable improvement in accuracy of the prediction. The experimental results show that, with the new prediction model, the presented DVS techniques achieve more energy reduction than previous works while delivering the same Quality of Service (QoS).

I. INTRODUCTION

The ever increasing computing power of battery operated portable devices opens a new era for mobile multimedia applications. It is important to develop techniques to reduce the energy dissipation of such applications so that the life time of the battery can be extended [1]-[5]. One of the representative examples of multimedia application is MPEG decoding. The processing time of MPEG decoding varies significantly due to different frame types and variation between scenes. We call this processing time as workload. *Dynamic Voltage Scaling (DVS)*, which allows the processor dynamically alter its speed and voltage at run time, is one of the most popular energy reduction techniques for the applications that have large workload variations [9].

Using DVS will impact the QoS of the MPEG decoder in several ways. The first to consider is the frame dropping rate. The decoder displays decoded frames at a constant rate. Each frame must be decoded before its display deadline. Otherwise, it will be dropped. The DVS algorithm has the potential to intensify frame dropping. Another impacting factor is the buffer size. Buffers are usually used with DVS to even the workload. Input buffers and output buffers can be inserted before and after the MPEG decoder. They provide the opportunity to "borrow" or "steal" processing time among adjacent frames so that a constant voltage can be used for decoding [1][4][5]. However, increasing the buffer adds the hardware cost. Careful trade-off decision should be made. Finally, the decoding time for each frame is different when using DVS, however, the frame input and display rate remain constant. To guarantee a smooth and continuous display, the decoding and displaying of the first frame need to be delayed, which we refer as *decoding latency*. Input buffers are needed to store the incoming frames before they entering the decoder. The buffer size is proportional to the length of the latency.

In this paper, we measure the quality of DVS strategy with its energy reduction, the buffer usage, the frame dropping rate and the decoding latency. Three DVS schemes for MPEG decoding are proposed, all of which achieve comparable energy reduction as the optimal solution.

Global-Grouping is an offline algorithm whose energy consumption is on average the most close to that of the optimal solution, i.e. decoding all the frames on the lowest possible and constant speed, among all three proposed approaches. With certain decoding latency and some input/output buffers, the Global-Grouping guarantees a continuous display at a constant rate without frame dropping, provided that the workload information of each frame is accurate. Two online heuristic algorithms, *Dynamic-Grouping* and *GOP-optimal*, are also proposed with different energy reductions and buffer requirements.

For most DVS techniques to achieve a good performance, it is important to predict the workload of each task as accurate as possible. In this paper we develop a linear model to predict the decoding workload of each frame. To the best of our knowledge, this is the first prediction model that penetrates into the layered structure of video stream and utilizes the information lying at block level instead of frame level or macro block level. It gives more than 50% reduction in prediction error compared with some of the best known approaches.

The rest of this paper is organized as follows. Section II describes the background of MPEG and related works in this area. In Section III we discuss our prediction model and its implementation in detail. Our scheduling methods are given in Section IV. We represent our experimental results and discussion in Section V. Finally, the conclusions are given in Section VI.

II. BACKGROUND AND RELATED WORKS

A. Background of MPEG

MPEG is a video compression standard which represents the video stream as a series of still images [3]. These images, also called frames, are displayed sequentially at constant rate (e.g. 25 fps or frames per second). There are three types of frames defined in MPEG standard. **I**-frames or *intra-coded* frames are encoded as a whole image i.e. it does not depend on any other picture. **P**-frames or *predictive coded* frames are encoded using past I or P frame as a reference. Finally there are **B**-frames also called as *bi-directionally predictive coded* frames which use both past and future I or P frames as references. The MPEG encoder always sends the encoded frames in a rearranged order so that the MPEG decoder can decompress the frames with minimum frame buffering [7]. For example, a movie with frame order of IBBPBBP will be rearranged in the output sequence as IPBBPBB.

The MPEG video stream has a hierarchical layered structure. From top to bottom, it can be divided into sequence, GOP, frame, slice, macro block and block layers. A video stream is a *sequence* of GOPs (*Group of Pictures*), each one of which comprises of several frames (ideally 12 to 15). Each frame is further divided into vertical strips called slices. Each slice contains several macro blocks which are a 16 by 16 pixel area of the image. There are six blocks per macro block amongst which four are luminance (Y) and two are chrominance (Cr and Cb) blocks.

There are different types of macro blocks similar as frames. *I macro blocks* are encoded without using any other macro block as a reference. *P, B* and *Bi macro* blocks are encoded with forward, backward and bi-directional references respectively. Further, there can be different types of macro blocks within a single frame. The I frame contains I macro blocks only. The P frame contains both I and P macro blocks and the B frame contains all of the four types of macro blocks.

Three major operations in MPEG decoding, which consumes most of the processing time, are Run Length Decoding, Inverse Discrete Cosine Transform (IDCT) and motion compensation. All of the four types of macro blocks require Run Length Decoding during their decoding. I macro blocks also require IDCT. P, B and Bi macro blocks may require IDCT and in addition also require motion compensation.

A detailed study of the MPEG coding algorithm shows that the matching process in motion estimation, which is the counter part of motion compensation at the encoder side, is done at *block* level. For example, in process of decoding a P macro block there can be a block which does not require IDCT and decoded only using motion compensation while the other blocks require both. This is the major motivating factor for our prediction model.

B. Related Works

Generally, previous works on DVS for MPEG decoding can be classified into two categories: prediction-based and non-prediction-based.

For the prediction-based scheduling, the accuracy of predicted workload plays a significant role in the performance of these techniques, either for energy saving or for QoS. Most prediction mechanisms utilize the correlation between the frame size and the frame decode time [2][8][9]. A linear relation is usually depicted between these two. The authors of [9] developed three predictors to predict decoding workload of each frame. The best one, which will be denoted as Frame_Type_Len in the rest of the paper, dynamically updates the average decode time for each frame type and then adds an offset using a weighted factor based on the slope of frame size vs. decode time curve. Our experimental results show that, by carefully analyzing the input video stream, our predictor gives more accurate results than the Fram_Type_Len predictor.

The prediction accuracy can be improved by considering other variables lying in a video stream apart from frame size and types. The authors of [1] divide the frame decoding time into two parts, frame-dependent (FD) part and frame-independent (FI) part. The time of the FD is predicted as the moving average of previous FD. The decoder in this work is assumed to decode only one frame in each display interval, which limits its energy saving.

There are also some improved DVS techniques that do not rely on workload prediction [4][5]. In both of the two works, buffers are used to avoid deadline missing. Reference [5] introduces an online DVS technique that fully utilizes the VST (workload-variation slack time) of each task. However, the worst case execution time is assumed to be known in advance, which is not very practical in real world. Another DVS technique using feedback control [4] to adjust the supply voltage based on the number of the frames in the output buffer. However, they assume that a frame is always ready for decoding. Furthermore, it is difficult to control the gain of the feedback controller and a slight change in the gain has a great impact on the entire performance. Despite of the above mentioned limitations, it is still the best existing technique we are aware of for both energy reduction and deadline missing control. In Section V, we will compare our algorithms with this approach.

III. WORKLOAD PREDICTION

The decode time of each frame varies primarily with the frame size. However, considering only frame size to estimate the workload leads to poor prediction accuracy. Another operation in decoding process which is responsible for introducing workload variation is IDCT. We experimentally developed a prediction model based on the number of IDCT computations required for each frame. This prediction scheme yields better results but still suffers from poor correlation.

We carefully analyzed the MPEG encoding/decoding algorithm and realized that different blocks inside the same macro block may require different decoding operations (i.e. IDCT and motion compensation). Some of them may need only motion compensation while others need both. It is because the matching process in motion estimation at the encoder side is done at block level and some blocks have a zero remaining energy after motion estimation (not requiring a DCT operation) while others have a non-zero remaining energy.

TABLE 1 RELATION BETWEEN THE MACRO BLOCKS AND THE BLOCKS

Block Type	Macro Block	I	P	B	Bi
IDCT only	M_1	X	X	X	X
IDCT + FW Motion	M_2		X		
FW Motion only	M_3		X		
IDCT + BW Motion	M_4			X	
BW Motion only	M_5			X	
IDCT + Bi Motion	M_6				X
Bi Motion only	M_7				X
No IDCT No Motion	M_8		X	X	X

The processing times for forward, backward and bi-directional motion compensation are different. For example, the bi-directional motion compensation takes the longest time because it needs to consider two references. Finally, in P, B and Bi macro blocks, there are a large number of skipped blocks which are copied directly from the reference block. No IDCT or motion compensation is needed for these blocks. The processing time for these skipped blocks is simply the time for memory read and write. Based on the above observations we divide the MPEG blocks into 8 different types given in the first column of TABLE 1. Different types of blocks require different processing during the decoding. Not all of the 8 types of blocks can co-exist in a macro block. The relation between the macro blocks and different types of blocks are summarized in TABLE 1. The variable M_i, $1 \leq i \leq 8$, is used to represent the number of type i blocks in a frame. The method to extract these values will be discussed later.

Our analysis shows that a considerable variation still exists for the decoding time of the I frames, although they have the same number of type 1 blocks. It means that the processing of the Run Length Decoder is not negligible. Further study shows that the processing time of the Run Length Decoder is proportional to the size of the data. Therefore, another variable, M_9, is introduced to account for the size of the frame.

Equation (1) shows our prediction model based on these nine variables. The coefficients w_0, w_1, ..., w_9 are obtained using linear regression analysis.

$$frame\_decode\_time = w_0 + \sum_{1 \leq i \leq 9} w_i \cdot M_i \qquad (1)$$

For the same MPEG decoder, the processing time of Run Length Decoding, IDCT and motion compensation on a single block is the same for different types of movies. With the above formulation of the frame decoding time, we can derive one set of regression coefficients that works for all types of movies. Hence, only one predictor is needed for each decoder. Most regression analysis based prediction model need to have several sets of regression coefficients for different types of movies.

We applied our prediction model to a variety of movies including animated movies, high motion and low motion scenes from actual movies. The length of these movie clips ranges from 150 frames up to 3000 frames. The target decoder is the Berkeley MPEG decoder [7] running on Pentium IV 2.6GHz processor. We simulated and compared our prediction results with that of two other predictors given by reference [1], which to the best of our knowledge are the most efficient amongst all contemporary approaches. The Frame_Avg approach predicts the decoding time as the moving average of previous decoding time for each type. The Frame_Type_Len approach improves the Frame_Avg approach by adding an offset to account for the frame size. The Frame_Type_Len approach is also trained with the same set of movies to obtain a fair comparison.

Fig. 1 depicts the comparison of the Prediction Errors (PE) of three predictors.

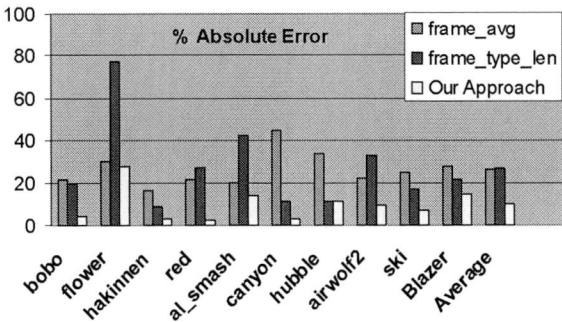

Fig. 1. Comparison of prediction errors

The PE is calculated as $PE = \frac{1}{n} \left| \frac{T_p - T_a}{T_a} \right|$, where T_p and T_a are predicted and actual decoding time respectively. As can be seen from the figure our predictor has an average of 66% improvement in the prediction error compared to both approaches.

In addition to the lower prediction error, the new predictor gives better correlation with the actual value. Fig. 2 gives the scatter plot of the predicted workload vs. actual workload for those three approaches. Fig. 3 gives the comparison of the correlation coefficients.

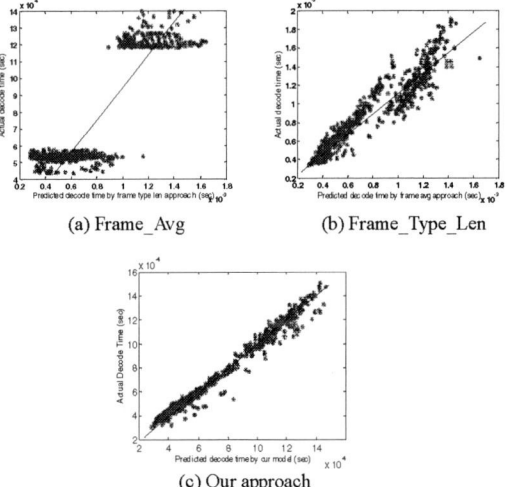

(a) Frame_Avg   (b) Frame_Type_Len

(c) Our approach

Fig. 2. Predicted workload vs. actual workload

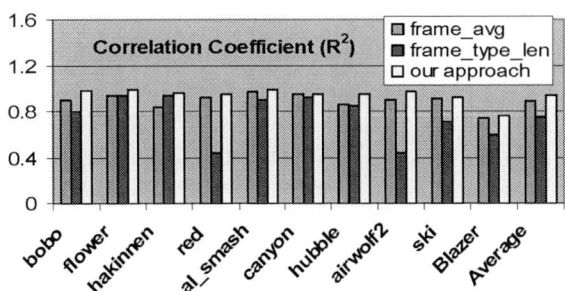

Fig. 3. Comparison of correlation coefficients

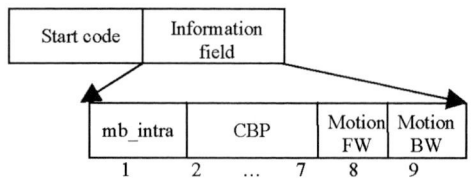

Fig. 4. Format of macro block header

At the last of this section, we go back to the layered structure of MPEG video stream and explain how to extract the variables M_1~M_8. There is a fix sized header at each hierarchy level of a MPEG stream, which contains important information needed for decoding. The information used by our model resides in the macro block header which contains a *start code* and 6 bits of information. The value of these bits is used as an index for a look up table stored in the decoder and each 9 bits entry in this table contains four fields as illustrated in Fig.4. The flag *mb_intra* indicates whether the given macro block is intra coded or not. The *CBP* (coded block pattern) is a six bit code that indicates whether the IDCT operation is needed for each block. The *Motion FW* and *Motion BW* flags indicate the existence of forward and backward motion compensations. By combining these parameters we can easily count the number of each categorized block for our model.

IV. DYNAMIC VOLTAGE SCALING

In this section, we introduce three DVS methods for MPEG decoding. For these approaches, continuous frequency/voltage scaling capabilities are assumed. Also, since the time needed to switch between different voltage settings is less than 1% of the time needed for decoding each frame [1], and some newer processors may even lower this percentage, we assume that the switch time is negligible. It is also assumed that the input and display of the MPEG decoder are at a constant rate whose period is T.

Given a MPEG stream with n frames, the optimal constant decoding voltage V_{opt} can be calculated as:

$$V_{opt} = \frac{\sum_{i=1}^{n} load_i}{i} \cdot \frac{V_{full}}{load_{max}},$$

where $load_i$ is the workload of the ith frame, $load_{max}$ is the largest workload and V_{full} is the level of supply voltage which finishes processing the $load_{max}$ within T. Using the V_{opt} the video stream can be decoded within $n*T$ time and the energy dissipation is minimal. We call this approach as optimal-VS.

There are several limitations with the optimal-VS. 1) It requires that the workload of each frame is known in advance 2) It decodes the frame at a constant speed without considering the frame incoming time or output deadline. As a result a large input/output buffer is needed; otherwise, there will be a QoS penalty. The following three DVS algorithms are developed to resolve these two problems.

A. GOP-optimal DVS

The GOP-optimal algorithm buffers all the frames in a GOP, estimates their workload, and decodes the entire GOP using a constant voltage that is calculated similar as V_{opt}. This is an on-line heuristic that is based on Optimal-VS. It does not need the workload information of the entire video stream before decoding. However, it does not consider the frame incoming time and display deadline either. The first frame in a GOP is always an I frame, which usually has larger workload than other frames in the GOP and needs longer decoding time.

To guarantee that there is always a frame ready to be displayed, the simplest way is to start displaying the first frame after the entire GOP has been decoded. As a result, the output buffer of the decoder should be large enough to hold all of the frames in a GOP. Furthermore, the input frames come in at a constant rate. Because the first frame in a GOP takes the longest decoding time, extra buffers are needed to store the incoming frames. In the worst case, the input buffer needs to be two GOP long. Note that these are only conservative estimations of the buffer size. The actual buffer usage could be less.

B. Global-Grouping

Let the display time of the ith frame denoted as D_i. The time to decode a n frame MPEG stream can be divided into n intervals $(0, D_1), (D_1, D_2), \ldots, (D_{n-1}, D_n)$. The Global-Grouping gathers consecutive intervals into groups. It divides the decoding time into m groups $G_1, G_2, \ldots, G_m$. Inside the jth group, a constant voltage V_j is used. V_j is selected such that all frames whose display time is within G_j can be decoded before its display deadline. Note that the Global-Grouping is applicable to the general applications with deadlines.

During the grouping procedure, the average workload in the time zones $(D_0, D_1), (D_0, D_2), \ldots, (D_0, D_n)$ will be tested and the intervals in the time zone that has the maximum average workload will be grouped together. After that a new search will start on the rest of the intervals. The pseudo code of the Global-Grouping algorithm is given in Fig. 5, where C_i is the decoding workload of frame i, which should be displayed at time D_i.

As mentioned in section II, a B frame has two reference frames. Both of them are received before the B frame and they also should be decoded before the B frame.

1.	Set *size* as the total number of frames;
2.	$index = 0, j = 0$;
3.	while $index < size$
4.	for $k = index+1$ to $k = size$
5.	find out the value of k, which makes $$\frac{\sum_{i=index}^{k} C_i}{D_k - D_i} \text{ maximum};$$
6.	make workload from $index+1$ to k group G_j;
7.	end
8.	$index = k, j++$, return to step 3;

Fig. 5. Global-Grouping algorithm

TABLE 2 WORKLOAD REARRANGEMENT

index	1	2	3	4	5	6	7	8
receive order	I_1	P_2	B_3	B_4	P_5	B_6	B_7	I_8
display order	I_1	B_3	B_4	P_2	B_6	B_7	P_5	I_8
Workload (C_i)	I_1	P_2+B_3	B_4	0	P_5+B_6	B_7	0	I_8

While the forward reference frame is displayed before the B frame, the backward reference frame is display after the B frame. Therefore, care should be used when calculating the workload C_i. TABLE 2 shows the relationship of receiving order, display order and rearranged decoding workload in each display interval. Note that the Global-Grouping enables the decoder to borrow the time from the previous interval; therefore the time to decode C_i could be larger than one T.

The result of Global-Grouping has some interesting characteristics as Theorem 1 and 2 states. The proof is straightforward and will be skipped due to the space limitation.

Theorem 1: For each group G_j, we can always find a constant voltage V_j, under which the decoding of each frame can be finished before its display deadline and the idle time of the processor is 0. V_j is monotonically decreasing as j increases and it is proportional to the average workload in this group: $C_{avg} = \dfrac{\sum_i C_i}{\sum_i (D_{i-1} - D_i)}$, where $(D_{i-1}, D_i) \in G_j$.

Theorem 2: The last frame of a group has the largest workload amongst the frames within this group. Its decoding is finished right at its display deadline. Other frames are decoded before their display deadline provided that the workload information is accurate.

Input and output buffers are needed for the Global-Grouping to guarantee that there is always a frame available for decoding or displaying. An example is illustrated in

Fig. 6. In order for each frame to be ready before their decoding starts, the receiving procedure must start $3T$ earlier than the decoding procedure. A buffer at the input side must be used to store the incoming frames during this $3T$ delay period. Therefore, the size of the buffer is 3 frames. Furthermore, since there is a *displaying lag*, output buffers are needed to store those that finish decoding before their display deadline. In the example, a buffer of 2 frames is needed. It is obvious that the input buffer size (*IB*) and the output buffer size (*OB*) always have the following relation $IB = OB \pm 1$.

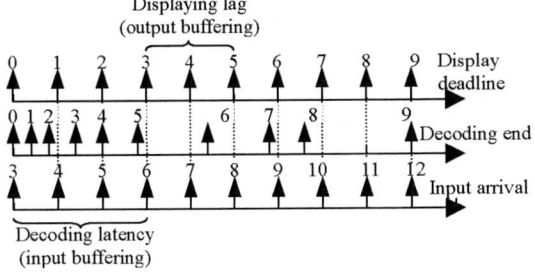

Fig. 6. Input and output buffering

The IB can be calculated as the following equation

$$IB = \max_j [\max_i (i - floor(\sum_{k=start\_of\_G_j}^{i} \frac{C_k}{C_{avg,j}}))] , \quad (2)$$

where $C_{avg,j}$ is the average workload of group G_j. The experiments show that, although the Global-Grouping needs fewer buffers than the Optimal-VS, the amount is still quite considerable.

The Global-Grouping is an offline strategy and it needs the workload profile of the entire video stream. Compared with the optimal solution, it requires less buffer while achieves a comparable energy reduction with less display latency. This algorithm is useful if the processor decodes and displays certain movie clips repeatedly. It can also be applied at the encoder side where the workload profile is available given the condition that the encoder can communicate with the decoder about the grouping and voltage selection results.

C. Dynamic-Grouping

The Dynamic-Grouping is an online heuristic based on Global-Grouping. It buffers the input frames up to a certain window size (e.g. a GOP size). The workload of each frame inside the window is predicted and grouping is applied within this window. When a new frame comes in, the decoder first predicts its workload *load_i* then updates the grouping dynamically as described in

Fig. 7. Here, M is the number of groups in current window and $C_{avg,j}$ is the average workload of the jth group.

The size of the input buffer for the Dynamic-Grouping is equal to the size of the window while the size of the output buffer can be calculated using equation (2). Compared with GOP-optimal, Dynamic-Grouping gives better trade-off between energy and buffer size.

```
1.   M = index of the last group in the current window;
2.   set the incoming frame to be group M+1;
3.   for l = M+1 to 1
4.       if (C_avg,l > C_avg,l-1)
5.           merge group l and l-1
6.           set the index of the new group as l − 1
7.       else
8.           stop
9.   end
```

Fig. 7. Dynamic group update

V. EXPERIMENTAL RESULTS AND DISCUSSION

The presented DVS algorithms are tested on several different movie clips. The statistics of these clips is given in TABLE 3.

We simulated and compared the proposed DVS algorithms with three other algorithms. *1)* Frame-based, which decodes one and only one frame in each display interval, *2)* Feedback control based [4] and *3)* Optimal-VS.

TABLE 3 CHARACTERISTICS OF MPEG CLIPS

MPEG clips		Frame type	# of frames	GOP size
name	index			
hakkinen	1	I,P,B	799	12
bobo	2	I,P,B	679	90
ski	3	I,P,B	1513	15
blazer	4	I,P,B	2998	12
wg	5	I,P	130	6

The energy values are reported as the percentage degradation over the optimal-VS approach. The buffer sizes are the size of display buffer in the unit of frames. Note that for the Global-Grouping and Optimal-VS, the input buffer size is the same as that of the output buffer plus or minus 1. For Dynamic-Grouping and GOP-optimal, input buffers of at most one GOP and two GOPs long are used, respectively. Some input buffers are also needed in feedback approach to guarantee that there is always a frame available to decode whenever the previous one finishes decoding. For all algorithms, the decoding latency is proportional to the input buffer size.

TABLE 4 gives the energy and buffer usage of different DVS algorithms when the workload prediction is perfect. It shows that while the Global-Grouping always gives similar energy reduction as the optimal one, it does not have much reduction in the buffer requirements. The Dynamic-Grouping gives the best balance between the energy reduction and the buffer requirements.

TABLE 5 gives the energy and the buffer usage when the workload prediction is not perfect. The workload prediction is given by our prediction model discussed in Section III. The accuracy of prediction has a great impact on the performance of DVS techniques, in terms of deadline missing and buffer usage, however, not so large impact on the energy dissipation. Since a frame will be dropped if it misses the decoding deadline, this causes unfair energy comparison. We scaled up the predicted workload by 5%, which is enough to make sure that the deadline miss rate is zero. For some cases, this scale increases the buffer usage however decrease the energy dissipation. The results show that both Global-Grouping and Dynamic-Grouping are pretty robust when working with our workload predictor.

TABLE 4 ENERGY AND BUFFER USAGE OF DVS ALGORITHMS WITH PERFECT WORKLOAD PREDICTION

MPEG clips		1	2	3	4	5
Frame-based	Energy (%)	200.3	65.1	88.6	118.2	55.8
	Buffer	1	1	1	1	1
Feedback	Energy (%)	3.3	18.1	20.2	13.4	17.7
	Buffer	9	10	9	10	9
GOP optimal	Energy (%)	1.4	4.5	11.8	5.6	10
	Buffer	3	9	3	5	2
Dynamic grouping	Energy (%)	2.2	2.4	10.1	4.1	10
	Buffer	3	12	4	2	2
Global grouping	Energy (%)	2.1	2.0	2.1	0.5	9.8
	Buffer	8	15	71	56	7
Optimal	Energy (%)	0	0	0	0	0
	Buffer	26	26	97	77	9

TABLE 5 ENERGY AND BUFFER USAGE OF DVS ALGORITHMS WITH IMPERFECT WORKLOAD PREDICTION

Clips	Global grouping		Dynamic grouping	
	Energy (%)	Buffer size	Energy (%)	Buffer size
1	0.78	32	2.8	3
2	0.99	18	1.4	12
3	2.7	79	11.8	5
4	0.3	66	10.7	7
5	0.4	9	4.6	4

VI. CONCLUSIONS

In conclusion of our work, we present a workload prediction model, which is motivated by detailed analysis of MPEG decoding procedure. The predictor utilizes the block level statistics of each MPEG frame and gives highly accurate prediction results. Three DVS algorithms are further presented. All of which gives comparable energy reduction as the optimal voltage scaling and work robustly with our predictor. The experimental results show that the Dynamic-Grouping algorithm gives the best trade-off between energy reduction and the quality of decoding.

REFERENCES

[1] K. Choi, K. Dantu, W. Cheng and M. Pedram, "Frame-based dynamic voltage scaling for a MPEG decoder," *ICCAD '02 – A give the CM/IEEE Int'l Conf. on Computer Aided Design, 2002, pp. 732-737*

[2] M. Mesarina and Y. Turner, "Reduced energy decoding of MPEG streams," *Proc. of Multimedia Computing and Networking, San Jose, CA 2002.*

[3] D. Son, C. Yu, and H. Kim, "Dynamic voltage scaling on MPEG decoding," *International Conference of Parallel and Distributed System (ICPADS), June 2001.*

[4] Z. Lu, J. Lach, M. Stan, and K. Skadron, "Reducing multimedia decode power using feedback control," In *Proceedings of the 21st International Conference on Computer Design (ICCD '03)*, 2003.

[5] C. Im, S. Ha, and H. Kim, "Dynamic voltage scheduling with buffers in low-power multimedia applications," *ACM Transactions on Embedded Computing Systems,* Vol. 3, pp 686-705, November 2004.

[6] Y. Lu, L. Benini, and G. D. Micheli, "Dynamic frequency scaling with buffer insertion for mixed workloads," *IEEE Transactions on computer-aided design of integrated circuits and systems,* 21(11), pp. 1284-1305, November 2002.

[7] http://bmrc.berkeley.edu/frame/research/mpeg/mpeg_overview.html

[8] E. Nurvitadhi, B. Lee, C. Yu and M. Kim, "A comparative study of dynamic voltage scaling for low-power video decoding," *Int'l Conf. on Embedded Systems and Applications,* June 23-26, 2003.

[9] A. Bavier, A. Montz, and L. Peterson, "Predicting MPEG execution times," SIGMETRICS / PERFORMANCE '98, *Int'l Conf. On Measurement and Modeling of Computer Systems, 1998, pp. 131-140.*

[10] L. Benini, G De Micheli, "System-level power optimization: techniques and tools," *International Symposium on Low Power Electronics and Design,* 1999.

Lazy BTB: Reduce BTB Energy Consumption Using Dynamic Profiling

Yen-Jen Chang

Department of Computer Science
National Chung-Hsing University, Taichung, 402 Taiwan
Tel : 886-4-22840497 ext.918
e-mail : ychang@cs.nchu.edu.tw

Abstract- In this paper, we propose an alternative BTB design, called *lazy BTB*, to reduce the BTB energy consumption by filtering out the redundant lookups. The most distinct feature of the *lazy BTB* is that it dynamically profiles the taken traces during program execution. Unlike the traditional design in which the BTB has to be looked up every instruction fetch, by introducing an additional field to record the trace information, our design can achieve the goal of one BTB lookup per taken trace. The experimental results show that with a negligible performance degradation the *lazy BTB* can reduce the BTB energy consumption by about 77% on average for the *MediaBench* applications.

I. Introduction

It is well known that the *control hazards* caused by the branch instructions are the major bottleneck in developing high performance processors. The most common solution to the control hazards is to introduce a specific hardware table, called *branch target buffer* (BTB). A BTB is a small associative memory that caches recently executed branch addresses and their target addresses. The purpose of the BTB is to provide early branch identification and its target address before the instruction is decoded. Thus, traditionally, the BTB has to be always looked up during instruction fetch stage. Because the BTB is actually a set-associative cache which is usually implemented using arrays of densely packed SRAM cells for high performance, the energy consumption of the BTB is considerable. For example, the Pentium Pro consumes about 5% of the total processor energy in the equipped 512-entry BTB [1].

The related techniques for BTB energy savings can be classified into two categories. One is to reduce the energy consumption per BTB lookup [2] [3], and the other is to reduce the number of BTB lookups [4] [5]. Based on the observation that reveals most BTB lookups are redundant, in this paper we propose an alternative BTB design, called *lazy BTB*, which aims to reduce the number of redundant BTB lookups. The key idea behind our design is to look up the BTB only when the instruction is likely to be a taken branch. We augment the conventional BTB organization with an additional field, called *taken trace size* (TTS) field, to store the instruction number between the predicted target and the next taken branch, referred to as a *taken trace*. We have developed a dynamic taken trace profiling technique which can collect the sufficient taken trace information during program execution. According to the profiled data from the previous runs, our design can conditionally skip the BTB lookup to reduce the energy consumption.

The distinct features of our design are summarized as follows. First, the lazy BTB is a software independent technique. Without any compiler instrument, it can dynamically profile the taken traces during program execution. Second, the lazy BTB can achieve the goal of one BTB lookup per taken trace. It is more energy efficient than other related work [4][5] that achieve one BTB lookup per basic block, because a taken trace contains more than one basic block. We use *SimpleScalar* [6] to perform the execution-driven simulation of *MediaBench* [7], and the BTB energy consumption are estimated by using *CACTI* [8] configured with 0.18μm technology. The results show that by eliminating a large amount of redundant lookups, our design can reduce the total energy consumption of BTB lookups by 56%~88% with a 1.7% IPC penalty.

The rest of this paper is organized as follows. Section 2 presents our motivation and the characteristics of BTB lookups, which reveals most BTB lookups are redundant. In Section 3, we describe the proposed lazy BTB in detail, including the necessary hardware augmentations. Then, the experimental results, including the impact of our design on energy reduction and performance, are given in Section 4, and Section 5 offers some brief conclusions.

II. Branch Target Buffer (BTB)

Pipelining is the key implementation technique to the high performance processors. As introduced in [9], for most RISC processors the widely used pipeline model is the typical five-stage pipeline, which is composed of instruction fetch (IF), instruction decode (ID), execution (EX), memory access (MEM), and write back (WB) stages. When a *taken* branch is executed, the branch target address is normally not determined until the end of ID. This implies that the pipelined processor needs to know the path of the branch (in order to fetch the next instruction) before it has been determined. There are two possible solutions to this problem. One is waiting for the branch to finish the target address calculation and the other is to continue fetching the instructions, possibly from the wrong path. Either solution would interrupt the steady pipeline flow, called *control hazard*, which has been shown to cause a great pipeline performance loss.

To eliminate the control hazard, the processor must perform the following jobs by the end of IF stage: identifying the instruction as a branch, deciding whether the branch is taken or not, and the target address calculation. This requirement can be achieved by using the *branch target buffer* (BTB). The BTB is a set-associative memory that caches several types of information, including recently executed branch addresses, their corresponding target addresses, and the prediction information. Fig. 1 shows a

9B-2

2006 Asia and South Pacific Design Automation Conference

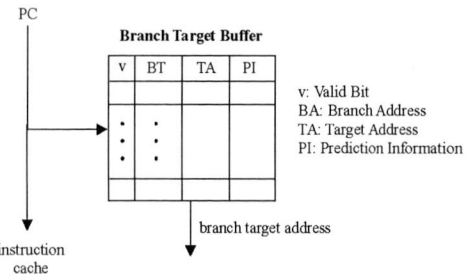

Fig. 1. A typical instruction fetch integrated with the BTB lookup.

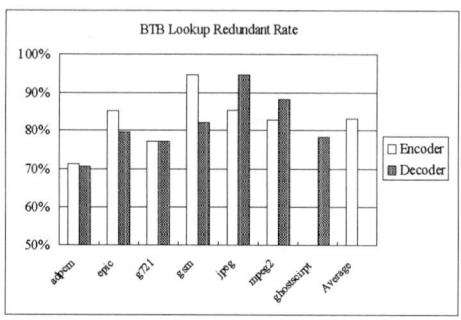

Fig. 2. BTB lookup redundant rate measured from *MediaBench*.

typical instruction fetch integrated with the BTB lookup. During IF stage, the instruction address, i.e., *program counter* (PC) value, is concurrently issued to the instruction cache and BTB. If a valid BTB entry is found for that address, then the instruction is a branch. According to the cached prediction information, if the branch is predicted taken, the BTB would output the corresponding target address to be used as the next PC. If the branch is predicted not taken, the processor continues fetching sequentially after the branch.

After the processor finishes executing the branch, it checks to see if the BTB correctly predicted the branch. If it has, all is well, and the processor can continue sequentially. If the branch was predicted incorrectly, the processor must flush the pipeline and begin fetching from the correct branch path. Then, the branch prediction information and branch target address (if changed) must be updated.

A. Characteristics of the BTB Lookups

Note that the BTB only caches the information regarding the recently executed branch instructions. Thus the BTB lookup is necessary only for the branch instructions. In the traditional BTB lookup mechanism, because the fetch engine has no sufficient information to distinguish the branch instructions, the BTB has to be looked up every instruction fetch, such that an overwhelming majority of the BTB lookups are redundant (or unnecessary). As indicated in [9], the branch instructions account for about 20% of the total executed instructions. It means that at least 80% of the BTB lookups are redundant. Fig. 2 shows the proportion of the non-branch instructions to the total executed instructions (referred to as *redundant rate*) measured from the execution traces of *MediaBench* benchmarks [7]. From this figure, the BTB lookup redundant rate is around 83% on average.

Unlike the conventional design where the BTB is always looked up every instruction fetch, motivated by most BTB lookups are redundant, we propose an alternative BTB design, called *lazy BTB*. The lazy BTB can dynamically profile sufficient information during program execution, and then use these profiled data to skip the BTB lookup conditionally. The goal is to look up the BTB only when the lookup is necessary. By filtering out most redundant BTB lookups, our design can effectively reduce the total energy consumption of the BTB.

B. Related Work

As described previously, we only survey the related work which target on reducing lookups to save energy

dissipated in BTB. Petrov and Orailoglu [4] proposed *application customizable branch target buffer* (ACBTB), which is a software profiling technique. By utilizing the precise control-flow information of the application, the ACBTB is accessed only when a branch instruction is to be executed. Because the control-flow information must be extracted during compile/link time, their method is static and not applicable to the existing executable programs. In addition, a large hardware modification is necessary.

We can use predecode technique to test if the instruction is a branch, but the drawback is that the predecode bits only become available at the end of the instruction fetch stage. This would result in a significant performance penalty. In [5], Parikh et al. proposed a small hardware table, called *prediction probe detector* (PPD), to reduce unnecessary predictor and BTB accesses. The PPD can use compiler hints and predecode bits to recognize when lookups to the direction-predictor and BTB can be avoided. The drawback of this approach is that the PPD lookup must be performed before accessing the predictor and BTB. That would result in the extra power consumption and possible performance penalty.

III. Lazy BTB

This section gives the detailed description of the proposed lazy BTB design. We first discuss the BTB management, and then develop a dynamic profiling technique, which is critical to the lazy BTB. In addition, the necessary hardware augmentations are also provided.

A. BTB Management

The BTB management is concerned with the issue of entry allocation and replacement. For most microprocessors, the BTB is a valuable resource with limited size. Thus, instead of allocating entry for each branch, we only cache the branches which have the potential for improving performance. Because caching the untaken branches does not improve the performance and they are unlikely to be taken in the future [10], the allocation policy used in our lazy BTB is that we only allocate a new entry for a branch on its first taken execution. If no entry is available, then the replacement is necessary. As indicated in [10], LRU is good enough. It achieves the similar performance gain to their proposed MPP algorithm which is an elaborate replacement policy. Thus, the entry replacement used in the lazy BTB is the simple LRU.

918

2006 Asia and South Pacific Design Automation Conference

9B-2

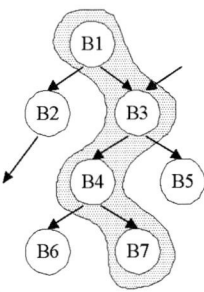

Fig. 3. An example of control flow graph (CFG). The shaded area is a taken trace.

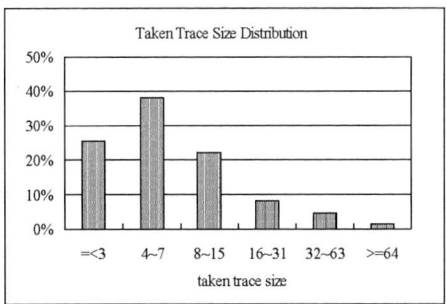

Fig. 4. Taken trace size (TTS) distribution measured from *MediaBench*.

B. Basic Block vs. Taken Trace

Fig. 3 shows a control flow graph (CFG), in which one node corresponds to one basic block. The *basic block*, by definition, is a sequential code that has no branch in except at the entry and no branch out except at the exit. Therefore, the branch instruction must be the last instruction of the basic block. Previous studies have shown that the average basic block size is usually small, especially for integer codes, it is around four to six instructions. As shown in Fig. 3, each basic block has two possible successors (caused by the *taken* and *untaken* path), but the correct path does not be determined until the codes are executed. Consequently, the basic block flow only depicts the static control structure of a program. It cannot reflect the dynamic behavior of a program.

In contrast to the basic block, we define a *taken trace* as the instruction stream between the two consecutive *taken* branches. A taken trace illustrates a snapshot of program execution. It can reflect the dynamic behavior of a program. A taken trace, by definition, contains more than one basic block. As shown in Fig. 3, the shaded area is a taken trace that is composed of basic blocks *B1, B3, B4* and *B7*. It means that the last instructions of *B1, B3* and *B4* are all untaken branches during program execution. Instead of *one BTB lookup per basic block*, the goal of our design is to achieve *one BTB lookup per taken trace*.

C. Hardware Augmentations

The lazy BTB design relies on the profiled taken trace from previous runs to skip the BTB lookup. A key issue in the realization of our design is how to profile the taken trace during program execution. Unlike the *ACBTB* technique presented in [4], which is based on the compiler profiling, our method is a hardware implementation without any software supports, including compiler. Before describing our design in detail, we first provide the necessary hardware augmentation.

(1) The conventional BTB has to be augmented with an extra field for each entry, called *taken trace size* (TTS) field, which is used to record the size of the following taken trace. The width of the TTS field must be large enough to accommodate most taken traces. Of course, the appropriate TTS field width depends on the dynamic behavior of the applications. Fig. 4 shows the average distribution of the TTS for *MediaBench* benchmark. For the best tradeoff between the energy reduction efficiency and hardware cost, the TTS field width is determined to be fixed 6-bit throughout this paper.

(2) Our design only performs the BTB lookup while the instruction is likely to be a taken branch. We need a counter, called *remainder trace length* (RTL), to indicate whether the currently fetched instruction locates within a taken trace or not. The initial RTL value is 0. When a BTB hit occurs, the RTL counter is set to the TTS value which is retrieved from the hit entry. Before looking up the BTB, if the RTL value is not equal to zero, then the currently fetched instruction is within a taken trace and is not a taken branch. Therefore, the BTB lookup can be skipped for energy saving. If the instruction is actually not a taken branch, then the RTL value is decreased by 1. In the other case, the RTL value is equal to zero, which implies that the currently fetched instruction is likely to be a taken branch. The BTB lookup is necessary for branch prediction and target address retrieval.

(3) An additional counter, called *trace size accumulator* (TSA), is needed to accumulate the taken trace size during program execution. The initial TSA value is 0 and increased by 1 every non-branch instruction execution. Until a taken branch is encountered, the TSA value is restored to the TTS field of the previous taken branch indexed by TE value (described below), and then it is reset to 0 to be accumulated until the next taken branch.

(4) Finally, in order to restore the TSA value to the corresponding taken branch, a temporal register, called *target entry* (TE), is needed to remember the index of the previous hit/allocated BTB entry during program execution. The initial TE value is 0. There are two cases where the TE value would be set. First, when we allocate a BTB entry for a new coming taken branch, the TE value has to be set to the allocated entry number. Second, if a BTB hit occurs and its prediction is correct, then the TE value has to be set to the hit entry number.

The hardware augmentations include an extra 6-bit field in BTB, three additional counters, and the necessary control circuitry. Except for the first one, the energy overheads caused by the remainder two are negligible to the energy consumption per BTB lookup.

D. Dynamic Taken Trace Profiling

Unlike the cache whose output must be accurate for correct program execution, the output of BTB is allowed to be inaccurate. The system can recover and continue by flushing any instructions fetched from the incorrect path before their results have been committed. This is the most important feature that guarantees our design can work well. Fig. 5 illustrates the dynamic taken trace profiling developed for the lazy BTB, which covers from the IF to EX stage. The

919

9B-2

2006 Asia and South Pacific Design Automation Conference

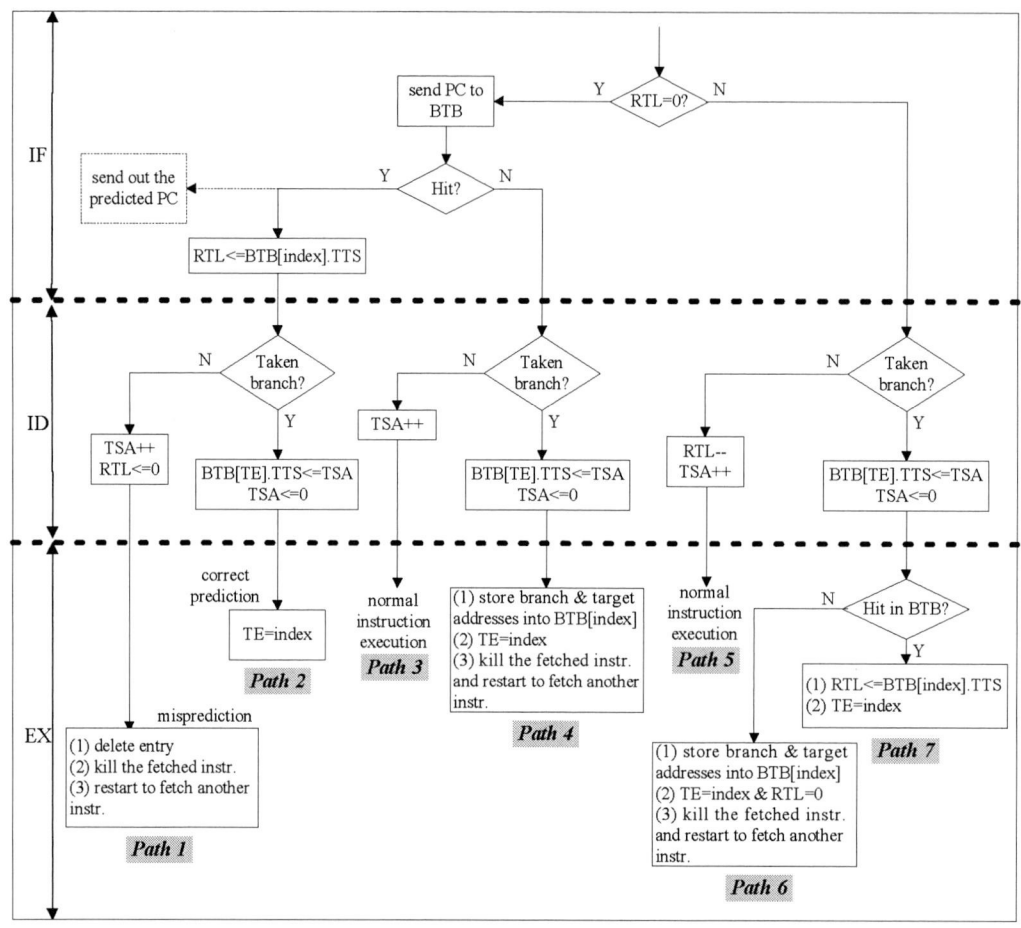

Fig. 5. The dynamic taken trace profiling technique developed for the lazy BTB design.

Table 1. The seven possible paths in the lazy BTB scheme.

Possible Paths	BTB Lookup	Hit/Miss	Prediction	Actual Branch	BTB Looup in EX	Penalty Cycles
Path 1	Y	Hit	taken	not taken	-	2
Path 2	Y	Hit	taken	taken	-	0
Path 3	Y	Miss	-	not taken	-	0
Path 4	Y	Miss	-	taken	-	2
Path 5	-	-	-	not taken	-	0
Path 6	-	-	-	taken	Y/Hit	3/4
Path 7	-	-	-	taken	Y/Miss	1/2

BTB lookup is performed (or skipped) during IF stage, and the actual branch result, i.e., the path and target address, would be determined in ID stage. If the prediction is correct, the execution continues with no stall. Otherwise, the recovery procedure for misprediction would be executed in the EX stage, which costs the performance penalty. From this figure, we can break the entire dynamic profiling scheme into seven possible paths. Their characteristics, including penalty cycles incurred by misprediction, are summarized in Table 1, and the detailed descriptions are provided as follow.

Path 1: In this path, because the instruction is found in the BTB and predicted taken, we can retrieve the corresponding TTS from the hit entry and set RTL to it during the IF stage. Next, in the ID stage, the branch is resolved and actually not taken. It is a misprediction case. The RTL value has to be reset to 0, and the TSA continues to accumulate the taken trace size. Note that the ID stage would be overlapped with the IF stage in the pipeline. In order to

avoid hardware conflict the RTL value changes only in the second phase of IF stage, and the first phase of ID stage. In the EX stage, due to the misprediction, we have to kill the fetched instruction, delete the BTB entry, and restart to fetch the instruction from the correct path. The penalty cycles are 2 for this path.

Path 2: Unlike the path 1 which is a misprediction, the BTB prediction is correct in this path. As shown in Fig. 5, the RTL is set to the retrieved TTS value during the IF stage. Next, in the ID stage, the TSA value has to be restored to the previous taken branch entry indexed by TE, and then be reset to 0 to accumulate the following taken trace size. Finally, the TE value is set to the index of the hit entry in the EX stage. Due to the correct prediction, the penalty cycle is 0.

Path 3: This path is the execution flow of the non-branch instructions. Thus, we only increase the TSA value by 1 to accumulate the taken trace size during the ID stage. Of course, the penalty cycle is 0.

Path 4: Due to the BTB miss, the instruction is predicted as non-branch (or not taken), but it is resolved as a taken branch in the ID stage. Consequently, the TSA value has to be restored to the previous taken branch entry indexed by TE, and then be reset to 0 to accumulate the next taken trace size. Next, in the EX stage, we allocate a BTB entry for this taken branch. After storing the branch address and its target addresses, the TE value is set to the index of the allocated entry. Finally, we have to kill the fetched

920

Table 2. Major processor and penalty parameters used in our processor model.

Processor Configuration	
Issue width	1 intr. per cycle
Intruction window	2-RUU, 2-LSQ
Function units	1 Int ALU, 1 Int Mult/Div
	1 FP ALU, 1 FP Mult/Div
L1 instruction cache	16KB, 32-way, 32B blocks
L1 data cache	16KB, 32-way, 32B blocks
TLB (iTLB & dTLB)	128-entry, 4-way
Branch perdictor	2-Level 1K-entry
BTB	512-entry, 4-way
Return address stack	8-entry
Penalty Parameters	
L1 hit latency	1 cycle
Branch misprediction	2 cycles
Memory access latency	8 cycles for the first chunk
	2 cycles for the rest of a burst access
TLB miss penalty	30 cycles

Table 3. Path distributions for each benchmark.

Benchmark	path 1~4	path 5	path 6~7
adpcm_en	37.53%	59.41%	3.06%
adpcm_de	32.83%	64.63%	2.54%
epic_en	13.89%	85.68%	0.43%
epic_de	15.95%	83.39%	0.66%
g721_en	18.00%	81.11%	0.89%
g721_de	17.72%	81.42%	0.86%
gsm_en	15.00%	84.45%	0.56%
gsm_de	11.35%	88.50%	0.15%
jpeg_en	14.57%	84.92%	0.51%
jpeg_de	14.44%	85.07%	0.49%
mpeg2_en	30.79%	66.90%	2.31%
mpeg2_de	17.37%	81.81%	0.82%
ghostsciprt	13.17%	86.48%	0.35%
Average	19.43%	79.52%	1.05%

instruction, and restart to fetch the instruction from the correct path. The penalty cycles are 2.

Path 5: Similar to the path 3, this path is also the execution flow of the non-branch instructions. The only difference between the paths 3 and 5 is that the BTB lookup can be skipped in this path due to RTL<>0. Note that, besides increasing the TSA value by 1, the RTL has to be decreased by 1 in the ID stage. The penalty cycle is also 0.

Path 6: Due to RTL<>0 the BTB lookup can be skipped in the IF stage, and then the instruction is resolved as a taken branch in the ID stage. Thus, we first restore the TSA value to the previous taken branch entry indexed by TE, and then reset TSA to 0 to accumulate the next taken trace size. Next, in the EX stage, before allocating a BTB entry for this taken branch, in order to avoid duplicated allocation we have to check whether it is already in the BTB or not. In the path 6, because this taken branch is not found in the BTB, we have to allocate a BTB entry for this taken branch as the steps in the path 4. Note that the RTL has to be reset to 0. Because a BTB lookup is unavoidable in the EX stage, in the worst case it may be overlapped with the BTB lookup in the IF stage. Thus, the penalty cycles are 3 for the normal case, and 4 for the worst case.

Path 7: This path is almost the same as the path 6. The only difference is that the taken branch is already in the BTB. Thus, instead of allocating a BTB entry for this taken branch, we can retrieve the corresponding TTS from the existing entry and set RTL to it in the EX stage. The penalty cycles are 1 for the normal case, and 2 for the worst case.

We summarize the important features of the new BTB design. (1) In paths 1~4, due to RTL=0 the lookup is necessary as the conventional BTB design. In contrast, because RTL<>0, the BTB lookup can be skipped in paths 5~7, as shown in shaded columns in Table 1. (2) Compared to the conventional BTB, a significant energy savings come from the path 5 in our design. This is because we have enough information profiled during program execution to skip the BTB lookup conditionally. (3) The lazy BTB achieves one BTB lookup per taken trace. It is more energy efficient than the ACBTB [6], which realizes one BTB lookup per basic block.

IV. Experimental Results

For the results presented in this study, we use *SimpleScalar* [6] toolset to model a baseline processor that closely resembles StrongARM processor [11]. It is a single-issue, in-order, pipelined machine with five stages. The major processor and penalty parameters are listed in Table 2. We use the execution-driven simulation to investigate the potential energy efficiency of the *lazy BTB* design, and its impact on performance.

A. Benchmarks

Because our baseline processor model is usually used in the embedded systems for multimedia or mobile applications, the input benchmark is *MediaBench* [7]. Unlike another popular benchmark, *SPEC2000*, which is a suit of general-purpose programs, the *MediaBench* is a suite of applications focus on multimedia and communications systems. Each benchmark of *MediaBench* has two separate programs: encoding and decoding.

B. Results and Discussions

Path Distributions: From Fig. 5, we know that the path distributions have a strong impact on the energy efficiency of the lazy BTB. Table 3 shows the path distributions for each benchmark. In this table, we divide the seven paths into three groups which are path 1~4, path 5 and path 6~7. Except for *adpcm_en*, *adpcm_de* and *mpeg2_en*, the percentage of path 5 is over 80% for all benchmark. This is because the BTB miss rates of *adpcm_en*, *adpcm_de* and *mpeg2_en* are higher than that of the other benchmarks. Because in our design the major energy savings come from the path 5, the large percentage of path 5 is preferred. In contrast, the BTB lookup is necessary in both path 1~4 and path 6~7, so their small percentages are favorable. Particularly, besides the energy consumption, path 6~7 further has a negative impact on the performance. This would be discussed below.

Total Energy Consumption of BTB Lookups: Because the organization of a BTB is essentially identical to that of a cache, we can use the *CACTI* tool [8], which is a widely accepted cache timing and power model, to estimate the BTB energy consumption. As listed in Table 2, the BTB organization is 512-entry 4-way. By using CACTI configured with 0.18μm technology, we obtained the energy

Table 4. Total energy consumption (measured in *mJ*) for both the conventional and lazy BTB designs.

	BTB$_{Conv}$	BTB$_{Lazy}$	Reduction
adpcm_en	290.8	126.9	56.35%
adpcm_de	239.2	90.7	62.09%
epic_en	25.3	3.7	85.25%
epic_de	3.2	0.6	82.73%
g721_en	131.9	26.1	80.22%
g721_de	128.5	25.0	80.56%
gsm_en	896.6	144.4	83.90%
gsm_de	305.7	35.6	88.35%
jpeg_en	48.6	7.6	84.41%
jpeg_de	12.3	1.9	84.58%
mpeg2_en	544.4	192.8	64.59%
mpeg2_de	82.2	15.6	80.99%
ghostscirpt	557.1	77.3	86.13%
Average	251.2	57.6	77.09%

consumption per lookup is about *0.484 nJ* and *0.492 nJ* for the conventional and lazy BTBs, respectively. Because in our design the BTB is augmented with an extra TTS (6-bit) field for each entry, the energy consumption per BTB lookup is slightly larger than that of the conventional BTB.

The metric used to evaluate the energy efficiency is the simple total energy consumption of BTB lookups. Table 4 shows the total energy consumption number in *mJ* for both the conventional and lazy BTB designs. Compared to the conventional BTB, one can immediately notice that the energy reduction would be an order of magnitudes. By filtering out most redundant BTB lookups, the lazy BTB can reduce the total energy consumption of BTB lookups by 56%~88% for *MediaBench*.

Performance Impact: The unit of performance measurement we use is *instructions per cycle* (IPC), calculated as the total number of execution instructions divided by execution cycles. Given that both the number of execution instructions and processor cycle time are constant, IPC is a direct measure of performance. Compared to the conventional BTB, from Fig. 5 we can see that only the paths 6 and 7 result in the extra penalty cycles. In this case, the BTB lookup is skipped, but the instruction is a taken branch actually. Thus, one BTB lookup has to be paid during the EX stage, that would decrease the overall performance. The paths 6 and 7 are, therefore, referred to as *unfavorable path*.

From previous discussion, we conclude that the negative impact of our design on the performance depends on the occurrence of unfavorable path. If most instructions follow the unfavorable path, then the lazy BTB would result in a significant decrease in IPC. Fortunately, the occurrence of unfavorable path is small enough. As shown in Table 3, the percentage of path 6~7 is about 1.05% on average, Therefore, the lazy BTB has a negligible degradation in performance. It can be seen from Fig. 6, which shows the IPC value for the conventional and lazy BTBs. Our design results in roughly 1.7% IPC degradation on average.

V. Conclusions

In this paper, we have proposed a low power BTB design, called *lazy BTB*. By using the developed dynamic

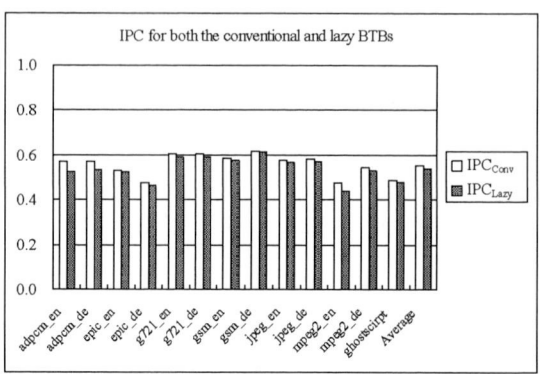

Fig. 6. The IPC value for the conventional and lazy BTBs.

taken trace profiling technique, the lazy BTB can achieve the goal of one BTB lookup per taken trace instead of one BTB lookup per basic block. The results show that without noticeable performance difference from the conventional BTB, our design can reduce the total energy dissipated in BTB lookups up to 88% for the *MediaBench* applications.

References

[1] S. Manne, A. Klauser, and D. Grunwald, "Pipeline Gating: Speculation Control for Energy Reduction," in Proc. of International Symposium on Computer Architecture, 1998, pp. 132-141.

[2] B. Fagin, "Partial Resolution in Branch Target Buffers," IEEE Transactions on Computers, Vol. 46, No. 10, 1997, pp. 1142-1145.

[3] D. H. Albonesi, "Selective Cache Ways: On-Demand Cache Resource Allocation," in Proc. of International Symposium on Microarchitecture, 1999, pp. 248-259.

[4] P. Petrov and A. Orailoglu, "Low-Power Branch Target Buffer for Application-Specific Embedded Processors," in Proc. of Euromicro Symposium on Digital System Design, 2003, pp. 158-165.

[5] D. Parikh, K. Shadron, Y. Zhang, and M. Stan, "Power-Aware Branch Prediction: Characterization and Design," IEEE Transactions on Computers, Vol. 53, No. 2, 2004, pp. 168-186.

[6] D.C. Burger and T. M. Austin, "The SimpleScalar Tool Set, Version 2.0," Computer Architecture News, 25 (3), pp. 13-25, June, 1997. Extended version appears as UW Computer Sciences Technical Report #1342, June 1997.

[7] C. Lee, M. Potkonjak and W. H. Mangione-Smith, "MediaBench: A Tool for Evaluating and Synthesizing Multimedia and Communications Systems," in Proc. of International Symposium on Microarchitecture, Dec. 1997, pp. 330-335.

[8] G. Reinman and N. P. Jouppi, "CACTI 2.0: An Integrated Cache Timing and Power Model," COMPAQ WRL Research Report, 2000.

[9] J. L. Hennessy and D. A. Patterson, "Computer Architecture: A Quantitative Approach," 3rd Ed., Morgan Kaufmann Publishers, Inc., 2003.

[10] C. H. Perleberg and A. J. Smith, "Branch Target Buffer Design and Optimization," IEEE Transactions on Computers, Vol. 42, No. 4, 1993, pp. 396-412.

[11] R. Witek and J. Montanaro, "StrongARM: A High-Performance ARM Processor," in Proc. of COMPCON, 1996, pp. 188-191.

Cache Size Selection for Performance, Energy and Reliability of Time-Constrained Systems *

Yuan Cai[1], Marcus T. Schmitz[2], Alireza Ejlali[2], Bashir M. Al-Hashimi[2], Sudhakar M. Reddy[1]

[1]Department of Electrical and Computer Engineering, University of Iowa
E-mail: {yucai, reddy}@engineering.uiowa.edu
[2]School of Electronics and Computer Science, University of Southampton
Email: {ms4, ae04v, bmah}@ecs.soton.ac.uk

Abstract— Improving performance, reducing energy consumption and enhancing reliability are three important objectives for embedded computing systems design. In this paper, we study the joint impact of cache size selection on these three objectives. For this purpose, we conduct extensive fault injection experiments on five benchmark examples using a cycle-accurate processor simulator. Performance and reliability are analyzed using the performability metric. Overall, our experiments demonstrate the importance of a careful cache size selection when designing energy-efficient and reliable systems. Furthermore, the experimental results show the existence of optimal or Pareto-optimal cache size selection to optimize the three design objectives.

I. INTRODUCTION

Cache memories are widely used in microprocessors to improve the system performance [1]. As small, fast on-chip memories, caches store frequently accessed instructions and data to avoid a large number of accesses to the slow, off-chip main memory. Depending on the way the cache blocks are mapped onto the main memory, we distinguish between direct-mapped caches (each main memory address is mapped to one and only one cache block) and n-way set-associative caches (each main memory address can be mapped to n possible cache blocks). As opposed to the slow dynamic main memory, cache memories are implemented as flip-flops using static logic. Though cache memories can improve the system performance dramatically, they are responsible for a large portion of the overall system's power dissipation [2]. To reduce the energy dissipation, several approaches of dynamic cache reconfiguration have been reported. Zhang et al. [3] proposed a technique called way concatenation that tunes the cache ways between one (direct-mapped), two and four. Accordingly, at a reduced number of ways (1 and 2) the corresponding unused cache ways are disable. This is carried out under software control, i.e., at application run-time. They report an average energy saving of 40% compared to a fixed four-way cache. Similarly, Dropsho et al. [4] introduced a cache design, called accounting cache based on the selective ways cache [5]. The number of active ways is dynamically change under hardware control. Powell et al. [6] applied way-prediction and selective direct-mapping to reduce the set-associative cache energy.

This is achieved by predicting the matching way and accessing only this matching one, instead of all ways. A dynamic online scheme that combines the processor voltage scaling and dynamic cache reconfiguration was proposed by Nacul et al. [7]. Their online algorithm adapts the processor speed and the cache subsystem to the workload requirements of the application. In a similar fashion, Zhang et al. [2] introduced an online heuristic that dynamically adjusts the cache size in order to minimize the cache energy. Their experiments point out that among all configurable cache parameters, cache size has the largest impact on cache performance and energy consumption. Yang et al. [8] investigated different design choices for resizable caches and evaluated their efficiency in reducing the system's energy dissipation. While these approaches are effective in reducing the energy dissipation, they neglect another important factor, namely the cache reliability.

Cache reliability is mainly threatened by transient faults, caused by strikes from alpha particles and energetic particles [9]. When a memory cell (flip-flop) is hit by such a particle, though the circuit itself is not damaged, the stored bit value can flip and cause an error. This problem is becoming more and more serious due to the ever-shrinking feature size and reduced supply voltage levels [10]. Hardware approaches that are dedicated to improve the cache reliability have been proposed. Such approaches make use of spatial redundancy to correct corrupted bits, for instance, data word parity [11] and Single Error Correct-Double Error Detect Error Correcting Codes (SEC-DED ECC) [12]. Li et al. [10] studied the impact of two leakage energy reduction approaches on the cache reliability. They also used the word parity and SEC to detect and correct the corrupted bits. However, the impact of cache size selection was not considered. Clearly, the spacial redundancy requires additional hardware and decreases the performance, hence is likely to increase the cache energy consumption.

Nevertheless, like cache energy and performance, the cache reliability is also affected by the cache size. The reason for this is threefold. Firstly, when the cache size is reduced (for instance, through disabling a portion of the cache), the probability of particle-hits in the smaller active area is also reduced and particles hitting the disabled part of the cache will not manifest in errors. Secondly, the execution time of the application generally increases as the cache size decreases. The probability of particle-hits during a longer execution time increases. Thirdly, if time redundancy techniques (e.g. rollback recovery)

---

*This work is supported in part by the EPSRC, U.K., under grant GR/S95770, EP/C512804

9B-3

2006 Asia and South Pacific Design Automation Conference

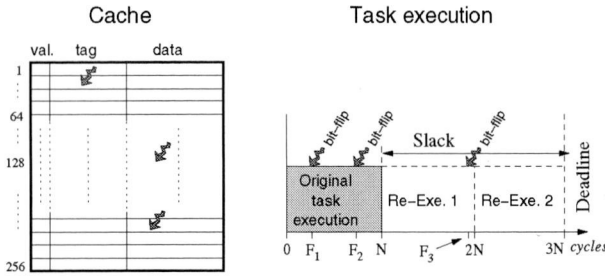

(a) Cache size set to 256 lines (slack for 2 re−executions)

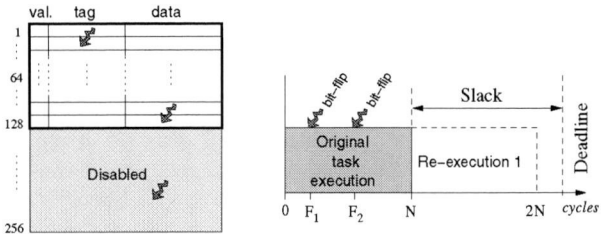

(b) Cache size set to 128 lines (slack for one re−execution)

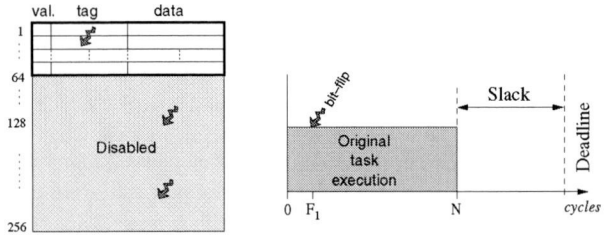

(c) Cache size set to 64 lines (insufficient slack for re−execution)

Fig. 1. Affection of the cache size on reliability

are used to correct faulty executions, the cache's influence on the task execution times will also affect the number of possible re-executions, hence affecting the system's reliability. For example, in Fig. 1(a), if we reduce the cache size from 256 lines to 128 and 64 lines (Fig. 1(b) and (c)), the number of faults in the active cache area decrease from 3 to 2 and 1, respectively. However, the smaller cache sizes result in prolonged task execution times, which, in turn, change the amount of slack left for re-executions.

The aim of this paper is to examine the combined effect of cache size selection on energy consumption, reliability and performance. To the best of our knowledge, this is the first investigation into the interaction between cache energy consumption and transient faults from an cache size perspective. We perform extensive fault inject simulations on five commonly used benchmarks, using a cycle-accurate microprocessor simulator. The experiments demonstrate that there exists a complex trade-off between the different objects. Ultimately, this trade-off can be exploited through dynamic cache resizing (enabling/disabling portions of the cache) at application runtime.

The remainder of this paper is organized as follows. We introduce the models of the transient faults, the cache performability and the cache energy in Section II. The simulation setup and results analysis are in Section III. Concluding remarks are given in Section IV.

II. PRELIMINARIES

A. Transient fault model

Transient faults within the cache are mainly caused by alpha particles hitting the flip-flops of the cache [9, 17]. The physical procedure of the particle-hits causing faults is complex and the effect depends on many factors, like the energy transferred from the particle into the circuit, the transistor size, etc. [18]. In this paper, however, we are not directly interested in the combined circuit and particle properties that can lead to transient faults, but rather in the effect of the transient faults within the cache on the task execution result with respect to performance, energy and reliability. Thus we will use the *bit-flip* as the transient fault model, i.e, when an alpha particle hits a flip-flop, the value stored in it changes its value from 1 to 0 or vice versa [19]. The arrival process of the transient faults is typically modeled as a Poisson process with an average fault rate λ_{fault} [13, 14]. With the arrival of a transient fault, each flip-flop in the cache has an equal probability to be hit since alpha particles are uniformly distributed over the circuit area. Of course, the transient faults can also occur in other parts of the processor (e.g. registers) and cause errors. However, the number of registers in modern microprocessor is far below the number of cache memory cells and it was pointed out in [21] that more than 90% of errors in a processor are originating from the transient faults in the cache. Therefore, we will focus throughout this paper on transient faults in the cache. When a transient fault occurs in the cache, though the value in a certain flip-flop is corrupted, this does not necessarily manifest in an error, that is, the computational result can still be correct. For example, when the transient fault happens in a tag array and the processor wants to read the corresponding cache line, the effect may be just one read miss causing a read from the external memory. Nevertheless, the overall correctness of the computation is not jeopardized. Even when a transient fault happens in a data array of the cache, it is possible that the corrupted value in the data array is overwritten due to a cache write from the processor or a data read from the memory to the cache before it is propagated into the data path of the processor so that the fault is masked. Hence, one important factor that characterizes the system's vulnerability is the ratio between the errors and transient faults. We define the *vulnerability factor* (VF) as this ratio. Further, we define the error rate λ_{error} as the product of VF and the fault rate, i.e., $\lambda_{error} = VF * \lambda_{fault}$. The error rate λ_{error} will be used in Section II.B to compute the performability of the cache. Though the fault rate λ_{fault} is independent of the cache size (3 faults appear in all cache configurations of Fig. 1), we can find that VF is actually a function of the cache size from the simulation results (see Section III.B), so the error rate as well as the performability both depend on the cache size. We will outline how to obtain VF through fault injection experiments in Section III.A.

B. Performability model

For real-time systems, the most important criterion of the system performance is whether the processor can finish executing a task within a given deadline. More specifically, suppose the processor needs N cycles to execute a task and the processor frequency is f, if the deadline before which the task should

924

complete is D, then the performance requirement is $N/f \leq D$. In the case that the task finishes execution before the deadline D, then there exists a slack. For example, Fig. 1 (b) shows the task execution for a cache size of 128 lines. As we can observe, it takes N cycles to execute the task, leaving a slack of $D - N/f$. This slack time can be utilized to increase the system's reliability against transient faults by performing roll-back recoveries (re-execution) when errors occur [14]. That is, in the presents of an error, the task is re-executed with the aim to achieve a non-faulty run. Nevertheless, since re-executions require time, the number of possible re-executions is limited by the amount of slack. Accordingly, the number of possible re-executions is given by:

$$k = \lfloor \frac{D}{N/f} \rfloor - 1 = \lfloor \frac{D \times f}{N} \rfloor - 1 \qquad (1)$$

For instance, in Fig. 1(a), the slack is large enough to perform two re-executions ($k = 2$), when needed. The number of possible re-execution, however, decreases as the cache size is reduced. For instance, in Fig. 1(b) and (c) the number of re-execution is $k = 1$ and $k = 0$, respectively.

Since the appearance of transient faults follows a Poisson distribution, the probability of at least one error during the execution of a task is [14]:

$$\rho_e = 1 - e^{\frac{-VF \times \lambda_{fault} \times N}{f}} = 1 - e^{\frac{-\lambda_{error} \times N}{f}} \qquad (2)$$

We use a combined metric called *performability* to measure the system performance and reliability together [13, 14]. Here, the performability is defined as the probability of finishing the task correctly within the deadline in the presence of faults [14].

Based on Eqs. (1) and 2, the performability can be expressed by [14]:

$$P = 1 - \rho_e^{k+1} = 1 - (1 - e^{\frac{-\lambda_{error} \times N}{f}})^{\lfloor \frac{D \times f}{N} \rfloor} \qquad (3)$$

The clock cycles N that the processor needs to execute a task is heavily impacted by the cache size. The direct result is that k, the number of possible re-executions, will be different for different cache sizes. Also, as outlined in Section III.A, the error rate λ_{error} is a function of the cache size. As a result, the performability is fundamentally impacted by the cache size. As our experimental results indicate (Section III), a careful selection of the cache size is of utmost importance to achieve the required performance and reliability.

C. Cache energy consumption

The energy dissipated in the cache is comprised by a static and a dynamic component. Static energy is caused by leakage currents in the CMOS circuit, while dynamic energy is mainly due to the charging and discharging of the load capacitance which are driven by switching gates. Dynamic energy consumption consists of most of the total energy dissipation in level one caches [16], hence we only consider the dynamic energy of the cache here. Cache energy is dissipated during read as well as write accesses. Write accesses include the normal write accesses and the cache line replacements after cache misses. Accordingly, the cache energy is given by:

$$E = E_{read} \times N_{read} + E_{write} \times N_{write} \qquad (4)$$

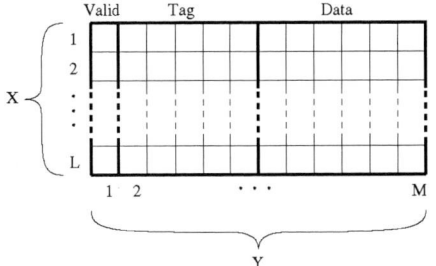

Fig. 2. Position of fault injection: instruction cache

where N_{read} and N_{write} are the numbers of the cache read access and write access respectively, while E_{read} and E_{write} are the energy consumed during one cache read access and one cache write access, respectively. Both N_{read} and N_{write} depend on the cache size since the cache hit rates are generally different for different cache sizes. Furthermore, the energy per access, E_{read} as well as E_{write}, is also cache size dependent. The main reason is that different memory wire lengths have different capacitances to be charged [23].

III. SIMULATIONS AND EXPERIMENTAL RESULTS

A. Simulation setup

In order to study the impacts of the cache size on performability and cache energy, we use a simulation-based approach. The experimental platform is MPARM [24], a cycle-accurate simulator that includes an ARM7 microprocessor model. The simulator reports detailed information regarding the number of clock cycles used to execute a benchmark as well as the energy consumed within the cache. The cache size of MPARM can be configured from 32 bytes up to 1M bytes (1024×1024) with the constraint that the size must be a power of 2. The cache line size is fixed to 16 bytes. In our simulation, we use separated data cache and instruction cache, and both have a maximum size of 256K bytes (256×1024). Cache memory accesses require one clock cycle, while accesses to the main DRAM memory are conservatively assumed to take 100 clock cycles [22]. Two way set-associativity is used for the data cache. The instruction cache is direct-mapped. Although the associativity of the cache is configurable, we do not change the cache associativity in our simulations, since our main focus is on the cache size.

We conduct the fault injection experiments to determine the vulnerability factor (Section II.A) of different cache configuration for five commonly used benchmark applications, namely, a fixed point FFT (FPFFT) with a 1024 points discrete sinewave as input, a cyclic redundancy check (CRC) with 300 ASCII characters as inputs, a 8×8 matrix multiplication (MM), a 12×12 matrix addition (MA) and a 100 integer quick sort algorithm (QSORT). In order to obtain accurate vulnerability factors, we inject 10^6 faults for each benchmarks. Each fault is injected in the following way: Two independent random variables X and Y are used to decide the position of the fault injection in the cache, as illustrated in Fig. 2. The random variable X is uniformly distributed between 1 and L, the number of lines in the cache. Accordingly, X decides in which line the fault will be injected. The random variable Y is uni-

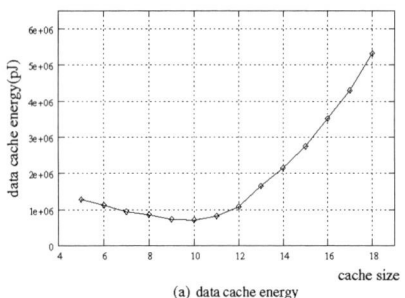

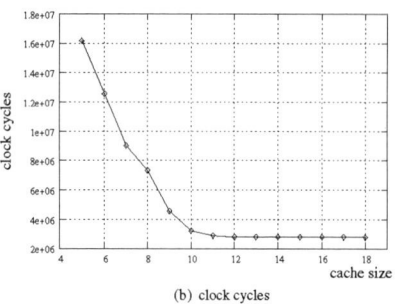

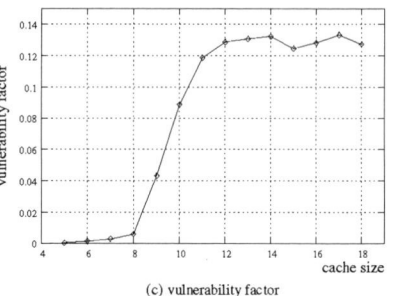

(a) data cache energy   (b) clock cycles   (c) vulnerability factor

Fig. 3. FPFFT simulation results: data cache

formly distributed between 1 and M, the number of bits in a line, to decide which bit in the line will flip its value. This determines if the flipped bit belongs to the valid bit, the tag array or the data array. Fig. 2 shows the case for the direct-mapped instruction cache, the fault position in the two way data cache is decided in a similar way, only that Y is uniformly distributed between 1 and $2M$. When Y is less than M, the fault is within set 0; otherwise, the fault is within set 1. According to the values of X and Y, we check whether the selected memory cell falls within the disabled cache region. If so, the fault injection will definitely not cause an error and a simulation run is unnecessary. On the other hand, if a memory cell within the enabled cache is selected, then the injected fault might manifest in an error, i.e., the simulation has to be performed to observe the effect of the injected fault. Before the simulation is performed, we also randomly determine the clock cycle I of the task execution during which the fault will be injected ($1 \leq I \leq N$). After a simulation has finished, we compare the outcome with the expected outcome to see if the injected fault resulted in an error. The number of errors is counted and divided by the number of injected faults (10^6) to obtain the vulnerability factor.

B. Experimental results

In the first set of experiments, we concentrate on the data cache and the results obtained for the FPFFT benchmark. Fig. 3 shows the outcomes of the experiments. The three graphs give the energy dissipation, the number of clock cycles and the vulnerability factor as a function of the cache size. Note that the cache size in the figures is the logarithm of the true cache size, e.g., cache size 10 means the true cache size is $2^{10} = 1024$ bytes. As we can observe from Fig. 3 (a), the optimal cache energy consumption for the FPFFT benchmark is obtained for a cache size of 2^{10} bytes. It is interesting to see that the cache energy increases for smaller sizes. The main reason for this behavior is the high miss rate for smaller caches, which consequently results in a large number of cache line replacements. This increases the number of cache accesses (read/write accesses + replacement accesses). Although the energy per cache access increases when the cache size gets larger, the number of cache accesses drops relatively faster and the overall effect is a decreasing energy consumption. When the cache size is above 2^{10} bytes, the number of the cache accesses reduces slower and becomes fixed after the cache size is greater than 2^{12} bytes. However, the energy per cache access continuously increases with the cache size increasing. This is mainly due to the fact that the address/data lines in the cache become longer and hence a larger capacitance has to be

TABLE I
PERFORMABILITY FOR FPFFT: DATA CACHE

cache size	number of 9's	digits after 9
5	6	89742
6	6	74106
7	6	59974
8	12	52998
9	16	69592
10	26	81057
11	26	52011
12	26	39275
13	26	34412
14	26	30301
15	26	48551
16	26	40212
17	26	27900
18	26	42730

charged for the accesses. As a result, the energy curve of the cache rises after the cache size is larger than 2^{10} bytes.

The processor clock cycles used to execute the benchmark is depicted in Fig. 3(b). As we can see, with increasing cache size the miss rate drops and the clock cycles decrease quickly. Clearly, less cache misses cause less time-consuming main memory accesses so that the total clock cycles are reduced. Nevertheless, once the cache size is above 2^{12} bytes, the clock cycles do not change any more. This is due to the fact that the application can not facilitate the extra cache and the increasing cache size will not further reduce the cache miss rate.

Fig. 3(c) shows the vulnerability factor (VF) as a function of the cache size. It shows an opposite trend when compared to the clock cycle curve of Fig. 3(b). This can be explained as follows. If the active cache portion is increasing, also an increasing number of the uniformly distributed transient faults will hit this active area, hence causing more errors. However, when the cache size has exceeded 2^{12} bytes, the FPFFT benchmark does not take advantage of additional cache and the processor will not access the additional cache lines. Hence, as the cache becomes larger and larger, although more and more transient faults fall into the cache, the number of those hitting the accessed lines of the cache does not change significantly. As a result, the VF curve becomes saturate when the cache size exceeds 2^{12} bytes.

Having obtained the clock cycles and vulnerability factor, we can now use Eq. 3 to compute the performability for each cache size. The deadline for the FPFFT benchmark is 80.9 ms, which is the execution time of the benchmark when the cache size is minimum (2^5 bytes). The performability results are given in Table I. Note that the performability is actually a probability, with a desired value of as close as possible to 1. To ease the comparison we report the results by the number of 9s after the decimal point (Column 2) and five more digits after the

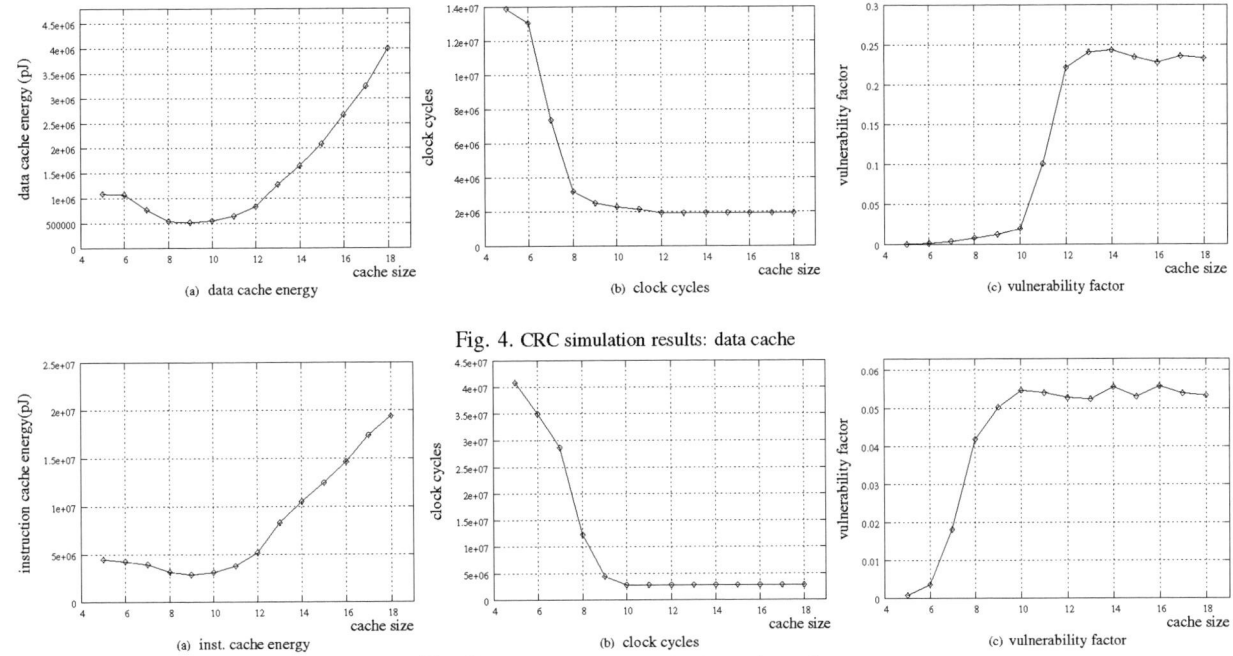

Fig. 4. CRC simulation results: data cache

Fig. 5. FPFFT simulation results: instruction cache

last 9 (Column 3). As we can observe, when the cache size is smaller than 2^8 bytes, though the VF value is small (Fig. 3(c)), the execution time is relatively long due to the large number of clock cycles (Fig. 3(b)). As a result, there is not enough slack for the re-execution so that the performability becomes low. With increasing cache size, the execution time reduces and there is more slack left for re-executions, even though the VF value increases, the performability still improves. This illustrates that compared to VF, slack time is more important to improve the performability for this benchmark. After the size reaches 2^{10} bytes, the difference between each performability value is marginal and the number of 9s remains the same. From Table I, it can be found that with the cache size of 2^{10} bytes, we can achieve the best performability. Taking the energy curve of Fig. 3(a) into consideration, we can find that selecting a cache size larger than 2^{10} is not a good choice because it is not energy efficient and the performability will not be improved. Clearly, for the FPFFT benchmark, 2^{10} bytes is the optimal cache size with which the cache energy is minimal and the performability is maximal.

The next set of experiments is concerned with the CRC benchmark, for which we performed the same evaluations as for the FPFFT benchmark. The deadline of the CRC benchmark is at 69.4 ms. Fig. 4 and Table II show the experimental results. When comparing Figs. 3 and 4 as well as Tables I and II, we can notice similar trends, however, we should note some important difference. For CRC the cache energy is minimal when the cache size is 2^9 bytes while the performability is maximal when the cache size is 2^{10} bytes. That is, as opposed to FPFFT, there is a Pareto-optimal set of the cache sizes: $\{2^9, 2^{10}\}$. The decision of the cache size selection can be made according to the system requirement. For safety-critical systems, we should select 2^{10} bytes to achieve the highest performability; for systems with tight energy budget, e.g. some battery powered systems, the cache size should be selected as

TABLE II
PERFORMABILITY FOR CRC: DATA CACHE

cache size	number of 9's	digits after 9
5	7	33991
6	6	75160
7	6	60276
8	25	78142
9	31	75245
10	36	87705
11	36	83265
12	36	31894
13	35	87930
14	35	86913
15	35	89884
16	35	17656
17	35	89517
18	35	03286

2^9 bytes.

Due to space limitations, we do not report here the detailed results of the matrix multiplication (MM), matrix addition (MA) and quick sort algorithm (QSORT).Nevertheless, the general trends of these benchmarks follow observations made for the FPFFT and CRC benchmarks. Overall these benchmarks have optimal data cache sizes for MM, MA and QSORT of 2^{10}, 2^9 and 2^9, respectively.

Since the above given experimental results concentrated on the data cache, we have conducted additional experiments for the instruction cache. The results of the FPFFT benchmark are given in Fig. 5 and Table III. The energy curve reaches the lowest point at the cache size of 2^9 bytes. In Table III, we can find the performability is maximum when the cache size is 2^{13} bytes. However, since the performability at size 2^{10} is very close to the maximum value and the number of 9s at size 2^{10} is the same as that of size 2^{13}, we can choose 2^9 and 2^{10} as the Pareto-optimal set of the cache size.

The experimental results for the other four benchmarks on the instruction cache follow a similar trend and details are omitted due to space limitations. Nevertheless, for the MA and QSORT benchmarks, there are Pareto-optimal sets: $\{2^7,$

TABLE III
PERFORMABILITY FOR FPFFT: INSTRUCTION CACHE

cache size	number of 9's	digits after 9
5	5	48157
6	4	80023
7	4	17800
8	12	45910
9	40	19101
10	64	75657
11	64	79140
12	64	85160
13	64	86659
14	64	69409
15	64	83935
16	64	67832
17	64	80194
18	64	83067

2^8} and {2^8, 2^9}, respectively. For the MM and CRC benchmarks the optimal cache sizes are 2^8 and 2^9, respectively.

Summarizing, we can draw the following conclusion from the above experimental results. There exist optimal or Pareto-optimal cache size choices with respect to performability and energy consumption. Depending on the application requirement a proper cache size should be selected to achieve the optimal energy and performability simultaneously or the best trade off. Furthermore, as the optimal cache sizes depend largely on the running application, dynamically changing the cache size to suit the particular application is not only beneficial from an energy point of view but also to improve the system's performability. For instance, when running the FPFFT benchmark after the MA benchmark, the processor's data cache should be adapted from 2^9 to 2^{10} bytes and the instruction cache should be changed from 2^8 to 2^9 bytes. This adaption would reduce the data cache energy by 2.6% and increasing the performability from sixteen 9s to twenty-six 9s. The instruction cache would reduce its energy by 8.8% and improve the performability from twelve 9s to forty 9s.

IV. CONCLUSIONS

In this paper, we studied the impact of the cache size selection on three important design objectives, namely, the system performance, the cache energy consumption and the cache reliability, which has not been addressed explicitly in previous work. Performability has been defined to combine the analysis of the performance and the reliability. We have conducted extensive experiments to analyze the interplay between the three objects. These experiments were performed using cycle-accurate processor simulations and it was found that the cache size selection affects not only the energy but also the performability. The results indicate that a careful cache size selection is needed, in order to take advantage of the found optimal energy/performability trade-off points.

REFERENCES

[1] J. L. Hennessy, D. A. Patterson, *Computer Architecture: A Quantitative Approach*, 2nd Edition, Morgan Kaufmann Publishing Co. 1996

[2] C. Zhang, F. Vahid and R. Lysecky, "A self-Tuning Cache Architecture for Embedded Systems", in Proc. of DATE, 2004.

[3] C.Zhang, F. Vahid, W. Najjar, "A Highly Configurable Cache Architecture for Embedded Systems", in Proc. of International Symposium on Computer Architecture, 2003.

[4] S. Dropsho et al., "Integrating Adaptive On-Chip Storage Structures for Reduced Dynamic Power", in Proc. of the International Conference on Parallel Architectures and Compilation Techniques, 2002.

[5] D. H. Albonesi, "Selective cache ways: On-demand cache resource allocation", in Proc. of International Symposium on Microarchitecutre, 1999.

[6] M. Powell A. Agaewal, T. Vijaykumar, B. Falsafi and K. Roy, "Reducing Set-Associative Cache Energy via Way-Prediction and Selective Direct Mapping", in Proc. of International Symposium on Microarchitecture, 2001.

[7] A. C. Nacul and T. Givargis, "Dynamic Voltage and Cache Reconfiguration for Low Power", in Proc. of DATE 04, March, 2004.

[8] S. Yang, M. D. Powell, B. Falsafi, T. N. Vijaykumar, "Exploiting Choice in Resizable Cache Design to Optimize Deep-Submicron Processor Energy-Delay", in Proc. of International Symposium on High-Performance Computer Architecture, 2002.

[9] G. Asadi, V. Sridharan, M. B. Tahoori, D. Kaeli, "Balancing Performacne and Reliability in the Memory Hierarchy", in Proc. of International Symposium on Performance Analysis of Systems and Software, 2005.

[10] L. Li, V. Degalahal, N. Vijaykrishnan, M. Kandemir, M. J. Irwin, "Soft Error and Energy Consumption interations: A Data Cache Perspective", in Proc. of ISLPED 04, Aug. 2004.

[11] S. Kim and A. K. Somani, "Area Efficient Architectures for Information Integrity in Cache Memories", in Proc. of International Symposium on Computer Architecture, 1999.

[12] W. Zhang, S. Gurumurthi, M. Kandemir, A. Sivasubramaniam, "ICR: in-cache replication for enhancing data cache reliability", in Proc. of International Conference on Dependable Systems and Networks, 2003.

[13] D. Zhu, R. Melhem and D. Mosse, "The Effecs of Energy Management on Reliability in Real-Time Embedded Systems", in Proc. of ICCAD 04, Nov. 2004.

[14] A. Ejlali, M. T. Schmitz, B. M. Al-Hashimi, S. G. Miremadi, "Energy Efficient SEU-Tolerance in DVS-Enabled Real-Time Systems through Information Redundancy", in Proc. of ISLPED 05, Aug. 2005

[15] R. Melhem, D. Mosse, E. Elnozahy, "The interplay of Power Management and Fault Recovery in Real-Time Systems", IEEE Transaction on Computers, Vol. 53, No. 2, February, 2004.

[16] H. Hanson, M. S. Hrishikesh, V. Agarwal, S. W. Keckler, D. Burger, "Static Energy Reduction Techniques for Microprocessor Caches", IEEE Transaction on VLSI systems, Vol. 11, No. 3, June, 2003.

[17] S. Mitra, N. Seifert, M. Zhang, Q. Shi, K. S. Kim, "Robust System Design with Built-In Soft-Error Resilience", IEEE Computer Magazine, Vol, 38, No. 2, Feburary, 2005.

[18] P. E. Dodd and F. W. Sexton, "Critical Charge Concepts for the CMOS SRAMS", IEEE Transactions on Nuclear Science, Vol. 42, No. 6, Dec. 1995.

[19] F. Faure, R. Velazco, M. Violante, M. Rebaudengo and M. Sonza Reorda, "Impact of Data Cache Memory on the Single Event Upset-Induced Error Rate of Microprocessors", IEEE Transactions on Nuclear Science, Vol. 50, No. 6, Dec. 2003.

[20] A. Maheshwari, W. Burleson, R. Tessier, "Trading off Transient Fault Tolerance and Power Consumption in Deep Submicron (DSM) VLSI Circuits", IEEE Transaction on VLSI systems, Vol. 12, No. 3, March 2004.

[21] M. Rebaudengo, M. S. Reorda and M. Violante, "An Accurate Analysis of the Effects of Soft Errors in the Instruction and Data Caches of a Pipelined Microprocessor", in Proc. of DATE 03, March, 2003.

[22] L. Li, I. Kadayif, Y-F. Tsai, N. Vijaykrishnan, M. Kandemir, M. J. Irwin and A. Sivsubramaniam, "Leakage Energy Management in Cache Hierarchies", in Proc. of International Conference on Parallel Architectures and Compilation Techniques, 2002.

[23] G. Reinmann and N. P. Jouppi, "CACTI2.0: An Integrated Cache Timing and Power Model", COMPAQ, Western Research Lab, Research Report, 2000.

[24] http://www-micrel.deis.unibo.it/sitonew/research/mparm.html

Reducing Dynamic Compilation Overhead by Overlapping Compilation and Execution *

P. Unnikrishnan, M. Kandemir, and F. Li
Computer Science and Engineering Department
Pennsylvania State University
e-mail: {unnikris,kandemir,feli}@cse.psu.edu

Abstract— An important problem in executing applications in energy-sensitive embedded environments is to tune their behavior based on dynamic variations in energy constraints. One option for achieving this is dynamic compilation — compiling code fragments on the fly to adapt to changing energy demands. While dynamic compilation can be very beneficial in many embedded environments where multiple criteria need to be satisfied during execution, it can also incur a significant performance overhead since compilation takes place at runtime. The goal in this work is to reduce this performance overhead of dynamic compilation by overlapping it with application execution. Specifically, provided that we have available hardware resources to perform dynamic compilation concurrently with application execution, our approach compiles the next code fragment to be executed while we are executing the current code fragment. The experimental results from our implementation indicate significant savings in execution times. Our experimental results also indicate that the proposed strategy performs consistently well under different parameters.

I. INTRODUCTION

Dynamic compilation and linking is an important technique for optimizing applications while they are executing. Most of the prior work in the area [1, 3, 4, 7] focused on using dynamic compilation for implementing performance-oriented compiler optimizations at runtime. For example, the value of a program variable may not be known statically (that is, at compile time), but once it is known at runtime it may enable several optimizations (e.g., constant propagation). Similarly, a code region turns out to be executed very frequently at runtime and thus deserves a more sophisticated compilation at runtime (anticipating frequent future executions). A recent study [11] also demonstrates how dynamic compilation and linking can be used for energy adaptation in battery-operated embedded environments; that is, when energy constraints (e.g., battery level) change, the application is recompiled (dynamically) to generate a more energy-efficient version. For example, when battery is low, some memory banks in the system may need to be turned off (to save power), and being able to work with fewer number of banks may demand recompilation of the application code. In contrast, when battery power is high, the application code can make use of all memory banks available in the system – this may require another dynamic compilation.

One of the most important problems in an energy-sensitive embedded platform that employs dynamic compilation is the extra time and energy taken by the dynamic compilation process itself. This is because this performance and energy overhead directly contributes to the execution energy and time (as compilation occurs at runtime). While it may not be possible to hide the compilation energy, it may be possible to hide some portion of the compilation overhead (time) by *overlapping* it with the application execution. In other words, provided that we have available hardware resources to perform dynamic compilation concurrently with application execution, we can compile the next code fragment to be executed while we are executing the current code fragment. The code fragment in question can be a loop nest, a subroutine, or several logically-related subroutines. Overlapping dynamic compilation with application execution is termed as *compilation parallelization* in this paper (since compilation occurs parallel with application execution). It should be emphasized that, while not quantified in this paper, reducing the time spent in dynamic compilation can also be beneficial from a leakage energy consumption viewpoint (as a side effect of reduction in execution cycles).

In this paper, we propose a strategy to hide the time spent in dynamic compilation. Our strategy is based on *predicting* the next code fragment to be executed and *pre-compiling* that fragment before it is actually needed. Krintz et al [8] propose an optimization strategy to reduce the performance overhead due to dynamic compilation. Their approach mainly targets Java applications and involves maintaining a global priority queue that determines the modules to be compiled ahead of time. The efficiency of this scheme depends heavily on the strategy employed to fill the priority queue and also the time taken to fill the queue. Our strategy makes use of a global history table to predict the next code fragment to compile, and is found to be 92.37% accurate in predicting the next code fragment that will be executed. Another difference between the two studies is that our target environment is a chip multiprocessor-based energy-sensitive embedded system and our idea is applicable to different programming environments. It should be observed that an on-chip multiprocessor platform is particularly suitable for parallelizing dynamic compilation since it has multiple processor cores, one of which can perform compilation while the others can execute the application. It is known that, in many multiprocessor applications, it may not be possible to fully utilize all the available processor cores. In such cases, using one of the cores for compilation does not affect the application execution in any significant way. We implemented the proposed optimization using Dyninst [2], a post-compiler program manipulation tool, and performed experiments using several applications. Our experimental results indicate that it is possible to hide a large percentage of the compilation time (32% on the average) if one employs one processor for dynamic compilation alone. In addition, our experimental results also show that allocating more processors for dynamic compilation can

---

*This work is supported in part by NSF Career Award #0093082 and by a grant from GSRC.

9B-4

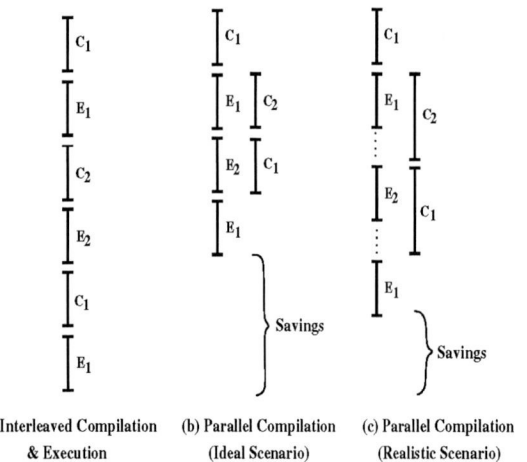

(a) Interleaved Compilation & Execution (b) Parallel Compilation (Ideal Scenario) (c) Parallel Compilation (Realistic Scenario)

Fig. 1. Different scenarios for dynamic compilation.

be beneficial until a threshold point is reached.

The remainder of this paper is organized as follows. Section II explains our approach and discusses our implementation. Section III presents the execution model in detail and discusses history-based next module (function/nest) prediction. Section IV introduces our benchmarks and gives experimental results. Section V concludes the paper with a summary.

II. OVERLAPPING COMPILATION WITH APPLICATION EXECUTION

A. Approach

Our goal in this paper is to hide the time spent in dynamic compilation as much as possible. Fig. 1 helps visualize why parallelizing compilation (i.e., overlapping compilation with application execution) might be useful in practice. In this figure, C_n denotes the dynamic compilation time for module (function/subprogram or loop nest) n before it is executed. E_n denotes the execution time for module n. We consider the worst case scenario where each module has to be optimized and recompiled before it can be executed (this may occur, for example, when constraints such as remaining battery power are constantly changing). Fig. 1(a) illustrates what happens when dynamic compilation is not parallelized. Since in this case the compilation and execution are interleaved over the lifetime of the application, the overall execution time of the application can be expressed as the sum of the compilation times and execution times of each module, i.e., $(C_1 + E_1 + C_2 + E_2 + \cdots + C_n + E_n)$. It is to be noted that this is the current state-of-the-art in dynamic compilation.

In comparison, Figures 1(b) and (c) illustrate the potential benefits of overlapping compilation with application execution. In Fig. 1(b), it is assumed that we are able to hide the entire time spent in dynamic compilation (except for the first module of course). Under this assumption, the overall execution time of the application can be expressed as $(C_1 + E_1 + E_2 + \cdots + E_n)$. It should be noted however that, depending on the actual values of C_i and E_i, this ideal scenario may not be achieved all the time. Fig. 1(c) depicts a more realistic scenario, where dynamic compilation takes place in parallel with application execution, however, the compilation is not always complete in time for execution. Consequently, the application has to wait until the compilation of the module is complete. In this case, the overall execution time of the application is given

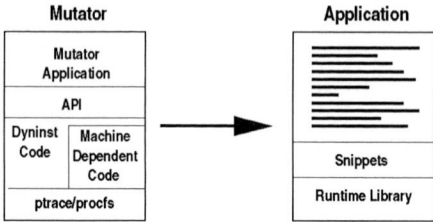

Fig. 2. Abstractions used in the Dyninst API.

by $(C_1 + E_1 + E_2 + E_3 + \cdots + E_n + TotalWaitingTime)$, where $TotalWaitingTime$ is the total time that the application has to wait for the compilation to complete. Under this scenario, the application in question incurs some compilation overhead but a significant saving can still be obtained when compared to the case in Fig. 1(a).

One might argue at this point that instead of utilizing extra resources for dynamic compilation, it may be a better idea to employ them in normal application execution. As a concrete example, instead of using one processor for application execution and another one for dynamic compilation, it may be a better option to use both these processors for application execution (and each can do dynamic compilation on a need basis). However, such an alternative may or may not be viable, depending on the specific case at hand. In particular, some embedded applications are not amenable to thread-level parallelization. As a result, they cannot take advantage of extra processors available in the system. However, as will be demonstrated in this paper, most such applications can still take advantage of extra resources if those resources are used for dynamic compilation.

B. Implementation Details

Our dynamic compilation infrastructure is implemented using the Dyninst software from the University of Maryland [2]. Dyninst is a post-compiler program manipulation tool which provides an Application Program Interface (API) called DyninstAPI for program instrumentation. Using the DyninstAPI library, it is possible to instrument and modify application programs during execution (i.e., as they are running). DyninstAPI is itself a C++ class library which can be included and directly called from a C++ program. With this interface, a program can create a new piece of code and insert it to another program while the latter is executing. The program being modified is able to continue execution and does not need to be recompiled entirely. While, as far as the applicability of our approach is concerned, it is not very important whether dynamic compilation is needed for performance or energy reasons, our current target environment in an energy-sensitive SoC platform where we have multiple cores and some of these cores can be used for parallelizing dynamic compilation. That is, in our environment, dynamic compilation is invoked as a result of some change in energy constraints at runtime. However, our approach is oriented towards reducing the performance impact of dynamic compilation rather than its energy consumption.

The overall structure of the Dyninst API and its implementation are shown in Fig. 2 (from [2]). There are two processes, called the mutator and the application (or mutatee). The left side of the figure shows the code for the mutator process that contains calls into the Dyninst API. It also contains the code that implements the runtime compiler and the utility routines to manipulate the application process (shown below the rectangle labeled API) as well as profiling/tracing tools. The right

930

half of the figure depicts the application process with the original code of the program shown in the top part of the figure. The bottom two parts of the application are the snippets that are inserted into the program, and the runtime library that supports the Dyninst API. To perform our experiments, we installed Dyninst on a Sun Solaris-based platform.

III. EXECUTION MODEL

A. Order of Events

Our execution model is depicted in Fig. 3. In this model, the application source code is augmented with "sensitivity lists." Each sensitivity list is attached to a module (a loop nest, subprogram, or function) and indicates the energy components that the module in question is sensitive to. E-Optimizer is a decision-making module which checks E-Script to determine the compilation strategy to choose, given the new energy constraint (e.g., remaining battery power). E-Script is a list that contains for each nest (or subprogram) the compilation strategies that should be activated based on energy constraints. After determining the optimization strategy, E-Optimizer asks E-Compiler whether there is already a compiled module in the "Compiled Code Repository," which corresponds to the new energy constraint. If there is, then E-Compiler supplies that module, which is subsequently inserted by E-Optimizer in the code and the execution resumes. If no such pre-compiled module exists in the repository, E-Compiler generates such a module and forwards it to E-Optimizer, which subsequently proceeds as explained above. It should be noted that, in our implementation, the variation in energy constraints is checked only when a sensitivity list check is being done (i.e., when a sensitivity list is reached during execution). This dynamic recompilation/linking-based execution environment has been implemented using Dyninst. In this implementation, E-Optimizer is a separate supervisory program that controls the code modifications to be performed on the target application.

We want to reiterate that our focus in this paper is on hiding as much compilation time as possible by overlapping it with application execution. Therefore, how the codes are optimized to adapt to changing energy constraints is beyond the scope of this paper (i.e., the approach proposed here is orthogonal to the set of compiler optimizations employed for reducing energy/execution cycles). It should be mentioned, however, that most of our optimizations applied during dynamic compilation target at memory banks, and increase/decrease the number of active banks to adapt to the energy constraints imposed. Our approach, however, can be made to work without any problem with any dynamic compilation framework and any set of compiler optimizations.

E-Optimizer comprises of two threads: E-CtrlThr (Controller Thread) and E-CompThr (Compilation Thread). E-CtrlThr monitors and controls the stopping/ starting and reconfiguration of the application. E-CompThr goes ahead of E-CtrlThr, predicts the next function/subroutine/nest that is energy sensitive, determines the appropriate compilation strategy for the new energy constraint, and compiles the new module and keeps it ready before the next sensitivity list for the energy sensitive region is reached. When the application reaches the next sensitivity list, it is stopped: If the dynamic compilation of the new module by E-CompThr is complete, it is inserted into the application and the application continues its execution. If the dynamic compilation is still in progress, the

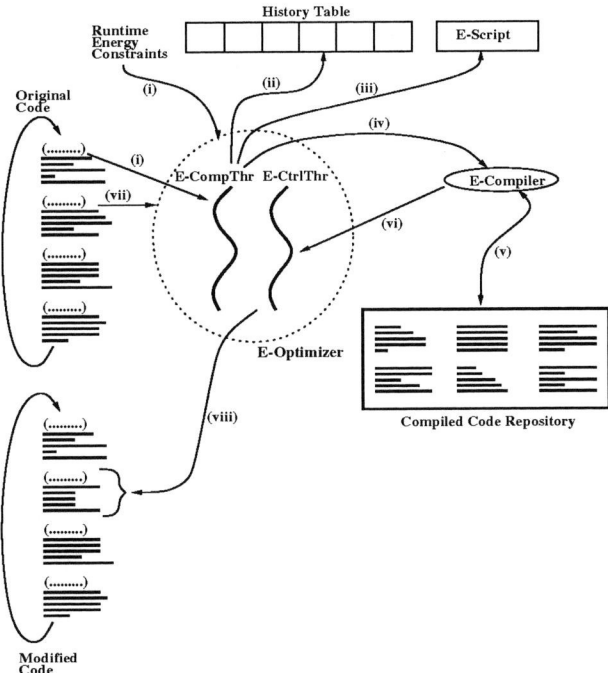

Fig. 3. Our execution model. The numbers attached to arrows indicate the order of events. This dynamic compilation framework has been implemented using Dyninst [2].

application waits until the compilation is complete before it can insert the new module and continue. In this framework, the only part that needs to be performed by the application programmer is to insert sensitivity lists at the appropriate places in the code. Our current implementation allows sensitivity lists to be added at the beginning of subprograms/functions and loops.

B. History-Based Next Module Prediction

The success of the proposed execution model hinges on the *prediction* of the next energy sensitive code region (module) to be executed in the application. On analyzing the function/subroutine/nest traces of several applications (that is, the order in which functions/subroutines/nests are executed), we observed regular patterns, which means that we can predict the next function/subroutine/nest to be executed with a reasonable accuracy. To benefit from this regularity, E-Optimizer maintains a history table (see Fig. 3). Every time the application program reaches an energy-sensitive region (annotated by a sensitivity list), it logs the function/subroutine/nest in the history table.[1] E-CompThr consults the history table before predicting the next energy sensitive region. Every time a sensitivity list is encountered, the compilation thread (E-CompThr) is activated to start the prediction and compilation of the next energy sensitive region. Savings in execution time are obtained when a correct prediction is made by E-CompThr. In the event of a misprediction, E-Optimizer invokes E-Compiler and compilation is done before the execution (incurring, of course, its performance overhead). Thus, the overall savings in compilation overhead are determined by the accuracy of the prediction and the history logged in the history

---

[1]*For array based applications, we focus on nests; that is, each nest is potentially an energy-sensitive module (code region). In contrast, for other applications, energy-sensitive modules are functions and subroutines. Unless stated otherwise, in our discussion, each nest (or function/subroutine) is considered as energy sensitive, and is augmented by a sensitivity list.*

9B-4

TABLE I
BENCHMARK CHARACTERISTICS.

Benchmark	Source/Type	Size/Input File
btrix	Spec95	721KB
tomcatv	Spec95	836KB
vpenta	Spec95	770KB
hier	Motion Est.	310KB
full_search	Motion Est.	310KB
epic	MediaBench	test_image.pgm
rasta	MediaBench	ex5_c1.wav
181.mcf	Spec2000	inp.in

TABLE II
PREDICTION ACCURACIES.

Benchmark	Prediction Accuracy
btrix	99.95%
tomcatv	88.38%
vpenta	99.96%
hier	100.00%
full_search	100.00%
epic	91.02%
rasta	72.13%
181.mcf	87.52%

TABLE III
THE BEST-CASE AND WORST-CASE TIMES FROM THE COMPILATION PERSPECTIVE (IN MILLISECONDS). THE VALUES WITHIN THE BRACKETS GIVE THE PERCENTAGE CONTRIBUTION TO THE TOTAL (COMPILATION PLUS EXECUTION) TIME.

Benchmark	Compilation Time (Worst Case)	Compilation Time (Best Case)	Execution Time (msec)
btrix	17651.69 [53.1%]	1810.51 [10.4%]	15599.96
tomcatv	22225.82 [80.3%]	1854.82 [25.4%]	5450.17
vpenta	20060.68 [36.5%]	1834.51 [5.0%]	34864.98
hier	3515.73 [36.7%]	1387.97 [18.7%]	6053.28
full_search	2468.43 [70.6%]	249.01 [19.4%]	1029.84
epic	90476.59 [82.0%]	1547.30 [7.2%]	19878.16
rasta	85515.07 [86.4%]	3138.66 [18.9%]	13436.02
181.mcf	5283755.26 [85.8%]	1490.7 [0.2%]	873999.61

table.

Our current prediction strategy makes use of the history table as follows. If the current module to execute is m1, we search the history table to find when was the last time m1 was called (invoked), and what was the method invoked following it. If this method (in the history list) is m2, we predict the next method to be called (after m1) as m2, and start pre-compiling it (if it needs to be recompiled). In other words, our approach assumes that if the most recent execution of m1 has been followed by an execution of m2, the current execution of m1 will also be followed by m2. Table II gives the prediction accuracies for the benchmarks used in this study. The prediction accuracies for array based codes are quite high due to the regular patterns in which loop nests are executed. That is, the loop nests in general are executed one after another with little (or no) control flow between them. For the other benchmarks, the prediction accuracy is dependent on the number of modules that are to be monitored and also on the regularity of their execution patterns. Overall, the prediction accuracies listed in Table II are very encouraging and imply that our pre-compilation based strategy can be successful in practice. However, we need to point out that a more sophisticated predictor can potentially generate even better predictions. In fact, we believe that it is even possible to parameterize the dynamic compiler to use different strategies at runtime based on a static profile.

IV. EXPERIMENTS AND RESULTS

A. Time Contribution of Dynamic Compilation

To evaluate our dynamic compilation based environment, we conducted experiments using eight benchmarks from different domains and benchmark suites. Table I gives important characteristics of these benchmark codes. The main reason for selecting these benchmarks is that we were able to run them through the Dyninst environment. The primary aim of our scheme is to hide the compilation time and thus reduce the overall execution time of applications in comparison to the default case, where dynamic compilation and execution are interleaved. Table III gives the best and the worst case (percentage) compilation times for each of the benchmarks in our suite when dynamic compilation and execution are fully interleaved (i.e., when no compilation time is hidden). To evaluate the worst case compilation time, each energy sensitive region is optimized and recompiled every time it is called before it can be executed. The compilation is based on the compilation strategies captured in the E-Script. In contrast, in the best case, each energy sensitive region is compiled only once during the course of entire execution, i.e., the first time the region is encountered. For every subsequent occurrence, the pre-compiled binaries from the Compiled Code Repository are used. Note that this best case represents the smallest amount of runtime that can be taken by the dynamic compiler. From the

results given in Table III, one can see that the compilation time constitutes more than 80% of the overall time in benchmarks tomcatv, epic, rasta and 181.mcf when the worst case scenario is considered. Even considering the best case scenario, it can be observed that, on the average, 13% of the total time goes to dynamic compilation. Consequently, an optimization strategy that hides this overhead (time) can be very useful in practice. In fact, the results in this table along with those in Table II provide a strong motivation for our research.

B. Reductions in Overall Execution Times

We recorded the total execution times with our strategy for various compilation probabilities (100%, 50%, and 25%) and compared them with the results when compilation and execution are interleaved. A 100% compilation probability implies that every module in the application has to be recompiled before it can be executed. On the other hand, a 50% (25%) compilation probability implies that only 50% (25%) of the modules in the application has to be recompiled before each time they need to be executed. Our experience with different applications indicate that most applications have compilation probabilities ranging from 25% to 80%. We performed experiments with these three different probabilities to cover a large spectrum. The experiments were executed on a Sun multi-processor machine running Solaris. The application process was bound to one of the processor (Processor 1) and the threads of E-Optimizer (i.e., E-CtrlThr and E-CompThr) were both bound to another processor (Processor 2). On analyzing the results obtained (see columns 2, 3, and 4 in Tables IV, V, and VI), one can make several observations. First, our approach reduces the time spent in dynamic compilation significantly. Specifically, with compilation probabilities of 100%, 50%, and 25%, the average reductions in the compilation time (the second column of the tables) are 32.06%, 20.45%, and 28.06%, respectively. However, to make a fair comparison, one needs to quantify the negative impact of our approach as well, that is, the increase in the execution time due to thread synchronization and contention. We can see from the third column in Tables IV, V, and VI that in some cases the increase in execution time is so much that it offsets all the benefits coming from parallelizing dynamic compilation (consider, for example, benchmarks such as rasta and 181.mcf with compilation probabilities of 50% and 25%). When one considers the total execution times (compilation time plus execution time) i.e., the fourth column in Tables IV, V, and VI, one sees an average of 10.35% and 4.07% reductions in total times with compilation probabilities 100% and 50%, respectively, and a 2.19% increase in the total time with a 25% compilation probability. Therefore, we can conclude that although we obtain

TABLE IV
PERCENTAGE IMPROVEMENTS DUE TO PARALLELIZING COMPILATION (WITH A 100% COMPILATION PROBABILITY).

Benchmark	2 Processors			3 Processors		
	Compl%	Excn%	Total%	Compl%	Exec%	Total%
btrix	23.11	-4.16	10.26	20.07	-0.24	10.50
tomcatv	30.80	-90.13	6.63	14.02	-5.32	10.15
vpenta	40.25	-2.38	13.17	38.69	-1.75	13.00
hier	63.75	-0.77	22.92	61.76	-0.68	22.25
full_search	27.06	-3.96	17.94	22.98	-2.28	15.55
epic	13.72	-39.36	4.01	28.66	-7.35	22.07
rasta	28.32	-162.50	2.00	36.71	-11.85	30.01
181.mcf	29.49	-129.02	5.93	38.25	-7.04	31.51

TABLE V
PERCENTAGE IMPROVEMENTS DUE TO PARALLELIZING COMPILATION (WITH A 50% COMPILATION PROBABILITY).

Benchmark	2 Processors			3 Processors		
	Compl%	Excn%	Total%	Compl%	Excn%	Total%
btrix	22.28	-0.67	8.34	15.99	0.38	5.65
tomcatv	8.48	-26.21	-2.25	9.92	-3.77	5.07
vpenta	33.19	-1.44	6.51	48.27	-1.09	6.64
hier	43.78	0.01	12.33	82.62	0.02	12.74
full_search	22.69	-2.72	12.17	33.17	-13.79	8.89
epic	11.63	-19.37	2.26	7.81	-4.51	3.69
rasta	12.50	-72.27	-6.42	16.03	-8.00	8.95
181.mcf	9.05	-28.50	-0.34	10.70	-9.65	4.84

TABLE VI
PERCENTAGE IMPROVEMENTS DUE TO PARALLELIZING COMPILATION (WITH A 25% COMPILATION PROBABILITY).

Benchmark	2 Processors			3 Processors		
	Compl%	Excn%	Total%	Compl%	Excn%	Total%
btrix	11.28	-0.51	2.52	4.45	0.22	1.26
tomcatv	14.54	-13.10	1.83	-6.12	-2.11	-4.49
vpenta	37.32	-0.47	5.89	63.42	-0.27	6.31
hier	37.27	0.43	9.22	60.86	0.32	9.27
full_search	-40.03	-4.85	-17.72	-9.55	-0.85	-4.40
epic	14.95	-11.19	2.87	8.87	-4.79	2.16
rasta	1.56	-49.92	-17.68	6.01	-6.32	1.19
181.mcf	-0.12	-10.11	-4.48	-8.59	1.08	-4.84

TABLE VII
VPENTA: IMPACT OF COMPILING ONLY CRITICAL LOOP NESTS.

Nests	Compile(ms)	Total(ms)
None	0	54375.86
Nest I	2874.97	55963.48
Nest I + Nest II	4824.32	58135.06
Nest I + Nest II + Nest IV	7319.83	46720.90
Nest I + Nest II + Nest IV + Next VIII	9692.83	59210.55
All Nests	20060.67	54989.12

some improvements, using only two processors does not bring impressive benefits.

These results led us to perform another set of experiments, where we used three processors. Specifically, the application process, as before, was bound to one processor (Processor 1), the controller thread (E-CtrlThr) was bound to a different processor (Processor 2), and the compilation thread (E-CompThr) was bound to a third processor (Processor 3). As can be seen from the columns 5, 6, and 7 of Tables IV, V, and VI, the results obtained using three processors are much better as compared to those obtained using two processors only. Specifically, with a compilation probability of 100%, the average reductions in the compilation time, execution time, and total time are 32.64%, -4.56%, and 19.38%, respectively. The corresponding values for compilation probabilities of 50% and 25% are 28.06%, -5.05%, and 7.05% and 14.91%, -1.59%, and 0.81%, respectively. These results indicate that our approach can make use of available processors effectively. It should also be emphasized that some of our applications, e.g., epic, rasta, and 181.mcf, are not amenable to chip-scale parallelism; so, we cannot use the available processors for reducing their execution times. However, it is still possible to use our approach to reduce their dynamic compilation times.

C. Impact of Compiling Critical Modules

Recall that in our experiments so far we considered all modules (functions/subroutines/ nests) as energy sensitive regions. In our next set of experiments, we analyzed the impact of dynamically compiling only a set of critical modules (instead of all modules). A critical module is the one that contributes to the overall execution time significantly. The experiments were performed assuming a 100% compilation probability. Table VII shows, for vpenta, the effect of compiling critical modules (nests) when execution and compilation are fully interleaved (i.e., when our scheme is not employed). The first column in this table lists the dynamically compiled critical nests in the application, the second column gives the compilation time, and the third column gives the overall time (compilation plus execution). It is easy to see that there is an optimum set of nests (Nest I + Nest II + Nest IV) such that, when com-

piled, lead to minimum overall time. In other words, for the minimum total time, it is important that only a specific set of critical modules need to be (predicted correctly and) dynamically compiled. In fact, as can be observed from this table, attempting to compile a larger set increases the total time. Similar results are presented in Table VIII when our approach is employed. Compiling only these critical modules (i.e., Nest I + Nest II + Nest IV) when execution and compilation are interleaved results in an improvement of 14.07% (for vpenta) in the total time. Overlapping execution and compilation yields an additional 13.98% improvement in the total time of the application, thereby producing a 28.05% overall performance improvement. Similar results have been observed with other benchmark codes as well, and those results are omitted here due to lack of space. One can conclude from these results that overlapping computation and compilation in addition with compiling only the critical modules yields the best results. In addition, we found that profiling can be of great help in determining the modules to compile. For example, in btrix, the seventh nest takes 95.2% of the time and is, therefore, a perfect candidate for compilation. Similarly, in 181.mcf, three subroutines, namely, refresh_potential, price_out_impl, and primal_bea_mpp take (together) 54.3% of the overall execution time. And, consequently, compiling only these critical modules generated between 20% and 30% improvements for these applications. As a result, a simple strategy for reducing the runtime overhead due to dynamic compilation would be (i) to profile the application to determine the critical modules and, (ii) to pre-compile these modules using the approach discussed in this paper.

D. Impact of Increasing the Number of Processors

In our experiments so far, we have used at most two processors for compilation purposes (in addition to the one that is executing the application itself). However, in some cases, allocating more processors to dynamic compilation can bring further reductions in overall execution time. Consider the example scenario depicted in Fig. 4. In this scenario, three processors are employed for performing dynamic compilation, whereas one processor is executing the application. Obviously, one might be able to get further benefits from even a larger number of processors depending on the application in question. It should be noted, however, that an opposite ar-

TABLE VIII
VPENTA: ABSOLUTE TIMES AND PERCENTAGE IMPROVEMENTS [DUE TO OVERLAPPED COMPUTATION AND COMPILATION OVER TABLE VII].

Nests	Compile (msec)	Total (msec)
Nest I	262.98 [90.85%]	52614.01 [5.99%]
Nest I + Nest II	472.14 [90.21%]	53585.28 [7.83%]
Nest I + Nest II + Nest IV	2652.98 [63.76%]	39124.29 [10.58%]
Nest I + Nest II + Nest IV + Next VIII	3169.78 [67.30%]	52648.76 [11.08%]
All Nests	11986.05[40.25%]	47747.07 [13.17%]

Fig. 4. An execution scenario where multiple processors are employed for dynamic compilation.

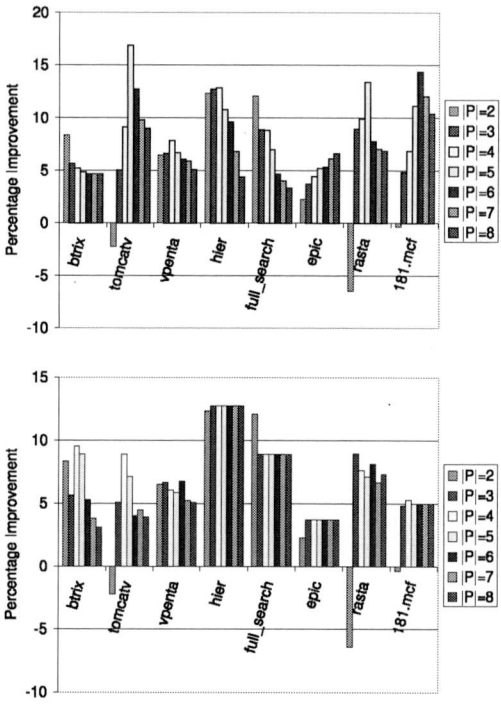

Fig. 5. Top: Percentage performance improvements with increasing number of processors employed for dynamic compilation. Bottom: Percentage performance improvements with increasing number of processors employed for application execution.

gument may defend using available processors for executing the application. As discussed earlier in the paper, such an approach may or may not be successful, depending on whether the application is amenable to module-level (loop nest-level or function/subroutine-level) parallelism. Specifically, if the application does not exhibit inherent parallelism, increasing the number of processors would not bring any benefits.

To study this tradeoff between parallelizing dynamic compilation and parallelizing application execution, we performed a set of experiments where we increased the number of processors and used them for compilation and execution. The results obtained are presented in Fig. 5 with a compilation probability of 50%. The top graph in Fig. 5 shows the results when the number of processors allocated for dynamic compilation is increased (and using only 1 processor for application execution). The bottom graph, on the other hand, gives the results when the number of processors allocated for application execution is increased (keeping the number of processors allocated for compilation at 2 except for the first bar where we have only 1 processor for compilation). We can make several observations from these two graphs. First, in general, employing the available processors for dynamic compilation (as opposed to application execution) generates better performance results. Specifically, the average performance improvements when using 5 processors are 9.49% and 7.31%, respectively, when the extra processors used for compilation and execution. Second, considering the top graph in Fig. 5, for each application, there is an optimum number of processors beyond which increasing the number of processors does not bring benefits. This is because of two main reasons. First, in some cases, accurately determining the multiple modules to pre-compile is not easy (that is, the prediction is difficult). Second, inter-thread communications can sometimes offset the benefits coming from employing multiple processors. Note that one can make a similar observation from the bottom graph in Fig. 5 as well. However, this time, the main reason for poor scalability is the inherent difficulty in parallelizing the application in question.

V. CONCLUDING REMARKS

The goal of this paper is to hide the time spent in dynamic compilation by overlapping it with application execution. We implemented a dynamic compilation/linking infrastructure that compiles/links program modules based on external energy constraints. Our strategy hides most of the dynamic compilation time by predicting the next module to be executed and by pre-compiling it. Our experimental results obtained using eight applications are very promising.

REFERENCES

[1] V. Bala, E. Duesterwald, and S. Banerjia. Transparent dynamic optimization: The design and implementation of Dynamo. *Technical Report HPL-1999-78*, HP Laboratories, 1999.

[2] B. R. Buck and J. K. Hollingsworth. An API for runtime code patching. *Journal of High Performance Computing Applications, 14(4):317–329*, Winter 1994.

[3] M. Burke, J. Choi, S. Fink, D. Grove, M. Hind, V. Sarkar, M. Serrano, V. Shreedhar, H. Srinivasan, and J. Whaley. The Jalapeno dynamically optimizing compiler for Java. In *Proc. the ACM Java Grande Conference*, June 1999.

[4] M. Cierniak, G. Lueh, and J. Stichnoth. Practicing JUDO: Java under dynamic optimizations. In *Proc. the ACM Conference on Programming Language Design and Implementation*, June 2000.

[5] K. Ebcioglu and E. R. Altman. DAISY: dynamic compilation for 100% architectural compatibility. In *Proc. the International Symposium on Computer Architecture*, 1997.

[6] D. R. Engler. VCODE: a retargettable, extensible, very fast dynamic code generation system. In *Proc. the 23rd ACM Conference on Programming Language Design and Implementation*, 1996.

[7] B. Grant, M. Philipose, M. Mock, C. Chambers, and S. J. Eggers. An evaluation of staged run-time optimizations in DyC. In *Proc. Conference on Programming Language Design and Implementation*, May 1999.

[8] C. Krintz, D. Grove, V. Sarkar, and B. Calder. Reducing the Overhead of Dynamic Compilation. *Software-Practice and Experience, 31(8):717–738*, 2001.

[9] J. R. Larus and E. Schnarr. EEL: machine-independent executable editing. In *Proc. SIGPLAN Conference on PLDI*, 1995.

[10] B. P. Miller, M. D. Callaghan, J. M. Cargille, J. K. Hollingsworth, R. B. Irvin, K. L. Karavanic, K. Kunchithapadam, and T. Newhall. The Paradyn parallel performance measurement tools. *IEEE Computer, 28(11), 1995, pp. 37–46*.

[11] P. Unnikrishnan, G. Chen, M. Kandemir, and D. R Mudgett. Dynamic compilation for energy adaptation. In *Proc. the International Conference on Computer Aided Design*, San Jose, CA, November, 2002.

2006 Asia and South Pacific Design Automation Conference

9B-5

Functional modeling techniques for efficient SW code generation of video codec applications

Sang-Il Han* **   Soo-Ik Chae*   Ahmed. A. Jerraya**

*Department of Electrical Engineering,
Seoul National Univ., Seoul, Korea
{sihan,chae}@sdgroup.snu.ac.kr

**SLS Group, TIMA Laboratory
Grenoble, France
{sang-il.han,ahmed.jerraya}@imag.fr

Abstract–Architectures with multiple programmable cores are becoming more attractive for video codec applications because they can provide highly concurrent computation and support multiple video standards and a shorter time-to-market. To find an efficient SW code for the multiple core architecture for a video codec application, it is very important to easily explore the design space by generating a SW code automatically from its functional model.

We introduce Abstract Clock Synchronous Model (ACSM) for functional modeling of video codec applications. The ACSM can easily represent both parallelism and conditionals, which are common in video codec applications. By applying ACSM to an H.264 baseline decoder on single core architecture, we reduced the execution time and the number of external memory accesses by 32 % and 46 % respectively compared to traditional dataflow model.

I. Introduction

Current video codec applications require a higher performance architecture that can process more complex algorithms for higher resolution images. To find an architecture that satisfies this requirement with a limited resource, it is essential to use functional modeling that can exploit deep pipelines and parallelism and maximize re-use of the limited resource. Furthermore, it is important to use a programmable architecture that supports multiple video formats and newly emerging standards. To improve the design productivity, both SW and HW codes should be generated automatically if possible because the time-to-market is getting shorter. Therefore, automatic SW code generation from a functional model is one of the essential technologies in system design. In this paper we focus on a functional modeling method to generate an efficient SW code from the functional model for a video codec application.

Four properties of the video codec applications should be addressed in functional modeling [1].

1) **Computation intensive operations**, such as motion compensation, sub-pel interpolation, and DCT transform, which should be executed within timing constraints.
2) **Massive data transfer operations**, e.g. for motion estimation and compensation.
3) **Data dependent operations** according to various image modes and macroblock (MB) modes.
4) **Iterative execution** for sub-macroblock, macroblock, and frame levels, which requires large memory buffers.

To generate an efficient SW code for the video codec applications with these properties, functional model should support the following requirements.

1) Parallelism and pipeline should be exploited using specific parallel architectures in order to perform computation-intensive operations within timing constraints. Therefore, a functional model should enable to represent intra- and inter-iteration

dependencies explicitly to exploit parallelism and pipeline respectively.
2) Communications should be expressed explicitly and the sizes of data transfers should be predictable in the functional model in order to efficiently use burst data transfers and data pre-fetches that are essential for the video applications.
3) A functional model should support conditionals such as if-then-else structure in order to efficiently represent conditional computation and communication of video codec applications.
4) A functional model should represent communication buffers explicitly in order to minimize a memory cost by allocating a minimal size of memories and reusing the buffers.

In this paper we propose Abstract Clock Synchronous Model (ACSM), which is an extension of the Clocked Synchronous Model for RTL modeling [2]. The ACSM employs a coarser clock to compose functional blocks, which will be mapped onto HW blocks or SW functions on a specific CPU core. By using the coarser clock, the ACSM can represent parallelism and conditionals of video codec applications while the existing data-driven models and event-driven models have difficulty to express conditionals and parallelism respectively. First we explain the basics of ACSM for video applications. Then we compare ACSM with previous functional modeling methods in detail. We will show its efficiency by comparing it with conventional dataflow methods on a single core architecture in terms of performance, and communication bandwidth.

The rest of the paper is organized as follows. In Section II we explain ACSM, a proposed solution suitable for modeling video codec applications. In Section III ACSM is compared with other functional models in detail. In Section IV we present several experimental results to check the efficiency of ACSM, which is followed by the conclusions in Section V.

II. Abstract Clock Synchronous Model

A. Assumptions

Fig. 1 shows the overall steps of automatic SW code generation from functional model. Three key design steps are 1) building a functional model of a target application, 2) mapping of the functional model to a target architecture, and 3) generation SW code automatically from the mapping result. The mapping step is based on an evaluation function where the inputs are a functional model and a target architecture and the outputs are evaluation metrics such as performance, power and cost. Therefore, to generate efficient SW code from a functional model, the target architecture should be taken into consideration in building a functional model.

In this paper we assume that a target architecture is composed of processor (and HW) subsystems for image processing, global memory subsystems for image store, and an interconnection for communication between subsystems, as shown in Fig. 2. A processor (or HW) subsystem is composed of a processor (or HW IP), a local memory, an interconnection interface, and a local bus.

0-7803-9451-8/06/$20.00 ©2006 IEEE.

935

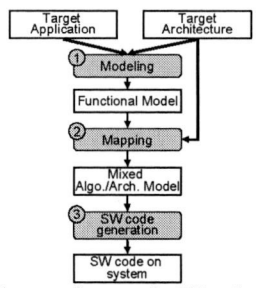

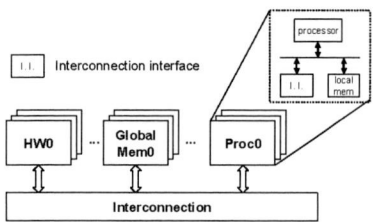

Fig. 1. The overall steps of automatic SW code generation.

Fig. 2. Generic target architecture.

Basically, the target architecture is a loosely coupled architecture, so that a processor (and HW) subsystem should copy necessary image data from its global memory subsystems to its local memory before processing it.

We limit the target applications of ACSM to macroblock-based video codecs, such as, MPEG-2, H.263, MPEG-4, and H.264 [1]. These video codecs combine inter-picture prediction to exploit temporal redundancy with transform-based codec of the prediction errors to exploit spatial redundancy. Fig. 3 shows a block diagram of an H.264 decoder that receives an encoded video bit stream from a network or a storage device and produces a frame sequence. Each frame is reconstructed by iterative executions of macroblock-level functions such as entropy decoding, inverse zigzag scan, inverse quantization, inverse transform, motion compensation, and deblocking filter.

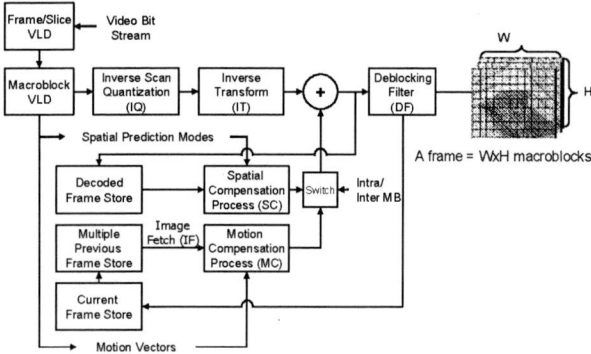

Fig. 3. Target application example: H.264 decoder block diagram.

In this paper, we focus on the functional modeling step for efficient SW code generation of video codec application.

B. Abstract clock synchronous model

The Clocked Synchronous model (CSM), which has been used in designing the hardware for clocked synchronous circuits, is

based on the **clock synchrony hypothesis** [2]: There is a global clock signal controlling the start of each computation in the system, and communication takes no time, and computation takes one clock cycle. This assumption makes it possible to describe the functionality of a circuit deterministically independent of the detailed timing of the gates in the circuit by separating each combinational logic block from others with clocked registers. In other words, the CSM is used to exploit the orthogonalization between functionality and timing in the synchronous design methodology [4]. In this paper we extend the CSM to the ACSM by using an abstract clock of larger granularity that is suitable for system-level design.

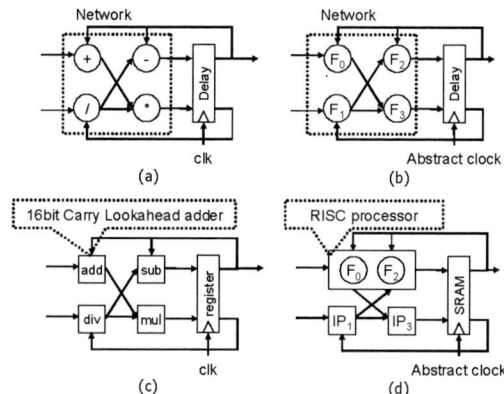

Fig. 4. Examples of (a) clocked synchronous model, (b) abstract clock synchronous model and their implementations (c-d).

Fig. 4 (a) shows an example of CSM for RTL modeling with a clock and Fig. 4 (b) shows an example of ACSM for functional modeling with an abstract clock. A CSM is composed of a network of combinational gates and delays. It is implemented by low level hardware as shown in Fig. 4 (c). For example, an addition and a delay in the CSM can be implemented by a 16bit carry lookahead adder and a register respectively. However, an ACSM is composed of a network of state-less functions and delays. It may be implemented by a combination of hardware and software as shown in Fig. 4 (d). For example, a function and a delay in the ACSM can be implemented by a SW code on RISC processor and an SRAM respectively. The major difference between the two models is the granularity of the clock and the components. Note that cyclic paths must contain at least one delay in both models.

C. Tagged signal model of ACSM

In order to describe ACSM and compare it with the existing functional modeling methods, we follow the tagged-signal model introduced in [3]. Given a set of *values V* and a set of *tags T*, an *event* e has a tag t and a value v, i.e. $e = (t, v) \in T \times V$. A *signal* s is a set of events. The tags are used to model time, precedence relationships, and synchronization points. The values represent the operands and results of computation.

In the ACSM, it is necessary to represent intra- and inter-iteration dependencies explicitly in order to exploit parallelism and pipeline as above mentioned. To do this, we use a set of tags $T \in \omega \times \omega$, where ω is the set of nonnegative integers with the usual numerical order. The set of the first components of all events is totally ordered. The first component is used to model data precedence between operations across an abstract clock boundary, i.e. inter-iteration dependencies. The set of the second components of all events is partially ordered. The second component is used to model data precedence between functions within an abstract clock interval. The set of all event tags is partially ordered because of the second component.

Let $e_{i,j}$ denote an event where the tag t is (i, j) and the value v is $e_{i,j}$. Given two events $e_{i,j}$ and $e_{n,m}$, $e_{i,j} < e_{n,m}$ if (i, j) < (n, m). Fig. 5 shows an example of precedence relationships between events in an ACSM. F_1 imposes a precedence constraint such that $e_{i,j1} < e_{i,j3}$. Delay imposes a precedence constraint such that $e_{i,j5} < e_{i+1,j2}$. But there is no precedence relationship between $e_{i,j3}$ and $e_{i,j4}$, and $e_{i,j1}$ and $e_{i+1,j0}$. The partially ordered events give rise to parallel and pipelined execution of functions in an ACSM

A data type D can be extended into a data type $D_\perp$ by adding the special value $\perp$ to model the absence of a value at a certain tag. Absent events are used to model the outputs of unselected operations in if-then-else structures in the ACSM.

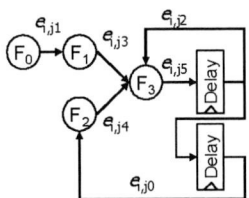

Fig. 5. An example of ACSM with events.

D. Abstract clock for video codec application

In the ACSM, it is important to select an abstract clock suitable for explicitly representing parallelism, communication, and communication buffers for efficient design space exploration. Video decoding (and encoding) processes are basically composed of four hierarchical iterations: sub-macroblock level, macroblock level, slice level, and image level. Each of these iteration indices can be a candidate for the abstract clock. Two iteration indices of slice and image levels are too coarse to represent parallelism, and communication explicitly between the essential functions of video codec applications such as motion estimation, motion compensation and inverse transform. Using the iteration index of 4x4 sub-macroblock level requires representing irregular delays due to the data dependencies among 4x4 blocks. For example, for a QCIF image as shown in Fig. 6, the delay between block 5 and its upper block is 166 delay units, but that between block 13 and its upper block 7 is 6 delay units. Therefore, we use the macroblock index as an abstract clock in an ACSM because the granularity of the macroblocks is good enough to represent parallelism, communication, and communication buffer explicitly and the decoding order of macroblocks as shown in Fig. 6 (a) is regular so that we can represent easily the data dependency.

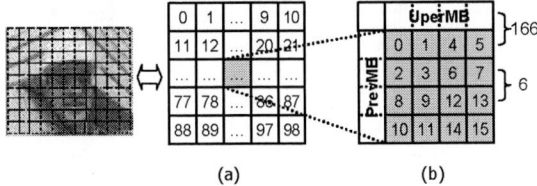

Fig. 6. Decoding orders of macroblock within an image (a), and 4x4 blocks within a macroblock (b).

E. Basic components and firing rules of ACSM

We decided to use Simulink [5] for a video codec application because it provides simulation and modeling environment of discrete-time systems that are sufficient to build an ACSM. It also includes Real-Time Workshop (RTW) [6], which generates a C code automatically from a Simulink model.

Fig. 7 shows basic components of ACSM that are expressed easily in Simulink.

- **Block** (process): A block, as shown in Fig. 7 (a), maps n input events on m output events: $(o_1, ..., o_m) = F_0(i_1, ..., i_n)$. It corresponds to S-function or pre-defined block with inherent sample rate in Simulink.

- **Delay**: A delay, as shown in Fig. 7 (b), represents that its output is delayed from its input by k abstract clock cycles. It corresponds to discrete delay in Simulink.

- **Arc** (edge): An arc carries events from one output port of a block or a delay to one or more input ports of one or more blocks and/or delays as shown in Fig. 7 (c). It corresponds to connecting line in Simulink.

We also defined two kinds of subsystems that are composed of blocks, delays, arcs, and other subsystems.

- **If-action subsystem (IAS)**: An IAS, as shown in Fig. 7 (d), represents an if-then-else structure. An IAS is enabled when its control input port, which is connected to an *if/else* block, has a present event, i.e. not absent. If an IAS is not enabled, its output ports have absent events. All output ports must be connected to a *merge* block and only one of them can have a present event at a time. It corresponds to "If-action subsystem" in Simulink.

- **For-iterator subsystem (FIS)**: A FIS, as shown in Fig. 7 (e), represents a for-loop structure. It is used to describe sequential or parallel repeated executions of blocks where the number of repetitions is known. It corresponds to "For-iterator subsystem" in Simulink. A FIS usually includes *Demux*s and *Mux*s. A *Demux* divides an event into several (sub-)events. A *Mux* integrates several (sub-)events into an event.

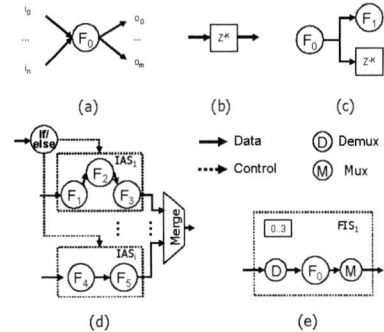

Fig. 7. Basic components in ACSM.

A block consumes one event from each input port and produces one event to each output port. This action is called *firing* and takes place under certain conditions called *firing rules*. A block except *merge* blocks in Fig. 7 (d) is fired when all input events are present. A *merge* block is fired when one of input events are present. A block except *if/else* blocks in Fig. 7 (d) produces one present event to each output port when fired. A *if/else* block produces one present event to one of output ports when fired.

F. An example of ACSM

Fig. 8 shows a simplified ACSM in Simulink for an H.264

baseline profile decoder. It includes two paths. A path consists of macroblock VLD (MB VLD), 8x8 sub-macroblock inverse quantization (8x8 IQ), and inverse quantization (8x8 IT) to compute a residual image from a video bit stream. The other path consists of MB VLD, spatial compensation (SC) or motion compensation (MC) from the current frame or previous frames. If four neighbor 4x4 sub-macroblocks, e.g. 0, 1, 2, 3 blocks in Fig. 6(b), have the same motion vector, it is possible to fetch less image data from previous frames (8x8 IF in IAS_2) and manipulate the image data more efficiently (8x8 MC in IAS_2) compared to 4x4 sub- macroblock based motion compensation (4x4 IF and 4x4 MC in IAS_3) by the elimination of common computation and communication. The ACSM consists of 83 S-functions, 286 arcs, 21 *if/else* blocks, 43 IASs, 5 FISs, and 24 delay blocks.

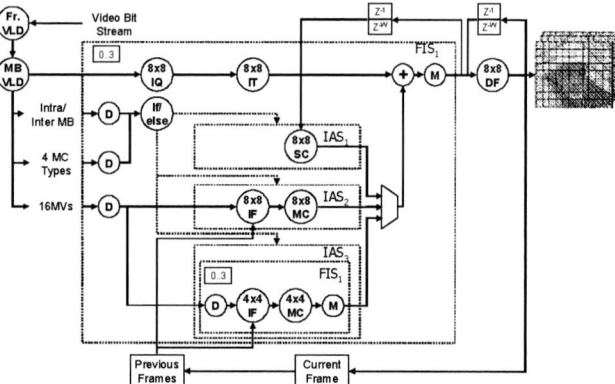

Fig. 8. A simplified ACSM of H.264 decoder in Simulink.

IV. Comparison with previous modeling styles

In order to obtain an efficient SW code from a functional model, it is important to select an appropriate functional modeling style according to the property of an application domain. In the following subsections, we will compare ACSM with the most popular previous modeling styles in respect of building functional models of video codec applications. The capabilities and the weakness of these models will be analyzed with regards to the requirements stated in section I.

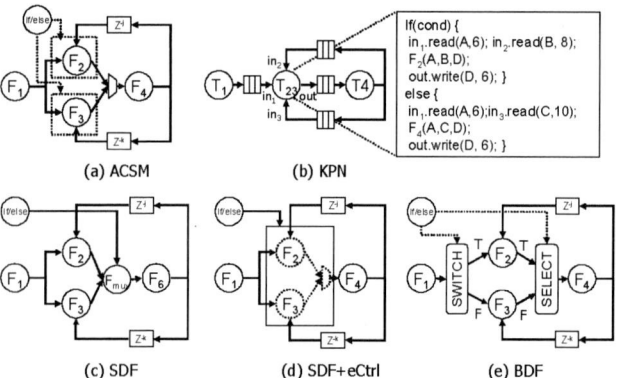

Fig. 9. Functional model examples of different styles.

A. Kahn Process networks

In the Kahn process networks [7], concurrent processes communicate through one-way FIFO channels with unbounded capacity. This means that writes to the channel always succeed immediately, while reads block until there is sufficient data in the channel to satisfy them. In particular, a process cannot test an input channel for the availability of data and then branch conditionally.

Fig. 9 (b) shows a KPN example corresponding to an ACSM model as shown in Fig. 9 (b). A process in a KPN model has a computation code mixed with a communication code. Therefore, The KPN doesn't support explicit and predictable communication that is required to efficiently use burst data transfers and data pre-fetchs. It also requires context switching to deal with blocked processes. To reduce the overhead of context switching, it is necessary to increase the granularity of process.

B. Synchronous Dataflow

In the Synchronous Dataflow (SDF) model [8], a process (actor) is fired when it has sufficient tokens (events) on its input ports. When an actor executes, it consumes a positive fixed number of data tokens from each input port, and produces a positive fixed number of tokens to each output port. If a SDF model is consistent, it is possible to execute the SDF model in bounded memory without context switching.

SDF cannot represent explicit conditional such as if-then-else structure, which is required for video codecs, because it doesn't allow absent token. To express an if-then-else structure, redundant computation and communication may be required as shown in Fig. 9 (c). According to the result of "if/else" actor, either F_2 or F_3 does not need to be executed because just only one of the outputs is used. Similarly, either communication between F_2 and Z^{-j} or that between F_3 and Z^{-k} is redundant.

Fig. 9 (d) shows an alternative to express an if-then-else structure with embedded controls in a SDF model, where blocks F_2, F_3, and an if-then-else structure are merged into a single block. It can remove unnecessary execution of F_2 or F_3, but redundant communication of F_2 with Z^{-j} or F_3 with Z^{-k} is still required. Furthermore, mapping F_2 and F_3 onto two different processors or HWs is not possible so that its design space is restricted. Therefore, it does not support explicit conditional well.

C. Boolean Dataflow

Boolean Dataflow (BDF) [9] is an extension of SDF to support conditionals. In the BDF model, an if-then-else structure is modeled with two actors, SWITCH and SELECT. The SWITCH actor reads one token from the control input port, and depending on whether the value of the control token is true or false, routes the input either to the output port marked T, or to the output marked F. It also produces an absent token to the other output port.

However, it is not guaranteed whether the execution of a BDF model is completed in a finite time or whether it requires a bounded memory [9]. In Fig. 9 (e), the tokens on the arc between F_2 and Z^{-j} and those on the arc between F_3 and Z^{-k} are accumulated because the execution ratios of F_2 and F_3 to F_4 are different depending on the control token produced by the if/else actor. To solve this problem, it is necessary to insert additional SWITCH actors both between F_2 and Z^{-j} and between F_3 and Z^{-k}. Building a BDF model of a video codec application at fine granularity will require many SWITCH actors, which is more error prone.

D. Synchronous model

Synchronous model (SM) [10] used in Esterel [11] and Lustre [12] is based on the perfect synchrony hypothesis assuming that the reaction to each set of the inputs is considered to be instantaneous. In the SM, a process is fired when it has at least one event on its input ports. When a process is fired, it can test absent events on its input ports and produce absent events on its output ports.

The synchronous assumption simplifies system specification and verification. However, difficulties arise especially for video codec applications if the target architecture is a distributed multi-core system, because it is very expensive to maintain a global clock for testing and producing absent events over a distributed system. The fully synchronous implementation based on time-triggered architecture [13] must be conservative, forcing the global clock to run as slow as the slowest computation and communication process. Therefore, SM is not suitable for video codec applications, which have large variations in computation and communication.

E. Analysis: comparing with ACSM

The ACSM overcomes all the restriction of the above mentioned models while providing the same scheduling facilities.

In data-driven models of KPN and SDF, the absent token is not defined. Therefore they have some difficulties in expressing an explicit if-then-else structure. In the BDF model the token overflow problem comes from the definition of the absent token without a global clock. However, the ACSM uses an abstract clock for the well-defined absent token. In the ACSM, an event on an arc is updated at every abstract clock tick, so there is no token overflow problem.

Although the dataflow model has difficulties in representing conditionals, it can express several valid schedules of a multi-rate algorithm with only a model. In a SDF model as shown in Fig. 10 (a), a schedule that requires minimal buffers is $F_0F_0F_1F_2F_2F_0F_1F_2F_2$ and another schedule with loop structures is $(3F_0)(2F_1(2F_2))$. Contrary to the dataflow model, the ACSM can express only one schedule with a model. Fig. 10 (b) shows an ACSM example that expresses a schedule that is $(3F_0)(2F_1(2F_2))$. However, it is possible to find other schedules by loop transformation techniques such as loop unrolling and loop split.

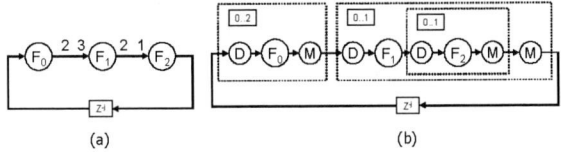

Fig. 10. Multi-rate modeling with SDF (a) and ACSM (b).

In the SM, there is no restriction on producing and testing absent events. It can cause a causality problem and requires a global clock to solve the problem. However, the ACSM allows only restricted absent events to express if-then-else structures. In other words, it is sufficient for a distributed multi-core system to replace absent events of unselected IAS outputs with present events only if an if-then-else structure is implemented over different processors.

V. Experiments

We performed several experiments to check the efficiency of the ACSM compared to the dataflow model, assuming that target architecture consists of a processor subsystem for image processing and an external global memory for image storage, as shown in Fig. 11, where a processor subsystem consists of a Tensilica Xtensa processor [14] with a default configuration, a local memory, and a DMA with a memory interface. We assumed that image fetch processes are executed in the DMA module.

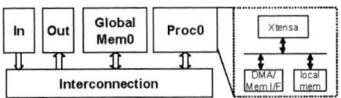

Fig. 11. Evaluation architecture.

The target application is an H.264 baseline decoder and its input bitstream is the Foreman sequence of QCIF format, which was encoded with QP=28 and IntraPeriod=5 when all MC block sizes were enabled.

Fig. 12 shows our experimental procedure. In the modeling step, we made three Simulink models from H.264 decoder reference C code as shown in Table I. In the Simulink model equivalent to a SDF model, an if-then-else structure is expressed by using a MUX block as shown in Fig. 9 (c) and the processing block size is 4x4. In Simulink model equivalent to an ACSM, an if-then-else structure is expressed by an *if/else* block and IASs and the processing block size is 4x4. In the optimized ACSM, the processing block sizes are either 4x4 or 8x8. For 8x8 block, an additional IAS is necessary as shown in Fig. 8. We limited the block size of the SDF model to 4x4 because the additional IAS causes redundant computation as explained in IV-B.

Then, we used Real-Time Workshop (RTW) to generate SW codes from the three different Simulink models. In the final step, we used Xtensa gdb and gprof to measure the execution cycles and the external memory access of the SW codes. We also added the reference C code to the test sets.

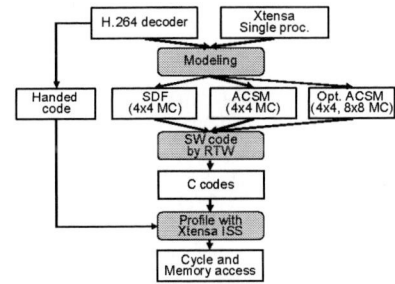

Fig. 12. Experimental procedure.

TABLE I four test sets with different configurations

Name	description
SDF	SDF model as shown in Fig. 9 (c)
ACSM	Proposed ACSM model as shown in Fig. 9 (a)
Optimized ACSM	Proposed ACSM model with 8x8 block size as shown in Fig. 8
Handed code	Reference C code

Fig. 13 shows the execution time and the number of external memory accesses for the four different configurations. According to the experimental results, the ACSM reduces both the execution time and the number of external memory accesses by 29 % and 22 % respectively compared to those of the SDF model. Furthermore, the optimized ACSM reduces them by 32 % and 46 % respectively compared to those of the SDF model.

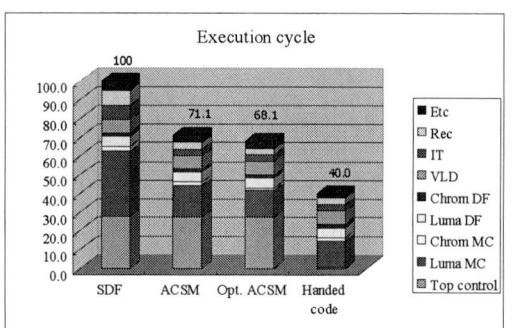 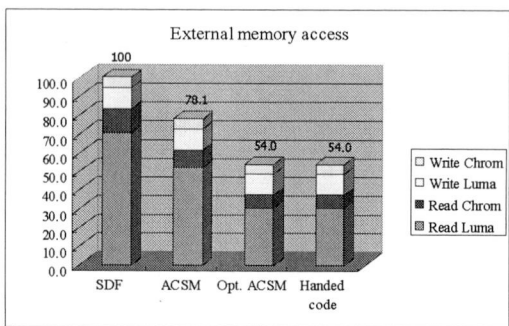

Fig. 13. Relative execution times and external memory accesses of different configurations.

According to the experimental results, it is necessary to express conditionals such as if-then-else structures explicitly for building a functional model of a video codec application. The optimized ACSM code from RTW requires 41% more execution time compared to the handed code because it includes many redundant copy operations.

Even if a SDF model with embedded controls can remove redundant computation, it cannot remove the external memory access as explained in section IV. It also limits the design space exploration due to increasing the granularity of blocks.

For the BDF model of the H.264 baseline decoder, 138 SWITCHes were required to resolve the token overflow problem mentioned in section IV. The number of the SWITCHes in the BDF model is larger than that of S-functions in the ACSM model, which means that although the BDF supports conditionals, it is not suitable in building a functional model for a video codec application because the BDF functional models get too complex.

VI. Conlusions

In this paper, we explained a functional modeling method for generating an efficient SW code from a functional model for video codec applications. It is based on ACSM, which is an extension of the clocked synchronous model by employing the macroblock index as an abstract clock. The ACSM can express conditionals easily by allowing absent events with the global abstract clock. It can also express parallelism and pipeline easily by partially ordered intra- and inter-dependencies. Therefore, the ACSM is suitable for functional modeling of video codec applications that require both parallelism and conditionals.

Experimental results with Simulink and RTW showed that the SW code generated from an ACSM of H.264 decoder is improved by up to 32% and 46 % compared to its SDF model in terms of the number of execution cycle and the number of external memory access respectively. We found that a BDF model of H.264 decoder requires many additional SWITCHes to express if-then-else structures. So the BDF model gets more complex. Therefore, the ACSM is more effective in building functional model for video codec applications because emerging standards such as H.264 require complex data-dependent operations.

For a H.264 baseline decoder, the SW code generated with the ACSM by RTW requires longer execution time and larger buffer memories compared to those of the handed code. We are in the process of developing a more efficient SW generation tool. Furthermore, we will extend this methodology for multi-processor systems because the current version of RTW can generate only a SW code for single-processor systems.

Acknowledgements

The authors are grateful to Xavier Guerin, Paul Amblard, Frédéric Pétrot and Nacer-Eddine Zergainoh from TIMA laboratory for their inputs on this paper. This work was supported by ITSoC Project and Brain Korea 21 Program.

References

[1] "Advanced video coding for generic audiovisual services," Int. Telecommum. Union-Telecommum. (ITU-T) and Int. Standards Org./Int. Electrotech. Comm. (ISO/IEC) JTC 1, Recommendation H.264 and ISO/IEC 14496-10 (MPEG-4) AVC, 2003

[2] Axel Jantcsh, "Modeling Embedded Systems and SoCs – Concurrency and Time in Models of Computation," Morgan Kaufmann, 2001.

[3] E. A. Lee and A. Sangiovanni-Vincentelli, "A Framework for Comparing Models of Computation," IEEE Trans. On CAD of Integrated Circuits and Systems. pp 1217-1229, December 1998.

[4] K. Keutzer et al, "System-level design: Orthogonalization of concerns and platform-based design," IEEE Trans. On CAD of Integrated Circuits and Systems.

[5] Simulink,http://www.mathworks.com/

[6] Real-Time Workshop, http://www.mathworks.com/

[7] G. Kahn and D.B. MacQueen, "Coroutines and Networks of Parallel Processes," In B. Gilchrist, editor, Information Processing 77, Proceedings, pp 993-998, Toronto, Canada.

[8] Lee, E. A., Parks, T. M. (1995), "Dataflow process networks," Proceedings of the IEEE83(5), 773-801

[9] J.T. Buck, "Scheduling Dynamic Dataflow Graphs with Bounded Memory using the Token Flow Model," PhD thesis, University of California, EECS Dept., Berkeley, CA, 1993. Technical Memorandum UCB/ERL M93/69

[10] Benveniste, A. et al, "The synchronous languages 12 years later," Proc. of the IEEE , Volume: 91 Issue: 1 , Jan 2003

[11] Gerard Berry, "The Foundations of Esterel," Proof, Language and Interaction: Essays in Honour of Robin Milner, G. Plotkin, C. Stirling and M. Tolfte, editors, MIT press, 1998

[12] N. Halbwachs, P. Caspi, P. Raymond and D. Pilaud. "The synchronous dataflow programming language Lustre,". Proc. of the IEEE, vol. 79, nr. 9. September 1991.

[13] H. Kopetz, "The time-triggered architecture," in Proc. First International Symposium on Object-Oriented Real-Time Distributed Computing (ISORC'98), Kyoto, Japan, 1998.

[14] Tensilica Xtensa V, http://www.tensilica.com/html/xtensa_v.html

Convergence-Provable Statistical Timing Analysis with Level-Sensitive Latches and Feedback Loops

Lizheng Zhang, Jengliang Tsai, Weijen Chen, Yuhen Hu, Charlie Chung-Ping Chen
ECE Department, University of Wisconsin, Madison, WI53706-1691, USA
Email:{lizhengz,weijen,jltsai}@cae.wisc.edu, {hu,chen}@engr.wisc.edu

ABSTRACT

Statistical timing analysis has been widely applied to predict the timing yield of VLSI circuits when process variations become significant. Existing statistical latch timing methods are either having exponential complexity or unable to treat the random variable's *self-dependence* caused by the coexistence of level-sensitive latches and feedback loops.

In this paper, an efficient iterative statistical timing algorithm with provable convergence is proposed for latch-based circuits with feedback loops. Based on a new notion of *iteration mean*, we prove that the algorithm converges unconditionally. Moreover, we show that the converged value of iteration mean can be used to predict the circuit yield during design time. Tested by ISCAS'89 benchmark circuits, the proposed algorithm shows an error of 1.1% and speedup of 303× on average when compared with the Monte Carlo simulation.

1. INTRODUCTION

With ever decreasing feature size of nano-scale integrated circuits, the variation of manufacturing parameters becomes more and more significant and must be considered during design. [1] Classical corner-based timing analysis produces timing predictions that are often too pessimistic and grossly conservative because we have only few chance to have parameters of all gates working on their corner values. Statistical static timing analysis (SSTA) that characterizes time variables as statistical random variables offers a better approach for more accurate and realistic timing prediction.

Correlated time variables due to spatial correlation or reconvergence fanout present themselves as a major challenge when applying SSTA to complicated circuits. Such correlations have been studied extensively in literatures [2–9]. In particular, an *extended canonical time model* has been proposed in [6] to represent these correlations in a compact form to be preserved during propagation of the timing random variables.

However, most of the existing SSTA methods only com-

pute the time variable distribution for combinational circuits. Although with some minor modifications, the existing SSTA method may be extended to deal with flip flop based sequential circuits, we have yet to find an effective SSTA method for sequential circuits consisting of level sensitive latches and feed-back loops. Specifically, with the presence of feedback loop, the timing-wise transparent level-sensitive latches may cause timing random variables to be *self-dependent*. In other words, a timing random variable will be dependent on a time variable with the same name, but instantiated in the previous iteration. Such a *self-dependence* presents itself as a new type of correlation that is caused by the coexistence of latches and feedback loops. An example of self-dependence is illustrated in figure 1.

Previously, a SSTA method for latch-based pipeline design have been proposed [10]. However, the issue of self-dependence is not addressed. In [11], a structural method is proposed to deal with the feedback loops by applying graph sorting algorithms. However, the computation complexity of these algorithms may grow exponentially [11].

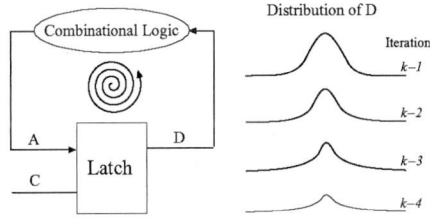

Figure 1: A simple latch circuit with a feedback loop and the possible divergence of the departure time distribution.

In classical deterministic timing analysis, an iterative latch timing algorithm, i.e. the SMO algorithm, has been proposed [12–14] to deal with the self dependence. In order to generalize the SMO algorithm to handle random time variables, one faces a convergence problem that can be briefly explained as follows: In deterministic timing analysis, each time variable assumes a deterministic value. Convergence is guaranteed if each time variable is bounded within a predefined, finite range. However, with SSTA, each time variable is modeled with a mean value and a standard deviation. Even the mean value can be bounded, the corresponding variance may still diverge. This is illustrated in the figure 1.

In this paper, we present a solution to such a convergence problem. Conceptually, we argue that the notion of *circuit convergence* should be differentiated from the overall *algo-*

Permission to make digital or hard copies of all or part of this work for personal or classroom use is granted without fee provided that copies are not made or distributed for profit or commercial advantage and that copies bear this notice and the full citation on the first page. To copy otherwise, to republish, to post on servers or to redistribute to lists, requires prior specific permission and/or a fee.
ASP-DAC 2005, Jan. 24-27, 2006, Yokohama, Japan
Copyright 2006 ACM X-XXXXX-XX-X/XX/XX ...$5.00.

rithm convergence. Therefore, even the actual data arrival time in the circuit may diverge, the convergence of the algorithm will not necessarily be affected. We proposed a novel SSTA algorithm, *StatITA*, for latch-based circuits with feedback loops based on a quantity of *iteration mean* which is the average latest data arrival time per iteration. We prove that *StatITA* converges unconditionally after sufficient number of iterations. Moreover, we show that the converged value of the iteration mean can be used to predict the circuit yield.

The rest of the paper is organized as following: Section 2 presents preliminary of the latch timing analysis and the graph model of the circuit with feedback loops; Section 3 introduces the theory of our iterative timing method; Section 4 summarizes the *StatITA* algorithm; Section 4 presents the C/C++ implementation and testing results; Section 5 gives the conclusions.

2. LATCH TIMING PRELIMINARY

$T = T_i^0 + T_i^1$: clock cycle time
T_i^0: clock low time at latch i
T_i^1: clock high time at latch i
H_i: hold time of latch i
S_i: setup time of latch i
C_i: rising clock edge arrival time at i^{th} latch
a_i: earliest data arrival time at i^{th} latch
A_i: latest data arrival time at i^{th} latch
d_i: earliest data departure time at i^{th} latch
D_i: latest data departure time at i^{th} latch
δ_{ij}: minimum combinational delay from latch i to j
Δ_{ij}: maximum combinational delay from latch i to j
$\lambda_{ij} = \delta_{ij} - T$: adjusted minimum delay
$\Lambda_{ij} = \Delta_{ij} - T$: adjusted maximum delay

Y: total circuit yield
N: total number of latches in the circuit
s_i: setup time violation at i^{th} latch
h_i: hold time violation i^{th} latch
s_c: critical setup time violation of the circuit
h_c: critical hold time violation of the circuit

---

p_m: number of latches in the feedback loop m
G_m: cycle mean of the feedback loop m
G_c: critical cycle mean of the circuit
O_i^k: iteration mean of latch i at k^{th} iteration

Figure 2: Notation used in this work

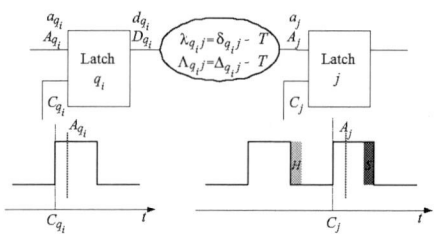

Figure 3: Latch Timing Diagram

It is common for high-end VLSI circuits to have feedback loops. If level-sensitive latches are used as the sequential elements, iterative methods will be applied for circuit timing due to the possible self-dependence issue.

Figure 3 shows a latch j and one of its input latch q_i that has combinational output paths to latch j. All latches are assumed to be active-high, but no generality is lost since no restriction is posted on clocks. So the data departure time of latch q_i at the kth iteration will be:

$$d_{q_i}^k = max(a_{q_i}^k, C_{q_i}) \qquad (1)$$
$$D_{q_i}^k = max(A_{q_i}^k, C_{q_i}) \qquad (2)$$

On the other hand, the data arrival time of latch j will be decided by its all input latches $q_1, q_2, ...$ as:

$$a_j^{k+1} = min(d_{q_1}^k + \lambda_{q_1 j}, d_{q_2}^k + \lambda_{q_2 j}, ...) \qquad (3)$$
$$A_j^{k+1} = max(D_{q_1}^k + \Lambda_{q_1 j}, D_{q_2}^k + \Lambda_{q_2 j}, ...) \qquad (4)$$

To make the circuit free from delay faults, the setup and hold time constraints must be satisfied at any latch $j = 1, 2, ..., N$ after sufficient iterations:

$$h_j^\infty = (C_j - T_j^0 + H_j) - a_j^\infty \le 0 \qquad (5)$$
$$s_j^\infty = A_j^\infty - (C_j + T_j^1 - S_j) \le 0 \qquad (6)$$

where s_j^∞ and h_j^∞ are the setup and hold time violations for latch j after sufficient iterations. It is more convenient to define the *critical setup time violation, s_c^∞* and *critical hold time violation, h_c^∞* as:

$$h_c^\infty = max(h_1^\infty, h_2^\infty, ..., h_N^\infty) \qquad (7)$$
$$s_c^\infty = max(s_1^\infty, s_2^\infty, ..., s_N^\infty) \qquad (8)$$

with which the setup and hold time constraints can be expressed compactly as:

$$h_c^\infty \le 0 \quad and \quad s_c^\infty \le 0 \qquad (9)$$

The above discussion, although it is intended to deal with the deterministic timing analysis, is also applicable when process variations are considered except that all time variables involved will become random variables.

2.1 Circuit Convergence

The major concern for an iterative latch timing method is the *circuit convergence*:

Definition 1. *A latch-based sequential circuit with feedback loops is said to* **converge** *during timing iterations if and only if both the latest and earliest data arrival times at the input of every latch are finite values after infinite number of iterations.*

According to the monotonicity of time variables involved in the iterative timing analysis, [13], it is impossible to have data arrival times, either the latest or the earliest, to be finite but oscillating among several values. So if a circuit converges as defined above, all of its time variables will converge to a fixed value or, in statistical case, to a fixed distribution.

Following theorem can simplify our convergence discussion by just focusing on the latest data arrival time only:

Theorem 1. *A latch-based circuit with feedback loops will converge during timing iterations if and only if every latest data arrival time in the circuit has an upper bound.*

PROOF. According to equations (1) and (2), both latest and earliest data departure time of any latch q_i will be lower bounded by the clock arrival time: $d_{q_i}^k \ge C_{q_i}$ and $D_{q_i}^k \ge C_{q_i}$.

So the latest and earliest data arrival times latch j will also have a finite lower bound:

$$a_j^{k+1} \geq min(\lambda_{q_1 j} + C_{q_1}, \lambda_{q_2 j} + C_{q_2}, ...)$$
$$A_j^{k+1} \geq max(\Lambda_{q_1 j} + C_{q_1}, \Lambda_{q_2 j} + C_{q_2}, ...)$$

Obviously, the earliest data arrival time of every latch is always upper bounded by the latest data arrival time of the same latch. So if all the latest arrival times in the circuit have upper bound, then circuit will converge as defined above. This proves the sufficiency. The necessity is trivial according to the definition of the convergence. □

2.2 Reduced Timing Graph

Timing iterations will be done at each latch in the circuit. To illustrate such iterations graphically, the entire circuit is partitioned into two parts: latches and combinational feedback sub-circuit. A *reduced timing graph*, $\{V, E\}$, is then constructed to represent the original circuit: latches are modeled by nodes of $n_i \in V$ and the combinational feedback sub-circuit is abstracted as directed edges of $e_{ij} \in E$ with weight of the adjusted maximum delay Λ_{ij}.

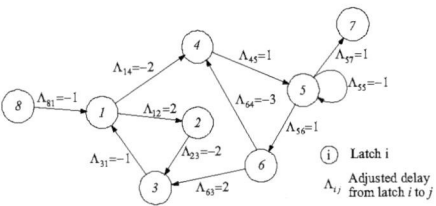

Figure 4: The reduced timing graph of an example circuit

A simple example of such reduced timing graph is shown in figure 4 where a circuit with 8 latches are modeled.

3. ITERATIVE TIMING THEORY

The main difficulty for iterative latch timing, as mentioned before, is the existence of feedback loops in the reduced timing graph. Every loop m with p_m latch nodes in it will have a *cycle mean(G_m)* defined as the average edge weight in the loop:

$$G_m = \frac{1}{p_m} \sum_{e_{ij} \in m} \Lambda_{ij} = \frac{1}{p_m} \sum_{e_{ij} \in m} \Delta_{ij} - T$$

and p_m is usually called *cycle length*.

For example, in the reduced timing graph shown in figure 4, latch nodes $1 \rightarrow 2 \rightarrow 3 \rightarrow 1$ will form a loop with length of 3, and the cycle mean of this loop is $G = (\Lambda_{12} + \Lambda_{23} + \Lambda_{31})/3 = -1/3$.

There will usually many loops existing in the reduced timing graph. Among them, the loop with the most positive cycle mean is with the most importance:

Definition 2. *The **critical cycle mean** of the reduce timing graph, G_c, is the larger value between 0 and the largest cycle mean among all possible loops:*

$$G_c = max(0, G_1, G_2, ...) \geq 0 \qquad (10)$$

where $G_1, G_2, ...$ are cycle means for all loops $1, 2, ...$ in the reduced timing graph.

For example, there are totally 4 loops in the example circuit shown in figure 4: loop $1(1 \rightarrow 2 \rightarrow 3 \rightarrow 1)$, loop $2(1 \rightarrow 4 \rightarrow 5 \rightarrow 6 \rightarrow 3 \rightarrow 1)$, loop $3(4 \rightarrow 5 \rightarrow 6 \rightarrow 4)$ and loop $4(5 \rightarrow 5)$. The cycle means of these loops are $G_1 = -1/3$, $G_2 = 1/5$, $G_3 - 1/3$ and $G_4 = -1$. So the critical cycle mean is $G_c = max(0, -1/3, 1/5, -1/3, -1) = 1/5$.

The critical cycle mean, G_c, revealed in later sections, is one of the most important circuit parameters determining the yield of the circuit with feedback loops. It is noticeable the similarity between the critical cycle mean G_c and the well-known concept of *maximum cycle mean(MCM)* in general graph theory. In deterministic cases, efficient algorithms are available to compute the MCM [15,16]. But these algorithms can not be directly applied when G_c becomes a random variable because of process variations. We here, instead, propose to compute G_c with an iterative method using the key idea of *iteration mean*:

Definition 3. *At every iteration k, each latch node i in the reduced timing graph will have an **iteration mean** defined as the latch's average latest data arrival time per iteration:*

$$O_i^k = \frac{A_i^k}{k+1} \qquad (11)$$

3.1 Graphical Imitation of Iterative Timing

Graphically, the update of the latest data arrival times at all latches, equations (2) and (4), can be imitated by one step of simultaneous "hop" of time variables at all nodes along all edges in the reduce timing graph with the following rules:

1. If a time variable hops along an edge, the edge weight is added.

2. If a time variable hops into a node, it will continue hopping only if it arrives later than the node's clock. Otherwise, it will "*die*" and a new time variable with the value of the clock arrival time at the node starts hopping.

3. If a time variable hops out of a node with multiple output edges, then multiple "*clones*" of the time variable will hop along all output edges.

4. If multiple time variables hop towards a node simultaneously, only the one with largest value will hop into the node.

It is clear that a time variable may not always hop along a given loop m in the reduce timing graph because it could possibly die. So it is meaningful to pickup those loops that are actually being followed by time variables and call them *timing loops*.

Definition 4. *If a time variable starts hopping at a node and it comes back to the same node later, then the loop through which the time variable passes is called a **timing loop**.*
*The relationship between a latch node and a timing loop can be one of the following three cases: (1) A node is **within** a timing loop if it is a node member of the timing loop; (2) node is **dominated** by a timing loop if it doesn't belong to the timing loop but the time variable hopping into it originates in the timing loop; (3) a node is **independent** on a timing loop if it is neither within nor dominated by the timing loop.*

For example, in figure 4, if the loop $1 \rightarrow 4 \rightarrow 5 \rightarrow 6 \rightarrow 3 \rightarrow 1$ becomes a timing loop, then node 8 will be independent to the timing loop and node 7 will be dominated by the timing loop.

3.2 Compute G_c from Iteration Mean

To prove the convergence of the iterative mean, we first prove a theorem which is valid for any loop in the reduce timing graph.

Theorem 2. *If there is a loop, m, with cycle mean of G_m and length of p_m, in the reduced timing graph, then for any iteration index of $k \geq p_m$ and any node $m_i (i = 1, 2, ..., p_m)$ in the loop, the latest data arrival time will satisfy the following inequality:*

$$A_{m_i}^k \geq A_{m_i}^{k-p_m} + p_m G_m \qquad (12)$$

where the equality holds if m becomes a timing loop.

PROOF. Assuming time variable $A_{m_1}^{k-p_m}$ of node m_1 at iteration $k - p_m$ tries to hop to node m_2, from iteration equations (2) and (4), we have:

$$A_{m_2}^{k-p_m+1} \geq max(A_{m_1}^{k-p_m}, C_{m_1}) + \Lambda_{m_1,m_2}$$
$$\geq A_{m_1}^{k-p_m} + \Lambda_{m_1,m_2}$$

where the equality holds if $A_{m_1}^{k-p_m}$ survives the hop from node m_1 to node m_2. Iteratively making such hops for p_m times along the loop of m, we will return back to node m_1 and

$$A_{m_1}^k \geq A_{m_1}^{k-p_m} + \Lambda_{m_1,m_2} + \Lambda_{m_2,m_3}, ..., + \Lambda_{m_{p_m},m_1}$$
$$= A_{m_1}^{k-p_m} + p_m G_m$$

where the equality holds only when the time variable survives every step of hopping which means the loop is a timing loop. $\square$

With this theorem in hand, we are then ready to present our first major contribution:

Theorem 3 (Convergence of Iteration Mean). *No matter what is the initial state of the reduced timing graph, the sequence of the iteration mean $O_j^0, O_j^1, O_j^2, ..., O_j^k, ...$ for any node j will always converge to one of the following two values after sufficient number of iterations:*

1. *$O_j^\infty = G_m$ if node j is within or dominated by a timing loop m with cycle mean G_m.*

2. *$O_j^\infty = 0$ if node j is independent on any timing loop.*

PROOF. In case 1, if node j is within the timing loop m whose length is p_m after r iterations, then for any index $k \geq r + p_m$ equality holds in theorem 2. Since there will exist an iteration index $r \leq t \leq r + p_m$ and an integer $n = 0, 1, 2, ...$ such that $k = np_m + t$, after applying theorem 2 for n times, the latest data arrival time of node j at iteration k will be $A_j^k = A_j^t + np_m G_m = A_j^t + (k - t)G_m$. So the iteration mean:

$$O_j^\infty = \lim_{k \to \infty} \frac{A_j^k}{k+1} = \lim_{k \to \infty} \frac{(k-t)G_m}{k+1} = G_m$$

If node j is not within but dominated by the timing loop m, then we define two constants Σ^+ and Σ^- as:

$$\Sigma^+ = \sum_{\Lambda_{i,j}>0} \Lambda_{i,j} \quad and \quad \Sigma^- = \sum_{\Lambda_{i,j}<0} \Lambda_{i,j}$$

which are clearly finite values.

With these two constants, if the time variable of node j is x hops away from node m_i in the timing loop m, then the latest data arrival time of node j at k^{th} iteration will be:

$$A_{m_i}^{k-x} + \Sigma^- \leq A_j^k \leq A_{m_i}^{k-x} + \Sigma^+$$

Since Σ^+ and Σ^- are finite constants, then

$$O_j^\infty = \lim_{k \to \infty} \frac{A_{m_i}^{k-x}}{k+1} = G_m$$

which proves the first case of the theorem.

For case 2, all time variables hopping into node j must start from a finite value such as clock arrival times or primary input arrival times. So the latest data arrival time at node j must be finite for any iteration index k. So the iteration mean of the node obviously converges to zero. $\square$

For the example graph shown in figure 4, the convergence of the iteration mean of some latch nodes is graphically illustrated in figure 5:

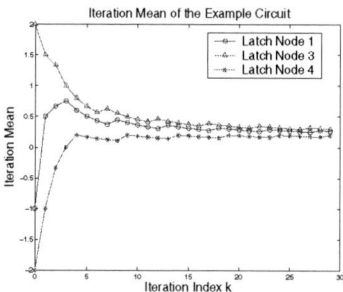

Figure 5: The iteration mean of three nodes in the example circuit of figure 4 and their convergence trend.

In [10], authors proposed to use the graph sorting algorithm which has exponential complexity in the worst cases. By substituting theorem 3 into equation (10), the critical cycle mean can be computed much simpler as:

$$G_c = max(O_1^\infty, O_2^\infty, ..., O_N^\infty) \qquad (13)$$

where $O_1^\infty, O_2^\infty, ..., O_N^\infty$ are converged iteration means for all N latches in the circuit.

3.3 Circuit Yield Prediction

In section 3, we defined a circuit parameter of *critical cycle mean* and claim that it is closely relating the circuit yield. We will establish this relationship solidly in this section.

Theorem 4 (Circuit Yield Computation). *A circuit will converge to a finite state if and only if its reduced timing graph has zero critical cycle mean: $G_c = 0$.*

PROOF. If $G_c > 0$, then according to the definition, there will be at least one loop whose cycle mean is G_c. Then applying equation 12 for enough times, the latest data arrival time in this loop can be arbitrarily large. Using theorem 1, the circuit will not converge in this case. Since $G_c \geq 0$ from definition, the necessity of the first assertion is proved.

If $G_c = 0$ then we have two cases. The first case is that there will be no timing loop in the reduced timing graph after sufficient number of iterations; This means that eventually all time variables hopping in the graph will come from

944

finite values and so that all latest data arrival times in the graph will be finite values. So circuit will converge by definition in section 2.1.

The second case is that there are some timing loops existing in the reduced timing graph after sufficient number of iterations. In this case, for any of such timing loop m, its cycle mean, G_m, must be non-positive since $G_m \leq G_c \leq 0$. So for any node m_i in m, applying theorem 2 for sufficiently large iteration index k:

$$A^k_{m_i} = A^q_{m_i} + (k-q)G_m \leq A^q_{m_i}$$

where q is a large but finite iteration index satisfies $k = q + np_m$ with loop length p_m and integer n. Since $A^q_{m_i}$ is obviously finite, the latest data arrival time must be upper bounded and the circuit converges in this case too. This proves the sufficiency. $\square$

Circuit will fail if it diverges since the latest data arrival time of some latches will go to infinity after sufficient number of iterations. But the circuit is not guaranteed to be functional even it converges in iterations. The setup and hold time constraints have to be additionally satisfied in order to be free of delay faults. So the overall timing yield of the circuit will be:

$$\begin{aligned} Y &= Pr\{G_c = 0 \cap s^\infty_c \leq 0 \cap h^\infty_c \leq 0\} \\ &= Pr\{max(G_c, s^\infty_c, h^\infty_c) = 0\} \end{aligned} \quad (14)$$

4. ITERATIVE TIMING ALGORITHM

The iterative latch timing algorithm, *StatITA*, is shown in figure 6 based on the theory in section 3.

StatITA takes the circuit as an input and compute the yield of the circuit at a given clock cycle time. The key iteration part of the algorithm is the repeat block from line 7 to line 19 where the convergence check is done in line 14.

StatITA has been implemented in C/C++ and tested on the ISCAS'89 benchmark circuits. The SSTA core is implemented based on the work from [6].

4.1 Accuracy of StatITA

All known tasks of statistical timing analysis can equivalently be accomplished by Monte Carlo simulations. The iterative timing analysis for latch-based circuits with feedback loops is not an exception either. So an iterative Monte Carlo timing analysis with 10,000 repetitions is also implemented in C/C++, *MontITA*, in parallel with *StatITA*.

Figure 7 shows the distributions of the critical cycle mean G_c computed from both *StatITA* and *MontITA* for circuit s526 at a clock cycle of $400ps$. The close match between *StatITA* and *MontITA* clearly shows the accuracy of the proposed iterative statistical latch timing algorithm.

The first application of a fast and accurate statistical latch timing algorithm is to predict the minimum clock cycle time at which the circuit will meet a yield goal. For this purpose, we define *the value of T_{97} as the minimum clock cycle time at which a given circuit will have a 97% timing yield.*

Table 1 shows the numerical T_{97} comparison between *MontITA* and *StatITA*. The average prediction error for the tested circuits is 1.1% which again demonstrates the accuracy of the proposed algorithm.

4.2 Performance of StatITA

```
1:  procedure StatITA(ClockCycle T)
2:      for (each latch i) do                    ▷ initialization
3:          C_i = setClockArrivalTimeForLatch(i)
4:          a_i^0 = 0; A_i^0 = 0; O_i^0 = 0;
5:      end for
6:      k = 1;
7:      repeat                                    ▷ iteration starts
8:          {a_i^k, A_i^k}=statisticalTiming(a_i^{k-1}, A_i^{k-1});
9:          done = true;
10:         for (each latch i ) do
11:             O_i^k = A_i^k/(k+1);
12:             mean = |μ_{O_i^{k-1}-O_i^k}|;
13:             std = σ_{O_i^{k-1}-O_i^k}
14:             if mean ≥ threshold ∪ std ≥ threshold then
15:                 done = false;              ▷ Not converged yet
16:             end if
17:         end for
18:         k = k + 1;
19:     until done                               ▷ iteration ends
20:     k = k-1;
21:     G_c = 0; s_c^∞ = -∞; h_c^∞ = -∞;
22:     for (each latch i) do
23:         G_c = max(O_i^k, G_c);
24:         s_c = max(A_i^k - (C_i + 0.5T - S), s_c^∞);
25:         h_c = max((C_i - 0.5T + H) - a_i^k, h_c^∞);
26:     end for
27:     Y = Pr{max(G_c, s_c^∞, h_c^∞) = 0};       ▷ circuit yield
28: end procedure
```

Figure 6: Iterative timing analysis algorithm **StatITA**

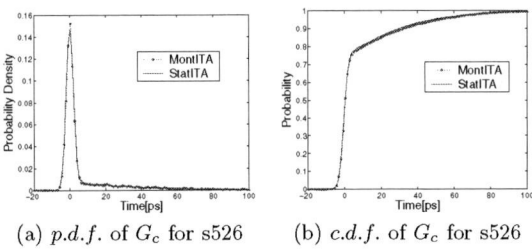

(a) *p.d.f.* of G_c for s526   (b) *c.d.f.* of G_c for s526

Figure 7: Critical cycle mean G_c for circuit s526 computed from *StatITA* and *MontITA* at clock cycle of 400ps

As reported in almost all SSTA works, Mont Carlo simulation is generally used as a "golden" method to evaluate the effect of process variations in the circuit timing. But the problem to directly apply Mont Carlo simulation in large circuit timing is the excessive CPU time it needs. This performance problem will be obviously more severe in iterative timing since each Monte Carlo sample will need many iterations to get the converged timing result.

From table 1, the excessive computation time needed by the MontITA is obvious. The average speedup of StatITA over MontITA is $303\times$.

The iteration core of statistical timing analysis is linear to the number of gates in the circuit according to [6]. The total number of iterations is determined by the convergence threshold and the circuit topology, not the size of the circuit. So the proposed StatITA algorithm will have linear complexity with respecting to the circuit size. This conclusion is clearly demonstrated in figure 8 where the linear trend of the run time with respecting to the gate number

			T_{97} [ps]			CPU Time[s]		
Circuits	Gates	Latches	StatITA	MontITA	Error	StatITA	MontITA	Speedup
s298	130	14	443	452	2.0%	2.14	320	150x
s526	196	21	465	469	0.9%	5.76	694	120x
s641	173	19	999	998	0.1%	1.17	372	320x
s820	279	5	777	788	1.4%	1.35	692	513x
s953	401	29	862	858	0.5%	3.32	1041	314x
s1423	616	74	2088	2051	1.8%	16.0	2083	130x
s5378	1517	179	764	780	2.1%	106	12372	117x
s9234	1827	211	859	858	0.1%	101	19073	189x
s13207	3516	638	1242	1246	0.3%	231	41571	180x
s15850	3889	534	1189	1199	0.8%	540	61044	113x
s38417	11543	1636	1544	–	–	1468	200hr*	490x*
s38584	12389	1426	1430	–	–	1209	303hr*	903x*
Average	–	–	–	–	1.1%	–	–	303x

Table 1: 97% yield clock cycle(T_{97}) and CPU time comparison between *StatITA* and *MontITA*.
(*)Estimation is from 100 repetitions and the accuracy of StatITA is not evaluated for these circuits.

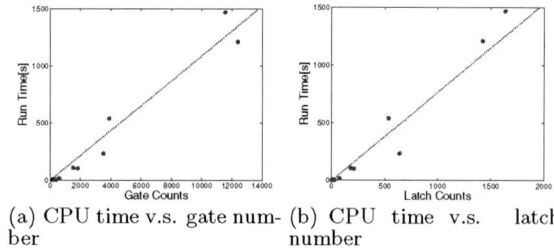

(a) CPU time v.s. gate number

(b) CPU time v.s. latch number

Figure 8: Run time of *StatITA* v.s. circuit size

and latch number in the circuit is shown.

5. CONCLUSIONS

A novel iterative timing statistical algorithm.*StatITA*, for latch-based circuits with feedback loops is proposed and its convergence during iterations is both theoretically proved and experimentally demonstrated. A novel concept of *iteration mean* is proposed to decide the convergence of the algorithm and the relationship between the converged iteration mean and circuit yield under process variations are founded both theoretically and experimentally.

Tested by the ISCAS'89 benchmark circuits, the proposed algorithm shows an error of 1.1% and 303× speedup on average when compared with Monte Carlo simulations.

6. ACKNOWLEDGEMENT

This work was partially funded by TSMC, UMC, Faraday, SpringSoft, National Science Foundation under grants CCR-0093309 & CCR-0204468 and National Science Council of Taiwan, R.O.C. under grant NSC 92-2218-E-002-030. Also great thanks to professor Barry D. Van Veen for the great discussions.

7. REFERENCES

[1] S. Nassif, "Within-chip variability analysis," *Electron Devices Meeting, 1998. IEDM '98 Technical Digest., International*, pp. 283 – 286, Dec 1998.

[2] C. S. Amin, N. Menezes, K. Killpack, F. Dartu, Y. Ismail, U. Choudhury, and N. Hakim, "Statistical static timing analysis: How simple can we get?" *42th Design Automation Conference, DAC'05*, 2005.

[3] J. Le, X. Li, and L. Pileggi, "Stac: statistical timing analysis with correlation," *Design Automation Conference, 2004. Proceedings. 41st*, pp. 343 – 348, June 2004.

[4] H. Chang and S. S. Sapatnekar, "Statistical timing analysis considering spatial correlations using a single pert-like traversal," *ICCAD'03*, pp. 621–625, Nov 2003.

[5] C. Visweswariah, K. Ravindran, and K. Kalafala, "First-order parameterized block-based statistical timing analysis," *TAU'04*, Feb 2004.

[6] L. Zhang, W. Chen, Y. Hu, and C. C. Chen, "Statistical timing analysis with extended pseudo-canonical timing model," *DATE'05*, March 2005.

[7] M. Orshansky, C. Spanos, and C. Hu, "Circuit performance variability decomposition," *4th International Workshop on Statistical Metrology, 1999. IWSM. 1999*, June 1999.

[8] S. Tsukiyama, M. Tanaka, and M. Fukui, "A statistical static timing analysis considering correlations between delays," *Proceedings of the 2001 conference on Asia South Pacific design automation*, January 2001.

[9] F. N. Najm and N. Menezes, "Yield estimation and optimization: Statistical timing analysis based on a timing yield model," *Proceedings of the 41st annual conference on Design automation*, 2004.

[10] M. C.-T. Chao, L.-C. Wang, K.-T. Cheng, and S. Kundu, "Static statistical timing analysis for latch-based pipeline designs," *IEEE/ACM International Conference on Computer Aided Design, 2004. ICCAD-2004*, pp. 468 – 472, 2004.

[11] R. Chen and H. Zhou, "Clock schedule verification under process variations," *IEEE/ACM International Conference on Computer Aided Design, 2004. ICCAD-2004*, pp. 619 – 625, Nov 2004.

[12] K. A. Sakallah, T. Mudge, and O. Olukotun, "$checkt_c$ and $mint_c$: timing verification and optimal clocking of synchronous digital circuits," *IEEE International Conference on Computer-Aided Design, 1990. ICCAD-90*, pp. 552 – 555, 1990.

[13] T. Szymanski and N. Shenoy, "Verifying clock schedules," *IEEE/ACM International Conference on Computer-Aided Design, 1992. ICCAD-92*, pp. 124 – 131, 1992.

[14] J.-F. Lee, D. Tang, and C. Wong, "A timing analysis algorithm for circuits with level-sensitive latches," *IEEE Transactions on Computer-Aided Design of Integrated Circuits and Systems*, vol. 15, pp. 535 – 543, May 1996.

[15] R. M. Karp, "A characterization of the minimum cycle mean in a digraph," *Discrete Mathematics*, vol. 23, pp. 309–311, 1978.

[16] S. M. Burns, "Performance analysis and optimization of asynchronous circuits," *PhD Thesis, California Institute of Technology*, 1991.

Parameterized Block-Based Non-Gaussian Statistical Gate Timing Analysis

Soroush Abbaspour, Hanif Fatemi, Massoud Pedram

Department of Electrical Engineering, University of Southern California

{sabbaspo, fatemi, pedram}@usc.edu

Abstract

As technology scales down, timing verification of digital integrated circuits becomes an increasingly challenging task due to the gate and wire variability. Therefore, statistical timing analysis (denoted by σTA) is becoming unavoidable. This paper introduces a new framework for performing statistical gate timing analysis for non-Gaussian sources of variation in block-based σTA. First, an approach is described to approximate a variational RC-π load by using a canonical first-order model. Next, an accurate variation-aware gate timing analysis based on statistical input transition, statistical gate timing library, and statistical RC-π load is presented. Finally, to achieve the aforementioned objective, a statistical effective capacitance calculation method is presented. Experimental results show an average error of 6% for gate delay and output transition time with respect to the Monte Carlo simulation with 10^4 samples while the runtime is nearly two orders of magnitude shorter.

1. Introduction

Process technology and environment-induced variability of gates and wires in VLSI circuits makes timing analysis of such circuits a challenging task [1]. More precisely, advanced analysis tools must be developed that are capable of verifying changes in the circuit timing which stem from various sources of variations [2]. In block-based statistical timing analysis (σTA), every timing quantity of interest (e.g., delay and slew, arrival time and required arrival time) is represented as a function of global sources of variation (denoted by X_i) and independent random sources of variation (denoted by S_i) in the canonical first-order (denoted by CFO) form. The advantages of such a formulation are that a) it can capture all correlations and b) it can produce delay sensitivities due to changes in various environmental and process-related parameters [2]. Sources of variations have often been assumed to be Gaussian, which in turn simplifies the block-based σTA. However, it has been recently reported that certain process parameters exhibit non-Gaussian probability distributions [3].

Block-based σTA breaks its analysis into two parts: 1) variational interconnect timing analysis [4][5] and 2) variational gate timing analysis. Unfortunately, block-based σTA is lacking in variation-aware gate timing analysis. The authors in [7] propose a modeling technique for gate delay variability considering multiple input switching. In [8], a model for calculating statistical gate delay variation caused by intra-chip and inter-chip variability is presented. Recent works do not provide an accurate means of analyzing the gate propagation delay and output slew as a function of variational input transition, variation-aware gate timing library, and variational gate load. In this paper a new framework is proposed for determining variational gate timing behavior. This is achieved by performing the following steps:

1. Given the variational resistive-capacitive load (where all resistances and capacitances are represented in the CFO form), an efficient and accurate algorithm is presented to calculate variation-aware RC-π load. To perform the analysis, we calculate the variation-aware admittance moments (cf. section 3), and as a result,

the resistance and capacitances in the RC-π load can be written in the CFO form.

2. Based on the statistical RC-π load obtained in step 1, we calculate the variation-aware effective capacitance in the CFO form. In order to achieve the aforementioned goal, a new approach for effective capacitance calculation in static timing analysis (STA) is proposed (cf. section 4.1.) This effective capacitance calculation method is used to calculate the variational effective capacitance considering non-Gaussian process and environmental sources of variation in the CFO form (cf. section 4.2.)

3. Given the variational input transition time, statistical gate timing library, and variational effective capacitance (c_{eff}) load in the CFO form, we calculate variational gate delay and output transition time in the CFO form (cf. sections 2.2.1)

We point out that although, in the remainder of this paper, we will mainly focus on the CFO random variables to represent process and environmental sources of variation as well as the performance quantities of interest; the work itself is not limited to the first-order approximation of these quantities. In fact, it is straightforward to extend the approach to more complex (e.g., second-order) forms regardless of considering Gaussian or non-Gaussian parameter variations.

The remainder of this paper is as follows. In section 2, we review the background of block-based σTA. We also show how to convert a quantity, which itself is a function of global and independent sources of variation, into a canonical first-order (CFO) form. The variation-aware RC-π calculation is presented in section 3. Section 4 explains the statistical gate timing analysis for the variational input rise time, variation-aware gate timing library, and variational RC-π load. In this section a new statistical effective capacitance calculation will be proposed and used for gate timing analysis, which is the key contribution of this paper. Section 5 presents experimental results. Finally, conclusions are discussed in section 6. We use the notation shown in Table 1 throughout the paper.

Table 1: Useful notation and descriptions

Notation	Description
A	A deterministic variable (does not take into account any statistical variation)
$\overset{p}{\underset{}{A}}$	An arbitrary (non-CFO) random variable, which is a function of m global and p independent random sources of variation
$\overset{\langle p \rangle}{A}$	A CFO random variable, which is a function of m global and p independent random sources of variation i.e., $\overset{\langle p \rangle}{A} = A_0 + \sum_{i=1}^{m} A_i \Delta X_i + \sum_{k=1}^{p} A_{m+j} \Delta S_j$

2. Background

As mentioned before, the sources of variation may exhibit non-Gaussian distributions. Therefore, in general, in addition to calculating the mean and variance of the electrical and timing parameters, we need to calculate the skewness of their distributions, i.e. using the first three moments of the parameters variations.

Definition: The degree of asymmetry of a distribution is called skewness (denoted by κ.) A distribution, or data set, is symmetric if it looks the same to the left and right of the center point. The skewness for a normal distribution is zero. Negative values for the skewness indicate data that are skewed left whereas positive values for the skewness indicate data that are skewed right. By skewed left (right), we mean that the left (right) tail is heavier than the right (left) tail.

The *skewness* of a distribution is defined to be $\kappa = \dfrac{\mu_3}{\sigma^3}$ where μ_3 is the 3rd central moment and σ^2 is the variance (second central moment.)

Lemma **1:** Suppose $\Delta S_1,...,\Delta S_n$ are n independent random variables with distribution $\Delta S_i \sim Dist_i(\mu=0, \sigma^2=1, \kappa_i)$. Then;

$$\sum_{i=1}^{n} a_i \Delta S_i = \sqrt{\sum_{i=1}^{n} a_i^2} \cdot \Delta S_{eq} \quad \text{where} \quad \Delta S_{eq} \sim Dist\left(\mu=0, \sigma^2=1, \kappa = \frac{\sum_{i=1}^{n} a_i^3 \kappa_i}{\left(\sum_{i=1}^{n} a_i^2\right)^{3/2}}\right)$$

Proof: It is omitted for brevity.

2.1 Canonical first-order (CFO) model for timing and electrical parameters

In block-based statistical timing analysis tool, a first-order variational model is employed for all timing quantities such as the gate and wire delays, arrival times, required arrival times, slacks and slews, i.e., all timing quantities are expressed in the CFO form as:

$$\overset{\triangleleft\triangleright}{a} = a_0 + \sum_{i=1}^{m} a_i \Delta X_i + a_{m+1} \Delta S_a$$

where a_0 is the nominal value; ΔX_i's represent the variation of m global sources of variation, X_i, from their nominal values, a_i's are the sensitivities to each of the global sources of variation, ΔS_a is the variation of independent random variable S_a and a_{m+1} is the sensitivity of the timing quantity to S_a. By scaling the sensitivity coefficients, we can assume that ΔX_i and ΔS_a have distributions with $\mu=0$ and $\sigma^2=1$ and skewness= κ denoted by $Dist(\mu=0,\sigma^2=1,\kappa)$.

Variation in the physical dimensions of the wire causes change in its resistance and capacitance, thereby, making the gate delay and slew as well as wire delay and slew to vary accordingly [9]. Therefore, we need to capture the effect of geometric variations on the electrical parameters. For instance, resistance and capacitance in the CFO form are calculated as follows:

$$\overset{\triangleleft\triangleright}{r} = r_0 + \sum_{i=1}^{m} r_i \Delta X_i + r_{m+1} \Delta S_r \qquad \overset{\triangleleft\triangleright}{c} = c_0 + \sum_{i=1}^{m} c_i \Delta X_i + c_{m+1} \Delta S_c$$

where r_0 and c_0 represent nominal resistance and capacitance values, computed when the wire dimensions are at their nominal or typical values. The other parameters are as explained above.

Observation: *Invariant Functional Form Property*: This property states that: $y = f(x) \Leftrightarrow \overset{\wp}{Y} = f(\overset{\wp}{X})$, which follows from the fact that form of function f is independent of its input type (deterministic or variational.)

2.2 Converting a variational function into CFO form

It is important to represent timing and electrical quantities in the CFO form. This in turn enables one to propagate first order sensitivities to different sources of variation through timing graph [2][9]. In addition, it makes statistical computations efficient and practical and provides timing diagnostics at a very small cost in run time. The remaining question is how to convert a quantity of interest (which itself is a function of different CFO variables) into the CFO form.

The following subsection presents a method to answer the above question. We use an example to show the procedure. The problem we address is how to convert the gate output transition time into the CFO form. However, this method can be easily applied to any other quantity of interest.

2.2.1 Gate timing analysis for lumped capacitive load

Problem Statement I: Given is a variational CMOS driver where its input rise time, t_{in}, is in the CFO form and drives an output capacitive load, also, in the CFO form. Note that the distribution characteristics of all global and independent sources of variation ($\mu=0$, $\sigma^2=1$, κ) are given. The objective is to calculate the output transition time, t_r, in the CFO form:

$$\overset{\triangleleft\triangleright}{t}_r = t_{r,0} + \sum_{i=1}^{m} t_{r,i} \Delta X_i + t_{r,m+1} \Delta S_{t_r}$$

i.e., calculate the nominal value ($t_{r,0}$) and the sensitivity coefficients ($t_{r,i}$ and $t_{r,m+1}$) as well as the skewness of distribution of ΔS_{t_r}.

The gate output transition time is a function of the input transition time, the logic gate characteristics (e.g., the W/L ratio, threshold voltage of transistors, V_{dd}, and temperature), and the output load. In commercial ASIC cell libraries, it is possible to characterize various output transition times (e.g. 10%, 50%, and 90%) as a function of above variables; i.e.;

$$t_r = TF(t_{in}, c_l, z) \quad \text{where} \quad z = \left\{\frac{W}{L}, V_T, V_{dd}, Temp, ...\right\} \quad (1)$$

where t_r is the output transition time and TF is the corresponding output transition time function. z captures the gate characteristics and environmental factors, t_{in} is the input transition time, and c_l is the output *capacitive* load. Based on the Invariant Functional Form Property, the form of function TF is independent of its input type (deterministic or variational.) Hence, we extend the above equation to the variational case. In block-based σTA, t_{in}, c_l, and every parameter z is given in the CFO form as a function of m global and exactly one independent random sources of variations. Therefore, t_r itself is a complex (non-CFO) random variable. Hence, to represent the complex t_r in the CFO form, we replace t_{in}, c_l, and z with their corresponding CFO models and collect terms. Hence, by differentiating with respect to global and independent random sources of variation, t_r as a function of m global sources of variation and p independent random sources of variation can be approximated as:

$$\overset{\wp}{t}_r = TF\left(\Delta X_1 ... \Delta X_m, \Delta S_1 ... \Delta S_p\right) \Rightarrow$$

$$\overset{\wp}{t}_r \cong TF\Big|_{\substack{\Delta X_i=0 \\ \Delta S_k=0}} + \sum_{i=1}^{m} \frac{\partial TF}{\partial \Delta X_i}\Big|_{\substack{\Delta X_i=0 \\ \Delta S_k=0}} \cdot \Delta X_i + \sum_{j=1}^{p} \frac{\partial TF}{\partial \Delta S_j}\Big|_{\substack{\Delta X_i=0 \\ \Delta S_k=0}} \cdot \Delta S_j \quad (2)$$

$$\text{where} \begin{cases} l = 1...m \\ k = 1...p \end{cases}$$

Considering ΔS_j's having $Dist_j(\mu=0,\ \sigma^2=1,\ \kappa_j)$, Eqn.(2) can be re-written as:

$$\overset{\triangleleft\triangleright}{t}_r = TF\Big|_{\substack{\Delta X_i=0 \\ \Delta S_k=0}} + \sum_{i=1}^{m} \frac{\partial TF}{\partial \Delta X_i}\Big|_{\substack{\Delta X_i=0 \\ \Delta S_k=0}} \cdot \Delta X_i + \sqrt{\sum_{j=1}^{p} \left(\frac{\partial TF}{\partial \Delta S_j}\Big|_{\substack{\Delta X_i=0 \\ \Delta S_k=0}}\right)^2} \cdot \Delta S_{t_r}$$

By using Lemma 1:

$$\Delta S_{t_r} \sim Dist\left(\mu=0, \sigma^2=1, \kappa = \sum_{j=1}^{p}\left(\frac{\partial TF}{\partial \Delta S_j}\Big|_{\substack{\Delta X_i=0 \\ \Delta S_k=0}}\right)^3 \cdot \kappa_j \middle/ \left(\sum_{j=1}^{p}\left(\frac{\partial TF}{\partial \Delta S_j}\Big|_{\substack{\Delta X_i=0 \\ \Delta S_k=0}}\right)^2\right)^{3/2}\right)$$

In Lemma 2, we present how to calculate addition, multiplication, and division of two CFO forms in a new CFO form.

Lemma 2: Suppose, a and b are two given CFO random variables as :

$$\overset{\triangleleft\triangleright}{a} = a_0 + \sum_{i=1}^{m} a_i \Delta X_i + a_{m+1} \Delta S_a \qquad \overset{\triangleleft\triangleright}{b} = b_0 + \sum_{i=1}^{m} b_i \Delta X_i + b_{m+1} \Delta S_b$$

Therefore, for addition, subtraction, multiplication and division of a and b, we have;

a) Addition and subtraction:

$$\overset{\triangleleft\triangleright}{c} = \overset{\triangleleft\triangleright}{a} \pm \overset{\triangleleft\triangleright}{b} = \left(a_0 \pm b_0\right) + \sum_{i=1}^{m} \left(a_i \pm b_i\right) \Delta X_i + \sqrt{a_{m+1}^2 + b_{m+1}^2} \Delta S_c$$

b) Multiplication:

$$\overset{\triangleleft\triangleright}{c} \cong \overset{\triangleleft\triangleright}{a} \times \overset{\triangleleft\triangleright}{b} = a_0 b_0 + \sum_{i=1}^{m} \left(a_0 b_i + a_i b_0\right) \Delta X_i + \sqrt{\left(a_0 b_{m+1}\right)^2 + \left(a_{m+1} b_0\right)^2} \Delta S_c$$

c) Division:

$$\overset{\triangleleft\triangleright}{c} \cong \frac{\overset{\triangleleft\triangleright}{a}}{\overset{\triangleleft\triangleright}{b}} = \frac{a_0}{b_0} + \sum_{i=1}^{m} \frac{a_i b_0 - a_0 b_i}{b_0^2} \Delta X_i + \sqrt{\left(\frac{a_{m+1}}{b_0}\right)^2 + \left(\frac{a_0 b_{m+1}}{b_0^2}\right)^2} \Delta S_c$$

Proof: It is omitted for brevity.

3. RC-π Load Calculation in the CFO Form

In VDSM technologies, one cannot neglect the effect of interconnect resistance of the load on the gate delay and output transition time. In STA, an adequate approximation of an nth order load seen by the gate (i.e., a load with n distributed capacitances to ground) is obtained by replacing the load by a second order RC-π model [10]. Equating the first, second, and third moments of the admittance of the real load with the first, second, and third moments of the RC-π load, one can compute c_n, r_π, and c_f as [11]:

$$c_n = Y_{1,in} - \frac{Y_{2,in}^2}{Y_{3,in}} \qquad r_\pi = -\frac{Y_{3,in}^2}{Y_{2,in}^3} \qquad c_f = \frac{Y_{2,in}^2}{Y_{3,in}} \qquad (3)$$

where $Y_{k,in}$ is the kth moment of the admittance of the real load. In σTA, it is required to consider the effect of variability of the load on the gate timing analysis, as detailed below.

Problem Statement II: Given is an RC network representation of the load of a logic gate in a design as exemplified in Figure 1(a), where each r and c is in the CFO form. Note that the distribution characteristics of all global and independent sources of variation ($\mu=0$, $\sigma^2=1$, κ) are given. The objective is to calculate an equivalent variational RC-π load (i.e., c_n, r_π, and c_f of Figure 1(b) are in the CFO form), while its admittance matches the admittance of the real load in the frequency range of interest.

c_n, r_π, and c_f are functions of the admittance moments as seen from Eqn. (3). Hence, by calculating the variational admittance moments, we can calculate the CFO parameters of RC-π load (using the technique explained in section 2.2.) This can be done by differentiating the expressions in Eqn. (3) with respect to the sources of variation (cf. section 2.2.) However, as it will be shown next, a recursive operation is utilized to calculate the variational admittance moments and since in each recursion step, we have a complex (non-CFO) random variable which will feeds in the next step and this may increase the complexity of the calculations;

We represent the admittance moments in the CFO form throughout the recursion. This helps us by controlling the complexity of presenting the moments as the recursive function proceeds. Following shows how to calculate the input admittance moments of the real load in the CFO form. Consider the RCY segment shown in

Figure 2. Assume that the admittances at nodes i and j are represented by infinite series using the admittance moments:

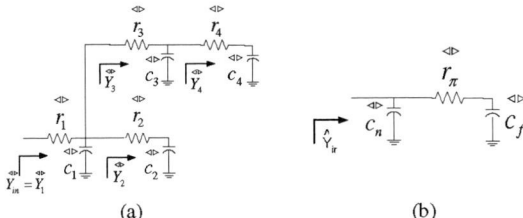

(a)   (b)

Figure 1: (a) a variational RC network representation of a net in a design. (b) the equivalent variational RC-π model.

$$Y_i(s) = sY_{1,i} + s^2 Y_{2,i} + \ldots + s^k Y_{k,i} + \ldots$$

$$Y_j(s) = sY_{1,j} + s^2 Y_{2,j} + \ldots + s^k Y_{k,j} + \ldots$$

where $Y_{k,l}$, which is the coefficient of s^k, denotes the kth moment of the admittance of node i. Thus, in STA, the admittance at node i is recursively computed in terms of the admittance at node j [11]:

$$Y_{1,i} = Y_{1,j} + c_i$$

$$Y_{k,i} = Y_{k,j} - r_i \sum_{l=1}^{k-1} Y_{l,i} Y_{k-l,i} - r_i c_i Y_{k-1,i} \quad \text{for } k \geq 2 \qquad (4)$$

Using the Invariant Functional Form Property, we extend the above equation to the variational case. Assume the admittance moments of node j are written in the CFO form. Thus, by differentiating $Y_{k,i}$ with respect to the sources of variations, the $Y_{k,i}$ moments can be also represented in the CFO form (cf. section 2.2.)

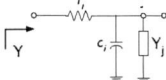

Figure 2: an RCY segment model for recursive admittance moment calculation.

By using the above recursive operations, we easily compute the moments of $Y_{in} = Y_1$ in the CFO form, and hence we calculate the values of c_n, r_π, and c_f in the CFO form using Eqn. (3).

4. Gate Timing Analysis for the RC-π Load in Block-Based σTA

Problem statement III: Given is a variational CMOS driver, whose input rise time, t_{in}, is in the CFO form and drives a variational RC-π load. The resistance and capacitances of this load are also in the CFO forms. The distribution characteristics of all global and independent sources of variation ($\mu=0$, $\sigma^2=1$, κ) are given. The objective is to calculate the output transition time, t_r, in the CFO form:

$$\overset{\triangleleft\triangleright}{t_r} = t_{r,0} + \sum_{i=1}^{m} t_{r,i} \Delta X_i + t_{r,m+1} \Delta S_{t_r}$$

i.e., calculate the nominal value ($t_{r,0}$) and the sensitivity coefficients ($t_{r,i}$ and $t_{r,m+1}$) as well as the skewness of distribution of ΔS_{t_r}.

Section 2.2.1 solves the same problem where the gate drives a variational purely-capacitive load in the CFO form. (cf. Eqn. (1)) Therefore, if we substitute the RC-π load with its equivalent variational effective capacitance, c_{eff}, in the CFO form, then the solution to problem statement I is an acceptable solution to problem statement III. Based on this reasoning, the following subsections propose a solution for calculating the effective capacitance in the CFO form. Section 4.1 presents a new effective capacitance calculation in static timing analysis. This approach is used in section 4.2 where statistical effective capacitance is calculated.

4.1 A new approach for effective capacitance calculation in static timing analysis

By definition, the effective capacitance is a pure capacitance that replaces an RC-π load and has the property that it gives the most accurate result from a timing model that is characterized with lumped capacitance. Typically, the effective capacitance stores the same amount of charge as the RC-π load until a certain point of the output voltage transition [11][12][13] (e.g., the 50% point of the output transition.) Figure 3(a) depicts a typical CMOS driver with its input waveform and RC-π load. The output voltage waveform may be modeled as a weighted linear sum of ramp and exponential waveforms as shown in Figure 3(b). We therefore assume that the *actual* c_{eff} can be obtained as a weighted average of that obtained for the ramp output waveform and that obtained for the exponential output waveform.

In the following, we calculate c_{eff} for ramp and exponential waveforms of the gate output voltage.

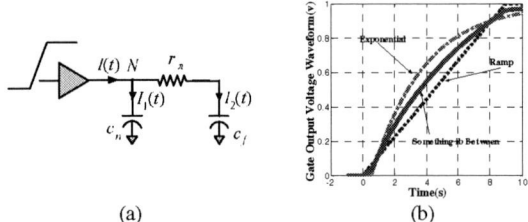

(a)   (b)

Figure 3: (a) A gate, which drives an RC-π calculated load.   (b) Gate output waveform is neither ramp nor exponential.

Theorem 1: Suppose that output voltage of a gate is approximated with an exponential waveform:

$$V_N(t) = V_{dd}\left(1 - e^{-pt}\right) \quad where \quad p = \frac{\ln\left(\frac{1-\alpha}{1-\beta}\right)}{t_r}$$

where $V_N(t)$ is the gate output voltage waveform in time domain and t_r is the output rise time from $\alpha\%$ transition to $\beta\%$ transition of this waveform. Note that t_r is a function of the input transition time (t_{in}) and the output load. Thus, the iterative effective capacitance equation for matching any $\theta\%$ point of the gate output transition time can be written as:

$$c_{eff}^{Exp}(\theta) = G(t_r, c_n, r_\pi, c_f) = c_n + k_{Exp}(\theta)c_f \quad where$$

$$k_{Exp}(\theta) = \left[1 + \frac{y}{\theta}\left(e^{\ln(1-\theta)/y} - 1\right)\right] \quad and \quad y = \ln\left(\frac{1-\alpha}{1-\beta}\right) \times \frac{r_\pi c_f}{t_r\left(t_{in}, c_{eff}^{Exp}(\theta)\right)}$$

Similarly for the ramp output voltage waveform, we have:

$$c_{eff}^{Ramp}(\theta) = H(t_r, c_n, r_\pi, c_f) = c_n + k_{Ramp}(\theta)c_f \quad where$$

$$k_{Ramp}(\theta) = \left[1 - \frac{x}{\theta}\left(1 - e^{-\theta/x}\right)\right] \quad and \quad x = (\beta - \alpha)\frac{r_\pi c_f}{t_r\left(t_{in}, c_{eff}^{Ramp}(\theta)\right)}$$

Proof: It is omitted for brevity.

Now, based on the assumption made above, an iterative equation for actual c_{eff} calculation for any $\theta\%$ point of the output transition time may be written as:

$$\left.\begin{aligned}c_{eff}^{Exp}(\theta) &= G\left(t_r, \left(t_{in}, c_{eff}^{Exp}(\theta)\right), c_n, r_\pi, c_f\right) \\ c_{eff}^{Ramp}(\theta) &= H\left(t_r, \left(t_{in}, c_{eff}^{Ramp}(\theta)\right), c_n, r_\pi, c_f\right)\end{aligned}\right\} \Rightarrow \tag{5}$$

$$c_{eff}(\theta) = F\left(t_r, \left(t_{in}, c_{eff}(\theta)\right), c_n, r_\pi, c_f\right) = \zeta \cdot G + (1-\zeta)H$$

where $0 \leq \zeta \leq 1$ is the weighting factor for the linear combination of exponential and ramp waveforms. However, we have observed that when $\theta\% = 50\%$, then $\zeta = 0.5$ results in the minimum error between the iterative c_{eff} equation in Eqn. (5) and the actual sign-off c_{eff} value.

4.2 Calculating c_{eff} in the CFO form

Suppose t_{in}, c_n, r_π, and c_f in the CFO form are given as:

$$\overset{\triangleleft\triangleright}{t_{in}} = t_{in,0} + \sum_{i=1}^{m} t_{in,i}\Delta X_i + t_{in,m+1}\Delta S_{t_{in}} \tag{6}$$

$$\overset{\triangleleft\triangleright}{c_n} = c_{n,0} + \sum_{i=1}^{m} c_{n,i}\Delta X_i + c_{n,m+1}\Delta S_{c_n} \tag{7}$$

$$\overset{\triangleleft\triangleright}{r_\pi} = r_{\pi,0} + \sum_{i=1}^{m} r_{\pi,i}\Delta X_i + r_{\pi,m+1}\Delta S_{r_\pi} \tag{8}$$

$$\overset{\triangleleft\triangleright}{c_f} = c_{f,0} + \sum_{i=1}^{m} c_{f,i}\Delta X_i + c_{f,m+1}\Delta S_{c_f} \tag{9}$$

$$\Delta S_{t_{in}} \sim Dist\left(\mu=0, \sigma^2=1, \kappa_{t_{in}}\right) \qquad \Delta S_{c_n} \sim Dist\left(\mu=0, \sigma^2=1, \kappa_{c_n}\right)$$
$$\Delta S_{r_\pi} \sim Dist\left(\mu=0, \sigma^2=1, \kappa_{r_\pi}\right) \qquad \Delta S_{c_f} \sim Dist\left(\mu=0, \sigma^2=1, \kappa_{c_f}\right) \tag{10}$$

The effective capacitance for this problem generally becomes a complex random variable, i.e. $\overset{p}{c_{eff}}$. Therefore, we approximate it with its CFO form and the objective becomes to calculate the coefficients of c_{eff} in the CFO form as well as the skewness of $\Delta S_{c_{eff}}$ as:

$$\overset{\triangleleft\triangleright}{c_{eff}} = c_{eff,0} + \sum_{i=1}^{m} c_{eff,i}\Delta X_i + c_{eff,m+1}\Delta S_{c_{eff}} \tag{11}$$

Such that $E\left[\left(\overset{\triangleleft\triangleright}{c_{eff}} - F\left(t_r, \left(\overset{\triangleleft\triangleright}{t_{in}}, \overset{\triangleleft\triangleright}{c_{eff}}\right), \overset{\triangleleft\triangleright}{c_n}, \overset{\triangleleft\triangleright}{r_\pi}, \overset{\triangleleft\triangleright}{c_f}\right)\right)^2\right]$ is minimized

where F is given in Eqn. (5). Theorem 2 presents the solution for calculating these unknown values.

Theorem 2: For a variational circuit, where t_{in}, c_n, r_π, and c_f in the CFO form are written as in Eqns. (6)-(10), the coefficients of c_{eff} in the CFO form (Eqn. (11)), can be calculated as:

$$c_{eff,0} = F\left(t_r, \left(t_{in,0}, c_{eff,0}\right), c_{n,0}, r_{\pi,0}, c_{f,0}\right) \tag{12}$$

$$c_{eff,i} = \frac{\left(\frac{\partial t_r}{\partial t_{in}}\right)^{nom} \cdot \left(\frac{\partial F}{\partial t_r}\right)^{nom} \cdot t_{in,i} + \left(\frac{\partial F}{\partial c_n}\right)^{nom} \cdot c_{n,i}}{1 - \left(\frac{\partial F}{\partial t_r}\right)^{nom} \cdot \left(\frac{\partial t_r}{\partial c_{eff}}\right)^{nom}}$$

$$+ \frac{\left(\frac{\partial F}{\partial r_\pi}\right)^{nom} \cdot r_{\pi,i} + \left(\frac{\partial F}{\partial c_f}\right)^{nom} \cdot c_{f,i}}{1 - \left(\frac{\partial F}{\partial t_r}\right)^{nom} \cdot \left(\frac{\partial t_r}{\partial c_{eff}}\right)^{nom}} \tag{13}$$

$$c_{eff,m+1} = \sqrt{\begin{array}{l}\left(c_{eff,m+1}^{t_{in}}\right)^2 + \left(c_{eff,m+1}^{c_n}\right)^2 \\ + \left(c_{eff,m+1}^{r_\pi}\right)^2 + \left(c_{eff,m+1}^{c_f}\right)^2\end{array}} \tag{14}$$

$$\Delta S_{c_{\text{eff}}} \sim Dist\left(\mu=0, \sigma^2=1, \kappa=\frac{\sum\limits_{u \in U}\left(c_{\text{eff},m+1}^u\right)^3 \kappa_u}{\left(\sum\limits_{u \in U}\left(c_{\text{eff},m+1}^u\right)^2\right)^{3/2}}\right) \quad (15)$$

$$\text{and } U=\left\{'t_{in}','c_n','r_\pi','c_f'\right\}$$

where;

$$c_{\text{eff},m+1}^{t_{in}}=\frac{\left(\frac{\partial F}{\partial t_r}\right)^{nom}\left(\frac{\partial t_r}{\partial t_{in}}\right)^{nom}t_{in,m+1}}{1-\left(\frac{\partial F}{\partial t_r}\right)^{nom}\left(\frac{\partial t_r}{\partial c_{\text{eff}}}\right)^{nom}} \qquad c_{\text{eff},m+1}^{c_n}=\frac{\left(\frac{\partial F}{\partial c_n}\right)^{nom}c_{n,m+1}}{1-\left(\frac{\partial F}{\partial t_r}\right)^{nom}\left(\frac{\partial t_r}{\partial c_{\text{eff}}}\right)^{nom}}$$

$$c_{\text{eff},m+1}^{r_\pi}=\frac{\left(\frac{\partial F}{\partial r_\pi}\right)^{nom}r_{\pi,m+1}}{1-\left(\frac{\partial F}{\partial t_r}\right)^{nom}\left(\frac{\partial t_r}{\partial c_{\text{eff}}}\right)^{nom}} \qquad c_{\text{eff},m+1}^{c_f}=\frac{\left(\frac{\partial F}{\partial c_f}\right)^{nom}c_{f,m+1}}{1-\left(\frac{\partial F}{\partial t_r}\right)^{nom}\left(\frac{\partial t_r}{\partial c_{\text{eff}}}\right)^{nom}}$$

Proof: It is omitted for brevity.

Eqn. (12) is the iterative c_{eff} calculation under the nominal conditions of the circuit. Hence, $c_{\text{eff},0}$ can be evaluated by using the conventional effective capacitance calculation [12][13].

$t_{in,i}$, $c_{n,i}$, $r_{\pi,i}$, $c_{f,i}$, are given (cf. Eqns. (6)-(9).) To evaluate Eqns. (13) and (14), we must calculate the derivatives of function F (function F is given in Eqn. (5)) with respect to t_r, c_n, r_π, c_f, and evaluate these derivatives for the nominal values of the circuit parameters (when all sources of variation are set to zero i.e., $(\partial F/\partial t_r)^{nom}$, $(\partial F/\partial c_n)^{nom}$, $(\partial F/\partial r_\pi)^{nom}$, and $(\partial F/\partial c_f)^{nom}$.) These terms are easy to evaluate. For the remaining terms, we need to calculate the derivatives of the output transition time (t_r) with respect to t_{in} and c_{eff} and evaluate them under the nominal condition of the circuit (i.e., $(\partial t_r/\partial t_{in})^{nom}$ and $(\partial t_r/\partial c_{\text{eff}})^{nom}$.) Therefore, we propose two different solutions:

1. Updating the gate library look-up table and utilizing the additional data during σTA: The revised tables now provide not only the timing quantity for each combination of t_{in} and c_l, but also the derivatives of the timing quantity (t_r) with respect to t_{in} and c_l for each combination of t_{in} and c_l.

2. Using the existing gate library look-up table, but performing additional calculations during σTA: To approximately calculate $(\partial t_r/\partial t_{in})^{nom}$, we read t_r (from the gate library) for $<t_{in,0}$; $c_{l,0}>$ and $<t_{in,0}+\delta$, $c_{l,0}>$. Next, we calculate $\Delta t_r/\delta$ as the approximation. $(\partial t_r/\partial c_{\text{eff}})^{nom}$ can be similarly calculated.

Using any of the above solutions, Eqns. (13) and (14) become closed form expressions, which can be evaluated in constant time. Note that we calculate $(\partial F/\partial t_r)^{nom}$, $(\partial F/\partial c_n)^{nom}$, $(\partial F/\partial r_\pi)^{nom}$, and $(\partial F/\partial c_f)^{nom}$ only once and in a constant time. Therefore, complexity of our method is dominated by the iterative effective capacitance calculation under the nominal conditions.

5. Experimental Results

Our experiments use 90nm CMOS process parameters to model gates and interconnect parasitics. We assumed two different configurations for the experimental setup. The first one consists of two inverters connected in series whereas the second one is a CMOS inverter followed by a 2-input NAND gate. For both configurations, we apply a ramp input to the first inverter while its nominal value is chosen

from the set $(t_{in})^{nom}=\{10ps,80ps,150ps,220ps,300ps\}$. For the first configuration, size of the first inverter is fixed at $W_p/W_n=30/15\mu m$ whereas size of the second inverter is chosen to be one of $W_p/W_n=\{20/10, 50/25, 70/35, 100/50\}\mu m$. For the second configuration, size of the first inverter is again fixed at $W_p/W_n=30/15\mu m$ whereas this time the size of the succeeding 2-input NAND gate is chosen to be one of $W_p/W_n=\{40/40, 50/50, 100/100\}\mu m$.

To characterize the timing behavior of the gate, a look-up table based library is employed which represents the gate delay and output transition time as a function of input rise time, output capacitive load, V_{dd}, and temperature. We apply different loading scenarios for the second-stage gate as explained in the following subsections, i.e., pure capacitive load, and general RC load. We have also considered four different global sources of variation (V_{dd}, temperature, Metal layer 1 width, and ILD) and one independent random sources of variation for each electrical parameter (i.e., r and c) and timing parameter (for instance t_{in}) in the circuit. The sensitivity of each given data to the sources of variation is chosen randomly, while the total σ variation for each data is chosen to be 10% and 15% of their nominal value. We also assumed that the sources of variation are skewed with different skewness values as explained in each subsection. Mean, variance, and skewness of effective capacitance, the gate 50% propagation delay, and 10%-90% output transition time (slew) are calculated using the approaches presented in this paper.

To compare the results, we ran Monte Carlo simulation with 10^4 samples on each test scenario and derived mean, variance, and skewness of the effective capacitance, gate 50% propagation delay, and 10%-90% output transition time. Average percentage errors for the mean, variance, and skewness of effective capacitance, the gate 50% propagation delay, and 10%-90% output transition time between the obtained results from the Monte Carlo and the calculated results based on using statistical gate timing analysis approach are reported.

A. Purely Capacitive Load

The load in this section is considered to be purely capacitive. Its nominal value is chosen to be $(C)^{nom}=\{400, 500, 800, 1400\}fF$. The scaled distribution of the sources of variation is considered to have a skewness of 0.4, 0.6, and 0.8. We performed our experiments on both circuit configurations explained above. The results for the first configuration (where the second gate is an inverter) are presented in Table 2 (the skewness of the given data is 0.4) and Table 3 (for the skewness of 0.8). The results for the second configuration are provided in Table 4 (for the skewness of 0.6). Experimental results indicate an average error of about 3% for two different σ values, i.e. 10% and 15%. As we increase the σ value (i.e. the total σ variation for each data; e.g. σ variation of t_{in}, and c_l) from 10% to 15%, the error in calculated mean, variance, and skewness of the delay and slew increase, but slightly. The sources of error can be mainly classified into two groups: 1) the inaccuracy of the gate library table lookup and 2) the linear first order approximation of the timing and electrical parameters with respect to the sources of variation. Note that, the runtime of the proposed algorithm in average is 89 times faster than the Monte Carlo based approach.

Table 2: Average error for the inverter driving pure capacitive load (Skewness=0.4)

Average error	σ=10%		σ=15%	
	Delay	Slew	Delay	Slew
Mean	1.5%	1.7%	2.2%	2.3%
Variance	1.2%	1.3%	1.8%	1.9%
Skewness	1.0%	1.1%	1.4%	1.3%

Table 3: Average error for the inverter driving pure capacitive load (Skewness=0.8)

Average error	σ=10%		σ=15%	
	Delay	Slew	Delay	Slew
Mean	1.9 %	2.3%	2.5%	2.9%
Variance	1.6%	1.7%	1.9%	2.1%
Skewness	1.4%	1.5%	1.5%	1.9%

Table 4: Average error for the 2-input NAND gate driving pure capacitive load (Skewness=0.6)

Average error	σ=10%		σ=15%	
	Delay	Slew	Delay	Slew
Mean	3.0 %	3.1%	3.2%	3.1%
Variance	2.5%	2.7%	2.8%	2.9%
Skewness	2.2%	2.3%	2.5%	2.6%

B. General *RC* Load

For this section, the load is considered to be an *RC* tree of varying topology. The nominal value of total load resistance is chosen from the set $(R)^{nom} = \{150, 260, 300, 710, 1000\}\Omega$ and the nominal value of the total capacitance of the load is chosen to be from the set $(C)^{nom} = \{400, 500, 800, 1400\}fF$. The scaled distribution of the sources of variation is considered to have a skewness of 0.5, 0.75, and 1.

Again, we performed the experiment on both circuit configurations as explained before. The results for the first configuration (where the second gate is an inverter) are presented in Table 5 (the skewness of the given data is 0.5) and Table 6 (the skewness of the given data is 0.75). The results for the second configuration are also provided in

Table 7 (the skewness of the given data is 1). Experimental results indicate an average error of about 6% for different σ values. As we increase the σ value (i.e. the total σ variation for each data; e.g. σ variation of t_{in}, c_n, r_π, and c_f from 10% to 15%, the error in calculated mean, variance, and skewness of c_{eff}, the gate delay, and output transition time increase, but slightly. Similarly, as skewness increases (e.g. skewness of t_{in}, c_n, r_π, and c_f) from 0.5 to 0.75, the error in calculated mean, variance, and skewness of the c_{eff}, as well as the error in delay and slew increases, but slightly. The sources of error can be mainly classified into four groups: 1) the inaccuracy of the gate library table lookup, 2) the linear first order approximation of the timing and electrical parameters with respect to the sources of variation, 3) the error in calculating the variational *RC*-π load and 4) the error in the effective capacitance iterative equation proposed in section 4.1. The runtime of the proposed algorithm is, on average, 95 times faster than the Monte Carlo based approach.

Table 5: Average error for the inverter driving general *RC* load (Skewness=0.5)

Average error	σ=10%			σ=15%		
	Ceff	Delay	Slew	Ceff	Delay	Slew
Mean	3.2%	3.5%	4.9%	3.5%	5.4%	5.8%
Variance	2.4%	3.3%	4.5%	2.6%	5.9%	5.2%
Skewness	2.5%	3.3%	4.9%	2.0%	5.5%	5.5%

Table 6: Average error for the inverter driving general *RC* load (Skewness=0.75)

Average error	σ=10%			σ=15%		
	Ceff	Delay	Slew	Ceff	Delay	Slew
Mean	3.5%	5.1 %	5.3%	3.8%	5.9%	6.1%
Variance	2.9%	4.3%	5.5%	3.6%	6.2%	6.2%
Skewness	2.8%	4.1%	4.9%	3.1%	5.9%	5.9%

Table 7: Average error for the 2-input NAND gate driving general *RC* load (Skewness=1)

Average error	σ=10%			σ=15%		
	Ceff	Delay	Slew	Ceff	Delay	Slew
Mean	4.1%	5.2 %	5.1%	4.2%	6.1%	6.7%
Variance	3.9%	5.4%	5.2%	4.3%	6.1%	6.1%
Skewness	4.0%	6.1%	5.6%	4.2%	6.5%	6.3%

6. Conclusion

In this paper we presented a framework to handle the variation-aware gate timing analysis in block-based σTA considering non-Gaussian sources of variation. First, we proposed an approach to calculate variational *RC*-π load, which can be utilized in place of the actual variational *RC* load for the gate timing analysis purposes. Next, we presented a new approach for calculating effective capacitance in STA. We used this technique to calculate the statistical c_{eff} in the CFO form, and thereby, calculated the gate delay and output slew in the that form. Experimental results show an average error of 6% with respect to Monte Carlo with 10^4 samples simulation.

7. References

[1] S. Nassif, "Modeling and Analysis of Manufacturing Variations," *CICC*, 2001, pp. 223-228.

[2] C. Visweswariah, K. Ravindran, K. Kalafala, S.G. Walker, S. Narayan, "First-order incremental block-based statistical timing analysis," *DAC*, 2004, pp. 331-336.

[3] H. Chang, V. Zolotov, S. Narayan, and C. Visweswariah "Parameterized Block-Based Statistical Timing Analysis with Non-Gaussian and Non-Linear Parameters," *Int'l Workshop on Timing Issues* (TAU), 2005.

[4] Y. Liu, L. T. Pileggi, and A. J. Strojwas, "Model Order Reduction of RC(L) Interconnect Including Variational Analysis," *DAC*, 1999, pp. 201-206.

[5] J.D. MA and R.A. Rutenbar, "Interval-Valued Reduced Order Statistical Interconnect Modeling," *ICCAD*, 2004, pp. 460-467.

[6] K. Agarwal, D. Sylvester, D. Blaauw, F. Liu, S. Nassif, and S. Vrudhula, "Variational delay metrics for interconnect timing analysis," DAC, 2004 pp. 381 – 384.

[7] A. Agarwal, F. Dartu, D.Blaauw, "Statistical Gate Delay Model Considering Multiple Input Switching," *DAC*, 2004, pp. 658 – 663.

[8] K. Okada, K. Yamaoka, and H. Onodera, "A statistical gate-delay model considering intra-gate variability" *ICCAD*, 2003, pp. 908 – 913.

[9] V. Mehrotra, S. Nassif, D. Boning, and J. Chung, "Modeling the Effects of Manufacturing Variation on High-Speed Microprocessor Interconnect Performance," *IEEE Electron Devices Meetings*, 1998, pp. 767-770.

[10] P.R. O'Brien and T. L. Savarino, "Modeling the Driving-Point Characteristics of Resistive Interconnect for Accurate Delay Estimation," *ICCAD*, 1989, pp.512-515.

[11] A.B. Kahng, S. Muddu, "Improved effective capacitance computations for use in logic and layout optimization," *VLSI Design*, 1999, pp. 578-582.

[12] F. Dartu, N. Menezes, and L. Pillegi, "Performance Computation for Precharacterized Gates with RC Loads," *IEEE Trans. On Computer Aided Design* 15(5):544-533, 1996.

[13] S. Abbaspour, M. Pedram, "Calculating the Effective Capacitance for the RC Interconnect in VDSM Technologies," *ASPDAC*, 2003.

Statistical Leakage Minimization through Joint Selection of Gate Sizes, Gate Lengths and Threshold Voltage

Sarvesh Bhardwaj[†], Yu Cao[†], Sarma Vrudhula[‡]
[†]Electrical Engineering, [‡]Computer Science and Engineering
Arizona State University, Tempe, AZ 85281
email:{sarvesh.bhardwaj, yu.cao, vrudhula}@asu.edu

ABSTRACT

This paper[1] proposes a novel methodology for statistical leakage minimization of digital circuits. A function of mean and variance of the circuit leakage is minimized with constraint on α-percentile of the delay using physical delay models. Since the leakage is a strong function of the threshold voltage and gate length, considering them as design variables can provide significant amount of power savings. The leakage minimization problem is formulated as a **multivariable convex optimization** problem. We demonstrate that statistical optimization can lead to more than 37% savings in nominal leakage compared to worst-case techniques that perform only gate sizing.

I. INTRODUCTION

The leakage power has become a major cause of concern during the design of high performance nano-scale circuits. For example, it was shown in [4] that for 30% variations in the circuit delay there can be up to 20X variations in the leakage current. Scaling has also resulted in significant increase in the variations of the process and design parameters [1, 5, 3]. The most important of these variations are the variations in the effective channel length L_e, and the threshold voltage V_{th} which are due to a lack of precise control in the lithography and channel doping steps [5]. Variations in these two parameters have a significant effect on the sub-threshold leakage of a gate because of its exponential dependency on these two parameters.

The problem of leakage reduction has been addressed at the design stage by various techniques such as transistor stacking [15], sleep transistor insertion [13], body biasing [24, 16] and driving the circuit into a minimum leakage sleep state. The power savings accrued by these techniques can be further supplemented by gate sizing, dual-threshold voltage (V_{th}) and supply voltage (V_{dd}) assignment [22, 12, 9, 11] and L_e biasing [10]. However, to the best of our knowledge, L_e

[1]Any opinions, findings, and conclusions or recommendations expressed in this material are those of the author(s) and do not necessarily reflect the views of the National Science Foundation. This work was carried out at the NSF's State/Industry/University Cooperative Research Centers' (NSF-S/IUCRC) Center for Low Power Electronics (CLPE). CLPE is supported by the NSF (Grant #EEC-9523338), the State of Arizona, and an industrial consortium. This work was also supported by NSF through grant #CCR-0205227. An extended version of this paper is available at:http://veda.eas.asu.edu/papers/bhardwaj-aspdac06.pdf

biasing has not been used in the past for leakage reduction in the presence of variations.

The traditional corner based design methodology treats the parameters as deterministic quantities and wastes expensive design resource in order to ensure a large guard-band on the design frequency as well as the power dissipation. Instead, a more effective methodology is to model the variations as random variables because of the stochastic nature of the underlying variations. Once this is done, the statistics (such as mean, variance etc.) of delay as well as leakage power of the circuit can be accurately estimated using the *probability density functions* (PDFs) of the parameters. A number of such statistical analysis techniques have been proposed recently [27, 20]. Although the proposed techniques predict the circuit delay or leakage accurately, their reliability with respect to what is manufactured greatly depends on the accuracy of the models used for the gate (and interconnect) delays and power [7]. While these models provide foundations for accurate statistical analysis, they are *necessary* for statistical optimization techniques.

A number of statistical optimization methods have been proposed recently [23, 19, 21, 14]. [23] uses a statistical timing analysis tool to check the satisfiability of the constraint on some percentile of the circuit delay. It then uses the *statistical sensitivities* to select the gates to be assigned high V_{th} as well as to be up-sized. In [19], a utility theoretic approach is used to identify a set of critical paths. The expected utility of the critical nodes (nodes on critical paths) is minimized subject to constraints on the expected delay and area. A robust circuit optimization technique is presented in [17] where the authors formulate the problem of maximizing the timing parametric yield as a geometric optimization problem with gate sizes as the decision variables. In [2], the problem of statistical leakage minimization using gate sizing is formulated as a geometric programming problem. The area minimization problem is solved in [21] by modeling the parametric variations using an uncertainty ellipsoid.

Introducing L_e and V_{th} as decision variables in the optimization problem instead of treating them fixed technological parameters, increases the size of the feasibility region. This can provide significant power savings that can not be achieved otherwise. Also, since V_{th} and its variance is dependent on L_e, L_e proves to be an extremely effective method of controlling the leakage variability. Instead of using a dual-V_{th} process, in which high V_{th} and low V_{th} are typically separated by about $50mV$ for speed improvement and power reduction, we use a **single** V_{th} for the circuit. The use of dual-V_{th} produces extra process corners and thus, exaggerates variability. Specifically, some high V_{th} gates may be faster (and more leaky) than some low V_{th} gates due to the variation in V_{th}. This phenomenon has led to many design failures, especially in low-power applications.

The major contributions of this work are as follows:

1. Both the **mean** and the **variance** of the leakage are minimized by formulating a statistical optimization problem with **gate sizes, gate lengths** and **threshold voltage** as decision variables.

2. Experimentally verified statistical models for the gate delay and gate leakage are used that take into account the variability in various device parameters,

3. The leakage minimization problem is formulated as a **multivariable convex optimization** problem and an optimal solution is obtained.

Section II formally describes the problem of statistical leakage minimization. Section III describes the models used for the gate delay and gate leakage. The transformation of the optimization problem into a convex optimization problem is described Section IV. The Experimental results and conclusions are outlined in Section V and VI respectively.

II. Problem Formulation

Let a circuit be represented using a Directed Acyclic Graph (DAG) $G = (N, E)$, where $N = \{1, 2, .., n\}$ is the set of nodes and $E = \{(i, j) : i, j \in N\}$ is the set of edges. The nodes correspond to the gates in the original circuit. An edge $e_{ij} = (i, j)$ represents that gate i fanouts to gate j.

Let the parameter space for each gate i be defined as $\hat{u}_i = (u_1^i, u_2^i, .., u_r^i)$, where r denotes the number of parameters. In the presence of process variations, each of these parameters is a random variable. Hence, if Ω denotes the space of manufacturing outcomes, $\hat{u}_i : \Omega \to \mathbb{R}^r$ is a function that maps every outcome $\omega \in \Omega$ to a point in an r-dimensional Euclidean space. Hence, the parameters for the manufacturing outcome ω are given by $\widehat{u_i}(\omega) = (u_1^i(\omega), u_2^i(\omega), .., u_r^i(\omega))$. The **random parameters** considered in this work include the **gate length** ($L_{e,i}$) and **threshold voltage** ($V_{th,i}$). Although all the gates in the circuit are assigned the **same** V_{th}, the dependence of V_{th} on $L_{e,i}$ as well as the random variations cause the threshold voltage of each gate to be different. As channel length L_e becomes shorter, V_{th} exhibits a greater dependence on L_e and drain bias (DIBL). Larger V_{dd} and smaller L_e usually lead to sharp degradation in V_{th} (i.e., V_{th} roll-off). The $V_{th,i}$ of a gate can then be represented as

$$V_{th,i} = V_{tho} + 0.05 - V_{dd}e^{-\delta L_{e,i}}. \quad (1)$$

where V_{tho} is the long channel V_{th} and δ is the DIBL coefficient. For simplicity, V_{tho} and $L_{e,i}$ are modeled as independent normal random variables. Also, w_i and V_{dd} are modeled as a deterministic quantities. Henceforth, the explicit dependency of $\hat{u}$ on the argument ω will not be shown.

The circuit leakage and delay under this variational model are also random variables. Let I_S denote the sub-threshold leakage of the circuit and D_p denote the delay of path $p \in \mathcal{P}$,

where $\mathcal{P}$ represents the set of paths in the circuit. The stochastic leakage minimization problem can now be formulated as follows

$$\min_{\omega \in \Omega} \quad I_S(\hat{u}_1, \hat{u}_2, .., \hat{u}_n, \omega) \quad (2)$$

$$\text{sub. to} \quad \mathbf{P}(D_p(\hat{u}_1, .., \hat{u}_n, \omega) \le T_{req}) \ge \alpha \quad \forall\, p \in \mathcal{P}. \quad (3)$$

where $\mathbf{P}(X \le x)$ denotes the probability that the random variable X is less than or equal to x. α can be considered to be a *confidence level*. As the number of manufacturing outcomes ω can be infinite, it does not make sense to solve the optimization problem for every ω. Hence, a more relevant objective would be some statistic (such as *mean* or *variance*) of the leakage current. Figure 1 shows the PDF of the leakage as a result of minimizing only the mean or the variance of the leakage. It can be seen that minimizing only the expected value of the leakage results in an increased number of chips having lower frequency (curve B). Whereas, minimizing just the variance without optimizing the mean leaves a scope of reduction in the leakage of the manufactured circuits (curve C). Hence the goal of maximizing the leakage yield can be achieved by minimizing a convex combination of the square of the mean and the variance of leakage. Thus, the new objective becomes $\lambda\mu^2(I_S) + (1-\lambda)\sigma^2(I_S)$, where μ and σ^2 are the mean and variance of leakage. $\lambda \in [0, 1]$ controls the relative weight of the mean and the variance of the leakage in the objective.

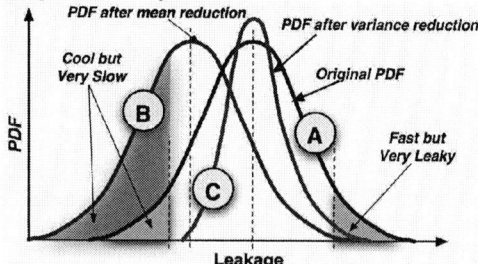

Fig. 1. Leakage Reduction

III. Leakage and Delay Models

A. Statistical Leakage Model

Let I_S represent the sub-threshold leakage of a circuit. As explained above, the I_S in the presence of the variations is random variable. In this work the sub-threshold leakage of a circuit is modeled using the model shown in (4) [25].

$$I_S = \sum_{i \in N} I_o \frac{w_i}{L_{e,i}^k} e^{\left(\frac{-V_{th,i}}{S}\right)}, \quad k > 1 \quad (4)$$

where $V_{th,i} = V_{tho} + 0.05 - V_{dd}(\delta_1 - \delta_2 L_{e,i})$ with $\delta_1 - \delta_2 L_{e,i}$ being approximation of $e^{-\delta L_{e,i}}$, $\delta_1, \delta_2 > 0$. I_o is th nominal sub-threshold leakage and k and S are positive fitting parameters. The summation in (4) is over all the nodes in the circuit. The above model captures the dependence of the sub-threshold leakage on the *all* the decision variables. Also, the dependence of V_{th} on L_e and V_{dd} has been taken into account. This model was fitted to the data from SPICE to obtain the parameters. From (4), we see that the sub-threshold leakage is inversely proportional to the gate length $L_{e,i}$. Since $L_{e,i}$ has been assumed to be a normally distributed random variable, the expectation of $L_{e,i}^{-k}$ does not

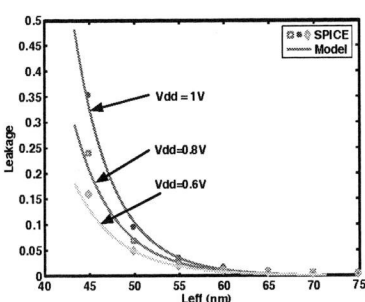

Fig. 2. Leakage Approximation

exist. Hence, we approximate the function $L_{e,i}^{-k}$ by writing it as $e^{-k \log L_{e,i}}$ and approximating $\log L_{e,i}$ by a quadratic function of $L_{e,i}$. Thus, the sub-threshold leakage is

$$I_S = \sum_{i \in N} I_o \frac{w_i}{L_{e,o}^k} e^{\left(\frac{-(V_{tho} + a_1 L_{e,i} + a_2 L_{e,i}^2)}{S} \right)}, \quad k > 1 \quad (5)$$

where a_1 and a_2 are some constants. The accuracy of the above approximation is shown in figure 2 which shows the variation of sub-threshold leakage with $L_{e,i}$. The parameters V_{tho} and $L_{e,i}$ are modeled as $V_{tho} = V_{To} + V_\xi$ where, V_{To} is the designer specified value of the threshold voltage and V_ξ is a zero mean normal random variable $N(0, \sigma^2(V_{To}))$. Notice, that the variation in the threshold voltage is dependent on the specified value of V_{To}. The variations in V_{tho} are modeled using the Pelgrom's model [18] as

$$\sigma^2(V_{th,i}) = \frac{k}{L_{e,i} w_i}. \quad (6)$$

Similarly, the gate length is also modeled as $L_{e,i} = L_{o,i} + L_{\xi,i}$ where $L_{\xi,i}$ is a zero mean normal random variable $N(0, \sigma_L^2)$. The variations in the gate length are assumed to be independent of the specified value of the gate length. Hence, the sub-threshold leakage of the circuit can be written as

$$I_S = \sum_{i \in N} I'_{o,i} e^{\left(\frac{-(V_\xi + (a_{11} + a_{12} V_{dd}) L_{\xi,i} + a_2 L_{\xi,i}^2)}{S} \right)}, \quad k > 1 \quad (7)$$

where

$$I'_{o,i} = I_o \frac{w_i}{L_{e,o}^k} e^{\left(\frac{-(V_{To} + (a_{11} + a_{12} V_{dd}) L_{o,i} + a_2 L_{o,i}^2)}{S} \right)}, \quad k > 1 \quad (8)$$

Now, $I'_{o,i}$ is a deterministic function of the assigned parameters. Also, as the underlying circuit parameters V_ξ and $L_{\xi,i}$ are assumed to be statistically independent, the mean of the leakage can be computed as shown in (9).

$$E[I_S] = \sum_{i \in N} I'_{o,i} E\left[e^{\left(\frac{-V_\xi}{S} \right)} \right] E\left[e^{\left(\frac{-((a_{11} + a_{12} V_{dd}) L_{\xi,i} + a_2 L_{\xi,i}^2)}{S} \right)} \right] \quad (9)$$

The expectation of the two functions dependent on the V_ξ and $L_{\xi,i}$ can be obtained by computing the expectation of a random variable U that is an exponential function of a zero mean normal random variable $W \sim N(0, \sigma_W^2)$. Thus the two functions in (9) have a general form $U = \exp(-(W + aW^2)/b)$. Table I gives the values of a and b for the functions having an exponential dependence on V_ξ and $L_{\xi,i}$ in (9).

Since the leakage has an exponential dependency on a linear function of V_ξ, for the V_ξ dependent term, $a = 0$ in the general form above. The mean of U can be computed using (10).

TABLE I
VALUES OF a AND b FOR THE FUNCTIONS DEPENDENT ON V_ξ AND $L_{\xi,i}$

Parameter	a	b
V_ξ	0	S
$L_{\xi,i}$	$\dfrac{a_2}{(a_{11} + a_{12} V_{dd})}$	$\dfrac{S}{(a_{11} + a_{12} V_{dd})}$

$$E[U] = \left(1 + \frac{2a}{b} \sigma_W^2 \right)^{-\frac{1}{2}} \cdot \exp\left(\frac{\sigma_W^2}{2b^2 + 4\sigma_W^2 ab} \right) \quad (10)$$

The second moment of U can also be computed from (10) by replacing b by $b/2$. Similarly, the second moment of the leakage can be computed by computing the expectation of

$$I_S^2 = \sum_{i,j \in N} I'_{o,i} I'_{o,j} e^{\left(\frac{-(2V_\xi + (a_{11} + a_{12} V_{dd})(L_{\xi,i} + L_{\xi,j}) + a_2 (L_{\xi,i}^2 + L_{\xi,j}^2))}{S} \right)}, \quad (11)$$

The procedure for computing the expectation of the above function $E[I_S^2]$ is analogous to the steps followed in computing the mean of the leakage. Using the second moment of the leakage, the variance can be computed using $\sigma^2(I_S) = E[I_S^2] - (E[I_S])^2$. The objective function of the leakage minimization problem can now be obtained using the mean and the variance of the leakage.

B. Statistical Delay Model

In the presence of process variations, the gate delays are random variables. For this work, we use the physical delay model proposed in [7]. Assuming that the transistors operate in the saturation mode, the proposed model can be simplified into the form shown in (12).

$$E[d_i] = \alpha \left(\frac{\beta_1}{w_i} + \beta_2 \right) \frac{L_{e,i} V_{dd}}{(V_{dd} - V_{th,i})^2} \left(1 + \frac{(V_{dd} - V_{th,i})}{\gamma L_{e,i}} \right) \quad (12)$$

The parameters for these models are obtained by performing SPICE simulation and by fitting these models to SPICE data. The accuracy of the mean of the delay is shown in Figure 3. The average error compared to the data from SPICE simulations is around 3-4% over ±25% range of $L_{e,i}$ for different values of the supply voltage and threshold voltage.

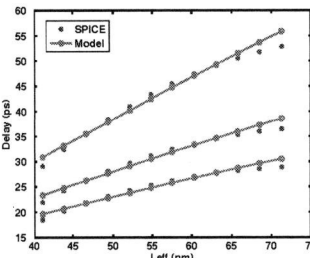

Fig. 3. Mean delay as a function of gate length for a NAND gate

Although the general form of the mean delay is not a linear function of the device parameters, for the values of the parameters in the saturation mode ($V_{dd} \in [0.8, 1.2]$) the delay can be safely assumed to have linear dependence of these parameters [7] (Figure 3 supports the linear dependence on L_e). Thus, the delay can be modeled as a normal random variable without having a significant impact on the accuracy.

9C-3

Under this assumption, the probabilistic constraint in (3) is equivalent to

$$z_\alpha(D_p) = E[D_p] + z_\alpha \sigma(D_p) \le T_{req} \qquad (13)$$

If the gate delays are totally correlated, the above constraint can be translated in terms of the *mean* and the *standard deviation* of the gate delays as shown in (14).

$$z_\alpha(D_p) = \sum_{i \in p} E[d_i] + z_\alpha \sum_{i \in p} \sigma(d_i) \le T_{req} \qquad (14)$$

where d_i is the delay of a gate on path $p \in \mathcal{P}$. Thus the problem reduces to obtaining the expressions for the variance of the individual gate delays. For a particular gate, at a higher value of V_{th} and fixed V_{dd}, the sensitivity of the gate delay to V_{th} is very high. Instead at a lower value of the V_{th}, the gate delay is more sensitive to the V_{dd} compared to V_{th}. Hence the variation in the delay is more for larger values of the V_{th} (higher delay), although the variations in V_{th} might be small at higher values of V_{th}. Thus, in this work, the variance of the delay is computed using the model shown in (15).

$$\frac{\sigma(d_i)}{E[d_i]} = k_\sigma (E[d_i])^\zeta \qquad (15)$$

where ζ and k_σ are fitting parameters. The α-percentile of the delay can now be computed as $z_\alpha(d_i) = E[d_i] + z_\alpha \sigma(d_i)$ using the models described above.

IV. CONVEX OPTIMIZATION

A function f of variable $\mathbf{x} \in \mathbb{R}^{+n}$ is a posynomial if it has the form

$$f(\mathbf{x}) = \sum_j \beta_j \prod_{i=1}^n x_i^{\alpha_{ij}} \qquad (16)$$

Posynomials have a useful property that they can be transformed into convex functions using the transformation $x_i = e^{y_i}$. Also, an exponential function of posynomials can be transformed into a convex function [6]. Since the objective function, which is a combination of the mean and the variance of the leakage, is a sum of exponential functions of *posynomials*, the objective function in the formulated optimization problem is convex.

In its original form, the expected delay is not a posynomial. However, it can be transformed into a posynomial by introducing the following inequality in the set of constraints

$$\frac{1}{V_{dd} - V_{th,i}} \le t_i \qquad (17)$$

and replacing $V_{dd} - V_{th,i}$ by t_i^{-1} in (12). The inequality given in (17) is equivalent to

$$t_i^{-1} + V_{tho} + \delta_2 V_{dd} L_{e,i} \le (1 + \delta_1) V_{dd} - 0.05. \qquad (18)$$

Since $\delta_1 > 0$ and $V_{dd} \ge 0.8$, the RHS of this inequality is positive. Hence, it is a valid posynomial inequality. As a result of this inequality being a valid posynomial inequality, the leakage minimization problem can be transformed into a convex optimization problem. Convex optimization problems are popular because efficient algorithms exist to solve them [26] as a locally optimal solution is also globally optimal. In this work, the convex optimization problem is solved using the optimization package LANCELOT [8]. The models used in the optimization correspond to the 90nm technological node. The range of the parameters, their nominal values and variances (where applicable) are given in Table II.

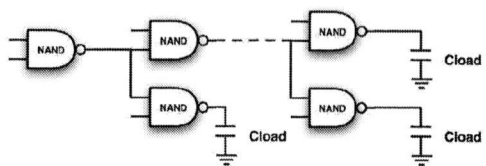

Fig. 4. Experimental circuit for the optimization problem

TABLE II
CIRCUIT PARAMETERS FOR THE 90NM TECHNOLOGICAL NODE

	$L_{e,i}$ (nm)	$V_{tho}(V)$	W (size)
Mean	55.0	0.30	-
Std.Dev	5.5	0.01	-
Upper Bound	70.0	0.40	10
Lower Bound	55.0	0.15	1

TABLE III
IMPACT OF INCLUDING V_{th} AS A DECISION VARIABLE. RESULTS FOR DELAY = 0.349 NS

	$V_{th}(V)$	Area	Leakage (fA/ns)	
			Mean	Var. ($X10^{-2}$)
A (Before)	0.30	32	0.19	8.74
B (After)	0.28	12.9	0.14	4.56
Diff (%)	-	59.0	26	47.8

V. EXPERIMENTAL RESULTS

A. Simultaneous V_{th} and Gate Sizing

In this section we discuss the savings that can be obtained by combining threshold voltage selection and gate sizing over simple gate sizing. For demonstration, we selected a chain of 10 NAND gates (similar to that in Figure 4) and performed gate sizing on the circuit with effective gate length and the threshold voltage fixed to their nominal values as shown in Table II. Also, V_{dd} was fixed to 1. The chain of 10 NAND gates is now optimized by minimizing the combination of mean and variance of leakage by considering V_{th} and gate sizes as decision variables. Compared to earlier works on gate sizing and threshold voltage assignment, this work performs statistical optimization and models the problem as a convex optimization and hence can guarantee the optimality of the solution while being efficient. V_{th} is treated as a continuous variable for the circuit as body biasing can provide significantly high level of granularity to achieve a given threshold voltage [24].

We start with a circuit optimized using only gate sizing with the rest of the parameters assigned to their nominal values given in Table II. We investigate the effect of changing the values of various parameters on different circuit attributes such as area and leakage with the delay constraint being fixed. It should be noted that if the delay constraint T_{req} is the minimum feasible delay $T_{req,min}$ at the nominal values of threshold voltage $V_{th,nom}$ and supply voltage $V_{dd,nom}$, then the optimal threshold voltage V_{th}^* in the circuit optimized with both V_{th} and w will be lower than $V_{th,nom}$ because decreasing the V_{th} decreases the delay.

Table III summarizes the value of the parameters and the circuit attributes before and after the optimization. The value of z_α is taken to be 3, which corresponds to 99% tim-

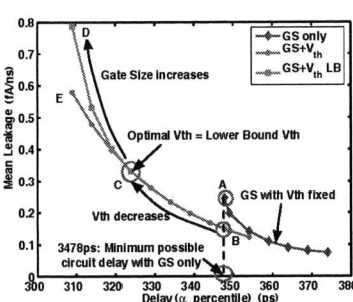

Fig. 5. Advantage of doing joint V_{th} and Gate Sizing

ing yield. The initial circuit A is obtained by performing only gate sizing with V_{th} and V_{dd} fixed to 0.3V and 1V respectively. The second circuit B is optimized by treating both the gate sizes and the threshold voltage as decision variables with V_{dd} fixed to 1V. As can be seen from the table, the circuit B has both lower area, lower mean leakage and lower variance of the leakage for the same value of the critical delay. By introducing V_{th} into the problem, we can get savings of up to 59% in the area and savings of 26% in the mean leakage. Thus we have significantly improved the leakage parametric yield without sacrificing the timing yield. Also, since the area of the circuit has reduced, it will lead to higher defect limited yield. Thus varying V_{th} is an effective method for optimizing leakage.

Figure 5 shows the trade-off of leakage and delay as a result of varying only sizes compared to simultaneous V_{th} selection and gate sizing. GS corresponds to the design methodology using only gate sizes as decision variables. As can be seen from the figure, point A corresponds to the minimum possible delay that can be achieved using GS only. However, if the V_{th} is introduced as an additional variable (with a lower bound = 0.25V) in the optimization, the feasible region is increased (because V_{th} can be changed) and we get a reduction in the objective function (point B). At point B, the threshold voltage is lower than that at point A and the area of the circuit corresponding to point B is lower than that of the circuit corresponding to point A. Also, from point B to point C, the threshold voltage proves to be the most effective method of reducing the circuit delay. For every optimal circuit between point B and point C, only the V_{th} is different, the area of all the circuits corresponding to points between B and C is the same.

At point C, the V_{th} reaches its lower bound of 0.25V. Since the threshold voltage cannot decrease any further, to achieve the critical delay, gate sizes have to be increased and thus the optimal circuits between C and D have only different area and their V_{th} is fixed to the minimum value. However, if the V_{th} were allowed to decrease further than 0.25V, we would have obtained optimal circuits having lower mean leakage as shown by the curve between the points C and E.

B. Simultaneous V_{th}, L_e and Gate Sizing

This section discusses the effect of introducing effective gate length and L_e as decision variables in the optimization along with V_{th} and gate sizes. Since the leakage has an exponential dependence on L_e, it is an effective factor for leakage

TABLE IV
CIRCUIT PARAMETERS FOR OPTIMIZED CIRCUIT WITH DIFFERENT
METHODS FOR 99-PERCENTILE CIRCUIT DELAY OF 349 PS

	V_{th}(V)	L_e (nm)	Area	Leak. (fA/ns)	
				Mean	Var. ($X10^{-2}$)
GS	3.0	54.8	32.0	0.19	8.74
GSV	2.8	54.8	12.9	0.14	4.56
GSVL	1.71	70.0	13.2	0.12	1.49

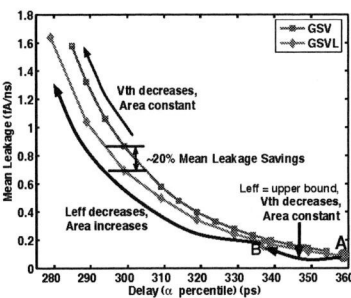

Fig. 6. Leakage-Delay tradeoffs for GSV and GSVL

reduction. We now show that L_e biasing provides significant leakage savings compared to the leakage savings obtained by simultaneous V_{th} and gate sizing. Table IV shows the parameters for the optimal circuit obtained by only gate sizing (GS), gate sizing and V_{th} assignment (GSV) and gate sizing, V_{th} and L_e biasing ($GSVL$) with a 99-percentile delay of 349ps and V_{dd} fixed to 1V. Changing the L_e from a fixed value to a variable causes the L_e to be assigned to its maximum value if the delay constraint is not very tight. The increase in L_e is shown in the third row and third column of the table. At this value of T_{req}, all the gates in the circuit have same L_e. Since, increasing the L_e increases the delay, the increase in the delay is compensated by decreasing the value of V_{th} as shown in the second column of GSVL. Thus the area of the circuit does not need to be increased. Hence, the overall effect is that the mean of the leakage reduces by around 15% with the area increasing by only 2%. Another important thing to be noticed is that the variance of the leakage decreases considerably. This is partly due to the fact that the variance of the threshold voltage reduces with increase in L_e as a result of Pelgrom's model. Thus the leakage parametric yield improves significantly without having a negative impact on the timing yield or the area of the circuit.

Figure 6 shows the leakage-delay trade-off for the two methods GSV and GSVL. In GSV, when the delay is reduced, the V_{th} has to be reduced to meet the timing constraint and thus the leakage increases. Instead, in GSVL, when the delay constraint is loose, the assignment of maximum value of L_e to all the gates in the circuit provides a leakage optimal design. Since increasing L_e increases the delay as well, V_{th} has to be reduced to satisfy the delay constraint. Hence, V_{th} for GSVL is lower than the V_{th} obtained after performing GSV for the same delay constraint. Thus, as shown in Figure 6, from point A to point B, the increase in L_e is compensated by the reduction in V_{th} and the area remains constant. At point B, the optimal design has the

TABLE V
AREA-LEAKAGE TRADE-OFF AS A RESULT OF INCLUDING L_e IN THE OPTIMIZATION

Delay (ps)	Area		Leakage (10^{-1}) (fA/ns)		Std. Dev. Leakage (10^{-1}) (fA/ns)	
	GSV	GSVL	GSV	GSVL	GSV	GSVL
289	13.54	24.83	13.2	10.3	18.73	12.00
299	13.42	24.89	8.67	6.95	12.25	7.64
309	13.42	23.94	5.81	4.98	8.21	5.25
319	13.21	18.32	3.98	3.41	5.63	3.53
329	13.12	14.78	2.78	2.41	3.93	2.45

minimum value of the $V_{th} = 0.15V$. Hence, beyond point B V_{th} cannot be decreased. Thus the decrease in circuit delay is achieved by reduction in L_e as well as increase in the gate sizes. The leakage savings in the mean leakage by including L_e in the optimization is around 20% more than that obtained using just GSV and around 37% compared to GS. The savings only increase as the delay constraint is tightened.

Table V compares the area, mean and the standard deviation of the leakage of the optimized circuit obtained by using GSV and GSVL for different values of the required time. We see from column 3 that the leakage reduction is obtained by using GSVL but at the cost of the increased area. From columns 5 and 7, we see that the mean of the leakage and the standard deviation of the leakage of the optimal circuit obtained using GSVL is much lower than that of a circuit obtained using only GSV.

VI. CONCLUSIONS

In this paper, we presented a novel methodology for simultaneously varying the threshold voltage, gate sizes and the gate length of a circuit to achieve a minimum leakage circuit. We included the effect of various process variations on different circuit parameters. A function of both mean and variance of the leakage was minimized with constraints on the α-percentile of the circuit delay. Also, to the best of our knowledge, this is the first work to include L_e as a method to reduce leakage variability. We showed that simultaneously using V_{th}, gate sizes and L_e provide significant improvement in the leakage parametric yield. We also demonstrated that we can obtain a considerably better circuit in terms of leakage and area by introducing L_e and V_{th} as decision variables in the optimization problem in addition to the gate sizes.

REFERENCES

[1] International technology roadmap for semiconductors. 2003.

[2] S. Bhardwaj and S. Vrudhula. Leakage minimizatio of nano-scale circuits in the presence of systematic and random variations. In *Proceedings Design Automation Conference (DAC)*, 2005.

[3] D. Boning and S. Nassif. *Models of process variations in device and interconnect, Design of High-Performance Microprocessor Circuits*, chapter 6. IEEE Press, 2000.

[4] S. Borkar et al. Parameteric variations and impact on circuits and microarchitecture. In *Proc. DAC*, 2003.

[5] K. A. Bowman, S. G. Duvall, and J. D. Meindl. Impact of die-to-die and within-die parameter fluctuations on the maximum clock frequency distribution for gigascale integration. *JSSC*, 37(2):183–190, Feb 2002.

[6] S. Boyd, S. J. Kim, L. Vandenberghe, and A. Hassibi. A tutorial on geometric programming. Technical report, www.stanford.edu/~boyd/gp_tutorial.html, 2004.

[7] Y. Cao and L. T. Clark. Mapping statistical process variations toward circuit performance variability: An analytical modeling approach. In *Proc. of DAC*, 2005.

[8] A. R. Conn, N. I. M. Gould, and P. L. Toint. *LANCELOT*. Springer-Verlag, 1992.

[9] W. H. et. al. Total power optimization through simultaneously multiple-v_{DD} multiple-v_{TH} assignment and device sizing with stack forcing. In *Proc. of ISLPED*, 2004.

[10] P. Gupta, A. B. Kahng, P. Sharma, and D. Sylvester. Selective gate-length biasing for cost-effective runtime leakage control. In *Proc. of DAC*, pages 327–330, 2004.

[11] M. Ketkar and S. S. Sapatnekar. Standby power optimization via transistor sizing and dual threshold voltage assignment. In *Proc. of ICCAD*, pages 375 – 378, 2002.

[12] D. Lee, H. Deogun, D. Blaauw, and D. Sylvester. Simultaneous state, Vt and Tox assignment for total standby power minimization. In *Proc. of DATE*, 2004.

[13] C. Long and L. He. Distributed sleep transistors network for power reduction. In *Proc of DAC*, pages 181 – 186, 2003.

[14] M. Mani, A. Devgan, and M. Orshansky. An efficient algorithm for statistical minimization of total power under timing yield constraints. In *ACM/IEEE Design Automation Conference*, 2005.

[15] S. Narendra et al. Full-chip subthreshold leakage power prediction and reduction techniques for sub-0.18-μ/m CMOS. *Journal of Solid-State Circuits*, 39(2):501–510, Feb 2004.

[16] C. Neau and K. Roy. Optimal body bias selection for leakage Improvement and Process Compensation over different technology generations. In *ISLPED*, pages 116–121, 2003.

[17] D. Patil et al. A new method for design of robust digital circuits. In *Proc. of ISQED*, 2005.

[18] M. J. M. Pelgrom, A. C. J. Duinmaijer, and A. P. G. Welbers. Matching properties of mos transistors. *IEEE Journal of Solid-State Circuits*, 24(5):1433–1439, Oct 1989.

[19] S. Raj, S. Vrudhula, and J. M. Wang. A methodology to improve timing yield in the presence of process variations. In *Proc. of DAC*, pages 448–453, 2004.

[20] R. Rao, A. Devgan, D. Blaauw, and D. Sylvester. Parametric yield estimation considering leakage variability. In *Proc. of DAC*, pages 442–447, 2004.

[21] J. Singh, V. Nookala, Z.-Q. Luo, and S. Sapatnekar. Robust gate sizing using geometric programming. In *ACM/IEEE Design Automation Conference*, 2005.

[22] S. Sirichotiyakul et al. Stand-by power minimization through simultaneous threshold voltage selection and circuit sizing. In *Proc. of DAC*, pages 436 – 441, 1999.

[23] A. Srivastava et al. Statistical Optimization of Leakage Power Considering Process Variations using Dual-Vth and Sizing. In *Proc. of DAC*, pages 773–778, 2004.

[24] J. Tschanz et al. Adaptive body bias for reducing impacts of die-to-die and within-die parameter variations on microprocessor frequency and leakage. *IEEE Journal of Solid-State Circuits*, 37(11):1396–1402, 2002.

[25] UC Berkeley Device Group. *BSIM 4.2.1 MOSFET Model - User's Manual*, 2004.

[26] P. M. Vaidya. A new algorithm for minimizing convex functions over convex sets. In *30th Annual Symposium on Foundations of Computer Science*, 1989.

[27] C. Visweswariah, K. Ravindran, K. Kalafala, S. G. Walker, and S. Narayan. First-order incremental block-based statistical timing analysis. In *Proc. of DAC*, pages 331–336, 2004.

2006 Asia and South Pacific Design Automation Conference

9C-4

Statistical Bellman-Ford Algorithm With An Application to Retiming

Mongkol Ekpanyapong Thaisiri Waterwai[†] Sung Kyu Lim

School of Electrical and Computer Engineering
Georgia Institute of Technology
{pop, limsk}@ece.gatech.edu

[†]Dept. of Industrial Engineering and Operations Research
University of California, Berkeley
thaisiri@uclink.berkeley.edu

Abstract— **Process variations in digital circuits make sequential circuit timing validation an extremely challenging task. In this paper, a Statistical Bellman-Ford (SBF) algorithm is proposed to compute the longest path length distribution for directed graphs with cycles. Our SBF algorithm efficiently computes the statistical longest path length distribution if there exist no positive cycles or detects one if the circuit is likely to have a positive cycle. An important application of SBF is Statistical Retiming-based Timing Analysis (SRTA), where SBF is used to check for the feasibility of a given target clock period distribution for retiming. Our gate and wire delay distribution model considers several high-impact intra-die process parameters and accurately captures the spatial and reconvergent path correlations. The Monte Carlo simulation is used to validate the accuracy of our SBF algorithm. To the best of our knowledge, this is the first paper that propose the statistic version of the longest path algorithm for sequential circuits.**

I. INTRODUCTION

Process variations in digital circuits make circuit timing validation an extremely challenging task. Variations on several high-impact intra-die process parameters such as effective gate length, wire width, and so forth, can easily invalidate the timing predictions made before the fabrication [1]. Therefore, statistical timing analysis tools that model gate and wire delay as probability distribution function became increasingly popular to tackle the timing validation under process variations [2], [3], [4]. However, most of the existing works focus on combinational circuits or sub-circuits (after FF removal) and fail to address sequential circuit timing validation directly. Partitioning circuit into sub-circuits and solving the problem on a sub-circuit by sub-circuit basis lead to a sub-optimal solution. By considering the sequential circuit, timing analysis can be done by using longest path algorithms that can handle graphs with negative cycles such as the Bellman-Ford algorithm. There are many CAD algorithms that adopt the Bellman-Ford algorithm including scheduling[5], clock scheduling[6], verification[7], and retiming [8]. A recent work on static timing analysis for sequential circuits [8] allows the users to model FFs and use them to predict the timing information *after* retiming. This work achieves a significant performance improvement by exploiting retiming-aware timing slack. Our goal in this paper is to develop the Statistical Bellman-Ford (SBF) algorithm. In addition, we show an application of SBF on global placement using retiming [8].

In this paper, we first develop a Statistical Bellman-Ford (SBF) algorithm to compute the longest path length distribution for directed graphs with negative cycles. We first prove that a statistical extension of the original Bellman-Ford algorithm correctly computes the longest path length distribution for the true distribution, but it requires infinite amount of time for the continuous distribution. Next, we show that two straightforward extensions of the Bellman-Ford algorithm for statistical analysis can not guarantee the correctness of the results. Lastly, we propose our SBF algorithm that closely approximates and efficiently computes the statistical longest path length distribution if there exists no positive cycles or detects one if the circuit is likely to have a positive cycle. Our SBF algorithm is integrated into SRTA, where SBF checks for the feasibility of a target clock period distribution for retiming. We show that the final critical path delay distribution after retiming is the statistical maximum among all primary outputs and all feedback vertices. The Monte Carlo simulation is used to validate the accuracy of our SRTA algorithm.

The remainder of the paper is organized as follows. Section II presents our statistical Bellman-Ford algorithm. Section III presents our statistical retiming-based algorithm and its application in retiming. We present the experimental results in Section IV and conclude in Section V.

II. STATISTICAL BELLMAN-FORD ALGORITHM

We first provide the analysis of statistical longest path algorithm for the sequential circuit including its properties. Next, we show that the simple modified Bellman-Ford algorithms can not compute the statistical longest path correctly. Finally, we propose a modified version of the Bellman-Ford algorithm that closely approximates the statistical longest path length distribution for the sequential circuit.

A. Statistical Longest Path Analysis

We first introduce a stochastic version of the Bellman-Ford algorithm that correctly solves the stochastic longest path problem for true distribution. Before we do so, we first introduce some quantities in the probability theory that are required to develop algorithms. For more precise definitions of the quantities, see [9], [10]. Let Ω be the set of outcomes of a fabrication process. A subset of Ω is called an event. Let $\mathbf{P}$ be a function that assigns a probability to each event. A random

0-7803-9451-8/06/$20.00 ©2006 IEEE.

959

variable $\mathbf{X} : \Omega \to \mathbb{R}^*$ maps each outcome $\varpi \in \Omega$ to a number in the extended real line $\mathbb{R}^* \triangleq [-\infty, \infty]$. The probability that a random variable $\mathbf{X}$ takes a value in a subset M of $\mathbb{R}^*$ is $\mathbf{P}[\varpi | \mathbf{X}(\varpi) \in M]$. Assume that the probability $\mathbf{P}$ determines the joint (and hence, marginal and conditional) distributions of all random variables of interest. Let $G = (V, E)$ be a directed graph with a source node s and a sink node t, and $w : E \times \Omega \to \mathbb{R}$ be an associated edge-length function, which is a random variable for each edge $(u, v) \in E$. We assume without loss of generality that there is no weight on nodes (since we can push the weights on nodes to their fan-in edges). Let K denote the number of directed *simple* (i.e., no cycles) $s - t$ paths in G, and $l_i : \Omega \to \mathbb{R}$ denote the length of the ith path, $i = 1, \dots, K$. Also, let $G(\varpi)$ be the graph G with length $w(u, v)(\varpi)$ on edge $(u, v) \in E$. If $\mathbf{X}$ is the longest path of G, it is defined as follows: for each $\varpi \in \Omega$, $\mathbf{X}(\varpi) = \max\{l_1(\varpi), \dots, l_K(\varpi)\}$, if there is no positive cycle in $G(\varpi)$, and $\mathbf{X}(\varpi) = \infty$, otherwise. The distribution of $\mathbf{X}$ is determined by the probability measure $\mathbf{P}$ as mentioned above. We define the *Statistical Longest Path Problem* as that of finding the distribution of the longest $s - t$ path in $G = (V, E)$ with edge length function $w : E \times \Omega \to \mathbb{R}$.

We extend the Bellman-Ford (BF) algorithm to obtain the outcome-by-outcome Statistical Bellman-Ford algorithm (oSBF). An illustration is shown in Figure 1. The algorithm is similar to the original Bellman-Ford algorithm but it is called for each outcome untill all possible outcomes are computed. The algorithm starts by first initialize all variables in the initialization step. The value of $a[v]$ represents the arrival time of node v. At the beginning, the arrival time of all nodes is set to $-\infty$, except the source node that has the value zero. After that, the relaxation step is called. Similar to the Bellman-Ford algorithm, the algorithm will stop when there is no update. During the relaxation, for a given outcome, the algorithm checks for all edges in the graph whether the value of sink node of each edge is greater than the summation of the gate delay of source node and the wire delay of that edge or not (the delay contraint). The algorithm stops when there is no update in the graph that is the value of sink node is greater than or equal to the summation of the delay of source node and the wire delay on the edge. During the positive cycles detection, the algorithm checks for each edge whether, is there any edge that violate the delay constraint. If there is an violation, the algorithm return false, otherwise the algorithm returns true with delay $a[t]$ of the sink node.

As opposed to updating a certain set of *constants* (such as arrival times $a[i]$) in the BF, at each step of the oSBF, we are required to update certain random variables that are *functions* on Ω by updating their values for each outcome $\varpi \in \Omega$. As a result, the complexity of the oSBF is high when the numbers of possible outcomes are large. In fact, when random variables are continuous such as uniform random variables, Ω has uncountably many elements, and the oSBF cannot be carried out in practice (require infinite runtime). However, its properties, which we prove here, are useful for our approximation algorithm, presented in the next section. Note that Monte Carlo simulation can be considered as an

---

oStatistical Bellman-Ford(G, w, s, t)

Initialization Step
for (each $v \in V$)
  $a[v](\varpi) \leftarrow -\infty, \forall \varpi \in \Omega$;
  $a[s](\varpi) \leftarrow 0, \forall \varpi \in \Omega$;
$Stop(\varpi) \leftarrow$ NO, $\forall \varpi \in \Omega$;
$g(\varpi) \leftarrow -1, \forall \varpi \in \Omega$;
$iter \leftarrow 1$;

Relaxation Step
while ($Stop(\varpi) = $ NO for some $\varpi \in \Omega$ and $iter < |V|$)
  $iter \leftarrow iter + 1$;
  $\widetilde{\Omega} \leftarrow \{\varpi \in \Omega | Stop(\varpi) = $ NO$\}$;
  $Stop(\varpi) \leftarrow$ YES, $\forall \varpi \in \widetilde{\Omega}$;
  for (each $v \in V$)
    for (each $\varpi \in \widetilde{\Omega}$)
      for (each edge $(u, v) \in E$)
        if $a[v](\varpi) < a[u](\varpi) + w(u,v)(\varpi)$;
        then $a[v](\varpi) \leftarrow a[u](\varpi) + w(u,v)(\varpi)$;
            $Stop(\varpi) \leftarrow$ NO;

Positive Cycles Detection Step
for (each $(u, v) \in E$)
  for (each $\varpi \in \Omega$)
    if ($a[v](\varpi) < a[u](\varpi) + w(u,v)(\varpi)$)
      $g(\varpi) \leftarrow +1$;

Output Step
if ($\mathbf{P}[\varpi | g(\varpi) = +1] > 0$)
  then return FALSE;
  else return TRUE and $a[t]$;

Fig. 1. A description of outcome-by-outcome Statistical Bellman-Ford (oSBF) algorithm

approximated version of the oSBF in which certain elements of Ω are sampled according to probability $\mathbf{P}$. The following theorem proves the correctness of the oSBF algorithm.

Theorem 1: If $\mathbf{P}[\varpi | G(\varpi)$ has a positive cycle$] = 0$, then $a[i]$ from the oSBF has the same distribution as that of the random variable representing the longest path from s to i, $i \in V$. Otherwise, the oSBF returns FALSE.

Proof: For each outcome $\varpi \in \Omega$, the oSBF is equivalent to the BF, which correctly calculates the lengths of the longest $s - i$ paths $a[i](\varpi), i \in V$ or correctly identifies a positive cycle in $G(\varpi)$. Hence, when the oSBF terminates, the function (random variable) $a[i]$ and the longest $s - i$ path are equal on $\Omega_1 \triangleq \{\varpi \in \Omega | G(\varpi)$ has no positive cycles$\}$ and are different on $\Omega_2 \triangleq \{\varpi \in \Omega | G(\varpi)$ has a positive cycle$\}$. If $\mathbf{P}[\Omega_2] = 0$, then they are equal with probability one, and hence, have the same distribution. Otherwise, there is a positive probability of having a positive cycle. Therefore, the oSBF returns FALSE. ∎

We define backward edges and nodes as follows:

Definition 1: For a given order of nodes P, $(u, v) \in E$ is a *forward edge* if u precedes v in P, and a *backward edge* otherwise. In the latter case, node u is called a *backward node*.

Lemma 1: If a node order $L = \{s, v_1, \dots, v_n, t\}$ is used in the relaxation step of the oSBF, after j relaxation iterations, $a[i]$ from oSBF is the random variable representing the longest (possibly not simple) $s - i$ path in G that contains $j - 1$ or fewer (possibly repeated) backward edges.[1]

---

[1] Similar proof was shown in [11]

Proof: Let B denote the set of all backward edges associated with the order L. Then the subgraph $G_B \triangleq (V, E \setminus B)$ is a directed acyclic graph (DAG). Now we recall the update step for node i in the relaxation step:

$$a[i] := \max_{u \in FI(i)} \Big[a[u] + w(u, i) \Big], \qquad (1)$$

where $FI(i) \triangleq \{u \in V | (u, i) \in E\}$ is the set of *fan-ins* of node i. At the first iteration of the relaxation step, when $a[i]$ is updated, $a[u] = -\infty$ for all *backward fan-ins* $u \in FI_b(i) \triangleq \{u \in FI(i) | (u, i) \in B\}$ because from Definition 1, nodes $u \in FI_b(i)$ come after node i in the order, and hence, have not been updated. Thus, at the first iteration of the relaxation step, it is sufficient to perform relaxation on G_B. Since G_B is a DAG, and L is a topological order of G_B, $a[i]$ represents the longest $s - i$ path that contains zero backward edge after the first relaxation step.

Now suppose that the result of Lemma 1 is true up to some $j \geq 1$ (Induction Hypothesis 1: IH 1). At iteration $j + 1$ of the relaxation step, we will show by induction that after $a[i]$ is updated using (1), it represents the longest $s - i$ path containing at most j backward edges. Consider the update for node v_1. Since s is the only node that precedes v_1 in L, all other nodes $u \in FI(v_1)$ are all updated after v_1. By IH 1, $a[u]$ represents the longest $s - u$ path with at most $j - 1$ backward edges for all $u \in FI(v_1)$. From (1), the updated $a[v_1]$ is the longest $s - v_1$ path with at most j backward edges because any $s - v_1$ path with $c > 0$ backward edges has the last edge being a backward edge, and removing such an edge results in a path with $c - 1$ backward edges.

Suppose that $a[v_i], i = 1, \ldots, r$ are now the longest $s - v_i$ paths with at most j backward edges for some $r \geq 1$ (Induction Hypothesis 2: IH 2). Similarly, from (1), the updated $a[v_{r+1}]$ is the longest $s - v_{r+1}$ path with at most j backward edges because any $s - v_{r+1}$ path with $c > 0$ backward edges either has the last edge being a backward edge and removing such an edge results in a path with $c - 1$ edges (this case corresponds to $u \in FI_b(v_{r+1})$, whose $a[u]$ are, from IH 1, the longest $s - u$ paths with at most $j - 1$ backward edges), or has the last edge being a forward edge and removing such edge results in a path with c backward edges (this case corresponds to $u \in FI_f(v_{r+1}) \triangleq \{u \in FI(v_{r+1}) | (u, v_{r+1}) \notin B\}$ whose $a[u]$ are, from IH 2, the longest $s - u$ path with at most j backward edges). This completes the proof. ∎

The following theorem improves the bound on the number of iterations of the relaxation step when there is no positive cycle. In the oSBF, the algorithm automatically terminates once the number of relaxation iterations reaches this bound (by the condition $Stop(\varpi) = \text{YES}$ for all $\varpi \in \Omega$). However, this is not true when the distribution of each $a[i]$ is approximately updated. Hence, the bound from this theorem will be useful for our approximation algorithm.

Theorem 2: If $\mathbf{P}[\varpi | G(\varpi)$ has a positive cycle$] = 0$, and a node order $L = \{s, v_1, \ldots, v_n, t\}$ is used in the relaxation step of the oSBF; then after $k + 1$ iterations, $a[i]$ from oSBF has the same distribution as that of the random variable representing the longest path from s to i, $i \in V$, where k is the maximum

number of connected backward nodes that can be in a simple $s - t$ path in G.

Proof: Since G has no positive cycles with probability 1, the longest $s - t$ path is a simple path with probability 1. According to the definition of k, the longest path has at most k backward nodes and hence, at most k backward edges. The theorem follows from Lemma 1. ∎

The result from Theorem 2 can be applied to approximation methods that are similar to the oSBF, except at each iteration $a[i], i \in V$ are *approximately* updated. More specifically, if $a[i]$ is a good approximation to the actual $a[i]$ obtained from the oSBF, then after $k + 1$ iterations, $a[i]$, obtained from the approximation method, is also a good approximation to the actual longest $s - i$ path. Hence, when there is no positive cycle with probability one, we can stop the approximation algorithm after $k + 1$ iterations.

B. Limitation of the Bellman-Ford Extensions

To the best of our knowledge, all proposed analytical models for statistical timing analysis suffer from the error introduced by the maximum function. This is because the output of the maximum function results in a new form of distribution. Unlike the true distribution, Bellman-Ford can return incorrect results because of the error from approximated distributions (such as the normal distribution approximation for the delay distribution). This is because the original Bellman-Ford algorithm is not designed to tolerate such error from stochastic computation. More precisely, it is typically assumed that the joint distribution of arrival times is fully characterized by vectors of parameters $\theta_i, i \in V$, which belong to a certain set Θ, which is closed under addition[2]. For example, when each node is assumed to be independently normally distributed, a two-dimensional vector $[\mu_i, \sigma_i^2]' \in \mathbb{R} \times [0, \infty)$ can be used to describe the mean and variance of the arrival time of node i. Let $f_{\max} : \Theta \times \Theta \to \Theta$ denote the maximum function that approximates the distribution of the maximum by a distribution characterized by a vector in Θ. In this section, two examples are used to demonstrate the drawback of simple extensions of Bellman-Ford algorithms.

The first extension is to use the Bellman-Ford algorithm to compute longest path length distribution as it is in a statistical timing analysis. Based on the Bellman-Ford algorithm, after all vertices are visited and there is still an updated edge, the algorithm will report a positive cycle. Because of the error from the approximation, it is possible that the algorithm can keep update the graph after $|V|$ iterations even when there is no positive cycle. Given two distributions of the same type, e.g., Gaussian distribution, the maximum function of two distributions is likely to exhibit the new distribution which is different from the input distributions. To make the timing analysis simple, the approximated maximum function is used instead that is assuming that the output of maximum distribution results in the same distribution as the input. Note that the good approximated maximum function should result in the

---

[2] A set A is closed under addition if for any $a, b \in A$, we have $a + b \in A$. This assumption leads to efficient propagation procedure which, however, can be relaxed.

9C-4   2006 Asia and South Pacific Design Automation Conference

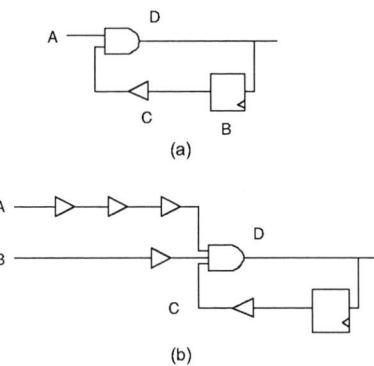

Fig. 2.   Illustration of Bellman-Ford update

new distribution that is greater than both input distributions. If the graph has at least a cycle, it is possible that the error from the approximated maximum distribution can result in the repeatly update of the graph over the cycle even though there is no new information propagated from that path or positive cycle. An illustration is shown in Figure 2(a). At the first iteration, an input arrives at node A with value a. Flip-flop B, buffer C, and gate D have delay b, c, and d, respectively. The output value of node D is $D^{(1)} = f_{\max}(a, C^{(0)}) + d = a + d$, where $C^{(0)}$ denotes the vector describing the distribution of the arrival time of gate C at iteration 0. After the signal propagates through flip-flop B and gate C, the new value of node D becomes $D^{(2)} = f_{\max}(a, a + d + b + c) + d$. Let $\triangle^{(2)} = D^{(2)} - D^{(1)}$ denote the change in the value of node D. Now if $d + b + c$ is originally negative with probability one, but its distribution is approximated by that of a negative-mean random variable with a small probability of being positive (for example, a normal random variable with mean -10 and variance 9), then we cannot guarantee that $\triangle^{(2)}$ will be a zero vector. Alternatively, even if the true distribution is initially used, the error from the maximum function could result in the same situation. After one more iteration of the Bellman-Ford, $D^{(3)} = f_{\max}(a, D^{(2)} + b + c) + d = D^{(2)} + \triangle^{(3)}$. A similar argument shows that $\triangle^{(3)}$ may not be a zero vector either. We observe from related experiments that $\triangle^{(i)}, i = 1, 2, \ldots$ are small but might not become zero vectors after the $|V|$ iterations; consequently the algorithm reports a positive cycle.

The second extension is to introduce error bound on longest path length distribution updates, which is named eSBF (error-bounded Statistical Bellman-Ford) algorithm.[3] Specifically, when the change in the distribution (for example the norm of $\triangle^{(i)}$) is less than such a bound, we consider it as no update. Although, imposing a positive error bound δ can help the Bellman-Ford algorithm terminate when the graph has no positive cycle, it could cause the Bellman-Ford to stop too early when δ is too large. Consequently, from Lemma 1, some paths are not considered if it stops before $k + 1$ iterations. Figure 2(b) shows an example in which the delay from path A is ignored as follows: Assume that the arrival time value of node D at iteration i is D^i. The total delay on path A,B, C, and gate D are a, b, c, and d respectively. If the change of the

---

[3]We use this algorithm in comparison with other Bellman-Ford extensions.

> Statistical Bellman-Ford$(G, w, s, t, \mathbf{P}a)$
>
> *Reachability Check Step*
> DFS(s); // find all backward edges
> $backward\_node \leftarrow 1$;
> for (each $v \in V$)
>   if (backward edge connected to v)
> $backward\_node \leftarrow backward\_node + 1$;
> $list\_backward\_node \leftarrow v$;
> $max\_k \leftarrow 0$;
> for (each $v \in list\_backward\_node$)
> $k \leftarrow$ DFS'(v); // backward nodes connected with v
>   if $(max\_k < k)$
> $k \leftarrow max\_k$;
>
> *Initialization Step*
> for (each $v \in V$)
> $a[v] \leftarrow -\infty$;
> $a[s] \leftarrow 0$;
>
> *Relaxation Step*
> for $(iter \leftarrow 1$ to $k + 1)$
>   for (each $v \in V$)
> $a[v] \leftarrow max_{u \in FI(v)} \Big[a[u] + w(u, v) \Big]$;
>
> *Checking Positive Cycles Step*
> $\mathbf{P}(cycle) \leftarrow check\_pos\_cycle()$;
>
> *Output Step*
> if $(\mathbf{P}(cycle) \leq \mathbf{P}a)$
>   then return FALSE;
>   else return TRUE and $a[t]$;

Fig. 3.   k-Statistical Bellman-Ford algorithm (kSBF) used in our SRTA (statistical retiming-based timing analysis)

arrival time at D resulting from delay propagated through A, which is $\triangle^{(i+1)} = f_{\max}(f_{\max}(D^{(i)} + c, b), a) + d - D^{(i)}$, is considered to be small with respect to the error bound δ, and the arrival time of node A has no further update, the algorithm could terminate without updating D. The information from path A is hence not propagated to the calculation of some other arrival times. As a consequence, the distribution of some $s - t$ paths that contain path A is not accounted for. Depending on the circuit structure, the total error due to this early termination could result in large error in the arrival times.

C. Statistical Bellman-Ford Algorithm

Our last extension of Bellman-Ford algorithm, named k-Statistical Bellman-Ford (kSBF), is shown in Figure 3. This is an algorithm that closely approximates and efficiently computes the longest path length distribution of directed graphs with negative cycles. We thus use kSBF in our SRTA (statistical retiming-based timing analysis) introduced in the next section. First, a depth first search algorithm is called to identify all backward edges and sort the nodes in a topological order. For each backward node, we call the depth first search DFS' by setting this backward node as a source node. DFS' returns the maximum number of connected backward nodes reachable by a simple path from the given source. The maximum number of connected backward nodes of the graph $(=k)$ is the largest number obtained by the DFS' algorithm. Note that this reachability algorithm needs to be performed only once. If all backward nodes are likely to be connected, the reachability step is not required, and instead, the total number

962

of backward nodes can be used.

After the maximum number of connected backward nodes of the graph is found, we initialize the arrival times of all nodes. Next, the relaxation step is called. For stochastic longest path, $k + 1$ iterations are required (from Theorem 2). After the relaxation is done, all simple paths from source to sink are considered according to Theorem 2. Then the stochastic positive cycle detection algorithm is used. We implement a stochastic positive cycle detection algorithm proposed in [12]. The positive cycle detection algorithm starts by first finding all backward edges. Then, it randomly pick a backward edge. Then, it creates a new graph G' by assigning sink node of the backward edge to be a new source node of G' and a source node of that backward edge to be a new sink node of G'. Then, depth first search is performed on G' to find the new set of backward edges and then remove this new set of backward edges from G'. After that, the maximum delay of G' is computed from source node to sink node of the graph (G'). The algorithm is randomly performed for M iterations, when M is an input parameter. Finally, the algorithm computes the probability of having positive cycle, the probability of the maximum of the delay distribution of all new M sink node greater than zero. If the probability of having no positive cycle is less than an acceptable probability, it returns FALSE, otherwise returns TRUE.

III. STATISTICAL RETIMING ANALYSIS

The reason that global placement is employed because the process variations can affect both gate and wire. With the increasingly important of wire delay, any optimization/modelling technique should target both gate and wire delay. In addition, spatial correlation information will be available only after placement. Note that Statistical Retiming based Timing Analysis (SRTA) is used to compute the timing solution of final placement only.

A. Modelling Delay Distribution

In this paper, the delay distribution model is based on the first order delay model from [4]. We assume that each gate and wire has the Gaussian distribution. Elmore delay model is used for wire delay computation based on the following equation (similar to [2]):

$$
\begin{aligned}
d_{int} = \ & d_{int}^0 + \sum_{i \in \Gamma_g} [\frac{\partial d}{\partial L_g^i}] \triangle L_g^i + \sum_{i \in \Gamma_g} [\frac{\partial d}{\partial W_g^i}] \triangle W_g^i \\
& + \sum_{i \in \Gamma_{int}} [\frac{\partial d}{\partial T_{int}^i}] \triangle T_{int}^i
\end{aligned}
$$

where d_{int}^0 is expected value of wire delay. Γ_g and Γ_{int} are the set of grids where all the receiver reside and the interconnect tree traverses, respectively. $\triangle L_g^i$, $\triangle W_g^i$, and $\triangle T_{int}^i$ are random variables representing the variation over the expected value of transistor length, transistor width, and metal thickness respectively. The differentiations are derived based on transistor and wire delay model from [13], [14].

Principal component analysis technique (PCA), similar to [2], is used to derived the first-order form for arrival time delay distribution. The basic idea of PCA is to classify input coefficient into orthogonal terms so that each coefficient term is uncorrelated. Reconvergent correlation can be efficiently handled by PCA. We use a grid hierarchical model for spatial correlation [2]. If two gates are located near each others, they are more correlated than putting them far apart.

There are four operations involved during statistical sequential arrival time computation: maximum, minimum, addition, and subtraction operations. The addition and subtraction of two Gaussian distributions result in another Gaussian distribution. The coefficient of each term in the first order model can be added and subtracted directly. Maximum and minimum functions require the tightness probability calculation [4], which is derived from [15]. Based on this model and the assumption that the maximum and/or minimum of two Gaussian distributions result in a new Gaussian distribution, the coefficient results can be expressed as the summation of product between input distributions and tightness probabilities.

In this paper, wire delay is computed based on Elmore delay model. Since the actual wire distance is not known until routing has been done, the approximated analytical model similar to [16] is used instead. We assume 10% variations in each process parameter terms.

B. Bounds on Target Clock Period

Here, we provide a theoretical result on the bounds of the target clock period, ϕ, which will be useful in the binary search procedure. Recall that the target clock period is set to the smallest value for which the graph $G = (V, E)$ has no positive cycle, and the arrival time of the sink node, $a[t]$, is no larger than ϕ. Let the delay of the ith directed simple $s - t$ path in G be represented by $l_i = \psi_i - \kappa_i \phi$, where ψ_i denotes the sum of gate and wire delays along path i, and κ_i denotes the number of flip-flops in path i. Let C denote the number of directed cycles in G, and $\zeta_j = \xi_j - \sigma_j \phi$ denote the total delay of the jth directed cycle, where ξ_j denotes the sum of gate and wire delays along cycle j. σ_j denotes the number of flip-flops in cycle j. Now ϕ is the smallest number that satisfies

$$\phi \geq a[t] = \max_{i=1,\dots,K} \{\psi_i - \kappa_i \phi\} \tag{2}$$

$$\zeta_j = \xi_j - \sigma_j \phi \leq 0 \qquad\qquad j = 1\dots,C. \tag{3}$$

Equivalently, the target clock period is given by

$$\phi = \max \left[\max_{i=1\dots,K} \left\{ \frac{\psi_i}{\kappa_i + 1} \right\}, \max_{j=1\dots,C} \left\{ \frac{\xi_j}{\sigma_j} \right\} \right] \tag{4}$$

Recall that each gate and wire delay is a random variable, and hence, ψ_i and ξ_j, which are sums of gate and wire delays, are random variables. Let $\phi_d^l, \phi_d^m, \phi_d^u$ denote the values of ϕ obtained from (4) when all gate and wire delays are replaced by their lower bounds (best case), means (average case), and upper bounds (worst case), respectively. It is obvious that ϕ, which is a random variable, is in $[\phi_d^l, \phi_d^u]$ with probability one. Moreover, as we show in the theorem below, the mean of ϕ is bounded below by ϕ_d^m.

Theorem 3: Let $\mathbf{E}[\phi]$ be the mean (expected value) of ϕ. Then $\phi_d^l \leq \phi \leq \phi_d^u$ with probability one, and $\phi_d^m \leq \mathbf{E}[\phi]$.

Proof: Only the mean case needs a proof. Equation (4) implies

$$\phi \geq \frac{\psi_i}{\kappa_i + 1}, \quad i = 1, \ldots, K \qquad \phi \geq \frac{\xi_j}{\sigma_j}, \quad j = 1, \ldots, C$$

Since the ψ_i and ξ_j are the sum of gate and wire delays, and κ_i and σ_j are constants, the expected values of the right-hand-side terms of the inequalities above can be obtained by replacing all gate and wire delays by their means. As a result, taking the expectation on both sides of the inequalities shows that the mean of ϕ is greater than each right hand side term under a deterministic average case. This implies that the mean of ϕ is greater than the maximum of all such terms (ϕ_d^m). ∎ Note that the bounds $\phi_d^l, \phi_d^m, \phi_d^u$ can be obtained by solving deterministic longest path problems.

IV. EXPERIMENTAL RESULTS

Our algorithms are implemented in C++/STL, compiled with gcc v3.2.2, and run on a Pentium IV 2.4 GHz machine. The benchmark set consists of six big circuits from ISCAS89 and five big circuits from ITC99 suites.

Table I shows a comparison of the results obtained by using Monte Carlo simulation, the modified Bellman-Ford algorithm with the error bound (eSBF), and the modified Bellman-Ford with $k + 1$ iterations (kSBF) on 8x8 dimension. Monte Carlo simulations are performed using 10,000 samples. We report the expectation (mean) and standard deviation (sigma) of the retiming delay distribution. We note that both eSBF and kSBF provide close results to Monte Carlo simulation results, especially in terms of the mean value. kSBF provides more accurate results than eSBF because, as pointed out in section II, eSBF can ignore some paths during its computation. Note that it is possible that eSBF can outperforms kSBF since kSBF may require more iterations than necessary. Because of the error arising from the analytical model, the higher the number of iterations, the more the error that is accumulated. However, most of the cases, kSBF is more accurate that eSBF. Both eSBF and kSBF substantially outperform Monte Carlo simulations in terms of runtime. The runtime reported is average runtime. Note that before using the bounds on target clock period, average runtime of kSBF is 9.4 hours. The bound can help reduce runtime substantially. Thus, we conclude kSBF, however, requires longer runtime than eSBF because of the higher the number of iterations required compared to eSBF. Figure 4 shows the comparison among Monte Carlo simulations, eSBF, and kSBF on s5378 benchmark. The solid, dotted, and dashed lines represent Monte Carlo simulation, eSBF, and kSBF, respectively. Results show that kSBF can provide a distribution similar to that of Monte Carlo simulations. In this case, it shows that kSBF is better since eSBF stops early and can ignore some paths.

V. CONCLUSIONS

In this paper, a Statistical Bellman-Ford (SBF) algorithm is proposed to compute the longest path length distribution for directed graphs with cycles. We use SBF in our Statistical Retiming-based Timing Analysis (SRTA), where SBF is used to check for the feasibility of a given target clock period

TABLE I

THE COMPARISON BETWEEN DETERMINISTIC MONTE CARLO
SIMULATION, ESBF, AND KSBF ON 8X8 DIMENSION

ckt	monte		eSBF		kSBF	
	mean	std.dev.	mean	std.dev.	mean	std.dev.
s5378	179.65	6.27	163.29	5.49	179.47	6.51
s9234	229.24	16.18	164.58	7.14	225.53	10.51
s13207	304.93	13.27	279.47	8.72	305.76	13.05
s15850	363.16	15.83	317.32	7.57	363.37	15.72
s38417	187.11	4.53	184.95	5.28	189.00	8.41
s38584	437.02	19.41	399.62	10.45	436.75	19.33
b14o	161.19	10.70	117.42	5.37	160.43	5.31
b15o	247.00	10.00	212.72	18.45	247.88	10.23
b20o	259.68	9.03	229.16	6.12	267.12	9.33
b21o	267.57	6.18	240.01	4.08	267.98	9.24
b22o	286.63	18.92	300.49	25.17	320.62	15.26
runtime	21 days		2.7 hours		3.7 hours	

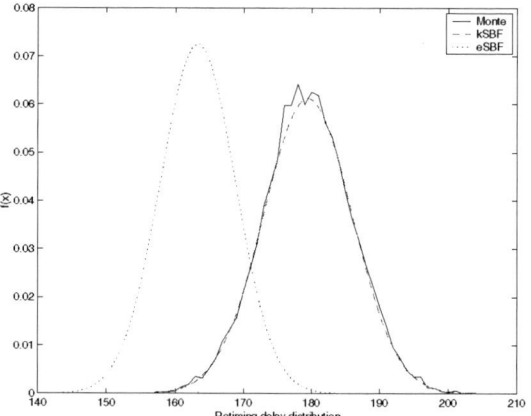

Fig. 4. The Distribution Comparison among Monte Carlo simulation (solid), eSBF (dotted), kSBF(dashed)

distribution for retiming. Our Monte Carlo simulation validates the accuracy of our SRTA algorithm.

REFERENCES

[1] K. Bowman, S. Duvall, and J. Meindl, "Impact of die-to-die and within-die paraemter fluctuations on ..." in *Proc. ISSCC*, 2001.

[2] H. Chang and S. Sapatnekar, "Statistical timing analysis considering spatial correlations using ..." in *Proc. ICCAD*, 2003.

[3] A. Agarwal, V. Zolotov, and D. Blaauw, "Statistical timing analysis using bounds and selective enumeration," in *IEEE TCAD*, 2003.

[4] C. Visweswariah, K. Ravindran, K. Kalafala, S. Walker, and S. Narayan, "First-order incremental block-based ..." in *Proc. DAC.*, 2004.

[5] G. D. Micheli, *Synthesis and Optimization of Digital Circuits*. McGraw-Hill, 1994.

[6] T. G. Szymanski, "Computing optimal clock schedules," in *Proc. DAC.*, 1992.

[7] Y. S. F-R. Boyer, El M. Aboulhamd, "An efficient verification method for a class of multi-phase sequential circuits," in *ICECS*, 2000.

[8] J. Cong and S. K. Lim, "Retiming-based timing analysis with an application to ..." *IEEE TCAD*, vol. 23, no. 12, 2004.

[9] R. Durrett, *Probability: Theory and Examples*. Duxbury Press, 1995.

[10] P. Billingsley, *Probability and Measure*. John Wiley & Sons, 1995.

[11] Y.-Z. Liao and C. K. Wong, "An algorithm to compact a vlsi symbolic layout with mixed constraints," *IEEE TCAD*, 1983.

[12] R. Chen and H. Zhou, "Clock schedule verification under process variations," in *Proc. ICCAD*, 2004.

[13] J. Rabaey, A. Chandrakasan, and B. Nikolic, *Digital Integrated Circuits*. Prentice Hall Electronics, 2003.

[14] SIA, "National Techonology Roadmap for Semiconductors," 2003.

[15] C. Clark, "The greatest of a finite set of random variables," in *Operations Research*, 1961.

[16] C. Ababei and K. Bazargan, "Statistical timing driven partitioning for vlsi circuits," in *Proc. DATE*, 2002.

An Exact Algorithm for the Statistical Shortest Path Problem *

Liang Deng
Dept. of Electrical and Computer Engineering
University of Illinois at Urbana-Champaign
ldeng@uiuc.edu

Martin D. F. Wong
Dept. of Electrical and Computer Engineering
University of Illinois at Urbana-Champaign
mdfwong@uiuc.edu

ABSTRACT

Graph algorithms are widely used in VLSI CAD. Traditional graph algorithms can handle graphs with deterministic edge weights. As VLSI technology continues to scale into nanometer designs, we need to use probability distributions for edge weights in order to model uncertainty due to parameter variations. In this paper, we consider the statistical shortest path (SSP) problem. Given a graph G, the edge weights of G are random variables. For each path P in G, let L_P be its length, which is the sum of all edge weights on P. Clearly L_P is a random variable and we let μ_P and σ_P^2 be its mean and variance, respectively. In the SSP problem, our goal is to find a path P connecting two given vertices to minimize the cost function $\mu_P + \Phi(\sigma_P^2)$ where Φ is an arbitrary function. (For example, if $\Phi(x) = 3\sqrt{x}$, the cost function is $\mu_P + 3\sigma_P$.) To minimize uncertainty in the final result, it is meaningful to look for paths with bounded variance, i.e., $\sigma_P^2 \leq B$ for a given fixed bound B. In this paper, we present an exact algorithm to solve the SSP problem in $O(B(V + E))$ time where V and E are the numbers of vertices and edges, respectively, in G. Our algorithm is superior to previous algorithms for SSP problem because we can handle: 1) *general graphs* (unlike previous works applicable only to directed acyclic graphs), 2) *arbitrary edge-weight distributions* (unlike previous algorithms designed only for specific distributions such as Gaussian), and 3) *general cost function* (none of the previous algorithms can even handle the cost function $\mu_P + 3\sigma_P$. Finally, we discuss applications of the SSP problem to maze routing, buffer insertions, and timing analysis under parameter variations.

1. INTRODUCTION

With continued technology scaling, parameter variations have become a major factor that affects circuit performance and could lead to excessive yield loss [1]. Variations in device and interconnect do not appear to be scaling at the same rate with the technology. For technology nodes of 65 nm or below, design methodologies that consider this important issue are needed to avoid over-pessimistic design or yield loss in manufacturing.

Graph algorithms are widely used in VLSI CAD and many CAD problems can be formulated as shortest path problems. Traditional shortest path algorithms can handle graphs only with deterministic edge weights. In or-

---

*This work was partially supported by the National Science Foundation under grant CCR-0306244

der to model uncertainty due to parameter variations, we need to use probability distributions for edge weights. In this paper, we consider the statistical shortest path (SSP) problem. Given a graph G in which edge weights are random variables. For each path P in G, let L_P be its length which is the sum of all edge weights on P. Clearly L_P is a random variable. Let μ_P and σ_P^2 be the mean and variance, respectively, of L_P. In the SSP problem, our goal is to find a path P to minimize the cost function $\mu_P + \Phi(\sigma_P^2)$ where Φ is an arbitrary function. For example, if $\Phi(x) = 3\sqrt{x}$, the cost function is $\mu_P + 3\sigma_P$. This cost function is widely used to measure the performance or yield when the distributions are Gaussian. To minimize uncertainty in the final result, it is meaningful to look for paths with bounded variance, i.e., $\sigma_P^2 \leq B$ for a given fixed bound B.

In this paper, we present an exact algorithm to solve the SSP problem in $O(B(V + E))$ time where V and E are the numbers of vertices and edges, respectively, in G. Our algorithm assumes all edge-weight variances are integers. (For graphs with non-integer edge-weight variances, we can simply discretized the range of real numbers for variances with desirable precisions and apply the algorithm designed for integer variances.) The main idea of our algorithm is to expand G into a larger graph G' by splitting each node into a number of nodes. New edges are added and edge weights (real numbers) are assigned intelligently. The resulting graph G' is guaranteed to be directed acyclic and that the deterministic shortest path in G' gives an optimal path in G. There were some previous efforts on the SSP problem [2–5]. Our algorithm is superior to previous algorithms because we can handle

- general graphs
- arbitrary edge-weight distributions
- general cost function

Note that previous algorithms for the SSP problem were designed for directed acyclic graphs with specific edge-weight distributions such as Gaussian. Moreover, none of the previous algorithms can handle the general cost function $\mu + \Phi(\sigma^2)$. In fact, none of them can handle the important cost function $\mu + 3\sigma$ even for Gaussian distributions.

The SSP problem has many applications in CAD for VLSI. We will briefly discuss its applications to timing analysis, maze routing and buffer insertion under parameter variations. For buffer insertion and timing analysis, the graphs are directed acyclic, but for maze routing the graphs are of general forms. We have applied our SSP algorithm to some problems in these applications and the results are encouraging. For example, in statistical timing analysis, our algorithm can find not only the worst $\mu_P + \Phi(\sigma_P^2)$ delay bound, but also the longest delay path candidates, even the delay distributions at outputs are not Gaussian.

The rest of paper is organized as follows. In Section 2, we will formally present the SSP problem. In Section 3, an efficient algorithm is proposed to find the optimal solution for SSP problem. In Section 4, some techniques are discussed to further improve its efficiency. We will present some CAD applications for the SSP problem in Section 5 and conclude the paper in Section 6.

2. STATISTICAL SHORTEST PATH PROBLEM

Given a directed graph G, we are interested to find a path from a vertex s (called source) to a vertex t (called sink). Such a path is called an $s - t$ path. As stated in the last section, the goal of the statistical shortest path (SSP) problem is to find a path P from s to t in G such that the cost function $\mu_P + \Phi(\sigma_P^2)$ is minimized, where L_P is the length of the path P, μ_P is the mean value of L_P, σ_P^2 is the variance of L_P, and Φ is an arbitrary function.

For each edge e in G, let X_e be its edge-weight distribution, μ_e be the mean of X_e, and σ_e^2 be the variance of X_e. All edge-weight distributions are assumed to be mutually independent. It is well known that if X and Y are independent random variables and $Z = X + Y$, then the mean and variances of Z can be obtained by adding the means and variances of X and Y, respectively.

Consider an $s - t$ path P in G. From the fact that means and variances are additive, it follows that:

$$\mu_P = \sum_{e \in P} \mu_e, \quad \sigma_P^2 = \sum_{e \in P} \sigma_e^2$$

Therefore, if the cost function is of the form $\mu_P + k\sigma_P^2$, then the SSP problem can be easily solved by assigning a real-valued weight $\mu_e + k\sigma_e^2$ to each edge and solving the traditional deterministic shortest path problem. Unfortunately, this approach would not work for the general cost function $\mu_P + \Phi(\sigma_P^2)$, since

$$\Phi(\sigma_P^2) \neq \sum_{e \in P} \Phi(\sigma_e^2)$$

For example, if $\Phi(x) = \sqrt{x}$, we have $\Phi(\sigma_P^2) = \sigma_P$, which is the standard deviation. It is well known that standard deviations are not additive. As we will see later that an important cost function is $\mu_P + k\sigma_P$ (i.e., $\Phi(x) = k\sqrt{x}$), we need to find a new approach to solve the problem.

We now discuss the importance of the cost function $\mu_P + k\sigma_P$. Typically, one would like to design for the worst case. First, let us assume all edge-weight distributions are Gaussian. Since adding two Gaussian distributions results in a Gaussian distribution, all path-length distributions are Gaussian. It is well known that with a Gaussian distribution for L_P, we have

$$P(|L_P - \mu_P| \leq 3\sigma_P) > 0.99$$

Therefore, one can use the cost function $\mu_P + 3\sigma_P$ to minimize for the worst case. As for general edge-weight

distributions, according to the Chebyshev's inequality,

$$P(|L_P - \mu_P| \leq k\sigma_P) > 1 - \frac{1}{k^2}$$

Clearly, the larger the k, the smaller is the "tail" probability. Therefore, we can fix a value for k to define what is considered to be the worst case. It follows that minimizing the cost function $\mu_P + k\sigma_P$ is for minimizing the statistical worst case path length.

As we mentioned in Section 1, we only consider $s - t$ paths P such that $\sigma_P^2 \leq B$ for a given fixed bound B. This is because we want to minimize uncertainty in the final result due to large variance σ_P^2. By bounding the variance of the path length of P, we have the following lemma.

LEMMA 1. *Let P be an $s - t$ path in G with $\sigma_P^2 \leq B$. Let u be a vertex on P and let $Q \subseteq P$ be an $s - u$ path. We have $\sigma_Q^2 < B$.*

PROOF. Let P be $< v_0, v_1, v_2, ..., v_n >$ where $v_0 = s$ and $v_n = t$. Let P_i be the subpath of P from v_0 to v_i. Let e_i be the edge (v_{i-1}, v_i). Note that P_{i+1} is obtained by concatenating P_i with e_{i+1}. Since variances are additive, it follows that $\sigma_{P_{i+1}}^2 = \sigma_{P_i}^2 + \sigma_{e_{i+1}}^2 > \sigma_{P_i}^2$. Thus the path length variances monotonically increase along the path P. Since the maximum path length variance is B which is at $v_n = t$, all other path length variances on the path P are strictly less than B. The lemma follows since u is on P. □

Finally, we assume all edge-weight variances are integers. This assumption allows us to design an algorithm to exactly solve the SSP problem. Note that for graphs with non-integer edge weights, we can discretize the range of real numbers for variances within desirable precision and then apply our algorithm.

3. ALGORITHM

We note that only μ and σ^2 are additive, not $\mu + \Phi(\sigma^2)$ with arbitrary function of Φ. So it is no longer true that the optimal path must consist of optimal subpaths. Without this *optimal-substructure property* [6], algorithms for classical shortest path problem are not valid for SSP problem.

Our approach to solve the SSP problem is to reconstruct the graph so that the optimal-substructure property is satisfied. We will construct a new graph G' from the original G. G' is a graph with deterministic edge weight, and thus the existing algorithms for classical shortest path problem can be used to find the shortest path in G'. Furthermore, the shortest path in G' must correspond to the shortest path in G whose path-length distribution has the minimum $\mu + \Phi(\sigma^2)$ value.

Note that the edge-weight distribution can be captured by μ and σ^2, and both of them are additive. We modify the graph so that only one of them is stored on the edge. Thus the edge weight becomes a deterministic number. The problem is reduced to how to modify the original graph G into a new graph G' so all edge weights become deterministic without losing the random variable information.

We use the node-splitting technique to achieve this objective. Since the SSP problem is to find optimal $\mu + \Phi(\sigma^2)$, which is linear in μ but non-linear in variance. We will use μ as the new edge weight in G'. Expanded vertices are used to preserve the variance. We split node

$u \in G$ into a set of nodes $\{u_1, u_2, \cdots, u_i, \cdots, u_B\}$ in G'. Note that each $s-u$ path in G has a corresponding $s-u_i$ path in G'. The variance of the length of an $s-u$ path in G must be an integer between 1 and B according to Lemma 1. Node u_i in G' represents the end point of an $s-u$ path in G with path length variance i. For each u_i, we call i the variance-index or just var-index, and write var-index$(u_i) = i$. According to Lemma 1, we only need to consider paths with path-length variances bounded by B in G. So for a node $u \in G$, we only create B nodes in G'. Of course, it is not necessary to expand the source node s. We create a vertex s_0 node in G' to represent s.

After vertex splitting, we create the edges in the new graph G'. We have three different cases. First of all, consider the edges from source node s_0. As shown in Figure 1, for any edge from s to a in G, we will create a corresponding edge in G'. It points from s_0 to a_i. Assume the edge-weight of (s, a) has mean μ_e and variance σ_e^2. Then, $i = \sigma_e^2$ and $w(s_0, a_i) = \mu_e$, where $w(s_0, a_i)$ is the deterministic edge weight of (s_0, a_i).

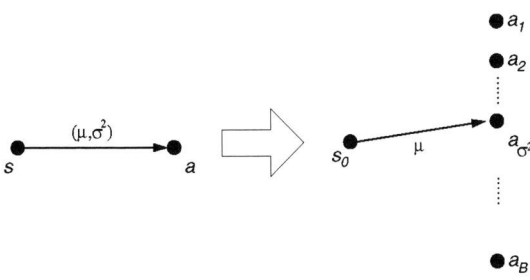

Figure 1: Illustration of node splitting for the source

Second, consider any edge $e = (u, v)$ in G where $u \neq s$ and $v \neq s$. μ_e and σ_e^2 are mean and variance of the edge weight of (u, v), respectively. As shown in Figure 2, u and v are divided into two sets of nodes in G'. Again, because of the additive property of variance, we will create the edge point from u_i to v_j, where i and j satisfy the following equation:

$$j = i + \sigma_{uv}^2$$

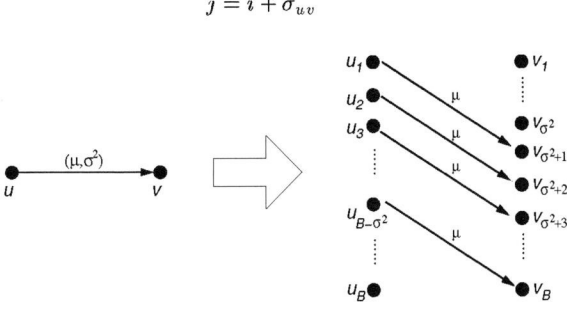

Figure 2: Illustration of node splitting

The edge weights are assigned as μ_e as illustrated. We also notice the created edges in G' are non-crossing. Thus, for each edge in the original graph G, there will be less than B edges created in G'

Finally, consider the sink node t in G. After vertex splitting, the sink node t in G is also divided into B

nodes in G'. To find the shortest path from s to t, we have to create a new dummy node in G' to represent t in G. As shown in Figure 3, t' is created as the sink node in G'.

According to the vertex splitting and edge creation procedures we discussed, any path from s_0 to t_i corresponds to a path from s to t in G with path-length variance i. To capture the $\Phi(\sigma^2)$ term in the cost function, we assign $\Phi(i)$ as the edge weight from t_i to t'.

Thus the new deterministic graph G' is constructed. The psudocode is shown as BUILDGRAPH. It is obvious that any path P' from s_0 to t' in G' corresponds to a path P from s to t in G. Because for any edge or vertex in the path Q from s_0 to t_i, it is mapped to one edge or vertex in G. So Q corresponds to a distinct path P in G. In G', t_i is connected to t by one edge, which implies the P' also corresponds to a distinct Q. Thus P' must corresponds to one P in G.

We can also prove that distinct P in G maps to distinct P' in G'. Let P be a path in G. For any sub-path of P from s to u, the path-length variance is known and fixed. So u corresponds to a vertex u_i in G'. Assume (u, v) is an edge in P, and v corresponds to vertex v_j in G'. Since only one edge (u_i, v_j) exists in G', for any $(u, v) \in P$, it maps to distinct (u_i, v_j). Thus the one-to-one correspondence between P and P' is proved.

BUILDGRAPH:
    create source node s_0 in G'
    s_0 corresponds to s in G
    for each node u in G except s
        Create B vertices $\{u_1, \cdots, u_B\}$ in G'
    for each edge $e = (u, v)$ in G
        for each node u_i
            $j = i + \sigma_e^2$
            if $j \leq B$
                create edge (u_i, v_j) in G'
                $w(u_i, v_j) = \mu_e$
    Create t' in G'
    for each t_i
        Connect t_i to t'
        $w(t_i, t') = \Phi i$

It is trivial to prove that P' has a path-length $\mu_P + \Phi(\sigma_P^2)$. We have the following theorem:

THEOREM 1. *The shortest path in G' which is created by* BUILDGRAPH *from G corresponds to the optimal path P in G which has the minimum $\mu_P + \Phi(\sigma_P^2)$ for the path-length distribution.*

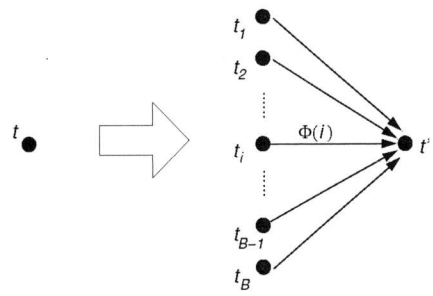

Figure 3: Illustration of node splitting for the sink

Theorem 2 states an important property of G'.

THEOREM 2. *The G' created by* BUILDGRAPH *from G is a directed acyclic graph (DAG). Moreover, G' has at most B levels.*

PROOF. Suppose there is a cycle $C = <v_0, v_1, ..., v_m>$ in G' where $v_0 = v_m$. According to Lemma 1,

$$\text{var-index}(v_0) < \text{var-index}(v_1) < \cdots < \text{var-index}(v_m)$$

This is a contradiction since $v_0 = v_m$. To see that G' has at most B levels, it suffices to show that every path P in G' has at most B edges. Note that the variance-index of the nodes on P is monotonically increasing (Lemma 1). If P has more than B edges, the var-index for the last node u on P has var-index$(u) > B$, contradicting that maximum var-index for every node is B. Therefore P has at most B edges. It then follows that G' has at most B levels. $\square$

It is well known that the shortest path problem in a directed acyclic graph (DAG) can be solved in $O(V + E)$ time. Moreover, existing linear time shortest path algorithms for DAG can handle positive and negative edge weights. Thus $\Phi(\sigma_P^2)$ can be arbitrary function with either positive or negative function values. In G', $E' < B \cdot E$ and $V' < B \cdot V$. So the SSP problem can be solved in $O(B(V + E))$ time.

4. IMPROVED IMPLEMENTATION

Since G' must be a DAG, we can create G' in a topological order. Thus, we don't have to create all B nodes for each vertex in G, which not only speeds up the runtime, but also saves the memory usage.

The first step is to create the s_0 vertex in G' and create those vertices adjacent to s_0 according to the structure of G. This step is similar to the procedure in BUILDGRAPH. For an edge (s, u) in G, we will create the u_i and connect the edge (s_0, u_i). We will also store some information in each created vertex u_i:

- Its parent $\pi(u_i) = s_0$.
- The path-length mean $m(u_i)$ from s to u_i
- The level of this vertex. It is used to represent the topological order. Now the level $l(u_i) = 1$. Here we assume the level of s_0 is 0 because any edge point to s_0 is redundant and can be ignored.

Then assume we already expand G' into level k. For each node u_i in level k, we will create v_j according to the edge $e = (u, v)$ in G, assuming $j = i + \sigma_e^2 \leq B$. If v_j doesn't exist in G', we create v_j, connect (u_i, v_j) and store the following in v_j:

- $\pi(v_j) = u_i$
- $m(v_j) = m(u_i) + \mu_e$
- $l(v_j) = k + 1$

If v_j already exists, we will compare $m(v_j)$ and $m(u_i) + \mu_e$. If $m(v_j) \leq m(u_i) + \mu_e$, it means the path from s through u to v is not a sub-path of optimal path from s to t. We don't do anything on v_j. Otherwise, the path from s through u to v is a better solution, so we update the information in v_j.

We will repeat this procedure to level n. If nodes in level n satisfy either one of the following conditions, we will terminate our algorithm:

- $n = B$. Now all vertices of level n must have the variance equals to B. According to Theorem 2, it cannot create any vertex in G' from these vertices.

- All vertices in level n corresponds to t in G.

We terminate this algorithm by connecting all t_i in G' to t' and calculate $m(t_i) + \Phi(i)$ to find the shortest path.

Figure 4 illustrates an example of our faster approach. Assume we use $B = 25$ to construct G' from G in Figure 4(a). Using BUILDGRAPH, 100 vertices need to be created by vertex splitting. By the improved implementation, the SSP problem can be solved very efficiently as shown in Figure 4(b). It only creates 10 vertices in G'.

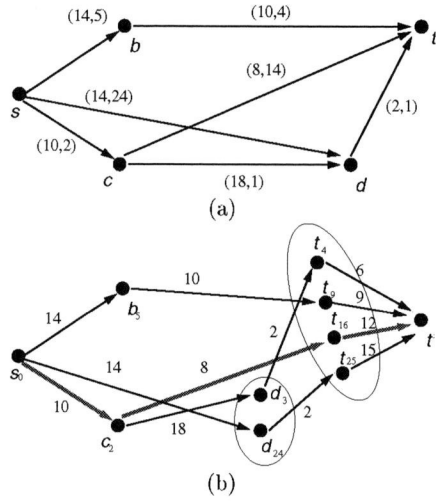

(a)

(b)

Figure 4: Improved implementation. (a) is the original graph G. The edges are annotated by (μ, σ^2) as the edge weights (b) is the expanded graph G'. The shortest path has a minimum $\mu + 3\sigma$ value.

5. APPLICATIONS

Many problems in CAD for VLSI can be formulated to SSP problem when variations become a concern. We will here briefly discuss its applications to maze routing, timing analysis and buffer insertion under parameter variations.

5.1 Maze Routing

Maze routing is to find the shortest path with minimum length in a grid routing problem. Traditionally, edge weights are assigned as real-valued number by cost function. However, parameter variations make it necessary to model the edge weights as random variables. The cost functions are often related to those parameters with variations. Considering the parameter variations, the maze routing problem can be formulated as a SSP problem.

Assume we want to find the optimal routing solution for a critical net. So we use the wire delay as the cost function. Considering the parameter variations, delay is not only the function of wire length. The geometric process variations come from various sources. For example, the 3σ value of wire width variation could reach 25% of nominal wire width [7]. Thus the delay value changes substantially. Because of the systemic variations, the nominal values of wire resistance and capacitance per unit length are no longer uniform within one die. Furthermore, temperature variation can also impact the performance.

Without considering the variation, the cost function is $D = f(\mathbf{P}, T)$, where $\mathbf{P}$ is a set of geometric parameters related to wire delay, and T is the temperature. Now, process variations become significant. $P_i \in \mathbf{P}$ are random variables. All these variations are assumed to have a distribution in Gaussian. We use first order approximation [8] to calculate the delay distribution:

$$D' = f(\mathbf{P_0}, T) + \sum_{P_i \in \mathbf{P}} \frac{\partial f}{\partial P_i} \Delta P_i \qquad (1)$$

where P_0 is a set of the nominal value of parameters. ΔP_i are parameter variables with zero mean. Now the new cost function D' will assign a Gaussian distribution to edge weight. Consider the systemic variation, we use the following equation to calculate the mean value of edge weight:

$$\mu = g(x, y, \mathbf{P_0}, T)$$

The variance can be calculated directly from Equation 1.

Since the path length is now obviously Gaussian, we use $\mu + 3\sigma$ to find the optimal path length. Now the maze routing problem is formulated as a SSP problem. It can be solved by our proposed algorithm.

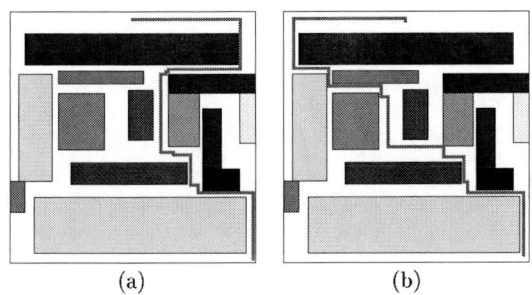

(a)   (b)

Figure 5: Compare the maze routing with or without consider the variations. (a) is the shortest path found by classical shortest path problem and (b) is the result by solving SSP problem.

We have applied our SSP algorithm to solve the maze routing problem. Physical parameters for 65nm technology from ITRS2001 [9] are used. 3σ values of these parameters are set to be 30% of their nominal values. Elmore's delay model is used to calculate the delay these physical parameters. Figure 5 shows one of the comparisons between the traditional approach and our method. The darker region in Figure 5 means higher temperature. To make the routing wire clear, the temperature profiles are only shown in the blocks. Figure 5(a) shows the results without considering variation. To make the comparison fair, we use temperature profile to calculate the mean delay value. Systemic variations are also considered. In Figure 5(b), all variations are considered and the shortest path is found by solving the SSP problem. It is clear that the path in Figure 5(b) intelligently avoids the hot spots.

5.2 Time Analysis

It is always a concern to predict the circuit performance accurately. Precise timing information is needed for circuit optimization to meet the yield or to avoid over design. Timing analysis considering variations is extensively studied recently. Path-based or block-based algorithms have been proposed to find the statistical longest paths and the delay distributions. Path-based approaches usually depend on static timer to find out a set of longest path candidates, and then the statistical approaches can be performed [10, 11]. Block based algorithms can get the delay distributions by propagating the random variables. However, they need path-based analysis to find out the longest paths [12, 13].

Our algorithm can be modified to find the earliest or latest arrival time. To perform timing analysis, a circuit is modeled as a DAG. The delay distributions for cells and interconnects are assigned as the weights of edges. Then our algorithm can be used to find the longest path with maximum $\mu + k\sigma$ value, or the shortest path with minimum $\mu - k\sigma$ value. Since this timing analysis is to find the extreme case of delays, the bound B could be large. With the pruning technologies similar to [14], our algorithm can find the longest or shortest path very efficiently, even if we set B to be infinity.

Another note on our proposed method is that the delay distribution for the longest path is not the delay distribution for the corresponding output. To find the true distribution of an output, we need to take MAX operation on delays of different paths to this output [10]. And even if the path delays are Gaussian, the delay distribution at the output is not necessarily Gaussian. However, we can use the $D_{MAX} = \mu + k\sigma$ value of the longest path as a good bound for delay distribution at output.

Our algorithm stated in Section 3 can also be modified to find longest path candidates. We will check all the edges to the sink node t'. If the delay distribution is not stochastically smaller than the longest path, we will treat it as a candidate of the statistical longest paths.

Table 5.2 shows some experimental results on ISCAS benchmark circuits. Our algorithm is performed to find the longest paths as well as the bound D_{MAX} ($k = 3$). The number of longest path candidates is labeled N_P in the table. Monte Carlo method is used for comparison. We perform 10000 runs of Monte Carlo analysis for each circuit. And the N_U shows the cases in Monte Carlo analysis which delays exceed the bound D_{MAX} (out of 10,000). We also show the mean μ_{mc} and variance σ_{mc}^2 of Monte Carlo analysis results. Since the delay distribution at an output is not Gaussian, $\mu_{mc} + 3\sigma_{mc}$ could lead to a worse bound for timing analysis. Column T shows the runtime for different testbench circuits.

bench	D_{MAX}	T	N_P	μ_{mc}	σ_{mc}	N_U
C432	75.0546	0.05s	95	73.73	0.47	19
C499	45.2776	0.02s	186	43.91	0.49	13
C880	72.6244	0.03s	13	71.26	0.48	20
C1355	61.6536	0.05s	93	60.46	0.43	9
C1908	109.891	0.77s	134	108.27	0.54	4
C2670	103.08	6.07s	2959	102.20	0.62	12
C3540	132.986	3.35s	585	131.05	0.64	13
C5315	121.185	0.71s	19	119.28	0.64	15
C6288	267.311	22.91s	33	265.22	0.775	21
C7552	103.156	0.64	4	101.83	0.445	8

Table 1: Timing analysis on ISCAS benchmark

5.3 Buffer Insertion

Buffer insertion is a widely used interconnection optimization method. It can also be formulated as a shortest path algorithm [15]. Figure 6 illustrates a simple example with three possible buffer locations. The input driver resistance is R_d and the output load capacitance is C_L.

We construct corresponding graph as shown in Figure 6(b), where s is the source node which represents the driver, t is the sink node of the graph which corresponds to the load of the wire. The remaining vertices a, b and c represent three possible buffer locations. Each edge corresponds to a wire segment between two possible buffers. We treat driver and load as buffers for convenience.

Edges are always directed along the signal transmission direction, The edge weight is defined as the delay t_d from one buffer input to the next buffer input. So $t_d = t_g + t_w$, where t_g is the delay of the buffer and t_w is the delay of the wire.

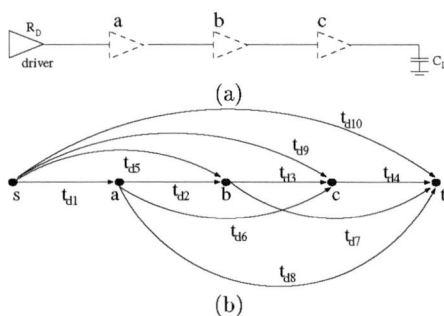

(a)

(b)

Figure 6: Formulate buffer insertion into shortest path problem. (a) is a small buffer insertion example with 3 possible locations. (b) Graph is built from (a).

Now consider the parameter variations. Delays of buffers and interconnects are modeled as random variables. All t_i in Figure 6(b) are now distributions instead of deterministic numbers. The buffer insertion becomes a SSP problem.

Similar to the maze routing problem, we use Elmore delay model. The parameter data are from ITRS. Table 5.3 shows some runtime results for our method. The wire length is set to be 10mm. 45nm technology parameters are used. And we set $B = 200$. It runs on a Linux box with 1GHz Pentium III CPU and 512MB memory. LMAX is the possible buffer locations uniformly distributed on the wire. BUF# is the number of inserted buffer. and μ_{delay} and σ_{delay} are mean and standard deviation of shortest path, respectively.

LMAX	BUF#	μ_{delay}	σ_{delay}	runtime
50	11	536	18.8	<0.01s
100	10	537	18.7	2.1s
150	11	536	18.8	17s

Table 2: Buffer Insertion Results

6. CONCLUSION

In this paper, we proposed a new algorithm to exactly solve the statistical shortest path problem. It can find the optimal solution in $O(B(V + E))$ time. Techniques are also proposed for improvement. This algorithm can be used in various applications in nanometer designs when the parameter variations become a concern.

7. REFERENCES

[1] Shekhar Borkar, Tanay Karnik, Siva Narendra, Jim Tschanz, Ali Keshavarzi, and Vivek De. Parameter variations and impact on circuits and microarchitecture. In *Proc. of the 40th Design Automation Conference*, pages 338–342. ACM Press, 2003.

[2] H. Frank. Shortest path in probabilistic graphs. *Oper. Res.*, 17:583–599, 1969.

[3] C. Elliott Sigal, A. Alan B. Pritsker, and James J. Solberg. The stochastic shortest route problem. *Opers. Res.*, 28:1122–1128, 1980.

[4] R. P. Loui. Optimal path in graphs with stochastic or multidimensional weights. *Comm. of ACM*, 26:670–676, 1983.

[5] Ishwar Murthy. Stochastic shortest path problems with piecewise-linear concave utility functions. *Management Science*, 44:125–136, 11 1998.

[6] Thomas H. Cormen, Charles E. Leiserson, Ronald L. Rivest, and Clifford Stein. *Introduction to Algorithms*. The MIT Press, 2001.

[7] Duane Boning and Sani Nassif. *Design of High-Performance Microprocessor Circuits*. 2002.

[8] Kanak Agarwal, Dennis Sylvester, David Blaauw, Frank Liu, Sani Nassif, and Sarma Vrudhula. Variational delay metrics for interconnect timing analysis. In *DAC 2004*, pages 381–384. ACM Press, 2004.

[9] Semiconductor Industry Association. *International Technology Roadmap for Semiconductors*, 2001.

[10] Michael Orshansky and Kurt Keutzer. A general probabilistic framework for worst case timing analysis. In *Proc. of the 39th Design Automation Conference*, pages 556–561. ACM Press, 2002.

[11] Hongliang Chang and S.S. Sapatnekar. Statistical timing analysis considering spatial correlations using a single pert-like traversal. In *Proc. International Conference on Computer Aided Design*, pages 621–625, 2003.

[12] A. Davgan and C. Kashyap. Block-based static timing analysis with uncertainty. In *Proc. of international conference on Computer Aided Design 2003*, pages 607– 614, 2003.

[13] C. Visweswariah, K. Ravindran, K. Kalafala, S. G. Walker, and S. Narayan. First-order incremental block-based statistical timing analysis. In *Proc. of the 41st Design Automation Conference*, pages 331–336, New York, NY, USA, 2004. ACM Press.

[14] Chirayu S. Amin, Noel Menezes, Kip Killpack, Florentin Dartu, Yehea Ismail, Umakanta Choudhury, and Nagib Hakim. Statistical static timing analysis: How simple can we get? In *Proc. of the 42nd Design Automation Conference*, pages 652–657, New York, NY, USA, 2005. ACM Press.

[15] Y. Gao and D. Wong. A graph based algorithm for optimal buffer insertion under accurate delay models. In *Proc. of the conference on Design, automation and test in Europe*, pages 535–539. IEEE Press, 2001.

Author Index

A

Abbaspour, Soroush	p. 947	(9C-2)
Abdi, Samar	p. 126	(1D-19)
Abraham, Jacob	p. 285	(3B-3)
Absar, Javed	p. 820	(8B-5)
Adachi, Hidekazu	p. 176	(2B-4)
Adachi, Hiroaki	p. 624	(6C-2)
Adachi, Kazunobu	p. 342	(4A-1)
Adamez, Jesus	p. 485	(5B-3)
Afzali-Kusha, Ali	p. 297	(3B-5)
Agarwal, Vineet	p. 718	(7C-3)
Agarwal, Vineet	p. 724	(7C-4)
Al-Hashimi, Bashir M.	p. 36	(1B-2)
Al-Hashimi, Bashir M.	p. 923	(9B-3)
Alizadeh, Bijan	p. 7	(1A-2)
Amirabadi, Amir	p. 297	(3B-5)
Anderson, Hans-Werner	p. 871	(8D-4)
Andraus, Zaher Semon	p. 19	(1A-4)
Asai, Hideki	p. 832	(8C-2)
Austin, Todd	p. 442	(5A-1)

B

Balakrishnan, Venkataramanan	p. 422	(4C-4)
Balakrishnan, Venkataramanan	p. 509	(5C-2)
Balasa, Florin	p. 802	(8B-2)
Banerjee, Kaustav	p. 223	(2D-2)
Banerjee, Sudarshan	p. 491	(5B-4)
Bansal, Aditya	p. 237	(2D-4)
Basu, Prasenjit	p. 13	(1A-3)
Behnen, Erwin	p. 871	(8D-4)
Ben-Romdhane, Mohamed	p. 30	(1B-1)
Bertacco, Valeria	p. 442	(5A-1)
Bhardwaj, Sarvesh	p. 953	(9C-3)
Bhunia, Swarup	p. 665	(7A-4)
Bhunia, Swarup	p. 712	(7C-2)
Blasco, Francisco	p. 485	(5B-3)
Boerstler, David	p. 856	(8D-1)
Bolliger, Mark	p. 871	(8D-4)
Bonaciu, Marius	p. 372	(4B-1)
Bouchhima, Aimen	p. 372	(4B-1)
Bozorgzadeh, Elaheh	p. 491	(5B-4)
Brockmeyer, Erik	p. 606	(6B-4)

Brokenshire, Daniel	p. 860	(8D-2)
Brown, Richard B.	p. 84	(1C-5)
Brown, Richard B.	p. 94	(1D-3)
Brown, Richard B.	p. 279	(3B-2)
Butler, Jon	p. 378	(4B-2)

C

Cai, Yici	p. 582	(6A-5)
Cai, Yici	p. 826	(8C-1)
Cai, Yuan	p. 923	(9B-3)
Cao, Yu	p. 953	(9C-3)
Cao, Zhen	p. 618	(6C-1)
Catthoor, Francky	p. 42	(1B-3)
Catthoor, Francky	p. 606	(6B-4)
Catthoor, Francky	p. 820	(8B-5)
Cauley, Stephen F	p. 422	(4C-4)
Cesario, Wander	p. 372	(4B-1)
Chae, Soo-Ik	p. 588	(6B-1)
Chae, Soo-Ik	p. 935	(9B-5)
Chakrabarti, Partha P	p. 13	(1A-3)
Chang, Cheng-Ru	p. 170	(2B-3)
Chang, Chia-Ming	p. 254	(3A-3)
Chang, Li-Pin	p. 334	(3D-2)
Chang, Shih-Chieh	p. 354	(4A-3)
Chang, Tian-Sheuan	p. 742	(7D-2)
Chang, Yao-Wen	p. 213	(2C-5)
Chang, Yao-Wen	p. 366	(4A-5)
Chang, Yao-Wen	p. 636	(6C-4)
Chang, Yen-Jen	p. 917	(9B-2)
Chang, Yuan-Hao	p. 334	(3D-2)
Chao, Chie-Min	p. 118	(1D-15)
Chao, Wen-Chang	p. 213	(2C-5)
Chaudhry, Rajat	p. 867	(8D-3)
Cheema, Muhammad Omer	p. 54	(1B-5)
Chen, Charlie Chungping	p. 941	(9C-1)
Chen, Chien-Chang	p. 777	(8A-3)
Chen, Chien-Hua	p. 600	(6B-3)
Chen, Guangyu	p. 128	(2A-1)
Chen, Guangyu	p. 140	(2A-3)
Chen, Guangyu	p. 808	(8B-3)
Chen, Guangyu	p. 814	(8B-4)
Chen, Guilin	p. 134	(2A-2)
Chen, Guilin	p. 140	(2A-3)

Chen, Hongyu . . . . p. 73 (1C-3)
Chen, Jian-Wen . . . . p. 736 (7D-1)
Chen, Jwu E . . . . p. 366 (4A-5)
Chen, Liang-Gee . . . . p. 750 (7D-3)
Chen, Po-Yuan . . . . p. 576 (6A-4)
Chen, Tai-Chen . . . . p. 636 (6C-4)
Chen, Tung-Chien . . . . p. 750 (7D-3)
Chen, Wei-Nien . . . . p. 742 (7D-2)
Chen, Weijen . . . . p. 941 (9C-1)
Chen, Xi . . . . p. 372 (4B-1)
Chen, Yiran . . . . p. 158 (2B-1)
Cheng, Chung-Kuan . . . . p. 73 (1C-3)
Cheng, Chung-Kuan . . . . p. 428 (4C-5)
Cheng, Kai-Ting . . . . p. 742 (7D-2)
Cheng, Kwang-Ting . . . . p. 25 (1A-5)
Cheng, Tim . . . . p. 360 (4A-4)
Cho, Hansu . . . . p. 126 (1D-19)
Cho, Minsik . . . . p. 765 (8A-1)
Choi, Jung Hwan . . . . p. 237 (2D-4)
Choi, Jung Hwan . . . . p. 712 (7C-2)
Choi, Kiyoung . . . . p. 905 (9A-5)
Choi, Kyu-Myung . . . . p. 551 (5D-4)
Choi, Kyu-Myung . . . . p. 905 (9A-5)
Chong, Philip . . . . p. 440 (4D-4)
Chu, Chris . . . . p. 195 (2C-2)
Chu, Fangqing . . . . p. 100 (1D-6)
Chung, Eui-Young . . . . p. 551 (5D-4)
Chung, Eui-Young . . . . p. 905 (9A-5)
Coenen, Martijn . . . . p. 146 (2A-4)
Cong, Jason . . . . p. 188 (2C-1)
Cong, Jason . . . . p. 384 (4B-3)
Cordone, Roberto . . . . p. 898 (9A-4)

D

Dai, Wayne Wei-Ming . . . . p. 730 (7C-5)
Das, Sayantan . . . . p. 13 (1A-3)
Dasgupta, Pallab . . . . p. 13 (1A-3)
Datta, Animesh . . . . p. 712 (7C-2)
De Micheli, Giovanni . . . . p. 146 (2A-4)
DeHon, Andre M. . . . . p. 789 (8A-5)
Deng, Liang . . . . p. 965 (9C-5)
Devgan, Anirudh . . . . p. 61 (1C-1)
Dhong, Sang . . . . p. 867 (8D-3)
Dick, Robert P. . . . . p. 879 (9A-1)
Dong, Sheqin . . . . p. 694 (7B-4)

Dou, Qingqi . . . . p. 285 (3B-3)
Du, Yu . . . . p. 521 (5C-4)
Dutt, Nikil . . . . p. 30 (1B-1)
Dutt, Nikil . . . . p. 491 (5B-4)
Dutt, Nikil . . . . p. 497 (5B-5)

E

Eda, Tsutomu . . . . p. 124 (1D-18)
Ejlali, Alireza . . . . p. 923 (9B-3)
Ekpanyapong, Mongkol . . . . p. 959 (9C-4)
Endoh, Chihiro . . . . p. 342 (4A-1)
Enomoto, Tadayoshi . . . . p. 90 (1D-1)
Eo, Soo-Kwan . . . . p. 551 (5D-4)

F

Fan, Yibo . . . . p. 122 (1D-17)
Fatemi, Hanif . . . . p. 947 (9C-2)
Feng, Xingguang . . . . p. 905 (9A-5)
Feng, Yan . . . . p. 771 (8A-2)
Feng, Zhe . . . . p. 630 (6C-3)
Ferrandi, Fabrizio . . . . p. 898 (9A-4)
Fujiwara, Hideo . . . . p. 671 (7A-5)
Fukazawa, Mitsuya . . . . p. 106 (1D-9)
Fukunaga, Masayasu . . . . p. 348 (4A-2)

G

Gajski, Daniel . . . . p. 116 (1D-14)
Gajski, Daniel . . . . p. 126 (1D-19)
Goel, Ashish . . . . p. 665 (7A-4)
Goossens, Kees . . . . p. 146 (2A-4)
Goplen, Brent . . . . p. 219 (2D-1)
Gorjiara, Bita . . . . p. 116 (1D-14)
Goto, Satoshi . . . . p. 112 (1D-12)
Goto, Yuichi . . . . p. 856 (8D-1)
Goyal, Prashant . . . . p. 291 (3B-4)
Gu, Zhenyu (Peter) . . . . p. 879 (9A-1)
Guiney, Michaela . . . . p. 434 (4D-1)
Gulati, Kanupriya . . . . p. 850 (8C-5)
Guo, Jin . . . . p. 42 (1B-3)
Gupta, Sanjay . . . . p. 871 (8D-4)
Guthaus, Matthew R. . . . . p. 84 (1C-5)

H

Ha, Soonhoi . . . . p. 152 (2A-5)

Ha, Soonhoi . . . . p. 497   (5B-5)
Ha, Yajun . . . . p. 886   (9A-2)
Hailu, Eskinder . . . . p. 856   (8D-1)
Hamada, Shuji . . . . p. 348   (4A-2)
Hammami, Omar . . . . p. 54   (1B-5)
Han, Jun . . . . p. 122   (1D-17)
Han, Sang-Il . . . . p. 935   (9B-5)
Hang, Hsueh-Ming . . . . p. 742   (7D-2)
Hanna, Ziyad . . . . p. 25   (1A-5)
Hara, Hiroyuki . . . . p. 533   (5D-1)
Harvey, Paul . . . . p. 871   (8D-4)
Hashimoto, Masanori . . . . p. 515   (5C-3)
He, Lei . . . . p. 207   (2C-4)
He, Lei . . . . p. 630   (6C-3)
Hedrich, Lars . . . . p. 700   (7B-5)
Higami, Yoshinobu . . . . p. 659   (7A-3)
Hiraide, Takahisa . . . . p. 342   (4A-1)
Hiraoka, Daisuke . . . . p. 860   (8D-2)
Hofstee, Peter . . . . p. 871   (8D-4)
Hon, Man Chung . . . . p. 326   (3C-5)
Hong, Xianlong . . . . p. 582   (6A-5)
Hong, Xianlong . . . . p. 618   (6C-1)
Hong, Xianlong . . . . p. 630   (6C-3)
Hong, Xianlong . . . . p. 694   (7B-4)
Hong, Xianlong . . . . p. 826   (8C-1)
Horikawa, Kazunari . . . . p. 1   (1A-1)
Hsieh, Cheng-Tao . . . . p. 354   (4A-3)
Hsieh, Jen-Wei . . . . p. 334   (3D-2)
Hu, Bo . . . . p. 92   (1D-2)
Hu, Qubo . . . . p. 606   (6B-4)
Hu, Xiaodong . . . . p. 618   (6C-1)
Hu, Yu . . . . p. 618   (6C-1)
Hu, Yuhen . . . . p. 941   (9C-1)
Huang, Chao-Wei . . . . p. 118   (1D-15)
Huang, Jiun-Lang . . . . p. 648   (7A-1)
Huang, Juinn-Dar . . . . p. 448   (5A-2)
Huang, Juinn-Dar . . . . p. 600   (6B-3)
Huang, Shih-Hsu . . . . p. 254   (3A-3)
Huang, Xuan-Lun . . . . p. 648   (7A-1)
Huang, Yu-Hui . . . . p. 576   (6A-4)
Hutton, Mike . . . . p. 73   (1C-3)
Hwang, Chanseok . . . . p. 201   (2C-3)
Hwang, Hyeyoung . . . . p. 152   (2A-5)
Hwang, TingTing . . . . p. 576   (6A-4)
Hyun, Chung Jin . . . . p. 758   (7D-4)

I

Ignjatovic, Aleksandar . . . . p. 612   (6B-5)
Ignjatovic, Aleksandar . . . . p. 796   (8B-1)
Ikeda, Hiroshi . . . . p. 624   (6C-2)
Ikenaga, Takeshi . . . . p. 112   (1D-12)
Imai, Satoshi . . . . p. 541   (5D-2)
Inoue, Atsuki . . . . p. 541   (5D-2)
Inoue, Yoshio . . . . p. 437   (4D-2)
Irwin, Mary Jane . . . . p. 140   (2A-3)
Ishida, Koichi . . . . p. 98   (1D-5)
Ishikawa, Tatsuyuki . . . . p. 112   (1D-12)
Ishizaka, Kinya . . . . p. 624   (6C-2)
Ismail, Yehea . . . . p. 231   (2D-3)
Isoda, Yutaka . . . . p. 342   (4A-1)
Isoda, Yutaka . . . . p. 624   (6C-2)
Ito, Noriyuki . . . . p. 342   (4A-1)
Ito, Noriyuki . . . . p. 624   (6C-2)

J

Jagannathan, Ashok . . . . p. 384   (4B-3)
Jain, Jitesh . . . . p. 422   (4C-4)
Janapsatya, Andhi . . . . p. 612   (6B-5)
Janapsatya, Andhi . . . . p. 796   (8B-1)
Jangkrajarng, Nuttorn . . . . p. 92   (1D-2)
Jen, Chein-Wei . . . . p. 118   (1D-15)
Jerraya, Ahmed . . . . p. 372   (4B-1)
Jerraya, Ahmed Amine . . . . p. 935   (9B-5)
Jia, Yongbo . . . . p. 730   (7C-5)
Jing, Tong . . . . p. 618   (6C-1)
Jing, Tong . . . . p. 630   (6C-3)
Johns, Charles . . . . p. 871   (8D-4)
Jou, Jing-Yang . . . . p. 448   (5A-2)
Jou, Jing-Yang . . . . p. 600   (6B-3)
Jung, Hyunuk . . . . p. 152   (2A-5)

K

Kadayif, Ismail . . . . p. 182   (2B-5)
Kahle, Jim . . . . p. 871   (8D-4)
Kajihara, Seiji . . . . p. 348   (4A-2)
Kameyama, Atsushi . . . . p. 871   (8D-4)
Kandemir, Mahmut . . . . p. 128   (2A-1)
Kandemir, Mahmut . . . . p. 134   (2A-2)
Kandemir, Mahmut . . . . p. 140   (2A-3)
Kandemir, Mahmut . . . . p. 182   (2B-5)

Kandemir, Mahmut . . . . p. 390 (4B-4)
Kandemir, Mahmut . . . . p. 808 (8B-3)
Kandemir, Mahmut . . . . p. 814 (8B-4)
Kandemir, Mahmut . . . . p. 929 (9B-4)
Kankani, Navneeth . . . . p. 724 (7C-4)
Kanuma, Akira . . . . p. 342 (4A-1)
Kao, Chao-Yang . . . . p. 736 (7D-1)
Katagiri, Hideaki . . . . p. 624 (6C-2)
Kato, Toshiyuki . . . . p. 124 (1D-18)
Katsuki, Kazuya . . . . p. 110 (1D-11)
Kawasaki, Kenichi . . . . p. 541 (5D-2)
Keaty, John . . . . p. 871 (8D-4)
Kettle, Neil . . . . p. 243 (3A-1)
Khatri, Sunil . . . . p. 850 (8C-5)
Ki, Wing-Hung . . . . p. 96 (1D-4)
Ki, Wing-Hung . . . . p. 102 (1D-7)
Ki, Wing-Hung . . . . p. 104 (1D-8)
Kim, Chris H . . . . p. 559 (6A-1)
Kim, Hyung-Ock . . . . p. 565 (6A-2)
Kim, Hyunsuk . . . . p. 551 (5D-4)
Kim, Seon Wook . . . . p. 120 (1D-16)
Kim, Suki . . . . p. 120 (1D-16)
Kim, Sung Dae . . . . p. 758 (7D-4)
Kimura, Shinji . . . . p. 1 (1A-1)
Kimura, Shinji . . . . p. 653 (7A-2)
King, Andy . . . . p. 243 (3A-1)
Kitahara, Takeshi . . . . p. 533 (5D-1)
Kjeldsberg, Per Gunnar . . . . p. 606 (6B-4)
Ko, Sung Jea . . . . p. 120 (1D-16)
Kobayashi, Fuminori . . . . p. 108 (1D-10)
Kobayashi, Kazutoshi . . . . p. 110 (1D-11)
Kobayashi, Masatsugu . . . . p. 124 (1D-18)
Kobayashi, Nobuaki . . . . p. 90 (1D-1)
Kobayashi, Shin-ya . . . . p. 659 (7A-3)
Kodakara, Sreekumar V. . . . . . . p. 61 (1C-1)
Koh, Cheng-Kok . . . . p. 158 (2B-1)
Koh, Cheng-Kok . . . . p. 422 (4C-4)
Koh, Cheng-Kok . . . . p. 509 (5C-2)
Kohara, Shunitsu . . . . p. 594 (6B-2)
Koide, Tetsushi . . . . p. 176 (2B-4)
Kolcu, Ibrahim . . . . p. 814 (8B-4)
Komatsu, Hiroaki . . . . p. 624 (6C-2)
Kong, Jeong-Taek . . . . p. 551 (5D-4)
Kono, Takeshi . . . . p. 342 (4A-1)
Koon, Suet-Chui . . . . p. 102 (1D-7)
Kosaka, Daisuke . . . . p. 677 (7B-1)

Kotani, Manabu . . . . p. 110 (1D-11)
Ku, Ja Chun . . . . p. 231 (2D-3)
Kubota, Hidemasa . . . . p. 832 (8C-2)
Kuh, Ernest S. . . . . p. 428 (4C-5)
Kulkarni, Sarvesh Hemchandra . . . . p. 838 (8C-3)
Kumar, Sanjay V . . . . p. 559 (6A-1)
Kuo, Ming-Hsine . . . . p. 783 (8A-4)
Kuo, Tei-Wei . . . . p. 334 (3D-2)
Kuo, Yu-Ting . . . . p. 118 (1D-15)
Kuroki, Wataru . . . . p. 408 (4C-2)

L

Lai, Xiaolue . . . . p. 273 (3B-1)
Lai, Xiaolue . . . . p. 291 (3B-4)
Lai, Xiaolue . . . . p. 527 (5C-5)
Lam, Yat-Hei . . . . p. 102 (1D-7)
Lam, Yat-Hei . . . . p. 104 (1D-8)
LaMeres, Brock . . . . p. 850 (8C-5)
Le, Bob . . . . p. 871 (8D-4)
Leavitt, Eric . . . . p. 434 (4D-1)
Lee, Chung-Len . . . . p. 366 (4A-5)
Lee, Geeng-Wei . . . . p. 600 (6B-3)
Lee, Ikhwan . . . . p. 551 (5D-4)
Lee, Jeong Hoo . . . . p. 758 (7D-4)
Lee, Junghee . . . . p. 460 (5A-4)
Lee, Kuang-Yao . . . . p. 303 (3C-1)
Lee, Sang . . . . p. 871 (8D-4)
Lee, Suh Ho . . . . p. 120 (1D-16)
Li, Fehui . . . . p. 128 (2A-1)
Li, Feihui . . . . p. 134 (2A-2)
Li, Feihui . . . . p. 182 (2B-5)
Li, Feihui . . . . p. 808 (8B-3)
Li, Feihui . . . . p. 929 (9B-4)
Li, Hai . . . . p. 158 (2B-1)
Li, Katherine Shu-Min . . . . p. 366 (4A-5)
Li, Wei . . . . p. 100 (1D-6)
Li, Zhao . . . . p. 402 (4C-1)
Li, Zhuo Robert . . . . p. 320 (3C-4)
Lian, Chung-Jr . . . . p. 750 (7D-3)
Liffiton, Mark Hammond . . . . p. 19 (1A-4)
Lim, Sung Kyu . . . . p. 959 (9C-4)
Lin, Hung-Chih . . . . p. 742 (7D-2)
Lin, Rung-Bin . . . . p. 783 (8A-4)
Lin, Sheng-Chih . . . . p. 223 (2D-2)
Lin, Shyh-Chang . . . . p. 636 (6C-4)

Lin, Tao . . . . p. 67 (1C-2)
Lin, Tay-Jyi . . . . p. 118 (1D-15)
Lin, Tsai-Ying . . . . p. 783 (8A-4)
Lin, Youn-Long . . . . p. 170 (2B-3)
Lin, Youn-Long . . . . p. 736 (7D-1)
Liu, Bin . . . . p. 582 (6A-5)
Liu, Chih-Wei . . . . p. 118 (1D-15)
Long, Di . . . . p. 694 (7B-4)
Luican, Ilie I. . . . . p. 802 (8B-2)
Luo, Fangyi . . . . p. 730 (7C-5)

M

Ma, Yuchun . . . . p. 384 (4B-3)
Maeda, Toshiyuki . . . . p. 348 (4A-2)
Mahmoodi, Hamid . . . . p. 665 (7A-4)
Mak, Wai-Kei . . . . p. 777 (8A-3)
Malani, Parth . . . . p. 911 (9B-1)
Marchal, Pol . . . . p. 42 (1B-3)
Markov, Igor . . . . p. 440 (4D-4)
Marsman, Eric D. . . . . p. 94 (1D-3)
Maruyama, Daisuke . . . . p. 342 (4A-1)
Matsumoto, Yuki . . . . p. 124 (1D-18)
Matsumura, Motoaki . . . . p. 541 (5D-2)
Matsuura, Munehiro . . . . p. 466 (5A-5)
Mattausch, Hans Juergen . . . . . . p. 176 (2B-4)
Mazumder, Pinaki . . . . p. 416 (4C-3)
McCann, James L. . . . . p. 279 (3B-2)
McCorquodale, Michael S. . . . . . . p. 94 (1D-3)
McCorquodale, Michael S. . . . . . . p. 279 (3B-2)
Mecha, Hortensia . . . . p. 396 (4B-5)
Meil, Gavin . . . . p. 860 (8D-2)
Meterelliyoz, Mesut . . . . p. 237 (2D-4)
Michael, Brad . . . . p. 860 (8D-2)
Miki, Kazuhiko . . . . p. 856 (8D-1)
Miki, Takuji . . . . p. 124 (1D-18)
Min, Sang Lyul . . . . p. 332 (3D-1)
Minami, Fumihiro . . . . p. 533 (5D-1)
Miyaoka, Yuichiro . . . . p. 594 (6B-2)
Miyazaki, Masahide . . . . p. 671 (7A-5)
Mochizuki, Isamu . . . . p. 547 (5D-3)
Mochizuki, Tsuyoshi . . . . p. 342 (4A-1)
Moondanos, John . . . . p. 25 (1A-5)
Moore, Simon . . . . p. 164 (2B-2)
Mori, Yutaka . . . . p. 624 (6C-2)
Morikawa, Kimihiro . . . . p. 547 (5D-3)

Morimoto, Takashi . . . . p. 176 (2B-4)
Mozos, Daniel . . . . p. 396 (4B-5)
Mukhopadhyay, Saibal . . . . p. 712 (7C-2)
Mullins, Robert . . . . p. 164 (2B-2)
Murali, Srinivasan . . . . p. 146 (2A-4)
Murthy, Jayathi . . . . p. 237 (2D-4)

N

Nagasawa, Shigeru . . . . p. 342 (4A-1)
Nagata, Makoto . . . . p. 106 (1D-9)
Nagata, Makoto . . . . p. 677 (7B-1)
Nagayama, Shinobu . . . . p. 378 (4B-2)
Nakahara, Hiroki . . . . p. 466 (5A-5)
Nam, Eyee Hyun . . . . p. 332 (3D-1)
Nazarian, Shahin . . . . p. 67 (1C-2)
Neto, Horácio . . . . p. 48 (1B-4)
Nguyen, Tuyen . . . . p. 871 (8D-4)
Nieh, Yow-Tyng . . . . p. 254 (3A-3)
Ninoi, Eizo . . . . p. 342 (4A-1)
Nishimura, Tsutomu . . . . p. 124 (1D-18)
Noguchi, Koichiro . . . . p. 106 (1D-9)
Nonomura, Kazuhiro . . . . p. 624 (6C-2)
Nugroho, Arif . . . . p. 114 (1D-13)

O

Obata, Koji . . . . p. 266 (3A-5)
Oda, Noriaki . . . . p. 706 (7C-1)
Ofek, Hillel . . . . p. 439 (4D-3)
Oh, Hyunok . . . . p. 497 (5B-5)
Oh, Taewook . . . . p. 152 (2A-5)
Ohba, Nobuyuki . . . . p. 454 (5A-3)
Ohkubo, Naoaki . . . . p. 570 (6A-3)
Ohtsuki, Tatsuo . . . . p. 594 (6B-2)
Ohtsuki, Tatsuo . . . . p. 653 (7A-2)
Onodera, Hidetoshi . . . . p. 110 (1D-11)
Onodera, Hidetoshi . . . . p. 515 (5C-3)
Onouchi, Masafumi . . . . p. 547 (5D-3)
Ou, Shih-Hao . . . . p. 118 (1D-15)
Ozturk, Ozcan . . . . p. 390 (4B-4)
Ozturk, Ozcan . . . . p. 814 (8B-4)

P

Palermo, Gianluca . . . . p. 898 (9A-4)
Palkovic, Martin . . . . p. 606 (6B-4)

Pan, David Z.	p. 61	(1C-1)
Pan, David Z.	p. 503	(5C-1)
Pan, David Z.	p. 765	(8A-1)
Pan, Min	p. 195	(2C-2)
Pan, Sung-Jui	p. 25	(1A-5)
Panda, Preeti Ranjan	p. 892	(9A-3)
Papa, David	p. 440	(4D-4)
Papanikolaou, Antonis	p. 42	(1B-3)
Parameswaran, Harindranath	p. 79	(1C-4)
Parameswaran, Sri	p. 612	(6B-5)
Parameswaran, Sri	p. 796	(8B-1)
Park, Ji Hwan	p. 120	(1D-16)
Park, Sanggyu	p. 588	(6B-1)
Pasricha, Sudeep	p. 30	(1B-1)
Pedram, Massoud	p. 67	(1C-2)
Pedram, Massoud	p. 201	(2C-3)
Pedram, Massoud	p. 473	(5B-1)
Pedram, Massoud	p. 947	(9C-2)
Peng, Chih-Yang	p. 213	(2C-5)
Petrovick, John	p. 871	(8D-4)
Pettengill, Sarah	p. 856	(8D-1)
Pham, Dac	p. 871	(8D-4)
Pham, Mydung	p. 871	(8D-4)
Pille, Juergen	p. 871	(8D-4)
Posadas, Hector	p. 485	(5B-3)
Posluszny, Stephen	p. 867	(8D-3)
Posluszny, Stephen	p. 871	(8D-4)
Pratap, Rajendra	p. 79	(1C-4)
Pu, Yu	p. 886	(9A-2)

Q

Qi, Jieming	p. 856	(8D-1)
Qiu, Qinru	p. 911	(9B-1)

R

Radulescu, Andrei	p. 146	(2A-4)
Rahmatullah, Nursani	p. 114	(1D-13)
Ramalingam, Anand	p. 61	(1C-1)
Rasouli, Seid Hadi	p. 297	(3B-5)
Reddy, Sudhakar M.	p. 923	(9B-3)
Reinman, Glenn	p. 384	(4B-3)
Ren, Junyan	p. 100	(1D-6)
Reshadi, Mehrdad	p. 116	(1D-14)
Riley, Mack	p. 871	(8D-4)
Rong, Peng	p. 473	(5B-1)

Rosdi, Bakhtiar Affendi	p. 260	(3A-4)
Roy, Kaushik	p. 158	(2B-1)
Roy, Kaushik	p. 237	(2D-4)
Roy, Kaushik	p. 665	(7A-4)
Roy, Kaushik	p. 712	(7C-2)
Roychowdhury, Jaijeet	p. 273	(3B-1)
Roychowdhury, Jaijeet	p. 291	(3B-4)
Roychowdhury, Jaijeet	p. 527	(5C-5)

S

Sakallah, Karem Ahmad	p. 19	(1A-4)
Sakurai, Takayasu	p. 98	(1D-5)
Sallay, Balazs	p. 860	(8D-2)
Saluja, Kewal K.	p. 659	(7A-3)
Sanchez, Pablo	p. 485	(5B-3)
Santambrogio, Marco Domenico	p. 898	(9A-4)
Sapatnekar, Sachin S.	p. 559	(6A-1)
Sapatnekar, Sachin S.	p. 219	(2D-1)
Sapatnekar, Sachin S.	p. 309	(3C-2)
Sapatnekar, Sachin S.	p. 771	(8A-2)
Sarto, Egino	p. 207	(2C-4)
Sasao, Tsutomu	p. 378	(4B-2)
Sasao, Tsutomu	p. 466	(5A-5)
Sato, Yasuo	p. 348	(4A-2)
Schmitz, Marcus T.	p. 36	(1B-2)
Schmitz, Marcus T.	p. 923	(9B-3)
Sciuto, Donatella	p. 898	(9A-4)
Sekine, Hidetoshi	p. 547	(5D-3)
Senger, Robert M.	p. 94	(1D-3)
Septien, Julio	p. 396	(4B-5)
Seyedi, Azam	p. 297	(3B-5)
Shafi, Hazim	p. 860	(8D-2)
Shang, Li	p. 879	(9A-1)
Shao, Hui	p. 96	(1D-4)
Shi, Jin	p. 826	(8C-1)
Shi, Richard	p. 92	(1D-2)
Shi, Richard	p. 402	(4C-1)
Shi, Rui	p. 428	(4C-5)
Shi, Sean X.	p. 503	(5C-1)
Shi, Weiping	p. 320	(3C-4)
Shi, Yiyu	p. 618	(6C-1)
Shi, Yiyu	p. 630	(6C-3)
Shi, Youhua	p. 653	(7A-2)
Shih, Che-Hua	p. 448	(5A-2)
Shih, Shen-Yu	p. 170	(2B-3)

Shimizu, Kazunori	p. 112	(1D-12)
Shin, Hongjoong	p. 765	(8A-1)
Shin, Youngsoo	p. 565	(6A-2)
Shiratake, Shinichiro	p. 533	(5D-1)
Shrivastava, Sachin	p. 79	(1C-4)
Singh, Siddharth	p. 237	(2D-4)
Son, Seung Woo	p. 128	(2A-1)
Srivastava, Navin	p. 223	(2D-2)
Stasiak, Daniel	p. 867	(8D-3)
Su, Chauchin	p. 366	(4A-5)
Su, Feng	p. 96	(1D-4)
Su, Man-Yun	p. 448	(5A-2)
Su, Ming-Hong	p. 249	(3A-2)
Suga, Atsuhiro	p. 541	(5D-2)
Sugawara, Osamu	p. 342	(4A-1)
Sugiura, Hiroaki	p. 124	(1D-18)
Sugiyama, Hiroyuki	p. 624	(6C-2)
Sugiyama, Yaroku	p. 342	(4A-1)
Sugiyama, Yaroku	p. 624	(6C-2)
Sunwoo, Myung Hoon	p. 758	(7D-4)
Sylvester, Dennis	p. 84	(1C-5)
Sylvester, Dennis	p. 838	(8C-3)

T

Tabero, Jesus	p. 396	(4B-5)
Takada, Hideyuki	p. 266	(3A-5)
Takagi, Kazuyoshi	p. 266	(3A-5)
Takahashi, Atsushi	p. 260	(3A-4)
Takahashi, Atsushi	p. 642	(6C-5)
Takahashi, Hiroshi	p. 659	(7A-3)
Takamatsu, Yuzo	p. 659	(7A-3)
Takano, Kohji	p. 454	(5A-3)
Taki, Kazuo	p. 106	(1D-9)
Tamtrakarn, Atit	p. 98	(1D-5)
Tan, Shelton X-D	p. 826	(8C-1)
Tan, Ying	p. 911	(9B-1)
Tanaka, Katsunori	p. 266	(3A-5)
Tanamura, Yoshiyasu	p. 624	(6C-2)
Tanji, Yuichi	p. 832	(8C-2)
Thiele, Lothar	p. 479	(5B-2)
Togawa, Nozomu	p. 594	(6B-2)
Togawa, Nozomu	p. 653	(7A-2)
Tomioka, Yoichi	p. 642	(6C-5)
Tomono, Naoki	p. 594	(6B-2)
Tsai, Chia-Yang	p. 742	(7D-2)

Tsai, Jengliang	p. 941	(9C-1)
Tsai, Shr-Cheng	p. 783	(8A-4)
Tseng, Wei-Chiu	p. 783	(8A-4)
Tsuchiya, Akira	p. 515	(5C-3)
Tsuchiya, Takehiko	p. 1	(1A-1)
Tsui, Chi-Ying	p. 96	(1D-4)
Tsui, Chi-Ying	p. 102	(1D-7)
Tsui, Chi-Ying	p. 104	(1D-8)
Tsukiboshi, Yoshiki	p. 533	(5D-1)
Tuncer, Emre	p. 67	(1C-2)

U

Uchida, Jumpei	p. 594	(6B-2)
Unnikrishnan, Priya	p. 929	(9B-4)
Usami, Kimiyoshi	p. 570	(6A-3)
Utsumi, Tetsuaki	p. 533	(5D-1)

V

Vandecappelle, Arnout	p. 606	(6B-4)
Verma, Manuj	p. 79	(1C-4)
Verock, Joseph	p. 871	(8D-4)
Véstias, Mário Pereira	p. 48	(1B-4)
Vijaykrishnan, Narayanan	p. 140	(2A-3)
Villar, Eugenio	p. 485	(5B-3)
Viswanathan, Natarajan	p. 195	(2C-2)
Vogel, Sebastian	p. 315	(3C-3)
Vrudhula, Sarma	p. 953	(9C-3)

W

Wagner, Ilya	p. 442	(5A-1)
Wakayama, Cherry	p. 92	(1D-2)
Wandeler, Ernesto	p. 479	(5B-2)
Wang, Baohua	p. 416	(4C-3)
Wang, Chun-Yao	p. 249	(3A-2)
Wang, Feng	p. 390	(4B-4)
Wang, J.-H.	p. 213	(2C-5)
Wang, Janet	p. 718	(7C-3)
Wang, Janet M	p. 724	(7C-4)
Wang, Jia	p. 879	(9A-1)
Wang, Ting-Chi	p. 303	(3C-1)
Wang, Xiaoying	p. 700	(7B-5)
Wang, Xiren	p. 683	(7B-2)
Wang, Yu-Jen	p. 742	(7D-2)
Wang, Zeyi	p. 521	(5C-4)

Wang, Zeyi . . . . p. 683 (7B-2)
Wang, Zeyi . . . . p. 844 (8C-4)
Warnock, James . . . . p. 871 (8D-4)
Watanabe, Minoru . . . . p. 108 (1D-10)
Watanabe, Takayuki . . . . p. 832 (8C-2)
Watanabe, Yukio . . . . p. 860 (8D-2)
Watewai, Thaisiri . . . . p. 959 (9C-4)
Wei, Jie . . . . p. 384 (4B-3)
Weitzel, Steve . . . . p. 871 (8D-4)
Wen, Xiaoqing . . . . p. 348 (4A-2)
Wendel, Dieter . . . . p. 871 (8D-4)
West, Andrew . . . . p. 164 (2B-2)
Wong, Martin D. F. . . . . p. 965 (9C-5)
Wong, Martin D.F. . . . . p. 315 (3C-3)
Wong, Yiu-Chung . . . . p. 207 (2C-4)
Wrighton, Michael G. . . . . p. 789 (8A-5)
Wu, Dong . . . . p. 36 (1B-2)
Wu, Kai-Chiang . . . . p. 354 (4A-3)
Wu, Meng-Chiou . . . . p. 783 (8A-4)
Wu, Min . . . . p. 122 (1D-17)
Wu, Qing . . . . p. 911 (9B-1)
Wu, Yongyi . . . . p. 122 (1D-17)

X

Xie, Min . . . . p. 188 (2C-1)
Xie, Yuan . . . . p. 390 (4B-4)
Xiong, Jinjun . . . . p. 207 (2C-4)
Xu, Xingwen . . . . p. 1 (1A-1)

Y

Yamada, Kenta . . . . p. 706 (7C-1)
Yamada, Tetsuya . . . . p. 547 (5D-3)
Yamamura, Kiyotaka . . . . p. 408 (4C-2)
Yamanaka, Hitoshi . . . . p. 342 (4A-1)
Yamaoka, Kousuke . . . . p. 176 (2B-4)
Yamashita, Ryoichi . . . . p. 624 (6C-2)
Yamashita, Shigeru . . . . p. 266 (3A-5)
Yamauchi, Hironori . . . . p. 124 (1D-18)
Yan, Changhao . . . . p. 844 (8C-4)

Yan, Guiying . . . . p. 618 (6C-1)
Yanagida, Masahiro . . . . p. 342 (4A-1)
Yanagisawa, Masao . . . . p. 594 (6B-2)
Yanagisawa, Masao . . . . p. 653 (7A-2)
Yang, Kai . . . . p. 360 (4A-4)
Yang, Peng . . . . p. 551 (5D-4)
Yang, Ya-Chi . . . . p. 509 (5C-2)
Yang, Yonghong . . . . p. 879 (9A-1)
Yao, Bo . . . . p. 73 (1C-3)
Ye, Zuochang . . . . p. 689 (7B-3)
Yeh, Shang-Yu . . . . p. 742 (7D-2)
Yi, Joonhwan . . . . p. 460 (5A-4)
Yoda, Tomoyuki . . . . p. 533 (5D-1)
Yoneda, Tomokazu . . . . p. 671 (7A-5)
Yoo, Jun-hee . . . . p. 905 (9A-5)
Yoo, Sungjoo . . . . p. 551 (5D-4)
Yoon, Sang-Yong . . . . p. 588 (6B-1)
Yoshitake, Akihiko . . . . p. 624 (6C-2)
Youssef, Wassim . . . . p. 372 (4B-1)
Yu, Wenjian . . . . p. 521 (5C-4)
Yu, Wenjian . . . . p. 683 (7B-2)
Yu, Wenjian . . . . p. 844 (8C-4)
Yu, Zhiping . . . . p. 689 (7B-3)

Z

Zeng, Xiaoyang . . . . p. 122 (1D-17)
Zhan, Yong . . . . p. 219 (2D-1)
Zhan, Yong . . . . p. 309 (3C-2)
Zhan, Yong . . . . p. 771 (8A-2)
Zhang, Lizheng . . . . p. 941 (9C-1)
Zhang, Mengsheng . . . . p. 521 (5C-4)
Zhang, Tianpei . . . . p. 309 (3C-2)
Zhang, Yan . . . . p. 384 (4B-3)
Zhou, Lili . . . . p. 92 (1D-2)
Zhou, Qiang . . . . p. 582 (6A-5)
Zhou, Shuo . . . . p. 73 (1C-3)
Zhu, Hongwei . . . . p. 802 (8B-2)
Zhu, Yi . . . . p. 73 (1C-3)
Zhu, Zhengyong . . . . p. 428 (4C-5)

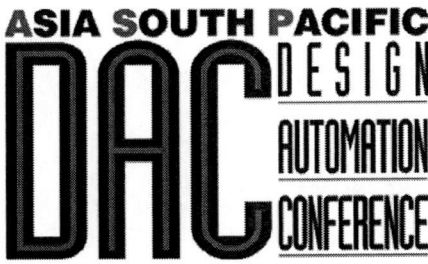

Call for Papers
ASP-DAC 2007
Asia and South Pacific Design Automation Conference 2007
http://www.aspdac.com/aspdac2007/
January 23-26, 2007
Pacifico Yokohama, Yokohama, JAPAN

Aims of the Conference:

ASP-DAC 2007 is the twelfth in a series of annual international conferences on VLSI design automation. Asia and South Pacific region is one of the most active regions of design and fabrication of silicon chips in the world. The conference aims are providing the Asian and South Pacific CAD/DA and Design community with opportunities of interchanging ideas and collaboratively discussing the directions of the technologies related to all of Electronic Design Automation (EDA). The goal of the conference is to provide a forum for presentation, discussion, and observation of the state-of-the-art of EDA technologies and design methodologies of electronic systems. The format of the meeting intends to cultivate and promote an instructive and productive interchange among EDA researchers/developers, and system/circuit/device designers. A wide variety of those scientists, engineers, and students who are interested in theoretical issues on EDA are also welcome.

Areas of Interest:
Original papers on, but not limited to, the following areas are invited.

[1] System Level Design:
System VLSI and SOC design methods, System specification, Specification languages, Design languages, Design reuse and IPs, Tools/methods for low power system design, Platform-based design, Network on chip design

[2] Embedded and Real-Time Systems:
Hardware-software co-design, Co-simulation, Co-verification, Real-time OS and middleware, Design language for embedded systems, Compilation techniques, ASIP synthesis

[3] Behavioral/Logic Synthesis and Optimization:
Behavioral/RTL synthesis, Technology independent optimization, Technology mapping, Interaction between logic design and layout, Sequential and asynchronous logic synthesis

[4] Validation and Verification for Behavioral/Logic Design:
Logic simulation, Symbolic simulation, Formal verification, Equivalence checking, Transaction-level/RTL and gate level modeling and validation

[5] Physical Design (Routing):
Routing, Repeater issues, Interconnect optimization, Interconnect planning, Module generation, Layout verification

[6] Physical Design (Placement):
Placement, Floorplanning, Partitioning, Hierarchical design

[7] Timing, Power, Signal/Power Integrity Analysis and Optimization:
Timing analysis, Power analysis, Signal/power integrity, Clock and global signal design

[8] Interconnect, Device and Circuit Modeling and Simulation:
Interconnect modeling, Interconnect extraction, Package modeling, Circuit simulation, Device modeling/simulation, Library design, Design fabrics, Design for manufacturability, Yield optimization, Reliability analysis, Emerging technologies

[9] Test and Design for Testability:
Test design, Fault modeling, ATPG, BIST and DFT, Memory, core and system test

[10] Analog, RF and Mixed Signal Design and CAD:
Analog/RF synthesis, Analog layout, Verification, Simulation techniques, Noise analysis, Analog circuit testing, Mixed signal design considerations

[11] Leading Edge Design Methodology for SOCs and SIPs:
Design methodology for Microprocessors, DSP, IP-core, multimedia processors, wireless communication systems, A/D mixed circuits, Memories, Sensors, MEMS chips, FPGAs, Novel reconfigurable systems, Rapid prototyping
Please note that ASP-DAC 2007 University LSI Design Contest encourages original papers on LSI design and implementation at universities and other educational organizations.

Submission of Papers:
Deadline for submission: 5 pm JST, July 10 (Mon), 2006
Notification of acceptance: September 29 (Fri), 2006
Deadline for final version: 5 pm JST, November 17 (Fri), 2006

Specification of the paper submission format will be available at our WEB site:
`http://www.aspdac.com/aspdac2007/`

Panels, Special Sessions and Tutorials:
Suggestions and proposals are welcome and have to be addressed to the Conference Secretariat (e-mail:aspdac2007@aspdac.com) no later than 5 pm JST, June 9 (Fri.), 2006.

Prospective Sponsors:
ACM SIGDA, IEEE Circuits and Systems Society, IEICE ESS (Institute of Electronics, Information and Communication Engineers, Engineering Sciences Society), IPSJ SIG-SLDM (Information Processing Society of Japan, SIG System LSI Design Methodology)

ASP-DAC2007 Chairs:
General Chair: Hidetoshi Onodera (Kyoto Univ.)
Technical Program Chair: Yusuke Matsunaga (Kyushu Univ.)
Technical Program Vice Chair: Kiyoung Choi (Seoul National Univ.)

Conference Secretariat:
Please contact Conference Secretariat (e-mail:aspdac2007@aspdac.com), if you have questions or comments.

ASIA SOUTH PACIFIC

DAC
DESIGN AUTOMATIOM CONFERENCE

University LSI Design Contest

Call for Designs
ASP-DAC 2007
University LSI Design Contest

http://www.aspdac.com/aspdac2007/
January 23-26, 2007
Pacifico Yokohama, Yokohama, JAPAN

Aims of the Contest:

As a unique feature of ASP-DAC 2007, the University LSI Design Contest will be held. The aim of the Contest is to encourage education and research on VLSI design at universities and other educational organizations. We solicit designs that fit in one or more of the following categories:

(1) Designed, and actually implemented on chips in universities or other educational organizations during the last two years.
(2) Designs that report actual measurements from implementations.
(3) Innovative design prototypes.

Interesting or excellent designs selected will be honored by providing the opportunities for presentation in a special session at the conference. Awards will be given to a few number of outstanding designs, selected from those presented at the conference.

Areas of Design:

Application areas, or types of circuits, of the original LSI circuit designs include (but are not limited to):
(1)Analog, RF and Mixed-Signal Circuits, (2) Digital Signal Processing, (3) Microprocessors, (4) Custom ASIC.
Methods, or technology, used for implementation include:
(a) Full Custom and Cell-Based LSIs, (b) Gate Arrays, (c) FPGA/PLDs

Submission of Design Descriptions:

A camera-ready summary is requested to be prepared within 2 pages including figures, tables, and references. It is strongly recommended that measured experimental results and a chip micrograph are included in the original LSI circuit design. If the experimental results and the chip micrograph have not been prepared before the deadline of submission, the authors can send the revised paper including them later. Please do not submit the same paper as a regular paper.

Specification of the submission format and the predetermined design tasks are available at
http://www.aspdac.com/aspdac2007/

Deadline for summary:	July	10(Mon),	2006
Notification of acceptance:	September	29(Fri),	2006
Deadline for camera-ready:	November	17(Fri),	2006

Review:

Submitted designs will be reviewed by the Design Contest Committee in a process similar to the review process for the technical papers. The following criteria will be applied in the selection of designs:
(1) Reliability of design and implementation, (2) Quality of implementation, (3) Performance of the design, (4) Novelty of application, algorithm, architecture, (5) Others.
Interesting or excellent designs selected will be presented at a special session of the conference.

Presentation:

An author of each selected design will be required to make a short presentation at a special session of ASP-DAC 2007. A digest of each design to be presented will be included in the conference proceedings.

ASP-DAC 2007 Chairs

General Chair:	Hidetoshi Onodera (Kyoto Univ.)
Technical Program Chair:	Yusuke Matsunaga (Kyushu Univ.)
Design Contest Chair:	Makoto Nagata (Kobe Univ.)